1999 Reorganization Highlights

210

Article 210 Reorganization

1996 NEC Section	Section Heading or Description	1999 NEC Section
220-4	Branch Circuits Required	210-11
220-4(a)	Number of Branch Circuits	210-11(a)
220-4(d)	Load Evenly Proportioned Among Branch Circuits	210-11(b)
	Dwelling Units	210-11(c)
220-4(b)	Small Appliance Branch Circuits	210-11(c)(1)
220-4(c)	Laundry Branch Circuits	210-11(c)(2)
210-52(d) (part)	Bathroom Branch Circuits	210-11(c)(3)
210-19	Conductors—Minimum Ampacity and Size	210-19
220-3(a), 210-22(c), 210-19 (part)	General (Continuous and Noncontinuous Loads)	210-19(a)
210-19(a) (part)	Multioutlet Branch Circuits	210-19(b)
210-19(b)	Household Ranges and Cooking Appliances	210-19(c)
210-19(c)	Other Loads	210-19(d)
210-20	Overcurrent Protection	210-20
210-22(c), 220-3(a)	Continuous and Noncontinuous Loads	210-20(a)
210-20(1)	Conductor Protection	210-20(b)
210-20(2)	Equipment	210-20(c)
210-20(3)	Outlet Devices	210-20(d)

220

Article 220 Reorganization

1996 NEC Section	Section Heading or Description	1999 NEC Section
220-3(b)	Lighting Load for Specified Occupancies	220-3(a)
220-3(c)	Other Loads—All Occupancies	220-3(b)
220-3(c)(1)	Specific Appliances or Loads	220-3(b)(1)
220-3(c)Exc.No. 2 & No. 5	Electric Dryers and Household Electric Cooking Appliances	220-3(b)(2)
220-3(c)(2)	Motor Loads	220-3(b)(3)
220-3(c)(3)	Recessed Lighting Fixtures	220-3(b)(4)
220-3(c)(4)	Heavy Duty Lampholders	220-3(b)(5)
220-3(c)(5)	Track Lighting (Deleted—See 410-102)	220-12(b)
220-3(c)(6)	Sign and Outline Lighting	220-3(b)(6)
220-3(c)Exc.No.3	Show Windows	220-3(b)(7)
220-3(c)Exc.No.1	Fixed Multioutlet Assemblies	220-3(b)(8)
220-3(c)(7)	Other Outlets	220-3(b)(9)
220-3(d)	Loads for Additions to Existing Installations	220-3(c)
Table 220-3(b) *Note	General use receptacles in dwellings and guest rooms of hotels and motels	220-3(b)(10)
Table 220-3(b)	General Lighting Loads by Occupancies	Table 220-3(a)
210-22	Maximum Loads	220-4
210-22(a)	Motor-Operated and Combination Loads	220-4(a)
210-22(b)	Inductive Lighting Loads	220-4(b)
210-22(c) (part)	Range Loads	220-4(c)

National Electrical Code® Handbook

EIGHTH EDITION

National Electrical Code®
Handbook

EIGHTH EDITION

Mark W. Earley, P.E.
Editor-in-Chief

Joseph V. Sheehan, P.E.
Editor

John M. Caloggero
Editor

With the complete text of the 1999 edition of the *National Electrical Code®*

NFPA® National Fire Protection Association, Quincy, Massachusetts

Product Manager: Jennifer Evans
Project Editor: Dana Richards
Developmental Editor: Jean Peck
Text Processing: Louise Grant
Composition: Modern Graphics Inc.
Art Coordinator: Nancy Maria
Illustrator: George Nichols
Cover Design: Corey & Company
Manufacturing Buyer: Ellen Glisker

Notice Concerning Liability: Publication of this handbook is for the purpose of circulating information and opinion among those concerned for fire and electrical safety and related subjects. While every effort has been made to achieve a work of high quality, neither the NFPA nor the contributors to this handbook guarantee the accuracy or completeness of or assume any liability in connection with the information and opinions contained in this handbook. The NFPA and the contributors shall in no event be liable for any personal injury, property, or other damages of any nature whatsoever, whether special, indirect, consequential, or compensatory, directly or indirectly resulting from the publication, use of, or reliance upon this handbook.

This handbook is published with the understanding that the NFPA and the contributors in this handbook are supplying information and opinion but are not attempting to render engineering or other professional services. If such services are required, the assistance of an appropriate professional should be sought.

Notice Concerning Code Interpretations: This eighth edition of the *National Electrical Code® Handbook* is based on the 1999 edition of NFPA 70, *National Electrical Code.* All NFPA codes, standards, recommended practices, and guides are developed in accordance with the published procedures of the NFPA by technical committees comprised of volunteers drawn from a broad array of relevant interests. The handbook contains the complete text of NFPA 70 and any applicable Formal Interpretations issued by the Association. These documents are accompanied by explanatory commentary.

The commentary in this handbook is not a part of the Code and does not constitute Formal Interpretations of the NFPA (which can be obtained only through requests processed by the responsible technical committees in accordance with the published procedures of the NFPA). The commentary, therefore, solely reflects the personal opinions of the editor or other contributors and does not necessarily represent the official position of the NFPA or its technical committees.

NFPA No.: 70HB99
ISBN: 0-87765-437-9
Library of Congress Card Catalog No: 89-63606

Printed in the United States of America RRD
03 02 01 00 5 4 3

Contents

Preface

On March 18–19, 1896, twenty-three persons, representing a wide variety of organizations, met at the headquarters of the American Society of Mechanical Engineers in New York City. Their purpose was to develop a national code of rules for electrical construction and operation. It is interesting to note that this first meeting took place a mere seventeen years after the invention of the incandescent light bulb. This was not the first attempt to establish consistent rules for electrical installations, but it was the first national effort. The number of electrical fires was on the increase, and the need was becoming urgent. One insurer reported 23 electrical fires in 65 insured textile mills in New England by 1881.

The major problem was the lack of electrical installation standards. As one of the early participants noted, "We were without standards and inspectors, while manufacturers were without experience and knowledge of real installation needs. The workmen frequently created the standards as they worked, and rarely did two men think and work alike."

Five electrical installation codes were in use in the United States by 1895. This caused considerable controversy and confusion. It was difficult to manufacture products that met the requirements of all five codes. Something had to be done to develop a single, national code. The committee that met in 1896 recognized that the five existing codes should collectively be used as the basis for the new code. In the first known instance of international harmonization, the group also referred to the German code, the code of the British Board of Trade, and the Phoenix Rules of England. The importance of industry consensus was immediately recognized. Therefore, the new code was reviewed by 1200 individuals in the United States and Europe before the committee met again in 1897. Shortly thereafter, the first U.S. electrical code was published.

The *National Electrical Code* has become the most widely adopted code in the United States. It is adopted in the territories and all 50 states. Moreover, it has grown well beyond the borders of the United States and is now used in several other countries. Because the code is a living document, constantly changing to reflect changes in technology, its use continues to grow.

Some things have not changed. The *National Electrical Code* continues to offer an open-consensus process. Anyone can submit a proposal for change or a public comment. The code is subject to a rigorous public review process. It still provides the best technical information, ensuring the practical safeguarding of persons and property from hazards arising from the use of electricity.

Throughout its history, the National Electrical Code Committee has been guided by giants in the electrical industry. The names are too numerous to mention. Certainly the first chairman, William J. Hammer, should be applauded for providing the leadership necessary to get the code started. Recently, the code has been chaired by outstanding leaders such as Richard L. Lloyd, Richard W. Osborne, Richard G. Biermann, and D. Harold Ware. Each of these men has devoted many years to the National Electrical Code Committee.

The editors wish to express special thanks to Paul Duks. Paul recently retired from Underwriters Laboratories Inc. after 42 years. He has made significant contributions to six editions of the *National Electrical Code Handbook.* The electrical industry has benefited from these contributions through greater understanding of the *NEC.* We wish him well in his retirement.

The editors have conferred closely with members of the National Electrical Code Committee in developing the revisions incorporated into the 1999 edition of the *Code*. The assistance and cooperation of code-making panel chairs and various committee members are herein gratefully acknowledged.

This edition of the handbook would not have been possible without the invaluable technical assistance of Merton W. Bunker, Chief Electrical Engineer, and Jeff Sargent, Electrical Specialist. We also wish to acknowledge the invaluable technical assistance of Mark C. Ode, Senior Electrical Specialist, and Lee F. Richardson, Senior Electrical Engineer. Their contributions are greatly appreciated.

The editors acknowledge with thanks the manufacturers and their representatives who have generously supplied photographs, drawings, and data upon request. Special thanks also to the editors and contributors to past editions. Their work has provided an excellent foundation on which to build.

Appreciation is expressed to the NFPA staff members who attended to the countless details that went into the preparation of this handbook, and especially to Dana Richards, Project Editor; Louise Grant, Composition; Nancy Maria, Art Coordinator; Jennifer Evans, Project Manager; George Nichols, Illustrator; Mary Warren-Pilson, Kathleen Stevens, and Kimberly L. Shea, Coordination; and Ellen Glisker, Manufacturing Buyer.

Finally, the editors wish to express their sincere appreciation to Richard Berman, Timothy Croushore, Philip H. Cox, George Gregory, Stanley Kaufman, James McDowell, Ray C. Mullin, James Pauley, Vince Saporita, Peter J. Schram, and John C. Wiles for special help on specific articles. We also wish to thank the following for contributing photos and graphics for this edition:

Ideal Industries, Inc.
Square D Co.
Cooper Bussman
Daniel Hurley, AEMC Instruments
Timothy Reeves, Thomas & Betts Co., Inc.
Jack D. Thomas, Thomas Lighting
Bill Childes, Midwest Electric Products, Inc.
William C. Boteler, Hubbell Incorporated
Bob Lester, Pass & Seymour/Legrand
Steven Benesh, Square D Co.
Debbie Coyne, Fluke Corporation
Terry Harmon, Thomas & Betts Co., Inc.
Joe Mogan, Lithonia Lighting
George Northrop, Dual-Lite, Inc.
M. Guzy, Automatic Switch Co.
Robert Hardin, S&C Electric Company
AT&T
Biddle Instruments
Underwriters Laboratories Inc.
Crouse-Hinds
Appleton Electric Co.
General Electric Co.
Neurodyne-Dempsey, Inc.
Daniel Woodhead Co.
Union Carbide Connector, Inc.
Electronic Theatre Controls, Inc.
Production Arts Lighting, Inc.
Kliegl Bros.
Colortran, Inc.
AMP Products Corp.
Anchor Electric
RACO
King Technology Inc.
3M Co., Electrical Products Division
AFC Cable Systems, Inc.
Alcoa
Alied Tube & Conduit, a Tyco International Co.
Amerace Corp.
Bargar Metal Fabricating Co.
Cable Tray Institute
Pyrotenax USA, Inc.
Carlon Electrical Products, a Lamson and Sessions Co.
Caterpillar
Ford Motor Co.
ITT Blackburn Co.
Lockwood Corp.
MP Husky Corp.
O.Z./Gedney Co.
Smart House
Solar Design Associates, Inc.
The Wiremold Co.
Thomas Industries Inc.
Tocco Division, Park Ohio Industries
Westinghouse Electric
Bose Corporation
Richard Bingham, Drametz - BMI
Rockbestos-Surprenant Cable Corp.

National Electrical Code®
Handbook

EIGHTH EDITION

Article 90

Introduction

Article 90 — Introduction

Contents

90-1. Purpose.

(a) Practical Safeguarding. The purpose of this *Code* is the practical safeguarding of persons and property from hazards arising from the use of electricity.

The *National Electrical Code®* *(NEC®)* is prepared by the NFPA *National Electrical Code* Committee. The scope of this committee is as follows.

Scope: This committee shall consist of a Correlating Committee and Code-Making Panels. It shall have primary responsibility for preparing documents on minimizing the risk of electricity as a source of electric shock and as an ignition source of fires and explosions. The Committee shall also be responsible for developing requirements to minimize the propagation of fire and explosions due to electrical installations.

The *NEC* is the most widely adopted set of electrical safety requirements in the world and is offered for use in law and for regulatory purposes in the interest of life and property protection. NFPA's model state law on inspection of electrical installations is entitled "The Model Law Providing for Inspection of Electrical Installations." The model law was prepared by the NFPA Advisory Committee on the Adoption and Use of the *National Electrical Code* and Related Documents as a guide for those jurisdictions that do not have formalized electrical inspection procedures or those that desire to amend their electrical inspection laws, as well as a guide to adoption of the *NEC*. The model law is intended for use by states and municipalities.

(b) Adequacy. This *Code* contains provisions that are considered necessary for safety. Compliance therewith and proper maintenance will result in an installation that is essentially free from hazard but not necessarily efficient, convenient, or adequate for good service or future expansion of electrical use.

FPN: Hazards often occur because of overloading of wiring systems by methods or usage not in conformity with this *Code*. This occurs because initial wiring did not provide for increases in the use of electricity. An initial adequate installation and reasonable provisions for system changes will provide for future increases in the use of electricity.

Consideration should always be given to future expansion in the use of electricity. Future expansion may be unlikely in some occupancies, but in others it is wise to plan an initial installation comprising service-entrance conductors and equipment, feeder conductors, and panelboards that will allow for future additions, alterations, designs, and so on.

(c) Intention. This *Code* is not intended as a design specification nor an instruction manual for untrained persons.

The *NEC* is intended for use by capable engineers and electrical contractors in the design and/or installation of electrical equipment; by inspection authorities exercising legal jurisdiction over electrical installations; by property insurance inspectors; by qualified industrial, commercial, and residential electricians; and by instructors of electrical apprentices or students.

90-2. Scope.

(a) Covered. This *Code* covers the following.

(1) Installations of electric conductors and equipment within or on public and private buildings or other structures, including mobile homes, recreational vehicles, and floating buildings; and other premises such as yards, carnival, parking, and other lots, and industrial substations.

FPN: For additional information concerning such installations in an industrial or multibuilding complex, see the *National Electrical Safety Code*, ANSI C2-1997.

Industrial and multibuilding complexes (e.g., some universities) often include substations and other installations that employ construction and wiring that is similar to electric utility installations. Although such nonutility installations are within the scope of the *NEC*, the *NEC* requirements may not always be sufficient or complete, for example, in such cases as clearances of conductors or clearance from buildings or structures for nominal voltages over 600 volts. In such cases, the user can find additional information in the *National Electrical Safety Code*, published by the Institute of Electrical and Electronics Engineers, Inc., P.O. Box 1331, 445 Hoes Lane, Piscataway, NJ 08855-1331.

(2) Installations of conductors and equipment that connect to the supply of electricity.

(3) Installations of other outside conductors and equipment on the premises.

(4) Installations of optical fiber cables and raceways.

See Article 770 for the installation requirements for optical fiber cables and raceways and Article 690 regarding solar photovoltaic systems.

(5) Installations in buildings used by the electric utility, such as office buildings, warehouses, garages, machine shops, and recreational buildings, that are not an integral part of a generating plant, substation, or control center.

Section 90-2(a)(5) was added in the 1996 *Code* to clarify what portions of electric utility facilities are covered by the *NEC*. See Section 90-2(b)(5) and the related commentary for information on facilities and specific lighting that is not covered by the *NEC*.

Figure 90.1 illustrates the distinction between electric utility facilities to which the *NEC* does and does not apply.

(b) Not Covered. This *Code* does not cover the following.

(1) Installations in ships, watercraft other than floating buildings, railway rolling stock, aircraft, or automotive vehicles other than mobile homes and recreational vehicles.

The requirements for floating buildings are found in Article 553.

FPN: While the scope of this *Code* indicates that the *Code* does not cover installations in ships, portions of this *Code* are incorporated by reference into Title 46, *Code of Federal Regulations*, Parts 110-113.

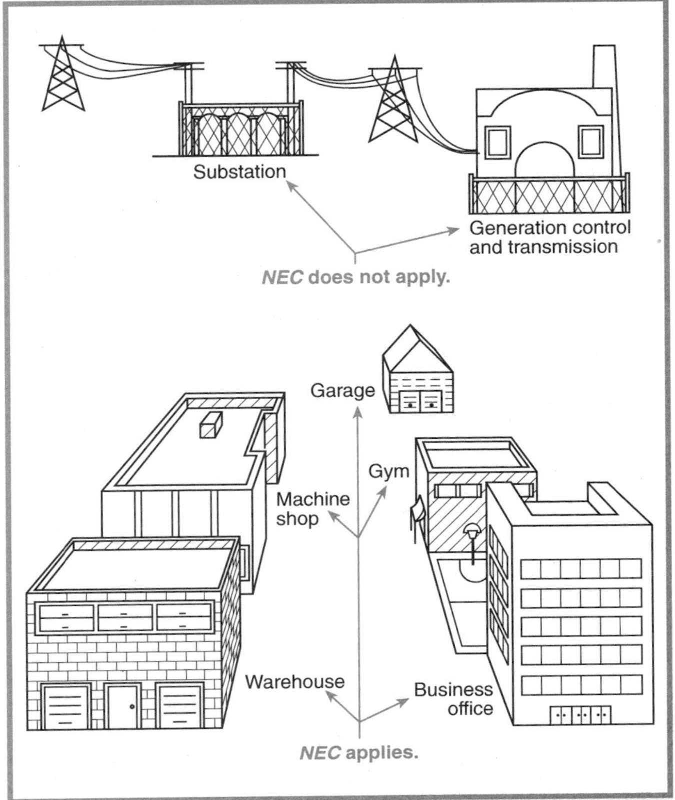

Figure 90.1 *A typical electric utility complex showing examples of facilities covered and not covered by the provisions of the NEC.*

This fine print note was added in the 1999 *NEC* to inform users that although the *NEC* does not specifically cover shipboard wiring, Title 46 of the *Code of Federal Regulations*, Parts 110–113, contains many specific *NEC* referenced requirements. These requirements, which originated in the *NEC*, are enforced by the U.S. Coast Guard.

(2) Installations underground in mines and self-propelled mobile surface mining machinery and its attendant electrical trailing cable.

(3) Installations of railways for generation, transformation, transmission, or distribution of power used exclusively for operation of rolling stock or installations used exclusively for signaling and communications purposes.

(4) Installations of communications equipment under the exclusive control of communications utilities located outdoors or in building spaces used exclusively for such installations.

(5) Installations, including associated lighting, under the exclusive control of electric utilities for the purpose of communications, metering, generation, control, transformation, transmission, or distribution of electric energy. Such installations shall be located in buildings used exclusively by utilities for such purposes; outdoors on property owned or leased by the utility; on or along public highways, streets, roads, etc.; or outdoors on private property by established rights such as easements.

Section 90-2(b)(5) was changed in the 1996 *Code* to clarify that, unless there are legal rights that allow otherwise, commercial parking lot lighting is covered by the *NEC*.

(c) Special Permission. The authority having jurisdiction for enforcing this *Code* may grant exception for the installation of conductors and equipment that are not under the exclusive control of the electric utilities and are used to connect the electric utility supply system to the service-entrance conductors of the premises served, provided such installations are outside a building or terminate immediately inside a building wall.

90-3. Code Arrangement. This *Code* is divided into the introduction and nine chapters. Chapters 1, 2, 3, and 4 apply generally; Chapters 5, 6, and 7 apply to special occupancies, special equipment, or other special conditions. These latter chapters supplement or modify the general rules. Chapters 1 through 4 apply except as amended by Chapters 5, 6, and 7 for the particular conditions.

Chapter 8 covers communications systems and is independent of the other chapters except where they are specifically referenced therein.

Chapter 9 consists of tables.

Material identified by the superscript letter "x" includes text extracted from other National Fire Protection Association (NFPA) documents as identified in Appendix A.

The reference to the Introduction is made with the intention that Article 90 be included in the application of this *Code*.

Chapters 1 through 4 apply generally, except as amended or specifically referenced in Chapters 5, 6, and 7 (Articles 500 through 780). For example, Section 300-22 (Chapter 3) is modified by Sections 725-3(b), 760-3(b), and 770-3(b) (Chapter 7) and is specifically referenced in Section 800-52(c) (Chapter 8).

90-4. Enforcement. This *Code* is intended to be suitable for mandatory application by governmental bodies that exercise legal jurisdiction over electrical installations and for use by insurance inspectors. The authority having jurisdiction for enforcement of the *Code* will have the responsibility for making interpretations of the rules, for deciding on the approval of equipment and materials, and for granting the special permission contemplated in a number of the rules.

Section 90-4 advises that all materials and equipment used under the requirements of this *Code* are subject to the approval of the authority having jurisdiction.

The text of Sections 90-7, 110-2, and 110-3, along with the definitions of the terms *approved, identified, labeled,* and *listed,* are intended to provide a basis for the authority having jurisdiction to make the judgments that fall within his or her area of responsibility. See also the commentary that follows Section 90-1(a).

The authority having jurisdiction may waive specific requirements in this *Code* or permit alternate methods where it is assured that equivalent objectives can be achieved by establishing and maintaining effective safety.

It is the responsibility of the local authority who is enforcing the *Code* to interpret the specific rules of the *Code*.

This paragraph of Section 90-4 is included to allow the authority having jurisdiction the option of permitting alternative methods where specific rules are not established in the *Code*. For example, the local authority may waive specific requirements in industrial occupancies, research and testing laboratories, and other occupancies where the specific type of installation is not covered in the *Code*.

Some localities do not adopt the *NEC*, but even in those localities installations that comply with the current *Code* are prima facie evidence that the electrical installation is safe.

This *Code* may require new products, constructions, or materials that may not yet be available at the time the *Code* is adopted. In such event, the authority having jurisdiction may permit the use of the products, constructions, or materials that comply with the most recent previous edition of this *Code* adopted by the jurisdiction.

This paragraph of Section 90-4 permits the authority having jurisdiction to waive a new *Code* requirement during the interim period between the acceptance of a new edition of the *NEC* and the availability of a new product, construction, or material redesigned to comply with the increased safety required by the latest edition. It is difficult to establish a viable future effective date within each section of the *NEC* because the time that is needed to change existing products and standards, as well as to develop new materials and test methods, is not usually known at the time the latest edition of the *Code* is adopted.

90-5. Mandatory Rules, Permissive Rules, and Explanatory Material.

(a) Mandatory Rules. Mandatory rules of this *Code* are those that identify actions that are specifically required or prohibited and are characterized by the use of the terms *shall* or *shall not.*

Section 90-5 was revised and reorganized in the 1999 *Code* to clarify that there are two distinctive types of rules stated in the *Code*. Mandatory rules are covered in (a) and are characterized by the term *shall* or *shall not*. Permissive rules are covered in (b).

(b) Permissive Rules. Permissive rules of this *Code* are those that identify actions that are allowed but not required, are normally used to describe options or alternative methods, and are characterized by the use of the terms *shall be permitted* or *shall not be required.*

Added in the 1999 *Code*, permissive rules are simply options or alternative methods of achieving equivalent safety — not requirements. A close reading of permissive terms is very important. Permissive rules can be misinterpreted. For example, the frequently used permissive term *shall be permitted* can be mistaken for a requirement. Substituting the phrase "the inspector must allow (item or method A)" for the phrase "(item A or method A) shall be permitted" generally clarifies the interpretation.

(c) Explanatory Material. Explanatory material, such as references to other standards, references to related sections of this *Code*, or information related to a *Code* rule, is included in this *Code* in the form of fine print notes (FPN). Fine print notes are informational only and are not enforceable as requirements of this *Code*.

Fine print notes (FPNs) do not contain "statements" of intent, nor do they contain recommendations. They do contain additional supplementary material that aids in the application of the requirement. In addition to printing explanatory material in fine print (small type), the material is further identified in the *Code* by the abbreviation FPN preceding the paragraph. Fine print notes are not requirements of the *NEC* and are not enforcable.

Footnotes to tables, although also in fine print, are not explanatory material unless they are identified by the abbreviation FPN. However, the footnotes are part of the tables and are necessary for the proper use of the tables. For example, see the footnotes to Table 310-13, which are necessary for the use of the table and are therefore mandatory and enforcable *Code* text.

Additional explanatory material is also found in Appendixes A, B, C, D, and E. Appendix A is a reference list of other NFPA documents from which text has been extracted. Appendix B provides guidance on the use of the general formula for ampacity found in Section 310-15(c). Appendix C consists of wire fill tables for conduit and tubing. Appendix D contains example calculations. Appendix E is a cross-reference list of section numbers to the revised Article 250.

FPN: The format and language used in this *Code* follows guidelines established by NFPA and published in the *NEC Style Manual*. Copies of this manual may be obtained from NFPA.

Added in the 1999 *Code*, this fine print note informs the user that a style manual is available for the *NEC*. A style manual is basically a "how-to" pamphlet for editors. The *NEC Style Manual* contains a list of rules and regulations that are used in the preparation of the *NEC*.

90-6. Formal Interpretations. To promote uniformity of interpretation and application of the provisions of this *Code*, formal interpretation procedures have been established and are found in the NFPA Regulations Governing Committee Projects.

The procedures for Formal Interpretations of the provisions of the *NEC* are outlined in "NFPA Regulations Governing Committee Projects." These regulations are included in the NFPA *Annual Directory*, which may be obtained from the Secretary of the Standards Council of the National Fire Protection Association. The Formal Interpretations procedure can be found in Section 6 of the Regulations.

The *National Electrical Code* Committee cannot be responsible for subsequent actions by authorities enforcing the *NEC* that accept or reject its findings. The authority having jurisdiction has responsibility for interpreting the *Code* rules and should attempt to resolve all disagreements at the local level.

Two general forms of Formal Interpretations are recognized: (1) those making an interpretation of the literal text and (2) those making an interpretation of the intent of the Committee at the time the particular text was issued.

Interpretations of the *NEC* not subject to processing are those that (a) involve a determination of compliance of a design, installation, product, or equivalency of protection; (b) involve a review of plans or specifications, or require judgment or knowledge that can be acquired only as a result of on-site inspection; (c) involve text that clearly and decisively provides the requested information; or (d) involve subjects that were not previously considered by the Technical Committee or that are not addressed in the document.

Formal Interpretations of *Code* rules are published in the NFPA *Fire News,* the NFPA Electrical Section News Bulletin, "Current Flashes," and the *National Fire Codes* subscription service and are sent to interested trade publications.

Most interpretations of the *NEC* are rendered as the personal opinion of NFPA Electrical Engineering staff or of the involved member of the *National Electrical Code* Committee, because the request for interpretation does not qualify for processing as a Formal Interpretation in accordance with the Regulations Governing Committee Projects. Such opinions are rendered in writing only in response to written requests. The correspondence contains a disclaimer indicating that it is not a Formal Interpretation issued pursuant to NFPA Regulations, and that any opinion expressed is the personal opinion of the author and does not necessarily represent the official position of NFPA or the *National Electrical Code* Committee.

90-7. Examination of Equipment for Safety. For specific items of equipment and materials referred to in this *Code*, examinations for safety made under standard conditions will provide a basis for approval where the record is made generally available through promulgation by organizations properly equipped and qualified for experimental testing, inspections of the run of goods at factories, and service-value determination through field inspections. This avoids the necessity for repetition of examinations by different examiners, frequently with inadequate facilities for such work, and the confusion that would result from conflicting reports as to the suitability of devices and materials examined for a given purpose.

It is the intent of this *Code* that factory-installed internal wiring or the construction of equipment need not be inspected at the time of installation of the equipment, except to detect alterations or damage, if the equipment has been listed by a qualified electrical testing laboratory that is recognized as having the facilities described above and that requires suitability for installation in accordance with this *Code*.

FPN No. 1: See requirements in Section 110-3.

FPN No. 2: *Listed* is defined in Article 100.

Testing laboratories, inspection agencies, or other organizations concerned with product evaluation publish lists of equipment or materials that have been tested and that meet nationally recognized standards or that have been found suitable for use in a specified manner. The *Code* does not contain detailed information on equipment or materials but refers to the products as "listed," "labeled," or "identified." See Article 100 for definitions of these terms.

The National Fire Protection Association does not approve, inspect, or certify any installations, procedures, equipment, or materials; nor does it approve or evaluate testing laboratories. In determining the acceptability of installations or procedures, equipment, or materials, the authority having jurisdiction may base acceptance on compliance with NFPA or other appropriate standards. In the absence of such standards, the authority may require evidence of proper installation, procedures, or use. The authority having jurisdiction may also refer to the listing or labeling practices of an organization concerned with product evaluations that is able to determine compliance with appropriate standards for the current production of listed items.

A publication entitled *Summary of Electrical and Building Code Requirements, Licensing Provisions and Laboratory Recognition at State and Local Levels* provides information on electrical testing laboratory recognition/accreditation programs and a list of those laboratories from which certain product certifications are accepted. The list is arranged according to jurisdiction. The publication is available from the International Association of Electrical Inspectors, 901 Waterfall Way, Suite 602, Richardson, TX, 75080-7702.

It is not intended that the *Code* apply to the internal factory-installed wiring or to the construction of listed equipment at the time of installation unless damage or alterations are detected.

90-8. Wiring Planning.

(a) Future Expansion and Convenience. Plans and specifications that provide ample space in raceways, spare raceways, and additional spaces will allow for future increases in the use of electricity. Distribution centers located in readily accessible locations will provide convenience and safety of operation.

The requirement for providing the exclusively dedicated space mandated by Section 110-26(f) supports the intent of Section 90-8(a) regarding future increases in the use of electricity.

Distribution centers should contain additional space and capacity for future additions and should be conveniently located for accessibility.

Where distribution equipment is installed so that easy access cannot be achieved, a spare raceway(s) or pull line(s) should be run at the initial installation, as illustrated in Figure 90.2.

(b) Number of Circuits in Enclosures. It is elsewhere provided in this *Code* that the number of wires and circuits

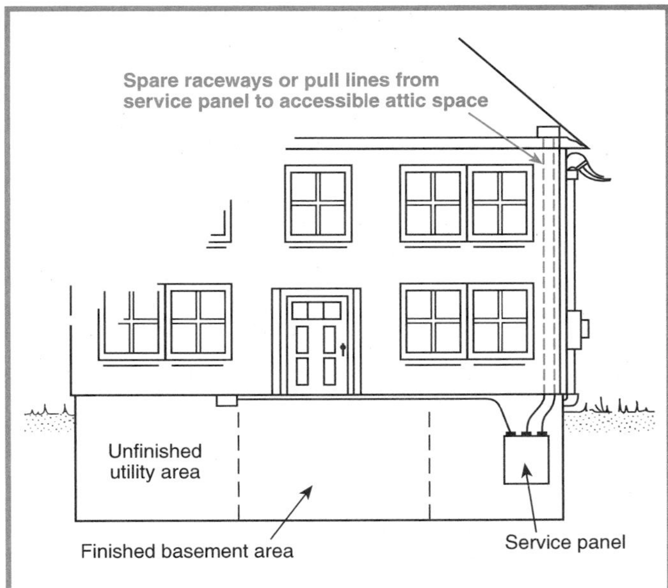

Figure 90.2 *A residential distribution system showing spare raceways or pull lines that allow for future circuits and loads.*

confined in a single enclosure be varyingly restricted. Limiting the number of circuits in a single enclosure will minimize the effects from a short circuit or ground fault in one circuit.

The heating effects inherently present wherever current-carrying conductors are grouped together will be minimized by these limitations. See Section 110-15 for restrictions on the number of overcurrent devices on one panelboard.

90-9. Metric Units of Measurement. For the purpose of this *Code*, metric units of measurement are in accordance with the modernized metric system known as the International System of Units (SI).

Metric trade sizes (numerical designations) of conduits were added in the 1996 *Code* as a fine print note after Sections 345-6(b), 346-5(b), and 348-7(b). An example using metric wire size was also added to the commentary following Table 8 of Chapter 9.

Generally, dimensions up to 36 in. are expressed in millimeters (mm), and those over 36 in. are expressed in meters (m). For example, the conversion of measurements in Table 110-26(a) to metric (SI) units of measurement is performed as follows:

$$3 \text{ ft} = 36 \text{ in.} \times 25.4 \text{ mm} = 914 \text{ mm}$$
$$3\tfrac{1}{2} \text{ ft} \times 0.3048 \text{ m} = 1.07 \text{ m}$$
$$4 \text{ ft} \times 0.3048 \text{ m} = 1.22 \text{ m}$$

Footnotes to tables involving dimensions in feet are expressed as "1 ft = 0.3048 m." However, throughout the *Code* text, the metric equivalent of 1 ft has been converted to millimeters and rounded off to 305 mm.

Note that warning signs stating specific clearances, such as required in Section 513-11(a), do not include metric equivalents.

For more information on metric conversion practices, see American National Standard (ANSI), *Standard for Use of the International System of Units (SI): The Modern Metric System,* IEEE/ASTM SI 10-1997.

Values of measurement in the *Code* text will be followed by an approximate equivalent value in SI units. Tables will have a footnote for SI conversion units used in the table.

Conduit size, wire size, horsepower designation for motors, and trade sizes that do not reflect actual measurements, e.g., box sizes, will not be assigned dual designation SI units.

CHAPTER 1

General

Article 100 — Definitions

Contents

A. General
B. Over 600 Volts, Nominal

Scope. This article contains only those definitions essential to the proper application of this *Code*. It is not intended to include commonly defined general terms or commonly defined technical terms from related codes and standards. In general, only those terms that are used in two or more articles are defined in Article 100. Other definitions are included in the article in which they are used but may be referenced in Article 100.

Part A of this article contains definitions intended to apply wherever the terms are used throughout this *Code*. Part B contains definitions applicable only to the parts of articles specifically covering installations and equipment operating at over 600 volts, nominal.

Commonly defined general terms include those terms defined in general English language dictionaries and terms that are not used in a unique or restricted manner in the *NEC*. Commonly defined technical terms such as *volt* (abbreviated V) and *ampere* (abbreviated A) are found in *The IEEE Standard Dictionary of Electrical and Electronic Terms*.

Definitions that are not listed in Article 100 are included in the following sections:

Term	Section
Alternating-Current Power Distribution Box (Motion Picture and Television Studios)	530-2
Alternating-Current Module (Solar Photovoltaic Systems)	690-2
Alternate Power Source (Health Care Facilities)	517-3
Ambulatory Health Care Center (Health Care Facilities)	517-3
Anesthetizing Location (Health Care Facilities)	517-3
Air-Conditioning or Comfort-Cooling Equipment (Recreational Vehicle)	551-2
Appliance, Fixed (Mobile Homes)	550-2
Appliance, Fixed (Recreational Vehicles)	551-2
Appliance, Portable (Mobile Homes)	550-2
Appliance, Portable (Recreational Vehicles)	551-2
Appliance, Stationary (Mobile Homes)	550-2
Appliance, Stationary (Recreational Vehicles)	551-2
Array (Solar Photovoltaic Systems)	690-2
Associated Apparatus (Intrinsically Safe Systems)	504-2
Audio Amplifier or Pre-Amplifier	640-2
Audio Auto Transformer	640-2
Audio Signal-Processing Equipment	640-2
Audio System	640-2
Audio Transformer	640-2
Block (Communications Circuits)	800-2
Block (Network-Powered Broadband Communications)	830-2
Blocking Diode (Solar Photovoltaic Systems)	690-2
Border Light (Theaters, Audience Areas)	520-2
Bottom Shield (Type FCC Cable)	328-2
Branch-Circuit Selection Current (Air-Conditioning and Refrigerating Equipment)	440-2
Breakout Assembly (Theaters, Audience Areas)	520-2
Building Component (Manufactured Buildings)	545-3
Building System (Manufactured Buildings)	545-3
Bulk Storage Plant	515-1
Bull Switch (Motion Picture and Television Studios)	530-2
Bundled (Theaters and Audience Areas)	520-2
Busway	364-2
Cable (Communications Circuits)	800-2
Cable Connector	328-2
Cable Sheath (Communication Circuits)	800-2
Cable Tray System	318-2
Cable, FCC System	328-2
Cable, Nonmetallic-Sheathed	336-2
Cable, Service-Entrance	338-1
Cable, Type AC (Armored Cable)	333-1
Cable, Type FCC (Flat Conductor Cable)	328-2
Cable, Type IGS (Integrated Gas Spacer Cable)	325-1
Cable, Type ITC (Instrumentation Tray Cable)	727-2
Cable, Type MC (Metal-Clad Cable)	334-1
Cable, Type MI (Mineral-Insulated, Metal-Sheathed Cable)	330-1
Cable, Type MV (Medium-Voltage Cable)	326-1
Cable, Type TC (Power and Control Tray Cable)	340-1
Cablebus	365-1
Camping Trailer (Recreational Vehicles)	551-2
Cell Line (Electrolytic Cells)	668-2
Cell (Cellular Concrete Floor Raceways)	358-2
Cell (Cellular Metal Floor Raceways)	356-1
Cellular Metal Floor Raceway	356-1
Center Pivot Irrigation Machines	675-2
Charge Controller (Solar Photovoltaic Systems)	690-2
Class 1 Circuit (Remote Control, Signaling)	725-2
Class 2 Circuit (Remote Control, Signaling)	725-2
Class 3 Circuit (Remote Control, Signaling)	725-2
Class I, Division 1 [Hazardous (Classified) Locations]	500-7(a)

Term	Section
Class I, Division 2 [Hazardous (Classified) Locations]	500-7(b)
Class I, Zone 0	505-9(a)
Class I, Zone 1	505-9(b)
Class I, Zone 2	505-9(c)
Class II, Division 1 [Hazardous (Classified) Locations]	500-8(a)
Class II, Division 2 [Hazardous (Classified) Locations]	500-8(b)
Class III, Division 1 [Hazardous (Classified) Locations]	500-9(a)
Class III, Division 2 [Hazardous (Classified) Locations]	500-9(b)
Closed Construction (Manufactured Buildings)	545-3
Collector Rings (Irrigation Machines)	675-2
Combustible Low-Density Cellulose Fiberboard	410-76(b), FPN
Concealed Knob-and-Tube Wiring	324-1
Connector Strip (Theaters, Audience Areas)	520-2
Control Drawing (Intrinsically Safe Systems)	504-2
Control System (Elevator)	620-2
Controller, Motion (Elevator)	620-2
Controller, Operation (Elevator)	620-2
Controller (Motor Controllers)	430-81(a)
Controller, Motor (Elevators)	620-2
Converter (Recreational Vehicles)	551-2
Coordination (Overcurrent Protection)	240-12
Cord- and Plug-Connected Lighting Assembly (Swimming Pools)	680-4
Cover (Wiring Methods)	Table 300-5
Critical Branch (Health Care Facilities)	517-3
Current-Limiting Overcurrent Protective Device	240-11
Dead Front (Recreational Vehicle)	551-2
Dielectric Heating (Induction and Dielectric Heating Equipment)	665-2
Different Intrinsically Safe Circuits (Intrinsically Safe Systems)	504-2
Disconnecting Means (Recreational Vehicles)	551-2
Distribution Panelboard (Mobile Homes)	550-2
Distribution Panelboard (Recreational Vehicles)	551-2
Distribution Point (Agricultural Buildings)	547-8(c)
Drop Box (Theaters, Audience Areas)	520-2
Dry-Niche Lighting Fixture (Swimming Pools)	680-4
Electric Vehicle	625-2
Electric-Discharge Lighting (Electric Signs)	600-2
Electrical Ducts	310-60(a)
Electrical Life-Support Equipment (Health Care Facilities)	517-3
Electrical Nonmetallic Tubing	331-1

Term	Section
Electrical Production and Distribution Network (Solar Photovoltaic Sysems)	690-2
Electrically Connected (Use and Identification of Grounded Conductors)	200-3
Electrically Connected (Electrolytic Cells)	668-2
Electrolytic Cell	668-2
Electrolytic Cell Line Working Zone	668-2
Emergency System (Health Care Facilities)	517-3
Emergency Systems	700-1
Equipment Rack (Audio)	640-2
Equipment System (Health Care Facilities)	517-3
Equipotential Plane (Agricultural Buildings)	547-9(a)
Essential Electrical System (Health Care Facilities)	517-3
Exposed (Communications Circuits)	800-2
Exposed (Optical Fiber Cables and Raceways)	770-2
Exposed (CATV)	820-2
Exposed to Accidental Contact with Electrical Light or Power Conductors	830-2
Exposed Conductive Surfaces (Health Care Facilities)	517-3
Fault-Protection Device	830-2
Feeder Assembly (Mobile Homes)	550-2
Finish Rating (Electrical Nonmetallic Tubing)	331-3(2), FPN
Fire Alarm Circuit	760-2
Fire Alarm Circuit Integrity (CI) Cable	760-2
Fire Resistant (Transformers and Transformer Vaults)	450-21(b)
First Floor (Nonmetallic-Sheathed Cable)	336-5(a)(1)
Flammable Anesthetics	517-3
Flammable Anesthetizing Location (Health Care Facilities)	517-3
Flexible Metal Conduit	350-2
Floating Building	553-2
Footlight (Theaters, Audience Areas)	520-2
Forming Shell (Swimming Pools)	680-4
Fountain (Swimming Pools)	680-4
Frame (Recreational Vehicle)	551-2
Gasoline Dispensing and Service Station	514-1
Group IIC	505-7(a)
Group IIB	505-7(b)
Group IIA	505-7(c)
Group A [Hazardous (Classified) Locations]	500-5(a)(1)
Group B [Hazardous (Classified) Locations]	500-5(a)(2)
Group C [Hazardous (Classified) Locations]	500-5(a)(3)
Group D [Hazardous (Classified) Locations]	500-5(a)(4)
Group E [Hazardous (Classified) Locations]	500-5(b)(1)

Term	Section
Group F [Hazardous (Classified) Locations]	500-5(b)(2)
Group G [Hazardous (Classified) Locations]	500-5(b)(3)
Grouped (Theaters, Audience Areas)	520-2
Hazard Current (Health Care Facilities)	517-3
Header (Cellular Concrete Floor Raceways)	358-2
Header (Cellular Metal Floor Raceways)	356-1
Health Care Facilities	517-3
Heating Equipment (Induction and Dielectric Heating Equipment)	665-2
Heating Panel (Space-Heating Equipment)	424-91(a)
Heating Panel Set (Space-Heating Equipment)	424-91(b)
Heating System (Deicing and Snow-Melting Equipment)	426-2
Hermetic Refrigerant Motor-Compressor (Air-Conditioning and Refrigerating Equipment)	440-2
High Voltage (Equipment, Over 600 Volts, Nominal)	490-2
Hospital (Health Care Facilities)	517-3
Hybrid System (Solar Photovoltaic Systems)	690-2
Hydromassage Bathtub (Swimming Pools)	680-4
Impedance Heating System (Deicing and Snow-Melting Equipment)	426-2
Impedance Heating System (Heating Equipment for Pipelines and Vessels)	427-2
Induction Heating (Induction and Dielectric Heating Equipment)	665-2
Induction Heating System (Heating Equipment for Pipelines and Vessels)	427-2
Industrial Machinery (Machine)	670-2
Industrial Manufacturing System (Industrial Machinery)	670-2
Insulating End (Type FCC Cable)	328-2
Insulation Level, 100 Percent and 133 Percent	Table 310-64
Integrated Heating System (Heating Equipment for Pipelines and Vessels)	427-2
Interactive System (Interconnected Electric Power Production Sources)	705-2
Interactive System (Solar Photovoltaic Systems)	690-2
Intermediate Metal Conduit	345-1
Intrinsically Safe Apparatus (Intrinsically Safe Systems)	504-2
Intrinsically Safe Circuit (Intrinsically Safe Systems)	504-2
Intrinsically Safe System	504-2
Inverter (Solar Photovoltaic Systems)	690-2
Inverter Input Circuit (Solar Photovoltaic Systems)	690-2
Inverter Output Circuit (Solar Photovoltaic Systems)	690-2

Term	Section
Irrigation Machines	675-2
Isolated Power System (Health Care Facilities)	517-3
Isolation Transformer (Health Care Facilities)	517-3
Laundry Area (Mobile Homes)	550-2
Life Safety Branch (Health Care Facilities)	517-3
Lighting and Appliance Branch-Circuit Panelboard	384-14(a)
Lighting Track	410-100
Limited Care Facility (Health Care Facilities)	517-3
Line Isolation Monitor (Health Care Facilities)	517-3
Liquidtight Flexible Metal Conduit	351-2
Liquidtight Flexible Nonmetallic Conduit	351-22
Location (Motion Picture and Television Studios)	530-2
Location Board (Motion Picture and Television Studios)	530-2
Loudspeaker	640-2
Low Voltage (Recreational Vehicle)	551-2
Luminaire (Lighting Fixtures)	410-1, FPN
Manufactured Building	545-3
Manufactured Home	550-2
Manufactured Phase (Phase Converters)	455-2
Manufactured Wiring System	604-2
Maximum Output Power (Audio)	640-2
Metal Shield Connections (Type FCC Cable)	328-2
Messenger Supported Wiring	321-1
Mixer (Audio)	640-2
Mixer-Amplifier (Audio)	640-2
Mobile Equipment	513-2
Mobile Home	550-2
Mobile Home Accessory Building or Structure	550-2
Mobile Home Lot	550-2
Mobile Home Park	550-2
Mobile Home Service Equipment	550-2
Module (Solar Photovoltaic Systems)	690-2
Motion Picture Studio (Lot) (Motion Picture and Television Studios)	530-2
Motor Control Circuit	430-71
Motor Home (Recreational Vehicles)	551-2
Network Interface Unit (NIU)	830-2
Network-Powered Broadband Communications Circuit	830-2
Neon Tubing (Electric Signs)	600-2
No-Niche Lighting Fixture (Swimming Pools)	680-4
Nominal Battery Voltage (Storage Batteries)	480-2
Nonflammable Dielectric Fluid (Transformers and Transformer Vaults)	450-24
Nonmetallic Extensions	342-1
Nonmetallic Underground Conduit with Conductors	343-1
Nonpower-Limited Fire Alarm Circuit (NPLFA)	760-2

Term	Section
Nursing Home (Health Care Facilities)	517-3
Nurses' Station (Health Care Facilities)	517-3
Open Wiring on Insulators	320-1
Optical Fiber Cables	770-4
Optical Fiber Raceway	770-2
Operating Device (Elevator)	620-2
Packaged Spa or Hot Tub Equipment Assembly (Swimming Pools)	680-4
Packaged Therapeutic Tub or Hydrotherapeutic Tank Equipment Assembly (Swimming Pools)	680-4
Panel (Solar Photovoltaic Systems)	690-2
Park Electrical Wiring Systems (Mobile Homes)	550-2
Park Trailer	552-2
Patient Bed Location (Health Care Facilities)	517-3
Patient Care Area (Health Care Facilities)	517-3
Patient Equipment Grounding Point (Health Care Facilities)	517-3
Patient Vicinity (Health Care Facilities)	517-3
Permanently Installed Decorative Fountains and Reflection Pools (Swimming Pools)	680-4
Permanently Installed Swimming, Wading, and Therapeutic Pools (Swimming Pools)	680-4
Phase Converter	455-2
Photovoltaic Power Source	690-2
Photovoltaic Source Circuit	690-2
Pipeline (Heating Equipment for Pipelines and Vessels)	427-2
Plugging Box (Motion Picture and Television Studios)	530-2
Point of Entrance (CATV)	820-3
Point of Entrance (Communications Circuits)	800-2
Point of Entrance (Optical Fiber Cables and Raceways)	770-2
Point of Entrance (Network-Powered Broadband Communications)	830-2
Pool Cover, Electrically Operated (Swimming Pools)	680-4
Pool (Swimming Pools)	680-4
Portable Equipment (Audio)	640-2
Portable Equipment (Aircraft Hangars)	513-2
Potable Equipment (Motion Picture and Television Studios)	530-2
Portable Equipment (Theaters, Audience Areas)	520-2
Plugging Box (Motion Picture and Television Studios)	530-2
Powered Loudspeaker (Audio)	640-2
Power-Limited Fire Alarm Circuit (PLFA)	760-2
Power Panelboard	384-14(b)
Power Supply Assembly (Recreational Vehicles)	551-2
Premises (Communications Circuits)	800-2
Premises (CATV)	820-2

Term	Section
Portable Power Distribution Unit (Theaters, Audience Areas)	520-2
Premises Wiring (Nework-Powered Broadband Communications)	830-2
Proscenium (Theaters, Audience Areas)	520-2
Psychiatric Hospital (Health Care Facilities)	517-3
Rated-Load Current (Air-Conditioning and Refrigerating Equipment)	440-2
Rated Load Impedance (Audio)	640-2
Rated Output Power	640-2
Rated Output Voltage	640-2
Rating of Service Disconnect	230-95
Recreational Vehicle	551-2
Recreational Vehicle Park	551-2
Recreational Vehicle Site	551-2
Recreational Vehicle Site Feeder Circuit Conductors	551-2
Recreational Vehicle Site Supply Equipment	551-2
Recreational Vehicle Stand	551-2
Reference Grounding Point (Health Care Facilities)	517-3
Resistance Heating Element (Deicing and Snow-Melting Equipment)	426-2
Resistance Heating Element (Heating Equipment for Pipelines and Vessels)	427-2
Rotary Phase Converter (Phase Converters)	455-2
Sealed Cell or Battery	480-2
Secondary Tie (Transformers and Transformer Vaults)	450-6
Selected Receptacles (Health Care Facilities)	517-3
Self-Contained Spa or Hot Tub (Swimming Pools)	680-4
Self-Contained Therapeutic Tubs or Hydrotherapeutic Tanks (Swimming Pools)	680-4
Set of Fuses (Service Equipment — Overcurrent Protection)	230-90(a)
Sign Body (Electric Signs)	600-2
Signal Equipment (Elevators)	620-2
Simple Apparatus (Intrinsically Safe Systems)	504-2
Single Grounding Electrode System	250-58
Single-Pole Separable Connector (Motion Picture and Television Studios)	530-2
Skeleton Tubing (Electric Signs)	600-2
Skin-Effect Heating System (Deicing and Snow-Melting Equipment)	426-2
Skin-Effect Heating System (Heating Equipment for Pipelines and Vessels)	427-2
Solar Cell (Solar Photovoltaic Systems)	690-2
Solar Photovoltaic System	690-2
Solidly Grounded (Service Equipment—Overcurrent Protection)	230-95
Spa or Hot Tub (Swimming Pools)	680-4
Spider (Cable Splicing Block) (Motion Picture and Television Studios)	530-2

Term	Section
Stage Effect (Special Effect) (Motion Picture and Television Studios)	530-2
Stage Property (Motion Picture and Television Studios)	530-2
Stage Set (Motion Picture and Television Studios)	530-2
Stand Lamp (Work Light) (Theaters, Audience Areas)	520-2
Stand Lamp (Work Light) (Motion Picture and Television Studios)	530-2
Stand-Alone System (Solar Photovoltaic Systems)	690-2
Standard Ampere Ratings (Overcurrent Protection)	240-6(a)
Static Phase Converter (Phase Converters)	455-2
Storable Swimming or Wading Pool (Swimming Pools)	680-4
Storage Battery	480-2
Storage Space (Lighting Fixtures)	410-8(a)
Strip Light (Theaters, Audience Areas)	520-2
Surge Arrester	280-2
System Voltage (Solar Photovoltaic Systems)	690-2
Task Illumination (Health Care Facilities)	517-3
Technical Power System (Audio)	640-2
Television Studio or Motion Picture Stage (Sound Stage)	530-2
Temperature Rating, Conductor	310-10, FPN
Temporary Equipment (Audio)	640-2
Therapeutic High-Frequency Diathermy Equipment	517-3
Thermal Resistivity	310-60(a)
Top Shield (Type FCC Cable)	328-2
Transformer	450-2, 450-6, 551-2
Travel Trailers (Recreational Vehicle)	551-2
Truck Camper	551-2
Transition Assembly (Type FCC Cable)	328-2
Two-fer (Theaters, Audience Areas)	520-2
Unfinished Basement (Branch Circuits—GFCI Protection)	210-8(a)(5)
Vessel (Heating Equipment for Pipelines and Vessels)	427-2
Wall Space (Branch Circuits—Dwelling Unit Receptacle Outlets)	210-52(a)(2)
Wet-Niche Lighting Fixture (Swimming Pools)	680-4
Wire (Communications Circuits)	800-2
Wireways	362-1
X-Ray Installations (Health Care Facilities)	517-3

A. General

Accessible (as applied to wiring methods). Capable of being removed or exposed without damaging the building structure or finish, or not permanently closed in by the structure or finish of the building.

Wiring methods located behind removable panels designed to allow access are not considered permanently enclosed and are considered exposed as applied to wiring methods. See Section 300-4(c) regarding cables through spaces behind panels designed to allow access.

Figure 100.1 illustrates examples of wiring methods and equipment that are considered accessible.

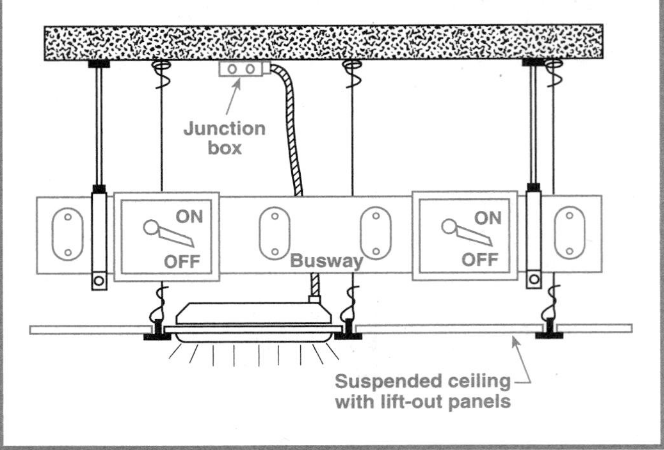

Figure 100.1 *Examples of busways and cables considered accessible even if located behind hung ceilings having lift-out panels.*

Accessible (as applied to equipment). Admitting close approach; not guarded by locked doors, elevation, or other effective means.

This definition of *accessible* does not preclude the use of externally operated fused switches and circuit breakers that are plugged into busways and mounted out of reach, provided they are located in accordance with the provisions of Section 380-8(a), Exception No. 1, and have suitable operating disconnecting means as required by Section 364-12.

Figure 100.2 illustrates some examples of equipment considered accessible under these provisions.

Accessible, Readily (Readily Accessible). Capable of being reached quickly for operation, renewal, or inspections, without requiring those to whom ready access is requisite to climb over or remove obstacles or to resort to portable ladders, etc.

The definition of *readily accessible* does not preclude the use of a locked door for service equipment or rooms containing service equipment, provided those for whom ready access is necessary have a key (or lock combination) available. For example, Sections 230-70(a) and 230-205(a) require service discon-

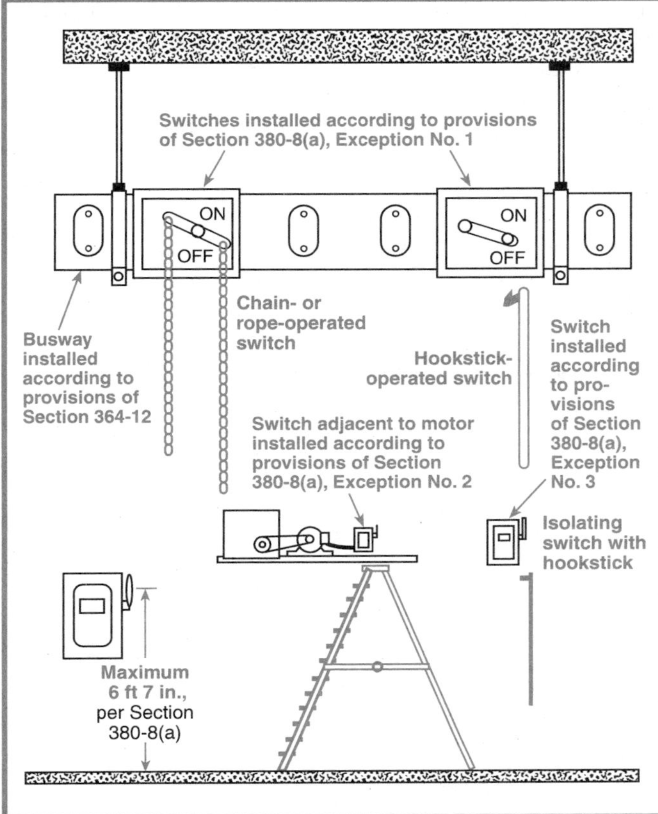

Figure 100.2 *Example of a busway and of switches considered accessible even if located above 6 ft 7 in.*

necting means to be readily accessible. Section 225-32 requires that feeder disconnecting means for separate buildings be readily accessible. It is common practice to locate this disconnecting means in the electrical equipment room of an office building or large apartment building and to keep the door to this room locked to prevent access by unauthorized persons. Section 240-24(a) requires that overcurrent devices be located as to be readily accessible also.

Ampacity. The current, in amperes, that a conductor can carry continuously under the conditions of use without exceeding its temperature rating.

The definition of the term *ampacity* states that the maximum current a conductor carries continuously will vary with the conditions of use as well as with the temperature rating of the conductor insulation. For example, ambient temperature is a condition of use. A conductor with insulation rated at 60°C, installed near a furnace where the ambient temperature is continuously maintained at 60°C, has no current-carrying capacity. Any current flowing through the conductor will raise its temperature above the 60°C insulation rating. Therefore, the ampacity of this conductor, regardless of its size, is zero. See the ampacity correction factors for temperature at the bottom of Tables 310-16 through 310-20, or see Appendix B.

Another condition of use is the number of conductors in a raceway or cable. *[See Section 310-15(b)(2).]*

Appliance. Utilization equipment, generally other than industrial, normally built in standardized sizes or types, that is installed or connected as a unit to perform one or more functions such as clothes washing, air conditioning, food mixing, deep frying, etc.

Approved. Acceptable to the authority having jurisdiction.

The phrase "authority having jurisdiction" is used in NFPA documents in a broad manner, since jurisdictions and approval agencies vary, as do their responsibilities. Where public safety is primary, the authority having jurisdiction may be a federal, state, local, or other regional department or individual such as a fire chief; fire marshal; chief of a fire prevention bureau, labor department, or health department; building official; electrical inspector; or others having statutory authority. For insurance purposes, an insurance inspection department, rating bureau, or other insurance company representative may be the authority having jurisdiction. In many circumstances, the property owner or his or her designated agent assumes the role of the authority having jurisdiction; at government installations, the commanding officer or departmental official may be the authority having jurisdiction.

Askarel. A generic term for a group of nonflammable synthetic chlorinated hydrocarbons used as electrical insulating media. Askarels of various compositional types are used. Under arcing conditions, the gases produced, while consisting predominantly of noncombustible hydrogen chloride, can include varying amounts of combustible gases depending on the askarel type.

Attachment Plug (Plug Cap) (Plug). A device that, by insertion in a receptacle, establishes a connection between the conductors of the attached flexible cord and the conductors connected permanently to the receptacle.

Standard attachment caps are also available with built-in options, such as switching, fuses, or even ground-fault circuit-interrupter protection.

Attachment plug contact blades have specific shapes, sizes, and configurations so that a receptacle or cord connector will not accept an attachment plug of a different voltage or current rating than that for which the device is intended. Configuration charts for general-purpose nonlocking and specific-purpose locking plugs and receptacles, respectively, are shown in Figures 210.7 and 210.8.

Automatic. Self-acting, operating by its own mechanism when actuated by some impersonal influence, as, for example, a change in current strength, pressure, temperature, or mechanical configuration.

Bathroom. An area including a basin with one or more of the following: a toilet, a tub, or a shower.

The definition of *bathroom* was moved to Article 100 in the 1996 *Code* because it is used in more than one article. Previously, this definition was located in Section 210-8(b)(1).

Bonding (Bonded). The permanent joining of metallic parts to form an electrically conductive path that will ensure electrical continuity and the capacity to conduct safely any current likely to be imposed.

The purpose of bonding, as explained in Sections 250-2(c) and (d) is to establish an effective path for fault current that, in turn, will facilitate the operation of the overcurrent protective device. Specific bonding requirements are found in Part E of Article 250 and in other sections of the *Code* as referenced in Section 250-4.

Bonding Jumper. A reliable conductor to ensure the required electrical conductivity between metal parts required to be electrically connected.

Figure 100.3 shows a bonding jumper used around a concentric knockout to ensure the required electrical conductivity between the metal parts. Bonding jumpers are also used where eccentric knockouts are encountered.

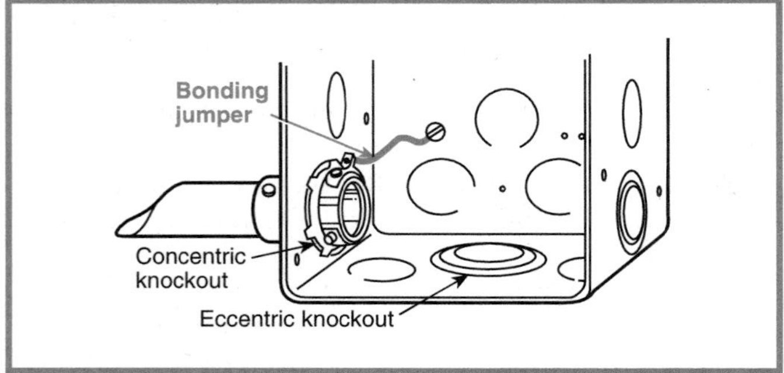

Figure 100.3 *A bonding jumper installed around a concentric knockout.*

Bonding Jumper, Equipment. The connection between two or more portions of the equipment grounding conductor.

Bonding Jumper, Main. The connection between the grounded circuit conductor and the equipment grounding conductor at the service.

Figure 100.4 shows a main bonding jumper used to provide the connection between the grounded circuit conductor of the service and the equipment ground. It may consist of a wire, busbar, or screw. Bonding jumpers may be located throughout the electrical system, but a main bonding jumper is only located at the service. Main bonding jumper requirements are found in Section 250-28.

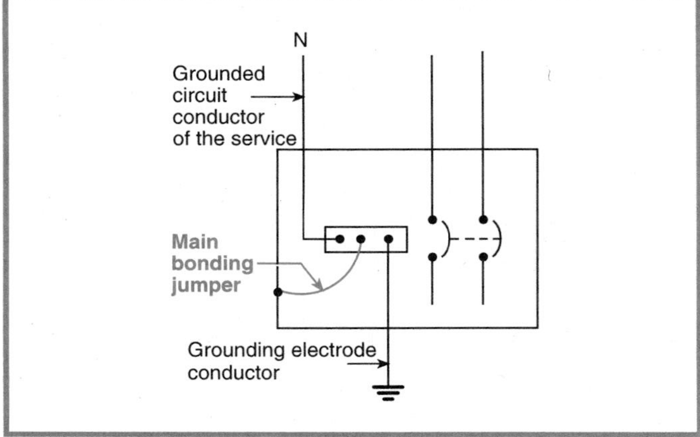

Figure 100.4 *A main bonding jumper installed between the grounded circuit conductor of the service and the equipment ground.*

Branch Circuit. The circuit conductors between the final overcurrent device protecting the circuit and the outlet(s).

Figure 100.5 shows the difference between branch circuits and feeders. Conductors between the overcurrent devices in the panelboards and the duplex receptacles are *branch-circuit conductors.* Conductors between the service equipment or source of separately derived systems and the panelboards are *feeders.*

Branch Circuit, Appliance. A branch circuit that supplies energy to one or more outlets to which appliances are to be connected, and that has no permanently connected lighting fixtures that are not a part of an appliance.

Two or more small-appliance branch circuits are required by Section 210-11(c)(1) for dwelling units. Section 210-52(b)(1) requires that these circuits supply

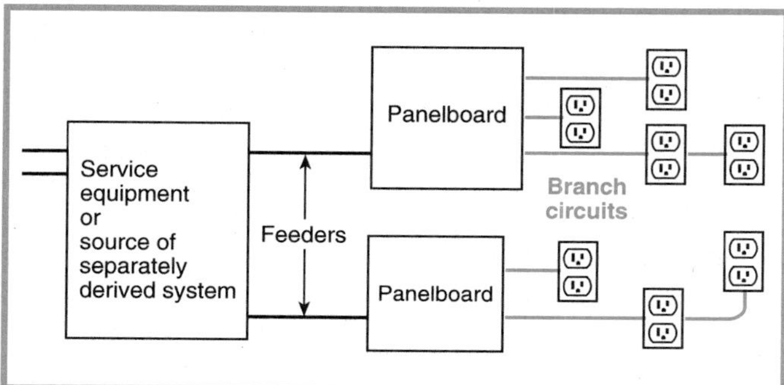

Figure 100.5 Branch circuits and feeders.

receptacle outlets located in such rooms as the kitchen, pantry, etc. These small-appliance branch circuits are not permitted to supply other outlets or permanently connected lighting fixtures. (See Section 210-52 for exact details.)

Branch Circuit, General Purpose. A branch circuit that supplies a number of outlets for lighting and appliances.

Branch Circuit, Individual. A branch circuit that supplies only one utilization equipment.

An *individual branch circuit* is a circuit that supplies *only* one piece of utilization equipment (e.g., one range, one space heater, or one motor). See Section 210-23 regarding permissible loads for branch circuits.

An individual branch circuit supplies *only* one single receptacle for the connection of a single attachment plug. This single receptacle is required to have an ampere rating not less than that of the branch circuit, as stated in Section 210-21(b)(1).

Figure 100.6 illustrates an individual branch circuit with a single receptacle intended for the connection of one piece of utilization equipment.

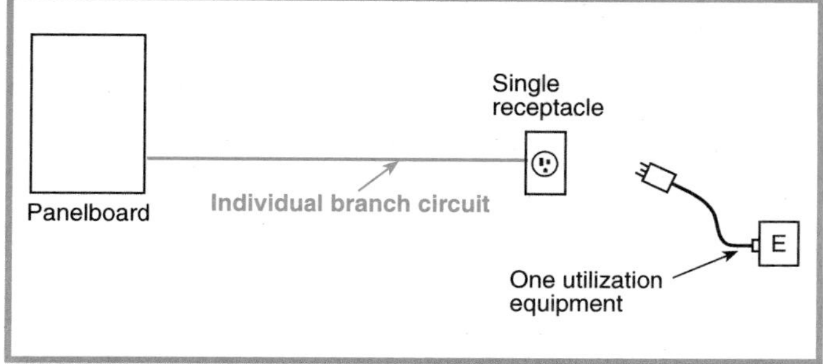

Figure 100.6 An individual branch circuit, which supplies only one utilization equipment via a single receptacle.

A branch circuit that supplies one duplex receptacle that can accommodate two cord- and plug-connected appliances or similar equipment is not an individual branch circuit.

Branch Circuit, Multiwire. A branch circuit that consists of two or more ungrounded conductors that have a potential difference between them, and a grounded conductor that has equal potential difference between it and each ungrounded conductor of the circuit and that is connected to the neutral or grounded conductor of the system.

See Sections 210-4, 240-20(b), and 300-13(b) for specific information about multiwire branch circuits.

Building. A structure that stands alone or that is cut off from adjoining structures by fire walls with all openings therein protected by approved fire doors.

A *building* is generally considered to be a roofed or walled structure that may be used or intended for supporting or sheltering any use or occupancy. However, it may also be a separate structure such as a pole, billboard sign, or water tower.

Definitions of the terms *fire walls* and *fire doors* are the responsibility of building codes.

Generically, a *fire wall* may be defined as a wall that separates buildings or subdivides a building to prevent the spread of fire and that has a fire resistance rating and structural stability. Fire doors (and fire windows) are used to protect openings in walls, floors, and ceilings against the spread of fire and smoke within, into, or out of buildings.

Cabinet. An enclosure designed either for surface mounting or flush mounting and is provided with a frame, mat, or trim in which a swinging door or doors are or can be hung.

Both cabinets and cutout boxes are covered in Article 373. *Cabinets* are designed for surface or flush mounting with a trim to which a swinging door(s) is hung. *Cutout boxes* are designed for surface mounting with a swinging door(s) secured directly to the box. *Panelboards* are electrical assemblies designed to be placed in a cabinet or cutout box. (See definitions of *cutout box* and *panelboard.*)

Circuit Breaker. A device designed to open and close a circuit by nonautomatic means and to open the circuit automatically on a predetermined overcurrent without damage to itself when properly applied within its rating.

FPN: The automatic opening means can be integral, direct acting with the circuit breaker, or remote from the circuit breaker.

Adjustable (as applied to circuit breakers). A qualifying term indicating that the circuit breaker can be set to trip at various values of current, time, or both, within a predetermined range.

Instantaneous Trip (as applied to circuit breakers). A qualifying term indicating that no delay is purposely introduced in the tripping action of the circuit breaker.

Inverse Time (as applied to circuit breakers). A qualifying term indicating that there is purposely introduced a delay in the tripping action of the circuit breaker, which delay decreases as the magnitude of the current increases.

Nonadjustable (as applied to circuit breakers). A qualifying term indicating that the circuit breaker does not have any adjustment to alter the value of current at which it will trip or the time required for its operation.

Setting (of circuit breakers). The value of current, time, or both, at which an adjustable circuit breaker is set to trip.

Concealed. Rendered inaccessible by the structure or finish of the building. Wires in concealed raceways are considered concealed, even though they may become accessible by withdrawing them.

Raceways and cables supported or located within hollow frames or permanently closed in by the finish of buildings are considered concealed. Open-type work — such as raceways and cables in exposed areas; in unfinished basements; in accessible underfloor areas or attics; attached to the surface of finished areas; or behind, above, or below panels designed to allow access and that may be removed without damage to the building structure or finish — is not considered concealed. [See definition of *exposed (as applied to wiring methods)*.]

Conductor.

Bare. A conductor having no covering or electrical insulation whatsoever.

Covered. A conductor encased within material of composition or thickness that is not recognized by this *Code* as electrical insulation.

Typical covered conductors are the green covered equipment grounding conductors in nonmetallic-sheathed cable or the uninsulated grounded system conductors in Type SE cable. Covered conductors should be treated as bare conductors for working clearances, since they are really uninsulated conductors.

Insulated. A conductor encased within material of composition and thickness that is recognized by this *Code* as electrical insulation.

In order for the covering on a conductor to be considered insulation, the conductor with the covering material is generally required to pass minimum testing required by a product standard. One such product standard is UL 83, *Thermoplastic-Insulated Wires and Cables*. In order to meet the requirements of UL 83, specimens of finished single conductor wires must pass specified tests that measure (1) resistance to flame propagation, (2) dielectric strength, even while immersed, and (3) resistance to abrasion, cracking, crushing, and impact. Only wires and cables that meet the minimum fire, electrical, and physical properties required by the applicable standards are permitted to be marked with letter designations found in Tables 310-13 and 310-62. See Section 310-13 for the exact requirements of insulated conductor construction and applications.

Conduit Body. A separate portion of a conduit or tubing system that provides access through a removable cover(s) to the interior of the system at a junction of two or more sections of the system or at a terminal point of the system.

Boxes such as FS and FD or larger cast or sheet metal boxes are not classified as conduit bodies.

Conduit bodies are a portion of a raceway system with removable covers to allow access to the interior of the system. They include the short-radius type as well as capped elbows and service-entrance elbows.

Some conduit bodies are referred to in the trade as "condulets" and include the LB, LL, LR, C, T, and X designs. (See Section 300-15 and Article 370 for rules on the usage of conduit bodies.)

Type FS or FD boxes are not classified as conduit bodies; they are listed with boxes in Table 370-16(a).

Connector, Pressure (Solderless). A device that establishes a connection between two or more conductors or between one or more conductors and a terminal by means of mechanical pressure and without the use of solder.

Continuous Load. A load where the maximum current is expected to continue for 3 hours or more.

Controller. A device or group of devices that serves to govern, in some predetermined manner, the electric power delivered to the apparatus to which it is connected.

A *controller* may be a remote-controlled magnetic contactor, switch, circuit breaker, or device that is normally used to start and stop motors and other apparatus and, in the case of motors, is required to be capable of interrupting the stalled-rotor current of the motor. Stop-and-start stations and similar control circuit components are not considered controllers.

Cooking Unit, Counter-Mounted. A cooking appliance designed for mounting in or on a counter and consisting of one or more heating elements, internal wiring, and built-in or separately mountable controls.

Copper-Clad Aluminum Conductors. Conductors drawn from a copper-clad aluminum rod with the copper metallurgically bonded to an aluminum core. The copper forms a minimum of 10 percent of the cross-sectional area of a solid conductor or each strand of a stranded conductor.

Cutout Box. An enclosure designed for surface mounting that has swinging doors or covers secured directly to and telescoping with the walls of the box proper.

Dead Front. Without live parts exposed to a person on the operating side of the equipment.

Demand Factor. The ratio of the maximum demand of a system, or part of a system, to the total connected load of a system or the part of the system under consideration.

Device. A unit of an electrical system that is intended to carry but not utilize electric energy.

Components (such as switches, circuit breakers, fuseholders, receptacles, attachment plugs, and lampholders) that distribute or control but do not consume electricity are considered devices. See the commentary following Section 210-4(b) for further explanation.

Disconnecting Means. A device, or group of devices, or other means by which the conductors of a circuit can be disconnected from their source of supply.

For disconnecting means for service equipment, see Part F of Article 230; for fuses, see Part D of Article 240; for circuit breakers, see Part G of Article 240; for appliances, see Part D of Article 422; for space-heating equipment, see Part C of Article 424; for motors and controllers, see Part J of Article 430; and for air-conditioning and refrigerating equipment, see Part B of Article 440. (See also references for *disconnecting means* in the Index.)

Dustproof. Constructed or protected so that dust will not interfere with its successful operation.

Dusttight. Constructed so that dust will not enter the enclosing case under specified test conditions.

Table 430-91, Motor Controller Enclosure Selection, provides a basis for selecting enclosure types that are dusttight. (See also the commentary following the definition of *enclosure*.)

Duty.

Continuous Duty. Operation at a substantially constant load for an indefinitely long time.

Intermittent Duty. Operation for alternate intervals of (1) load and no load; or (2) load and rest; or (3) load, no load, and rest.

Periodic Duty. Intermittent operation in which the load conditions are regularly recurrent.

Short-Time Duty. Operation at a substantially constant load for a short and definitely specified time.

Varying Duty. Operation at loads, and for intervals of time, both of which may be subject to wide variation.

Information on the protection of intermittent, periodic, short-time, and varying-duty motors against overload can be found in Section 430-33.

Dwelling.

Dwelling Unit. One or more rooms for the use of one or more persons as a housekeeping unit with space for eating, living, and sleeping, and permanent provisions for cooking and sanitation.

A mobile home may be considered to be a dwelling unit. Where dwelling units are referenced throughout the *Code*, it is important to note that rooms in motels, hotels, and similar occupancies could be classified as dwelling units if they satisfy the requirements of the definition. For example, the motel or hotel room illustrated in Figure 100.7 clearly meets the definition because it has eating, living, and sleeping space and permanent areas for cooking and sanitation.

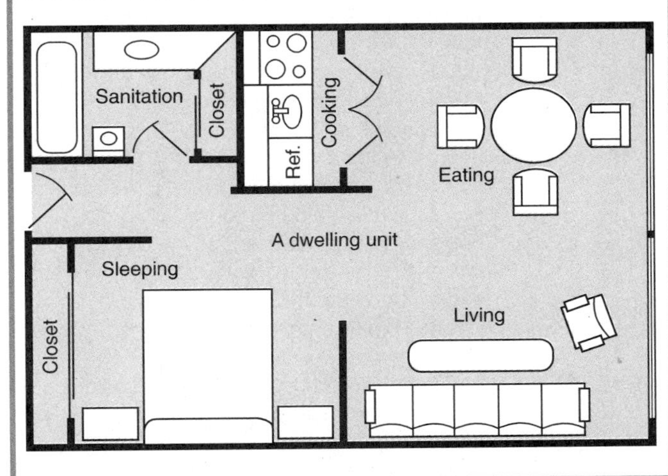

Figure 100.7 *Example of motel or hotel room considered to be a dwelling unit.*

Multifamily Dwelling. A building that contains three or more dwelling units.

One-Family Dwelling. A building that consists solely of one dwelling unit.

Two-Family Dwelling. A building that consists solely of two dwelling units.

Electric Sign. A fixed, stationary, or portable self-contained, electrically illuminated utilization equipment with words or symbols designed to convey information or attract attention.

Enclosed. Surrounded by a case, housing, fence, or walls that prevent persons from accidentally contacting energized parts.

Enclosure. The case or housing of apparatus, or the fence or walls surrounding an installation to prevent personnel from accidentally contacting energized parts, or to protect the equipment from physical damage.

FPN: See Table 430-91 for examples of enclosure types.

The following information is taken from the 1998 UL *General Information Directory* (White Book), category AALZ, "Electrical Equipment for Use in Ordinary Locations." It summarizes the intended uses of the various types of enclosures for nonhazardous locations.

Enclosure Type Number	Provides a Degree of Protection Against the Following Environmental Conditions[1]
1	Indoor use
2	Indoor use, limited amounts of falling water
3R	Outdoor use, undamaged by the formation of ice on the enclosure[2]
3	Same as 3R plus windblown dust
3S	Same as 3R plus windblown dust, with external mechanisms remaining operable while ice laden
4	Outdoor use, splashing water, windblown dust, hose-directed water, undamaged by the formation of ice on the enclosure[2]
4X	Same as 4 plus resists corrosion
5	Indoor use to provide a degree of protection against settling airborne dust, falling dirt, and dripping noncorrosive liquids
6	Same as 3R plus entry of water during temporary submersion at a limited depth
6P	Same as 3R plus entry of water during prolonged submersion at a limited depth
12, 12K	Indoor use, dust, dripping noncorrosive liquids
13	Indoor use, dust, spraying of water, oil, and noncorrosive coolants

[1]All types of enclosures provide a degree of protection against ordinary corrosion and against accidental contact with the enclosed equipment when doors or covers are closed and in place. All types of enclosures provide protection against a limited amount of falling dirt.

[2]All outdoor-type enclosures provide a degree of protection against rain, snow, and sleet. Outdoor enclosures are also suitable for use indoors if they meet the environmental conditions present.

An enclosure that complies with the requirements for more than one type of enclosure may be marked with multiple designations. Enclosures marked with a Type may also be marked as follows.

Type 1 may be marked "Indoor Use Only."

Type 3, 3S, 4, 4X, 6, or 6P may be marked "Raintight."

Type 3R may be marked "Rainproof."

Type 4, 4X, 6, or 6P may be marked "Watertight."

Type 4X or 6P may be marked "Corrosion resistant."

Type 2, 5, 12, 12K, or 13 may be marked "Driptight."

Type 3, 3S, 5, 12K, or 13 may be marked "Dusttight."

For equipment designated "Raintight," testing designed to simulate exposure to a beating rain will not result in entrance of water. For equipment designated "Rainproof," testing designed to simulate exposure to a beating rain will not interfere with the operation of the apparatus or result in wetting of live parts and wiring within the enclosure. "Watertight" equipment is constructed so that water does not enter the enclosure when subjected to a stream of water. "Corrosion resistant" equipment is constructed so that it provides a degree of protection against exposure to corrosive agents such as salt spray. "Driptight" equipment is constructed so that falling moisture or dirt does not enter the enclosure. "Dusttight" equipment is constructed so that circulating or airborne dust does not enter the enclosure.

Energized. Electrically connected to a source of potential difference.

The terms *energized* and *live parts* were added to Article 100 for the 1996 *Code.* [See also the definitions of *exposed (as applied to live parts)* and *live parts.*]

Equipment. A general term including material, fittings, devices, appliances, fixtures, apparatus, and the like used as a part of, or in connection with, an electrical installation.

Explosionproof Apparatus. Apparatus enclosed in a case that is capable of withstanding an explosion of a specified gas or vapor that may occur within it and of preventing the ignition of a specified gas or vapor surrounding the enclosure by sparks, flashes, or explosion of the gas or vapor within, and that operates at such an external temperature that a surrounding flammable atmosphere will not be ignited thereby.

FPN: For further information, see *Explosion-Proof and Dust-Ignition-Proof Electrical Equipment for Use in Hazardous (Classified) Locations,* ANSI/UL 1203-1994.

Exposed (as applied to live parts). Capable of being inadvertently touched or approached nearer than a safe distance by a person. It is applied to parts that are not suitably guarded, isolated, or insulated.

See Section 110-27 for guarding of live parts. (See also definitions of *energized* and *live parts*.)

Exposed (as applied to wiring methods). On or attached to the surface or behind panels designed to allow access.

See Figure 100.1, where wiring methods located behind a suspended ceiling with lift-out panels are considered exposed (as applied to wiring methods).

Externally Operable. Capable of being operated without exposing the operator to contact with live parts.

Feeder. All circuit conductors between the service equipment, the source of a separately derived system, or other power supply source and the final branch-circuit overcurrent device.

Figure 100.5 illustrates the difference between branch circuits and feeders. See also the commentary following the definition of *branch circuit.*

Festoon Lighting. A string of outdoor lights that is suspended between two points.

The definition of *festoon lighting* moved into Article 100 for the 1999 *Code* because it is used in more than one article. Previously, this definition was located in Section 225-6(b).

Fitting. An accessory such as a locknut, bushing, or other part of a wiring system that is intended primarily to perform a mechanical rather than an electrical function.

Items such as condulets, conduit couplings, EMT connectors and couplings, and threadless connectors are considered fittings.

Garage. A building or portion of a building in which one or more self-propelled vehicles carrying volatile flammable liquid for fuel or power are kept for use, sale, storage, rental, repair, exhibition, or demonstrating purposes, and all that portion of a building that is on or below the floor or floors in which such vehicles are kept and that is not separated therefrom by suitable cutoffs.

FPN: For commercial garages, repair, and storage, see Section 511-1.

Ground. A conducting connection, whether intentional or accidental, between an electrical circuit or equipment and the earth, or to some conducting body that serves in place of the earth.

Grounded. Connected to earth or to some conducting body that serves in place of the earth.

Grounded, Effectively. Intentionally connected to earth through a ground connection or connections of sufficiently low impedance and having sufficient current-carrying capacity to prevent the buildup of voltages that may result in undue hazards to connected equipment or to persons.

Grounded Conductor. A system or circuit conductor that is intentionally grounded.

Grounding Conductor. A conductor used to connect equipment or the grounded circuit of a wiring system to a grounding electrode or electrodes.

Grounding Conductor, Equipment. The conductor used to connect the noncurrent-carrying metal parts of equipment, raceways, and other enclosures to the system grounded conductor, the grounding electrode conductor, or both, at the service equipment or at the source of a separately derived system.

See Section 250-118 for types of equipment grounding conductors.

Grounding Electrode Conductor. The conductor used to connect the grounding electrode to the equipment grounding conductor, to the grounded conductor, or to both, of the circuit at the service equipment or at the source of a separately derived system.

The grounding electrode conductor is required to be copper, aluminum, or copper-clad aluminum. It is used to connect the equipment grounding conductor or the grounded conductor (at the service equipment or at the separately derived system) to the grounding electrode for either grounded or ungrounded systems. (See Figures 100.4 and 250.1, where the grounding electrode conductor is shown in a typical grounding system for a single-phase, 3-wire service.)

The grounding electrode conductor is sized according to Table 250-66. (See also Article 250, Part C.)

Ground-Fault Circuit Interrupter. A device intended for the protection of personnel that functions to de-energize a circuit or portion thereof within an established period of time when a current to ground exceeds some predetermined value that is less than that required to operate the overcurrent protective device of the supply circuit.

The commentary following Section 210-8 contains a list of applicable cross references for ground-fault circuit interrupters (GFCIs). Figures 210.9 through 210.17 contain specific information regarding the requirements for GFCIs.

A *ground-fault circuit interrupter* is a device that is designed to protect personnel against electrocution. The device immediately shuts off the electric current when a difference of 5 milliamperes, plus or minus 1 milliampere, between the two circuit conductors is detected.

Ground-Fault Protection of Equipment. A system intended to provide protection of equipment from damaging line-to-ground fault currents by operating to cause a disconnecting means to open all ungrounded conductors of the faulted circuit. This protection is provided at current levels less than those required to protect conductors from damage through the operation of a supply circuit overcurrent device.

See the commentary on Sections 230-95, 426-28, and 427-22.

Guarded. Covered, shielded, fenced, enclosed, or otherwise protected by means of suitable covers, casings, barriers, rails, screens, mats, or platforms to remove the likelihood of approach or contact by persons or objects to a point of danger.

Hoistway. Any shaftway, hatchway, well hole, or other vertical opening or space in which an elevator or dumbwaiter is designed to operate.

See Article 620 for the installation of electrical equipment and wiring methods in hoistways.

Identified (as applied to equipment). Recognizable as suitable for the specific purpose, function, use, environment, application, etc., where described in a particular *Code* requirement.

FPN: Suitability of equipment for a specific purpose, environment, or application may be determined by a qualified testing laboratory, inspection agency, or other organization concerned with product evaluation. Such identification may include labeling or listing. (See definitions of *Labeled* and *Listed*.)

In Sight From (Within Sight From, Within Sight). Where this *Code* specifies that one equipment shall be in sight from, within sight from, or within sight, etc., of another equipment, the specified equipment is to be visible and not more than 50 ft (15.24 m) distant from the other.

Figures 430.20, 430.21, and 600.1 depict requirements for the placement of a disconnecting means that is not in sight.

Interrupting Rating. The highest current at rated voltage that a device is intended to interrupt under standard test conditions.

FPN: Equipment intended to interrupt current at other than fault levels may have its interrupting rating implied in other ratings, such as horsepower or locked rotor current.

Interrupting ratings are essential in the coordination of electrical systems so that available fault currents can be properly controlled. Other sections specifically dealing with interrupting ratings are Sections 110-9, 240-60(c), 240-83(c), and 240-86.

Isolated (as applied to location). Not readily accessible to persons unless special means for access are used.

See the definition of *accessible, readily.*

Labeled. Equipment or materials to which has been attached a label, symbol, or other identifying mark of an organization that is acceptable to the authority having jurisdiction and concerned with product evaluation, that maintains periodic inspection of production of labeled equipment or materials, and by whose labeling the manufacturer indicates compliance with appropriate standards or performance in a specified manner.

Equipment and conductors required or permitted by this *Code* are acceptable only if they have been approved for a specific environment or application by the authority having jurisdiction, as stated in Section 110-2.

See Section 90-7 regarding the examination of equipment for safety. Listing or labeling by a qualified testing laboratory will provide a basis for approval.

Lighting Outlet. An outlet intended for the direct connection of a lampholder, a lighting fixture, or a pendant cord terminating in a lampholder.

Listed. Equipment, materials, or services included in a list published by an organization that is acceptable to the authority having jurisdiction and concerned with evaluation of products or services, that maintains periodic inspection of production of listed equipment or materials or periodic evaluation of services, and whose listing states that either the equipment, material, or services meets identified standards or has been tested and found suitable for a specified purpose.

FPN: The means for identifying listed equipment may vary for each organization concerned with product evaluation, some of which do not recognize equipment as listed unless it is also labeled. Use of the system employed by the listing organization allows the authority having jurisdiction to identify a listed product.

The definition of *listed* and the accompanying FPN were revised for the 1999 *Code*. This slightly reworded definition is now more in line with the definition of *listed* found in the NFPA Regulations Governing Committee Projects.

See the commentary following the definition of *labeled*.

Live Parts. Electric conductors, buses, terminals, or components that are uninsulated or exposed and a shock hazard exists.

The term *live parts* was added to Article 100 in the 1996 *Code*. The term *live parts* indicates that a shock hazard exists due to electrical components such as terminals, busbars, and fuseholders that are not insulated or protected by barriers.

Location.

Damp Location. Partially protected locations under canopies, marquees, roofed open porches, and like locations, and interior locations subject to moderate degrees of moisture, such as some basements, some barns, and some cold-storage warehouses.

Dry Location. A location not normally subject to dampness or wetness. A location classified as dry may be temporarily subject to dampness or wetness, as in the case of a building under construction.

Wet Location. Installations underground or in concrete slabs or masonry in direct contact with the earth, and locations subject to saturation with water or other liquids, such as vehicle washing areas, and locations exposed to weather and unprotected.

It is intended that the inside of a raceway in a wet location or a raceway installed underground be considered a wet location. Therefore, any conductors contained therein would be required to be suitable for wet locations.

See Section 300-6(c) for some examples of wet locations and Section 410-4(a) for information on fixtures installed in wet locations.

See "patient care area" in Section 517-3 for a definition of wet locations in a patient care area.

Motor Control Center. An assembly of one or more enclosed sections having a common power bus and principally containing motor control units.

Multioutlet Assembly. A type of surface, flush, or freestanding raceway; designed to hold conductors and receptacles, assembled in the field or at the factory.

Revised for the 1999 *Code*, the definition of *multioutlet assembly* now includes a reference to a freestanding assembly with multiple outlets, commonly called a *power pole*.

In dry locations, metallic and nonmetallic multioutlet assemblies are permitted; however, they are not permitted to be installed if concealed. See Article 353 for details on recessing these assemblies.

Nonautomatic. Action requiring personal intervention for its control. As applied to an electric controller, nonautomatic control does not necessarily imply a manual controller, but only that personal intervention is necessary.

Nonincendive Circuit. A circuit, other than field wiring, in which any arc or thermal effect produced under intended operating conditions of the equipment, is not capable, under specified test conditions, of igniting the flammable gas–, vapor–, or dust–air mixture.

A nonincendive circuit employs a protection technique that prevents electrical circuits from causing a fire or explosion in a hazardous (classified) location. A careful reading of Section 500-4(f) points out that this protection technique is only permitted for Division 2 areas of Class I or Class II locations. Nonincendive circuits are not permitted in Division 1 areas of Class I or Class II locations. Nonincendive circuits are sometimes referred to as intrinsically safe circuits for Division 2 locations only. Nonincendive circuits and equipment are tested or evaluated essentially in the same way that intrinsically safe circuits and equipment are tested and evaluated, except that abnormal conditions are not considered. Because of its definition, a nonincendive circuit is a low-energy circuit.

FPN: For test conditions, see *Nonincendive Electrical Equipment for Use in Class I and II, Division 2 and Class III, Divisions 1 and 2 Hazardous (Classified) Locations,* ANSI/ISA-S12.12-1994.

Nonincendive Field Wiring. Wiring that enters or leaves an equipment enclosure and, under normal operating conditions of the equipment, is not capable, due to arcing or thermal effects, of igniting the flammable gas–, vapor–, or dust–air mixture. Normal operation includes opening, shorting, or grounding the field wiring.

The definition of *nonincendive field wiring* was added to the 1999 *Code* to alert users that although the circuits and equipment may have been evaluated and approved as nonincendive, field wiring is not generally approved by a testing laboratory. Field wiring meeting this definition would require limitations of energy on the wiring under conditions such as opening, shorting, or grounding. For example, stored energy in the form of mutual inductance or capacitance could be released during an opening, shorting, or grounding of nonincendive field wiring, thus defeating the purpose of this protection technique. Further information regarding the nonincendive protection techniques may be found in the definition of *nonincendive circuit* in Article 100 and also in Section 500-4(f).

Nonlinear Load. A load where the wave shape of the steady-state current does not follow the wave shape of the applied voltage.

FPN: Electronic equipment, electronic/electric-discharge lighting, adjustable-speed drive systems, and similar equipment may be nonlinear loads.

The definition of *nonlinear load* was added to Article 100 in the 1996 *Code*. Nonlinear loads are a major cause of harmonic currents in modern circuits. Information on the undesirable operational effects that are often associated with harmonic currents, including additional heating, were also added in the 1996 *Code*. The FPN following Section 310-10 points out that harmonic current, as well as fundamental current, should be used in determining the heat generated internally in a conductor. Also, limited use of parallel neutrals is permitted in Section 310-4, Exception No. 4.

Actual circuit measurements of current for nonlinear loads should be made using only true rms-measuring ammeter instruments. Averaging ammeters will produce inaccurate values if used to measure nonlinear loads. [See Section 310-15(b)(4)(c) and associated commentary.]

Outlet. A point on the wiring system at which current is taken to supply utilization equipment.

An example is a lighting outlet or a receptacle outlet.

Outline Lighting. An arrangement of incandescent lamps or electric-discharge lighting to outline or call attention to certain features such as the shape of a building or the decoration of a window.

See Articles 410 and 600 for details on outline lighting.

Oven, Wall-Mounted. An oven for cooking purposes and consisting of one or more heating elements, internal wiring, and built-in or separately mountable controls.

Overcurrent. Any current in excess of the rated current of equipment or the ampacity of a conductor. It may result from overload, short circuit, or ground fault.

FPN: A current in excess of rating may be accommodated by certain equipment and conductors for a given set of conditions. Therefore the rules for overcurrent protection are specific for particular situations.

Overload. Operation of equipment in excess of normal, full-load rating, or of a conductor in excess of rated ampacity that, when it persists for a sufficient length of time, would cause damage or dangerous overheating. A fault, such as a short circuit or ground fault, is not an overload.

Panelboard. A single panel or group of panel units designed for assembly in the form of a single panel; including buses, automatic overcurrent devices, and equipped with or without switches for the control of light, heat, or power circuits; designed to be placed in a cabinet or cutout box placed in or against a wall or partition and accessible only from the front.

See Article 384 for details on panelboards.

Plenum. A compartment or chamber to which one or more air ducts are connected and that forms part of the air distribution system.

The definition of *plenum* is essentially the same as the definition of *plenum* in *Standard for the Installation of Air Conditioning and Ventilating Systems,* NFPA 90A-1996. For information on wiring methods permitted within plenums, see Section 300-22(b).

The definition of *plenum* is not intended to apply to the space above a suspended ceiling that is used for environmental air referred to in Section 300-22(c).

Power Outlet. An enclosed assembly that may include receptacles, circuit breakers, fuseholders, fused switches, buses, and watt-hour meter mounting means; intended to supply and control power to mobile homes, recreational vehicles, park trailers, or boats; or to serve as a means for distributing power required to operate mobile or temporarily installed equipment.

See Figure 550.1 for an example of a power outlet assembly.

Premises Wiring (System). That interior and exterior wiring, including power, lighting, control, and signal circuit wiring together with all of their associated hardware, fittings, and wiring devices, both permanently and temporarily installed, that extends from the service point of utility conductors or source of power such as a battery, a solar photovoltaic system, or a generator, transformer, or converter windings, to the outlet(s). Such wiring does not include wiring internal to appliances, fixtures, motors, controllers, motor control centers, and similar equipment.

The definition of *premises wiring system* was revised for the 1999 *Code* by removing the limiting term *separately derived system* and expanding the definition to be more inclusive. The definition now specifically includes sources of power such as a battery, a solar photovoltaic system, a generator, a transformer, and converter windings.

It is important to the understanding of this definition to know that a premises wiring system does not include the internal wiring of electrical equipment, including that of the transformer.

Qualified Person. One familiar with the construction and operation of the equipment and the hazards involved.

Raceway. An enclosed channel of metal or nonmetallic materials designed expressly for holding wires, cables, or busbars, with additional functions as permitted in this *Code*. Raceways include, but are not limited to, rigid metal conduit, rigid nonmetallic conduit, intermediate metal conduit, liquidtight flexible conduit, flexible metallic tubing, flexible metal conduit, electrical nonmetallic tubing, electrical metallic tubing, underfloor raceways, cellular concrete floor raceways, cellular metal floor raceways, surface raceways, wireways, and busways.

Cable trays (see Article 318) are not considered raceways in the *NEC*.

Rainproof. Constructed, protected, or treated so as to prevent rain from interfering with the successful operation of the apparatus under specified test conditions.

See the commentary following the definition of *enclosure*.

Raintight. Constructed or protected so that exposure to a beating rain will not result in the entrance of water under specified test conditions.

Raceways on exterior surfaces of buildings are required to be made raintight (see Sections 225-22 and 230-53). For boxes and cabinets, see Section 300-6. Also see the commentary following the definitions of *wet location* and *enclosure*.

Receptacle. A receptacle is a contact device installed at the outlet for the connection of an attachment plug. A single receptacle is a single contact device with no other contact device on the same yoke. A multiple receptacle is two or more contact devices on the same yoke.

The definition of *receptacle* was editorially revised for the 1999 *Code*. Figure 100.8 shows one single and two multiple receptacles.

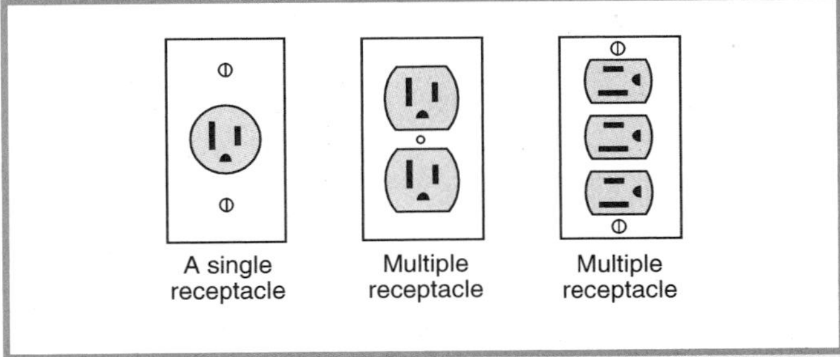

Figure 100.8 *Receptacles.*

Receptacle Outlet. An outlet where one or more receptacles are installed.

See Figure 100.8 and the commentary following Section 220-3(c).

Remote-Control Circuit. Any electric circuit that controls any other circuit through a relay or an equivalent device.

Figure 100.9 illustrates a remote-control circuit that starts and stops an electric motor.

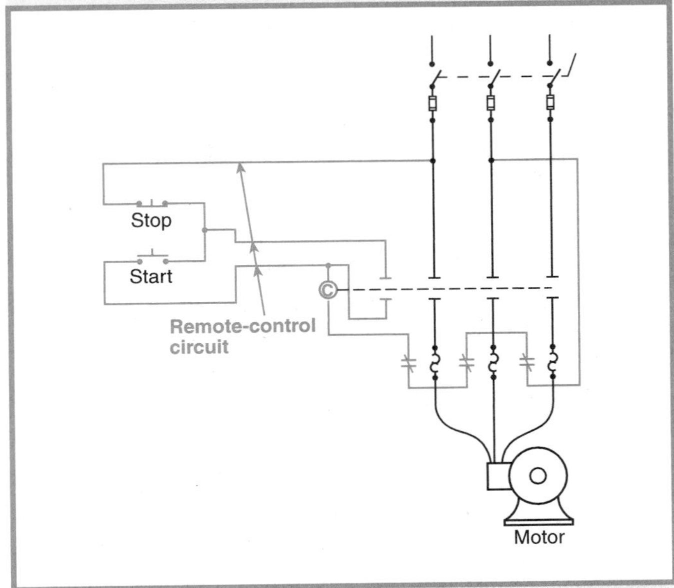

Figure 100.9 *A remote-control circuit for starting and stopping an electric motor.*

Sealable Equipment. Equipment enclosed in a case or cabinet that is provided with a means of sealing or locking so that live parts cannot be made accessible without opening the enclosure. The equipment may or may not be operable without opening the enclosure.

Separately Derived System. A premises wiring system whose power is derived from a battery, a solar photovoltaic system, or from a generator, transformer, or converter windings, and that has no direct electrical connection, including a solidly connected grounded circuit conductor, to supply conductors originating in another system.

Service. The conductors and equipment for delivering electric energy from the serving utility to the wiring system of the premises served.

The definition of *service* was modified for the 1999 *Code* to state that electric energy to a service can only be supplied by the serving utility. If electric energy is supplied by other than the serving utility, the supplied conductors and equipment could not be considered a service.

Service Cable. Service conductors made up in the form of a cable.

Service Conductors. The conductors from the service point to the service disconnecting means.

The definition of *service conductor* was revised in the 1999 *Code* to be more precise. The phrase "or other source of power" was deleted. The service conductors originate at the service point (where the serving utility ends) and they end at the service disconnect. Due to the associated 1999 *Code* change in the definition of *service,* these service conductors may originate only from the serving utility.

Service conductors is a broad term and may include service drops, service laterals, and service-entrance conductors. But, this term specifically excludes any wiring on the supply side (serving utility side) of the service point.

If the utility has specified that the service point is at the utility pole, then the service conductors from an overhead distribution system originate at the utility pole and terminate at the service disconnecting means.

If the utility has specified that the service point is at the utility manhole, then the service conductors from an underground distribution system originate at the utility manhole and terminate at the service disconnecting means. Where utility-owned primary conductors are extended to outdoor pad-mounted transformers on private property, the service conductors originate at the secondary connections of the transformers, only if the utility has specified that the service point is at the secondary connections.

See Article 230, Part H, and the commentary following Section 230-200 for service conductors exceeding 600 volts, nominal.

Service Drop. The overhead service conductors from the last pole or other aerial support to and including the splices, if any, connecting to the service-entrance conductors at the building or other structure.

In Figure 100.10, the overhead service-drop conductors run from the utility pole and connect to the service-entrance conductors at the service point. Conductors on the utility side of the service point are not covered by the *NEC.* The utility specifies the location of the service point. Exact locations of the service point may vary from utility to utility, as well as from occupancy to occupancy.

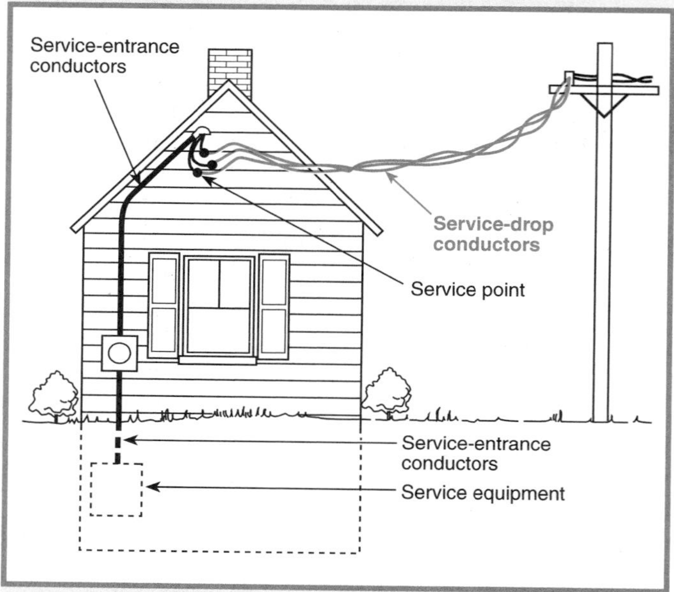

Figure 100.10 *Overhead system showing a service drop from a utility pole to attachment on a house and service-entrance conductors from point of attachment (spliced to service-drop conductors), down the side of the house, through the meter socket, and terminating within the service equipment.*

Service-Entrance Conductors, Overhead System. The service conductors between the terminals of the service equipment and a point usually outside the building, clear of building walls, where joined by tap or splice to the service drop.

See Figure 100.10 for an illustration of service-entrance conductors in an overhead system.

Service-Entrance Conductors, Underground System. The service conductors between the terminals of the service equipment and the point of connection to the service lateral.

See Figure 100.11 for an illustration of service-entrance conductors in an underground system.

FPN: Where service equipment is located outside the building walls, there may be no service-entrance conductors, or they may be entirely outside the building.

Service Equipment. The necessary equipment, usually consisting of a circuit breaker(s) or switch(es) and fuse(s) and their accessories, connected to the load end of service conductors to a building or other structure, or an otherwise designated area, and intended to constitute the main control and cutoff of the supply.

The definition of *service equipment* was revised for the 1999 *Code*. In addition to editorial corrections, the phrase "connected to the load end of service conductors" was substituted for the phrase "located near the point of entrance." This change is in concert with the other associated changes made to the definitions of *service* and *service conductors*.

Service equipment may consist of circuit breakers or fused switches provided to disconnect all ungrounded conductors in a building or other structure from the service-entrance conductors.

The disconnecting means at any *one* location is not allowed to consist of more than six circuit breakers or six switches and is required to be readily accessible either outside the building or structure or inside nearest the point of entrance of the service-entrance conductors. See Section 230-6 and Article 230, Part F, for service conductors outside the building and service disconnecting means, respectively.

Service Lateral. The underground service conductors between the street main, including any risers at a pole or other structure or from transformers, and the first point of connection to the service-entrance conductors in a terminal box or meter or other enclosure, inside or outside the building wall. Where there is no terminal box, meter, or other enclosure, the point of connection shall be considered to be the point of entrance of the service conductors into the building.

The definition of *service lateral* was revised for the 1999 *Code* eliminating the references to adequate space.

As Figure 100.11 shows, the underground service laterals may be run from poles or from transformers and with or without terminal boxes, provided they begin at the service point. Conductors on the utility side of the service point are not covered by the *NEC*. The utility specifies the location of the service point. Exact locations of the service point may vary from utility to utility, as well as from occupancy to occupancy.

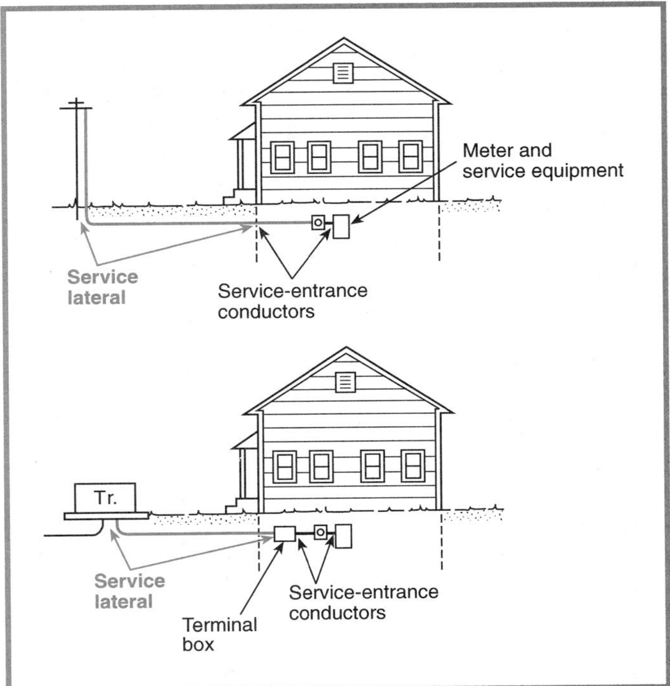

Figure 100.11 *Underground systems showing service laterals run from a pole and from a transformer.*

Service Point. The point of connection between the facilities of the serving utility and the premises wiring.

The *service point* is the point of demarcation between the serving utility and the premises wiring. The *service point* is the point on the wiring system where the serving utility ends and where the premises wiring begins. The serving utility generally specifies the location of the service point.

Because the location of the service point is generally determined by the utility, the service-drop conductors and the service-lateral conductors may or may not be part of the service as covered by the *NEC*. In order for these types of conductors to be covered, they must be physically located on the premises wiring side of the service point. If these conductors are located on the utility side of the service point, they are not covered in the definition of *service conductors* and, therefore, are not covered by the *NEC*.

Generally, based on the definitions of the terms *service point* and *service conductors*, any conductor on the serving utility side of the service point is not covered by the *NEC*. For example, a typical suburban residence has an overhead service drop from the utility pole to the house. If the utility specifies that the service point is at the point of attachment of the service drop to the house, then the service-drop conductors are not considered service conductors in this case because the service drop is not on the premises wiring side of the service point. Alternately, if the service point is specified as "at the pole" by the utility, then the service-drop conductors are considered service conductors, and the *NEC* would apply to the service drop.

Exact locations for a service point may vary from utility to utility, as well as from occupancy to occupancy.

Show Window. Any window used or designed to be used for the display of goods or advertising material, whether it is fully or partly enclosed or entirely open at the rear and whether or not it has a platform raised higher than the street floor level.

See Section 220-12(a) and Figure 220.1 for show-window lighting load requirements.

Signaling Circuit. Any electric circuit that energizes signaling equipment.

Solar Photovoltaic System. The total components and subsystems that, in combination, convert solar energy into electrical energy suitable for connection to a utilization load.

See Article 690 for solar photovoltaic system requirements.

Special Permission. The written consent of the authority having jurisdiction.

The authority having jurisdiction for enforcement of the *Code* is responsible for making interpretations and granting special permission contemplated in a number of the rules. For examples, see Sections 110-26(a)(1), Exception No. 2; 230-2(b); and 426-14.

Switchboard. A large single panel, frame, or assembly of panels on which are mounted, on the face or back, or both, switches, overcurrent and other protective devices, buses, and usually instruments. Switchboards are generally accessible from the rear as well as from the front and are not intended to be installed in cabinets.

Busbars are to be arranged to avoid inductive overheating. Service busbars are required to be isolated by barriers from the remainder of the switchboard. Most modern switchboards are totally enclosed to minimize the probability of spreading fire to adjacent combustible materials and to guard live parts.

Switches.

Bypass Isolation Switch. A manually operated device used in conjunction with a transfer switch to provide a means of directly connecting load conductors to a power source, and of disconnecting the transfer switch.

See Sections 700-6 and 701-7(b) for further information on bypass isolation switches.

General-Use Snap Switch. A form of general-use switch constructed so that it can be installed in device boxes or on box covers, or otherwise used in conjunction with wiring systems recognized by this *Code*.

General-Use Switch. A switch intended for use in general distribution and branch circuits. It is rated in amperes, and it is capable of interrupting its rated current at its rated voltage.

Isolating Switch. A switch intended for isolating an electric circuit from the source of power. It has no interrupting rating, and it is intended to be operated only after the circuit has been opened by some other means.

Motor-Circuit Switch. A switch rated in horsepower that is capable of interrupting the maximum operating overload current of a motor of the same horsepower rating as the switch at the rated voltage.

Transfer Switch. An automatic or nonautomatic device for transferring one or more load conductor connections from one power source to another.

Thermally Protected (as applied to motors). The words "Thermally Protected" appearing on the nameplate of a motor or motor-compressor indicate that the motor is provided with a thermal protector.

Thermal Protector (as applied to motors). A protective device for assembly as an integral part of a motor or motor-compressor that, when properly applied, protects the motor against dangerous overheating due to overload and failure to start.

FPN: The thermal protector may consist of one or more sensing elements integral with the motor or motor-compressor and an external control device.

Utilization Equipment. Equipment that utilizes electric energy for electronic, electromechanical, chemical, heating, lighting, or similar purposes.

Ventilated. Provided with a means to permit circulation of air sufficient to remove an excess of heat, fumes, or vapors.

See the commentary following Section 110-13(b).

Volatile Flammable Liquid. A flammable liquid having a flash point below 38°C (100°F), or a flammable liquid whose temperature is above its flash point, or a Class II combustible liquid having a vapor pressure not exceeding 40 psia (276 kPa) at 38°C (100°F) whose temperature is above its flash point.

The *flash point* of a liquid is the minimum temperature at which it gives off sufficient vapor to form an ignitible mixture with the air near the surface of the liquid or within the vessel used to contain the liquid. An *ignitible mixture* is a mixture within the explosive or flammable range (between upper and lower limits) that is capable of the propagation of flame away from the source of ignition when ignited. Some emission of vapors takes place below the flash point but not in sufficient quantities to form an ignitible mixture.

Voltage (of a circuit). The greatest root-mean-square (rms) (effective) difference of potential between any two conductors of the circuit concerned.

FPN: Some systems, such as 3-phase 4-wire, single-phase 3-wire, and 3-wire direct current, may have various circuits of various voltages.

Common 3-phase, 4-wire wye systems are 480/277 volts and 208/120 volts. The *voltage of the circuit* is the highest voltage between any two conductors (i.e., 480 volts or 208 volts). The voltage of the circuit of a 2-wire feeder or branch circuit (single phase and the grounded conductor) derived from these systems would be the voltage between the two conductors at the lower voltage (i.e., 277 volts or 120 volts). The same applies to dc or single-phase, 3-wire systems where there are two voltages.

Voltage, Nominal. A nominal value assigned to a circuit or system for the purpose of conveniently designating its voltage class (e.g., 120/240 volts, 480Y/277 volts, 600 volts). The actual voltage at which a circuit operates can vary from the nominal within a range that permits satisfactory operation of equipment.

See Section 220-2, which lists the nominal voltages used in computing branch-circuit and feeder loads.

FPN: See *Voltage Ratings for Electric Power Systems and Equipment (60 Hz),* ANSI C84.1-1995.

Voltage to Ground. For grounded circuits, the voltage between the given conductor and that point or conductor of the circuit that is grounded; for ungrounded circuits, the greatest voltage between the given conductor and any other conductor of the circuit.

The voltage to ground of a 277/480-volt wye system would be 277 volts; of a 120/208-volt wye system, 120 volts; and of a 3-phase, 3-wire ungrounded 480-volt system, 480 volts.

For a 3-phase, 4-wire delta system, with the center of one leg grounded, there are two voltages to ground. For example, on a 240-volt system, two legs would each have 120 volts to ground and the third, or "high" leg, would have 208 volts to ground. See Sections 215-8, 230-56, and 384-3(e) for special marking on such circuits.

Watertight. Constructed so that moisture will not enter the enclosure under specified test conditions.

Unless the enclosure is hermetically sealed, it is possible for moisture to enter the enclosure. See the commentary after the definition of *enclosure* and after Table 430-91.

Weatherproof. Constructed or protected so that exposure to the weather will not interfere with successful operation.

See the commentary after the definition of *enclosure* and after Table 430-91.

FPN: Rainproof, raintight, or watertight equipment can fulfill the requirements for weatherproof where varying weather conditions other than wetness, such as snow, ice, dust, or temperature extremes, are not a factor.

B. Over 600 Volts, Nominal

Whereas the preceding definitions are intended to apply wherever the terms are used throughout this *Code*, the following definitions are applicable only to parts of the article specifically covering installations and equipment operating at over 600 volts, nominal.

Electronically Actuated Fuse. An overcurrent protective device that generally consists of a control module that provides current sensing, electronically derived time-current characteristics, energy to initiate tripping, and an interrupting module that interrupts current when an overcurrent occurs. Electronically actuated fuses may or may not operate in a current-limiting fashion, depending on the type of control selected.

Although they are called fuses because they interrupt current by melting a fusible element, electronically actuated fuses respond to a signal from an electronic control rather than from the heat generated by actual current passing through a fusible element. Electronically actuated fuses have controls similar to electronic circuit breakers.

Fuse. An overcurrent protective device with a circuit-opening fusible part that is heated and severed by the passage of overcurrent through it.

FPN: A fuse comprises all the parts that form a unit capable of performing the prescribed functions. It may or may not be the complete device necessary to connect it into an electrical circuit.

Controlled Vented Power Fuse. A fuse with provision for controlling discharge circuit interruption such that no solid material may be exhausted into the surrounding atmosphere.

FPN: The fuse is designed so that discharged gases will not ignite or damage insulation in the path of the discharge or propagate a flashover to or between grounded members or conduction members in the path of the discharge where the distance between the vent and such insulation or conduction members conforms to manufacturer's recommendations.

Expulsion Fuse Unit (Expulsion Fuse). A vented fuse unit in which the expulsion effect of gases produced by the arc and lining of the fuseholder, either alone or aided by a spring, extinguishes the arc.

Nonvented Power Fuse. A fuse without intentional provision for the escape of arc gases, liquids, or solid particles to the atmosphere during circuit interruption.

Power Fuse Unit. A vented, nonvented, or controlled vented fuse unit in which the arc is extinguished by being drawn through solid material, granular material, or liquid, either alone or aided by a spring.

Vented Power Fuse. A fuse with provision for the escape of arc gases, liquids, or solid particles to the surrounding atmosphere during circuit interruption.

Multiple Fuse. An assembly of two or more single-pole fuses.

Switching Device. A device designed to close, open, or both, one or more electric circuits.

Switching Devices.

Circuit Breaker. A switching device capable of making, carrying, and breaking currents under normal circuit conditions, and also making, carrying for a specified time, and breaking currents under specified abnormal circuit conditions, such as those of short circuit.

Cutout. An assembly of a fuse support with either a fuseholder, fuse carrier, or disconnecting blade. The fuseholder or fuse carrier may include a conducting element (fuse link), or may act as the disconnecting blade by the inclusion of a nonfusible member.

Disconnecting (or Isolating) Switch (Disconnector, Isolator). A mechanical switching device used for isolating a circuit or equipment from a source of power.

Disconnecting Means. A device, group of devices, or other means whereby the conductors of a circuit can be disconnected from their source of supply.

Interrupter Switch. A switch capable of making, carrying, and interrupting specified currents.

Oil Cutout (Oil-Filled Cutout). A cutout in which all or part of the fuse support and its fuse link or disconnecting blade is mounted in oil with complete immersion of the contacts and the fusible portion of the conducting element (fuse link) so that arc interruption by severing of the fuse link or by opening of the contacts will occur under oil.

Oil Switch. A switch having contacts that operate under oil (or askarel or other suitable liquid).

Regulator Bypass Switch. A specific device or combination of devices designed to bypass a regulator.

Article 110 — Requirements for Electrical Installations

Contents

A. General

110-2. Approval. The conductors and equipment required or permitted by this *Code* shall be acceptable only if approved.

FPN: See Examination of Equipment for Safety, Section 90-7, and Examination, Identification, Installation, and Use of Equipment, Section 110-3. See definitions of *Approved, Identified, Labeled,* and *Listed.*

Section 110-2 of the *Code* requires that all equipment be approved as defined in Article 100 and, as such, be acceptable to the authority having jurisdiction. Section 110-3 provides guidance for the evaluation of equipment and recognizes listing or labeling as a means of establishing suitability.

Approval of equipment is the responsibility of the electrical inspection authority, and many such approvals are based on tests and listings of testing laboratories.

110-3. Examination, Identification, Installation, and Use of Equipment.

(a) Examination. In judging equipment, considerations such as the following shall be evaluated:

(1) Suitability for installation and use in conformity with the provisions of this *Code*

FPN: Suitability of equipment use may be identified by a description marked on or provided with a product to identify the suitability of the product for a specific purpose, environment, or application. Suitability of equipment may be evidenced by listing or labeling.

(2) Mechanical strength and durability, including, for parts designed to enclose and protect other equipment, the adequacy of the protection thus provided
(3) Wire-bending and connection space
(4) Electrical insulation
(5) Heating effects under normal conditions of use and also under abnormal conditions likely to arise in service
(6) Arcing effects
(7) Classification by type, size, voltage, current capacity, and specific use
(8) Other factors that contribute to the practical safeguarding of persons using or likely to come in contact with the equipment

For wire-bending and connection space in cabinets and cutout boxes, see Section 373-6, Tables 373-6(a) and (b), and Sections 373-7, 373-9, and 373-11. For wire-bending and connection space in other equipment, see the appropriate *NEC* article and section. For example, see Sections 370-16 and 370-28 for outlet, device, pull, and junction boxes, as well as conduit bodies; Sections 380-3 and 380-18 for switches; Section 384-3(g) for switchboards and panelboards; and Section 430-10 for motors and motor controllers.

(b) Installation and Use. Listed or labeled equipment shall be installed and used in accordance with any instructions included in the listing or labeling.

Manufacturers usually supply installation instructions with equipment for use by general contractors, erectors, electrical contractors, electrical inspectors, and others concerned with an installation. It is most important to follow the listing or labeling installation instructions. For example, Section 210-52, second paragraph, permits permanently installed electric baseboard heaters to be equipped with receptacle outlets that meet the requirements for the wall space utilized by such heaters. The installation instructions for such permanent baseboard heaters indicate that these heaters should not be mounted beneath a receptacle. In dwelling units, it is very common to use low-density heat units that may measure in excess of 12 ft in length. Therefore, to meet the provisions of Section 210-52(a), and also the installation instructions, a receptacle must either be part of the heating unit or be installed in the floor close to the wall, but not above the heating unit. (See Section 210-52, FPN, and Figure 210.24 for more specific details.)

Section 110-3(b) does not, in itself, require listing or labeling. It does, however, require considerable evaluation of equipment. Section 110-2 requires that equipment be acceptable only if approved. The term *approved* is defined in Article 100 as acceptable to the authority having jurisdiction. Before issuing approval, the authority having jurisdiction may require evidence of compliance with Section 110-3(a). The most common form of this evidence that is considered acceptable by authorities having jurisdiction is a listing or labeling by a third party.

There are some sections in the *Code* that require listed or labeled equipment, such as Section 250-8, which includes the phrase "listed pressure connectors, listed clamps, or other listed means."

110-4. Voltages. Throughout this *Code*, the voltage considered shall be that at which the circuit operates. The voltage rating of electrical equipment shall not be less than the nominal voltage of a circuit to which it is connected.

Voltages used for computing branch-circuit and feeder loads are nominal voltages as listed in Section 220-2. See the definitions of *voltage (of a circuit); voltage, nominal;* and *voltage to ground* in Article 100. See also Sections 300-2 and 300-3, which specify the voltage limitations of conductors of circuits rated 600 volts, nominal, or less, and over 600 volts, nominal.

110-5. Conductors. Conductors normally used to carry current shall be of copper unless otherwise provided in this *Code*. Where the conductor material is not specified, the material and the sizes given in this *Code* shall apply to copper conductors. Where other materials are used, the size shall be changed accordingly.

FPN: For aluminum and copper-clad aluminum conductors, see Section 310-15.

See Section 310-14 for aluminum conductor material.

110-6. Conductor Sizes. Conductor sizes are expressed in American Wire Gage (AWG) or in circular mils.

For copper, aluminum, or copper-clad aluminum conductors up to size No. 4/0, this *Code* uses the American Wire Gage (AWG) for size identification, which is the same as the Brown and Sharpe Gage (BS).

Conductors larger than 4/0 are sized in circular mils, beginning with 250,000 circular mils. Prior to the 1990 edition, a 250,000-circular mil conductor was labeled "250 MCM." The term *MCM* was defined as 1000 circular mils (the first "M" being the Roman numeral designation for 1000). In the 1990 edition, the notation was changed to "250 kcmil" to recognize the accepted convention that "k" indicates 1000. Recent UL standards and IEEE standards now use the notation "kcmil" rather than "MCM."

The circular mil area of a conductor is equal to its diameter in mils squared (1 in. = 1000 mils). For example, the circular mil area of a No. 8 solid conductor that has a 0.1285-in. diameter is calulated as follows:

$$0.1285 \text{ in.} \times 1000 = 128.5 \text{ mils}$$
$$128.5 \times 128.5 = 16{,}512.25 \text{ circular mils,}$$
$$\text{or } 16{,}510 \text{ circular mils (rounded off)}$$

According to Table 8 in Chapter 9, this rounded value represents the circular mil area for one conductor. Where stranded conductors are used, the circular mil area of each strand must be multiplied by the number of strands to determine the circular mil area of the conductor (see Table 8 in Chapter 9).

110-7. Insulation Integrity. Completed wiring installations shall be free from short circuits and from grounds other than as required or permitted in Article 250.

Insulation is the material that prevents the flow of electricity between points of different potential in an electrical system. Failure of the insulation system is one of the most common causes of problems in electrical installations. This is true in both high-voltage and low-voltage systems.

Insulation tests are performed on new or existing

installations to determine the quality or condition of the insulation of conductors and equipment.

The principal causes of insulation failures are heat, moisture, dirt, and physical damage (abrasion or nicks) occurring during and after installation. Insulation can also fail due to chemical attack, sunlight, and excessive voltage stresses.

Insulation integrity must be maintained during overcurrent conditions. Overcurrent protective devices must be selected and coordinated using tables of insulation thermal-withstand ability in order to ensure that the damage point of an insulated conductor is never reached. These tables, entitled "Allowable Short-Circuit Currents for Insulated Copper (or Aluminum) Conductors" are contained in the Insulated Cable Engineers Association's publication ICEA P-32-382. See Section 110-10 for other circuit components.

In an insulation resistance test, an applied voltage ranging from 100 to 5000 volts (usually 500 to 1000 volts for systems of 600 volts or less), supplied from a source of constant potential, is applied across the insulation. A megohmmeter is usually the potential source, and it indicates the insulation resistance directly on a scale calibrated in megohms (MΩ). The quality of the insulation is evaluated based on the level of the insulation resistance.

The insulation resistance of many types of insulation is quite variable with temperature, so the field data obtained should be corrected to the standard temperature for the class of equipment being tested.

The megohm value of insulation resistance obtained will be inversely proportional to the volume of insulation being tested. For example, a cable 1000 ft long would be expected to have one-tenth the insulation resistance of a cable that is 100 ft long, if all other conditions were identical.

The insulation resistance test is relatively easy to perform and is useful on all types and classes of electrical equipment. Its main value lies in the charting of data from periodic tests, corrected for temperature, over a long period so that deteriorative trends can be detected.

Manuals on this subject are available from instrument manufacturers. Thorough knowledge in the use of insulation testers is essential if the test results are to be meaningful. Figure 110.1 shows a typical megohmmeter insulation tester.

110-8. Wiring Methods. Only wiring methods recognized as suitable are included in this *Code*. The recognized methods of wiring shall be permitted to be installed in any type of building or occupancy, except as otherwise provided in this *Code*.

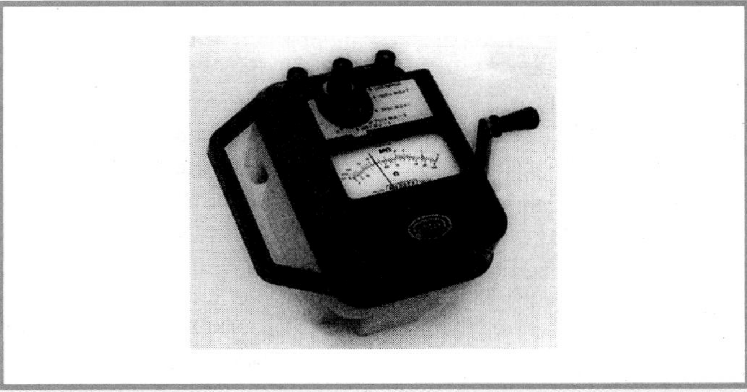

Figure 110.1 *A multivoltage multirange insulation tester. (Biddle Instruments)*

The scope of Article 300 applies generally to all wiring methods, except as amended, modified, or supplemented by Chapter 5 (Special Occupancies), Chapter 6 (Special Equipment), and Chapter 7 (Special Conditions).

Chapter 8 (Communications Systems) is independent of the other chapters, except where some other section is specifically referenced in Chapter 8 of the *Code*. For example, see Section 800-40(b).

110-9. Interrupting Rating. Equipment intended to interrupt current at fault levels shall have an interrupting rating sufficient for the nominal circuit voltage and the current that is available at the line terminals of the equipment.

Equipment intended to interrupt current at other than fault levels shall have an interrupting rating at nominal circuit voltage sufficient for the current that must be interrupted.

Section 110-9 was changed in the 1999 *Code* by substituting the word *interrupt* for the word *break* in two places.

The interrupting rating of overcurrent protective devices is determined under standard test conditions. It is important that the test conditions match the actual installation needs. Section 110-9 states that all fuses and circuit breakers intended to interrupt the circuit at fault levels must have an adequate interrupting rating wherever they are used in the electrical system. Fuses or circuit breakers that do not have adequate interrupting ratings could rupture while attempting to clear a short circuit.

Interrupting ratings should not be confused with short-circuit current ratings. Short-circuit current ratings are further explained in the commentary following Section 110-10.

110-10. Circuit Impedance and Other Characteristics. The overcurrent protective devices, the total impedance, the component short-circuit current ratings, and other characteristics of the circuit to be protected shall be selected and coordinated to permit the circuit-protective devices used to clear a fault to do so without extensive damage to the electrical components of the circuit. This fault shall be assumed to be either between two or more of the circuit conductors, or between any circuit conductor and the grounding conductor or enclosing metal raceway. Listed products applied in accordance with their listing shall be considered to meet the requirements of this section.

Section 110-10 was changed in the 1999 *Code* by substituting the new word *current* for *withstand*. This change correlates the *Code* language with the standard marking language used on equipment. Withstand ratings are not marked on equipment; short-circuit current ratings are marked on the equipment. This marking appears on many pieces of equipment, such as panelboards, switchboards, busways, contactors, and starters. Additionally, a new last sentence was added in the 1999 *Code* to address concerns of what exactly constitutes "extensive damage." Since, under product safety requirements, electrical equipment is evaluated for indications of extensive damage, listed products used within their ratings are considered to meet the requirements of this section.

The basic purpose of overcurrent protection is to open the circuit before conductors or conductor insulation is damaged when an overcurrent condition occurs. An overcurrent condition can be the result of an overload, a ground fault, or a short circuit and must be eliminated before the conductor insulation damage point is reached.

Overcurrent protective devices (such as fuses and circuit breakers) should be selected to ensure that the short-circuit current rating of the system components will not be exceeded should a short circuit or high-level ground fault occur.

System components include wire, bus structures, switching, protection and disconnect devices, and distribution equipment, all of which have limited short-circuit ratings and would be damaged or destroyed if these short-circuit ratings were exceeded. Merely providing overcurrent protective devices with sufficient interrupting rating will not ensure adequate short-circuit protection for the system components. When the available short-circuit current exceeds the short-circuit current rating of an electrical component, the overcurrent protective device must limit the let-through energy to within the rating of that electrical component.

Utility companies usually determine and provide information on available short-circuit current levels at the service equipment. Literature on how to calculate short-circuit currents at each point in any distribution can generally be obtained by contacting the manufacturers of overcurrent protective devices or by referring to IEEE 141-1993, *IEEE Recommended Practice for Electric Power Distribution for Industrial Plants* (Red Book).

For a typical one-family dwelling with a 100-ampere service using No. 2 aluminum supplied by a 1.72 percent, 37$^1/_2$ kVA transformer at a distance of 25 ft, the available short-circuit current would be approximately 6000 amperes.

Available short-circuit current to multifamily structures, where pad-mounted transformers are located very close to the multimetering location, can be relatively high. For example, the line-to-line fault current values close to a low-impedance transformer could exceed 22,000 amperes. At the secondary of a single-phase, center-tapped transformer, the line-to-neutral fault current is approximately 1$^1/_2$ times that of the line-to-line fault current. The short-circuit current rating of utilization equipment located and connected near the service equipment should be known. For example, HVAC equipment is tested at 3500 amperes through a 40-ampere load rating and at 5000 amperes for loads rated more than 40 amperes.

Adequate short-circuit protection can be provided by fuses, molded-case circuit breakers, and low-voltage power circuit breakers depending on specific circuit and installation requirements.

Fuses for Branch-Circuit and Service Overcurrent Protection

The following tables and explanations are based on information for fuses contained in the 1997 UL *Electrical Construction Materials Directory* and the applicable UL standards.

> Several classes of fuse are available that provide branch-circuit, feeder, and/or service overcurrent protection. These fuses and their ratings are tabulated in Table 1.1.
>
> The interrupting rating of a fuse refers to the highest prospective current that the fuse will interrupt. Fuses are tested for operation at any available current equal to or less than their interrupting rating.
>
> A current-limiting fuse limits the clearing time at rated voltage to an interval equal to or less than the first major or symmetrical current loop duration ($^1/_2$ cycle) and limits the peak current to a value that is less than the available peak current. Because the time required for a fuse to melt depends on the current in the circuit, a fuse that is current limiting when subjected to a specific short-

Table 1.1 Description of Fuse Class Designations

Class Designation	Subclasses	Voltage Rating, V_{ac}	Ampere Ratings	Interrupting Rating, rms Symmetrical Amperes	Current Limitation
CC		600	0–30	200,000	Current limiting
G		480	0–60	100,000	Current limiting
H		250 or 600	0–600	10,000	None
J		600	0–600	200,000	Current limiting
K	K1, K5, and K9	250 or 600	0–600	200,000	Not marked current limiting
L		600	601–6000	200,000	Current limiting
Plug Fuse	Edison Base	125	0–30	10,000	Not current limiting
	Type S	125	0–30	10,000	Not current limiting
R	RK1	250 or 600	0–600	200,000	High degree of current limitation
	RK5	250 or 600	0–600	200,000	Moderate degree of current limitation
T		300	0–1200	200,000	Current limiting
		600	0–1200	200,000	Current limiting

circuit current may not be current limiting on a circuit of lower current.

Equipment, such as switches, motor starters, and panelboards, that has been investigated and found suitable for use with the above current-limiting fuses is marked with the class of fuse that is intended to be used in the equipment and with the short-circuit rating. The equipment is suitable for use on circuits that have a maximum available fault current up to the short-circuit rating of the equipment or the interrupting rating of the fuse, whichever is lower. The interrupting rating of a fuse does not automatically qualify the equipment use on circuits with higher available currents than the rating of the equipment itself.

Fuses of a given class, voltage rating, and current-rating range are intended for use only in fuseholders that are designed and marked for that particular fuse. Class L fuses are designed for use in equipment where the line and load connections are made by means of solid busbars.

Fuses designated as Class H are available in both a nonrenewable and renewable form. Renewable fuses allow the replacement of the fuse element after it has cleared the overcurrent. The renewable elements have been investigated for use only with the same line of fuses from the same manufacturer.

Fuses designated as Class K are classified as to interrupting rating and in terms of maximum clearing ampere-squared-seconds and maximum peak let-through current. Class K1 fuses have a high degree of current limitation, Class K5 fuses have a moderate degree of current limitation, and Class K9 fuses have a minimal degree of current limitation. Class K fuses can be placed into fuseholders intended for Class H fuses. As such, they should not be used where greater than 10,000-ampere capacity is available because the Class K fuse might easily be replaced with a Class H fuse. For this reason, Class K fuses are not marked "Current Limiting."

Fuses designated as Class R are provided with dimensional features that allow the insertion into rejection-type fuseholders designed to accept only Class R fuses. In addition, Class R fuses can be inserted into fuseholders intended to accept Class H fuses. This substitution is acceptable since the Class R fuse has interrupting ratings above that of the Class H fuse.

Fuses that meet certain requirements may

be marked "Time Delay," indicating that they are intended for use with motors or other inductive circuits.

Fuses are tested for operation at any voltage equal to or less than their voltage rating.

Fuses may have dc ratings in addition to their ac ratings. If so, they will be marked with their dc voltage and interrupting rating. When dc ratings are required, care must be taken during fuse replacement to substitute a fuse of sufficient dc voltage and interrupting rating. Class K or R fuses that have been investigated for use in protecting trailing cables in dc circuits in mines are marked "Mine Duty" and have an interrupting rating of 20,000 amperes dc.

The maximum permissible let-through values that may be obtained when the fuses are connected to a circuit that has the indicated available current are shown in Table 1.2.

Some fuses are listed as special-purpose fuses because they have an interrupting rating of 300,000 amperes rms symmetrical. Standard industry values have not yet been established for I_p and I^2t at this level. Fuses that are physically identical to another class of listed fuse and that meet all the performance requirements for that listed class of fuse may be marked to indicate their allowable substitution for that class.

The following standards are used to evaluate products in this category:

UL 248-1/CSA-C22.2 No. 248.1, *Low Voltage Fuses, General Requirements*

UL 248-2/CSA-22.2 No. 248.2, *Class C Fuses*

UL 248-3/CSA-22.2 No. 248.3, *Class CA and CB Fuses*

UL 248-4/CSA-C22.2 No. 248.4, *Class CC Fuses*

UL 248-5/CSA-22.2 No. 248.5, *Class G Fuses*

UL 248-6/CSA-22.2 No. 248.6, *Class H Non-Renewable Fuses*

UL 248-7/CSA-22.2 No. 248.7, *Class H Renewable Fuses*

UL 248-8/CSA-C22.2 No. 248.8, *Class J Fuses*

UL 248-9/CSA-22.2 No. 248.9, *Class K Fuses*

UL 248-10/CSA-C22.2 No. 248.10, *Class L Fuses*

UL 248-11/CSA-22.2 No. 248.11, *Plug Fuses*

UL 248-12/CSA-C22.2 No. 248.12, *Class R Fuses*

UL 248-13/CSA-22.2 No. 248.13, *Semiconductor Fuses*

UL 248-15/CSA-C22.2 No. 248.15, *Class T Fuses*

UL 198L, *DC Fuses for Industrial Use*

UL 198M, *Mine-Duty Fuses*

Fuses, Miscellaneous, Miniature, and Micro (JDYX)

This category covers supplemental fuses, which are also described as miscellaneous, miniature, and micro fuses. These fuses provide supplemental protection in appliances or utilization equipment to provide individual protection for components or internal circuits. They are not suitable for branch circuit use. Specific physical dimensions are not specified, but dimensional limitations apply to prevent insertion of supplementary protection fuses into branch circuit fuse holders intended to accommodate branch-circuit fuses of the Class CC, H, J, K, L, R, or T type.

The voltage rating of supplemental fuses is 125 volts ac or greater. Supplemental fuses (except for micro fuses) rated 250 volts may have short circuit ratings at 250 volts, as shown in the table below, but have a minimum 10,000 ampere short-circuit rating at 125 volts. All other supplemental fuses (except for micro fuses) have a minimum short-circuit rating of 10,000 amperes at their rated voltage, with an optional short-circuit rating of 50,000 or 100,000 amperes.

Fuse Rating (amperes)	Short Circuit Rating (amperes)
1 or less	35
More than 1 to 3.5	100
More than 3.5 to 10	200
More than 10 to 15	750
More than 15 to 30	1500

Short circuit ratings are marked on supplemental fuses or on the smallest packaging container.

Micro fuses are supplemental fuses with no principal dimension exceeding 10 mm (excluding leads). They are rated 125 volts ac or greater, with a 50-ampere short-circuit rating.

Supplemental fuses may have a time delay characteristic. This information may be marked on the fuse or on the smallest packaging container.

Fuses covered under this category are not marked "Current Limiting."

Table 1.2 Maximum Peak Let-Through Current (I_p-amperes) and Clearing I^2t (ampere-squared-seconds)

Fuse Class	Current Rating Range, A	Between Threshold and 50 kA		100 kA		200 kA	
		$I_p \times 10^3$	$I^2t \times 10^3$	$I_p \times 10^3$	$I^2t \times 10^3$	$I_p \times 10^3$	$I^2t \times 10^3$
CC	0–15	3	2	3	2	4	3
	16–20	3	2	4	3	5	3
	21–30	6	7	7.5	7	12	7
G	0–15	—	—	4	3.8	n/a	n/a
	16–20	—	—	5	5	n/a	n/a
	21–30	—	—	7	7	n/a	n/a
	31–60	—	—	10.5	25	n/a	n/a
J and 600 V Class T	0–30	6	7	7.5	7	12	7
	31–60	8	30	10	30	16	30
	61–100	12	60	14	80	20	80
	101–200	16	200	20	300	30	300
	201–400	25	1000	30	1100	45	1100
	401–600	35	2500	45	2500	70	2500
	601–800	50	4000	55	4000	75	4000
Class T only	801–1200	55	7500	70	7500	88	7500
300 V Class T	0–30	5	3.5	7	3.5	9	3.5
	31–60	7	15	9	15	12	15
	61–100	9	40	12	40	15	40
	101–200	13	150	16	150	150	20
	201–400	22	550	28	550	35	550
	401–600	29	1000	37	1000	46	1000
	601–800	37	1500	50	1500	65	1500
	801–1200	50	3500	65	3500	80	4000
L	601–800	80	10	80	10	80	10
	801–1200	80	12	80	12	120	15
	1201–1600	100	22	100	22	150	30
	1601–2000	110	35	120	35	165	40
	2001–2500	—	—	165	75	180	75
	2501–3000	—	—	175	100	200	100
	3001–4000	—	—	220	150	250	150
	4001–5000	—	—	—	350	300	350
	5001–6000	—	—	—	350	350	500
RK1	0–30	6	10	10	10	12	11
	31–60	10	40	12	40	16	50
	61–100	14	100	16	100	20	100
	101–200	18	400	22	400	30	400
	201–400	33	1200	35	1200	50	1600
	401–600	43	3000	50	3000	70	4000
RK5	0–30	11	50	11	50	14	50
	31–60	20	200	21	200	26	200
	61–100	22	500	25	500	32	500
	101–200	32	1600	40	1600	50	2000
	201–400	50	5000	60	5000	75	6000
	401–600	65	10,000	80	10,000	100	12,000

UL 248-1/CSA-C22.2 No. 248.1, *Low Voltage Fuses, General Requirements*
UL 248-14/CSA-C22.2 No. 248.14, *Supplemental Fuses*

Circuit Breakers

When using circuit breakers, "current limiting" can be accomplished by using a current-limiting molded-case circuit breaker, a fused circuit breaker that has integrally mounted current-limiting fuses, or by using current limiters designed to be used in conjunction with specific circuit breakers.

The following information on circuit breakers is based on the 1997 UL *General Information Directory* (White Book), "Circuit Breakers, Molded-Case, and Circuit Breaker Enclosures (DIVQ)."

> Listed circuit breakers are intended for use in listed enclosures, or as part of other listed equipment or without enclosures where acceptable.
>
> Listed circuit breakers are rated 600 volts or less. A circuit breaker is marked ac or dc or both ac and dc. A symbol (~), where used, represents ac. The frequency is included if other than 60 Hz.
>
> Class CTL circuit breakers may be identified by the words *Class CTL* or *CTL* on the circuit breaker as part of the marking. Class CTL circuit breakers have physical size, configuration, or other means that, in conjunction with the physical means provided in a Class CTL assembly, are designed to prevent the installation of more circuit breaker poles than the number for which the assembly is designed and rated.
>
> Some circuit breakers are not provided with a means to prevent their installation in Class CTL assemblies. These circuit breakers are for use in old-style, non-Class CTL equipment and are marked "For Replacement Use Only, Not CTL Assemblies."
>
> Circuit breakers marked "SWD" are suitable for switching 120-, 277-, or 347-volt fluorescent lighting on a regular basis at their rated voltage. Circuit breakers marked H1D have been evaluated for switching high-intensity discharge lighting loads on a regular basis at rated voltage.
>
> Three-pole circuit breakers having provision for two poles to be connected to a bus structure and a third isolated pole (commonly referred to as delta breakers) are marked "For Replacement Use Only."
>
> Some circuit breakers include a pole intended to disconnect the grounded circuit conductor of a branch circuit. All poles of these circuit breakers open simultaneously.
>
> Single-pole circuit breakers with handle ties rated 120/240 volts ac and multipole independent trip circuit breakers rated 120/240 volts ac are suitable for use in a single-phase multiwire circuit where the voltage to ground does not exceed 120 volts.
>
> Single-pole circuit breakers rated 125 volts, 125 volts dc, 125/250 volts, or 125/250 volts dc and multipole independent trip circuit breakers rated 125/250 volts or 125/250 volts dc are suitable for use in a single-phase and dc multiwire circuit where the voltage to ground does not exceed 125 volts.
>
> Some independent trip circuit breakers are marked "Independent Trip," "No Common Trip," or with equivalent wording.
>
> Multipole common trip circuit breakers rated 120/240 volts ac are suitable for use in a single-phase multiwire circuit with or without the neutral connected to the load where the voltage to ground does not exceed 120 volts.
>
> Multipole common trip circuit breakers rated 125/250 volts or 125/250 volts dc are suitable for use in a single-phase and a dc multiwire circuit with or without the neutral connected to the load where the voltage to ground does not exceed 125 volts.
>
> Circuit breakers, the performance of which may be affected by a 40°C ambient temperature within the enclosure and that have been evaluated for this application, are marked "40°C."
>
> Unless otherwise marked, circuit breakers should not be loaded in excess of 80 percent of their current rating where in normal operation the load will continue for 3 hours or more.
>
> Circuit breakers that have an interrupting rating higher than 5000 amperes are marked to indicate the higher rating(s).
>
> Circuit breakers are tested under overload conditions at six times rating to cover motor circuit applications and are suitable for use as motor circuit disconnects per Section 430-109 of the *NEC*.
>
> An interrupting rating on a circuit breaker included in a piece of equipment does not automatically qualify the equipment in which

the circuit breaker is installed for use on circuits with higher available currents than the rating of the equipment itself.

A circuit breaker that includes an accessory device, whether attached to the circuit breaker by the manufacturer of the circuit breaker, or, by others, is marked to indicate the presence of that accessory.

Where the accessory is a shunt trip device that is suitable for operation with ground-fault sensing and relaying equipment, such suitability is indicated in the marking of the circuit breaker.

Two-pole circuit breakers that are marked to indicate use on 3-phase circuits have been investigated and found suitable for controlling 3-phase, delta corner grounded circuits.

Three-pole circuit breakers are suitable for use only on 3-phase circuits unless marked to indicate otherwise.

Circuit breakers rated 50 amperes or less and 125/250 volts or less are evaluated for use with tungsten-filament lamp loads.

Listed circuit breakers may be mounted in any position unless marked to indicate otherwise. If, however, the circuit breaker is mounted so that the handle is operated vertically rather than rotationally or horizontally, the up position of the handle is to be the "on" position.

Line and load markings on a circuit breaker are intended to limit connections thereto as marked.

Circuit breaker enclosures that are suitable for use as service equipment are marked accordingly.

Circuit breaker enclosures marked for service equipment use may also be used to provide the main control and means of cutoff for a separately derived system or a second building.

Circuit breakers that have been found suitable for use with heating, air- conditioning, and refrigeration equipment comprising multimotor or combination loads are marked "HACR Type" in conjunction with the listing mark. Such circuit breakers are suitable for use with heating, air-conditioning, and refrigerating equipment marked for use with an HACR-type circuit breaker. Use of these circuit breakers with heating, air-conditioning, and refrigerating equipment is limited to installations where the equipment is marked as suitable for use with any properly sized circuit breaker, or is marked for use with an HACR-type circuit breaker, or is not limited by any marking as to the type of branch-circuit, short-circuit, and ground-fault protective device.

A current-limiting circuit breaker is one that does not employ a fusible element and that, when operating within its current-limiting range, limits the let-through I^2t to a value that is less than the I^2t of a $^1/_2$-cycle wave of the symmetrical prospective current.

Current-limiting circuit breakers are marked "Current Limiting" and are marked either to indicate the let-through characteristics or to indicate where such information may be obtained.

UL 489, *Molded-Case Circuit Breakers, Molded-Case Switches, and Circuit-Breaker Enclosures,* is used to investigate products in this category.

Circuit-Breaker Current Limiters

The following information on circuit breakers is based on the 1997 UL *General Information Directory* (White Book), "Circuit-Breaker Current Limiters (DIRW)."

> Circuit-breaker current limiters are designed to be used in conjunction with specific circuit breakers and to be directly connected to the load terminals of the circuit breakers. They contain fusible elements that function only to increase the fault-current interrupting ability of the combination that is intended for use in the same manner as circuit breakers where installed at the service and as branch-circuit protection. The limiters are rated 600 volts or less.
>
> The fusible elements in circuit-breaker current limiters are coordinated so that they function at currents below those specified in short-circuit test requirements for circuit breakers. Except for this feature of short-circuit operation, combinations of circuit breakers and circuit-breaker current limiters meet all requirements applicable to branch-circuit and service circuit breakers and, in addition, are required to clear circuits up to and including 25 times their ampere rating, and circuits of 1000 amperes or less regardless of ampere rating, without causing operation of the fusible elements in the current limiter.

Circuit-breaker current limiters are marked to indicate the breakers with which they are intended to be used.

Circuit-breaker current limiters that are marked as follows have been investigated in conjunction with the circuit breaker and found suitable for the marked interrupting rating.

Current-Interrupting Rating(s),
MAXIMUM RMS SYM,
AMPERES ______
VOLTS ______

An interrupting rating on a circuit-breaker current limiter included in a piece of equipment does not automatically qualify the equipment in which the combination is installed as suitable for use on circuits with higher available currents than the rating of the equipment itself. The combination of circuit breaker and circuit-breaker current limiter is intended to be mounted in listed enclosures.

Equipment (such as panelboards, service equipment, and dead-front switchboards) that has been investigated and found suitable for use with the combination of circuit breaker and circuit-breaker current limiter is marked to indicate that both may be used.

Circuit-breaker current limiters as listed herein are for use with copper conductors unless marked to indicate which terminals are suitable for use with aluminum conductors. Such marking shall be independent of any marking on terminal connectors and shall be readily visible.

Unless the circuit-breaker current limiter is marked to indicate otherwise, the wiring space and current-carrying capacity are based on the use of 60°C wire in circuits rated 100 amperes or less, and the use of 75°C wire for higher ampere-rated circuits.

UL 489, *Molded-Case Circuit Breakers, Molded-Case Switches, and Circuit Breaker Enclosures,* is used to investigate products in this category.

110-11. Deteriorating Agents. Unless identified for use in the operating environment, no conductors or equipment shall be located in damp or wet locations; where exposed to gases, fumes, vapors, liquids, or other agents that have a deteriorating effect on the conductors or equipment; or where exposed to excessive temperatures.

FPN No. 1: See Section 300-6 for protection against corrosion.

FPN No. 2: Some cleaning and lubricating compounds can cause severe deterioration of many plastic materials used for insulating and structural applications in equipment.

Equipment identified only as "dry locations," "Type 1," or "indoor use only" shall be protected against permanent damage from the weather during building construction.

110-12. Mechanical Execution of Work. Electrical equipment shall be installed in a neat and workmanlike manner.

The regulation is Section 110-12 calling for "neat and workmanlike" installations has appeared in the *NEC* as currently worded for over a half century. It stands as a basis for pride in one's work and has been emphasized by persons involved in the training of apprentice electricians for many years.

Many *Code* conflicts or violations have been cited by the authority having jurisdiction based on the authority's interpretation of "neat and workmanlike manner."

Many electrical inspection authorities use their own experience, or precedents in their local areas, as the basis for their judgments.

Examples of installations that do not qualify as "neat and workmanlike" are exposed runs of cables or raceways that are improperly supported (i.e., sagging between supports or using improper support methods); field-bent and kinked, flattened, or poorly measured raceways; or cabinets, cutout boxes, and enclosures that are not plumb or that are not properly secured.

(a) Unused Openings. Unused openings in boxes, raceways, auxiliary gutters, cabinets, equipment cases, or housings shall be effectively closed to afford protection substantially equivalent to the wall of the equipment.

(b) Subsurface Enclosures. Conductors shall be racked to provide ready and safe access in underground and subsurface enclosures, into which persons enter for installation and maintenance.

(c) Integrity of Electrical Equipment and Connections. Internal parts of electrical equipment, including busbars, wiring terminals, insulators, and other surfaces, shall not be damaged or contaminated by foreign materials such as paint, plaster, cleaners, abrasives, or corrosive residues. There shall be no damaged parts that may adversely affect safe operation or mechanical strength of the equipment such as parts that are broken; bent; cut; or deteriorated by corrosion, chemical action, or overheating.

110-13. Mounting and Cooling of Equipment.

(a) Mounting. Electrical equipment shall be firmly secured to the surface on which it is mounted. Wooden plugs driven into holes in masonry, concrete, plaster, or similar materials shall not be used.

(b) Cooling. Electrical equipment that depends upon the natural circulation of air and convection principles for cooling of exposed surfaces shall be installed so that room airflow over such surfaces is not prevented by walls or by adjacent installed equipment. For equipment designed for floor mounting, clearance between top surfaces and adjacent surfaces shall be provided to dissipate rising warm air.

See Sections 430-14(a) and 430-16 for motor locations, and Sections 450-9 and 450-45 for transformer locations.

Electrical equipment provided with ventilating openings shall be installed so that walls or other obstructions do not prevent the free circulation of air through the equipment.

For example, a ventilated busway must be located where there are no walls or other objects that might interfere with the natural circulation of air and convection principles for cooling. See the definition of *ventilated* in Article 100.

Some types of equipment, such as panelboards and transformers, are adversely affected if enclosure surfaces normally exposed to room air are covered or tightly enclosed. Ventilating openings in equipment are provided to allow the circulation of room air around internal components of the equipment, and the blocking of such openings can cause dangerous overheating.

110-14. Electrical Connections. Because of different characteristics of dissimilar metals, devices such as pressure terminal or pressure splicing connectors and soldering lugs shall be identified for the material of the conductor and shall be properly installed and used. Conductors of dissimilar metals shall not be intermixed in a terminal or splicing connector where physical contact occurs between dissimilar conductors (such as copper and aluminum, copper and copper-clad aluminum, or aluminum and copper-clad aluminum), unless the device is identified for the purpose and conditions of use. Materials such as solder, fluxes, inhibitors, and compounds, where employed, shall be suitable for the use and shall be of a type that will not adversely affect the conductors, installation, or equipment.

FPN: Many terminations and equipment are marked with a tightening torque.

Section 110-3(b) applies where terminations and equipment are marked with tightening torques.

Tightening Torques

Tables 1.3 through 1.6 provide information on the tightening torques used by Underwriters Laboratories Inc. when testing wire connectors, unless the manufacturer assigns another value appropriate for the design. This information should be used for guidance only if no tightening information on the specific wire connector is available. It should not be used to replace the manufacturer's instructions, which should always be followed. This information was taken from the edition of UL Standard 486B that was in effect at the time of the printing of the 1999 edition of this handbook. Similar information can be found in UL 486A and UL 486C.

(a) Terminals. Connection of conductors to terminal parts shall ensure a thoroughly good connection without damaging the conductors and shall be made by means of pressure connectors (including set-screw type), solder lugs, or splices to flexible leads. Connection by means of wire-binding screws or studs and nuts that have upturned lugs or the equivalent shall be permitted for No. 10 or smaller conductors. Terminals for more than one conductor and terminals used to connect aluminum shall be so identified.

(b) Splices. Conductors shall be spliced or joined with splicing devices identified for the use or by brazing, welding, or soldering with a fusible metal or alloy. Soldered splices shall first be spliced or joined so as to be mechanically and electrically secure without solder and then soldered. All splices and joints and the free ends of conductors shall be covered with an insulation equivalent to that of the conductors or with an insulating device identified for the purpose.

Wire connectors or splicing means installed on conductors for direct burial shall be listed for such use.

Field observations and trade magazine articles indicate that failures of electrical connections have been determined as the cause of many equipment burnouts and fires. Many of these failures are attributable to improper terminations, poor workmanship, the differing characteristics of dissimilar metals, and improper binding screws or splicing devices.

UL's requirements for listing solid aluminum conductors in sizes No. 12 and No. 10 and for listing snap switches and receptacles for use on 15- and 20-ampere branch circuits incorporate stringent tests that take the factors listed in the previous paragraph into account. For further information regarding receptacles and switches using CO/ALR-rated terminals, refer to Sections 380-14(c) and 410-56(b).

Table 1.3 Tightening Torques for Screws* in Pound-Inches

Wire Size (AWG or kcmil)	Slotted Head No. 10 and Larger		Hexagonal Head-External Drive Socket Wrench	
	Slot Width to 3/64 or Slot Length to 1/4 in.**	Slot Width Over 3/64 or Slot Length Over 1/4 in.**	Split-Bolt Connectors	Other Connectors
30–10	20	35	80	75
8	25	40	80	75
6	35	45	165	110
4	35	45	165	110
3	35	50	275	150
2	40	50	275	150
1	—	50	275	150
1/0	—	50	385	180
2/0	—	50	385	180
3/0	—	50	500	250
4/0	—	50	500	250
250	—	50	650	325
300	—	50	650	325
350	—	50	650	325
400	—	50	825	325
500	—	50	825	375
600	—	50	1000	375
700	—	50	1000	375
750	—	50	1000	375
800	—	50	1100	500
900	—	50	1100	500
1000	—	50	1100	500
1250	—	—	1100	600
1500	—	—	1100	600
1750	—	—	1100	600
2000	—	—	1100	600

*Clamping screws with multiple tightening means. For example, for a slotted hexagonal head screw, use the torque value associated with the tool used in the installation. UL uses both values when testing.

**For values of slot width or length other than those specified, select the largest torque value associated with conductor size.

Screwless pressure terminal connectors of the conductor push-in type are for use with copper and copper-clad aluminum conductors only.

Instructions describing proper installation techniques and emphasizing the need to follow these techniques and practice good workmanship are required to be included with each coil of No. 12 and No. 10 insulated aluminum wire or cable. See also the commentary on tightening torque that follows Section 110-14(c)(3), FPN.

New product and material designs that provide for increased levels of safety of aluminum wire terminations have been developed by the electrical industry.

To assist all concerned parties in the proper and safe use of solid aluminum wire in making connections to wiring devices used on 15- and 20-ampere branch circuits, the following information is presented. Understanding and using this information is essential for proper application of materials and devices that are now available.

For New Installations

The following is based on a report that was prepared by the Ad Hoc Committee on Aluminum Terminations prior to the 1975 *Code*. This information is still pertinent today, and needed in order to comply with Section 110-14(a) where aluminum wire is used in new installations.

New Materials and Devices For direct connection, only 15- and 20-ampere receptacles and switches marked "CO/ALR" and connected as described in Installation Method should be used.

The "CO/ALR" marking is on the device mounting yoke or strap. The "CO/ALR" marking means the

Table 1.4 Torques in Pound-Inches for Slotted Head Screws* Smaller than No. 10, for Use with No. 8 Smaller Conductors

Screw-Slot Length — in.**	Screw-Slot Width Less than 3/64 in.	Screw-Slot Width 3/64 in. and Larger
to 5/32	7	9
5/32	7	12
3/16	7	12
7/32	7	12
1/4	9	12
9/32	—	15
above 9/32	—	20

*Clamping screws with multiple tightening means. For example, for a slotted hexagonal head screw, use the torque value associated with the tool used in the installation. UL uses both values when testing.
**For slot lengths of intermediate values, select torques pertaining to next shorter slot length.

Table 1.5 Torques for Recessed Allen Head Screws

Socket Size Across Flats—in. Pound-Inches	Torque, Pound-Inches
1/8	45
5/32	100
3/16	120
7/32	150
1/4	200
5/16	275
3/8	375
1/2	500
9/16	600

Table 1.6 Lug Bolting Torques for Connection of Wire Connectors to Busbars, etc.

Bolt Diameter—Inches	Tightening Torque, Pound-Feet
No. 8 or smaller	1.5
No. 10	2.0
1/4 or less	6
5/16	11
3/8	19
7/16	30
1/2	40
9/16 or larger	55

devices have been tested to stringent heat-cycling requirements to determine their suitability for use with UL-labeled aluminum, copper, or copper-clad aluminum wire.

Listed solid aluminum wire, No. 12 or No. 10, marked with the aluminum insulated wire label should be used. The installation instructions that are packaged with the wire should be used.

Installation Method Figure 110.2 illustrates the correct method of connection to be used.

1. The freshly stripped end of the wire should be wrapped two-thirds to three-quarters of the distance around the wire-binding screw post, as shown in Step A of Figure 110.2. The loop is made so that rotation of the screw while tightening will tend to wrap the wire around the post rather than unwrap it.

2. The screw should be tightened until the wire is snugly in contact with the underside of the screw head and with the contact plate on the wiring device, as shown in Step B of Figure 110.2.

3. The screw should be tightened an additional half-turn, thereby providing a firm connection, as shown in Step C of Figure 110.2. Where torque screwdrivers are used, the screw should be tightened to 12 pound-inches, as shown in Step C of Figure 110.2.

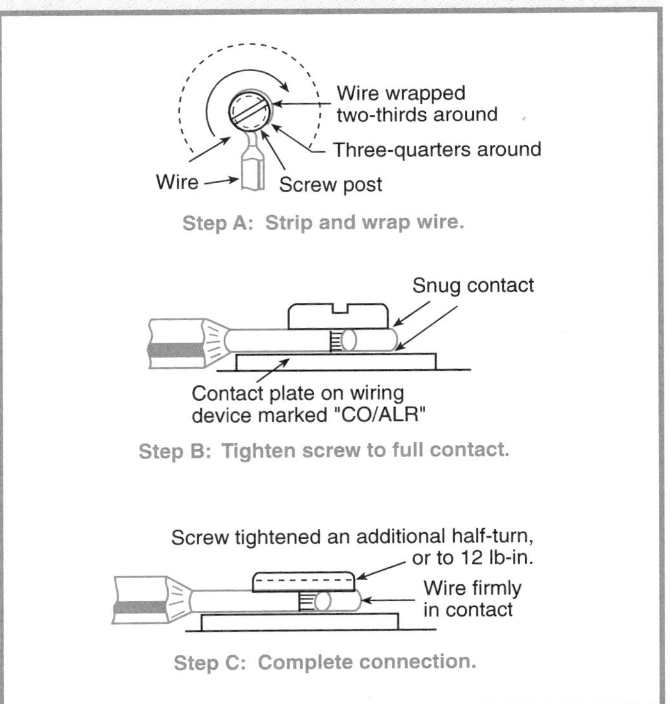

Figure 110.2 *Correct method of terminating aluminum wire at wire-binding screw terminals of receptacles and snap switches. (Underwriters Laboratories Inc.)*

4. The wires should be positioned behind the wiring device so as to decrease the likelihood of the terminal screws loosening when the device is positioned into the outlet box.

Figure 110.3 illustrates incorrect methods for connection. These methods should *not* be used.

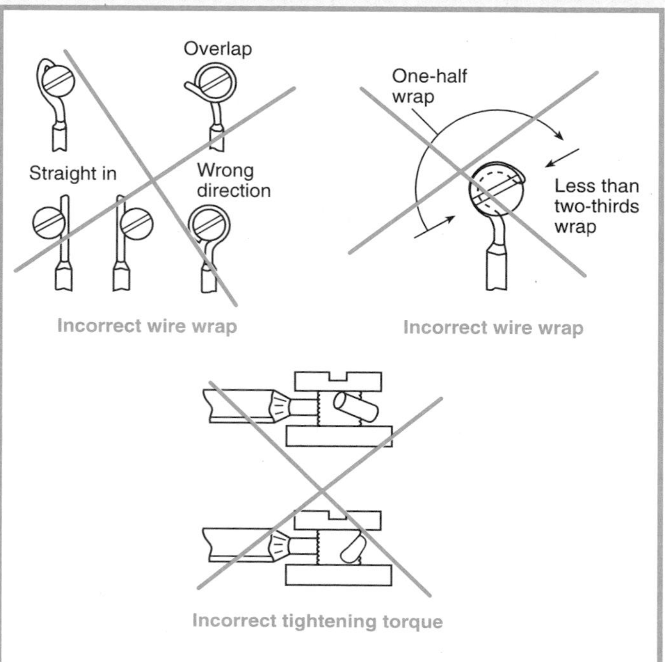

Figure 110.3 *Incorrect methods of terminating aluminum wire at wire-binding screw terminals of receptacles and snap switches. (Underwriters Laboratories Inc.)*

Existing Inventory When labeled, solid aluminum wire No. 12 and No. 10 that does not bear the new aluminum wire label is used, it should be used with wiring devices that are marked "CO/ALR" and connected as described in Installation Method. This is the preferred and recommended method for using such wire.

In the following types of devices, the terminals should not be directly connected to aluminum conductors but may be used with labeled copper or copper-clad conductors:

1. Receptacles and snap switches marked "AL-CU"
2. Receptacles and snap switches having no conductor marking
3. Receptacles and snap switches that have back-wired terminals or screwless terminals of the push-in type

For Existing Installations

If examination discloses overheating or loose connections, the recommendations described under Existing Inventory should be followed.

Twist-On Wire Connectors

Because Section 110-14(b) requires conductors to be spliced with "splicing devices identified for the use," wire connectors are required to be marked for conductor suitability. Twist-on wire connectors are not suitable for splicing aluminum conductors or copper-clad aluminum to copper conductors unless it is so stated and marked as such on the shipping carton. The marking is typically "AL-CU (dry locations)." Presently, one style of wire nut has been listed and one style of crimp-type connector has been listed as having met these requirements.

On February 2, 1995, Underwriters Laboratories Inc. announced the listing of a twist-on wire connector suitable for use with UL 486C aluminum-to-copper conductors, in accordance with *Splicing Wire Connectors*. This was the first listing of a twist-on type of connector for aluminum-to-copper conductors since 1987. The UL listing does *not* cover aluminum-to-aluminum combinations. However, more than one aluminum or copper conductor is allowed when used in combination with each other.

These listed wire-connecting devices are available for pigtailing short lengths of copper conductors to the original aluminum branch-circuit conductors, as shown in Figure 110.4. Primarily, these pigtailed con-

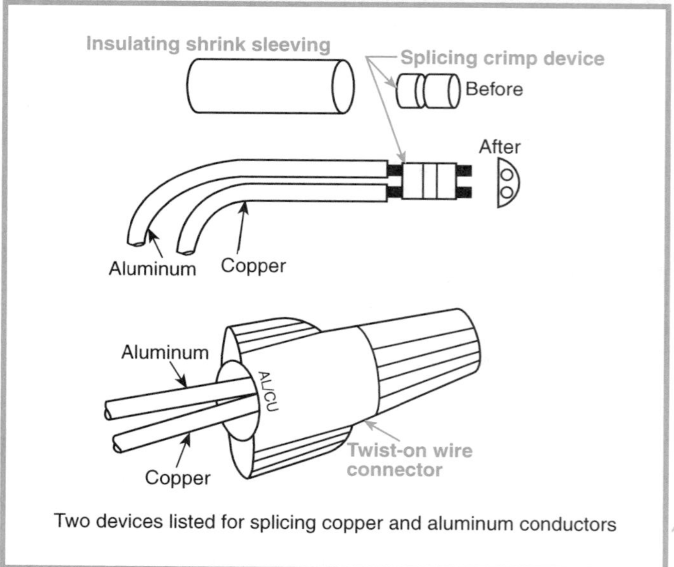

Figure 110.4 *Pigtailing copper to aluminum conductors.*

ductors supply 15- and 20-ampere wiring devices. Pigtailing is permitted provided there is suitable space within the enclosure.

(c) Temperature Limitations. The temperature rating associated with the ampacity of a conductor shall be selected and coordinated so as not to exceed the lowest temperature rating of any connected termination, conductor, or device. Conductors with temperature ratings higher than specified for terminations shall be permitted to be used for ampacity adjustment, correction, or both.

(1) Termination provisions of equipment for circuits rated 100 amperes or less, or marked for Nos. 14 through 1 conductors, shall be used only for one of the following.

(a) Conductors rated 60°C (140°F), or
(b) Conductors with higher temperature ratings, provided the ampacity of such conductors is determined based on the 60°C (140°F) ampacity of the conductor size used, or
(c) Conductors with higher temperature ratings if the equipment is listed and identified for use with such conductors, or
(d) For motors marked with design letters B, C, D, or E, conductors having an insulation rating of 75°C (167°F) or higher shall be permitted to be used provided the ampacity of such conductors does not exceed the 75°C (167°F) ampacity.

(2) Termination provisions of equipment for circuits rated over 100 amperes, or marked for conductors larger than No. 1, shall be used only for

(a) Conductors rated 75°C (167°F), or
(b) Conductors with higher temperature ratings provided the ampacity of such conductors does not exceed the 75°C (167°F) ampacity of the conductor size used, or up to their ampacity if the equipment is listed and identified for use with such conductors.

(3) Separately installed pressure connectors shall be used with conductors at the ampacities not exceeding the ampacity at the listed and identified temperature rating of the connector.

FPN: With respect to Sections 110-14(c)(1), (2), and (3), equipment markings or listing information may additionally restrict the sizing and temperature ratings of connected conductors.

Revised for the 1999 *Code,* the exceptions in this section were reworded into positive language and placed in the main body of the text. Also, Section 110-14(c)(1)(d) was added to grant permission for motors to be connected using the 75°C ampacity for a conductor smaller than No. 1. This change was necessary because motors most likely would not meet the requirements of Section 110-14(c)(1)(c), that is, motors are generally not listed nor do motors have temperature ratings for the supply conductor terminations.

Section 110-14(c)(3) requires that conductor terminations, as well as conductors, be rated for the operating temperature of the circuit. For example, the load on a No. 8 THHN, 90°C copper wire would be limited to 40 amperes if it was connected to a disconnect switch that had its terminals rated at 60°C. Also, this same No. 8 THHN, 90°C wire would be limited to 50 amperes if it was connected to a fusible switch that had its terminals rated at 75°C. These ampacities are from Table 310-16. Not only does this requirement apply to conductor terminations of breakers and fusible switches, but the equipment enclosure must also permit terminations above 60°C. See Figure 110.5 for an example of termination temperature markings.

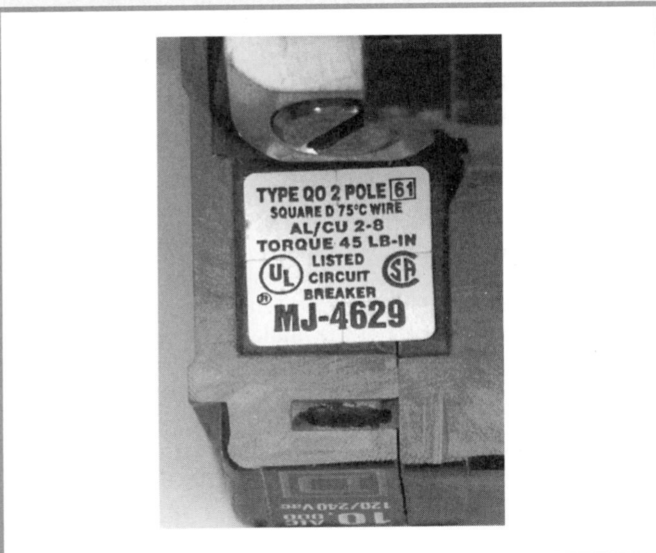

Figure 110.5 *An example of termination temperature markings on a main circuit breaker. (Square D Co.)*

110-18. Arcing Parts. Parts of electric equipment, which in ordinary operation produce arcs, sparks, flames, or molten metal, shall be enclosed or separated and isolated from all combustible material.

FPN: For hazardous (classified) locations, see Articles 500 through 517. For motors, see Section 430-14.

Examples of electrical equipment that may produce sparks during ordinary operation are open motors having a centrifugal starting switch, or open motors with commutators, or collector rings. Adequate separation from combustible material is essential if open motors with these features are used.

110-19. Light and Power from Railway Conductors. Circuits for lighting and power shall not be connected to any system that contains trolley wires with a ground return.

Exception: Car houses, power houses, or passenger and freight stations operated in connection with electric railways.

110-21. Marking. The manufacturer's name, trademark, or other descriptive marking by which the organization responsible for the product can be identified shall be placed on all electric equipment. Other markings that indicate voltage, current, wattage, or other ratings shall be provided as specified elsewhere in this *Code*. The marking shall be of sufficient durability to withstand the environment involved.

The *Code* requires that the rating of equipment be marked on the equipment and that such markings be located so as to be visible or easily accessible during or after installation.

110-22. Identification of Disconnecting Means. Each disconnecting means required by this *Code* for motors and appliances, and each service, feeder, or branch circuit at the point where it originates, shall be legibly marked to indicate its purpose unless located and arranged so the purpose is evident. The marking shall be of sufficient durability to withstand the environment involved.

Where circuit breakers or fuses are applied in compliance with the series combination ratings marked on the equipment by the manufacturer, the equipment enclosure(s) shall be legibly marked in the field to indicate the equipment has been applied with a series combination rating. The marking shall be readily visible and state the following:

CAUTION — SERIES COMBINATION SYSTEM
RATED _____ AMPERES. IDENTIFIED
REPLACEMENT COMPONENTS REQUIRED.

FPN: See Section 240-83(c) for interrupting rating marking for end-use equipment.

Proper identification needs to be specific. For example, the marking should not merely indicate "motor," but rather "motor, water pump"; not merely "lights," but rather "lights, front lobby." Consideration should also be given to the form of identification. Marking often fades or is covered by paint after installation.

The second paragraph of Section 110-22 requires series-rated overcurrent devices to be legibly marked. The equipment manufacturer can mark the equipment that may be used with series combination ratings. If the equipment is installed in the field at its marked series combination rating, the equipment must have an additional label, as specified in Section 110-22, to indicate that the series combination rating has been used.

B. 600 Volts, Nominal, or Less

110-26. Spaces About Electrical Equipment. Sufficient access and working space shall be provided and maintained about all electric equipment to permit ready and safe operation and maintenance of such equipment. Enclosures housing electrical apparatus that are controlled by lock and key shall be considered accessible to qualified persons.

For the 1999 *Code,* Section 110-26 was revised and reorganized. Previously found in Section 110-16, this entire section was retitled and placed in a new Part B, 600 Volts, Nominal, or Less, of Article 110. Significant organizational changes were made to make it easier to use. In Section 110-26(a), the three parts of a safe working space are now clearly separated into depth, width, and height sections. Installation requirements for switchboards, panelboards, and motor control centers, previously found in Section 384-4, were moved to a new Section 110-26(f), Dedicated Equipment Space. This change provides the *Code* user with a single section in which to access all of the requirements that pertain to spaces about electrical equipment.

Key to the understanding of Section 110-26 is that requirements for spaces about electrical equipment are divided into two separate and very distinct catagories, that is, working space and dedicated equipment space. Working space generally applies to the protection of the worker, and dedicated equipment space applies to the protection of electrical equipment.

A new last sentence was added to the first paragraph of Section 110-26 in the 1999 *Code,* which should clarify that this equipment remains accessible even if it is protected by lock and key.

(a) Working Space. Working space for equipment operating at 600 volts, nominal, or less to ground and likely to require examination, adjustment, servicing, or maintenance while energized shall comply with the dimensions of (1), (2), and (3) or as required or permitted elsewhere in this *Code*.

The intent of Section 110-26(a) is to provide enough space for personnel to perform any of the operations listed without jeopardizing the worker's safety. These operations include examination, adjustment, servicing, or maintenance of equipment. Some examples of such equipment include panelboards, switches, circuit breakers, controllers, and controls on heating and air-conditioning equipment. It is important to understand that the word *examination,* as used in Section 110-26(a), includes such tasks as checking for the presence of voltage using a portable voltmeter.

Minimum working clearances are not required, however, if the equipment is such that it is not likely to require examination, adjustment, servicing, or maintenance while energized. But, "sufficient" access and working space are still required by the first paragraph of Section 110-26.

(1) Depth of Working Space. The depth of the working space in the direction of access to live parts shall not be less than indicated in Table 110-26(a). Distances shall be measured from the live parts if such are exposed or from the enclosure front or opening if such are enclosed.

Table 110-26(a). Working Spaces

	Minimum Clear Distance (ft)		
Nominal Voltage to Ground	**Condition 1**	**Condition 2**	**Condition 3**
0–150	3	3	3
151–600	3	3½	4

Notes:
1. For SI units, 1 ft = 0.3048 m.
2. Where the conditions are as follows:
Condition 1 — Exposed live parts on one side and no live or grounded parts on the other side of the working space, or exposed live parts on both sides effectively guarded by suitable wood or other insulating materials. Insulated wire or insulated busbars operating at not over 300 volts to ground shall not be considered live parts.
Condition 2 — Exposed live parts on one side and grounded parts on the other side. Concrete, brick, or tile walls shall be considered as grounded.
Condition 3 — Exposed live parts on both sides of the work space (not guarded as provided in Condition 1) with the operator between.

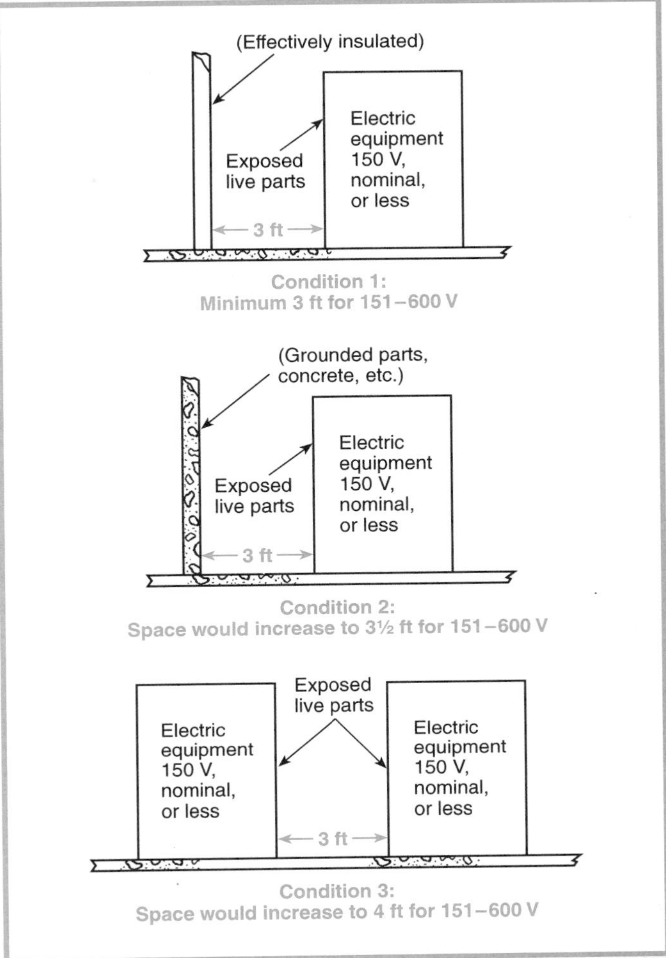

Figure 110.6 *Distances are measured from the live parts if the live parts are exposed, or from the enclosure front if live parts are enclosed. If any assemblies, such as switchboards or motor-control centers, are accessible from the back and expose live parts, the working clearance dimensions would be required at the rear of the equipment, as illustrated. Note that for Condition 3, where there is an enclosure on opposite sides of the working space, the clearance for only one working space is required.*

Included in these clearance requirements is the step-back distance from the face of the equipment. Table 110-26(a) provides requirements for clearances away from the equipment, based on the circuit voltage to ground, and whether there are grounded or ungrounded objects in the step-back space, or if there are exposed live parts across from each other. The voltages to ground consist of two groups: 0 to 150 and 151 to 600, inclusive. Remember, where an ungrounded system is utilized, the voltage to ground will be the greatest voltage between the given conductor and any other conductor of the circuit. For example, the voltage to ground for a 480-volt ungrounded delta system is 480 volts. See Figure 110.6 for general working clearance requirements.

Exception No. 1: Working space shall not be required in back or sides of assemblies, such as dead-front switchboards or motor control centers, where there are no renewable or adjustable parts, such as fuses or switches, on the back or sides and where all connections are accessible from locations other than the back or sides. Where rear access is required to work on de-energized parts on the back of enclosed equipment, a minimum working space of 30 in. (762 mm) horizontally shall be provided.

Generally, the working clearance requirements for the sides and the rear of equipment are the same as for the front of equipment. Exception No. 1 only applies to equipment if it actually requires rear or side access. According to Table 110-26(a), the minimum clear working space required is 3 ft for 0 to 600 volts. However, if rear or side access is required to work on de-energized parts only, then the 3-ft requirement may be reduced to 30 in. by this exception.

Equipment that does not require rear or side access is not required to have any working clearances at the sides or in the rear of the equipment.

Exception No. 1 was modified for the 1999 *Code* by including the sides of equipment in addition to the rear of equipment. (See Figure 110.7.)

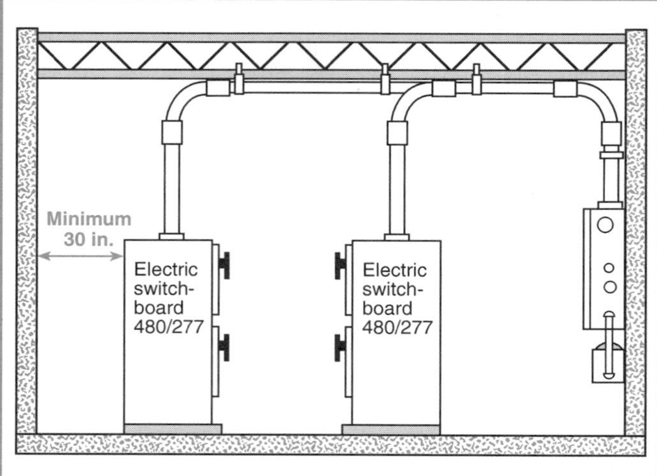

Figure 110.7 *An example of the 30-in. minimum working space that may be required in the rear of equipment.*

Exception No. 2: By special permission, smaller spaces shall be permitted where all uninsulated parts are at a voltage no greater than 30 volts rms, 42 volts peak, or 60 volts dc.

Exception No. 3: In existing buildings where electrical equipment is being replaced, Condition 2 working clearance shall be permitted between dead-front switchboards, panelboards, or motor control centers located across the aisle from each other where conditions of maintenance and supervision ensure that written procedures have been adopted to prohibit equipment on both sides of the aisle from being open at the same time and qualified persons who are authorized will service the installation.

Exception No. 3 permits some relief for installations being upgraded. Where dead-front switchboards, panelboards, or motor-control centers are being replaced in an existing building, the working clearance allowed is that required by Table 110-26(a), Condition 2. This clearance is only allowed where there is written procedure prohibiting facing doors of equipment from being open at the same time, and where only authorized and qualified persons will service the installation. (See Figure 110.8.)

(2) Width of Working Space. The width of the working space in front of the electric equipment shall be the width of the equipment or 30 in. (762 mm), whichever is greater.

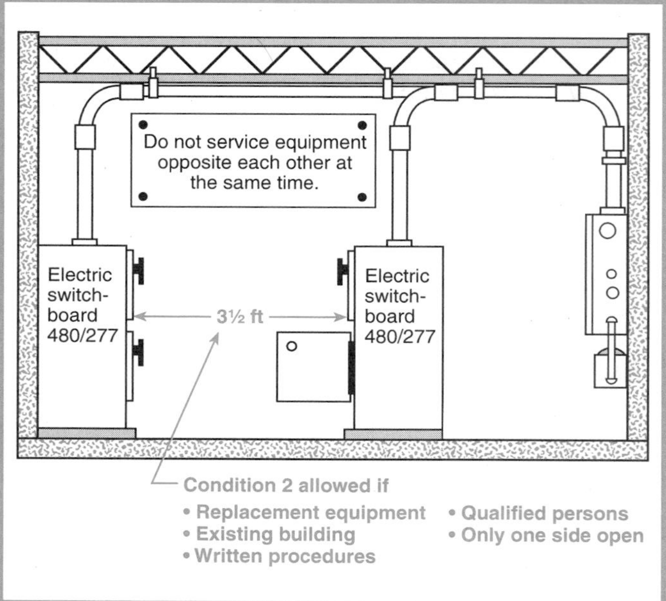

Figure 110.8 *An example of Section 110-26(a)(1), Exception No. 3.*

In all cases, the work space shall permit at least a 90 degree opening of equipment doors or hinged panels.

Regardless of the width of the electrical equipment, the working space cannot be less than 30 in. wide. Also, according to Section 110-26(e), the working space must be clear from the floor up to a height of 6½ ft or the height of the equipment, whichever is greater. This allows an individual to have at least shoulder-width space when in front of the equipment. This 30-in. measurement can be made from either the left or the right edge of the equipment and can overlap other electrical equipment, provided the other equipment does not extend beyond the clearance required by Table 110-26(a). If the equipment is greater than 30 in. wide, the left-to-right space must be equal to the width of the equipment. See Figure 110.9 for an explanation of the 30-in. width requirement.

In addition to the width requirements, sufficient depth in the working space must also be provided to allow a panel or door to open at least 90 degrees. If doors or hinged panels are larger than 3 ft, more than a 3-ft-deep working space must be provided to allow a full 90 degree opening. (See Figure 110.10.)

(3) Height of Working Space. The work space shall be clear and extend from the grade, floor, or platform to the height required by Section 110-26(e). Within the height requirements of this section, other equipment associated with

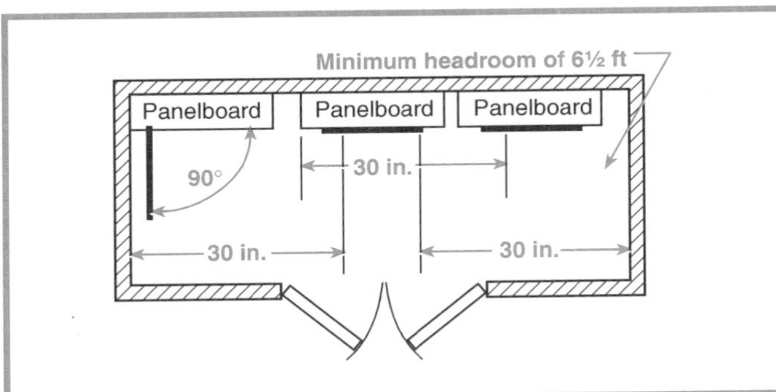

Figure 110.9. The 30-in.-wide front working space is not required to be directly centered on the electrical equipment if it can be ensured that the space is sufficient for safe operation and maintenance of such equipment.

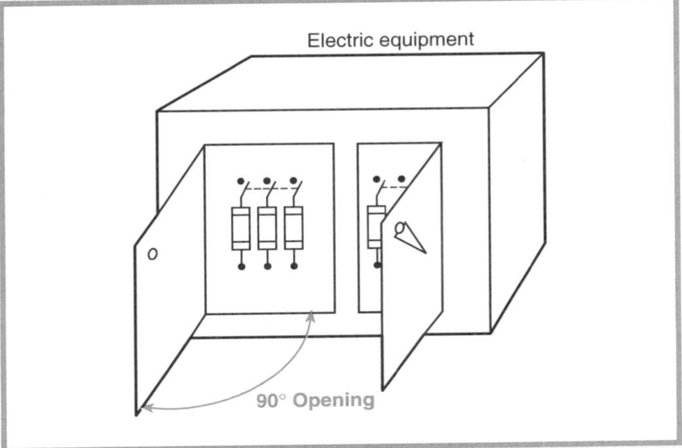

Figure 110.10 Equipment doors are required to open a full 90 degrees to ensure a safe working space.

the electrical installation located above or below the electrical equipment shall be permitted to extend not more than 6 in. (153 mm) beyond the front of the electrical equipment.

Revised in the 1999 *Code*, Section 110-26(a)(3) now permits electrical equipment located above or below other electrical equipment to extend into the "working space" not more than 6 in. This revised requirement permits a 12-in. by 12-in. wireway to be placed directly above or below a 6-in.-deep panelboard without impinging into the working space or compromising practical working clearances. The requirement continues to prohibit large differences in depth of equipment below or above other equipment that specifically require working space. For equipment that produces heat or that otherwise requires ventilation refer to Sections 110-3(b) and 110-13.

(b) Clear Spaces. Working space required by this section shall not be used for storage. When normally enclosed live parts are exposed for inspection or servicing, the working space, if in a passageway or general open space, shall be suitably guarded.

Section 110-26(b) as well as the rest of Section 110-26 does not prohibit the placement of panelboards in corridors or passageways.

(c) Access and Entrance to Working Space. At least one entrance of sufficient area shall be provided to give access to the working space about electric equipment.

For equipment rated 1200 amperes or more and over 6 ft (1.83 m) wide that contains overcurrent devices, switching devices, or control devices, there shall be one entrance not less than 24 in. (610 mm) wide and 6½ ft (1.98 m) high at each end of the working space.

For a graphical explanation of access and entrance requirements to a working space, see Figures 110.11 and 110.12. Notice the unacceptable and hazardous situation shown in Figure 110.13.

Exception No. 1: Where the location permits a continuous and unobstructed way of exit travel, one means of access shall be permitted.

Exception No. 2: Where the work space required by Section 110-26(a) is doubled, only one entrance to the working space is required. It shall be located so the edge of the entrance nearest the equipment is the minimum clear distance given in Table 110-26(a) away from such equipment.

For an explanation of the two exceptions to Section 110-26(c), see Figures 110.14 and 110.15.

(d) Illumination. Illumination shall be provided for all working spaces about service equipment, switchboards, panelboards, or motor control centers installed indoors. Additional lighting fixtures shall not be required where the work space is illuminated by an adjacent light source. In electrical equipment rooms, the illumination shall not be controlled by automatic means only.

(e) Headroom. The minimum headroom of working spaces about service equipment, switchboards, panelboards, or motor control centers shall be 6½ ft (1.98 m). Where the electrical equipment exceeds 6½ ft (1.98 m) in height, the minimum headroom shall not be less than the height of the equipment.

Exception: Service equipment or panelboards, in existing dwelling units, that do not exceed 200 amperes.

(f) Dedicated Equipment Space. Equipment within the scope of Article 384, and motor control centers, shall be

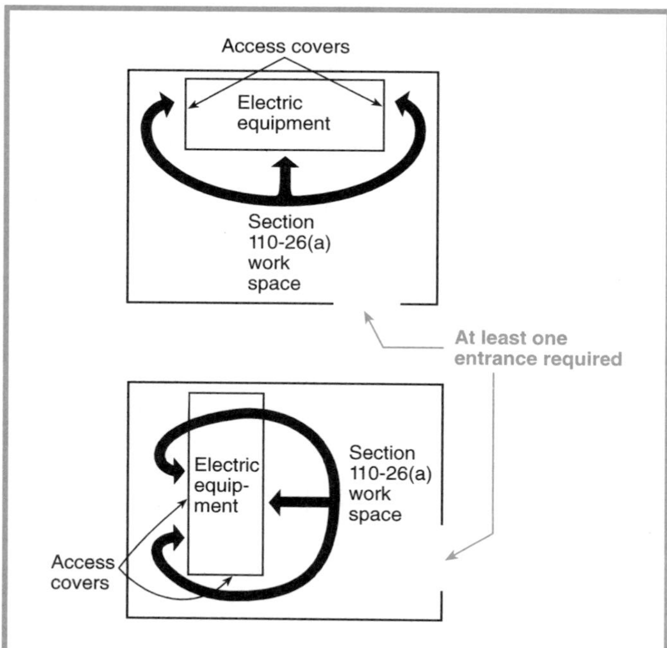

Figure 110.11 Section 110-26(c), Basic Rule, first paragraph. At least one entrance is required to provide access to the working space around electrical equipment. The installation shown on the bottom would not be acceptable if the electrical equipment was a switchboard over 6 ft wide and rated 1200 amperes or more.

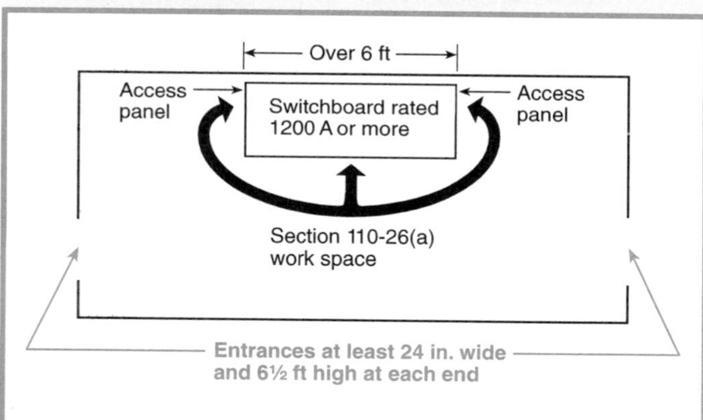

Figure 110.12 Section 110-26(c), Basic Rule, second paragraph. For equipment rated 1200 amperes or more and over 6 ft wide, one entrance not less than 24 in. wide and 6½ ft high is required at each end.

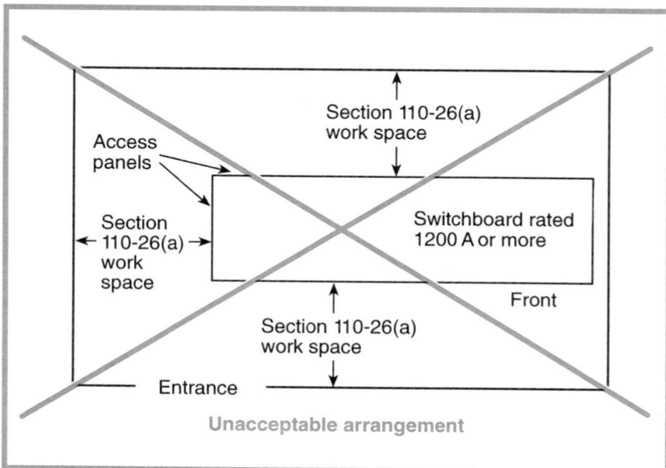

Figure 110.13 Unacceptable arrangement of large switchboard—a person could be trapped behind arcing electrical equipment.

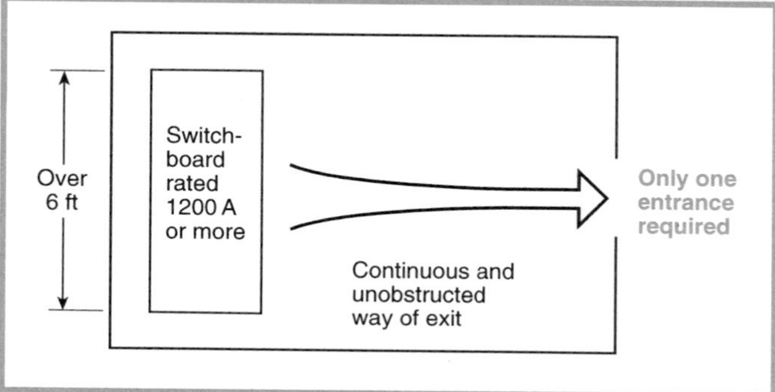

Figure 110.14 The equipment location permits a continuous and unobstructed way of exit travel.

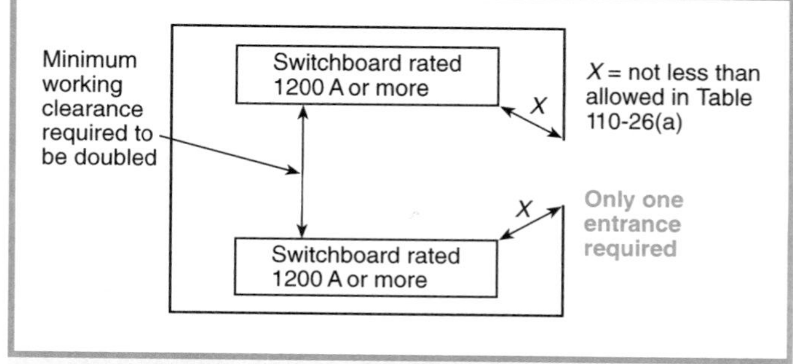

Figure 110.15 If the working space required by Section 110-26(a) is doubled, only one entrance to the working space is required.

located in dedicated spaces and protected from damage as covered in (1) and (2).

Exception: Control equipment that by its very nature or because of other rules of the Code must be adjacent to or within sight of its operating machinery shall be permitted in those locations.

The former Section 384-4 was moved and became a new Section 110-26(f) in the 1999 *Code*. This action was part of the *NEC* Technical Correlating Committee efforts toward making the *Code* a more usable document. See the commentary immediately following the first paragraph of Section 110-26.

Section 110-26(f) only applies to equipment spaces for items such as panelboards, switchboards, and motor control centers. Refer to Section 384-1 for the list of specific equipment that requires dedicated spaces.

(1) Indoor. For indoor installations, the dedicated space shall comply with the following.

(a) *Dedicated Electrical Space.* The space equal to the width and depth of the equipment and extending from the floor to a height of 6 ft (1.83 m) above the equipment or to the structural ceiling, whichever is lower, shall be dedicated to the electrical installation. No piping, ducts, or equipment foreign to the electrical installation shall be located in this zone.

Exception: Equipment that is isolated from the foreign equipment by height or physical enclosures or covers that will afford adequate mechanical protection from vehicular traffic or accidental contact by unauthorized personnel or that complies with (b), shall be permitted in areas that do not have the dedicated space described in this rule.

(b) *Foreign Systems.* The space equal to the width and depth of the equipment shall be kept clear of foreign systems unless protection is provided to avoid damage from condensation, leaks, or breaks in such foreign systems. This zone shall extend from the top of the electrical equipment to the structural ceiling.

(c) *Sprinkler Protection.* Sprinkler protection shall be permitted for the dedicated space where the piping complies with this section.

(d) *Suspended Ceilings.* A dropped, suspended, or similar ceiling that does not add strength to the building structure shall not be considered a structural ceiling.

The dedicated electrical space includes the space defined by extending the footprint of the switchboard or panelboard from the floor to a height of 6 ft above the height of the equipment or to the structural ceiling, whichever is lower. This reserved space permits busways, conduits, raceways, and cables to enter the equipment. The dedicated electrical space must be clear of any piping, ducts, or equipment that is foreign to the electrical installation. Plumbing, heating, ventilation, or air-conditioning piping, ducts, and equipment must be installed outside the width and depth zone. Where foreign systems are installed directly above the footprint area of electrical equipment (from top of equipment to the structural ceiling), that foreign system must include protective equipment that will ensure that occurances such as leaks, condensation, and even breaks will not damage the electrical equipment located below.

Sprinkler protection is permitted for the dedicated spaces as long as it complies with this section. A dropped, suspended, or similar ceiling is permitted to be located directly in the dedicated space, as are building structural members.

In addition to being located in a dedicated space, the electrical equipment must also be protected from physical damage. Damage can be caused by activities near this equipment, such as material handling by personnel or the operation of a fork lift or other mobile equipment. See Section 110-27(b) for other provisions relating to the protection of electrical equipment.

Example Figures

Figures 110.16, 110.17, and 110.18 illustrate the two distinct indoor installation spaces required in Sections 110-26(a) and (f), that is, the working space and the equipment space.

Refer to Figure 110.16 and note that Section 110-26(f) is intended to require that the space outlined by the footprint area (the space equal to the width and depth of the equipment) of the panelboard or switchboard be reserved for the installation of electrical equipment or for the installation of conduits, cable trays, and so on, entering or exiting the equipment. The working space [required by Section 110-26(a)] is the outlined area in front of the switchboard. Note that sprinkler protection is afforded the entire dedicated space and working space without actually entering either space. Also, note that the exhaust duct is not located in or directly above the dedicated electrical equipment space. Although not specifically required to be located here, this duct location may be a cost effective solution that avoids the substantial physical protection requirements of Section 110-26(f)(1)(b).

Figure 110.17 illustrates the working space required in front of the panelboard by Section 110-26(a). No equipment, electrical or otherwise, is allowed in the working space.

Figure 110.18 illustrates the dedicated electrical space required over and under the panelboard by Section 110-26(f)(1). This space is for the cables, raceways, and so on, that run to and from the panelboard.

(2) Outdoor. Outdoor electrical equipment shall be installed in suitable enclosures and shall be protected from accidental contact by unauthorized personnel, or by vehicu-

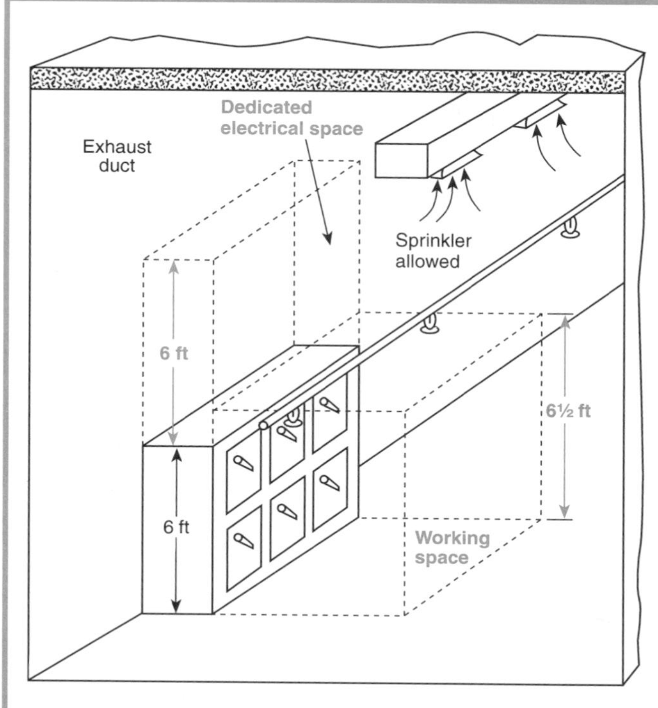

Figure 110.16 The two distinct indoor installation spaces required by Sections 110-26(a) and 110-26(f), that is, the "working space" and the "dedicated electrical space."

lar traffic, or by accidental spillage or leakage from piping systems. The working clearance space shall include the zone described in Section 110-26(a). No architectural appurtenance or other equipment shall be located in this zone.

Extreme care should be taken where protection from unauthorized personnel or vehicle traffic is added to existing installations in order to comply with this requirement. Any excavation or the driving of steel into the ground required for the placement of fencing, vehicle stops, or bollards should only be done after a thorough investigation of the below-grade wiring.

110-27. Guarding of Live Parts.

(a) Live Parts Guarded Against Accidental Contact. Except as elsewhere required or permitted by this *Code,* live parts of electric equipment operating at 50 volts or more shall be guarded against accidental contact by approved enclosures or by any of the following means.

(1) By location in a room, vault, or similar enclosure that is accessible only to qualified persons.
(2) By suitable permanent, substantial partitions or screens arranged so that only qualified persons will have access to the space within reach of the live parts. Any openings in such partitions or screens shall be sized and located so that persons are not likely to come into accidental contact with the live parts or to bring conducting objects into contact with them.
(3) By location on a suitable balcony, gallery, or platform elevated and arranged so as to exclude unqualified persons.
(4) By elevation of 8 ft (2.44 m) or more above the floor or other working surface.

For more information, see the requirements of Sections 610-13(b) and 610-21(a). Although contact conductors obviously have to be bare in order for contact shoes on the moving member to make contact with the conductor, it is possible to place guards near the conductor to prevent its accidental contact with persons and still have slots or spaces through which the moving contacts can operate. Note that the *Code* also recognizes the guarding of live parts by elevation.

(b) Prevent Physical Damage. In locations where electric equipment is likely to be exposed to physical damage, enclosures or guards shall be so arranged and of such strength as to prevent such damage.

(c) Warning Signs. Entrances to rooms and other guarded locations that contain exposed live parts shall be marked with conspicuous warning signs forbidding unqualified persons to enter.

FPN: For motors, see Sections 430-132 and 430-133. For over 600 volts, see Section 110-34.

Live parts of electrical equipment should be covered, shielded, enclosed, or otherwise protected by covers, barriers, mats, or platforms to prevent the likelihood of contact by persons or objects. See the definitions of *dead front, guarded,* and *isolated* in Article 100.

Although not changed, Section 110-27(c) was relocated for the 1999 *Code.* This was previously Section 110-17(c).

C. Over 600 Volts, Nominal

110-30. General. Conductors and equipment used on circuits over 600 volts, nominal, shall comply with Part A of this article and with the following sections, which supplement or modify Part A. In no case shall the provisions of this part apply to equipment on the supply side of the service point.

See "Over 600 Volts" in the Index for various articles, sections, and parts that include requirements for installations over 600 volts.

Equipment on the supply side of the service point is outside of the scope of the *NEC.* Such equipment

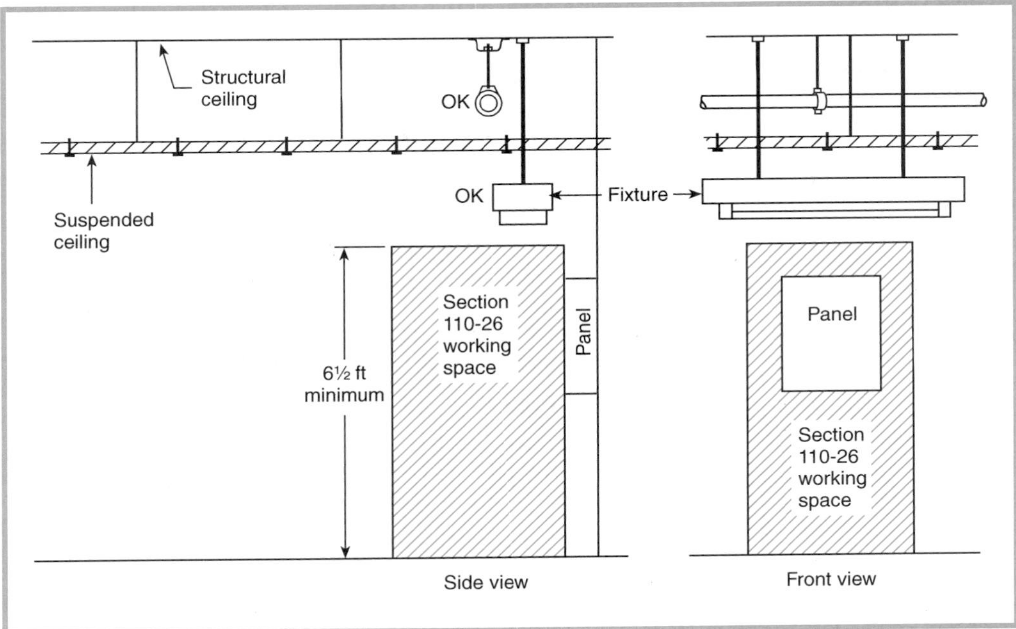

Figure 110.17 The working space in front of a panelboard as required by Section 110-26. This illustration supplements the dedicated equipment space shown in Figure 110.18.

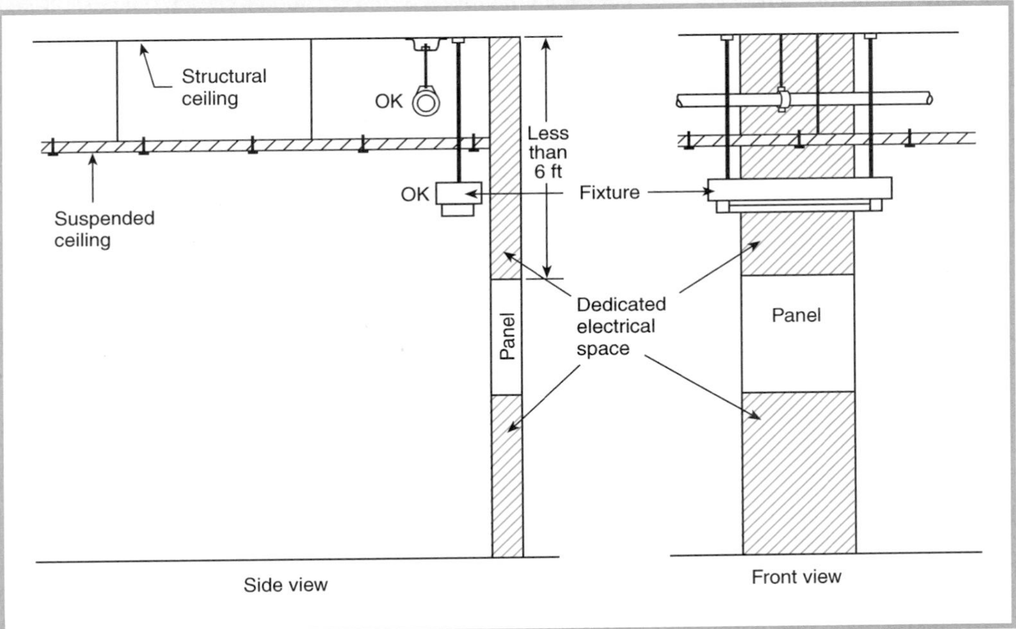

Figure 110.18 The dedicated electrical space over and under a panelboard, as required by Section 110-26(f)(1).

is covered by the *National Electrical Safety Code* (ANSI C2), published by the Institute of Electrical and Electronic Engineers (IEEE).

110-31. Enclosure for Electrical Installations. Electrical installations in a vault, room, or closet or in an area surrounded by a wall, screen, or fence, access to which is controlled by lock and key or other approved means, shall be considered to be accessible to qualified persons only. The type of enclosure used in a given case shall be designed and constructed according to the nature and degree of the hazard(s) associated with the installation.

For installations other than equipment as described in Section 110-31(c), a wall, screen, or fence shall be used to enclose an outdoor electrical installation to deter access by persons who are not qualified. A fence shall not be less than 7 ft (2.13 m) in height or a combination of 6 ft (1.80 m) or more of fence fabric and a 1-ft (305-mm) or more extension utilizing three or more strands of barbed wire or equivalent.

FPN: See Article 450 for construction requirements for transformer vaults.

(a) Indoor Installations.

(1) In Places Accessible to Unqualified Persons. Indoor electrical installations that are open to unqualified persons shall be made with metal-enclosed equipment or shall be enclosed in a vault or in an area to which access is controlled by a lock. Metal-enclosed switchgear, unit substations, transformers, pull boxes, connection boxes, and other similar associated equipment shall be marked with appropriate caution signs. Openings in ventilated dry-type transformers or similar openings in other equipment shall be designed so that foreign objects inserted through these openings will be deflected from energized parts.

(2) In Places Accessible to Qualified Persons Only. Indoor electrical installations considered accessible only to qualified persons in accordance with this section shall comply with Sections 110-34, 490-24, and 110-36.

(b) Outdoor Installations.

(1) In Places Accessible to Unqualified Persons. Outdoor electrical installations that are open to unqualified persons shall comply with Article 225.

FPN: For clearances of conductors for system voltages over 600 volts, nominal, see *National Electrical Safety Code*, ANSI C2-1997.

(2) In Places Accessible to Qualified Persons Only. Outdoor electrical installations that have exposed live parts shall be accessible to qualified persons only in accordance with the first paragraph of this section and shall comply with Sections 110-34, 490-24, and 110-36.

(c) Enclosed Equipment Accessible to Unqualified Persons. Ventilating or similar openings in equipment shall be designed so that foreign objects inserted through these openings will be deflected from energized parts. Where exposed to physical damage from vehicular traffic, suitable guards shall be provided. Nonmetallic or metal-enclosed equipment located outdoors and accessible to the general public shall be designed so that exposed nuts or bolts cannot be readily removed, permitting access to live parts. Where nonmetallic or metal-enclosed equipment is accessible to the general public and the bottom of the enclosure is less than 8 ft (2.44 m) above the floor or grade level, the enclosure door or hinged cover shall be kept locked. Doors and covers of enclosures used solely as pull boxes, splice boxes, or junction boxes shall be locked, bolted, or screwed on. Underground box covers that weigh over 100 lb (45.4 kg) shall be considered as meeting this requirement.

110-32. Work Space About Equipment. Sufficient space shall be provided and maintained about electric equipment to permit ready and safe operation and maintenance of such equipment. Where energized parts are exposed, the minimum clear work space shall not be less than 6½ ft (1.98 m) high (measured vertically from the floor or platform), or less than 3 ft (914 mm) wide (measured parallel to the equipment). The depth shall be as required in Section 110-34(a). In all cases, the work space shall be adequate to permit at least a 90 degree opening of doors or hinged panels.

110-33. Entrance and Access to Work Space.

(a) Entrance. At least one entrance not less than 24 in. (610 mm) wide and 6½ ft (1.98 m) high shall be provided to give access to the working space about electric equipment.

(1) On switchboard and control panels exceeding 6 ft (1.83 m) in width, there shall be one entrance at each end of such boards unless the location of the switchboards and control panels permits a continuous and unobstructed way of exit travel, or unless the work space required in Section 110-34(a) is doubled.

(2) Where one entrance to the working space is permitted under the conditions described in (1), the entrance shall be located so that the edge of the entrance nearest the switchboards and control panels is the minimum clear distance given in Table 110-34(a) away from such equipment.

(3) Where bare energized parts at any voltage or insulated energized parts above 600 volts, nominal, to ground are located adjacent to such entrance, they shall be suitably guarded.

Section 110-33(a) contains similar requirements to Section 110-26(c). Because of the high voltages involved, Section 110-33(a) differs slightly in that a minimum working space entrance dimension is required even for switchboards for 6 ft or less in width. For further information, see the commentary following Section 110-26(c) since most of it remains valid for over 600-volt installations as well.

(b) Access. Permanent ladders or stairways shall be provided to give safe access to the working space around electric equipment installed on platforms, balconies, mezzanine floors, or in attic or roof rooms or spaces.

110-34. Work Space and Guarding.

(a) Working Space. Except as elsewhere required or permitted in this *Code,* the minimum clear working space in the direction of access to live parts of electrical equipment shall not be less than specified in Table 110-34(a). Distances shall be measured from the live parts, if such are exposed, or from the enclosure front or opening if such are enclosed.

Exception: Working space shall not be required in back of equipment such as dead-front switchboards or control assemblies where there are no renewable or adjustable parts (such as fuses or switches) on the back and where all connections are accessible from locations other than the back. Where rear access is required to work on de-energized parts on the back of enclosed equipment, a minimum working space of 30 in. (762 mm) horizontally shall be provided.

Table 110-34(a). Minimum Depth of Clear Working Space at Electrical Equipment

	Minimum Clear Distance (ft)		
Nominal Voltage to Ground	**Condition 1**	**Condition 2**	**Condition 3**
601–2500 V	3	4	5
2501–9000 V	4	5	6
9001–25,000 V	5	6	9
25,001–75 kV	6	8	10
Above 75 kV	8	10	12

Notes:
1. For SI units, 1 ft = 0.3048 m.
2. Where the conditions are as follows:

Condition 1 — Exposed live parts on one side and no live or grounded parts on the other side of the working space, or exposed live parts on both sides effectively guarded by suitable wood or other insulating materials. Insulated wire or insulated busbars operating at not over 300 volts shall not be considered live parts.
Condition 2 — Exposed live parts on one side and grounded parts on the other side. Concrete, brick, or tile walls will be considered as grounded surfaces.
Condition 3 — Exposed live parts on both sides of the work space (not guarded as provided in Condition 1) with the operator between.

(b) Separation from Low-Voltage Equipment. Where switches, cutouts, or other equipment operating at 600 volts, nominal, or less, are installed in a room or enclosure where there are exposed live parts or exposed wiring operating at over 600 volts, nominal, the high-voltage equipment shall be effectively separated from the space occupied by the low-voltage equipment by a suitable partition, fence, or screen.

Exception: Switches or other equipment operating at 600 volts, nominal, or less, and serving only equipment within the high-voltage vault, room, or enclosure shall be permitted to be installed in the high-voltage enclosure, room, or vault if accessible to qualified persons only.

(c) Locked Rooms or Enclosures. The entrances to all buildings, rooms, or enclosures containing exposed live parts or exposed conductors operating at over 600 volts, nominal, shall be kept locked unless such entrances are under the observation of a qualified person at all times.

Where the voltage exceeds 600 volts, nominal, permanent and conspicuous warning signs shall be provided, reading as follows:

DANGER — HIGH VOLTAGE — KEEP OUT

If equipment is used on circuits over 600 volts, nominal, and it contains exposed live parts or exposed conductors, then the equipment is required to be located in a locked room or an enclosure. The provisions for locking are not required if the location is under observation at all times, as in the case of some engine rooms.

(d) Illumination. Illumination shall be provided for all working spaces about electrical equipment. The lighting outlets shall be arranged so that persons changing lamps or making repairs on the lighting system will not be endangered by live parts or other equipment.

The points of control shall be located so that persons are not likely to come in contact with any live part or moving part of the equipment while turning on the lights.

(e) Elevation of Unguarded Live Parts. Unguarded live parts above working space shall be maintained at elevations not less than required by Table 110-34(e).

Table 110-34(e). Elevation of Unguarded Live Parts Above Working Space

Nominal Voltage Between Phases	Elevation
601–7500 V	8 ft 6 in.
7501–35,000 V	9 ft
Over 35 kV	9 ft + 0.37 in./kV above 35

Note: For SI units, 1 in. = 25.4 mm; 1 ft = 0.3048 m.

(f) Protection of Service Equipment, Metal-Enclosed Power Switchgear, and Industrial Control Assemblies. Pipes or ducts foreign to the electrical installation that require periodic maintenance or whose malfunction would endanger the operation of the electrical system shall not be located in the vicinity of the service equipment, metal-enclosed power switchgear, or industrial control assemblies. Protection shall be provided where necessary to avoid damage from condensation leaks and breaks in such foreign systems. Piping and other facilities shall not be considered foreign if provided for fire protection of the electrical installation.

110-36. Circuit Conductors. Circuit conductors shall be permitted to be installed in raceways, in cable trays, as metal-clad cable, as bare wire, cable, and busbars, or as Type MV cables, or conductors as provided in Sections 300-37, 300-39, 300-40, and 300-50. Bare live conductors shall conform with Section 490-24.

Insulators, together with their mounting and conductor attachments, where used as supports for wires, single-conductor cables, or busbars, shall be capable of safely withstanding the maximum magnetic forces that would prevail when two or more conductors of a circuit were subjected to short-circuit current.

Open runs of insulated wires and cables that have a bare lead sheath or a braided outer covering shall be supported in a manner designed to prevent physical damage to the braid or sheath. Supports for lead-covered cables shall be designed to prevent electrolysis of the sheath.

110-40. Temperature Limitations at Terminations. Conductors shall be permitted to be terminated based on the 90°C (194°F) temperature rating and ampacity as given in Tables 310-67 through 310-86, unless otherwise identified.

D. Tunnel Installations Over 600 Volts, Nominal

Because the requirements for systems operating over 600 volts, nominal, appeared to be somewhat fragmented and placed throughout the *NEC*, housekeeping of these requirements was determined to be in order for the 1999 *Code*. Therefore, tunnel installation requirements for systems operating over 600 volts, nominal, were moved from Article 710, Part F, to Article 110, Part D.

110-51. General.

(a) Covered. The provisions of this part shall apply to installation and use of high-voltage power distribution and utilization equipment that is portable and/or mobile, such as substations, trailers, or cars, mobile shovels, draglines, hoists, drills, dredges, compressors, pumps, conveyors, underground excavators, and the like.

(b) Other Articles. The requirements of this part shall be additional to, or amendatory of, those prescribed in Articles 100 through 490 of this *Code*. Special attention shall be paid to Article 250.

(c) Protection Against Physical Damage. Conductors and cables in tunnels shall be located above the tunnel floor and so placed or guarded to protect them from physical damage.

110-52. Overcurrent Protection. Motor-operated equipment shall be protected from overcurrent in accordance with Article 430. Transformers shall be protected from overcurrent in accordance with Article 450.

110-53. Conductors. High-voltage conductors in tunnels shall be installed in metal conduit or other metal raceway, Type MC cable, or other approved multiconductor cable. Multiconductor portable cable shall be permitted to supply mobile equipment.

110-54. Bonding and Equipment Grounding Conductors.

(a) Grounded and Bonded. All noncurrent-carrying metal parts of electric equipment and all metal raceways and cable sheaths shall be effectively grounded and bonded to all metal pipes and rails at the portal and at intervals not exceeding 1000 ft (305 m) throughout the tunnel.

(b) Equipment Grounding Conductors. An equipment grounding conductor shall be run with circuit conductors inside the metal raceway or inside the multiconductor cable jacket. The equipment grounding conductor shall be permitted to be insulated or bare.

110-55. Transformers, Switches, and Electrical Equipment. All transformers, switches, motor controllers, motors, rectifiers, and other equipment installed below ground shall be protected from physical damage by location or guarding.

110-56. Energized Parts. Bare terminals of transformers, switches, motor controllers, and other equipment shall be enclosed to prevent accidental contact with energized parts.

110-57. Ventilation System Controls. Electrical controls for the ventilation system shall be arranged so that the airflow can be reversed.

110-58. Disconnecting Means. A switching device, meeting the requirements of Article 430 or 450, shall be installed at each transformer or motor location for disconnecting the transformer or motor. The switching device shall open all ungrounded conductors of a circuit simultaneously.

110-59. Enclosures. Enclosures for use in tunnels shall be dripproof, weatherproof, or submersible as required by the environmental conditions. Switch or contactor enclosures shall not be used as junction boxes or raceways for conductors feeding through or tapping off to other switches, unless special designs are used to provide adequate space for this purpose.

CHAPTER 2

Wiring and Protection

Article 200 — Use and Identification of Grounded Conductors

Contents

200-1. Scope. This article provides requirements for the following:

(1) Identification of terminals
(2) Grounded conductors in premises wiring systems
(3) Identification of grounded conductors

FPN: See Article 100 for definitions of *Grounded Conductor* and *Grounding Conductor.*

Article 200 contains requirements for grounded conductor use and identification. The grounded circuit conductor is referred to throughout the *Code* as the "grounded conductor." The grounded conductor is often, but not always (as in corner-grounded delta systems), the neutral conductor.

The 1999 *Code* provides a new alternative for identifying the grounded conductor in Section 200-6. The new method allows the use of three continuous white stripes along the entire length of conductor insulation that is colored other than green (instead of a continuous white or natural gray coloring) as a means to identify it as the grounded conductor. The three white stripes method of identification is permitted for all conductor sizes.

200-2. General. All premises wiring systems, other than circuits and systems exempted or prohibited by Sections 210-10, 215-7, 250-21, 250-22, 250-162, 503-13, 517-63, 668-11, 668-21, and Section 690-41, Exception, shall have a grounded conductor that is identified in accordance with Section 200-6.

The grounded conductor, where insulated, shall have insulation that is (1) suitable, other than color, for any ungrounded conductor of the same circuit on circuits of less than 1000 volts or impedance grounded neutral systems of 1 kV and over, or (2) rated not less than 600 volts for solidly grounded neutral systems of 1 kV and over as described in Section 250-184(a).

200-3. Connection to Grounded System. Premises wiring shall not be electrically connected to a supply system unless the latter contains, for any grounded conductor of the interior system, a corresponding conductor that is grounded. For the purpose of this section, *electrically connected* shall mean connected so as to be capable of carrying current, as distinguished from connection through electromagnetic induction.

Grounded conductors of premises wiring (other than separately derived systems) are required to be connected to the supply system grounded conductor to ensure a common, continuous, grounded system.

200-6. Means of Identifying Grounded Conductors.

(a) Sizes No. 6 or Smaller. An insulated grounded conductor of No. 6 or smaller shall be identified by a continuous white or natural gray outer finish or by three continuous white stripes on other than green insulation along its entire length. Wires that have their outer covering finished to show a white or natural gray color but have colored tracer threads in the braid identifying the source of manufacture shall be considered as meeting the provisions of this section. Insulated grounded conductors shall also be permitted to be identified as follows.

(1) The grounded conductor of a mineral-insulated, metal-sheathed cable shall be identified at the time of installation by distinctive marking at its terminations.

(2) A single-conductor, sunlight-resistant, outdoor-rated cable used as a grounded conductor in photovoltaic power systems as permitted by Section 690-31 shall be identified at the time of installation by distinctive white marking at all terminations.

(3) Fixture wire shall comply with the requirements for grounded conductor identification as specified in Section 402-8.

(4) For aerial cable, the identification shall be as above, or by means of a ridge located on the exterior of the cable so as to identify it.

The general rule of Section 200-6(a) requires insulated conductors to be white or natural gray for their entire length. Added for the 1999 *Code,* the general requirement now includes a new concept of marking the grounded conductor, that is, three continuous white stripes along the entire length of the insulated conductor. Other methods of identification are also permitted within Section 200-6(a). For example, the grounded conductor of MI cable, due to its unique construction, is permitted to be identified at the time of installation. Aerial cable is permitted to have the grounded conductor identified by a ridge along its insulated surface. Fixture wires are permitted to have the grounded conductor identified by many different methods. Those methods include colored insulation, stripes on the insulation, colored braid, colored separator, and tinned conductors. These identification methods are explained in detail in Section 400-22. Distinctive white markings at the time of installation is only permitted for conductors No. 6 and smaller in photovoltaic power installations that are located outdoors. Identification of the grounded conductor for sizes No. 6 or smaller solely by distinctive white marking at the time of installation is not permitted except as described for multiconductor cables and cords in Section 200-6(c) and (e).

(b) Sizes Larger than No. 6. An insulated grounded conductor larger than No. 6 shall be identified either by a continuous white or natural gray outer finish or by three continuous white stripes on other than green insulation along its entire length or at the time of installation by a distinctive white marking at its terminations. This marking shall encircle the conductor or insulation.

The general rule of Section 200-6(b) requires the insulated conductors to be white or natural gray for their entire length. Added for the 1999 *Code,* the general requirement now includes a new concept of marking the grounded conductor, that is, three continuous white stripes along the entire length of the insulated conductor. Another permitted method for these larger conductors is field applied distinctive white markings applied at the time of installation at all of the conductor termination points. Where field applied, the white marking must completely encircle the conductor so as to be clearly visible. This method of identification is shown in Figure 200.1.

(c) Flexible Cords. An insulated conductor that is intended for use as a grounded conductor, where contained within a flexible cord, shall be identified by a white or natural gray outer finish or by three continuous white stripes on other than green insulation or by methods permitted by Section 400-22.

(d) Grounded Conductors of Different Systems. Where conductors of different systems are installed in the same raceway, cable, box, auxiliary gutter, or other type of enclosure, one system grounded conductor, if required, shall have an outer covering conforming to Section 200-6(a) or 200-6(b). Each other system grounded conductor shall have an outer covering of white with a readily distinguishable different colored stripe (not green) running along the insulation, or other and different means of identification as allowed by Section 200-6(a) or (b) that will distinguish each system grounded conductor.

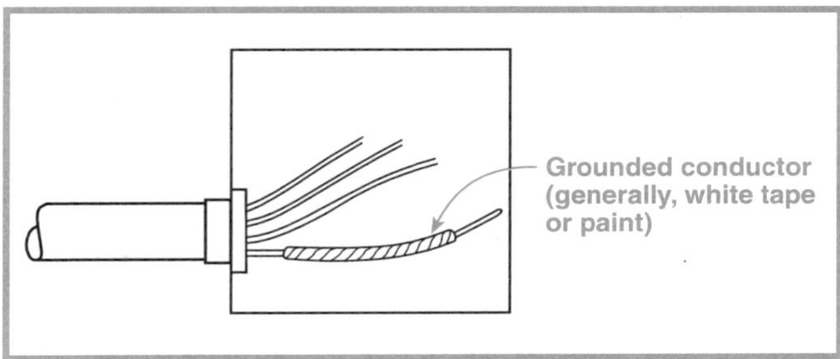

Figure 200.1 *Field-applied identification to a No. 4 conductor identifying it as the grounded conductor as permitted by Section 200-6(b).*

The requirements found in Section 200-6(d) have been there since the 1987 edition of the *NEC.* However, these requirements are often misapplied. As the Figure 200.2 shows, if grounded conductors of different systems are present in the same enclosure, these grounded conductors must be designated by different colors. Careful study of Section 200-6(d) reveals what Figure 200.2 shows, that is, that one system uses a white or natural gray insulation for the grounded conductor, and the other system must use an identification different than a solid white or natural gray colored insulation. Often the second system is white with a stripe. This section no longer permits the past long time practice of identifying the grounded conductor of one system white and the other system gray. See Figure 200.2, which illustrates

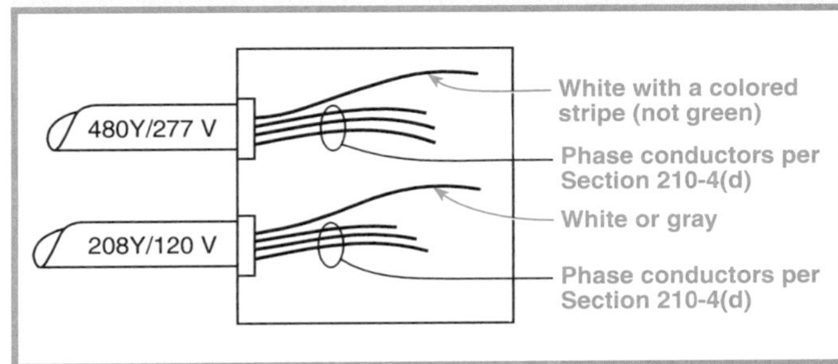

Figure 200.2 *Grounded conductors of different systems in the same enclosure. The insulation of the grounded conductors are of a color as prescribed by Section 200-6(d).*

branch-circuit grounded conductors of different systems in the same enclosure.

(e) Grounded Conductors of Multiconductor Cables. The insulated grounded conductors in a multiconductor cable shall be identified by a continuous white or natural gray outer finish or by three continuous white stripes on other than green insulation along its entire length. Multiconductor flat cable No. 4 or larger shall be permitted to employ an external ridge on the grounded conductor.

Exception No. 1: Where the conditions of maintenance and supervision ensure that only qualified persons will service the installation, grounded conductors in multiconductor cables shall be permitted to be permanently identified at their terminations at the time of installation by a distinctive white marking or other equally effective means.

Sections 200-6(e), Exception No. 1, introduces a concept for identifying grounded conductors of multiconductor cables. This exceptions allow identification of a grounded conductor of a multiconductor cable, as illustrated in Figure 200.3, at the time of installation by use of a distinctive white marking or other equally effective means, such as, for example, numbering, lettering, or tagging. This exception is intended to apply in locations where a regulated system of maintenance and supervision ensures that only qualified persons will service the installation.

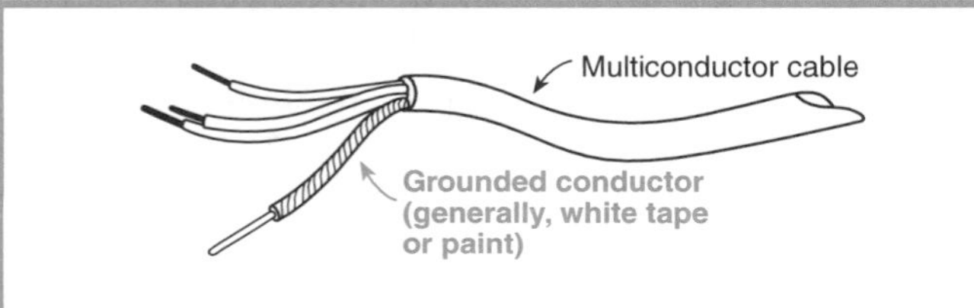

Figure 200.3 *Field-applied identification to the conductor of a multiconductor cable that will be used as the grounded conductor as permitted by Section 200-6(e), Exception No. 1.*

Exception No. 2: The grounded conductor of a multiconductor varnished-cloth-insulated cable shall be permitted to be identified at its terminations at the time of installation by a distinctive white marking or other equally effective means.

200-7. Use of Insulation of a White or Natural Gray Color or with Three Continuous White Stripes.

(a) General. The following shall be used only for the grounded circuit conductor, unless otherwise permitted in (b) and (c):

(1) A conductor with continuous white or natural gray covering
(2) A conductor with three continuous white stripes on other than green insulation
(3) A marking at the termination of white or natural gray color

(b) Circuits of Less than 50 Volts. A conductor with white or natural gray color insulation or three continuous white stripes or having a marking of white or natural gray at the termination for circuits of less than 50 volts shall be required to be grounded only as required by Section 250-20(a).

(c) Circuits of 50 Volts or More. The use of insulation that is white or natural gray or that has three continuous white stripes for other than a grounded conductor for circuits of 50 volts or more shall be permitted only as in (1) through (3).

(1) If part of a cable assembly and where the insulation is permanently re-identified to indicate its use as an ungrounded conductor, by painting or other effective means at its termination, and at each location where the conductor is visible and accessible.

(2) Where a cable contains an insulated conductor for single-pole, 3-way, or 4-way switch loops, and the conductor with white or natural gray insulation or a marking of three continuous white stripes is used for the supply to the switch, but not as a return conductor from the switch to the switched outlet. In these applications, the conductor with white or natural gray insulation or with three continuous white stripes shall be permanently re-identified to indicate its use by painting or other effective means at its terminations and at each location where the conductor is visible and accessible.

Section 200-7(c)(2) was formerly Section 200-7, Exception No. 2. Previous editions of the *Code* permitted switch loops containing a white insulated conductor that was used to supply the switch and not used to supply the light fixtures, to remain white. Re-identification of this particular ungrounded conductor was not required. However, revised for the 1999 *Code,* re-identification of all ungrounded conductors that are white or any permitted white coloring, is now required at each and every termination point. The required re-identifcation must be effective, permanent and suitable for the environment, so as to clearly identify the insulated conductor as an ungrounded conductor. The reason for this re-identification requirement is that many electronic automation devices requiring a grounded conductor are now available for installation into switch outlets. Where re-identification is done properly, this may eliminate the possible mis-wiring of these new automation devices during future installations.

(3) Where a flexible cord, having one conductor identified by a white or natural gray outer finish or three continuous white stripes or by any other means permitted by Section

400-22, is used for connecting an appliance or equipment permitted by Section 400-7. This shall apply to flexible cords connected to outlets whether or not the outlet is supplied by a circuit that has a grounded conductor.

200-9. Means of Identification of Terminals. The identification of terminals to which a grounded conductor is to be connected shall be substantially white in color. The identification of other terminals shall be of a readily distinguishable different color.

Exception: Where the conditions of maintenance and supervision ensure that only qualified persons will service the installations, terminals for grounded conductors shall be permitted to be permanently identified at the time of installation by a distinctive white marking or other equally effective means.

See the commentary following Section 200-6(e), Exception No. 1, which elaborates on the means for identifying grounded conductors for these qualified locations.

200-10. Identification of Terminals.

(a) Device Terminals. All devices, excluding lighting and appliance branch-circuit panelboards, provided with terminals for the attachment of conductors and intended for connection to more than one side of the circuit shall have terminals properly marked for identification, unless the electrical connection of the terminal intended to be connected to the grounded conductor is clearly evident.

Exception: Terminal identification shall not be required for devices that have a normal current rating of over 30 amperes, other than polarized attachment plugs and polarized receptacles for attachment plugs as required in Section 200-10(b).

(b) Receptacles, Plugs, and Connectors. Receptacles, polarized attachment plugs, and cord connectors for plugs and polarized plugs shall have the terminal intended for connection to the grounded conductor identified.

Identification shall be by a metal or metal coating that is substantially white in color or by the word *white* or the letter *W* located adjacent to the identified terminal.

If the terminal is not visible, the conductor entrance hole for the connection shall be colored white or marked with the word *white* or the letter *W*.

Section 200-10(b) requires that terminals of receptacles, plugs, and connectors intended for the connection of the grounded conductor be marked by one of several methods. Those methods include the word *white*, the letter *W*, and a terminal identified by a distinctive white color. This variety of methods will allow plating of all screws and terminals to meet other requirements of specific applications, such as corrosion-resistant devices.

FPN: See Section 250-126 for identification of wiring device equipment grounding conductor terminals.

(c) Screw Shells. For devices with screw shells, the terminal for the grounded conductor shall be the one connected to the screw shell.

(d) Screw Shell Devices with Leads. For screw shell devices with attached leads, the conductor attached to the screw shell shall have a white or natural gray finish. The outer finish of the other conductor shall be of a solid color that will not be confused with the white or natural gray finish used to identify the grounded conductor.

(e) Appliances. Appliances that have a single-pole switch or a single-pole overcurrent device in the line or any line-connected screw shell lampholders, and that are to be connected by (1) a permanent wiring method or (2) field-installed attachment plugs and cords with three or more wires (including the equipment grounding conductor), shall have means to identify the terminal for the grounded circuit conductor (if any).

200-11. Polarity of Connections. No grounded conductor shall be attached to any terminal or lead so as to reverse the designated polarity.

Article 210 — Branch Circuits

Contents

A. General Provisions

210-1. Scope. This article covers branch circuits except for branch circuits that supply only motor loads, which are covered in Article 430. Provisions of this article and Article 430 apply to branch circuits with combination loads.

The exception covering branch circuits for electrolytic cells was removed for the 1999 *Code*. However, Sections 668-3(c)(1) through (4), indicate that electrolytic cell line conductors, cells, cell line attachments, and the wiring of auxiliary equipment and devices within the cell line working zone are not required to comply with the provisions of Article 210.

210-2. Other Articles for Specific-Purpose Branch Circuits. Branch circuits shall comply with this article and also with the applicable provisions of other articles of this *Code*. The provisions for branch circuits supplying equipment in the following list amend or supplement the provisions in this article and shall apply to branch circuits referred to therein.

	Article	Section
Air-conditioning and refrigerating equipment		440-6 440-31 440-32
Busways		364-9
Circuits and equipment operating at less than 50 volts	720	
Central heating equipment other than fixed electric space-heating equipment		422-12
Class 1, Class 2, and Class 3 remote-control, signaling, and power-limited circuits	725	
Closed-loop and programmed power distribution	780	
Cranes and hoists		610-42
Electric signs and outline lighting		600-6
Electric welders	630	
Elevators, dumbwaiters, escalators, moving walks, wheel chairlifts, and stairway chair lifts		620-61
Fire alarm systems	760	
Fixed electric heating equipment for pipelines and vessels		427-4
Fixed electric space-heating equipment		424-3
Fixed outdoor electric deicing and snow-melting equipment		426-4
Information technology equipment		645-5
Infrared lamp industrial heating equipment		422-48 424-3
Induction and dielectric heating equipment	665	
Marinas and boatyards		555-5

	Article	Section
Mobile homes, manufactured homes, and mobile home parks	550	
Motion picture and television studios and similar locations	530	
Motors, motor circuits, and controllers	430	
Pipe organs		650-7
Recreational vehicles and recreational vehicle parks	551	
Sound-recording and similar equipment		640-6
Switchboards and panelboards		384-32
Theaters, audience areas of motion picture and television studios, and similar locations		520-41 520-52 520-62
X-ray equipment		660-2 517-73

210-3. Rating. Branch circuits recognized by this article shall be rated in accordance with the maximum permitted ampere rating or setting of the overcurrent device. The rating for other than individual branch circuits shall be 15, 20, 30, 40, and 50 amperes. Where conductors of higher ampacity are used for any reason, the ampere rating or setting of the specified overcurrent device shall determine the circuit rating.

To compensate for voltage drop in a long circuit, larger conductors with a higher ampacity are commonly used. For example, a branch circuit of No. 10 Type TW copper conductors has a 30-ampere ampacity. However, where a 20-ampere overcurrent device protects this branch circuit, it is *rated* as a 20-ampere branch circuit.

Exception: Multioutlet branch circuits greater than 50 amperes shall be permitted to supply nonlighting outlet loads on industrial premises where maintenance and supervision indicate that qualified persons will service the equipment.

It is common in industrial establishments to provide several single receptacles of 50-ampere or higher rating on a single branch circuit to allow quick relocation of equipment for production and/or maintenance use, such as in the case of electric welders. Generally, only one piece of equipment is operated at a time. The type of receptacle used is generally of the non-ANSI-type configuration known as a pin-and-sleeve receptacle, although the *Code* does not so limit the design. These receptacles may or may not be horsepower rated.

210-4. Multiwire Branch Circuits.

(a) General. Branch circuits recognized by this article shall be permitted as multiwire circuits. A multiwire branch circuit shall be permitted to be considered as multiple circuits. All conductors shall originate from the same panelboard.

FPN: A 3-phase, 4-wire, wye-connected power system used to supply power to nonlinear loads may necessitate that the power system design allow for the possibility of high harmonic neutral currents.

The power supplies for equipment such as computers, printers, and adjustable-speed motor drives may introduce harmonic currents in the system neutral conductor. The total harmonic distortion current that results can exceed the load current of the device itself. See the commentary following Section 310-15(b)(4) for a discussion of the *NEC* Correlating Committee Ad Hoc Subcommittee on Nonlinear Loads regarding neutral conductor ampacity.

(b) Dwelling Units. In dwelling units, a multiwire branch circuit supplying more than one device or equipment on the same yoke shall be provided with a means to disconnect simultaneously all ungrounded conductors at the panelboard where the branch circuit originated.

A *device* is defined in Article 100 as a unit of an electrical system that is intended to carry but not utilize electric energy. Devices include receptacles, switches, and lampholders. Wiring devices or equipment include such items as dimmers, pilot lights, and home automation controls that use electrical energy.

Many 125-volt, 15- and 20-ampere duplex receptacles have a break-off tab, as shown in Figure 210.1 (top), that permits each of the two receptacles to be supplied from different circuits or a 3-wire (multiwire) branch circuit. This is commonly called a split-wired receptacle (i.e., one circuit supplies half of the duplex receptacle and another circuit supplies the other half). The simultaneous opening of both "hot" conductors at the panelboard effectively protects personnel from inadvertent contact during servicing with an energized conductor or device terminal. The simultaneous disconnection can be achieved by a 2-pole circuit breaker or by two single-pole circuit breakers with an approved handle tie, as shown in Figure 210.1 (bottom). Where fuses are used, a 2-pole disconnect switch is required.

(c) Line-to-Neutral Loads. Multiwire branch circuits shall supply only line-to-neutral loads.

Exception No. 1: A multiwire branch circuit that supplies only one utilization equipment.

Exception No. 2: Where all ungrounded conductors of the multiwire branch circuit are opened simultaneously by the branch-circuit overcurrent device.

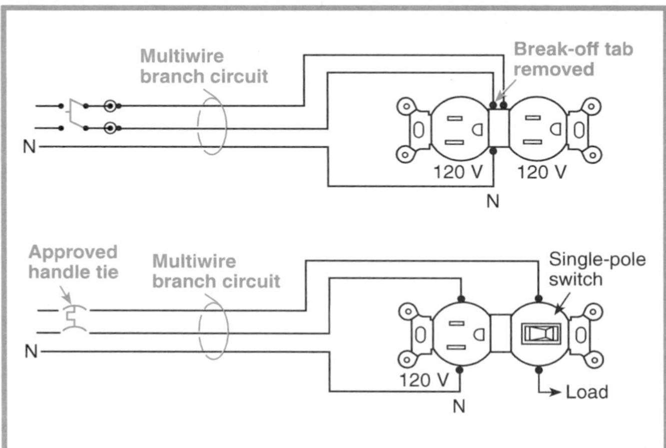

Figure 210.1 *Where a multiwire branch circuit supplies more than one device or equipment on the same yoke, Section 210-4(b) requires the simultaneous disconnection of all ungrounded conductors. This is accomplished at the dwelling unit panelboard by using a 2-pole circuit breaker, or approved handle ties on two single-pole circuit breakers, or a 2-pole disconnect.*

FPN: See Section 300-13(b) for continuity of grounded conductor on multiwire circuits.

The term *branch circuit, multiwire* is defined in Article 100 as a branch circuit that consists of two or more ungrounded conductors that have a potential difference between them, and a grounded conductor that has equal potential difference between it and each ungrounded conductor of the circuit and that is connected to the neutral or grounded conductor of the system.

The circuit most commonly used as a multiwire branch circuit, as illustrated in Figure 210.2, consists of two ungrounded conductors and one grounded conductor supplied from a 120/240-volt, single-phase, 3-wire system. Multiwire branch circuits have many advantages, such as three wires doing the work of four (in place of two 2-wire circuits), less raceway fill, easier balancing and phasing of a system, and less voltage drop. See the commentary following Section 215-2(d), FPN No. 3, for further information on voltage drop for branch circuits.

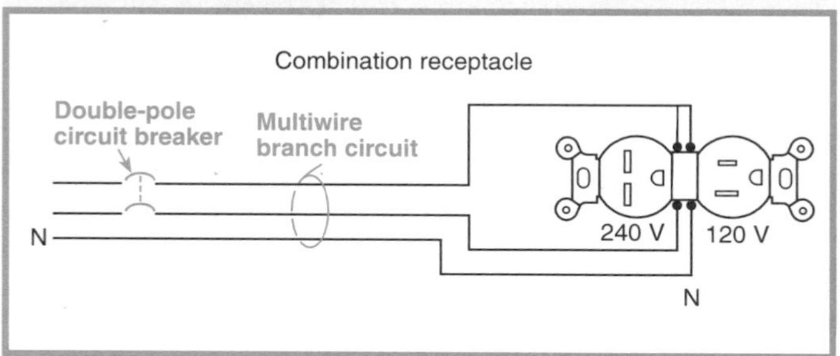

Figure 210.2 *Section 210-4(c), Exception No. 2, permits a multiwire branch circuit to supply line-to-neutral loads where the ungrounded conductors are opened simultaneously by the branch-circuit overcurrent device.*

Multiwire branch circuits may be derived from a 120/240-volt, single-phase; 208Y/120-volt, 3-phase, 4-wire; or a 240/120-volt, 3-phase, 4-wire delta system. Section 220-4(d) requires multiwire branch circuits to be properly balanced. If two ungrounded conductors and a common neutral are used as a multiwire branch circuit supplied from a 208Y/120-volt, 3-phase, 4-wire system, the neutral carries the same current as the phase conductor with the highest current and, therefore, should be the same size. The neutral for a 2-phase, 3-wire or a 2-phase, 5-wire circuit must be sized to carry 140 percent of the ampere rating of the circuit as required by See Section 220-22. See the commentary following Section 210-4(a), FPN, for further information on 3-phase, 4-wire system neutral conductors.

If loads are connected line-to-line (i.e., utilization equipment connected between 2 or 3 phases), 2-pole or 3-pole circuit breakers are required to disconnect all ungrounded conductors simultaneously. In testing 240-volt equipment, it is quite possible not to realize that the circuit is still energized with 120 volts if one pole of the overcurrent device is open. See Sections 210-10 and 240-20(b) for further information on circuit breaker overcurrent protection of ungrounded conductors. Other precautions concerning device removal on multiwire branch circuits are found in the commentary following Section 300-13(b).

(d) Identification of Ungrounded Conductors. Where more than one nominal voltage system exists in a building, each ungrounded conductor of a multiwire branch circuit, where accessible, shall be identified by phase and system. This means of identification shall be permitted to be by separate color coding, marking tape, tagging, or other approved means and shall be permanently posted at each branch-circuit panelboard.

Figure 210.3 shows an example where two different nominal voltage systems exist in a building. Each ungrounded system conductor has been identified with color-coded marking tape. A notice indicating the means of the identification has been permanently located at each panelboard. It should be noted that this requirement applies only to multiwire branch circuits.

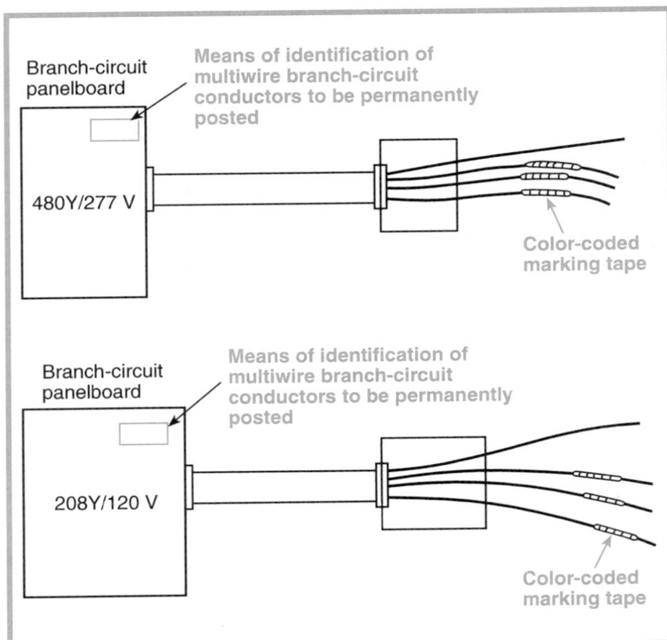

Figure 210.3 An example of accessible (ungrounded) phase conductors identified by marking tape.

210-5. Identification for Branch Circuits.

(a) Grounded Conductor. The grounded conductor of a branch circuit shall be identified in accordance with Section 200-6.

(b) Equipment Grounding Conductor. The equipment grounding conductor shall be identified in accordance with Section 250-119.

210-6. Branch-Circuit Voltage Limitations.

(a) Occupancy Limitation. In dwelling units and guest rooms of hotels, motels, and similar occupancies, the voltage shall not exceed 120 volts, nominal, between conductors that supply the terminals of the following:

(1) Lighting fixtures
(2) Cord- and plug-connected loads 1440 volt-amperes, nominal, or less, or less than ¼ hp

Similar occupancies include sleeping rooms in dormitories, fraternities, sororities, nursing homes, and other such facilities.

Small loads such as those of 1440 volt-amperes or less and motors of less than ¼ horsepower are limited to 120-volt circuits. High-wattage cord- and plug-connected loads such as electric ranges, clothes dryers, and some window air conditioners are permitted to be connected to a 208-volt or 240-volt circuit.

(b) 120 Volts Between Conductors. Circuits not exceeding 120 volts, nominal, between conductors shall be permitted to supply the following:

(1) The terminals of lampholders applied within their voltage ratings

Section 210-6(b)(1) allows lampholders to be used only within their voltage ratings. See the commentary following Section 210-6(c)(2) for details on voltage limitations for listed incandescent lighting fixtures.

(2) Auxiliary equipment of electric-discharge lamps

This includes ballasts for fluorescent and high-intensity-discharge (e.g., mercury vapor, metal halide, and sodium) lamps.

(3) Cord- and plug-connected or permanently connected utilization equipment

(c) 277 Volts to Ground. Circuits exceeding 120 volts, nominal, between conductors and not exceeding 277 volts, nominal, to ground shall be permitted to supply the following:

(1) Listed electric-discharge lighting fixtures

Section 210-6(c)(1) allows listed electric-discharge lighting fixtures to be used only within their ratings. See Sections 225-7(c) and (d) for additional restrictions for installation of outdoor lighting fixtures on 277-volt and 480-volt circuits.

Figure 210.4 shows some examples of lighting units permitted to be connected to branch circuits. Medium-base screw shell lampholders are not permitted to be directly connected to 277-volt branch circuits. Other types of lampholders may be connected to 277-volt circuits, but only if the lampholders have a 277-volt rating. Connecting the 277-volt branch circuit to a listed electric-discharge fixture or a listed autotransformer-type incandescent fixture with a medium-base screw shell lampholder is permitted.

(2) Listed incandescent lighting fixtures, where supplied at 120 volts or less from the output of a stepdown autotransformer that is an integral component of the fixture and the outer shell terminal is electrically connected to a grounded conductor of the branch circuit

Section 210-6(c)(2) permits an incandescent fixture on a 277-volt circuit only if it is a listed fixture with an integral autotransformer and an output to the lampholder not exceeding 120 volts. In this application, the autotransformer supplies 120 volts to the lampholder, and the grounded conductor is connected to the screw shell of the lampholder. This application is similar to a branch circuit derived from

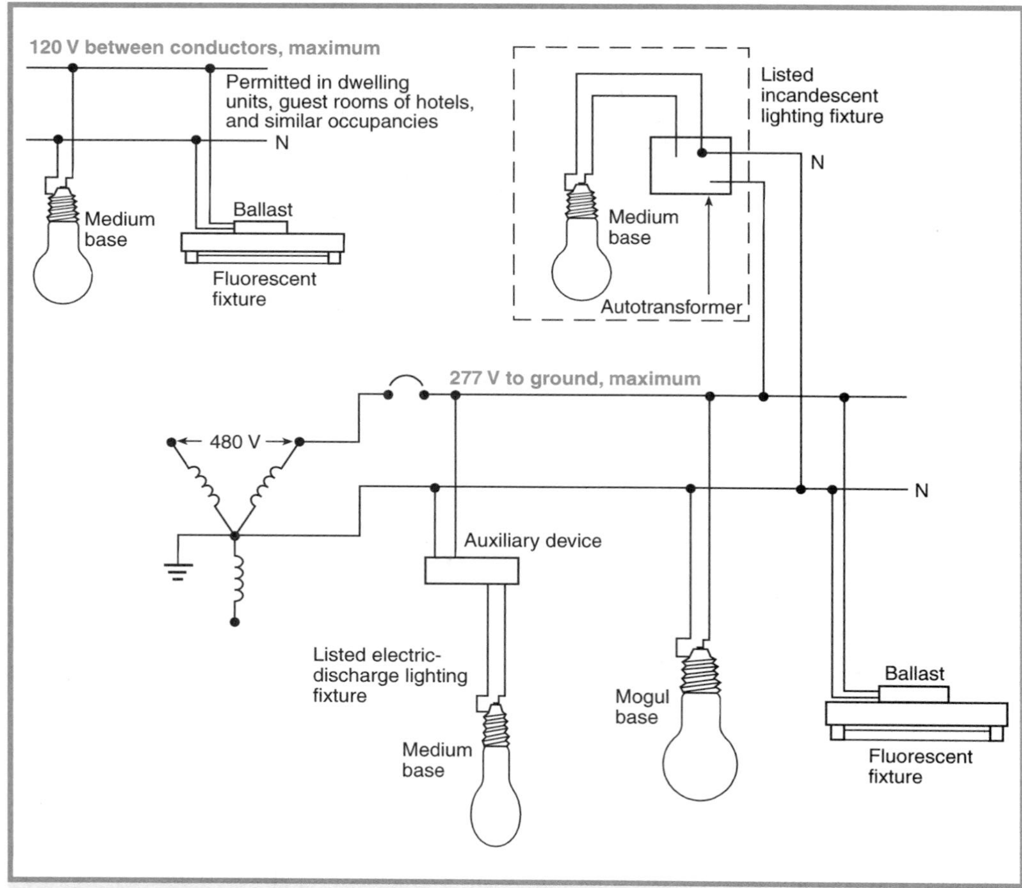

Figure 210.4 *Examples of lighting units permitted by Section 210-6 to be connected to branch circuits.*

an autotransformer, except that the 120-volt circuit is the internal wiring of the fixture.

(3) Lighting fixtures equipped with mogul-base screw shell lampholders
(4) Lampholders, other than the screw shell type, applied within their voltage ratings
(5) Auxiliary equipment of electric-discharge lamps
(6) Cord- and plug-connected or permanently connected utilization equipment

The equipment listed in Section 210-6(c) is permitted to be connected to circuits not exceeding 277 volts to ground. This includes 480-volt grounded systems in general; however, 480-volt corner-grounded delta systems do not qualify. Typical examples of cord- and plug-connected equipment listed under Section 210-6(c)(6) are through-the-wall heating and air-conditioning units and restaurant deep fat fryers operating at 480 volts, 3 phase, from a grounded wye system.

(d) 600 Volts Between Conductors. Circuits exceeding 277 volts, nominal, to ground and not exceeding 600 volts, nominal, between conductors shall be permitted to supply the following:

(1) The auxiliary equipment of electric-discharge lamps mounted in permanently installed fixtures where the fixtures are mounted in accordance with one of the following:
   (a) Not less than a height of 22 ft (6.71 m) on poles or similar structures for the illumination of outdoor areas such as highways, roads, bridges, athletic fields, or parking lots
   (b) Not less than a height of 18 ft (5.49 m) on other structures such as tunnels

Minimum mounting heights required by Section 210-6(d)(1) are for circuits that exceed 277 volts to ground and do not exceed 600 volts phase-to-phase. These circuits supply the auxiliary equipment of electric-discharge lighting. Figure 210.5 (top) indicates the

minimum mounting height of 22 ft for fixtures in outdoor areas such as parking lots. Figure 210.5 (bottom) shows the minimum mounting height of 18 ft for fixtures installed in tunnels and similar structures.

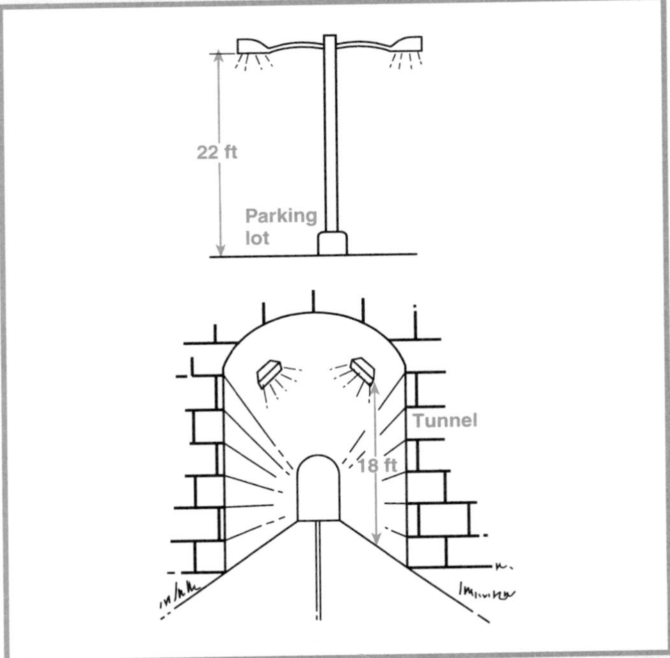

Figure 210.5 *Minimum mounting heights are required by Section 210-6(d)(1) for circuits exceeding 277 volts to ground and not exceeding 600 volts between conductors where supplying auxiliary equipment of electric-discharge lampholders.*

(2) Cord- and plug-connected or permanently connected utilization equipment

FPN: See Section 410-78 for auxiliary equipment limitations.

Exception No. 1 to (b), (c), and (d): For lampholders of infrared industrial heating appliances as provided in Section 422-14.

Exception No. 2 to (b), (c), and (d): For railway properties as described in Section 110-19.

210-7. Receptacles and Cord Connectors.

(a) Grounding Type. Receptacles installed on 15- and 20-ampere branch circuits shall be of the grounding type. Grounding-type receptacles shall be installed only on circuits of the voltage class and current for which they are rated, except as provided in Tables 210-21(b)(2) and (b)(3).

Exception: Nongrounding-type receptacles installed in accordance with Section 210-7(d).

(b) To Be Grounded. Receptacles and cord connectors that have grounding contacts shall have those contacts effectively grounded.

Exception No. 1: Receptacles mounted on portable and vehicle-mounted generators in accordance with Section 250-34.

Exception No. 2: Replacement receptacles as permitted by Section 210-7(d)

(c) Methods of Grounding. The grounding contacts of receptacles and cord connectors shall be grounded by connection to the equipment grounding conductor of the circuit supplying the receptacle or cord connector.

FPN: For installation requirements for the reduction of electrical noise, see Section 250-146(d).

The branch-circuit wiring method shall include or provide an equipment grounding conductor to which the grounding contacts of the receptacle or cord connector shall be connected.

FPN No. 1: Section 250-118 describes acceptable grounding means.

FPN No. 2: For extensions of existing branch circuits, see Section 250-130.

(d) Replacements. Replacement of receptacles shall comply with (1), (2), and (3) as applicable.

(1) Where a grounding means exists in the receptacle enclosure or a grounding conductor is installed in accordance with Section 250-130(c), grounding-type receptacles shall be used and shall be connected to the grounding conductor in accordance with Sections 210-7(c) or 250-130(c).

Section 210-7(d)(1) states the basic rule for replacing an existing receptacle. If a grounding means exists or is added, a grounding-type receptacle must be installed and connected to the equipment grounding conductor. Section 250-130(c) lists five options for connecting the equipment grounding conductor when receptacles are replaced.

(2) Ground-fault circuit-interrupter protected receptacles shall be provided where replacements are made at receptacle outlets that are required to be so protected elsewhere in this *Code*.

Section 210-7(d)(2) requires replacement receptacles to be GFCI protected if GFCI protection is required in the current *Code*. For example, if a kitchen was designed in accordance with the 1984 *NEC*, GFCI-protected receptacles were not required by that edition of the *Code*. However, if an existing receptacle that serves the countertop surfaces is replaced, the replacement receptacle would be required to be a GFCI type or connected to the load side of a GFCI-type receptacle or to be protected by a GFCI-type

circuit breaker. See Section 210-8(a)(6) for this GFCI requirement.

(3) Where a grounding means does not exist in the receptacle enclosure, the installation shall comply with (a), (b), or (c).

(a) A nongrounding-type receptacle(s) shall be permitted to be replaced with another nongrounding-type receptacle(s).
(b) A nongrounding-type receptacle(s) shall be permitted to be replaced with a ground-fault circuit interrupter-type of receptacle(s). These receptacles shall be marked "No Equipment Ground." An equipment grounding conductor shall not be connected from the ground-fault circuit interrupter-type receptacle to any outlet supplied from the ground-fault circuit interrupter receptacle.
(c) A nongrounding-type receptacle(s) shall be permitted to be replaced with a grounding-type receptacle(s) where supplied through a ground-fault circuit interrupter. Grounding-type receptacles supplied through the ground-fault circuit interrupter shall be marked "GFCI Protected" and "No Equipment Ground." An equipment grounding conductor shall not be connected between the grounding-type receptacles.

If existing nongrounding-type receptacles are replaced, only grounding-type receptacles are permitted if a grounding means exists in the receptacle enclosure. If a grounding means does not exist in the receptacle enclosure, then there are four choices (as follows) for the replacement.

1. A nongrounding-type receptacle can be used, which indicates to the user that a grounding means for an appliance is not available.

2. A grounding-type receptacle can be used as permitted by Section 250-130(c).

3. A GFCI-type receptacle can be used if visibly marked "No Equipment Ground."

4. A GFCI-type circuit breaker may be used to supply the branch circuit if visibly marked "GFCI Protected" and "No Equipment Ground."

If a GFCI-type receptacle is used, it can be either the non-feed-through type or a feed-through type wired so that the equipment grounding conductor is not connected to any outlet supplied by this receptacle. For circuits without an equipment grounding conductor, a GFCI receptacle is permitted to supply downstream outlets connected to grounding-type receptacles if they are visibly marked "GFCI Protected" and "No Equipment Ground."

The ability of the GFCI to provide shock hazard protection is independent of whether the tool, appliance, and such, to which it is connected is grounded. A nongrounding-type receptacle provides no shock hazard protection.

If the ungrounded GFCI-type receptacle is used, care should be exercised in the installation to ensure that an equipment grounding conductor is not connected between the GFCI grounding terminal and any grounding-type receptacle on the load side of the GFCI-type receptacle. Proper labeling of the replaced device as well as any downstream devices must also be addressed.

(e) Cord- and Plug-Connected Equipment. The installation of grounding-type receptacles shall not be used as a requirement that all cord- and plug-connected equipment be of the grounded type.

FPN: See Section 250-114 for types of cord- and plug-connected equipment to be grounded.

Over the years, Section 210-7(d) has required grounding-type receptacles to replace nongrounding-type receptacles. Grounding-type receptacles should be conveniently located for use with utilization equipment that is required to be grounded. For cord- and plug-connected equipment that is required to be grounded, see Section 250-114.

Many appliances are not required to be grounded — for example, table lamps, toasters, flatirons, TVs, VCRs, and some heating equipment. Also, distinctively marked, listed, double-insulated tools and appliances are not required to be grounded.

(f) Noninterchangeable Types. Receptacles connected to circuits that have different voltages, frequencies, or types of current (ac or dc) on the same premises shall be of such design that the attachment plugs used on these circuits are not interchangeable.

Configuration charts for 15-, 20-, 30-, 40-, 50-, and 60-ampere general-purpose nonlocking and specific-purpose locking plugs and receptacles are given in Figures 210.6 and 210.7.

210-8. Ground-Fault Circuit-Interrupter Protection for Personnel.

FPN: See Section 215-9 for ground-fault circuit-interrupter protection for personnel on feeders.

(a) Dwelling Units. All 125-volt, single-phase, 15- and 20-ampere receptacles installed in the locations specified below shall have ground-fault circuit-interrupter protection for personnel.

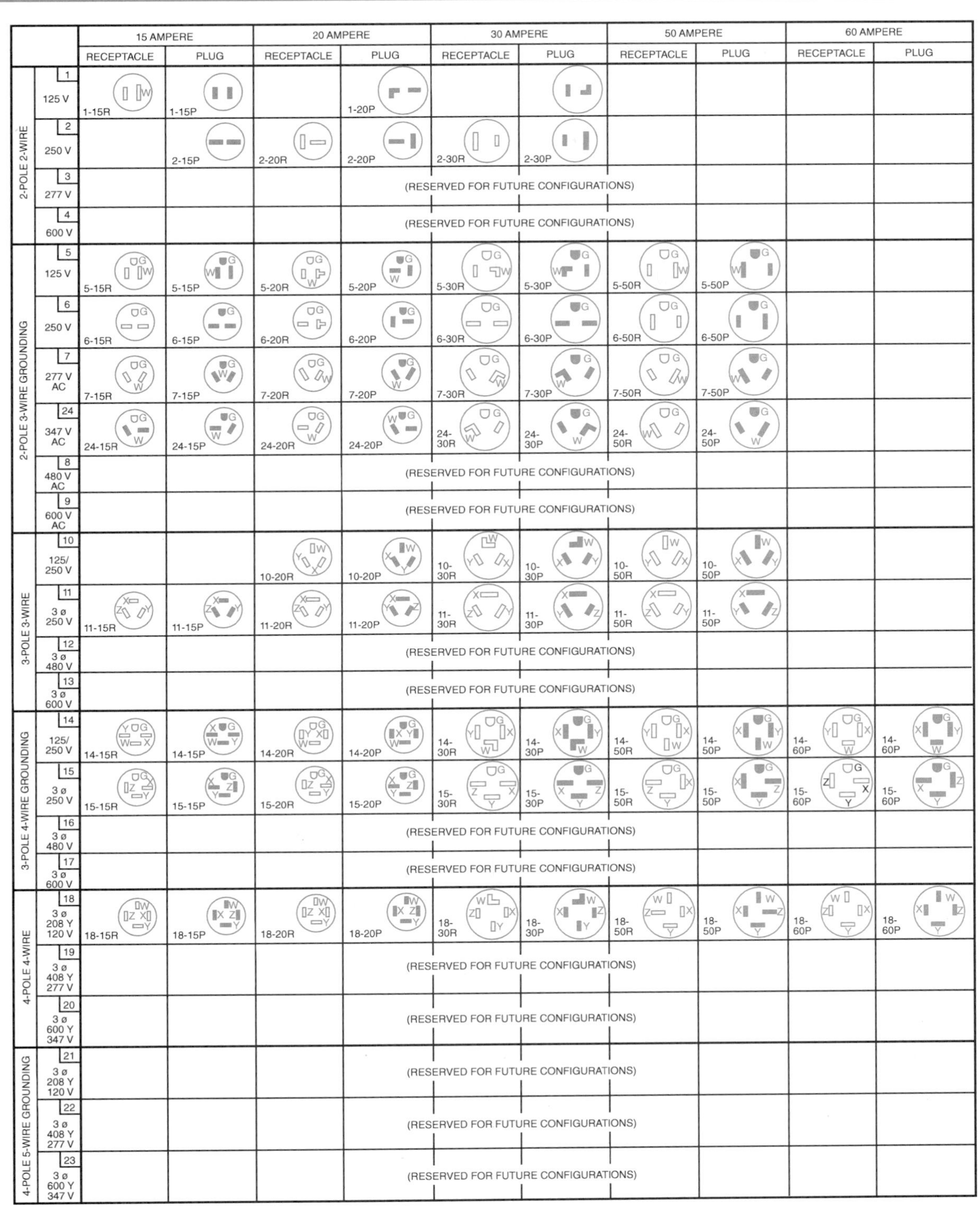

Figure 210.6 *Configuration chart for general-purpose nonlocking plugs and receptacles. Reproduced from Wiring Devices — Dimensional Requirements, NEMA WD 6-88 (revision and redesignation of ANSI C73-73).*

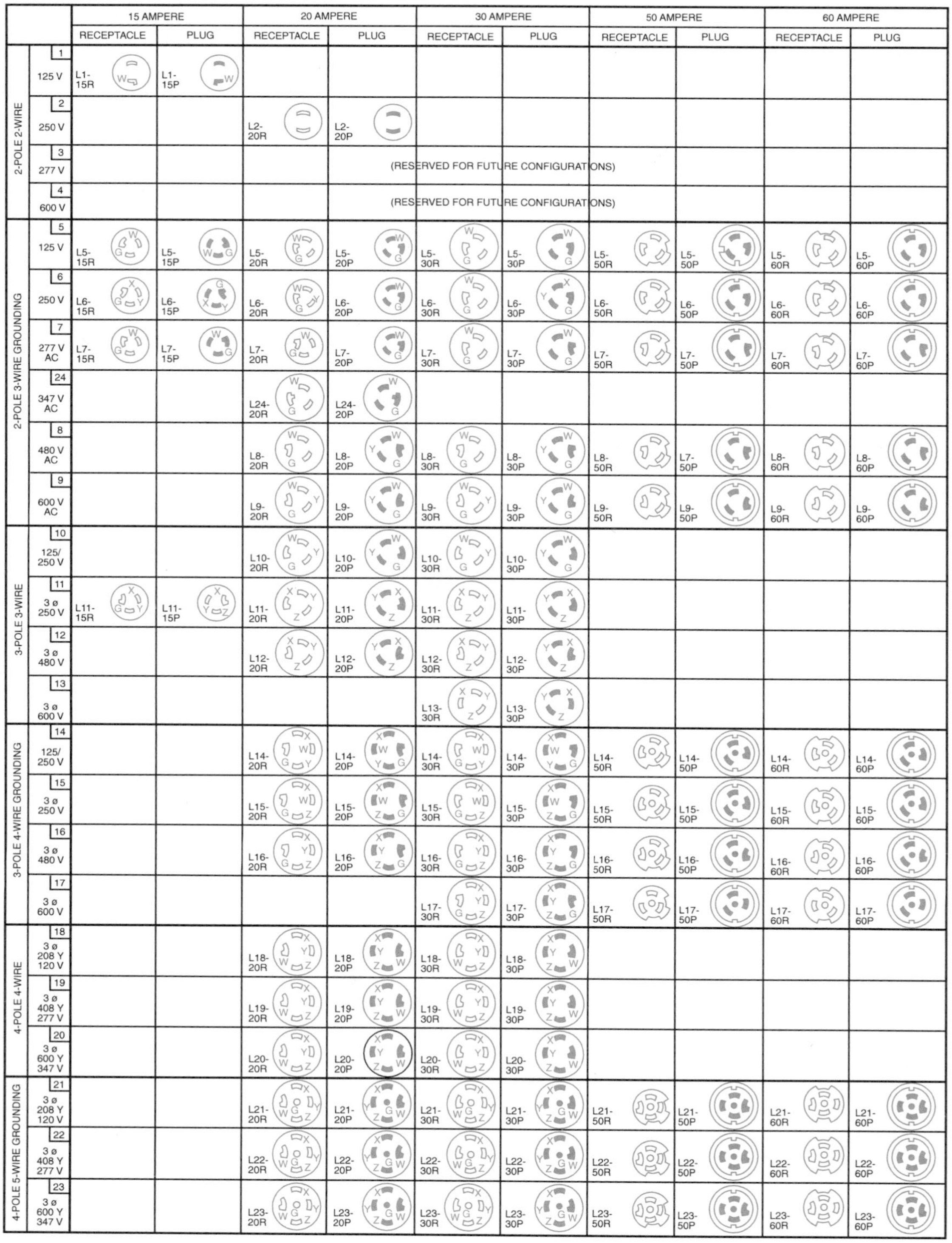

		15 AMPERE		20 AMPERE		30 AMPERE		50 AMPERE		60 AMPERE	
		RECEPTACLE	PLUG	RECEPTACLE	PLUG	RECEPTACLE	PLUG	RECEPTACLE	PLUG	RECEPTACLE	PLUG
2-POLE 2-WIRE	1 125 V	L1-15R	L1-15P								
	2 250 V			L2-20R	L2-20P						
	3 277 V					(RESERVED FOR FUTURE CONFIGURATIONS)					
	4 600 V					(RESERVED FOR FUTURE CONFIGURATIONS)					
2-POLE 3-WIRE GROUNDING	5 125 V	L5-15R	L5-15P	L5-20R	L5-20P	L5-30R	L5-30P	L5-50R	L5-50P	L5-60R	L5-60P
	6 250 V	L6-15R	L6-15P	L6-20R	L6-20P	L6-30R	L6-30P	L6-50R	L6-50P	L6-60R	L6-60P
	7 277 V AC	L7-15R	L7-15P	L7-20R	L7-20P	L7-30R	L7-30P	L7-50R	L7-50P	L7-60R	L7-60P
	24 347 V AC			L24-20R	L24-20P						
	8 480 V AC			L8-20R	L8-20P	L8-30R	L8-30P	L8-50R	L7-50P	L8-60R	L8-60P
	9 600 V AC			L9-20R	L9-20P	L9-30R	L9-30P	L9-50R	L9-50P	L9-60R	L9-60P
3-POLE 3-WIRE	10 125/250 V			L10-20R	L10-20P	L10-30R	L10-30P				
	11 3 ø 250 V	L11-15R	L11-15P	L11-20R	L11-20P	L11-30R	L11-30P				
	12 3 ø 480 V			L12-20R	L12-20P	L12-30R	L12-30P				
	13 3 ø 600 V					L13-30R	L13-30P				
3-POLE 4-WIRE GROUNDING	14 125/250 V			L14-20R	L14-20P	L14-30R	L14-30P	L14-50R	L14-50P	L14-60R	L14-60P
	15 3 ø 250 V			L15-20R	L15-20P	L15-30R	L15-30P	L15-50R	L15-50P	L15-60R	L15-60P
	16 3 ø 480 V			L16-20R	L16-20P	L16-30R	L16-30P	L16-50R	L16-50P	L16-60R	L16-60P
	17 3 ø 600 V					L17-30R	L17-30P	L17-50R	L17-50P	L17-60R	L17-60P
4-POLE 4-WIRE	18 3 ø 208 Y 120 V			L18-20R	L18-20P	L18-30R	L18-30P				
	19 3 ø 408 Y 277 V			L19-20R	L19-20P	L19-30R	L19-30P				
	20 3 ø 600 Y 347 V			L20-20R	L20-20P	L20-30R	L20-30P				
4-POLE 5-WIRE GROUNDING	21 3 ø 208 Y 120 V			L21-20R	L21-20P	L21-30R	L21-30P	L21-50R	L21-50P	L21-60R	L21-60P
	22 3 ø 408 Y 277 V			L22-20R	L22-20P	L22-30R	L22-30P	L22-50R	L22-50P	L22-60R	L22-60P
	23 3 ø 600 Y 347 V			L23-20R	L23-20P	L23-30R	L23-30P	L23-50R	L23-50P	L23-60R	L23-60P

Figure 210.7 *Configuration chart for specific-purpose locking plugs and receptacles. Reproduced from Wiring Devices — Dimensional Requirements, NEMA WD 6-88 (revision and redesignation of ANSI C73-73).*

Section 210-8 is the main rule for the application of ground-fault circuit interrupters (GFCIs). Since the original introduction of the GFCI in the 1971 *Code,* these devices have proven to their users and the electrical community that they are worth the added cost during construction or remodeling. Published data from the Consumer Product Safety Commission show a decreasing trend in the number of electrocutions in the United States since the introduction of GFCI devices. Unfortunately, no statistics are available for the actual number of lives saved by GFCI devices or the actual number of injuries prevented by GFCI devices. However, most would agree that the number of saved lives and prevented injuries is substantial.

Figure 210.8 shows a typical circuit arrangement of a GFCI for personnel protection. The line conductors are passed through a toroidal coil and connected to a shunt-trip device.

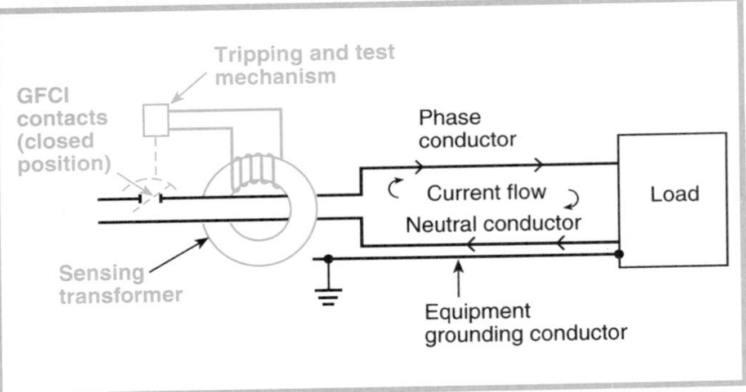

Figure 210.8 *The circuitry and components of a typical GFCI.*

As long as the current in each conductor remains equal, the device remains in a closed position. If one of the conductors comes in contact with a grounded object, either directly or through a person's body, some of the current returns by an alternative path, resulting in an unbalanced current. The toroidal coil senses the unbalanced current, and a circuit is established to the shunt-trip mechanism that reacts and opens the circuit. Note that the circuit design does not require the presence of an equipment grounding conductor, which is the reason Section 210-7(d) permits the use of GFCIs as replacements for receptacles where a grounding means does not exist. GFCIs operate on currents of 5 mA. Listing standards permit a differential of 4 to 6 mA. At trip levels of 5 mA (the instantaneous current could be much higher), a shock can be felt during the time of the fault. The shock can lead to involuntary reactions that may cause secondary accidents such as falls. GFCIs will not protect personnel from shock hazards where contact is between phase and neutral or phase-to-phase conductors.

A variety of GFCIs are available, including portable and plug-in types and circuit-breaker types, types built into attachment plug caps, and receptacle types. Each has a test switch so that the unit can be checked periodically to ensure proper operation. See Figures 210.9 and 210.10.

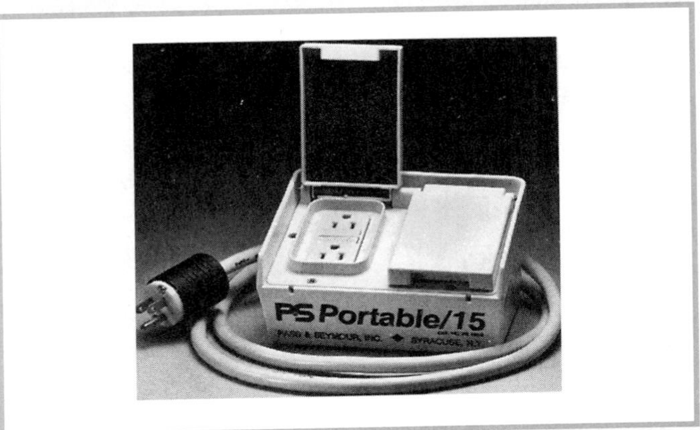

Figure 210.9 *A portable plug-in type of GFCI. (Pass & Seymour/ Legrand)*

Figure 210.10 *A 15-ampere duplex receptacle with integral GFCI that also protects downstream loads. (Pass & Seymour/ Legrand)*

The 1999 *Code* has revised and clarified the requirements for the use of ground-fault circuit interrupters (GFCI) in garages and other accessory buildings. While Section 210-8 is the main rule for GFCIs, other specific applications require the use of GFCIs. These specific applications are noted in the Table 210.1.

Table 210.1 Additional Requirements for Location of Ground-Fault Circuit-Interrupter Protection for Personnel

Location	Section
Audio System Equipment	640-10(a)
Boathouses	555-3
Commercial Garages	511-10
Construction Sites	305-6
Electric Vehicle Charging Systems	625-22
Elevators, Escalators, and Moving Walkways	620-85
Feeders	215-9
Fountains	680-51(a)
Health Care Facilities	517-20
	517-21
High-Pressure Spray Washing Appliances	422-8(d)(3)
Hydromassage Bathtubs	680-70
Marinas	555-3
Mobile Homes	550-8(b)
	550-23(e)
Motion Picture and Television Studios	530-73(a)(1)
Park Trailers	552-41(c)
Pools, Permanently Installed	680-6(a)
Pools, Storable	680-31
Signs, Mobile or Portable	600-10(c)(2)
Signs with Fountains	680-57
Recreational Vehicles	551-41
Recreational Vehicle Parks	551-71
Replacement Receptacles	210-7(d)

(1) Bathrooms.

The requirement for GFCIs for receptacles in bathrooms was originally included in the *Code* because data supplied with the *Code* proposals for GFCIs indicated that they could prevent a number of accidents occurring in bathrooms.

Section 210-8(a)(1) requires that all 125-volt, single-phase, 15- and 20-ampere receptacles in bathrooms are to have GFCI protection for personnel. This includes any such receptacles located for a clothes washer, clothes dryer (gas), integral with a lighting fixture, and, of course, the wall-mounted receptacles adjacent to the basin.

A *bathroom* is defined in Article 100 as an area that includes a basin with one or more of the following: a toilet, a tub, or a shower. The term applies to the entire area, whether a separating door, as illustrated in Figure 210.11, is present or not. Note that Section 210-52(d) requires that a receptacle be located adjacent to each basin location. However, basins are adjacent and in close proximity, then one receptacle outlet may satisfy the requirement.

(2) Garages, and also accessory buildings that have a floor located at or below grade level not intended as habitable rooms and limited to storage areas, work areas, and areas of similar use.

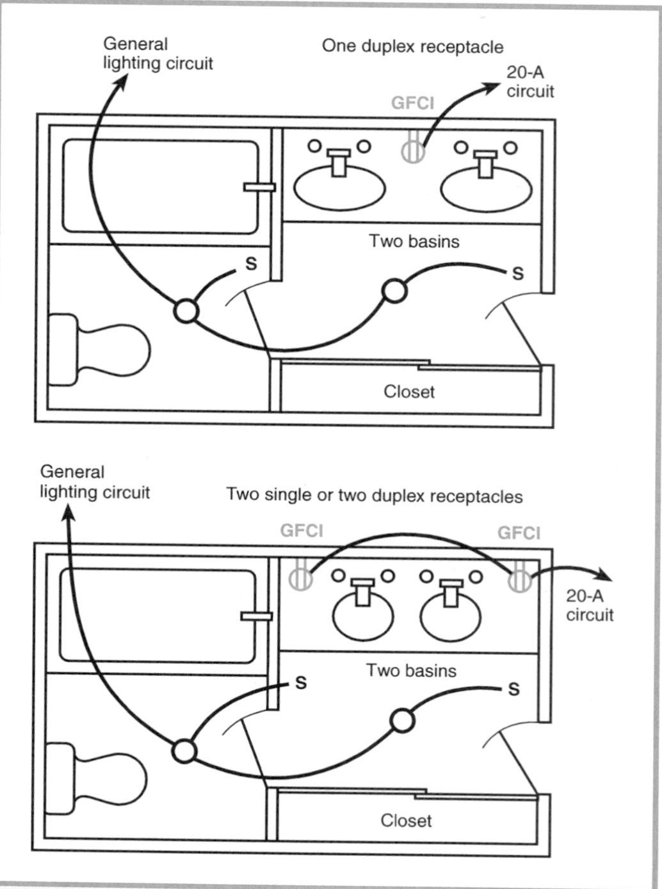

Figure 210.11 *GFCI-protected receptacles in accordance with Section 210-8(a)(1) in bathrooms where a separating door is present.*

Exception No. 1: Receptacles that are not readily accessible.

Exception No. 2: A single receptacle or a duplex receptacle for two appliances located within dedicated space for each appliance that, in normal use, is not easily moved from one place to another, and that is cord- and plug-connected in accordance with Section 400-7(a)(6), (a)(7), or (a)(8).

The requirement for GFCI receptacles in garages and sheds, as illustrated in Figure 210.12, provides a degree of safety for persons using portable hand-held tools, gardening appliances, lawn mowers, string trimmers, snow blowers, and so on, that might be connected to these receptacles, which are often the closest available. Also, GFCI protection is required in garage areas where auto repair work and general workshop electrical tools are used.

Exception No. 1 permits a ceiling-mounted receptacle installed for connection of a garage door opener

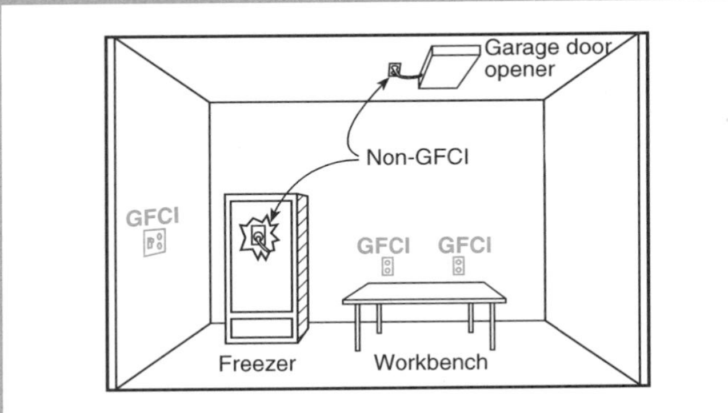

Figure 210.12 *Examples of receptacles in a garage that are required Section 210-8(a)(2) to have GFCI protection. Some receptacles are exempt because they are not readily accessible or are located for an appliance occupying dedicated space.*

to be exempt from the GFCI requirement. Exception No. 2 allows a duplex receptacle located where two cord- and plug-connected appliances occupy a dedicated space to be exempt from the GFCI requirement. If only a single cord- and plug-connected appliance, such as a food freezer, occupies the dedicated space, then a single receptacle must be used.

Receptacles installed under the exceptions to Section 210-8(a)(2) shall not be considered as meeting the requirements of Section 210-52(g).

(3) Outdoors.

Exception: Receptacles that are not readily accessible and are supplied by a dedicated branch circuit for electric snow-melting or deicing equipment shall be permitted to be installed in accordance with the applicable provisions of Article 426.

The dwelling unit shown in Figure 210.13 has four outdoor receptacles. Three of these receptacles are considered to be at direct grade level access and must have GFCI protection for personnel. The fourth receptacle located adjacent to the gutter for the roof-mounted snow-melting cable is not readily accessible and, therefore, is exempt from the GFCI requirements of Section 210-8(a)(3). However, this receptacle is covered by the equipment protection requirements of Section 426-28. See the commentary following Sections 210-52(e) and 410-57 regarding the installation of outdoor receptacles subject to moisture.

(4) Crawl spaces. Where the crawl space is at or below grade level.

(5) Unfinished basements. For purposes of this section, unfinished basements are defined as portions or areas of the basement not intended as habitable rooms and limited to storage areas, work areas, and the like.

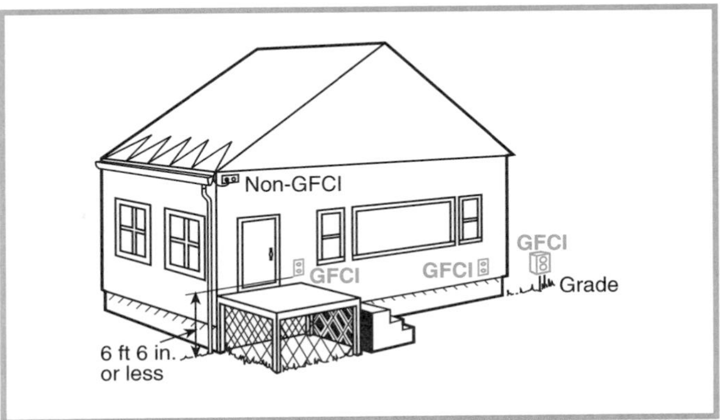

Figure 210.13 *A dwelling unit with three receptacles that are required by Section 210-8(a)(3) to have GFCI protection and one that is exempt because it is not readily accessible.*

Exception No. 1: Receptacles that are not readily accessible.

Exception No. 2: A single receptacle or a duplex receptacle for two appliances located within dedicated space for each appliance that, in normal use, is not easily moved from one place to another, and that is cord- and plug-connected in accordance with Section 400-7(a)(6), (a)(7), or (a)(8).

An unfinished portion of a basement is limited to storage areas, work areas, and the like. The receptacles in the work area of the basement shown in Figure 210.14 must have GFCI protection. Section 210-8(a)(5) does not apply to finished areas in basements, such

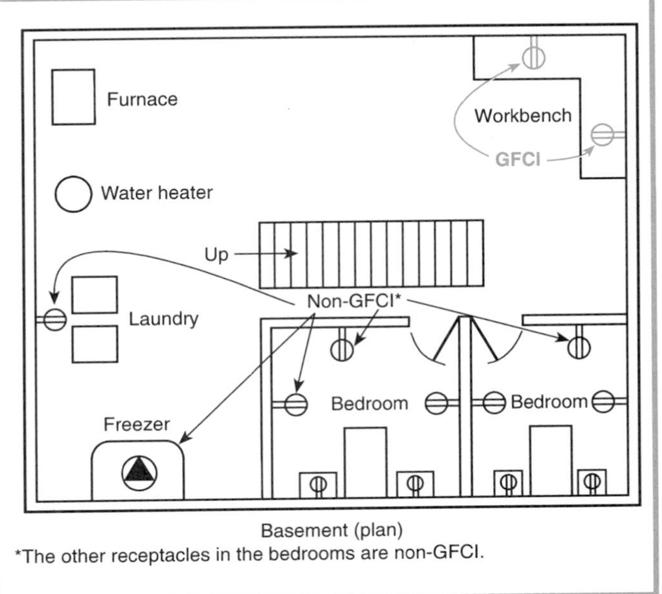

Figure 210.14 *A basement with GFCI-protected receptacles in accordance with Section 210-8(a)(5) in the work area and non-GFCI receptacles elsewhere.*

as sleeping rooms or family rooms. GFCI protection of these receptacles is not required. Freezer and laundry receptacles do not require GFCI protection in accordance with Section 210-8(a)(5), Exception No. 2.

Receptacles installed under the exceptions to Section 210-8(a)(5) shall not be considered as meeting the requirements of Section 210-52(g).

(6) Kitchens. Where the receptacles are installed to serve the countertop surfaces.

Many countertop kitchen appliances are ungrounded, and the presence of water and grounded surfaces contributes to a hazardous environment, leading to this requirement for GFCI protection around a kitchen sink. See Figure 210.26 for kitchens. The requirement is intended for receptacles serving the countertop. Receptacles installed for disposals, dishwashers, and trash compactors are not required to be protected by a GFCI.

(7) Wet bar sinks. Where the receptacles are installed to serve the countertop surfaces and are located within 6 ft (1.83 m) of the outside edge of the wet bar sink. Receptacle outlets shall not be installed in a face-up position in the work surfaces or countertops.

Many countertop wet bar appliances are ungrounded, and the presence of water and grounded surfaces contributes to a hazardous environment, leading to this requirement for GFCI protection around a wet bar sink. As illustrated in Figure 210.15, receptacles within 6 ft of a wet bar sink are required to be GFCI protected. These receptacles cannot be installed in the countertop in the face-up position because liquid, dirt, and other foreign material can enter the receptacle.

(b) Other than Dwelling Units. All 125-volt, single-phase, 15- and 20-ampere receptacles installed in the locations specified below shall have ground-fault circuit-interrupter protection for personnel.

(1) Bathrooms

Where receptacles are provided in bathroom areas of hotels and motels, GFCI-protected receptacles are required. Lavatories in airports, commercial buildings, industrial facilities, and other nondwelling occupancies are required to have *all* their receptacles GFCI protected. The only exception to this rule is found in Section 517-21, which permits receptacles in hospital critical care areas to be non-GFCI where the toilet and basin are installed in the patient room

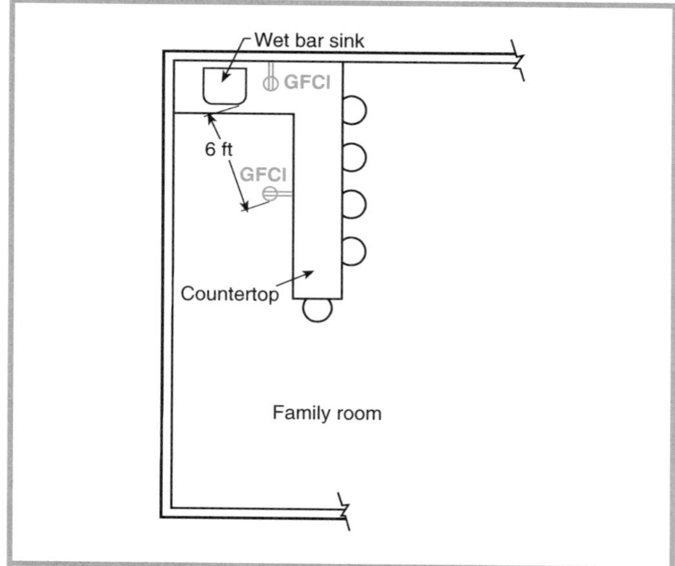

Figure 210.15 *GFCI-protected receptacles in accordance with Section 210-8(a)(7) that are located near a wet bar and serve the countertop.*

rather than a separate bathroom. Some motel and hotel bathrooms, like the one shown in Figure 210.16, have the washbasin located outside the door to the room containing the tub, toilet, or another washbasin. The definition of *bathroom* as found in Article 100 applies, as does the GFCI requirement of Section 210-8(b)(1).

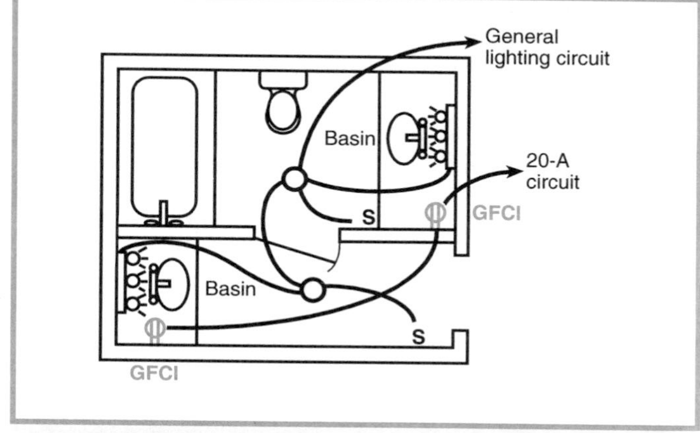

Figure 210.16 *GFCI-protected receptacles shown in accordance with Section 210-8(b)(1) in a motel/hotel bathroom where one basin is located outside the door to the rest of the bathroom area.*

(2) Rooftops

Section 210-8(b)(2) requires all rooftop 15- and 20-ampere receptacles in nondwelling occupancies to be

GFCI protected. See Section 210-63 for placement of rooftop receptacles.

Exception: Receptacles that are not readily accessible and are supplied from a dedicated branch circuit for electric snow-melting or deicing equipment shall be permitted to be installed in accordance with the applicable provisions of Article 426.

210-9. Circuits Derived from Autotransformers. Branch circuits shall not be derived from autotransformers unless the circuit supplied has a grounded conductor that is electrically connected to a grounded conductor of the system supplying the autotransformer.

Figures 210.17 through 210.20 illustrate typical applications of autotransformers. In Figure 210.17, a 120-volt supply is derived from a 240-volt system. The grounded conductor of the primary system is electrically connected to the grounded conductor of the secondary system.

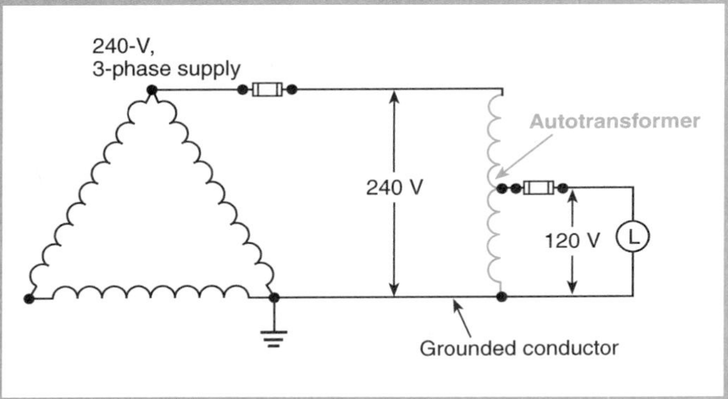

Figure 210.17 *The circuitry for an autotransformer used to derive a 2-wire, 120-volt system for lighting or convenience receptacles from a 240-volt corner-grounded delta-powered system.*

A buck-boost transformer is classified as an autotransformer. A buck-boost transformer provides a means of raising or lowering (boosting or bucking) a supply line voltage by a small amount (usually no more than ±20 percent). It is an insulating transformer with two primary windings (H1-H2 and H3-H4) and two secondary windings (X1-X2 and X3-X4). Its primary and secondary windings are connected together so that the electrical characteristics are changed from an insulating transformer to those of a boosting or bucking autotransformer, correcting voltage up to ±20 percent.

A single unit is used to boost or buck single-phase voltage, but two or three units are used to boost or buck 3-phase voltage. An autotransformer requires little physical space, is economical, and, above all, is efficient.

One common application of a boost transformer is to derive a single-phase, 240-volt supply system for ranges, air conditioners, heating elements, and motors from a 3-phase, 208Y/120-volt source system. The boosted leg should not be used to supply line-to-neutral loads because the boosted line-to-neutral voltage will be higher than 120 volts.

Another common boost transformer application is to increase a single-phase, 240-volt source to a single-phase, 277-volt supply for lighting systems. One common 3-phase application is to boost 440 volts to 550 volts for power equipment. Transforming 240 volts to 208 volts for use with 208-volt appliances is a common application of a buck transformer. Another common application of a buck transformer is to convert a 480Y/277-volt source to a 416Y/240-volt supply system.

Literature containing diagrams for connection and application of autotransformers is available from manufacturers.

Exception No. 1: An autotransformer shall be permitted without the connection to a grounded conductor where transforming from a nominal 208 volts to a nominal 240-volt supply or similarly from 240 volts to 208 volts.

Exception No. 1 allows an autotransformer (without an electrical connection to a grounded conductor) to extend or add an individual branch circuit in an existing installation where transforming (boosting) 208 volts to 240 volts, as shown in Figure 210.18. Figures 210.19 and 210.20 illustrate typical single-phase and 3-phase buck and boost transformers connected as autotransformers to change 240 volts to 208 volts or vice versa.

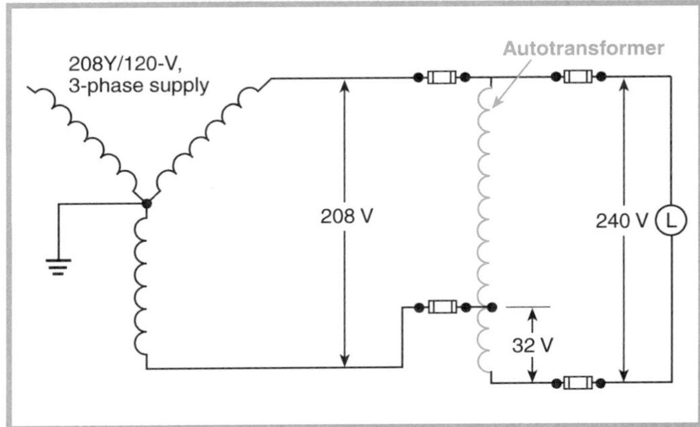

Figure 210.18 *The circuitry for an autotransformer used to derive a 240-volt system for appliances from a 208Y/120-volt source.*

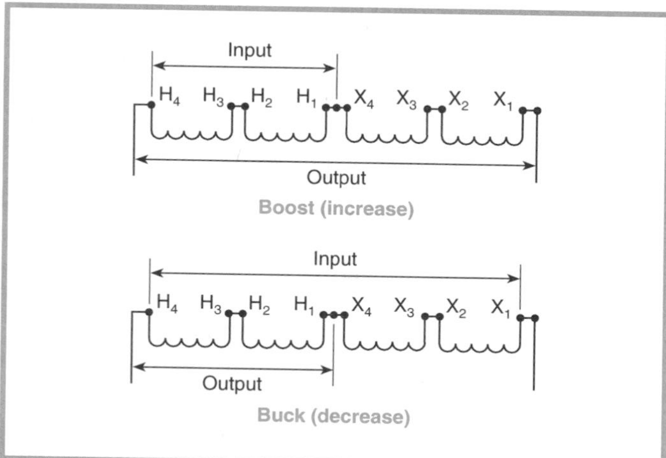

Figure 210.19 *Typical single-phase connection diagrams for buck or boost transformers connected as autotransformers to change 240 volts single-phase to 208 volts or vice versa.*

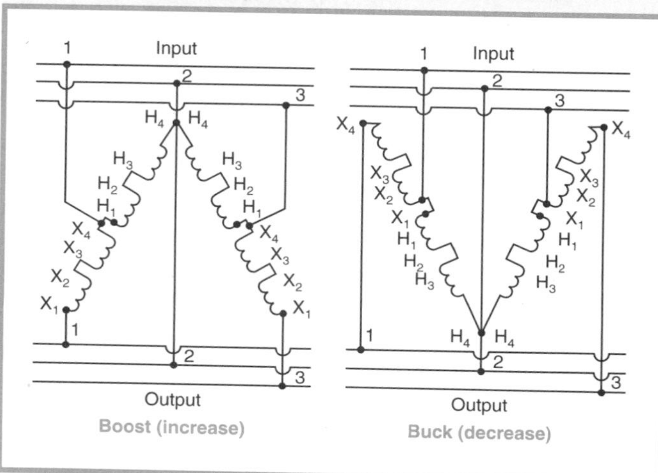

Figure 210.20 *Typical connection diagrams for buck or boost transformers connected in 3-phase open delta as autotransformers to change 240 volts to 208 volts or vice versa.*

Exception No. 2: In industrial occupancies, where conditions of maintenance and supervision ensure that only qualified persons will service the installation, autotransformers shall be permitted to supply nominal 600-volt loads from nominal 480-volt systems, and 480-volt loads from nominal 600-volt systems, without the connection to a similar grounded conductor.

In industrial locations, Exception No. 2 allows the use of an autotransformer to supply a 600-volt load from 480-volt systems, provided there are qualified personnel to service the installation. It also allows 480-volt loads to be supplied through an autotransformer supplied by a 600-volt system.

210-10. Ungrounded Conductors Tapped from Grounded Systems. Two-wire dc circuits and ac circuits of two or more ungrounded conductors shall be permitted to be tapped from the ungrounded conductors of circuits that have a grounded neutral conductor. Switching devices in each tapped circuit shall have a pole in each ungrounded conductor. All poles of multipole switching devices shall manually switch together where such switching devices also serve as a disconnecting means as required by the following:

(1) Section 410-48 for double-pole switched lampholders
(2) Section 410-54(b) for electric-discharge lamp auxiliary equipment switching devices
(3) Section 422-31(b) for an appliance
(4) Section 424-20 for a fixed electric space-heating unit
(5) Section 426-51 for electric deicing and snow-melting equipment
(6) Section 430-85 for a motor controller
(7) Section 430-103 for a motor

Two-wire ungrounded branch circuits may be tapped from ac or dc circuits of two or more ungrounded conductors that have a grounded neutral conductor. Figure 210.21 (top) illustrates an ungrounded 2-wire branch circuit tapped from the ungrounded conductors of a dc or single-phase system to supply a small motor. Figure 210.21 (bottom) illustrates a 3-phase, 4-wire wye system.

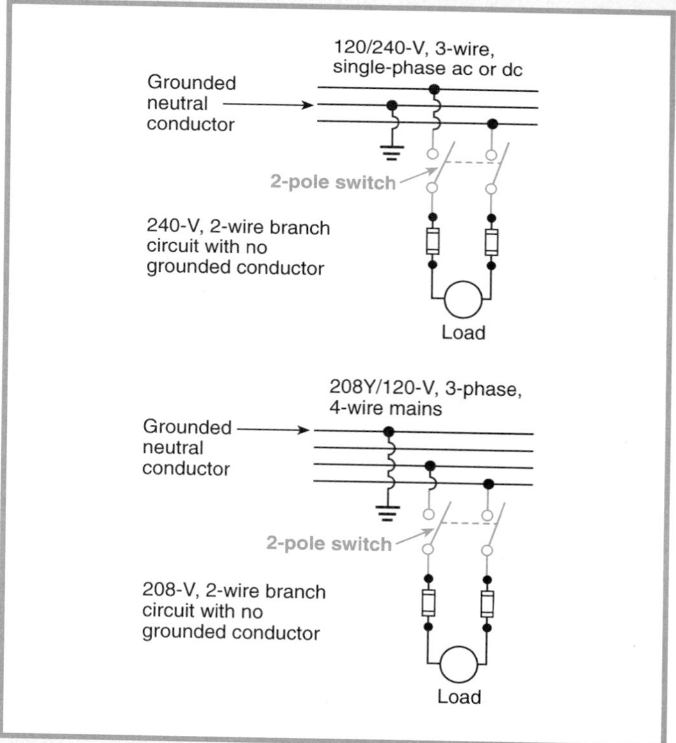

Figure 210.21 *Branch circuits tapped from ungrounded conductors of multiwire systems.*

Circuit breakers or switches that are used as the disconnecting means for a branch circuit are required to open all poles simultaneously. This requirement involves only the manual operation of the disconnecting means; therefore, if switches and fuses are used and one fuse blows, or if circuit breakers (two single-pole circuit breakers with a handle tie) are used and one breaker trips, one pole could remain closed. The intention is not to provide a common trip of fuses or circuit breakers but to disconnect (manually) the ungrounded conductors of the branch circuit with one manual operation. See Section 240-20(b) for information on handle ties.

210-11. Branch Circuits Required. Branch circuits for lighting and for appliances, including motor-operated appliances, shall be provided to supply the loads computed in accordance with Section 220-3. In addition, branch circuits shall be provided for specific loads not covered by Section 220-3 where required elsewhere in this *Code* and for dwelling unit loads as specified in (c).

Section 210-11 is new for the 1999 *NEC*. It is derived from Section 220-4 of the 1996 *NEC*. Since this text contains requirements that deal with the number of circuits and the required branch circuits, those requirements have been moved from Article 220 to Article 210. In addition, a new subsection, (c) Dwelling Units, was created. It includes the dwelling unit requirements previously located in Sections 220-4(b) and (c). Also, the required branch circuit for the bathroom receptacle outlets is specified in this section in lieu of Section 210-52(d) since it is a required branch circuit and is consistent with the requirements in Sections 210-52(b) and 210-52(f). A summary of the relocated text is as follows:

New Section	1996 *NEC* Corresponding Section
210-11	220-4
210-11(a)	220-4(a)
210-11(b)	220-4(d)
210-11(c)	220-4(b) and (c)
210-11(c)(1)	220-4(b)
210-11(c)(2)	220-4(c)
210-11(c)(3)	210-52(d), 2nd and 3rd sentences

(a) Number of Branch Circuits. The minimum number of branch circuits shall be determined from the total computed load and the size or rating of the circuits used. In all installations, the number of circuits shall be sufficient to supply the load served. In no case shall the load on any circuit exceed the maximum specified by Section 220-4.

(b) Load Evenly Proportioned Among Branch Circuits. Where the load is computed on a volt-amperes/square foot (0.093 m^2) basis, the wiring system up to and including the branch-circuit panelboard(s) shall be provided to serve not less than the calculated load. This load shall be evenly proportioned among multioutlet branch circuits within the panelboard(s). Branch-circuit overcurrent devices and circuits need only be installed to serve the connected load.

(c) Dwelling Units.

(1) Small-Appliance Branch Circuits. In addition to the number of branch circuits required by other parts of this section, two or more 20-ampere small-appliance branch circuits shall be provided for all receptacle outlets specified by Section 210-52(b).

(2) Laundry Branch Circuits. In addition to the number of branch circuits required by other parts of this section, at least one additional 20-ampere branch circuit shall be provided to supply the laundry receptacle outlet(s) required by Section 210-52(f). This circuit shall have no other outlets.

(3) Bathroom Branch Circuits. In addition to the number of branch circuits required by other parts of this section, at least one 20-ampere branch circuit shall be provided to supply the bathroom receptacle outlet(s). Such circuits shall have no other outlets.

Exception: Where the 20-ampere circuit supplies a single bathroom, outlets for other equipment within the same bathroom shall be permitted to be supplied in accordance with Section 210-23(a).

FPN: See Examples D1(a), D1(b), D2(b), and D4(a) in Appendix D.

210-12. Arc-Fault Circuit-Interrupter Protection.

(a) Definition. An *arc-fault circuit interrupter* is a device intended to provide protection from the effects of arc faults by recognizing characteristics unique to arcing and by functioning to de-energize the circuit when an arc fault is detected.

(b) Dwelling Unit Bedrooms. All branch circuits that supply 125-volt, single-phase, 15- and 20-ampere receptacle outlets installed in dwelling unit bedrooms shall be protected by an arc-fault circuit interrupter(s). This requirement shall become effective January 1, 2002.

The definition given in Section 210-12(a) explains the function of an AFCI. These special devices are intended to mitigate the effects of arcing faults that may pose risk of fire ignition under certain conditions if the arcing persists. They are tested using methods that create or simulate arcing conditions to determine their ability to recognize and react to arcing faults. These devices are also tested to verify that operation is not unduly inhibited by the presence of loads and circuit characteristics that may mask or attenuate symptoms of unwanted arcing. In addition, these devices are evaluated to determine resistance to unwanted tripping because of the presence of arcing that occurs in control and utilization equipment under normal operating conditions or a loading condition which closely mimics an arcing fault, such as the current waveform produced by some solid-state loads, e.g., electronic ballasts.

UL has announced the development of a standard for devices that are intended to mitigate the effects of arcing faults that may pose a risk of fire ignition. The standard will be designated UL 1699. The scope of this standard will cover devices having a maximum rating of 20 amperes and intended for use in 120-volt ac, 60-Hz circuits. These devices may also include capability to perform other functions such as overcurrent protection, ground-fault circuit interruption, surge suppression, etc. The draft UL 1699 standard describes a Branch Circuit Arcing Fault Circuit Interrupter as a device intended to be installed at the origin of a branch circuit or feeder, such as at a panelboard. This device is intended to provide protection for the branch circuit against the unwanted effects of arcing. This device also provides limited protection to branch circuit extension wiring.

With no real field history of this newly available technology, the *NEC* requirement is drafted to implement branch-circuit protection only for bedroom receptacle outlet circuits. Restricting the requirement to bedroom circuits reflects the desire to gain field experience in a limited application before mandating installation of devices in other unit circuits. Bedrooms contain readily ignitible cloth and cotton materials, and occupants may be sleeping when ignition occurs and not likely able to take protective action rapidly.

B. Branch-Circuit Ratings

210-19. Conductors — Minimum Ampacity and Size.

(a) General. Branch-circuit conductors shall have an ampacity not less than the maximum load to be served. Where a branch circuit supplies continuous loads or any combination of continuous and noncontinuous loads, the minimum branch-circuit conductor size, before the application of any adjustment or correction factors, shall have an allowable ampacity equal to or greater than the noncontinuous load plus 125 percent of the continuous load.

Section 210-19 was revised for the 1999 *NEC* to include the conductor sizing rules previously found in Sections 220-3(a) and 210-22(c) for branch circuits. Section 210-19 is the logical location for this requirement because it deals with the ampacity and size requirements for the conductors. The text previously in Section 210-22(c) was consolidated into Section 210-19.

Exception: Where the assembly, including the overcurrent devices protecting the branch circuit(s), is listed for operation at 100 percent of its rating, the ampacity of the branch circuit conductors shall be permitted to be not less than the sum of the continuous load plus the noncontinuous load.

FPN No. 1: See Section 310-15 for ampacity ratings of conductors.

FPN No. 2: See Part B of Article 430 for minimum rating of motor branch-circuit conductors.

FPN No. 3: See Section 310-10 for temperature limitation of conductors.

FPN No. 4: Conductors for branch circuits as defined in Article 100, sized to prevent a voltage drop exceeding 3 percent at the farthest outlet of power, heating, and lighting loads, or combinations of such loads, and where the maximum total voltage drop on both feeders and branch circuits to the farthest outlet does not exceed 5 percent, will provide reasonable efficiency of operation. See Section 215-2 for voltage drop on feeder conductors.

Improper voltage due to a voltage drop in supply conductors is a major source of trouble and inefficient operation in electrical equipment. Undervoltage conditions reduce the capability and reliability of motors, lighting sources, heaters, and solid-state equipment. Sample voltage-drop calculations may be found in the commentary following Section 215-2(b), FPN No. 3, and following Table 9 in Chapter 9.

(b) Multioutlet Branch Circuits. Conductors of branch circuits supplying more than one receptacle for cord- and plug-connected portable loads shall have an ampacity of not less than the rating of the branch circuit.

Section 210-19(b) is a new section for the 1999 *NEC* that incorporates the multioutlet requirement previously found in Section 210-19(a) (second sentence). Since this is a critical requirement for these branch circuits, a dedicated subsection was created to clearly indicate the rule.

Only multioutlet branch-circuit conductors supplying receptacles for cord- and plug-connected portable loads are required to have an ampacity that is not less than the rating of the circuit (the rating of the overcurrent device per Section 210-3), because the loading of such circuits is unpredictable.

(c) Household Ranges and Cooking Appliances. Branch-circuit conductors supplying household ranges, wall-mounted ovens, counter-mounted cooking units, and other household cooking appliances shall have an ampacity not less than the rating of the branch circuit and not less than the maximum load to be served. For ranges of 8¾ kW or more rating, the minimum branch-circuit rating shall be 40 amperes.

A minimum 40-ampere branch-circuit rating would be, for example, No. 8 Type TW copper or No. 6 Type TW aluminum. See Table 310-16 for other applications.

Exception No. 1: Tap conductors supplying electric ranges, wall-mounted electric ovens, and counter-mounted electric cooking units from a 50-ampere branch circuit shall have an ampacity of not less than 20 and shall be sufficient for the load to be served. The taps shall not be longer than necessary for servicing the appliance.

As illustrated in Figure 210.22, Exception No. 1 permits a 20-ampere tap conductor from a range, oven, or cooking unit to be connected to a 50-ampere branch circuit, if the following four conditions are met.

1. The taps are not longer than necessary to service or permit access to the junction box.
2. The taps to each unit are properly spliced.
3. The junction box is adjacent to each unit.
4. The taps are of sufficient size for the load to be served.

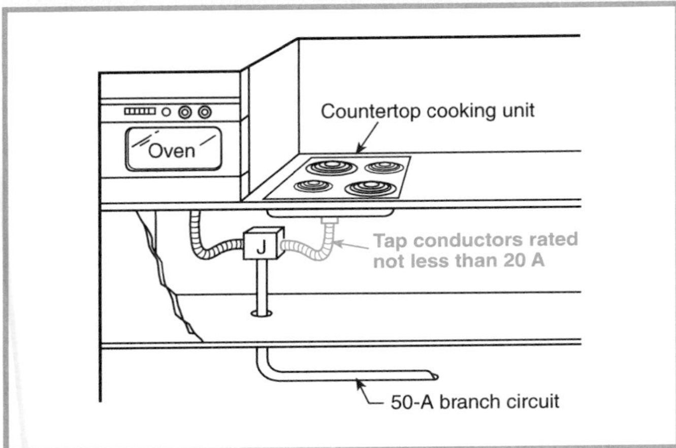

Figure 210.22 *Tap conductors permitted by Section 210-19(c), Exception No. 1, to be sized smaller than the branch-circuit conductors but not longer than necessary for servicing.*

Exception No. 2: The neutral conductor of a 3-wire branch circuit supplying a household electric range, a wall-mounted oven, or a counter-mounted cooking unit shall be permitted to be smaller than the ungrounded conductors where the maximum demand of a range of 8¾ kW or more rating has been computed according to Column A of Table 220-19, but shall have an ampacity of not less than 70 percent of the branch-circuit rating and shall not be smaller than No. 10.

Column A of Table 220-19 and Section 220-19 indicate that the maximum demand for one range (not over 12 kW rating) is 8 kW (8 kW = 8000 volt-amperes; 8000 volt-amperes ÷ 240 volts = 33.3 amperes). The allowable ampacity of a No. 8 TW copper conductor is 40 amperes (see Table 310-16) and may be used for the range branch circuit. According to this computation, the neutral of this 3-wire circuit can be smaller than No. 8 but not smaller than No. 10, which has an allowable ampacity of 30 amperes (30 amperes is more than 70 percent of 40 amperes, as per the exception). The maximum demand for a neutral of an 8-kW range circuit seldom exceeds 25 amperes, since current is drawn from the neutral only for lights, clocks, timers, and some heating elements when in the low-heating position.

(d) Other Loads. Branch-circuit conductors that supply loads other than those specified in Section 210-2 and other than cooking appliances as covered above shall have an ampacity sufficient for the loads served and shall not be smaller than No. 14.

Exception No. 1: Tap conductors for such loads shall have an ampacity of not less than 15 for circuits rated less than 40 amperes and of not less than 20 for circuits rated at 40 or 50 amperes and only where these tap conductors supply any of the following loads:

(a) Individual lampholders or fixtures with taps extending not longer than 18 in. (457 mm) beyond any portion of the lampholder or fixture
(b) A fixture having tap conductors as provided in Section 410-67
(c) Individual outlets, other than receptacle outlets, with taps not over 18 in. (457 mm) long
(d) Infrared lamp industrial heating appliances
(e) Nonheating leads of deicing and snow-melting cables and mats

Tap conductors are generally required to have the same ampacity as the branch circuit overcurrent device. Exception No. 1 lists specific applications in subparts (a) through (e) where the tap conductors are permitted with reduced ampacities. These tap conductors are required to have an ampacity of 15 amperes or more (No. 14 copper conductors) for circuits rated less than 40 amperes. They are required to have an ampacity of 20 amperes or more (No. 12 copper conductors) for circuits rated 40 or 50 amperes. See Section 240-4 for special requirements for the overcurrent protection of flexible cords and fixture wires as indicated in Exception No. 2.

Exception No. 2: Fixture wires and flexible cords shall be permitted to be smaller than No. 14 as permitted by Section 240-4.

210-20. Overcurrent Protection. Branch-circuit conductors and equipment shall be protected by overcurrent protective devices that have a rating or setting that complies with (a) through (d).

(a) Continuous and Noncontinuous Loads. Where a branch circuit supplies continuous loads or any combination of continuous and noncontinuous loads, the rating of the overcurrent device shall not be less than the noncontinuous load plus 125 percent of the continuous load.

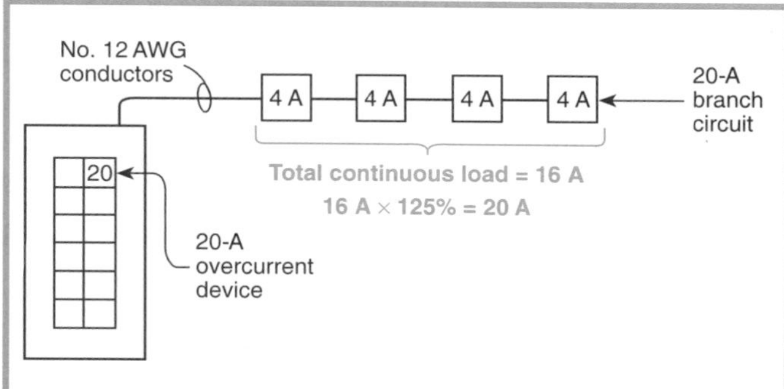

Figure 210.23 A continuous load (store lighting) calculated at 125 percent to determine the ampacity of the conductor and the branch-circuit size.

Exception: Where the assembly, including the overcurrent devices protecting the branch circuit(s), is listed for operation at 100 percent of its rating, the ampere rating of the overcurrent device shall be permitted to be not less than the sum of the continuous load plus the noncontinuous load

According to Section 210-20, an overcurrent device that supplies noncontinuous and continuous loads must have a rating that is not less than the sum of 100 percent of the noncontinuous load plus 125 percent of the continuous load, calculated in accordance with Article 210.

In addition, Section 210-19 requires that the circuit conductors, chosen from the ampacity tables, must have an initial ampacity of not less than the sum of 100 percent of the noncontinuous load plus 125 percent of the continuous load, the same as calculated for the overcurrent device.

The rating of the overcurrent device cannot exceed the final ampacity of the circuit conductors after all the derating or correction factors have been applied, such as for temperature or number of conductors.

Example

Determine the minimum-size overcurrent protective device and the minimum conductor size for the following circuit:

1. 25-ampere continuous load
2. 60°C overcurrent device terminal rating
3. Type THWN conductors
4. Four current-carrying conductors in a raceway

Overcurrent Protective Device (OCPD) Size Selection

Referring to Section 210-20(a), 1.25 × 25 amperes = 31.25 amperes minimum. The minimum standard-size OCPD, according to Section 240-6, is 35 amperes.

Conductor Size Selection Before Derating

1.25 × 25 amperes = 31.25 amperes minimum

Due to the 60°C rating of the overcurrent device terminal, it is necessary to choose a conductor based on ampacities in the 60°C column. The calculated load must not exceed the conductor ampacity. Therefore, a No. 8 THWN conductor with an ampacity of 40 amperes is selected. The selected conductor has a 75°C rating. The higher temperature rating was chosen because further derating will be necessary. The maximum ampacity of a THWN copper conductor in the 60°C column of Table 310-16 is 40 amperes, which results in a No. 8 size.

Application of Derating Factors

Since there are four conductors in the raceway, Section 310-15(b)(2)(a) and Table 310-15(b)(2) require an 80 percent derating factor. According to Table 310-16, No. 8 THWN has an ampacity of 50 amperes in the 75°C column, to which the derating factor is applied as follows:

0.80 × 50 amperes = 40 amperes

A conductor with an ampacity of 40 amperes protected by a 35-ampere OCPD complies with Section 240-3(b).

Circuit Evaluation

The previous calculation results in four No. 8 THWN copper conductors in one raceway, each with an ampacity of 50 amperes, supplying a 25-ampere continuous load and protected by a 35-ampere OCPD.

The rating of a branch circuit is based on the size of the overcurrent protective device. Even though the conductor in this example has an ampacity of 50 amperes, the branch circuit is still a 35-ampere branch circuit (see Section 210-3). A 40-ampere overcurrent device may be installed in this circuit to utilize the current-carrying capacity of the conductor.

(b) Conductor Protection. Conductors shall be protected in accordance with Section 240-3.

Exception No. 1: Tap conductors as permitted in Section 210-19(d) shall be permitted to be protected by the branch-circuit overcurrent device.

Exception No. 2: Fixture wires and flexible cords shall be permitted to be protected in accordance with Section 240-4.

(c) Equipment. The rating or setting of the overcurrent protective device shall not exceed that specified in the applicable articles referenced in Section 240-2 for equipment.

(d) Outlet Devices. The rating or setting shall not exceed that specified in Section 210-21 for outlet devices.

210-21. Outlet Devices. Outlet devices shall have an ampere rating that is not less than the load to be served and shall comply with (a) and (b).

(a) Lampholders. Where connected to a branch circuit having a rating in excess of 20 amperes, lampholders shall be of the heavy-duty type. A heavy-duty lampholder shall have a rating of not less than 660 watts if of the admedium type and not less than 750 watts if of any other type.

The intent of Section 210-21(a) is to restrict a fluorescent lighting branch-circuit rating to not more than 20 amperes, because most lampholders manufactured for use with fluorescent lights are not of the heavy-duty type and are rated at 660 watts or 250 watts.

Branch-circuit conductors for fluorescent electric-discharge lighting are usually connected to a ballast rather than to lampholders, and, by specifying a wattage rating for these lampholders, a limit of 20 amperes is applied to ballast circuits.

It is only the admedium-base lampholder that is recognized as heavy duty at the rating of 660 watts. Other lampholders are required to have a rating of not less than 750 watts to be recognized as heavy duty. The requirement of Section 210-21(a) prohibits the use of medium-base screw shell lampholders on branch circuits that are in excess of 20 amperes.

(b) Receptacles.

(1) A single receptacle installed on an individual branch circuit shall have an ampere rating of not less than that of the branch circuit.

Exception No. 1: Where installed in accordance with Section 430-81(c).

Exception No. 2: A receptacle installed exclusively for the use of a cord- and plug-connected arc welder shall be permitted to have an ampere rating not less than the minimum branch-circuit conductor ampacity determined by Section 630-11(a) for arc welders.

FPN: See definition of *Receptacle* in Article 100.

(2) Where connected to a branch circuit supplying two or more receptacles or outlets, a receptacle shall not supply a total cord- and plug-connected load in excess of the maximum specified in Table 210-21(b)(2).

Table 210-21(b)(2). Maximum Cord- and Plug-Connected Load to Receptacle

Circuit Rating (Amperes)	Receptacle Rating (Amperes)	Maximum Load (Amperes)
15 or 20	15	12
20	20	16
30	30	24

(3) Where connected to a branch circuit supplying two or more receptacles or outlets, receptacle ratings shall conform to the values listed in Table 210-21(b)(3), or, where larger than 50 amperes, the receptacle rating shall not be less than the branch-circuit rating.

Table 210-21(b)(3). Receptacle Ratings for Various Size Circuits

Circuit Rating (Amperes)	Receptacle Rating (Amperes)
15	Not over 15
20	15 or 20
30	30
40	40 or 50
50	50

A single receptacle installed on an individual branch circuit is required to have an ampere rating of not less than that of the branch circuit. For example, a single receptacle on a 20-ampere branch circuit must be rated at 20 amperes; however, two or more 15-ampere receptacles are permitted on a 20-ampere general-purpose branch circuit. This requirement does not apply to specific types of cord- and plug-connected arc welders.

Exception: Receptacles for one or more cord- and plug-connected arc welders shall be permitted to have ampere ratings not less than the minimum branch-circuit conductor ampacity permitted by Sections 630-11(a) or (b) as applicable for arc welders.

(4) The ampere rating of a range receptacle shall be permitted to be based on a single range demand load as specified in Table 220-19.

210-23. Permissible Loads. In no case shall the load exceed the branch-circuit ampere rating. An individual branch circuit shall be permitted to supply any load for which it is rated. A branch circuit supplying two or more outlets or receptacles shall supply only the loads specified according to its size as specified in (a) through (d) and as summarized in Section 210-24 and Table 210-24.

(a) 15- and 20-Ampere Branch Circuits. A 15- or 20-ampere branch circuit shall be permitted to supply lighting units or other utilization equipment, or a combination of both. The rating of any one cord- and plug-connected utilization equipment shall not exceed 80 percent of the branch-circuit ampere rating. The total rating of utilization equipment fastened in place, other than lighting fixtures, shall not exceed 50 percent of the branch-circuit ampere rating where lighting units, cord- and plug-connected utilization equipment not fastened in place, or both, are also supplied.

Exception: The small appliance branch circuits, laundry branch circuits, and bathroom branch circuits required in a dwelling unit(s) by Sections 210-11(c)(1), (2), and (3) shall supply only the receptacle outlets specified in that section.

Section 210-23(a) permits a 15- or 20-ampere branch circuit for lighting to also supply utilization equipment fastened in place, such as an air conditioner. The equipment load must not exceed 50 percent of the branch-circuit ampere rating (7.5 amperes on a 15-ampere circuit and 10 amperes on a 20-ampere circuit). Such equipment, fastened in place, is not permitted on the small-appliance branch circuits required in the kitchen, dining room, and so on, by Section 220-4(b).

(b) 30-Ampere Branch Circuits. A 30-ampere branch circuit shall be permitted to supply fixed lighting units with heavy-duty lampholders in other than a dwelling unit(s) or utilization equipment in any occupancy. A rating of any one cord- and plug-connected utilization equipment shall not exceed 80 percent of the branch-circuit ampere rating.

(c) 40- and 50-Ampere Branch Circuits. A 40- or 50-ampere branch circuit shall be permitted to supply cooking appliances that are fastened in place in any occupancy. In other than dwelling units, such circuits shall be permitted to supply fixed lighting units with heavy-duty lampholders, infrared heating units, or other utilization equipment.

A branch circuit that supplies two or more outlets is permitted to supply only the loads specified according to its size, in accordance with Sections 210-23(a) through (c) and as summarized in Section 210-24 and Table 210-24. Other circuits are not permitted to have more than one outlet and are considered individual branch circuits. However, Sections 517-71 and 660-4(b) do not require individual branch circuits for portable, mobile, and transportable medical X-ray equipment requiring a capacity of not over 60 amperes.

(d) Branch Circuits Larger than 50 Amperes. Branch circuits larger than 50 amperes shall supply only nonlighting outlet loads.

See the commentary following Section 210-3, Exception, regarding multioutlet branch circuits greater than 50 amperes that are permitted to supply nonlighting outlet loads in industrial establishments.

210-24. Branch-Circuit Requirements — Summary. The requirements for circuits that have two or more outlets or receptacles, other than the receptacle circuits of Sections 210-11(c)(1) and (2) as specifically provided for above, are summarized in Table 210-24.

Table 210-24. Summary of Branch-Circuit Requirements

Circuit Rating	**15 A**	**20 A**	**30 A**	**40 A**	**50 A**
Conductors (min. size):					
Circuit wires[1]	14	12	10	8	6
Taps	14	14	14	12	12
Fixture wires and cords — See Section 240-4					
Overcurrent Protection	**15 A**	**20 A**	**30 A**	**40 A**	**50 A**
Outlet Devices:					
Lampholders permitted	Any type	Any type	Heavy duty	Heavy duty	Heavy duty
Receptacle rating[2]	15 max. A	15 or 20 A	30 A	40 or 50 A	50 A
Maximum Load	**15 A**	**20 A**	**30 A**	**40 A**	**50 A**
Permissible load	See Section 210-23(a)	See Section 210-23(a)	See Section 210-23(b)	See Section 210-23(c)	See Section 210-23(c)

[1]These gauges are for copper conductors.

[2]For receptacle rating of cord-connected electric-discharge lighting fixtures, see Section 410-30(c).

Table 210-24 summarizes branch-circuit requirements of conductors, overcurrent protection, outlet devices, maximum load, and permissible loads where two or more outlets are supplied.

If the branch circuit serves a fixture load and supplies two or more fixture outlets, Section 210-23 requires the branch circuit to have a specific ampere rating that is also the rating of the overcurrent device, as stated in Section 210-3. Thus, if the circuit breaker that protects the branch circuit is rated 20 amperes, the conductors supplying this circuit must have an ampacity of 20 amperes. Note that in accordance with the Article 100 definition of *ampacity*, the ampacity is determined after applying all derating factors, such as those in Section 310-15(b)(2)(a). Where seven to nine such conductors are in one conduit, a No. 12 Type THHN copper conductor (30 amperes, per Table 310-16) derated to 70 percent, per Table 310-15(b)(2), would have an ampacity of 21 amperes and would be suitable for a load of 20 amperes. Thus, this conductor would be acceptable for use on the 20-ampere multioutlet branch circuit.

210-25. Common Area Branch Circuits. Branch circuits in dwelling units shall supply only loads within that dwelling unit or loads associated only with that dwelling unit. Branch circuits required for the purpose of lighting, central alarm, signal, communications, or other needs for public or common areas of a two-family or multifamily dwelling shall not be supplied from equipment that supplies an individual dwelling unit.

Not only does Section 210-25 prohibit branch circuits from feeding more than one dwelling unit, it also prohibits the sharing of systems, equipment, or common lighting if that equipment is fed from any of the dwelling units. The systems, equipment, or lighting for public or common areas is required to be supplied from a separate "house load" panelboard. This requirement permits access to the branch-circuit disconnecting means without the need to enter the space of any tenants. It is also intended to prevent a tenant from turning off important circuits that may affect other tenants.

C. Required Outlets

210-50. General. Receptacle outlets shall be installed as specified in Sections 210-52 through 210-63.

(a) Cord Pendants. A cord connector that is supported by a permanently installed cord pendant shall be considered a receptacle outlet.

(b) Cord Connections. A receptacle outlet shall be installed wherever flexible cords with attachment plugs are used. Where flexible cords are permitted to be permanently connected, receptacles shall be permitted to be omitted for such cords.

Flexible cords are permitted to be permanently connected to boxes or fittings where specifically permitted by the *Code*. However, plugging a cord into a lampholder by inserting a screw-plug adapter is not permissible. See Section 410-47 for permitted use of screw-shell lampholders.

(c) Appliance Outlets. Appliance receptacle outlets installed in a dwelling unit for specific appliances, such as laundry equipment, shall be installed within 6 ft (1.83 m) of the intended location of the appliance.

See Sections 210-52(f) and 210-11(c)(2) regarding laundry receptacle outlets and branch circuits.

210-52. Dwelling Unit Receptacle Outlets. Receptacle outlets required by this section shall be in addition to any receptacle that is part of a lighting fixture or appliance, located within cabinets or cupboards, or located more than 5½ ft (1.68 m) above the floor.

Permanently installed electric baseboard heaters equipped with factory-installed receptacle outlets or outlets provided as a separate assembly by the manufacturer shall be permitted as the required outlet or outlets for the wall space utilized by such permanently installed heaters. Such receptacle outlets shall not be connected to the heater circuits.

FPN: Listed baseboard heaters include instructions that may not permit their installation below receptacle outlets.

Section 210-52 was completely rewritten and reorganized to improve clarity in the 1999 *NEC*. The rules for dwelling unit receptacle outlets are basically the same as previous editions of the *Code*. The receptacle spacing requirements are for receptacles that are normally used to supply lighting and general-purpose electrical equipment. These receptacles are in addition to the ones that are 5½ ft above the floor and within cupboards and cabinets.

According to listing requirements [see Section 110-3(b)], permanent electric baseboard heaters are not permitted to be located beneath wall receptacles. If the receptacle is a part of the heater, appliance or lamp cords are less apt to be exposed to the heating elements, as might occur should they fall into convector slots. Many electric baseboard heaters are of the low-density type and are designed to be longer than 12 ft. To meet the spacing requirements of Section 210-52(a), the required receptacle must be located as a part of the heater unit, as shown as Figure 210.24.

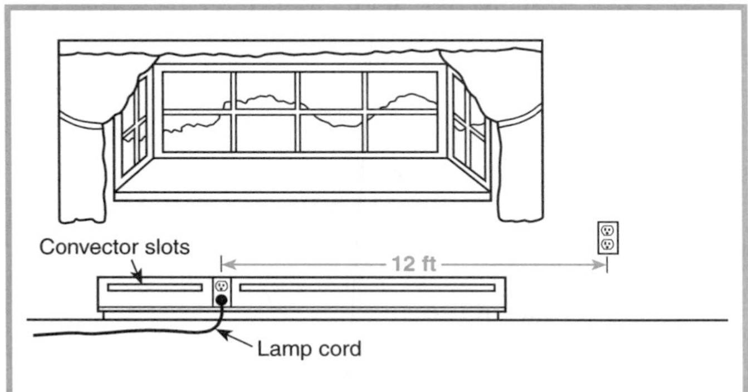

Figure 210.24 Permanent electric baseboard heater equipped with a receptacle outlet to meet the spacing requirements of Section 210-52(a).

(a) General Provisions. In every kitchen, family room, dining room, living room, parlor, library, den, sunroom, bedroom, recreation room, or similar room or area of dwelling units, receptacle outlets shall be installed in accordance with the general provisions specified in (1) through (3).

(1) Spacing. Receptacles shall be installed so that no point along the floor line in any wall space is more than 6 ft (1.83 m), measured horizontally, from an outlet in that space. Receptacle outlets shall, insofar as practicable, be spaced equal distances apart.

Receptacles are required to be located so that no "point" in any wall space is more than 6 ft from a receptacle. This rule intends that an appliance or lamp with a flexible cord attached may be placed anywhere in the room near a wall and be within 6 ft of a receptacle, thus eliminating the need for extension cords. Receptacles should be placed equal distances apart where there is no specific room layout for the general use of electrical equipment. However, the last sentence of Section 210-52(a)(1) does not prohibit a receptacle layout designed for intended utilization equipment or practical room use. For example, receptacles in a living room, family room, or den that are intended to serve home entertainment equipment or home office equipment may be placed in corners, may be grouped, or may be placed in a convenient location. Receptacles that are intended for window-type holiday lighting may be placed under windows. In any event, even if more receptacles than the minimum are installed in a room, no point in any wall space is permitted to be more than 6 ft from a receptacle.

(2) Wall Space. As used in this section, a wall space shall include the following:

(a) Any space 2 ft (610 mm) or more in width (including space measured around corners) and unbroken along the floor line by doorways, fireplaces, and similar openings

(b) The space occupied by fixed panels in exterior walls, excluding sliding panels

(c) The space afforded by fixed room dividers such as free-standing bar-type counters or railings

A *wall space* is a wall unbroken along the floor line by doorways, fireplaces, archways, and similar openings and may include two or more walls of a room (around corners), as illustrated in Figure 210.25.

Fixed room dividers, such as bar-type counters and railings, are to be included in the 6-ft measurement. Fixed panels in exterior walls are counted as regular wall space, and a floor-type receptacle close to the wall can be used to meet the required spacing. Isolated, individual wall spaces 2 ft or more in width are considered usable for the location of a lamp or appliance, and a receptacle outlet is required to be provided.

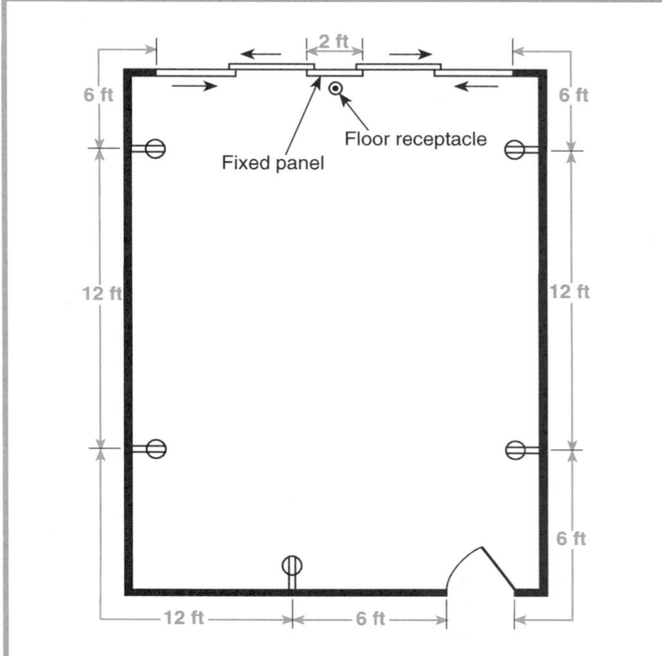

Figure 210.25 Typical room plan view of the location of dwelling unit receptacles meeting the requirements of Section 210-52(a).

(3) Floor Receptacles. Receptacle outlets in floors shall not be counted as part of the required number of receptacle outlets unless located within 18 in. (457 mm) of the wall.

(b) Small Appliances.

Section 210-52(b) requires two or more 20-ampere circuits for all receptacle outlets for the small-appliance loads, including refrigeration equipment, in the kitchen, dining room, pantry, and breakfast room of a dwelling unit. The countertop receptacle outlets in

kitchens must be supplied by no fewer than two small-appliance branch circuits. These circuits may also supply receptacle outlets in the pantry, dining room, and breakfast room, as well as an electric clock receptacle and electric loads associated with gas-fired appliances; but, these circuits are to have no other outlets. See Sections 210-8(a)(6) and (7) for GFCI requirements.

No restriction is placed on the number of outlets connected to a general lighting or small-appliance branch circuit. The minimum number of receptacle outlets in a room is determined by Section 210-52(a). It is desirable to provide more than the minimum number of receptacle outlets required, thereby further reducing the need for extension cords.

Figure 210.26 illustrates the application of the requirements of Sections 210-52(b)(1), (2), and (3). The circuits illustrated are not permitted to serve any other outlets, such as might be connected to exhaust hoods or fans, disposals, or dishwashers. The countertop receptacles are also required to be supplied by these two circuits. Receptacles installed to serve countertop surfaces are required to be GFCI protected in accordance with Section 210-8(a)(6). The dining room switched receptacle on a 15-ampere general-purpose branch circuit is permitted according to Section 210-52(b)(1), Exception No. 1. The refrigerator receptacle supplied by a 15-ampere individual branch circuit (Figure 210.26, bottom) is permitted by Section 210-52(b)(1), Exception No. 2.

(1) In the kitchen, pantry, breakfast room, dining room, or similar area of a dwelling unit, the two or more 20-ampere small-appliance branch circuits required by Section 210-11(c)(1) shall serve all receptacle outlets covered by Sections 210-52(a) and (c) and receptacle outlets for refrigeration equipment.

Exception No. 1: In addition to the required receptacles specified by Section 210-52, switched receptacles supplied from a general-purpose branch circuit as defined in Section 210-70(a)(1), Exception No. 1, shall be permitted.

Exception No. 1 permits switched receptacles supplied from general-purpose 15-ampere branch circuits to be located in kitchens, pantries, breakfast rooms, and similar areas. See Section 210-70(a) and Figure 210.26 for details and an illustration.

Exception No. 2: The receptacle outlet for refrigeration equipment shall be permitted to be supplied from an individual branch circuit rated 15 amperes or greater.

Exception No. 2 is intended to allow a choice for refrigeration equipment receptacle outlets located in a kitchen or similar area. An individual 15-ampere or larger branch circuit may serve this equipment, or it may be included in the 20-ampere small-appliance branch circuit. Refrigeration equipment is also exempt from the GFCI requirements of Section 210-8. See Figure 210.26 for an illustration.

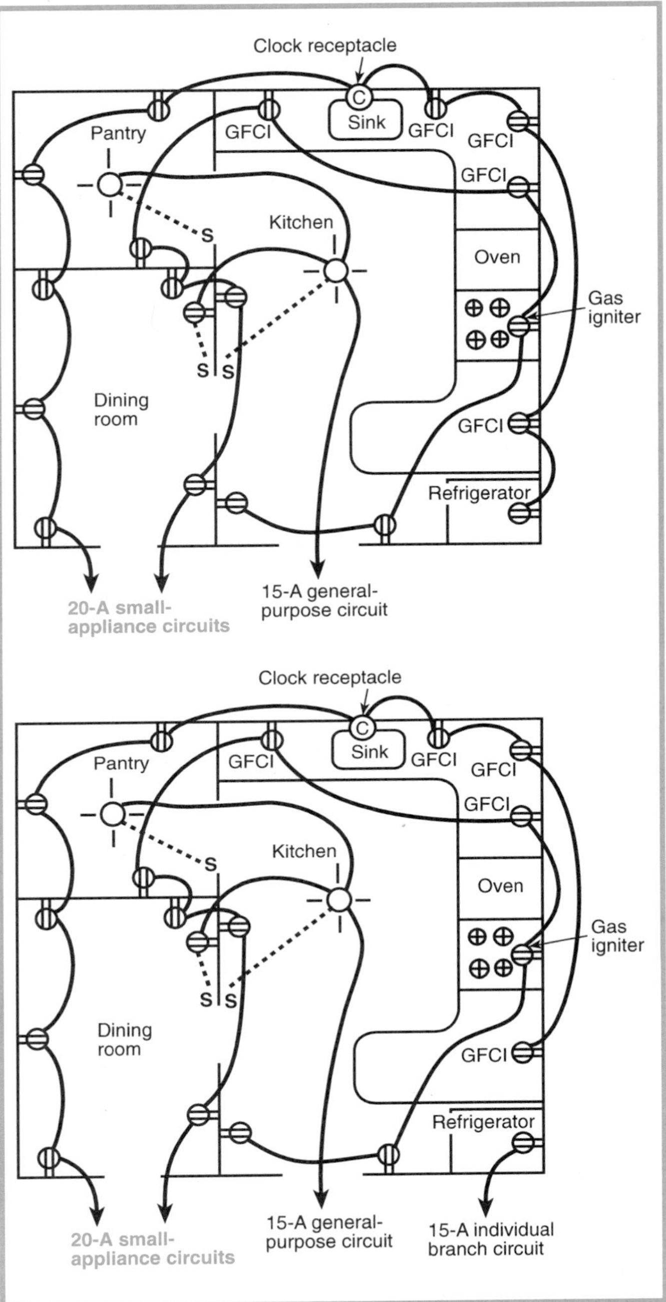

***Figure 210.26** Small-appliance branch circuits as applied to the requirements of Sections 210-52(b)(1), (2), and (3) for all receptacle outlets in the kitchen (including refrigerator), pantry, and dining room.*

(2) The two or more small-appliance branch circuits specified in (b)(1) shall have no other outlets.

Exception No. 1: A receptacle installed solely for the electrical supply to and support of an electric clock in any of the rooms specified above.

Exception No. 2: Receptacles installed to provide power for supplemental equipment and lighting on gas-fired ranges, ovens, or counter-mounted cooking units.

Exception No. 2 is intended to allow the small electrical loads associated with gas-fired appliances to be connected to small-appliance branch circuits. See Figure 210.26 for an illustration.

(3) Receptacles installed in a kitchen to serve countertop surfaces shall be supplied by not less than two small-appliance branch circuits, either or both of which shall also be permitted to supply receptacle outlets in the same kitchen and in other rooms specified in Section 210-52(b)(1). Additional small-appliance branch circuits shall be permitted to supply receptacle outlets in the kitchen and other rooms specified in Section 210-52(b)(1). No small-appliance branch circuit shall serve more than one kitchen.

(c) Countertops. In kitchens and dining rooms of dwelling units, receptacle outlets for counter spaces shall be installed in accordance with (1) through (5).

(1) Wall Counter Spaces. A receptacle outlet shall be installed at each wall counter space that is 12 in. (305 mm) or wider. Receptacle outlets shall be installed so that no point along the wall line is more than 24 in. (610 mm), measured horizontally from a receptacle outlet in that space.

(2) Island Counter Spaces. At least one receptacle outlet shall be installed at each island counter space with a long dimension of 24 in. (610 mm) or greater and a short dimension of 12 in. (305 mm) or greater.

(3) Peninsular Counter Spaces. At least one receptacle outlet shall be installed at each peninsular counter space with a long dimension of 24 in. (610 mm) or greater and a short dimension of 12 in. (305 mm) or greater. A peninsular countertop is measured from the connecting edge.

(4) Separate Spaces. Countertop spaces separated by range tops, refrigerators, or sinks shall be considered as separate countertop spaces in applying the requirements of (1), (2), and (3).

(5) Receptacle Outlet Location. Receptacle outlets shall be located above, but not more than 18 in. (458 mm) above the countertop. Receptacle outlets shall not be installed in a face-up position in the work surfaces or countertops. Receptacle outlets rendered not readily accessible by appliances fastened in place or appliances occupying dedicated space shall not be considered as these required outlets.

Dwelling unit receptacles serving countertop spaces in kitchens, dining areas, and similar rooms, as illustrated in Figure 210.27, are required to be installed as follows:

1. Installed in each wall space wider than 12 in. and spaced so that no point along the wall line is more than 24 in. from a receptacle
2. Installed not more than 18 in. above the countertop and not in a face-up position (Receptacles installed in a face-up position in a countertop could collect crumbs, liquids, and other debris, resulting in a potential fire or shock hazard.)
3. Installed at each countertop island and peninsular countertop with a short dimension of at least 12 in. and a long dimension of at least 24 in. (The measurement of a peninsular-type countertop is from the edge connecting to the nonpeninsular counter.)

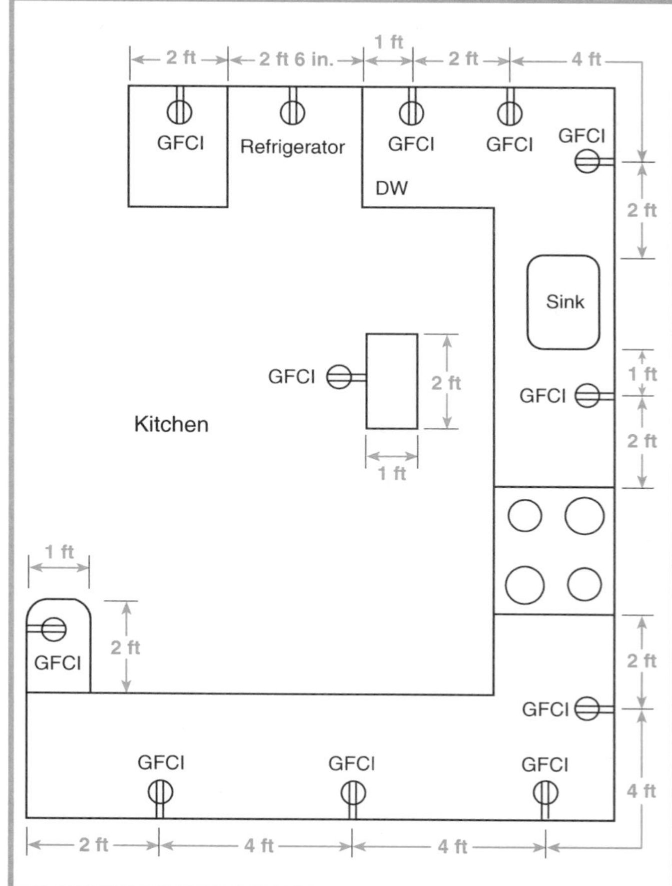

Figure 210.27 Dwelling unit receptacles serving countertop spaces in a kitchen and installed in accordance with Section 210-52(c).

4. Accessible for use and not blocked by appliances occupying dedicated space or fastened in place
5. Fed from two or more of the required 20-ampere small-appliance branch circuits and GFCI protected according to Section 210-8(a)(6)

Exception: To comply with the conditions as specified in (a) or (b), receptacle outlets shall be permitted to be mounted not more than 12 in. (305 mm) below the countertop. Receptacles mounted below the countertop in accordance with this exception shall not be located where the countertop extends more than 6 in. (153 mm) beyond its support base.

(a) Construction for the physically impaired
(b) On island and peninsular countertops where the countertop is flat across its entire surface (no backsplashes, dividers, etc.) and there are no means to mount a receptacle within 18 in. (458 mm) above the countertop, such as an overhead cabinet

(d) Bathrooms. In dwelling units, at least one wall receptacle outlet shall be installed in bathrooms within 36 in. (914 mm) of the outside edge of each basin. The receptacle outlet shall be located on a wall that is adjacent to the basin location. See Section 210-8(a)(1).

Receptacle outlets shall not be installed in a face-up position in the work surfaces or countertops in a bathroom basin location.

Section 210-52(d) requires one wall receptacle in each bathroom of a dwelling unit to be installed adjacent (within 36 in.) to the washbasin. This receptacle is required in addition to any receptacle that may be part of any lighting fixture or medicine cabinet. If there is more than one basin, a receptacle outlet is required adjacent to each basin location. If the basins are in close proximity, one receptacle outlet installed between the two basins may satisfy this requirement. See Section 410-57(c), which prohibits installation of a receptacle inside bathtub and shower spaces. See Figure 210.11 for a typical electrical layout of a bathroom.

Section 210-11(c)(3) requires the receptacle outlets to be supplied from a 20-ampere branch circuit with no other outlets. However, this circuit is permitted to supply the required receptacles in more than one bathroom. This receptacle is also required to be GFCI protected according to Section 210-8(a)(1).

(e) Outdoor Outlets. For a one-family dwelling and each unit of a two-family dwelling that is at grade level, at least one receptacle outlet accessible at grade level and not more than 6½ ft (1.98 m) above grade shall be installed at the front and back of the dwelling. See Section 210-8(a)(3).

The rule for one- and two-family dwellings requires two receptacle outlets for each dwelling unit, as shown in Figure 210.28. Installation of outdoor receptacles requires that care be taken to ensure that the receptacle faceplate rests securely on the supporting surface to prevent moisture from entering the enclosure. If uneven surfaces, such as brick, stone, or stucco, are encountered, it may be necessary to close openings with caulking compound or mastic. See Section 410-57 for further information on receptacles in damp or wet locations.

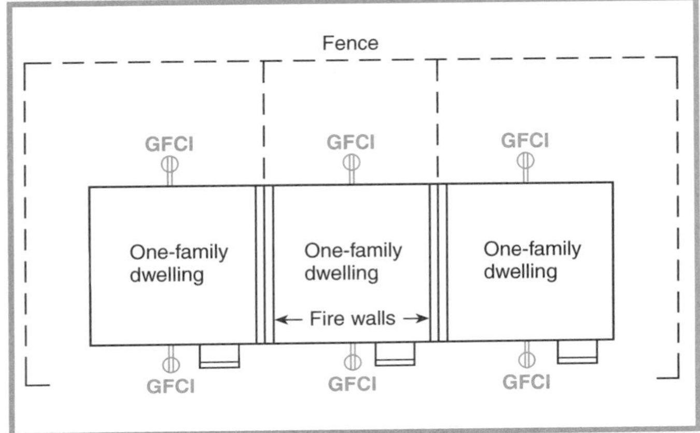

***Figure 210.28** An example of "row" housing with GFCI-protected receptacles located as required by Section 210-52(e) at the front and back of each one-family dwelling.*

(f) Laundry Areas. In dwelling units, at least one receptacle outlet shall be installed for the laundry.

A laundry receptacle outlet(s) is supplied by a 20-ampere branch circuit. This circuit is to have no other outlets. See Section 210-11(c)(2) for further information.

Exception No. 1: In a dwelling unit that is an apartment or living area in a multifamily building where laundry facilities are provided on the premises that are available to all building occupants, a laundry receptacle shall not be required.

Exception No. 2: In other than one-family dwellings where laundry facilities are not to be installed or permitted, a laundry receptacle shall not be required.

(g) Basements and Garages. For a one-family dwelling, at least one receptacle outlet, in addition to any provided for laundry equipment, shall be installed in each basement and in each attached garage, and in each detached garage with electric power. See Sections 210-8(a)(2) and (a)(5). Where a portion of the basement is finished into a habitable room(s), the receptacle outlet required by this section shall be installed in the unfinished portion.

In a one-family dwelling, it is mandatory to install a receptacle in the basement (in addition to the laundry receptacle), in each attached garage, and in each detached garage with electric power. Section 210-8(a)(5) requires receptacles in unfinished basements to be protected by a GFCI. Section 210-8(a)(2) requires receptacles that are installed in garages to be protected by a GFCI. If no electric power is provided, it is not mandatory to install receptacles in detached garages.

(h) Hallways. In dwelling units, hallways of 10 ft (3.05 m) or more in length shall have at least one receptacle outlet.

As used in this subsection, the hall length shall be considered the length along the centerline of the hall without passing through a doorway.

This requirement is intended to minimize strain or damage to cords and receptacles. The requirement is for dwelling unit receptacles and is not applicable to common hallways of hotels, motels, apartment buildings, condominiums, and so on.

210-60. Guest Rooms.

(a) General. Guest rooms in hotels, motels, and similar occupancies shall have receptacle outlets installed in accordance with Section 210-52. See Section 210-8(b)(1).

(b) Receptacle Placement. In applying the provisions of Section 210-52(a), the total number of receptacle outlets shall not be less than the minimum number that would comply with the provisions of that section. These receptacle outlets shall be permitted to be located conveniently for permanent furniture layout. At least two receptacle outlets shall be readily accessible. Where receptacles are installed behind the bed, the receptacle shall be located to prevent the bed from contacting any attachment plug that may be installed, or the receptacle shall be provided with a suitable guard.

Section 210-60(b) permits the receptacles in guest rooms of hotels and motels to be placed in accessible locations that are compatible with the existence of permanent furniture. However, the minimum number of receptacles required in accordance with Section 210-52 is not permitted to be reduced. The minimum number of receptacle outlets should be determined by assuming there is no furniture in the room. The location of this minimum number of receptacles is then determined based on the permanent furniture layout.

Hotel and motel rooms are commonly being used as remote offices for business people carrying laptop computers and other plug-in devices. The 1999 *NEC* requires *two* receptacles to be available without moving furniture to access the receptacles. For receptacles that are located behind beds, guards must be installed to prevent the bed from contacting the attachment plugs.

Bathroom areas for guest rooms in hotels and motels are required to be provided with a receptacle outlet adjacent to the basin location, in accordance with Section 210-8(b)(1).

In Figure 210.29, the receptacle outlets in the hotel guest room shown are located conveniently with respect to the permanent furniture layout. Some spaces that are 2 ft or more in width have no receptacle outlets because the exception permits the required number of outlets to be placed in convenient locations that are compatible with the permanent furniture layout. The receptacle outlet adjacent to the permanent dresser is needed because the exception applies only to the location of receptacle outlets, not to the minimum number of receptacle outlets.

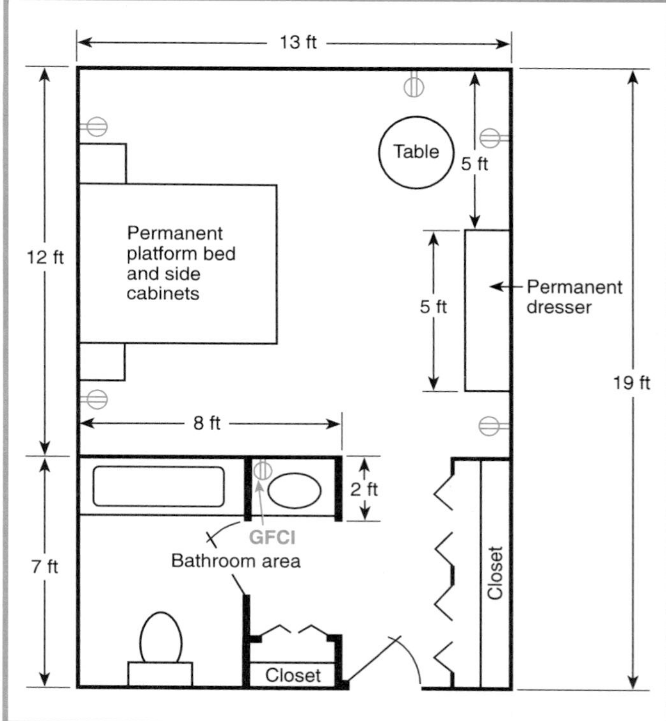

Figure 210.29 An example of a hotel guest room with receptacles located as permitted by Section 210-60(b), with respect to permanent furniture.

210-62. Show Windows. At least one receptacle outlet shall be installed directly above a show window for each

12 linear ft (3.66 m) or major fraction thereof of show window area measured horizontally at its maximum width.

Show windows are usually designed from floor to ceiling for maximum display. To discourage floor receptacles and unsightly extension cords likely to cause physical injury, receptacles must be installed "directly above" a show window, and one receptacle is required for every 12 linear ft or "major fraction thereof" (6 ft or more). See Sections 220-3(b)(7) and 220-12(a) for information regarding load computations for show windows.

210-63. Heating, Air-Conditioning, and Refrigeration Equipment Outlet. A 125-volt, single-phase, 15- or 20-ampere–rated receptacle outlet shall be installed at an accessible location for the servicing of heating, air-conditioning, and refrigeration equipment on rooftops and in attics and crawl spaces. The receptacle shall be located on the same level and within 25 ft (7.62 m) of the heating, air-conditioning, and refrigeration equipment. The receptacle outlet shall not be connected to the load side of the equipment disconnecting means.

Exception: Rooftop equipment on one- and two-family dwellings.

Section 210-63 is intended to help prevent makeshift methods of obtaining power for servicing and troubleshooting heating, air-conditioning, and refrigeration equipment in attics, in crawl spaces, and on rooftops. Note the exception for rooftop equipment on one- and two-family dwellings.

FPN: See Section 210-8 for ground-fault circuit-interrupter requirements.

210-70. Lighting Outlets Required. Lighting outlets shall be installed where specified in (a), (b), and (c).

(a) Dwelling Units. In dwelling units, lighting outlets shall be installed in accordance with (1), (2), and (3).

(1) Habitable Rooms. At least one wall switch-controlled lighting outlet shall be installed in every habitable room and bathroom.

Exception No. 1: In other than kitchens and bathrooms, one or more receptacles controlled by a wall switch shall be permitted in lieu of lighting outlets.

Section 210-70 was rewritten and reorganized to improve clarity for the 1999 *NEC*. A receptacle is not permitted to be switched as a lighting outlet on a small-appliance branch circuit. A receptacle can be switched as a lighting outlet (in the dining room, for example) supplied by a branch circuit other than a small-appliance branch circuit. See Figure 210-26, which shows a dining room switched receptacle on a 15-ampere general-purpose branch circuit.

Exception No. 2: Lighting outlets shall be permitted to be controlled by occupancy sensors that are (1) in addition to wall switches or (2) located at a customary wall switch location and equipped with a manual override that will allow the sensor to function as a wall switch.

(2) Additional Locations. At least one wall switch-controlled lighting outlet shall be installed in hallways, stairways, attached garages, and detached garages with electric power; and to provide illumination on the exterior side of outdoor entrances or exits with grade level access. A vehicle door in a garage shall not be considered as an outdoor entrance or exit. Where lighting outlets are installed in interior stairways, there shall be a wall switch at each floor level to control the lighting outlet where the difference between floor levels is six steps or more.

Section 210-70 points out that adequate lighting and proper control and location of switching is as essential to the safety of occupants of dwelling units, hotels, motels, and so on, as are proper wiring requirements. Proper illumination ensures safe movement for persons of all ages; thus, many accidents are avoided.

Although the requirement calls for a switched lighting outlet at outdoor entrances and exits, it does not prohibit a single lighting outlet, if suitably located, from serving more than one door.

A wall-switch-controlled lighting outlet is required in the kitchen and bathroom. A receptacle outlet controlled by a wall switch is not permitted to serve as a lighting outlet in these rooms. Occupancy sensors are permitted to be used for switching these lighting outlets, provided they are equipped with a manual override or are used in addition to regular switches.

Exception: In hallways, stairways, and at outdoor entrances, remote, central, or automatic control of lighting shall be permitted.

(3) Storage or Equipment Spaces. For attics, underfloor spaces, utility rooms, and basements, at least one lighting outlet containing a switch or controlled by a wall switch shall be installed where these spaces are used for storage or contain equipment requiring servicing. At least one point of control shall be at the usual point of entry to these spaces. The lighting outlet shall be provided at or near the equipment requiring servicing.

Installation of lighting outlets in attics, under floor spaces or crawl areas, and in utility rooms or basements is required *only* where these spaces are used for storage (e.g., holiday decorations and luggage).

If such spaces contain equipment that requires

servicing (e.g., air-handling units, cooling and heating equipment, water pumps, and sump pumps), Section 210-70(c) requires that a lighting outlet be installed in these spaces.

(b) Guest Rooms. At least one wall switch-controlled lighting outlet or wall switch-controlled receptacle shall be installed in guest rooms in hotels, motels, or similar occupancies.

(c) Other Locations. For attics and underfloor spaces containing equipment requiring servicing, such as heating, air-conditioning, and refrigeration equipment, at least one lighting outlet containing a switch or controlled by a wall switch shall be installed in such spaces. At least one point of control shall be at the usual point of entry to these spaces. The lighting outlet shall be provided at or near the equipment requiring servicing.

Article 215 — Feeders

Contents

215-1. Scope. This article covers the installation requirements, overcurrent protection requirements, minimum size, and ampacity of conductors for feeders supplying branch-circuit loads as computed in accordance with Article 220.

Exception: Feeders for electrolytic cells as covered in Section 668-3(c)(1) and (4).

The scope statement for Article 215 has been revised for the 1999 *NEC* to include a reference to overcurrent protection requirements for feeders. This change encompasses both the fact that Article 215 references Article 240 for overcurrent protection and the fact that Section 215-3 includes the 125 percent sizing rules for overcurrent devices.

A thorough calculation of the total connected load to be supplied by the feeder is required to accurately determine feeder conductor ampacity. The sum of the computed and connected loads supplied by a feeder is multiplied by the "demand factor" to determine the load that the feeder conductors must be sized to serve. See Article 100 for the definition of *demand factor.*

When the total connected load is operated simultaneously, the demand factor is 100 percent; that is, the maximum demand is equal to the total connected load. Due to diversity, the maximum operating load carried at any time may be only three-quarters of the total connected load; thus, the demand factor is 75 percent.

On a new installation, a minimum value for the demand factor can be determined by applying the requirements and tables of Article 220, Branch Circuit, Feeder, and Service Calculations.

Feeder conductor sizes are determined by calculating the total volt-amperes (VA) of the feeder load at the nominal voltage of the feeder circuit. See Section 220-2 for the nominal system voltages used in computing branch-circuit and feeder loads. See Section 310-15 for allowable ampacities and sizes of insulated conductors.

Feeder circuits are required to have sufficient ampacity for safety. Overloading of a wiring system that does not provide for increases in the use of electricity often creates hazards. It is a good practice to allow for future expansion and convenience increases, as stated in Section 90-8.

215-2. Minimum Rating and Size.

(a) General. Feeder conductors shall have an ampacity not less than required to supply the load as computed in Parts B, C, and D of Article 220. The minimum feeder-circuit conductor size, before the application of any adjustment or correction factors, shall have an allowable ampacity equal to or greater than the noncontinuous load plus 125 percent of the continuous load.

Exception: Where the assembly, including the overcurrent devices protecting the feeder(s), is listed for operation at 100 percent of its rating, the ampacity of the feeder conductors shall be permitted to be not less than the sum of the continuous load plus the noncontinuous load.

Additional minimum sizes shall be as specified in (b), (c), and (d) under the conditions stipulated.

Section 215-2 has been revised and rewritten for the 1999 *NEC* to include the conductor sizing requirement previously found in Section 220-10(b). Section 215-2 is the most appropriate location for this material because it deals with conductor ampacity and size and Section 220-10 deals only with load calculation.

(b) For Specified Circuits. The ampacity of feeder conductors shall not be less than 30 amperes where the load supplied consists of any of the following number and types of circuits:

(1) Two or more 2-wire branch circuits supplied by a 2-wire feeder
(2) More than two 2-wire branch circuits supplied by a 3-wire feeder
(3) Two or more 3-wire branch circuits supplied by a 3-wire feeder
(4) Two or more 4-wire branch circuits supplied by a 3-phase, 4-wire feeder

(c) Ampacity Relative to Service-Entrance Conductors. The feeder conductor ampacity shall not be less than that of the service-entrance conductors where the feeder conductors carry the total load supplied by service-entrance conductors with an ampacity of 55 amperes or less.

Section 215-2(b) was previously Section 215-2(a) in the 1996 *Code* and has been renumbered to allow for the creation of Section 215-2(a), General. Section 215-2(c) was previously Section 215-2(b) in the 1996 *Code*.

(d) Individual Dwelling Unit or Mobile Home Conductors. Feeder conductors for individual dwelling units or mobile homes need not be larger than service-entrance conductors. Section 310-15(b)(6) shall be permitted to be used for conductor size.

Section 215-2(d) was reorganized for the 1999 *NEC*. It incorporates the permission for applying Section 310-15(b)(6) allowing feeder conductors to be the same size as the service-entrance conductors for dwelling units. This permission was found in the last two sentences of Section 215-2 in the 1996 *Code*.

For example, Table 310-16 allows 200 amperes for a 3/0 Type THW copper wire. However, for a 3-wire, single-phase dwelling service, as shown in Figure 215.1, Table 310-15(b)(6) permits 200 amperes for a 2/0 Type THW copper conductor or 200 amperes for a 4/0 Type THW aluminum conductor. Feeder conductors carrying the total load supplied by the service are not required to be sized larger than the service-entrance conductors.

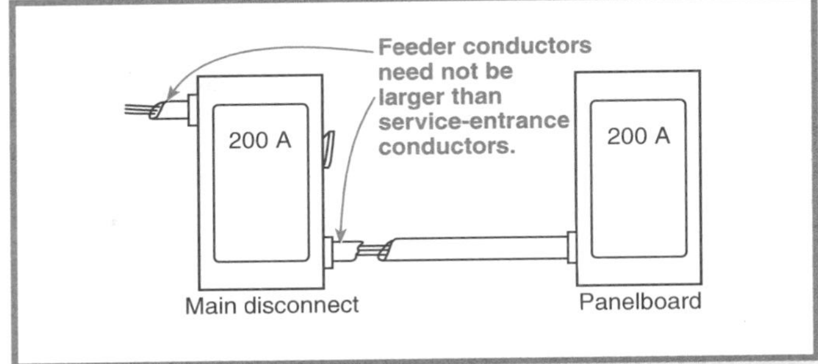

Figure 215.1 *A 3-wire, single-phase dwelling service with an ampacity of 200 amperes for 2/0 copper or 4/0 aluminum conductors used as service-entrance conductors and feeder conductors.*

FPN No. 1: See Examples D1 through D10 in Appendix D.

FPN No. 2: Conductors for feeders as defined in Article 100, sized to prevent a voltage drop exceeding 3 percent at the farthest outlet of power, heating, and lighting loads, or combinations of such loads, and where the maximum total voltage drop on both feeders and branch circuits to the farthest outlet does not exceed 5 percent, will provide reasonable efficiency of operation.

FPN No. 3: See Section 210-19(a), FPN No. 4, for voltage drop for branch circuits.

Reasonable operating efficiency is achieved if the voltage drop of a feeder or the voltage drop of a branch circuit is limited to 3 percent. However, the total voltage drop of a branch circuit plus a feeder can reach 5 percent and still achieve reasonable operating efficiency. See Article 100 for definitions of *feeder* and *branch circuit*.

The 5 percent voltage-drop value is explanatory material, and, as such, appears as a fine print note. Fine print notes are not mandatory (see Section 90-5). However, Section 250-122(b) does require an increase in circular mil area for equipment grounding conductors where circuit conductors are increased due to voltage drop.

The resistance or impedance of conductors may cause a substantial difference between voltage values at service equipment and voltage values at the point-of-utilization equipment. Excessive voltage drop impairs the starting and operation of electrical equipment. Undervoltage can result in inefficient operation of heating, lighting, and motor loads. An applied voltage of 10 percent below rating can result in a decrease in efficiency of substantially more than 10 percent. For example, fluorescent light output would be reduced by 15 percent, while incandescent light output would be reduced by 30 percent. Induction motors would run hotter and produce less torque. The running current would increase 11 percent, while the operating temperature would increase by 12 percent. At the same time, torque would be reduced by 19 percent.

In addition to resistance or impedance, the type of raceway or cable enclosure, the type of circuit (ac, dc, single-phase, 3-phase), and the power factor should be considered to determine voltage drop.

The following basic formula can be used to determine the voltage drop in a 2-wire dc circuit, a 2-wire ac circuit, or a 3-wire ac single-phase circuit, all with a balanced load at 100 percent power factor and where reactance can be neglected.

$$VD = \frac{2 \times L \times R \times I}{1000}$$

where:

VD = voltage drop (based on conductor temperature of 75°F)

L = one-way length of circuit (ft)

R = conductor resistance in ohms (Ω) per thousand feet (from Chapter 9, Table 8)

I = load current (amperes)

For 3-phase circuits (at 100 percent power factor), the voltage drop between any two phase conductors is 0.866 times the voltage drop calculated by the preceding formula.

Example

Determine the voltage drop in a 240-volt, 2-wire heating circuit with a load of 50 amperes. The circuit size is No. 6 Type THHN copper, and the one-way circuit length is 100 ft.

First, find the conductor resistance in Chapter 9, Table 8. Then, substitute values into the following voltage-drop formula:

$$VD = \frac{2 \times L \times R \times I}{1000} = \frac{2 \times 100 \times 0.491 \times 50}{1000}$$

$$= \frac{4910}{1000} = 4.91\text{-volt drop}$$

A 12-volt drop on a 240-volt circuit is a 5 percent drop. A 4.91-volt drop falls within this percentage. Should the total voltage drop exceed 5 percent, or 12 volts, larger-size conductors should be used, the circuit length should be shortened, or the circuit load should be reduced.

See the commentary following Chapter 9, Table 9, for an example of voltage-drop calculation using ac reactance and resistance. Voltage-drop tables and calculations are also available from various manufacturers.

215-3. Overcurrent Protection. Feeders shall be protected against overcurrent in accordance with the provisions of Part A of Article 240. Where a feeder supplies continuous loads or any combination of continuous and noncontinuous loads, the rating of the overcurrent device shall not be less than the noncontinuous load plus 125 percent of the continuous load.

Section 215-3 on overcurrent protection requirements has been revised for the 1999 *NEC* to include the overcurrent protection (OCP) sizing rule previously found in Section 220-10(b). Article 215 is the appropriate location for this requirement since Section 215-3 deals with overcurrent protection requirements for feeders and Article 220 deals only with computed loads.

Exception: Where the assembly, including the overcurrent devices protecting the feeder(s), is listed for operation at 100 percent of its rating, the ampere rating of the overcurrent device shall be permitted to be not less than the sum of the continuous load plus the noncontinuous load.

215-4. Feeders with Common Neutral.

(a) Feeders with Common Neutral. Feeders containing a common neutral shall be permitted to supply two or three sets of 3-wire feeders, or two sets of 4-wire or 5-wire feeders.

(b) In Metal Raceway or Enclosure. Where installed in a metal raceway or other metal enclosure, all conductors of all feeders using a common neutral shall be enclosed within the same raceway or other enclosure as required in Section 300-20.

If feeder conductors carrying ac current, including the neutral, are installed in metal raceways, they are required to be grouped together to avoid induction heating of the surrounding metal. If it is necessary to run parallel conductors through multiple raceways, conductors from each phase plus the neutral must be run in each raceway. See Sections 250-102(e), 250-134(b), 300-3, 300-5(i), and 300-20(a) for feeder conductor installation details.

A 3-phase, 4-wire (208Y/120-volt, 480Y/277-volt) system is often used to supply both lighting and motor loads. The 3-phase motor loads are typically not connected to the neutral and, therefore, will not cause current to flow in the neutral conductor. The maximum current on the neutral, therefore, is due to lighting loads or circuits where the neutral is used. On this type of system (3-phase, 4-wire), a demand factor of 70 percent is permitted by Section 220-22 for that portion of the neutral load in excess of 200 amperes. For example, if the maximum possible unbalanced load is 500 amperes, the neutral would have to be large enough to carry 410 amperes (200 amperes plus 70 percent of 300 amperes, or 410 amperes). No reduction of the neutral capacity for that portion of the load consisting of electric-discharge lighting is permitted. See Section 310-15(b)(4) for other loads with harmonic currents. See Section 220-22 for other systems where the 70 percent demand factor may be applied. The maximum unbalanced load for feeders supplying clothes dryers, household ranges, wall-mounted ovens, and counter-mounted cooking units is required to be considered 70 percent of the load on the ungrounded conductors. See Examples D1(a) through D5(b) of Appendix D. See Examples D5(a) and D5(b) of Appendix D for a multifamily dwelling served at 208Y/120 volts, 3 phase, 4 wire.

215-5. Diagrams of Feeders. If required by the authority having jurisdiction, a diagram showing feeder details shall be provided prior to the installation of the feeders. Such a diagram shall show the area in square feet of the building or other structure supplied by each feeder, the total connected load before applying demand factors, the demand factors used, the computed load after applying demand factors, and the size and type of conductors to be used.

215-6. Feeder Conductor Grounding Means. Where a feeder supplies branch circuits in which equipment grounding conductors are required, the feeder shall include or provide a grounding means, in accordance with the provisions of Section 250-134, to which the equipment grounding conductors of the branch circuits shall be connected.

215-7. Ungrounded Conductors Tapped from Grounded Systems. Two-wire dc circuits and ac circuits of two or more ungrounded conductors shall be permitted to be tapped from the ungrounded conductors of circuits having a grounded neutral conductor. Switching devices in each tapped circuit shall have a pole in each ungrounded conductor.

Section 215-7 does not require a common trip or simultaneous opening of circuit breakers or fuses but, rather, requires a switching device to "manually" disconnect the ungrounded conductors of the feeder. See Section 210-10, which relates to the ungrounded conductors of the branch circuit.

215-8. Means of Identifying Conductor with the Higher Voltage to Ground. On a 4-wire, delta-connected secondary where the midpoint of one phase winding is grounded to supply lighting and similar loads, the phase conductor having the higher voltage to ground shall be identified by an outer finish that is orange in color or by tagging or other effective means. Such identification shall be placed at each point where a connection is made if the grounded conductor is also present.

Where the midpoint of one phase winding is grounded in order to supply 120-volt lighting and similar loads from a delta-connected, 3-phase secondary, one phase conductor will have a higher voltage to ground. An orange finish, orange tape, or other effective means identifies this phase conductor at any point, such as junction or pull boxes or panelboards, where connections may be made and the grounded conductor is also present. The orange high leg of a 3-phase, 4-wire (120/240 240/120-volt) delta system is 208 volts to ground (120 volts multiplied by 1.73 equals 208 volts) and should obviously not be used for 120-volt circuits. See Sections 230-56 and 384-3(e) and (f) for details on high-leg marking and phase arrangement.

With a 4-wire, 240/120-volt, delta-connected secondary in which the midpoint of one phase is grounded, one phase conductor has a higher voltage to ground (208 volts) than the other two. The conductor with the higher voltage to ground is identified in Figure 215.2 as having an orange finish. The identification must be visible at every point where a connection is made if the grounded conductor (neutral) is present.

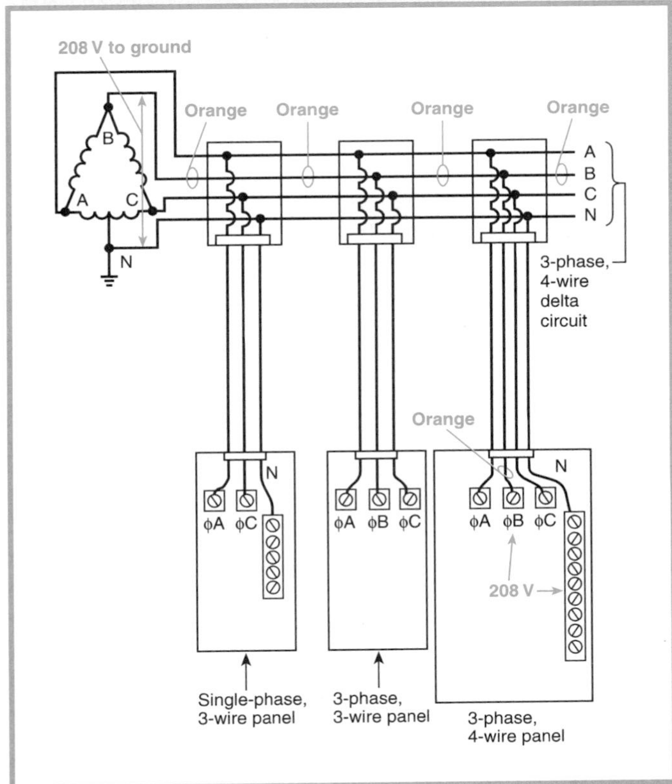

Figure 215.2 *Identification of the high-leg (orange) conductor of a 3-phase, 4-wire, 240/120-volt delta system.*

215-9. Ground-Fault Circuit-Interrupter Protection for Personnel. Feeders supplying 15- and 20-ampere receptacle branch circuits shall be permitted to be protected by a ground-fault circuit interrupter in lieu of the provisions for such interrupters as specified in Section 210-8 and Article 305.

Several manufacturers offer double-pole 120/240-volt circuit-breaker-type ground-fault circuit interrupters (GFCIs) for application to a feeder, thereby protecting all branch circuits supplied by that feeder. This installation is in lieu of provisions of Section 210-8 for outdoor, bathroom, garage, kitchen, basement, and boathouse receptacles. They can also be used to protect construction-site receptacles, as covered in Section 305-6(a), provided that the feeder supplies no lighting branch circuits.

It may be more economical or convenient to install GFCIs for feeders. However, consideration should be given to the fact that a ground-fault circuit interrupter may be monitoring several branch circuits and will de-energize all branch circuits in response to a line-to-ground fault from one branch circuit.

215-10. Ground-Fault Protection of Equipment. Each feeder disconnect rated 1000 amperes or more and installed on solidly grounded wye electrical systems of more than 150 volts to ground, but not exceeding 600 volts phase-to-phase, shall be provided with ground-fault protection of equipment in accordance with the provisions of Section 230-95.

Exception No. 1: The provisions of this section shall not apply to a disconnecting means for a continuous industrial process where a nonorderly shutdown will introduce additional or increased hazards.

Exception No. 2: The provisions of this section shall not apply to fire pumps.

Exception No. 3: The provisions of this section shall not apply if ground-fault protection of equipment is provided on the supply side of the feeder.

The intent of Section 215-10 is to require ground-fault protection of equipment for feeder disconnects that are rated 1000 amperes or more at 480Y/277 volts. A similar requirement for services is found in Section 230-95. The reason for the requirement is the unusually high number of burndowns reported on feeders and services in this voltage range.

It should be noted that ground-fault protection of feeder equipment is not required if protection is provided on an upstream feeder or at the service. However, it may be desirable to have additional levels of ground-fault protection on feeders so that a single ground fault does not de-energize the whole electrical system. See Section 230-95 for further commentary on ground-fault protection of services. Also, see Section 517-17(a), which requires an additional level of ground-fault protection for health care facilities.

For emergency feeders, according to Article 700, the ground-fault protection requirements are different. See Section 700-26 for further details.

215-11. Circuits Derived from Autotransformers. Feeders shall not be derived from autotransformers unless the system supplied has a grounded conductor that is electrically connected to a grounded conductor of the system supplying the autotransformer.

Section 215-11 addresses autotransformers for feeders. This section is similar to the requirements in Section 210-9 for branch circuits. See the commentary following Section 210-9 for further information on autotransformers supplying branch circuits.

Exception No. 1: An autotransformer shall be permitted without the connection to a grounded conductor where transforming from a nominal 208 volts to a nominal 240-volt supply or similarly from 240 volts to 208 volts.

Exception No. 2: In industrial occupancies, where conditions of maintenance and supervision ensure that only qualified persons service the installation, autotransformers shall be permitted to supply nominal 600-volt loads from nominal 480-volt systems, and 480-volt loads from nominal 600-volt systems, without the connection to a similar grounded conductor.

Article 220 — Branch-Circuit, Feeder, and Service Calculations

Contents

A. General

220-1. Scope. This article provides requirements for computing branch-circuit, feeder, and service loads.

The scope of Article 220 has been revised for the 1999 *NEC* to clearly indicate that the article deals with load computation for branch-circuit, feeder, and service loads and not with requirements for determining the number of branch circuits needed. The determination for the number of branch circuits is now contained in Article 210.

Exception: Branch-circuit and feeder calculations for electrolytic cells as covered in Section 668-3(c)(1) and (4).

220-2. Computations.

(a) Voltages. Unless other voltages are specified, for purposes of computing branch-circuit and feeder loads, nominal system voltages of 120, 120/240, 208Y/120, 240, 347, 480Y/277, 480, 600Y/347, and 600 volts shall be used.

(b) Fractions of an Ampere. Except where computations result in a fraction of an ampere 0.5 or larger, such fractions shall be permitted to be dropped.

For uniform calculation of load, nominal voltages, as listed in Section 220-2(a), are required to be used in computing the ampere load on the conductors. To select conductor sizes, refer to Sections 310-15(a) and (b).

Loads are computed on the basis of volt-amperes (VA) or kilovolt-amperes (kVA) rather than watts or kilowatts (kW). However, the rating of equipment is given in watts or kilowatts. Such ratings are considered to be the equivalent of the same rating in volt-amperes or kilovolt-amperes. See, for example, Section 220-19. This concept recognizes that load calculations are to determine conductor and circuit sizes, that the power factor of the load is often unknown, and that the conductor "sees" the circuit volt-amperes only, not the circuit power (watts).

See Examples D1(a) through D5(b) of Appendix D. The results of these examples are generally expressed in amperes. Unless the computations result in a major fraction of an ampere (0.5 or larger), such fractions (less than 0.5) are permitted to be dropped.

220-3. Computation of Branch Circuit Loads. Branch-circuit loads shall be computed as shown in (a) through (c).

(a) Lighting Load for Specified Occupancies. A unit load of not less than that specified in Table 220-3(a) for occupancies specified therein shall constitute the minimum lighting load for each square foot (0.093 m^2) of floor area. The floor area for each floor shall be computed from the outside dimensions of the building, dwelling unit, or other area involved. For dwelling units, the computed floor area shall not include open porches, garages, or unused or unfinished spaces not adaptable for future use.

Section 220-3 was significantly revised for the 1999 *NEC* so that it correlates with the changes in Article 210. Also, former Section 220-3(c) was revised to incorporate all of the exceptions into *Code* rules. This simple arrangement allows *Code* users to find the specific text that applies to the load to be calculated.

Former Section 220-3(a) was relocated in the 1999 *NEC* to Sections 210-19 and 210-20, which incorporate

Table 220-3(a). General Lighting Loads by Occupancies

Type of Occupancy	Unit Load per Square Foot (Volt-Amperes)
Armories and auditoriums	1
Banks	3½[b]
Barber shops and beauty parlors	3
Churches	1
Clubs	2
Court rooms	2
Dwelling units[a]	3
Garages — commercial (storage)	½
Hospitals	2
Hotels and motels, including apartment houses without provision for cooking by tenants[a]	2
Industrial commercial (loft) buildings	2
Lodge rooms	1½
Office buildings	3½[b]
Restaurants	2
Schools	3
Stores	3
Warehouses (storage)	¼
In any of the above occupancies except one-family dwellings and individual dwelling units of two-family and multifamily dwellings:	
Assembly halls and auditoriums	1
Halls, corridors, closets, stairways	½
Storage spaces	¼

Note: For SI units, 1 ft^2 = 0.093 m^2.
[a]See Section 220-3(b)(10).
[b]In addition, a unit load of 1 volt-ampere per square foot shall be included for general-purpose receptacle outlets where the actual number of general-purpose receptacle outlets is unknown.

the sizing rules for the conductor and overcurrent protection.

Examples of unused or unfinished spaces are some attics, cellars, and crawl spaces.

FPN: The unit values herein are based on minimum load conditions and 100 percent power factor, and may not provide sufficient capacity for the installation contemplated.

(b) Other Loads — All Occupancies. In all occupancies, the minimum load for each outlet for general-use receptacles and outlets not used for general illumination shall not be less than that computed in (1) through (11), the loads shown being based on nominal branch-circuit voltages.

Exception: The loads of outlets serving switchboards and switching frames in telephone exchanges shall be waived from the computations.

(1) Specific Appliances or Loads. An outlet for a specific appliance or other load not covered in (2) through (11) shall be computed based on the ampere rating of the appliance or load served.

(2) Electric Dryers and Household Electric Cooking Appliances. Load computations shall be permitted as specified in Section 220-18 for electric dryers and Section 220-19 for electric ranges and other cooking appliances.

(3) Motor Loads. Outlets for motor loads shall be computed in accordance with the requirements in Sections 430-22 and 430-24 and Article 440.

(4) Recessed Lighting Fixtures. An outlet supplying recessed lighting fixture(s) shall be computed based on the maximum volt-ampere rating of the equipment and lamps for which the fixture(s) is rated.

Recessed lighting fixtures are included in the 3 volt-amperes per square foot calculation used for dwellings if they are used for general lighting.

(5) Heavy Duty Lampholders. Outlets for heavy-duty lampholders shall be computed at a minimum of 600 volt-amperes.

(6) Sign and Outline Lighting. Sign and outline lighting outlets shall be computed at a minimum of 1200 volt-amperes for each required branch circuit specified in Section 600-5(a).

Section 220-3(b)(6) assigns 1200 volt-amperes as a minimum circuit load for the signs and outline lighting outlets required by Section 600-5. If the specific load is known to be larger, then, according to Section 220-3(b)(1), the actual load is used.

(7) Show Windows. Show windows shall be computed in accordance with either (a) or (b).

The following two options are permitted for the load calculations for branch circuits serving show windows:

1. 180 volt-amperes per receptacle (Note that Section 210-62 requires one receptacle per 12 linear ft.)
2. 200 volt-amperes per linear ft [See Section 220-3(c), Exception No. 3.]

As shown in Figure 220.1, the linear-foot calculation method is permitted in lieu of the specified unit load per outlet for branch circuits serving show windows.

(a) The unit load per outlet as required in other provisions of this section
(b) At 200 volt-amperes per linear foot of show window

(8) Fixed Multioutlet Assemblies. Fixed multioutlet assemblies used in other than dwelling units or the guest rooms of hotels or motels shall be computed in accordance with (a) or (b).

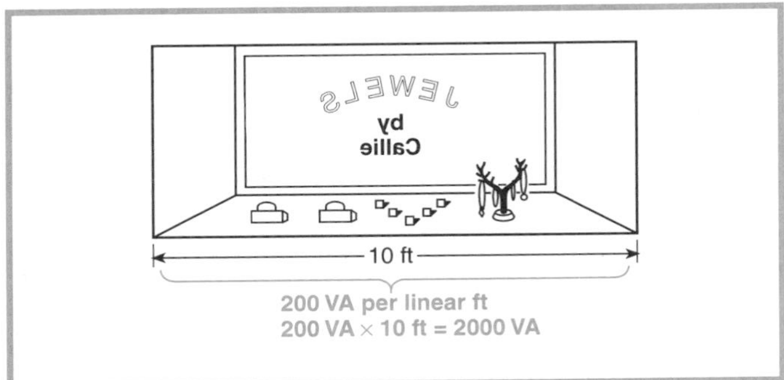

Figure 220.1 An example of the linear-foot load calculation method for branch circuits serving a show window.

(a) Where appliances are unlikely to be used simultaneously, each 5 ft (1.52 m) or fraction thereof of each separate and continuous length shall be considered as one outlet of not less than 180 volt-amperes.

(b) Where appliances are likely to be used simultaneously, each 1 ft (305 mm) or fraction thereof shall be considered as an outlet of not less than 180 volt-amperes.

Fixed multioutlet assemblies are commonly used in commercial or industrial locations and may have been selected to provide a number of receptacles along a given work area (light use) or to allow simultaneous connection and use of a number of appliances (heavy use). As shown in Figure 220.2, the requirements of Section 220-3(b)(8) state that each 5 ft of a fixed multioutlet assembly be considered as on outlet rated 180 volt-amperes and that, if appliances are likely to be used simultaneously, each 1 ft be considered as one outlet rated 180 volt-amperes.

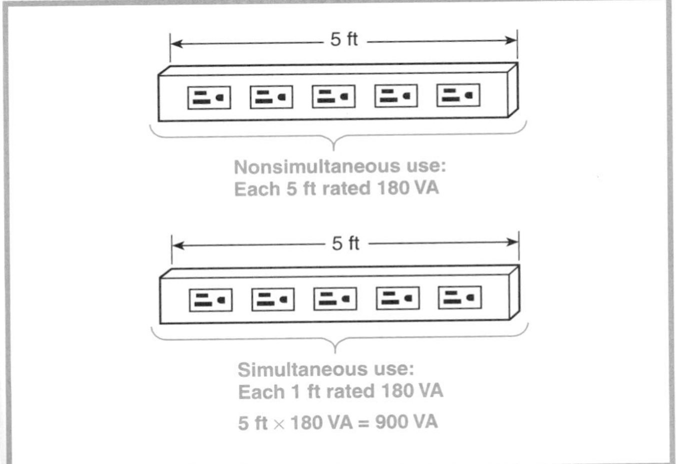

Figure 220.2 The requirements of Section 220-3(b)(8) as applied to fixed multioutlet assemblies.

(9) Receptacle Outlets. Except as covered in (10), receptacle outlets shall be computed at not less than 180 volt-amperes for each single or for each multiple receptacle on one strap. A single piece of equipment consisting of a multiple receptacle comprised of four or more receptacles shall be computed at not less than 90 volt-amperes per receptacle.

This provision shall not be applicable to the receptacle outlets specified in Sections 210-11(c)(1) and (2).

As illustrated in Figure 220.3, the 180-volt-ampere load is applied to single and multiple receptacles mounted on a single yoke or strap. These are considered "receptacle outlets," in accordance with Section 220-3(b)(9). The receptacle outlets are not the lighting outlets installed for general illumination or the small-appliance branch circuits, as indicated in Section 220-3(b)(10). The receptacle load for outlets for general illumination in one- and two-family and multifamily dwellings and in guest rooms of hotels and motels is included in Table 220-3(a). The load requirement for the small-appliance branch circuits is 1500 volt-amperes per circuit, as described in Section 220-16(a).

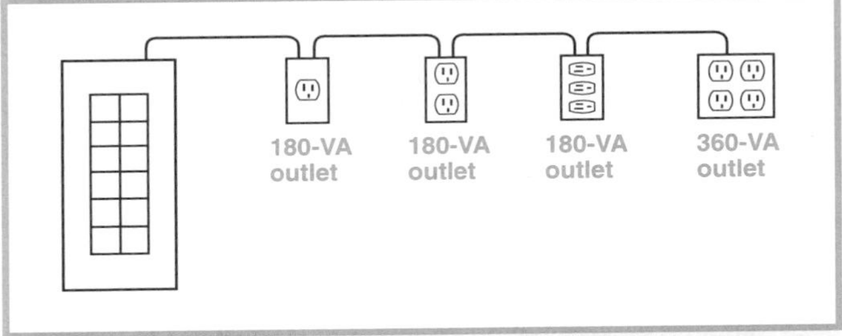

Figure 220.3 The 180-volt-ampere load requirement of Section 220-3(b)(9) as applied to single- and multiple-receptacle outlets on a single strap.

Note in Figure 220.3 that the last outlet consists of two duplex receptacles on separate straps. Therefore, that outlet would be calculated at 360 volt-amperes. New in the 1999 *NEC,* each single receptacle of a multiple receptacle, comprised of four or more receptacles, is calculated at 90 volt-amperes per receptacle. As an example, a single-strap device or a multiple-receptacle device is calculated as follows:

Single-Strap Device	Computed Load
Duplex receptacle	180 VA
Triplex receptacle	180 VA
Quad or four-plex-type receptacle	90 × 4 = 360 VA
Five-plex-type receptacle	90 × 5 = 450 VA

A load of 180 volt-amperes is not required to be considered for outlets supplying recessed lighting fixtures, lighting outlets for general illumination, and small-appliance branch circuits. To apply the 180-

volt-ampere requirement in this case would be unrealistic, as it would unnecessarily restrict the number of lighting or receptacle outlets on a branch circuit. See the note below Table 220-3(a) that references Section 220-3(b)(10). This note indicates that the 180-volt-ampere requirement does not apply to most receptacle outlets in dwellings.

In Figure 220.4, the maximum number of outlets permitted on 15- and 20-ampere branch circuits is 10 and 13 outlets, respectively. This restriction does not apply to outlets connected to general lighting or small-appliance branch circuits.

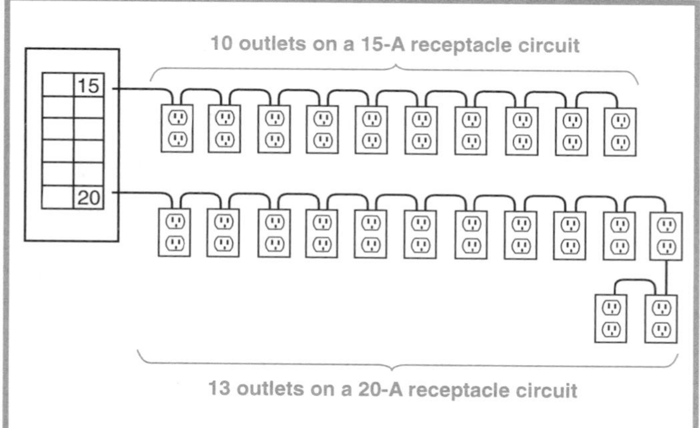

Figure 220.4 Computation of the maximum number of outlets permitted on 15- and 20-ampere branch circuits.

(10) Dwelling Occupancies. In one-family, two-family, and multifamily dwellings and in guest rooms of hotels and motels, the outlets specified in (a), (b), and (c) are included in the general lighting load calculations of Section 220-3(a). No additional load calculations shall be required for such outlets.

In the example shown in Figure 220.5, each panel serves a computed load of 80 amperes. The main feeder is sized to carry a minimum of 240 amperes (3 multiplied by 80 amperes). The feeders from the meter enclosure to the panelboards are sized to provide a minimum of 80 amperes. The main feeder is not intended to be sized to carry 300 amperes.

See Figure 230.13 for a similar example for service conductors. The ungrounded service conductors are no longer required to be sized for the sum of the main overcurrent device rating of 300 amperes. Service conductors are required to have sufficient ampacity to carry the loads computed in accordance with Article 220, with appropriate demand factors applied.

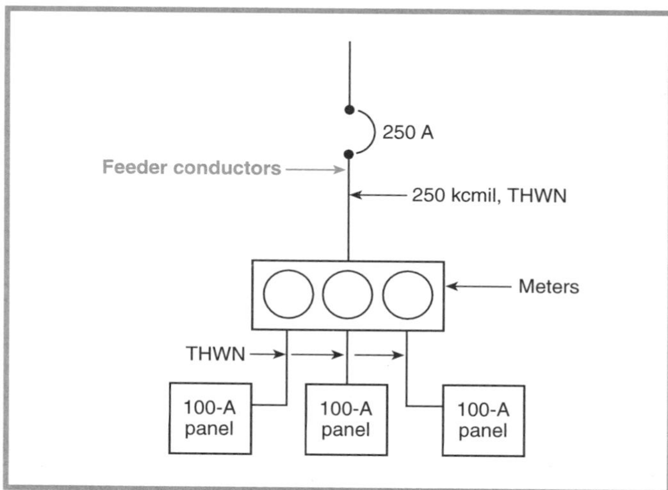

Figure 220.5 Feeder conductors sized in accordance with Section 220-10.

See Sections 230-23, 230-31, and 230-42 for specifics on size and rating of conductors.

Part B of Article 220 contains the requirements for calculating feeder and service loads. Parts C and D provide optional methods for calculating feeder and service loads in dwelling units and multifamily dwellings.

Except as permitted in Section 240-3, the rating of the overcurrent device cannot exceed the final ampacity of the circuit conductors after all the derating or correction factors have been applied, such as for temperature or number of conductors.

Example

Determine the minimum-size overcurrent protective device and the minimum conductor size for a circuit with the following characteristics:

1. A 3-phase, 4-wire feeder (full-size neutral)
2. A 125-ampere noncontinuous load
3. A 200-ampere continuous load
4. A 75°C overcurrent device terminal rating
5. Type THWN conductors
6. Four current-carrying conductors in a raceway
7. A major portion of the load is nonlinear

Answer

Step 1. Sum of continuous and noncontinuous loads

200 A × 125% [Section 220-10(b)] = 250 A computed continuous load

125 A noncontinuous + 250 A continuous = 375 A

Step 2. Overcurrent protective device (OCPD) size selection

The minimum standard-size OCPD, according to Section 240-6, is 400 amperes.

Step 3. Conductor size selection before derating

200 A continuous load × 125% = 250 A

125 A noncontinuous + 250 A continuous = 375 A

Due to the 75°C rating of the overcurrent device terminal, the minimum-size Type THWN copper conductor, in the 75°C column of Table 310-16, that can supply a computed load of 375 amperes is a 500-kcmil copper with an ampacity of 380 amperes.

Step 4. Application of derating factors to conductors

Since a major portion of the load consists of fluorescent lighting and HID lighting fixtures, the 3-phase conductors and the neutral conductor constitute four current-carrying conductors in the same raceway. Therefore, Sections 310-15(b)(2) and (4) require an 80 percent derating factor. According to Table 310-16, 500-kcmil Type THWN conductors have an ampacity of 380 amperes. The derating factors are applied to this ampacity as follows:

380 A × 80% = 304 A

According to Sections 240-3 and 240-6, a conductor with a computed ampacity of 304 amperes is not allowed to be protected by a 400-ampere OCPD. Therefore, the 500-kcmil THWN copper conductor cannot be used. The alternative is to choose a 600-kcmil conductor in the 90°C column.

Step 5. Conductor selection and check

Where higher-temperature insulations are used, adjustment factors can be applied to the higher ampacity. Since THWN is a 75°C insulation, a 90°C THHN is selected.

Where a 600-kcmil THHN copper conductor is used, the check is performed as follows:

1. According to Table 310-16, a 600-kcmil conductor has an ampacity of 475 amperes in the THHN 90°C column:

   475 A × 80% = 380 A

2. A conductor with a computed ampacity of 380 amperes is allowed to be protected by a 400-ampere OCPD, in accordance with Section 240-3(b). However, since the OCPD terminations are rated 75°C, the load current cannot exceed the ampacity of a 600-kcmil conductor in the 75°C column, which has a value of 420 amperes.

Step 6. Circuit evaluation

The previous calculation results in four 600-kcmil Type THHN copper conductors in one raceway, each with an ampacity of 380 amperes, supplying a 375-ampere continuous load and protected by a 400-ampere OCPD.

Section 110-14(c)(2) permits a 90°C 600-kcmil copper conductor with a final computed ampacity of 380 amperes to terminate on a 75°C-rated terminal.

(a) All general-use receptacle outlets of 20-ampere rating or less, including receptacles connected to the circuits in Section 210-11(c)(3)
(b) The receptacle outlets specified in Sections 210-52(e) and (g)
(c) The lighting outlets specified in Sections 210-70(a) and (b)

(11) Other Outlets. Other outlets not covered in (1) through (10) shall be computed based on 180 volt-amperes per outlet.

(c) Loads for Additions to Existing Installations.

(1) Dwelling Units. Loads for structural additions to an existing dwelling unit or to a previously unwired portion of an existing dwelling unit, either of which exceeds 500 ft^2 (46.5 m^2), shall be computed in accordance with (b). Loads for new circuits or extended circuits in previously wired dwelling units shall be computed in accordance with either (a) or (b).

(2) Other than Dwelling Units. Loads for new circuits or extended circuits in other than dwelling units shall be computed in accordance with either (a) or (b).

220-4. Maximum Loads. The total load shall not exceed the rating of the branch circuit, and it shall not exceed the maximum loads specified in (a) through (c) under the conditions specified therein.

(a) Motor-Operated and Combination Loads. Where a circuit supplies only motor-operated loads, Article 430 shall apply. Where a circuit supplies only air-conditioning equipment, refrigerating equipment, or both, Article 440 shall apply. For circuits supplying loads consisting of motor-operated utilization equipment that is fastened in place and that has a motor larger than ⅛ hp in combination with other loads, the total computed load shall be based on 125 percent of the largest motor load plus the sum of the other loads.

(b) Inductive Lighting Loads. For circuits supplying lighting units that have ballasts, transformers, or autotransformers, the computed load shall be based on the total ampere ratings of such units and not on the total watts of the lamps.

(c) Range Loads. It shall be acceptable to apply demand factors for range loads in accordance with Table 220-19, including Note 4.

B. Feeders and Services

220-10. General. The computed load of a feeder or service shall not be less than the sum of the loads on the branch circuits supplied, as determined by Part A of this article, after any applicable demand factors permitted by Parts B, C, or D have been applied.

FPN: See Examples D1(a) through D10 in Appendix D. See Section 220-4(b) for the maximum load in amperes permitted for lighting units operating at less than 100 percent power factor.

220-11. General Lighting. The demand factors specified in Table 220-11 shall apply to that portion of the total branch-circuit load computed for general illumination. They shall not be applied in determining the number of branch circuits for general illumination.

Table 220-11. Lighting Load Demand Factors

Type of Occupancy	Portion of Lighting Load to Which Demand Factor Applies (Volt-Amperes)	Demand Factor (Percent)
Dwelling units	First 3000 or less at	100
	From 3001 to 120,000 at	35
	Remainder over 120,000 at	25
Hospitals*	First 50,000 or less at	40
	Remainder over 50,000 at	20
Hotels and motels, including apartment houses without provision for cooking by tenants*	First 20,000 or less at	50
	From 20,001 to 100,000 at	40
	Remainder over 100,000 at	30
Warehouses (storage)	First 12,500 or less at	100
	Remainder over 12,500 at	50
All others	Total volt-amperes	100

*The demand factors of this table shall not apply to the computed load of feeders or services supplying areas in hospitals, hotels, and motels where the entire lighting is likely to be used at one time, as in operating rooms, ballrooms, or dining rooms.

220-12. Show-Window and Track Lighting.

(a) Show Windows. For show-window lighting, a load of not less than 200 volt-amperes shall be included for each linear foot (305 mm) of show window, measured horizontally along its base.

FPN: See Section 220-3(b)(7) for branch circuits supplying show windows.

The 200-volt-ampere calculation for each linear foot of a show window is required to calculate the *feeder* load. See the commentary following Section 220-3(b)(7) for load calculations for branch circuits.

(b) Track Lighting. For track lighting in other than dwelling units or guest rooms of hotels or motels, an additional load of 150 volt-amperes shall be included for every 2 ft (610 mm) of lighting track or fraction thereof.

The requirements of Section 220-12(b) previously appeared in Section 410-102 but they were moved to Article 220 for the 1999 *Code* since they deal primarily with load calculations.

Example

Suppose a lighting plan shows 62.5 linear ft of single-circuit track lighting for a small department store featuring clothing. Since the actual track lighting fixtures are owner supplied, neither the quantity of track lighting fixtures or the lamp size are specified. What is the minimum calculated load associated with the track lighting that must be added to the service or feeder supplying this store?

Answer

According to Section 220-12(b), the minimum calculated load to be added to the service or feeder supplying this track light installation is calculated as follows:

$$62.5 \text{ ft} \div 2 \text{ ft} = 31.25 \text{ ft}$$

31.25 must be rounded up to 32

$$32 \times 150 \text{ VA} = 4800 \text{ VA}$$

4800 VA is the minimum load that must be added to the service or feeder calculation.

It is important to note that the branch circuits supplying this installation are covered in Section 410-101(b). For the track lighting branch-circuit load, the maximum load on the track cannot exceed the rating of the branch circuit supplying the track. Also, the track must be supplied by a branch circuit that has a rating not exceeding the rating of the track. The track length does not enter into the branch-circuit calculation.

Section 220-12(b) is not intended to limit the number of feet of track on a single branch circuit nor is it intended to limit the number of fixtures on an individual track. Rather, this section is intended to be used solely for load calculations of feeders and services.

220-13. Receptacle Loads—Nondwelling Units. In other than dwelling units, receptacle loads computed at not more than 180 volt-amperes per outlet in accordance with Section 220-3(b)(9) and fixed multi-outlet assemblies computed in accordance with Section 220-3(b)(8) shall be permitted to be added to the lighting loads and made subject to the demand factors given in Table 220-11, or they shall be permitted to be made subject to the demand factors given in Table 220-13.

Table 220-13. Demand Factors for Nondwelling Receptacle Loads

Portion of Receptacle Load to Which Demand Factor Applies (Volt-Amperes)	Demand Factor (Percent)
First 10 kVA or less at	100
Remainder over 10 kVA at	50

Section 220-13 permits receptacle loads, calculated at not more than 180 volt-amperes per strap, to be computed by either of the following two methods.

1. **The receptacle loads are added to the lighting load. The demand factors in Table 220-11 are then applied to the combined load.**
2. **The receptacle loads are calculated (without the lighting load) with demand factors from Table 220-13 applied.**

220-14. Motors. Motor loads shall be computed in accordance with Sections 430-24, 430-25, and 430-26.

220-15. Fixed Electric Space Heating. Fixed electric space heating loads shall be computed at 100 percent of the total connected load; however, in no case shall a feeder or service load current rating be less than the rating of the largest branch circuit supplied.

Exception: Where reduced loading of the conductors results from units operating on duty-cycle, intermittently, or from all units not operating at one time, the authority having jurisdiction may grant permission for feeder conductors to have an ampacity less than 100 percent, provided the conductors have an ampacity for the load so determined.

220-16. Small Appliance and Laundry Loads — Dwelling Unit.

(a) Small Appliance Circuit Load. In each dwelling unit, the load shall be computed at 1500 volt-amperes for each 2-wire small-appliance branch circuit required by Section 210-11(c)(1). Where the load is subdivided through two or more feeders, the computed load for each shall include not less than 1500 volt-amperes for each 2-wire small-appliance branch circuit. These loads shall be permitted to be included with the general lighting load and subjected to the demand factors provided in Table 220-11.

See the commentary following Section 210-52(b) regarding required receptacle outlets for small-appliance branch circuits.

Exception: The individual branch circuit permitted by Section 210-52(b)(1), Exception No. 2 shall be permitted to be excluded from the calculation required by Section 220-16.

(b) Laundry Circuit Load. A load of not less than 1500 volt-amperes shall be included for each 2-wire laundry branch circuit installed as required by Section 210-11(c)(2). This load shall be permitted to be included with the general lighting load and subjected to the demand factors provided in Table 220-11.

In each dwelling unit, the feeder load is required to be calculated at 1500 volt-amperes for each of the two or more (2-wire) small-appliance branch circuits and at 1500 volt-amperes for each (2-wire) laundry branch circuit. These loads are permitted to be totaled and then added to the general lighting load. The total load (i.e., small appliance, laundry, and general lighting) is subjected to the demand factors provided in Table 220-11.

220-17. Appliance Load — Dwelling Unit(s). It shall be permissible to apply a demand factor of 75 percent to the nameplate rating load of four or more appliances fastened in place, other than electric ranges, clothes dryers, space-heating equipment, or air-conditioning equipment, that are served by the same feeder or service in a one-family, two-family, or multifamily dwelling.

For appliances fastened in place (other than ranges, clothes dryers, space-heating and air-conditioning equipment), feeder capacity must be provided for the sum of these loads, and the total load of four or more such appliances may be reduced by a demand factor of 75 percent. Seventy-five percent of 8944 volt-amperes equals 6708 volt-amperes, which is the load to be added to the other determined loads where calculating the size of service and feeder conductors.

Example

A 120/240-volt fastened-in-place appliance load — dwelling unit(s) — for the following:

Appliance	Rating	Load
Water heater	4000 W, 240 V	4000 VA
Kitchen disposal	½ hp, 120 V	1176 VA
Dishwasher	1200 W, 120 V	1200 VA
Furnace motor	¼ hp, 120 V	696 VA
Attic fan	¼ hp, 120 V	696 VA
Water pump	½ hp, 240 V	1176 VA
Total		8944 VA

See Table 430-148 for the full-load current, in amperes, for single-phase ac motors, in accordance with Section 220-14.

220-18. Electric Clothes Dryers — Dwelling Unit(s). The load for household electric clothes dryers in a dwelling unit(s) shall be 5000 watts (volt-amperes) or the nameplate rating, whichever is larger, for each dryer served. The use of the demand factors in Table 220-18 shall be permitted. Where two or more single-phase dryers are supplied by a 3-phase, 4-wire feeder or service, the total load shall be computed on the basis of twice the maximum number connected between any two phases.

Table 220-18. Demand Factors for Household Electric Clothes Dryers

Number of Dryers	Demand Factor (Percent)
1	100
2	100
3	100
4	100
5	80
6	70
7	65
8	60
9	55
10	50
11–13	45
14–19	40
20–24	35
25–29	32.5
30–34	30
35–39	27.5
40 and over	25

To compute the load of household electric dryers, this requirement specifies a minimum demand of 5 kVA for the calculation of feeder conductors. If the nameplate rating is known and exceeds 5 kW, the larger rating is applied.

220-19. Electric Ranges and Other Cooking Appliances — Dwelling Unit(s). The demand load for household electric ranges, wall-mounted ovens, counter-mounted cooking units, and other household cooking appliances individually rated in excess of 1¾ kW shall be permitted to be computed in accordance with Table 220-19. Where two or more single-phase ranges are supplied by a 3-phase, 4-wire feeder or service, the total load shall be computed on the basis of twice the maximum number connected between any two phases. Kilovolt-amperes (kVA) shall be considered equivalent to kilowatts (kW) for loads computed under this section.

FPN: See Example D5(a) in Appendix D.

Table 220-19. Demand Loads for Household Electric Ranges, Wall-Mounted Ovens, Counter-Mounted Cooking Units, and Other Household Cooking Appliances over 1¾ kW Rating (Column A to be used in all cases except as otherwise permitted in Note 3.)

	Maximum Demand (kW) (See Notes)	Demand Factor (Percent) (See Note 3)	
Number of Appliances	Column A (Not over 12 kW Rating)	Column B (Less than 3½ kW Rating)	Column C (3½ kW to 8¾ kW Rating)
1	8	80	80
2	11	75	65
3	14	70	55
4	17	66	50
5	20	62	45
6	21	59	43
7	22	56	40
8	23	53	36
9	24	51	35
10	25	49	34
11	26	47	32
12	27	45	32
13	28	43	32
14	29	41	32
15	30	40	32
16	31	39	28
17	32	38	28
18	33	37	28
19	34	36	28
20	35	35	28
21	36	34	26
22	37	33	26
23	38	32	26
24	39	31	26
25	40	30	26
26–30	15 kW + 1 kW for each range	30	24
31–40		30	22
41–50	25 kW + ¾ kW for each range	30	20
51–60		30	18
61 and over		30	16

Notes:
1. Over 12 kW through 27 kW ranges all of same rating. For ranges individually rated more than 12 kW but not more than 27 kW, the maximum demand in Column A shall be increased 5 percent for each additional kilowatt of rating or major fraction thereof by which the rating of individual ranges exceeds 12 kW.

Table 220-19 Notes (Continued)

The size of the conductors is required to be determined by the rating of the range. According to Table 220-19, for a range rated not over 12 kW, the demand load is 8 kW (8 kVA per Section 220-19), and No. 8 copper conductors with 60°C insulation would suffice.

Note that Section 210-19(c) does not permit the branch-circuit rating of a circuit supplying household ranges of $8^3/_4$-kW nameplate rating to be less than 40 amperes.

2. Over 8¾ kW through 27 kW ranges of unequal ratings. For ranges individually rated more than 8¾ kW and of different ratings, but none exceeding 27 kW, an average value of rating shall be computed by adding together the ratings of all ranges to obtain the total connected load (using 12 kW for any range rated less than 12 kW) and dividing by the total number of ranges. Then the maximum demand in Column A shall be increased 5 percent for each kilowatt or major fraction thereof by which this average value exceeds 12 kW.

Note 2 provides for ranges larger than $8^3/_4$ kW, and Note 4 covers installations where the circuit supplies multiple cooking components, which are combined and treated as a single range.

3. Over 1¾ kW through 8¾ kW. In lieu of the method provided in Column A, it shall be permissible to add the nameplate ratings of all household cooking appliances rated more than 1¾ kW but not more than 8¾ kW and multiply the sum by the demand factors specified in Column B or C for the given number of appliances. Where the rating of cooking appliances falls under both Column B and Column C, the demand factors for each column shall be applied to the appliances for that column, and the results added together.

The branch-circuit load for one range is permitted to be computed by using the nameplate rating of the appliance or Table 220-19. Where a single branch circuit supplies a counter-mounted cooking unit and not more than two wall-mounted ovens, all of which are located in the same room, the nameplate ratings of these appliances can be added together and this total can be treated as equivalent to one range.

Example

A single branch circuit supplies the following loads:

Appliance	Rating
One counter-mounted cooking unit	8 kW
One wall-mounted oven	7 kW
One wall-mounted oven	6 kW
Total	21 kW

Table 220-19 Notes (Continued)

This can be treated as one range. The maximum demand from Column A of Table 220-19 for one range not over 12 kW is 8 kW.

The total of 21 kW is less than 27 kW and exceeds 12 kW by 9 kW (see Note 1). The maximum demand in Column A (8 kW) is increased 5 percent for each additional kW exceeding 12 kW (9 kW) as follows:

$$5\% \text{ per kW} \times 9 \text{ kW} = 45\% \text{ increase}$$

$$45\% \times 8 \text{ kW} = 3.6\text{-kW increase}$$

$$8 \text{ kW} + 3.6 \text{ kW} = 11.6 \text{ kW}$$

$$11{,}600 \text{ W} = 11{,}600 \text{ VA} \div 240 \text{ V} = 48.3 \text{ A}$$

4. Branch-Circuit Load. It shall be permissible to compute the branch-circuit load for one range in accordance with Table 220-19. The branch-circuit load for one wall-mounted oven or one counter-mounted cooking unit shall be the nameplate rating of the appliance. The branch-circuit load for a counter-mounted cooking unit and not more than two wall-mounted ovens, all supplied from a single branch circuit and located in the same room, shall be computed by adding the nameplate rating of the individual appliances and treating this total as equivalent to one range.

5. This table also applies to household cooking appliances rated over 1¾ kW and used in instructional programs.

It is permissible to add the nameplate ratings of all household cooking appliances rated more than 1¾ kW but not more than 8¾ kW and multiply the sum by the demand factors specified in Column B or C of Table 220-19 for the given number of appliances.

For feeder demand factors other than for dwelling units, that is, commercial electric cooking equipment, dishwasher booster heaters, water heaters, and so on, see Table 220-20.

The demand factors of this *Code* are based on the diversified use of appliances, because it is unlikely that all appliances will be used simultaneously or that all cooking units and the oven of a range will be at maximum heat for any length of time.

FPN No. 1: See Table 220-20 for commercial cooking equipment.

FPN No. 2: See the examples in Appendix D.

220-20. Kitchen Equipment — Other Than Dwelling Unit(s). It shall be permissible to compute the load for commercial electric cooking equipment, dishwasher booster heaters, water heaters, and other kitchen equipment in accordance with Table 220-20. These demand factors shall be applied to all equipment that has either thermostatic control or intermittent use as kitchen equipment. They shall not apply to space-heating, ventilating, or air-conditioning equipment.

However, in no case shall the feeder or service demand

be less than the sum of the largest two kitchen equipment loads.

Table 220-20. Demand Factors for Kitchen Equipment — Other than Dwelling Unit(s)

Number of Units of Equipment	Demand Factor (Percent)
1	100
2	100
3	90
4	80
5	70
6 and over	65

220-21. Noncoincident Loads. Where it is unlikely that two or more noncoincident loads will be in use simultaneously, it shall be permissible to use only the largest load(s) that will be used at one time, in computing the total load of a feeder or service.

220-22. Feeder or Service Neutral Load. The feeder or service neutral load shall be the maximum unbalance of the load determined by this article. The maximum unbalanced load shall be the maximum net computed load between the neutral and any one ungrounded conductor, except that the load thus obtained shall be multiplied by 140 percent for 3-wire, 2-phase or 5-wire, 2-phase systems. For a feeder or service supplying household electric ranges, wall-mounted ovens, counter-mounted cooking units, and electric dryers, the maximum unbalanced load shall be considered as 70 percent of the load on the ungrounded conductors, as determined in accordance with Table 220-19 for ranges and Table 220-18 for dryers. For 3-wire dc or single-phase ac; 4-wire, 3-phase; 3-wire, 2-phase; or 5-wire, 2-phase systems, a further demand factor of 70 percent shall be permitted for that portion of the unbalanced load in excess of 200 amperes. There shall be no reduction of the neutral capacity for that portion of the load that consists of nonlinear loads supplied from a 4-wire, wye-connected, 3-phase system nor the grounded conductor of a 3-wire circuit consisting of two phase wires and the neutral of a 4-wire, 3-phase, wye-connected system.

FPN No. 1: See Examples D1(a), D1(b), D2(b), D4(a), and D5(a) in Appendix D.

FPN No. 2: A 3-phase, 4-wire, wye-connected power system used to supply power to nonlinear loads may necessitate that the power system design allow for the possibility of high harmonic neutral currents.

Section 220-22 describes the basis for calculating the neutral load of feeders or services as the maximum unbalanced load that can occur between the neutral and any other ungrounded conductor.

For a household electric range or clothes dryer, the maximum unbalanced load may be assumed to be 70 percent, so the neutral may be sized on this basis. Although Section 220-22 permits the reduction of the feeder neutral conductor size under specific conditions of use, Section 310-10, Temperature Limitations of Conductors, contains the principal determinates that affect a conductor operating temperature. Principal determinate No. 2 requires that heating due to harmonic current also be taken into consideration. In certain cases, this may actually prohibit the reduction of a neutral feeder and require that the neutral feeder conductors be adjusted to a larger size.

Where the system also supplies nonlinear loads such as electric-discharge lighting, including fluorescent and HID, or data-processing or similar equipment, the neutral is considered a current-carrying conductor if the electric-discharge lighting, data-processing, and the like, load on the feeder neutral consists of more than half the total load. Electric-discharge lighting and data-processing equipment may have harmonic currents in the neutral that may exceed the load current in the ungrounded conductors. It would be appropriate to require a full-size or larger feeder neutral conductor, depending on the total harmonic distortion contributed by the equipment to be supplied (see Section 220-22, FPN No. 2).

In some instances, the neutral current may exceed the current in the phase conductors. See the commentary following Section 310-15(b)(4) regarding neutral conductor ampacity.

C. Optional Calculations for Computing Feeder and Service Loads

220-30. Optional Calculation — Dwelling Unit.

(a) Feeder and Service Load. For a dwelling unit having the total connected load served by a single 3-wire, 120/240-volt or 208Y/120-volt set of service-entrance or feeder conductors with an ampacity of 100 or greater, it shall be permissible to compute the feeder and service loads in accor dance with this section instead of the method specified in Part B of this article. The calculated load shall be the result of adding the loads from (b) and (c). Feeder and service-entrance conductors whose demand load is determined by this optional calculation shall be permitted to have the neutral load determined by Section 220-22.

The optional method given in Section 220-30 is applicable to a single-dwelling unit, whether it is a separate building or located in a multifamily dwelling. The optional calculation permitted by Section 220-30 may be used only where the service-entrance or feeder conductors have an ampacity of at least 100 amperes.

Examples of the optional calculation for a dwelling unit are given in Examples D2(a), D2(b), D2(c), and D4(b) of Appendix D.

See Article 100 for the definition of *dwelling unit.*

(b) General Loads. The general calculated load shall be not less than 100 percent of the first 10 kVA plus 40 percent of the remainder of the following loads:

(1) 1500 volt-amperes for each 2-wire, 20-ampere small-appliance branch circuit and each laundry branch circuit specified in Section 220-16
(2) 3 volt-amperes per square foot (0.093 m^2) for general lighting and general-use receptacles
(3) The nameplate rating of all appliances that are fastened in place, permanently connected, or located to be on a specific circuit, ranges, wall-mounted ovens, counter-mounted cooking units, clothes dryers, and water heaters

Section 220-30(b)(3) includes appliances that may not be "fastened in place" but may be permanently connected or on a specific circuit, such as clothes dryers, dishwashers, and freezers.

(4) The nameplate ampere or kVA rating of all motors and of all low-power-factor loads

(c) Heating and Air-Conditioning Load. Include the largest of the following six selections (load in kVA).

(1) 100 percent of the nameplate rating(s) of the air conditioning and cooling.
(2) 100 percent of the nameplate ratings of the heat pump compressors and supplemental heating unless the controller prevents the compressor and supplemental heating from operating at the same time.
(3) 100 percent of the nameplate ratings of electric thermal storage and other heating systems where the usual load is expected to be continuous at the full nameplate value. Systems qualifying under this selection shall not be calculated under any other selection in (c).
(4) 65 percent of the nameplate rating(s) of the central electric space heating, including integral supplemental heating in heat pumps where the controller prevents the compressor and supplemental heating from operating at the same time.
(5) 65 percent of the nameplate rating(s) of electric space heating if less than four separately controlled units.
(6) 40 percent of the nameplate rating(s) of electric space heating if four or more separately controlled units.

Section 220-21 states that when considering loads that do not operate simultaneously, the largest load being considered is used. In concert with Section 220-21, Section 220-30(c) requires that only the largest of the six choices has to be included in the feeder or service calculation. Examples of calculations using air conditioning and heating are found in Appendix D, Examples D2(b) and D2(c).

220-31. Optional Calculation for Additional Loads in Existing Dwelling Unit. For an existing dwelling unit presently being served by an existing 120/240-volt or 208Y/120-volt, 3-wire service, it shall be permissible to compute load calculations as follows:

Load (kVa)	Percent of Load
First 8 kVA of load at	100
Remainder of load at	40

Load calculations shall include lighting at 3 volt-amperes/ft^2 (0.093 m^2); 1500 volt-amperes for each 2-wire, small-appliance branch circuit and each laundry branch circuit as specified in Section 220-16; range or wall-mounted oven and counter-mounted cooking unit; and other appliances that are permanently connected or fastened in place, at nameplate rating.

If air-conditioning equipment or electric space-heating equipment is to be installed, the following formula shall be applied to determine if the existing service is of sufficient size.

Air-conditioning equipment*	100%
Central electric space heating*	100%
Less than four separately controlled space-heating units*	100%
First 8 kVA of all other loads	100%
Remainder of all other loads	40%

Other loads shall include the following:

(1) 1500 volt-amperes for each 20-ampere appliance circuit
(2) Lighting and portable appliances at 3 volt-amperes/ft^2 (0.093 m^2)
(3) Household range or wall-mounted oven and counter-mounted cooking unit
(4) All other appliances fastened in place, including four or more separately controlled space-heating units, at nameplate rating

*Use larger connected load of air conditioning and space heating, but not both.

The optional method described in Section 220-31(4) allows an additional load to be supplied by an existing service.

Example

An existing dwelling unit is served by a 100-ampere service. An additional load of 5 kVA, 240-volt air con-

ditioning is to be installed. The existing load consists of the following:

Load	Volt-Amperes
General lighting	
24 ft × 40 ft = 960 ft² × 3 VA per ft²	2,880 VA
Small-appliance circuits (3 × 1500 VA)	4,500 VA
Laundry circuit at 1500 VA	1,500 VA
Electric range rated 10.5 kW	10,500 VA
Electric water heater rated 3.0 kW	3,000 VA
Total existing load	22,380 VA
Computation of existing load + additional	
load of air conditioning (5 kVA × 100%)	5,000 VA
First 8 kVA of other load at 100%	8,000 VA
Remainder of other load at 40% (22,380 – 8000) = 14,380 × 40%	5,752 VA
Total load	18,752 VA
18,752 VA ÷ 240 V = 78.1 A	

The additional load contributed by the 5-kVA air conditioning does not exceed the allowable load permitted on a 100-ampere service.

220-32. Optional Calculation — Multifamily Dwelling.

(a) Feeder or Service Load. It shall be permissible to compute the feeder or service load of a multifamily dwelling in accordance with Table 220-32 instead of Part B of this article where all the following conditions are met.

(1) No dwelling unit is supplied by more than one feeder.
(2) Each dwelling unit is equipped with electric cooking equipment.

The method of calculation of load under Section 220-32 is optional and applies only where one service or feeder supplies the entire load of a dwelling unit. Where all the stated conditions prevail, the optional calculations in Section 220-32 may be used instead of those in Part B of Article 220.

Exception: When the computed load for multifamily dwellings without electric cooking in Part B of this article exceeds that computed under Part C for the identical load plus electric cooking (based on 8 kW per unit), the lesser of the two loads shall be permitted to be used.

Section 220-32(a)(2) requires that each dwelling unit to be equipped with electric cooking equipment in order to use this method of calculation. The exception to Section 220-32(a)(2) permits calculation for dwelling units that do not have electric cooking equipment by calculating a simulated electric cooking equipment load of 8 kW per unit and selecting the lesser of the two loads.

(3) Each dwelling unit is equipped with either electric space heating or air conditioning or both. Feeders and service-entrance conductors whose demand load is determined by this optional calculation shall be permitted to have the neutral load determined by Section 220-22.

(b) House Loads. House loads shall be computed in accordance with Part B of this article and shall be in addition to the dwelling unit loads computed in accordance with Table 220-32.

(c) Connected Loads. The connected load to which the demand factors of Table 220-32 apply shall include the following.

(1) 1500 volt-amperes for each 2-wire, 20-ampere small-appliance branch circuit and each laundry branch circuit specified in Section 220-16.
(2) 3 volt-amperes/ft^2 (0.093 m^2) for general lighting and general-use receptacles.
(3) The nameplate rating of all appliances that are fastened in place, permanently connected or located to be on a specific circuit, ranges, wall-mounted ovens, counter-mounted cooking units, clothes dryers, water heaters, and space heaters. If water heater elements are interlocked so that all elements cannot be used at the same time, the maximum possible load shall be considered the nameplate load.
(4) The nameplate ampere or kilovolt-ampere rating of all motors and of all low-power-factor loads.
(5) The larger of the air-conditioning load or the space-heating load.

220-33. Optional Calculation — Two Dwelling Units. Where two dwelling units are supplied by a single feeder and the computed load under Part B of this article exceeds that for three identical units computed under Section 220-32, the lesser of the two loads shall be permitted to be used.

220-34. Optional Method — Schools. The calculation of a feeder or service load for schools shall be permitted in accordance with Table 220-34 in lieu of Part B of this article where equipped with electric space heating or air conditioning, or both. The connected load to which the demand factors of Table 220-34 apply shall include all of the interior and exterior lighting, power, water heating, cooking, other loads, and the larger of the air-conditioning load or space-heating load within the building or structure.

Feeders and service-entrance conductors whose demand load is determined by this optional calculation shall be permitted to have the neutral load determined by Section 220-22. Where the building or structure load is calculated by this optional method, feeders within the building or structure shall have ampacity as permitted in Part B of this article; however, the ampacity of an individual feeder shall not be required to be larger than the ampacity for the entire building.

This section shall not apply to portable classroom buildings.

Many schools add small, portable classroom buildings. The air-conditioning load must comply with Ar-

Table 220-32. Optional Calculations — Demand Factors for Three or More Multifamily Dwelling Units

Number of Dwelling Units	Demand Factor (Percent)
3–5	45
6–7	44
8–10	43
11	42
12–13	41
14–15	40
16–17	39
18–20	38
21	37
22–23	36
24–25	35
26–27	34
28–30	33
31	32
32–33	31
34–36	30
37–38	29
39–42	28
43–45	27
46–50	26
51–55	25
56–61	24
62 and over	23

Table 220-34. Optional Method — Demand Factors for Feeders and Service-Entrance Conductors for Schools

Connected Load (VA/ft^2)	Demand Factor (Percent)
The first 3 VA/ft^2 at	100
Plus,	
Over 3 to 20 VA/ft^2 at	75
Plus,	
Remainder over 20 VA/ft^2 at	25

Note: For SI units, $1\ ft^2 = 0.093\ m^2$.

ticle 440, and the lighting load must be considered continuous. The demand factors of Table 220-34 do not apply to portable classrooms because this would decrease the feeder or service size to below that required for the connected continuous load.

220-35. Optional Calculations for Determining Existing Loads. The calculation of a feeder or service load for existing installations shall be permitted to use actual maximum demand to determine the existing load under the following conditions.

(1) The maximum demand data is available for a 1-year period.

Exception: If the maximum demand data for a 1-year period is not available, the calculated load shall be permitted to be based on the maximum demand (measure of average power demand over a 15-minute period) continuously recorded over a minimum 30-day period using a recording ammeter or power meter connected to the highest loaded phase of the feeder or service, based on the initial loading at the start of the recording. The recording shall reflect the maximum demand of the feeder or service by being taken when the building or space is occupied and shall include by measurement or calculation the larger of the heating or cooling equipment load, and other loads that may be periodic in nature due to seasonal or similar conditions.

(2) The maximum demand at 125 percent plus the new load does not exceed the ampacity of the feeder or rating of the service.

(3) The feeder has overcurrent protection in accordance with Section 240-3, and the service has overload protection in accordance with Section 230-90.

Where existing installations have been checked, the maximum demand kVA data for a minimum one-year period is available (or the 30-day alternate method from the exception), and the installation complies with Sections 220-35(2) and (3), the connection of additional loads to existing services and feeders is permitted.

220-36. Optional Calculation — New Restaurants. Calculation of a service load or feeder, where the feeder serves the total load, for a new restaurant shall be permitted in accordance with Table 220-36 in lieu of Part B of this article.

The overload protection of the service-entrance conductors shall be in accordance with Sections 230-90 and 240-3.

Feeder conductors shall not be required to be of greater ampacity than the service-entrance conductors.

Service-entrance or feeder conductors whose demand load is determined by this optional calculation shall be permitted to have the neutral load determined by Section 220-22.

Section 220-36 recognizes the effects of load diversity typical of restaurant occupancies. It also recognizes the amount of continuous loads as a percentage of the total connected load. The National Restaurant Association, the Edison Electric Institute, and the Electric Power Research Institute based the data for this change on load studies of 262 restaurants. These studies have shown that the demand factors decline

Table 220-36. Optional Method — Demand Factors for Service-Entrance and Feeder Conductors for New Restaurants

Total Connected Load (kVA)	All Electric Demand Factor (Percent)	Not All Electric Demand Factor (Percent)
0–250	80	100
251–280	70	90
281–325	60	80
326–375	50	70
376–800	50	65
Over 800	50	50

Note: Add all electrical loads, including both heating and cooling loads, to compute the total connected load. Select the one demand factor that applies from the table and multiply the total connected load by this single demand factor.

as the connected loads increase. Based on this information, it was determined that demand factors would be appropriate.

In using this optional method, it is very important to notice that all loads are added together, even heating and air conditioning, and then the appropriate demand factor from the table is applied.

Example 1

An all-electric restaurant with a total connected load of 325 kVA at 208Y/120 volts would have a demand factor of 60 percent. Therefore, the service would be sized as follows:

$$\text{Amperes} = \frac{\text{kVA} \times 1000}{\text{volts} \times \sqrt{3}} \times \text{demand factor}$$
$$= \frac{325{,}000}{208 \times \sqrt{3}} \times 60\%$$
$$= 541.27, \text{ or } 541 \text{ A}$$

The next larger standard-size overcurrent device is 600 amperes. The minimum size of the conductors must be adequate to handle the load, but Section 240-3(b) would permit the next larger standard-rated overcurrent device to be used.

Example 2

A restaurant with gas cooking appliances and a total connected load of 325 kVA at 208Y/120 volts would have a demand factor of 80 percent. Therefore, the service would be sized as follows:

$$\text{Amperes} = \frac{\text{kVA} \times 1000}{\text{volts} \times \sqrt{3}} \times \text{demand factor}$$
$$= \frac{325{,}000}{208 \times \sqrt{3}} \times 80\%$$
$$= 721.68, \text{ or } 722 \text{ A}$$

Section 230-79 requires that the service disconnecting means have a rating that is not less than the computed load (722 amperes). The next higher standard rating for an overcurrent device, according to Section 240-6, is 800 amperes.

D. Method for Computing Farm Loads

220-40. Farm Loads — Buildings and Other Loads.

(a) Dwelling Unit. The feeder or service load of a farm dwelling unit shall be computed in accordance with the provisions for dwellings in Part B or C of this article. Where the dwelling has electric heat and the farm has electric grain-drying systems, Part C of this article shall not be used to compute the dwelling load.

(b) Other than Dwelling Unit. For each farm building or load supplied by two or more separate branch circuits, the load for feeders, service-entrance conductors, and service equipment shall be computed in accordance with demand factors not less than indicated in Table 220-40.

FPN: See Section 230-21 for overhead conductors from a pole to a building or other structure.

Table 220-40. Method for Computing Farm Loads for Other than Dwelling Unit

Ampere Load at 240 Volts Maximum	Demand Factor (Percent)
Loads expected to operate without diversity, but not less than 125 percent full-load current of the largest motor and not less than the first 60 amperes of load	100
Next 60 amperes of all other loads	50
Remainder of other load	25

Section 230-21 requires such overhead conductors to be considered a service drop and to be installed accordingly.

220-41. Farm Loads — Total. The total load of the farm for service-entrance conductors and service equipment shall be computed in accordance with the farm dwelling unit load and demand factors specified in Table 220-41. Where there is equipment in two or more farm equipment buildings or for loads having the same function, such loads shall be computed in accordance with Table 220-40 and shall be permitted to be combined as a single load in Table 220-41 for computing the total load.

FPN: See Section 230-21 for overhead conductors from a pole to a building or other structure.

Table 220-41. Method for Computing Total Farm Load

Individual Loads Computed in Accordance with Table 220-40	Demand Factor (Percent)
Largest load	100
Second largest load	75
Third largest load	65
Remaining loads	50

Note: To this total load, add the load of the farm dwelling unit computed in accordance with Part B or C of this article. Where the dwelling has electric heat and the farm has electric grain-drying systems, Part C of this article shall not be used to compute the dwelling load.

Article 225 — Outside Branch Circuits and Feeders

Contents

225-1. Scope. This article covers requirements for outside branch circuits and feeders run on or between buildings, structures, or poles on the premises; and electric equipment and wiring for the supply of utilization equipment that is located on or attached to the outside of buildings, structures, or poles.

FPN: For additional information on wiring over 600 volts, see *National Electrical Safety Code*, ANSI C2-1997.

225-2. Other Articles. Application of other articles, including additional requirements to specific cases of equipment and conductors, is as follows:

	Article
Branch circuits	210
Class 1, Class 2, and Class 3 remote-control, signaling, and power-limited circuits	725
Communications circuits	800
Community antenna television and radio distribution systems	820
Conductors for general wiring	310
Electrically driven or controlled irrigation machines	675
Electric signs and outline lighting	600
Feeders	215
Fire alarm systems	760
Fixed outdoor electric deicing and snow-melting equipment	426
Floating buildings	553
Grounding	250
Hazardous (classified) locations	500
Hazardous (classified) locations — specific	510
Marinas and boatyards	555
Messenger supported wiring	321
Open wiring on insulators	320
Over 600 volts, general	490
Overcurrent protection	240
Radio and television equipment	810
Services	230
Solar photovoltaic systems	690
Swimming pools, fountains, and similar installations	680
Use and identification of grounded conductors	200

A. General

Article 225 was reorganized into three parts for the 1999 *NEC*. Part A contains the general requirements for outside branch circuits and feeders. Part B contains the requirements from former Section 225-8 for more than one building or other structure. Part C contains the requirements for over 600 volts, nominal. Former Section 225-8 was deleted in the 1999 *NEC*.

225-3. Calculation of Load.

(a) Branch Circuits. The load on outdoor branch circuits shall be as determined by Section 220-3.

(b) Feeders. The load on outdoor feeders shall be as determined by Part B of Article 220.

225-4. Conductor Covering. Where within 10 ft (3.05 m) of any building or structure other than supporting poles or towers, open individual (aerial) overhead conductors shall be insulated or covered. Conductors in cables or raceways, except Type MI cable, shall be of the rubber-covered type or thermoplastic type and, in wet locations, shall comply with Section 310-8. Conductors for festoon lighting shall be of the rubber-covered or thermoplastic type.

Exception: Equipment grounding conductors and grounded circuit conductors shall be permitted to be bare or covered as specifically permitted elsewhere in this Code.

The exception to Section 225-4 and Exception No. 2 to Section 250-184(a) clarify that bare neutral messenger wire cable assemblies are permitted to be regrounded at additional buildings.

225-5. Size of Conductors. The ampacity of outdoor branch-circuit and feeder conductors shall be in accordance with Section 310-15 based on loads as determined under Section 220-3 and Part B of Article 220.

225-6. Conductor Size and Support.

(a) Overhead Spans. Open individual conductors shall not be smaller than the following:

(1) For 600 volts, nominal, or less, No. 10 copper or No. 8 aluminum for spans up to 50 ft (15.2 m) in length and No. 8 copper or No. 6 aluminum for a longer span, unless supported by a messenger wire

(2) For over 600 volts, nominal, No. 6 copper or No. 4 aluminum where open individual conductors and No. 8 copper or No. 6 aluminum where in cable

The size limitation of copper and aluminum conductors is based on the need for adequate mechanical strength to support the weight of the conductors and withstand wind, ice, and other similar conditions. Figure 225.1 is an illustration of overhead spans that are not messenger supported and are run between buildings, structures, or poles, and that are 600 volts or less. If the conductors are supported on a messenger, the messenger cable provides the mechanical strength. See Section 321-3 for wiring methods permitted to be messenger supported.

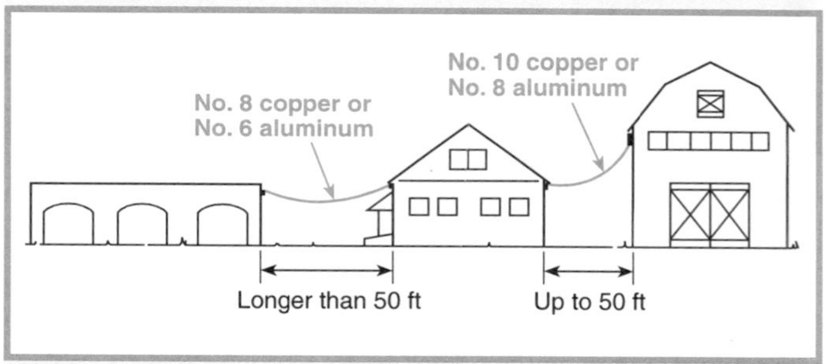

***Figure 225.1** Minimum sizes of conductors in overhead spans as specified by Section 225-6(a)(1) for 600 volts, nominal, or less.*

(b) Festoon Lighting. Overhead conductors for festoon lighting shall not be smaller than No. 12 unless the conductors are supported by messenger wires. In all spans exceeding

40 ft (12.2 m), the conductors shall be supported by messenger wire. The messenger wire shall be supported by strain insulators. Conductors or messenger wires shall not be attached to any fire escape, down spout, or plumbing equipment.

Article 100 defines *festoon lighting* as a string of outdoor lights suspended between two points. The conductors are not permitted to be smaller than No. 12 unless a messenger wire supports them. On all spans of festoon lighting exceeding 40 ft, messenger wire is required and is to be supported by strain insulators. If no messenger wire is required, the No. 12 or larger conductors are required to be supported by strain insulators. Figure 225.2 shows an example where messenger wire is required for festoon lighting conductors in a span exceeding 40 ft.

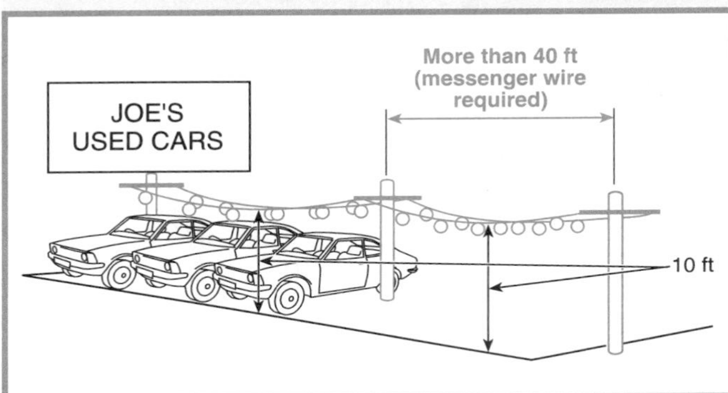

Figure 225.2 *Messenger wire required by Section 225-6(b) for festoon lighting conductors in a span exceeding 40 ft.*

Attachment to fire escapes, plumbing equipment, or drain spouts is prohibited because it could provide a path to ground. Moreover, such methods of attachment could not be relied on for a permanent or secure means of support.

These requirements previously appeared in Section 225-13, but for the 1999 *Code,* were moved here in Section 225-6(b).

225-7. Lighting Equipment Installed Outdoors.

(a) General. For the supply of lighting equipment installed outdoors, the branch circuits shall comply with Article 210 and (b) through (d).

(b) Common Neutral. The ampacity of the neutral conductor shall not be less than the maximum net computed load current between the neutral and all ungrounded conductors connected to any one phase of the circuit.

Multiwire branch circuits consisting of a neutral and two or more ungrounded conductors are permitted, provided the neutral capacity is not less than the total load of all ungrounded conductors connected to any one phase of the circuit.

Figures 225.3 and 225.4 show an example using a 120/240-volt, single-phase, 3-wire system and an example using a 208Y/120-volt, 3-phase, 4-wire system, respectively. In Figure 225.3, all branch circuits are rated at 20 amperes. The maximum unbalanced current that can occur will be four times 20 amperes, which is 80 amperes. In Figure 225.4, all branch circuits are rated at 20 amperes. The maximum unbalanced current that can occur on a 3-phase system with the load connected as shown is 80 amperes, due to the load on phase A.

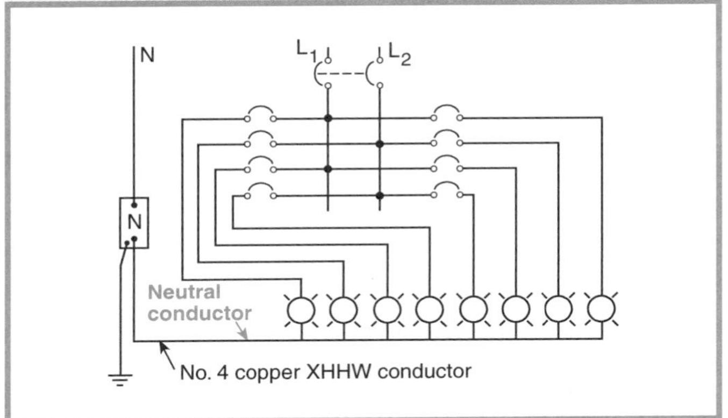

Figure 225.3 *An example using a 120/240-volt, single-phase, 3-wire system (branch circuits rated at 20 amperes; maximum unbalance current of 80 amperes).*

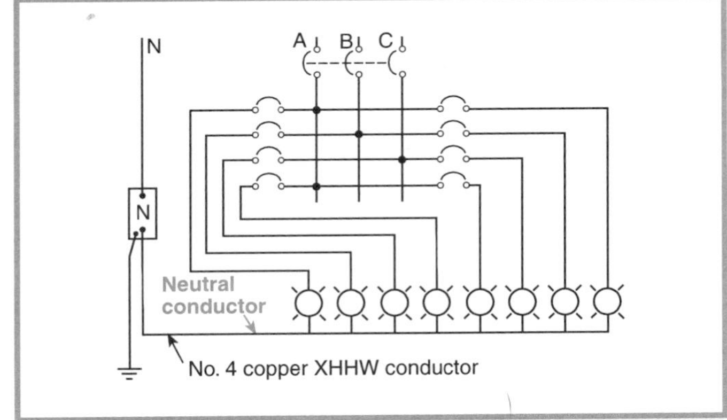

Figure 225.4 *An example using a 208Y/120-volt, 3-phase, 4-wire system (branch circuits rated at 20 amperes; maximum unbalance current of 80 amperes).*

(c) 277 Volts to Ground. Circuits exceeding 120 volts, nominal, between conductors and not exceeding 277 volts, nominal, to ground shall be permitted to supply lighting fixtures for illumination of outdoor areas of industrial establishments, office buildings, schools, stores, and other commercial or public buildings where the fixtures are not less than 3 ft (914 mm) from windows, platforms, fire escapes, and the like.

Branch circuits for the outdoor illumination of industrial establishments, office buildings, schools, stores, and other commercial or public buildings are permitted to operate with a voltage to ground of 277 volts. See Section 210-6(c)(1) for voltages greater than 277 volts to ground.

Many designers of outdoor lighting installations on poles prefer using 130-volt-rated lamps on 120-volt circuits (a nominal 10 percent undervoltage). Although this decreases the light intensity somewhat, it prolongs the life of the lamps, minimizing pole lamp maintenance, which can be difficult.

The restrictions outlined in Section 225-7(c) are in addition to those in Section 210-6(c).

(d) 600 Volts Between Conductors. Circuits exceeding 277 volts, nominal, to ground and not exceeding 600 volts, nominal, between conductors shall be permitted to supply the auxiliary equipment of electric-discharge lamps in accordance with Section 210-6(d)(1).

Section 210-6(d)(1) contains the minimum height requirements for circuits exceeding 277 volts, nominal, to ground, but not exceeding 600 volts, nominal, between conductors for circuits that supply the auxiliary equipment of electric-discharge lamps.

225-9. Overcurrent Protection. Overcurrent protection shall be in accordance with Section 210-20 for branch circuits and Article 240 for feeders.

Section 225-9 ensures that feeder overcurrent devices meet the same general accessibility requirements as service overcurrent devices.

225-10. Wiring on Buildings. The installation of outside wiring on surfaces of buildings shall be permitted for circuits of not over 600 volts, nominal, as open wiring on insulators, as multiconductor cable, as Type MC cable, as Type MI cable, as messenger supported wiring, in rigid metal conduit, in intermediate metal conduit, in rigid nonmetallic conduit, in cable trays, as cablebus, in wireways, in auxiliary gutters, in electrical metallic tubing, in flexible metal conduit, in liquidtight flexible metal conduit, in liquidtight flexible nonmetallic conduit, and in busways. Circuits of over 600 volts, nominal, shall be installed as provided in Section 300-37. Circuits for signs and outline lighting shall be installed in accordance with Article 600.

225-11. Circuit Exits and Entrances. Where outside branch and feeder circuits leave or enter a building, the requirements of Sections 230-52 and 230-54 shall apply.

Section 225-11 references Section 230-52, Individual Conductors Entering Buildings or Other Structures, and Section 230-54, Overhead Service Locations. Section 225-18 covers the requirements for clearances from ground (not over 600 volts) and Section 225-19 covers the requirements for clearances from buildings for conductors not over 600 volts. See Section 225-19(d) for clearances from windows, doors, fire escapes, and so on. Figure 225.5 shows an example where these requirements apply.

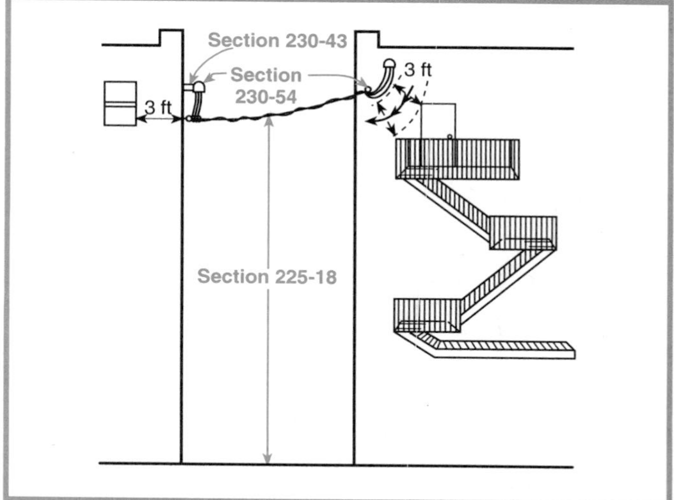

Figure 225.5 Section 225-11 references Sections 230-52, Individual Conductors Entering Buildings or Other Structures, and Section 230-54, Connections at Service Head. Section 225-18 covers the requirements for clearances from ground (not over 600 volts) and Section 225-19 covers the requirements for clearances from buildings (not over 600 volts). See Section 225-19(d) for clearances from windows, doors, fire escapes, and so on.

225-12. Open-Conductor Supports. Open conductors shall be supported on glass or porcelain knobs, racks, brackets, or strain insulators.

225-14. Open-Conductor Spacings.

(a) 600 Volts, Nominal, or Less. Conductors of 600 volts, nominal, or less, shall comply with the spacings provided in Table 230-51(c).

(b) Over 600 Volts, Nominal. Conductors of over 600 volts, nominal, shall comply with the spacings provided in Sections 110-36 and 490-24.

(c) Separation from Other Circuits. Open conductors shall be separated from open conductors of other circuits or systems by not less than 4 in. (102 mm).

(d) Conductors on Poles. Conductors on poles shall have a separation of not less than 1 ft (305 mm) where not placed on racks or brackets. Conductors supported on poles shall provide a horizontal climbing space not less than the following:

Ample space is required for linemen to climb over or through conductors to safely work with conductors on the pole.

(1) Power conductors below communications conductors — 30 in. (762 mm)
(2) Power conductors alone or above communications conductors:
300 volts or less — 24 in. (610 mm)
Over 300 volts — 30 in. (762 mm)
(3) Communications conductors below power conductors — same as power conductors
(4) Communications conductors alone — no requirement

225-15. Supports Over Buildings. Supports over a building shall be in accordance with Section 230-29.

225-16. Point of Attachment to Buildings. The point of attachment to a building shall be in accordance with Section 230-26.

225-17. Means of Attachment to Buildings. The means of attachment to a building shall be in accordance with Section 230-27.

225-18. Clearance from Ground. Overhead spans of open conductors and open multiconductor cables of not over 600 volts, nominal, shall conform to the following:

10 ft (3.05 m) — above finished grade, sidewalks, or from any platform or projection from which they might be reached where the voltage does not exceed 150 volts to ground and accessible to pedestrians only

12 ft (3.66 m) — over residential property and driveways, and those commercial areas not subject to truck traffic where the voltage does not exceed 300 volts to ground

15 ft (4.57 m) — for those areas listed in the 12-ft (3.66-m) classification where the voltage exceeds 300 volts to ground

18 ft (5.49 m) — over public streets, alleys, roads, parking areas subject to truck traffic, driveways on other than residential property, and other land traversed by vehicles such as cultivated, grazing, forest, and orchard

FPN: For clearances of conductors of over 600 volts, see *National Electrical Safety Code*, ANSI C2-1997.

225-19. Clearances from Buildings for Conductors of Not Over 600 Volts, Nominal.

(a) Above Roofs. Overhead spans of open conductors and open multiconductor cables shall have a vertical clearance of not less than 8 ft (2.44 m) above the roof surface. The vertical clearance above the roof level shall be maintained for a distance not less than 3 ft (914 mm) in all directions from the edge of the roof.

Exception No. 1: The area above a roof surface subject to pedestrian or vehicular traffic shall have a vertical clearance from the roof surface in accordance with the clearance requirements of Section 225-18.

Exception No. 2: Where the voltage between conductors does not exceed 300, and the roof has a slope of 4 in. (102 mm) in 12 in. (305 mm) or greater, a reduction in clearance to 3 ft (914 mm) shall be permitted.

Exception No. 3: Where the voltage between conductors does not exceed 300, a reduction in clearance above only the overhanging portion of the roof to not less than 18 in. (457 mm) shall be permitted if (1) not more than 6 ft (1.83 m) of the conductors, 4 ft (1.22 m) horizontally, pass above the roof overhang, and (2) they are terminated at a through-the-roof raceway or approved support.

Exception No. 4: The requirement for maintaining the vertical clearance 3 ft (914 mm) from the edge of the roof shall not apply to the final conductor span where the conductors are attached to the side of a building.

(b) From Nonbuilding or Nonbridge Structures. From signs, chimneys, radio and television antennas, tanks, and other nonbuilding or nonbridge structures, clearances — vertical, diagonal, and horizontal — shall not be less than 3 ft (914 mm).

(c) Horizontal Clearances. Clearances shall not be less than 3 ft (914 mm).

(d) Final Spans. Final spans of feeders or branch circuits to a building they supply or from which they are fed shall be permitted to be attached to the building, but they shall be kept not less than 3 ft (914 mm) from windows that are designed to be opened, doors, porches, balconies, ladders, stairs, fire escapes, or similar locations. Vertical clearance of final spans above, or within 3 ft (914 mm) measured horizontally of, platforms, projections, or surfaces from which they might be reached shall be maintained in accordance with Section 225-18.

Exception: Conductors run above the top level of a window shall be permitted to be less than the 3 ft (914 mm) requirement above.

Overhead branch-circuit and feeder conductors shall not be installed beneath openings through which materials may be moved, such as openings in farm and commercial buildings, and shall not be installed where they will obstruct entrance to these building openings.

(e) Zone for Fire Ladders. Where buildings exceed three stories or 50 ft (15.2 m) in height, overhead lines shall be arranged, where practicable, so that a clear space (or zone) at least 6 ft (1.83 m) wide will be left either adjacent to the buildings or beginning not over 8 ft (2.44 m) from them to facilitate the raising of ladders when necessary for fire fighting.

FPN: For clearance of conductors over 600 volts, see *National Electrical Safety Code*, ANSI C2-1997.

225-20. Mechanical Protection of Conductors. Mechanical protection of conductors on buildings, structures, or poles shall be as provided for services in Section 230-50.

225-21. Multiconductor Cables on Exterior Surfaces of Buildings. Supports for multiconductor cables on exterior surfaces of buildings shall be as provided in Section 230-51.

225-22. Raceways on Exterior Surfaces of Buildings. Raceways on exterior surfaces of buildings shall be raintight and arranged to drain.

Exception: Flexible metal conduit, where permitted in Section 350-5(1), shall not be required to be raintight.

Raintight is defined in Article 100 as so constructed or protected that exposure to a beating rain will not result in the entrance of water under specified test conditions. To ensure this, all conduit bodies, fittings, and boxes used in wet locations are required to be provided with threaded hubs or other approved means. Threadless couplings and connectors used with metal conduit or electrical metallic tubing are required to be of the raintight type [see Sections 345-9(a), 346-9(a), and 348-10].

If raceways are exposed to weather or rain through weatherhead openings, condensation is likely to occur, causing moisture to accumulate within raceways at low points of the installation and in junction boxes. Raceways should thus be installed to permit drainage through drain holes at appropriate locations.

225-24. Outdoor Lampholders. Where outdoor lampholders are attached as pendants, the connections to the circuit wires shall be staggered. Where such lampholders have terminals of a type that puncture the insulation and make contact with the conductors, they shall be attached only to conductors of the stranded type.

Splices to branch-circuit conductors for outdoor lampholders of the Edison-base type or "pigtail sockets" are required to be staggered so that splices will not be in close proximity to each other.

"Pin-type" terminal sockets are required to be attached to *stranded* conductors only and are intended for installations for temporary lighting or decorations, signs, or specifically approved applications.

225-25. Location of Outdoor Lamps. Locations of lamps for outdoor lighting shall be below all energized conductors, transformers, or other electric utilization equipment, unless

(1) Clearances or other safeguards are provided for relamping operations, or

(2) Equipment is controlled by a disconnecting means that can be locked in the open position.

Because Section 225-18 requires a minimum clearance for open conductors of 10 ft above grade or platforms, it would be difficult to keep all electrical equipment above the lamps. Section 225-25(1), allows other clearances or safeguards to permit safe relamping, and Section 225-25(2) permits the use of a disconnecting means to de-energize the circuit.

225-26. Vegetation. Vegetation such as trees shall not be used for support of overhead conductor spans.

Exception: For temporary wiring in accordance with Article 305.

Where overhead conductor spans are attached to a tree, normal tree growth around the attachment device causes the mounting insulators to break and the conductor insulation to be degraded. The Section 225-26 requirement also reduces the likelihood of chafing of the conductor insulation and the danger of shock to tree trimmers and tree climbers. Trees are permitted as support of overhead conductor spans on a temporary basis only. Outdoor lighting fixtures and associated equipment are permitted by Section 410-16(h) to be supported by trees. In order to prevent the chafing damage, conductors are run up the tree from an underground wiring method. See Section 300-5(d) for protection of conductors.

B. More than One Building or Other Structure

Old Section 225-8 and Section 225-9 were expanded and relocated to a new Part B of Article 225 for the 1999 *NEC*. Part B covers outside branch circuits and feeders on single managed properties with more than one building or structure. This new part covers the feeders that supply buildings on college campuses, multi-building industrial facilities, multi-building commercial facilities, and so on, provided that these buildings are under one management.

Many of the requirements for the disconnecting means are similar to the requirements for services and are included in this part.

225-30. Number of Supplies. Where more than one building or other structure is on the same property and under single management, each building or other structure served shall be supplied by one feeder or branch circuit unless permitted in (a) through (e). For the purpose of this section, a multiwire branch circuit shall be considered a single circuit.

(a) Special Conditions. Additional feeders or branch circuits shall be permitted to supply the following:

(1) Fire pumps
(2) Emergency systems
(3) Legally required standby systems
(4) Optional standby systems
(5) Parallel power production systems

(b) Special Occupancies. By special permission, additional feeders or branch circuits shall be permitted for

(1) Multiple-occupancy buildings where there is no available space for supply equipment accessible to all occupants, or
(2) A single building or other structure sufficiently large to make two or more supplies necessary.

(c) Capacity Requirements. Additional feeders or branch circuits shall be permitted where the capacity requirements are in excess of 2000 amperes at a supply voltage of 600 volts or less.

(d) Different Characteristics. Additional feeders or branch circuits shall be permitted for different voltages, frequencies, or phases, or for different uses, such as control of outside lighting from multiple locations.

(e) Documented Switching Procedures. Additional feeders or branch circuits shall be permitted to supply installations under single management where documented safe switching procedures are established and maintained for disconnection.

Buildings on college campuses, multi-building industrial facilities, and multi-building commercial facilities are permitted to be supplied by secondary loop supply networks provided that documented switching procedures are established. These switching procedures must establish a method to safely operate the switch for the facility during maintenance, alternate supply and emergency supply conditions. Keyed interlock systems are often used to reduce the likelihood of inappropriate switching procedures resulting in a hazardous condition.

225-31. Disconnecting Means. Means shall be provided for disconnecting all ungrounded conductors that supply or pass through the building or structure.

225-32. Location. The disconnecting means shall be installed either inside or outside of the building or structure served or where the conductors pass through the building or structure. The disconnecting means shall be at a readily accessible location nearest the point of entrance of the conductors. For the purposes of this section, the requirements in Section 230-6 shall be permitted to be utilized.

Exception No. 1: For installations under single management, where documented safe switching procedures are established and maintained for disconnection, the disconnecting means shall be permitted to be located elsewhere on the premises.

Exception No. 2: For buildings or other structures qualifying under the provisions of Article 685, the disconnecting means shall be permitted to be located elsewhere on the premises.

Exception No. 3: For towers or poles used as lighting standards, the disconnecting means shall be permitted to be located elsewhere on the premises.

Exception No. 4: For poles or similar structures used only for support of signs installed in accordance with Article 600, the disconnecting means shall be permitted to be located elsewhere on the premises.

225-33. Maximum Number of Disconnects.

(a) General. The disconnecting means for each supply permitted by Section 225-30 shall consist of not more than six switches or six circuit breakers mounted in a single enclosure, in a group of separate enclosures, or in or on a switchboard. There shall be no more than six disconnects per supply grouped in any one location.

Exception: For the purpose of this section, disconnecting means used solely for the control circuit of the ground-fault protection system, installed as part of the listed equipment, shall not be considered a supply disconnecting means.

(b) Single-Pole Units. Two or three single-pole switches or breakers, capable of individual operation, shall be permitted on multiwire circuits, one pole for each ungrounded conductor, as one multipole disconnect, provided they are equipped with handle ties or a master handle to disconnect all ungrounded conductors with no more than six operations of the hand.

225-34. Grouping of Disconnects.

(a) General. The two to six disconnects as permitted in Section 225-33 shall be grouped. Each disconnect shall be marked to indicate the load served.

Exception: One of the two to six disconnecting means permitted in Section 225-33, where used only for a water pump also intended to provide fire protection, shall be permitted to be located remote from the other disconnecting means.

(b) Additional Disconnecting Means. The one or more additional disconnecting means for fire pumps or for emergency, legally required standby, or optional standby system permitted by Section 225-30 shall be installed sufficiently remote from the one to six disconnecting means for normal supply to minimize the possibility of simultaneous interruption of supply.

225-35. Access to Occupants. In a multiple-occupancy building, each occupant shall have access to the occupant's supply disconnecting means.

Exception: In a multiple-occupancy building where electric supply and electrical maintenance are provided by the building management and where these are under continuous building management supervision, the supply disconnecting means supplying more than one occupancy shall be permit-

ted to be accessible to authorized management personnel only.

225-36. Suitable for Service Equipment. The disconnecting means specified in Section 225-31 shall be suitable for use as service equipment.

Exception: For garages and outbuildings on residential property, a snap switch or a set of 3-way or 4-way snap switches shall be permitted as the disconnecting means.

225-37. Identification. Where a building or structure has any combination of feeders, branch circuits, or services passing through it or supplying it, a permanent plaque or directory shall be installed at each feeder and branch-circuit disconnect location denoting all other services, feeders, or branch circuits supplying that building or structure or passing through that building or structure and the area served by each. See Section 230-2(e).

Exception No. 1: A plaque or directory shall not be required for large capacity multibuilding industrial installations under single management, where it is ensured that disconnection can be accomplished by establishing and maintaining safe switching procedures.

Exception No. 2: This identification shall not be required for branch circuits installed from a dwelling unit to a second building or structure.

225-38. Disconnect Construction.

(a) Manually or Power Operable. The disconnecting means shall consist of either (1) a manually operable switch or a circuit breaker equipped with a handle or other suitable operating means or (2) a power-operable switch or circuit breaker, provided the switch or circuit breaker can be opened by hand in the event of a power failure.

(b) Simultaneous Opening of Poles. Each building or structure disconnecting means shall simultaneously disconnect all ungrounded supply conductors that it controls from the building or structure wiring system.

(c) Disconnection of Grounded Conductor. Where the building or structure disconnecting means does not disconnect the grounded conductor from the grounded conductors in the building or structure wiring, other means shall be provided for this purpose at the location of disconnecting means. A terminal or bus to which all grounded conductors can be attached by means of pressure connectors shall be permitted for this purpose.

In a multisection switchboard, disconnects for the grounded conductor shall be permitted to be in any section of the switchboard, provided any such switchboard section is marked.

(d) Indicating. The building or structure disconnecting means shall plainly indicate whether it is in the open or closed position.

225-39. Rating of Disconnect. The feeder or branch-circuit disconnecting means shall have a rating of not less than the load to be carried, determined in accordance with Article 220. In no case shall the rating be lower than specified in (a), (b), (c), or (d).

(a) One-Circuit Installation. For installations to supply only limited loads of a single branch-circuit, the branch-circuit disconnecting means shall have a rating of not less than 15 amperes.

(b) Two-Circuit Installations. For installations consisting of not more than two 2-wire branch circuits, the feeder or branch-circuit disconnecting means shall have a rating of not less than 30 amperes.

(c) One-Family Dwelling. For a one-family dwelling, the feeder disconnecting means shall have a rating of not less than 100 amperes, 3-wire.

(d) All Others. For all other installations, the feeder or branch-circuit disconnecting means shall have a rating of not less than 60 amperes.

225-40. Access to Overcurrent Protective Devices. Where a feeder overcurrent device is not readily accessible, branch-circuit overcurrent devices shall be installed on the load side, shall be mounted in a readily accessible location, and shall be of a lower ampere rating than the feeder overcurrent device.

C. Over 600 Volts

225-50. Warning Signs. Signs with the words "WARNING — HIGH VOLTAGE — KEEP OUT" shall be posted in plain view where unauthorized persons might come in contact with live parts.

225-51. Isolating Switches. Where oil switches or air, oil, vacuum, or sulfur hexafluoride circuit breakers constitute a building disconnecting means, an isolating switch with visible break contacts and meeting the requirements of Section 230-204(b), (c), and (d) shall be installed on the supply side of the disconnecting means and all associated equipment.

Exception: The isolating switch shall not be required where the disconnecting means is mounted on removable truck panels or metal-enclosed switchgear units that cannot be opened unless the circuit is disconnected and that, when removed from the normal operating position, automatically disconnect the circuit breaker or switch from all energized parts.

225-52. Location. A building or structure disconnecting means shall be located in accordance with Section 225-32, or it shall be electrically operated by a similarly located remote-control device.

225-53. Type. Each building or structure disconnect shall simultaneously disconnect all ungrounded supply conductors it controls and shall have a fault-closing rating not less than the maximum available short-circuit current available at its supply terminals.

Where fused switches or separately mounted fuses are installed, the fuse characteristics shall be permitted to contribute to the fault closing rating of the disconnecting means.

The requirement for a disconnecting means for over 600 volt feeders to buildings or structures is similar to the requirements found in Section 230-205(b). This disconnect can be an air, oil, SF6, or vacuum breaker. It also can be an air, oil, SF6 or vacuum switch. Where a switch is used, fuses are permitted to help with the fault closing capability of the switch. The common practice of using fused load-break cutouts to switch sections of overhead lines or using load-break elbows to switch sections of underground lines are permitted. However, the building disconnecting means must be gang-operated to simultaneously open and close all ungrounded supply conductors. Load-break elbows and fused cutouts cannot be used as the building disconnection means.

Article 230 — Services

Contents

230-1. Scope. This article covers service conductors and equipment for control and protection of services and their installation requirements.

FPN: See Figure 230-1.

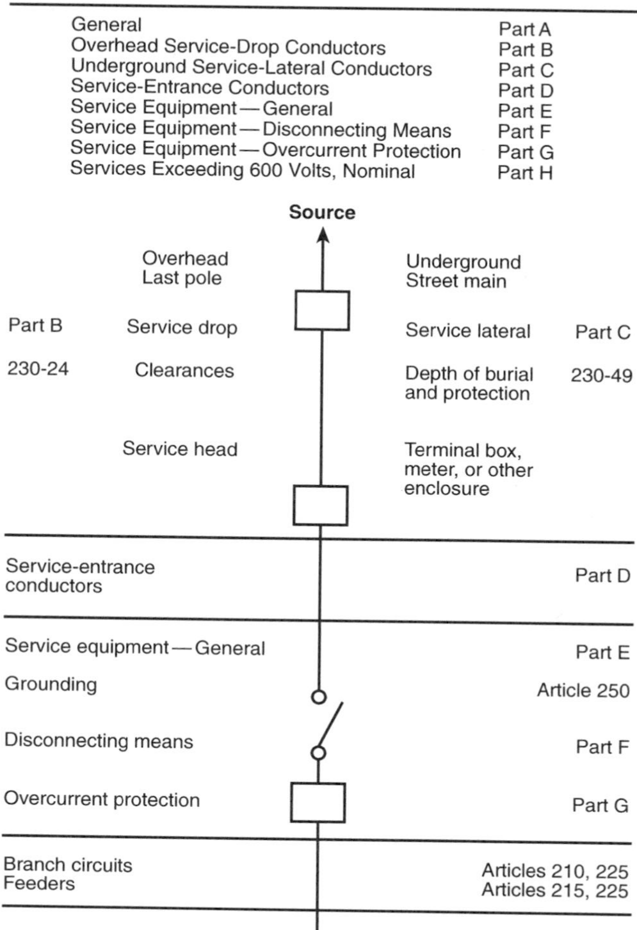

Figure 230-1 Services.

The requirements for the subjects covered in Parts A through H of Article 230 are arranged as follows:

Subject	Part	Section
General	A	230-2 – 230-9
Overhead Service-Drop Conductors	B	230-21 – 230-29
Underground Service-Lateral Conductors	C	230-30 – 230-32
Service-Entrance Conductors	D	230-40 – 230-56
Service-Equipment — General	E	230-62 – 230-66
Service-Equipment — Disconnecting Means	F	230-70 – 230-82
Service-Equipment — Overcurrent Protection	G	230-90 – 230-95
Services Exceeding 600 Volts, Nominal	H	230-200 – 230-212

A. General

230-2. Number of Services. A building or other structure served shall be supplied by only one service unless permitted in (a) through (d). For the purpose of Section 230-40, Exception No. 2 only, underground sets of conductors, size 1/0 and larger, running to the same location and connected together at their supply end, but not connected together at their load end, shall be considered to be supplying one service.

The basic requirement of Section 230-2 is that a building or other structure can be supplied by only one service. If more than one service is installed, as permitted by (a) through (d), a permanent plaque or directory is required by (e). The plaque or directory must be installed at each service drop or lateral or at each service-equipment location. All the other services on or in that building or structure and the area served by each must be noted on the plaque or directory. This plaque or directory should be constructed of sufficient durability to withstand the ambient environment. Figure 230.1 shows an example where more than one service is installed.

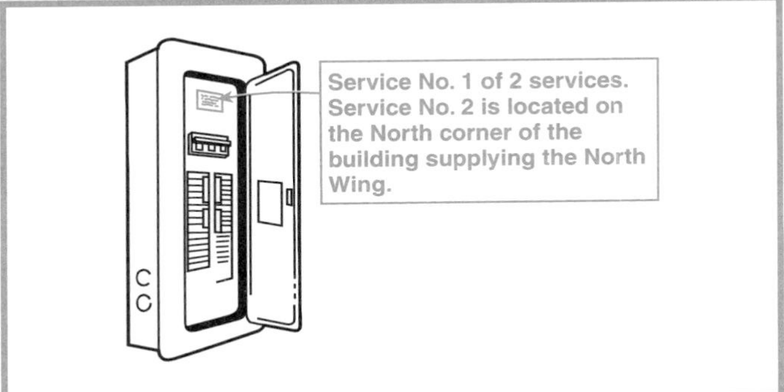

Figure 230.1 An example where more than one service is installed and a permanent plaque or directory is required, denoting all other services and the area served by each.

(a) Special Conditions. Additional services shall be permitted to supply

(1) Fire pumps
(2) Emergency systems
(3) Legally required standby systems
(4) Optional standby systems
(5) Parallel power production systems

The intent here is that a disruption of the main building service should not disconnect fire pump equipment or emergency, legally required standby, optional standby, or parallel power production systems.

(b) Special Occupancies. By special permission, additional services shall be permitted for

(1) Multiple-occupancy buildings where there is no available space for service equipment accessible to all occupants, or
(2) A single building or other structure sufficiently large to make two or more services necessary.

Section 230-2(b) permits additional services for certain occupancies by special permission (the written consent of the authority having jurisdiction; see the definition of *authority having jurisdiction* in Article 100). A separate service in multiple-occupancy buildings may be necessary if no space is available for service equipment accessible to all occupants. Also, if a building is large, more than one service is permitted. The expansion of buildings, shopping centers, or industrial plants often necessitates the addition of one or more services. It may, for example, be impractical or impossible to install one service for an industrial plant with the capacity requirements to compensate for any and all future load requirements. It is also impractical to run feeders for extremely long distances due to high costs and voltage drop problems. The serving utility and the authority having jurisdiction should be consulted before contemplating the use of this exception.

(c) Capacity Requirements.

(1) Additional services shall be permitted where the capacity requirements are in excess of 2000 amperes at a supply voltage of 600 volts or less.

(2) Additional services shall be permitted where the load requirements of a single-phase installation are greater than the serving agency normally supplies through one service.

(3) Additional services shall be permitted by special permission.

Section 230-2(c) permits two or more services if capacity requirements are in excess of 2000 amperes where supplied by 600 volts or less for large single-phase services or for lesser loads by special permission.

Many electric power companies have specifications and have adopted special regulations covering certain types of electrical loads and service equipment that may be energized from their lines. It is advisable to consult with the serving utility to determine line and transformer capacities before designing electrical services for large buildings.

(d) Different Characteristics. Additional services shall be permitted for different voltages, frequencies, or phases, or for different uses, such as for different rate schedules.

Section 230-2(d) permits the installation of more than one service for different characteristics, such as different voltages, frequencies, single-phase or 3-phase services, or different utility rate schedules. For example, different service characteristics exist between a 3-wire, 120/240-volt, single-phase service and a 3-phase, 4-wire, 480Y/277-volt service. For different applications, such as different rate schedules, this requirement allows a second service for supplying a second meter on a different rate, for equipment such as an electric water heater.

(e) Identification. Where a building or structure is supplied by more than one service, or any combination of branch circuits, feeders, and services, a permanent plaque or directory shall be installed at each service disconnect location denoting all other services, feeders, and branch circuits supplying that building or structure and the area served by each. See Section 225-37.

Section 230-2(e) states that where any combination of branch circuits, feeders, and services supplies power to a building or structure, a permanent plaque or directory must be installed at each service disconnect location to indicate where the other disconnects that feed the building are located.

Example Figures

Figures 230.2A through 230.2L illustrate some examples of permitted service configurations. The figures are intended to clarify some of the *Code* rules that affect services and are often misunderstood. No attempt has been made to include every type of service arrangement. It should be understood that the term *one location,* as applied to services, is determined by the authority having jurisdiction. Circuit breakers are shown in all diagrams for consistency; however, the service overcurrent device may consist of fuses.

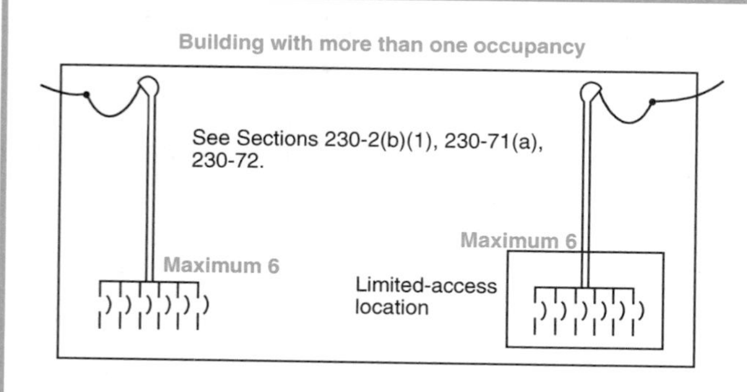

Figure 230.2A *More than one service and no available space accessible to all occupants.*

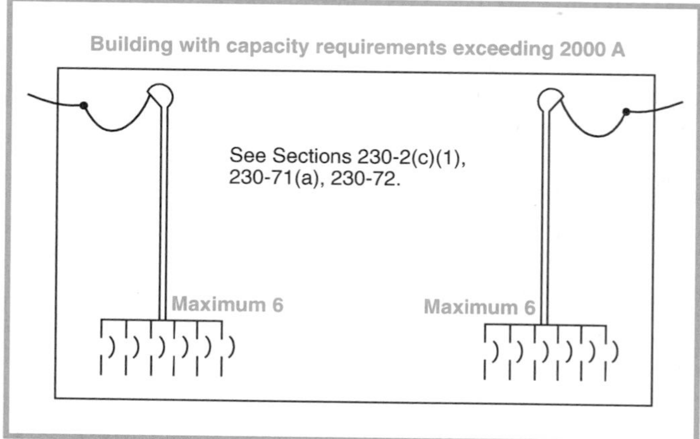

Figure 230.2B *More than one service.*

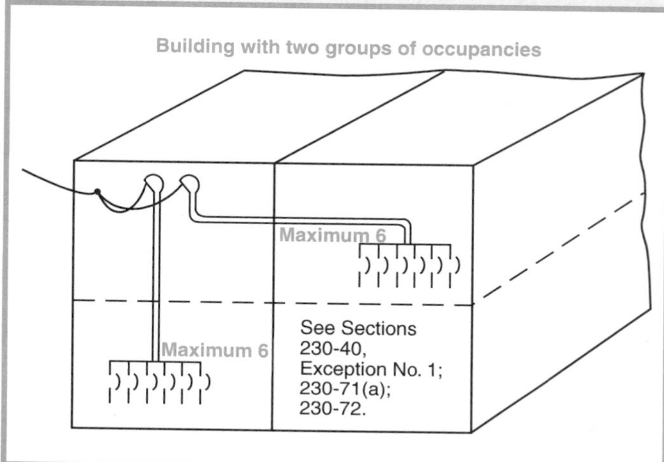

Figure 230.2C *One service with two sets of service-entrance conductors in different locations.*

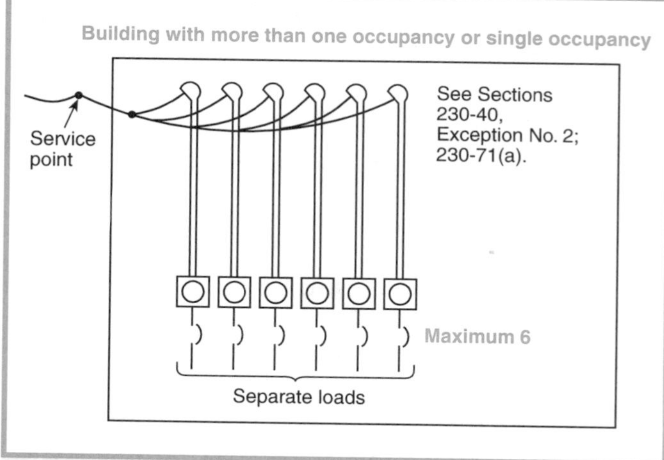

Figure 230.2D *One service with six sets of service-entrance conductors connected to separate enclosures supplying separate loads in a group at one location.*

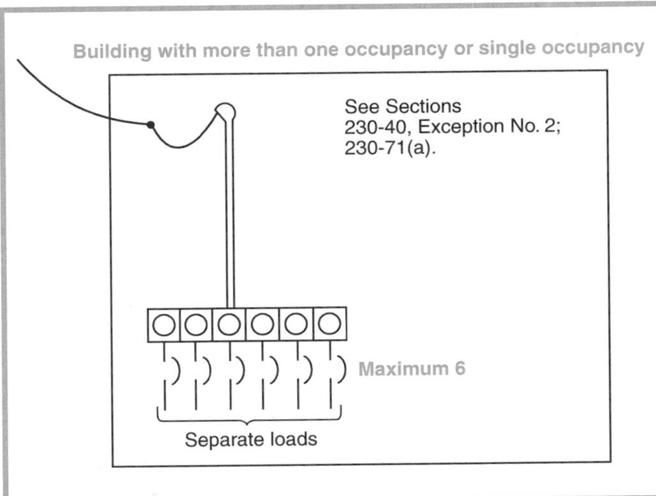

Figure 230.2E *Same as Figure 230.2(D) in an optional arrangement.*

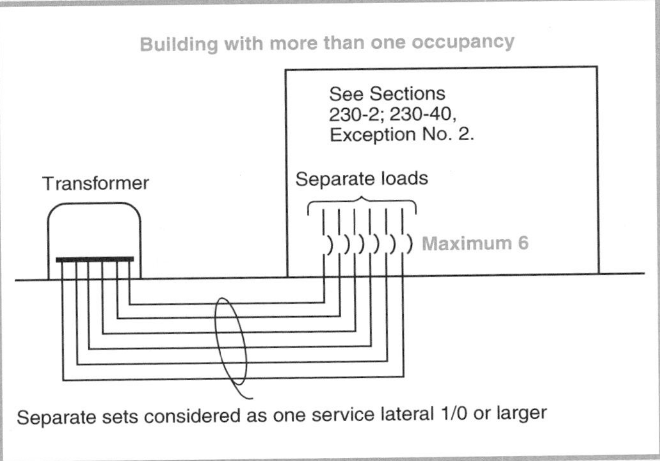

Figure 230.2F *One service lateral in one location.*

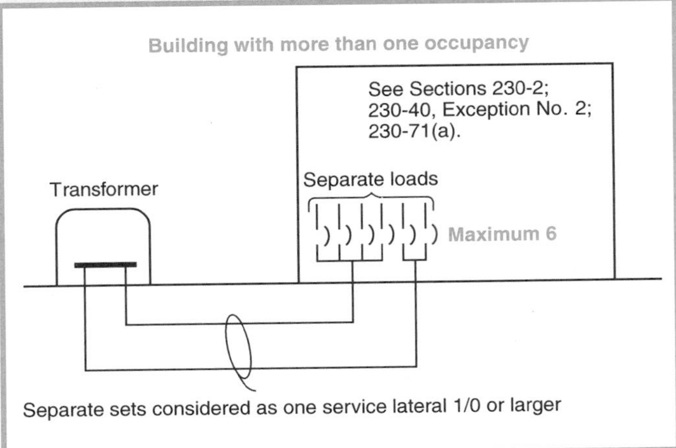

Figure 230.2G *One service lateral in one location.*

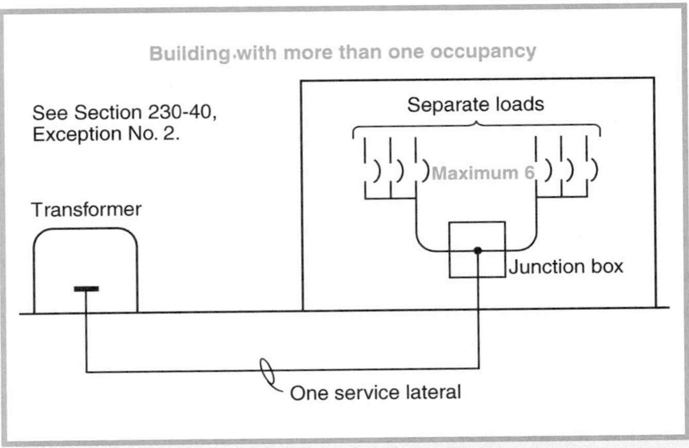

Figure 230.2H *Two sets of service-entrance conductors, each supplying three disconnects in one location.*

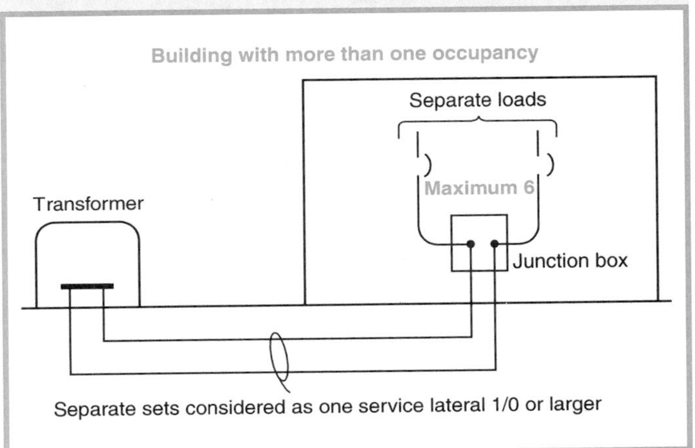

Figure 230.2I *One service lateral with one set of service-entrance conductors tapped from each service-lateral set supplying one disconnect in one location.*

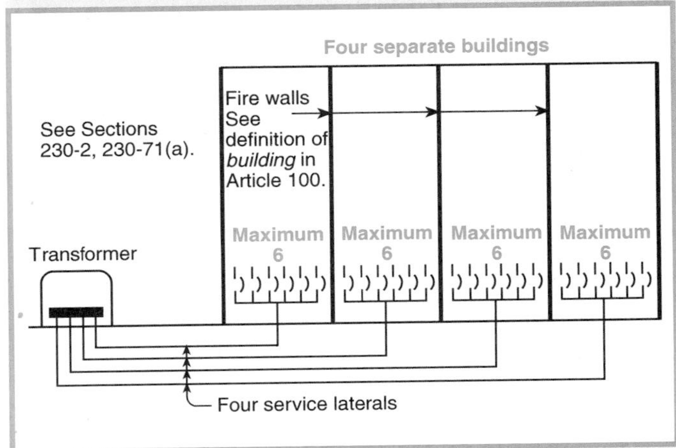

Figure 230.2J *Four service laterals in four different locations.*

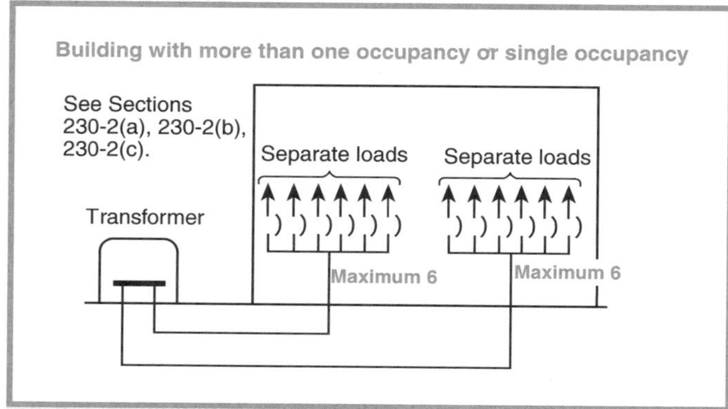

Figure 230.2K Two service laterals in two different locations.

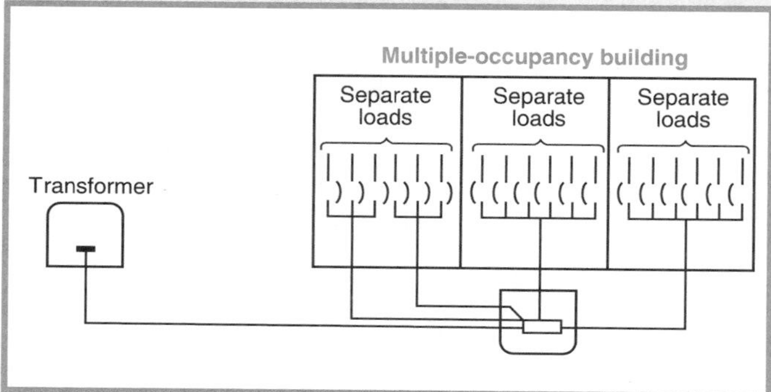

Figure 230.2L One service lateral with four sets of service-entrance conductors.

230-3. One Building or Other Structure Not to Be Supplied Through Another. Service conductors supplying a building or other structure shall not pass through the interior of another building or other structure.

Service conductors are permitted to be installed along the "exterior" of one building to supply another building. However, service conductors supplying a building are not permitted to pass through the "interior" of a building. Each building served in this manner is required to be provided with a disconnecting means for all ungrounded conductors, in accordance with Part F, Service Equipment — Disconnecting Means. For example, in Figure 230.3, Building No. 2 is *not* to be supplied through Building No. 1. The disconnecting means shown for Building Nos. 1 and 2 are located on the exterior walls. A disconnecting means suitable for use as service equipment is required to be provided for each building.

230-6. Conductors Considered Outside the Building. Conductors shall be considered outside of a building or other structure under any of the following conditions:

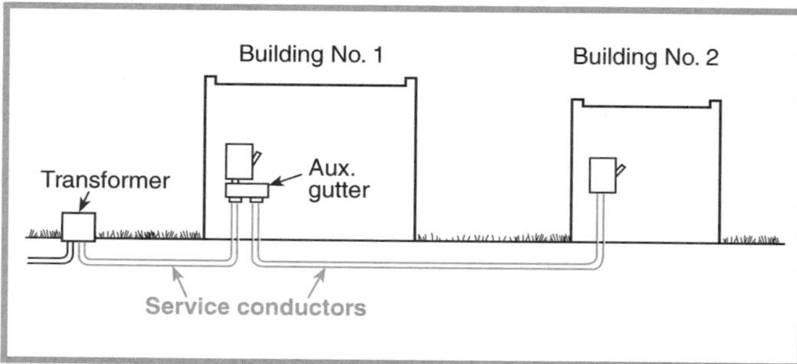

Figure 230.3 Service conductors installed in accordance with Section 230-3 so as not to pass through the interior of Building No. 1 to supply Building No. 2.

(1) Where installed under not less than 2 in. (50.8 mm) of concrete beneath a building or other structure
(2) Where installed within a building or other structure in a raceway that is encased in concrete or brick not less than 2 in. (50.8 mm) thick

Service-entrance conductors, as illustrated in Figure 230.4, are considered outside a building if they are installed as illustrated in Figure 230.4. Figure 230.4 shows service-entrance conductors installed beneath the building under not less than 2 in. of concrete or concealed in a raceway within the building and encased by not less than 2 in. of concrete or brick. They are also considered outside the building if they are installed in a transformer vault. See Sections 450-41 through 450-48 for transformer vaults.

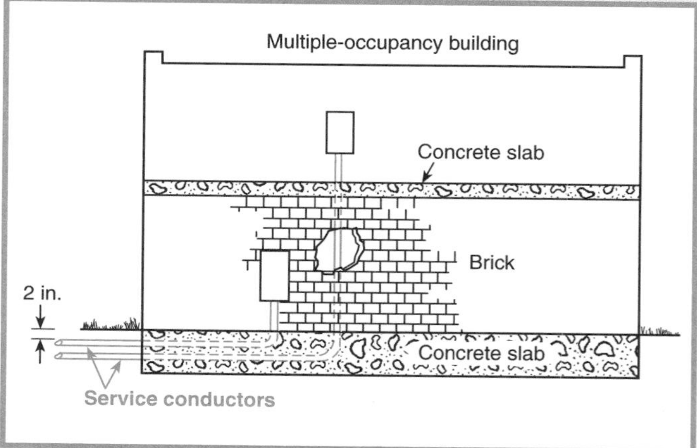

Figure 230.4 Service conductors considered outside a building where installed under not less than 2 in. of concrete beneath the building or in a raceway encased by 2 in. of concrete or brick within the building.

(3) Where installed in a transformer vault conforming to the requirements of Article 450, Part C

230-7. Other Conductors in Raceway or Cable. Conductors other than service conductors shall not be installed in the same service raceway or service cable.

All feeder and branch-circuit conductors must be separated from service conductors. Service conductors are not provided with overcurrent protection where they receive their supply; they are protected against overload conditions at their load end by the service disconnect fuses or circuit breakers. The amount of current that could be imposed on feeder or branch-circuit conductors, should they be in the same raceway and a fault occur, would be much higher than the ampacity of the feeder or branch-circuit conductors.

Exception No. 1: Grounding conductors and bonding jumpers.

Exception No. 2: Load management control conductors having overcurrent protection.

Since load management control conductors, control circuit, or switch leg conductors for use with special rate meters are usually short, they are allowed in the service raceway or cable. See Figure 230.11 for an example.

230-8. Raceway Seal. Where a service raceway enters a building or structure from an underground distribution system, it shall be sealed in accordance with Section 300-5. Spare or unused raceways shall also be sealed. Sealants shall be identified for use with the cable insulation, shield, or other components.

Sealant, such as duct seal or a bushing incorporating the physical characteristics of a seal, is required to be used to seal the ends of service raceways. The intent of this requirement is to prevent water, usually the result of condensation due to temperature differences, from entering the service equipment via the raceway. The sealant material should be compatible with the conductor insulation and should not cause deterioration over time. For underground services over 600 volts, nominal, refer to Section 300-50(e) for raceway seal requirements. See Figure 300.8 for an example.

230-9. Clearance from Building Openings. Service conductors installed as open conductors or multiconductor cable without an overall outer jacket shall have a clearance of not less than 3 ft (914 mm) from windows that are designed to be opened, doors, porches, balconies, ladders, stairs, fire escapes, or similar locations. Vertical clearance of final spans above, or within 3 ft (914 mm) measured horizontally of, platforms, projections, or surfaces from which they might be reached shall be maintained in accordance with Section 230-24(b).

Exception: Conductors run above the top level of a window shall be permitted to be less than the 3 ft (914 mm) requirement above.

As illustrated in Figure 230.5, the clearance of 3 ft applies to open conductors, not to a raceway or cable assembly with an overall outer jacket approved for use as a service conductor. The intent is to protect the conductors from physical damage and/or accidental contact by personnel. The exception permits service conductors, including drip loops and service-drop conductors, to be located just above window openings, because they are considered "out of reach." Section 230-9 was expanded for the 1999 *Code* to clarify that the clearance requirements also apply to stairs, ladders, and balconies. If service conductors are within 3 ft measured horizontally from the balcony, stair landing, or other platform, clearance to the platform of at least 10 ft must be maintained.

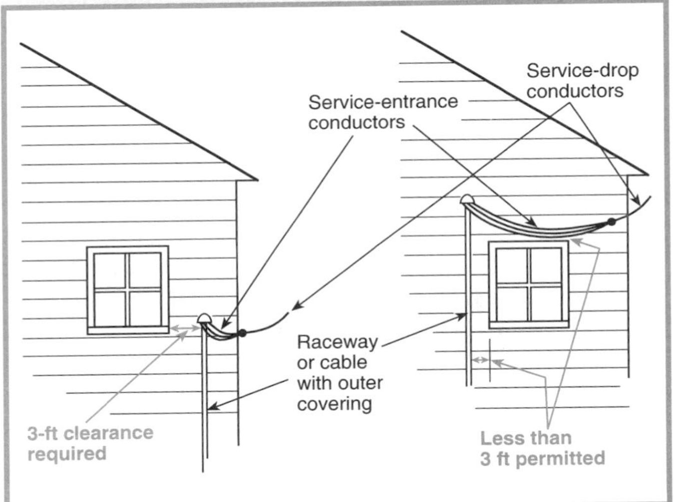

Figure 230.5 *Required dimensions for service conductors located alongside a window and service conductors above the top level of a window designed to be opened.*

Overhead service conductors shall not be installed beneath openings through which materials may be moved, such as openings in farm and commercial buildings, and shall not be installed where they will obstruct entrance to these building openings.

Elevated building openings through which materials may be moved are often high enough for the service

conductors to be installed below the opening. This is prohibited by this requirement so as to reduce the likelihood of damage to the service conductors and the potential for electric shock to persons using these high building openings.

B. Overhead Service-Drop Conductors

230-21. Overhead Supply. Overhead service conductors to a building or other structure (such as a pole) on which a meter or disconnecting means is installed shall be considered as a service drop and installed accordingly.

FPN: For example, see farm loads in Part D of Article 220.

230-22. Insulation or Covering. Individual conductors shall be insulated or covered with an extruded thermoplastic or thermosetting insulating material.

Exception: The grounded conductor of a multiconductor cable shall be permitted to be bare.

The intent of Section 230-22 is to prevent problems created by weather and abrasion and other deleterious effects that reduce the insulating quality of the covering or insulation.

230-23. Size and Rating.

(a) General. Conductors shall have sufficient ampacity to carry the current for the load as computed in accordance with Article 220 and shall have adequate mechanical strength.

Where adding load to any service, it is the installer's responsibility to be aware of all existing loads. The potential for overloading the service conductors must be governed by installer responsibility and inspector awareness, not by the usual method of overcurrent protection. The serving electric utility should be alerted to added load to ensure that adequate facilities are available.

(b) Minimum Size. The conductors shall not be smaller than No. 8 copper or No. 6 aluminum or copper-clad aluminum.

Exception: Conductors supplying only limited loads of a single branch circuit — such as small polyphase power, controlled water heaters, and similar loads — shall not be smaller than No. 12 hard-drawn copper or equivalent.

(c) Grounded Conductors. The grounded conductor shall not be less than the minimum size as required by Section 250-24(b).

230-24. Clearances. The vertical clearances of all service-drop conductors shall be based on conductor temperature of 60°F (15°C), no wind, with final unloaded sag in the wire, conductor, or cable.

Service-drop conductors shall not be readily accessible and shall comply with (a) through (d) for services not over 600 volts, nominal.

(a) Above Roofs. Conductors shall have a vertical clearance of not less than 8 ft (2.44 m) above the roof surface. The vertical clearance above the roof level shall be maintained for a distance of not less than 3 ft (914 mm) in all directions from the edge of the roof.

Service-drop conductors are not permitted to be readily accessible. This main rule applies to services of up to 600 volts. An 8-ft vertical clearance is required over the roof surface, extending 3 ft in all directions from the edge. Note that Exception No. 4 allows penetration of this space for the final span of the service drop.

Exception No. 1: The area above a roof surface subject to pedestrian or vehicular traffic shall have a vertical clearance from the roof surface in accordance with the clearance requirements of Section 230-24(b).

Exception No. 1 requires service-drop conductor clearance above a roof surface subject to vehicular or pedestrian traffic, such as the rooftop parking area shown in Figure 230.6A, to meet the clearance requirements of Section 230-24(b).

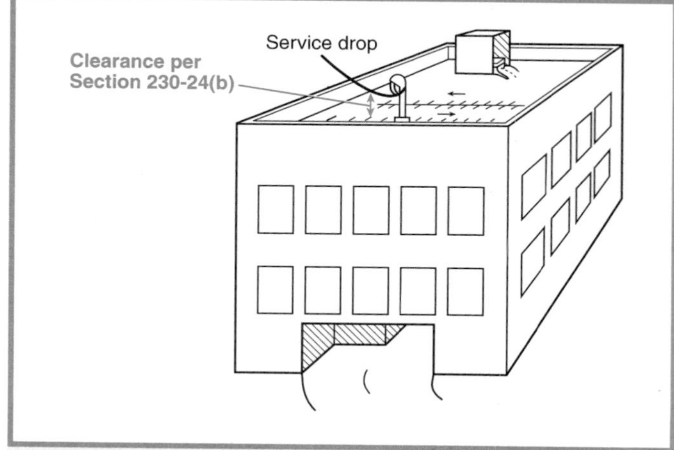

Figure 230.6A *Service-drop conductor clearance required by Section 230-24(a), Exception No. 1.*

Exception No. 2: Where the voltage between conductors does not exceed 300 and the roof has a slope of 4 in. (102 mm) in 12 in. (305 mm), or greater, a reduction in clearance to 3 ft (914 mm) shall be permitted.

Exception No. 2 permits a reduction in service-drop conductor clearance above the roof from 8 ft to 3 ft, as illustrated in Figure 230.6B, where the voltage between conductors does not exceed 300 volts and the roof is sloped at not less than 4 in. vertically in 12 in. horizontally. This exception applies to steeply sloped roofs that are less likely to be walked on. The roof slope is to be equal to or greater than 4 in. in 12 in. There are no restrictions on the length of the conductors over the roof or the horizontal distance over the roof.

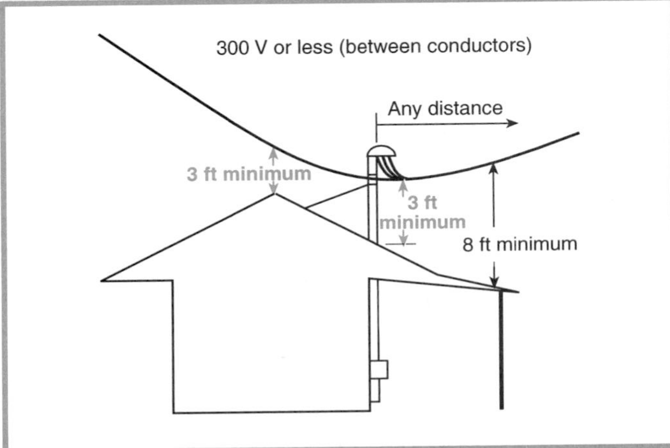

Figure 230.6B *Reduction in clearance as permitted by Section 230-24(a), Exception No. 2.*

Exception No. 3: Where the voltage between conductors does not exceed 300, a reduction in clearance above only the overhanging portion of the roof to not less than 18 in. (457 mm) shall be permitted if (1) not more than 6 ft (1.83 m) of service-drop conductors, 4 ft (1.22 m) horizontally, pass above the roof overhang, and (2) they are terminated at a through-the-roof raceway or approved support.

FPN: See Section 230-28 for mast supports.

Exception No. 3 permits a reduction of service-drop conductor clearances to 18 in. above the roof as illustrated in Figure 230.6C. This reduction is for service mast (through-the-roof) installations where the voltage between conductors does not exceed 300 volts and the mast is located within 4 ft of the edge of the roof, measured horizontally. This exception applies to either sloped or flat roofs that are easily walked on. Not more than 6 ft of conductors is permitted to pass over the roof.

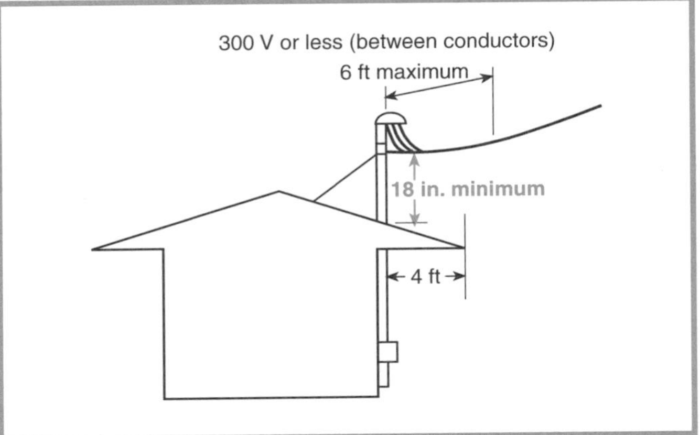

Figure 230.6C *Reduction in clearance as permitted by Section 230-24(a), Exception No. 3.*

Exception No. 4: The requirement for maintaining the vertical clearance 3 ft (914 mm) from the edge of the roof shall not apply to the final conductor span where the service drop is attached to the side of a building.

Section 230-24(a) applies to the vertical clearance above roofs for service-drop conductors up to 600 volts. This main rule requires a vertical clearance of 8 ft above the roof, including those areas 3 ft in all directions beyond the edge of the roof.

Exception No. 4 exempts the final span of a service drop attached to the side of a building, from the 8-ft requirement as illustrated in Figure 230.6D. Exception Nos. 2 and 3 permit lesser clearances for service drops of 300 volts or less. See Figures 230.6B and 230.6C for illustrations.

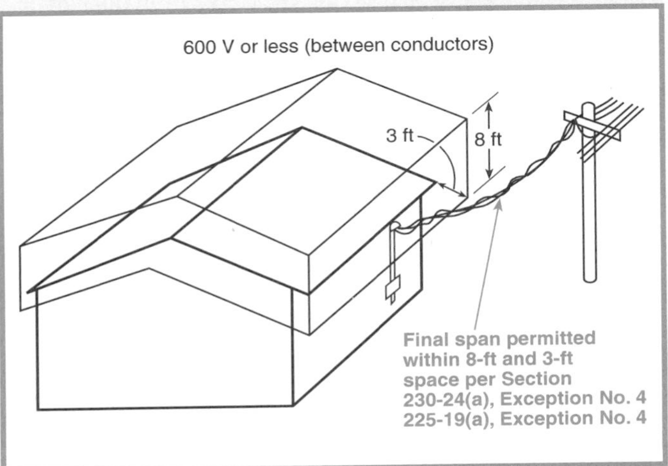

Figure 230.6D *Final span clearance as permitted by Section 230-24(a), Exception No. 4.*

If the roof is subject to pedestrian or vehicular traffic, the vertical clearance of the service drop must be the same as the vertical clearance from the ground, in accordance with Section 230-24(b).

(b) Vertical Clearance from Ground. Service-drop conductors where not in excess of 600 volts, nominal, shall have the following minimum clearance from final grade:

Figure 230.7 illustrates the 10-ft, 12-ft, 15-ft, and 18-ft vertical clearance from ground for service-drop conductors up to 600 volts, as specified by Section 230-24(b).

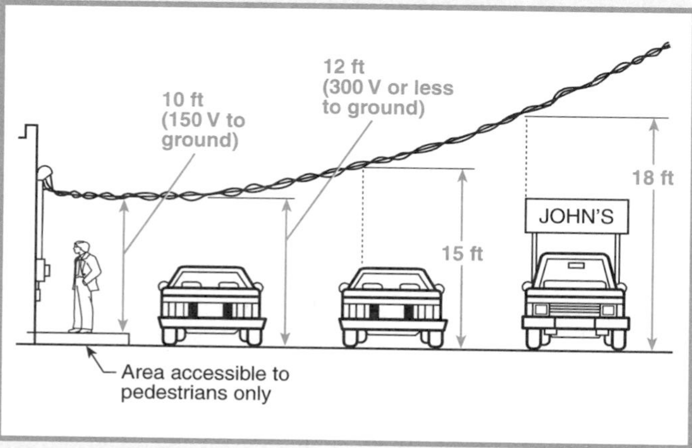

***Figure 230.7** Clearances in accordance with Section 230-24(b).*

10 ft (3.05 m) — at the electric service entrance to buildings, also at the lowest point of the drip loop of the building electric entrance, and above areas or sidewalks accessible only to pedestrians, measured from final grade or other accessible surface only for service-drop cables supported on and cabled together with a grounded bare messenger where the voltage does not exceed 150 volts to ground

12 ft (3.66 m) — over residential property and driveways, and those commercial areas not subject to truck traffic where the voltage does not exceed 300 volts to ground

15 ft (4.57 m) — for those areas listed in the 12-ft (3.66-m) classification where the voltage exceeds 300 volts to ground

18 ft (5.49 m) — over public streets, alleys, roads, parking areas subject to truck traffic, driveways on other than residential property, and other land such as cultivated, grazing, forest, and orchard

(c) Clearance from Building Openings. See Section 230-9.

(d) Clearance from Swimming Pools. See Section 680-8.

230-26. Point of Attachment. The point of attachment of the service-drop conductors to a building or other structure shall provide the minimum clearances as specified in Section 230-24. In no case shall this point of attachment be less than 10 ft (3.05 m) above finished grade.

230-27. Means of Attachment. Multiconductor cables used for service drops shall be attached to buildings or other structures by fittings identified for use with service conductors. Open conductors shall be attached to fittings identified for use with service conductors or to noncombustible, nonabsorbent insulators securely attached to the building or other structure.

See Section 230-51 for mounting and supporting of service cables or individual open service conductors and Section 230-54 for connections at service heads.

230-28. Service Masts as Supports. Where a service mast is used for the support of service-drop conductors, it shall be of adequate strength or be supported by braces or guys to withstand safely the strain imposed by the service drop. Where raceway-type service masts are used, all raceway fittings shall be identified for use with service masts. Only power service-drop conductors shall be permitted to be attached to a service mast.

Where the service drop is secured to the mast, guying may be necessary to provide adequate mechanical strength to support the service drop.

230-29. Supports Over Buildings. Service-drop conductors passing over a roof shall be securely supported by substantial structures. Where practicable, such supports shall be independent of the building.

C. Underground Service-Lateral Conductors

230-30. Insulation. Service-lateral conductors shall be insulated for the applied voltage.

Exception: A grounded conductor shall be permitted to be uninsulated as follows:

(a) Bare copper used in a raceway
(b) Bare copper for direct burial where bare copper is judged to be suitable for the soil conditions
(c) Bare copper for direct burial without regard to soil conditions where part of a cable assembly identified for underground use
(d) Aluminum or copper-clad aluminum without individual insulation or covering where part of a cable assembly identified for underground use in a raceway or for direct burial

Figure 230.8 shows various applications of bare, grounded service-lateral and service-entrance conductors for underground locations. Aluminum or

copper-clad aluminum conductors must be insulated if run in a raceway or direct buried, unless they are part of a cable assembly identified for the use.

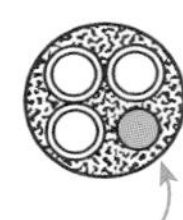

Figure 230.8 *Bare, grounded service-lateral and service-entrance conductors for underground locations.*

230-31. Size and Rating.

(a) General. Service-lateral conductors shall have sufficient ampacity to carry the current for the load as computed in accordance with Article 220 and shall have adequate mechanical strength.

(b) Minimum Size. The conductors shall not be smaller than No. 8 copper or No. 6 aluminum or copper-clad aluminum.

Exception: Conductors supplying only limited loads of a single branch circuit — such as small polyphase power, controlled water heaters, and similar loads — shall not be smaller than No. 12 copper or No. 10 aluminum or copper-clad aluminum.

(c) Grounded Conductors. The grounded conductor shall not be less than the minimum size required by Section 250-24(b).

FPN: Reasonable efficiency of operation can be provided when voltage drop is taken into consideration in sizing the service-lateral conductors.

See Sections 310-15(b)(3) and (b)(6) and the commentary following Section 230-42(c) for further information on the sizing of grounded conductors.

230-32. Protection Against Damage. Underground service-lateral conductors shall be protected against damage in accordance with Section 300-5. Service-lateral conductors entering a building shall be installed in accordance with Section 230-6 or protected by a raceway wiring method identified in Section 230-43.

D. Service-Entrance Conductors

230-40. Number of Service-Entrance Conductor Sets. Each service drop or lateral shall supply only one set of service-entrance conductors.

Exception No. 1: Buildings with one or more than one occupancy shall be permitted to have one set of service-entrance conductors for each class of service run to each occupancy or group of occupancies.

Exception No. 2: Where two to six service disconnecting means in separate enclosures are grouped at one location and supply separate loads from one service drop or lateral, one set of service-entrance conductors shall be permitted to supply each or several such service equipment enclosures.

See Figures 230.2A through 230.2L for examples of permitted service configurations. Note that in such cases the lateral conductors are considered *one* service lateral.

Exception No. 3: A single-family dwelling unit and a separate structure shall be permitted to have one set of service-entrance conductors run to each from a single service drop or lateral.

Each set of service-drop or service-lateral conductors is allowed to supply only one set of service-entrance conductors. However, if they supply buildings with more than one occupancy, such as multifamily dwellings, strip malls, and office buildings, they are allowed to supply more than one set of service-entrance conductors, provided that they run to each occupancy or group of occupancies.

Exception No. 3 allows a second set of service-entrance conductors supplied by a single service drop or lateral at a single-family dwelling unit to also supply a second building on the premises, such as a garage or storage shed.

Exception No. 4: A two-family dwelling or a multifamily dwelling shall be permitted to have one set of service-entrance conductors installed to supply the circuits covered in Section 210-25.

Exception No. 5: One set of service-entrance conductors connected to the supply side of the normal service disconnecting means shall be permitted to supply each or several systems covered by Section 230-82(4).

230-41. Insulation of Service-Entrance Conductors. Service-entrance conductors entering or on the exterior of buildings or other structures shall be insulated.

Exception: A grounded conductor shall be permitted to be uninsulated as follows:

(a) Bare copper used in a raceway or part of a service cable assembly

(b) Bare copper for direct burial where bare copper is judged to be suitable for the soil conditions

(c) *Bare copper for direct burial without regard to soil conditions where part of a cable assembly identified for underground use*
(d) *Aluminum or copper-clad aluminum without individual insulation or covering where part of a cable assembly or identified for underground use in a raceway, or for direct burial*

Service-entrance conductors are required to be insulated; however, bare grounded conductors are permitted under the same conditions as underground service laterals. See Figure 230.8, which shows examples of bare grounded conductors for underground locations.

230-42. Minimum Size and Rating.

(a) General. The ampacity of the service-entrance conductors before the application of any adjustment or correction factors shall not be less than either (1) or (2). Loads shall be determined in accordance with Article 220. Ampacity shall be determined from Section 310-15. The maximum allowable current of busways shall be that value for which the busway has been listed or labeled.

(1) The sum of the noncontinuous loads plus 125 percent of continuous loads
(2) The sum of noncontinuous load plus the continuous load if the service-entrance conductors terminate in an overcurrent device where both the overcurrent device and its assembly are listed for operation at 100 percent of their rating

(b) Ungrounded Conductors. Ungrounded conductors shall have an ampacity of not less than the minimum rating of the disconnecting means specified in Section 230-79.

The basic rule in Section 230-42(b) requires the *ungrounded* service conductors to be sized large enough to carry the load.

(c) Grounded Conductors. The grounded conductor shall not be less than the minimum size as required by Section 250-24(b).

The maximum unbalanced load determines the size of the *grounded* service conductor. Section 220-22 allows this value to be reduced, based on the maximum unbalanced load that can occur between the ungrounded conductors and the grounded (neutral) conductor. However, the minimum size of the grounded service conductor cannot be smaller than that required by Section 250-24(b)(1).

The additional heating effect of harmonic currents, due to nonlinear loads when sizing the neutral conductor of a 3-phase, 4-wire wye system should be considered. If the service to a building is a single-phase, 3-wire, 120/240-volt system with no 240-volt loads, the maximum current in the neutral would be the same as the maximum current in the ungrounded conductor. If all loads connected to one leg are "on" then all of the other loads are "off," creating the maximum current on the neutral. In such cases, the service neutral size could not be reduced but would be sized the same as the ungrounded conductors. See Section 310-15(b)(4) for more information on sizing the ungrounded conductor.

230-43. Wiring Methods for 600 Volts, Nominal, or Less. Service-entrance conductors shall be installed in accordance with the applicable requirements of this *Code* covering the type of wiring method used and shall be limited to the following methods:

(1) Open wiring on insulators
(2) Type IGS cable
(3) Rigid metal conduit
(4) Intermediate metal conduit
(5) Electrical metallic tubing
(6) Electrical nonmetallic tubing (ENT)
(7) Service-entrance cables
(8) Wireways
(9) Busways
(10) Auxiliary gutters
(11) Rigid nonmetallic conduit
(12) Cablebus
(13) Type MC cable
(14) Mineral-insulated, metal-sheathed cable
(15) Flexible metal conduit not over 6 ft (1.83 m) long or liquidtight flexible metal conduit not over 6 ft (1.83 m) long between raceways, or between raceway and service equipment, with equipment bonding jumper routed with the flexible metal conduit or the liquidtight flexible metal conduit according to the provisions of Section 250-102(a), (b), (c), and (e)
(16) Liquidtight flexible nonmetallic conduit

Where flexible metal conduit or liquidtight flexible metal conduit is installed for services, a bonding jumper is required to be installed between both ends within the raceway. The bonding jumper is not to be wrapped or spiraled around the conduit.

Cable tray systems shall be permitted to support cables for use as service-entrance conductors in accordance with Article 318.

230-46. Spliced Conductors. Service-entrance conductors shall be permitted to be spliced or tapped by clamped or bolted connections. Splices shall be made in enclosures or, if directly buried, with a listed underground splice kit.

Splices of conductors shall be made in accordance with Sections 110-14, 300-5(e), 300-13, and 300-15.

New for the 1999 *NEC,* splices are permitted in service-entrance conductors if the splice meets the requirements of Section 230-46. Splices must be in an enclosure or direct buried using a listed underground splice kit. It is common to have an underground service lateral terminate at a terminal box either inside or outside the building. At this point, service conductors may be spliced or run directly to the service equipment. Splices are permitted where, for example, the cable enters a terminal box and a different wiring method, such as conduit, continues to the service equipment. Splices are most common where metering equipment is located on the line side of service equipment, service busways, taps for supplying up to six disconnecting means, and where repairing damaged existing underground conductors. See Section 230-43 for wiring methods for service conductors for 600 volts, nominal, or less.

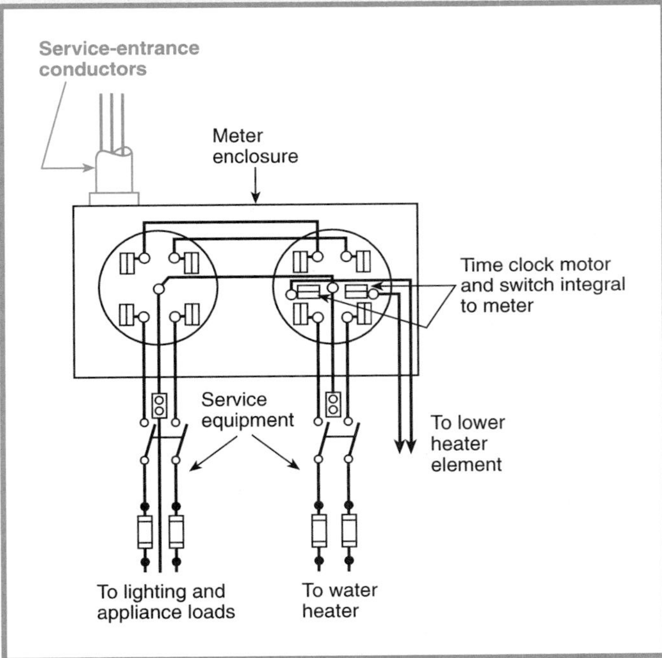

Figure 230.9 *Time clock and control switch integral to a meter for use, generally, with water heaters.*

230-49. Protection Against Physical Damage — Underground. Underground service-entrance conductors shall be protected against physical damage in accordance with Section 300-5.

230-50. Protection of Open Conductors and Cables Against Damage — Above Ground. Service-entrance conductors installed above ground shall be protected against physical damage as specified in (a) or (b).

(a) Service Cables. Service cables, where subject to physical damage, shall be protected by any of the following:

(1) Rigid metal conduit
(2) Intermediate metal conduit
(3) Rigid nonmetallic conduit suitable for the location
(4) Electrical metallic tubing
(5) Other approved means

(b) Other than Service Cable. Individual open conductors and cables other than service cables shall not be installed within 10 ft (3.05 m) of grade level or where exposed to physical damage.

Exception: Type MI and Type MC cable shall be permitted within 10 ft (3.05 m) of grade level where not exposed to physical damage or where protected in accordance with Section 300-5(d).

230-51. Mounting Supports. Cables or individual open service conductors shall be supported as specified in (a), (b), or (c).

(a) Service Cables. Service cables shall be supported by straps or other approved means within 12 in. (305 mm) of every service head, gooseneck, or connection to a raceway or enclosure and at intervals not exceeding 30 in. (762 mm).

(b) Other Cables. Cables that are not approved for mounting in contact with a building or other structure shall be mounted on insulating supports installed at intervals not exceeding 15 ft (4.57 m) and in a manner that will maintain a clearance of not less than 2 in. (50.8 mm) from the surface over which they pass.

(c) Individual Open Conductors. Individual open conductors shall be installed in accordance with Table 230-51(c). Where exposed to the weather, the conductors shall be mounted on insulators or on insulating supports attached to racks, brackets, or other approved means. Where not exposed to the weather, the conductors shall be mounted on glass or porcelain knobs.

Table 230-51(c). Supports and Clearances for Individual Open Service Conductors

Maximum Volts	Maximum Distance Between Supports (ft)	Minimum Clearance (in.) Between Conductors	Minimum Clearance (in.) From Surface
600	9	6	2
600	15	12	2
300	4½	3	2
600*	4½*	2½*	1*

Note: For SI units, 1 in. = 25.4 mm; 1 ft = 0.3048 m.
*Where not exposed to weather.

230-52. Individual Conductors Entering Buildings or Other Structures. Where individual open conductors enter a building or other structure, they shall enter through roof bushings or through the wall in an upward slant through individual, noncombustible, nonabsorbent insulating tubes. Drip loops shall be formed on the conductors before they enter the tubes.

230-53. Raceways to Drain. Where exposed to the weather, raceways enclosing service-entrance conductors shall be raintight and arranged to drain. Where embedded in masonry, raceways shall be arranged to drain.

Service raceways exposed to the weather are required to have raintight fittings and drain holes. During the installation of raceways in masonry, it is nearly impossible to prevent the entrance of surface water, rain, or water from poured concrete.

Exception: As permitted in Section 350-5.

230-54. Overhead Service Locations.

(a) Raintight Service Head. Service raceways shall be equipped with a raintight service head at the point of connection to service-drop conductors.

(b) Service Cable Equipped with Raintight Service Head or Gooseneck. Service cables shall be equipped with a raintight service head.

Exception: Type SE cable shall be permitted to be formed into a gooseneck and taped with a self-sealing weather-resistant thermoplastic.

(c) Service Heads Above Service-Drop Attachment. Service heads and goosenecks in service-entrance cables shall be located above the point of attachment of the service-drop conductors to the building or other structure.

Exception: Where it is impracticable to locate the service head above the point of attachment, the service head location shall be permitted not farther than 24 in. (610 mm) from the point of attachment.

(d) Secured. Service cables shall be held securely in place.

(e) Separately Bushed Openings. Service heads shall have conductors of different potential brought out through separately bushed openings.

Exception: For jacketed multiconductor service cable without splice.

(f) Drip Loops. Drip loops shall be formed on individual conductors. To prevent the entrance of moisture, service-entrance conductors shall be connected to the service-drop conductors either (1) below the level of the service head or (2) below the level of the termination of the service-entrance cable sheath.

(g) Arranged that Water Will Not Enter Service Raceway or Equipment. Service-drop conductors and service-entrance conductors shall be arranged so that water will not enter service raceway or equipment.

Service raceways and service cables are required to be equipped with a raintight service (weather) head. Type SE service-entrance cables, however, may be run continuously from a utility pole to metering or service equipment or, if shaped in a downward direction or "gooseneck" and sealed by taping and painting, as shown in Figure 230.10, may be installed without a service head.

Service heads and goosenecks are required to be located above the service-drop point of attachment to the building or structure. Individual conductors should extend in a downward direction, as shown in Figure 230.10, or drip loops should be formed so that, where splices are made, they are at the lowest point of the drip loop.

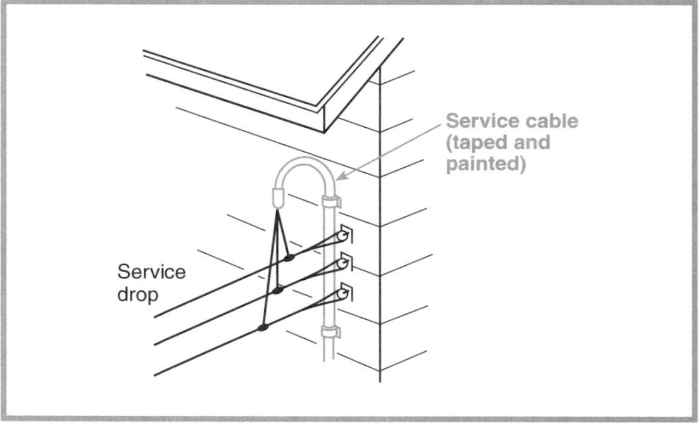

Figure 230.10 *A service-entrance cable that terminates in a "gooseneck" without a raintight service weatherhead.*

230-56. Service Conductor with the Higher Voltage to Ground. On a 4-wire, delta-connected service where the midpoint of one phase winding is grounded, the service conductor having the higher phase voltage to ground shall be durably and permanently marked by an outer finish that is orange in color, or by other effective means, at each termination or junction point.

Proper service connections require the service conductors having the higher voltage to ground to be durably marked by an outer finish of orange, such as by painting, colored adhesive tagging, or taping. Marking should be at both the point of connection to the service-drop conductors and the point of connection to the service disconnect. See Sections 215-8 and 384-3(e) and (f) for high-leg marking requirements.

E. Service Equipment — General

230-62. Service Equipment — Enclosed or Guarded. Energized parts of service equipment shall be enclosed as specified in (a), or guarded as specified in (b).

(a) Enclosed. Energized parts shall be enclosed so that they will not be exposed to accidental contact or shall be guarded as in (b).

(b) Guarded. Energized parts that are not enclosed shall be installed on a switchboard, panelboard, or control board and guarded in accordance with Sections 110-18 and 110-27. Where energized parts are guarded as provided in Sections 110-27(a)(1) and (a)(2), a means for locking or sealing doors providing access to energized parts shall be provided.

230-66. Marking. Service equipment rated at 600 volts or less shall be marked to identify it as being suitable for use as service equipment. Individual meter socket enclosures shall not be considered service equipment.

According to the listing information, panelboards with the neutral "factory bonded" to the enclosure will be marked "Suitable Only for Use as Service Equipment." Other types of equipment intended for optional use, either as service equipment or as feeders on the load side of the service disconnect, are required by Section 230-66 to be marked suitable for use as service equipment. Section 225-36 requires the feeder disconnecting means to be suitable for use as service equipment.

F. Service Equipment — Disconnecting Means

230-70. General. Means shall be provided to disconnect all conductors in a building or other structure from the service-entrance conductors.

(a) Location. The service disconnecting means shall be installed at a readily accessible location either outside of a building or structure or inside nearest the point of entrance of the service conductors.

Service disconnecting means shall not be installed in bathrooms.

No maximum distance is specified from the point of entrance of service conductors to a readily accessible location for the installation of a service disconnecting means. The authority enforcing this *Code* has the responsibility for, and is charged with, making a decision as to how far inside the building the service-entrance conductors are allowed to travel to the main disconnecting means. The length of service-entrance conductors should be kept to a minimum inside buildings, since power utilities provide limited overcurrent protection and, in the event of a fault, the service conductors could ignite nearby combustible materials.

Some local jurisdictions have ordinances that allow service-entrance conductors to run within the building up to a specified length to terminate at the disconnecting means. The authority having jurisdiction may permit service conductors to bypass fuel storage tanks or gas meters, and the like, permitting the service disconnecting means to be located in a readily accessible location. However, if the authority judges the distance as being excessive, the disconnecting means may be required to be located on the outside of the building or near the building at a readily accessible location that is not necessarily nearest the point of entrance of the conductors. See also Section 230-6 and Figure 230.4 for conductors considered to be outside a building.

See Section 380-8(a) for mounting height restrictions for switches and for circuit breakers used as switches.

(b) Marking. Each service disconnect shall be permanently marked to identify it as a service disconnect.

(c) Suitable for Use. Each service disconnecting means shall be suitable for the prevailing conditions. Service equipment installed in hazardous (classified) locations shall comply with the requirements of Articles 500 through 517.

230-71. Maximum Number of Disconnects.

(a) General. The service disconnecting means for each service permitted by Section 230-2, or for each set of service-entrance conductors permitted by Section 230-40, Exception Nos. 1 or 3, shall consist of not more than six switches or six circuit breakers mounted in a single enclosure, in a group of separate enclosures, or in or on a switchboard. There shall be no more than six disconnects per service grouped in any one location. For the purpose of this section, disconnecting means used solely for power monitoring equipment or the control circuit of the ground-fault protection system, installed as part of the listed equipment, shall not be considered a service disconnecting means.

One set of service-entrance conductors, either overhead or underground, is permitted to supply two to six service disconnecting means in lieu of a single main disconnect. A single-occupancy building can have up to six disconnects for each set of service-entrance conductors. Multiple-occupancy buildings (residential or other than residential) can be provided with one main service disconnect or up to six main disconnects for each set of service-entrance conductors.

Multiple-occupancy buildings may have service-entrance conductors run to each occupancy, and each such set of service-entrance conductors may have from one to six disconnects (see Section 230-40, Exception No. 1).

Where service-entrance conductors are routed outside the building (see Section 230-6 and Figure 230.4), each set of service-entrance conductors is permitted to supply not more than six disconnecting means at each occupancy of a multiple-occupancy building. See Figures 230.2A through 230.2L for examples of permitted service configurations.

Figure 230.11 shows a single enclosure for grouping service equipment that consists of six circuit breakers or six fused switches. This arrangement does not require a main switch. Six separate enclosures would also be permitted as the service equipment. The last sentence of Section 230-71(a) makes clear that although the disconnect for the control circuit of a ground-fault protector may be a seventh disconnect switch, it is *not* considered one of the one to six service disconnects. It is permitted even where six disconnects are in use.

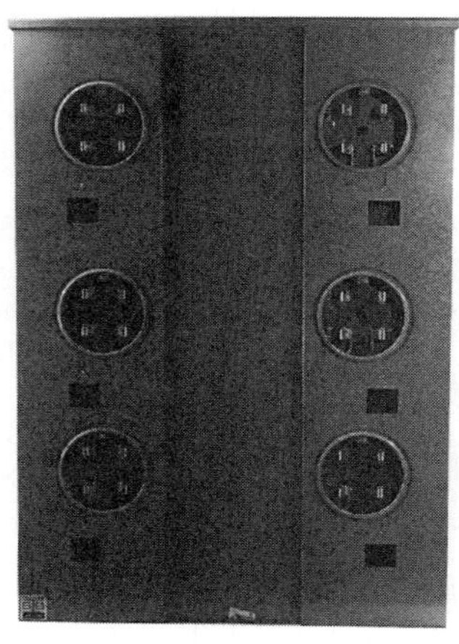

Figure 230.11 *An enclosure for grouping service equipment consisting of six circuit breakers or six fused switches. (Anchor Electric)*

(b) Single-Pole Units. Two or three single-pole switches or breakers, capable of individual operation, shall be permitted on multiwire circuits, one pole for each ungrounded conductor, as one multipole disconnect, provided they are equipped with handle ties or a master handle to disconnect all conductors of the service with no more than six operations of the hand.

FPN: See Section 384-16(a) for service equipment in panelboards, and see Section 430-95 for service equipment in motor control centers.

230-72. Grouping of Disconnects.

(a) General. The two to six disconnects as permitted in Section 230-71 shall be grouped. Each disconnect shall be marked to indicate the load served.

Exception: One of the two to six service disconnecting means permitted in Section 230-71, where used only for a water pump also intended to provide fire protection, shall be permitted to be located remote from the other disconnecting means.

This water pump is not the fire pump covered by Article 695.

(b) Additional Service Disconnecting Means. The one or more additional service disconnecting means for fire pumps, for legally required standby, or for optional standby services permitted by Section 230-2 shall be installed sufficiently remote from the one to six service disconnecting means for normal service to minimize the possibility of simultaneous interruption of supply.

The intent of Section 230-2(a) is to permit separate services where necessary for fire pumps (with one to six disconnects) or for emergency, legally required standby or optional standby systems (with one to six disconnects), in addition to the one to six disconnects for the normal building service. Article 230 recognizes that a disruption of the normal building service should not disconnect the fire pump, emergency, or other exempted systems. Because these services are in addition to the normal services, the one to six disconnects allowed for them are not included as one of the six disconnects for the normal supply.

(c) Access to Occupants. In a multiple-occupancy building, each occupant shall have access to the occupant's service disconnecting means.

Exception: In a multiple-occupancy building where electric service and electrical maintenance are provided by the building management and where these are under continuous building management supervision, the service disconnecting means supplying more than one occupancy shall be permitted to be accessible to authorized management personnel only.

A multiple-occupancy building is a building that may have any number of dwelling units, offices, and the like that are independent of each other. Unless electric service and maintenance are provided by and under continuous supervision of the building management, the occupants must have ready access to their disconnecting means as required by Section 240-24(b).

230-74. Simultaneous Opening of Poles. Each service disconnect shall simultaneously disconnect all ungrounded service conductors that it controls from the premises wiring system.

230-75. Disconnection of Grounded Conductor. Where the service disconnecting means does not disconnect the grounded conductor from the premises wiring, other means shall be provided for this purpose in the service equipment. A terminal or bus to which all grounded conductors can be attached by means of pressure connectors shall be permitted for this purpose.

In a multisection switchboard, disconnects for the grounded conductor shall be permitted to be in any section of the switchboard, provided any such switchboard section is marked.

Provisions are required at the service equipment for disconnecting the grounded conductor from the premises wiring. This disconnection does not have to be by operation of the service disconnecting means. Disconnection can be, and most commonly is, accomplished by manually removing the grounded conductor from the bus or terminal bar to which it is lugged or bolted. This is often referred to as the *neutral disconnect link.*

Manufacturers design neutral terminal bars for service equipment so that grounded conductors must be cut to be attached; that is, the grounded conductor cannot be run straight through the service equipment without means of disconnection from the premises wiring.

230-76. Manually or Power Operable. The service disconnecting means for ungrounded service conductors shall consist of either (1) a manually operable switch or circuit breaker equipped with a handle or other suitable operating means or (2) a power-operated switch or circuit breaker provided the switch or circuit breaker can be opened by hand in the event of a power supply failure.

230-77. Indicating. The service disconnecting means shall plainly indicate whether it is in the open or closed position.

230-79. Rating of Service Disconnecting Means. The service disconnecting means shall have a rating not less than the load to be carried, determined in accordance with Article 220. In no case shall the rating be lower than specified in (a), (b), (c), or (d).

Three-wire services supplying one-family dwellings are required to be installed using wire with an ampacity of at least 100 amperes. Section 230-79 was relocated from Section 230-42 in the 1996 *NEC* and modified in the 1999 *NEC* to require a 100-ampere service for all single-family dwellings.

A conductor ampacity of 60 amperes is still permitted for other loads. Smaller sizes are permitted down to No. 14 copper (No. 12 aluminum) for installations with one circuit. Two-circuit installations must have a rating of at least 30 amperes. Figure 230.12 illustrates the conductor sizing requirements of Section 230-79 for ungrounded service-entrance conductors.

A single service disconnecting means is required to have a rating of not less than the load to be carried. See Example D2(a) of Appendix D and Figure 230.12.

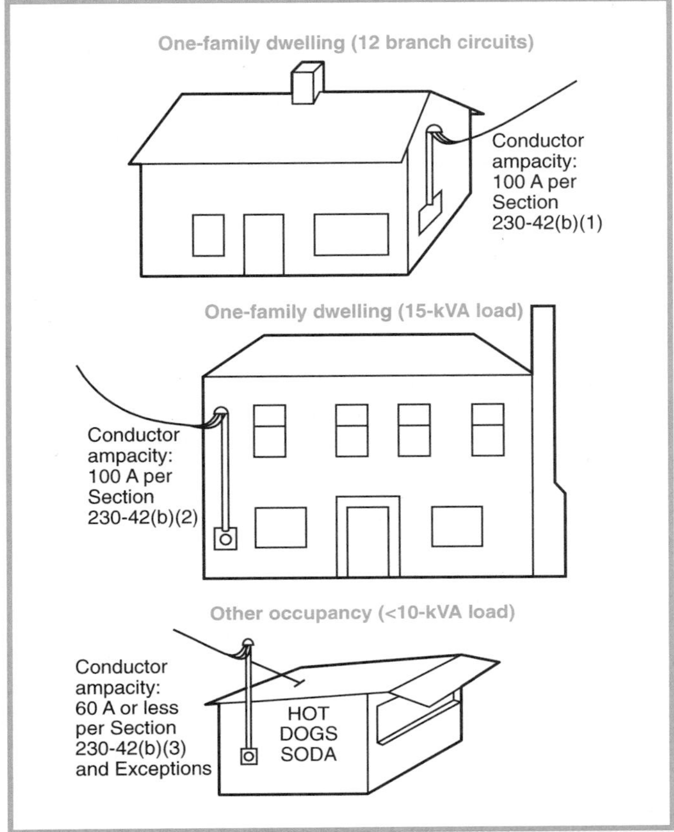

Figure 230.12 *Section 230-79 as applies to ungrounded conductors required to have an ampacity of at least 100 amperes. Conductors with an ampacity of 60 amperes are permissible for other occupancies.*

(a) One-Circuit Installation. For installations to supply only limited loads of a single branch circuit, the service disconnecting means shall have a rating of not less than 15 amperes.

(b) Two-Circuit Installations. For installations consisting of not more than two 2-wire branch circuits, the service disconnecting means shall have a rating of not less than 30 amperes.

(c) One-Family Dwelling. For a one-family dwelling, the service disconnecting means shall have a rating of not less than 100 amperes, 3-wire.

(d) All Others. For all other installations, the service disconnecting means shall have a rating of not less than 60 amperes.

230-80. Combined Rating of Disconnects. Where the service disconnecting means consists of more than one switch or circuit breaker, as permitted by Section 230-71, the combined ratings of all the switches or circuit breakers used shall not be less than the rating required by Section 230-79.

Section 230-71(a) permits up to six individual switches or circuit breakers, mounted in a single enclosure, in a group of separate enclosures or in or on a switchboard or several switchboards, to serve as the required service disconnecting means at any one location. Section 230-80 refers to situations where more than one switch or circuit breaker is used as the disconnecting means and indicates that the combined rating of all the switches or circuit breakers used cannot be less than the rating required for a single switch or circuit breaker.

Section 230-90 requires that overcurrent protection be provided in each ungrounded service conductor. This overcurrent protection must have a rating or setting that is not higher than the allowable ampacity of the service conductors. However, Exception No. 3 to Section 230-90(a) allows not more than six circuit breakers or six sets of fuses to be considered the overcurrent device. None of these individual overcurrent devices can have a rating or setting higher than the ampacity of the service conductors.

In complying with these rules, it is possible for the total of the six overcurrent devices to be greater than the rating of the service-entrance conductors. However, the size of the service-entrance conductors is required to be adequate for the computed load only, and each individual service disconnecting means is required to be large enough for the individual loads supplied. See the commentary following Section 230-90(a), Exception No. 3.

230-81. Connection to Terminals. The service conductors shall be connected to the service disconnecting means by pressure connectors, clamps, or other approved means. Connections that depend on solder shall not be used.

230-82. Equipment Connected to the Supply Side of Service Disconnect. Only the following equipment shall be permitted to be connected to the supply side of the service disconnecting means:

(1) Cable limiters or other current-limiting devices
(2) Meters nominally rated not in excess of 600 volts, provided all metal housings and service enclosures are grounded in accordance with Article 250
(3) Instrument transformers (current and voltage), high-impedance shunts, surge-protective devices identified for use on the supply side of the service disconnect, load management devices, and surge arresters
(4) Taps used only to supply load management devices, circuits for stand-by power systems, fire pump equipment, and fire and sprinkler alarms, if provided with service equipment and installed in accordance with requirements for service-entrance conductors

Systems such as fire alarms, fire pumps, standby power, and sprinkler alarms are permitted to be connected ahead of the normal service disconnecting means only if such systems are provided with a separate disconnecting means and overcurrent protection.

(5) Solar photovoltaic systems or interconnected electric power production sources (See Articles 690 or 705 as applicable.)
(6) Control circuits for power-operable service disconnecting means, if suitable overcurrent protection and disconnecting means are provided
(7) Ground-fault protection systems where installed as part of listed equipment, if suitable overcurrent protection and disconnecting means are provided

G. Service Equipment — Overcurrent Protection

230-90. Where Required. Each ungrounded service conductor shall have overload protection.

Service-entrance conductors, overhead or underground, are the supply conductors between the point of connection to the service-drop or service-lateral conductors and the service equipment. Service equipment is intended to constitute the main control and means of cutoff of the electrical supply to the premises wiring system. At this point, an overcurrent device, usually a circuit breaker or a fuse, is required to be installed in series with each ungrounded service conductor to provide overload protection only.

The service overcurrent device will not protect the service conductors under conditions of short circuit or ground fault on the line side of the disconnect. A degree of protection under these overcurrent conditions is provided by the special requirements for service conductor physical protection and location.

(a) Ungrounded Conductor. Such protection shall be provided by an overcurrent device in series with each ungrounded service conductor that has a rating or setting not higher than the allowable ampacity of the conductor.

Exception No. 1: For motor-starting currents, ratings that conform with Sections 430-52, 430-62, and 430-63 shall be permitted.

If a service supplies a motor load as well as lighting or a lighting and appliance load, then the overcurrent protective device is required to have a rating that is sufficient for the lighting and/or appliance load, as determined in accordance with Articles 210 and 220. For an individual motor, the rating is specified by Section 430-52 and, for two or more motors, the rating is specified by Section 430-62.

Example

Determine the minimum-size service conductors to supply a 100-ampere lighting and appliance load plus three squirrel-cage induction motors rated 460 volts, 3 phase, code letter F, service factor 1.15, 40°C, full-voltage starting: one 100-hp and two 25-hp motors on a 480-volt, 3-phase system.

Answer

Step 1. Conductor Loads

The full-load current of the 100-hp motor is 124 amperes (see Table 430-150). The full-load current of each 25-hp motor is 34 amperes (see Table 430-150). The service-entrance conductors are calculated at 125 percent of 124 amperes (155 amperes) plus two motors at 34 amperes each (68 amperes) for a total of 223 amperes (see Section 430-24). The motor load of 223 amperes plus the lighting and appliance load of 100 amperes equals a total load of 323 amperes. Based on this calculation, the service-entrance conductors cannot be smaller than 400-kcmil copper or 600-kcmil aluminum (see Table 310-16).

Step 2. Overcurrent Protection

The maximum rating of the service overcurrent protective device is based on the lighting and appliance loads calculated in accordance with Article 220, plus the largest motor branch-circuit overcurrent device, plus the sum of the other full-load motor currents. Using an inverse-time circuit breaker (see Table 430-152), 250 percent of 124 amperes (100-hp motor) equals 310 amperes. The next standard size allowed is 350 amperes, plus 2 times 34 amperes, for a total of 418 amperes, plus the lighting and appliance loads, for a total of 518 amperes. Since there is no permission to go up to the next standard-size overcurrent device, the next lower standard size is 450 amperes. See Sections 240-6 and 430-63.

Exception No. 2: Fuses and circuit breakers with a rating or setting that conform with Section 240-3(b) or (c) and Section 240-6 shall be permitted.

Exception No. 3: Two to six circuit breakers or sets of fuses shall be permitted as the overcurrent device to provide the overload protection. The sum of the ratings of the circuit breakers or fuses shall be permitted to exceed the ampacity of the service conductors, provided the calculated load in accordance with Article 220 does not exceed the ampacity of the service conductors.

Circuit breaker or fuse ampere ratings are permitted to be greater than the ampacity of the service conductors if the load consists of motors or motors and other loads and the ground-fault short-circuit device does not exceed the value calculated in accordance with Sections 430-62 and 430-63.

If the conductor rating does not correspond to the standard ampere rating of a circuit breaker or fuse and does not exceed the value calculated by Sections 430-62 and 430-63, the next larger-size circuit breaker or fuse may be used, provided its rating does not exceed 800 amperes, as permitted in Section 240-3(b)(3). See Section 240-6 for standard ampere ratings of fuses and circuit breakers. If multiple disconnects are used as the disconnecting means, the ampacity of the service conductors must be equal to or greater than the load calculated in accordance with Article 220; however, they are not required to be sized equal to or greater than the sum of the multiple disconnects.

Example

The computed load for a service is 350 amperes. The ampacity of a 500-kcmil Type XHHW copper conductor is 380 amperes (see Table 310-16) and the conductor is allowed to be protected by a 400-ampere fuse or circuit breaker. The rating of the fuse or circuit breaker is based on the ampacity of the service conductor and not the rating of the service disconnect switch.

In this example, a 400-ampere fuse or circuit breaker may be considered properly sized for the protection of 500-kcmil XHHW copper service conductors. If the service disconnecting means [see Exception No. 3 to Section 230-90(a)] consists of six circuit breakers or six sets of fuses, the combined ratings must not be less than the rating required for a single switch or circuit breaker, in accordance with Section 230-80. See the commentary following Section 230-80 and Figure 230.13.

As Figure 230.13 shows, the combined ratings of the overcurrent devices equals the size of the overcurrent device required by Section 240-3 and Section 230-90(a), Exception No. 4. However, the combined ratings of the overcurrent devices are not required to be equal to or less than this value. For example, all disconnects could have been rated 100 amperes. See the commentary following Section 230-23(a).

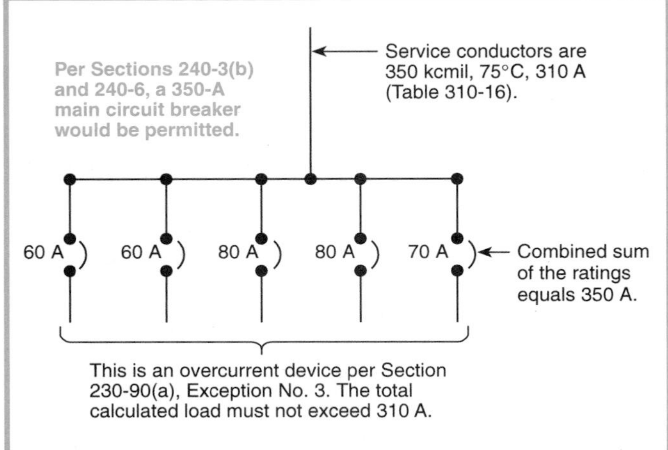

Figure 230.13 *An example where the combined ratings of the overcurrent devices are permitted to exceed the ampacity of the service conductors.*

Exception No. 4: Overload protection for fire pump supply conductors shall conform with Section 695-4(b)(1).

Exception No. 5: Overload protection for 120/240-volt, 3-wire, single-phase dwelling services shall be permitted in accordance with the requirements of Section 310-15(b)(6).

FPN: See *Standard for the Installation of Centrifugal Fire Pumps*, NFPA 20-1996.

A set of fuses shall be considered all the fuses required to protect all the ungrounded conductors of a circuit. Single-pole circuit breakers, grouped in accordance with Section 230-71(b), shall be considered as one protective device.

On multiwire circuits, two or three single-pole switches or circuit breakers that are capable of individual operation are permitted as one protective device. This allowance is acceptable, provided the switches or circuit breakers are equipped with handle ties or a master handle, so that all ungrounded conductors of a service can be disconnected with not more than six operations of the hand per Section 230-71(b).

(b) Not in Grounded Conductor. No overcurrent device shall be inserted in a grounded service conductor except a circuit breaker that simultaneously opens all conductors of the circuit.

230-91. Location. The service overcurrent device shall be an integral part of the service disconnecting means or shall be located immediately adjacent thereto.

230-92. Locked Service Overcurrent Devices. Where the service overcurrent devices are locked or sealed, or not readily accessible to the occupant, branch-circuit overcurrent devices shall be installed on the load side, shall be mounted in a readily accessible location, and shall be of lower ampere rating than the service overcurrent device.

230-93. Protection of Specific Circuits. Where necessary to prevent tampering, an automatic overcurrent device that protects service conductors supplying only a specific load, such as a water heater, shall be permitted to be locked or sealed where located so as to be accessible.

230-94. Relative Location of Overcurrent Device and Other Service Equipment. The overcurrent device shall protect all circuits and devices.

Exception No. 1: The service switch shall be permitted on the supply side.

Exception No. 2: High-impedance shunt circuits, surge arresters, surge-protective capacitors, and instrument transformers (current and voltage) shall be permitted to be connected and installed on the supply side of the service disconnecting means as permitted in Section 230-82.

Exception No. 3: Circuits for load management devices shall be permitted to be connected on the supply side of the service overcurrent device where separately provided with overcurrent protection.

Exception No. 4: Circuits used only for the operation of fire alarm, other protective signaling systems, or the supply to fire pump equipment shall be permitted to be connected on the supply side of the service overcurrent device where separately provided with overcurrent protection.

Exception No. 5: Meters nominally rated not in excess of 600 volts, provided all metal housings and service enclosures are grounded in accordance with Article 250.

Exception No. 6: Where service equipment is power operable, the control circuit shall be permitted to be connected ahead of the service equipment if suitable overcurrent protection and disconnecting means are provided.

230-95. Ground-Fault Protection of Equipment. Ground-fault protection of equipment shall be provided for solidly grounded wye electrical services of more than 150 volts to ground, but not exceeding 600 volts phase-to-phase for each service disconnect rated 1000 amperes or more.

The rating of the service disconnect shall be considered to be the rating of the largest fuse that can be installed or the highest continuous current trip setting for which the actual overcurrent device installed in a circuit breaker is rated or can be adjusted.

See the definition of *ground-fault protection of equipment* in Article 100. Ground-fault protection of equipment on services rated 1000 amperes or more operating at 480Y/277 volts was first required in the 1971 *Code* because of the unusually high number of burndowns reported on this type of service. Ground-fault protection of services will not protect the conductors on the supply side of the service disconnecting means but is designed to provide protection from line-to-ground faults that occur on the load side

of the service disconnecting means. An alternative to installing ground-fault protection may be to provide multiple disconnects rated less than 1000 amperes. For instance, up to six 800-ampere disconnecting means may be used and, in this case, ground-fault protection would not be required. FPN No. 2 in Section 230-95(c) recognizes that ground-fault protection may be desirable at lesser amperages on solidly grounded systems for voltages exceeding 150 volts to ground but not exceeding 600 volts phase-to-phase.

In addition to providing ground-fault protection, engineering studies are recommended to determine the circuit impedance and short-circuit currents that would be available at the supply terminals so that equipment and overcurrent protection of the proper interrupting rating are used. See Sections 110-9 and 110-10 for details on interrupting rating and circuit impedance.

There are two basic types of ground-fault equipment protectors, as illustrated in Figures 230.14 and 230.15. In Figure 230.14, the ground-fault sensor is installed around all the circuit conductors, and a stray current on a line-to-ground fault sets up an unbalance of the currents flowing in individual conductors installed through the ground-fault sensor. When this current exceeds the setting of the ground-fault sensor, the shunt trip will operate and open the circuit breakers.

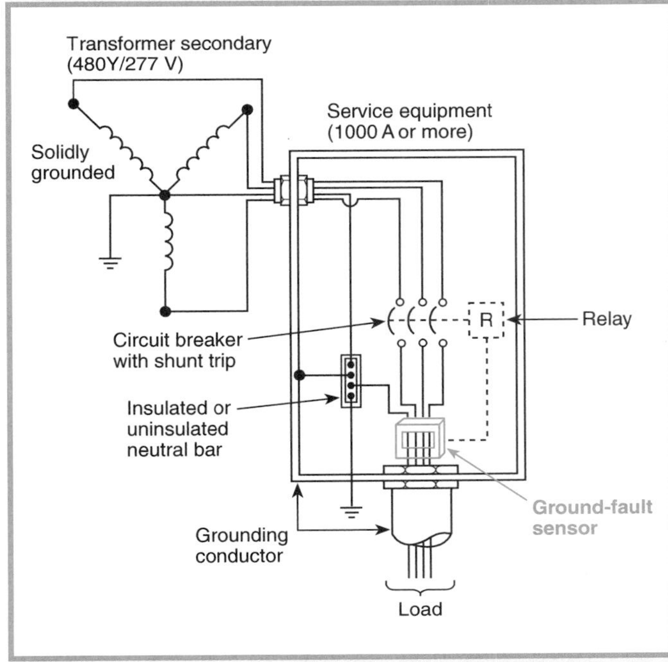

Figure 230.14 *A ground-fault sensor encircling all circuit conductors, including the neutral.*

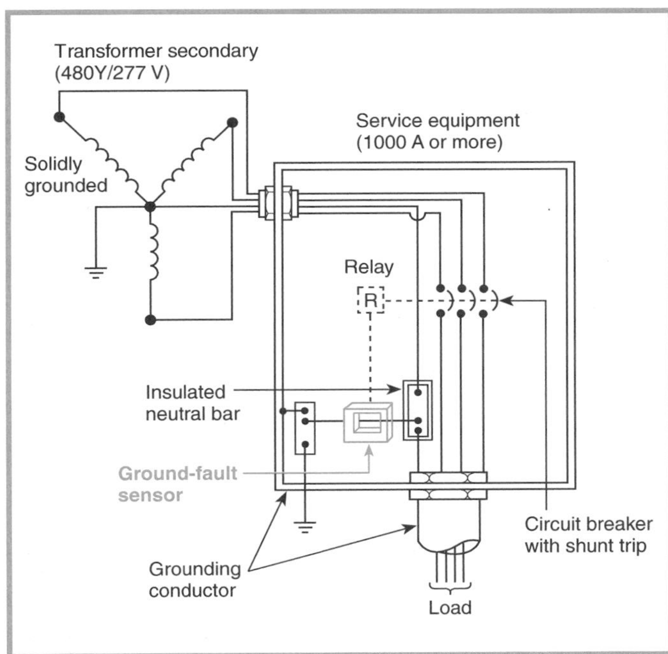

Figure 230.15 *A ground-fault sensor encircling the bonding jumper conductor only.*

The ground-fault sensor illustrated in Figure 230.15 is installed around the bonding jumper only. When an unbalanced current from a line-to-ground fault occurs, the current flows through the bonding jumper, and the shunt trip causes the circuit breaker to operate, removing the load from the line. See also Section 250-24(a),(4), which permits a grounding electrode conductor connection to the equipment grounding terminal bar or bus.

The maximum setting for ground-fault sensors is 1200 amperes; however, there is no minimum, and it should be noted that setting at lower levels increases the likelihood of unwanted shutdowns. The requirements of Section 230-95 place a restriction on fault currents greater than 3000 amperes and limit the duration of the fault to not more than 1 second. This minimizes the amount of damage done by an arcing fault, which is directly proportional to the time the arcing fault is allowed to burn.

Care should be taken to ensure that interconnecting multiple supply systems does not negate proper sensing by the ground-fault protection equipment. A careful engineering study must be made to ensure that fault currents do not take parallel paths to the supply system, thereby bypassing the ground-fault detection device. See Sections 215-10, 240-13, 517-17, and 705-32 for further information on ground-fault protection of equipment.

Definition. *Solidly grounded* means that the grounded conductor is grounded without inserting any resistor or impedance device.

Exception No. 1: The ground-fault protection provisions of this section shall not apply to a service disconnect for a continuous industrial process where a nonorderly shutdown will introduce additional or increased hazards.

Exception No. 2: The ground-fault protection provisions of this section shall not apply to fire pumps.

Most fire pumps rated 100 hp and over would require a disconnecting means rated at 1000 amperes or more. However, due to the emergency nature of their use, fire pumps are exempt from the provisions of Section 230-95.

The requirement for ground-fault protection system performance testing is a result of numerous reports of ground-fault protection systems that were improperly wired and could not or did not perform the function for which they were intended. The *Code* and qualified testing laboratories require a set of performance testing instructions to be supplied with the equipment. Evaluation and listing of the instructions fall under the jurisdiction of those best qualified to make such judgments: the qualified electrical testing laboratory (see Section 90-7). If listed equipment is not installed in accordance with the instructions provided, the installation does not comply with Section 110-3(b). Figure 230.16 shows a ground-fault protection system that does comply.

(a) Setting. The ground-fault protection system shall operate to cause the service disconnect to open all ungrounded conductors of the faulted circuit. The maximum setting of the ground-fault protection shall be 1200 amperes, and the maximum time delay shall be one second for ground-fault currents equal to or greater than 3000 amperes.

(b) Fuses. If a switch and fuse combination is used, the fuses employed shall be capable of interrupting any current higher than the interrupting capacity of the switch during a time when the ground-fault protective system will not cause the switch to open.

(c) Performance Testing. The ground-fault protection system shall be performance tested when first installed on site. The test shall be conducted in accordance with instructions that shall be provided with the equipment. A written record of this test shall be made and shall be available to the authority having jurisdiction.

FPN No. 1: Ground-fault protection that functions to open the service disconnect will afford no protection from faults on the line side of the protective element. It serves only to limit damage to conductors and equipment on the load side in the event of an arcing ground fault on the load side of the protective element.

FPN No. 2: This added protective equipment at the service equipment may make it necessary to review the overall wiring system for proper selective overcurrent protection coordination. Additional installations of ground-fault protective equipment may be needed on feeders and branch circuits where maximum continuity of electrical service is necessary.

FPN No. 3: Where ground-fault protection is provided for the service disconnect and interconnection is made with another supply system by a transfer device, means or devices may be needed to ensure

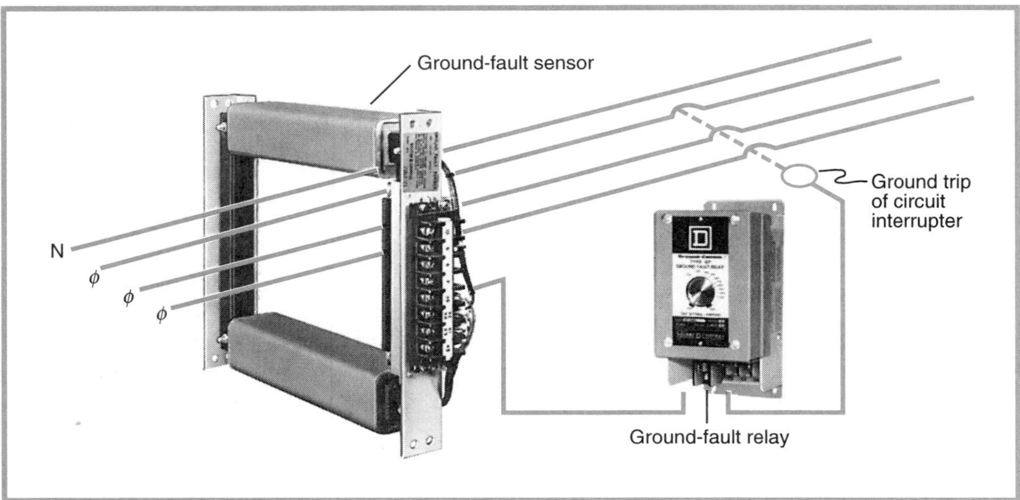

Figure 230.16 *A ground-fault protection system. (Square D Co.)*

HL-7

ROOM 129
SURFACE MOUNT
BOTTOM FEED
INTEGRATED SHORT CIRCUIT CURRENT RATING
(IER) - 18,000 RMS

480/277 VOLT, 3 PHASE, 4 WIRE, WITH GROUND BUS
400 A BUS
MLO
SUPPLIED FROM MDS

DESCRIPTION	KVA	LOAD QTY	LOAD TYPE	BREAKER AMP/POLE	#		#	BREAKER AMP/POLE	LOAD TYPE	LOAD QTY	KVA	DESCRIPTION
RTU-1	5.9	(1)	MS	25/3	1	A	2	45/3	MS	(1)	9.5	RTU-5
	5.9		MS		3	B	4		MS		9.5	
	5.9		MS		5	C	6		MS		9.5	
RTU-2	9.5	(1)	MS	40/3	7	A	8	45/3	MS	(1)	11.1	RTU-7
	9.5		MS		9	B	10		MS		11.1	
	9.5		MS		11	C	12		MS		11.1	
RTU-3	9.5	(1)	MS	40/3	13	A	14	45/3	MS	(1)	9.5	RTU-8
	9.5		MS		15	B	16		MS		9.5	
	9.5		MS		17	C	18		MS		9.5	
RTU-4	9.7	(1)	MS	40/3	19	A	20	60/3	MS	(1)	9.5	RTU-9
	9.7		MS		21	B	22		MS		9.5	
	9.7		MS		23	C	24		MS		9.5	
VAV-4-1	4.5	(1)	PW	25/1	25	A	26		SC			SPACE
VAV-4-2	4	(1)	PW	20	27	B	28		SC			SPACE
VAV-4-3	7	(1)	PW	35/1	29	C	30		SC			SPACE
VAV-4-4	6.5	(1)	PW	30/1	31	A	32	25/3	PW	(1)	4.67	VAV-5-1
VAV-4-5	3	(1)	PW	20/1	33	B	34		PW		4.67	
VAV-4-8	7.5	(1)	PW	35/1	35	C	36		PW		4.67	
VAV-5-2	8	(1)	PW	40/1	37	A	38		SC			SPACE
VAV-5-3	9.5	(1)	PW	45/1	39	B	40		SC			SPACE
VAV-5-4	3	(1)	PW	20/1	41	C	42		SC			SPACE

Connected Load	KVA	AMPS
Phase A	97.9	353.3
Phase B	95.4	344.3
Phase C	96.4	347.9
Total	289.6	
Average		348.5

*Spares not included

Type	KVA	Factor	KVA
L	0.0	x 1.25	0.0
R	0.0	x 0.5	0.0
PN	0.0	x 1 =	0.0
PW	67.0	x 1 =	67.0
MN	0.0	x 1 =	0.0
MNI	0.0	x 1 =	0.0
MS	222.6	x 1 =	222.6
MW	0.0	x 1 =	0.0
KT	0.0	x 0.65	0.0
Total	289.6		289.6

*100% of 1st 10kva, plus 50% of remainder.

289.6	Total Connected KVA
348.5	Total Connected Amps
289.6	Total Demand KVA
348.5	Total Demand Amps

Al, please see me re: this panel schedule. Thanks. Gary

proper ground-fault sensing by the ground-fault protection equipment.

H. Services Exceeding 600 Volts, Nominal

230-200. General. Service conductors and equipment used on circuits exceeding 600 volts, nominal, shall comply with all the applicable provisions of the preceding sections of this article and with the following sections, which supplement or modify the preceding sections. In no case shall the provisions of Part H apply to equipment on the supply side of the service point.

Where services rated over 600 volts supply utility-owned and -maintained transformers, the conductors on the line side and load side of the service point are service-lateral conductors; however, only those conductors on the load side of the service point come under the requirements of the *NEC.* The "service point" is a specific location where the supply conductors of the electrical utility and the customer-owned (premises wiring) conductors connect.

In the installations depicted in Figures 230.17A and 230.17B, the main service disconnecting means is located as shown at the building location, in accordance with Section 230-70(a) or 230-205(a). The conductors between the service disconnect and the point of connection to the service lateral are service-entrance conductors. The connection point may be in a belowground or aboveground junction box, where a change in the wiring method might occur. Conductors on the load side of the service disconnecting means are feeders. Each building or structure is required to have a disconnecting means, in accordance with Section 225-31.

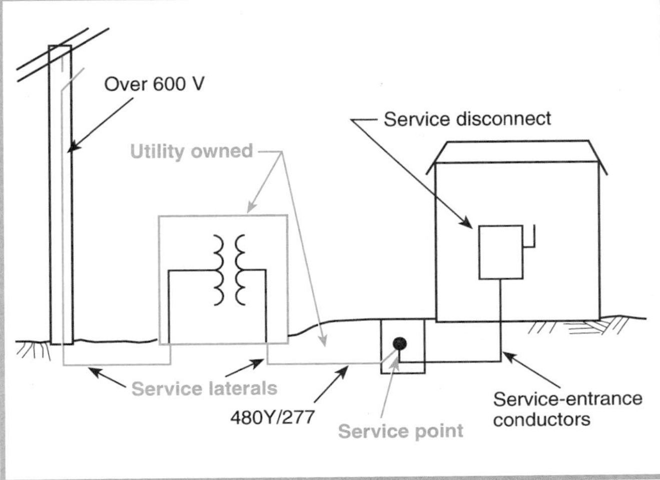

Figure 230.17A *Service rated over 600 volts supplying utility-owned transformer.*

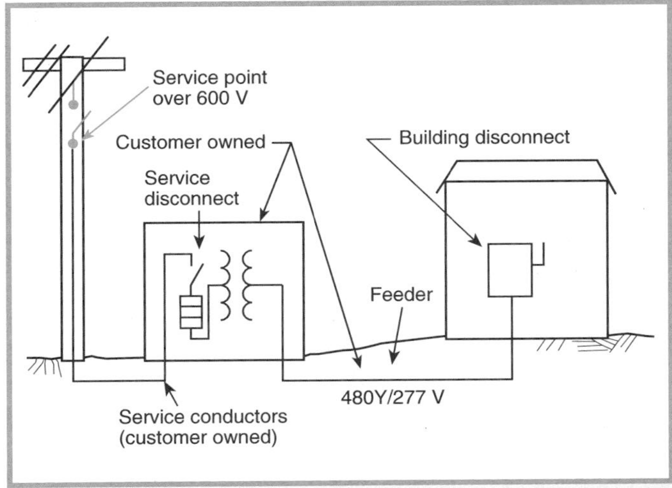

Figure 230.17B *Service rated over 600 volts supplying customer-owned transformer.*

FPN: For clearances of conductors of over 600 volts, nominal, see *National Electrical Safety Code,* ANSI C2-1997.

230-202. Service-Entrance Conductors. Service-entrance conductors to buildings or enclosures shall be installed to conform to the following.

(a) Conductor Size. Service-entrance conductors shall not be smaller than No. 6 unless in multiconductor cable. Multiconductor cable shall not be smaller than No. 8.

(b) Wiring Methods. Service-entrance conductors shall be installed by one of the wiring methods covered in Sections 300-37 and 300-50.

230-203. Warning Signs. Signs with the words "DANGER — HIGH VOLTAGE — KEEP OUT" shall be posted in plain view where unauthorized persons might come in contact with energized parts.

230-204. Isolating Switches.

(a) Where Required. Where oil switches or air, oil, vacuum, or sulfur hexafluoride circuit breakers constitute the service disconnecting means, an isolating switch with visible break contacts shall be installed on the supply side of the disconnecting means and all associated service equipment.

Exception: An isolating switch shall not be required where the circuit breaker or switch is mounted on removable truck panels or metal-enclosed switchgear units, that

(a) Cannot be opened unless the circuit is disconnected, and
(b) Where all energized parts are automatically disconnected when the circuit breaker or switch is removed from the normal operating position.

(b) Fuses as Isolating Switch. Where fuses are of the type that can be operated as a disconnecting switch, a set of such fuses shall be permitted as the isolating switch.

(c) Accessible to Qualified Persons Only. The isolating switch shall be accessible to qualified persons only.

(d) Grounding Connection. Isolating switches shall be provided with a means for readily connecting the load side conductors to ground when disconnected from the source of supply.

A means for grounding the load side conductors shall not be required for any duplicate isolating switch installed and maintained by the electric supply company.

Figure 230.18 illustrates two-position isolating switches for grounding load-side conductors when they are disconnected from high-voltage line buses.

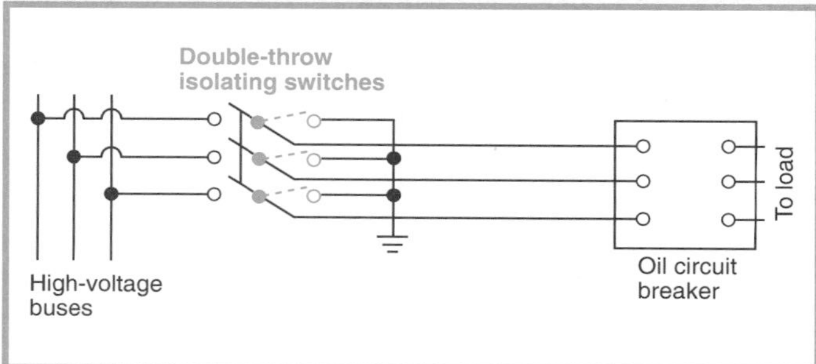

Figure 230.18 Two-position isolating switches for grounding load-side conductors when disconnected from high-voltage line buses.

230-205. Disconnecting Means.

(a) Location. The service disconnecting means shall be located in accordance with Section 230-70.

(b) Type. Each service disconnect shall simultaneously disconnect all ungrounded service conductors that it controls and shall have a fault-closing rating that is not less than the maximum short-circuit current available at its supply terminals.

Where fused switches or separately mounted fuses are installed, the fuse characteristics shall be permitted to contribute to the fault-closing rating of the disconnecting means.

(c) Remote Control. For multibuilding, industrial installations under single management, the service disconnecting means shall be permitted to be located at a separate building or structure. In such cases, the service disconnecting means shall be permitted to be electrically operated by a readily accessible, remote-control device.

230-206. Overcurrent Devices as Disconnecting Means. Where the circuit breaker or alternative for it, as specified in Section 230-208 for service overcurrent devices, meets the requirements specified in Section 230-205, they shall constitute the service disconnecting means.

230-208. Protection Requirements. A short-circuit protective device shall be provided on the load side of, or as an integral part of, the service disconnect, and shall protect all ungrounded conductors that it supplies. The protective device shall be capable of detecting and interrupting all values of current, in excess of its trip setting or melting point, that can occur at its location. A fuse rated in continuous amperes not to exceed three times the ampacity of the conductor, or a circuit breaker with a trip setting of not more than six times the ampacity of the conductors, shall be considered as providing the required short-circuit protection.

FPN: See Tables 310-67 through 310-86 for ampacities of conductors rated 2001 volts and above.

Overcurrent devices shall conform to the following:

(a) Equipment Type. Equipment used to protect service-entrance conductors shall meet the requirements of Article 490, Part B.

(b) Enclosed Overcurrent Devices. The restriction to 80 percent of the rating for an enclosed overcurrent device on continuous loads shall not apply to overcurrent devices installed in services operating at over 600 volts.

230-209. Surge Arresters (Lightning Arresters). Surge arresters installed in accordance with the requirements of Article 280 shall be permitted on each ungrounded overhead service conductor.

230-210. Service Equipment — General Provisions. Service equipment, including instrument transformers, shall conform to Article 490, Part A.

230-211. Metal-Enclosed Switchgear. Metal-enclosed switchgear shall consist of a substantial metal structure and a sheet metal enclosure. Where installed over a combustible floor, suitable protection thereto shall be provided.

Figure 230.19 shows an assembly of metal-enclosed switchgear.

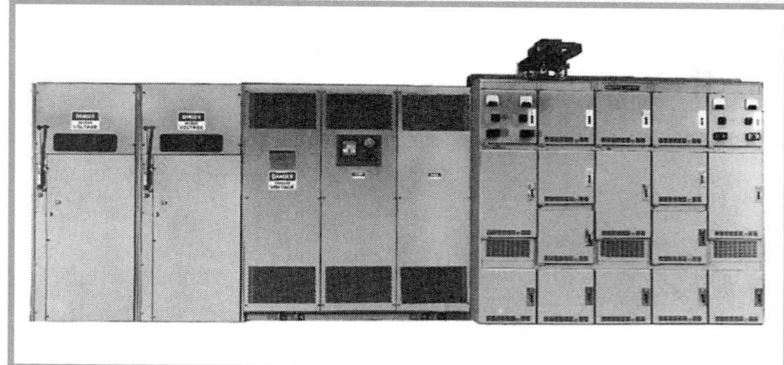

Figure 230.19 An assembly of metal-enclosed switchgear. (Square D Co.)

230-212. Services Over 15,000 Volts. Where the voltage exceeds 15,000 volts between conductors, they shall enter either metal-enclosed switchgear or a transformer vault conforming to the requirements of Sections 450-41 through 450-48.

Article 240 — Overcurrent Protection

Contents

240-1. Scope. Parts A through G of this article provide the general requirements for overcurrent protection and overcurrent protective devices not more than 600 volts, nominal. Part H covers overcurrent protection for those portions of supervised industrial installations operating at voltages of not more than 600 volts, nominal. Part I covers overcurrent protection over 600 volts, nominal.

FPN: Overcurrent protection for conductors and equipment is provided to open the circuit if the current reaches a value that will cause an excessive or dangerous temperature in conductors or conductor insulation. See also Section 110-9 for requirements for interrupting ratings and Section 110-10 for requirements for protection against fault currents.

A. General

240-2. Protection of Equipment. Equipment shall be protected against overcurrent in accordance with the article in this *Code* that covers the type of equipment as specified in the following list:

	Article
Air-conditioning and refrigerating equipment	440
Appliances	422
Audio signal processing, amplification, and reproduction equipment	640
Branch circuits	210
Busways	364
Capacitors	460
Class 1, Class 2, and Class 3 remote-control, signaling, and power-limited circuits	725
Closed-loop and programmed power distribution	780
Cranes and hoists	610
Electric signs and outline lighting	600
Electric welders	630
Electrolytic cells	668
Elevators, dumbwaiters, escalators, moving walks, wheelchair lifts, and stairway chair lifts	620
Emergency systems	700
Fire alarm systems	760
Fire pumps	695
Fixed electric heating equipment for pipelines and vessels	427
Fixed electric space-heating equipment	424
Fixed outdoor electric deicing and snow-melting equipment	426
Generators	445
Health care facilities	517
Induction and dielectric heating equipment	665
Industrial machinery	670
Lighting fixtures, lampholders, lamps, and receptacles	410
Motion picture and television studios and similar locations	530
Motors, motor circuits, and controllers	430
Phase converters	455
Pipe organs	650
Places of assembly	518
Services	230
Solar photovoltaic systems	690
Switchboards and panelboards	384
Theaters, audience areas of motion picture and television studios, and similar locations	520
Transformers and transformer vaults	450
X-ray equipment	660

240-3. Protection of Conductors. Conductors, other than flexible cords and fixture wires, shall be protected against overcurrent in accordance with their ampacities as specified in Section 310-15, unless otherwise permitted or required in (a) through (g).

(a) Power Loss Hazard. Conductor overload protection shall not be required where the interruption of the circuit would create a hazard, such as in a material handling magnet circuit or fire pump circuit. Short-circuit protection shall be provided.

FPN: See *Standard for the Installation of Centrifugal Fire Pumps*, NFPA 20-1996.

(b) Devices Rated 800 Amperes or Less. The next higher standard overcurrent device rating (above the ampacity of the conductors being protected) shall be permitted to be used, provided all of the following conditions are met.

(1) The conductors being protected are not part of a multioutlet branch circuit supplying receptacles for cord- and plug-connected portable loads.
(2) The ampacity of the conductors does not correspond with the standard ampere rating of a fuse or a circuit breaker without overload trip adjustments above its rating (but that shall be permitted to have other trip or rating adjustments).
(3) The next higher standard rating selected does not exceed 800 amperes.

Table 210-24 summarizes the requirements for the size of conductors where two or more outlets are required. The first footnote also indicates that the wire sizes are for copper conductors. Section 210-3 indicates that branch-circuit conductors rated 15, 20, 30, 40, and 50 amperes, with two or more outlets, must be protected at their ratings. Section 210-19(a) requires that branch-circuit conductors have an ampacity of not less than the rating of the branch circuit and not less than the maximum load to be served.

These are specific requirements that take precedence over Section 240-3(b), which applies generally. Where the conductor is not part of a multioutlet branch circuit supplying receptacles for cord- and plug-connected portable loads, and the conductor ampacities do not correspond to standard ampere ratings of fuses or circuit breakers, the next standard device rating is permitted if it does not exceed 800 amperes.

(c) Devices Rated Over 800 Amperes. Where the overcurrent device is rated over 800 amperes, the ampacity of the conductors it protects shall be equal to or greater than the rating of the overcurrent device as defined in Section 240-6.

(d) Small Conductors. Unless specifically permitted in (e) through (g), the overcurrent protection shall not exceed 15 amperes for No. 14, 20 amperes for No. 12, and 30 amperes for No. 10 copper; or 15 amperes for No. 12 and 25 amperes for No. 10 aluminum and copper-clad aluminum after any correction factors for ambient temperature and number of conductors have been applied.

(e) Tap Conductors. Tap conductors shall be permitted to be protected against overcurrent in accordance with Sections 210-19(d), 240-21, 364-11, 364-12, and 430-53(d).

As used in this article, a *tap conductor* is defined as a conductor, other than a service conductor that has overcurrent protection ahead of its point of supply, that exceeds the value permitted for similar conductors that are protected as described elsewhere in this section.

(f) Transformer Secondary Conductors. Single-phase (other than 2-wire) and multiphase (other than delta-delta, 3-wire) transformer secondary conductors shall not be considered to be protected by the primary overcurrent protective device. Conductors supplied by the secondary side of a single-phase transformer having a 2-wire (single-voltage) secondary, or a three-phase, delta-delta connected transformer having a 3-wire (single-voltage) secondary, shall be permitted to be protected by overcurrent protection provided on the primary (supply) side of the transformer, provided this protection is in accordance with Section 450-3 and does not exceed the value determined by multiplying the secondary conductor ampacity by the secondary to primary transformer voltage ratio.

The first paragraph of Section 240-3 requires that conductors be protected against overcurrent in accordance with their ampacity. Section 240-3(f) permits the secondary circuit conductors from a transformer to be protected by overcurrent devices in the primary circuit conductors of the transformer only in the following two special cases:

1. A transformer with a 2-wire primary and a 2-wire secondary, provided the transformer primary is protected in accordance with Section 450-3.
2. A 3-phase, delta-delta-connected transformer having a 3-wire, single-voltage secondary, provided its primary is protected in accordance with Section 450-3.

Example

The overcurrent device in the primary of a transformer is allowed to protect the conductors supplied by the secondary if a single-phase transformer with a 2-wire secondary has a 50-ampere overcurrent device in the primary, and the secondary conductors have an ampacity of at least 100 amperes.

The secondary-to-primary voltage ratio is 50 ÷ 100, a ratio of 0.5. Multiplying the secondary conductor ampacity (100) by 0.5 yields 50, the maximum rating of the overcurrent device allowed on the primary of the transformer so as to also provide overcurrent protection for the secondary conductors. These secondary conductors are not tap conductors, are not limited in length, and do not require overcurrent protection where they receive their supply, which is at the transformer secondary terminals.

However, where the secondary consists of a 3-wire, 240/120-volt system, a 120-volt line-to-neutral load could draw up to 200 amperes before the overcurrent device in the primary actuated. This is the result of the 1:4 secondary-to-primary voltage ratio of the 120-volt winding of the transformer secondary, which can cause dangerous overloading of the secondary conductors.

Except for these two special cases, transformer secondary conductors must be protected by the use of overcurrent devices, since the primary overcurrent devices do not provide this protection.

(g) Overcurrent Protection for Specific Conductor Applications. Overcurrent protection for the specific conductors shall be permitted to be provided as referenced in the following list:

	Article	Section
Air-conditioning and refrigeration equipment circuit conductors	440, Parts C, F	
Capacitor circuit conductors	460	460-8(b) and 460-25(a)–(d)
Control and instrumentation circuit conductors (Type ITC)	727	727-9

	Article	Section
Electric welder circuit conductors	630	630-12 and 630-32
Fire alarm system circuit conductors	760	760-23, 760-24, 760-41, and Chapter 9, Tables 12(a) and 12(b)
Motor-operated appliance circuit conductors	422, Part B	
Motor and motor-control circuit conductors	430, Parts C, D, E, F, H	
Phase converter supply conductors	455	455-7
Remote-control, signaling, and power-limited circuit conductors	725	725-23, 725-24, 725-41, and Chapter 9, Tables 11(a) and 11(b)
Secondary tie conductors	450	450-6

240-4. Protection of Flexible Cords and Fixture Wires. Flexible cord, including tinsel cord and extension cords, and fixture wires shall be protected against overcurrent by either (a) or (b).

(a) Ampacities. Flexible cord shall be protected by an overcurrent device in accordance with its ampacity as specified in Tables 400-5(A) and (B). Fixture wire shall be protected against overcurrent in accordance with its ampacity as specified in Table 402-5. Supplementary overcurrent protection, as in Section 240-10, shall be permitted to be an acceptable means for providing this protection.

(b) Branch Circuit Overcurrent Device. Flexible cord shall be protected where supplied by a branch circuit in accordance with one of the methods described below.

(1) Supply Cord of Listed Appliance or Portable Lamps. Where flexible cord or tinsel cord is approved for and used with a specific listed appliance or portable lamp, it shall be permitted to be supplied by a branch circuit of Article 210 in accordance with the following:

20-ampere circuits — tinsel cord or No. 18 cord and larger

30-ampere circuits — No. 16 cord and larger

40-ampere circuits — cord of 20-ampere capacity and over

50-ampere circuits — cord of 20-ampere capacity and over

Section 240-4 was rewritten for the 1999 *NEC* to clarify the requirements for overcurrent protection of flexible cords and fixture wire. Section 240-4(a) references Tables 400-5(A) and 400-5(B) for flexible cords and Table 402-5 for fixture wire ampacity. Supplementary protection, as described in Section 240-10, is also acceptable as an alternate for protection of either flexible cord or fixture wire.

Section 240-4(b) permits smaller conductors to be connected to branch circuits of a greater rating where the smaller conductors are approved for and used with a specific listed appliance, portable lamp, or extension cord, or where fixture wire is protected as listed in the exceptions. Paragraph (3) also permits listed cord sets No. 16 or larger to be protected by 20-ampere or less branch-circuit overcurrent devices.

(2) Fixture Wire. Fixture wire shall be permitted to be tapped to the branch circuit conductor of a branch circuit of Article 210 in accordance with the following:

20-ampere circuits — No. 18, up to 50 ft (15.2 m) of run length

20-ampere circuits — No. 16, up to 100 ft (30.5 m) of run length

20-ampere circuits — No. 14 and larger

30-ampere circuits — No. 14 and larger

40-ampere circuits — No. 12 and larger

50-ampere circuits — No. 12 and larger

(3) Extension Cord Sets. Flexible cord used in listed extension cord sets, or in extension cords made with separately listed and installed components, shall be permitted to be supplied by a branch circuit of Article 210 in accordance with the following:

20-ampere circuits — No. 16 and larger

240-6. Standard Ampere Ratings.

(a) Fuses and Fixed-Trip Circuit Breakers. The standard ampere ratings for fuses and inverse time circuit breakers shall be considered 15, 20, 25, 30, 35, 40, 45, 50, 60, 70, 80, 90, 100, 110, 125, 150, 175, 200, 225, 250, 300, 350, 400, 450, 500, 600, 700, 800, 1000, 1200, 1600, 2000, 2500, 3000, 4000, 5000, and 6000 amperes. The use of fuses and inverse time circuit breakers with nonstandard ampere ratings shall be permitted. The standard ampere rating for fuses shall be considered 1, 3, 6, 10, and 601.

(b) Adjustable-Trip Circuit Breakers. The rating of adjustable-trip circuit breakers having external means for adjusting the current setting (long-time pickup setting), not meeting the requirements of (c), shall be the maximum setting possible.

(c) Restricted Access Adjustable-Trip Circuit Breakers. A circuit breaker(s) that has restricted access to the adjusting means shall be permitted to have an ampere rating(s) that is equal to the adjusted current setting (long-time pickup setting). Restricted access shall be defined as located behind one of the following:

(1) Removable and sealable covers over the adjusting means
(2) Bolted equipment enclosure doors
(3) Locked doors accessible only to qualified personnel

The set long-time pickup rating (as opposed to the instantaneous trip rating) of an adjustable-trip circuit breaker can be considered the circuit breaker rating if access to the adjustment means is limited. Limitation is due to location of the adjustment means behind sealable covers shown in Figure 240.1 or behind locked doors accessible only to qualified personnel.

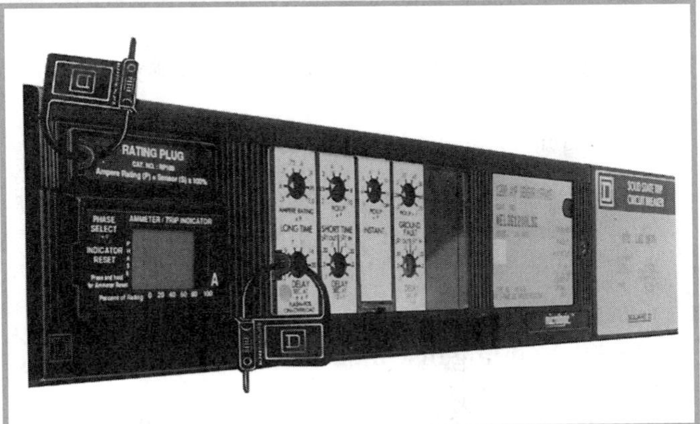

Figure 240.1 *An adjustable-trip circuit breaker with a transparent, removable and sealable cover. (Square D Co.)*

Manufacturers of both fuses and inverse time circuit breakers have or can make available products with ampere ratings other than those listed in Section 240-6 or the exception. Selection of such nonstandard ratings is not required by the *Code* but may permit better protection for conductors.

240-8. Fuses or Circuit Breakers in Parallel. Fuses and circuit breakers shall be permitted to be connected in parallel where they are factory assembled in parallel and listed as a unit. Individual fuses, circuit breakers, or combinations thereof shall not otherwise be connected in parallel.

Section 240-8 prohibits the use of fuses or circuit breakers in parallel unless they are factory assembled in parallel and listed as a unit. Section 380-17 prohibits the use of fuses in parallel in fused switches.

It is not the intent of Section 240-8 to restore the use of standard fuses in parallel in disconnect switches. This section, however, gives recognition to parallel low-voltage circuit breakers or fuses and parallel high-voltage circuit breakers or fuses where tested and factory-assembled in parallel and listed as a unit.

High-voltage fuses have long been recognized in parallel where assembled in an identified common mounting.

240-9. Thermal Devices. Thermal relays and other devices not designed to open short circuits shall not be used for the protection of conductors against overcurrent due to short circuits or grounds, but the use of such devices shall be permitted to protect motor branch-circuit conductors from overload if protected in accordance with Section 430-40.

240-10. Supplementary Overcurrent Protection. Where supplementary overcurrent protection is used for lighting fixtures, appliances, and other equipment or for internal circuits and components of equipment, it shall not be used as a substitute for branch-circuit overcurrent devices or in place of the branch-circuit protection specified in Article 210. Supplementary overcurrent devices shall not be required to be readily accessible.

240-11. Definition of Current-Limiting Overcurrent Protective Device. A *current-limiting overcurrent protective device* is a device that, when interrupting currents in its current-limiting range, will reduce the current flowing in the faulted circuit to a magnitude substantially less than that obtainable in the same circuit if the device were replaced with a solid conductor having comparable impedance.

Most electrical distribution systems can deliver high ground-fault or short-circuit currents to components such as conductors, service equipment, and so on. These components may not be able to handle short-circuit currents; they may be damaged or destroyed, and serious "burndowns" and fires could result. Properly selected current-limiting overcurrent protective devices, such as the ones shown in Figures 240.2 and 240.3, limit the let-through energy to within

Figure 240.2 *A Class R current-limiting fuse with rejection feature to prohibit the installation of noncurrent-limiting fuses. (Bussmann Division, Cooper Industries)*

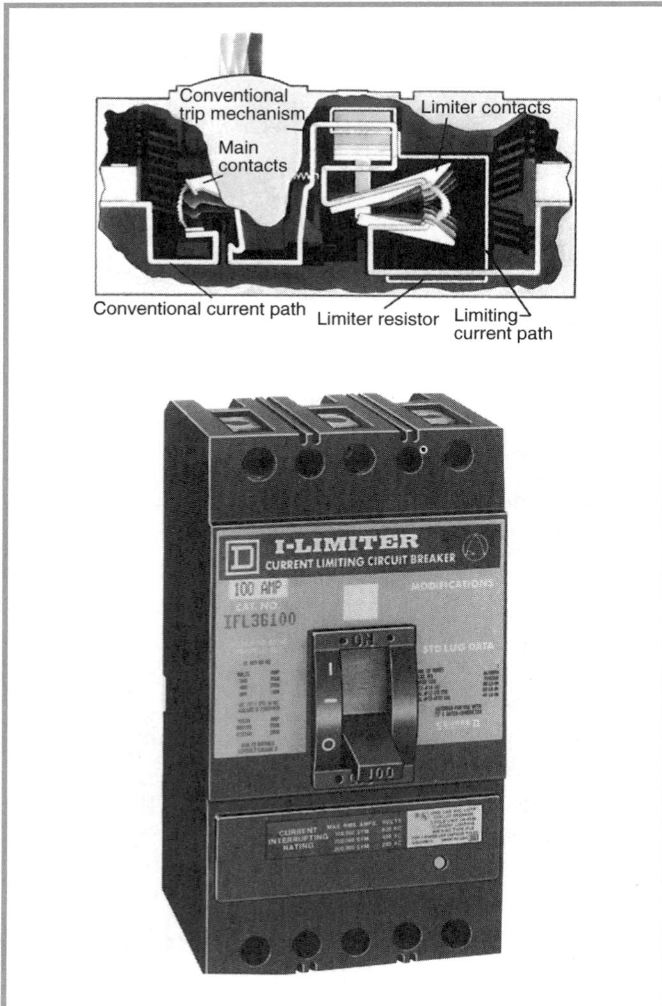

Figure 240.3 An example of a current-limiting circuit breaker. (Square D Co.)

the rating of the components, in spite of high available short-circuit currents. See the commentary following Section 240-83(c). See Sections 110-9 and 110-10 for details on interrupting ratings, circuit impedance, and other characteristics.

240-12. Electrical System Coordination. Where an orderly shutdown is required to minimize the hazard(s) to personnel and equipment, a system of coordination based on the following two conditions shall be permitted:

(1) Coordinated short-circuit protection
(2) Overload indication based on monitoring systems or devices

For the purposes of this section, *coordination* is defined as properly localizing a fault condition to restrict outages to the equipment affected, accomplished by the choice of selective fault-protective devices.

FPN: The monitoring system may cause the condition to go to alarm, allowing corrective action or an orderly shutdown, thereby minimizing personnel hazard and equipment damage.

240-13. Ground-Fault Protection of Equipment. Ground-fault protection of equipment shall be provided in accordance with the provisions of Section 230-95 for solidly grounded wye electrical systems of more than 150 volts to ground, but not exceeding 600 volts phase-to-phase for each individual device used as a building or structure main disconnecting means rated 1000 amperes or more.

The provisions of this section shall not apply to the disconnecting means for the following:

(1) Continuous industrial processes where a nonorderly shutdown will introduce additional or increased hazards
(2) Installations where ground-fault protection is provided by other requirements for services or feeders
(3) Fire pumps installed in accordance with Article 695

Section 240-13 extends the requirement of Section 230-95 to building disconnects, regardless of how the disconnects are classified (service disconnects or building disconnects for feeders or even branch circuits). See Sections 215-10 and Article 225 Part B for the requirement for building disconnects not on the utility service.

Section 240-13 requires each building or structure disconnect that is rated 1000 amperes or more, on a solidly grounded system of more than 150 volts to ground (e.g., a 480Y/277-volt system), to be provided with ground-fault protection for equipment. Provisions are included for continuous industrial processes where a nonorderly shutdown will introduce additional hazards, and for fire pumps.

Where ground-fault protection for equipment is installed at the service equipment, and other buildings or structures are supplied by feeders or branch circuits, Section 250-24 requires regrounding of the grounded conductor if an equipment grounding conductor is not included in the run. However, Section 240-13(2) exempts the grounded conductor from being regrounded downstream from the ground-fault protected service. Regrounding of the neutral at the second building may nullify the ground-fault protection of the second building that would otherwise be provided by ground-fault protection at the main service.

B. Location

240-20. Ungrounded Conductors.

(a) Overcurrent Device Required. A fuse or an overcurrent trip unit of a circuit breaker shall be connected in series with each ungrounded conductor. A combination of a current transformer and overcurrent relay shall be considered equivalent to an overcurrent trip unit.

FPN: For motor circuits, see Parts C, D, F, and K of Article 430.

(b) Circuit Breaker as Overcurrent Device. Circuit breakers shall open all ungrounded conductors of the circuit unless otherwise permitted in (1), (2), or (3).

(1) Except where limited by Section 210-4(b), individual single-pole circuit breakers, with or without approved handle ties, shall be permitted as the protection for each ungrounded conductor of multiwire branch circuits that serve only single-phase, line-to-neutral loads.

(2) In grounded systems, individual single-pole circuit breakers with approved handle ties shall be permitted as the protection for each ungrounded conductor for line-to-line connected loads for single-phase circuits or 3-wire, direct-current circuits.

(3) For line-to-line loads in 4-wire, 3-phase systems or 5-wire, 2-phase systems having a grounded neutral and no conductor operating at a voltage greater than permitted in Section 210-6, individual single-pole circuit breakers with approved handle ties shall be permitted as the protection for each ungrounded conductor.

Before discussing handle ties it is important to understand the Article 100 definition of the term *branch circuit, multiwire,* as well as Section 210-4(c) and its two exceptions. Multiwire branch circuits are permitted to supply line-to-line loads that consist of one piece of utilization equipment or where all ungrounded conductors are opened simultaneously by the branch-circuit overcurrent device. See the commentary following Section 210-4(c) for additional information.

Section 240-20(b) requires that if a circuit breaker is used, it must open all ungrounded conductors of the circuit when it trips or is manually operated. For 2-wire circuits with one conductor grounded, this rule is simple and needs no further explanation. For multiwire branch circuits of 600 volts or less, however, there are three methods of implementing this rule.

The first, and certainly the most common, method is to use a multipole circuit breaker with an internal common trip mechanism. This breaker is operated by an external single lever internally attached to two or three poles of a circuit breaker, or the external lever may be attached to multiple handles operated as one, provided the breaker is a factory-assembled unit. Underwriters Laboratories refers to these devices as multipole common trip circuit breakers. These circuit breakers are required to be used for multiwire branch circuits fed from both 3-phase and single-phase ungrounded systems. Of course, they are permitted to be used on any multiwire branch circuit within their rating.

The second method is to use two or three single-pole circuit breakers and add an approved handle tie to function as a common operating handle. This multipole circuit breaker is field assembled by externally attaching an approved common lever (handle tie) onto the two or three individual circuit breakers. It is important to understand that handle ties do not cause the circuit breaker to serve as a common trip mechanism, but rather allow for common switching only. Handle tie mechanism circuit breakers are permitted as a substitute for internal common trip mechanism circuit breakers only for limited applications. Unless specifically prohibited elsewhere, circuit breakers with approved handle ties are permitted for multiwire branch circuits only if the circuit is supplied from grounded 3-phase or grounded single-phase systems. The single-pole circuit breakers used together in this fashion must be rated for the dual voltage encountered, such as 120/240 volts.

The third method is to use individual single-pole circuit breakers without common trip mechanisms or without handle ties for multiwire branch circuits. Unless limited by other sections of the *Code,* this method is permitted for multiwire circuits, provided the multiwire branch circuit supplies only single-phase line-to-neutral loads.

Example Figures

See Figures 240.4, 240.5, and 240.6 illustrate some examples of how these requirements are applied. In

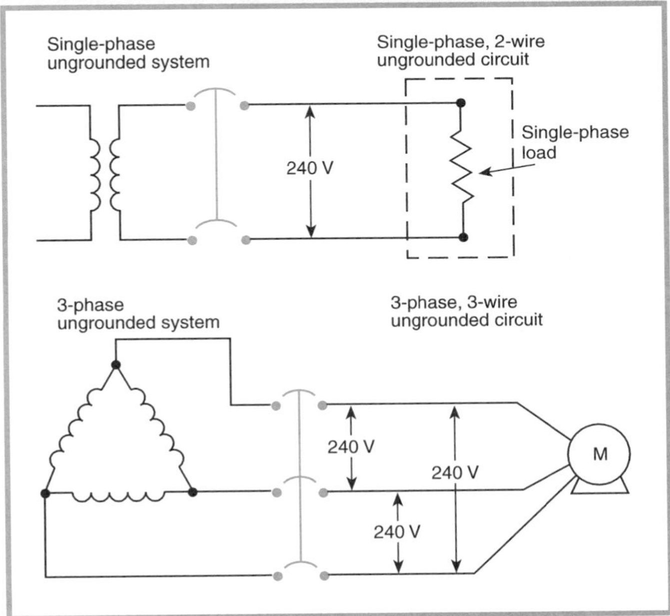

Figure 240.4 *Examples of circuits that require multipole common trip-type circuit breakers, in accordance with Section 240-20(b).*

Figure 240.4, where multipole circuit breakers are required, handle ties are not permitted because the circuits are supplied from ungrounded systems. In Figure 240.5, where single-pole circuit breakers are permitted, handle ties are not permitted since the circuits supply line-to-neutral loads. In Figure 240.6, where line-to-line loads are supplied from single-phase or 4-wire, 3-phase systems, approved handle ties are permitted.

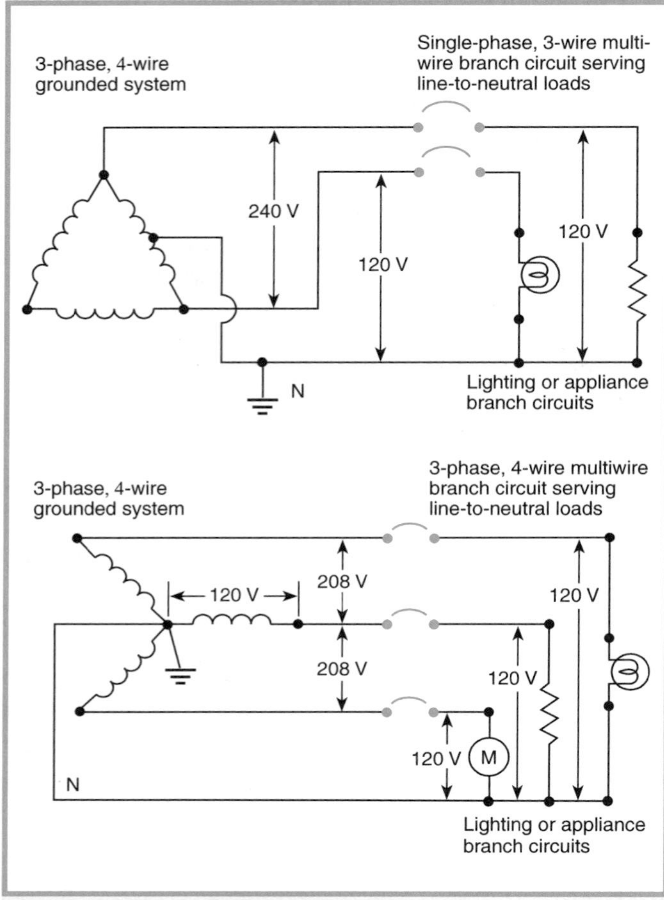

Figure 240.5 Examples of circuits where single-pole circuit breakers are permitted, in accordance with Section 240-20(b)(1), because they open the ungrounded conductor of the circuit.

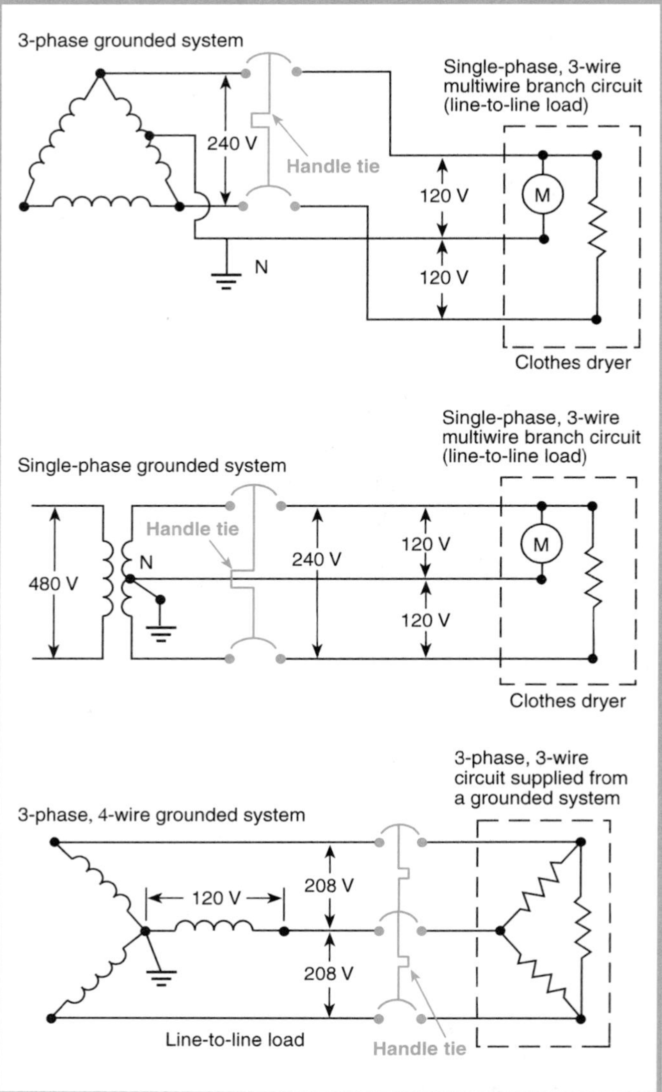

Figure 240.6 Examples of circuits where approved handle ties are permitted according to Section 240-20(b)(2) or 240-20(b)(3).

(c) Closed-Loop Power Distribution Systems. Listed devices that provide equivalent overcurrent protection in closed-loop power distribution systems shall be permitted as a substitute for fuses or circuit breakers.

240-21. Location in Circuit. Overcurrent protection shall be provided in each ungrounded circuit conductor and shall be located at the point where the conductors receive their supply except as specified in (a) through (g). No conductor supplied under the provisions of (a) through (g) shall supply another conductor under those provisions, except through an overcurrent protective device meeting the requirements of Section 240-3.

(a) Branch-Circuit Conductors. Branch-circuit tap conductors meeting the requirements specified in Section 210-19 shall be permitted to have overcurrent protection located as specified in that Section.

(b) Feeder Taps. Conductors shall be permitted to be tapped, without overcurrent protection at the tap, to a feeder as specified in (1) through (5).

Figure 240.7 illustrates how a smaller 1/0 THW conductor (150 amperes) is tapped from a larger 3/0 THW feeder conductor (200 amperes) that is in turn protected by a 150-ampere circuit breaker that is equal

to the ampacity of the 1/0 tap conductor. The circuit breaker protecting the feeder conductors also protects the tap conductors. ***Additional overcurrent protection is not required.***

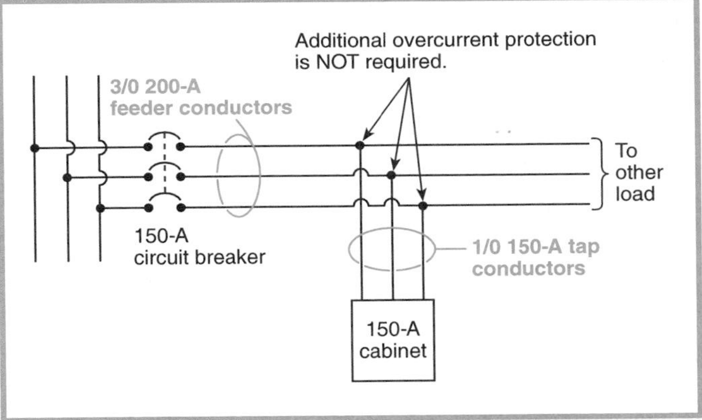

Figure 240.7 *An example where the circuit breaker protecting the feeder conductors is permitted by Section 240-21(a) to protect the tap conductors to the cabinet.*

(1) Taps Not Over 10 ft (3.05 m) Long. Where the length of the tap conductors does not exceed 10 ft (3.05 m) and the tap conductors comply with all of the following.

(a) The ampacity of the tap conductors is

(1) Not less than the combined computed loads on the circuits supplied by the tap conductors, and

(2) Not less than the rating of the device supplied by the tap conductors or not less than the rating of the overcurrent-protective device at the termination of the tap conductors.

(b) The tap conductors do not extend beyond the switchboard, panelboard, disconnecting means, or control devices they supply.

(c) Except at the point of connection to the feeder, the tap conductors are enclosed in a raceway, which shall extend from the tap to the enclosure of an enclosed switchboard, panelboard, or control devices, or to the back of an open switchboard.

(d) For field installations where the tap conductors leave the enclosure or vault in which the tap is made, the rating of the overcurrent device on the line side of the tap conductors shall not exceed 10 times the ampacity of the tap conductor.

FPN: For overcurrent protection requirements for lighting and appliance branch-circuit panelboards, see Sections 384-16(a) and (e).

(2) Taps Not Over 25 ft (7.62 m) Long. Where the length of the tap conductors does not exceed 25 ft (7.62 m) and the tap conductors comply with all of the following.

(a) The ampacity of the tap conductors is not less than one-third of the rating of the overcurrent device protecting the feeder conductors.

(b) The tap conductors terminate in a single circuit breaker or a single set of fuses that will limit the load to the ampacity of the tap conductors. This device shall be permitted to supply any number of additional overcurrent devices on its load side.

(c) The tap conductors are suitably protected from physical damage or are enclosed in a raceway.

Figure 240.8 illustrates the conditions of Section 240-21(b)(2), where three No. 3/0 THW copper tap conductors are protected from physical damage in a raceway. The lengths of the tap conductors are not more than 25 ft between terminations, and the conductors are tapped from 500-kcmil THW copper feeders and terminate in a circuit breaker.

Note that a No. 3/0 THW copper conductor (200 amperes) is more than one-third the ampacity of a 500-kcmil THW copper conductor (380 amperes). See Table 310-16 for the ampacity of copper conductors in conduit.

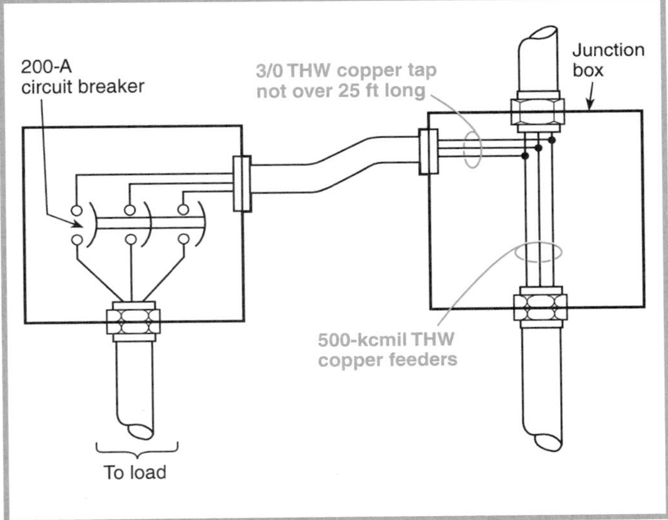

Figure 240.8 *An example where the feeder taps terminate in a single circuit breaker, as per Section 240-21(b)(2).*

(3) Taps Supplying a Transformer [Primary Plus Secondary Not Over 25 ft (7.62 m) Long]. Where the tap conductors supply a transformer and comply with all the following conditions.

(a) The conductors supplying the primary of a transformer have an ampacity at least one-third of the rating of the overcurrent device protecting the feeder conductors.

(b) The conductors supplied by the secondary of the transformer shall have an ampacity that, when multiplied by the ratio of the secondary-to-primary voltage, is at least one-third of the rating of the overcurrent device protecting the feeder conductors.

(c) The total length of one primary plus one secondary conductor, excluding any portion of the primary conductor that is protected at its ampacity, is not over 25 ft (7.62 m).

(d) The primary and secondary conductors are suitably protected from physical damage.

(e) The secondary conductors terminate in a single circuit breaker or set of fuses that will limit the load current to not more than the conductor ampacity that is permitted by Section 310-15.

Figure 240.9 illustrates the conditions of Section 240-21(b)(3). The overcurrent protection requirements of Sections 384-16(a) and 450-3(b) also apply.

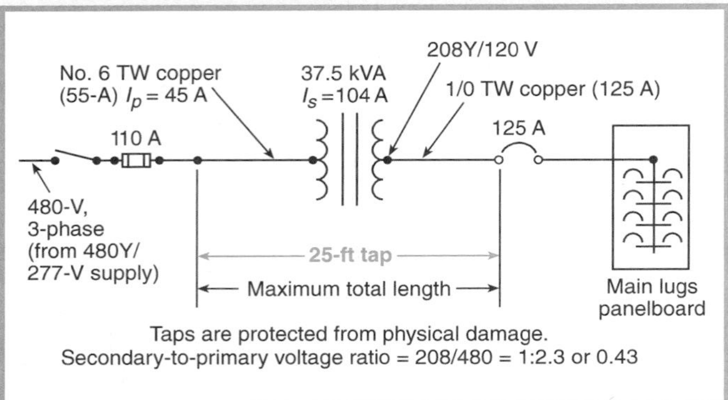

Figure 240.9 *An example where the transformer feeder taps (primary plus secondary) are not over 25 ft long, as per Section 240-21(b)(3).*

(4) Taps Over 25 ft (7.62 m) Long. Where the feeder is in a high bay manufacturing building over 35 ft (10.67 m) high at walls, and the installation complies with all of the following conditions.

(a) Conditions of maintenance and supervision ensure that only qualified persons will service the systems.

(b) The tap conductors are not over 25 ft (7.62 m) long horizontally and not over 100 ft (30.5 m) total length.

(c) The ampacity of the tap conductors is not less than one-third of the rating of the overcurrent device protecting the feeder conductors.

(d) The tap conductors terminate at a single circuit breaker or a single set of fuses that will limit the load to the ampacity of the tap conductors. This single overcurrent device shall be permitted to supply any number of additional overcurrent devices on its load side.

(e) The tap conductors are suitably protected from physical damage or are enclosed in a raceway.

(f) The tap conductors are continuous from end-to-end and contain no splices.

(g) The tap conductors are sized No. 6 copper or No. 4 aluminum or larger.

(h) The tap conductors do not penetrate walls, floors, or ceilings.

(i) The tap is made no less than 30 ft (9.14 m) from the floor.

Figure 240.10 illustrates the conditions of Section 240-21(b)(4). It permits a tap of 100 ft for manufacturing buildings with walls that are over 35 ft high, if the tap connection is not less than 30 ft from the floor, and conditions of maintenance and supervision ensure that only qualified persons will service these systems.

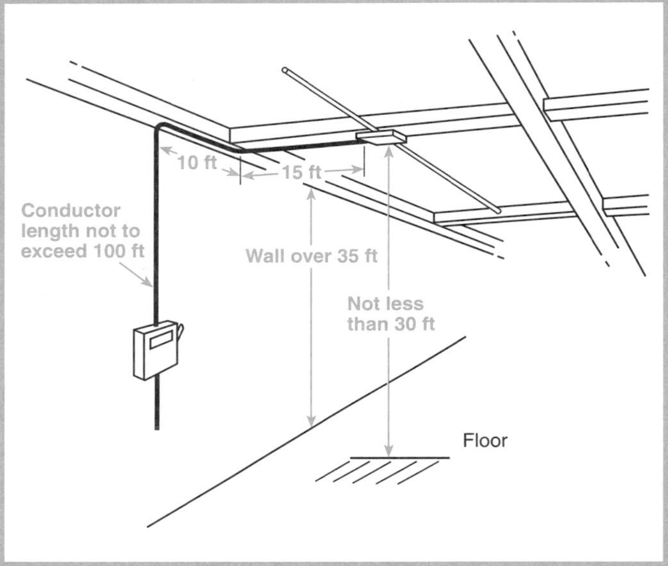

Figure 240.10 *An example where the feeder taps are over 25 ft long, the tap connection being not less than 30 ft from the floor, as per Section 240-21(b)(4).*

(5) Outside Taps of Unlimited Length. Where the conductors are located outdoors, except at the point of termination, and comply with all of the following conditions.

(a) The conductors are suitably protected from physical damage.

(b) The conductors terminate at a single circuit breaker or a single set of fuses that will limit the load to the ampacity of the conductors. This single overcurrent device shall be permitted to supply any number of additional overcurrent devices on its load side.

(c) The overcurrent device for the conductors is an integral part of a disconnecting means or shall be located immediately adjacent thereto.

(d) The disconnecting means for the conductors is installed at a readily accessible location either outside of a building or structure, or inside nearest the point of entrance of the conductors.

Section 240-21(b)(5) permits outside conductors to be tapped from a feeder without any limitations on the length of the tap conductors. The tap conductors must be protected and must terminate in a single, fused disconnect or single circuit breaker rated at not more than the ampacity of the tap conductors. Also, this fused disconnect or circuit breaker must be installed at a readily accessible location either inside or outside a building or structure. Furthermore, if this fused disconnect or circuit breaker is installed inside a building or structure, it must be located nearest the point of entrance of the tap conductors.

(c) Transformer Secondary Conductors. Conductors shall be permitted to be connected to a transformer secondary, without overcurrent protection at the secondary, as specified in (1) through (4).

Transformer secondary conductors are permitted without an overcurrent protective device in the secondary under the following four separate conditions.

1. **The primary overcurrent protective device, as described in Section 240-21(c)(1) can protect single-phase (2-wire) and 3-phase (delta-delta) transformer secondary conductors.**
2. **The transformer secondary conductors do not exceed 10 ft.**
3. **The transformer secondary conductors do not exceed 25 ft.**
4. **The transformer secondary conductors are located outdoors.**

Typical transformers, dry type and oil filled, must have their overcurrent protection within 10 ft (conductor length) of the transformer (25 ft for industrial establishments). Outdoor secondary conductors are not limited to a particular distance. However, the maximum number of overcurrent protective devices is similar to services.

FPN: For overcurrent protection requirements for transformers, see Section 450-3.

(1) Protection by Primary Overcurrent Device. Single-phase (other than 2-wire) and multiphase (other than delta-delta, 3-wire) transformer secondary conductors are not considered to be protected by the primary overcurrent protective device. Conductors supplied by the secondary side of a single-phase transformer having a 2-wire (single-voltage) secondary, or a three-phase, delta-delta connected transformer having a 3-wire (single-voltage) secondary, shall be permitted to be protected by overcurrent protection provided on the primary (supply) side of the transformer, provided this protection is in accordance with Section 450-3 and does not exceed the value determined by multiplying the secondary conductor ampacity by the secondary to primary transformer voltage ratio.

(2) Transformer Secondary Conductors Not Over 10 ft (3.05 m) Long. Where the length of secondary conductor does not exceed 10 ft (3.05 m) and complies with all of the following.

(a) The ampacity of the secondary conductors is
  (1) Not less than the combined computed loads on the circuits supplied by the secondary conductors, and
  (2) Not less than the rating of the device supplied by the secondary conductors or not less than the rating of the overcurrent-protective device at the termination of the secondary conductors.

(b) The secondary conductors do not extend beyond the switchboard, panelboard, disconnecting means, or control devices they supply.

(c) The secondary conductors are enclosed in a raceway, which shall extend from the transformer to the enclosure of an enclosed switchboard, panelboard, or control devices, or to the back of an open switchboard.

FPN: For overcurrent protection requirements for lighting and appliance branch-circuit panelboards, see Sections 384-16(a) and (e).

(3) Secondary Conductors Not Over 25 ft (7.62 m) Long. For industrial installations only, where the length of the secondary conductors does not exceed 25 ft (7.62 m) and complies with all of the following.

(a) The ampacity of the secondary conductors is not less than the secondary current rating of the transformer, and the sum of the ratings of the overcurrent devices does not exceed the ampacity of the secondary conductors.

(b) All overcurrent devices are grouped.

(c) The secondary conductors are suitably protected from physical damage.

(4) Outside Secondary Conductors. Where the conductors are located outdoors, except at the point of termination, and comply with all of the following conditions.

(a) The conductors are suitably protected from physical damage.

(b) The conductors terminate at a single circuit breaker or a single set of fuses that will limit the load to the ampacity of the conductors. This single overcurrent device shall be permitted to supply any number of additional overcurrent devices on its load side.

(c) The overcurrent device for the conductors is an integral part of a disconnecting means or shall be located immediately adjacent thereto.

(d) The disconnecting means for the conductors are installed at a readily accessible location either outside of a building or structure, or inside nearest the point of entrance of the conductors.

(5) Secondary Conductors from a Feeder Tapped Transformer. Transformer secondary conductors installed in accordance with Section 240-21(b)(3) shall be permitted to have overcurrent protection as specified in that section.

(d) Service Conductors. Service-entrance conductors shall be permitted to be protected by overcurrent devices in accordance with Section 230-91.

(e) Busway Taps. Busways and busway taps shall be permitted to be protected against overcurrent in accordance with Sections 364-10 through 364-13.

(f) Motor Circuit Taps. Motor-feeder and branch-circuit conductors shall be permitted to be protected against overcurrent in accordance with Sections 430-28 and 430-53 respectively.

(g) Conductors from Generator Terminals. Conductors from generator terminals that meet the size requirement in Section 445-5 shall be permitted to be protected against overload by the generator overload protective device(s) required by Section 445-4.

240-22. Grounded Conductor. No overcurrent device shall be connected in series with any conductor that is intentionally grounded, unless one of the following two conditions are met:

(1) The overcurrent device opens all conductors of the circuit, including the grounded conductor, and is designed so that no pole can operate independently
(2) Where required by Sections 430-36 or 430-37 for motor overload protection

240-23. Change in Size of Grounded Conductor. Where a change occurs in the size of the ungrounded conductor, a similar change shall be permitted to be made in the size of the grounded conductor.

Section 240-23 acknowledges that the size of a grounded conductor may be increased (because of voltage-drop problems, for example) or reduced to correspond to a change made in the size of an ungrounded conductor, as in the case of tap conductors, where all are of the same circuit.

240-24. Location in or on Premises.

(a) Accessibility. Overcurrent devices shall be readily accessible unless one of the following applies.

Section 240-24(a) is intended to correlate with Section 230-91, that is, the service overcurrent device must be an integral part of the service disconnecting means or must be located immediately adjacent thereto.

(1) For busways, as provided in Section 364-12.
(2) For supplementary overcurrent protection, as described in Section 240-10.
(3) For overcurrent devices, as described in Sections 225-40 and 230-92.
(4) For overcurrent devices adjacent to utilization equipment that they supply, access shall be permitted to be by portable means.

Section 240-24(a)(4) recognizes the need for overcurrent protection in locations that are not readily accessible, such as above suspended ceilings. It permits overcurrent devices to be located so that they are not readily accessible, as long as they are located next to the appliance, motor, or other equipment they supply and can be reached by using a ladder. See also Sections 240-10 and 380-8(a) regarding the accessibility of supplementary overcurrent devices and switches.

(b) Occupancy. Each occupant shall have ready access to all overcurrent devices protecting the conductors supplying that occupancy. Where electric service and electrical maintenance are provided by the building management and where these are under continuous building management supervision, the service overcurrent devices and feeder overcurrent devices supplying more than one occupancy shall be permitted to be accessible to only authorized management personnel in the following:

(1) In multiple occupancy buildings
(2) For guest rooms of hotels and motels that are intended for transient occupancy

(c) Not Exposed to Physical Damage. Overcurrent devices shall be located where they will not be exposed to physical damage.

FPN: See Section 110-11, Deteriorating Agents.

(d) Not in Vicinity of Easily Ignitible Material. Overcurrent devices shall not be located in the vicinity of easily ignitible material, such as in clothes closets.

Examples of locations where combustibles may be stored are linen closets, paper storage closets, and clothes closets.

(e) Not Located in Bathrooms. In dwelling units and guest rooms of hotels and motels, overcurrent devices, other than supplementary overcurrent protection, shall not be located in bathrooms as defined in Article 100.

C. Enclosures

240-30. General.

(a) Overcurrent devices shall be protected from physical damage by one of the following:

(1) Installation in enclosures, cabinets, cutout boxes, or equipment assemblies
(2) Mounting on open-type switchboards, panelboards, or control boards that are in rooms or enclosures free from dampness and easily ignitible material, and are accessible only to qualified personnel

(b) The operating handle of a circuit breaker shall be permitted to be accessible without opening a door or cover.

Properly selected overcurrent protective devices are designed to open a circuit before an overcurrent condition can seriously damage conductor insulation. Requirements that overcurrent devices be enclosed in cabinets or cutout boxes ensure that hot metal particles will not be ejected in the vicinity of combustibles. Also, use of an enclosure prevents contact with live parts by personnel.

Overcurrent devices mounted on open-type switchboards, panelboards, or control boards and having exposed energized parts are to be located where accessible only to qualified persons.

240-32. Damp or Wet Locations. Enclosures for overcurrent devices in damp or wet locations shall comply with Section 373-2(a).

240-33. Vertical Position. Enclosures for overcurrent devices shall be mounted in a vertical position. Circuit breaker enclosures shall be permitted to be installed horizontally where the circuit breaker is installed in accordance with Section 240-81. Listed busway plug-in units shall be permitted to be mounted in orientations corresponding to the busway mounting position.

A wall-mounted vertical position is desirable to achieve easier access, natural hand operation, normal swing or closing of doors or covers, and legibility of the manufacturer's markings.

D. Disconnecting and Guarding

240-40. Disconnecting Means for Fuses. A disconnecting means shall be provided on the supply side of all fuses in circuits over 150 volts to ground and cartridge fuses in circuits of any voltage where accessible to other than qualified persons so that each individual circuit containing fuses can be independently disconnected from the source of power. A current-limiting device without a disconnecting means shall be permitted on the supply side of the service disconnecting means as permitted by Section 230-82. A single disconnecting means shall be permitted on the supply side of more than one set of fuses as permitted by Section 430-112, Exception, for group operation of motors and Section 424-22(c) for fixed electric space-heating equipment.

A single disconnect switch is allowed to serve more than one set of fuses, such as in multimotor installations or for electric space-heating equipment where the heating element load is required to be subdivided, each element with its own set of fuses.

240-41. Arcing or Suddenly Moving Parts. Arcing or suddenly moving parts shall comply with (a) and (b).

(a) Location. Fuses and circuit breakers shall be located or shielded so that persons will not be burned or otherwise injured by their operation.

(b) Suddenly Moving Parts. Handles or levers of circuit breakers, and similar parts that may move suddenly in such a way that persons in the vicinity are likely to be injured by being struck by them, shall be guarded or isolated.

Arcing or suddenly moving parts are usually associated with switchboards or control boards that may be of the open type. They should be under competent supervision and accessible only to qualified persons. Fuses or circuit breakers are required to be located or shielded so that under an abnormal condition, the subsequent arc across the opening device will not injure persons in the vicinity.

Guardrails may be provided in the vicinity of disconnecting means because sudden-moving handles may be capable of causing injury. Modern switchboards, for example, are equipped with removable handles. See Article 100 for the definition of *guarded.* See also Section 110-27 for the guarding of live parts (600 volts, nominal, or less).

E. Plug Fuses, Fuseholders, and Adapters

240-50. General.

(a) Maximum Voltage. Plug fuses shall be permitted to be used in the following circuits:

(1) Circuits not exceeding 125 volts between conductors
(2) Circuits supplied by a system having a grounded neutral where the line-to-neutral voltage does not exceed 150 volts

Plug fuses can be installed in circuits supplied by 120/240-volt, single-phase, 3-wire systems and 208Y/120-volt, 3-phase, 4-wire systems.

(b) Marking. Each fuse, fuseholder, and adapter shall be marked with its ampere rating.

(c) Hexagonal Configuration. Plug fuses of 15-ampere and lower rating shall be identified by a hexagonal configuration of the window, cap, or other prominent part to distinguish them from fuses of higher ampere ratings.

Figure 240.11 shows some examples of plug fuses and Type S fuses. Note the hexagonal feature on the 15-ampere fuses.

Figure 240.11 Plug fuses and Type S fuses. (Bussmann Division, Cooper Industries)

(d) No Energized Parts. Plug fuses, fuseholders, and adapters shall have no exposed energized parts after fuses or fuses and adapters have been installed.

(e) Screw Shell. The screw shell of a plug-type fuseholder shall be connected to the load side of the circuit.

240-51. Edison-Base Fuses.

(a) Classification. Plug fuses of the Edison-base type shall be classified at not over 125 volts and 30 amperes and below.

(b) Replacement Only. Plug fuses of the Edison-base type shall be used only for replacements in existing installations where there is no evidence of overfusing or tampering.

240-52. Edison-Base Fuseholders. Fuseholders of the Edison-base type shall be installed only where they are made to accept Type S fuses by the use of adapters.

240-53. Type S Fuses. Type S fuses shall be of the plug type and shall comply with (a) and (b).

(a) Classification. Type S fuses shall be classified at not over 125 volts and 0 to 15 amperes, 16 to 20 amperes, and 21 to 30 amperes.

(b) Noninterchangeable. Type S fuses of an ampere classification as specified in (a) shall not be interchangeable with a lower ampere classification. They shall be designed so that they cannot be used in any fuseholder other than a Type S fuseholder or a fuseholder with a Type S adapter inserted.

240-54. Type S Fuses, Adapters, and Fuseholders.

(a) To Fit Edison-Base Fuseholders. Type S adapters shall fit Edison-base fuseholders.

(b) To Fit Type S Fuses Only. Type S fuseholders and adapters shall be designed so that either the fuseholder itself or the fuseholder with a Type S adapter inserted cannot be used for any fuse other than a Type S fuse.

(c) Nonremovable. Type S adapters shall be designed so that once inserted in a fuseholder, they cannot be removed.

(d) Nontamperable. Type S fuses, fuseholders, and adapters shall be designed so that tampering or shunting (bridging) would be difficult.

(e) Interchangeability. Dimensions of Type S fuses, fuseholders, and adapters shall be standardized to permit interchangeability regardless of the manufacturer.

Figure 240.12 shows a Type S nonrenewable plug fuse and adapter.

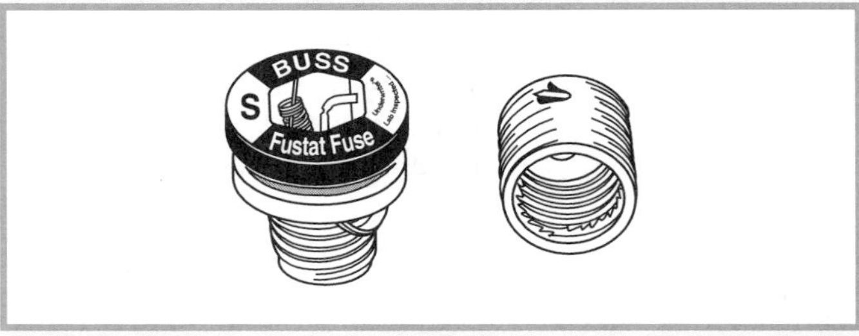

Figure 240.12 A Type S nonrenewable plug fuse and adapter. (Bussmann Division, Cooper Industries)

F. Cartridge Fuses and Fuseholders

240-60. General.

(a) Maximum Voltage — 300-Volt Type. Cartridge fuses and fuseholders of the 300-volt type shall be permitted to be used in the following circuits:

(1) Circuits not exceeding 300 volts between conductors
(2) Single-phase line-to-neutral circuits supplied from a 3-phase, 4-wire, solidly grounded neutral source where the line-to-neutral voltage does not exceed 300 volts

(b) Noninterchangeable — 0–6000-Ampere Cartridge Fuseholders. Fuseholders shall be designed so that it will be difficult to put a fuse of any given class into a fuseholder that is designed for a current lower, or voltage higher, than that of the class to which the fuse belongs. Fuseholders for current-limiting fuses shall not permit insertion of fuses that are not current limiting.

(c) Marking. Fuses shall be plainly marked, either by printing on the fuse barrel or by a label attached to the barrel showing the following:

(1) Ampere rating
(2) Voltage rating
(3) Interrupting rating where other than 10,000 amperes
(4) Current limiting where applicable
(5) The name or trademark of the manufacturer

The interrupting rating shall not be required to be marked on fuses used for supplementary protection.

Figure 240.13 shows examples of Class G fuses rated 300 volts. Note the plainly marked barrels.

Figure 240.13 *Various Class G fuses rated 300 volts. (Bussmann Division, Cooper Industries)*

Class H-type cartridge fuses have an interrupting capacity (IC) rating of 10,000 amperes, and this rating need not be marked on the fuse. However, Class CC, G, J, K, L, R, and T cartridge fuses exceed the 10,000-ampere IC rating and must be marked with the IC rating.

Section 240-83(c) requires that the IC rating for circuit breakers for other than 5000 amperes be indicated on the circuit breaker.

Fuses or circuit breakers used for supplementary protection of fluorescent fixtures, semiconductor rectifiers, motor-operated appliances, and so on, need not be marked for IC. See also the commentary on circuit breakers following Section 110-10.

240-61. Classification. Cartridge fuses and fuseholders shall be classified according to voltage and amperage ranges. Fuses rated 600 volts, nominal or less, shall be permitted to be used for voltages at or below their ratings.

See Section 490-21(b) for application of high-voltage fuses.

G. Circuit Breakers

240-80. Method of Operation. Circuit breakers shall be trip free and capable of being closed and opened by manual operation. Their normal method of operation by other than manual means, such as electrical or pneumatic, shall be permitted if means for manual operation is also provided.

240-81. Indicating. Circuit breakers shall clearly indicate whether they are in the open "off" or closed "on" position.

Where circuit breaker handles are operated vertically rather than rotationally or horizontally, the "up" position of the handle shall be the "on" position.

See Sections 240-83(d), 380-11, and 410-81(a) for requirements on circuit breakers used as switches.

Section 240-81 prohibits a circuit breaker from being inverted.

240-82. Nontamperable. A circuit breaker shall be of such design that any alteration of its trip point (calibration) or the time required for its operation will require dismantling of the device or breaking of a seal for other than intended adjustments.

240-83. Marking.

(a) Durable and Visible. Circuit breakers shall be marked with their ampere rating in a manner that will be durable and visible after installation. Such marking shall be permitted to be made visible by removal of a trim or cover.

(b) Location. Circuit breakers rated at 100 amperes or less and 600 volts or less shall have the ampere rating molded, stamped, etched, or similarly marked into their handles or escutcheon areas.

(c) Interrupting Rating. Every circuit breaker having an interrupting rating other than 5000 amperes shall have its interrupting rating shown on the circuit breaker. The interrupting rating shall not be required to be marked on circuit breakers used for supplementary protection.

This requirement recognizes series-rated circuit breakers and requires that the end-use equipment be marked with the series combination rating. For example, a circuit breaker with an interrupting rating of 10,000 amperes may perform safely on a circuit with an available fault current that is greater than 10,000 amperes under the following conditions.

1. If it is protected on its line side by a circuit breaker with a suitable interrupting rating.
2. If the series combination has been tested and demonstrated to safely open a short-circuit current higher than the 10,000 amperes on the load side of the breaker.

In the 1995 and 1998 editions of the UL *General Information Directory* (White Book), under the category "Switchboards, Dead Front Type (WEVZ)," the following information is provided.

> *Short Circuit Ratings:* Dead-front switchboard sections or interiors are marked with their dc or rms symmetrical short-circuit current rating in amperes. The marking states that short-circuit ratings are limited to the lowest short-circuit rating of (1) any switchboard section connected in series or (2) the lowest short-circuit rating of any device installed or intended to be installed therein. However, for

combination series-connected devices, the short-circuit current rating marked on the switchboard may be higher than the short-circuit current rating of a specific circuit breaker installed or to be installed in the switchboard. This higher rating is valid only if the specific overcurrent devices identified in the marking are used within or ahead of the switchboard in accordance with the marked instructions. In many cases, the short-circuit ratings are associated with instructions for securing supply wiring within the switchboard. [1998]

Since the line side (protecting) circuit breaker may be field installed and the series rating may or may not be needed, depending on the equipment installed and the available fault current, only the system installer or designer can determine the need for this newly required marking.

Most often, the series rating of devices is specific to individual manufacturers. It is especially important to consult listing information or product literature for retrofit installations. [1995]

(d) Used as Switches. Circuit breakers used as switches in 120-volt and 277-volt fluorescent lighting circuits shall be listed and shall be marked "SWD."

Circuit breakers marked "SWD" are 15- or 20-ampere breakers that have been subjected to additional endurance and temperature testing.

(e) Voltage Marking. Circuit breakers shall be marked with a voltage rating not less than the nominal system voltage that is indicative of their capability to interrupt fault currents between phases or phase to ground.

240-85. Applications. A circuit breaker with a straight voltage rating, such as 240V or 480V, shall be permitted to be applied in a circuit in which the nominal voltage between any two conductors does not exceed the circuit breaker's voltage rating. A two-pole circuit breaker shall not be used for protecting a 3-phase, corner-grounded delta circuit unless the circuit breaker is marked 1ϕ–3ϕ to indicate such suitability.

A circuit breaker with a slash rating, such as 120/240V or 480Y/277V, shall be permitted to be applied in a circuit where the nominal voltage of any conductor to ground does not exceed the lower of the two values of the circuit breaker's voltage rating and the nominal voltage between any two conductors does not exceed the higher value of the circuit breaker's voltage rating.

These requirements were considered explanatory material in the past and as such were contained in the FPN following Section 240-83(e). In the 1996 *Code*, these same requirements were moved into Section 240-85.

A circuit breaker marked "480Y/277V" is not intended for use on a 480-volt system with up to 480 volts to ground, such as a 480-volt circuit derived from a delta-connected system. A circuit breaker marked "480V" or "600V" should be used on such a system. In like manner, a circuit breaker marked "120/240V" is not intended for use on a delta-connected 240-volt circuit. A 240-volt, 480-volt, or 600-volt circuit breaker should be used on such a circuit. The slash "/" between the lower and higher voltage ratings in the marking indicates that the circuit breaker has been tested for use on a circuit with the higher voltage between phases and with the lower voltage to ground.

240-86. Series Ratings. Where a circuit breaker is used on a circuit having an available fault current higher than its marked interrupting rating by being connected on the load side of an acceptable overcurrent protective device having the higher rating, the following shall apply.

(a) Marking. The additional series combination interrupting rating shall be marked on the end use equipment, such as switchboards and panelboards.

(b) Motor Contribution. Series ratings shall not be used where

(1) Motors are connected on the load side of the higher-rated overcurrent device and on the line side of the lower-rated overcurrent device, and
(2) The sum of the motor full-load currents exceeds 1 percent of the interrupting rating of the lower-rated circuit breaker.

H. Supervised Industrial Installations

240-90. General. Overcurrent protection in areas of supervised industrial installations shall comply with all the applicable provisions of the other sections of this article, except as provided in Part H. The provisions of Part H shall only be permitted to apply to those portions of the electrical system in the supervised industrial installation used exclusively for manufacturing or process control activities.

Part H, Supervised Industrial Installations, relaxes some of the location requirements for low voltage (600 volts, nominal, and under) overcurrent protection on transformer secondary conductors and outside feeder taps for large manufacturing or process industries. Normally, Section 240-21 provides the requirements for the location of circuit overcurrent protection. However, Part H can apply to the electrical

systems serving the process or manufacturing areas that meet the criteria of Section 240-91.

240-91. Definition of Supervised Industrial Installation. For the purposes of Part H, *supervised industrial installation* is defined as the industrial portions of a facility where all of the following conditions are met.

(1) Conditions of maintenance and engineering supervision ensure that only qualified persons will monitor and service the system.
(2) The premises wiring system has 2500 kVA or greater of load used in industrial process(es), manufacturing activities, or both, as calculated in accordance with Article 220.
(3) The premises has at least one service that is more than 150 volts to ground and more than 300 volts phase-to-phase.

This definition shall not apply to those installations in buildings used by the industrial facility for offices, warehouses, garages, machine shops, and recreational facilities that are not an integral part of the industrial plant, substation, or control center.

An industrial establishment must have a combined process and manufacturing load greater than 2500 kVA; have at least one service, 480Y/277V, nominal, or greater; and have maintenance and engineering supervision that will ensure that only qualified persons will monitor and service the electrical system. All process or manufacturing loads from each low-, medium-, and high-voltage system can be added together to satisfy the load requirements of Part H. Loads are calculated in accordance with Article 220. However, loads not associated with manufacturing or processing cannot be used to meet the 2500-kVA minimum requirement.

The provisions of Part H only apply to the low voltage (600 volts, nominal, or less) electrical systems used for process or manufacturing. Part H does not apply to electrical systems operating at over 600 volts, nominal, or for electrical systems that serve nonmanufacturing or nonprocess facilities such as offices, warehouses, garages, machine shops, or recreational facilities. However, if a part of the process or manufacturing electrical system is used to serve an office, warehouse, garage, machine shop, or recreational facility that is an integral part of the industrial plant, control center, or substation, Part H can still apply to the total process or manufacturing electrical system.

240-92. Location in Circuit. An overcurrent device shall be connected in each ungrounded circuit conductor as follows.

(a) Feeder and Branch-Circuit Conductors. Feeder and branch-circuit conductors shall be protected at the point the conductors receive their supply, as permitted in Section 240-21, or as otherwise permitted in (b) or (c).

(b) Transformer Secondary Conductors of Separately Derived Systems. Conductors shall be permitted to be connected to a transformer secondary of a separately derived system, without overcurrent protection at the connection, where the conditions of (1), (2), and (3) are met.

(1) Short-Circuit and Ground-Fault Protection. The conductors shall be protected from short-circuit and ground-fault conditions by complying with one of the following conditions.

(a) The length of the secondary conductors does not exceed 50 ft (15.24 m) and the transformer primary overcurrent device has a rating or setting that does not exceed 150 percent of the value determined by multiplying the secondary conductor ampacity by the secondary-to-primary transformer voltage ratio.
(b) The length of the secondary conductors does not exceed 75 ft (22.86 m) and the conductors are protected by a differential relay with a trip setting equal to or less than the conductor ampacity.
(c) The conductors shall be considered to be protected if the length of the secondary conductors does not exceed 75 ft (22.86 m) and if calculations, made under engineering supervision, determine that the system overcurrent devices will protect the conductors within recognized time vs. current limits for all short-circuit and ground-fault conditions.

(2) Overload Protection. The conductors shall be protected against overload conditions by complying with one of the following.

(a) The conductors terminate in a single overcurrent device that will limit the load to the conductor ampacity.
(b) The sum of the overcurrent devices at the conductor termination limits the load to the conductor ampacity. The overcurrent devices shall consist of not more than six circuit breakers or sets of fuses, mounted in a single enclosure, in a group of separate enclosures, or in or on a switchboard. There shall be no more than six overcurrent devices grouped in any one location.
(c) Overcurrent relaying is connected [with a current transformer(s), if needed] to sense all of the secondary conductor current and limit the load to the conductor ampacity by opening upstream or downstream devices.
(d) Conductors shall be considered to be protected if calculations, made under engineering supervision, determine that the system overcurrent devices will protect the conductors from overload conditions.

(3) Physical Protection. The secondary conductors shall be suitably protected from physical damage.

(c) Outside Feeder Taps. Outside conductors shall be permitted to be tapped to a feeder or to be connected at a transformer secondary, without overcurrent protection at the tap or connection, where all the following conditions are met.

(1) The conductors are suitably protected from physical damage.
(2) The sum of the overcurrent devices at the conductor termination limits the load to the conductor ampacity. The overcurrent devices shall consist of not more than six circuit breakers or sets of fuses mounted in a single enclosure, in a group of separate enclosures, or in or on a switchboard. There shall be no more than six overcurrent devices grouped in any one location.
(3) The tap conductors are installed outdoors, except at the point of termination.
(4) The overcurrent device for the conductors is an integral part of a disconnecting means or shall be located immediately adjacent thereto.
(5) The disconnecting means for the conductors are installed at a readily accessible location either outside of a building or structure or inside nearest the point of entrance of the conductors.

I. Overcurrent Protection Over 600 Volts, Nominal

240-100. Feeders and Branch Circuits.

(a) Feeder and branch-circuit conductors shall have overcurrent protection in each ungrounded conductor located at the point where the conductor receives its supply or at a location in the circuit determined under engineering supervision. The overcurrent protection shall be permitted to be provided by one of the following.

(1) Overcurrent Relays and Current Transformers. Circuit breakers used for overcurrent protection of 3-phase circuits shall have a minimum of three overcurrent relays operated from three current transformers. On 3-phase, 3-wire circuits, an overcurrent relay in the residual circuit of the current transformers shall be permitted to replace one of the phase relays.

An overcurrent relay, operated from a current transformer that links all phases of a 3-phase, 3-wire circuit, shall be permitted to replace the residual relay and one of the phase-conductor current transformers. Where the neutral is not regrounded on the load side of the circuit as permitted in Section 250-184(b), the current transformer shall be permitted to link all 3-phase conductors and the grounded circuit conductor (neutral).

(2) Fuses. A fuse shall be connected in series with each ungrounded conductor.

(b) Protective Devices. The protective device(s) shall be capable of detecting and interrupting all values of current that can occur at their location in excess of their trip setting or melting point.

(c) Conductor Protection. The operating time of the protective device, the available short-circuit current, and the conductor used shall be coordinated to prevent damaging or dangerous temperatures in conductors or conductor insulation under short-circuit conditions.

240-101. Additional Requirements for Feeders.

(a) Rating or Setting of Overcurrent Protective Devices. The continuous ampere rating of a fuse shall not exceed three times the ampacity of the conductors. The long-time trip element setting of a breaker or the minimum trip setting of an electronically actuated fuse shall not exceed six times the ampacity of the conductor. For fire pumps, conductors shall be permitted to be protected for overcurrent in accordance with Section 695-4(b).

(b) Feeder Taps. Conductors tapped to a feeder shall be permitted to be protected by the feeder overcurrent device where that overcurrent device also protects the tap conductor.

Article 250 — Grounding

Contents

See Appendix E for a cross reference list of section numbers between the 1996 Article 250 and the 1999 Article 250.

The complete revision of Article 250 is one of the most significant changes to the 1999 *Code.* The revision was a collective effort of the NEC Usability Task Group, Code-Making Panel 5, and *NEC* users who submitted proposals and comments. Similar requirements that formerly appeared in different parts of Article 250 have been grouped together in the same part and exceptions have been converted into positive code language. An overall new approach to the layout has provided a more "user friendly" Article 250. Appendix E provides two cross-reference lists. Section E1 compares the 1999 sections to the 1996 sections and Section E2 compares the 1996 sections to the 1999 sections.

A. General

250-1. Scope. This article covers general requirements for grounding and bonding of electrical installations, and specific requirements in (1) through (6).

(1) Systems, circuits, and equipment required, permitted, or not permitted to be grounded
(2) Circuit conductor to be grounded on grounded systems
(3) Location of grounding connections
(4) Types and sizes of grounding and bonding conductors and electrodes
(5) Methods of grounding and bonding
(6) Conditions under which guards, isolation, or insulation may be substituted for grounding

250-2. General Requirements for Grounding and Bonding. The following general requirements identify what grounding and bonding of electrical systems are required to accomplish. The prescriptive methods contained in Article 250 shall be followed to comply with the performance requirements of this section.

Fine print notes 1 and 2 to Section 250-1 in the 1996 *Code* now appear in the text of Section 250-2 in the 1999 *Code.* Section 250-2 provides the performance requirements for grounding and bonding of electrical systems and equipment. Performance-based requirements provide an overall objective without stating the specifics for accomplishing that objective. The first paragraph of Section 250-2 indicates that the performance objectives stated in (a), (b), (c), and (d) are accomplished by complying with the prescriptive requirements found in the rest of Article 250. Section 250-51 of the 1996 *Code* was considered to be a performance requirement for the grounding path. The requirements of that section did not provide a specific rule for the sizing or connection of grounding conductors, rather it stated overall performance considerations for grounding conductors. In the 1999 *Code,* Section 250-2(d) contains the fault current path objectives that were contained in Section 250-51 of previous editions of the *Code.*

(a) Grounding of Electrical Systems. Electrical systems that are required to be grounded shall be connected to earth in a manner that will limit the voltage imposed by lightning, line surges, or unintentional contact with higher voltage lines and that will stabilize the voltage to earth during normal operation.

(b) Grounding of Electrical Equipment. Conductive materials enclosing electrical conductors or equipment, or forming part of such equipment, shall be connected to earth so as to limit the voltage to ground on these materials. Where the electrical system is required to be grounded, these materials shall be connected together and to the supply system grounded conductor as specified by this article. Where the electrical system is not solidly grounded, these materials shall be connected together in a manner that establishes an effective path for fault current.

(c) Bonding of Electrically Conductive Materials and Other Equipment. Electrically conductive materials, such as metal water piping, metal gas piping, and structural steel members, that are likely to become energized shall be bonded as specified by this article to the supply system grounded conductor or, in the case of an ungrounded electrical system, to the electrical system grounded equipment, in a manner that establishes an effective path for fault current.

(d) Performance of Fault Current Path. The fault current path shall be permanent and electrically continuous, shall be capable of safely carrying the maximum fault likely to be imposed on it, and shall have sufficiently low impedance to facilitate the operation of overcurrent devices under fault conditions.

The earth shall not be used as the sole equipment grounding conductor or fault current path.

FPN: See Figure 250-2 for information on the organization of Article 250.

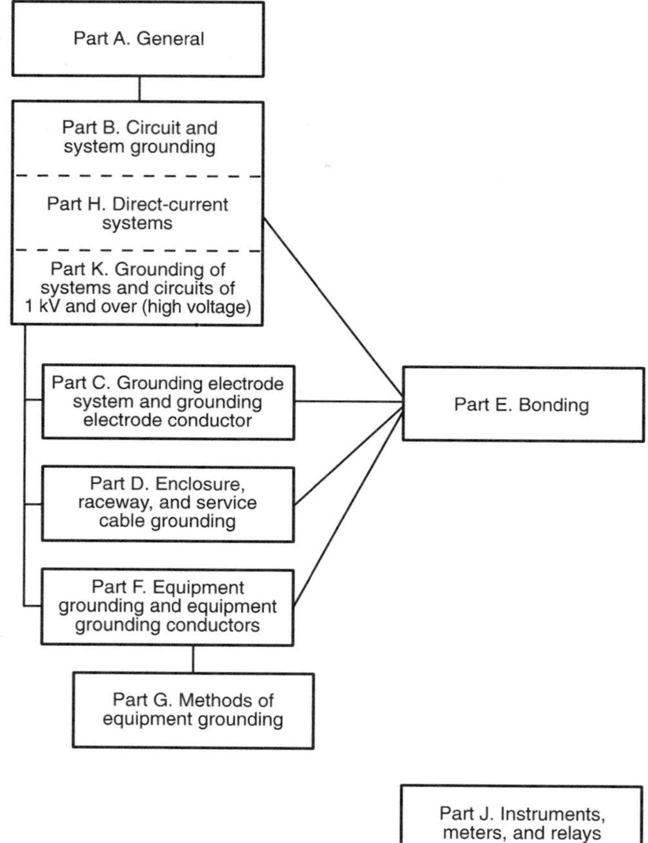

Figure 250-2 Grounding.

It is imperative that *Code* users be familiar with the definitions in Article 100, especially those terms associated with Article 250. Specific terms to be aware of are *grounded conductor, equipment grounding conductor,* and *grounding electrode conductor.*

Grounding can be divided into two areas: system grounding and equipment grounding. These two areas are kept separate from each other except at the point where they receive their source of power, such as at the service equipment or at a separately derived system.

Grounding is the intentional connection of a current-carrying conductor to ground or something that serves in place of ground. In most instances, this connection is made at the supply source, such as a transformer, and at the main service disconnecting means of the premises using the energy.

There are three basic reasons for grounding. They are as follows:

1. To limit the voltages caused by lightning or by accidental contact of the supply conductors with conductors of higher voltage
2. To stabilize the voltage under normal operating conditions (This maintains the voltage at one level relative to ground, so that any equipment connected to the system will be subject only to that potential difference.)
3. To facilitate the operation of overcurrent devices, such as fuses, circuit breakers, or relays, under ground-fault conditions

Figure 250.1 is a diagram of a typical grounding system for a single-phase, 3-wire service. One conductor of the system is intentionally connected to ground (via the grounding electrode conductor). Bonding the equipment ground bus to the neutral bus via the main bonding jumper provides a ground reference for exposed, noncurrent-carrying parts of the electrical system and a circuit through the grounded service conductor for ground-fault current.

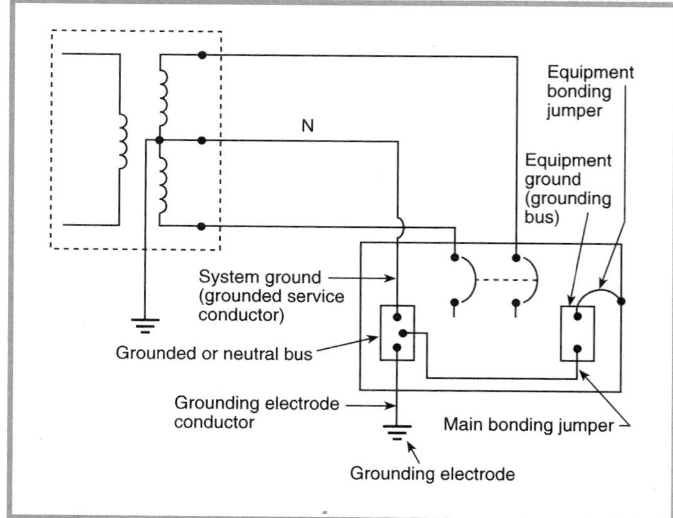

Figure 250.1 A typical grounding system for a single-phase, 3-wire service.

250-4. Application of Other Articles. In other articles applying to particular cases of installation of conductors and equipment, there are requirements that are in addition to those of this article or are modifications of them.

	Article	Section
Agricultural buildings		547-8
Audio signal processing, amplification, and reproduction equipment		640-7
Branch circuits		210-5 210-6 210-7
Cablebus		365-9
Capacitors		460-10 460-27
Circuits and equipment operating at less than 50 volts	720	
Class 1, Class 2, and Class 3 remote-control, signaling, and power-limited circuits		725-6
Closed-loop and programmed power distribution		780-3
Communications circuits	800	
Community antenna television and radio distribution systems		820-33 820-40 820-41
Conductors for general wiring	310	
Cranes and hoists	610	
Electrically driven or controlled irrigation machines		675-11(c) 675-12 675-13 675-14 675-15
Electric signs and outline lighting	600	
Electrolytic cells	668	
Elevators, dumbwaiters, escalators, moving walks, wheelchair lifts, and stairway chair lifts	620	
Fire alarm systems		760-6
Fixed electric heating equipment for pipelines and vessels		427-29 427-48
Fixed outdoor electric deicing and snow-melting equipment		426-27
Fixtures and lighting equipment		410-17 410-18 410-20 410-21 410-105(b)
Flexible cords and cables		400-22 400-23
Floating buildings		553-8 553-10 553-11
Grounding-type receptacles, adapters, cord connectors, and attachment plugs		410-58
Hazardous (classified) locations	500-517	
Health care facilities	517	
Induction and dielectric heating equipment	665	
Industrial machinery	670	
Information technology equipment		645-15
Intrinsically safe systems		504-50
Lighting fixtures, lampholders, lamps, and receptacles	410	
Marinas and boatyards		555-8
Mobile homes and mobile home park	550	
Motion picture and television studios and similar locations		530-20 530-66
Motors, motor circuits, and controllers	430	
Outlet, device, pull and junction boxes, conduit bodies and fittings		370-4 370-25
Over 600 volts, nominal, underground wiring methods		300-50(b)
Panelboards		384-20
Pipe organs	650	
Radio and television equipment	810	
Receptacles and cord connectors		210-7
Recreational vehicles and recreational vehicle parks	551	
Services	230	
Solar photovoltaic systems		690-41 690-42 690-43 690-45 690-47
Swimming pools, fountains, and similar installations	680	
Switchboards and panelboards		384-3(d)
Switches		380-12
Theaters, audience areas of motion picture and television studios, and similar locations		520-81
Transformers and transformer vaults		450-10
Use and identification of grounded conductors	200	
X-ray equipment	660	517-78

250-6. Objectionable Current Over Grounding Conductors.

(a) Arrangement to Prevent Objectionable Current. The grounding of electrical systems, circuit conductors, surge arresters, and conductive noncurrent-carrying materials and equipment shall be installed and arranged in a manner that will prevent an objectionable flow of current over the grounding conductors or grounding paths.

(b) Alterations to Stop Objectionable Current. If the use of multiple grounding connections results in an objectionable flow of current, one or more of the following alterations shall be permitted to be made, provided that the requirements of Section 250-2(d) are met.

(1) Discontinue one or more but not all of such grounding connections.
(2) Change the locations of the grounding connections.
(3) Interrupt the continuity of the conductor or conductive path interconnecting the grounding connections.
(4) Take other suitable remedial action satisfactory to the authority having jurisdiction.

An increase in the use of electronic controls and computer equipment, that is sensitive to stray currents has caused installation designers to look for ways to isolate electronic equipment from the effects of such stray circulating currents. Circulating currents on equipment grounding conductors, metal raceways, and building steel develop potential differences between ground and the electronic equipment.

One of the first solutions recommended by designers is to isolate the electronic equipment from all other power equipment by disconnecting it from the power ground. This is achieved by removing the equipment grounding means or installing nonmetallic spacers in the raceway system. The electronic equipment is then grounded to an earth ground isolated from the power ground. Isolating the equipment in this manner creates a potential difference that is a shock hazard and does not establish a low-impedance ground-fault return path to the power source, which is necessary to actuate the overcurrent protection device. Section 250-6(b) is not intended to allow disconnection of all power grounding connections to the electronic equipment. See also the commentary following Section 250-6(d).

(c) Temporary Currents Not Classified as Objectionable Currents. Temporary currents resulting from accidental conditions, such as ground-fault currents, that occur only while the grounding conductors are performing their intended protective functions shall not be classified as objectionable current for the purposes specified in (a) and (b).

(d) Limitations to Permissible Alterations. The provisions of this section shall not be considered as permitting electronic equipment from being operated on ac systems or branch circuits that are not grounded as required by this article. Currents that introduce noise or data errors in electronic equipment shall not be considered the objectionable currents addressed in this section.

Section 250-6(d) indicates that currents that result in noise or data errors in electronic equipment are not considered to be the "objectionable currents" referred to in Section 250-6. This section limits the alterations permitted by Section 250-6(b). See Section 250-96(b), which provides methods to eliminate noise and data errors.

(e) Isolation of Objectionable Direct-Current Ground Currents. Where isolation of objectionable dc ground currents from cathodic protection systems is required, a listed ac coupling/dc isolating device shall be permitted in the equipment grounding path to provide an effective return path for ac ground-fault current while blocking dc current.

The dc ground current on grounding conductors as a result of a cathodic protection system may be considered objectionable. Due to the required grounding and bonding connections associated with metal piping systems, it is inevitable that where cathodic protection for the piping system is provided, dc current will be present on grounding and bonding conductors. Section 250-6(e) allows the use of a listed ac coupling/dc isolating device. This device prevents the direct-current current on grounding and bonding conductors and allows the ground-fault return path to function properly. As part of the product testing, these devices are evaluated for proper performance under ground-fault conditions.

250-8. Connection of Grounding and Bonding Equipment. Grounding conductors and bonding jumpers shall be connected by exothermic welding, listed pressure connectors, listed clamps, or other listed means. Connection devices or fittings that depend solely on solder shall not be used. Sheet metal screws shall not be used to connect grounding conductors to enclosures.

Section 250-8 prohibits the use of sheet metal screws as a means for attaching equipment grounding conductors to equipment. Connection means that are listed or are part of listed equipment and exothermic welding are required to ensure a permanent and low resistance connection. Figures 250.2 and 250.3 illustrate two methods of attaching an equipment grounding conductor to a metal box.

250-10. Protection of Ground Clamps and Fittings. Ground clamps or other fittings shall be approved for general use without protection or shall be protected from physical damage as indicated in (1) or (2).

(1) In installations where they are not likely to be damaged
(2) Where enclosed in metal, wood, or equivalent protective covering

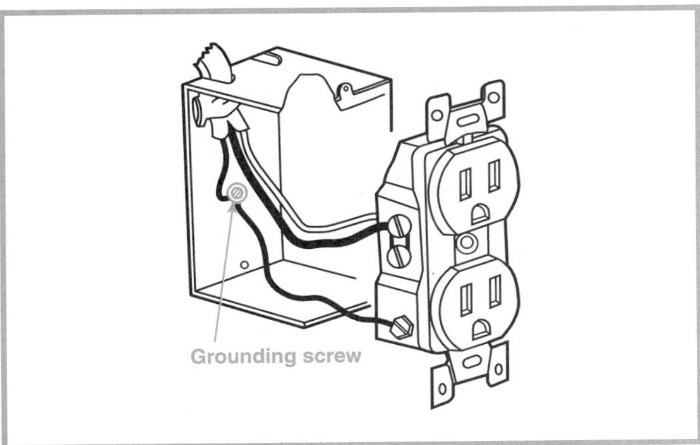

Figure 250.2 An application of a grounding screw.

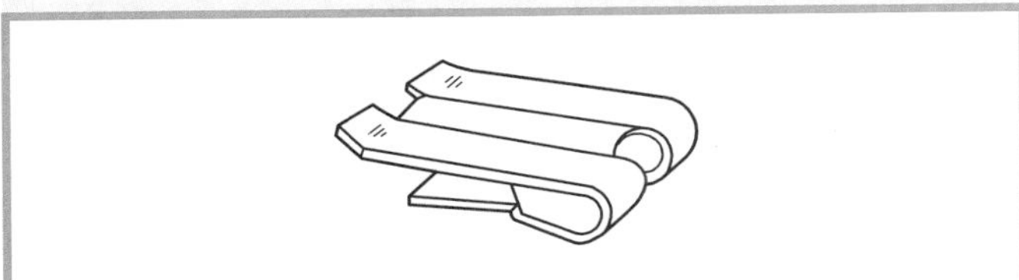

Figure 250.3 An application of a grounding clip.

250-12. Clean Surfaces. Nonconductive coatings (such as paint, lacquer, and enamel) on equipment to be grounded shall be removed from threads and other contact surfaces to ensure good electrical continuity or be connected by means of fittings designed so as to make such removal unnecessary.

B. Circuit and System Grounding

250-20. Alternating-Current Circuits and Systems to Be Grounded. Alternating-current circuits and systems shall be grounded as provided for in (a), (b), (c), or (d). Other circuits and systems shall be permitted to be grounded.

> FPN: An example of a system permitted to be grounded is a corner-grounded delta transformer connection. See Section 250-26(4) for conductor to be grounded.

(a) Alternating-Current Circuits of Less than 50 Volts. Alternating-current circuits of less than 50 volts shall be grounded under any of the following conditions:

(1) Where supplied by transformers, if the transformer supply system exceeds 150 volts to ground
(2) Where supplied by transformers, if the transformer supply system is ungrounded
(3) Where installed as overhead conductors outside of buildings

(b) Alternating-Current Systems of 50 Volts to 1000 Volts. Alternating-current systems of 50 volts to 1000 volts that supply premises wiring and premises wiring systems shall be grounded under any of the following conditions:

(1) Where the system can be grounded so that the maximum voltage to ground on the ungrounded conductors does not exceed 150 volts

Figure 250.4 illustrates the grounding requirements of Section 250-20(b)(1) as applied to a 120-volt, single-phase, 2-wire system and to a 120/240-volt, single-phase, 3-wire system.

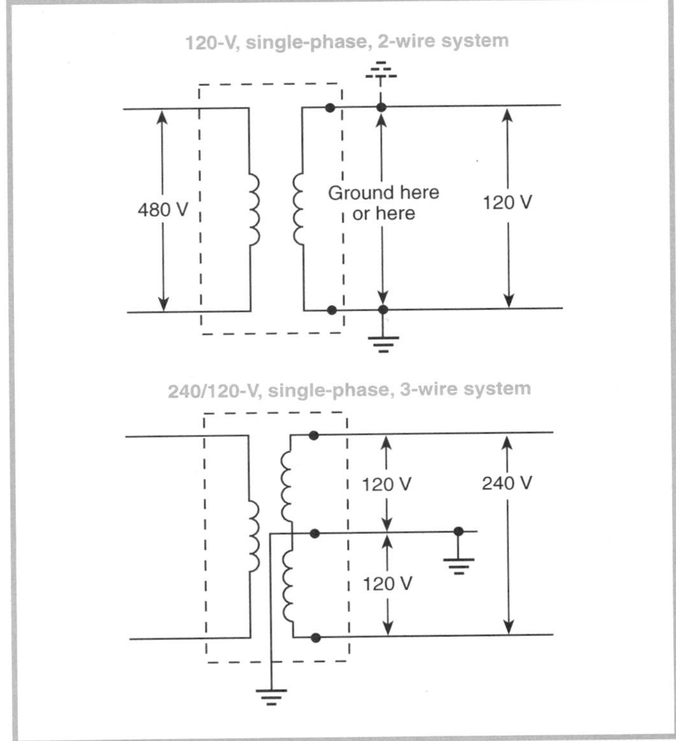

Figure 250.4 Typical systems required to be grounded by Section 250-20(b)(1).

(2) Where the system is 3-phase, 4-wire, wye connected in which the neutral is used as a circuit conductor
(3) Where the system is 3-phase, 4-wire, delta connected in which the midpoint of one phase winding is used as a circuit conductor

As Figure 250.5 illustrates, grounding is required for all wye systems if the neutral is used as a grounded circuit conductor and for all 3-phase, 4-wire delta systems if the midpoint of one phase is used as a grounded circuit conductor. See Sections 215-8, 230-56, and 384-3(e) for identification of the "high leg."

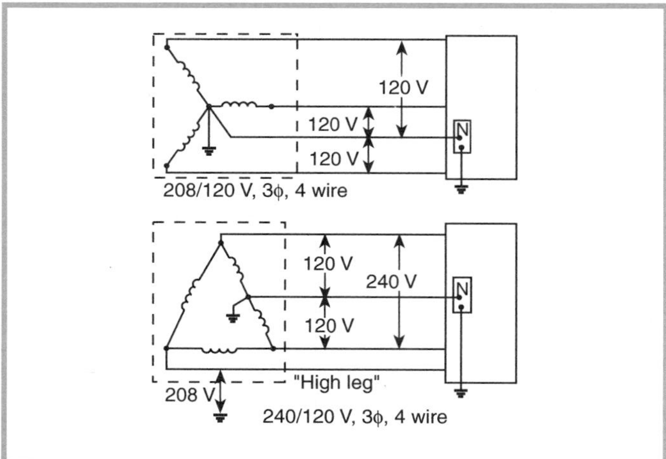

Figure 250.5 Typical systems required to be grounded by Sections 250-20(b)(2) and (3).

(c) Alternating-Current Systems of 1 kV and Over. Alternating-current systems supplying mobile or portable equipment shall be grounded as specified in Section 250-188. Where supplying other than mobile or portable equipment, such systems shall be permitted to be grounded. Where such systems are grounded, they shall comply with the applicable provisions of this article.

(d) Separately Derived Systems. If required to be grounded as in (a) or (b), separately derived systems shall be grounded as specified in Section 250-30.

Two of the most common sources of separately derived systems in premises wiring are transformers and generators. An autotransformer or step-down transformer that is part of electrical equipment and does not supply premises wiring is not the source of a separately derived system. See the definition of *premises wiring* in Article 100.

FPN No. 1: An alternate ac power source such as an on-site generator is not a separately derived system if the neutral is solidly interconnected to a service-supplied system neutral.

Figures 250.6 and 250.7 depict a 208Y/120-volt, 3-phase, 4-wire electrical service supplying a service disconnecting means to a building. The system is fed through a transfer switch connected to a generator intended to provide power for an emergency or standby system.

In Figure 250.6, the neutral conductor from the generator to the load is not disconnected by the transfer switch. There is a direct electrical connection between the normal grounded system conductor (neutral) and the generator neutral through the neutral bus in the transfer switch, thereby grounding the generator neutral. Since the generator is grounded by connection to the normal system ground, it is not a separately derived system, and there are no requirements for grounding the neutral at the generator. Under these conditions, it is necessary to run an equipment grounding conductor from the service equipment to the 3-pole transfer switch and from the 3-pole transfer switch to the generator. This can be in the form of any one of the items listed in Section 250-118.

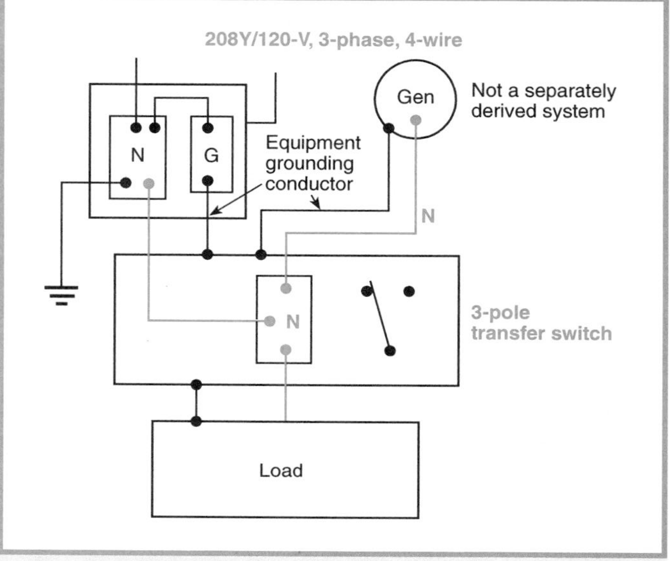

Figure 250.6 A 208Y/120-volt, 3-phase, 4-wire system that has a direct electrical connection of the grounded circuit conductor (neutral) to the generator.

In Figure 250.7, the grounded conductor (neutral) is connected to the switching contacts of a 4-pole transfer switch. Therefore, the generator system does not have a direct electrical connection to the other supply system grounded conductor (neutral), and the system supplied by the generator is considered separately derived. This separately derived system (3-phase, 4-wire, wye-connected system that supplies line to neutral loads) is required to be grounded in accordance with Sections 250-20(b) and (d). The methods for grounding the system are specified in Section 250-30(a).

Section 250-30(a)(1) requires separately derived systems to have a bonding jumper connected between the generator frame and the grounded circuit conductor (neutral). The grounding electrode conductor from the generator is required to be connected

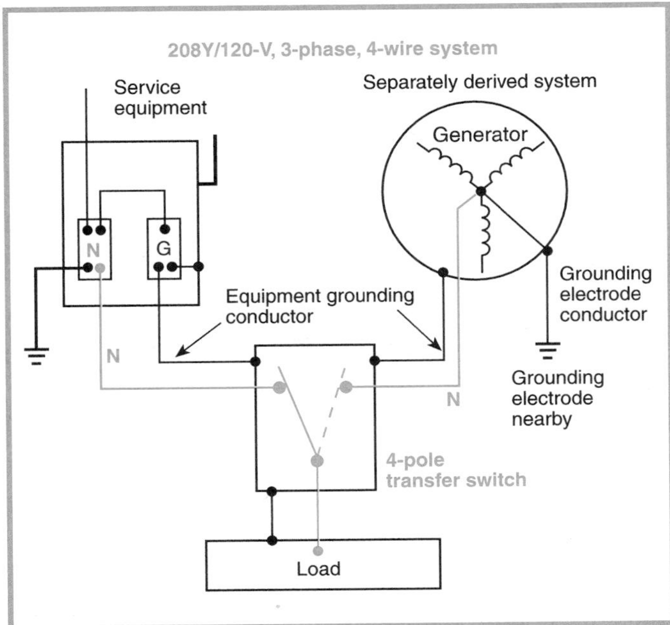

Figure 250.7 *A 208Y/120-volt, 3-phase, 4-wire system that does not have a direct electrical connection of the grounded circuit conductor (neutral) to the generator.*

to a grounding electrode. This conductor should be located as close to the generator as practicable. If the generator is in a building, the preferred grounding electrode is required to be one of the following, depending on which grounding electrode conductor is nearest to the generator: (1) effectively grounded structural metal member or (2) the first 5 ft of water pipe into a building where this piping is effectively grounded. (There is an exception with specific conditions that permits the grounding connection to the water piping beyond the first 5 ft.) For buildings or structures in which the preferred electrodes are not available, the choice can be made from any of the grounding electrodes specified in Sections 250-50 and 250-52.

FPN No. 2: For systems that are not separately derived and are not required to be grounded as specified in Section 250-30, see Section 445-5 for minimum size of conductors that must carry fault current.

250-21. Alternating-Current Systems of 50 Volts to 1000 Volts Not Required to Be Grounded. The following ac systems of 50 volts to 1000 volts shall be permitted to be grounded but shall not be required to be grounded:

(1) Electric systems used exclusively to supply industrial electric furnaces for melting, refining, tempering, and the like
(2) Separately derived systems used exclusively for rectifiers that supply only adjustable speed industrial drives
(3) Separately derived systems supplied by transformers that have a primary voltage rating less than 1000 volts, provided that all of the following conditions are met:
   (a) The system is used exclusively for control circuits.
   (b) The conditions of maintenance and supervision ensure that only qualified persons will service the installation.
   (c) Continuity of control power is required.
   (d) Ground detectors are installed on the control system.
(4) Isolated systems as permitted or required in Articles 517 and 668

FPN: The proper use of suitable ground detectors on ungrounded systems can provide additional protection.

(5) High-impedance grounded neutral systems as specified in Section 250-36

See Section 250-36 for the requirements for high-impedance grounded neutral systems.

250-22. Circuits Not to Be Grounded. The following circuits shall not be grounded:

(1) Cranes (circuits for electric cranes operating over combustible fibers in Class III locations, as provided in Section 503-13)
(2) Health care facilities (circuits as provided in Article 517)
(3) Electrolytic cells (circuits as provided in Article 668)

250-24. Grounding Service-Supplied Alternating-Current Systems.

(a) System Grounding Connections. A premises wiring system that is supplied by an ac service that is grounded shall have at each service a grounding electrode conductor connected to the grounding electrode(s) required by Part C of this article. The grounding electrode conductor shall be connected to the grounded service conductor in accordance with (1) through (5).

(1) General. The connection shall be made at any accessible point from the load end of the service drop or service lateral to and including the terminal or bus to which the grounded service conductor is connected at the service disconnecting means.

FPN: See definition of *Service Drop* and *Service Lateral* in Article 100; see also Section 230-21 for overhead supply.

(2) Outdoor Transformer. Where the transformer supplying the service is located outside the building, at least one additional grounding connection shall be made from the grounded service conductor to a grounding electrode, either at the transformer or elsewhere outside the building.

Exception: The additional grounding connection shall not be made on high-impedance grounded neutral systems. The system shall meet the requirements of Section 250-36.

See Figure 250.8 for an illustration of a distribution system transformer connected to a grounding electrode.

(3) Dual Fed Services. For services that are dual fed (double ended) in a common enclosure or grouped together in separate enclosures and employing a secondary tie, a single grounding electrode connection to the tie point of the grounded circuit conductors from each power source shall be permitted.

(4) Main Bonding Jumper as Wire or Busbar. Where the main bonding jumper specified in Section 250-28 is a wire or busbar, and is installed from the neutral bar or bus to the equipment grounding terminal bar or bus in the service equipment, the grounding electrode conductor shall be permitted to be connected to the equipment grounding terminal bar or bus to which the main bonding jumper is connected.

Where a ground-return-type sensor is used for ground-fault protection of equipment (see Article 100 for the definition of *ground-fault protection of equipment*), the sensor is required to be installed on the main bonding jumper, and the grounding electrode conductor is required to be connected to the equipment grounding bus or terminal bar, so that ground-fault current can be accurately sensed. Where the service equipment is power switchgear, it is often more practical to connect the grounding electrode conductor to the grounding bus inside the switchgear. See Figures 230.14, 230.15, and 230.16 for illustrations of ground-fault protection system connections.

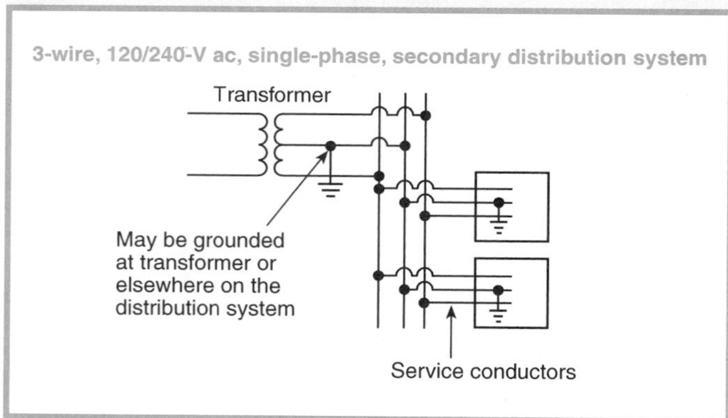

Figure 250.8 *An example of a 3-wire, 120/240-volt ac, single-phase, secondary distribution system where grounding connections are required on the secondary side of the transformer and the supply side of the service disconnecting means.*

(5) Load-Side Grounding Connections. A grounding connection shall not be made to any grounded circuit conductor on the load side of the service disconnecting means except as otherwise permitted in this article.

FPN: See Section 250-30(b) for separately derived systems, Section 250-32 for connections at separate buildings or structures, and Section 250-142 for use of the grounded circuit conductor for grounding equipment.

The power for ac premises wiring systems is either separately derived, in accordance with Section 250-20(d), or supplied by the service. See the definition of *service* in Article 100. Section 250-30 covers grounding requirements for separately derived ac systems. Section 250-24(a) covers system grounding requirements for service-supplied ac systems.

According to Section 250-24, a premises wiring system supplied by an ac service that is required to be grounded shall have a grounding electrode conductor at each service connected to the grounding electrode(s) that meets the requirements in Part C of Article 250. Note that the grounding electrode requirements for a separately derived system are specified in Section 250-30(a)(3).

The grounding electrode conductor connection to the grounded conductor is very specific. The *Code* requires that the connection be made to the grounded service conductor and describes where this connection is permitted. If a transformer is installed outdoors on the load side of the service point and supplies power to a building or structure, a grounding connection must be made at the transformer secondary under the conditions listed in Section 250-20. In addition, the conductor grounded at the transformer is required to be grounded again at the building or structure.

The main rule is that a grounding connection shall not be made to any grounded circuit conductor on the load side of the service disconnecting means. See Section 250-24(a)(5).

(b) Grounded Conductor Brought to Service Equipment. Where an ac system operating at less than 1000 volts is grounded at any point, the grounded conductor(s) shall be run to each service disconnecting means, and shall be bonded to each disconnecting means enclosure. The grounded conductor(s) shall be installed in accordance with (1) through (3).

If the utility service supplying premises wiring is grounded, the grounded conductor, whether used to supply a load or not, must be run to the service equipment, be bonded to the equipment, and be connected

to a grounding electrode system. Figure 250.9 shows an example.

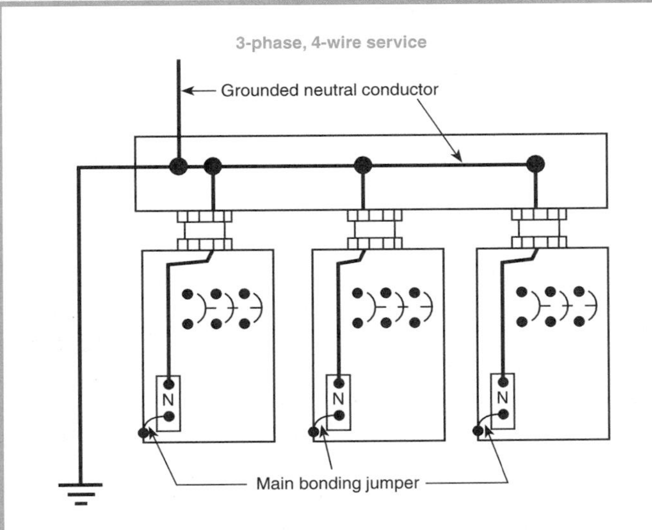

Figure 250.9 *An example of a grounded system where the grounded conductor is brought into each 3-phase, 4-wire service equipment enclosure and to the 3-phase, 3-wire service equipment enclosure, where it is bonded to the enclosure.*

Exception: Where more than one service disconnecting means is located in an assembly listed for use as service equipment, it shall be permitted to run the grounded conductor(s) to the assembly, and the conductor(s) shall be bonded to the assembly enclosure.

The main rule in Section 250-24(b) requires the grounded conductor to be brought in and bonded to each service disconnecting means enclosure. See Figure 250.9 for an example. The exception to Section 250-24(b) permits one bonding connection to a listed service assembly containing more than one service disconnecting means as illustrated in Figure 250.10.

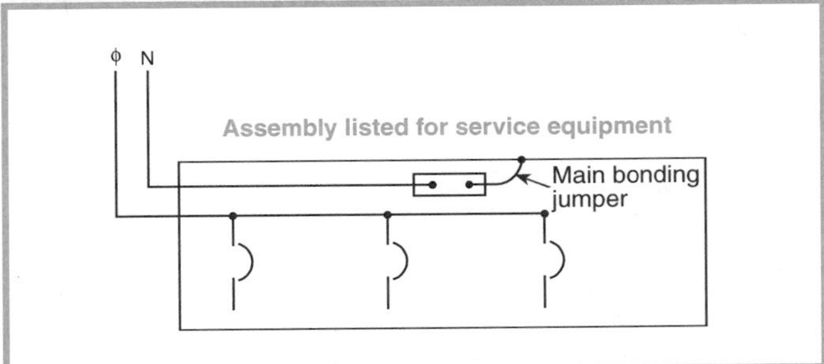

Figure 250.10 *One bonding connection to a listed service assembly containing multiple service disconnecting means, as per Section 250-24(b), Exception.*

(1) Routing. This conductor shall be routed with the phase conductors and shall not be smaller than the required grounding electrode conductor specified in Table 250-66, but shall not be required to be larger than the largest ungrounded service-entrance phase conductor. In addition, for service-entrance phase conductors larger than 1100 kcmil copper or 1750 kcmil aluminum, the grounded conductor shall not be smaller than 12½ percent of the area of the largest service-entrance phase conductor.

(2) Parallel Conductors. Where the service-entrance phase conductors are installed in parallel, the size of the grounded conductor shall be based on the total circular mil area of the parallel conductors as indicated in this section. Where installed in two or more raceways, the size of the grounded conductor in each raceway shall be based on the size of the ungrounded service-entrance conductor in the raceway but not smaller than No. 1/0.

For a multiple raceway or cable service installation, the minimum size for the grounded conductor in each raceway or cable cannot be less than No. 1/0. Although the cumulative size of the parallel grounded conductors may be larger than is required by Section 250-24(b)(1), the minimum 1/0 per raceway or cable correlates with the requirements for parallel conductors contained in Section 310-4.

FPN: See Section 310-4 for grounded conductors connected in parallel.

(3) High Impedance. The grounded conductor on a high-impedance grounded neutral system shall be grounded in accordance with Section 250-36.

(c) Grounding Electrode Conductor. A grounding electrode conductor shall be used to connect the equipment grounding conductors, the service-equipment enclosures, and, where the system is grounded, the grounded service conductor to the grounding electrode(s) required by Part C of this article.

High-impedance grounded neutral system connections shall be made as covered in Section 250-36.

FPN: See Section 250-24(a) for ac system grounding connections.

(d) Ungrounded System Grounding Connections. A premises wiring system that is supplied by an ac service that is ungrounded shall have, at each service, a grounding electrode conductor connected to the grounding electrode(s) required by Part C of this article. The grounding electrode conductor shall be connected to a metal enclosure of the service conductors at any accessible point from the load end of the service drop or service lateral to the service disconnecting means.

250-26. Conductor to Be Grounded—Alternating-Current Systems. For ac premises wiring systems, the conductor to be grounded shall be as specified in the following:

Section 250-26 works in conjunction with Section 250-20(b), which identifies ac systems that are required to be grounded. Section 250-26 identifies which conductor of the systems in Section 250-20(b) must be grounded.

(1) Single-phase, 2-wire — one conductor
(2) Single-phase, 3-wire — the neutral conductor
(3) Multiphase systems having one wire common to all phases — the common conductor
(4) Multiphase systems requiring one grounded phase — one phase conductor
(5) Multiphase systems in which one phase is used as in (2) — the neutral conductor

250-28. Main Bonding Jumper. For a grounded system, an unspliced main bonding jumper shall be used to connect the equipment grounding conductor(s) and the service-disconnect enclosure to the grounded conductor of the system within the enclosure for each service disconnect.

Where the service equipment of a grounded system consists of multiple disconnects, a main bonding jumper for each service disconnect is required to connect the grounded service conductor, the equipment grounding conductor, and the service equipment enclosure. See Figure 250.9.

Exception No. 1: Where more than one service disconnecting means is located in an assembly listed for use as service equipment, an unspliced main bonding jumper shall bond the grounded conductor(s) to the assembly enclosure.

If multiple service disconnects are part of an assembly listed as service equipment, all grounded service conductors are required to be run to and bonded to the assembly. However, only one section is required to have the main bonding jumper connection. See Figure 250.10.

Exception No. 2: Impedance grounded neutral systems shall be permitted to be connected as provided in Sections 250-36 and 250-186.

(a) Material. Main bonding jumpers shall be of copper or other corrosion-resistant material. A main bonding jumper shall be a wire, bus, screw, or similar suitable conductor.

(b) Construction. Where a main bonding jumper is a screw only, the screw shall be identified with a green finish that shall be visible with the screw installed.

The requirement of Section 250-28(b) makes it possible to readily distinguish the main bonding jumper screw for inspection.

(c) Attachment. Main bonding jumpers shall be attached in the manner specified by the applicable provisions of Section 250-8.

(d) Size. The main bonding jumper shall not be smaller than the sizes shown in Table 250-66 for grounding electrode conductors. Where the service-entrance phase conductors are larger than 1100 kcmil copper or 1750 kcmil aluminum, the bonding jumper shall have an area that is not less than 12½ percent of the area of the largest phase conductor except that where the phase conductors and the bonding jumper are of different materials (copper or aluminum), the minimum size of the bonding jumper shall be based on the assumed use of phase conductors of the same material as the bonding jumper and with an ampacity equivalent to that of the installed phase conductors.

The size of the equipment bonding jumper and the main bonding jumper on the *supply* side of a service is based on the size of the supply-phase conductors. Section 250-28(d) uses Table 250-66 for sizing the main and supply side bonding jumpers. The title of this table states that it is used for sizing the grounding electrode conductor, however, it is also used to size the main and supply side bonding jumpers. Unlike the grounding electrode conductor, bonding jumpers may be required to be larger than a No. 3/0 copper or 250-kcmil aluminum conductor. Where the size of the phase conductors exceeds 1100 kcmil copper or 1750 kcmil aluminum, the bonding jumpers cannot be less than 12½ percent of the cross-sectional area of the phase conductors.

To apply the bonding jumper requirements, each switch is treated as separate service equipment. Figure 250.11 provides an example.

Example

In applying the bonding requirements to Figure 250.11, the grounded conductor is sized for the maximum unbalance that can occur, but is not smaller than allowed by Section 250-24(b)(1), which has similar requirements to this section.

Assuming a computed load of 450 amperes for the main service, the service-entrance conductors are sized at 750 kcmil copper. The bonding jumpers for the metal service conduit and trough are based on the size of the main service-entrance conductors and cannot be less than No. 2/0 copper, based on Table 250-66.

The service-entrance conductors to the enclosures are sized No. 3/0 and 500 kcmil copper, based on their loads. The bonding jumpers for the disconnects and short nipples are sized based on the size of the phase conductors supplying each disconnect. In this case, No. 3/0 and 500 kcmil require No. 4 and No. 1/0 copper bonding jumpers, respectively.

In some instances, the bonding jumper may be required to be larger than the grounding electrode conductor. Section 250-28(d) indicates that where the

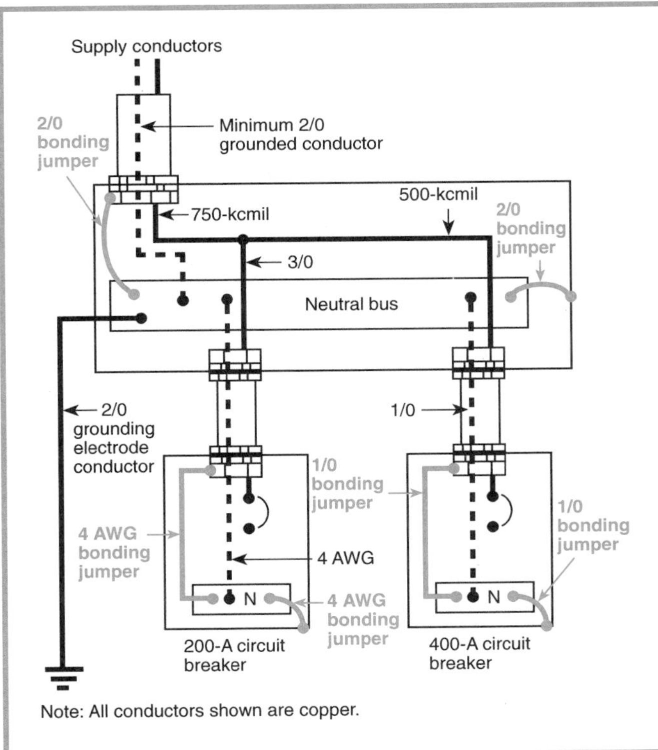

Figure 250.11 *An example of the bonding requirements for service equipment.*

service-entrance conductors are larger than 1100 kcmil copper or 1750 kcmil aluminum, the bonding jumper is required to have a cross-sectional area of not less than 12½ percent of the cross-sectional area of the largest phase conductor. For example, if a service is supplied by four 500-kcmil conductors in parallel for each phase, the minimum cross-sectional area of the bonding jumper is calculated as follows: 4 × 500 kcmil = 2000 kcmil. Therefore, the bonding jumper cannot be less than 12½ percent of 2000 kcmil, which results in a 250-kcmil copper conductor.

250-30. Grounding Separately Derived Alternating-Current Systems.

(a) Grounded Systems. A separately derived ac system that is grounded shall comply with (1) through (4).

Section 250-30(a) provides the requirements for bonding and grounding the separately derived systems described in Section 250-20(d).

A *separately derived system* is a premises wiring system whose power is derived from a battery, a solar photovoltaic system, or from a generator, transformer, or converter windings, and that has no direct electrical connection, including a solidly connected grounded circuit conductor, to supply conductors originating in another system.

The requirements of Section 250-30 are most commonly applied to 480Y/277-volt transformers that transform a 480-volt supply to a 208Y/120-volt system for lighting and appliance loads.

These requirements provide for a low-impedance path to ground so that line-to-ground faults on circuits supplied by the transformer result in sufficient current flow to operate the overcurrent devices. These requirements also apply to generators or systems derived from converter windings, although these systems do not have the same wide use as separately derived systems that are derived from transformers.

Exception: High-impedance grounded neutral system grounding connection requirements shall not be required to comply with (1) and (2) and shall be made as specified in Sections 250-36 and 250-186.

(1) Bonding Jumper. A bonding jumper in compliance with Sections 250-28(a) through (d), that is sized for the derived phase conductors, shall be used to connect the equipment grounding conductors of the separately derived system to the grounded conductor. Except as permitted by Section 250-24(a)(4), this connection shall be made at any point on the separately derived system from the source to the first system disconnecting means or overcurrent device, or it shall be made at the source of a separately derived system that has no disconnecting means or overcurrent devices. The point of connection shall be the same as the grounding electrode conductor as required in Section 250-30(a)(2).

Where a separately derived system provides a grounded conductor, a bonding jumper must be installed to connect the equipment grounding conductors to the grounded conductor. They are connected to the grounding electrode system by the grounding electrode conductor. The bonding jumper is sized according to Section 250-28(d) and may be located at any point between the source terminals (transformer, generator, etc.) and the first disconnecting means or overcurrent device.

Exception No. 1: A bonding jumper at both the source and the first disconnecting means shall be permitted where doing so does not establish a parallel path for the grounded circuit conductor. Where a grounded conductor is used in this manner, it shall not be smaller than the size specified for the bonding jumper but shall not be required to be larger than the ungrounded conductor(s). For the purposes of this exception, connection through the earth is not considered as providing a parallel path.

Exception No. 2: The size of the bonding jumper for a system that supplies a Class 1, Class 2, or Class 3 circuit, and is derived from a transformer rated not more than 1000 volt-

amperes, shall not be smaller than the derived phase conductors and shall not be smaller than No. 14 copper or No. 12 aluminum.

Section 250-30(a)(1) requires the bonding jumper to be not smaller than the sizes given in Table 250-66, that is, not smaller than No. 8 copper. This exception permits a bonding jumper for a Class 1, Class 2, or Class 3 circuit to be not smaller than No. 14 copper or No. 12 aluminum.

(2) Grounding Electrode Conductor. A grounding electrode conductor, sized in accordance with Section 250-66 for the derived phase conductors, shall be used to connect the grounded conductor of the derived system to the grounding electrode as specified in (3). Except as permitted by Sections 250-24(a)(3) or (a)(4), this connection shall be made at the same point on the separately derived system where the bonding jumper is installed.

Where a separately derived system is required to be grounded, the conductor to be grounded is allowed to be connected to the grounding electrode system at any location between the source terminals (transformer, generator, etc.) and the first disconnecting means or overcurrent device. The 1999 *Code* requires that the location of the grounding electrode conductor connection to the grounded conductor be at the same point as where the bonding jumper is connected to the grounded conductor. By establishing a common point of connection, normal neutral current will be carried only on the system grounded conductor and metal raceways. Piping systems and structural steel will not provide a parallel circuit. Figures 250.12A and 250.12B illustrate examples of grounding electrode connections for separately derived systems.

Exception: A grounding electrode conductor shall not be required for a system that supplies a Class 1, Class 2, or Class 3 circuit and is derived from a transformer rated not more than 1000 volt-amperes, provided the system grounded conductor is bonded to the transformer frame or enclosure by a jumper sized in accordance with Section 250-30(a)(1), Exception No. 2, and the transformer frame or enclosure is grounded by one of the means specified in Section 250-134.

(3) Grounding Electrode. The grounding electrode shall be as near as practicable to and preferably in the same area as the grounding electrode conductor connection to the system. The grounding electrode shall be the nearest one of the following:

(a) An effectively grounded structural metal member of the structure
(b) An effectively grounded metal water pipe within 5 ft (1.52 m) from the point of entrance into the building

Exception: In industrial and commercial buildings where conditions of maintenance and supervision ensure that only

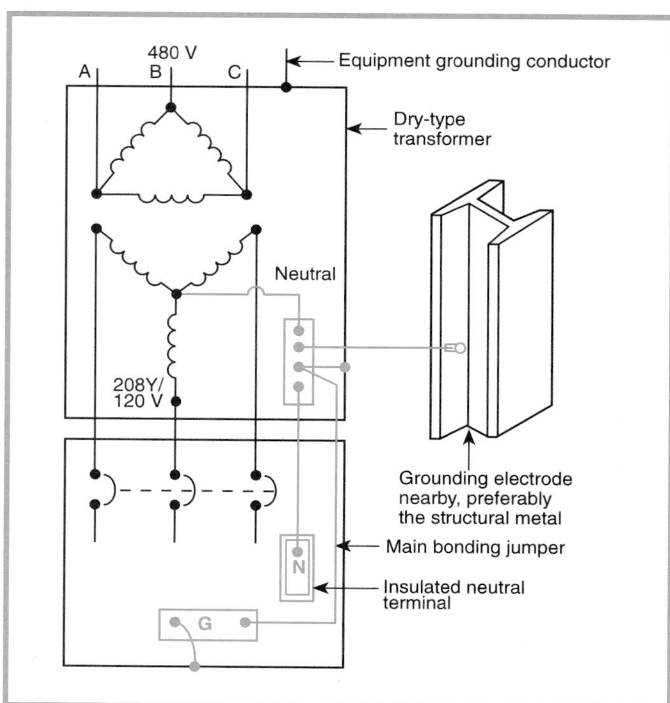

Figure 250.12A *A grounding arrangement for a separately derived system where the grounding electrode conductor connection is made at the transformer.*

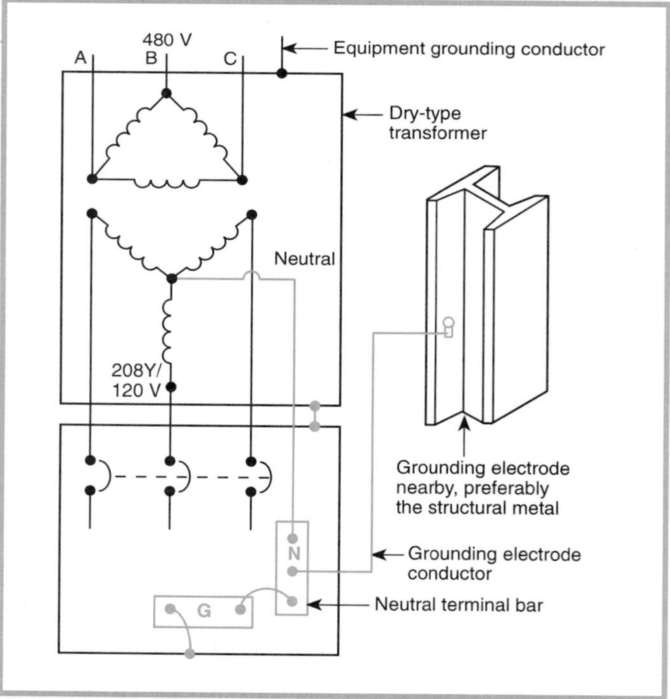

Figure 250.12B *A grounding arrangement for a separately derived system where the grounding electrode conductor connection is made at the first disconnect.*

qualified persons will service the installation and the entire length of the interior metal water pipe that is being used for the grounding electrode is exposed, the connection shall be permitted at any point on the water pipe system.

(c) Other electrodes as specified in Sections 250-50 and 250-52 where the electrodes specified by (a) or (b) are not available.

Exception for (a), (b), and (c): Where a separately derived system originates in listed equipment also used as service equipment, the grounding electrode used for the service equipment shall be permitted as the grounding electrode for the separately derived system provided the grounding electrode conductor from the service equipment to the grounding electrode is of sufficient size for the separately derived system. Where the equipment ground bus internal to the service equipment is not smaller than the required grounding electrode conductor, the grounding electrode connection for the separately derived system shall be permitted to be made to the bus.

FPN: See Section 250-104(a)(4) for bonding requirements of interior metal water piping in the area served by separately derived systems.

Section 250-30(a)(3) requires that the grounding electrode be as near as practicable to the grounding conductor connection to the system to minimize the impedance to ground. If an effectively grounded structural metal member of the building structure or an effectively grounded metal water pipe is available nearby, Section 250-30(a)(3) requires that this grounding electrode be used. For example, where a transformer is installed on the fiftieth floor, the grounding electrode conductor is not required to be run to the service grounding electrode system. However, where an effectively grounded metal water pipe is used as an electrode for a separately derived system, the 1999 *Code* requires that the grounding electrode conductor connection be made within 5 ft of where the piping system enters the building. Concern over the use of nonmetallic piping or fittings is the basis for this new requirement. Where the piping system is located in an industrial or commercial building that is serviced only by qualified persons and the entire length that will be used as an electrode is exposed, the connection is permitted to be made at any point on the piping system.

The practice of grounding the secondary of an isolating transformer to a ground rod or running the grounding electrode conductor back to the service ground (usually to reduce electrical noise on data processing systems) is prohibited if an item in (a) or (b) of Section 250-30(a)(3) is available. However, an isolation transformer that is part of a listed power supply for a data processing room is not required to be grounded in accordance with Section 250-30(a)(3) but must be grounded in accordance with the manufacturer's instructions. See Figures 250.12A and 250.12B.

(4) Grounding Methods. In all other respects, grounding methods shall comply with requirements prescribed in other parts of this *Code*.

Figures 250.12A and 250.12B are typical wiring diagrams for dry-type transformers supplied from a 480-volt, 3-phase feeder to derive a 208Y/120-volt or 480Y/277-volt secondary. As indicated in Section 250-30(a)(1), the bonding jumper connection is required to be sized according to Section 250-28(d). In Figure 250.12A, this connection is made at the source of the separately derived system, in the transformer enclosure. In Figure 250.12B, the bonding jumper connection is made at the disconnect. With the grounding electrode conductor, the bonding jumper, and the bonding of the grounded circuit conductor (neutral) as shown, line-to-ground fault currents are able to return to the supply source rather than through a long, high-impedance path. A path of lower impedance is provided that facilitates the operation of overcurrent devices, in accordance with Section 250-2(d). The grounding electrode conductor from the secondary grounded circuit conductor is sized according to Section 250-66.

(b) Ungrounded Systems. The equipment of an ungrounded separately derived system shall be grounded as specified in (1) through (3).

(1) Grounding Electrode Conductor. A grounding electrode conductor, sized in accordance with Section 250-66 for the derived phase conductors, shall be used to connect the metal enclosures of the derived system to the grounding electrode as specified in (2). This connection shall be made at any point on the separately derived system from the source to the first system disconnecting means.

In a separately derived system that is ungrounded, a bonding jumper must be installed to connect the disconnect enclosure and equipment grounding conductors to the grounding electrode system.

(2) Grounding Electrode. Except as permitted by Section 250-34 for portable and vehicle mounted generators, the grounding electrode shall comply with Section 250-30(a)(3).

(3) Grounding Methods. In all other respects, grounding methods shall comply with requirements prescribed in other parts of this *Code*.

250-32. Two or More Buildings or Structures Supplied from a Common Service.

(a) Grounding Electrode. Where two or more buildings or structures are supplied from a common ac service by a feeder(s) or branch circuit(s), the grounding electrode(s) required in Part C of this article at each building or structure shall be connected in the manner specified in (b) or (c). Where there are no existing grounding electrodes, the grounding electrode(s) required in Part C of this article shall be installed.

Exception: A grounding electrode at separate buildings or structures shall not be required where only one branch circuit supplies the building or structure and the branch circuit includes an equipment grounding conductor for grounding the noncurrent-carrying parts of all equipment.

If a building is supplied by only one branch circuit with an equipment grounding conductor grounding equipment, it is not required to establish a grounding electrode system or connect to one if one exists. See Figure 250.13 for an example of such a system.

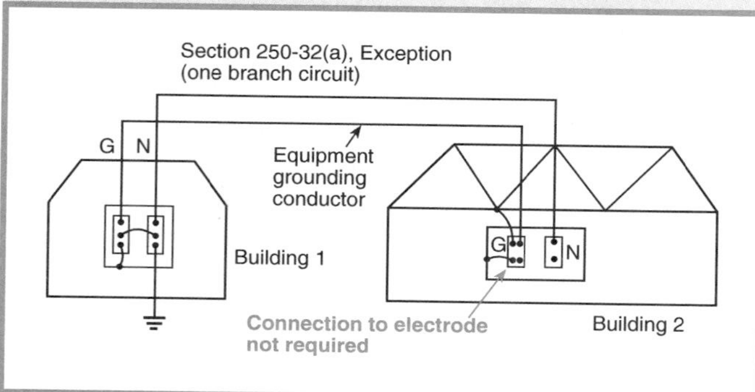

Figure 250.13 An installation where connection from the enclosure of the building disconnecting means to the electrode is not required since an equipment grounding conductor is run with the circuit conductors.

(b) Grounded Systems. For a grounded system at the separate building or structure, the connection to the grounding electrode and grounding or bonding of equipment, structures, or frames required to be grounded or bonded shall comply with either (1) or (2).

(1) Equipment Grounding Conductor. An equipment grounding conductor as described in Section 250-118 shall be run with the supply conductors and connected to the building or structure disconnecting means and to the grounding electrode(s). The equipment grounding conductor shall be used for grounding or bonding of equipment, structures, or frames required to be grounded or bonded. Any installed grounded conductor shall not be connected to the equipment grounding conductor or to the grounding electrode(s).

If a single service supplies more than one building, such as illustrated in Figure 250.15, and the feeder is installed with an equipment grounding conductor, Section 250-32(a) requires that a grounding electrode system be established, unless one already exists. The equipment grounding bus must be bonded to the grounding electrode system. The disconnecting means, building steel, and interior metal water piping must be bonded to the grounding electrode system. All noncurrent-carrying metal parts of electrical equipment are required to be grounded by connection to the equipment grounding bus.

If a feeder supplies another building from the same service and an equipment grounding conductor is run with the feeder, the grounded conductor (neutral) is not permitted to be connected to the equipment grounding conductor or the grounding electrode system as illustrated in Figure 250.15.

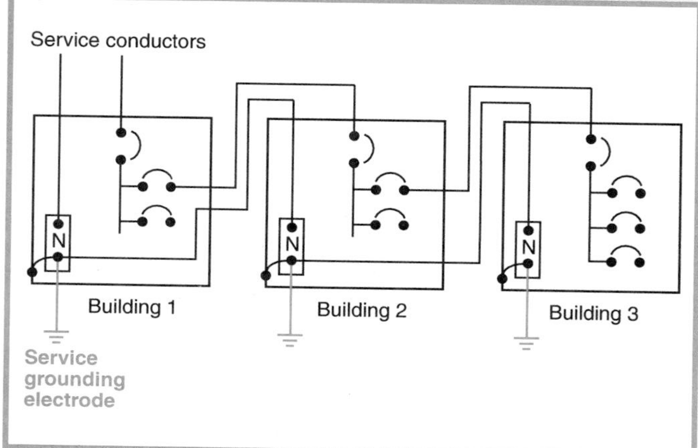

Figure 250.14 A single service (grounded system) supplying three buildings. Provided there are no parallel paths for neutral current, buildings 2 and 3 are permitted to have the grounding electrode conductor connected to the grounded conductor per Section 250-32(b)(2).

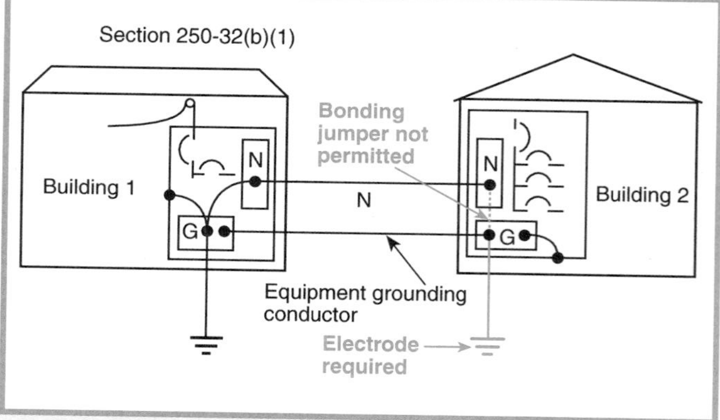

Figure 250.15 An installation where a connection between the gounded conductor (N) and the equipment grounding conductor (G) is not permitted at Building 2.

(2) Grounded Conductor. Where (1) an equipment grounding conductor is not run with the supply to the building or structure, and (2) there are no continuous metallic paths bonded to the grounding system in both buildings or structures involved, and (3) ground-fault protection of equipment has not been installed on the common ac service, the grounded circuit conductor run with the supply to the building or structure shall be connected to the building or structure disconnecting means and to the grounding electrode(s) and shall be used for grounding or bonding of equipment, structures, or frames required to be grounded or bonded.

Similar to the provisions of Section 250-30(a)(2), Section 250-32(b)(2) also was revised to eliminate the creation of parallel paths for normal neutral current on grounding conductors, metal raceways, metal piping, and other metal structures. In previous editions of the *Code,* connecting the grounding electrode conductor and equipment grounding conductors to the grounded conductor at a separate building or structure was permitted. This multiple location grounding arrangement could provide parallel paths for neutral current along the electrical system and also other continuous metallic piping and mechanical systems. Connection of the grounded conductor to a grounding electrode system at a separate building or structure is only permitted where these parallel paths are not created and where there is no common GFPE protection provided at the service where the feeder or branch circuit originates. See Figure 250.14.

(c) Ungrounded Systems. The grounding electrode(s) shall be connected to the building or structure disconnecting means.

(d) Disconnecting Means Located in Separate Building or Structure on the Same Premises. Where one or more disconnecting means supply one or more additional buildings or structures under single management, and where these disconnecting means are located remote from those buildings or structures in accordance with the provisions of Section 225-32, Exception Nos. 1 and 2, all of the following conditions shall be met.

(1) The connection of the grounded circuit conductor to the grounding electrode at a separate building or structure shall not be made.
(2) An equipment grounding conductor for grounding any noncurrent-carrying equipment, interior metal piping systems, and building or structural metal frames is run with the circuit conductors to a separate building or structure and bonded to existing grounding electrode(s) required in Part C of this article, or, where there are no existing electrodes, the grounding electrode(s) required in Part C of this article shall be installed where a separate building or structure is supplied by more than one branch circuit.
(3) Bonding the equipment grounding conductor to the grounding electrode at a separate building or structure shall be made in a junction box, panelboard, or similar enclosure located immediately inside or outside the separate building or structure.

Figure 250.16 illustrates an installation where the disconnect for Building 2 is located in Building 1. The rules are applicable to separate buildings or structures that do not have a disconnect, as permitted by Exception No. 1 to Section 225-32. The feeder conductors must terminate in a panelboard, junction box, or similar enclosure that is located immediately inside or outside the building.

Note that an equipment grounding conductor (see Section 250-118) must be run with the feeder conductors, the grounded conductor (neutral) is not bonded to the enclosure or equipment grounding bus, and the equipment grounding bus is connected to a new or existing grounding electrode system at the second building. All noncurrent-carrying metal parts of equipment, building steel, and interior metal piping systems must be connected to the grounding electrode system.

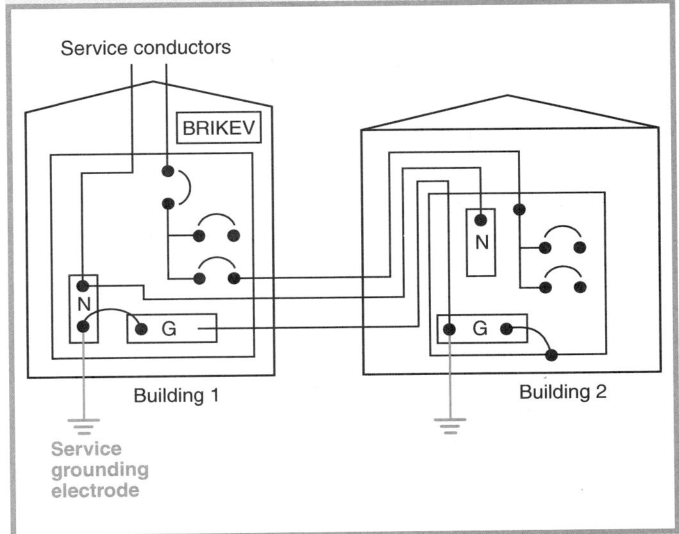

Figure 250.16 *Grounding and bonding requirements for a separate building under single management with the disconnect remotely located from the building.*

(e) Agricultural Buildings or Structures. Where livestock is housed, any portion of the equipment grounding conductor run underground to the building or structure disconnecting means shall be insulated or covered copper.

FPN: See Section 547-8 for special grounding requirements for agricultural buildings.

(f) Grounding Conductor. The size of the grounding conductor to the grounding electrode(s) shall not be less than given in Table 250-122, and shall not be required to be larger than the largest ungrounded supply conductor. The installation shall comply with Part C of this article.

Section 250-32(f) recognizes the difference between a grounding electrode conductor, sized in accordance with Table 250-66, and an equipment grounding conductor connected to a grounding electrode, sized in accordance with Table 250-122. The equipment grounding conductor must be sized based on the rating of the overcurrent protective device protecting the feeder to the building. If the feeder conductor size is increased for voltage drop consideration, the equipment grounding conductor run with the feeder must be increased proportionately.

250-34. Portable and Vehicle-Mounted Generators.

(a) Portable Generators. The frame of a portable generator shall not be required to be grounded and shall be permitted to serve as the grounding electrode for a system supplied by the generator under the following conditions:

(1) The generator supplies only equipment mounted on the generator or cord- and plug-connected equipment through receptacles mounted on the generator, or both, and
(2) The noncurrent-carrying metal parts of equipment and the equipment grounding conductor terminals of the receptacles are bonded to the generator frame.

Portable describes equipment that is easily carried from one location to another. *Mobile* describes equipment that is capable of being moved, as on wheels or rollers, such as vehicle-mounted or on a trailer.

The frame of a portable generator is not required to be connected to ground (ground rod, water pipe, etc.) if the generator has receptacles mounted on the generator panel and the receptacles have equipment grounding terminals bonded to the generator frame.

(b) Vehicle-Mounted Generators. The frame of a vehicle shall be permitted to serve as the grounding electrode for a system supplied by a generator located on the vehicle under the following conditions:

(1) The frame of the generator is bonded to the vehicle frame, and
(2) The generator supplies only equipment located on the vehicle or cord- and plug-connected equipment through receptacles mounted on the vehicle, or both equipment located on the vehicle and cord- and plug-connected equipment through receptacles mounted on the vehicle or on the generator, and
(3) The noncurrent-carrying metal parts of equipment and the equipment grounding conductor terminals of the receptacles are bonded to the generator frame, and
(4) The system complies with all other provisions of this article.

Vehicle-mounted generators that provide a neutral conductor and are installed as separately derived systems supplying equipment and receptacles on the vehicle are required to have the neutral conductor bonded to the generator frame and vehicle frame. The noncurrent-carrying parts of the equipment are required to be bonded to the generator frame.

(c) Grounded Conductor Bonding. A system conductor that is required to be grounded by Section 250-26 shall be bonded to the generator frame where the generator is a component of a separately derived system.

Portable and vehicle-mounted generators installed as separately derived systems and that provide a neutral conductor, such as 3-phase, 4-wire wye or single-phase, 240/120-volt or 3-phase, 4-wire delta-connected, are required to have the neutral conductor bonded to the generator frame.

FPN: For grounding portable generators supplying fixed wiring systems, see Section 250-20(d).

250-36. High-Impedance Grounded Neutral Systems. High-impedance grounded neutral systems in which a grounding impedance, usually a resistor, limits the ground-fault current to a low value shall be permitted for 3-phase ac systems of 480 volts to 1000 volts where all of the following conditions are met.

(1) The conditions of maintenance and supervision ensure that only qualified persons will service the installation.
(2) Continuity of power is required.
(3) Ground detectors are installed on the system.
(4) Line-to-neutral loads are not served.

Section 250-36 covers high-impedance grounded neutral systems of 50 to 1000 volts. Systems rated over 1000 volts are covered in Section 250-186. For information on the differences between solidly grounded systems and high-impedance grounded neutral systems, see "Grounding for Emergency and Standby Power Systems," by Robert B. West, *IEEE Transactions on Industry Applications,* Vol. IA-15, No. 2, March/April 1979.

As the schematic diagram of Figure 250.17 shows, a high-impedance grounded neutral system is designed to minimize the amount of fault current that

can flow during a ground fault. The grounding impedance is usually selected to limit fault current to a value that is slightly greater than or equal to the capacitive discharge current. This system is used where continuity of power is required. Therefore, a ground fault results in an alarm condition rather than in the tripping of a circuit breaker. This allows for a safe and orderly shutdown.

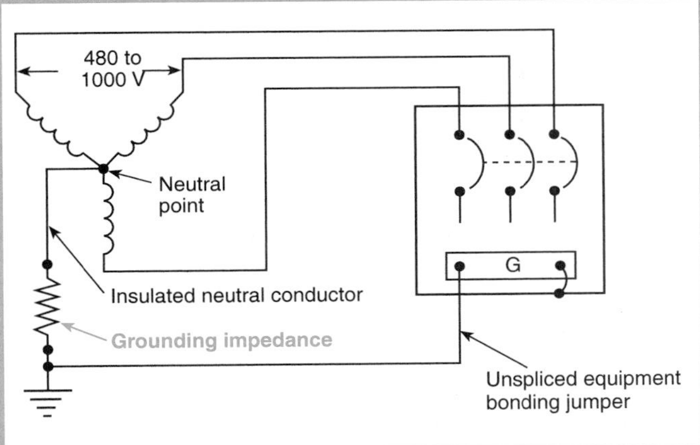

Figure 250.17 Schematic diagram of a high-impedance grounded neutral system.

High-impedance grounded neutral systems shall comply with the provisions of (a) through (f).

(a) Grounding Impedance Location. The grounding impedance shall be installed between the grounding electrode conductor and the system neutral. Where a neutral is not available, the grounding impedance shall be installed between the grounding electrode conductor and the neutral derived from a grounding transformer.

(b) Neutral Conductor. The neutral conductor from the neutral point of the transformer or generator to its connection point to the grounding impedance shall be fully insulated.

The neutral conductor shall have an ampacity of not less than the maximum current rating of the grounding impedance. In no case shall the neutral conductor be smaller than No. 8 copper or No. 6 aluminum or copper-clad aluminum.

The current through the neutral conductor is limited by the grounding impedance. Therefore, the neutral conductor is not required to be sized to carry high fault current. The neutral conductor cannot be smaller than No. 8 copper or No. 6 aluminum.

(c) System Neutral Connection. The system neutral conductor shall not be connected to ground except through the grounding impedance.

FPN: The impedance is normally selected to limit the ground-fault current to a value slightly greater than or equal to the capacitive charging current of the system. This value of impedance will also limit transient overvoltages to safe values. For guidance, refer to criteria for limiting transient overvoltages in *Recommended Practice for Grounding of Industrial and Commercial Power Systems*, ANSI/IEEE 142-1991.

Additional information may be found in "Charging Current Data for Guesswork-Free Design of High-Resistance Grounded Systems," by D.S. Baker, *IEEE Transactions on Industry Applications*, Vol. IA-15, No. 2, March/April 1979; and "High-Resistance Grounding," by Baldwin Bridger, Jr., *IEEE Transactions on Industry Applications*, Vol. IA-19, No. 1, January/February 1983.

(d) Neutral Conductor Routing. The conductor connecting the neutral point of the transformer or generator to the grounding impedance shall be permitted to be installed in a separate raceway. It shall not be required to run this conductor with the phase conductors to the first system disconnecting means or overcurrent device.

(e) Equipment Bonding Jumper. The equipment bonding jumper (the connection between the equipment grounding conductors and the grounding impedance) shall be an unspliced conductor run from the first system disconnecting means or overcurrent device to the grounded side of the grounding impedance.

(f) Grounding Electrode Conductor Location. The grounding electrode conductor shall be attached at any point from the grounded side of the grounding impedance to the equipment grounding connection at the service equipment or first system disconnecting means.

C. Grounding Electrode System and Grounding Electrode Conductor

250-50. Grounding Electrode System. If available on the premises at each building or structure served, each item (a) through (d), and any made electrodes in accordance with Sections 250-52(c) and (d), shall be bonded together to form the grounding electrode system. The bonding jumper(s) shall be installed in accordance with Sections 250-64(a), (b), and (e), shall be sized in accordance with Section 250-66, and shall be connected in the manner specified in Section 250-70.

An unspliced grounding electrode conductor shall be permitted to be run to any convenient grounding electrode available in the grounding electrode system or to one or more grounding electrode(s) individually. It shall be sized for the largest grounding electrode conductor required among all the electrodes connected to it.

Figure 250.18 shows an example of a grounding electrode system. Note the location of the grounding electrode conductor.

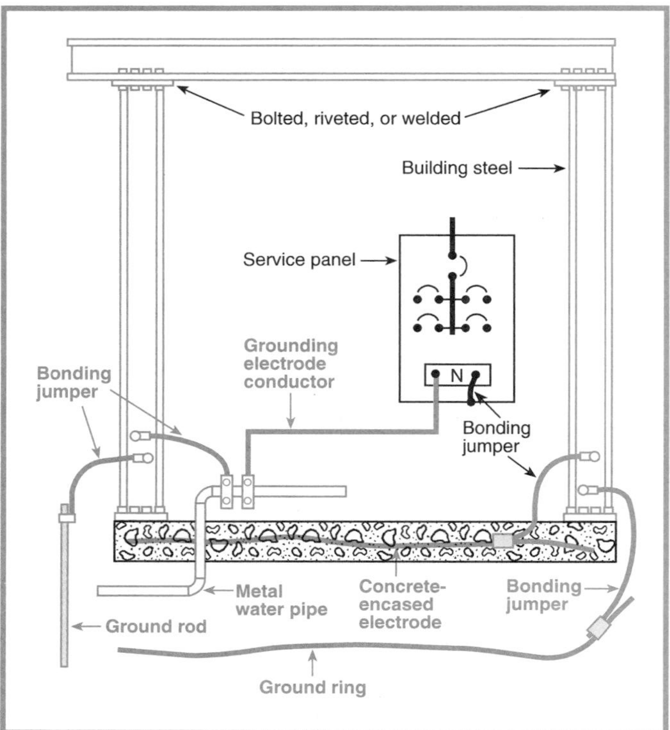

Figure 250.18 An example of bonding electrodes together to form the grounding electrode system.

The grounding electrode conductor shall be unspliced or spliced by means of irreversible compression-type connectors listed for the purpose or by the exothermic welding process.

Interior metal water piping located more than 5 ft (1.52 m) from the point of entrance to the building shall not be used as a part of the grounding electrode system or as a conductor to interconnect electrodes that are part of the grounding electrode system.

Exception: In industrial and commercial buildings where conditions of maintenance and supervision ensure that only qualified persons will service the installation and the entire length of the interior metal water pipe that is being used for the conductor is exposed.

Section 250-50 prohibits the use of that portion of the interior metal water piping system that extends more than 5 ft beyond the point of entrance into the building to interconnect grounding electrodes and the grounding electrode conductor. The exception, however, permits this practice, provided there is qualified maintenance and the entire length of the pipe is exposed. See the commentary following Section 250-130(c) for further explanation.

FPN: See Sections 547-8 and 547-9 for special grounding and bonding requirements for agricultural buildings.

(a) Metal Underground Water Pipe. A metal underground water pipe in direct contact with the earth for 10 ft (3.05 m) or more (including any metal well casing effectively bonded to the pipe) and electrically continuous (or made electrically continuous by bonding around insulating joints or sections or insulating pipe) to the points of connection of the grounding electrode conductor and the bonding conductors.

There has always been uncertainty as to whether metal water piping systems should be used as grounding electrodes, so a number of years ago the electrical industry and the waterworks industry formed a committee of all interested parties to evaluate the use of metal underground water piping systems as grounding electrodes. Based on their findings, the committee issued an authoritative report on this subject. The International Association of Electrical Inspectors published the report, *Interim Report of the American Research Committee on Grounding,* in January 1944 (reprinted March 1949).

The National Institute of Standards and Technology has monitored the electrolysis of metal systems, because a flow of current at a grounding electrode on dc systems can cause displacement of metal. The results of this monitoring have shown that problems are minimal.

(1) Continuity. Continuity of the grounding path or the bonding connection to interior piping shall not rely on water meters or filtering devices and similar equipment.

(2) Supplemental Electrode Required. A metal underground water pipe shall be supplemented by an additional electrode of a type specified in Sections 250-50 or 250-52. Where the supplemental electrode is a made electrode of the rod, pipe, or plate type, it shall comply with Section 250-56. The supplemental electrode shall be permitted to be bonded to the grounding electrode conductor, the grounded service-entrance conductor, the nonflexible grounded service raceway, or any grounded service enclosure.

Section 250-50(a)(2) specifically requires that rod, pipe, and plate electrodes used to supplement metal water piping be installed in accordance with Section 250-56. This clarifies that the supplemental electrode system must be installed as if it were the sole grounding electrode for the system. Where 25 ohms earth resistance cannot be achieved with one rod, pipe, or plate, another electrode (other than the metal piping that is being supplemented) shall be provided. One of the permitted methods of bonding a supplemental grounding electrode conductor to the primary electrode system is to connect it to the service enclosure.

Exception: The supplemental electrode shall be permitted to be bonded to the interior metal water piping at any convenient point as covered in Section 250-50, Exception.

Where the supplemental electrode is a made electrode as in Section 250-52(c) or (d), that portion of the bonding jumper that is the sole connection to the supplemental grounding electrode shall not be required to be larger than No. 6 copper wire or No. 4 aluminum wire.

This paragraph correlates with Section 250-52(c) or (d) and Section 250-66(a). For example, if a "made electrode" is used, the size of the "sole" bonding jumper between the service equipment grounded conductor and a ground rod is not required to be larger than No. 6 copper or No. 4 aluminum.

The requirement to supplement the metal water pipe is based on the practice of using a plastic pipe for replacement when the original metal water pipe fails. This leaves the system without a grounding electrode unless a supplementary electrode is provided.

(b) Metal Frame of the Building or Structure. The metal frame of the building or structure, where effectively grounded.

(c) Concrete-Encased Electrode. An electrode encased by at least 2 in. (50.8 mm) of concrete, located within and near the bottom of a concrete foundation or footing that is in direct contact with the earth, consisting of at least 20 ft (6.1 m) of one or more bare or zinc galvanized or other electrically conductive coated steel reinforcing bars or rods of not less than ½-in. (12.7-mm) diameter, or consisting of at least 20 ft (6.1 m) of bare copper conductor not smaller than No. 4. Reinforcing bars shall be permitted to be bonded together by the usual steel tie wires or other effective means.

Figure 250.19 provides an example showing a concrete-encased electrode.

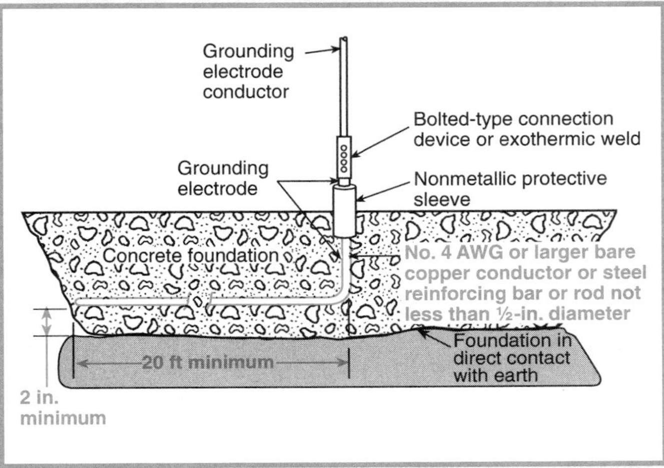

Figure 250.19 *A concrete-encased electrode.*

(d) Ground Ring. A ground ring encircling the building or structure, in direct contact with the earth at a depth below the earth's surface of not less than 2½ ft (762 mm), consisting of at least 20 ft (6.1 m) of bare copper conductor not smaller than No. 2.

Section 250-50(d) requires the metal underground water pipe, the metal frame of the building, a concrete-encased electrode, and a ground-ring type of electrode to be bonded together only where they are available to form the grounding electrode system, as shown in Figure 250.20. If a metal underground water pipe is the only grounding electrode available, it must be supplemented by one of the grounding electrodes specified in Sections 250-50 or 250-52.

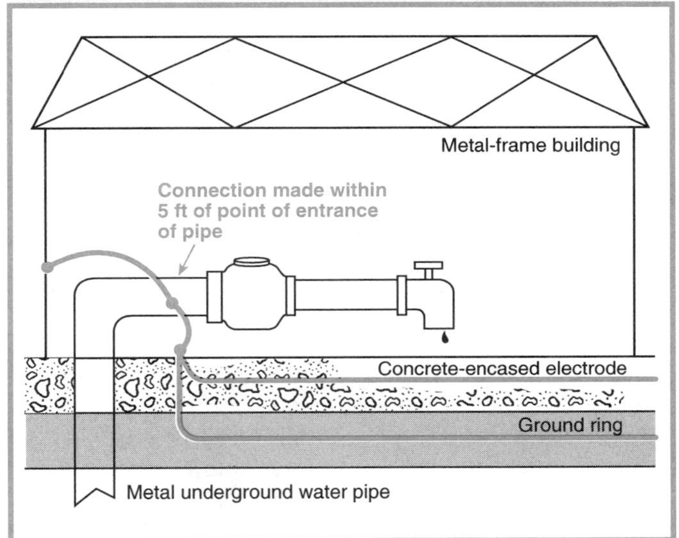

Figure 250.20 *A grounding electrode system.*

250-52. Made and Other Electrodes. Where none of the electrodes specified in Section 250-50 is available, one or more of the electrodes specified in (b) through (d) shall be used. Where practicable, made electrodes shall be embedded below permanent moisture level. Made electrodes shall be free from nonconductive coatings such as paint or enamel. Where more than one electrode is used, each electrode of one grounding system (including that used for air terminals) shall not be less than 6 ft (1.83 m) from any other electrode of another grounding system. Two or more grounding electrodes that are effectively bonded together shall be considered a single grounding electrode system.

(a) Metal Underground Gas Piping System. A metal underground gas piping system shall not be used as a grounding electrode.

(b) Other Local Metal Underground Systems or Structures. Other local metal underground systems or structures such as piping systems and underground tanks.

(c) Rod and Pipe Electrodes. Rod and pipe electrodes shall not be less than 8 ft (2.44 m) in length, shall consist of the following materials, and shall be installed in the following manner.

(1) Electrodes of pipe or conduit shall not be smaller than ¾ in. trade size and, where of iron or steel, shall have the outer surface galvanized or otherwise metal-coated for corrosion protection.

(2) Electrodes of rods of iron or steel shall be at least ⅝ in. (15.87 mm) in diameter. Stainless steel rods less than ⅝ in. (15.87 mm) in diameter, nonferrous rods, or their equivalent shall be listed and shall not be less than ½ in. (12.7 mm) in diameter.

(3) The electrode shall be installed such that at least 8 ft (2.44 m) of length is in contact with the soil. It shall be driven to a depth of not less than 8 ft (2.44 m) except that, where rock bottom is encountered, the electrode shall be driven at an oblique angle not to exceed 45 degrees from the vertical or shall be buried in a trench that is at least 2½ ft (762 mm) deep. The upper end of the electrode shall be flush with or below ground level unless the aboveground end and the grounding electrode conductor attachment are protected against physical damage as specified in Section 250-10.

All pipe and rod electrodes must have 8 ft of length in contact with the soil, regardless of rock bottom. Where rock bottom is encountered, the electrodes must either be driven at not more than a 45 degree angle or be buried in a 2½-ft deep trench. Ground clamps used on buried electrodes must be listed for direct earth burial. Ground clamps installed above ground must be protected where subject to physical damage. Figure 250.21 illustrates these requirements.

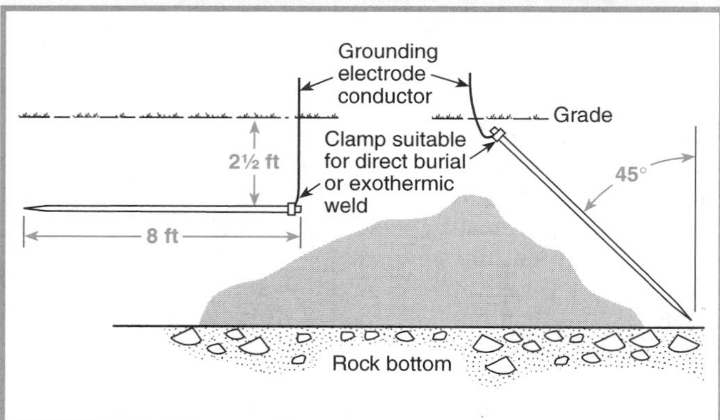

Figure 250.21 The installation requirements of rod and pipe electrodes as specified by Section 250-52(c)(3).

(d) Plate Electrodes. Each plate electrode shall expose not less than 2 ft² (0.186 m²) of surface to exterior soil. Electrodes of iron or steel plates shall be at least ¼ in. (6.35 mm) in thickness. Electrodes of nonferrous metal shall be at least 0.06 in. (1.52 mm) in thickness. Plate electrodes shall be installed not less than 2½ ft (762 mm) below the surface of the earth.

(e) Aluminum Electrodes. Aluminum electrodes shall not be permitted.

250-54. Supplementary Grounding Electrodes. Supplementary grounding electrodes shall be permitted to be connected to the equipment grounding conductors specified in Section 250-118, but the earth shall not be used as the sole equipment grounding conductor.

Grounding electrodes, such as ground rods, connected to equipment are not permitted to be used in lieu of the equipment grounding conductor but may be used for supplementary protection. For example, they may be used for lightning protection or to equalize potentials in the area of the equipment. The second paragraph of Section 250-2(d) also specifies that the earth must not be used as the sole equipment grounding conductor.

250-56. Resistance of Made Electrodes. A single electrode consisting of a rod, pipe, or plate that does not have a resistance to ground of 25 ohms or less shall be augmented by one additional electrode of any of the types specified in Sections 250-50 or 250-52. Where multiple rod, pipe, or plate electrodes are installed to meet the requirements of this section, they shall not be less than 6 ft (1.83 m) apart.

FPN: The paralleling efficiency of rods longer than 8 ft (2.44 m) is improved by spacing greater than 6 ft (1.83 m).

A supplementary made electrode must be spaced at least 6 ft from another made electrode. See Figure 250.22.

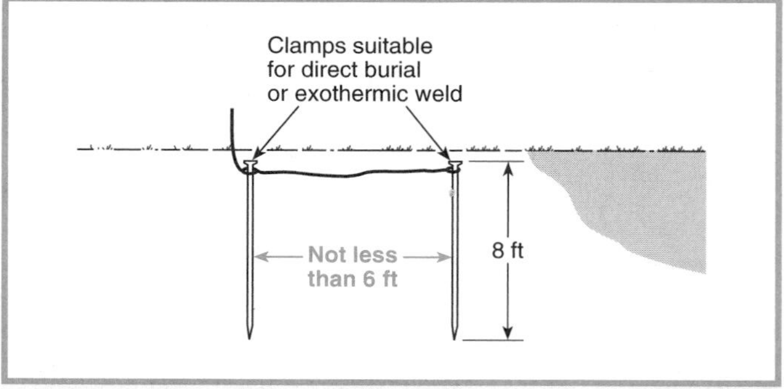

Figure 250.22 The 6-ft spacing between electrodes, as required by Section 250-52 or 250-56.

The resistance to ground of a driven grounding electrode can be measured by a ground tester used in the manner shown in Figure 250.23.

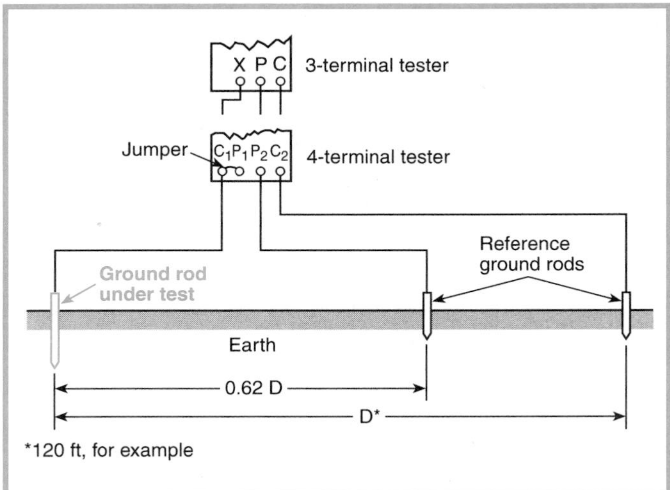

Figure 250.23 *The resistance to ground of a ground rod being measured by a ground tester.*

250-58. Common Grounding Electrode. Where an ac system is connected to a grounding electrode in or at a building as specified in Sections 250-24 and 250-32, the same electrode shall be used to ground conductor enclosures and equipment in or on that building. Where separate services supply a building and are required to be connected to a grounding electrode, the same grounding electrode shall be used.

Two or more grounding electrodes that are effectively bonded together shall be considered as a single grounding electrode system in this sense.

250-60. Use of Air Terminals. Air terminal conductors and driven pipes, rods, or other made electrodes used for grounding air terminals shall not be used in lieu of the made grounding electrodes required by Section 250-52 for grounding wiring systems and equipment. This provision shall not prohibit the required bonding together of grounding electrodes of different systems.

FPN No. 1: See Section 250-106 for spacing from air terminals. See Sections 800-40(d), 810-21(j), and 820-40(d) for bonding of electrodes.

FPN No. 2: Bonding together of all separate grounding electrodes will limit potential differences between them and between their associated wiring systems.

250-62. Grounding Electrode Conductor Material. The grounding electrode conductor shall be of copper, aluminum, or copper-clad aluminum. The material selected shall be resistant to any corrosive condition existing at the installation or shall be suitably protected against corrosion. The conductor shall be solid or stranded, insulated, covered, or bare.

250-64. Grounding Electrode Conductor Installation. Grounding electrode conductors shall be installed as specified in (a) through (e).

(a) Aluminum or Copper-Clad Aluminum Conductors. Insulated or bare aluminum or copper-clad aluminum grounding conductors shall not be used where in direct contact with masonry or the earth or where subject to corrosive conditions. Where used outside, aluminum or copper-clad aluminum grounding conductors shall not be installed within 18 in. (457 mm) of the earth.

(b) Grounding Electrode Conductor. A grounding electrode conductor or its enclosure shall be securely fastened to the surface on which it is carried. A No. 4 copper or aluminum, or larger conductor shall be protected if exposed to severe physical damage. A No. 6 grounding conductor that is free from exposure to physical damage shall be permitted to be run along the surface of the building construction without metal covering or protection where it is securely fastened to the construction; otherwise, it shall be in rigid metal conduit, intermediate metal conduit, rigid nonmetallic conduit, electrical metallic tubing, or cable armor. Grounding conductors smaller than No. 6 shall be in rigid metal conduit, intermediate metal conduit, rigid nonmetallic conduit, electrical metallic tubing, or cable armor.

(c) Continuous. The grounding electrode conductor shall be installed in one continuous length without a splice or joint, unless spliced only by irreversible compression-type connectors listed for the purpose or by the exothermic welding process.

It is frequently necessary to splice the grounding electrode conductor to add additional equipment. Section 250-64(c) permits splicing the grounding electrode conductor by irreversible compression-type fittings or exothermic welding. The intent is to require a permanent connection that will not become loose after installation. This permission is no longer limited to commercial and industrial locations.

Exception: Busbars shall be permitted to have splices.

(d) Grounding Electrode Conductor Taps. Where a service consists of more than a single enclosure as permitted in Section 230-40, Exception No. 2, it shall be permitted to connect taps to the grounding electrode conductor. Each such tap conductor shall extend to the inside of each such enclosure. The grounding electrode conductor shall be sized in accordance with Section 250-66, but the tap conductors shall be permitted to be sized in accordance with the grounding electrode conductors specified in Section 250-66 for the largest conductor serving the respective enclosures. The tap conductors shall be connected to the grounding electrode conductor in such a manner that the grounding electrode conductor remains without a splice or joint.

The grounding electrode (tap) conductors are required to be sized using Table 250-66 and are based on the size of the largest phase conductors serving each enclosure. The grounding electrode conductor is determined by the size of the largest service-entrance conductor or equivalent cross-sectional area for parallel conductors, per Table 250-66. As illustrated in

Figure 250.24, the tap method eliminates the difficulties found in looping grounding electrode conductors from one enclosure to another. The No. 2 grounding electrode conductor is required to be installed without a splice or joint, except as permitted in Section 250-64(c).

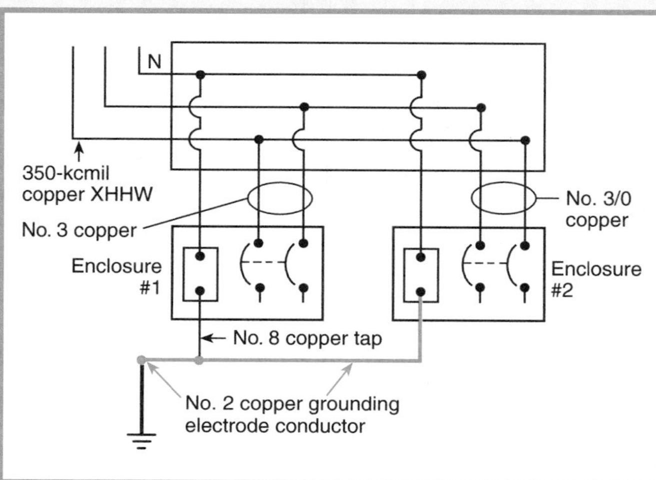

Figure 250.24 The tap method of connecting grounding electrode conductors from one enclosure to another.

(e) Enclosures for Grounding Electrode Conductors. Metal enclosures for grounding electrode conductors shall be electrically continuous from the point of attachment to cabinets or equipment to the grounding electrode, and shall be securely fastened to the ground clamp or fitting. Metal enclosures that are not physically continuous from cabinet or equipment to the grounding electrode shall be made electrically continuous by bonding each end to the grounding conductor. Where a raceway is used as protection for a grounding conductor, the installation shall comply with the requirements of the appropriate raceway article.

Bonding jumpers installed to ensure the electrical continuity of metal enclosures are required to be sized in accordance with Section 250-102(c). See Figure 250.29, which shows the bonding of a metal raceway to a grounding electrode conductor at both ends to ensure that the raceway and conductor are in parallel.

250-66. Size of Alternating-Current Grounding Electrode Conductor. The size of the grounding electrode conductor of a grounded or ungrounded ac system shall not be less than given in Table 250-66, except as permitted in (a) through (c).

Table 250-66. Grounding Electrode Conductor for Alternating-Current Systems

Size of Largest Service-Entrance Conductor or Equivalent Area for Parallel Conductors[1]		Size of Grounding Electrode Conductor	
Copper	**Aluminum or Copper-Clad Aluminum**	**Copper**	**Aluminum or Copper-Clad Aluminum[2]**
2 or smaller	1/0 or smaller	8	6
1 or 1/0	2/0 or 3/0	6	4
2/0 or 3/0	4/0 or 250 kcmil	4	2
Over 3/0 through 350 kcmil	Over 250 kcmil through 500 kcmil	2	1/0
Over 350 kcmil through 600 kcmil	Over 500 kcmil through 900 kcmil	1/0	3/0
Over 600 kcmil through 1100 kcmil	Over 900 kcmil through 1750 kcmil	2/0	4/0
Over 1100 kcmil	Over 1750 kcmil	3/0	250 kcmil

Notes:

1. Where multiple sets of service-entrance conductors are used as permitted in Section 230-40, Exception No. 2, the equivalent size of the largest service-entrance conductor shall be determined by the largest sum of the areas of the corresponding conductors of each set.

2. Where there are no service-entrance conductors, the grounding electrode conductor size shall be determined by the equivalent size of the largest service-entrance conductor required for the load to be served.

[1]This table also applies to the derived conductors of separately derived ac systems.

[2]See installation restrictions in Section 250-64(a).

Example

In applying the sizing requirements to Figure 250.25, the grounding electrode conductor is sized as No. 2 copper, in accordance with Table 250-66, by the following calculation:

No. 3 AWG	=	52,620	circular mil area, Table 8
No. 3/0 AWG	=	167,800	circular mil area, Table 8
		220,420	Sum of areas

From Table 8, nearest size = 250 kcmil

From Table 250-94 = No. 2 AWG grounding electrode conductor

Note that the size of the tap to the grounding electrode conductor from the disconnect on the left in Figure 250.25 is based on the size of the service-entrance conductors supplying the enclosure.

FPN: See Section 250-24(b) for size of ac system conductor brought to service equipment.

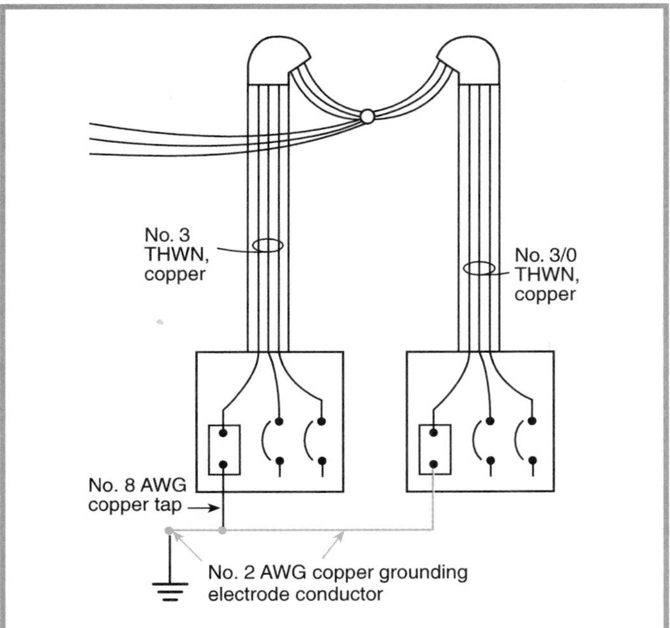

Figure 250.25 An example of sizing the grounding electrode conductor with multiple sets of service-conductors per Table 250-66, Note 1.

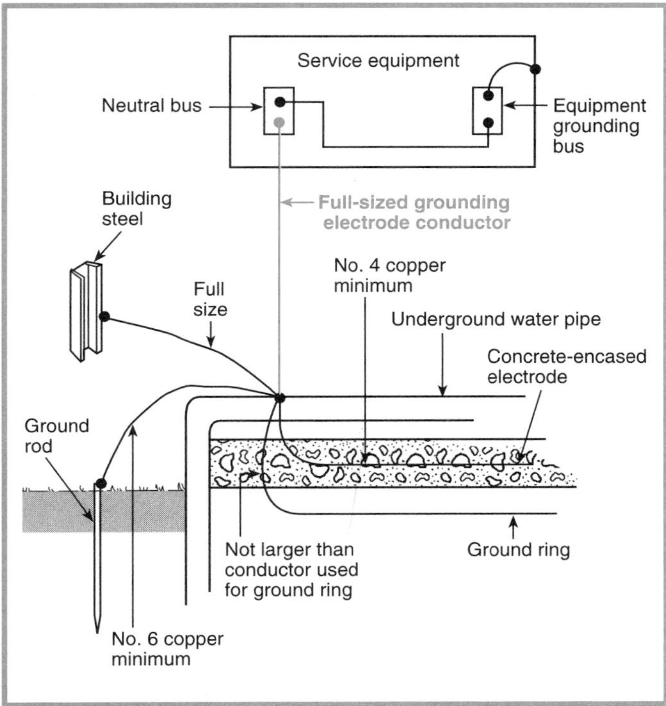

Figure 250.26 Illustration of sizing the grounding electrode conductor in accordance with Sections 250-66(a), (b), and (c).

(a) Connections to Made Electrodes. Where the grounding electrode conductor is connected to made electrodes as permitted in Section 250-52(c) or (d), that portion of the conductor that is the sole connection to the grounding electrode shall not be required to be larger than No. 6 copper wire or No. 4 aluminum wire.

(b) Connections to Concrete-Encased Electrodes. Where the grounding electrode conductor is connected to a concrete-encased electrode as permitted in Section 250-50(c), that portion of the conductor that is the sole connection to the grounding electrode shall not be required to be larger than No. 4 copper wire.

(c) Connections to Ground Rings. Where the grounding electrode conductor is connected to a ground ring as permitted in Section 250-50(d), that portion of the conductor that is the sole connection to the grounding electrode shall not be required to be larger than the conductor used for the ground ring.

As illustrated in Figure 250.26, if a grounding electrode conductor is run from the service equipment or separately derived system to a water pipe or structural metal building member and from that point to one of the electrodes mentioned in Section 250-66(a), that portion of the grounding electrode between the service equipment or separately derived system and the water pipe or structural metal building member must be a full-sized conductor, per Table 250-66. If the grounding electrode conductor from the service equipment were run, for example, to the ground rod first and then to the water pipe, the conductor to the ground rod would also have to be full-sized, per Table 250-66. Note that Figure 250.26 is not intended to show the physical routing and connection of the bonding jumpers.

250-68. Grounding Electrode Conductor Connection to Grounding Electrodes.

(a) Accessibility. The connection of a grounding electrode conductor to a grounding electrode shall be accessible.

Exception: An encased or buried connection to a concrete-encased, driven, or buried grounding electrode shall not be required to be accessible.

If the exposed portion of an encased, driven, or buried electrode is used for the termination of a grounding electrode conductor, the terminations are required to be accessible. However, if the connection is buried or encased, it is not required to be accessible. Note that ground clamps and other connectors suitable for use where buried in earth or embedded in concrete must be listed for such use, either by a marking on the connector or by a tag attached to the connector.

See Figures 250.19 and 250.22 for illustrations of encased and buried electrodes.

(b) Effective Grounding Path. The connection of a grounding electrode conductor shall be made in a manner that will ensure a permanent and effective grounding path. Where necessary to ensure the grounding path for a metal piping system used as a grounding electrode, effective bonding shall be provided around insulated joints and sections and around any equipment that is likely to be disconnected for repairs or replacement. Bonding conductors shall be of sufficient length to permit removal of such equipment while retaining the integrity of the bond.

A water meter or water filter system is typical of equipment likely to be disconnected for repairs or replacement.

250-70. Methods of Grounding Conductor Connection to Electrodes. The grounding conductor shall be connected to the grounding electrode by exothermic welding, listed lugs, listed pressure connectors, listed clamps, or other listed means. Connections depending on solder shall not be used. Ground clamps shall be listed for the materials of the grounding electrode and the grounding electrode conductor and, where used on pipe, rod, or other buried electrodes, shall also be listed for direct soil burial. Not more than one conductor shall be connected to the grounding electrode by a single clamp or fitting unless the clamp or fitting is listed for multiple conductors. One of the following methods shall be used:

(1) A listed bolted clamp of cast bronze or brass, or plain or malleable iron
(2) A pipe fitting, pipe plug, or other approved device screwed into a pipe or pipe fitting
(3) For indoor telecommunications purposes only, a listed sheet metal strap-type ground clamp having a rigid metal base that seats on the electrode and having a strap of such material and dimensions that it is not likely to stretch during or after installation
(4) An equally substantial approved means

Where a ground clamp is used and terminates on a galvanized water pipe, for example, it is required to be of a material that is compatible with steel so as to prevent corrosion. The same type of compatibility requirement applies to ground clamps on copper water pipe.

Figure 250.27 shows a listed ground clamp generally used with No. 8 through No. 4 grounding electrode conductors. Exothermic welds are also acceptable. Kits are commercially available for this purpose.

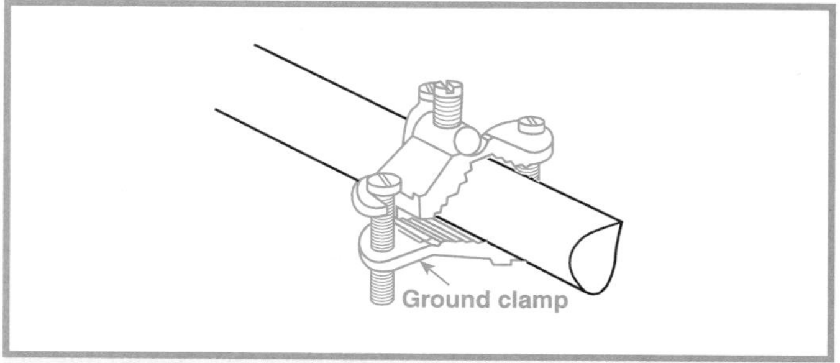

Figure 250.27 *An application of a listed ground clamp.*

Figure 250.28 shows a listed U-bolt ground clamp. These clamps are available for all pipe sizes and all grounding electrode conductor sizes. Where grounding electrode conductors are run in conduit, conduit hubs may be bolted to the threaded portion of the U-bolt.

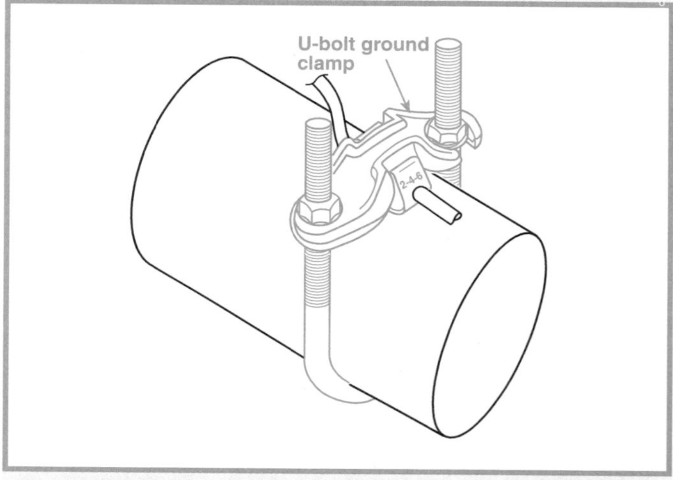

Figure 250.28 *An application of a listed U-bolt ground clamp.*

D. Enclosure, Raceway, and Service Cable Grounding

250-80. Service Raceways and Enclosures. Metal enclosures and raceways for service conductors and equipment shall be grounded.

Exception: A metal elbow that is installed in an underground installation of rigid nonmetallic conduit and is isolated from possible contact by a minimum cover of 18 in. (457 mm) to any part of the elbow shall not be required to be grounded.

The exception to Section 250-80 recognizes that metal sweep elbows are often installed in underground installations of rigid nonmetallic conduit. The metal elbows are installed because nonmetallic elbows can

be damaged by friction from the pulling ropes during conductor installation. The elbows are isolated from physical contact by burial so that no part of the elbow is less than 18 in. below grade.

250-84. Underground Service Cable or Conduit.

(a) Underground Service Cable. The sheath or armor of a continuous underground metal-sheathed service cable system that is metallically connected to the underground system shall not be required to be grounded at the building. The sheath or armor shall be permitted to be insulated from the interior conduit or piping.

(b) Underground Service Conduit Containing Cable. An underground service conduit that contains a metal-sheathed cable bonded to the underground system shall not be required to be grounded at the building. The sheath or armor shall be permitted to be insulated from the interior conduit or piping.

250-86. Other Conductor Enclosures and Raceways. Except as permitted by Section 250-112(i), metal enclosures and raceways for other than service conductors shall be grounded.

Section 250-86 requires grounding, bonding, and ensurance of electrical continuity of all enclosures and metal raceways. Connectors, couplings, or other similar fittings that perform mechanical and electrical functions must ensure bonding and grounding continuity between the fitting, the metal raceway, and the enclosure. Metal enclosures are required to be grounded so that when a fault occurs between an ungrounded (hot) conductor and ground, the potential difference between the noncurrent-carrying parts of the electrical installation will be minimized, thereby reducing the shock hazard to personnel.

Exception No. 1: Metal enclosures and raceways for conductors added to existing installations of open wire, knob and tube wiring, and nonmetallic-sheathed cable shall not be required to be grounded where these enclosures or wiring methods

(a) Do not provide an equipment ground;
(b) Are in runs of less than 25 ft (7.62 m);
(c) Are free from probable contact with ground, grounded metal, metal lath, or other conductive material; and
(d) Are guarded against contact by persons.

Exception No. 2: Short sections of metal enclosures or raceways used to provide support or protection of cable assemblies from physical damage shall not be required to be grounded.

Exception No. 3: A metal elbow that is installed in an underground installation of rigid nonmetallic conduit and is isolated from possible contact by a minimum cover of 18 in. (457 mm) to any part of the elbow shall not be required to be grounded.

E. Bonding

250-90. General. Bonding shall be provided where necessary to ensure electrical continuity and the capacity to conduct safely any fault current likely to be imposed.

250-92. Services.

(a) Bonding of Services. The noncurrent-carrying metal parts of equipment indicated in (1), (2), and (3) shall be effectively bonded together.

(1) The service raceways, cable trays, cablebus framework, or service cable armor or sheath except as permitted in Section 250-84.
(2) All service enclosures containing service conductors, including meter fittings, boxes, or the like, interposed in the service raceway or armor.
(3) Any metallic raceway or armor enclosing a grounding electrode conductor as specified in Section 250-64(b). Bonding shall apply at each end and to all intervening raceways, boxes, and enclosures between the service equipment and the grounding electrode.

Section 250-92(a)(3) is intended to clarify that where metal raceways, boxes, or enclosures contain a grounding electrode conductor, both ends of the raceway, box, or enclosure must be bonded to the grounding electrode conductor as illustrated in Figure 250.29. Bonding the raceway to the conductor reduces the impedance and minimizes the potential difference between the electrical equipment and ground. See also Sections 250-64(e) and 250-102(a).

(b) Bonding to Other Systems. An accessible means external to enclosures for connecting intersystem bonding and grounding conductors shall be provided at the service by at least one of the following means:

(1) Exposed nonflexible metallic service raceways
(2) Exposed grounding electrode conductor
(3) Approved means for the external connection of a copper or other corrosion-resistant bonding or grounding conductor to the service raceway or equipment

For the purposes of providing an accessible means for intersystem bonding, the disconnecting means at a separate building or structure as permitted in Section 250-32 and the disconnecting means at a mobile home as permitted in Section 550-23(a) shall be considered the service equipment.

Section 250-92(b) clarifies that the approved means for an external connection for bonding other systems must be copper or another approved corrosion-resistant material. For example, aluminum would not be suitable for connection to electrodes near earth or in concrete.

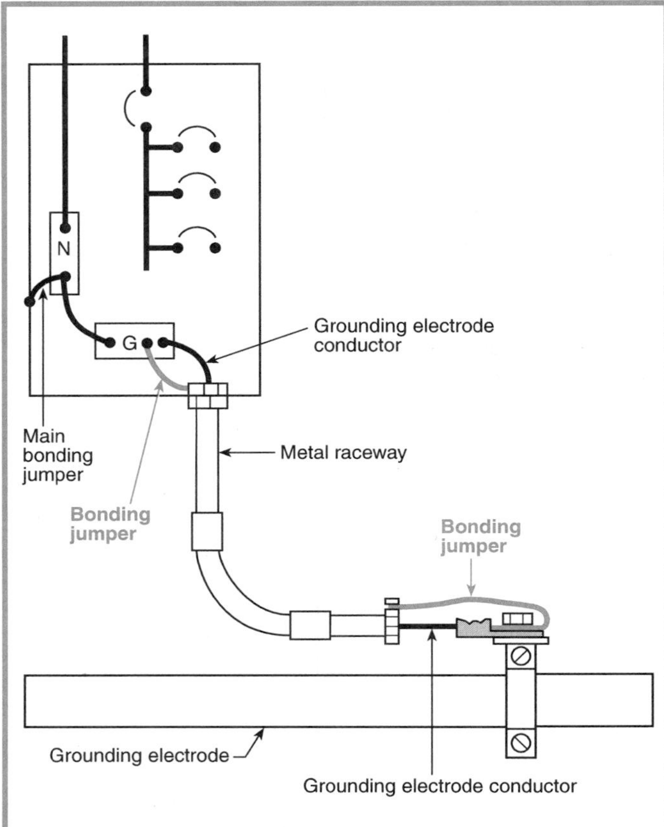

Figure 250.29 *Bonding of metal raceway containing a grounding electrode conductor to the conductor at both ends as required by Section 250-64(e).*

The need for an external accessible bonding means is equally important for separate buildings and mobile homes. In these occupancies, the disconnecting means enclosure on the load side of the service can be considered the equivalent of the service equipment for the purpose of intersystem bonding.

Two different grounding and bonding arrangements are shown in Figure 250.30. On the left is a grounding electrode conductor at the meter housing with the service equipment enclosure bonded to the grounded service conductor. On the right is the grounding electrode conductor at the service equipment with the service equipment enclosure and the meter housing bonded to the grounded service conductor.

FPN No. 1: A No. 6 copper conductor with one end bonded to the service raceway or equipment and with 6 in. (152 mm) or more of the other end made accessible on the outside wall is an example of the approved means covered in (b)(3).

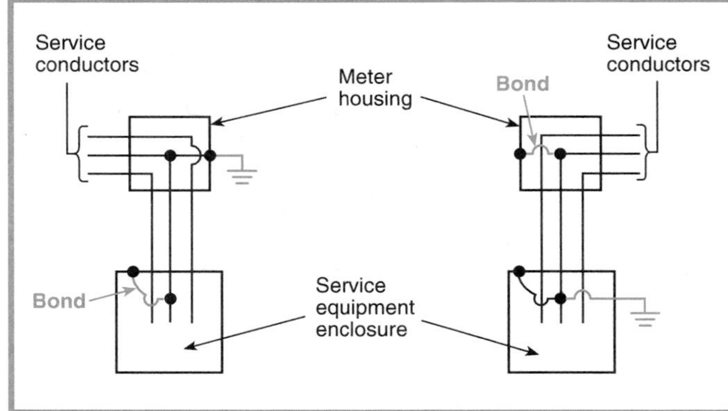

Figure 250.30 *Two different grounding and bonding arrangements.*

Other means are illustrated in Figure 250.31. On the left is an illustration of accessible means for the connection. On the right is a method of providing the required bonding means when the panelboard is a flush type.

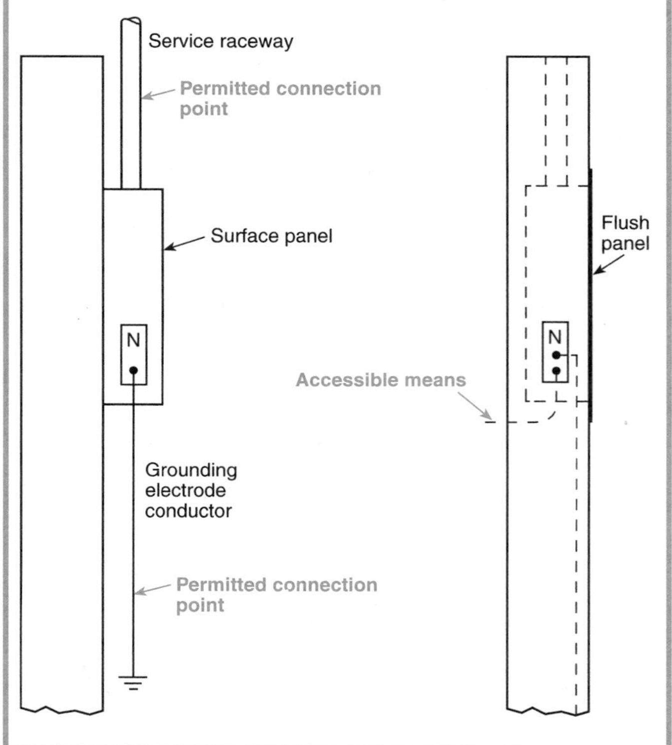

Figure 250.31 *Examples of accessible external means for intersystem bonding, as required by Section 250-92(b).*

FPN No. 2: See Sections 800-40, 810-21, and 820-40 for bonding and grounding requirements for communications circuits, radio and television equipment, and CATV circuits.

The *Code* requires that separate systems be bonded together to reduce the differences of potential between them due to lightning or accidental contact with power lines. Lightning protection systems, communications, radio and TV, and CATV systems are required to be bonded together to minimize the potential differences between the systems. Lack of interconnection can result in a severe shock and fire hazard.

The reason for this potential hazard is illustrated by Figure 250.32, which shows a CATV cable with its jacket grounded to a separate ground rod and not bonded to the power ground. This cable is connected to the tuner of a television set inside a home. Also connected to the tuner is the 120-volt supply, with one conductor grounded at the service (the power ground). In each case, a dc resistance to ground will be present at the grounding electrode, as shown in the equivalent circuit. This resistance to ground will vary widely, depending on soil conditions and the type of grounding electrode. The resistance at the CATV ground (R_{TV}) is likely to be higher than the power ground resistance (R_p), because the power ground is often an underground metal water piping system or concrete-encased electrode, whereas the CATV ground is commonly a ground rod.

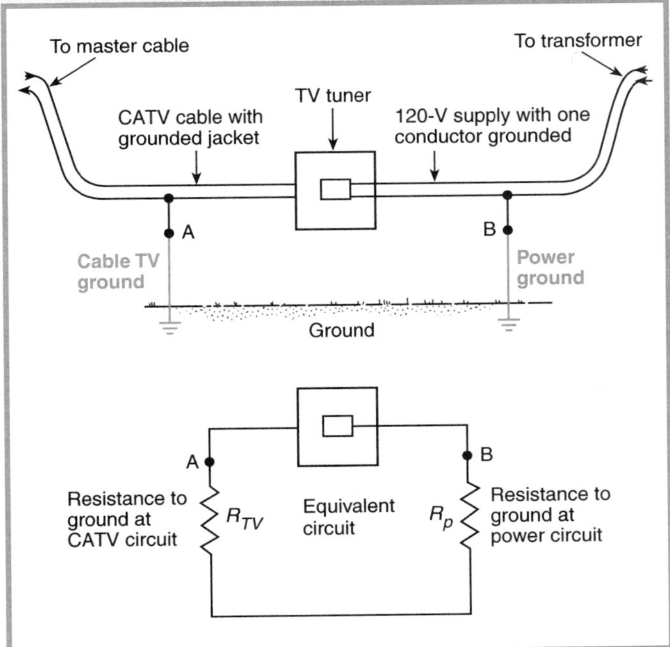

Figure 250.32 *A cable TV installation that does not comply with the Code and illustrates why bonding between different systems is necessary.*

Example

For the cable TV installation shown in Figure 250.32, assume that a current is induced in the power line by a switching surge or a nearby lightning strike, so that a momentary current of 1000 amperes flows over the power line to the power-line ground. This amount of current is not unusual under such circumstances. It could be, and often is, considerably higher. Also assume that the power ground has a resistance (R_p) of 10 ohms, a *very* low value in most circumstances. A single ground rod in average soil has a resistance to ground in the neighborhood of 40 ohms. According to Ohm's law, point B will be raised momentarily to a potential of $1000 \times 10 = 10{,}000$ volts, resulting in a potential of 10,000 volts between points A and B and anything connected to them — in this case, parts within the tuner. This voltage would also exist between the grounded conductor within the CATV cable and grounded surfaces in the walls of the home, such as water pipes (which are connected to the power ground), over which the cable runs. This potential could also appear across a person with one hand on the CATV cable and the other on a metal surface connected to the power ground (a radiator or refrigerator, for example). Actual voltage is likely to be many times the 10,000 volts calculated, as extremely low (below-normal) values were assumed for both resistance to ground and current. Most insulation systems, however, are not designed to withstand even 10,000 volts. Even if the insulation system does withstand this voltage surge, it is likely to be damaged. Breakdown of the insulation system will result in sparking.

The same situation would exist if the current surge were on the CATV cable or on a telephone line. The only difference would be the voltage involved, depending on the individual resistance to ground of the grounding electrodes.

The solution is to bond between points A and B or to connect the CATV cable jacket to the power ground, and this is exactly what the *Code* requires. Thus, even though point A is raised to some very high voltage above ground, so is point B, and no voltage exists between the two grounding systems.

These rules are provided to address the difficulties encountered by communications and CATV installers in complying with *Code* grounding and bonding requirements arising from the increasing use of plastic for water pipe, fittings, water meters, and service conduit. In the past, bonding between communications,

CATV, and power systems was usually achieved by connecting the communications protector grounds or cable shield to an interior metallic water pipe, because the pipe was often used as the power grounding electrode. Thus, the requirement that the power, communications, CATV cable shield, and metallic water piping systems be bonded together was easily satisfied. If the power was grounded to one of the other electrodes permitted by the *Code,* usually by a made electrode such as a ground rod, the bond was connected to the power grounding electrode conductor or to a metallic service raceway, since at least one of these was usually accessible. With the proliferation of plastic water pipe and the increasing tendency for service equipment (often flush-mounted) to be installed in finished areas, where the grounding electrode conductor is often concealed, as well as the increased use of plastic service-entrance conduit, communications and CATV installers no longer have access to a point for connecting bonding jumpers or grounding conductors. See Figure 250.31 and also the commentary to Section 820-40(d), FPN No. 2.

250-94. Method of Bonding at the Service. Electrical continuity at service equipment, service raceways, and service conductor enclosures shall be ensured by one of the following methods:

(1) Bonding equipment to the grounded service conductor in a manner provided in Section 250-8

Figure 250.33 is an illustration of the grounding and bonding at an individual service. Figure 250.34 shows a grounding and bonding arrangement for up to six switches that serve as the service disconnecting means for an individual service. Section 250-24(b) clarifies that the grounded service conductor is required to be run to each service disconnect and bonded to the enclosure. Section 250-94(1) permits the bonding of service equipment to be accomplished by bonding to the grounded service conductor.

(2) Connections utilizing threaded couplings or threaded bosses on enclosures where made up wrenchtight
(3) Threadless couplings and connectors where made up tight for metal raceways and metal-clad cables
(4) Other approved devices, such as bonding-type locknuts and bushings

Note that method No. 4 requires bonding-type locknuts or bushings. Standard locknuts or sealing locknuts are not acceptable as the "sole means" for bonding on the line side of service equipment.

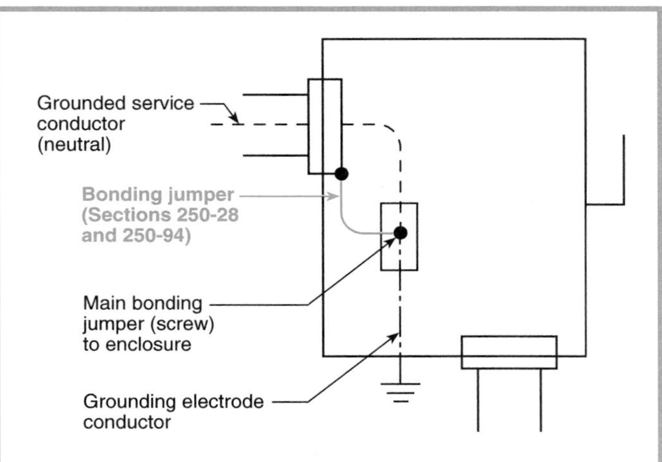

Figure 250.33 *Grounding and bonding for a service with one disconnecting means.*

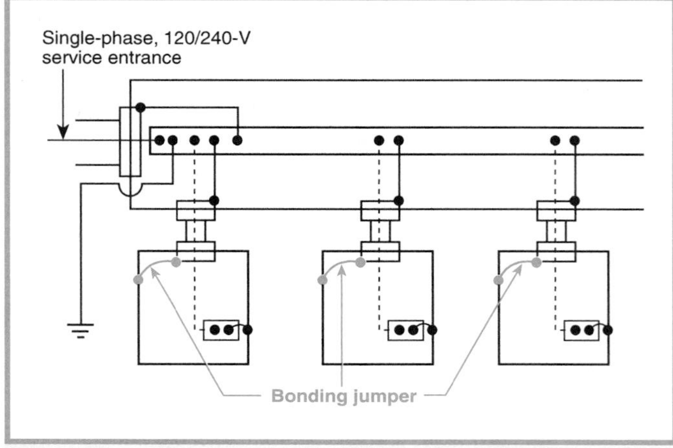

Figure 250.34 *A grounding and bonding arrangement for up to six switches that serve as the service disconnecting means for an individual service.*

Bonding bushings for use with rigid or intermediate metal conduit are provided with means (usually one or more set screws) for reliably bonding the bushing and the conduit on which it is threaded to the metal equipment enclosure or box. If means for connecting a grounding conductor or bonding jumper are not provided and such a conductor is needed, a grounding-type bushing is required to be used.

Grounding-type bushings used with rigid or intermediate metal conduit such as those shown in Figures 250.35 and 250.37 have provisions for connecting a bonding jumper or have means provided by the manufacturer for use in mounting a wire connector. This type of bushing may also have means (usually one or more set screws) to reliably bond the bushing to the metal equipment enclosure or box in the same

manner as with a bonding jumper. See Figure 250.36 for a bonding type wedge lug used to connect a conduit to a box.

Figure 250.35 *A grounding bushing used to connect a copper bonding or grounding wire to a conduit. (The Thomas & Betts Co. Inc.)*

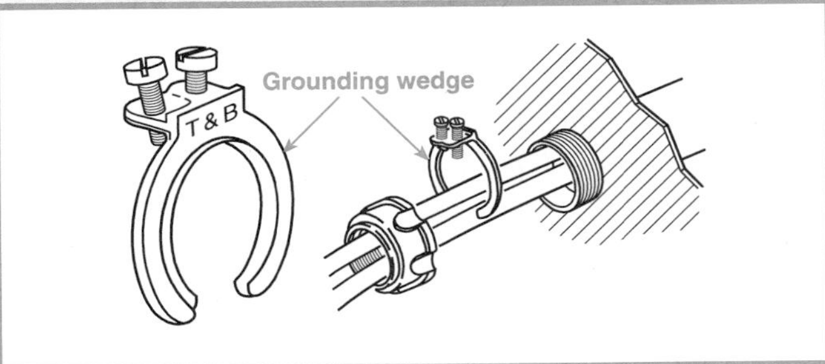

Figure 250.36 *A grounding wedge lug used to provide an electrical connection between a conduit and a box. (The Thomas & Betts Co. Inc.)*

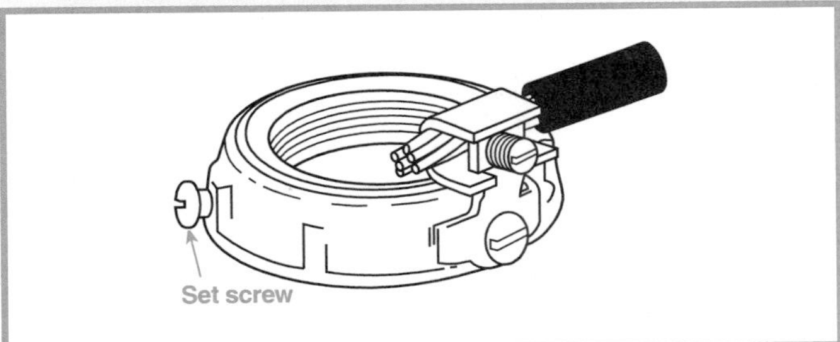

Figure 250.37 *A threaded grounding bushing with set screws used to ensure electrical and mechanical connection. (General Electric Co.)*

Bonding jumpers meeting the other requirements of this article shall be used around concentric or eccentric knockouts that are punched or otherwise formed so as to impair the electrical connection to ground. Standard locknuts or bushings shall not be the sole means for the bonding required by this section.

250-96. Bonding Other Enclosures.

(a) General. Metal raceways, cable trays, cable armor, cable sheath, enclosures, frames, fittings, and other metal noncurrent-carrying parts that are to serve as grounding conductors, with or without the use of supplementary equipment grounding conductors, shall be effectively bonded where necessary to ensure electrical continuity and the capacity to conduct safely any fault current likely to be imposed on them. Any nonconductive paint, enamel, or similar coating shall be removed at threads, contact points, and contact surfaces or be connected by means of fittings designed so as to make such removal unnecessary.

(b) Isolated Grounding Circuits. Where required for the reduction of electrical noise (electromagnetic interference) on the grounding circuit, an equipment enclosure supplied by a branch circuit shall be permitted to be isolated from a raceway containing circuits supplying only that equipment by one or more listed nonmetallic raceway fittings located at the point of attachment of the raceway to the equipment enclosure. The metal raceway shall comply with provisions of this article and shall be supplemented by an internal insulated equipment grounding conductor installed in accordance with Section 250-146(d) to ground the equipment enclosure.

FPN: Use of an isolated equipment grounding conductor does not relieve the requirement for grounding the raceway system.

Section 250-96(b) permits electronic equipment to be isolated from the raceway in a manner similar to that for cord- and plug-connected equipment, to reduce electromagnetic interference. It specifies that an equipment enclosure supplied by a branch circuit is the subject of the requirement and that subsequent wiring, raceways, or other equipment beyond the insulating fitting is not permitted.

Figures 250.38 and 250.39 show examples of installations. In Figure 250.38 note that the metal raceway

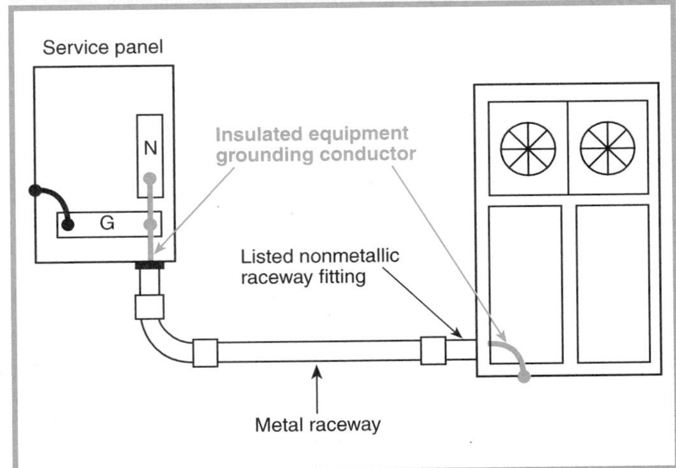

Figure 250.38 *An installation where the electronic equipment is grounded through the isolated grounding conductor.*

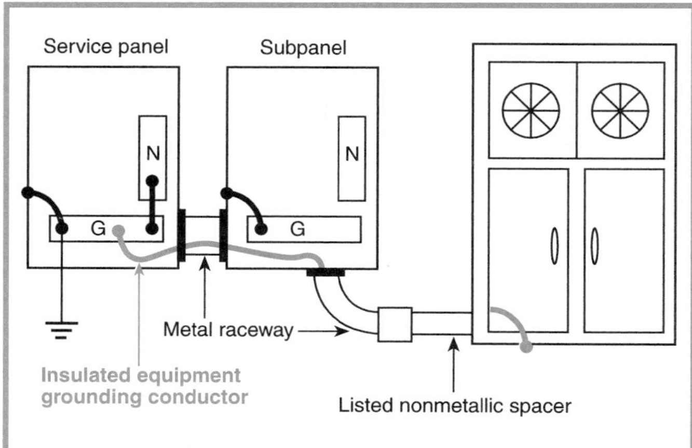

Figure 250.39 An installation where the insulated equipment grounding conductor is allowed to pass through the subpanel without connecting to the grounding bus to terminate at the service grounding bus.

is grounded in the usual manner, by attachment to the grounded service enclosure, satisfying the concern mentioned in the fine print note. In Figure 250.39, note that Section 384-20, Exception, permits the isolated equipment grounding conductor (it is required to be insulated) to pass through the panelboard.

250-97. Bonding for Over 250 Volts. For circuits of over 250 volts to ground, the electrical continuity of metal raceways and cables with metal sheaths that contain any conductor other than service conductors shall be ensured by one or more of the methods specified for services in Section 250-94, except for (1).

Exception: Where oversized, concentric, or eccentric knockouts are not encountered, or where a box or enclosure with concentric or eccentric knockouts is listed for the purpose, the following methods shall be permitted:

(a) Threadless couplings and connectors for cables with metal sheaths
(b) Two locknuts, on rigid metal conduit or intermediate metal conduit, one inside and one outside of boxes and cabinets
(c) Fittings with shoulders that seat firmly against the box or cabinet, such as electrical metallic tubing connectors, flexible metal conduit connectors, and cable connectors, with one locknut on the inside of boxes and cabinets
(d) Listed fittings

Bonding around prepunched concentric or eccentric knockouts is not required if the enclosure containing the knockouts has been tested and listed as suitable for bonding.

The methods in (a), (b), and (c) of the exception to Section 250-97, are permitted for circuits over 250 volts to ground only where there are no oversize, concentric, or eccentric knockouts. Note that method (c) permits fittings, such as EMT connectors, cable connectors, and similar fittings with shoulders that seat firmly against the metal of a box or cabinet, to be installed with only one locknut, located on the inside of the box.

250-98. Bonding Loosely Jointed Metal Raceways. Expansion fittings and telescoping sections of metal raceways shall be made electrically continuous by equipment bonding jumpers or other means.

250-100. Bonding in Hazardous (Classified) Locations. Regardless of the voltage of the electrical system, the electrical continuity of noncurrent-carrying metal parts of equipment, raceways, and other enclosures in any hazardous (classified) location as defined in Article 500 shall be ensured by any of the methods specified for services in Section 250-94 that are approved for the wiring method used.

250-102. Equipment Bonding Jumpers.

(a) Material. Equipment bonding jumpers shall be of copper or other corrosion-resistant material. A bonding jumper shall be a wire, bus, screw, or similar suitable conductor.

(b) Attachment. Equipment bonding jumpers shall be attached in the manner specified by the applicable provisions of Section 250-8 for circuits and equipment and by Section 250-70 for grounding electrodes.

(c) Size — Equipment Bonding Jumper on Supply Side of Service. The bonding jumper shall not be smaller than the sizes shown in Table 250-66 for grounding electrode conductors. Where the service-entrance phase conductors are larger than 1100 kcmil copper or 1750 kcmil aluminum, the bonding jumper shall have an area not less than 12½ percent of the area of the largest phase conductor except that, where the phase conductors and the bonding jumper are of different materials (copper or aluminum), the minimum size of the bonding jumper shall be based on the assumed use of phase conductors of the same material as the bonding jumper and with an ampacity equivalent to that of the installed phase conductors. Where the service-entrance conductors are paralleled in two or more raceways or cables, the equipment bonding jumper, where routed with the raceways or cables, shall be run in parallel. The size of the bonding jumper for each raceway or cable shall be based on the size of the service-entrance conductors in each raceway or cable.

The bonding jumper for a grounding electrode conductor raceway or cable armor as covered in Section 250-64(d) shall be the same size or larger than the required enclosed grounding electrode conductor.

(d) Size — Equipment Bonding Jumper on Load Side of Service. The equipment bonding jumper on the load side of the service overcurrent devices shall be sized, as a minimum, in accordance with the sizes listed in Table 250-

122, but not required to be larger than the circuit conductors supplying the equipment, and not smaller than No. 14.

A single common continuous equipment bonding jumper shall be permitted to bond two or more raceways or cables where the bonding jumper is sized in accordance with Table 250-122 for the largest overcurrent device supplying circuits therein.

(e) Installation. The equipment bonding jumper shall be permitted to be installed inside or outside of a raceway or enclosure. Where installed on the outside, the length of the equipment bonding jumper shall not exceed 6 ft (1.83 m) and shall be routed with the raceway or enclosure. Where installed inside of a raceway, the equipment bonding jumper shall comply with the requirements of Sections 250-119 and 250-148.

In many applications, it is necessary to install equipment bonding jumpers on the outside of metal raceways and enclosures. For example, it would be impractical to install the bonding jumper for a conduit expansion joint on the inside of the conduit. For some metal raceway and rigid conduit systems and conduit systems in hazardous (classified) locations, it is desirable to install the bonding jumper where it is visible and accessible for inspection and maintenance. An external bonding jumper will have a higher impedance than an internal bonding jumper. However, by limiting the length to 6 ft and routing it with the raceway, the increase in the impedance of the equipment grounding circuit will be insignificant. For example, this rule permits a maximum length of 6 ft for a bonding jumper run outside a length of flexible metal conduit, as illustrated in Figure 250.40. Since the function of a bonding jumper is readily apparent, color identification is not necessary.

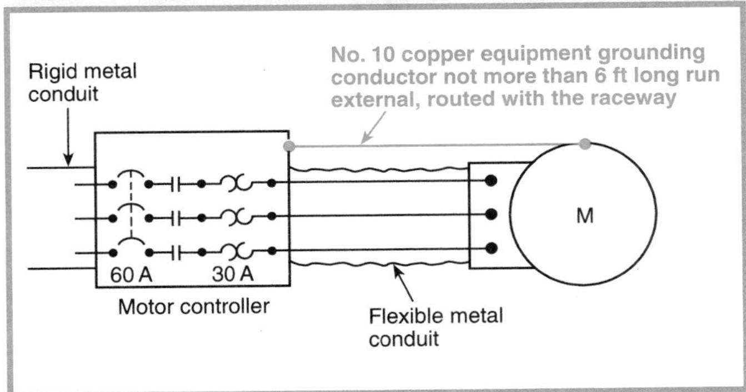

Figure 250.40 *Bonding around the outside of flexible metal conduit.*

250-104. Bonding of Piping Systems and Exposed Structural Steel.

(a) Metal Water Piping. The interior metal water piping system shall be bonded as required in (1), (2), (3), or (4) of this section. The bonding jumper shall be installed in accordance with Section 250-64(a), (b), and (e). The points of attachment of the bonding jumper(s) shall be accessible.

(1) General. The interior metal water piping system shall be bonded to the service equipment enclosure, the grounded conductor at the service, the grounding electrode conductor where of sufficient size, or to the one or more grounding electrodes used. The bonding jumper shall be sized in accordance with Table 250-66 except as permitted in (2) and (3).

Bonding the interior metal water piping system is not the same as using the metal water piping system as a grounding electrode. Bonding to the grounding electrode system places the bonded components at the same voltage level. For example, a current of 2000 amperes across a 25-ft-long No. 6 copper conductor produces a voltage differential of approximately 26 volts. Section 250-104(a)(1) requires the interior metal water piping system and any other metal piping systems likely to become energized to be bonded to the service equipment or grounding electrode conductor. The bonding jumper must be sized not smaller than required by Table 250-66.

Some judgment must be exercised in each case. Where it cannot reasonably be concluded that the hot and cold water pipes are reliably interconnected, an electrical bonding jumper is required to ensure that this connection is made. The special installation requirements provided in Sections 250-64(a), (b), and (e) are also required for the water piping bonding jumper.

(2) Buildings of Multiple Occupancy. In buildings of multiple occupancy, where the interior metal water piping system for the individual occupancies is metallically isolated from all other occupancies by use of nonmetallic water piping, the interior metal water piping system for each occupancy shall be permitted to be bonded to the equipment grounding terminal of the panelboard or switchboard enclosure (other than service equipment) supplying that occupancy. The bonding jumper shall be sized in accordance with Table 250-122.

The intent of Section 250-104(a)(2) is to recognize that the increased use of nonmetallic water piping mains causes the interior metal piping system of a multiple-occupancy building to be isolated from ground. Therefore, the water pipe is permitted to be bonded to the panelboard serving only that particular occupancy. The bonding jumper in this case is permitted to be sized according to Table 250-122, based on the size of the main overcurrent device supplying the occupancy.

(3) Multiple Buildings or Structures Supplied from a Common Service. The interior metal water piping system shall be bonded to the building or structure disconnecting means enclosure where located at the building or structure, or to the equipment grounding conductor run with the supply conductors, or to the one or more grounding electrodes used. The bonding jumper shall be sized in accordance with Section 250-122 based on the rating or setting of the largest overcurrent device protecting the feeder(s) or branch circuit(s) that supply the building.

(4) Separately Derived Systems. The grounded conductor of the separately derived system shall be bonded to the nearest available point of the interior metal water piping system in the area served by the separately derived system. This connection shall be made at the same point on the separately derived system where the grounding electrode conductor is connected. The bonding jumper shall be sized in accordance with Table 250-66.

Section 250-104(a)(4) requires that where a separately derived system supplies the power, the interior metal piping system in the area must be bonded to the grounded conductor at the point nearest the derived system, and this connection must be accessible.

x**(b) Metal Gas Piping.** Each aboveground portion of a gas piping system upstream from the equipment shutoff valve shall be electrically continuous and bonded to the grounding electrode system.

(c) Other Metal Piping. Interior metal piping that may become energized shall be bonded to the service equipment enclosure, the grounded conductor at the service, the grounding electrode conductor where of sufficient size, or to the one or more grounding electrodes used. The bonding jumper shall be sized in accordance with Table 250-122 using the rating of the circuit that may energize the piping.

The equipment grounding conductor for the circuit that may energize the piping shall be permitted to serve as the bonding means.

FPN: Bonding all piping and metal air ducts within the premises will provide additional safety.

(d) Structural Steel. Exposed interior structural steel that is interconnected to form a steel building frame and is not intentionally grounded and may become energized shall be bonded to the service equipment enclosure, the grounded conductor at the service, the grounding electrode conductor where of sufficient size, or to the one or more grounding electrodes used. The bonding jumper shall be sized in accordance with Table 250-66 and installed in accordance with Sections 250-64(a), (b), and (e). The points of attachment of the bonding jumpers shall be accessible.

Section 250-104(d) requires exposed metal building framework not intentionally or inherently grounded to be bonded to the service equipment or grounding electrode system.

250-106. Lightning Protection Systems. The lightning protection system ground terminals shall be bonded to the building or structure grounding electrode system.

In previous editions of the *Code,* all electrical equipment located within 6 ft of downleads of the lightning protection system were required to be bonded to the downlead at that location. The 1999 *Code* specifies that the ground terminals (electrode system) of the lightning protection system be bonded to the electrical service grounding electrode system as shown in Figure 250.41. A similar requirement is found in Section 3-14 of the 1997 edition of NFPA 780, *Standard for the Installation of Lightning Protection Systems.* Additional bonding between the lightning protection system and the electrical system may be necessary based on proximity and whether separation between the systems is through air or building materials. A new fine print note references NFPA 780 for guidance on determining the need for additional bonding connections. A method of calculating flashover distances is contained in Section 3-21.2 of NFPA 780-1997.

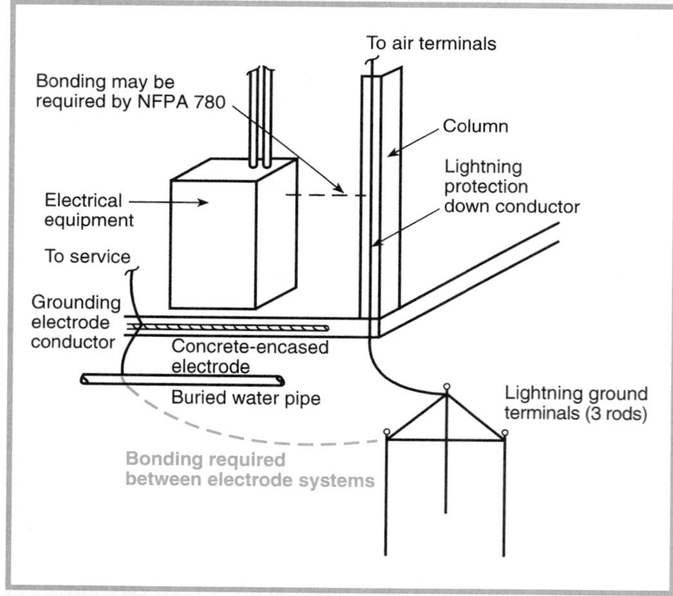

Figure 250.41 *Bonding between lightning protection ground terminal and electrical service grounding electrode system in accordance with Section 250-106.*

Exposed, noncurrent-carrying metal parts of fixed equipment that are not likely to become energized are not required to be grounded. These include some metal nameplates on nonmetallic enclosures and small parts, such as bolts and screws, if they are located so that they are not likely to become energized.

FPN No. 1: See Section 250-60 for use of air terminals. For further information, see *Standard for the Installation of Lightning Protection Systems*, NFPA 780-1997, which contains detailed information on grounding, bonding, and spacing from lightning protection systems.

FPN No. 2: Metal raceways, enclosures, frames, and other noncurrent-carrying metal parts of electric equipment installed on a building equipped with a lightning protection system may require bonding or spacing from the lightning protection conductors in accordance with *Standard for the Installation of Lightning Protection Systems*, NFPA 780-1997. Separation from lightning protection conductors is typically 6 ft (1.83 m) through air or 3 ft (0.92 m) through dense materials such as concrete, brick, or wood.

F. Equipment Grounding and Equipment Grounding Conductors

250-110. Equipment Fastened in Place or Connected by Permanent Wiring Methods (Fixed). Exposed noncurrent-carrying metal parts of fixed equipment likely to become energized shall be grounded under any of the following conditions:

(1) Where within 8 ft (2.44 m) vertically or 5 ft (1.52 m) horizontally of ground or grounded metal objects and subject to contact by persons
(2) Where located in a wet or damp location and not isolated
(3) Where in electrical contact with metal
(4) Where in a hazardous (classified) location as covered by Articles 500 through 517
(5) Where supplied by a metal-clad, metal-sheathed, metal-raceway, or other wiring method that provides an equipment ground, except as permitted by Section 250-86, Exception No. 2 for short sections of metal enclosures
(6) Where equipment operates with any terminal at over 150 volts to ground

Exception No. 1 to (1) through (6): Metal frames of electrically heated appliances, exempted by special permission, in which case the frames shall be permanently and effectively insulated from ground.

Exception No. 2 to (1) through (6): Distribution apparatus, such as transformer and capacitor cases, mounted on wooden poles, at a height exceeding 8 ft (2.44 m) above ground or grade level.

Exception No. 3 to (1) through (6): Listed equipment protected by a system of double insulation, or its equivalent, shall not be required to be grounded. Where such a system is employed, the equipment shall be distinctively marked.

250-112. Fastened in Place or Connected by Permanent Wiring Methods (Fixed) — Specific. Exposed, noncurrent-carrying metal parts of the kinds of equipment described in (a) through (k), and noncurrent-carrying metal parts of equipment and enclosures described in (l) and (m), shall be grounded regardless of voltage.

(a) Switchboard Frames and Structures. Switchboard frames and structures supporting switching equipment, except frames of 2-wire dc switchboards where effectively insulated from ground.

Section 250-112(a) clarifies that dc switchboards insulated from ground are not required to be grounded.

(b) Pipe Organs. Generator and motor frames in an electrically operated pipe organ, unless effectively insulated from ground and the motor driving it.

(c) Motor Frames. Motor frames, as provided by Section 430-142.

(d) Enclosures for Motor Controllers. Enclosures for motor controllers unless attached to ungrounded portable equipment.

(e) Elevators and Cranes. Electric equipment for elevators and cranes.

(f) Garages, Theaters, and Motion Picture Studios. Electric equipment in commercial garages, theaters, and motion picture studios, except pendant lampholders supplied by circuits not over 150 volts to ground.

(g) Electric Signs. Electric signs, outline lighting, and associated equipment as provided in Article 600.

(h) Motion Picture Projection Equipment. Motion picture projection equipment.

(i) Power-Limited Remote-Control, Signaling, and Fire Alarm Circuits. Equipment supplied by Class 1 power-limited circuits and Class 1, Class 2, and Class 3 remote-control and signaling circuits, and by fire alarm circuits, shall be grounded where system grounding is required by Part B of this article.

If a fire alarm circuit operates at any system voltage required to be grounded by Article 250, Part B, then the noncurrent-carrying metal parts of the equipment are required to be grounded by an equipment grounding conductor, in accordance with Part F of Article 250.

(j) Lighting Fixtures. Lighting fixtures as provided in Part E of Article 410.

(k) Skid-Mounted Equipment. Permanently mounted electrical equipment and skids shall be grounded with an equipment bonding jumper sized as required by Section 250-122.

(l) Motor-Operated Water Pumps. Motor-operated water pumps including the submersible type.

The requirement of Section 250-112(l) is intended to reduce stray voltages and minimize shock hazard during maintenance, where the pump is hauled out of the well casing and might be tested in a barrel of water.

(m) Metal Well Casings. Where a submersible pump is used in a metal well casing, the well casing shall be bonded to the pump circuit equipment grounding conductor.

Section 250-112(m) is intended to prevent a shock hazard that could exist due to a potential difference between the pump, which is grounded to the system ground, and the metal well casing.

250-114. Equipment Connected by Cord and Plug. Under any of the conditions described in (1) through (4), exposed noncurrent-carrying metal parts of cord- and plug-connected equipment likely to become energized shall be grounded.

Exception: Listed tools, listed appliances, and listed equipment covered in (2) through (4) shall not be required to be grounded where protected by a system of double insulation or its equivalent. Double insulated equipment shall be distinctively marked.

The exception to Section 250-114 recognizes listed double-insulated appliances, motor-operated hand-held tools, stationary and fixed motor-operated tools, and light industrial motor-operated tools as not requiring equipment grounding connections.

(1) In hazardous (classified) locations (see Articles 500 through 517)
(2) Where operated at over 150 volts to ground

Exception No. 1: Motors, where guarded, shall not be required to be grounded.

Exception No. 2: Metal frames of electrically heated appliances, exempted by special permission, shall not be required to be grounded, in which case the frames shall be permanently and effectively insulated from ground.

(3) In residential occupancies
  (a) Refrigerators, freezers, and air conditioners
  (b) Clothes-washing, clothes-drying, dish-washing machines; information technology equipment; sump pumps and electrical aquarium equipment
  (c) Hand-held motor-operated tools, stationary and fixed motor-operated tools, light industrial motor-operated tools
  (d) Motor-operated appliances of the following types: hedge clippers, lawn mowers, snow blowers, and wet scrubbers
  (e) Portable handlamps
(4) In other than residential occupancies
  (a) Refrigerators, freezers, and air conditioners
  (b) Clothes-washing, clothes-drying, dish-washing machines; information technology equipment; sump pumps and electrical aquarium equipment
  (c) Hand-held motor-operated tools, stationary and fixed motor-operated tools, light industrial motor-operated tools
  (d) Motor-operated appliances of the following types: hedge clippers, lawn mowers, snow blowers, and wet scrubbers
  (e) Cord- and plug-connected appliances used in damp or wet locations or by persons standing on the ground or on metal floors or working inside of metal tanks or boilers
  (f) Tools likely to be used in wet or conductive locations
  (g) Portable handlamps

Exception: Tools and portable handlamps likely to be used in wet or conductive locations shall not be required to be grounded where supplied through an isolating transformer with an ungrounded secondary of not over 50 volts.

Tools are required to be grounded by an equipment grounding conductor within the cord or cable supplying the tool, except where the tool is supplied by an isolating transformer, as permitted by the exception. Portable tools and appliances protected by an approved system of double insulation are required to be listed by a qualified electrical testing laboratory as being suitable for the purpose, and the equipment is required to be distinctively marked as double insulated.

Cord-connected portable tools or appliances are not intended to be used in damp, wet, or conductive locations unless they are grounded, supplied by an isolation transformer with a secondary of not more than 50 volts, or protected by an approved system of double insulation.

Figure 250.42 shows an example of lighting equipment supplied through an isolating transformer operating at 6 or 12 volts that provides safe illumination for work inside boilers, tanks, and similar locations that may be metal or wet. (Daniel Woodhead Co.)

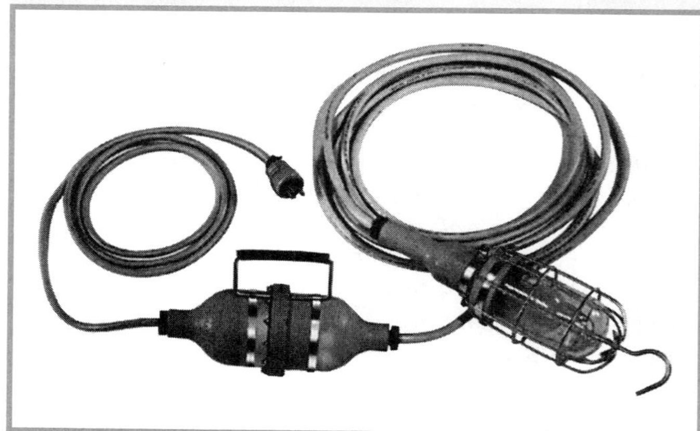

Figure 250.42 *Lighting equipment not required to be grounded where supplied through an isolating transformer operating at 6 or 12 volts.*

250-116. Nonelectric Equipment. The metal parts of nonelectric equipment described in this section shall be grounded.

(1) Frames and tracks of electrically operated cranes and hoists
(2) Frames of nonelectrically driven elevator cars to which electric conductors are attached
(3) Hand-operated metal shifting ropes or cables of electric elevators

FPN: Where extensive metal in or on buildings may become energized and is subject to personal contact, adequate bonding and grounding will provide additional safety.

Because metal siding on buildings is not electrical equipment, it is outside the scope of the *Code* [see Section 90-2(a)], and the *Code* cannot require that it be grounded.

Quite often, lighting fixtures, signs, or receptacles are installed on buildings with metal siding that could become energized and is subject to contact by persons. Therefore, grounding of metal siding reduces the shock hazard to personnel.

See Sections 547-8 and 547-9 for additional rules on grounding and bonding of agricultural buildings.

250-118. Types of Equipment Grounding Conductors. The equipment grounding conductor run with or enclosing the circuit conductors shall be one or more or a combination of the following.

Figure 250.43 provides an example of the various sizes of metal conduits enclosing the circuit conductors used as equipment grounding conductors, as they apply to a service and feeder system.

(1) A copper or other corrosion-resistant conductor. This conductor shall be solid or stranded; insulated, covered, or bare; and in the form of a wire or a busbar of any shape.
(2) Rigid metal conduit.
(3) Intermediate metal conduit.
(4) Electrical metallic tubing.
(5) Flexible metal conduit where both the conduit and fittings are listed for grounding.
(6) Listed flexible metal conduit that is not listed for grounding, meeting all the following conditions.
   (a) The conduit is terminated in fittings listed for grounding.
   (b) The circuit conductors contained in the conduit are protected by overcurrent devices rated at 20 amperes or less.
   (c) The combined length of flexible metal conduit and flexible metallic tubing and liquidtight flexible metal conduit in the same ground return path does not exceed 6 ft (1.83 m).
   (d) The conduit is not installed for flexibility.
(7) Listed liquidight flexible metal conduit meeting all the following conditions.

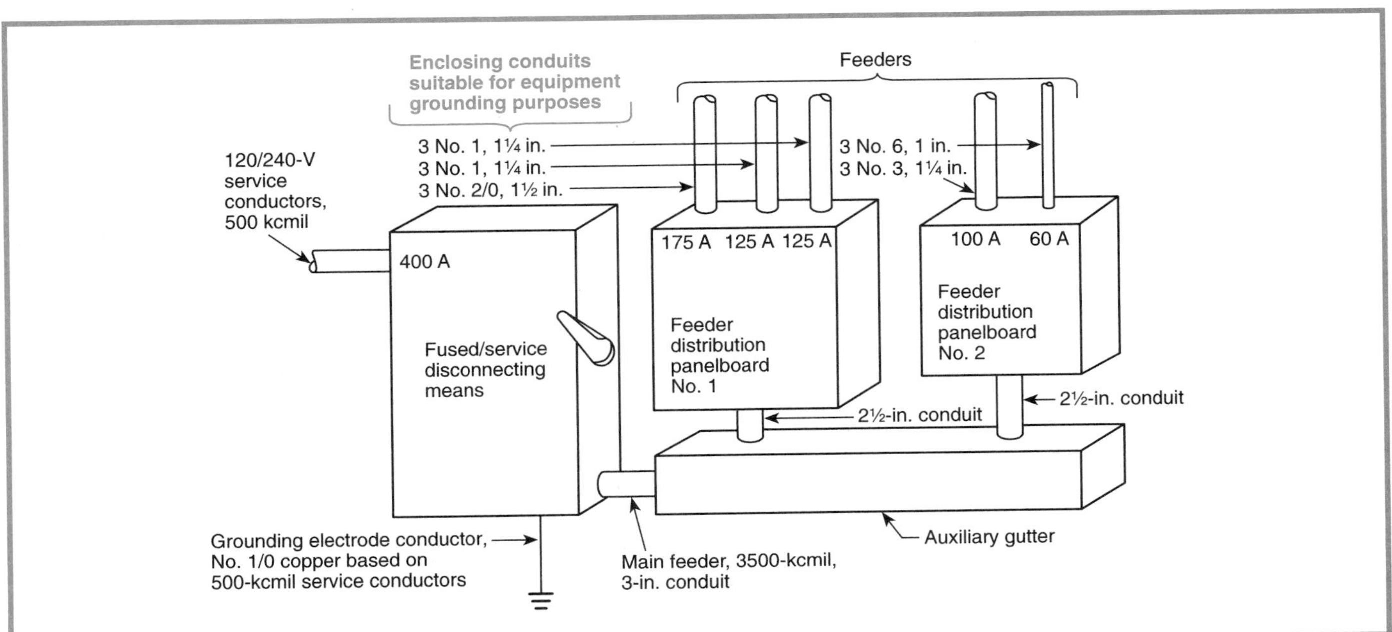

Figure 250.43 *An example of the various sizes of enclosing metal conduits used as equipment grounding conductors, as they apply to a service and feeder system.*

(a) The conduit is terminated in fittings listed for grounding.
(b) For trade sizes ⅜ in. through ½ in., the circuit conductors contained in the conduit are protected by overcurrent devices rated at 20 amperes or less.
(c) For trade sizes ¾ in. through 1¼ in., the circuit conductors contained in the conduit are protected by overcurrent devices rated not more than 60 amperes and there is no flexible metal conduit, flexible metallic tubing, or liquidtight flexible metal conduit in trade sizes ⅜ in. or ½ in. in the grounding path.
(d) The combined length of flexible metal conduit and flexible metallic tubing and liquidtight flexible metal conduit in the same ground return path does not exceed 6 ft (1.83 m).
(e) The conduit is not installed for flexibility.

(8) Flexible metallic tubing where the tubing is terminated in fittings listed for grounding and meeting all the following conditions.

(a) The circuit conductors contained in the tubing are protected by overcurrent devices rated at 20 amperes or less.
(b) The combined length of flexible metal conduit and flexible metallic tubing and liquidtight flexible metal conduit in the same ground return path does not exceed 6 ft (1.83 m).

(9) Armor of Type AC cable as provided in Section 333-21.
(10) The copper sheath of mineral-insulated, metal-sheathed cable.
(11) The metallic sheath or the combined metallic sheath and grounding conductors of Type MC cable.
(12) Cable trays as permitted in Sections 318-3(c) and 318-7.
(13) Cablebus framework as permitted in Section 365-2(a).
(14) Other electrically continuous metal raceways listed for grounding.

250-119. Identification of Equipment Grounding Conductors. Unless required elsewhere in this *Code,* equipment grounding conductors shall be permitted to be bare, covered, or insulated. Individually covered or insulated equipment grounding conductors shall have a continuous outer finish that is either green or green with one or more yellow stripes except as permitted in this section.

(a) Conductors Larger than No. 6. An insulated or covered conductor larger than No. 6 copper or aluminum shall be permitted, at the time of installation, to be permanently identified as an equipment grounding conductor at each end and at every point where the conductor is accessible. Identification shall be accomplished by one of the following:

(1) Stripping the insulation or covering from the entire exposed length
(2) Coloring the exposed insulation or covering green
(3) Marking the exposed insulation or covering with green tape or green adhesive labels

(b) Multiconductor Cable. Where the conditions of maintenance and supervision ensure that only qualified persons will service the installation, one or more insulated conductors in a multiconductor cable, at the time of installation, shall be permitted to be permanently identified as equipment grounding conductors at each end and at every point where the conductors are accessible by one of the following means:

(1) Stripping the insulation from the entire exposed length
(2) Coloring the exposed insulation green
(3) Marking the exposed insulation with green tape or green adhesive labels

(c) Flexible Cord. An uninsulated equipment grounding conductor shall be permitted, but, if individually covered, the covering shall have a continuous outer finish that is either green or green with one or more yellow stripes.

250-120. Equipment Grounding Conductor Installation. An equipment grounding conductor shall be installed as follows.

(a) Raceway, Cable Trays, Cable Armor, or Cable Sheaths. Where it consists of a raceway, cable tray, cable armor, or cable sheath or where it is a wire within a raceway or cable, it shall be installed in accordance with the applicable provisions in this *Code* using fittings for joints and terminations approved for use with the type raceway or cable used. All connections, joints, and fittings shall be made tight using suitable tools.

(b) Aluminum and Copper-Clad Aluminum Conductors. Aluminum and copper-clad aluminum conductors shall be installed in accordance with the restrictions of Section 250-64.

(c) Equipment Grounding Conductors Smaller than No. 6. Equipment grounding conductors smaller than No. 6 shall be protected from physical damage by a raceway or cable armor except where run in hollow spaces of walls or partitions, where not subject to physical damage, or where protected from physical damage.

250-122. Size of Equipment Grounding Conductors.

(a) General. Copper, aluminum, or copper-clad aluminum equipment grounding conductors of the wire type shall not be smaller than shown in Table 250-122, but shall not be required to be larger than the circuit conductors supplying the equipment. Where a raceway or a cable armor or sheath is used as the equipment grounding conductor, as provided in Sections 250-118 and 250-134(a), it shall comply with Section 250-2(d).

The last sentence of Section 250-122(a) alerts users that some of the wiring methods allowed for grounding purposes must be evaluated further where a long distance exists between a power source and utilization equipment.

(b) Adjustment for Voltage Drop. Where conductors are adjusted in size to compensate for voltage drop, equipment grounding conductors, where installed, shall be adjusted proportionately according to circular mil area.

Equipment grounding conductors on the load side of the service disconnecting means overcurrent devices are sized based on the size of the overcurrent devices ahead of them. Where the ungrounded conductors are increased in size to compensate for voltage drop, the equipment grounding conductors must also be increased proportionately.

Example

A 240-volt, single-phase, 250-ampere load is supplied from a 300-ampere breaker located in a panelboard 500 ft away. The conductors are 250 kcmil copper, installed in rigid nonmetallic conduit with a No. 4 copper equipment grounding conductor. The conductors are increased to 350 kcmil; therefore, the equipment grounding conductor must be increased as follows:

$$\frac{350{,}000 \text{ circular mils}}{250{,}000 \text{ circular mils}} = 1.4$$

Referring to Chapter 9, Table 8, we find that No. 4 has a cross-sectional area of 41,740 circular mils. $1.4 \times 41{,}740$ circular mils $= 58{,}436$ circular mils

Again referring to Chapter 9, Table 8, we find that 58,436 circular mils converts to a No. 2 copper equipment grounding conductor.

(c) Multiple Circuits. Where a single equipment grounding conductor is run with multiple circuits in the same raceway or cable, it shall be sized for the largest overcurrent device protecting conductors in the raceway or cable.

A single equipment grounding conductor is required to be sized for the largest overcurrent device and is not required to be sized for the composite of all the circuits in the raceway. It is not anticipated that all circuits will develop faults at the same time.

Example

Three 3-phase circuits in the same raceway, protected by overcurrent devices rated 30, 60, and 100 amperes, would require only one equipment grounding conductor, sized according to the largest overcurrent device (in this case, 100 amperes). Therefore, a No. 8 copper or No. 6 aluminum conductor or copper-clad aluminum conductor is required by Table 250-122.

(d) Motor Circuits. Where the overcurrent device consists of an instantaneous trip circuit breaker or a motor short-circuit protector, as allowed in Section 430-52, the equipment grounding conductor size shall be permitted to be based on the rating of the motor overload protective device but not less than the size shown in Table 250-122.

(e) Flexible Cord and Fixture Wire. Equipment grounding conductors that are part of flexible cords or used with fixture wires in accordance with Section 240-4 shall be not smaller than No. 18 copper and not smaller than the circuit conductors.

(f) Conductors in Parallel. Where conductors are run in parallel in multiple raceways or cables as permitted in Section 310-4, the equipment grounding conductors, where used, shall be run in parallel in each raceway or cable. One of the following methods shall be used to ensure the equipment grounding conductors are protected.

(1) Each parallel equipment grounding conductor shall be sized on the basis of the ampere rating of the overcurrent device protecting the circuit conductors in the raceway or cable in accordance with Table 250-122.

A full-sized equipment grounding conductor is needed in each raceway of an array of raceways enclosing paralleled phase conductors, to prevent overloading and possible burn-out of the grounding conductor should a ground fault occur along one of the parallel branches. The installation conditions for paralleled conductors prescribed in Section 310-4 result in proportional distribution of the current-time duty among the several paralleled grounding conductors only for overcurrent conditions downstream of the paralleled array.

This requirement is also applicable to cables that are installed in parallel. Since cable assemblies are manufactured in standard conductor size configurations, the equipment grounding conductor in a cable is properly sized for some circuit arrangements. However, where the cable is used in large capacity parallel circuits, the equipment grounding conductor in each cable may not be sized in accordance with Table 250-122. To address this problem, Section 250-122(f)(2) permits the sizing of the equipment grounding conductor within a multiconductor cable to be based on the trip setting of an equipment ground-fault device. The use of this method of protection is only permitted where the installation will be serviced by qualified personnel and the ground-fault device is specifically listed for this purpose. The trip setting of the GFPE cannot exceed the ampacity of a single ungrounded conductor installed in one of the parallel cables. This will preclude the need to manufacture multiconductor cables with equipment grounding conductors that are sized for a specific parallel circuit configuration.

Figure 250.44 shows a parallel array of two nonmetallic conduits installed underground. For clarity, a one-line diagram with equipment grounding conductors is shown. A ground fault at the enclosure will cause the equipment grounding conductor in conduit No. 1 to carry more than its proportionate share of fault current. Note that the fault is fed by two different conductors of the same phase, one from the left and two from the right. The shortest and lowest-impedance path to ground from the fault to the supply panelboard is through the equipment grounding conductor in conduit No. 1. The grounding path from the fault through conduit No. 2 is longer and of higher impedance. Therefore, the equipment grounding conductor in each raceway must be capable of carrying a major portion of the fault current without burning open.

Figure 250.45 illustrates the provisions of Section 250-122(f)(2) for multiconductor cables used in a parallel circuit arrangement. The equipment grounding conductor in each cable is sized from Table 250-122 based on the trip setting of the GFPE device. The setting cannot exceed the ampacity of a single ungrounded conductor contained within the multiconductor cable.

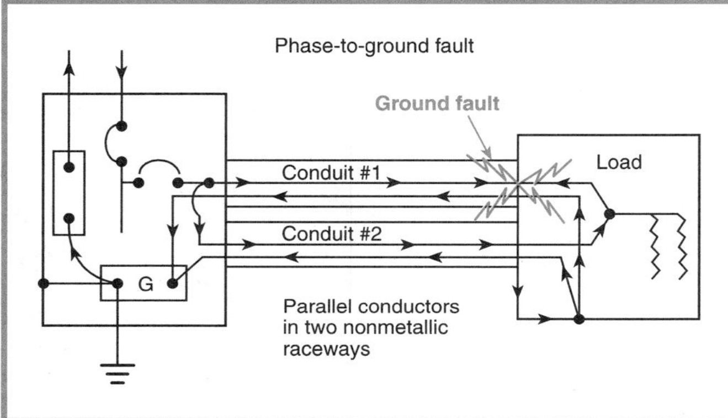

Figure 250.44 Grounding paths for ground fault at the load supplied by parallel conductors in two nonmetallic raceways, illustrating the reason for the requirement of Section 250-122(f)(1).

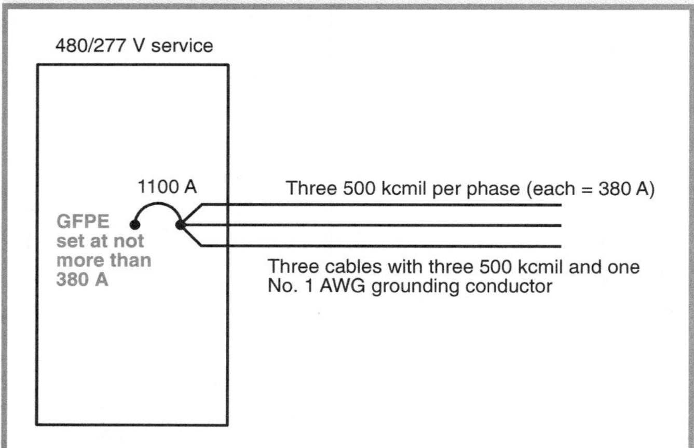

Figure 250.45 Use of ground fault protection of equipment device with multiconductor cables installed in parallel.

(2) Where ground-fault protection of equipment is installed, each parallel equipment grounding conductor in a multiconductor cable shall be permitted to be sized in accordance with Table 250-122 on the basis of the trip rating of the ground-fault protection where the following conditions are met.

(1) Conditions of maintenance and supervision ensure that only qualified persons will service the installation.
(2) The ground-fault protection equipment is set to trip at not more than the ampacity of a single ungrounded conductor of one of the cables in parallel.
(3) The ground-fault protection is listed for the purpose.

Table 250-122. Minimum Size Equipment Grounding Conductors for Grounding Raceway and Equipment

Rating or Setting of Automatic Overcurrent Device in Circuit Ahead of Equipment, Conduit, etc., Not Exceeding (Amperes)	Size (AWG or kcmil)	
	Copper	Aluminum or Copper-Clad Aluminum*
15	14	12
20	12	10
30	10	8
40	10	8
60	10	8
100	8	6
200	6	4
300	4	2
400	3	1
500	2	1/0
600	1	2/0
800	1/0	3/0
1000	2/0	4/0
1200	3/0	250
1600	4/0	350
2000	250	400
2500	350	600
3000	400	600
4000	500	800
5000	700	1200
6000	800	1200

Note: Where necessary to comply with Section 250-2(d), the equipment grounding conductor shall be sized larger than this table.

*See installation restrictions in Section 250-120.

250-124. Equipment Grounding Conductor Continuity.

(a) Separable Connections. Separable connections such as those provided in drawout equipment or attachment plugs and mating connectors and receptacles shall provide for first-make, last-break of the equipment grounding conductor. First-make, last-break shall not be required where interlocked equipment, plugs, receptacles, and connectors preclude energization without grounding continuity.

(b) Switches. No automatic cutout or switch shall be placed in the equipment grounding conductor of a premises wiring system unless the opening of the cutout or switch disconnects all sources of energy.

250-126. Identification of Wiring Device Terminals. The terminal for the connection of the equipment grounding conductor shall be identified by one of the following.

(1) A green, not readily removable terminal screw with a hexagonal head.
(2) A green, hexagonal, not readily removable terminal nut.
(3) A green pressure wire connector. If the terminal for the grounding conductor is not visible, the conductor entrance hole shall be marked with the word *green* or *ground*, the letters *G* or *GR* or the grounding symbol shown in Figure 250-126, or otherwise identified by a distinctive green color. If the terminal for the equipment grounding conductor is readily removable, the area adjacent to the terminal shall be similarly marked.

Figure 250-126 Grounding symbol.

G. Methods of Equipment Grounding

250-130. Equipment Grounding Conductor Connections. Equipment grounding conductor connections at the source of separately derived systems shall be made in accordance with Section 250-30(a)(1). Equipment grounding conductor connections at service equipment shall be made as indicated in (a) or (b). For replacement of nongrounding-type receptacles with grounding type receptacles and for branch-circuit extensions only in existing installations that do not have an equipment grounding conductor in the branch circuit, connections shall be permitted as indicated in (c).

(a) For Grounded Systems. The connection shall be made by bonding the equipment grounding conductor to the grounded service conductor and the grounding electrode conductor.

The grounding arrangement for a grounded system is illustrated in Figure 250.46.

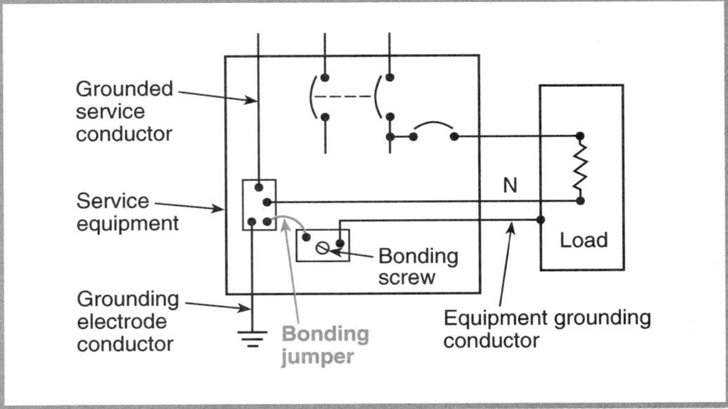

Figure 250.46 *Grounding for grounded systems, per Section 250-130(a), illustrating connection of the equipment grounding bus to the enclosures and the grounded service conductor.*

(b) For Ungrounded Systems. The connection shall be made by bonding the equipment grounding conductor to the grounding electrode conductor.

(c) Nongrounding Receptacle Replacement or Branch Circuit Extensions. The equipment grounding conductor of a grounding-type receptacle or a branch-circuit extension shall be permitted to be connected to any of the following:

Section 250-130(c) applies to both ungrounded and grounded systems. It permits a nongrounding-type receptacle to be replaced with a grounding-type receptacle under the following conditions.

1. The branch circuit does not contain an equipment ground.
2. An existing branch circuit is being extended for additional receptacle outlets.
3. An equipment grounding conductor is connected between the receptacle grounding terminal and any accessible point on the grounding electrode system.

Due to changes made in the 1993 *Code* to Section 250-50 (formerly 250-81), an interior metal water pipe more than 5 ft from the point of entrance of the water pipe into the building is no longer allowed to serve as a conductor. This revision was due to the increasing use of nonmetallic piping and fittings.

Figure 250.47 shows a branch circuit extension made from an existing installation. This method is also permitted to ground a replacement 3-wire receptacle in the existing ungrounded box on the left, where no grounding conductor is available.

(1) Any accessible point on the grounding electrode system as described in Section 250-50
(2) Any accessible point on the grounding electrode conductor
(3) The equipment grounding terminal bar within the enclosure where the branch circuit for the receptacle or branch circuit originates
(4) For grounded systems, the grounded service conductor within the service equipment enclosure

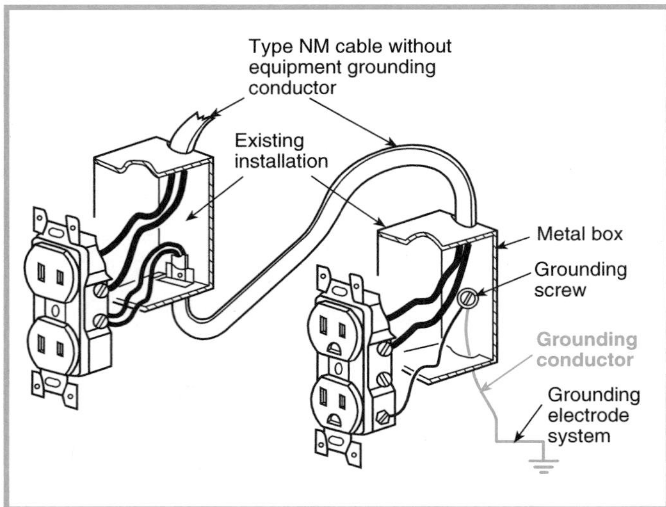

Figure 250.47 Branch-circuit extension to an existing installation, per Section 250-130(c), illustrating a separate grounding conductor connected to the main grounding electrode.

(5) For ungrounded systems, the grounding terminal bar within the service equipment enclosure

FPN: See Section 210-7(d) for the use of a ground-fault circuit-interrupting type of receptacle.

250-132. Short Sections of Raceway. Isolated sections of metal raceway or cable armor, where required to be grounded, shall be grounded in accordance with Section 250-134.

250-134. Equipment Fastened in Place or Connected by Permanent Wiring Methods (Fixed) — Grounding. Unless grounded by connection to the grounded circuit conductor as permitted by Sections 250-32, 250-140, and 250-142, noncurrent-carrying metal parts of equipment, raceways, and other enclosures, if grounded, shall be grounded by one of the following methods.

Section 250-134 eliminates any conflict between Section 250-134(a), which requires an equipment grounding conductor to be used for equipment grounding, and Sections 250-32, 250-140, and 250-142, which permit the grounded circuit conductor to be used for equipment grounding if certain specified conditions are met.

(a) Equipment Grounding Conductor Types. By any of the equipment grounding conductors permitted by Section 250-118.

(b) With Circuit Conductors. By an equipment grounding conductor contained within the same raceway, cable, or otherwise run with the circuit conductors.

One of the functions of an equipment grounding conductor is to provide a low-impedance ground-fault path between a ground fault and the electrical source, which allows the overcurrent protective device to actuate, interrupting the current. To keep the impedance at a minimum, it is necessary to run the equipment grounding conductor within the same raceway or cable as the circuit conductor(s). This allows the magnetic field developed by the circuit conductor and the equipment grounding conductor to cancel, reducing their impedance. Magnetic flux strength is inversely proportional to the square of the distance between the two conductors. By placing an equipment conductor away from the conductor delivering the fault current, the magnetic flux cancellation decreases. This increases the impedance of the fault path and delays operation of the protective device.

Exception No. 1: As provided in Section 250-130(c), the equipment grounding conductor shall be permitted to be run separately from the circuit conductors.

Exception No. 1 to Section 250-134(b) permits an equipment grounding conductor to be run to the grounding electrode separately from the other conductors of the circuit. This applies only where a grounding-type receptacle is used on a circuit that does not include an equipment grounding conductor. See the commentary following Section 250-130(c) for further explanation.

Exception No. 2: For dc circuits, the equipment grounding conductor shall be permitted to be run separately from the circuit conductors.

FPN No. 1: See Sections 250-102 and 250-168 for equipment bonding jumper requirements.

FPN No. 2: See Section 400-7 for use of cords for fixed equipment.

250-136. Equipment Considered Effectively Grounded. Under the conditions specified in (a) and (b), the noncurrent-carrying metal parts of the equipment shall be considered effectively grounded.

(a) Equipment Secured to Grounded Metal Supports. Electric equipment secured to and in electrical contact with a metal rack or structure provided for its support and grounded by one of the means indicated in Section 250-134. The structural metal frame of a building shall not be used as the required equipment grounding conductor for ac equipment.

Figure 250.48 shows an example of electrical equipment secured to and in electrical contact with a metal rack that is effectively grounded in accordance with Section 250-136(a).

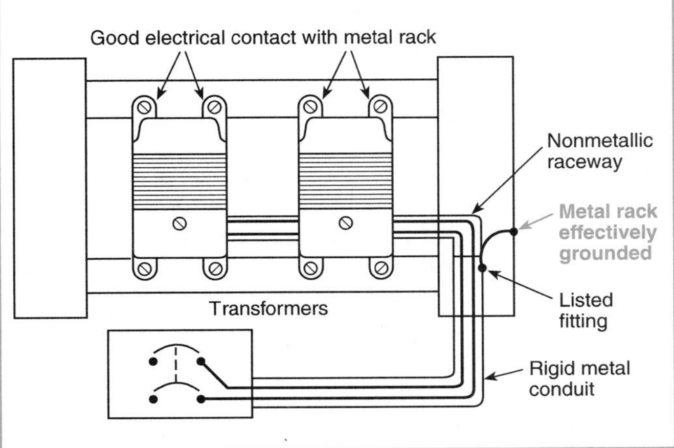

Figure 250.48 *An example of equipment grounding on a metal rack.*

(b) Metal Car Frames. Metal car frames supported by metal hoisting cables attached to or running over metal sheaves or drums of elevator machines that are grounded by one of the methods indicated in Section 250-134.

250-138. Cord- and Plug-Connected Equipment. Non-current-carrying metal parts of cord- and plug-connected equipment if grounded, shall be grounded by one of the following methods.

(a) By Means of an Equipment Grounding Conductor. By means of an equipment grounding conductor run with the power supply conductors in a cable assembly or flexible cord properly terminated in a grounding-type attachment plug with one fixed grounding contact.

Exception: The grounding contacting pole of grounding-type plug-in ground-fault circuit interrupters shall be permitted to be of the movable, self-restoring type on circuits operating at not over 150 volts between any two conductors, or over 150 volts between any conductor and ground.

(b) By Means of a Separate Flexible Wire or Strap. By means of a separate flexible wire or strap, insulated or bare, protected as well as practicable against physical damage, where part of equipment.

250-140. Frames of Ranges and Clothes Dryers. This section shall apply to existing branch-circuit installations only. New branch-circuit installations shall comply with Sections 250-134 and 250-138. Frames of electric ranges, wall-mounted ovens, counter-mounted cooking units, clothes dryers, and outlet or junction boxes that are part of the circuit for these appliances shall be grounded in the manner specified by Section 250-134 or 250-138; or, except for mobile homes and recreational vehicles, shall be permitted to be grounded to the grounded circuit conductor if all of the following conditions are met.

(1) The supply circuit is 120/240-volt, single-phase, 3-wire; or 208Y/120-volt derived from a 3-phase, 4-wire wye-connected system.
(2) The grounded conductor is not smaller than No. 10 copper or No. 8 aluminum.
(3) The grounded conductor is insulated, or the grounded conductor is uninsulated and part of a Type SE service-entrance cable and the branch circuit originates at the service equipment.
(4) Grounding contacts of receptacles furnished as part of the equipment are bonded to the equipment.

Section 250-140(4) applies only to existing branch circuits supplying the appliances specified in this section. The grounded conductor (neutral) is no longer allowed to be used for grounding the metal noncurrent-carrying parts of the appliances listed in this section. Branch circuits installed for new appliance installations are required to provide an equipment grounding conductor for grounding the noncurrent-carrying parts.

Caution should be exercised to ensure that new appliances connected to an existing branch circuit are properly grounded. An older appliance connected to a new branch circuit must have its 3-wire cord and plug replaced with a 4-conductor cord, one being an equipment grounding conductor. The bonding jumper between the neutral and frame of the appliance must be removed. If a new appliance is connected to an existing branch circuit where the neutral conductor was previously used for grounding the appliance, a bonding jumper must be installed at the appliance terminals to connect the frame to the neutral.

The grounded circuit conductor of an existing branch circuit is still permitted to be used to ground the frame of an electric range, wall-mounted oven, or counter-mounted cooking unit, provided all the conditions of (1) through (4) are met. The grounded circuit conductor is also permitted to be used to ground any junction boxes in the circuit supplying the appliance, and a 3-wire pigtail and range receptacle are permitted to be used, regardless of whether the circuit to the receptacle contains a separate equipment grounding conductor.

Where service-entrance cable was previously installed, an uninsulated neutral was allowed. However, the circuit was required to originate at the service equipment to avoid neutral current from downstream panelboards flowing on metal objects, such as pipes or ducts. Figure 250.49 shows an existing installation where Type SE service-entrance cable was used for ranges, dryers, wall-mounted ovens, and counter-mounted cooking units. Junction boxes in the supply circuit were also permitted to be grounded from the grounded neutral conductor.

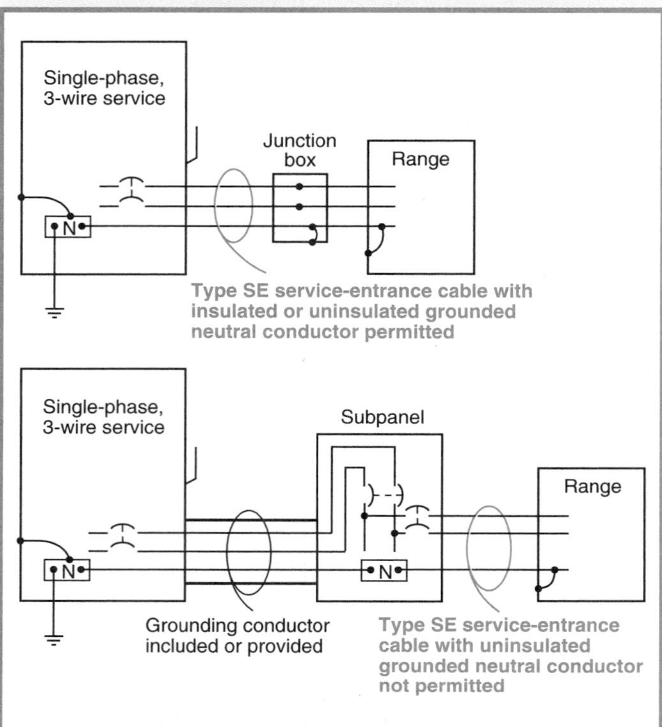

Figure 250.49 *An existing installation where Type SE service-entrance cable was used for appliances specified in Section 250-140.*

250-142. Use of Grounded Circuit Conductor for Grounding Equipment.

(a) Supply-Side Equipment. A grounded circuit conductor shall be permitted to ground noncurrent-carrying metal parts of equipment, raceways, and other enclosures at any of the following locations:

(1) On the supply side or within the enclosure of the ac service-disconnecting means

(2) On the supply side or within the enclosure of the main disconnecting means for separate buildings as provided in Section 250-32(b)

(3) On the supply side or within the enclosure of the main disconnecting means or overcurrent devices of a separately derived system where permitted by Section 250-30(a)(1)

In separately derived systems, the grounded circuit conductor is permitted to ground noncurrent-carrying metal parts of equipment, raceways, and other enclosures only on the supply side of the main disconnecting means.

(b) Load-Side Equipment. Except as permitted in Sections 250-30(a)(1) and 250-32(b), a grounded circuit conductor shall not be used for grounding noncurrent-carrying metal parts of equipment on the load side of the service disconnecting means or on the load side of a separately derived system disconnecting means or the overcurrent devices for a separately derived system not having a main disconnecting means.

One major reason the grounded circuit conductor is not permitted to be grounded on the load side of the service (except as allowed in Section 250-140) is that, should the grounded service conductor become disconnected at any point on the line side of the ground, the equipment grounding conductor and all conductive parts connected to it will carry the neutral current, raising the potential to ground of exposed metal parts not normally intended to carry current. This could result in arcing in concealed spaces and could pose a severe shock hazard, particularly if the path should be opened inadvertently by a worker. Even without an open grounded conductor (usually referred to as an open neutral), the equipment grounding conductor path will become a parallel path with the grounded conductor, and there will be some potential drop on exposed and concealed dead metal parts. The magnitude of this potential difference will be determined by the relative impedances of the equipment grounding path and grounded conductor circuits. Not only would the equipment grounding conductor path be affected, but all parallel paths not intended as equipment grounding conductors would be affected as well. This could involve current flowing through metal building structures, piping, and ducts.

Exception No. 1: The frames of ranges, wall-mounted ovens, counter-mounted cooking units, and clothes dryers under the conditions permitted for existing installations by Section 250-140 shall be permitted to be grounded by a grounded circuit conductor.

Exception No. 2: It shall be permissible to ground meter enclosures by connection to the grounded circuit conductor on the load-side of the service disconnect if

(a) No service ground-fault protection is installed, and

(b) All meter enclosures are located near the service disconnecting means, and

(c) The size of the grounded circuit conductor is not smaller than the size specified in Table 250-122 for equipment grounding conductors.

Exception No. 3: Direct-current systems shall be permitted to be grounded on the load side of the disconnecting means or overcurrent device in accordance with Section 250-164.

250-144. Multiple Circuit Connections. Where equipment is required to be grounded, and is supplied by separate connection to more than one circuit or grounded premises wiring system, a means for grounding shall be provided for each such connection as specified in Sections 250-134 and 250-138.

250-146. Connecting Receptacle Grounding Terminal to Box. An equipment bonding jumper shall be used to connect the grounding terminal of a grounding-type receptacle to a grounded box unless grounded as in (a) through (d).

(a) Surface Mounted Box. Where the box is mounted on or at the surface, direct metal-to-metal contact between the device yoke and the box shall be permitted to ground the receptacle to the box. This provision shall not apply to cover-mounted receptacles unless the box and cover combination are listed as providing satisfactory ground continuity between the box and the receptacle.

The main rule of Section 250-146 requires an equipment bonding jumper to be installed between the device box and the receptacle grounding terminal. However, Section 250-146(a) permits the equipment bonding jumper to be omitted if the metal yoke of the device is in direct contact with the metal device box ears as illustrated in Figure 250.50.

Cover-mounted wiring devices, such as on 4-in. square covers, are not considered grounded. Section 250-146(a) is not applicable to cover-mounted receptacles, as illustrated in Figure 250.51. Box-cover and device combinations listed as providing grounding continuity are permitted.

(b) Contact Devices or Yokes. Contact devices or yokes designed and listed for the purpose shall be permitted in conjunction with the supporting screws to establish the grounding circuit between the device yoke and flush-type boxes.

Section 250-146(b) is illustrated by Figure 250.52, which shows a receptacle designed with a spring-type grounding strap for holding the mounting screw and establishing the grounding circuit so that no bonding jumper is required.

(c) Floor Boxes. Floor boxes designed for and listed as providing satisfactory ground continuity between the box and the device shall be permitted.

(d) Isolated Receptacles. Where required for the reduction of electrical noise (electromagnetic interference) on the grounding circuit, a receptacle in which the grounding terminal is purposely insulated from the receptacle mounting means shall be permitted. The receptacle grounding terminal

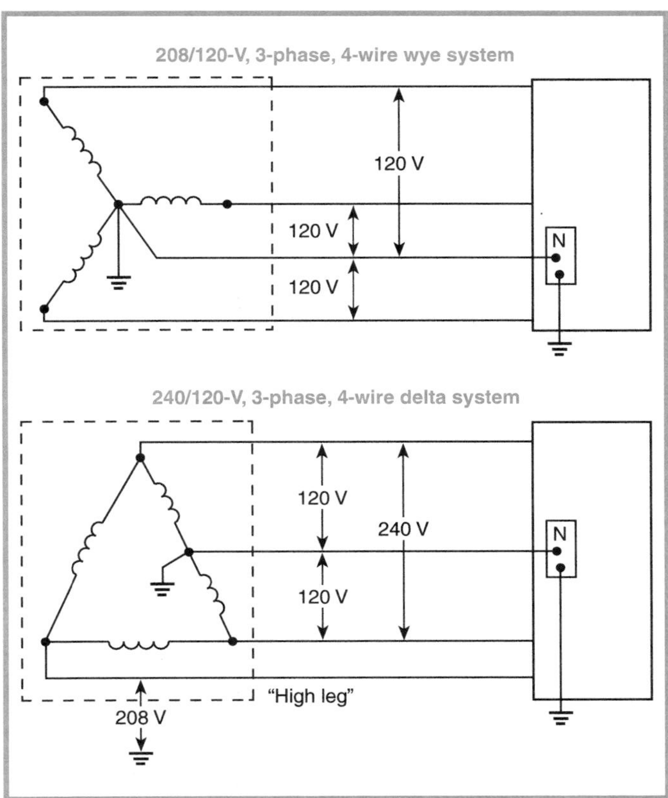

Figure 250.50 *An example of a box-mounted receptacle attached to a surface box where bonding jumper is not required.*

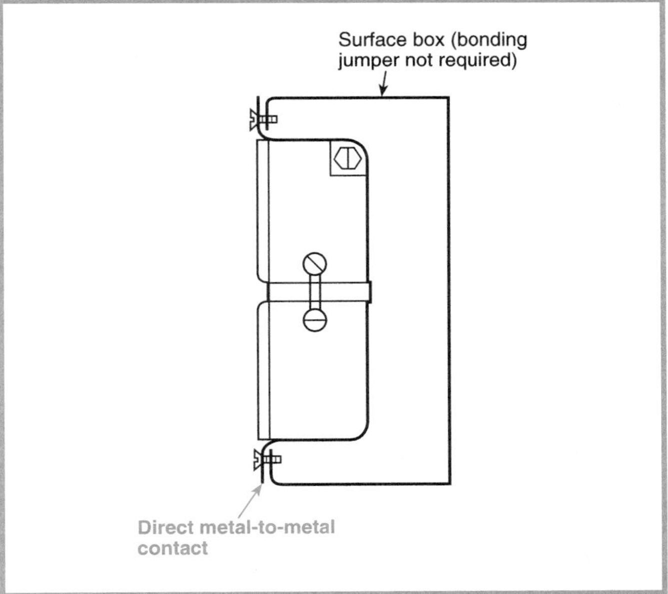

Figure 250.51 *An example of a cover-mounted receptacle attached to a surface box where bonding jumper is not required.*

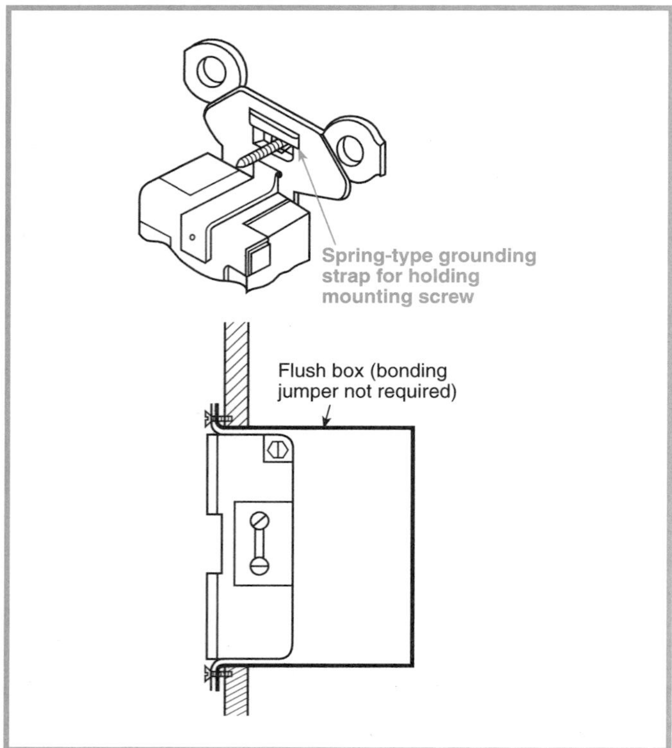

Figure 250.52 *A receptacle designed with a spring-type grounding strap that holds the mounting screw captive and eliminates a bonding jumper to the box, in accordance with Section 250-146(b).*

shall be grounded by an insulated equipment grounding conductor run with the circuit conductors. This grounding conductor shall be permitted to pass through one or more panelboards without connection to the panelboard grounding terminal as permitted in Section 384-20, Exception, so as to terminate within the same building or structure directly at an equipment grounding conductor terminal of the applicable derived system or service.

Section 250-146(d) allows an isolated-ground-type receptacle to be installed without a bonding jumper between the metal device box and the receptacle grounding terminal. An insulated equipment grounding conductor as shown in Figure 250.53 is installed with the branch-circuit conductors. This conductor may originate in the service panel, pass through any number of subpanels without being connected to the equipment grounding bus, and terminate at the isolated-ground-type receptacle ground terminal. However, this does not exempt the metal device box from being grounded. The metal device box must be grounded either by an equipment grounding conductor run with the circuit conductors or by a wiring method that serves as an equipment grounding conductor. See Section 250-118 for types of equipment grounding conductors.

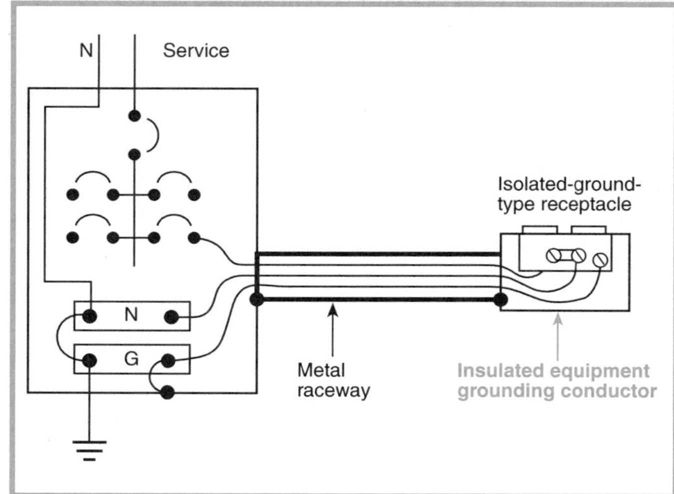

Figure 250.53 *An isolated-ground-type receptacle with an insulated equipment grounding conductor and with the device box grounded through the metal raceway.*

Section 250-146(d) requires that where isolated-ground-type receptacles are used, the isolated equipment grounding conductor must terminate at an equipment grounding terminal of the applicable service or derived system within the same building as the receptacle. If the isolated equipment grounding conductor terminates at a separate building, a large voltage difference may exist between buildings during lightning transients. Such transients could cause damage to equipment connected to an isolated-ground-type receptacle and present a shock hazard between the isolated equipment frame and other grounded surfaces.

FPN: Use of an isolated equipment grounding conductor does not relieve the requirement for grounding the raceway system and outlet box.

The fine print note to Section 250-146(d) is a reminder that metallic raceways and boxes are still required to be grounded by one of the usual required methods. This could require a separate grounding conductor, for example, to ground a metal box in a nonmetallic raceway system or to ground a metal box supplied by flexible metal conduit. Where replacing an ordinary grounding-type receptacle with an isolated-ground type receptacle, use of an existing equipment grounding conductor as the isolated receptacle grounding conductor could effectively defeat or seriously com-

promise the required box or raceway equipment ground.

250-148. Continuity and Attachment of Equipment Grounding Conductors to Boxes. Where more than one equipment grounding conductor enters a box, all such conductors shall be spliced or joined within the box or to the box with devices suitable for the use. Connections depending solely on solder shall not be used. Splices shall be made in accordance with Section 110-14(b) except that insulation shall not be required. The arrangement of grounding connections shall be such that the disconnection or the removal of a receptacle, fixture, or other device fed from the box will not interfere with or interrupt the grounding continuity.

Exception: The equipment grounding conductor permitted in Section 250-146(d) shall not be required to be connected to the other equipment grounding conductors or to the box.

(a) Metal Boxes. A connection shall be made between the one or more equipment grounding conductors and a metal box by means of a grounding screw that shall be used for no other purpose or a listed grounding device.

(b) Nonmetallic Boxes. One or more equipment grounding conductors brought into a nonmetallic outlet box shall be arranged so that a connection can be made to any fitting or device in that box requiring grounding.

H. Direct-Current Systems

250-160. General. Direct-current systems shall comply with Part H and other sections of Article 250 not specifically intended for ac systems.

250-162. Direct-Current Circuits and Systems to Be Grounded. Direct-current circuits and systems shall be grounded as provided for in (a) and (b).

(a) Two-Wire, Direct-Current Systems. A two-wire, dc system supplying premises wiring and operating at greater than 50 volts but not greater than 300 volts, shall be grounded.

Exception No. 1: A system equipped with a ground detector and supplying only industrial equipment in limited areas shall not be required to be grounded.

Exception No. 2: A rectifier-derived dc system supplied from an ac system complying with Section 250-20 shall not be required to be grounded.

Exception No. 3: Direct-current fire alarm circuits having a maximum current of 0.030 amperes as specified in Article 760, Part C, shall not be required to be grounded.

(b) Three-Wire, Direct-Current Systems. The neutral conductor of all 3-wire, dc systems supplying premises wiring shall be grounded.

250-164. Point of Connection for Direct-Current Systems.

(a) Off-Premises Source. Direct-current systems to be grounded and supplied from an off-premises source shall have the grounding connection made at one or more supply stations. A grounding connection shall not be made at individual services or at any point on the premises wiring.

As shown in the 3-wire dc distribution system in Figure 250.54, the neutral is grounded at the off-premises generator site. Grounding of a 2-wire dc system would be accomplished in the same manner. For an on-premises generator, a grounding connection is required and is to be located at the source of the first system disconnecting means or overcurrent device. Other equivalent means that utilize equipment listed and identified for such use are permitted.

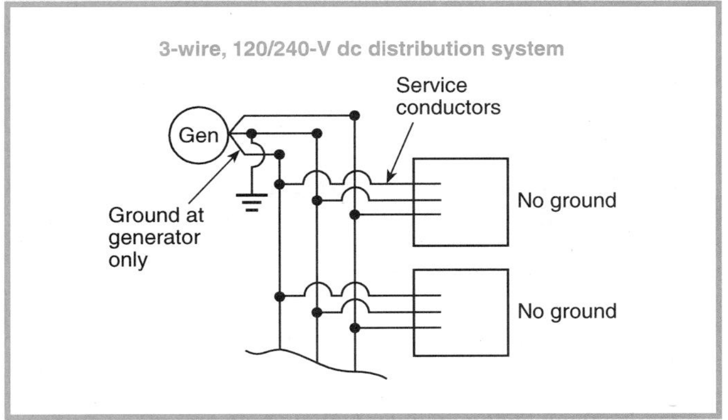

Figure 250.54 *A 3-wire, 120/240-volt dc distribution system that has the neutral grounded at the off-premises generator site.*

(b) On-Premises Source. Where the dc system source is located on the premises, a grounding connection shall be made at one of the following:

(1) The source
(2) The first system disconnection means or overcurrent device
(3) By other means that accomplish equivalent system protection and that utilize equipment listed and identified for the use

250-166. Size of Direct-Current Grounding Electrode Conductor. The size of the grounding electrode conductor for a dc system shall be as specified in (a) through (e).

(a) Not Smaller than the Neutral Conductor. Where the dc system consists of a 3-wire balancer set or a balancer winding with overcurrent protection as provided in Section 445-4(d), the grounding electrode conductor shall not be smaller than the neutral conductor, and not smaller than No. 8 copper or No. 6 aluminum.

(b) Not Smaller than the Largest Conductor. Where the dc system is other than as in (a), the grounding electrode conductor shall not be smaller than the largest conductor supplied by the system, and not smaller than No. 8 copper or No. 6 aluminum.

(c) Connected to Made Electrodes. Where connected to made electrodes as in Section 250-52(c) or (d), that portion of the grounding electrode conductor that is the sole connection to the grounding electrode shall not be required to be larger than No. 6 copper wire or No. 4 aluminum wire.

(d) Connected to a Concrete-Encased Electrode. Where connected to a concrete-encased electrode as in Section 250-50(c), that portion of the grounding electrode conductor that is the sole connection to the grounding electrode shall not be required to be larger than No. 4 copper wire.

(e) Connected to a Ground Ring. Where connected to a ground ring as in Section 250-50(d), that portion of the grounding electrode conductor that is the sole connection to the grounding electrode shall not be required to be larger than the conductor used for the ground ring.

250-168. Direct-Current Bonding Jumper. For dc systems, the size of the bonding jumper shall not be smaller than the system grounding conductor specified in Section 250-166.

250-169. Ungrounded Direct-Current Separately Derived Systems. Except as otherwise permitted in Section 250-34 for portable and vehicle mounted generators, an ungrounded dc separately derived system supplied from a stand-alone power source (such as an engine–generator set) shall have a grounding electrode conductor connected to an electrode that complies with Part C to provide for grounding of metal enclosures, raceways, cables, and exposed noncurrent-carrying metal parts of equipment. The grounding electrode conductor connection shall be to the metal enclosure at any point on the separately derived system from the source to the first system disconnecting means or overcurrent device, or it shall be made at the source of a separately derived system that has no disconnecting means or overcurrent devices.

The size of the grounding electrode conductor shall be in accordance with Section 250-166.

J. Instruments, Meters, and Relays

250-170. Instrument Transformer Circuits. Secondary circuits of current and potential instrument transformers shall be grounded where the primary windings are connected to circuits of 300 volts or more to ground and, where on switchboards, shall be grounded irrespective of voltage.

Exception: Circuits where the primary windings are connected to circuits of less than 1000 volts with no live parts or wiring exposed or accessible to other than qualified persons.

250-172. Instrument Transformer Cases. Cases or frames of instrument transformers shall be grounded where accessible to other than qualified persons.

Exception: Cases or frames of current transformers, the primaries of which are not over 150 volts to ground and that are used exclusively to supply current to meters.

250-174. Cases of Instruments, Meters, and Relays Operating at Less than 1000 Volts. Instruments, meters, and relays operating with windings or working parts at less than 1000 volts shall be grounded as specified in (a), (b), or (c).

(a) Not on Switchboards. Instruments, meters, and relays not located on switchboards, operating with windings or working parts at 300 volts or more to ground, and accessible to other than qualified persons, shall have the cases and other exposed metal parts grounded.

(b) On Dead-Front Switchboards. Instruments, meters, and relays (whether operated from current and potential transformers or connected directly in the circuit) on switchboards having no live parts on the front of the panels shall have the cases grounded.

(c) On Live-Front Switchboards. Instruments, meters, and relays (whether operated from current and potential transformers or connected directly in the circuit) on switchboards having exposed live parts on the front of panels shall not have their cases grounded. Mats of insulating rubber or other suitable floor insulation shall be provided for the operator where the voltage to ground exceeds 150.

250-176. Cases of Instruments, Meters, and Relays — Operating Voltage 1 kV and Over. Where instruments, meters, and relays have current-carrying parts of 1 kV and over to ground, they shall be isolated by elevation or protected by suitable barriers, grounded metal, or insulating covers or guards. Their cases shall not be grounded.

Exception: Cases of electrostatic ground detectors where the internal ground segments of the instrument are connected to the instrument case and grounded and the ground detector is isolated by elevation.

250-178. Instrument Grounding Conductor. The grounding conductor for secondary circuits of instrument transformers and for instrument cases shall not be smaller than No. 12 copper or No. 10 aluminum. Cases of instrument transformers, instruments, meters, and relays that are mounted directly on grounded metal surfaces of enclosures or grounded metal switchboard panels shall be considered to be grounded, and no additional grounding conductor shall be required.

K. Grounding of Systems and Circuits of 1 kV and Over (High Voltage)

250-180. General. Where high-voltage systems are grounded, they shall comply with all applicable provisions of the preceding sections of this article and with the following sections, which supplement and modify the preceding sections.

250-182. Derived Neutral Systems. A system neutral derived from a grounding transformer shall be permitted to be used for grounding high-voltage systems.

250-184. Solidly Grounded Neutral Systems.

(a) Neutral Conductor. The minimum insulation level for neutral conductors of solidly grounded systems shall be 600 volts.

Exception No. 1: Bare copper conductors shall be permitted to be used for the neutral of service entrances and the neutral of direct-buried portions of feeders.

Exception No. 2: Bare conductors shall be permitted for the neutral of overhead portions installed outdoors.

FPN: See Section 225-4 for conductor covering where within 10 ft (3.05 m) of any building or other structure.

(b) Multiple Grounding. The neutral of a solidly grounded neutral system shall be permitted to be grounded at more than one point for the following:

(1) Services
(2) Direct-buried portions of feeders employing a bare copper neutral
(3) Overhead portion installed outdoors

(c) Neutral Grounding Conductor. The neutral grounding conductor shall be permitted to be a bare conductor if isolated from phase conductors and protected from physical damage.

250-186. Impedance Grounded Neutral Systems. Impedance grounded neutral systems shall comply with the provisions of (a) through (d).

(a) Location. The grounding impedance shall be inserted in the grounding conductor between the grounding electrode of the supply system and the neutral point of the supply transformer or generator.

(b) Identified and Insulated. Where the neutral conductor of an impedance grounded neutral system is used, it shall be identified, as well as fully insulated with the same insulation as the phase conductors.

(c) System Neutral Connection. The system neutral shall not be connected to ground, except through the neutral grounding impedance.

(d) Equipment Grounding Conductors. Equipment grounding conductors shall be permitted to be bare and shall be connected to the ground bus and grounding electrode conductor at the service-entrance equipment or the disconnecting means for a separately derived system and extended to the system ground.

250-188. Grounding of Systems Supplying Portable or Mobile Equipment. Systems supplying portable or mobile high-voltage equipment, other than substations installed on a temporary basis, shall comply with (a) through (f).

Portable describes equipment that is easily carried from one location to another. *Mobile* describes equipment that is easily moved, as on wheels, treads, and so on.

(a) Portable or Mobile Equipment. Portable or mobile high-voltage equipment shall be supplied from a system having its neutral grounded through an impedance. Where a delta-connected high-voltage system is used to supply portable or mobile equipment, a system neutral shall be derived.

(b) Exposed Noncurrent-Carrying Metal Parts. Exposed noncurrent-carrying metal parts of portable or mobile equipment shall be connected by an equipment grounding conductor to the point at which the system neutral impedance is grounded.

(c) Ground-Fault Current. The voltage developed between the portable or mobile equipment frame and ground by the flow of maximum ground-fault current shall not exceed 100 volts.

(d) Ground-Fault Detection and Relaying. Ground-fault detection and relaying shall be provided to automatically de-energize any high-voltage system component that has developed a ground fault. The continuity of the equipment grounding conductor shall be continuously monitored so as to de-energize automatically the high-voltage circuit to the portable or mobile equipment upon loss of continuity of the equipment grounding conductor.

(e) Isolation. The grounding electrode to which the portable or mobile equipment system neutral impedance is connected shall be isolated from and separated in the ground by at least 20 ft (6.1 m) from any other system or equipment grounding electrode, and there shall be no direct connection between the grounding electrodes, such as buried pipe, fence, etc.

(f) Trailing Cable and Couplers. High-voltage trailing cable and couplers for interconnection of portable or mobile equipment shall meet the requirements of Part C of Article 400 for cables and Section 490-55 for couplers.

250-190. Grounding of Equipment. All noncurrent-carrying metal parts of fixed, portable, and mobile equipment and associated fences, housings, enclosures, and supporting structures shall be grounded.

Exception: Where isolated from ground and located so as to prevent any person who can make contact with ground from contacting such metal parts when the equipment is energized.

Grounding conductors not an integral part of a cable assembly shall not be smaller than No. 6 copper or No. 4 aluminum.

FPN: See Section 250-110, Exception No. 2 for pole-mounted distribution apparatus.

Article 280 — Surge Arresters

Contents

A. General

280-1. Scope. This article covers general requirements, installation requirements, and connection requirements for surge arresters installed on premises wiring systems.

Voltage surges with peaks of several thousand volts, even on 120-volt circuits, are not uncommon. These surges occur because of induced voltages in power and transmission lines resulting from lightning strikes in the vicinity of the line. They also occur as a result of switching inductive circuits on the premises. Surge arresters for installation as part of an electric service, such as the one shown in Figure 280.1 and for use with cord- and plug-connected solid-state electronic equipment are commercially available. The basic standards used to investigate surge arresters are ANSI/IEEE C62.1-1989, *Standard for Gapped Silicon-Carbide Surge Arresters for AC Power Circuits,* and ANSI/IEEE C62.11, *Standard for Metal-Oxide Surge Arresters for AC Power Circuits.*

Figure 280.1 *A lightning surge protector suitable for service-entrance installation and for mounting in a panel knockout. (General Electric Co.)*

280-2. Definition. A *surge arrester* is a protective device for limiting surge voltages by discharging or bypassing surge current, and it also prevents continued flow of follow current while remaining capable of repeating these functions.

280-3. Number Required. Where used at a point on a circuit, a surge arrester shall be connected to each ungrounded conductor. A single installation of such surge arresters shall be permitted to protect a number of interconnected circuits, provided that no circuit is exposed to surges while disconnected from the surge arresters.

Means are required to be provided for protection of circuits that may be disconnected from the generating station bus. A switch with double-throw action used to disconnect the outside circuits from the station generator and alternatively connect these circuits to ground would satisfy the condition of a single set of arresters protecting more than one circuit.

Surge arresters are required to be installed on circuits in buildings that house explosives. See Chapter 6 of NFPA 495-1996, *Explosive Materials Code,* for details.

280-4. Surge Arrester Selection.

(a) Circuits of Less than 1000 Volts. The rating of the surge arrester shall be equal to or greater than the maximum continuous phase-to-ground power frequency voltage available at the point of application.

Surge arresters installed on circuits of less than 1000 volts shall be listed for the purpose.

(b) Circuits of 1 kV and Over — Silicon Carbide Types. The rating of a silicon carbide-type surge arrester shall be not less than 125 percent of the maximum continuous phase-to-ground voltage available at the point of application.

FPN No. 1: For further information on surge arresters, see *Standard for Gapped Silicon-Carbide Surge Arresters for AC Power Circuits*, ANSI/IEEE C62.1-1989; *Guide for the Application of Gapped Silicon-Carbide Surge Arresters for Alternating-Current Systems*, ANSI/IEEE C62.2-1987; *Standard for Metal-Oxide Surge Arresters for Alternating-Current Power Circuits*, ANSI/IEEE C62.11-1993; and *Guide for the Application of Metal-Oxide Surge Arresters for Alternating-Current Systems*, ANSI/IEEE C62.22-1991.

FPN No. 2: The selection of a properly rated metal oxide arrester is based on considerations of maximum continuous operating voltage and the magnitude and duration of overvoltages at the arrester location as affected by phase-to-ground faults, system grounding techniques, switching surges, and other causes. See the manufacturer's application rules for selection of the specific arrester to be used at a particular location.

B. Installation

280-11. Location. Surge arresters shall be permitted to be located indoors or outdoors. Surge arresters shall be made inaccessible to unqualified persons, unless listed for installation in accessible locations.

Maximum protection is achieved where the surge protective device is located as close as practicable to the equipment to be protected. When a surge passes through an arrester, a wave is reflected in both directions on the conductors connected to the surge arrester. The magnitude of the reflected wave increases as the distance from the arrester increases. If the length of the conductor between the protected equipment and the surge arrester is short, the magnitude of the wave reflected through the equipment is minimized.

280-12. Routing of Surge Arrester Connections. The conductor used to connect the surge arrester to line or bus and to ground shall not be any longer than necessary and shall avoid unnecessary bends.

Arrester conductors should be as short and be run as straight as practicable, avoiding any sharp bends and turns, which would increase the impedance. High-frequency currents, such as those common to lightning discharges, tend to reduce the effectiveness of a grounding conductor.

C. Connecting Surge Arresters

280-21. Installed at Services of Less than 1000 Volts. Line and ground connecting conductors shall not be smaller than No. 14 copper or No. 12 aluminum. The arrester grounding conductor shall be connected to one of the following:

(1) Grounded service conductor
(2) Grounding electrode conductor
(3) Grounding electrode for the service
(4) Equipment grounding terminal in the service equipment

Single-phase or 3-phase grounded or ungrounded services are permitted to have the surge arrester grounded to the equipment grounding terminal in the service equipment. Figure 280.2 shows three methods of grounding the ground terminals of surge arresters at service entrances. On the top is an arrester connected to a neutral service conductor. In the middle is an arrester connected to a grounding electrode conductor. On the bottom is an arrester connected to a grounding electrode conductor of an ungrounded system.

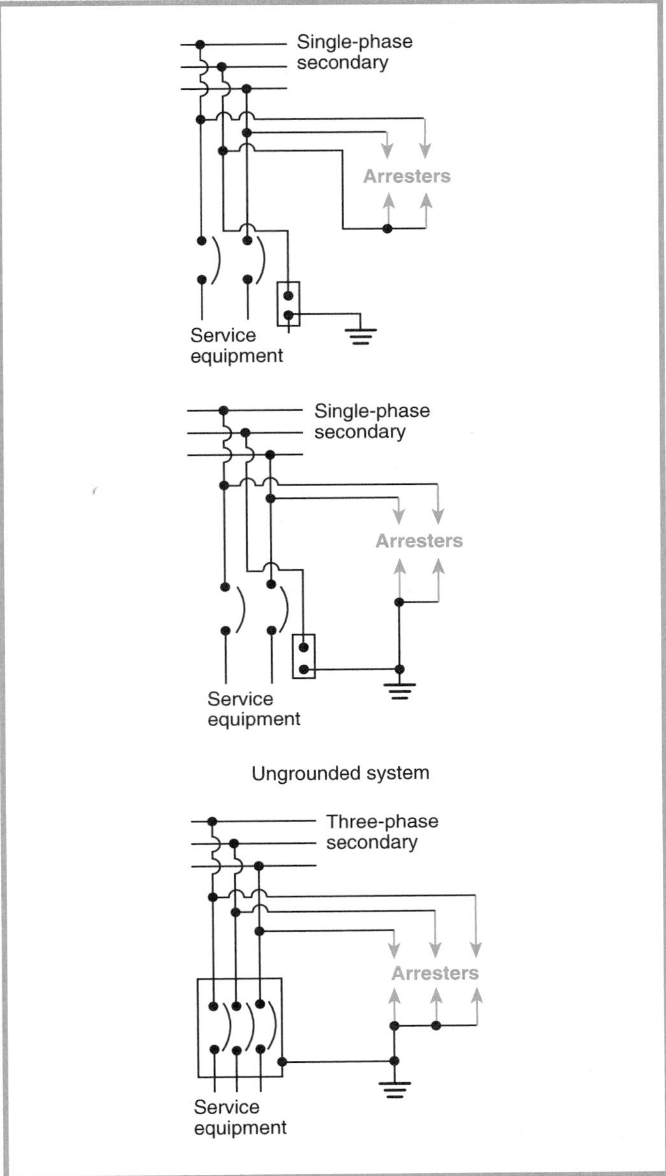

Figure 280.2 Three locations of grounding surge arresters at service entrances.

280-22. Installed on the Load Side Services of Less than 1000 Volts. Line and ground connecting conductors shall not be smaller than No. 14 copper or No. 12 aluminum. A surge arrester shall be permitted to be connected between any two conductors—ungrounded conductor(s), grounded conductor, grounding conductor. The grounded conductor and the grounding conductor shall be interconnected only by the normal operation of the surge arrester during a surge.

280-23. Circuits of 1 kV and Over — Surge-Arrester Conductors. The conductor between the surge arrester and the line and the surge arrester and the grounding connection shall not be smaller than No. 6 copper or aluminum.

280-24. Circuits of 1 kV and Over — Interconnections. The grounding conductor of a surge arrester protecting a transformer that supplies a secondary distribution system shall be interconnected as specified in (a), (b), or (c).

(a) Metallic Interconnections. A metallic interconnection shall be made to the secondary grounded circuit conductor or the secondary circuit grounding conductor provided that, in addition to the direct grounding connection at the surge arrester, the following occurs.

(1) The grounded conductor of the secondary has elsewhere a grounding connection to a continuous metal underground water piping system. However, in urban water-pipe areas where there are at least four water-pipe connections on the neutral and not fewer than four such connections in each mile of neutral, the metallic interconnection shall be permitted to be made to the secondary neutral with omission of the direct grounding connection at the surge arrester.

(2) The grounded conductor of the secondary system is a part of a multiground neutral system of which the primary neutral has at least four ground connections in each mile of line in addition to a ground at each service.

(b) Through Spark Gap or Device. Where the surge arrester grounding conductor is not connected as in (a) or where the secondary is not grounded as in (a) but is otherwise grounded as in Sections 250-50 and 250-52, an interconnection shall be made through a spark gap or listed device as follows.

(1) For ungrounded or unigrounded primary systems, the spark gap or listed device shall have a 60-Hz breakdown voltage of at least twice the primary circuit voltage but not necessarily more than 10 kV, and there shall be at least one other ground on the grounded conductor of the secondary that is not less than 20 ft (6.1 m) distant from the surge arrester grounding electrode.

(2) For multigrounded neutral primary systems, the spark gap or listed device shall have a 60-Hz breakdown of not more than 3 kV, and there shall be at least one other ground on the grounded conductor of the secondary that is not less than 20 ft (6.1 m) distant from the surge arrester grounding electrode.

(c) By Special Permission. An interconnection of the surge arrester ground and the secondary neutral, other than as provided in (a) or (b), shall be permitted to be made only by special permission.

280-25. Grounding. Except as indicated in this article, surge arrester grounding connections shall be made as specified in Article 250. Grounding conductors shall not be run in metal enclosures unless bonded to both ends of such enclosure.

CHAPTER 3

Wiring Methods and Materials

Article 300 — Wiring Methods

Contents

A. General Requirements

300-1. Scope.

(a) All Wiring Installations. This article covers wiring methods for all wiring installations unless modified by other articles.

This scope statement was simplified for the 1999 *Code* without changing the intent. The seven previous exceptions (1996 *Code*) were replaced with "unless modified by other articles" because Section 90-3 allows other chapters to modify Chapter 3. The articles that were previously listed included Articles 504, 725, 760, 770, 800, 810, 820, and 830. The modifications to Article 300 are stated within the individual articles.

(b) Integral Parts of Equipment. The provisions of this article are not intended to apply to the conductors that form an integral part of equipment, such as motors, controllers, motor control centers, or factory assembled control equipment or listed utilization equipment.

The *Electrical Standard for Industrial Machinery*, NFPA 79-1997, is one example of where the *NEC* does not apply.

300-2. Limitations.

(a) Voltage. Wiring methods specified in Chapter 3 shall be used for voltages 600 volts, nominal, or less where not specifically limited in some section of Chapter 3. They shall be permitted for voltages over 600 volts, nominal, where specifically permitted elsewhere in this *Code*.

(b) Temperature. Temperature limitation of conductors shall be in accordance with Section 310-10.

See Section 110-14(c) and the commentary on temperature limitations of conductor terminations.

300-3. Conductors.

(a) Single Conductors. Single conductors specified in Table 310-13 shall only be installed where part of a recognized wiring method of Chapter 3.

Section 300-3(a) makes it clear that building wire, such as individual insulated conductors identified as THHN, is prohibited from use outside of a recognized wiring method.

(b) Conductors of the Same Circuit. All conductors of the same circuit and, where used, the grounded conductor and all equipment grounding conductors shall be contained within the same raceway, auxiliary gutter, cable tray, trench, cable, or cord, unless otherwise permitted in accordance with (1) through (4).

For the 1999 *Code*, Section 300-3(b) was arranged in a more logical order. There are, however, no technical changes to this section. The general rule remains consistent with previous editions of the *Code*, that is, that circuit conductors should be grouped to avoid increases in the overall circuit impedance and to reduce inductive heating. Similar requirements are found in Section 300-5(i).

(1) Paralleled Installations. Conductors shall be permitted to be run in parallel in accordance with the provisions of Section 310-4. The requirement to run all circuit conductors within the same raceway, auxiliary gutter, cable tray, trench, cable, or cord shall apply separately to each portion of the paralleled installation, and the equipment grounding conductors shall comply with the provisions of Section 250-122. Parallel runs in cable tray shall comply with the provisions of Section 318-8(d).

Exception: Conductors installed in nonmetallic raceways run underground shall be permitted to be arranged as isolated phase installations. The raceways shall be installed in close proximity and the conductors shall comply with the provisions of Section 300-20(b).

(2) Grounding Conductors. Equipment grounding conductors shall be permitted to be installed outside a raceway or cable assembly where in accordance with the provisions of Section 250-130(c) for certain existing installations, or in accordance with Section 250-134(b), Exception No. 2, for dc circuits. Equipment bonding conductors shall be permitted to be installed on the outside of raceways in accordance with Section 250-102(e).

Section 300-3(b)(2) recognizes that some types of grounding and bonding conductors can be run as single conductors on the exterior of the raceway or outside of a cable assembly.

(3) **Nonferrous Wiring Methods.** Conductors in wiring methods with a nonmetallic or other nonmagnetic sheath shall, where run in different raceways, auxiliary gutters, cable trays, trenches, cables, or cords, comply with the provisions of Section 300-20(b). Conductors in single-conductor Type MI cable with a nonmagnetic sheath shall comply with the provisions of Section 330-16.

(4) **Enclosures.** Where an auxiliary gutter runs between a column-width panelboard and a pull box, and the pull box includes neutral terminations, the neutral conductors of circuits supplied from the panelboard shall be permitted to originate in the pull box.

Section 300-3(b)(4) recognizes the practice of supplying narrow column-type panelboard through an auxiliary gutter from an overhead pull box and running only the ungrounded conductors down from the pull box to the panelboard. As shown in Figure 300.1, the feeder and branch-circuit neutral conductors are terminated in the overhead pull box and are not carried with the ungrounded conductors. Inductive heating does not occur, since the load-carrying conductors extend down and back up within the same enclosure.

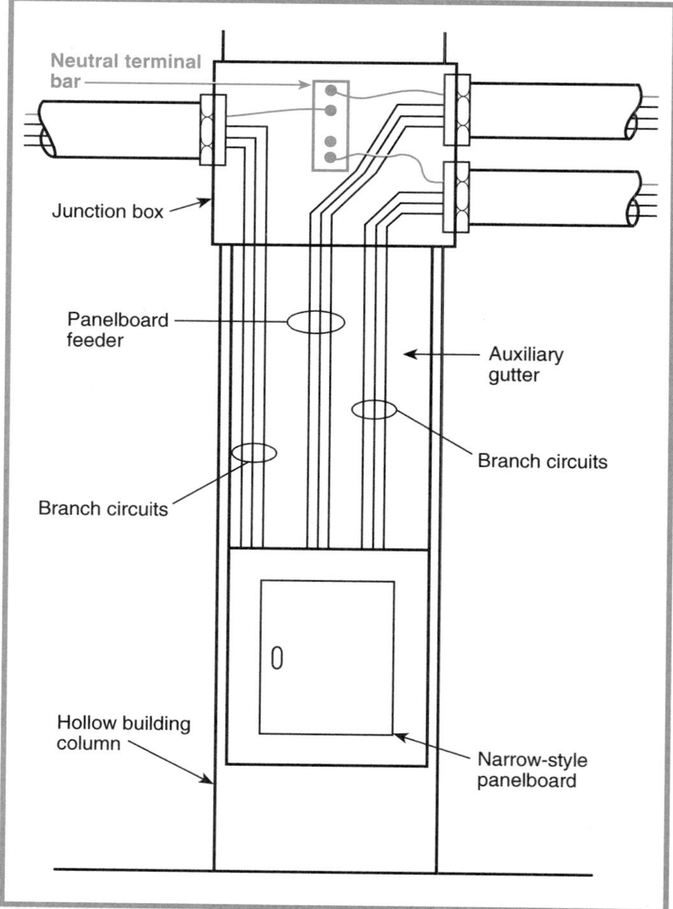

Figure 300.1 *An installation where an auxiliary gutter extends from the panelboard up to a pull box that is used as a termination point for the feeder and branch-circuit grounded conductors (neutrals).*

(c) **Conductors of Different Systems.**

(1) **600 Volts, Nominal, or Less.** Conductors of circuits rated 600 volts, nominal, or less, ac circuits, and dc circuits shall be permitted to occupy the same equipment wiring enclosure, cable, or raceway. All conductors shall have an insulation rating equal to at least the maximum circuit voltage applied to any conductor within the enclosure, cable, or raceway.

Section 300-3(c)(1) makes it clear that it is the *maximum circuit voltage* in the raceway, not the maximum insulation voltage rating of the conductors in the raceway, that determines the minimum voltage rating required for the insulation of conductors for systems of 600 volts or less.

The conductors of a 3-phase, 4-wire, 208Y/120-volt ac circuit; a 3-phase, 4-wire, 480Y/277-volt ac circuit; and a 3-wire, 120/240-volt dc circuit may occupy the same equipment wiring enclosure, cable, or raceway if all of the conductors are insulated for the maximum circuit voltage of any conductor. In this case, the maximum circuit voltage would be 480 volts, and 600-volt insulation would be suitable for all of the conductors.

If a 2-wire, 120-volt circuit is included in the same raceway with a 3-wire, 120/240-volt circuit having 600-volt conductors, then the 2-wire, 120-volt circuit conductors could use 300-volt insulation because the maximum circuit voltage is only 240 volts.

Exception: For solar photovoltaic systems in accordance with Section 690-4(b).

Section 690-4(b) prohibits solar photovoltaic circuits to be located within the same enclosure as conductors of other systems unless the conductors are separated by a partition or are connected together.

FPN: See Section 725-54(a)(1) for Class 2 and Class 3 circuit conductors.

Section 725-54(a)(1) prohibits Class 2 and Class 3 circuit conductors from occupying the same enclosure, cable, or raceway as Class 1, electric light, and power conductors, with some exceptions.

(2) Over 600 Volts, Nominal. Conductors of circuits rated over 600 volts, nominal, shall not occupy the same equipment wiring enclosure, cable, or raceway with conductors of circuits rated 600 volts, nominal, or less unless otherwise permitted in (a) through (e).

(a) Secondary wiring to electric-discharge lamps of 1000 volts or less, if insulated for the secondary voltage involved, shall be permitted to occupy the same fixture, sign, or outline lighting enclosure as the branch-circuit conductors.

(b) Primary leads of electric-discharge lamp ballasts, insulated for the primary voltage of the ballast, where contained within the individual wiring enclosure, shall be permitted to occupy the same fixture, sign, or outline lighting enclosure as the branch-circuit conductors.

(c) Excitation, control, relay, and ammeter conductors used in connection with any individual motor or starter shall be permitted to occupy the same enclosure as the motor-circuit conductors.

(d) In motors, switchgear and control assemblies, and similar equipment, conductors of different voltage ratings shall be permitted.

(e) In manholes, if the conductors of each system are permanently and effectively separated from the conductors of the other systems and securely fastened to racks, insulators, or other approved supports, conductors of different voltage ratings shall be permitted.

Conductors having nonshielded insulation and operating at different voltage levels shall not occupy the same enclosure, cable, or raceway.

300-4. Protection Against Physical Damage. Where subject to physical damage, conductors shall be adequately protected.

(a) Cables and Raceways Through Wood Members.

(1) Bored Holes. In both exposed and concealed locations, where a cable or raceway-type wiring method is installed through bored holes in joists, rafters, or wood members, holes shall be bored so that the edge of the hole is not less than 1¼ in. (31.8 mm) from the nearest edge of the wood member. Where this distance cannot be maintained, the cable or raceway shall be protected from penetration by screws or nails by a steel plate or bushing, at least 1⁄16 in. (1.59 mm) thick, and of appropriate length and width installed to cover the area of the wiring.

Exception: Steel plates shall not be required to protect rigid metal conduit, intermediate metal conduit, rigid nonmetallic conduit, or electrical metallic tubing.

(2) Notches in Wood. Where there is no objection because of weakening the building structure, in both exposed and concealed locations, cables or raceways shall be permitted to be laid in notches in wood studs, joists, rafters, or other wood members where the cable or raceway at those points is protected against nails or screws by a steel plate at least 1⁄16 in. (1.59 mm) thick installed before the building finish is applied.

Exception: Steel plates shall not be required to protect rigid metal conduit, intermediate metal conduit, rigid nonmetallic conduit, or electrical metallic tubing.

The intent of Section 300-4(a)(1) is to prevent nails and screws from being driven into cables and raceways. By keeping the edge of a drilled hole 1¼ in. from the nearest edge of a stud, as shown in Figure 300.2, nails would not likely penetrate the stud far enough to injure a cable. Building codes limit the maximum size of bored or notched holes in studs, and Section 300-4(a)(2) indicates that consideration should be given to the size of notches in studs, so as not to affect the strength of the structure.

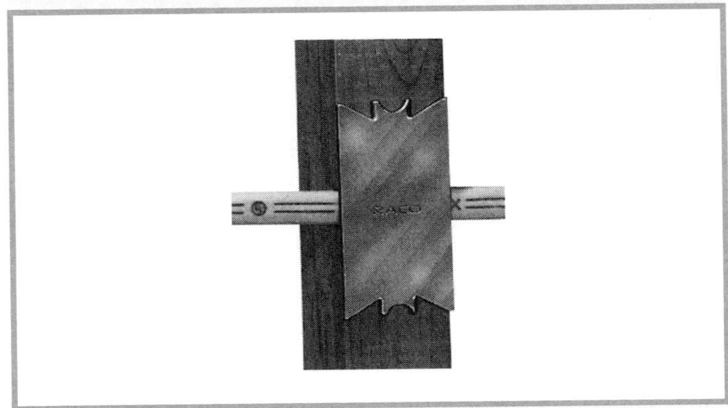

Figure 300.2 *A steel plate used to protect a nonmetallic-sheathed cable within 1¼ in. of the edge of a wood stud. (RACO)*

The exceptions to Sections 300-4(a)(1) and 300-4(a)(2) permit intermediate metal conduit, rigid metal conduit, rigid nonmetallic conduit, and electrical metallic tubing to be installed through bored holes or laid in notches less than 1¼ in. from the nearest edge of the stud, without a steel plate or bushing.

(b) Nonmetallic-Sheathed Cables and Electrical Nonmetallic Tubing Through Metal Framing Members.

(1) Nonmetallic-Sheathed Cable. In both exposed and concealed locations where nonmetallic-sheathed cables pass through either factory or field punched, cut, or drilled slots or holes in metal members, the cable shall be protected by bushings or grommets covering all metal edges and securely fastened in the opening prior to installation of the cable.

Section 300-4(b)(1) was revised for the 1999 *Code* by adding the phrase "covering all metal edges." This change prohibits the use of grommets that do not completely encircle the NM type of cables as they

pass through a hole in the metal studs. This new requirement affords better physical protection for nonmetallic cables as they are "pulled" through the openings in metal studs.

(2) Nonmetallic-Sheathed Cable and Electrical Nonmetallic Tubing. Where nails or screws are likely to penetrate nonmetallic-sheathed cable or electrical nonmetallic tubing, a steel sleeve, steel plate, or steel clip not less than 1/16 in. (1.59 mm) in thickness shall be used to protect the cable or tubing.

(c) Cables Through Spaces Behind Panels Designed to Allow Access. Cables or raceway-type wiring methods, installed behind panels designed to allow access, shall be supported according to their applicable articles.

Cable- or raceway-type wiring methods installed above suspended ceilings with lift-up panels cannot be laid on the suspended ceiling. They are required to be supported according to Section 300-11(a), Section 300-23, and the requirements of the article applicable to the wiring method involved. Similarly, low-voltage and communications cables are not permitted to block access to equipment above the suspended ceiling. Examples of this requirement are also found in Sections 725-5, 760-5, 800-5, and 820-5. For supporting of low-voltage cables, see Sections 720-11, 725-7, 760-8, 800-6, and 820-6.

(d) Cables and Raceways Parallel to Framing Members. In both exposed and concealed locations, where a cable- or raceway-type wiring method is installed parallel to framing members, such as joists, rafters, or studs, the cable or raceway shall be installed and supported so that the nearest outside surface of the cable or raceway is not less than 1¼ in. (31.8 mm) from the nearest edge of the framing member where nails or screws are likely to penetrate. Where this distance cannot be maintained, the cable or raceway shall be protected from penetration by nails or screws by a steel plate, sleeve, or equivalent at least 1/16 in. (1.59 mm) thick.

Exception No. 1: Steel plates, sleeves, or the equivalent shall not be required to protect rigid metal conduit, intermediate metal conduit, rigid nonmetallic conduit, or electrical metallic tubing.

The intent of Section 300-4(d) is to prevent mechanical damage to cables and raceways from nails and screws. The *Code* offers two means of protection. The first method is to fasten the cable or raceway so that it is 1¼ in. from the edge of the framing member, as illustrated in Figure 300.3. This requirement generally applies to exposed and concealed work. The second method permits the cable or raceway to be installed closer than 1¼ in. from the edge of the framing member if physical protection, such as a steel plate or a sleeve, is provided. As stated in Exception No. 1, this requirement does not apply to rigid metal conduit, rigid nonmetallic conduit, intermediate metal conduit, or electrical metallic tubing wiring methods, as these methods provide physical protection for the conductors.

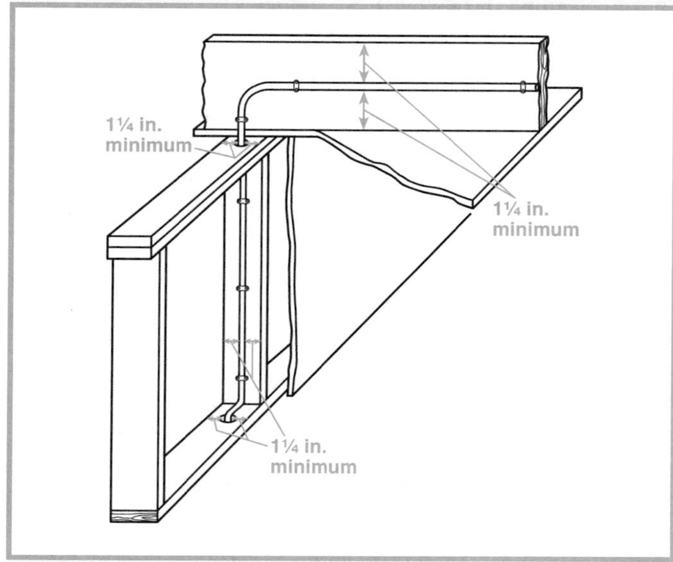

***Figure 300.3** Cables and raceways installed parallel to framing members in accordance with Section* 300-4(d).

Exception No. 2: For concealed work in finished buildings, or finished panels for prefabricated buildings where such supporting is impracticable, it shall be permissible to fish the cables between access points.

Exception No. 3: Steel plates, sleeves, or the equivalent shall not be required to protect cables or raceways in mobile homes and recreational vehicles.

(e) Cables and Raceways Installed in Shallow Grooves. Cable- or raceway-type wiring methods installed in a groove, to be covered by wallboard, siding, paneling, carpeting, or similar finish, shall be protected by 1/16 in. (1.59 mm) thick steel plate, sleeve, or equivalent or by not less than 1¼ in. (31.8 mm) free space for the full length of the groove in which the cable or raceway is installed.

Exception: Steel plates, sleeves, or the equivalent shall not be required to protect rigid metal conduit, intermediate metal conduit, rigid nonmetallic conduit, or electrical metallic tubing.

Added in the 1996 *Code,* Section 300-4(e) requires supplemental mechanical protection for flexible conduit and cable installed in shallow grooves of rigid insulation intended to be covered by wallboard, siding, paneling, and the like, unless the cable is set back at least 1¼ in. from the surface of the framing member.

(f) Insulated Fittings. Where raceways containing ungrounded conductors No. 4 or larger enter a cabinet, box enclosure, or raceway, the conductors shall be protected by a substantial fitting providing a smoothly rounded insulating surface, unless the conductors are separated from the fitting or raceway by substantial insulating material that is securely fastened in place.

These requirements, originally found in Sections 370-17 and 373-6(c), were placed here for the 1996 *Code* to ensure that they will apply to all raceway wiring methods.

Exception: Where threaded hubs or bosses that are an integral part of a cabinet, box enclosure, or raceway provide a smoothly rounded or flared entry for conductors.

Conduit bushings constructed wholly of insulating material shall not be used to secure a fitting or raceway. The insulating fitting or insulating material shall have a temperature rating not less than the insulation temperature rating of the installed conductors.

300-5. Underground Installations.

(a) Minimum Cover Requirements. Direct-buried cable or conduit or other raceways shall be installed to meet the minimum cover requirements of Table 300-5.

Conductors under residential driveways must be at least 18 in. below grade. But, if the conductors are protected by an overcurrent device rated at not more than 20 amperes and provided with ground-fault circuit-interrupter (GFCI) protection for personnel, the burial depth may be reduced to 12 in. Figures 300.4 and 300.5 provide examples showing underground installations of 18 in. and 12 in., respectively.

(b) Grounding. All underground installations shall be grounded and bonded in accordance with Article 250.

Rigid nonmetallic conduit elbows installed as part of a long run of conduit are often damaged in the process of pulling the conductors, due to friction at the bend. For service raceways, Section 250-80, Exception, permits a metal elbow to be installed without being grounded, provided it is isolated from possible contact by at least 18 in. of cover to any part of the elbow, as shown in Figure 300.6. For other than service raceways, Exception No. 3 to Section 250-86 applies.

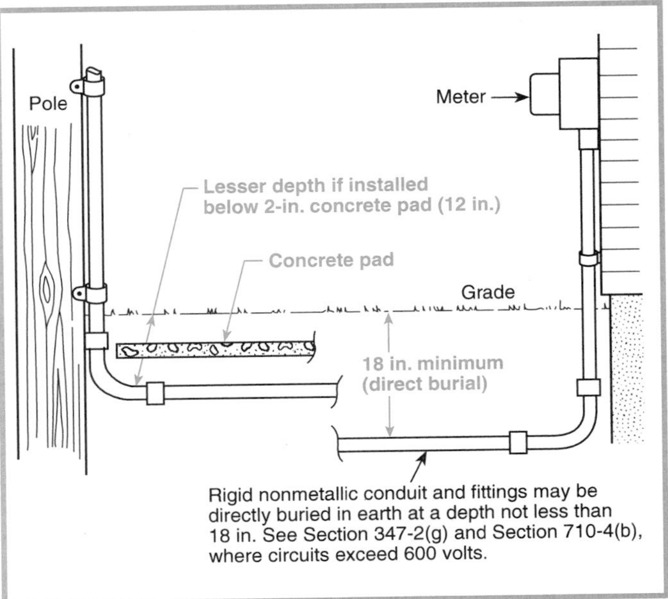

Figure 300.4 *PVC rigid nonmetallic conduit buried in compliance with Table 300-5 and installed in accordance with Section 300-5(a).*

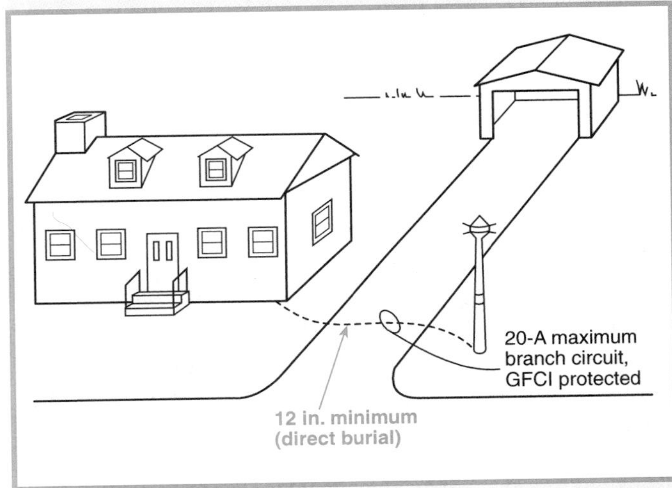

Figure 300.5 *A 20-ampere, GFCI-protected residential branch circuit installed with a minimum burial depth of 12 in. beneath a residential driveway.*

(c) Underground Cables Under Buildings. Underground cable installed under a building shall be in a raceway that is extended beyond the outside walls of the building.

(d) Protection from Damage. Direct-buried conductors and cables emerging from the ground shall be protected by enclosures or raceways extending from the minimum cover distance required by Section 300-5(a) below grade to a point at least 8 ft (2.44 m) above finished grade. In no case shall the protection be required to exceed 18 in. (457 mm) below finished grade.

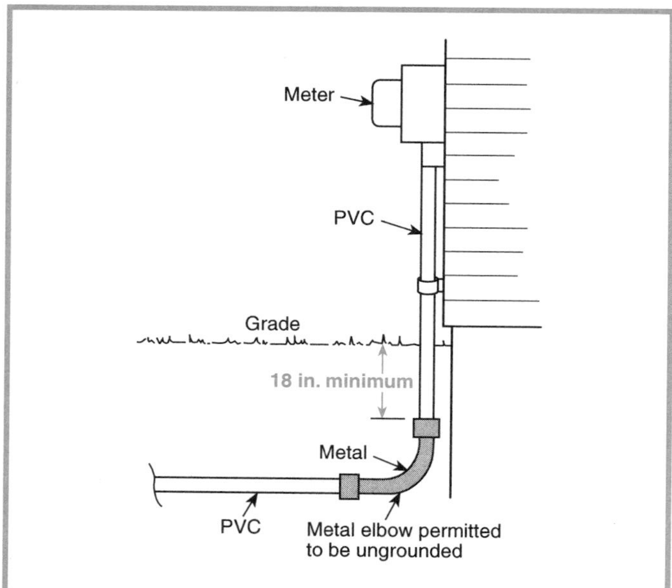

Figure 300.6 *An application of Section 250-80, Exception, which permits the metal elbow to be ungrounded, provided it is isolated from contact by a minimum cover of 18 in. to any part of the elbow.*

Service laterals that are not encased in concrete and that are buried 18 in. (457 mm) or more below grade shall have their location identified by a warning ribbon that is placed in the trench at least 12 in. (305 mm) above the underground installation.

Warning ribbon, or tape, was added for the 1999 *Code* to reduce the risk of an accident or electrocution during excavation near underground service lateral conductors and to correlate the *NEC®* with utility company practices. This requirement did not extend to feeders and branch circuits because those circuits contain short-circuit protection.

Conductors entering a building shall be protected to the point of entrance.

Where the enclosure or raceway is subject to physical damage, the conductors shall be installed in rigid metal conduit, intermediate metal conduit, Schedule 80 rigid nonmetallic conduit, or equivalent.

(e) Splices and Taps. Direct-buried conductors or cables shall be permitted to be spliced or tapped without the use of splice boxes. The splices or taps shall be made in accordance with Section 110-14(b).

There is a difference between multiconductor cables labeled for direct burial and single conductors labeled for direct burial. Since direct-burial multiconductor cables may or may not contain individual conductors that are labeled for direct burial, the overall cable jacket may be the only underground protection technique for the contained conductors. Although the direct-burial splicing techniques used on multiconductor cables can be much different than the techniques used on direct-burial single-conductor cables, the *Code* requirements are generally the same. The splicing technique should be listed for the cable type and listed for direct burial, due to the identified requirements and the listing requirements of Sections 110-14(b) and 250-8.

Examples of underground splicing methods used with single-conductor direct-burial cables are shown in Figure 300.7.

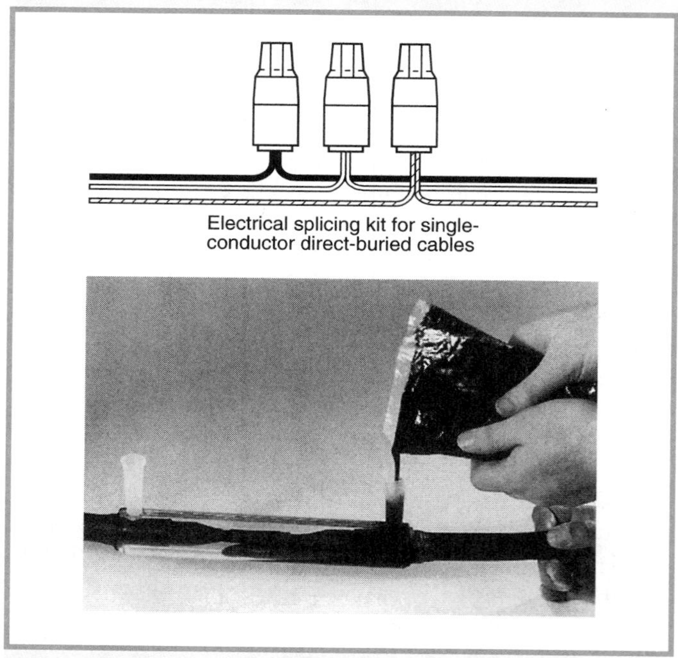

Figure 300.7 *Examples of underground splicing methods. (Top: King Technology Inc.; Bottom: 3M Co., Electrical Products Division)*

(f) Backfill. Backfill that contains large rocks, paving materials, cinders, large or sharply angular substances, or corrosive material shall not be placed in an excavation where materials may damage raceways, cables, or other substructures or prevent adequate compaction of fill or contribute to corrosion of raceways, cables, or other substructures.

Where necessary to prevent physical damage to the raceway or cable, protection shall be provided in the form of granular or selected material, suitable running boards, suitable sleeves, or other approved means.

(g) Raceway Seals. Conduits or raceways through which moisture may contact energized live parts shall be sealed or plugged at either or both ends.

FPN: Presence of hazardous gases or vapors may also necessitate sealing of underground conduits or raceways entering buildings.

Figure 300.8 shows a conduit sealing bushing used to prevent the entrance of gas or moisture. See Section 230-8 for sealing service raceways.

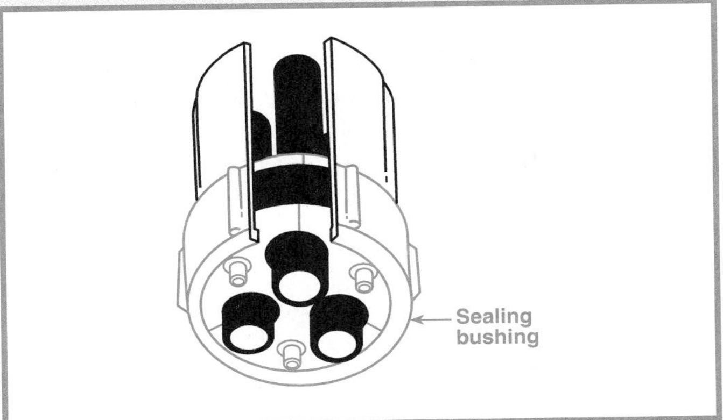

Figure 300.8 *A conduit sealing bushing used to prevent the entrance of gas or moisture. (O.Z./Gedney Co.)*

(h) Bushing. A bushing, or terminal fitting, with an integral bushed opening shall be used at the end of a conduit or other raceway that terminates underground where the conductors or cables emerge as a direct burial wiring method. A seal incorporating the physical protection characteristics of a bushing shall be permitted to be used in lieu of a bushing.

Figure 300.9 shows a Type UF cable buried in compliance with Table 300-5. Note the protective bushing where the cable is used with metal conduit.

(i) Conductors of the Same Circuit. All conductors of the same circuit and, where used, the grounded conductor and all equipment grounding conductors shall be installed in the same raceway or shall be installed in close proximity in the same trench.

Exception No. 1: Conductors in parallel in raceways shall be permitted, but each raceway shall contain all conductors of the same circuit including grounding conductors.

Exception No. 1 permits the installation of paralleled conductors in different raceways. See Section 310-4 for conductors in parallel.

Exception No. 2: Isolated phase installations shall be permitted in nonmetallic raceways in close proximity where conductors are paralleled as permitted in Section 310-4, and where the conditions of Section 300-20 are met.

Isolated phase installations are those that contain only one phase per raceway. Such an installation may have some advantages where there are many large conductors in parallel, as it avoids crossing conduc-

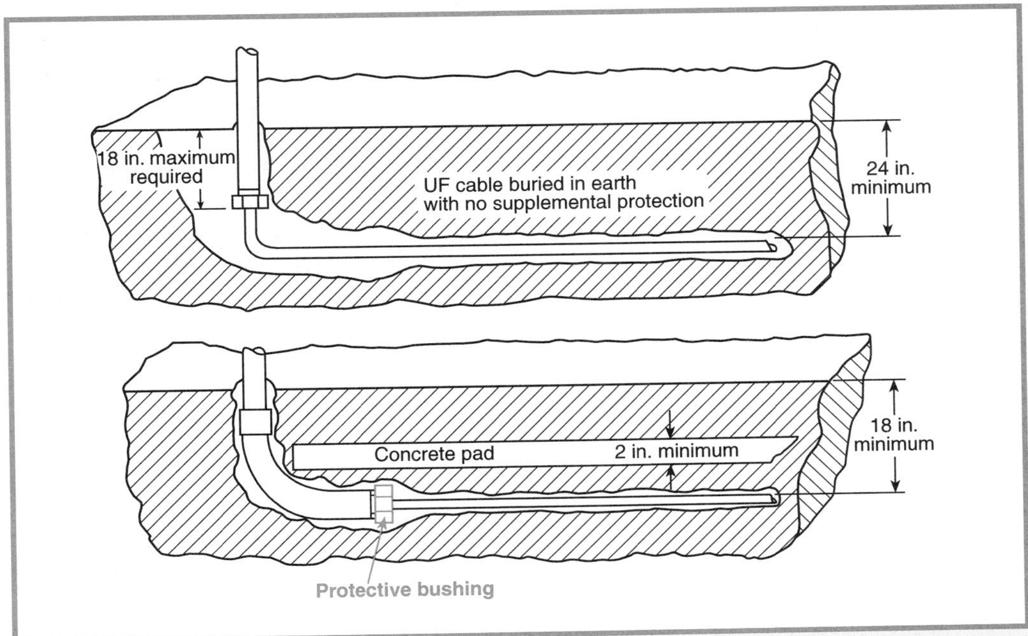

Figure 300.9 *A Type UF cable buried in compliance with Table 300-5.*

tors in the close quarters of the termination compartment of switchboards, unit substations, or transformer enclosures. The spacing between isolated phase raceways should be as small as possible and the length of the run limited, in order to avoid the increased circuit impedance and resulting increase in voltage drop inherent in an installation involving alternating-current circuits.

Isolated phase installations present an inherent hazard of overheating that must be understood and carefully controlled. This hazard is a result of induced currents in metal surrounding a raceway that contains only one phase conductor. See Sections 300-20(a) and 300-20(b). The surrounding metal acts as a shorted transformer turn. In underground installations, it is unlikely that a single conductor will be installed in a metal raceway or, if it were, that it would present a fire hazard. This is not true, however, for aboveground raceways, and it is the reason isolated phase installations are not permitted above ground.

See Sections 300-3(b)(3) and 330-16, which recognize single-conductor Type MI cable.

(j) Ground Movement. Where direct-buried conductors, raceways, or cables are subject to movement by settlement or frost, direct-buried conductors, raceways, or cables shall be arranged to prevent damage to the enclosed conductors or to equipment connected to the raceways.

FPN: This section recognizes "S" loops in underground direct burial to raceway transitions, expansion joints in raceway risers to fixed equipment, and, generally, the provision of flexible connections to equipment subject to settlement or frost heaves.

Section 300-5(j), added in the 1996 *Code,* points out the need for installers to allow for movement of direct-buried equipment, cables, and raceways. Slack must be allowed in cables or expansion joints, or other measures must be taken, if ground movement due to frost or settlement is anticipated.

300-6. Protection Against Corrosion. Metal raceways, cable trays, cablebus, auxiliary gutters, cable armor, boxes, cable sheathing, cabinets, elbows, couplings, fittings, supports, and support hardware shall be of materials suitable for the environment in which they are to be installed.

Section 300-6 applies generally. For specific applications, see the particular article covering the appropriate cables, raceways, or enclosures.

(a) General. Ferrous raceways, cable trays, cablebus, auxiliary gutters, cable armor, boxes, cable sheathing, cabinets, metal elbows, couplings, fittings, supports, and support hardware shall be suitably protected against corrosion inside and outside (except threads at joints) by a coating of approved corrosion-resistant material such as zinc, cadmium, or enamel. Where protected from corrosion solely by enamel, they shall not be used outdoors or in wet locations as described in (c). Where boxes or cabinets have an approved system of organic coatings and are marked "Raintight," "Rainproof," or "Outdoor Type," they shall be permitted outdoors.

Exception: Threads at joints shall be permitted to be coated with an identified electrically conductive compound.

Zinc chromate paste is one type of electrically conductive compound that can be used.

(b) In Concrete or in Direct Contact with the Earth. Ferrous or nonferrous metal raceways, cable armor, boxes, cable sheathing, cabinets, elbows, couplings, fittings, supports, and support hardware shall be permitted to be installed in concrete or in direct contact with the earth, or in areas subject to severe corrosive influences where made of material judged suitable for the condition, or where provided with corrosion protection approved for the condition.

Where ferrous or nonferrous metal conduit has corrosion protection and is judged suitable for the condition, it may be installed in concrete, in contact with the earth, or in areas exposed to severe corrosive influence. Special precautions are normally necessary for installing aluminum conduits in concrete, and specific approval by the authority having jurisdiction may be necessary.

Metal raceways installed in the earth can be coated with an asphalt compound, plastic sheath, or other equivalent protection to help prevent deterioration. Also, metallic raceways are available with a bonded PVC coating.

Galvanized rigid steel conduit and steel intermediate metal conduit do not generally require supplementary corrosion protection.

(c) Indoor Wet Locations. In portions of dairies, laundries, canneries, and other indoor wet locations, and in locations where walls are frequently washed or where there are surfaces of absorbent materials, such as damp paper or wood, the entire wiring system, where installed exposed, including all boxes, fittings, conduits, and cable used therewith, shall be mounted so that there is at least a ¼-in. (6.35-mm) airspace between it and the wall or supporting surface.

Exception: Nonmetallic raceways, boxes, and fittings shall be permitted to be installed without the airspace on a concrete, masonry, tile, or similar surface.

Table 300-5. Minimum Cover Requirements, 0 to 600 Volts, Nominal, Burial in Inches (Cover is defined as the shortest distance in inches measured between a point on the top surface of any direct-buried conductor, cable, conduit, or other raceway and the top surface of finished grade, concrete, or similar cover.)

	Type of Wiring Method or Circuit				
Location of Wiring Method or Circuit	**Column 1 Direct Burial Cables or Conductors**	**Column 2 Rigid Metal Conduit or Intermediate Metal Conduit**	**Column 3 Nonmetallic Raceways Listed for Direct Burial Without Concrete Encasement or Other Approved Raceways**	**Column 4 Residential Branch Circuits Rated 120 Volts or Less with GFCI Protection and Maximum Overcurrent Protection of 20 Amperes**	**Column 5 Circuits for Control of Irrigation and Landscape Lighting Limited to Not More than 30 Volts and Installed with Type UF or in Other Identified Cable or Raceway**
All locations not specified below	24	6	18	12	6
In trench below 2-in. thick concrete or equivalent	18	6	12	6	6
Under a building	0 (in raceway only)	0	0	0 (in raceway only)	0 (in raceway only)
Under minimum of 4-in. thick concrete exterior slab with no vehicular traffic and the slab extending not less than 6 in. beyond the underground installation	18	4	4	6 (direct burial) 4 (in raceway)	6 (direct burial) 4 (in raceway)
Under streets, highways, roads, alleys, driveways, and parking lots	24	24	24	24	24
One- and two-family dwelling driveways and outdoor parking areas, and used only for dwelling-related purposes	18	18	18	12	18
In or under airport runways, including adjacent areas where trespassing prohibited	18	18	18	18	18

Notes:
1. For SI units, 1 in. = 25.4 mm.
2. Raceways approved for burial only where concrete encased shall require concrete envelope not less than 2 in. thick.
3. Lesser depths shall be permitted where cables and conductors rise for terminations or splices or where access is otherwise required.
4. Where one of the wiring method types listed in Columns 1–3 is used for one of the circuit types in Columns 4 and 5, the shallower depth of burial shall be permitted.
5. Where solid rock prevents compliance with the cover depths specified in this table, the wiring shall be installed in metal or nonmetallic raceway permitted for direct burial. The raceways shall be covered by a minimum of 2 in. of concrete extending down to rock.

The exception to Section 300-6(c) is in harmony with Section 547-4, permitting nonmetallic boxes, fittings, conduit, and cables to be installed without the airspace in corrosive locations of agricultural buildings.

FPN: In general, areas where acids and alkali chemicals are handled and stored may present such corrosive conditions, particularly when wet or damp. Severe corrosive conditions may also be present in portions of meatpacking plants, tanneries, glue houses, and some stables; installations immediately adjacent to a seashore and swimming pool areas; areas where chemical deicers are used; and storage cellars or rooms for hides, casings, fertilizer, salt, and bulk chemicals.

300-7. Raceways Exposed to Different Temperatures.

(a) Sealing. Where portions of an interior raceway system are exposed to widely different temperatures, as in refrigerating or cold-storage plants, circulation of air from a warmer to a colder section through the raceway shall be prevented.

Where a raceway is used to enclose the lighting and refrigeration branch-circuit conductors within a walk-in chest, the circulation of air through the raceway from a warmer to a colder section could cause condensation within the raceway. This can be prevented by sealing the raceway with a suitable pliable compound at a conduit body or junction box, usually installed in the raceway before it enters the colder section. Special sealing fittings, such as those used in hazardous (classified) locations, are not necessary.

(b) Expansion Joints. Raceways shall be provided with expansion joints where necessary to compensate for thermal expansion and contraction.

FPN: Table 347-9(A) provides the expansion information for polyvinyl chloride (PVC). A nominal number for steel conduit can be determined by multiplying the expansion length in this table by 0.20. The coefficient of expansion for steel electrical metallic tubing, intermediate metal conduit, and rigid conduit is 6.50×10^{-6} (0.0000065 in. per inch of conduit for each °F in temperature change).

This fine print note, added in the 1996 *Code*, provides the relationship of linear expansion of PVC rigid nonmetallic conduit to steel conduit. For example, if a calculation indicated a linear expansion of 1¼ in. for PVC conduit, the steel conduit equivalent of expansion would be only ¼ in.

300-8. Installation of Conductors with Other Systems. Raceways or cable trays containing electric conductors shall not contain any pipe, tube, or equal for steam, water, air, gas, drainage, or any service other than electrical.

Unless modified by other articles as stated in Section 300-1, this section specifically prohibits installation of an electrical conductor in a raceway or cable tray that includes a drain, water, or air pipe, etc.

300-10. Electrical Continuity of Metal Raceways and Enclosures. Metal raceways, cable armor, and other metal enclosures for conductors shall be metallically joined together into a continuous electric conductor and shall be connected to all boxes, fittings, and cabinets so as to provide effective electrical continuity. Unless specifically permitted elsewhere in this *Code*, raceways and cable assemblies shall be mechanically secured to boxes, fittings, cabinets, and other enclosures.

Section 250-2 sets forth in detail what must be accomplished by grounding and bonding metal parts of the electrical system. These metal parts must form an effective low-impedance path to ground in order to safely conduct any fault current and facilitate the operation of overcurrent devices protecting the enclosed circuit conductors.

Exception: Short sections of raceways used to provide support or protection of cable assemblies from physical damage shall not be required to be made electrically continuous.

300-11. Securing and Supporting.

(a) Secured in Place. Raceways, cable assemblies, boxes, cabinets, and fittings shall be securely fastened in place. Support wires that do not provide secure support shall not be permitted as the sole support. Support wires and associated fittings that provide secure support and that are installed in addition to the ceiling grid support wires, shall be permitted as the sole support. Where independent support wires are used, they shall be secured at both ends. Cables and raceways shall not be supported by ceiling grids.

(1) Wiring located within the cavity of a fire-rated floor–ceiling or roof–ceiling assembly shall not be secured to, or supported by, the ceiling assembly, including the ceiling support wires. An independent means of secure support shall be provided. Where independent support wires are used, they shall be distinguishable by color, tagging, or other effective means from those that are part of the fire-rated design.

Exception: The ceiling support system shall be permitted to support wiring and equipment that have been tested as part of the fire-rated assembly.

Wiring methods of any type or lighting fixtures are not allowed to be supported or secured to the support wires or T bars of a fire-rated ceiling assembly unless the assembly has been tested and listed for that use.

Section 300-11(a)(1) dealing with fire-rated floor–ceiling and roof–ceiling assemblies was revised for the 1999 *Code.* Now, if support wires are selected as the supporting means for the electrical system within the fire-rated ceiling cavity, they must be distinguishable from the ceiling support wires and they must be secured at both ends.

Generally, the rule for supporting electrical equipment is "securely fastened in place." This phrase means not only that vertical support for the weight of the equipment must be provided, but also that the equipment must be secured to prevented horizontal movement or sway. The intention is to prevent the loss of grounding continuity provided by the raceway that could result from horizontal movement.

Parts (1) and (2) of this section are quite similar. Unless the exceptions apply, these sections make clear that all types of wiring are prohibited from being attached in any way to the support wires of the ceiling assembly. Unless ceiling grids are part of the building structure, they, too, are prohibited from furnishing support for cables and raceways.

However, it should be noted that if wiring and equipment are located within the ceiling cavity and rigidly supported independent of the ceiling, without the use of ceiling-type hanger wire, then the requirements of this section are met.

Refer to the appropriate wiring method article in Chapter 3 for cable and raceway supporting requirements. See Sections 410-15(a) and 410-16 for proper support of lighting fixtures. See Sections 370-23 for support of outlet boxes. And see Sections 725-7, 760-8, 770-8 for various low voltage, fire alarm and optical fiber cable supports. See Chapter 8 for communications cable supports.

FPN: One method of determining fire rating is testing in accordance with *Standard Methods of Tests of Fire Endurance of Building Construction and Materials,* NFPA 251-1995.

(2) Wiring located within the cavity of a nonfire-rated floor–ceiling or roof–ceiling assembly shall not be secured to, or supported by, the ceiling assembly, including the ceiling support wires. An independent means of secure support shall be provided.

Exception: The ceiling support system shall be permitted to support branch-circuit wiring and associated equipment where installed in accordance with the ceiling system manufacturer's instructions.

(b) Raceways Used as Means of Support. Raceways shall only be used as a means of support for other raceways, cables, or nonelectric equipment under the following conditions:

(1) Where the raceway or means of support is identified for the purpose; or
(2) Where the raceway contains power supply conductors for electrically controlled equipment and is used to support Class 2 circuit conductors or cables that are solely for the purpose of connection to the equipment control circuits; or
(3) Where the raceway is used to support boxes or conduit-bodies in accordance with Section 370-23 or to support fixtures in accordance with Section 410-16(f)

The intent of Section 300-11(b)(3) is to prevent cables from being attached to the exterior of a raceway. Electrical, telephone, and computer cables wrapped around a raceway can prevent dissipation of heat from the raceway and affect the temperature of the conductors therein. This section also prohibits the use of a raceway as a means of support for nonelectric equipment, such as suspended ceilings, water pipes, nonelectric signs, and the like, which could cause a mechanical failure of the raceway.

However, Section 300-11(b)(2) does allow the installation of Class 2 thermostat conductors for a boiler or air conditioner unit to be supported by the conduit supplying power to the unit, as shown in Figure 300.10. These Class 2 circuits are functionally associated with the branch-circuit wiring method.

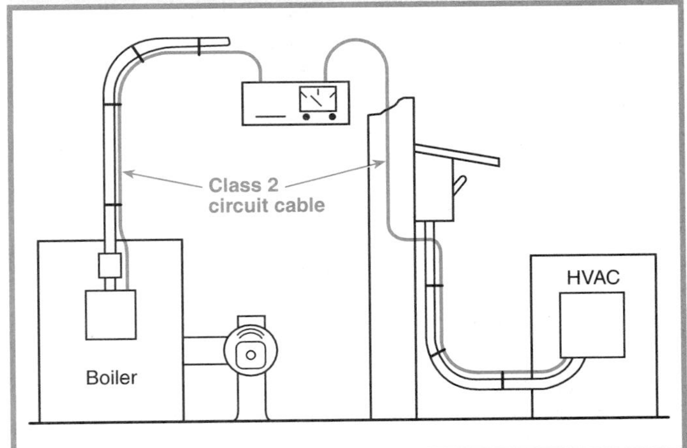

Figure 300.10 *Raceways used to support Class 2 thermostat wires.*

300-12. Mechanical Continuity — Raceways and Cables. Metal or nonmetallic raceways, cable armors, and cable sheaths shall be continuous between cabinets, boxes, fittings, or other enclosures or outlets.

Exception: Short sections of raceways used to provide support or protection of cable assemblies from physical damage shall not be required to be mechanically continuous.

300-13. Mechanical and Electrical Continuity — Conductors.

(a) General. Conductors in raceways shall be continuous between outlets, boxes, devices, etc. There shall be no splice or tap within a raceway unless permitted by Sections 300-15; 352-7; 352-29; 354-6, Exception; 362-7; 362-21; or 364-8(a).

Splices or taps are prohibited within raceways unless the raceways are equipped with hinged or removable covers. Busway conductors are exempt from this requirement.

(b) Device Removal. In multiwire branch circuits, the continuity of a grounded conductor shall not depend on device connections such as lampholders, receptacles, etc., where the removal of such devices would interrupt the continuity.

Grounded conductors (neutrals) of multiwire branch circuits supplying receptacles, lampholders, or other such devices are not permitted to depend on terminal connections for continuity between devices. For such installations (3- or 4-wire circuits), a splice is made and a jumper is connected to the terminal, unless the neutral is "looped"; that is, a receptacle or lampholder could be replaced without interrupting the continuity of energized downstream line-to-neutral loads (see commentary to Section 300-14). Opening the neutral could cause unbalanced voltages, and a considerably higher voltage would be impressed on one part of a multiwire branch circuit, especially if the downstream line-to-neutral loads were appreciably unbalanced. This requirement does not apply to individual 2-wire circuits or other circuits that do not contain a grounded (neutral) conductor.

300-14. Length of Free Conductors at Outlets, Junctions, and Switch Points. At least 6 in. (152 mm) of free conductor, measured from the point in the box where it emerges from its raceway or cable sheath, shall be left at each outlet, junction, and switch point for splices or the connection of fixtures or devices. Where the opening to an outlet, junction, or switch point is less than 8 in. (203 mm) in any dimension, each conductor shall be long enough to extend at least 3 in. (76.2 mm) outside the opening.

Exception: Conductors that are not spliced or terminated at the outlet, junction, or switch point.

A conductor looping through an outlet box and intended for connection to receptacles, switches, lampholders, or other such devices requires slack so that terminal connections may be made easily.

Conductors running through a box should have sufficient slack to prevent physical damage from the insertion of devices or from the use of fixture studs, hickeys, or other fixture supports within the box.

Revised for the 1999 *Code,* this section is much more specific about the exact measurements of free conductor length required at each splice point or device outlet. For these free conductor length measurements, see Figure 300.11.

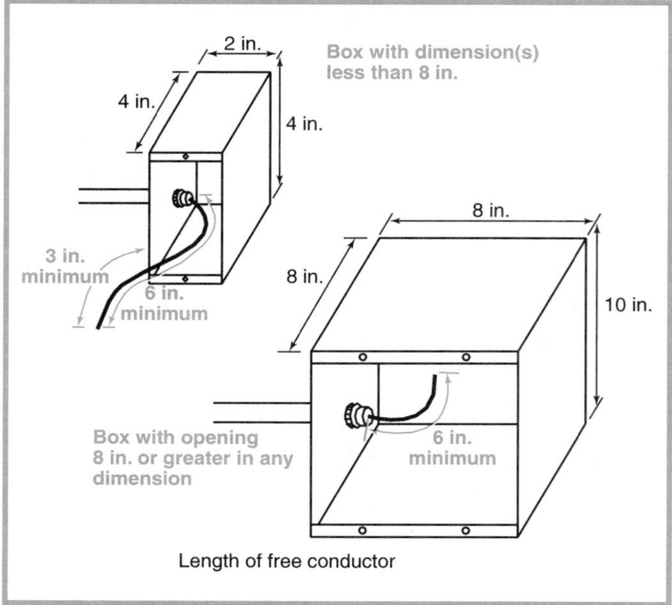

Figure 300.11 *Outlet box with free conductor lengths shown.*

300-15. Boxes, Conduit Bodies, or Fittings — Where Required.

(a) Box or Conduit Body. Where the wiring method is conduit, electrical metallic tubing, Type AC cable, Type MC cable, Type MI cable, nonmetallic-sheathed cable, or other cables, a box or conduit body complying with Article 370 shall be installed at each conductor splice point, outlet, switch point, junction point, or pull point, unless otherwise permitted in (b) through (n). A box shall be installed at each outlet and switch point for concealed knob-and-tube wiring.

Fittings and connectors shall be used only with the specific wiring methods for which they are designed and listed.

(b) Equipment. An integral junction box or wiring compartment as part of approved equipment shall be permitted.

(c) Protection. Where cables enter or exit from conduit or tubing that is used to provide cable support or protection against physical damage. A fitting shall be provided on the end(s) of the conduit or tubing to protect the cable from abrasion.

Section 300-15(c) permits conduit or tubing to be used as support and protection against physical damage without terminating in a box. It also permits conduit

or tubing to be used as physical protection for underground cables that exit from buildings or that are located outdoors on poles, without a box being required on the end of the conduit. A fitting to protect the wires or cables against physical damage is required on the ends of the conduit or tubing.

(d) Type MI Cable. Where accessible fittings are used for straight-through splices in mineral-insulated metal-sheathed cable.

(e) Integral Enclosure. A wiring device with integral enclosure identified for the use, having brackets that securely fasten the device to walls or ceilings of conventional onsite frame construction, for use with nonmetallic-sheathed cable, shall be permitted in lieu of a box or conduit body.

FPN: See Sections 336-18, Exception No. 2; 545-10; 550-10(i); and 551-47(e), Exception No. 1.

Section 300-15(e) applies to a device with an integral enclosure (boxless device) such as the one shown in Figure 300.12.

Figure 300.12 *A self-contained device (SCD) receptacle. (Pass & Seymour/Legrand)*

(f) Fitting. A fitting identified for the use shall be permitted in lieu of a box or conduit body where accessible after installation and not containing spliced or terminated conductors.

Where a cable system makes a transition to a raceway to provide mechanical protection against damage. Section 300-15(f) permits the use of a fitting instead of a box. For example, where nonmetallic-sheathed cable that is run overhead on floor joists and drops down on a concrete wall to supply a receptacle needs to be protected from physical damage, a short length of raceway is installed to the outlet device box. The cable sheath is removed for the length of the raceway. The cable is then inserted in the raceway and secured by a combination fitting that is fastened to the raceway.

(g) Buried Conductors. As permitted in Section 300-5(e) for splices and taps in buried conductors and cables.

(h) Insulated Devices. As permitted in Section 336-21 for insulated devices supplied by nonmetallic-sheathed cable.

(i) Enclosures. As permitted in Section 373-8 for switches and overcurrent devices, and Section 430-10(a) for motor controllers.

(j) Fixtures. As permitted in Section 410-31 where a fixture is used as a raceway.

(k) Embedded. Where conductors are embedded as covered in Sections 424-40, 424-41(d), 426-22(b), 426-24(a), and 427-19(a).

(l) Manufactured Wiring System. Where manufactured wiring systems in accordance with Article 604 are used.

(m) Closed Loop. Where a device identified and listed as suitable for installation without a box is used with a closed-loop power distribution system.

See Article 780, Closed-Loop and Programmed Power Distribution.

(n) Manholes. A box or conduit body shall not be required for conductors in manholes where accessible only to qualified persons.

300-16. Raceway or Cable to Open or Concealed Wiring.

(a) Box or Fitting. A box or terminal fitting having a separately bushed hole for each conductor shall be used wherever a change is made from conduit, electrical metallic tubing, electrical nonmetallic tubing, nonmetallic-sheathed cable, Type AC cable, Type MC cable, or mineral-insulated, metal-sheathed cable and surface raceway wiring to open wiring or to concealed knob-and-tube wiring. A fitting used for this purpose shall contain no taps or splices and shall not be used at fixture outlets.

(b) Bushing. A bushing shall be permitted in lieu of a box or terminal fitting where the conductors emerge from a raceway or conduit and enter or terminate at equipment, such as open switchboards, unenclosed control equipment, or similar equipment. The bushing shall be of the insulating type for other than lead-sheathed conductors.

300-17. Number and Size of Conductors in Raceway. The number and size of conductors in any raceway shall not be more than will permit dissipation of the heat and ready installation or withdrawal of the conductors without damage to the conductors or to their insulation.

FPN: See the following sections of this *Code*: electrical nonmetallic tubing, Section 331-6; intermediate metal conduit, Section 345-7; rigid metal conduit, Section 346-7; rigid nonmetallic conduit, Section 347-11; electrical metallic tubing, Section 348-8; flexible metallic tubing, Section 349-12; flexible metal conduit, Section 350-12; liquidtight flexible metal conduit, Section 351-6; liquidtight nonmetallic flexible conduit, Section 351-25; surface raceways, Sections 352-4 and 352-25; underfloor raceways, Section 354-5; cellular metal floor raceways, Section 356-5; cellular concrete floor raceways, Section 358-11; wireways, Section 362-5; fixture wire, Section 402-7; theaters, Section 520-6; signs, Section 600-31(c); elevators, Section 620-33; audio signal processing, amplification, and reproduction equipment, Sections 640-23(a) and 640-24; Class 1, Class 2, and Class 3 circuits, Article 725; fire alarm circuits, Article 760, and optical fiber cables and raceways, Article 770.

Listed wire-pulling compounds are available to assist in the process of pulling wires into raceways.

300-18. Raceway Installations.

(a) Complete Runs. Raceways, other than busways or exposed raceways having hinged or removable covers, shall be installed complete between outlet, junction, or splicing points prior to the installation of conductors. Where required to facilitate the installation of utilization equipment, the raceway shall be permitted to be initially installed without a terminating connection at the equipment. Prewired raceway assemblies shall be permitted only where specifically permitted in this *Code* for the applicable wiring method.

One of the primary functions of a raceway is to provide physical protection for conductors. If raceways are incomplete at the time of installation, a greater possibility exists for damage to the conductors.

Section 300-18(a) does, however, permit the installation of conductors in a raceway prior to the complete installation of the raceway, up to the point of utilization. The motor installation shown in Figure 300.13 is a typical application, where the motor is supplied through liquidtight flexible metal conduit that terminates in the motor terminal box through a 90 degree angle connector. Wiring a fixture whip prior to connecting a lighting fixture is also permitted by this section.

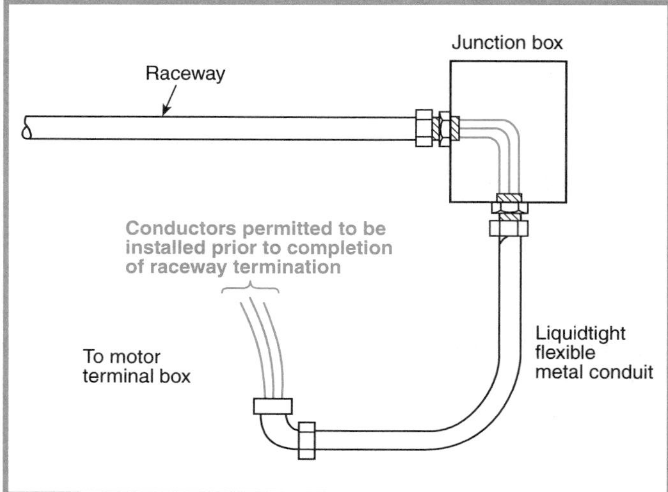

***Figure 300.13** An application of Section 300-18(a), which permits the conductors supplying a motor through liquidtight flexible metal conduit to be installed prior to the connection of the raceway to the motor terminal box.*

(b) Welding. Metal raceways shall not be supported, terminated, or connected by welding to the raceway unless specifically designed to be or otherwise specifically permitted to be in this *Code*.

300-19. Supporting Conductors in Vertical Raceways.

(a) Spacing Intervals — Maximum. Conductors in vertical raceways shall be supported if the vertical rise exceeds the values in Table 300-19(a). One cable support shall be provided at the top of the vertical raceway or as close to the top as practical. Intermediate supports shall be provided as necessary to limit supported conductor lengths to not greater than those values specified in Table 300-19(a).

Table 300-19(a). Spacings for Conductor Supports

Size of Wire	Support of Conductors in Vertical Raceways	Conductors: Aluminum or Copper-Clad Aluminum	Conductors: Copper
18 AWG through 8 AWG	Not greater than	100 ft	100 ft
6 AWG through 1/0 AWG	Not greater than	200 ft	100 ft
2/0 AWG through 4/0 AWG	Not greater than	180 ft	80 ft
Over 4/0 AWG through 350 kcmil	Not greater than	135 ft	60 ft
Over 350 kcmil through 500 kcmil	Not greater than	120 ft	50 ft
Over 500 kcmil through 750 kcmil	Not greater than	95 ft	40 ft
Over 750 kcmil	Not greater than	85 ft	35 ft

Note: For SI units, 1 ft = 0.3048 m.

Exception: Steel wire armor cable shall be supported at the top of the riser with a cable support that clamps the steel wire armor. A safety device shall be permitted at the lower end of the riser to hold the cable in the event there is slippage of the cable in the wire-armored cable support. Additional wedge-type supports shall be permitted to relieve the strain on the equipment terminals caused by expansion of the cable under load.

(b) Support Methods. One of the following methods of support shall be used.

(1) By clamping devices constructed of or employing insulating wedges inserted in the ends of the raceways. Where clamping of insulation does not adequately support the cable, the conductor also shall be clamped.

(2) By inserting boxes at the required intervals in which insulating supports are installed and secured in a satisfactory manner to withstand the weight of the conductors attached thereto, the boxes being provided with covers.

(3) In junction boxes, by deflecting the cables not less than 90 degrees and carrying them horizontally to a distance not less than twice the diameter of the cable, the cables being carried on two or more insulating supports and additionally secured thereto by tie wires if desired. Where this method is used, cables shall be supported at intervals not greater than 20 percent of those mentioned in the preceding tabulation.

(4) By a method of equal effectiveness.

Conductors in long vertical runs are required to be supported if the vertical rise exceeds the values in Table 300-19(a), to prevent the weight of the conductors from damaging the insulation where they leave the conduit and to prevent the conductors from being pulled out of the terminals. Supports such as those shown in Figures 300.14 and 300.15 may be used, in addition to many other types of grips manufactured for this purpose.

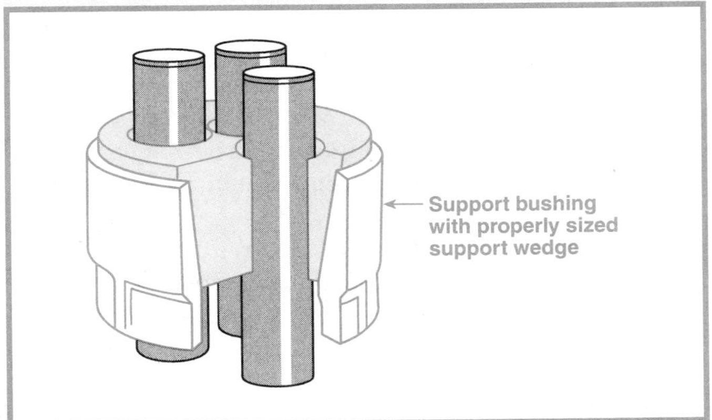

Figure 300.14 *A support bushing, located at the top of a vertical conduit at a cabinet or pull box, used to prevent the weight of the conductors from damaging insulation or placing strain on termination points. (O.Z./Gedney Co.)*

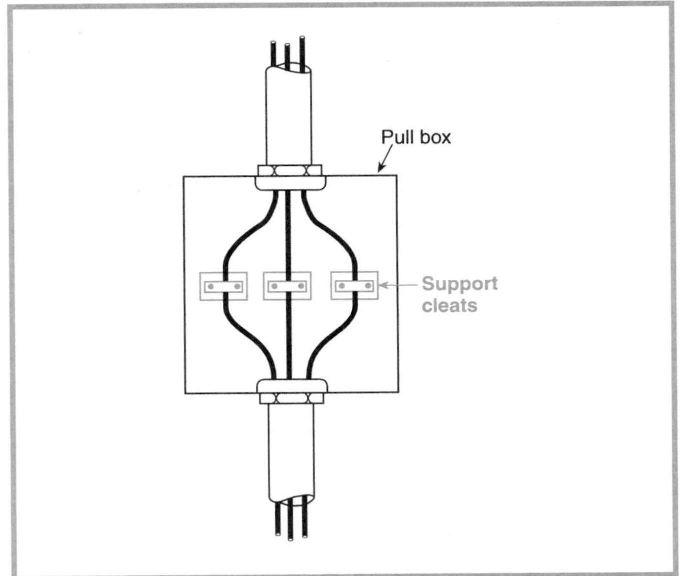

Figure 300.15 *Support cleats used to prevent the weight of vertical conductors from damaging insulation or placing strain on termination points.*

Example

A vertical raceway contains No. 1/0 copper conductors. One cable support near the top of the run would be required if the vertical run exceeds 100 ft. Intermediate supports may be required to limit the supported length to the table values. If the vertical run was less than 100 ft, cable supports would not be required.

300-20. Induced Currents in Metal Enclosures or Metal Raceways.

(a) Conductors Grouped Together. Where conductors carrying alternating current are installed in metal enclosures or metal raceways, they shall be arranged so as to avoid heating the surrounding metal by induction. To accomplish this, all phase conductors and, where used, the grounded conductor and all equipment grounding conductors shall be grouped together.

Exception No. 1: Equipment grounding conductors for certain existing installations shall be permitted to be installed separate from their associated circuit conductors where run in accordance with the provisions of Section 250-130(c).

Exception No. 2: A single conductor shall be permitted to be installed in a ferromagnetic enclosure and used for skin-effect heating in accordance with the provisions of Sections 426-42 and 427-47.

(b) Individual Conductors. Where a single conductor carrying alternating current passes through metal with magnetic properties, the inductive effect shall be minimized by (1) cutting slots in the metal between the individual holes through which the individual conductors pass, or (2) passing all the conductors in the circuit through an insulating wall sufficiently large for all of the conductors of the circuit.

Exception: In the case of circuits supplying vacuum or electric-discharge lighting systems or signs, or X-ray apparatus, the currents carried by the conductors are so small that the inductive heating effect can be ignored where these conductors are placed in metal enclosures or pass through metal.

FPN: Because aluminum is not a magnetic metal, there will be no heating due to hysteresis; however, induced currents will be present. They will not be of sufficient magnitude to require grouping of conductors or special treatment in passing conductors through aluminum wall sections.

Section 300-3(b)(3) permits *single*-conductor Type MI cable, provided it complies with Section 330-16. This section requires the installation to conform to Section 300-20 regarding inductive effects.

300-21. Spread of Fire or Products of Combustion. Electrical installations in hollow spaces, vertical shafts, and ventilation or air-handling ducts shall be made so that the possible spread of fire or products of combustion will not be substantially increased. Openings around electrical penetrations through fire-resistant-rated walls, partitions, floors, or ceilings shall be firestopped using approved methods to maintain the fire resistance rating.

FPN: Directories of electrical construction materials published by qualified testing laboratories contain many listing installation restrictions necessary to maintain the fire-resistive rating of assemblies where penetrations or openings are made. An example is the 24-in. (610-mm) minimum horizontal separation that usually applies between boxes on opposite sides of the wall. Assistance in complying with Section 300-21 can be found in these directories and product listings.

It is the intent of Section 300-21 that cables, cable trays, and raceways be installed in such a manner through rated wall, floor and ceiling assemblies that they do not contribute to the spread of fire or the products of combustion. According to NFPA 221-1997, *Standard for Fire Walls and Fire Barrier Walls, fire resistance rating* is defined as the time, in minutes or hours, that materials or assemblies have withstood a fire test exposure as established in accordance with the test procedures of NFPA 251, *Standard Methods of Tests of Fire Endurance of Building Construction and Materials.* [Note: ASTM E119, *Standard Test Methods for Fire Tests of Building Construction and Materials,* and UL 263, *Fire Tests of Building Construction and Materials,* are similar to NFPA 251, *Standard Methods of Tests of Fire Endurance of Building Construction and Materials.*]

Further, NFPA 221, Section 4-2, Penetration Seals, states the following:

All through-penetration protection systems shall be tested and rated in accordance with ASTM E814, *Standard Test Method for Fire Tests of Through-Penetration Fire Stops.* The positive pressure difference between the exposed and unexposed surfaces of the test assembly shall not be less than 0.01 in. (2.5 Pa) water gauge. A through-penetration protection system shall have an F rating (as defined by ASTM E814) not less than the required fire resistance rating of the fire wall or fire barrier wall.

Exception: Concrete, mortar, or grout shall be permitted with maximum 6-in. (153-mm) nominal diameter steel or copper pipe or steel conduit. Concrete, mortar, or grout shall be the thickness required to maintain the required fire resistance rating of the wall being penetrated. The maximum opening size shall be 144 in.^{2} (0.094 m^2).

According to 1998 UL *Fire Resistance Directory — Volume 2, category XHEX, Through-Penetration Firestop Systems,* a firestop system is a specific construction consisting of a wall or floor assembly, a penetrating item passing through an opening in the wall or floor assembly, and the materials designed to prevent the spread of fire through the openings. The specifications for materials in a firestop system and the assembly of the materials are details which directly relate to the established ratings. Information concerning these details is described in the individual systems. The hourly ratings apply only to the complete systems. Individual components are designated for use in a specific system to achieve specified ratings. The individual components are not assigned ratings and are not intended to be interchanged between systems.

The basic standard used to investigate products in this category is ANSI/UL 1479, *Fire Tests of Through-Penetration Firestops.* This standard defines the criteria for hourly F, T, and L ratings for firestop systems. The F rating criteria prohibits flame passage through the system and requires acceptable hose stream test performance. The T rating criteria prohibits flame passage through the system and requires the maximum temperature rise on the unexposed surface of the wall or floor assembly, on the penetrating item, and on the fill material not to exceed 325°F (181°C) above ambient and requires acceptable hose stream test performance. The L rating criteria determines the amount of air leakage, in cubic feet per minute per square foot of opening (CFM/sq ft), through the firestop system at ambient and/or 400°F air temperatures at an air pressure differential of 0.30 in. W.C. The L ratings are intended to assist authorities having jurisdiction, and others, in determining the suitability of firestop systems for the protection of penetrations and miscellaneous openings in floors and smoke barriers for the purpose of restricting the movement of smoke in accordance with the *Life Safety Code,* NFPA *101*.

Materials used in the firestop systems are to be installed in accordance with the manufacturer's instructions provided with the materials. The structural integrity of the floor or wall assembly needs to be evaluated when providing openings for the penetrating items.

Firestop systems such as shown in Figure 300.16 and Figure 300.17 may be used to meet the requirements of Section 300-21.

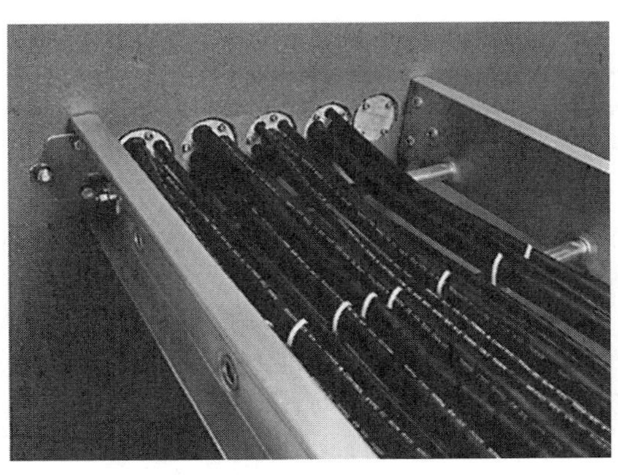

Figure 300.16 *Fire seals used in a through-penetration firestop system to maintain the fire resistance rating of the wall, as required by Section 300-21. (O.Z./Gedney Co.)*

Figure 300.17 *Fire-barrier caulking used as part of a through-penetration firestop system to meet the requirements of Section 300-21. (3M Co., Electrical Products Division)*

300-22. Wiring in Ducts, Plenums, and Other Air-Handling Spaces. The provisions of this section apply to the installation and uses of electric wiring and equipment in ducts, plenums, and other air-handling spaces.

FPN: Article 424, Part F.

(a) Ducts for Dust, Loose Stock, or Vapor Removal. No wiring systems of any type shall be installed in ducts used to transport dust, loose stock, or flammable vapors. No wiring system of any type shall be installed in any duct, or shaft containing only such ducts, used for vapor removal or for ventilation of commercial-type cooking equipment.

(b) Ducts or Plenums Used for Environmental Air. Only wiring methods consisting of Type MI cable, Type MC cable employing a smooth or corrugated impervious metal sheath without an overall nonmetallic covering, electrical metallic tubing, flexible metallic tubing, intermediate metal conduit, or rigid metal conduit shall be installed in ducts or plenums specifically fabricated to transport environmental air. Flexible metal conduit and liquidtight flexible metal conduit shall be permitted, in lengths not to exceed 4 ft (1.22 m), to connect physically adjustable equipment and devices permitted to be in these ducts and plenum chambers. The connectors used with flexible metal conduit shall effectively close any openings in the connection. Equipment and devices shall be permitted within such ducts or plenum chambers only if necessary for their direct action upon, or sensing of, the contained air. Where equipment or devices are installed and illumination is necessary to facilitate maintenance and repair, enclosed gasketed-type fixtures shall be permitted.

The intent of Section 300-22 is to limit the use of materials that would contribute smoke and products of combustion during a fire in an area that handles environmental air and, in the case of paragraph (b), to provide an effective barrier against the spread of products of combustion into the ducts or plenums.

Paragraph (b) applies to ducts and plenums, such as sheet-metal ducts, specifically constructed to transport environmental air. Equipment and devices such as lighting fixtures and motors are not normally permitted in ducts or plenums; therefore, wiring methods in paragraph (b) are different from those permitted in paragraph (c).

(c) Other Space Used for Environmental Air. This section applies to space used for environmental air-handling purposes other than ducts and plenums as specified in (a) and (b). It does not include habitable rooms or areas of buildings, the prime purpose of which is not air handling.

FPN: The space over a hung ceiling used for environmental air-handling purposes is an example of the type of other space to which this section applies.

Section 300-22(c) applies to other spaces used to transport environmental air that are not specifically manufactured as ducts or plenums, such as the space or cavity between a structural floor or roof and a suspended (hung) ceiling. Many spaces above suspended ceilings are intended to transport return air. Some are also used for supply air, but these are not nearly so common as those used for return air.

This section does not apply to habitable rooms and other areas whose prime purpose is *not* air handling. Such an area is shown in Figure 300.18 and designated as area A.

If the prime purpose of the room or space *is* air handling, such as areas B and C of Figure 300.18, then the restrictions in Section 300-22(c) apply, whether or not electrical equipment is located in the room.

Exception: This section shall not apply to the joist or stud spaces of dwelling units where the wiring passes through such spaces perpendicular to the long dimension of such spaces.

The exception to Section 300-22(c) permits cable to pass through joist or stud spaces of a dwelling unit, as illustrated in Figure 300.19. The joist space is covered with sheet metal and used as a cold-air return for a forced warm-air central heating system. Equipment such as junction boxes or device enclosures is not permitted in this location.

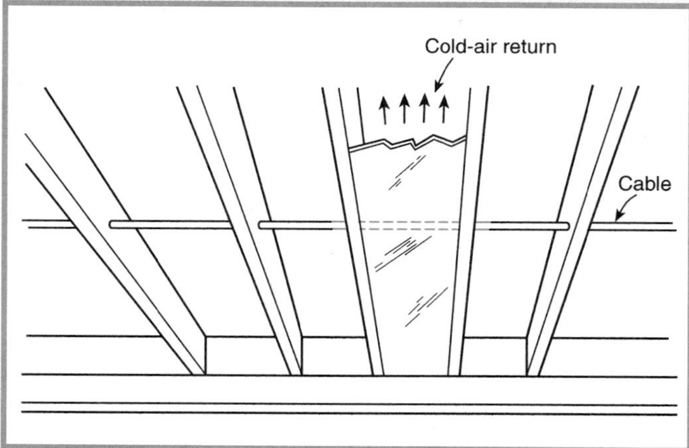

Figure 300.19 *A cable passing through joist spaces of a dwelling unit, as permitted by Section 300-22(c), Exception.*

(1) Wiring Methods. The wiring methods for such other space shall be limited to totally enclosed, nonventilated,

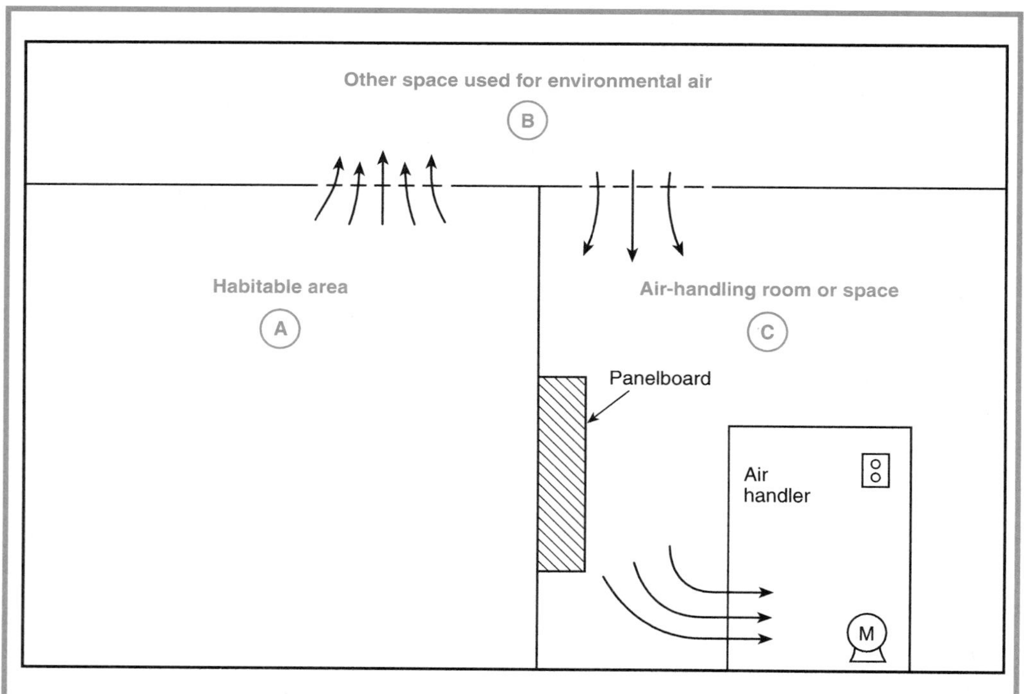

Figure 300.18 *An application of Section 300-22(c). Ordinary wiring methods are permitted in habitable areas that are not used primarily for air handling, such as area A. Only wiring methods described in Section 300-22(c)(1) are permitted in rooms or spaces that are primarily used for air handling, such as area B and area C.*

insulated busway having no provisions for plug-in connections, Type MI cable, Type MC cable without an overall nonmetallic covering, Type AC cable, or other factory-assembled multiconductor control or power cable that is specifically listed for the use, or listed prefabricated cable assemblies of metallic manufactured wiring systems without nonmetallic sheath. Other type cables and conductors shall be installed in electrical metallic tubing, flexible metallic tubing, intermediate metal conduit, rigid metal conduit, flexible metal conduit, or, where accessible, surface metal raceway or metal wireway with metal covers or solid bottom metal cable tray with solid metal covers.

Note that the cable assemblies of metallic manufactured wiring systems are required to be listed for use in space used for environmental air. See Article 604, Manufactured Wiring Systems.

Exception: Liquidtight flexible metal conduit shall be permitted in single lengths not exceeding 6 ft (1.83 m).

(2) Equipment. Electrical equipment with a metal enclosure, or with a nonmetallic enclosure listed for the use and having adequate fire-resistant and low-smoke-producing characteristics, and associated wiring material suitable for the ambient temperature shall be permitted to be installed in such other space unless prohibited elsewhere in this *Code.*

Electrical equipment with metal enclosures is allowed within spaces used for environmental air. However, nonmetallic enclosures must be listed for this use.

Exception: Integral fan systems shall be permitted where specifically identified for such use.

(d) Information Technology Equipment. Electric wiring in air-handling areas beneath raised floors for information technology equipment shall be permitted in accordance with Article 645.

It is not intended that the requirements of Sections 300-22(b) or 300-22(c) apply to air-handling areas beneath raised floors in information technology rooms. See Article 645, Information Technology Equipment.

300-23. Panels Designed to Allow Access. Cables, raceways, and equipment installed behind panels designed to allow access, including suspended ceiling panels, shall be arranged and secured so as to allow the removal of panels and access to the equipment.

Section 300-23 is intended to prevent the excess accumulation of wires and cables that could limit access to electrical equipment by preventing the removal of access panels.

B. Requirements for Over 600 Volts, Nominal

Part B of this article has been changed for the 1999 *Code.* Some of the requirements previously located within Article 710, Over 600 Volts, Nominal, in the 1996 *Code* have been relocated to Article 300, Part B. The flow chart shown in Figure 300.20 shows the relocated and rearranged sections.

300-31. Covers Required. Suitable covers shall be installed on all boxes, fittings, and similar enclosures to prevent accidental contact with energized parts or physical damage to parts or insulation.

300-32. Conductors of Different Systems. See Section 300-3(c)(2).

300-34. Conductor Bending Radius. The conductor shall not be bent to a radius less than 8 times the overall diameter for nonshielded conductors or 12 times the diameter for shielded or lead-covered conductors during or after installation. For multiconductor or multiplexed single conductor cables having individually shielded conductors, the minimum bending radius is 12 times the diameter of the individually shielded conductors or 7 times the overall diameter, whichever is greater.

300-35. Protection Against Induction Heating. Metallic raceways and associated conductors shall be arranged so as to avoid heating of the raceway in accordance with the applicable provisions of Section 300-20.

300-37. Aboveground Wiring Methods. Aboveground conductors shall be installed in rigid metal conduit, in intermediate metal conduit, in electrical metallic tubing, in rigid nonmetallic conduit, in cable trays, as busways, as cablebus, in other identified raceways, or as open runs of metal-clad cable suitable for the use and purpose. In locations accessible to qualified persons only, open runs of Type MV cables, bare conductors, and bare busbars shall also be permitted. Busbars shall be permitted to be either copper or aluminum.

In transformer vaults, switch rooms, and similar areas restricted to qualified personnel, any suitable wiring method may be used. Open wiring using bare or insulated conductors on insulators is commonly employed, as is rigid metal conduit and rigid nonmetallic conduit.

300-39. Braid-Covered Insulated Conductors — Open Installation. Open runs of braid-covered insulated conductors shall have a flame-retardant braid. If the conductors used do not have this protection, a flame-retardant saturant shall be applied to the braid covering after installation. This treated braid covering shall be stripped back a safe distance at conductor terminals, according to the operating voltage. This distance shall not be less than 1 in. (25.4 mm) for each kilovolt of the conductor-to-ground voltage of the circuit, where practicable.

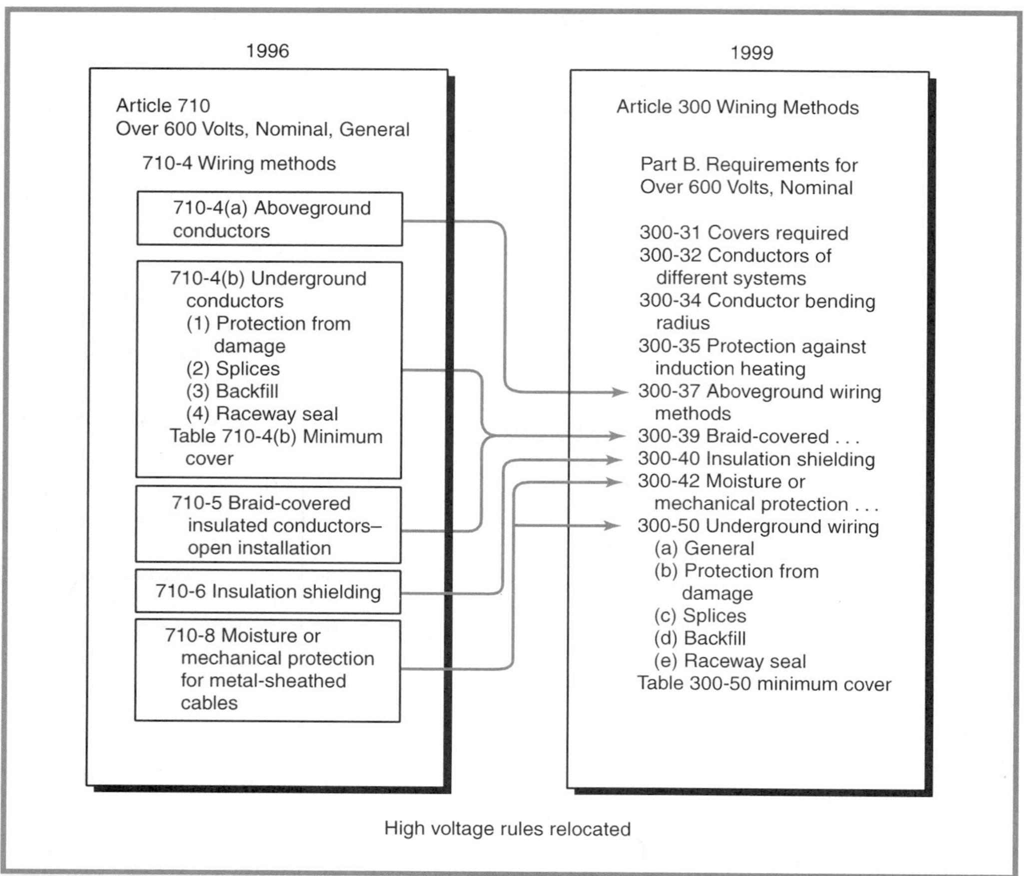

Figure 300.20 *The relocation of Article 710, Over 600 Volts, Nominal, (1996 Code) into Article 300, Part B.*

300-40. Insulation Shielding. Metallic and semiconducting insulation shielding components of shielded cables shall be removed for a distance dependent on the circuit voltage and insulation. Stress reduction means shall be provided at all terminations of factory-applied shielding.

Metallic shielding components such as tapes, wires, or braids, or combinations thereof, and their associated conducting or semiconducting components shall be grounded.

Special kits are available from several manufacturers that permit a quick and easy means of providing the required stress reductions when terminating solid dielectric cables.

300-42. Moisture or Mechanical Protection for Metal-Sheathed Cables. Where cable conductors emerge from a metal sheath and where protection against moisture or physical damage is necessary, the insulation of the conductors shall be protected by a cable sheath terminating device.

300-50. Underground Installations.

(a) General. Underground conductors shall be identified for the voltage and conditions under which they are installed. Direct burial cables shall comply with the provisions of Section 310-7. Underground cables shall be installed in accordance with (1) or (2), and the installation shall meet the depth requirements of Table 300-50.

(1) Shielded Cables and Nonshielded Cables in Metal-Sheathed Cable Assemblies. Underground cables, including nonshielded, Type MC and moisture-impervious metal sheath cables shall have those sheaths grounded through an effective grounding path meeting the requirements of Section 250-2(d). They shall be direct buried or installed in raceways identified for the use.

(2) Other Nonshielded Cables. Other nonshielded cables not covered in (1) shall be installed in rigid metal conduit, intermediate metal conduit, or rigid nonmetallic conduit encased in not less than 3 in. (76 mm) of concrete.

(b) Protection from Damage. Conductors emerging from the ground shall be enclosed in listed raceways. Raceways installed on poles shall be of rigid metal conduit, intermediate metal conduit, PVC Schedule 80, or equivalent, extending from the minimum cover depth specified in Table 300-50 to a point 8 ft (2.44 m) above finished grade. Conductors entering a building shall be protected by an approved enclosure or raceway from the minimum cover depth to the point of entrance. Where direct-buried conductors, raceways, or cables are subject to movement by settlement or frost, they shall be installed to prevent damage to the enclosed conductors or to the equipment connected to the raceways. Metallic enclosures shall be grounded.

Table 300-50. Minimum Cover Requirements (*Cover* is defined as the shortest distance in inches measured between a point on the top surface of any direct-buried conductor, cable, conduit, or other raceway and the top surface of finished grade, concrete, or similar cover.)

Circuit Voltage	Direct-Buried Cables	Rigid Nonmetallic Conduit Approved for Direct Burial*	Rigid Metal Conduit and Intermediate Metal Conduit
Over 600 V through 22 kV	30	18	6
Over 22 kV through 40 kV	36	24	6
Over 40 kV	42	30	6

Note: For SI units, 1 in. = 25.4 mm.

*Listed by a qualified testing agency as suitable for direct burial without encasement. All other nonmetallic systems shall require 2 in. (50.8 mm) of concrete or equivalent above conduit in addition to above depth.

Exception No. 1: Areas subject to vehicular traffic, such as thoroughfares or commercial parking areas, shall have a minimum cover of 24 in. (610 mm).

Exception No. 2: The minimum cover requirements for other than rigid metal conduit and intermediate metal conduit shall be permitted to be reduced 6 in. (152 mm) for each 2 in. (50.8 mm) of concrete or equivalent protection placed in the trench over the underground installation.

Exception No. 3: The minimum cover requirements shall not apply to conduits or other raceways that are located under a building or exterior concrete slab not less than 4 in. (102 mm) in thickness and extending not less than 6 in. (152 mm) beyond the underground installation. A warning ribbon or other effective means suitable for the conditions shall be placed above the underground installation.

Exception No. 4: Lesser depths shall be permitted where cables and conductors rise for terminations or splices or where access is otherwise required.

Exception No. 5: In airport runways, including adjacent defined areas where trespass is prohibited, cable shall be permitted to be buried not less than 18 in. (457 mm) deep and without raceways, concrete enclosement, or equivalent.

Exception No. 6: Raceways installed in solid rock shall be permitted to be buried at lesser depth where covered by 2 in. (50.8 mm) of concrete, which shall be permitted to extend to the rock surface.

(c) Splices. Direct burial cables shall be permitted to be spliced or tapped without the use of splice boxes, provided they are installed using materials suitable for the application. The taps and splices shall be watertight and protected from mechanical damage. Where cables are shielded, the shielding shall be continuous across the splice or tap.

Exception: At splices of an engineered cabling system, metallic shields of direct-buried single-conductor cables with maintained spacing between phases shall be permitted to be interrupted and overlapped. Where shields are interrupted and overlapped, each shield section shall be grounded at one point.

(d) Backfill. Backfill containing large rocks, paving materials, cinders, large or sharply angular substances, or corrosive materials shall not be placed in an excavation where materials can damage raceways, cables, or other substructures, or prevent adequate compaction of fill, or contribute to corrosion of raceways, cables, or other substructures.

Protection in the form of granular or selected material or suitable sleeves shall be provided to prevent physical damage to the raceway or cable.

(e) Raceway Seal. Where a raceway enters from an underground system, the end within the building shall be sealed with an identified compound so as to prevent the entrance of moisture or gases, or it shall be so arranged to prevent moisture from contacting live parts.

Article 305 — Temporary Wiring

Contents

305-1. Scope. The provisions of this article apply to temporary electrical power and lighting wiring methods that may be of a class less than would be required for a permanent installation.

305-2. All Wiring Installations.

(a) Other Articles. Except as specifically modified in this article, all other requirements of this *Code* for permanent wiring shall apply to temporary wiring installations.

Wiring methods installed as temporary wiring are required to be installed in accordance with the rules of the article governing that wiring method, except where modified by the rules in this article.

(b) Approval. Temporary wiring methods shall be acceptable only if approved based on the conditions of use and any special requirements of the temporary installation.

Section 305-2(b) requires that all temporary wiring methods be approved based on criteria such as (1) length of time in service, (2) severity of physical abuse, (3) exposure to weather, and (4) other special requirements. Special requirements may range from tunnel construction projects and tent cities constructed after a natural disaster to flammable hazardous material reclamation projects.

305-3. Time Constraints.

(a) During the Period of Construction. Temporary electrical power and lighting installations shall be permitted during the period of construction, remodeling, maintenance, repair, or demolition of buildings, structures, equipment, or similar activities.

(b) 90 Days. Temporary electrical power and lighting installations shall be permitted for a period not to exceed 90 days for Christmas decorative lighting and similar purposes.

Note that the temporary wiring applications of Section 305-3(a) and (c) are *not* restricted to 90 days.

(c) Emergencies and Tests. Temporary electrical power and lighting installations shall be permitted during emergencies and for tests, experiments, and developmental work.

(d) Removal. Temporary wiring shall be removed immediately upon completion of construction or purpose for which the wiring was installed.

All temporary wiring must be not only disconnected but also removed.

305-4. General.

(a) Services. Services shall be installed in conformance with Article 230.

(b) Feeders. Feeders shall be protected as provided in Article 240. They shall originate in an approved distribution center. Conductors shall be permitted within cable assemblies, or within cords or cables of a type identified in Table 400-4 for hard usage or extra-hard usage. For the purpose of this section, Type NM and Type NMC cables shall be permitted to be used in any dwelling, building, or structure without any height limitation.

Exception: Single insulated conductors shall be permitted where installed for the purpose(s) specified in Section 305-3(c), where accessible only to qualified persons.

Temporary feeders are permitted to be (1) cable assemblies, (2) multiconductor cords, or (3) single-conductor cords. Cords used as feeders must be identified for hard or extra-hard usage according to Table 400-4. Individual conductors, as described in Section 310-13, are not permitted as open conductors but, rather, must be part of a cable assembly or used in a raceway system. Open or individual conductor feeders are permitted only during emergencies or tests.

All temporary wiring methods must be approved by the authority having jurisdiction. *[See Section 305-2(b)].*

(c) Branch Circuits. All branch circuits shall originate in an approved power outlet or panelboard. Conductors shall be permitted within cable assemblies, or within multiconductor cord or cable of a type identified in Table 400-4 for hard usage or extra-hard usage. All conductors shall be protected as provided in Article 240. For the purposes of this section, Type NM and Type NMC cables shall be permitted to be used in any dwelling, building, or structure without any height limitation.

Branch circuits installed for the purposes specified in Sections 305-3(b) or (c) shall be permitted to be run as single insulated conductors. Where the wiring is installed in accordance with Section 305-3(b), the voltage to ground shall not exceed 150 volts, the wiring shall not be subject to physical damage, and the conductors shall be supported on insulators at intervals of not more than 10 ft (3.05 m); or, for festoon lighting, the conductors shall be arranged so that excessive strain is not transmitted to the lampholders.

The basic requirement for safety in Section 305-4(c) is that temporary wiring be located and installed so that it will not be physically damaged. Section 305-2(a) requires wiring to be installed in accordance with the appropriate article (unless modified in Article 305).

Note that hard-usage or extra-hard-usage extension cords are permitted to be laid on the floor.

(d) Receptacles. All receptacles shall be of the grounding type. Unless installed in a continuous grounded metal raceway or metal-covered cable, all branch circuits shall contain a separate equipment grounding conductor, and all receptacles shall be electrically connected to the equipment grounding conductors. Receptacles on construction sites shall not be installed on branch circuits that supply temporary lighting. Receptacles shall not be connected to the same ungrounded conductor of multiwire circuits that supply temporary lighting.

The intent of Section 305-4(d) is to require separate ungrounded conductors for lighting and receptacle loads, so that the activation of a fuse, circuit breaker, or ground-fault circuit interrupter due to a fault or equipment overload will not de-energize the lighting circuit.

(e) Disconnecting Means. Suitable disconnecting switches or plug connectors shall be installed to permit the disconnection of all ungrounded conductors of each temporary circuit. Multiwire branch circuits shall be provided with a means to disconnect simultaneously all ungrounded conductors at the power outlet or panelboard where the branch circuit originated. Approved handle ties shall be permitted.

(f) Lamp Protection. All lamps for general illumination shall be protected from accidental contact or breakage by a suitable fixture or lampholder with a guard.

Brass shell, paper-lined sockets, or other metal-cased sockets shall not be used unless the shell is grounded.

(g) Splices. On construction sites, a box shall not be required for splices or junction connections where the circuit conductors are multiconductor cord or cable assemblies. See Sections 110-14(b) and 400-9. A box, conduit body, or terminal fitting having a separately bushed hole for each conductor shall be used wherever a change is made to a conduit or tubing system or a metal-sheathed cable system.

(h) Protection from Accidental Damage. Flexible cords and cables shall be protected from accidental damage. Sharp corners and projections shall be avoided. Where passing through doorways or other pinch points, protection shall be provided to avoid damage.

Note that, unlike Section 400-8, Section 305-4(h) permits flexible cords and cables, because of the nature of their use, to pass through doorways.

(i) Termination(s) at Devices. Flexible cords and cables entering enclosures containing devices requiring termination shall be secured to the box with fittings designed for the purpose.

(j) Support. Cable assemblies and flexible cords and cables shall be supported in place at intervals that ensure that they will be protected from physical damage. Support shall be in the form of staples, cable ties, straps, or similar type fittings installed so as not to cause damage.

Section 305-4(j) was added in the 1999 *Code* to clarify that temporary wiring methods should not have to be supported in the same manner afforded a permanent installation.

305-6. Ground-Fault Protection for Personnel. Ground-fault protection for personnel for all temporary wiring installations shall be provided to comply with (a) and (b). This section shall apply only to temporary wiring installations used to supply temporary power to equipment used by personnel during construction, remodeling, maintenance, repair, or demolition of buildings, structures, equipment, or similar activities.

(a) Receptacle Outlets. All 125-volt, single-phase, 15-, 20-, and 30-ampere receptacle outlets that are not a part of the permanent wiring of the building or structure and that are in use by personnel shall have ground-fault circuit interrupter protection for personnel. If a receptacle(s) is installed or exists as part of the permanent wiring of the building or structure and is used for temporary electric power, ground-fault circuit-interrupter protection for personnel shall be provided. For the purposes of this section, cord sets or devices incorporating listed ground-fault circuit interrupter protection for personnel identified for portable use shall be permitted.

Exception No. 1: Receptacles on a 2-wire, single-phase portable or vehicle-mounted generator rated not more than 5 kW, where the circuit conductors of the generator are insulated from the generator frame and all other grounded surfaces, shall be permitted without ground-fault protection for personnel.

Exception No. 2: In industrial establishments only, where conditions of maintenance and supervision ensure that only qualified personnel are involved, an assured equipment grounding conductor program as specified in Section 305-6(b)(2) shall be permitted to be utilized for all receptacle outlets.

(b) Use of Other Outlets. Receptacles other than 125-volt, single-phase, 15-, 20-, and 30-ampere receptacles shall have protection in accordance with (1) or, the assured equipment grounding conductor program in accordance with (2).

(1) Ground-fault circuit interrupter protection for personnel.
(2) A written assured equipment grounding conductor program continuously enforced at the site by one or more designated persons to ensure that equipment grounding conductors for all cord sets, receptacles that are not a part of the permanent wiring of the building or structure, and equipment connected by cord and plug are installed and maintained in accordance with the applicable requirements of Sections 210-7(c), 250-114, 250-138, and 305-4(d).

Personnel using temporary wiring while performing activities such as construction, remodeling, maintenance, repair, and demolition must have electrical

shock or electrocution protection. This personnel protection may be provided by either ground-fault circuit-interrupter (GFCI) protection or according to the "Assured Equipment Grounding Conductor Program." However, for all 125-volt, single-phase, 15-, 20-, and 30-ampere receptacle outlets, there simply is no option. For these specific receptacle outlets, personnel protection can only be afforded by ground-fault circuit-interrupter (GFCI) protection. The assured equipment grounding conductor program may not be used for 125-volt, single-phase, 15-, 20-, and 30-ampere receptacles, unless the user qualifies under Section 305-6(a), Exception No. 2.

According to OSHA 29CFR 1926.404(b)(1)(iii),

> The employer shall establish and implement an assured equipment grounding conductor program on construction sites covering all cord sets, receptacles which are not a part of the building or structure, and equipment connected by cord and plug which are available for use or used by employees. This program shall comply with the following minimum requirements:
>
> (A) A written description of the program, including the specific procedures adopted by the employer, shall be available at the jobsite for inspection and copying by the Assistant Secretary and any affected employee.
>
> (B) The employer shall designate one or more competent persons. . .

These OSHA requirements are very similar to the present *NEC* requirements for an assured grounding program.

Receptacle outlets, other than the 125-volt, single-phase, 15-, 20- and 30-ampere outlets, are not required to be GFCI protected, provided, of course, they are installed and maintained in accordance with an assured equipment grounding conductor program of Section 305-6(b)(2). Otherwise, they must be GFCI protected.

GFCI protection for construction or maintenance personnel using receptacles that are part of the permanent wiring that are not GFCI protected may be provided by using cord sets or listed portable GFCIs identified for portable use. An example of a GFCI cord set that is identified for portable use is shown in Figure 305.1.

Example Figures

Figures 305.1 through 305.4 show some examples of implementing the temporary wiring requirements of Section 305-6.

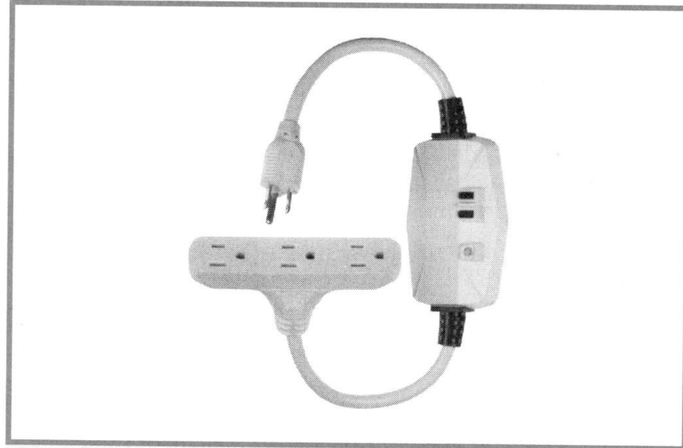

Figure 305.1 *A raintight GFCI with open neutral protection that is designed for use on the line end of a flexible cord. (Pass & Seymour/Legrand)*

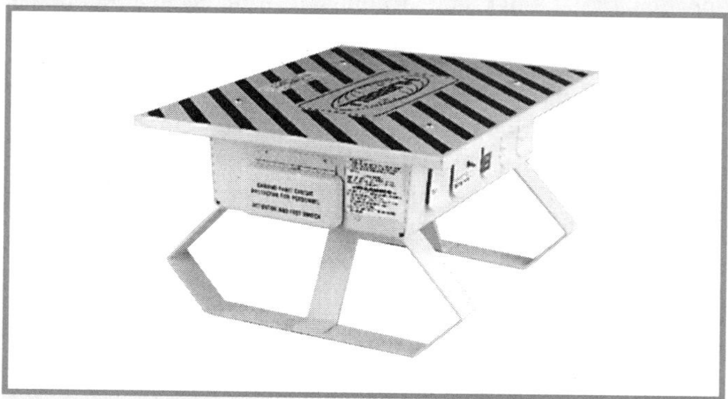

Figure 305.2 *A temporary power outlet unit commonly used on construction sites with a variety of configurations, including GFCI protection. (Hubbell)*

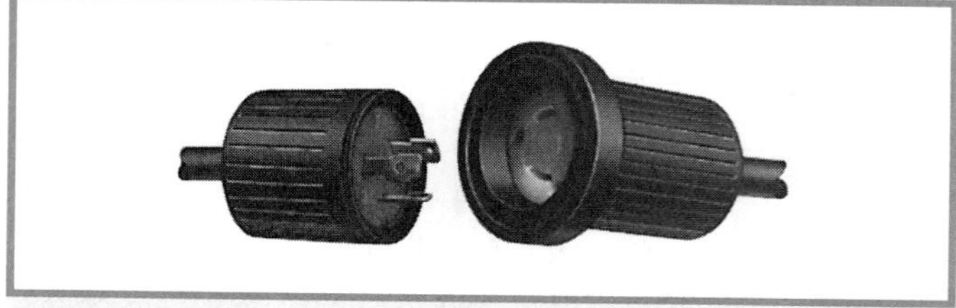

Figure 305.3 *A watertight plug and connector used to prevent tripping of GFCI protective devices in wet or damp weather. (Hubbell)*

(a) The following tests shall be performed on all cord sets, receptacles that are not part of the permanent wiring of the building or structure, and cord- and plug-connected equipment required to be grounded.

(1) All equipment grounding conductors shall be tested for continuity and shall be electrically continuous.

Figure 305.4 *A 15-ampere duplex receptacle with integral GFCI that also protects downstream loads. (Pass & Seymour/Legrand)*

(2) Each receptacle and attachment plug shall be tested for correct attachment of the equipment grounding conductor. The equipment grounding conductor shall be connected to its proper terminal.

(3) All required tests shall be performed

(a) Before first use on site,
(b) When there is evidence of damage,
(c) Before equipment is returned to service following any repairs,
(d) At intervals not exceeding 3 months.

(b) The tests required in (2)(a) shall be recorded and made available to the authority having jurisdiction.

305-7. Guarding. For wiring over 600 volts, nominal, suitable fencing, barriers, or other effective means shall be provided to limit access only to authorized and qualified personnel.

Article 310 — Conductors for General Wiring

Contents

310-1. Scope. This article covers general requirements for conductors and their type designations, insulations, markings, mechanical strengths, ampacity ratings, and uses. These requirements do not apply to conductors that form an integral part of equipment, such as motors, motor controllers, and similar equipment, or to conductors specifically provided for elsewhere in this *Code*.

FPN: For flexible cords and cables, see Article 400. For fixture wires, see Article 402.

310-2. Conductors.

As the electrical industry moves more toward global suppliers of insulated conductors, it is important for the user to understand that the all "insulated" conductors are not equal. Only wires and cables that meet the minimum fire, electrical, and physical properties required by the applicable standards are permitted to be marked with the letter designations found in Tables 310-13 and 310-62. See Section 310-13 for the exact requirements of insulated conductor construction and applications.

(a) Insulated. Conductors shall be insulated.

Exception: Where covered or bare conductors are specifically permitted elsewhere in this Code.

FPN: See Section 250-184 for insulation of neutral conductors of a solidly grounded high-voltage system.

(b) Conductor Material. Conductors in this article shall be of aluminum, copper-clad aluminum, or copper unless otherwise specified.

310-3. Stranded Conductors. Where installed in raceways, conductors of size No. 8 and larger shall be stranded.

Larger-size conductors are required to be stranded to provide greater flexibility. This requirement does not apply to busbars and the conductors of Type MI mineral-insulated metal-sheathed cable. In addition, the bonding conductors of a common bonding grid of a permanently installed swimming pool are required to be solid, nonstranded conductors of size No. 8 or larger.

Exception: As permitted or required elsewhere in this Code.

310-4. Conductors in Parallel. Aluminum, copper-clad aluminum, or copper conductors of size No. 1/0 and larger, comprising each phase, neutral, or grounded circuit conductor, shall be permitted to be connected in parallel (electrically joined at both ends to form a single conductor).

Conductors connected in parallel, in accordance with Section 310-4, are considered a single conductor with a total cross-sectional area of all conductors in parallel. Therefore, if individual conductors are tapped from conductors in parallel, the tap connection must include all the conductors in parallel for that particular phase.

Tapping into only one of the parallel conductors will result in an unbalance in the distribution of the tap load current between parallel conductors. This will result in one of the conductors carrying more than its share of the load, which can cause overheating and conductor insulation failure.

For example, if a 250-kcmil conductor is tapped from a set of two 500-kcmil conductors in parallel, the splicing device must include both 500-kcmil conductors and the single 250-kcmil tap conductor.

Exception No. 1: As permitted in Section 620-12(a)(1).

Exception No. 2: Conductors in sizes smaller than No. 1/0 shall be permitted to be run in parallel to supply control power to indicating instruments, contactors, relays, solenoids, and similar control devices provided

(a) They are contained within the same raceway or cable,
(b) The ampacity of each individual conductor is sufficient to carry the entire load current shared by the parallel conductors, and
(c) The overcurrent protection is such that the ampacity of each individual conductor will not be exceeded if one or more of the parallel conductors become inadvertently disconnected.

Exception No. 3: Conductors in sizes smaller than No. 1/0 shall be permitted to be run in parallel for frequencies of 360 Hz and higher where conditions (a), (b), and (c) of Exception No. 2 are met.

For example, in control wiring and circuits that operate at frequencies greater than 360 Hz, it may be necessary to reduce cable capacitance or voltage drop in long lengths of wire. A No. 14 conductor may have more than sufficient capacity to carry the load, but by installing two conductors in parallel, the voltage drop can be reduced to acceptable limits. This method is permissible, provided the safeguards listed in Exception No. 2 are followed.

Exception No. 4: Under engineering supervision, grounded neutral conductors in sizes No. 2 and larger shall be permitted to be run in parallel for existing installations.

FPN: Exception No. 4 can be utilized to alleviate overheating of neutral conductors in existing installations due to high content of triplen harmonic currents.

The exception and fine print note were added to Section 310-4 in the 1996 *Code*. The word *triplen* refers to a third-order harmonic current, such as the third, sixth, ninth, and so on.

The paralleled conductors in each phase, neutral, or grounded circuit conductor shall

(1) Be the same length,
(2) Have the same conductor material,
(3) Be the same size in circular mil area,
(4) Have the same insulation type,
(5) Be terminated in the same manner.

Where run in separate raceways or cables, the raceways or cables shall have the same physical characteristics. Conductors of one phase, neutral, or grounded circuit conductor shall not be required to have the same physical characteristics as those of another phase, neutral, or grounded circuit conductor to achieve balance.

For example, the conductors in phases A and B may be copper, and those in phase C may be aluminum. However, all raceways or cables must be of the same size, material, and length. Cables, in this case, means cables such as Type MC.

FPN: Differences in inductive reactance and unequal division of current can be minimized by choice of materials, methods of construction, and orientation of conductors.

Where equipment grounding conductors are used with conductors in parallel, they shall comply with the requirements of this section except that they shall be sized in accordance with Section 250-122.

Where conductors are used in parallel, space in enclosures shall be given consideration (see Articles 370 and 373).

Conductors installed in parallel shall comply with the provisions of Section 310-15(b)(2)(a).

Section 310-4 permits a practical means for installing large-capacity conductors for feeders or services. The paralleling of two or more conductors in place of one large conductor to ensure equal division of current depends on a number of factors. Therefore, several conditions must be satisfied so as not to overload any of the individual paralleled conductors. Other than as permitted in Section 250-122 and the exceptions to Section 310-4, there does not appear to be any practical need to parallel conductors smaller than No. 1/0.

To avoid excessive voltage drop and also to ensure equal division of current, it is essential that different phase conductors be located close together and that each phase conductor, grounded conductor, and the grounding conductor (if used) be grouped together in each raceway or cable. However, isolated phase installations are permitted underground where the phase conductors are run in nonmetallic raceways that are in close proximity.

The impedance of a circuit in an aluminum raceway or aluminum-sheathed cable will be different from the impedance of the same circuit in a steel raceway or steel-sheathed cable; therefore, it is required that separate raceways and cables have the same physical characteristics. See Section 300-20 regarding induced currents in metal enclosures or raceways.

Note that all conductors of the same phase or neutral are required to be of the same conductor material. For example, if 12 conductors are paralleled for a 3-phase, 4-wire, 480Y/277-volt ac circuit, four conductors could be installed in each of three raceways. The *Code* does not intend that all 12 conductors be copper or aluminum but does intend that the individual conductors in parallel for each phase, grounded conductor, and neutral be the same material, insulation type, length, and so on. Also, it is intended that the three raceways have the same physical characteristics (e.g., three rigid aluminum conduits, three steel IMC conduits, three EMTs, or three nonmetallic conduits), not a mixture (e.g., two rigid aluminum conduits and one rigid steel conduit).

It is neither economical nor practical to use conductors larger than 1000 kcmil in raceways unless the conductor size is governed by voltage drop. The ampacity of larger sizes would increase very little in proportion to the increase in the size of the conductor. Where the cross-sectional area of a conductor increases 50 percent (e.g., from 1000 to 1500 kcmil), a Type THW conductor ampacity increases only 80 amperes (less than 15 percent). A 100 percent increase (from 1000 to 2000 kcmil) causes an increase of only 120 amperes (approximately 2 percent). Generally, in situations where cost is a factor, installation of two (or more) paralleled conductors per phase may be beneficial.

310-5. Minimum Size of Conductors. The minimum size of conductors shall be as shown in Table 310-5.

Table 310-5. Minimum Size of Conductors

Conductor Voltage Rating (Volts)	Minimum Conductor Size (AWG) Copper	Aluminum or Copper-Clad Aluminum
0–2000	14	12
2001–8000	8	8
8001–15,000	2	2
15,001–28,000	1	1
28,001–35,000	1/0	1/0

Exception No. 1: For flexible cords as permitted by Section 400-12.

Exception No. 2: For fixture wire as permitted by Section 410-24, FPN.

Exception No. 3: For motors rated 1 hp or less as permitted by Section 430-22(c).

Exception No. 4: For cranes and hoists as permitted by Section 610-14.

Exception No. 5: For elevator control and signaling circuits as permitted by Section 620-12.

Exception No. 6: For Class 1, Class 2, and Class 3 circuits as permitted by Sections 725-27(a) and 725-51, Exception.

Exception No. 7: For fire alarm circuits as permitted by Sections 760-27(a), 760-51, Exception, and 760-71(b).

Exception No. 8: For motor-control circuits as permitted by Section 430-72.

Exception No. 9: For control and instrumentation circuits as permitted by Section 727-6.

310-6. Shielding. Solid dielectric insulated conductors operated above 2000 volts in permanent installations shall have ozone-resistant insulation and shall be shielded. All metallic insulation shields shall be grounded through an effective grounding path meeting the requirements of Section 250-2(d). Shielding shall be for the purpose of confining the voltage stresses to the insulation.

Exception: Nonshielded insulated conductors listed by a qualified testing laboratory shall be permitted for use up to 8000 volts under the following conditions:

(a) Conductors shall have insulation resistant to electric discharge and surface tracking, or the insulated conductor(s) shall be covered with a material resistant to ozone, electric discharge, and surface tracking.

(b) Where used in wet locations, the insulated conductor(s) shall have an overall nonmetallic jacket or a continuous metallic sheath.

(c) Where operated at 5001 to 8000 volts, the insulated conductor(s) shall have a nonmetallic jacket over the insulation. The insulation shall have a specific inductive capacity not greater than 3.6, and the jacket shall have a specific inductive capacity not greater than 10 and not less than 6.

(d) Insulation and jacket thicknesses shall be in accordance with Table 310-63.

Solid dielectric insulated conductors that are permanently installed and that operate at greater than 2000 volts are required to have ozone-resistant insulation and must be shielded with a grounded metallic shield (note exception).

Shielding is accomplished by applying a metal tape or nonmetallic semiconducting tape around the conductor surface, which prevents corona from forming and reduces high-voltage stresses. Corona is a faint glow adjacent to the surface of the electrical conductor at high voltage. If high-voltage stresses and a charging current are flowing between the conductor and ground (usually due to moisture), the surrounding atmosphere is ionized, and ozone — generated by an electric discharge in ordinary oxygen or air — is formed, which will attack the conductor jacket and insulation and may eventually break them down. The shield is at ground potential; therefore, no voltage above ground is present on the jacket outside the shield, thus preventing a discharge from the jacket and the subsequent formation of ozone.

Specialized training and close adherence to manufacturers' instructions are absolutely essential for high-voltage cable installations.

Example Figures

Figures 310.1, 310.2, and 310.3 show some examples of shielded cable installations: a three-conductor cable of the shielded type, a stress-relief cone for an indoor cable terminator, and a stress cone on a single-conductor shielded cable terminating inside a pothead. Notice, in Figure 310.3, that a clamping ring provides a grounding connection between the copper shielding tape and the shield to the metallic base of the pothead.

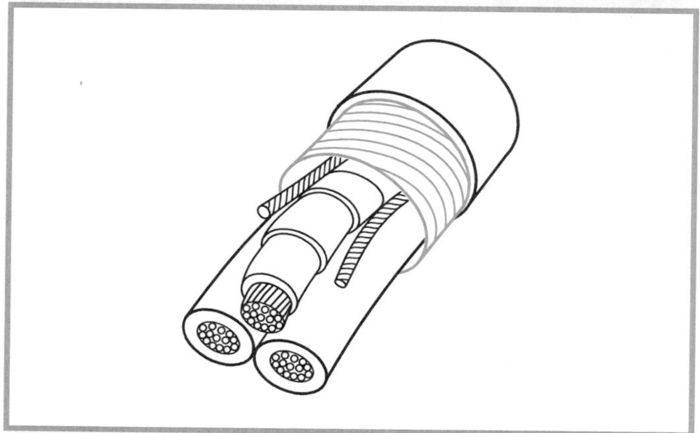

Figure 310.1 *A three-conductor cable of the shielded type.*

Figure 310.2 *A one-piece, premolded stress-relief cone for indoor cable terminations of up to 35 kV phase-to-phase. (ITT Blackburn Co.)*

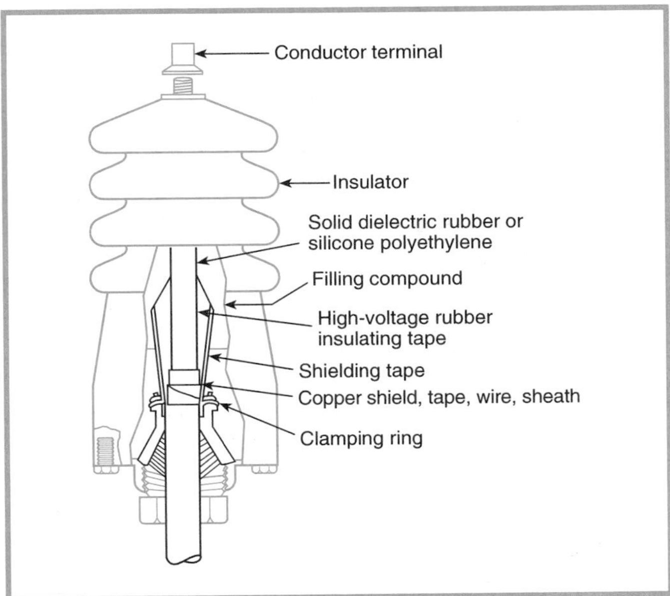

Figure 310.3 *A stress cone on a single-conductor shielded cable terminating inside a pothead.*

310-7. Direct Burial Conductors. Conductors used for direct burial applications shall be of a type identified for such use.

Cables rated above 2000 volts shall be shielded.

Exception: Nonshielded multiconductor cables rated 2001–5000 volts shall be permitted if the cable has an overall metallic sheath or armor.

The metallic shield, sheath, or armor shall be grounded through an effective grounding path meeting the requirements of Section 250-2(d).

FPN No. 1: See Section 300-5 for installation requirements for conductors rated 600 volts or less.

FPN No. 2: See Section 300-50 for installation requirements for conductors rated over 600 volts.

310-8. Locations.

(a) Dry Locations. Insulated conductors and cables used in dry locations shall be any of the types identified in this *Code*.

(b) Dry and Damp Locations. Insulated conductors and cables used in dry and damp locations shall be Types FEP, FEPB, MTW, PFA, RH, RHH, RHW, RHW-2, SA, THHN, THW, THW-2, THHW, THHW-2, THWN, THWN-2, TW, XHH, XHHW, XHHW-2, Z, or ZW.

(c) Wet Locations. Insulated conductors and cables used in wet locations shall be

(1) Moisture-impervious metal-sheathed;
(2) Types MTW, RHW, RHW-2, TW, THW, THW-2, THHW, THHW-2, THWN, THWN-2, XHHW, XHHW-2, ZW; or
(3) Of a type listed for use in wet locations.

(d) Locations Exposed to Direct Sunlight. Insulated conductors and cables used where exposed to direct rays of the sun shall be of a type listed or marked "sunlight resistant."

Section 310-8 was revised for the 1999 *Code* by organizing each location type and stipulating a list of identified insulation types for each location. Additionally, a new paragraph (d), Locations Exposed to Direct Sunlight, was added.

310-9. Corrosive Conditions. Conductors exposed to oils, greases, vapors, gases, fumes, liquids, or other substances having a deleterious effect on the conductor or insulation shall be of a type suitable for the application.

See the commentary following Section 501-13 regarding gasoline-resistant conductors.

310-10. Temperature Limitation of Conductors. No conductor shall be used in such a manner that its operating temperature will exceed that designated for the type of insulated conductor involved. In no case shall conductors be associated together in such a way with respect to type of circuit, the wiring method employed, or the number of conductors that the limiting temperature of any conductor is exceeded.

Most terminations are normally designed for 60°C (140°F) or 75°C (167°F) maximum temperatures, although some are now being designed for 90°C (194°F). Therefore, the higher-rated ampacities for conductors of 90°C (194°F), 110°C (230°F), and so on, cannot be used unless the terminals at which the conductors terminate have comparable ratings.

Tables 310-16 through 310-20 have ampacity correction factors for ambient temperatures greater or less than the ambient temperature identified in the table heading. To assign the proper ampacity to a conductor in an ambient above 30°C, the appropriate correction factor must be used. This correction factor is applied in addition to any adjustment factor, such as in Section 310-15(b)(2)(a).

Example

No. 2 THHN copper conductors are to be installed in a raceway in an ambient temperature of 50°C (122°F). According to Table 310-16, the allowable ampacity of the conductor at 30°C is 130 amperes, which is multiplied by 0.82 (taken from the correction factors at the bottom of the table). Thus, the allowable ampacity of the No. 2 conductor at 50°C is reduced to 106.6 amperes (130 amperes × 0.82 = 106.6 amperes).

If six of these conductors were run in a raceway, Section 310-15(b)(2)(a) would require the allowable ampacity to be further reduced to 80 percent, which, in this case, would be 106.6 amperes × 0.8 = 85.28 amperes. Under these conditions, the No. 2 conductors would be suitable for an 80-ampere circuit.

The basis for determining the ampacities of conductors for Tables 310-16 and 310-17 was the NEMA "Report of Determination of Maximum Permissible Current-Carrying Capacity of Code Insulated Wires and Cables for Building Purposes," dated June 27, 1938. The basis for determining the ampacities of conductors for Tables 310-18 and 310-19 and the ampacity tables in Appendix B was the Neher-McGrath method. See the commentary following Section 310-15(c) for further explanation.

Conductors should be chosen that have a rating above the anticipated maximum ambient temperature. The operating temperature of conductors should be controlled at or below the conductor rating by coordinating conductor size, number of associated conductors, and ampacity for the particular conductor rating and ambient temperature. All tabulations should be corrected for the anticipated ambient temperature, using the correction factors at the bottom of the ampacity tables. If more than three conductors are associated together, the additional adjustment shown in Section 310-15(b)(2)(a) must be applied.

FPN: The temperature rating of a conductor (see Tables 310-13 and 310-61) is the maximum temperature, at any location along its length, that the conductor can withstand over a prolonged time period without serious degradation. The allowable ampacity tables, the ampacity tables of Article 310 and the ampacity tables of Appendix B, the correction factors at the bottom of these tables, and the notes to the tables provide guidance for coordinating conductor sizes, types, allowable ampacities, ampacities, ambient temperatures, and number of associated conductors.

The principal determinants of operating temperature are as follows:

(1) Ambient temperature — ambient temperature may vary along the conductor length as well as from time to time.
(2) Heat generated internally in the conductor as the result of load current flow, including fundamental and harmonic currents.
(3) The rate at which generated heat dissipates into the ambient medium. Thermal insulation that covers or surrounds conductors will affect the rate of heat dissipation.
(4) Adjacent load-carrying conductors — adjacent conductors have the dual effect of raising the ambient temperature and impeding heat dissipation.

The fine print note focuses attention on the necessity for derating conductors where high ambient temperatures are encountered and provides users with helpful information in coordinating ampacities, ambient temperatures, conductor size and number, and so on, to ensure operation at or below rating.

The second principal determinant explains that heating due to harmonic current should also be taken into consideration. In certain cases, this may require conductors to be adjusted to a larger size. For existing installations, see Section 310-4, Exception No. 4.

310-11. Marking.

(a) Required Information. All conductors and cables shall be marked to indicate the following information, using the applicable method described in (b):

(1) The maximum rated voltage for which the conductor was listed
(2) The proper type letter or letters for the type of wire or cable as specified elsewhere in this *Code*
(3) The manufacturer's name, trademark, or other distinctive marking by which the organization responsible for the product can be readily identified
(4) The AWG size or circular mil area
(5) Cable assemblies where the neutral conductor is smaller than the ungrounded conductors shall be so marked

(b) Method of Marking.

(1) Surface Marking. The following conductors and cables shall be durably marked on the surface. The AWG size or circular mil area shall be repeated at intervals not exceeding 24 in. (610 mm). All other markings shall be repeated at intervals not exceeding 40 in. (1.02 m).

(a) Single- and multiconductor rubber- and thermoplastic-insulated wire and cable
(b) Nonmetallic-sheathed cable
(c) Service-entrance cable
(d) Underground feeder and branch-circuit cable
(e) Tray cable
(f) Irrigation cable
(g) Power-limited tray cable
(h) Instrumentation tray cable

(2) Marker Tape. Metal-covered multiconductor cables shall employ a marker tape located within the cable and running for its complete length.

Exception No. 1: Mineral-insulated, metal-sheathed cable.

Exception No. 2: Type AC cable.

Exception No. 3: The information required in Section 310-11(a) shall be permitted to be durably marked on the outer nonmetallic covering of Type MC, Type ITC, or Type PLTC cables at intervals not exceeding 40 in. (1.02 m).

Exception No. 4: The information required in Section 310-11(a) shall be permitted to be durably marked on a nonmetallic covering under the metallic sheath of Type ITC or Type PLTC cable at intervals not exceeding 40 in. (1.02 m).

Type PLTC cable is permitted to have a metallic sheath or armor over a nonmetallic jacketed cable. A second nonmetallic jacket covering the metallic sheath is optional. Exception Nos. 3 and 4 define the marking requirements for either case.

FPN: Included in the group of metal-covered cables are Type AC cable (Article 333), Type MC cable (Article 334), and lead-sheathed cable.

(3) Tag Marking. The following conductors and cables shall be marked by means of a printed tag attached to the coil, reel, or carton:

(a) Mineral-insulated, metal-sheathed cable
(b) Switchboard wires
(c) Metal-covered, single-conductor cables
(d) Conductors that have an outer surface of asbestos
(e) Type AC cable

(4) Optional Marking of Wire Size. The information required in (a)(4) shall be permitted to be marked on the surface of the individual insulated conductors for the following multiconductor cables:

(a) Type MC cable
(b) Tray cable
(c) Irrigation cable
(d) Power-limited tray cable
(e) Power-limited fire alarm cable
(f) Instrumentation tray cable

(c) Suffixes to Designate Number of Conductors. A type letter or letters used alone shall indicate a single insulated conductor. The letter suffixes shall be indicated as follows:

D — For two insulated conductors laid parallel within an outer nonmetallic covering

M — For an assembly of two or more insulated conductors twisted spirally within an outer nonmetallic covering

(d) Optional Markings. All conductors and cables contained in Chapter 3 shall be permitted to be surface marked to indicate special characteristics of the cable materials.

FPN: Examples of these markings include but are not limited to "LS" for limited smoke and markings such as "sunlight resistant."

New cable insulations that have special characteristics are frequently developed. An example is the family of limited-smoke cables, which are permitted to be marked "LS." Other materials have been developed or are in development that have other characteristics such as sunlight resistance and low corrosivity. Section 310-11(d) allows these developments without identifying each characteristic in the *Code*.

310-12. Conductor Identification.

(a) Grounded Conductors. Insulated or covered grounded conductors shall be identified in accordance with Section 200-6.

Sections 200-6(a) and (b) were revised for the 1999 *Code* to permit an alternate method of identifying a grounded conductor. This new method of identification is described as "three continuous white stripes on other than green insulation along the conductor's entire length."

(b) Equipment Grounding Conductors. Equipment grounding conductors shall be in accordance with Section 250-119.

(c) Ungrounded Conductors. Conductors that are intended for use as ungrounded conductors, whether used as single conductors or in multiconductor cables, shall be finished to be clearly distinguishable from grounded and grounding conductors. Ungrounded conductors shall be distinguished by colors other than white, natural gray, or green; or by a combination of color plus distinguishing marking. Distinguishing markings shall also be in a color other than white, natural gray, or green, and shall consist of a stripe or stripes or a regularly spaced series of identical marks. Distinguishing markings shall not conflict in any manner with the surface markings required by Section 310-11(b)(1).

Exception: As permitted by Section 200-7.

Ungrounded conductors with white or natural gray insulation are permitted if the conductors are permanently re-identified at termination points and if the conductor is visible and accessible. The normal methods of re-identification include colored tape, tagging, or paint. Other applications where white conductors are permitted include flexible cords and circuits less than 50 volts. Revised in the 1999 *Code*, a white conductor used in single-pole, 3-way and 4-way switch loops now also requires re-identification (a color other than white, natural gray, or green) if it uses an ungrounded conductor. See Section 200-7(c) for exact details.

310-13. Conductor Constructions and Applications. Insulated conductors shall comply with the applicable provisions of one or more of the following: Tables 310-13, 310-61, 310-62, 310-63, and 310-64.

These conductors shall be permitted for use in any of the

wiring methods recognized in Chapter 3 and as specified in their respective tables.

FPN: Thermoplastic insulation may stiffen at temperatures colder than minus 10°C (plus 14°F). Thermoplastic insulation may also be deformed at normal temperatures where subjected to pressure, such as at points of support. Thermoplastic insulation, where used on dc circuits in wet locations, may result in electroendosmosis between conductor and insulation.

Table 310-13 lists the various types of insulated conductors covered by the requirements of this *Code*. More detailed wire classification information from sizes No. 14 through 2000 kcmil may be obtained from standards or directories, such as those published by Underwriters Laboratories Inc.

Table 310-13 also includes conductor applications and maximum operating temperatures. Some conductors have dual ratings. For example, Type XHHW is rated 90°C (194°F) for dry and damp locations and 75°C (167°F) for wet locations; Type THW is rated 75°C (167°F) for dry and wet locations and 90°C (194°F) for special applications within electric-discharge lighting equipment.

Types RHW-2, XHHW-2, and other types identified by the suffix "-2" are rated 90°C (194°F) for wet locations as well as dry and damp locations. Conductors permitted to be identified by the suffix "-2," other than Types RHW-2 and XHHW-2, are identified in the table by a footnote 5.

The maximum continuous ampacities for copper, copper-clad aluminum, and aluminum conductors, rated 0 through 2000 volts, are listed in Tables 310-16 through 310-20 and accompanying adjustment factors of Sections 310-15(b)(1) through (7), or they can be calculated in accordance with Section 310-15(c).

Copper-clad aluminum conductors are drawn from a copper-clad aluminum rod with the copper metallurgically bonded to an aluminum core. The copper forms a minimum of 10 percent of the cross-sectional area of a solid conductor or of each strand of a stranded conductor. See the additional commentary

Table 310-13. Conductor Application and Insulations

Trade Name	Type Letter	Maximum Operating Temperature	Application Provisions	Insulation	Thickness of Insulation: AWG or kcmil	Thickness of Insulation: Mils	Outer Covering[1]
Fluorinated ethylene propylene	FEP or FEPB	90°C 194°F	Dry and damp locations	Fluorinated ethylene propylene	14–10 8–2	20 30	None
		200°C 392°F	Dry locations — special applications[2]	Fluorinated ethylene propylene	14–8	14	Glass braid
					6–2	14	Asbestos or other suitable braid material
Mineral insulation (metal sheathed)	MI	90°C 194°F 250°C 482°F	Dry and wet locations For special applications[2]	Magnesium oxide	18–16[3] 16–10 9–4 3–500	23 36 50 55	Copper or alloy steel
Moisture-, heat-, and oil-resistant thermoplastic	MTW	60°C 140°F 90°C 194°F	Machine tool wiring in wet locations as permitted in NFPA 79 (see Article 670) Machine tool wiring in dry locations as permitted in NFPA 79 (see Article 670)	Flame-retardant moisture-, heat-, and oil-resistant thermoplastic	 22–12 10 8 6 4–2 1–4/0 213–500 591–1000	(A) (B) 30 15 30 20 45 30 60 30 60 40 80 50 95 60 110 70	(A) None (B) Nylon jacket or equivalent
Paper		85°C 185°F	For underground service conductors, or by special permission	Paper			Lead sheath

[1]Some insulations do not require an outer covering.
[2]Where design conditions require maximum conductor operating temperatures above 90°C (194°F).
[3]For signaling circuits permitting 300-volt insulation.

Table 310-13. (Continued)

Trade Name	Type Letter	Maximum Operating Temperature	Application Provisions	Insulation	Thickness of Insulation: AWG or kcmil	Thickness of Insulation: Mils	Outer Covering[1]
Perfluoroalkoxy	PFA	90°C 194°F 200°C 392°F	Dry and damp locations Dry locations — special applications[2]	Perfluoroalkoxy	14–10 8–2 1–4/0	20 30 45	None
Perfluoroalkoxy	PFAH	250°C 482°F	Dry locations only. Only for leads within apparatus or within raceways connected to apparatus (Nickel or nickel-coated copper only)	Perfluoroalkoxy	14–10 8–2 1–4/0	20 30 45	None
Thermoset Thermoset	RH RHH	75°C 167°F 90°C 194°F	Dry and damp locations Dry and damp locations	Flame-retardant thermoset	14–12[4] 10 8–2 1–4/0 213–500 501–1000 1001–2000 For 601–2000, see Table 310-62	30 45 60 80 95 110 125	Moisture-resistant, flame-retardant, non-metallic covering[1]
Moisture-resistant thermoset	RHW[5]	75°C 167°F	Dry and wet locations Where over 2000 volts insulation, shall be ozone-resistant	Flame-retardant, moisture-resistant thermoset	14–10 8–2 1–4/0 213–500 501–1000 1001–2000 For 601–2000 volts, see Table 310-62	45 60 80 95 110 125	Moisture-resistant, flame-retardant, non-metallic covering[6]
Moisture-resistant thermoset	RHW-2	90°C 194°F	Dry and wet locations	Flame-retardant moisture-resistant thermoset	14–10 8–2 1–4/0 213–500 501–1000 1001–2000 For 601–2000 volts, see Table 310-62	45 60 80 95 110 125	Moisture-resistant, flame-retardant, non-metallic covering[6]
Silicone	SA	90°C 194°F 200°C 392°F	Dry and damp locations For special application[2]	Silicone rubber	14–10 8–2 1–4/0 213–500 501–1000 1001–2000	45 60 80 95 110 125	Glass or other suitable braid material
Thermoset	SIS	90°C 194°F	Switchboard wiring only	Flame-retardant thermoset	14–10 8–2 1–4/0	30 45 95	None
Thermoplastic and fibrous outer braid	TBS	90°C 194°F	Switchboard wiring only	Thermoplastic	14–10 8 6–2 1–4/0	30 45 60 80	Flame-retardant, nonmetallic covering

[1]Some insulations do not require an outer covering.

[2]Where design conditions require maximum conductor operating temperatures above 90°C (194°F).

[4]For size Nos. 14-12, RHH insulation shall be 45 mils thickness.

[5]Listed wire types designated with the suffix "-2," such as RHW-2, shall be permitted to be used at a continuous 90°C (194°F) operating temperature, wet or dry.

[6]Some rubber insulations do not require an outer covering.

(continues)

Table 310-13. (Continued)

Trade Name	Type Letter	Maximum Operating Temperature	Application Provisions	Insulation	Thickness of Insulation: AWG or kcmil	Thickness of Insulation: Mils	Outer Covering[1]
Extended polytetra-fluoroethylene	TFE	250°C 482°F	Dry locations only. Only for leads within apparatus or within raceways connected to apparatus, or as open wiring (Nickel or nickel-coated copper only)	Extruded poly-tetrafluoro-ethylene	14–10 8–2 1–4/0	20 30 45	None
Heat-resistant thermoplastic	THHN	90°C 194°F	Dry and damp locations	Flame-retardant, heat-resistant thermoplastic	14–12 10 8–6 4–2 1–4/0 250–500 501–1000	15 20 30 40 50 60 70	Nylon jacket or equivalent
Moisture- and heat-resistant thermoplastic	THHW	75°C 167°F 90°C 194°F	Wet location Dry location	Flame-retardant, moisture- and heat-resistant thermoplastic	14–10 8 6–2 1–4/0 213–500 501–1000	30 45 60 80 95 110	None
Moisture- and heat-resistant thermoplastic	THW[5]	75°C 167°F 90°C 194°F	Dry and wet locations Special applications within electric discharge lighting equipment. Limited to 1000 open-circuit volts or less (Size 14-8 only as permitted in Section 410-31)	Flame-retardant, moisture- and heat-resistant thermoplastic	14–10 8 6–2 1–4/0 213–500 501–1000 1001–2000	30 45 60 80 95 110 125	None
Moisture- and heat-resistant thermoplastic	THWN[5]	75°C 167°F	Dry and wet locations	Flame-retardant, moisture- and heat-resistant thermoplastic	14–12 10 8–6 4–2 1–4/0 250–500 501–1000	15 20 30 40 50 60 70	Nylon jacket or equivalent
Moisture-resistant thermoplastic	TW	60°C 140°F	Dry and wet locations	Flame-retardant, moisture-resistant thermoplastic	14–10 8 6–2 1–4/0 213–500 501–1000 1001–2000	30 45 60 80 95 110 125	None
Underground feeder and branch-circuit cable — single conductor (For Type UF cable employing more than one conductor, see Article 339)	UF	60°C 140°F	See Article 339	Moisture-resistant	14–10 8–2 1–4/0	60[7] 80[7] 95[7]	Integral with insulation
		75°C 167°F[8]		Moisture- and heat-resistant			

[1]Some insulations do not require an outer covering.
[5]Listed wire types designated with the suffix "-2," such as RHW-2, shall be permitted to be used at a continuous 90°C (194°F) operating temperature, wet or dry.
[7]Includes integral jacket.
[8]For ampacity limitation, see Section 339-5.

Table 310-13. (Continued)

Trade Name	Type Letter	Maximum Operating Temperature	Application Provisions	Insulation	Thickness of Insulation AWG or kcmil	Mils	Outer Covering[1]
Underground service-entrance cable — single conductor (For Type USE cable employing more than one conductor, see Article 338)	USE[5]	75°C 167°F	See Article 338	Heat- and moisture-resistant	14–10 8–2 1–4/0 213–500 501–1000 1001–2000	45 60 80 95[9] 110 125	Moisture-resistant nonmetallic covering [see Section 338-1(b)]
Thermoset	XHH	90°C 194°F	Dry and damp locations	Flame-retardant thermoset	14–10 8–2 1–4/0 213–500 501–1000 1001–2000	30 45 55 65 80 95	None
Moisture-resistant thermoset	XHHW[5]	90°C 194°F 75°C 167°F	Dry and damp locations Wet locations	Flame-retardant, moisture-resistant thermoset	14–10 8–2 1–4/0 213–500 501–1000 1001–2000	30 45 55 65 80 95	None
Moisture-resistant thermoset	XHHW-2	90°C 194°F	Dry and wet locations	Flame-retardant, moisture-resistant thermoset	14–10 8–2 1–4/0 213–500 501–1000 1001–2000	30 45 55 65 80 95	None
Modified ethylene tetrafluoroethylene	Z	90°C 194°F 150°C 302°F	Dry and damp locations Dry locations — special applications[2]	Modified ethylene-tetrafluoroethylene	14–12 10 8–4 3–1 1/0–4/0	15 20 25 35 45	None
Modified ethylene tetrafluoroethylene	ZW[5]	75°C 167°F 90°C 194°F 150°C 302°F	Wet locations Dry and damp locations Dry locations — special applications[2]	Modified ethylene tetrafluoroethylene	14–10 8–2	30 45	None

[1]Some insulations do not require an outer covering.

[2]Where design conditions require maximum conductor operating temperatures above 90°C (194°F).

[5]Listed wire types designated with the suffix "-2," such as RHW-2, shall be permitted to be used at a continuous 90°C (194°F) operating temperature, wet or dry.

[9]Insulation thickness shall be permitted to be 80 mils for listed Type USE conductors that have been subjected to special investigations. The nonmetallic covering over individual rubber-covered conductors of aluminum-sheathed cable and of lead-sheathed or multiconductor cable shall not be required to be flame retardant. For Type MC cable, see Section 334-20. For nonmetallic-sheathed cable, see Section 336-30. For Type UF cable, see Section 339-1.

following Section 110-14 for making electrical connections with different types of conductor material.

The characteristics of copper, copper-clad aluminum, and aluminum conductors may be compared as follows in Table 3.1.

310-14. Aluminum Conductor Material. Solid aluminum conductors No. 8, 10, and 12 shall be made of an AA-8000 series electrical grade aluminum alloy conductor material. Stranded aluminum conductors No. 8 through 1000 kcmil marked as Type XHHW, THW, THHW, THWN, THHN, service-entrance Type SE Style U and SE Style R shall be

Table 3.1 Conductor Characteristics

Characteristic	Copper	Copper-Clad Aluminum	Aluminum
Density (lb/in.³)	0.323	0.121	0.098
Density (g/cm³)	8.91	3.34	2.71
Resistivity ohms/CMF	10.37	16.08	16.78
Resistivity Microhm — CM	1.724	2.673	2.790
Conductivity (IACS %)	100	61–63	61.0
Weight % Copper	100	26.8	—
Tensile K psi — Hard	65.0	30.0	27.0
Tensile kg/mm² — Hard	45.7	21.1	19.0
Tensile K psi — Annealed	35.0	17.0	17.0*
Tensile kg/mm² — Annealed	24.6	12.0	12.0
Specific Gravity	8.91	3.34	2.71

*Semi-annealed

made of an AA-8000 series electrical grade aluminum alloy conductor material.

Section 310-14 provides proper recognition of approved AA-8000 series electrical-grade aluminum alloy conductor materials for wire and cable products. It also provides coordination with the UL listing requirements for testing terminations, such as CO/ALR devices and other connectors, suitable for use with aluminum conductors. AA-8000 series aluminum alloy materials and the connectors suitable for use with aluminum conductors have been developed by the electrical industry to provide for safe and stable connections. Connections suitable for use with aluminum conductors are generally listed as suitable for use with copper conductors also and are marked accordingly, such as AL7CU or AL9CU. Numbers 7 and 9 identify the temperature ratings of 75°C and 90°C, respectively, for these connectors.

310-15. Ampacities for Conductors Rated 0–2000 Volts.

(a) General.

Section 310-15 was reorganized for the 1999 *Code*. The notes affecting the ampacity of conductors, previously entitled Article 310, Notes to Ampacity Tables 0 to 2000 Volts, now appear as adjustment factors in Section 310-15(b)(1) through (7). Although this reorganization was a significant change, no substantive changes were made to the requirements.

(1) Tables or Engineering Supervision. Ampacities for conductors shall be permitted to be determined by tables or under engineering supervision, as provided in (b) and (c).

FPN No. 1: Ampacities provided by this section do not take voltage drop into consideration. See Section 210-19(a), FPN No. 4, for branch circuits and Section 215-2(d), FPN No. 2, for feeders.

FPN No. 2: For the allowable ampacities of Type MTW wire, see Table 11 in the *Electrical Standard for Industrial Machinery*, NFPA 79-1997.

Section 310-15(a)(1)permits either of two methods of determining conductor ampacity for conductors rated 0 through 2000 volts. The methods are (1) to select the ampacity from a table, using correction factors in the table or notes where necessary, or (2) to calculate the ampacity. The latter method can be complex and time consuming and requires engineering supervision. It may, however, result in lower installation costs in some cases, and if the calculation is done properly, it provides a mathematically exact ampacity. See the commentary following Section 310-15(c) and accompanying Appendix B for further explanation.

(2) Selection of Ampacity. Where more than one calculated or tabulated ampacity could apply for a given circuit length, the lowest value shall be used.

Exception: Where two different ampacities apply to adjacent portions of a circuit, the higher ampacity shall be permitted to be used beyond the point of transition, a distance equal to 10 ft (3.05 m) or 10 percent of the circuit length figured at the higher ampacity, whichever is less.

Example

Three 500-kcmil THW conductors in a rigid conduit are run from a motor control center for 12 ft past a heat-treating furnace to a pump motor located 150 ft from the motor control center. Where run in a 78°F to 86°F ambient, the conductors have an ampacity of 380 amperes, per Table 310-16.

The ambient temperature near the furnace, where the conduit is run, is found to be 113°F, and the length of this particular part of the run is greater than 10 ft and more than 10 percent of the total length of the run at the 78°F to 86°F ambient.

In accordance with the correction factors for temperature at the bottom of Table 310-16, the ampacity is 0.82 × 380 amperes, or 311.6 amperes. This, therefore, is the ampacity of the total run in accordance with Section 310-15(a)(2).

Had the run near the furnace at the 113°F ambient been 10 ft or less in length, the ampacity of the entire run would have been 380 amperes, in accordance with the exception to Section 310-15(a)(2). The heat-sinking effect of the run at the lower ambient temperature would have been sufficient to reduce the temperature of the conductor near the furnace.

FPN: See Section 110-14(c) for conductor temperature limitations due to termination provisions.

(b) Tables. Ampacities for conductors rated 0 to 2000 volts shall be as specified in the Allowable Ampacity Tables 310-16 through 310-19 and Ampacity Tables 310-20 and 310-21 as modified by (1) through (7).

FPN: Tables 310-16 through 310-19 are application tables for use in determining conductor sizes on loads calculated in accordance with Article 220. Allowable ampacities result from consideration of one or more of the following:

(1) Temperature compatibility with connected equipment, especially at the connection points.
(2) Coordination with circuit and system overcurrent protection.
(3) Compliance with the requirements of product listings or certifications. See Section 110-3(b).
(4) Preservation of the safety benefits of established industry practices and standardized procedures.

The ampacities in Table 310-16, in particular, do not take into account all of the many factors affecting ampacity. See the commentary following Section 310-15(c) and accompanying Appendix B for further explanation. However, experience over many years has proved the table values to be adequate where loads are calculated in accordance with Article 220. The reason is that not all of the diversity factors and load factors in most actual installations are specifically provided for in Article 220. If loads are not calculated in accordance with the requirements of Article 220, the table ampacities, even when corrected in accordance with ambient correction factors and the notes to the tables, may be too high. This result can be particularly true where many cables or raceways are routed close to each other underground.

(1) General. For explanation of type letters used in tables and for recognized sizes of conductors for the various conductor insulations, see Section 310-13. For installation requirements, see Sections 310-1 through 310-10 and the various articles of this *Code*. For flexible cords, see Tables 400-4, 400-5(A), and 400-5(B).

(2) Adjustment Factors.

(a) *More than Three Current-Carrying Conductors in a Raceway or Cable.* Where the number of current-carrying conductors in a raceway or cable exceeds three, or where single conductors or multiconductor cables are stacked or bundled longer than 24 in. (610 mm) without maintaining spacing and are not installed in raceways, the allowable ampacity of each conductor shall be reduced as shown in Table 310-15(b)(2)(a).

Table 310-15(b)(2)(a). Adjustment Factors for More than Three Current-Carrying Conductors in a Raceway or Cable

Number of Current-Carrying Conductors	Percent of Values in Tables 310-16 through 310-19 as Adjusted for Ambient Temperature if Necessary
4–6	80
7–9	70
10–20	50
21–30	45
31–40	40
41 and above	35

FPN: See Appendix B, Table B-310-11, for adjustment factors for more than three current-carrying conductors in a raceway or cable with load diversity.

The factors in the second column of Table 310-15(b)(2)(a) are based on no diversity. This means that all conductors in the raceway or cable are loaded to their maximum rated load. For load diversity, the user is directed to Appendix B.

Specific cross references for raceway fill and adjustment factors of Section 310-15(b)(2) may be found in the fine print note following Section 300-17. For Class 1 conductors, see Section 725-28(a). For fire alarm systems, see Sections 760-28 and 760-52. For optical fiber cables and raceways, see Section 770-52.

Exception No. 1: Where conductors of different systems, as provided in Section 300-3, are installed in a common race-

way or cable, the derating factors shown in Table 310-15(b)(2)(a) shall apply to the number of power and lighting conductors only (Articles 210, 215, 220, and 230).

Exception No. 1 assumes that the watt loss (heating) from any control and signal conductors in the same raceway or cable will not be enough to significantly increase the temperature of the power and lighting conductors. See Sections 725-26 and 725-54 for limitations on the installation of control and signal conductors in the same raceway or cable as power and lighting conductors.

Exception No. 2: For conductors installed in cable trays, the provisions of Section 318-11 shall apply.

Exception No. 3: Derating factors shall not apply to conductors in nipples having a length not exceeding 24 in. (610 mm).

Exception No. 4: Derating factors shall not apply to underground conductors entering or leaving an outdoor trench if those conductors have physical protection in the form of rigid metal conduit, intermediate metal conduit, or rigid nonmetallic conduit having a length not exceeding 10 ft (3.05 m) and the number of conductors does not exceed four.

Exception No. 4 was relocated here as part of the 1999 *Code* reorganization of Section 310-15. It previously appeared in the 1996 *Code* as Note 8 of Notes to Ampacity Tables of 0 to 2000 Volts after Table 310-19.

Figure 310.4 illustrates Section 310-15(b)(2)(a), Exception No. 4, in that derating factors do not apply to the conductors because they have physical protection (conduit) that does not exceed 10 ft in length and the number of conductors does not exceed four.

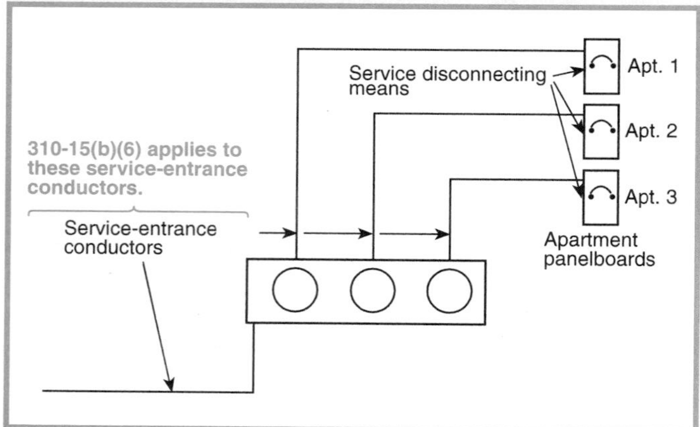

Figure 310.4 *An application of Section 310-15(b)(2)(a) Exception No. 4.*

(b) *More than One Conduit, Tube, or Raceway.* Spacing between conduits, tubing, or raceways shall be maintained.

Spacing is normally maintained between individual conduits in groups of conduit runs from junction box to junction box because of the need to separate the conduits where they enter the junction box, to allow room for locknuts and bushings. Field experience has indicated this degree of spacing between runs has not caused any problems.

(3) Bare or Covered Conductors. Where bare or covered conductors are used with insulated conductors, their allowable ampacities shall be limited to those permitted for the adjacent insulated conductors.

(4) Neutral Conductor.

(a) A neutral conductor that carries only the unbalanced current from other conductors of the same circuit shall not be required to be counted when applying the provisions of Section 310-15(b)(2)(a).

(b) In a 3-wire circuit consisting of two phase wires and the neutral of a 4-wire, 3-phase wye-connected system, a common conductor carries approximately the same current as the line-to-neutral load currents of the other conductors and shall be counted when applying the provisions of Section 310-15(b)(2)(a).

(c) On a 4-wire, 3-phase wye circuit where the major portion of the load consists of nonlinear loads, harmonic currents are present in the neutral conductor; the neutral shall therefore be considered a current-carrying conductor.

During the cycle for the 1996 *NEC*, a task group composed of interested parties was created to recommend to the *National Electrical Code* Committee what direction should be taken in its standards that would result in improving the safeguarding of persons and property from conditions that may be introduced by nonlinear loads. This group was designated the *NEC* Correlating Committee Ad Hoc Subcommittee on Nonlinear Loads. The scope of this subcommittee was as follows:

1. To study the effects of electrical loads producing substantial current distortion upon electrical system distribution components including, but not limited to
   a. Distribution transformers, current transformers, and others
   b. Switchboards and panelboards
   c. Phase and neutral feeder conductors
   d. Phase and neutral branch-circuit conductors
   e. Proximate data and communications conductors
2. To study harmful effects, if any, to the system components from overheating resulting from these load characteristics

3. To make recommendations for methods to minimize the harmful effects of nonlinear loads considering all means, including compensating methods at load sources
4. To prepare proposals, if necessary, to amend the 1996 *National Electrical Code,* where amelioration to fire safety may be achieved

The subcommittee reviewed technical literature and electrical theory on the fundamental nature of harmonic distortion. They reviewed the requirements in and proposals for the 1993 *NEC* regarding nonlinear loads. The conclusion of the subcommittee was that while nonlinear loads can cause undesirable operational effects, including additional heating, no significant threat to persons and property has been adequately substantiated.

The subcommittee agreed with the existing *Code* text regarding nonlinear loads. However, the subcommittee submitted many proposals for the 1996 *NEC.* These proposals included a definition of nonlinear load, revised text reflecting that definition, fine print notes calling attention to the effects of nonlinear loads, and permitting the paralleling of neutral conductors in existing installations under engineering supervision.

As part of the subcommittee's final report, nine proposals for changes to the 1993 *NEC* were submitted. All were accepted without modification as changes to the 1996 *NEC.*

Also included in this report and pertinent to the 1999 *NEC* Section 310-15(b)(4)(c) is the following discussion.

Should Neutral Conductors Be Oversized?

There is concern that because the theoretical maximum neutral current is 1.73 times the balanced phase conductor current, there is a potential for neutral conductor overheating in 3-phase, 4-wire, wye-connected power systems. The subcommittee acknowledges this theoretical basis; however, in reviewing documented information, fires attributed to the use of nonlinear loads could not be identified.

The subcommittee reviewed all the data that was made available to the subcommittee regarding measurements of circuits that contain nonlinear loads. This data was obtained from consultants, equipment manufacturers, and testing laboratories, and included hundreds of feeder and branch circuits involving 3-phase, 4-wire, wye-connected systems with nonlinear loads. This data revealed that many circuits had neutral conductor current greater than the phase conductor current, and approximately 5 percent of all circuits reported had neutral conductor current exceeding 125 percent of the highest phase conductor current. One documented survey with data collected in 1988 from 146 three-phase computer power system sites determined that 3.4 percent of the sites had neutral current in excess of the rated system full-load current.

According to Section 384-16(c) of the 1993 *NEC* the total continuous load on any overcurrent device located in a panelboard shall not exceed 80 percent of its rating (the exception being assemblies listed for continuous operation at 100 percent of its rating). Since the neutral conductor is usually not connected to an overcurrent device, derating for continuous operation is not necessary. Therefore, neutral conductor ampacity is usually 125 percent of the maximum continuous current allowed by the overcurrent device.

Also important for gathering electrically measured data from existing installations is the following excerpt continuing in this report.

Measurement of Nonsinusoidal Voltages and Currents

The measurement of nonsinusoidal voltages and currents may require instruments different from the conventional meters used to measure sinusoidal waveforms. Many voltage and current meters respond only to the peak value of a waveform, and indicate a value that is equivalent to the rms value of a sinusoidal waveform. For a sinusoidal waveform the rms value will be 70.7 percent of the peak value. Meters of this type are known as "average responding meters" and will only give a true indication if the waveform being measured is sinusoidal. Both analog and digital meters may be average responding instruments. Voltages and currents that are nonsinusoidal, such as those with harmonic frequencies, cannot be accurately measured using an average responding meter. Only a meter that measures "true rms" can be used to correctly measure the rms value of a nonsinusoidal waveform.

Figure 310.5 shows an example of a clamp-on ammeter that uses true rms measurements. Figure 310.6 shows an example of a portable diagnostic analyser used for more sophisticated power measurements including measuring harmonic distortion.

(5) Grounding or Bonding Conductor. A grounding or bonding conductor shall not be counted when applying the provisions of Section 310-15(b)(2)(a).

(6) 120/240-Volt, 3-Wire, Single-Phase Dwelling Services and Feeders. For dwelling units, conductors, as listed in Table 310-15(b)(6) shall be permitted as 120/240-volt, 3-wire, single-phase service-entrance conductors,

Figure 310.5 *A clamp-on ammeter that uses true rms measurements. (Fluke Corp.)*

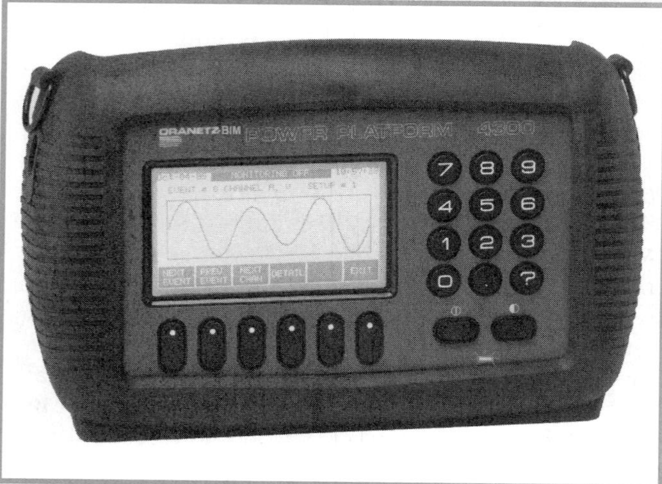

Figure 310.6 *A portable tool for such tasks as diagnostic power analysis including harmonic distortion. (Dranetz-BMI)*

service lateral conductors, and feeder conductors that serve as the main power feeder to a dwelling unit and are installed in raceway or cable with or without an equipment grounding conductor. For application of this section, the main power feeder shall be the feeder(s) between the main disconnect and the lighting and appliance branch-circuit panelboard(s), and the feeder conductors to a dwelling unit shall not be required to be larger than their service-entrance conductors. The grounded conductor shall be permitted to be smaller, than the ungrounded conductors, provided the requirements of Sections 215-2, 220-22, and 230-42 are met.

Table 310-15(b)(6). Conductor Types and Sizes for 120/240-Volt, 3-Wire, Single-Phase Dwelling Services and Feeders. Conductor Types RH, RHH, RHW, RHW-2, THHN, THHW, THW, THW-2, THWN, THWN-2, XHHW, XHHW-2, SE, USE, USE-2

Conductor (AWG or kcmil)		
Copper	Aluminum or Copper-Clad Aluminum	Service or Feeder Rating (Amperes)
4	2	100
3	1	110
2	1/0	125
1	2/0	150
1/0	3/0	175
2/0	4/0	200
3/0	250	225
4/0	300	250
250	350	300
350	500	350
400	600	400

Section 310-15(b)(6) was placed here as part of the 1999 *Code* reorganization of Section 310-15. It previously appeared in the 1996 *Code* as Note 3 of Notes to Ampacity Tables of 0 to 2000 Volts after Table 310-19.

If a single set of 3-wire, single-phase, service-entrance conductors in raceway or cable supplies a one-family, two-family, or multifamily dwelling, the reduced conductor size permitted by Section 310-15(b)(6) is applicable to the service-entrance conductors, service-lateral conductors, or any feeder conductors that supply the main power feeder to a dwelling unit.

This section permits the main feeder to a dwelling unit to be sized based on the conductor sizes in Table 310-15(b)(6) even if other loads, such as ac units and pool loads, are fed from the same service. The feeder conductors to a dwelling unit are not required to be larger than its service-entrance conductors.

Figures 310.7 and 310.8 illustrate the application of Section 310-15(b)(6). In Figure 310.7, the reduced conductor size permitted is applicable to the service-entrance conductors run to each apartment from the meters. In Figure 310.8, the reduced conductor size permitted is also applicable to the feeder conductors run to each apartment from the service disconnecting means, because these feeders carry the entire load to each apartment.

Previously limited to a two-wire size reduction but changed in the 1996 *Code,* the grounded conductor is now permitted to be reduced more than two sizes.

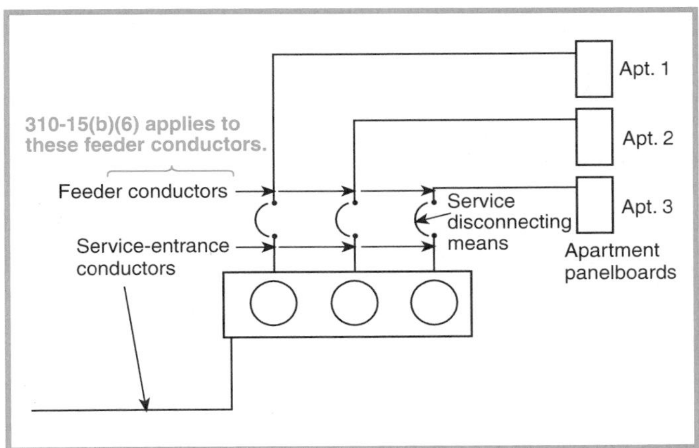

Figure 310.7 *An application of Section 310-15(b)(6).*

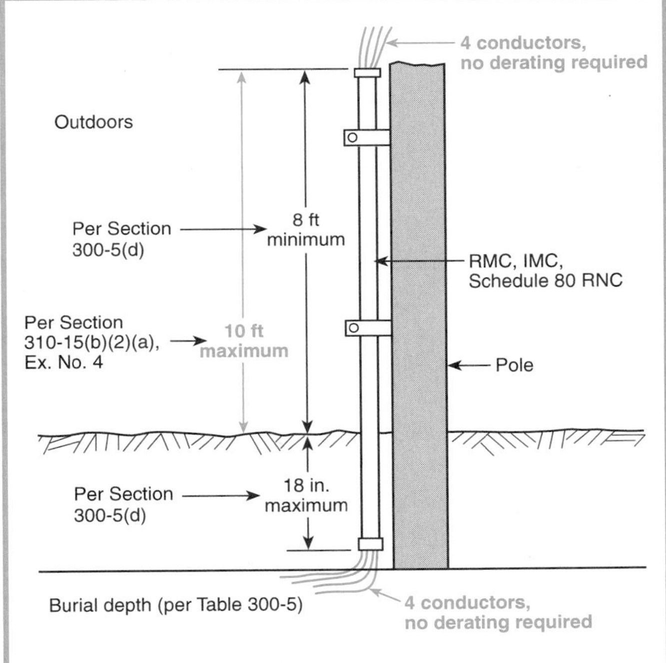

Figure 310.8 *Another application of Section 310-15(b)(6).*

The stipulation is that the requirements from the other applicable *Code* sections are observed, especially Section 250-24(b).

Other sections of the *Code* must also be applied for determining the size of service or feeder conductors. Section 230-42 requires service conductors to be of sufficient size to carry the load calculated in accordance with Article 220. It should not be taken for granted that the grounded (neutral) conductor can be automatically reduced. The grounded (neutral) conductor must provide a low-impedance fault path capable of conducting fault current back to the transformer. Section 250-24(b) provides the requirements for determining the minimum-size grounded (neutral) conductor.

Section 250-24(b) is not applicable to the grounded (neutral) conductor of *feeders;* therefore, the minimum size is governed by Section 220-22. This section requires the grounded (neutral) conductor to be large enough to carry the maximum unbalance of the net computed load connected to the neutral and any one ungrounded conductor. In the event that there are no 240-volt loads, the neutral, under severe unbalanced conditions, carries the same current as the ungrounded conductor supplying the load.

(7) Mineral-Insulated, Metal-Sheathed Cable. The temperature limitations on which the ampacities of mineral-insulated, metal-sheathed cable are based shall be determined by the insulating materials used in the end seal. Termination fittings incorporating unimpregnated, organic, insulating materials shall be limited to 90°C (194°F) operation.

(c) Engineering Supervision. Under engineering supervision, conductor ampacities shall be permitted to be calculated by means of the following general formula:

$$I = \sqrt{\frac{TC - (TA + DeltaTD)}{RDC(1 + YC)RCA}}$$

Where:

TC = Conductor temperature in degrees Celsius (°C)

TA = Ambient temperature in degrees Celsius (°C)

$Delta\ TD$ = Dielectric loss temperature rise

RDC = dc resistance of conductor at temperature TC

YC = Component ac resistance resulting from skin effect and proximity effect

RCA = Effective thermal resistance between conductor and surrounding ambient

FPN: See Appendix B for examples of formula applications.

The formula in Section 310-15(c) was developed by J.H. Neher and M.H. McGrath to determine conductor ampacity. It is actually a composite of a number of separate formulas. A description of this method of calculation was given in AIEE paper No. 5-660, "The Calculation of the Temperature Rise and Load Capability of Cable Systems," by J. H. Neher and M. H. McGrath. This paper was presented to the AIEE general meeting in Montreal, Quebec, on June 24–28, 1956, and was published in *AIEE Transactions, Part III*

(*Power Apparatus and Systems*), Vol. 76, October 1957, pp. 752–772. AIEE (American Institute of Electrical Engineers) is now IEEE (Institute of Electrical and Electronic Engineers).

The Neher-McGrath formula in Section 310-15(c) is a heat-transfer formula, composed of a series of heat-transfer calculations, that takes into account all heat sources and the thermal resistances between the heat sources and free air. The most common use for the Neher-McGrath formula is to calculate the ampacity of conductors in underground electrical ducts (raceways), although the formula is applicable to all conductor installations.

It is not the intent of the following discussion to provide instruction on the use of the Neher-McGrath method of calculation. The intent is to identify the many factors affecting the calculations. It is because of these many variables and the complexities of the many formulas involved that the *Code* requires the calculation to be made under engineering supervision.

Current passing through a conductor produces I^2R losses in the form of heat, which results from conductor losses and appears as a temperature rise in the conductor. This heat must pass through the cable insulation, the air in the raceway, and the raceway itself to the surrounding medium, usually earth or concrete, where it is dissipated into the air by radiation and convection. Unless the heat is dissipated, the temperature in the conductor will exceed the rating of the conductor insulation.

The conductor's ampacity is based on the rate of heat dissipation through the thermal resistances surrounding the conductor. Current traveling through a material with a specific resistance at a specified temperature generates this heat. Additional heat is caused by skin and proximity effects, as usually the current is ac and there are other conductors in the same duct.

For conductors in underground electrical ducts, there are several heat sources, as follows, and as illustrated in Figure 310.9.

1. *Conductor losses due to the load current I^2R.* These losses vary with the load current, conductor material, and conductor cross-sectional area (conductor size).

2. *Skin-effect heating if the current is alternating current.* The heat developed by the skin effect is due to the shape of the conductor and is based on the configuration of the conductors (i.e., solid, stranded, or compact).

3. *Hysteresis losses if the duct is steel or other magnetic material.* These losses are dependent upon the magnetic properties of the electrical duct and the shape of the duct.

4. *Heating from other conductors in the duct.* This heating is based on the number, location, and proximity of other conductors as well as the losses in the other conductors. The more conductors in the raceway, the greater the heating effect from these conductors is likely to be. This factor replaces the adjustment factors in Section 310-15(b)(2)(a) to the ampacity tables.

5. *Mutual heating from other ducts, cables, etc., in the vicinity.* The closer the other heat sources and the more they surround the duct for which calculations are being made, the greater the heating effect. For example, in the case of a symmetrical 9-duct bank, 3 ducts high and 3 ducts wide, the center duct will receive the most heat as a result of mutual heating.

Heat generated by the following various types of losses is conducted through the different thermal barriers or resistances, as illustrated in Figure 310.10.

1. *Conductor insulation.* The conductor insulation, which is designed to perform as a good electrical

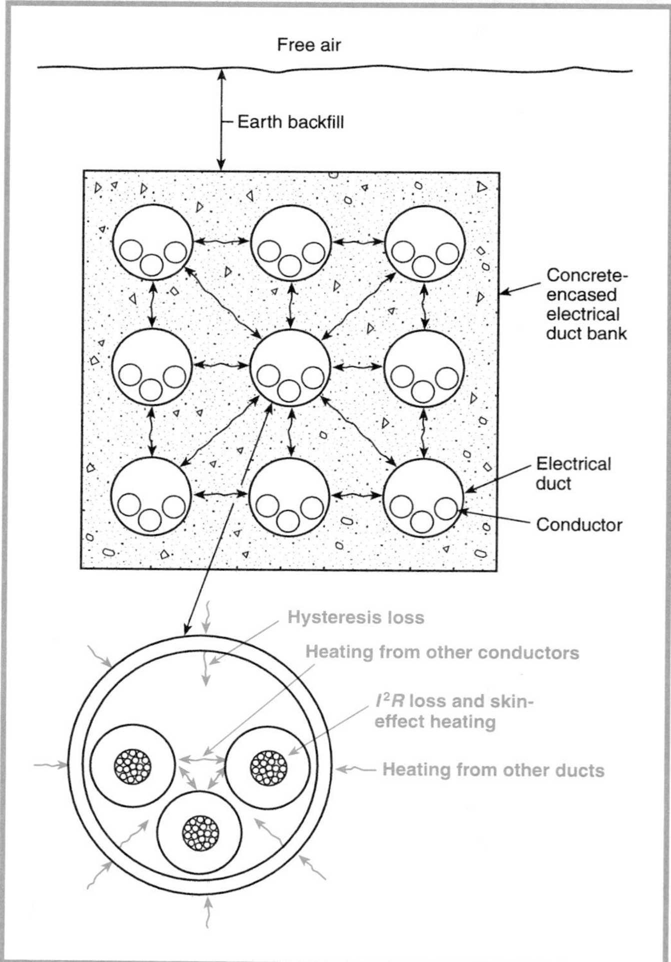

Figure 310.9 *Heat sources.*

insulator, also serves as a good thermal insulator. It presents a thermal resistance to heat generated by the conductor due to the I^2R losses, including any dielectric losses. This thermal resistance value depends on the thickness of the insulation and the type of insulating material used. Materials such as polyvinyl chloride, used in Type THW and other conductors; cross-linked polyethylene, used in Type XHHW and other conductors; and rubber, used in Type RHW and other conductors, have different thermal resistivities. In addition, the thickness of the conductor insulation varies from one type of insulation to another, even for the same size conductor.

2. *Airspace.* The next thermal barrier encountered by the heat flow generated in the conductor is the airspace between the conductor insulation and the surrounding wall or raceway. The thermal resistance of this airspace is based on the number of conductors in the duct, the assumed mean value of the temperature of the air in the duct, and the constants provided

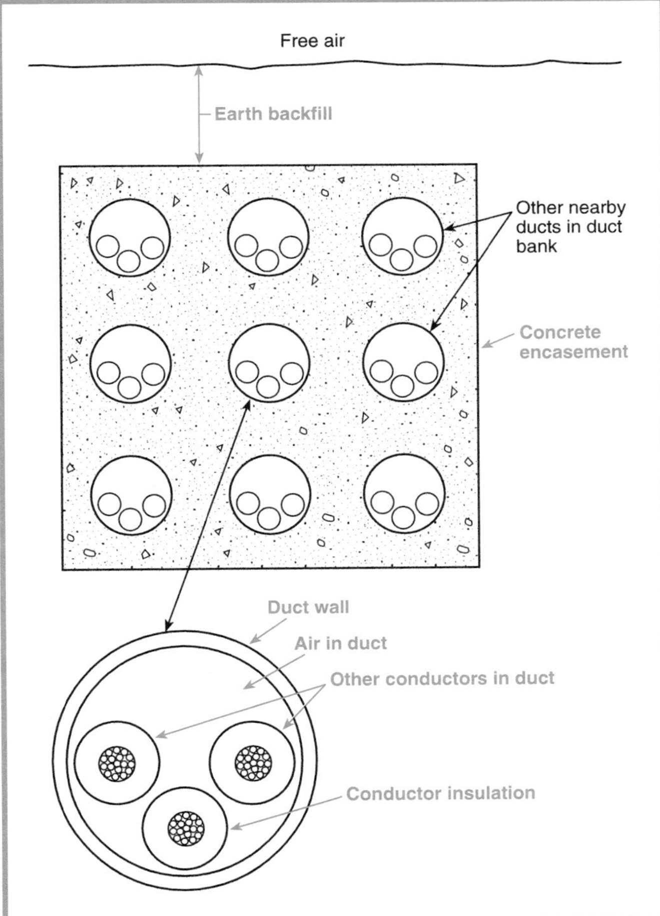

Figure 310.10 *Thermal barriers (resistances).*

in the Neher-McGrath paper, which were determined from experimental data.

3. *Duct wall.* After it passes through the airspace around the conductors, the heat encounters the thermal resistance of the duct wall. This thermal resistance is based on the thermal resistivity of the type of material used and the thickness of the duct wall. Metallic materials have less thermal resistance than nonmetallic materials. The thicker the wall, the greater the thermal resistance.

4. *Earth backfill.* The thermal resistance that must be considered next is that offered by the earth or other backfill material above the duct. This incorporates not only the thermal resistivity and ambient temperature of the earth but also the number of current-carrying conductors within the duct, the outside diameter of the duct, the burial depth, a loss factor, and the mutual heating factor caused by other nearby ducts. The deeper the duct is buried, the greater the thermal resistance.

To prevent the temperature of the conductors from exceeding the rated temperature of the insulation, heat dissipation through these thermal resistances must be equal to or greater than the heat developed. Thus, the thermal resistances of all the components of a conductor must be determined, and the allowable temperature differential above the ambient temperature and between the conductors and the surface of the earth must be known.

In addition to the Neher-McGrath paper itself, as described in the first paragraph of this commentary, the following references provide more detailed information on the use of this method of calculation:

"Power Cable Ampacity Tables," Insulated Power Cable Engineers Association, IPCEA P-46-426 (IEEE S-135-1).

"IEEE Standard Power Cable Ampacity Tables," Institute of Electrical and Electronic Engineers, IEEE Standard 835-1994.

"Neher-McGrath Calculations for Insulated Power Cables," Peter Pollak, IEEE Paper No. CH2040-4/84/0000-0172, presented at the 1984 IEEE Industrial and Commercial Power Systems Technical Conference, Atlanta, GA, May 9, 1984.

"Cable Ampacities, the *NEC* and Computerized Applications," M. T. Brown et al., IEEE Paper No. CH2207-9/85/00000-0323.

"How to Use the Neher-McGrath Method to Calculate Ampacity of Underground Conductors," John M. Caloggero, NFPA *Fire Journal*, pp. 17–18, May/June 1988.

Table 310-16. Allowable Ampacities of Insulated Conductors Rated 0 through 2000 Volts, 60°C through 90°C (140°F through 194°F) Not More than Three Current-Carrying Conductors in Raceway, Cable, or Earth (Directly Buried), Based on Ambient Temperature of 30°C (86°F)

Size	Temperature Rating of Conductor (See Table 310-13)						Size
	60°C (140°F)	75°C (167°F)	90°C (194°F)	60°C (140°F)	75°C (167°F)	90°C (194°F)	
AWG or kcmil	Types TW, UF	Types FEPW, RH, RHW, THHW, THW, THWN, XHHW, USE, ZW	Types TBS, SA, SIS, FEP, FEPB, MI, RHH, RHW-2, THHN, THHW, THW-2, THWN-2, USE-2, XHH, XHHW, XHHW-2, ZW-2	Types TW, UF	Types RH, RHW, THHW, THW, THWN, XHHW, USE	Types TBS, SA, SIS, THHN, THHW, THW-2, THWN-2, RHH, RHW-2, USE-2, XHH, XHHW, XHHW-2, ZW-2	AWG or kcmil
	COPPER			ALUMINUM OR COPPER-CLAD ALUMINUM			
18	—	—	14	—	—	—	—
16	—	—	18	—	—	—	—
14*	20	20	25	—	—	—	—
12*	25	25	30	20	20	25	12*
10*	30	35	40	25	30	35	10*
8	40	50	55	30	40	45	8
6	55	65	75	40	50	60	6
4	70	85	95	55	65	75	4
3	85	100	110	65	75	85	3
2	95	115	130	75	90	100	2
1	110	130	150	85	100	115	1
1/0	125	150	170	100	120	135	1/0
2/0	145	175	195	115	135	150	2/0
3/0	165	200	225	130	155	175	3/0
4/0	195	230	260	150	180	205	4/0
250	215	255	290	170	205	230	250
300	240	285	320	190	230	255	300
350	260	310	350	210	250	280	350
400	280	335	380	225	270	305	400
500	320	380	430	260	310	350	500
600	355	420	475	285	340	385	600
700	385	460	520	310	375	420	700
750	400	475	535	320	385	435	750
800	410	490	555	330	395	450	800
900	435	520	585	355	425	480	900
1000	455	545	615	375	445	500	1000
1250	495	590	665	405	485	545	1250
1500	520	625	705	435	520	585	1500
1750	545	650	735	455	545	615	1750
2000	560	665	750	470	560	630	2000
	CORRECTION FACTORS						
Ambient Temp. (°C)	For ambient temperatures other than 30°C (86°F), multiply the allowable ampacities shown above by the appropriate factor shown below.						Ambient Temp. (°F)
21–25	1.08	1.05	1.04	1.08	1.05	1.04	70–77
26–30	1.00	1.00	1.00	1.00	1.00	1.00	78–86
31–35	0.91	0.94	0.96	0.91	0.94	0.96	87–95
36–40	0.82	0.88	0.91	0.82	0.88	0.91	96–104
41–45	0.71	0.82	0.87	0.71	0.82	0.87	105–113
46–50	0.58	0.75	0.82	0.58	0.75	0.82	114–122
51–55	0.41	0.67	0.76	0.41	0.67	0.76	123–131
56–60	—	0.58	0.71	—	0.58	0.71	132–140
61–70	—	0.33	0.58	—	0.33	0.58	141–158
71–80	—	—	0.41	—	—	0.41	159–176

*See Section 240-3.

Table 310-17. Allowable Ampacities of Single-Insulated Conductors Rated 0 Through 2000 Volts in Free Air, Based on Ambient Air Temperature of 30°C (86°F)

Size	Temperature Rating of Conductor (See Table 310-13)						Size
	60°C (140°F)	75°C (167°F)	90°C (194°F)	60°C (140°F)	75°C (167°F)	90°C (194°F)	
AWG or kcmil	Types TW, UF	Types FEPW, RH, RHW, THHW, THW, THWN, XHHW, ZW	Types TBS, SA, SIS, FEP, FEPB, MI, RHH, RHW-2, THHN, THHW, THW-2, THWN-2, USE-2, XHH, XHHW, XHHW-2, ZW-2	Types TW, UF	Types RH, RHW, THHW, THW, THWN, XHHW	Types TBS, SA, SIS, THHN, THHW, THW-2, THWN-2, RHH, RHW-2, USE-2, XHH, XHHW, XHHW-2, ZW-2	AWG or kcmil
	COPPER			ALUMINUM OR COPPER-CLAD ALUMINUM			
18	—	—	18	—	—	—	—
16	—	—	24	—	—	—	—
14*	25	30	35	—	—	—	—
12*	30	35	40	25	30	35	12*
10*	40	50	55	35	40	40	10*
8	60	70	80	45	55	60	8
6	80	95	105	60	75	80	6
4	105	125	140	80	100	110	4
3	120	145	165	95	115	130	3
2	140	170	190	110	135	150	2
1	165	195	220	130	155	175	1
1/0	195	230	260	150	180	205	1/0
2/0	225	265	300	175	210	235	2/0
3/0	260	310	350	200	240	275	3/0
4/0	300	360	405	235	280	315	4/0
250	340	405	455	265	315	355	250
300	375	445	505	290	350	395	300
350	420	505	570	330	395	445	350
400	455	545	615	355	425	480	400
500	515	620	700	405	485	545	500
600	575	690	780	455	540	615	600
700	630	755	855	500	595	675	700
750	655	785	885	515	620	700	750
800	680	815	920	535	645	725	800
900	730	870	985	580	700	785	900
1000	780	935	1055	625	750	845	1000
1250	890	1065	1200	710	855	960	1250
1500	980	1175	1325	795	950	1075	1500
1750	1070	1280	1445	875	1050	1185	1750
2000	1155	1385	1560	960	1150	1335	2000
			CORRECTION FACTORS				
Ambient Temp. (°C)	For ambient temperatures other than 30°C (86°F), multiply the allowable ampacities shown above by the appropriate factor shown below.						Ambient Temp. (°F)
21–25	1.08	1.05	1.04	1.08	1.05	1.04	70–77
26–30	1.00	1.00	1.00	1.00	1.00	1.00	78–86
31–35	0.91	0.94	0.96	0.91	0.94	0.96	87–95
36–40	0.82	0.88	0.91	0.82	0.88	0.91	96–104
41–45	0.71	0.82	0.87	0.71	0.82	0.87	105–113
46–50	0.58	0.75	0.82	0.58	0.75	0.82	114–122
51–55	0.41	0.67	0.76	0.41	0.67	0.76	123–131
56–60	—	0.58	0.71	—	0.58	0.71	132–140
61–70	—	0.33	0.58	—	0.33	0.58	141–158
71–80	—	—	0.41	—	—	0.41	159–176

*See Section 240-3.

Table 310-18. Allowable Ampacities of Three Single-Insulated Conductors, Rated 0 Through 2000 Volts, 150°C Through 250°C (302°F Through 482°F), in Raceway or Cable, Based on Ambient Air Temperature of 40°C (104°F)

Size	Temperature Rating of Conductor (See Table 310-13)				Size
	150°C (302°F)	200°C (392°F)	250°C (482°F)	150°C (302°F)	
AWG or kcmil	Type Z	Types FEP, FEPB, PFA	Types PFAH, TFE	Type Z	AWG or kcmil
	COPPER		NICKEL OR NICKEL-COATED COPPER	ALUMINUM OR COPPER-CLAD ALUMINUM	
14	34	36	39	—	14
12	43	45	54	30	12
10	55	60	73	44	10
8	76	83	93	57	8
6	96	110	117	75	6
4	120	125	148	94	4
3	143	152	166	109	3
2	160	171	191	124	2
1	186	197	215	145	1
1/0	215	229	244	169	1/0
2/0	251	260	273	198	2/0
3/0	288	297	308	227	3/0
4/0	332	346	361	260	4/0
CORRECTION FACTORS					
Ambient Temp. (°C)	For ambient temperatures other than 40°C (104°F), multiply the allowable ampacities shown above by the appropriate factor shown below.				Ambient Temp. (°F)
41–50	0.95	0.97	0.98	0.95	105–122
51–60	0.90	0.94	0.95	0.90	123–140
61–70	0.85	0.90	0.93	0.85	141–158
71–80	0.80	0.87	0.90	0.80	159–176
81–90	0.74	0.83	0.87	0.74	177–194
91–100	0.67	0.79	0.85	0.67	195–212
101–120	0.52	0.71	0.79	0.52	213–248
121–140	0.30	0.61	0.72	0.30	249–284
141–160	—	0.50	0.65	—	285–320
161–180	—	0.35	0.58	—	321–356
181–200	—	—	0.49	—	357–392
201–225	—	—	0.35	—	393–437

310-60. Conductors Rated 2001 to 35,000 Volts.

(a) Definitions.

Electrical Ducts. As used in Article 310, electrical ducts shall include any of the electrical conduits recognized in Chapter 3 as suitable for use underground; and other raceways round in cross section, listed for underground use, and embedded in earth or concrete.

The term *electrical duct* is used to differentiate from ducts used for air handling, for example. It is intended to include nonmetallic electrical ducts commonly used for underground wiring, as well as other raceways (e.g., rigid metal conduit, intermediate metal conduit, rigid nonmetallic conduit, etc.) listed for use underground in earth or concrete.

Thermal Resistivity. As used in this *Code, thermal resistivity* refers to the heat transfer capability through a substance by conduction. It is the reciprocal of thermal conductivity and is designated Rho and expressed in the units °C-cm/watt.

(b) Ampacities of Conductors Rated 2001 to 35,000 Volts. Ampacities for solid dielectric-insulated conductors shall be permitted to be determined by tables or under engineering supervision, as provided in (c) and (d).

Table 310-19. Allowable Ampacities of Single-Insulated Conductors, Rated 0 Through 2000 Volts, 150°C Through 250°C (302°F Through 482°F), in Free Air, Based on Ambient Air Temperature of 40°C (104°F)

Size	Temperature Rating of Conductor (See Table 310-13)				Size
AWG or kcmil	150°C (302°F)	200°C (392°F)	250°C (482°F)	150°C (302°F)	AWG or kcmil
	Type Z	Types FEP, FEPB, PFA	Types PFAH, TFE	Type Z	
	COPPER		NICKEL, OR NICKEL-COATED COPPER	ALUMINUM OR COPPER-CLAD ALUMINUM	
14	46	54	59	—	14
12	60	68	78	47	12
10	80	90	107	63	10
8	106	124	142	83	8
6	155	165	205	112	6
4	190	220	278	148	4
3	214	252	327	170	3
2	255	293	381	198	2
1	293	344	440	228	1
1/0	339	399	532	263	1/0
2/0	390	467	591	305	2/0
3/0	451	546	708	351	3/0
4/0	529	629	830	411	4/0
CORRECTION FACTORS					
Ambient Temp. (°C)	For ambient temperatures other than 40°C (104°F), multiply the allowable ampacities shown above by the appropriate factor shown below.				Ambient Temp. (°F)
41–50	0.95	0.97	0.98	0.95	105–122
51–60	0.90	0.94	0.95	0.90	123–140
61–70	0.85	0.90	0.93	0.85	141–158
71–80	0.80	0.87	0.90	0.80	159–176
81–90	0.74	0.83	0.87	0.74	177–194
91–100	0.67	0.79	0.85	0.67	195–212
101–120	0.52	0.71	0.79	0.52	213–248
121–140	0.30	0.61	0.72	0.30	249–284
141–160	—	0.50	0.65	—	285–320
161–180	—	0.35	0.58	—	321–356
181–200	—	—	0.49	—	357–392
201–225	—	—	0.35	—	393–437

(1) Selection of Ampacity. Where more than one calculated or tabulated ampacity could apply for a given circuit length, the lowest value shall be used.

Exception: Where two different ampacities apply to adjacent portions of a circuit, the higher ampacity shall be permitted to be used beyond the point of transition, a distance equal to 10 ft (3.05 m) or 10 percent of the circuit length figured at the higher ampacity, whichever is less.

FPN: See Section 110-40 for conductor temperature limitations due to termination provisions.

(c) Tables. Ampacities for conductors rated 2001 to 35,000 volts shall be as specified in the Ampacity Tables 310-67 through 310-86. Ampacities at ambient temperatures other than those shown in the tables shall be determined by the formula in (4).

FPN No. 1: For ampacities calculated in accordance with Section 310-60(b), reference IEEE *Standard Power Cable Ampacity Tables,* IEEE 835-1994 (IPCEA Pub. No. P-46-426) and the references therein for availability of all factors and constants.

FPN No. 2: Ampacities provided by this section do not take voltage drop into consideration. See Section 210-19(a), FPN No. 4, for branch circuits and Section 215-2(d), FPN No. 2, for feeders.

(1) Grounded Shields. Ampacities shown in Tables 310-69, 310-70, 310-81, and 310-82 are for cable with shields grounded at one point only. Where shields are grounded at more than one point, ampacities shall be adjusted to take into consideration the heating due to shield currents.

Table 310-20. Ampacities of Two or Three Single-Insulated Conductors, Rated 0 Through 2000 Volts, Supported on a Messenger, Based on Ambient Air Temperature of 40°C (104°F)

Size	Temperature Rating of Conductor (See Table 310-13)				Size
	75°C (167°F)	90°C (194°F)	75°C (167°F)	90°C (194°F)	
AWG or kcmil	Types RH, RHW, THHW, THW, THWN, XHHW, ZW	Types THHN, THHW, THW-2, THWN-2, RHH, RWH-2, USE-2, XHHW, XHHW-2, ZW-2	Types RH, RHW, THW, THWN, THHW, XHHW	Types THHN, THHW, RHH, XHHW, RHW-2, XHHW-2, THW-2, THWN-2, USE-2, ZW-2	AWG or kcmil
	COPPER		ALUMINUM OR COPPER-CLAD ALUMINUM		
8	57	66	44	51	8
6	76	89	59	69	6
4	101	117	78	91	4
3	118	138	92	107	3
2	135	158	106	123	2
1	158	185	123	144	1
1/0	183	214	143	167	1/0
2/0	212	247	165	193	2/0
3/0	245	287	192	224	3/0
4/0	287	335	224	262	4/0
250	320	374	251	292	250
300	359	419	282	328	300
350	397	464	312	364	350
400	430	503	339	395	400
500	496	580	392	458	500
600	553	647	440	514	600
700	610	714	488	570	700
750	638	747	512	598	750
800	660	773	532	622	800
900	704	826	572	669	900
1000	748	879	612	716	1000
Ambient Temp. (°C)	For ambient temperatures other than 40°C (104°F), multiply the allowable ampacities shown above by the appropriate factor shown below.				Ambient Temp. (°F)
21–25	1.20	1.14	1.20	1.14	70–77
26–30	1.13	1.10	1.13	1.10	79–86
31–35	1.07	1.05	1.07	1.05	88–95
36–40	1.00	1.00	1.00	1.00	97–104
41–45	0.93	0.95	0.93	0.95	106–113
46–50	0.85	0.89	0.85	0.89	115–122
51–55	0.76	0.84	0.76	0.84	124–131
56–60	0.65	0.77	0.65	0.77	133–140
61–70	0.38	0.63	0.38	0.63	142–158
71–80	—	0.45	—	0.45	160–176

Table 310-21. Ampacities of Bare or Covered Conductors, Based on 40°C (104°F) Ambient, 80°C (176°F) Total Conductor Temperature, 2 ft/sec (610 mm/sec) Wind Velocity

Copper Conductors			
Bare		Covered	
AWG or kcmil	Amperes	AWG or kcmil	Amperes
8	98	8	103
6	124	6	130
4	155	4	163
2	209	2	219
1/0	282	1/0	297
2/0	329	2/0	344
3/0	382	3/0	401
4/0	444	4/0	466
250	494	250	519
300	556	300	584
500	773	500	812
750	1000	750	1050
1000	1193	1000	1253
—	—	—	—
—	—	—	—
—	—	—	—
AAC Aluminum Conductors			
8	76	8	80
6	96	6	101
4	121	4	127
2	163	2	171
1/0	220	1/0	231
2/0	255	2/0	268
3/0	297	3/0	312
4/0	346	4/0	364
266.8	403	266.8	423
336.4	468	336.4	492
397.5	522	397.5	548
477.0	588	477.0	617
556.5	650	556.5	682
636.0	709	636.0	744
795.0	819	795.0	860
954.0	920	—	—
1033.5	968	1033.5	1017
1272	1103	1272	1201
1590	1267	1590	1381
2000	1454	2000	1527

(2) Burial Depth of Underground Circuits. Where the burial depth of direct burial or electrical duct bank circuits is modified from the values shown in a figure or table, ampacities shall be permitted to be modified as indicated in (a) and (b).

(a) Where burial depths are increased in part(s) of an electrical duct run, no decrease in ampacity of the conductors is needed, provided the total length of parts of the duct run increased in depth is less than 25 percent of the total run length.

(b) Where burial depths are deeper than shown in a specific underground ampacity table or figure, an ampacity derating factor of 6 percent per increased foot (305 mm) of depth for all values of Rho shall be permitted.

No rating change is needed where the burial depth is decreased.

Table 310-61. Conductor Application and Insulation

Trade Name	Type Letter	Maximum Operating Temperature	Application Provision	Insulation	Outer Covering
Medium voltage solid dielectric	MV-90 MV-105*	90°C 105°C	Dry or wet locations rated 2001 volts and higher	Thermo-plastic or thermo-setting	Jacket, sheath, or armor

*Where design conditions require maximum conductor temperatures above 90°C.

Table 310-62. Thickness of Insulation for 601 to 2000-Volt Nonshielded Types RHH and RHW, in Mils

Conductor Size (AWG or kcmil)	Column A[1]	Column B[2]
14–10	80	60
8	80	70
6–2	95	70
1–2/0	110	90
3/0–4/0	110	90
213–500	125	105
501–1000	140	120

[1]Column A insulations are limited to natural, SBR, and butyl rubbers.
[2]Column B insulations are materials such as cross-linked polyethylene, ethylene propylene rubber, and composites thereof.

(3) Electrical Ducts in Figure 310-60. At locations where electrical ducts enter equipment enclosures from underground, spacing between such ducts, as shown in Figure 310-60, shall be permitted to be reduced without requiring the ampacity of conductors therein to be reduced.

(4) Ambients Not in Tables. Ampacities at ambient temperatures other than those shown in the tables shall be determined by means of the following formula:

$$I_2 = I_1 \sqrt{\frac{TC - TA_2 - Delta\ TD}{TC - TA_1 - Delta\ TD}}$$

Where:

I_1 = ampacity from tables at ambient TA_1
I_2 = ampacity at desired ambient TA_2
TC = conductor temperature in degrees Celsius (°C)
TA_1 = surrounding ambient from tables in degrees Celsius (°C)
TA_2 = desired ambient in degrees Celsius (°C)
$Delta\ TD$ = dielectric loss temperature rise

(d) Engineering Supervision. Under engineering supervision, conductor ampacities shall be permitted to be calculated by means of the following general formula:

$$I = \sqrt{\frac{TC - (TA + Delta\ TD)}{RDC\ (1 + YC)RCA}}$$

Where:

TC = conductor temperature in °C
TA = ambient temperature in °C
$Delta\ TD$ = dielectric loss temperature rise
RDC = dc resistance of conductor at temperature TC
YC = component ac resistance resulting from skin effect and proximity effect
RCA = effective thermal resistance between conductor and surrounding ambient

FPN: See Appendix B for examples of formula applications.

Table 310-63. Thickness of Insulation and Jacket for Nonshielded Solid Dielectric Insulated Conductors Rated 2001 to 8000 Volts, in Mils

	2001–5000 Volts						5001–8000 Volts 100 Percent Insulation Level Wet or Dry Locations		
	Dry Locations, Single Conductor			Wet or Dry Locations					
	Without Jacket	With Jacket		Single Conductor		Multi-conductor*	Single Conductor		Multi-conductor*
Conductor Size (AWG or kcmil)	Insulation	Insulation	Jacket	Insulation	Jacket	Insulation	Insulation	Jacket	Insulation
8	110	90	30	125	80	90	180	80	180
6	110	90	30	125	80	90	180	80	180
4–2	110	90	45	125	80	90	180	95	180
1–2/0	110	90	45	125	80	90	180	95	180
3/0–4/0	110	90	65	125	95	90	180	110	180
213–500	120	90	65	140	110	90	210	110	210
501–750	130	90	65	155	125	90	235	125	235
751–1000	130	90	65	155	125	90	250	140	250

*Under a common overall covering such as a jacket, sheath, or armor.

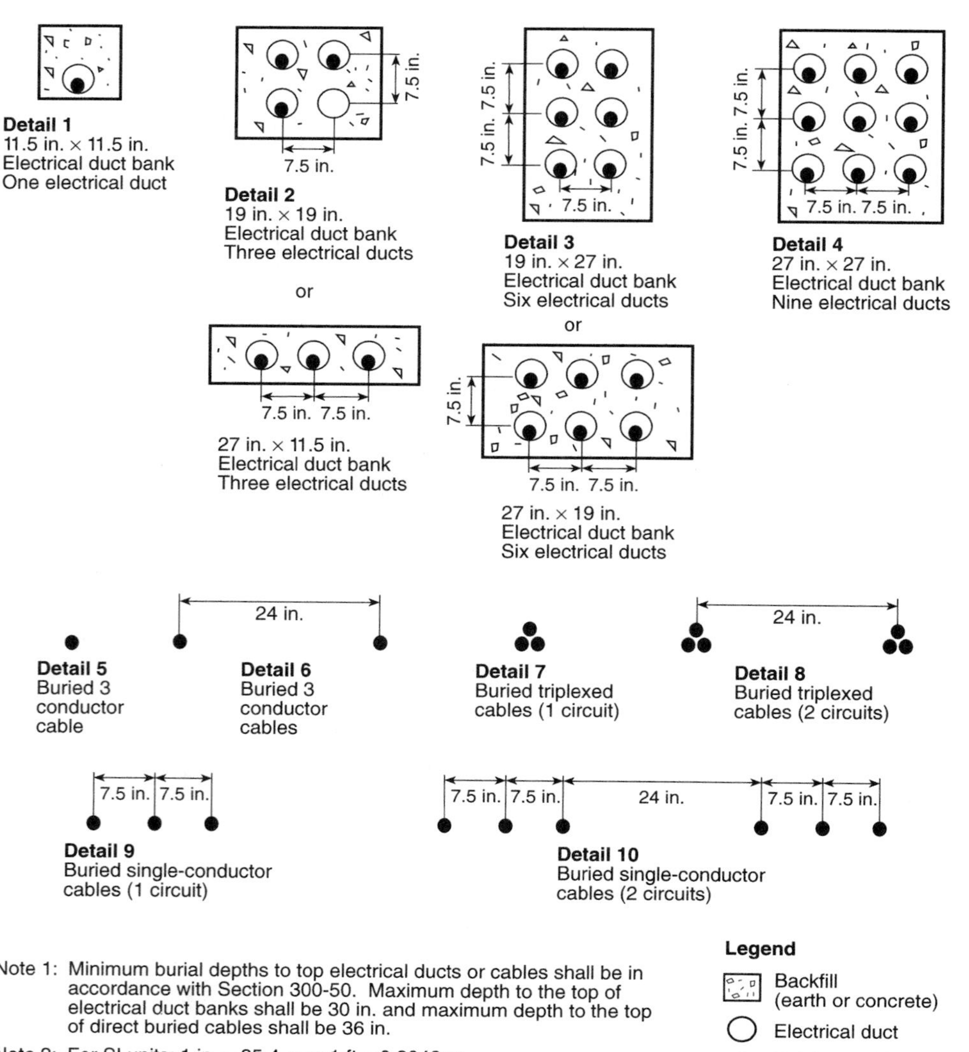

Figure 310-60 Cable installation dimensions for use with Tables 310-77 through 310-86.

Table 310-64. Thickness of Insulation for Shielded Solid Dielectric Insulated Conductors Rated 2001 to 35,000 Volts, in Mils

Conductor Size (AWG or kcmil)	2001–5000 Volts	5001–8000 Volts		8001–15,000 Volts		15,001–25,000 Volts		25,001–28,000 Volts		28,001–35,000 Volts	
		100 Percent Insulation Level[1]	133 Percent Insulation Level[2]	100 Percent Insulation Level[1]	133 Percent Insulation Level[2]	100 Percent Insulation Level[1]	133 Percent Insulation Level[2]	100 Percent Insulation Level[1]	133 Percent Insulation Level[2]	100 Percent Insulation Level[1]	133 Percent Insulation Level[2]
8	90	—	—	—	—	—	—	—	—	—	—
6–4	90	115	140	—	—	—	—	—	—	—	—
2	90	115	140	175	215	—	—	—	—	—	—
1	90	115	140	175	215	260	345	280	345	—	—
1/0–2000	90	115	140	175	215	260	345	280	345	345	420

[1]**100 Percent Insulation Level.** Cables in this category shall be permitted to be applied where the system is provided with relay protection such that ground faults will be cleared as rapidly as possible but, in any case, within 1 minute. While these cables are applicable to the great majority of cable installations that are on grounded systems, they shall be permitted to be used also on other systems for which the application of cables is acceptable, provided the above clearing requirements are met in completely de-energizing the faulted section.

[2]**133 Percent Insulation Level.** This insulation level corresponds to that formerly designated for ungrounded systems. Cables in this category shall be permitted to be applied in situations where the clearing time requirements of the 100 percent level category cannot be met, and yet there is adequate assurance that the faulted section will be de-energized in a time not exceeding 1 hour. Also, they shall be permitted to be used where additional insulation strength over the 100 percent level category is desirable.

Table 310-67. Ampacities of Insulated Single Copper Conductor Cables Triplexed in Air Based on Conductor Temperatures of 90°C (194°F) and 105°C (221°F) and Ambient Air Temperature of 40°C (104°F)

Conductor Size (AWG or kcmil)	Temperature Rating of Conductor (See Table 310-61)			
	2001–5000 Volts Ampacity		5001–35,000 Volts Ampacity	
	90°C (194°F) Type MV-90	105°C (221°F) Type MV-105	90°C (194°F) Type MV-90	105°C (221°F) Type MV-105
8	65	74	—	—
6	90	99	100	110
4	120	130	130	140
2	160	175	170	195
1	185	205	195	225
1/0	215	240	225	255
2/0	250	275	260	295
3/0	290	320	300	340
4/0	335	375	345	390
250	375	415	380	430
350	465	515	470	525
500	580	645	580	650
750	750	835	730	820
1000	880	980	850	950

Table 310-68. Ampacities of Insulated Single Aluminum Conductor Cables Triplexed in Air Based on Conductor Temperatures of 90°C (194°F) and 105°C (221°F) and Ambient Air Temperature of 40°C (104°F)

Conductor Size (AWG or kcmil)	Temperature Rating of Conductor (See Table 310-61)			
	2001–5000 Volts Ampacity		5001–35,000 Volts Ampacity	
	90°C (194°F) Type MV-90	105°C (221°F) Type MV-105	90°C (194°F) Type MV-90	105°C (221°F) Type MV-105
8	50	57	—	—
6	70	77	75	84
4	90	100	100	110
2	125	135	130	150
1	145	160	150	175
1/0	170	185	175	200
2/0	195	215	200	230
3/0	225	250	230	265
4/0	265	290	270	305
250	295	325	300	335
350	365	405	370	415
500	460	510	460	515
750	600	665	590	660
1000	715	800	700	780

Table 310-69. Ampacities of Insulated Single Copper Conductor Isolated in Air Based on Conductor Temperatures of 90°C (194°F) and 105°C (221°F) and Ambient Air Temperature of 40°C (104°F)

Conductor Size (AWG or kcmil)	Temperature Rating of Conductor (See Table 310-61)					
	2001–5000 Volts Ampacity		5001–15,000 Volts Ampacity		15,001–35,000 Volts Ampacity	
	90°C (194°F) Type MV-90	105°C (221°F) Type MV-105	90°C (194°F) Type MV-90	105°C (221°F) Type MV-105	90°C (194°F) Type MV-90	105°C (221°F) Type MV-105
8	83	93	—	—	—	—
6	110	120	110	125	—	—
4	145	160	150	165	—	—
2	190	215	195	215	—	—
1	225	250	225	250	225	250
1/0	260	290	260	290	260	290
2/0	300	330	300	335	300	330
3/0	345	385	345	385	345	380
4/0	400	445	400	445	395	445
250	445	495	445	495	440	490
350	550	615	550	610	545	605
500	695	775	685	765	680	755
750	900	1000	885	990	870	970
1000	1075	1200	1060	1185	1040	1160
1250	1230	1370	1210	1350	1185	1320
1500	1365	1525	1345	1500	1315	1465
1750	1495	1665	1470	1640	1430	1595
2000	1605	1790	1575	1755	1535	1710

Table 310-70. Ampacities of Insulated Single Aluminum Conductor Isolated in Air Based on Conductor Temperatures of 90°C (194°F) and 105°C (221°F) and Ambient Air Temperature of 40°C (104°F)

Conductor Size (AWG or kcmil)	Temperature Rating of Conductor (See Table 310-61)					
	2001–5000 Volts Ampacity		5001–15,000 Volts Ampacity		15,001–35,000 Volts Ampacity	
	90°C (194°F) Type MV-90	105°C (221°F) Type MV-105	90°C (194°F) Type MV-90	105°C (221°F) Type MV-105	90°C (194°F) Type MV-90	105°C (221°F) Type MV-105
8	64	71	—	—	—	—
6	85	95	87	97	—	—
4	115	125	115	130	—	—
2	150	165	150	170	—	—
1	175	195	175	195	175	195
1/0	200	225	200	225	200	225
2/0	230	260	235	260	230	260
3/0	270	300	270	300	270	300
4/0	310	350	310	350	310	345
250	345	385	345	385	345	380
350	430	480	430	480	430	475
500	545	605	535	600	530	590
750	710	790	700	780	685	765
1000	855	950	840	940	825	920
1250	980	1095	970	1080	950	1055
1500	1105	1230	1085	1215	1060	1180
1750	1215	1355	1195	1335	1165	1300
2000	1320	1475	1295	1445	1265	1410

Table 310-71. Ampacities of an Insulated Three-Conductor Copper Cable Isolated in Air Based on Conductor Temperatures of 90°C (194°F) and 105°C (221°F) and Ambient Air Temperature of 40°C (104°F)

	Temperature Rating of Conductor (See Table 310-61)			
	2001–5000 Volts Ampacity		5001–35,000 Volts Ampacity	
Conductor Size (AWG or kcmil)	90°C (194°F) Type MV-90	105°C (221°F) Type MV-105	90°C (194°F) Type MV-90	105°C (221°F) Type MV-105
8	59	66	—	—
6	79	88	93	105
4	105	115	120	135
2	140	154	165	185
1	160	180	185	210
1/0	185	205	215	240
2/0	215	240	245	275
3/0	250	280	285	315
4/0	285	320	325	360
250	320	355	360	400
350	395	440	435	490
500	485	545	535	600
750	615	685	670	745
1000	705	790	770	860

Table 310-72. Ampacities of Insulated Three-Conductor Aluminum Cable Isolated in Air Based on Conductor Temperatures of 90°C (194°F) and 105°C (221°F) and Ambient Air Temperature of 40°C (104°F)

	Temperature Rating of Conductor (See Table 310-61)			
	2001–5000 Volts Ampacity		5001–35,000 Volts Ampacity	
Conductor Size (AWG or kcmil)	90°C (194°F) Type MV-90	105°C (221°F) Type MV-105	90°C (194°F) Type MV-90	105°C (221°F) Type MV-105
8	46	51	—	—
6	61	68	72	80
4	81	90	95	105
2	110	120	125	145
1	125	140	145	165
1/0	145	160	170	185
2/0	170	185	190	215
3/0	195	215	220	245
4/0	225	250	255	285
250	250	280	280	315
350	310	345	345	385
500	385	430	425	475
750	495	550	540	600
1000	585	650	635	705

Table 310-73. Ampacities of an Insulated Triplexed or Three Single-Conductor Copper Cables in Isolated Conduit in Air Based on Conductor Temperatures of 90°C (194°F) and 105°C (221°F) and Ambient Air Temperature of 40°C (104°F)

	Temperature Rating of Conductor (See Table 310-61)			
	2001–5000 Volts Ampacity		5001–35,000 Volts Ampacity	
Conductor Size (AWG or kcmil)	90°C (194°F) Type MV-90	105°C (221°F) Type MV-105	90°C (194°F) Type MV-90	105°C (221°F) Type MV-105
8	55	61	—	—
6	75	84	83	93
4	97	110	110	120
2	130	145	150	165
1	155	175	170	190
1/0	180	200	195	215
2/0	205	225	225	255
3/0	240	270	260	290
4/0	280	305	295	330
250	315	355	330	365
350	385	430	395	440
500	475	530	480	535
750	600	665	585	655
1000	690	770	675	755

Table 310-74. Ampacities of an Insulated Triplexed or Three Single-Conductor Aluminum Cables in Isolated Conduit in Air Based on Conductor Temperatures of 90°C (194°F) and 105°C (221°F) and Ambient Air Temperature of 40°C (104°F)

	Temperature Rating of Conductor (See Table 310-61)			
	2001–5000 Volts Ampacity		5001–35,000 Volts Ampacity	
Conductor Size (AWG or kcmil)	90°C (194°F) Type MV-90	105°C (221°F) Type MV-105	90°C (194°F) Type MV-90	105°C (221°F) Type MV-105
8	43	48	—	—
6	58	65	65	72
4	76	85	84	94
2	100	115	115	130
1	120	135	130	150
1/0	140	155	150	170
2/0	160	175	175	200
3/0	190	210	200	225
4/0	215	240	230	260
250	250	280	255	290
350	305	340	310	350
500	380	425	385	430
750	490	545	485	540
1000	580	645	565	640

Table 310-75. Ampacities of an Insulated Three-Conductor Copper Cable in Isolated Conduit in Air Based on Conductor Temperatures of 90°C (194°F) and 105°C (221°F) and Ambient Air Temperature of 40°C (104°F)

	Temperature Rating of Conductor (See Table 310-61)			
	2001–5000 Volts Ampacity		5001–35,000 Volts Ampacity	
Conductor Size (AWG or kcmil)	90°C (194°F) Type MV-90	105°C (221°F) Type MV-105	90°C (194°F) Type MV-90	105°C (221°F) Type MV-105
8	52	58	—	—
6	69	77	83	92
4	91	100	105	120
2	125	135	145	165
1	140	155	165	185
1/0	165	185	195	215
2/0	190	210	220	245
3/0	220	245	250	280
4/0	255	285	290	320
250	280	315	315	350
350	350	390	385	430
500	425	475	470	525
750	525	585	570	635
1000	590	660	650	725

Table 310-76. Ampacities of an Insulated Three-Conductor Aluminum Cable in Isolated Conduit in Air Based on Conductor Temperatures of 90°C (194°F) and 105°C (221°F) and Ambient Air Temperature of 40°C (104°F)

	Temperature Rating of Conductor (See Table 310-61)			
	2001–5000 Volts Ampacity		5001–35,000 Volts Ampacity	
Conductor Size (AWG or kcmil)	90°C (194°F) Type MV-90	105°C (221°F) Type MV-105	90°C (194°F) Type MV-90	105°C (221°F) Type MV-105
8	41	46	—	—
6	53	59	64	71
4	71	79	84	94
2	96	105	115	125
1	110	125	130	145
1/0	130	145	150	170
2/0	150	165	170	190
3/0	170	190	195	220
4/0	200	225	225	255
250	220	245	250	280
350	275	305	305	340
500	340	380	380	425
750	430	480	470	520
1000	505	560	550	615

Table 310-77. Ampacities of Three Single-Insulated Copper Conductors in Underground Electrical Ducts (Three Conductors per Electrical Duct) Based on Ambient Earth Temperature of 20°C (68°F), Electrical Duct Arrangement per Figure 310-60, 100 Percent Load Factor, Thermal Resistance (RHO) of 90, Conductor Temperatures of 90°C (194°F) and 105°C (221°F)

	Temperature Rating of Conductor (See Table 310-61)			
	2001–5000 Volts Ampacity		5001–35,000 Volts Ampacity	
Conductor Size (AWG or kcmil)	90°C (194°F) Type MV-90	105°C (221°F) Type MV-105	90°C (194°F) Type MV-90	105°C (221°F) Type MV-105
One Circuit (See Figure 310-60, Detail 1)				
8	64	69	—	—
6	85	92	90	97
4	110	120	115	125
2	145	155	155	165
1	170	180	175	185
1/0	195	210	200	215
2/0	220	235	230	245
3/0	250	270	260	275
4/0	290	310	295	315
250	320	345	325	345
350	385	415	390	415
500	470	505	465	500
750	585	630	565	610
1000	670	720	640	690
Three Circuits (See Figure 310-60, Detail 2)				
8	56	60	—	—
6	73	79	77	83
4	95	100	99	105
2	125	130	130	135
1	140	150	145	155
1/0	160	175	165	175
2/0	185	195	185	200
3/0	210	225	210	225
4/0	235	255	240	255
250	260	280	260	280
350	315	335	310	330
500	375	405	370	395
750	460	495	440	475
1000	525	565	495	535
Six Circuits (See Figure 310-60, Detail 3)				
8	48	52	—	—
6	62	67	64	68
4	80	86	82	88
2	105	110	105	115
1	115	125	120	125
1/0	135	145	135	145
2/0	150	160	150	165
3/0	170	185	170	185
4/0	195	210	190	205
250	210	225	210	225
350	250	270	245	265
500	300	325	290	310
750	365	395	350	375
1000	410	445	390	415

Table 310-78. Ampacities of Three Single-Insulated Aluminum Conductors in Underground Electrical Ducts (Three Conductors per Electrical Duct) Based on Ambient Earth Temperature of 20°C (68°F), Electrical Duct Arrangement per Figure 310-60, 100 Percent Load Factor, Thermal Resistance (RHO) of 90, Conductor Temperatures of 90°C (194°F) and 105°C (221°F)

	Temperature Rating of Conductor (See Table 310-61)			
	2001–5000 Volts Ampacity		5001–35,000 Volts Ampacity	
Conductor Size (AWG or kcmil)	90°C (194°F) Type MV-90	105°C (221°F) Type MV-105	90°C (194°F) Type MV-90	105°C (221°F) Type MV-105
One Circuit (See Figure 310-60, Detail 1)				
8	50	54	—	—
6	66	71	70	75
4	86	93	91	98
2	115	125	120	130
1	130	140	135	145
1/0	150	160	155	165
2/0	170	185	175	190
3/0	195	210	200	215
4/0	225	245	230	245
250	250	270	250	270
350	305	325	305	330
500	370	400	370	400
750	470	505	455	490
1000	545	590	525	565
Three Circuits (See Figure 310-60, Detail 2)				
8	44	47	—	—
6	57	61	60	65
4	74	80	77	83
2	96	105	100	105
1	110	120	110	120
1/0	125	135	125	140
2/0	145	155	145	155
3/0	160	175	165	175
4/0	185	200	185	200
250	205	220	200	220
350	245	265	245	260
500	295	320	290	315
750	370	395	355	385
1000	425	460	405	440
Six Circuits (See Figure 310-60, Detail 3)				
8	38	41	—	—
6	48	52	50	54
4	62	67	64	69
2	80	86	80	88
1	91	98	90	99
1/0	105	110	105	110
2/0	115	125	115	125
3/0	135	145	130	145
4/0	150	165	150	160
250	165	180	165	175
350	195	210	195	210
500	240	255	230	250
750	290	315	280	305
1000	335	360	320	345

Table 310-79. Ampacities of Three Insulated Copper Conductors Cabled within an Overall Covering (Three-Conductor Cable) in Underground Electrical Ducts (One Cable per Electrical Duct) Based on Ambient Earth Temperature of 20°C (68°F), Electrical Duct Arrangement per Figure 310-60, 100 Percent Load Factor, Thermal Resistance (RHO) of 90, Conductor Temperatures of 90°C (194°F) and 105°C (221°C)

	Temperature Rating of Conductor (See Table 310-61)			
	2001–5000 Volts Ampacity		5001–35,000 Volts Ampacity	
Conductor Size (AWG or kcmil)	90°C (194°F) Type MV-90	105°C (221°F) Type MV-105	90°C (194°F) Type MV-90	105°C (221°F) Type MV-105
One Circuit (See Figure 310-60, Detail 1)				
8	59	64	—	—
6	78	84	88	95
4	100	110	115	125
2	135	145	150	160
1	155	165	170	185
1/0	175	190	195	210
2/0	200	220	220	235
3/0	230	250	250	270
4/0	265	285	285	305
250	290	315	310	335
350	355	380	375	400
500	430	460	450	485
750	530	570	545	585
1000	600	645	615	660
Three Circuits (See Figure 310-60, Detail 2)				
8	53	57	—	—
6	69	74	75	81
4	89	96	97	105
2	115	125	125	135
1	135	145	140	155
1/0	150	165	160	175
2/0	170	185	185	195
3/0	195	210	205	220
4/0	225	240	230	250
250	245	265	255	270
350	295	315	305	325
500	355	380	360	385
750	430	465	430	465
1000	485	520	485	515
Six Circuits (See Figure 310-60, Detail 3)				
8	46	50	—	—
6	60	65	63	68
4	77	83	81	87
2	98	105	105	110
1	110	120	115	125
1/0	125	135	130	145
2/0	145	155	150	160
3/0	165	175	170	180
4/0	185	200	190	200
250	200	220	205	220
350	240	270	245	275
500	290	310	290	305
750	350	375	340	365
1000	390	420	380	405

Table 310-80. Ampacities of Three Insulated Aluminum Conductors Cabled within an Overall Covering (Three-Conductor Cable) in Underground Electrical Ducts (One Cable per Electrical Duct) Based on Ambient Earth Temperature of 20°C (68°F), Electrical Duct Arrangement per Figure 310-60, 100 Percent Load Factor, Thermal Resistance (RHO) of 90, Conductor Temperatures of 90°C (194°F) and 105°C (221°C)

	Temperature Rating of Conductor (See Table 310-61)			
	2001–5000 Volts Ampacity		5001–35,000 Volts Ampacity	
Conductor Size (AWG or kcmil)	90°C (194°F) Type MV-90	105°C (221°F) Type MV-105	90°C (194°F) Type MV-90	105°C (221°F) Type MV-105
One Circuit (See Figure 310-60, Detail 1)				
8	46	50	—	—
6	61	66	69	74
4	80	86	89	96
2	105	110	115	125
1	120	130	135	145
1/0	140	150	150	165
2/0	160	170	170	185
3/0	180	195	195	210
4/0	205	220	220	240
250	230	245	245	265
350	280	310	295	315
500	340	365	355	385
750	425	460	440	475
1000	495	535	510	545
Three Circuits (See Figure 310-60, Detail 2)				
8	41	44	—	—
6	54	58	59	64
4	70	75	75	81
2	90	97	100	105
1	105	110	110	120
1/0	120	125	125	135
2/0	135	145	140	155
3/0	155	165	160	175
4/0	175	185	180	195
250	190	205	200	215
350	230	250	240	255
500	280	300	285	305
750	345	375	350	375
1000	400	430	400	430
Six Circuits (See Figure 310-60, Detail 3)				
8	36	39	—	—
6	46	50	49	53
4	60	65	63	68
2	77	83	80	86
1	87	94	90	98
1/0	99	105	105	110
2/0	110	120	115	125
3/0	130	140	130	140
4/0	145	155	150	160
250	160	170	160	170
350	190	205	190	205
500	230	245	230	245
750	280	305	275	295
1000	320	345	315	335

Table 310-81. Ampacities of Single Insulated Copper Conductors Directly Buried in Earth Based on Ambient Earth Temperature of 20°C (68°F), Arrangement per Figure 310-60, 100 Percent Load Factor, Thermal Resistance (RHO) of 90, Conductor Temperatures of 90°C (194°F) and 105°C (221°C)

	Temperature Rating of Conductor (See Table 310-61)			
	2001–5000 Volts Ampacity		5001–35,000 Volts Ampacity	
Conductor Size (AWG or kcmil)	90°C (194°F) Type MV-90	105°C (221°F) Type MV-105	90°C (194°F) Type MV-90	105°C (221°F) Type MV-105
One Circuit, Three Conductors (See Figure 310-60, Detail 9)				
8	110	115	—	—
6	140	150	130	140
4	180	195	170	180
2	230	250	210	225
1	260	280	240	260
1/0	295	320	275	295
2/0	335	365	310	335
3/0	385	415	355	380
4/0	435	465	405	435
250	470	510	440	475
350	570	615	535	575
500	690	745	650	700
750	845	910	805	865
1000	980	1055	930	1005
Two Circuits, Six Conductors (See Figure 310-60, Detail 10)				
8	100	110	—	—
6	130	140	120	130
4	165	180	160	170
2	215	230	195	210
1	240	260	225	240
1/0	275	295	255	275
2/0	310	335	290	315
3/0	355	380	330	355
4/0	400	430	375	405
250	435	470	410	440
350	520	560	495	530
500	630	680	600	645
750	775	835	740	795
1000	890	960	855	920

Table 310-82. Ampacities of Single Insulated Aluminum Conductors Directly Buried in Earth Based on Ambient Earth Temperature of 20°C (68°F), Arrangement per Figure 310-60, 100 Percent Load Factor, Thermal Resistance (RHO) of 90, Conductor Temperatures of 90°C (194°F) and 105°C (221°F)

	Temperature Rating of Conductor (See Table 310-61)			
	2001–5000 Volts Ampacity		5001–35,000 Volts Ampacity	
Conductor Size (AWG or kcmil)	90°C (194°F) Type MV-90	105°C (221°F) Type MV-105	90°C (194°F) Type MV-90	105°C (221°F) Type MV-105
One Circuit, Three Conductors (See Figure 310-60, Detail 9)				
8	85	90	—	—
6	110	115	100	110
4	140	150	130	140
2	180	195	165	175
1	205	220	185	200
1/0	230	250	215	230
2/0	265	285	245	260
3/0	300	320	275	295
4/0	340	365	315	340
250	370	395	345	370
350	445	480	415	450
500	540	580	510	545
750	665	720	635	680
1000	780	840	740	795
Two Circuits, Six Conductors (See Figure 310-60, Detail 10)				
8	80	85	—	—
6	100	110	95	100
4	130	140	125	130
2	165	180	155	165
1	190	200	175	190
1/0	215	230	200	215
2/0	245	260	225	245
3/0	275	295	255	275
4/0	310	335	290	315
250	340	365	320	345
350	410	440	385	415
500	495	530	470	505
750	610	655	580	625
1000	710	765	680	730

Table 310-83. Ampacities of Three Insulated Copper Conductors Cabled within an Overall Covering (Three-Conductor Cable), Directly Buried in Earth Based on Ambient Earth Temperature of 20°C (68°F), Arrangement per Figure 310-60, 100 Percent Load Factor, Thermal Resistance (RHO) of 90, Conductor Temperatures of 90°C (194°F) and 105°C (221°F)

	Temperature Rating of Conductor (See Table 310-61)			
	2001–5000 Volts Ampacity		5001–35,000 Volts Ampacity	
Conductor Size (AWG or kcmil)	90°C (194°F) Type MV-90	105°C (221°F) Type MV-105	90°C (194°F) Type MV-90	105°C (221°F) Type MV-105
One Circuit (See Figure 310-60, Detail 5)				
8	85	89	—	—
6	105	115	115	120
4	135	150	145	155
2	180	190	185	200
1	200	215	210	225
1/0	230	245	240	255
2/0	260	280	270	290
3/0	295	320	305	330
4/0	335	360	350	375
250	365	395	380	410
350	440	475	460	495
500	530	570	550	590
750	650	700	665	720
1000	730	785	750	810
Two Circuits (See Figure 310-60, Detail 10)				
8	80	84	—	—
6	100	105	105	115
4	130	140	135	145
2	165	180	170	185
1	185	200	195	210
1/0	215	230	220	235
2/0	240	260	250	270
3/0	275	295	280	305
4/0	310	335	320	345
250	340	365	350	375
350	410	440	420	450
500	490	525	500	535
750	595	640	605	650
1000	665	715	675	730

Table 310-84. Ampacities of Three Insulated Aluminum Conductors Cabled within an Overall Covering (Three-Conductor Cable), Directly Buried in Earth Based on Ambient Earth Temperature of 20°C (68°F), Arrangement per Figure 310-60, 100 Percent Load Factor, Thermal Resistance (RHO) of 90, Conductor Temperatures of 90°C (194°F) and 105°C (221°F)

	Temperature Rating of Conductor (See Table 310-61)			
	2001–5000 Volts Ampacity		5001–35,000 Volts Ampacity	
Conductor Size (AWG or kcmil)	90°C (194°F) Type MV-90	105°C (221°F) Type MV-105	90°C (194°F) Type MV-90	105°C (221°F) Type MV-105
One Circuit (See Figure 310-60, Detail 5)				
8	65	70	—	—
6	80	88	90	95
4	105	115	115	125
2	140	150	145	155
1	155	170	165	175
1/0	180	190	185	200
2/0	205	220	210	225
3/0	230	250	240	260
4/0	260	280	270	295
250	285	310	300	320
350	345	375	360	390
500	420	450	435	470
750	520	560	540	580
1000	600	650	620	665
Two Circuits (See Figure 310-60, Detail 6)				
8	60	66	—	—
6	75	83	80	95
4	100	110	105	115
2	130	140	135	145
1	145	155	150	165
1/0	165	180	170	185
2/0	190	205	195	210
3/0	215	230	220	240
4/0	245	260	250	270
250	265	285	275	295
350	320	345	330	355
500	385	415	395	425
750	480	515	485	525
1000	550	590	560	600

Table 310-85. Ampacities of Three Triplexed Single Insulated Copper Conductors Directly Buried in Earth Based on Ambient Earth Temperature of 20°C (68°F), Arrangement per Figure 310-60, 100 Percent Load Factor, Thermal Resistance (RHO) of 90, Conductor Temperatures 90°C (194°F) and 105°C (221°F)

	Temperature Rating of Conductor (See Table 310-61)			
	2001–5000 Volts Ampacity		5001–35,000 Volts Ampacity	
Conductor Size (AWG or kcmil)	90°C (194°F) Type MV-90	105°C (221°F) Type MV-105	90°C (194°F) Type MV-90	105°C (221°F) Type MV-105
One Circuit, Three Conductors (See Figure 310-60, Detail 7)				
8	90	95	—	—
6	120	130	115	120
4	150	165	150	160
2	195	205	190	205
1	225	240	215	230
1/0	255	270	245	260
2/0	290	310	275	295
3/0	330	360	315	340
4/0	375	405	360	385
250	410	445	390	410
350	490	580	470	505
500	590	635	565	605
750	725	780	685	740
1000	825	885	770	830
Two Circuits, Six Conductors (See Figure 310-60, Detail 8)				
8	85	90	—	—
6	110	115	105	115
4	140	150	140	150
2	180	195	175	190
1	205	220	200	215
1/0	235	250	225	240
2/0	265	285	255	275
3/0	300	320	290	315
4/0	340	365	325	350
250	370	395	355	380
350	445	480	425	455
500	535	575	510	545
750	650	700	615	660
1000	740	795	690	745

Table 310-86. Ampacities of Three Triplexed Single Insulated Aluminum Conductors Directly Buried in Earth Based on Ambient Earth Temperature of 20°C (68°F), Arrangement per Figure 310-60, 100 Percent Load Factor, Thermal Resistance (RHO) of 90, Conductor Temperatures 90°C (194°F) and 105°C (221°F)

	Temperature Rating of Conductor (See Table 310-61)			
	2001–5000 Volts Ampacity		5001–35,000 Volts Ampacity	
Conductor Size (AWG or kcmil)	90°C (194°F) Type MV-90	105°C (221°F) Type MV-105	90°C (194°F) Type MV-90	105°C (221°F) Type MV-105
One Circuit, Three Conductors (See Figure 310-60, Detail 7)				
8	70	75	—	—
6	90	100	90	95
4	120	130	115	125
2	155	165	145	155
1	175	190	165	175
1/0	200	210	190	205
2/0	225	240	215	230
3/0	255	275	245	265
4/0	290	310	280	305
250	320	350	305	325
350	385	420	370	400
500	465	500	445	480
750	580	625	550	590
1000	670	725	635	680
Two Circuits, Six Conductors (See Figure 310-60, Detail 8)				
8	65	70	—	—
6	85	95	85	90
4	110	120	105	115
2	140	150	135	145
1	160	170	155	170
1/0	180	195	175	190
2/0	205	220	200	215
3/0	235	250	225	245
4/0	265	285	255	275
250	290	310	280	300
350	350	375	335	360
500	420	455	405	435
750	520	560	485	525
1000	600	645	565	605

Article 318 — Cable Trays

Contents

(d) Solid Bottom Cable Tray — Multiconductor Control and/or Signal Cables Only
(e) Ventilated Channel Cable Trays
318-10. Number of Single Conductor Cables, Rated 2000 Volts or Less, in Cable Trays
(a) Ladder or Ventilated Trough Cable Trays
(b) Ventilated Channel Cable Trays
318-11. Ampacity of Cables, Rated 2000 Volts or Less, in Cable Trays
(a) Multiconductor Cables
(b) Single Conductor Cables
318-12. Number of Type MV and Type MC Cables (2001 Volts or Over) in Cable Trays
318-13. Ampacity of Type MV and Type MC Cables (2001 Volts or Over) in Cable Trays
(a) Multiconductor Cables (2001 Volts or Over)
(b) Single Conductor Cables (2001 Volts or Over)

318-1. Scope. This article covers cable tray systems, including ladder, ventilated trough, ventilated channel, solid bottom, and other similar structures.

Cable trays are mechanical support systems. Cable trays are not raceways. See definition of *raceway* in Article 100.

The requirements of Article 318, Cable Trays, cover four types of cable trays: ladder, ventilated trough, ventilated channel, and solid bottom.

318-2. Definition.

Cable Tray System. A unit or assembly of units or sections and associated fittings forming a structural system used to securely fasten or support cables and raceways.

318-3. Uses Permitted. Cable tray installations shall not be limited to industrial establishments.

Cable tray installations are typically an industrial-type wiring method. However, cable tray installations have never been restricted to just industrial installations. Cable tray installations are being applied more in commercial installations than ever before, especially as a wire and cable management system for tel/data installations.

(a) Wiring Methods. The following shall be permitted to be installed in cable tray systems under the conditions described in their respective articles and sections:

	Section	Article
Armored cable		333
Electrical metallic tubing		348
Electrical nonmetallic tubing		331
Fire alarm cables		760
Flexible metal conduit		350
Flexible metallic tubing		349
Instrumentation tray cable		727
Intermediate metal conduit		345
Liquidtight flexible metal conduit and liquidtight flexible nonmetallic conduit		351
Metal-clad cable		334
Mineral-insulated, metal-sheathed cable		330
Multiconductor service-entrance cable		338
Multiconductor underground feeder and branch-circuit cable		339
Multipurpose and communications cables		800
Nonmetallic-sheathed cable		336
Power and control tray cable		340
Power-limited tray cable	725-61(c) and 725-71(e)	
Optical fiber cables		770
Other factory-assembled, multiconductor control, signal, or power cables that are specifically approved for installation in cable trays		
Rigid metal conduit		346
Rigid nonmetallic conduit		347

Section 318-3(a) identifies the raceways and many of the cable types that may be supported in commercial and industrial cable tray installations. Cable tray is rarely used as a major raceway support system. For raceway support systems, the versatility of strut systems exceeds that of cable tray support systems.

Section 318-3(a) does not identify *all* the specific cable types that may be installed in commercial and industrial cable tray systems. According to Section 318-3(a)(11), other factory-assembled, multiconductor control, signal, or power cables that are specifically approved for installation in cable trays are permitted as well.

(b) In Industrial Establishments. The wiring methods in Section 318-3(a) shall be permitted to be used in any industrial establishment under the conditions described in their respective articles. In industrial establishments only, where conditions of maintenance and supervision ensure that only qualified persons will service the installed cable tray system, any of the cables in (1) and (2) shall be permitted to be installed in ladder, ventilated trough, or ventilated channel cable trays.

In the 1996 *Code*, this section was revised to positively state that all of the cable types in Section 318-3(a) are permitted to be installed in cable trays in industrial facilities.

Section 318-3(b) also permits single-conductor cables (rated 0 to 2000 volts) and Type MV cables to be installed in ladder, ventilated-trough, or ventilated-channel cable trays, provided the installation is located in a qualifying industrial facility. Single conductors and Type MV cable installations are not permitted in solid-bottom cable trays.

(1) Single Conductors. Single conductor cables shall be permitted to be installed in accordance with the following.

(a) Single conductor cable shall be No. 1/0 or larger and shall be of a type listed and marked on the surface for use in cable trays. Where Nos. 1/0 through 4/0 single conductor cables are installed in ladder cable tray, the maximum allowable rung spacing for the ladder cable tray shall be 9 in. (229 mm). Where exposed to direct rays of the sun, cables shall be identified as being sunlight resistant.

(b) Welding cables shall comply with the provisions of Article 630, Part D.

Cable trays used to support welding cables are required to be dedicated for welding cable installation. See Section 630-42 for installation details.

(c) Single conductors used as equipment grounding conductors shall be insulated, covered, or bare and they shall be No. 4 or larger.

(2) Multiconductor. Multiconductor cables, Type MV (Article 326) where exposed to direct rays of the sun, shall be identified as being sunlight resistant.

(c) Equipment Grounding Conductors. Metallic cable trays shall be permitted to be used as equipment grounding conductors where continuous maintenance and supervision ensure that qualified persons will service the installed cable tray system and the cable tray complies with provisions of Section 318-7.

Section 318-3(c) expanded the optional use of the cable tray as the equipment grounding conductor. No longer is this practice limited just to qualifying industrial installations. In order for the cable tray system to qualify as an equipment grounding conductor, it must meet all four requirements of Section 318-7(b).

(d) Hazardous (Classified) Locations. Cable trays in hazardous (classified) locations shall contain only the cable types permitted in Sections 501-4, 502-4, 503-3, and 504-20.

(e) Nonmetallic Cable Tray. Nonmetallic cable tray shall be permitted in corrosive areas and in areas requiring voltage isolation.

Fiberglass cable trays are often used to support cables in corrosive environments or in electrolytic cell rooms where voltage isolation is required. See Article 668, Electrolytic Cells.

318-4. Uses Not Permitted. Cable tray systems shall not be used in hoistways or where subject to severe physical damage. Cable tray systems shall not be used in environmental airspaces, except as permitted in Section 300-22, to support wiring methods recognized for use in such spaces.

Section 300-22(c) specifically limits the types of wiring methods that may be used within other spaces used for environmental air. Metallic cable trays may be used within these spaces to support only the recognized wiring methods permitted in these spaces. The cable tray types may be ladder, ventilated trough, ventilated channel, or solid bottom. Metal cable trays are not the limiting factor; rather, the cable or wiring method is the limiting factor.

318-5. Construction Specifications.

(a) Strength and Rigidity. Cable trays shall have suitable strength and rigidity to provide adequate support for all contained wiring.

(b) Smooth Edges. Cable trays shall not have sharp edges, burrs, or projections that may damage the insulation or jackets of the wiring.

(c) Corrosion Protection. Cable tray systems shall be corrosion resistant. If made of ferrous material, the system shall be protected from corrosion as required by Section 300-6.

(d) Side Rails. Cable trays shall have side rails or equivalent structural members.

(e) Fittings. Cable trays shall include fittings or other suitable means for changes in direction and elevation of runs.

(f) Nonmetallic Cable Tray. Nonmetallic cable trays shall be made of flame-retardant material.

318-6. Installation.

(a) Complete System. Cable trays shall be installed as a complete system. Field bends or modifications shall be made so that the electrical continuity of the cable tray system and support for the cables shall be maintained. Cable tray systems shall be permitted to have mechanically discontinuous segments between cable tray runs or between cable tray runs and equipment. The system shall provide for the support of the cables in accordance with their corresponding articles.

Where cable trays support individual conductors and where the conductors pass from one cable tray to another, or from a cable tray to raceways or to equipment where the conductors are terminated, the support distance between cable trays or between the cable tray and the equipment shall not exceed 6 ft (1.83 m). The conductors shall be secured to the cable tray(s) at the transition and they shall be protected, by guarding or by location, from physical damage.

A bonding jumper sized in accordance with Section 250-102 shall connect the two sections of cable tray, or the cable tray and the raceway or equipment. Bonding shall be in accordance with Section 250-96.

Runs of cable tray are not required to be totally mechanically continuous from the equipment source to the equipment termination. Breaks in the mechanical continuity of cable tray systems are permitted and often occur at tees, crossovers, elevation changes, fire stops, or for thermal contraction and expansion. Also, cable tray systems are not required to be mechanically connected to the equipment they serve.

Revised in the 1999 *Code*, a distance limit of 6 ft has been added concerning mechanically discontinuous cable tray segments for individual conductors. However, this distance limit does not apply to trays containing multiconductor cables.

Most important, especially for discontinuous cable tray segments, is the bonding of the entire cable tray system. According to Section 250-96, properly sized and installed bonding conductors are required to be installed across any mechanical discontinuities in the cable tray system and across any space between the cable tray and the conductor termination equipment enclosure or its equipment ground bus.

Of course, the cables installed within cable tray systems must always be supported to the minimum requirements of the applicable article. This requirement either limits the gap distance in cable tray runs and between the cable tray and the equipment enclosures or it requires intermediate cable supports at the appropriate distances in place of the cable tray.

(b) Completed Before Installation. Each run of cable tray shall be completed before the installation of cables.

(c) Supports. Supports shall be provided to prevent stress on cables where they enter raceways or other enclosures from cable tray systems.

(d) Covers. In portions of runs where additional protection is required, covers or enclosures providing the required protection shall be of a material that is compatible with the cable tray.

(e) Multiconductor Cables Rated 600 Volts or Less. Multiconductor cables rated 600 volts or less shall be permitted to be installed in the same cable tray.

(f) Cables Rated Over 600 Volts. Cables rated over 600 volts and those rated 600 volts or less installed in the same cable tray shall comply with either (1) or (2).

(1) The cables rated over 600 volts are Type MC.

(2) The cables rated over 600 volts are separated from the cables rated 600 volts or less by a solid fixed barrier of a material compatible with the cable tray.

(g) Through Partitions and Walls. Cable trays shall be permitted to extend transversely through partitions and walls or vertically through platforms and floors in wet or dry locations where the installations, complete with installed cables, are made in accordance with the requirements of Section 300-21.

Figure 318.1 shows an example where fire wall penetration seals are used to maintain the integrity of fire-rated walls, floors, or ceilings, as required by Section 300-21.

Figure 318.1 *Fire wall penetration seals used to maintain the integrity of fire-rated walls, floors, partitions, or ceilings, as required by Section 300-21. (O.Z./Gedney Co.)*

(h) Exposed and Accessible. Cable trays shall be exposed and accessible except as permitted by Section 318-6(g).

(i) Adequate Access. Sufficient space shall be provided and maintained about cable trays to permit adequate access for installing and maintaining the cables.

(j) Raceways, Cables, and Outlet Boxes Supported from Cable Trays. In industrial facilities where conditions of maintenance and supervision ensure only qualified persons will service the installation and where cable trays are designed to support the load, raceways, cables, and outlet boxes shall be permitted to be supported from cable trays. For raceway terminating at the tray, a listed cable tray clamp or adapter shall be used and no nearby support, such as a support within 3 ft (914 mm), shall be required.

For raceway or cable running parallel to, but under or

beside, a tray, support shall be in accordance with the requirements of the appropriate raceway or cable article.

For outlet boxes located under or beside a tray, support shall be in accordance with the requirements of Article 370.

Changed in the 1999 *Code,* Section 318-6(j) permits conduit and cable termination supports as well as outlet boxes supported solely by the cable tray in qualifying industrial facilities only. These items are not permitted to be supported solely by the cable tray in commercial installations.

For commercial installations (and nonqualifying industrial facilities), conduits must be supported within 3 ft of the cable tray or within 5 ft if structural members do not permit fastening within 3 ft of the cable tray. Cables connecting to equipment outside the cable tray system must be supported according to their respective article. For example, Type MC cable in the larger sizes is required to be supported outside a cable tray system at intervals not exceeding 6 ft according to Section 334-10.

318-7. Grounding.

(a) Metallic Cable Trays. Metallic cable trays that support electrical conductors shall be grounded as required for conductor enclosures in Article 250.

Section 318-7(a), together with Section 250-96, requires all cable tray systems that support electrical conductors (whether mechanically continuous or with isolated segments) to be electrically continuous and effectively bonded and grounded.

This requirement applies whether or not the cable tray is used as an equipment grounding conductor.

(b) Steel or Aluminum Cable Tray Systems. Steel or aluminum cable tray systems shall be permitted to be used as equipment grounding conductors provided that all the following requirements are met.

(1) The cable tray sections and fittings shall be identified for grounding purposes.

(2) The minimum cross-sectional area of cable trays shall conform to the requirements in Table 318-7(b)(2).

(3) All cable tray sections and fittings shall be legibly and durably marked to show the cross-sectional area of metal in channel cable trays, or cable trays of one-piece construction, and the total cross-sectional area of both side rails for ladder or trough cable trays.

(4) Cable tray sections, fittings, and connected raceways shall be bonded in accordance with Section 250-96 using bolted mechanical connectors or bonding jumpers sized and installed in accordance with Section 250-102.

Table 318-7(b)(2). Metal Area Requirements for Cable Trays Used as Equipment Grounding Conductors

Maximum Fuse Ampere Rating, Circuit Breaker Ampere Trip Setting, or Circuit Breaker Protective Relay Ampere Trip Setting for Ground-Fault Protection of Any Cable Circuit in the Cable Tray System	Minimum Cross-Sectional Area of Metal[a] (in.^{2})	
	Steel Cable Trays	Aluminum Cable Trays
60	0.20	0.20
100	0.40	0.20
200	0.70	0.20
400	1.00	0.40
600	1.50[b]	0.40
1000	—	0.60
1200	—	1.00
1600	—	1.50
2000	—	2.00[b]

Note: For SI units, 1 in.^{2} = 645 sq mm^2.

[a]Total cross-sectional area of both side rails for ladder or trough cable trays; or the minimum cross-sectional area of metal in channel cable trays or cable trays of one-piece construction.

[b]Steel cable trays shall not be used as equipment grounding conductors for circuits with ground-fault protection above 600 amperes. Aluminum cable trays shall not be used as equipment grounding conductors for circuits with ground-fault protection above 2000 amperes.

For cable tray systems in commercial occupancies, designers are afforded the option to specify multiconductor cables without equipment grounding conductors (EGCs) and use the cable tray system as the required equipment grounding conductor, provided the cable tray system meets the requirements of Sections 318-7(a) and 318-7(b). Figure 318.2 shows an example of the grounding and bonding of multiconductor cables in cable trays with conduit runs to power equipment.

For cable tray systems in industrial establishments that qualify, the designer is also afforded the option to specify single-conductor cables without a cable equipment grounding conductor and to use the cable tray as the required equipment grounding conductor. Again, this option is available only if the cable tray system meets the requirements of Sections 318-7(a) and 318-7(b).

318-8. Cable Installation.

(a) Cable Splices. Cable splices made and insulated by approved methods shall be permitted to be located within a cable tray provided they are accessible and do not project above the side rails.

(b) Fastened Securely. In other than horizontal runs, the cables shall be fastened securely to transverse members of the cable trays.

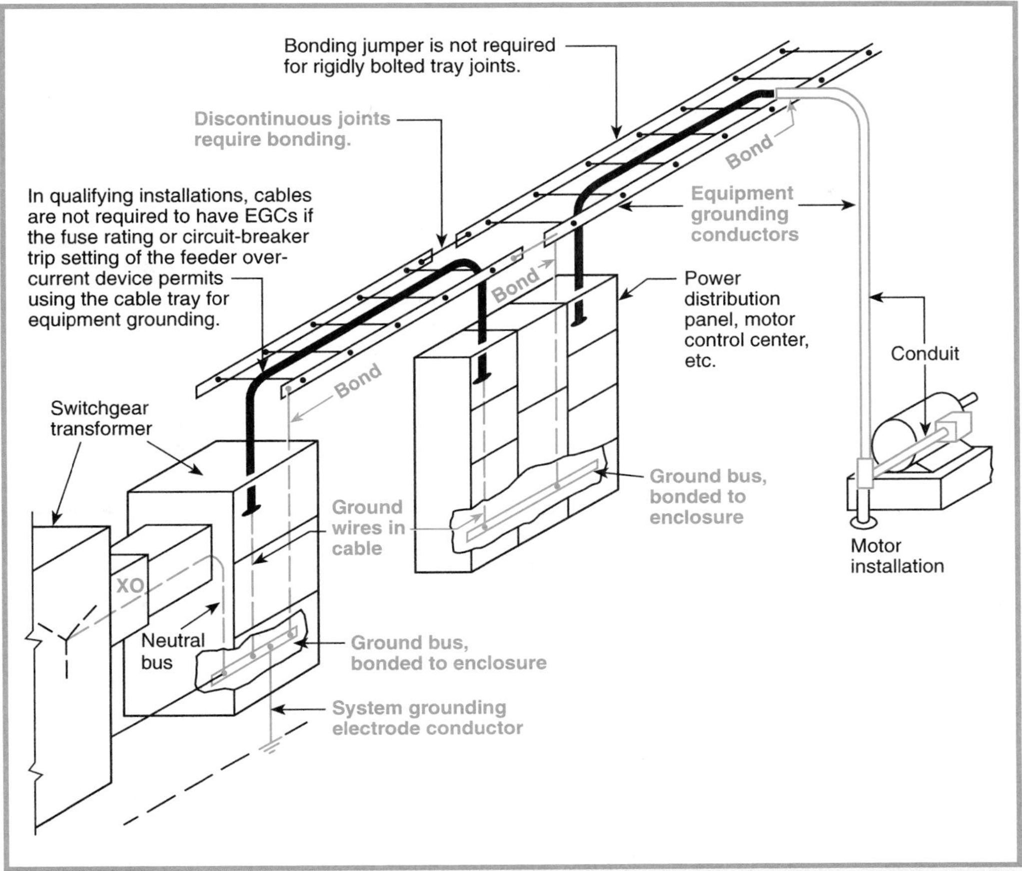

Figure 318.2 *An example of multiconductor cables in cable trays with conduit runs to power equipment where bonding is provided in accordance with Section 318-7(b)(4). (Cable Tray Institute)*

Other fastening requirements are found in Sections 318-6(a) and (c).

(c) Bushed Conduit and Tubing. A box shall not be required where cables or conductors are installed in bushed conduit and tubing used for support or for protection against physical damage.

(d) Connected in Parallel. Where single conductor cables comprising each phase or neutral of a circuit are connected in parallel as permitted in Section 310-4, the conductors shall be installed in groups consisting of not more than one conductor per phase or neutral to prevent current unbalance in the paralleled conductors due to inductive reactance.

Single conductors shall be securely bound in circuit groups to prevent excessive movement due to fault-current magnetic forces unless single conductors are cabled together, such as triplexed assemblies.

The binding or otherwise grouping of 3-phase circuits is good engineering practice. If properly done, it results in the phase reactances of the conductors being balanced, which reduces the voltage unbalance between the phases of the 3-phase circuit.

(e) Single Conductors. Where any of the single conductors installed in ladder or ventilated trough cable trays are Nos. 1/0 through 4/0, all single conductors shall be installed in a single layer. Conductors that are bound together to comprise each circuit group shall be permitted to be installed in other than a single layer.

318-9. Number of Multiconductor Cables, Rated 2000 Volts or Less, in Cable Trays. The number of multiconductor cables, rated 2000 volts or less, permitted in a single cable tray shall not exceed the requirements of this section. The conductor sizes herein apply to both aluminum and copper conductors.

(a) Any Mixture of Cables. Where ladder or ventilated trough cable trays contain multiconductor power or lighting cables, or any mixture of multiconductor power, lighting, control, and signal cables, the maximum number of cables shall conform to the following.

(1) Where all of the cables are No. 4/0 or larger, the sum of the diameters of all cables shall not exceed the cable tray width, and the cables shall be installed in a single layer.

(2) Where all of the cables are smaller than No. 4/0, the sum of the cross-sectional areas of all cables shall not exceed the maximum allowable cable fill area in Column 1 of Table 318-9 for the appropriate cable tray width.

(3) Where No. 4/0 or larger cables are installed in the same cable tray with cables smaller than No. 4/0, the sum of the cross-sectional areas of all cables smaller than No. 4/0 shall not exceed the maximum allowable fill area resulting from the computation in Column 2 of Table 318-9 for the appropriate cable tray width. The No. 4/0 and larger cables shall be installed in a single layer, and no other cables shall be placed on them.

Table 318-9. Allowable Cable Fill Area for Multiconductor Cables in Ladder, Ventilated Trough, or Solid Bottom Cable Trays for Cables Rated 2000 Volts or Less

	Maximum Allowable Fill Area for Multiconductor Cables			
	Ladder or Ventilated Trough Cable Trays, Section 318-9(a)		Solid Bottom Cable Trays, Section 318-9(c)	
Inside Width of Cable Tray (in.)	Column 1 Applicable for Section 318-9(a)(2) Only (in.^{2})	Column 2[a] Applicable for Section 318-9(a)(3) Only (in.^{2})	Column 3 Applicable for Section 318-9(c)(2) Only (in.^{2})	Column 4[a] Applicable for Section 318-9(c)(3) Only (in.^{2})
6.0	7.0	7–(1.2 Sd)[b]	5.5	5.5–Sd[b]
9.0	10.5	10.5–(1.2 Sd)	8.0	8.0–Sd
12.0	14.0	14–(1.2 Sd)	11.0	11.0–Sd
18.0	21.0	21–(1.2 Sd)	16.5	16.5–Sd
24.0	28.0	28–(1.2 Sd)	22.0	22.0–Sd
30.0	35.0	35–(1.2 Sd)	27.5	27.5–Sd
36.0	42.0	42–(1.2 Sd)	33.0	33.0–Sd

Note: For SI units, 1 in.^{2} = 645 mm^2.

[a]The maximum allowable fill areas in Columns 2 and 4 shall be computed. For example, the maximum allowable fill, in square inches, for a 6-in. (152-mm) wide cable tray in Column 2 shall be 7 minus (1.2 multiplied by Sd).

[b]The term *Sd* in Columns 2 and 4 is equal to the sum of the diameters, in inches, of all Nos. 4/0 and larger multiconductor cables in the same cable tray with smaller cables.

(b) Multiconductor Control and/or Signal Cables Only. Where a ladder or ventilated trough cable tray, having a usable inside depth of 6 in. (152 mm) or less, contains multiconductor control and/or signal cables only, the sum of the cross-sectional areas of all cables at any cross section shall not exceed 50 percent of the interior cross-sectional area of the cable tray. A depth of 6 in. (152 mm) shall be used to compute the allowable interior cross-sectional area of any cable tray that has a usable inside depth of more than 6 in. (152 mm).

(c) Solid Bottom Cable Trays Containing Any Mixture. Where solid bottom cable trays contain multiconductor power or lighting cables, or any mixture of multiconductor power, lighting, control, and signal cables, the maximum number of cables shall conform to the following.

(1) Where all of the cables are No. 4/0 or larger, the sum of the diameters of all cables shall not exceed 90 percent of the cable tray width, and the cables shall be installed in a single layer.

(2) Where all of the cables are smaller than No. 4/0, the sum of the cross-sectional areas of all cables shall not exceed the maximum allowable cable fill area in Column 3 of Table 318-9 for the appropriate cable tray width.

(3) Where No. 4/0 or larger cables are installed in the same cable tray with cables smaller than No. 4/0, the sum of the cross-sectional areas of all cables smaller than No. 4/0 shall not exceed the maximum allowable fill area resulting from the computation in Column 4 of Table 318-9 for the appropriate cable tray width. The No. 4/0 and larger cables shall be installed in a single layer, and no other cables shall be placed on them.

(d) Solid Bottom Cable Tray — Multiconductor Control and/or Signal Cables Only. Where a solid bottom cable tray, having a usable inside depth of 6 in. (152 mm) or less, contains multiconductor control and/or signal cables only, the sum of the cross-sectional areas of all cables at any cross section shall not exceed 40 percent of the interior cross-sectional area of the cable tray. A depth of 6 in. (152 mm) shall be used to compute the allowable interior cross-sectional area of any cable tray that has a usable inside depth of more than 6 in. (152 mm).

(e) Ventilated Channel Cable Trays. Where ventilated channel cable trays contain multiconductor cables of any type, the following shall apply.

(1) Where only one multiconductor cable is installed, the cross-sectional area shall not exceed the value specified in Column 1 of Table 318-9(e).

(2) Where more than one multiconductor cable is installed, the sum of the cross-sectional area of all cables shall not exceed the value specified in Column 2 of Table 318-9(e).

Table 318-9(e). Allowable Cable Fill Area for Multiconductor Cables in Ventilated Channel Cable Trays for Cables Rated 2000 Volts or Less

	Maximum Allowable Fill Area for Multiconductor Cables (in.^{2})	
Inside Width of Cable Tray (in.)	Column 1 One Cable	Column 2 More than One Cable
3	2.3	1.3
4	4.5	2.5
6	7.0	3.8

Note: For SI units, 1 in.^{2} = 645 mm^2.

Cable trays may be ladder, ventilated trough, ventilated channel, or solid bottom of various widths. The depth of a cable tray is a structural consideration. The depth (up to 6 in.) is related to fill only where the cable tray contains signal and control cables or if the cable tray contains large cable splices. Heat dissipation is generally not a problem for signal and control cables. Inspectors or contractors, therefore, are not expected to compute the various combinations of cable tray fill in the field. Installation handbooks for cable tray applications are available from various cable tray manufacturers.

318-10. Number of Single Conductor Cables, Rated 2000 Volts or Less, in Cable Trays. The number of single conductor cables, rated 2000 volts or less, permitted in a single cable tray section shall not exceed the requirements of this section. The single conductors, or conductor assemblies, shall be evenly distributed across the cable tray. The conductor sizes herein apply to both aluminum and copper conductors.

(a) Ladder or Ventilated Trough Cable Trays. Where ladder or ventilated trough cable trays contain single conductor cables, the maximum number of single conductors shall conform to the following.

(1) Where all of the cables are 1000 kcmil or larger, the sum of the diameters of all single conductor cables shall not exceed the cable tray width.

(2) Where all of the cables are from 250 kcmil up to 1000 kcmil, the sum of the cross-sectional areas of all single conductor cables shall not exceed the maximum allowable cable fill area in Column 1 of Table 318-10, for the appropriate cable tray width.

(3) Where 1000 kcmil or larger single conductor cables are installed in the same cable tray with single conductor cables smaller than 1000 kcmil, the sum of the cross-sectional areas of all cables smaller than 1000 kcmil shall not exceed the maximum allowable fill area resulting from the computation in Column 2 of Table 318-10, for the appropriate cable tray width.

(4) Where any of the single conductor cables are Nos. 1/0 through 4/0, the sum of the diameters of all single conductor cables shall not exceed the cable tray width.

(b) Ventilated Channel Cable Trays. Where 3-in. (76-mm), 4-in. (102-mm), or 6-in. (152-mm) wide ventilated channel cable trays contain single conductor cables, the sum of the diameters of all single conductors shall not exceed the inside width of the channel.

318-11. Ampacity of Cables, Rated 2000 Volts or Less, in Cable Trays.

(a) Multiconductor Cables. The allowable ampacity of multiconductor cables, nominally rated 2000 volts or less, installed according to the requirements of Section 318-9,

Table 318-10. Allowable Cable Fill Area for Single Conductor Cables in Ladder or Ventilated Trough Cable Trays for Cables Rated 2000 Volts or Less

	Maximum Allowable Fill Area for Single Conductor Cables in Ladder or Ventilated Trough Cable Trays	
Inside Width of Cable Tray (in.)	**Column 1 Applicable for Section 318-10(a)(2) Only ($in.^2$)**	**Column 2[a] Applicable for Section 318-10(a)(3) Only ($in.^2$)**
6	6.5	6.5–(1.1 Sd)[b]
9	9.5	9.5–(1.1 Sd)
12	13.0	13.0–(1.1 Sd)
18	19.5	19.5–(1.1 Sd)
24	26.0	26.0–(1.1 Sd)
30	32.5	32.5–(1.1 Sd)
36	39.0	39.0–(1.1 Sd)

Note: For SI units, 1 $in.^2$ = 645 mm^2.

[a]The maximum allowable fill areas in Column 2 shall be computed. For example, the maximum allowable fill, in square inches, for a 6-in. (152-mm) wide cable tray shall be 6.5 minus (1.1 multiplied by Sd).

[b]The term *Sd* in Column 2 is equal to the sum of the diameters, in inches, of all 1000 kcmil and larger single conductor cables in the same ladder or ventilated trough cable tray with small cables.

shall be as given in Tables 310-16 and 310-18, subject to the provisions of (1), (2), and (3).

(1) The derating factors of Section 310-15(b)(2)(a) shall apply only to multiconductor cables with more than three current-carrying conductors. Derating shall be limited to the number of current-carrying conductors in the cable and not to the number of conductors in the cable tray.

(2) Where cable trays are continuously covered for more than 6 ft (1.83 m) with solid unventilated covers, not over 95 percent of the allowable ampacities of Tables 310-16 and 310-18 shall be permitted for multiconductor cables.

(3) Where multiconductor cables are installed in a single layer in uncovered trays, with a maintained spacing of not less than one cable diameter between cables, the ampacity shall not exceed the allowable ambient temperature-corrected ampacities of multiconductor cables, with not more than three insulated conductors rated 0 through 2000 volts in free air, in accordance with Section 310-15(c).

Figure 318.3 illustrates this requirement. Note that the cables, rated 2000 volts or less, are installed in a single layer in an uncovered tray, with not less than one cable diameter between cables and not more than three conductors per cable. Refer to Table B-310-3 in Appendix B for the ampacity of the conductors in this configuration.

FPN: See Table B-310-3 in Appendix B.

(b) Single Conductor Cables. The derating factors of Section 310-15(b)(2)(a), shall not apply to the ampacity of cables

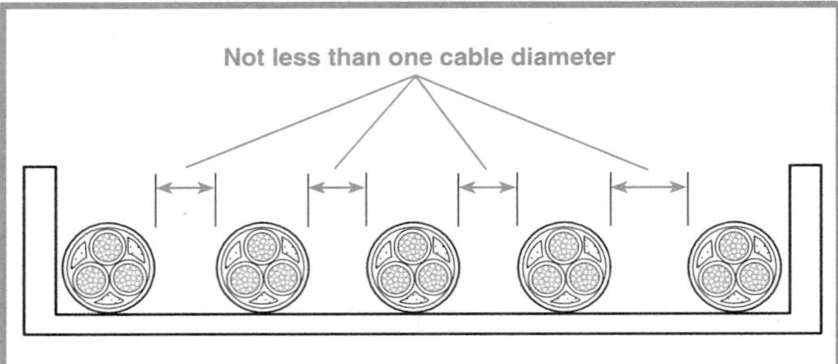

Figure 318.3 An illustration of Section 318-11(a)(3) for multiconductor cables, 2000 volts or less, with not more than three conductors per cable (ampacity to be determined from Table B-310-3 in Appendix B).

in cable trays. The ampacity of single conductor cables, or single conductors cabled together (triplexed, quadruplexed, etc.), nominally rated 2000 volts or less, shall comply with the following.

(1) Where installed according to the requirements of Section 318-10, the ampacities for 600 kcmil and larger single conductor cables in uncovered cable trays shall not exceed 75 percent of the allowable ampacities in Tables 310-17 and 310-19. Where cable trays are continuously covered for more than 6 ft (1.83 m) with solid unventilated covers, the ampacities for 600 kcmil and larger cables shall not exceed 70 percent of the allowable ampacities in Tables 310-17 and 310-19.

(2) Where installed according to the requirements of Section 318-10, the ampacities for No. 1/0 through 500 kcmil single conductor cables in uncovered cable trays shall not exceed 65 percent of the allowable ampacities in Tables 310-17 and 310-19. Where cable trays are continuously covered for more than 6 ft (1.83 m) with solid unventilated covers, the ampacities for No. 1/0 through 500 kcmil cables shall not exceed 60 percent of the allowable ampacities in Tables 310-17 and 310-19.

(3) Where single conductors are installed in a single layer in uncovered cable trays, with a maintained space of not less than one cable diameter between individual conductors, the ampacity of Nos. 1/0 and larger cables shall not exceed the allowable ampacities in Tables 310-17 and 310-19.

(4) Where single conductors are installed in a triangular or square configuration in uncovered cable trays, with a maintained free air space of not less than 2.15 times one conductor diameter (2.15 × OD) of the largest conductor contained within the configuration and adjacent conductor configurations or cables, the ampacity of Nos. 1/0 and larger cables shall not exceed the allowable ampacities of two or three single insulated conductors rated 0 through 2000 volts supported on a messenger in accordance with Section 310-15(b).

Section 318-11(b)(4) recognizes single conductors in a triangular configuration installed in a cable tray with maintained spacing as having the same ampacity as three single insulated conductors on a messenger. The maintained spacing allows air to circulate around the cable.

Where three single conductors, nominally rated 2000 volts or less, are cabled together in a triangular configuration, with not less than 2.15 times the conductor diameter (2.15 × OD) between groups, as illustrated in Figure 318.4, the ampacity of the conductors is determined in accordance with Table 310-20. Prior to the 1999 *Code,* Table 310-20 was located in Appendix B as Table B-310-2. This change allows the user to determine conductor ampacity for this application without engineering supervision.

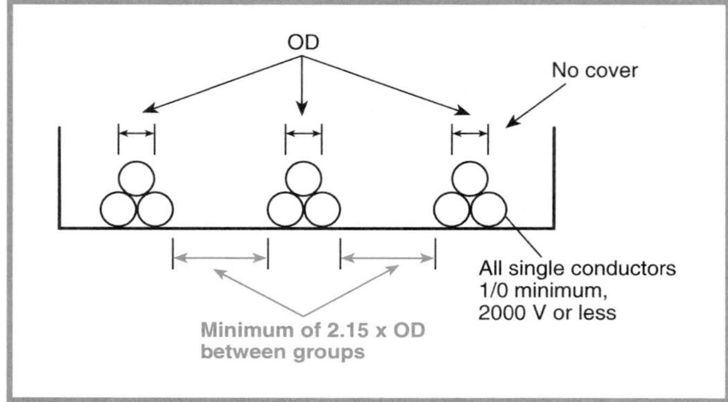

Figure 318.4 An illustration of Section 318-11(b)(4), for three single conductors installed in a triangular configuration with spacing between groups of not less than 2.15 times the conductor diameter (ampacities to be determined from Table 310-20).

FPN: See Table 310-20.

318-12. Number of Type MV and Type MC Cables (2001 Volts or Over) in Cable Trays. The number of cables, rated 2001 volts or over, permitted in a single cable tray shall not exceed the requirements of this section.

The sum of the diameters of single conductor and multiconductor cables shall not exceed the cable tray width, and the cables shall be installed in a single layer. Where single conductor cables are triplexed, quadruplexed, or bound together in circuit groups, the sum of the diameters of the single conductors shall not exceed the cable tray width, and these groups shall be installed in single layer arrangement.

318-13. Ampacity of Type MV and Type MC Cables (2001 Volts or Over) in Cable Trays. The ampacity of cables, rated 2001 volts, nominal, or over, installed according to Section 318-12 shall not exceed the requirements of this section.

(a) Multiconductor Cables (2001 Volts or Over). The allowable ampacity of multiconductor cables shall be as given in Tables 310-75 and 310-76, subject to the provisions of (1) and (2).

(1) Where cable trays are continuously covered for more than 6 ft (1.83 m) with solid unventilated covers, not more than 95 percent of the allowable ampacities of Tables 310-75 and 310-76 shall be permitted for multiconductor cables.

(2) Where multiconductor cables are installed in a single layer in uncovered cable trays, with maintained spacing of not less than one cable diameter between cables, the ampacity shall not exceed the allowable ampacities of Tables 310-71 and 310-72.

(b) Single Conductor Cables (2001 Volts or Over). The ampacity of single conductor cables, or single conductors cabled together (triplexed, quadruplexed, etc.), shall comply with the following.

(1) The ampacities for Nos. 1/0 and larger single conductor cables in uncovered cable trays shall not exceed 75 percent of the allowable ampacities in Tables 310-69 and 310-70. Where the cable trays are covered for more than 6 ft (1.83 m) with solid unventilated covers, the ampacities for Nos. 1/0 and larger single conductor cables shall not exceed 70 percent of the allowable ampacities in Tables 310-69 and 310-70.

(2) Where single conductor cables are installed in a single layer in uncovered cable trays, with a maintained space of not less than one cable diameter between individual conductors, the ampacity of Nos. 1/0 and larger cables shall not exceed the allowable ampacities in Tables 310-69 and 310-70.

(3) Where single conductors are installed in a triangular or square configuration in uncovered cable trays, with a maintained free air space of not less than 2.15 times the diameter (2.15 × OD) of the largest conductor contained within the configuration and adjacent conductor configurations or cables, the ampacity of Nos. 1/0 and larger cables shall not exceed the allowable ampacities in Tables 310-67 and 310-68.

Article 320 — Open Wiring on Insulators

Contents

320-1. Definition. Open wiring on insulators is an exposed wiring method using cleats, knobs, tubes, and flexible tubing for the protection and support of single insulated conductors run in or on buildings, and not concealed by the building structure.

Open wiring on insulators is an exposed wiring method that is not permitted to be concealed by the structure or finish of the building. It is permitted indoors or outdoors, in dry or wet locations, and where subject to corrosive vapors provided the insulation choice from Table 310-13 is suitable for use in a corrosive environment.

This method of wiring is no longer permitted for temporary lighting and power circuits on construction sites but is permitted for lighting and power circuits in agricultural buildings (see Section 547-4). It may also be used for services (see Section 230-43).

320-2. Other Articles. Open wiring on insulators shall comply with this article and also with the applicable provisions of other articles in this *Code*, especially Articles 225 and 300.

320-3. Uses Permitted. Open wiring on insulators shall be permitted on systems of 600 volts, nominal, or less, only for industrial or agricultural establishments, indoors or outdoors, in wet or dry locations, where subject to corrosive vapors, and for services.

320-5. Conductors.

(a) Type. Conductors shall be of a type specified by Article 310.

(b) Ampacity. The ampacity shall comply with Section 310-15.

See Tables 310-17 and 310-19 for ampacities of conductors.

320-6. Conductor Supports.

(a) Conductor Sizes Smaller than No. 8. Conductors smaller than No. 8 shall be rigidly supported on noncombustible, nonabsorbent insulating materials and shall not contact any other objects.

Supports shall be installed as follows:

(1) Within 6 in. (152 mm) from a tap or splice
(2) Within 12 in. (305 mm) of a dead-end connection to a lampholder or receptacle
(3) At intervals not exceeding 4½ ft (1.37 m) and at closer intervals sufficient to provide adequate support where likely to be disturbed

(b) Conductor Sizes No. 8 and Larger. Supports for conductors No. 8 or larger installed across open spaces shall be permitted up to 15 ft (4.57 m) apart if noncombustible, nonabsorbent insulating spacers are used at least every 4½ ft (1.37 m) to maintain at least 2½ in. (64 mm) between conductors.

Where not likely to be disturbed in buildings of mill construction, No. 8 and larger conductors shall be permitted to be run across open spaces if supported from each wood cross member on approved insulators maintaining 6 in. (152 mm) between conductors.

Mill construction is generally considered to be a building where the floors and ceilings are supported by wood timbers or beams or wood cross members spaced approximately 15 ft apart. This type of construction is sometimes referred to as plank-on-timber construction. This section permits No. 8 and larger conductors to span this distance where the ceilings are high and free of obstructions and the conductors are unlikely to come into contact with other objects.

(c) Industrial Establishments. In industrial establishments only, where conditions of maintenance and supervision ensure that only qualified persons will service the system, conductors of sizes 250 kcmil and larger shall be permitted to be run across open spaces where supported on intervals up to 30 ft (9.1 m) apart.

It is common practice in industrial buildings to install open feeders on insulators, which are mounted on the bottom of roof trusses at every bay location. Many bays are more than 15 ft wide. Therefore, this section permits size 250 kcmil and larger conductors to be supported at 30-ft intervals where it is assured that qualified persons will service the system.

In addition to the ease and economy of installation or alteration of open wiring, it is to be noted that, by close spacing of conductors, the reactance of a circuit is reduced; hence, the voltage drop is reduced.

320-7. Mounting of Conductor Supports. Where nails are used to mount knobs, they shall not be smaller than tenpenny. Where screws are used to mount knobs, or where nails or screws are used to mount cleats, they shall be of a length sufficient to penetrate the wood to a depth equal to at least one-half the height of the knob and the full thickness of the cleat. Cushion washers shall be used with nails.

320-8. Tie Wires. No. 8 or larger conductors supported on solid knobs shall be securely tied thereto by tie wires having an insulation equivalent to that of the conductor.

320-10. Flexible Nonmetallic Tubing. In dry locations, where not exposed to severe physical damage, conductors shall be permitted to be separately enclosed in flexible nonmetallic tubing. The tubing shall be in continuous lengths not exceeding 15 ft (4.57 m) and secured to the surface by straps at intervals not exceeding 4½ ft (1.37 m).

320-11. Through Walls, Floors, Wood Cross Members, etc. Open conductors shall be separated from contact with walls, floors, wood cross members, or partitions through which they pass by tubes or bushings of noncombustible, nonabsorbent insulating material. Where the bushing is shorter than the hole, a waterproof sleeve of noninductive material shall be inserted in the hole and an insulating bushing slipped into the sleeve at each end in such a manner as to keep the conductors absolutely out of contact with the sleeve. Each conductor shall be carried through a separate tube or sleeve.

FPN: See Section 310-10 for temperature limitation of conductors.

320-12. Clearance from Piping, Exposed Conductors, etc. Open conductors shall be separated at least 2 in. (50.8 mm) from metal raceways, piping, or other conducting material, and from any exposed lighting, power, or signaling conductor, or shall be separated therefrom by a continuous and firmly fixed nonconductor in addition to the insulation of the conductor. Where any insulating tube is used, it shall be secured at the ends. Where practicable, conductors shall pass over rather than under any piping subject to leakage or accumulations of moisture.

The provision for additional protective insulation on open wiring is to prevent contact with metal piping, metal objects, or exposed conductors of other circuits.

320-13. Entering Spaces Subject to Dampness, Wetness, or Corrosive Vapors. Conductors entering or leaving locations subject to dampness, wetness, or corrosive vapors shall have drip loops formed on them and shall then pass upward and inward from the outside of the buildings, or from the damp, wet, or corrosive location, through noncombustible, nonabsorbent insulating tubes.

FPN: See Section 230-52 for individual conductors entering buildings or other structures.

320-14. Protection from Physical Damage. Conductors within 7 ft (2.13 m) from the floor shall be considered exposed to physical damage. Where open conductors cross ceiling joists and wall studs and are exposed to physical damage, they shall be protected by one of the following methods:

(1) Guard strips not less than 1 in. (25.4 mm) nominal in thickness and at least as high as the insulating supports, placed on each side of and close to the wiring
(2) A substantial running board at least ½ in. (12.7 mm) thick in back of the conductors with side protections. Running boards shall extend at least 1 in. (25.4 mm) outside the conductors, but not more than 2 in. (50.8 mm), and the protecting sides shall be at least 2 in. (50.8 mm) high and at least 1 in. (25.4 mm) nominal in thickness
(3) Boxing made as above and furnished with a cover kept at least 1 in. (25.4 mm) away from the conductors within. Where protecting vertical conductors on side walls, the boxing shall be closed at the top and the holes through which the conductors pass shall be bushed.
(4) Rigid metal conduit, intermediate metal conduit, rigid nonmetallic conduit, or electrical metallic tubing, in which case the rules of Articles 345, 346, 347, or 348 shall apply; or by metal piping, in which case the conductors shall be encased in continuous lengths of approved flexible tubing.

320-15. Unfinished Attics and Roof Spaces. Conductors in unfinished attics and roof spaces shall comply with (a) or (b).

(a) Accessible by Stairway or Permanent Ladder. Conductors shall be installed along the side of or through bored holes in floor joists, studs, or rafters. Where run through bored holes, conductors in the joists and in studs or rafters to a height of not less than 7 ft (2.13 m) above the floor or floor joists shall be protected by substantial running boards extending not less than 1 in. (25.4 mm) on each side of the conductors. Running boards shall be securely fastened in place. Running boards and guard strips shall not be required for conductors installed along the sides of joists, studs, or rafters.

(b) Not Accessible by Stairway or Permanent Ladder. Conductors shall be installed along the sides of or through bored holes in floor joists, studs, or rafters.

Exception: In buildings completed before the wiring is installed, attic and roof spaces that are not accessible by stairway or permanent ladder and have headroom at all points less than 3 ft (914 mm), the wiring shall be permitted to be installed on the edges of rafters or joists facing the attic or roof space.

320-16. Switches. Surface-type snap switches shall be mounted in accordance with Section 380-10(a), and boxes shall not be required. Other type switches shall be installed in accordance with Section 380-4.

Article 321 — Messenger Supported Wiring

Contents

321-1. Definition. *Messenger supported wiring* is an exposed wiring support system using a messenger wire to support insulated conductors by any one of the following:

(1) A messenger with rings and saddles for conductor support
(2) A messenger with a field-installed lashing material for conductor support
(3) Factory-assembled aerial cable
(4) Multiplex cables utilizing a bare conductor, factory assembled and twisted with one or more insulated conductors, such as duplex, triplex, or quadruplex type of construction

Messenger supported wiring systems have been manufactured and successfully used in industrial installations for many years. They have also been used for many years as service drops by utilities for commercial and residential installations.

See references to messenger supported wiring in Sections 225-6(a)(1) and 225-6(b).

321-2. Other Articles. Messenger supported wiring shall comply with this article and also with the applicable provisions of other articles in this *Code*, especially Articles 225 and 300.

321-3. Uses Permitted.

(a) Cable Types. The following shall be permitted to be installed in messenger supported wiring under the conditions described in the article or section referenced for each:

	Section	Article
Metal-clad cable		334
Mineral-insulated, metal-sheathed cable		330
Multiconductor service-entrance cable		338
Multiconductor underground feeder and branch-circuit cable		339
Other factory-assembled, multi-conductor control, signal, or power cables that are identified for the use		
Power and control tray cable		340
Power-limited tray cable	725-61(c) and 725-71(e)	

(b) In Industrial Establishments. In industrial establishments only, where conditions of maintenance and supervision ensure that only qualified persons will service the installed messenger supported wiring, the following shall be permitted:

(1) Any of the conductor types shown in Table 310-13 or Table 310-62

Some of the triplex and quadruplex cable used by utilities as service-drop cable does not use conductors recognized in Table 310-13 and does not meet the requirements of Article 310. Such triplex and quadruplex cable would, therefore, be acceptable only where approved by the authority having jurisdiction.

(2) MV cable

Where exposed to weather, conductors shall be listed for use in wet locations. Where exposed to direct rays of the sun, conductors or cables shall be sunlight resistant.

(c) Hazardous (Classified) Locations. Messenger supported wiring shall be permitted to be used in hazardous (classified) locations where the contained cables are permitted for such use in Sections 501-4, 502-4, 503-3, and 504-20.

321-4. Uses Not Permitted. Messenger supported wiring shall not be used in hoistways or where subject to severe physical damage.

321-5. Ampacity. The ampacity shall be determined by Section 310-15.

See Sections 310-15(b) and Table 310-20 for two or three single-insulated conductors supported on a messenger. See Section 310-15(c) and Appendix B, Table B-310-3, for ampacities of conductors for other cable types.

321-6. Messenger Support. The messenger shall be supported at dead ends and at intermediate locations so as to eliminate tension on the conductors. The conductors shall not be permitted to come into contact with the messenger supports or any structural members, walls, or pipes.

321-7. Grounding. The messenger shall be grounded as required by Sections 250-80 and 250-86 for enclosure grounding.

321-8. Conductor Splices and Taps. Conductor splices and taps made and insulated by approved methods shall be permitted in messenger supported wiring.

Article 324 — Concealed Knob-and-Tube Wiring

Contents

324-1. Definition. *Concealed knob-and-tube wiring* is a wiring method using knobs, tubes, and flexible nonmetallic tubing for the protection and support of single insulated conductors.

Open wiring on insulators (Article 320) is required to be exposed, whereas knob-and-tube wiring is allowed to be concealed. Conductors used for knob-and-tube work may be of any general-use type specified by Article 310.

324-2. Other Articles. Concealed knob-and-tube wiring shall comply with this article and also with the applicable provisions of other articles in this *Code*, especially Article 300.

324-3. Uses Permitted. Concealed knob-and-tube wiring shall be permitted to be installed in the hollow spaces of walls and ceilings or in unfinished attics and roof spaces as provided in Section 324-11 only as follows:

(1) For extensions of existing installations, or
(2) Elsewhere by special permission.

Concealed knob-and-tube wiring is permitted to be installed only for extensions of existing installations or where special permission is granted by the authority having jurisdiction of enforcement of the *Code*. See definition of *special permission* in Article 100.

324-4. Uses Not Permitted. Concealed knob-and-tube wiring shall not be used in commercial garages, theaters and similar locations, motion picture studios, hazardous (classified) locations, or in the hollow spaces of walls, ceilings, and attics where such spaces are insulated by loose, rolled, or foamed-in-place insulating material that envelops the conductors.

Concealed knob-and-tube wiring is designed for use in hollow spaces of walls, ceilings, and attics and utilizes the free air in such spaces for heat dissipation. Weatherization of hollow spaces by blown-in, foamed-in, or rolled insulation prevents the dissipation of heat into the free air space. This will result in higher conductor temperature, which could cause insulation breakdown and possible ignition of the insulation.

324-5. Conductors.

(a) Type. Conductors shall be of a type specified by Article 310.

(b) Ampacity. The ampacity shall be determined by Section 310-15.

See Tables 310-17 and 310-19 for ampacities of conductors.

324-6. Conductor Supports. Conductors shall be rigidly supported on noncombustible, nonabsorbent insulating materials and shall not contact any other objects. Supports shall be installed as follows:

(1) Within 6 in. (152 mm) of each side of each tap or splice
(2) At intervals not exceeding 4½ ft (1.37 m)

Where it is impracticable to provide supports, conductors shall be permitted to be fished through hollow spaces in dry locations, provided each conductor is individually enclosed in flexible nonmetallic tubing that is in continuous lengths between supports, between boxes, or between a support and a box.

324-7. Tie Wires. Where solid knobs are used, conductors shall be securely tied thereto by tie wires having insulation equivalent to that of the conductor.

324-8. Conductor Clearances.

(a) General. A clearance of not less than 3 in. (76 mm) shall be maintained between conductors and a clearance of not less than 1 in. (25.4 mm) between the conductor and the surface over which it passes.

(b) Limited Conductor Space. Where space is too limited to provide these minimum clearances, such as at meters, panelboards, outlets, and switch points, the individual conductors shall be enclosed in flexible nonmetallic tubing, which shall be continuous in length between the last support and the enclosure or terminal point.

324-9. Through Walls, Floors, Wood Cross Members, etc. Conductors shall comply with Section 320-11 where passing through holes in structural members. Where passing through wood cross members in plastered partitions, conductors shall be protected by noncombustible, nonabsorbent, insulating tubes extending not less than 3 in. (76 mm) beyond the wood member.

The provision for insulated tubes where knob-and-tube wiring passes through wood cross members in plastered partitions is to protect the wire from contact with plaster that is likely to accumulate on horizontal wood members.

324-10. Clearance from Piping, Exposed Conductors, etc. Conductors shall comply with Section 320-12 for clearances from other exposed conductors, piping, etc.

324-11. Unfinished Attics and Roof Spaces. Conductors in unfinished attics and roof spaces shall comply with (a) or (b).

FPN: See Section 310-10 for temperature limitation of conductors.

(a) Accessible by Stairway or Permanent Ladder. Conductors shall be installed along the side of or through bored holes in floor joists, studs, or rafters. Where run through bored holes, conductors in the joists and in studs or rafters to a height of not less than 7 ft (2.13 m) above the floor or floor joists shall be protected by substantial running boards extending not less than 1 in. (25.4 mm) on each side of the conductors. Running boards shall be securely fastened in place. Running boards and guard strips shall not be required where conductors are installed along the sides of joists, studs, or rafters.

Figure 324.1 illustrates the "running board" method of protecting open-type conductors where they are installed at a height less than 7 ft above the floor or floor joists in an accessible attic. This method is applied in attics that are accessible by stairways or permanent ladders and where such spaces are generally used for storage.

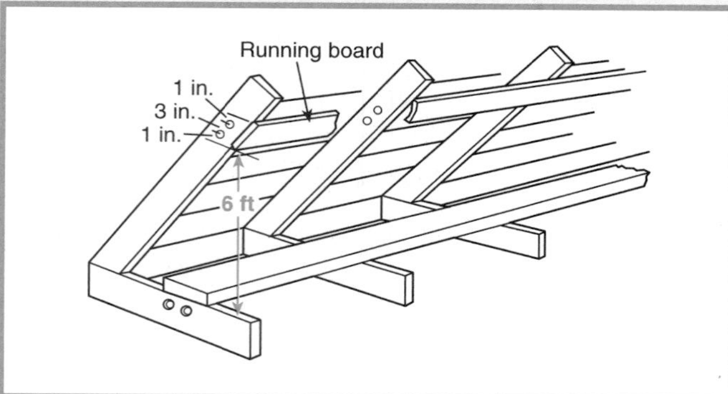

Figure 324.1 Open wiring in an accessible attic protected by running boards, as required by Section 324-11(a).

(b) Not Accessible by Stairway or Permanent Ladder. Conductors shall be installed along the sides of or through bored holes in floor joists, studs, or rafters.

Exception: In buildings completed before the wiring is installed, attic and roof spaces that are not accessible by stairway or permanent ladder and have headroom at all points less than 3 ft (914 mm), the wiring shall be permitted to be installed on the edges of rafters or joists facing the attic or roof space.

324-12. Splices. Splices shall be soldered unless approved splicing devices are used. In-line or strain splices shall not be used.

324-13. Boxes. Outlet boxes shall comply with Article 370.

324-14. Switches. Switches shall comply with Sections 380-4 and 380-10(b).

Article 325 — Integrated Gas Spacer Cable: Type IGS

Contents

A. General

325-1. Definition. *Type IGS cable* is a factory assembly of one or more conductors, each individually insulated and enclosed in a loose fit, nonmetallic flexible conduit as an integrated gas spacer cable rated 0 through 600 volts.

As illustrated in Figure 325.1, Type IGS (integrated gas spacer) cable consists of solid aluminum rod conductors, 250-kcmil minimum size. These conductors are insulated with dry kraft paper and are factory installed in a medium-density polyethylene gas pipe, 2-in. minimum trade size, which is then filled with sulfur hexafluoride (SF_6) gas at a pressure of approximately 20 psi.

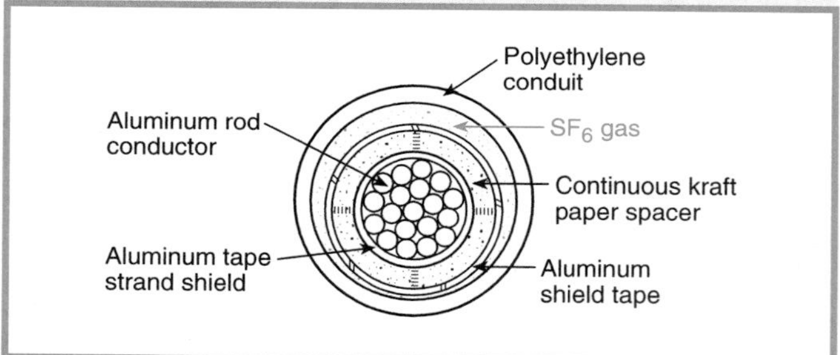

Figure 325.1 Cross section of single-conductor, 4750-kcmil Type IGS cable.

325-2. Other Articles. Type IGS cable shall comply with this article and also with the applicable provisions of other articles in this *Code.*

325-3. Uses Permitted. Type IGS cable shall be permitted for use underground, including direct burial in the earth, as service-entrance conductors, or as feeder or branch-circuit conductors.

325-4. Uses Not Permitted. Type IGS cable shall not be used as interior wiring or be exposed in contact with buildings.

B. Installation

325-11. Bending Radius. Where the coilable nonmetallic conduit and cable is bent for installation purposes or is flexed or bent during shipment or installation, the radii of bends measured to the inside of the bend shall not be less than specified in Table 325-11.

Table 325-11. Minimum Radii of Bends

Conduit Trade Size (in.)	Minimum Radii	
	in.	mm
2	24	610
3	35	889
4	45	1143

325-12. Bends. A run of Type IGS cable between pull boxes or terminations shall not contain more than the equivalent of four quarter bends (360 degrees total), including those bends located immediately at the pull box or terminations.

325-13. Fittings. Terminations and splices for Type IGS cable shall be identified as a type that is suitable for maintaining the gas pressure within the conduit. A valve and cap shall be provided for each length of the cable and conduit to check the gas pressure or to inject gas into the conduit.

325-14. Ampacity. The ampacity of Type IGS cable and conduit shall not exceed values shown in Table 325-14 for single conductor or multiconductor cable.

Table 325-14. Ampacity of Type IGS Cable

Size (kcmil)	Amperes	Size (kcmil)	Amperes
250	119	2500	376
500	168	3000	412
750	206	3250	429
1000	238	3500	445
1250	266	3750	461
1500	292	4000	476
1750	344	4250	491
2000	336	4500	505
2250	357	4750	519

C. Construction Specifications

325-20. Conductors. The conductors shall be solid aluminum rods, laid parallel, consisting of one to nineteen ½-in. (12.7-mm) diameter rods.

The minimum conductor size shall be 250 kcmil and the maximum size shall be 4750 kcmil.

325-21. Insulation. The insulation shall be dry kraft paper tapes and a pressurized sulfur hexafluoride gas (SF_6), both approved for electrical use. The nominal gas pressure shall be 20 pounds per square inch gauge (psig) (138 kPa gauge).

The thickness of the paper spacer shall be as specified in Table 325-21.

325-22. Conduit. The conduit shall be a medium density polyethylene identified as suitable for use with natural gas rated pipe in 2-in., 3-in., or 4-in. trade size. The percent fill dimensions for the conduit are shown in Table 325-22.

The size of the conduit permitted for each conductor size shall be calculated for a percent fill not to exceed those found in Table 1, Chapter 9.

Table 325-21. Paper Spacer Thickness

Size (kcmil)	Thickness	
	in.	mm
250–1000	0.040	1.02
1250–4750	0.060	1.52

Table 325-22. Conduit Dimensions

Conduit Trade Size (in.)	Outside Diameter		Inside Diameter	
	in.	mm	in.	mm
2	2.375	60	1.947	49.46
3	3.500	89	2.886	73.30
4	4.500	114	3.710	94.23

325-24. Marking. The cable shall be marked in accordance with Sections 310-11(a), 310-11(b)(1), and 310-11(d).

Article 326 — Medium Voltage Cable: Type MV

Contents

326-1. Definition. Type MV cable is a single or multiconductor solid dielectric insulated cable rated 2001 volts or higher.

Type MV (medium voltage) cables are rated 2001 to 35,000 volts. Cables rated 2001 to 8000 volts may be shielded or nonshielded. All insulated conductors 8001 volts and higher have electrostatic shielding. When nonshielded cables are used, they must comply with Section 310-6, Exception. If these cables are installed in underground conduits, they must be encased in 3 in. of concrete to comply with Section 300-50(a)(2).

326-2. Other Articles. In addition to the provisions of this article, Type MV cable shall comply with the applicable provisions of this *Code*, especially those of Articles 300, 305, 310, 318, and 490.

326-3. Uses Permitted. Type MV cables shall be permitted for use on power systems rated up to 35,000 volts, nominal, in wet or dry locations, in raceways, in cable trays as specified in Section 318-3(b)(1) or directly buried in accordance with Section 300-50, and in messenger supported wiring.

326-4. Uses Not Permitted. Type MV cable shall not be used unless identified for the use (1) where exposed to direct sunlight, and (2) in cable trays.

Cables intended for installation in cable trays in accordance with Article 318 are marked "For CT Use" or "For Use in Cable Trays."

326-5. Construction. Type MV cables shall have copper, aluminum, or copper-clad aluminum conductors and shall be constructed in accordance with Article 310.

Cables with aluminum conductors are marked with the word *Aluminum* or the letters *AL*.

326-6. Ampacity. The ampacity of Type MV cable shall be in accordance with Section 310-60.

Exception: The ampacity of Type MV cable installed in cable tray shall be in accordance with Section 318-13.

The exception to Section 326-6 points the user to corrected ampacities for conductors and multiconductor cables where installed in cable trays.

326-7. Marking. Medium voltage cable shall be marked as required in Section 310-11.

Cables are marked with their conductor size, voltage rating, and insulation level (100 percent or 133 percent).

Article 328 — Flat Conductor Cable: Type FCC

Contents

A. General

328-1. Scope. This article covers a field-installed wiring system for branch circuits incorporating Type FCC cable and associated accessories as defined by the article. The wiring system is designed for installation under carpet squares.

The FCC (flat conductor cable) system is designed to provide a completely accessible, flexible power system. As shown in Figure 328.1, it also provides an easy method for reworking obsolete wiring systems currently in use in many office facilities. The carpet squares are not permitted to be larger than 36 in. by 36 in. to comply with Section 328-10. This limitation is judged necessary to provide ready access to the cable by lifting a carpet square. It also reduces the likelihood of damage to the cable by an individual cutting through the carpet above the cable with a knife or razor blade and possibly penetrating the top shield of the cable in the process.

Figure 328.1 *Installing Type FCC cable. (AMP Products Corp.)*

328-2. Definitions.

Type FCC Cable. *Type FCC cable* consists of three or more flat copper conductors placed edge-to-edge and separated and enclosed within an insulating assembly.

FCC System. A complete wiring system for branch circuits that is designed for installation under carpet squares. The FCC system includes Type FCC cable and associated shielding, connectors, terminators, adapters, boxes, and receptacles.

Cable Connector. A connector designed to join Type FCC cables without using a junction box.

Insulating End. An insulator designed to electrically insulate the end of a Type FCC cable.

Top Shield. A grounded metal shield covering undercarpet components of the FCC system for the purposes of providing protection against physical damage.

Bottom Shield. A protective layer that is installed between the floor and Type FCC flat conductor cable to protect the cable from physical damage and may or may not be incorporated as an integral part of the cable.

Transition Assembly. An assembly to facilitate connection of the FCC system to other wiring systems, incorporating (1) a means of electrical interconnection, and (2) a suitable box or covering for providing electrical safety and protection against physical damage.

Metal Shield Connections. Means of connection designed to electrically and mechanically connect a metal shield to another metal shield, to a receptacle housing or self-contained device, or to a transition assembly.

328-3. Other Articles. The FCC systems shall conform with the applicable provisions of Articles 210, 220, 240, 250, and 300.

328-4. Uses Permitted.

(a) Branch Circuits. Use of FCC systems shall be permitted both for general-purpose and appliance branch circuits and for individual branch circuits.

(b) Floors. Use of FCC systems shall be permitted on hard, sound, smooth, continuous floor surfaces made of concrete, ceramic, or composition flooring, wood, and similar materials.

(c) Walls. Use of FCC systems shall be permitted on wall surfaces in surface metal raceways.

(d) Damp Locations. Use of FCC systems in damp locations shall be permitted.

(e) Heated Floors. Materials used for floors heated in excess of 30°C (86°F) shall be identified as suitable for use at these temperatures.

328-5. Uses Not Permitted. FCC systems shall not be used in the following:

(1) Outdoors or in wet locations
(2) Where subject to corrosive vapors
(3) In any hazardous (classified) location
(4) In residential, school, and hospital buildings

Section 328-5(4) prohibits Type FCC wiring systems throughout school and hospital buildings, even though parts of these buildings may be office or administrative spaces.

328-6. Branch-Circuit Ratings.

(a) Voltage. Voltage between ungrounded conductors shall not exceed 300 volts. Voltage between ungrounded conductors and the grounded conductor shall not exceed 150 volts.

(b) Current. General-purpose and appliance branch circuits shall have ratings not exceeding 20 amperes. Individual branch circuits shall have ratings not exceeding 30 amperes.

B. Installation

328-10. Coverings. Floor-mounted Type FCC cable, cable connectors, and insulating ends shall be covered with carpet squares not larger than 36 in. (914 mm) square. Those carpet squares that are adhered to the floor shall be attached with release-type adhesives.

328-11. Cable Connections and Insulating Ends. All Type FCC cable connections shall use connectors identified for their use, installed such that electrical continuity, insulation, and sealing against dampness and liquid spillage are provided. All bare cable ends shall be insulated and sealed against dampness and liquid spillage using listed insulating ends.

328-12. Shields.

(a) Top Shield. A metal top shield shall be installed over all floor-mounted Type FCC cable, connectors, and insulating ends. The top shield shall completely cover all cable runs, corners, connectors, and ends.

(b) Bottom Shield. A bottom shield shall be installed beneath all Type FCC cable, connectors, and insulating ends.

328-13. Enclosure and Shield Connections. All metal shields, boxes, receptacle housings, and self-contained devices shall be electrically continuous to the equipment grounding conductor of the supplying branch circuit. All such electrical connections shall be made with connectors identified for this use. The electrical resistivity of such shield system shall not be more than that of one conductor of the Type FCC cable used in the installation.

328-14. Receptacles. All receptacles, receptacle housings, and self-contained devices used with the FCC system shall be identified for this use and shall be connected to the Type FCC cable and metal shields. Connection from any grounding conductor of the Type FCC cable shall be made to the shield system at each receptacle.

328-15. Connection to Other Systems. Power feed, grounding connection, and shield system connection between the FCC system and other wiring systems shall be accomplished in a transition assembly identified for this use.

328-16. Anchoring. All FCC system components shall be firmly anchored to the floor or wall using an adhesive or mechanical anchoring system identified for this use. Floors shall be prepared to ensure adherence of the FCC system to the floor until the carpet squares are placed.

328-17. Crossings. Crossings of more than two Type FCC cable runs shall not be permitted at any one point. Crossings of a Type FCC cable over or under a flat communications or signal cable shall be permitted. In each case, a grounded layer of metal shielding shall separate the two cables, and crossings of more than two flat cables shall not be permitted at any one point.

328-18. System Height. Any portion of an FCC system with a height above floor level exceeding 0.090 in. (2.29 mm) shall be tapered or feathered at the edges to floor level.

328-19. FCC Systems Alterations. Alterations to FCC systems shall be permitted. New cable connectors shall be used at new connection points to make alterations. It shall be permitted to leave unused cable runs and associated cable connectors in place and energized. All cable ends shall be covered with insulating ends.

328-20. Polarization of Connections. All receptacles and connections shall be constructed and installed so as to maintain proper polarization of the system.

C. Construction

328-30. Type FCC Cable. Type FCC cable shall be listed for use with the FCC system and shall consist of three, four, or five flat copper conductors, one of which shall be an equipment grounding conductor. The insulating material of the cable shall be moisture resistant and flame retardant.

Section 328-30 requires all FCC cables to be listed. Previous to the 1996 *Code,* there was no listing requirement.

328-31. Markings. Type FCC cable shall be clearly and durably marked on both sides at intervals of not more than 24 in. (610 mm) with the information required by Section 310-11(a) and with the following additional information:

(1) Material of conductors
(2) Maximum temperature rating
(3) Ampacity

328-32. Conductor Identification.

(a) Colors. Conductors shall be clearly and durably marked on both sides throughout their length as specified in Section 310-12.

(b) Order. For a 2-wire FCC system with grounding, the grounding conductor shall be central.

328-33. Corrosion Resistance. Metal components of the system shall be either corrosion resistant, coated with corrosion-resistant materials, or insulated from contact with corrosive substances.

328-34. Insulation. All insulating materials in the FCC systems shall be identified for their use.

328-35. Shields.

(a) Materials and Dimensions. All top and bottom shields shall be of designs and materials identified for their use. Top shields shall be metal. Both metallic and nonmetallic materials shall be permitted for bottom shields.

(b) Resistivity. Metal shields shall have cross-sectional areas that provide for electrical resistivity of not more than that of one conductor of the Type FCC cable used in the installation.

(c) Metal-Shield Connectors. Metal shields shall be connected to each other and to boxes, receptacle housings, self-contained devices, and transition assemblies using metal-shield connectors.

328-36. Receptacles and Housings. Receptacle housings and self-contained devices designed either for floor mounting or for in- or on-wall mounting shall be permitted for use with the FCC system. Receptacle housings and self-contained devices shall incorporate means for facilitating entry and termination of Type FCC cable and for electrically connecting the housing or device with the metal shield. Receptacles and self-contained devices shall comply with Section 210-7. Power and communications outlets installed together in common housing shall be permitted in accordance with Section 800-52(a)(2), Exception No. 1.

328-37. Transition Assemblies. All transition assemblies shall be identified for their use. Each assembly shall incorporate means for facilitating entry of the Type FCC cable into the assembly, for connecting the Type FCC cable to grounded conductors, and for electrically connecting the assembly to the metal cable shields and to equipment grounding conductors.

Article 330 — Mineral-Insulated, Metal-Sheathed Cable: Type MI

Contents

A. General

330-1. Definition. *Type MI mineral-insulated, metal-sheathed cable* is a factory assembly of one or more conductors insulated with a highly compressed refractory mineral insulation and enclosed in a liquidtight and gastight continuous copper or alloy steel sheath.

Type MI mineral-insulated, metal-sheathed cable consists of one or more solid copper conductors insulated with highly compressed magnesium oxide and enclosed in a continuous copper or alloy steel (e.g., stainless steel) sheath. It is manufactured in size Nos. 16 to 500 kcmil, one conductor; Nos. 16 to 4, two and three conductor; Nos. 16 to 6, four conductor; and Nos. 16 to 10, seven conductor. The cable is rated 600 volts.

Terminations specifically investigated for use with this cable are listed as "mineral-insulated cable fittings."

Supplementary nonmetallic coatings presently used have not been investigated for resistance to corrosion.

Fittings for use on Type MI cable are suitable for use at a maximum operating temperature of 90°C (194°F) in dry locations and 60°C (140°F) in wet locations. As shown in Figure 330.1, a complete box connector consists of a connector body and a screw-on potting fitting that may be used separately as an end fitting for change to open wiring. The screw-on potting fitting is to be assembled with a special tool and consists of a screw-on pot, insulating cap, insulating sleeving, anchoring bead, and sealing compound.

330-2. Other Articles. Type MI cable shall comply with this article and also with the applicable provisions of other articles in this *Code*, especially Article 300.

330-3. Uses Permitted. Type MI cable shall be permitted as follows:

(1) For services, feeders, and branch circuits
(2) For power, lighting, control, and signal circuits
(3) In dry, wet, or continuously moist locations
(4) Indoors or outdoors
(5) Where exposed or concealed

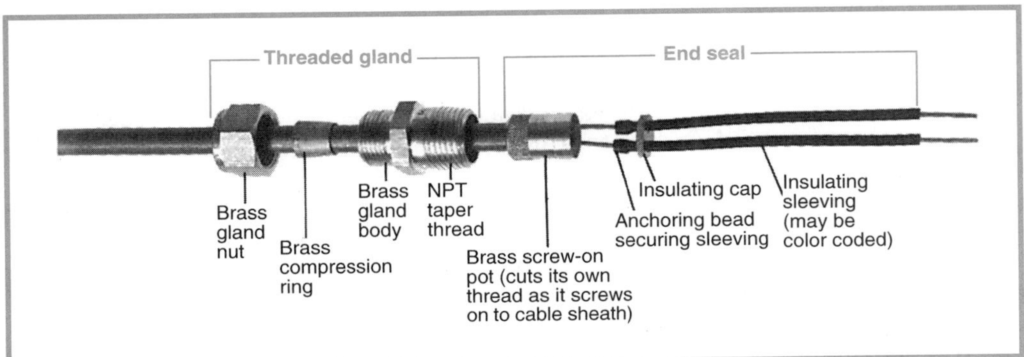

Figure 330.1 *A Type MI cable fitting used for terminating cable to an enclosure, a box, or directly to equipment. (Pyrotenax USA, Inc.)*

(6) Embedded in plaster, concrete, fill, or other masonry, whether above or below grade
(7) In any hazardous (classified) location
(8) Where exposed to oil and gasoline
(9) Where exposed to corrosive conditions not deteriorating to its sheath
(10) In underground runs where suitably protected against physical damage and corrosive conditions

Type MI cable is suitable for all power and control circuits up to 600 volts and may be used for services, feeders, and branch circuits in exposed and concealed work. It is permitted to be installed in dry and wet locations; in plaster, masonry, concrete, or fill; or underground. It may be installed where exposed to weather, continuous moisture, oil, or gasoline and in any hazardous (classified) location or in other conditions not having a deteriorating effect on the metal sheath.

330-4. Uses Not Permitted. Type MI cable shall not be used where exposed to destructive corrosive conditions, unless protected by materials suitable for the conditions.

B. Installation

330-10. Wet Locations. Where installed in wet locations, Type MI cable shall comply with Section 300-6(c).

330-11. Through Joists, Studs, or Rafters. Type MI cable shall comply with Section 300-4 where installed through studs, joists, rafters, or similar wood members.

330-12. Supports. Type MI cable shall be supported by one of the following means:

(1) Supported securely at intervals not exceeding 6 ft (1.83 m) by straps, staples, hangers, or similar fittings designed and installed so as not to damage the cable

Exception: Where cable is fished, support shall not be required.

(2) Supported in accordance with Section 318-8(b) where installed in cable trays

330-13. Bends. Bends in Type MI cable shall be made so as not to damage the cable. The radius of the inner edge of any bend shall not be less than shown as follows:

(1) Five times the external diameter of the metallic sheath for cable not more than ¾ in. (19 mm) in external diameter
(2) Ten times the external diameter of the metallic sheath for cable greater than ¾ in. (19 mm) but not more than 1 in. (25.4 mm) in external diameter

The minimum bending radius is intended to prevent mechanical damage to the conductor insulation or the sheath that could result in cracking or a hot spot at the point of damage, or both.

As illustrated in Figure 330.2, for cables with an external diameter (OD) not greater than ¾ in., the minimum bending radius *(R)* is 5 times the cable OD. For cables greater than ¾ in. but not greater than 1 in. in diameter, the minimum bending radius is 10 times the cable OD.

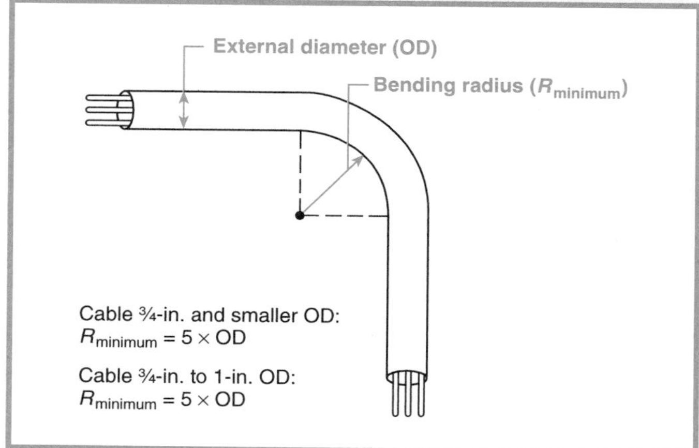

Figure 330.2 *An illustration of Section 330-13, for bends in Type MI cable.*

330-14. Fittings. Fittings used for connecting Type MI cable to boxes, cabinets, or other equipment shall be identified for such use. Where single-conductor cables enter ferrous metal boxes or cabinets, the installation shall comply with Section 300-20 to prevent inductive heating.

330-15. Terminal Seals. Where Type MI cable terminates, a seal shall be provided immediately after stripping to prevent the entrance of moisture into the insulation. The conductors extending beyond the sheath shall be individually provided with an insulating material.

330-16. Single Conductors. Where single-conductor cables are used, all phase conductors and, where used, the neutral conductor shall be grouped together to minimize induced voltage on the sheath. Where single-conductor cables enter ferrous enclosures, the installation shall comply with Section 300-20 to prevent heating from induction.

Section 330-16 permits Type MI cables to be used as single conductors. The larger sizes of Type MI cable are available only as single-conductor cables. Single conductors in a metal sheath can result in induced voltage on the sheath. Therefore, this section requires all conductors of the circuit to be grouped together to minimize the voltage on the sheath.

Where single conductors enter a ferrous metal enclosure, inductive heating can occur. This is due to hysteresis loss caused by the magnetic flux that occurs in ferrous metals and I^2R losses from the currents induced by the conductor. Section 300-20 requires additional measures to minimize this magnetic heating of enclosures. The additional measures may consist of slots cut in the metal between the individual holes for each conductor connector. Available from the cable manufacturer are nonferrous connecting plates that accept individual threaded connections of all circuit conductors, thereby eliminating circulating currents and fully complying with Section 300-20.

C. Construction Specifications

330-20. Conductors. Type MI cable conductors shall be of solid copper or nickel-clad copper with a resistance corresponding to standard AWG sizes.

330-21. Insulation. The conductor insulation in Type MI cable shall be a highly compressed refractory mineral that provides proper spacing for all conductors.

330-22. Outer Sheath. The outer sheath shall be of a continuous construction to provide mechanical protection and moisture seal. Where made of copper, it shall provide an adequate path for equipment grounding purposes. Where made of steel, an equipment grounding conductor in accordance with Article 250 shall be provided.

Article 331 — Electrical Nonmetallic Tubing

Contents

A. General

331-1. Definition. *Electrical nonmetallic tubing* is a pliable corrugated raceway of circular cross section with integral or associated couplings, connectors, and fittings listed for the installation of electric conductors. It is composed of a material that is resistant to moisture and chemical atmospheres and is flame retardant.

A pliable raceway is a raceway that can be bent by hand with a reasonable force, but without other assistance.

Electrical nonmetallic tubing shall be made of material that does not exceed the ignitibility, flammability, smoke generation, and toxicity characteristics of rigid (nonplasticized) polyvinyl chloride.

Electrical nonmetallic tubing (ENT) is made of the same material (PVC) used for rigid nonmetallic conduit (Article 347) suitable for aboveground use. The outside diameters of ENT (½-in. through 2-in. trade sizes only) are such that standard couplings and other fittings for rigid PVC conduit can be used.

Because of the corrugations, the raceway can be bent by hand and has some degree of flexibility, but it is not intended for use where flexibility is necessary, as at motor terminations to prevent transmission of noise and vibration, or for connection of adjustable fixtures or moving parts. ENT is suitable for the installation of conductors having a temperature rating as indicated on the product. The maximum allowable ambient temperature is 50°C (122°F). Figure 331.1 shows an example of electrical nonmetallic tubing.

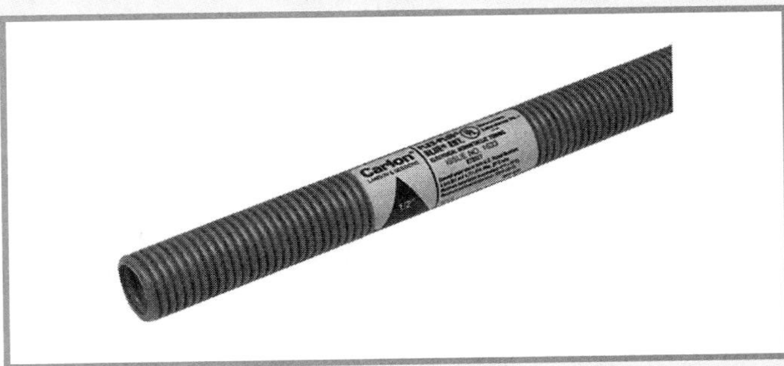

Figure 331.1 *An example of electrical nonmetallic tubing (ENT). (Carlon, a Lamson & Sessions Co.)*

331-2. Other Articles. Installations for electrical nonmetallic tubing shall comply with the provisions of the applicable sections of Article 300. Where equipment grounding is required by Article 250, a separate equipment grounding conductor shall be installed in the raceway.

331-3. Uses Permitted. The use of electrical nonmetallic tubing and fittings shall be permitted in the following:

(1) In any building not exceeding three floors above grade

(a) For exposed work, where not subject to physical damage

(b) Concealed within walls, floors, and ceilings

FPN: See Section 336-5(a)(1) for definition of first floor.

Electrical nonmetallic tubing (ENT) is permitted to be installed, either concealed or exposed. Where exposed and subject to physical damage, ENT is required to be protected and is limited to use in buildings not exceeding three floors above grade, as defined in Section 336-5(a). See definition of *exposed (as applied to wiring methods)* in Article 100. Where concealed or above a suspended ceiling, ENT is permitted to be installed within walls, floors, or ceilings in buildings of three floors or less without the need for fire-rated construction. The three-floor limitation is based on the likelihood that only a small quantity of ENT will be exposed to fire and that the occupants will have adequate time to exit the building before the products of combustion make the building untenable. Figure 331.2A shows some examples of permitted uses of ENT in a building of three floors or less.

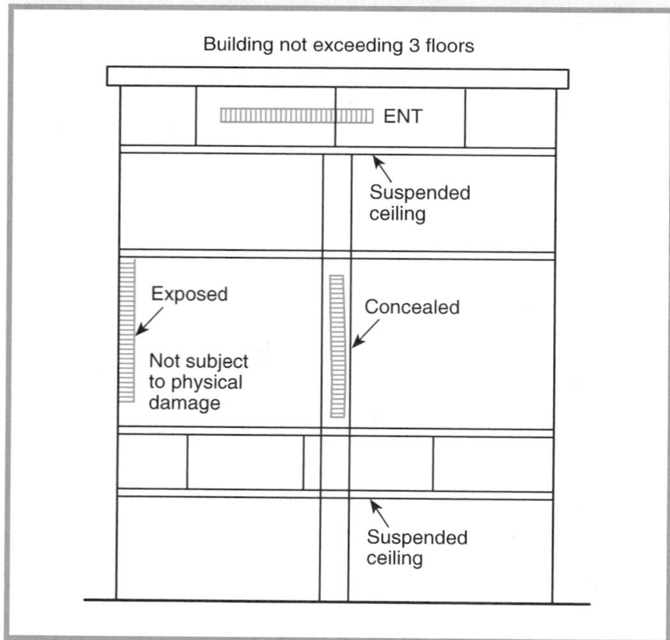

***Figure 331.2A** Examples of permitted uses of ENT in a building not exceeding three floors.*

(2) In any building exceeding three floors above grade, electrical nonmetallic tubing shall be concealed within walls, floors, and ceilings where the walls, floors, and ceilings provide a thermal barrier of material that has at least a 15-minute finish rating as identified in listings of fire-rated assemblies. The 15-minute finish rated thermal barrier shall be permitted to be used for combustible or noncombustible walls, floors, and ceilings.

Electrical nonmetallic tubing (ENT) is permitted to be installed within the walls, floors, or ceilings of a building of any height where the walls, floors, or ceilings provide a thermal barrier of material that has at least a 15-minute finish rating. See the commentary to the fine print note following Section 331-3(2). It is not permitted or intended that ENT be used exposed in the first three floors of a building that exceeds three floors except as permitted in Section 331-3(5). Where installed in a building over three floors, ENT must be installed behind the 15-minute thermal barrier on all floors. Figure 331.2B shows some examples of permitted uses of ENT in a building exceeding three floors.

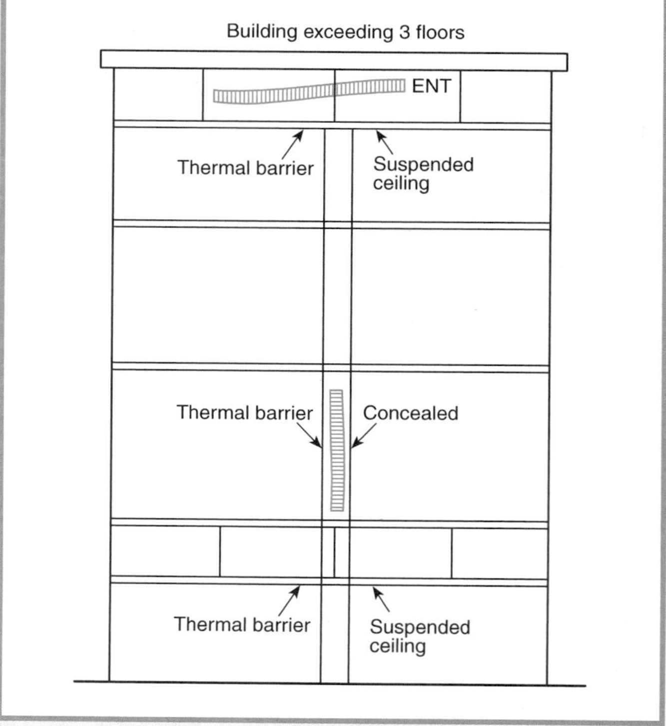

***Figure 331.2B** Examples of permitted uses of ENT in a building exceeding three floors.*

FPN: A finish rating is established for assemblies containing combustible (wood) supports. The finish rating is defined as the time at which the wood stud or wood joist reaches an average temperature rise of 121°C (250°F) or an individual temperature of 163°C (325°F) as measured on the plane of the wood nearest the fire. A finish rating is not intended to represent a rating for a membrane ceiling.

Interior finish is generally considered to consist of those materials or combinations of materials that form the exposed interior surface of walls and ceilings in a building. Common interior finish materials include plaster, gypsum wallboard, wood, plywood paneling, fibrous ceiling tiles, and a variety of wall coverings. Ordinary paint, wallpaper, or other similar wall coverings not exceeding 1/28 in. in thickness are generally considered incidental to interior finish, except where the authority having jurisdiction deems them a hazard.

The finish rating of a wall or ceiling finish material is the time required for the unexposed surface of the finish membrane to reach an average temperature rise of 121°C (250°F) above ambient or an individual temperature rise at any one point not exceeding 163°C (325°F) when the assembly is tested in accordance with *Standard Methods of Tests of Fire Endurance of Building Construction and Materials*, NFPA 251 (also known as ANSI/UL 263 or ASTM E119).

The finish rating of wall and ceiling finish materials tested and rated by UL as part of wall and ceiling assemblies can be found in the UL *Fire Resistance Directory*, immediately following the assembly rating and just below the design number. Only assemblies containing combustible support members, however, have published finish ratings. Obviously, it is not the intent to limit ENT to constructions consisting of combustible support members. This section is intended to provide a 15-minute thermal barrier as a minimum threshold of acceptability.

The following chart, reproduced from the NFPA *Fire Protection Handbook* (18th edition), Chapter 7, Table 7-4O, page 7-68, provides ratings for common finish materials. If the finish rating concealing the ENT is unknown or is less than 15 minutes, the ENT can still be used if the installation meets the criteria in Section 331-3, including the three-floor limitation, where required, and the installation is not prohibited by Section 331-4. For finish materials not tested and rated, use Table 331.1.

(3) In locations subject to severe corrosive influences as covered in Section 300-6 and where subject to chemicals for which the materials are specifically approved

(4) In concealed, dry, and damp locations not prohibited by Section 331-4

Table 331.1 Various Finishes Over Wood Framing, One Side (Combustible) with Exposure on Finish Side

Finish Material	Fire Resistance Rating[1] (minutes)
Fiberboard, ½ in. thick	5
Fiberboard, flameproofed, ½ in. thick	10
Fiberboard, ½ in. thick, with ½-in. 1:2, 1:2 gypsum-sand plaster	15
Gypsum wallboard, ⅜ in. thick	10
Gypsum wallboard, ½ in. thick	15
Gypsum wallboard, ⅝ in. thick	20
Gypsum wallboards, laminated, two ⅜ in. thick	28
Gypsum wallboards, laminated, one ⅜ in. plus one ½ in. thick	37
Gypsum wallboards, laminated, two ½ in. thick	47
Gypsum wallboards, laminated, two ⅝ in. thick	60
Gypsum lath, plain or indented, ⅜ in. thick, with ½-in. 1:2, 1:2 gypsum-sand plaster	20
Gypsum lath, perforated, ⅜ in. thick, with ½-in. 1:2, 1:2 gypsum-sand plaster	30
Gypsum-sand plaster, 1:2, 1:3, ½ in. thick, on wood lath	15
Lime-sand plaster, 1:5, 1:7.5, ½ in. thick, on wood lath	15
Gypsum-sand plaster, 1:2, 1:2, ¾ in. thick, on metal lath (no paper backing)	15
Neat gypsum plaster, ¾ in. thick on metal lath (no paper backing)[2]	15
Neat gypsum plaster, 1 in. thick on metal lath (no paper backing)[2]	35
Lime-sand plaster, 1:5, 1:7.5, ¾ in. thick, on metal lath (no paper backing)	10
Portland cement plaster, ¾ in. thick, on metal lath (no paper backing)	10
Gypsum-sand plaster, 1:2, 1:3, ¾ in. thick, on paper-backed metal lath	20

Note: For SI units, 1 in. = 25.4 mm.

[1]From National Bureau of Standards, now known as the National Institute for Standards and Technology, BMS-92.

[2]Unsanded wood-fiber plaster.

(5) Above suspended ceilings where the suspended ceilings provide a thermal barrier of material that has at least a 15-minute finish rating as identified in listings of fire-rated assemblies, except as permitted in Section 331-3(1)(a).
(6) Encased in poured concrete, or embedded in a concrete slab on grade where ENT is placed on sand or approved screenings, provided fittings identified for this purpose are used for connections
(7) For wet locations indoors as permitted in this section or in a concrete slab on or below grade, with fittings listed for the purpose
(8) ½ in. through 1 in. as listed manufactured prewired assembly

Prewired ENT is a listed assembly where the conductors are required to be installed at the manufacturing facility where controlled conditions will prevent damage to the conductor insulation. Special cutting tools are required to be used when cutting prewired ENT to prevent "nicking" the conductor installation. This prewired assembly is shown in Figure 331.3.

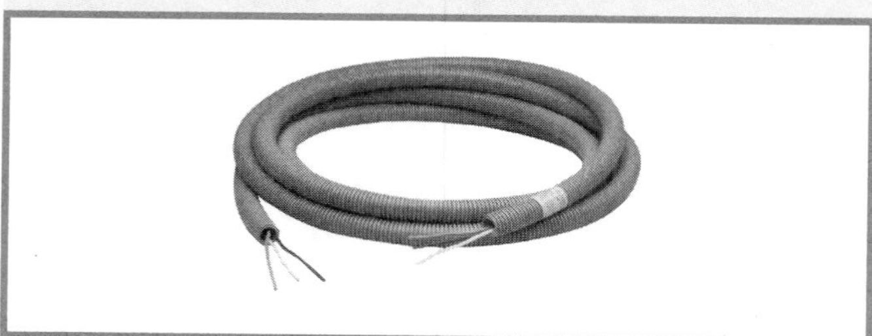

Figure 331.3 *Listed manufactured prewired ENT assembly. (Carlon, a Lamson & Sessions Co.)*

FPN: Extreme cold may cause some types of nonmetallic conduits to become brittle and, therefore, more susceptible to damage from physical contact.

331-4. Uses Not Permitted. Electrical nonmetallic tubing shall not be used in the following:

(1) In hazardous (classified) locations, except as permitted by Section 504-20 and Section 505-15(a)(1)
(2) For the support of fixtures and other equipment
(3) Where subject to ambient temperatures in excess of 50°C (122°F) unless listed otherwise
(4) For conductors whose insulation temperature limitations would exceed those for which the tubing is listed
(5) For direct earth burial
(6) Where the voltage is over 600 volts
(7) In exposed locations, except as permitted by Sections 331-3(1), 331-3(5), and 331-3(7)
(8) In theaters and similar locations, except as provided in Articles 518 and 520

See Sections 518-4 and 520-5 for specific wiring methods permitted.

(9) Where exposed to the direct rays of the sun, unless identified as sunlight resistant

B. Installation

331-5. Size.

(a) Minimum. Tubing smaller than ½-in. electrical trade size shall not be used.

(b) Maximum. Tubing larger than 2-in. electrical trade size shall not be used.

FPN: Metric trade numerical designations for electrical nonmetallic tubing are ½ = 16, ¾ = 21, 1 = 27, 1¼ = 35, 1½ = 41, and 2 = 53.

331-6. Number of Conductors in Tubing. The number of conductors in a single tubing shall not exceed that permitted by the percentage fill in Table 1, Chapter 9.

Table 1 of Chapter 9 specifies the maximum percent fill of a conduit or tubing. Table 4 (expanded and revised for the 1996 *Code*) provides the usable area within the selected conduit or tubing, and Table 5 provides the required area for each of the conductors. *Examples* using these tables to calculate a conduit or tubing size are provided following Chapter 9, *Table 1, Notes to Tables, Note 6.*

If the conductors are of the same wire size, Appendix C, added in the 1996 *Code,* may be consulted. Appendix C, which contains 12 sets of tables, very accurately indicates the maximum number of conductors permitted in a conduit or tubing. This appendix replaced Tables 3A, 3B, and 3C in the 1993 *Code.* Examples using this appendix to select a conduit or tubing size are provided following the introduction in Appendix C.

In order to select the proper trade size electrical nonmetallic tubing, the section entitled "Electrical Nonmetallic Tubing" in Table 4 of Chapter 9 should be followed. Appendix C Tables C2 and C2A for electrical nonmetallic tubing are also permissible.

331-7. Trimming. All cut ends of tubing shall be trimmed inside and outside to remove rough edges.

331-8. Joints. All joints between lengths of tubing and between tubing and couplings, fittings, and boxes shall be by an approved method.

331-9. Bends — How Made. Bends of electrical nonmetallic tubing shall be made so that the tubing will not be damaged and that the internal diameter of the tubing will not be effectively reduced. Bends shall be permitted to be made manually without auxiliary equipment, and the radius of the curve of the inner edge of such bends shall not be less than shown in Table 346-10.

331-10. Bends — Number in One Run. There shall not be more than the equivalent of four quarter bends (360 degrees total) between pull points, e.g., conduit bodies and boxes.

331-11. Supports. Electrical nonmetallic tubing shall be installed as a complete system as provided in Article 300 and shall be secured at intervals not exceeding 3 ft (914 mm) or, for horizontal runs, it shall be permitted to be supported by openings in framing members at intervals not exceeding 3 ft (914 mm). In addition, it shall be securely fastened in place within 3 ft (914 mm) of each outlet box, device box, junction box, cabinet, or fitting where it terminates.

As illustrated in Figure 331.4, where ENT is run on the surface of framing members, it is required to be fastened to the framing member every 3 ft and within 3 ft of every box. See Section 300-4(d) for protection against physical damage.

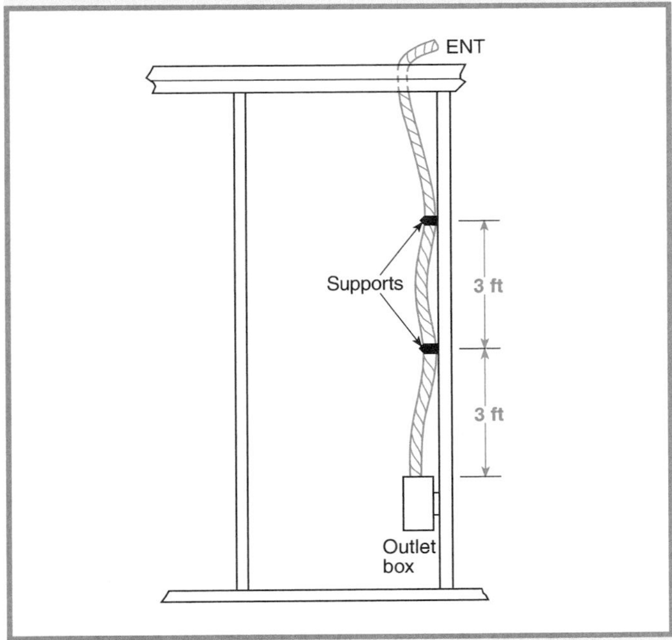

Figure 331.4 An application of Section 331-11, showing ENT supported every 3 ft and within 3 ft of the box.

Exception: Lengths not exceeding a distance of 6 ft (1.83 m) from a fixture terminal connection for tap connections to lighting fixtures shall be permitted without being secured.

As the example in Figure 331.5 shows, ENT is permitted to be used as fixture whip without support for lengths not exceeding 6 ft. See Section 410-67(c) for tap conductor wiring details.

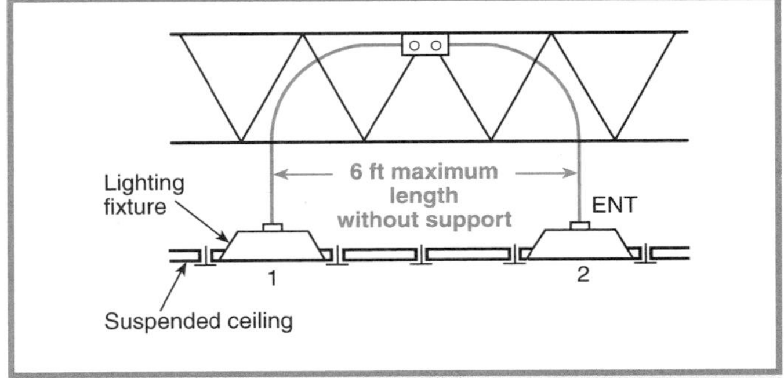

Figure 331.5 An application of Section 331-11, Exception showing ENT unsupported in lengths not exceeding 6 ft.

331-12. Boxes and Fittings. Electrical nonmetallic tubing shall be used only with listed fittings. Boxes and fittings shall comply with the applicable provisions of Article 370.

331-13. Splices and Taps. Splices and taps shall be made only in junction boxes, outlet boxes, device boxes, or conduit bodies. See Article 370 for rules on the installation and use of boxes and conduit bodies.

331-14. Bushings. Where a tubing enters a box, fitting, or other enclosure, a bushing or adapter shall be provided to protect the wire from abrasion unless the box, fitting, or enclosure design provides equivalent protection.

FPN: See Section 300-4(f) for the protection of conductors size No. 4 or larger.

C. Construction Specifications

331-15. General. Electrical nonmetallic tubing shall be clearly and durably marked at least every 10 ft (3.05 m) as required in the first sentence of Section 110-21. The type of material shall also be included in the marking. Tubing that has limited smoke-producing characteristics shall be permitted to be identified with the suffix LS. The type, size, and quantity of conductors used in prewired manufactured assemblies shall be identified by means of a printed tag or label attached to each end of the manufactured assembly and either the carton, coil, or reel. The enclosed conductors shall be marked in accordance with Section 310-11.

ENT, as a prewired manufactured assembly, shall be provided in continuous lengths capable of being shipped in a coil, reel, or carton without damage.

Article 333 — Armored Cable: Type AC

Contents

A. General

333-1. Definition. *Type AC cable* is a fabricated assembly of insulated conductors in a flexible metallic enclosure. See Section 333-19.

Type AC cable (armored cable) is listed by Underwriters Laboratories Inc. in sizes No. 14 through No. 1 copper and No. 12 through No. 1 aluminum or copper-clad aluminum and is rated at 600 volts or less. Figure 333.1 shows an example of Type AC cable.

Figure 333.1 An example of Type AC cable. (AFC Cable Systems, Inc.)

333-2. Other Articles. Type AC cable shall comply with this article and also with the applicable provisions of other articles in this *Code*, especially Article 300.

333-3. Uses Permitted. Except where otherwise specified in this *Code* and where not subject to physical damage, Type AC cable shall be permitted for branch circuits and feeders in both exposed and concealed work and in cable trays where identified for such use.

Type AC cable shall be permitted in dry locations and embedded in plaster finish on brick or other masonry, except in damp or wet locations. It shall be permissible to run or fish this cable in the air voids of masonry block or tile walls where such walls are not exposed or subject to excessive moisture or dampness.

333-4. Uses Not Permitted. Type AC cable shall not be used where prohibited elsewhere in this *Code*, including the following:

(1) In theaters and similar locations, except as provided in Article 518, Places of Assembly
(2) In motion picture studios
(3) In any hazardous (classified) location except as permitted by Sections 501-4(b), Exception, 502-4(b), Exception No. 1, and 504-20
(4) Where exposed to corrosive fumes or vapors
(5) On cranes or hoists, except as provided in Section 610-11(c)
(6) In storage battery rooms
(7) In hoistways or on elevators, except as provided in Section 620-21
(8) In commercial garages where prohibited in Article 511

B. Installation

333-7. Support. Type AC cable shall be secured by staples, cable ties, straps, hangers, or similar fittings designed and installed so as not to damage the cable at intervals not exceeding 4½ ft (1.37 m) and within 12 in. (305 mm) of every outlet box, junction box, cabinet, or fitting.

Section 333-7 requires that Type AC cable be *secured.* Simply draping the cable over air ducts or lower members of bar joists, pipes, and ceiling grid members is not permitted. Section 333-7(a) permits Type AC cable, where run horizontally through framing members, to be passed through bored or punched holes in framing members without additional securing provided the cable is secured within 12 in. of the outlet and the framing members are less than 54 in. apart. See Section 300-4(c) for support requirements of cables through space behind panels designed to allow access.

(a) Horizontal Runs. Type AC cable installed in other than vertical runs through bored or punched holes in wood or metal framing members, or through notches in wooden framing members and protected by a steel plate at least 1/16 in. (1.59 mm) thick, shall be considered secured where the support intervals do not exceed 4½ ft (1.37 m) and the armored cable is securely fastened in place by an approved means within 12 in. (505 mm) of each box, cabinet, conduit body, or other armored cable termination.

(b) Unsupported. Type AC cable shall be permitted to be unsupported where the cable is

(1) Fished between access points, where concealed in finished buildings or structures and supporting is impracticable;
(2) Not more than 2 ft (610 mm) in length at terminals where flexibility is necessary; or
(3) Not more than 6 ft (1.83 m) in length from an outlet for connections within an accessible ceiling to lighting fixtures or equipment.

(c) Cable Tray Installations. Type AC cable installed in cable trays shall comply with Section 318-8(b).

333-8. Bending Radius. All bends shall be made so that the cable will not be damaged, and the radius of the curve of the inner edge of any bend shall not be less than five times the diameter of the Type AC cable.

333-9. Boxes and Fittings. At all points where the armor of AC cable terminates, a fitting shall be provided to protect wires from abrasion, unless the design of the outlet boxes or fittings is such as to afford equivalent protection, and, in addition, an insulating bushing or its equivalent protection shall be provided between the conductors and the armor. The connector or clamp by which the Type AC cable is fastened to boxes or cabinets shall be of such design that the insulating bushing or its equivalent will be visible for inspection. Where change is made from Type AC cable to other cable or raceway wiring methods, a box, fitting, or conduit body shall be installed at junction points as required in Section 300-15.

Armored cable connectors are considered suitable for equipment grounding if installed in accordance with Section 300-10.

333-10. Through or Parallel to Framing Members. Type AC cable shall comply with Section 300-4 where installed through or parallel to studs, joists, rafters, or similar wood or metal members.

333-11. Exposed Work. Exposed runs of cable shall closely follow the surface of the building finish or of running boards. Exposed runs shall also be permitted to be installed on the underside of joists where supported at each joist and located so as not to be subject to physical damage.

333-12. In Accessible Attics. Type AC cables in accessible attics or roof spaces shall be installed as specified in (a) and (b).

(a) Where Run Across the Top of Floor Joists. Where run across the top of floor joists, or within 7 ft (2.13 m) of floor or floor joists across the face of rafters or studding, in attics and roof spaces that are accessible, the cable shall be protected by substantial guard strips that are at least as high as the cable. Where this space is not accessible by permanent stairs or ladders, protection shall only be required within 6 ft (1.83 m) of the nearest edge of the scuttle hole or attic entrance.

(b) Cable Installed Parallel to Framing Members. Where the cable is installed parallel to the sides of rafters, studs, or floor joists, neither guard strips nor running boards shall be required, and the installation shall also comply with Section 300-4(d).

C. Construction Specifications

333-19. Construction. Type AC cable shall have an armor of flexible metal tape. The insulated conductors shall be in accordance with Section 333-20. Cables of the AC type shall have an internal bonding strip of copper or aluminum in intimate contact with the armor for its entire length.

The armor of Type AC cable is recognized as an equipment grounding conductor by Section 250-118. The required internal bonding strip can be simply cut off at the termination of the armored cable, or it can be bent back on the armor. It is not necessary to connect it to an equipment grounding terminal. It reduces the inductive reactance of the spiral armor and increases the armor's effectiveness as an equipment ground. Many installers use this strip to help prevent the insulating bushing required by Section 333-9 (the "red head") from falling out during rough wiring.

333-20. Conductors. Insulated conductors shall be of a type listed in Table 310-13 or those identified for use in this cable. In addition, the conductors shall have an overall moisture-resistant and fire-retardant fibrous covering. For Type ACT, a moisture-resistant fibrous covering shall be required only on the individual conductors. The ampacity shall be determined by Section 310-15.

Armored cable installed in thermal insulation shall have conductors rated at 90°C (194°F). The ampacity of cable installed in these applications shall be that of 60°C (140°F) conductors.

Cable marked "ACTH" indicates an armored cable rated 75°C employing conductors having *thermoplastic* insulation. Cable marked "ACTHH" indicates an armored cable rated 90°C employing conductors having *thermoplastic* insulation. Cable marked "ACHH" indicates armored cable rated 90°C employing conductors having *thermosetting* insulation.

The requirements for armored cable installed in thermal insulation recognizes the decrease in heat dissipation capability of cables.

333-21. Grounding. Type AC cable shall provide an adequate path for equipment grounding as required by Section 250-2(d).

333-22. Marking. The cable shall be marked in accordance with Section 310-11, except that Type AC shall have ready identification of the manufacturer by distinctive external markings on the cable sheath throughout its entire length.

Article 334 — Metal-Clad Cable: Type MC

Contents

A. General

334-1. Definition. *Type MC cable* is a factory assembly of one or more insulated circuit conductors with or without optical fiber members enclosed in an armor of interlocking metal tape, or a smooth or corrugated metallic sheath.

Type MC (metal-clad) cable is available in three designs: interlocked metal tape, corrugated tube, and smooth tube. All are intended for aboveground use, except where marked for direct burial. Cables suitable for use in cable trays, direct sunlight, or when directly buried are so marked. Type MC cable includes Type CS (copper sheath) and Type ALS (aluminum sheath). Figure 334.1 shows some examples of Type MC cable.

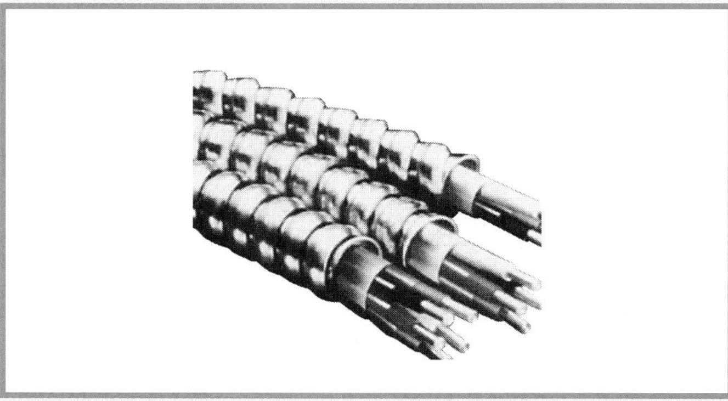

Figure 334.1 *Some examples of Type MC cable. (AFC Cable Systems, Inc.)*

334-2. Other Articles. Metal-clad cable shall comply with this article and also with the applicable provisions of other articles in this *Code*, especially Article 300.

Type MC cable shall be permitted for systems in excess of 600 volts, nominal. See Section 300-2(a).

Type MC cable is rated for use up to 2000 volts and listed in sizes No. 18 and larger for copper and No. 12 and larger for aluminum or copper-clad aluminum, and it employs thermoset or thermoplastic insulated conductors. Composite electrical MC and optical fiber cables are permitted by Section 770-5(c) and are marked "MC-OF." See Figure 770.2 for an example.

334-3. Uses Permitted. Unless specifically prohibited elsewhere in this *Code* and where not subject to physical damage, Type MC cables shall be permitted as follows:

(1) For services, feeders, and branch circuits
(2) For power, lighting, control, and signal circuits
(3) Indoors or outdoors
(4) Where exposed or concealed
(5) Direct buried where identified for such use
(6) In cable tray
(7) In any raceway
(8) As open runs of cable
(9) As aerial cable on a messenger
(10) In hazardous (classified) locations as permitted in Articles 501, 502, 503, 504, and 505
(11) In dry locations and embedded in plaster finish on brick or other masonry except in damp or wet locations
(12) In wet locations where any of the following conditions are met:
  - (a) The metallic covering is impervious to moisture.
  - (b) A lead sheath or moisture-impervious jacket is provided under the metal covering.

(c) The insulated conductors under the metallic covering are listed for use in wet locations.

334-4. Uses Not Permitted. Type MC cable shall not be used where exposed to destructive corrosive conditions, such as direct burial in the earth, in concrete, or where exposed to cinder fills, strong chlorides, caustic alkalis, or vapors of chlorine or of hydrochloric acids, unless the metallic sheath is suitable for the conditions or is protected by material suitable for the conditions.

B. Installation

334-10. Installation. Type MC cable shall be installed in compliance with Articles 300, 490, 725, and Section 770-52 as applicable and in accordance with the following.

(a) Supported Cables. Type MC cable shall be supported and secured at intervals not exceeding 6 ft (1.83 m). Cables containing four or fewer conductors, sized no larger than No. 10 shall be secured within 12 in. (305 mm) of every box, cabinet, fitting, or other cable termination.

Section 334-10(a) contains the general support requirements for Type MC cable, that is, supported and secured at least every 6 ft. Additionally, MC cable containing four or fewer conductors of size No. 10 or less are required to be secured within 12 in. from every box, cabinet, or fitting, as illustrated in Figure 334.2.

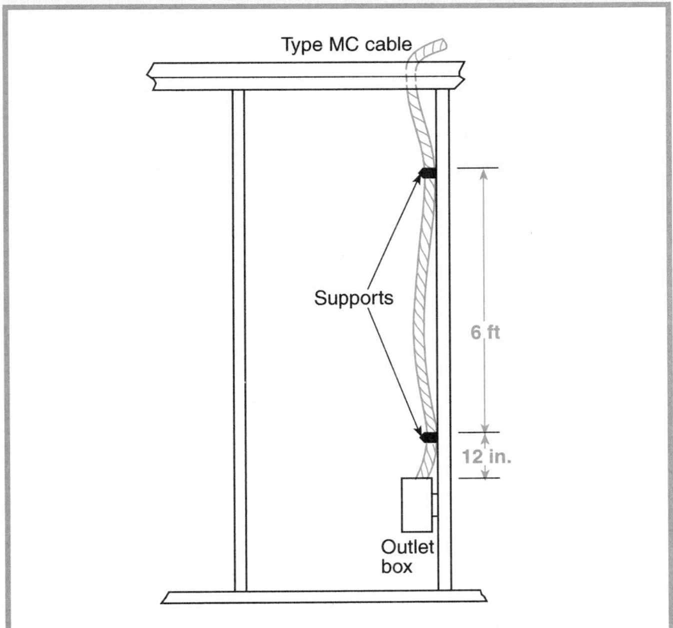

Figure 334.2 *An application of Section 334-10(a), showing Type MC cable supported and secured at intervals not exceeding 6 ft and within 12 in. of the box.*

(1) Horizontal Runs. Cables installed in other than vertical runs through bored or punched holes in wood or metal framing members, or through notches in wooden framing members and protected by a steel plate at least 1/16 in. (1.59 mm) thick shall be considered supported and secured where such support does not exceed 6-ft (1.83-m) intervals.

(2) At Terminations. Cables containing four or fewer conductors, sized not larger than No. 10 shall be secured within 12 in. (305 mm) of every box, cabinet, fitting, or other cable termination.

Type MC cable that is run horizontally through framing members (spaced less than 6 ft apart) and passes through bored or punched holes in framing members without additional securing is considered supported by the framing members. Cable ties are not required as the cable passes through these members. But, the MC cable must be secured (fastened in place) within 12 in. of the outlet box. See Section 300-4(c) for support requirements of cables through spaces behind panels designed to allow access.

(b) Unsupported Cables. Type MC cable shall not be required to be supported and secured where the cable is fished between access points, where concealed in finished buildings or structures and supporting is impracticable, or where used in lengths not more than 6 ft (1.83 m) from an outlet for connections within an accessible ceiling to lighting fixture(s) or equipment.

Section 334-10(b) permits Type MC cable to be fished, as shown in Figure 334.3.

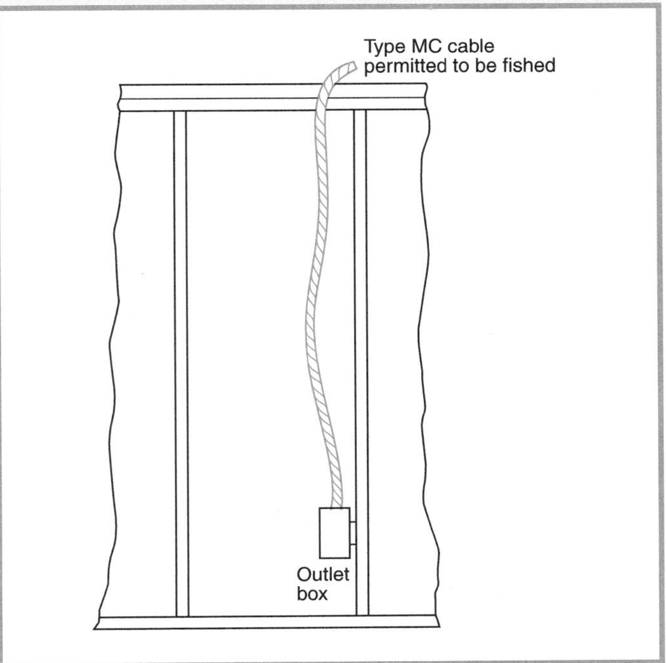

Figure 334.3 *An application of Section 334-10(b), which permits Type MC cable to be fished in walls, floors, or ceilings.*

(c) Cable Tray. Type MC cable installed in cable tray shall comply with Article 318.

(d) Direct Buried. Direct-buried cable shall comply with Sections 300-5 or 300-50, as appropriate.

(e) Installed as Service-Entrance Cable. Type MC cable installed as service-entrance cable shall comply with Article 230.

(f) Installed Outside of Buildings or as Aerial Cable. Type MC cable installed outside of buildings or as aerial cable shall comply with Article 225 and Article 321.

(g) Through or Parallel to Joists, Studs, and Rafters. Type MC cable shall comply with Section 300-4 where installed through or parallel to joists, studs, rafters, or similar wood or metal members.

(h) In Accessible Attics. The installation of Type MC cable in accessible attics or roof spaces shall also comply with Section 333-12.

In accessible attics, Type MC cable installed across the top of floor joists or within 7 ft of the floor or floor joists across the face of rafters or studs must be protected by guard strips. Where the attic is not accessible by a permanent ladder or stairs, guard strips are required only within 6 ft of the scuttle hole or opening.

334-11. Bending Radius. All bends shall be made so that the cable will not be damaged, and the radius of the curve of the inner edge of any bend shall not be less than shown in (a) through (c).

(a) Smooth Sheath.

(1) Ten times the external diameter of the metallic sheath for cable not more than ¾ in. (19 mm) in external diameter

(2) Twelve times the external diameter of the metallic sheath for cable more than ¾ in. (19 mm) but not more than 1½ in. (38 mm) in external diameter

See the commentary following Section 334-11(c).

(3) Fifteen times the external diameter of the metallic sheath for cable more than 1½ in. (38 mm) in external diameter

(b) Interlocked-Type Armor or Corrugated Sheath. Seven times the external diameter of the metallic sheath.

(c) Shielded Conductors. Twelve times the overall diameter of one of the individual conductors or seven times the overall diameter of the multiconductor cable, whichever is greater.

The minimum bending radius of 12 times the outside diameter (OD) of a single shielded conductor [Section 334-11(a)(2)] is consistent with ICEA (formerly IPCEA) requirements and good engineering practice. The same minimum on a multiconductor cable, however, would be excessive.

Example

Consider 500-kcmil, 15-kV, 100 percent insulation level (0.175-in. insulation):

Cable	OD	12 × OD	7 × OD
Single-conductor	1.50 in.	18 in.	—
Three-conductor	3.15–3.50 in.	38–42 in.	22–24 in.

334-12. Fittings. Fittings used for connecting Type MC cable to boxes, cabinets, or other equipment shall be listed and identified for such use. Where single-conductor cables enter ferrous metal boxes or cabinets, the installation shall comply with Section 300-20 to prevent inductive heating.

Connectors should be selected in accordance with the size and type of cable for which they are designated. Bronze connectors are intended for use only with cable employing corrugated copper armor. Some Type MC cable connectors are also acceptable for Type AC cable when specifically indicated on the device or the shipping carton.

334-13. Ampacity. The ampacity of Type MC cable shall be in accordance with Sections 310-15 or 310-60.

Exception No. 1: The ampacities for Type MC cable installed in cable tray shall be determined in accordance with Sections 318-11 and 318-13.

Exception No. 2: The ampacities of No. 18 and No. 16 conductors shall be in accordance with Table 402-5.

FPN: See Section 310-10 for temperature limitation of conductors.

C. Construction Specifications

334-20. Conductors. The conductors shall be of copper, aluminum, or copper-clad aluminum, solid or stranded.

The minimum conductor size shall be No. 18 copper and No. 12 aluminum or copper-clad aluminum.

334-21. Insulation. The insulated conductors shall comply with (a) or (b).

(a) 600 Volts. Insulated conductors in sizes No. 18 and No. 16 shall be of a type listed in Table 402-3, with a maximum operating temperature not less than 90°C (194°F), and as permitted by Section 725-27. Conductors larger than No. 16 shall be of a type listed in Table 310-13 or of a type identified for use in Type MC cable.

(b) Over 600 Volts. Insulated conductors shall be of a type listed in Tables 310-61 through 310-64.

334-22. Metallic Sheath. The metallic covering shall be one of the following types: smooth metallic sheath, corrugated metallic sheath, interlocking metal tape armor. The metallic sheath shall be continuous and close fitting.

Supplemental protection of an outer covering of corrosion-resistant material shall be permitted and shall be required where such protection is needed. The sheath shall not be used as a current-carrying conductor.

FPN: See Section 300-6 for protection against corrosion.

334-23. Grounding. Type MC cable shall provide an adequate path for equipment grounding as required by Article 250.

The armor of interlocking Type MC cable is not recognized by UL as the sole means of providing an equipment grounding circuit but may be used to supplement the internal grounding conductor.

334-24. Marking. The cable shall be marked in accordance with Section 310-11.

Type MC cable is required to be marked with the maximum rated voltage, the proper insulation type letter or letters, and the AWG size or circular mil area. This marking may be on a marker tape located within the cable and running its complete length, or, if the metallic covering is of smooth construction that permits surface marking, the MC cable may be durably marked on the outer covering at intervals not exceeding 24 in. For MC cable with an outer nonmetallic covering, the marking is permitted on the surface of the nonmetallic jacket. See Section 310-11 for marking requirements.

Article 336 — Nonmetallic-Sheathed Cable: Types NM, NMC, and NMS

Contents

336-1. Scope. This article covers the use, installation, and construction specifications of nonmetallic-sheathed cable.

A. General

336-2. Definition. *Nonmetallic-sheathed cable* is a factory assembly of two or more insulated conductors having an outer sheath of moisture-resistant, flame-retardant, nonmetallic material.

Types NM, NMC, and NMS nonmetallic-sheathed cable may be used for either exposed or concealed wiring. Where exposed, it should not be subject to physical damage. It was first recognized in the 1928 *NEC* as a substitute for concealed knob-and-tube wiring (Article 324) and open wiring on insulators (Article 320). The basic advantages of nonmetallic-sheathed cable are that the outer sheath provides continuous protection in addition to the insulation applied to the conductors; the cable is easily fished in partitions of finished buildings; no insulating supports are required; and only one hole need be bored that can accommodate more than one cable passing through a wood cross member.

336-3. Other Articles. Installations of nonmetallic-sheathed cable shall comply with the other applicable provisions of this *Code*, especially Articles 300 and 310.

Two examples (one from Article 300 and one from Article 310) of the many applicable requirements are as follows.

1. In accordance with Section 300-4(b)(1), where the cable passes through factory- or field-punched holes in metal studs or similar members, it is required to be protected by bushings or grommets covering all metal edges and securely fastened in the opening prior to installation of the cable.

2. Section 310-15(b)(2)(a) states in part: "or where single conductors or multiconductor cables are stacked or bundled longer than 24 in. (610 mm) without maintaining spacing and are not installed in raceways, the allowable ampacity of each conductor shall be reduced as shown in Table 310-15(b)(2)." Failure to comply with the appropriate adjustment ampacity derating called for by this table, where nonmetallic-sheathed cables may be stacked or bundled, can lead to overheating of conductors.

336-4. Uses Permitted. Type NM, Type NMC, and Type NMS cables shall be permitted to be used in the following:

(1) One- and two-family dwellings
(2) Multifamily dwellings and other structures, except as prohibited in Section 336-5
(3) Cable trays, where the cables are identified for the use

FPN: See Section 310-10 for temperature limitation of conductors.

(a) Type NM. Type NM cable shall be permitted for both exposed and concealed work in normally dry locations. It shall be permissible to install or fish Type NM cable in air voids in masonry block or tile walls where such walls are not exposed or subject to excessive moisture or dampness.

For concealed work, nonmetallic-sheathed cable should be installed where it is protected from physical damage often caused by nails or screws. Where practical, care should be taken to avoid areas where trim, door and window casings, baseboards, moldings, and so on, are likely to be nailed. See Section 300-4 for details on protection against physical damage.

(b) Type NMC. Type NMC cable shall be permitted as follows:

(1) For both exposed and concealed work in dry, moist, damp, or corrosive locations
(2) In outside and inside walls of masonry block or tile
(3) In a shallow chase in masonry, concrete, or adobe protected against nails or screws by a steel plate at least 1/16-in. (1.59-mm) thick, and covered with plaster, adobe, or similar finish

Type NMC (corrosion-resistant) cable is required for installations in dairy barns and similar farm buildings (see Article 547), where cable will be exposed to fumes, vapors, or liquids such as ammonia and barnyard acids. Under such circumstances, ordinary types of nonmetallic-sheathed cable have in some cases deteriorated rapidly due to ammonia fumes or the growth of fungus or mold.

In addition to insulated conductors, nonmetallic-sheathed cable may have an insulated or bare conductor for equipment grounding purposes only. See Section 250-119 and Table 250-122 for identification requirements and minimum conductor sizes.

(c) Type NMS. Type NMS cable shall be permitted for both exposed and concealed work in normally dry locations. It shall be permissible to install or fish Type NMS cable in air voids in masonry block or tile walls where such walls are not exposed or subject to excessive moisture or dampness. Type NMS cable shall be used as permitted in Article 780.

The construction of this hybrid cable, Type NMS, is fully described in Sections 336-30(a)(3) and 780-6(a). Type NMS nonmetallic-sheathed cable is intended for use with "smart house" circuits as permitted in Article 780.

336-5. Uses Not Permitted.

(a) Types NM, NMC, and NMS. Types NM, NMC, and NMS cables shall not be used in the following:

(1) In any multifamily dwelling or other structure exceeding three floors above grade

Nonmetallic-sheathed cable is permitted for dwellings and other structures such as stores, professional offices, motels, industrial occupancies, and the like, provided these dwellings or structures do not exceed three floors above grade. The word *floor,* as used in Section 336-5, means the entire story of the building, from floor to ceiling. It is required that the calculations for determining the first story be based on 50 percent of the area of the lower story exterior wall surfaces, as illustrated in Figure 336.1. One additional level "not designed for human habitation" is permitted, as depicted in Figure 336.2.

For the purpose of this article, the first floor of a building shall be that floor that has 50 percent or more of the exterior wall surface area level with or above finished grade. One additional level that is the first level and not designed for human habitation and used only for vehicle parking, storage, or similar use shall be permitted.

(2) As service-entrance cable
(3) In commercial garages having hazardous (classified) locations as provided in Section 511-3

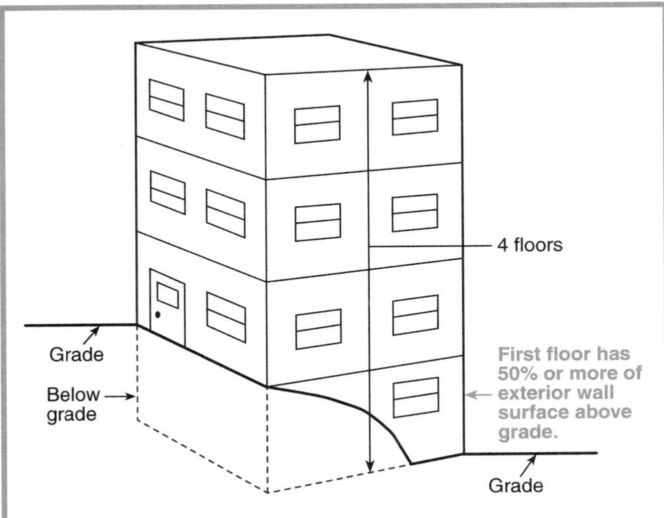

Figure 336.1 *A representation of a first floor of a building that is level with or above finished grade for 50 percent or more of its exterior wall surface area between grade and the floor above.*

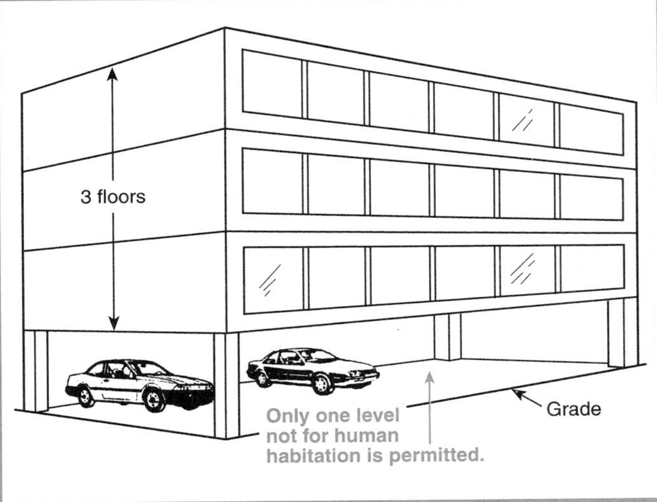

Figure 336.2 *A representation of a building where the first level is a parking area so that, for the purposes of Section 336-5(a)(1), the first floor is the first level above the parking area.*

(4) In theaters and similar locations, except as provided in Article 518, Places of Assembly
(5) In motion picture studios
(6) In storage battery rooms
(7) In hoistways
(8) Embedded in poured cement, concrete, or aggregate
(9) In any hazardous (classified) location, except as permitted by Sections 501-4(b), Exception, 502-4(b), Exception, and 504-20

(b) Types NM and NMS. Types NM and NMS cable shall not be installed in the following:

(1) Where exposed to corrosive fumes or vapors
(2) Where embedded in masonry, concrete, adobe, fill, or plaster
(3) In a shallow chase in masonry, concrete, or adobe and covered with plaster, adobe, or similar finish

B. Installation

336-6. Exposed Work — General. In exposed work, except as provided in Section 300-11(a), the cable shall be installed as specified in (a) through (d).

(a) To Follow Surface. The cable shall closely follow the surface of the building finish or of running boards.

(b) Protection from Physical Damage. The cable shall be protected from physical damage where necessary by conduit, electrical metallic tubing, Schedule 80 PVC rigid nonmetallic conduit, pipe, guard strips, listed surface metal or nonmetallic raceway, or other means. Where passing through a floor, the cable shall be enclosed in rigid metal conduit, intermediate metal conduit, electrical metallic tubing, Schedule 80 PVC rigid nonmetallic conduit, listed surface metal or nonmetallic raceway, or other metal pipe extending at least 6 in. (152 mm) above the floor.

(c) In Unfinished Basements. Where the cable is run at angles with joists in unfinished basements, it shall be permissible to secure cables not smaller than two No. 6 or three No. 8 conductors directly to the lower edges of the joists. Smaller cables shall either be run through bored holes in joists or on running boards.

As illustrated in Figure 336.3, nonmetallic-sheathed cables installed in an unfinished basement can be run through joists (A) and attached to the side or face of joists or beams (B) and running boards (C). Section 300-4(d) requires cables that are run parallel to framing members be installed at least 1¼ in. from the nearest edge of studs, joists, or rafters.

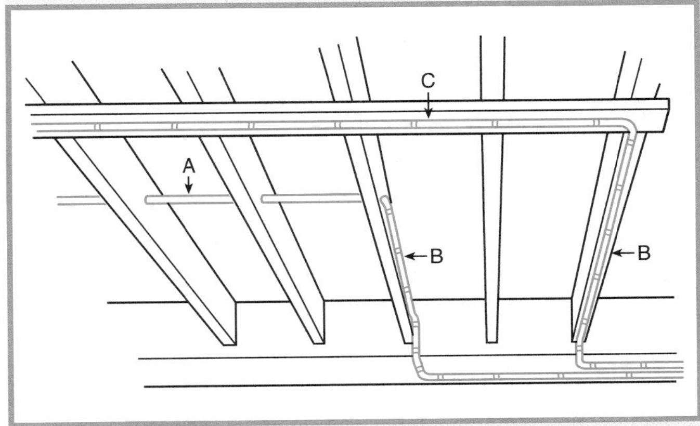

Figure 336.3 *Nonmetallic-sheathed cables installed in an unfinished basement.*

(d) In Accessible Attics. The installation of cable in accessible attics or roof spaces shall also comply with Section 333-12.

336-9. Through or Parallel to Framing Members. Types NM, NMC, or NMS cable shall comply with Section 300-4 where installed through or parallel to joists, studs, rafters, or similar wood or metal members.

336-16. Bends. Bends in cable shall be made so, and other handling shall be such, that the cable will not be damaged and the radius of the curve of the inner edge of any bend shall not be less than five times the diameter of the cable.

336-18. Supports. Nonmetallic-sheathed cable shall be secured by staples, cable ties, straps, or similar fittings designed and installed so as not to damage the cable. Cable shall be secured in place at intervals not exceeding 4½ ft (1.37 m) and within 12 in. (305 mm) from every cabinet, box, or fitting. Flat cables shall not be stapled on edge. Cables run through holes in wood or metal joists, rafters, or studs shall be considered to be supported and secured.

FPN: See Section 370-17(c) for support where nonmetallic boxes are used.

Exception No. 1: For concealed work in finished buildings or finished panels for prefabricated buildings where such supporting is impracticable, it shall be permissible to fish cable between access points.

Exception No. 2: A wiring device identified for the use, without a separate outlet box, incorporating an integral cable clamp shall be permitted where the cable is secured in place at intervals not exceeding 4½ ft (1.37 m) and within 12 in. (305 mm) from the wiring device wall opening, and there shall be at least a 12-in. (305-mm) loop of unbroken cable or 6 in. (152 mm) of a cable end available on the interior side of the finished wall to permit replacement.

Section 336-18 requires that the cable be *secured*. Simply draping the cable over air ducts or lower members of bar joists, pipes, and ceiling grid members is not permitted, except where fished as allowed by Exception No. 1.

It also prohibits two-conductor nonmetallic-sheathed cable (or other flat configurations) from being stapled on edge. The intent is to prohibit the cable from being installed with its short dimension against a wood joist. When stapled in this manner, two cables are usually placed side by side under the staple. If the staple is driven too far into the stud, damage to the insulation and conductors can occur. See Section 300-4(c) for support requirements of cables through spaces behind panels designed to allow access. For an example of Exception No. 2, see Figure 300.12.

Exception No. 3: Lengths not more than 4½ ft (1.37 m) from an outlet for connection within an accessible ceiling to lighting fixture(s) or equipment.

Added for the 1999 *Code,* Exception No. 3 permits short unsupported lengths of nonmetallic-sheathed cable for fixture and equipment connections.

336-20. Boxes of Insulating Material. Nonmetallic outlet boxes shall be permitted as provided in Section 370-3.

Nonmetallic boxes and nonmetallic wiring systems are recommended for use in corrosive atmospheres. See Sections 370-3, 370-17(c), and Article 547 for details.

336-21. Devices of Insulating Material. Switch, outlet, and tap devices of insulating material shall be permitted to be used without boxes in exposed cable wiring and for rewiring in existing buildings where the cable is concealed and fished. Openings in such devices shall form a close fit around the outer covering of the cable and the device shall fully enclose the part of the cable from which any part of the covering has been removed.

Where connections to conductors are by binding-screw terminals, there shall be available as many terminals as conductors.

336-25. Devices with Integral Enclosures. Wiring devices with integral enclosures identified for such use shall be permitted as provided in Section 300-15(e).

336-26. Ampacity. The ampacity of Types NM, NMC, and NMS cable shall be that of 60°C (140°F) conductors and shall comply with Section 310-15.

The 90°C (194°F) rating shall be permitted to be used for ampacity derating purposes provided the final derated ampacity does not exceed that for a 60°C (140°F) rated conductor.

C. Construction Specifications

336-30. General. Nonmetallic-sheathed cable shall comply with (a) and (b).

(a) Construction. The outer cable sheath shall be a nonmetallic material.

(1) Type NM. The overall covering shall be flame retardant and moisture resistant.

(2) Type NMC. The overall covering shall be flame retardant, moisture resistant, fungus resistant, and corrosion resistant.

(3) Type NMS. Type NMS cable is a factory assembly of insulated power, communications, and signaling conductors enclosed within a common sheath of moisture-resistant, flame-retardant, nonmetallic material. The sheath shall be applied so as to separate the power conductors from the communications and signaling conductors. The signaling conductors shall be permitted to be shielded. An optional outer jacket shall be permitted.

FPN: For composite optical cable, see Sections 770-4 and 770-52.

(b) Conductors. The insulated power conductors shall be one of the types listed in Table 310-13 that is suitable for branch-circuit wiring or one that is identified for use in these cables.

The power conductors shall be sizes No. 14 through No. 2 with copper conductors or sizes No. 12 through No. 2 with aluminum or copper-clad aluminum conductors.

The signaling conductors shall comply with Section 780-5.

In addition to the insulated conductors, the cable shall be permitted to have an insulated or bare conductor for equipment grounding purposes only. Where provided, the grounding conductor shall be sized in accordance with Article 250.

Conductors shall be rated at 90°C (194°F).

FPN: Types NM, NMC, and NMS cable identified by the markings NM-B, NMC-B, and NMS-B meet this requirement.

336-31. Marking. The cable shall be marked in accordance with Section 310-11.

Article 338 — Service-Entrance Cable: Types SE and USE

Contents

338-1. Definition. *Service-entrance cable* is a single conductor or multiconductor assembly provided with or without an overall covering, primarily used for services, and is of the following types.

(a) Type SE. Type SE, having a flame-retardant, moisture-resistant covering.

(b) Type USE. Type USE, identified for underground use, having a moisture-resistant covering, but not required to have a flame-retardant covering.

Cabled, single-conductor, Type USE constructions recognized for underground use may have a bare copper conductor cabled with the assembly. Type USE single, parallel, or cabled conductor assemblies recognized for underground use may have a bare copper concentric conductor applied. These constructions do not require an outer overall covering.

FPN: See Section 230-41, Exception, item (b) for directly buried, uninsulated service-entrance conductors.

(c) One Uninsulated Conductor. If Type SE or USE cable consists of two or more conductors, one shall be permitted to be uninsulated.

According to the 1998 UL *Electrical Construction Materials Directory,* category TXKT, service cable and category TYLZ service-entrance cable, rated 600 volts, is listed in sizes No. 12 and larger for copper and No. 10 and larger for aluminum or copper-clad aluminum. Type SE cable contains Type RHW, XHHW, or THWN conductors. Type USE cable contains conductors with insulation equivalent to RHW or XHHW. The type designation of the conductors may be marked on the surface of the cable. When used, this marking indicates the temperature rating for the cable corresponding to the temperature rating of the conductors. When this marking does not appear, the temperature rating of the cable is 75°C. Type USE-2 contains insulation equivalent to RHW-2 or XHHW-2 and is rated 90°C wet or dry.

The cables are designated as follows:

1. Type SE — cable for aboveground installation. The outer jacket or finish of Type SE is suitable for use where exposed to sun.

2. Type USE or USE-2 — cable for underground installation, including burial directly in the earth. Cable in sizes No. 4/0 AWG and smaller and having all conductors insulated is suitable for all of the underground uses for which Type UF cable is permitted by the *Code*. Types USE and USE-2 are not suitable for use in premises or above ground except to terminate at the service equipment or metering equipment.

3. Submersible water pump cable — indicates a multiconductor cable in which 2, 3, or 4 single-conductor, Type USE or USE-2 cables are provided in a flat or twisted assembly. The cable is listed in sizes No. 12 to 4/0 inclusive for copper and No. 10 to 4/0 inclusive for aluminum or copper-clad aluminum. The cable is tag-marked "For Use Within the Well Casing for Wiring Deep-Well Water Pumps Where

the Cable Is Not Subject to Repetitive Handling Caused by Frequent Servicing of the Pump Units." The insulation may also be surface-marked "Pump Cable." The cable may be directly buried in the earth in conjunction with this use.

The equipment grounding conductor, where required, is permitted to be bare. See Section 230-41 for service-entrance conductor insulation requirements.

338-2. Uses Permitted as Service-Entrance Conductors. Service-entrance cable used as service-entrance conductors shall be installed as required by Article 230.

Type USE used for service laterals shall be permitted to emerge above ground outside at terminations in meter bases or other enclosures where protected in accordance with Section 300-5(d).

Section 338-2 recognizes that Type USE cable may rise out of the ground to terminate in a meter base above ground outside. Since Type USE service-entrance cable may not be flame retardant, it is not permitted indoors.

338-3. Uses Permitted as Branch Circuits or Feeders.

(a) Grounded Conductor Insulated. Type SE service-entrance cables shall be permitted in interior wiring systems where all of the circuit conductors of the cable are of the rubber-covered or thermoplastic type.

(b) Grounded Conductor Not Insulated. Type SE service-entrance cables without individual insulation on the grounded circuit conductor shall not be used as a branch circuit or as a feeder within a building, except a cable that has a final nonmetallic outer covering and is supplied by alternating current at not over 150 volts to ground shall be permitted as a feeder to supply only other buildings on the same premises.

Type SE service-entrance cable shall be permitted for use where the fully insulated conductors are used for circuit wiring and the uninsulated conductor is used for equipment grounding purposes.

(c) Temperature Limitations. Type SE service-entrance cable used to supply appliances shall not be subject to conductor temperatures in excess of the temperature specified for the type of insulation involved.

Branch circuits using service-entrance cable as a wiring method are permitted only if all circuit conductors within the cable are fully insulated according to Section 310-13. The equipment grounding conductor is the only conductor permitted to be bare within service-entrance cable used for branch circuits.

Omitted from this section and not permitted for new installations is service-entrance cable containing a grounded conductor that is not insulated (a bare neutral conductor). If used as a branch circuit, service-entrance cable cannot serve appliances such as ranges, wall-mounted ovens, counter-mounted cooking units, or clothes dryers. This is coordinated with Sections 250-140 and 250-142 of the 1999 *Code.*

According to the 1998 UL *Electrical Construction Materials Directory,* Category TYLZ, service entrance cable "Based upon tests which have been made involving the maximum heating that can be produced, an uninsulated conductor employed in a service cable assembly is considered to have the same current-carrying capacity as the insulated conductors, even though it may be smaller in size."

338-4. Installation Methods for Branch Circuits and Feeders.

(a) Interior Installations. In addition to the provisions of this article, Type SE service-entrance cable used for interior wiring shall comply with the installation requirements of Parts A and B of Article 336 and shall comply with the applicable provisions of Article 300.

FPN: See Section 310-10 for temperature limitation of conductors.

Section 336-5 prohibits the use of nonmetallic-sheathed cable in dwellings and other structures exceeding three floors above grade (plus an additional level, if it is the first level not for human habitation and is used only for vehicle storage, parking, etc.). Type SE cable, which may be similar in construction to Types NM and NMC, where used for interior branch circuits and feeders, should also meet the provisions of Sections 336-4 and 336-5.

A similar restriction, as indicated in Section 339-3(a)(4), applies to Type UF cable where used for interior wiring.

Note, however, that service-entrance cables do not have to meet the same temperature requirements that are required of nonmetallic-sheathed cables, since the requirements do not extend to Part C of Article 336.

(b) Exterior Installations. In addition to the provisions of this article, service-entrance cable used for feeders or branch circuits, where installed as exterior wiring, shall be installed as required by Article 225. The cable shall be supported in accordance with Section 336-18, unless used as messenger-supported wiring as allowed by Article 321.

Type USE cable shall be installed outside in accordance with the provisions of Article 339. Where Type USE cable emerges above ground at terminations, it shall be protected in accordance with Section 300-5(d).

Multiconductor service-entrance cable shall be permitted

to be installed as messenger-supported wiring in accordance with Articles 225 and 321.

338-5. Marking. Service-entrance cable shall be marked as required in Section 310-11. Cable with the neutral conductor smaller than the ungrounded conductors shall be so marked.

338-6. Bends. Bends in cable shall be made, and other handlings shall be such, that the protective coverings of the cable will not be damaged, and the radius of the curve of the inner edge of any bend shall not be less than five times the diameter of the cable.

Article 339 — Underground Feeder and Branch-Circuit Cable: Type UF

Contents

339-1. Description and Marking.

According to the 1998 UL *Electrical Construction Materials Directory,* category YDUX, underground feeder and branch-circuit cable rated 600 volts is listed in sizes No. 14 to 4/0 AWG inclusive for copper and No. 12 to 4/0 AWG inclusive for aluminum or copper-clad aluminum, for single and multiconductor cable.

Some multiconductor cable is surface-marked with the suffix "B" immediately following the type letters to indicate conductors employing 90°C rated insulation.

Submersible water pump cable indicates a multiconductor cable in which 2, 3, or 4 single-conductor, Type UF cables are provided in a flat or twisted assembly. The cable is listed in sizes No. 14 to 4/0 AWG inclusive for copper and No. 12 to 4/0 AWG inclusive for aluminum or copper-clad aluminum. The cable is tag-marked "For Use Within the Well Casing for Wiring Deep-Well Water Pumps Where the Cable Is Not Subject to Repetitive Handling Caused by Frequent Servicing of the Pump Units." The insulation may be surface-marked "Pump Cable." The cable may be directly buried in the earth in conjunction with this use.

This cable may employ copper, aluminum, or copper-clad aluminum conductors. Cable with copper-clad aluminum conductors is surface-printed "AL (CU-CLAD)" or "Cu-Clad Al." Cable with aluminum conductors is surface-printed "AL."

This cable may be terminated at boxes and other enclosures by using nonmetallic-sheathed cable connectors.

If single-conductor Type UF cable is terminated with a fitting not specifically recognized for use with single-conductor cable, special care should be taken to ensure that it is properly secured and not subject to damage.

(a) Description. Underground feeder and branch-circuit cable shall be a listed Type UF cable in sizes No. 14 copper or No. 12 aluminum or copper-clad aluminum through No. 4/0. The conductors of Type UF shall be one of the moisture-resistant types listed in Table 310-13 that is suitable for branch-circuit wiring or one that is identified for such use. In addition to the insulated conductors, the cable shall be permitted to have an insulated or bare conductor for equipment grounding purposes only. The overall covering shall be flame retardant; moisture, fungus, and corrosion resistant; and suitable for direct burial in the earth.

(b) Marking. The cable shall be marked in accordance with Section 310-11.

339-2. Other Articles. In addition to the provisions of this article, installations of underground feeder and branch-circuit cable (Type UF) shall comply with other applicable provisions of this *Code*, especially Article 300 and Section 310-13.

339-3. Use.

(a) Uses Permitted.

(1) Type UF cable shall be permitted for use underground, including direct burial in the earth, as feeder or branch-circuit cable where provided with overcurrent protection of the rated ampacity as required in Section 339-4.

(2) Where single-conductor cables are installed, all cables of the feeder circuit or branch circuit, including the neutral and equipment grounding conductor, if any, shall be run together in the same trench or raceway.

(3) For underground requirements, see Section 300-5.

(4) Type UF cable shall be permitted for interior wiring in wet, dry, or corrosive locations under the recognized wiring methods of this *Code*, and, where installed as nonmetallic-sheathed cable, the installation and conductor requirements shall comply with the provisions of Article 336 and shall be of the multiconductor type.

Where UF cable is installed as nonmetallic-sheathed cable, the ampacity of Type UF cable is determined according to Section 336-26. See the commentary following Sections 336-4, 336-5, and 338-4 regarding installation requirements as applied to Type UF cable where used for interior wiring.

(5) For solar photovoltaic systems in accordance with Section 690-31.

(6) Single-conductor cables shall be permitted as the nonheating leads for heating cables as provided in Section 424-43.

Type UF cable supported by cable trays shall be of the multiconductor type.

FPN: See Section 310-10 for temperature limitation of conductors.

(b) Uses Not Permitted. Type UF cable shall not be used in the following:

(1) As service-entrance cables
(2) In commercial garages
(3) In theaters
(4) In motion picture studios
(5) In storage battery rooms
(6) In hoistways
(7) In any hazardous (classified) location
(8) Embedded in poured cement, concrete, or aggregate, except where embedded in plaster as nonheating leads as provided in Article 424
(9) Where exposed to direct rays of the sun, unless identified as sunlight resistant

Type UF cable suitable for exposure to the direct rays of the sun is indicated by tag marking and marking on the surface of the cable with the designation "Sunlight Resistant."

(10) Where subject to physical damage

Added in the 1999 *Code,* this physical protection requirement ensures that Type UF cable, as it emerges from underground, will be protected from physical damage.

339-4. Overcurrent Protection. Overcurrent protection shall be provided in accordance with the provisions of Section 240-3.

339-5. Ampacity. The ampacity of Type UF cable shall be that of 60°C (140°F) conductors in accordance with Section 310-15.

Article 340 — Power and Control Tray Cable: Type TC

Contents

340-1. Definition. *Type TC power and control tray cable* is a factory assembly of two or more insulated conductors, with or without associated bare or covered grounding conductors under a nonmetallic sheath, for installation in cable trays, in raceways, or where supported by a messenger wire.

340-2. Other Articles. In addition to the provisions of this article, installations of Type TC tray cable shall comply with other applicable articles of this *Code*, especially Articles 300 and 318.

340-3. Construction. The insulated conductors of Type TC tray cable shall be in sizes No. 18 through 1000 kcmil copper and sizes No. 12 through 1000 kcmil aluminum or copper-clad aluminum. Insulated conductors of sizes No. 14 and larger copper and sizes No. 12 and larger aluminum or copper-clad aluminum shall be one of the types listed in Tables 310-13 or 310-62 that is suitable for branch circuit and feeder circuits or one that is identified for such use. The outer sheath shall be a flame-retardant, nonmetallic material. A metallic sheath shall not be permitted either under or over the nonmetallic sheath.

(a) Wet Locations. Where installed in wet locations, Type TC cable shall also be resistant to moisture and corrosive agents.

(b) Fire Alarm Systems. Where used for fire alarm systems, conductors shall also be in accordance with Section 760-27.

(c) Thermocouple Circuits. Conductors in Type TC cables used for thermocouple circuits in accordance with Article 725 shall also be permitted to be any of the materials used for thermocouple extension wire.

There shall be no voltage marking on a Type TC cable employing thermocouple extension wire.

(d) Class 1 Circuit Conductors. Insulated conductors of sizes No. 18 and No. 16 copper shall also be in accordance with Section 725-27.

340-4. Use Permitted. Type TC tray cable shall be permitted to be used in the following.

(1) For power, lighting, control, and signal circuits.
(2) In cable trays, or in raceways, or where supported in outdoor locations by a messenger wire.
(3) In cable trays in hazardous (classified) locations as permitted in Articles 318, 501, 502, 504, and 505 in industrial establishments where the conditions of maintenance and supervision ensure that only qualified persons will service the installation.
(4) For Class 1 circuits as permitted in Article 725.
(5) For nonpower-limited fire alarm circuits if conductors comply with the requirements of Section 760-27.
(6) In industrial establishments where the conditions of maintenance and supervision ensure that only qualified persons will service the installation, and where the cable is not subject to physical damage, Type TC tray cable that complies with the crush and impact requirements of Type MC cable and is identified for such use shall be permitted as open wiring in lengths not to exceed a total of 50 ft (15.24 m) between a cable tray and the utilization equipment or device.

    The cable shall be supported and secured at intervals not exceeding 6 ft (1.83 m).

    Equipment grounding for the utilization equipment shall be provided by an equipment grounding conductor within the cable.

    FPN: See Section 310-10 for temperature limitation of conductors.

Section 340-4 permits Type TC tray cable to be used for nonpower-limited fire alarm circuits. According to Section 760-27, the conductor material must be copper. Aluminum and copper-clad aluminum conductors are not permitted for fire alarm circuits.

Specific types of TC tray cable used in qualifying occupancies are permitted to extend from a cable tray to a piece of equipment without the use of conduit.

Type TC cable is permitted to be installed in a hazardous location only if that location is in an industrial establishment where conditions of maintenance and supervision ensure that only qualified persons will service the installation.

340-5. Uses Not Permitted. Type TC tray cable shall not be used in the following:

(1) Installed where it will be exposed to physical damage
(2) Installed as open cable on brackets or cleats
(3) Used where exposed to direct rays of the sun, unless identified as sunlight resistant
(4) Direct buried, unless identified for such use

Type TC cable is not generally permitted to be installed as open cable. However, as stated in Section 340-4(6), Type TC cable is permitted in qualifying occupancies to be run as open wiring in lengths of 50 ft or less, where the TC cable complies with crush and impact requirements.

340-6. Marking. The cable shall be marked in accordance with Section 310-11.

340-7. Ampacity. The ampacities of the conductors of Type TC tray cable shall be determined from Section 402-5 for conductors smaller than No. 14 and from Section 318-11.

340-8. Bends. Bends in Type TC cable shall be made so as not to damage the cable.

Article 342 — Nonmetallic Extensions

Contents

342-1. Definition. *Nonmetallic extensions* are an assembly of two insulated conductors within a nonmetallic jacket or an extruded thermoplastic covering. The classification includes both surface extensions, intended for mounting directly on the surface of walls or ceilings, and aerial cable containing a supporting messenger cable as an integral part of the cable assembly.

342-2. Other Articles. In addition to the provisions of this article, nonmetallic extensions shall be installed in accordance with the applicable provisions of this *Code*.

342-3. Uses Permitted. Nonmetallic extensions shall be permitted only where all of the following conditions are met:

(a) From an Existing Outlet. The extension is from an existing outlet on a 15- or 20-ampere branch circuit in conformity with the requirements of Article 210.

(b) Exposed and in a Dry Location. The extension is run exposed and in a dry location.

(c) Nonmetallic Surface Extensions. For nonmetallic surface extensions, the building is occupied for residential or office purposes and does not exceed the height limitations specified in Section 336-5(a)(1).

(c1) [Alternate to (c)]. For aerial cable, the building is occupied for industrial purposes, and the nature of the occupancy requires a highly flexible means for connecting equipment.

FPN: See Section 310-10 for temperature limitation of conductors.

Nonmetallic surface extensions are permitted to be installed only in residential or office buildings no higher than three floors above grade, in accordance with Section 336-5(a)(1). An extension is permitted, however, as an aerial cable in industrial occupancies where a highly flexible means for connecting equipment is necessary. See Sections 342-4(a) and 342-7(b) for aerial cable installation details.

A nonmetallic extension is an assembly of two insulated circuit conductors with or without an equipment grounding conductor within a nonmetallic jacket or extruded thermoplastic covering. Assemblies without a grounding conductor are marked "Intended for Replacement Use Only."

342-4. Uses Not Permitted. Nonmetallic extensions shall not be used as follows.

(a) Aerial Cable. As aerial cable to substitute for one of the general wiring methods specified by this *Code*.

(b) Unfinished Areas. In unfinished basements, attics, or roof spaces.

(c) Voltage Between Conductors. Where the voltage between conductors exceeds 150 volts for nonmetallic surface extension and 300 volts for aerial cable.

(d) Corrosive Vapors. Where subject to corrosive vapors.

(e) Through a Floor or Partition. Where run through a floor or partition, or outside the room in which it originates.

342-5. Splices and Taps. Extensions shall consist of a continuous unbroken length of the assembly, without splices, and without exposed conductors between fittings. Taps shall be permitted where approved fittings completely covering the tap connections are used. Aerial cable and its tap connectors shall be provided with an approved means for polarization. Receptacle-type tap connectors shall be of the locking type.

342-6. Fittings. Each run shall terminate in a fitting that covers the end of the assembly. All fittings and devices shall be of a type identified for the use.

342-7. Installation. Nonmetallic extensions shall be installed as specified in (a) and (b).

(a) Nonmetallic Surface Extensions.

(1) One or more extensions shall be permitted to be run in any direction from an existing outlet, but not on the floor or within 2 in. (50.8 mm) from the floor.

(2) Nonmetallic surface extensions shall be secured in place by approved means at intervals not exceeding 8 in. (203 mm), with an allowance for 12 in. (305 mm) to the first fastening where the connection to the supplying outlet is by means of an attachment plug. There shall be at least one fastening between each two adjacent outlets supplied. An extension shall be attached to only woodwork or plaster finish, and shall not be in contact with any metal work or other conductive material other than with metal plates on receptacles.

(3) A bend that reduces the normal spacing between the conductors shall be covered with a cap to protect the assembly from physical damage.

(b) Aerial Cable.

(1) Aerial cable shall be supported by its messenger cable and securely attached at each end with clamps and turnbuckles. Intermediate supports shall be provided at not more than 20-ft (6.1-m) intervals. Cable tension shall be adjusted to eliminate excessive sag. The cable shall have a clearance of not less than 2 in. (50.8 mm) from steel structural members or other conductive material.

(2) Aerial cable shall have a clearance of not less than 10 ft (3.05 m) above floor areas accessible to pedestrian traffic, and not less than 14 ft (4.27 m) above floor areas accessible to vehicular traffic.

(3) Cable suspended over work benches, not accessible to pedestrian traffic, shall have a clearance of not less than 8 ft (2.44 m) above the floor.

(4) Aerial cables shall be permitted as a means to support lighting fixtures where the total load on the supporting messenger cable does not exceed that for which the assembly is intended.

(5) The supporting messenger cable, where installed in conformity with the applicable provisions of Article 250 and if properly identified as an equipment grounding conductor, shall be permitted to ground equipment. The messenger cable shall not be used as a branch-circuit conductor.

342-8. Marking. Nonmetallic extensions shall be marked in accordance with Section 110-21.

Article 343 — Nonmetallic Underground Conduit with Conductors

Contents

A. General

343-1. Description. *Nonmetallic underground conduit with conductors* is a factory assembly of conductors or cables inside a nonmetallic, smooth wall conduit with a circular cross section.

The nonmetallic conduit shall be composed of a material that is resistant to moisture and corrosive agents. It shall also be capable of being supplied on reels without damage or distortion and shall be of sufficient strength to withstand abuse, such as impact or crushing, in handling and during installation without damage to conduit or conductors.

Nonmetallic underground conduit with conductors (preassembled conductors in conduit) has been used by electric utilities for outdoor lighting for several years. It is supplied in continuous lengths on coils or reels, or in cartons. It consists of nonmetallic conduit with the conductors preinstalled by the manufacturer. The intent of preassembly is to minimize the damage to conductors and raceways that can result from pulling. The product is designed to allow conductors to be removed and reinserted; therefore, maintenance is not precluded.

The use of nonmetallic underground conduit with conductors is restricted and is not permitted inside buildings. The conductors may enter the building for termination, but the raceway may not. See Sections 343-3 and 343-4 for uses permitted and not permitted.

343-2. Other Articles. Installations for nonmetallic underground conduit with conductors shall comply with the provisions of the applicable sections of Article 300. Where equipment grounding is required by Article 250, an assembly containing a separate equipment grounding conductor shall be used.

343-3. Uses Permitted. The use of listed nonmetallic underground conduit with conductors and fittings shall be permitted in the following.

(1) For direct burial underground installation. For minimum cover requirements, see Tables 300-5 and 300-50 under rigid nonmetallic conduit.
(2) Encased or embedded in concrete.
(3) In cinder fill.
(4) In underground locations subject to severe corrosive influences as covered in Section 300-6 and where subject to chemicals for which the assembly is specifically approved.

Nonmetallic underground conduit with conductors is required to be listed.

343-4. Uses Not Permitted. Nonmetallic underground conduit with conductors shall not be used in the following:

(1) In exposed locations
(2) Inside buildings

Exception: The conductor or the cable portion of the assembly, where suitable, shall be permitted to extend within the building for termination purposes in accordance with Section 300-3.

(3) In hazardous (classified) locations except as permitted by Sections 503-3(a), 504-20, 514-8, and 515-5, and in Class I, Division 2 locations as permitted in Section 501-4(b), Exception

B. Installation

343-5. Size.

(a) Minimum. Nonmetallic underground conduit with conductors smaller than ½-in. electrical trade size shall not be used.

(b) Maximum. Nonmetallic underground conduit with conductors larger than 4-in. electrical trade size shall not be used.

FPN: Metric trade numerical designations for nonmetallic underground conduit with conductors are ½ = 16, ¾ = 21, 1 = 27, 1¼ = 35, 1½ = 41, 2 = 53, 2½ = 63, 3 = 78, 3½ = 91, and 4 = 103.

343-6. Trimming. For termination, the conduit shall be trimmed away from the conductors or cables using an approved method that will not damage the conductor or cable insulation or jacket. All ends shall be trimmed inside and out to remove rough edges.

343-7. Joints. All joints between conduit, fittings, and boxes shall be made by an approved method.

343-8. Conductor Terminations. All terminations between the conductors or cables and equipment shall be made by an approved method for that type of conductor or cable.

343-9. Bushings. Where the nonmetallic underground conduit with conductors enters a box, fitting, or other enclosure, a bushing or adapter shall be provided to protect the conductor or cable from abrasion unless the design of the box, fitting, or enclosure provides equivalent protection.

FPN: See Section 300-4(f) for the protection of conductors size No. 4 or larger.

343-10. Bends — How Made. Bends of nonmetallic underground conduit with conductors shall be manually made so that the conduit will not be damaged and the internal diameter of the conduit will not be effectively reduced. The radius of the curve of the centerline of such bends shall not be less than shown in Table 343-10.

Table 343-10. Radius of Conduit Bends

Trade Size (in.)	Minimum Bending Radius (in.)
½	10
¾	12
1	14
1¼	18
1½	20
2	26
2½	36
3	48
4	60

Note: For SI units, in. = 25.4 mm (radius).

Added for the 1999 *Code*, Article 343 now contains the minimum bending radius for nonmetallic underground conduit with conductors as described in Table 343-10.

343-11. Bends — Number in One Run. There shall not be more than the equivalent of four quarter bends (360 degrees total) between termination points.

343-12. Splices and Taps. Splices and taps shall be made in junction boxes or other enclosures. See Article 370 for rules on the installation and use of boxes and conduit bodies.

C. Construction

343-13. General. Nonmetallic underground conduit with conductors is an assembly that is provided in continuous lengths shipped in a coil, reel, or carton.

343-14. Conductors and Cables. Conductors and cables used in nonmetallic underground conduit with conductors shall be listed, shall be suitable for use in wet locations, and shall be as follows.

(a) 600 Volts or Less. Alternating-current and direct-current circuits shall be permitted. All conductors shall have an insulation rating equal to at least the maximum nominal circuit voltage of any conductor or cable within the conduit.

(b) Over 600 Volts. Conductors or cables rated over 600 volts shall not occupy the same conduit with conductors or cables of circuits rated 600 volts or less.

343-15. Conductor Fill. The maximum number of conductors or cables in nonmetallic underground conduit with conductors shall not exceed that permitted by the percentage fill in Table 1, Chapter 9.

343-16. Marking. Nonmetallic underground conduit with conductors shall be clearly and durably marked at least every 10 ft (3.05 m) as required by Section 110-21. The type of conduit material shall also be included in the marking.

Identification of conductors or cables used in the assembly shall be provided on a tag attached to each end of the assembly or to the side of a reel. Enclosed conductors or cables shall be marked in accordance with Section 310-11.

Article 345 — Intermediate Metal Conduit

Contents

A. General

345-1. Definition. *Intermediate metal conduit* is a listed steel raceway of circular cross section with integral or associated couplings, approved for the installation of electrical conductors and used with listed fittings to provide electrical continuity.

Intermediate metal conduit (IMC) is a listed thinner-walled rigid metal conduit that is satisfactory for use in all locations where rigid metal conduit is permitted to be used. Also, threaded fittings, couplings, conectors, and so on, are interchangeable for either IMC or rigid metal conduit. Threadless fittings for IMC are suitable only for the type of conduit indicated by the marking on the carton. Galvanized IMC installed in concrete does not require supplementary corrosion protection. Galvanized IMC installed in contact with soil does not generally require supplementary corrosion protection. As a guide in the absence of experience with the corrosive effects of soil in a specific location, soils producing severe corrosive effects are generally characterized by low resistivity, less than 2000 ohm-cm. Wherever ferrous metal conduit runs directly from concrete encasement to soil burial, the metal in contact with the soil can be severely corroded.

345-2. Other Articles. Installations for intermediate metal conduit shall comply with the provisions of the applicable sections of Article 300.

345-3. Uses Permitted.

(a) All Atmospheric Conditions and Occupancies. Use of intermediate metal conduit shall be permitted under all atmospheric conditions and occupancies. Where practicable, dissimilar metals in contact anywhere in the system shall be avoided to eliminate the possibility of galvanic action. Intermediate metal conduit shall be permitted as an equipment grounding conductor.

Exception: Aluminum fittings and enclosures shall be permitted to be used with steel intermediate metal conduit.

(b) Corrosion Protection. Intermediate metal conduit, elbows, couplings, and fittings shall be permitted to be installed in concrete, in direct contact with the earth, or in areas subject to severe corrosive influences where protected by corrosion protection and judged suitable for the condition.

FPN: See Section 300-6 for protection against corrosion.

(c) Cinder Fill. Intermediate metal conduit shall be permitted to be installed in or under cinder fill where subject to permanent moisture where protected on all sides by a layer of noncinder concrete not less than 2 in. (50.8 mm) thick; where the conduit is not less than 18 in. (457 mm) under the fill; or where protected by corrosion protection and judged suitable for the condition.

FPN: See Section 300-6 for protection against corrosion.

B. Installation

345-5. Wet Locations. All supports, bolts, straps, screws, etc., shall be of corrosion-resistant materials or protected against corrosion by corrosion-resistant materials.

FPN: See Section 300-6 for protection against corrosion.

345-6. Size.

(a) Minimum. Conduit smaller than ½-in. electrical trade size shall not be used.

(b) Maximum. Conduit larger than 4-in. electrical trade size shall not be used.

FPN: Metric trade numerical designations for intermediate metal conduit are the same as those found in *Extra-heavy Duty Rigid Steel Conduits for Electrical Installations*, IEC 981-1989; namely, ½ = 16, ¾ = 21, 1 = 27, 1¼ = 35, 1½ = 41, 2 = 53, 2½ = 63, 3 = 78, 3½ = 91, and 4 = 103.

The metric trade size designations for the various sizes of electrical conduit have been provided to assist the *Code* user.

345-7. Number of Conductors in Conduit. The number of conductors in a single conduit shall not exceed that permitted by the percentage fill specified in Table 1, Chapter 9, using the conduit dimensions of Table 4, Chapter 9.

Table 1 of Chapter 9 specifies the maximum percent fill of a conduit or tubing. Table 4 provides the usable area within the selected conduit or tubing, and Table 5 provides the required area for each of the conductors. Examples using these tables to calculate a conduit or tubing size are provided following Chapter 9, Table 1, Notes to Tables, Note 6.

If the conductors are of the same wire size, instead of doing the calculations, the tables of Appendix C may be used. Appendix C, which contains 12 sets of tables, very accurately indicates the maximum number of conductors permitted in a conduit or tubing. This appendix replaced Tables 3A, 3B, and 3C from the 1993 *Code*. Examples using this appendix to select a conduit or tubing size are provided following the introduction in Appendix C.

In order to select the proper trade size intermediate metal conduit, the section entitled "Intermediate Metal Conduit" in Table 4 of Chapter 9 should be

followed. Appendix C Tables C4 and C4A for intermediate metal conduit are also permissible.

345-8. Reaming and Threading. All cut ends of conduits shall be reamed or otherwise finished to remove rough edges. Where conduit is threaded in the field, a standard cutting die with a ¾-in. taper per foot (1 in 16) shall be used.

FPN: See *Standards for Pipe Threads, General Purpose (Inch),* ANSI/ASME B.1.20.1-1983.

See the commentary following Section 348-11 for reaming considerations.

345-9. Couplings and Connectors.

(a) Threadless. Threadless couplings and connectors used with conduit shall be made tight. Where buried in masonry or concrete, they shall be the concretetight type. Where installed in wet locations, they shall be the raintight type.

(b) Running Threads. Running threads shall not be used on conduit for connection at couplings.

See the commentary following Section 346-9 for examples of threadless fittings.

345-10. Bends—How Made. Bends of intermediate metal conduit shall be made so that the conduit will not be damaged and so that the internal diameter of the conduit will not be effectively reduced. The radius of the curve of the inner edge of any field bend shall not be less than indicated in Table 346-10.

Exception: For field bends for conductors without lead sheath and made with a single operation (one shot) bending machine designed for the purpose, the minimum radius shall not be less than that indicated in Table 346-10, Exception.

See the commentary following Section 346-10 for a definition of the term *field bend.*

345-11. Bends — Number in One Run. There shall not be more than the equivalent of four quarter bends (360 degrees total) between pull points, e.g., conduit bodies and boxes.

See the commentary following Section 346-11 for the rationale behind limiting the number of bends.

345-12. Securing and Supporting. Intermediate metal conduit shall be installed as a complete system as provided in Article 300 and shall be securely fastened in place and supported in accordance with (a) and (b).

(a) Securely Fastened. Each intermediate metal conduit shall be securely fastened within 3 ft (914 mm) of each outlet box, junction box, device box, cabinet, conduit body, or other conduit termination. Fastening shall be permitted to be increased to a distance of 5 ft (1.52 m) where structural members do not readily permit fastening within 3 ft (914 mm). Where approved, conduit shall not be required to be securely fastened within 3 ft (914 mm) of the service head for above-the-roof termination of a mast.

As illustrated in Figure 345.1, intermediate metal conduit (IMC) is required to be securely fastened at least every 10 ft. Fastening is also required within 3 ft of outlet boxes, junction boxes, cabinets, and conduit bodies. However, where structural support members do not permit fastening within 3 ft, the support may be located up to 5 ft away.

(b) Supports. Intermediate metal conduit shall be supported in accordance with one of the following.

(1) Conduit shall be supported at intervals not exceeding 10 ft (3.05 m).

(2) The distance between supports for straight runs of conduit shall be permitted in accordance with Table 346-12, provided the conduit is made up with threaded couplings, and such supports prevent transmission of stresses to termination where conduit is deflected between supports.

(3) Exposed vertical risers from industrial machinery shall be permitted to be supported at intervals not exceeding 20 ft (6.1 m), provided the conduit is made up with threaded couplings, is firmly supported at the top and bottom of the

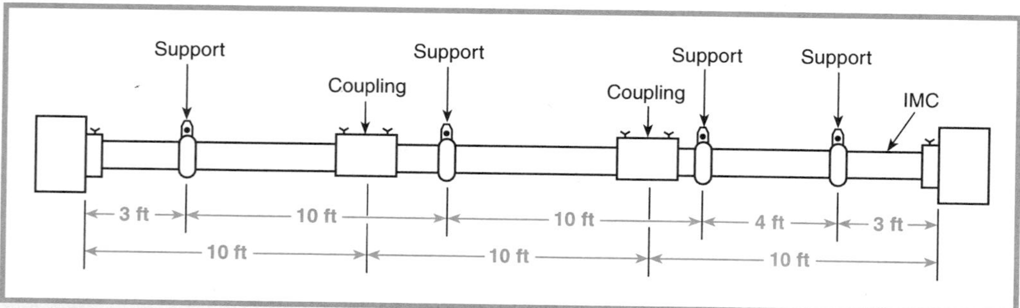

Figure 345.1 *Minimum fastening requirements for intermediate metal conduit according to Section 345-12(a).*

riser, and no other means of intermediate support is readily available.

(4) Horizontal runs of intermediate metal conduit supported by openings through framing members at intervals not exceeding 10 ft (3.05 m) and securely fastened within 3 ft (914 mm) of termination points shall be permitted.

Section 345-13(b)(4) permits lengths of intermediate metal conduit (IMC) to be supported (but not necessarily secured) by framing members at 10-ft intervals, provided the IMC is secured and supported at least 3 ft from the box or enclosure. Installations where the IMC is installed through the bar joists is just one example and is shown in Figure 345.2.

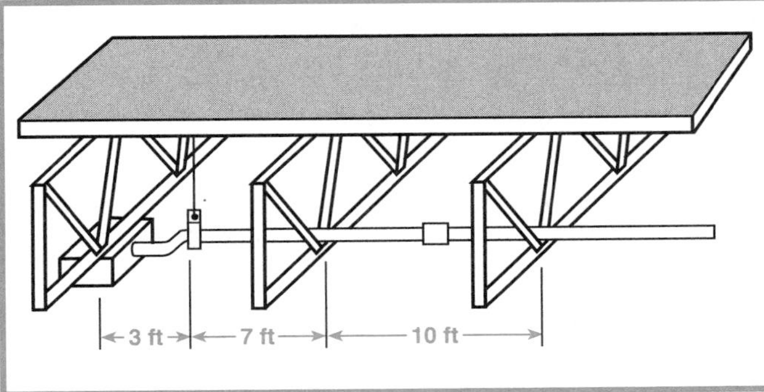

Figure 345.2 *An example of intermediate metal conduit supported by framing members and securely fastened at the 3-ft distance from the box required by Section 345-12(b)(4).*

345-13. Boxes and Fittings. See Article 370.

345-14. Splices and Taps. Splices and taps shall be made in accordance with Section 300-15. See Article 370 for rules on the installation and use of boxes and conduit bodies.

345-15. Bushings. Where a conduit enters a box, fitting, or other enclosure, a bushing shall be provided to protect the wire from abrasion unless the design of the box, fitting, or enclosure is such as to afford equivalent protection.

FPN: See Section 300-4(f) for the protection of conductors sizes No. 4 and larger at bushings.

C. Construction Specifications

345-16. General. Intermediate metal conduit shall comply with (a) through (c).

Revised for the 1999 *Code,* Section 345-16 reduces the marking requirements from every 2½ ft to every 5 ft. Also changed, lengths other than 10 ft are no longer limited for specific applications or use. Now, longer or shorter conduit lengths are permitted.

(a) Standard Lengths. The standard length of intermediate metal conduit shall be 10 ft (3.05 m), including an attached coupling, and each end shall be threaded. Longer or shorter lengths with or without coupling and threaded or unthreaded shall be permitted.

(b) Corrosion-Resistant Material. Nonferrous conduit of corrosion-resistant material shall have suitable markings.

(c) Marking. Each length shall be clearly and durably marked at least every 5 ft (1.52 m) with the letters *IMC.* Each length shall be marked as required in the first sentence of Section 110-21.

Article 346 — Rigid Metal Conduit

Contents

A. General

346-1. Definition. *Rigid metal conduit* is a listed metal raceway of circular cross section with integral or associated couplings, approved for the installation of electrical conductors and used with listed fittings to provide electrical continuity.

346-2. Other Articles. Installations for rigid metal conduit shall comply with the provisions of the applicable sections of Article 300.

346-3. Uses Permitted.

(a) All Atmospheric Conditions and Occupancies. Use of rigid metal conduit shall be permitted under all atmospheric conditions and occupancies. Where practicable, dissimilar metals in contact anywhere in the system shall be avoided to eliminate the possibility of galvanic action. Rigid metal conduit shall be permitted as an equipment grounding conductor. Ferrous raceways and fittings protected from corrosion solely by enamel shall be permitted only indoors and in occupancies not subject to severe corrosive influences.

Aluminum fittings and enclosures shall be permitted to be used with steel rigid metal conduit, and steel fittings and enclosures shall be permitted to be used with aluminum rigid metal conduit.

Section 346-3(a) makes it clear that aluminum rigid conduit can be used with steel fittings and enclosures, as can aluminum fittings and enclosures with steel rigid conduit. Tests have shown that the galvanic corrosion at steel and aluminum interfaces is minor compared to the natural corrosion on the combination of steel and steel or aluminum and aluminum.

(b) Corrosion Protection. Rigid metal conduit, elbows, couplings, and fittings shall be permitted to be installed in concrete, in direct contact with the earth, or in areas subject to severe corrosive influences where protected by corrosion protection and judged suitable for the condition.

FPN: See Section 300-6 for protection against corrosion.

Section 346-3(b) indicates the permitted uses for listed ferrous and nonferrous conduit, including their use in concrete, in direct contact with the earth, and in corrosive areas. The fine print note references Section 300-6 for additional information on protection against corrosion and specific types of corrosion-resistant materials.

It is advisable to consult with the authority enforcing this *Code* for approval of corrosion-resistant materials or for requirements prior to the installation of nonferrous metal (aluminum) conduit in concrete, since chloride additives in the concrete mix may cause corrosion.

(c) Cinder Fill. Rigid metal conduit shall be permitted to be installed in or under cinder fill where subject to permanent moisture where protected on all sides by a layer of noncinder concrete not less than 2 in. (50.8 mm) thick; where the conduit is not less than 18 in. (457 mm) under the fill; or where protected by corrosion protection and judged suitable for the condition.

FPN: See Section 300-6 for protection against corrosion.

Although cinder fill is not commonly used in modern construction, it is still encountered at older building sites. Care should be taken to ensure the proper installation of rigid metal conduit as permitted by this section. Where cinders have been used as fill, they may contain sulfur, and where they have combined with moisture, sulfuric acid is formed, which can corrode metal raceways.

B. Installation

346-5. Wet Locations. All supports, bolts, straps, screws, etc., shall be of corrosion-resistant materials or protected against corrosion by corrosion-resistant materials.

FPN: See Section 300-6 for protection against corrosion.

346-6. Size.

(a) Minimum. Rigid metal conduit smaller than ½-in. electrical trade size shall not be used.

Exception: For enclosing the leads of motors as permitted in Section 430-145(b).

(b) Maximum. Rigid metal conduit larger than 6-in. electrical trade size shall not be used.

FPN: Metric trade numerical designations for rigid metal conduit are the same as those found in *Extra-heavy Duty Rigid Steel Conduits for Electrical Installations*, IEC 981-1989; namely, ½ = 16, ¾ = 21, 1 = 27, 1¼ = 35, 1½ = 41, 2 = 53, 2½ = 63, 3 = 78, 3½ = 91, 4 = 103, 5 = 129, and 6 = 155.

The metric trade size designations for the various sizes of electrical conduit have been provided to assist the *Code* user.

346-7. Number of Conductors in Conduit. The number of conductors in a single conduit shall not exceed that permitted by the percentage fill specified in Table 1, Chapter 9, using the conduit dimensions of Table 4, Chapter 9.

Table 1 of Chapter 9 specifies the maximum percent fill of a conduit or tubing. Table 4 provides the usable area within the selected conduit or tubing, and Table 5 provides the required area for each of the conductors. Examples using these tables to calculate a conduit or tubing size are provided following Chapter 9, Table 1, Notes to Tables, Note 6.

If the conductors are of the same wire size, instead of doing the calculations, Appendix C may be used. Appendix C, which contains 12 sets of tables, very accurately indicates the maximum number of conductors permitted in a conduit or tubing. This appendix replaced Tables 3A, 3B, and 3C from the 1993 *Code*. Examples using this appendix to select a conduit or tubing size are provided following the introduction in Appendix C.

In order to select the proper trade size rigid metal conduit, the section entitled "Rigid Metal Conduit" in Table 4 of Chapter 9 should be followed. Appendix C Tables C8 and C8A for rigid metal conduit are also permissible.

346-8. Reaming and Threading. All cut ends of conduits shall be reamed or otherwise finished to remove rough edges. Where conduit is threaded in the field, a standard cutting die with a ¾-in. taper per foot (1 in 16) shall be used.

FPN: See *Standards for Pipe Threads, General Purpose (Inch)*, ANSI/ASME B.1.20.1-1983.

See the commentary following Section 348-11 for reaming considerations.

346-9. Couplings and Connectors.

(a) Threadless. Threadless couplings and connectors used with conduit shall be made tight. Where buried in masonry or concrete, they shall be the concretetight type. Where installed in wet locations, they shall be the raintight type.

(b) Running Threads. Running threads shall not be used on conduit for connection at couplings.

Figure 346.1 illustrates a threadless connection integral to a conduit body, FS box, and so on. This type of connection may be separate from the conduit body or box as an individual fitting of the compression type (raintight) suitable for wet locations, or it may be of the set-screw type.

Threadless fittings are not intended for use over threads, as the fitting will not seat properly. The threaded end of the conduit should be cut off and reamed before installation.

Figure 346.2 illustrates a three-piece coupling (the electrical equivalent of a pipe union), which is used to join two lengths of conduit where it is impossible to turn either length, such as in underground or concrete slab construction. Another fitting for joining conduit is a bolted split coupling. Running threads are not permitted to join two conduits but may be permitted to join two boxes if electrical and mechanical connections are assured (locknuts and bushings).

Figure 346.1 *Conduit body with a threadless connector. (Appleton Electric Co.)*

Figure 346.2 *A three-piece-type (union-type) coupling. (Appleton Electric Co.)*

346-10. Bends — How Made. Bends of rigid metal conduit shall be made so that the conduit will not be damaged and that the internal diameter of the conduit will not be effectively reduced. The radius of the curve of the inner edge of any field bend shall not be less than indicated in Table 346-10.

Exception: For field bends for conductors without lead sheath and made with a single operation (one shot) bending machine designed for the purpose, the minimum radius shall not be less than that indicated in Table 346-10, Exception.

The term *field bend* means any bend or offset made by installers, using proper tools and equipment, during the installation of conduit systems.

346-11. Bends — Number in One Run. There shall not be more than the equivalent of four quarter bends (360 degrees total) between pull points, e.g., conduit bodies and boxes.

Table 346-10. Radius of Conduit Bends

Size of Conduit (in.)	Conductors Without Lead Sheath (in.)
½	4
¾	5
1	6
1¼	8
1½	10
2	12
2½	15
3	18
3½	21
4	24
5	30
6	36

Note: For SI units, 1 in. = 25.4 mm (radius).

Table 346-10. Exception, Radius of Conduit Bends

Size of Conduit (in.)	*Radius to Center of Conduit (in.)*
½	*4*
¾	*4½*
1	*5¾*
1¼	*7¼*
1½	*8¼*
2	*9½*
2½	*10½*
3	*13*
3½	*15*
4	*16*
5	*24*
6	*30*

Note: For SI units, 1 in. = 25.4 mm (radius).

Limiting the number of bends in a conduit run reduces pulling tension on conductors and helps ensure easy insertion or removal of conductors during later phases of construction, when the conduit may be permanently enclosed by the finish of the building. Adjustments during that time are often impossible.

The *Code* does not limit the pull points to conduit bodies and boxes; these are only examples of pull points.

346-12. Securing and Supporting. Rigid metal conduit shall be installed as a complete system as provided in Article 300 and shall be securely fastened in place and supported in accordance with (a) and (b).

(a) Securely Fastened. Each rigid metal conduit shall be securely fastened within 3 ft (914 mm) of each outlet box, junction box, device box, cabinet, conduit body, or other conduit termination. Fastening shall be permitted to be increased to a distance of 5 ft (1.52 m) where structural members do not readily permit fastening within 3 ft (914 mm). Where approved, conduit shall not be required to be securely fastened within 3 ft (914 mm) of the service head for above-the-roof termination of a mast.

As illustrated in Figure 346.3, rigid metal conduit is required to be securely fastened at least every 10 ft. Secure fastening is also required within 3 ft of outlet boxes, junction boxes, cabinets, and conduit bodies. However, where structural support members do not permit fastening within 3 ft, secure fastening may be located up to 5 ft away.

(b) Supports. Rigid metal conduit shall be supported in accordance with one of the following.

(1) Conduit shall be supported at intervals not exceeding 10 ft (3.05 m).

(2) The distance between supports for straight runs of conduit shall be permitted in accordance with Table 346-12(b)(2), provided the conduit is made up with threaded couplings, and such supports prevent transmission of stresses to termination where conduit is deflected between supports.

(3) Exposed vertical risers from stationary equipment or fixtures shall be permitted to be supported at intervals not exceeding 20 ft (6.1 m), provided the conduit is made up with threaded couplings, is firmly supported at the top and bottom of the riser, and no other means of intermediate support is readily available.

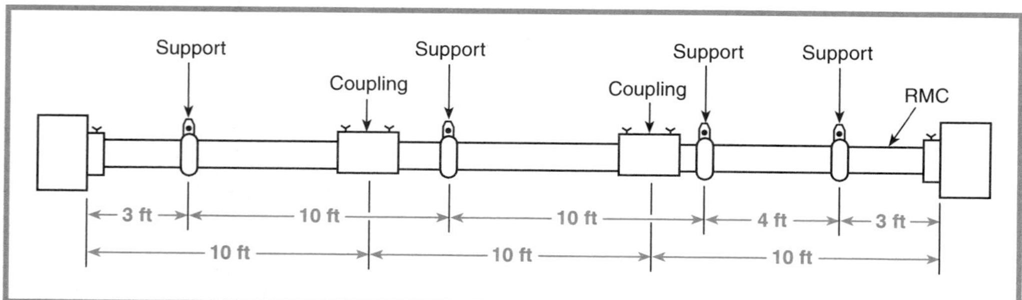

Figure 346.3 *Minimum support required for rigid metal conduit according to Section 346-12(a).*

Table 346-12(b)(2). Supports for Rigid Metal Conduit

Conduit Size (in.)	Maximum Distance Between Rigid Metal Conduit Supports (ft)
½–¾	10
1	12
1¼–1½	14
2–2½	16
3 and larger	20

Note: For SI units, 1 ft = 0.3048 m (supports).

(4) Horizontal runs of rigid metal conduit supported by openings through framing members at intervals not exceeding 10 ft (3.05 m) and securely fastened within 3 ft (914 mm) of termination points shall be permitted.

Section 346-12(b)(4) permits lengths of rigid metal conduit to be supported (but not necessarily secured) by framing members at 10-ft intervals, provided the rigid metal conduit is secured and supported at least 3 ft from the box or enclosure. Installations where the rigid metal conduit is installed through the bar joists is just one example and is shown in Figure 346.4.

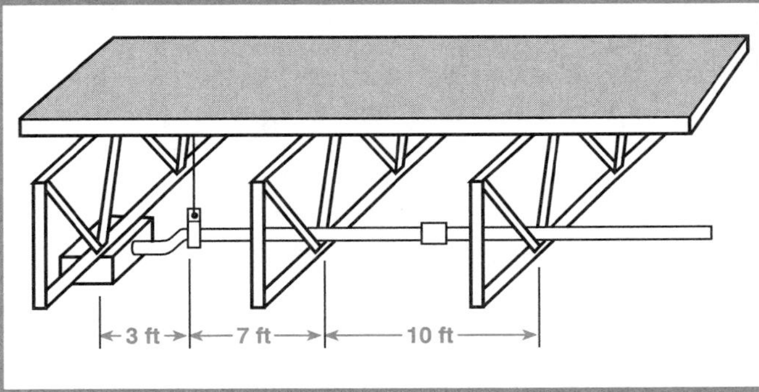

Figure 346.4 *An example of rigid metal conduit supported by framing members and securely fastened at the 3-ft distance from the box required by Section 346-12(b)(4).*

346-13. Boxes and Fittings. See Article 370.

346-14. Splices and Taps. Splices and taps shall be made in accordance with Section 300-15. See Article 370 for rules on the installation and use of boxes and conduit bodies.

346-15. Bushings. Where a conduit enters a box, fitting, or other enclosure, a bushing shall be provided to protect the wire from abrasion unless the design of the box, fitting, or enclosure is such as to afford equivalent protection.

FPN: See Section 300-4(f) for the protection of conductors sizes No. 4 and larger at bushings.

C. Construction Specifications

346-16. General. Rigid metal conduit shall comply with (a) through (c).

(a) Standard Lengths. The standard length of rigid metal conduit shall be 10 ft (3.05 m), including an attached coupling, and each end shall be threaded. Longer or shorter lengths with or without coupling and threaded or unthreaded shall be permitted.

Revised for the 1999 *Code,* lengths other than 10 ft are no longer limited for specific applications or use. Now, longer or shorter lengths of conduit are permitted.

(b) Corrosion-Resistant Material. Nonferrous conduit of corrosion-resistant material shall have suitable markings.

(c) Marking. Each length shall be clearly and durably identified in every 10 ft (3.05 m) as required in the first sentence of Section 110-21.

Article 347 — Rigid Nonmetallic Conduit

Contents

347-1. Description. This article shall apply to a type of conduit and fittings of suitable nonmetallic material that is resistant to moisture and chemical atmospheres. For use aboveground, it shall also be flame retardant, resistant to impact and crushing, resistant to distortion from heat under conditions likely to be encountered in service, and resistant to low temperature and sunlight effects. For use underground, the material shall be acceptably resistant to moisture and corrosive agents and shall be of sufficient strength to withstand abuse, such as by impact and crushing, in handling and during installation. Conduits listed for the purpose shall be permitted to be installed underground in continuous lengths from a reel. Where intended for direct burial, without encasement in concrete, the material shall also be capable of withstanding continued loading that is likely to be encountered after installation.

The 1998 UL *General Information Directory* (White Book) describes three types of rigid nonmetallic conduit recognized for use in Article 347 and Article 300, Part B:

1. Rigid nonmetallic Schedule 40 and Schedule 80 PVC conduit (DZYR)
2. Rigid nonmetallic underground conduit, plastic (EAZY)
3. Rigid nonmetallic fiberglass conduit (DZKT)

Rigid Nonmetallic Schedule 40 and Schedule 80 PVC Conduit (DZYR)

Rigid nonmetallic PVC conduit (Schedule 40 and Schedule 80) is suitable for use aboveground, underground, and for direct burial without encasement in concrete. Unless marked for higher temperature, rigid nonmetallic conduit is intended for use with wires rated 75°C or less, including where it is encased in concrete within buildings and where ambient temperature is 50°C or less. Where encased in concrete in trenches outside buildings, it is suitable for use with wires rated 90°C or less.

Schedule 40 conduit is suitable for exposed work where it will not be subject to physical damage.

The marking "Schedule 80 PVC" identifies conduit suitable for use where it may be exposed to physical damage to comply with Section 347-3(c).

Schedule 80 rigid PVC conduit has a reduced cross-sectional area available for wiring space. The actual cross-sectional area and the need for reference to the *National Electrical Code,* Chapter 9, Table 1, for wire fill capacity are prominently marked on the conduit surface.

PVC conduit is designed for connection to couplings, fittings, and boxes by the use of a suitable solvent-type cement. Instructions supplied by the solvent-type cement manufacturer describe the method of assembly and precautions to be followed.

Listed PVC conduit is inherently resistant to atmosphere containing common industrial corrosive agents and will also withstand vapors or mist of caustic, pickling, plating bath, hydrofluoric, and chromic acids.

PVC conduit, elbows, and bends (including couplings) that have been investigated for direct exposure to other reagents may be identified by the designation "Reagent Resistant" printed on the surface of the product. Further information on reagent resistance may be found in the 1998 UL *Electrical Construction Materials Directory* and in UL 651, "Schedule 40 and 80 Rigid PVC Conduit."

Rigid Nonmetallic Underground Conduit, Plastic (EAZX)

Plastic types of rigid nonmetallic conduit, for use only when installed underground, may be (1) polyvinyl chloride (PVC) Type A or Type EB, or (2) high-density polyethylene (HDPE) Schedule 40. The various conduit types differ in their inside and outside diameters.

The conduit is designed for use in underground work under the following conditions, as indicated on the listing mark: (1) when laid with its entire length in concrete (Type A); (2) when laid with its entire length in concrete in outdoor trenches (Type EB); and (3) direct burial with or without being encased in concrete (HDPE Schedule 40). The conduit is intended for use in ambient temperatures of 50°C or less and, unless marked otherwise,

Type A and HDPE Schedule 40 conduit are intended for use with wires rated 75°C or less. Type EB conduit and Type A conduit encased in concrete in trenches outside buildings may be used with wires rated 90°C or less. HDPE Schedule 40 conduit, when directly buried or encased in concrete, may be used with wires rated 90°C or less.

Where conduit emerges from underground installation, the wiring method shall be of a type recognized for the purpose.

PVC conduit is designed for joining with PVC couplings by the use of a solvent-type cement. HDPE conduit is designed for joining by threaded couplings, drive-on couplings, or a butt fusing process. Instructions supplied by the solvent-type cement manufacturer describe the method of assembly and precautions to be followed.

Reinforced Thermosetting Resin Conduit (DZKT) [formally referred to as Rigid Nonmetallic Fiberglass Conduit and sometimes Fiberglass Reinforced Epoxy Conduit (FRE) conduit.]

Reinforced thermosetting resin conduit has been evaluated for use with wires rated 90°C or less.

Conduit marked "RTRC Type BG" or "Rigid Nonmetallic Conduit Underground (Reinforced Thermosetting Resin Conduit)" has been evaluated for underground use only — for direct burial, with or without encasement in concrete.

Conduit marked "RTRC Type AG" or "Rigid Nonmetallic Conduit (Reinforced Thermosetting Resin Conduit)" has been evaluated for use aboveground, underground, and for direct burial with or without encasement in concrete. This conduit has been evaluated for concealed or exposed work where not subject to physical damage.

Reinforced thermosetting resin conduit is listed in sizes ½ to 6 in. in IPS, ID, RTRC 40 and RTRC 80 dimensions, as marked on the product. Listing includes straight conduit, elbows, bends, and other fittings, unless otherwise noted.

Reinforced thermosetting resin conduit, elbows, bends, and other fittings, which have been investigated for direct exposure to reagents, are identified by the designation "Reagent Resistant" and are marked to indicate the specific reagents.

Reinforced thermosetting resin conduit is designed for connection to couplings, fittings, and boxes by use of a suitable epoxy-type cement or drive-on bell and spigot. Instructions supplied by the epoxy-type cement manufacturer describe the method of assembly and precautions to be followed.

For use of Schedule 80, see Sections 300-5(d), 551-80(b), and 300-50(b).

347-2. Uses Permitted. The use of listed rigid nonmetallic conduit shall be permitted under the following conditions.

FPN: Extreme cold may cause some nonmetallic conduits to become brittle and therefore more susceptible to damage from physical contact.

(a) Concealed. In walls, floors, and ceilings.

(b) Corrosive Influences. In locations subject to severe corrosive influences as covered in Section 300-6 and where subject to chemicals for which the materials are specifically approved.

(c) Cinders. In cinder fill.

(d) Wet Locations. In portions of dairies, laundries, canneries, or other wet locations and in locations where walls are frequently washed, the entire conduit system including boxes and fittings used therewith shall be installed and equipped so as to prevent water from entering the conduit. All supports, bolts, straps, screws, etc., shall be of corrosion-resistant materials or be protected against corrosion by approved corrosion-resistant materials.

(e) Dry and Damp Locations. In dry and damp locations not prohibited by Section 347-3.

(f) Exposed. For exposed work where not subject to physical damage if identified for such use.

Nonmetallic conduits are not permitted to be installed in ducts, plenums, and other air-handling spaces. See Section 300-22, which limits the use of materials in ducts, plenums, and other air-handling spaces that may contribute smoke and products of combustion during a fire.

Also, rigid nonmetallic conduit is not permitted in places of assembly or theaters except as permitted in Sections 518-4 and 520-5. See these sections for specific details.

(g) Underground Installations. For underground installations, see Sections 300-5 and 300-50.

(h) Support of Conduit Bodies. Rigid nonmetallic conduit shall be permitted to support nonmetallic conduit bodies not larger than the largest trade size of an entering raceway. The conduit bodies shall not contain devices or support fixtures or other equipment.

347-3. Uses Not Permitted. Rigid nonmetallic conduit shall not be used in the following locations.

(a) Hazardous (Classified) Locations.

(1) In hazardous (classified) locations, except as covered in Sections 503-3(a), 504-20, 514-8, and 515-5
(2) In Class I, Division 2 locations, except as permitted in Section 501-4(b), Exception

(b) Support of Fixtures. For the support of fixtures or other equipment not described in Section 347-2(h).

(c) Physical Damage. Where subject to physical damage unless identified for such use.

(d) Ambient Temperatures. Where subject to ambient temperatures in excess of 50°C (122°F) unless listed otherwise.

(e) Insulation Temperature Limitations. For conductors whose insulation temperature limitations would exceed those for which the conduit is listed.

(f) Theaters and Similar Locations. In theaters and similar locations, except as provided in Articles 518 and 520.

347-4. Other Articles. Installation of rigid nonmetallic conduit shall comply with the applicable provisions of Article 300. Where equipment grounding is required by Article 250, a separate equipment grounding conductor shall be installed in the conduit.

Exception No. 1: As permitted in Section 250-134(b), Exception No. 2, for dc circuits and Section 250-134(b), Exception No. 1, for separately run equipment grounding conductors.

Exception No. 2: Where the grounded conductor is used to ground equipment as permitted in Section 250-142.

A. Installations

347-5. Trimming. All cut ends shall be trimmed inside and outside to remove rough edges.

347-6. Joints. All joints between lengths of conduit, and between conduit and couplings, fittings, and boxes, shall be made by an approved method.

347-8. Securing and Supporting. Rigid nonmetallic conduit shall be installed as a complete system as provided in Section 300-18 and shall be fastened so that movement from thermal expansion or contraction will be permitted. Rigid nonmetallic conduit shall be securely fastened and supported in accordance with (a) and (b).

The requirements of Section 347-8 are fairly stringent because they are based on ambient temperatures higher than normally encountered and use horizontal-support tests only. Expansion can cause damage to the raceway or its supports. Expansion fittings, therefore, should be used, and the supports must allow expansion/contraction cycles without damage. See the commentary on expansion fittings following Section 347-9.

(a) Securely Fastened. Each rigid nonmetallic conduit shall be securely fastened within 3 ft (914 mm) of each outlet box, junction box, device box, conduit body, or other conduit termination. Conduit listed for securing at other than 3 ft (914 mm) shall be permitted to be installed in accordance with the listing.

(b) Supports. Rigid nonmetallic conduit shall be supported as required in Table 347-8. Conduit listed for support at spacings other than as shown in Table 347-8 shall be permitted to be installed in accordance with the listing. Horizontal runs of rigid nonmetallic conduit supported by openings through framing members at intervals not exceeding those in Table 347-8 and securely fastened within 3 ft (914 mm) of termination points shall be permitted.

Table 347-8. Support of Rigid Nonmetallic Conduit

Conduit Size (in.)	Maximum Spacing Between Supports (ft)
½–1	3
1¼–2	5
2½–3	6
3½–5	7
6	8

Note: For SI units, 1 ft = 0.3048 m (supports).

347-9. Expansion Fittings. Expansion fittings for rigid nonmetallic conduit shall be provided to compensate for thermal expansion and contraction where the length change, in accordance with Tables 347-9(A) and (B), is expected to be ¼ in. (6.36 mm) or greater in a straight run between securely mounted items such as boxes, cabinets, elbows, or other conduit terminations.

Expansion fittings are generally provided in exposed runs of rigid nonmetallic conduit where (1) the run is long, or (2) the run is subjected to large temperature variations during installation or afterward, or (3) expansion and contraction measures are provided for the building or other structures. Rigid nonmetallic conduit exhibits a considerably greater change in length per degree change in temperature than metal raceway systems.

In some parts of the United States and other countries, outdoor temperature variations of over 38°C (100°F) are common. According to Chapter 9, Table 10, a 100-ft run of PVC rigid nonmetallic conduit will change 4.1 in. in length if the temperature change is 38°C (100°F).

The normal expansion range of most larger sizes

Table 347-9(A). Expansion Characteristics of PVC Rigid Nonmetallic Conduit Coefficient of Thermal Expansion = 3.38×10^{-5} in./in./°F

Temperature Change (°F)	Length Change of PVC Conduit (in./100 ft)	Temperature Change (°F)	Length Change of PVC Conduit (in./100 ft)
5	0.2	105	4.2
10	0.4	110	4.5
15	0.6	115	4.7
20	0.8	120	4.9
25	1.0	125	5.1
30	1.2	130	5.3
35	1.4	135	5.5
40	1.6	140	5.7
45	1.8	145	5.9
50	2.0	150	6.1
55	2.2	155	6.3
60	2.4	160	6.5
65	2.6	165	6.7
70	2.8	170	6.9
75	3.0	175	7.1
80	3.2	180	7.3
85	3.4	185	7.5
90	3.6	190	7.7
95	3.8	195	7.9
100	4.1	200	8.1

Table 347-9(B). Expansion Characteristics of Fiberglass Reinforced Conduit (Rigid Nonmetallic Conduit) Coefficient of Thermal Expansion = 1.5×10^{-5} in./in./°F

Temperature Change (°F)	Length Change of Fiberglass Conduit (in./100 ft)	Temperature Change (°F)	Length Change of Fiberglass Conduit (in./100 ft)
5	0.1	105	1.9
10	0.2	110	2.0
15	0.3	115	2.1
20	0.4	120	2.2
25	0.5	125	2.3
30	0.5	130	2.3
35	0.6	135	2.4
40	0.7	140	2.5
45	0.8	145	2.6
50	0.9	150	2.7
55	1.0	155	2.8
60	1.1	160	2.9
65	1.2	165	3.0
70	1.3	170	3.1
75	1.4	175	3.2
80	1.4	180	3.2
85	1.5	185	3.3
90	1.6	190	3.4
95	1.7	195	3.5
100	1.8	200	3.6

of rigid nonmetallic conduit expansion couplings is generally 6 in. Information concerning installation and application of this type of coupling may be obtained from manufacturers' instructions.

Expansion fittings are seldom used underground, where temperatures are relatively constant. If rigid nonmetallic conduit is buried or covered immediately, expansion and contraction are not a problem.

347-10. Size.

(a) Minimum. Rigid nonmetallic conduit smaller than ½-in. electrical trade size shall not be used.

(b) Maximum. Rigid nonmetallic conduit larger than 6-in. electrical trade size shall not be used.

FPN: Metric trade numerical designations for rigid nonmetallic conduit are ½ = 16, ¾ = 21, 1 = 27, 1¼ = 35, 1½ = 41, 2 = 53, 2½ = 63, 3 = 78, 3½ = 91, 4 = 103, 5 = 129, and 6 = 155.

347-11. Number of Conductors. The number of conductors permitted in a single conduit shall not exceed the percentage fill specified in Table 1, Chapter 9.

Because of the extra wall thickness of Schedule 80 PVC conduit, the cross-sectional area available for wires is less than that of other raceways of the same trade size.

Table 1 of Chapter 9 specifies the maximum percent fill of a conduit or tubing. Table 4 provides the usable area within the selected conduit or tubing, and Table 5 provides the required area for each of the conductors. Examples using these tables to calculate a conduit or tubing size are provided following Chapter 9, Table 1, Notes to Tables, Note 6.

If the conductors are of the same wire size, instead of doing the calculations, the tables of Appendix C may be used. Appendix C, which contains 12 sets of tables, very accurately indicates the maximum number of conductors permitted in a conduit or tubing. This appendix replaced Tables 3A, 3B, and 3C from the 1993 *Code*. Examples using this appendix to select a conduit or tubing size are provided following the introduction in Appendix C.

To permit selection of the proper trade size rigid nonmetallic conduit, Table 4 of Chapter 9 contains four separate subtables, one for each type of rigid nonmetallic conduit. The appropriate table for the given type of rigid nonmetallic conduit should be followed. Appendix C Tables C9 and C9A through C12 and C12A are also permissible, provided the appropriate table for the given type of rigid nonmetallic conduit is used.

347-12. Bushings. Where a conduit enters a box, fitting, or other enclosure, a bushing or adapter shall be provided to protect the wire from abrasion unless the box, fitting, or enclosure design provides equivalent protection.

FPN: See Section 300-4(f) for the protection of conductors No. 4 and larger at bushings.

347-13. Bends — How Made. Bends of rigid nonmetallic conduit shall be made so that the conduit will not be damaged and that the internal diameter of the conduit will not be effectively reduced. Field bends shall be made only with bending equipment identified for the purpose, and the radius of the curve of the inner edge of such bends shall not be less than shown in Table 346-10.

See the commentary following Section 346-10 for a definition of the term *field bend.*

347-14. Bends — Number in One Run. There shall not be more than the equivalent of four quarter bends (360 degrees total) between pull points, e.g., conduit bodies and boxes.

See the commentary following Section 346-11 for the rationale behind limiting the number of bends.

347-15. Boxes and Fittings. Rigid nonmetallic conduit shall be used only with listed fittings. Boxes and fittings shall comply with the applicable provisions of Article 370.

347-16. Splices and Taps. Splices and taps shall be made in accordance with Section 300-15. See Article 370 for rules on the installation and use of boxes and conduit bodies.

B. Construction Specifications

347-17. General. Rigid nonmetallic conduit shall comply with the following.

Marking. Each length of nonmetallic conduit shall be clearly and durably marked at least every 10 ft (3.05 m) as required in the first sentence of Section 110-21. The type of material shall also be included in the marking unless it is visually identifiable. For conduit recognized for use above ground, these markings shall be permanent. For conduit limited to underground use only, these markings shall be sufficiently durable to remain legible until the material is installed. Conduit shall be permitted to be surface marked to indicate special characteristics of the material.

FPN: Examples of these optional markings include but are not limited to "LS" for limited-smoke and markings such as "sunlight resistant."

Article 348 — Electrical Metallic Tubing

Contents

A. General

348-1. Definition. *Electrical metallic tubing* is a listed metallic tubing of circular cross section approved for the installation of electrical conductors when joined together with listed fittings.

The requirement for listed electrical metallic tubing (EMT) is not a new requirement. However, a new definition of electrical metallic tubing was added for the 1999 *Code*.

348-2. Other Articles. Installations of electrical metallic tubing shall comply with the applicable provisions of Article 300.

348-4. Uses Permitted.

(a) Exposed and Concealed. The use of listed electrical metallic tubing shall be permitted for both exposed and concealed work.

(b) Corrosion Protection. Ferrous or nonferrous electrical metallic tubing, elbows, couplings, and fittings shall be permitted to be installed in concrete, in direct contact with the earth, or in areas subject to severe corrosive influences where protected by corrosion protection and judged suitable for the condition.

FPN: See Section 300-6 for information on protection against corrosion.

According to the 1998 UL *General Information for Electrical Equipment Directory* (White Book), category FJMX, galvanized steel electrical metallic tubing (EMT) installed in concrete, on grade or above, generally requires no supplementary corrosion protection. Galvanized steel electrical metallic tubing in concrete slab below grade level may require supplementary corrosion protection. In general, galvanized steel EMT in contact with soil requires supplementary corrosion protection. Where galvanized steel EMT without supplementary corrosion protection extends directly from concrete encasement to soil burial, severe corrosive effects are likely to occur on the metal in contact with the soil.

348-5. Uses Not Permitted. Electrical metallic tubing shall not be used

(1) Where, during installation or afterward, it will be subject to severe physical damage.
(2) Where protected from corrosion solely by enamel.
(3) In cinder concrete or cinder fill where subject to permanent moisture unless protected on all sides by a layer of noncinder concrete at least 2 in. (50.8 mm) thick or unless the tubing is at least 18 in. (457 mm) under the fill.
(4) In any hazardous (classified) location except as permitted by Sections 502-4, 503-3, and 504-20.
(5) For the support of fixtures or other equipment except conduit bodies no larger than the largest trade size of the tubing. Where practicable, dissimilar metals in contact anywhere in the system shall be avoided to eliminate the possibility of galvanic action.

Exception: Aluminum fittings and enclosures shall be permitted to be used with steel electrical metallic tubing.

B. Installation

348-6. Wet Locations. All supports, bolts, straps, screws, etc., shall be of corrosion-resistant materials or protected against corrosion by corrosion-resistant materials.

Table 1 of Chapter 9 specifies the maximum percent fill of a conduit or tubing. Table 4 provides the usable area within the selected conduit or tubing, and Table 5 provides the required area for each of the conductors. Examples using these tables to calculate a conduit or tubing size are provided following Chapter 9, Table 1, Notes to Tables, Note 6.

If the conductors are of the same wire size, instead of doing the calculations, the tables of Appendix C may be used. Appendix C, which contains 12 sets of tables, very accurately indicates the maximum number of conductors permitted in a conduit or tubing. This appendix replaced Tables 3A, 3B, and 3C from the 1993 *Code*. Examples using this appendix to select a conduit or tubing size are provided following the introduction in Appendix C.

In order to select the proper trade size electrical metallic tubing, the section entitled "Electrical Metallic Tubing" in Table 4 of Chapter 9 should be followed. Appendix C Tables C1 and C1A for electrical metallic tubing are also permissible.

FPN: See Section 300-6 for information on protection against corrosion.

348-7. Size.

(a) Minimum. Tubing smaller than ½-in. electrical trade size shall not be used.

Exception: For enclosing the leads of motors as permitted in Section 430-145(b).

(b) Maximum. The maximum size of tubing shall be the 4-in. electrical trade size.

FPN: Metric trade numerical designations for electrical metallic tubing are the same as those found in *Extra-heavy Duty Rigid Steel Conduits for Electrical Installations*, IEC 981-1989; namely, ½ = 16, ¾ = 21, 1 = 27, 1¼ = 35, 1½ = 41, 2 = 53, 2½ = 63, 3 = 78, 3½ = 91, and 4 = 103.

The metric trade size designations for the various sizes of electrical conduit have been provided to assist the *Code* user.

348-8. Number of Conductors in Tubing. The number of conductors permitted in a single tubing shall not exceed the percentage fill specified in Table 1, Chapter 9.

348-9. Reaming and Threading. Electrical metallic tubing shall not be threaded. Where integral couplings are utilized, such couplings shall be permitted to be factory threaded.

All cut ends of electrical metallic tubing shall be reamed or otherwise finished to remove rough edges.

In addition to a reamer, a half-round file has proved practical for removing rough edges. The handle of a pair of pump pliers, the nose of side-cutting pliers, or an electrician's knife can be effective as well.

348-10. Couplings and Connectors. Couplings and connectors used with tubing shall be made up tight. Where buried in masonry or concrete, they shall be concretetight type. Where installed in wet locations, they shall be of the raintight type.

UL-listed fittings that are suitable for use in poured concrete or where exposed to rain are so indicated on the fitting or carton. The term *raintight* or the equivalent on the carton indicates suitability for use where directly exposed to rain. The term *concretetight* or equivalent on the carton indicates suitability for use in poured concrete. See Sections 225-22 and 230-54(a) for raintight requirements as applied to raceways on exterior surfaces of buildings and to service raceways.

Fittings have been tested for use only with steel tubing unless marked on the device or carton to indicate suitability for use with aluminum or other material.

Indentor-type fittings are for use with metallic-coated electrical metallic tubing only and require a special tool supplied by the manufacturer for proper installation. Diametrically opposed indentor-type tools require two sets of indentations nominally 90 degrees apart. Triple-indent tools require one set of indentations.

348-11. Bends — How Made. Bends in the tubing shall be made so that the tubing will not be damaged and the internal diameter of the tubing will not be effectively reduced. The radius of the curve of the inner edge of any field bend shall not be less than shown in Table 346-10.

Exception: For field bends made with a bending machine designed for the purpose, the minimum radius shall not be less than indicated in Table 346-10, Exception.

348-12. Bends — Number in One Run. There shall not be more than the equivalent of four quarter bends (360 degrees total) between pull points, e.g., conduit bodies and boxes.

348-13. Supports. Electrical metallic tubing shall be installed as a complete system as provided in Article 300. Each tubing length shall be securely fastened in place at least every 10 ft (3.05 m). In addition, each tube shall be securely fastened within 3 ft (914 mm) of each outlet box, junction box, device box, cabinet, conduit body, or other tubing terminations.

"Securely fastened in place" means the EMT must be supported and secured at the prescribed intervals, as illustrated in Figure 348.1.

Exception No. 1: Fastening of unbroken lengths shall be permitted to be increased to a distance of 5 ft (1.52 m) where structural members do not readily permit fastening within 3 ft (914 mm).

As the example in Figure 348.2 shows, Exception No. 1 permits boxes secured to ceiling or roof support structural members that are spaced not more than 5 ft apart to serve as support for runs of EMT perpendicular to the axis of the ceiling or roof support members.

Exception No. 2: For concealed work in finished buildings or prefinished wall panels where such securing is impracticable, unbroken lengths (without coupling) of electrical metallic tubing shall be permitted to be fished.

Horizontal runs of electrical metallic tubing supported by openings through framing members at intervals not greater than 10 ft (3.05 m) and securely fastened within 3 ft (914 mm) of termination points shall be permitted.

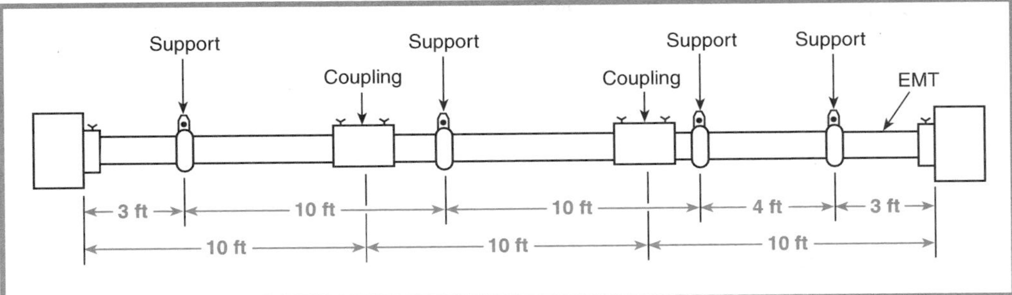

Figure 348.1 *Minimum support required (unless exception applies) for electrical metallic tubing.*

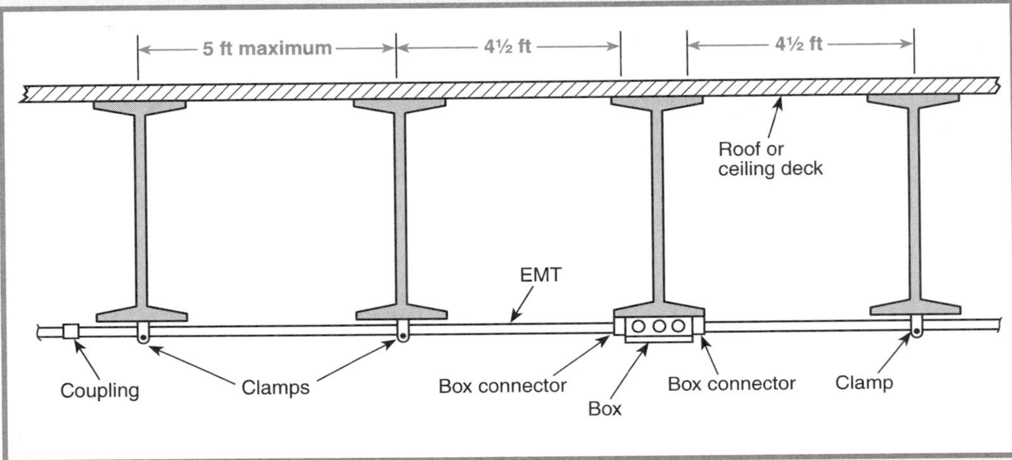

Figure 348.2 *An example of an EMT installation complying with Section 348-13, Exception No. 1.*

See the commentary and example following Section 345-12(b)(4), which apply to horizontal runs of intermediate metal conduit.

348-14. Boxes and Fittings. Boxes and fittings shall comply with the applicable provisions of Article 370.

348-15. Splices and Taps. Splices and taps shall be made in accordance with Section 300-15. See Article 370 for rules on the installation and use of boxes and conduit bodies.

C. Construction Specifications

348-16. General. Electrical metallic tubing shall comply with (a) through (d).

(a) Cross Section. The tubing, and elbows and bends for use with the tubing, shall have a circular cross section.

(b) Finish. Tubing shall have such a finish or treatment of outer surfaces as will provide an approved durable means of readily distinguishing it, after installation, from rigid metal conduit.

(c) Connectors. Where the tubing is coupled together by threads, the connector shall be designed so as to prevent bending of the tubing at any part of the thread.

(d) Marking. Electrical metallic tubing shall be clearly and durably marked at least every 10 ft (3.05 m) as required in the first sentence of Section 110-21.

Article 349 — Flexible Metallic Tubing

Contents

A. General

Flexible metallic tubing is a type of raceway used for certain specific applications, particularly under the requirements of Sections 300-22(b) and (c) for wiring in ducts, plenums, and other air-handling spaces. Flexible metallic tubing is very flexible and is rarely affected by vibration or other movement. It is an effective barrier to gases and products of combustion if it is installed with matching listed fittings and is of adequate mechanical strength for use where not exposed to physical damage.

Flexible metallic tubing not greater than 6 ft in length is suitable for use as a raceway for branch-circuit tap conductors that supply lighting fixtures where outlet boxes are located not less than 1 ft from the fixture.

349-1. Definition. Flexible metallic tubing is a listed tubing that is circular in cross section, flexible, metallic, and liquidtight without a nonmetallic jacket.

349-3. Other Articles. Installations of flexible metallic tubing shall comply with the provisions of the applicable sections of Article 300 and Section 110-21.

349-4. Uses Permitted. Flexible metallic tubing shall be permitted to be used for branch circuits

A common application of flexible metallic tubing is as a branch-circuit wiring method for equipment or lighting fixtures mounted on or above suspended ceilings.

The 1000-volt limitation prohibits flexible metallic tubing from being used for the secondary circuits of sign ballasts, sign transformers, electronic sign power supplies, or oil burner ignition transformers unless these circuits are less than 1000 volts.

(1) In dry locations,
(2) Where concealed,
(3) In accessible locations, and
(4) For system voltages of 1000 volts maximum.

349-5. Uses Not Permitted. Flexible metallic tubing shall not be used

(1) In hoistways;
(2) In storage battery rooms;
(3) In hazardous (classified) locations unless otherwise permitted under other articles in this *Code;*
(4) Underground for direct earth burial, or embedded in poured concrete or aggregate;
(5) Where subject to physical damage; and
(6) In lengths over 6 ft (1.83 m).

Unlike flexible metal conduit or liquidtight flexible metal conduit, flexible metallic tubing is limited in use to 6-ft lengths.

B. Construction and Installation

349-10. Size.

(a) Minimum. Flexible metallic tubing smaller than ½-in. electrical trade size shall not be used.

Exception No. 1: Flexible metallic tubing of ⅜-in. trade size shall be permitted to be installed in accordance with Sections 300-22(b) and (c).

Exception No. 2: Flexible metallic tubing ⅜-in. trade size shall be permitted in lengths not in excess of 6 ft (1.83 m) as part of an approved assembly or for lighting fixtures. See Section 410-67(c).

(b) Maximum. The maximum size of flexible metallic tubing shall be the ¾-in. trade size.

FPN: Metric trade numerical designations for flexible metallic tubing are ⅜ = 12, ½ = 16, and ¾ = 21.

349-12. Number of Conductors.

(a) Flexible Metallic Tubing — ½-in. and ¾-in. Trade Size. The number of conductors permitted in ½-in. and ¾-in. trade sizes of flexible metallic tubing shall not exceed the percentage of fill specified in Table 1, Chapter 9.

(b) Flexible Metallic Tubing — ⅜-in. Trade Size. The number of conductors permitted in ⅜-in. trade size flexible metallic tubing shall not exceed that permitted in Table 350-12.

Table 1 of Chapter 9 specifies the maximum percent fill of a conduit or tubing. Table 4 provides the usable area within the selected conduit or tubing, and Table 5 provides the required area for each of the conductors. Examples using these tables to calculate a conduit or tubing size are provided following Chapter 9, Table 1, Notes to Tables, Note 6.

If the conductors are of the same wire size, instead of doing the calculations, the tables of Appendix C may be used. Appendix C, which contains 12 sets of tables, very accurately indicates the maximum number of conductors permitted in a conduit or tubing. This appendix replaced Tables 3A, 3B, and 3C from the 1993 *Code*. Examples using this appendix to select a conduit or tubing size are provided following the introduction in Appendix C.

In order to select the proper trade size flexible metallic tubing, the section entitled "Flexible Metal Conduit" in Table 4 of Chapter 9 or the manufacturer's instruction should be followed. Appendix C Tables C3 and C3A for flexible metal conduit for sizes ½ in. and ¾ in. are also permissible.

349-16. Grounding. See Section 250-118(8) for rules on the use of flexible metallic tubing as an equipment grounding conductor.

349-17. Splices and Taps. Splices and taps shall be made in accordance with Section 300-15. See Article 370 for rules on the installation and use of boxes and conduit bodies.

349-18. Fittings. Flexible metallic tubing shall be used only with listed terminal fittings. Fittings shall effectively close any openings in the connection.

349-20. Bends.

(a) Infrequent Flexing Use. Where the flexible metallic tubing shall be infrequently flexed in service after installation, the radii of bends measured to the inside of the bend shall not be less than specified in Table 349-20(a).

Table 349-20(a). Minimum Radii for Flexing Use

Trade Size (in.)	Minimum Radii (in.)
⅜	10
½	12½
¾	17½

Note: For SI units, 1 in. = 25.4 mm (radii).

(b) Fixed Bends. Where the flexible metallic tubing is bent for installation purposes and is not flexed or bent as required by use after installation, the radii of bends measured to the inside of the bend shall not be less than specified in Table 349-20(b).

Table 349-20(b). Minimum Radii for Fixed Bends

Trade Size (in.)	Minimum Radii (in.)
⅜	3½
½	4
¾	5

Note: For SI units, 1 in. = 25.4 mm (radii).

Article 350 — Flexible Metal Conduit

Contents

A. General

350-1. Scope. This article covers the use and installation of flexible metal conduit and associated fittings.

350-2. Definition. Flexible metal conduit is a raceway of circular cross section made of helically wound, formed, interlocked metal strip.

350-3. Other Articles. Installations of flexible metal conduit shall comply with the applicable provisions of Article 300.

350-4. Uses Permitted. Flexible metal conduit shall be listed and shall be permitted to be used in exposed and concealed locations.

All flexible metal conduit as well as all flexible metal conduit fittings are required to be listed.

As the example in Figure 350.1 shows, listed flexible metal conduit (FMC) is permitted for use in wet locations provided the completed installation prevents water from entering enclosures or other raceways to which the conduit is connected. Also, for this application, the conductors must be suitable for wet locations. Listed flexible metal conduit ½ in. and larger may be installed in unlimited lengths where an equipment grounding conductor is installed with the circuit conductors. See Sections 250-118(5) and (6) for specific requirements related to its use as a grounding conductor.

350-5. Uses Not Permitted. Flexible metal conduit shall not be used in the following:

(1) In wet locations unless the conductors are approved for the specific conditions and the installation is such that liquid is not likely to enter raceways or enclosures to which the conduit is connected
(2) In hoistways, other than as permitted in Section 620-21(a)(1)

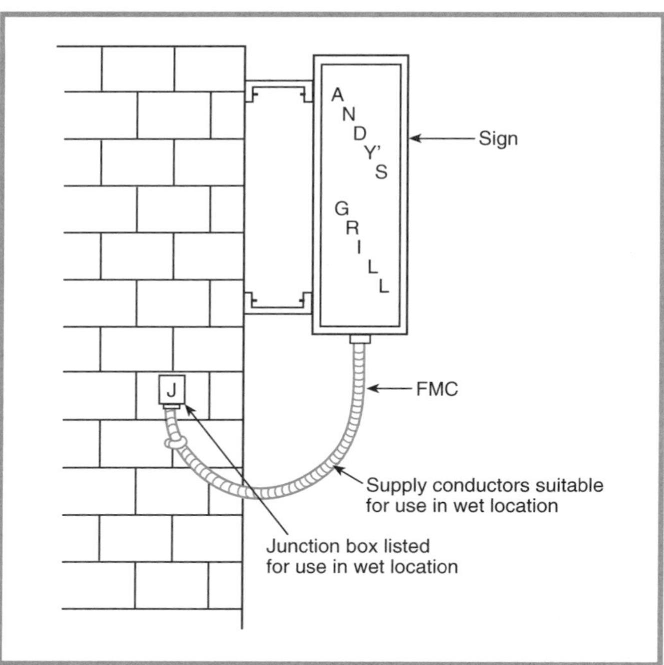

Figure 350.1 *An example of an acceptable application of flexible metal conduit on the exterior of a building. The branch-circuit supply conductors are suitable for use in a wet location, and the junction box is listed for use in a wet location.*

(3) In storage-battery rooms
(4) In any hazardous (classified) location other than as permitted in Sections 501-4(b) and 504-20
(5) Where exposed to materials having a deteriorating effect on the installed conductors, such as oil or gasoline
(6) Underground or embedded in poured concrete or aggregate
(7) Where subject to physical damage

B. Installation

350-10. Size.

(a) Minimum. Flexible metal conduit less than ½-in. electrical trade size shall not be used unless permitted in (1) through (5) below for ⅜-in. electrical trade size.

(1) For enclosing the leads of motors as permitted in Section 430-145(b)
(2) In lengths not in excess of 6 ft (1.83 m)
  (a) For utilization equipment, or
  (b) As part of a listed assembly, or
  (c) For tap connections to lighting fixtures as permitted in Section 410-67(c)

Section 350-10(a)(2) makes it clear that ⅜-in. flexible metal conduit is permitted to be used as the manufactured or field-installed metal raceway (4 ft to 6 ft in length) to enclose tap conductors between the outlet box and the fixture terminal housing of recessed lighting fixtures.

Flexible metal conduit is also permitted to be used as a 6-ft fixture whip from an outlet box to a lighting fixture.

(3) For manufactured wiring systems as permitted in Section 604-6(a)

Section 604-6 permits a smaller minimum size for manufactured wiring systems because the conductors are not as prone to physical damage when assembled under factory-controlled conditions.

(4) In hoistways, as permitted in Section 620-21(a)(1)
(5) As part of a listed assembly to connect wired fixture sections as permitted in Section 410-77(c)

(b) Maximum. Flexible metal conduit larger than 4-in. electrical trade size shall not be used.

FPN: Metric trade numerical designations for flexible metal conduit are ⅜ = 12, ½ = 16, ¾ = 21, 1 = 27, 1¼ = 35, 1½ = 41, 2 = 53, 2½ = 63, 3 = 78, 3½ = 91, and 4 = 103.

350-12. Number of Conductors. The number of conductors permitted in a flexible metal conduit shall not exceed the percentage of fill specified in Table 1, Chapter 9, or as permitted in Table 350-12 for ⅜-in. flexible metal conduit.

Table 350-12. Maximum Number of Insulated Conductors in ⅜-in. Flexible Metal Conduit*

	Types RFH-2, SF-2		Types TF, XHHW, AF, TW		Types TFN, THHN, THWN		Types FEP, FEPB, PF, PGF	
Size (AWG)	Fittings Inside Conduit	Fittings Outside Conduit	Fittings Inside Conduit	Fittings Outside Conduit	Fittings Inside Conduit	Fittings Outside Conduit	Fittings Inside Conduit	Fittings Outside Conduit
18	2	3	3	5	5	8	5	8
16	1	2	3	4	4	6	4	6
14	1	2	2	3	3	4	3	4
12	—	—	1	2	2	3	2	3
10	—	—	1	1	1	1	1	2

*In addition, one covered or bare equipment grounding conductor of the same size shall be permitted.

Table 1 of Chapter 9 specifies the maximum percent fill of a conduit or tubing. Table 4 provides the usable area within the selected conduit or tubing, and Table 5 provides the required area for each of the conductors. Examples using these tables to calculate a conduit or tubing size are provided following Chapter 9, Table 1, Notes to Tables, Note 6.

If the conductors are of the same wire size, instead of doing the calculations, the tables of Appendix C may be used. Appendix C, which contains 12 sets of tables, very accurately indicates the maximum number of conductors permitted in a conduit or tubing. This appendix has replaced Tables 3A, 3B, and 3C from the 1993 *Code*. Examples using this appendix to select a conduit or tubing size are provided following the introduction in Appendix C.

In order to select the proper trade size flexible metal conduit, the section entitled "Flexible Metal Conduit" in Table 4 of Chapter 9 should be followed. Appendix C Tables C3 and C3A for flexible metal conduit are also permissible.

350-14. Grounding. Flexible metal conduit shall be permitted as a grounding means as covered in Section 250-118. Where an equipment bonding jumper is required around flexible metal conduit, it shall be installed in accordance with Section 250-102.

In general, flexible metal conduit (FMC) requires the inclusion of an equipment grounding conductor. The exception permits short lengths (6 ft or less) of flexible metal conduit to serve as the equipment grounding conductor.

Where the length of the total ground-fault return path exceeds 6 ft or the circuit overcurrent exceeds 20 amperes, a separate equipment grounding conductor must be installed with the circuit conductors. Section 250-102(e), however, permits the equipment grounding conductor to be routed on the outside of the raceway in lengths that are no longer than 6 ft and bonded at each end.

Where used in hazardous (classified) locations, a bonding jumper is specifically required. See Sections 501-16(b), 502-16(b), and 503-16(b) for types of equipment grounding conductors.

The sketch on the right in Figure 350.2 shows an acceptable application of flexible metal conduit where the total length of any ground return path is limited to 6 ft. The sketch on the left shows an application that is unacceptable, as the grounding return path for fixture 2 exceeds the permitted maximum of 6 ft.

Where used to connect equipment where flexibility is required, an equipment grounding conductor shall be installed.

Where flexible metal conduit is installed to minimize the transmission of vibration from equipment such as motors or to provide flexibility for equipment that requires adjustment, such as floodlights or spotlights, a separate equipment grounding conductor is required to be installed regardless of the length of the flexible conduit.

350-16. Bends. There shall not be more than the equivalent of four quarter bends (360 degrees total) between pull points, e.g., conduit bodies and boxes. Bends in the conduit shall be made so that the conduit will not be damaged and the internal diameter of the conduit will not be effectively reduced. The radius of the curve of the inner edge of any field bend shall not be less than shown in Table 346-10.

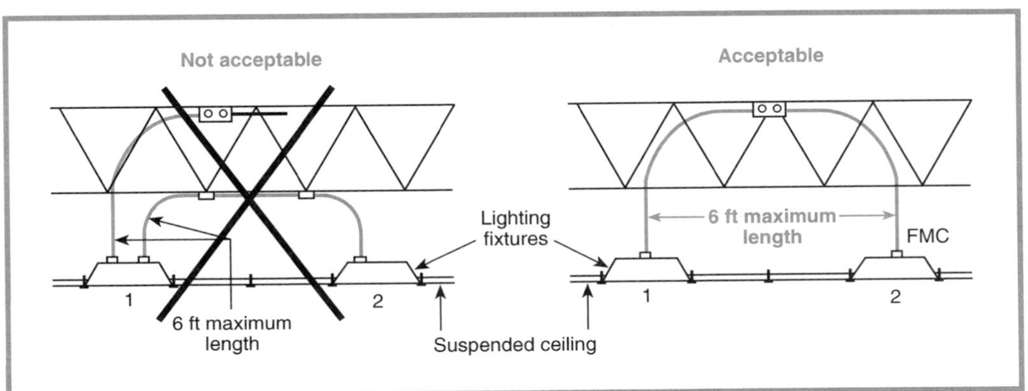

Figure 350.2 *An example of acceptable and unacceptable applications of flexible metal conduit without separate equipment grounding conductors used as a fixture whip.*

As with other raceways, a run of flexible metal conduit installed between boxes, conduit bodies, and so on, is not permitted to contain more than the equivalent of four quarter bends (360 degrees total). Proper shaping and support of this flexible wiring method will ensure that conductors can be easily installed or withdrawn at any time.

350-18. Supports. Flexible metal conduit shall be securely fastened in place by an approved means within 12 in. (305 mm) of each box, cabinet, conduit body, or other conduit termination and shall be supported and secured at intervals not to exceed 4½ ft (1.37 m).

Exception No. 1: Where flexible metal conduit is fished.

Exception No. 2: Lengths not exceeding 3 ft (914 mm) at terminals where flexibility is required.

Exception No. 3: Lengths not exceeding 6 ft (1.83 m) from a fixture terminal connection for tap connections to light fixtures as permitted in Section 410-67(c).

Horizontal runs of flexible metal conduit supported by openings through framing members at intervals not greater than 4½ ft (1.37 m) and securely fastened within 12 in. (305 mm) of termination points shall be permitted.

350-20. Fittings. Fittings used with flexible metal conduit shall be listed. Angle connectors shall not be used for concealed raceway installations.

350-22. Trimming. All cut ends of flexible metal conduit shall be trimmed or otherwise finished to remove rough edges, except where fittings that thread into the convolutions are used.

350-24. Splices and Taps. Splices and taps shall be made in accordance with Section 300-15. See Article 370 for rules on the installation and use of boxes and conduit bodies.

Article 351 — Liquidtight Flexible Metal Conduit and Liquidtight Flexible Nonmetallic Conduit

Contents

351-1. Scope. This article covers liquidtight flexible metal conduit and liquidtight flexible nonmetallic conduit.

Liquidtight flexible metal and liquidtight flexible nonmetallic conduit are intended for use in wet locations or where exposed to mineral oil, both at a maximum temperature of 60°C (140°F). They are not intended for use where exposed to gasoline or similar light petroleum solvents unless so marked on the product.

Liquidtight flexible nonmetallic conduit is used extensively in the machine tool and related industries. See clause 17-8 in *Electrical Standard for Industrial Machinery*, NFPA 79-1997, for the uses permitted on an industrial machine.

A. Liquidtight Flexible Metal Conduit

351-2. Definition. *Liquidtight flexible metal conduit* is a listed raceway of circular cross section having an outer liquidtight, nonmetallic, sunlight-resistant jacket over an inner flexible metal core with associated couplings, connectors, and fittings and approved for the installation of electric conductors.

351-3. Other Articles. Installations of liquidtight flexible metal conduit shall comply with the applicable provisions of Article 300 and with the specific sections of Articles 350, 501, 502, 503, and 553 referenced below.

FPN: For marking requirements, see Section 110-21.

351-4. Use.

(a) Permitted. Listed liquidtight flexible metal conduit shall be permitted to be used in exposed or concealed locations as follows:

(1) Where conditions of installation, operation, or maintenance require flexibility or protection from liquids, vapors, or solids
(2) As permitted by Sections 501-4(b), 502-4, 503-3, and 504-20 and in other hazardous (classified) locations where specifically approved, and by Section 553-7(b)
(3) For direct burial where listed and marked for the purpose

Liquidtight flexible metal conduit is required to be a listed product. If properly marked for the application, liquidtight flexible metal conduit is permitted for direct burial in the earth. Note that the requirements of Section 300-5 are applicable to liquidtight flexible metal conduit if installed underground.

Liquidtight flexible metal conduit may be installed in unlimited lengths provided it meets the other requirements of this article and a separate equipment grounding conductor is installed with the circuit conductors.

Liquidtight flexible metal conduit is on the permitted list of wiring methods allowed for services (Section 230-43), provided the length does not exceed 6 ft and an equipment bonding jumper is installed in accordance with Section 250-102.

(b) Not Permitted. Liquidtight flexible metal conduit shall not be used as follows:

(1) Where subject to physical damage
(2) Where any combination of ambient and conductor temperature will produce an operating temperature in excess of that for which the material is approved

351-5. Size.

(a) Minimum. Liquidtight flexible metal conduit smaller than ½-in. electrical trade size shall not be used.

Exception: Liquidtight flexible metal conduit of ⅜-in. size shall be permitted as covered in Section 350-10(a).

(b) Maximum. The maximum size of liquidtight flexible metal conduit shall be the 4-in. trade size.

FPN: Metric trade numerical designations for liquidtight flexible metal conduit are ⅜ = 12, ½ = 16, ¾ = 21, 1 = 27, 1¼ = 35, 1½ = 41, 2 = 53, 2½ = 63, 3 = 78, 3½ = 91, and 4 = 103.

351-6. Number of Conductors.

(a) Single Conduit. The number of conductors permitted in a single conduit, ½-in. through 4-in. trade sizes, shall not exceed the percentage of fill specified in Table 1, Chapter 9.

(b) Liquidtight Flexible Metal Conduit — ⅜-in. Size. The number of conductors permitted in ⅜-in. liquidtight flexible metal conduit shall not exceed that permitted in Table 350-12.

Table 1 of Chapter 9 specifies the maximum percent fill of a conduit or tubing. Table 4 provides the usable area within the selected conduit or tubing, and Table 5 provides the required area for each of the conductors. Examples using these tables to calculate a conduit or tubing size are provided following Chapter 9, Table 1, Notes to Tables, Note 6.

If the conductors are of the same wire size, instead of doing the calculations, the tables of Appendix C may be used. Appendix C, which contains 12 sets of tables, very accurately indicates the maximum number of conductors permitted in a conduit or tubing. This appendix replaced Tables 3A, 3B, and 3C from the 1993 *Code*. Examples using this appendix to select a conduit or tubing size are provided following the introduction in Appendix C.

In order to select the proper trade size liquidtight flexible metal conduit, the section entitled "Liquidtight Flexible Metal Conduit" in Table 4 of Chapter 9 should be followed. Appendix C Tables C7 and C7A for liquidtight flexible metal conduit are also permissible.

351-7. Fittings. Liquidtight flexible metal conduit shall be used only with listed terminal fittings. Angle connectors shall not be used for concealed raceway installations.

351-8. Supports. Liquidtight flexible metal conduit shall be securely fastened in place by an approved means within 12 in. (305 mm) of each box, cabinet, conduit body, or other conduit termination and shall be supported and secured at intervals not to exceed 4½ ft (1.37 m).

Exception No. 1: Where liquidtight flexible metal conduit is fished.

Exception No. 2: Lengths not exceeding 3 ft (914 mm) at terminals where flexibility is necessary.

Exception No. 3: Lengths not exceeding 6 ft (1.83 m) from a fixture terminal connection for tap conductors to lighting fixtures as permitted in Section 410-67(c).

The support requirements are modified if the liquidtight flexible metal conduit encloses tap conductors. Section 410-67(c) is part of the special provisions for flush and recessed fixtures.

Horizontal runs of liquidtight flexible metal conduit supported by openings through framing members at intervals not greater than 4½ ft (1.37 m) and securely fastened within 12 in. (305 mm) of termination points shall be permitted.

351-9. Grounding. Liquidtight flexible metal conduit shall be permitted as a grounding means as covered in Section

250-118. Where an equipment bonding jumper is required around liquidtight flexible metal conduit, it shall be installed in accordance with Section 250-102.

Where used to connect equipment where flexibility is required, an equipment grounding conductor shall be installed.

FPN: See Sections 501-16(b), 502-16(b), and 503-16(b) for types of equipment grounding conductors.

Where liquidtight flexible metal conduit is installed in hazardous (classified) locations, a bonding jumper may be required. There are some exceptions, however, as indicated in the exceptions to Sections 501-16(b), 502-16(b), and 503-16(b).

351-10. Bends — Number in One Run. There shall not be more than the equivalent of four quarter bends (360 degrees total) between pull points, e.g., conduit bodies and boxes.

351-11. Splices and Taps. Splices and taps shall be made in accordance with Section 300-15. See Article 370 for rules on the installation and use of boxes and conduit bodies.

B. Liquidtight Flexible Nonmetallic Conduit

351-22. Definition. Liquidtight flexible nonmetallic conduit is a listed raceway of circular cross section of various types as follows:

(1) A smooth seamless inner core and cover bonded together and having one or more reinforcement layers between the core and cover, designated as Type LFNC-A
(2) A smooth inner surface with integral reinforcement within the conduit wall, designated as Type LFNC-B
(3) A corrugated internal and external surface without integral reinforcement within the conduit wall, designated as Type LFNC-C

This conduit is flame resistant and, with fittings, is approved for the installation of electrical conductors.

351-23. Use.

(a) Permitted. Listed liquidtight flexible nonmetallic conduit shall be permitted to be used in exposed or concealed locations for the following purposes.

FPN: Extreme cold may cause some types of nonmetallic conduits to become brittle and therefore more susceptible to damage from physical contact.

(1) Where flexibility is required for installation, operation, or maintenance.
(2) Where protection of the contained conductors is required from vapors, liquids, or solids.
(3) For outdoor locations where listed and marked as suitable for the purpose.

FPN: For marking requirements, see Section 110-21.

(4) For direct burial where listed and marked for the purpose.
(5) Liquidtight flexible nonmetallic conduit as defined in Section 351-22(2) shall be permitted to be installed in lengths longer than 6 ft (1.83 m) where secured in accordance with Section 351-27.
(6) As a listed manufactured prewired assembly, ½-in. through 1-in. conduit, as defined in Section 351-22(2).

Prewired Type LFNC-B is a listed assembly where the conductors are required to be installed at the manufacturing facility where controlled conditions prevent damage to the conductor insulation. Special cutting tools are required to be used when cutting prewired Type LFNC-B to prevent nicking the conductor installation. This prewired assembly is shown in Figure 351.1.

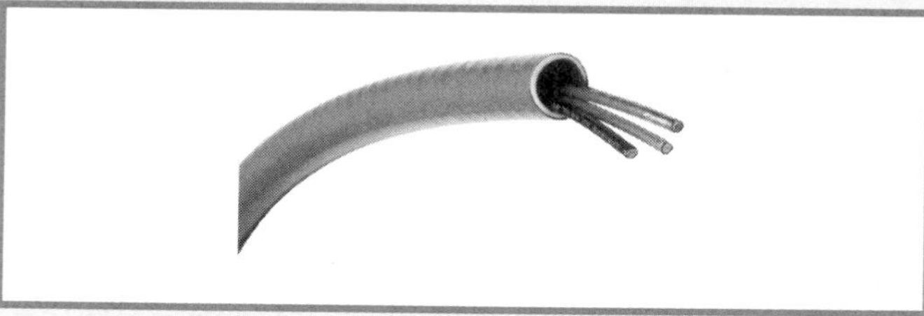

Figure 351.1 *Listed manufactured prewired assembly of liquidtight flexible nonmetallic conduit, Type B. (Carlon®, a Lamson & Sessions Company)*

(b) Not Permitted. Liquidtight flexible nonmetallic conduit shall not be used in the following:

(1) Where subject to physical damage
(2) Where any combination of ambient and conductor temperatures is in excess of that for which the liquidtight flexible nonmetallic conduit is approved
(3) In lengths longer than 6 ft (1.83 m), except as permitted by Section 351-23(a)(5) or where a longer length is approved as essential for a required degree of flexibility
(4) Where voltage of the contained conductors is in excess of 600 volts, nominal

Exception: As permitted in Section 600-32(a) for electric signs over 600 volts.

351-24. Size. The electrical trade sizes of liquidtight flexible nonmetallic conduit shall be in accordance with (a) or (b):

(a) ½ in. to 4 in. inclusive
(b) ⅜ in. as permitted below
   (1) For enclosing the leads of motors as permitted in Section 430-145(b)
   (2) In lengths not exceeding 6 ft (1.83 m) as part of a listed assembly for tap connections to lighting fix-

tures as required in Section 410-67(c), or for utilization equipment

(3) For electric sign conductors in accordance with Section 600-32(a)

FPN: Metric trade numerical designations for liquidtight flexible nonmetallic conduit are ⅜ = 12, ½ = 16, ¾ = 21, 1 = 27, 1¼ = 35, 1½ = 41, 2 = 53, 2½ = 63, 3 = 78, 3½ = 91, and 4 = 103.

351-25. Number of Conductors. The number of conductors permitted in a single conduit shall be in accordance with the percentage fill specified in Table 1, Chapter 9.

Table 1 of Chapter 9 specifies the maximum percent fill of a conduit or tubing. Table 4 provides the usable area within the selected conduit or tubing, and Table 5 provides the required area for each of the conductors. Examples using these tables to calculate a conduit or tubing size are provided following Chapter 9, Table 1, Notes to Tables, Note 6.

If the conductors are of the same wire size, instead of doing the calculations, the tables of Appendix C may be used. Appendix C, which contains 12 sets of tables, very accurately indicates the maximum number of conductors permitted in a conduit or tubing. This appendix replaced Tables 3A, 3B, and 3C from the 1993 *Code*. Examples using this appendix to select a conduit or tubing size are provided following the introduction in Appendix C.

To permit selection of the proper trade size liquidtight flexible nonmetallic conduit, Table 4 of Chapter 9 contains two sections entitled "Liquidtight Flexible Nonmetallic Conduit" (one for FNMC-A and one for FNMC-B). The correct table according to Section 351-22 should be followed. Appendix C Tables C5 and C5A or C6 and C6A for liquidtight flexible nonmetallic conduit (FNMC-A or FNMC-B) are also permissible.

351-26. Fittings. Liquidtight flexible nonmetallic conduit shall be used only with listed terminal fittings. Angle connectors shall not be used for concealed raceway installations.

351-27. Securing and Supporting. Liquidtight flexible nonmetallic conduit, as defined in Section 351-22(2), shall be securely fastened and supported in accordance with one of the following.

(a) The conduit shall be securely fastened at intervals not exceeding 3 ft (914 mm) and within 12 in. (305 mm) on each side of every outlet box, junction box, cabinet, or fitting.

(b) Securing and supporting of the conduit shall not be required where it is fished, installed in lengths not exceeding 3 ft (914 mm) at terminals where flexibility is required, or where installed in lengths not exceeding 6 ft (1.83 m) from a fixture terminal connection for tap conductors to lighting fixtures as permitted in Section 410-67(c).

(c) Horizontal runs of liquidtight flexible nonmetallic conduit supported by openings through framing members at intervals not exceeding 3 ft (914 mm) and securely fastened within 12 in. (305 mm) of termination points shall be permitted.

351-28. Equipment Grounding. Where an equipment grounding conductor is required for the circuits installed in liquidtight flexible nonmetallic conduit, it shall be permitted to be installed on the inside or outside of the conduit. Where installed on the outside, the length of the equipment grounding conductor shall not exceed 6 ft (1.83 m) and shall be routed with the raceway or enclosure. Fittings and boxes shall be bonded or grounded in accordance with Article 250.

351-29. Splices and Taps. Splices and taps shall be made in accordance with Section 300-15. See Article 370 for rules on the installation and use of boxes and conduit bodies.

351-30. Bends — Number in One Run. There shall not be more than the equivalent of four quarter bends (360 degrees total) between pull points, e.g., conduit bodies and boxes.

Article 352 — Surface Metal Raceways and Surface Nonmetallic Raceways

Contents

A. Surface Metal Raceways

352-1. Uses.

(a) Permitted. The use of surface metal raceways shall be permitted in the following:

(1) In dry locations
(2) In Class I, Division 2 hazardous (classified) locations as permitted in Section 501-4(b), Exception
(3) Under raised floors, as permitted in Section 645-5(d)(2)

(b) Not Permitted. The use of surface metal raceways shall not be permitted in the following:

(1) Where subject to severe physical damage, unless otherwise approved
(2) Where the voltage is 300 volts or more between conductors, unless the metal has a thickness of not less than 0.040 in. (1.016 mm) nominal
(3) Where subject to corrosive vapors
(4) In hoistways
(5) Where concealed, except as permitted in Section 352-1(a)(3)

The installation shown in Figure 352.1 is typical of how a surface metal raceway can be used.

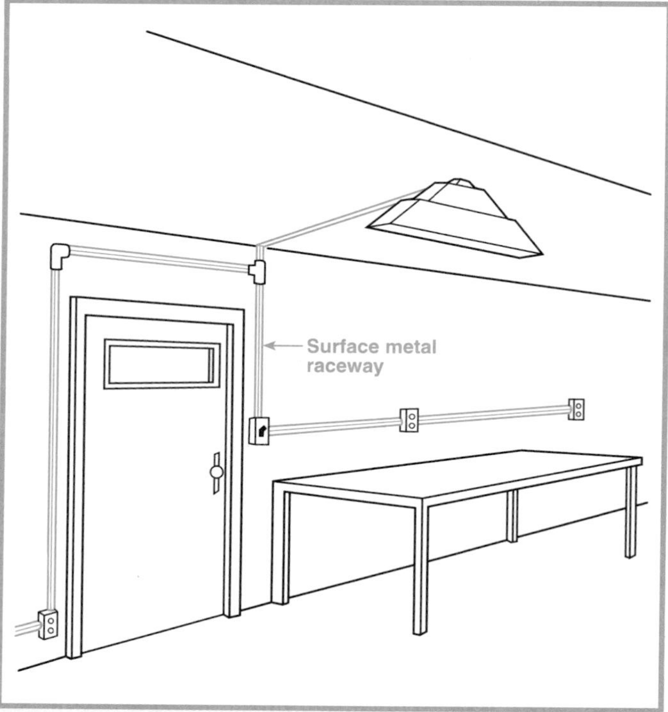

Figure 352.1 An example of a surface metal raceway extending from an existing receptacle outlet.

352-2. Other Articles. Surface metal raceways shall comply with the applicable provisions of Article 300.

352-3. Size of Conductors. No conductor larger than that for which the raceway is designed shall be installed in surface metal raceway.

352-4. Number of Conductors in Raceways. The number of conductors installed in any raceway shall not be greater than the number for which the raceway is designed.

The derating factors of Section 310-15(b)(2)(a) shall not apply to conductors installed in surface metal raceways where all of the following conditions are met:

(1) The cross-sectional area of the raceway exceeds 4 in.^{2} (2580 mm^2)
(2) The current-carrying conductors do not exceed 30 in number
(3) The sum of the cross-sectional areas of all contained conductors does not exceed 20 percent of the interior cross-sectional area of the surface metal raceway

The number, type, and sizes of conductors permitted to be installed in a listed surface metal raceway are marked on the raceway or on the package in which they are shipped. Typically, this information is available in detail in the manufacturer's catalog. Figure 352.2 provides conductor fill information for specific surface metal raceways.

352-5. Extension Through Walls and Floors. Surface metal raceways shall be permitted to pass transversely through dry walls, dry partitions, and dry floors if the length passing through is unbroken. Access to the conductors shall be maintained on both sides of the wall, partition, or floor.

352-6. Combination Raceways. Where combination surface metal raceways are used both for signaling and for lighting and power circuits, the different systems shall be run in separate compartments identified by sharply contrasting colors of the interior finish, and the same relative position of compartments shall be maintained throughout the premises.

Type of Raceway	Wire Size Gauge No.	Number of Wires: Types RHH, RHW	Number of Wires: Type THW	Number of Wires: Type TW	Number of Wires: Types THHN, THWN
No. 200 (½ in. × 11/32 in.)	12	—	2	3	3
	14	—	2	3	5
No. 500 (¾ in. × 17/32 in.)	8	—	—	2	2
	10	2	2	3	4
	12	2	3	4	7
	14	2	4	6	9
No. 700 (¾ in. × 21/32 in.)	6	—	—	—	2
	8	—	2	2	3
	10	2	3	4	5
	12	2	4	6	8
	14	3	5	7	11
No. 1500 (1 9/16 in. × 11/32 in.)	6	—	—	—	2
	8	—	—	2	3
	10	2	3	4	5
	12	2	3	5	7
	14	2	4	6	10
No. 2000[a] (1 9/16 in. × ¾ in.)	12	—	—	7	7
	14	—	—	7	7
No. 2100[a] (1¼ in. × 7/8 in.)	6	2	4	4	6
	8	4	6	8	10
	10	7	10	14	17
	12	8	13	19	28
	14	10	15	24	37
No. 2200[a] (2 3/8 in. × ¾ in.)	6	5	7	3[b] 7	11
	8	8	11	7[b] 14	19
	10	13	19	10[b] 26	32
	12	15	23	10[b] 34	51
	14	18	29	10[b] 44	69
No. 2600 (2 7/32 in. × 23/32 in.)	6	2	3	3	5
	8	4	5	7	9
	10	6	9	12	15
	12	7	11	16	24
	14	9	14	21	33
G-3000 (2¾ in. × 1 17/32 in.)	6	4[b] 11	6[b] 19	6[b] 17	6[b] 27
	8	6[b] 18	8[b] 26	8[b] 34	8[b] 44
	10	10[b] 30	10[b] 45	10[b] 62	10[b] 76
	12	14[b] 36	18[b] 55	18[b] 81	18[b] 119
	14	16[b] 42	26[b] 67	26[b] 103	26[b] 160
G-4000, with divider (4¾ in. × 1¾ in.)	2	— 7	— 10	— 10	— 12
	3	— 8	— 11	— 11	— 15
	4	— 9	— 13	— 13	— 17
	6	4[b] 12	4[b] 18	4[b] 18	7[b] 28
	8	7[b] 19	7[b] 28	7[b] 36	8[b] 47
	10	11[b] 32	11[b] 48	11[b] 66	15[b] 81
	12	15[b] 39	15[b] 59	15[b] 86	24[b] 128
	14	17[b] 45	17[b] 72	17[b] 110	32[b] 171
G-4000, without divider (4¾ in. × 1¾ in.)	2	— 14	— 20	— 20	— 25
	3	— 16	— 23	— 23	— 30
	4	— 18	— 27	— 27	— 35
	6	8[b] 24	8[b] 36	8[b] 36	10[b] 57
	8	10[b] 39	10[b] 57	10[b] 78	15[b] 94
	10	15[b] 65	15[b] 96	12[b] 133	18[b] 163
	12	21[b] 78	21[b] 119	16[b] 174	34[b] 256
	14	21[b] 91	21[b] 145	17[b] 222	34[b] 344
G-6000 (4¾ in. × 3 9/16 in.)	2/0	10[b] 17	12[b] 22	12[b] 22	15[b] 27
	1/0	11[b] 20	14[b] 26	14[b] 26	18[b] 33
	1	12[b] 23	17[b] 31	17[b] 31	21[b] 39
	2	16[b] 30	23[b] 43	23[b] 43	29[b] 53
	3	19[b] 34	27[b] 50	27[b] 50	34[b] 63
	4	21[b] 39	32[b] 58	32[b] 58	40[b] 74
	6	27[b] 51	42[b] 77	42[b] 77	66[b] 122
	8	40[b] 74	57[b] 106	73[b] 134	92[b] 169
	10	75[b] 137	111[b] 203	154[b] 282	187[b] 343
	12	90[b] 164	137[b] 252	200[b] 386	295[b] 540
	14	105[b] 193	167[b] 307	255[b] 469	396[b] 726

[a]Figures for Nos. 2000, 2100, 2200, G-3000, G-4000, and G-6000 are without receptacles, except where noted.
[b]With receptacles.

Figure 352.2 *Conductor fill table for various surface metal raceways. (The Wiremold Co.)*

352-7. Splices and Taps. Splices and taps shall be permitted in surface metal raceways having a removable cover that is accessible after installation. The conductors, including splices and taps, shall not fill the raceway to more than 75 percent of its area at that point. Splices and taps in surface metal raceways without removable covers shall be made only in junction boxes. All splices and taps shall be made by approved methods.

Taps of Type FC cable installed in surface metal raceway shall be made in accordance with Section 363-10.

352-8. General. Surface metal raceways shall be of such construction as will distinguish them from other raceways. Surface metal raceways and their elbows, couplings, and similar fittings shall be designed so that the sections can be electrically and mechanically coupled together and installed without subjecting the wires to abrasion.

Where covers and accessories of nonmetallic materials are used on surface metal raceways, they shall be identified for such use.

352-9. Grounding. Surface metal raceway enclosures providing a transition from other wiring methods shall have a means for connecting an equipment grounding conductor.

As the example in Figure 352.3 shows, where a surface metal raceway is supplied by Type MC or NM cable, a means (e.g., grounding terminal screw or lug) must be available at the surface metal raceway for terminating the equipment grounding conductor provided within the cable.

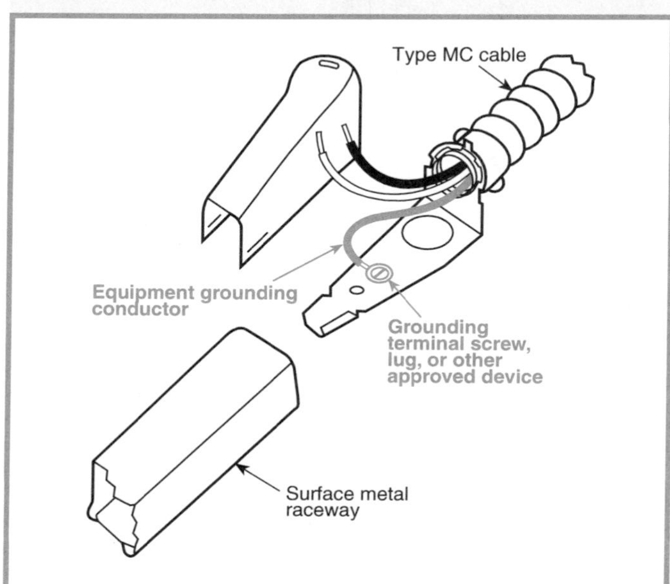

Figure 352.3 *An example of providing a means for terminating an equipment grounding conductor at a surface metal raceway.*

B. Surface Nonmetallic Raceways

352-21. Description. Part B of this article shall apply to a type of surface nonmetallic raceway and fittings of suitable nonmetallic material that is resistant to moisture and chemical atmospheres. It shall also be flame retardant, resistant to impact and crushing, resistant to distortion from heat under conditions likely to be encountered in service, and resistant to low-temperature effects. Surface nonmetallic raceways that have limited smoke-producing characteristics shall be permitted to be identified with the suffix *LS.*

352-22. Use.

(a) Permitted. The use of surface nonmetallic raceways shall be permitted in dry locations.

(b) Not Permitted. Surface nonmetallic raceways shall not be used as follows:

(1) Where concealed
(2) Where subject to severe physical damage
(3) Where the voltage is 300 volts or more between conductors, unless listed for higher voltage
(4) In hoistways
(5) In any hazardous (classified) location except Class I, Division 2 locations as permitted in Section 501-4(b), Exception
(6) Where subject to ambient temperatures exceeding those for which the nonmetallic raceway is listed
(7) For conductors whose insulation temperature limitations would exceed those for which the nonmetallic raceway is listed

The installations shown in Figures 352.4 and 352.5 are typical of how a surface nonmetallic raceway can be used.

Figure 352.4 *An example of a surface nonmetallic raceway extending from an existing receptacle outlet. (The Wiremold Co.)*

Figure 352.5 *An example of a surface nonmetallic raceway supplying a lighting outlet. (The Wiremold Co.)*

352-23. Other Articles. Surface nonmetallic raceways shall comply with the applicable provisions of Article 300. Where equipment grounding is required by Article 250, a separate equipment grounding conductor shall be installed in the raceway.

352-24. Size of Conductors. No conductor larger than that for which the raceway is designed shall be installed in surface nonmetallic raceway.

352-25. Number of Conductors in Raceways. The number of conductors installed in any raceway shall not be greater than the number for which the raceway is designed.

352-26. Combination Raceways. Where combination surface nonmetallic raceways are used both for signaling and for lighting and power circuits, the different systems shall be run in separate compartments, identified by printed legend or by sharply contrasting colors of the interior finish, and the same relative position of compartments shall be maintained throughout the premises.

352-27. General. Surface nonmetallic raceways shall be of such construction as will distinguish them from other raceways. Surface nonmetallic raceways and their elbows, couplings, and similar fittings shall be designed so that the sections can be mechanically coupled together and installed without subjecting the wires to abrasion.

352-28. Extension Through Walls and Floors. Surface nonmetallic raceways shall be permitted to pass transversely through dry walls, dry partitions, and dry floors if the length passing through is unbroken. Access to the conductors shall be maintained on both sides of the wall, partition, or floor.

352-29. Splices and Taps. Splices and taps shall be permitted in surface nonmetallic raceways having a removable cover that is accessible after installation. The conductors, including splices and taps, shall not fill the raceway to more than 75 percent of its area at that point. Splices and taps in surface nonmetallic raceways without removable covers shall be made only in junction boxes. All splices and taps shall be made by approved methods.

C. Strut-Type Channel Raceway

352-40. Description. Part C of this article shall apply to strut-type channel raceways and accessories formed of metal that are resistant to moisture or protected by corrosion protection and judged suitable for the condition. These channel raceways shall be permitted to be galvanized, stainless, enameled, or PVC-coated steel, or aluminum. Covers shall be either metallic or nonmetallic.

352-41. Uses Permitted. The installation of listed strut-type channel raceways shall be permitted:

(1) Where exposed
(2) In dry locations
(3) In locations subject to corrosive vapors where protected by finishes judged suitable for the condition
(4) Where the voltage is 600 volts or less
(5) As power poles
(6) In Class I, Division 2 hazardous (classified) locations as permitted in Section 501-4(b), Exception

The installation shown in Figure 352.6 is typical of how a strut-type channel raceway can be used.

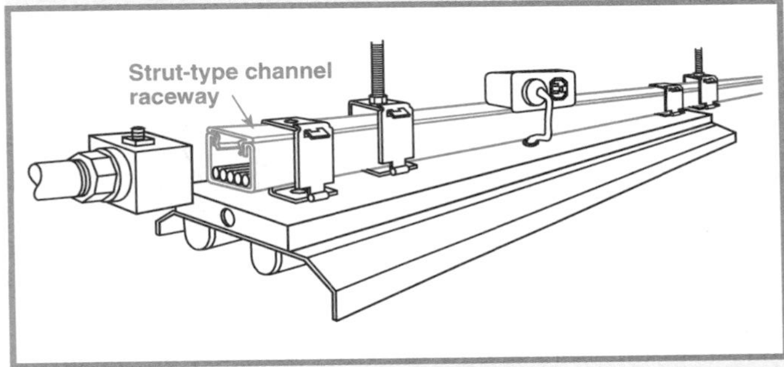

Figure 352.6 *An example of a strut-type channel raceway using accessories to support and supply power to lighting fixtures. (Allied Tube & Conduit, a Tyco International Co.)*

352-42. Uses Not Permitted.

(a) General. Strut-type channel raceways shall not be permitted to be used where concealed.

(b) Ferrous Metal. Ferrous channel raceways and fittings protected from corrosion solely by enamel shall be permitted only indoors and in occupancies not subject to severe corrosive influences.

352-43. Other Articles. Installation of strut-type channel raceways shall comply with the applicable provisions of Articles 250 and 300.

352-44. Size of Conductors. No conductor larger than that for which the raceway is listed shall be installed in strut-type channel raceways.

352-45. Number of Conductors in Raceways. The number of conductors permitted in strut-type channel raceway shall not exceed the percentage fill using Table 352-45 and applicable outside diameter (O.D.) dimensions of specific types and sizes of wire given in the Tables in Chapter 9.

The derating factors of Section 310-15(b)(2)(a) shall not apply to conductors installed in strut-type channel raceways where all of the following conditions are met:

(1) The cross-sectional area of the raceway exceeds 4 in.^{2} (2580 mm^2)
(2) The current-carrying conductors do not exceed 30 in number
(3) The sum of the cross-sectional areas of all contained conductors does not exceed 20 percent of the interior cross-sectional area of the strut-type channel raceway

$$\text{Formula for wire fill: } N = \frac{CA}{WA}$$

Where:

N = number of wires
CA = channel area in square inches
WA = wire area

Table 352-45. Channel Size and Inside Diameter Area

Channel Size	Area		40% Area*		25% Area**	
	in.^{2}	mm^2	in.^{2}	mm^2	in.^{2}	mm^2
1⅝ × 13⁄16	0.887	572	0.355	229	0.222	143
1⅝ × 1	1.151	743	0.460	297	0.288	186
1⅝ × 1⅜	1.677	1076	0.671	433	0.419	270
1⅝ × 1⅝	2.028	1308	0.811	523	0.507	327
1⅝ × 2 7⁄16	3.169	2045	1.267	817	0.792	511
1⅝ × 3¼	4.308	2780	1.723	1112	1.077	695
1½ × ¾	0.849	548	0.340	219	0.212	137
1½ × 1½	1.828	1179	0.731	472	0.457	295
1½ × 1⅞	2.301	1485	0.920	594	0.575	371
1½ × 3	3.854	2487	1.542	995	0.964	622

*Raceways with external joiners shall use a 40 percent wire fill calculation to determine the number of conductors permitted.

**Raceways with internal joiners shall use a 25 percent wire fill calculation to determine the number of conductors permitted.

The adjustment factors of Table 310-15(b)(2) are applicable to strut-type channel raceways because Table 352-45 does not contain any raceways with cross-sectional areas greater than 4 in.^{2}

Example

Calculate the maximum quantity of No. 10 Type THWN-2 copper conductors permitted in a normal 1½-in. by 1½-in. strut-type channel raceway where the joiners are mounted internally. (Generally, ordinary-duty strut-type channel raceway couplings or joiners are of the internal type, and heavy-duty strut raceway couplings are of the external type.)

Answer (working in inches)

Step 1. Since the strut-type channel raceway joiners are mounted on the internal surface of the raceway, Note 2 of Table 352-45 requires the use of the "25% Area" column for the maximum usable internal area of the raceway.

Step 2. According to Table 352-45, and using the 25% Area column, the usable internal area of the raceway for a 1½-in. by 1½-in. strut-type channel is 0.457 in.^{2}

Step 3. According to Chapter 9, Table 5, a No. 10 Type THWN-2 copper conductor has an area of 0.0211 in.^{2}

Step 4. Using the formula for wire fill (located directly below Table 352-45),

$$N = \frac{CA}{WA}$$

where:

N = number of wires
CA = channel area (in.^{2})
WA = wire area (in.^{2})

and substituting the table values,

$$N = \frac{0.457 \text{ in.}^2}{0.0211 \text{ in.}^2} = 21.66$$

or not more than 21 No. 10 Type THWN-2 copper conductors.

However, the adjustment factors of Table 310-15(b)(2) are applicable where a raceway contains more than three current-carrying conductors.

352-46. Extensions Through Walls and Floors. It shall be permitted to extend unbroken lengths of strut-type channel raceway through walls, partitions, and floors where closure strips are removeable from either side and the portion within the wall, partition, or floor remains covered.

352-47. Support of Strut-Type Channel Raceways.

(a) A surface mount strut-type channel raceway shall be secured to the mounting surface with retention straps external to the channel at intervals not exceeding 10 ft (3.05 m) and within 3 ft (914 mm) of each outlet box, cabinet, junction box, or other channel raceway termination.

(b) Suspension Mount. Strut-type channel raceways shall be permitted to be suspension mounted in air with approved appropriate methods designed for the purpose at intervals not to exceed 10 ft (3.05 m).

352-48. Splices and Taps. Splices and taps shall be permitted in raceways that are accessible after installation by having a removable cover. The conductors, including splices and taps, shall not fill the raceway to more than 75 percent of its area at that point. All splices and taps shall be made by approved methods.

352-49. General. Strut-type channel raceways shall be of a construction that distinguishes them from other raceways. Raceways and their elbows, couplings, and other fittings shall be designed so that the sections can be electrically and mechanically coupled together and installed without subjecting the wires to abrasion.

Where closure strips and accessories of nonmetallic materials are used on metallic strut-type channel raceways, they shall be listed and identified for such use.

352-50. Grounding. Strut-type channel raceway enclosures providing a transition to or from other wiring methods shall have a means for connecting an equipment grounding conductor. Strut-type channel raceway shall be permitted as an equipment grounding conductor in accordance with Section 250-118(14). Where a snap-fit metal cover for strut-type channel raceway is used to achieve electrical continuity in accordance with the listing, this cover shall not be permitted as the means for providing electrical continuity for a receptacle mounted in the cover.

352-51. Marking. Each length shall be clearly and durably identified as required in the first sentence of Section 110-21.

Article 353 — Multioutlet Assembly

Contents

353-1. Other Articles. A multioutlet assembly shall comply with applicable provisions of Article 300.

FPN: See the definition of *multioutlet assembly* in Article 100.

Multioutlet assemblies are metal and nonmetallic raceways that are usually surface mounted and designed to contain branch-circuit conductors and receptacles. Receptacles may be spaced at desired intervals and may be assembled at the factory or in the field. See Section 220-3(b)(8) and Figure 220.4 for load calculations.

353-2. Use.

(a) Permitted. The use of a multioutlet assembly shall be permitted in dry locations.

(b) Not Permitted. A multioutlet assembly shall not be installed as follows:

(1) Where concealed, except that it shall be permissible to surround the back and sides of a metal multioutlet assembly by the building finish or recess a nonmetallic multioutlet assembly in a baseboard
(2) Where subject to severe physical damage
(3) Where the voltage is 300 volts or more between conductors unless the assembly is of metal having a thickness of not less than 0.040 in. (1.02 mm)
(4) Where subject to corrosive vapors
(5) In hoistways
(6) In any hazardous (classified) locations except Class I, Division 2, locations as permitted in Section 501-4(b), Exception.

353-3. Metal Multioutlet Assembly Through Dry Partitions. It shall be permissible to extend a metal multioutlet assembly through (not run within) dry partitions, if arrangements are made for removing the cap or cover on all exposed portions and no outlet is located within the partitions.

Article 354 — Underfloor Raceways

Contents

354-1. Other Articles. Underfloor raceways shall comply with the applicable provisions of Article 300.

354-2. Use.

An underfloor raceway is a practical means of bringing light, power, and signal and communications systems to desks, work benches, or tables that are not lo-

cated adjacent to wall space. This wiring method offers flexibility in layout where used with movable partitions and is commonly used in large retail stores and office buildings to supply power at any desired location.

Underfloor raceways are permitted beneath the surface of concrete, wood, or other flooring material. The wiring method between cabinets, raceway junction boxes, and outlet boxes may be rigid metal conduit, intermediate metal conduit, rigid nonmetallic conduit, liquidtight flexible nonmetallic conduit, electrical nonmetallic tubing, or electrical metallic tubing. Flexible metal conduit may be used where not installed in concrete.

(a) Permitted. The installation of underfloor raceways shall be permitted beneath the surface of concrete or other flooring material or in office occupancies, where laid flush with the concrete floor and covered with linoleum or equivalent floor covering.

(b) Not Permitted. Underfloor raceways shall not be installed (1) where subject to corrosive vapors or (2) in any hazardous (classified) locations, except as permitted by Section 504-20 and in Class I, Division 2 locations as permitted in Section 501-4(b), Exception. Unless made of a material judged suitable for the condition or unless corrosion protection approved for the condition is provided, ferrous or nonferrous metal underfloor raceways, junction boxes, and fittings shall not be installed in concrete or in areas subject to severe corrosive influences.

354-3. Covering. Raceway coverings shall comply with (a) through (d).

(a) Raceways Not Over 4 in. (102 mm) Wide. Half-round and flat-top raceways not over 4 in. (102 mm) in width shall have not less than ¾ in. (19 mm) of concrete or wood above the raceway.

Exception: As permitted in (c) and (d) for flat-top raceways.

As Figure 354.1 illustrates, a ¾-in. concrete or wood covering is required over underfloor raceways not over 4 in. wide, except for trench-type and other raceways that are flush with concrete.

(b) Raceways Over 4 in. (102 mm) Wide but Not Over 8 in. (203 mm) Wide. Flat-top raceways over 4 in. (102 mm) but not over 8 in. (203 mm) wide with a minimum of 1 in. (25.4 mm) spacing between raceways shall be covered with concrete to a depth of not less than 1 in. (25.4 mm). Raceways spaced less than 1 in. (25.4 mm) apart shall be covered with concrete to a depth of 1½ in. (38 mm).

As Figure 354.2 illustrates, flat-top underfloor raceways over 4 in. wide and spaced less than 1 in. apart must be covered with at least 1½ in. of concrete.

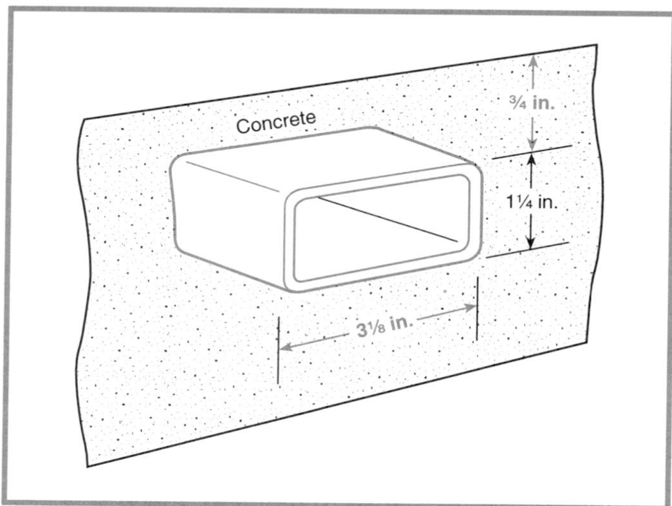

Figure 354.1 *An underfloor raceway not over 4 in. installed in compliance with Section 354-3(a). (Walker Systems, a Wiremold Co.)*

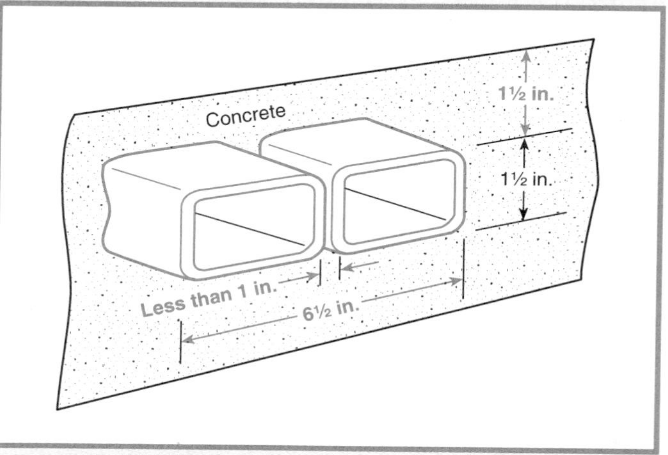

Figure 354.2 *Two side-by-side underfloor raceways over 4 in. installed in compliance with Section 354-3(b). (Walker Systems, a Wiremold Co.)*

(c) Trench-Type Raceways Flush with Concrete. Trench-type flush raceways with removable covers shall be permitted to be laid flush with the floor surface. Such approved raceways shall be designed so that the cover plates will provide adequate mechanical protection and rigidity equivalent to junction box covers.

Approved flush-type underfloor raceways may be installed flush with the floor surface provided they have covers at least equal to those of junction box covers.

(d) Other Raceways Flush with Concrete. In office occupancies, approved metal flat-top raceways, if not over 4 in. (102 mm) in width, shall be permitted to be laid flush with the concrete floor surface, provided they are covered with substantial linoleum that is not less than 1/16 in. (1.59 mm) in thickness or with equivalent floor covering. Where more than one and not more than three single raceways are each installed flush with the concrete, they shall be contiguous with each other and joined to form a rigid assembly.

354-4. Size of Conductors. No conductor larger than that for which the raceway is designed shall be installed in underfloor raceways.

354-5. Maximum Number of Conductors in Raceway. The combined cross-sectional area of all conductors or cables shall not exceed 40 percent of the interior cross-sectional area of the raceway.

354-6. Splices and Taps. Splices and taps shall be made only in junction boxes.

For the purposes of this section, so-called loop wiring (continuous, unbroken conductor connecting the individual outlets) shall not be considered to be a splice or tap.

Exception: Splices and taps shall be permitted in trench-type flush raceway having a removable cover that is accessible after installation. The conductors, including splices and taps, shall not fill the raceway more than 75 percent of its area at that point.

354-7. Discontinued Outlets. When an outlet is abandoned, discontinued, or removed, the sections of circuit conductors supplying the outlet shall be removed from the raceway. No splices or reinsulated conductors, such as would be the case with abandoned outlets on loop wiring, shall be allowed in raceways.

Loop wiring (continuous, unbroken conductors) is recognized where it runs from the underfloor raceway up to the terminals of attached receptacles, back into the raceway, and then on to the next device, as illustrated in Figure 354.3. When an outlet is removed, the sections of conductors supplying the outlet must be removed from the raceway as well. As is the case with abandoned outlets on loop wiring, reinsulated or spliced conductors are not allowed in raceways, except trench-type raceways as covered in the exception to Section 354-6.

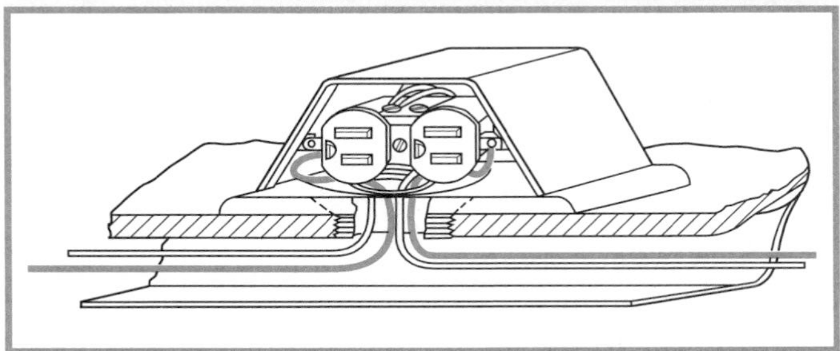

Figure 354.3 *A receptacle outlet supplied from an underfloor raceway by the loop method of wiring.*

354-8. Laid in Straight Lines. Underfloor raceways shall be laid so that a straight line from the center of one junction box to the center of the next junction box will coincide with the centerline of the raceway system. Raceways shall be firmly held in place to prevent disturbing this alignment during construction.

354-9. Markers at Ends. A suitable marker shall be installed at or near each end of each straight run of raceways to locate the last insert.

354-10. Dead Ends. Dead ends of raceways shall be closed.

354-13. Junction Boxes. Junction boxes shall be leveled to the floor grade and sealed to prevent the free entrance of water or concrete. Junction boxes used with metal raceways shall be metal and shall be electrically continuous with the raceways.

354-14. Inserts. Inserts shall be leveled and sealed to prevent the entrance of concrete. Inserts used with metal raceways shall be metal and shall be electrically continuous with the raceway. Inserts set in or on fiber raceways before the floor is laid shall be mechanically secured to the raceway. Inserts set in fiber raceways after the floor is laid shall be screwed into the raceway. When cutting through the raceway wall and setting inserts, chips and other dirt shall not be allowed to remain in the raceway, and tools shall be used that are designed so as to prevent the tool from entering the raceway and damaging conductors that may be in place.

354-15. Connections to Cabinets and Wall Outlets. Connections between raceways and distribution centers and wall outlets shall be made by means of flexible metal conduit where not installed in concrete, rigid metal conduit, intermediate metal conduit, electrical metallic tubing, or approved fittings. Where a metallic underfloor raceway system provides for the termination of an equipment grounding conductor, rigid nonmetallic conduit, electrical nonmetallic tubing, or liquidtight flexible nonmetallic conduit where not installed in concrete, shall be permitted.

Article 356 — Cellular Metal Floor Raceways

Contents

356-1. Definitions. For the purposes of this article, a *cellular metal floor raceway* shall be defined as the hollow spaces of cellular metal floors, together with suitable fittings, that may be approved as enclosures for electric conductors. A *cell* shall be defined as a single, enclosed tubular space in a cellular metal floor member, the axis of the cell being parallel to the axis of the metal floor member. A *header* shall be defined as a transverse raceway for electric conductors, providing access to predetermined cells of a cellular metal floor, thereby permitting the installation of electric conductors from a distribution center to the cells.

Cellular metal floor raceways are a form of metal floor deck construction designed for use in steel-frame buildings and consisting of sheet metal formed into shapes that are combined to form cells or raceways. The cells extend across the building and, depending on the structural strength required, can have various shapes and sizes.

One type of cellular metal floor is shown in Figure 356.1. Figure 356.2 shows the installation of header ducts (one for power conductors and one for telephone conductors) prior to the concrete being poured.

Figure 356.1 *One type of cellular metal floor construction, as viewed in the "slab" stage of installation. (Square D Co.)*

Figure 356.2 *Power and telephone header ducts feeding a cellular metal floor distribution system. (Square D Co.)*

356-2. Uses Not Permitted. Conductors shall not be installed in cellular metal floor raceways as follows:

(1) Where subject to corrosive vapor
(2) In any hazardous (classified) location except as permitted by Section 504-20, and in Class I, Division 2, locations as permitted in Section 501-4(b), Exception
(3) In commercial garages, other than for supplying ceiling outlets or extensions to the area below the floor but not above

FPN: See Section 300-8 for installation of conductors with other systems.

Section 300-8 prohibits the installation of electric conductors in raceways or cable trays containing any pipes, tubes, and so on, for steam, water, air, gas, drainage, or any service other than electrical.

356-3. Other Articles. Cellular metal floor raceways shall comply with the applicable provisions of Article 300.

A. Installation

356-4. Size of Conductors. No conductor larger than No. 1/0 shall be installed, except by special permission.

356-5. Maximum Number of Conductors in Raceway. The combined cross-sectional area of all conductors or cables shall not exceed 40 percent of the interior cross-sectional area of the cell or header.

Connections to the cells are made by means of headers extending across the cells and connecting only to those cells that are to be used as raceways for the conductors. Two or three separate headers, connecting to different sets of cells, may be used for different systems, such as light and power, signaling, and communications systems.

Figure 356.3 shows the cells with a header in place. The header is extended up to a cabinet or distribution center on a wall or column by means of a special elbow fitting. A junction box or access fitting is provided at each point where the header crosses a cell to which it connects. See the commentary following Section 354-7 and Figure 354.3 for discussion of loop wiring and discontinued outlets.

Figure 356.3 A typical cellular metal floor raceway installation showing cells, header ducts, junction boxes, and special elbow fittings. (Square D Co.)

356-6. Splices and Taps. Splices and taps shall be made only in header access units or junction boxes.

For the purposes of this section, so-called loop wiring (continuous unbroken conductor connecting the individual outlets) shall not be considered to be a splice or tap.

356-7. Discontinued Outlets. When an outlet is abandoned, discontinued, or removed, the sections of circuit conductors supplying the outlet shall be removed from the raceway. No splices or reinsulated conductors, such as would be the case with abandoned outlets on loop wiring, shall be allowed in raceways.

356-8. Markers. A suitable number of markers shall be installed for locating cells in the future.

Markers are brass flat-head screws set into the top side of the cells and adjusted so that their heads are flush with the floor finish and are exposed in order to aid in the location of cells for future installations.

356-9. Junction Boxes. Junction boxes shall be leveled to the floor grade and sealed against the free entrance of water or concrete. Junction boxes used with these raceways shall be of metal and shall be electrically continuous with the raceway.

See Figure 356.3, which illustrates cells connected to headers with junction boxes for future access and a special elbow fitting for connecting the header to a cabinet. Connections to wall outlets are to be made with metal raceways unless there are provisions for equipment grounding termination, as required by Section 356-11.

Installation instructions are supplied by the manufacturer for use by the general contractor, erector, electrical contractor, inspector, and others concerned with the installation.

356-10. Inserts. Inserts shall be leveled to the floor grade and sealed against the entrance of concrete. Inserts shall be of metal and shall be electrically continuous with the raceway. In cutting through the cell wall and setting inserts, chips and other dirt shall not be allowed to remain in the raceway, and tools shall be used that are designed to prevent the tool from entering the cell and damaging the conductors.

356-11. Connection to Cabinets and Extensions from Cells. Connections between raceways and distribution centers and wall outlets shall be made by means of flexible metal conduit where not installed in concrete, rigid metal conduit, intermediate metal conduit, electrical metallic tubing, or approved fittings. Where there are provisions for the termination of an equipment grounding conductor, nonmetallic conduit, electrical nonmetallic tubing, or liquidtight flexible nonmetallic conduit where not installed in concrete, shall be permitted.

B. Construction Specifications

356-12. General. Cellular metal floor raceways shall be constructed so that adequate electrical and mechanical continuity of the complete system will be secured. They shall provide a complete enclosure for the conductors. The interior surfaces shall be free from burrs and sharp edges, and surfaces over which conductors are drawn shall be smooth. Suitable bushings or fittings having smooth rounded edges shall be provided where conductors pass.

Article 358 — Cellular Concrete Floor Raceways

Contents

358-1. Scope. This article covers cellular concrete floor raceways, the hollow spaces in floors constructed of precast cellular concrete slabs, together with suitable metal fittings designed to provide access to the floor cells.

Cellular concrete floor raceways are a form of floor deck construction commonly used in high-rise office buildings. This method is very similar in design, application, and adaptation to cellular metal floor raceways.

Basically, this wiring method consists of floor cells (that are part of the structural floor system), header ducts laid at right angles to the cells and used to carry conductors from cabinets to cells, and junction boxes, as shown in Figure 358.1.

Figure 358.1 *A standard underfloor duct used on a precast cellular concrete floor raceway installation. (Square D Co.)*

358-2. Definitions. A *cell* shall be defined as a single, enclosed tubular space in a floor made of precast cellular concrete slabs, the direction of the cell being parallel to the direction of the floor member. A *header* shall be defined as transverse metal raceways for electric conductors, providing access to predetermined cells of a precast cellular concrete floor, thereby permitting the installation of electric conductors from a distribution center to the floor cells.

358-3. Other Articles. Cellular concrete floor raceways shall comply with the applicable provisions of Article 300.

358-4. Uses Not Permitted. Conductors shall not be installed in precast cellular concrete floor raceways as follows:

(1) Where subject to corrosive vapor
(2) In any hazardous (classified) locations except as permitted by Section 504-20, and in Class I, Division 2, locations as permitted in Section 501-4(b), Exception
(3) In commercial garages, other than for supplying ceiling outlets or extensions to the area below the floor but not above

FPN: See Section 300-8 for installation of conductors with other systems.

Section 300-8 prohibits the installation of electric conductors in raceways or cable trays containing any pipes, tubes, and so on, for steam, water, air, gas, drainage, or any service other than electrical.

358-5. Header. The header shall be installed in a straight line at right angles to the cells. The header shall be mechanically secured to the top of the precast cellular concrete floor. The end joints shall be closed by a metal closure fitting and sealed against the entrance of concrete. The header shall be electrically continuous throughout its entire length and shall be electrically bonded to the enclosure of the distribution center.

358-6. Connection to Cabinets and Other Enclosures. Connections from headers to cabinets and other enclosures shall be made by means of listed metal raceways and listed fittings.

358-7. Junction Boxes. Junction boxes shall be leveled to the floor grade and sealed against the free entrance of water or concrete. Junction boxes shall be of metal and shall be mechanically and electrically continuous with the header.

Figure 358.2 illustrates a trench-type raceway with a rigid cover plate extending at a right angle across the cells, with access to predetermined cells.

358-8. Markers. A suitable number of markers shall be installed for the future location of cells.

358-9. Inserts. Inserts shall be leveled and sealed against the entrance of concrete. Inserts shall be of metal and shall be fitted with grounded-type receptacles. A grounding conductor shall connect the insert receptacles to a positive ground connection provided on the header. Where cutting through the cell wall for setting inserts or other purposes (such as providing access openings between header and cells), chips and other dirt shall not be allowed to remain in the raceway, and the tool used shall be designed so as to prevent the tool from entering the cell and damaging the conductors.

358-10. Size of Conductors. No conductor larger than No. 1/0 shall be installed, except by special permission.

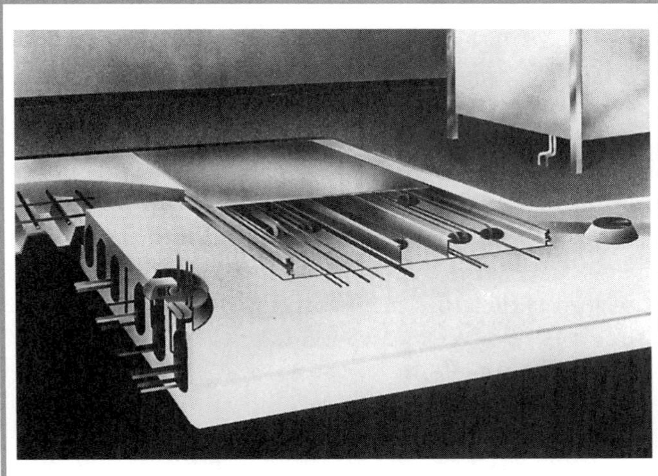

Figure 358.2 A trench-type raceway with removable cover shown extending at right angles across the cells of cellular concrete floor construction. (Bargar Metal Fabricating Co.)

358-11. Maximum Number of Conductors. The combined cross-sectional area of all conductors or cables shall not exceed 40 percent of the cross-sectional area of the cell or header.

358-12. Splices and Taps. Splices and taps shall be made only in header access units or junction boxes.

For the purposes of this section, so-called loop wiring (continuous unbroken conductor connecting the individual outlets) shall not be considered to be a splice or tap.

358-13. Discontinued Outlets. When an outlet is abandoned, discontinued, or removed, the sections of circuit conductors supplying the outlet shall be removed from the raceway. No splices or reinsulated conductors, such as would be the case of abandoned outlets on loop wiring, shall be allowed in raceways.

Article 362 — Metal Wireways and Nonmetallic Wireways

Contents

A. Metal Wireways

362-1. Definition. *Wireways* are sheet metal troughs with hinged or removable covers for housing and protecting electric wires and cable and in which conductors are laid in place after the wireway has been installed as a complete system.

Wireways are sheet-metal enclosures equipped with hinged or removable covers and are manufactured in 1-ft to 10-ft lengths and various widths and depths. Couplings, elbows, end plates, and accessories such as "T" and "X" fittings are available.

Unlike auxiliary gutters, which are not permitted to extend more than 30 ft from the equipment they supplement, wireways may be run throughout an entire area.

362-2. Uses.

(a) Permitted. The use of wireways shall be permitted as follows:

(1) For exposed work
(2) In concealed spaces only in accordance with Section 640-24

(3) In hazardous (classified) locations as permitted by Section 501-4(b) for Class I, Division 2, locations; Section 502-4(b) for Class II, Division 2, locations; and Section 504-20 for intrinsically safe wiring

Where installed in wet locations, wireways shall be listed for the purpose.

(b) Not Permitted. The use of wireways shall not be permitted where subject to severe physical damage or corrosive vapor.

362-3. Other Articles. Installations of wireways shall comply with the applicable provisions of Article 300.

362-4. Size of Conductors. No conductor larger than that for which the wireway is designed shall be installed in any wireway.

362-5. Number of Conductors. Wireways shall not contain more than 30 current-carrying conductors at any cross section. Conductors for signaling circuits or controller conductors between a motor and its starter and used only for starting duty shall not be considered as current-carrying conductors.

The sum of cross-sectional areas of all contained conductors at any cross section of the wireway shall not exceed 20 percent of the interior cross-sectional area of the wireway.

The derating factors specified in Section 310-15(b)(2)(a) shall not be applicable to the 30 current-carrying conductors at 20 percent fill specified above.

The total of the cross-sectional areas of all conductors is not permitted to exceed 20 percent of the interior cross-sectional area of the wireway. Provided the quantity of conductors does not exceed 30, the adjustment factors of Section 310-15(b)(2) do not apply.

Example

A wireway contains 26 conductors: 10 No. 3/0 XHHW-2, 3 No. 6 THWN, 3 No. 8 THHN, and 10 No. 12 THHN. Find the minimum standard-size wireway required according to Article 362.

Answer

Conductor Type and Size	Quantity	Individual Area* (in.^{2})	Total Area (in.^{2})
No. 3/0 XHHW-2	10 ×	0.2642	= 2.6420
No. 6 THWN	3 ×	0.0507	= 0.1521
No. 8 THHN	3 ×	0.0366	= 0.1098
No. 12 THHN	10 ×	0.0133	= 0.1330
Total area occupied by conductors			= 3.0369
Minimum wireway area required:	= 3.0369 ÷ 20% fill		= 15.1845
Minimum size square wireway required:	15.1845 = 3.9 in. × 3.9 in. or 4 in. × 4 in. wireway		

*Individual area dimensions of conductors are from Chapter 9, Table 5.

Exception No. 1: Where the derating factors specified in Section 310-15(b)(2)(a) are applied, the number of current-carrying conductors shall not be limited, but the sum of the cross-sectional areas of all contained conductors at any cross section of the wireway shall not exceed 20 percent of the interior cross-sectional area of the wireway.

The exception permits the number of current-carrying conductors in a wireway to exceed 30 provided the sum of the cross-sectional areas of all the conductors does not exceed 20 percent of the interior cross-sectional area of the wireway and the adjustment factors of Section 310-15(b)(2) are applied.

Exception No. 2: As provided in Section 520-6, the 30-conductor limitation shall not apply to theaters and similar locations.

Exception No. 3: As provided in Section 620-32, the 20 percent fill limitation shall not apply to elevators and dumbwaiters.

362-6. Deflected Insulated Conductors. Where insulated conductors are deflected within a wireway, either at the ends or where conduits, fittings, or other raceways or cables enter or leave the wireway, or where the direction of the wireway is deflected greater than 30 degrees, dimensions corresponding to Section 373-6 shall apply. Where insulated conductors No. 4 or larger enter a wireway through a raceway or cable, the distance between those raceway and cable entries shall not be less than six times the trade diameter of the larger raceway or cable connector.

The intent of Section 362-6 is to provide adequate space for bending conductors without damage to the insulation.

362-7. Splices and Taps. Splices and taps shall be permitted within a wireway provided they are accessible. The conductors, including splices and taps, shall not fill the wireway to more than 75 percent of its area at that point.

Conductors in wireways are accessible through hinged or removable covers. Circuits, taps, or splices may be added or altered if necessary.

See Section 362-5 and the associated example in the commentary regarding the number of conductors permitted.

362-8. Supports. Wireways shall be supported in accordance with the following.

(a) Horizontal Support. Wireways shall be supported where run horizontally at each end and at intervals not to exceed 5 ft (1.52 m) or for individual lengths longer than 5 ft (1.52 m) at each end or joint, unless listed for other support intervals. The distance between supports shall not exceed 10 ft (3.05 m).

(b) Vertical Support. Vertical runs of wireways shall be securely supported at intervals not exceeding 15 ft (4.57 m) and shall not have more than one joint between supports. Adjoining wireway sections shall be securely fastened together to provide a rigid joint.

362-9. Extension Through Walls. Wireways shall be permitted to pass transversely through walls if the length passing through the wall is unbroken. Access to the conductors shall be maintained on both sides of the wall.

362-10. Dead Ends. Dead ends of wireways shall be closed.

362-11. Extensions from Wireways. Extensions from wireways shall be made with cord pendants installed in accordance with Section 400-10 or any wiring method in Chapter 3 that includes a means for equipment grounding. Where a separate equipment grounding conductor is employed, connection of the equipment grounding conductors in the wiring method to the wireway shall comply with Sections 250-8 and 250-12. Where rigid nonmetallic conduit, electrical nonmetallic tubing, or liquidtight flexible nonmetallic conduit is used, connection of the equipment grounding conductor in the nonmetallic raceway to a metal wireway shall comply with Sections 250-8 and 250-12.

Extensions from wireways using metal raceways, metal-sheathed cables, and nonmetallic-sheathed cables are made through knockouts provided on the wireway or field punched. Rigid nonmetallic conduit, electrical nonmetallic tubing, and liquidtight flexible nonmetallic conduit may also be used. Cables and nonmetallic raceways as well as the wireway are required to include a means for ensuring an effective continuation of the equipment grounding conductor.

Sections of wireways, including accessory fittings (elbows, endplates, flanges, etc.), are bolted together, assuring a rigid mechanical and electrical connection.

362-12. Marking. Wireways shall be marked so that their manufacturer's name or trademark will be visible after installation.

362-13. Grounding. Grounding shall be in accordance with the provisions of Article 250.

B. Nonmetallic Wireways

362-14. Definition. Nonmetallic wireways are flame-retardant, nonmetallic troughs with removable covers for housing and protecting electric wires and cables in which conductors are laid in place after the wireway has been installed as a complete system.

Part B was added to Article 362 to provide installation requirements for nonmetallic wireways.

Nonmetallic wireways are troughs with removable covers in which conductors are laid after the wireway has been installed as a complete system. The wireway must be installed on the surface of the structure, not concealed. It is allowed to pass through walls, provided that an unbroken length passes through the wall.

362-15. Uses Permitted. The use of listed nonmetallic wireways shall be permitted as follows:

(1) Only for exposed work, except as permitted in accordance with Section 640-24
(2) Where subject to corrosive vapors
(3) In wet locations where listed for the purpose

FPN: Extreme cold may cause nonmetallic wireways to become brittle and, therefore, more susceptible to damage from physical contact.

362-16. Uses Not Permitted. Nonmetallic wireways shall not be used as follows:

(1) Where subject to physical damage
(2) In any hazardous (classified) location, except as permitted in Section 504-20
(3) Where exposed to sunlight unless listed and marked as suitable for the purpose
(4) Where subject to ambient temperatures other than those for which nonmetallic wireway is listed
(5) For conductors whose insulation temperature limitations would exceed those for which the nonmetallic wireway is listed

362-17. Other Articles. Installations of nonmetallic wireways shall comply with the applicable provisions of Article 300. Where equipment grounding is required by Article 250, a separate equipment grounding conductor shall be installed in the nonmetallic wireway.

Exception: Where the grounded conductor is used to ground equipment as permitted in Section 250-142.

362-18. Size of Conductors. No conductor larger than that for which the nonmetallic wireway is designed shall be installed in any nonmetallic wireway.

362-19. Number of Conductors. The sum of cross-sectional areas of all contained conductors at any cross section of the nonmetallic wireway shall not exceed 20 percent of the interior cross-sectional area of the nonmetallic wireway. Conductors for signaling circuits or controller conductors between a motor and its starter and used only for starting duty shall not be considered as current-carrying conductors.

The derating factors specified in Section 310-15(b)(2)(a) shall be applicable to the current-carrying conductors up to and including the 20 percent fill specified above.

362-20. Deflected Insulated Conductors. Where insulated conductors are deflected within a nonmetallic wireway, either at the ends or where conduits, fittings, or other raceways or cables enter or leave the nonmetallic wireway, or where the direction of the nonmetallic wireway is deflected greater than 30 degrees, dimensions corresponding to Section 373-6 shall apply.

362-21. Splices and Taps. Splices and taps shall be permitted within a nonmetallic wireway provided they are accessible. The conductors, including splices and taps, shall not fill the nonmetallic wireway to more than 75 percent of its area at that point.

362-22. Supports. Nonmetallic wireway shall be supported in accordance with (a) and (b).

(a) Horizontal Support. Nonmetallic wireways shall be supported where run horizontally at intervals not to exceed 3 ft (914 mm), and at each end or joint, unless listed for other support intervals. In no case shall the distance between supports exceed 10 ft (3.05 m).

(b) Vertical Support. Vertical runs of nonmetallic wireway shall be securely supported at intervals not exceeding 4 ft (1.22 m), unless listed for other support intervals, and shall not have more than one joint between supports. Adjoining nonmetallic wireway sections shall be securely fastened together to provide a rigid joint.

362-23. Expansion Fittings. Expansion fittings for nonmetallic wireway shall be provided to compensate for thermal expansion and contraction where the length change is expected to be 0.25 in. (6.36 mm) or greater in a straight run.

FPN: See Table 347-9(A) for expansion characteristics of PVC rigid nonmetallic conduit. The expansion characteristics of PVC nonmetallic wireway are identical.

362-24. Extension Through Walls. Nonmetallic wireways shall be permitted to pass transversely through walls if the length passing through the wall is unbroken. Access to the conductors shall be maintained on both sides of the wall.

362-25. Dead Ends. Dead ends of nonmetallic wireways shall be closed using listed fittings.

362-26. Extensions from Nonmetallic Wireways. Extensions from nonmetallic wireways shall be made with cord pendants or any wiring method of Chapter 3. A separate equipment grounding conductor shall be installed in, or an equipment grounding connection shall be made to, any of the wiring methods used for the extension.

362-27. Marking. Nonmetallic wireways shall be marked so that the manufacturer's name or trademark and interior cross-sectional area in square inches shall be visible after installation. Nonmetallic wireways that have limited smoke-producing characteristics shall be permitted to be identified with the suffix *LS*.

Article 363 — Flat Cable Assemblies: Type FC

Contents

363-1. Definition. *Type FC, a flat cable assembly,* is an assembly of parallel conductors formed integrally with an insulating material web specifically designed for field installation in surface metal raceway.

Type FC (flat) cable is an assembly of three or four parallel No. 10 special stranded copper wires formed integrally with an insulating material web. The cable is marked with the size of the maximum branch circuit to which it may be connected, the cable type designation, manufacturer's identification, maximum working voltage, conductor size, and temperature rating. A marking accompanying the cable on a tag or reel indicates the special metal raceways and specific FC cable fittings with which the cable is intended to be used. Figures 363.1 and 363.2 show the basic components of this wiring method.

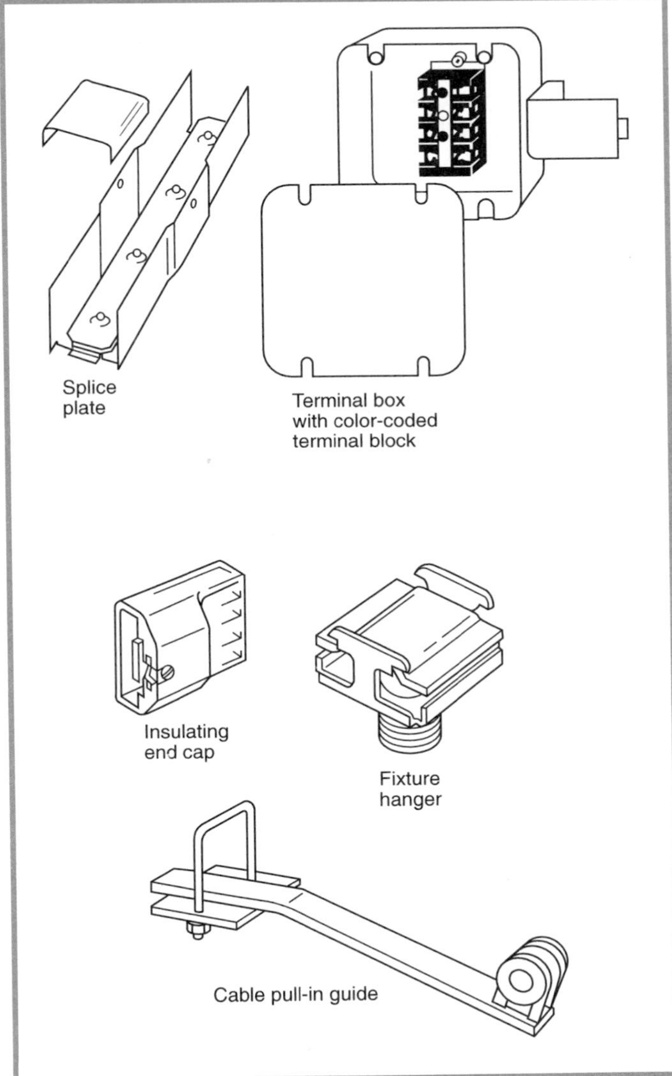

Figure 363.1 *Basic components and accessories used for an installation of Type FC cable assembly. (The Wiremold Co.)*

363-2. Other Articles. In addition to the provisions of this article, installation of Type FC cable shall conform with the applicable provisions of Articles 210, 220, 250, 300, 310, and 352.

363-3. Uses Permitted. Flat cable assemblies shall be permitted only as branch circuits to supply suitable tap devices

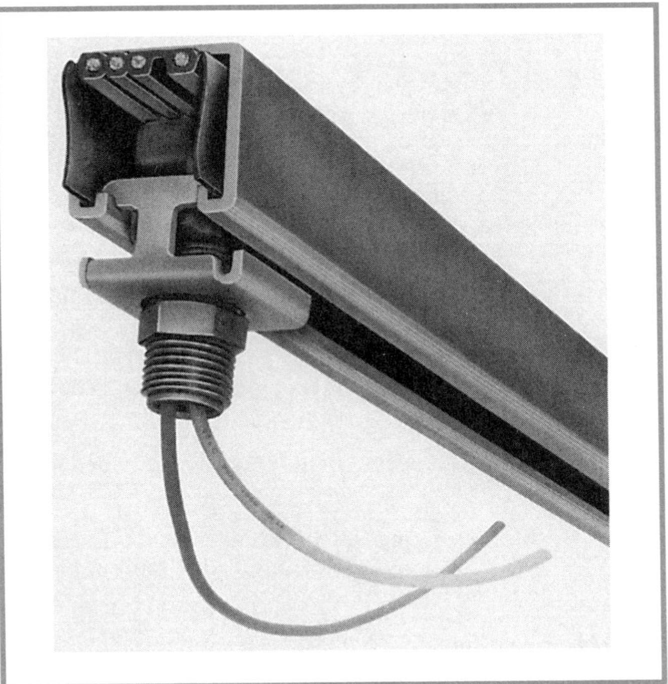

Figure 363.2 *A fixture hanger used with Type FC cable assembly. (The Wiremold Co.)*

for lighting, small appliances, or small power loads. Flat cable assemblies shall be installed for exposed work only. Flat cable assemblies shall be installed in locations where they will not be subjected to severe physical damage.

363-4. Uses Not Permitted. Flat cable assemblies shall not be installed in the following:

(1) Where subject to corrosive vapors unless suitable for the application
(2) In hoistways
(3) In any hazardous (classified) location
(4) Outdoors or in wet or damp locations unless identified for use in wet locations

363-5. Installation. Flat cable assemblies shall be installed in the field only in surface metal raceways identified for the use. The channel portion of the surface metal raceway systems shall be installed as complete systems before the flat cable assemblies are pulled into the raceways.

363-6. Number of Conductors. The flat cable assemblies shall consist of either two, three, or four conductors.

363-7. Size of Conductors. Flat cable assemblies shall have conductors of No. 10 special stranded copper wires.

363-8. Conductor Insulation. The entire flat cable assembly shall be formed to provide a suitable insulation covering all of the conductors and using one of the materials recognized in Table 310-13 for general branch-circuit wiring.

363-9. Splices. Splices shall be made in listed junction boxes.

363-10. Taps. Taps shall be made between any phase conductor and the grounded conductor or any other phase conductor by means of devices and fittings identified for the use. Tap devices shall be rated at not less than 15 amperes, or more than 300 volts to ground, and they shall be color-coded in accordance with the requirements of Section 363-20.

363-11. Dead Ends. Each flat cable assembly dead end shall be terminated in an end-cap device identified for the use.

The dead-end fitting for the enclosing surface metal raceway shall be identified for the use.

363-12. Fixture Hangers. Fixture hangers installed with the flat cable assemblies shall be identified for the use.

363-13. Fittings. Fittings to be installed with flat cable assemblies shall be designed and installed to prevent physical damage to the cable assemblies.

363-14. Extensions. All extensions from flat cable assemblies shall be made by approved wiring methods, within the junction boxes, installed at either end of the flat cable assembly runs.

363-15. Supports. The flat cable assemblies shall be supported by means of their special design features, within the surface metal raceways.

The surface metal raceways shall be supported as required for the specific raceway to be installed.

363-16. Rating. The rating of the branch circuit shall not exceed 30 amperes.

363-17. Marking. In addition to the provisions of Section 310-11, Type FC cable shall have the temperature rating durably marked on the surface at intervals not exceeding 24 in. (610 mm).

363-18. Protective Covers. Where a flat cable assembly is installed less than 8 ft (2.44 m) above the floor or fixed working platform, it shall be protected by a metal cover identified for the use.

363-19. Identification. The grounded conductor shall be identified throughout its length by means of a distinctive and durable white or natural gray marking.

363-20. Terminal Block Identification. Terminal blocks identified for the use shall have distinctive and durable markings for color or word coding. The grounded conductor section shall have a white marking or other suitable designation. The next adjacent section of the terminal block shall have a black marking or other suitable designation. The next section shall have a red marking or other suitable designation. The final or outer section, opposite the grounded conductor section of the terminal block, shall have a blue marking or other suitable designation.

Article 364 — Busways

Contents

A. General Requirements

364-1. Scope. This article covers service-entrance, feeder, and branch-circuit busways and associated fittings.

364-2. Definition. For the purpose of this article, a *busway* is considered to be a grounded metal enclosure containing factory-mounted, bare or insulated conductors, which are usually copper or aluminum bars, rods, or tubes.

FPN: For cablebus, refer to Article 365.

According to the 1997 UL *Electrical Construction Materials Directory,* category CWFT, busways and short run-run busways are provided with metal enclosures. These enclosures are in some cases an additional ground bus, and are intended for use as equipment grounding conductors. Some busways are not intended for use ahead of service equipment and are marked with the maximum rating of overcurrent protection to be used on the supply side of the busway. Busway that has been investigated to determine its suitability for installation in a specified position, or for use in vertical runs, or for support at intervals greater than 5 ft, or for outdoor use, is so marked. This marking is on or contiguous with the nameplate incorporating the manufacturer's name and electrical rating. A busway or fitting containing a vapor seal is so marked, but unless marked otherwise, the busway or fitting has not been investigated for passage through a fire wall.

Short-run busways are marked to limit the run to 30 ft or less, and no more than 10 ft vertically. They are intended primarily for a feed to switchboards. Except for transformer stubs, short-run busways are not intended to have intermediate taps. Short-run busways are not ventilated and may be marked for outdoor use.

Busways and associated fittings marked "Short Circuit Current Rating(s) Maximum rms Symmetrical Amps ________ Volts ________" have been investigated for the rating indicated.

Busways that are intended to supply and support industrial and commercial lighting fixtures are classified as "Lighting Busway" and are so marked. Trolley busway is marked "Trolley Busway" and is additionally marked "Lighting Busway" if intended to supply and support industrial and commercial lighting fixtures. A busway with provision for insertion of plug-in devices at any point along its length and intended for general use is classified as "Continuous Plug-in Busway" and is so marked. The marking is contiguous with the marking of the manufacturer's name and the electrical rating.

Busway marked "Lighting Busway" and protected by overcurrent devices rated in excess of 20 amperes is intended for use only with fixtures employing heavy-duty lampholders unless additional overcurrent protection is provided for the fixture in accordance with this *Code*.

"Trolley Busway" should be installed out of the reach of people, or it should be otherwise installed to prevent accidental contact with exposed conductors.

Figures 364.1 and 364.2 show examples of trolley busways. Note the busway in Figure 364.2. has provision for insertion of a plug-in device for stationary use.

Figure 364.1 *A trolley busway with the trolley in place. (Amerace Corp.)*

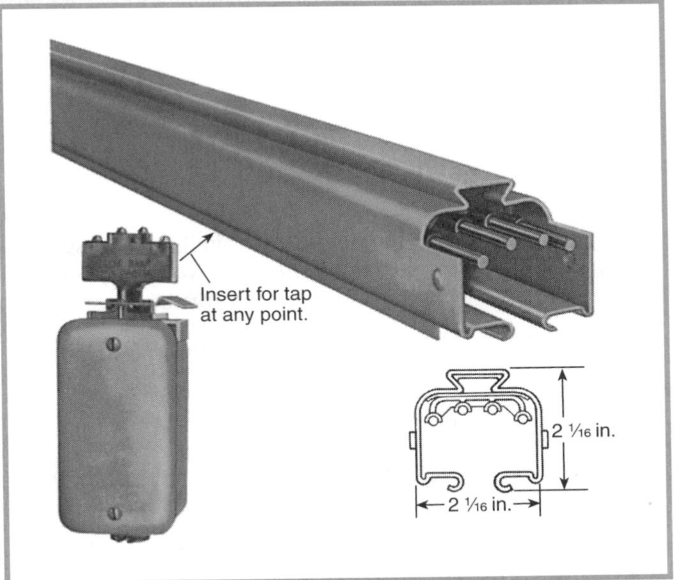

Figure 364.2 *A plug-in device for stationary use on a trolley busway. (Amerace Corp.)*

364-3. Other Articles. Installations of busways shall comply with the applicable provisions of Article 300.

364-4. Use.

(a) Uses Permitted. Busways shall be permitted to be installed where they are located as follows:

(1) Located in the open and are visible, or
(2) Installed behind access panels, provided the busways are totally enclosed, of nonventilating-type construction, and installed so that the joints between sections and at fittings are accessible for maintenance purposes. Where installed behind access panels, means of access shall be provided, and the following conditions shall be met:
 (a) The space behind the access panels shall not be used for air-handling purposes, or
 (b) Where the space behind the access panels is used for environmental air, other than ducts and plenums, there shall be no provisions for plug-in connections, and the conductors shall be insulated.

Unless busways are mounted in the open and visible, the installation must comply with Section 364-4(a)(2).

Busways are commonly used as feeders, mounted horizontally in industrial buildings or mounted vertically in high-rise buildings. See Figure 100.1 for an example of a busway mounted above a hung ceiling.

(b) Uses Not Permitted. Busways shall not be installed as follows:

(1) Where subject to severe physical damage or corrosive vapors
(2) In hoistways
(3) In any hazardous (classified) location, unless specifically approved for such use

FPN: See Section 501-4(b).

(4) Outdoors or in wet or damp locations unless identified for such use

Lighting busway and trolley busway shall not be installed less than 8 ft (2.44 m) above the floor or working platform unless provided with a cover identified for the purpose.

364-5. Support. Busways shall be securely supported at intervals not exceeding 5 ft (1.52 m) unless otherwise designed and marked.

364-6. Through Walls and Floors.

(a) Walls. Unbroken lengths of busway shall be permitted to be extended through dry walls.

(b) Floors. Floor penetrations shall comply with (1) and (2).

(1) Busways shall be permitted to be extended vertically through dry floors if totally enclosed (unventilated) where passing through and for a minimum distance of 6 ft (1.83 m) above the floor to provide adequate protection from physical damage.

(2) In other than industrial establishments, where a vertical riser penetrates two or more dry floors, a minimum 4-in. (102-mm) high curb shall be installed around all floor openings for riser busways to prevent liquids from entering the opening. The curb shall be installed within 12 in. (304.8 mm) of the floor opening. Electrical equipment shall be located so that it will not be damaged by liquids that are retained by the curb.

FPN: See Section 300-21 for information concerning the spread of fire or products of combustion.

A busway or fitting containing a vapor seal is so marked, but, unless marked otherwise, the busway or fitting has not been investigated for passage through a fire-rated wall. The requirements of Section 300-21 are most important in order to confine a fire and the products of combustion at their origin.

Revised for the 1999 *Code,* this section now requires that a curb be placed around a busway if the busway penetrates two or more dry floors. Experience has shown that when liquid spills occur on upper floors of normally dry buildings, the spilled liquid often flows to the busway floor penetration and down the vertical rise of the busway. Spills may cause extensive damage to the busway and the building's electrical system. The addition of a 4-in. curb encircling the busway will help eliminate this possibly dangerous situation.

364-7. Dead Ends. A dead end of a busway shall be closed.

364-8. Branches from Busways. Branches from busways shall be permitted to be made by the following.

(a) Branches from busways shall be made in accordance with Articles 331, 334, 345, 346, 347, 348, 350, 351, 352, and 364. Where a nonmetallic raceway is used, connection of equipment grounding conductors in the nonmetallic raceway to the busway shall comply with Sections 250-8 and 250-12.

(b) Suitable cord and cable assemblies approved for extra-hard usage or hard usage and listed bus drop cable shall be permitted as branches from busways for the connection of portable equipment or the connection of stationary equipment to facilitate their interchange in accordance with Sections 400-7 and 400-8 and the following conditions.

Section 364-8(b) very clearly sets forth the permitted raceway, cable, and cord extensions from a busway and the appropriate installation methods to accomplish these branches.

(1) The cord or cable shall be attached to the building by an approved means.
(2) The length of the cord or cable from a busway plug-in device to a suitable tension take-up support device shall not exceed 6 ft (1.83 m).

Figure 364.3 shows an example of a cable or cord branch from a busway installed according to the provisions of Section 364-8(b)(2).

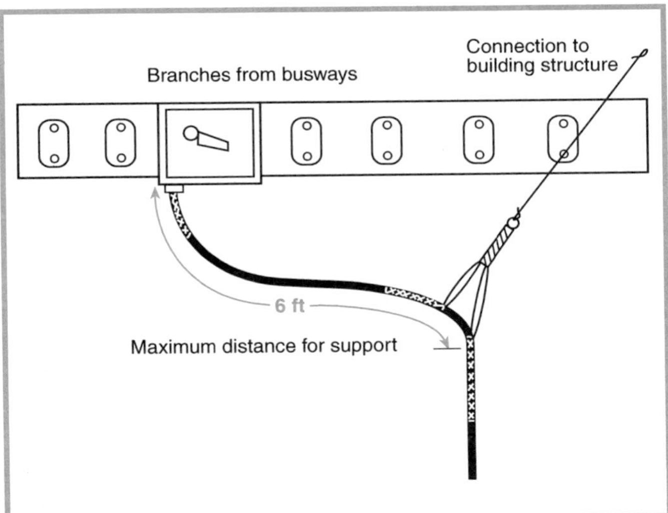

Figure 364.3 An example of a cable or cord branch from a busway according to Section 364-8(b).

Exception: In industrial establishments only, where the conditions of maintenance and supervision ensure that only qualified persons will service the installation, lengths exceeding 6 ft. (1.83 m) shall be permitted between the busway plug-in device and the tension take-up support device where the cord or cable is supported at intervals not exceeding 8 ft (2.4 m).

Section 400-7(a) specifically prohibits the installation of spliced cords. Figure 364.4 shows an example of a cable or cord branch from a busway installed according to Section 364-8(b)(2), Exception.

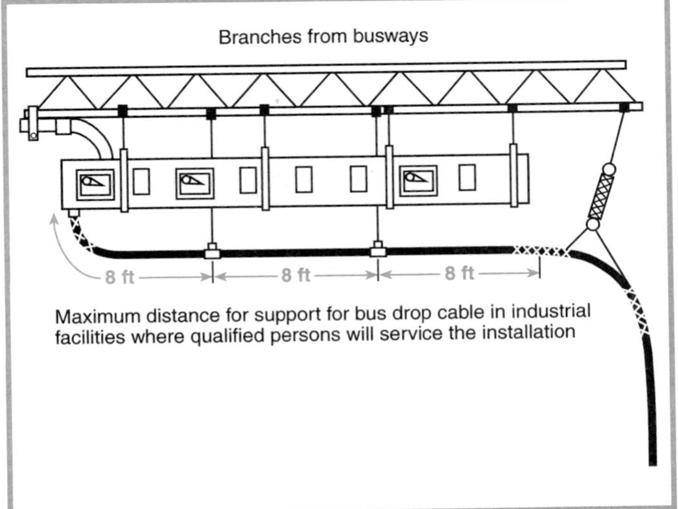

Figure 364.4 An example of an installation permitted only in industrial occupancies with other restrictions according to Section 364-8(b)(2), Exception.

(3) The cord or cable shall be installed as a vertical riser from the tension take-up support device to the equipment served.

(4) Strain relief cable grips shall be provided for the cord or cable at the busway plug-in device and equipment terminations.

(c) Suitable cord and cable assemblies approved for extra-hard usage or hard usage and listed bus drop cable shall be permitted as branches from trolley-type busways for the connection of movable equipment in accordance with Sections 400-7 and 400-8.

364-9. Overcurrent Protection. Overcurrent protection shall be provided in accordance with Sections 364-10 through 364-13.

364-10. Rating of Overcurrent Protection — Feeders. A busway shall be protected against overcurrent in accordance with the allowable current rating of the busway.

The rated ampacity of a busway is based on the allowable temperature rise of the conductors and can be determined in the field only by reference to the nameplate data.

The requirements of Sections 240-3(b) and 240-3(c) are applicable for busways.

Exception No. 1: The applicable provisions of Section 240-3 shall be permitted.

Exception No. 2: Where used as transformer secondary ties, the provisions of Section 450-6(a)(3) shall be permitted.

364-11. Reduction in Ampacity Size of Busway. Overcurrent protection shall be required where busways are reduced in ampacity.

Exception: For industrial establishments only, omission of overcurrent protection shall be permitted at points where busways are reduced in ampacity, provided that the length of the busway having the smaller ampacity does not exceed 50 ft (15.2 m) and has an ampacity at least equal to one-third the rating or setting of the overcurrent device next back on the line, and provided that such busway is free from contact with combustible material.

In industrial establishments, where the size of the smaller busway is kept within the specified limits, the additional cost of providing overcurrent protection at the point where the size is changed is not warranted. For example, busway protected by a 1200-ampere overcurrent device may be reduced in size, provided the smaller busway has a current rating of 400 amperes (⅓ of 1200 amperes) and does not extend more than 50 ft. In this case, overcurrent protection would be required if the smaller busway were rated less than 400 amperes (e.g., 200 amperes, 300 amperes, etc.).

364-12. Feeder or Branch Circuits. Where a busway is used as a feeder, devices or plug-in connections for tapping off feeder or branch circuits from the busway shall contain the overcurrent devices required for the protection of the feeder or branch circuits. The plug-in device shall consist of an externally operable circuit breaker or an externally operable fusible switch. Where such devices are mounted out of reach and contain disconnecting means, suitable means such as ropes, chains, or sticks shall be provided for operating the disconnecting means from the floor.

Externally operated fused switches and circuit breakers that are plugged into busways and mounted out of reach are to be considered accessible if operated by means such as ropes, chains, or hooksticks. Figure 364.5 shows a 10-ft section of feeder busway. See Figure 100.2 for other example illustrations.

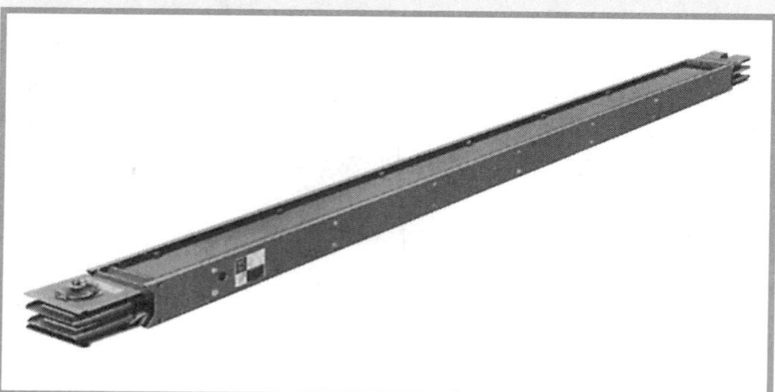

Figure 364.5 *A 10-ft section of feeder busway. (Square D Co.)*

Exception No. 1: As permitted in Section 240-21.

Exception No. 2: For fixed or semifixed lighting fixtures, where the branch-circuit overcurrent device is part of the fixture cord plug on cord-connected fixtures.

Exception No. 3: Where fixtures without cords are plugged directly into the busway and the overcurrent device is mounted on the fixture.

364-13. Rating of Overcurrent Protection — Branch Circuits. A busway used as a branch circuit shall be protected against overcurrent in accordance with Section 210-20. Where so used, the circuit shall comply with the applicable requirements of Articles 210, 430, and 440.

364-15. Marking. Busways shall be marked with the voltage and current rating for which they are designed, and with the manufacturer's name or trademark in such manner as to be visible after installation.

B. Requirements for Over 600 Volts, Nominal

364-21. Identification. Each bus run shall be provided with a permanent nameplate on which the following information shall be provided:

(1) Rated voltage
(2) Rated continuous current; if bus is forced-cooled, both the normal forced-cooled rating and the self-cooled (not forced-cooled) rating for the same temperature rise shall be given
(3) Rated frequency
(4) Rated impulse withstand voltage
(5) Rated 60-Hz withstand voltage (dry)
(6) Rated momentary current
(7) Manufacturer's name or trademark

FPN: See *Guide for Metal-Enclosed Bus and Calculating Losses in Isolated-Phase Bus,* ANSI C37.23-1987 (R1991), for construction and testing requirements for metal-enclosed buses.

364-22. Grounding. Metal-enclosed bus shall be grounded in accordance with Article 250.

364-23. Adjacent and Supporting Structures. Metal-enclosed busways shall be installed so that temperature rise from induced circulating currents in any adjacent metallic parts will not be hazardous to personnel or constitute a fire hazard.

364-24. Neutral. Neutral bus, where required, shall be sized to carry all neutral load current, including harmonic currents, and shall have adequate momentary and short-circuit rating consistent with system requirements.

364-25. Barriers and Seals. Bus runs that have sections located both inside and outside of buildings shall have a vapor seal at the building wall to prevent interchange of air between indoor and outdoor sections.

Exception: Vapor seals shall not be required in forced-cooled bus.

Fire barriers shall be provided where fire walls, floors, or ceilings are penetrated.

FPN: See Section 300-21 for information concerning the spread of fire or products of combustion.

364-26. Drain Facilities. Drain plugs, filter drains, or similar methods shall be provided to remove condensed moisture from low points in bus run.

364-27. Ventilated Bus Enclosures. Ventilated bus enclosures shall be installed in accordance with Article 110, Part C, and Section 490-24, unless designed so that foreign objects inserted through any opening will be deflected from energized parts.

364-28. Terminations and Connections. Where bus enclosures terminate at machines cooled by flammable gas, seal-off bushings, baffles, or other means shall be provided to prevent accumulation of flammable gas in the bus enclosures.

Flexible or expansion connections shall be provided in long, straight runs of bus to allow for temperature expansion or contraction, or where the bus run crosses building vibration insulation joints.

All conductor termination and connection hardware shall be accessible for installation, connection, and maintenance.

364-29. Switches. Switching devices or disconnecting links provided in the bus run shall have the same momentary rating as the bus. Disconnecting links shall be plainly marked to be removable only when bus is de-energized. Switching devices that are not load break shall be interlocked to prevent operation under load, and disconnecting link enclosures shall be interlocked to prevent access to energized parts.

364-30. Wiring 600 Volts or Less, Nominal. Secondary control devices and wiring that are provided as part of the metal-enclosed bus run shall be insulated by fire-retardant barriers from all primary circuit elements with the exception of short lengths of wire, such as at instrument transformer terminals.

Article 365 — Cablebus

Contents

365-1. Definition. *Cablebus* is an assembly of insulated conductors with fittings and conductor terminations in a completely enclosed, ventilated protective metal housing. Cablebus is ordinarily assembled at the point of installation from the components furnished or specified by the manufacturer in accordance with instructions for the specific job. This assembly is designed to carry fault current and to withstand the magnetic forces of such current.

As shown in Figure 365.1, cablebus consists of a metal structure or framework installed in a manner similar to a cable tray support system. Insulated conductors, No. 1/0 or larger, are field installed within the framework on special insulating blocks at specified intervals to provide controlled spacing between conductors. To completely enclose the conductors, a ventilated top cover is attached to the framework.

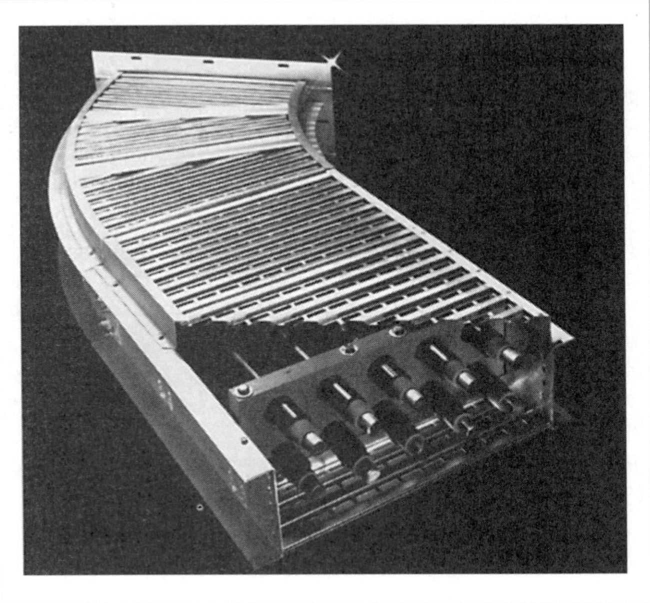

Figure 365.1 A section of cablebus with conductors in place and the ventilated top cover ready to be attached to the busway frame. (MP Husky Corp.)

365-2. Use.

(a) 600 Volts or Less. Approved cablebus shall be permitted at any voltage or current for which spaced conductors are rated and shall be installed for exposed work only. Cablebus installed outdoors or in corrosive, wet, or damp locations shall be identified for such use. Cablebus shall not be installed in hoistways or hazardous (classified) locations unless specifically approved for such use. Cablebus shall be permitted to be used for branch circuits, feeders, and services.

Cablebus framework, where bonded as required by Article 250, shall be permitted as the equipment grounding conductor for branch circuits and feeders.

(b) Over 600 Volts. Approved cablebus shall be permitted for systems in excess of 600 volts, nominal. See Section 300-37.

365-3. Conductors.

(a) Types of Conductors. The current-carrying conductors in cablebus shall have an insulation rating of 75°C (167°F) or higher of an approved type and suitable for the application in accordance with Articles 310 and 490.

(b) Ampacity of Conductors. The ampacity of conductors in cablebus shall be in accordance with Tables 310-17 and

310-19, or Tables 310-69 and 310-70 for installations over 600 volts.

(c) Size and Number of Conductors. The size and number of conductors shall be that for which the cablebus is designed, and in no case smaller than No. 1/0.

(d) Conductor Supports. The insulated conductors shall be supported on blocks or other mounting means designed for the purpose.

The individual conductors in a cablebus shall be supported at intervals not greater than 3 ft (914 mm) for horizontal runs and 1½ ft (457 mm) for vertical runs. Vertical and horizontal spacing between supported conductors shall not be less than one conductor diameter at the points of support.

365-5. Overcurrent Protection. Cablebus shall be protected against overcurrent in accordance with the allowable ampacity of the cablebus conductors in accordance with Section 240-3.

Exception: Overcurrent protection shall be permitted in accordance with Sections 240-100 and 240-101 for over 600 volts, nominal.

365-6. Support and Extension Through Walls and Floors.

(a) Support. Cablebus shall be securely supported at intervals not exceeding 12 ft (3.66 m).

Exception: Where spans longer than 12 ft (3.66 m) are required, the structure shall be specifically designed for the required span length.

(b) Transversely Routed. Cablebus shall be permitted to extend transversely through partitions or walls, other than fire walls, provided the section within the wall is continuous, protected against physical damage, and unventilated.

(c) Through Dry Floors and Platforms. Except where firestops are required, cablebus shall be permitted to extend vertically through dry floors and platforms, provided the cablebus is totally enclosed at the point where it passes through the floor or platform and for a distance of 6 ft (1.83 m) above the floor or platform.

(d) Through Floors and Platforms in Wet Locations. Except where firestops are required, cablebus shall be permitted to extend vertically through floors and platforms in wet locations where (1) there are curbs or other suitable means to prevent water flow through the floor or platform opening, and (2) where the cablebus is totally enclosed at the point where it passes through the floor or platform and for a distance of 6 ft (1.83 m) above the floor or platform.

365-7. Fittings. A cablebus system shall include approved fittings for the following:

(1) Changes in horizontal or vertical direction of the run
(2) Dead ends
(3) Terminations in or on connected apparatus or equipment or the enclosures for such equipment
(4) Additional physical protection where required, such as guards where subject to severe physical damage

365-8. Conductor Terminations. Approved terminating means shall be used for connections to cablebus conductors.

365-9. Grounding. A cablebus installation shall be grounded and bonded in accordance with Article 250, excluding Section 250-86, Exception No. 2.

365-10. Marking. Each section of cablebus shall be marked with the manufacturer's name or trade designation and the maximum diameter, number, voltage rating, and ampacity of the conductors to be installed. Markings shall be located so as to be visible after installation.

Article 370 — Outlet, Device, Pull and Junction Boxes, Conduit Bodies and Fittings

Contents

A. Scope and General

370-1. Scope. This article covers the installation and use of all boxes and conduit bodies used as outlet, junction, or pull boxes, depending on their use, and manholes and other electric enclosures intended for personnel entry. Cast, sheet metal, nonmetallic, and other boxes such as FS, FD, and larger boxes are not classified as conduit bodies. This article also includes installation requirements for fittings used to join raceways and to connect raceways and cables to boxes and conduit bodies.

370-2. Round Boxes. Round boxes shall not be used where conduits or connectors requiring the use of locknuts or bushings are to be connected to the side of the box.

Section 370-2 requires the use of rectangular or octagonal boxes having a flat bearing surface at each knockout for locknuts and bushings to ensure effective grounding continuity. Round boxes, however, can be used if the conduit or cable is secured by clamps within the box or if the cable does not need attachment to the box, as permitted by Section 370-17(c), Exception.

370-3. Nonmetallic Boxes. Nonmetallic boxes shall be permitted only with open wiring on insulators, concealed knob-and-tube wiring, nonmetallic-sheathed cable, and nonmetallic raceways.

Exception No. 1: Where internal bonding means are provided between all entries, nonmetallic boxes shall be permitted to be used with metal raceways or metal-armored cables.

Exception No. 2: Where integral bonding means with a provision for attaching an equipment grounding jumper inside the box are provided between all threaded entries in nonmetallic boxes listed for the purpose, nonmetallic boxes shall be permitted to be used with metal raceways or metal-armored cables.

Exception No. 1 applies to nonmetallic boxes without threaded entries and permits the use of metal raceways and metal-armored cables with nonmetallic boxes. Internal bonding means is required to be installed to ensure ground continuity between the metal raceways or metal-armored cables. For the purposes of this exception the term *metal-armored cable* includes cables with a metal covering such as mineral-insulated, metal-sheathed cable (Type MI), metal-clad cable (Type MC), and armored cable (Type AC).

Exception No. 2 permits the use of metal raceways and metal-armored cables with listed nonmetallic boxes equipped with threaded entries. An integral bonding means is required to ensure ground continuity between the threaded entries. The requirement for means to attach a grounding jumper is intended to accommodate devices or equipment attached to the box. For the purposes of this exception the term *metal-armored cable* includes cables with a metal covering such as mineral-insulated, metal-sheathed cable (Type MI), metal-clad cable (Type MC), and armored cable (Type AC).

370-4. Metal Boxes. All metal boxes shall be grounded in accordance with the provisions of Article 250.

370-5. Short Radius Conduit Bodies. Conduit bodies such as capped elbows and service-entrance elbows enclosing conductors No. 6 or smaller, and that are only intended to enable the installation of the raceway and the contained conductors, shall not contain splices, taps, or devices and shall be of sufficient size to provide free space for all conductors enclosed in the conduit body.

Short radius conduit bodies are not permitted to contain splices.

B. Installation

370-15. Damp, Wet, or Hazardous (Classified) Locations.

(a) Damp or Wet Locations. In damp or wet locations, boxes, conduit bodies, and fittings shall be placed or equipped so as to prevent moisture from entering or accumulating within the box, conduit body, or fitting. Boxes, conduit bodies, and fittings installed in wet locations shall be listed for use in wet locations.

FPN No. 1: For boxes in floors, see Section 370-27(b).

FPN No. 2: For protection against corrosion, see Section 300-6.

Article 100 defines the term *weatherproof* as "constructed or protected so that exposure to the weather will not interfere with successful operation." Rainproof, raintight, or watertight equipment can fulfill the requirements for this definition where varying weather conditions other than wetness, such as snow, ice, dust, or temperature extremes, are not a factor.

A weatherhead fitting is considered to be weatherproof because the openings for the conductors are placed in a downward position so that rain or snow cannot enter the fitting.

See the definitions of *damp location* and *wet location* under *location* in Article 100, as well as the commentary following the definition of *enclosure,* for further explanation.

(b) Hazardous (Classified) Locations. Installations in hazardous (classified) locations shall conform to Articles 500 through 517.

370-16. Number of Conductors in Outlet, Device, and Junction Boxes, and Conduit Bodies. Boxes and conduit bodies shall be of sufficient size to provide free space for all enclosed conductors. In no case shall the volume of the box, as calculated in (a), be less than the fill calculation as calculated in (b). The minimum volume for conduit bodies shall be as calculated in (c).

The provisions of this section shall not apply to terminal housings supplied with motors. (See Section 430-12.)

Boxes and conduit bodies enclosing conductors, size No. 4 or larger, shall also comply with the provisions of Section 370-28.

(a) Box Volume Calculations. The volume of a wiring enclosure (box) shall be the total volume of the assembled sections, and, where used, the space provided by plaster rings, domed covers, extension rings, etc., that are marked with their volume in cubic inches or are made from boxes the dimensions of which are listed in Table 370-16(a).

(1) Standard Boxes. The volumes of standard boxes that are not marked with a cubic inch capacity shall be as given in Table 370-16(a).

(2) Other Boxes. Boxes 100 in.^{3} (1640 cm^3) or less, other than those described in Table 370-16(a), and nonmetallic boxes shall be durably and legibly marked by the manufacturer with their cubic inch capacity. Boxes described in Table 370-16(a) that have a larger cubic inch capacity than is designated in the table shall be permitted to have their cubic inch capacity marked as required by this section.

(b) Box Fill Calculations. The volumes in paragraphs (1) through (5), as applicable, shall be added together. No allowance shall be required for small fittings such as locknuts and bushings.

(1) Conductor Fill. Each conductor that originates outside the box and terminates or is spliced within the box shall be counted once, and each conductor that passes through

Table 370-16(a). Metal Boxes

Box Dimension in Inches, Trade Size, or Type	Minimum Capacity (in.^{3})	Maximum Number of Conductors*						
		No. 18	No. 16	No. 14	No. 12	No. 10	No. 8	No. 6
4 × 1¼ round or octagonal	12.5	8	7	6	5	5	4	2
4 × 1½ round or octagonal	15.5	10	8	7	6	6	5	3
4 × 2⅛ round or octagonal	21.5	14	12	10	9	8	7	4
4 × 1¼ square	18.0	12	10	9	8	7	6	3
4 × 1½ square	21.0	14	12	10	9	8	7	4
4 × 2⅛ square	30.3	20	17	15	13	12	10	6
4 11/16 × 1¼ square	25.5	17	14	12	11	10	8	5
4 11/16 × 1½ square	29.5	19	16	14	13	11	9	5
4 11/16 × 2⅛ square	42.0	28	24	21	18	16	14	8
3 × 2 × 1½ device	7.5	5	4	3	3	3	2	1
3 × 2 × 2 device	10.0	6	5	5	4	4	3	2
3 × 2 × 2¼ device	10.5	7	6	5	4	4	3	2
3 × 2 × 2½ device	12.5	8	7	6	5	5	4	2
3 × 2 × 2¾ device	14.0	9	8	7	6	5	4	2
3 × 2 × 3½ device	18.0	12	10	9	8	7	6	3
4 × 2⅛ × 1½ device	10.3	6	5	5	4	4	3	2
4 × 2⅛ × 1⅞ device	13.0	8	7	6	5	5	4	2
4 × 2⅛ × 2⅛ device	14.5	9	8	7	6	5	4	2
3¾ × 2 × 2½ masonry box/gang	14.0	9	8	7	6	5	4	2
3¾ × 2 × 3½ masonry box/gang	21.0	14	12	10	9	8	7	4
FS — Minimum internal depth 1¾ single cover/gang	13.5	9	7	6	6	5	4	2
FD — Minimum internal depth 2⅜ single cover/gang	18.0	12	10	9	8	7	6	3
FS — Minimum internal depth 1¾ multiple cover/gang	18.0	12	10	9	8	7	6	3
FD — Minimum internal depth 2⅜ multiple cover/gang	24.0	16	13	12	10	9	8	4

Note: For SI units, 1 in.^{3} = 16.4 cm^3.
*Where no volume allowances are required by Sections 370-16(b)(2) through 370-16(b)(5).

the box without splice or termination shall be counted once. The conductor fill, in cubic inches, shall be computed using Table 370-16(b). A conductor, no part of which leaves the box, shall not be counted.

Table 370-16(b). Volume Allowance Required per Conductor

Size of Conductor (AWG)	Free Space Within Box for Each Conductor (in.^{3})
18	1.50
16	1.75
14	2.00
12	2.25
10	2.50
8	3.00
6	5.00

Note: For SI units, 1 in.^{3} = 16.4 cm^3.

Section 370-16 provides the requirements and identifies the allowances for the number of conductors permitted to be enclosed within a box. This section continues to require that the *total box volume* be equal to or greater than the *total box fill.*

General Requirements

The *total box volume* is determined by adding the individual volumes of the box components. The components include the box itself plus any attachments to the box, such as a plaster ring, an extension ring, or a dome cover. The volume of each box component is determined either from the volume marking on the component itself or from the standard volumes listed in Table 370-16(a). If a box is marked with a larger volume than listed in Table 370-16(a), the larger volume can be used instead of the table value.

Adding all of the volume allowances for all items contributing to box fill determines the *total box fill.* The volume allowance for each fill item is based on the volume listed in Table 370-16(b) for the conductor size indicated. See Table 370.1, Summary of Components Contributing to Box Fill.

Exception: An equipment grounding conductor or conductors or not over four fixture wires smaller than No. 14, or both, shall be permitted to be omitted from the calculations where they enter a box from a domed fixture or similar canopy and terminate within that box.

(2) Clamp Fill. Where one or more internal cable clamps, whether factory or field supplied, are present in the box, a single volume allowance in accordance with Table 370-16(b) shall be made based on the largest conductor present in the box. No allowance shall be required for a cable connector with its clamping mechanism outside the box.

(3) Support Fittings Fill. Where one or more fixture studs or hickeys are present in the box, a single volume allowance in accordance with Table 370-16(b) shall be made for each type of fitting based on the largest conductor present in the box.

(4) Device or Equipment Fill. For each yoke or strap containing one or more devices or equipment, a double volume allowance in accordance with Table 370-16(b) shall be made for each yoke or strap based on the largest conductor connected to a device(s) or equipment supported by that yoke or strap.

(5) Equipment Grounding Conductor Fill. Where one or more equipment grounding conductors or equipment bonding jumpers enters a box, a single volume allowance in accordance with Table 370-16(b) shall be made based on the largest equipment grounding conductor or equipment bonding jumper present in the box. Where an additional set of equipment grounding conductors, as permitted by Section 250-146(d), is present in the box, an additional volume allowance shall be made based on the largest equipment grounding conductor in the additional set.

Specific Examples

The next three examples are presented to identify the applicable requirements of Section 370-16 and accompanying tables.

Example 1

The simple method [according to the footnote of Table 370-16(a)] to select a standard-size box can be used if all the conductors are the same size and the box does not contain any cable clamps, support fittings, devices, or equipment grounding conductors. Refer to Figure 370.1A as an example. To determine the number of conductors permitted in this standard 4 × 1½ square box (21.0 in.³) is to count the conductors in the box and compare the total to the maximum number of conductors permitted by Table 370-16(a). Each unspliced conductor running through the box is counted as one conductor, and each other conduc-

Table 370.1 Summary of Items Contributing to Box Fill

Items Contained Within Box	Volume Allowance	Based on [see Table 370-16(b)]
Conductors that originate outside box	One for each conductor	Actual conductor size
Conductors that pass through box without splice or connection	One for each conductor	Actual conductor size
Conductors that originate within box and do not leave box	None (These conductors are not counted.)	NA
Fixture conductors [per Section 370-16(b)(1), Exception]	None (These conductors are not counted.)	NA
Internal cable clamps (one or more)	One only	Largest-size conductor present
Support fittings (such as fixture studs, hickeys, etc.)	One for each type of support fitting	Largest-size conductor present
Devices (such as receptacles, switches, etc.)	Two for each yoke or mounting strap	Largest-size conductor connected to device or equipment
Equipment grounding conductor (one or more)	One only	Largest equipment grounding conductor present
Isolated equipment grounding conductor (one or more) (see Section 250-74, Exception No. 4)	One only	Largest isolated and insulated equipment grounding conductor present

Table 370.2 Total Box Fill for Example No. 2

Items Contained Within Box	Volume Allowance	Unit Volume Based on Table 370-16(b) (in.^{3})	Total Box Fill (in.^{3})
4 conductors	4 volume allowances for No. 14 conductors	2.00	8.00
1 clamp	1 volume allowance (based on No. 14 conductors)	2.00	2.00
1 device	2 volume allowances (based on No. 14 conductors)	2.00	4.00
Equipment grounding conductors (all)	1 volume allowance (based on No. 14 conductors)	2.00	2.00
Total			16.00 in.^{3}

tor is counted as one conductor. Therefore, the total conductor count for this box is nine conductors. Table 370-16(a) indicates that the maximum fill for this box is nine No. 12 conductors, so the box is adequately sized.

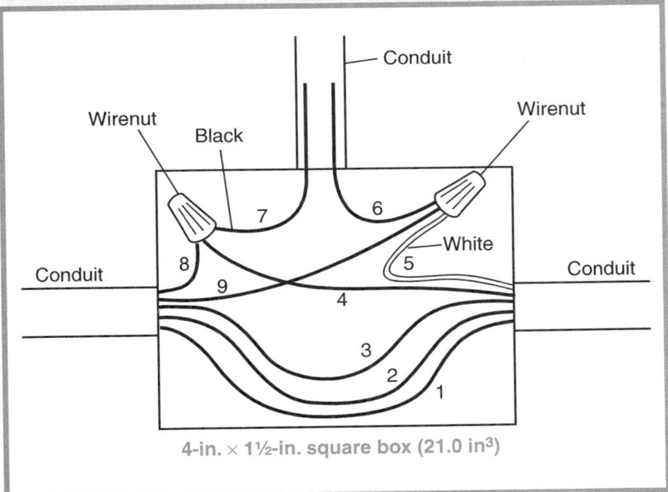

Figure 370.1A *Example 1: A standard-size 4-in. × 1½-in. [square box (21.0 in.^{3})] containing no fittings or devices, such as fixture studs, cable clamps, switches, receptacles, or equipment grounding conductors.*

Example 2

The standard method is used by adding the total box volume and then subtracting the total box fill to ensure compliance. As an example, refer to Figure 370.1B. For a standard 3 × 2 × 3½ device box (18 in.^{3}) Table 370-16(a) allows up to a maximum of nine No. 14 conductors. The box fill for this situation is as given in the Table 370.2. Therefore, since the total box fill of 16 in.^{3} is less than the 18 in.^{3} total box volume permitted, the box is adequately sized.

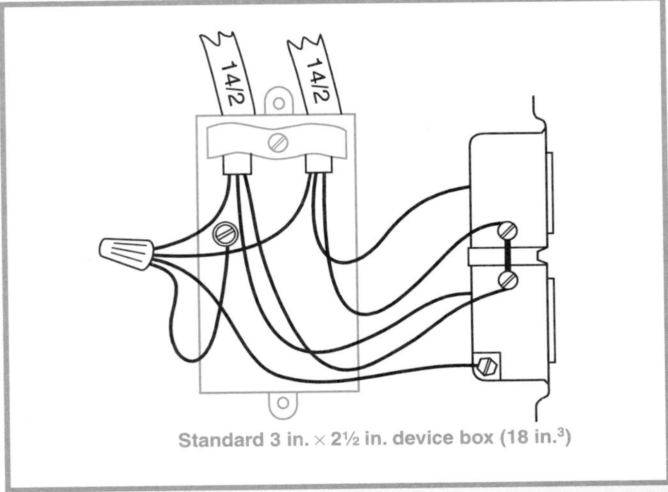

Figure 370.1B *Example 2: A device box that contains components and conductors requiring deductions in accordance with Section 370-16.*

Example 3

This standard-method example uses Figure 370.2. The figure illustrates two 3 × 2 × 3½-in. device boxes assembled to configure a single box. The total box volume, referring to Table 370-16(a), is 36 in.^{3} (2 × 18 in.^{3}). The total box fill, referring to Table 370-16(b), is determined as given in Table 370.3. With only 26 in.^{3} of the 36 in.^{3} filled, the box is adequately sized.

(c) Conduit Bodies.

(1) General. Conduit bodies enclosing No. 6 conductors or smaller, other than short radius conduit bodies as described in Section 370-5, shall have a cross-sectional area not less than twice the cross-sectional area of the largest conduit or tubing to which it is attached. The maximum number of conductors permitted shall be the maximum number permitted by Table 1 of Chapter 9 for the conduit or tubing to which it is attached.

Table 370.3 Total Box Fill for Example No. 3

Items Contained Within Box	Volume Allowance	Unit Volume Based on Table 370-16(b) (in.³)	Total Box Fill (in.³)
6 conductors	2 volume allowances for No. 14 conductors	2.00	4.00
	4 volume allowances for No. 12 conductors	2.25	9.00
2 clamps	1 volume allowance (based on No. 12 conductors)	2.25	2.25
2 devices	2 volume allowances (based on No. 14 conductors)	2.00	4.00
	2 volume allowances (based on No. 12 conductors)	2.25	4.50
Equipment grounding conductors (all)	1 volume allowance (based on No. 12 conductors)	2.25	2.25
Total			26.00 in.³

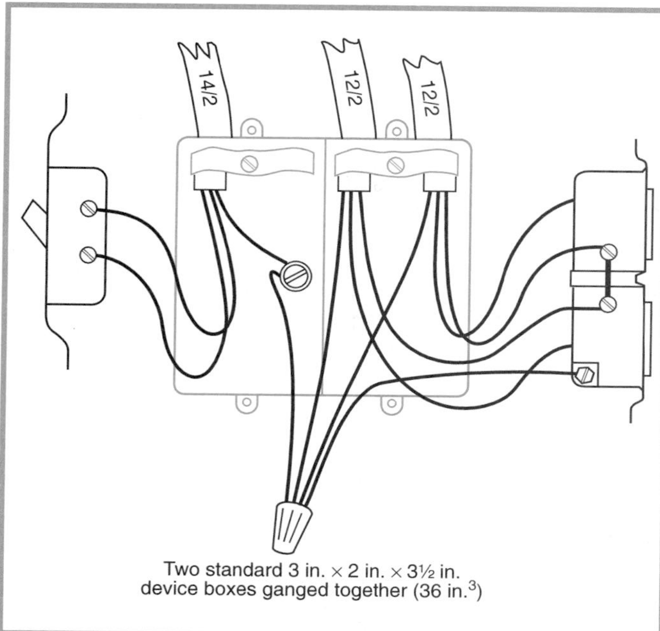

Figure 370.2 *Example 3: Two standard gangable device boxes containing conductors of different sizes.*

(2) With Splices, Taps, or Devices. Only those conduit bodies that are durably and legibly marked by the manufacturer with their cubic inch capacity shall be permitted to contain splices, taps, or devices. The maximum number of conductors shall be computed in accordance with Section 370-16(b). Conduit bodies shall be supported in a rigid and secure manner.

As illustrated in Figure 370.3, conduit bodies other than the short-radius type are permitted to contain splices or taps, provided the conduit bodies are marked with their cubic inch capacity. Such conduit bodies are required to have a cross-sectional area not less than twice that of the conduit to which they are attached and are not permitted to contain more conductors than the attached raceway. The volume requirements for splicing or tapping are provided in Section 370-16(c).

Conduit bodies must be rigidly supported. See Section 370-23(d)(2) and the exception to Section 370-23(e), and Exception No. 1 of Section 370-23(f), which permit the raceway to support the conduit body, provided the conduit body is not larger than the attached raceway.

See Section 370-28 for requirements where conduit bodies are used as pull and junction boxes.

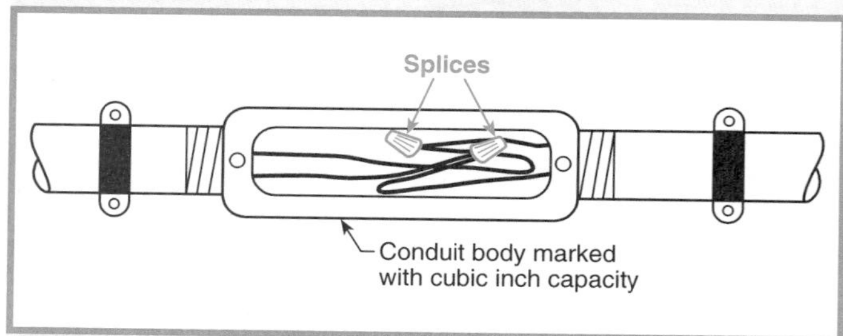

Figure 370.3 *An example of splices in a raceway-supported conduit body.*

370-17. Conductors Entering Boxes, Conduit Bodies, or Fittings. Conductors entering boxes, conduit bodies, or fittings shall be protected from abrasion and shall comply with (a) through (d).

(a) Openings to Be Closed. Openings through which conductors enter shall be adequately closed.

(b) Metal Boxes and Conduit Bodies. Where metal boxes or conduit bodies are installed with open wiring or concealed knob-and-tube wiring, conductors shall enter through insulating bushings or, in dry locations, through flexible tubing extending from the last insulating support and firmly secured to the box or conduit body. Where raceway or cable is installed with metal boxes or conduit bodies, the raceway or cable shall be secured to such boxes and conduit bodies.

(c) Nonmetallic Boxes. Nonmetallic boxes shall be suitable for the lowest temperature-rated conductor entering the box. Where nonmetallic boxes are used with open wiring or concealed knob-and-tube wiring, the conductors shall enter the box through individual holes. Where flexible tubing is used to encase the conductors, the tubing shall extend from the last insulating support to no less than ¼ in. (6.35 mm) inside the box. Where nonmetallic-sheathed cable is used, the cable assembly, including the sheath, shall extend into the box no less than ¼ in. (6.35 mm) through a nonmetallic-sheathed cable knockout opening. In all instances, all permitted wiring methods shall be secured to the boxes.

Standard nonmetallic boxes are permitted for use with 90°C (194°F) insulated conductors. A nonmetallic box used for splicing a conductor of a higher temperature rating to a conductor of a lower temperature rating is required to be identified as suitable for the temperature rating of the lower-rated conductor. The intent is to avoid the necessity of giving a high temperature rating to boxes in a normal temperature location simply because high-temperature conductors enter the box from, or exit to, a high-temperature location. However, where insulated conductors rated at higher temperatures are necessary in a high-temperature environment, the box is required to be suitably identified by marking on the box or in the listing of the box to comply with Section 110-3(b).

Exception: Where nonmetallic-sheathed cable or underground feeder and branch-circuit cable is used with single gang boxes no larger than a nominal size 2¼ in. × 4 in., mounted in walls or ceilings, and where the cable is fastened within 8 in. (203 mm) of the box measured along the sheath and where the sheath extends through a cable knockout no less than ¼ in. (6.35 mm), securing the cable to the box shall not be required. Multiple cable entries shall be permitted in a single cable knockout opening.

Some cable-securing means is required in all single gang boxes larger than 2¼ in. by 4 in. The requirement is based on the width of the box and the likelihood that the cable will be pushed back out of the box when the conductors and device, if any, are folded back into the box during installation of receptacles, switches, dimmers, and so on. The last sentence of this exception permits multiple cable entries in a single knockout opening.

(d) Conductors No. 4 or Larger. Installation shall comply with Section 300-4(f).

370-18. Unused Openings. Unused cable or raceway openings in boxes and conduit bodies shall be effectively closed to afford protection substantially equivalent to that of the wall of the box or conduit body. Metal plugs or plates used with nonmetallic boxes or conduit bodies shall be recessed at least ¼ in. (6.35 mm) from the outer surface of the box.

370-19. Boxes Enclosing Flush Devices. Boxes used to enclose flush devices shall be of such design that the devices will be completely enclosed on back and sides, and that substantial support for the devices will be provided. Screws for supporting the box shall not be used in attachment of the device contained therein.

370-20. In Wall or Ceiling. In walls or ceilings of concrete, tile, or other noncombustible material, boxes shall be installed so that the front edge of the box will not be set back of the finished surface more than ¼ in. (6.35 mm). In walls and ceilings constructed of wood or other combustible material, boxes shall be flush with the finished surface or project therefrom.

Drywall, plaster, and gypsum board are "other noncombustible materials." A wall with metal or wood studs covered by drywall, plaster, or gypsum board is also considered a "wall of noncombustible material" for the purposes of Section 370-20.

370-21. Repairing Plaster and Drywall or Plasterboard. Plaster, drywall, or plasterboard surfaces that are broken or incomplete shall be repaired so there will be no gaps or open spaces greater than ⅛ in. (3.18 mm) at the edge of the box or fitting.

370-22. Exposed Surface Extensions. Surface extensions from a box of a concealed wiring system shall be made by mounting and mechanically securing a box or extension ring over the concealed box. Where required, equipment grounding shall be in accordance with Article 250.

Exception: A surface extension shall be permitted to be made from the cover of a concealed box where the cover is designed so it is unlikely to fall off, or be removed if its securing means becomes loose. The wiring method shall be flexible and arranged so that any required grounding continuity is independent of the connection between the box and cover.

Figure 370.4 shows an example of a flexible surface extension from a concealed outlet box. Section 370-22, Exception, which permits this technique, requires the use of a cover that will not fall off if the securing screws become loose.

370-23. Supports. Enclosures within the scope of this article shall be supported in accordance with one or more of the provisions in (a) through (h).

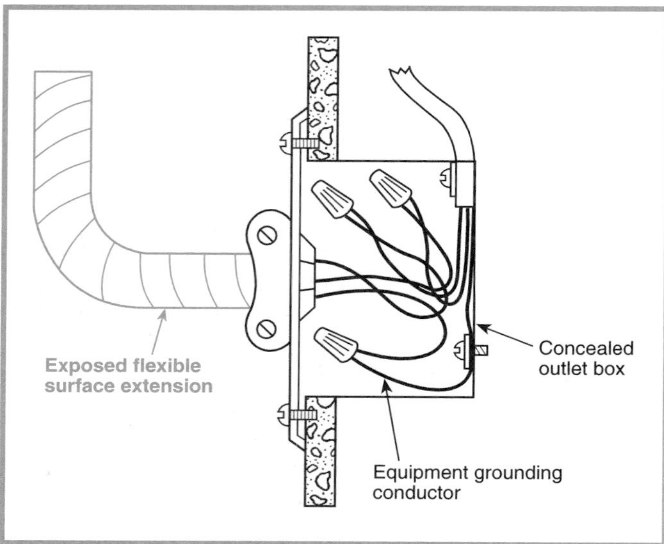

Figure 370.4 *An example of a flexible surface extension from a concealed outlet box.*

(a) Surface Mounting. An enclosure mounted on a building or other surface shall be rigidly and securely fastened in place. If the surface does not provide rigid and secure support, additional support in accordance with other provisions of this section shall be provided.

Figure 370.5 illustrates one method of supporting a box where the ceiling construction affords no support. See also Section 410-16 regarding means of support for lighting fixtures.

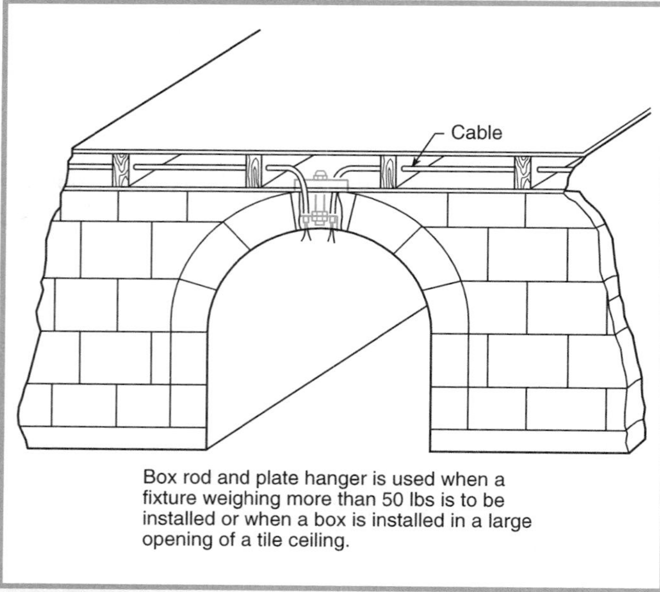

Figure 370.5 *A plate and threaded-rod hanger used to support an outlet box in a tile arch ceiling.*

(b) Structural Mounting. An enclosure supported from a structural member of a building or from grade shall be rigidly supported either directly, or by using a metal, polymeric, or wood brace.

(1) Nails. Nails, where used as a fastening means, shall be attached by using brackets on the outside of the enclosure, or they shall pass through the interior within ¼ in. (6.35 mm) of the back or ends of the enclosure.

This rule prevents the nails from interfering with the installation of devices. Permitting nails inside the box within ¼ in. of the ends reduces splitting of the smaller wooden studs used in some frame-type construction. However, splitting sometimes occurs where nails are within ¼ in. of the back of the box.

(2) Braces. Metal braces shall be protected against corrosion and formed from metal that is not less than 0.020 in. (508 μm) thick uncoated. Wood braces shall have a cross section not less than nominal 1 in. × 2 in. Wood braces in wet locations shall be treated for the conditions. Polymeric braces shall be identified as being suitable for the use.

(c) Mounting in Finished Surfaces. An enclosure mounted in a finished surface shall be rigidly secured thereto by clamps, anchors, or fittings identified for the application.

Where there are no structural members, or where boxes are cut into existing walls, boxes are permitted to be secured by clamps or anchors. Figure 370.6 shows one example of an acceptable mounting method. Shown in the bottom portion of the figure

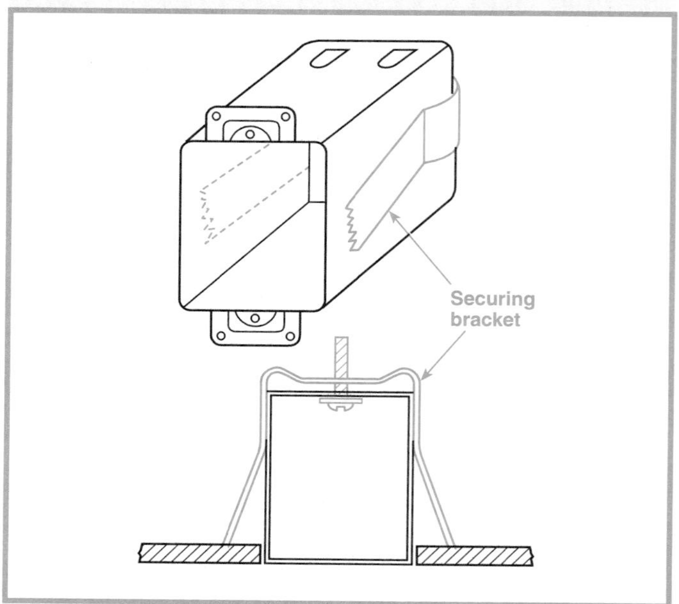

Figure 370.6 *One type of device box used for old work.*

is a device box mounted in an opening in an existing wall by means of a securing bracket that is part of the box.

(d) Suspended Ceilings. An enclosure mounted to structural or supporting elements of a suspended ceiling shall be not more than 100 in.^{3} (1640 cm^3) in size and shall be securely fastened in place in one of the following ways.

Section 370-23(d) was modified for the 1999 *NEC* to clarify that suspended ceilings could be used to support boxes that are 100 in.^{3} or less.

(1) Framing Members. An enclosure shall be fastened to the framing members by mechanical means such as bolts, screws, or rivets, or by the use of clips or other securing means identified for use with the type of ceiling framing member(s) and enclosure(s) employed. The framing members shall be adequately supported and securely fastened to each other and to the building structure.

(2) Support Wires. The installation shall comply with the provisions of Section 300-11(a). The enclosure shall be secured, using methods identified for the purpose, to ceiling support wires, including any additional support wires installed for that purpose. Support wires used for enclosure support shall be fastened at each end so as to be taut within the ceiling cavity.

(e) Raceway Supported Enclosure, Without Devices or Fixtures. An enclosure that does not contain a device(s) or support a fixture(s) or other equipment, and is supported by entering raceways shall not exceed 100 in.^{3} (1640 cm^3) in size. It shall have threaded entries or have hubs identified for the purpose. It shall be supported by two or more conduits threaded wrenchtight into the enclosure or hubs. Each conduit shall be secured within 3 ft (914 mm) of the enclosure, or within 18 in. (457 mm) of the enclosure if all entries are on the same side.

Boxes are not permitted to be supported by rigid raceways using locknuts and bushings. Enclosures without devices or fixtures, however, are considered to be adequately supported, provided the conduit is connected to the enclosure using threaded hubs and the threaded conduits enter the box on two or more sides and are supported within 3 ft of the enclosure. A box is not permitted to be supported by a single raceway.

Exception: Rigid metal, intermediate metal, or rigid nonmetallic conduit or electrical metallic tubing shall be permitted to support a conduit body of any size, including a conduit body constructed with only one conduit entry, provided the conduit body is not larger than the largest trade size of the conduit or electrical metallic tubing.

(f) Raceway Supported Enclosures, with Devices or Fixtures. An enclosure that contains a device(s) or supports a fixture(s) or other equipment and is supported by entering raceways shall not exceed 100 in.^{3} (1640 cm^3) in size. It shall have threaded entries or have hubs identified for the purpose. It shall be supported by two or more conduits threaded wrenchtight into the enclosure or hubs. Each conduit shall be secured within 18 in. (457 mm) of the enclosure.

The conduit is required to be supported within 18 in. if the enclosure contains devices or fixtures. See the commentary following Section 370-23(e) regarding enclosures without devices or fixtures.

Exception No. 1: Rigid metal or intermediate metal conduit shall be permitted to support a conduit body of any size, including a conduit body constructed with only one conduit entry, provided the conduit bodies are not larger than the largest trade size of the conduit.

Exception No. 2: An unbroken length(s) of rigid or intermediate metal conduit shall be permitted to support a box used for fixture support, or to support a wiring enclosure within a fixture and used in lieu of a box in accordance with Section 300-15(b), where all of the following conditions are met.

(a) The conduit is securely fastened at a point so that the length of conduit beyond the last point of conduit support does not exceed 3 ft (914 mm).

(b) The unbroken conduit length before the last point of conduit support is 12 in. (305 mm) or greater, and that portion of the conduit is securely fastened at some point not less than 12 in. (305 mm) from its last point of support.

(c) Where accessible to unqualified persons, the fixture, measured to its lowest point, is at least 8 ft (2.44 m) above grade or standing area and at least 3 ft (914 mm) measured horizontally to the 8 ft (2.44 m) elevation from windows, doors, porches, fire escapes, or similar locations.

(d) A fixture supported by a single conduit does not exceed 12 in. (305 mm) in any direction from the point of conduit entry.

(e) The weight supported by any single conduit does not exceed 20 lb (9.08 kg).

(f) At the fixture end, the conduit(s) is threaded wrenchtight into the box or wiring enclosure, or into hubs identified for the purpose.

(g) Enclosures in Concrete or Masonry. An enclosure supported by embedment shall be identified as suitably protected from corrosion and securely embedded in concrete or masonry.

Boxes are permitted to be embedded in masonry or concrete, provided they are rigid and secure. Figure 370.7 shows a mud box installed in a concrete ceiling. Additional support is not required.

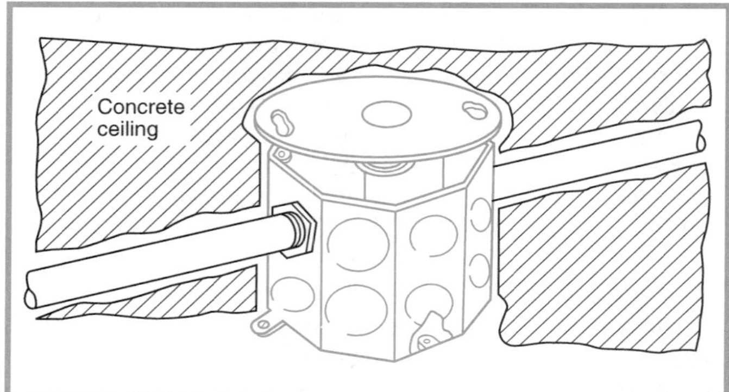

Figure 370.7 *A mud box installed in a concrete ceiling.*

(h) Pendant Boxes. An enclosure supported by a pendant shall comply with (1) or (2).

(1) Flexible Cord. A box shall be supported from a multiconductor cord or cable in an approved manner that protects the conductors against strain, such as a strain-relief connector threaded into a box with a hub.

(2) Conduit. A box supporting lampholders or lighting fixtures, or wiring enclosures within fixtures used in lieu of boxes in accordance with Section 300-15(b), shall be supported by rigid or intermediate metal conduit stems. For stems longer than 18 in. (457 mm), the stems shall be connected to the wiring system with flexible fittings suitable for the location. At the fixture end, the conduit(s) shall be threaded wrenchtight into the box or wiring enclosure, or into hubs identified for the purpose.

Where supported by only a single conduit, the threaded joints shall be prevented from loosening by the use of setscrews or other effective means, or the fixture, at any point, shall be at least 8 ft (2.44 m) above grade or standing area and at least 3 ft (914 mm) measured horizontally to the 8 ft (2.44 m) elevation from windows, doors, porches, fire escapes, or similar locations. A fixture supported by a single conduit shall not exceed 12 in. (305 mm) in any horizontal direction from the point of conduit entry.

370-24. Depth of Outlet Boxes. No box shall have an internal depth of less than ½ in. (12.7 mm). Boxes intended to enclose flush devices shall have an internal depth of not less than 15⁄16 in. (23.8 mm).

The use of a shallow box may become necessary because of old work or existing construction where very shallow partitions, plumbing pipes, or ductwork is encountered within the partition, for example. The selection of a box used in this type of situation is required to be based on its having sufficient cubic-inch capacity.

370-25. Covers and Canopies. In completed installations, each box shall have a cover, faceplate, or fixture canopy.

(a) Nonmetallic or Metal Covers and Plates. Nonmetallic or metal covers and plates shall be permitted. Where metal covers or plates are used, they shall comply with the grounding requirements of Section 250-110.

FPN: For additional grounding requirements, see Section 410-18(a) for metal fixture canopies, and Sections 380-12 and 410-56(d) for metal faceplates.

(b) Exposed Combustible Wall or Ceiling Finish. Where a fixture canopy or pan is used, any combustible wall or ceiling finish exposed between the edge of the canopy or pan and the outlet box shall be covered with noncombustible material.

Because heat from a short circuit, ground fault or from overlamping could create a fire hazard within a fixture canopy or pan, any exposed combustible wall or ceiling space between the edge of the outlet box and the perimeter of the fixture is required to be covered with noncombustible material. The noncombustible material need not be metal. Glass fiber pads commonly provided as thermal barriers on ceiling pan fixtures can be used to meet this requirement. Where the wall or ceiling finish is concrete, tile, plaster, or other noncombustible material, the requirements of this section do not apply (see Section 370-20).

(c) Flexible Cord Pendants. Covers of outlet boxes and conduit bodies having holes through which flexible cord pendants pass shall be provided with bushings designed for the purpose or shall have smooth, well-rounded surfaces on which the cords may bear. So-called hard rubber or composition bushings shall not be used.

370-27. Outlet Boxes.

(a) Boxes at Lighting Fixture Outlets. Boxes used at lighting fixture outlets shall be designed for the purpose. At every outlet used exclusively for lighting, the box shall be designed or installed so that a lighting fixture may be attached.

Exception: A wall-mounted fixture weighing not more than 6 lb (2.72 kg) and not exceeding 16 in. (406 mm) in any dimension shall be permitted to be supported on other boxes, provided the fixture or its supporting yoke is secured to the box with no fewer than two No. 6 or larger screws.

Device boxes are designed for the mounting of devices such as snap switches and receptacles, usually with 6-32 screws (No. 6 screws with 32 threads per inch). These boxes are not suitable for the support of other than lightweight wall-mounted lighting fixtures. The exception to Section 370-27(a) permits fixtures such as wall-bracket types or sconces weighing less than 6 lb to be supported by the No. 6 or larger

screws. For heavier or ceiling-mounted lighting fixtures, see Section 410-16, Means of Support. The outlet box is required to provide "adequate" support.

(b) Floor Boxes. Boxes listed specifically for this application shall be used for receptacles located in the floor.

Exception: Where the authority having jurisdiction judges them free from likely exposure to physical damage, moisture, and dirt, boxes located in elevated floors of show windows and similar locations shall be permitted to be other than those listed for floor applications. Receptacles and covers shall be listed as an assembly for this type of location.

(c) Boxes at Ceiling-Suspended (Paddle) Fan Outlets. Outlet boxes shall not be used as the sole support for ceiling-suspended (paddle) fans.

The change for the 1999 *NEC* clarified that this section applied to paddle-type fans and not other ceiling support fans such as a bathroom ventilation fan. One method of supporting a ceiling fan so that the box does not serve as the sole support is shown in Figure 370.8. See also Section 422-18, Support of Ceiling-Suspended (Paddle) Fans. The exception to this section clarifies that listed boxes are available that provide a suitable means for support. However, there are several alternative and retrofit methods that can provide suitable support for a paddle fan.

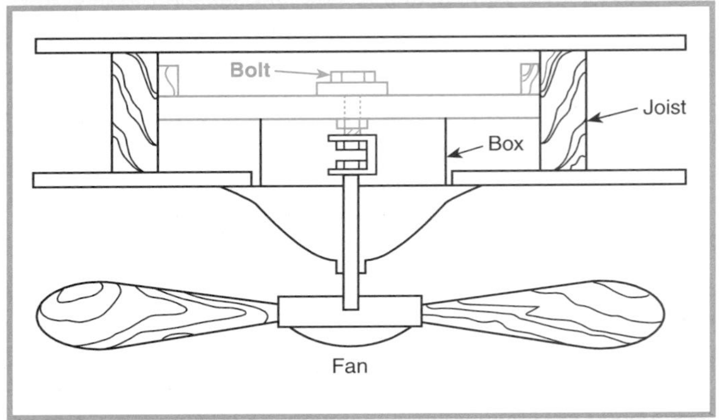

Figure 370.8 *A ceiling fan supported without depending on the box for sole support.*

Exception: Boxes listed for the application shall be permitted as the sole means of support.

370-28. Pull and Junction Boxes. Boxes and conduit bodies used as pull or junction boxes shall comply with (a) through (d).

Exception: Terminal housings supplied with motors shall comply with the provisions of Section 430-12.

(a) Minimum Size. For raceways containing conductors of No. 4 or larger, and for cables containing conductors of No. 4 or larger, the minimum dimensions of pull or junction boxes installed in a raceway or cable run shall comply with the following.

(1) Straight Pulls. In straight pulls, the length of the box shall not be less than eight times the trade diameter of the largest raceway.

Section 370-28(a)(1) applies to minimum dimensions of pull and junction boxes or conduit bodies used with raceways or cables containing conductors No. 4 or larger.

For straight pulls, for example, a 2-in. conduit containing four No. 4/0 Type THHW conductors (see Appendix C, Table C8) requires a 16-in.-long pull box (8 × 2 in. = 16 in.). It should be understood that 16 in. is the required minimum length; however, for maximum ease in handling this size conductor, a longer pull box may be desired.

(2) Angle or U Pulls. Where splices, or where angle or U pulls are made, the distance between each raceway entry inside the box and the opposite wall of the box shall not be less than six times the trade diameter of the largest raceway in a row. This distance shall be increased for additional entries by the amount of the sum of the diameters of all other raceway entries in the same row on the same wall of the box. Each row shall be calculated individually, and the single row that provides the maximum distance shall be used.

Exception: Where a raceway or cable entry is in the wall of a box or conduit body opposite a removable cover, the distance from that wall to the cover shall be permitted to comply with the distance required for one wire per terminal in Table 373-6(a).

The distance between raceway entries enclosing the same conductor shall not be less than six times the trade diameter of the larger raceway.

When transposing cable size into raceway size in (a)(1) and (a)(2), the minimum trade size raceway required for the number and size of conductors in the cable shall be used.

Figures 370.9A and 370.9B provide examples of calculations required by Section 370-28(a)(2). As the example in Figure 370.9A illustrates, where splices, angle pulls, or U pulls are made. The distance between each raceway entry inside the box and the opposite wall of the box must not be less than six times the trade diameter of the largest raceway, plus the distance for additional raceway entries. This additional distance is calculated by adding the diameters of the other raceway entries in one row on the same side of the box. The example in Figure 370.9B shows that raceway entries enclosing the same conductor are

required to have a minimum separation between them. The intent is to provide adequate space for the conductor to make the bend.

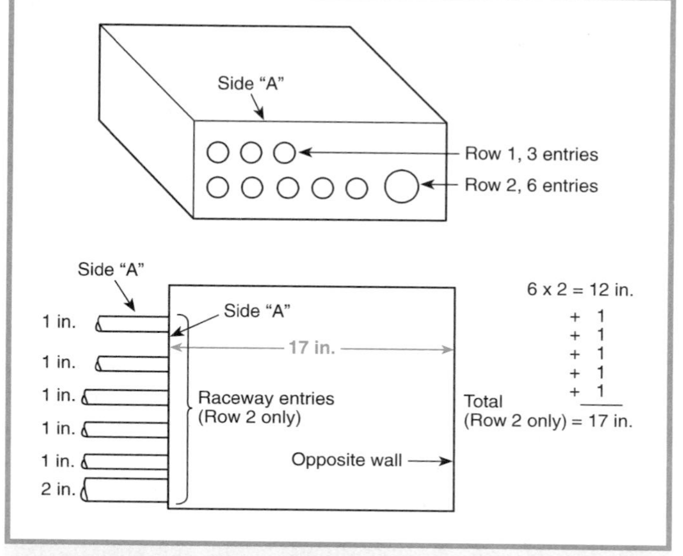

Figure 370.9A *An example showing calculations required by Section 370-28(a)(2) for splices, angle pulls, or U pulls.*

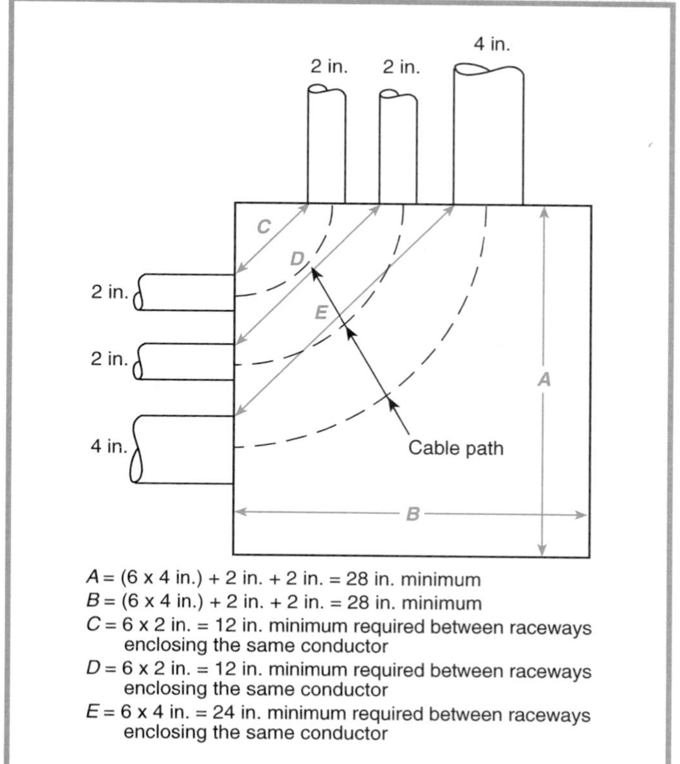

Figure 370.9B *An example showing calculations required by Section 370-28(a)(2) for raceways enclosing the same conductor.*

(3) Boxes or conduit bodies of dimensions less than those required in (a)(1) and (a)(2) shall be permitted for installations of combinations of conductors that are less than the maximum conduit or tubing fill (of conduits or tubing being used) permitted by Table 1 of Chapter 9, provided the box or conduit body has been approved for and is permanently marked with the maximum number and maximum size of conductors permitted.

(b) Conductors in Pull or Junction Boxes. In pull boxes or junction boxes having any dimension over 6 ft (1.83 m), all conductors shall be cabled or racked up in an approved manner.

(c) Covers. All pull boxes, junction boxes, and conduit bodies shall be provided with covers compatible with the box or conduit body construction and suitable for the conditions of use. Where metal covers are used, they shall comply with the grounding requirements of Section 250-110. An extension from the cover of an exposed box shall comply with Section 370-22, Exception.

(d) Permanent Barriers. Where permanent barriers are installed in a box, each section shall be considered as a separate box.

370-29. Conduit Bodies, Junction, Pull, and Outlet Boxes to Be Accessible. Conduit bodies, junction, pull, and outlet boxes shall be installed so that the wiring contained in them can be rendered accessible without removing any part of the building or, in underground circuits, without excavating sidewalks, paving, earth, or other substance that is to be used to establish the finished grade.

Exception: Listed boxes shall be permitted where covered by gravel, light aggregate, or noncohesive granulated soil if their location is effectively identified and accessible for excavation.

Consideration should also be given to the accessibility of junction boxes installed on a structural ceiling above a suspended ceiling.

A box is permitted to be used at any point for the connection of conduit, tubing, or cable, provided it is not rendered inaccessible. See Article 100 for the definition of *accessible (as applied to wiring methods)*. See Section 300-15 for other requirements for boxes, conduit bodies, or fittings.

C. Construction Specifications

370-40. Metal Boxes, Conduit Bodies, and Fittings.

(a) Corrosion Resistant. Metal boxes, conduit bodies, and fittings shall be corrosion resistant or shall be well-galvanized, enameled, or otherwise properly coated inside and out to prevent corrosion.

FPN: See Section 300-6 for limitation in the use of boxes and fittings protected from corrosion solely by enamel.

(b) Thickness of Metal. Sheet steel boxes not over 100 $in.^3$ (1640 cm^3) in size shall be made from steel not less than

0.0625 in. (1.59 mm) thick. The wall of a malleable iron box or conduit body and a die-cast or permanent-mold cast aluminum, brass, bronze, or zinc box or conduit body shall not be less than 3⁄32 in. (2.38 mm) thick. Other cast metal boxes or conduit bodies shall have a wall thickness not less than 1⁄8 in. (3.17 mm).

Exception No. 1: Listed boxes and conduit bodies shown to have equivalent strength and characteristics shall be permitted to be made of thinner or other metals.

Exception No. 2: The walls of listed short radius conduit bodies, as covered in Section 370-5, shall be permitted to be made of thinner metal.

(c) Metal Boxes Over 100 in.^{3} Metal boxes over 100 in.^{3} (1640 cm^3) in size shall be constructed so as to be of ample strength and rigidity. If of sheet steel, the metal thickness shall not be less than 0.053 in. (1.35 mm) uncoated.

(d) Grounding Provisions. A means shall be provided in each metal box for the connection of an equipment grounding conductor. The means shall be permitted to be a tapped hole or equivalent.

The "means provided" by the box manufacturer is usually in the form of a 10-32 tapped hole marked "GR" or "GRD," or the equivalent, next to the hole. It should be noted, however, that the "means provided" may not necessarily be used. See the commentary following Section 250-148 regarding two arrangements for the termination of equipment grounding conductors.

370-41. Covers. Metal covers shall be of the same material as the box or conduit body with which they are used, or they shall be lined with firmly attached insulating material that is not less than 1⁄32 in. (0.79 mm) thick, or they shall be listed for the purpose. Metal covers shall be the same thickness as the boxes or conduit body for which they are used, or they shall be listed for the purpose. Covers of porcelain or other approved insulating materials shall be permitted if of such form and thickness as to afford the required protection and strength.

370-42. Bushings. Covers of outlet boxes and conduit bodies having holes through which flexible cord pendants may pass shall be provided with approved bushings or shall have smooth, well-rounded surfaces on which the cord may bear. Where individual conductors pass through a metal cover, a separate hole equipped with a bushing of suitable insulating material shall be provided for each conductor. Such separate holes shall be connected by a slot as required by Section 300-20.

370-43. Nonmetallic Boxes. Provisions for supports or other mounting means for nonmetallic boxes shall be outside of the box, or the box shall be constructed so as to prevent contact between the conductors in the box and the supporting screws.

370-44. Marking. All boxes and conduit bodies, covers, extension rings, plaster rings, and the like shall be durably and legibly marked with the manufacturer's name or trademark.

D. Manholes and Other Electric Enclosures Intended for Personnel Entry

New for the 1999 *NEC*, Part D of Article 370 covers manholes and other electrical enclosures intended for personnel entry. However, general electrical equipment installation requirements within the manhole or large enclosure are still covered by Article 110.

370-50. General. Electric enclosures intended for personnel entry and specifically fabricated for this purpose shall be of sufficient size to provide safe work space about electric equipment with live parts that is likely to require examination, adjustment, servicing, or maintenance while energized. They shall have sufficient size to permit ready installation or withdrawal of the conductors employed without damage to the conductors or to their insulation. They shall comply with the provisions of this part.

Exception: Where electric enclosures covered by Part D of this article are part of an industrial wiring system operating under conditions of maintenance and supervision that ensure only qualified persons will monitor and supervise the system, they shall be permitted to be designed and installed in accordance with appropriate engineering practice. If required by the authority having jurisdiction, design documentation shall be provided.

370-51. Strength. Manholes, vaults, and their means of access shall be designed under qualified engineering supervision and shall withstand all loads likely to be imposed on the structures.

FPN: See *National Electrical Safety Code*, ANSI C2-1997, for additional information on the loading that can be expected to bear on underground enclosures.

370-52. Cabling Work Space. A clear work space not less than 3 ft (914 mm) wide shall be provided where cables are located on both sides, and not less than 2½ ft (762 mm) where cables are only on one side. The vertical headroom shall not be less than 6 ft (1.83 m) unless the opening is within 1 ft (305 mm), measured horizontally, of the adjacent interior side wall of the enclosure.

Exception: A manhole containing only one or more of the following shall be permitted to have one of the horizontal work space dimensions reduced to 2 ft (608 mm) where the other horizontal clear work space is increased so the sum of the two dimensions is not less than 6 ft (1.83 m):

(a) Optical fiber cables as covered in Article 770
(b) Power-limited fire alarm circuits supplied in accordance with Section 760-41
(c) Class 2 or Class 3 remote-control and signaling circuits, or both, supplied in accordance with Section 725-41

370-53. Equipment Work Space. Where electric equipment with live parts that is likely to require examination, adjustment, servicing, or maintenance while energized is installed in a manhole, vault, or other enclosure designed for personnel access, the work space and associated requirements in Section 110-26 shall be met for installations operating at 600 volts or less. Where the installation is over 600 volts, the work space and associated requirements in Section 110-34 shall be met. A manhole access cover that weighs over 100 lb (45.4 kg) shall be considered as meeting the requirements of Section 110-34(c).

370-54. Bending Space for Conductors. Bending space for conductors operating at 600 volts or below shall be provided in accordance with the requirements of Section 370-28(a). Conductors operating over 600 volts shall be provided with bending space in accordance with Sections 370-71(a) and 370-71(b) as applicable. Where any horizontal dimension exceeds 6 ft (1.83 m), all conductors shall be cabled or racked up in an approved manner.

Exception: Where Section 370-71(b) applies, each row or column of ducts on one wall of the enclosure shall be calculated individually, and the single row or column that provides the maximum distance shall be used.

370-55. Access to Manholes.

(a) Dimensions. Rectangular access openings shall not be less than 26 in. × 22 in. (659 mm × 557 mm). Round access openings in a manhole shall not be less than 26 in. (659 mm) in diameter.

Exception: A manhole that has a fixed ladder that does not obstruct the opening, or that contains only one or more of the following shall be permitted to reduce the minimum cover diameter to 2 ft (608 mm):

(a) Optical fiber cables as covered in Article 770
(b) Power-limited fire alarm circuits supplied in accordance with Section 760-41
(c) Class 2 or Class 3 remote-control and signaling circuits, or both, supplied in accordance with Section 725-41

(b) Obstructions. Manhole openings shall be free of protrusions that could injure personnel or prevent ready egress.

(c) Location. Manhole openings for personnel shall be located where they are not directly above electric equipment or conductors in the enclosure. Where this is not practicable, either a protective barrier or a fixed ladder shall be provided.

(d) Covers. Covers shall be over 100 lb (45.4 kg) or otherwise designed to require the use of tools to open. They shall be designed or restrained so they cannot fall into the manhole or protrude sufficiently to contact electrical conductors or equipment within the manhole.

(e) Marking. Manhole covers shall have an identifying mark or logo that prominently indicates their function, such as "electric."

370-56. Access to Vaults and Tunnels.

(a) Location. Access openings for personnel shall be located where they are not directly above electric equipment or conductors in the enclosure. Other openings shall be permitted over equipment to facilitate installation, maintenance, or replacement of equipment.

(b) Locks. In addition to compliance with the requirements of Section 110-34(c), if applicable, access openings for personnel shall be arranged so that a person on the inside can exit when the access door is locked from the outside, or in the case of normally locking by padlock, the locking arrangement shall be such that the padlock can be closed on the locking system to prevent locking from the outside.

370-57. Ventilation. Where manholes, tunnels, and vaults have communicating openings into enclosed areas used by the public, ventilation to open air shall be provided wherever practicable.

370-58. Guarding. Where conductors or equipment, or both, could be contacted by objects falling or being pushed through a ventilating grating, both conductors and live parts shall be protected in accordance with the requirements of Sections 110-27(a)(2) or 110-31(a)(1), depending on the voltage.

370-59. Fixed Ladders. Fixed ladders shall be corrosion resistant.

E. Pull and Junction Boxes for Use on Systems Over 600 Volts, Nominal

370-70. General. Where pull and junction boxes are used on systems over 600 volts, the installation shall comply with the provisions of Part E and also the following general provisions of this article:

(1) In Part A, Sections 370-2, 370-3, and 370-4
(2) In Part B, Sections 370-15; 370-17; 370-18; 370-20; 370-23(a), (b), or (g); and 370-29
(3) In Part C, Sections 370-40(a) and (c) and 370-41

370-71. Size of Pull and Junction Boxes. Pull and junction boxes shall provide adequate space and dimensions for the installation of conductors, and they shall comply with the specific requirements of this section.

Exception: Terminal housings supplied with motors shall comply with the provisions of Section 430-12.

(a) For Straight Pulls. The length of the box shall not be less than 48 times the outside diameter, over sheath, of the largest shielded or lead-covered conductor or cable entering the box. The length shall not be less than 32 times the outside diameter of the largest nonshielded conductor or cable.

(b) For Angle or U Pulls.

(1) The distance between each cable or conductor entry inside the box and the opposite wall of the box shall not be less than 36 times the outside diameter, over sheath, of the largest cable or conductor. This distance shall be increased for additional entries by the amount of the sum of the outside diameters, over sheath, of all other cables or conductor entries through the same wall of the box.

Exception No. 1: Where a conductor or cable entry is in the wall of a box opposite a removable cover, the distance from that wall to the cover shall be permitted to be not less than the bending radius for the conductors as provided in Section 300-34.

Exception No. 2: Where cables are nonshielded and not lead covered, the distance of 36 times the outside diameter shall be permitted to be reduced to 24 times the outside diameter.

(2) The distance between a cable or conductor entry and its exit from the box shall not be less than 36 times the outside diameter, over sheath, of that cable or conductor.

Exception: Where cables are nonshielded and not lead covered, the distance of 36 times the outside diameter shall be permitted to be reduced to 24 times the outside diameter.

(c) Removable Sides. One or more sides of any pull box shall be removable.

370-72. Construction and Installation Requirements.

(a) Corrosion Protection. Boxes shall be made of material inherently resistant to corrosion or shall be suitably protected, both internally and externally, by enameling, galvanizing, plating, or other means.

(b) Passing Through Partitions. Suitable bushings, shields, or fittings having smooth, rounded edges shall be provided where conductors or cables pass through partitions and at other locations where necessary.

(c) Complete Enclosure. Boxes shall provide a complete enclosure for the contained conductors or cables.

(d) Wiring Is Accessible. Boxes shall be installed so that the wiring is accessible without removing any part of the building. Working space shall be provided in accordance with Section 110-34.

(e) Suitable Covers. Boxes shall be closed by suitable covers securely fastened in place. Underground box covers that weigh over 100 lb (45.4 kg) shall be considered meeting this requirement. Covers for boxes shall be permanently marked "DANGER — HIGH VOLTAGE — KEEP OUT." The marking shall be on the outside of the box cover and shall be readily visible. Letters shall be block type and at least ½ in. (12.7 mm) in height.

(f) Suitable for Expected Handling. Boxes and their covers shall be capable of withstanding the handling to which they may likely be subjected.

Article 373 — Cabinets, Cutout Boxes, and Meter Socket Enclosures

Contents

373-1. Scope. This article covers the installation and construction specifications of cabinets, cutout boxes, and meter socket enclosures.

See the definitions of *cabinet* and *cutout box* in Article 100. Cabinets and cutout boxes are designed with a swinging door(s) to enclose potential transformers, current transformers, switches, overcurrent devices, meters, or control equipment. Some cabinets for circuit breaker panelboards or load centers do not have doors, as permitted by Section 240-30(a)(2). Cabinets and cutout boxes are required to be of sufficient size to accommodate all devices and conductors without overcrowding or jamming. This condition can be prevented by the use of auxiliary gutters (Article 374).

The serving electric utility often has equipment specifications or service requirements beyond the *Code* for meter sockets, metering cabinets, and metering compartments within switchgear, switchboards, and panelboards. Consultation with the local electric utility on these requirements will help identify suitable equipment for the installation. One organization, the Electric Utility Service Equipment Requirements Committee (EUSERC), promotes uniform electric utility metering service requirements

for these enclosures that meet the requirements of the *Code* and the serving utility. Many electric equipment manufacturers identify their equipment that meets the EUSERC metering space requirements.

A. Installation

373-2. Damp, Wet, or Hazardous (Classified) Locations.

(a) Damp and Wet Locations. In damp or wet locations, surface-type enclosures within the scope of this article shall be placed or equipped so as to prevent moisture or water from entering and accumulating within the cabinet or cutout box, and shall be mounted so there is at least ¼-in. (6.35-mm) airspace between the enclosure and the wall or other supporting surface. Enclosures installed in wet locations shall be weatherproof.

Exception: Nonmetallic enclosures shall be permitted to be installed without the airspace on a concrete, masonry, tile, or similar surface.

The exception to Section 373-2(a) waives the airspace requirement behind nonmetallic cabinets and cutout boxes mounted on specific surfaces.

FPN: For protection against corrosion, see Section 300-6.

(b) Hazardous (Classified) Locations. Installations in hazardous (classified) locations shall conform to Articles 500 through 517.

373-3. Position in Wall. In walls of concrete, tile, or other noncombustible material, cabinets shall be installed so that the front edge of the cabinet will not set back of the finished surface more than ¼ in. (6.35 mm). In walls constructed of wood or other combustible material, cabinets shall be flush with the finished surface or project therefrom.

373-4. Unused Openings. Unused openings in enclosures within the scope of this article shall be effectively closed to afford protection substantially equivalent to that of the enclosures within the scope of this article. Where metal plugs or plates are used with nonmetallic cabinets or cutout boxes, they shall be recessed at least ¼ in. (6.35 mm) from the outer surface.

373-5. Cabinets, Cutout Boxes, and Meter Socket Enclosures. Conductors entering enclosures within the scope of this article shall be protected from abrasion and shall comply with (a) through (c).

See Section 300-4(f) on insulated fittings.

(a) Openings to Be Closed. Openings through which conductors enter shall be adequately closed.

(b) Metal Cabinets, Cutout Boxes, and Meter Socket Enclosures. Where metal enclosures within the scope of this article are installed with open wiring or concealed knob-and-tube wiring, conductors shall enter through insulating bushings or, in dry locations, through flexible tubing extending from the last insulating support and firmly secured to the enclosure.

(c) Cables. Where cable is used, each cable shall be secured to the cabinet, cutout box, or meter socket enclosure.

Section 373-5(c) prohibits the installation of several cables bunched together and run through a knockout or chase nipple. Individual cable clamps or connectors are required to be used with only one cable per clamp or connector, unless the clamp or connector is approved for more than a single cable.

Exception: Cables with entirely nonmetallic sheaths shall be permitted to enter the top of a surface-mounted enclosure through one or more nonflexible raceways not less than 18 in. (457 mm) or more than 10 ft (3.05 m) in length, provided all the following conditions are met.

(a) Each cable is fastened within 12 in. (305 mm), measured along the sheath, of the outer end of the raceway.
(b) The raceway extends directly above the enclosure and does not penetrate a structural ceiling.
(c) A fitting is provided on each end of the raceway to protect the cable(s) from abrasion and the fittings remain accessible after installation.
(d) The raceway is sealed or plugged at the outer end using approved means so as to prevent access to the enclosure through the raceway.
(e) The cable sheath is continuous through the raceway and extends into the enclosure beyond the fitting not less than ¼ in. (6.35 mm).
(f) The raceway is fastened at its outer end and at other points in accordance with the applicable article.
(g) Where installed as conduit or tubing, the allowable cable fill does not exceed that permitted for complete conduit or tubing systems by Table 1 of Chapter 9 of this Code, and all applicable notes thereto.

FPN: See Table 1 in Chapter 9, including Note 9, for allowable cable fill in circular raceways. See Section 310-15(b)(2)(a) for required ampacity reductions for multiple cables installed in a common raceway.

The exception, which is new for the 1999 *NEC*, spells out the requirements that allow nonmetallic sheathed and UF cables to enter the top of a surface-mounted enclosure through a nonflexible raceway nipple. Nipples are permitted to be 18 in. to 10 ft in length. However, if the nipple length exceeds 24 in., ampacity adjustment factors as specified in Section 310-15(b)(2) apply.

373-6. Deflection of Conductors. Conductors at terminals or conductors entering or leaving cabinets or cutout boxes and the like shall comply with (a) through (c).

Exception: Wire-bending space in enclosures for motor controllers with provisions for one or two wires per terminal shall comply with Section 430-10(b).

(a) Width of Wiring Gutters. Conductors shall not be deflected within a cabinet or cutout box unless a gutter having a width in accordance with Table 373-6(a) is provided. Conductors in parallel in accordance with Section 310-4 shall be judged on the basis of the number of conductors in parallel.

(b) Wire-Bending Space at Terminals. Wire-bending space at each terminal shall be provided in accordance with (1) or (2).

(1) Table 373-6(a) shall apply where the conductor does not enter or leave the enclosure through the wall opposite its terminal.

(2) Table 373-6(b) shall apply where the conductor does enter or leave the enclosure through the wall opposite its terminal.

Section 373-6(b)(2) and Table 373-6(b) provide the requirements for wire-bending space where straight-in wiring or offset (double bends) is employed at terminals. Section 373-6(b)(1) is for use where only 90 degree bends are involved.

The notes to Table 373-6(b) are intended to permit a reduction in required bending space for removable and lay-in wire terminals. The removable terminal wire connectors can be either compression type or set screw type. However, connectors are required to be of the type intended for a single conductor (single barrel). Removable connectors designed for multiple wires are not permitted to have a reduction in bending space.

To make wiring easier, a terminal may be placed on the stripped end of the conductor, which has been cut to the proper length. The terminal is crimped or lightly torqued on the wire as intended. The wire is bent and routed to facilitate mounting onto the stud or landing pad for proper connection. All mechanical screws, bolts, and nuts involved should then be torqued to the proper value. See the commentary following the fine print note under Section 110-14 regarding tightening torques.

Note that in accordance with the notes to Tables 373-6(a) and 373-6(b), when using Table 373-6(a), bending space is measured in the direction in which the wire leaves the terminal, and when using Table 373-6(b), it is measured in a direction perpendicular to the enclosure wall.

A lay-in-type terminal is a pressure wire connector in which part of the connector is removable or swings away so that the stripped end of the conductor can be laid into the fixed portion of the connector. The removable or swing-away portion is then put back in place and the connector tightened down on the conductor.

In conjunction with the following listing, Figure 373.1 shows the application of the rules of Sections 373-6(b)(1) and 373-6(b)(2) and Tables 373-6(a) and 373-6(b) to the wiring for a lay-in-type terminal.

T_1, Section 373-6(b)(2): Table 373-6(b) applies for conductors *M*.

T_2, Section 373-6(b)(2): Table 373-6(b) applies for conductors BR_2 unless Exception No. 2 of Section 373-6(b)(1) applies. This exception allows Table 373-6(a) to apply to T_2 as long as BR_2 enters or leaves the enclosure where gutter G_2 joins gutter G_1, and gutter G_1 has a width conforming to Table 373-6(b) for BR_2.

T_3, Section 373-6(b)(2): Table 373-6(b) applies for conductors BR_3.

T_4, Section 373-6(b)(1): Table 373-6(a) applies for conductor *N*.

G_1, Section 373-6(a): Table 373-6(a) applies for conductors *M*. Table 373-6(b) applies for conductors BR_2 where T_2 does not comply with Table 373-6(b).

G_2, Section 373-6(a): Table 373-6(a) applies for conductors BR_2.

G_3, Section 373-6(a): Table 373-6(a) applies for conductors BR_3.

G_4, Section 373-6(a): Table 373-6(a) applies for conductor *N*.

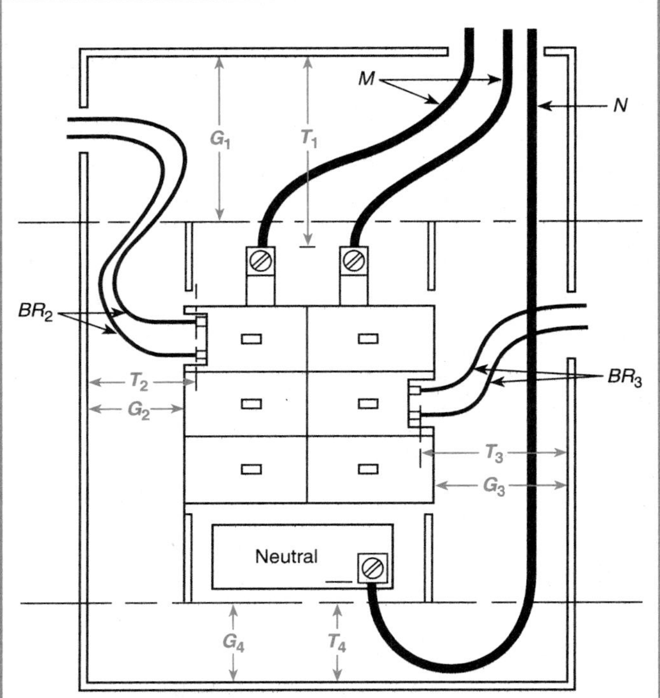

Figure 373.1 *Wiring for a lay-in type terminal to which the accompanying list of rules apply.*

Figure 373.2 is an illustration of the conditions under which Section 373-6(b)(1), Exception No. 2, is applicable. See also Figure 430.1, which shows an example of wire-bending space in enclosures for motor controllers.

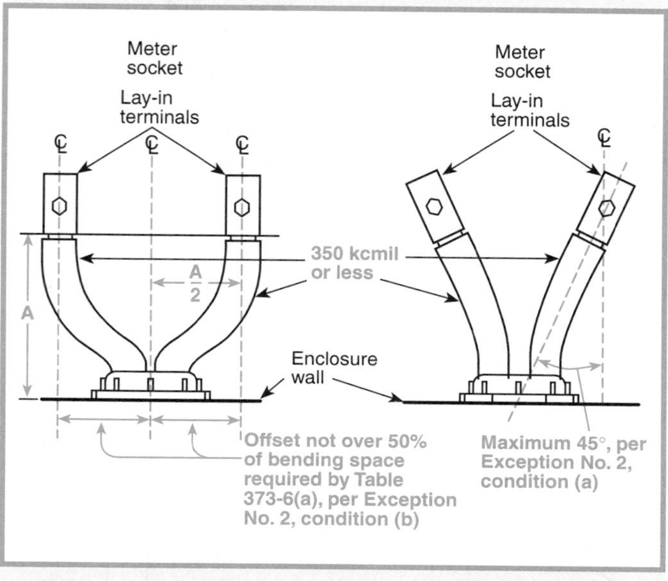

Figure 373.2 *Conditions under which Section 373-6(b)(1), Exception No. 2, is applicable.*

Exception No. 1: Where the distance between the wall and its terminal is in accordance with Table 373-6(a), a conductor shall be permitted to enter or leave an enclosure through the wall opposite its terminal provided the conductor enters or leaves the enclosure where the gutter joins an adjacent gutter that has a width that conforms to Table 373-6(b) for that conductor.

Exception No. 2: A conductor not larger than 350 kcmil shall be permitted to enter or leave an enclosure containing only a meter socket(s) through the wall opposite its terminal, provided the distance between the terminal and the opposite wall is not less than that specified in Table 373-6(a) and the terminal is a lay-in type where the terminal is either of the following:

(a) Directed toward the opening in the enclosure and is within a 45 degree angle of directly facing the enclosure wall

(b) Directly facing the enclosure wall and offset not greater than 50 percent of the bending space specified in Table 373-6(a)

FPN: *Offset* is the distance measured along the enclosure wall from the axis of the centerline of the terminal to a line passing through the center of the opening in the enclosure.

Table 373-6(a). Minimum Wire-Bending Space at Terminals and Minimum Width of Wiring Gutters in Inches

Wire Size (AWG or kcmil)	Wires per Terminal				
	1	2	3	4	5
14–10	Not specified	—	—	—	—
8–6	1½	—	—	—	—
4–3	2	—	—	—	—
2	2½	—	—	—	—
1	3	—	—	—	—
1/0–2/0	3½	5	7	—	—
3/0–4/0	4	6	8	—	—
250	4½	6	8	10	—
300–350	5	8	10	12	—
400–500	6	8	10	12	14
600–700	8	10	12	14	16
750–900	8	12	14	16	18
1000–1250	10	—	—	—	—
1500–2000	12	—	—	—	—

Notes:
1. For SI units, 1 in. = 25.4 mm.
2. Bending space at terminals shall be measured in a straight line from the end of the lug or wire connector (in the direction that the wire leaves the terminal) to the wall, barrier, or obstruction.

(c) Conductors No. 4 or Larger. Installation shall comply with Section 300-4(f).

Where No. 4 or larger ungrounded conductors enter a cabinet, box enclosure, meter socket, or raceway, the conductors shall be protected by a substantial fitting providing a smoothly rounded insulating surface to protect the conductors from abrasion during and after installation. This requirement was moved from Section 373-6(c) to Section 300-4(f) and therefore applies generally to all wiring methods and all enclosures. See also Sections 345-15, 346-15, and 347-12 relating to bushings.

Metal conduit bushings or fittings provided with insulated sleeves or linings are commonly used. It is also possible to use a separate insulating sleeve or lining to separate the conductors from the raceway fitting.

Listed insulating bushings provided separately or as part of a fitting are colored black or brown if they are suitable for a temperature of 150°C and any other color for 90°C, unless specifically marked for a higher temperature. Figure 373.3 shows an insulated thermoplastic or fiber bushing that is used to protect the conductors from chafing against a metal conduit fitting. Note the use of a double locknut.

373-7. Space in Enclosures. Cabinets and cutout boxes shall have sufficient space to accommodate all conductors installed in them without crowding.

373-8. Enclosures for Switches or Overcurrent Devices. Enclosures for switches or overcurrent devices shall not be

Table 373-6(b). Minimum Wire-Bending Space at Terminals in Inches

Wire Size (AWG or kcmil)	Wires per Termial							
	1		2		3		4 or More	
14–10	Not specified		—		—		—	
8	1½		—		—		—	
6	2		—		—		—	
4	3		—		—		—	
3	3		—		—		—	
2	3½		—		—		—	
1	4½		—		—		—	
1/0	5½		5½		7		—	
2/0	6		6		7½		—	
3/0	6½	(½)	6½	(½)	8		—	
4/0	7	(1)	7½	(1½)	8½	(½)	—	
250	8½	(2)	8½	(2)	9	(1)	10	
300	10	(3)	10	(2)	11	(1)	12	
350	12	(3)	12	(3)	13	(3)	14	(2)
400	13	(3)	13	(3)	14	(3)	15	(3)
500	14	(3)	14	(3)	15	(3)	16	(3)
600	15	(3)	16	(3)	18	(3)	19	(3)
700	16	(3)	18	(3)	20	(3)	22	(3)
750	17	(3)	19	(3)	22	(3)	24	(3)
800	18		20		22		24	
900	19		22		24		24	
1000	20		—		—		—	
1250	22		—		—		—	
1500	24		—		—		—	
1750	24		—		—		—	
2000	24		—		—		—	

Notes:
1. For SI units, 1 in. = 25.4 mm.
2. Bending space at terminals shall be measured in a straight line from the end of the lug or wire connector in a direction perpendicular to the enclosure wall.
3. For removable and lay-in wire terminals intended for only one wire, bending space shall be permitted to be reduced by the number of inches shown in parentheses.

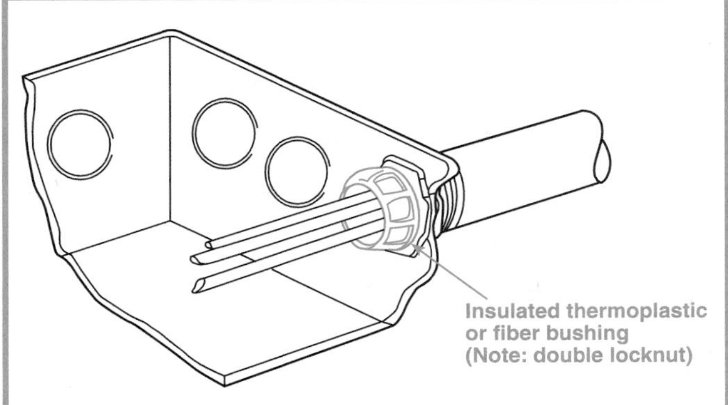

Figure 373.3 *An insulating bushing used to protect conductors from chafing against a metal conduit fitting.*

used as junction boxes, auxiliary gutters, or raceways for conductors feeding through or tapping off to other switches or overcurrent devices, unless adequate space for this purpose is provided. The conductors shall not fill the wiring space at any cross section to more than 40 percent of the cross-sectional area of the space, and the conductors, splices, and taps shall not fill the wiring space at any cross section to more than 75 percent of the cross-sectional area of that space.

Most enclosures are intended to accommodate only those conductors that will be connected to terminals for switches or overcurrent devices within the enclosures themselves. Where adequate space is provided for additional conductors, such as control circuits, the total conductor fill in the enclosure is not permitted to exceed 40 percent of the cross section of the wiring space in the enclosure, and no more than 75 percent if splices or taps are necessary.

Example

If an enclosure has a wiring space of 4 in. by 3 in., the cross-sectional area is 12 in.^{2} Thus, the total conductor fill (see Chapter 9, Table 5 for dimensions of conductors) at any cross section cannot exceed 4.8 in.^{2} (40 percent of 12 in.^{2}), and the maximum space for conductors and splices or taps at any cross section cannot exceed 9 in.^{2} (75 percent of 12 in.^{2}).

In general, the best way to avoid overcrowding enclosures is to use properly sized auxiliary gutters (Sections 374-5 and 374-8) or junction boxes (Sections 370-16 and 370-28). See Section 430-10 and commentary for wiring space in enclosures for motor controllers and disconnecting means. See also Section 710-59 for installations over 600 volts.

373-9. Side or Back Wiring Spaces or Gutters. Cabinets and cutout boxes shall be provided with back wiring spaces, gutters, or wiring compartments as required by Section 373-11(c) and (d).

B. Construction Specifications

373-10. Material. Cabinets, cutout boxes, and meter socket enclosures shall comply with (a) through (c).

(a) Metal Cabinets and Cutout Boxes. Metal enclosures within the scope of this article shall be protected both inside and outside against corrosion.

FPN: For information on protection against corrosion, see Section 300-6.

(b) Strength. The design and construction of enclosures within the scope of this article shall be such as to secure ample strength and rigidity. If constructed of sheet steel, the metal thickness shall not be less than 0.053 in. (1.35 mm) uncoated.

(c) Nonmetallic Cabinets. Nonmetallic cabinets shall be listed or they shall be submitted for approval prior to installation.

373-11. Spacing. The spacing within cabinets and cutout boxes shall comply with (a) through (d).

(a) General. Spacing within cabinets and cutout boxes shall be sufficient to provide ample room for the distribution of wires and cables placed in them, and for a separation between metal parts of devices and apparatus mounted within them as follows.

(1) Base. Other than at points of support, there shall be an airspace of at least 1/16 in. (1.59 mm) between the base of the device and the wall of any metal cabinet or cutout box in which the device is mounted.

(2) Doors. There shall be an airspace of at least 1 in. (25.4 mm) between any live metal part, including live metal parts of enclosed fuses, and the door.

Exception: Where the door is lined with an approved insulating material or is of a thickness of metal not less than 0.093 in. (2.36 mm) uncoated, the airspace shall not be less than 1/2 in. (12.7 mm).

(3) Live Parts. There shall be an airspace of at least 1/2 in. (12.7 mm) between the walls, back, gutter partition, if of metal, or door of any cabinet or cutout box and the nearest exposed current-carrying part of devices mounted within the cabinet where the voltage does not exceed 250. This spacing shall be increased to at least 1 in. (25.4 mm) for voltages of 251 to 600, nominal.

Exception: Where the conditions in Section 373-11(a)(2), Exception, are met, the airspace for nominal voltages from 251 to 600 shall be permitted to be not less than 1/2 in. (12.7 mm).

(b) Switch Clearance. Cabinets and cutout boxes shall be deep enough to allow the closing of the doors when 30-ampere branch-circuit panelboard switches are in any position, when combination cutout switches are in any position, or when other single-throw switches are opened as far as their construction will permit.

(c) Wiring Space. Cabinets and cutout boxes that contain devices or apparatus connected within the cabinet or box to more than eight conductors, including those of branch circuits, meter loops, feeder circuits, power circuits, and similar circuits, but not including the supply circuit or a continuation thereof, shall have back-wiring spaces or one or more side-wiring spaces, side gutters, or wiring compartments.

(d) Wiring Space — Enclosure. Side-wiring spaces, side gutters, or side-wiring compartments of cabinets and cutout boxes shall be made tight enclosures by means of covers, barriers, or partitions extending from the bases of the devices contained in the cabinet, to the door, frame, or sides of the cabinet.

Exception: Side-wiring spaces, side gutters, and side-wiring compartments of cabinets shall not be required to be made tight enclosures where those side spaces contain only conductors that enter the cabinet directly opposite to the devices where they terminate.

Partially enclosed back-wiring spaces shall be provided with covers to complete the enclosure. Wiring spaces that are required by (c), and that are exposed when doors are open, shall be provided with covers to complete the enclosure. Where adequate space is provided for feed-through conductors and for splices as required in Section 373-8, additional barriers shall not be required.

Article 374 — Auxiliary Gutters

Contents

374-1. Use. Auxiliary gutters shall be permitted to supplement wiring spaces at meter centers, distribution centers, switchboards, and similar points of wiring systems and may enclose conductors or busbars but shall not be used to enclose switches, overcurrent devices, appliances, or other similar equipment.

Auxiliary gutter sections and associated fittings are identical to those of wireways, and if listed by UL, bear the UL listing mark "Listed Wireway" or "Auxiliary Gutter." They differ only in their intended use. See the commentary following Section 362-1 for comparative discussion. Gutters (and wireways) are required to be constructed and installed to ensure

adequate electrical and mechanical continuity of the complete system per Section 250-118(14).

Auxiliary gutters installed in wet locations are required to be suitable for such locations. See Section 620-35 for less restrictive requirements where auxiliary gutters are used for elevators, dumbwaiters, escalators, and moving walks.

374-2. Extension Beyond Equipment. An auxiliary gutter shall not extend a greater distance than 30 ft (9.14 m) beyond the equipment that it supplements.

Exception: As permitted in Section 620-35 for elevators, an auxiliary gutter shall be permitted to extend a distance greater than 30 ft (9.14 m) beyond the equipment that it supplements.

FPN: For wireways, see Article 362. For busways, see Article 364.

374-3. Supports.

(a) Sheet Metal Auxiliary Gutters. Sheet metal auxiliary gutters shall be supported throughout their entire length at intervals not exceeding 5 ft (1.52 m).

(b) Nonmetallic Auxiliary Gutters. Nonmetallic auxiliary gutters shall be supported at intervals not to exceed 3 ft (914 mm) and at each end or joint, unless listed for other support intervals. In no case shall the distance between supports exceed 10 ft (3.05 m).

374-4. Covers. Covers shall be securely fastened to the gutter.

374-5. Number of Conductors.

(a) Sheet Metal Auxiliary Gutters. The number of conductors permitted in a sheet metal auxiliary gutter shall be in accordance with (1) through (4).

(1) Sheet metal auxiliary gutters shall not contain more than 30 current-carrying conductors at any cross section. Conductors for signaling circuits or controller conductors between a motor and its starter and used only for starting duty shall not be considered as current-carrying conductors.

(2) As provided in Section 620-35 for elevators, the 30 conductor limitation shall not apply.

(3) The sum of the cross-sectional areas of all contained conductors at any cross-section of a sheet metal auxiliary gutter shall not exceed 20 percent of the interior cross-sectional area of the sheet metal auxiliary gutter.

(4) Where the 20 percent fill specified in (3) is not exceeded and the derating factors specified in Section 310-15(b)(2)(a) is applied, the number of current-carrying conductors shall not be limited.

(b) Nonmetallic Auxiliary Gutters. The sum of cross-sectional areas of all contained conductors at any cross section of the nonmetallic auxiliary gutter shall not exceed 20 percent of the interior cross-section area of the nonmetallic auxiliary gutter.

Section 374-5 calls out the requirements for metal and nonmetallic auxiliary gutters. A common requirement for both types of auxiliary gutters is that all of the contained conductors are not permitted to exceed 20 percent fill of the interior cross-sectional area of the gutter. The dimensions of insulated conductors, found in Tables 5 and 5A of Chapter 9, may be used to compute the size of auxiliary gutters.

Where sheet metal auxiliary gutters contain 30 or fewer current-carrying conductors, the correction factors in Section 310-15(b)(2) do not apply. However, if more than 30 conductors are installed in a sheet metal auxiliary gutter, the ampacity adjustment factors of Section 310-15(b)(2) apply and there is no limit on the number of current-carrying conductors up to the 20 percent fill.

The requirements for nonmetallic auxiliary gutters limits the cross-sectional area of all conductors to 20 percent. There is no 30-conductor allowance. The derating factors specified in Section 310-15(b)(2) must be applied.

See the example for calculating the size of a wireway in the commentary following Section 362-5. This calculation method is also applicable to auxiliary gutters.

No limit is placed on the size of conductors that may be installed in an auxiliary gutter; however, see Section 374-6 for ampacity limitations of bare copper or aluminum busbars enclosed in gutters.

374-6. Ampacity of Conductors.

(a) Sheet Metal Auxiliary Gutters. Where the number of current-carrying conductors contained in the sheet metal auxiliary gutter is 30 or less, the correction factors specified in Section 310-15(b)(2)(a) shall not apply. The current carried continuously in bare copper bars in sheet metal auxiliary gutters shall not exceed 1000 amperes/in.^{2} (645 mm^2) of cross section of the conductor. For aluminum bars, the current carried continuously shall not exceed 700 amperes/in.^{2} (645 mm^2) of cross section of the conductor.

(b) Nonmetallic Auxiliary Gutters. The derating factors specified in Section 310-15(b)(2)(a) shall be applicable to the current-carrying conductors in the nonmetallic auxiliary gutter.

374-7. Clearance of Bare Live Parts. Bare conductors shall be securely and rigidly supported so that the minimum clearance between bare current-carrying metal parts of different potential mounted on the same surface will not be less than 2 in. (50.8 mm), nor less than 1 in. (25.4 mm) for parts that are held free in the air. A clearance not less than 1 in. (25.4 mm) shall be secured between bare current-carrying metal parts and any metal surface. Adequate provisions shall be made for the expansion and contraction of busbars.

374-8. Splices and Taps. Splices and taps shall comply with (a) through (d).

(a) Within Gutters. Splices or taps shall be permitted within gutters where they are accessible by means of removable covers or doors. The conductors, including splices and taps, shall not fill the gutter to more than 75 percent of its area.

(b) Bare Conductors. Taps from bare conductors shall leave the gutter opposite their terminal connections, and conductors shall not be brought in contact with uninsulated current-carrying parts of different potential.

(c) Suitably Identified. All taps shall be suitably identified at the gutter as to the circuit or equipment that they supply.

(d) Overcurrent Protection. Tap connections from conductors in auxiliary gutters shall be provided with overcurrent protection as required in Section 240-21.

Suitable bushings, shields, and so on, must be provided where conductors pass around bends or between gutters and cabinets and other locations to prevent abrasion of the conductor insulation.

Per Sections 374-8(c) and (d), all taps from gutters are required to be identified (as to circuits or equipment) and be protected with overcurrent devices per Section 240-21.

374-9. Construction and Installation. Auxiliary gutters shall comply with (a) through (f).

(a) Electrical and Mechanical Continuity. Gutters shall be constructed and installed so that adequate electrical and mechanical continuity of the complete system will be secured.

(b) Substantial Construction. Gutters shall be of substantial construction and shall provide a complete enclosure for the contained conductors. All surfaces, both interior and exterior, shall be suitably protected from corrosion. Corner joints shall be made tight, and where the assembly is held together by rivets, bolts, or screws, such fasteners shall be spaced not more than 12 in. (305 mm) apart.

(c) Smooth Rounded Edges. Suitable bushings, shields, or fittings having smooth, rounded edges shall be provided where conductors pass between gutters, through partitions, around bends, between gutters and cabinets or junction boxes, and at other locations where necessary to prevent abrasion of the insulation of the conductors.

(d) Deflected Insulated Conductors. Where insulated conductors are deflected within an auxiliary gutter, either at the ends or where conduits, fittings, or other raceways or cables enter or leave the gutter, or where the direction of the gutter is deflected greater than 30 degrees, dimensions corresponding to Section 373-6 shall apply.

Also, conductors are required to be shaped or formed in a permanent manner so that they are not in contact with bare busbars within the gutter.

(e) Indoor and Outdoor Use.

(1) Sheet Metal Auxiliary Gutters. Sheet metal auxiliary gutters installed in wet locations shall be suitable for such locations.

(2) Nonmetallic Auxiliary Gutters.

(a) Nonmetallic auxiliary gutters installed outdoors shall

(1) Be listed and marked as suitable for exposure to sunlight;
(2) Be listed and marked as suitable for use in wet locations;
(3) Be listed for the maximum ambient temperature of the installation, and marked for the installed conductor insulation temperature rating; and
(4) Have expansion fittings installed where the expected length change due to expansion and contraction due to temperature change is more than 0.25 in. (6.35 mm).

(b) Nonmetallic auxiliary gutters installed indoors shall

(1) Be listed for the maximum ambient temperature of the installation and marked for the installed conductor insulation temperature rating, and
(2) Have expansion fittings installed where expected length change, due to expansion and contraction due to temperature change, is more than ¼ in. (6.35 mm).

This section provides requirements for both indoor and outdoor installations. Nonmetallic gutters must have expansion fittings where temperature changes are expected to change the length more than ¼ in. See the fine print note following Section 362-23 regarding expansion characteristic of PVC rigid nonmetallic conduit and PVC nonmetallic wireway.

FPN: Extreme cold may cause nonmetallic auxiliary gutter to become brittle and, therefore, more susceptible to damage from physical contact.

(f) Grounding. Grounding shall be in accordance with the provisions of Article 250.

Article 380 — Switches

Contents

A. Installation

380-1. Scope. The provisions of this article shall apply to all switches, switching devices, and circuit breakers where used as switches.

380-2. Switch Connections.

(a) Three-Way and Four-Way Switches. Three-way and four-way switches shall be wired so that all switching is done only in the ungrounded circuit conductor. Where in metal raceways or metal-armored cables, wiring between switches and outlets shall be in accordance with Section 300-20(a).

Exception: Switch loops shall not require a grounded conductor.

The *NEC* does not specifically prohibit the use of *two* 2-conductor nonmetallic-sheathed cables instead of a *single* 3-conductor cable for wiring 3-way and 4-way switches. However, the use of *two* 2-conductor cables could easily result in a violation of Section 300-20 if metal boxes are used and the cables enter the box through separate knockouts. Also, use of the same clamp or section of a clamp for both cables would, in most cases, be in violation of Section 110-3(b), because clamps have been tested for only one cable per clamp or section of clamp.

The grounded conductor is not needed in a switch loop *[see Section 300-20(a)]* because the ungrounded conductor both enters and leaves the enclosure in the same cable or raceway, thus avoiding inductive heating.

(b) Grounded Conductors. Switches or circuit breakers shall not disconnect the grounded conductor of a circuit.

Exception: A switch or circuit breaker shall be permitted to disconnect a grounded circuit conductor where all circuit conductors are disconnected simultaneously, or where the device is arranged so that the grounded conductor cannot be disconnected until all the ungrounded conductors of the circuit have been disconnected.

380-3. Enclosure.

(a) General. Switches and circuit breakers shall be of the externally operable type mounted in an enclosure listed for the intended use. The minimum wire-bending space at terminals and minimum gutter space provided in switch enclosures shall be as required in Section 373-6.

Exception No. 1: Pendant- and surface-type snap switches and knife switches mounted on an open-face switchboard or panelboard shall be permitted without enclosures.

Exception No. 2: Switches and circuit breakers installed in accordance with Sections 110-27(a)(1), (2), (3), or (4) shall be permitted without enclosures.

Exception No. 2 to Section 380-3 was added to the *Code* to recognize the variety of means allowed by Section 110-27(a) for guarding of live parts. Switches and circuit breakers guarded by these means are permitted without enclosures.

(b) Used as a Raceway. Enclosures shall not be used as junction boxes, auxiliary gutters, or raceways for conductors feeding through or tapping off to other switches or overcurrent devices, unless the enclosure complies with Section 373-8.

380-4. Wet Locations. A switch or circuit breaker in a wet location or outside of a building shall be enclosed in a weatherproof enclosure or cabinet that shall comply with

Section 373-2(a). Switches shall not be installed within wet locations in tub or shower spaces unless installed as part of a listed tub or shower assembly.

380-5. Time Switches, Flashers, and Similar Devices. Time switches, flashers, and similar devices shall be of the enclosed type or shall be mounted in cabinets or boxes or equipment enclosures. Energized parts shall be barriered to prevent operator exposure when making manual adjustments or switching.

Exception: Devices mounted so they are accessible only to qualified persons shall be permitted without barriers, provided they are located within an enclosure such that any energized parts within 6 in. (152 mm) of the manual adjustment or switch are covered by suitable barriers.

380-6. Position and Connection of Switches.

(a) Single-Throw Knife Switches. Single-throw knife switches shall be placed so that gravity will not tend to close them. Single-throw knife switches, approved for use in the inverted position, shall be provided with a locking device that will ensure that the blades remain in the open position when so set.

(b) Double-Throw Knife Switches. Double-throw knife switches shall be permitted to be mounted so that the throw will be either vertical or horizontal. Where the throw is vertical, a locking device shall be provided to hold the blades in the open position when so set.

(c) Connection of Switches. Single-throw knife switches and switches with butt contacts shall be connected so that the blades are de-energized when the switch is in the open position. Single-throw knife switches, molded-case switches, switches with butt contacts, and circuit breakers used as switches shall be connected so that the terminals supplying the load are de-energized when the switch is in the open position.

Exception: The blades and terminals supplying the load of a switch shall be permitted to be energized when the switch is in the open position where the switch is connected to circuits or equipment inherently capable of providing a backfeed source of power. For such installations, a permanent sign shall be installed on the switch enclosure or immediately adjacent to open switches that reads:

WARNING — LOAD SIDE TERMINALS MAY BE ENERGIZED BY BACKFEED.

Typical backfeed sources are batteries, generators, and double-ended switchboard ties. These may cause the load side of the switch or circuit breaker to be energized when it is in the open position, a condition inherent to the circuitry.

380-7. Indicating. General-use and motor-circuit switches and circuit breakers, where mounted in an enclosure as described in Section 380-3, shall clearly indicate whether they are in the open (off) or closed (on) position.

Where these switch or circuit breaker handles are operated vertically rather than rotationally or horizontally, the up position of the handle shall be the (on) position.

Exception: Vertically-operated double-throw switches shall be permitted to be in the closed (on) position with the handle in either the up or down position.

380-8. Accessibility and Grouping.

(a) Location. All switches and circuit breakers used as switches shall be located so that they may be operated from a readily accessible place. They shall be installed so that the center of the grip of the operating handle of the switch or circuit breaker, when in its highest position, will not be more than 6 ft 7 in. (2.0 m) above the floor or working platform.

Exception No. 1: On busway installations, fused switches and circuit breakers shall be permitted to be located at the same level as the busway. Suitable means shall be provided to operate the handle of the device from the floor.

Exception No. 2: Switches installed adjacent to motors, appliances, or other equipment that they supply shall be permitted to be located higher than specified in the foregoing and to be accessible by portable means.

Exception No. 3: Hookstick operable isolating switches shall be permitted at greater heights.

(b) Voltage Between Adjacent Switches. Snap switches shall not be grouped or ganged in enclosures unless they can be arranged so that the voltage between adjacent switches does not exceed 300, or unless they are installed in enclosures equipped with permanently installed barriers between adjacent switches.

380-9. Provisions for Snap Switch Faceplates.

(a) Position. Snap switches mounted in boxes shall have faceplates installed so as to completely cover the opening and seat against the finished surface.

(b) Grounding. Snap switches, including dimmer switches, shall be effectively grounded and shall provide a means to ground metal faceplates, whether or not a metal faceplate is installed. Snap switches shall be considered effectively grounded if either of the following conditions are met.

(1) The switch is mounted with metal screws to a metal box or to a nonmetallic box with integral means for grounding devices.

(2) An equipment grounding conductor or equipment bonding jumper is connected to an equipment grounding termination of the snap switch.

Exception to (b): Where no grounding means exists within the snap-switch enclosure or where the wiring method does not include or provide an equipment ground, a snap switch without a grounding connection shall be permitted for replacement purposes only. A snap switch wired under the provisions of this exception and located within reach of conducting floors or other conducting surfaces shall be provided with a faceplate of nonconducting, noncombustible material.

Section 380-9 was revised for the 1999 *NEC* to require all snap switches and dimmers to ground faceplates. This provision is necessary whether a metal or nonconducting faceplate is used. The requirements in (1) or (2) of this section describe the provisions to satisfy the main requirement. Switch plates in existing installations attached to switches in boxes without an equipment grounding conductor are required to be made of insulating material.

(c) Construction. Metal faceplates shall be of ferrous metal not less than 0.030 in. (0.762 mm) in thickness or of nonferrous metal not less than 0.040 in. (1.016 mm) in thickness. Faceplates of insulating material shall be noncombustible and not less than 0.10 in. (2.54 mm) in thickness, but they shall be permitted to be less than 0.10 in. (2.54 mm) in thickness if formed or reinforced to provide adequate mechanical strength.

380-10. Mounting of Snap Switches.

(a) Surface-Type. Snap switches used with open wiring on insulators shall be mounted on insulating material that will separate the conductors at least ½ in. (12.7 mm) from the surface wired over.

(b) Box Mounted. Flush-type snap switches mounted in boxes that are set back of the wall surface as permitted in Section 370-20 shall be installed so that the extension plaster ears are seated against the surface of the wall. Flush-type snap switches mounted in boxes that are flush with the wall surface or project therefrom shall be installed so that the mounting yoke or strap of the switch is seated against the box.

Cooperation is necessary among the building trades (carpenters, drywall installers, plasterers, etc.) in order for electricians to properly set device boxes flush with the finish surface, thereby ensuring a secure seating of the switch yoke and permitting the maximum projection of switch handles through the installed switch plate.

380-11. Circuit Breakers as Switches. A hand-operable circuit breaker equipped with a lever or handle, or a power-operated circuit breaker capable of being opened by hand in the event of a power failure, shall be permitted to serve as a switch if it has the required number of poles.

Circuit breakers capable of being hand operated are required to clearly indicate whether they are in the open (off) or closed (on) position. See Section 380-7 for details on handle positions.

See Section 240-83(d) for marking (SWD) for circuit breakers used as switches for 120-volt and 277-volt fluorescent lighting circuits.

FPN: See the provisions contained in Sections 240-81 and 240-83.

380-12. Grounding of Enclosures. Metal enclosures for switches or circuit breakers shall be grounded as specified in Article 250. Where nonmetallic enclosures are used with metal raceways or metal-armored cables, provisions shall be made for grounding continuity.

Metal boxes for switches shall be effectively grounded. Nonmetallic boxes for switches shall be installed with a wiring method that provides or includes an equipment ground.

Figure 380.1 illustrates how the effective grounding of metal faceplate can be accomplished by connecting the equipment grounding conductor provided with the wiring method to a grounding terminal on a metal yoke or strap.

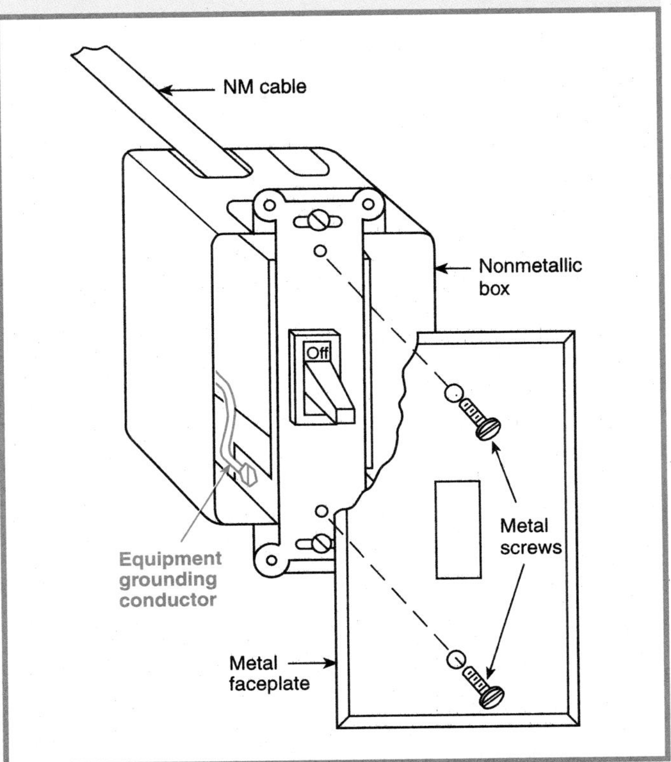

Figure 380.1 *Grounding of a metal faceplate accomplished by connecting the equipment grounding conductor to a grounding terminal on a metal yoke or strap.*

380-13. Knife Switches.

(a) Isolating Switches. Knife switches rated at over 1200 amperes at 250 volts or less, and at over 600 amperes at 251 to 600 volts, shall be used only as isolating switches and shall not be opened under load.

(b) To Interrupt Currents. To interrupt currents over 1200 amperes at 250 volts, nominal, or less, or over 600 amperes at 251 to 600 volts, nominal, a circuit breaker or a switch of special design listed for such purpose shall be used.

(c) General-Use Switches. Knife switches of ratings less than specified in (a) and (b) shall be considered general-use switches.

FPN: See definition of *general-use switch* in Article 100.

(d) Motor-Circuit Switches. Motor-circuit switches shall be permitted to be of the knife-switch type.

FPN: See definition of a *motor-circuit switch* in Article 100.

380-14. Rating and Use of Snap Switches. Snap switches shall be used within their ratings and as indicated in (a) through (d).

FPN No. 1: For switches on signs and outline lighting, see Section 600-6.

FPN No. 2: For switches controlling motors, see Sections 430-83, 430-109, and 430-110.

(a) Alternating Current General-Use Snap Switch. A form of general-use snap switch suitable only for use on ac circuits for controlling the following:

(1) Resistive and inductive loads, including electric-discharge lamps, not exceeding the ampere rating of the switch at the voltage involved
(2) Tungsten-filament lamp loads not exceeding the ampere rating of the switch at 120 volts
(3) Motor loads not exceeding 80 percent of the ampere rating of the switch at its rated voltage

(b) Alternating-Current or Direct-Current General-Use Snap Switch. A form of general-use snap switch suitable for use on either ac or dc circuits for controlling the following.

(1) Resistive loads not exceeding the ampere rating of the switch at the voltage applied.
(2) Inductive loads not exceeding 50 percent of the ampere rating of the switch at the applied voltage. Switches rated in horsepower are suitable for controlling motor loads within their rating at the voltage applied.
(3) Tungsten-filament lamp loads not exceeding the ampere rating of the switch at the applied voltage if T-rated.

(c) CO/ALR Snap Switches. Snap switches rated 20 amperes or less directly connected to aluminum conductors shall be listed and marked CO/ALR.

(d) Alternating-Current Specific-Use Snap Switches Rated for 347 Volts. Snap switches rated 347 volts ac shall be listed and shall be used only for controlling the following.

(1) Noninductive loads other than tungsten-filament lamps not exceeding the ampere and voltage ratings of the switch.

(2) Inductive loads not exceeding the ampere and voltage ratings of the switch. Where particular load characteristics or limitations are specified as a condition of the listing, those restrictions shall be observed regardless of the ampere rating of the load.

Section 380-14(d)(2) permits a relatively new type of AC specific-use snap switch that is 347-volt rated. These switches, unless specifically restricted, are permitted to be used on circuits of a lower voltage, such as 277- and 120-volt circuits.

The ampere rating of the switch shall not be less than 15 amperes at a voltage rating of 347 volts ac. Flush-type snap switches rated 347 volts ac shall not be readily interchangeable in box mounting with switches identified in Section 380-14(a) and (b).

B. Construction Specifications

380-15. Marking. Switches shall be marked with the current and voltage and, if horsepower rated, the maximum rating for which they are designed.

380-16. 600-Volt Knife Switches. Auxiliary contacts of a renewable or quick-break type or the equivalent shall be provided on all knife switches rated 600 volts designed for use in breaking current over 200 amperes.

380-17. Fused Switches. A fused switch shall not have fuses in parallel except as permitted in Section 240-8.

380-18. Wire-Bending Space. The wire-bending space required by Section 380-3 shall meet Table 373-6(b) spacings to the enclosure wall opposite the line and load terminals.

Article 384 — Switchboards and Panelboards

Contents

A. General

384-1. Scope. This article covers the following:

(1) All switchboards, panelboards, and distribution boards installed for the control of light and power circuits
(2) Battery-charging panels supplied from light or power circuits

384-2. Other Articles. Switches, circuit breakers, and overcurrent devices used on switchboards, panelboards, and distribution boards, and their enclosures, shall comply with this article and also with the requirements of Articles 240, 250, 370, 373, 380, and other articles that apply. Switchboards and panelboards in hazardous (classified) locations shall comply with the requirements of Articles 500 through 517.

See the definitions of *panelboard* and *switchboard* in Article 100.

384-3. Support and Arrangement of Busbars and Conductors.

(a) Conductors and Busbars on a Switchboard or Panelboard. Conductors and busbars on a switchboard or panelboard shall comply with (1), (2), and (3) as applicable.

(1) Location. Conductors and busbars shall be located so as to be free from physical damage and shall be held firmly in place.

(2) Service Switchboards. Barriers shall be placed in all service switchboards such that no uninsulated, ungrounded service busbar or service terminal will be exposed to inadvertent contact by persons or maintenance equipment while servicing load terminations.

It is often impractical to disconnect or de-energize the service conductors supplying a service switchboard. For this reason, qualified electricians work on these switchboards with the service bus energized. Barriers are required in all service switchboards to isolate the service busbars and terminals from the remainder of the switchboard, thus providing some measure of safety against contact with line-energized parts during maintenance and installation of new feeders or branch circuits. It must be clearly understood that de-energizing the load side of a switchboard, by operation of the disconnecting means, does not de-energize the ungrounded service conductors. Installers should also be careful to remove tools that may have been left on the inside of the gear in order to prevent ground faults or short circuits due to tools coming in contact with busbars.

(3) Same Vertical Section. Other than the required interconnections and control wiring, only those conductors that are intended for termination in a vertical section of a switchboard shall be located in that section.

Exception: Conductors shall be permitted to travel horizontally through vertical sections of switchboards where such conductors are isolated from busbars by a barrier.

The exception to Section 384-3(a)(3) permits conductors to travel horizontally through vertical sections of a switchboard where barriers are provided to isolate the conductors from the busbars.

(b) Overheating and Inductive Effects. The arrangement of busbars and conductors shall be such as to avoid overheating due to inductive effects.

(c) Used as Service Equipment. Each switchboard or panelboard, if used as service equipment, shall be provided with a main bonding jumper sized in accordance with Section 250-28(d) or the equivalent placed within the panelboard or one of the sections of the switchboard for connecting the grounded service conductor on its supply side to the switchboard or panelboard frame. All sections of a switchboard shall be bonded together using an equipment grounding conductor sized in accordance with Table 250-122.

Exception: Switchboards and panelboards used as service equipment on high-impedance grounded-neutral systems in accordance with Section 250-36 shall not be required to be provided with a main bonding jumper.

(d) Terminals. In switchboards and panelboards, load terminals for field wiring, including grounded circuit conductor load terminals and connections to the ground bus for load equipment grounding conductors, shall be located so that it will not be necessary to reach across or beyond an uninsulated ungrounded line bus in order to make connections.

(e) High-Leg Marking. On a switchboard or panelboard supplied from a 4-wire, delta-connected system where the midpoint of one phase winding is grounded, that phase busbar or conductor having the higher voltage to ground shall be durably and permanently marked by an outer finish that is orange in color or by other effective means.

The high leg is common on a 240/120-volt 3-phase, 4-wire delta system. It is typically designated as "B phase." The high-leg marking is required to be the color orange or other similar effective means and is intended to prevent problems due to the lack of complete standardization where metered and nonmetered equipment are installed in the same installation. Electricians should always test each phase to ground with suitable equipment in order to know exactly where this high leg is located in the system.

(f) Phase Arrangement. The phase arrangement on 3-phase buses shall be A, B, C from front to back, top to bottom, or left to right, as viewed from the front of the switchboard or panelboard. The B phase shall be that phase having the higher voltage to ground on 3-phase, 4-wire, delta-connected systems. Other busbar arrangements shall be permitted for additions to existing installations and shall be marked.

Exception: Equipment within the same single section or multisection switchboard or panelboard as the meter on 3-phase, 4-wire, delta-connected systems shall be permitted to have the same phase configuration as the metering equipment.

The exception to Section 384-3(f) permits the phase leg having the higher voltage to ground to be located at the right-hand position (C phase). The exception recognizes the fact that metering compartments have been standardized with the high leg at the right position (C phase) rather than in the center on B phase. This exception makes it unnecessary to transpose the panelboard or switchboard busbar arrangement ahead of and beyond a metering compartment. See also Sections 215-8 and 230-56 for further information on identifying conductors with the higher voltage to ground. Other busbar arrangements for addition to existing installations are permitted by the main rule.

(g) Minimum Wire-Bending Space. The minimum wire-bending space at terminals and minimum gutter space provided in panelboards and switchboards shall be as required in Section 373-6.

Section 384-3(g) requires that installations in the field comply with Section 373-6. See also the commentary following Section 384-35, which covers the size of the enclosure.

384-4. Installation.

FPN: For the dedicated space requirement, see Section 110-26(f).

B. Switchboards

384-5. Location of Switchboards. Switchboards that have any exposed live parts shall be located in permanently dry locations and then only where under competent supervision and accessible only to qualified persons. Switchboards shall be located so that the probability of damage from equipment or processes is reduced to a minimum.

384-6. Switchboards in Damp or Wet Locations. Switchboards in damp or wet locations shall be installed to comply with Section 373-2(a).

384-7. Location Relative to Easily Ignitible Material. Switchboards shall be placed so as to reduce to a minimum the probability of communicating fire to adjacent combustible materials. Where installed over a combustible floor, suitable protection thereto shall be provided.

One method of compliance with this rule is to form and set a piece of sheet steel or other suitable noncombustible material on the floor under the equipment.

384-8. Clearances.

(a) From Ceiling. For other than a totally enclosed switchboard, a space not less than 3 ft (914 mm) shall be provided between the top of the switchboard and any combustible ceiling, unless a noncombustible shield is provided between the switchboard and the ceiling.

(b) Around Switchboards. Clearances around switch boards shall comply with the provisions of Section 110-26.

Sufficient access and working space is required to permit ready and safe operation and maintenance of such equipment. Table 110-26(a) indicates minimum working clearances from 0 to 600 volts and Table 110-34(a) is used for voltages over 600 volts.

384-9. Conductor Insulation. An insulated conductor used within a switchboard shall be listed, flame retardant, and shall be rated not less than the voltage applied to it and not less than the voltage applied to other conductors or busbars with which it may come in contact.

384-10. Clearance for Conductors Entering Bus Enclosures. Where conduits or other raceways enter a switchboard, floor-standing panelboard, or similar enclosure at the bottom, sufficient space shall be provided to permit installation of conductors in the enclosure. The wiring space shall not be less than shown in Table 384-10 where the conduit or raceways enter or leave the enclosure below the busbars, their supports, or other obstructions. The conduit or raceways, including their end fittings, shall not rise more than 3 in. (76 mm) above the bottom of the enclosure.

Table 384-10. Clearance for Conductors Entering Bus Enclosures

	Minimum Spacing Between Bottom of Enclosure and Busbars, Their Supports, or Other Obstructions	
Conductor	**in.**	**mm**
Insulated busbars, their supports, or other obstructions	8	203
Noninsulated busbars	10	254

Section 384-10 should be carefully considered where installing underground conduit or raceways that terminate in the bottom of an open switchboard. For example, larger sizes of conduit used for service laterals or feeders and extending more than 3 in. above the bottom of the enclosure are difficult to shorten. On the other hand, conduits or raceways should not be installed flush with the finished floor under switchboards that are located on the outside of buildings or in other locations where water could enter the raceways.

384-12. Grounding of Instruments, Relays, Meters, and Instrument Transformers on Switchboards. Instruments, relays, meters, and instrument transformers located on switchboards shall be grounded as specified in Sections 250-170 through 250-178.

C. Panelboards

384-13. General. All panelboards shall have a rating not less than the minimum feeder capacity required for the load computed in accordance with Article 220. Panelboards shall be durably marked by the manufacturer with the voltage and the current rating and the number of phases for which they are designed and with the manufacturer's name or trademark in such a manner so as to be visible after installation, without disturbing the interior parts or wiring. All panelboard circuits and circuit modifications shall be legibly identified as to purpose or use on a circuit directory located on the face or inside of the panel doors.

FPN: See Section 110-22 for additional requirements.

Some panelboards are suitable for use as service equipment and are so marked.

Listed panelboards are for use with copper conductors, unless marked to indicate which terminals are suitable for use with aluminum conductors. Such marking is required to be independent of any marking on terminal connectors and is required to appear on a wiring diagram or other readily visible location. If all terminals are suitable for use with aluminum conductors as well as with copper conductors, the panelboard will be marked "Use Copper or Aluminum Wire." A panelboard employing terminals or main or branch-circuit units individually marked "AL-CU" will be marked as noted above or will be marked "Use Copper Wire Only." The latter marking indicates that wiring space or other factors make the panelboard unsuitable for aluminum conductors. [See Section 110-14(c).]

Panelboards to which units (circuit breakers, switches, etc.) may be added in the field are marked with the name or trademark of the manufacturer and the catalog number or equivalent of those units intended for installation in the field.

Unless the panelboard is marked to indicate otherwise, the termination provisions are based on the use of 60°C (140°F) ampacities for wire sizes No. 14 through 1 and 75°C (167°F) ampacities for wire sizes No. 1/0 and larger.

384-14. Classification of Panelboards. Panelboards shall be classified for the purposes of this article as either lighting and appliance branch-circuit panelboards or power panelboards.

(a) Lighting and Appliance Branch-Circuit Panelboard. A lighting and appliance branch-circuit panelboard is one having more than 10 percent of its overcurrent devices protecting lighting and appliance branch circuits. A lighting and appliance branch circuit is a branch circuit that has a connection to the neutral of the panelboard and that has overcurrent protection of 30 amperes or less in one or more conductors.

A lighting and appliance branch-circuit panelboard is a panelboard having more than 10 percent of the installed overcurrent devices supplying circuits with neutrals. At least one circuit in the panelboard must have an overcurrent device (typically a breaker) rated at 30 amperes or less. One common example is a 30 position, 120/240 volt, residential panelboard. This panelboard must have at least 4 circuits (10 percent of 30 = 3) at 120 volts supplying lighting and appliance branch circuits. At least one of these circuits must be rated 15, 20, or 30 amperes.

(b) Power Panelboard. A power panelboard is one having 10 percent or fewer of its overcurrent devices protecting lighting and appliance branch circuits.

A power panelboard is a panelboard having 10 percent or less of the installed overcurrent devices supplying lighting and appliance branch circuits. Any panelboard that is not classified as a lighting and appliance panelboard is a power panelboard.

384-15. Number of Overcurrent Devices on One Panelboard. Not more than 42 overcurrent devices (other than those provided for in the mains) of a lighting and appliance branch-circuit panelboard shall be installed in any one cabinet or cutout box.

A lighting and appliance branch-circuit panelboard shall be provided with physical means to prevent the installation of more overcurrent devices than that number for which the panelboard was designed, rated, and approved.

For the purposes of this article, a 2-pole circuit breaker shall be considered two overcurrent devices; a 3-pole circuit breaker shall be considered three overcurrent devices.

Class CTL panelboards are identified by the words "Class CTL."

Class CTL panelboards incorporate physical features that, in conjunction with the physical size, configuration, or other means provided in Class CTL circuit breakers, fuseholders, or fusible switches, are designed to prevent the installation of more overcurrent protective poles than the number for which the panelboard is designed and rated.

"Class CTL" is the Underwriters Laboratories Inc. designation for the *Code* requirement for circuit limitation within a lighting and appliance branch-circuit panelboard and means "circuit limiting."

384-16. Overcurrent Protection.

(a) Lighting and Appliance Branch-Circuit Panelboard Individually Protected. Each lighting and appliance branch-circuit panelboard shall be individually protected on the supply side by not more than two main circuit breakers or two sets of fuses having a combined rating not greater than that of the panelboard.

Exception No. 1: Individual protection for a lighting and appliance panelboard shall not be required if the panelboard feeder has overcurrent protection not greater than the rating of the panelboard.

Main overcurrent protection may be an integral part of a panelboard or located remote from the panelboard. See also the commentary following Section 384-15.

Figure 384.1 shows a panelboard with a 200-ampere main circuit breaker. Figure 384.2 shows a panelboard without main overcurrent protection, but it may be protected by overcurrent protection as illustrated in Figure 384.3. Note that in this latter figure that the panelboard feeders have overcurrent protection not greater than the rating of the panelboard.

Exception No. 2: For existing installations, individual protection for lighting and appliance branch-circuit panelboards shall not be required where such panelboards are used as service equipment in supplying an individual residential occupancy.

The phrase "for existing installations" in Exception No. 2 means the existing panelboard. It is not intended that a split-bus panelboard used in an individual dwelling occupancy be replaced if a circuit is added to the existing panelboard. It does mean, however, that for installation of new panelboards in new or existing residential occupancies, a split-bus six-disconnect panelboard (with more than two circuit breakers or sets of fuses protecting the panelboard) is not permitted for the service equipment.

An individual residential occupancy could be a dwelling unit in a multifamily dwelling where the panelboard is used as service equipment. See the definition of *dwelling unit* in Article 100.

Figure 384.4 shows the split-bus circuitry for a 200-ampere (*top*) and a 150-ampere (*bottom*) panelboard. The upper panel has two 100-ampere main breakers installed as disconnecting means, and 200-ampere main lugs. The lower is a split-bus panel with 150-ampere main lugs and six main breaker disconnecting means. The 150-ampere panelboard is suitable for use as service equipment only if it is not a lighting and appliance panelboard or if it presently exists in an individual residential occupancy.

Figure 384.1 A panelboard with main circuit breaker disconnect, suitable for use as service equipment. (Square D Co.)

Figure 384.2 A panelboard with main lugs only. (Square D Co.)

(b) Power Panelboard Protection. In addition to the requirements of Section 384-13, a power panelboard with supply conductors that include a neutral and having more than 10 percent of its overcurrent devices protecting branch circuits rated 30 amperes or less shall be protected on the supply side by an overcurrent protective device having a rating not greater than that of the panelboard.

Exception: This individual protection shall not be required for a power panelboard used as service equipment with multiple disconnecting means in accordance with Section 230-71.

(c) Snap Switches Rated at 30 Amperes or Less. Panelboards equipped with snap switches rated at 30 amperes or less shall have overcurrent protection not in excess of 200 amperes.

The requirement of Section 384-16(c) is limited to snap switches; it does not apply to panelboards equipped with circuit breakers.

(d) Continuous Load. The total load on any overcurrent device located in a panelboard shall not exceed 80 percent of its rating where, in normal operation, the load will continue for three hours or more.

Exception: An assembly, including the overcurrent device, shall be permitted to be used for continuous operation at 100 percent of its rating where it is listed for this purpose.

(e) Supplied Through a Transformer. Where a panelboard is supplied through a transformer, the overcurrent protection in 384-16(a), (b), and (c) shall be located on the secondary side of the transformer.

Exception: A panelboard supplied by the secondary side of a transformer shall be considered as protected by the overcurrent protection provided on the primary side of the transformer where that protection is in accordance with Section 240-21(c)(1).

(f) Delta Breakers. A 3-phase disconnect or overcurrent device shall not be connected to the bus of any panelboard that has less than 3-phase buses. Delta breakers shall not be installed in panelboards.

(g) Back-Fed Devices. Plug-in-type overcurrent protection devices or plug-in type main lug assemblies that are back fed and used to terminate field-installed ungrounded supply conductors shall be secured in place by an additional fastener that requires other than a pull to release the device from the mounting means on the panel.

384-17. Panelboards in Damp or Wet Locations. Panelboards in damp or wet locations shall be installed to comply with Section 373-2(a).

384-18. Enclosure. Panelboards shall be mounted in cabinets, cutout boxes, or enclosures designed for the purpose and shall be dead front.

Exception: Panelboards other than of the dead-front externally operable type shall be permitted where accessible only to qualified persons.

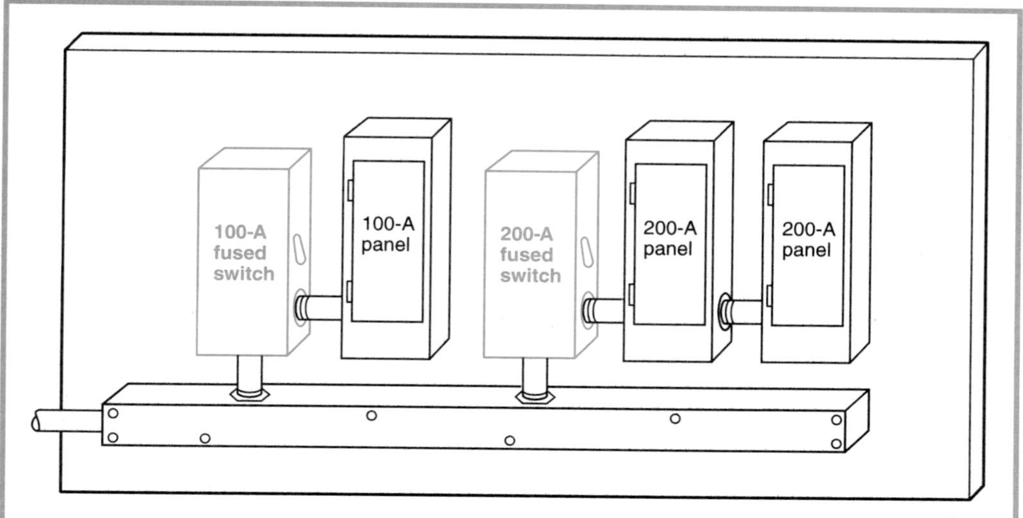

Figure 384.3 *An arrangement of three individual lighting and appliance branch-circuit panelboards with main overcurrent protection remote from the panelboards.*

384-19. Relative Arrangement of Switches and Fuses. In panelboards, fuses of any type shall be installed on the load side of any switches.

Exception: Fuses installed as part of service equipment in accordance with the provisions of Section 230-94 shall be permitted on the line side of the service switch.

Section 230-94 permits the service switch on either the supply side or load side of fuses. Where fuses of panelboards are accessible to other than qualified persons, such as occupants of a multifamily dwelling, Section 240-40 requires that disconnecting means be located on the supply side of all fuses in circuits of over 150 volts to ground, and in cartridge-type fuses in circuits of any voltage. Thus, when the disconnect switch is opened, the fuses are de-energized, and danger from shock is reduced.

384-20. Grounding of Panelboards. Panelboard cabinets and panelboard frames, if of metal, shall be in physical contact with each other and shall be grounded. Where the panelboard is used with nonmetallic raceway or cable or where separate grounding conductors are provided, a terminal bar for the grounding conductors shall be secured inside the cabinet. The terminal bar shall be bonded to the cabinet and panelboard frame, if of metal, otherwise it shall be connected to the grounding conductor that is run with the conductors feeding the panelboard.

A separate equipment grounding conductor terminal bar is required to be installed and bonded to the panelboard for the termination of feeder and branch-circuit equipment grounding conductors. Where installed within service equipment, this terminal is bonded to the neutral terminal bar, as illustrated in Figure 384.5. Any other connection between the equipment grounding terminal bar and the neutral bar, other than allowed in Section 250-32 is not permitted. If this downstream connection occurs, current flow in the neutral or grounded conductor would take parallel paths through the equipment grounding conductors (the raceway, for example) back to the service equipment. Normal load currents flowing on the equipment grounding conductors could create a shock hazard. Exposed metal parts of equipment could have a potential difference of several volts created by the load current on the grounding conductors. Another safety hazard could be created by this effect where subpanels are used, since arcing or loose connections at connectors and raceway fittings, etc., could serve to create a potential fire hazard.

Exception: Where an isolated equipment grounding conductor is provided as permitted by Section 250-146(d), the insulated equipment grounding conductor that is run with the circuit conductors shall be permitted to pass through the panelboard without being connected to the panelboard's equipment grounding terminal bar.

Grounding conductors shall not be connected to a terminal bar provided for grounded conductors (may be a neutral) unless the bar is identified for the purpose and is located where interconnection between equipment grounding conductors and grounded circuit conductors is permitted or required by Article 250.

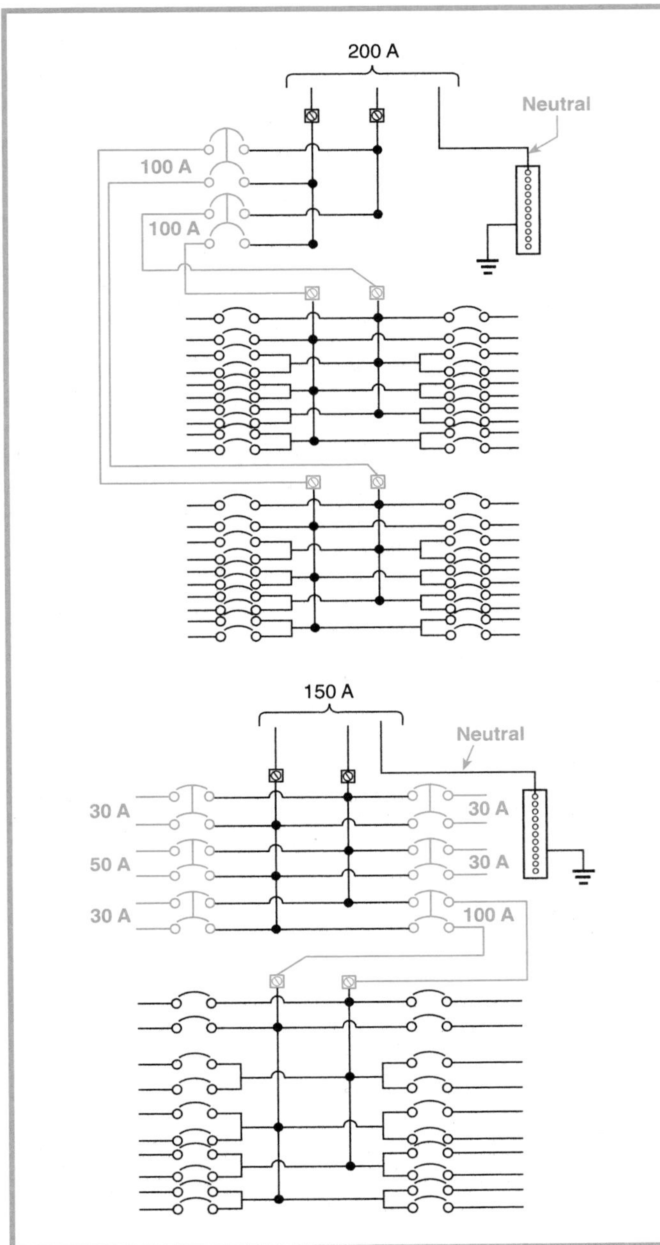

Figure 384.4 Circuitry for a 200-ampere (top) and a 150-ampere (bottom) split-bus panelboard.

The grounding of electronic equipment as well as overall power quality is of concern to the electrical industry. Sensitive electronic equipment used in industrial and commercial power systems may fail to perform properly where electrical noise is present in the equipment grounding conductor.

Where required for the reduction of electrical noise on the grounding circuit, an isolated equipment grounding terminal is permitted. This equipment

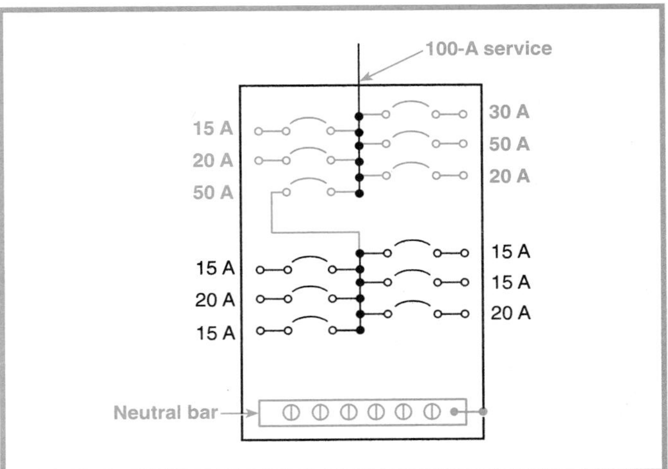

Figure 384.5 A split-bus lighting and appliance branch-circuit panelboard supplying an individual residential occupancy.

grounding terminal is required to be grounded by an insulated equipment grounding conductor that is run with the circuit conductors. The isolated equipment grounding conductor is also permitted to pass through one or more panelboards (without connection to the panelboard grounding terminal), but it is very important that it terminate directly at the applicable separately derived system or service grounding terminal. If the isolated equipment grounding conductor is run in a separate building, however, Section 250-146(d), requires the isolated equipment grounding conductor to terminate at a panelboard within the same building.

A connection to only a separate grounding electrode that places the earth in the fault return path may prevent sufficient current for opening overcurrent protection when a ground fault occurs. See the commentary following Section 250-146(d) FPN, and Section 250-54.

D. Construction Specifications

384-30. Panels. The panels of switchboards shall be made of moisture-resistant, noncombustible material.

384-31. Busbars. Insulated or bare busbars shall be rigidly mounted.

384-32. Protection of Instrument Circuits. Instruments, pilot lights, potential transformers, and other switchboard devices with potential coils shall be supplied by a circuit that is protected by standard overcurrent devices rated 15 amperes or less.

Exception No. 1: Overcurrent devices rated more than 15 amperes shall be permitted where the interruption of the circuit could create a hazard. Short-circuit protection shall be provided.

Exception No. 2: For ratings of 2 amperes or less, special types of enclosed fuses shall be permitted.

384-33. Component Parts. Switches, fuses, and fuseholders used on panelboards shall comply with the applicable requirements of Articles 240 and 380.

384-34. Knife Switches. Exposed blades of knife switches shall be de-energized when open.

FPN: See Section 380-6(c), Exception, for installation.

384-35. Wire-Bending Space in Panelboards. The enclosure for a panelboard shall have the top and bottom wire-bending space sized in accordance with Table 373-6(b) for the largest conductor entering or leaving the enclosure. Side wire-bending space shall be in accordance with Table 373-6(a) for the largest conductor to be terminated in that space.

Exception No. 1: Either the top or bottom wire-bending space shall be permitted to be sized in accordance with Table 373-6(a) for a lighting and appliance branch-circuit panelboard rated 225 amperes or less.

Exception No. 2: Either the top or bottom wire-bending space for any panelboard shall be permitted to be sized in accordance with Table 373-6(a) where at least one side wire-bending space is sized in accordance with Table 373-6(b) for the largest conductor to be terminated in any side wire-bending space.

Exception No. 3: The top and bottom wire-bending space shall be permitted to be sized in accordance with Table 373-6(a) spacings if the panelboard is designed and constructed for wiring using only one single 90 degree bend for each conductor, including the grounded circuit conductor, and the wiring diagram shows and specifies the method of wiring that shall be used.

Exception No. 4: Either the top or the bottom wire-bending space, but not both, shall be permitted to be sized in accordance with Table 373-6(a) where there are no conductors terminated in that space.

Section 384-35 covers the size of the enclosure for a panelboard. With reference to Figure 373.1 (see Section 373-6), the general rule calls for wire-bending space T_1 and T_4 to be in accordance with Table 373-6(b) for size *M* conductors (assuming these are the largest conductors entering the enclosure). Side wire-bending space T_2 must be in accordance with Table 373-6(a) for the wire size to be used with the largest-rated unit facing that side space, and T_3 must be similarly sized for the largest-rated unit facing the right side of the enclosure.

Exception No. 1 of Section 384-35 permits either T_1 or T_4 (not both) to be reduced to the space required by Table 373-6(a) for size *M* conductors for a panelboard rated 225 amperes or less.

Exception No. 2 of Section 384-35 permits either T_1 or T_4 (not both) to be reduced to the space required by Table 373-6(a) for size *M* conductors for any panelboard. Exception No. 2 is valid where either T_2 or T_3 (or both) is sized in accordance with Table 373-6(b) for the largest conductor to be terminated in either the left or right side spaces. Under the construction rules of Section 384-35, a panelboard enclosure might not be of adequate size for all manner of wiring, and Section 373-6 must be considered when wiring is planned.

Exception No. 3 of Section 384-35 permits both the top and bottom wire-bending space to be reduced as noted. A single 90 degree bend, meaning one and only one 90 degree bend, is required to be present for the ungrounded conductors. A grounded conductor is permitted to be wired straight in if Table 373-6(b) spacing is provided for the grounded conductor.

Exception No. 4 of Section 384-35 permits a reduction to the Table 373-6(a) spacing for the top or bottom space where no terminals face that space. In this case, the space is a gutter space, and measurement is on a line perpendicular to the wall of the enclosure and to the closest barrier post or side of a switch, fuse, or circuit breaker unit that is, or may be, installed.

384-36. Minimum Spacings. The distance between bare metal parts, busbars, etc., shall not be less than specified in Table 384-36.

Table 384-36. Minimum Spacings Between Bare Metal Parts

Voltage	Opposite Polarity Where Mounted on the Same Surface (in.)	Opposite Polarity Where Held Free in Air (in.)	Live Parts to Ground* (in.)
Not over 125 volts, nominal	¾	½	½
Not over 250 volts, nominal	1¼	¾	½
Not over 600 volts, nominal	2	1	1

Note: For SI units, 1 in. = 25.4 mm.

*For spacing between live parts and doors of cabinets, see Sections 373-11(a)(1), (2), and (3).

Where close proximity does not cause excessive heating, parts of the same polarity at switches, enclosed fuses, etc., shall be permitted to be placed as close together as convenience in handling will allow.

Exception: The distance shall be permitted to be less than that specified in Table 384-36 at circuit breakers and switches and in listed components installed in switchboards and panelboards.

CHAPTER 4

Equipment for General Use

Article 400 — Flexible Cords and Cables

Contents

A. General

400-1. Scope. This article covers general requirements, applications, and construction specifications for flexible cords and flexible cables.

Flexible cords and cables, as covered by Article 400, because of the nature of their use, are not considered a wiring method. Wiring methods are covered in Chapter 3 of the *Code*. Careful study of Sections 400-7 and 400-8 is required before choosing flexible cords or cables for a specific application.

400-2. Other Articles. Flexible cords and flexible cables shall comply with this article and with the applicable provisions of other articles of this *Code*.

400-3. Suitability. Flexible cords and cables and their associated fittings shall be suitable for the conditions of use and location.

400-4. Types. Flexible cords and flexible cables shall conform to the description in Table 400-4. Types of flexible cords and flexible cables other than those listed in the table shall be the subject of special investigation.

400-5. Ampacities for Flexible Cords and Cables. Table 400-5(A) provides the allowable ampacities and Table 400-5(B) provides the ampacities for flexible cords and cables with not more than three current-carrying conductors. These tables shall be used in conjunction with applicable end-use product standards to ensure selection of the proper size and type. If the number of current-carrying conductors exceeds three, the allowable ampacity or the ampacity of each conductor shall be reduced from the 3-conductor rating as shown in the following table.

Number of Conductors	Percent of Value in Tables 400-5(A) and 400-5(B)
4–6	80
7–9	70
10–20	50
21–30	45
31–40	40
41 and above	35

Ultimate Insulation Temperature. In no case shall conductors be associated together in such a way with respect to the kind of circuit, the wiring method used, or the number of conductors such that the limiting temperature of the conductors is exceeded.

A neutral conductor that carries only the unbalanced current from other conductors of the same circuit need not be considered as a current-carrying conductor.

In a 3-wire circuit consisting of two phase wires and the neutral of a 4-wire, 3-phase wye-connected system, a common conductor carries approximately the same current as the line-to-neutral currents of the other conductors and shall be considered to be a current-carrying conductor.

On a 4-wire, 3-phase wye circuit where the major portion

Table 400-4. Flexible Cords and Cables (See Section 400-4)

Trade Name	Type Letter	Size (AWG)	Number of Conductors	Insulation	Nominal Insulation Thickness[1] AWG	Nominal Insulation Thickness[1] Mils	Braid on Each Conductor	Outer Covering	Use		
Lamp Cord	C	18–10	2 or more	Thermset or thermo-plastic	18–16 14–10	30 45	Cotton	None	Pendant or portable	Dry locations	Not hard usage
Elevator cable	E See Note 5. See Note 9. See Note 10.	20–2	2 or more	Thermoset	20–16 14–12 12–10 8–2	20 30 45 60	Cotton	Three cotton, Outer one flame-retardant & moisture-resistant. See Note 3.	Elevator lighting and control	Nonhazardous locations	
					20–16 14–12 12–10 8–2	20 30 45 60	Flexible nylon jacket				
Elevator cable	EO See Note 5. See Note 10.	20–2	2 or more	Thermoset	20–16 14–12 12–10 8–2	20 30 45 60	Cotton	Three cotton, Outer one flame-retardant & moisture-resistant. See Note 3.	Elevator lighting and control	Nonhaz-ardous locations	
								One cotton and a neo-prene jacket. See Note 3.		Hazardous (classified) locations	
Elevator cable	ET See Note 5. See Note 10.	20–2	2 or more	Thermo-plastic	20–16 14–12 12–10 8–2	20 30 45 60	Rayon	Three cotton or equivalent, Outer one flame-retardant & moisture-resistant See Note 3.	Nonhazardous locations		
	ETLB See Note 5. See Note 10.						None				
	ETP See Note 5. See Note 10.						Rayon	Thermo-plastic	Hazardous (classified) locations		
	ETT See Note 5. See Note 10.						None	One cotton or equiva-lent and a thermo-plastic jacket			
Portable power cable	G	8–500 kcmil	2–6 plus grounding conduc-tor(s)	Thermoset	8–2 1–4/0 250 kcmil–500 kcmil	60 80 95		Oil-resistant thermoset	Portable and extra hard usage		
	G-GC	8–500 kcmil	3 plus 2 grounding conductors and 1 ground check conductor	Thermoset	8–2 1–4/0 250 kcmil–500 kcmil	60 80 95		Oil-resistant thermoset			

(continues)

Table 400-4. (Continued)

Trade Name	Type Letter	Size (AWG)	Number of Conductors	Insulation	Nominal Insulation Thickness[1] AWG	Nominal Insulation Thickness[1] Mils	Braid on Each Conductor	Outer Covering	Use		
Heater cord	HPD	18–12	2, 3, or 4	Thermoset	18–16 14-12	15 30	None	Cotton or rayon	Portable heaters	Dry locations	Not hard usage
Parallel heater cord	HPN See Note 6.	18–12	2 or 3	Oil-resistant thermoset	18–16 14 12	45 60 95	None	Oil-resistant thermoset	Portable	Damp locations	Not hard usage
Thermoset jacketed heater cords	HS	14–12	2, 3, or 4	Thermoset	18–16	30	None	Cotton and thermoset	Portable or portable heater	Damp	Extra hard usage
	HSJ	18–12									Hard usage
	HSO	14–12						Cotton and oil-resistant thermoset			Extra hard usage
	HSJO	18–12		Oil-resistant thermoset	14–12	45					Hard usage
	HSOO	14–12									Extra hard usage
	HSJOO	18–12									Hard usage
Twisted portable cord	PD	18–10	2 or more	Thermoset or thermoplastic	18–16 14–10	30 45	Cotton	Cotton or rayon	Pendant or portable	Dry locations	Not hard usage
Portable power cable	PPE	8–500 kcmil	1–6 plus optional grounding conductor(s)	Thermoplastic elastomer	8–2 1–4/0 250 kcmil–500 kcmil	60 80 95		Oil-resistant thermoplastic elastomer	Portable, extra hard usage		
Hard service cord	S See Note 4.	18–12	2 or more	Thermoset	18–16 14–10 8–2	30 45 60	None	Thermoset	Pendant or portable	Damp locations	Extra hard usage
Flexible stage and lighting power cable	SC	8–250 kcmil	1 or more		8–2 1–4/0 250 kcmil	60 80 95		Thermoset[2]	Portable, extra hard usage		
	SCE			Thermoplastic elastomer				Thermoplastic elastomer[2]			
	SCT			Thermoplastic				Thermoplastic[2]			
Hard service cord	SE See Note 4.	18–2	2 or more	Thermoplastic elastomer	18–16 14–10 8–2	30 45 60	None	Thermoplastic elastomer	Pendant or portable	Damp locations	Extra hard usage
	SEO See Note 4.							Oil-resistant thermoplastic elastomer			
	SEOO See Note 4.			Oil-resistant thermoplastic elastomer							
Junior hard service cord	SJ	18–10	2, 3, 4, or 5	Thermoset	18–12	30	None	Thermoset	Pendant or portable	Damp locations	Hard usage
	SJE			Thermoplastic elastomer				Thermoplastic elastomer			

Table 400-4. (Continued)

Trade Name	Type Letter	Size (AWG)	Number of Conductors	Insulation	Nominal Insulation Thickness[1] AWG	Nominal Insulation Thickness[1] Mils	Braid on Each Conductor	Outer Covering	Use		
Junior hard service cord	SJEO	18–10	2, 3, 4, or 5	Thermoset	18–12	30	None	Oil-resistant thermoplastic elastomer	Pendant or portable	Damp locations	Hard usage
	SJEOO			Oil-resistant thermoplastic elastomer							
	SJO			Thermoset				Oil-resistant thermoset			
	SJOO			Oil-resistant thermoset							
	SJT			Thermoplastic	10	45		Thermoplastic			
	SJTO			Thermoplastic	18–12	30		Oil-resistant thermoplastic			
	SJTOO			Oil-resistant thermoplastic	10	45					
Hard service cord	SO See Note 4.	18–2	2 or more	Thermoset	18–16	30		Oil-resistant thermoset	Pendant or portable	Damp locations	Extra hard usage
	SOO See Note 4.			Oil-resistant thermoset	14–10 8–2	45 60					
All thermoset parallel cord	SP-1 See Note 6.	20–18	2 or 3	Thermoset	20–18	30	None	Thermoset	Pendant or portable	Damp locations	Not hard usage
	SP-2 See Note 6.	18–16			18–16	45					
	SP-3 See Note 6.	18–10			18–16 14 12 10	60 80 95 110			Refrigerators, room air conditioners, and as permitted in Section 422-16(b)		
All elastomer (thermoplastic) parallel cord	SPE-1 See Note 6.	20–18	2 or 3	Thermoplastic elastomer	20–18	30	None	Thermoplastic elastomer	Pendant or portable	Damp locations	Not hard usage
	SPE-2 See Note 6.	18–16			18–16	45					

(continues)

Table 400-4. (Continued)

Trade Name	Type Letter	Size (AWG)	Number of Conductors	Insulation	Nominal Insulation Thickness[1] AWG	Mils	Braid on Each Conductor	Outer Covering	Use		
All elastomer (thermoplastic) parallel cord	SPE-3 See Note 6.	18–10	2 or 3	Thermoplastic elastomer	18–16 14 12 10	60 80 95 110	None	Thermoplastic elastomer	Refrigerators, room air conditioners, and as permitted in Section 422-16(b)	Damp locations	Not hard usage
All plastic parallel cord	SPT-1 See Note 6.	20–18	2 or 3	Thermoplastic	20–18	30	None	Thermoplastic	Pendant or portable	Damp locations	Not hard usage
	SPT-2 See Note 6.	18–16			18–16	45					
	SPT-3 See Note 6.	18–10			18–16 14 12 10	60 80 95 110		Thermoplastic	Refrigerators, room air conditioners, and as permitted in Section 422-16(6)	Damp locations	Not hard usage
Range, dryer cable	SRD	10–4	3 or 4	Thermoset	10–4	45	None	Thermoset	Portable	Damp locations	Ranges, dryers
	SRDE	10–4	3 or 4	Thermoplastic elastomer			None	Thermoplastic elastomer			
	SRDT	10–4	3 or 4	Thermoplastic			None	Thermoplastic			
Hard service cord	ST See Note 4.	18–2	2 or More	Thermoplastic	18–16 14–10 8–2	30 45 60	None	Thermoplastic	Pendant or portable	Damp locations	Extra hard usage
	STO See Note 4.							Oil-resistant thermoplastic			
	STOO See Note 4.			Oil-resistant thermoplastic							
Vacuum cleaner cord	SV See Note 6.	18–16	2 or 3	Thermoset	18–16	15	None	Thermoset	Pendant or portable	Damp locations	Not hard usage
	SVE See Note 6.			Thermoplastic elastomer				Thermoplastic elastomer			
	SVEO See Note 6.							Oil-resistant thermoplastic elastomer			

Table 400-4. (Continued)

Trade Name	Type Letter	Size (AWG)	Number of Conductors	Insulation	Nominal Insulation Thickness[1] AWG	Nominal Insulation Thickness[1] Mils	Braid on Each Conductor	Outer Covering	Use		
Vacuum cleaner cord	SVEOO See Note 6.	18–16	2 or 3	Oil-resistant thermo-plastic elastomer	18–16	15	None	Oil-resistant thermo-plastic elastomer	Pendant or portable	Damp locations	Not hard usage
	SVO			Thermoset				Oil-resistant thermoset			
	SVOO			Oil-resistant thermoset				Oil-resistant thermoset			
	SVT See Note 6.			Thermo-plastic				Thermo-plastic			
	SVTO See Note 6.			Thermo-plastic				Oil-resistant thermo-plastic			
	SVTOO			Oil-resistant thermo-plastic							
Parallel tinsel cord	TPT See Note 2.	27	2	Thermo-plastic	27	30	None	Thermo-plastic	At-tached to an appli-ance	Damp locations	Not hard usage
Jacketed tinsel cord	TS See Note 2.	27	2	Thermoset	27	15	None	Thermoset	At-tached to an appli-ance	Damp locations	Not hard usage
	TST See Note 2.			Thermo-plastic				Thermo-plastic			
Portable power cable	W	8–500 kcmil	1–6	Thermoset	8–2 1–4/0 250 kcmil–500 kcmil	60 80 95		Oil-resistant thermoset	Portable, extra hard usage		
Electric vehicle cable	EV	18–500 kcmil See Note 11.	2 or more plus ground-ing conduc-tor(s), plus optional hybrid data, signal, com-muni-cations, and optical fiber cables	Thermoset with optional nylon See Note 12.	18–16 14–10 8–2 1–4/0 250 kcmil–500 kcmil	30 (20) 45 (30) 60 (45) 80 (60) 95 (75) See Note 12.	Optional	Thermoset	Electric vehicle charg-ing	Wet locations	Extra hard usage
	EVJ	18–12 See Note 11.			18–12	30 (20) See Note 12.					Hard usage

(continues)

Table 400-4. (Continued)

Trade Name	Type Letter	Size (AWG)	Number of Conductors	Insulation	Nominal Insulation Thickness[1] AWG	Mils	Braid on Each Conductor	Outer Covering	Use		
Electric vehicle cable	EVE	18–500 kcmil See Note 11.	2 or more plus grounding conductor(s), plus optional hybrid data, signal, communi-	Thermoplastic elastomer with optional nylon See Note 12.	18–16 14–10 8–2 1–4/0 250 kcmil–500 kcmil	30 (20) 45 (30) 60 (45) 80 (60) 95 (75) See Note 12.	Optional	Thermoplastic elastomer	Electric vehicle charging	Wet locations	Extra hard usage
	EVJE	18–12 See Note 11.			18–12	30 (20) See Note 12.					Hard usage
Electric vehicle cable	EVT	18–500 kcmil See Note 11.	2 or more plus grounding conductor(s), plus optional hybrid data, signal, communications, and optical fiber cables	Thermoplastic with optional nylon See Note 12.	18–16 14–10 8–2 1–4/0 250 kcmil–500 kcmil	30 (20) 45 (30) 60 (45) 80 (60) 95 (75) See Note 12.	Optional	Thermoplastic	Electric vehicle charging	Wet locations	Extra hard usage
	EVJT	18–12 See Note 11.			18–12	30 (20) See Note 12.					Hard usage

[1]See Note 8.

[2]The required outer covering on some single conductor cables may be integral with the insulation.

Notes:

1. Except for Types HPN, SP-1, SP-2, SP-3, SPE-1, SPE-2, SPE-3, SPT-1, SPT-2, SPT-3, TPT, and three-conductor parallel versions of SRD, SRDE, SRDT, individual conductors are twisted together.

2. Types TPT, TS, and TST shall be permitted in lengths not exceeding 8 ft (2.44 m) where attached directly, or by means of a special type of plug, to a portable appliance rated at 50 watts or less and of such nature that extreme flexibility of the cord is essential.

3. Rubber-filled or varnished cambric tapes shall be permitted as a substitute for the inner braids.

4. Types G, G-GC, S, SC, SCE, SCT, SE, SEO, SEOO, SO, SOO, ST, STO, STOO, PPE, and W shall be permitted for use on theater stages, in garages, and elsewhere where flexible cords are permitted by this *Code*.

5. Elevator traveling cables for operating control and signal circuits shall contain nonmetallic fillers as necessary to maintain concentricity. Cables shall have steel supporting members as required for suspension by Section 620-41. In locations subject to excessive moisture or corrosive vapors or gases, supporting members of other materials shall be permitted. Where steel supporting members are used, they shall run straight through the center of the cable assembly and shall not be cabled with the copper strands of any conductor.

In addition to conductors used for control and signaling circuits, Tyes E, EO, ET, ETLB, ETP, and ETT elevator cables shall be permitted to incorporate in the construction, one or more No. 20 telephone conductor pairs, one or more coaxial cables, or one or more optical fibers. The No. 20 conductor pairs shall be permitted to be covered with suitable shielding for telephone, audio, or higher frequency communications circuits; the coaxial cables consist of a center conductor, insulation, and shield for use in video or other radio frequency communications circuits. The optical fiber shall be suitably covered with flame-retardant thermoplastic. The insulation of the conductors shall be rubber or thermoplastic of thickness not less than specified for the other conductors of the particular type of cable. Metallic shields shall have their own protective covering. Where used, these components shall be permitted to be incorporated in any layer of the cable assembly but shall not run straight through the center.

6. The third conductor in these cables shall be used for equipment grounding purposes only. The insulation of the grounding conductor for Types SPE-1, SPE-2, SPE-3, SPT-1, SPT-2, and SPT-3 shall be permitted to be thermoset polymer.

7. The individual conductors of all cords, except those of heat-resistant cords, shall have a thermoset or thermoplastic insulation, except that the equipment grounding conductor where used shall be in accordance with Section 400-23(b).

Table 400-4. (Continued)

8. Where the voltage between any two conductors exceeds 300, but does not exceed 600, flexible cord of Nos. 10 and smaller shall have thermoset or thermoplastic insulation on the individual conductors at least 45 mils in thickness, unless Type S, SE, SEO, SEOO, SO, SOO, ST, STO, or STOO cord is used.

9. Insulations and outer coverings that meet the requirements as flame retardant, limited smoke, and are so listed, shall be permitted to be designated limited smoke with the suffix *LS* after the code type designation.

10. Elevator cables in sizes No. 20 through 14 are rated 300 volts, and sizes 10 through 2 are rated 600 volts. No. 12 is rated 300 volts with a 30 mil-insulation thickness and 600 volts with a 45-mil insulation thickness.

11. Conductor size for Types EV, EVJ, EVE, EVJE, EVT, and EVJT cables apply to nonpower-limited circuits only. Conductors of power-limited (data, signal, or communications) circuits may extend beyond the stated AWG size range. All conductors shall be insulated for the same cable voltage rating.

12. Insulation thickness for Types EV, EVJ, EVE, EVJE, EVT, and EVJT cables of nylon construction is indicated in parentheses.

A new type of cable, Type G-CG, was added to Table 400-4 for the 1999 *Code*. This cable is similar to Type G, except it also incorporates an insulated ground-check (GC) conductor. The ground-check conductor is used as part of a low-voltage circuit that monitors the grounding conductor continuity.

Notes 11 and 12 following Table 400-4 coordinate these cable types with Article 625, Electric Vehicle Charging System Equipment.

Table 400-5(A). Allowable Ampacity for Flexible Cords and Cables [Based on Ambient Temperature of 30°C (86°F). See Section 400-13 and Table 400-4.]

Size (AWG)	Thermoset Type TS / Thermoplastic Types TPT, TST	Thermoset Types C, E, EO, PD, S, SJ, SJO, SJOO, SO, SOO, SP-1, SP-2, SP-3, SRD, SV, SVO, SVOO / Thermoplastic Types ET, ETLB, ETP, ETT, SE, SEO, SJE, SJEO, SJT, SJTO, SJTOO, SPE-1, SPE-2, SPE-3, SPT-1, SPT-2, SPT-3, ST, SRDE, SRDT, STO, STOO, SVE, SVEO, SVT, SVTO, SVTOO		Types HPD, HPN, HS, HSJ, HSO, HSJO, HSOO, HSJOO
		A†	B†	
27*	0.5	—	—	—
20	—	5**	***	—
18	—	7	10	10
17	—	—	12	—
16	—	10	13	15
15	—	—	—	17
14	—	15	18	20
12	—	20	25	30
10	—	25	30	35
8	—	35	40	—
6	—	45	55	—
4	—	60	70	—
2	—	80	95	—

*Tinsel cord

**Elevator cables only

***7 amperes for elevator cables only; 2 amperes for other types

†The allowable currents under subheading A apply to 3-conductor cords and other multiconductor cords connected to utilization equipment so that only 3 conductors are current-carrying. The allowable currents under subheading B apply to 2-conductor cords and other multiconductor cords connected to utilization equipment so that only 2 conductors are current carrying.

Table 400-5(B). Ampacity of Cable Types SC, SCE, SCT, PPE, G, G-GC, and W. [Based on Ambient Temperature of 30°C (86°F). See Table 400-4.]

Size (AWG or kcmil)	Temperature Rating of Cable								
	60°C (140°F)			75°C (167°F)			90°C (194°F)		
	D[1]	E[2]	F[3]	D[1]	E[2]	F[3]	D[1]	E[2]	F[3]
8	60	55	48	70	65	57	80	74	65
6	80	72	63	95	88	77	105	99	87
4	105	96	84	125	115	101	140	130	114
3	120	113	99	145	135	118	165	152	133
2	140	128	112	170	152	133	190	174	152
1	165	150	131	195	178	156	220	202	177
1/0	195	173	151	230	207	181	260	234	205
2/0	225	199	174	265	238	208	300	271	237
3/0	260	230	201	310	275	241	350	313	274
4/0	300	265	232	360	317	277	405	361	316
250	340	296	259	405	354	310	455	402	352
300	375	330	289	445	395	346	505	449	393
350	420	363	318	505	435	381	570	495	433
400	455	392	343	545	469	410	615	535	468
500	515	448	392	620	537	470	700	613	536

[1]The ampacities under subheading D shall be permitted for single-conductor Types SC, SCE, SCT, PPE, and W cable only where the individual conductors are not installed in raceways and are not in physical contact with each other except in lengths not to exceed 24 in. (610 mm) where passing through the wall of an enclosure.

[2]The ampacities under subheading E apply to two-conductor cables and other multiconductor cables connected to utilization equipment so that only two conductors are current carrying.

[3]The ampacities under subheading F apply to three-conductor cables and other multiconductor cables connected to utilization equipment so that only three conductors are current carrying.

of the load consists of nonlinear loads, there are harmonic currents present in the neutral conductor and the neutral shall be considered to be a current-carrying conductor.

An equipment grounding conductor shall not be considered a current-carrying conductor.

Where a single conductor is used for both equipment grounding and to carry unbalanced current from other conductors, as provided for in Section 250-140 for electric ranges and electric clothes dryers, it shall not be considered as a current-carrying conductor.

Exception: For other loading conditions, adjustment factors shall be permitted to be calculated under Section 310-15(c).

FPN: See Appendix B, Table B-310-11 for adjustment factors for more than three current-carrying conductors in a raceway or cable with load diversity.

400-6. Markings.

(a) Standard Markings. Flexible cords and cables shall be marked by means of a printed tag attached to the coil reel or carton. The tag shall contain the information required in Section 310-11(a). Types S, SC, SCE, SCT, SE, SEO, SEOO, SJ, SJE, SJEO, SJEOO, SJO, SJT, SJTO, SJTOO, SO, SOO, ST, STO, and STOO flexible cords and G, G-GC, PPE, and W flexible cables shall be durably marked on the surface at intervals not exceeding 24 in. (610 mm) with the type designation, size, and number of conductors.

(b) Optional Markings. Flexible cords and cable types listed in Table 400-4 shall be permitted to be surface marked to indicate special characteristics of the cable materials.

FPN: Examples of these markings include, but are not limited to, "LS" for limited smoke and markings such as "sunlight resistant."

The 1998 UL *Electrical Construction Materials Directory*, under the category Flexible Cord (ZJCZ), indicates that additional markings may include the following:

"Water Resistant" — indicates the cord is suitable for immersion in water.

"For Mobile Home Use" or "For Recreational Vehicle Use" or "For Mobile Home and Recreational Vehicle Use," followed by current rating in amperes — indicates suitability for use in mobile homes or recreational vehicles.

"Outdoor" or "W-A" — indicates suitability for use outdoors. The minimum temperature rating for these cords is −40°C, unless otherwise marked on the cord.

"W" — indicates suitability for use outdoors and for immersion in water. The low temperature rat-

ing for these cords is −40°C, unless otherwise marked on the cord with optional ratings of –50, –60, or –70°C. The low temperature ratings are determined by means of a bend test (not a suppleness test) at the given temperature.

"VW-1" — indicates the cord complies with a vertical flame test.

400-7. Uses Permitted.

(a) Uses. Flexible cords and cables shall be used only for the following:

(1) Pendants
(2) Wiring of fixtures
(3) Connection of portable lamps, portable and mobile signs, or appliances
(4) Elevator cables
(5) Wiring of cranes and hoists
(6) Connection of stationary equipment to facilitate their frequent interchange
(7) Prevention of the transmission of noise or vibration
(8) Appliances where the fastening means and mechanical connections are specifically designed to permit ready removal for maintenance and repair, and the appliance is intended or identified for flexible cord connection
(9) Data processing cables as permitted by Section 645-5
(10) Connection of moving parts
(11) Temporary wiring as permitted in Sections 305-4(b) and 305-4(c)

Flexible cords are permitted to be hard-wired into a junction box when the cord is used for pendant lighting fixtures; lighting fixtures where the cord and canopy are part of the listed fixture; fixed lighting fixtures when supplied with an attachment plug; supplies to pendant pushbutton stations for cranes; and portable lamp (droplight) connections. Revised for the 1999 *Code*, Section 400-7(a)(3) now permits cords to supply portable and mobile signs.

(b) Attachment Plugs. Where used as permitted in subsections (a)(3), (a)(6), and (a)(8), each flexible cord shall be equipped with an attachment plug and shall be energized from a receptacle outlet.

Exception: As permitted in Section 364-8.

400-8. Uses Not Permitted. Unless specifically permitted in Section 400-7, flexible cords and cables shall not be used for the following:

(1) As a substitute for the fixed wiring of a structure
(2) Where run through holes in walls, structural ceilings suspended ceilings, dropped ceilings, or floors
(3) Where run through doorways, windows, or similar openings
(4) Where attached to building surfaces

Exception: Flexible cord and cable shall be permitted to be attached to building surfaces in accordance with the provisions of Section 364-8.

(5) Where concealed behind building walls, structural ceilings, suspended ceilings, dropped ceilings, or floors
(6) Where installed in raceways, except as otherwise permitted in this *Code*

The flexible cords and cables referred to in Article 400 are not limited to use with portable equipment. They are not permitted to be used, however, as a substitute for the fixed wiring of a structure or where concealed behind building walls, ceilings (including structural, suspended, or dropped types), or floors. See Section 410-30 for cord-connected lighting fixtures. Also, see Article 305 for the requirement for the use of extension cords as temporary wiring, and see Section 240-4 for overcurrent protection requirements for flexible cord.

400-9. Splices. Flexible cord shall be used only in continuous lengths without splice or tap where initially installed in applications permitted by Section 400-7(a). The repair of hard-service cord and junior hard-service cord (see Trade Name column in Table 400-4) No. 14 and larger shall be permitted if conductors are spliced in accordance with Section 110-14(b) and the completed splice retains the insulation, outer sheath properties, and usage characteristics of the cord being spliced.

The initial installation of a cord containing a splice is prohibited by Section 400-9. Also, if a cord contains a splice, it is not permitted to be reused in a different location or for a different purpose. This applies to all uses of flexible cords covered under Section 400-7(a).

400-10. Pull at Joints and Terminals. Flexible cords and cables shall be connected to devices and to fittings so that tension will not be transmitted to joints or terminals.

Exception: Listed portable single pole devices that are intended to accommodate such tension at their terminals, shall be permitted to be used with single conductor flexible cable.

FPN: Some methods of preventing pull on a cord from being transmitted to joints or terminals are knotting the cord, winding with tape, and fittings designed for the purpose.

400-11. In Show Windows and Show Cases. Flexible cords used in show windows and show cases shall be Type S, SE, SEO, SEOO, SJ, SJE, SJEO, SJEOO, SJO, SJOO, SJT, SJTO, SJTOO, SO, SOO, ST, STO, or STOO.

Exception No. 1: For the wiring of chain-supported lighting fixtures.

Exception No. 2: As supply cords for portable lamps and other merchandise being displayed or exhibited.

Flexible cords listed for hard usage or extra-hard usage should be used in show windows and show cases. Precautions should be taken to ensure that these cords are maintained in good condition, because they may come in contact with combustible materials usually present at these locations and because they are exposed to wear and tear from continual housekeeping and display changes.

400-12. Minimum Size. The individual conductors of a flexible cord or cable shall not be smaller than the sizes in Table 400-4.

Exception: The size of the insulated ground-check conductor of Type G-GC cables shall be not smaller than No. 10.

Added in the 1999 *Code,* the exception to Section 400-12 correlates with the introduction of the new cable Type G-CG added to Table 400-4. This cable is similar to Type G, except it also incorporates an insulated ground-check (GC) conductor. The ground-check conductor is used as part of a low-voltage circuit that monitors the grounding conductor continuity.

400-13. Overcurrent Protection. Flexible cords not smaller than No. 18, and tinsel cords or cords having equivalent characteristics of smaller size approved for use with specific appliances, shall be considered as protected against overcurrent by the overcurrent devices described in Section 240-4.

400-14. Protection from Damage. Flexible cords and cables shall be protected by bushings or fittings where passing through holes in covers, outlet boxes, or similar enclosures.

A variety of bushings and fittings are available for this purpose, both insulated and noninsulated. Some include pull-relief means, as required in Section 400-10. Many insulating bushings are listed by Underwriters Laboratories Inc. in the following product categories:

1. Conduit Fittings (bushings and fittings for use on the ends of conduit in boxes, gutters)
2. Insulating Devices and Materials
3. Bushings (bushings for the protection of cords where they pass through walls or barriers of metal)
4. Outlet Bushings and Fittings (bushings and fittings for use on the ends of conduit, EMT, or armored cable, where a change to open wiring is made)

B. Construction Specifications

400-20. Labels. Flexible cords shall be examined and tested at the factory and labeled before shipment.

See the definition of *labeled* in Article 100.

400-21. Nominal Insulation Thickness. The nominal thickness of insulation for conductors of flexible cords and cables shall not be less than specified in Table 400-4.

Exception: The nominal insulation thickness for the ground-check conductors of Type G-GC cables shall not be less than 45 mils for No. 8 and not less than 30 mils for No. 10.

400-22. Grounded-Conductor Identification. One conductor of flexible cords that is intended to be used as a grounded circuit conductor shall have a continuous marker that readily distinguishes it from the other conductor or conductors. The identification shall consist of one of the methods indicated in (a) through (f).

(a) Colored Braid. A braid finished to show a white or natural gray color and the braid on the other conductor or conductors finished to show a readily distinguishable solid color or colors.

(b) Tracer in Braid. A tracer in a braid of any color contrasting with that of the braid and no tracer in the braid of the other conductor or conductors. No tracer shall be used in the braid of any conductor of a flexible cord that contains a conductor having a braid finished to show white or natural gray.

Exception: In the case of Types C and PD and cords having the braids on the individual conductors finished to show white or natural gray. In such cords, the identifying marker shall be permitted to consist of the solid white or natural gray finish on one conductor provided there is a colored tracer in the braid of each other conductor.

(c) Colored Insulation. A white or natural gray insulation on one conductor and insulation of a readily distinguishable color or colors on the other conductor or conductors for cords having no braids on the individual conductors.

For jacketed cords furnished with appliances, one conductor having its insulation colored light blue, with the other conductors having their insulation of a readily distinguishable color other than white or natural gray.

Exception: Cords that have insulation on the individual conductors integral with the jacket.

The insulation shall be permitted to be covered with an outer finish to provide the desired color.

(d) Colored Separator. A white or natural gray separator on one conductor and a separator of a readily distinguishable solid color on the other conductor or conductors of cords having insulation on the individual conductors integral with the jacket.

(e) Tinned Conductors. One conductor having the individual strands tinned and the other conductor or conductors having the individual strands untinned for cords having insulation on the individual conductors integral with the jacket.

(f) Surface Marking. One or more stripes, ridges, or grooves located on the exterior of the cord so as to identify one conductor for cords having insulation on the individual conductors integral with the jacket.

400-23. Equipment Grounding Conductor Identification. A conductor intended to be used as an equipment grounding conductor shall have a continuous identifying marker readily distinguishing it from the other conductor or conductors. Conductors having a continuous green color or a continuous green color with one or more yellow stripes shall not be used for other than equipment grounding purposes. The identifying marker shall consist of one of the methods in (a) or (b).

(a) Colored Braid. A braid finished to show a continuous green color or a continuous green color with one or more yellow stripes.

(b) Colored Insulation or Covering. For cords having no braids on the individual conductors, an insulation of a continuous green color or a continuous green color with one or more yellow stripes.

400-24. Attachment Plugs. Where a flexible cord is provided with an equipment grounding conductor and equipped with an attachment plug, the attachment plug shall comply with Sections 250-138(a) and (b).

C. Portable Cables Over 600 Volts, Nominal

400-30. Scope. This part applies to multiconductor portable cables used to connect mobile equipment and machinery.

400-31. Construction.

(a) Conductors. The conductors shall be No. 8 copper or larger and shall employ flexible stranding.

Exception: The size of the insulated ground-check conductor of Type G-GC cables shall be not smaller than No. 10.

(b) Shields. Cables operated at over 2000 volts shall be shielded. Shielding shall be for the purpose of confining the voltage stresses to the insulation.

(c) Equipment Grounding Conductor(s). An equipment grounding conductor(s) shall be provided. The total area shall not be less than that of the size of the equipment grounding conductor required in Section 250-122.

400-32. Shielding. All shields shall be grounded.

400-33. Grounding. Grounding conductors shall be connected in accordance with Part E of Article 250.

400-34. Minimum Bending Radii. The minimum bending radii for portable cables during installation and handling in service shall be adequate to prevent damage to the cable.

400-35. Fittings. Connectors used to connect lengths of cable in a run shall be of a type that lock firmly together. Provisions shall be made to prevent opening or closing these connectors while energized. Suitable means shall be used to eliminate tension at connectors and terminations.

400-36. Splices and Terminations. Portable cables shall not contain splices unless the splices are of the permanent molded, vulcanized types in accordance with Section 110-14(b). Terminations on portable cables rated over 600 volts, nominal, shall be accessible only to authorized and qualified personnel.

Article 402 — Fixture Wires

Contents

402-1. Scope. This article covers general requirements and construction specifications for fixture wires.

402-2. Other Articles. Fixture wires shall comply with this article and also with the applicable provisions of other articles of this *Code*.

FPN: For application in lighting fixtures, see Article 410.

402-3. Types. Fixture wires shall be of a type listed in Table 402-3, and they shall comply with all requirements of that table. The fixture wires listed in Table 402-3 are all suitable for service at 600 volts, nominal, unless otherwise specified.

FPN: Thermoplastic insulation may stiffen at temperatures colder than -10°C (+14°F), requiring that care be exercised during installation at such temperatures. Thermoplastic insulation may also be deformed at normal temperatures where subjected to pressure, requiring care be exercised during installation and at points of support.

Table 402-3. Fixture Wires

Trade Name	Type Letter	Insulation	Thickness of Insulation: AWG	Mils	Mils	Outer Covering	Maximum Operating Temperature	Application Provisions
Asbestos Covered Heat-resistant Fixture Wire	AF	Impregnated asbestos or moisture-resistant insulation and impregnated asbestos	 18–14 12–10	Thickness of Moisture-resistant Insulation Mils — 20 — 25	Thickness of Asbestos Mils 30 10 45 20	None	150°C 302°F	Fixture wiring — limited to 300 volts and indoor dry locations
Heat-resistant Rubber-Covered Fixture Wire — Flexible Stranding	FFH-2	Heat-resistant rubber Cross-linked synthetic polymer	18–16 18–16	— —	30 30	Nonmetallic covering	75°C 167°F	Fixture wiring
ECTFE — Solid or 7-Strand	HF	Ethylene chloro-trifluoro-ethylene	18–14	—	15	None	150°C 302°F	Fixture wiring
ECTFE — Flexible Stranding	HFF	Ethylene chlorotrifluoro-ethylene	18–14	—	15	None	150°C 302°F	Fixture wiring
Tape Insulated Fixture Wire — Solid or 7-Strand	KF-1	Aromatic polyimide tape	18–10	—	5.5	None	200°C 392°F	Fixture wiring — limited to 300 volts
	KF-2	Aromatic polyimide tape	18–10	—	8.4	None	200°C 392°F	Fixture wiring
Tape Insulated Fixture Wire — Flexible Stranding	KFF-1	Aromatic polyimide tape	18–10	—	5.5	None	200°C 392°F	Fixture wiring — limited to 300 volts
	KFF-2	Aromatic polyimide tape	18–10	—	8.4	None	200°C 392°F	Fixture wiring
Perfluoroalkoxy — Solid or 7-Strand (Nickel or Nickel-Coated Copper)	PAF	Perfluoroalkoxy	18–14	—	20	None	250°C 482°F	Fixture wiring (nickel or nickel-coated copper)
Perfluoroalkoxy — Flexible Stranding	PAFF	Perfluoroalkoxy	18–14	—	20	None	150°C 302°F	Fixture wiring
Fluorinated Ethylene Propylene Fixture Wire — Solid or 7-Strand	PF	Fluorinated ethylene propylene	18–14	—	20	None	200°C 392°F	Fixture wiring
Fluorinated Ethylene Propylene Fixture Wire — Flexible Stranding	PFF	Fluorinated ethylene propylene	18–14	—	20	None	150°C 302°F	Fixture wiring
Fluorinated Ethylene Propylene Fixture Wire — Solid or 7-Strand	PGF	Fluorinated ethylene propylene	18–14	—	14	Glass braid	200°C 392°F	Fixture wiring
Fluorinated Ethylene Propylene Fixture Wire — Flexible Stranding	PGFF	Fluorinated ethylene propylene	18–14	—	14	Glass braid	150°C 302°F	Fixture wiring

Table 402-3. (Continued)

Trade Name	Type Letter	Insulation	Thickness of Insulation AWG		Mils	Outer Covering	Maximum Operating Temperature	Application Provisions
Extruded Polytetra-Fluoroethylene — Solid or 7-Strand (Nickel or Nickel-Coated Copper)	PTF	Extruded polytetra-fluoroethylene	18–14	—	20	None	250°C 482°F	Fixture wiring (nickel or nickel-coated copper)
Extruded Polytetra-Fluoroethylene — Flexible Stranding 26-36 AWG (Silver or Nickel-Coated Copper)	PTFF	Extruded polytetra-fluoroethylene	18–14	—	20	None	150°C 302°F	Fixture wiring (silver or nickel-coated copper)
Heat-resistant Rubber-Covered Fixture Wire — Solid or 7-Strand	RFH-1	Heat-resistant rubber	18	—	15	Nonmetallic covering	75°C 167°F	Fixture wiring — limited to 300 volts
	RFH-2	Heat-resistant rubber Cross-linked synthetic polymer	18–16	—	30	None or nonmetallic covering	75°C 167°F	Fixture wiring
Heat-resistant Cross-Linked Synthetic Polymer-Insulated Fixture Wire — Solid or Stranded	RFHH-2* RFHH-3*	Cross-linked synthetic polymer	18–16 18–16	— —	30 45	None or nonmetallic covering	90°C 194°F	Fixture wiring — multiconductor cable
Silicone Insulated Fixture Wire — Solid or 7-Strand	SF-1	Silicone rubber	18	—	15	Nonmetallic covering	200°C 392°F	Fixture wiring — limited to 300 volts
	SF-2	Silicone rubber	18–14	—	30	Nonmetallic covering	200°C 392°F	Fixture wiring
Silicone Insulated Fixture Wire — Flexible Stranding	SFF-1	Silicone rubber	18	—	15	Nonmetallic covering	150°C 302°F	Fixture wiring — limited to 300 volts
	SFF-2	Silicone rubber	18–14	—	30	Nonmetallic covering	150°C 302°F	Fixture wiring
Thermoplastic Covered Fixture Wire — Solid or 7-Strand	TF*	Thermoplastic	18–16	—	30	None	60°C 140°F	Fixture wiring
Thermoplastic Covered Fixture Wire — Flexible Stranding	TFF*	Thermoplastic	18–16	—	30	None	60°C 140°F	Fixture wiring
Heat-resistant Thermoplastic-Covered Fixture Wire — Solid or 7-Strand	TFN*	Thermoplastic	18–16	—	15	Nylon-jacketed or equivalent	90°C 194°F	Fixture wiring
Heat-resistant Thermoplastic Covered Fixture Wire — Flexible Stranded	TFFN*	Thermoplastic	18–16	—	15	Nylon-jacketed or equivalent	90°C 194°F	Fixture wiring

(continues)

Table 402-3. (Continued)

Trade Name	Type Letter	Insulation	Thickness of Insulation AWG		Mils	Outer Covering	Maximum Operating Temperature	Application Provisions
Cross-Linked Polyolefin Insulated Fixture Wire — Solid or 7-Strand	XF*	Cross-linked polyolefin	18–14 12–10	— —	30 45	None	150°C 302°F	Fixture wiring — limited to 300 volts
Cross-Linked Polyolefin Insulated Fixture Wire — Flexible Stranded	XFF*	Cross-linked polyolefin	18–14 12–10	— —	30 45	None	150°C 302°F	Fixture wiring — limited to 300 volts
Modified ETFE — Solid or 7-Strand	ZF	Modified ethylene tetrafluoro-ethylene	18–14	—	15	None	150°C 302°F	Fixture wiring
Flexible Stranding	ZFF	Modified ethylene tetrafluoro-ethylene	18–14	—	15	None	150°C 302°F	Fixture wiring
High Temp. Modified ETFE — Solid or 7-Strand	ZHF	Modified ethylene tetrafluoro-ethylene	18–14	—	15	None	200°C 392°F	Fixture wiring

*Insulations and outer coverings that meet the requirements of flame retardant, limited smoke and are so listed shall be permitted to be designated limited smoke with the suffix */LS* after the code type designation.

402-5. Allowable Ampacities for Fixture Wires. The allowable ampacity of fixture wire shall be as specified in Table 402-5.

No conductor shall be used under such conditions that its operating temperature will exceed the temperature specified in Table 402-3 for the type of insulation involved.

FPN: See Section 310-10 for temperature limitation of conductors.

Table 402-5. Allowable Ampacity for Fixture Wires

Size (AWG)	Allowable Ampacity
18	6
16	8
14	17
12	23
10	28

402-6. Minimum Size. Fixture wires shall not be smaller than No. 18.

402-7. Number of Conductors in Conduit or Tubing. The number of fixture wires permitted in a single conduit or tubing shall not exceed the percentage fill specified in Table 1, Chapter 9.

Table 1 of Chapter 9 specifies the maximum percent fill of a conduit or tubing. Table 4 provides the usable area within the selected conduit or tubing, and Table 5 provides the required area for each of the conductors. Examples using these tables to calculate a conduit or tubing size are provided following Chapter 9, Table 1, Note 6.

Example 1

A remote ballast installation requires a single flexible metal conduit to contain 14 No. 16 TFFN fixture wires and three No. 12 THHN conductors. What size flexible metal conduit will be required?

Answer

The solution is found by using Table 1 of Chapter 9 and the accompanying Note 6 following Table 1. Table 1 sets the maximum percentage of conduit and tubing fill based on the internal cross-sectional area of the raceway in question. Note 6 states, "For (calculating) combinations of conductors of different sizes, use Tables 5 and 5A for dimensions of conductors and Table 4 for the applicable conduit or tubing dimensions."

Step 1. Using Table 1, look up the maximum percent of cross section of conduit permitted for conductors. Table 1 sets the limit of conductor fill for over two conductors at 40 percent of the total cross-sectional area of the raceway.

Step 2. Look up the individual conductor cross-sectional areas in Chapter 9, Table 5.

No. 16 TFFN = 0.0072 in.^{2}

No. 12 THHN = 0.0133 in.^{2}

Step 3. Calculate the total area occupied by the wires as follows:

No. 16 TFFN 14	×	0.0072	=	0.1008 in.^{2}
No. 12 TFFN 3	×	0.0133	=	0.0399 in.^{2}
Total area				0.1407 in.^{2}

Step 4. Using the 40% column of Table 4, look up the appropriate flexible metal conduit size based on 40% fill and a total conductor area fill of 0.1407 in.^{2} Since 0.1407 in.^{2} is greater than 0.127 and less than 0.213, select $^3/_4$-in. trade size flexible metal conduit.

The answer is $^3/_4$-in. flexible metal conduit.

Instead of doing the calculations, the tables of Appendix C may be used if the conductors in a raceway or tubing are all of the same wire size. Using Tables C1 and C1A of Appendix C for electrical metallic tubing, the following example is provided.

Example 2

A fire alarm system requires a riser to contain 36 No. 16 TFF conductors. What size electrical metallic tubing will be required?

Answer

According to Appendix C, Table C1, a 1-in. EMT is required.

402-8. Grounded Conductor Identification. One conductor of fixture wires that is intended to be used as a grounded conductor shall be identified by means of stripes or by the means described in Sections 400-22(a) through (e).

402-9. Marking.

(a) Required Information. All fixture wires shall be marked to indicate the information required in Section 310-11(a).

(b) Method of Marking. Thermoplastic insulated fixture wire shall be durably marked on the surface at intervals not exceeding 24 in. (610 mm). All other fixture wire shall be marked by means of a printed tag attached to the coil, reel, or carton.

(c) Optional Marking. Fixture wire types listed in Table 402-3 shall be permitted to be surface marked to indicate special characteristics of the cable materials.

FPN: Examples of these markings include, but are not limited to, "LS" for limited smoke or markings such as "sunlight resistant."

402-10. Uses Permitted. Fixture wires shall be permitted (1) for installation in lighting fixtures and in similar equipment where enclosed or protected and not subject to bending or twisting in use, or (2) for connecting lighting fixtures to the branch-circuit conductors supplying the fixtures.

402-11. Uses Not Permitted. Fixture wires shall not be used as branch-circuit conductors.

Exception: As permitted by Section 725-27 for Class 1 circuits and Section 760-27 for fire alarm circuits.

402-12. Overcurrent Protection. Overcurrent protection for fixture wires shall be as specified in Section 240-4.

Article 410 — Lighting Fixtures, Lampholders, Lamps, and Receptacles

Contents

D. Fixture Supports
- 410-15. Supports
  - (a) General
  - (b) Metal Poles Supporting Lighting Fixtures
- 410-16. Means of Support
  - (a) Outlet Boxes
  - (b) Inspection
  - (c) Suspended Ceilings
  - (d) Fixture Studs
  - (e) Insulating Joints
  - (f) Raceway Fittings
  - (g) Busways
  - (h) Trees

E. Grounding
- 410-17. General
- 410-18. Exposed Fixture Parts
  - (a) Exposed Conductive Parts
  - (b) Made of Insulating Material
- 410-20. Equipment Grounding Conductor Attachment
- 410-21. Methods of Grounding

F. Wiring of Fixtures
- 410-22. Fixture Wiring — General
- 410-23. Polarization of Fixtures
- 410-24. Conductor Insulation
- 410-27. Pendant Conductors for Incandescent Filament Lamps
  - (a) Support
  - (b) Size
  - (c) Twisted or Cabled
- 410-28. Protection of Conductors and Insulation
  - (a) Properly Secured
  - (b) Protection Through Metal
  - (c) Fixture Stems
  - (d) Splices and Taps
  - (e) Stranding
  - (f) Tension
- 410-29. Cord-Connected Showcases
  - (a) Cord Requirements
  - (b) Receptacles, Connectors, and Attachment Plugs
  - (c) Support
  - (d) No Other Equipment
  - (e) Secondary Circuit(s)
- 410-30. Cord-Connected Lampholders and Fixtures
  - (a) Lampholders
  - (b) Adjustable Fixtures
  - (c) Electric-Discharge Fixtures
- 410-31. Fixtures as Raceways

G. Construction of Fixtures
- 410-34. Combustible Shades and Enclosures
- 410-35. Fixture Rating
  - (a) Marking
  - (b) Electrical Rating
- 410-36. Design and Material
- 410-37. Nonmetallic Fixtures
- 410-38. Mechanical Strength
  - (a) Tubing for Arms
  - (b) Metal Canopies
  - (c) Canopy Switches
- 410-39. Wiring Space
- 410-42. Portable Lamps
  - (a) General
  - (b) Portable Handlamps
- 410-44. Cord Bushings
- 410-45. Tests
- 410-46. Live Parts

H. Installation of Lampholders
- 410-47. Screw-Shell Type
- 410-48. Double-Pole Switched Lampholders
- 410-49. Lampholders in Wet or Damp Locations

J. Construction of Lampholders
- 410-50. Insulation
- 410-52. Switched Lampholders

K. Lamps and Auxiliary Equipment
- 410-53. Bases, Incandescent Lamps
- 410-54. Electric-Discharge Lamp Auxiliary Equipment
  - (a) Enclosures
  - (b) Switching

L. Receptacles, Cord Connectors, and Attachment Plugs (Caps)
- 410-56. Rating and Type
  - (a) Receptacles
  - (b) CO/ALR Receptacles
  - (c) Isolated Ground Receptacles
  - (d) Faceplates
  - (e) Position of Receptacle Faces
  - (f) Receptacle Mounting
  - (g) Attachment Plugs
  - (h) Attachment Plug Ejector Mechanisms
  - (i) Noninterchangeability
- 410-57. Receptacles in Damp or Wet Locations
  - (a) Damp Locations
  - (b) Wet Locations
  - (c) Bathtub and Shower Space
  - (d) Protection for Floor Receptacles
  - (e) Flush Mounting with Faceplate
  - (f) Installation
- 410-58. Grounding-Type Receptacles, Adapters, Cord Connectors, and Attachment Plugs
  - (a) Grounding Poles
  - (b) Grounding-Pole Identification
  - (c) Grounding Terminal Use

A. General

410-1. Scope. This article covers lighting fixtures, lampholders, pendants, receptacles, incandescent filament lamps, arc lamps, electric-discharge lamps, the wiring and equipment forming part of such lamps, fixtures, and lighting installations.

FPN: The international term for a lighting fixture is *luminaire* and is defined as a complete lighting unit consisting of a lamp or lamps together with the parts designed to distribute the light, to position and protect the lamps, and to connect the lamps to the power supply.

410-2. Application of Other Articles. Equipment for use in hazardous (classified) locations shall conform to Articles 500 through 517. Lighting systems operating at 30 volts or less shall conform to Article 411. Arc lamps used in theaters shall comply with Section 520-61, and arc lamps used in projection machines shall comply with Section 540-20. Arc lamps used on constant-current systems shall comply with the general requirements of Article 490.

410-3. Live Parts. Fixtures, lampholders, lamps, and receptacles shall have no live parts normally exposed to contact. Exposed accessible terminals in lampholders, receptacles, and switches shall not be installed in metal fixture canopies or in open bases of portable table or floor lamps.

See the definition of *live parts* in Article 100. Notice that two conditions are necessary to have *live parts* present. The first condition is exposed or uninsulated components, terminals, or conductors. The second condition is that a shock hazard exists.

Exception: Cleat-type lampholders and receptacles located at least 8 ft (2.44 m) above the floor shall be permitted to have exposed terminals.

B. Fixture Locations

410-4. Fixtures in Specific Locations.

A pamphlet entitled *Fixture Marking Guide,* available from Underwriters Laboratories Inc., was developed to help the authority having jurisdiction quickly determine whether common types of UL-listed fluorescent, high-intensity discharge, and incandescent fixtures are installed correctly.

(a) Wet and Damp Locations. Fixtures installed in wet or damp locations shall be installed so that water cannot enter or accumulate in wiring compartments, lampholders, or other electrical parts. All fixtures installed in wet locations shall be marked, "Suitable for Wet Locations." All fixtures installed in damp locations shall be marked, "Suitable for Wet Locations" or "Suitable for Damp Locations."

Fixtures marked "Suitable for Wet Locations" are required where exposed to the weather or subject to water saturation. Construction, design, and installation are to prevent the entrance of rain, snow, ice, and dust. Outdoor parks and parking lots, outdoor recreational areas (tennis, golf, baseball, etc.), car wash areas, and building exteriors are examples of wet locations.

Locations protected from the weather and not subject to water saturation but still exposed to moisture, such as the following, may be considered damp locations:

1. The underside of store or gasoline-station canopies or theater marquees
2. Some cold-storage warehouses
3. Some agricultural buildings
4. Some basements
5. Roofed open porches and carports

Fixtures used in these locations are required to be marked "Suitable for Damp Locations." See the definitions of *location, damp location, dry location,* and *wet location* in Article 100.

(b) Corrosive Locations. Fixtures installed in corrosive locations shall be of a type suitable for such locations.

(c) In Ducts or Hoods. Fixtures shall be permitted to be installed in commercial cooking hoods where all of the following conditions are met.

(1) The fixture shall be identified for use within commercial cooking hoods and installed so that the temperature limits of the materials used are not exceeded.

(2) The fixture shall be constructed so that all exhaust vapors, grease, oil, or cooking vapors are excluded from the lamp and wiring compartment. Diffusers shall be resistant to thermal shock.

(3) Parts of the fixture exposed within the hood shall be corrosion resistant or protected against corrosion, and the surface shall be smooth so as not to collect deposits and facilitate cleaning.

(4) Wiring methods and materials supplying the fixture(s) shall not be exposed within the cooking hood.

FPN: See Section 110-11 for conductors and equipment exposed to deteriorating agents.

The requirements for Section 410-4(c)(4) were initially taken from NFPA 96, *Standard for Ventilation Control and Fire Protection of Commercial Cooking Operations.* NFPA 96 provides the minimum fire safety requirements (preventative and operative) related to the design, installation, operation, inspection, and maintenance of all public and private cooking operations, except single-family residential usage. This includes, but is not limited to, all manner of cooking equipment, exhaust hoods, grease removal devices, exhaust ductwork, exhaust fans, dampers, fire-extinguishing equipment, and all other auxiliary or ancillary components or systems that are involved in the capture, containment, and control of grease-laden cooking effluent. Also, NFPA 96 is intended to include residential cooking equipment where used for purposes other than residential family use, such as employee kitchens or break areas and church and meeting hall kitchens, regardless of frequency of use. As stated here, NFPA 96 does not apply to single-family residential dwellings.

Grease may cause the deterioration of conductor insulation resulting in short circuits or ground faults in wiring, hence the requirement prohibiting wiring methods and materials (raceways, cables, lampholders) within ducts or hoods. Conventional enclosed and gasketed fixtures located in the path of travel of exhaust products are not permitted because a fire could result from the high temperatures on grease-coated glass bowls or globes enclosing the lamps. Recessed or surface fixtures intended for location within hoods are required to be identified as suitable for the specific purpose and should be installed with the required clearances maintained. Note that wiring systems, including rigid metal conduit, are not permitted to be run exposed within the cooking hood.

For further information, refer also to UL 710-1995, *Safety Standard for Exhaust Hoods for Commercial Cooking Equipment.*

(d) Bathtub and Shower Areas. No parts of cord-connected fixtures, hanging fixtures, lighting track, pendants, or ceiling-suspended (paddle) fans shall be located within a zone measured 3 ft (914 mm) horizontally and 8 ft (2.44 m) vertically from the top of the bathtub rim or shower stall threshold. This zone is all encompassing and includes the zone directly over the tub or shower stall.

The intent of Section 410-4(d) is to keep cord-connected, hanging, or pendant fixtures and suspended fans out of the reach of an individual standing on a bathtub rim. This list implies that the same risk of electric shock is present for each prohibited item.

Figure 410.1 illustrates the restricted zone in which the specified fixtures, lighting track, and fans are prohibited. This requirement applies to hydromassage bathtubs, as defined in Section 680-4, as well as other bathtub types and shower areas. See Section 680-41 for installation requirements for spas and hot tubs (as defined in Section 680-4) installed indoors.

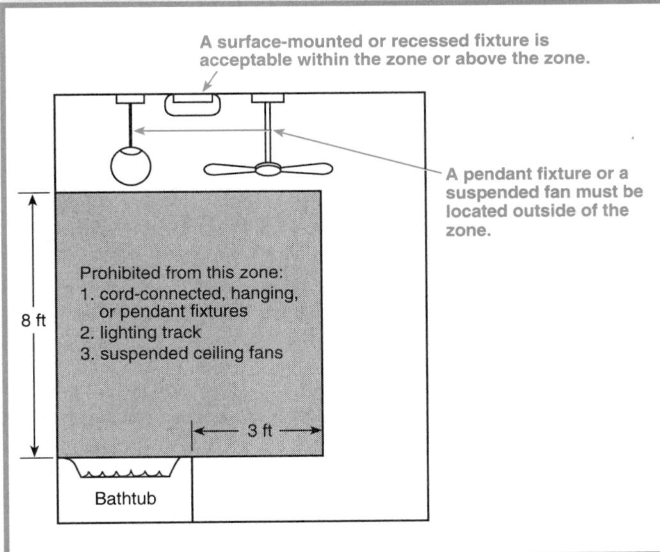

Figure 410.1 *Fixtures, lighting track, and suspended fans located near bathtubs.*

410-5. Fixtures Near Combustible Material. Fixtures shall be constructed, or installed, or equipped with shades or guards so that combustible material will not be subjected to temperatures in excess of 90°C (194°F).

Nearly every fire requires an initial heat source, an initial fuel source, and something to bring them together. Sections 410-5, 410-6, 410-7, and 410-8 regulate only the placement of the heat source. It is important to remember that successful fire prevention is most likely to come about if all three components are treated with due care.

Test have also shown that hot particles from broken incandescent lamps can ignite combustibles below the lamps.

410-6. Fixtures Over Combustible Material. Lampholders installed over highly combustible material shall be of the unswitched type. Unless an individual switch is provided for each fixture, lampholders shall be located at least 8 ft (2.44 m) above the floor, or shall be located or guarded so that the lamps cannot be readily removed or damaged.

Section 410-6 refers to pendants and fixed lighting equipment installed above highly combustible material. If the lamp cannot be located out of reach, the requirement can be met by equipping the lamp with a suitable guard. Section 410-6 is not applicable to portable lamps.

410-7. Fixtures in Show Windows. Chain-supported fixtures used in a show window shall be permitted to be externally wired. No other externally wired fixtures shall be used.

410-8. Fixtures in Clothes Closets.

(a) Definition.

Storage Space. *Storage space* shall be defined as a volume bounded by the sides and back closet walls and planes extending from the closet floor vertically to a height of 6 ft (1.83 m) or the highest clothes-hanging rod and parallel to the walls at a horizontal distance of 24 in. (610 mm) from the sides and back of the closet walls respectively, and continuing vertically to the closet ceiling parallel to the walls at a horizontal distance of 12 in. (305 mm) or the width of the shelf, whichever is greater.

FPN: See Figure 410-8.

For a closet that permits access to both sides of a hanging rod, the storage space shall include the volume below the highest rod extending 12 in. (305 mm) on either side of the rod on a plane horizontal to the floor extending the entire length of the rod.

The 24-in. rule is intended to cover the clothes-hanging space, even if no clothes-hanging rod is installed. If a clothes-hanging rod is installed, the space extends from the floor to the top of the highest rod. If no clothes-hanging rod is installed, it extends from the floor to a height of 6 ft.

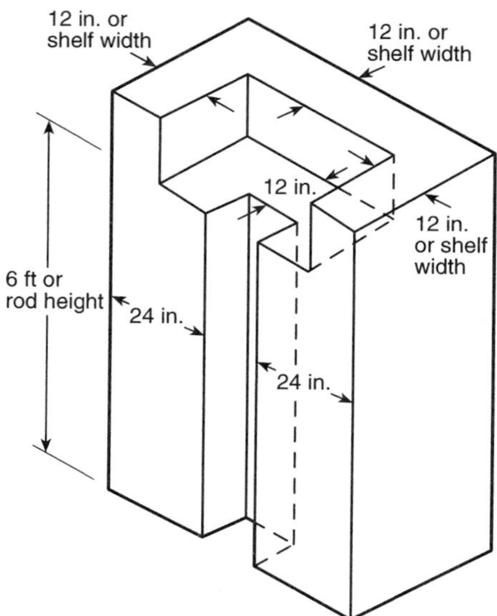

Figure 410-8 Closet storage space.

The 12-in. rule in the first paragraph is intended to cover the shelf space if shelving is not installed at the time of fixture installation. If shelving is installed and the shelves are wider than 12 in., the greater width applies.

The 12-in. rule in the second paragraph is intended to cover the clothes-hanging space in closets where there is access to both sides of a hanging rod.

(b) Fixture Types Permitted. Listed fixtures of the following types shall be permitted to be installed in a closet:

(1) A surface-mounted or recessed incandescent fixture with a completely enclosed lamp
(2) A surface-mounted or recessed fluorescent fixture

(c) Fixture Types Not Permitted. Incandescent fixtures with open or partially enclosed lamps and pendant fixtures or lampholders shall not be permitted.

See the commentary following Section 410-8(d)(3).

(d) Location. Fixtures in clothes closets shall be permitted to be installed as follows.

(1) Surface-mounted incandescent fixtures installed on the wall above the door or on the ceiling, provided there is a minimum clearance of 12 in. (305 mm) between the fixture and the nearest point of a storage space.

(2) Surface-mounted fluorescent fixtures installed on the wall above the door or on the ceiling, provided there is a minimum clearance of 6 in. (152 mm) between the fixture and the nearest point of a storage space.

(3) Recessed incandescent fixtures with a completely enclosed lamp installed in the wall or the ceiling, provided there is a minimum clearance of 6 in. (152 mm) between the fixture and the nearest point of a storage space.

The requirement of Section 410-8(d)(3) is the result of tests that have shown that a hot filament falling from a broken incandescent lamp can ignite combustible material below the fixture in which the lamp is installed.

(4) Recessed fluorescent fixtures installed in the wall or on the ceiling, provided there is a minimum clearance of 6 in. (152 mm) between the fixture and the nearest point of a storage space.

Note that the clearance measurement for each requirement here is to the fixture, not to the lamp itself.

It is not mandatory to install a lighting fixture in a clothes closet; if one is installed, however, the conditions for installation are as required by Section 410-8(d).

The requirements of Section 410-8(d) apply to incandescent and fluorescent lighting in clothes closets of various kinds of occupancies. The intent is to prevent hot lamps or parts of broken lamps from coming in contact with cartons, blankets, and the like, stored on shelves, and clothing hung in closets.

410-9. Space for Cove Lighting. Coves shall have adequate space and shall be located so that lamps and equipment can be properly installed and maintained.

Adequate space is necessary for easy access for relamping fixtures or replacing lampholders, ballasts, and so on; adequate space also improves ventilation.

C. Provisions at Fixture Outlet Boxes, Canopies, and Pans

410-10. Space for Conductors. Canopies and outlet boxes taken together shall provide adequate space so that fixture conductors and their connecting devices can be properly installed.

410-11. Temperature Limit of Conductors in Outlet Boxes. Fixtures shall be of such construction or installed so that the conductors in outlet boxes shall not be subjected to temperatures greater than that for which the conductors are rated.

Branch-circuit wiring, other than 2-wire or multiwire branch-circuits supplying power to fixtures connected together, shall not be passed through an outlet box that is an integral part of a fixture unless the fixture is identified for through-wiring.

Branch-circuit conductors run to a lighting outlet box are not permitted to be subjected to greater temperatures than those for which they are rated. For example, conductors are rated 75°C and are to supply a ceiling outlet box for the connection of a surface-mounted fixture or attached outlet box of a recessed fixture. The design and installation of the fixture should be such that the heat of the lamps does not subject the conductors to a greater temperature than 75°C. These types of fixtures are listed by Underwriters Laboratories Inc. based on a heat-contributing factor of the supply conductors of not more than the maximum permitted lamp wattage of the fixture.

Figure 410.2 illustrates recessed fixtures listed by Underwriters Laboratories Inc. for one set of supply conductors, and Figure 410.3 illustrates fixtures listed for a feed-through installation.

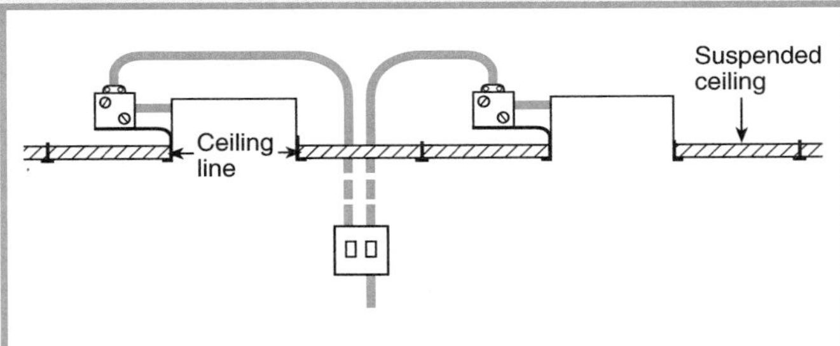

Figure 410.2 *Recessed lighting fixtures designed for branch-circuit conductors terminating at each fixture (no feed-through).*

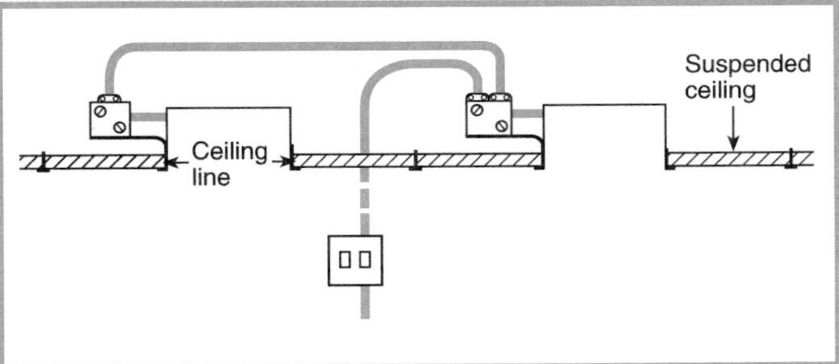

Figure 410.3 *Recessed lighting fixtures designed for feed-through branch-circuit conductors.*

FPN: See Section 410-31, Exceptions No. 2 and 3 for circuits supplying power to fixtures connected together.

410-12. Outlet Boxes to Be Covered. In a completed installation, each outlet box shall be provided with a cover unless covered by means of a fixture canopy, lampholder, receptacle, or similar device.

410-13. Covering of Combustible Material at Outlet Boxes. Any combustible wall or ceiling finish exposed between the edge of a fixture canopy or pan and an outlet box shall be covered with noncombustible material.

Lighting fixtures are required to be designed and installed not only to prevent overheating of conductors but also to prevent overheating of adjacent combustible wall or ceiling finishes. Hence, it is required that any combustible finish between the edge of a fixture canopy and an outlet box be covered with a noncombustible material or fixture accessory. See Section 370-20 for the requirements covering combustible finishes.

Where lighting fixtures are not directly mounted on outlet boxes, suitable outlet box covers are required.

410-14. Connection of Electric-Discharge Lighting Fixtures.

(a) Independent of the Outlet Box. Electric-discharge lighting fixtures supported independent of the outlet box shall be connected to the branch circuit through metal raceway, nonmetallic raceway, Type MC cable, Type AC cable, Type MI cable, nonmetallic sheathed cable, or by flexible cord as permitted in Section 410-30(b) or (c).

(b) Access to Boxes. Electric-discharge lighting fixtures surface mounted over concealed outlet, pull, or junction boxes shall be installed with suitable openings in back of the fixture to provide access to the boxes.

D. Fixture Supports

410-15. Supports.

(a) General. Fixtures, lampholders, and receptacles shall be securely supported. A fixture that weighs more than 6 lb (2.72 kg) or exceeds 16 in. (406 mm) in any dimension shall not be supported by the screw shell of a lampholder.

(b) Metal Poles Supporting Lighting Fixtures. Metal poles shall be permitted to be used to support lighting fixtures and as a raceway to enclose supply conductors, provided the following conditions are met.

(1) A metal pole shall have a handhole not less than 2 in. × 4 in. (50.8 mm × 102 mm) with a raintight cover to provide access to the supply terminations within the pole or pole base.

Exception No 1: No handhole shall be required in a pole 8 ft (2.44 m) or less in height above grade where the supply wiring method continues without splice or pull point, and where the interior of the pole and any splices are accessible by removing the fixture.

Exception No. 1 is typically used for both landscape (bollard type) lighting and pole lights at dwellings.

Exception No. 2: No handhole shall be required in a metal pole 20 ft (6.10 m) or less in height above grade that is provided with a hinged base.

Exception No. 2 recognizes metal light poles that do not have a handhole, but rather use a hinged-base pole to permit access to splices made in the pole base. The height of the pole is limited to 20 ft. Both parts of the pole must be bonded in accordance with Section 250-96 (systems operating at 250 volts or less) or Section 250-97 (circuits operating at over 250 volts), depending on the voltage of the system.

Figure 410.4 illustrates a metal light pole with a hinged baseplate that meets the requirements of Exception No. 2. A handhole is not necessary because the pole can be tilted to allow access to terminations in the base.

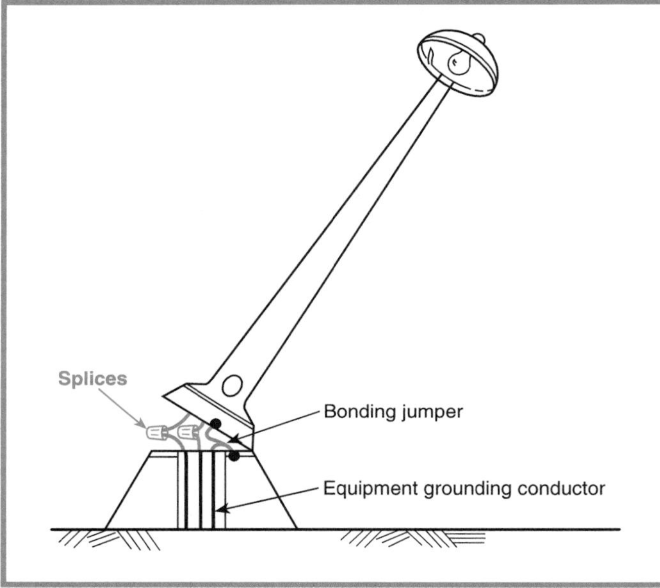

Figure 410.4 A hinged-base metal pole supporting a lighting fixture.

(2) Where raceway risers or cable is not installed withinthe pole, a threaded fitting or nipple shall be brazed or welded to the pole opposite the handhole for the supply connection.

(3) A metal pole shall be provided with a grounding terminal.

(a) A pole with a handhole shall have the grounding terminal accessible from the handhole.
(b) A pole with a hinged base shall have the grounding terminal accessible within the base.

Exception: No grounding terminal shall be required in a pole 8 ft (2.44 m) or less in height above grade where the supply wiring method continues without splice or pull, and where the interior of the pole and any splices are accessible by removing the fixture.

(4) A pole with a hinged base shall have the hinged base and pole bonded together.

(5) Metal raceways or other equipment grounding conductors shall be bonded to the pole with an equipment grounding conductor recognized by Section 250-118 and sized in accordance with Section 250-122.

(6) Conductors in vertical metal poles used as raceway shall be supported as provided in Section 300-19.

The changes to Section 410-15 for the 1999 *Code* were basically editorial. However, the change made in Section 410-15(b) by adding the word *raceway* was not editorial. By adding the word *raceway,* the user is reminded to maintain the required separation between power wiring and communications, signaling, and power-limited circuits.

For example, where a light pole supports a lighting fixture and a security camera, the security camera signaling and power-limited wiring may be installed within the pole cavity. Since the pole contains open circuit conductors (e.g., THW or XHHW conductors) to supply the lighting fixture already, the separation requirements can be fulfilled by enclosing the camera conductors within a flexible raceway and installing that raceway within the pole. Of course, the use of a cable or cord assembly for the lighting circuit conductors remains an optional choice for the user. Section 410-15(b) is not intended to necessarily require the placement of a raceway for communications cables on the exterior of a lighting pole.

410-16. Means of Support.

(a) Outlet Boxes. Outlet boxes or fittings installed as required by Section 370-23 shall be permitted to support fixtures weighing 50 lb (22.7 kg) or less. A fixture that weighs more than 50 lb (22.7 kg) shall be supported independent of the outlet box unless the outlet box is listed for the weight to be supported.

Regardless of whether a lighting fixture is attached to an outlet box or supported independently of the outlet box, care should be taken to securely fasten the outlet box or support the independent rod or hanger.

(b) Inspection. Fixtures shall be installed so that the connections between the fixture conductors and the circuit conductors can be inspected without requiring the disconnection of any part of the wiring unless the fixtures are connected by attachment plugs and receptacles.

(c) Suspended Ceilings. Framing members of suspended ceiling systems used to support fixtures shall be securely fastened to each other and shall be securely attached to the building structure at appropriate intervals. Fixtures shall be securely fastened to the ceiling framing member by mechanical means such as bolts, screws, or rivets. Listed clips identified for use with the type of ceiling framing member(s) and fixture(s) shall also be permitted.

Section 410-16(c) provides requirements for fixtures installations where the fixture are supported by a suspended or "hung" ceiling. Where fixtures are supported independent of a suspended ceiling, Section 410-16(c) does not apply.

Revised for the 1999 *Code,* Section 410-16(c) now requires that if clips are used to support fixtures to the framing members of a suspended ceiling, those clips must be of a type listed for the application. However, the use of listed clips for fixture support does not complete the requirements of this section. Additionally, the ceiling framing members must be securely attached to each other and to the building structure. For the support of wiring that is located within the cavity of floor–ceiling assemblies, see Section 300-11(a).

(d) Fixture Studs. Fixture studs that are not a part of outlet boxes, hickeys, tripods, and crowfeet shall be made of steel, malleable iron, or other material suitable for the application.

(e) Insulating Joints. Insulating joints that are not designed to be mounted with screws or bolts shall have an exterior metal casing, insulated from both screw connections.

(f) Raceway Fittings. Raceway fittings used to support a lighting fixture(s) shall be capable of supporting the weight of the complete fixture assembly and lamp(s).

(g) Busways. Fixtures shall be permitted to be connected to busways in accordance with Section 364-12.

(h) Trees. Outdoor lighting fixtures and associated equipment shall be permitted to be supported by trees.

FPN No. 1: See Section 225-26 for restrictions for support of overhead conductors.

Section 225-26 prohibits the support of overhead conductor spans on trees.

FPN No. 2: See Section 300-5(d) for protection of conductors.

Section 300-5(d) requires buried conductors and cables to be protected from physical damage by the use of raceways where the conductors emerge from the ground to a point at least 8 ft above finish grade.

E. Grounding

410-17. General. Fixtures and lighting equipment shall be grounded as required in Article 250 and Part E of this article.

410-18. Exposed Fixture Parts.

(a) Exposed Conductive Parts. Exposed metal parts shall be grounded or insulated from ground and other conducting surfaces or inaccessible to unqualified personnel. Lamp tie wires, mounting screws, clips, and decorative bands on glass spaced at least 1½ in. (38 mm) from lamp terminals shall not be required to be grounded.

(b) Made of Insulating Material. Fixtures directly wired or attached to outlets supplied by a wiring method that does not provide a ready means for grounding shall be made of insulating material and shall have no exposed conductive parts.

410-20. Equipment Grounding Conductor Attachment. Fixtures with exposed metal parts shall be provided with a means for connecting an equipment grounding conductor for such fixtures.

410-21. Methods of Grounding. Fixtures and equipment shall be considered grounded where mechanically connected to an equipment grounding conductor as specified in Section 250-118 and sized in accordance with Section 250-122.

F. Wiring of Fixtures

410-22. Fixture Wiring — General. Wiring on or within fixtures shall be neatly arranged and shall not be exposed to physical damage. Excess wiring shall be avoided. Conductors shall be arranged so that they shall not be subjected to temperatures above those for which they are rated.

410-23. Polarization of Fixtures. Fixtures shall be wired so that the screw shells of lampholders will be connected to the same fixture or circuit conductor or terminal. The grounded conductor, where connected to a screw-shell lampholder, shall be connected to the screw shell.

410-24. Conductor Insulation. Fixtures shall be wired with conductors having insulation suitable for the environmental conditions, current, voltage, and temperature to which the conductors will be subjected.

FPN: For ampacity of fixture wire, maximum operating temperature, voltage limitations, minimum wire size, etc., see Article 402.

410-27. Pendant Conductors for Incandescent Filament Lamps.

(a) Support. Pendant lampholders with permanently attached leads, where used for other than festoon wiring, shall be hung from separate stranded rubber-covered conductors that are soldered directly to the circuit conductors but supported independently thereof.

(b) Size. Unless part of listed decorative lighting assemblies, pendant conductors shall not be smaller than No. 14 for mogul-base or medium-base screw-shell lampholders, nor smaller than No. 18 for intermediate or candelabra-base lampholders.

(c) Twisted or Cabled. Pendant conductors longer than 3 ft (914 mm) shall be twisted together where not cabled in a listed assembly.

410-28. Protection of Conductors and Insulation.

(a) Properly Secured. Conductors shall be secured in a manner that will not tend to cut or abrade the insulation.

(b) Protection Through Metal. Conductor insulation shall be protected from abrasion where it passes through metal.

(c) Fixture Stems. Splices and taps shall not be located within fixture arms or stems.

(d) Splices and Taps. No unnecessary splices or taps shall be made within or on a fixture.

FPN: For approved means of making connections, see Section 110-14.

(e) Stranding. Stranded conductors shall be used for wiring on fixture chains and on other movable or flexible parts.

(f) Tension. Conductors shall be arranged so that the weight of the fixture or movable parts will not put a tension on the conductors.

410-29. Cord-Connected Showcases. Individual showcases, other than fixed, shall be permitted to be connected by flexible cord to permanently installed receptacles, and groups of not more than six such showcases shall be permitted to be coupled together by flexible cord and separable locking-type connectors with one of the group connected by flexible cord to a permanently installed receptacle.

The installation shall comply with (a) through (e) of this section.

(a) Cord Requirements. Flexible cord shall be of the hard-service type, having conductors not smaller than the branch-circuit conductors, having ampacity at least equal to the branch-circuit overcurrent device, and having an equipment grounding conductor.

FPN: See Table 250-122 for size of equipment grounding conductor.

(b) Receptacles, Connectors, and Attachment Plugs. Receptacles, connectors, and attachment plugs shall be of a listed grounding type rated 15 or 20 amperes.

(c) Support. Flexible cords shall be secured to the undersides of showcases so that (1) wiring will not be exposed to mechanical damage; (2) a separation between cases not in excess of 2 in. (50.8 mm), nor more than 12 in. (305 mm) between the first case and the supply receptacle, will be ensured; and (3) the free lead at the end of a group of showcases will have a female fitting not extending beyond the case.

(d) No Other Equipment. Equipment other than showcases shall not be electrically connected to showcases.

(e) Secondary Circuit(s). Where showcases are cord-connected, the secondary circuit(s) of each electric-discharge lighting ballast shall be limited to one showcase.

410-30. Cord-Connected Lampholders and Fixtures.

(a) Lampholders. Where a metal lampholder is attached to a flexible cord, the inlet shall be equipped with an insulating bushing that, if threaded, shall not be smaller than nominal 3/8-in. pipe size. The cord hole shall be of a size appropriate for the cord, and all burrs and fins shall be removed in order to provide a smooth bearing surface for the cord.

Bushing having holes 9/32 in. (7.14 mm) in diameter shall be permitted for use with plain pendant cord and holes 13/32 in. (10.3 mm) in diameter with reinforced cord.

Metal lampholders (brass- and aluminum-shell type) used with flexible-cord pendants should be equipped with smooth and permanently secured insulating bushings. Nonmetallic-type lampholders do not require a bushing because the material and design afford equivalent protection.

(b) Adjustable Fixtures. Fixtures that require adjusting or aiming after installation shall not be required to be equipped with an attachment plug or cord connector provided the exposed cord is of the hard usage or extra-hard usage type and is not longer than that required for maximum adjustment. The cord shall not be subject to strain or physical damage.

(c) Electric-Discharge Fixtures.

(1) A listed fixture or a listed assembly shall be permitted to be cord connected if

(a) The fixture is located directly below the outlet box or busway, and

(b) The flexible cord is

- (1) Visible for its entire length outside the fixture,
- (2) Not subject to strain or physical damage, and
- (3) Terminated in a grounding-type attachment plug cap, busway plug or have a fixture assembly with a strain relief and canopy.

(2) Electric-discharge lighting fixtures provided with mogul-base, screw-shell lampholders shall be permitted to be connected to branch circuits of 50 amperes or less by cords complying with Section 240-4. Receptacles and attachment plugs shall be permitted to be of lower ampere rating than the branch circuit but not less than 125 percent of the fixture full-load current.

(3) Electric-discharge lighting fixtures equipped with a flanged surface inlet shall be permitted to be supplied by cord pendants equipped with cord connectors. Inlets and connectors shall be permitted to be of lower ampere rating than the branch circuit but not less than 125 percent of the fixture load current.

Section 410-30(c)(1) applies to listed cord- and plug-connected electric-discharge lighting fixtures, such as the fixture illustrated in Figure 410.5, or electric-discharge lighting fixture assemblies. The supply cord is not permitted to penetrate a suspended ceiling, since it is required to be continuously visible along its entire length.

Supply cords cannot be used as a supporting means, and the fixtures are suspended directly below

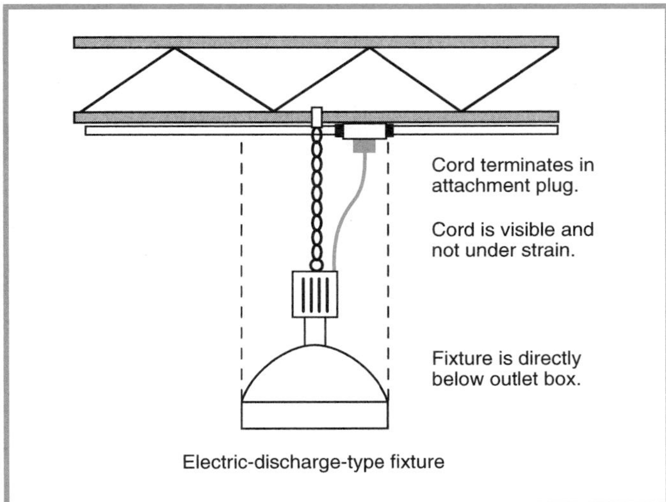

Figure 410.5 *A listed fluorescent lighting fixture that is cord- and plug-connected.*

the outlet boxes supplying each fixture. Electric-discharge lighting fixtures are not permitted to be supplied by cord if they are installed in lift-out-type suspended ceilings. If the electric-discharge fixture is suspended below the lift-out-type ceiling, the cord is not permitted to penetrate the ceiling. Section 400-8 further explains the uses not permitted for cords. According to Section 364-12, electric-discharge fixtures are permitted to be connected to busways by cords plugged directly into the busway or suspended from the busway.

410-31. Fixtures as Raceways. Fixtures shall not be used as a raceway for circuit conductors.

According to the UL *Marking Guide for Fixtures,* fixtures listed for use as raceways are marked "Suitable for use as a raceway" and also "Maximum of ___°C permitted in raceway."

Section 410-31, Exception No. 2 permits an additional 2-wire circuit to be carried through the fixtures to supply switched night lighting commonly used to conserve energy.

Exception No. 1: Fixtures listed for use as a raceway.

Exception No. 2: Fixtures designed for end-to-end assembly to form a continuous raceway or fixtures connected together by recognized wiring methods shall be permitted to carry through conductors of a 2-wire or multiwire branch circuit supplying the fixtures.

Exception No. 3: One additional 2-wire branch circuit separately supplying one or more of the connected fixtures described in Exception No. 2 shall be permitted to be carried through the fixtures.

FPN: See Article 100 for definition of "Multiwire Branch Circuit."

Branch-circuit conductors within 3 in. (76 mm) of a ballast within the ballast compartment shall have an insulation temperature rating not lower than 90°C (194°F), such as Types RHH, THW, THHN, THHW, FEP, FEPB, SA, and XHHW.

Temperature ratings along with other insulated conductor specifications are found in Table 310-13. Note the 90°C rating that may be applied to Type THW conductors for special application in electric-discharge lighting equipment.

G. Construction of Fixtures

410-34. Combustible Shades and Enclosures. Adequate airspace shall be provided between lamps and shades or other enclosures of combustible material.

410-35. Fixture Rating.

(a) Marking. All fixtures shall be marked with the maximum lamp wattage or electrical rating, manufacturer's name, trademark, or other suitable means of identification. A fixture requiring supply wire rated higher than 60°C (140°F) shall be marked in letters not smaller than ¼ in. (6.35 mm) high, prominently displayed on the fixture and shipping carton or equivalent.

(b) Electrical Rating. The electrical rating shall include the voltage and frequency and shall indicate the current rating of the unit, including the ballast, transformer, or autotransformer.

410-36. Design and Material. Fixtures shall be constructed of metal, wood, or other material suitable for the application and shall be designed and assembled so as to secure requisite mechanical strength and rigidity. Wiring compartments, including their entrances, shall be such that conductors may be drawn in and withdrawn without physical damage.

410-37. Nonmetallic Fixtures. When fixture wiring compartments are constructed from combustible material, armored or lead-covered conductors with suitable fittings shall be used or the wiring compartment shall be lined with metal.

410-38. Mechanical Strength.

(a) Tubing for Arms. Tubing used for arms and stems where provided with cut threads shall not be less than 0.040 in. (1.016 mm) in thickness and where provided with rolled (pressed) threads shall not be less than 0.025 in. (0.635 mm) in thickness. Arms and other parts shall be fastened to prevent turning.

(b) Metal Canopies. Metal canopies supporting lampholders, shades, etc., exceeding 8 lb (3.63 kg), or incorporating attachment-plug receptacles, shall not be less than

0.020 in. (0.508 mm) in thickness. Other canopies shall not be less than 0.016 in. (0.4064 mm) if made of steel and not less than 0.020 in. (0.508 mm) if of other metals.

(c) Canopy Switches. Pull-type canopy switches shall not be inserted in the rims of metal canopies that are less than 0.025 in. (0.635 mm) in thickness unless the rims are reinforced by the turning of a bead or the equivalent. Pull-type canopy switches, whether mounted in the rims or elsewhere in sheet metal canopies, shall not be located more than 3½ in. (89 mm) from the center of the canopy. Double set-screws, double canopy rings, a screw ring, or equal method shall be used where the canopy supports a pull-type switch or pendant receptacle.

The above thickness requirements shall apply to measurements made on finished (formed) canopies.

410-39. Wiring Space. Bodies of fixtures, including portable lamps, shall provide ample space for splices and taps and for the installation of devices, if any. Splice compartments shall be of nonabsorbent, noncombustible material.

410-42. Portable Lamps.

(a) General. Portable lamps shall be wired with flexible cord recognized by Section 400-4 and an attachment plug of the polarized or grounding type. Where used with Edison-base lampholders, the grounded conductor shall be identified and attached to the screw shell and the identified blade of the attachment plug.

(b) Portable Handlamps. In addition to the provisions of Section 410-42(a), portable handlamps shall comply with the following.

(1) Metal shell, paper-lined lampholders shall not be used.
(2) Handlamps shall be equipped with a handle of molded composition or other insulating material.
(3) Handlamps shall be equipped with a substantial guard attached to the lampholder or handle.
(4) Metallic guards shall be grounded by the means of an equipment grounding conductor run with circuit conductors within the power-supply cord.
(5) Portable handlamps shall not be required to be grounded where supplied through an isolating transformer with an ungrounded secondary of not over 50 volts.

See Figure 410.6 for an example of a portable handlamp (drop light) meeting the requirements of Section 410-42.

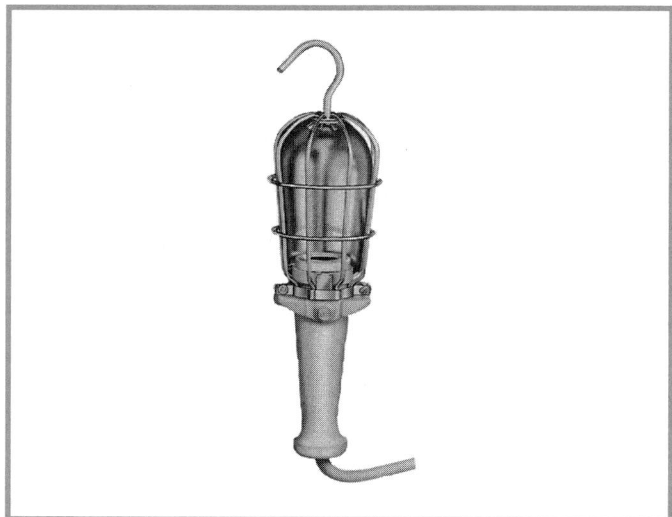

Figure 410.6 *A portable handlamp with a grounded metal guard and reflector and a swivel-type hook that permits the lamp to be adjusted in different positions. (Daniel Woodhead Co.)*

410-44. Cord Bushings. A bushing or the equivalent shall be provided where flexible cord enters the base or stem of a portable lamp. The bushing shall be of insulating material unless a jacketed type of cord is used.

410-45. Tests. All wiring shall be free from short circuits and grounds and shall be tested for these defects prior to being connected to the circuit.

410-46. Live Parts. Exposed live parts within porcelain fixtures shall be suitably recessed and located so as to make it improbable that wires will come in contact with them. There shall be a spacing of at least ½ in. (12.7 mm) between live parts and the mounting plane of the fixture.

H. Installation of Lampholders

410-47. Screw-Shell Type. Lampholders of the screw-shell type shall be installed for use as lampholders only. Where supplied by a circuit having a grounded conductor, the grounded conductor shall be connected to the screw shell.

Many years ago it was common practice to install screw-shell lampholders with screw-shell adapters in baseboards and walls to connect cord-connected appliances and lighting equipment. This now-prohibited practice permitted exposed live parts to be contacted by persons when the adapters were removed.

See Section 410-56(a) for permitted uses of receptacles.

410-48. Double-Pole Switched Lampholders. Where supplied by the ungrounded conductors of a circuit, the switching device of lampholders of the switched type shall simultaneously disconnect both conductors of the circuit.

Single-pole switching may be used to interrupt the ungrounded conductor of a 2-wire circuit having one conductor grounded. The grounded conductor is required to be connected to the screw shell of the lampholder.

Where a 2-wire circuit is derived from the two ungrounded conductors of a multiwire circuit (3- or 4-wire system) or from the two ungrounded conductors of a 2-wire circuit (3-wire system) and is used with switched lampholders, the switching device is required to be double-pole and to simultaneously disconnect both ungrounded conductors of the circuit. See Section 410-52 on the construction of switched lampholders.

410-49. Lampholders in Wet or Damp Locations. Lampholders installed in wet or damp locations shall be of the weatherproof type.

J. Construction of Lampholders

410-50. Insulation. The outer metal shell and the cap shall be lined with insulating material that shall prevent the shell and cap from becoming a part of the circuit. The lining shall not extend beyond the metal shell more than ⅛ in. (3.17 mm) but shall prevent any current-carrying part of the lamp base from being exposed when a lamp is in the lampholding device.

410-52. Switched Lampholders. Switched lampholders shall be of such construction that the switching mechanism interrupts the electrical connection to the center contact. The switching mechanism shall also be permitted to interrupt the electrical connection to the screw shell if the connection to the center contact is simultaneously interrupted.

K. Lamps and Auxiliary Equipment

410-53. Bases, Incandescent Lamps. An incandescent lamp for general use on lighting branch circuits shall not be equipped with a medium base if rated over 300 watts, nor with a mogul base if rated over 1500 watts. Special bases or other devices shall be used for over 1500 watts.

410-54. Electric-Discharge Lamp Auxiliary Equipment.

(a) Enclosures. Auxiliary equipment for electric-discharge lamps shall be enclosed in noncombustible cases and treated as sources of heat.

The UL *General Information Electrical Equipment Directory*—1998, contains two categories for ballasts under Electric Discharge Lamp Control Equipment (FKOT): fluorescent ballasts (FKVS) and HID (high intensity discharge) ballasts (FLCR).

Fluorescent ballast enclosures are categorized by UL as open, indoor, outdoor (both Type 1 and Type 2), and weatherproof. HID ballasts are categorized the same, except there is no open-type ballast for HID ballasts.

Open Type. Open core and coil constructions (i.e., ballasts without complete metal enclosures) are intended for use within suitable enclosures.

Indoor Ballasts. Indoor ballasts are suitable for use in an indoor, dry location only.

Outdoor Ballasts. Type 1 outdoor ballasts are suitable for use in (1) outdoor equipment, (2) fixtures intended for wet or damp locations, or (3) an outdoor sign if the ballasts are within an overall electrical enclosure. Ballasts of this type are marked "Type 1 Outdoor" or "Type 1." Type 2 outdoor ballasts are suitable for use in (1) outdoor equipment, (2) fixtures intended for wet or damp locations, or (3) an outdoor sign if the ballasts, in addition to its own enclosure, are within an overall enclosure. Ballasts of this type are marked "Type 2 Outdoor" or "Type 2."

Weatherproof Ballasts. Weatherproof ballasts are suitable for use where completely exposed to the weather without an additional enclosure and are marked "Weatherproof" or "WP."

(b) Switching. Where supplied by the ungrounded conductors of a circuit, the switching device of auxiliary equipment shall simultaneously disconnect all conductors.

L. Receptacles, Cord Connectors, and Attachment Plugs (Caps)

410-56. Rating and Type.

(a) Receptacles. Receptacles installed for the attachment of portable cords shall be rated at not less than 15 amperes, 125 volts, or 15 amperes, 250 volts, and shall be of a type not suitable for use as lampholders.

(b) CO/ALR Receptacles. Receptacles rated 20 amperes or less and directly connected to aluminum conductors shall be marked CO/ALR.

Section 410-56(b) requires that 15- and 20-ampere receptacles directly connected to aluminum conductors be suitable for such use. If the receptacle is not of the CO/ALR type, it can be connected with a copper pigtail to an aluminum branch-circuit conductor only if the wire connector is suitable for such a connection and is marked with the letters "AL" and "CU." The commentary following Section 110-14(b) further explains the suitability of wire connectors used to join copper and aluminum conductors.

(c) Isolated Ground Receptacles. Receptacles intended for the reduction of electrical noise (electromagnetic interference) as permitted in Section 250-146(d) shall be identified by an orange triangle located on the face of the receptacle. Receptacles so identified shall be used only with grounding conductors that are isolated in accordance with Section 250-146(d). Isolated ground receptacles installed in nonmetallic boxes shall be covered with a nonmetallic faceplate.

Exception: Where an isolated ground receptacle is installed in a nonmetallic box, a metal faceplate shall be permitted if the box contains a feature or accessory that permits the effective grounding of the faceplate.

Section 410-56(d) requires that metal receptacle faceplates be grounded. Generally, this requirement is easily met by grounding the metal box. However, isolated ground receptacles installed in nonmetallic boxes are problematic, since grounding the receptacle in this case does not ground the faceplate. Section 410-56(c) contains two solutions concerning the receptacle faceplate. The general solution is to use only nonmetallic faceplates. A change for the 1999 *Code* provides an alternate solution — a new exception that allows a nonmetallic box manufacturer to add a feature or accessory to accomplish effective grounding of a metal faceplate.

(d) Faceplates. Metal faceplates shall be of ferrous metal not less than 0.030 in. (0.762 mm) in thickness or of nonferrous metal not less than 0.040 in. (1.016 mm) in thickness. Metal faceplates shall be grounded. Faceplates of insulating material shall be noncombustible and not less than 0.10 in. (2.54 mm) in thickness but shall be permitted to be less than 0.10 in. (2.54 mm) in thickness if formed or reinforced to provide adequate mechanical strength.

(e) Position of Receptacle Faces. After installation, receptacle faces shall be flush with or project from faceplates of insulating material and shall project a minimum of 0.015 in. (0.381 mm) from metal faceplates. Faceplates shall be installed so as to completely cover the opening and seat against the mounting surface.

The reason for requiring receptacles to project from metal faceplates is to prevent faults between the blades of attachment plugs and metal faceplates. The proper mounting of faceplates will help to ensure that attachment plugs can be fully inserted, thus providing a better contact.

(f) Receptacle Mounting.

(1) Receptacles mounted in boxes that are set back of the wall surface, as permitted in Section 370-20, shall be installed so that the mounting yoke or strap of the receptacle is held rigidly at the surface of the wall.

(2) Receptacles mounted in boxes that are flush with the wall surface or project therefrom shall be installed so that the mounting yoke or strap of the receptacle is seated against the box or raised box cover.

Prior to achieving compliance with Section 410-56(f)(2), the outlet box used to enclose a receptacle must be rigidly and securely supported according to Section 370-23(b) or (c). In addition, mounting outlet boxes with the proper set back, according to Section 370-20, requires the cooperation of other construction trades (drywall applicators, plasterers, and carpenters) as well as the building designers.

The intent of Sections 410-56(f)(1), (2), and (3) is to allow attachment plugs to be inserted or removed without movement of the receptacle. Additionally, by restricting movement of the receptacle, effective grounding continuity can be maintained for contact devices or receptacle yokes where the box is installed flush with the wall surface or projects therefrom. The proper installation of receptacles helps ensure that attachment plugs can be fully inserted, thus providing a better contact. Sections 410-56(e) and (f) have been reorganized for the 1999 *Code* without adding any new requirements.

(3) **Receptacles Mounted on Covers.** Receptacles mounted to and supported by a cover shall be secured by more than one screw or shall be a device assembly or box cover listed and identified for securing by a single screw.

Receptacles mounted on raised covers, such as the receptacle illustrated in Figure 410.7, are not permitted to be secured by a single screw unless listed and identified for the use.

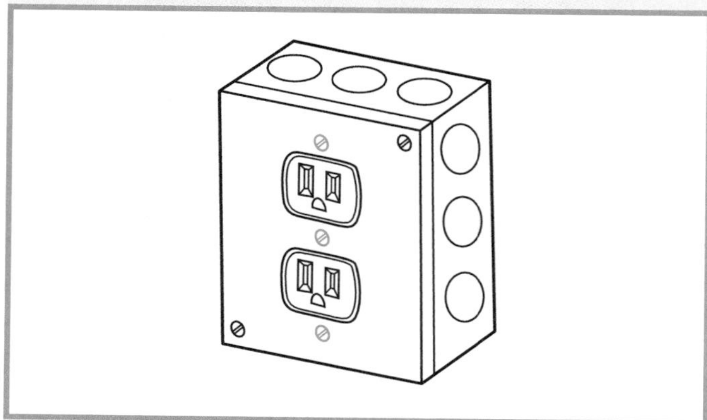

Figure 410.7 *A receptacle mounted on a raised cover.*

(g) Attachment Plugs. All 15- and 20-ampere attachment plugs and connectors shall be constructed so that there are no exposed current-carrying parts except the prongs, blades, or pins. The cover for wire terminations shall be a part, which is essential for the operation of an attachment plug or connector (dead-front construction). Attachment plugs shall be installed so that their prongs, blades, or pins are not energized unless inserted into an energized receptacle. No receptacle shall be installed so as to require an energized attachment plug as its source of supply.

The design requirements found in Section 410-56(g) (referred to as dead-front construction) have minimized the occurrence of electrical faults between metal plates and attachment plugs with terminal screws exposed on the face of the plug.

Added for the 1999 *Code,* the requirements stated in the last two sentences of Section 410-56(g) were originally found within product information only. However, as an aid to the inspection community, these requirements are now clearly stated in the *NEC.* A live attachment cap can be a most dangerous situation. Never should attachment caps be installed so as to allow the blades to be energized without being plugged into a device.

(h) Attachment Plug Ejector Mechanisms. Attachment plug ejector mechanisms shall not adversely affect engagement of the blades of the attachment plug with the contacts of the receptacle.

Section 410-56(h) permits a device that reduces the likelihood of damage to the cord when the cord is pulled to remove the plug. This device is designed for use by persons with mobility or visual impairments.

(i) Noninterchangeability. Receptacles, cord connectors, and attachment plugs shall be constructed so that receptacle or cord connectors will not accept an attachment plug with a different voltage or current rating than that for which the device is intended; however, a 20-ampere T-slot receptacle or cord connector shall be permitted to accept a 15-ampere attachment plug of the same voltage rating. Nongrounding-type receptacles and connectors shall not accept grounding-type attachment plugs.

See Figures 210.6 and 210.7 for ANSI C73 standard configuration chart.

410-57. Receptacles in Damp or Wet Locations.

(a) Damp Locations. A receptacle installed outdoors in a location protected from the weather or in other damp locations shall have an enclosure for the receptacle that is weatherproof when the receptacle is covered (attachment plug cap not inserted and receptacle covers closed).

An installation suitable for wet locations shall also be considered suitable for damp locations.

A receptacle shall be considered to be in a location protected from the weather where located under roofed open porches, canopies, marquees, and the like, and will not be subjected to a beating rain or water runoff.

(b) Wet Locations.

(1) A receptacle installed in a wet location where the product intended to be plugged into it is not attended while in use (e.g., sprinkler system controllers, landscape lighting, holiday lights, etc.) shall have an enclosure that is weatherproof with the attachment plug cap inserted or removed.

Section 410-57(b)(1) applies to receptacles for cord- and plug-connected equipment likely to be used outdoors or in a wet location for long periods of time. Examples include pump motors, vending machines, and so on. The intent is that receptacles for this application remain weatherproof while they are in use.

(2) A receptacle installed in a wet location where the product intended to be plugged into it will be attended while in use (e.g., portable tools, etc.) shall have an enclosure that is weatherproof when the attachment plug cap is removed.

Section 410-57(b)(2) applies to receptacles for cord- and plug-connected portable tools or other portable equipment likely to be used outdoors for a specific purpose and then removed. Examples include gardening tools, television sets, radios, and so on. Figures 410.8 and 410.9 show receptacles considered suitable for wet locations. The receptacle on the left in Figure 410.8 is permitted for portable tools and appliances normally used only when attended. The receptacle on the right is required for equipment likely to be used when unattended.

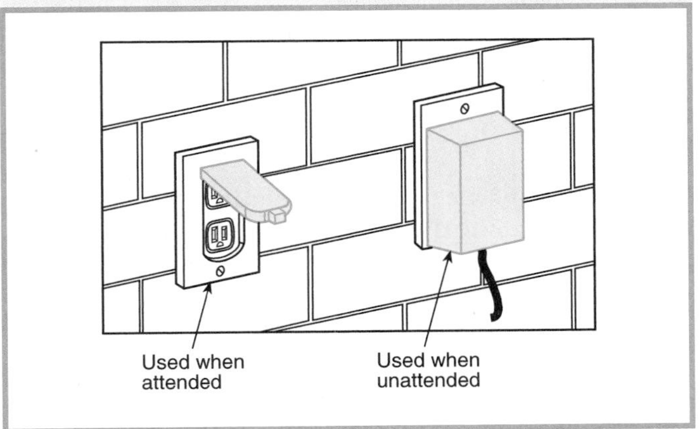

Figure 410.8 *Weatherproof receptacles installed in a wet location.*

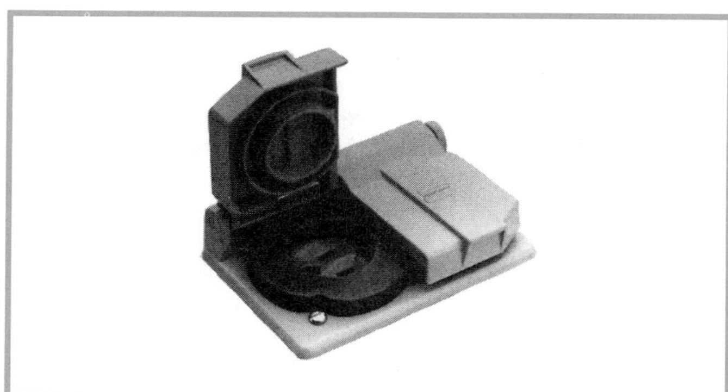

Figure 410.9 Weatherproof receptacle cover suitable for wet locations. (Crouse-Hinds)

(c) Bathtub and Shower Space. A receptacle shall not be installed within a bathtub or shower space.

Section 410-57(c) was added for the 1996 *Code* to prohibit the installation of receptacles inside bathtub and shower spaces, even if installed in a weatherproof enclosure. This helps prevent the use of shavers, radios, hair dryers, and so on, in these areas. The unprotected line side of GFCI-protected receptacles installed in these areas could possibly become wet and could, therefore, create a shock hazard by energizing surrounding wet surfaces.

(d) Protection for Floor Receptacles. Standpipes of floor receptacles shall allow floor-cleaning equipment to be operated without damage to receptacles.

(e) Flush Mounting with Faceplate. The enclosure for a receptacle installed in an outlet box flush-mounted on a wall surface shall be made weatherproof by means of a weatherproof faceplate assembly that provides a watertight connection between the plate and the wall surface.

(f) Installation. A receptacle outlet installed outdoors shall be located so that water accumulation is not likely to touch the outlet cover or plate.

Outdoor receptacle outlets are required for one-family and two-family dwellings and for swimming pool areas. They are also very common at other locations, such as shopping centers (for children's rides and decorative lighting), truck terminals (for refrigeration units and engine heaters), marinas, mobile home and recreational vehicle sites, and so on.

The enclosure for an outdoor receptacle may be flush or surface mounted or supported by conduits, in accordance with Section 370-23, and used with a variety of receptacles and covers; however, it is also the installer's responsibility to provide a watertight seal.

410-58. Grounding-Type Receptacles, Adapters, Cord Connectors, and Attachment Plugs.

(a) Grounding Poles. Grounding-type receptacles, cord connectors, and attachment plugs shall be provided with one fixed grounding pole in addition to the circuit poles. The grounding contacting pole of grounding-type plug-in ground-fault circuit interrupters shall be permitted to be of the movable, self-restoring type on circuits operating at not over 150 volts between and to conductors nor over 150 volts between any conductor and ground.

(b) Grounding-Pole Identification. Grounding-type receptacles, adapters, cord connections, and attachment plugs shall have a means for connection of a grounding conductor to the grounding pole.

A terminal for connection to the grounding pole shall be designated by one of the following.

(1) A green-colored hexagonal-headed or shaped terminal screw or nut, not readily removable.

(2) A green-colored pressure wire connector body (a wire barrel).

(3) A similar green-colored connection device, in the case of adapters. The grounding terminal of a grounding adapter shall be a green-colored rigid ear, lug, or similar device. The grounding connection shall be designed so that it cannot make contact with current-carrying parts of the receptacle, adapter, or attachment plug. The adapter shall be polarized.

Section 410-58(b)(3) requires the grounding terminal of an adapter to be a green-colored ear, lug, or similar device, thereby prohibiting use of an adapter with an attached pigtail grounding wire, which had been used for many years.

(4) If the terminal for the equipment grounding conductor is not visible, the conductor entrance hole shall be marked with the word "green" or "ground," the letters "G" or "GR" or the grounding symbol, as shown in Figure 410-58(b)(4), or otherwise identified by a distinctive green color. If the terminal for the equipment grounding conductor is readily removable, the area adjacent to the terminal shall be similarly marked.

Figure 410-58(b)(4) Grounding symbol.

(c) Grounding Terminal Use. A grounding terminal or grounding-type device shall not be used for purposes other than grounding.

(d) Grounding-Pole Requirements. Grounding-type attachment plugs and mating cord connectors and receptacles shall be designed so that the grounding connection is made before the current-carrying connections. Grounding-type devices shall be designed so grounding poles of attachment plugs cannot be brought into contact with current-carrying parts of receptacles or cord connectors.

The grounding blade of the attachment plug cap of most grounding-type combinations is longer to ensure a "make-first, break-last" grounding connection. In some non-ANSI pin-and-sleeve types, the grounding contact of the receptacle is closer to the face of the receptacle than other contacts, serving the same purpose.

(e) Use. Grounding-type attachment plugs shall be used only with a cord having an equipment grounding conductor.

M. Special Provisions for Flush and Recessed Fixtures

410-64. General. Fixtures installed in recessed cavities in walls or ceilings shall comply with Sections 410-65 through 410-72.

Formal Interpretation 81-6

Reference: Article 410, Part M

Question: Is it intended that fixtures installed in suspended ceilings be subject to the requirements of Part M of Article 410?

Answer. Yes.

Issue Edition: 1981

Reference: 410, Part M

Date: October 1982

410-65. Temperature.

(a) Combustible Material. Fixtures shall be installed so that adjacent combustible material will not be subjected to temperatures in excess of 90°C (194°F).

(b) Fire-Resistant Construction. Where a fixture is recessed in fire-resistant material in a building of fire-resistant construction, a temperature higher than 90°C (194°F) but not higher than 150°C (302°F), shall be considered acceptable if the fixture is plainly marked that it is listed for that service.

(c) Recessed Incandescent Fixtures. Incandescent fixtures shall have thermal protection and shall be identified as thermally protected.

Because many recessed incandescent fixtures are suitable for a wide variety of lamp sizes and types and finish trims, the temperature close to the lamp can vary widely. Therefore, many manufacturers have chosen to locate thermal protectors away from the source of heat, such as in the outlet box, and to design the protector so that it will detect a change in temperature resulting from the addition of thermal insulation around the fixture. This prevents nuisance tripping of the protector (as a result of changing lamp wattage, for example) but still provides protection against overheating arising from thermal insulation around a recessed fixture not designed for such use.

Exception No. 1: Thermal protection shall not be required in a recessed fixture identified for use and installed in poured concrete.

Exception No. 2: Thermal protection shall not be required in a recessed fixture whose design, construction, and thermal performance characteristics are equivalent to a thermally protected fixture, and are identified as inherently protected.

Recessed incandescent fixtures without thermal protection are not permitted unless they are listed and identified as providing equivalent temperature protection by construction design.

Figure 410.10 illustrates a listed recessed fixture installed in direct contact with thermal insulation. Thermal protection is provided to deactivate the lamp should the fixture be mislamped so that it overheats.

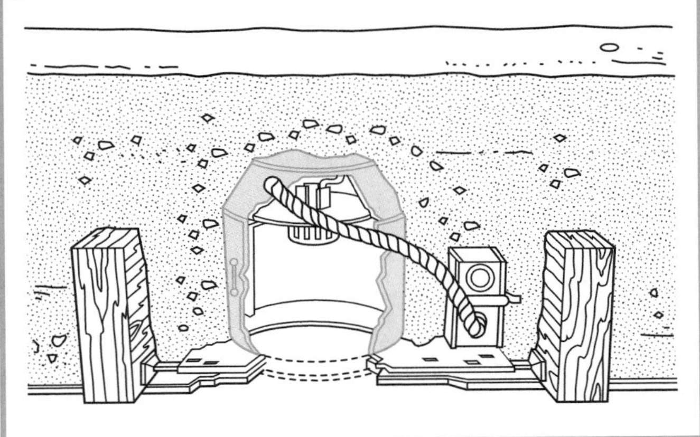

Figure 410.10 *A listed Type IC recessed fixture suitable for use in insulated ceilings installed in direct contact with thermal insulation. (Thomas Industries Inc.)*

410-66. Clearance and Installation.

(a) Clearance.

(1) A recessed fixture that is not identified for contact with insulation shall have all recessed parts spaced at least

½ in. (12.7 mm) from combustible materials. The points of support and the trim finishing off the opening in the ceiling or wall surface shall be permitted to be in contact with combustible materials.

(2) A recessed fixture that is identified for contact with insulation, Type IC, shall be permitted to be in contact with combustible materials at recessed parts, points of support, and portions passing through or finishing off the opening in the building structure.

(b) Installation. Thermal insulation shall not be installed above a recessed fixture or within 3 in. (76 mm) of the recessed fixture's enclosure, wiring compartment, or ballast unless it is identified for contact with insulation, Type IC.

410-67. Wiring.

(a) General. Conductors that have insulation suitable for the temperature encountered shall be used.

(b) Circuit Conductors. Branch-circuit conductors that have an insulation suitable for the temperature encountered shall be permitted to terminate in the fixture.

(c) Tap Conductors. Tap conductors of a type suitable for the temperature encountered shall be permitted to run from the fixture terminal connection to an outlet box placed at least 1 ft (305 mm) from the fixture. Such tap conductors shall be in suitable raceway or Type AC or MC cable of at least 18 in. (450 mm) but not more than 6 ft (1.83 m) in length.

N. Construction of Flush and Recessed Fixtures

410-68. Temperature. Fixtures shall be constructed so that adjacent combustible material will not be subject to temperatures in excess of 90°C (194°F).

410-70. Lamp Wattage Marking. Incandescent lamp fixtures shall be marked to indicate the maximum allowable wattage of lamps. The markings shall be permanently installed, in letters at least ¼ in. (6.35 mm) high, and shall be located where visible during relamping.

410-71. Solder Prohibited. No solder shall be used in the construction of a fixture box.

410-72. Lampholders. Lampholders of the screw-shell type shall be of porcelain or other suitable insulating materials. Where used, cements shall be of the high-heat type.

P. Special Provisions for Electric-Discharge Lighting Systems of 1000 Volts or Less

410-73. General.

(a) Open-Circuit Voltage of 1000 Volts or Less. Equipment for use with electric-discharge lighting systems and designed for an open-circuit voltage of 1000 volts or less shall be of a type intended for such service.

(b) Considered as Energized. The terminals of an electric-discharge lamp shall be considered as energized where any lamp terminal is connected to a circuit of over 300 volts.

(c) Transformers of the Oil-Filled Type. Transformers of the oil-filled type shall not be used.

(d) Additional Requirements. In addition to complying with the general requirements for lighting fixtures, such equipment shall comply with Part P of this article.

(e) Thermal Protection.

(1) The ballast of a fluorescent fixture installed indoors shall have integral thermal protection. Replacement ballasts shall also have thermal protection integral with the ballast.

(2) A simple reactance ballast in a fluorescent fixture with straight tubular lamps shall not be required to be thermally protected.

(3) A ballast in a fluorescent exit fixture shall not have thermal protection.

(4) A ballast in a fluorescent fixture that is used for egress lighting and energized only during an emergency shall not have thermal protection.

Thermal protection integral with the ballast is required for fluorescent fixtures installed indoors. Thermally protected ballasts are required as replacements for nonthermally protected ballasts in older fixtures. Thermally protected fluorescent lamp ballasts intended for use in accordance with Section 410-73(e) are marked "Class P."

As different Class P ballasts have different heating characteristics, the heating characteristics should be considered when selecting replacements for nonthermally protected ballasts. This type of ballast protection is set to open the circuit at a predetermined temperature, to prevent abnormal ballast heat buildup caused by a fault in one or more of the ballast components or by some lampholder or wiring fault.

Section 410-73(e)(3) exempts exit sign fixtures from the thermal protection requirement, because overheating during high ambient conditions could cause the thermal protection to operate. This could impair evacuation during a fire.

Section 410-73(e)(4) exempts egress lighting from the thermal protection requirement for the same reason. However, this applies to egress lighting that is energized only during the emergency condition.

Figure 410.11 illustrates a reactance-type ballast used in series with a 30-watt or less preheat-type fluorescent lamp. This type of ballast does not require thermal protection, and the fixture may be equipped with automatic-type starters (such as used with medicine cabinet fixtures) or a manual momentary contact starter (such as used with desk lamps and some small under-cabinet fixtures).

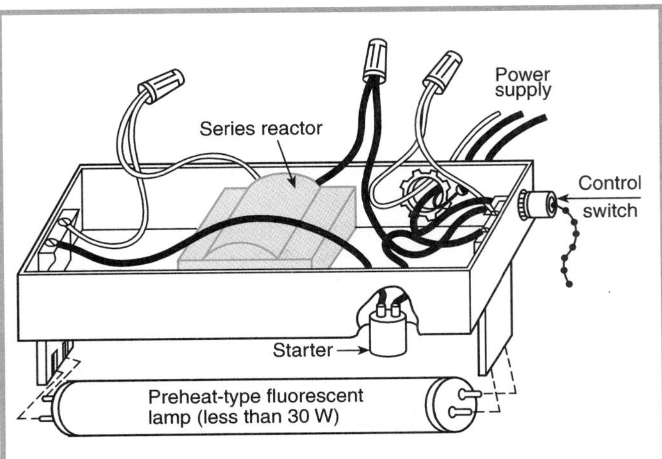

Figure 410.11 *The circuitry for a simple reactance-type ballast for fluorescent lighting.*

(f) High-Intensity Discharge Fixtures.

(1) Recessed high-intensity fixtures designed to be installed in wall or ceiling cavities shall have thermal protection and be identified as thermally protected.

(2) Thermal protection shall not be required in a recessed high-intensity fixture whose design, construction, and thermal performance characteristics are equivalent to a thermally protected fixture and are identified as inherently protected.

(3) Thermal protection shall not be required in a recessed high-intensity discharge fixture identified for use and installed in poured concrete.

(4) A recessed remote ballast for a high-intensity discharge fixture shall have thermal protection that is integral with the ballast and be identified as thermally protected.

410-74. Direct-Current Equipment. Fixtures installed on dc circuits shall be equipped with auxiliary equipment and resistors especially designed and for dc operation, and the fixtures shall be so marked.

410-75. Open-Circuit Voltage Exceeding 300 Volts. Equipment having an open-circuit voltage exceeding 300 volts shall not be installed in dwelling occupancies unless such equipment is designed so that there will be no exposed live parts when lamps are being inserted, are in place, or are being removed.

Fixtures intended for use in other-than-dwelling occupancies are so marked. This usually indicates that the fixture has maintenance features considered beyond the capabilities of the ordinary homeowner or involves voltages in excess of those permitted by this *Code* for dwelling occupancies. For other references to voltage limitations within dwelling units, see Sections 210-6 and 410-80(b).

410-76. Fixture Mounting.

(a) Exposed Ballasts. Fixtures that have exposed ballasts or transformers shall be installed so that such ballasts or transformers will not be in contact with combustible material.

(b) Combustible Low-Density Cellulose Fiberboard. Where a surface-mounted fixture containing a ballast is to be installed on combustible low-density cellulose fiberboard, it shall be listed for this condition or shall be spaced not less than 1½ in. (38 mm) from the surface of the fiberboard. Where such fixtures are partially or wholly recessed, the provisions of Sections 410-64 through 410-72 shall apply.

FPN: Combustible low-density cellulose fiberboard includes sheets, panels, and tiles that have a density of 20 lb/ft^3 (320.36 kg/m^3) or less, and that are formed of bonded plant fiber material but does not include solid or laminated wood, nor fiberboard that has a density in excess of 20 lb/ft^3 (320.36 kg/m^3) or is a material that has been integrally treated with fire-retarding chemicals to the degree that the flame spread in any plane of the material will not exceed 25, determined in accordance with tests for surface burning characteristics of building materials. See *Test Method for Surface Burning Characteristics of Building Materials*, ANSI/ASTM E84-1997.

Fluorescent lamp fixtures intended for mounting on combustible low-density cellulose fiberboard ceilings have been evaluated with thermal insulation above the ceiling in the vicinity of the fixture and are marked "Suitable for Surface Mounting on Combustible Low-Density Cellulose Fiberboard."

Fluorescent lamp fixtures not so marked may be directly mounted against a ceiling surface of *other-than-combustible low-density fiberboard* or may be spaced not less than 1½ in. from the surface of the low-density fiberboard.

410-77. Equipment Not Integral with Fixture.

(a) Metal Cabinets. Auxiliary equipment, including reactors, capacitors, resistors, and similar equipment, where not installed as part of a lighting fixture assembly, shall be enclosed in accessible, permanently installed metal cabinets.

(b) Separate Mounting. Separately mounted ballasts that are intended for direct connection to a wiring system shall not be required to be separately enclosed.

(c) Wired Fixture Sections. Wired fixture sections are paired, with a ballast(s) supplying a lamp or lamps in both. For interconnection between paired units, it shall be permissible to use ⅜-in. flexible metal conduit in lengths not exceeding 25 ft (7.62 m) in conformance with Article 350. Fixture wire operating at line voltage, supplying only the ballast(s) of one of the paired fixtures, shall be permitted in the same raceway as the lamp supply wires of the paired fixtures.

Wired fixture sections are shipped in pairs and marked for use in pairs. Each individual unit includes lamps in odd-numbered quantities (1 or 3 is most common), with the odd lamp in each fixture supplied by a two-lamp ballast located in one fixture of the pair. Two-lamp ballasts are more energy efficient than single-lamp or three-lamp ballasts.

410-78. Autotransformers. An autotransformer that is used to raise the voltage to more than 300 volts, as part of a ballast for supplying lighting units, shall be supplied only by a grounded system.

410-79. Switches. Snap switches shall comply with Section 380-14.

Q. Special Provisions for Electric-Discharge Lighting Systems of More than 1000 Volts

Sections 410-80 through 410-92 apply to interior electric-discharge neon tube-type lighting that contains neon, helium, or argon gas, with or without mercury, at low vapor pressure; long-length fluorescent tube lighting requiring more than 1000 volts; and cold-cathode fluorescent lamp installations arranged to operate with several tubes in series.

410-80. General.

(a) Open-Circuit Voltage Exceeding 1000 Volts. Equipment for use with electric-discharge lighting systems and designed for an open-circuit voltage exceeding 1000 volts shall be of a type intended for such service.

(b) Dwelling Occupancies. Equipment that has an open-circuit voltage exceeding 1000 volts shall not be installed in or on dwelling occupancies.

Section 410-80(b) specifically prohibits electric-discharge lighting (such as neon tube, high-intensity discharge, or fluorescent lighting) that has an open-circuit voltage greater than 1000 volts to be installed in dwelling occupancies. Such lighting systems are often used as decorative lighting as well as for outline lighting and signs. Where used as outline lighting or signs in nonresidential occupancies, see also Article 600.

(c) Live Parts. The terminal of an electric-discharge lamp shall be considered as a live part where any lamp terminal is connected to a circuit of over 300 volts.

(d) Additional Requirements. In addition to complying with the general requirements for lighting fixtures, such equipment shall comply with Part Q of this article.

FPN: For signs and outline lighting, see Article 600.

410-81. Control.

(a) Disconnection. Fixtures or lamp installations shall be controlled either singly or in groups by an externally operable switch or circuit breaker that opens all ungrounded primary conductors.

(b) Within Sight or Locked Type. The switch or circuit breaker shall be located within sight from the fixtures or lamps, or it shall be permitted elsewhere if it is provided with a means for locking in the open position.

The requirement in Section 410-81(b) is intended to help protect the service person against the disconnecting means being turned on or closed while the equipment is being serviced.

410-82. Lamp Terminals and Lampholders. Parts that must be removed for lamp replacement shall be hinged or held captive. Lamps or lampholders will be designed so that there shall be no exposed live parts when lamps are being inserted or are being removed.

410-83. Transformer Ratings. Transformers and ballasts shall have a secondary open-circuit voltage of not over 15,000 volts with an allowance on test of 1000 volts additional. The secondary-current rating shall not be more than 120 milliamperes if the open-circuit voltage is over 7500 volts, and not more than 240 milliamperes if the open-circuit voltage is 7500 volts or less.

410-84. Transformer Type. Transformers shall be enclosed and listed.

410-85. Transformers and Secondary Connections. The high-voltage windings of transformers shall not be connected in series or parallel.

410-86. Transformer Locations.

(a) Accessible. Transformers shall be accessible after installation.

(b) Secondary Conductors. Transformers shall be installed as near to the lamps as practicable to keep the secondary conductors as short as possible.

(c) Adjacent to Combustible Materials. Transformers shall be located so that adjacent combustible materials will not be subjected to temperatures in excess of 90°C (194°F).

410-87. Transformer Loading. The lamps connected to any transformer shall be of such length and characteristics so as not to cause a condition of continuous overvoltage on the transformer.

Transformers are required to be enclosed and listed, and they should be rated to supply the proper current and voltage for the lamp or tube. For field-installed skeleton tubing, this requirement is restated in Section 600-41(a).

410-88. Wiring Method — Secondary Conductors. Conductors shall be installed in accordance with Section 600-32.

Table 310-13, which lists various types of insulated conductors, does not include the type of cable required in Sections 410-88 and 600-31(b). There are, however, standards for this type of cable. The following information is an excerpt from the 1998 UL *General Information for Electrical Equipment Directory* (UL White Book), Gas Tube Sign and Ignition Cable (ZJQX).

> Gas tube sign and ignition cable is listed as single conductor Type GTO-5 (5000 volts), GTO-10 (10,000 volts), or GTO-15 (15,000 volts), in sizes Nos. 18-10 AWG copper and Nos. 12–10 aluminum and copper-clad aluminum. This material is intended for use with gas tube signs, oil burners, and inside lighting.
>
> "L" used as a suffix in combination with any of the preceding type letter designations indicates that an outer covering of lead has been applied.
>
> Wires with copper-clad aluminum are surface printed "AL (CU-CLAD)" or "CU-CLAD AL." Wires with aluminum conductors are surface printed "AL."

410-89. Lamp Supports. Lamps shall be adequately supported as required in Section 600-41.

410-90. Exposure to Damage. Lamps shall not be located where normally exposed to physical damage.

410-91. Marking. Each fixture or each secondary circuit of tubing having an open-circuit voltage of over 1000 volts shall have a clearly legible marking in letters not less than ¼ in. (6.35 mm) high reading: "Caution . . . volts." The voltage indicated shall be the rated open-circuit voltage.

410-92. Switches. Snap switches shall comply with Section 380-14.

R. Lighting Track

410-100. Definition. *Lighting track* is a manufactured assembly designed to support and energize lighting fixtures that are capable of being readily repositioned on the track. Its length may be altered by the addition or subtraction of sections of track.

410-101. Installation.

(a) Lighting Track. Lighting track shall be permanently installed and permanently connected to a branch circuit. Only lighting track fittings shall be installed on lighting track. Lighting track fittings shall not be equipped with general-purpose receptacles.

A lighting track fitting differs from a fitting as defined in Article 100 in that it usually performs both an electrical and a mechanical function.

It is not the intent that such assemblies be a system for locating convenience receptacles or an alternative for required receptacle outlets such as those required in Section 210-62. Lighting track can be removed and relocated and therefore is not a substitute for required outlets.

(b) Connected Load. The connected load on lighting track shall not exceed the rating of the track. Lighting track shall be supplied by a branch circuit having a rating not more than that of the track.

(c) Locations Not Permitted. Lighting track shall not be installed in the following locations:

(1) Where likely to be subjected to physical damage
(2) In wet or damp locations
(3) Where subject to corrosive vapors
(4) In storage battery rooms
(5) In hazardous (classified) locations
(6) Where concealed
(7) Where extended through walls or partitions
(8) Less than 5 ft (1.52 m) above the finished floor except where protected from physical damage or track operating at less than 30 volts rms open-circuit voltage
(9) Within the zone measured 3 ft (914 mm) horizontally and 8 ft (2.44 m) vertically from the top of the bathtub rim

Low-voltage lighting track operating at less than 30 volts is permitted to be installed less than 5 ft above the floor.

(d) Support. Fittings identified for use on lighting track shall be designed specifically for the track on which they are to be installed. They shall be securely fastened to the track, maintain polarization and grounding, and shall be designed to be suspended directly from the track.

Moved for the 1999 *Code,* previous Section 410-102 that addressed track lighting loads was relocated to Section 220-12(b).

The volt-ampere (VA) load for 2 ft of track was reduced from 180 volt-amperes to 150 volt-amperes for the 1996 *Code* because a value of 150 VA is more consistant with standard lamp values for 2 ft of track. It should be understood that this section is not intended to limit the number of feet of track on a single branch circuit nor is it intended to limit the number of fixtures on an individual track. Rather, this section is intended to be used for load calculations of feeders and services.

Example

Suppose a lighting plan shows 62.5 linear ft of single circuit track lighting for a small department store featuring clothing. Since the actual track lighting fixtures are owner supplied, neither the quantity of track lighting fixtures nor the lamp size are specified. What is the minimum calculated load associated with the track lighting that must be added to the service or feeder supplying this store?

Answer

According to Section 220-12(b), the minimum calculated load to be added to the service or feeder supplying this track light installation is determined as follows:

62.5 ft ÷ 2 ft = 31.25

31.25 must be rounded up to 32

32 × 150 VA = 4800 VA

4800 VA is the minimum load that must be added to the service or feeder calculation.

It is important to note that the branch circuits supplying this installation are covered in Section 410-101(b). For the track lighting branch-circuit load, the maximum load on the track cannot exceed the rating of the branch circuit supplying the track. Also, the track must be supplied by a branch circuit having a rating not exceeding the rating of the track. The track length does not enter into the branch-circuit calculation.

410-103. Heavy-Duty Lighting Track. Heavy-duty lighting track is lighting track identified for use exceeding 20 amperes. Each fitting attached to a heavy-duty lighting track shall have individual overcurrent protection.

410-104. Fastening. Lighting track shall be securely mounted so that each fastening will be suitable for supporting the maximum weight of fixtures that can be installed. Unless identified for supports at greater intervals, a single section 4 ft (1.22 m) or shorter in length shall have two supports, and, where installed in a continuous row, each individual section of not more than 4 ft (1.22 m) in length shall have one additional support.

410-105. Construction Requirements.

(a) Construction. The housing for the lighting track system shall be of substantial construction to maintain rigidity. The conductors shall be installed within the track housing permitting insertion of a fixture, and designed to prevent tampering and accidental contact with live parts. Components of lighting track systems of different voltages shall not be interchangeable. The track conductors shall be a minimum No. 12 or equal, and shall be copper. The track system ends shall be insulated and capped.

(b) Grounding. Lighting track shall be grounded in accordance with Article 250, and the track sections shall be securely coupled to maintain continuity of the circuitry, polarization, and grounding throughout.

Article 411 — Lighting Systems Operating at 30 Volts or Less

Contents

411-1. Scope. This article covers lighting systems operating at 30 volts or less and their associated components.

Article 411 was added for the 1996 *Code* to address low-voltage lighting systems. Article 411 is intended to cover low-voltage interior lighting and low-voltage exterior (landscape) lighting installations. It covers systems operating at 30 volts rms or 42.4 volts peak or less, having a maximum output of 600 volt-amperes.

411-2. Lighting Systems Operating at 30 Volts or Less. A lighting system consisting of an isolating power supply operating at 30 volts (42.4 volts peak) or less, under any load condition, with one or more secondary circuits, each limited to 25 amperes maximum, supplying lighting fixtures and associated equipment identified for the use.

411-3. Listing Required. Lighting systems operating at 30 volts or less shall be listed for the purpose.

411-4. Locations Not Permitted. Lighting systems operating at 30 volts or less shall not be installed (1) where concealed or extended through a building wall, unless using a wiring method specified in Chapter 3, or (2) within 10 ft (3.05 m) of pools, spas, fountains, or similar locations, except as permitted by Article 680.

The language contained in Section 411-4 recognizes that shock and fire hazards still exist, even with low-voltage systems.

411-5. Secondary Circuits.

(a) Grounding. Secondary circuits shall not be grounded.

(b) Isolation. The secondary circuit shall be insulated from the branch circuit by an isolating transformer.

(c) **Bare Conductors.** Exposed bare conductors and current-carrying parts shall be permitted. Bare conductors shall not be installed less than 7 ft (2.2 m) above the finished floor, unless specifically listed for a lower installation height.

411-6. Branch Circuit. Lighting systems operating at 30 volts or less shall be supplied from a maximum 20-ampere branch circuit.

411-7. Hazardous (Classified) Locations. Where installed in hazardous (classified) locations, these systems shall conform with Articles 500 through 517 in addition to this article.

Article 422 — Appliances

Contents

A. General

422-1. Scope. This article covers electric appliances used in any occupancy.

Article 422 was completely reorganized for the 1999 *Code*. Except for new Section 422-15 on central vacuum outlet assemblies, the reorganization did not significantly alter the 1996 requirements for appliances.

Article 422 covers electric appliances that may be used in a dwelling unit or in commercial and industrial locations. It also covers appliances that may be fastened in place or cord- and plug-connected, such as air-conditioning units, dishwashers, heating appliances, water heaters, infrared heating lamps, and so on. See Section 422-3 for the requirements of other articles. Also see Article 100 for the definition of *appliance.*

422-3. Other Articles. Appliances for use in hazardous (classified) locations shall comply with Articles 500 through 517.

The requirements of Article 430 shall apply to the installation of motor-operated appliances and the requirements of Article 440 shall apply to the installation of appliances containing a hermetic refrigerant motor-compressor(s), except as specifically amended in this article.

422-4. Live Parts. Appliances shall have no live parts normally exposed to contact other than those parts functioning as open-resistance heating elements, such as the heating element of a toaster, which are necessarily exposed.

B. Installation

422-10. Branch-Circuit Rating. This section specifies the ratings of branch circuits capable of carrying appliance current without overheating under the conditions specified.

Conductors that form integral parts of appliances are tested as part of the listing or labeling process.

(a) Individual Circuits. The rating of an individual branch circuit shall not be less than the marked rating of the appliance or the marked rating of an appliance having combined loads as provided in Section 422-62.

The rating of an individual branch-circuit for motor-operated appliances not having a marked rating shall be in accordance with Part B of Article 430.

The branch-circuit rating for an appliance that is continuously loaded, other than a motor-operated appliance, shall not be less than 125 percent of the marked rating; or not less than 100 percent of the marked rating if the branch-circuit device and its assembly are listed for continuous loading at 100 percent of its rating.

Branch circuits for household cooking appliances shall be permitted to be in accordance with Table 220-19.

(b) Circuits Supplying Two or More Loads. For branch circuits supplying appliance and other loads, the rating shall be determined in accordance with Section 210-23.

422-11. Overcurrent Protection. Appliances shall be protected against overcurrent in accordance with (a) through (g) and Section 422-10.

(a) Branch-Circuit Overcurrent Protection. Branch circuits shall be protected in accordance with Section 240-3.

If a protective device rating is marked on an appliance, the branch-circuit overcurrent device rating shall not exceed the protective device rating marked on the appliance.

If a labeled or listed appliance is provided with installation instructions from the manufacturer, the branch-circuit size is not permitted to be less than the minimum size stated in the installation instructions. See Section 110-3(b) regarding installation and use of listed or labeled equipment.

(b) Household-Type Appliance with Surface Heating Elements. A household-type appliance with surface heating elements having a maximum demand of more than 60 amperes computed in accordance with Table 220-19 shall have its power supply subdivided into two or more circuits, each of which is provided with overcurrent protection rated at not over 50 amperes.

(c) Infrared Lamp Commercial and Industrial Heating Appliances. Infrared lamp commercial and industrial heating appliances shall have overcurrent protection not exceeding 50 amperes.

(d) Open-Coil or Exposed Sheathed-Coil Types of Surface Heating Elements in Commercial-Type Heating Appliances. Open-coil or exposed sheathed-coil types of surface heating elements in commercial-type heating appliances shall be protected by overcurrent protective devices rated at not over 50 amperes.

(e) Single Nonmotor-Operated Appliance. If the branch circuit supplies a single nonmotor-operated appliance, the rating of overcurrent protection shall

(1) Not exceed that marked on the appliance;
(2) If the overcurrent protection rating is not marked and the appliance is rated 13.3 amperes or less, not exceed 20 amperes; or
(3) If the overcurrent protection rating is not marked and the appliance is rated over 13.3 amperes, not exceed 150 percent of the appliance rated current. Where 150 percent of the appliance rating does not correspond to a standard overcurrent device ampere rating, the next higher standard rating shall be permitted.

(f) Electric Heating Appliances Employing Resistance-Type Heating Elements Rated More than 48 Amperes.

(1) Electric heating appliances employing resistance-type heating elements rated more than 48 amperes, other than household appliances with surface heating elements covered by Section 422-11(b), and commercial-type heating appliances covered by Section 422-11(d), shall have the heating elements subdivided. Each subdivided load shall not exceed 48 amperes and shall be protected at not more than 60 amperes.

These supplementary overcurrent protective devices shall be (1) factory-installed within or on the heater enclosure or provided as a separate assembly by the heater manufacturer; (2) accessible; and (3) suitable for branch-circuit protection.

The main conductors supplying these overcurrent protective devices shall be considered branch-circuit conductors.

(2) Commercial kitchen and cooking appliances using sheathed-type heating elements not covered in Section 422-11(d) shall be permitted to be subdivided into circuits not exceeding 120 amperes and protected at not more than 150 amperes where one of the following is met:

(a) Elements are integral with and enclosed within a cooking surface
(b) Elements are completely contained within an enclosure identified as suitable for this use
(c) Elements are contained within an ASME-rated and stamped vessel

(3) Water heaters and steam boilers employing resistance-type immersion electric heating elements contained in an ASME-rated and stamped vessel shall be permitted to be subdivided into circuits not exceeding 120 amperes and protected at not more than 150 amperes.

(g) Motor-Operated Appliances. Motors of motor-operated appliances shall be provided with overload protection in accordance with Part C of Article 430. Hermetic refrigerant motor-compressors in air-conditioning or refrigerating equipment shall be provided with overload protection in accordance with Part F of Article 440. Where appliance overcurrent protective devices that are separate from the appliance are required, data for selection of these devices shall be marked on the appliance. The minimum marking shall be that specified in Sections 430-7 and 440-4.

422-12. Central Heating Equipment. Central heating equipment other than fixed electric space-heating equipment shall be supplied by an individual branch circuit.

Exception: Auxiliary equipment, such as a pump, valve, humidifier, or electrostatic air cleaner directly associated with the heating equipment, shall be permitted to be connected to the same branch circuit.

The exception to Section 422-12 permits the electric motors, ignition systems, controls, and so on, of fossil-fuel–fired central heating equipment to be connected to the same individual branch circuit, as defined in Article 100 under *branch circuit, individual.*

422-13. Storage-Type Water Heaters. A branch circuit supplying a fixed storage-type water heater that has a capacity of 120 gal (454.2 L) or less shall have a rating not less than 125 percent of the nameplate rating of the water heater.

FPN: For branch-circuit rating, see Section 422-10.

422-14. Infrared Lamp Industrial Heating Appliances. Infrared industrial heating appliance lampholders shall be permitted to be connected to any of the branch circuits in Article 210 and, in industrial occupancies, shall be permitted to be operated in series on circuits of over 150 volts to ground provided the voltage rating of the lampholders is not less than the circuit voltage.

Each section, panel, or strip carrying a number of infrared lampholders (including the internal wiring of such section, panel, or strip) shall be considered an appliance. The terminal connection block of each such assembly shall be considered an individual outlet.

422-15. Central Vacuum Outlet Assemblies.

(a) Listed central vacuum outlet assemblies shall be permitted to be connected to a branch circuit in accordance with Section 210-23(a).

(b) The ampacity of the connecting conductors shall not be less than the ampacity of the branch circuit conductors to which they are connected.

(c) An equipment grounding conductor shall be used where the central vacuum outlet assembly has accessible noncurrent-carrying metal parts.

Added for the 1999 *Code,* Section 422-15 permits listed central vacuum outlet devices to be connected to the ordinary 15- or 20-ampere general-purpose branch circuits that may be located in the same area as the vacuum outlet is installed. Starting and stopping of the central vacuum system is achieved by a Class 2 control circuit that originates at the main unit of the central vacuum system. The circuit is switched at each outlet by the insertion or removal of the matching vacuum hose in the outlet.

422-16. Flexible Cords.

(a) General. Flexible cord shall be permitted (1) for the connection of appliances to facilitate their frequent interchange or to prevent the transmission of noise or vibration or (2) to facilitate the removal or disconnection of appliances that are fastened in place, where the fastening means and mechanical connections are specifically designed to permit ready removal for maintenance or repair, and the appliance is intended or identified for flexible cord connection.

It should be understood that a cord-connected appliance is required to be specifically designed to be readily removable for maintenance and repair, mechanically as well as electrically.

(b) Specific Appliances.

(1) Electrically operated kitchen waste disposers shall be permitted to be cord- and plug-connected with a flexible cord identified as suitable for the purpose in the installation instructions of the appliance manufacturer, where all of the following conditions are met.

(a) The flexible cord shall be terminated with a grounding type attachment plug.

Exception: A listed kitchen waste disposer distinctly marked to identify it as protected by a system of double insulation, or its equivalent, shall not be required to be terminated with a grounding-type attachment plug.

(b) The length of the cord shall not be less than 18 in. (457 mm) and not over 36 in. (914 mm).
(c) Receptacles shall be located to avoid physical damage to the flexible cord.
(d) The receptacle shall be accessible.

The kitchen waste disposer illustrated in Figure 422.1 is an example of a cord- and plug-connected appliance with mechanical connections designed to permit removal. The cord and receptacle are designed and installed in accordance with Section 422-16(b)(1).

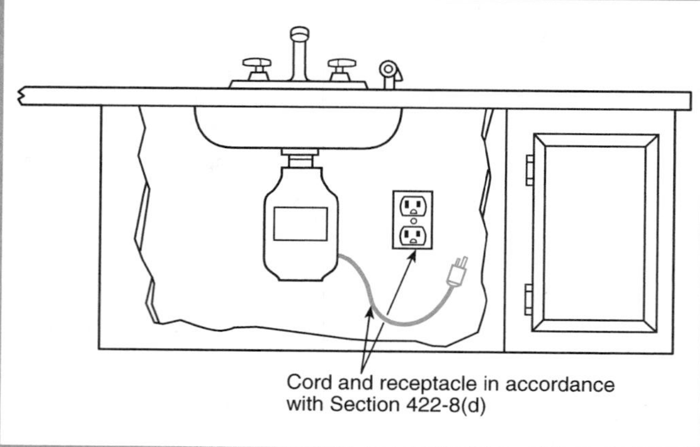

Figure 422.1 A cord- and plug-connected kitchen waste disposer.

(2) Built-in dishwashers and trash compactors shall be permitted to be cord- and plug-connected with a flexible cord identified as suitable for the purpose in the installation instructions of the appliance manufacturer where all of the following conditions are met.

(a) The flexible cord shall be terminated with a grounding type attachment plug.

Exception: A listed dishwasher or trash compactor distinctly marked to identify it as protected by a system of double insulation, or its equivalent, shall not be required to be terminated with a grounding-type attachment plug.

(b) The length of the cord shall be 3 ft to 4 ft (0.914 m to 1.22 m) measured from the face of the attachment plug to the plane of the rear of the appliance.
(c) Receptacles shall be located to avoid physical damage to the flexible cord.
(d) The receptacle shall be located in the space occupied by the appliance or adjacent thereto.
(e) The receptacle shall be accessible.

(3) Wall-mounted ovens and counter-mounted cooking units complete with provisions for mounting and for making electrical connections shall be permitted to be permanently connected or, only for ease in servicing or for installation, cord- and plug-connected.

A separable connector or a plug and receptacle combination in the supply line to an oven or cooking unit shall (1) not be installed as the disconnecting means required by Section 422-30 and (2) be approved for the temperature of the space in which it is located.

422-17. Protection of Combustible Material. Each electrically heated appliance that is intended by size, weight, and service to be located in a fixed position shall be placed so as to provide ample protection between the appliance and adjacent combustible material.

422-18. Support of Ceiling-Suspended (Paddle) Fans.

(a) Ceiling-Suspended (Paddle) Fans 35 lb (15.88 kg) or Less. Ceiling-suspended (paddle) fans that do not exceed 35 lb (15.88 kg) in weight, with or without accessories, shall be permitted to be supported by outlet boxes identified for such use and supported in accordance with Sections 370-23 and 370-27.

Section 370-27(c) does not permit standard-type boxes to support ceiling fans unless provided with supplemental support as shown in Figure 370.8. However, Section 370-27(c), Exception, does permit boxes listed for the application as the sole support for the fan. Section 422-18(a) permits boxes identified for fan support to be used to support ceiling fans that do not exceed 35 lb, such as the box and fan shown in Figure 422.2.

(b) Ceiling-Suspended (Paddle) Fans Exceeding 35 lb (15.88 kg). Ceiling-suspended (paddle) fans exceeding 35 lb (15.88 kg) in weight, with or without accessories, shall be supported independently of the outlet box. See Section 370-23.

Section 422-18(b) requires listed fans exceeding 35 lb, with or without accessories, to be supported independently from the outlet box as shown in Figure 370.8. Additionally, Section 370-23 requires boxes to be rigid and securely fastened in place. A new exception was added to Section 422-18(b) for the 1999 *Code,* allowing a box identified or listed for the application to provide the sole support for this heavier class of fan.

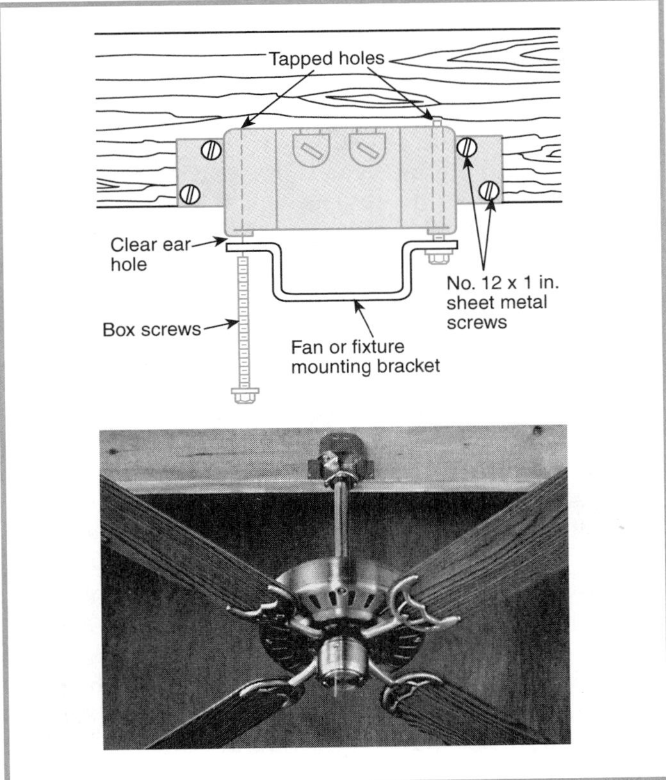

Figure 422.2 An example of supporting a ceiling fan (35 lb or less) with a box identified for such use. (Raco Inc.)

Exception: Listed outlet boxes or outlet box systems that are identified for the purpose shall be permitted to support ceiling-suspended fans, with or without asccessories, that weigh 70 lb (31.76 kg) or less.

422-20. Other Installation Methods. Appliances employing methods of installation other than covered by this article shall be permitted to be used only by special permission.

C. Disconnecting Means

422-30. General. A means shall be provided to disconnect each appliance from all ungrounded conductors in accordance with the following sections of Part C. If an appliance is supplied by more than one source, the disconnecting means shall be grouped and identified.

422-31. Disconnection of Permanently Connected Appliances.

(a) Rated at Not Over 300 Volt-Amperes or ⅛ Horsepower. For permanently connected appliances rated at not over 300 volt-amperes or ⅛ hp, the branch-circuit overcurrent device shall be permitted to serve as the disconnecting means.

(b) Appliances Rated Over 300 Volt-Amperes or ⅛ Horsepower. For permanently connected appliances rated over 300 volt-amperes or ⅛ hp, the branch-circuit switch or circuit breaker shall be permitted to serve as the disconnecting means where the switch or circuit breaker is within sight from the appliance or is capable of being locked in the open position.

FPN No. 1: For motor-driven appliances of more than ⅛ hp, see Section 422-35.

FPN No. 2: For appliances employing unit switches, see Section 422-33.

422-32. Disconnection of Cord- and Plug-Connected Appliances.

(a) Separable Connector or an Attachment Plug and Receptacle. For cord- and plug-connected appliances, an accessible separable connector or an accessible plug and receptacle shall be permitted to serve as the disconnecting means. Where the separable connector or plug and receptacle are not accessible, cord- and plug-connected appliances shall be provided with disconnecting means in accordance with Section 422-31.

(b) Connection at the Rear Base of a Range. For cord- and plug-connected household electric ranges, an attachment plug and receptacle connection at the rear base of a range, if it is accessible from the front by removal of a drawer, shall be considered as meeting the intent of Section 422-32(a).

(c) Rating. The rating of a receptacle or of a separable connector shall not be less than the rating of any appliance connected thereto.

Exception: Demand factors authorized elsewhere in this Code shall be permitted to be applied to the rating of a receptacle or of a separable connector.

422-33. Unit Switch(es) as Disconnecting Means. A unit switch(es) with a marked off position that is a part of an appliance and disconnects all ungrounded conductors shall be permitted as the disconnecting means required by this article where other means for disconnection are provided in the following types of occupancies.

(a) Multifamily Dwellings. In multifamily dwellings, the other disconnecting means shall be within the dwelling unit, or on the same floor as the dwelling unit in which the appliance is installed, and shall be permitted to control lamps and other appliances.

(b) Two-Family Dwellings. In two-family dwellings, the other disconnecting means shall be permitted either inside or outside of the dwelling unit in which the appliance is installed. In this case, an individual switch or circuit breaker for the dwelling unit shall be permitted and shall also be permitted to control lamps and other appliances.

(c) One-Family Dwellings. In one-family dwellings, the service disconnecting means shall be permitted to be the other disconnecting means.

(d) Other Occupancies. In other occupancies, the branch-circuit switch or circuit breaker, where readily accessible for servicing of the appliance, shall be permitted as the other disconnecting means.

422-34. Switch and Circuit Breaker to Be Indicating. Switches and circuit breakers used as disconnecting means shall be of the indicating type.

422-35. Disconnecting Means for Motor-Driven Appliance. If a switch or circuit breaker serves as the disconnecting means for a permanently connected motor-driven appliance of more than ⅛ hp, it shall be located within sight from the motor controller and shall comply with Part J of Article 430.

Exception: If a motor-driven appliance of more than ⅛ hp is provided with a unit switch that complies with Section 422-33(a), (b), (c), or (d), the switch or circuit breaker serving as the other disconnecting means shall be permitted to be out of sight from the motor controller.

D. Construction

422-40. Polarity in Cord- and Plug-Connected Appliances. If the appliance is provided with a manually operated, line-connected, single-pole switch for appliance on-off operation, an Edison-base lampholder, or a 15- or 20-ampere receptacle, the attachment plug shall be of the polarized or grounding type.

A 2-wire, nonpolarized attachment plug shall be permitted to be used on a listed double-insulated shaver.

FPN: For polarity of Edison-base lampholders, see Section 410-42(a).

422-41. Cord- and Plug-Connected Appliances Subject to Immersion. Cord- and plug-connected portable freestanding hydromassage units and hand-held hair dryers shall be constructed to provide protection for personnel against electrocution when immersed while in the "on" or "off" position.

Although receptacles in bathrooms of dwelling units have been required to be protected by ground-fault circuit interrupters since the 1975 edition of the *Code,* many receptacles in existing bathrooms are not so protected. Cord- and plug-connected appliances such as hand-held hair dryers, curling irons, and so on, which can and have accidentally fallen into bathtubs, causing fatalities, are required to be provided with some form of protective device that is part of the appliance. Three types of protectors comply with this requirement: ALCIs, IDCIs, and GFCIs.

Appliance-leakage circuit interrupters (ALCIs) are intended to de-energize the supply to the appliance when leakage current exceeds a predetermined value. Immersion-detector circuit interrupters (IDCIs) de-energize the supply when a liquid causes a conductive path between a live part and a sensor. Ground-fault circuit interrupters (GFCIs) de-energize the supply when the current to ground exceeds a predetermined value.

422-42. Signals for Heated Appliances. In other than dwelling-type occupancies, each electrically heated appliance or group of appliances intended to be applied to combustible material shall be provided with a signal or an integral temperature-limiting device.

A common way to provide a signal light for electrically heated appliances in commercial or industrial locations is to use a red light connected to and within sight of the appliance so as to indicate that the appliance is energized and operating.

No signal is required for an electrically heated appliance provided with an integral thermostat or equivalent that limits the temperature to which it can heat.

422-43. Flexible Cords.

(a) Heater Cords. All cord- and plug-connected smoothing irons and electrically heated appliances that are rated at more than 50 watts and produce temperatures in excess of 121°C (250°F) on surfaces with which the cord is likely to be in contact shall be provided with one of the types of approved heater cords listed in Table 400-4.

(b) Other Heating Appliances. All other cord- and plug-connected electrically heated appliances shall be connected with one of the approved types of cord listed in Table 400-4, selected in accordance with the usage specified in that table.

422-44. Cord- and Plug-Connected Immersion Heaters. Electric heaters of the cord- and plug-connected immersion type shall be constructed and installed so that current-carrying parts are effectively insulated from electrical contact with the substance in which they are immersed.

422-45. Stands for Cord- and Plug-Connected Appliances. Each smoothing iron and other cord- and plug-connected electrically heated appliance intended to be applied to combustible material shall be equipped with an approved stand, which shall be permitted to be a separate piece of equipment or a part of the appliance.

422-46. Flatirons. Electrically heated smoothing irons shall be equipped with an identified temperature-limiting means.

422-47. Water Heater Controls. All storage or instantaneous-type water heaters other than

(a) Storage water heaters that are identified as being suitable for use with supply water temperature of 82°C (180°F) or above and a capacity of 60 kW or above, or

(b) Instantaneous-type water heaters that are identified as being suitable for such use, with a capacity of 1 gal (3.785 L) or less shall be equipped with a temperature-limiting means in addition to its control thermostat to disconnect all ungrounded conductors, and such means shall be (1) installed to sense maximum water temperature and (2) be either a trip-free, manually reset type or a type having a replacement element. Such water heaters shall be marked to require the installation of a temperature and pressure relief valve.

FPN: See *Relief Valves and Automatic Gas Shutoff Devices for Hot Water Supply Systems,* ANSI Z21.22-1986.

422-48. Infrared Lamp Industrial Heating Appliances.

(a) 300 Watts or Less. Infrared heating lamps rated at 300 watts or less shall be permitted with lampholders of the medium-base, unswitched porcelain type or other types identified as suitable for use with infrared heating lamps rated 300 watts or less.

(b) Over 300 Watts. Screw-shell lampholders shall not be used with infrared lamps rated over 300 watts, unless the lampholders are identified as being suitable for use with infrared heating lamps rated over 300 watts.

Infrared (heat) radiation lamps are tungsten-filament incandescent lamps similar in appearance to lighting lamps; however, they are designed to operate at a lower temperature, thus transferring more heat radiation and less light intensity. Infrared lamps are used for a variety of heating and drying purposes in industrial locations.

422-49. High-Pressure Spray Washers. All single-phase cord- and plug-connected high-pressure spray washing machines rated at 250 volts or less shall be provided with factory-installed ground-fault circuit-interrupter protection for personnel. The ground-fault circuit interrupter shall be an integral part of the attachment plug or shall be located in the supply cord within 12 in. (305 mm) of the attachment plug.

Section 422-49 permits high-pressure spray washers to be used without a ground-fault circuit interrupter as part of the supply cord for washers rated 3-phase or over 250 volts.

422-50. Cord- and Plug-Connected Pipe Heating Assemblies. Cord- and plug-connected pipe heating assemblies intended to prevent freezing of piping shall be listed.

The *listing* requirement was added in the 1993 *Code.* It was included as a result of data submitted that substantiated numerous fires initiated by heat tapes. Additional requirements for ground-fault protection are in Section 427-22.

E. Marking

422-60. Nameplate.

(a) Nameplate Marking. Each electric appliance shall be provided with a nameplate giving the identifying name and the rating in volts and amperes, or in volts and watts. If the appliance is to be used on a specific frequency or frequencies, it shall be so marked.

Where motor overload protection external to the appliance is required, the appliance shall be so marked.

FPN: See Section 422-11 for overcurrent protection requirements.

(b) To Be Visible. Marking shall be located so as to be visible or easily accessible after installation.

422-61. Marking of Heating Elements. All heating elements that are rated over one ampere, replaceable in the field, and a part of an appliance shall be legibly marked with the ratings in volts and amperes, or in volts and watts, or with the manufacturer's part number.

422-62. Appliances Consisting of Motors and Other Loads. Appliances shall be marked in accordance with (a) or (b).

(a) Marking. In addition to the marking required in Section 422-60, the marking on an appliance that consists of a motor with other load(s) or motors with or without other load(s), not provided with factory-equipped cords and attachment plugs, or where both the minimum supply circuit conductor ampacity and maximum rating of the circuit overcurrent protective device are not more than 15 amperes, shall specify the minimum supply circuit conductor ampacity and the maximum rating of the circuit overcurrent protective device.

(b) Alternate Marking Method. For appliances, other than those factory-equipped with cords and attachment plugs, complying with Section 422-60, an alternate marking method shall be permitted to specify the rating of the largest motor in volts and amperes, and the additional load(s) in volts and amperes, or volts and watts in addition to the marking required in Section 422-60. The ampere rating of a motor ⅛ hp or less or a nonmotor load 1 ampere or less shall not be required to be marked unless such loads constitute the principal load.

Article 424 — Fixed Electric Space-Heating Equipment

Contents

A. General

424-1. Scope. This article covers fixed electric equipment used for space heating. For the purpose of this article, heating equipment shall include heating cable, unit heaters, boilers, central systems, or other approved fixed electric space-heating equipment. This article shall not apply to process heating and room air conditioning.

The electric furnace shown in Figure 424.1 is one example of the type of fixed electric equipment used for space heating that this article covers.

Figure 424.1 *An electric furnace with cooling coils for air conditioning.*

424-2. Other Articles. All requirements of this *Code* shall apply where applicable. Fixed electric space-heating equipment for use in hazardous (classified) locations shall comply with Articles 500 through 517. Fixed electric space-heating equipment incorporating a hermetic refrigerant motor-compressor shall also comply with Article 440.

424-3. Branch Circuits.

(a) Branch-Circuit Requirements. Individual branch circuits shall be permitted to supply any size fixed electric space-heating equipment.

Branch circuits supplying two or more outlets for fixed electric space-heating equipment shall be rated 15, 20, or 30 amperes. In other than residential occupancies, fixed infrared heating equipment shall be permitted to be supplied from branch circuits rated not over 50 amperes.

(b) Branch-Circuit Sizing. The ampacity of the branch-circuit conductors and the rating or setting of overcurrent protective devices supplying fixed electric space-heating equipment consisting of resistance elements with or without a motor shall not be less than 125 percent of the total load of the motors and the heaters. The rating or setting of overcurrent protective devices shall be permitted in accordance with Section 240-3(b). A contactor, thermostat, relay, or similar device, listed for continuous operation at 100 percent of its rating, shall be permitted to supply its full-rated load as provided in Section 210-19(a), Exception.

The size of the branch-circuit conductors and overcurrent protective devices supplying fixed electric space-heating equipment, including a hermetic refrigerant motor-compressor with or without resistance units, shall be computed in accordance with Sections 440-34 and 440-35. The provisions of this section shall not apply to conductors that form an integral part of approved fixed electric space-heating equipment.

The sizing of branch-circuit conductors supplying fixed electric space-heating equipment at 125 percent of the total load of the heaters (and motors) is predicated on the need to protect overcurrent devices, particularly in panelboards, from overheating during periods of prolonged operation. See Section 384-16(d). Since the branch-circuit conductors and overcurrent devices are required to be rated at 125 percent of the total load, an additional increase of 125 percent is not required for continuous loads. The requirement in Section 424-3(b) is phrased to avoid arguments as to whether or not the heating equipment is a "continuous load" as defined in Article 100.

B. Installation

424-9. General. All fixed electric space-heating equipment shall be installed in an approved manner.

Permanently installed electric baseboard heaters equipped with factory-installed receptacle outlets, or outlets provided as a separate listed assembly, shall be permitted in lieu of a receptacle outlet(s) that is required by Section 210-50(b). Such receptacle outlets shall not be connected to the heater circuits.

FPN: Listed baseboard heaters include instructions that may not permit their installation below receptacle outlets.

The second paragraph of Section 424-9 restates the permission granted in Section 210-52, second paragraph, that is, it allows factory-installed receptacle outlets in electric baseboard heaters to satisfy the spacing requirements for receptacle outlets in dwelling units according to Section 210-52(a).

Heating equipment and systems often have special installation instructions for spacings, types of supply wires, or special control equipment, which must be considered in determining the suitability of the installation.

424-10. Special Permission. Fixed electric space-heating equipment and systems installed by methods other than covered by this article shall be permitted only by special permission.

424-11. Supply Conductors. Fixed electric space-heating equipment requiring supply conductors with over 60°C insulation shall be clearly and permanently marked. This marking shall be plainly visible after installation and shall be permitted to be adjacent to the field connection box.

Fixed electric space-heating equipment may require supply conductors with a temperature rating greater than 60°C.

424-12. Locations.

(a) Exposed to Physical Damage. Where subject to physical damage, fixed electric space-heating equipment shall be protected in an approved manner.

(b) Damp or Wet Locations. Heaters and related equipment installed in damp or wet locations shall be approved for such locations and shall be constructed and installed so that water or other liquids cannot enter or accumulate in or on wired sections, electrical components, or ductwork.

FPN No. 1: See Section 110-11 for equipment exposed to deteriorating agents.

FPN No. 2: See Section 680-27 for pool deck areas.

424-13. Spacing from Combustible Materials. Fixed electric space-heating equipment shall be installed to provide the required spacing between the equipment and adjacent combustible material, unless it has been found to be acceptable where installed in direct contact with combustible material.

C. Control and Protection of Fixed Electric Space-Heating Equipment

424-19. Disconnecting Means. Means shall be provided to disconnect the heater, motor controller(s), and supplementary overcurrent protective device(s) of all fixed electric space-heating equipment from all ungrounded conductors. Where heating equipment is supplied by more than one source, the disconnecting means shall be grouped and marked.

(a) Heating Equipment with Supplementary Overcurrent Protection. The disconnecting means for fixed electric space-heating equipment with supplementary overcurrent protection shall be within sight from the supplementary overcurrent protective device(s), on the supply side of these devices, if fuses, and, in addition, shall comply with either (1) or (2).

(1) Heater Containing No Motor Rated Over ⅛ Horsepower. The above disconnecting means or unit switches complying with Section 424-19(c) shall be permitted to serve as the required disconnecting means for both the motor controller(s) and heater under either item (a) or (b).

(a) The disconnecting means provided is also within sight from the motor controller(s) and the heater.

(b) The disconnecting means provided shall be capable of being locked in the open position.

(2) Heater Containing a Motor(s) Rated Over ⅛ Horsepower. The above disconnecting means shall be permitted to serve as the required disconnecting means for both the motor controller(s) and heater by one of the means specified in items (a) through (d).

(a) Where the disconnecting means is also in sight from the motor controller(s) and the heater.

(b) Where the disconnecting means is not within sight from the heater, a separate disconnecting means shall be installed, or the disconnecting means shall be capable of being locked in the open position, or unit switches complying with Section 424-19(c) shall be permitted.

(c) Where the disconnecting means is not within sight from the motor controller location, a disconnecting means complying with Section 430-102 shall be provided.

(d) Where the motor is not in sight from the motor controller location, Section 430-102(b) shall apply.

(b) Heating Equipment Without Supplementary Overcurrent Protection.

(1) Without Motor or with Motor Not Over ⅛ Horsepower. For fixed electric space-heating equipment without a motor rated over ⅛ hp, the branch-circuit switch or circuit breaker shall be permitted to serve as the disconnecting means where the switch or circuit breaker is within sight from the heater or is capable of being locked in the open position.

(2) Over ⅛ Horsepower. For motor-driven electric space-heating equipment with a motor rated over ⅛ hp, a disconnecting means shall be located within sight from the motor controller, or shall be permitted to comply with the requirements in Section 424-19(a)(2).

(c) Unit Switch(es) as Disconnecting Means. A unit switch(es) with a marked "off" position that is part of a fixed heater and disconnects all ungrounded conductors shall be permitted as the disconnecting means required by this article where other means for disconnection are provided in the following types of occupancies.

Section 424-19(c) permits a unit switch to serve as the disconnecting means, provided it has a marked "off" position and disconnects all ungrounded conductors and that other means are also provided in accordance with paragraphs (1), (2), (3), and (4). Such *other means* are not required to be capable of being locked in the "open" position as required by Section 424-19(b)(1).

See Section 424-20 for thermostatically controlled switching devices.

(1) Multifamily Dwellings. In multifamily dwellings, the other disconnecting means shall be within the dwelling unit, or on the same floor as the dwelling unit in which the fixed heater is installed, and shall also be permitted to control lamps and appliances.

(2) Two-Family Dwellings. In two-family dwellings, the other disconnecting means shall be permitted either inside or outside of the dwelling unit in which the fixed heater is installed. In this case, an individual switch or circuit breaker for the dwelling unit shall be permitted and shall also be permitted to control lamps and appliances.

(3) One-Family Dwellings. In one-family dwellings, the service disconnecting means shall be permitted to be the other disconnecting means.

(4) Other Occupancies. In other occupancies, the branch-circuit switch or circuit breaker, where readily accessible for servicing of the fixed heater, shall be permitted as the other disconnecting means.

424-20. Thermostatically Controlled Switching Devices.

(a) Serving as Both Controllers and Disconnecting Means. Thermostatically controlled switching devices and combination thermostats and manually controlled switches shall be permitted to serve as both controllers and disconnecting means, provided all of the following conditions are met:

(1) Provided with a marked "off" position
(2) Directly open all ungrounded conductors when manually placed in the "off" position
(3) Designed so that the circuit cannot be energized automatically after the device has been manually placed in the "off" position
(4) Located as specified in Section 424-19

(b) Thermostats that Do Not Directly Interrupt All Ungrounded Conductors. Thermostats that do not directly interrupt all ungrounded conductors and thermostats that operate remote-control circuits shall not be required to meet the requirements of (a). These devices shall not be permitted as the disconnecting means.

424-21. Switch and Circuit Breaker to Be Indicating. Switches and circuit breakers used as disconnecting means shall be of the indicating type.

424-22. Overcurrent Protection.

(a) Branch-Circuit Devices. Electric space-heating equipment, other than such motor-operated equipment as required by Articles 430 and 440 to have additional overcurrent protection, shall be permitted to be protected against overcurrent where supplied by one of the branch circuits in Article 210.

(b) Resistance Elements. Resistance-type heating elements in electric space-heating equipment shall be protected at not more than 60 amperes. Equipment rated more than 48 amperes and employing such elements shall have the heating elements subdivided, and each subdivided load shall not exceed 48 amperes. Where a subdivided load is less than 48 amperes, the rating of the supplementary overcurrent protective device shall comply with Section 424-3(b). A boiler employing resistance-type immersion heating elements contained in an ASME rated and stamped vessel shall be permitted to comply with Section 424-72(a).

The reason for this subdivision requirement is to minimize the amount of damaging energy released into the heating elements during a short circuit, thereby reducing the risk of fire. In addition to safety, a second benefit may be partial continuity of service.

When a short circuit occurs, large amounts of damaging energy are released. This damage comes in the form of both heat and magnetic energy. By limiting the size of the overcurrent device protecting the individual heating elements, the damaging short-circuit energy released at the element is greatly reduced, thereby greatly reducing the risk of fire.

Historically, it has been stated that the subdivision size of 60 amperes was originally selected to utilize the maximum fuseholder size of 60 amperes while maintaining up to a 48-ampere heating element size (48 amperes × 125% = 60 amperes).

The following example makes a strong case for the 60-ampere subdivision requirement when a short circuit occurs at the element level. Although this example uses fuses, the same case can be made using circuit breakers.

Example

A 200 kW, 3-phase, 480-volt resistance-type unit heater has a phase current of 240.6 amperes. If the present *Code* rule for subdivision was not observed and this load were to be protected by just one device, the selected overcurrent device could be sized by multiplying 240.6 amperes times 125 percent and selecting a maximum overcurrent protective device sized 350 amperes. If the subdivision requirement were followed, the heater would probably contain six separately protected internal circuits limited in size to 60 amperes.

Answer

By using the 1998 UL *Electrical Construction Equipment Directory* (Green Book), the energy let-through of a 350-ampere fuse can be compared to the energy let-through of a 60-ampere fuse. In the fuse section (JCQR), the let-through energy, approximated by the current squared and then multiplied by the time, or I^2t, is provided for various fuse classes (UL). For this example, a 600-volt, 60-ampere Class T fuse could have a let-through I^2t as high as 30,000 ampere squared seconds. But, a 600-volt, 350-ampere Class T fuse could have a let-through I^2t as high as 1,100,000 ampere squared seconds. That means the 350-ampere fuse could let through 36.67 times as much damaging energy as the 60-ampere fuse during a short circuit. The difference in energy let-through between these two overcurrent devices (the single 350-ampere device and the group of 60-ampere devices) is significant enough to make the difference between replacing a single element or replacing a good portion of the entire system.

It should be obvious from this example that subdivision of a circuit greatly reduces the risk of fire.

(c) Overcurrent Protective Devices. The supplementary overcurrent protective devices for the subdivided loads specified in (b) shall be (1) factory-installed within or on the heater enclosure or supplied for use with the heater as a separate assembly by the heater manufacturer; (2) accessible, but shall not be required to be readily accessible; and (3) suitable for branch-circuit protection.

FPN: See Section 240-10.

Where cartridge fuses are used to provide this overcurrent protection, a single disconnecting means shall be permitted to be used for the several subdivided loads.

FPN No. 1: For supplementary overcurrent protection, see Section 240-10.

FPN No. 2: For disconnecting means for cartridge fuses in circuits of any voltage, see Section 240-40.

Where supplementary overcurrent protection is required, the heating equipment manufacturer is required to furnish the necessary overcurrent protective devices.

(d) Branch-Circuit Conductors. The conductors supplying the supplementary overcurrent protective devices shall be considered branch-circuit conductors.

Where the heaters are rated 50 kW or more, the conductors supplying the supplementary overcurrent protective devices specified in (c) shall be permitted to be sized at not less than 100 percent of the nameplate rating of the heater, provided all of the following conditions are met.

(1) The heater is marked with a minimum conductor size.
(2) The conductors are not smaller than the marked minimum size.
(3) A temperature-actuated device controls the cyclic operation of the equipment.

(e) Conductors for Subdivided Loads. Field-wired conductors between the heater and the supplementary overcurrent protective devices shall be sized at not less than 125 percent of the load served. The supplementary overcurrent protective devices specified in (c) shall protect these conductors in accordance with Section 240-3.

Where the heaters are rated 50 kW or more, the ampacity of field-wired conductors between the heater and the supplementary overcurrent protective devices shall be permitted to be not less than 100 percent of the load of their respective subdivided circuits, provided all of the following conditions are met.

(1) The heater is marked with a minimum conductor size.
(2) The conductors are not smaller than the marked minimum size.
(3) A temperature-activated device controls the cyclic operation of the equipment.

D. Marking of Heating Equipment

424-28. Nameplate.

(a) Marking Required. Each unit of fixed electric space-heating equipment shall be provided with a nameplate giving the identifying name and the normal rating in volts and watts, or in volts and amperes.

Electric space-heating equipment intended for use on alternating current only or direct current only shall be marked to so indicate. The marking of equipment consisting of motors over ⅛ hp and other loads shall specify the rating of the motor in volts, amperes, and frequency, and the heating load in volts and watts, or in volts and amperes.

(b) Location. This nameplate shall be located so as to be visible or easily accessible after installation.

424-29. Marking of Heating Elements. All heating elements that are replaceable in the field and are a part of an electric heater shall be legibly marked with the ratings in volts and watts, or in volts and amperes.

E. Electric Space-Heating Cables

424-34. Heating Cable Construction. Heating cables shall be furnished complete with factory-assembled nonheating leads at least 7 ft (2.13 m) in length.

424-35. Marking of Heating Cables. Each unit shall be marked with the identifying name or identification symbol, catalog number, ratings in volts and watts, or in volts and amperes.

Each unit length of heating cable shall have a permanent legible marking on each nonheating lead located within 3 in. (76 mm) of the terminal end. The lead wire shall have the following color identification to indicate the circuit voltage on which it is to be used:

120 volt, nominal — yellow

208 volt, nominal — blue

240 volt, nominal — red

277 volt, nominal — brown

480 volt, nominal — orange

424-36. Clearances of Wiring in Ceilings. Wiring located above heated ceilings shall be spaced not less than 2 in. (50.8 mm) above the heated ceiling and shall be considered as operating at an ambient temperature of 50°C (122°F). The ampacity of conductors shall be computed on the basis of the correction factors shown in the 0–2000 volt ampacity tables of Article 310. If this wiring is located above thermal insulation having a minimum thickness of 2 in. (50.8 mm), the wiring shall not require correction for temperature.

424-37. Location of Branch-Circuit and Feeder Wiring in Exterior Walls. Wiring methods shall comply with Article 300 and Section 310-10.

424-38. Area Restrictions.

(a) Shall Not Extend Beyond the Room or Area. Heating cables shall not extend beyond the room or area in which they originate.

(b) Uses Prohibited. Heating cables shall not be installed in the following:

(1) In closets
(2) Over walls
(3) Over partitions that extend to the ceiling, unless they are isolated single runs of imbedded cable
(4) Over cabinets whose clearance from the ceiling is less than the minimum horizontal dimension of the cabinet to the nearest cabinet edge that is open to the room or area

(c) In Closet Ceilings as Low-Temperature Heat Sources to Control Relative Humidity. The provisions of (b) shall not prevent the use of cable in closet ceilings as low-temperature heat sources to control relative humidity, provided they are used only in those portions of the ceiling that are unobstructed to the floor by shelves or other permanent fixtures.

424-39. Clearance from Other Objects and Openings. Heating elements of cables shall be separated at least 8 in. (203 mm) from the edge of outlet boxes and junction boxes that are to be used for mounting surface lighting fixtures. A clearance of not less than 2 in. (50.8 mm) shall be provided from recessed fixtures and their trims, ventilating openings, and other such openings in room surfaces. Sufficient area shall be provided to ensure that no heating cable will be covered by any surface-mounted units.

424-40. Splices. Embedded cables shall be spliced only where necessary and only by approved means, and in no case shall the length of the heating cable be altered.

424-41. Installation of Heating Cables on Dry Board, in Plaster, and on Concrete Ceilings.

(a) In Walls. Cables shall not be installed in walls unless it is necessary for an isolated single run of cable to be installed down a vertical surface to reach a dropped ceiling.

(b) Adjacent Runs. Adjacent runs of cable not exceeding 2¾ watts/ft (2¾ watts/305 mm) shall not be installed less than 1½ in. (38 mm) on centers.

(c) Surfaces to Be Applied. Heating cables shall be applied only to gypsum board, plaster lath, or other fire-resistant material. With metal lath or other electrically conductive surfaces, a coat of plaster shall be applied to completely separate the metal lath or conductive surface from the cable.

FPN: See also Section 424-41(f).

(d) Splices. All heating cables, the splice between the heating cable and nonheating leads, and 3-in. (76-mm) minimum of the nonheating lead at the splice shall be embedded in plaster or dry board in the same manner as the heating cable.

(e) Ceiling Surface. The entire ceiling surface shall have a finish of thermally noninsulating sand plaster that has a nominal thickness of ½ in. (12.7 mm), or other noninsulating material identified as suitable for this use and applied according to specified thickness and directions.

(f) Secured. Cables shall be secured by means of approved stapling, tape, plaster, nonmetallic spreaders, or other approved means at either intervals not exceeding 16 in. (406 mm) or at intervals not exceeding 6 ft (1.83 m) for cables identified for such use. Staples or metal fasteners that straddle the cable shall not be used with metal lath or other electrically conductive surfaces.

(g) Dry Board Installations. In dry board installations, the entire ceiling below the heating cable shall be covered with gypsum board not exceeding ½ in. (12.7 mm) thickness. The void between the upper layer of gypsum board, plaster lath, or other fire-resistant material and the surface layer of gypsum board shall be completely filled with thermally conductive, nonshrinking plaster or other approved material or equivalent thermal conductivity.

(h) Free from Contact with Conductive Surfaces. Cables shall be kept free from contact with metal or other electrically conductive surfaces.

(i) Joists. In dry board applications, cable shall be installed parallel to the joist, leaving a clear space centered under the joist of 2½ in. (64 mm) (width) between centers of adjacent runs of cable. A surface layer of gypsum board shall be mounted so that the nails or other fasteners do not pierce the heating cable.

(j) Crossing Joists. Cables shall cross joists only at the ends of the room unless the cable is required to cross joists elsewhere in order to satisfy the manufacturer's instructions that the installer avoid placing the cable too close to ceiling penetrations and light fixtures.

424-42. Finished Ceilings. Finished ceilings shall not be covered with decorative panels or beams constructed of materials that have thermal insulating properties, such as wood, fiber, or plastic. Finished ceilings shall be permitted to be covered with paint, wallpaper, or other approved surface finishes.

424-43. Installation of Nonheating Leads of Cables.

(a) Free Nonheating Leads. Free nonheating leads of cables shall be installed in accordance with approved wiring methods from the junction box to a location within the ceiling. Such installations shall be permitted to be single conductors in approved raceways, single or multiconductor Type UF, Type NMC, Type MI, or other approved conductors.

(b) Leads in Junction Box. Not less than 6 in. (152 mm) of free nonheating lead shall be within the junction box. The marking of the leads shall be visible in the junction box.

(c) Excess Leads. Excess leads of heating cables shall not be cut but shall be secured to the underside of the ceiling and embedded in plaster or other approved material, leaving only a length sufficient to reach the junction box with not less than 6 in. (152 mm) of free lead within the box.

424-44. Installation of Cables in Concrete or Poured Masonry Floors.

(a) Watts per Linear Foot. Constant wattage heating cables shall not exceed 16½ watts/linear foot (305 mm) of cable.

(b) Spacing Between Adjacent Runs. The spacing between adjacent runs of cable shall not be less than 1 in. (25.4 mm) on centers.

(c) Secured in Place. Cables shall be secured in place by nonmetallic frames or spreaders or other approved means while the concrete or other finish is applied.

Cables shall not be installed where they bridge expansion joints unless protected from expansion and contraction.

(d) Spacings Between Heating Cable and Metal Embedded in the Floor. Spacings shall be maintained between the heating cable and metal embedded in the floor, unless the cable is a grounded metal-clad cable.

(e) Leads Protected. Leads shall be protected where they leave the floor by rigid metal conduit, intermediate metal conduit, rigid nonmetallic conduit, electrical metallic tubing, or by other approved means.

(f) Bushings or Approved Fittings. Bushings or approved fittings shall be used where the leads emerge within the floor slab.

(g) Ground-Fault Circuit-Interrupter Protection for Conductive Heated Floors of Bathrooms, Hydromassage Bathtub, Spa, and Hot Tub Locations. Ground-fault circuit-interrupter protection for personnel shall be provided for electrically heated floors in bathrooms, hydromassage bathtub, spa, and hot tub locations. This shall apply to all systems used with conductive floor coverings, whether cable, panel, or other approved types.

Added for the 1999 *Code,* this new Section 424-44(g) requires the use of GFCI protection for electrically heated conductive floors and serves to reduce shock hazards to persons with bare feet in these area.

424-45. Inspection and Tests. Cable installations shall be made with due care to prevent damage to the cable assembly and shall be inspected and approved before cables are covered or concealed.

F. Duct Heaters

424-57. General. Part F shall apply to any heater mounted in the airstream of a forced-air system where the air-moving unit is not provided as an integral part of the equipment.

The insert-type electric duct heater shown in Figure 424.2 is typical of the type of heater covered by Part F of Article 424.

Figure 424.2 *An insert-type electric duct heater.*

424-58. Identification. Heaters installed in an air duct shall be identified as suitable for the installation.

424-59. Airflow. Means shall be provided to ensure uniform and adequate airflow over the face of the heater in accordance with the manufacturer's instructions.

> FPN: Heaters installed within 4 ft (1.22 m) of the outlet of an air-moving device, heat pump, air conditioner, elbows, baffle plates, or other obstructions in ductwork may require turning vanes, pressure plates, or other devices on the inlet side of the duct heater to ensure an even distribution of air over the face of the heater.

424-60. Elevated Inlet Temperature. Duct heaters intended for use with elevated inlet air temperature shall be identified as suitable for use at the elevated temperatures.

424-61. Installation of Duct Heaters with Heat Pumps and Air Conditioners. Heat pumps and air conditioners having duct heaters closer than 4 ft (1.22 m) to the heat pump or air conditioner shall have both the duct heater and heat pump or air conditioner identified as suitable for such installation and so marked.

424-62. Condensation. Duct heaters used with air conditioners or other air-cooling equipment that may result in condensation of moisture shall be identified as suitable for use with air conditioners.

424-63. Fan Circuit Interlock. Means shall be provided to ensure that the fan circuit is energized when any heater circuit is energized. However, time- or temperature-controlled delay in energizing the fan motor shall be permitted.

424-64. Limit Controls. Each duct heater shall be provided with an approved, integral, automatic-reset temperature-limiting control or controllers to de-energize the circuit or circuits.

In addition, an integral independent supplementary control or controllers shall be provided in each duct heater that will disconnect a sufficient number of conductors to interrupt current flow. This device shall be manually resettable or replaceable.

424-65. Location of Disconnecting Means. Duct heater controller equipment shall be either accessible with the disconnecting means installed at or within sight from the controller or as permitted by Section 424-19(a).

424-66. Installation. Duct heaters shall be installed in accordance with the manufacturer's instructions in a manner so that operation will not create a hazard to persons or property. Furthermore, duct heaters shall be located with respect to building construction and other equipment so as to permit access to the heater. Sufficient clearance shall be maintained to permit replacement of controls and heating elements and for adjusting and cleaning of controls and other parts requiring such attention. See Section 110-26.

> FPN: For additional installation information, see *Standard for the Installation of Air Conditioning and Ventilating Systems*, NFPA 90A-1996, and *Standard for the Installation of Warm Air Heating and Air Conditioning Systems*, NFPA 90B-1996.

G. Resistance-Type Boilers

424-70. Scope. The provisions in Part G of this article shall apply to boilers employing resistance-type heating elements. Electrode-type boilers shall not be considered as employing resistance-type heating elements. See Part H of this article.

424-71. Identification. Resistance-type boilers shall be identified as suitable for the installation.

424-72. Overcurrent Protection.

(a) Boiler Employing Resistance-Type Immersion Heating Elements in an ASME Rated and Stamped Vessel. A boiler employing resistance-type immersion heating elements contained in an ASME rated and stamped vessel shall have the heating elements protected at not more than 150 amperes. Such a boiler rated more than 120 amperes shall have the heating elements subdivided into loads not exceeding 120 amperes.

Where a subdivided load is less than 120 amperes, the rating of the overcurrent protective device shall comply with Section 424-3(b).

(b) Boiler Employing Resistance-Type Heating Elements Rated More than 48 Amperes and Not Contained in an ASME Rated and Stamped Vessel. A boiler employing resistance-type heating elements not contained in an ASME rated and stamped vessel shall have the heating elements protected at not more than 60 amperes. Such a boiler rated more than 48 amperes shall have the heating elements subdivided into loads not exceeding 48 amperes.

Where a subdivided load is less than 48 amperes, the rating of the overcurrent protective device shall comply with Section 424-3(b).

See the commentary following Section 424-22(b) for an explanation of the subdivision requirement.

(c) Supplementary Overcurrent Protective Devices. The supplementary overcurrent protective devices for the subdivided loads as required by Sections 424-72(a) and (b) shall be (1) factory-installed within or on the boiler enclosure or provided as a separate assembly by the boiler manufacturer; and (2) accessible, but need not be readily accessible; and (3) suitable for branch-circuit protection.

Where cartridge fuses are used to provide this overcurrent protection, a single disconnecting means shall be permitted for the several subdivided circuits. See Section 240-40.

(d) Conductors Supplying Supplementary Overcurrent Protective Devices. The conductors supplying these supplementary overcurrent protective devices shall be considered branch-circuit conductors.

Where the heaters are rated 50 kW or more, the conductors supplying the overcurrent protective device specified in (c) shall be permitted to be sized at not less than 100 percent of the nameplate rating of the heater, provided all of the following conditions are met.

(1) The heater is marked with a minimum conductor size.
(2) The conductors are not smaller than the marked minimum size.
(3) A temperature- or pressure-actuated device controls the cyclic operation of the equipment.

(e) Conductors for Subdivided Loads. Field-wired conductors between the heater and the supplementary overcurrent protective devices shall be sized at not less than 125 percent of the load served. The supplementary overcurrent protective devices specified in (c) shall protect these conductors in accordance with Section 240-3.

Where the heaters are rated 50 kW or more, the ampacity of field-wired conductors between the heater and the supplementary overcurrent protective devices shall be permitted to be not less than 100 percent of the load of their respective subdivided circuits, provided all of the following conditions are met.

(1) The heater is marked with a minimum conductor size.
(2) The conductors are not smaller than the marked minimum size.
(3) A temperature-activated device controls the cyclic operation of the equipment.

424-73. Over-Temperature Limit Control. Each boiler designed so that in normal operation there is no change in state of the heat transfer medium shall be equipped with a temperature-sensitive limiting means. It shall be installed to limit maximum liquid temperature and shall directly or indirectly disconnect all ungrounded conductors to the heating elements. Such means shall be in addition to a temperature regulating system and other devices protecting the tank against excessive pressure.

424-74. Over-Pressure Limit Control. Each boiler designed so that in normal operation there is a change in state of the heat transfer medium from liquid to vapor shall be equipped with a pressure-sensitive limiting means. It shall be installed to limit maximum pressure and shall directly or indirectly disconnect all ungrounded conductors to the heating elements. Such means shall be in addition to a pressure regulating system and other devices protecting the tank against excessive pressure.

H. Electrode-Type Boilers

424-80. Scope. The provisions in Part H of this article shall apply to boilers for operation at 600 volts, nominal, or less, in which heat is generated by the passage of current between electrodes through the liquid being heated.

FPN: For over 600 volts, see Part E of Article 490.

424-81. Identification. Electrode-type boilers shall be identified as suitable for the installation.

424-82. Branch-Circuit Requirements. The size of branch-circuit conductors and overcurrent protective devices shall be calculated on the basis of 125 percent of the total load (motors not included). A contactor, relay, or other device, approved for continuous operation at 100 percent of its rating, shall be permitted to supply its full-rated load. See Section 210-19(a), Exception. The provisions of this section shall not apply to conductors that form an integral part of an approved boiler.

Where an electrode boiler is rated 50 kW or more, the conductors supplying the boiler electrode(s) shall be permitted to be sized at not less than 100 percent of the nameplate rating of the electrode boiler, provided all the following conditions are met.

(1) The electrode boiler is marked with a minimum conductor size.
(2) The conductors are not smaller than the marked minimum size.
(3) A temperature- or pressure-actuated device controls the cyclic operation of the equipment.

424-83. Over-Temperature Limit Control. Each boiler designed so that in normal operation there is no change in state of the heat transfer medium shall be equipped with a temperature-sensitive limiting means. It shall be installed to limit maximum liquid temperature and shall directly or indirectly interrupt all current flow through the electrodes. Such means shall be in addition to the temperature regulating system and other devices protecting the tank against excessive pressure.

424-84. Over-Pressure Limit Control. Each boiler designed so that in normal operation there is a change in state of the heat transfer medium from liquid to vapor shall be equipped with a pressure-sensitive limiting means. It shall be installed to limit maximum pressure and shall directly or indirectly interrupt all current flow through the electrodes. Such means shall be in addition to a pressure regulating system and other devices protecting the tank against excessive pressure.

424-85. Grounding. For those boilers designed such that fault currents do not pass through the pressure vessel, and the pressure vessel is electrically isolated from the electrodes, all exposed noncurrent-carrying metal parts, including the pressure vessel, supply, and return connecting piping, shall be grounded in accordance with Article 250.

For all other designs, the pressure vessel containing the electrodes shall be isolated and electrically insulated from ground.

424-86. Markings. All electrode-type boilers shall be marked to show the following:

(1) The manufacturer's name
(2) The normal rating in volts, amperes, and kilowatts
(3) The electrical supply required specifying frequency, number of phases, and number of wires
(4) The marking: "Electrode-Type Boiler"
(5) A warning marking: "All Power Supplies Shall Be Disconnected Before Servicing, Including Servicing the Pressure Vessel"

The nameplate shall be located so as to be visible after installation.

J. Electric Radiant Heating Panels and Heating Panel Sets

424-90. Scope. The provisions of Part J of this article shall apply to radiant heating panels and heating panel sets.

424-91. Definitions.

(a) Heating Panel. A *heating panel* is a complete assembly provided with a junction box or a length of flexible conduit for connection to a branch circuit.

(b) Heating Panel Set. A *heating panel set* is a rigid or nonrigid assembly provided with nonheating leads or a terminal junction assembly identified as being suitable for connection to a wiring system.

424-92. Markings.

(a) Markings shall be permanent and in a location that is visible prior to application of panel finish.

(b) Each unit shall be identified as suitable for the installation.

(c) Each unit shall be marked with the identifying name or identification symbol, catalog number, and rating in volts and watts, or in volts and amperes.

(d) The manufacturers of heating panels or heating panel sets shall provide marking labels that indicate that the space-heating installation incorporates heating panels or heating panel sets and instructions that the labels shall be affixed to the panelboards to identify which branch circuits supply the circuits to those space-heating installations. If the heating panels and heating panel set installations are visible and distinguishable after installation, the labels shall not be required to be provided and affixed to the panelboards.

424-93. Installation.

(a) General.

(1) Heating panels and heating panel sets shall be installed in accordance with the manufacturer's instructions.

(2) The heating portion shall not

(a) Be installed in or behind surfaces where subject to physical damage.
(b) Be run through or above walls, partitions, cupboards, or similar portions of structures that extend to the ceiling.
(c) Be run in or through thermal insulation, but shall be permitted to be in contact with the surface of thermal insulation.

(3) Edges of panels and panel sets shall be separated by not less than 8 in. (203 mm) from the edges of any outlet boxes and junction boxes that are to be used for mounting surface lighting fixtures. A clearance of not less than 2 in. (50.8 mm) shall be provided from recessed fixtures and their trims, ventilating openings, and other such openings in room surfaces, unless the heating panels and panel sets are listed and marked for lesser clearances, in which case, they shall be permitted to be installed at the marked clearances. Sufficient area shall be provided to ensure that no heating panel or heating panel set is to be covered by any surface-mounted units.

(4) After the heating panels or heating panel sets are installed and inspected, it shall be permitted to install a surface that has been identified by the manufacturer's instructions as being suitable for the installation. The surface shall be secured so that the nails or other fastenings do not pierce the heating panels or heating panel sets.

(5) Surfaces permitted by Section 424-93(a)(4) shall be permitted to be covered with paint, wallpaper, or other approved surfaces identified in the manufacturer's instructions as being suitable.

(b) Heating Panel Sets.

(1) Heating panel sets shall be permitted to be secured to the lower face of joists or mounted in between joists, headers, or nailing strips.

(2) Heating panel sets shall be installed parallel to joists or nailing strips.

(3) Nailing or stapling of heating panel sets shall be done only through the unheated portions provided for this purpose. Heating panel sets shall not be cut through or nailed through any point closer than ¼ in. (6.35 mm) to the element. Nails, staples, or other fasteners shall not be used where they penetrate current-carrying parts.

(4) Heating panel sets shall be installed as complete units unless identified as suitable for field cutting in an approved manner.

424-94. Clearances of Wiring in Ceilings. Wiring located above heated ceilings shall be spaced not less than 2 in. (50.8 mm) above the heated ceiling and shall be considered as operating at an ambient of 50°C (122°F). The ampacity shall be computed on the basis of the correction factors given in the 0–2000 volt ampacity tables of Article 310. If this wiring is located above thermal insulations having a minimum thickness of 2 in. (50.8 mm), the wiring shall not require correction for temperature.

424-95. Location of Branch-Circuit and Feeder Wiring in Walls.

(a) Exterior Walls. Wiring methods shall comply with Article 300 and Section 310-10.

(b) Interior Walls. Any wiring behind heating panels or heating panel sets located in interior walls or partitions shall be considered as operating at an ambient temperature of 40°C (104°F), and the ampacity shall be computed on the basis of the correction factors given in the 0–2000 volt ampacity tables of Article 310.

424-96. Connection to Branch-Circuit Conductors.

(a) General. Heating panels or heating panel sets assembled together in the field to form a heating installation in one room or area shall be connected in accordance with the manufacturer's instructions.

(b) Heating Panels. Heating panels shall be connected to branch-circuit wiring by an approved wiring method.

(c) Heating Panel Sets.

(1) Heating panel sets shall be connected to branch-circuit wiring by a method identified as being suitable for the purpose.

(2) A heating panel set provided with terminal junction assembly shall be permitted to have the nonheating leads attached at the time of installation in accordance with the manufacturer's instructions.

424-97. Nonheating Leads. Excess nonheating leads of heating panels or heating panel sets shall be permitted to be cut to the required length. They shall meet the installation requirements of the wiring method employed in accordance with Section 424-96. Nonheating leads shall be an integral part of a heating panel and a heating panel set and shall not be subjected to the ampacity requirements of Section 424-3(b) for branch circuits.

424-98. Installation in Concrete or Poured Masonry.

(a) Maximum Heated Area. Heating panels or heating panel sets shall not exceed 33 watts/ft^2 (33 watts/ 0.093 m^2) of heated area.

(b) Secured in Place and Identified as Suitable. Heating panels or heating panel sets shall be secured in place by means specified in the manufacturer's instructions and identified as suitable for the installation.

(c) Expansion Joints. Heating panels or heating panel sets shall not be installed where they bridge expansion joints unless provision is made for expansion and contraction.

(d) Spacings. Spacings shall be maintained between heating panels or heating panel sets and metal embedded in the floor. Grounded metal-clad heating panels shall be permitted to be in contact with metal embedded in the floor.

(e) Protection of Leads. Leads shall be protected where they leave the floor by rigid metal conduit, intermediate metal conduit, rigid nonmetallic conduit, electrical metallic tubing, or by other approved means.

(f) Bushings or Fittings Required. Bushings or approved fittings shall be used where the leads emerge within the floor slabs.

424-99. Installation Under Floor Covering.

(a) Identification. Heating panels or heating panel sets for installation under floor covering shall be identified as suitable for installation under floor covering.

(b) Maximum Heated Area. Heating panels or panel sets, installed under floor covering, shall not exceed 15 watts/ft^2 (15 watts/0.093 m^2) of heated area.

(c) Installation. Listed heating panels or panel sets, if installed under floor covering, shall be installed on floor surfaces that are smooth and flat in accordance with the manufacturer's instructions and shall also comply with the following.

(1) Expansion Joints. Heating panels or heating panel sets shall not be installed where they bridge expansion joints unless protected from expansion and contraction.

(2) Connection to Conductors. Heating panels and heating panel sets shall be connected to branch-circuit and supply wiring by wiring methods recognized in Chapter 3.

(3) Anchoring. Heating panels and heating panel sets shall be firmly anchored to the floor using an adhesive or anchoring system identified for this use.

(4) Coverings. After heating panels or heating panel sets are installed and inspected, they shall be permitted to be covered by a floor covering that has been identified by the manufacturer as being suitable for the installation. The covering shall be secured to the heating panel or heating panel sets with release-type adhesives or by means identified for this use.

(5) Fault Protection. A device to open all ungrounded conductors supplying the heating panels or heating panel sets, provided by the manufacturer, shall function when a low- or high-resistance line-to-line, line-to-grounded conductor, or line-to-ground fault occurs, such as the result of a penetration of the element or element assembly.

FPN: An integral grounding shield may be required to provide this protection.

An example of a heating system that could be installed in this fashion is a system using conductive-film heating elements.

Article 426 — Fixed Outdoor Electric Deicing and Snow-Melting Equipment

Contents

A. General

426-1. Scope. The requirements of this article shall apply to electrically energized heating systems and the installation of these systems.

(a) Embedded. Embedded in driveways, walks, steps, and other areas.

(b) Exposed. Exposed on drainage systems, bridge structures, roofs, and other structures.

Article 426 includes requirements for resistance heating elements, impedance heating systems, or skin-effect heating systems used for deicing and snow melting. These systems are defined in Section 426-2.

426-2. Definitions. For the purpose of this article:

Heating System. A complete system consisting of components such as heating elements, fastening devices, nonheating circuit wiring, leads, temperature controllers, safety signs, junction boxes, raceways, and fittings.

Impedance Heating System. A system in which heat is generated in a pipe or rod, or combination of pipes and rods, by causing current to flow through the pipe or rod by direct connection to an ac voltage source from a dual-winding transformer. The pipe or rod shall be permitted to be embedded in the surface to be heated, or constitute the exposed components to be heated.

Resistance Heating Element. A specific separate element to generate heat that is embedded in or fastened to the surface to be heated.

FPN: Tubular heaters, strip heaters, heating cable, heating tape, and heating panels are examples of resistance heaters.

Skin-Effect Heating System. A system in which heat is generated on the inner surface of a ferromagnetic envelope embedded in or fastened to the surface to be heated.

FPN: Typically, an electrically insulated conductor is routed through and connected to the envelope at the other end. The envelope and the electrically insulated conductor are connected to an ac voltage source from a dual-winding transformer.

426-3. Application of Other Articles. All requirements of this *Code* shall apply except as specifically amended in this article. Cord- and plug-connected fixed outdoor electric deicing and snow-melting equipment intended for specific use and identified as suitable for this use shall be installed according to Article 422. Fixed outdoor electric deicing and snow-melting equipment for use in hazardous (classified) locations shall comply with Articles 500 through 516.

426-4. Branch-Circuit Sizing. The ampacity of branch-circuit conductors and the rating or setting of overcurrent protective devices supplying fixed outdoor electric deicing and snow-melting equipment shall not be less than 125 percent of the total load of the heaters. The rating or setting of overcurrent protective devices shall be permitted in accordance with Section 240-3(b).

B. Installation

426-10. General. Equipment for outdoor electric deicing and snow melting shall be identified as being suitable for the following:

(1) The chemical, thermal, and physical environment
(2) Installation in accordance with the manufacturer's drawings and instructions

426-11. Use. Electrical heating equipment shall be installed in such a manner as to be afforded protection from physical damage.

The instructions required by Underwriters Laboratories Inc. for UL-listed mat or cable deicing and snow-melting equipment intended for burial in concrete specifically indicate that the slab must be a double pour (poured in two parts) if that is the only acceptable means of installation. If such a limitation is not specifically mentioned in the installation instructions, either a single or double pour may be used. See Section 110-3(b) regarding the installation and use of listed or labeled equipment.

426-12. Thermal Protection. External surfaces of outdoor electric deicing and snow-melting equipment that operate at temperatures exceeding 60°C (140°F) shall be physically guarded, isolated, or thermally insulated to protect against contact by personnel in the area.

426-13. Identification. The presence of outdoor electric deicing and snow-melting equipment shall be evident by the posting of appropriate caution signs or markings where clearly visible.

426-14. Special Permission. Fixed outdoor deicing and snow-melting equipment employing methods of construction or installation other than covered by this article shall be permitted only by special permission.

See the definition of *special permission* in Article 100.

C. Resistance Heating Elements

426-20. Embedded Deicing and Snow-Melting Equipment.

(a) Watt Density. Panels or units shall not exceed 120 watts/ft^2 (120 watts/0.093 m^2) of heated area.

(b) Spacing. The spacing between adjacent cable runs is dependent upon the rating of the cable, and shall be not less than 1 in. (25.4 mm) on centers.

(c) Cover. Units, panels, or cables shall be installed as follows:

(1) On a substantial asphalt or masonry base at least 2 in. (50.8 mm) thick and have at least 1½ in. (38 mm) of asphalt or masonry applied over the units, panels, or cables; or
(2) They shall be permitted to be installed over other approved bases and embedded within 3½ in. (89 mm) of masonry or asphalt but not less than 1½ in. (38 mm) from the top surface; or
(3) Equipment that has been specially investigated for other forms of installation shall be installed only in the manner for which it has been investigated.

(d) Secured. Cables, units, and panels shall be secured in place by frames or spreaders or other approved means while the masonry or asphalt finish is applied.

(e) Expansion and Contraction. Cables, units, and panels shall not be installed where they bridge expansion joints unless provision is made for expansion and contraction.

426-21. Exposed Deicing and Snow-Melting Equipment.

(a) Secured. Heating element assemblies shall be secured to the surface being heated by approved means.

(b) Over-Temperature. Where the heating element is not in direct contact with the surface being heated, the design of the heater assembly shall be such that its temperature limitations shall not be exceeded.

(c) Expansion and Contraction. Heating elements and assemblies shall not be installed where they bridge expansion joints unless provision is made for expansion and contraction.

(d) Flexural Capability. Where installed on flexible structures, the heating elements and assemblies shall have a flexural capability that is compatible with the structure.

426-22. Installation of Nonheating Leads for Embedded Equipment.

(a) Grounding Sheath or Braid. Nonheating leads having a grounding sheath or braid shall be permitted to be embedded in the masonry or asphalt in the same manner as the heating cable without additional physical protection.

(b) Raceways. All but 1 in. to 6 in. (25.4 mm to 152 mm) of nonheating leads of Type TW and other approved types not having a grounding sheath shall be enclosed in a rigid conduit, electrical metallic tubing, intermediate metal conduit, or other raceways within asphalt or masonry; and the distance from the factory splice to raceway shall not be less than 1 in. (25.4 mm) or more than 6 in. (152 mm).

(c) Bushings. Insulating bushings shall be used in the asphalt or masonry where leads enter conduit or tubing.

See Section 300-4(f) and the associated commentary for further information regarding insulating bushings.

(d) Expansion and Contraction. Leads shall be protected in expansion joints and where they emerge from masonry or asphalt by rigid conduit, electrical metallic tubing, intermediate metal conduit, other raceways, or other approved means.

(e) Leads in Junction Boxes. Not less than 6 in. (152 mm) of free nonheating lead shall be within the junction box.

426-23. Installation of Nonheating Leads for Exposed Equipment.

(a) Nonheating Leads. Power supply nonheating leads (cold leads) for resistance elements shall be suitable for the temperature encountered. Preassembled nonheating leads on approved heaters shall be permitted to be shortened if the markings specified in Section 426-25 are retained. Not less than 6 in. (152 mm) of nonheating leads shall be provided within the junction box.

(b) Protection. Nonheating power supply leads shall be enclosed in a rigid conduit, intermediate metal conduit, electrical metallic tubing, or other approved means.

426-24. Electrical Connection.

(a) Heating Element Connections. Electrical connections, other than factory connections of heating elements to nonheating elements embedded in masonry or asphalt or on exposed surfaces, shall be made with insulated connectors identified for the use.

(b) Circuit Connections. Splices and terminations at the end of the nonheating leads, other than the heating element end, shall be installed in a box or fitting in accordance with Sections 110-14 and 300-15.

426-25. Marking. Each factory-assembled heating unit shall be legibly marked within 3 in. (76 mm) of each end of the nonheating leads with the permanent identification symbol, catalog number, and ratings in volts and watts, or in volts and amperes.

426-26. Corrosion Protection. Ferrous and nonferrous metal raceways, cable armor, cable sheaths, boxes, fittings, supports, and support hardware shall be permitted to be installed in concrete or in direct contact with the earth, or in areas subject to severe corrosive influences, where made of material suitable for the condition, or where provided with corrosion protection identified as suitable for the condition.

426-27. Grounding Braid or Sheath. Grounding means, such as copper braid, metal sheath, or other approved means, shall be provided as part of the heated section of the cable, panel, or unit.

426-28. Equipment Protection. Ground-fault protection of equipment shall be provided for fixed outdoor electric deicing and snow-melting equipment, except for equipment that employs mineral-insulated, metal-sheathed cable embedded in a noncombustible medium.

Section 426-28 has been revised for the 1999 *Code*. Rather than protecting the entire branch circuit, the ground-fault protection requirement is now focused on protecting just the equipment itself. This change affords the manufacturer and the user an option of providing both circuit and equipment protection, or just the required equipment protection. This required protection for fixed outdoor deicing and snow-melting equipment may be accomplished by using circuit breakers equipped with ground-fault equipment protection (GFEP) or an integral device supplied as part of the deicing or snow-melting equipment that is sensitive to leakage currents in the magnitude of 6 milliamperes to 50 milliamperes [referred to by UL as ground-fault equipment protection circuit interrupters (GFEPCI)]. These protection devices, if applied properly, will substantially reduce the risk of a fire being started by low-level electrical arcing.

It is important to understand that this required equipment protection is not the same as a ground-fault circuit interrupter used for personnel protection that trips at 5 milliamperes (±1 milliampere).

For further information regarding ground-fault equipment protection used to comply with Sections 426-28 and 427-22, refer to the 1998 UL *General Information for Electrical Equipment Directory* (White Book), category KCZI.

D. Impedance Heating

426-30. Personnel Protection. Exposed elements of impedance heating systems shall be physically guarded, isolated, or thermally insulated with a weatherproof jacket to protect against contact by personnel in the area.

426-31. Isolation Transformer. A dual-winding transformer with a grounded shield between the primary and secondary windings shall be used to isolate the distribution system from the heating system.

426-32. Voltage Limitations. Unless protected by a ground-fault circuit-interrupter protection for personnel, the secondary winding of the isolation transformer connected to the impedance heating elements shall not have an output voltage greater than 30 volts ac.

Where ground-fault circuit-interrupter protection for personnel is provided, the voltage shall be permitted to be greater than 30 but not more than 80 volts.

426-33. Induced Currents. All current-carrying components shall be installed in accordance with Section 300-20.

426-34. Grounding. An impedance heating system that is operating at a voltage greater than 30, but not more than 80, shall be grounded at a designated point(s).

E. Skin-Effect Heating

426-40. Conductor Ampacity. The current through the electrically insulated conductor inside the ferromagnetic envelope shall be permitted to exceed the ampacity values shown in Article 310, provided it is identified as suitable for this use.

426-41. Pull Boxes. Where pull boxes are used, they shall be accessible without excavation by location in suitable vaults or above grade. Outdoor pull boxes shall be of watertight construction.

426-42. Single Conductor in Enclosure. The provisions of Section 300-20 shall not apply to the installation of a single conductor in a ferromagnetic envelope (metal enclosure).

426-43. Corrosion Protection. Ferromagnetic envelopes, ferrous or nonferrous metal raceways, boxes, fittings, supports, and support hardware shall be permitted to be installed in concrete or in direct contact with the earth, or in areas subjected to severe corrosive influences, where made of material suitable for the condition, or where provided with corrosion protection identified as suitable for the condition. Corrosion protection shall maintain the original wall thickness of the ferromagnetic envelope.

426-44. Grounding. The ferromagnetic envelope shall be grounded at both ends; and, in addition, it shall be permitted to be grounded at intermediate points as required by its design.

The provisions of Section 250-30 shall not apply to the installation of skin-effect heating systems.

FPN: For grounding methods, see Sections 250-30(a)(4) and (b)(3).

F. Control and Protection

426-50. Disconnecting Means.

(a) Disconnection. All fixed outdoor deicing and snow-melting equipment shall be provided with a means for disconnection from all ungrounded conductors. Where readily accessible to the user of the equipment, the branch-circuit switch or circuit breaker shall be permitted to serve as the disconnecting means. Switches used as the disconnecting means shall be of the indicating type.

(b) Cord- and Plug-Connected Equipment. The factory-installed attachment plug of cord- and plug-connected equipment rated 20 amperes or less and 150 volts or less to ground shall be permitted to be the disconnecting means.

426-51. Controllers.

(a) Temperature Controller with "Off" Position. Temperature controlled switching devices that indicate an "off" position and that interrupt line current shall open all ungrounded conductors when the control device is in the "off" position. These devices shall not be permitted to serve as the disconnecting means unless provided with a positive lockout in the "off" position.

(b) Temperature Controller Without "Off" Position. Temperature controlled switching devices that do not have an "off" position shall not be required to open all ungrounded conductors and shall not be permitted to serve as the disconnecting means.

(c) Remote Temperature Controller. Remote controlled temperature-actuated devices shall not be required to meet the requirements of Section 426-51(a). These devices shall not be permitted to serve as the disconnecting means.

(d) Combined Switching Devices. Switching devices consisting of combined temperature-actuated devices and manually controlled switches that serve both as the controller and the disconnecting means shall comply with all of the following conditions:

(1) Open all ungrounded conductors when manually placed in the "off" position
(2) Be so designed that the circuit cannot be energized automatically if the device has been manually placed in the "off" position
(3) Be provided with a positive lockout in the "off" position.

426-52. Overcurrent Protection. Fixed outdoor electric deicing and snow-melting equipment shall be permitted to be protected against overcurrent where supplied by a branch circuit as specified in Section 426-4.

426-54. Cord- and Plug-Connected Deicing and Snow-Melting Equipment. Cord- and plug-connected deicing and snow-melting equipment shall be listed.

According to the 1998 UL *General Information for Electrical Equipment Directory* (White Book), category KOBQ, UL listed deicing and snow-melting equipment is provided with means for permanent wiring connection, except, the equipment rated 20 amperes or less and 150 volts or less to ground may be of cord- and plug-connected construction. See the definition of listed in Article 100.

Article 427 — Fixed Electric Heating Equipment for Pipelines and Vessels

Contents

A. General

427-1. Scope. The requirements of this article shall apply to electrically energized heating systems and the installation of these systems used with pipelines or vessels or both.

Article 427 includes requirements for impedance heating, induction heating, and skin-effect heating, in addition to resistance heating elements. Definitions of the various systems are provided in Section 427-2.

427-2. Definitions. For the purpose of this article:

Impedance Heating System. A system in which heat is generated in a pipeline or vessel wall by causing current to flow through the pipeline or vessel wall by direct connection to an ac voltage source from a dual-winding transformer.

Induction Heating System. A system in which heat is generated in a pipeline or vessel wall by inducing current and hysteresis effect in the pipeline or vessel wall from an external isolated ac field source.

Integrated Heating System. A complete system consisting of components such as pipelines, vessels, heating elements, heat transfer medium, thermal insulation, moisture barrier, nonheating leads, temperature controllers, safety signs, junction boxes, raceways, and fittings.

Pipeline. A length of pipe including pumps, valves, flanges, control devices, strainers, and/or similar equipment for conveying fluids.

Resistance Heating Element. A specific separate element to generate heat that is applied to the pipeline or vessel externally or internally.

FPN: Tubular heaters, strip heaters, heating cable, heating tape, heating blankets, and immersion heaters are examples of resistance heaters.

Skin-Effect Heating System. A system in which heat is generated on the inner surface of a ferromagnetic envelope attached to a pipeline or vessel, or both.

FPN: Typically, an electrically insulated conductor is routed through and connected to the envelope at the other end. The envelope and the electrically insulated conductor are connected to an ac voltage source from a dual-winding transformer.

Vessel. A container such as a barrel, drum, or tank for holding fluids or other material.

427-3. Application of Other Articles. All requirements of this *Code* shall apply except as specifically amended in this article. Cord-connected pipe heating assemblies intended for specific use and identified as suitable for this use shall be installed according to Article 422. Fixed electric pipeline and vessel heating equipment for use in hazardous (classified) locations shall comply with Articles 500 through 516.

427-4. Branch-Circuit Sizing. The ampacity of branch-circuit conductors and the rating or setting of overcurrent protective devices that supply fixed electric heating equipment for pipelines and vessels shall be not less than 125 percent of the total load of the heaters. The rating or setting of overcurrent protective devices shall be permitted in accordance with Section 240-3(b).

B. Installation

427-10. General. Equipment for pipeline and vessel electrical heating shall be identified as being suitable for (1) the chemical, thermal, and physical environment; and (2) installation in accordance with the manufacturer's drawings and instructions.

427-11. Use. Electrical heating equipment shall be installed in such a manner as to be afforded protection from physical damage.

427-12. Thermal Protection. External surfaces of pipeline and vessel heating equipment that operate at temperatures exceeding 60°C (140°F) shall be physically guarded, isolated, or thermally insulated to protect against contact by personnel in the area.

427-13. Identification. The presence of electrically heated pipelines or vessels, or both, shall be evident by the posting of appropriate caution signs or markings at frequent intervals along the pipeline or vessel.

C. Resistance Heating Elements

427-14. Secured. Heating element assemblies shall be secured to the surface being heated by means other than the thermal insulation.

427-15. Not in Direct Contact. Where the heating element is not in direct contact with the pipeline or vessel being heated, means shall be provided to prevent overtemperature of the heating element unless the design of the heater assembly is such that its temperature limitations will not be exceeded.

427-16. Expansion and Contraction. Heating elements and assemblies shall not be installed where they bridge expansion joints unless provisions are made for expansion and contraction.

427-17. Flexural Capability. Where installed on flexible pipelines, the heating elements and assemblies shall have a flexural capability that is compatible with the pipeline.

427-18. Power Supply Leads.

(a) Nonheating Leads. Power supply nonheating leads (cold leads) for resistance elements shall be suitable for the temperature encountered. Preassembled nonheating leads on approved heaters shall be permitted to be shortened if the markings specified in Section 427-20 are retained. Not less than 6 in. (152 mm) of nonheating leads shall be provided within the junction box.

(b) Power Supply Leads Protection. Nonheating power supply leads shall be protected where they emerge from electrically heated pipeline or vessel heating units by rigid metal conduit, intermediate metal conduit, electrical metallic tubing, or other raceways identified as suitable for the application.

(c) Interconnecting Leads. Interconnecting nonheating leads connecting portions of the heating system shall be permitted to be covered by thermal insulation in the same manner as the heaters.

427-19. Electrical Connections.

(a) Nonheating Interconnections. Nonheating interconnections, where required under thermal insulation, shall be made with insulated connectors identified as suitable for this use.

(b) Circuit Connections. Splices and terminations outside the thermal insulation shall be installed in a box or fitting in accordance with Sections 110-14 and 300-15.

427-20. Marking. Each factory-assembled heating unit shall be legibly marked within 3 in. (76 mm) of each end of the nonheating leads with the permanent identification symbol, catalog number, and ratings in volts and watts, or in volts and amperes.

427-22. Equipment Protection. Ground-fault protection of equipment shall be provided for electric heat tracing and heating panels. This requirement shall not apply in industrial establishments where there is alarm indication of ground faults and

The words "not having a metal covering" were removed from Section 427-22 for the 1996 *Code*. Revised for the 1999 *Code*, the requirement was rewritten to emphasize that the required ground-fault protection was for the equipment only.

Rather than protecting the entire branch circuit, the ground-fault protection requirement is now focused on protecting just the equipment itself. This change affords the manufacturer and the user an option of providing both circuit and equipment protection, or just the required equipment protection. This required protection may be accomplished by using circuit breakers equipped with ground-fault equipment protection (GFEP) or an integral device supplied as part of the pipeline or vessel heating equipment that is sensitive to leakage currents in the magnitude of 6 milliamperes to 50 milliamperes [referred to by UL as ground-fault equipment protection circuit-interrupters (GFEPCI)]. These protection devices, if applied properly, will substantially reduce the risk of a fire being started by low-level electrical arcing.

It is important to understand that this required equipment protection is not the same as a ground-fault circuit interrupter used for personnel protection that trips at 5 milliamperes (±1 milliampere).

For further information regarding ground-fault equipment protection used to comply with Sections 426-28 and 427-22, refer to the 1998 UL *General Information for Electrical Equipment Directory* (White Book), category KCZI.

(1) Conditions of maintenance and supervision ensure that only qualified persons will service the installed systems.

(2) Continued circuit operation is necessary for safe operation of equipment or processes.

427-23. Metal Covering. Electric heating equipment shall have a grounded metal covering in accordance with (a) or (b). The metal covering shall provide an effective ground path for equipment protection.

This requirement was changed for the 1996 *Code.* It requires a grounded metal covering on all heaters. This metal covering is intended to provide a path to ground for fault current, in order to trip phase or ground-fault protective devices, thus reducing the potential for fire and electric shock. It also provides added mechanical protection of the heating cable or panel. This requirement became effective July 1, 1996.

(a) Heating Wires or Cables. Heating wires or cables shall have a grounded metal covering that surrounds the heating element and bus wires, if any, and their electrical insulation.

(b) Heating Panels. Heating panels shall have a grounded metal covering over the heating element and its electrical insulation on the side opposite the side attached to the surface to be heated.

D. Impedance Heating

427-25. Personnel Protection. All accessible external surfaces of the pipeline or vessel, or both, being heated shall be physically guarded, isolated, or thermally insulated (with a weatherproof jacket for outside installations) to protect against contact by personnel in the area.

427-26. Isolation Transformer. A dual-winding transformer with a grounded shield between the primary and secondary windings shall be used to isolate the distribution system from the heating system.

427-27. Voltage Limitations. Unless protected by ground-fault circuit-interrupter protection for personnel, the secondary winding of the isolation transformer connected to the pipeline or vessel being heated shall not have an output voltage greater than 30 volts ac.

Where ground-fault circuit-interrupter protection for personnel is provided, the voltage shall be permitted to be greater than 30 but not more than 80 volts.

427-28. Induced Currents. All current-carrying components shall be installed in accordance with Section 300-20.

427-29. Grounding. The pipeline or vessel, or both, being heated that is operating at a voltage greater than 30 but not more than 80 shall be grounded at designated points.

427-30. Secondary Conductor Sizing. The ampacity of the conductors connected to the secondary of the transformer shall be at least 100 percent of the total load of the heater.

E. Induction Heating

427-35. Scope. This part covers the installation of line frequency induction heating equipment and accessories for pipelines and vessels.

FPN: See Article 665 for other applications.

427-36. Personnel Protection. Induction coils that operate or may operate at a voltage greater than 30 volts ac shall be enclosed in a nonmetallic or split metallic enclosure, isolated or made inaccessible by location to protect personnel in the area.

427-37. Induced Current. Induction coils shall be prevented from inducing circulating currents in surrounding metallic equipment, supports, or structures by shielding, isolation, or insulation of the current paths. Stray current paths shall be bonded to prevent arcing.

F. Skin-Effect Heating

427-45. Conductor Ampacity. The ampacity of the electrically insulated conductor inside the ferromagnetic envelope shall be permitted to exceed the values given in Article 310, provided it is identified as suitable for this use.

427-46. Pull Boxes. Pull boxes for pulling the electrically insulated conductor in the ferromagnetic envelope shall be permitted to be buried under the thermal insulation, provided their locations are indicated by permanent markings on the insulation jacket surface and on drawings. For outdoor installations, pull boxes shall be of watertight construction.

427-47. Single Conductor in Enclosure. The provisions of Section 300-20 shall not apply to the installation of a single conductor in a ferromagnetic envelope (metal enclosure).

427-48. Grounding. The ferromagnetic envelope shall be grounded at both ends and, in addition, it shall be permitted to be grounded at intermediate points as required by its design. The ferromagnetic envelope shall be bonded at all joints to ensure electrical continuity.

The provisions of Section 250-30 shall not apply to the installation of skin-effect heating systems.

FPN: See Sections 250-30(a)(4) and (b)(3) for grounding methods.

G. Control and Protection

427-55. Disconnecting Means.

(a) Switch or Circuit Breaker. Means shall be provided to disconnect all fixed electric pipeline or vessel heating equipment from all ungrounded conductors. The branch-circuit switch or circuit breaker, where readily accessible to the user of the equipment, shall be permitted to serve as the disconnecting means. The disconnecting means shall be of the indicating type, and shall be provided with a positive lockout in the "off" position.

(b) Cord- and Plug-Connected Equipment. The factory-installed attachment plug of cord- and plug-connected equipment rated 20 amperes or less and 150 volts or less to ground shall be permitted to be the disconnecting means.

427-56. Controls.

(a) Temperature Control with "Off" Position. Temperature controlled switching devices that indicate an "off" position and that interrupt line current shall open all ungrounded conductors when the control device is in this "off" position. These devices shall not be permitted to serve as the disconnecting means unless provided with a positive lockout in the "off" position.

(b) Temperature Control Without "Off" Position. Temperature controlled switching devices that do not have an "off" position shall not be required to open all ungrounded conductors and shall not be permitted to serve as the disconnecting means.

(c) Remote Temperature Controller. Remote controlled temperature-actuated devices shall not be required to meet the requirements of Sections 427-56(a) and (b). These devices shall not be permitted to serve as the disconnecting means.

(d) Combined Switching Devices. Switching devices consisting of combined temperature-actuated devices and manually controlled switches that serve both as the controllers and the disconnecting means shall comply with all the following conditions:

(1) Open all ungrounded conductors when manually placed in the "off" position
(2) Be designed so that the circuit cannot be energized automatically if the device has been manually placed in the "off" position
(3) Be provided with a positive lockout in the "off" position

427-57. Overcurrent Protection. Heating equipment shall be considered as protected against overcurrent where supplied by a branch circuit as specified in Section 427-4.

Article 430 — Motors, Motor Circuits, and Controllers

Contents

A. General

Most electrical equipment is rated in volt-amperes (VA) or watt input. Basic to the understanding of Article 430 is the fact that motors have traditionally been rated in horsepower output. Circuits supplying motors are sized according to the input to the motor (input equals output plus losses of the motor). The losses are not the type of information found on the nameplate of a motor. Tables 430-147 through 430-152 contain very accurate industry-wide input ampere ratings for motors.

However, some motors are available with their output ratings expressed in watts and kilowatts. (One horsepower equals approximately 746 watts.) It is important to understand that circuits that supply motors not rated in horsepower must still be sized according to the input of the motor, rated in amperes. Sizing circuits based solely on kilowatt output results in seriously undersized conductors and the improper application of overcurrent devices. See Section 430-6 for ampacity and motor rating determination.

430-1. Scope. This article covers motors, motor branch-circuit and feeder conductors and their protection, motor overload protection, motor control circuits, motor controllers, and motor control centers.

FPN No. 1: Installation requirements for motor control centers are covered in Section 110-26(f). Air conditioning and refrigerating equipment are covered in Article 440.

In previous editions of the *NEC,* Section 384-4 contained the dedicated space requirements for motor control centers. For the 1999 edition, these requirements were moved to Section 110-26(f).

FPN No. 2: Figure 430-1 is for information only.

Figure 430-1 in the *Code* text is intended to assist the user in following the provisions of Article 430. It is explanatory material and not a *Code* requirement.

430-2. Adjustable-Speed Drive Systems. The incoming branch circuit or feeder to power conversion equipment included as a part of an adjustable-speed drive system shall be based on the rated input to the power conversion equipment. Where the power conversion equipment is marked to indicate that overload protection is included, additional overload protection shall not be required.

The disconnecting means shall be permitted to be in the incoming line to the conversion equipment and shall have a rating not less than 115 percent of the rated input current of the conversion unit.

FPN: Electrical resonance can result from the interaction of the nonsinusoidal currents from this type of load with power factor correction capacitors.

The characteristics of the adjustable-speed drive systems may create resonance circuit conditions if capacitors are improperly applied for power factor correction. Power factor correction with harmonic filtering can safely be applied to this type of load

through proper design. For specific details of the harmonic effects related to capacitors, the capacitor manufacturer should be consulted.

General, Sections 430-1 through 430-18	Part A
Motor Circuit Conductors, Sections 430-21 through 430-29	Part B
Motor and Branch-Circuit Overload Protection, Sections 430-31 through 430-44	Part C
Motor Branch-Circuit Short-Circuit and Ground-Fault Protection, Sections 430-51 through 430-58	Part D
Motor Feeder Short-Circuit and Ground-Fault Protection, Sections 430-61 through 430-63	Part E
Motor Control Circuits, Sections 430-71 through 430-74	Part F
Motor Controllers, Sections 430-81 through 430-91	Part G
Motor Control Centers, Sections 430-92 through 430-98	Part H
Disconnecting Means, Sections 430-101 through 430-113	Part J
Over 600 Volts, Nominal, Sections 430-121 through 430-127	Part K
Protection of Live Parts—All Voltages, Sections 430-131 through 430-133	Park L
Grounding—All Voltages, Sections 430-141 through 430-145	Part M
Tables, Tables 430-147 through 430-152	Part N

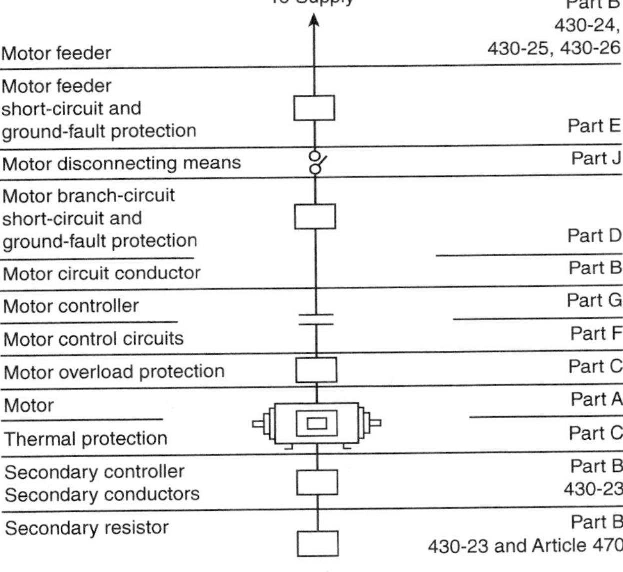

FPN: Figure 430-1.

430-3. Part-Winding Motors. A part-winding start induction or synchronous motor is one that is arranged for starting by first energizing part of its primary (armature) winding and, subsequently, energizing the remainder of this winding in one or more steps. A standard part-winding start induction motor is arranged so that one-half of its primary winding can be energized initially, and, subsequently, the remaining half can be energized, both halves then carrying equal current. A hermetic refrigerant compressor motor shall not be considered a standard part-winding start induction motor.

Where separate overload devices are used with a standard part-winding start induction motor, each half of the motor winding shall be individually protected in accordance with Sections 430-32 and 430-37 with a trip current one-half that specified.

Each motor-winding connection shall have branch-circuit short-circuit and ground-fault protection rated at not more than one-half that specified by Section 430-52.

Exception: A short-circuit and ground-fault protective device shall be permitted for both windings if the device will allow the motor to start. Where time-delay (dual-element) fuses are used, they shall be permitted to have a rating not exceeding 150 percent of the motor full-load current.

430-5. Other Articles. Motors and controllers shall also comply with the applicable provisions of the following:

	Article	Section
Air-conditioning and refrigerating equipment	440	
Capacitors		460-8, 460-9
Commercial garages, aircraft hangars, gasoline dispensing and service stations, bulk storage plants, spray application, dipping, and coating processes, and inhalation anesthetizing locations	511, 513, 514, 515, 516, and 517, Part D	
Cranes and hoists	610	
Electrically driven or controlled irrigation machines	675	
Elevators, dumbwaiters, escalators, moving walks, wheelchair lifts, and stairway chair lifts	620	
Fire pumps	695	
Hazardous (classified) locations	500–503 and 505	
Industrial machinery	670	
Motion picture projectors		540-11 and 540-20
Motion picture and television studios and similar locations	530	
Resistors and reactors	470	
Theaters, audience areas of motion picture and television studios, and similar locations		520-48
Transformers and transformer vaults	450	

430-6. Ampacity and Motor Rating Determination. The size of conductors supplying equipment covered by Article 430 shall be selected from the allowable ampacity tables in accordance with Section 310-15(b) or shall be calculated in accordance with Section 310-15(c). Where flexible cord is used, the size of the conductor shall be selected in accordance with Section 400-5. The required ampacity and motor ratings shall be determined as specified in (a), (b), and (c).

(a) General Motor Applications. For general motor applications, current ratings shall be determined based on (1) and (2).

(1) Table Values. The values given in Tables 430-147, 430-148, 430-149, and 430-150, including notes, shall be used to determine the ampacity of conductors or ampere ratings of switches, branch-circuit short-circuit and ground-fault protection, instead of the actual current rating marked on the motor nameplate. Where a motor is marked in amperes, but not horsepower, the horsepower rating shall be assumed to be that corresponding to the value given in Tables 430-147, 430-148, 430-149, and 430-150, interpolated if necessary.

Exception No. 1: Multispeed motors shall be in accordance with Sections 430-22(a) and 430-52.

Exception No. 2: For equipment that employs a shaded-pole or permanent-split capacitor-type fan or blower motor that is marked with the motor type, the full load current for such motor marked on the nameplate of the equipment in which the fan or blower motor is employed shall be used instead of the horsepower rating to determine the ampacity or rating of the disconnecting means, the branch-circuit conductors, the controller, the branch-circuit short-circuit and ground-fault protection, and the separate overload protection. This marking on the equipment nameplate shall not be less than the current marked on the fan or blower motor nameplate.

Exception No. 3: For a listed motor-operated appliance that is marked with both motor horsepower and full-load current, the motor full-load current marked on the nameplate of the appliance shall be used instead of the horsepower rating on the appliance nameplate to determine the ampacity or rating of the disconnecting means, the branch-circuit conductors, the controller, the branch-circuit short-circuit and ground-fault protection, and any separate overload protection.

Added for the 1999 *Code*, this new Exception No. 3 is intended to resolve confusion that may result when motor-operated appliances are labeled with both horsepower and ampere ratings. The nameplate current rating in amperes is closely evaluated by the testing laboratories and is considered more accurate than the marked horsepower rating for motor-operated appliances.

(2) Nameplate Values. Separate motor overload protection shall be based on the motor nameplate current rating.

(b) Torque Motors. For torque motors, the rated current shall be locked-rotor current, and this nameplate current shall be used to determine the ampacity of the branch-circuit conductors covered in Sections 430-22 and 430-24, the ampere rating of the motor overload protection, and the ampere rating of motor branch-circuit short-circuit and ground-fault protection in accordance with Section 430-52(b).

FPN: For motor controllers and disconnecting means, see Sections 430-83(d) and 430-110.

(c) Alternating-Current Adjustable Voltage Motors. For motors used in alternating-current, adjustable voltage, variable torque drive systems, the ampacity of conductors, or ampere ratings of switches, branch-circuit short-circuit and ground-fault protection, etc., shall be based on the maximum operating current marked on the motor or control nameplate, or both. If the maximum operating current does not appear on the nameplate, the ampacity determination shall be based on 150 percent of the values given in Tables 430-149 and 430-150.

For general motor applications other than ac adjustable voltage or torque motors, the ampacity of motor branch-circuit conductors, branch-circuit and ground-fault protection, and ampere rating of the motor disconnecting means are determined by the ampere values listed in Tables 430-147 through 430-150. The ampere values are based on the horsepower rating and nominal voltage listed on the motor nameplate.

The ampere rating provided on the motor nameplate is used to size the overload protective devices intended to protect the motor, motor control apparatus, and motor branch-circuit conductors.

430-7. Marking on Motors and Multimotor Equipment.

(a) Usual Motor Applications. A motor shall be marked with the following information.

(1) Manufacturer's name.

(2) Rated volts and full-load amperes. For a multispeed motor, full-load amperes for each speed, except shaded-pole and permanent-split capacitor motors where amperes are required only for maximum speed.

(3) Rated frequency and number of phases, if an ac motor.

(4) Rated full-load speed.

(5) Rated temperature rise or the insulation system class and rated ambient temperature.

(6) Time rating. The time rating shall be 5, 15, 30, or 60 minutes, or continuous.

(7) Rated horsepower if ⅛ hp or more. For a multispeed motor ⅛ hp or more, rated horsepower for each speed, except shaded-pole and permanent-split capacitor motors ⅛ hp or more where rated horsepower is required only for maximum speed. Motors of arc welders are not required to be marked with the horsepower rating.

(8) Code letter or locked-rotor amperes if an alternating-current motor rated ½ hp or more. On polyphase wound-rotor motors, the code letter shall be omitted.

FPN: See Section 430-7(b).

(9) Design letter for design B, C, D, or E motors.

FPN: Motor design letter definitions are found in *Motors and Generators, Part 1, Definitions*, ANSI/NEMA MG 1-1993, and in *Standard Dictionary of Electrical and Electronic Terms*, ANSI/IEEE 100-1996.

Design letters are not to be confused with code letters. For technical accuracy, code letters should be referred to as "locked-rotor indicating code letters" and are explained in Section 430-7(b). Design letters reflect characteristics inherent in motor design, such as locked-rotor current, slip at rated load, and locked-rotor and breakdown torque. Design E motors have significant differences from the older and very common Design B motors. For example, Design E motors are very energy efficient (lower I^2R losses) and have high locked-rotor currents (see Table 430-151B). Some Design E motors may even have higher full-load currents than their equivalent Design B motors. It is imperative that these electrical differences be understood before a Design E motor is used for an initial installation or as a replacement.

Where existing motors are being considered for replacement with Design E motors, the existing motor circuits must be evaluated for minimum code compliance (i.e., disconnect rating, controller rating, short-circuit and ground-fault protection, and overload protection) as well as standard design compliance issues such as starting voltage-drop or power quality impact on the facility. Of course it is prudent, where energy conservation is concerned, to also upgrade the mechanical portion of the system, since the electric motor does not reflect all the losses of a typical mechanical system.

As a result of the basic differences of Design E motors, various sections in Article 430 had to be adjusted to safely permit the installation and proper use of Design E motors for this edition as well as in previous editions of the *NEC*.

(10) Secondary volts and full-load amperes if a wound-rotor induction motor.

(11) Field current and voltage for dc excited synchronous motors.

(12) Winding — straight shunt, stabilized shunt, compound, or series, if a dc motor. Fractional horsepower dc motors 7 in. (178 mm) or less in diameter shall not be required to be marked.

(13) A motor provided with a thermal protector complying with Sections 430-32(a)(2) or (c)(2) shall be marked "Thermally Protected." Thermally protected motors rated 100 watts or less and complying with Section 430-32(c)(2) shall be permitted to use the abbreviated marking "T.P."

(14) A motor complying with Section 430-32(c)(4) shall be marked "Impedance Protected." Impedance protected motors rated 100 watts or less and complying with Section 430-32(c)(4) shall be permitted to use the abbreviated marking "Z.P."

(b) Locked-Rotor Indicating Code Letters. Code letters marked on motor nameplates to show motor input with locked rotor shall be in accordance with Table 430-7(b).

The code letter indicating motor input with locked rotor shall be in an individual block on the nameplate, properly designated.

Table 430-7(b). Locked-Rotor Indicating Code Letters

Code Letter	Kilovolt-Amperes per Horsepower with Locked Rotor	Code Letter	Kilovolt-Amperes per Horsepower with Locked Rotor
A	0–3.14	L	9.0–9.99
B	3.15–3.54	M	10.0–11.19
C	3.55–3.99	N	11.2–12.49
D	4.0–4.49	P	12.5–13.99
E	4.5–4.99	R	14.0–15.99
F	5.0–5.59	S	16.0–17.00
G	5.6–6.29	T	18.0–19.99
H	6.3–7.09	U	29.0–22.39
J	7.1–7.99	V	22.4 and up
K	8.0–8.99		

(1) Multispeed motors shall be marked with the code letter designating the locked-rotor kilovolt-ampere (kVA) per horsepower for the highest speed at which the motor can be started.

Exception: Constant horsepower multispeed motors shall be marked with the code letter giving the highest locked-rotor kilovolt-ampere (kVA) per horsepower.

(2) Single-speed motors starting on Y connection and running on delta connections shall be marked with a code letter corresponding to the locked-rotor kilovolt-ampere (kVA) per horsepower for the Y connection.

(3) Dual-voltage motors that have a different locked-rotor kilovolt-ampere (kVA) per horsepower on the two voltages shall be marked with the code letter for the voltage giving the highest locked-rotor kilovolt-ampere (kVA) per horsepower.

(4) Motors with 60- and 50-hz ratings shall be marked with a code letter designating the locked-rotor kilovolt-ampere (kVA) per horsepower on 60 hz.

(5) Part-winding start motors shall be marked with a code letter designating the locked-rotor kilovolt-ampere (kVA) per horsepower that is based on the locked-rotor current for the full winding of the motor.

Example

Using Table 430-7(b), find the maximum locked-rotor current for a 20-hp, 460-volt, 3-phase motor with a nameplate kilovolt-ampere code letter "G."

Answer

Since the example requests the maximum value and the table lists a range of values, select the maximum value for code letter "G" range, or 6.29 kilovolt-amperes per horsepower. Then, use the following formula to find the maximum locked-rotor current.

$$\text{Locked-rotor amperes} = \text{motor hp} \times \text{maximum code letter value } \frac{\text{kVA}}{\text{hp}} \times \frac{1000}{\text{volts} \times 1.73}$$

Substitute as follows:

$$\text{Locked-rotor amperes} = 20 \text{ hp} \times 6.29 \frac{\text{kVA}}{\text{hp}} \times \frac{1000}{460 \text{ volts} \times 1.73} = 158 \text{ amperes}$$

Therefore, the maximum locked-rotor current for a 20-hp, 460-volt motor with a code letter "G" is 158 amperes.

(c) Torque Motors. Torque motors are rated for operation at standstill and shall be marked in accordance with section 430-7 (a), except that locked-rotor torque shall replace horsepower.

(d) Multimotor and Combination-Load Equipment.

(1) Multimotor and combination-load equipment shall be provided with a visible nameplate marked with the manufacturer's name, the rating in volts, frequency, number of phases, minimum supply circuit conductor ampacity, and the maximum ampere rating of the circuit short-circuit and ground-fault protective device. The conductor ampacity shall be computed in accordance with Section 430-24 and counting all of the motors and other loads that will be operated at the same time. The short-circuit and ground-fault protective device rating shall not exceed the value computed in accordance with Section 430-53. Multimotor equipment for use on two or more circuits shall be marked with the above information for each circuit.

The nameplate marking for the maximum ampere rating of the branch-circuit short-circuit and ground-fault protective device may limit the type of protective device to a fuse by stipulating "fuse" without reference to a circuit breaker. This means that the circuit to the equipment is required to be protected by fuses, such as by a fused disconnect switch. The fused switch may be supplied from a circuit breaker in a panelboard.

(2) Where the equipment is not factory-wired and the individual nameplates of motors and other loads are visible after assembly of the equipment, the individual nameplates shall be permitted to serve as the required marking.

430-8. Marking on Controllers. A controller shall be marked with the manufacturer's name or identification, the voltage, the current or horsepower rating, and such other necessary data to properly indicate the motors for which it is suitable. A controller that includes motor overload protection suitable for group motor application shall be marked with the motor overload protection and the maximum branch-circuit short-circuit and ground-fault protection for such applications.

Combination controllers that employ adjustable instantaneous trip circuit breakers shall be clearly marked to indicate the ampere settings of the adjustable trip element.

Where a controller is built-in as an integral part of a motor or of a motor-generator set, individual marking of the controller shall not be required if the necessary data are on the nameplate. For controllers that are an integral part of equipment approved as a unit, the above marking shall be permitted on the equipment nameplate.

FPN: See Section 110-10 for information on circuit impedance and other characteristics.

430-9. Terminals.

(a) Markings. Terminals of motors and controllers shall be suitably marked or colored where necessary to indicate the proper connections.

(b) Conductors. Motor controllers and terminals of control circuit devices shall be connected with copper conductors unless identified for use with a different conductor.

Terminals for motor controllers are tested, designed, and listed using copper conductors, unless marked for other conductors. Section 430-9(b) highlights this limitation while permitting other conductors to be used if they are determined to be suitable and are identified as such.

(c) Torque Requirements. Control circuit devices with screw-type pressure terminals used with No. 14 or smaller copper conductors shall be torqued to a minimum of 7 lb-in. (0.79 N-m) unless identified for a different torque value.

Proper torque is essential for safe and reliable connections. Section 430-9(c) enhances safety by providing a minimum torque value for screw-type pressure terminals.

430-10. Wiring Space in Enclosures.

(a) General. Enclosures for motor controllers and disconnecting means shall not be used as junction boxes, auxiliary gutters, or raceways for conductors feeding through or tapping off to the other apparatus unless designs are employed that provide adequate space for this purpose.

FPN: See Section 373-8 for switch and overcurrent-device enclosures.

During the planning stages of a motor(s) installation, consideration should be given to location and proper working spaces (as required by Sections 110-26 and 110-34) for motor controllers and disconnects. This includes provisions for the use of auxiliary gutters or junction boxes, to ensure space for conductors feeding through or tapping off to other apparatus. For switch and overcurrent device enclosures, see Section 373-8.

(b) Wire-Bending Space in Enclosures. Minimum wire-bending space within the enclosures for motor controllers shall be in accordance with Table 430-10(b) where measured in a straight line from the end of the lug or wire connector (in the direction the wire leaves the terminal) to the wall or barrier. Where alternate wire termination means are substituted for that supplied by the manufacturer of the controller, they shall be of a type identified by the manufacturer for use with the controller and shall not reduce the minimum wire-bending space.

Table 430-10(b). Minimum Wire-Bending Space at the Terminals of Enclosed Motor Controllers (in.)

Size of Wire (AWG or kcmil)	Wires per Terminal*	
	1	2
14–10	Not specified	—
8–6	1½	—
4–3	2	—
2	2½	—
1	3	—
1/0	5	5
2/0	6	6
3/0–4/0	7	7
250	8	8
300	10	10
350–500	12	12
600–700	14	16
750–900	18	19

*Where provision for three or more wires per terminal exists, the minimum wire-bending space shall be in accordance with the requirements of Article 373.

See Figure 430.1 for an illustration of measurement, in accordance with Section 430-10(b) or 373-6(b).

430-11. Protection Against Liquids. Suitable guards or enclosures shall be provided to protect exposed current-carrying parts of motors and the insulation of motor leads where installed directly under equipment, or in other locations where dripping or spraying oil, water, or other injurious liquid may occur, unless the motor is designed for the existing conditions.

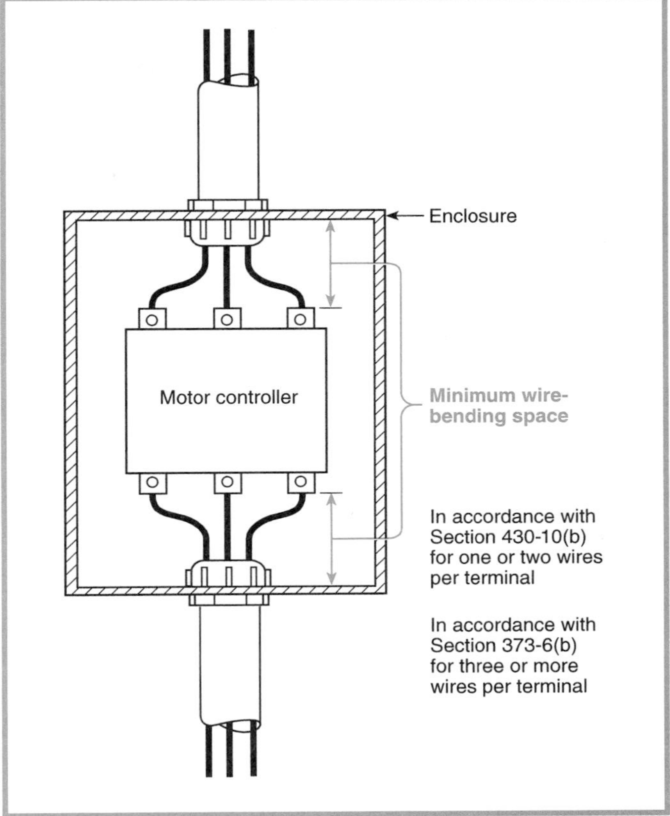

Figure 430.1 *Wire-bending space in enclosures for motor controllers.*

Exposed current-carrying parts and insulated leads of motors should be suitably protected from liquids (dripping or spraying oil or other lubricants, water, or excessive moisture) whose presence may cause deterioration and insulation breakdown.

430-12. Motor Terminal Housings.

(a) Material. Where motors are provided with terminal housings, the housings shall be of metal and of substantial construction.

Exception: In other than hazardous (classified) locations, substantial, nonmetallic, nonburning housings shall be permitted provided an internal grounding means between the motor frame and the equipment grounding connection is incorporated within the housing.

Nonmetallic terminal housings are permitted on motors of any size, provided the housing material has been determined to be nonburning, with, for example, a flammability rating of at least 94-5V in accordance with UL 746C, *Polymeric Materials — Use in Electrical Equipment Evaluations.*

(b) Dimensions and Space — Wire-to-Wire Connections. Where these terminal housings enclose wire-to-wire connections, they shall have minimum dimensions and usable volumes in accordance with Table 430-12(b).

Table 430-12(b). Terminal Housings — Wire-to-Wire Connections

Motors 11 in. in Diameter or Less

Horsepower	Cover Opening Minimum Dimension (in.)	Usable Volume Minimum (in.^{3})
1 and smaller[a]	1⅝	10.5
1½, 2, and 3[b]	1¾	16.8
5 and 7½	2	22.4
10 and 15	2½	36.4

Motors Over 11 in. in Diameter Alternating-Current Motors

Maximum Full Load Current for 3-Phase Motors with Maximum of 12 Leads (Amperes)	Terminal Box Cover Opening Minimum Dimension (in.)	Usable Volume Minimum (in.^{3})	Typical Maximum Horsepower 3-Phase	
			230 Volt	460 Volt
45	2.5	36.4	15	30
70	3.3	77	25	50
110	4.0	140	40	75
160	5.0	252	60	125
250	6.0	450	100	200
400	7.0	840	150	300
600	8.0	1540	250	500

Direct-Current Motors

Maximum Full-Load Current for Motors with Maximum of Six Leads (Amperes)	Terminal Box Minimum Dimensions (in.)	Usable Volume Minimum (in.^{3})
68	2.5	26
105	3.3	55
165	4.0	100
240	5.0	180
375	6.0	330
600	7.0	600
900	8.0	1100

Notes:

1. For SI units, 1 in. = 25.4 mm.
2. Auxiliary leads for such items as brakes, thermostats, space heaters, exciting fields, etc., shall be permitted to be neglected if their current-carrying area does not exceed 25 percent of the current-carrying area of the machine power leads.

[a]For motors rated 1 hp and smaller and with the terminal housing partially or wholly integral with the frame or end shield, the volume of the terminal housing shall not be less than 1.1 in.^{3} per wire-to-wire connection. The minimum cover opening dimension is not specified.

[b]For motors rated 1½, 2, and 3 hp and with the terminal housing partially or wholly integral with the frame or end shield, the volume of the terminal housing shall not be less than 1.4 in.^{3} per wire-to-wire connection. The minimum cover opening dimension is not specified.

Changed in the 1999 *Code,* the minimum cover opening dimensions given in Table 430-12(b) have been reduced to reflect more practical values in use today without changing the minimum usable volumes.

(c) Dimensions and Space — Fixed Terminal Connections. Where these terminal housings enclose rigidly mounted motor terminals, the terminal housing shall be of sufficient size to provide minimum terminal spacings and usable volumes in accordance with Table 430-12(c)(1) and Table 430-12(c)(2).

Table 430-12(c)(1). Terminal Spacings — Fixed Terminals

Nominal Volts	Minimum Spacing (in.) Between Line Terminals	Minimum Spacing (in.) Between Line Terminals and Other Uninsulated Metal Parts
240 or less	¼	¼
Over 250–600	⅜	⅜

Note: For SI units, 1 in. = 25.4 mm.

Table 430-12(c)(2). Usable Volumes — Fixed Terminals

Power-Supply Conductor Size (AWG)	Minimum Usable Volume per Power-Supply Conductor (in.^{3})
14	1
12 and 10	1¼
8 and 6	2¼

Note: For SI units, 1 in. = 25.4 mm.

(d) Large Wire or Factory Connections. For motors with larger ratings, greater number of leads, or larger wire sizes, or where motors are installed as a part of factory-wired equipment, without additional connection being required at the motor terminal housing during equipment installation, the terminal housing shall be of ample size to make connections, but the foregoing provisions for the volumes of terminal housings shall not be considered applicable.

(e) Equipment Grounding Connections. A means for attachment of an equipment grounding conductor termination in accordance with Section 250-8 shall be provided at motor terminal housings for wire-to-wire connections or fixed terminal connections. The means for such connections shall be permitted to be located either inside or outside the motor terminal housing.

Exception: Where a motor is installed as a part of factory-wired equipment that is required to be grounded and without additional connection being required at the motor terminal housing during equipment installation, a separate means for motor grounding at the motor terminal housing shall not be required.

430-13. Bushing. Where wires pass through an opening in an enclosure, conduit box, or barrier, a bushing shall be used to protect the conductors from the edges of openings having sharp edges. The bushing shall have smooth, well-rounded surfaces where it may be in contact with the conductors. If used where oils, greases, or other contaminants may be present, the bushing shall be made of material not deleteriously affected.

FPN: For conductors exposed to deteriorating agents, see Section 310-9.

430-14. Location of Motors.

(a) Ventilation and Maintenance. Motors shall be located so that adequate ventilation is provided and so that maintenance, such as lubrication of bearings and replacing of brushes, can be readily accomplished.

(b) Open Motors. Open motors that have commutators or collector rings shall be located or protected so that sparks cannot reach adjacent combustible material.

Exception: Installation of these motors on wooden floors or supports shall be permitted.

430-16. Exposure to Dust Accumulations. In locations where dust or flying material will collect on or in motors in such quantities as to seriously interfere with the ventilation or cooling of motors and thereby cause dangerous temperatures, suitable types of enclosed motors that will not overheat under the prevailing conditions shall be used.

FPN: Especially severe conditions may require the use of enclosed pipe-ventilated motors, or enclosure in separate dusttight rooms, properly ventilated from a source of clean air.

For motors exposed to combustible dust or readily ignitible flying material, see the requirements of Sections 502-8 and 502-9 (Class II, Divisions 1 and 2) and Sections 503-6 and 503-7 (Class III, Divisions 1 and 2). For classification of locations, see Sections 500-8 (Class II locations) and 500-9 (Class III locations).

430-17. Highest Rated or Smallest Rated Motor. In determining compliance with Sections 430-24, 430-53(b), and 430-53(c), the highest rated or smallest rated motor shall be based on the rated full-load current as selected from Tables 430-147, 430-148, 430-149, and 430-150.

430-18. Nominal Voltage of Rectifier Systems. The nominal value of the ac voltage being rectified shall be used to determine the voltage of a rectifier derived system.

Exception: The nominal dc voltage of the rectifier shall be used if it exceeds the peak value of the ac voltage being rectified.

B. Motor Circuit Conductors

430-21. General. Part B specifies sizes of conductors that are capable of carrying the motor current without overheating under the conditions specified.

The provisions of Part B shall not apply to motor circuits rated over 600 volts, nominal. See Part K.

The provisions of Articles 250, 300, and 310 shall not apply to conductors that form an integral part of equipment, such as motors, motor controllers, motor control centers, or other factory-assembled control equipment.

FPN No. 1: See Sections 300-1(b) and 310-1 for similar requirements.

FPN No. 2: See Section 430-9(b) for equipment device terminal requirements.

430-22. Single Motor.

(a) General. Branch-circuit conductors that supply a single motor used in a continuous duty application shall have an ampacity of not less than 125 percent of the motor's full-load current rating as determined by Section 430-6(a)(1).

The provision for a conductor with an ampacity of at least 125 percent of the motor full-load current rating does not constitute a conductor derating. It is based on the need to provide for a sustained running current that is greater than the rated full-load current and for protection of the conductors by the motor overload protective device set above the motor full-load current rating.

The ampacity of the motor branch-circuit conductors is based on the full-load current rating values provided in Tables 430-147 through 430-150.

For a multispeed motor, the selection of branch-circuit conductors on the line side of the controller shall be based on the highest of the full-load current ratings shown on the motor nameplate. The selection of branch-circuit conductors between the controller and the motor shall be based on the current rating of the winding(s) that the conductors energize.

Exception No. 1: For dc motors operating from a rectified single-phase power supply, the conductors between the field wiring terminals of the rectifier and the motor shall have an ampacity of not less than the following percent of the motor full-load current rating:

(a) Where a rectifier bridge of the single-phase half-wave type is used, 190 percent
(b) Where a rectifier bridge of the single-phase full-wave type is used, 150 percent

Exception No. 2: Circuit conductors supplying power conversion equipment included as part of an adjustable-speed drive system shall have an ampacity not less than 125 percent of the rated input to the power conversion equipment

FPN: See Appendix D, Example No. D8, and Figure 430-1.

Figure 430.2 illustrates each motor on an individual branch circuit with branch-circuit short-circuit and ground-fault protective devices and disconnecting means in one location, and controllers and overload protection at the motor locations.

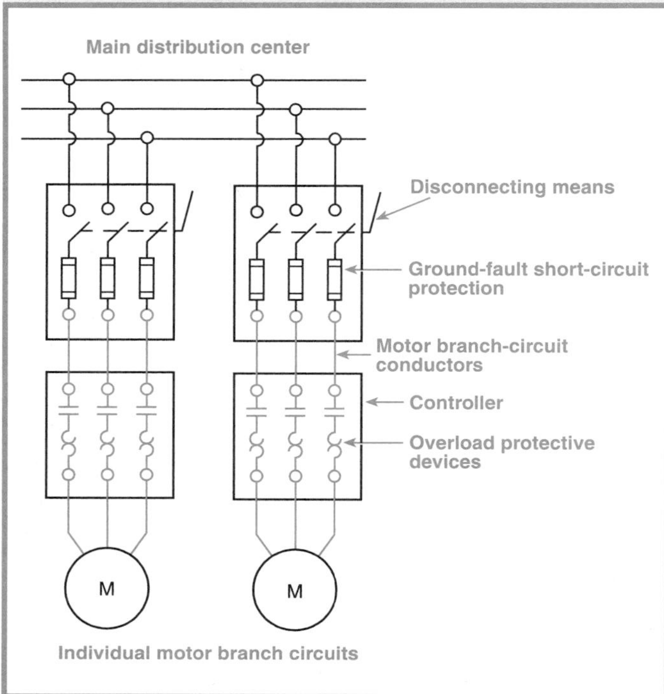

Figure 430.2 *A distribution center supplying individual branch circuits to each motor (branch-circuit short-circuit and ground-fault protective devices).*

Figure 430.3 also illustrates each motor on an individual branch circuit, but, unlike Figure 430.2, the branch circuits are tapped from a feeder at a convenient location, such as a junction box or wireway, or from open wiring. The tap conductors are required to terminate in a branch-circuit protective device located not more than 25 ft from where the taps are connected to the feeder, in accordance with Section 430-28. Also see Section 430-28, Exception, which permits a 100-ft tap under some conditions.

If motors are connected to a 15- or 20-ampere branch circuit that also supplies lighting or other appliance loads, as illustrated in Figure 430.4, the provisions of Articles 210 and 430 apply. Motors rated less than 1 hp may be connected to these circuits and are required to be provided with overload protective devices unless the motors are not permanently installed, are started manually, and are within sight from the controller location. For additional information on the installation of motors (1 hp or less), see Sections 430-32(b) and (c) and 430-53(a).

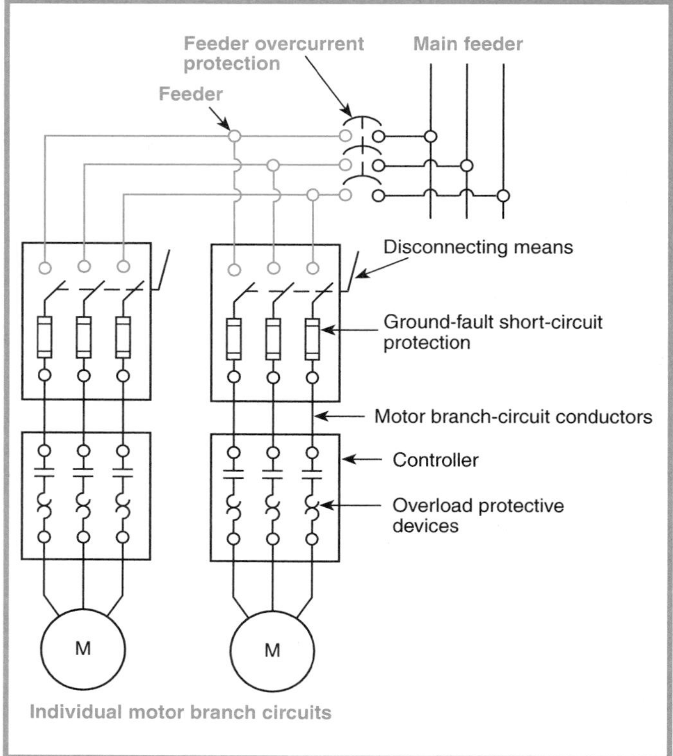

Figure 430.3 *A feeder supplying individual branch circuits to each motor.*

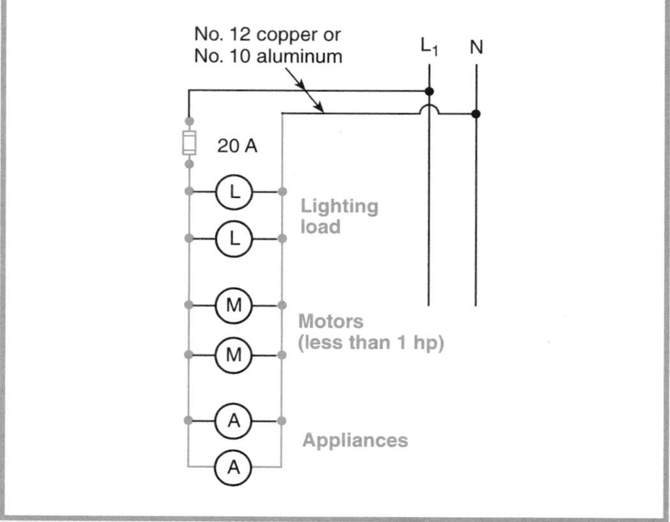

Figure 430.4 *A 20-ampere branch circuit supplying small motors, lighting, and appliances.*

Figure 430.5 illustrates the following essential parts of a motor branch circuit:

1. The branch-circuit conductors
2. The disconnecting means
3. The branch-circuit short-circuit and ground-fault protective devices
4. The motor overload protective devices.

The branch-circuit short-circuit and ground-fault protective device may be a fuse or a circuit breaker and is required to be capable of carrying the starting current of the motor without opening the circuit. See Table 430-152.

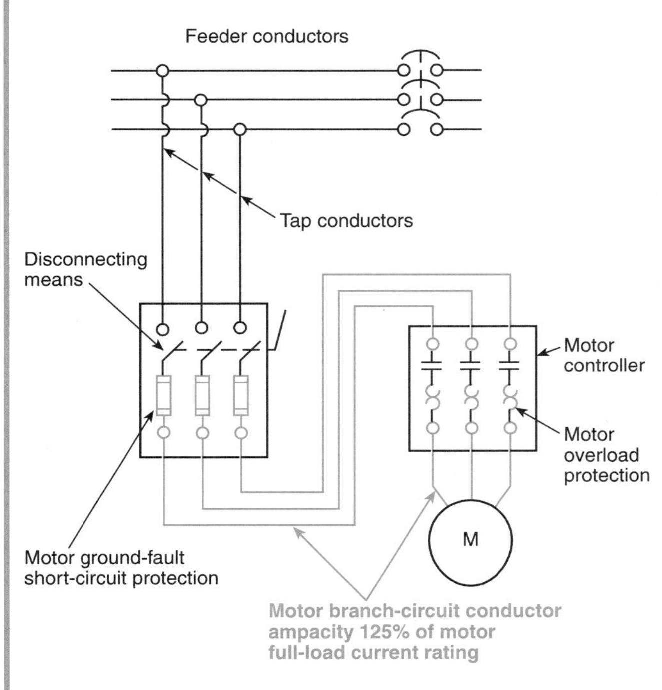

Figure 430.5 *A motor branch circuit showing the essential parts: branch-circuit conductors, disconnecting means, branch-circuit overcurrent protection, and motor overload devices.*

In general, it is required that every motor be provided with overload protective devices intended to protect the motor windings, motor-control apparatus, and motor branch-circuit conductors against excessive heating due to motor overloads and failure to start. Overload in equipment is defined as operation in excess of normal full-load rating, which, when it persists for a sufficient length of time, will cause damage or dangerous overheating. Overload in a motor includes a stalled rotor but does not include fault currents due to short circuits or ground faults. See Section 430-44 for conditions where providing automatic opening of a motor circuit due to overload may be objectionable.

For a wye-start, delta-run connected motor, the selection of branch-circuit conductors on the line side of the controller shall be based on the motor full-load current. The selection of conductors between the controller and the motor shall be based on 58 percent of the motor full-load current.

Wye start, delta run is a method of providing reduced-voltage starting for a polyphase induction motor. It requires a specific type of motor controller and a delta-wired motor with all leads brought out to the terminal box. This method of starting finds wide application in compressors used for air conditioning, where the machine is allowed to start unloaded. Although the three ungrounded conductors that supply the total load to the motor controller carry 100 percent of the full-load current, the six conductors from the motor controller to the motor share this full-load current, with a mathematical ratio of $1/\sqrt{3} = 0.5774$, or 58 percent.

Figure 430.6, conductors from terminals T_1, T_2, and T_3 to the motor, as well as the conductors from terminals T_4, T_5, and T_6 to the motor, are all sized at 58 percent of the full-load current used to size the conductors supplying L_1, L_2, and L_3.

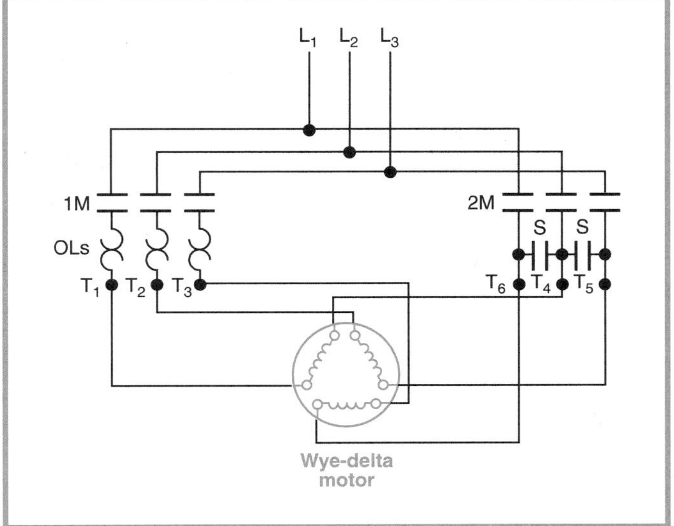

Figure 430.6 *An elementary wiring diagram of a typical wye-start, delta-run motor and controller. During START, contacts 1M and S are closed and contacts 2M are open. During RUN, contacts 1M and 2M are closed and contacts S are open. (Square D Co.)*

(b) Other than Continuous Duty. Conductors for a motor used in a short-time, intermittent, periodic, or varying duty application shall have an ampacity of not less than the percentage of the motor nameplate current rating shown in

Table 430-22(b), unless the authority having jurisdiction grants special permission for conductors of lower ampacity.

Table 430-22(b). Duty-Cycle Service

	Nameplate Current Rating Percentages			
Classification of Service	**5-Minute Rated Motor**	**15-Minute Rated Motor**	**30- & 60-Minute Rated Motor**	**Continuous Rated Motor**
Short-time duty operating valves, raising or lowering rolls, etc.	110	120	150	—
Intermittent duty freight and passenger elevators, tool heads, pumps, drawbridges, turntables, etc. (for arc welders, see Section 630-11)	85	85	90	140
Periodic duty rolls, ore- and coal-handling machines, etc.	85	90	95	140
Varying duty	110	120	150	200

Note: Any motor application shall be considered as continuous duty unless the nature of the apparatus it drives is such that the motor will not operate continuously with load under any condition of use.

A motor is considered to be for continuous duty unless the nature of the apparatus it drives is such that the motor cannot operate continuously with load under any condition of use. Conductors for a motor used for short-time, intermittent, periodic, or varying duty are required to have an ampacity in accordance with Table 430-22(b). Branch-circuit conductors for a motor with a rated horsepower used for 5-minute short-time duty service are permitted to be sized smaller than for the same motor with a 60-minute rating, due to the cooling intervals between operating periods. See Article 100 for the definition of *duty.*

(c) Separate Terminal Enclosure. The conductors between a stationary motor rated 1 hp or less and the separate terminal enclosure permitted in Section 430-145(b) shall be permitted to be smaller than No. 14 but not smaller than No. 18, provided they have an ampacity as specified in Section 430-22(a).

430-23. Wound-Rotor Secondary.

(a) Continuous Duty. For continuous duty, the conductors connecting the secondary of a wound-rotor ac motor to its controller shall have an ampacity not less than 125 percent of the full-load secondary current of the motor.

(b) Other than Continuous Duty. For other than continuous duty, these conductors shall have an ampacity, in percent of full-load secondary current, not less than that specified in Table 430-22(b).

(c) Resistor Separate from Controller. Where the secondary resistor is separate from the controller, the ampacity of the conductors between controller and resistor shall not be less than that shown in Table 430-23(c).

Table 430-23(c). Secondary Conductor

Resistor Duty Classification	Ampacity of Conductor in Percent of Full-Load Secondary Current
Light starting duty	35
Heavy starting duty	45
Extra-heavy starting duty	55
Light intermittent duty	65
Medium intermittent duty	75
Heavy intermittent duty	85
Continuous duty	110

Before the advent of variable frequency drives, wound-rotor ac motors were generally used where speed control was desired; where high starting torque for a rapid, smooth acceleration to full load was required; for frequent starting; and for low starting current. These motors are also known as slip-ring motors, because three slip rings are mounted on the shaft, and brushes in contact with the slip rings are connected to field-installed external resistance units and a controller, as Figure 430.7 illustrates. The resistors are a part of the rotor circuit, and the full value of the external resistance is in the circuit when the motor starts. This value is gradually reduced until the motor attains the desired speed.

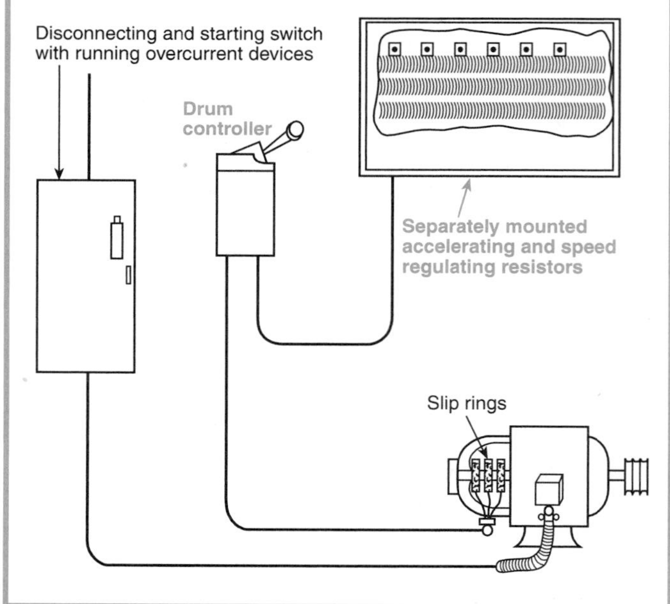

Figure 430.7 *A branch circuit to a wound-rotor induction motor showing a drum controller and separate bank of resistors for motor starting and speed regulation.*

The selection of a controller used for speed regulation, usually a dial-type or a drum-type switch, is basically for two types of loads: constant torque (machine loads) and variable torque (fan loads).

The ampacities of the conductors between the controller and the resistance units are the allowable percentages of Table 430-23(c) for the resistor duty classification.

430-24. Several Motors or a Motor(s) and Other Load(s). Conductors supplying several motors, or a motor(s) and other load(s), shall have an ampacity at least equal to the sum of the full-load current rating as determined by Section 430-6(a)(1) of all the motors, plus 25 percent of the highest rated motor in the group, plus the ampere rating of other loads determined in accordance with Article 220 and other applicable articles.

FPN: See Appendix D, Example No. D8.

Where the conductors are feeders, the highest rating or setting of the feeder short-circuit and ground-fault protective devices for the minimum-size feeder conductor permitted by Section 430-24 is specified in Section 430-62.

Where the selection of a feeder protective device of higher rating or setting is based on the simultaneous starting of two or more motors, the size of the feeder conductors is required to be increased accordingly.

These requirements, and those of Section 430-62 for the short-circuit and ground-fault protection of power feeders, are based on the principle that the power feeder conductors should be sized to have an ampacity equal to 125 percent of the full-load current of the largest motor, plus the full-load currents of all other motors supplied by the feeder.

Except where two or more motors may be started simultaneously, the heaviest load that a power feeder will ever be required to carry occurs when the largest motor is started and all the other motors supplied by the same feeder are running and delivering their full-rated horsepower.

Where the conductors are branch-circuit conductors to multimotor equipment, Section 430-53 specifies the maximum rating of the branch-circuit short-circuit and ground-fault protective device Section 430-7(d)(1) requires the maximum ampere rating of the short-circuit and ground-fault protective device to be marked on multimotor equipment.

See Section 430-62(b) and the associated commentary if the size of the feeder conductors is sized larger than the minimum size.

Exception No. 1: Where one or more of the motors of the group are used for short-time, intermittent, periodic, or varying duty, the ampere rating of such motors to be used in the summation shall be determined in accordance with Section 430-22(b). For the highest rated motor, the greater of either the ampere rating from 430-22(b) or the largest continuous duty motor full-load current multiplied by 1.25 shall be used in the summation.

Exception No. 2: The ampacity of conductors supplying motor-operated fixed electric space-heating equipment shall conform with Section 424-3(b).

Exception No. 3: Where the circuitry is interlocked so as to prevent operation of selected motors or other loads at the same time, the conductor ampacity shall be permitted to be based on the summation of the currents of the motors and other loads to be operated at the same time that results in the highest total current.

430-25. Multimotor and Combination-Load Equipment. The ampacity of the conductors supplying multimotor and combination-load equipment shall not be less than the minimum circuit ampacity marked on the equipment in accordance with Section 430-7(d). Where the equipment is not factory-wired and the individual nameplates are visible in accordance with Section 430-7(d)(2), the conductor ampacity shall be determined in accordance with Section 430-24.

To compute the load for the minimum allowable conductor size for a combination lighting (or lighting and appliance) load and motor load, the capacity for the lighting load is determined in accordance with Article 220 (and other applicable sections, Article 422, etc.) plus the sum of the motor load, determined in accordance with Section 430-22 (single motor) or Section 430-24 (two or more motors).

430-26. Feeder Demand Factor. Where reduced heating of the conductors results from motors operating on duty-cycle, intermittently, or from all motors not operating at one time, the authority having jurisdiction may grant permission for feeder conductors to have an ampacity less than specified in Section 430-24, provided the conductors have sufficient ampacity for the maximum load determined in accordance with the sizes and number of motors supplied and the character of their loads and duties.

The authority having jurisdiction may grant permission to allow a demand factor of less than 100 percent if operational procedures, production demands, or the nature of the work is such that not all the motors are running at one time.

430-27. Capacitors with Motors. Where capacitors are installed in motor circuits, conductors shall comply with Sections 460-8 and 460-9.

430-28. Feeder Taps. Feeder tap conductors shall have an ampacity not less than that required by Part B, shall terminate in a branch-circuit protective device and, in addition, shall meet one of the following requirements:

(1) Be enclosed by either an enclosed controller or by a raceway, be not more than 10 ft (3.05 m) in length, and, for field installation, be protected by an overcurrent device on the line side of the tap conductor, the rating or setting of which shall not exceed 1000 percent of the tap conductor ampacity
(2) Have an ampacity of at least one-third that of the feeder conductors, be suitably protected from physical damage or enclosed in a raceway, and be not more than 25 ft (7.62 m) in length
(3) Have the same ampacity as the feeder conductors

Exception: Feeder Taps Over 25 ft (7.62 m) Long. In high-bay manufacturing buildings [over 35 ft (10.67 m) high at walls], where conditions of maintenance and supervision ensure that only qualified persons will service the systems, conductors tapped to a feeder shall be permitted to be not over 25 ft (7.62 m) long horizontally and not over 100 ft (30.5 m) in total length where all of the following conditions are met.

(a) The ampacity of the tap conductors is not less than one-third that of the feeder conductors.
(b) The tap conductors terminate with a single circuit breaker or a single set of fuses conforming with (1) Part D, where the load-side conductors are a branch circuit or (2) Part E, where the load-side conductors are a feeder.
(c) The tap conductors are suitably protected from physical damage and are installed in raceways.
(d) The tap conductors are continuous from end-to-end and contain no splices.
(e) The tap conductors shall be No. 6 copper or No. 4 aluminum or larger.
(f) The tap conductors shall not penetrate walls, floors, or ceilings.
(g) The tap shall not be made less than 30 ft (9.14 m) from the floor.

Section 430-28 contains three basic requirements for feeder taps that supply motor circuits.

First, the tap conductor from the feeder to the motor overcurrent device must be sized according to Part B of Article 430. For a single motor load, the tap conductors are sized the same as the motor branch-circuit conductors, that is, according to Section 430-22. Section 430-22 requires that motor branch-circuit conductors be sized at least 125 percent of the table value of full-load current for the motor. The table value, rather than the nameplate value, is the full-load current used for conductor sizing according to Section 430-6(a).

Second, the tap conductors must terminate in a set of fuses or a circuit breaker, thus limiting the load on the tap conductors. It is important to point out here that reduced size tap conductors are protected from overload by the terminal overcurrent device, but protected from short-circuit (and ground-fault) only from the feeder overcurrent device.

Third, where the tap conductor ampacity is less than the ampacity of the feeder, the tap conductor installation must meet the additional requirements associated with their tap conductor distance limits, that is, 10 ft, 25 ft, or by exception, 100 ft.

The requirements for tap conductors that supply motor loads are somewhat similar to the basic tap requirements found in Section 240-21. For example, where tap conductors supply a motor load and do not exceed 10 ft, the tap conductors shall be sized for the load, terminate in a set of fuses or a circuit breaker, be enclosed by a controller or a raceway, and be protected by a feeder overcurrent device not exceeding 10 times the tap conductor ampacity.

Where the tap conductors supply a motor load and do not exceed 25 ft, the tap conductors shall be sized for the load, terminate in a set of fuses or a circuit breaker, be protected from physical damage or be enclosed a raceway, and have an ampacity at least one-third that of the feeder conductor.

In a high-bay manufacturing building, feeder taps up to 100 ft long are conditionally permitted under Section 430-28, Exception.

Example

A 15-hp, 230-volt, 3-phase, NEMA design B, squirrel cage induction motor with a service factor of 1.15 is to be supplied by a tap from a 250-kcmil feeder. Assuming three conductors in an individual raceway, all Type THWN copper, and no ambient correction factor, the feeder has an ampacity of 255 amperes

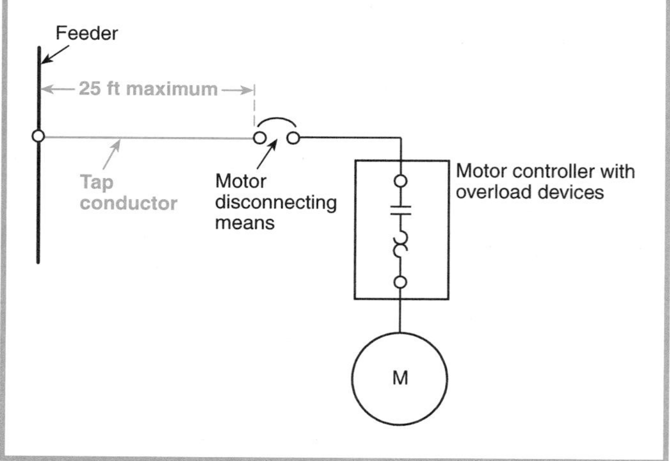

Figure 430.8 *Protective devices (branch-circuit short-circuit and ground-fault) for a motor branch circuit located not more than 25 ft from the point where the conductors are tapped to a feeder.*

(from Table 310-16, 75°C column). Where the tap conductors are not over 25 ft long (see Figure 430.8), No. 4 conductors with an ampacity of 85 amperes are permitted ($^1/_3$ × 255 amperes = 85 amperes).

Answer

Based on Section 430-6(a) and Table 430-150, the full-load current of the 15-hp motor is 42 amperes. According to Section 430-52(c)(1), the motor branch-circuit short-circuit and ground-fault protective device cannot exceed the values given in Table 430-152. The maximum time-delay fuse value is 42 × 1.75 = 73.5 amperes. The maximum inverse time circuit breaker value is 42 × 2.50 = 105 amperes. Section 430-52(c)(1), Exception No. 1, allows the next higher standard size — 80 and 110 amperes, respectively. A higher size, based on Exception No. 2, is allowed if the 80- or 110-ampere size is not adequate to start the motor. Based on Section 430-32, the motor overload protective devices (heaters) are required to be set at a value not greater than 125 percent of the full-load ampere rating marked on the motor nameplate. A higher size motor overload protective device setting of up to 140 percent may be used according to the permissive rules set forth in Section 430-34. Regardless of the exact setting, with the motor overload protection set at approximately 50 amperes, the No. 4 THWN copper motor branch-circuit tap conductors are well protected from overload.

For additional information concerning taps supplying motor circuits for group installations, refer to Section 430-53(d) and the associated commentary.

430-29. Constant Voltage Direct-Current Motors — Power Resistors. Conductors connecting the motor controller to separately mounted power accelerating and dynamic braking resistors in the armature circuit shall have an ampacity not less than the value calculated from Table 430-29 using motor full-load current. If an armature shunt resistor is used, the power accelerating resistor conductor ampacity shall be calculated using the total of motor full-load current and armature shunt resistor current.

Armature shunt resistor conductors shall have an ampacity of not less than that calculated from Table 430-29 using rated shunt resistor current as full-load current.

C. Motor and Branch-Circuit Overload Protection

430-31. General. Part C specifies overload devices intended to protect motors, motor-control apparatus, and motor branch-circuit conductors against excessive heating due to motor overloads and failure to start.

Table 430-29. Conductor Rating Factors for Power Resistors

Time in Seconds		Ampacity of Conductor in Percent of Full-Load Current
On	Off	
5	75	35
10	70	45
15	75	55
15	45	65
15	30	75
15	15	85
Continuous Duty		110

Overload in electrical apparatus is an operating overcurrent that, when it persists for a sufficient length of time, would cause damage or dangerous overheating of the apparatus. It does not include short circuits or ground faults.

These provisions shall not be interpreted as requiring overload protection where it might introduce additional or increased hazards, as in the case of fire pumps.

FPN: For protection of fire pump supply conductors, see Section 695-6.

The provisions of Part C shall not apply to motor circuits rated over 600 volts, nominal. See Part K.

FPN: See Appendix D, Example No. D8.

430-32. Continuous-Duty Motors.

(a) More than 1 Horsepower. Each continuous-duty motor rated more than 1 hp shall be protected against overload by one of the following means.

(1) A separate overload device that is responsive to motor current. This device shall be selected to trip or shall be rated at no more than the following percent of the motor nameplate full-load current rating:

Motors with a marked service factor not less than 1.15	125%
Motors with a marked temperature rise not over 40°C	125%
All other motors	115%

Motors are required to be protected from overloads. Shown in Figure 430.9 is a magnetic motor starter equipped with an adjustable overload device. The maximum setting of this device is covered in Section 430-32(a)(1). Failure to select and maintain the proper setting of this overload device could result in an overload condition of the motor, premature failure of the motor, or worst, could result in a fire.

Figure 430.9 Line-voltage magnetic starter equipped with an adjustable overload device. (Square D Co.)

Modification of this value shall be permitted as provided in Section 430-34. For a multispeed motor, each winding connection shall be considered separately.

Where a separate motor overload device is connected so that it does not carry the total current designated on the motor nameplate, such as for wye-delta starting, the proper percentage of nameplate current applying to the selection or setting of the overload device shall be clearly designated on the equipment, or the manufacturer's selection table shall take this into account.

FPN: Where power factor correction capacitors are installed on the load side of the motor overload device, see Section 460-9.

(2) A thermal protector integral with the motor, approved for use with the motor it protects on the basis that it will prevent dangerous overheating of the motor due to overload and failure to start. The ultimate trip current of a thermally protected motor shall not exceed the following percentage of motor full-load current given in Tables 430-148, 430-149, and 430-150:

Motor full-load current not exceeding 9 amperes	170%
Motor full-load current from 9.1 to, and including, 20 amperes	156%
Motor full-load current greater than 20 amperes	140%

If the motor current-interrupting device is separate from the motor and its control circuit is operated by a protective device integral with the motor, it shall be arranged so that the opening of the control circuit will result in interruption of current to the motor.

(3) A protective device integral with a motor that will protect the motor against damage due to failure to start shall be permitted if the motor is part of an approved assembly that does not normally subject the motor to overloads.

(4) For motors larger than 1500 hp, a protective device having embedded temperature detectors that cause current to the motor to be interrupted when the motor attains a temperature rise greater than marked on the nameplate in an ambient temperature of 40°C.

Continuous-duty-rated motors of more than 1 hp can be protected against overload conditions by any one of the following four methods in accordance with Section 430-32(a):

1. An overload device located in the motor controller, such as a bimetallic element or eutectic material
2. A thermal protector located in the motor that senses excessive current or temperature
3. A protective device in the motor, where the motor is part of an assembly that does not normally subject the motor to overloads
4. For motors larger than 1500 hp, a temperature-sensitive device embedded in the motor windings that will de-energize the motor

(b) One Horsepower or Less, Nonautomatically Started.

(1) Each continuous-duty motor rated at 1 hp or less that is not permanently installed, is nonautomatically started, and is within sight from the controller location shall be permitted to be protected against overload by the branch-circuit short-circuit and ground-fault protective device. This branch-circuit protective device shall not be larger than that specified in Part D of Article 430.

Exception: Any such motor shall be permitted on a nominal 120-volt branch circuit protected at not over 20 amperes.

Motors rated 1 hp or less that are not permanently installed and not automatically started, such as bench grinders, drill presses, and portable electric tools, are not required to have overload protection and may be protected by the branch-circuit short-circuit fuse or circuit breaker. This type of equipment is usually attended by the operator, who can immediately shut off power to the motor should it overheat and start smoking.

(2) Any such motor that is not in sight from the controller location shall be protected as specified in Section 430-32(c). Any motor rated at 1 hp or less that is permanently installed shall be protected in accordance with Section 430-32(c).

(c) One Horsepower or Less, Automatically Started. Any motor of 1 hp or less that is started automatically shall be protected against overload by one of the following means.

(1) A separate overload device that is responsive to motor current. This device shall be selected to trip or shall be rated at not more than the following percentage of the motor nameplate full-load current rating:

Motors with a marked service factor not less than 1.15	125%
Motors with a marked temperature rise not over 40°C	125%
All other motors	115%

For a multispeed motor, each winding connection shall be considered separately. Modification of this value shall be permitted as provided in Section 430-34.

(2) A thermal protector integral with the motor, approved for use with the motor that it protects on the basis that it will prevent dangerous overheating of the motor due to overload and failure to start. Where the motor current-interrupting device is separate from the motor and its control circuit is operated by a protective device integral with the motor, it shall be arranged so that the opening of the control circuit will result in interruption of current to the motor.

(3) A protective device integral with a motor that will protect the motor against damage due to failure to start shall be permitted (1) if the motor is part of an approved assembly that does not subject the motor to overloads, or (2) if the assembly is also equipped with other safety controls (such as the safety combustion controls on a domestic oil burner) that protect the motor against damage due to failure to start. Where the assembly has safety controls that protect the motor, it shall be so indicated on the nameplate of the assembly where it will be visible after installation.

(4) In case the impedance of the motor windings is sufficient to prevent overheating due to failure to start, the motor shall be permitted to be protected as specified in Section 430-32(b)(1) for manually started motors if the motor is part of an approved assembly in which the motor will limit itself so that it will not be dangerously overheated.

> FPN: Many ac motors of less than 1/20 hp, such as clock motors, series motors, etc., and also some larger motors such as torque motors, come within this classification. It does not include split-phase motors having automatic switches that disconnect the starting windings.

(d) Wound-Rotor Secondaries. The secondary circuits of wound-rotor ac motors, including conductors, controllers, resistors, etc., shall be permitted to be protected against overload by the motor-overload device.

The basic premise behind Sections 430-32(a) through (d) is that the operation of a motor in excess of its normal full-load rating for a prolonged period of time causes damage or dangerous overheating that may start a fire. Overload protection is intended to protect the motor and the system components from damaging overload currents.

A continuous-duty motor with a marked service factor of not less than 1.15 or with a marked temperature rise of not over 40°C can carry a 25 percent overload for an extended period without damage to the motor. Other such types of motors are those with a service factor of less than 1.15 or those with a marked temperature rise of greater than 40°C that are incapable of withstanding a prolonged overload, where the motor overload protective device opens the circuit if the motor continues to draw 115 percent of its rated full-load current.

A thermal protector located inside the motor housing, as shown in Figure 430.10 is connected in series with the motor winding. This protective device commonly consists of a set of normally closed contacts attached to a bimetallic disk through which the circuit is normally closed. The thermal protector heating coil (in series with the motor winding) causes the disk to heat rapidly. The heat-actuated disk snaps the contacts open to protect the motor windings from overheating due to failure to start, a sudden heavy overload, or a prolonged overload.

After the circuit opens and the motor has cooled to a normal temperature, the contacts will automatically close and restart the motor. In some cases, this may not be desirable. For such applications, the protective device is designed so that it must be returned to the closed position by a manually controlled reset, as required by Section 430-43. For larger motors (usually over 1 hp), a similar device is used. This device, upon abnormal overload, acts as a control-circuit switch and operates the control circuit of a motor current-interrupting device, usually a motor contactor or starter, located separately from the motor. A thermal protector and circuit-interrupting device should be approved for use with the motor it protects and is required to open the circuit on an overcurrent, as specified in Section 430-32(a)(2).

430-33. Intermittent and Similar Duty. A motor used for a condition of service that is inherently short-time, intermittent, periodic, or varying duty, as illustrated by Table 430-22(b), shall be permitted to be protected against overload by the branch-circuit short-circuit and ground-fault protective device, provided the protective device rating or setting does not exceed that specified in Table 430-152.

Any motor application shall be considered to be for continuous duty unless the nature of the apparatus it drives is such that the motor cannot operate continuously with load under any condition of use.

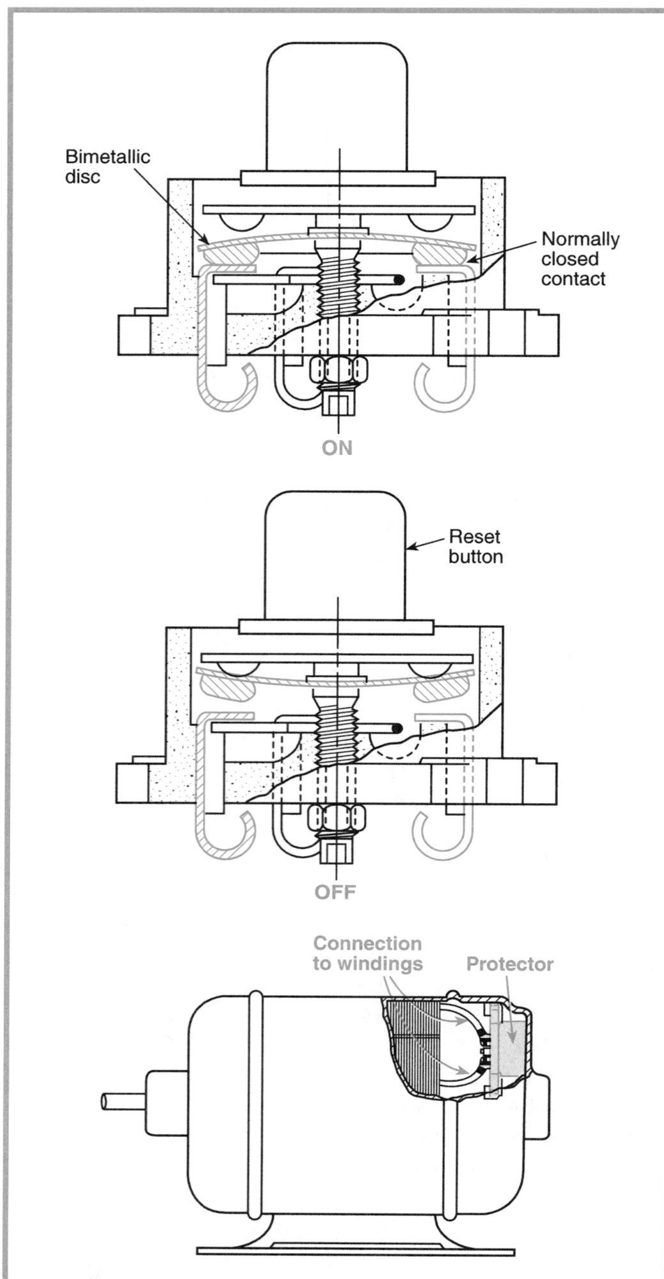

Figure 430.10 A thermal protector for a motor, in which a heat-sensitive snap-action disk opens contacts and protects the motor against dangerous overheating. This device is integrally mounted within the motor. (Texas Instruments, Inc.)

Where a motor is selected for duty-cycle service (short-time, intermittent, periodic, or varying), it can be assumed that the motor will not operate continuously, due to the nature of the apparatus or machinery it drives. Therefore, prolonged overloads are not likely to occur unless mechanical failure in the driven apparatus stalls the motor; in this case, however, the branch-circuit protective device would open the circuit. The omission of overload protective devices for such motors is based on the type of duty and not on the time rating of the motor.

430-34. Selection of Overload Relay. Where the overload relay selected in accordance with Sections 430-32(a)(1) and (c)(1) is not sufficient to start the motor or to carry the load, the next higher size overload relay shall be permitted to be used, provided the trip current of the overload relay does not exceed the following percentage of motor nameplate full-load current rating:

Motors with marked service factor not less than 1.15	140%
Motors with a marked temperature rise not over 40°C	140%
All other motors	130%

If not shunted during the starting period of the motor as provided in Section 430-35, the overload device shall have sufficient time delay to permit the motor to start and accelerate its load.

FPN: A Class 20 or 30 overload relay will provide a longer motor acceleration time than a Class 10 or 20, respectively. Use of a higher class overload relay may preclude the need for selection of a higher trip current.

430-35. Shunting During Starting Period.

(a) Nonautomatically Started. For a nonautomatically started motor, the overload protection shall be permitted to be shunted or cut out of the circuit during the starting period of the motor if the device by which the overload protection is shunted or cut out cannot be left in the starting position and if fuses or inverse time circuit breakers rated or set at not over 400 percent of the full-load current of the motor are located in the circuit so as to be operative during the starting period of the motor.

(b) Automatically Started. The motor overload protection shall not be shunted or cut out during the starting period if the motor is automatically started.

If not shunted during the starting period of the motor, the overload device is required to have sufficient time delay to start and accelerate its load; whereas, if shunting is employed, the overload protection is permitted to be bypassed only during the starting period of the motor. See Figure 430.11, which illustrates this type of bypass during motor starting. When the switch is thrown momentarily in one direction (starting position), the overload protective fuses are shunted or cut out of the circuit. The switch is then thrown in the opposite direction (running position) and is required to be designed so that it cannot be left in the starting position.

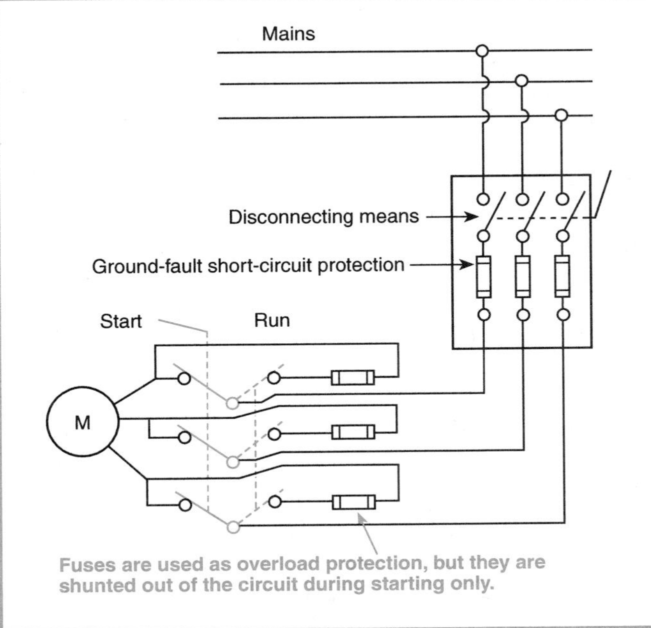

Figure 430.11 *Arrangement for across-the-line starting of a motor.*

Where fuses are used as overload protection, they may be shunted or cut out of the circuit during the starting period by a device (in this case a double-throw switch) designed so that it cannot be left in the starting position. Therefore, during the starting period, the motor is protected only by the branch-circuit fuses that are always rated within the limits of Section 430-35(b). If there are no branch-circuit fuses, as permitted by Section 430-53, then a starter (shunting) device is not allowed during the starting period unless the feeder protection is within the limits of Section 430-35(b) (not over 400 percent of the full-load motor current).

Exception: The motor overload protection shall be permitted to be shunted or cut out during the starting period on an automatically started motor where

(a) The motor starting period exceeds the time delay of available motor overload protective devices, and

(b) Listed means are provided to

(1) Sense motor rotation and to automatically prevent the shunting or cutout in the event that the motor fails to start, and

(2) Limit the time of overload protection shunting or cutout to less than the locked rotor time rating of the protected motor, and

(3) Provide for shutdown and manual restart if motor running condition is not reached.

430-36. Fuses — In Which Conductor. Where fuses are used for motor overload protection, a fuse shall be inserted in each ungrounded conductor and also in the grounded conductor if the supply system is 3-wire, 3-phase ac with one conductor grounded.

430-37. Devices Other than Fuses — In Which Conductor. Where devices other than fuses are used for motor overload protection, Table 430-37 shall govern the minimum allowable number and location of overload units such as trip coils or relays.

Table 430-37. Overload Units

Kind of Motor	Supply System	Number and Location of Overload Units, Such as Trip Coils or Relays
1-phase ac or dc	2-wire, 1-phase ac or dc ungrounded	1 in either conductor
1-phase ac or dc	2-wire, 1-phase ac or dc, one conductor grounded	1 in ungrounded conductor
1-phase ac or dc	3-wire, 1-phase ac or dc, grounded neutral	1 in either ungrounded conductor
1-phase ac	Any 3-phase	1 in ungrounded conductor
2-phase ac	3-wire, 2-phase ac, ungrounded	2, one in each phase
2-phase ac	3-wire, 2-phase ac, one conductor grounded	2 in ungrounded conductors
2-phase ac	4-wire, 2-phase ac, grounded or ungrounded	2, one per phase in ungrounded conductors
2-phase ac	5-wire, 2-phase ac, grounded neutral or ungrounded	2, one per phase in any ungrounded phase wire
3-phase ac	Any 3-phase	3, one in each phase*

**Exception: An overload unit in each phase shall not be required where overload protection is provided by other approved means.*

All 3-phase motors are required to be provided with three overload units, one in each phase. The exceptions are those protected by other approved means. Specially designed or integral-type detectors, with or without supplementary external protective devices, are some exceptions. See also Section 430-36 for instances where fuses used as overloads are required even in the grounded conductor.

430-38. Number of Conductors Opened by Overload Device. Motor overload devices, other than fuses or thermal protectors, shall simultaneously open a sufficient number of ungrounded conductors to interrupt current flow to the motor.

430-39. Motor Controller as Overload Protection. A motor controller shall also be permitted to serve as an overload device if the number of overload units complies with Table 430-37 and if these units are operative in both the starting and running position in the case of a dc motor, and in the running position in the case of an ac motor.

For the purpose of Article 430, a controller may be a switch, a circuit breaker, a contactor, or other device to start and stop a motor by making and breaking the motor circuit current. The controller is required to be capable of interrupting the stalled-rotor current of the motor and must have a horsepower rating that is not lower than the horsepower rating of the motor. Motor controllers are covered in Part G of Article 430.

Dual-element fuses can be sized to provide motor overload protection (see Section 430-36). Automatically operated contactors or circuit breakers (with trip units) are governed by the requirements of Sections 430-37 and 430-38 where these devices are used to provide overload protection.

430-40. Overload Relays. Overload relays and other devices for motor overload protection that are not capable of opening short circuits shall be protected by fuses or circuit breakers with ratings or settings in accordance with Section 430-52 or by a motor short-circuit protector in accordance with Section 430-52.

Some overload devices are marked with a maximum short-circuit and ground-fault protective device rating or setting. This rating sets the limit on the maximum fuse or breaker size that may be upstream from the overload device. It also notifies the user that coordination between the overload device and the short-circuit and ground-fault device is required. This is most often the case for group motor installation.

Exception: Where approved for group installation and marked to indicate the maximum size of fuse or inverse time circuit breaker by which they must be protected, the overload devices shall be protected in accordance with this marking.

FPN: For instantaneous trip circuit breakers or motor short-circuit protectors, see Section 430-52.

430-42. Motors on General-Purpose Branch Circuits. Overload protection for motors used on general-purpose branch circuits as permitted in Article 210 shall be provided as specified in (a), (b), (c), or (d).

(a) Not Over 1 Horsepower. One or more motors without individual overload protection shall be permitted to be connected to a general-purpose branch circuit only where the installation complies with the limiting conditions specified in Sections 430-32(b) and (c) and Sections 430-53(a)(1) and (a)(2).

(b) Over 1 Horsepower. Motors of larger ratings than specified in Section 430-53(a) shall be permitted to be connected to general-purpose branch circuits only where each motor is protected by overload protection selected to protect the motor as specified in Section 430-32. Both the controller and the motor overload device shall be approved for group installation with the short-circuit and ground-fault protective device selected in accordance with Section 430-53.

(c) Cord- and Plug-Connected. Where a motor is connected to a branch circuit by means of an attachment plug and receptacle and individual overload protection is omitted as provided in Section 430-42(a), the rating of the attachment plug and receptacle shall not exceed 15 amperes at 125 volts or 250 volts. Where individual overload protection is required as provided in Section 430-42(b) for a motor or motor-operated appliance that is attached to the branch circuit through an attachment plug and receptacle, the overload device shall be an integral part of the motor or of the appliance. The rating of the attachment plug and receptacle shall determine the rating of the circuit to which the motor may be connected, as provided in Article 210.

(d) Time Delay. The branch-circuit short-circuit and ground- fault protective device protecting a circuit to which a motor or motor-operated appliance is connected shall have sufficient time delay to permit the motor to start and accelerate its load.

430-43. Automatic Restarting. A motor overload device that can restart a motor automatically after overload tripping shall not be installed unless approved for use with the motor it protects. A motor overload device that can restart a motor automatically after overload tripping shall not be installed if automatic restarting of the motor can result in injury to persons.

An integral motor overload protective device may be of the type that, after tripping and sufficiently cooling, will automatically restart the motor, or it may be of the type that, after tripping, is closed by use of a manually operated reset button. See the commentary following Section 430-32(d).

430-44. Orderly Shutdown. If immediate automatic shutdown of a motor by a motor overload protective device(s) would introduce additional or increased hazard(s) to a person(s) and continued motor operation is necessary for safe shutdown of equipment or process, a motor overload sensing device(s) conforming with the provisions of Part C of this article shall be permitted to be connected to a supervised alarm instead of causing immediate interruption of the motor circuit, so that corrective action or an orderly shutdown can be initiated.

D. Motor Branch-Circuit Short-Circuit and Ground-Fault Protection

430-51. General. Part D specifies devices intended to protect the motor branch-circuit conductors, the motor control apparatus, and the motors against overcurrent due to short circuits or grounds. These rules add to or amend the provisions of Article 240. The devices specified in Part D do not include the types of devices required by Sections 210-8, 230-95, and 305-6.

The provisions of Part D do not apply to motor circuits rated over 600 volts, nominal. See Part K.

FPN: See Appendix D, Example No. D8.

430-52. Rating or Setting for Individual Motor Circuit.

(a) General. The motor branch-circuit short-circuit and ground-fault protective device shall comply with (b) and either (c) or (d), as applicable.

Section 430-52(a) establishes the maximum allowable ratings or settings of devices (fuses or circuit breakers) acceptable for motor branch-circuit short-circuit and ground-fault protection and states that these devices are expected to carry the starting current of the motor and provide short-circuit and ground-fault protection. For certain exceptions to the maximum rating or setting of these motor branch-circuit protective devices, as specified in Table 430-152, see Sections 430-52, 430-53, and 430-54.

Section 430-6 requires that where the current rating of a motor is used to determine the ampacity of conductors or ampere ratings of switches, branch-circuit overcurrent devices, and so on, the values given in Tables 430-147 through 430-150 (including notes) are required to be used instead of the actual motor nameplate current rating. Separate motor overload protection is required to be based on the motor nameplate current rating.

Figure 430.5 illustrates a typical motor circuit where the branch-circuit short-circuit and ground-fault protective fuse or circuit breaker rating must carry the starting current and may be sized 150 to 300 percent of the motor full-load current (depending on the type of motor, but excluding Design E). It should be noted that it is not necessary to size the branch-circuit conductors to the percentages (150 to 300) permitted for the branch-circuit short-circuit and ground-fault protective devices.

The rules for short-circuit and ground-fault protection are specific for particular situations. A *short circuit* is a fault between two conductors or between phases. A *ground fault* is a fault between an ungrounded conductor and ground. During a short-circuit or phase-to-ground condition, the extremely high current causes the protective fuses or circuit breakers to open the circuit. Excess current flow caused by an overload condition passes through the overload protective device at the motor controller, thereby causing the device to open the control circuit or motor circuit conductors. Branch-circuit conductors with an ampacity of 125 percent (not 150 to 300 percent) of the motor full-load current are reasonably protected by motor-protective devices set to operate at nearly the same current as the ampacity of the conductors. Branch-circuit short-circuit and ground-fault protective devices will open the circuit under short-circuit conditions and thereby provide short-circuit and ground-fault protection for both the motor and overload protective device; however, the overload protective device is not intended to open short circuits or ground faults.

The rating or setting of the branch-circuit short-circuit and ground-fault protective device should be selected as low as possible for maximum protection. However, where the rating or setting specified in Table 430-152 or permitted by Section 430-52(c)(1), Exception No. 1, is not sufficient for the starting current of the motor, such as in the case of severe starting conditions where the motor and its driven machinery require an extended period of time to reach the desired speed, it is allowable to use a higher rating or setting, as permitted in Section 430-52(c)(1), Exception No. 2.

(b) All Motors. The motor branch-circuit short-circuit and ground-fault protective device shall be capable of carrying the starting current of the motor.

(c) Rating or Setting.

(1) A protective device that has a rating or setting not exceeding the value calculated according to the values given in Table 430-152 shall be used.

Exception No. 1: Where the values for branch-circuit short-circuit and ground-fault protective devices determined by Table 430-152 do not correspond to the standard sizes or ratings of fuses, nonadjustable circuit breakers, thermal protective devices, or possible settings of adjustable circuit breakers, the next higher standard size, rating, or possible setting shall be permitted.

Exception No. 2: Where the rating specified in Table 430-152, as modified by Exception No. 1, is not sufficient for the starting current of the motor:

(a) The rating of a nontime-delay fuse not exceeding 600 amperes or a time-delay Class CC fuse shall be permit-

ted to be increased but shall in no case exceed 400 percent of the full-load current.

(b) The rating of a time-delay (dual-element) fuse shall be permitted to be increased but shall in no case exceed 225 percent of the full-load current.

(c) The rating of an inverse time circuit breaker shall be permitted to be increased but shall in no case exceed 400 percent for full-load currents of 100 amperes or less or 300 percent for full-load currents greater than 100 amperes.

(d) The rating of a fuse of 601–6000 ampere classification shall be permitted to be increased but shall in no case exceed 300 percent of the full-load current.

FPN: See Appendix D, Example No. D8, and Figure 430-1.

(2) Where maximum branch-circuit short-circuit and ground-fault protective device ratings are shown in the manufacturer's overload relay table for use with a motor controller or are otherwise marked on the equipment, they shall not be exceeded even if higher values are allowed as shown above.

(3) An instantaneous trip circuit breaker shall be used only if adjustable and if part of a listed combination motor controller having coordinated motor overload and short-circuit and ground-fault protection in each conductor, and the setting is adjusted to no more than the value specified in Table 430-152.

FPN: For the purpose of this article, instantaneous-trip circuit breakers may include a damping means to accommodate a transient motor inrush current without nuisance tripping of the circuit breaker.

Exception No. 1: Where the setting specified in Table 430-152 is not sufficient for the starting current of the motor, the setting of an instantaneous trip circuit breaker shall be permitted to be increased but shall in no case exceed 1300 percent of the motor full-load current for other than Design E motors or Design B energy efficient motors and no more than 1700 percent of full-load motor current for Design E motors or Design B energy efficient motors. Trip settings above 800 percent for other than Design E motors or Design B energy efficient motors and above 1100 percent for Design E motors or Design B energy efficient motors shall be permitted where the need has been demonstrated by engineering evaluation. In such cases, it shall not be necessary to first apply an instantaneous-trip circuit breaker at 800 percent or 1100 percent.

FPN: For additional information on the requirements for a motor to be classified "energy efficient," see *Motors and Generators,* NEMA Standards Publication No. MG1-1993, Revision 1, Part 12.59.

Exception No. 2: Where the motor full-load current is 8 amperes or less, the setting of the instantaneous-trip circuit breaker with a continuous current rating of 15 amperes or less in a listed combination motor controller that provides coordinated motor branch-circuit overload and short-circuit and ground-fault protection shall be permitted to be increased to the value marked on the controller.

(4) For a multispeed motor, a single short-circuit and ground-fault protective device shall be permitted for two or more windings of the motor, provided the rating of the protective device does not exceed the above applicable percentage of the nameplate rating of the smallest winding protected.

Exception: For a multispeed motor, a single short-circuit and ground-fault protective device shall be permitted to be used and sized according to the full-load current of the highest current winding, where all of the following conditions are met.

(a) Each winding is equipped with individual overload protection sized according to its full-load current.

(b) The branch-circuit conductors supplying each winding are sized according to the full-load current of the highest full-load current winding.

(c) The controller for each winding has a horsepower rating not less than that required for the winding having the highest horsepower rating.

(5) Suitable fuses shall be permitted in lieu of devices listed in Table 430-152 for power electronic devices in a solid state motor controller system provided that the marking for replacement fuses is provided adjacent to the fuses.

(6) A listed self-protected combination controller shall be permitted in lieu of the devices specified in Table 430-152. Adjustable instantaneous-trip settings shall not exceed 1300 percent of full-load motor current for other than Design E motors or Design B energy efficient motors and not more than 1700 percent of full-load motor current for Design E motors or Design B energy efficient motors.

Revised for the 1999 *Code,* Section 430-52 recognizes a new product referred to as a "self-protected combination controller." This unit combines the functions of short-circuit protection, disconnect, controller, and overload protection into one single unit. Since only listed units are permitted, they must be applied within their ratings.

(7) A motor short-circuit protector shall be permitted in lieu of devices listed in Table 430-152 if the motor short-circuit protector is part of a listed combination motor controller having coordinated motor overload protection and short-circuit and ground-fault protection in each conductor and it will open the circuit at currents exceeding 1300 percent of motor full-load current for other than Design E motors or Design B energy efficient motors and 1700 percent of motor full-load motor current for Design E motors or Design B energy efficient motors.

(d) Torque Motors. Torque motor branch circuits shall be protected at the motor nameplate current rating in accordance with Section 240-3(b).

430-53. Several Motors or Loads on One Branch Circuit. Two or more motors or one or more motors and other loads shall be permitted to be connected to the same branch circuit under conditions specified in (d) and in (a), (b), or (c).

(a) Not Over 1 Horsepower. Several motors, each not exceeding 1 hp in rating, shall be permitted on a nominal 120-volt branch circuit protected at not over 20 amperes or a branch circuit of 600 volts, nominal, or less, protected at not over 15 amperes, if all of the following conditions are met.

(1) The full-load rating of each motor does not exceed 6 amperes.
(2) The rating of the branch-circuit short-circuit and ground-fault protective device marked on any of the controllers is not exceeded.
(3) Individual overload protection conforms to Section 430-32.

Two or more motors or one or more motors and other loads may be connected to the same 120-volt, 15- or 20-ampere, single-phase lighting circuit. The provision is that each motor is to be rated not more than 1 hp, the full-load rating of each motor is not to exceed 6 amperes, and the rating of the branch-circuit protective device is not permitted to be exceeded.

It should be understood that the requirements for overload protection, as provided in Section 430-32, are required to be applied in all cases, regardless of the number (one or more) of motors or the type of branch circuit.

(b) If Smallest Rated Motor Protected. If the branch-circuit short-circuit and ground-fault protective device is selected not to exceed that allowed by Section 430-52 for the smallest rated motor, two or more motors or one or more motors and other load(s), with each motor having individual overload protection, shall be permitted to be connected to a branch circuit where it can be determined that the branch-circuit short-circuit and ground-fault protective device will not open under the most severe normal conditions of service that might be encountered.

(c) Other Group Installations. Two or more motors of any rating or one or more motors and other load(s), with each motor having individual overload protection, shall be permitted to be connected to one branch circuit where the motor controller(s) and overload device(s) are (1) installed as a listed factory assembly and the motor branch-circuit short-circuit and ground-fault protective device is either provided as part of the assembly or is specified by a marking on the assembly, or (2) the motor branch-circuit short-circuit and ground-fault protective device, the motor controller(s), and overload device(s) are field-installed as separate assemblies listed for such use and provided with manufacturers' instructions for use with each other, and (3) all of the following conditions are complied with.

(1) Each motor overload device is listed for group installation with a specified maximum rating of fuse or inverse time circuit breaker, or both.
(2) Each motor controller is listed for group installation with a specified maximum rating of fuse or circuit breaker, or both.
(3) Each circuit breaker is one of the inverse time type and listed for group installation.
(4) The branch circuit shall be protected by fuses or inverse time circuit breakers having a rating not exceeding that specified in Section 430-52 for the highest rated motor connected to the branch circuit plus an amount equal to the sum of the full-load current ratings of all other motors and the ratings of other loads connected to the circuit. Where this calculation results in a rating less than the ampacity of the supply conductors, it shall be permitted to increase the maximum rating of the fuses or circuit breaker to a value not exceeding that permitted by Section 240-3(b).
(5) The branch-circuit fuses or inverse time circuit breakers are not larger than allowed by Section 430-40 for the overload relay protecting the smallest rated motor of the group.

FPN: See Section 110-10 for circuit impedance and other characteristics.

Section 110-10 addresses characteristics of components such as impedance and short-circuit current ratings. It should be noted that devices with the same ampere rating may have significantly different short-circuit current ratings. A proper selection of components includes consideration of the characteristics of all components so that the occurrence of a fault will not cause unacceptable damage.

(d) Single Motor Taps. For group installations described above, the conductors of any tap supplying a single motor shall not be required to have an individual branch-circuit short-circuit and ground-fault protective device, provided they comply with either of the following.

(1) No conductor to the motor shall have an ampacity less than that of the branch-circuit conductors.
(2) No conductor to the motor shall have an ampacity less than one-third that of the branch-circuit conductors, with a minimum in accordance with Section 430-22; the conductors to the motor overload device being not more than 25 ft (7.62 m) long and being protected from physical damage.

The two examples illustrated in Figures 430.12 and 430.13 describe installations that permit the omission of individual motor branch-circuit short-circuit and ground-fault protective devices.

Figure 430.12 illustrates main branch-circuit conductors supplying a motor that is part of a group

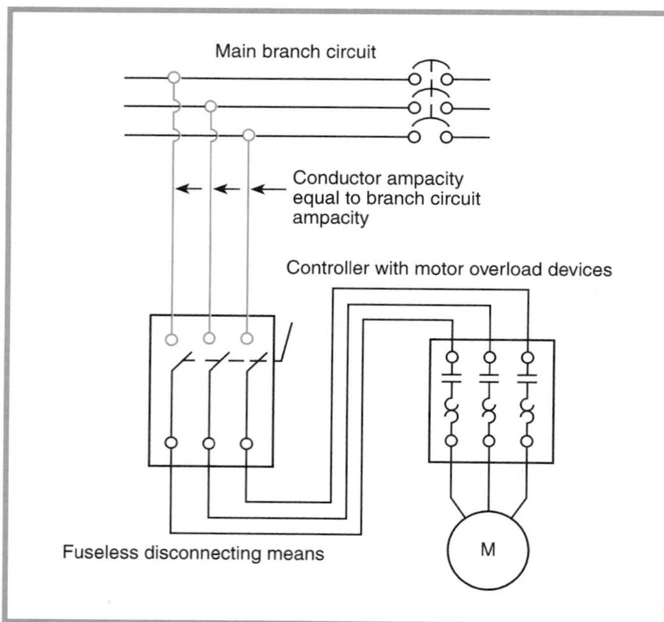

Figure 430.12 *An example of the permissible omission of motor branch-circuit protective devices for tap conductors that have the same ampacity as the main conductors.*

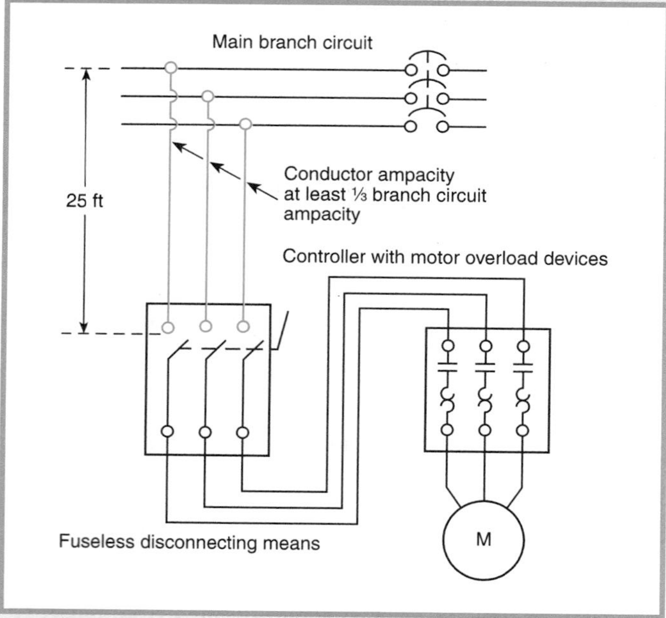

Figure 430.13 *An example of the permissible omission of motor branch-circuit protective devices for tap conductors that have at least one-third the ampacity of the main conductors, are not over 25 ft long, and are protected from physical damage.*

installation. The tap conductors have an ampacity equal to the ampacity of the main branch-circuit conductors; therefore, branch-circuit short-circuit and ground-fault protective devices, fuses, or circuit breakers for the conductors in the tap are not required at the point of connection of the tap conductors to the main conductors, provided that the motor controller and motor overload protective device are listed for group installation with the size of the main branch-circuit short-circuit and ground-fault protective device used.

Figure 430.13 also illustrates main branch-circuit conductors supplying a motor that is part of a group installation. Here, the tap conductors have an ampacity at least one-third the ampacity of the main branch-circuit conductors, are not more than 25 ft in length, and are suitably protected from physical damage. The motor controller and motor overload protective device are required to be listed for group installation with the size of the main branch-circuit short-circuit and ground-fault protective device used.

In both examples, the main branch-circuit fuses or circuit breakers would operate in the event of a short circuit, and the overload protective device would operate to protect the motor and tap conductors under overload conditions.

It should be noted that the tap conductors should never be of a smaller size and ampacity than the branch-circuit conductors required by Section 430-22. That is, a tap conductor (25 ft or less) may be one-third the ampacity of the main branch-circuit conductor to which it is connected; however, this ampacity is required to be equal to or larger than 125 percent of the motor's full-load current rating (see Section 430-22).

Example

A feeder sized at No. 2/0 copper THW typically has an ampacity of 175 amperes (see Table 310-16). A tap conductor (25 ft or less) would normally be permitted to be sized at No. 6 copper THW (65 amperes). But one-third of 175 amperes is 58 amperes, and the motor circuit conductors must have an ampacity of at least 85 amperes. If a 25-hp, 230-volt, 3-phase squirrel-cage motor is to be supplied from this feeder, a No. 6 tap conductor would not meet the requirements of Section 430-22. That is, 125 percent of the full-load current of the motor (68 amperes from Table 430-150) is 85 amperes (1.25 × 68 amperes = 85 amperes). Therefore, the branch-circuit tap conductors are not permitted to be smaller than No. 4 copper THW, with a normal ampacity of 85 amperes (see Table 310-16). Note that the ampacities in Table 310-16 are reduced for ambient temperatures above 30°C and for more than three conductors in the raceway or cable.

430-54. Multimotor and Combination-Load Equipment. The rating of the branch-circuit short-circuit and ground-fault protective device for multimotor and combination-load equipment shall not exceed the rating marked on the equipment in accordance with Section 430-7(d).

430-55. Combined Overcurrent Protection. Motor branch-circuit short-circuit and ground-fault protection and motor overload protection shall be permitted to be combined in a single protective device where the rating or setting of the device provides the overload protection specified in Section 430-32.

Fuses and circuit breakers are not permitted to be sized as overload protection according to the values of Section 430-34. Rather fuses are only permitted to be sized as overload protection according to the values found in Section 430-32(a)(1), (b)(1), and (c)(1).

Either a circuit breaker with inverse time characteristics or a dual-element (time-delay) fuse may serve as both motor overload protection and also as the branch-circuit short-circuit and ground-fault protection.

One-time, time-delay dual-element and Type S dual-element fuses and adapters are available with up to a 30-ampere rating. Type S fuses are designed to prevent oversize fusing. See Sections 240-50 through 240-54 for more information about these fuses and adapters.

Figures 430.14 and 430.15 are examples of time-delay, cartridge-type dual-element fuses able to withstand the normal motor starting current where sized at or near the motor full-load rating but that open when subjected to prolonged overload or blow quickly during a short circuit or ground fault. The dual-element characteristics are the thermal cutout element, which permits harmless high-inrush currents to flow for short periods (but which would open the circuit during a prolonged period), and the fuse link element, which has current-limiting ability for short-circuit currents (and which would blow quickly). Dual-element fuses may be used in larger sizes to provide only short-circuit and ground-fault protection.

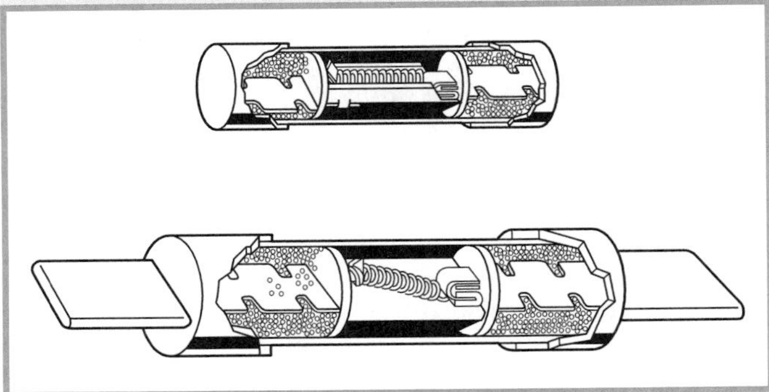

Figure 430.14 *A Fusetron cartridge-type fuse. (Bussmann Division, Cooper Industries)*

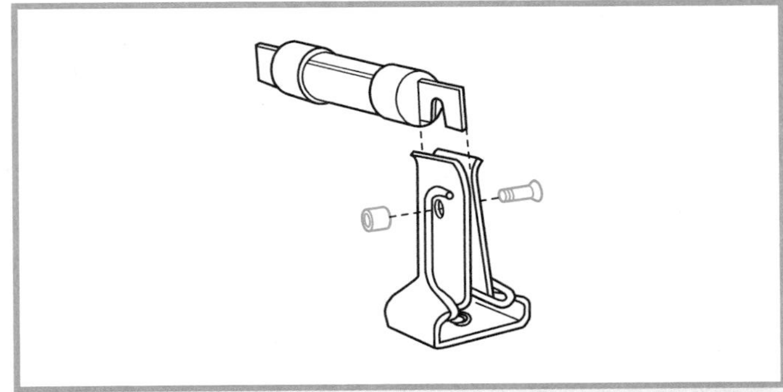

Figure 430.15 *A Class R dual-element fuse with physical rejection feature to prevent interchangeability. (Bussmann Division, Cooper Industries)*

430-56. Branch-Circuit Protective Devices — In Which Conductor. Branch-circuit protective devices shall comply with the provisions of Section 240-20.

430-57. Size of Fuseholder. Where fuses are used for motor branch-circuit short-circuit and ground-fault protection, the fuseholders shall not be of a smaller size than required to accommodate the fuses specified by Table 430-152.

Exception: Where fuses having time delay appropriate for the starting characteristics of the motor are used, it shall be permitted to use fuseholders sized to fit the fuses that are used.

The use of dual-element (time-delay) fuses makes it possible to use smaller fuses, thereby providing better protection because of the lower ratings. Dual-element fuses also save in installation cost by allowing smaller-size switches and panels, and they allow for easier arrangement of equipment where space is at a premium at motor control centers.

430-58. Rating of Circuit Breaker. A circuit breaker for motor branch-circuit short-circuit and ground-fault protection shall have a current rating in accordance with Sections 430-52 and 430-110.

E. Motor Feeder Short-Circuit and Ground-Fault Protection

430-61. General. Part E specifies protective devices intended to protect feeder conductors supplying motors against overcurrents due to short circuits or grounds.

FPN: See Appendix D, Example No. D8.

430-62. Rating or Setting — Motor Load.

(a) Specific Load. A feeder supplying a specific fixed motor load(s) and consisting of conductor sizes based on Section 430-24 shall be provided with a protective device having a rating or setting not greater than the largest rating or setting of the branch-circuit short-circuit and ground-fault protective device for any motor supplied by the feeder [based on the maximum permitted value for the specific type of a protective device shown in Table 430-152, or Section 440-22(a) for hermetic refrigerant motor-compressors], plus the sum of the full-load currents of the other motors of the group.

Where the same rating or setting of the branch-circuit short-circuit and ground-fault protective device is used on two or more of the branch circuits supplied by the feeder, one of the protective devices shall be considered the largest for the above calculations.

Section 430-62(a) recognizes the lower setting for motor overload devices required for hermetic refrigerant motor-compressors.

Exception: Where one or more instantaneous trip circuit breakers or motor short-circuit protectors are used for motor branch-circuit short-circuit and ground-fault protection as permitted in Section 430-52(c), the procedure provided above for determining the maximum rating of the feeder protective device shall apply with the following provision. For the purpose of the calculation, each instantaneous trip circuit breaker or motor short-circuit protector shall be assumed to have a rating not exceeding the maximum percentage of motor full-load current permitted by Table 430-152 for the type of feeder protective device employed.

FPN: See Appendix D, Example No. D8.

(b) Other Installations. Where feeder conductors have an ampacity greater than required by Section 430-24, the rating or setting of the feeder overcurrent protective device shall be permitted to be based on the ampacity of the feeder conductors.

Section 430-62(b) states the obvious solution of how to size a motor feeder larger than the minimum size required by the *Code*. If the motor feeder conductors are sized larger than the minimum required, then the size of the overcurrent device for the feeder is based on the size of the feeder conductors.

430-63. Rating or Setting — Power and Light Loads. Where a feeder supplies a motor load and, in addition, a lighting or a lighting and appliance load, the feeder protective device shall be permitted to have a rating or setting that is sufficient to carry the lighting or the lighting and appliance load as determined in accordance with Articles 210 and 220 plus, for a single motor, the rating permitted by Section 430-52, and, for two or more motors, the rating permitted by Section 430-62.

F. Motor Control Circuits

430-71. General. Part F contains modifications of the general requirements and applies to the particular conditions of motor control circuits.

FPN: See Section 430-9(b) for equipment device terminal requirements.

Definition of Motor Control Circuit. The circuit of a control apparatus or system that carries the electric signals directing the performance of the controller, but does not carry the main power current.

430-72. Overcurrent Protection.

(a) General. A motor control circuit tapped from the load side of a motor branch-circuit short-circuit and ground-fault protective device(s) and functioning to control the motor(s) connected to that branch circuit shall be protected against overcurrent in accordance with Section 430-72. Such a tapped control circuit shall not be considered to be a branch circuit and shall be permitted to be protected by either a supplementary or branch-circuit overcurrent protective device(s). A motor control circuit other than such a tapped control circuit shall be protected against overcurrent in accordance with Section 725-23 or the notes to Tables 11(a) and (b) in Chapter 9, as applicable.

(b) Conductor Protection. The overcurrent protection for conductors shall be provided as specified in (1) or (2).

Exception No. 1: Where the opening of the control circuit would create a hazard as, for example, the control circuit of a fire pump motor, and the like, conductors of control circuits shall require only short-circuit and ground-fault protection and shall be permitted to be protected by the motor branch-circuit short-circuit and ground-fault protective device(s).

Exception No. 2: Conductors supplied by the secondary side of a single-phase transformer having only a two-wire (single-voltage) secondary shall be permitted to be protected by overcurrent protection provided on the primary (supply) side of the transformer, provided this protection does not exceed the value determined by multiplying the appropriate maximum rating of the overcurrent device for the secondary conductor from Table 430-72(b) by the secondary-to-primary voltage ratio. Transformer secondary conductors (other than two-wire) shall not be considered to be protected by the primary overcurrent protection.

(1) Where the motor branch-circuit short-circuit and ground-fault protective device does not provide protection in accordance with Section 430-72(b)(2), separate overcurrent protection shall be provided. The overcurrent protection shall not exceed the values specified in Column A of Table 430-72(b).

(2) Conductors shall be permitted to be protected by the motor branch-circuit short-circuit and ground-fault protective device and shall require only short-circuit and ground-fault protection. Where the conductors do not extend beyond the motor control equipment enclosure, the rating of the protective device(s) shall not exceed the value specified in Column B of Table 430-72(b). Where the conductors extend beyond the motor control equipment enclosure, the rating of the protective device(s) shall not exceed the value specified in Column C of Table 430-72(b).

(c) Control Circuit Transformer. Where a motor control circuit transformer is provided, the transformer shall be protected in accordance with (1), (2), (3), (4), or (5).

Exception: Overcurrent protection shall be omitted where the opening of the control circuit would create a hazard as, for example, the control circuit of a fire pump motor and the like.

(1) Where the transformer supplies a Class 1 power-limited circuit, Class 2, or Class 3 remote-control circuit conforming with the requirements of Article 725, the protection shall comply with Article 725.

(2) Protection shall be permitted to be provided in accordance with Section 450-3.

(3) Control circuit transformers rated less than 50 volt-amperes (VA) and that are an integral part of the motor controller and located within the motor controller enclosure shall be permitted to be protected by primary overcurrent devices, impedance limiting means, or other inherent protective means.

(4) Where the control circuit transformer rated primary current is less than 2 amperes, an overcurrent device rated or set at not more than 500 percent of the rated primary current shall be permitted in the primary circuit.

(5) Protection shall be permitted to be provided by other approved means.

Motor control circuits are allowed to receive their power from either the load side of the motor short-circuit and ground-fault protective device or from a separate source, such as a panelboard.

Motor control circuits that receive their power from a separate source are protected against overcurrent in accordance with Section 725-23 for Class 1 circuits. Conductor sizes No. 14 and larger are protected according to their ampacity listed in Tables 310-16 through 310-20. Conductor sizes No. 16 and 18 must be protected at not more than 10 and 7 amperes, respectively, as specified in Table 430-72(b).

If a motor control circuit is tapped from the load side of the motor branch-circuit short-circuit and ground-fault protective device, the size of the tapped conductor and rating of the overcurrent device are based on whether the conductor stays within the motor control enclosure or leaves it. The load on a motor control circuit is similar to a motor branch-circuit load in that there is a predetermined connected load. There is also an initial high inrush of current, until the armature of the relay is seated and the current decreases to a steady state. Therefore, the overcurrent protection is similar to the short-circuit and ground-fault protection provided for a motor and is allowed to be greater than the ampacity of the control circuit conductor.

430-73. Mechanical Protection of Conductor. Where damage to a motor control circuit would constitute a hazard, all conductors of such a remote motor control circuit that are outside the control device itself shall be installed in a

Table 430-72(b). Maximum Rating of Overcurrent Protective Device in Amperes

			Protection Provided by Motor Branch-Circuit Protective Device(s)			
	Separate Protection Provided Column A		Conductors Within Enclosure Column B		Conductors Extend Beyond Enclosure Column C	
Control Circuit Conductor Size (AWG)	Copper	Aluminum or Copper-Clad Aluminum	Copper	Aluminum or Copper-Clad Aluminum	Copper	Aluminum or Copper-Clad Aluminum
18	7	—	25	—	7	—
16	10	—	40	—	10	—
14	(Note 1)	—	100	—	45	—
12	(Note 1)	(Note 1)	120	100	60	45
10	(Note 1)	(Note 1)	160	140	90	75
larger than 10	(Note 1)	(Note 1)	(Note 2)	(Note 2)	(Note 3)	(Note 3)

Notes:

1. Value specified in Section 310-15 as applicable.
2. 400 percent of value specified in Table 310-17 for 60°C conductors.
3. 300 percent of value specified in Table 310-16 for 60°C conductors.

raceway or be otherwise suitably protected from physical damage.

Where one side of the motor control circuit is grounded, the motor control circuit shall be arranged so that an accidental ground in the control circuit remote from the motor controller will (1) not start the motor, and (2) not bypass manually operated shutdown devices or automatic safety shutdown devices.

If damage to the motor control circuit conductors would constitute a fire or accident hazard, physical protection of the motor control circuit conductors is extremely important. If damage to the control circuit conductors could result in an accidental ground fault or short circuit, causing the device to operate or rendering the device inoperative (either condition constituting a hazard to persons or property), conductors are required to be installed in a raceway. Where boilers or furnaces are equipped with an automatic safety control device, damage to the conductors of the low-voltage (Article 725, Part C) control circuit (thermostat, etc.) is not considered to constitute a hazard.

The second paragraph of Section 430-73 requires that if one side of the motor control circuit is grounded, the circuit be arranged so that an accidental ground in the remote control device will not start the motor. For example, see the control wiring illustrated in Figure 430.16. If the control circuit is a 120-volt, single-phase circuit derived from a 208-volt, 3-phase wye system supplying the motor, one side of the control circuit will be the grounded neutral. If the start button of the motor control circuit is in the grounded neutral, a ground fault on the coil side of the start button can short-circuit the start circuit and start the motor. The same condition exists if the ground fault is in the wiring rather than in the control device itself. By locating the start button in the ungrounded side of the control circuit, this hazardous condition is avoided. See Figure 430.17.

Combinations of ground faults in motor and motor control circuits can produce the same problem. If the circuit is ungrounded, the first fault may go undetected. One solution is to use double-pole control devices, with one pole in each of the two control lines.

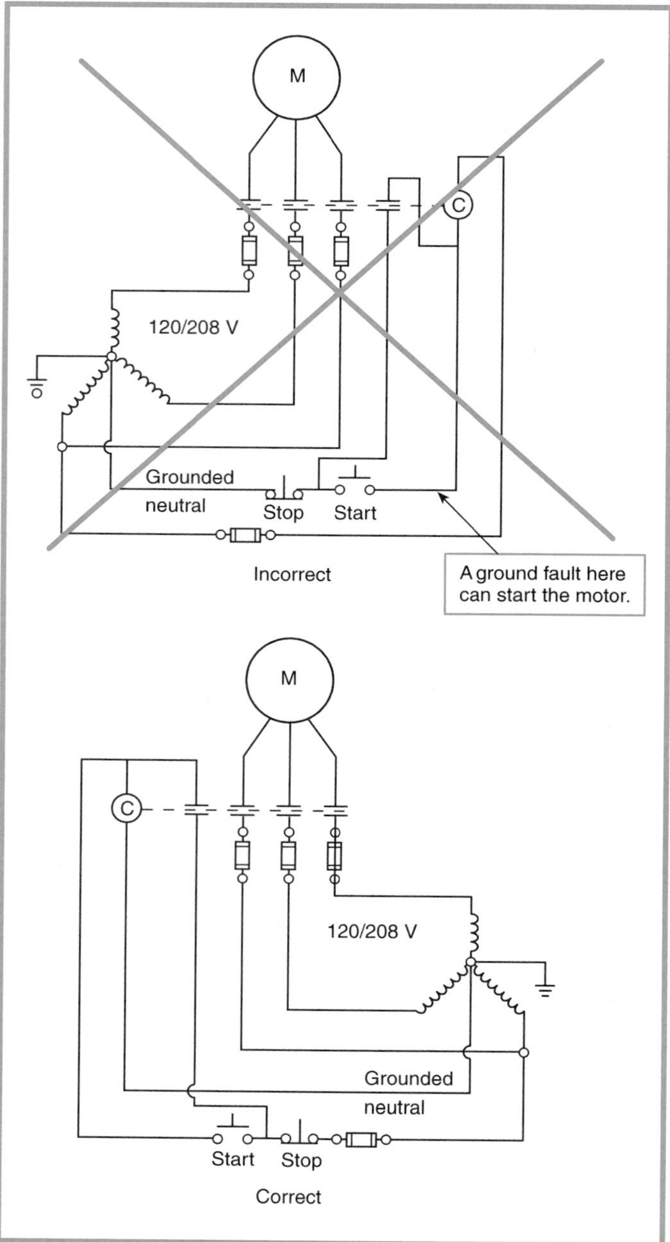

Figure 430.16 *An example of control wiring in violation of Section 430-73, paragraph 2 (top) and in compliance with Section 430-73, paragraph 2 (bottom). (For simplification, motor overload elements and disconnecting means are not shown.)*

430-74. Disconnection.

(a) General. Motor control circuits shall be arranged so that they will be disconnected from all sources of supply when the disconnecting means is in the open position. The disconnecting means shall be permitted to consist of two or more separate devices, one of which disconnects the motor and the controller from the source(s) of power supply for the motor, and the other(s), the motor control circuit(s) from its power supply. Where separate devices are used, they shall be located immediately adjacent to each other.

Exception No. 1: Where more than 12 motor control circuit conductors are required to be disconnected, the disconnecting means shall be permitted to be located other than immediately adjacent to each other where all of the following conditions are complied with.

(a) Access to energized parts is limited to qualified persons in accordance with Part L of this article.

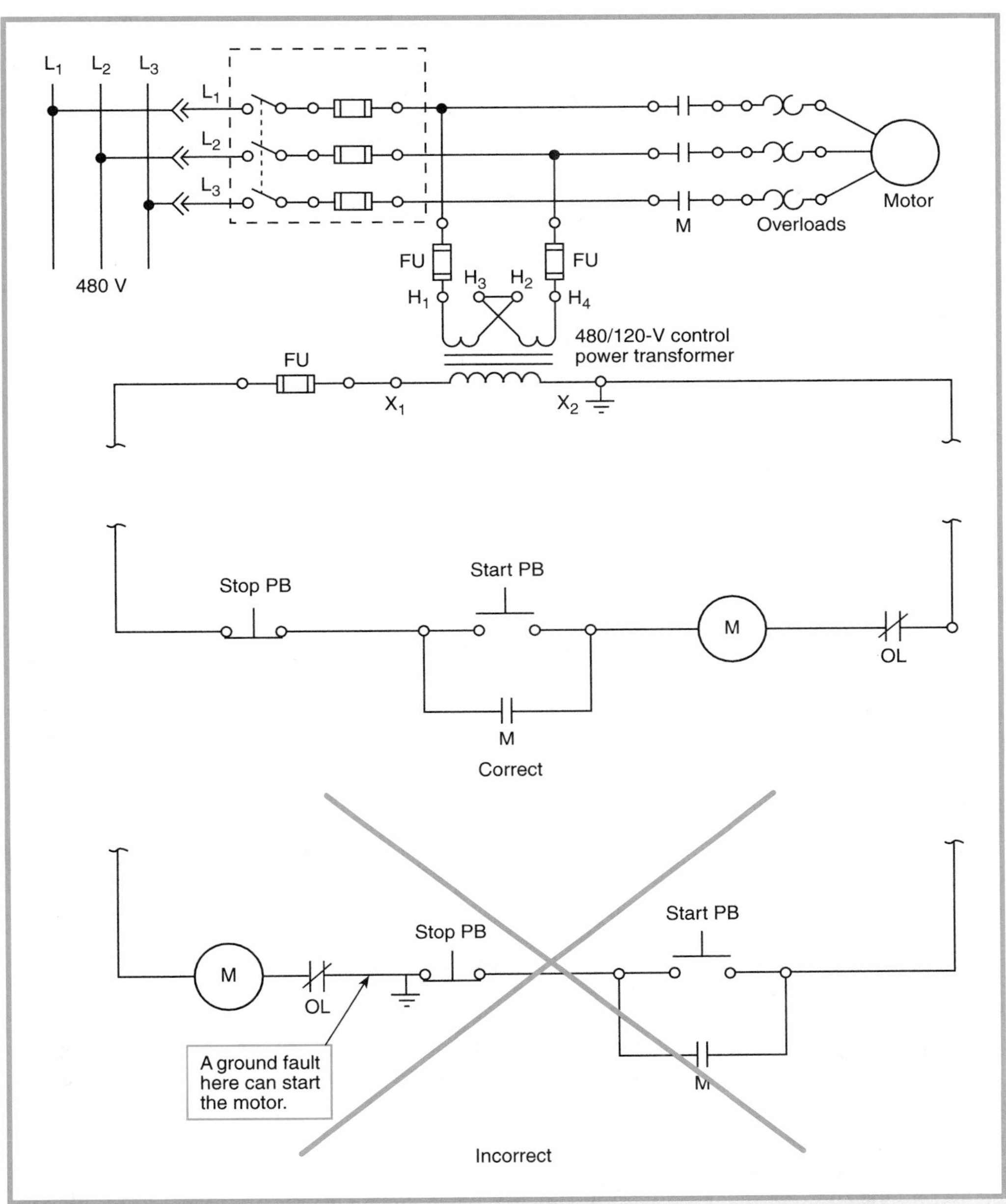

Figure 430.17 *An example of control wiring utilizing a 480/120-volt control power transformer. (The lower control circuit is not in compliance with Section 430-73, paragraph 2.)*

(b) A warning sign is permanently located on the outside of each equipment enclosure door or cover permitting access to the live parts in the motor control circuit(s), warning that motor control circuit disconnecting means are remotely located and specifying the location and identification of each disconnect. Where energized parts are not in an equipment enclosure as permitted by Sections 430-132 and 430-133, an additional warning sign(s) shall be located where visible to persons who may be working in the area of the energized parts.

Exception No. 2: The motor control circuit disconnecting means shall be permitted to be remote from the motor controller power supply disconnecting means where the opening of one or more motor control circuit disconnect means may result in potentially unsafe conditions for personnel or property and the conditions of items (a) and (b) of Exception No. 1 above are complied with.

(b) Control Transformer in Controller Enclosure. Where a transformer or other device is used to obtain a reduced voltage for the motor control circuit and is located in the controller enclosure, such transformer or other device shall be connected to the load side of the disconnecting means for the motor control circuit.

G. Motor Controllers

430-81. General. Part G is intended to require suitable controllers for all motors.

(a) Definition. For the definition of *Controller*, see Article 100. For the purpose of this article, a controller is any switch or device that is normally used to start and stop a motor by making and breaking the motor circuit current.

(b) Stationary Motor of ⅛ Horsepower or Less. For a stationary motor rated at ⅛ hp or less that is normally left running and is constructed so that it cannot be damaged by overload or failure to start, such as clock motors and the like, the branch-circuit protective device shall be permitted to serve as the controller.

(c) Portable Motor of ⅓ Horsepower or Less. For a portable motor rated at ⅓ hp or less, the controller shall be permitted to be an attachment plug and receptacle.

430-82. Controller Design.

(a) Starting and Stopping. Each controller shall be capable of starting and stopping the motor it controls and shall be capable of interrupting the locked-rotor current of the motor.

(b) Autotransformer. An autotransformer starter shall provide an "off" position, a running position, and at least one starting position. It shall be designed so that it cannot rest in the starting position or in any position that will render the overload device in the circuit inoperative.

(c) Rheostats. Rheostats shall be in compliance with the following.

(1) Motor-starting rheostats shall be designed so that the contact arm cannot be left on intermediate segments. The point or plate on which the arm rests when in the starting position shall have no electrical connection with the resistor.

(2) Motor-starting rheostats for dc motors operated from a constant voltage supply shall be equipped with automatic devices that will interrupt the supply before the speed of the motor has fallen to less than one-third its normal rate.

430-83. Ratings. The controller shall have a rating as specified in (a) of this section, unless otherwise permitted in (b) or (c), or as specified in (d) of this section, under the conditions specified.

Section 430-83 was reorganized for the 1999 *Code* without changing the basic requirements.

(a) General.

(1) Controllers, other than inverse time circuit breakers and molded case switches, shall have horsepower ratings at the application voltage not lower than the horsepower rating of the motor. A controller for a Design E motor rated more than 2 hp shall (1) be marked as rated for use with a Design E motor, or (2) have a horsepower rating not less than 1.4 times the rating of a motor rated 3 through 100 hp, or not less than 1.3 times the rating of a motor rated over 100 hp.

(2) A branch-circuit inverse time circuit breaker rated in amperes shall be permitted as a controller for all motors, including Design E. Where this circuit breaker is also used for overload protection, it shall conform to the appropriate provisions of this article governing overload protection.

(b) Small Motors. Devices as specified in Section 430-81(b) and (c) shall be permitted as a controller.

(c) Stationary Motors of 2 Horsepower or Less. For stationary motors rated at 2 hp or less and 300 volts or less, the controller shall be permitted to be either of the following:

(1) A general-use switch having an ampere rating not less than twice the full-load current rating of the motor
(2) On ac circuits, a general-use snap switch suitable only for use on ac (not general-use ac-dc snap switches) where the motor full-load current rating is not more than 80 percent of the ampere rating of the switch

(d) Torque Motors. For torque motors, the controller shall have a continuous-duty, full-load current rating not less than the nameplate current rating of the motor. For a motor controller rated in horsepower but not marked with the foregoing current rating, the equivalent current rating shall be determined from the horsepower rating by using Tables 430-147, 430-148, 430-149, or 430-150.

(e) Voltage Rating. A controller with a straight voltage rating, e.g., 240 volts or 480 volts, shall be permitted to be applied in a circuit in which the nominal voltage between any two conductors does not exceed the controller's voltage rating. A controller with a slash rating, e.g., 120/240 volts or 480/277 volts, shall only be applied in a circuit in which the nominal voltage to ground from any conductor does not exceed the lower of the two values of the controller's voltage rating and the nominal voltage between any two conductors does not exceed the higher value of the controller's voltage rating.

Section 430-83(e) requires that controllers identified with slash voltage ratings, such as 120/240-volt and 480/277-volt grounded systems, are allowed to be used only on electrical systems in which the nominal voltage to ground does not exceed the lower voltage rating of the controller, and the nominal voltage between any two phases of the electrical system is not greater than the higher value of the controller voltage rating.

430-84. Need Not Open All Conductors. The controller shall not be required to open all conductors to the motor.

Exception: Where the controller serves also as a disconnecting means, it shall open all ungrounded conductors to the motor as provided in Section 430-111.

A controller that does not also serve as a disconnecting means is required to open only as many motor circuit conductors as may be necessary to stop the motor. That is, one conductor for a dc or single-phase motor circuit, two conductors for a 3-phase motor circuit, and three conductors for a 2-phase motor circuit.

430-85. In Grounded Conductors. One pole of the controller shall be permitted to be placed in a permanently grounded conductor, provided the controller is designed so that the pole in the grounded conductor cannot be opened without simultaneously opening all conductors of the circuit.

Generally, one conductor of a 120-volt circuit is grounded, and a single-pole device is required to be connected in the ungrounded conductor to serve as a controller. A 2-pole controller is permitted for such a circuit, where both conductors (grounded and ungrounded) are opened simultaneously. The same requirement can be applied to other circuits, such as 240-volt, 3-wire circuits with one conductor grounded.

430-87. Number of Motors Served by Each Controller. Each motor shall be provided with an individual controller.

Exception: For motors rated 600 volts or less, a single controller rated at not less than the equivalent horsepower, as determined in accordance with Section 430-110(c)(1), of all the motors in the group shall be permitted to serve the group under any of the following conditions:

(a) Where a number of motors drive several parts of a single machine or piece of apparatus, such as metal and woodworking machines, cranes, hoists, and similar apparatus
(b) Where a group of motors is under the protection of one overcurrent device as permitted in Section 430-53(a)
(c) Where a group of motors is located in a single room within sight from the controller location

The conditions stated in the exception to Section 430-87 are similar those specified in Section 430-112, which permits the use of a single disconnecting means for a group of motors.

430-88. Adjustable-Speed Motors. Adjustable-speed motors that are controlled by means of field regulation shall be equipped and connected so that they cannot be started under a weakened field.

Exception: Starting under a weakened field shall be permitted where the motor is designed for such starting.

The torque and speed of a motor depend on the amount of current passing through the armature. This current is a function of shunt field strength and rpm of the armature. A reduction of the shunt field magnetic flux causes a reduction of the counterelectromotive force in the armature, resulting in an increase in armature current, thereby increasing torque and thus increasing speed.

Because of excessive armature starting currents, this type of motor is not permitted to be started under a weakened field condition unless some means is provided to limit the speed to within safe limits.

430-89. Speed Limitation. Machines of the following types shall be provided with speed limiting devices or other speed limiting means:

(1) Separately excited dc motors
(2) Series motors
(3) Motor-generators and converters that can be driven at excessive speed from the dc end, as by a reversal of current or decrease in load

Exception: Separate speed-limiting devices or means shall not be required under either of the following conditions:

(a) Where the inherent characteristics of the machines, the system, or the load and the mechanical connection thereto are such as to safely limit the speed
(b) Where the machine is always under the manual control of a qualified operator

It is still fairly common for dc motors to be used where speed control is essential, such as in the case of electric railways and elevators, where a smooth start, controlled acceleration, and a smooth stop are necessary.

If the load is removed from a series motor when it is running, the speed of the motor will increase until it is dangerously high. To produce the necessary counterelectromotive force with a weakened field, the armature must turn correspondingly faster. Series motors are commonly used as gear-drive traction motors of electric locomotives and, thus, are continuously loaded.

The Ward Leonard speed control system is widely used to control a separately excited dc motor for the operation of electric elevators or hoists. Figure 430.18 is a simplified diagram of the speed control system, where the armatures of two generators (G_1 and G_2) are mounted on a shaft driven by a motor (not shown). M is the motor drive for the elevator. The fields of G_1 and M are excited by G_1. By varying the rheostat R position, the voltage generated by G_2 is also varied,

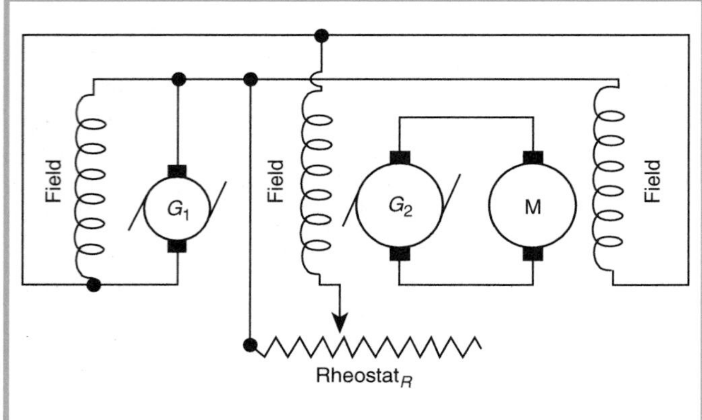

Figure 430.18 *A schematic diagram of the Ward Leonard speed control system.*

which in turn controls the speed of *M*. If the field circuit of *M* were to be accidentally opened during a light load, the motor would reach an excessive speed; however, no speed-limiting device is required, as the motor is always loaded.

Separately excited dc motors, series motors, and motor (compound-wound dc) generators and (synchronous) converters are required to be provided with speed-limiting devices (note exceptions), such as a centrifugal device on the shaft of the machine or a remotely located overspeed device, which may be set at a predetermined speed to operate a set of contacts and thereby trip a circuit breaker and deenergize the machine.

430-90. Combination Fuseholder and Switch as Controller. The rating of a combination fuseholder and switch used as a motor controller shall be such that the fuseholder will accommodate the size of the fuse specified in Part C of this article for motor overload protection.

Time-delay (dual-element) fuses can commonly be used for both motor overload and branch-circuit short-circuit and ground-fault protection and can be sized in accordance with Section 430-32. See also Sections 430-36, 430-55, and 430-57 for other requirements regarding fuses and fuseholders.

Exception: Where fuses having time delay appropriate for the starting characteristics of the motor are used, fuseholders of smaller size than specified in Part C of this article shall be permitted.

430-91. Motor Controller Enclosure Types. Table 430-91 provides the basis for selecting enclosures for use in specific locations, other than hazardous (classified) locations. The enclosures are not intended to protect against conditions such as condensation, icing, corrosion, or contamination that may occur within the enclosure or enter via the conduit or unsealed openings. These internal conditions shall require special consideration by the installer and user.

See the commentary following the definition of *enclosure* in Article 100. Enclosure type numbers are described in greater detail in industry standards such as ANSI/NEMA ICS6-1988, NEMA Pub. 250-179, UL 508, and the controller manufacturer's literature. For other than general-use Type No. 1 enclosures, the type number is required to be marked on the motor-controller enclosure.

H. Motor Control Centers

430-92. General. Part H covers motor control centers installed for the control of motors, lighting, and power circuits.

Part H provides the requirements for the installation of motor control centers. Section 430-1, FPN No. 1, states that the installation requirements for motor control centers are covered in Section 110-26(f).

Motor control centers are made up of a number of motor starters, controls, and disconnect switches assembled in one large enclosure. Motor control centers are allowed to be used as service equipment if provided with a single main disconnecting means. A second service disconnecting means, however, is permitted in the motor control center if it is provided to serve other loads.

Access and working space clearances as well as dedicated space requirements of Section 110-26 are applicable to motor control centers.

430-94. Overcurrent Protection. Motor control centers shall be provided with overcurrent protection in accordance with Article 240 based on the rating of the common power bus. This protection shall be provided by (1) an overcurrent protective device located ahead of the motor control center or (2) a main overcurrent protective device located within the motor control center.

430-95. Service-Entrance Equipment. Where used as service equipment, each motor control center shall be provided with a single main disconnecting means to disconnect all ungrounded service conductors.

Exception: A second service disconnect shall be permitted to supply additional equipment.

Where a grounded conductor is provided, the motor control center shall be provided with a main bonding jumper, sized in accordance with Section 250-28(d), within one of the sections for connecting the grounded conductor, on its supply side, to the motor control center equipment ground bus.

Table 430-91. Motor Controller Enclosure Selection

For Outdoor Use

Provides a Degree of Protection Against the Following Environmental Conditions	Enclosure Type Number[1]						
	3	3R	3S	4	4X	6	6P
Incidental contact with the enclosed equipment	X	X	X	X	X	X	X
Rain, snow, and sleet	X	X	X	X	X	X	X
Sleet[2]	—	—	X	—	—	—	—
Windblown dust	X	—	X	X	X	X	X
Hosedown	—	—	—	X	X	X	X
Corrosive agents	—	—	—	—	X	—	X
Occasional temporary submersion	—	—	—	—	—	X	X
Occasional prolonged submersion	—	—	—	—	—	—	X

For Indoor Use

Provides a Degree of Protection Against the Following Environmental Conditions	Enclosure Type Number[1]									
	1	2	4	4X	5	6	6P	12	12K	13
Incidental contact with the enclosed equipment	X	X	X	X	X	X	X	X	X	X
Falling dirt	X	X	X	X	X	X	X	X	X	X
Falling liquids and light splashing	—	X	X	X	X	X	X	X	X	X
Circulating dust, lint, fibers, and flyings	—	—	X	X	—	X	X	X	X	X
Settling airborne dust, lint, fibers, and flyings	—	—	X	X	X	X	X	X	X	X
Hosedown and splashing water	—	—	X	X	—	X	X	—	—	—
Oil and coolant seepage	—	—	—	—	—	—	—	X	X	X
Oil or coolant spraying and splashing	—	—	—	—	—	—	—	—	—	X
Corrosive agents	—	—	—	X	—	—	X	—	—	—
Occasional temporary submersion	—	—	—	—	—	X	X	—	—	—
Occasional prolonged submersion	—	—	—	—	—	—	X	—	—	—

[1]Enclosure type number shall be marked on the motor controller enclosure.

[2]Mechanism shall be operable when ice covered.

430-96. Grounding. Multisection motor control centers shall be bonded together with an equipment grounding conductor or an equivalent grounding bus sized in accordance with Table 250-122. Equipment grounding conductors shall terminate on this grounding bus or to a grounding termination point provided in a single-section motor control center.

430-97. Busbars and Conductors.

(a) Support and Arrangement. Busbars shall be protected from physical damage and be held firmly in place. Other than for required interconnections and control wiring, only those conductors that are intended for termination in a vertical section shall be located in that section.

Exception: Conductors shall be permitted to travel horizontally through vertical sections where such conductors are isolated from the busbars by a barrier.

(b) Phase Arrangement. The phase arrangement on 3-phase horizontal common power and vertical buses shall be A, B, C from front to back, top to bottom, or left to right, as viewed from the front of the motor control center. The B phase shall be that phase having the higher voltage to ground on 3-phase, 4-wire, delta-connected systems. Other busbar arrangements shall be permitted for additions to existing installations and shall be marked.

Exception: Rear-mounted units connected to a vertical bus that is common to front-mounted units shall be permitted to have a C, B, A phase arrangement where properly identified.

(c) Minimum Wire-Bending Space. The minimum wire-bending space at the motor control center terminals and minimum gutter space shall be as required in Article 373.

(d) Spacings. Spacings between motor control center bus terminals and other bare metal parts shall not be less than specified in Table 430-97.

(e) Barriers. Barriers shall be placed in all service-entrance motor control centers to isolate service busbars and terminals from the remainder of the motor control center.

430-98. Marking.

(a) Motor Control Centers. Motor control centers shall be marked according to Section 110-21, and such marking shall be plainly visible after installation. Marking shall also include common power bus current rating and motor control center short-circuit rating.

Table 430-97. Minimum Spacing Between Bare Metal Parts

	Opposite Polarity Where Mounted on the Same Surface (in.)	Opposite Polarity Where Held Free in Air (in.)	Live Parts to Ground (in.)
Not over 125 volts, nominal	¾	½	½
Not over 250 volts, nominal	1¼	¾	½
Not over 600 volts, nominal	2	1	1

Note: For SI units, 1 in. = 25.4 mm.

(b) Motor Control Units. Motor control units in a motor control center shall comply with Section 430-8.

J. Disconnecting Means

430-101. General. Part J is intended to require disconnecting means capable of disconnecting motors and controllers from the circuit.

FPN No. 1: See Figure 430-1.

FPN No. 2: See Section 110-22 for identification of disconnecting means.

430-102. Location.

(a) Controller. An individual disconnecting means shall be provided for each controller and shall disconnect the controller. The disconnecting means shall be located in sight from the controller location.

Exception No. 1: For motor circuits over 600 volts, nominal, a controller disconnecting means capable of being locked in the open position shall be permitted to be out of sight of the controller, provided the controller is marked with a warning label giving the location of the disconnecting means.

Exception No. 2: A single disconnecting means shall be permitted for a group of coordinated controllers that drive several parts of a single machine or piece of apparatus. The disconnecting means and the controllers shall be located in sight from the machine or apparatus.

(b) Motor. A disconnecting means shall be located in sight from the motor location and the driven machinery location.

Exception: A disconnecting means, in addition to the controller disconnecting means as required in accordance with Section 430-102(a), shall not be required for the motor where the disconnecting means for the controller is individually capable of being locked in the open position.

FPN: For information on lockout/tagout procedures, see *Standard for Electrical Safety Requirements for Employee Workplaces,* NFPA 70E-1995.

The main rules of Sections 430-102(a) and (b) require that the disconnecting means be in sight from the controller, the motor location, and the driven machinery location. For motors over 600 volts, the controller disconnecting means may be out of sight of the controller, as illustrated in Figure 430.19, provided that the controller has a warning label indicating the location and identification of the disconnecting means, which must be capable of being locked in the open position.

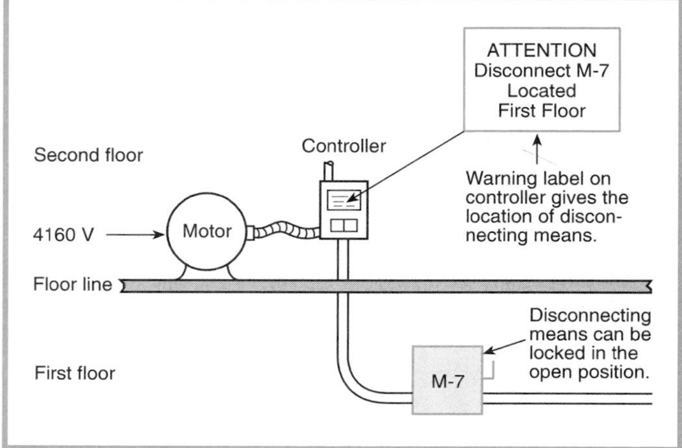

Figure 430.19 A controller (disconnecting means) that is out of sight of the controller (for motors over 600 volts).

The disconnecting means may be out of sight of the motor, as illustrated in Figure 430.20, if the disconnecting means complying with Section 430-102(a) is individually capable of being locked in the open position.

A single disconnecting means may be located adjacent to a group of coordinated controllers, as illustrated in Figure 430.21, where the controllers are mounted on a multimotor continuous process machine.

Section 430-102 clearly requires that individual disconnect switches or circuit breakers must be capable of being locked in the open position. Disconnect switches or circuit breakers located behind the locked door of a panelboard or located within locked rooms do not comply with the requirements of this section.

The fine print note points out an out an important consideration and reference standard for employee safety in the workplace. Part II, Chapter 5, of NFPA 70E-1995, *Standard for Electrical Safety Requirements for Employee Workplaces,* requires in part that, "All electrical circuit conductors and circuit parts shall not be considered to be in an electrically safe condition until

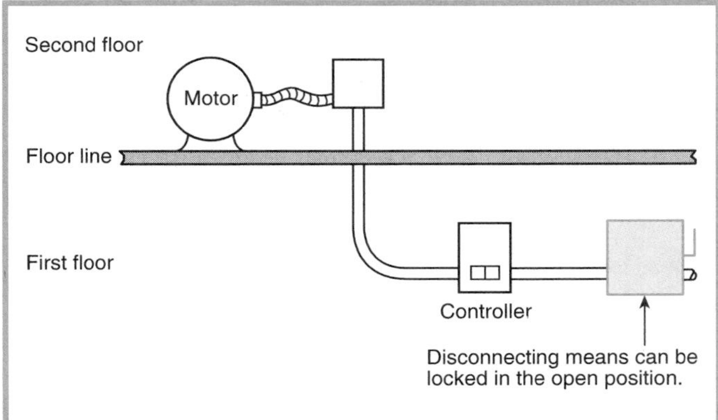

Figure 430.20 A controller disconnecting means that is out of sight of the motor.

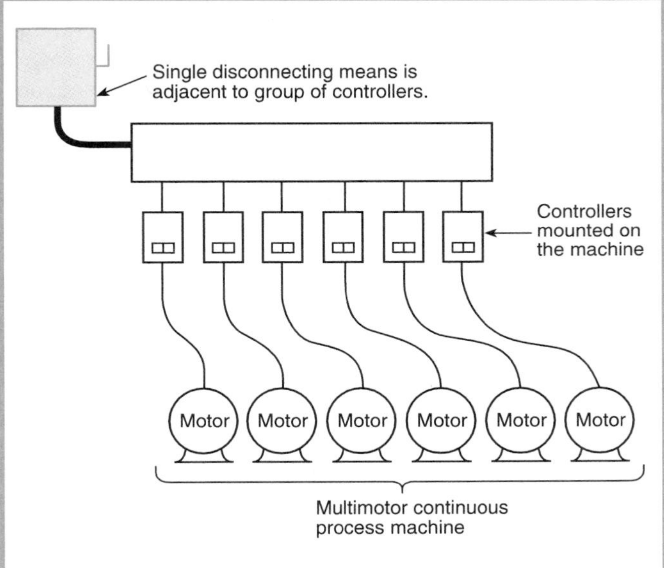

Figure 430.21 A single disconnecting means located adjacent to a group of coordinated controllers mounted on a multimotor continuous process machine.

all sources of energy are removed, the disconnecting means is under lockout/tagout, the absence of voltage is verified by an approved voltage testing device, etc." Further, it states that, "Lockout/tagout requirements shall apply to fixed permanently installed equipment, temporarily installed equipment, and to portable equipment." Principles and procedures are set forth in NFPA 70E establishing strict work rules requiring locking-off and tagging-out of disconnect switches.

430-103. Operation. The disconnecting means shall open all ungrounded supply conductors and shall be designed so that no pole can be operated independently. The disconnecting means shall be permitted in the same enclosure with the controller.

FPN: See Section 430-113 for equipment receiving energy from more than one source.

The *Code* requires that a switch, circuit breaker, or other device serve as a disconnecting means for both the controller and the motor, thereby providing safety during maintenance and inspection shutdown periods. The disconnecting means also disconnects the controller; therefore, it cannot be a part of the controller. However, separate disconnects and controllers may be mounted on the same panel or be contained in the same enclosure, such as combination fused-switch, magnetic-starter units.

Depending on the size of the motor and other conditions, the type of disconnecting means required may be a motor circuit switch, a circuit breaker, a general-use switch, an isolating switch, an attachment plug and receptacle, or a branch-circuit short-circuit and ground-fault protective device, as specified in Section 430-109.

If a motor is stalled or under heavy overload and the motor controller fails to properly open the circuit, the disconnecting means, which is required to be rated to interrupt locked-rotor current, can be used to open the circuit. For motors larger than 100 hp ac or 40 hp dc, the disconnecting means is, in accordance with Section 430-109(e), permitted to be a general-use or an isolating switch where plainly marked "Do not operate under load."

430-104. To Be Indicating. The disconnecting means shall plainly indicate whether it is in the open (off) or closed (on) position.

430-105. Grounded Conductors. One pole of the disconnecting means shall be permitted to disconnect a permanently grounded conductor, provided the disconnecting means is designed so that the pole in the grounded conductor cannot be opened without simultaneously disconnecting all conductors of the circuit.

430-107. Readily Accessible. One of the disconnecting means shall be readily accessible.

430-108. Every Switch. Every disconnecting means in the motor circuit between the point of attachment to the feeder and the point of connection to the motor shall comply with the requirements of Sections 430-109 and 430-110.

Figure 430.22 represents an example of a listed disconnect switch rated in horsepower meeting the requirements of Sections 430-109 and 430-110.

Figure 430.22 Heavy-duty safety switches, UL-listed for use on systems up to 200,000 amperes, fault current rms symmetrical with Class J or Class R fuses installed. (Square D Co.)

430-109. Type. The disconnecting means shall be a type specified in (a), unless otherwise permitted in (b) through (g), under the conditions specified.

(a) General.

(1) A listed motor-circuit switch rated in horsepower. For Design E motors rated greater than 2 hp, the motor circuit switch shall be either (a) marked as rated for use with Design E motors or (b) have a horsepower rating not less than 1.4 times the rating of a motor rated 3–100 hp, or not less than 1.3 times the rating of a motor rated over 100 hp.

(2) A listed molded case circuit breaker.

(3) A listed molded case switch.

(4) An instantaneous trip circuit breaker that is part of a listed combination motor controller.

(5) Listed self-protected combination controller.

(6) Listed manual motor controllers additionally marked "Suitable as Motor Disconnect" shall be permitted as a disconnecting means where installed between the final motor branch-circuit short-circuit and ground-fault protective device and the motor.

(b) Stationary Motors of ⅛ Horsepower or Less. For stationary motors of ⅛ hp or less, the branch-circuit overcurrent device shall be permitted to serve as the disconnecting means.

(c) Stationary Motors of 2 Horsepower or Less. For stationary motors rated at 2 hp or less and 300 volts or less, the disconnecting means shall be permitted to be one of the devices specified in (1), (2), or (3).

(1) A general-use switch having an ampere rating not less than twice the full-load current rating of the motor
(2) On ac circuits, a general-use snap switch suitable only for use on ac (not general-use ac-dc snap switches) where the motor full-load current rating is not more than 80 percent of the ampere rating of the switch
(3) A listed manual motor controller having a horsepower rating not less than the rating of the motor and marked "Suitable as Motor Disconnect"

(d) Autotransformer-Type Controlled Motors. For motors of over 2 hp to and including 100 hp, the separate disconnecting means required for a motor with an autotransformer-type controller shall be permitted to be a general-use switch where all of the following provisions are met.

(1) The motor drives a generator that is provided with overload protection.
(2) The controller is capable of interrupting the locked-rotor current of the motors, is provided with a no voltage release, and is provided with running overload protection not exceeding 125 percent of the motor full-load current rating.
(3) Separate fuses or an inverse time circuit breaker rated or set at not more than 150 percent of the motor full-load current are provided in the motor branch circuit.

(e) Isolating Switches. For stationary motors rated at more than 40 hp dc or 100 hp ac, the disconnecting means shall be permitted to be a general-use or isolating switch where plainly marked "Do not operate under load."

(f) Cord- and Plug-Connected Motors. For a cord- and plug-connected motor, a horsepower-rated attachment plug and receptacle having ratings no less than the motor ratings shall be permitted to serve as the disconnecting means for other than a Design E motor, and for a Design E motor rated 2 hp or less. For a Design E motor rated more than 2 hp, an attachment plug and receptacle used as the disconnecting means shall have a horsepower rating not less than 1.4 times the motor rating. A horsepower-rated attachment plug and receptacle shall not be required for a cord- and plug-connected appliance in accordance with Section 422-32, a room air conditioner in accordance with Section 440-63, or a portable motor rated ⅓ hp or less.

Section 430-109 requires the disconnecting means to be a motor circuit switch, a circuit breaker, or a molded-case switch (nonautomatic circuit interrupter). A motor circuit switch is a horsepower-rated switch capable of interrupting the maximum overload current of a motor (see the definition of *switches, motor-circuit switch* in Article 100). A molded-case switch (nonautomatic circuit interrupter) is a circuit-breaker-like device without the overcurrent element and automatic-trip mechanism. It is rated in amperes and is suitable for use as a motor circuit disconnectbased on its ampere rating, as is a circuit breaker. The disconnecting means is required to be listed.

Figures 430.23, 430.24, 430.25, and 430.26 illustrate Sections 430-109(b), (c), (e), and (f), respectively, to this general requirement.

Section 430-109(a) recognizes that Design E motors have a high starting (locked-rotor) current. Disconnect switches used with Design E motors over 2 hp must be identified for use with these design motors, or the size of the disconnect switch must be adjusted to compensate for this high starting current.

Where horsepower-rated fused switches are required, it should be noted that marking within the enclosure usually permits a dual horsepower rating. The standard horsepower rating is based on the largest non-time-delay (non-dual-element) fuse rating that can be used in the switch and that will permit the motor to start. The maximum horsepower rating is based on the largest rated time-delay (dual-element) fuse that can be used in the switch and that will permit the motor to start. Thus, where time-delay fuses are used, smaller-size switches and fuseholders can be used (see Section 430-57, Exception).

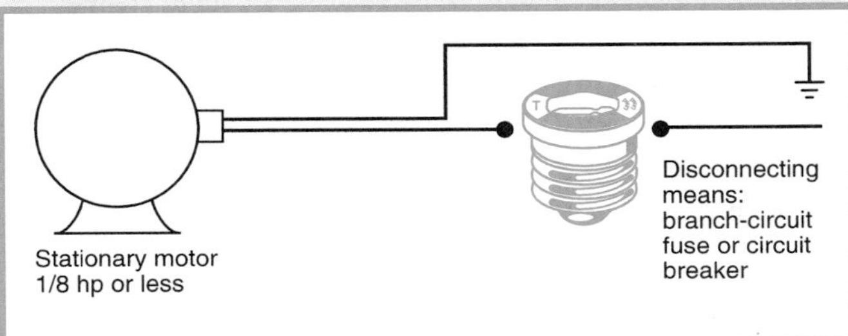

Figure 430.23 A branch-circuit overcurrent device serving as the disconnecting means for a stationary motor of 1/8 hp or less according to Section 430-109(b).

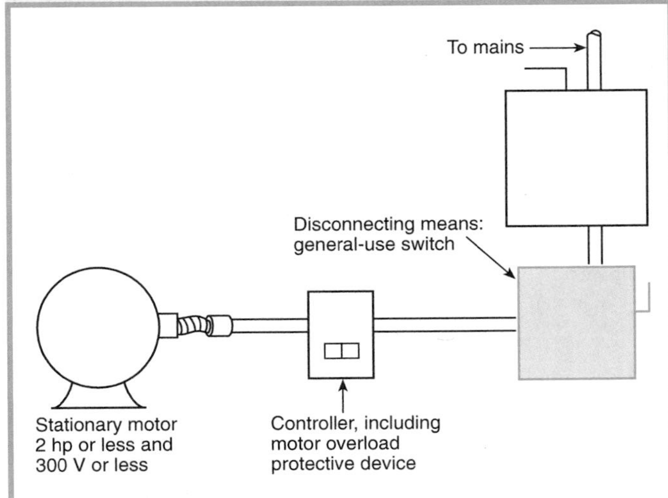

Figure 430.24 A general-use switch having an ampere rating not less than twice the motor full-load current serving as the disconnecting means for a stationary motor rated at 2 hp or less and at 300 volts or less according to Section 430-109(c).

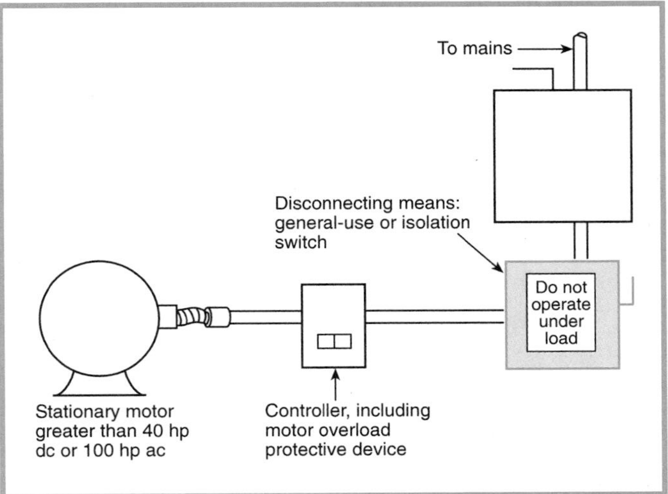

Figure 430.25 A general-use or an isolation switch serving as the disconnecting means for a stationary motor rated at more than 40 hp dc or 100 hp ac according to Section 430-109(e).

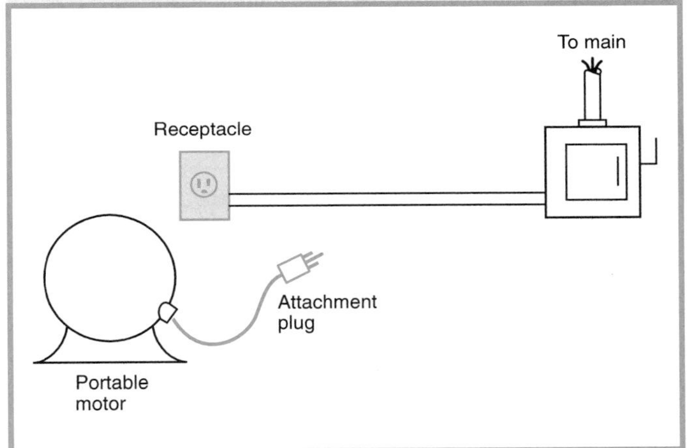

Figure 430.26 An attachment plug and receptacle of the proper rating serving as the disconnecting means for a certain cord- and plug-connected motor according to Section 430-109(f).

(g) Torque Motors. For torque motors, the disconnecting means shall be permitted to be a general-use switch.

430-110. Ampere Rating and Interrupting Capacity.

(a) General. The disconnecting means for motor circuits rated 600 volts, nominal, or less, shall have an ampere rating of at least 115 percent of the full-load current rating of the motor.

Exception: A listed nonfused motor-circuit switch having a horsepower rating equal to or greater than the motor horsepower shall be permitted to have an ampere rating less than 115 percent of the full-load current rating of the motor.

(b) For Torque Motors. Disconnecting means for a torque motor shall have an ampere rating of at least 115 percent of the motor nameplate current.

(c) For Combination Loads. Where two or more motors are used together or where one or more motors are used in combination with other loads, such as resistance heaters, and where the combined load may be simultaneous on a single disconnecting means, the ampere and horsepower ratings of the combined load shall be determined as follows.

(1) The rating of the disconnecting means shall be determined from the sum of all currents, including resistance loads, at the full-load condition and also at the locked-rotor condition. The combined full-load current and the combined locked-rotor current so obtained shall be considered as a single motor for the purpose of this requirement as follows.

The full-load current equivalent to the horsepower rating of each motor shall be selected from Table 430-147, 430-148, 430-149, or 430-150. These full-load currents shall be added to the rating in amperes of other loads to obtain an equivalent full-load current for the combined load.

The locked-rotor current equivalent to the horsepower rating of each motor shall be selected from Table 430-151(A) or 430-151(B). The locked-rotor currents shall be added to the rating in amperes of other loads to obtain an equivalent locked-rotor current for the combined load. Where two or more motors or other loads cannot be started simultaneously, the largest sum of locked rotor currents of a motor or group of motors that can be started simultaneously and the full load currents of other concurrent loads shall be permitted to be used to determine the equivalent locked-rotor current for the simultaneous combined loads.

Exception: Where part of the concurrent load is resistance load, and where the disconnecting means is a switch rated in horsepower and amperes, the switch used shall be permitted to have a horsepower rating that is not less than the combined load of the motor(s), if the ampere rating of the switch is not less than the locked-rotor current of the motor(s) plus the resistance load.

(2) The ampere rating of the disconnecting means shall not be less than 115 percent of the sum of all currents at the full-load condition determined in accordance with Section 430-110(c)(1).

Exception: A listed nonfused motor-circuit switch having a horsepower rating equal to or greater than the equivalent horsepower of the combined loads, determined in accordance with Section 430-110(c)(1), shall be permitted to have an ampere rating less than 115 percent of the sum of all currents at the full-load condition.

(3) For small motors not covered by Table 430-147, 430-148, 430-149, or 430-150, the locked-rotor current shall be assumed to be six times the full-load current.

A general-use switch, fuse, circuit breaker, molded-case switch (nonautomatic circuit interrupter), or attachment plug and receptacle used as a disconnecting means is required to have an ampere rating of not less than 115 percent of the motor full-load current.

Following is an example of how to determine the size of a disconnect for a combination load.

Example

Assume the load consists of one 5-hp, one 3-hp, and two $^{1}/_{2}$-hp motors, plus a 10-kilowatt heater, all rated 240 volts, 3-phase. All motors are Design B motors. The full-load and locked-rotor current equivalents are calculated, using the appropriate tables, as follows:

Calculation of Equivalent Full-Load Current Rating

Motor or Other Load	Table	Amperes
5-hp motor	430-150	15.2
3-hp motor	430-150	9.6
½-hp motor	430-150	2.2
½-hp motor	430-150	2.2
10-kW heater	$\frac{10 \times 1000}{240 \times 1.732}$	24.1
Total		53.3

Calculation of Equivalent Locked-Rotor Current Rating

Motor or Other Load	Table	Amperes
5-hp motor	430-151B	92
3-hp motor	430-151B	64
½-hp motor	430-151B	20
½-hp motor	430-151B	20
10-kW heater	$\frac{10 \times 1000}{240 \times 1.732}$	24.1
Total		220.1

Answer

The disconnecting means for a motor must have an ampere rating that is not less than 115 percent and also have a horsepower rating that is not less than the combined load. Therefore, the minimum ampere rating of the disconnecting means is 61.3 amperes (1.15 × 53.3 amperes = 61.3 amperes). Using Table

430-151B to obtain an equivalent horsepower from a locked-rotor current rating, the closest value equal to or greater than 220.1 amperes under the 230-volt column is 232 amperes, which equates to 15 hp. A 15-hp general-purpose or heavy-duty switch satisfies the minimum horsepower requirement but fails to satisfy the minimum current requirement of 61.3 amperes. Therefore, the next larger size disconnect switch must be used, which in horsepower ratings is 20 hp. This is because the locked-rotor current for a 20-hp, 3-phase, 230-volt motor is 290 amperes (over the needed 220.1 amperes) and the full-load current is 54 amperes (over the full-load rating of 53.3 amperes).

Listed circuit breakers and molded-case switches are tested under overload conditions at six times their rating, to cover motor circuit applications, and are suitable for use as a motor disconnecting means.

430-111. Switch or Circuit Breaker as Both Controller and Disconnecting Means. A switch or circuit breaker shall be permitted to be used as both the controller and disconnecting means if it complies with (a) and is one of the types specified in (b).

(a) General. The switch or circuit breaker complies with the requirements for controllers specified in Section 430-83, opens all ungrounded conductors to the motor, and is protected by an overcurrent device in each ungrounded conductor (which shall be permitted to be the branch-circuit fuses). The overcurrent device protecting the controller shall be permitted to be part of the controller assembly or shall be permitted to be separate. An autotransformer-type controller shall be provided with a separate disconnecting means.

(b) Type. The device shall be one of the types specified in (1), (2), or (3).

(1) Air-Break Switch. An air-break switch, operable directly by applying the hand to a lever or handle.

(2) Inverse Time Circuit Breaker. An inverse time circuit breaker operable directly by applying the hand to a lever or handle. The circuit breaker shall be permitted to be both power and manually operable.

(3) Oil Switch. An oil switch used on a circuit whose rating does not exceed 600 volts or 100 amperes, or by special permission on a circuit exceeding this capacity where under expert supervision. The oil switch shall be permitted to be both power and manually operable.

If used as a controller, a switch or circuit breaker is required to meet all the requirements for controllers and to be protected by branch-circuit short-circuit and ground-fault protective devices (fuses or a circuit breaker), which ensure that all ungrounded conductors will be opened.

If the controller consists of a manually operable air-break switch, an inverse time circuit breaker, or a 100-ampere maximum oil switch (higher rating by special permission), the controller is considered a satisfactory disconnecting means. It is the intent of Section 430-11 to permit omission of an additional device to serve as a disconnecting means.

It should be noted that a separate disconnecting means is required to be provided if the controller is of the autotransformer or compensator type. (This switch may be combined in the same enclosure with a motor overload protective device.)

430-112. Motors Served by Single Disconnecting Means. Each motor shall be provided with an individual disconnecting means.

Exception: A single disconnecting means shall be permitted to serve a group of motors under any one of the conditions of (a), (b), and (c). The single disconnecting means shall be rated in accordance with Section 430-110(c).

(a) Where a number of motors drive several parts of a single machine or piece of apparatus, such as metal and woodworking machines, cranes, and hoists

(b) Where a group of motors is under the protection of one set of branch-circuit protective devices as permitted by Section 430-53(a)

(c) Where a group of motors is in a single room within sight from the location of the disconnecting means

The exception after Section 430-112 permits a single disconnecting means to serve a group of motors. The disconnecting means is required to have a rating equal to the sum of the horsepower or current of each motor in the group. If the sum is over 2 hp, a motor circuit switch (horsepower-rated) is required to be used; thus, for five 2-hp motors, the disconnecting means should be a motor circuit switch rated at not less than 10 hp.

Part (a) of the exception to Section 430-112 indicates that a single disconnecting means may be used where a number of motors drive several parts of a single machine, such as cranes (see Sections 610-31 and 610-32), metal or woodworking machines, steel rolling mill machinery, and so on. The single disconnecting means for multimotor machinery provides a positive means of simultaneously de-energizing all motor branch circuits, including remote control circuits, interlocking circuits, limit-switch circuits, and operator control stations.

Part (b) of the exception to Section 430-112 refers to Section 430-53(a), which permits a group of motors under the protection of the same branch-circuit device, provided the device is rated not more than 20 amperes at 125 volts or 15 amperes at more than 125 volts but not more than 600 volts. The motors are required to be rated 1 hp or less, and the full-load current for each motor is not permitted to exceed 6 amperes. A single disconnecting means is both practical and economical for a group of such small motors.

Part (c) of the exception to Section 430-112 covers the common situation where a group of motors is located in one room, such as a pump room, compressor room, mixer room, and so on. It is therefore possible to design the layout of a single disconnecting means with an unobstructed view (not more than 50 ft) from each motor.

These conditions for an individual disconnecting means are similar to those specified in Section 430-87, which permits the use of a single controller for a group of motors.

430-113. Energy from More than One Source. Motor and motor-operated equipment receiving electrical energy from more than one source shall be provided with disconnecting means from each source of electrical energy immediately adjacent to the equipment served. Each source shall be permitted to have a separate disconnecting means.

Exception No. 1: Where a motor receives electrical energy from more than one source, the disconnecting means for the main power supply to the motor shall not be required to be immediately adjacent to the motor, provided the controller disconnecting means is capable of being locked in the open position.

Exception No. 2: A separate disconnecting means shall not be required for a Class 2 remote-control circuit conforming with Article 725, rated not more than 30 volts, and that is isolated and ungrounded.

Certain motors may require multiple separate sources of power to operate properly. For example, large synchronous motors commonly receive electrical energy from more than one source. Section 430-113 could apply as well to control circuits that supply power to sensors mounted within or otherwise attached to the motor or the driven machine. Exception No. 2 to Section 430-113 relieves the user from the disconnect requirement only for Class 2 circuits.

K. Over 600 Volts, Nominal

430-121. General. Part K recognizes the additional hazard due to the use of higher voltages. It adds to or amends the other provisions of this article. Other requirements for circuits and equipment operating at over 600 volts, nominal, are in Article 490.

430-122. Marking on Controllers. In addition to the marking required by Section 430-8, a controller shall be marked with the control voltage.

430-123. Conductor Enclosures Adjacent to Motors. Flexible metal conduit or liquidtight flexible metal conduit not exceeding 6 ft (1.83 m) in length shall be permitted to be employed for raceway connection to a motor terminal enclosure.

430-124. Size of Conductors. Conductors supplying motors shall have an ampacity not less than the current at which the motor overload protective device(s) is selected to trip.

430-125. Motor-Circuit Overcurrent Protection.

(a) General. Each motor circuit shall include coordinated protection to automatically interrupt overload and fault currents in the motor, the motor-circuit conductors, and the motor control apparatus.

Exception: Where a motor is vital to operation of the plant and the motor should operate to failure if necessary to prevent a greater hazard to persons, the sensing device(s) is permitted to be connected to a supervised annunciator or alarm instead of interrupting the motor circuit.

(b) Overload Protection.

(1) Each motor shall be protected against dangerous heating due to motor overloads and failure to start by a thermal protector integral with the motor or external current-sensing devices, or both.

(2) The secondary circuits of wound-rotor ac motors including conductors, controllers, and resistors rated for the application shall be considered as protected against overcurrent by the motor overload protection means.

(3) Operation of the overload interrupting device shall simultaneously disconnect all ungrounded conductors.

(4) Overload sensing devices shall not automatically reset after trip unless resetting of the overload sensing device does not cause automatic restarting of the motor or there is no hazard to persons created by automatic restarting of the motor and its connected machinery.

(c) Fault-Current Protection.

(1) Fault-current protection shall be provided in each motor circuit by one of the following means.

(a) A circuit breaker of suitable type and rating arranged so that it can be serviced without hazard. The circuit breaker shall simultaneously disconnect all ungrounded conductors. The circuit breaker shall be permitted to sense the fault current by means of integral or external sensing elements.

(b) Fuses of a suitable type and rating placed in each ungrounded conductor. Fuses shall be used with suitable disconnecting means or they shall be of a type that can also serve as the disconnecting means. They shall be arranged so that they cannot be serviced while they are energized.

(2) Fault-current interrupting devices shall not automatically reclose the circuit.

Exception: Automatic reclosing of a circuit shall be permitted where the circuit is exposed to transient faults and where such automatic reclosing does not create a hazard to persons.

(3) Overload protection and fault-current protection shall be permitted to be provided by the same device.

430-126. Rating of Motor Control Apparatus. The ultimate trip current of overcurrent (overload) relays or other motor-protective devices used shall not exceed 115 percent of the controller's continuous current rating. Where the motor branch-circuit disconnecting means is separate from the controller, the disconnecting means current rating shall not be less than the ultimate trip setting of the overcurrent relays in the circuit.

430-127. Disconnecting Means. The controller disconnecting means shall be capable of being locked in the open position.

L. Protection of Live Parts — All Voltages

430-131. General. Part L specifies that live parts shall be protected in a manner judged adequate for the hazard involved.

430-132. Where Required. Exposed live parts of motors and controllers operating at 50 volts or more between terminals shall be guarded against accidental contact by enclosure or by location as follows:

(1) By installation in a room or enclosure that is accessible only to qualified persons
(2) By installation on a suitable balcony, gallery, or platform, elevated and arranged so as to exclude unqualified persons.
(3) By elevation 8 ft (2.44 m) or more above the floor

Exception: Live parts of motors operating at more than 50 volts between terminals shall not require additional guarding for stationary motors that have commutators, collectors, and brush rigging located inside of motor-end brackets and not conductively connected to supply circuits operating at more than 150 volts to ground.

430-133. Guards for Attendants. Where live parts of motors or controllers operating at over 150 volts to ground are guarded against accidental contact only by location as specified in Section 430-132, and where adjustment or other attendance may be necessary during the operation of the apparatus, suitable insulating mats or platforms shall be provided so that the attendant cannot readily touch live parts unless standing on the mats or platforms.

FPN: For working space, see Sections 110-26 and 110-34.

M. Grounding — All Voltages

430-141. General. Part M specifies the grounding of exposed noncurrent-carrying metal parts, likely to become energized, of motor and controller frames to prevent a voltage above ground in the event of accidental contact between energized parts and frames. Insulation, isolation, or guarding are suitable alternatives to grounding of motors under certain conditions.

430-142. Stationary Motors. The frames of stationary motors shall be grounded under any of the following conditions:

(1) Where supplied by metal-enclosed wiring
(2) Where in a wet location and not isolated or guarded
(3) If in a hazardous (classified) location as covered in Articles 500 through 517
(4) If the motor operates with any terminal at over 150 volts to ground

Any motor in a wet location and subject to contact by personnel constitutes a serious hazard and, unless it is isolated, elevated, or guarded from reach, is required to be grounded.

Stationary motors are usually supplied by wiring that is enclosed in metal raceways, flexible metal conduit, or cables with metal sheaths. When effectively attached to the motor junction box or frame, the metal raceway or cable armor serves as the equipment grounding conductor. See Section 250-118 for more information on types of equipment grounding conductors.

Where the frame of the motor is not grounded, it shall be permanently and effectively insulated from the ground.

430-143. Portable Motors. The frames of portable motors that operate at over 150 volts to ground shall be guarded or grounded.

FPN No. 1: See Section 250-114(4) for grounding of portable appliances in other than residential occupancies.

FPN No. 2: See Section 250-138(b) for color of equipment grounding conductor.

430-144. Controllers. Controller enclosures shall be grounded regardless of voltage. Controller enclosures shall have means for attachment of an equipment grounding conductor termination in accordance with Section 250-8.

Exception: Enclosures attached to ungrounded portable equipment shall not be required to be grounded.

430-145. Method of Grounding. Where required, grounding shall be done in the manner specified in Article 250.

(a) Grounding Through Terminal Housings. Where the wiring to fixed motors is metal-enclosed cable or in metal raceways, junction boxes to house motor terminals shall be provided, and the armor of the cable or the metal raceways shall be connected to them in the manner specified in Article 250.

FPN: See Section 430-12(e) for grounding connection means required at motor terminal housings.

(b) Separation of Junction Box from Motor. The junction box required by (a) shall be permitted to be separated from the motor by not more than 6 ft (1.83 m), provided the leads to the motor are Type AC cable or armored cord or are stranded leads enclosed in liquidtight flexible metal conduit, flexible metal conduit, intermediate metal conduit, rigid metal conduit, or electrical metallic tubing not smaller than 3⁄8-in. electrical trade size, the armor or raceway being connected both to the motor and to the box.

Liquidtight flexible nonmetallic conduit and rigid nonmetallic conduit shall be permitted to enclose the leads to the motor provided the leads are stranded and the required equipment grounding conductor is connected to both the motor and to the box.

Where stranded leads are used, protected as specified above, they shall not be larger than No. 10 and shall comply with other requirements of this *Code* for conductors to be used in raceways.

(c) Grounding of Controller Mounted Devices. Instrument transformer secondaries and exposed noncurrent-carrying metal or other conductive parts or cases of instrument transformers, meters, instruments, and relays shall be grounded as specified in Sections 250-170 through 250-178.

Most motors are subject to vibration, and good wiring practice requires that, in nearly all cases, the wiring to motors that are fixed be installed with a short section (not more than 6 ft) of liquidtight flexible metal or nonmetallic conduit or of flexible metal conduit to the motor terminal housing. Such use of flexible conduit requires an equipment grounding conductor.

N. Tables

Tables 430-147 through 430-150 accurately reflect the typical and most used 4-pole and 2-pole induction motors in use.

Table 430-151 in the 1993 *Code* was revised for the 1996 *Code*. It was divided into two tables, Table 430-151(A) for single-phase motors and Table 430-151(B) for 2- and 3-phase motors, and was changed to accommodate newly developed Design E motors. [Table 430-151(B) now includes two columns: one for Design B, C, and D motors and one for Design E motors.]

Table 430-147. Full-Load Current in Amperes, Direct-Current Motors

The following values of full-load currents* are for motors running at base speed.

	Armature Voltage Rating*					
Horsepower	90 Volts	120 Volts	180 Volts	240 Volts	500 Volts	550 Volts
1⁄4	4.0	3.1	2.0	1.6	—	—
1⁄3	5.2	4.1	2.6	2.0	—	—
1⁄2	6.8	5.4	3.4	2.7	—	—
3⁄4	9.6	7.6	4.8	3.8	—	—
1	12.2	9.5	6.1	4.7	—	—
1 1⁄2	—	13.2	8.3	6.6	—	—
2	—	17	10.8	8.5	—	—
3	—	25	16	12.2	—	—
5	—	40	27	20	—	—
7 1⁄2	—	58	—	29	13.6	12.2
10	—	76	—	38	18	16
15	—	—	—	55	27	24
20	—	—	—	72	34	31
25	—	—	—	89	43	38
30	—	—	—	106	51	46
40	—	—	—	140	67	61
50	—	—	—	173	83	75
60	—	—	—	206	99	90
75	—	—	—	255	123	111
100	—	—	—	341	164	148
125	—	—	—	425	205	185
150	—	—	—	506	246	222
200	—	—	—	675	330	294

*These are average dc quantities.

Table 430-148. Full-Load Currents in Amperes, Single-Phase Alternating-Current Motors

The following values of full-load currents are for motors running at usual speeds and motors with normal torque characteristics. Motors built for especially low speeds or high torques may have higher full-load currents, and multispeed motors will have full-load current varying with speed, in which case the namplate current ratings shall be used.

The voltages listed are rated motor voltages. The currents listed shall be permitted for system voltage ranges of 110 to 120 and 220 to 240 volts.

Horsepower	115 Volts	200 Volts	208 Volts	230 Volts
⅙	4.4	2.5	2.4	2.2
¼	5.8	3.3	3.2	2.9
⅓	7.2	4.1	4.0	3.6
½	9.8	5.6	5.4	4.9
¾	13.8	7.9	7.6	6.9
1	16	9.2	8.8	8.0
1½	20	11.5	11.0	10
2	24	13.8	13.2	12
3	34	19.6	18.7	17
5	56	32.2	30.8	28
7½	80	46.0	44.0	40
10	100	57.5	55.0	50

Table 430-149. Full-Load Current, Two-Phase Alternating-Current Motors (4-Wire)

The following values of full-load current are for motors running at speeds usual for belted motors and motors with normal torque characteristics. Motors built for especially low speeds or high torques may require more running current, and multispeed motors will have full-load current varying with speed, in which case the nameplate current rating shall be used. Current in the common conductor of a 2-phase, 3-wire system will be 1.41 times the value given.

The voltages listed are rated motor voltages. The currents listed shall be permitted for system voltage ranges of 110 to 120, 220 to 240, 440 to 480, and 550–600 volts.

	Induction Type Squirrel Cage and Wound Rotor (Amperes)				
Horsepower	115 Volts	230 Volts	460 Volts	575 Volts	2300 Volts
½	4.0	2.0	1.0	0.8	—
¾	4.8	2.4	1.2	1.0	—
1	6.4	3.2	1.6	1.3	—
1½	9.0	4.5	2.3	1.8	—
2	11.8	5.9	3.0	2.4	—
3	—	8.3	4.2	3.3	—
5	—	13.2	6.6	5.3	—
7½	—	19	9.0	8.0	—
10	—	24	12	10	—
15	—	36	18	14	—
20	—	47	23	19	—
25	—	59	29	24	—
30	—	69	35	28	—
40	—	90	45	36	—
50	—	113	56	45	—
60	—	133	67	53	14
75	—	166	83	66	18
100	—	218	109	87	23
125	—	270	135	108	28
150	—	312	156	125	32
200	—	416	208	167	43

Table 430-150. Full-Load Current Three-Phase Alternating-Current Motors

The following values of full-load currents are typical for motors running at speeds usual for belted motors and motors with normal torque characteristics.

Motors built for low speeds (1200 rpm or less) or high torques may require more running current, and multispeed motors will have full-load current varying with speed. In these cases, the nameplate current rating shall be used.

The voltages listed are rated motor voltages. The currents listed shall be permitted for system voltage ranges of 110 to 120, 220 to 240, 440 to 480, and 550 to 600 volts.

	Induction Type Squirrel Cage and Wound Rotor (Amperes)							Synchronous-Type Unity Power Factor* (Amperes)			
Horsepower	115 Volts	200 Volts	208 Volts	230 Volts	460 Volts	575 Volts	2300 Volts	230 Volts	460 Volts	575 Volts	2300 Volts
½	4.4	2.5	2.4	2.2	1.1	0.9	—	—	—	—	—
¾	6.4	3.7	3.5	3.2	1.6	1.3	—	—	—	—	—
1	8.4	4.8	4.6	4.2	2.1	1.7	—	—	—	—	—
1½	12.0	6.9	6.6	6.0	3.0	2.4	—	—	—	—	—
2	13.6	7.8	7.5	6.8	3.4	2.7	—	—	—	—	—
3	—	11.0	10.6	9.6	4.8	3.9	—	—	—	—	—
5	—	17.5	16.7	15.2	7.6	6.1	—	—	—	—	—
7½	—	25.3	24.2	22	11	9	—	—	—	—	—
10	—	32.2	30.8	28	14	11	—	—	—	—	—
15	—	48.3	46.2	42	21	17	—	—	—	—	—
20	—	62.1	59.4	54	27	22	—	—	—	—	—
25	—	78.2	74.8	68	34	27	—	53	26	21	—
30	—	92	88	80	40	32	—	63	32	26	—
40	—	120	114	104	52	41	—	83	41	33	—
50	—	150	143	130	65	52	—	104	52	42	—
60	—	177	169	154	77	62	16	123	61	49	12
75	—	221	211	192	96	77	20	155	78	62	15
100	—	285	273	248	124	99	26	202	101	81	20
125	—	359	343	312	156	125	31	253	126	101	25
150	—	414	396	360	180	144	37	302	151	121	30
200	—	552	528	480	240	192	49	400	201	161	40
250	—	—	—	—	302	242	60	—	—	—	—
300	—	—	—	—	361	289	72	—	—	—	—
350	—	—	—	—	414	336	83	—	—	—	—
400	—	—	—	—	477	382	95	—	—	—	—
450	—	—	—	—	515	412	103	—	—	—	—
500	—	—	—	—	590	472	118	—	—	—	—

*For 90 and 80 percent power factor, the figures shall be multiplied by 1.1 and 1.25, respectively.

Table 430-151(A). Conversion Table of Single-Phase Locked-Rotor Currents for Selection of Disconnecting Means and Controllers as Determined from Horsepower and Voltage Rating
For use only with Sections 430-110, 440-12, 440-41, and 455-8(c).

Rated Horsepower	Maximum Locked-Rotor Current in Amperes, Single Phase		
	115 Volts	208 Volts	230 Volts
½	58.8	32.5	29.4
¾	82.8	45.8	41.4
1	96	53	48
1½	120	66	60
2	144	80	72
3	204	113	102
5	336	186	168
7½	480	265	240
10	600	332	300

Table 430-151(B). Conversion Table of Polyphase Design B, C, D, and E Maximum Locked-Rotor Currents for Selection of Disconnecting Means and Controllers as Determined from Horsepower and Voltage Rating and Design Letter
For use only with Sections 430-110, 440-12,* 440-41,* and 455-8(c).

Rated Horsepower	Maximum Motor Locked-Rotor Current in Amperes Two- and Three-Phase, Design B, C, D and E											
	115 Volts		200 Volts		208 Volts		230 Volts		460 Volts		575 Volts	
	B, C, D	E	B, C, D	E	B, C, D	E	B, C, D	E	B, C, D	E	B, C, D	E
½	40	40	23	23	22.1	22.1	20	20	10	10	8	8
¾	50	50	28.8	28.8	27.6	27.6	25	25	12.5	12.5	10	10
1	60	60	34.5	34.5	33	33	30	30	15	15	12	12
1½	80	80	46	46	44	44	40	40	20	20	16	16
2	100	100	57.5	57.5	55	55	50	50	25	25	20	20
3	—	—	73.6	84	71	81	64	73	32	36.5	25.6	29.2
5	—	—	105.8	140	102	135	92	122	46	61	36.8	48.8
7½	—	—	146	210	140	202	127	183	63.5	91.5	50.8	73.2
10	—	—	186.3	259	179	249	162	225	81	113	64.8	90
15	—	—	267	388	257	373	232	337	116	169	93	135
20	—	—	334	516	321	497	290	449	145	225	116	180
25	—	—	420	646	404	621	365	562	183	281	146	225
30	—	—	500	775	481	745	435	674	218	337	174	270
40	—	—	667	948	641	911	580	824	290	412	232	330
50	—	—	834	1185	802	1139	725	1030	363	515	290	412
60	—	—	1001	1421	962	1367	870	1236	435	618	348	494
75	—	—	1248	1777	1200	1708	1085	1545	543	773	434	618
100	—	—	1668	2154	1603	2071	1450	1873	725	937	580	749
125	—	—	2087	2692	2007	2589	1815	2341	908	1171	726	936
150	—	—	2496	3230	2400	3106	2170	2809	1085	1405	868	1124
200	—	—	3335	4307	3207	4141	2900	3745	1450	1873	1160	1498
250	—	—	—	—	—	—	—	—	1825	2344	1460	1875
300	—	—	—	—	—	—	—	—	2200	2809	1760	2247
350	—	—	—	—	—	—	—	—	2550	3277	2040	2622
400	—	—	—	—	—	—	—	—	2900	3745	2320	2996
450	—	—	—	—	—	—	—	—	3250	4214	2600	3371
500	—	—	—	—	—	—	—	—	3625	4682	2900	3746

*In determining compliance with Sections 440-12 and 440-41, the values in the B, C, D columns shall be used.

Table 430-152. Maximum Rating or Setting of Motor Branch-Circuit Short-Circuit and Ground-Fault Protective Devices

	Percentage of Full-Load Current			
Type of Motor	Nontime Delay Fuse[1]	Dual Element (Time-Delay) Fuse[1]	Instantaneous Trip Breaker	Inverse Time Breaker[2]
Single-phase motors	300	175	800	250
AC polyphase motors other than wound-rotor				
Squirrel cage—				
Other than Design E	300	175	800	250
Design E	300	175	1100	250
Synchronous[3]	300	175	800	250
Wound rotor	150	150	800	150
Direct current (constant voltage)	150	150	250	150

Note: For certain exceptions to the values specified, see Sections 430-52 through 430-54.

[1]The values in the Nontime Delay Fuse column apply to Time-Delay Class CC fuses.

[2]The values given in the last column also cover the ratings of nonadjustable inverse time types of circuit breakers that may be modified as in Section 430-52.

[3]Synchronous motors of the low-torque, low-speed type (usually 450 rpm or lower), such as are used to drive reciprocating compressors, pumps, etc., that start unloaded, do not require a fuse rating or circuit-breaker setting in excess of 200 percent of full-load current.

Although Class CC fuses are rated as time delay, because they are so fast acting, they are permitted to be sized according to the requirements of non-time-delay rated fuses. Examples of Class CC fuses are shown in Figure 430.27.

Figure 430.27 *Class CC fuses. (Bussmann Division, Cooper Industries)*

Article 440 — Air-Conditioning and Refrigerating Equipment

Contents

A. General

440-1. Scope. The provisions of this article apply to electric motor-driven air-conditioning and refrigerating equipment, and to the branch circuits and controllers for such equipment. It provides for the special considerations necessary for circuits supplying hermetic refrigerant motor-compressors and for any air-conditioning or refrigerating equipment that is supplied from a branch circuit that supplies a hermetic refrigerant motor-compressor.

440-2. Definitions.

Branch-Circuit Selection Current. *Branch-circuit selection current* is the value in amperes to be used instead of the rated-load current in determining the ratings of motor branch-circuit conductors, disconnecting means, controllers, and branch-circuit short-circuit and ground-fault protective devices wherever the running overload protective device permits a sustained current greater than the specified percentage of the rated-load current. The value of branch-circuit selection current will always be equal to or greater than the marked rated-load current.

Hermetic Refrigerant Motor-Compressor. A combination consisting of a compressor and motor, both of which are enclosed in the same housing, with no external shaft or shaft seals, the motor operating in the refrigerant.

Rated-Load Current. The rated-load current for a hermetic refrigerant motor-compressor is the current resulting when the motor-compressor is operated at the rated load, rated voltage, and rated frequency of the equipment it serves.

440-3. Other Articles.

(a) Article 430. These provisions are in addition to, or amendatory of, the provisions of Article 430 and other articles in this *Code*, which apply except as modified in this article.

(b) Articles 422, 424, or 430. The rules of Articles 422, 424, or 430, as applicable, shall apply to air-conditioning and refrigerating equipment that does not incorporate a hermetic refrigerant motor-compressor. This equipment includes devices that employ refrigeration compressors driven by conventional motors, furnaces with air-conditioning evaporator coils installed, fan-coil units, remote forced air-cooled condensers, remote commercial refrigerators, etc.

(c) Article 422. Devices such as room air conditioners, household refrigerators and freezers, drinking water coolers, and beverage dispensers shall be considered appliances, and the provisions of Article 422 shall also apply.

(d) Other Applicable Articles. Hermetic refrigerant motor-compressors, circuits, controllers, and equipment shall also comply with the applicable provisions of the following:

	Article	Section
Capacitors		460-9
Commercial garages, aircraft hangars, gasoline dispensing and service stations, bulk storage plants, spray application, dipping, and coating processes, and inhalation anesthetizing locations	511, 513, 514, 515, 516, and 517, Part D	
Hazardous (classified) locations	500–503 and 505	
Motion picture and television studios and similar locations	530	
Resistors and reactors	470	

Article 440 provides special considerations necessary for circuits supplying *hermetic* refrigerant motor-compressors and is in addition to or amendatory of the provisions of Article 430 and other applicable articles. However, many requirements, such as disconnecting means, controllers, single or group installations, and sizing of conductors, are the same as or similar to those applied in Article 430.

Article 440 does not apply unless a hermetic refrigerant motor-compressor is supplied. Article 440 must be applied in conjunction with Article 430.

Note the terms *rated-load current* and *branch-circuit selection current,* defined in Section 440-2. When a branch-circuit selection current is marked on a nameplate, it is required to be used instead of the rated-load current to determine the size of the disconnecting means, the controller, the motor branch-circuit conductors, and the overcurrent protective devices for the branch-circuit conductors and the motor. The value of branch-circuit selection current will always be greater than the marked rated-load current.

440-4. Marking on Hermetic Refrigerant Motor-Compressors and Equipment.

(a) Hermetic Refrigerant Motor-Compressor Nameplate. A hermetic refrigerant motor-compressor shall be provided with a nameplate that shall indicate the manufacturer's name, trademark, or symbol; identifying designation; phase; voltage; and frequency. The rated-load current in amperes of the motor-compressor shall be marked by the equipment manufacturer on either or both the motor-compressor nameplate and the nameplate of the equipment in which the motor-compressor is used. The locked-rotor current of each single-phase motor-compressor having a rated-load current of more than 9 amperes at 115 volts, or more than 4.5 amperes at 230 volts, and each polyphase motor-compressor shall be marked on the motor-compressor nameplate. Where a thermal protector complying with Sections 440-52(a)(2) and (b)(2) is used, the motor-compressor nameplate or the equipment nameplate shall be marked with the words "thermally protected." Where a protective system complying with Sections 440-52(a)(4) and (b)(4) is used and is furnished with the equipment, the equipment nameplate shall be marked with the words, "thermally protected system." Where a protective system complying with Sections 440-52(a)(4) and (b)(4) is specified, the equipment nameplate shall be appropriately marked.

(b) Multimotor and Combination-Load Equipment. Multimotor and combination-load equipment shall be provided with a visible nameplate marked with the maker's name, the rating in volts, frequency and number of phases, minimum supply circuit conductor ampacity, and the maximum rating of the branch-circuit short-circuit and ground-fault protective device. The ampacity shall be calculated by using Part D and counting all the motors and other loads that will be operated at the same time. The branch-circuit short-circuit and ground-fault protective device rating shall not exceed the value calculated by using Part C. Multimotor or combination-load equipment for use on two or more circuits shall be marked with the above information for each circuit.

Exception No. 1: Multimotor and combination-load equipment that is suitable under the provisions of this article for connection to a single 15- or 20-ampere, 120-volt, or a 15-ampere, 208- or 240-volt, single-phase branch circuit shall be permitted to be marked as a single load.

Exception No. 2: The minimum supply circuit conductor ampacity and the maximum rating of the branch-circuit short-circuit and ground-fault protective device shall not be required to be marked on a room air conditioner conforming with Section 440-62(a).

(c) Branch-Circuit Selection Current. A hermetic refrigerant motor-compressor, or equipment containing such a compressor, having a protection system that is approved for use with the motor-compressor that it protects and that permits continuous current in excess of the specified percentage of nameplate rated-load current given in Section 440-52(b)(2) or (b)(4), shall also be marked with a branch-circuit selection current that complies with Section 440-52(b)(2) or (b)(4). This marking shall be provided by the equipment manufacturer and shall be on the nameplate(s) where the rated-load current(s) appears.

440-5. Marking on Controllers. A controller shall be marked with the manufacturer's name, trademark, or symbol; identifying designation; voltage; phase; full-load and locked-rotor current (or horsepower) rating; and such other data as may be needed to properly indicate the motor-compressor for which it is suitable.

440-6. Ampacity and Rating. The size of conductors for equipment covered by this article shall be selected from Tables 310-16 through 310-19 or calculated in accordance with Section 310-15 as applicable. The required ampacity of conductors and rating of equipment shall be determined as follows.

(a) Hermetic Refrigerant Motor-Compressor. For a hermetic refrigerant motor-compressor, the rated-load current marked on the nameplate of the equipment in which the motor-compressor is employed shall be used in determining the rating or ampacity of the disconnecting means, the branch-circuit conductors, the controller, the branch-circuit short-circuit and ground-fault protection, and the separate motor overload protection. Where no rated-load current is shown on the equipment nameplate, the rated-load current shown on the compressor nameplate shall be used. For disconnecting means and controllers, see also Sections 440-12 and 440-41.

Exception No. 1: Where so marked, the branch-circuit selection current shall be used instead of the rated-load current to determine the rating or ampacity of the disconnecting means, the branch-circuit conductors, the controller, and the branch-circuit short-circuit and ground-fault protection.

Exception No. 2: For cord- and plug-connected equipment, the nameplate marking shall be used in accordance with Section 440-22(b), Exception No. 2.

(b) Multimotor Equipment. For multimotor equipment employing a shaded-pole or permanent split-capacitor-type fan or blower motor, the full-load current for such motor marked on the nameplate of the equipment in which the fan or blower motor is employed shall be used instead of the horsepower rating to determine the ampacity or rating of the disconnecting means, the branch-circuit conductors, the controller, the branch-circuit short-circuit and ground-fault protection, and the separate overload protection. This marking on the equipment nameplate shall not be less than the current marked on the fan or blower motor nameplate.

440-7. Highest Rated (Largest) Motor. In determining compliance with this article and with Sections 430-24, 430-53(b) and (c), and 430-62(a), the highest rated (largest) motor shall be considered to be the motor that has the highest rated-load current. Where two or more motors have the same highest rated-load current, only one of them shall be considered as the highest rated (largest) motor. For other than hermetic refrigerant motor-compressors, and fan or blower motors as covered in Section 440-6(b), the full-load current used to determine the highest rated motor shall be the equivalent value corresponding to the motor horsepower rating selected from Tables 430-148, 430-149, or 430-150.

Exception: Where so marked, the branch-circuit selection current shall be used instead of the rated-load current in determining the highest rated (largest) motor-compressor.

440-8. Single Machine. An air-conditioning or refrigerating system shall be considered to be a single machine under the provisions of Section 430-87, Exception, and Section 430-112, Exception. The motors shall be permitted to be located remotely from each other.

B. Disconnecting Means

440-11. General. The provisions of Part B are intended to require disconnecting means capable of disconnecting air-conditioning and refrigerating equipment including motor-compressors and controllers from the circuit conductors. See Figure 430-1.

440-12. Rating and Interrupting Capacity.

(a) Hermetic Refrigerant Motor-Compressor. A disconnecting means serving a hermetic refrigerant motor-compressor shall be selected on the basis of the nameplate rated-load current or branch-circuit selection current, whichever is greater, and locked-rotor current, respectively, of the motor-compressor as follows.

(1) The ampere rating shall be at least 115 percent of the nameplate rated-load current or branch-circuit selection current, whichever is greater.

(2) To determine the equivalent horsepower in complying with the requirements of Section 430-109, the horsepower rating shall be selected from Tables 430-148, 430-149, or 430-150 corresponding to the rated-load current or branch-circuit selection current, whichever is greater, and also the horsepower rating from Tables 430-151(A) or 430-151(B) corresponding to the locked-rotor current. In case the nameplate rated-load current or branch-circuit selection current and locked-rotor current do not correspond to the currents shown in Tables 430-148, 430-149, 430-150, 430-151(A), or 430-151(B), the horsepower rating corresponding to the next higher value shall be selected. In case different horsepower ratings are obtained when applying these tables, a horsepower rating at least equal to the larger of the values obtained shall be selected.

(b) Combination Loads. Where one or more hermetic refrigerant motor-compressors are used together or are used in combination with other motors or loads, and where the combined load may be simultaneous on a single disconnecting means, the rating for the combined load shall be determined as follows.

(1) The horsepower rating of the disconnecting means shall be determined from the sum of all currents, including resistance loads, at the rated-load condition and also at the locked-rotor condition. The combined rated-load current and the combined locked-rotor current so obtained shall be considered as a single motor for the purpose of this requirement as follows.

(a) The full-load current equivalent to the horsepower rating of each motor, other than a hermetic refrigerant motor-compressor, and fan or blower motors as covered in Section 440-6(b) shall be selected from Tables 430-148, 430-149, or 430-150. These full-load currents shall be added to the motor-compressor rated-load current(s) or branch-circuit selection current(s), whichever is greater, and to the rating in amperes of other loads to obtain an equivalent full-load current for the combined load.

(b) The locked-rotor current equivalent to the horsepower rating of each motor, other than a hermetic refrigerant motor-compressor, shall be selected from Table 430-151(A) or 430-151(B), and, for fan and blower motors of the shaded-pole or permanent split-capacitor type marked with the locked-rotor current, the marked value shall be used. The locked-rotor currents shall be added to the motor-compressor locked-rotor current(s) and to the rating in amperes of other loads to obtain an equivalent locked-rotor current for the combined load. Where two or more motors or other loads such as resistance heaters, or both, cannot be started simultaneously, appropriate combinations of locked-rotor and rated-load current or branch-circuit selection current, whichever is greater, shall be an acceptable means of determining the equivalent locked-rotor current for the simultaneous combined load.

Exception: Where part of the concurrent load is a resistance load and the disconnecting means is a switch rated in horsepower and amperes, the switch used shall be permitted to have a horsepower rating not less than the combined load to the motor-compressor(s) and other motor(s) at the locked-

rotor condition, if the ampere rating of the switch is not less than this locked-rotor load plus the resistance load.

(2) The ampere rating of the disconnecting means shall be at least 115 percent of the sum of all currents at the rated-load condition determined in accordance with Section 440-12(b)(1).

(c) Small Motor-Compressors. For small motor-compressors not having the locked-rotor current marked on the nameplate, or for small motors not covered by Tables 430-147, 430-148, 430-149, or 430-150, the locked-rotor current shall be assumed to be six times the rated-load current. See Section 440-3(a).

(d) Every Switch. Every disconnecting means in the refrigerant motor-compressor circuit between the point of attachment to the feeder and the point of connection to the refrigerant motor-compressor shall comply with the requirements of Section 440-12.

(e) Disconnecting Means Rated in Excess of 100 Horsepower. Where the rated-load or locked-rotor current as determined above would indicate a disconnecting means rated in excess of 100 hp, the provisions of Section 430-109(e) shall apply.

440-13. Cord-Connected Equipment. For cord-connected equipment such as room air conditioners, household refrigerators and freezers, drinking water coolers, and beverage dispensers, a separable connector or an attachment plug and receptacle shall be permitted to serve as the disconnecting means. See also Section 440-63.

440-14. Location. Disconnecting means shall be located within sight from and readily accessible from the air-conditioning or refrigerating equipment. The disconnecting means shall be permitted to be installed on or within the air-conditioning or refrigerating equipment.

Exception No. 1: Where the disconnecting means provided in accordance with Section 430-102(a) is capable of being locked in the open position, and the refrigerating or air conditioning equipment is essential to an industrial process in a facility where the conditions of maintenance and the supervision ensure that only qualified persons will service the equipment, a disconnecting means within sight from the equipment shall not be required.

Exception No. 1 was added to the 1996 *Code* to accommodate special conditions associated with process refrigeration equipment. Typically, this equipment is very large, so rated disconnects may not be available. Additionally, this equipment may be in hazardous locations, and locating disconnecting means within sight of the motor may introduce additional hazards.

Exception No. 2: Where an attachment plug and receptacle serve as the disconnecting means in accordance with Section 440-13, their location shall be accessible, but shall not be required to be readily accessible.

FPN: See Parts G and J of Article 430 for additional requirements.

The reference to Parts G and J of Article 430 in the fine print note is intended to call attention to the additional disconnect location requirements in Sections 430-102, 430-107, and 430-113. Since Section 440-3(a) makes the requirements in Article 440 in addition to or amendatory of the provisions of Article 430, the requirement of Section 440-14 mandates that the equipment disconnecting means be within sight from and readily accessible from the equipment, even if there is also a remote disconnect capable of being locked in the open position under the provision of Section 430-102(b), Exception. This special requirement for air-conditioning and refrigeration equipment covered by Article 440 is more stringent than the provisions in Article 430 to provide protection for service personnel working on equipment located in attics, on roofs, or outside in a remote location where it is difficult to gain access to a remote lockable disconnect. See Section 440-14, Exception No. 1.

C. Branch-Circuit Short-Circuit and Ground-Fault Protection

440-21. General. The provisions of Part C specify devices intended to protect the branch-circuit conductors, control apparatus, and motors in circuits supplying hermetic refrigerant motor-compressors against overcurrent due to short circuits and grounds. They are in addition to or amendatory of the provisions of Article 240.

Where an air conditioner is listed by a qualified electrical testing laboratory with a nameplate that reads "maximum fuse size," the listing restricts the use of this unit to fuse protection only and does not cover its use with circuit breakers. If the air conditioner has been evaluated for both fuses and ordinary circuit breakers, or both fuses and HACR-type circuit breakers, it may be so marked. UL-listed circuit breakers that have been found suitable for use with heating, air-conditioning, and refrigeration equipment comprising multimotor or combination loads are marked "Listed HACR Type." It is the intent of Section 110-3(b) to require that the manufacturer's installation specifications be closely followed and that any restriction of the listing be applied to the installation of the equipment, in order to comply with the *Code.*

The 1998 UL *Electrical Appliance and Utilization Equipment Directory* states the following under Air-Conditioners, Central Cooling (ACAV): "This marked protective device rating is the maximum for which the equipment has been investigated and found acceptable. Where the marking specifies fuses, or 'HACR Type' circuit breakers, the circuit is intended to be protected only by the type of protective device specified." Figure 440.1 illustrates three wiring configurations where the equipment is under fuse protection only.

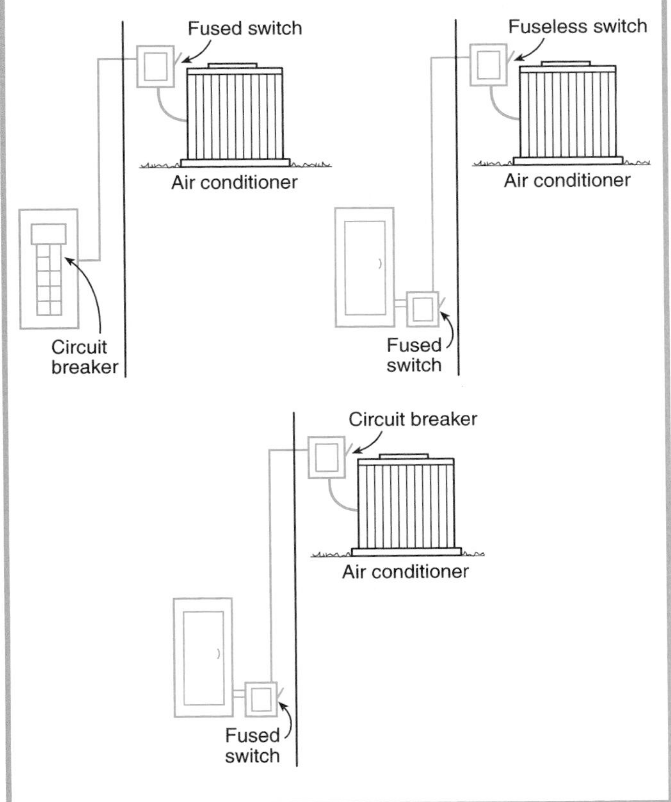

Figure 440.1 *Three correct alternate wiring configurations satisfying a nameplate that specifies fuses, thus restricting the equipment to protection by fuses only.*

The 1998 UL *Electrical Appliance and Utilization Equipment Directory* further indicates that "in air conditioners employing two or more motors or a motor(s) and other loads operating from a single supply circuit, the motor overload protective devices (including thermal protectors for motors) and other factory-installed motor circuit components and wiring are investigated on the basis of compliance with the motor branch-circuit short-circuit and ground-fault protection requirements of Section 430-53(c), Other Group Installations. Such multimotor and combination load equipment is to be connected only to a circuit protected by fuses or a circuit breaker with a rating that does not exceed the value marked on the data plate of the equipment.

Current-limiting overcurrent devices, which may reduce the amount of fault current to which the equipment will be subjected, can be installed in the branch circuit supplying the equipment. See Section 240-11 for the definition of *current-limiting device* and commentary explaining short-circuit damage. This is important particularly for the larger commercial and industrial installations. See also Sections 110-3(b) and 110-10 regarding the installation and use of listed or labeled equipment and the selection of overcurrent protective devices (such as fuses and circuit breakers).

440-22. Application and Selection.

(a) Rating or Setting for Individual Motor-Compressor. The motor-compressor branch-circuit short-circuit and ground-fault protective device shall be capable of carrying the starting current of the motor. A protective device having a rating or setting not exceeding 175 percent of the motor-compressor rated-load current or branch-circuit selection current, whichever is greater, shall be permitted, provided that, where the protection specified is not sufficient for the starting current of the motor, the rating or setting shall be permitted to be increased, but shall not exceed 225 percent of the motor rated-load current or branch-circuit selection current, whichever is greater.

Exception: The rating of the branch-circuit short-circuit and ground-fault protective device shall not be required to be less than 15 amperes.

(b) Rating or Setting for Equipment. The equipment branch-circuit short-circuit and ground-fault protective device shall be capable of carrying the starting current of the equipment. Where the hermetic refrigerant motor-compressor is the only load on the circuit, the protection shall conform with Section 440-22(a). Where the equipment incorporates more than one hermetic refrigerant motor-compressor or a hermetic refrigerant motor-compressor and other motors or other loads, the equipment short-circuit and ground-fault protection shall conform with Section 430-53 and the following.

(1) Where a hermetic refrigerant motor-compressor is the largest load connected to the circuit, the rating or setting of the branch-circuit short-circuit and ground-fault protective device shall not exceed the value specified in Section 440-22(a) for the largest motor-compressor plus the sum of the rated-load current or branch-circuit selection current, whichever is greater, of the other motor-compressor(s) and the ratings of the other loads supplied.

(2) Where a hermetic refrigerant motor-compressor is not the largest load connected to the circuit, the rating or setting of the branch-circuit short-circuit and ground-fault protective device shall not exceed a value equal to the sum of the rated-load current or branch-circuit selection current, whichever is greater, rating(s) for the motor-compressor(s) plus the value specified in Section 430-53(c)(4) where other motor loads are supplied, or the value specified in Section 240-3 where only nonmotor loads are supplied in addition to the motor-compressor(s).

Exception No. 1: Equipment that will start and operate on a 15- or 20-ampere 120-volt, or 15-ampere 208- or 240-volt single-phase branch circuit, shall be permitted to be protected by the 15- or 20-ampere overcurrent device protecting the branch circuit, but if the maximum branch-circuit short-circuit and ground-fault protective device rating marked on the equipment is less than these values, the circuit protective device shall not exceed the value marked on the equipment nameplate.

Exception No. 2: The nameplate marking of cord- and plug-connected equipment rated not greater than 250 volts, single-phase, such as household refrigerators and freezers, drinking water coolers, and beverage dispensers, shall be used in determining the branch-circuit requirements, and each unit shall be considered as a single motor unless the nameplate is marked otherwise.

(c) Protective Device Rating Not to Exceed the Manufacturer's Values. Where maximum protective device ratings shown on a manufacturer's heater table for use with a motor controller are less than the rating or setting selected in accordance with Sections 440-22(a) and (b), the protective device rating shall not exceed the manufacturer's values marked on the equipment.

D. Branch-Circuit Conductors

440-31. General. The provisions of Part D and Article 310 specify ampacities of conductors required to carry the motor current without overheating under the conditions specified, except as modified in Section 440-6(a), Exception No. 1.

The provisions of these articles shall not apply to integral conductors of motors, motor controllers and the like, or to conductors that form an integral part of approved equipment.

FPN: See Sections 300-1(b) and 310-1 for similar requirements.

440-32. Single Motor-Compressor. Branch-circuit conductors supplying a single motor-compressor shall have an ampacity not less than 125 percent of either the motor-compressor rated-load current or the branch-circuit selection current, whichever is greater.

440-33. Motor-Compressor(s) With or Without Additional Motor Loads. Conductors supplying one or more motor-compressor(s) with or without an additional load(s) shall have an ampacity not less than the sum of the rated-load or branch-circuit selection current ratings, whichever is larger, of all the motor-compressors plus the full-load currents of the other motors, plus 25 percent of the highest motor or motor-compressor rating in the group.

Exception No. 1: Where the circuitry is interlocked so as to prevent the starting and running of a second motor-compressor or group of motor-compressors, the conductor size shall be determined from the largest motor-compressor or group of motor-compressors that is to be operated at a given time.

Exception No. 2: The branch circuit conductors for room air conditioners shall be in accordance with Part G of Article 440.

Branch circuits for listed air-conditioning and refrigeration equipment that have a nameplate marked with the branch-circuit conductor size and branch-circuit short-circuit protective device size are not required to have the branch-circuit conductors sized in accordance with Section 440-33. The testing laboratory standard includes the 25 percent increase for the largest motor or compressor in the group plus the other nonmotor or noncompressor load; therefore, the actual nameplate full-load amperes for the complete assembly can be used to size the branch-circuit conductors.

440-34. Combination Load. Conductors supplying a motor-compressor load in addition to a lighting or appliance load as computed from Article 220 and other applicable articles shall have an ampacity sufficient for the lighting or appliance load plus the required ampacity for the motor-compressor load determined in accordance with Section 440-33, or, for a single motor-compressor, in accordance with Section 440-32.

Exception: Where the circuitry is interlocked so as to prevent simultaneous operation of the motor-compressor(s) and all other loads connected, the conductor size shall be determined from the largest size required for the motor-compressor(s) and other loads to be operated at a given time.

440-35. Multimotor and Combination-Load Equipment. The ampacity of the conductors supplying multimotor and combination-load equipment shall not be less than the minimum circuit ampacity marked on the equipment in accordance with Section 440-4(b).

E. Controllers for Motor-Compressors

440-41. Rating.

(a) Motor-Compressor Controller. A motor-compressor controller shall have both a continuous-duty full-load current rating and a locked-rotor current rating, not less than the nameplate rated-load current or branch-circuit selection current, whichever is greater, and locked-rotor current, respectively of the compressor (see Sections 440-6 and 440-7). In case the motor controller is rated in horsepower, but is without one or both of the foregoing current ratings, equivalent currents shall be determined from the ratings as follows. Use Tables 430-148, 430-149, or 430-150 to determine the equivalent full-load current rating. Use Tables 430-151(A) or 430-151(B) to determine the equivalent locked-rotor current ratings.

(b) Controller Serving More than One Load. A controller, serving more than one motor-compressor or a motor-compressor and other loads, shall have a continuous-duty full-load current rating and a locked-rotor current rating not less than the combined load as determined in accordance with Section 440-12(b).

F. Motor-Compressor and Branch-Circuit Overload Protection

440-51. General. The provisions of Part F specify devices intended to protect the motor-compressor, the motor-control apparatus, and the branch-circuit conductors against excessive heating due to motor overload and failure to start. See Sections 240-3(g) and 430-31.

440-52. Application and Selection.

(a) Protection of Motor-Compressor. Each motor-compressor shall be protected against overload and failure to start by one of the following means.

(1) A separate overload relay that is responsive to motor-compressor current. This device shall be selected to trip at not more than 140 percent of the motor-compressor rated-load current.

(2) A thermal protector integral with the motor-compressor, approved for use with the motor-compressor that it protects on the basis that it will prevent dangerous overheating of the motor-compressor due to overload and failure to start. If the current-interrupting device is separate from the motor-compressor and its control circuit is operated by a protective device integral with the motor-compressor, it shall be arranged so that the opening of the control circuit will result in interruption of current to the motor-compressor.

(3) A fuse or inverse time circuit breaker responsive to motor current, which shall also be permitted to serve as the branch-circuit short-circuit and ground-fault protective device. This device shall be rated at not more than 125 percent of the motor-compressor rated-load current. It shall have sufficient time delay to permit the motor-compressor to start and accelerate its load. The equipment or the motor-compressor shall be marked with this maximum branch-circuit fuse or inverse time circuit breaker rating.

(4) A protective system, furnished or specified and approved for use with the motor-compressor that it protects on the basis that it will prevent dangerous overheating of the motor-compressor due to overload and failure to start. If the current-interrupting device is separate from the motor-compressor and its control circuit is operated by a protective device that is not integral with the current-interrupting device, it shall be arranged so that the opening of the control circuit will result in interruption of current to the motor-compressor.

(b) Protection of Motor-Compressor Control Apparatus and Branch-Circuit Conductors. The motor-compressor controller(s), the disconnecting means, and the branch-circuit conductors shall be protected against overcurrent due to motor overload and failure to start by one of the following means, which shall be permitted to be the same device or system protecting the motor-compressor in accordance with Section 440-52(a).

Exception: Overload protection of motor-compressors and equipment on 15- and 20-ampere, single-phase, branch circuits shall be permitted to be in accordance with Sections 440-54 and 440-55.

(1) An overload relay selected in accordance with Section 440-52(a)(1)
(2) A thermal protector applied in accordance with Section 440-52(a)(2) and that will not permit a continuous current in excess of 156 percent of the marked rated-load current or branch-circuit selection current
(3) A fuse or inverse time circuit breaker selected in accordance with Section 440-52(a)(3)
(4) A protective system in accordance with Section 440-52(a)(4) and that will not permit a continuous current in excess of 156 percent of the marked rated-load current or branch-circuit selection current

440-53. Overload Relays. Overload relays and other devices for motor overload protection that are not capable of opening short circuits shall be protected by fuses or inverse time circuit breakers with ratings or settings in accordance with Part C unless approved for group installation or for part-winding motors and marked to indicate the maximum size of fuse or inverse time circuit breaker by which they shall be protected.

Exception: The fuse or inverse time circuit breaker size marking shall be permitted on the nameplate of approved equipment in which the overload relay or other overload device is used.

440-54. Motor-Compressors and Equipment on 15- or 20-Ampere Branch Circuits — Not Cord- and Attachment Plug-Connected. Overload protection for motor-compressors and equipment used on 15- or 20-ampere 120-volt, or 15-ampere 208- or 240-volt single-phase branch circuits as permitted in Article 210 shall be permitted as indicated in (a) and (b).

(a) Overload Protection. The motor-compressor shall be provided with overload protection selected as specified in Section 440-52(a). Both the controller and motor overload protective device shall be approved for installation with the short-circuit and ground-fault protective device for the branch circuit to which the equipment is connected.

(b) Time Delay. The short-circuit and ground-fault protective device protecting the branch circuit shall have sufficient time delay to permit the motor-compressor and other motors to start and accelerate their loads.

440-55. Cord- and Attachment Plug-Connected Motor-Compressors and Equipment on 15- or 20-Ampere Branch Circuits. Overload protection for motor-compressors and equipment that are cord- and attachment plug-connected and used on 15- or 20-ampere 120-volt, or 15-ampere 208- or 240-volt single-phase branch circuits as permitted in Article 210 shall be permitted as indicated in (a), (b), and (c).

(a) Overload Protection. The motor-compressor shall be provided with overload protection as specified in Section

440-52(a). Both the controller and the motor overload protective device shall be approved for installation with the short-circuit and ground-fault protective device for the branch circuit to which the equipment is connected.

(b) Attachment Plug and Receptacle Rating. The rating of the attachment plug and receptacle shall not exceed 20 amperes at 125 volts or 15 amperes at 250 volts.

(c) Time Delay. The short-circuit and ground-fault protective device protecting the branch circuit shall have sufficient time delay to permit the motor-compressor and other motors to start and accelerate their loads.

G. Provisions for Room Air Conditioners

440-60. General. The provisions of Part G shall apply to electrically energized room air conditioners that control temperature and humidity. For the purpose of Part G, a room air conditioner (with or without provisions for heating) shall be considered as an ac appliance of the air-cooled window, console, or in-wall type that is installed in the conditioned room and that incorporates a hermetic refrigerant motor-compressor(s). The provisions of Part G cover equipment rated not over 250 volts, single phase, and such equipment shall be permitted to be cord- and attachment plug-connected.

A room air conditioner that is rated three phase or rated over 250 volts shall be directly connected to a wiring method recognized in Chapter 3, and provisions of Part G shall not apply.

440-61. Grounding. Room air conditioners shall be grounded in accordance with Sections 250-110, 250-112, and 250-114.

440-62. Branch-Circuit Requirements.

(a) Room Air Conditioner as a Single Motor Unit. A room air conditioner shall be considered as a single motor unit in determining its branch-circuit requirements where all the following conditions are met.

(1) It is cord- and attachment plug-connected.
(2) Its rating is not more than 40 amperes and 250 volts, single phase.
(3) Total rated-load current is shown on the room air-conditioner nameplate rather than individual motor currents.
(4) The rating of the branch-circuit short-circuit and ground-fault protective device does not exceed the ampacity of the branch-circuit conductors or the rating of the receptacle, whichever is less.

(b) Where No Other Loads Are Supplied. The total marked rating of a cord- and attachment plug-connected room air conditioner shall not exceed 80 percent of the rating of a branch circuit where no other loads are supplied.

(c) Where Lighting Units or Other Appliances Are Also Supplied. The total marked rating of a cord- and attachment plug-connected room air conditioner shall not exceed 50 percent of the rating of a branch circuit where lighting units or other appliances are also supplied.

440-63. Disconnecting Means. An attachment plug and receptacle shall be permitted to serve as the disconnecting means for a single-phase room air conditioner rated 250 volts or less if (1) the manual controls on the room air conditioner are readily accessible and located within 6 ft (1.83 m) of the floor or (2) an approved manually operable switch is installed in a readily accessible location within sight from the room air conditioner.

440-64. Supply Cords. Where a flexible cord is used to supply a room air conditioner, the length of such cord shall not exceed 10 ft (3.05 m) for a nominal, 120-volt rating or 6 ft (1.83 m) for a nominal, 208- or 240-volt rating.

Article 445 — Generators

Contents

445-1. General. Generators and their associated wiring and equipment shall also comply with the applicable provisions of Articles 695, 700, 701, 702, and 705.

445-2. Location. Generators shall be of a type suitable for the locations in which they are installed. They shall also meet the requirements for motors in Section 430-14. Generators installed in hazardous (classified) locations as described in Articles 500 through 503 and Article 505, or in other locations as described in Articles 510 through 517, and in Articles 518, 520, 525, 530, 665, and 695 shall also comply with the applicable provisions of those articles.

445-3. Marking. Each generator shall be provided with a nameplate giving the manufacturer's name, the rated frequency, power factor, number of phases if of alternating current, the rating in kilowatts or kilovolt amperes, the normal volts and amperes corresponding to the rating, rated revolutions per minute, insulation system class and rated ambient temperature or rated temperature rise, and time rating.

445-4. Overcurrent Protection.

(a) Constant-Voltage Generators. Constant-voltage generators, except ac generator exciters, shall be protected from overloads by inherent design, circuit breakers, fuses, or other acceptable overcurrent protective means suitable for the conditions of use.

(b) Two-Wire Generators. Two-wire, dc generators shall be permitted to have overcurrent protection in one conductor only if the overcurrent device is actuated by the entire current generated other than the current in the shunt field. The overcurrent device shall not open the shunt field.

(c) 65 Volts or Less. Generators operating at 65 volts or less and driven by individual motors shall be considered as protected by the overcurrent device protecting the motor if these devices will operate when the generators are delivering not more than 150 percent of their full-load rated current.

(d) Balancer Sets. Two-wire, dc generators used in conjunction with balancer sets to obtain neutrals for 3-wire systems shall be equipped with overcurrent devices that will disconnect the 3-wire system in case of excessive unbalancing of voltages or currents.

(e) Three-Wire, Direct-Current Generators. Three-wire, dc generators, whether compound or shunt wound, shall be equipped with overcurrent devices, one in each armature lead, and connected so as to be actuated by the entire current from the armature. Such overcurrent devices shall consist either of a double-pole, double-coil circuit breaker, or of a 4-pole circuit breaker connected in the main and equalizer leads and tripped by two overcurrent devices, one in each armature lead. Such protective devices shall be interlocked so that no one pole can be opened without simultaneously disconnecting both leads of the armature from the system.

Exception to (a) through (e): Where deemed by the authority having jurisdiction, a generator is vital to the operation of an electrical system and the generator should operate to failure to prevent a greater hazard to persons. The overload sensing device(s) shall be permitted to be connected to an annunciator or alarm supervised by authorized personnel instead of interrupting the generator circuit.

Alternating-current generators can be designed so that during short periods of time, when the generator may carry an excessive overload, the voltage will fall off sufficiently to limit the current and power output to values that will not damage the generator.

The connection of a 2-wire generator, protected by a single-pole circuit breaker, is illustrated in Figure 445.1. Where two or more dc generators are operated in parallel or multiple, an equalizer conductor lead is connected to the positive terminal of each generator and, in effect, connects the series fields in parallel so as to maintain equal output voltage for each generator. The current could divide at the positive terminal, some flowing through the series field and positive lead and some flowing through the equalizer lead. The entire current generated flows through the negative lead; therefore, the fuse or circuit breaker (or at least the operating coil of the circuit breaker) must be placed in the negative lead. Overcurrent devices are required to be connected so as to be actuated by the entire armature output current.

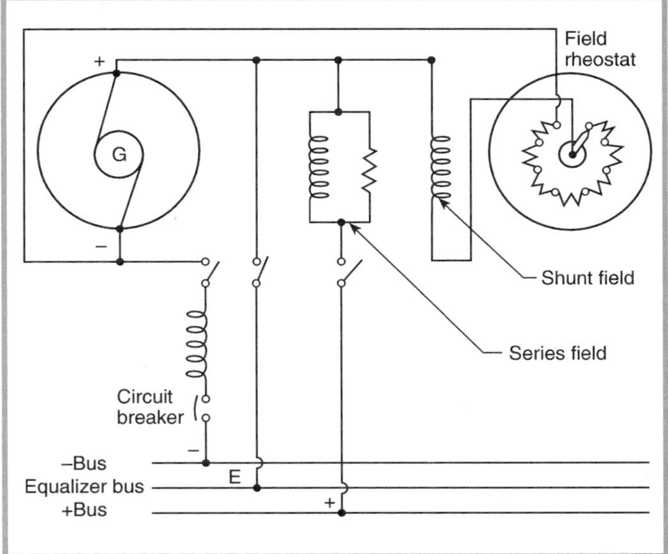

Figure 445.1 *A 2-wire dc generator protected by a single-pole circuit breaker.*

An overcurrent device should not be placed in the shunt field circuit because, if the circuit were to open when the field was at full strength, an extremely high voltage would be induced that could damage the field winding insulation and the generator.

Section 445-4(c) indicates that generators operating at 65 volts or less are to be thought of as protected by the overcurrent devices that also protect the drive motor, provided these devices operate when the generator delivers 150 percent of its full-load rated current.

Figure 445.2 illustrates a double-pole circuit breaker with one pole connected in each lead of the main generator and with the operating coil properly designed to be connected in the neutral lead from the balancer, and arranged so as to be operated by either of the "A" coils or by the "B" coil. Each of the two generators used as a balancer set carries approximately half the unbalanced load and, thus, is always smaller than the main generator. During an excessive imbalance of the load, the balancer set would be overloaded, with no overload on the main generator; hence, a double-pole circuit breaker is connected (as noted) to guard against this condition.

It should be noted that the authority having juris-

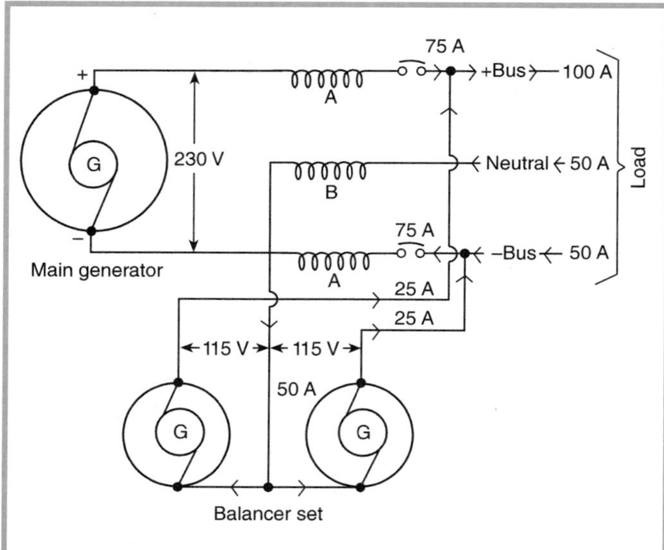

Figure 445.2 A double-pole circuit breaker (one pole connected in each lead of the main generator) with the opening coil connected in the neutral of the balancer set.

diction may judge that having the generator operate to failure is preferable to providing automatic means to shut it down, which, in many cases, could present a greater hazard to personnel. An overload sensing device(s) would be permitted to be connected to an annunciator or an alarm (instead of interrupting the generator) and allow operating personnel to shut down load-side equipment in a safe and orderly fashion.

445-5. Ampacity of Conductors. The ampacity of the conductors from the generator terminals to the first overcurrent device shall not be less than 115 percent of the nameplate current rating of the generator. It shall be permitted to size the neutral conductors in accordance with Section 220-22. Conductors that must carry ground-fault currents shall not be smaller than required by Section 250-24(b). Neutral conductors of dc generators that must carry ground-fault currents shall not be smaller than the minimum required size of the largest conductor.

Exception: Where the design and operation of the generator prevent overloading, the ampacity of the conductors shall not be less than 100 percent of the nameplate current rating of the generator.

445-6. Protection of Live Parts. Live parts of generators operated at more than 50 volts to ground shall not be exposed to accidental contact where accessible to unqualified persons.

445-7. Guards for Attendants. Where necessary for the safety of attendants, the requirements of Section 430-133 shall apply.

445-8. Bushings. Where wires pass through an opening in an enclosure, conduit box, or barrier, a bushing shall be used to protect the conductors from the edges of an opening having sharp edges. The bushing shall have smooth, well-rounded surfaces where it may be in contact with the conductors. If used where oils, grease, or other contaminants may be present, the bushing shall be made of a material not deleteriously affected.

445-9. Generator Terminal Housings. Generator terminal housings shall comply with Section 430-12.

445-10. Disconnecting Means Required for Generators. Generators shall be equipped with a disconnect by means of which the generator and all protective devices and control apparatus are able to be disconnected entirely from the circuits supplied by the generator except where

(1) The driving means for the generator can be readily shut down; and
(2) The generator is not arranged to operate in parallel with another generator or other source of voltage.

Added for the 1999 *Code,* Section 445-10 requires that generators be equipped with a disconnect switch.

Article 450 — Transformers and Transformer Vaults (Including Secondary Ties)

Contents

450-1. Scope. This article covers the installation of all transformers.

Exception No. 1: Current transformers.

Exception No. 2: Dry-type transformers that constitute a component part of other apparatus and comply with the requirements for such apparatus.

Exception No. 3: Transformers that are an integral part of an X-ray, high-frequency, or electrostatic-coating apparatus.

Exception No. 4: Transformers used with Class 2 and Class 3 circuits that comply with Article 725.

Exception No. 5: Transformers for sign and outline lighting that comply with Article 600.

Exception No. 6: Transformers for electric-discharge lighting that comply with Article 410.

Exception No. 7: Transformers used for power-limited fire alarm circuits that comply with Part C of Article 760.

Exception No. 8: Transformers used for research, development, or testing, where effective arrangements are provided to safeguard persons from contacting energized parts.

This article covers the installation of transformers dedicated to supplying power to a fire pump installation as modified by Article 695.

This article also covers the installation of transformers in hazardous (classified) locations as modified by Articles 501 through 504.

A. General Provisions

450-2. Definitions. For the purpose of this article:

Transformer. The word *transformer* is intended to mean an individual transformer, single- or polyphase, identified by a single nameplate, unless otherwise indicated in this article.

450-3. Overcurrent Protection. Overcurrent protection of transformers shall comply with (a), (b), or (c). As used in this section, the word *transformer* shall mean a transformer or polyphase bank of two or more single-phase transformers operating as a unit.

Section 450-3 was reorganized for the 1999 *Code* in order to present transformer protection requirements in a simpler and more user friendly table format. The fundamental protection requirements have not changed, only the presentation of the requirements is different. Instead of 11 paragraphs, 10 exceptions and 2 tables, Section 450-3 now contains two basic tables with notes and one exception. Remember, a note to a table is part of the requirements of the table, and, as stated in Section 90-5(c), fine print notes (FPNs), are explanatory in nature and thus are not enforceable.

FPN No. 1: See Sections 240-3, 240-21, 240-100, and 240-101 for overcurrent protection of conductors.

The requirements for protection of transformer secondaries in Article 450 apply only to the protection of transformers, not to the protection of conductors. Article 240 applies only to the protection of conductors, not to the protection of transformers. It is possible that the overcurrent protection required by Article 450 also satisfies the requirements in Article

240 for protection of the conductors, and vice versa, but it is also possible that they will not.

FPN No. 2: Nonlinear loads can increase heat in a transformer without operating its overcurrent protective device.

The increased heating effects of nonlinear load currents must be taken into account when determining the load on a transformer. There are several methods for dealing with the heating effects of nonlinear loads, including derating equipment, oversizing equipment, increasing insulation ratings, installing thermal protection systems, and using K-factor transformers. The optimum method for dealing with transformer overheating will vary, depending on several technical and economic factors, and should be considered during the design phase of the electrical system.

(a) Transformers Over 600 Volts, Nominal. Overcurrent protection shall be provided in accordance with Table 450-3(a).

The ratings or settings obtained from Table 450-3(a) are based on the type of protective device (fuse, electronic fuse, or circuit breaker), transformer rated current and impedance, and primary and secondary voltages. According to this table, the maximum ratings or settings of an overcurrent protective device for transformers rated over 600 volts is separated into two broad categories: *Any location* (or unsupervised) and *supervised locations only.*

The first category for over 600-volt transformers is not limited by location and it is referred to as *any location.* The maximum ratings or settings for overcurrent devices permitted in the *any location* row are applicable for all unsupervised locations. An *any location* transformer installation must be provided with overcurrent protection in both the primary and secondary circuit. Of course, the user may select the *any location* row even if the location qualifies as a *supervised locations only* since the requirements for the *any location* row are the same or exceed the *supervised locations only* row. See Figure 450.1 for an example of an installation using circuit breakers on the primary and the secondary for an over 600-volt transformer with 6 percent impedance.

The second category for over 600-volt transformers is *supervised locations only.* The maximum ratings or settings for overcurrent devices permitted in the *supervised locations only* rows are strictly limited to the

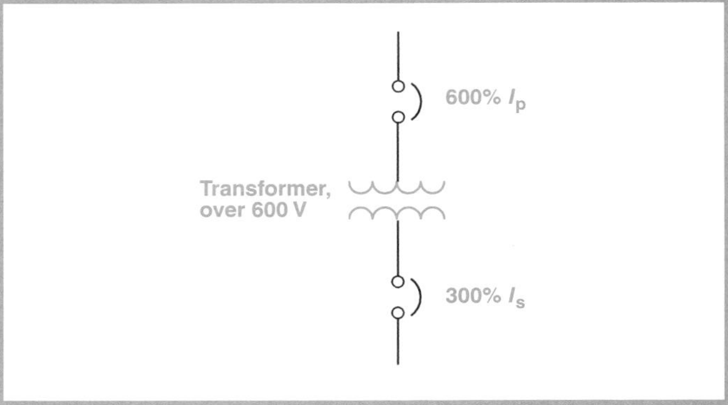

Figure 450.1 *A transformer with 6 percent impedance and rated over 600 volts using circuit-breaker protection for both the primary and the secondary. For the transformer in this figure, both the primary and the secondary voltages are over 600 volts.*

conditions explained in Note 3 of the table. If the location of a transformer does not qualify as a *supervised* location, then it is necessary to select the general rows of *any location* in Table 450-3(a). See Figure 450.1 for an example of an installation using circuit breakers on the primary and the secondary for an over 600-volt transformer with 6 percent impedance. Notice that the installation shown fulfills the requirements of both *any location* and *supervised locations only.*

For the 1999 *Code,* Article 240, Overcurrent Protection, contains a new Part H, Supervised Industrial Installations. Part H contains many revised overcurrent protection requirements for feeders and feeder taps associated with transformers. Also, requirements for the overcurrent protection of transformer secondary conductors are found in Section 240-3(f).

(b) Transformers 600 Volts, Nominal, or Less. Overcurrent protection shall be provided in accordance with Table 450-3(b).

The ratings or settings of the overcurrent protective device obtained from Table 450-3(b) is based on the transformer-rated current and whether or not secondary protection is provided. According to Table 450-3(b), the maximum ratings or settings of overcurrent protective devices for transformers rated 600 volts and less is separated into two categories: *primary only protection* and *primary and secondary protection.*

According to Table 450-3(b), transformers with currents of 9 amperes or more must be protected by either of two methods. Method 1 requires primary protection only and is set at not more than 125 percent of the primary side rating. Method 1 does not require secondary side overcurrent protection. Method 2 requires secondary side overcurrent protection to be

Table 450-3(a). Maximum Rating or Setting of Overcurrent Protection for Transformers Over 600 Volts (as a Percentage of Transformer-Rated Current)

		Primary Protection Over 600 Volts		Secondary Protection (see Note 2)		
				Over 600 Volts		600 Volts or Below
Location Limitations	**Transformer Rated Impedance**	**Circuit Breaker (see Note 4)**	**Fuse Rating**	**Circuit Breaker (see Note 4)**	**Fuse Rating**	**Circuit Breaker or Fuse Rating**
Any location	Not more than 6%	600% (see Note 1)	300% (see Note 1)	300% (see Note 1)	250% (see Note 1)	125% (see Note 1)
	More than 6% and not more than 10%	400% (see Note 1)	300% (see Note 1)	250% (see Note 1)	225% (see Note 1)	125% (see Note 1)
Supervised locations only (see Note 3)	Any	300% (see Note 1)	250% (see Note 1)	Not required	Not required	Not required
	Not more than 6%	600%	300%	300% (see Note 5)	250% (see Note 5)	250% (see Note 5)
	More than 6% and not more than 10%	400%	300%	250% (see Note 5)	225% (see Note 5)	250% (see Note 5)

Notes:

1. Where the required fuse rating or circuit breaker setting does not correspond to a standard rating or setting, the next higher standard rating or setting shall be permitted.

2. Where secondary overcurrent protection is required, the secondary overcurrent device shall be permitted to consist of not more than six circuit breakers or six sets of fuses grouped in one location. Where multiple overcurrent devices are utilized, the total of all the device ratings shall not exceed the allowed value of a single overcurrent device. If both breakers and fuses are utilized as the overcurrent device, the total of the device ratings shall not exceed that allowed for fuses.

3. A supervised location where conditions of maintenance and supervision ensure that only qualified persons will monitor and service the transformer installation.

4. Electronically actuated fuses that may be set to open at a specific current shall be set in accordance with settings for circuit breakers.

5. A transformer equipped with a coordinated thermal overload protection by the manufacturer shall be permitted to have separate secondary protection omitted.

For Note 1, for standard ratings of circuit breakers and fuses, see Section 240-6.

For Note 2, overcurrent protection to protect the secondary of a transformer is allowed to consist of not more than six sets of fuses or six circuit breakers, under the following conditions.

1. They must be grouped in one location.
2. The sum of the overcurrent devices must not exceed the maximum value of a single device.
3. Where a combination of fuses and circuit breakers is used, the maximum value will be that of a single set of fuses.

See Figures 450.2 and 450.3 for an application of Note 2. This note also appears in Table 450-3(b) and applies to transformers rated 600 volts or less as well.

For Note 3, a supervised location is where maintenance of the equipment is performed by personnel familiar with proper operation of the equipment and aware of the hazards associated with it.

For Note 4, *electronically actuated fuse* is defined in Article 100, Part B. Over 600 Volts, Nominal.

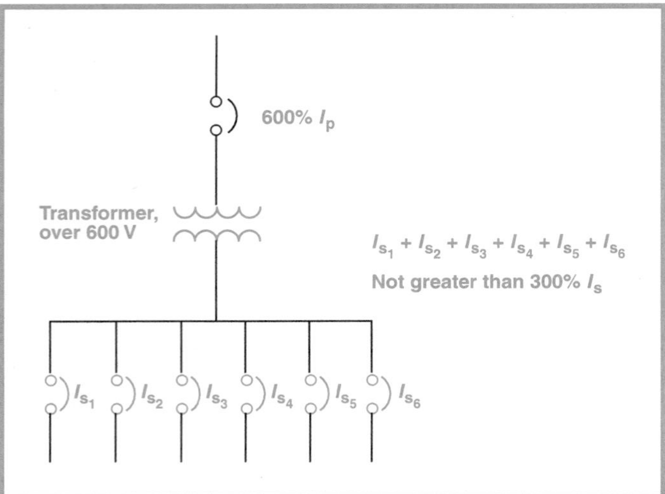

Figure 450.2 *A transformer rated over 600 volts with a secondary rated over 600 volts, with secondary protection consisting of six circuit breakers. The sum of the ratings of the circuit breakers is not permitted to exceed 300 percent of the rated secondary current.*

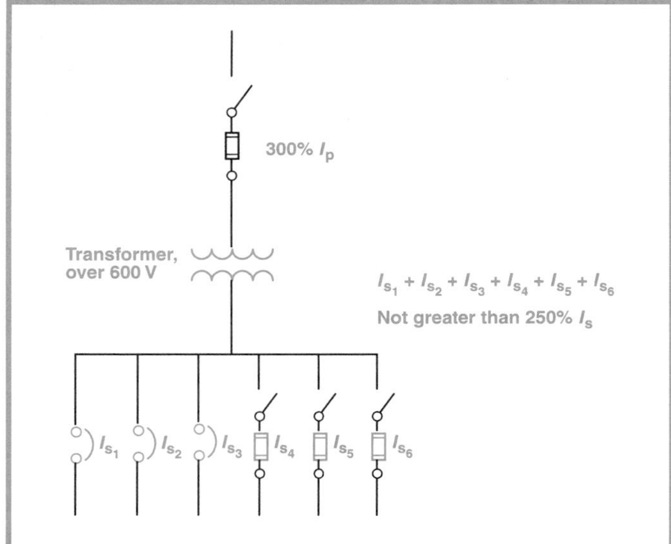

Figure 450.3 *A transformer rated over 600 volts with a secondary rated over 600 volts, with secondary protection consisting of fuses and circuit breakers. The sum of the ratings of all the overcurrent devices is not permitted to exceed the rating permitted for fuses.*

set at not more than 125 percent, provided the primary side overcurrent protection is set at not more than 250 percent of the primary side rating. Although not required, following either protection method will free the user from any further protection requirements of Table 450-3(b). According to this table, smaller transformers have protection requirements that are less restrictive. For overcurrent protection of motor control circuit transformers, see Section 430-72. An example of *primary only protection* is found in

Table 450-3(b). Maximum Rating or Setting of Overcurrent Protection for Transformers 600 Volts and Less (as a Percentage of Transformer-Rated Current)

	Primary Protection			Secondary Protection (see Note 2)	
Protection Method	**Currents of 9 Amperes or More**	**Currents Less than 9 Amperes**	**Currents Less than 2 Amperes**	**Currents of 9 Amperes or More**	**Currents Less than 9 Amperes**
Primary only protection	125% (see Note 1)	167%	300%	Not required	Not required
Primary and secondary protection	250% (see Note 3)	250% (see Note 3)	250% (see Note 3)	125% (see Note 1)	167%

Notes:

1. Where 125 percent of this current does not correspond to a standard rating of a fuse or nonadjustable circuit breaker, the next higher standard rating described in Section 240-6 shall be permitted.

2. Where secondary overcurrent protection is required, the secondary overcurrent device shall be permitted to consist of not more than six circuit breakers or six sets of fuses grouped in one location. Where multiple overcurrent devices are utilized, the total of all the device ratings shall not exceed that allowed value of a single overcurrent device. If both breakers and fuses are utilized as the overcurrent device, the total of the device ratings shall not exceed that allowed for fuses.

3. A transformer equipped with coordinated thermal overload protection by the manufacturer and arranged to interrupt the primary current, shall be permitted to have primary overcurrent protection rated or set at a current value that is not more than six times the rated current of the transformer for transformers having not more than 6 percent impedance, and not more than four times the rated current of the transformer for transformers having more than 6 percent but not more than 10 percent impedence.

Figure 450.4. An example of *primary and secondary protection* is found in Figure 450.5.

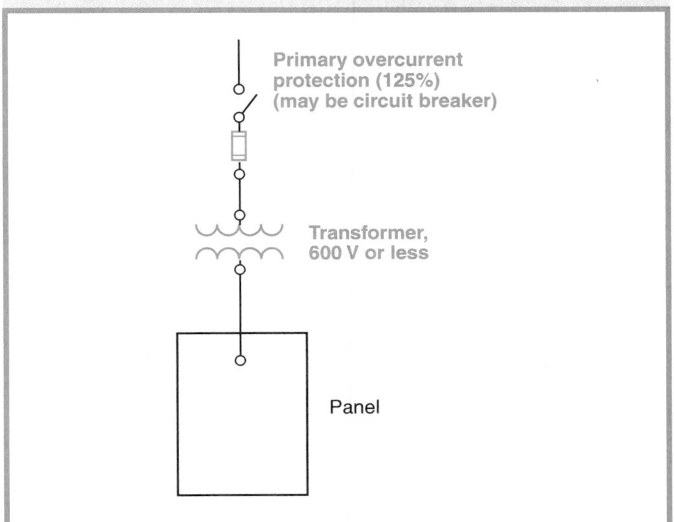

Figure 450.4 *A transformer (9 amperes or more) of 600 volts and less protected by an individual overcurrent device on the primary side according to Table 450-3(b), primary only protection.*

Questions frequently arise as to whether the overcurrent protection required for transformers, as specified in Section 450-3, provides satisfactory protection for the primary and secondary conductors. Where polyphase transformers are involved, primary and secondary conductors will usually not be properly protected. The rules in Section 450-3 are intended to protect the transformer alone. The primary overcurrent device provides short-circuit protection for the primary conductors and a degree of overload protection for the transformer, and secondary overcurrent devices prevent the transformer and secondary conductors from being overloaded. The transformer is considered the point of supply, and the conductors it supplies are required to be protected in accordance with their ampacity.

Section 240-3(f) permits the secondary circuit conductors from a transformer to be protected by overcurrent devices in the primary circuit conductors of the transformer only in two special cases. The first case is a transformer with a 2-wire primary and a 2-wire secondary, provided the transformer primary is protected in accordance with Section 450-3. The second case is a 3-phase, delta-delta-connected transformer having a 3-wire, single-voltage secondary, provided its primary is protected in accordance with Section 450-3. In cases where the primary feeder to the transformer incorporates overcurrent protective devices rated (or set) at a level not to exceed those prescribed herein, it is not necessary to duplicate them at the transformer.

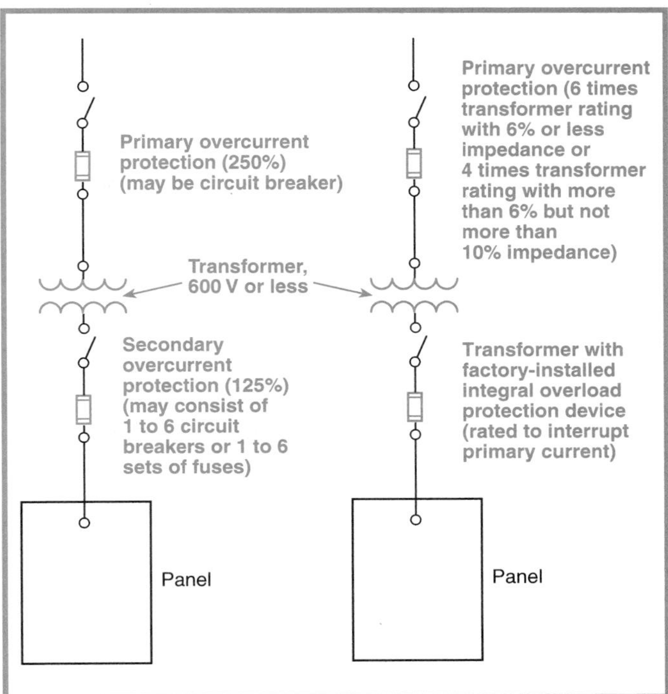

Figure 450.5 *Two transformers (with currents of 9 amperes or more) of 600 volts or less with overcurrent protection according to Table 450-3(b), primary and secondary protection. The left side of the diagram shows a standard protection technique; the right side diagram shows a protection technique according to Note 3 of Table 450-3(b).*

Exception: Where the transformer is installed as a motor-control circuit transformer in accordance with Sections 430-72(c)(1) through (5).

(c) Voltage Transformers. Voltage transformers installed indoors or enclosed shall be protected with primary fuses.

FPN: For protection of instrument circuits including voltage transformers, see Section 384-32.

450-4. Autotransformers 600 Volts, Nominal, or Less.

(a) Overcurrent Protection. Each autotransformer 600 volts, nominal, or less, shall be protected by an individual overcurrent device installed in series with each ungrounded input conductor. Such overcurrent device shall be rated or set at not more than 125 percent of the rated full-load input current of the autotransformer. Where this calculation does not correspond to a standard rating of a fuse or nonadjustable circuit breaker and the rated input current is 9 amperes or more, the next higher standard rating described in Section 240-6 shall be permitted. An overcurrent device shall not be installed in series with the shunt winding (the winding common to both the input and the output circuits) of the autotransformer between Points A and B as shown in Figure 450-4.

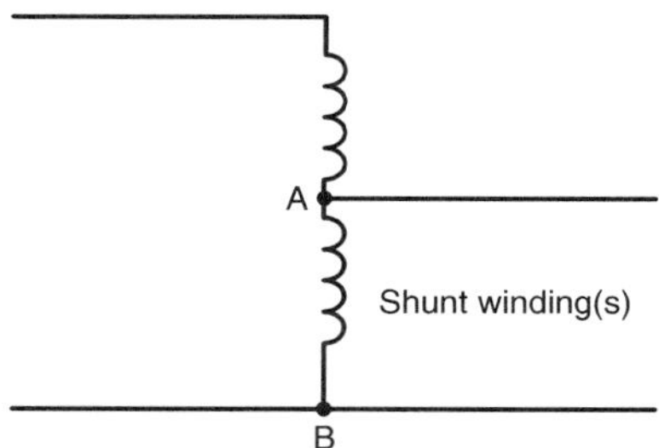

Figure 450-4 Autotransformer.

Exception: Where the rated input current of the autotransformer is less than 9 amperes, an overcurrent device rated or set at not more than 167 percent of the input current shall be permitted.

Because of the voltage feedback problem that may occur, an overcurrent device is not permitted between points A and B in Figure 450-4, Autotransformer.

Figure 450.6 provides an example of overcurrent protection for an autotransformer. It shows a two-winding, single-phase transformer connected to boost a 208-volt supply to 240 volts. It is provided with a two-pole disconnect switch with both overcurrent devices (OC-1a and OC-1b) located on the supply side of the autotransformer. If an overcurrent device were located between points A and B and this overcurrent device opened, the full 208-volt supply voltage would be applied across the 32-volt secondary winding in series with the load. Under these conditions, a higher-than-normal voltage would appear across the primary winding. If the load impedance were very low, this voltage could approach 208/32 × 208 = 1352 volts.

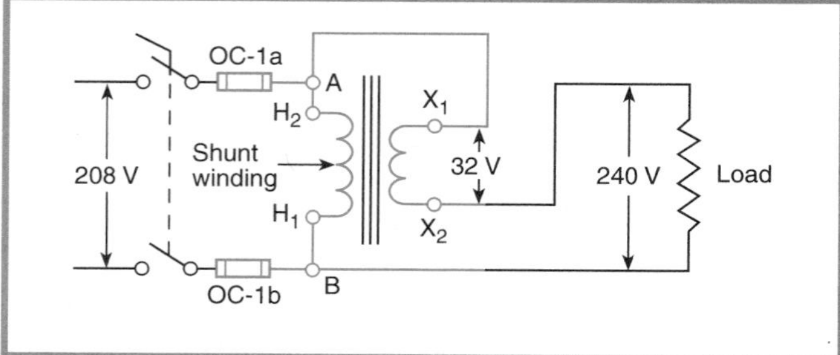

Figure 450.6 A disconnect switch with overcurrent devices properly connected to protect an autotransformer and located to meet the requirements of Section 450-4(a), last sentence.

(b) Transformer Field-Connected as an Autotransformer. A transformer field-connected as an autotransformer shall be identified for use at elevated voltage.

The requirement in Section 450-4(b) is necessary because of the dielectric voltage withstand test requirements applied to transformers. The test is conducted at 2500 volts for windings rated 250 volts or less, and 4000 volts for higher-rated windings. A transformer intended for buck or boost operation would require that the test for the low-voltage winding be based on the sum of the primary and secondary voltage ratings.

FPN: For information on permitted uses of autotransformers, see Section 210-9 and Section 215-11.

450-5. Grounding Autotransformers. Grounding autotransformers covered in this section are zigzag or T-connected transformers connected to 3-phase, 3-wire ungrounded systems for the purpose of creating a 3-phase, 4-wire distribution system or to provide a neutral reference for grounding purposes. Such transformers shall have a continuous per phase current rating and a continuous neutral current rating.

FPN: The phase current in a grounding autotransformer is one-third the neutral current.

(a) Three-Phase, 4-Wire System. A grounding autotransformer used to create a 3-phase, 4-wire distribution system from a 3-phase, 3-wire ungrounded system shall conform to the following.

(1) Connections. The transformer shall be directly connected to the ungrounded phase conductors and shall not be switched or provided with overcurrent protection that is independent of the main switch and common-trip overcurrent protection for the 3-phase, 4-wire system.

(2) Overcurrent Protection. An overcurrent sensing device shall be provided that will cause the main switch or common-trip overcurrent protection referred to in (a)(1) to open if the load on the autotransformer reaches or exceeds 125 percent of its continuous current per phase or neutral rating. Delayed tripping for temporary overcurrents sensed at the autotransformer overcurrent device shall be permitted for the purpose of allowing proper operation of branch or feeder protective devices on the 4-wire system.

(3) Transformer Fault Sensing. A fault-sensing system that will cause the opening of a main switch or common-trip overcurrent device for the 3-phase, 4-wire system shall be provided to guard against single-phasing or internal faults.

FPN: This can be accomplished by the use of two subtractive-connected donut-type current transformers installed to sense and signal when an unbalance occurs in the line current to the autotransformer of 50 percent or more of rated current.

(4) Rating. The autotransformer shall have a continuous neutral-current rating that is sufficient to handle the maximum possible neutral unbalanced load current of the 4-wire system.

Figure 450.7 shows the proper method of protecting a grounding autotransformer where it is used to provide a neutral for a 3-phase system where necessary to supply a group of single-phase, line-to-neutral loads. Separate overcurrent protection is not provided for the autotransformer, because there will be no control of the system line-to-neutral voltages if the autotransformer becomes disconnected. Consequently, simultaneous interruption of the power supply to all the line-to-neutral loads is necessary whenever the grounding autotransformer is switched off.

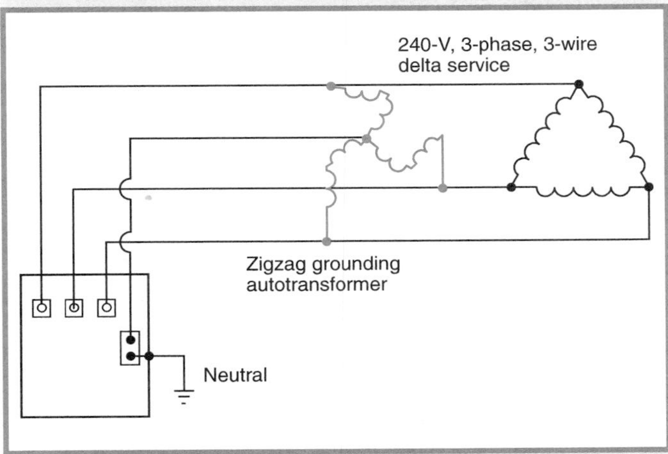

Figure 450.7 *A zigzag autotransformer used to create a 3-phase, 4-wire distribution system or to provide a neutral reference for grounding purposes.*

The donut-type current transformers CT-1, CT-2, and CT-3 shown in Figure 450.8 are required to be arranged to trip the circuit breaker located upstream of both the autotransformer and line-to-neutral connected loads, to satisfy the requirements of Section 450-5(a)(1). All three relays are intended to trip the main breaker if the current in any phase or the neutral conductor exceeds 125 percent of the rated current, as specified in Section 450-5(a)(2). The current transformers CT-2 and CT-3 are also differentially connected, to protect against an internal failure of the autotransformer, as required by Section 450-5(a)(3).

(b) Ground Reference for Fault Protection Devices. A grounding autotransformer used to make available a specified magnitude of ground-fault current for operation of a ground-responsive protective device on a 3-phase, 3-wire ungrounded system shall conform to the following requirements.

(1) Rating. The autotransformer shall have a continuous neutral-current rating sufficient for the specified ground-fault current.

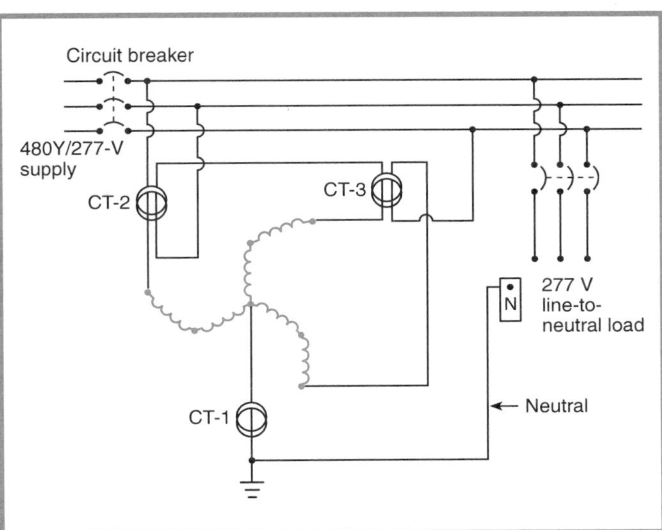

Figure 450.8 *A zigzag autotransformer used to establish a neutral connection for a 480Y/277-volt, 3-phase ungrounded system to supply single-phase line-to-neutral loads.*

(2) Overcurrent Protection. An overcurrent protective device of adequate short-circuit rating that will open simultaneously all ungrounded conductors when it operates shall be applied in the grounding autotransformer branch circuit and shall be rated or set at a current not exceeding 125 percent of the autotransformer continuous per phase-current rating or 42 percent of the continuous-current rating of any series connected devices in the autotransformer neutral connection. Delayed tripping for temporary overcurrents to permit the proper operation of ground-responsive tripping devices on the main system shall be permitted, but shall not exceed values that would be more than the short-time current rating of the grounding autotransformer or any series connected devices in the neutral connection thereto.

Figure 450.9 shows the proper method of protecting a grounding autotransformer where it is used as a ground reference for fault protection devices. The overcurrent protective device is to have a rating (or setting) not in excess of 125 percent of the rated phase current of the autotransformer (42 percent of the neutral current rating) and not more than 42 percent of the continuous current rating of the neutral grounding resistor or other current-carrying device in the neutral connection, as specified in Section 450-5(b)(2).

(c) Ground Reference for Damping Transitory Overvoltages. A grounding autotransformer used to limit transitory overvoltages shall be of suitable rating and connected in accordance with Section 450-5(a)(1).

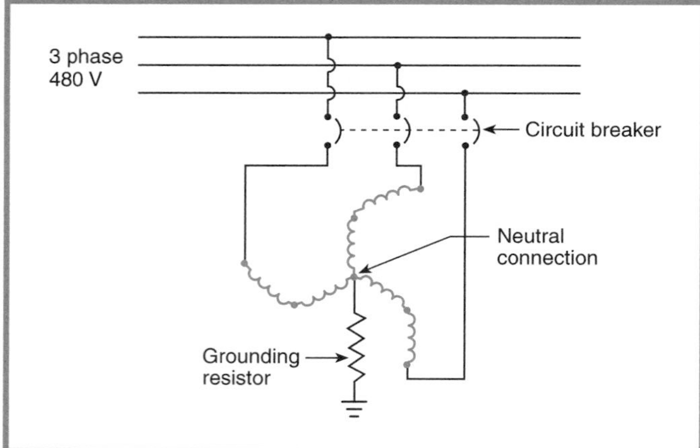

Figure 450.9 *A zigzag autotransformer used to establish a reference ground-fault current for fault-protective-device operation or for damping-transitory-overvoltage surges.*

For installations involving a high-resistance grounding package, the functional performance of the installation parallels that described in Section 450-5(b), differing only in that the magnitude of available ground-fault current would likely be a lower value. It would be appropriate to employ the connections displayed in Figure 450.9 and conform with the overcurrent protection requirements prescribed in Section 450-5(b)(2).

With any of the grounding autotransformer applications covered by Sections 450-5(a), (b), or (c), it is important to emphasize the use of a ganged 3-pole switching interrupter for connecting and disconnecting the autotransformer, in order to accomplish simultaneous connection (and disconnection) of the three line terminals. If, at any time, one or two of the line connections to the autotransformer should open, which could occur if the protective devices were single pole, the grounding autotransformer would cease to function in the desired fashion and would act as a high-inductive-reactance connection between the electrical system and ground. The latter connection is prone to create high-value transitory overvoltages, line-to-ground.

450-6. Secondary Ties. A secondary tie is a circuit operating at 600 volts, nominal, or less, between phases that connects two power sources or power supply points, such as the secondaries of two transformers. The tie shall be permitted to consist of one or more conductors per phase.

As used in this section, the word *transformer* means a transformer or a bank of transformers operating as a unit.

(a) Tie Circuits. Tie circuits shall be provided with overcurrent protection at each end as required in Article 240.

Under the conditions described in (a)(1) and (a)(2), the overcurrent protection shall be permitted to be in accordance with (a)(3).

(1) Loads at Transformer Supply Points Only. Where all loads are connected at the transformer supply points at each end of the tie and overcurrent protection is not provided in accordance with Article 240, the rated ampacity of the tie shall not be less than 67 percent of the rated secondary current of the largest transformer connected to the secondary tie system.

(2) Loads Connected Between Transformer Supply Points. Where load is connected to the tie at any point between transformer supply points and overcurrent protection is not provided in accordance with Article 240, the rated ampacity of the tie shall not be less than 100 percent of the rated secondary current of the largest transformer connected to the secondary tie system.

Exception: As otherwise provided in Section 450-6(a)(4).

(3) Tie Circuit Protection. Under the conditions described in (a)(1) and (a)(2), both ends of each tie conductor shall be equipped with a protective device that will open at a predetermined temperature of the tie conductor under short-circuit conditions. This protection shall consist of one of the following: (1) a fusible link cable connector, terminal, or lug, commonly known as a limiter, each being of a size corresponding with that of the conductor and of construction and characteristics according to the operating voltage and the type of insulation on the tie conductors or (2) automatic circuit breakers actuated by devices having comparable current-time characteristics.

(4) Interconnection of Phase Conductors Between Transformer Supply Points. Where the tie consists of more than one conductor per phase, the conductors of each phase shall comply with one of the following provisions.

(a) *Interconnected.* The conductors shall be interconnected in order to establish a load supply point, and the protection specified in (a)(3) shall be provided in each tie conductor at this point.

(b) *Not Interconnected.* The loads shall be connected to one or more individual conductors of a paralleled conductor tie without interconnecting the conductors of each phase and without the protection specified in (a)(3) at load connection points. Where this is done, the tie conductors of each phase shall have a combined capacity of not less than 133 percent of the rated secondary current of the largest transformer connected to the secondary tie system, the total load of such taps shall not exceed the rated secondary current of the largest transformer, and the loads shall be equally divided on each phase and on the individual conductors of each phase as far as practicable.

(5) Tie Circuit Control. Where the operating voltage exceeds 150 volts to ground, secondary ties provided with limiters shall have a switch at each end that, when open, will de-energize the associated tie conductors and limiters. The current rating of the switch shall not be less than the rated current of the conductors connected to the switch. It shall be capable of opening its rated current, and it shall be constructed so that it will not open under the magnetic forces resulting from short-circuit current.

(b) Overcurrent Protection for Secondary Connections. Where secondary ties are used, an overcurrent device rated or set at not more than 250 percent of the rated secondary current of the transformers shall be provided in the secondary connections of each transformer. In addition, an automatic circuit breaker actuated by a reverse-current relay set to open the circuit at not more than the rated secondary current of the transformer shall be provided in the secondary connection of each transformer.

The requirements of Section 450-6 apply specifically to network systems for power distribution commonly employed where the load density is high and reliability of service is important. Such a system is illustrated in Figure 450.10. This type of distribution system introduces a variety of problems not encountered in the more common radial-type distribution system and must be designed by experienced electrical engineers. Figure 450.10 shows a typical 3-phase network system for an industrial plant fed by two primary feeders, preferably from separate substations, energized at any standard voltage up to 34,500 volts. Each of the transformers is supplied by the two primary feeders, which are arranged by means of a double-throw switch at the transformer so that the transformer may be supplied by either feeder.

Each of the network transformers is rated in the range of 300 to 1000 kilovolt-amperes (kVA) and is required to be protected as illustrated in Figure 450.11. The primary and secondary protection is in accordance with Section 450-3, but an additional protective device is also required to be provided on the secondary side. This protective device is known as a *network protector,* consisting of a circuit breaker and a reverse-current relay. This operates on reverse current to prevent power from being fed back into the transformer through the secondary ties should a fault occur in the transformer or a primary feeder. The reverse-current relay is set to trip the circuit breaker at a current value not more than the rated secondary current of the transformer. The relay is not designed to trip the circuit breaker in the event of an overload on the secondary of the transformer.

The secondary ties shown in Figure 450.10 are required to be protected at each end with an overcurrent device, in accordance with Section 450-6(a)(3). The overcurrent device most commonly provided for this purpose is a special type of fuse known as a *current limiter.* Such a device is illustrated in Figure 450.12. This high-interrupting-capacity device is de-

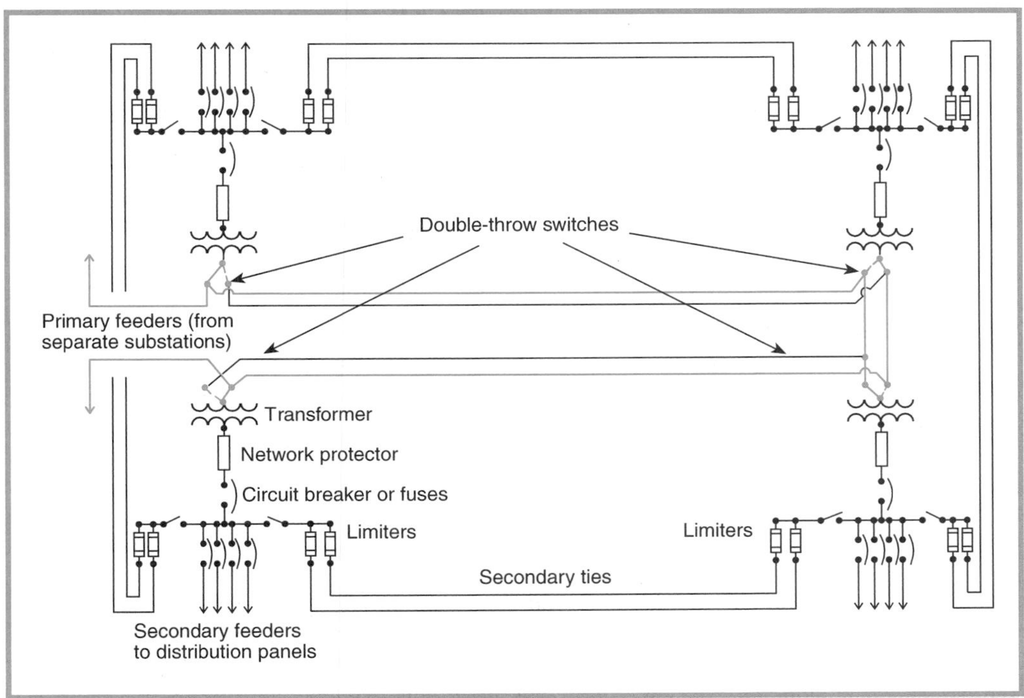

Figure 450.10 *A typical 3-phase network system for an industrial plant fed by two primary feeders.*

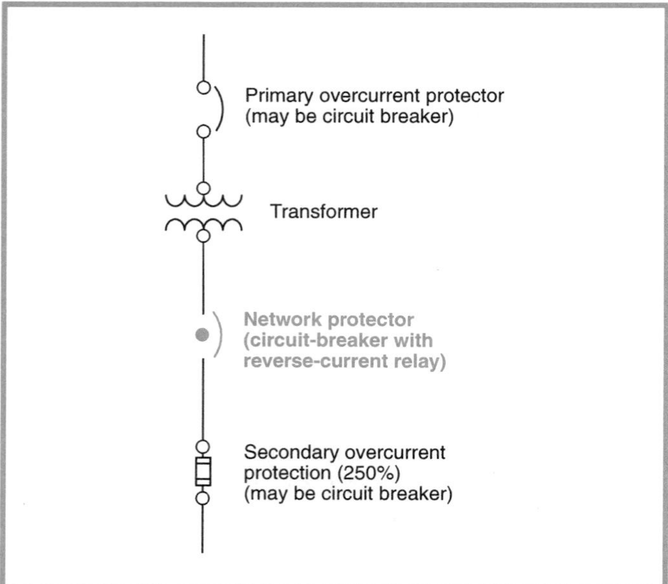

Figure 450.11 *Primary and secondary overcurrent protection for a transformer in a network system, showing a network protector (an automatic circuit breaker actuated by a reverse-current relay).*

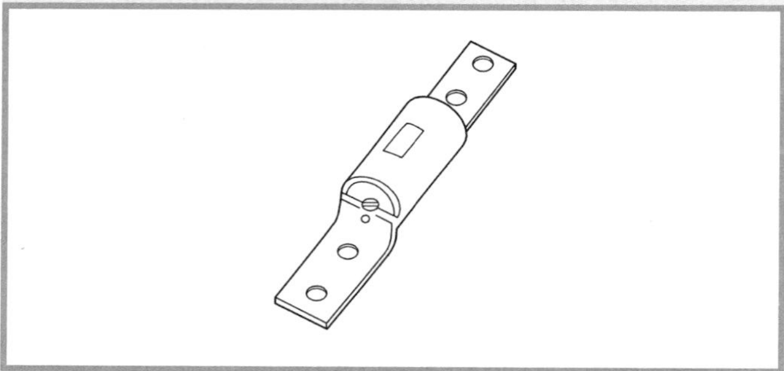

Figure 450.12 *A current limiter (a special type of high-interrupting-capacity fuse).*

signed to provide short-circuit protection only for the secondary ties, which will open safely before temperatures damaging to the cable insulation are reached. See Section 240-11 for the definition of *current-limiting overcurrent protective device* and its accompanying commentary. The secondary ties form a closed loop equipped with switching devices so that any part of the loop may be isolated when repairs are needed or a current limiter must be replaced.

450-7. Parallel Operation. Transformers shall be permitted to be operated in parallel and switched as a unit provided the overcurrent protection for each transformer meets the requirements of Section 450-3(a) for primary and secondary protective devices over 600 volts or Section 450-3(b) for primary and secondary protective devices 600 volts or less.

Parallel operation of transformers that are not switched as a unit can present dangerous backfeed situations for workers performing electrical maintenance. Appropriate lockout/tagout procedures must be implemented during maintenance of electrical equipment operated or connected in parallel. See NFPA 70E-1995, *Standard for Electrical Safety Requirements for Employee Workplaces,* for safety-related work practices and appropriate lockout/tagout procedures.

450-8. Guarding. Transformers shall be guarded as specified in (a) through (d).

(a) Mechanical Protection. Appropriate provisions shall be made to minimize the possibility of damage to transformers from external causes where the transformers are exposed to physical damage.

One method of providing mechanical protection is to strategically place bollards around the transformer. This provides a degree of protection from vehicles.

(b) Case or Enclosure. Dry-type transformers shall be provided with a noncombustible moisture-resistant case or enclosure that will provide protection against the accidental insertion of foreign objects.

(c) Exposed Energized Parts. Switches or other equipment operating at 600 volts, nominal, or less, and serving only equipment within a transformer enclosure shall be permitted to be installed in the transformer enclosure if accessible to qualified persons only. All energized parts shall be guarded in accordance with Sections 110-27 and 110-34.

(d) Voltage Warning. The operating voltage of exposed live parts of transformer installations shall be indicated by signs or visible markings on the equipment or structures.

450-9. Ventilation. The ventilation shall be adequate to dispose of the transformer full-load losses without creating a temperature rise that is in excess of the transformer rating.

FPN No. 1: See *General Requirements for Liquid-Immersed Distribution, Power, and Regulating Transformers*, ANSI/IEEE C57.12.00-1993, and *General Requirements for Dry-Type Distribution and Power Transformers*, ANSI/IEEE C57.12.01-1989.

FPN No. 2: Additional losses may occur in some transformers where nonsinusoidal currents are present, resulting in increased heat in the transformer above its rating. See *Recommended Practice for Establishing Transformer Capability When Supplying Nonsinusoidal Load Currents*, ANSI/ IEEEC57.110-1993, where transformers are utilized with nonlinear loads.

Transformers with ventilating openings shall be installed so that the ventilating openings are not blocked by walls or other obstructions. The required clearances shall be clearly marked on the transformer.

Section 450-9 is intended to clarify that transformers are not permitted to be installed directly against walls or other obstructions that block openings for ventilation and that the required clearances should be clearly marked on the transformer (see Section 450-11).

Fine Print Note No. 2 of Section 450-9 warns of increased heating of transformers. See the commentary following Section 450-3, FPN No. 2, and the commentary following Section 310-15(b)(4) for additional information concerning nonlinear loads.

450-10. Grounding. Exposed noncurrent-carrying metal parts of transformer installations, including fences, guards, etc., shall be grounded where required under the conditions and in the manner specified for electric equipment and other exposed metal parts in Article 250.

450-11. Marking. Each transformer shall be provided with a nameplate giving the name of the manufacturer, rated kilovolt-amperes, frequency, primary and secondary voltage, impedance of transformers 25 kVA and larger, required clearances for transformers with ventilating openings, and the amount and kind of insulating liquid where used. In addition, the nameplate of each dry-type transformer shall include the temperature class for the insulation system.

The information given on a transformer nameplate is necessary to determine whether special precautions must be used pertaining to clearances for ventilation, overcurrent protection, or liquid confinement.

450-12. Terminal Wiring Space. The minimum wire-bending space at fixed, 600-volt and below terminals of transformer line and load connections shall be as required in Section 373-6. Wiring space for pigtail connections shall conform to Table 370-16(b).

The requirement in Section 450-12 ensures adequate wire bending space at fixed terminals of transformer line and load connections rated 600 volts or less, as this is a point of maximum mechanical and electrical stress on the conductor insulation.

450-13. Accessibility. All transformers and transformer vaults shall be readily accessible to qualified personnel for inspection and maintenance, or meet the requirements of (a) or (b).

Transformers are not accessible if wiring methods or other equipment obstruct the access of a worker or prevent removal of the covers for inspection or maintenance. Practical clearance considerations required for removal and replacement of the transformer are also important.

(a) Open Installations. Dry-type transformers 600 volts, nominal, or less, located in the open on walls, columns, or structures, shall not be required to be readily accessible.

(b) Hollow Space Installations. Dry-type transformers 600 volts, nominal, or less and not exceeding 50 kVA shall be permitted in hollow spaces of buildings not permanently closed in by structure, and provided they meet the ventilation requirements of Section 450-9 and separation from combustible materials requirements of Section 450-21(a). Transformers so installed shall not be required to be readily accessible.

Section 450-13(b) continues to permit the installation of dry-type transformers rated 600 volts or less and not exceeding 50 kVA in hollow spaces of hung ceiling areas, provided these spaces are fire resistant, ventilated, and accessible. According to Section 300-22(c)(2), transformers are permitted to be installed in hollow spaces where the space is used for environmental air provided the transformer is in a metal enclosure (ventilated or non-ventilated) and the transformer is suitable for the ambient air temperature within the hollow space. Of course, the requirement of Section 450-13(b) applies to transformer installations in "other spaces used for environmental air."

B. Specific Provisions Applicable to Different Types of Transformers

450-21. Dry-Type Transformers Installed Indoors.

(a) Not Over 112½ kVA. Dry-type transformers installed indoors and rated 112½ kVA or less shall have a separation of at least 12 in. (305 mm) from combustible material unless separated from the combustible material by a fire-resistant, heat-insulated barrier.

Exception: This rule shall not apply to transformers rated for 600 volts, nominal, or less, completely enclosed, with or without ventilating openings.

(b) Over 112½ kVA. Individual dry-type transformers of more than 112½ kVA rating shall be installed in a transformer room of fire-resistant construction. Unless specified otherwise in this article, the term *fire resistant* means a construction having a minimum fire rating of 1 hour.

Exception No. 1: Transformers with Class 155 or higher insulation systems and separated from combustible material by a fire-resistant, heat-insulating barrier or by not less than 6 ft (1.83 m) horizontally and 12 ft (3.66 m) vertically.

Exception No. 2: Transformers with Class 155 or higher insulation systems and completely enclosed except for ventilating openings.

Dry-type transformers with a Class 155 or higher insulation system rating are not required to be installed in transformer rooms or vaults if space separation or a fire-resistant heat-insulating barrier is provided. Although these units are designed for higher operating temperatures, the need for a transformer vault is mitigated by the fire-resistant characteristics of high-temperature insulations.

The two exceptions to Section 450-21(b) were revised for the 1999 *Code* by eliminating the reference to the specific temperature rating (80°C rise or higher rating) and substituting the class insulation system (Class 155 or higher.) The transformer class insulation system provides a more complete reference than simply using the permitted temperature rise. Further information on specific transformer class insulation systems may be found in UL 1561, *Dry-Type General Purpose and Power Transformers.*

FPN: See *Method for Fire Tests of Building Construction and Materials*, ANSI/ASTM E119-1995, and *Standard Methods of Tests of Fire Endurance of Building Construction and Materials*, NFPA 251-1995.

(c) Over 35,000 Volts. Dry-type transformers rated over 35,000 volts shall be installed in a vault complying with Part C of this article.

Dry-type transformers depend on the surrounding air for adequate ventilation and, where rated $112^1/_2$ kilovolt-amperes or less, are not required to be installed in a fire-resistant transformer room but must comply with Section 450-9.

Dry-type transformers, or gas-filled or less-flammable liquid-insulated transformers (see Section 450-23), installed indoors with a primary voltage of not more than 35,000 volts are commonly used because a transformer vault is not required.

For the same reason, askarel-filled transformers have been extensively used indoors in the past. Askarel, which contains a polychlorinated biphenyl (PCB), is no longer being manufactured. Acceptable substitutes that comply with Section 450-23 are readily available.

Figure 450.13 shows a 1500-kilovolt-ampere gas-filled dry-type transformer, available in ratings through 5000 kilovolt-amperes, $34^1/_2$ kilovolts, for use indoors or outdoors. These units are built using the conventional ventilated dry-type core and coils but sealed in a heavy-gauge steel tank and filled with fluorocarbon gas for dielectric strength as well as cooling. Since the transformers are completely hermetically sealed in a heavy-gauge steel tank, they are considered safe indoors and outdoors, and there is little likelihood of flames or gases escaping in the event of a short circuit or transformer failure. See Sections 501-2, 502-2, and 503-2 for requirements for transformers in hazardous (classified) locations.

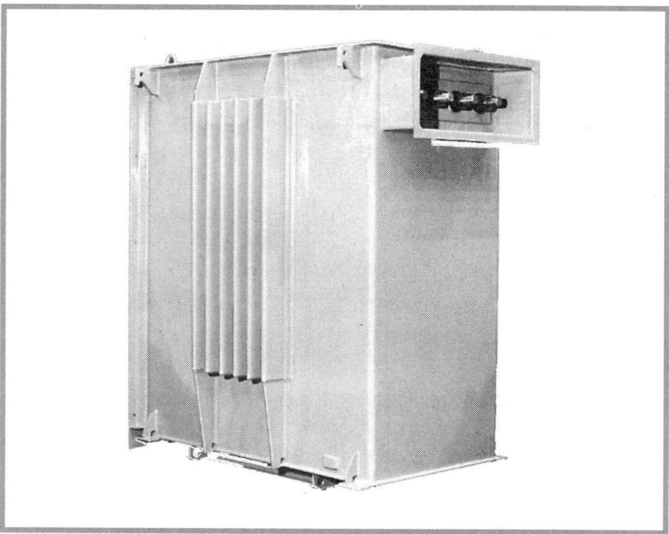

Figure 450.13 *A 1500-kVA gas-filled dry-type transformer, available in ratings through 5000 kVA, 341/2 kV, for use indoors or outdoors. (Westinghouse Electric Corp.)*

Figure 450.14 shows a dry-type transformer with the outside casing in place and with the latest core and coil design for a typical dry-type power transformer rated at 1000 kilovolt-amperes, 13,800 volts to 480 volts, 3-phase, 60 Hz. This transformer has a high-voltage and low-voltage flange for connection to switchgear and a high-voltage, 2-position (double-throw), 3-pole-load air-break switch that may be attached to the case and arranged as a selector switch for the connection of the transformer primary to either of two feeder sources.

Dry-type transformers rated $112^1/_2$ kilovolt-amperes or less require 12 in. of separation from combustible material or separation by fire-resistant barriers. Transformers rated less than 600 volts and completely enclosed, except for ventilating openings, are exempt from this requirement unless the manufacturer's installation instructions specify clearance distances. Noncombustible insulations used in transformers, such as mica, porcelain, and glass, which can withstand high temperatures, have permitted the application of larger dry-type transformers. Combustible materials, however, such as varnishes, may have been used with those insulations, and, under short-circuit conditions, flames can escape from the transformer enclosure. Transformers rated over $112^1/_2$ kilovolt-amperes are required to be located in fire-

Figure 450.14 A dry-type transformer with a core and coil design rated at 1000 kVA, 13,800 volts to 480 volts, 3-phase, 60 Hz. (Westinghouse Electric Corp.)

resistant transformer rooms or vaults unless either of the exceptions apply.

450-22. Dry-Type Transformers Installed Outdoors. Dry-type transformers installed outdoors shall have a weatherproof enclosure.

Transformers exceeding 112½ kVA shall not be located within 12 in. (305 mm) of combustible materials of buildings unless the transformer has Class 155 insulation systems or higher and is completely enclosed except for ventilating openings.

See the commentary following 450-21, Exception No. 2, for an explanation of the change from specific temperature rating (80°C rise or higher rating) to the Class insulation system (Class 155 or higher.)

450-23. Less-Flammable Liquid-Insulated Transformers. Transformers insulated with listed less-flammable liquids that have a fire point of not less than 300°C shall be permitted to be installed in accordance with (a) or (b).

(a) Indoor Installations. In accordance with one of the following.

(1) In Type I or Type II buildings, in areas where all of the following requirements are met:

(a) The transformer is rated 35,000 volts or less.
(b) No combustible materials are stored.
(c) A liquid confinement area is provided.
(d) The installation complies with all restrictions provided for in the listing of the liquid.

(2) With an automatic fire extinguishing system and a liquid confinement area, provided the transformer is rated 35,000 volts or less.
(3) In accordance with Section 450-26.

(b) Outdoor Installations. Less-flammable liquid-filled transformers shall be permitted to be installed outdoors attached to, adjacent to, or on the roof of buildings, where installed in accordance with (1) or (2).

(1) For Type I and Type II buildings, the installation shall comply with all restrictions provided for in the listing of the liquid.

FPN: Installations adjacent to combustible material, fire escapes, or door and window openings may require additional safeguards such as those listed in Section 450-27.

(2) In accordance with Section 450-27.

FPN No. 1: As used in this section, *Type I and Type II buildings* refers to Type I and Type II building construction as defined in *Standard on Types of Building Construction,* NFPA 220-1995. *Combustible materials* refers to those materials not classified as noncombustible or limited-combustible as defined in *Standard on Types of Building Construction,* NFPA 220-1995.

NFPA 220-1995, *Standard on Types of Building Construction,* defines Type I building construction as "that type in which the structural members, including walls, columns, beams, girders, trusses, arches, floors, and roofs, are of approved noncombustible or limited-combustible materials and have fire resistance ratings not less than those specified in Table 3-1."

Type II building construction is defined as "that type not qualifying as Type I construction in which the structural members, including walls, columns, beams, girders, trusses, arches, floors, and roofs, are of approved noncombustible or limited-combustible materials and shall have fire resistance ratings not less than those specified in Table 3-1."

FPN No. 2: See definition of *Listed* in Article 100.

Two listing agencies, Factory Mutual Research Corp. and Underwriters Laboratories Inc., list less-flammable liquids for transformers. These liquids have a fire point of at least 300°C. Figure 450.15 shows an example of a liquid-insulated transformer.

The Factory Mutual Research Corp. listing is based on the use of a Factory Mutual approved less-flammable fluid in a transformer tank that meets certain criteria. Pressure-relief devices must be provided. Factory Mutual also recommends the

Table 3-1 Fire Resistance Ratings (in Hours) for Type I and Type II Construction*

	Type I		Type II	
Exterior Bearing Walls –				
Supporting more than one floor, columns, or other bearing walls	4	3	2	1
Supporting one floor only	4	3	2	1
Supporting a roof only	4	3	1	1
Interior Bearing Walls –				
Supporting more than one floor, columns, or other bearing walls	4	3	2	1
Supporting one floor only	3	2	2	1
Supporting roofs only	3	2	1	1
Columns –				
Supporting more than one floor, columns, or other bearing walls	4	3	2	1
Supporting one floor only	3	2	2	1
Supporting roofs only	3	2	1	1
Beams, Girders, Trusses & Arches –				
Supporting more than one floor, columns, or other bearing walls	4	3	2	1
Supporting one floor only	3	2	2	1
Supporting roofs only	3	2	1	1
Floor Construction	3	2	2	1
Roof Construction	2	1½	1	1
Exterior Nonbearing Walls	0	0	0	0

*For further information, see NFPA 220-1995, *Standard on Types of Building Construction.*

Figure 450.15 *A liquid-insulated transformer filled with a listed less-flammable liquid having fire point of at least 300°C. (Square D Co.)*

use of enhanced electrical protection. Spacing from adjacent combustibles must be provided, based on the fluid capacity of the transformer tank, as illustrated in Figure 450.16. In the event of a leak, the liquid confinement area is intended to prevent transformer dielectric fluid from spreading beyond the vicinity of the transformer, as illustrated in Figure 450.17. Further information on applications may be found in the Factory Mutual Loss Prevention Data Sheet 5-4/14-8.

The Underwriters Laboratories Inc. listing is based on UL requirements that no tank rupture or noted fluid leakage occur during low- and high-current arcing fault tests. Further information may be obtained from the 1997 UL *Gas and Oil Equipment Directory,* under Transformer Fluids (EOVK), or from the manufacturer.

450-24. Nonflammable Fluid-Insulated Transformers. Transformers insulated with a dielectric fluid identified as nonflammable shall be permitted to be installed indoors or outdoors. Such transformers installed indoors and rated over 35,000 volts shall be installed in a vault. Such transformers installed indoors shall be furnished with a liquid confinement area and a pressure-relief vent. The transformers shall be furnished with a means for absorbing any gases generated

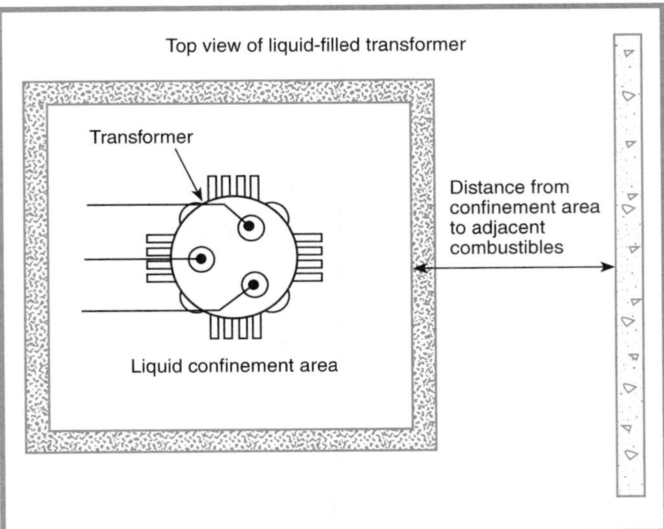

Figure 450.16 *A transformer tank containing a Factory Mutual listed less-flammable fluid, where the spacing from adjacent combustibles to the liquid confinement area is based on the capacity of the tank.*

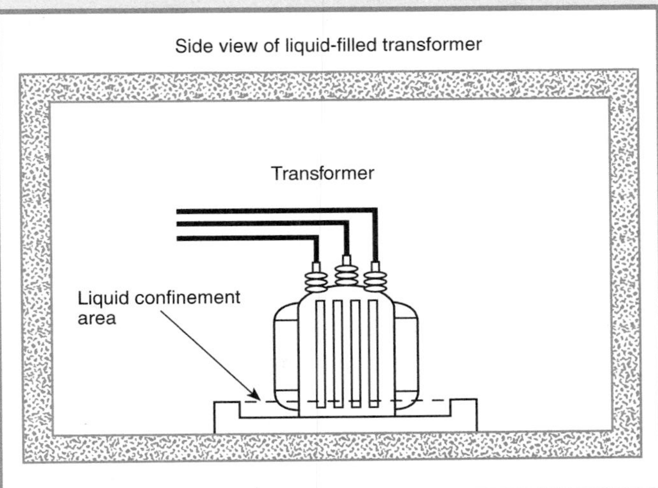

Figure 450.17 *A transformer tank containing a Factory Mutual listed less-flammable fluid, where the volume of the confinement area is based on the capacity of the tank.*

by arcing inside the tank, or the pressure-relief vent shall be connected to a chimney or flue that will carry such gases to an environmentally safe area.

FPN: Safety may be increased if fire hazard analyses are performed for such transformer installations.

For the purposes of this section, a nonflammable dielectric fluid is one that does not have a flash point or fire point, and is not flammable in air.

Section 450-24 was revised for the 1990 *Code* to require a liquid confinement area and a pressure-relief vent. The liquid confinement area is intended to limit the extent of a spill if the tank leaks or ruptures. If a means for absorbing gases generated by arcing within the transformer is not provided, the pressure-relief vent must be connected to a chimney or flue that vents to an environmentally safe area.

The need for a gas absorption system or a chimney or flue that vents to an environmentally safe area is due to concerns about products generated during arcing. The high arc temperatures may cause the insulating medium to break down, resulting in the evolution of toxic or corrosive compounds.

450-25. Askarel-Insulated Transformers Installed Indoors. Askarel-insulated transformers installed indoors and rated over 25 kVA shall be furnished with a pressure-relief vent. Where installed in a poorly ventilated place, they shall be furnished with a means for absorbing any gases generated by arcing inside the case, or the pressure-relief vent shall be connected to a chimney or flue that will carry such gases outside the building. Askarel-insulated transformers rated over 35,000 volts shall be installed in a vault.

450-26. Oil-Insulated Transformers Installed Indoors. Oil-insulated transformers installed indoors shall be installed in a vault constructed as specified in Part C of this article.

Exception No. 1: Where the total capacity does not exceed 112½ kVA, the vault specified in Part C of this article shall be permitted to be constructed of reinforced concrete that is not less than 4 in. (102 mm) thick.

Exception No. 2: Where the nominal voltage does not exceed 600, a vault shall not be required if suitable arrangements are made to prevent a transformer oil fire from igniting other materials, and the total capacity in one location does not exceed 10 kVA in a section of the building classified as combustible, or 75 kVA where the surrounding structure is classified as fire-resistant construction.

Exception No. 3: Electric furnace transformers that have a total rating not exceeding 75 kVA shall be permitted to be installed without a vault in a building or room of fire-resistant construction, provided suitable arrangements are made to prevent a transformer oil fire from spreading to other combustible material.

Exception No. 4: A transformer that has a total rating not exceeding 75 kVA and a supply voltage of 600 volts or less that is an integral part of charged particle accelerating equipment shall be permitted to be installed without a vault in a building or room of noncombustible or fire-resistant construction, provided suitable arrangements are made to prevent a transformer oil fire from spreading to other combustible material.

Exception No. 5: Transformers shall be permitted to be installed in a detached building that does not comply with Part C of this article if neither the building nor its contents present a fire hazard to any other building or property, and if the building is used only in supplying electric service and the interior is accessible only to qualified persons.

Exception No. 6: Oil-insulated transformers shall be permitted to be used without a vault in portable and mobile surface mining equipment (such as electric excavators) if each of the following conditions is met.

(a) Provision is made for draining leaking fluid to the ground.
(b) Safe egress is provided for personnel.
(c) A minimum ¼-in. (6.35-mm) steel barrier is provided for personnel protection.

450-27. Oil-Insulated Transformers Installed Outdoors. Combustible material, combustible buildings, and parts of buildings, fire escapes, and door and window openings shall be safeguarded from fires originating in oil-insulated transformers installed on roofs, attached to or adjacent to a building or combustible material.

Space separations, fire-resistant barriers, automatic water spray systems, and enclosures that confine the oil of a ruptured transformer tank are recognized safeguards. One or more of these safeguards shall be applied according to the degree of hazard involved in cases where the transformer installation presents a fire hazard.

Oil enclosures shall be permitted to consist of fire-resistant dikes, curbed areas or basins, or trenches filled with coarse, crushed stone. Oil enclosures shall be provided with trapped drains where the exposure and the quantity of oil involved are such that removal of oil is important.

FPN: For additional information on transformers installed on poles or structures or underground, see *National Electrical Safety Code*, ANSI C2-1997.

450-28. Modification of Transformers. When modifications are made to a transformer in an existing installation that change the type of the transformer with respect to Part B of this article, such transformer shall be marked to show the type of insulating liquid installed, and the modified transformer installation shall comply with the applicable requirements for that type of transformer.

Askarel-insulated transformers are permitted to be modified by replacing the askarel with either oil or a less-flammable liquid. Where such a modification takes place, the completed installation is required to have the same degree of safety as a new installation. For example, replacement of askarel with oil in an indoor installation without a vault may not be acceptable (see Section 450-26 and its exceptions). The same is true if the replacement liquid is a less-flammable liquid (see Section 450-23).

C. Transformer Vaults

450-41. Location. Vaults shall be located where they can be ventilated to the outside air without using flues or ducts wherever such an arrangement is practicable.

450-42. Walls, Roofs, and Floors. The walls and roofs of vaults shall be constructed of materials that have adequate structural strength for the conditions with a minimum fire resistance of 3 hours. The floors of vaults in contact with the earth shall be of concrete that is not less than 4 in. (102 mm) thick, but where the vault is constructed with a vacant space or other stories below it, the floor shall have adequate structural strength for the load imposed thereon and a minimum fire resistance of 3 hours. For the purposes of this section, studs and wallboard construction shall not be acceptable.

Vaults are intended primarily as passive fire protection. The need for vaults is dictated by the combustibility of the dielectric media and the size of the transformer. Transformers insulated with mineral oil have the greatest need for passive protection, to prevent the spread of burning oil to other combustible materials.

In past editions of the *Code*, Section 450-42 prohibited the use of studs and wallboard to construct a 3-hour-rated fire wall. This restriction was removed for the 1999 *Code*. Now, vault walls and roofs are permitted to be constructed of any fire resistive material including studs and wallboard, provided, of course, the wall or roof assembly has a 3-hour fire resistance rating.

There is less need for a vault around a dry-type transformer of less than 35,000 volts. If the transformer has adequate clearance from combustible construction and storage (see Section 450-21), a fire would be confined to the transformer.

Askarel-insulated transformers of less than 35,000 volts do not require vaults, since askarel is considered a noncombustible fluid. Transformers with a listed less-flammable liquid insulation may be installed without a vault, as permitted in Section 450-23. See the commentary following Section 450-23(b)(2) that relates to Type I and Type II building construction.

Exception: Where transformers are protected with automatic sprinkler, water spray, carbon dioxide, or halon, construction of 1-hour rating shall be permitted.

FPN No. 1: For additional information, see *Method for Fire Tests of Building Construction and Materials*, ANSI/ASTM E119-1995, and *Standard Methods of Tests of Fire Endurance of Building Construction and Materials*, NFPA 251-1995.

FPN No. 2: A typical 3-hour construction is 6 in. (152-mm) thick reinforced concrete.

450-43. Doorways. Vault doorways shall be protected as follows.

(a) Type of Door. Each doorway leading into a vault from the building interior shall be provided with a tight-fitting door that has a minimum fire rating of 3 hours. The authority having jurisdiction shall be permitted to require such a door for an exterior wall opening where conditions warrant.

Exception: Where transformers are protected with automatic sprinkler, water spray, carbon dioxide, or halon, construction of 1-hour rating shall be permitted.

FPN: For additional information, see *Standard for Fire Doors and Fire Windows*, NFPA 80-1995.

(b) Sills. A door sill or curb that is of sufficient height to confine the oil from the largest transformer within the vault shall be provided, and in no case shall the height be less than 4 in. (102 mm).

(c) Locks. Doors shall be equipped with locks, and doors shall be kept locked, access being allowed only to qualified persons. Personnel doors shall swing out and be equipped with panic bars, pressure plates, or other devices that are normally latched but open under simple pressure.

Section 450-43 prohibits the use of conventional rotation-type door knobs on transformer vault doors. It is believed that an injured worker attempting to escape from a transformer vault may not be able to operate a rotating-type door knob but would be able to operate panic-type door hardware.

450-45. Ventilation Openings. Where required by Section 450-9, openings for ventilation shall be provided in accordance with (a) through (f).

(a) Location. Ventilation openings shall be located as far away as possible from doors, windows, fire escapes, and combustible material.

(b) Arrangement. A vault ventilated by natural circulation of air shall be permitted to have roughly half of the total area of openings required for ventilation in one or more openings near the floor and the remainder in one or more openings in the roof or in the sidewalls near the roof, or all of the area required for ventilation shall be permitted in one or more openings in or near the roof.

(c) Size. For a vault ventilated by natural circulation of air to an outdoor area, the combined net area of all ventilating openings, after deducting the area occupied by screens, gratings, or louvers, shall not be less than 3 in.^{2} (1936 mm^2) per kVA of transformer capacity in service, and in no case shall the net area be less than 1 ft^2 (0.093 m^2) for any capacity under 50 kVA.

(d) Covering. Ventilation openings shall be covered with durable gratings, screens, or louvers, according to the treatment required in order to avoid unsafe conditions.

(e) Dampers. All ventilation openings to the indoors shall be provided with automatic closing fire dampers that operate in response to a vault fire. Such dampers shall possess a standard fire rating of not less than 1½ hours.

FPN: See *Standard for Fire Dampers*, ANSI/UL 555-1995.

(f) Ducts. Ventilating ducts shall be constructed of fire-resistant material.

450-46. Drainage. Where practicable, vaults containing more than 100 kVA transformer capacity shall be provided with a drain or other means that will carry off any accumulation of oil or water in the vault unless local conditions make this impracticable. The floor shall be pitched to the drain where provided.

450-47. Water Pipes and Accessories. Any pipe or duct system foreign to the electrical installation shall not enter or pass through a transformer vault. Piping or other facilities provided for vault fire protection, or for transformer cooling, shall not be considered foreign to the electrical installation.

Section 450-47 permits automatic sprinkler protection for transformer vaults. Piping or ductwork for cooling of the transformer is also permitted to be installed in a transformer vault. No other piping or ductwork is permitted to enter or pass through a transformer vault.

450-48. Storage in Vaults. Materials shall not be stored in transformer vaults.

Article 455 — Phase Converters

Contents

A. General

455-1. Scope. This article covers the installation and use of phase converters.

The requirements for phase converters were relocated from Article 430 to Article 455 for the 1993 *Code*. A *phase converter* is an electrical device that converts single-phase electrical power to 3-phase, for the operation of equipment that normally operates from a 3-phase electrical supply. Phase converters are of two types: *static*, with no moving parts, and *rotary*, with an internal rotor that is required to be rotating before a load is applied (see Section 455-2 for definitions).

Phase converters are most commonly used to supply 3-phase motor loads in locations where only single-phase power is available from the local utility. Electrical installations on farms and in other rural areas are examples of such locations. Although their most common loads are motors, phase converters are increasingly used to supply such loads as cellular telephone transmitter sites.

455-2. Definitions.

Manufactured Phase. The manufactured or derived phase originates at the phase converter and is not solidly connected to either of the single-phase input conductors.

Phase Converter. An electrical device that converts single-phase power to 3-phase electrical power.

FPN: Phase converters have characteristics that modify the starting torque and locked-rotor current of motors served, and consideration is required in selecting a phase converter for a specific load.

Rotary-Phase Converter. A device that consists of a rotary transformer and capacitor panel(s) that permits the operation of 3-phase loads from a single-phase supply.

Static-Phase Converter. A device without rotating parts, sized for a given 3-phase load to permit operation from a single-phase supply.

455-3. Other Articles. All applicable requirements of this *Code* shall apply to phase converters except as amended by this article.

455-4. Marking. Each phase converter shall be provided with a permanent nameplate indicating the following:

(1) Manufacturer's name
(2) Rated input and output voltages
(3) Frequency
(4) Rated single-phase input full-load amperes
(5) Rated minimum and maximum single load in kilovolt-amperes (kVA) or horsepower
(6) Maximum total load in kilovolt-ampere (kVA) or horsepower
(7) For a rotary-phase converter, 3-phase amperes at full load

455-5. Equipment Grounding Connection. A means for attachment of an equipment grounding conductor termination in accordance with Section 250-8 shall be provided.

455-6. Conductors.

(a) Ampacity. The ampacity of the single-phase supply conductors shall be determined by (1) or (2).

FPN: Single-phase conductors sized to prevent a voltage drop not exceeding 3 percent from the source of supply to the phase converter may help ensure proper starting and operation of motor loads.

(1) Variable Loads. Where the loads to be supplied are variable, the conductor ampacity shall not be less than 125 percent of the phase converter nameplate single-phase input full-load amperes.

(2) Fixed Loads. Where the phase converter supplies specific fixed loads, and the conductor ampacity is less than 125 percent of the phase converter nameplate single-phase input full-load amperes, the conductors shall have an ampacity not less than 250 percent of the sum of the full-load, 3-phase current rating of the motors and other loads served where the input and output voltages of the phase converter are identical. Where the input and output voltages of the phase converter are different, the current as determined by this section shall be multiplied by the ratio of output to input voltage.

(b) Manufactured Phase Marking. The manufactured phase conductors shall be identified in all accessible locations with a distinctive marking.

455-7. Overcurrent Protection. The single-phase supply conductors and phase converter shall be protected from overcurrent by (a) or (b). Where the required fuse rating or circuit breaker setting does not correspond to a standard rating or setting, the next higher standard rating or setting shall be permitted.

(a) Variable Loads. Where the loads to be supplied are variable, overcurrent protection shall be set at not more than 125 percent of the phase converter nameplate single-phase input full-load amperes.

(b) Fixed Loads. Where the phase converter supplies specific fixed loads, and the conductors are sized in accordance with Section 455-6(a)(2), the conductors shall be protected in accordance with their ampacity. The overcurrent protection determined from this section shall not exceed 125 percent of the phase converter nameplate single-phase input amperes.

455-8. Disconnecting Means. Means shall be provided to disconnect simultaneously all ungrounded single-phase supply conductors to the phase converter.

(a) Location. The disconnecting means shall be readily accessible and located in sight from the phase converter.

(b) Type. The disconnecting means shall be a switch rated in horsepower, a circuit breaker, or a molded-case switch. Where only nonmotor loads are served, an ampere-rated switch shall be permitted.

(c) Rating. The ampere rating of the disconnecting means shall not be less than 115 percent of the rated maximum single-phase input full-load amperes or, for specific fixed loads, shall be permitted to be selected from (1) or (2).

(1) Current Rated Disconnect. The disconnecting means shall be a circuit breaker or molded-case switch with an ampere rating not less than 250 percent of the sum of the following:

(a) Full-load, 3-phase current ratings of the motors
(b) Other loads served

(2) Horsepower Rated Disconnect. The disconnecting means shall be a switch with a horsepower rating. The equivalent locked rotor current of the horsepower rating of the switch shall not be less than 200 percent of the sum of the following:

(a) Nonmotor loads
(b) The 3-phase, locked-rotor current of the largest motor as determined from Table 430-151(B) and
(c) The full-load current of all other 3-phase motors operating at the same time

(d) Voltage Ratios. The calculations in (c) shall apply directly where the input and output voltages of the phase converter are identical. Where the input and output voltages of the phase converter are different, the current shall be multiplied by the ratio of the output to input voltage.

455-9. Connection of Single-Phase Loads. Where single-phase loads are connected on the load side of a phase converter, they shall not be connected to the manufactured phase.

455-10. Terminal Housings. A terminal housing shall be provided on a phase converter, and the terminal housing shall be in accordance with the provisions of Section 430-12.

B. Specific Provisions Applicable to Different Types of Phase Converters

455-20. Disconnecting Means. The single-phase disconnecting means for the input of a static phase converter shall be permitted to serve as the disconnecting means for the phase converter and a single load if the load is within sight of the disconnecting means.

455-21. Start-Up. Power to the utilization equipment shall not be supplied until the rotary-phase converter has been started.

455-22. Power Interruption. Utilization equipment supplied by a rotary-phase converter shall be controlled in such a manner that power to the equipment will be disconnected in the event of a power interruption.

FPN: Magnetic motor starters, magnetic contactors, and similar devices, with manual or time delay restarting for the load, will provide restarting after power interruption.

455-23. Capacitors. Capacitors that are not an integral part of the rotary-phase conversion system but are installed for a motor load shall be connected to the line side of that motor overload protective device.

Article 460 — Capacitors

Contents

460-1. Scope. This article covers the installation of capacitors on electric circuits.

Surge capacitors or capacitors included as a component part of other apparatus and conforming with the requirements of such apparatus are excluded from these requirements.

This article also covers the installation of capacitors in hazardous (classified) locations as modified by Articles 501 through 503.

460-2. Enclosing and Guarding.

(a) Containing More than 3 gal (11.36 L) of Flammable Liquid. Capacitors containing more than 3 gal (11.36 L) of flammable liquid shall be enclosed in vaults or outdoor fenced enclosures complying with Article 110, Part C. This limit shall apply to any single unit in an installation of capacitors.

(b) Accidental Contact. Where capacitors are accessible to unauthorized and unqualified persons, they shall be enclosed, located, or guarded so that persons cannot come into accidental contact or bring conducting materials into accidental contact with exposed energized parts, terminals, or buses associated with them. However, no additional guarding is required for enclosures accessible only to authorized and qualified persons.

Means are required to drain off the stored charge in a capacitor after the supply circuit has been opened. Otherwise, a person servicing the equipment could receive a severe shock, or damage may occur to the equipment.

Figure 460.1, diagram (a), shows a method in which capacitors are connected in a motor circuit so that they may be switched with the motor. In this arrangement, the stored charge will drain off through the windings when the circuit is opened. Diagram (b) shows another arrangement where the capacitor is connected to the line side of the motor starter contacts. An automatic discharge device and a separate disconnecting means are required.

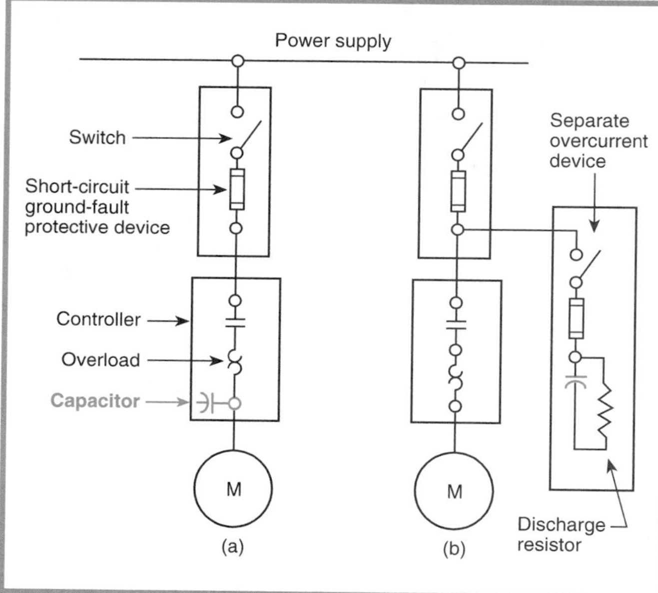

Figure 460.1 *Methods of connecting capacitors in induction motor circuit for power factor correction.*

As shown in Figure 460.2, capacitors are often equipped with built-in resistors to drain off the stored charge, although this type is not needed where connected as shown in Figure 460.1, diagram (a).

A. 600 Volts, Nominal, and Under

460-6. Discharge of Stored Energy. Capacitors shall be provided with a means of discharging stored energy.

(a) Time of Discharge. The residual voltage of a capacitor shall be reduced to 50 volts, nominal, or less, within 1 minute after the capacitor is disconnected from the source of supply.

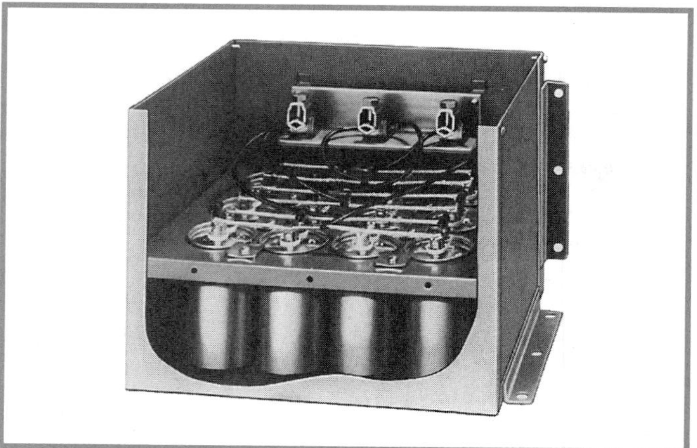

Figure 460.2 *Power factor correction capacitors with discharge resistors. (Square D Co.)*

(b) Means of Discharge. The discharge circuit shall be either permanently connected to the terminals of the capacitor or capacitor bank, or provided with automatic means of connecting it to the terminals of the capacitor bank on removal of voltage from the line. Manual means of switching or connecting the discharge circuit shall not be used.

460-8. Conductors.

(a) Ampacity. The ampacity of capacitor circuit conductors shall not be less than 135 percent of the rated current of the capacitor. The ampacity of conductors that connect a capacitor to the terminals of a motor or to motor circuit conductors shall not be less than one-third the ampacity of the motor circuit conductors and in no case less than 135 percent of the rated current of the capacitor.

(b) Overcurrent Protection.

(1) An overcurrent device shall be provided in each ungrounded conductor for each capacitor bank.

Except as permitted in the exception to Section 460-8(b)(1), it is intended that the overcurrent device be separate from the overcurrent device protecting any other equipment or conductor. See Figure 460.1, (a) and (b).

Exception: A separate overcurrent device shall not be required for a capacitor connected on the load side of a motor overload protective device.

(2) The rating or setting of the overcurrent device shall be as low as practicable.

(c) Disconnecting Means.

(1) A disconnecting means shall be provided in each ungrounded conductor for each capacitor bank.

Except as permitted in the exception to Section 460-8(c)(1), it is intended that the disconnecting means be separate from the disconnecting means for any other equipment. See Figure 460.1, (a) and (b).

Exception: A separate disconnecting means shall not be required where a capacitor is connected on the load side of a motor controller.

(2) The disconnecting means shall open all ungrounded conductors simultaneously.

(3) The disconnecting means shall be permitted to disconnect the capacitor from the line as a regular operating procedure.

(4) The rating of the disconnecting means shall not be less than 135 percent of the rated current of the capacitor.

Capacitors are rated in kilovars, which is abbreviated "kVAr" and stands for reactive kilovolt-amperes. The kVAr rating shows how many reactive kilovolt-amperes the capacitor will supply in order to cancel out the reactive kilovolt-amperes caused by inductance. For example, a 20-kVAr capacitor will cancel out 20 kilovolt-amperes of inductive reactive kilovolt-amperes.

The basic unit is 3-phase and delta-connected internally, but single-phase and 2-phase units are also available. They are constructed with built-in fuses for short-circuit protection and discharge resistors that reduce the voltage to a 50-volt crest or less when disconnected from the power supply. This will occur within 1 minute on 600-volt units and within 5 minutes on 2400- and 4160-volt units.

The capacitor circuit conductors and disconnecting means are required to have an ampacity not less than 135 percent of the rated current of the capacitor. The reason is that all capacitors are manufactured with a tolerance of zero percent to +15 percent, so a 100-kVAr capacitor may actually draw a current equivalent to a 115-kVAr capacitor. In addition, the current drawn by a capacitor varies directly with the line voltage, and any variation in the line voltage from a pure sine wave form causes the capacitor to draw an increased current. Considering these several factors, the increased current can amount to 135 percent of the rated current of the capacitor.

The current corresponding to the kVAr rating of a 3-phase capacitor, I_c, is computed from the following formula:

$$I_c = \frac{\text{kVAr} \times 1000}{1.73 \times \text{volts}}$$

The ampacity of the conductors and the switching device is then determined by multiplying I_c by 1.35. The most effective power factor correction is obtained where the individual capacitors are connected directly to the terminals of the motors, transformers, and other inductive machinery.

Where capacitors are connected together and operated as a unit, no complicated calculations are needed to determine the proper size capacitor to use. Capacitor manufacturers publish tables in which the required capacitor value is obtained by referring to the speed and horsepower of the motor. These values will improve the motor power factor to approximately 95 percent. To improve a plant power factor, capacitor manufacturers also publish tables to assist in calculating the total kVAr rating of capacitors required to improve the power factor to any desired value.

460-9. Rating or Setting of Motor Overload Device. Where a motor installation includes a capacitor connected on the load side of the motor overload device, the rating or setting of the motor overload device shall be based on the improved power factor of the motor circuit.

The effect of the capacitor shall be disregarded in determining the motor circuit conductor rating in accordance with Section 430-22.

Where a capacitor is connected on the load side of the overload relays, as shown in Figure 460.1, diagram (a), consideration is required to be given when selecting the rating or setting of the motor overload device because the line current will be reduced due to an improved power factor. A value lower than indicated in Section 430-32 should be used for proper protection of the motor.

460-10. Grounding. Capacitor cases shall be grounded in accordance with Article 250.

Exception: Capacitor cases shall not be grounded where the capacitor units are supported on a structure designed to operate at other than ground potential.

460-12. Marking. Each capacitor shall be provided with a nameplate giving the name of the manufacturer, rated voltage, frequency, kilovar or amperes, number of phases, and, if filled with a combustible liquid, the amount of liquid in gallons. Where filled with a nonflammable liquid, the nameplate shall so state. The nameplate shall also indicate if a capacitor has a discharge device inside the case.

B. Over 600 Volts, Nominal

460-24. Switching.

(a) Load Current. Group-operated switches shall be used for capacitor switching and shall be capable of the following:

(1) Carrying continuously not less than 135 percent of the rated current of the capacitor installation
(2) Interrupting the maximum continuous load current of each capacitor, capacitor bank, or capacitor installation that will be switched as a unit
(3) Withstanding the maximum inrush current, including contributions from adjacent capacitor installations
(4) Carrying currents due to faults on capacitor side of switch

(b) Isolation.

(1) A means shall be installed to isolate from all sources of voltage each capacitor, capacitor bank, or capacitor installation that will be removed from service as a unit.

(2) The isolating means shall provide a visible gap in the electrical circuit adequate for the operating voltage.

(3) Isolating or disconnecting switches (with no interrupting rating) shall be interlocked with the load-interrupting device or shall be provided with prominently displayed caution signs in accordance with Section 490-22 to prevent switching load current.

(c) Additional Requirements for Series Capacitors. The proper switching sequence shall be ensured by use of one of the following:

(1) Mechanically sequenced isolating and bypass switches
(2) Interlocks
(3) Switching procedure prominently displayed at the switching location

460-25. Overcurrent Protection.

(a) Provided to Detect and Interrupt Fault Current. A means shall be provided to detect and interrupt fault current likely to cause dangerous pressure within an individual capacitor.

(b) Single-Phase or Multiphase Devices. Single-phase or multiphase devices shall be permitted for this purpose.

(c) Protected Individually or in Groups. Capacitors shall be permitted to be protected individually or in groups.

(d) Protective Devices Rated or Adjusted. Protective devices for capacitors or capacitor equipment shall be rated or adjusted to operate within the limits of the safe zone for individual capacitors. If the protective devices are rated or adjusted to operate within the limits for Zone 1 or Zone 2, the capacitors shall be enclosed or isolated.

In no event shall the rating or adjustment of the protective devices exceed the maximum limit of Zone 2.

FPN: For definitions of *Safe Zone, Zone 1,* and *Zone 2,* see *Shunt Power Capacitors,* ANSI/IEEE 18-1992.

The reference to Zones 1 and 2 of ANSI/IEEE 18-1992 pertains to the performance of the capacitors under fault conditions. If a fault current exceeds the limit established for Zone 2, the capacitor tank may burst.

460-26. Identification. Each capacitor shall be provided with a permanent nameplate giving the manufacturer's name, rated voltage, frequency, kilovar or amperes, number of phases, and the amount of liquid in gallons identified as flammable, if such is the case.

460-27. Grounding. Capacitor neutrals and cases, if grounded, shall be grounded in accordance with Article 250.

Exception: Where the capacitor units are supported on a structure that is designed to operate at other than ground potential.

460-28. Means for Discharge.

(a) Means to Reduce the Residual Voltage. A means shall be provided to reduce the residual voltage of a capacitor to 50 volts or less within 5 minutes after the capacitor is disconnected from the source of supply.

(b) Connection to Terminals. A discharge circuit shall be either permanently connected to the terminals of the capacitor or provided with automatic means of connecting it to the terminals of the capacitor bank after disconnection of the capacitor from the source of supply. The windings of motors, or transformers, or of other equipment directly connected to capacitors without a switch or overcurrent device interposed shall meet the requirements of (a).

Article 470 — Resistors and Reactors (For Rheostats, See Section 430-82.)

Contents

A. 600 Volts, Nominal, and Under

470-1. Scope. This article covers the installation of separate resistors and reactors on electric circuits.

Exception: Resistors and reactors that are component parts of other apparatus.

This article also covers the installation of resistors and reactors in hazardous (classified) locations as modified by Articles 501 through 504.

470-2. Location. Resistors and reactors shall not be placed where exposed to physical damage.

470-3. Space Separation. A thermal barrier shall be required if the space between the resistors and reactors and any combustible material is less than 12 in. (305 mm).

470-4. Conductor Insulation. Insulated conductors used for connections between resistance elements and controllers shall be suitable for an operating temperature of not less than 90°C (194°F).

Exception: Other conductor insulations shall be permitted for motor starting service.

Resistors are made in many sizes and shapes and for different purposes. They may be wire or ribbon wound, form wound, edgewise wound, cast grid, punched steel grid, or box resistors. They may be mounted in the open or in ventilated metal boxes or cabinets, depending on their use and location. Since they give off heat, they are required to be guarded and located at safe distances from combustible materials. Where mounted on switchboards or installed in control panels, they are not required to have additional guards.

Reactors are installed in a circuit to introduce inductance for motor starting, controlling the current, and paralleling transformers. Current-limiting reactors are installed to limit the amount of current that can flow in a circuit when a short circuit occurs. Reactors can be divided into two classes: those with iron cores and those with no magnetic materials in the windings. Either type may be air cooled or oil immersed.

Mechanical stresses exist between adjacent air-core reactors due to their external fields, and the manufacturer's recommendations should be followed in spacing and bracing units and fastening supporting insulators.

Saturable reactors may be used for theater dimming (see Section 520-25 and commentary). These have, in addition to the ac winding, an auxiliary winding connected line-to-line or line-to-ground, in order to neutralize charging current and prevent a voltage rise. Those used on high-voltage systems may be oil immersed.

B. Over 600 Volts, Nominal

470-18. General.

(a) Protected Against Physical Damage. Resistors and reactors shall be protected against physical damage.

(b) Isolated by Enclosure or Elevation. Resistors and reactors shall be isolated by enclosure or elevation to protect personnel from accidental contact with energized parts.

(c) Combustible Materials. Resistors and reactors shall not be installed in close enough proximity to combustible materials to constitute a fire hazard and shall have a clearance of not less than 1 ft (305 mm) from combustible materials.

(d) Clearances. Clearances from resistors and reactors to grounded surfaces shall be adequate for the voltage involved.

FPN: See Article 490.

(e) Temperature Rise from Induced Circulating Currents. Metallic enclosures of reactors and adjacent metal parts shall be installed so that the temperature rise from induced circulating currents will not be hazardous to personnel or constitute a fire hazard.

470-19. Grounding. Resistor and reactor cases or enclosures shall be grounded in accordance with Article 250.

Exception: Resistor or reactor cases or enclosures supported on a structure designed to operate at other than ground potential shall not be grounded.

470-20. Oil-Filled Reactors. Installation of oil-filled reactors, in addition to the above requirements, shall comply with applicable requirements of Article 450.

Article 480 — Storage Batteries

Contents

480-1. Scope. The provisions of this article shall apply to all stationary installations of storage batteries.

There are two general types of storage cells: the *lead-acid* type and the *alkali* (nickel-cadmium) type. Basically, a lead-acid cell consists of a positive plate, usually lead peroxide (a semisolid compound) mounted on a framework or grid for support, and a negative plate, made of sponge lead mounted on a grid. Grids are generally made of a lead alloy, such as lead-calcium, lead-antimony, or lead-selenium. The electrolyte is sulfuric acid and distilled water.

Lead-acid cells may be of the vented or sealed (valve-regulated) type. Under normal charging conditions, the vented type will liberate gases, hydrogen at the negative plate and oxygen at the positive plate. The valve-regulated type provides a means to recombine this gas, thus minimizing emissions from the cell.

In the nickel-cadmium battery, the principal active material in the positive plate is nickelous hydroxide; in the negative plate, it is cadmium hydroxide. The electrolyte is potassium hydroxide (an alkali).

In stationary installations, nickel-cadmium cells are generally of the vented type and will liberate hydrogen and oxygen during normal charging. Hermetically sealed nickel-cadmium cells are sometimes used, but they require special charging equipment to prevent gas emissions.

480-2. Definitions.

Nominal Battery Voltage. The voltage computed on the basis of 2 volts per cell for the lead-acid type and 1.2 volts per cell for the alkali type.

Sealed Cell or Battery. A sealed cell or battery is one that has no provision for the addition of water or electrolyte or for external measurement of electrolyte specific gravity. The individual cells shall be permitted to contain a venting arrangement as described in Section 480-9(b).

Storage Battery. A battery comprised of one or more rechargeable cells of the lead-acid, nickel-cadmium, or other rechargeable electrochemical types.

480-3. Wiring and Equipment Supplied from Batteries. Wiring and equipment supplied from storage batteries shall be subject to the requirements of this *Code* applying to wiring and equipment operating at the same voltage.

480-4. Grounding. The requirements of Article 250 shall apply.

480-5. Insulation of Batteries Not Over 250 Volts. This section shall apply to storage batteries having cells connected so as to operate at a nominal battery voltage of not over 250 volts.

(a) Vented Lead-Acid Batteries. Cells and multicompartment batteries with covers sealed to containers of nonconductive, heat-resistant material shall not require additional insulating support.

(b) Vented Alkaline-Type Batteries. Cells with covers sealed to jars of nonconductive, heat-resistant material shall require no additional insulation support. Cells in jars of conductive material shall be installed in trays of nonconductive material with not more than 20 cells (24 volts, nominal) in the series circuit in any one tray.

(c) Rubber Jars. Cells in rubber or composition containers shall require no additional insulating support where the total nominal voltage of all cells in series does not exceed 150 volts. Where the total voltage exceeds 150 volts, batteries shall be sectionalized into groups of 150 volts or less, and each group shall have the individual cells installed in trays or on racks.

(d) Sealed Cells or Batteries. Sealed cells and multicompartment sealed batteries constructed of nonconductive, heat-resistant material shall not require additional insulating support. Batteries constructed of a conducting container shall have insulating support if a voltage is present between the container and ground.

480-6. Insulation of Batteries of Over 250 Volts. The provisions of Section 480-5 shall apply to storage batteries having the cells connected so as to operate at a nominal voltage exceeding 250 volts, and, in addition, the provisions of this section shall also apply to such batteries. Cells shall be installed in groups having a total nominal voltage of not over 250 volts. Insulation, which can be air, shall be provided between groups and shall have a minimum separation between live battery parts of opposite polarity of 2 in. (50.8 mm) for battery voltages not exceeding 600 volts.

480-7. Racks and Trays. Racks and trays shall comply with (a) and (b).

(a) Racks. Racks, as required in this article, are rigid frames designed to support cells or trays. They shall be substantial and made of the following:

(1) Metal, treated so as to be resistant to deteriorating action by the electrolyte and provided with nonconducting members directly supporting the cells or with continuous insulating material other than paint on conducting members, or
(2) Other construction such as fiberglass or other suitable nonconductive materials

(b) Trays. Trays are frames, such as crates or shallow boxes usually of wood or other nonconductive material, constructed or treated so as to be resistant to deteriorating action by the electrolyte.

480-8. Battery Locations. Battery locations shall conform to (a), (b), and (c).

(a) Ventilation. Provisions shall be made for sufficient diffusion and ventilation of the gases from the battery to prevent the accumulation of an explosive mixture.

Compliance with Section 480-8(a) is necessary to prevent classification of a battery location as a hazardous (classified) location, in accordance with Article 500.

It is not the intent of Section 480-8(a) to mandate mechanical ventilation. Hydrogen disperses rapidly and requires little air movement to prevent accumulation. Unrestricted natural air movement in the vicinity of the battery, together with normal air changes for occupied spaces or heat removal, will normally be sufficient. If the space is confined, mechanical ventilation may be required in the vicinity of the battery.

Hydrogen is lighter than air and will tend to concentrate at ceiling level, so some form of ventilation should be provided at the upper portion of the structure. Ventilation can be a fan, roof ridge vent, or louvered area.

Although valve-regulated batteries are often referred to as "sealed," they actually emit very small quantities of hydrogen gas under normal operation, and are capable of liberating large quantities of explosive gases if overcharged. These batteries therefore require the same amount of ventilation as their vented counterparts.

(b) Live Parts. Guarding of live parts shall comply with Section 110-27.

Batteries should be located in clean, dry rooms. Batteries must be arranged to provide sufficient work space for inspection and maintenance. Provisions must also be made for adequate ventilation, in order to prevent an accumulation of an explosive mixture of the gases from the batteries.

The fumes given off by storage batteries are very corrosive; therefore, wiring and its insulation are required to be of a type that will withstand corrosive action (see Section 310-9). Special precautions are necessary to ensure that all metalwork (metal raceways, metal racks, etc.) is designed or treated so as to be corrosion resistant. Manufacturers sometimes suggest that aluminum or plastic conduit be used to withstand the corrosive battery fumes, or, if steel conduit is used, that it be zinc coated and corrosion protected with a coating of an asphaltum-type paint (see Section 300-6).

Overcharging heats a battery and causes gassing and loss of water. A battery should not be allowed to reach temperatures over 110°F (43.3°C), because heat causes a shedding of active materials from the plates that will eventually form a sediment buildup in the bottom of the case and short circuit the plates and the cell. Because mixtures of oxygen and hydrogen are highly explosive, flame or sparks should never be allowed near a cell, especially if the filler cap is removed.

(c) Working Space. Working space about the battery systems shall comply with Section 110-26. Working clearance shall be measured from the edge of the battery rack.

480-9. Vents.

(a) Vented Cells. Each vented cell shall be equipped with a flame arrester that is designed to prevent destruction of the cell due to ignition of gases within the cell by an external spark or flame under normal operating conditions.

(b) Sealed Cells. Sealed battery or cells shall be equipped with a pressure-release vent to prevent excessive accumulation of gas pressure, or the battery or cell shall be designed to prevent scatter of cell parts in event of a cell explosion.

Article 490 — Equipment, Over 600 Volts, Nominal

Contents

D. Mobile and Portable Equipment
- 490-51. General
  - (a) Covered
  - (b) Other Requirements
  - (c) Protection
  - (d) Disconnecting Means
- 490-52. Overcurrent Protection
- 490-53. Enclosures
- 490-54. Collector Rings
- 490-55. Power Cable Connections to Mobile Machines
- 490-56. High-Voltage Portable Cable for Main Power Supply

E. Electrode-Type Boilers
- 490-70. General
- 490-71. Electric Supply System
- 490-72. Branch-Circuit Requirements
  - (a) Rating
  - (b) Common-Trip Fault Interrupting Device
  - (c) Phase-Fault Protection
  - (d) Ground Current Detection
  - (e) Grounded Neutral Conductor
- 490-73. Pressure and Temperature Limit Control
- 490-74. Grounding

A. General

490-1. Scope. This article covers the general requirements for equipment operating at more than 600 volts, nominal.

For the 1999 *Code*, the high-voltage requirements previously found in the 1996 *NEC* Article 710, Over 600 Volts, Nominal, General, were distributed to Article 110, General Requirements; Article 240, Overcurrent Protection; Article 300, Wiring Methods; and this new Article 490. Due to this reorganization of Article 710, only the high-voltage requirements for equipment operating at over 600 volts, nominal, are contained in Article 490.

In order to determine the current location of previous requirements from Article 710, the following table is provided:

1996 NEC	1999 NEC
Article 710 — Over 600 Volts, Nominal, General	Article 490— Equipment, Over 600 Volts, Nominal
Part A. General	Article 490, Part A, General
710-1. Scope	490-1
710-2. Definition	490-2
710-3. Other Articles	Deleted
710-4. Wiring Methods	Article 300, Part B
710-4(a). Aboveground Conductors	300-37
710-4(b). Underground Conductors	300-50
Table 710-4(b)	Table 300-50
710-4(c). Busbars	300-37
710-5. Braid-Covered Insulated Conductors — Open Installation	300-39
710-6. Insulation Shielding	300-40
710-7. Grounding	Deleted
710-8. Moisture or Mechanical Protection for Metal-Sheathed Cables	300-42
710-9. Protection of Service Equipment, Metal-Enclosed Power Switchgear, and Industrial Control Assemblies	110-31
Part B. Equipment — General Provisions	Deleted title
710-11. Indoor Installations	Deleted
710-12. Outdoor Installations	Deleted
710-13. Metal-Enclosed Equipment	Deleted
710-14. Oil-Filled Equipment	490-3
Part C. Equipment — Specific Provisions	Title, Article 490, Part B
710-20. Overcurrent Protection	240-100
710-21. Circuit-Interrupting Devices	490-21
710-22. Isolating Means	490-22
710-23. Voltage Regulators	490-23
710-24. Metal-Enclosed Power Switchgear and Industrial Control Assemblies	Article 490, Part C
710-24(a). Scope	490-30
710-24(b). Arrangement of Devices in Assemblies	490-31
710-24(c). Guarding of High-Voltage Energized Parts Within a Compartment	490-32
710-24(d). Guarding of Low-Voltage Energized Parts Within a Compartment	490-33
710-24(e). Clearance for Cable Conductors Entering Enclosure	490-34
710-24(f). Accessibility of Energized Parts	490-35
710-24(g). Grounding	490-36
710-24(h). Grounding of Devices	490-37
710-24(i). Door Stops and Cover Plates	490-38
710-24(j). Gas Discharge from Interrupting Devices	490-39
710-24(k). Inspection Windows	490-40
710-24(l). Location of Devices	490-41
710-24(m). Interlocks — Interrupter Switches	490-42
710-24(n). Stored Energy for Opening	490-43
710-24(o). Fused Interrupter Switches	490-44
710-24(p). Interlocks — Circuit Breakers	490-45
Part D. Installations Accessible to Qualified Persons Only	Deleted
710-31. Enclosure for Electrical Installations	Deleted
710-32. Circuit Conductors	110-36
710-33. Minimum Space Separation	490-46
Table 710-33. Minimum Clearance of Live Parts	Table 490-46

1996 NEC	1999 NEC
710-34. Work Space and Guarding	Deleted
Part E. Mobile and Portable Equipment	Title, Article 490, Part D
710-41. General	490-51
710-42. Overcurrent Protection	490-52
710-43. Enclosures	490-53
710-44. Collector Rings	490-54
710-45. Power Cable Connections to Mobile Machines	490-55
710-46. High-Voltage Portable Cable for Main Power Supply	490-56
710-47. Grounding	Deleted
Part F. Tunnel Installations	Title, Article 110, Part D
710-51. General	110-51
710-52. Overcurrent Protection	110-52
710-53. Conductors	110-53
710-54. Bonding and Equipment Grounding Conductor	110-54
710-55. Transformers, Switches, and Electric Equipment	110-55
710-56. Energized Parts	110-56
710-57. Ventilation System Controls	110-57
710-58. Disconnecting Means	110-58
710-59. Enclosures	110-59
710-60. Grounding	Deleted
Part G. Electrode-Type Boilers	Title, Article 490, Part E
710-70. General	490-70
710-71. Electric Supply System	490-71
710-72. Branch Circuit Requirements	490-72
710-73. Pressure and Temperature Limit Control	490-73
710-74. Grounding	490-74

FPN: See *Standard for Electrical Safety Requirements for Employee Workplaces*, NFPA 70E-1995, for electrical safety requirements for employee workplaces.

490-2. Definition. For the purposes of this article, *high voltage* shall be defined as more than 600 volts, nominal.

490-3. Oil-Filled Equipment. Installation of electrical equipment, other than transformers covered in Article 450, containing more than 10 gal (37.85 L) of flammable oil per unit shall meet the requirements of Parts B and C of Article 450.

B. Equipment — Specific Provisions

490-21. Circuit-Interrupting Devices.

(a) Circuit Breakers.

(1) Circuit breakers installed indoors shall be mounted either in metal-enclosed units or fire-resistant cell-mounted units, or they shall be permitted to be open mounted in locations accessible to qualified persons only.

(2) Circuit breakers used to control oil-filled transformers shall be located either outside the transformer vault or be capable of operation from outside the vault.

(3) Oil circuit breakers shall be arranged or located so that adjacent readily combustible structures or materials are safeguarded in an approved manner.

(4) Circuit breakers shall have the following equipment or operating characteristics.

(a) An accessible mechanical or other approved means for manual tripping, independent of control power.
(b) Be release free (trip free).
(c) If capable of being opened or closed manually while energized, the main contacts shall operate independently of the speed of the manual operation.
(d) A mechanical position indicator at the circuit breaker to show the open or closed position of the main contacts.
(e) A means of indicating the open and closed position of the breaker at the point(s) from which they may be operated.
(f) A permanent and legible nameplate showing manufacturer's name or trademark, manufacturer's type or identification number, continuous current rating, interrupting rating in megavolt-amperes (MVA) or amperes, and maximum voltage rating. Modification of a circuit breaker affecting its rating(s) shall be accompanied by an appropriate change of nameplate information.

(5) The continuous current rating of a circuit breaker shall not be less than the maximum continuous current through the circuit breaker.

(6) The interrupting rating of a circuit breaker shall not be less than the maximum fault current the circuit breaker will be required to interrupt, including contributions from all connected sources of energy.

(7) The closing rating of a circuit breaker shall not be less than the maximum asymmetrical fault current into which the circuit breaker can be closed.

(8) The momentary rating of a circuit breaker shall not be less than the maximum asymmetrical fault current at the point of installation.

(9) The rated maximum voltage of a circuit breaker shall not be less than the maximum circuit voltage.

(b) Power Fuses and Fuseholders.

(1) Use. Where fuses are used to protect conductors and equipment, a fuse shall be placed in each ungrounded conductor. Two power fuses shall be permitted to be used in parallel to protect the same load, if both fuses have identical ratings, and both fuses are installed in an identified common mounting with electrical connections that will divide the current equally. Power fuses of the vented type shall not be used indoors, underground, or in metal enclosures unless identified for the use.

(2) Interrupting Rating. The interrupting rating of power fuses shall not be less than the maximum fault current the fuse will be required to interrupt, including contributions from all connected sources of energy.

(3) Voltage Rating. The maximum voltage rating of power fuses shall not be less than the maximum circuit voltage. Fuses having a minimum recommended operating voltage shall not be applied below this voltage.

(4) Identification of Fuse Mountings and Fuse Units. Fuse mountings and fuse units shall have permanent and legible nameplates showing the manufacturer's type or designation, continuous current rating, interrupting current rating, and maximum voltage rating.

(5) Fuses. Fuses that expel flame in opening the circuit shall be designed or arranged so that they will function properly without hazard to persons or property.

(6) Fuseholders. Fuseholders shall be designed or installed so that they will be de-energized while replacing a fuse.

Exception: Fuses and fuseholders designed to permit fuse replacement by qualified persons using equipment designed for the purpose without de-energizing the fuseholder shall be permitted.

(7) High-Voltage Fuses. Metal-enclosed switchgear and substations that utilize high-voltage fuses shall be provided with a gang-operated disconnecting switch. Isolation of the fuses from the circuit shall be provided by either connecting a switch between the source and the fuses or providing roll-out switch and fuse-type construction. The switch shall be of the load-interrupter type, unless mechanically or electrically interlocked with a load-interrupting device arranged to reduce the load to the interrupting capability of the switch.

Exception: More than one switch shall be permitted as the disconnecting means for one set of fuses where the switches are installed to provide connection to more than one set of supply conductors. The switches shall be mechanically or electrically interlocked to permit access to the fuses only when all switches are open. A conspicuous sign shall be placed at the fuses reading

WARNING — FUSES MAY BE ENERGIZED FROM MORE THAN ONE SOURCE.

(c) Distribution Cutouts and Fuse Links — Expulsion Type.

(1) Installation. Cutouts shall be located so that they may be readily and safely operated and re-fused, and so that the exhaust of the fuses will not endanger persons. Distribution cutouts shall not be used indoors, underground, or in metal enclosures.

(2) Operation. Where fused cutouts are not suitable to interrupt the circuit manually while carrying full load, an approved means shall be installed to interrupt the entire load. Unless the fused cutouts are interlocked with the switch to prevent opening of the cutouts under load, a conspicuous sign shall be placed at such cutouts reading:

WARNING — DO NOT OPEN UNDER LOAD.

(3) Interrupting Rating. The interrupting rating of distribution cutouts shall not be less than the maximum fault current the cutout will be required to interrupt, including contributions from all connected sources of energy.

(4) Voltage Rating. The maximum voltage rating of cutouts shall not be less than the maximum circuit voltage.

(5) Identification. Distribution cutouts shall have on their body, door, or fuse tube a permanent and legible nameplate or identification showing the manufacturer's type or designation, continuous current rating, maximum voltage rating, and interrupting rating.

(6) Fuse Links. Fuse links shall have a permanent and legible identification showing continuous current rating and type.

(7) Structure Mounted Outdoors. The height of cutouts mounted outdoors on structures shall provide safe clearance between lowest energized parts (open or closed position) and standing surfaces, in accordance with Section 110-34(e).

(d) Oil-Filled Cutouts.

(1) Continuous Current Rating. The continuous current rating of oil-filled cutouts shall not be less than the maximum continuous current through the cutout.

(2) Interrupting Rating. The interrupting rating of oil-filled cutouts shall not be less than the maximum fault current the oil-filled cutout will be required to interrupt, including contributions from all connected sources of energy.

(3) Voltage Rating. The maximum voltage rating of oil-filled cutouts shall not be less than the maximum circuit voltage.

(4) Fault Closing Rating. Oil-filled cutouts shall have a fault closing rating not less than the maximum asymmetrical fault current that can occur at the cutout location, unless suitable interlocks or operating procedures preclude the possibility of closing into a fault.

(5) Identification. Oil-filled cutouts shall have a permanent and legible nameplate showing the rated continuous current, rated maximum voltage, and rated interrupting current.

(6) Fuse Links. Fuse links shall have a permanent and legible identification showing the rated continuous current.

(7) Location. Cutouts shall be located so that they will be readily and safely accessible for re-fusing, with the top of the cutout not over 5 ft (1.52 m) above the floor or platform.

(8) Enclosure. Suitable barriers or enclosures shall be provided to prevent contact with nonshielded cables or energized parts of oil-filled cutouts.

(e) Load Interrupters. Load-interrupter switches shall be permitted if suitable fuses or circuit breakers are used in conjunction with these devices to interrupt fault currents. Where these devices are used in combination, they shall be coordinated electrically so that they will safely withstand the effects of closing, carrying, or interrupting all possible currents up to the assigned maximum short-circuit rating.

Where more than one switch is installed with interconnected load terminals to provide for alternate connection to different supply conductors, each switch shall be provided with a conspicuous sign reading

WARNING — SWITCH MAY BE ENERGIZED BY BACKFEED.

(1) Continuous Current Rating. The continuous current rating of interrupter switches shall equal or exceed the maximum continuous current at the point of installation.

(2) Voltage Rating. The maximum voltage rating of interrupter switches shall equal or exceed the maximum circuit voltage.

(3) Identification. Interrupter switches shall have a permanent and legible nameplate including the following information: manufacturer's type or designation, continuous current rating, interrupting current rating, fault closing rating, maximum voltage rating.

(4) Switching of Conductors. The switching mechanism shall be arranged to be operated from a location where the operator is not exposed to energized parts and shall be arranged to open all ungrounded conductors of the circuit simultaneously with one operation. Switches shall be arranged to be locked in the open position. Metal-enclosed switches shall be operable from outside the enclosure.

(5) Stored Energy for Opening. The stored-energy operator shall be permitted to be left in the uncharged position after the switch has been closed if a single movement of the operating handle charges the operator and opens the switch.

(6) Supply Terminals. The supply terminals of fused interrupter switches shall be installed at the top of the switch enclosure or, if the terminals are located elsewhere, the equipment shall have barriers installed so as to prevent persons from accidentally contacting energized parts or dropping tools or fuses into energized parts.

See Figures 490.1 and 490.2 for an example of a fused interrupter switch and the fuseholder components.

490-22. Isolating Means. Means shall be provided to completely isolate an item of equipment. The use of isolating switches shall not be required where there are other ways of de-energizing the equipment for inspection and repairs, such as drawout-type metal-enclosed switchgear units and removable truck panels.

Isolating switches not interlocked with an approved circuit-interrupting device shall be provided with a sign warning against opening them under load.

A fuseholder and fuse, designed for the purpose, shall be permitted as an isolating switch.

490-23. Voltage Regulators. Proper switching sequence for regulators shall be ensured by use of one of the following:

(1) Mechanically sequenced regulator bypass switch(es)
(2) Mechanical interlocks
(3) Switching procedure prominently displayed at the switching location

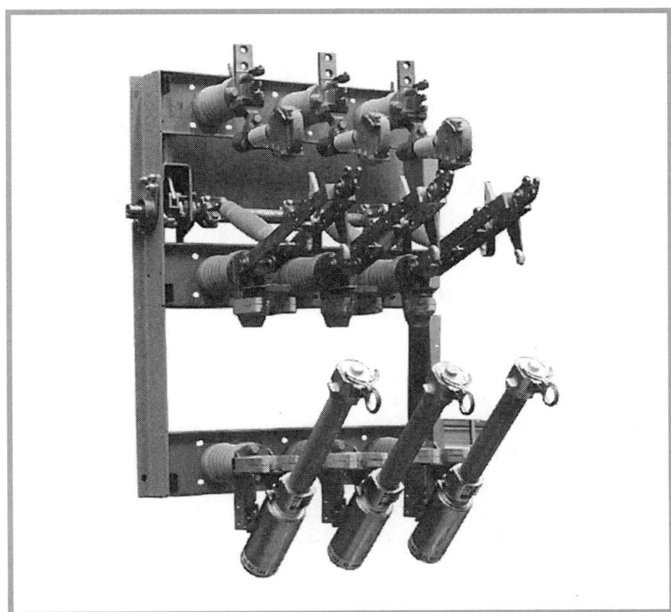

Figure 490.1 *Group-operated interrupter-switch and powerfuse combination rated at 13.8 kV, 600 amperes continuous and interrupting, 40,000 amperes momentary, 40,000 amperes fault closing.*

Figure 490.2 *Components of the indoor solid-material (SM) power fuseholder (borix-acid arc-extinguishing type) with a 14.4 kV, 400E-ampere maximum, 40,000-ampere rms asymmetrical interrupting rating. Shown here are the spring and cable assembly, refill unit, holder, and snuffler. (S & C Electric Co.)*

490-24. Minimum Space Separation. In field-fabricated installations, the minimum air separation between bare live conductors and between such conductors and adjacent grounded surfaces shall not be less than the values given in Table 490-24. These values shall not apply to interior portions or exterior terminals of equipment designed, manufactured, and tested in accordance with accepted national standards.

C. Equipment — Metal-Enclosed Power Switchgear and Industrial Control Assemblies

490-30. General. This part covers assemblies of metal-enclosed power switchgear and industrial control, including but not limited to switches, interrupting devices and their control, metering, protection and regulating equipment, where an integral part of the assembly, with associated interconnections and supporting structures. This part also includes metal-enclosed power switchgear assemblies that form a part of unit substations, power centers, or similar equipment.

For the control and protection of feeders leaving a substation, Figure 490.3 [old Figure 710-1] illustrates a typical example of modern, metal-enclosed switchgear. This industrial unit substation includes a high-voltage disconnect switch, transformer, and low-voltage switchgear with a fully functioning ground-fault relay protection system.

Indicator instruments, such as voltmeters, ammeters, wattmeters, and protective relays, may be mounted on the panel doors as desired. This switchgear affords a high degree of safety because all live parts are metal-enclosed, and interlocks are provided for safe operation.

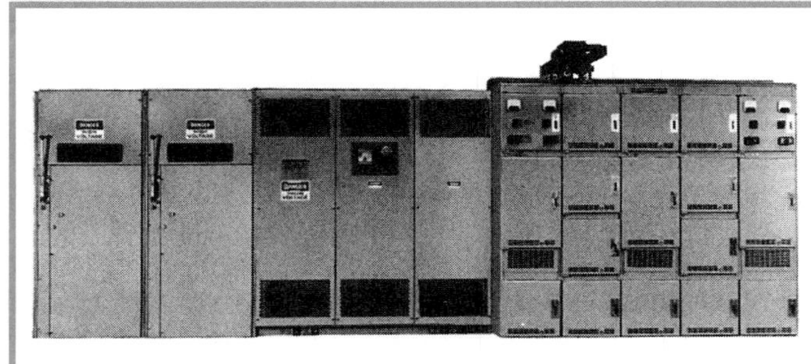

Figure 490.3 *An assembly of metal-enclosed switchgear. (Square D Co.)*

An example of a high-voltage pad-mounted transformer and enclosure that may contain primary and

Table 490-24. Minimum Clearance of Live Parts*

Nominal Voltage Rating (kV)	Impulse Withstand, B.I.L (kV)		Minimum Clearance of Live Parts (in.)			
			Phase-to-Phase		Phase-to-Ground	
	Indoors	Outdoors	Indoors	Outdoors	Indoors	Outdoors
2.4–4.16	60	95	4.5	7	3.0	6
7.2	75	95	5.5	7	4.0	6
13.8	95	110	7.5	12	5.0	7
14.4	110	110	9.0	12	6.5	7
23	125	150	10.5	15	7.5	10
34.5	150	150	12.5	15	9.5	10
	200	200	18.0	18	13.0	13
46	—	200	—	18	—	13
	—	250	—	21	—	17
69	—	250	—	21	—	17
	—	350	—	31	—	25
115	—	550	—	53	—	42
138	—	550	—	53	—	42
	—	650	—	63	—	50
161	—	650	—	63	—	50
	—	750	—	72	—	58
230	—	750	—	72	—	58
	—	900	—	89	—	71
	—	1050	—	105	—	83

Note: For SI units, 1 in. = 25.4 mm.

*The values given are the minimum clearance for rigid parts and bare conductors under favorable service conditions. They shall be increased for conductor movement or under unfavorable service conditions or wherever space limitations permit. The selection of the associated impulse withstand voltage for a particular system voltage is determined by the characteristics of the surge protective equipment.

secondary switches or circuit breakers is shown in Figure 490.4.

Figure 490.4 A 300-kVA, 15-kV pad-mounted transformer integral unit containing a primary hookstick-operated switch with a limited number of secondary breakers or switches. (Square D Co.)

An example of a high voltage indoor transformer for placement into a unit substation is shown in Figure 490.5.

Figure 490.5 A 1500-kVA, 15-kV unit substation transformer or power-center-type transformer adaptable to a lineup of a high-voltage switch, transformer, and low-voltage secondary. (Square D Co.)

490-31. Arrangement of Devices in Assemblies. Arrangement of devices in assemblies shall be such that individual components can safely perform their intended function without adversely affecting the safe operation of other components in the assembly.

490-32. Guarding of High-Voltage Energized Parts Within a Compartment. Where access for other than visual inspection is required to a compartment that contains energized high-voltage parts, barriers shall be provided to prevent accidental contact by persons, tools, or other equipment with energized parts. Exposed live parts shall only be permitted in compartments accessible to qualified persons. Fuses and fuseholders designed to enable future replacement without de-energizing the fuse holder shall only be permitted for use by qualified persons.

490-33. Guarding of Low Voltage Energized Parts Within a Compartment. Energized bare parts mounted on doors shall be guarded where the door must be opened for maintenance of equipment or removal of draw-out equipment.

490-34. Clearance for Cable Conductors Entering Enclosure. The unobstructed space opposite terminals or opposite raceways or cables entering a switchgear or control assembly shall be adequate for the type of conductor and method of termination.

490-35. Accessibility of Energized Parts.

(a) Doors that would provide unqualified persons access to high-voltage energized parts shall be locked.

(b) Low-voltage control equipment, relays, motors, and the like shall not be installed in compartments with exposed high-voltage energized parts or high-voltage wiring unless the access means is interlocked with the high-voltage switch or disconnecting means to prevent the access means from being opened or removed unless the high-voltage switch or disconnecting means is in the isolating position.

Exception: High-voltage instrument or control transformers and space heaters shall be permitted to be installed in the high-voltage compartment without access restrictions beyond those that apply to the high-voltage compartment generally.

490-36. Grounding. Frames of switchgear and control assemblies shall be grounded.

490-37. Grounding of Devices. Devices with metal cases or frames, or both, such as instruments, relays, meters, and instrument and control transformers, located in or on switchgear or control, shall have the frame or case grounded.

490-38. Door Stops and Cover Plates. External hinged doors or covers shall be provided with stops to hold them in the open position. Cover plates intended to be removed for inspection of energized parts or wiring shall be equipped with lifting handles and shall not exceed 12 ft^2(1.11 m^2) in area or 60 lb (27.22 kg) in weight, unless they are hinged and bolted or locked.

490-39. Gas Discharge from Interrupting Devices. Gas discharged during operating of interrupting devices shall be directed so as not to endanger personnel.

490-40. Inspection Windows. Windows intended for inspection of disconnecting switches or other devices shall be of suitable transparent material.

490-41. Location of Devices.

(a) Control and instrument transfer switch handles or push buttons other than those covered in (b), shall be in a readily accessible location at an elevation of not over 78 in. (1.98 m).

Exception: Operating handles requiring more than 50 lb (22.68 kg) of force shall be located no higher than 66 in. (1.68 m) in either the open or closed position.

(b) Operating handles for infrequently operated devices, such as drawout fuses, fused potential or control transformers and their primary disconnects, and bus transfer switches, shall be permitted to be located where they are safely operable and serviceable from a portable platform.

490-42. Interlocks — Interrupter Switches. Interrupter switches equipped with stored energy mechanisms shall have mechanical interlocks to prevent access to the switch compartment unless the stored energy mechanism is in the discharged or blocked position.

490-43. Stored Energy for Opening. The stored energy operator shall be permitted to be left in the uncharged position after the switch has been closed if a single movement of the operating handle charges the operator and opens the switch.

490-44. Fused Interrupter Switches.

(a) Supply Terminals. The supply terminals of fused interrupter switches shall be installed at the top of the switch enclosure or, if the terminals are located elsewhere, the equipment shall have barriers installed so as to prevent persons from accidentally contacting energized parts or dropping tools or fuses into energized parts.

(b) Backfeed. Where fuses can be energized by backfeed, a sign shall be placed on the enclosure door reading

WARNING — FUSES MAY BE ENERGIZED
BY BACKFEED.

(c) Switching Mechanism. The switching mechanism shall be arranged to be operated from a location outside the enclosure where the operator is not exposed to energized parts and shall be arranged to open all ungrounded conductors of the circuit simultaneously with one operation. Switches shall be capable of being locked in the open position.

490-45. Interlocks — Circuit Breakers.

(a) Circuit breakers equipped with stored energy mechanisms shall be designed to prevent the release of the stored energy unless the mechanism has been fully charged.

(b) Mechanical interlocks shall be provided in the housing to prevent the complete withdrawal of the circuit breaker from the housing when the stored energy mechanism is in the fully charged position, unless a suitable device is provided to block the closing function of the circuit breaker before complete withdrawal.

D. Mobile and Portable Equipment

490-51. General.

(a) Covered. The provisions of this part shall apply to installations and use of high-voltage power distribution and utilization equipment that is portable or mobile, or both, such as substations and switch houses mounted on skids, trailers, or cars; mobile shovels; draglines; cranes; hoists; drills; dredges; compressors; pumps; conveyors; underground excavators; and the like.

(b) Other Requirements. The requirements of this part shall be additional to, or amendatory of, those prescribed in Articles 100 through 725 of this *Code*. Special attention shall be paid to Article 250.

(c) Protection. Adequate enclosures or guarding, or both, shall be provided to protect portable and mobile equipment from physical damage.

(d) Disconnecting Means. Disconnecting means shall be installed for mobile and portable high-voltage equipment according to the requirements of Part H of Article 230 and shall disconnect all ungrounded conductors.

490-52. Overcurrent Protection. Motors driving single or multiple dc generators supplying a system operating on a cyclic load basis do not require overload protection, provided that the thermal rating of the ac drive motor cannot be exceeded under any operating condition. The branch-circuit protective device(s) shall provide short-circuit and locked-rotor protection, and shall be permitted to be external to the equipment.

490-53. Enclosures. All energized switching and control parts shall be enclosed in effectively grounded metal cabinets or enclosures. These cabinets or enclosures shall be marked "DANGER — HIGH VOLTAGE — KEEP OUT" and shall be locked so that only authorized and qualified persons can enter. Circuit breakers and protective equipment shall have the operating means projecting through the metal cabinet or enclosure so these units can be reset without opening locked doors. With doors closed, reasonable safe access for normal operation of these units shall be provided.

490-54. Collector Rings. The collector ring assemblies on revolving-type machines (shovels, draglines, etc.) shall be guarded to prevent accidental contact with energized parts by personnel on or off the machine.

490-55. Power Cable Connections to Mobile Machines. A metallic enclosure shall be provided on the mobile machine for enclosing the terminals of the power cable. The enclosure shall include provisions for a solid connection for the ground wire(s) terminal to effectively ground the machine frame. Ungrounded conductors shall be attached to insulators or be terminated in approved high-voltage cable couplers (which include ground wire connectors) of proper voltage and ampere rating. The method of cable termination used shall prevent any strain or pull on the cable from stressing the electrical connections. The enclosure shall have provision for locking so only authorized and qualified persons may open it and shall be marked

DANGER — HIGH VOLTAGE — KEEP OUT.

490-56. High-Voltage Portable Cable for Main Power Supply. Flexible high-voltage cable supplying power to portable or mobile equipment shall comply with Article 250 and Article 400, Part C.

E. Electrode-Type Boilers

490-70. General. The provisions of this part shall apply to boilers operating over 600 volts, nominal, in which heat is generated by the passage of current between electrodes through the liquid being heated.

490-71. Electric Supply System. Electrode-type boilers shall be supplied only from a 3-phase, 4-wire solidly grounded wye system, or from isolating transformers arranged to provide such a system. Control circuit voltages shall not exceed 150 volts, shall be supplied from a grounded system, and shall have the controls in the ungrounded conductor.

490-72. Branch-Circuit Requirements.

(a) Rating. Each boiler shall be supplied from an individual branch circuit rated not less than 100 percent of the total load.

(b) Common-Trip Fault Interrupting Device. The circuit shall be protected by a 3-phase, common-trip fault-interrupting device, which shall be permitted to automatically reclose the circuit upon removal of an overload condition but shall not reclose after a fault condition.

(c) Phase-Fault Protection. Phase-fault protection shall be provided in each phase, consisting of a separate phase-overcurrent relay connected to a separate current transformer in the phase.

(d) Ground Current Detection. Means shall be provided for detection of the sum of the neutral and ground currents and shall trip the circuit-interrupting device if the sum of those currents exceeds the greater of 5 amperes or 7½ percent of the boiler full-load current for 10 seconds or exceeds an instantaneous value of 25 percent of the boiler full-load current.

(e) Grounded Neutral Conductor. The grounded neutral conductor shall be as follows:

(1) Connected to the pressure vessel containing the electrodes
(2) Insulated for not less than 600 volts
(3) Have not less than the ampacity of the largest ungrounded branch-circuit conductor
(4) Installed with the ungrounded conductors in the same raceway, cable, or cable tray, or where installed as open conductors, in close proximity to the ungrounded conductors
(5) Not be used for any other circuit

490-73. Pressure and Temperature Limit Control. Each boiler shall be equipped with a means to limit the maximum temperature or pressure, or both, by directly or indirectly interrupting all current flow through the electrodes. Such means shall be in addition to the temperature or pressure, or both, regulating systems and pressure relief or safety valves.

490-74. Grounding. All exposed noncurrent-carrying metal parts of the boiler and associated exposed grounded structures or equipment shall be bonded to the pressure vessel or to the neutral conductor to which the vessel is connected in accordance with Section 250-102, except the ampacity of the bonding jumper shall not be less than the ampacity of the neutral conductor.

CHAPTER 5

Special Occupancies

Article 500 — Hazardous (Classified) Locations, Classes I, II, and III, Divisions 1 and 2

Contents

500-1. Scope — Articles 500 through 504. Articles 500 through 504 cover the requirements for electrical and electronic equipment and wiring for all voltages in Class I, Divisions 1 and 2; Class II, Divisions 1 and 2; and Class III, Divisions 1 and 2 locations where fire or explosion hazards may exist due to flammable gases or vapors, flammable liquids, combustible dust, or ignitible fibers or flyings.

Article 500 was revised and reorganized for the 1999 *Code.* Article 500 is limited to those locations classified as Class I, II, or III, Divisions 1 or 2. The new method of classification using zones is for locations classified in accordance the International Electrotechnical Commission (IEC) (see Article 505).

The *Code* does not classify areas for the manufacture, transportation, storage, and use of explosive materials, such as ammunition, dynamite, and blasting powder. Areas where such materials are present are not considered hazardous (classified) locations in accordance with Article 500. However, many organizations responsible for the safety of such areas require equipment and wiring methods suitable for hazardous locations as part of many safety precautions, even though the equipment and wiring have not been investigated for such locations. The logic is that hazardous location equipment and wiring methods are safer than ordinary location equipment and wiring methods. Further information on these locations can be found in NFPA 495-1996, *Explosive Materials Code.*

The *National Electrical Code* does not classify the area. Determination of the parameters and distances and degrees of hazard, is determined by other NFPA technical committees that have the experience and expertise of working with the flammable liquids, gases, vapors, dusts, and flyings that may be present. However, some of this information has been extracted from other NFPA documents and is included in Articles 511 through 517 of the *Code.*

For information on hazardous locations in general, including background on the classification of areas, equipment protection systems, ignition sources, static electricity and lightning, and requirements that apply outside the United States, see *Electrical Installations in Hazardous Locations,* by Peter J. Schram and Mark W. Earley. This book is available from NFPA.

FPN: For the requirements for electrical and electronic equipment and wiring for all voltages in Class I, Zone 0, Zone 1, and Zone 2 hazardous (classified) locations where fire or explosion hazards may exist due to flammable gases or vapors, or flammable liquids, refer to Article 505.

500-2. Other Articles. Except as modified in Articles 500 through 504, all other applicable rules contained in this Code shall apply to electrical equipment and wiring installed in hazardous (classified) locations.

The first four chapters of this *Code* apply to the electrical equipment and wiring in Articles 500 through 505, except for any amendments that appear within these articles. This means that the materials and equipment must be suitable for environmental conditions such as rain, snow, ice, altitude, and heat; deteriorating effect on the conductors and equipment; and interrupting rating sufficient for the nominal circuit voltage and available fault current, just to name a few. Installation of any wiring methods allowed by Articles

500 through 505 must be in accordance with the article governing that particular wiring method and, any amendments to the rules.

500-3. General.

(a) Classifications of Locations. Locations shall be classified depending on the properties of the flammable vapors, liquids, or gases, or combustible dusts or fibers that may be present and the likelihood that a flammable or combustible concentration or quantity is present. Where pyrophoric materials are the only materials used or handled, these locations shall not be classified.

Each room, section, or area shall be considered individually in determining its classification.

Pyrophoric materials ignite spontaneously upon contact with the air. The use of electrical equipment suitable for a hazardous (classified) location will not prevent ignition of the material. The process containment system should be designed to prevent contact between the pyrophoric material and the air.

FPN: Through the exercise of ingenuity in the layout of electrical installations for hazardous (classified) locations, it is frequently possible to locate much of the equipment in less hazardous or in nonhazardous locations and, thus, to reduce the amount of special equipment required.

(b) Documentation. All areas designated as hazardous (classified) locations shall be properly documented. This documentation shall be available to those authorized to design, install, inspect, maintain, or operate electrical equipment at the location.

A new Section 500-3(b) has been added to require documentation of the areas that have been designated hazardous (classified) locations and the degree of classification. One type of documentation consists of area classification drawings. See the FPN to Section 505-10(c). This documentation provides the necessary information for installers and service personnel to ensure that any electrical equipment installed or maintained in those classified areas is of the proper type.

(c) Reference Standards.

FPN No. 1: It is important that the authority having jurisdiction be familiar with recorded industrial experience as well as with the standards of the National Fire Protection Association, the American Petroleum Institute, and the Instrument Society of America that may be of use in the classification of various locations, the determination of adequate ventilation, and the protection against static electricity and lightning hazards.

FPN No. 2: For further information on the classification of locations, see *Flammable and Combustible Liquids Code,* NFPA 30-1996; *Standard for Drycleaning Plants,* NFPA 32-1996; *Standard for Spray Application Using Flammable or Combustible Materials*, NFPA 33-1995; *Standard for Dipping and Coating Processes Using Flammable or Combustible Liquids*, NFPA 34-1995; *Standard for the Manufacture of Organic Coatings,* NFPA 35-1995; *Standard for Solvent Extraction Plants,* NFPA 36-1997; *Standard on Fire Protection for Laboratories Using Chemicals*, NFPA 45-1996; *Standard for Gaseous Hydrogen Systems at Consumer Sites*, NFPA 50A-1994; *Standard for Liquefied Hydrogen Systems at Consumer Sites*, NFPA 50B-1994; *Liquefied Petroleum Gas Code,* NFPA 58-1998; *Standard for the Storage and Handling of Liquefied Petroleum Gases at Utility Gas Plants,* NFPA 59-1998; *Recommended Practice for the Classification of Flammable Liquids, Gases, or Vapors and of Hazardous (Classified) Locations for Electrical Installations in Chemical Process Areas*, NFPA 497-1997; *Recommended Practice for the Classification of Combustible Dusts and of Hazardous (Classified) Locations for Electrical Installations in Chemical Process Areas,* NFPA 499-1997; *Standard for Fire Protection in Wastewater Treatment and Collection Facilities,* NFPA 820-1995; *Recommended Practice for Classification of Locations for Electrical Installations at Petroleum Facilities Classified as Class I, Division 1 and Division 2,* ANSI/API RP500-1997; *Area Classification In Hazardous (Classified) Dust Locations,* ANSI/ISAS12.10-1988.

FPN No. 3: For further information on protection against static electricity and lightning hazards in hazardous (classified) locations, see *Recommended Practice on Static Electricity*, NFPA 77-1993; *Standard for the Installation of Lightning Protection Systems*, NFPA 780-1997; and *Protection Against Ignitions Arising Out of Static Lightning and Stray Currents,* API RP 2003-1991.

FPN No. 4: For further information on ventilation, see *Flammable and Combustible Liquids Code,* NFPA 30-1996; and *Recommended Practice for Classification of Locations for Electrical Installations at Petroleum Facilities Classified as Class I, Division 1 and Division 2,* API RP 500-1997, Section 4.6.

FPN No. 5: For further information on electrical systems for hazardous (classified) locations on offshore oil and gas producing platforms, see *Recommended Practice for Design and Installation of Electrical Systems for Offshore Production Platforms,* ANSI/API RP 14F-1991.

(d) Threaded Conduit. All threaded conduit referred to herein shall be threaded with an NPT standard conduit cutting die that provides ¾-in. taper per foot. Such conduit shall be made wrenchtight to prevent sparking when fault current flows through the conduit system, and ensure the explosionproof or dust-ignitionproof integrity of the conduit system where applicable.

For equipment provided with metric threaded entries, such entries shall be identified as being metric, or listed adapters to permit connection to conduit or NPT-threaded fittings shall be provided with the equipment. Adapters shall be used for connection to conduit or NPT-threaded fittings. Listed cable fittings that have metric threads shall be permitted to be used.

FPN: Threading specifications for metric threaded entries are located in *Metric Screw Threads,* ISO 965/1:1980 and *Metric Screw Threads,* ISO 965/3:1980.

All conduit joints are required to be made up wrenchtight. This is intended to prevent arcing of the conduit under fault conditions. The use of a bonding jumper in lieu of a wrenchtight connection is no longer permitted. Ground continuity, without arcing or sparking, is critical in hazardous locations.

The information on metric threads was expanded in the 1999 *Code.* Equipment with metric threaded entries will have to be identified or be provided with suitable adapters that permit the connection of conduit and fittings that are NPT threaded.

(e) Fiber Optic Cable Assembly. Where a fiber optic cable assembly contains conductors that are capable of carrying current, the fiber optic cable assembly shall be installed in accordance with the requirements of Articles 500, 501, 502, or 503, as applicable.

The requirements of Article 500, 501, 502, or 503 are applicable even if the conductor is grounded.

500-4. Protection Techniques. The following shall be acceptable protection techniques for electrical and electronic equipment in hazardous (classified) locations.

(a) Explosionproof Apparatus. This protection technique shall be permitted for equipment in those Class I, Division 1 and 2 locations for which it is approved.

FPN: Explosionproof apparatus is defined in Article 100. For further information, see *Explosionproof and DustIgnitionproof Electrical Equipment for Use in Hazardous (Classified) Locations,* ANSI/UL 1203-1994.

(b) Dust Ignitionproof. This protection technique shall be permitted for equipment in those Class II, Division 1 and 2 locations for which it is approved.

FPN: Dust-ignitionproof equipment is defined in Section 502-1. For further information, see *Explosionproof and DustIgnitionproof Electrical Equipment for Use in Hazardous (Classified) Locations,* ANSI/UL 1203-1994.

(c) Dusttight. This protection technique shall be permitted for equipment in those Class II, Division 2 and Class III locations for which it is approved.

FPN No. 1: Dusttight enclosures are constructed so that dust will not enter the enclosing cases under specified test conditions.

FPN No. 2: For further information, see *Nonincendive Electrical Equipment for Use in Class I and II, Division 2, and Class III, Divisions 1 and 2 Hazardous (Classified) Locations,* ANSI/ISA S12.12-1994, and *Electrical Equipment for Use in Class I and II, Division 2 and Class III Hazardous (Classified) Locations,* UL 1604-1994.

FPN No. 3: For further information on test conditions for equipment other than rotating equipment, see *Enclosures for Electrical Equipment (1000 volts Maximum),* ANSI/NEMA 250-1991.

(d) Purged and Pressurized. This protection technique shall be permitted for equipment in any hazardous (classified) location for which it is approved.

NFPA 496-1998, *Standard for Purged and Pressurized Enclosures for Electrical Equipment,* covers purged and pressurized enclosures for electrical equipment in Class I hazardous (classified) locations and pressurized enclosures for electrical equipment in Class II hazardous (classified) locations.

In Class I, the purpose of purged and pressurized enclosures is to eliminate or reduce within the enclosure a Class I hazardous (classified) location classification, as defined in Article 500 of the *Code.* By use of purged and pressurized enclosures, equipment not otherwise acceptable for hazardous (classified) locations may be used for these locations, in accordance with the *Code.* In the 1993 edition of NFPA 496, the three types of purged enclosures were revised slightly to refer to three types of pressurizing.

Purging is defined as the process of supplying an enclosure with a protective gas, at a sufficient flow and positive pressure to reduce the concentration of any flammable gas or vapor initially present to an acceptable level. The types of pressurizing are as follows.

1. Type X pressurizing reduces the classification within a protected enclosure from Division 1 to nonclassified.
2. Type Y pressurizing reduces the classification within a protected enclosure from Division 1 to Division 2.
3. Type Z pressurizing reduces the classification within a protected enclosure from Division 2 to nonclassified.

In Class II hazardous (classified) locations, the purpose of pressurized enclosures is to prevent the entrance of dusts into an enclosure. By use of pressurized enclosures, equipment not otherwise acceptable for hazardous (classified) locations may be used for these locations, in accordance with the *Code*. Pressurization, for the purposes of NFPA 496, is defined as the process of supplying an enclosure with a protective gas, with or without continuous flow, at sufficient pressure to prevent the entrance of a flammable gas or vapor, a combustible dust, or an ignitible fiber.

It should be noted that an atmosphere that is made hazardous by combustible dust inside an enclosure cannot be reduced to a safe level by supplying a flow of protective gas in the same manner as with gases or vapors. The enclosure must be opened and the dust removed. Visual inspection can determine if the dust has been removed. Positive pressure will prevent entrance of dust into a clean enclosure.

Field-installed devices and equipment, such as push-button controls and pilot lights, are permitted in purged or pressurized enclosures, provided the conductor terminals are within the purged or pressurized atmosphere.

Section 5-3.4.1 of NFPA 30-1996, *Flammable and Combustible Liquids Code,* and Section 4.6 of API RP500 provide guidelines on ventilation.

FPN No. 1: In some cases, hazards may be reduced or hazardous (classified) locations limited or eliminated by adequate positive-pressure ventilation from a source of clean air in conjunction with effective safeguards against ventilation failure.

FPN No. 2: For further information, see *Standard for Purged and Pressurized Enclosures for Electrical Equipment,* NFPA 496-1998.

(e) Intrinsically Safe Systems. Intrinsically safe apparatus and wiring shall be permitted in any hazardous (classified) location for which it is approved, and the provisions of Articles 501 through 503 and 510 through 516 shall not be considered applicable to such installations, except as required by Article 504.

Installation of intrinsically safe apparatus and wiring shall be in accordance with the requirements of Article 504.

FPN: For further information, see *Intrinsically Safe Apparatus and Associated Apparatus for Use in Class I, II, and III, Division 1, Hazardous Locations,* ANSI/UL 913-1997.

(f) Nonincendive. A protective technique where, under normal operating conditions, any arcing or thermal effects are not capable of igniting the flammable gas, vapor, or dust-in-air mixture. This protection technique shall be permitted for equipment in those Class I, Division 2; Class II, Division 2; and Class III locations for which it is approved.

(1) Nonincendive Circuit. A circuit in which any arc or thermal effect produced, under intended operating conditions of the equipment or due to opening, shorting, or grounding of field wiring, is not capable, under specified test conditions, of igniting the flammable gas–, vapor–, or dust–air mixture.

FPN No. 1: For test conditions, see *Nonincendive Electrical Equipment for Use in Class I and II, Division 2, and Class III, Division 1 and 2 Hazardous (Classified) Locations,* ANSI/ISA S12.12-1994.

FPN No. 2: Nonincendive circuit is defined in Article 100. For further information, see *Nonincendive Electrical Equipment for Use in Class I and II, Division 2, and Class III, Division 1 and 2 Hazardous (Classified) Locations,* ANSI/ISA S12.12-1994.

(2) Nonincendive Equipment. Equipment having electrical/electronic circuitry that is incapable, under normal operating conditions, of causing ignition of a specified flammable gas–, vapor–, or dust–air mixture due to arcing or thermal means.

FPN: For further information, see *Nonincendive Electrical Equipment for Use in Class I and II, Division 2, and Class III, Divisions 1 and 2 Hazardous (Classified) Locations,* ANSI/ISA-S12.12-1994.

(a) *Nonincendive Component.* A component having contacts for making or breaking an incendive circuit and the contacting mechanism shall be constructed so that the component is incapable of igniting the specified flammable gas– or vapor–air mixture. The housing of a nonincendive component is not intended to exclude the flammable atmosphere or contain an explosion.

This protection technique shall be permitted for current-interrupting contacts in those Class I, Division 2 locations for which the component is approved.

FPN: For further information, see *Electrical Equipment for Use in Class I and II, Division 2, and Class III Hazardous (Classified) Locations,* UL 1604-1994.

(g) Oil Immersion. This protection technique shall be permitted for current-interrupting contacts in Class I, Division 2 locations as described in Section 501-6(b)(1)(b).

FPN: See Sections 501-3(b)(1), Exception (a); 501-5(a)(1), Exception (b), 501-6(b)(1); 501-14(b)(1), Exception (a); 502-14(a)(2), Exception; and 502-14(a)(3), Exception. For further information, see *Industrial Control Equipment for Use in Hazardous (Classified) Locations,* ANSI/UL 698-1995.

(h) Hermetically Sealed. A hermetically sealed device shall be sealed against the entrance of an external atmosphere and the seal shall be made by fusion, e.g., soldering, brazing, welding, or the fusion of glass to metal.

This protection technique shall be permitted for current-interrupting contacts in Class I, Division 2 locations.

FPN: See Sections 501-3(b)(1), Exception (b); 501-5(a)(1), Exception (a); 501-6(b)(1)(a); and 501-14(b)(1), Exception (b). For further information, see *Nonincendive Electrical Equipment for Use in Class I and II, Division 2, and Class III, Division 1 and 2 Hazardous (Classified) Locations,* ANSI/ISA S12.12-1994.

(i) Other Protection Techniques. Other protection techniques used in equipment listed for use in hazardous (classified) locations.

Some listed equipment employs unique or a combination of protection techniques. Examples are listed attachment plugs for use in hazardous locations and listed battery-operated flashlights and lanterns.

500-5. Special Precaution. Articles 500 through 504 require equipment construction and installation that will ensure safe performance under conditions of proper use and maintenance.

FPN No. 1: It is important that inspection authorities and users exercise more than ordinary care with regard to installation and maintenance.

FPN No. 2: Low ambient conditions require special consideration. Explosionproof or dust-ignitionproof equipment may not be suitable for use at temperatures lower than −25°C (−13°F) unless they are identified for lowtemperature service. However, at low ambient temperatures, flammable concentrations of vapors may not exist in a location classified Class I, Division 1 at normal ambient temperature.

At low ambient temperatures, such as those encountered in the Arctic, explosion pressures increase at very low temperatures. The strengths of materials change, and the explosion pressure within the explosionproof equipment may increase beyond the safe operating strength of the material. In addition, some sealing materials for sealing fittings may become brittle. However, the extent of the hazardous (classified) location may also change under low ambient conditions. The material may be used in a location where the temperature range is so low that no vapors are produced based on the flash point of the material involved.

For purposes of testing, approval, and area classification, various air mixtures (not oxygen-enriched) shall be grouped in accordance with Sections 500-5(a) and 500-5(b).

Oxygen enrichment can drastically change the explosion characteristics of materials. It lowers the minimum ignition energies, increases explosion pressures, and can reduce the maximum experimental safe gap, rendering both intrinsically safe and explosionproof equipment unsafe unless the equipment has been tested for the conditions involved.

Exception: Equipment approved for a specific gas, vapor, or dust.

Determining the proper group classification for flammable gases and vapors involves a determination of explosion pressures and maximum safe clearance between parts of a clamped joint under several conditions, and comparison of the values obtained with those obtained for presently classified materials under the same test conditions. Although some work has been done on the classification of flammable materials on the basis of chemical structure, the method is not sufficiently refined or accurate to ensure proper classification of all flammable materials.

For additional information on the rationale for classification, reference may be made to the following:

1. *Rationale for Classification of Combustible Gases, Vapors, and Dusts with Reference to the National Electrical Code,* Publication NMAB 353-6, 1982, a report of the Committee on Evaluation of Industrial Hazards, The National Materials Advisory Board, Commission on Engineering and Technical Systems, National Research Council. This publication is available from the National Technical Information Service (NTIS), Springfield, VA 22151.
2. *An Investigation of Fifteen Flammable Gases or Vapors with Respect to Explosion-Proof Electrical Equipment,* Bulletin of Research No. 58 by Underwriters Laboratories Inc., August 1969 (also Bulletin of Research Nos. 58A and 58B, which supplement No. 58). These are available from Underwriters Laboratories Inc., Publications Stock, 333 Pfingsten Rd., Northbrook, IL 60062.
3. *Electrical Installations in Hazardous Locations,* Chapter 2, Peter J. Schram and Mark W. Earley. This book is available from NFPA.

FPN: This grouping is based on the characteristics of the materials. Facilities are available for testing and identifying equipment for use in the various atmospheric groups.

(a) Class I Group Classifications. Class I groups shall be as follows.

x**(1) Group A.** Acetylene.

x**(2) Group B.** Flammable gas, flammable liquid–produced vapor, or combustible liquid–produced vapor mixed with air that may burn or explode, having either a maximum experimental safe gap (MESG) value less than or equal to 0.45 mm or a minimum igniting current ratio (MIC ratio) less than or equal to 0.40.

FPN: A typical Class I, Group B material is hydrogen.

Exception No. 1: Group D equipment shall be permitted to be used for atmospheres containing butadiene provided all conduit runs into explosionproof equipment are provided with explosionproof seals installed within 18 in. (457 mm) of the enclosure.

Exception No. 2: Group C equipment shall be permitted to be used for atmospheres containing allyl glycidyl ether, n-butyl glycidyl ether, ethylene oxide, propylene oxide, and acrolein provided all conduit runs into explosionproof equipment are provided with explosionproof seals installed within 18 in. (457 mm) of the enclosure.

The materials identified in Exception Nos. 1 and 2 produce high pressures as a result of pressure piling in unsealed conduits. If all conduits are sealed, however, this problem is eliminated.

x**(3) Group C.** Flammable gas, flammable liquid–produced vapor, or combustible liquid–produced vapor mixed with air that may burn or explode, having either a maximum experimental safe gap (MESG) value greater than 0.45 mm and less than or equal to 0.75 mm, or a minimum igniting current ratio (MIC ratio) greater than 0.40 and less than or equal to 0.80.

FPN: A typical Class I, Group C material is ethylene.

x**(4) Group D.** Flammable gas, flammable liquid–produced vapor, or combustible liquid–produced vapor mixed with air that may burn or explode, having either a maximum experimental safe gap (MESG) value greater than 0.75 mm or a minimum igniting current ratio (MIC ratio) greater than 0.80.

FPN: A typical Class I, Group D material is propane.

Exception: For atmospheres containing ammonia, the authority having jurisdiction for enforcement of this Code shall be permitted to reclassify the location to a less hazardous location or a nonhazardous location.

FPN No. 1: For additional information on the properties and group classification of Class I materials, see *Recommended Practice for the Classification of Flammable Liquids, Gases, or Vapors and of Hazardous (Classified) Locations for Electrical Installations in Chemical Process Areas*, NFPA 497-1997, and *Guide to Fire Hazard Properties of Flammable Liquids, Gases, and Volatile Solids,* NFPA 325-1994.

FPN No. 2: The explosion characteristics of air mixtures of gases or vapors vary with the specific material involved. For Class I locations, Groups A, B, C, and D, the classification involves determinations of maximum explosion pressure and maximum safe clearance between parts of a clamped joint in an enclosure. It is necessary, therefore, that equipment be approved not only for class but also for the specific group of the gas or vapor that will be present.

FPN No. 3: Certain chemical atmospheres may have characteristics that require safeguards beyond those required for any of the above groups. Carbon disulfide is one of these chemicals because of its low ignition temperature [100°C (212°F)] and the small joint clearance permitted to arrest its flame.

FPN No. 4: For classification of areas involving ammonia atmosphere, see *Safety Code for Mechanical Refrigeration,* ANSI/ASHRAE 15-1994, and *Safety Requirements for the Storage and Handling of Anhydrous Ammonia,* ANSI/CGA G2.1-1989.

Table 2-1 is taken directly from NFPA 497-1997, *Classification of Flammable Liquids, Gases, or Vapors and of Hazardous (Classified) Locations for Electrical Installations in Chemical Process Areas.*

Combustible materials are classified into four Class I, Division Groups, A, B, C, and D, or three Class I, Zone Groups, IIC, IIB, and IIA, depending on their properties. Table 2-1 is an alphabetical listing of selected combustible materials with their group classification and relevant physical properties. For the definitions of terms used in this table, refer to the original document, NFPA 497-1997.

Publication NMAB 353-4, *Classifications of Dusts Relative to Electrical Equipment in Class II Hazardous Locations,* published in 1982 by the Committee on Evaluation of Industrial Hazards, National Materials Advisory Board, Commission on Engineering and Technical Systems, National Research Council, and National Academy of Sciences, includes a description of test methods to determine the ignition temperatures and electrical resistivity of combustible dusts. The publication is available from the National Technical Information Service (NTIS), Springfield, VA 22151.

(b) Class II Group Classifications. Class II groups shall be as follows.

x**(1) Group E.** Atmospheres containing combustible metal dusts, including aluminum, magnesium, and their commercial alloys, or other combustible dusts whose particle size, abrasiveness, and conductivity present similar hazards in the use of electrical equipment.

FPN: Certain metal dusts may have characteristics that require safeguards beyond those required for atmospheres containing the dusts of aluminum, magnesium, and their commercial alloys. For example, zirconium, thorium, and uranium dusts have extremely low ignition temperatures [as low as 20°C (68°F)] and

Table 2-1 Selected Chemicals

Chemical	CAS No.	*NEC* Group	Type[6]	Flash Point (°C)	AIT (°C)	%LFL	%UFL	Vapor Density (Air=1)	Vapor Pressure[7] (mm Hg)	Class I Zone Group[3]	MIE (mJ)	MIC Ratio	MESG (mm)
Acetaldehyde	75-07-0	C*	I	-38	175	4.0	60.0	1.5	874.9	IIA	0.37	0.98	0.92
Acetic Acid	64-19-7	D*	II	43	464	4.0	19.9	2.1	15.6	IIA		2.67	1.76
Acetic Acid-Tert.-Butyl Ester	540-88-5	D	II			1.7	9.8	4.0	40.6				
Acetic Anhydride	108-24-7	D	II	54	316	2.7	10.3	3.5	4.9				
Acetone	67-64-1	D*	I		465	2.5	12.8	2.0	230.7	IIA	1.15	1.00	1.02
Acetone Cyanohydrin	75-86-5	D	IIIA	74	688	2.2	12.0	2.9	0.3				
Acetonitrile	75-05-8	D	I	6	524	3.0	16.0	1.4	91.1	IIA			1.50
Acetylene	74-86-2	A*	GAS		305	2.5	99.9	0.9	36600	IIC	0.017	0.28	0.25
Acrolein (Inhibited)	107-02-8	B(C)*	I		235	2.8	31.0	1.9	274.1	IIB	0.13		
Acrylic Acid	79-10-7	D	II	54	438	2.4	8.0	2.5	4.3				
Acrylonitrile	107-13-1	D*	I	-26	481	3.0	17.0	1.8	108.5	IIB	0.16	0.78	0.87
Adiponitrile	111-69-3	D	IIIA	93	550			1.0	0.002				
Allyl Alcohol	107-18-6	C*	I	22	378	2.5	18.0	2.0	25.4				0.84
Allyl Chloride	107-05-1	D	I	-32	485	2.9	11.1	2.6	366			1.33	1.17
Allyl Glycidyl Ether	106-92-3	B(C)[1]	II		57			3.9					
Alpha-Methyl Styrene	98-83-9	D	II		574	0.8	11.0	4.1	2.7				
n-Amyl Acetate	628-63-7	D	I	25	360	1.1	7.5	4.5	4.2				1.02
sec-Amyl Acetate	626-38-0	D	I	23		1.1	7.5	4.5		IIA			
Ammonia	7664-41-7	D*[2]	I		498	15.0	28.0	0.6	7498.0	IIA	680	6.85	3.17
Aniline	62-53-3	D	IIIA	70	615	1.3	11.0	3.2	0.7	IIA			
Benzene	71-43-2	D*	I	-11	498	1.2	7.8	2.8	94.8	IIA	0.20	1.00	0.99
Benzyl Chloride	98-87-3	D	IIIA		585	1.1		4.4	0.5				
Bromopropyne	106-96-7	D	I	10	324	3.0							
n-Butane	3583-47-9	D*[5]	GAS		288	1.9	8.5	2.0			0.25	0.94	1.07
1,3-Butadiene	106-99-0	B(D)*[1]	GAS	-76	420	2	12	1.9		IIB	0.13	0.76	0.79
1-Butanol	71-36-3	D*	I	36	343	1.4	11.2	2.6	7.0	IIA			0.91
2-Butanol	71-36-5	D*	I	36	405	1.7	9.8	2.6		IIA			
Butylamine	109-73-9	D	GAS	-12	312	1.7	9.8	2.5	92.9			1.13	
Butylene	25167-67-3	D	I		385	1.6	10.0	1.9	2214.6				
n-Butyraldehyde	123-72-8	C*	I	-12	218	1.9	12.5	2.5	112.2				0.92
n-Butyl Acetate	123-86-4	D*	I	22	421	1.7	7.6	4.0	11.5	IIA		1.08	1.04
sec-Butyl Acetate	105-46-4	D	II	-8		1.7	9.8	4.0	22.2				
tert.-Butyl Acetate	540-88-5	D	II			1.7	9.8	4.0	40.6				
n-Butyl Acrylate (Inhibited)	141-32-2	D	II	49	293	1.7	9.9	4.4	5.5				
n-Butyl Glycidyl Ether	2426-08-6	B(C)[1]	II										
n-Butyl Formal	110-62-3	C	IIIA						34.3				
Butyl Mercaptan	109-79-5	C	I	2				3.1	46.4				
Butyl-2-Propenoate	141-32-2	D	II	49		1.7	9.9	4.4	5.5				
para tert.-Butyl Toluene	98-51-1	D	IIIA										
n-Butyric Acid	107-92-6	D	IIIA	72	443	2.0	10.0	3.0	0.8				
Carbon Disulfide	75-15-0	*3	I	-30	90	1.3	50.0	2.6	358.8	IIC	0.009	0.39	0.20
Carbon Monoxide	630-08-0	C*	GAS	609	700	12.5	74.	0.97		IIA			
Chloroacetaldehyde	107-20-0	C	IIIA	88					63.1				
Chlorobenzene	108-90-7	D	I	29	593	1.3	9.6	3.9	11.9				
1-Chloro-1-Nitropropane	2425-66-3	C	IIIA										
Chloroprene	126-99-8	D	GAS	-20		4.0	20.0	3.0					
Cresol	1319-77-3	D	IIIA	81	559	1.1		3.7					
Crotonaldehyde	4170-30-3	C*	I	13	232	2.1	15.5	2.4	33.1	IIB			0.81
Cumene	98-82-8	D	I	36	424	0.9	6.5	4.1	4.6	IIA			
Cyclohexane	110-82-7	D	I	-17	245	1.3	8.0	2.9	98.8	IIA	0.22	1.0	0.94

Table 2-1 (Continued)

Chemical	CAS No.	*NEC* Group	Type[6]	Flash Point (°C)	AIT (°C)	%LFL	%UFL	Vapor Density (Air=1)	Vapor Pressure[7] (mm Hg)	Class I Zone Group[3]	MIE (mJ)	MIC Ratio	MESG (mm)
Cyclohexanol	108-93-0	D	IIIA	68	300			3.5	0.7	IIA			
Cyclohexanone	108-94-1	D	II	44	245	1.1	9.4	3.4	4.3	IIA			0.98
Cyclohexene	110-83-8	D	I	-6	244	1.2		2.8	89.4			0.97	
Cyclopropane	75-19-4	D*	I		503	2.4	10.4	1.5	5430	IIB	0.17	0.84	0.91
p-Cymene	99-87-6	D	II	47	436	0.7	5.6	4.6	1.5	IIA			
Decene	872-05-9	D	II		235			4.8	1.7				
n-Decaldehyde	112-31-2	C	IIIA						0.09				
n-Decanol	112-30-1	D	IIIA	82	288			5.3	0.008				
Decyl Alcohol	112-30-1	D	IIIA	82	288			5.3	0.008				
Diacetone Alcohol	123-42-2	D	IIIA	64	603	1.8	6.9	4.0	1.4				
Di-Isobutylene	25167-70-8	D*	I	2	391	0.8	4.8	3.8			0.96		
Di-Isobutyl Ketone	108-83-8	D	II	60	396	0.8	7.1	4.9	1.7				
o-Dichlorobenzene	955-50-1	D	IIIA	66	647	2.2	9.2	5.1		IIA			
1,4-Dichloro-2,3 Epoxybutane	3583-47-9	D*	I			1.9	8.5	2.0			0.25	0.98	1.07
1,1-Dichloroethane	1300-21-6	D	I		438	6.2	16.0	3.4	227				1.82
1,2-Dichloroethylene	156-59-2	D	I	97	460	5.6	12.8	3.4	204	IIA			
1,1-Dichloro-1-Nitroethane	594-72-9	C	IIIA	76				5.0					
1,3-Dichloropropene	10061-02-6	D	I	35		5.3	14.5	3.8					
Dicyclopentadiene	77-73-6	C	I	32	503				2.8				0.91
Diethylamine	109-87-9	C*	I	-28	312	1.8	10.1	2.5		IIA			1.15
Diethylaminoethanol	100-37-8	C	IIIA	60	320			4.0	1.6	IIA			
Diethyl Benzene	25340-17-4	D	II	57	395			4.6					
Diethyl Ether	60-29-7	C*	I	12	160	1.9	36.0	2.6	38.2	IIB	0.19	0.88	0.83
Diethylene Glycol Monobutyl Ether	112-34-5	C	IIIA	78	228	0.9	24.6	5.6	0.02				
Diethylene Glycol Monomethyl Ether	111-77-3	C	IIIA	93	241				0.2				
n-n-Dimethyl Aniline	121-69-7	C	IIIA	63	371	1.0		4.2	0.7				
Dimethyl Formamide	68-12-2	D	II	58	455	2.2	15.2	2.5	4.1				1.08
Dimethyl Sulfate	77-78-1	D	IIIA	83	188			4.4	0.7				
Dimethylamine	124-40-3	C	GAS		400	2.8	14.4	1.6		IIA			
2,2-Dimethylbutane	75-83-2	D[5]	I	-48	405				319.3				
2,3-Dimethylbutane		D[5]	I		396								
3,3-Dimethylheptane	1071-26-7	D[5]	I		325				10.8				
2,3-Dimethylhexane	31394-54-4	D[5]	I		438								
2,3-Dimethylpentane	107-83-5	D[5]	I		335				211.7				
Di-N-Propylamine	142-84-7	C	I	17	299				27.1				
1,4-Dioxane	123-91-1	C*	I	12	180	2.0	22.0	3.0	38.2	IIB	0.19		0.70
Dipentene	138-86-3	D	II	45	237	0.7	6.1	4.7					1.18
Dipropylene Glycol Methyl Ether	34590-94-8	C	IIIA	85		1.1	3.0	5.1	0.5				
Diisopropylamine	108-18-9	C	GAS	-6	316	1.1	7.1	3.5					
Dodecene	6842-15-5	D	IIIA	100	255								
Epichlorohydrin	3132-64-7	C*	I	33	411	3.8	21.0	3.2	13.0				
Ethane	74-84-0	D*	GAS	-29	472	3.0	12.5	1.0		IIA	0.24	0.82	0.91
Ethanol	64-17-5	D*	I	13	363	3.3	19.0	1.6	59.5	IIA		0.88	0.89
Ethylamine	75-04-7	D*	I	-18	385	3.5	14.0	1.6	1048		2.4		
Ethylene	74-85-1	C*	GAS	0	450	2.7	36.0	1.0		IIB	0.070	0.53	0.65
Ethylenediamine	107-15-3	D*	I	33	385	2.5	12.0	2.1	12.5				
Ethylenimine	151-56-4	C*	I	-11	320	3.3	54.8	1.5	211		0.48		
Ethylene Chlorohydrin	107-07-3	D	IIIA	59	425	4.9	15.9	2.8	7.2				

(continues)

Table 2-1 (Continued)

Chemical	CAS No.	*NEC* Group	Type[6]	Flash Point (°C)	AIT (°C)	%LFL	%UFL	Vapor Density (Air=1)	Vapor Pressure[7] (mm Hg)	Class I Zone Group[3]	MIE (mJ)	MIC Ratio	MESG (mm)
Ethylene Dichloride	107-06-2	D*	I	13	413	6.2	16.0	3.4	79.7				
Ethylene Glycol Monoethyl Ether Acetate	111-15-9	C	II	47	379	1.7		4.7	2.3			0.53	0.97
Ethylene Glycol Monobutyl Ether Acetate	112-07-2	C	IIIA		340	0.9	8.5		0.9				
Ethylene Glycol Monobutyl Ether	111-76-2	C	IIIA		238	1.1	12.7	4.1	1.0				
Ethylene Glycol Monoethyl Ether	110-80-5	C	II		235	1.7	15.6	3.0	5.4				0.84
Ethylene Glycol Monomethyl Ether	109-86-4	D	II		285	1.8	14.0	2.6	9.2				0.85
Ethylene Oxide	75-21-8	B(C)*[1]	I	-20	429	3.0	99.9	1.5	1314	IIB	0.065	0.47	0.59
2-Ethylhexaldehyde	123-05-7	C	II	52	191	0.8	7.2	4.4	1.9				
2-Ethylhexanol	104-76-7	D	IIIA	81		0.9	9.7	4.5	0.2				
2-Ethylhexyl Acrylate	103-09-3	D	IIIA	88	252				0.3				
Ethyl Acetate	141-78-6	D*	I	-4	427	2.0	11.5	3.0	93.2		0.46		0.99
Ethyl Acrylate (Inhibited)	140-88-5	D*	I	9	372	1.4	14.0	3.5	37.5	IIA			0.86
Ethyl Alcohol	64-17-5	D*	I	13	363	3.3	19.0	1.6	59.5				0.89
Ethyl Sec-Amyl Ketone	541-85-5	D	II	59									
Ethyl Benzene	100-41-4	D	I	21	432	0.8	6.7	3.7	9.6				
Ethyl Butanol	97-95-0	D	II	57		1.2	7.7	3.5	1.5				
Ethyl Butyl Ketone	106-35-4	D	II	46				4.0	3.6				
Ethyl Chloride	75-00-3	D	GAS	-50	519	3.8	15.4	2.2					
Ethyl Ether	60-29-7	C*	I	-45	160	1.9	36.0	2.6	538		0.19	0.88	0.84
Ethyl Formate	109-94-4	D	GAS	-20	455	2.8	16.0	2.6		IIA			0.94
Ethyl Mercaptan	75-08-1	C*	I	-18	300	2.8	18.0	2.1	527.4			0.90	0.90
n-Ethyl Morpholine	100-74-3	C	I	32				4.0					
2-Ethyl-3-Propyl Acrolein	645-62-5	C	IIIA	68				4.4					
Ethyl Silicate	78-10-4	D	II					7.2					
Formaldehyde (Gas)	50-00-0	B	GAS	60	429	7.0	73.0	1.0					0.57
Formic Acid	64-18-6	D	II	50	434	18.0	57.0	1.6	42.7				1.86
Fuel Oil 1	8008-20-6	D	II	72	210	0.7	5.0						
Furfural	98-01-1	C	IIIA	60	316	2.1	19.3	3.3	2.3				0.94
Furfuryl Alcohol	98-00-0	C	IIIA	75	490	1.8	16.3	3.4	0.6				
Gasoline	8006-61-9	D*	I	-46	280	1.4	7.6	3.0					
n-Heptane	142-82-5	D*	I	-4	204	1.0	6.7	3.5	45.5	IIA	0.24	0.88	0.91
n-Heptene	81624-04-6	D[5]	I	-1	204			3.4					0.97
n-Hexane	110-54-3	D*[5]	I	-23	225	1.1	7.5	3.0	152	IIA	0.24	0.88	0.93
Hexanol	111-27-3	D	IIIA	63				3.5	0.8	IIA			0.98
2-Hexanone	591-78-6	D	I	35	424	1.2	8.0	3.5	10.6				
Hexene	592-41-6	D	I	-26	245	1.2	6.9		186				
sec-Hexyl Acetate	108-84-9	D	II	45				5.0					
Hydrazine	302-01-2	C	II	38	23		98.0	1.1	14.4				
Hydrogen	1333-74-0	B*	GAS		520	4.0	75.0	0.1		IIC	0.019	0.25	0.28
Hydrogen Cyanide	74-90-8	C*	GAS	-18	538	5.6	40.0	0.9		IIB			0.80
Hydrogen Selenide	7783-07-5	C	I						7793				
Hydrogen Sulfide	7783-06-4	C*	GAS		260	4.0	44.0	1.2			0.068		0.90
Isoamyl Acetate	123-92-2	D	I	25	360	1.0	7.5	4.5	6.1				
Isoamyl Alcohol	123-51-3	D	II	43	350	1.2	9.0	3.0	3.2				1.02
Isobutane	75-28-5	D[5]	GAS		460	1.8	8.4	2.0					
Isobutyl Acetate	110-19-0	D*	I	18	421	2.4	10.5	4.0	17.8				
Isobutyl Acrylate	106-63-8	D	I		427			4.4	7.1				
Isobutyl Alcohol	78-83-1	D*	I	-40	416	1.2	10.9	2.5	10.5			0.92	0.98

Table 2-1 (Continued)

Chemical	CAS No.	*NEC* Group	Type[6]	Flash Point (°C)	AIT (°C)	%LFL	%UFL	Vapor Density (Air=1)	Vapor Pressure[7] (mm Hg)	Class I Zone Group[3]	MIE (mJ)	MIC Ratio	MESG (mm)
Isobutyraldehyde	78-84-2	C	GAS	-40	196	1.6	10.6	2.5					
Isodecaldehyde	112-31-2	C	IIIA					5.4	0.09				
Isohexane	107-83-5	D[5]			264				211.7			1.00	
Isopentane	78-78-4	D[5]			420				688.6				
Isooctyl Aldehyde	123-05-7	C	II		197				1.9				
Isophorone	78-59-1	D		84	460	0.8	3.8	4.8	0.4				
Isoprene	78-79-5	D*	I	-54	220	1.5	8.9	2.4	550.6				
Isopropyl Acetate	108-21-4	D	I		460	1.8	8.0	3.5	60.4				
Isopropyl Ether	108-20-3	D*	I	-28	443	1.4	7.9	3.5	148.7		1.14		0.94
Isopropyl Glycidyl Ether	4016-14-2	C	I										
Isopropylamine	75-31-0	D	GAS	-26	402	2.3	10.4	2.0			2.0		
Kerosene	8008-20-6	D	II	72	210	0.7	5.0			IIA			
Liquified Petroleum Gas	68476-85-7	D	I		405								
Mesityl Oxide	141-97-9	D*	I	31	344	1.4	7.2	3.4	47.6				
Methane	74-82-8	D*	GAS	-223	630	5.0	15.0	0.6		IIA	0.28	1.00	1.12
Methanol	67-56-1	D*	I	12	385	6.0	36.0	1.1	126.3	IIA	0.14	0.82	0.92
Methyl Acetate	79-20-9	D	GAS	-10	454	3.1	16.0	2.6		IIB		1.08	0.99
Methyl Acrylate	96-33-3	D	GAS	-3	468	2.8	25.0	3.0				0.98	0.85
Methyl Alcohol	67-56-1	D*	I		385	6.0	36.0	1.1	126.3				0.91
Methyl Amyl Alcohol	108-11-2	D	II	41		1.0	5.5	3.5	5.3				1.01
Methyl Chloride	74-87-3	D	GAS	-46	632	8.1	17.4	1.7					1.00
Methyl Ether	115-10-6	C*	GAS	-41	350	3.4	27.0	1.6				0.85	0.84
Methyl Ethyl Ketone	78-93-3	D*	I	-6	404	1.4	11.4	2.5	92.4		0.53	0.92	0.84
Methyl Formal	534-15-6	C*	I	1	238			3.1					
Methyl Formate	107-31-3	D	GAS	-19	449	4.5	23.0	2.1					0.94
2-Methylhexane	31394-54-4	D[5]	I		280								
Methyl Isobutyl Ketone	141-79-7	D*	I	31	440	1.2	8.0	3.5	11				
Methyl Isocyanate	624-83-9	D	GAS	-15	534	5.3	26.0	2.0					
Methyl Mercapatan	74-93-1	C	GAS	-18		3.9	21.8	1.7					
Methyl Methacrylate	80-62-6	D	I	10	422	1.7	8.2	3.6	37.2	IIA			0.95
Methyl N-Amyl Ketone	110-43-0	D	II	49	393	1.1	7.9	3.9	3.8				
Methyl Tertiary Butyl Ether	1634-04-4	D	I	-80	435	1.6	8.4	0.2	250.1				
2-Methyloctane	3221-61-2				220				6.3				
2-Methylpropane	75-28-5	D[5]	I		460				2639				
Methyl-1-Propanol	78-83-1	D*	I	-40	416	1.2	10.9	2.5	10.1				0.98
Methyl-2-Propanol	75-65-0	D*	I	10	360	2.4	8.0	2.6	42.2				
2-Methyl-5-Ethyl Pyridine	104-90-5	D		74		1.1	6.6	4.2					
Methylacetylene	74-99-7	C*	I			1.7		1.4	4306		0.11		
Methylacetylene-Propadiene	27846-30-6	C	I										0.74
Methylal	109-87-5	C	I	-18	237	1.6	17.6	2.6	398				
Methylamine	74-89-5	D	GAS		430	4.9	20.7	1.0		IIA			1.10
2-Methylbutane	78-78-4	D[5]		-56	420	1.4	8.3	2.6	688.6				
Methylcyclohexane	208-87-2	D	I	-4	250	1.2	6.7	3.4			0.27		
Methylcyclohexanol	25630-42-3	D		68	296			3.9					
2-Methycyclohexanone	583-60-8	D	II					3.9					
2-Methylheptane		D[5]			420								
3-Methylhexane	589-34-4	D[5]			280				61.5				
3-Methylpentane	94-14-0	D[5]			278								
2-Methylpropane	75-28-5	D[5]	I		460				2639				
2-Methyl-1-Propanol	78-83-1	D*	I	-40	223	1.2	10.9	2.5	10.5				

(continues)

Table 2-1 (Continued)

Chemical	CAS No.	*NEC* Group	Type[6]	Flash Point (°C)	AIT (°C)	%LFL	%UFL	Vapor Density (Air=1)	Vapor Pressure[7] (mm Hg)	Class I Zone Group[3]	MIE (mJ)	MIC Ratio	MESG (mm)
2-Methyl-2-Propanol	75-65-0	D*	I		478	2.4	8.0	2.6	42.2				
2-Methyloctane	2216-32-2	D[5]			220								
3-Methyloctane	2216-33-3	D[5]			220				6.3				
4-Methyloctane	2216-34-4	D[5]			225				6.8				
Monoethanolamine	141-43-5	D		85	410			2.1	0.4	IIA			
Monoisopropanolamine	78-96-6	D		77	374			2.6	1.1				
Monomethyl Aniline	100-61-8	C			482				0.5				
Monomethyl Hydrazine	60-34-4	C	I	23	194	2.5	92.0	1.6					
Morpholine	110-91-8	C*	II	35	310	1.4	11.2	3.0	10.1				0.95
Naphtha (Coal Tar)	8030-30-6	D	II	42	277					IIA			
Naphtha (Petroleum)	8030-30-6	D*[4]	I	42	288	1.1	5.9	2.5		IIA			
Neopentane	463-82-1	D[5]		-65	450	1.4	8.3	2.6	1286				
Nitrobenzene	98-95-3	D		88	482	1.8		4.3	0.3				0.94
Nitroethane	79-24-3	C	I	28	414	3.4		2.6	20.7	IIA			0.87
Nitromethane	75-52-5	C	I	35	418	7.3		2.1	36.1	IIA		0.92	1.17
1-Nitropropane	108-03-2	C	I	34	421	2.2		3.1	10.1				0.84
2-Nitropropane	79-46-9	C*	I	28	428	2.6	11.0	3.1	17.1				
n-Nonane	111-84-2	D[5]	I	31	205	0.8	2.9	4.4	4.4	IIA			
Nonene	27214-95-8	D	I			0.8		4.4					
Nonyl Alcohol	143-08-8	D				0.8	6.1	5.0	0.02	IIA			
n-Octane	111-65-9	D*[5]	I	13	206	1.0	6.5	3.9	14.0	IIA			0.94
Octene	25377-83-7	D	I	8	230	0.9		3.9					
n-Octyl Alcohol	111-87-5	D						4.5	0.08	IIA			1.05
n-Pentane	109-66-0	D*[5]	I	-40	243	1.5	7.8	2.5	513		0.28	0.97	0.93
1-Pentanol	71-41-0	D*	I	33	300	1.2	10.0	3.0	2.5	IIA			
2-Pentanone	107-87-9	D	I	7	452	1.5	8.2	3.0	35.6				0.99
1-Pentene	109-67-1	D	I	-18	275	1.5	8.7	2.4	639.7				
2-Pentene	109-68-2	D	I	-18				2.4					
2-Pentyl Acetate	626-38-0	D	I	23		1.1	7.5	4.5					
Phenylhydrazine	100-63-0	D		89				3.7	0.03				
Process Gas > 30% H2	1333-74-0	B**	GAS		520	4.0	75.0	0.1			0.019	0.45	
Propane	74-98-6	D*	GAS	-104	450	2.1	9.5	1.6		IIA	0.25	0.82	0.97
1-Propanol	71-23-8	D*	I	15	413	2.2	13.7	2.1	20.7	IIA			0.89
2-Propanol	67-63-0	D*	I	12	399	2.0	12.7	2.1	45.4		0.65		1.00
Propiolactone	57-57-8	D				2.9		2.5	2.2				
Propionaldehyde	123-38-6	C	I	-9	207	2.6	17.0	2.0	318.5				
Propionic Acid	79-09-4	D	II	54	466	2.9	12.1	2.5	3.7				
Propionic Anhydride	123-62-6	D		74	285	1.3	9.5	4.5	1.4				
n-Propyl Acetate	109-60-4	D	I	14	450	1.7	8.0	3.5	33.4				1.05
n-Propyl Ether	111-43-3	C*	I	21	215	1.3	7.0	3.5	62.3				
Propyl Nitrate	627-13-4	B*	I	20	175	2.0	100.0						
Propylene	115-07-1	D*	GAS	-108	455	2.0	11.1	1.5			0.28		0.91
Propylene Dichloride	78-87-5	D	I	16	557	3.4	14.5	3.9	51.7				1.32
Propylene Oxide	75-56-9	B(C)*[1]	I	-37	449	2.3	36.0	2.0	534.4		0.13		0.70
Pyridine	110-86-1	D*	I	20	482	1.8	12.4	2.7	20.8	IIA			
Styrene	100-42-5	D*	I	31	490	0.9	6.8	3.6	6.1	IIA		1.21	
Tetrahydrofuran	109-99-9	C*	I	-14	321	2.0	11.8	2.5	161.6	IIB	0.54		0.87
Tetrahydronaphthalene	119-64-2	D	IIIA		385	0.8	5.0	4.6	0.4				
Tetramethyl Lead	75-74-1	C	II	38				9.2					
Toluene	108-88-3	D*	I	4	480	1.1	7.1	3.1	28.53	IIA	0.24		
n-Tridecene	2437-56-1	D	IIIA			0.6		6.4	593.4				

Table 2-1 (Continued)

Chemical	CAS No.	*NEC* Group	Type[6]	Flash Point (°C)	AIT (°C)	%LFL	%UFL	Vapor Density (Air=1)	Vapor Pressure[7] (mm Hg)	Class I Zone Group[3]	MIE (mJ)	MIC Ratio	MESG (mm)
Triethylamine	121-44-8	C*	I	-9	249	1.2	8.0	3.5	68.5	IIA	0.75		
Triethylbenzene	25340-18-5	D		83			56.0	5.6					
2,2,3-Trimethylbutane		D[5]			442								
2,2,4-Trimethylbutane		D[5]			407								
2,2,3-Trimethylpentane		D[5]			396								
2,2,4-Trimethylpentane		D[5]			415								
2,3,3-Trimethylpentane		D[5]			425								
Tripropylamine	102-69-2	D	II	41				4.9	1.5				1.13
Turpentine	8006-64-2	D	I	35	253	0.8			4.8				
n-Undecene	28761-27-5	D	IIIA			0.7		5.5					
Unsymmetrical Dimethyl Hydrazine	57-14-7	C*	I	-15	249	2.0	95.0	1.9					0.85
Valeraldehyde	110-62-3	C	I	280	222			3.0	34.3				
Vinyl Acetate	108-05-4	D*	I	-6	402	2.6	13.4	3.0	113.4	IIA	0.70		0.94
Vinyl Chloride	75-01-4	D*	GAS	-78	472	3.6	33.0	2.2					0.96
Vinyl Toluene	25013-15-4	D		52	494	0.8	11.0	4.1					
Vinylidene Chloride	75-35-4	D	I		570	6.5	15.5	3.4	599.4				3.91
Xylene	1330-20-7	D*	I	25	464	0.9	7.0	3.7		IIA	0.2		
Xylidine	121-69-7	C	IIIA	63	371	1.0		4.2	0.7				

* Material has been classified by test.

** Fuel and process gas mixtures found by test not to present hazards similar to those of hydrogen, may be grouped based on the test results.

NOTES:

[1]If explosionproof equipment is isolated by sealing all conduits 1/2 in. or larger, in accordance with Section 501-5(a) of NFPA 70, *National Electrical Code,* equipment for the group classification shown in parentheses is permitted.

[2]For classification of areas involving ammonia, see *Safety Code for Mechanical Refrigeration,* ANSI/ASHRAE 15, and *Safety Requirements for the Storage and Handling of Anhydrous Ammonia,* ANSI/CGA G2.1.

[3]Certain chemicals may have characteristics that require safeguards beyond those required for any of the above groups. Carbon disulfide is one of these chemicals because of its low autoignition temperature and the small joint clearance necessary to arrest its flame propagation.

[4]Petroleum naphtha is a saturated hydrocarbon mixture whose boiling range is 68°F to 275°F (20°C to 135°C). It is also known as benzine, ligroin, petroleum ether, and naphtha.

[5]Commercial grades of aliphatic hydrocarbon solvents are mixtures of several isomers of the same chemical formula (or molecular weight). The autoignition temperatures of the individual isomers are significantly different. The electrical equipment should be suitable for the AIT of the solvent mixture.

[6]Type is used to designate if the material is a gas, flammable liquid, or combustible liquid.

[7]Vapor pressure reflected in units of mm Hg at 77°F (25°C) unless stated otherwise.

[8]Class I, Zone Groups are based upon "Electrical apparatus for explosive gas atmospheres — Part 20: Data for flammable gases and vapors, relating to the use of electrical apparatus, IEC 79-20 (1996)."

minimum ignition energies lower than any material classified in any of the Class I or Class II Groups.

x**(2) Group F.** Atmospheres containing combustible carbonaceous dusts that have more than 8 percent total entrapped volatiles (see *Standard Test Method for Volatile Material in the Analysis Sample for Coal and Coke,* ASTM D3175-89, for coal and coke dusts) or that have been sensitized by other materials so that they present an explosion hazard. Coal, carbon black, charcoal, and coke dusts are examples of carbonaceous dusts.

(3) Group G. Atmospheres containing combustible dusts not included in Group E or F, including flour, grain, wood, plastic, and chemicals.

FPN No. 1: For additional information on group classification of Class II materials, see *Recommended Practice for the Classification of Combustible Dusts and of Hazardous (Classified) Locations for Electrical Installations in Chemical Process Areas,* NFPA 499-1997.

FPN No. 2: The explosion characteristics of air mixtures of dust vary with the materials involved. For Class II locations, Groups E, F, and G, the classification involves the tightness of the joints of assembly and shaft openings to prevent the entrance of dust in the dust-ignitionproof enclosure, the blanketing effect of layers of dust on the equipment that may cause overheating, and the ignition temperature of the dust. It is necessary, therefore, that equipment be approved not only for the class, but also for the specific group of dust that will be present.

FPN No. 3: Certain dusts may require additional precautions due to chemical phenomena that can result in the generation of ignitible gases. See *National Electrical Safety Code,* ANSI C2-1997, Section 127A, Coal Handling Areas.

Table 2-5 is taken directly from NFPA 499-1997, *Recommended Practice for the Classification of Combustible Dusts and of Hazardous (Classified) Locations for Electrical Installations in Chemical Process Areas.*

Combustible dusts are classified into three Class II, Division Groups, E, F, and G, depending on their properties. Table 2-5 is an alphabetical listing of selected combustible materials with their group classification and relevant physical properties. For the definitions of terms used in the table, refer to the original document, NFPA 499-1997.

Table 2-5 Selected Combustible Materials

Chemical Name	CAS No.	*NEC* Group	Code	Layer or Cloud Ignition Temp °C
Acetal, Linear		G	NL	440
Acetoacet-p-phenetidide	122-82-7	G	NL	560
Acetoacetanilide	102-01-2	G	M	440
Acetylamino-t-nitrothiazole		G		450
Acrylamide Polymer		G		240
Acrylonitrile Polymer		G		460
Acrylonitrile-Vinyl Chloride-Vinylidene chloride copolymer (70-20-10)		G		210
Acrylonitrile-Vinyl Pyridine Copolymer		G		240
Adipic Acid	124-04-9	G	M	550
Alfalfa Meal		G		200
Alkyl Ketone Dimer Sizing Compound		G		160
Allyl Alcohol Derivative (CR-39)		G	NL	500
Almond Shell		G		200
Aluminum, A422 Flake	7429-90-5	E		320
Aluminum, Atomized Collector Fines		E	CL	550
Aluminum—cobalt alloy (60-40)		E		570
Aluminum—copper alloy (50-50)		E		830
Aluminum—lithium alloy (15% Li)		E		400
Aluminum—magnesium alloy (Dowmetal)		E	CL	430
Aluminum—nickel alloy (58-42)		E		540
Aluminum—silicon alloy (12% Si)		E	NL	670
Amino-5-nitrothiazole	121-66-4	G		460
Anthranilic Acid	118-92-3	G	M	580
Apricot Pit		G		230
Aryl-nitrosomethylamide		G	NL	490
Asphalt	8052-42-4	F		510
Aspirin [acetol (2)]	50-78-2	G	M	660
Azelaic Acid	109-31-9	G	M	610
Azo-bis-butyronitrile	78-67-1	G		350
Benzethonium Chloride		G	CL	380
Benzoic Acid	65-85-0	G	M	440
Benzotriazole	95-14-7	G	M	440
Beta-naphthalene-axo-dimethylaniline		G		175
Bis(2-hydroxy-5-chlorophenyl) Methane	97-23-4	G	NL	570
Bisphenol-A	80-05-7	G	M	570
Boron, Commercial Amorphous (85% B)	7440-42-8	E		400
Calcium Silicide		E		540
Carbon Black (More Than 8% Total Entrapped Volatiles)		F		
Carboxymethyl Cellulose	9000-11-7	G		290
Carboxypolymethylene		G	NL	520
Cashew Oil, Phenolic, Hard		G		180
Cellulose		G		260
Cellulose Acetate		G		340
Cellulose Acetate Butyrate		G	NL	370
Cellulose Triacetate		G	NL	430
Charcoal (Activated)	64365-11-3	F		180
Charcoal (More Than 8% Total Entrapped Volatiles)		F		
Cherry Pit		G		220
Chlorinated Phenol		G	NL	570
Chlorinated Polyether Alcohol		G		460
Chloroacetoacetanilide	101-92-8	G	M	640
Chromium (97%) Electrolytic, Milled	7440-47-3	E		400
Cinnamon		G		230
Citrus Peel		G		270
Coal, Kentucky Bituminous		F		180
Coal, Pittsburgh Experimental		F		170
Coal, Wyoming		F		

Table 2-5 (Continued)

Chemical Name	CAS No.	NEC Group	Code	Layer or Cloud Ignition Temp °C
Cocoa Bean Shell		G		370
Cocoa, Natural, 19% Fat		G		240
Coconut Shell		G		220
Coke (More Than 8% Total Entrapped Volatiles)		F		
Cork		G		210
Corn		G		250
Corn Dextrine		G		370
Corncob Grit		G		240
Cornstarch, Commercial		G		330
Cornstarch, Modified		G		200
Cottonseed Meal		G		200
Coumarone-Indene, Hard		G	NL	520
Crag No. 974	533-74-4	G	CL	310
Cube Root, South America	83-79-4	G		230
Di-alphacumyl Peroxide, 40-60 on CA	80-43-3	G		180
Diallyl Phthalate	131-17-9	G	M	480
Dicyclopentadiene Dioxide		G	NL	420
Dieldrin (20%)	60-57-1	G	NL	550
Dihydroacetic Acid		G	NL	430
Dimethyl Isophthalate	1459-93-4	G	M	580
Dimethyl Terephthalate	120-61-6	G	M	570
Dinitro-o-toluamide	148-01-6	G	NL	500
Dinitrobenzoic Acid		G	NL	460
Diphenyl	92-52-4	G	M	630
Ditertiary-butyl-paracresol	128-37-0	G	NL	420
Dithane m-45	8018-01-7	G		180
Epoxy		G	NL	540
Epoxy-bisphenol A		G	NL	510
Ethyl Cellulose		G	CL	320
Ethyl Hydroxyethyl Cellulose		G	NL	390
Ethylene Oxide Polymer		G	NL	350
Ethylene-maleic Anhydride Copolymer		G	NL	540
Ferbam	14484-64-1	G		150
Ferromanganese, Medium Carbon	12604-53-4	E		290
Ferrosilicon (88% Si, 9% Fe)	8049-17-0	E		800
Ferrotitanium (19% Ti, 74.1% Fe, 0.06% C)		E	CL	380
Flax Shive		G		230
Fumaric Acid	110-17-8	G	M	520
Garlic, Dehydrated		G	NL	360
Gilsonite	12002-43-6	F		500
Green Base Harmon Dye		G		175
Guar Seed		G	NL	500
Gulasonic Acid, Diacetone		G	NL	420
Gum, Arabic		G		260
Gum, Karaya		G		240
Gum, Manila		G	CL	360
Gum, Tragacanth	9000-65-1	G		260
Hemp Hurd		G		220
Hexamethylene Tetramine	100-97-0	G	S	410
Hydroxyethyl Cellulose		G	NL	410
Iron, 98% H_2 Reduced		E		290
Iron, 99% Carbonyl	13463-40-6	E		310
Isotoic Anhydride		G	NL	700
L-sorbose		G	M	370
Lignin, Hydrolized, Wood-type, Fine		G	NL	450
Lignite, California		F		180
Lycopodium		G		190
Malt Barley		G		250
Manganese	7439-96-5	E		240
Magnesium, Grade B, Milled		E		430
Manganese Vancide		G		120
Mannitol	69-65-8	G	M	460
Methacrylic Acid Polymer		G		290
Methionine (l-methionine)	63-68-3	G		360
Methyl Cellulose		G		340
Methyl Methacrylate Polymer	9011-14-7	G	NL	440
Methyl Methacrylate-ethyl Acrylate		G	NL	440
Methyl Methacrylate-styrene-butadiene		G	NL	480
Milk, Skimmed		G		200
N,N-Dimethylthio-formamide		G		230
Nitropyridone	100703-82-0	G	M	430
Nitrosamine		G	NL	270
Nylon Polymer	63428-84-2	G		430
Para-oxy-benzaldehyde	123-08-0	G	CL	380
Paraphenylene Diamine	106-50-3	G	M	620
Paratertiary Butyl Benzoic Acid	98-73-7	G	M	560
Pea Flour		G		260
Peach Pit Shell		G		210
Peanut Hull		G		210
Peat, Sphagnum	94114-14-4	G		240
Pecan Nut Shell	8002-03-7	G		210
Pectin	5328-37-0	G		200
Pentaerythritol	115-77-5	G	M	400
Petrin Acrylate Monomer	7659-34-9	G	NL	220
Petroleum Coke (More Than 8% Total Entrapped Volatiles)		F		
Petroleum Resin	64742-16-1	G		500
Phenol Formaldehyde	9003-35-4	G	NL	580
Phenol Formaldehyde, Polyalkylene-p	9003-35-4	G		290
Phenol Furfural	26338-61-4	G		310
Phenylbetanaphthylamine	135-88-6	G	NL	680
Phthalic Anydride	85-44-9	G	M	650

(continues)

Table 2-5 (Continued)

Chemical Name	CAS No.	NEC Group	Code	Layer or Cloud Ignition Temp °C
Phthalimide	85-41-6	G	M	630
Pitch, Coal Tar	65996-93-2	F	NL	710
Pitch, Petroleum	68187-58-6	F	NL	630
Polycarbonate		G	NL	710
Polyethylene, High Pressure Process	9002-88-4	G		380
Polyethylene, Low Pressure Process	9002-88-4	G	NL	420
Polyethylene Terephthalate	25038-59-9	G	NL	500
Polyethylene Wax	68441-04-8	G	NL	400
Polypropylene (no antioxidant)	9003-07-0	G	NL	420
Polystyrene Latex	9003-53-6	G		500
Polystyrene Molding Compound	9003-53-6	G	NL	560
Polyurethane Foam, Fire Retardant	9009-54-5	G		390
Polyurethane Foam, No Fire Retardant	9009-54-5	G		440
Polyvinyl Acetate	9003-20-7	G	NL	550
Polyvinyl Acetate/Alcohol	9002-89-5	G		440
Polyvinyl Butyral	63148-65-2	G		390
Polyvinyl Chloride-dioctyl Phthalate		G	NL	320
Potato Starch, Dextrinated	9005-25-8	G	NL	440
Pyrethrum	8003-34-7	G		210
Rayon (Viscose) Flock	61788-77-0	G		250
Red Dye Intermediate		G		175
Rice		G		220
Rice Bran		G	NL	490
Rice Hull		G		220
Rosin, DK	8050-09-7	G	NL	390
Rubber, Crude, Hard	9006-04-6	G	NL	350
Rubber, Synthetic, Hard (33% S)	64706-29-2	G	NL	320
Safflower Meal		G		210
Salicylanilide	87-17-2	G	M	610
Sevin	63-25-2	G		140
Shale, Oil	68308-34-9	F		
Shellac	9000-59-3	G	NL	400
Sodium Resinate	61790-51-0	G		220
Sorbic Acid (Copper Sorbate or Potash)	110-44-1	G		460
Soy Flour	68513-95-1	G		190
Soy Protein	9010-10-0	G		260
Stearic Acid, Aluminum Salt	637-12-7	G		300
Stearic Acid, Zinc Salt	557-05-1	G	M	510
Styrene Modified Polyester-Glass Fiber	100-42-5	G		360
Styrene-acrylonitrile (70-30)	9003-54-7	G	NL	500
Styrene-butadiene Latex (75% styrene)	903-55-8	G	NL	440
Styrene-maleic Anhydride Copolymer	9011-13-6	G	CL	470
Sucrose	57-50-1	G	CL	350
Sugar, Powdered	57-50-1	G	CL	370
Sulfur	7704-34-9	G		220
Tantalum	7440-25-7	E		300
Terephthalic Acid	100-21-0	G	NL	680
Thorium, 1.2% O_2	7440-29-1	E	CL	280
Tin, 96%, Atomized, (2% Pb)	7440-31-5	E		430
Titanium, 99% Ti	7440-32-6	E	CL	330
Titanium Hydride (95% Ti, 3.8% H_2)	7704-98-5	E	CL	480
Trithiobisdimethylthio-formamide		G		230
Tung, Kernels, Oil-free	8001-20-5	G		240
Urea Formaldehyde Molding Compound	9011-05-6	G	NL	460
Urea Formaldehyde-phenol Formaldehyde	25104-55-6	G		240
Vanadium, 86.4%	7440-62-2	E		490
Vinyl Chloride-acrylonitrile Copolymer	9003-00-3	G		470
Vinyl Toluene-acrylonitrile Butadiene	76404-69-8	G	NL	530
Violet 200 Dye		G		175
Vitamin B1, Mononitrate	59-43-8	G	NL	360
Vitamin C	50-81-7	G		280
Walnut Shell, Black		G		220
Wheat		G		220
Wheat Flour	130498-22-5	G		360
Wheat Gluten, Gum	100684-25-1	G	NL	520
Wheat Starch		G	NL	380
Wheat Straw		G		220
Wood Flour		G		260
Woodbark, Ground		G		250
Yeast, Torula	68602-94-8	G		260
Zirconium Hydride	7704-99-6	E		270
Zirconium		E	CL	330

Notes:
1. Normally, the minimum ignition temperature of a layer of a specific dust is lower than the minimum ignition temperature of a cloud of that dust. Since this is not universally true, the lower of the two minimum ignition temperatures is listed. If no symbol appears between the two temperature columns, then the layer ignition temperature is shown. "CL" means the cloud ignition temperature is shown. "NL" means that no layer ignition temperature is available, and the cloud ignition temperature is shown. "M" signifies that the dust layer melts before it ignites; the cloud ignition temperature is shown. "S" signifies that the dust layer sublimes before it ignites; the cloud ignition temperature is shown.
2. Certain metal dusts may have characteristics that require safeguards beyond those required for atmospheres containing the dusts of aluminum, magnesium, and their commercial alloys. For example, zirconium, thorium, and uranium dusts have extremely low ignition temperatures [as low as 68°F (20°C)] and minimum ignition energies lower than any material classified in any of the Class I or Class II groups.

(c) Approval for Class and Properties. Equipment, regardless of the classification of the location in which it is installed, that depends on a single compression seal, diaphragm, or tube to prevent flammable or combustible fluids from entering the equipment, shall be approved for a Class I, Division 2 location.

Exception: Equipment installed in a Class I, Division 1 location shall be suitable for the Division 1 location.

FPN: See Section 501-5(f)(3) for additional requirements.

Equipment shall be approved not only for the class of location but also for the explosive, combustible, or ignitible properties of the specific gas, vapor, dust, fiber, or flyings that will be present. In addition, Class I equipment shall not have any exposed surface that operates at a temperature in excess of the ignition temperature of the specific gas or vapor. Class II equipment shall not have an external temperature higher than that specified in Section 500-5(f). Class III equipment shall not exceed the maximum surface temperatures specified in Section 503-1.

Equipment that has been approved for a Division 1 location shall be permitted in a Division 2 location of the same class and group.

Where specifically permitted in Articles 501 through 503, general-purpose equipment or equipment in general-purpose enclosures shall be permitted to be installed in Division 2 locations if the equipment does not constitute a source of ignition under normal operating conditions.

Unless otherwise specified, normal operating conditions for motors shall be assumed to be rated full-load steady conditions.

It is not intended that locked-rotor or other motor overload conditions, such as single phasing, be considered when evaluating motor operating temperatures (internal and external) in Class I, Division 2 locations. However, such abnormal load conditions are considered when evaluating external temperatures of explosionproof motors for Class I, Division 1 locations and motors for Class II, Division 1 locations, such as dust-ignitionproof motors.

Where flammable gases or combustible dusts are or may be present at the same time, the simultaneous presence of both shall be considered when determining the safe operating temperature of the electrical equipment.

A coal-handling facility is an example of a location where methane gas and coal dust may be present at the same time.

FPN: The characteristics of various atmospheric mixtures of gases, vapors, and dusts depend on the specific material involved.

(d) Marking. Approved equipment shall be marked to show the class, group, and operating temperature or temperature range referenced to a 40°C ambient.

The marked operating temperature or temperature range is normally referenced to a 40°C (104°F) ambient. Unless the equipment is provided with thermally actuated sensors that limit the temperature to that marked on the equipment, operation in ambient temperatures higher than 40°C (104°F) will increase the operating temperature of the equipment. Many explosionproof and dust-ignitionproof motors are equipped with thermal protectors. In like manner, operation in ambient temperatures lower than 40°C (104°F) will usually reduce the operating temperature.

Exception No. 1: Equipment of the non-heat-producing type, such as junction boxes, conduit, and fittings, and equipment of the heat-producing type having a maximum temperature not more than 100°C (212°F) shall not be required to have a marked operating temperature or temperature range.

Exception No. 2: Fixed lighting fixtures marked for use in Class I, Division 2 or Class II, Division 2 locations only shall not be required to be marked to indicate the group.

Exception No. 3: Fixed general-purpose equipment in Class I locations, other than fixed lighting fixtures, that is acceptable for use in Class I, Division 2 locations shall not be required to be marked with the class, group, division, or operating temperature.

A squirrel-cage induction motor without brushes, switching mechanisms, or similar arc-producing devices is an example of fixed general-purpose equipment. See the second paragraph of Section 501-8(b).

Exception No. 4: Fixed dusttight equipment other than fixed lighting fixtures that are acceptable for use in Class II, Division 2 and Class III locations shall not be required to be marked with the class, group, division, or operating temperature.

Exception No. 5: Electric equipment suitable for ambient temperatures exceeding 40°C (104°F) shall be marked with both the maximum ambient temperature and the operating temperature or temperature range at that ambient temperature.

The ignition temperature of a solid, liquid, or gaseous substance is the minimum temperature required to initiate or cause self-sustained combustion independent of the heating or heated element. The ignition temperature and the flash point are unrelated properties, except that the flash point is always lower than the ignition temperature.

FPN: Equipment not marked to indicate a division, or marked "Division 1" or "Div. 1," is suitable for both Division 1 and 2 locations. Equipment marked "Division 2" or "Div. 2" is suitable for Division 2 locations only.

The temperature range, if provided, shall be indicated in identification numbers, as shown in Table 500-5(d).

Identification numbers marked on equipment nameplates shall be in accordance with Table 500-5(d).

Table 500-5(d). Identification Numbers

Maximum Temperature °C	Maximum Temperature °F	Identification Number
450	842	T1
300	572	T2
280	536	T2A
260	500	T2B
230	446	T2C
215	419	T2D
200	392	T3
180	356	T3A
165	329	T3B
160	320	T3C
135	275	T4
120	248	T4A
100	212	T5
85	185	T6

Equipment that is approved for Class I and Class II shall be marked with the maximum safe operating temperature, as determined by simultaneous exposure to the combinations of Class I and Class II conditions.

FPN: Since there is no consistent relationship between explosion properties and ignition temperature, the two are independent requirements.

(e) Class I Temperature. The temperature marking specified in (d) shall not exceed the ignition temperature of the specific gas or vapor to be encountered.

Flammable gases or vapors are separated into four different atmospheric groups in Article 500: A, B, C, and D. The *Code* requirements for Class I locations do not vary for different kinds of gas or vapor contained in the atmosphere, except in those cases where seals may be used in all conduits to change the group classification. See Exception Nos. 1 and 2 to Section 500-5(a)(2) for materials such as butadiene and ethylene oxide. It is necessary to select equipment designed for use in the particular group involved. The reason for designating the groups this way is that explosive mixtures have different explosion pressures and maximum safe clearances between parts of a joint in an enclosure.

Underwriters Laboratories Inc. and Factory Mutual Research Corp. list or approve electrical equipment suitable for use in all groups of Class I locations. Further information is available from the UL *Hazardous Locations Equipment Directory* and the FM *Approval Guide*. It should be noted (from the UL Directory) that "only those products bearing the appropriate listing mark and the company's name, trade name, trademark, or other recognized identification should be considered as covered by UL's Listing and Follow-Up Service." Other testing laboratories may also list equipment for hazardous locations for use in the United States. The acceptance of the listing agency is the responsibility of the authority having jurisdiction.

Several testing laboratories outside the United States also provide listing, approval, or certification of equipment for use in hazardous locations. However, they may not be testing and investigating the equipment for use in hazardous locations, as defined in Article 500. Some international laboratories certify equipment for installation where the classification is according to the IEC classification scheme covered in Article 505.

FPN: For information regarding ignition temperatures of gases and vapors, see *Recommended Practice for the Classification of Combustible Dusts and of Hazardous (Classified) Locations for Electrical Installations in Chemical Process Areas,* NFPA 499-1997, and *Guide to Fire Hazard Properties of Flammable Liquids, Gases, and Volatile Solids,* NFPA 325-1994.

(f) Class II Temperature. The temperature marking specified in (d) shall be less than the ignition temperature of the specific dust to be encountered. For organic dusts that may dehydrate or carbonize, the temperature marking shall not exceed the lower of either the ignition temperature or 165°C (329°F).

As in Class I locations, equipment must be approved not only for the class but also for the specific group. It is important that, in addition to the proper selection of equipment, high standards of installation be maintained for subsequent additions or alterations.

The NFPA and ANSI standards referenced in Articles 500 through 517 are essential for proper application of these articles.

The following NFPA standards and recommended practices include information on hazardous (classified) locations and the extent of hazardous (classified) locations in specific occupancies or industries.

NFPA 30, *Flammable and Combustible Liquids Code*

NFPA 30A, *Automotive and Marine Service Station Code*

NFPA 32, *Standard for Drycleaning Plants*

NFPA 33, *Standard for Spray Application Using Flammable and Combustible Materials*

NFPA 34, *Standard for Dipping and Coating Processes Using Flammable or Combustible Liquids*

NFPA 35, *Standard for the Manufacture of Organic Coatings*

NFPA 36, *Standard for Solvent Extraction Plants*

NFPA 45, *Standard for Fire Protection for Laboratories Using Chemicals*

NFPA 50A, *Standard for Gaseous Hydrogen Systems at Consumer Sites*

NFPA 50B, *Standard for Liquefied Hydrogen Systems at Consumer Sites*

NFPA 51, *Standard for the Design and Installation of Oxygen-Fuel Gas Systems for Welding, Cutting, and Allied Processes*

NFPA 51A, *Standard for Acetylene Cylinder Charging Plant*

NFPA 52, *Compressed Natural Gas (CNG) Vehicular Fuel Systems Code*

NFPA 54, *Natural Fuel Gas Code*

NFPA 58, *Liquefied Petroleum Gas Code*

NFPA 59, *Standard for the Storage and Handling of Liquefied Petroleum Gases at Utility Gas Plants*

NFPA 59A, *Standard for the Production, Storage, and Handling of Liquefied Natural Gas (LNG)*

NFPA 61, *Standard for the Prevention of Fire and Dust Explosions in Agricultural and Food Products Facilities*

NFPA 88A, *Standard for Parking Structures*

NFPA 88B, *Standard for Repair Garages*

NFPA 99, *Standard for Health Care Facilities*

NFPA 407, *Standard for Aircraft Fuel Servicing*

NFPA 409, *Standard for Aircraft Hangars*

NFPA 480, *Standard for the Storage, Handling, and Processing of Magnesium Solids and Powders*

NFPA 481, *Standard for the Production, Processing, Handling, and Storage of Titanium*

NFPA 495, *Standard for Explosive Materials Code*

NFPA 496, *Standard for Purged and Pressurized Enclosures for Electrical Equipment*

NFPA 497, *Recommended Practice for the Classification of Flammable Liquids, Gases, or Vapors and of Hazardous (Classified) Locations for Electrical Installations in Chemical Process Areas*

NFPA 499, *Recommended Practice for the Classification of Combustible Dusts and of Hazardous (Classified) Locations for Electrical Installations in Chemical Process Areas*

NFPA 651, *Standard for the Machining and Finishing of Aluminum and the Production and Handling of Aluminum Powders*

NFPA 654, *Standard for the Prevention of Fire and Dust Explosions from the Manufacturing, Processing, and Handling of Combustible Particulate Solids*

NFPA 655, *Standard for Prevention of Sulfur Fires and Explosions*

NFPA 8502, *Standard for the Prevention of Furnace Explosions/Implosions in Multiple Burner Boilers*

NFPA 8503, *Standard for Pulverized Fuel Systems*

NFPA 8504, *Standard on Atmospheric Fluidized-Bed Boiler Operation*

NFPA 8505, *Standard for Stoker Operation*

FPN: See *Recommended Practice for the Classification of Combustible Dusts and of Hazardous (Classified) Locations for Electrical Installations in Chemical Process Areas,* NFPA 499-1997, for minimum ignition temperatures of specific dusts.

The ignition temperature for which equipment was approved prior to this requirement shall be assumed to be as shown in Table 500-5(f).

Table 500-5(f).

Class II Group	Equipment that Is Not Subject to Overloading		Equipment (Such as Motors or Power Transformers) that May Be Overloaded			
			Normal Operation		Abnormal Operation	
	°C	°F	°C	°F	°C	°F
E	200	392	200	392	200	392
F	200	392	150	302	200	392
G	165	329	120	248	165	329

500-6. Specific Occupancies. Articles 510 through 517 cover garages, aircraft hangars, gasoline dispensing and service stations, bulk storage plants, spray application, dipping and coating processes, and health care facilities.

500-7. Class I Locations. Class I locations are those in which flammable gases or vapors are or may be present in the air in quantities sufficient to produce explosive or ignitible mixtures. Class I locations shall include those specified in (a) and (b).

(a) Class I, Division 1. A Class I, Division 1 location is a location

(1) In which ignitible concentrations of flammable gases or vapors can exist under normal operating conditions, or
(2) In which ignitible concentrations of such gases or vapors

may exist frequently because of repair or maintenance operations or because of leakage, or

(3) In which breakdown or faulty operation of equipment or processes might release ignitible concentrations of flammable gases or vapors, and might also cause simultaneous failure of electrical equipment in such a way as to directly cause the electrical equipment to become a source of ignition.

FPN No. 1: This classification usually includes the following locations:

(1) Where volatile flammable liquids or liquefied flammable gases are transferred from one container to another
(2) Interiors of spray booths and areas in the vicinity of spraying and painting operations where volatile flammable solvents are used
(3) Locations containing open tanks or vats of volatile flammable liquids
(4) Drying rooms or compartments for the evaporation of flammable solvents
(5) Locations containing fat and oil extraction equipment using volatile flammable solvents
(6) Portions of cleaning and dyeing plants where flammable liquids are used
(7) Gas generator rooms and other portions of gas manufacturing plants where flammable gas may escape
(8) Inadequately ventilated pump rooms for flammable gas or for volatile flammable liquids
(9) The interiors of refrigerators and freezers in which volatile flammable materials are stored in open, lightly stoppered, or easily ruptured containers
(10) All other locations where ignitible concentrations of flammable vapors or gases are likely to occur in the course of normal operations

FPN No. 2: In some Division 1 locations, ignitible concentrations of flammable gases or vapors may be present continuously or for long periods of time. Examples include the following:

(1) The inside of inadequately vented enclosures containing instruments normally venting flammable gases or vapors to the interior of the enclosure
(2) The inside of vented tanks containing volatile flammable liquids
(3) The area between the inner and outer roof sections of a floating roof tank containing volatile flammable fluids
(4) Inadequately ventilated areas within spraying or coating operations using volatile flammable fluids
(5) The interior of an exhaust duct that is used to vent ignitible concentrations of gases or vapors

Experience has demonstrated the prudence of avoiding the installation of instrumentation or other electric equipment in these particular areas altogether or where it cannot be avoided because it is essential to the process and other locations are not feasible [see Section 500-3(a), FPN], using electric equipment or instrumentation approved for the specific application or consisting of intrinsically safe systems as described in Article 504.

Fine Print Note No. 2 describes locations that are defined in Section 505-9(a) as Class I, Zone 0 locations. Only intrinsically safe equipment and wiring suitable for Zone 0 locations is permitted in such locations. The intent here is to warn against using other than intrinsically safe systems in these parts of Class I, Division 1 locations, based on experience.

(b) Class I, Division 2. A Class I, Division 2 location is a location

(1) In which volatile flammable liquids or flammable gases are handled, processed, or used, but in which the liquids, vapors, or gases will normally be confined within closed containers or closed systems from which they can escape only in case of accidental rupture or breakdown of such containers or systems, or in case of abnormal operation of equipment, or
(2) In which ignitible concentrations of gases or vapors are normally prevented by positive mechanical ventilation, and which might become hazardous through failure or abnormal operation of the ventilating equipment, or
(3) That is adjacent to a Class I, Division 1 location, and to which ignitible concentrations of gases or vapors might occasionally be communicated unless such communication is prevented by adequate positivepressure ventilation from a source of clean air, and effective safeguards against ventilation failure are provided.

FPN No. 1: This classification usually includes locations where volatile flammable liquids or flammable gases or vapors are used but that, in the judgment of the authority having jurisdiction, would become hazardous only in case of an accident or of some unusual operating condition. The quantity of flammable material that might escape in case of accident, the adequacy of ventilating equipment, the total area involved, and the record of the industry or business with respect to explosions or fires are all factors that merit consideration in determining the classification and extent of each location.

FPN No. 2: Piping without valves, checks, meters, and similar devices would not ordinarily introduce a hazardous condition even though used for flammable liquids or gases. Depending on factors such as the quantity and size of the containers and ventilation, locations used for the storage of flammable liquids or liquefied or compressed gases in sealed containers may either be considered hazardous (classified) or unclassified locations. See *Flammable and Combusti-*

ble Liquids Code, NFPA 30-1996, and *Liquefied Petroleum Gas Code,* NFPA 58-1998.

500-8. Class II Locations. Class II locations are those that are hazardous because of the presence of combustible dust. Class II locations shall include those specified in (a) and (b).

(a) Class II, Division 1. A Class II, Division 1 location is a location

(1) In which combustible dust is in the air under normal operating conditions in quantities sufficient to produce explosive or ignitible mixtures, or
(2) Where mechanical failure or abnormal operation of machinery or equipment might cause such explosive or ignitible mixtures to be produced, and might also provide a source of ignition through simultaneous failure of electric equipment, operation of protection devices, or from other causes, or
(3) In which combustible dusts of an electrically conductive nature may be present in hazardous quantities.

FPN: Combustible dusts that are electrically nonconductive include dusts produced in the handling and processing of grain and grain products, pulverized sugar and cocoa, dried egg and milk powders, pulverized spices, starch and pastes, potato and wood-flour, oil meal from beans and seed, dried hay, and other organic materials that may produce combustible dusts when processed or handled. Only Group E dusts are considered to be electrically conductive for classification purposes. Dusts containing magnesium or aluminum are particularly hazardous, and the use of extreme precaution will be necessary to avoid ignition and explosion.

(b) Class II, Division 2. A Class II, Division 2 location is a location

(1) Where combustible dust is not normally in the air in quantities sufficient to produce explosive or ignitible mixtures, and dust accumulations are normally insufficient to interfere with the normal operation of electrical equipment or other apparatus, but combustible dust may be in suspension in the air as a result of infrequent malfunctioning of handling or processing equipment and
(2) Where combustible dust accumulations on, in, or in the vicinity of the electrical equipment may be sufficient to interfere with the safe dissipation of heat from electrical equipment or may be ignitible by abnormal operation or failure of electrical equipment.

FPN No. 1: The quantity of combustible dust that may be present and the adequacy of dust removal systems are factors that merit consideration in determining the classification and may result in an unclassified area.

FPN No. 2: Where products such as seed are handled in a manner that produces low quantities of dust, the amount of dust deposited may not warrant classification.

500-9. Class III Locations. Class III locations are those that are hazardous because of the presence of easily ignitible fibers or flyings, but in which such fibers or flyings are not likely to be in suspension in the air in quantities sufficient to produce ignitible mixtures. Class III locations shall include those specified in (a) and (b).

(a) Class III, Division 1. A Class III, Division 1 location is a location in which easily ignitible fibers or materials producing combustible flyings are handled, manufactured, or used.

FPN No. 1: Such locations usually include some parts of rayon, cotton, and other textile mills; combustible fiber manufacturing and processing plants; cotton gins and cotton-seed mills; flax-processing plants; clothing manufacturing plants; woodworking plants; and establishments and industries involving similar hazardous processes or conditions.

FPN No. 2: Easily ignitible fibers and flyings include rayon, cotton (including cotton linters and cotton waste), sisal or henequen, istle, jute, hemp, tow, cocoa fiber, oakum, baled waste kapok, Spanish moss, excelsior, and other materials of similar nature.

(b) Class III, Division 2. A Class III, Division 2 location is a location in which easily ignitible fibers are stored or handled other than in the process of manufacture.

Sections 500-7, 500-8, and 500-9 recognize three classes of hazardous (classified) locations, based on the type of material involved. Within each class there are varying degrees of hazard, so each class is subdivided into two divisions. The classification by division is based on the likelihood the material will be present. The requirements for Division 1 of each class are more stringent than those for Division 2.

The materials in the three classes are defined as follows: Class I, flammable gases or vapors; Class II, combustible dust; and Class III, combustible fibers or flyings.

Where a given location is classified as hazardous, it should not be difficult to determine in which of the three classes (it may belong in more than one class) it belongs. Common sense and good judgment must prevail in classifying an area that is likely to become hazardous and in determining those portions of the premises to be classed Division 1 or Division 2.

Article 501 — Class I Locations

Contents

501-1. General. The general rules of this *Code* shall apply to the electric wiring and equipment in locations classified as Class I in Section 500-7.

Equipment listed and marked in accordance with Section 505-10 for use in Class I, Zone 0, 1, or 2 locations shall be permitted in Class I, Division 2 locations for the same gas and with a suitable temperature rating.

Exception: As modified by this article.

The most common Class I locations are those areas involved in the handling or processing of volatile flammable liquids such as gasoline, naphtha, benzene, diethyl ether, and acetone, or flammable gases such as hydrogen, methane, and propane.

Where ignitible concentrations (concentrations within the flammable or explosive limits) of flammable gases or vapors are present, atmospheres exist that are explosive when ignited by an arc, a spark, or high temperature. NFPA 325-1994, *Guide to Fire Hazard Properties of Flammable Liquids, Gases, and Volatile Solids,* includes information on the explosive limits of flammable liquids, gases, and volatile solids. All electrical equipment that may cause ignition-capable arcs or sparks should be kept out of Class I locations where practicable. If this is not practicable, such apparatus must be approved for the purpose and installed properly. The arc produced at the contacts of listed or labeled intrinsically safe equipment is not ignition-capable because the energy available is insufficient to cause ignition.

Hermetic sealing of all electrical equipment is impractical, because equipment such as motors, conventional switches, and circuit breakers have movable parts that must be operated through the enclosing case; that is, the lever of a switch or the shaft of a motor must have sufficient clearance to operate freely. In addition, in many cases, it is necessary to have access to the inside of enclosures for installation, servicing, or alterations.

It is practically impossible to make threaded conduit joints gastight. The conduit system and apparatus enclosure "breathe" due to temperature changes, and any flammable gases or vapors in the room may slowly enter the conduit or enclosure, creating an explosive mixture. Should an arc occur, an explosion could take place.

When an explosion occurs within the enclosure or conduit system, the burning mixture or hot gases must be sufficiently confined within the system to prevent ignition of any explosive mixture that might be present in the area outside the enclosures or conduit system. An apparatus enclosure must be designed with sufficient strength to withstand the maximum pressure generated by an internal explosion in order to prevent rupture and the release of burning or hot gases. Enclosures have been designed to withstand such internal explosions. The ability to withstand the internal explosion is one criteria by which "explosionproof" enclosures are evaluated.

It has been found that during an explosion within an enclosure, gases will escape through any paths or openings that exist, but the gases will be sufficiently cooled if they are carried out through an opening that is long in proportion to its width; that is, the spiral path of at least five fully engaged threads of a screwed-on junction box cover, as illustrated in Figure 501.1. This principle is also applied in the design of explosionproof enclosures for apparatus where a wide machined flange on the body of the enclosure and a similar machined flange on the cover are provided as illustrated in Figure 501.2. These machined flanges are ground so that when the cover is seated in place, the clearance between the two surfaces will at no point exceed, for example, 0.0015 in. If an explosion occurs within the enclosure, escaping gas travels a considerable distance through a very small opening. It is, therefore, sufficiently cooled, when it enters and mixes with the surrounding atmosphere, that ignition of the external explosive mixture cannot occur.

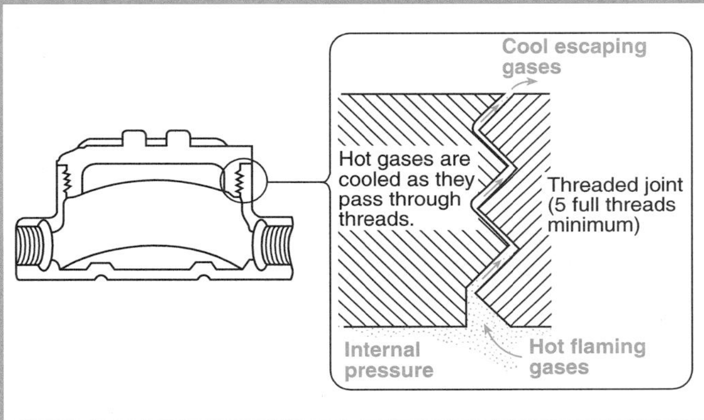

Figure 501.1 *Hot gases are cooled as they pass through the threads of a screw-type cover of an explosionproof junction box.*

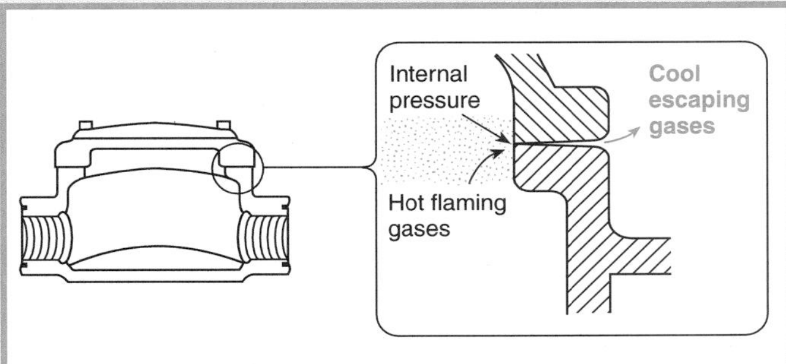

Figure 501.2 *Hot gases are cooled as they pass across a machined-flanged joint. The clearance between the machined surfaces is kept very small.*

The clearance between flat surfaces may increase somewhat under explosion conditions, because the internal pressures created by the explosion tend to force the surfaces apart. The amount of increase in the joint clearance depends on the "stiffness" of the enclosure parts; the size, strength, and spacing of the bolts; and the explosion pressure (see Figure 501.3). Simply measuring the joint width and clearance when there are no internal pressures will not indicate the actual clearances under the dynamic conditions of an explosion. Explosion tests are usually needed to demonstrate the acceptability of the design.

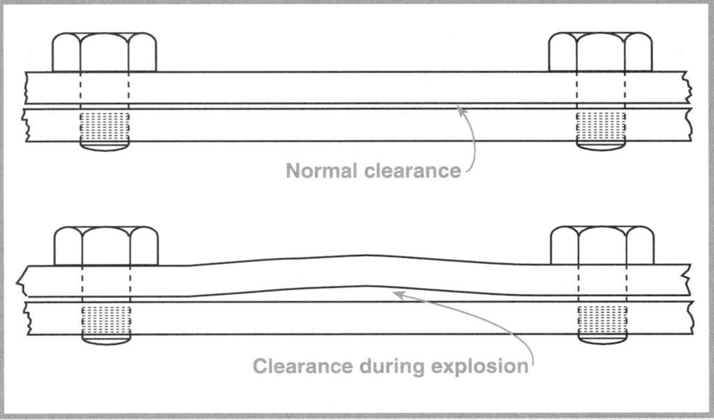

Figure 501.3 *Effect of internal explosion on cover-to-body joint clearance in explosionproof enclosure. (Underwriters Laboratories Inc.)*

501-2. Transformers and Capacitors.

(a) Class I, Division 1. In Class I, Division 1 locations, transformers and capacitors shall comply with the following.

(1) Containing Liquid that Will Burn. Transformers and capacitors containing a liquid that will burn shall be installed only in approved vaults that comply with Sections 450-41 through 450-48, and, in addition,

(a) There shall be no door or other communicating opening between the vault and the Division 1 location, and
(b) Ample ventilation shall be provided for the continuous removal of flammable gases or vapors, and
(c) Vent openings or ducts shall lead to a safe location outside of buildings, and
(d) Vent ducts and openings shall be of sufficient area to relieve explosion pressures within the vault, and all portions of vent ducts within the buildings shall be of reinforced concrete construction.

(2) Not Containing Liquid that Will Burn. Transformers and capacitors that do not contain a liquid that will burn shall be installed in vaults complying with (a)(1) or be approved for Class I locations.

(b) Class I, Division 2. In Class I, Division 2 locations, transformers and capacitors shall comply with Sections 450-21 through 450-27.

501-3. Meters, Instruments, and Relays.

(a) Class I, Division 1. In Class I, Division 1 locations, meters, instruments, and relays, including kilowatt-hour meters, instrument transformers, resistors, rectifiers, and thermionic tubes, shall be provided with enclosures approved for Class I, Division 1 locations.

Enclosures approved for Class I, Division 1 locations include explosionproof enclosures and purged and pressurized enclosures.

FPN: See *Standard for Purged and Pressurized Enclosures for Electrical Equipment*, NFPA 496-1998.

See the commentary on purged and pressurized enclosures for electrical equipment in hazardous (classified) locations following Section 500-4(d) and the commentary on explosionproof enclosures following the exception to Section 501-1.

(b) Class I, Division 2. In Class I, Division 2 locations, meters, instruments, and relays shall comply with the following.

(1) Contacts. Switches, circuit breakers, and make-and-break contacts of pushbuttons, relays, alarm bells, and horns shall have enclosures approved for Class I, Division 1 locations in accordance with Section 501-3(a).

Exception: General-purpose enclosures shall be permitted, if current-interrupting contacts are

(a) Immersed in oil, or

(b) Enclosed within a chamber that is hermetically sealed against the entrance of gases or vapors, or

There are several types of hermetic seals, including fusion seals, such as the glass-to-metal seals in mercury-tube switches and some reed switches, welded seals, soldered seals, and seals made with gaskets. Seals of the glass-to-metal-fusion type are usually the most reliable. Soft solder seals can be relatively porous, and their effectiveness is highly dependent on workmanship. Although gasketed seals can be very effective, depending on the gasket material used, gasket materials can be damaged and can deteriorate rapidly where exposed to atmospheres containing solvent vapors. Gasketed enclosures are not considered as being hermetically sealed in accordance with Section 500-4(h).

(c) In nonincendive circuits, or

See the definitions of *nonincendive circuit, nonincendive equipment,* and *nonincendive component* in Section 500-4(h). *Nonincendive* is similar to *intrinsically safe,* which is defined in Article 504 and ANSI/UL 913, but does not include consideration of faults and all the abnormal conditions inherent in the definition of intrinsically safe. A circuit may be nonincendive in a Group D atmosphere but not in a Group C atmosphere, as the minimum ignition energies for the various flammable materials differ.

(d) Part of a listed nonincendive component.

(2) Resistors and Similar Equipment. Resistors, resistance devices, thermionic tubes, rectifiers, and similar equipment that are used in or in connection with meters, instruments, and relays shall comply with Section 501-3(a).

Exception: General-purpose–type enclosures shall be permitted if such equipment is without make-and-break or sliding contacts [other than as provided in (b)(1)] and if the maximum operating temperature of any exposed surface will not exceed 80 percent of the ignition temperature in degrees Celsius of the gas or vapor involved or has been tested and found incapable of igniting the gas or vapor. This exception shall not apply to thermionic tubes.

The intent of the phrase "or has been tested and found incapable of igniting the gas or vapor" is to permit listed equipment with operating temperatures higher than 80 percent of the ignition temperature. If the equipment has been tested, the safety factor inherent in this 80 percent rule is not needed. The system of temperature measurement must be specified, as 80 percent of a temperature in degrees Celsius is not the same temperature as 80 percent of that temperature in degrees Fahrenheit.

The last sentence of this exception concerns ionization of the air from thermionic tubes, such as cathode-ray tubes.

(3) Without Make-or-Break Contacts. Transformer windings, impedance coils, solenoids, and other windings that do not incorporate sliding or make-or-break contacts shall be provided with enclosures. General-purpose–type enclosures shall be permitted.

(4) General-Purpose Assemblies. Where an assembly is made up of components for which general-purpose enclosures are acceptable as provided in (b)(1), (b)(2), and (b)(3), a single general-purpose enclosure shall be acceptable for the assembly. Where such an assembly includes any of the equipment described in (b)(2), the maximum obtainable surface temperature of any component of the assembly shall be clearly and permanently indicated on the outside of the enclosure. Alternatively, approved equipment shall be permitted to be marked to indicate the temperature range for which it is suitable, using the identification numbers of Table 500-5(d).

(5) Fuses. Where general-purpose enclosures are permitted in (b)(1), (b)(2), (b)(3), and (b)(4), fuses for overcurrent protection of instrument circuits not subject to overloading in normal use shall be permitted to be mounted in general-purpose enclosures if each such fuse is preceded by a switch complying with (b)(1).

(6) Connections. To facilitate replacements, process control instruments shall be permitted to be connected through flexible cord, attachment plug, and receptacle, provided the following:

(1) A switch complying with (b)(1) is provided so that the attachment plug is not depended on to interrupt current; and
(2) The current does not exceed 3 amperes at 120 volts, nominal; and
(3) The power-supply cord does not exceed 3 ft (914 mm), is of a type approved for extra-hard usage or for hard usage if protected by location, and is supplied through an attachment plug and receptacle of the locking and grounding type; and
(4) Only necessary receptacles are provided; and
(5) The receptacle carries a label warning against unplugging under load.

501-4. Wiring Methods. Wiring methods shall comply with (a) and (b).

(a) Class I, Division 1.

Section 501-4(a) indicates that termination fittings used with Type MI cable are required to be approved for the specific purpose. It is intended that Type MI cable fittings approved and marked [see Section 500-5(d)] for the hazardous (classified) location class and group are to be used. Type MI cable fittings have a clamp-type joint that must be investigated to determine that it is explosionproof. Type MI cable fittings suitable for nonhazardous locations may not be suitable for Class I, Division 1 hazardous (classified) locations (see Figure 501.4).

Rigid metal conduit and intermediate metal conduit are required to be threaded with an (NPT) standard conduit cutting die that provides a $^{3}/_{4}$-in. taper per foot, and five full threads must be engaged. The *Code* recognizes electrical equipment with metric threaded entries [see Section 500-3(d)]. Equipment with metric threaded entries must be identified to indicate that metric threads are provided or be provided with listed adapters to allow the connection of NPT-threaded conduit or fittings to the equipment. Each joint is required to be made up wrenchtight at couplings and unions, threaded hubs of junction boxes, device boxes, conduit bodies, and so on.

Figure 501.5 shows an explosionproof junction box having three hubs and a threaded opening for the screw-type cover. Unused openings must be effectively closed by inserting threaded metal plugs engaging at least five full threads and affording protection equivalent to that of the wall of the box. Figure 501.6 shows a larger-type explosionproof junction box with a bolted flanged cover.

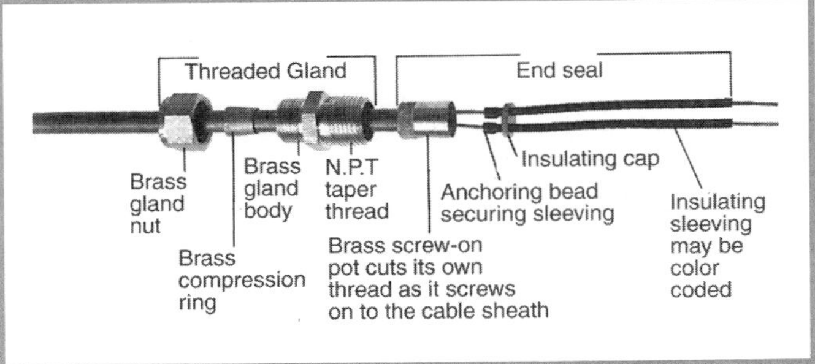

Figure 501.4 *Type MI cable and fitting. The flat joint between the gland body and compression ring and between the compression ring and gland nut must be explosionproof. Also, the threaded connections must be explosionproof. Type MI cable fittings not investigated for use in hazardous locations may not be explosionproof.*

Figure 501.5 *An explosionproof junction box with a screw-type cover. (Crouse-Hinds)*

Figure 501.6 *Cutaway view of explosionproof junction box and bolted flanged cover. (Appleton Electric Co.)*

Figure 501.7 shows a flexible fitting, available in lengths up to 3 ft, for use in Class I, Division 1 locations. The fitting consists of a deeply corrugated bronze tube with an internal nonmetallic tubular protective liner and an outer cover of braided fine bronze wires. A threaded fitting is securely attached to each end of the flexible tube. The flexible fitting is commonly used at motor connections, can withstand continuous vibration for long periods, is explosionproof, and affords maximum protection to any enclosed conductors.

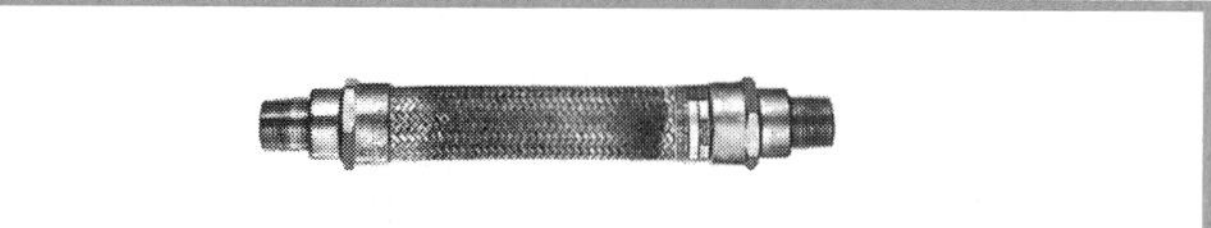

Figure 501.7 *A flexible explosionproof fitting. (Crouse-Hinds)*

(1) Fixed Wiring. In Class I, Division 1 locations, threaded rigid metal conduit, threaded steel intermediate metal conduit, or Type MI cable with termination fittings approved for the location shall be the wiring method employed. All boxes, fittings, and joints shall be threaded for connection to conduit or cable terminations and shall be explosionproof. Threaded joints shall be made up with at least five threads fully engaged. Type MI cable shall be installed and supported in a manner to avoid tensile stress at the termination fittings.

Exception No. 1: Rigid nonmetallic conduit complying with Article 347 shall be permitted where encased in a concrete envelope a minimum of 2 in. (50.8 mm) thick and provided with not less than 24 in. (610 mm) of cover measured from the top of the conduit to grade. The concrete encasement shall be permitted to be omitted where subject to the provisions of Section 511-4, Exception; 514-8, Exception No. 2; and Section 515-5(a).

Where rigid nonmetallic conduit is used, threaded rigid metal conduit or threaded steel intermediate metal conduit shall be used for the last 24 in. (610 mm) of the underground run to emergence or to the point of connection to the aboveground raceway. An equipment grounding conductor shall be included to provide for electrical continuity of the raceway system and for grounding of noncurrent-carrying metal parts.

Exception No. 1 allows the use of rigid nonmetallic conduit in some underground installations. If rigid nonmetallic conduit is used for underground wiring, threaded rigid metal conduit or threaded steel intermediate metal conduit must be used for the last 2 ft of the underground run to the point of emergence or the point of connection to the aboveground raceway. The rigid nonmetallic conduit, including rigid nonmetallic conduit elbows and fittings, is required to be located not less than 2 ft below grade. The conduit must also be encased in not less than 2 in. of concrete.

The exceptions in Sections 511-4, 514-8, and 515-5 do not require concrete encasement for rigid nonmetallic conduit. These exceptions are for specific occupancies where there has been considerable experience with underground nonmetallic conduit. Exception No. 1 to Section 501-4(a) applies to other occupancies with underground Class I, Division 1 locations.

Where rigid nonmetallic conduit is used, an equipment grounding conductor must be included and must be bonded to the metal raceway system.

Exception No. 2: In industrial establishments with restricted public access, where the conditions of maintenance and supervision ensure that only qualified persons will service the installation, Type MC cable, listed for use in Class I, Division 1 locations, with a gas/vaportight continuous corrugated aluminum sheath, an overall jacket of suitable polymeric material, separate grounding conductors in accordance with Section 250-122, and provided with termination fittings listed for the application shall be permitted.

Due to the potential for physical damage to Type MC cable, its use in Class I, Division 1 locations is limited to cable that is listed specifically for use in Class I, Division 1 locations and installed at facilities that have full-time, qualified maintenance personnel who would be doing regular maintenance, might notice if cables were damaged, would know the hazards, and could de-energize the circuit to repair the installation.

FPN: See Sections 334-3 and 334-4 for restrictions on use of Type MC cable.

Exception No. 3: In industrial establishments with restricted public access, where the conditions of maintenance and supervision ensure that only qualified persons will service the installation, Type ITC cable, listed for use in Class I, Division 1 locations, with a gas/vaportight continuous corrugated aluminum sheath, an overall jacket of suitable polymeric material and provided with termination fittings listed for the application shall be permitted.

A new Exception No. 3 was added to the 1999 *Code* that allows the use of Type ITC cable with a gas/vaportight corrugated aluminum sheath and polymeric jacket, listed for use in Class I, Division 1 locations in industrial establishments where all of the prequisites have been satisfied.

(2) Flexible Connections. Where necessary to employ flexible connections, as at motor terminals, flexible fittings listed for Class I locations shall be used.

Exception: Flexible cord installed in accordance with the provisions of Section 501-11 shall be permitted.

(b) Class I, Division 2. In Class I, Division 2 locations, threaded rigid metal conduit, threaded steel intermediate metal conduit, enclosed gasketed busways, enclosed gasketed wireways, or Type PLTC cable in accordance with the provisions of Article 725, or Type ITC cable in cable trays, in raceways, supported by messenger wire, or directly buried where the cable is listed for this use; Type MI, MC, MV, or TC cable with approved termination fittings shall be the wiring method employed. Type ITC, PLTC, MI, MC, MV, or TC cable shall be permitted to be installed in cable tray systems and shall be installed in a manner to avoid tensile stress at the termination fittings. Boxes, fittings, and joints shall not be required to be explosionproof except as required by Sections 501-3(b)(1), 501-6(b)(1), and 501-14(b)(1). Where provision must be made for limited flexibility, as at motor terminals, flexible metal fittings, flexible metal conduit with approved fittings, liquidtight flexible metal conduit with approved fittings, liquidtight flexible nonmetallic conduit with approved fittings, or flexible cord approved for extra-hard usage and provided with approved bushed fittings shall be used. An additional conductor for grounding shall be included in the flexible cord.

In Class I, Division 2 locations, boxes, fittings, and joints are not required to be explosionproof at lighting outlets or at enclosures containing no arcing devices. Where general-purpose enclosures are permitted by Section 501-4(b), rigid or intermediate metal conduit may be used with locknuts and bushings. However, a bonding jumper with proper fittings is required to be used between the enclosure and the raceway to ensure adequate bonding from the hazardous area to the point of grounding for the service equipment. See Section 501-16(a).

Where limited flexibility is necessary and approved fittings are required for use with flexible metal conduit, liquidtight flexible conduit, and extra-hard-usage flexible cord, the fittings are not required to be specifically approved for Class I locations. Also, where flexible conduit or liquidtight flexible conduit is used, internal or external bonding jumpers with proper fittings are required to be provided, in accordance with Section 501-16(b) and the exception to Section 501-4(b).

Section 501-4(b) also permits a variety of cables, cable tray systems in accordance with Section 318-3(d), enclosed gasketed wireways, and enclosed gasketed busways. The cable and cable fittings, cable trays, wireways, and busways are not required to be specifically listed or labeled for Class I locations. For example, if Type ITC or MC cable is used, neither the cable nor the fittings needs to be listed for use in hazardous (classified) locations. Type AC cable is not a permitted wiring method in Section 501-4(b). This is due to concern about arcing between convolutions during ground-fault conditions. This prohibition was introduced in the 1993 *Code*.

Any wiring method suitable for ordinary locations may be used in a nonincendive circuit. See Section 501-4(b), Exception.

FPN: See Section 501-16(b) for grounding requirements where flexible conduit is used.

Exception: Nonincendive field wiring shall be permitted using any of the methods suitable for wiring in ordinary locations.

The exception to Section 501-4(b) is intended to permit what have been termed "nonincendive field circuits." See the comments on "nonincendive" following Section 501-3(b)(1), Exception (c) and the definition of *nonincendive circuit* in Section 500-4(f)(1). Many low-voltage, low-energy circuits are of the nonincendive type. However, a Class 2 circuit, as defined in Article 725, is not necessarily nonincendive. Considerable equipment listed by such testing laboratories as Factory Mutual Research Corp. and Underwriters Laboratories Inc. has nonincendive circuits intended for field wiring in the various groups in Class I, Division 2 locations. Some common telephone circuits and thermocouple circuits are also nonincendive.

501-5. Sealing and Drainage. Seals in conduit and cable systems shall comply with (a) through (f). Sealing compound shall be of a type approved for the conditions and use. Sealing compound shall be used in Type MI cable termination fittings to exclude moisture and other fluids from the cable insulation.

FPN No. 1: Seals are provided in conduit and cable systems to minimize the passage of gases and vapors and prevent the passage of flames from one portion of the electrical installation to another through the conduit. Such communication through Type MI cable is inherently prevented by construction of the cable. Unless specifically designed and tested for the purpose, conduit and cable seals are not intended to prevent the passage of liquids, gases, or vapors at a continuous pressure differential across the seal. Even at differences in pressure across the seal equivalent to a few inches of water, there may be a slow passage of gas or vapor through a seal, and through conductors passing through the seal. See Section 501-5(e)(2). Temperature extremes and highly corrosive liquids and vapors can affect the ability of seals to perform their intended function. See Section 501-5(c)(2).

FPN No. 2: Gas or vapor leakage and propagation of flames may occur through the interstices between the strands of standard stranded conductors larger than No. 2. Special conductor constructions, e.g., compacted strands or sealing of the individual strands, are means of reducing leakage and preventing the propagation of flames.

The sealing compound used in conduit seal fittings is somewhat porous, so that gases, particularly those under slight pressure, and gases with small molecules, such as hydrogen, can pass slowly through the sealing compound. Also, the seal is around the insulation on the conductor, and gases can be transmitted slowly through the air spaces (the interstices) between strands of stranded conductors. See Section 501-5(e)(2), FPN No. 2. Experience has shown, however, that under normal conditions for the smaller conductors, and with only normal atmospheric pressure differentials across the seal, the passage of gas through a seal is not sufficient to result in a hazard. For larger conductors, however, see Section 501-5, FPN No. 2. Sealing fittings should be used only with the sealing compound or compounds recommended by the fitting manufacturer. Different sealing compounds have different rates of expansion and contraction that may affect their function within a given fitting.

The use of teflon tapes or joint compounds on conduit threads may weaken the seal fitting and interrupt the equipment grounding path. Cracks have occurred during hydrostatic testing of fittings where these materials have been used.

(a) Conduit Seals, Class I, Division 1. In Class I, Division 1 locations, conduit seals shall be located as follows.

(1) In each conduit entry into an explosionproof enclosure where either (a) the enclosure contains apparatus, such as switches, circuit breakers, fuses, relays, or resistors, that may produce arcs, sparks, or high temperatures that are considered to be an ignition source in normal operation, or (b) the entry is 2-in. size or larger and the enclosure contains terminals, splices, or taps. For the purposes of this section high temperatures shall be considered to be any temperatures exceeding 80 percent of the autoignition temperature in degrees Celsius of the gas or vapor involved.

Exception to (a)(1)(a): Conduit entering an enclosure where such switches, circuit breakers, fuses, relays, or resistors are

(a) Enclosed within a chamber hermetically sealed against the entrance of gases or vapors, or

(b) Immersed in oil in accordance with Section 501-6(b)(1)(b), or

(c) Enclosed within a factory-sealed explosionproof chamber located within the enclosure, approved for the location, and marked "factory sealed" or equivalent.

Factory-sealed enclosures shall not be considered to serve as a seal for another adjacent explosionproof enclosure that is required to have a conduit seal.

Conduit seals shall be installed within 18 in. (457 mm) from the enclosure. Only explosionproof unions, couplings, reducers, elbows, capped elbows, and conduit bodies similar to L, T, and Cross types that are not larger than the trade size of the conduit shall be permitted between the sealing fitting and the explosionproof enclosure.

(2) In each conduit entry into a pressurized enclosure where the conduit is not pressurized as part of the protection system. Conduit seals shall be installed within 18 in. (457 mm) from the pressurized enclosure.

FPN No. 1: Installing the seal as close as possible to the enclosure will reduce problems with purging the dead airspace in the pressurized conduit.

FPN No. 2: For further information, see *Standard for Purged and Pressurized Enclosures for Electrical Equipment,* NFPA 496-1998.

(3) Where two or more explosionproof enclosures for which conduit seals are required under (a)(1) are connected by nipples or by runs of conduit not more than 36 in. (914 mm) long, a single conduit seal in each such nipple connection or run of conduit shall be considered sufficient if located not more than 18 in. (457 mm) from either enclosure.

An example of Section 501-5(a) requirements is illustrated in Figure 501.8.

(4) In each conduit run leaving a Class I, Division 1 location. The sealing fitting shall be permitted on either side of the boundary of such location within 10 ft (3.05 m) of the boundary, and shall be designed and installed so to minimize the amount of gas or vapor within the Division 1 portion of the conduit from being communicated to the conduit beyond the seal. Except for approved explosionproof reducers at the conduit seal, there shall be no union, coupling, box, or fitting between the conduit seal and the point at which the conduit leaves the Division 1 location.

Exception: Metal conduit containing no unions, couplings, boxes, or fittings that passes completely through a Class I, Division 1 location with no fittings less than 12 in. (305 mm) beyond each boundary shall not require a conduit seal if the termination points of the unbroken conduit are in unclassified locations.

(b) Conduit Seals, Class I, Division 2. In Class I, Division 2 locations, conduit seals shall be located as follows.

(1) For connections to enclosures that are required to be explosionproof, a conduit seal shall be provided in accor-

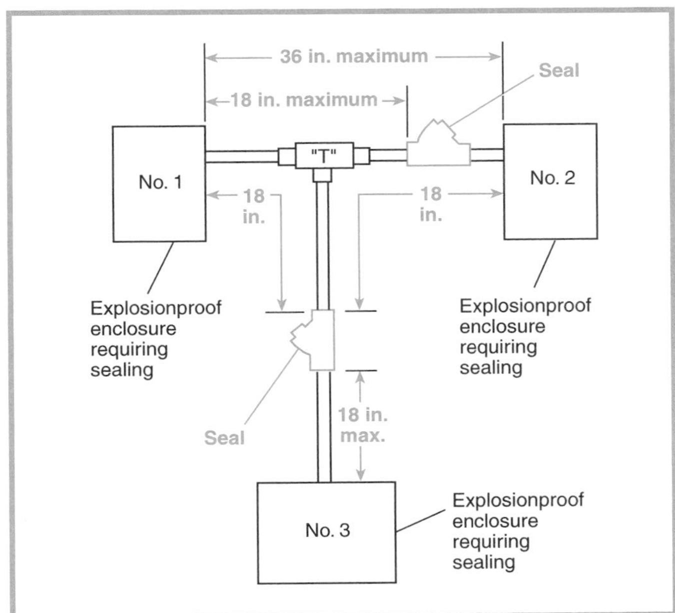

Figure 501.8 *Two seals are required here, so that the run of conduit between Enclosure No. 1 and Enclosure No. 2 is sealed. Even if Enclosure No. 3 were not required to be sealed, the seal in the vertical run of conduit to Enclosure No. 3 would be required to be sealed within 18 in. of Enclosure No. 1, because the vertical conduit run to the "T" fitting is a conduit run to Enclosure No. 1.*

dance with (a)(1)(a) and (a)(3). All portions of the conduit run or nipple between the seal and such enclosure shall comply with Section 501-4(a).

(2) In each conduit run passing from a Class I, Division 2 location into an unclassified location. The sealing fitting shall be permitted on either side of the boundary of such location within 10 ft (3.05 m) of the boundary, and shall be designed and installed so to minimize the amount of gas or vapor within the Division 2 portion of the conduit from being communicated to the conduit beyond the seal. Rigid metal conduit or threaded steel intermediate metal conduit shall be used between the sealing fitting and the point at which the conduit leaves the Division 2 location, and a threaded connection shall be used at the sealing fitting. Except for approved explosionproof reducers at the conduit seal, there shall be no union, coupling, box, or fitting between the conduit seal and the point at which the conduit leaves the Division 2 location.

Exception No. 1: Metal conduit containing no unions, couplings, boxes or fittings that passes completely through a Class I, Division 2 location with no fittings less than 12 in. (305 mm) beyond each boundary shall not be required to be sealed if the termination points of the unbroken conduit are in unclassified locations.

Exception No. 2: Conduit systems terminating at an unclassified location where a wiring method transition is made to cable tray, cablebus, ventilated busway, Type MI cable, or open wiring shall not be required to be sealed where passing from the Class I, Division 2 location into the unclassified location. The unclassified location shall be outdoors or, if the conduit system is all in one room, it shall be permitted to be indoors. The conduits shall not terminate at an enclosure containing an ignition source in normal operation.

Exception No. 3: Conduit systems passing from an enclosure or room that is unclassified as a result of pressurization into a Class I, Division 2 location shall not require a seal at the boundary.

FPN: For further information, refer to *Standard for Purged and Pressurized Enclosures for Electrical Equipment,* NFPA 496-1998.

Exception No. 4: Segments of aboveground conduit systems shall not be required to be sealed where passing from a Class I, Division 2 location into an unclassified location if the following conditions are met:

(a) No part of the conduit system segment passes through a Class I, Division I location where the conduit contains unions, couplings, boxes, or fittings within 12 in. (305 mm) of the Class I, Division 1 location; and
(b) The conduit system segment is located entirely in outdoor locations; and
(c) The conduit system segment is not directly connected to canned pumps, process or service connections for flow, pressure, or analysis measurement, etc., that depend on a single compression seal, diaphragm, or tube to prevent flammable or combustible fluids from entering the conduit system; and
(d) The conduit system segment contains only threaded metal conduit, unions, couplings, conduit bodies, and fittings in the unclassified location; and
(e) The conduit system segment is sealed at its entry to each enclosure or fitting housing terminals, splices, or taps in Class I, Division 2 locations.

(c) Class I, Divisions 1 and 2. Where required, seals in Class I, Division 1 and 2 locations shall comply with the following.

(1) Fittings. Enclosures for connections or equipment shall be provided with an approved integral means for sealing, or sealing fittings approved for Class I locations shall be used. Sealing fittings shall be accessible.

(2) Compound. Sealing compound shall be approved and shall provide a seal against passage of gas or vapors through the seal fitting, shall not be affected by the surrounding atmosphere or liquids, and shall not have a melting point of less than 93°C (200°F).

(3) Thickness of Compounds. In a completed seal, the minimum thickness of the sealing compound shall not be less than the trade size of the sealing fitting and, in no case, less than ⅝ in. (16 mm).

Exception: Listed cable sealing fittings shall not be required to have a minimum thickness equal to the trade size of the fitting.

(4) Splices and Taps. Splices and taps shall not be made in fittings intended only for sealing with compound, nor shall other fittings in which splices or taps are made be filled with compound.

(5) Assemblies. In an assembly where equipment that may produce arcs, sparks, or high temperatures is located in a compartment separate from the compartment containing splices or taps, and an integral seal is provided where conductors pass from one compartment to the other, the entire assembly shall be approved for Class I locations. Seals in conduit connections to the compartment containing splices or taps shall be provided in Class I, Division 1 locations where required by (a)(1)(b).

(6) Conductor Fill. The cross-sectional area of the conductors permitted in a seal shall not exceed 25 percent of the cross-sectional area of a rigid metal conduit of the same trade size unless it is specifically approved for a higher percentage of fill.

(d) Cable Seals, Class I, Division 1. In Class I, Division 1 locations, cable seals shall be located as follows.

(1) Cable shall be sealed at all terminations. The sealing fitting shall comply with (c). Multiconductor Type MC cables with a gas/vaportight continuous corrugated aluminum sheath and an overall jacket of suitable polymeric material shall be sealed with an approved fitting after removing the jacket and any other covering so that the sealing compound will surround each individual insulated conductor in such a manner as to minimize the passage of gases and vapors.

Exception: Shielded cables and twisted pair cables shall not require the removal of the shielding material or separation of the twisted pairs provided the termination is by an approved means to minimize the entrance of gases or vapors and prevent propagation of flame into the cable core.

A new exception to Section 501-5(d)(1) has been added to the 1999 *Code* that allows the installation of shielded and twisted pair cables commonly used for signalling and instrumentation, to be sealed without removal of the outer sheathing nor the separation of the twisted conductors.

(2) Cables in conduit with a gas/vaportight continuous sheath capable of transmitting gases or vapors through the cable core shall be sealed in the Division 1 location after removing the jacket and any other coverings so that the sealing compound will surround each individual insulated conductor and the outer jacket.

Exception: Multiconductor cables with a gas/vaportight continuous sheath capable of transmitting gases or vapors through the cable core shall be permitted to be considered as a single conductor by sealing the cable in the conduit within 18 in. (457 mm) of the enclosure and the cable end within the enclosure by an approved means to minimize the entrance of gases or vapors and prevent the propagation of flame into the cable core, or by other approved methods. For shielded cables and twisted pair cables, it shall not be required to remove the shielding material or separate the twisted pair.

The intent of the exception to Section 501-5(d)(2) is to permit flat computer cables, coaxial cables, and twisted pairs to be treated as a single conductor where installed in conduit, since separating the individual conductors or removing the outer jacket (of a coaxial cable or a twisted pair, for example) is impractical and can destroy the electrical properties of the cable. In addition to the cable seal, the end of the cable within the enclosure must also be sealed.

(3) Each multiconductor cable in conduit shall be considered as a single conductor if the cable is incapable of transmitting gases or vapors through the cable core. These cables shall be sealed in accordance with (a).

(e) Cable Seals, Class I, Division 2. In Class I, Division 2 locations, cable seals shall be located as follows.

(1) Cables entering enclosures that are required to be approved for Class I locations shall be sealed at the point of entrance. The sealing fitting shall comply with (b)(1). Multiconductor cables with a gas/vaportight continuous sheath capable of transmitting gases or vapors through the cable core shall be sealed in an approved fitting in the Division 2 location after removing the jacket and any other coverings so that the sealing compound will surround each individual insulated conductor in such a manner as to minimize the passage of gases and vapors. Multiconductor cables in conduit shall be sealed as described in (d).

Exception No. 1: Cables passing from an enclosure or room that is unclassified as a result of Type Z pressurization into a Class I, Division 2 location shall not require a seal at the boundary.

Exception No. 2: Shielded cables and twisted pair cables shall not require the removal of the shielding material or separation of the twisted pairs provided the termination is by an approved means to minimize the entrance of gases or vapors and prevent propagation of flame into the cable core.

A new exception to Section 501-5(e)(1) was added in the 1999 *Code* to allow cables that go from a Type Z pressurized room or enclosure into a Class 1, Division 2 location to be installed without a sealing fitting at the boundary. This will correlate with the requirement for conduit found in Section 501-5(b), Exception No. 1.

(2) Cables with a gas/vaportight continuous sheath and that will not transmit gases or vapors through the cable core in excess of the quantity permitted for seal fittings shall not be required to be sealed except as required in (e)(1). The minimum length of such cable run shall not be less than that length that limits gas or vapor flow through the cable core to the rate permitted for seal fittings [0.007 ft^3/hour (198 cm^3/hour) of air at a pressure of 6 in. of water (1493 pascals)].

The ability of a cable to transmit gases or vapors through the core (primarily between insulated conductors) depends not only on how tightly packed the conductors are within the outer sheaths, and the location and composition of fillers, but also on how the cable has been handled and the geometry of the cable run. If there is any question whether or not the cable run is capable of transmitting gases or vapors through the core, sealing is suggested. See Section 501-5, FPN No. 2.

FPN No. 1: See *Outlet Boxes and Fittings for Use in Hazardous (Classified) Locations,* ANSI/UL 886-1994.

FPN No. 2: The cable core does not include the interstices of the conductor strands.

The intent of this fine print note is that the conductors themselves be individually sealed, such as by dipping the ends in wax, before measuring the rate of flow. If this is done, however, the wax should be removed before making electrical connections or putting the system into service.

(3) Cables with a gas/vaportight continuous sheath capable of transmitting gases or vapors through the cable core shall not be required to be sealed except as required in (e)(1), unless the cable is attached to process equipment or devices that may cause a pressure in excess of 6 in. of water (1493 pascals) to be exerted at a cable end, in which case a seal, barrier, or other means shall be provided to prevent migration of flammables into an unclassified area.

Exception: Cables with an unbroken gas/vaportight continuous sheath shall be permitted to pass through a Class I, Division 2 location without seals.

(4) Cables that do not have gas/vaportight continuous sheath shall be sealed at the boundary of the Division 2 and unclassified location in such a manner as to minimize the passage of gases or vapors into an unclassified location.

FPN: The sheath mentioned in (d) and (e) may be either metal or a nonmetallic material.

See Figures 501.9, 501.10, and 501.11. Also see the commentary following Section 501-5(f)(3).

Table 5.1 summarizes the seal requirements of Section 501-5.

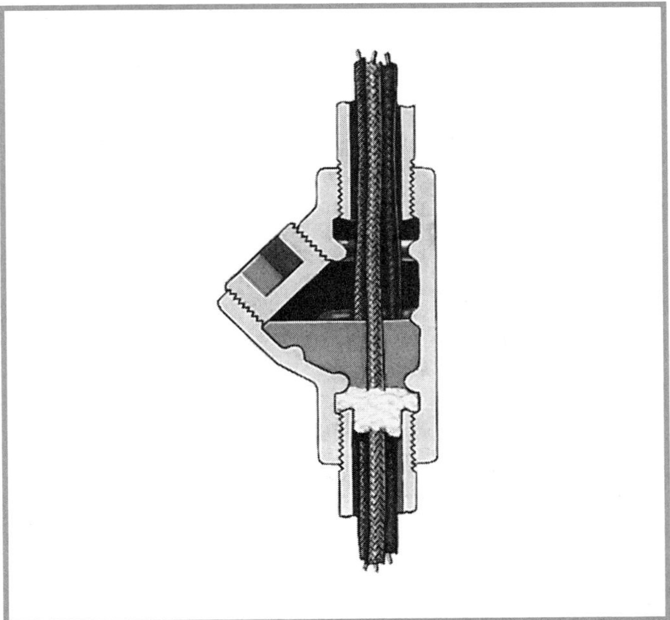

Figure 501.9 *A sealing fitting placed in a run of conduit to minimize the passage of gases from one portion of the electrical installation to another. (Crouse-Hinds)*

Figure 501.10 *A sealing fitting for a horizontal or vertical conduit run with a removable screw-type fitting for draining purposes. (Appleton Electric Co.)*

(f) Drainage.

(1) Control Equipment. Where there is a probability that liquid or other condensed vapor may be trapped within enclosures for control equipment or at any point in the raceway system, approved means shall be provided to prevent accumulation or to permit periodic draining of such liquid or condensed vapor.

(2) Motors and Generators. Where the authority having jurisdiction judges that there is a probability that liquid or condensed vapor may accumulate within motors or gener-

Table 5.1 Conduit and Cable Seal Requirements

Classification	Application	Location of Seal
Conduit Seals		
Class I, Division 1	Switch enclosure Circuit breaker enclosure Fuse enclosure Relay enclosure Resistor enclosure Arcing or sparking apparatus High-temperature apparatus	In conduit run within 18 in. of enclosure
	Explosionproof enclosure containing arcing or sparking contacts that are hermetically sealed against gas or vapor entry Explosionproof enclosure containing arcing or sparking contacts that are immersed in oil, in accordance with Section 501-6(b)(1)(2)	In conduit runs of 1½ in. and smaller, no seal is required. Where conduit is larger than 1½ in., in conduit run within 18 in. of enclosure.
	Enclosure containing terminals, splices, or taps fitting containing terminals, splices or taps	In conduit runs smaller than 2 in. trade size, no seal is required. Where conduit is 2 in. or larger, in conduit run within 18 in. of enclosure.
	Two explosionproof enclosures with a conduit run between them of 36 in. or less	In conduit run within 18 in. of each enclosure. Permitted to use a single seal in each run as long as the seal is within 18 in. of each enclosure.
	Two explosionproof enclosures with a conduit run between them greater than 36 in.	In conduit run within 18 in. of each enclosure.
	Conduit run leaving Division 1 location	On either side of boundary. No unions, couplings, boxes, or fittings permitted between the seal fitting and the point where the conduit leaves the Division 1 location.
	Metal conduit containing no unions, couplings, boxes, or fittings that passes completely through a Class I, Division 1 location, with no fittings less than 12 in. (305 mm) beyond each boundary.	Not required to be sealed if the termination points of the unbroken conduit are in unclassified locations.
Class I, Division 2	Enclosure required to be explosionproof	Seal as required for similar equipment in Division 1 location.
	Conduit run leaving Division 2 location	On either side of boundary. No unions, couplings, boxes or fittings permitted between the seal fitting and the point where the conduit leaves the Division 2 location.
	Metal conduit containing no unions, couplings, boxes, or fittings that passes completely through a Division 2 location with no fittings less than 12 in. (305 mm) beyond each boundary.	Not required to be sealed if the termination points of the unbroken conduit are in unclassified locations.
	Conduit systems terminating at an unclassified location where a wiring method transition is made to cable tray, cablebus, ventilated busway, Type MI cable, or open wiring.	Not required to be sealed where passing from the Class I, Division 2 location into an outdoor unclassified location or an indoor location if the conduit system is all in one room. The conduits shall not terminate at an enclosure containing an ignition source in normal operation.
Cable Seals		
Class I, Division 1	Enclosure with integral seal	Conduit seal fitting not required
	Multiconductor Type MC cables with a gas/vaportight continuous corrugated aluminum sheath and an overall jacket of suitable polymeric material.	Seal at all terminations with an approved fitting after removing the jacket and any other covering, so that the sealing compound will surround each individual insulated conductor.

Table 5.1 (Continued)

Classification	Application	Location of Seal
	Cables in conduit with a gas/vaportight continuous sheath capable of transmitting gases or vapors through the cable core.	Seal in the Division 1 location after removing the jacket and any other coverings, so that the sealing compound will surround each individual insulated conductor and the outer jacket.
	Multiconductor cables with a gas/vaportight continuous sheath capable of transmitting gases or vapors through the cable core.	Permitted to be considered a single conductor by sealing the cable in the conduit within 18 in. (457 mm) of the enclosure and the cable end within the enclosure by an approved means, to minimize the entrance of gases or vapors and prevent the propagation of flame into the cable core, or by other approved methods.
	For shielded cables and twisted pair cables	Removal of the shielding material or separation of the twisted pair is not required. Sealing the cable in the conduit within 18 in. (457 mm) of the enclosure and the cable end within the enclosure by an approved means, to minimize the entrance of gases or vapors and prevent the propagation of flame into the cable core, or by other approved methods.
	Each multiconductor cable in conduit if the cable is incapable of transmitting gases or vapors through the cable core.	Considered a single conductor. These cables and sealed in accordance with Section 501-5(a).
Class I, Division 2	Cables entering enclosures that are required to be approved for Class I locations.	Sealed at the point of entrance.
	Multiconductor cables in conduit	Sealed in accordance with the requirements for Division 1 locations.
	Multiconductor cables with a gas/vaportight continuous sheath capable of transmitting gases or vapors through the cable core.	Sealed in an approved fitting in the Division 2 location after removing the jacket and any other coverings, so that the sealing compound will surround each individual insulated conductor.
	Cables with a gas/vaportight continuous sheath and that will not transmit gases or vapors through the cable core in excess of the quantity permitted for seal fittings. The minimum length of such cable run shall not be less than that length that limits gas or vapor flow through the cable core to the rate permitted for seal fittings [0.007 cu ft per hour (198 cu cm per hour) of air at a pressure of 6 in. of water (1493 pascals)].	Not required to be sealed unless entering an enclosure that is required to be approved for Class I locations.
	Cables with a gas/vaportight continuous sheath capable of transmitting gases or vapors through the cable core.	Not required to be sealed unless entering an enclosure that is required to be approved for Class I locations or unless the cable is attached to process equipment or devices that may cause a pressure in excess of 6 in. of water (1493 pascals) to be exerted at a cable end, in which case a seal, barrier, or other means shall be provided to prevent migration of flammables into an unclassified area.
	Cables with an unbroken gas/vaportight continuous sheath that pass through a Class I, Division 2 location.	No seal required
	Cables that do not have a gas/vaportight continuous sheath.	Sealed at the boundary of the Division 2 and unclassified location in such a manner as to minimize the passage of gases or vapors into an unclassified location.

ators, joints and conduit systems shall be arranged to minimize entrance of liquid. If means to prevent accumulation or to permit periodic draining are judged necessary, such means shall be provided at the time of manufacture and shall be considered an integral part of the machine.

(3) Canned Pumps, Process or Service Connections, etc. For canned pumps, process or service connections for flow, pressure, or analysis measurement, etc., that depend on a single compression seal, diaphragm, or tube to prevent flammable or combustible fluids from entering the electrical raceway or cable system capable of transmitting fluids, an additional approved seal, barrier, or other means shall be provided to prevent the flammable or combustible fluid from entering the raceway or cable system capable of transmitting fluids beyond the additional devices or means, if the primary seal fails.

Other means of preventing flammable or combustible fluid from entering the conduit system employed by the liquefied natural gas industry is the use of a vented enclosure containing busbars or specifically designed process-connected instrumentation. The conduit from the canned pump terminates at the enclosure, and the conductors are connected to one end of the solid busbars. The circuit continues through the short busbar section to another set of conductors and a sealed conduit.

The additional approved seal or barrier and the interconnecting enclosure shall meet the temperature and pressure conditions to which they will be subjected upon failure of the primary seal, unless other approved means are provided to accomplish the purpose above.

Drains, vents, or other devices shall be provided so that primary seal leakage will be obvious.

FPN: See also the fine print notes to Section 501-5.

Seals in conduits are used to prevent an explosion from traveling through the conduit to another enclosure and to minimize the passage of gases or vapors from a hazardous (classified) location to a nonhazardous location. Sealing compound must be used as soon as possible on Type MI cable terminations to exclude moisture from cable insulation.

Where the conduit enters an enclosure containing arcing or high-temperature equipment, a sealing fitting is required to be placed within 18 in. of the enclosure it isolates; conduit bodies ("L," "T," etc.), couplings, unions, and elbows are the only enclosures or fittings permitted between the seal and the enclosure. See Figure 501.13 for an approved type of union. If two enclosures are spaced not more than 36 in. apart, a single seal may be placed between two connecting nipples if the seal is located not more than 18 in. from either enclosure.

In each 2-in. or larger conduit, a sealing fitting is required to be placed within 18 in. of the entrance to an explosionproof enclosure that houses terminals, splices, or taps.

A sealing fitting is required where the conduit leaves a Division 1 location or passes from a Division 2 location to a nonhazardous location. The sealing fitting is permitted on either side of the boundary, and there is to be no union, coupling, box, and such, between the seal and the boundary. However, approved explosionproof reducers are permitted to be installed in conduit seals.

The maximum permitted fill for the conduit is 40 percent; the maximum permitted fill of most conduit seals is 25 percent. If the conduit fill exceeds 25 percent of the cross-sectional area of the sealing fitting, the use of a larger trade size seal may be required. The use of reducers is allowed for connection of a larger trade size sealing fitting to the conduit.

It is preferable to locate the fitting on the nonhazardous-location side. This sealing fitting serves two purposes: it completes the explosionproof wiring method and it completes the explosionproof enclosure system. Note that a ½-in. conduit connected to an explosionproof box containing only splices, even in a Division 1 location, is not required to be sealed within 18 in. of the box. The seal at the boundary of the Division 1 location serves to complete the explosionproof system. The sealing fitting at the boundary also prevents the conduit system from serving as a pipe to transmit flammable mixtures from a Division 1 or Division 2 location to a nonhazardous location.

In Class I, Division 2 locations, a seal is required in each conduit entering an enclosure that is required to be explosionproof, in order to complete the explosionproof enclosure.

Figure 501.9 illustrates the proper sealing of a fitting. A dam must be provided to prevent the sealing material from running out of the fitting while still in the liquid state. All conductors must be separated to permit the sealing material to run between them. The sealing compound is required to have a minimum thickness of not less than the trade size of the conduit, and in no case less than ⅝ in. Conduit fittings for sealing are for use only with sealing compound supplied with the fitting and specified by the manufacturer in instructions furnished with the fitting.

Unless the additional seal or barrier, described in Section 501-5(f)(3), and interconnecting enclosures meet the performance requirements of the primary seal, the application of pressure or exposure to extreme temperatures must be prevented at the additional seal or barrier so that the process fluid will not enter the conduit system if the primary seal fails. If the process fluid is a gas or can become a gas under ordinary atmospheric conditions (liquefied natural gas, for example), the drain mentioned in Section 501-5(f)(3) should be a vent. See the commentary following Section 501-5(f)(3).

The necessary sealing may be accomplished by a sealing fitting and compound. To eliminate the time-consuming task of field-poured seals, a factory-sealed device with the seal designed into the device is permissible. A wide selection of factory-sealed devices are available for a variety of installations in hazardous (classified) locations. Factory-sealed devices are usually marked as such. Explosionproof motors are normally factory sealed, and no other seal is required. Where a conduit terminates in a motor, however, and if the conduit is 2 in. or larger, a seal is required to be placed within 18 in. of the motor terminal housing.

Figure 501.11 shows a sealing fitting designed for use in a vertical run of conduit, to provide drainage for any condensation of moisture trapped by the seal above an enclosure. Any accumulation of water runs down over the surface of the sealing compound, flowing through an explosionproof drain. Figure 501.10 illustrates a sealing fitting designed for either a horizontal or vertical conduit run. Figure 501.12 shows a combination drain and breather fitting. These fittings are specially designed to serve as a water drain and air vent while providing positive explosionproof protection. The fitting permits the escape of accumulated water through its drain, and the breather allows the continuous circulation of air, preventing condensation of any moisture that may be present. Individual drain or breather fittings are also available. It is good practice to consider the installation of drain, breather, or combination fittings to guard against water accumulation, which can cause future insulation failures, even though prevalent conditions may not indicate a need.

Figure 501.11 *A sealing fitting with an automatic drain plug. (Appleton Electric Co.)*

Figure 501.12 *A combination breather-drainage fitting. (Appleton Electric Co.)*

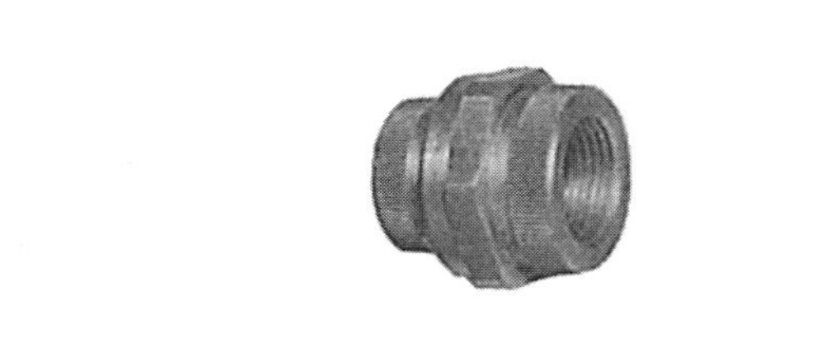

Figure 501.13 *An explosionproof union. (Appleton Electric Co.)*

Figure 501.14 illustrates a Class I, Division 1 location using threaded rigid metal conduit or threaded intermediate metal conduit and explosionproof fittings and equipment, including motors, motor controllers, push-button stations, lighting outlets, and junction boxes. The enclosures for the disconnecting means and motor controller for the motor (right portion of the drawing) are placed in a nonhazardous location and are thus not required to be explosionproof.

Each of the three conduits is sealed on the nonhazardous side before passing into the hazardous (classified) location. The pigtail leads of both motors are factory sealed at the motor-terminal housing, and, unless the size of the flexible fitting entering the motor-terminal housing is 2-in. trade size or larger, no other seals are needed at this point. Because the push-button control station and the motor controller and disconnect (left portion of the drawing) are considered arc-producing devices, conduits are sealed within 18 in. of the entrance to these enclosures. Seals are required even though the contacts may be immersed in oil.

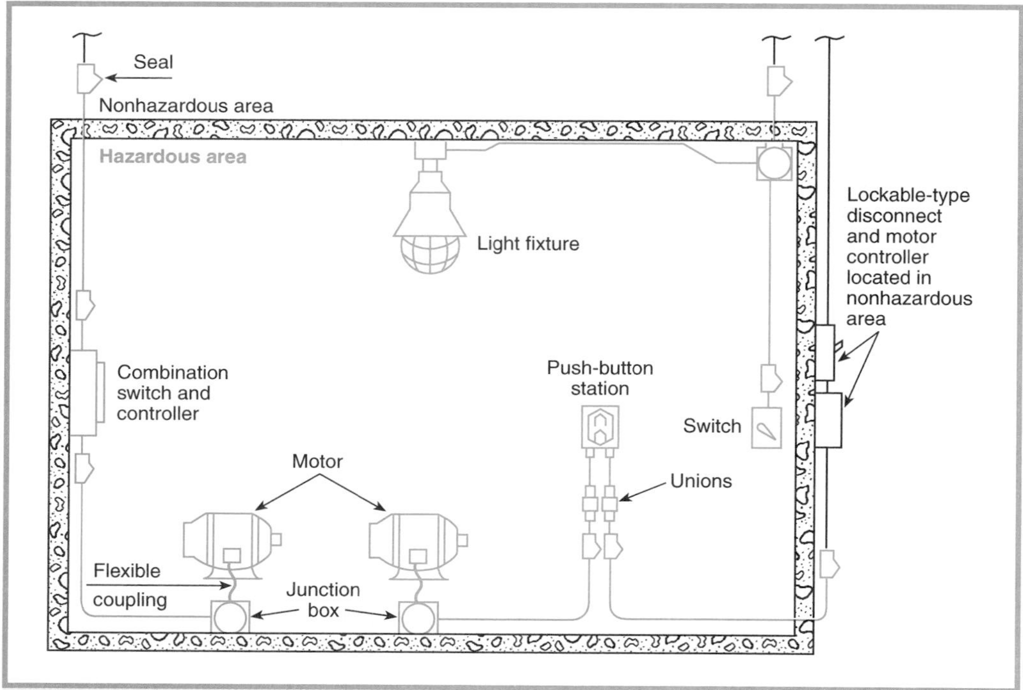

Figure 501.14 A Class I, Division 1 location where threaded metal conduits, sealing fittings, explosionproof fittings, and equipment for power and lights are used.

A seal is provided within 18 in. of the lighting switch. The design of the lighting fixture, as required by ANSI/UL 844, is such that the explosionproof chamber for the wiring must be separated or sealed from the lamp compartment; hence, a separate seal is not required adjacent to lighting fixtures complying with ANSI/UL 844. The lighting fixture is suspended on a conduit stem threaded into the cover of an explosionproof ceiling box. (See Section 501-9.)

501-6. Switches, Circuit Breakers, Motor Controllers, and Fuses.

(a) Class I, Division 1. In Class I, Division 1 locations, switches, circuit breakers, motor controllers, and fuses, including pushbuttons, relays, and similar devices, shall be provided with enclosures, and the enclosure in each case, together with the enclosed apparatus, shall be approved as a complete assembly for use in Class I locations.

(b) Class I, Division 2. Switches, circuit breakers, motor controllers, and fuses in Class I, Division 2 locations shall comply with the following:

(1) Type Required. Circuit breakers, motor controllers, and switches intended to interrupt current in the normal performance of the function for which they are installed shall be provided with enclosures approved for Class I, Division 1 locations in accordance with Section 501-3(a), unless general-purpose enclosures are provided and

(a) The interruption of current occurs within a chamber hermetically sealed against the entrance of gases and vapors, or

(b) The current make-and-break contacts are oil-immersed and of the general-purpose type having a 2-in. (50.8-mm) minimum immersion for power contacts and a 1-in. (25.4-mm) minimum immersion for control contacts, or

(c) The interruption of current occurs within a factory-sealed explosionproof chamber approved for the location, or

(d) The device is a solid state, switching control without contacts, where the surface temperature does not exceed 80 percent of the ignition temperature in degrees Celsius of the gas or vapor involved.

(2) Isolating Switches. Fused or unfused disconnect and isolating switches for transformers or capacitor banks that are not intended to interrupt current in the normal performance of the function for which they are installed shall be permitted to be installed in general-purpose enclosures.

(3) Fuses. For the protection of motors, appliances, and lamps, other than as provided in (b)(4), standard plug or cartridge fuses shall be permitted, provided they are placed within enclosures approved for the location; or fuses shall be permitted if they are within general-purpose enclosures, and if they are of a type in which the operating element is immersed in oil or other approved liquid, or the operating element is enclosed within a chamber hermetically sealed against the entrance of gases and vapors, or the fuse is a nonindicating, filled, current-limiting type.

(4) Fuses Internal to Lighting Fixtures. Approved cartridge fuses shall be permitted as supplementary protection within lighting fixtures.

Figure 501.15 shows an explosionproof panelboard consisting of an assembly of branch-circuit devices enclosed in a cast metal explosionproof housing. These panelboards are provided with bolted access covers and threaded conduit-entry hubs designed to withstand the force of an internal explosion.

Figure 501.15 *An explosionproof panelboard. (Appleton Electric Co.)*

Figure 501.16 shows an open and closed view of a cylindrical-type ("spin-top") combination motor controller, motor control starter, and circuit breaker in an explosionproof enclosure. The top and bottom covers are threaded on for quick removal for installation and servicing. Figure 501.17 shows the same type of equipment in a rectangular enclosure with a hinged, bolted-on cover. These types of housings are designed to accommodate a wide range of manually or magnetically operated across-the-line types of motor starters in a variety of ratings.

In Class I, Division 2 locations, it is assumed that fuses or circuit breakers will seldom open the circuit where used to protect feeders or branch circuits supplying lamps in fixed positions only. Division 2 locations are not normally hazardous but may become so [see Section 500-7(b)(3)], and since it is unlikely that the fuse or circuit breaker in such a circuit will operate simultaneously with the occurrence of an explosive mixture inside the enclosure, general-purpose enclosures are permitted for such overcurrent devices.

Section 501-6(b)(4) permits fuses, often used for ballast protection in high-intensity-discharge and outdoor fluorescent fixtures.

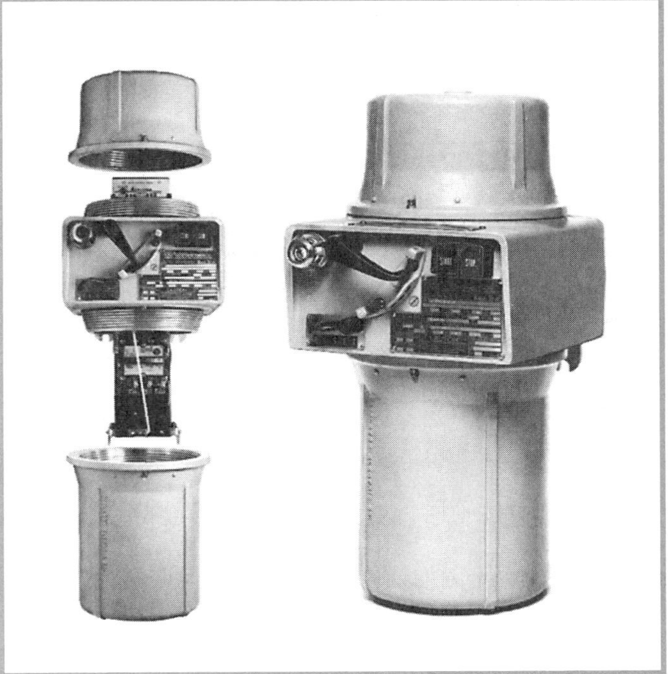

Figure 501.16 *An explosionproof enclosure for a motor control starter and circuit breaker (open and closed views). (Appleton Electric Co.)*

Figure 501.18 illustrates a standard toggle switch in an explosionproof enclosure.

501-7. Control Transformers and Resistors. Transformers, impedance coils, and resistors used as, or in conjunction with, control equipment for motors, generators, and appliances shall comply with (a) and (b).

(a) Class I, Division 1. In Class I, Division 1 locations, transformers, impedance coils, and resistors, together with any switching mechanism associated with them, shall be provided with enclosures approved for Class I, Division 1 locations in accordance with Section 501-3(a).

(b) Class I, Division 2. In Class I, Division 2 locations, control transformers and resistors shall comply with the following.

Figure 501.17 A magnetic motor starter for use in a Class I, Group D location. Note the number of securing bolts and the width of the flange. (General Electric Co.)

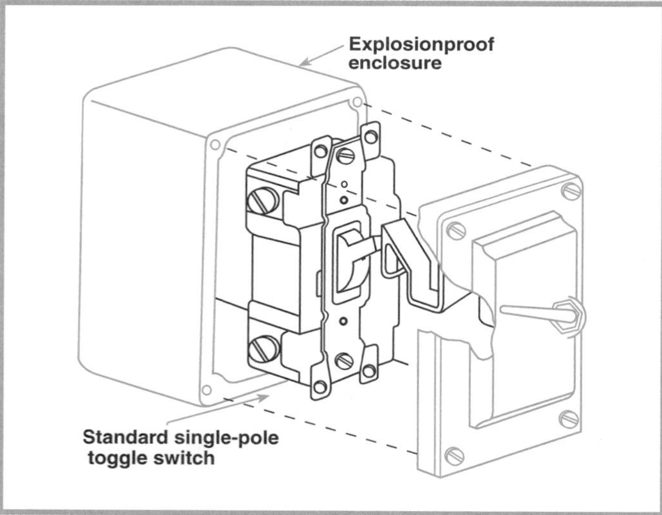

Figure 501.18 A standard toggle switch in an explosionproof enclosure.

(1) Switching Mechanisms. Switching mechanisms used in conjunction with transformers, impedance coils, and resistors shall comply with Section 501-6(b).

(2) Coils and Windings. Enclosures for windings of transformers, solenoids, or impedance coils shall be permitted to be of the general-purpose type.

(3) Resistors. Resistors shall be provided with enclosures; and the assembly shall be approved for Class I locations, unless resistance is nonvariable and maximum operating temperature, in degrees Celsius, will not exceed 80 percent of the ignition temperature of the gas or vapor involved, or has been tested and found incapable of igniting the gas or vapor.

501-8. Motors and Generators.

(a) Class I, Division 1. In Class I, Division 1 locations, motors, generators, and other rotating electric machinery shall be as follows:

(1) Approved for Class I, Division 1 locations; or
(2) Of the totally enclosed type supplied with positivepressure ventilation from a source of clean air with discharge to a safe area, so arranged to prevent energizing of the machine until ventilation has been established and the enclosure has been purged with at least 10 volumes of air, and also arranged to automatically de-energize the equipment when the air supply fails; or
(3) Of the totally enclosed inert gas-filled type supplied with a suitable reliable source of inert gas for pressuring the enclosure, with devices provided to ensure a positive pressure in the enclosure and arranged to automatically de-energize the equipment when the gas supply fails; or
(4) Of a type designed to be submerged in a liquid that is flammable only when vaporized and mixed with air, or in a gas or vapor at a pressure greater than atmospheric and that is flammable only when mixed with air; and the machine is arranged so to prevent energizing it until it has been purged with the liquid or gas to exclude air, and also arranged to automatically de-energize the equipment when the supply of liquid or gas or vapor fails or the pressure is reduced to atmospheric.

Totally enclosed motors of Types (2) or (3) shall have no external surface with an operating temperature in degrees Celsius in excess of 80 percent of the ignition temperature of the gas or vapor involved. Appropriate devices shall be provided to detect and automatically de-energize the motor or provide an adequate alarm if there is any increase in temperature of the motor beyond designed limits. Auxiliary equipment shall be of a type approved for the location in which it is installed.

FPN: See ASTM *Test Procedure,* D 2155-69.

The intent of Section 501-8(a), item (4), is to permit nonexplosionproof motors submerged in liquefied natural gas (LNG), liquefied petroleum gas (LPG), and so on. It does not permit nonexplosionproof motors under water, such as in wet pits, unless the motors are provided with some other system of explosion protection, for example, purged and pressurized per NFPA 496-1998, *Standard for Purged and Pressurized Enclosures for Electrical Equipment.*

The ASTM test procedure is used to determine the

ignition temperature of some flammable and combustible liquids.

(b) Class I, Division 2. In Class I, Division 2 locations, motors, generators, and other rotating electric machinery in which are employed sliding contacts, centrifugal or other types of switching mechanism (including motor overcurrent, overloading, and overtemperature devices), or integral resistance devices, either while starting or while running, shall be approved for Class I, Division 1 locations, unless such sliding contacts, switching mechanisms, and resistance devices are provided with enclosures approved for Class I, Division 2 locations in accordance with Section 501-3(b). The exposed surface of space heaters used to prevent condensation of moisture during shutdown periods shall not exceed 80 percent of the ignition temperature in degrees Celsius of the gas or vapor involved when operated at rated voltage, and the maximum surface temperature [based on a 40°C (104°F) ambient] shall be permanently marked on a visible nameplate mounted on the motor. Otherwise, space heaters shall be approved for Class I, Division 2 locations.

In Class I, Division 2 locations, the installation of open or nonexplosionproof enclosed motors, such as squirrel-cage induction motors without brushes, switching mechanisms, or similar arc-producing devices that are not identified for use in a Class I, Division 2 location, shall be permitted.

Many motor heaters are de-energized automatically when the motor is running. However, the heater ratings are usually low when compared with the normal heat generated during motor operation. Unless otherwise indicated on the motor wiring diagram or in instructions provided with the motor, there is no need to de-energize the heater except to save energy. Note the requirement for marking the heater temperature on the motor or the use of heaters approved for the location.

FPN No. 1: It is important to consider the temperature of internal and external surfaces that may be exposed to the flammable atmosphere.

FPN No. 2: It is important to consider the risk of ignition due to currents arcing across discontinuities and overheating of parts in multisection enclosures of large motors and generators. Such motors and generators may need equipotential bonding jumpers across joints in the enclosure and from enclosure to ground. Where the presence of ignitible gases or vapors is suspected, clean-air purging may be needed immediately prior to and during start-up periods.

It is intended that the phrase "other rotating electric machinery" include electric brakes. Listed and labeled electric brakes are available for Class I, Division 1, Group C and D locations.

Figures 501.19 and 501.20 show closed and opened views of a totally enclosed fan-cooled motor listed for use in explosive atmospheres. The main frame and end-bells are designed with sufficient strength to withstand an internal explosion. Flames or hot gases are cooled while escaping because of the wide metal-to-metal joints between the frame and end-bells and the long, close-tolerance clearance provided for the free turn of the shaft. Air circulation outside the motor is maintained by a nonsparking (aluminum, bronze, or non-static-generating-type plastic) fan on the end opposite the shaft end of the motor.

Figure 501.19 Terminal housing of a motor listed for use in specific hazardous locations. Note integral sealing of the motor. (General Electric Co.)

Figure 501.20 View showing internal fan of motor in Figure 501.19.

A sheet metal housing surrounds this fan to reduce the likelihood of an individual or object coming into contact with the moving blades and to direct the flow of air. Internal fans on the shaft circulate air around the windings.

Motors that have arcing or sparking devices, such as commutators, internal switches, or other control devices, must be explosionproof. General-purpose squirrel-cage induction motors without such devices may be used in Division 2 locations.

Since some open-type motors are permitted in Class I, Division 2 locations, care should be exercised in selecting the motor types where flammable gases or vapors with very low ignition temperatures may be present. Modern motors with high-temperature insulation systems, such as Class H [180°C (356°F)], may operate close to or above the ignition temperature of the flammable mixture.

501-9. Lighting Fixtures. Lighting fixtures shall comply with (a) or (b).

(a) Class I, Division 1. In Class I, Division 1 locations, lighting fixtures shall comply with the following.

(1) Approved Fixtures. Each fixture shall be approved as a complete assembly for the Class I, Division 1 location and shall be clearly marked to indicate the maximum wattage of lamps for which it is approved. Fixtures intended for portable use shall be specifically approved as a complete assembly for that use.

(2) Physical Damage. Each fixture shall be protected against physical damage by a suitable guard or by location.

(3) Pendant Fixtures. Pendant fixtures shall be suspended by and supplied through threaded rigid metal conduit stems or threaded steel intermediate conduit stems, and threaded joints shall be provided with set-screws or other effective means to prevent loosening. For stems longer than 12 in. (305 mm), permanent and effective bracing against lateral displacement shall be provided at a level not more than 12 in. (305 mm) above the lower end of the stem, or flexibility in the form of a fitting or flexible connector approved for the Class I, Division 1 location shall be provided not more than 12 in. (305 mm) from the point of attachment to the supporting box or fitting.

(4) Supports. Boxes, box assemblies, or fittings used for the support of lighting fixtures shall be approved for Class I locations.

(b) Class I, Division 2. In Class I, Division 2 locations, lighting fixtures shall comply with the following.

(1) Portable Lighting Equipment. Portable lighting equipment shall comply with (a)(1).

Exception: Where portable lighting equipment are mounted on movable stands and are connected by flexible cords, as covered in Section 501-11, they shall be permitted, where mounted in any position, if they conform to Section 501-9(b)(2).

(2) Fixed Lighting. Lighting fixtures for fixed lighting shall be protected from physical damage by suitable guards or by location. Where there is danger that falling sparks or hot metal from lamps or fixtures might ignite localized concentrations of flammable vapors or gases, suitable enclosures or other effective protective means shall be provided. Where lamps are of a size or type that may, under normal operating conditions, reach surface temperatures exceeding 80 percent of the ignition temperature in degrees Celsius of the gas or vapor involved, fixtures shall comply with (a)(1) or shall be of a type that has been tested in order to determine the marked operating temperature or temperature range.

(3) Pendant Fixtures. Pendant fixtures shall be suspended by threaded rigid metal conduit stems, threaded steel intermediate metal conduit stems, or by other approved means. For rigid stems longer than 12 in. (305 mm), permanent and effective bracing against lateral displacement shall be provided at a level not more than 12 in. (305 mm) above the lower end of the stem, or flexibility in the form of an approved fitting or flexible connector shall be provided not more than 12 in. (305 mm) from the point of attachment to the supporting box or fitting.

(4) Switches. Switches that are a part of an assembled fixture or of an individual lampholder shall comply with Section 501-6(b)(1).

(5) Starting Equipment. Starting and control equipment for electric-discharge lamps shall comply with Section 501-7(b).

Exception: A thermal protector potted into a thermally protected fluorescent lamp ballast if the lighting fixture is approved for locations of this class and division.

Figures 501.21 and 501.22 show typical lighting fixtures for Class I, Group C and D locations and a variety of parts of a complete lighting fixture assembly. The outlet boxes have an internally threaded opening designed to receive the cover. A pendant fixture is attached to the cover by threaded rigid metal conduit or threaded intermediate metal conduit. To prevent loosening from vibration or lamp changing, threaded joints are required to be provided with setscrews. The setscrews should not interrupt the explosionproof joint. Rigid metal conduit or intermediate metal conduit stems longer than 12 in. require effective bracing or a flexible fitting approved for the purpose, placed not more than 12 in. from the point of attachment to the supporting box, cover, or fitting.

A globe holder is threaded onto the body of the fixture housing and supports a heavy glass globe, guard, and reflector. It is available in sizes suitable for lamps from 40 watts through 500 watts. In designing any hazardous (classified) location lighting sys-

Figure 501.21 A typical lighting fixture for use in Class I, Group C and D locations. (Crouse-Hinds)

tem, operating temperatures must be considered. Therefore, if the area is Class I, Division 1, fixtures approved for this location that are properly marked must be used. Generally, enclosed and gasketed fixtures (previously called vaportight fixtures) without guards, if breakage is unlikely, or fixtures approved for Class I, Division 2 locations, are required in Division 2 locations. Fixtures listed by Underwriters Laboratories Inc. for use in any of the groups under Class I, either Division 1 or 2 locations or both, are designed to operate without causing ignition of surrounding flammable gas or vapor atmospheres and are marked with the operating temperature or temperature range code. See Table 500-5(d).

Figure 501.23 shows an explosionproof hand lamp. It is required that lamp compartments be sealed from the terminal compartment. Provisions are required

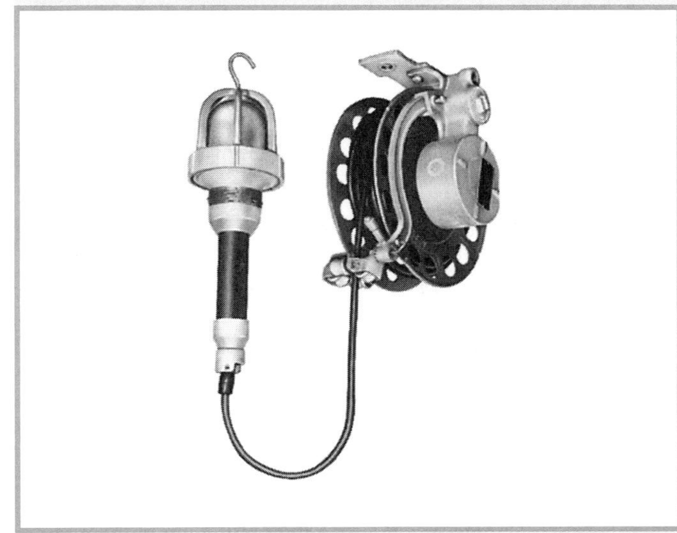

Figure 501.23 An explosionproof hand lamp for use in Class I locations. (Appleton Electric Co.)

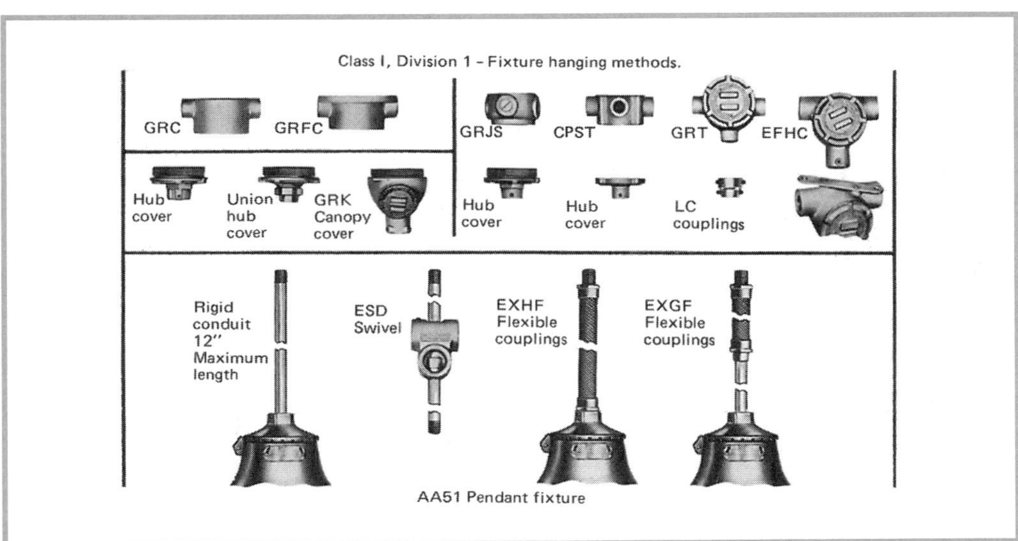

Figure 501.22 Various components used in an explosionproof lighting fixture installation. (Appleton Electric Co.)

to be made for the connection of 3-conductor (one must be a grounding conductor) flexible, extra-hard-usage cord. See Section 501-11.

501-10. Utilization Equipment.

(a) Class I, Division 1. In Class I, Division 1 locations, all utilization equipment shall be approved for Class I, Division 1 locations.

(b) Class I, Division 2. In Class I, Division 2 locations, all utilization equipment shall comply with the following.

(1) Heaters. Electrically heated utilization equipment shall conform with either item (a) or (b).

(a) The heater shall not exceed 80 percent of the ignition temperature in degrees Celsius of the gas or vapor involved on any surface that is exposed to the gas or vapor when continuously energized at the maximum rated ambient temperature. If a temperature controller is not provided, these conditions shall apply when the heater is operated at 120 percent of rated voltage.

Exception No. 1: For motor-mounted anticondensation space heaters, see Section 501-8(b).

Exception No. 2: A current-limiting device is applied to the circuit serving the heater that will limit the current in the heater to a value less than that required to raise the heater surface temperature to 80 percent of the ignition temperature.

(b) The heater shall be approved for Class I, Division 1 locations.

Exception: Electrical resistance heat tracing approved for Class I, Division 2 locations.

(2) Motors. Motors of motor-driven utilization equipment shall comply with Section 501-8(b).

(3) Switches, Circuit Breakers, and Fuses. Switches, circuit breakers, and fuses shall comply with Section 501-6(b).

The requirements for utilization equipment in Class I locations are virtually identical for Division 1 and 2 locations, except for heaters. Electric pipe heat-tracing systems listed for Class I, Division 2 locations and complying with Section 501-10(b)(1)(b), Exception, are available.

501-11. Flexible Cords, Class I, Divisions 1 and 2. A flexible cord shall be permitted for connection between portable lighting equipment or other portable utilization equipment and the fixed portion of their supply circuit. Flexible cord shall also be permitted for that portion of the circuit where the fixed wiring methods of Section 501-4(a) cannot provide the necessary degree of movement for fixed and mobile electrical utilization equipment, in an industrial establishment where conditions of maintenance and engineering supervision ensure that only qualified persons will install and service the installation, and the flexible cord is protected by location or by a suitable guard from damage. The length of the flexible cord shall be continuous. Where flexible cords are used, the cords shall be as follows:

(1) Of a type approved for extra-hard usage;
(2) Contain, in addition to the conductors of the circuit, a grounding conductor complying with Section 400-23;
(3) Connected to terminals or to supply conductors in an approved manner;
(4) Be supported by clamps or by other suitable means in such a manner that there will be no tension on the terminal connections; and
(5) Be provided with suitable seals where the flexible cord enters boxes, fittings, or enclosures of the explosionproof type.

Exception: As provided in Sections 501-3(b)(6) and 501-4(b).

Electric submersible pumps with means for removal without entering the wet-pit shall be considered portable utilization equipment. The extension of the flexible cord within a suitable raceway between the wet-pit and the power source shall be permitted.

Electric mixers intended for travel into and out of open-type mixing tanks or vats shall be considered portable utilization equipment.

FPN: See Section 501-13 for flexible cords exposed to liquids having a deleterious effect on the conductor insulation.

The second paragraph of Section 501-11 recognizes a wet-pit type of installation that is finding increasing acceptance for use in waste-water systems. This section was revised for the 1993 *Code* to permit a flexible cord to be run in raceway to a junction box outside the wet-pit.

Mixers that travel into and out of open-type mixing tanks are considered portable utilization equipment and may be wired with flexible cord. See the commentary following Section 501-8(a).

501-12. Receptacles and Attachment Plugs, Class I, Divisions 1 and 2. Receptacles and attachment plugs shall be of the type providing for connection to the grounding conductor of a flexible cord and shall be approved for the location.

Figure 501.24 shows an explosionproof receptacle and attachment plug with an interlocking switch. The design of this device is such that when the switch is in the "on" position, the plug cannot be removed. Also, the switch cannot be placed in the "on" position when the plug has been removed; that is, the plug cannot be inserted or removed unless the switch is in the "off" position. The receptacle is factory sealed, with a provision for threaded-conduit entry to the switch compartment; the plug is to be used with Type S or equivalent extra-hard-service flexible cord having a grounding conductor.

Figure 501.24 *A receptacle and attachment plug of the explosionproof type with an interlocking switch. The switch must be in the "off" position to remove the attachment plug. (Appleton Electric Co.)*

Figure 501.25 shows a 30-ampere, 4-pole receptacle and attachment plug assembly suitable for use without a switch. The design is such that the mating parts of the receptacle and plug are enclosed in a chamber that seals the arc and, by delayed-action construction, prevents complete removal of the plug until the arc or hot metal has cooled. The receptacle is factory sealed, and the attachment plug is designed for use with a 4-conductor cord (3-conductor, 3-phase circuit with one grounding conductor) or a 3-conductor cord

Figure 501.25 *A 4-pole (delayed action) explosionproof receptacle and attachment plug suitable for use without a switch. (Appleton Electric Co.)*

(two circuit conductors and one grounding conductor).

Exception: As provided in Section 501-3(b)(6).

501-13. Conductor Insulation, Class I, Divisions 1 and 2. Where condensed vapors or liquids may collect on, or come in contact with, the insulation on conductors, such insulation shall be of a type approved for use under such conditions; or the insulation shall be protected by a sheath of lead or by other approved means.

Nylon-jacketed conductors, such as Types THWN and TW, suitable for use where exposed to gasoline, have gained widespread acceptance because of their ease in handling and application and for economic reasons.

Underwriters Laboratories Inc.'s 1998 *Electrical Construction Materials Directory* states the following under "Wires, Thermoplastic-Insulated."

> Gasoline resistant wire has been tested at 23°C when immersed in gasoline. It is considered inherently resistant to gasoline vapors within the limits of the temperature rating of the wire type.
>
> Gasoline resistant THWN indicates a THWN conductor suitable for use in wet locations, and for exposure to mineral oil, and to liquid gasoline and gasoline vapors at ordinary ambient temperature. It is identified by tag marking and by printing on the insulation or jacket with the designation "Type THWN Gasoline and Oil Resistant I" if suitable for exposure to mineral oil at 60°C or "Type THWN Gasoline and Oil Resistant II" if the compound is suitable for exposure to mineral oil at 75°C.
>
> Gasoline resistant TW indicates a TW conductor with a jacket of extruded nylon or equivalent material suitable for use in wet locations, and for exposure to mineral oil, and to liquid gasoline and gasoline vapors at ordinary ambient temperature. It is identified by a tag marking and by the designation on the insulation or jacket "Type TW Gasoline and Oil Resistant I."
>
> The conductor itself must bear the marking legend designating its use as suitable for gasoline exposure. Such designation on the tag alone is not sufficient.

501-14. Signaling, Alarm, Remote-Control, and Communications Systems.

(a) Class I, Division 1. In Class I, Division 1 locations, all apparatus and equipment of signaling, alarm, remote-control, and communications systems, regardless of voltage, shall be approved for Class I, Division 1 locations, and all wiring shall comply with Sections 501-4(a) and 501-5(a) and (c).

(b) Class I, Division 2. In Class I, Division 2 locations, signaling, alarm, remote-control, and communications systems shall comply with the following.

(1) Contacts. Switches, circuit breakers, and make-and-break contacts of pushbuttons, relays, alarm bells, and horns shall have enclosures approved for Class I, Division 1 locations in accordance with Section 501-3(a).

Exception: General-purpose enclosures shall be permitted if current-interrupting contacts are one of the following:

(a) Immersed in oil, or
(b) Enclosed within a chamber hermetically sealed against the entrance of gases or vapors, or
(c) In nonincendive circuits, or

See the commentary following item (c) of Section 501-3(b)(1), Exception.

(d) Part of a listed nonincendive component

(2) Resistors and Similar Equipment. Resistors, resistance devices, thermionic tubes, rectifiers, and similar equipment shall comply with Section 501-3(b)(2).

(3) Protectors. Enclosures shall be provided for lightning protective devices and for fuses. Such enclosures shall be permitted to be of the general-purpose type.

(4) Wiring and Sealing. All wiring shall comply with Sections 501-4(b) and 501-5(b) and (c).

Audible signaling devices, such as bells, sirens, and horns, other than electronic types, usually involve make-and-break contacts capable of producing a spark of sufficient energy to cause ignition of a hazardous atmospheric mixture. Where used in Class I locations, therefore, this type of equipment is required to be contained in explosionproof or purged and pressurized enclosures, wiring methods are required to comply with Section 501-4, and sealing fittings are required to be provided in accordance with Section 501-5. Figure 501.26 shows a signal siren for use in Class I locations.

Explosionproof devices or explosionproof enclosures may prove more practical than oil-immersed contacts, because maintaining the condition and level of the oil can be a problem. Hermetically sealed enclosures, such as float-operated mercury-tube switches, are available for some applications. Electronic signal devices without make-and-break contacts will usually not require explosionproof enclosures in Division 2 locations.

Figure 501.26 *A signal siren mounted on an explosionproof enclosure, for use in hazardous areas. (Crouse-Hinds)*

501-15. Live Parts, Class I, Divisions 1 and 2. There shall be no exposed live parts.

Contact with the circuit could produce sparks that could cause an explosion in a hazardous (classified) location.

501-16. Grounding, Class I, Divisions 1 and 2. Wiring and equipment in Class I, Division 1 and 2 locations shall be grounded as specified in Article 250 and with the following additional requirements.

(a) Bonding. The locknut-bushing and double-locknut types of contacts shall not be depended on for bonding purposes, but bonding jumpers with proper fittings or other approved means of bonding shall be used. Such means of bonding shall apply to all intervening raceways, fittings, boxes, enclosures, etc., between Class I locations and the point of grounding for service equipment or point of grounding of a separately derived system.

Exception: The specific bonding means shall only be required to the nearest point where the grounded circuit conductor and the grounding electrode are connected together on the line side of the building or structure disconnecting means as specified in Sections 250-32(a), (b), and (c), provided the branch-circuit overcurrent protection is located on the load side of the disconnecting means.

The exception to Section 501-16(a) was new in the 1993 *Code* and revised in the 1999 *Code*. Where the building in which the hazardous location exists is supplied through a service in another building, it is not necessary to apply the bonding requirement from the hazardous location back to the service. It is only necessary to apply the bonding requirement from the hazardous location back to the grounding electrode on the line side of the building or structure disconnecting means. This connection must be ahead of the branch circuits that are on the load side of the disconnecting means for the building or structure.

FPN: See Section 250-100 for additional bonding requirements in hazardous (classified) locations.

(b) Types of Equipment Grounding Conductors. Where flexible metal conduit or liquidtight flexible metal conduit is used as permitted in Section 501-4(b) and is to be relied on to complete a sole equipment grounding path, it shall be installed with internal or external bonding jumpers in parallel with each conduit and complying with Section 250-102.

Special consideration is necessary in the grounding and bonding of exposed noncurrent-carrying metal parts of equipment, such as the frames or metal exteriors of motors, fixed or portable lamps, lighting fixtures, enclosures, and raceways, to ensure permanent and effective mechanical and electrical connections, in order to prevent the possibility of arcs or sparks caused by ineffective or poor grounding methods. One example of an external bonding jumper is shown in Figure 501.27. To be effective, proper grounding and bonding applies to all interconnected raceways, fittings, enclosures, and so on, between hazardous (classified) locations and the point of grounding for service equipment or the point of grounding of building disconnecting means that supplies the branch-circuit overcurrent protection. Where conduit is used in hazardous (classified) locations, it is preferable that threaded connections also be employed in the nonhazardous location.

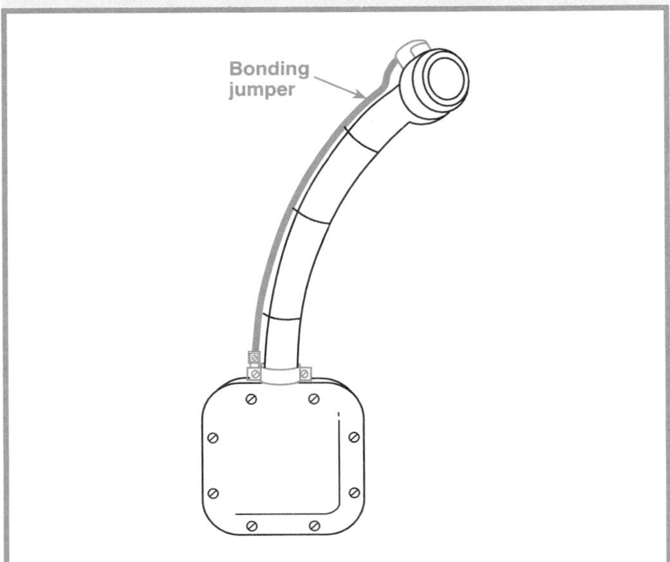

Figure 501.27 *A fitting for the connection of an external bonding jumper used with liquidtight flexible metal conduit.*

Exception: In Class I, Division 2 locations, the bonding jumper shall be permitted to be deleted where all the following conditions are met.

(a) Listed liquidtight flexible metal conduit 6 ft (1.83 m) or less in length, with fittings listed for grounding, is used.
(b) Overcurrent protection in the circuit is limited to 10 amperes or less.
(c) The load is not a power utilization load.

501-17. Surge Protection.

(a) Class I, Division 1. Surge arresters, including their installation and connection, shall comply with Article 280. The surge arresters and capacitors shall be installed in enclosures approved for Class I, Division 1 locations. Surge-protective capacitors shall be of a type designed for specific duty.

(b) Class I, Division 2. Surge arresters shall be nonarcing, such as metal-oxide varistor (MOV), sealed type, and surge-protective capacitors shall be of a type designed for specific duty. Installation and connection shall comply with Article 280.

Enclosures shall be permitted to be of the general-purpose type. Surge protection of types other than described above shall be installed in enclosures approved for Class I, Division 1 locations.

Some surge arresters, such as the older style lightning arresters, are spark-producing devices. Others, such as solid-state types, are not. Surge arresters should be connected to the service conductors outside the building and be bonded to the service-entrance raceway system. For services less than 1000 volts, the arrester grounding conductor is connected as provided in Article 280, Part C.

Where the service voltage is less than 600 volts, the supply system is a secondary system; therefore, the grounded service conductor should always be bonded to the equipment grounding conductor, as required by Article 250.

In Class I, Division 1 locations, all surge arresters are required to be installed in explosionproof or purged enclosures. In Class I, Division 2 locations, only the spark-producing types of surge arresters need such protection. They can also be installed in oil-filled enclosures or have the arcing or sparking contacts enclosed in hermetically sealed chambers. The nonsparking type of surge arrester needs no special enclosure in a Class I, Division 2 location.

501-18. Multiwire Branch Circuits. In a Class I, Division 1 location, a multiwire branch circuit shall not be permitted.

Exception: Where the disconnect device(s) for the circuit opens all ungrounded conductors of the multiwire circuit simultaneously.

Section 501-18 was added in the 1993 *Code* to indicate that multiwire branch circuits (see the definition of *branch circuit, multiwire* in Article 100) are not permitted in Class I hazardous locations unless the disconnecting device opens all the ungrounded conductors of the multiwire branch circuit simultaneously.

Article 502 — Class II Locations

Contents

502-1. General. The general rules of this *Code* shall apply to the electric wiring and equipment in locations classified as Class II locations in Section 500-8.

A Class II, Division 1 (combustible dust) location is an area where two different types of situations may exist. One is where a dust cloud of combustible dust is likely to be present continuously or intermittently under normal operating conditions, repair, maintentance, or leakage. The other situation is where a dust layer is likely to accumulate to a depth of greater than $^{1}/_{8}$ in. over a period of time, usually a 24-hour period. The size of the dust particle is a factor in determining whether the area should be classified as Class II or Class III. Combustible dust, as defined in NFPA 499-1997, *Recommended Practice for the Classification of Combustible Dusts and of Hazardous (Classified) Locations for Electrical Installations in Chemical Process Areas,* is any finely divided solid material 420 microns (0.0165 in.) or less in diameter (i.e., material passing through a U.S. No. 40 standard sieve) that presents a fire or explosion hazard when dispersed.

Exception: As modified by this article.

Dust-ignitionproof, as used in this article, shall mean enclosed in a manner that will exclude dusts and, where installed and protected in accordance with this *Code*, will not permit arcs, sparks, or heat otherwise generated or liberated inside of the enclosure to cause ignition of exterior accumulations or atmospheric suspensions of a specified dust on or in the vicinity of the enclosure.

FPN: For further information on dust-ignitionproof enclosures, see Type 9 enclosure in *Enclosures for Electrical Equipment,* ANSI/NEMA 250-1991, and *Explosionproof and Dust-Ignitionproof Electrical Equipment for Hazardous (Classified) Locations,* ANSI/UL 1203-1994.

Equipment installed in Class II locations shall be able to function at full rating without developing surface temperatures high enough to cause excessive dehydration or gradual carbonization of any organic dust deposits that may occur.

FPN: Dust that is carbonized or excessively dry is highly susceptible to spontaneous ignition.

Equipment and wiring of the type defined in Article 100 as explosionproof shall not be required and shall not be acceptable in Class II locations unless approved for such locations.

Where Class II, Group E dusts are present in hazardous quantities, there are only Division 1 locations.

Class II, Division 1 and 2 locations are defined in Section 500-8 as hazardous due to the presence of combustible dust. These locations are separated into three groups: Group E, Group F, and Group G [see Section 500-5(b)].

It should be noted that equipment suitable for one class and group is not necessarily suitable for any other class and group. Class I equipment is not necessarily "better" for, or even suitable for, a Class II location. The hazard contemplated in the equipment design is different. Class II equipment is designed to prevent the ignition of layers of dust, which may increase the operating temperature of the equipment. Class I equipment is not designed for dust layering unless it is also designed and approved for Class II locations. To protect against explosions in hazardous (classified) locations, all electrical equipment exposed to the hazardous atmosphere is required to be suitable for such locations. Grain dust, for example, will ignite at a temperature lower than that of most flammable vapors. Motors listed for use in Class I locations may not have dust shields on the bearings to prevent entrance of dust into the bearing race, thereby causing overheating of the bearing resulting in ignition of dust on the motor.

One or more of the following four hazards may be present in a Class II location:

1. An explosive mixture of air and dust in suspension
2. Accumulation of dust that acts as a thermal blanket and interferes with the safe dissipation of heat from electrical equipment
3. Accumulation of electrically conductive dust lodging between terminals that have a difference of potential, thereby causing tracking and glowing hot particles, short-circuits, or ground faults that may ignite dust accumulated in the vicinity
4. Deposits of dust that could be ignited by arcs or sparks

In the layout of electrical installations for hazardous (classified) locations, it is preferable to locate service equipment, switchboards, panelboards, and much of the electrical equipment in less hazardous areas, usually in a separate room. The use of pressurized rooms, as described in NFPA 496-1998, *Standard for Purged and Pressurized Enclosures for Electrical Equipment,* is a common method of protecting panelboards and switchboards in grain elevators and similar locations.

502-2. Transformers and Capacitors.

(a) Class II, Division 1. In Class II, Division 1 locations, transformers and capacitors shall comply with the following.

(1) Containing Liquid that Will Burn. Transformers and capacitors containing a liquid that will burn shall be installed only in approved vaults complying with Sections 450-41 through 450-48, and, in addition, the following shall apply.

(a) Doors or other openings communicating with the Division 1 location shall have self-closing fire doors on both sides of the wall, and the doors shall be carefully fitted and provided with suitable seals (such as weather stripping) to minimize the entrance of dust into the vault.
(b) Vent openings and ducts shall communicate only with the outside air.
(c) Suitable pressure-relief openings communicating with the outside air shall be provided.

(2) Not Containing Liquid that Will Burn. Transformers and capacitors that do not contain a liquid that will burn shall be installed in vaults complying with Sections 450-41 through 450-48 or be approved as a complete assembly, including terminal connections for Class II locations.

(3) Metal Dusts. No transformer or capacitor shall be installed in a location where dust from magnesium, aluminum, aluminum bronze powders, or other metals of similarly hazardous characteristics may be present.

(b) Class II, Division 2. In Class II, Division 2 locations, transformers and capacitors shall comply with the following.

(1) Containing Liquid that Will Burn. Transformers and capacitors containing a liquid that will burn shall be installed in vaults that comply with Sections 450-41 through 450-48.

(2) Containing Askarel. Transformers containing askarel and rated in excess of 25 kVA shall be as follows:

(a) Provided with pressure-relief vents
(b) Provided with a means for absorbing any gases generated by arcing inside the case, or the pressure-relief vents shall be connected to a chimney or flue that will carry such gases outside the building
(c) Have an airspace of not less than 6 in. (152 mm) between the transformer cases and any adjacent combustible material

(3) Dry-Type Transformers. Dry-type transformers shall be installed in vaults or shall have their windings and terminal connections enclosed in tight metal housings without ventilating or other openings and shall operate at not over 600 volts, nominal.

Where it is necessary to install a transformer, it may be possible to use a small, low-voltage, dusttight (without ventilating openings), dry-type transformer, but transformers that have a primary voltage rating of over 600 volts are required to be either less flammable liquid-insulated or installed in a vault. In almost all cases, transformers can be remotely located from dust atmospheres.

Capacitors used for power-factor correction of individual motors are of sealed construction but, if installed in Class II, Division 1 locations, are also required to be approved as a complete assembly, including dusttight terminal enclosures. The only special requirement for capacitors in Division 2 locations is that they are not permitted to contain oil or any other liquid that will burn; otherwise, they are to be installed in vaults.

Capacitor and transformer oils (such as askarel) that do not burn have, in the past, contained PCBs. Production of such oil has been stopped.

502-4. Wiring Methods. Wiring methods shall comply with (a) and (b).

(a) Class II, Division 1. In Class II, Division 1 locations, threaded rigid metal conduit, threaded steel intermediate metal conduit, or Type MI cable with termination fittings approved for the location shall be the wiring method employed. Type MI cable shall be installed and supported in a manner to avoid tensile stress at the termination fittings.

Exception: In industrial establishments with limited public access, where the conditions of maintenance and supervision ensure that only qualified persons will service the installation, Type MC cable, listed for use in Class II, Division 1 locations, with a gas/vaportight continuous corrugated aluminum sheath, an overall jacket of suitable polymeric material, separate grounding conductors in accordance with Section 250-122, and provided with termination fittings listed for the application shall be permitted.

Due to the potential for physical damage to MC cable, its use in Class II, Division 1 locations is limited to cable listed for Class II, Division 1 locations installed at facilities that have full-time, qualified maintenance personnel who would be doing regular maintenance, might notice if cables were damaged, would know the hazards, and could de-energize the circuit to repair the installation.

(1) Fittings and Boxes. Fittings and boxes shall be provided with threaded bosses for connection to conduit or cable terminations, shall have close-fitting covers, and shall have no openings (such as holes for attachment screws) through which dust might enter or through which sparks or burning material might escape. Fittings and boxes in which taps, joints, or terminal connections are made, or that are used in locations where dusts are of a combustible, electrically conductive nature, shall be approved for Class II locations.

Boxes or fittings such as conduit "L," "T," or "C" fittings installed in a Class II, Division 1 location, where splices, taps, or terminations are made within them, must be of the threaded type, with close-fitting covers. There can be no holes in the box or fitting, such as mounting holes, that would allow dust to enter or sparks or burning material to be ejected from the enclosure. Where used as stated above, the equipment must be approved for Class II locations, usually of the dust-ignitionproof or pressurized type.

Where boxes or fittings of the type described in the preceding paragraph are installed with no splices, taps, or terminations made within them, they are required to be dusttight, with threaded bosses for connection of conduit or cable connectors.

(2) Flexible Connections. Where necessary to employ flexible connections, dusttight flexible connectors, liquidtight flexible metal conduit with approved fittings, liquidtight flexible nonmetallic conduit with approved fittings, or flexible cord approved for extra-hard usage and provided with bushed fittings shall be used. Where flexible cords are used, they shall comply with Section 502-12. Where flexible connections are subject to oil or other corrosive conditions, the insulation of the conductors shall be of a type approved for the condition or shall be protected by means of a suitable sheath.

FPN: See Section 502-16(b) for grounding requirements where flexible conduit is used.

(b) Class II, Division 2. In Class II, Division 2 locations, rigid metal conduit, intermediate metal conduit, electrical metallic tubing, dusttight wireways, Type MC or MI cable with approved termination fittings, Type PLTC in cable trays, Type ITC in cable trays, or Type MC, MI, or TC cable installed in ladder, ventilated trough, or ventilated channel cable trays in a single layer, with a space not less than the larger cable diameter between the two adjacent cables, shall be the wiring method employed.

Exception No. 1: Nonincendive field wiring shall be permitted using any of the methods suitable for wiring in ordinary locations.

Exception No. 2: Type MC cable listed for use in Class II, Division 1 locations shall be permitted to be installed without the above required spacings.

(1) Wireways, Fittings, and Boxes. Wireways, fittings, and boxes in which taps, joints, or terminal connections are made shall be designed to minimize the entrance of dust and (1) shall be provided with telescoping or close-fitting covers or other effective means to prevent the escape of sparks or burning material and (2) shall have no openings (such as holes for attachment screws) through which, after installation, sparks or burning material might escape or through which adjacent combustible material might be ignited.

(2) Flexible Connections. Where flexible connections are necessary, (a)(2) shall apply.

Where it is necessary to use flexible connections, liquidtight flexible conduit or extra-hard-usage flexible cord is permitted. Another method would be to use a flexible fitting, as shown in Figure 501.7. Where liquidtight flexible conduit is used, a bonding jumper (internal or external) is required to be provided around such conduit. See Section 502-16(b). An additional conductor for grounding is required to be provided where flexible cord is used.

In Division 1 locations, boxes containing taps, joints, or terminal connections must be dust-ignitionproof and must be provided with threaded hubs, as shown in Figure 502.1.

To provide adequate bonding in Division 2 locations, a threaded hub, as shown in Figure 502.1, is a method of achieving this requirement. Figure 502.1 also shows a close-fitting cover, as required for Class II locations. Standard pressed-steel boxes are permitted where they do not contain taps, joints, or terminal connections, and where a bonding jumper is provided around the box.

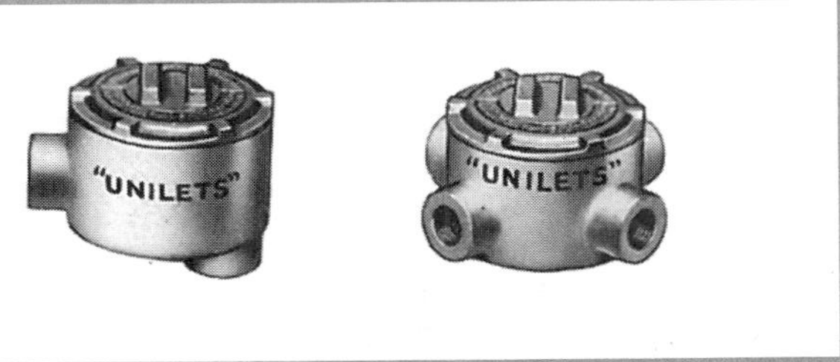

Figure 502.1 *Junction boxes with threaded hubs, suitable for use in Class II, Group E hazardous atmospheres. (Appleton Electric Co.)*

502-5. Sealing, Class II, Divisions 1 and 2. Where a raceway provides communication between an enclosure that is required to be dust-ignitionproof and one that is not, suitable means shall be provided to prevent the entrance of dust into the dust-ignitionproof enclosure through the raceway. One of the following means shall be permitted:

(1) A permanent and effective seal,
(2) A horizontal raceway not less than 10 ft (3.05 m) long, or
(3) A vertical raceway not less than 5 ft (1.52 m) long and extending downward from the dust-ignitionproof enclosure

Where a raceway provides communication between an enclosure that is required to be dust-ignitionproof and an enclosure in an unclassified location, seals shall not be required.

Sealing fittings shall be accessible.

Seals shall not be required to be explosionproof.

FPN: Electrical sealing putty is a method of sealing.

Section 502-5 provides three suitable ways to prevent dust from entering the dust-ignitionproof enclosure through the raceway, where a raceway connects an enclosure required to be dust-ignitionproof to one that is not. Where sealing fittings are used, any of those designed for use in Class I locations can be used. However, seals are not required to be explosionproof in Class II locations. No sealing method is needed in the special, but not unusual, situation in which no dust can enter the raceway in the hazardous (classified) location. See Figure 502.2.

Figure 502.2 *Four methods permitted for preventing the entrance of dust into the dust-ignitionproof enclosure through the raceway.*

502-6. Switches, Circuit Breakers, Motor Controllers, and Fuses.

(a) Class II, Division 1. In Class II, Division 1 locations, switches, circuit breakers, motor controllers, and fuses shall comply with the following:

(1) Type Required. Switches, circuit breakers, motor controllers, and fuses, including pushbuttons, relays, and similar devices that are intended to interrupt current during normal operation or that are installed where combustible dusts of an electrically conductive nature may be present, shall be provided with approved dust-ignitionproof enclosures.

(2) Isolating Switches. Disconnecting and isolating switches containing no fuses and not intended to interrupt current and not installed where dusts may be of an electrically conductive nature shall be provided with tight metal enclosures that shall be designed to minimize the entrance of dust and that shall (1) be equipped with telescoping or close-fitting covers or with other effective means to prevent the escape of sparks or burning material and (2) have no openings (such as holes for attachment screws) through which, after installation, sparks or burning material might escape or through which exterior accumulations of dust or adjacent combustible material might be ignited.

(3) **Metal Dusts.** In locations where dust from magnesium, aluminum, aluminum bronze powders, or other metals of similarly hazardous characteristics may be present, fuses, switches, motor controllers, and circuit breakers shall have enclosures specifically approved for such locations.

(b) Class II, Division 2. In Class II, Division 2 locations, enclosures for fuses, switches, circuit breakers, and motor controllers, including pushbuttons, relays, and similar devices, shall be dusttight.

The electrical equipment required in Class II locations is different from that for Class I locations. Dust-ignitionproof enclosures for Class II locations are not required to be explosionproof. However, explosionproof equipment is allowed to be used in Class II locations if the equipment is dual rated and identified as suitable for the Class II division and group. Explosionproof enclosures are not necessarily dust-ignitionproof.

Explosionproof enclosures, where used in a Class II environment, are not required to be sealed within 18 in. of the enclosure to complete the explosionproof assembly, as required in a Class I environment. They must, however, be provided with seals to prevent the entrance of dust into the dust-ignitionproof enclosure where a raceway provides a means for dust to enter the system, such as in a conduit run between a dust-ignitionproof enclosure and a general-purpose junction box.

Where equipment is located in a Class II, Division 1 location and is likely to produce arcs or sparks during normal operation, it must be installed in a dust-ignitionproof or pressurized enclosure. In addition, heat-generating equipment, such as control transformers, solenoids, impedance coils, resistors, and any associated overcurrent devices or switching mechanisms, must have dust-ignitionproof enclosures or pressurized enclosures installed in accordance with NFPA 496-1998, *Standard for Purged and Pressurized Enclosures for Electrical Equipment.* Caution should be used where metal dusts such as magnesium, aluminum, or other metals with similar hazardous characteristics may be present. Enclosures must be approved specifically for that environment (suitable for Class II, Division 1, Group E).

Figure 502.3 shows a pushbutton station with pilot light, suitable for both Class I and Class II hazardous (classified) locations. Figure 502.4 shows a panelboard suitable for use in Class II locations only. Many, but not all, of the switches or circuit breakers approved for Class I, Division 1 locations are also approved for Class II locations. It is important to

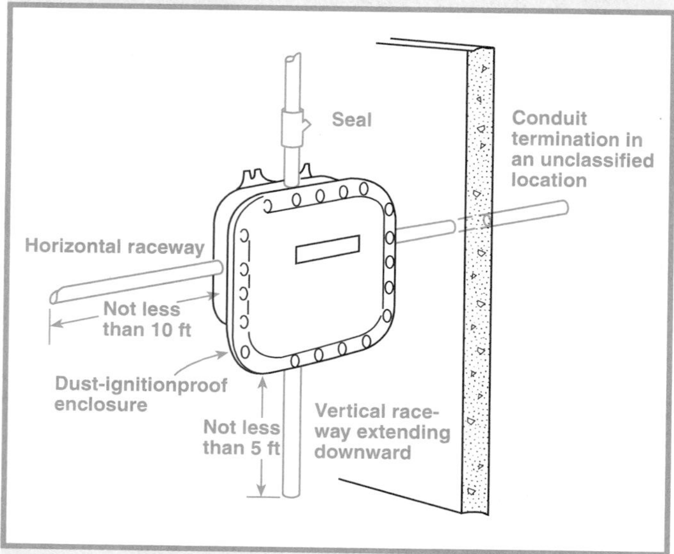

Figure 502.3 *An explosionproof and dusttight pushbutton control station suitable for use in Class I, Group C and D, and Class II, Group E, F, and G locations. (Appleton Electric Co.)*

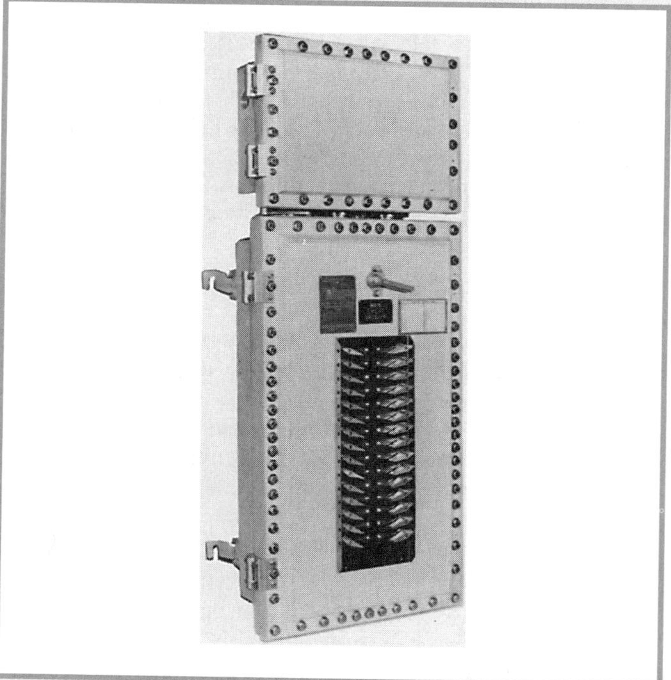

Figure 502.4 *A dust-ignitionproof panelboard for use in Class II, Group E, F, and G locations. (Crouse-Hinds)*

always look for the listing and identification of the hazardous (classified) locations for which the listing has been given.

502-7. Control Transformers and Resistors.

(a) Class II, Division 1. In Class II, Division 1 locations, control transformers, solenoids, impedance coils, resistors, and any overcurrent devices or switching mechanisms associated with them shall have dust-ignitionproof enclosures approved for Class II locations. No control transformer, impedance coil, or resistor shall be installed in a location where dust from magnesium, aluminum, aluminum bronze powders, or other metals of similarly hazardous characteristics may be present unless provided with an enclosure approved for the specific location.

(b) Class II, Division 2. In Class II, Division 2 locations, transformers and resistors shall comply with the following.

(1) Switching Mechanisms. Switching mechanisms (including overcurrent devices) associated with control transformers, solenoids, impedance coils, and resistors shall be provided with dusttight enclosures.

(2) Coils and Windings. Where not located in the same enclosure with switching mechanisms, control transformers, solenoids, and impedance coils shall be provided with tight metal housings without ventilating openings.

(3) Resistors. Resistors and resistance devices shall have dust-ignitionproof enclosures approved for Class II locations.

Exception: Where the maximum normal operating temperature of the resistor will not exceed 120°C (248°F), nonadjustable resistors or resistors that are part of an automatically timed starting sequence shall be permitted to have enclosures complying with (b)(2).

502-8. Motors and Generators.

(a) Class II, Division 1. In Class II, Division 1 locations, motors, generators, and other rotating electrical machinery shall be

(1) Approved for Class II, Division 1 locations, or
(2) Totally enclosed pipe-ventilated, meeting temperature limitations in Section 502-1.

It is intended that the phrase "other rotating electrical machinery" include electric brakes. Listed and labeled electric brakes are available for Class II, Group E, F, and G locations.

Although some explosionproof (Class I, Division 1) motors are dust-ignitionproof and approved for both Class I and II locations, this is by no means true of all motors. Always look for the marking to be sure the motor is designed and tested for the Class II location involved. If control wiring to the motor is necessary (see motor installation instructions), be sure the control circuit is properly installed and connected. Most motors for Class II locations require internal thermal protection to comply with the temperature limitations in Section 500-5(f), and integral horsepower Class II motors may require both power and control circuit wiring from the motor controller to the motor.

(b) Class II, Division 2. In Class II, Division 2 locations, motors, generators, and other rotating electrical equipment shall be totally enclosed nonventilated, totally enclosed pipe-ventilated, totally enclosed water–air cooled, totally enclosed fan-cooled or dust-ignitionproof for which maximum full-load external temperature shall be in accordance with Section 500-5(f) for normal operation when operating in free air (not dust blanketed) and shall have no external openings.

Exception: If the authority having jurisdiction believes accumulations of nonconductive, nonabrasive dust will be moderate and if machines can be easily reached for routine cleaning and maintenance, the following shall be permitted to be installed:

(a) Standard open-type machines without sliding contacts, centrifugal or other types of switching mechanism (including motor overcurrent, overloading, and overtemperature devices), or integral resistance devices
(b) Standard open-type machines with such contacts, switching mechanisms, or resistance devices enclosed within dusttight housings without ventilating or other openings
(c) Self-cleaning textile motors of the squirrel-cage type

Section 502-8(b) permits all types of totally enclosed motors in Class II, Division 2 locations if the external surface temperatures, without a dust blanket, do not exceed the temperatures indicated under maximum full-load (normal operation) conditions in Section 500-5(f). Totally enclosed fan-cooled (TEFC) motors are specifically mentioned. The motor should be examined carefully to be sure there are no external openings, even though the motor may be marked "TEFC."

Figure 502.5 shows a totally enclosed pipe-ventilated motor illustrating intake piping through which cool, clean air is delivered to the motor from a fan or blower. The exhaust opening is connected to a pipe discharging to the outside of the building, to prevent dust accumulation inside the motor.

Totally enclosed motors with no special provision for cooling may be used in Class II, Division 2 locations, but to deliver the same horsepower, they must be considerably larger than an open-type, fan-cooled, or pipe-ventilated motor.

Figure 502.5 A pipe-ventilated motor that meets the temperature limitations of Section 502-1. (General Electric Co.)

502-9. Ventilating Piping. Ventilating pipes for motors, generators, or other rotating electric machinery, or for enclosures for electric equipment, shall be of metal not less than 0.021 in. (533 μm) in thickness, or of equally substantial noncombustible material, and shall comply with the following:

(1) Lead directly to a source of clean air outside of buildings,
(2) Be screened at the outer ends to prevent the entrance of small animals or birds, and
(3) Be protected against physical damage and against rusting or other corrosive influences

Ventilating pipes shall also comply with (a) and (b).

(a) Class II, Division 1. In Class II, Division 1 locations, ventilating pipes, including their connections to motors or to the dust-ignitionproof enclosures for other equipment, shall be dusttight throughout their length. For metal pipes, seams and joints shall comply with one of the following:

(1) Be riveted and soldered,
(2) Be bolted and soldered,
(3) Be welded, or
(4) Be rendered dusttight by some other equally effective means

(b) Class II, Division 2. In Class II, Division 2 locations, ventilating pipes and their connections shall be sufficiently tight to prevent the entrance of appreciable quantities of dust into the ventilated equipment or enclosure and to prevent the escape of sparks, flame, or burning material that might ignite dust accumulations or combustible material in the vicinity. For metal pipes, lock seams and riveted or welded joints shall be permitted; and tight-fitting slip joints shall be permitted where some flexibility is necessary, as at connections to motors.

502-10. Utilization Equipment.

(a) Class II, Division 1. In Class II, Division 1 locations, all utilization equipment shall be approved for Class II locations. Where dust from magnesium, aluminum, aluminum bronze powders, or other metals of similarly hazardous characteristics may be present, such equipment shall be approved for the specific location.

(b) Class II, Division 2. In Class II, Division 2 locations, all utilization equipment shall comply with the following.

(1) Heaters. Electrically heated utilization equipment shall be approved for Class II locations.

Exception: Metal-enclosed radiant heating panel equipment shall be dusttight and marked in accordance with Section 500-5(d).

(2) Motors. Motors of motor-driven utilization equipment shall comply with Section 502-8(b).

(3) Switches, Circuit Breakers, and Fuses. Enclosures for switches, circuit breakers, and fuses shall be dusttight.

(4) Transformers, Solenoids, Impedance Coils, and Resistors. Transformers, solenoids, impedance coils, and resistors shall comply with Section 502-7(b).

502-11. Lighting Fixtures. Lighting fixtures shall comply with (a) and (b).

(a) Class II, Division 1. In Class II, Division 1 locations, lighting fixtures for fixed and portable lighting shall comply with the following.

(1) Approved Fixtures. Each fixture shall be approved for Class II locations and shall be clearly marked to indicate the maximum wattage of the lamp for which it is approved. In locations where dust from magnesium, aluminum, aluminum bronze powders, or other metals of similarly hazardous characteristics may be present, fixtures for fixed or portable lighting and all auxiliary equipment shall be approved for the specific location.

(2) Physical Damage. Each fixture shall be protected against physical damage by a suitable guard or by location.

(3) Pendant Fixtures. Pendant fixtures shall be suspended by threaded rigid metal conduit stems, threaded steel intermediate metal conduit stems, by chains with approved fittings, or by other approved means. For rigid stems longer than 12 in. (305 mm), permanent and effective bracing against lateral displacement shall be provided at a level not more than 12 in. (305 mm) above the lower end of the stem, or flexibility in the form of a fitting or a flexible connector approved for the location shall be provided not more than 12 in. (305 mm) from the point of attachment to the supporting box or fitting. Threaded joints shall be provided with set-screws or other effective means to prevent loosening. Where wiring between an outlet box or fitting and a pendant fixture is not enclosed in conduit, flexible cord approved for hard usage shall be used, and suitable seals shall be provided where the cord enters the fixture and the outlet box or fitting. Flexible cord shall not serve as the supporting means for a fixture.

(4) Supports. Boxes, box assemblies, or fittings used for the support of lighting fixtures shall be approved for Class II locations.

(b) Class II, Division 2. In Class II, Division 2 locations, lighting fixtures shall comply with the following.

(1) Portable Lighting Equipment. Portable lighting equipment shall be approved for Class II locations. They shall be clearly marked to indicate the maximum wattage of lamps for which they are approved.

(2) Fixed Lighting. Lighting fixtures for fixed lighting, where not of a type approved for Class II locations, shall provide enclosures for lamps and lampholders that shall be designed to minimize the deposit of dust on lamps and to prevent the escape of sparks, burning material, or hot metal. Each fixture shall be clearly marked to indicate the maximum wattage of the lamp that shall be permitted without exceeding an exposed surface temperature in accordance with Section 500-5(f) under normal conditions of use.

(3) Physical Damage. Lighting fixtures for fixed lighting shall be protected from physical damage by suitable guards or by location.

(4) Pendant Fixtures. Pendant fixtures shall be suspended by threaded rigid metal conduit stems, threaded steel intermediate metal conduit stems, by chains with approved fittings, or by other approved means. For rigid stems longer than 12 in. (305 mm), permanent and effective bracing against lateral displacement shall be provided at a level not more than 12 in. (305 mm) above the lower end of the stem, or flexibility in the form of an approved fitting or a flexible connector shall be provided not more than 12 in. (305 mm) from the point of attachment to the supporting box or fitting. Where wiring between an outlet box or fitting and a pendant fixture is not enclosed in conduit, flexible cord approved for hard usage shall be used. Flexible cord shall not serve as the supporting means for a fixture.

(5) Electric-Discharge Lamps. Starting and control equipment for electric-discharge lamps shall comply with the requirements of Section 502-7(b).

Figure 502.6 shows a listed fixture suitable for use in Class II, Group E, F, and G locations. Lighting fixtures, fixed or portable, and auxiliary equipment (such as ballasts) are required to be approved for use in Group E atmospheres if metal dusts are present.

Other than the requirement that the fixture be marked to indicate maximum lamp wattage, the only requirements for fixtures in Division 2 locations are that lamps be enclosed in suitable globes to minimize dust deposits on the lamps and prevent the escape of sparks or burning material. Guards are required to be provided, unless, of course, globe breakage is unlikely.

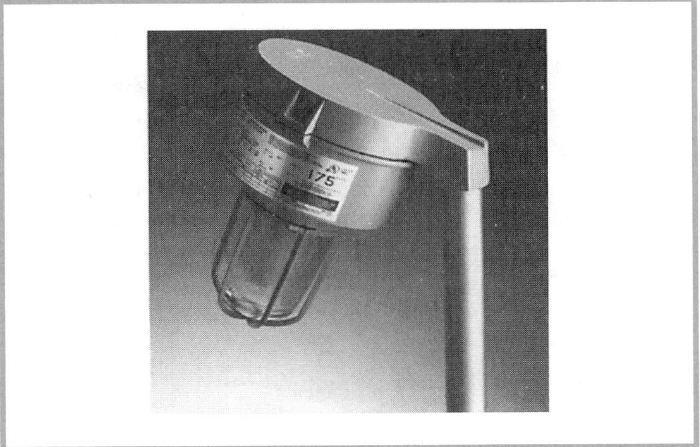

Figure 502.6 *A typical lighting fixture for use in Class II, Division 1 locations. Where breakage is unlikely, a metal guard is not required. (Crouse-Hinds)*

Flexible cord of the hard-usage type is permitted with approved sealed connections for the wiring of chain-suspended or hook-and-eye-suspended fixtures. Flexible cords are not intended to be used as a cord pendant or drop cord.

The portable hand lamp shown in Figure 501.23 is approved as a complete assembly for use in Class I locations and also in any Class II, Group F or G location.

502-12. Flexible Cords — Class II, Divisions 1 and 2. Flexible cords used in Class II locations shall comply with the following:

(1) Be of a type approved for extra-hard usage,
(2) Contain, in addition to the conductors of the circuit, a grounding conductor complying with Section 400-23,
(3) Be connected to terminals or to supply conductors in an approved manner,
(4) Be supported by clamps or by other suitable means in such a manner that there will be no tension on the terminal connections, and
(5) Be provided with suitable seals to prevent the entrance of dust where the flexible cord enters boxes or fittings that are required to be dust-ignitionproof

502-13. Receptacles and Attachment Plugs.

(a) Class II, Division 1. In Class II, Division 1 locations, receptacles and attachment plugs shall be of the type providing for connection to the grounding conductor of the flexible cord and shall be approved for Class II locations.

(b) Class II, Division 2. In Class II, Division 2 locations, receptacles and attachment plugs shall be of the type that provide for connection to the grounding conductor of the flexible cord and shall be designed so that connection to the supply circuit cannot be made or broken while live parts are exposed.

502-14. Signaling, Alarm, Remote-Control, and Communications Systems; and Meters, Instruments, and Relays.

FPN: See Article 800 for rules governing the installation of communications circuits.

(a) Class II, Division 1. In Class II, Division 1 locations, signaling, alarm, remote-control, and communications systems; and meters, instruments, and relays shall comply with the following:

(1) Wiring Methods. The wiring method shall comply with Section 502-4(a).

(2) Contacts. Switches, circuit breakers, relays, contactors, fuses and current-breaking contacts for bells, horns, howlers, sirens, and other devices in which sparks or arcs may be produced shall be provided with enclosures approved for a Class II location.

Exception: Where current-breaking contacts are immersed in oil or where the interruption of current occurs within a chamber sealed against the entrance of dust, enclosures shall be permitted to be of the general-purpose type.

(3) Resistors and Similar Equipment. Resistors, transformers, choke coils, rectifiers, thermionic tubes, and other heat-generating equipment shall be provided with enclosures approved for Class II locations.

Exception: Where resistors or similar equipment are immersed in oil or enclosed in a chamber sealed against the entrance of dust, enclosures shall be permitted to be of the general-purpose type.

(4) Rotating Machinery. Motors, generators, and other rotating electric machinery shall comply with Section 502-8(a).

(5) Combustible, Electrically Conductive Dusts. Where dusts are of a combustible, electrically conductive nature, all wiring and equipment shall be approved for Class II locations.

(6) Metal Dusts. Where dust from magnesium, aluminum, aluminum bronze powders, or other metals of similarly hazardous characteristics may be present, all apparatus and equipment shall be approved for the specific conditions.

(b) Class II, Division 2. In Class II, Division 2 locations, signaling, alarm, remote-control, and communications systems; and meters, instruments, and relays shall comply with the following.

(1) Contacts. Enclosures shall comply with (a)(2), or contacts shall have tight metal enclosures designed to minimize the entrance of dust and shall have telescoping or tight-fitting covers and no openings through which, after installation, sparks or burning material might escape.

Exception: In nonincendive circuits, enclosures shall be permitted to be of the general-purpose type.

(2) Transformers and Similar Equipment. The windings and terminal connections of transformers, choke coils, and similar equipment shall be provided with tight metal enclosures without ventilating openings.

(3) Resistors and Similar Equipment. Resistors, resistance devices, thermionic tubes, rectifiers, and similar equipment shall comply with (a)(3).

Exception: Enclosures for thermionic tubes, nonadjustable resistors, or rectifiers for which maximum operating temperature will not exceed 120°C (248°F) shall be permitted to be of the general-purpose type.

(4) Rotating Machinery. Motors, generators, and other rotating electric machinery shall comply with Section 502-8(b).

(5) Wiring Methods. The wiring method shall comply with Section 502-4(b).

502-15. Live Parts, Class II, Divisions 1 and 2. Live parts shall not be exposed.

502-16. Grounding, Class II, Divisions 1 and 2. Wiring and equipment in Class II, Divisions 1 and 2 locations shall be grounded as specified in Article 250 and with the following additional requirements.

(a) Bonding. The locknut-bushing and double-locknut types of contact shall not be depended on for bonding purposes, but bonding jumpers with proper fittings or other approved means of bonding shall be used. Such means of bonding shall apply to all intervening raceways, fittings, boxes, enclosures, etc., between Class II locations and the point of grounding for service equipment or point of grounding of a separately derived system.

Exception: The specific bonding means shall only be required to the nearest point where the grounded circuit conductor and the grounding electrode conductor are connected together on the line side of the building or structure disconnecting means as specified in Sections 250-32(a), (b), and (c), if the branch-circuit overcurrent protection is located on the load side of the disconnecting means.

FPN: See Section 250-100 for additional bonding requirements in hazardous (classified) locations.

(b) Types of Equipment Grounding Conductors. Where flexible conduit is used as permitted in Section 502-4, it shall be installed with internal or external bonding jumpers in parallel with each conduit and complying with Section 250-102.

Exception: In Class II, Division 2 locations, the bonding jumper shall be permitted to be deleted where all the following conditions are met.

(a) Listed liquidtight flexible metal conduit 6 ft (1.83 m) or less in length, with fittings listed for grounding, is used.
(b) Overcurrent protection in the circuit is limited to 10 amperes or less.
(c) The load is not a power utilization load.

Single locknuts or double locknuts and bushings are not to be depended on for bonding purposes. Bonding jumpers or other approved means with proper fittings are required for the interconnection of all raceways, junction boxes, fittings, enclosures, and so on, between the hazardous area and the grounding electrode conductor connection point at the service equipment, the grounding electrode system of the building disconnecting means of separate buildings, or the source of a separately derived system. Where installed outside the raceway or enclosure, the grounding conductor is not to exceed 6 ft and is to be routed with the raceway or enclosure. See Section 250-102(e).

502-17. Surge Protection — Class II, Divisions 1 and 2. Surge arresters, including their installation and connection, shall comply with Article 280. In addition, surge arresters, if installed in a Class II, Division 1 location, shall be in suitable enclosures.

Surge-protective capacitors shall be of a type designed for specific duty.

502-18. Multiwire Branch Circuits. In a Class II, Division 1 location, a multiwire branch circuit shall not be permitted.

Exception: Where the disconnect device(s) for the circuit opens all ungrounded conductors of the multiwire circuit simultaneously.

Article 503 — Class III Locations

Contents

503-1. General. The general rules of this *Code* shall apply to electric wiring and equipment in locations classified as Class III locations in Section 500-9.

Exception: As modified by this article.

Equipment installed in Class III locations shall be able to function at full rating without developing surface temperatures high enough to cause excessive dehydration or gradual carbonization of accumulated fibers or flyings. Organic material that is carbonized or excessively dry is highly susceptible to spontaneous ignition. The maximum surface temperatures under operating conditions shall not exceed 165°C (329°F) for equipment that is not subject to overloading, and 120°C (248°F) for equipment (such as motors or power transformers) that may be overloaded.

FPN: For electric trucks, see *Fire Safety Standard for Powered Industrial Trucks Including Type Designations, Areas of Use, Conversions, Maintenance, and Operation,* NFPA 505-1996.

Class III locations usually include textile mills (cotton, rayon, etc.) where easily ignitible fibers or combustible flyings are present in the manufacturing process. Sawmills and other woodworking plants, where sawdust, wood shavings, and combustible fibers or flyings are present, may also become hazardous. If wood flour (dust) is present, the location is a Class II, Group G location, not a Class III location.

Fibers or flyings are hazardous not only because they are easily ignited, but also because flames spread through them quickly. Such fires travel with a rapidity approaching an explosion and are commonly called "flash fires."

Class III, Division 1 applies to locations where ma-

terial is handled, manufactured, or used. Division 2 applies to locations where material is stored or handled but where no manufacturing processes are performed. There are no group designations in Class III locations.

503-2. Transformers and Capacitors — Class III, Divisions 1 and 2. Transformers and capacitors shall comply with Section 502-2(b).

503-3. Wiring Methods. Wiring methods shall comply with (a) and (b).

(a) Class III, Division 1. In Class III, Division 1 locations, the wiring method shall be rigid metal conduit, rigid nonmetallic conduit, intermediate metal conduit, electrical metallic tubing, dusttight wireways, or Type MC or MI cable with approved termination fittings.

(1) Boxes and Fittings. All boxes and fittings shall be dusttight.

(2) Flexible Connections. Where necessary to employ flexible connections, dusttight flexible connectors, liquidtight flexible metal conduit with approved fittings, liquidtight flexible nonmetallic conduit with approved fittings, or flexible cord in conformance with Section 503-10 shall be used.

FPN: See Section 503-16(b) for grounding requirements where flexible conduit is used.

(b) Class III, Division 2. In Class III, Division 2 locations, the wiring method shall comply with (a).

Exception: In sections, compartments, or areas used solely for storage and containing no machinery, open wiring on insulators shall be permitted where installed in accordance with Article 320, but only on condition that protection as required by Section 320-14 be provided where conductors are not run in roof spaces and are well out of reach of sources of physical damage.

503-4. Switches, Circuit Breakers, Motor Controllers, and Fuses — Class III, Divisions 1 and 2. Switches, circuit breakers, motor controllers, and fuses, including pushbuttons, relays, and similar devices, shall be provided with dusttight enclosures.

See the definition of *dusttight* in Article 100.

503-5. Control Transformers and Resistors — Class III, Divisions 1 and 2. Transformers, impedance coils, and resistors used as or in conjunction with control equipment for motors, generators, and appliances shall be provided with dusttight enclosures complying with the temperature limitations in Section 503-1.

See the definition of *dusttight* in Article 100.

503-6. Motors and Generators — Class III, Divisions 1 and 2. In Class III, Divisions 1 and 2 locations, motors, generators, and other rotating machinery shall be totally enclosed nonventilated, totally enclosed pipe ventilated, or totally enclosed fan cooled.

Exception: In locations where, in the judgment of the authority having jurisdiction, only moderate accumulations of lint or flyings will be likely to collect on, in, or in the vicinity of a rotating electric machine and where such machine is readily accessible for routine cleaning and maintenance, one of the following shall be permitted:

(a) Self-cleaning textile motors of the squirrel-cage type;
(b) Standard open-type machines without sliding contacts, centrifugal or other types of switching mechanism, including motor overload devices; or
(c) Standard open-type machines having such contacts, switching mechanisms, or resistance devices enclosed within tight housings without ventilating or other openings

It is intended that the phrase "other rotating machinery" include electric brakes. Listed and labeled electric brakes are available for Class II, Group G locations, and, according to the Underwriters Laboratories Inc. *Hazardous Location Equipment Directory*, such brakes are suitable for Class III locations.

503-7. Ventilating Piping — Class III, Divisions 1 and 2. Ventilating pipes for motors, generators, or other rotating electric machinery, or for enclosures for electric equipment, shall be of metal not less than 0.021 in. (533 μm) in thickness, or of equally substantial noncombustible material, and shall comply with the following:

(1) Lead directly to a source of clean air outside of buildings,
(2) Be screened at the outer ends to prevent the entrance of small animals or birds, and
(3) Be protected against physical damage and against rusting or other corrosive influences

Ventilating pipes shall be sufficiently tight, including their connections, to prevent the entrance of appreciable quantities of fibers or flyings into the ventilated equipment or enclosure and to prevent the escape of sparks, flame, or burning material that might ignite accumulations of fibers or flyings or combustible material in the vicinity. For metal pipes, lock seams and riveted or welded joints shall be permitted; and tight-fitting slip joints shall be permitted where some flexibility is necessary, as at connections to motors.

503-8. Utilization Equipment — Class III, Divisions 1 and 2.

(a) Heaters. Electrically heated utilization equipment shall be approved for Class III locations.

(b) Motors. Motors of motor-driven utilization equipment shall comply with Section 503-6.

(c) Switches, Circuit Breakers, Motor Controllers, and Fuses. Switches, circuit breakers, motor controllers, and fuses shall comply with Section 503-4.

503-9. Lighting Fixtures — Class III, Divisions 1 and 2.

(a) Fixed Lighting. Lighting fixtures for fixed lighting shall provide enclosures for lamps and lampholders that are designed to minimize entrance of fibers and flyings and to prevent the escape of sparks, burning material, or hot metal. Each fixture shall be clearly marked to show the maximum wattage of the lamps that shall be permitted without exceeding an exposed surface temperature of 165°C (329°F) under normal conditions of use.

(b) Physical Damage. A fixture that may be exposed to physical damage shall be protected by a suitable guard.

(c) Pendant Fixtures. Pendant fixtures shall be suspended by stems of threaded rigid metal conduit, threaded intermediate metal conduit, threaded metal tubing of equivalent thickness, or by chains with approved fittings. For stems longer than 12 in. (305 mm), permanent and effective bracing against lateral displacement shall be provided at a level not more than 12 in. (305 mm) above the lower end of the stem, or flexibility in the form of an approved fitting or a flexible connector shall be provided not more than 12 in. (305 mm) from the point of attachment to the supporting box or fitting.

(d) Portable Lighting Equipment. Portable lighting equipment shall be equipped with handles and protected with substantial guards. Lampholders shall be of the unswitched type with no provision for receiving attachment plugs. There shall be no exposed current-carrying metal parts, and all exposed noncurrent-carrying metal parts shall be grounded. In all other respects, portable lighting equipment shall comply with (a).

503-10. Flexible Cords — Class III, Divisions 1 and 2. Flexible cords shall comply with the following:

(1) Be of a type approved for extra-hard usage;
(2) Contain, in addition to the conductors of the circuit, a grounding conductor complying with Section 400-23;
(3) Be connected to terminals or to supply conductors in an approved manner;
(4) Be supported by clamps or other suitable means in such a manner that there will be no tension on the terminal connections; and
(5) Be provided with suitable means to prevent the entrance of fibers or flyings where the cord enters boxes or fittings

503-11. Receptacles and Attachment Plugs — Class III, Divisions 1 and 2. Receptacles and attachment plugs shall be of the grounding type and shall be designed so to minimize the accumulation or the entry of fibers or flyings, and shall prevent the escape of sparks or molten particles.

Exception: In locations where, in the judgment of the authority having jurisdiction, only moderate accumulations of lint or flyings will be likely to collect in the vicinity of a receptacle, and where such receptacle is readily accessible for routine cleaning, general-purpose grounding-type receptacles mounted so as to minimize the entry of fibers or flyings shall be permitted.

503-12. Signaling, Alarm, Remote-Control, and Local Loudspeaker Intercommunications Systems — Class III, Divisions 1 and 2. Signaling, alarm, remote-control, and local loudspeaker intercommunications systems shall comply with the requirements of Article 503 regarding wiring methods, switches, transformers, resistors, motors, lighting fixtures, and related components.

503-13. Electric Cranes, Hoists, and Similar Equipment — Class III, Divisions 1 and 2. Where installed for operation over combustible fibers or accumulations of flyings, traveling cranes and hoists for material handling, traveling cleaners for textile machinery, and similar equipment shall comply with (a) through (d).

(a) Power Supply. Power supply to contact conductors shall be isolated from all other systems and shall be equipped with an acceptable ground detector that will give an alarm and automatically de-energize the contact conductors in case of a fault to ground or will give a visual and audible alarm as long as power is supplied to the contact conductors and the ground fault remains.

(b) Contact Conductors. Contact conductors shall be located or guarded so as to be inaccessible to other than authorized persons and shall be protected against accidental contact with foreign objects.

(c) Current Collectors. Current collectors shall be arranged or guarded so as to confine normal sparking and prevent escape of sparks or hot particles. To reduce sparking, two or more separate surfaces of contact shall be provided for each contact conductor. Reliable means shall be provided to keep contact conductors and current collectors free of accumulations of lint or flyings.

(d) Control Equipment. Control equipment shall comply with Sections 503-4 and 503-5.

In a Class III location, cranes installed over accumulations of fibers or flyings and equipped with rolling or sliding collectors making contact with bare conductors introduce the following two hazards.

1. Any arcing between a conductor and a collector rail may ignite combustible fibers or lint accumulated on or near the bare conductor. This hazard may be prevented by maintaining the proper alignment of the bare conductor, by using a collector designed so that proper contact is always maintained, and by using guards or shields to confine hot metal particles that may be produced by arcing.

2. If enough moisture is present, fibers and flyings accumulating on the insulating supports of the bare conductors may form a conductive path between the conductors or from one conductor to ground, permitting enough current to flow to ignite the fibers. If the system is ungrounded, a current flow to ground is

unlikely to start a fire. A suitable recording ground detector will sound an alarm and automatically de-energize contact conductors when the insulation resistance is lowered by an accumulation of fibers on the insulators or in case of a fault to ground. A ground-fault indicator is permitted that will maintain an alarm until the system is de-energized or the ground fault is cleared.

503-14. Storage Battery Charging Equipment — Class III, Divisions 1 and 2. Storage battery charging equipment shall be located in separate rooms built or lined with substantial noncombustible materials. The rooms shall be constructed to prevent ignitible amounts of flyings or lint and shall be well ventilated.

503-15. Live Parts — Class III, Divisions 1 and 2. Live parts shall not be exposed.

Exception: As provided in Section 503-13.

503-16. Grounding — Class III, Divisions 1 and 2. Wiring and equipment in Class III, Divisions 1 and 2 locations shall be grounded as specified in Article 250 and with the following additional requirements.

(a) Bonding. The locknut-bushing and double-locknut types of contacts shall not be depended on for bonding purposes, but bonding jumpers with proper fittings or other approved means of bonding shall be used. Such means of bonding shall apply to all intervening raceways, fittings, boxes, enclosures, etc., between Class III locations and the point of grounding for service equipment or point of grounding of a separately derived system.

Exception: The specific bonding means shall only be required to the nearest point where the grounded circuit conductor and the grounding electrode conductor are connected together on the line side of the building or structure disconnecting means as specified in Sections 250-32(a), (b), and (c), if the branch-circuit overcurrent protection is located on the load side of the disconnecting means.

FPN: See Section 250-100 for additional bonding requirements in hazardous (classified) locations.

(b) Types of Equipment Grounding Conductors. Where flexible conduit is used as permitted in Section 503-3, it shall be installed with internal or external bonding jumpers in parallel with each conduit and complying with Section 250-102.

Exception: In Class III, Division 1 and 2 locations, the bonding jumper shall be permitted to be deleted where all the following conditions are met.

(a) Listed liquidtight flexible metal 6 ft (1.83 m) or less in length, with fittings listed for grounding, is used;
(b) Overcurrent protection in the circuit is limited to 10 amperes or less; and
(c) The load is not a power utilization load.

Article 504 — Intrinsically Safe Systems

Contents

504-1. Scope. This article covers the installation of intrinsically safe (I.S.) apparatus, wiring, and systems for Class I, II, and III locations.

FPN: For further information, see *Wiring Practices for Hazardous (Classified) Locations Instrumentation Part 1: Intrinsic Safety,* ANSI/ISA RP 12.6-1995.

Article 504 was first included in the 1990 *NEC.* Previously, the installation requirements were in ANSI/ISA RP 12.6-1987 and its earlier 1976 edition.

504-2. Definitions.

The standard used in the United States for construction and performance requirements is ANSI/UL 913-1997, *Intrinsically Safe Apparatus and Associated Apparatus for Use in Class I, II, and III, Division 1 Hazardous (Classified) Locations.* This standard is similar to standards in other countries; all standards are based on the IEC (International Electrotechnical Committee) standard. The *National Electrical Code* now offers the choice of designating hazardous (classified) locations into two divisions (1 and 2) or three zones (0, 1, and

2); but, ANSI/UL 913 requirements are based on the IEC Zone 0 requirements, which are the most stringent. Equipment certified by a testing laboratory for Zone 1 would not necessarily meet ANSI/UL 913 requirements for Division 1.

Associated Apparatus. Apparatus in which the circuits are not necessarily intrinsically safe themselves, but that affect the energy in the intrinsically safe circuits and are relied on to maintain intrinsic safety. Associated apparatus may be either of the following:

(1) Electrical apparatus that has an alternative-type protection for use in the appropriate hazardous (classified) location, or
(2) Electrical apparatus not so protected that shall not be used within a hazardous (classified) location

FPN No. 1: Associated apparatus has identified intrinsically safe connections for intrinsically safe apparatus and also may have connections for nonintrinsically safe apparatus.

FPN No. 2: An example of associated apparatus is an intrinsic safety barrier, which is a network designed to limit the energy (voltage and current) available to the protected circuit in the hazardous (classified) location, under specified fault conditions.

For an illustration of an intrinsic safety barrier, see Figure 504.1.

Figure 504.1 *A typical intrinsic safety barrier that limits energy to the hazardous location. (Crouse-Hinds)*

Control Drawing. A drawing or other document provided by the manufacturer of the intrinsically safe or associated apparatus that details the allowed interconnections between the intrinsically safe and associated apparatus.

Different Intrinsically Safe Circuits. Different intrinsically safe circuits are intrinsically safe circuits in which the possible interconnections have not been evaluated and approved as intriniscally safe.

Intrinsically Safe Apparatus. Apparatus in which all the circuits are intrinsically safe.

Intrinsically Safe Circuit. A circuit in which any spark or thermal effect is incapable of causing ignition of a mixture of flammable or combustible material in air under prescribed test conditions.

Due to its physical and electrical characteristics, an intrinsically safe circuit will not develop sufficient electrical energy (millijoules) in an arc or spark to cause ignition or sufficient thermal energy due to an overload condition to cause the temperature of the installed circuit to exceed the ignition temperature of a specified gas or vapor under normal or abnormal operating conditions.

An abnormal condition may be due to accidental damage, failure of electrical components, excessive voltage, or improper adjustment or maintenance of the equipment.

FPN: Test conditions are described in *Standard for Safety, Intrinsically Safe Apparatus and Associated Apparatus for Use in Class I, II, and III, Division 1, Hazardous (Classified) Locations,* ANSI/UL 913-1997.

Intrinsically Safe System. An assembly of interconnected intrinsically safe apparatus, associated apparatus, and interconnecting cables in that those parts of the system that may be used in hazardous (classified) locations are intrinsically safe circuits.

Although low-energy devices, such as thermocouples, crystal strain, or pressure transducers, generate millivolts and currents in the microampere range, this does not ensure that they are intrinsically safe. These devices are normally connected to amplifiers and power supplies connected to 120-volt or higher circuits. Should a fault occur within the amplifier or power supply, or a voltage surge occur in the electrical supply systems, high-energy arcing, sparking, or overheating of the low-energy portion of the circuit could occur.

FPN: An intrinsically safe system may include more than one intrinsically safe circuit.

Simple Apparatus. A device that will neither generate nor store more than 1.2 volts, 0.1 ampere, 25 milliwatts, or 20 microjoules.

FPN: Examples are switches, thermocouples, light-emitting diodes (LEDs), connectors, and resistance temperature devices (RTDs).

The definition of *simple apparatus* was added in the 1996 *Code* to clarify the use of the term in Sections 504-4, Exception, and 504-10, Exception. The intent is to permit the use of apparatus that stores little or no energy without requiring the apparatus to be approved or to comply with the control drawing. See the fine print note for examples of the apparatus contemplated.

504-3. Application of Other Articles. Except as modified by this article, all applicable articles of this *Code* shall apply.

Because intrinsically safe wiring must be low-energy wiring in order to be intrinsically safe, the wiring itself is most likely to be Class 2, in accordance with Section 725-41, or, in a fire-protective signaling system, power-limited in accordance with Section 760-41. See Article 725 or 760, as appropriate, for the requirements for such wiring. If the installation falls under the scope of Article 800, that is the applicable article. The intrinsically safe apparatus and associated apparatus, on the other hand, may be supplied by ordinary power circuits, and other *Code* rules may be applicable. It is common for intrinsically safe apparatus or associated apparatus supplied by a power circuit to be located in a hazardous (classified) location, with the apparatus protected by one of the protection systems required by Articles 500 through 503 or 505, that is, explosionproof, dust-ignitionproof, purged, and pressurized. In this case, Articles 500 through 503 and 505 are also applicable. Also, intrinsically safe systems are not exempt from the grounding and bonding requirements of Sections 501-16, 502-16, 503-16, and 505-25.

504-4. Equipment Approval. All intrinsically safe apparatus and associated apparatus shall be approved.

Exception: Simple apparatus, as described on the control drawing, shall not be required to be approved.

504-10. Equipment Installation.

(a) Control Drawing. Intrinsically safe apparatus, associated apparatus, and other equipment shall be installed in accordance with the control drawing(s).

The control drawing may put limitations on cables and separation of circuits within an intrinsically safe system. The control drawing also illustrates what is permitted to be connected in the system. Compliance with all the provisions in the control drawing is essential if intrinsic safety is to be maintained. The investigation of the equipment by third-party testing laboratories is based on installation in accordance with the control drawing.

For an illustration of intrinsic safety barriers that are not suitable for Class I, Division 1 locations, see Figure 504.2.

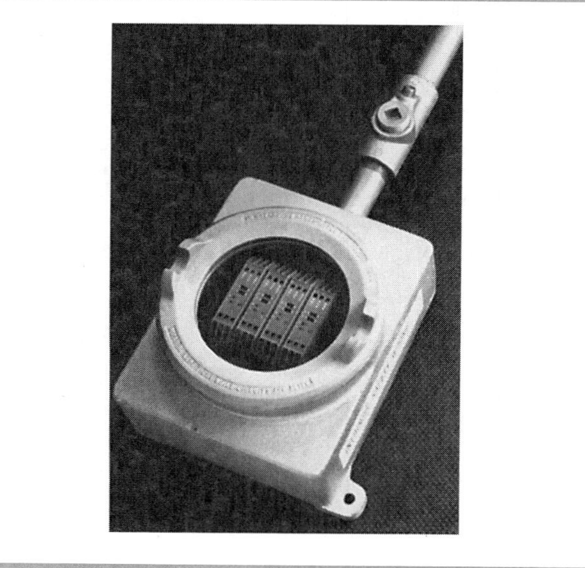

Figure 504.2 *Intrinsic safety barriers are not suitable for Class I, Division 1 locations. Where the barriers may need to be in the Division 1 location, additional protective measures, such as installation in an explosionproof enclosure, are required. (Crouse-Hinds)*

Exception: A simple apparatus that does not interconnect intrinsically safe circuits.

FPN: The control drawing identification is marked on the apparatus.

(b) Location. Intrinsically safe apparatus shall be permitted to be installed in any hazardous (classified) locations for which it has been approved. General-purpose enclosures shall be permitted for intrinsically safe apparatus.

Associated apparatus shall be permitted to be installed in any hazardous (classified) location for which it has been approved, or if protected by other means permitted by Articles 501 through 503 and 505.

504-20. Wiring Methods. Intrinsically safe apparatus and wiring shall be permitted to be installed using any of the wiring methods suitable for unclassified locations, including Chapter 7 and Chapter 8. Sealing shall be as provided in Section 504-70, and separation shall be as provided in Section 504-30.

See the commentary following Section 504-3.

504-30. Separation of Intrinsically Safe Conductors.

(a) From Nonintrinsically Safe Circuit Conductors.

(1) Open Wiring. Conductors and cables of intrinsically safe circuits not in raceways or cable trays shall be separated at least 1.97 in. (50 mm) and secured from conductors and cables of any nonintrinsically safe circuits.

Exception: Where either (1) all of the intrinsically safe circuit conductors are in Type MI or MC cables or (2) all of the nonintrinsically safe circuit conductors are in raceways or Type MI or MC cables where the sheathing or cladding is capable of carrying fault current to ground.

(2) In Raceways, Cable Trays, and Cables. Conductors of intrinsically safe circuits shall not be placed in any raceway, cable tray, or cable with conductors of any nonintrinsically safe circuit.

Exception No. 1: Where conductors of intrinsically safe circuits are separated from conductors of nonintrinsically safe circuits by a distance of at least 1.97 in. (50 mm) and secured, or by a grounded metal partition or an approved insulating partition.

FPN: No. 20 gauge sheet metal partitions 0.0359 in. (912 μm) or thicker are generally considered acceptable.

Exception No. 2: Where either (1) all of the intrinsically safe circuit conductors or (2) all of the nonintrinsically safe circuit conductors are in grounded metal-sheathed or metal-clad cables where the sheathing or cladding is capable of carrying fault current to ground.

FPN: Cables meeting the requirements of Articles 330 and 334 are typical of those considered acceptable.

The cable types mentioned are Type MI cable and Type MC cable, with an interlocked-tape-type outer sheath (as opposed to the smooth or corrugated continuous outer sheath) has not usually been investigated for suitability as an equipment grounding conductor and, therefore, may not be capable of carrying a fault current to ground as required by Exception No. 2.

(3) Within Enclosures.

(a) Conductors of intrinsically safe circuits shall be separated at least 1.97 in. (50 mm) from conductors of any nonintrinsically safe circuits, or as specified in Section 504-30(a)(2).

(b) All conductors shall be secured so that any conductor that might come loose from a terminal cannot come in contact with another terminal.

FPN No. 1: The use of separate wiring compartments for the intrinsically safe and nonintrinsically safe terminals is the preferred method of complying with this requirement.

FPN No. 2: Physical barriers such as grounded metal partitions or approved insulating partitions or approved restricted access wiring ducts separated from other such ducts by at least ¾ in. (19 mm) can be used to help ensure the required separation of the wiring.

The intent of Section 504-30(a) is to prevent intrusion of unsafe energy into the intrinsically safe system as a result of a wiring fault. Since low-voltage, low-energy, but nonintrinsically safe circuits are permitted by Articles 725 and 800 to be minimally insulated and may not be protected by being installed in raceways or cables, particularly in nonhazardous locations, it is essential that nonintrinsically safe circuits and intrinsically safe circuits be physically and electrically separated.

(b) From Different Intrinsically Safe Circuit Conductors. Different intrinsically safe circuits shall be in separate cables or shall be separated from each other by one of the following means.

(1) The conductors of each circuit are within a grounded metal shield.
(2) The conductors of each circuit have insulation with a minimum thickness of 0.01 in. (254 μm).

Exception: Unless otherwise approved.

504-50. Grounding.

(a) Intrinsically Safe Apparatus, Associated Apparatus, and Raceways. Intrinsically safe apparatus, associated apparatus, cable shields, enclosures, and raceways, if of metal, shall be grounded.

FPN: Supplementary bonding to the grounding electrode may be needed for some associated apparatus, e.g., zener diode barriers, if specified in the control drawing. See *Wiring Practices for Hazardous (Classified) Locations Instrumentation Part 1: Intrinsic Safety*, ANSI/ISA RP 12.6-1995.

It is very important to maintain a low-impedance path to ground for zener diode barrier systems, since such systems shunt fault currents to ground.

Figure 504.3 illustrates a common type of zener diode barrier system. F_1 limits the duration of the current in Z_1 and Z_2 to the power rating of the zener diodes under fault conditions, so that the diodes need not be excessively large. Although the diodes themselves can be considered subject to open-circuit fault, they are redundant components, and either one alone will provide the necessary protection in the event of a first fault, that is, high voltage across input terminals 1 and 2. Resistor R_3 is the current-limiting resistor and has been investigated as a protective component not subject to short-circuit fault. Resistor R_1 is used primarily to permit testing to determine that the diodes are intact. Terminals 2 and 4 and the ends of the two diodes are connected to a ground bus that, in turn, is connected to a ground system to which all grounds in the intrinsically safe system are connected. A very low impedance (1 ohm or less is usually recommended) is necessary so that the voltage level on the ground bus will not be raised to an unsafe level under high-current fault conditions.

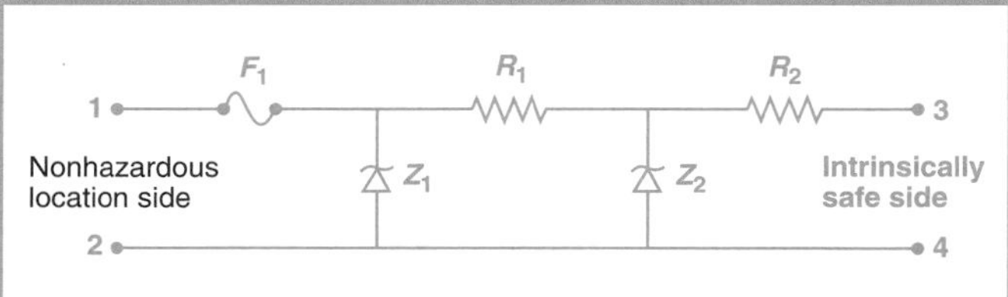

Figure 504.3 *Fused zener diode barrier.*

A shunt diode barrier system is commonly tested with 250 volts ac across terminals 1 and 2, representing a fault in associated apparatus on the nonhazardous location side of the barrier, which imposes a voltage up to 250 volts ac on these terminals. Diode Z_1 conducts at its rated voltage, typically 14 or 15 volts for a barrier designed for a 12-volt (normal operation) system, thus limiting the output voltage of the circuit to 14 or 15 volts. The voltage at which this diode conducts is designed to be higher than the rated input voltage (under no-fault conditions) at terminals 1 and 2. The fuse is selected so that it will open before the power rating of Z_1 is exceeded. Diode Z_2 usually conducts within one or two volts of Z_1 and serves as a backup to Z_1 in the event Z_1 fails in an open-circuit condition for any reason.

Resistor R_2 limits the current in the intrinsically safe circuit to the required value at the output voltage of Z_1 and Z_2. Therefore, even though up to 250 volts ac may be applied to the input of the circuit (terminals 1 and 2) as a result of a fault in the equipment or circuit connected to terminals 1 and 2, the output at terminals 3 and 4 cannot exceed the voltage and current permitted by the diodes and resistor.

Shunt diode barriers normally have maximum inductance and capacitance ratings on the intrinsically safe side because, even though both the voltage and current are limited at terminals 3 and 4, too much inductance in the circuit connected to terminals 3 and 4 could result in the release of an ignition-capable spark when the circuit is opened. In a like manner, too much capacitance between terminals 3 and 4 could result in the release of an ignition-capable spark if there were a short-circuit between wiring connected to terminals 3 and 4 or between wiring connected to terminal 3 and ground.

Wiring always has inductance and capacitance associated with it, depending on the spacing between conductors, size of conductors, and length of conductors. It is necessary, therefore, to limit the length of conductors connected to terminals 3 and 4, just as it is necessary to limit the inductance and capacitance of connected equipment. The length limitation is usually on the order of thousands of feet. The manufacturer's instructions provide the information on installation limitations.

By adjusting the values of its components, the barrier can be designed for a variety of uses, including different types of instrument systems and thermocouples. Barriers such as described are available commercially from a number of manufacturers.

(b) Connection to Grounding Electrodes. Where connection to a grounding electrode is required, the grounding electrode shall be as specified in Sections 250-50(a), (b), (c), and (d) and shall comply with Section 250-30(a)(3). Section 250-52 shall not be used if electrodes specified in Section 250-50 are available.

Electrodes as specified in Section 250-50 are metal underground water pipe, metal frame of building, concrete-encased electrode, and ground ring. These electrodes usually provide lower resistance grounds than electrodes covered in Section 250-52, such as ground rods and plate electrodes.

(c) Shields. Where shielded conductors or cables are used, shields shall be grounded.

Exception: Where a shield is part of an intrinsically safe circuit.

504-60. Bonding.

(a) Hazardous Locations. In hazardous (classified) locations, intrinsically safe apparatus shall be bonded in the hazardous (classified) location in accordance with Section 250-100.

(b) Nonhazardous Locations. In nonhazardous locations, where metal raceways are used for intrinsically safe system wiring in hazardous locations, associated apparatus shall be bonded in accordance with Sections 501-16(a), 502-16(a), 503-16(a), or 505-25, as applicable.

504-70. Sealing. Conduits and cables that are required to be sealed by Sections 501-5 and 502-5 shall be sealed to minimize the passage of gases, vapors, or dusts. Such seals shall not be required to be explosionproof or flameproof.

Exception: Seals shall not be required for enclosures that contain only intrinsically safe apparatus, except as required by Section 501-5(f)(3).

504-80. Identification. Labels required by this section shall be suitable for the environment where they are installed with consideration given to exposure to chemicals and sunlight.

(a) Terminals. Intrinsically safe circuits shall be identified at terminal and junction locations in a manner that will prevent unintentional interference with the circuits during testing and servicing.

(b) Wiring. Raceways, cable trays, and open wiring for intrinsically safe system wiring shall be identified with permanently affixed labels with the wording "Intrinsic Safety Wiring" or equivalent. The labels shall be located so as to be visible after installation and placed so that they may be readily traced through the entire length of the installation. Spacing between labels shall not be more than 25 ft (7.62 m).

Exception: Circuits run underground shall be permitted to be identified where they become accessible after emergence from the ground.

FPN No. 1: Wiring methods permitted in unclassified locations may be used for intrinsically safe systems in hazardous (classified) locations. Without labels to identify the application of the wiring, enforcement authorities cannot determine that an installation is in compliance with the *Code*.

FPN No. 2: In unclassified locations, the identification is necessary to ensure that nonintrinsically safe wire will not be inadvertently added to existing raceways at a later date.

(c) Color Coding. Color coding shall be permitted to identify intrinsically safe conductors where they are colored light blue and where no other conductors colored light blue are used. Likewise, color coding shall be permitted to identify raceways, cable trays, and junction boxes where they are colored light blue and contain only intrinsically safe wiring.

Article 505 — Class I, Zone 0, 1, and 2 Locations

Contents

505-1. Scope. This article covers the requirements for the zone classification system as an alternative to the division classification system covered in Article 500 for electrical and electronic equipment and wiring for all voltages in Class I, Zone 0, Zone 1, and Zone 2 hazardous (classified) locations where fire or explosion hazards may exist due to flammable gases, vapors, or liquids.

Article 505 was introduced in the 1996 *Code* and greatly expanded in the 1999 *Code*. It offers an alternative method of classifying hazardous locations. This classification method is based on the standards for area classification used by the International Electrotechnical Commission (IEC).

The IEC classification scheme includes underground mines. In the United States, mines are under the jurisdiction of the Mine Safety and Health Administration (MSHA) and are outside of the scope of the *Code*.

FPN: For the requirements for electrical and electronic equipment and wiring for all voltages in Class I, Division 1 or Division 2; Class II, Division 1 or Division 2; and Class III, Division 1 or Division 2 hazardous (classified) locations where fire or explosion hazards may exist due to flammable gases or vapors, flammable liquids, or combustible dusts or fibers, refer to Articles 500 through 504.

505-2. Other Articles. All other applicable rules contained in this *Code* shall apply to electrical equipment and wiring installed in hazardous (classified) locations.

Exception: As modified by Article 504 and this article.

See the commentary following Section 500-2.

505-3. Location and General Requirements.

(a) Classification of Locations. Locations shall be classified depending on the properties of the flammable vapors, liquids, or gases that may be present and the likelihood that a flammable or combustible concentration or quantity is present. Where pyrophoric materials are the only materials used or handled, these locations shall not be classified.

Each room, section, or area shall be considered individually in determining its classification.

FPN No. 1: See Section 505-6 for restrictions on area classification.

FPN No. 2: Through the exercise of ingenuity in the layout of electrical installations for hazardous (classified) locations, it is frequently possible to locate much of the equipment in less hazardous or in nonhazardous locations and, thus, to reduce the amount of special equipment required.

(b) Threading. All threaded conduit referred to herein shall be threaded with an NPT standard conduit cutting die that provides ¾-in. taper per foot. Such conduit shall be made wrenchtight to prevent sparking when fault current flows through the conduit system, and to ensure the explosionproof or flameproof integrity of the conduit system where applicable.

Equipment provided with threaded entries for field wiring connections shall be installed in accordance with (1) or (2).

(1) Equipment Provided with Threaded Entries for NPT Threaded Conduit or Fittings. For equipment provided with threaded entries for NPT threaded conduit or fittings, listed conduit, conduit fittings, or cable fittings shall be used.

(2) Equipment Provided with Threaded Entries for Metric Threaded Conduit or Fittings. For equipment with metric threaded entries, such entries shall be identified as being metric, or listed adapters to permit connection to conduit or NPT-threaded fittings shall be provided with the equipment. Adapters shall be used for connection to conduit or NPT-threaded fittings. Listed cable fittings that have metric threads shall be permitted to be used.

FPN: Threading specifications for metric threaded entries are located in *Metric Screw Threads,* ISO 965/1:1980, and *Metric Screw Threads,* ISO 965/3:1980.

See the commentary on Section 500-3.

505-4. Protection Techniques. The following shall be acceptable protection techniques for electrical and electronic equipment in hazardous (classified) locations.

FPN: For additional information, see *Electrical Apparatus for Use in Class I, Zone 0, 1 Hazardous (Classified) Locations General Requirements,* ISA S12.0.01-1997; *Electrical Equipment for Use in Class I, Zone 0, 1, and 2 Hazardous (Classified) Locations,* ANSI/UL 2279, 1997; and *Electrical Apparatus for Explosive Gas Atmospheres - Part 0: General Requirements,* IEC 79-0-1983, Amendment No. 1 (1987), and Amendment No. 2 (1991).

Electrical and electronic equipment located in accordance with the zone method may be protected by the following methods:

1. In Class I, Zone 1, equipment approved as flameproof "d," which is very similar to explosionproof equipment in the USA
2. Purged and pressurized equipment for Class I, Zone 1 or Zone 2 locations for which it is approved
3. Intrinsic safety techniques for the Class I, Zone 0 or Zone 1 for which it is approved
4. In Class I, Zone 2, equipment approved as type "n," which under normal operation is not capable of igniting a surrounding explosive gas atmosphere, and is not likely to create a fault that is capable of causing ignition
5. Oil immersion "o" technique for Class I, Zone 1, where the equipment or part of the equipment is immersed in a protective liquid so that an explosive gases above or outside the enclosure cannot be ignited
6. A type of protection technique of increased safety "e" approved for Class I, Zone 1, where the electrical equipment involved does not produce arcs or sparks or excessive temperature under normal operation, as well as under abnormal conditions, under specified conditions
7. Encapsulation "m" technique in which the arcing, sparking, or hot parts are completely surrounded in a compound in such a way that an explosive gas or vapor cannot be ignited in a Class I, Zone 1 area
8. A technique using powder filling "q" in which the arcing, sparking, or hot parts of the equipment are surrounded by a material such as glass or quartz powder to prevent ignition of an external explosive atmosphere in a Class I, Zone 1 area

The letter in quotation marks following the protection method is the letter used in the marking on the equipment. See Section 505-10(b).

Of the many protection techniques, intrinsic safety, flameproof, and increased safety are the most common for Zone 1 locations.

(a) Flameproof "d." This protection technique shall be permitted for equipment in those Class I, Zone 1 locations for which it is approved.

FPN No. 1: *Flameproof* is a type of protection of electrical equipment in which the enclosure will withstand an internal explosion of a flammable mixture that has penetrated into the interior, without suffering damage and without causing ignition, through any joints or structural openings in the enclosure, of an external explosive atmosphere consisting of one or more of the gases or vapors for which it is designed.

FPN No. 2: For further information, see *Electrical Apparatus for Use In Class I, Zone 1 and 2 Hazardous (Classified) Locations, Type of Protection — Flameproof "d,"* ISA S12.22.01-1996; *Electrical Apparatus for Explosive Gas Atmospheres, Part 1 — Construction and Verification Test of Flameproof Enclosures of Electrical Apparatus,* IEC 79-1-1990 and Amendment No. 1 (1993).

(b) Purged and Pressurized. This protection technique shall be permitted for equipment in those Class I, Zone 1 or Zone 2 locations for which it is approved.

FPN No. 1: In some cases, hazards may be reduced or hazardous (classified) locations limited or eliminated by adequate positive-pressure ventilation from a source of clean air in conjunction with effective safeguards against ventilation failure.

FPN No. 2: For further information, see *Standard for Purged and Pressurized Enclosures for Electrical Equipment,* NFPA 496-1998.

Flameproof protection is commonly combined with increased safety protection. For example, motor control and other switching contacts are commonly protected by flameproof enclosures with the field wiring terminals protected in a separate but attached enclosure by increased safety. The conductors between the enclosures are protected by flameproof feed-through insulators.

FPN No. 3: Pressurized "p" is a type of protection of electrical equipment that uses the technique of guarding against the ingress of the external atmosphere, which may be explosive, into an enclosure by maintaining a protective gas therein at a pressure above that of the external atmosphere. For further information, see *Electrical Apparatus for Explosive Gas Atmospheres - Part 2: Electrical Apparatus, Type of Protection "p,"* IEC 79-2-1983; and *Electrical Apparatus for Explosive Gas Atmospheres - Part 13: Construction and Use of Rooms or Buildings Protected by Pressurization,* IEC 79-13-1982.

(c) Intrinsic Safety. This protection technique shall be permitted for equipment in those Class I, Zone 0 or Zone 1 locations for which it is approved.

The identifying letter for intrinsic safety is "i" followed by either "a" or "b," identifying whether the equipment is suitable for Zone 0 (ia) or Zone 1 (ib). The associated apparatus is identified by the same letters in brackets, that is, [ia] or [ib].

FPN No. 1: *Intrinsic safety* is designated type of protection "ia" by IEC 79-11 for use in Zone 0 locations. *Intrinsic safety* is designated type of protection "ib" by IEC 79-11 for use in Zone 1 locations.

FPN No. 2: For further information, see *Intrinsically Safe Apparatus and Associated Apparatus for Use in Class I, II, and III, Hazardous Locations,* ANSI/UL 913-1997; *Electrical Apparatus for Explosive Gas Atmospheres — Part 11: Intrinsic Safety "i,"* IEC 79-11-1991; and *Electrical Apparatus for Explosive Gas Atmospheres — Part 3: Spark-test Apparatus for Intrinsically-safe Circuits,* IEC 79-3-1990.

FPN No. 3: Intrinsically safe associated apparatus, designated by [ia] or [ib], is connected to intrinsically safe equipment ("ia" or "ib" respectively), but is located outside the hazardous (classified) location unless also protected by another type of protection (such as flameproof).

(d) Type of Protection "n." This protection technique shall be permitted for equipment in those Class I, Zone 2 locations for which it is approved. Type of protection "n" is further subdivided into nA, nC, and nR.

FPN No. 1: See Table 505-10(b) 1 for the descriptions of subdivisions for type of protection "n."

FPN No. 2: *Type "n" protection* is a type of protection applied to electrical equipment such that, in normal operation, the electrical equipment is not capable of igniting a surrounding explosive gas atmosphere and a fault capable of causing ignition is not likely to occur.

FPN No. 3: For further information, see *Electrical Apparatus for Explosive Gas Atmospheres, Part 15 — Electrical Apparatus with Type of Protection "n,"* IEC 79-15-1987.

(e) Oil Immersion "o." This protection technique shall be permitted for equipment in those Class I, Zone 1 locations for which it is approved.

FPN No. 1: *Oil immersion* is a type of protection in which the electrical equipment or parts of the electrical equipment are immersed in a protective liquid in such a way that an explosive atmosphere that may be above the liquid or outside the enclosure cannot be ignited.

FPN No. 2: For further information, see *Electrical Apparatus for Use in Class I, Zone 1 Hazardous (Classified) Locations, Type of Protection — Oil-Immersion "o,"* ISA S12.26.01 — 1996; and *Electrical Apparatus for Explosive Gas Atmospheres, Part 6 — Oil-Immersion "o,"* IEC 79-6-1995.

(f) Increased Safety "e." This protection technique shall be permitted for equipment in those Class I, Zone 1 locations for which it is approved.

FPN No. 1: *Increased safety* is a type of protection applied to electrical equipment that does not produce arcs or sparks in normal service and under specified abnormal conditions, in which additional measures are applied so as to give increased security against the possibility of excessive temperatures and of the occurrence of arcs and sparks.

FPN No. 2: For further information, see *Electrical Apparatus for Use in Class I, Zone 1 Hazardous (Classified) Locations, Type of Protection — Increased Safety "e,"* ISA S12.16.01-1996; and *Electrical Apparatus for Explosive Gas Atmospheres — Part 7: Increased Safety "e,"* IEC 79-7-1990, Amendment No. 1 (1991), and Amendment No. 2 (1993).

The increased safety protection technique is commonly used for motors and generators (see Section 505-21) and fluorescent lighting fixtures. It is also used for terminal boxes. See the commentary following Section 505-4(b).

(g) Encapsulation "m." This protection technique shall be permitted for equipment in those Class I, Zone 1 locations for which it is approved.

FPN No. 1: *Encapsulation* is a type of protection in which the parts that could ignite an explosive atmosphere by either sparking or heating are enclosed in a compound in such a way that this explosive atmosphere cannot be ignited.

FPN No. 2: For further information, see *Electrical Apparatus for Use in Class I, Zone 1 Hazardous (Classified) Locations Type of Protection — Encapsulation "m,"* ISA S12.23.01-1996, and *Electrical Apparatus for Explosive Gas Atmospheres — Part 18: Encapsulation "m,"* IEC 79-18-1992.

(h) Powder Filling "q." This protection technique shall be permitted for equipment in those Class I, Zone 1 locations for which it is approved.

FPN No. 1: *Powder filling* is a type of protection in which the parts capable of igniting an explosive atmosphere are fixed in position and completely surrounded by filling material (glass or quartz powder) to prevent the ignition of an external explosive atmosphere.

FPN No. 2: For further information, see *Electrical Apparatus for Use in Class I, Zone 1 Hazardous (Classified) Locations Type of Protection — Powder Filling "q,"* ISA S12.25.01-1996, and *Electrical Apparatus for Explosive Gas Atmospheres — Part 5: Powder Filling, Type of Protection "q,"* IEC 79-5-1967.

505-5. Reference Standards.

FPN No. 1: It is important that the authority having jurisdiction be familiar with recorded industrial experience as well as with standards of the National Fire Protection Association, the American Petroleum Institute, and the International Society for Measurement and Control (ISA), that may be of use in the classification of various locations, the determination of adequate ventilation, and the protection against static electricity and lightning hazards.

FPN No. 2: For further information on the classification of locations, see *Electrical Apparatus for Explosive Gas Atmospheres, Classification of Hazardous Areas,* IEC 79-10-1995; *Classification of Locations for Electrical Installations at Petroleum Facilities Classified as Class I, Zone 0, Zone 1, or Zone 2,* API RP 505-1996; *Electrical Apparatus for Explosive Gas Atmospheres, Classifications of Hazardous (Classified) Locations,* ISA S12.24.01-1997; and *Model Code of Safe Practice in the Petroleum Industry, Part 15: Area Classification Code for Petroleum Installations,* IP 15, The Institute of Petroleum, London.

FPN No. 3: For further information on protection against static electricity and lightning hazards in hazardous (classified) locations, see *Recommended Practice on Static Electricity,* NFPA 77-1993; *Standard for the Installation of Lightning Protection Systems,* NFPA 780-1997; and *Protection Against Ignitions Arising Out of Static Lightning and Stray Currents,* API RP 2003-1991.

FPN No. 4: For further information on ventilation, see *Flammable and Combustible Liquids Code,* NFPA 30-1996, and *Recommended Practice for Classification of Locations for Electrical Installations at Petroleum Facilities Classified as Class I, Division 1 or Division 2,* API RP 500-1997, Section 6.3.

FPN No. 5: For further information on electrical systems for hazardous (classified) locations on offshore oil and gas producing platforms, see *Design and Installation of Electrical Systems for Offshore Production Platforms,* ANSI/API RP 14F-1991.

FPN No. 6: For further information on the installation of electrical equipment in hazardous (classified) locations in general, see *Electrical Apparatus for Explosive Gas Atmospheres — Part 14: Electrical Installations in Explosive Gas Atmospheres (Other than Mines),* IEC 79-14-1996, and *Electrical Apparatus for Explosive Gas Atmospheres — Part 16: Artifi-*

cial Ventilation for the Protection of Analyzer(s) Houses, IEC 79-16-1990.

505-6. Special Precaution. Article 505 requires equipment construction and installation that will ensure safe performance under conditions of proper use and maintenance.

FPN No. 1: It is important that inspection authorities and users exercise more than ordinary care with regard to the installation and maintenance of electrical equipment in hazardous (classified) locations.

FPN No. 2: Low ambient conditions require special consideration. Electrical equipment depending on the protection techniques described by Section 505-4(a) may not be suitable for use at temperatures lower than −20°C (−13°F) unless they are approved for use at lower temperatures. However, at low ambient temperatures, flammable concentrations of vapors may not exist in a location classified Class I, Zones 0, 1, or 2 at normal ambient temperature.

(a) Supervision of Work. Classification of areas and selection of equipment and wiring methods shall be under the supervision of a qualified Registered Professional Engineer.

It should be noted that Section 505-6(a) requires area classification, wiring, and equipment selection to be under the supervision of a qualified Registered Professional Engineer for installations in Class I, Zone 0, 1, and 2.

(b) Dual Classification. In instances of areas within the same facility classified separately, Class I, Zone 2 locations shall be permitted to abut, but not overlap, Class I, Division 2 locations. Class I, Zone 0 or Zone 1 locations shall not abut Class I, Division 1 or Division 2 locations.

It is the intent that an installation be designed using either the classification of Article 500 or the classification scheme of Article 505. Both schemes cannot be used for classifying the same area. In the situation of areas within the same facility, Class I, Zone 2 locations are allowed to be adjacent to and share the same border, but not overlap Class I, Division 2 locations. However, Class I, Zone 0 or Zone 1 locations are not allowed to be adjacent to and share the same border with Class I, Division 1 or Division 2 locations.

(c) Reclassification Permitted. A Class I, Division 1 or Division 2 location shall be permitted to be reclassified as a Class I, Zone 0, Zone 1, or Zone 2 location provided all of the space that is classified because of a single flammable gas or vapor source is reclassified under the requirements of this article.

505-7. Grouping and Classification. For purposes of testing, approval, and area classification, various air mixtures (not oxygen enriched) shall be grouped as required in (a), (b), and (c).

FPN: Group I is intended for use in describing atmospheres that contain firedamp (a mixture of gases, composed mostly of methane, found underground, usually in mines). This *Code* does not apply to installations underground in mines. See Section 90-2(b).

Group II shall be subdivided into IIC, IIB, and IIA, as noted in (a), (b), and (c), according to the nature of the gas or vapor, for protection techniques "d," "ia," "ib," "[ia]," and "[ib]," and, where applicable, "n" and "o."

FPN No. 1: The gas and vapor subdivision as described above is based on the maximum experimental safe gap (MESG), minimum igniting current (MIC), or both. Test equipment for determining the MESG is described in *Construction and Verification Tests of Flameproof Enclosures of Electrical Apparatus*, IEC 79-1A-1975, Amendment No. 1 (1993) and UL *Technical Report No. 58* (1993). The test equipment for determining MIC is described in *Spark-Test Apparatus for Intrinsically-Safe Circuits*, IEC 79-3-1990. The classification of gases or vapors according to their maximum experimental safe gaps and minimum igniting currents is described in *Classification of Mixtures of Gases or Vapours with Air According to Their Maximum Experimental Safe Gaps and Minimum Igniting Currents*, IEC 79-12-1978.

FPN No. 2: Verification of electrical equipment utilizing protection techniques "e," "m," "p," and "q," due to design technique, does not require tests involving MESG or MIC. Therefore, Group II is not required to be subdivided for these protection techniques.

FPN No. 3: It is necessary that the meanings of the different equipment markings and Group II classifications be carefully observed to avoid confusion with Class I, Divisions 1 and 2, Groups A, B, C, and D.

The group order is the approximate inverse of the groups specified in Article 500. Group IIC includes Article 500, Groups A and B. This classification scheme includes the evaluation of maximum safe experimental gap as well as minimum igniting current. Although the maximum safe experimental gap for Group A is less than that for Group B in some circumstances, the minimum igniting current is less for hydrogen (Group B) than it is for acetylene (Group A). This difference has been accounted for in ANSI/UL 913-1997, *UL Standard for Safety Intrinsically Safe Apparatus and Associated Apparatus for Use in Class I, II, III, Division 1, Hazardous (Classified) Locations,* since it is a consideration for the evaluation of intrinsically safe apparatus.

x**(a) Group IIC.** Atmospheres containing acetylene, hydrogen, or flammable gas, flammable liquid-produced vapor, or combustible liquid-produced vapor mixed with air that may burn or explode, having either a maximum experimental safe gap (MESG) value less than or equal to 0.50 mm or minimum igniting current ratio (MIC ratio) less than or equal to 0.45.

FPN: Group IIC is equivalent to a combination of Class I, Group A, and Class I, Group B, as described in Sections 500-5(a)(1) and (a)(2).

ˣ**(b) Group IIB.** Atmospheres containing acetaldehyde, ethylene, or flammable gas, flammable liquid-produced vapor, or combustible liquid-produced vapor mixed with air that may burn or explode, having either maximum experimental safe gap (MESG) values greater than 0.50 mm and less than or equal to 0.90 mm or minimum igniting current ratio (MIC ratio) greater than 0.45 and less than or equal to 0.80.

FPN: Group IIB is equivalent to Class I, Group C, as described in Section 500-5(a)(3).

ˣ**(c) Group IIA.** Atmospheres containing acetone, ammonia, ethyl alcohol, gasoline, methane, propane, or flammable gas, flammable liquid-produced vapor, or combustible liquid-produced vapor mixed with air that may burn or explode, having either a maximum experiment safe gap (MESG) value greater than 0.90 mm or minimum igniting current ratio (MIC ratio) greater than 0.80.

FPN: Group IIA is equivalent to Class I, Group D as described in Section 500-5(a)(4).

(d) Other. Equipment shall be permitted to be listed for a specific gas or vapor, specific mixtures of gases or vapors, or any specific combination of gases or vapors.

FPN: One common example is equipment marked for "IIB + H_2."

505-8. Class I Temperature. The temperature marking specified in Section 505-10(b) shall not exceed the ignition temperature of the specific gas or vapor to be encountered.

FPN: For information regarding ignition temperatures of gases and vapors, see *Recommended Practice for the Classification of Flammable Liquids, Gases, or Vapors and of Hazardous (Classified) Locations for Electrical Installations in Chemical Process Areas,* NFPA 497-1997; *Guide to Fire Hazard Properties of Flammable Liquids, Gases, and Volatile Solids,* NFPA 325-1994; and *Electrical Apparatus for Explosive Gas Atmospheres, Data for Flammable Gases and Vapours, Relating to the Use of Electrical Apparatus,* IEC 79-20-1996.

505-9. Zone Classification. The classification into zones shall be in accordance with the following.

(a) Class I, Zone 0. A Class I, Zone 0 location is a location

(1) In which ignitible concentrations of flammable gases or vapors are present continuously, or
(2) In which ignitible concentrations of flammable gases or vapors are present for long periods of time.

FPN No. 1: As a guide in determining when flammable gases or vapors are present continuously or for long periods of time, refer to *Recommended Practice for Classification of Locations for Electrical Installations of Petroleum Facilities Classified as Class I, Zone 0, Zone 1 or Zone 2,* API RP 505-1996; *Electrical Apparatus for Explosive Gas Atmospheres, Classifications of Hazardous Areas,* IEC 79-10-1995; and *Area Classification Code for Petroleum Installations, Model Code, Part 15, Institute of Petroleum;* and *Electrical Apparatus for Explosive Gas Atmospheres, Classifications of Hazardous (Classified) Locations,* ISA S12.24.01-1997.

FPN No. 2: This classification includes locations inside vented tanks or vessels that contain volatile flammable liquids; inside inadequately vented spraying or coating enclosures, where volatile flammable solvents are used; between the inner and outer roof sections of a floating roof tank containing volatile flammable liquids; inside open vessels, tanks and pits containing volatile flammable liquids; the interior of an exhaust duct that is used to vent ignitible concentrations of gases or vapors; and inside inadequately ventilated enclosures that contain normally venting instruments utilizing or analyzing flammable fluids and venting to the inside of the enclosures.

FPN No. 3: It is not good practice to install electrical equipment in Zone 0 locations except when the equipment is essential to the process or when other locations are not feasible. [See Section 505-3(a) FPN No. 2.] If it is necessary to install electrical systems in a Zone 0 location, it is good practice to install intrinsically safe systems as described by Article 504.

(b) Class I, Zone 1. A Class I, Zone 1 location is a location

(1) In which ignitible concentrations of flammable gases or vapors are likely to exist under normal operating conditions; or
(2) In which ignitible concentrations of flammable gases or vapors may exist frequently because of repair or maintenance operations or because of leakage; or
(3) In which equipment is operated or processes are carried on, of such a nature that equipment breakdown or faulty operations could result in the release of ignitible concentrations of flammable gases or vapors and also cause simultaneous failure of electrical equipment in a mode to cause the electrical equipment to become a source of ignition; or
(4) That is adjacent to a Class I, Zone 0 location from which ignitible concentrations of vapors could be communicated, unless communication is prevented by adequate positive pressure ventilation from a source of clean air and effective safeguards against ventilation failure are provided.

FPN No. 1: Normal operations is considered the situation when plant equipment is operating within its design parameters. Minor releases of flammable material may be part of normal operations. Minor releases include the releases from mechanical packings on pumps. Failures that involve repair or shutdown (such as the breakdown of pump seats and flange gaskets, and spillage caused by accidents) are not considered normal operation.

FPN No. 2: This classification usually includes locations where volatile flammable liquids or liquefied flammable gases are transferred from one container to another. In areas in the vicinity of spraying and painting operations where flammable solvents are used; adequately ventilated drying rooms or compartments for evaporation of flammable solvents; adequately ventilated locations containing fat and oil extraction equipment using volatile flammable solvents; portions of cleaning and dyeing plants where volatile flammable liquids are used; adequately ventilated gas generator rooms and other portions of gas manufacturing plants where flammable gas may escape; inadequately ventilated pump rooms for flammable gas or for volatile flammable liquids; the interiors of refrigerators and freezers in which volatile flammable materials are stored in the open, lightly stoppered, or easily ruptured containers; and other locations where ignitible concentrations of flammable vapors or gases are likely to occur in the course of normal operation, but not classified Zone 0.

(c) Class I, Zone 2. A Class I, Zone 2 location is a location

(1) In which ignitible concentrations of flammable gases or vapors are not likely to occur in normal operation and if they do occur will exist only for a short period; or
(2) In which volatile flammable liquids, flammable gases, or flammable vapors are handled, processed, or used, but in which the liquids, gases, or vapors normally are confined within closed containers of closed systems from which they can escape, only as a result of accidental rupture or breakdown of the containers or system, or as a result of the abnormal operation of the equipment with which the liquids or gases are handled, processed, or used; or
(3) In which ignitible concentrations of flammable gases or vapors normally are prevented by positive mechanical ventilation, but which may become hazardous as a result of failure or abnormal operation of the ventilation equipment; or
(4) That is adjacent to a Class I, Zone 1 location, from which ignitible concentrations of flammable gases or vapors could be communicated, unless such communication is prevented by adequate positive-pressure ventilation from a source of clean air, and effective safeguards against ventilation failure are provided.

FPN: The Zone 2 classification usually includes locations where volatile flammable liquids or flammable gases or vapors are used, but which would become hazardous only in case of an accident or of some unusual operating condition.

505-10. Listing, Marking, and Documentation.

(a) Listing. Equipment that is listed for a Zone 0 location shall be permitted in a Zone 1 or Zone 2 location of the same gas or vapor. Equipment that is listed for a Zone 1 location shall be permitted in a Zone 2 location of the same gas or vapor.

(b) Marking. Equipment shall be marked in accordance with (1) or (2).

(1) Division Equipment. Equipment approved for Class I, Division 1 or Class I, Division 2 shall, in addition to being marked in accordance with Section 500-5(d), be permitted to be marked with the following:

(a) Class I, Zone 1 or Class I, Zone 2 (as applicable), and
(b) Applicable gas classification group(s) in accordance with Table 505-10(b)(2), and
(c) Temperature classification in accordance with Section 505-10(b)(3)

Table 505-10(b)(1). Types of Protection Designation

Designation	Technique	Zone*
d	Flameproof enclosure	1
e	Increased safety	1
ia	Intrinsic safety	0
ib	Intrinsic safety	1
[ia]	Intrinsically safe associated apparatus	Nonhazardous
[ib]	Intrinsically safe associated apparatus	Nonhazardous
m	Encapsulation	1
nA	Nonsparking equipment	2
nC	Sparking equipment in which the contacts are suitably protected other than by restricted breathing enclosure	2
nR	Restricted breathing enclosure	2
o	Oil immersion	1
p	Purged and pressurized	1 or 2
q	Powder filled	1

*Does not address use where a combination of techniques is used.

(2) Zone Equipment. Equipment meeting one or more of the protection techniques described in Section 505-4 shall be marked with the following in the order shown:

(a) Class
(b) Zone
(c) Symbol "AEx"
(d) Protection technique(s) in accordance with Table 505-10(b)(1)
(e) Applicable gas classification group(s) in accordance with Table 505-10(b)(2)

(f) Temperature classification in accordance with Section 505-10(b)(3)

Exception: Intrinsically safe associated apparatus shall be required to be marked only with (c), (d), and (e).

FPN No. 1: An example of such a required marking is "Class I, Zone 0, AEx ia IIC T6."

The symbol "AEx" identifies the equipment as meeting United States (American) standards. In European Common Market countries the symbol is "EEx." In the IEC standards on which both are based, the symbol is "Ex."

Electrical equipment of types of protection "e," "m," "p," or "q," shall be marked Group II. Electrical equipment of types of protection "d," "ia," "ib," "[ia]," or "[ib]" shall be marked Group IIA, or IIB, or IIC, or for a specific gas or vapor. Electrical equipment of types of protection "n" shall be marked Group II unless it contains enclosed-break devices, nonincendive components, or energy-limited equipment or circuits, in which case it shall be marked Group IIA, IIB, or IIC, or a specific gas or vapor. Electrical equipment of other types of protection shall be marked Group II unless the type of protection utilized by the equipment requires that it shall be marked Group IIA, IIB, or IIC, or a specific gas or vapor.

FPN No. 2: An explanation of the marking that is required follows.

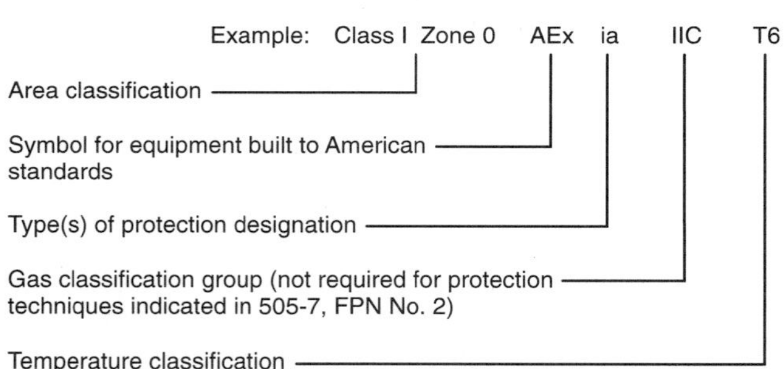

Figure 505-10(b)(1).

Table 505-10(b)(2). Gas Classification Groups

Gas Group	Comment
IIC	See Section 505-7(a)
IIB	See Section 505-7(b)
IIA	See Section 505-7(c)

(3) Temperature Classifications. Approved equipment shall be marked to show the operating temperature or temperature range referenced to a 40°C (104°F) ambient. The temperature range, if provided, shall be indicated in identification numbers, as shown in Table 505-10(b)(3).

Table 505-10(b)(3). Classification of Maximum Surface Temperature for Group II Electrical Equipment

Temperature Class	Maximum Surface Temperature (°C)
T1	≤450
T2	≤300
T3	≤200
T4	≤135
T5	≤100
T6	≤85

Electrical equipment designed for use in the ambient temperature range between −20°C and +40°C shall require no additional temperature marking.

Electrical equipment that is designed for use in a range of ambient temperatures other than −20°C and +40°C is considered to be special; and the ambient temperature range shall then be marked on the equipment, including either the symbol "Ta" or "Tamb" together with the special range of ambient temperatures. As an example, such a marking might be "−30°C Ta + 40°C."

Electrical equipment suitable for ambient temperatures exceeding 40°C (104°F) shall be marked with both the maximum ambient temperature and the operating temperature or temperature range at that ambient temperature.

Exception No. 1: Equipment of the nonheat-producing type, such as conduit fittings, and equipment of the heat-producing type having a maximum temperature of not more than 100°C (212°F) shall not be required to have a marked operating temperature or temperature range.

Exception No. 2: Equipment approved for Class I, Division 1 or Division 2 locations as permitted by Sections 505-20(b) and (c) shall be permitted to be marked in accordance with Section 500-5(d) and Table 500-5(d).

(c) Documentation for Industrial Occupancies. All areas in industrial occupancies designated as hazardous (classified) locations shall be properly documented. This documentation shall be available to those authorized to design, install, inspect, maintain, or operate electrical equipment at the location.

FPN: For examples of area classification drawings, see *Electrical Apparatus for Explosive Gas Atmospheres, Classification of Hazardous Areas,* IEC 79-10-1995; *Classification of Locations for Electrical Installations at Petroleum Facilities Classified as Class I, Zone 0, Zone 1, or Zone 2,* API RP 505-1997; *Electrical Apparatus for Explosive Gas Atmospheres, Classifications of Hazardous (Classified) Locations,* ISA S12.24.01-1997; *Recommended Practice for Classification of Locations for Electrical Installations at Petroleum Facilities Classified as Class I, Division I or Division 2,* API RP 500-1997, *Model Code of*

Safe Practice in the Petroleum Industry, Part 15: Area Classification Code for Petroleum Installations, IP 15, The Institute of Petroleum, London.

505-15. Wiring Methods.

(a) Zone 0. In Class I, Zone 0 locations, only the following wiring methods shall be permitted.

(1) Intrinsically safe wiring in accordance with Article 504.

FPN: Article 504 only includes protection technique "ia."

Section 505-15(a)(1) no longer permits any wiring method other than intrinsically safe wiring in Zone 0. The 1996 *Code* permitted intrinsically safe and nonincendive circuits in a Zone 0 location.

(2) Seals shall be provided within 10 ft (3.05 m) of where a conduit leaves a Zone 0 location. There shall be no unions, couplings, boxes, or fittings, except reducers at the seal, in the conduit run between the seal and the point at which the conduit leaves the location.

Exception: A rigid unbroken conduit that passes completely through the Zone 0 location with no fittings less than 12 in. (305 mm) beyond each boundary, shall not be required to be sealed, if the termination points of the unbroken conduit are in unclassified locations.

(3) Seals shall be provided on cables at the first point of termination after entry into the Zone 0 location.

(4) Seals shall not be required to be explosionproof or flameproof.

(b) Zone 1. In Class I, Zone 1 locations, all wiring methods permitted for Class I, Division 1 locations shall be permitted.

Where Class I, Division 1 wiring methods are used, sealing and drainage shall be provided in accordance with Sections 501-5(a), (c), (d), and (f), except where the term "Division 1" is used "Zone 1" shall be substituted.

An explosionproof seal, constructed in accordance with Section 501-5(c), shall be provided for each conduit entering an enclosure having type of protection "e" or "d," except where the type of protection "d" enclosure is marked to indicate that a seal is not required.

Wiring methods shall maintain the integrity of protection techniques.

FPN No. 1: For example, equipment with type of protection "e" requires that conduit seals or cable fittings incorporate suitable methods to maintain the "ingress protection" (minimum IP54) of the enclosure; and, for conduit, serve to maintain the explosionproof integrity of the conduit system.

FPN No. 2: Different electrical enclosures provide different degrees of "ingress protection." The measures applied to enclosures of electrical apparatus include

(1) The protection of persons against contact with or approach to live parts and against contact with moving parts (other than smooth rotating shafts and the like) inside the enclosure;
(2) The protection of the apparatus inside the enclosure against ingress of solid foreign bodies; and
(3) The protection of the apparatus inside the enclosure against harmful ingress of water.

(c) Zone 2. In Class I, Zone 2 locations, all wiring methods permitted for Class I, Division 2 locations shall be permitted. Sealing and drainage shall be provided in accordance with Sections 501-5(b), (c), (e), and (f), except where the term "Division 2" is used, "Zone 2" shall be substituted and where the term "Division 1" is used, "Zone 1" shall be substituted.

Wiring methods shall maintain the integrity of protection techniques.

(d) Solid Obstacles. Flameproof equipment with flanged joints shall not be installed such that the flange openings are closer than the distances shown in Table 505-15 to any solid obstacle that is not a part of the equipment (such as steelworks, walls, weather guards, mounting brackets, pipes, or other electrical equipment) unless the equipment is listed for a smaller distance of separation.

Table 505-15. Minimum Distance of Obstructions from Flameproof "d" Flange Openings

	Minimum Distance	
Gas Group	in.	mm
IIC	1 37/64	40
IIB	1 3/16	30
IIA	25/64	10

505-20. Equipment.

(a) Zone 0. In Class I, Zone 0 locations, only equipment specifically listed and marked as suitable for the location shall be permitted.

Exception: Intrinsically safe equipment listed for use in Class I, Division 1 locations for the same gas, or as permitted by Section 505-7(d), and with a suitable temperature rating shall be permitted.

The exception to Section 505-20(a) is based on the fact that ANSI/UL 913-1997, the standard used to evaluate intrinsically safe systems for Class I, Division 1 locations, is based on the IEC requirements for intrinsically safe equipment for Class I, Zone 0 locations.

(b) Zone 1. In Class I, Zone 1 locations, only equipment specifically listed and marked as suitable for the location shall be permitted.

Exception: Equipment approved for use in Class I, Division 1 or listed for use in Class I, Zone 0 locations for the same gas, or as permitted by Section 505-7(d) and with a suitable temperature rating shall be permitted.

(c) Zone 2. In Class I, Zone 2 locations, only equipment specifically listed and marked as suitable for the location shall be permitted.

Exception No. 1: Equipment listed for use in Class I, Zone 0 or Zone 1 locations for the same gas, or as permitted by Section 505-7(d), and with a suitable temperature rating, shall be permitted.

Exception No. 2: Equipment approved for use in Class I, Division 1 or Division 2 locations for the same gas, or as permitted by Section 505-7(d), and with a suitable temperature rating shall be permitted.

Exception No. 3: In Class I, Zone 2 locations, the installation of open or nonexplosionproof or nonflameproof enclosed motors, such as squirrel-cage induction motors without brushes, switching mechanisms, or similar arc-producing devices that are not identified for use in a Class I, Zone 2 location shall be permitted.

FPN No. 1: It is important to consider the temperature of internal and external surfaces that may be exposed to the flammable atmosphere.

FPN No. 2: It is important to consider the risk of ignition due to currents arcing across discontinuities and overheating of parts in multisection enclosures of large motors and generators. Such motors and generators may need equipotential bonding jumpers across joints in the enclosure and from enclosure to ground. Where the presence of ignitible gases or vapors is suspected, clean air purging may be needed immediately prior to and during start-up periods.

See the commentary following Section 501-8(b), FPN No. 2.

(d) Manufacturer's Instructions. Electrical equipment installed in hazardous (classified) locations shall be installed in accordance with the instructions (if any) provided by the manufacturer.

505-21. Increased Safety "e" Motors and Generators. In Class I, Zone 1 locations, Increased Safety "e" motors and generators of all voltage ratings shall be listed for Class I, Zone 1 locations, and shall comply with the following.

(1) Motors shall be marked with the current ratio, I_A/I_N, and time, t_E;
(2) Motors shall have controllers marked with the model or identification number, output rating (horsepower or kilowatt), full-load amperes, starting current ratio (I_A/I_N), and time (t_E) of the motors that they are intended to protect; the controller marking shall also include the specific overload protection type (and setting, if applicable) that is listed with the motor or generator;
(3) Connections shall be made with the specific terminals listed with the motor or generator;
(4) Terminal housings shall be permitted to be of substantial, nonmetallic, nonburning material provided an internal grounding means between the motor frame and the equipment grounding connection is incorporated within the housing;
(5) The provisions of Part C of Article 430 shall apply regardless of the voltage rating of the motor;
(6) The motors shall be protected against overload by a separate overload device that is responsive to motor current. This device shall be selected to trip or shall be rated in accordance with the listing of the motor and its overload protection;
(7) Sections 430-34 and 430-44 shall not apply to such motors; and
(8) The motor overload protection shall not be shunted or cut out during the starting period.

505-25. Grounding and Bonding. Grounding and bonding shall comply with Article 250 and Section 501-16.

Article 510 — Hazardous (Classified) Locations — Specific

Contents

510-1. Scope. Articles 511 through 517 cover occupancies or parts of occupancies that are or may be hazardous because of atmospheric concentrations of flammable liquids, gases, or vapors, or because of deposits or accumulations of materials that may be readily ignitible.

510-2. General. The general rules of this *Code* and the provisions of Articles 500 through 504, shall apply to electric wiring and equipment in occupancies within the scope of Articles 511 through 517, except as such rules are modified in Articles 511 through 517. Where unusual conditions exist in a specific occupancy, the authority having jurisdiction shall judge with respect to the application of specific rules.

NFPA publishes a number of standards and recommended practices that provide requirements or guidance on the classification of hazardous locations in specific occupancies. Information and copies of standards may be obtained from the National Fire Protection Association (NFPA), 1 Batterymarch Park, P.O. Box 9101, Quincy, MA 02269-9101. If calling to order standards, the number is 1-800-344-3555. For information, call 617-770-3000.

Article 511 — Commercial Garages, Repair and Storage

Contents

511-1. Scope. These occupancies shall include locations used for service and repair operations in connection with self-propelled vehicles (including, but not limited to, passenger automobiles, buses, trucks, and tractors) in which volatile flammable liquids are used for fuel or power.

Article 100 defines *garage* as a building or portion of a building in which one or more self-propelled vehicles carrying volatile flammable liquid for fuel or power are kept for use, sale, storage, rental, repair, exhibition, or demonstrating purposes, and all that portion of a building that is on or below the floor or floors in which such vehicles are kept and that is not separated therefrom by suitable cutoffs.

A mechanical ventilating system that is capable of continuously providing at least six air changes per hour is required for all enclosed, basement, and underground parking garages.

Operations involving open flame or electric arcs, including fusion gas and electric welding, are required to be restricted to areas specifically provided for such purposes.

Approved suspended unit heaters may be used, provided they are located not less than 8 ft above the floor and are installed in accordance with the conditions of their approval.

For the requirements for locations used for the storage or repair of boats, see NFPA 303-1995, *Fire Protection Standard for Marinas and Boatyards.*

511-2. Locations. Areas in which flammable fuel is transferred to vehicle fuel tanks shall conform to Article 514. Parking garages used for parking or storage and where no repair work is done except exchange of parts and routine maintenance requiring no use of electrical equipment, open flame, welding, or the use of volatile flammable liquids are not classified.

FPN: For further information, see *Standard for Parking Structures,* NFPA 88A-1995, and *Standard for Repair Garages,* NFPA 88B-1997.

511-3. Class I Locations. Classification under Article 500.

(a) Up to a Level of 18 in. (457 mm) Above the Floor. For each floor, the entire area up to a level of 18 in. (457 mm) above the floor shall be considered to be a Class I, Division 2 location.

Exception: Where the enforcing agency determines that there is mechanical ventilation providing a minimum of four air changes per hour.

(b) Any Pit or Depression Below Floor Level. Any pit or depression below floor level shall be considered to be a Class I, Division 1 location, which shall extend up to said floor level, except that any pit or depression in which six air changes per hour are exhausted at the floor level of the pit shall be permitted to be judged by the enforcing agency to be a Class I, Division 2 location.

Exception: Lubrication and service rooms without dispensing shall be classified in accordance with Table 514-2.

The exception to Section 511-3(b) (which first appeared in the 1993 *Code*) provides guidance for determining the classification of facilities primarily offering oil and filter change and lubrication-type service, not the dispensing of fuel. Where the lower level work area of a lubritorium is provided with exhaust ventilation at a rate specified in Table 514-2 under "Lubrication or Service Room — Without Dispensing," the lower level is not a hazardous (classified) location. The referenced table in Article 514 is extracted from NFPA 30A-1996, *Automotive and Marine Service Station Code.*

(c) Areas Adjacent to Defined Locations or with Positive-Pressure Ventilation. Areas adjacent to defined locations in which flammable vapors are not likely to be released, such as stock rooms, switchboard rooms, and other similar locations, shall not be classified where mechanically ventilated at a rate of four or more air changes per hour or where effectively cut off by walls or partitions.

(d) Adjacent Areas by Special Permission. Adjacent areas that by reason of ventilation, air pressure differentials, or physical spacing are such that, in the opinion of the authority enforcing this *Code*, no ignition hazard exists, shall be unclassified.

(e) Fuel-Dispensing Units. Where fuel-dispensing units (other than liquid petroleum gas, which is prohibited) are located within buildings, the requirements of Article 514 shall govern.

Where mechanical ventilation is provided in the dispensing area, the controls shall be interlocked so that the dispenser cannot operate without ventilation as prescribed in Section 500-7(b).

See Figure 514-2 in the *Code* and commentary Figure 514.1.

(f) Portable Lighting Equipment. Portable lighting equipment shall be equipped with handle, lampholder, hook, and substantial guard attached to the lampholder or handle. All exterior surfaces that might come in contact with battery terminals, wiring terminals, or other objects shall be of nonconducting material or shall be effectively protected with insulation. Lampholders shall be of an unswitched type and shall not provide means for plug-in of attachment plugs. The outer shell shall be of molded composition or other suitable material. Unless the lamp and its cord are supported or arranged in such a manner that they cannot be used in the locations classified in Section 511-3, they shall be of a type approved for Class I, Division 1 locations.

Often, a reel-type portable hand lamp is used for supplemental lighting while performing vehicle service. See Figure 511.1 for an illustration of a cord reel assembly.

Where there is no transfer of flammable liquids to vehicle fuel tanks, but there is use of electrical diagnostic equipment, electrically operated tools or machinery, or open flames such as used for welding or cutting, the area must be classified in accordance with Section 511-3. See Figure 511.2. Fuel dispensing units located within the garage are governed by the requirements of Article 514.

The Class I, Division 2 location above grade within a commercial garage extends 18 in. above floor level, unless the authority having jurisdiction determines otherwise because mechanical ventilation provides at least four air changes per hour.

The Class I, Division 1 location below grade extends from the floor of the pit or depression to floor level, unless the authority having jurisdiction permits

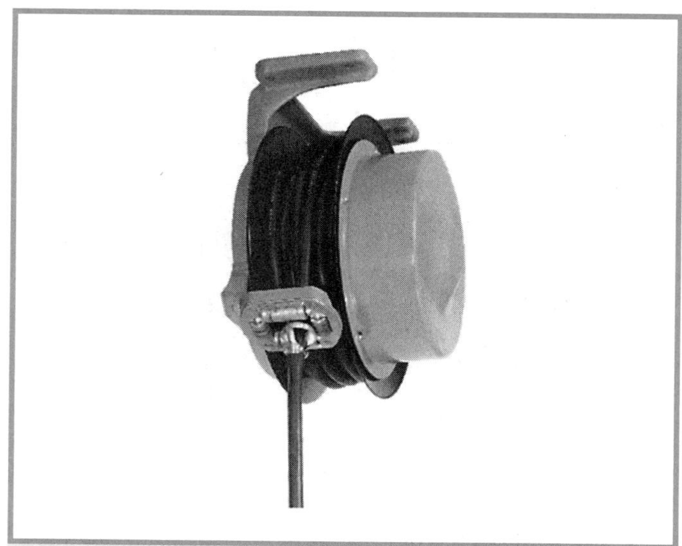

Figure 511.1 *This cord, which is part of a portable lamp assembly, is required to be arranged so that the lamp cannot be used in a Class I location. Otherwise, the lamp must be of a type approved for Class I, Division 1 hazardous locations (explosionproof). (Appleton Electric Co.)*

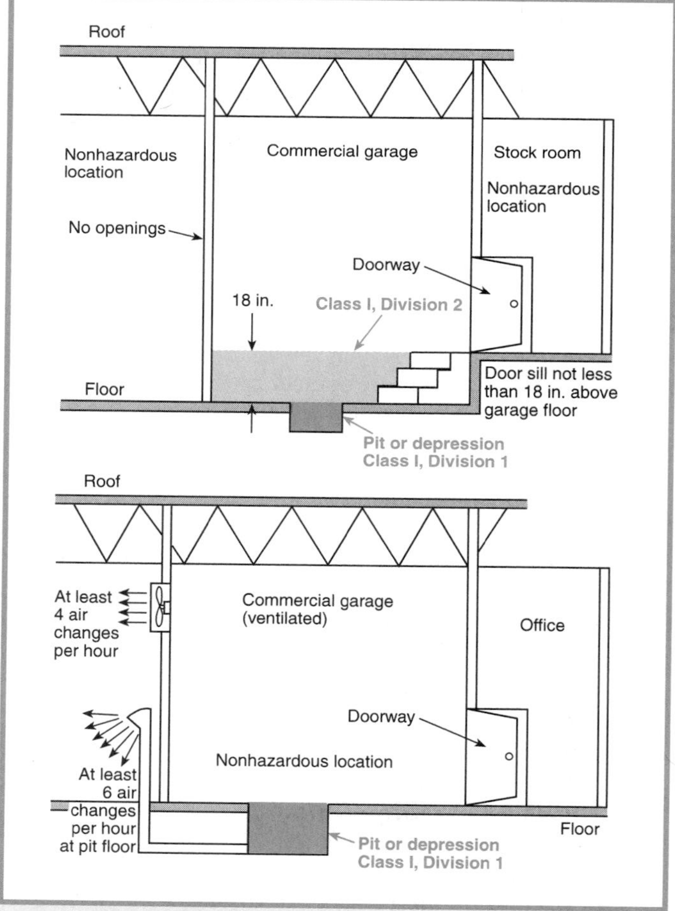

Figure 511.2 *Classification of locations in commercial garages.*

the pit or depression to be classified as Class I, Division 2 because ventilation providing at least six air changes per hour exhausts air at the floor level of the pit or depression. Areas suitably cut off and areas adjacent to nonhazardous ventilated garages are not classified hazardous.

511-4. Wiring and Equipment in Class I Locations. Within Class I locations as defined in Section 511-3, wiring and equipment shall conform to applicable provisions of Article 501. Raceways embedded in a masonry wall or buried beneath a floor shall be considered to be within the Class I location above the floor if any connections or extensions lead into or through such areas.

Section 511-4 applies to raceways located in walls and below floors of Class I locations in commercial garages. A raceway in a masonry wall or buried beneath a floor is not permitted to have any connections or extensions leading into or through a Class I, Division 1 or 2 location if it is to be considered located in a nonhazardous location and not subject to the provisions of Article 501. See Figure 511.3. Article 501 applies to the raceway if any part of it is not embedded in the wall or if the wall is not of masonry. Article 501 applies to the raceway if any part of it is not buried beneath the floor. The floor material is not specified.

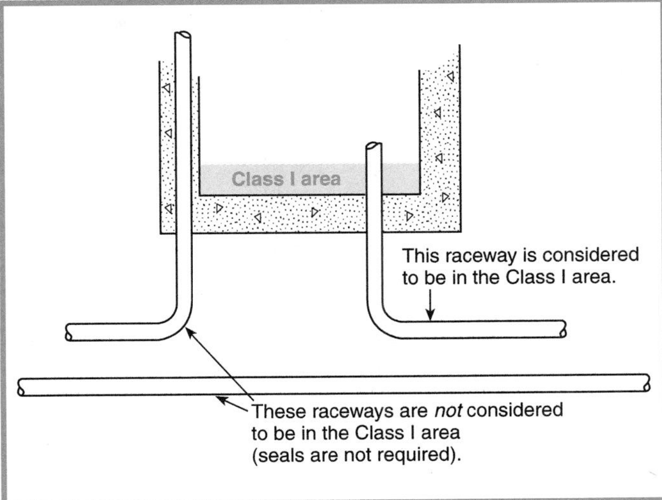

Figure 511.3 *Raceways embedded in a masonry wall or buried beneath the floor are not considered located in the Class I location only under conditions where there are no extensions into the Class I location.*

Exception: Rigid nonmetallic conduit that complies with Article 347 shall be permitted where buried under not less than 2 ft (610 mm) of cover. Where rigid nonmetallic conduit is used, threaded rigid metal conduit or threaded steel intermediate metal conduit shall be used for the last 2 ft (610 mm) of the underground run to emergence or to the point of connection to the aboveground raceway and an equipment grounding conductor shall be included to provide electrical continuity of the raceway system and for grounding of noncurrent-carrying metal parts.

The exception to Section 511-4 was new in the 1996 *Code*. It permits the use of rigid nonmetallic conduit underground in commercial garages. Underground conduit in similar facilities has been damaged by corrosion. Where nonmetallic raceways are used, the bonding and grounding requirements of Section 501-16 must be complied with.

511-5. Sealing. Approved seals conforming to the requirements of Section 501-5 shall be provided, and Section 501-5(b)(2) shall apply to horizontal as well as vertical boundaries of the defined Class I locations.

Seals are required if any part of the raceway is in, or passes through, a Class I, Division 2 location. See the commentary on seals in Section 501-5.

511-6. Wiring in Spaces Above Class I Locations.

For the installation of electrical wiring in areas above those designated by Section 511-3 as Class I locations, Section 511-6 permits a number of raceway and cable systems.

(a) Fixed Wiring Above Class I Locations. All fixed wiring above Class I locations shall be in metal raceways, rigid nonmetallic conduit, electrical nonmetallic tubing, flexible metal conduit, liquidtight flexible metal conduit, or liquidtight flexible nonmetallic conduit or shall be Type MC, MI, manufactured wiring systems, or PLTC cable in accordance with Article 725, or Type TC cable. Cellular metal floor raceways or cellular concrete floor raceways shall be permitted to be used only for supplying ceiling outlets or extensions to the area below the floor, but such raceways shall have no connections leading into or through any Class I location above the floor.

(b) Pendants. For pendants, flexible cord suitable for the type of service and approved for hard usage shall be used.

(c) Grounded and Grounding Conductors. Where a circuit supplies portables or pendants and includes a grounded conductor as provided in Article 200, receptacles, attachment plugs, connectors, and similar devices shall be of the grounding type, and the grounded conductor of the flexible cord shall be connected to the screw shell of any lampholder or to the grounded terminal of any utilization equipment supplied. Approved means shall be provided for maintaining continuity of the grounding conductor between the fixed wiring system and the noncurrent-carrying metal portions of pendant fixtures, portable lamps, and portable utilization equipment.

(d) Attachment Plug Receptacles. Attachment plug receptacles in a fixed position shall be located above the level of any defined Class I location or shall be approved for the location.

511-7. Equipment Above Class I Locations.

(a) Arcing Equipment. Equipment that is less than 12 ft (3.66 m) above the floor level and that may produce arcs, sparks, or particles of hot metal, such as cutouts, switches, charging panels, generators, motors, or other equipment (excluding receptacles, lamps, and lampholders) having make-and-break or sliding contacts, shall be of the totally enclosed type or constructed so as to prevent the escape of sparks or hot metal particles.

(b) Fixed Lighting. Lamps and lampholders for fixed lighting that is located over lanes through which vehicles are commonly driven or that may otherwise be exposed to physical damage shall be located not less than 12 ft (3.66 m) above floor level, unless of the totally enclosed type or constructed so as to prevent escape of sparks or hot metal particles.

511-8. Battery Charging Equipment. Battery chargers and their control equipment, and batteries being charged, shall not be located within locations classified in Section 511-3.

511-9. Electric Vehicle Charging.

(a) General. All electrical equipment and wiring shall be installed in accordance with Article 625, except as noted in Section 511-9(b) and (c). Flexible cords shall be of a type approved for extra-hard usage.

(b) Connector Location. No connector shall be located within a Class I location as defined in Section 511-3.

(c) Plug Connections to Vehicles. Where plugs are provided for direct connection to vehicles, the point of connection shall not be within a Class I location as defined in Section 511-3, and, where the cord is suspended from overhead, it shall be arranged so that the lowest point of sag is at least 6 in. (152 mm) above the floor. Where an automatic arrangement is provided to pull both cord and plug beyond the range of physical damage, no additional connector shall be required in the cable or at the outlet.

511-10. Ground-Fault Circuit-Interrupter Protection for Personnel. All 125-volt, single-phase, 15- and 20-ampere receptacles installed in areas where electrical diagnostic equipment, electrical hand tools, or portable lighting equipment are to be used shall have ground-fault circuit-interrupter protection for personnel.

Ground-fault circuit interrupters (GFCIs) intended to protect personnel from shock hazards are designed to trip when a ground-fault current of 5 milliamperes (plus or minus 1 mA) or greater is detected. Ground-fault protectors intended to protect equipment and prevent fires may trip at a much higher level. The GFCI necessary to comply with this requirement may be contained within the receptacle or may be a circuit-breaker-type GFCI.

See the definition of *ground-fault circuit interrupter* in Article 100 and see Figures 210.13 and 210.14.

511-16. Grounding. All metal raceways, the metal armor or metallic sheath on cables, and all noncurrent-carrying metal parts of fixed or portable electrical equipment, regardless of voltage, shall be grounded as provided in Article 250. Grounding in Class I locations shall comply with Section 501-16.

Article 513 — Aircraft Hangars

Contents

513-1. Scope. This article shall apply to buildings or structures inside any part of which aircraft containing Class I (flammable) liquids or Class II (combustible) liquids whose temperatures are above their flash points are housed or stored and in which aircraft might undergo service, repairs, or alterations. It shall not apply to locations used exclusively for aircraft that have never contained fuel or unfueled aircraft.

The scope of Article 513 appears as a new section, but is actually a revision of the 1996 *Code,* Section 513-1. It clarifies that it is not necessary to classify areas in which aircraft fuel is a Class II combustible liquid, unless the liquid is going to be used or stored above its flashpoint.

For further information, see NFPA 409-1995, *Standard on Aircraft Hangars.*

FPN No. 1: For definitions of aircraft hangar and unfueled aircraft, see *Standard on Aircraft Hangars,* NFPA 409-1995.

FPN No. 2: For further information on fuel classification, see *Flammable and Combustible Liquids Code,* NFPA 30-1996, and *Guide to Fire Hazard Properties of Flammable Liquids, Gases, and Volatile Solids,* NFPA 325-1994.

513-2. Definitions. For the purpose of this article, the following definitions shall apply.

The definitions in Section 513-2 apply only to the portable and mobile equipment covered in Article 513.

Mobile Equipment. Equipment with electric components suitable to be moved only with mechanical aids or which are provided with wheels for movements by person(s) or powered devices.

Portable Equipment. Equipment with electric components suitable to be moved by a single person without mechanical aids.

513-3. Classification of Locations.

(a) Below Floor Level. Any pit or depression below the level of the hangar floor shall be classified as a Class I, Division 1 location that shall extend up to said floor level.

(b) Areas Not Cut Off or Ventilated. The entire area of the hangar, including any adjacent and communicating areas not suitably cut off from the hangar, shall be classified as a Class I, Division 2 location up to a level 18 in. (457 mm) above the floor.

(c) Vicinity of Aircraft. The area within 5 ft (1.52 m) horizontally from aircraft power plants or aircraft fuel tanks shall be classified as a Class I, Division 2 location that shall extend upward from the floor to a level 5 ft (1.52 m) above the upper surface of wings and of engine enclosures.

In order to properly classify the area in accordance with Section 513-3(c), it is necessary to obtain information on the aircraft parking patterns, the types of aircraft, and the operations to be performed in the hangar. See Figure 513.1 for area classification in aircraft hangars.

Consideration of future changes in aircraft types and locations is appropriate to avoid the need for costly wiring and equipment alterations as a result of changes in the area classification.

(d) Areas Suitably Cut Off and Ventilated. Adjacent areas in which flammable liquids or vapors are not likely to be released, such as stock rooms, electrical control rooms, and other similar locations, shall not be classified where adequately ventilated and where effectively cut off from the hangar itself by walls or partitions.

513-4. Wiring and Equipment in Class I Locations. All wiring and equipment that is or may be installed or operated within any of the Class I locations defined in Section 513-3 shall comply with the applicable provisions of Article 501. All wiring installed in or under the hangar floor shall comply with the requirements for Class I, Division 1 locations. Where such wiring is located in vaults, pits, or ducts, adequate drainage shall be provided.

Attachment plugs and receptacles in Class I locations shall be approved for Class I locations or shall be designed so that they cannot be energized while the connections are being made or broken.

513-5. Wiring Not Within Class I Locations.

(a) Fixed Wiring. All fixed wiring in a hangar, but not within a Class I location as defined in Section 513-3, shall be installed in metal raceways or shall be Type MI, TC, or MC cable.

Exception: Wiring in unclassified locations, as defined in Section 513-3(d), shall be of a type recognized in Chapter 3.

(b) Pendants. For pendants, flexible cord suitable for the type of service and approved for hard usage shall be used. Each such cord shall include a separate equipment grounding conductor.

(c) Portable Equipment. For portable utilization equipment and lamps, flexible cord suitable for the type of service and approved for extra-hard usage shall be used. Each such cord shall include a separate equipment grounding conductor.

(d) Grounded and Grounding Conductors. Where a circuit supplies portables or pendants and includes a grounded conductor as provided in Article 200, receptacles, attachment plugs, connectors, and similar devices shall be of the grounding type, and the grounded conductor of the flexible cord shall be connected to the screw shell of any lampholder or to the grounded terminal of any utilization equipment supplied. Approved means shall be provided for maintaining continuity of the grounding conductor between the fixed wiring system and the noncurrent-carrying metal portions

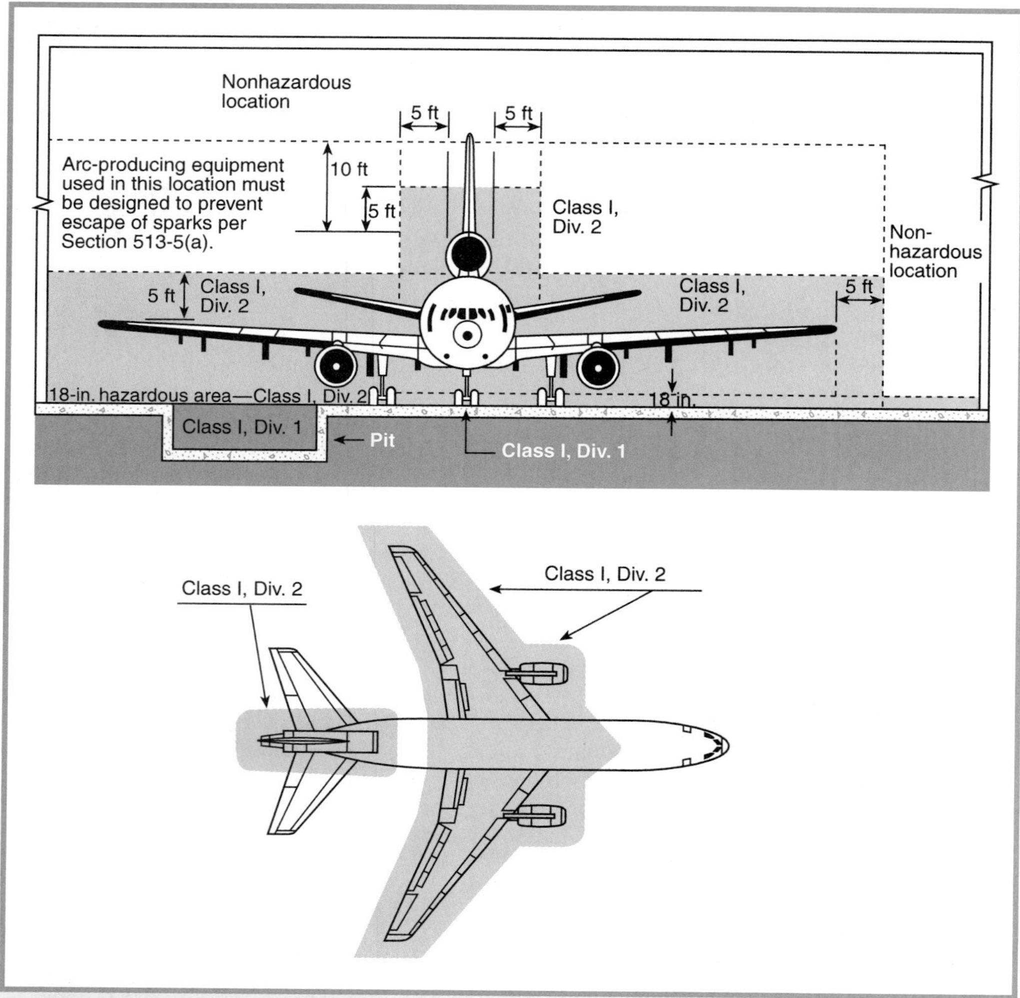

Figure 513.1 Area classification in aircraft hangars.

of pendant fixtures, portable lamps, and portable utilization equipment.

513-6. Equipment Not Within Class I Locations.

(a) Arcing Equipment. In locations other than those described in Section 513-3, equipment that is less than 10 ft (3.05 m) above wings and engine enclosures of aircraft and that may produce arcs, sparks, or particles of hot metal, such as lamps and lampholders for fixed lighting, cutouts, switches, receptacles, charging panels, generators, motors, or other equipment having make-and-break or sliding contacts, shall be of the totally enclosed type or constructed so as to prevent the escape of sparks or hot metal particles.

Exception: Equipment in areas described in Section 513-3(d) shall be permitted to be of the general-purpose type.

(b) Lampholders. Lampholders of metal-shell, fiber-lined types shall not be used for fixed incandescent lighting.

(c) Portable Lighting Equipment. Portable lighting equipment that are used within a hangar shall be approved for the location in which they are used.

(d) Portable Equipment. Portable utilization equipment that is or may be used within a hangar shall be of a type suitable for use in Class I, Division 2 locations.

513-7. Stanchions, Rostrums, and Docks.

(a) In Class I Location. Electric wiring, outlets, and equipment (including lamps) on or attached to stanchions, rostrums, or docks that are located or likely to be located in a Class I location, as defined in Section 513-3(c), shall comply with the requirements for Class I, Division 2 locations.

(b) Not in Class I Location. Where stanchions, rostrums, or docks are not located or likely to be located in a Class I location, as defined in Section 513-3(c), wiring and equipment shall comply with Sections 513-5 and 513-6, except that such wiring and equipment not more than 18 in.

(457 mm) above the floor in any position shall comply with Section 513-7(a). Receptacles and attachment plugs shall be of a locking type that will not readily disconnect.

(c) Mobile Type. Mobile stanchions with electric equipment complying with (b) above shall carry at least one permanently affixed warning sign to read

WARNING — KEEP 5 FT CLEAR OF AIRCRAFT ENGINES AND FUEL TANK AREAS.

513-8. Sealing. Approved seals shall be provided in accordance with Section 501-5. Sealing requirements specified in Sections 501-5(a)(4) and (b)(2) shall apply to horizontal as well as to vertical boundaries of the defined Class I locations. Raceways embedded in a concrete floor or buried beneath a floor shall be considered to be within the Class I location above the floor.

513-9. Aircraft Electrical Systems. Aircraft electrical systems shall be de-energized when the aircraft is stored in a hangar and, whenever possible, while the aircraft is undergoing maintenance.

513-10. Aircraft Batteries — Charging and Equipment. Aircraft batteries shall not be charged where installed in an aircraft located inside or partially inside a hangar.

Battery chargers and their control equipment shall not be located or operated within any of the Class I locations defined in Section 513-3 and shall preferably be located in a separate building or in an area such as defined in Section 513-3(d). Mobile chargers shall carry at least one permanently affixed warning sign to read "WARNING — KEEP 5 FT CLEAR OF AIRCRAFT ENGINES AND FUEL TANK AREAS." Tables, racks, trays, and wiring shall not be located within a Class I location and, in addition, shall comply with Article 480.

513-11. External Power Sources for Energizing Aircraft.

(a) Not Less than 18 in. (457 mm) Above Floor. Aircraft energizers shall be designed and mounted so that all electric equipment and fixed wiring will be at least 18 in. (457 mm) above floor level and shall not be operated in a Class I location as defined in Section 513-3(c).

(b) Marking for Mobile Units. Mobile energizers shall carry at least one permanently affixed warning sign to read

WARNING — KEEP 5 FT CLEAR OF AIRCRAFT ENGINES AND FUEL TANK AREAS.

(c) Cords. Flexible cords for aircraft energizers and ground support equipment shall be approved for the type of service and extra-hard usage and shall include an equipment grounding conductor.

513-12. Mobile Servicing Equipment with Electric Components.

(a) General. Mobile servicing equipment (such as vacuum cleaners, air compressors, air movers, etc.) having electric wiring and equipment not suitable for Class I, Division 2 locations shall be designed and mounted so that all such fixed wiring and equipment will be at least 18 in. (457 mm) above the floor. Such mobile equipment shall not be operated within the Class I location defined in Section 513-3(c) and shall carry at least one permanently affixed warning sign to read

WARNING — KEEP 5 FT CLEAR OF AIRCRAFT ENGINES AND FUEL TANK AREAS.

(b) Cords and Connectors. Flexible cords for mobile equipment shall be suitable for the type of service and approved for extra-hard usage and shall include an equipment grounding conductor. Attachment plugs and receptacles shall be approved for the location in which they are installed and shall provide for connection of the equipment grounding conductor.

(c) Restricted Use. Equipment that is not identified as suitable for Class I, Division 2 locations shall not be operated in locations where maintenance operations likely to release flammable liquids or vapors are in progress.

513-16. Grounding. All metal raceways, the metal armor or metallic sheath on cables, and all noncurrent-carrying metal parts of fixed or portable electrical equipment, regardless of voltage, shall be grounded as provided in Article 250. Grounding in Class I locations shall comply with Section 501-16.

Article 514 — Gasoline Dispensing and Service Stations

Contents

514-1. Definition. A *gasoline dispensing and service station* is a location where gasoline or other volatile flammable liquids or liquefied flammable gases are transferred to the fuel tanks (including auxiliary fuel tanks) of self-propelled vehicles or approved containers.

Other areas used as lubritoriums, service rooms, repair rooms, offices, salesrooms, compressor rooms, and similar locations shall comply with Articles 510 and 511 with respect to electric wiring and equipment.

Where the authority having jurisdiction can satisfactorily

determine that flammable liquids having a flash point below 38°C (100°F), such as gasoline, will not be handled, such location shall not be required to be classified.

The filling of approved containers was included in the scope of Article 514 in the 1993 *Code*. Article 514 applies to dispensing locations for liquefied petroleum gas (LPG), even if no vehicles are served.

FPN No. 1: For further information regarding safeguards for gasoline dispensing and service stations, see *Automotive and Marine Service Station Code,* NFPA 30A-1996.

FPN No. 2: For information on classified areas pertaining to LP-Gas systems other than residential or commercial, see *Standard for the Storage and Handling of Liquefied Petroleum Gases,* NFPA 58-1995, and *Standard for the Storage and Handling of Lique-*

ˣTable 514-2. Class I Locations — Service Stations

Location	Class I, Group D Division	Extent of Classified Location
Underground Tank		
Fill Opening	1	Any pit, box, or space below grade level, any part of which is within the Division 1 or 2 classified location.
	2	Up to 18 in. above grade level within a horizontal radius of 10 ft from a loose fill connection and within a horizontal radius of 5 ft from a tight fill connection.
Vent — Discharging Upward	1	Within 3 ft of open end of vent, extending in all directions.
	2	Space between 3 ft and 5 ft of open end of vent, extending in all directions.
Dispensing Device[1,4]		
(except overhead type)[2]		
Pits	1	Any pit, box, or space below grade level, any part of which is within the Division 1 or 2 classified location.
Dispenser		FPN: Space classification inside the dispenser enclosure is covered in *Power Operated Dispensing Devices for Petroleum Products,* ANSI/UL 87-1995.
	2	Within 18 in. horizontally in all directions extending to grade from the dispenser enclosure or that portion of the dispenser enclosure containing liquid handling components.
		FPN: Space classification inside the dispenser enclosure is covered in *Power Operated Dispensing Devices for Petroleum Products,* ANSI/UL 87-1995.
Outdoor	2	Up to 18 in. above grade level within 20 ft horizontally of any edge of enclosure.
Indoor		
with Mechanical Ventilation	2	Up to 18 in. above grade or floor level within 20 ft horizontally of any edge of enclosure.
with Gravity Ventilation	2	Up to 18 in. above grade or floor level within 25 ft horizontally of any edge of enclosure.
Dispensing Device[4]		
Overhead Type[2]	1	The space within the dispenser enclosure, and all electrical equipment integral with the dispensing hose or nozzle.
	2	A space extending 18 in. horizontally in all directions beyond the enclosure and extending to grade.
	2	Up to 18 in. above grade level within 20 ft horizontally measured from a point vertically below the edge of any dispenser enclosure.
Remote Pump — Outdoor	1	Any pit, box, or space below grade level if any part is within a horizontal distance of 10 ft from any edge of pump.
	2	Within 3 ft of any edge of pump, extending in all directions. Also up to 18 in. above grade level within 10 ft horizontally from any edge of pump.

Table 514-2. Continued

Location	Class I, Group D Division	Extent of Classified Location
Remote Pump — Indoor	1	Entire space within any pit.
	2	Within 5 ft of any edge of pump, extending in all directions. Also up to 3 ft above grade level within 25 ft horizontally from any edge of pump.
Lubrication or Service Room — with Dispensing	1	Any pit within any unventilated space.
	2	Any pit with ventilation.
	2	Space up to 18 in. above floor or grade level and 3 ft horizontally from a lubrication pit.
Dispenser for Class I Liquids	2	Within 3 ft of any fill or dispensing point, extending in all directions.
Lubrication or Service Room — Without Dispensing	2	Entire area within any pit used for lubrication or similar services where Class I liquids may be released.
	2	Area up to 18 in. above any such pit, and extending a distance of 3 ft horizontally from any edge of the pit.
	2	Entire unventilated area within any pit, below grade area, or subfloor area.
	2	Area up to 18 in. above any such unventilated pit, below grade work area, or subfloor work area and extending a distance of 3 ft horizontally from the edge of any such pit, belowgrade work area, or subfloor work area.
	Nonclassified	Any pit, belowgrade work area, or subfloor work area that is provided with exhaust ventilation at a rate of not less than 1 cfm/ft^2 (0.3 m^3/minute/m^2) of floor area at all times that the building is occupied or when vehicles are parked in or over this area and where exhaust air is taken from a point within 12 in. (0.3 m) of the floor of the pit, belowgrade work area, or subfloor work area.
Special Enclosure Inside Building[3]	1	Entire enclosure.
Sales, Storage, and Rest Rooms	Nonclassified	If there is any opening to these rooms within the extent of a Division 1 location, the entire room shall be classified as Division 1.
Vapor Processing Systems Pits	1	Any pit, box, or space below grade level, any part of which is within a Division 1 or 2 classified location or that houses any equipment used to transport or process vapors.
Vapor Processing Equipment Located Within Protective Enclosures FPN: See *Automotive and Marine Service Station Code,* NFPA 30A-1996, Section 4-5.7	2	Within any protective enclosure housing vapor processing equipment.
Vapor Processing Equipment Not Within Protective Enclosures (excluding piping and combustion devices)	2	The space within 18 in. in all directions of equipment containing flammable vapor or liquid extending to grade level. Up to 18 in. above grade level within 10 ft horizontally of the vapor processing equipment.
Equipment Enclosures	1	Any space within the enclosure where vapor or liquid is present under normal operating conditions.
Vacuum-Assist Blowers	2	The space within 18 in. in all directions extending to grade level. Up to 18 in. above grade level within 10 ft horizontally.

Note: For SI units, 1 in. = 2.5 cm; 1 ft = 0.3048 m.

[1]Refer to Figure 514-2 for an illustration of classified location around dispensing devices.

[2]Ceiling mounted hose reel.

[3]FPN: See *Automotive and Marine Service Station Code,* NFPA 30A-1996, Section 2-2.

[4]FPN: Area classification inside the dispenser enclosure is covered in *Power-Operated Dispensing Devices for Petroleum Products,* ANSI/UL 87-1995.

fied Petroleum Gases at Utility Gas Plants, NFPA 59-1995.

FPN No. 3: See Section 555-10 for gasoline dispensing stations in marinas and boatyards.

x514-2. Class I Locations. Table 514-2 shall be applied where Class I liquids are stored, handled, or dispensed and shall be used to delineate and classify service stations. A Class I location shall not extend beyond an unpierced wall, roof, or other solid partition.

Table 514-2 was extracted from NFPA 30A-1996, *Automotive and Marine Service Station Code.* See the commentary following Section 511-3(b), Exception, regarding the classification of facilities that primarily offer lubrication-type service. See Figure 514.1.

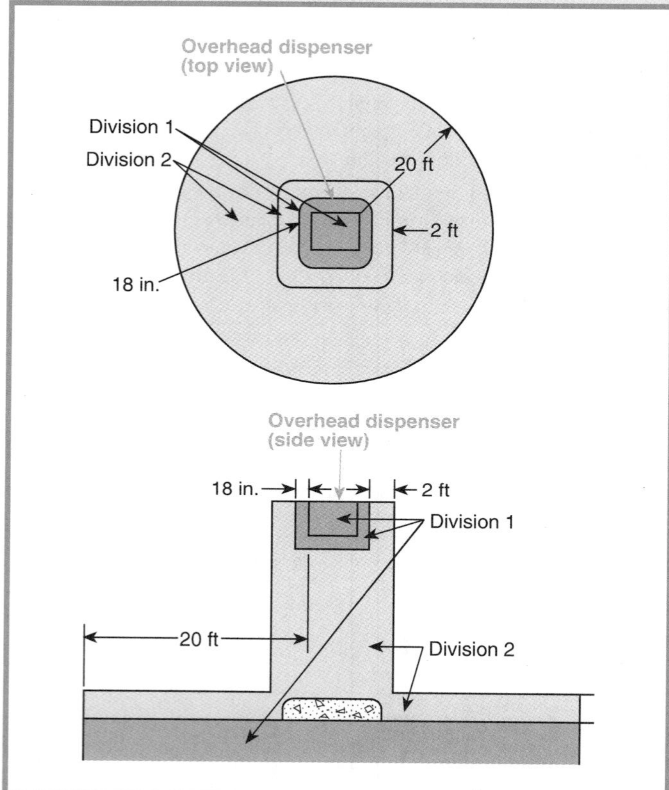

Figure 514.1 *Extent of Class I location around overhead gasoline dispensing units, in accordance with Table 514-2 in NFPA 30A.*

FPN: For information on area classification where liquefied petroleum gases are dispensed, see *Standard for the Storage and Handling of Liquefied Petroleum Gases,* NFPA 58-1995.

514-3. Wiring and Equipment Within Class I Locations. All electrical equipment and wiring within Class I locations defined in Section 514-2 shall comply with the applicable provisions of Article 501.

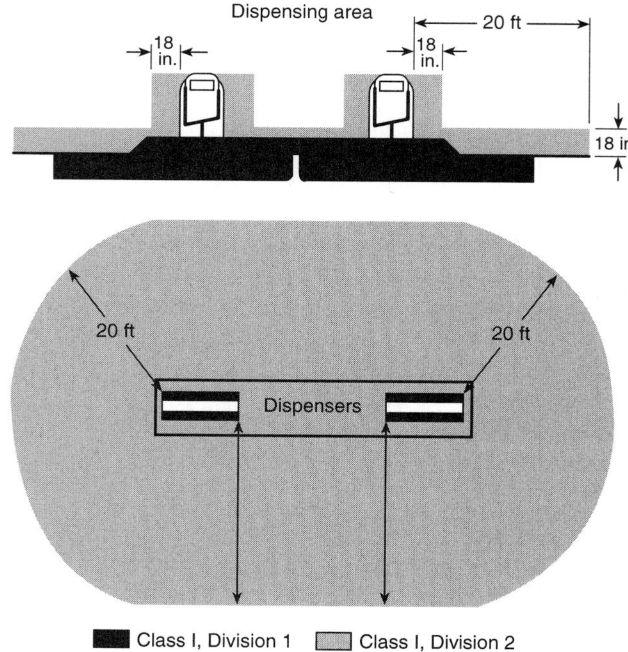

xFigure 514-2 Classified locations adjacent to dispensers as detailed in Table 514-2.

For gasoline- and oil-resistant insulated conductors, see the commentary following Section 501-13.

Exception: As permitted in Section 514-8.

FPN: For special requirements for conductor insulation, see Section 501-13.

514-4. Wiring and Equipment Above Class I Locations. Wiring and equipment above the Class I locations defined in Section 514-2 shall comply with Sections 511-6 and 511-7.

514-5. Circuit Disconnects.

(a) General. Each circuit leading to or through dispensing equipment, including equipment for remote pumping systems, shall be provided with a clearly identified and readily accessible switch or other acceptable means, located remote from the dispensing devices, to disconnect simultaneously from the source of supply, all conductors of the circuit, including the grounded conductor, if any.

Single-pole breakers utilizing handle ties shall not be permitted.

For 1996, Section 514-5(a) was revised to require that the disconnecting means be clearly marked and readily accessible. The disconnecting means is also required to be remote from the dispensing device, so that if an emergency requires rapid disconnection, the person operating the disconnecting means will not be exposed to the hazard.

It is important to note that all conductors of a circuit, including the grounded conductor, that may be present within a dispensing device are required to be provided with a switch or special-type circuit breaker that will simultaneously disconnect all conductors. Handle ties on single-pole circuit breakers are not permitted. The intent is that no energized conductors be in the dispenser vicinity during maintenance or alteration. Considering possible accidental reversal of the polarities of conductors at panelboards, the grounded conductor must be able to be switched to the open or "off" position. Grounded conductors may be present in old-style pump motors, or they may pass through a dispenser as part of a circuit for the island lighting.

[x]**(b) Attended Self-Service Stations.** Emergency controls as specified in Section 514-5(a) shall be installed at a location acceptable to the authority having jurisdiction, but controls shall not be more than 100 ft (30 m) from dispensers.

[x]**(c) Unattended Self-Service Stations.** Emergency controls as specified in Section 514-5(a) shall be installed at a location acceptable to the authority having jurisdiction, but the controls shall be more than 20 ft (7 m) but less than 100 ft (30 m) from the dispensers. Additional emergency controls shall be installed on each group of dispensers or the outdoor equipment used to control the dispensers. Emergency controls shall shut off all power to all dispensing equipment at the station. Controls shall be manually reset only in a manner approved by the authority having jurisdiction.

FPN: For additional information, see 9-4.5 and 9-5.3 of *Automotive and Marine Service Station Code,* NFPA 30A1996.

Since a fire or large gasoline spill at the dispensing island may make it impossible to operate the switches on the dispensing island that shut off the flow of gasoline, Section 4-1.2 of NFPA 30A-1996, *Automotive and Marine Service Station Code,* requires an easily accessible and clearly identified emergency power cutoff to be provided at a location remote from the dispensing device. Sections 514-5(b) and 514-5(c), added in the 1993 *Code,* are extracted from NFPA 30A and provide the maximum and minimum distances for the location of the emergency control for attended and unattended self-service stations. The term *clearly identified* means that a sign is to be posted indicating where the cutoff switch is located. This emergency power cutoff should be readily accessible and not blocked by the storage of such things as tires or cases of lubricating oil. All service station operators as well as responding fire fighters should know the location of the emergency power cutoff.

514-6. Provisions for Maintenance and Service of Dispensing Equipment. Each dispensing device shall be provided with a means to remove all external voltage sources, including feedback, during periods of maintenance and service of the dispensing equipment.

A new Section 514-6, Provisions for Maintenance and Service of Dispensing Equipment, was added to the 1999 *Code.* It requires a means of removing all external voltages sources from each dispensing device, including sources that may backfeed into the dispenser. This will provide additional safety to personnel servicing the equipment.

514-7. Sealing.

(a) At Dispenser. An approved seal shall be provided in each conduit run entering or leaving a dispenser or any cavities or enclosures in direct communication therewith. The sealing fitting shall be the first fitting after the conduit emerges from the earth or concrete.

(b) At Boundary. Additional seals shall be provided in accordance with Section 501-5. Sections 501-5(a)(4) and (b)(2) shall apply to horizontal as well as to vertical boundaries of the defined Class I locations.

Sealing fittings are required in all conduits leaving a Class I location. All conduits passing under the boundaries of the hazardous (classified) locations (20-ft radius from dispenser) or the tank fill-pipe (10-ft radius from a loose-fill connection and 5-ft radius from a tight-fill connection) are considered in a Class I, Division 1 location (see Section 514-8), and the seal is to be the first fitting at the point of emergence. A seal is required to be provided in each conduit run entering or leaving a dispenser; therefore, even though a conduit runs from dispenser to dispenser and does not leave the hazardous (classified) location, a seal is necessary where leaving, and again where entering, the dispenser. Panelboards are generally located in a room classified as a nonhazardous location; however, any conduit coming from the dispenser or passing under the hazardous (classified) location boundaries from the dispenser or tank fill-pipe would require a seal at the panelboard location to minimize the likelihood of gas migration into the remote location. Where the panelboard is located in the lube or repair room, all conduits emerging into the 18-in. hazardous (classified) location would require seals. See Figures 514.2 and 514.3.

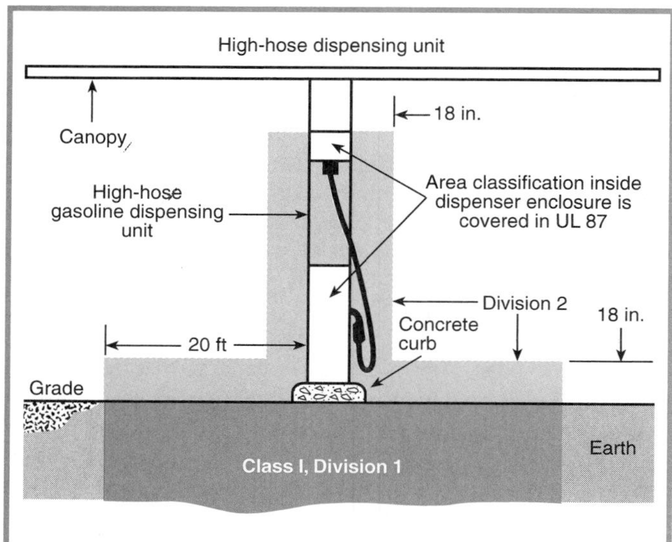

Figure 514.2 A gasoline dispensing installation indicating locations for sealing fittings. Emergency controls are required for self-service stations. See Sections 514-5(a) and (b).

514-8. Underground Wiring. Underground wiring shall be installed in threaded rigid metal conduit or threaded steel intermediate metal conduit. Any portion of electrical wiring or equipment that is below the surface of a Class I, Division 1 or Division 2 location (as defined in Table 514-2) shall be considered to be in a Class I, Division 1 location, which shall extend at least to the point of emergence above grade. Refer to Table 300-5.

Experience has shown that the fuel spilled in the vicinity of gasoline pumps in service stations tends to accumulate underground, where it can enter electrical conduits and accumulate in voids. Therefore, Section 514-8 classifies the space below surface areas subject to fuel spills as Class I, Division 1 locations.

Exception No. 1: Type MI cable shall be permitted where it is installed in accordance with Article 330.

Exception No. 2: Rigid nonmetallic conduit complying with Article 347 shall be permitted where buried under not less than 2 ft (610 mm) of cover. Where rigid nonmetallic conduit is used, threaded rigid metal conduit or threaded steel intermediate metal conduit shall be used for the last 2 ft (610 mm) of the underground run to emergence or to the point of connection to the aboveground raceway, and an equipment grounding conductor shall be included to provide electrical continuity of the raceway system and for grounding of non-current-carrying metal parts.

Exception No. 2 to Section 514-8 makes it clear that if rigid nonmetallic conduit is used for underground wiring, threaded rigid metal conduit or threaded steel intermediate metal conduit must be used for the last 2 ft of the underground run to the point of emergence or to the point of connection to the aboveground raceway. The rigid nonmetallic conduit, including rigid nonmetallic conduit elbows and fittings, is required to be located not less than 2 ft below grade. See Figure 514.4.

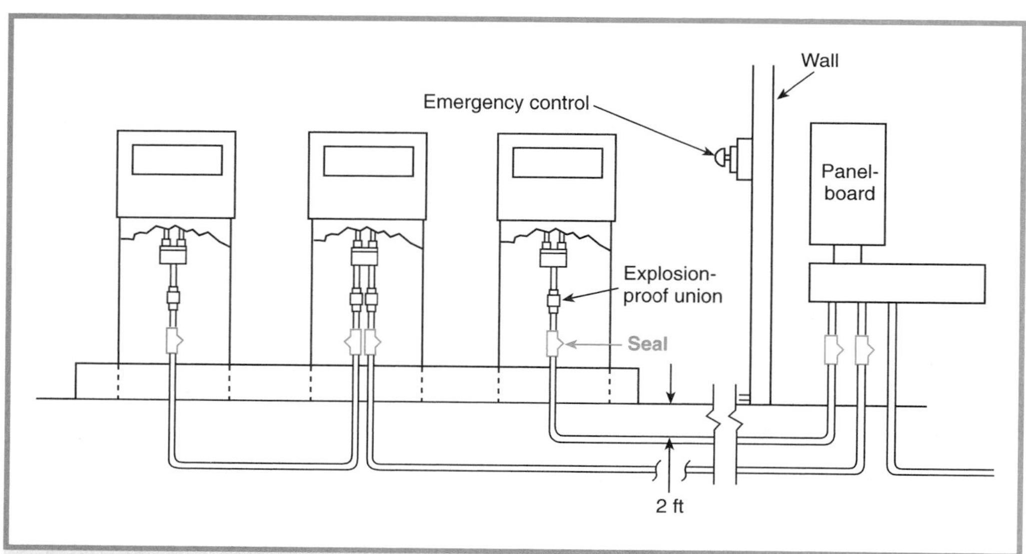

Figure 514.3 Seals are required at points marked "S." Seals are not required at the sign and two of the lights because conduit runs do not pass through a hazardous location.

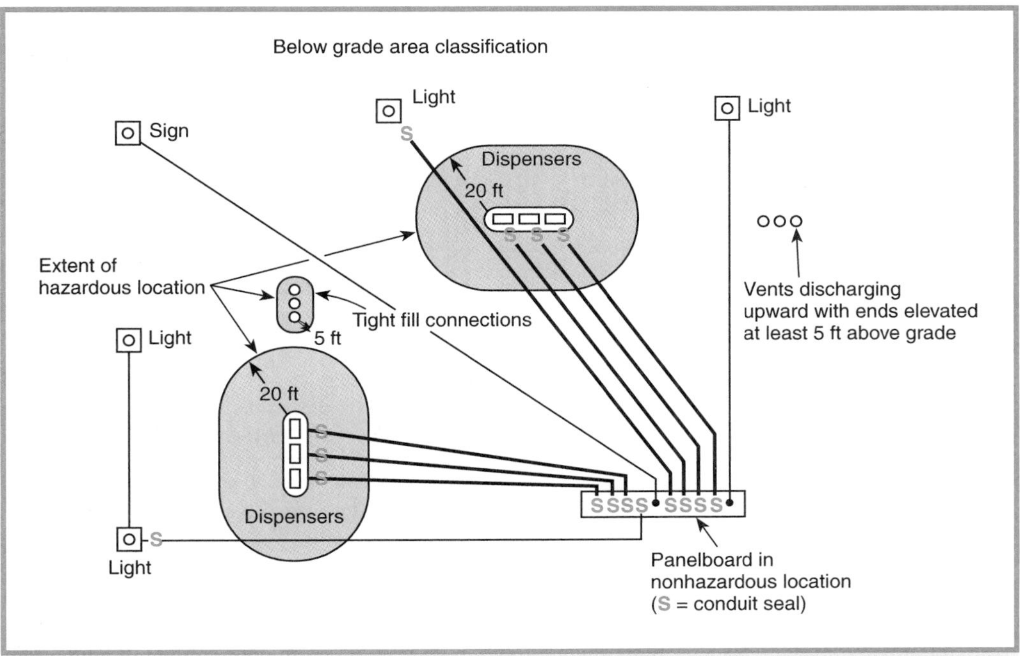

Figure 514.4 *Use of rigid nonmetallic conduit in accordance with Exception No. 2 to Section 514-8.*

If rigid nonmetallic conduit is used, an equipment grounding conductor must be included and must be bonded to the explosionproof raceway system inside the dispenser. This is accomplished by terminating the equipment grounding conductor on the ground screw (or other means) provided in the dispenser junction box.

514-16. Grounding. All metal raceways, the metal armor or metallic sheath on cables, and all noncurrent-carrying metal parts of fixed or portable electrical equipment, regardless of voltage, shall be grounded as provided in Article 250. Grounding in Class I locations shall comply with Section 501-16.

Article 515 — Bulk Storage Plants

Contents

515-1. Definition. A *bulk storage plant* is that portion of a property where flammable liquids are received by tank vessel, pipelines, tank car, or tank vehicle, and are stored or blended in bulk for the purpose of distributing such liquids by tank vessel, pipeline, tank car, tank vehicle, portable tank, or container.

FPN: For further information, see *Flammable and Combustible Liquids Code,* NFPA 30-1996.

[x]**515-2. Class I Locations.** Table 515-2 shall be applied where Class I liquids are stored, handled, or dispensed and shall be used to delineate and classify bulk storage plants. The Class I location shall not extend beyond a floor, wall, roof, or other solid partition that has no communicating openings.

FPN: The area classifications listed in Table 515-2 are based on the premise that the installation meets the applicable requirements of *Flammable and Combustible Liquids Code,* NFPA 30-1996, Chapter 5, in all respects. Should this not be the case, the authority having jurisdiction has the authority to classify the extent of the classified space.

Table 515-2 is essentially the same as Table 5-9.5.3 in NFPA 30-1996, *Flammable and Combustible Liquids*

[x]Table 515-2. Class I Locations — Bulk Plants

Location	Class I, Division	Extent of Classified Location
Indoor equipment installed in accordance with *Flammable and Combustible Liquids Code,* NFPA 30-1996, Section 5-3.4.5, where flammable vapor-air mixtures may exist under normal operation	1	Space within 5 ft of any edge of such equipment, extending in all directions.
	2	Space between 5 ft and 8 ft of any edge of such equipment, extending in all directions. Also, space up to 3 ft above floor or grade level within 5 ft to 25 ft horizontally from any edge of such equipment.[1]
Outdoor equipment of the type covered in *Flammable and Combustible Liquids Code,* NFPA 30-1996, Section 5-3.4.5, where flammable vapor-air mixtures may exist under normal operation	1	Space within 3 ft of any edge of such equipment, extending in all directions.
	2	Space between 3 ft and 8 ft of any edge of such equipment, extending in all directions. Also, space up to 3 ft above floor or grade level within 3 ft to 10 ft horizontally from any edge of such equipment.
Tank — Aboveground[2]	1	Space inside dike where dike height is greater than the distance from the tank to the dike for more than 50 percent of the tank circumference.
Shell, Ends, or Roof and Dike Space	2	Within 10 ft from shell, ends, or roof of tank. Space inside dikes to level of top of dike.
Vent	1	Within 5 ft of open end of vent, extending in all directions.
	2	Space between 5 ft and 10 ft from open end of vent, extending in all directions.
Floating Roof	1	Space above the roof and within the shell.
Underground Tank Fill Opening	1	Any pit, box, or space below grade level, if any part is within a Division 1 or 2 classified location.
	2	Up to 18 in. above grade level within a horizontal radius of 10 ft from a loose fill connection, and within a horizontal radius of 5 ft from a tight fill connection.
Vent — Discharging Upward	1	Within 3 ft of open end of vent, extending in all directions.
	2	Space between 3 ft and 5 ft of open end of vent, extending in all directions.
Drum and Container Filling	1	Within 3 ft of vent and fill openings, extending in all directions.
Outdoors, or Indoors with Adequate Ventilation	2	Space between 3 ft and 5 ft from vent or fill opening, extending in all directions. Also, up to 18 in. above floor or grade level within a horizontal radius of 10 ft from vent or fill openings.
Pumps, Bleeders, Withdrawal Fittings, Meters, and Similar Devices		
Indoors	2	Within 5 ft of any edge of such devices, extending in all directions. Also up to 3 ft above floor or grade level within 25 ft horizontally from any edge of such devices.
Outdoors	2	Within 3 ft of any edge of such devices, extending in all directions. Also up to 18 in. above grade level within 10 ft horizontally from any edge of such devices.
Pits		
Without Mechanical Ventilation	1	Entire space within pit if any part is within a Division 1 or 2 classified location.
With Adequate Mechanical Ventilation	2	Entire space within pit if any part is within a Division 1 or 2 classified location.
Containing Valves, Fittings, or Piping, and Not Within a Division 1 or 2 Classified Location	2	Entire pit.

[x]**Table 515-2. (Continued)**

Location	Class I, Division	Extent of Classified Location
Drainage Ditches, Separators, Impounding Basins		
Outdoor	2	Space up to 18 in. above ditch, separator, or basin. Also up to 18 in. above grade within 15 ft horizontally from any edge.
Indoor		Same as pits.
Tank Vehicle and Tank Car[3]		
Loading Through Open Dome	1	Within 3 ft of edge of dome, extending in all directions.
	2	Space between 3 ft and 15 ft from edge of dome, extending in all directions.
Loading through Bottom Connections with Atmospheric Venting	1	Within 3 ft of point of venting to atmosphere, extending in all directions.
	2	Space between 3 ft and 15 ft from point of venting to atmosphere, extending in all directions. Also up to 18 in. above grade within a horizontal radius of 10 ft from point of loading connection.
Office and Rest Rooms	Unclassified	If there is any opening to these rooms within the extent of an indoor classified location, the room shall be classified the same as if the wall, curb, or partition did not exist.
Loading through Closed Dome with Atmospheric Venting	1	Within 3 ft of open end of vent, extending in all directions.
	2	Space between 3 ft and 15 ft from open end of vent, extending in all directions. Also within 3 ft of edge of dome, extending in all directions.
Loading through Closed Dome with Vapor Control	2	Within 3 ft of point of connection of both fill and vapor lines, extending in all directions.
Bottom Loading with Vapor Control Any Bottom Unloading	2	Within 3 ft of point of connections, extending in all directions. Also up to 18 in. above grade within a horizontal radius of 10 ft from point of connections.
Storage and Repair Garage for Tank Vehicles	1	All pits or spaces below floor level.
	2	Space up to 18 in. above floor or grade level for entire storage or repair garage.
Garages for other than Tank Vehicles	Unclassified	If there is any opening to these rooms within the extent of an outdoor classified location, the entire room shall be classified the same as the space classification at the point of the opening.
Outdoor Drum Storage	Unclassified	
Indoor Warehousing Where There Is No Flammable Liquid Transfer	Unclassified	If there is any opening to these rooms within the extent of an indoor classified location, the room shall be classified the same as if the wall, curb, or partition did not exist.
Piers and Wharves		See Figure 515-2.

Note: For SI units, 1 in. = 25.4 mm; 1 ft = 0.3048 m.

[1]The release of Class I liquids may generate vapors to the extent that the entire building, and possibly a zone surrounding it, should be considered a Class I, Division 2 location.

[2]For Tanks — Underground, see Section 514-2.

[3]When classifying extent of space, consideration shall be given to the fact that tank cars or tank vehicles may be spotted at varying points. Therefore, the extremities of the loading or unloading positions shall be used.

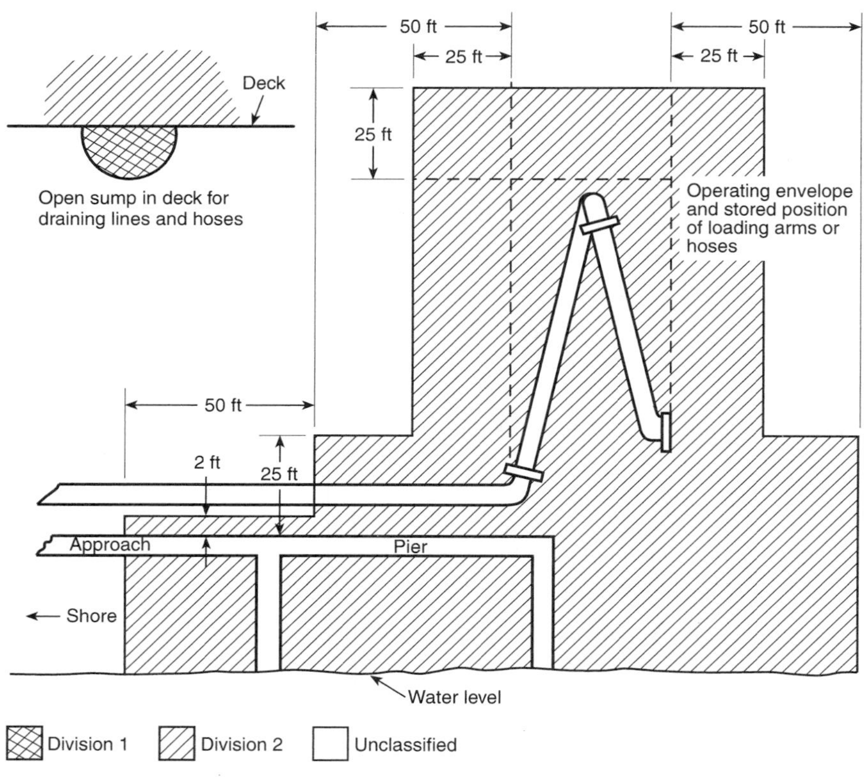

[x]Figure 515-2 Marine terminal handling flammable liquids.

Code. Section 5-3.4.5 of NFPA 30-1996, as referenced in the first two items in column 1 of Table 515-2, is as follows:

> 5-3.4.5 Equipment used in a building and the ventilation of the building shall be designed to limit flammable vapor–air mixtures under normal operating conditions to the interior of equipment, and to not more than 5 ft (1.5 m) from equipment that exposes Class I liquids to the air. Examples of such equipment are dispensing stations, open centrifuges, plate and frame filters, open vacuum filters, and surfaces of open equipment.

515-3. Wiring and Equipment Within Class I Locations. All electric wiring and equipment within the Class I locations defined in Section 515-2 shall comply with the applicable provisions of Article 501.

Exception: As permitted in Section 515-5.

515-4. Wiring and Equipment Above Class I Locations. All fixed wiring above Class I locations shall be in metal raceways or PVC Schedule 80 rigid nonmetallic conduit, or equivalent, or be Type MI, TC, or MC cable. Fixed equipment that may produce arcs, sparks, or particles of hot metal, such as lamps and lampholders for fixed lighting, cutouts, switches, receptacles, motors, or other equipment having make-and-break or sliding contacts, shall be of the totally enclosed type or be constructed so as to prevent the escape of sparks or hot metal particles. Portable lamps or other utilization equipment and their flexible cords shall comply with the provisions of Article 501 for the class of location above which they are connected or used.

515-5. Underground Wiring.

(a) Wiring Method. Underground wiring shall be installed in threaded rigid metal conduit or threaded steel intermediate metal conduit or, where buried under not less than 2 ft (610 mm) of cover, shall be permitted in rigid nonmetallic conduit or an approved cable. Where rigid nonmetallic conduit is used, threaded rigid metal conduit or threaded steel intermediate metal conduit shall be used for the last 2 ft (610 mm) of the conduit run to emergence or to the point of connection to the aboveground raceway. Where cable is used, it shall be enclosed in threaded rigid metal conduit or threaded steel intermediate metal conduit from the point of lowest buried cable level to the point of connection to the aboveground raceway.

See the commentary following Section 514-8.

(b) Insulation. Conductor insulation shall comply with Section 501-13.

(c) Nonmetallic Wiring. Where rigid nonmetallic conduit or cable with a nonmetallic sheath is used, an equipment grounding conductor shall be included to provide for electrical continuity of the raceway system and for grounding of noncurrent-carrying metal parts.

515-6. Sealing. Sealing requirements in Sections 501-5(a)(4) and (b)(2) shall apply to horizontal as well as to vertical boundaries of the defined Class I locations. Buried raceways under defined Class I locations shall be considered to be within a Class I, Division 1 location.

515-7. Gasoline Dispensing. Where gasoline or other volatile flammable liquids or liquefied flammable gases are dispensed at bulk stations, the applicable provisions of Article 514 shall apply.

515-16. Grounding. All metal raceways, the metal armor or metallic sheath on cables, and all noncurrent-carrying metal parts of fixed or portable electrical equipment, regardless of voltage, shall be grounded as provided in Article 250. Grounding in Class I locations shall comply with Section 501-16.

FPN: For information on grounding for static protection, see 5-6.3.4 and 5-6.3.5 of *Flammable and Combustible Liquids Code,* NFPA 30-1996.

See Figures 515.1 through 515.7.

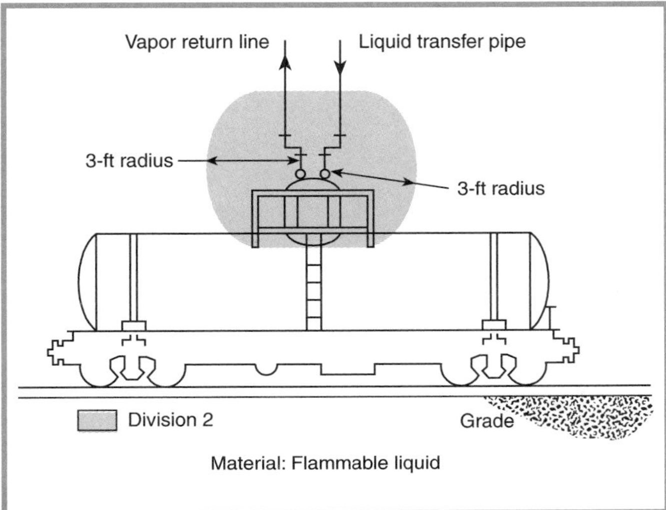

Figure 515.1 *Tank car/tank truck loading and unloading via closed system. Transfer through dome only.*

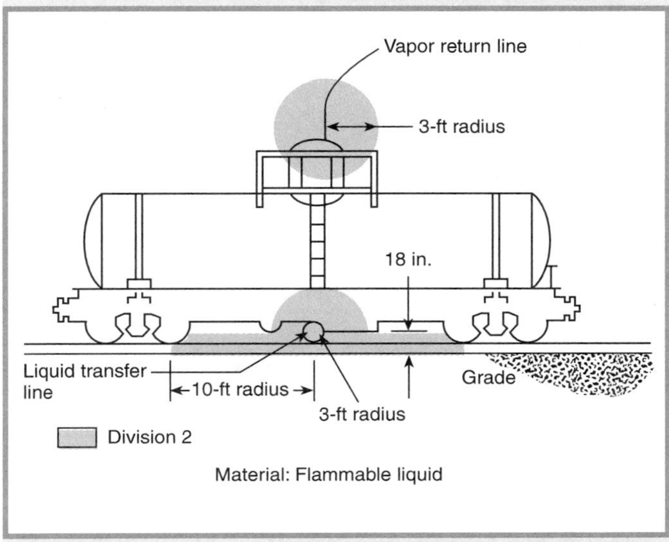

Figure 515.2 *Tank car/tank truck loading and unloading via closed system. Bottom product transfer only.*

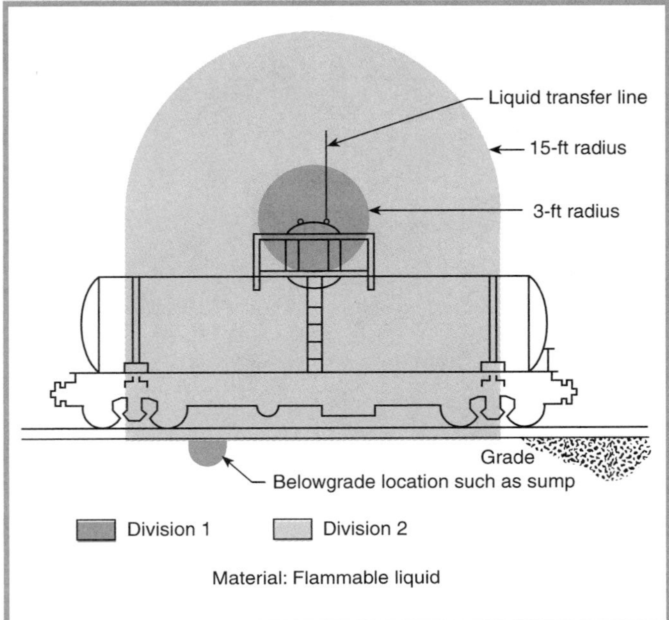

Figure 515.3 *Tank car/tank truck loading and unloading via open system. Top or bottom product transfer.*

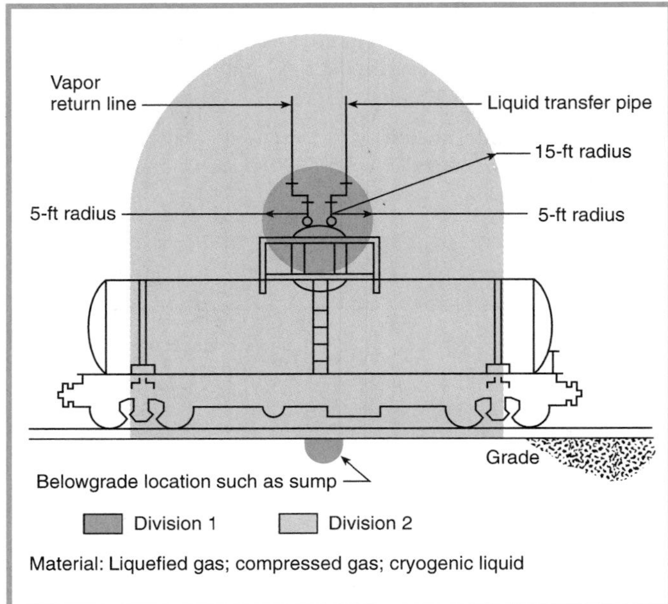

Figure 515.4 *Tank car/tank truck loading and unloading via closed system. Transfer through dome only.*

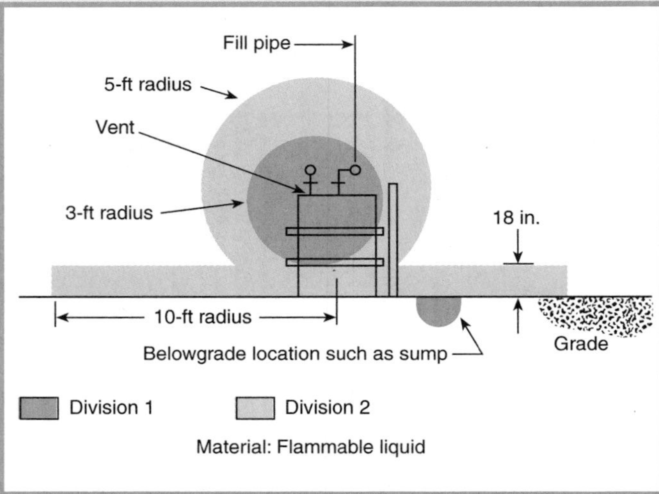

Figure 515.5 *Drum filling station, outdoors or indoors, with adequate ventilation.*

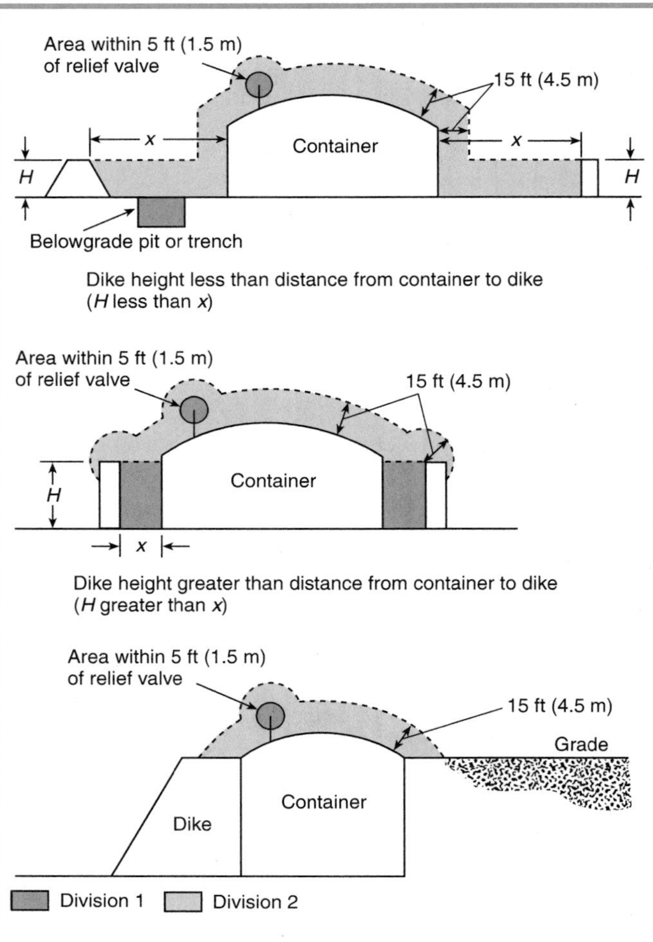

Figure 515.6 *Storage tanks for cryogenic liquids. [From NFPA 59A-1996, Standard for the Production, Storage, and Handling of Liquefied Natural Gas (LNG).]*

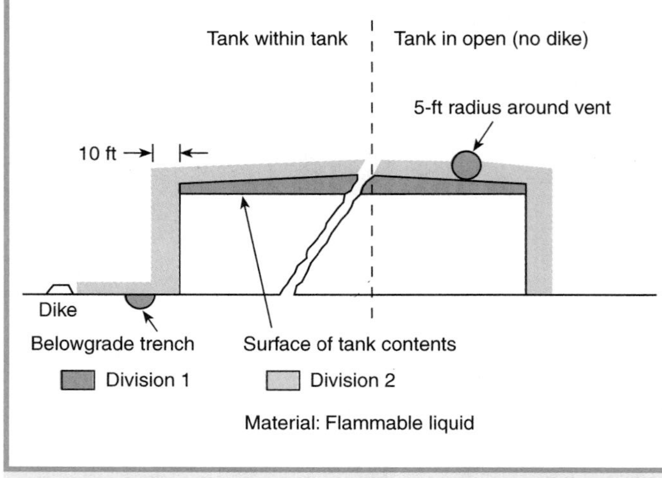

Figure 515.7 *Storage tanks, outdoors at grade. (From API Recommended Practice 500A).*

Article 516 — Spray Application, Dipping, and Coating Processes

Contents

516-1. Scope. This article covers the regular or frequent application of flammable liquids, combustible liquids, and combustible powders by spray operations and the application of flammable liquids, or combustible liquids at temperatures above their flashpoint, by dipping, coating, or other means.

FPN: For further information regarding safeguards for these processes, such as fire protection, posting of warning signs, and maintenance, see *Standard for Spray Application Using Flammable and Combustible Materials*, NFPA 33-1995, and *Standard for Dipping and Coating Processes Using Flammable or Combustible Liquids*, NFPA 34-1995. For additional information regarding ventilation, see *Standard for Exhaust Systems for Air Conveying of Materials,* NFPA 91-1995.

516-2. Classification of Locations. Classification is based on dangerous quantities of flammable vapors, combustible mists, residues, dusts, or deposits.

(a) Class I or Class II, Division 1 Locations. The following spaces shall be considered Class I or Class II, Division 1 locations, as applicable.

[x]**(1)** The interiors of spray booths and rooms except as specifically provided in Section 516-3(d).

[x]**(2)** The interior of exhaust ducts.

[x]**(3)** Any area in the direct path of spray operations.

[x]**(4)** For dipping and coating operations, all space within a 5-ft (1.52-m) radial distance from the vapor sources extending from these surfaces to the floor. The vapor source shall be the liquid exposed in the process and the drainboard, and any dipped or coated object from which it is possible to measure vapor concentrations exceeding 25 percent of the lower flammable limit at a distance of 1 ft (305 mm), in any direction, from the object.

[x]**(5)** Sumps, pits, or below-grade channels within 25 ft (7.625 m) horizontally of a vapor source. If the sump, pit, or channel extends beyond 25 ft (7.625 m) from the vapor source, it shall be provided with a vapor stop or it shall be classified as Class I, Division 1 for its entire length.

[x]**(6)** The interior of any enclosed dipping or coating process or apparatus.

(b) Class I or Class II, Division 2 Locations. The following spaces shall be considered Class I or Class II, Division 2 as applicable.

[x]**(1)** For open spraying, all space outside of but within 20 ft (6.10 m) horizontally and 10 ft (3.05 m) vertically of the Class I, Division 1 location as defined in Section 516-2(a), and not separated from it by partitions. See Figure 516-2(b)(1).

[x]**(2)** If spray application operations are conducted within a closed-top, open-face, or open-front booth or room, any electrical wiring or utilization equipment located outside of the booth or room but within the boundaries designated as Division 2 in Figure 516-2(b)(2) shall be suitable for Class I, Division 2 or Class II, Division 2 locations, whichever is applicable.

The Class I, Division 2 or Class II, Division 2 locations shown in Figure 516-2(b)(2) shall extend from the edges of the open face or open front of the booth or room in accordance with the following.

(a) If the exhaust ventilation system is interlocked with the spray application equipment, then the Division 2 location shall extend 5 ft (1525 mm) horizontally and 3 ft (915 mm)

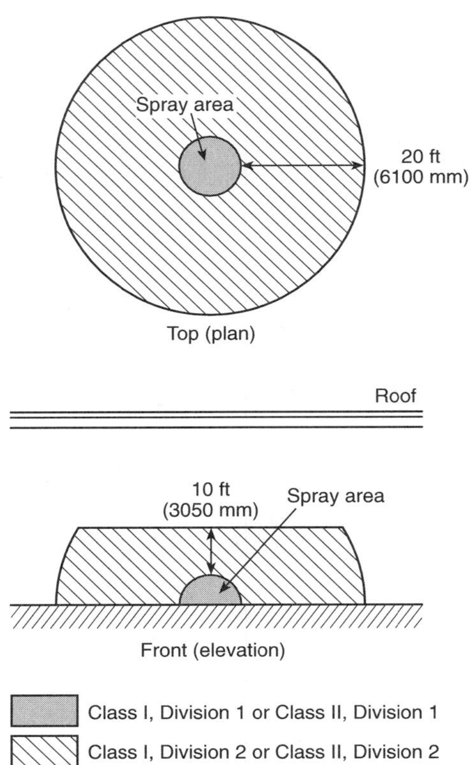

ˣFigure 516-2(b)(1) Electrical area classification for open spray areas.

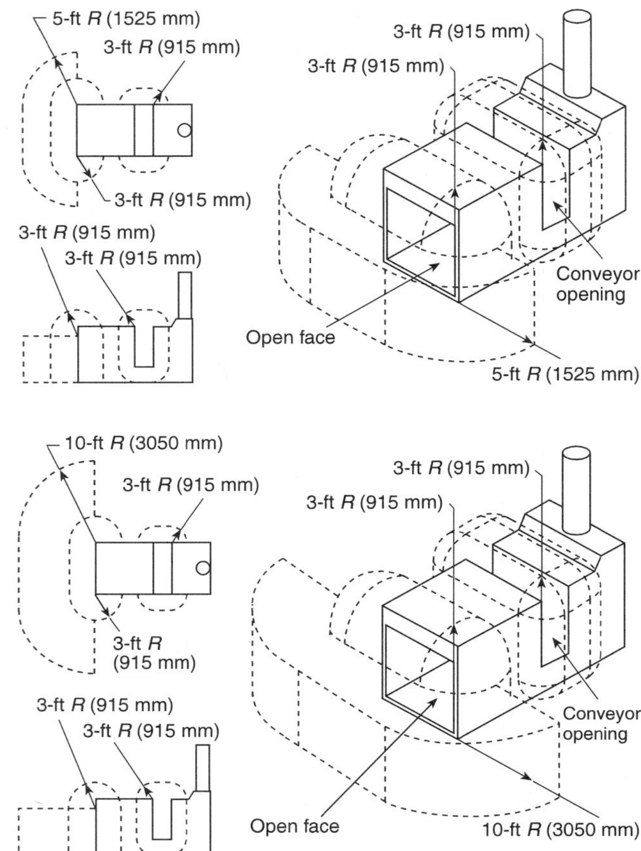

ˣFigure 516-2(b)(2) Class I or Class II, Division 2 locations adjacent to a closed top, open face, or open front spray booth or room.

vertically from the open face or open front of the booth or room, as shown in Figure 516-2(b)(2) top.

(b) If the exhaust ventilation system is not interlocked with the spray application equipment, then the Division 2 location shall extend 10 ft (3050 mm) horizontally and 3 ft (915 mm) vertically from the open face or open front of the booth or room, as shown in Figure 516-2(b)(2) bottom.

For the purposes of this subsection, *interlocked* shall mean that the spray application equipment cannot be operated unless the exhaust ventilation system is operating and functioning properly and spray application is automatically stopped if the exhaust ventilation system fails.

ˣ**(3)** For spraying operations conducted within an open top spray booth, the space 3 ft (914 mm) vertically above the booth and within 3 ft (914 mm) of other booth openings shall be considered Class I or Class II, Division 2.

ˣ**(4)** For spraying operations confined to an enclosed spray booth or room, the space within 3 ft (914 mm) in all directions from any openings shall be considered Class I or Class II, Division 2 as shown in Figure 516-2(b)(4).

ˣ**(5)** For dip tanks and drain boards, the 3-ft (914-mm) space surrounding the Class I, Division 1 location as defined in Section 516-2(a)(4) and as shown in Figure 516-2(b)(5).

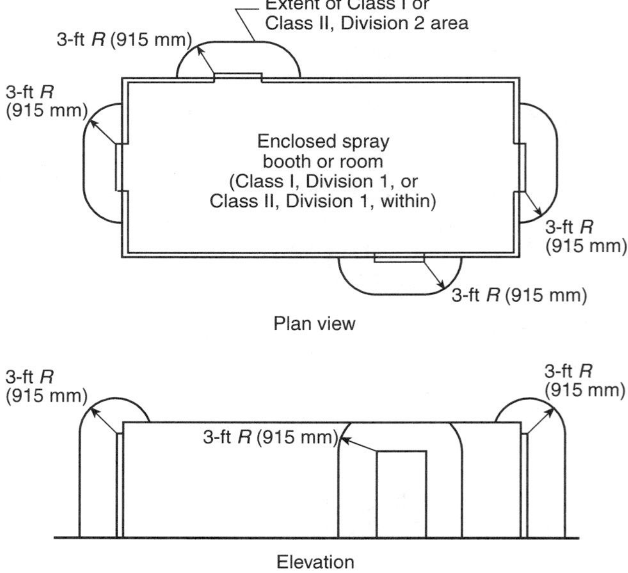

ˣFigure 516-2(b)(4) Class I (or Class II), Division 2 locations adjacent to an enclosed spray booth or spray room.

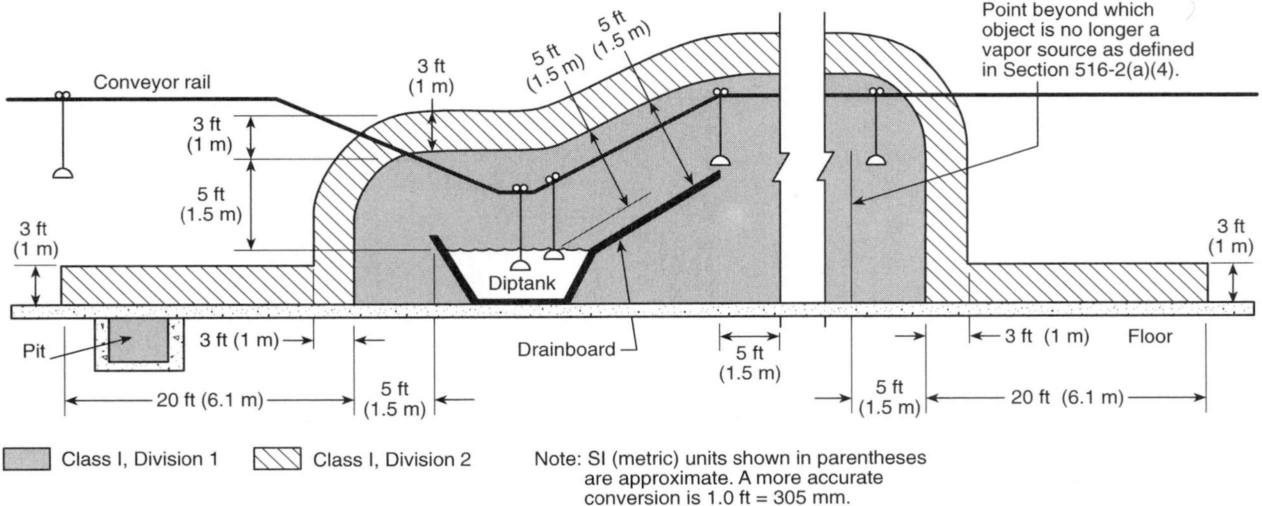

ˣFigure 516-2(b)(5) Electrical area classification for open processes without vapor containment or ventilation.

ˣ(6) For dip tanks and drain boards, the space 3 ft (914 mm) above the floor and extending 20 ft (6.1 m) horizontally in all directions from the Class I, Division 1 location.

Exception: This space shall not be required to be considered a hazardous (classified) location where the vapor source area is 5 ft² (0.46 m²) or less, and where the contents of the open tank, trough, or container do not exceed 5 gal (18.9 L). In addition, the vapor concentration during operation and shutdown periods shall not exceed 25 percent of the lower flammable limit outside the Class I location specified in Section 516-2(a)(4).

ˣ**(c) Enclosed Coating and Dipping Operations.** The space adjacent to an enclosed dipping or coating process or apparatus shall be considered unclassified.

Exception: The space within 3 ft (914 mm) in all directions from any opening in the enclosure shall be classified as Class I, Division 2.

(d) Adjacent Locations. Adjacent locations that are cut off from the defined Class I or Class II locations by tight partitions without communicating openings, and within which flammable vapors or combustible powders are not likely to be released, shall be unclassified.

(e) Unclassified Locations. Locations using drying, curing, or fusion apparatus and provided with positive mechanical ventilation adequate to prevent accumulation of flammable concentrations of vapors, and provided with effective interlocks to de-energize all electrical equipment (other than equipment approved for Class I locations) in case the ventilating equipment is inoperative, shall be permitted to be unclassified where the authority having jurisdiction so judges.

FPN: For further information regarding safeguards, see *Standard for Ovens and Furnaces*, NFPA 86-1995.

516-3. Wiring and Equipment in Class I Locations.

(a) Wiring and Equipment — Vapors. All electric wiring and equipment within the Class I location (containing vapor only — not residues) defined in Section 516-2 shall comply with the applicable provisions of Article 501.

ˣ**(b) Wiring and Equipment — Vapors and Residues.** Unless specifically listed for locations containing deposits of dangerous quantities of flammable or combustible vapors, mists, residues, dusts, or deposits (as applicable), there shall be no electrical equipment in any spray area as herein defined whereon deposits of combustible residue may readily accumulate, except wiring in rigid metal conduit, intermediate metal conduit, Type MI cable, or in metal boxes or fittings containing no taps, splices, or terminal connections.

Only electrical equipment specifically listed for the location is permitted where deposits of combustible residue may accumulate, as stated in Section 516-3(b). The latter part of the requirement, however, permits wiring in rigid metal conduit, intermediate metal conduit, Type MI cable, or standard metal boxes without splices or taps.

(c) Illumination. Illumination of readily ignitible areas through panels of glass or other transparent or translucent material shall be permitted only if it complies with the following.

(1) Fixed lighting units are used as the source of illumination;
(2) The panel effectively isolates the Class I location from the area in which the lighting unit is located;
(3) The lighting unit is approved for its specific location;

(4) The panel is of a material or is protected so that breakage will be unlikely; and
(5) The arrangement is such that normal accumulations of hazardous residue on the surface of the panel will not be raised to a dangerous temperature by radiation or conduction from the source of illumination.

Underwriters Laboratories Inc.'s 1998 *Hazardous Locations Equipment Directory* indicates that UL-listed fixtures suitable for locations having deposits of readily combustible paint residue are so marked.

ˣ**(d) Portable Equipment.** Portable electric lamps or other utilization equipment shall not be used in a spray area during spray operations.

Exception No. 1: Where portable electric lamps are required for operations in spaces not readily illuminated by fixed lighting within the spraying area, they shall be of the type approved for Class I, Division 1 locations where readily ignitible residues may be present.

Exception No. 2: Where portable electric drying apparatus are used in automobile refinishing spray booths and the following requirements are met.

(a) The apparatus and its electrical connections are not located within the spray enclosure during spray operations.
(b) Electrical equipment within 18 in. (45.7 cm) of the floor is approved for Class I, Division 2 locations.
(c) All metallic parts of the drying apparatus are electrically bonded and grounded.
(d) Interlocks are provided to prevent the operation of spray equipment while drying apparatus is within the spray enclosure, to allow for a 3-minute purge of the enclosure before energizing the drying apparatus, and to shut off drying apparatus on failure of ventilation system.

(e) Electrostatic Equipment. Electrostatic spraying or detearing equipment shall be installed and used only as provided in Section 516-4.

FPN: For further information, see *Standard for Spray Application Using Flammable and Combustible Materials,* NFPA 33-1995.

ˣ**516-4. Fixed Electrostatic Equipment.** This section shall apply to any equipment using electrostatically charged elements for the atomization, charging, and/or precipitation of hazardous materials for coatings on articles or for other similar purposes in which the charging or atomizing device is attached to a mechanical support or manipulator. This shall include robotic devices. This section shall not apply to devices that are held or manipulated by hand. Where robotic programming procedures involve manual manipulation of the robot arm while spraying with the high voltage on, the provisions of Section 516-5 shall apply. The installation of electrostatic spraying equipment shall comply with Sections 516-4(a) through (j). Spray equipment shall be listed or approved.

All automatic electrostatic equipment systems shall comply with Sections 516-4(a) through (i).

(a) Power and Control Equipment. Transformers, high-voltage supplies, control apparatus, and all other electric portions of the equipment shall be installed outside of the Class I location as defined in Section 516-2 or be of a type approved for the location.

Exception: High-voltage grids, electrodes, electrostatic atomizing heads, and their connections shall be permitted within the Class I location.

(b) Electrostatic Equipment. Electrodes and electrostatic atomizing heads shall be adequately supported in permanent locations and shall be effectively insulated from ground. Electrodes and electrostatic atomizing heads that are permanently attached to their bases, supports, reciprocators, or robots shall be deemed to comply with this section.

(c) High-Voltage Leads. High-voltage leads shall be properly insulated and protected from mechanical damage or exposure to destructive chemicals. Any exposed element at high voltage shall be effectively and permanently supported on suitable insulators and shall be effectively guarded against accidental contact or grounding.

(d) Support of Goods. Goods being coated using this process shall be supported on conveyors or hangers. The conveyors or hangers shall be arranged to (1) ensure that the parts being coated are electrically connected to ground with a resistance of 1 megohm or less and (2) to prevent parts from swinging.

(e) Automatic Controls. Electrostatic apparatus shall be equipped with automatic means that will rapidly de-energize the high-voltage elements under any of the following conditions:

(1) Stoppage of ventilating fans or failure of ventilating equipment from any cause
(2) Stoppage of the conveyor carrying goods through the high-voltage field unless stoppage is required by the spray process
(3) Occurrence of excessive current leakage at any point in the high-voltage system
(4) De-energizing the primary voltage input to the power supply

(f) Grounding. All electrically conductive objects in the spray area, except those objects required by the process to be at high voltage, shall be adequately grounded. This requirement shall apply to paint containers, wash cans, guards, hose connectors, brackets, and any other electrically conductive objects or devices in the area.

(g) Isolation. Safeguards such as adequate booths, fencing, railings, interlocks, or other means shall be placed about the equipment or incorporated therein so that they, either by their location or character, or both, ensure that a safe separation of the process is maintained.

(h) Signs. Signs shall be conspicuously posted to

(1) Designate the process zone as dangerous with regard to fire and accident,

(2) Identify the grounding requirements for all electrically conductive objects in the spray area, and
(3) Restrict access to qualified personnel only.

(i) Insulators. All insulators shall be kept clean and dry.

(j) Other Than Nonincendive Equipment. Spray equipment that cannot be classified as nonincendive shall comply with (1) and (2).

(1) Conveyors or hangers shall be arranged so to maintain a safe distance of at least twice the sparking distance between goods being painted and electrodes, electrostatic atomizing heads, or charged conductors. Warnings defining this safe distance shall be posted.

(2) The equipment shall provide an automatic means of rapidly de-energizing the high-voltage elements in the event the distance between the goods being painted and the electrodes or electrostatic atomizing heads falls below that specified in (1).

[x]516-5. Electrostatic Hand-Spraying Equipment. This section shall apply to any equipment using electrostatically charged elements for the atomization, charging, and/or precipitation of materials for coatings on articles, or for other similar purposes in which the atomizing device is hand held or manipulated during the spraying operation. Electrostatic hand-spraying equipment and devices used in connection with paint-spraying operations shall be of approved types and shall comply with Sections 516-5(a) through (e).

(a) General. The high-voltage circuits shall be designed so as not to produce a spark of sufficient intensity to ignite the most readily ignitible of those vapor–air mixtures likely to be encountered, nor result in appreciable shock hazard upon coming in contact with a grounded object under all normal operating conditions. The electrostatically charged exposed elements of the hand gun shall be capable of being energized only by an actuator that also controls the coating material supply.

(b) Power Equipment. Transformers, power packs, control apparatus, and all other electric portions of the equipment shall be located outside of the Class I location or be approved for the location.

Exception: The hand gun itself and its connections to the power supply shall be permitted within the Class I location.

(c) Handle. The handle of the spraying gun shall be electrically connected to ground by a metallic connection and be constructed so that the operator in normal operating position is in intimate electrical contact with the grounded handle to prevent buildup of a static charge on the operator's body. Signs indicating the necessity for grounding other persons entering the spray area shall be conspicuously posted.

(d) Electrostatic Equipment. All electrically conductive objects in the spraying area shall be adequately grounded. This requirement shall apply to paint containers, wash cans, and any other electrically conductive objects or devices in the area. The equipment shall carry a prominent, permanently installed warning regarding the necessity for this grounding feature.

(e) Support of Objects. Objects being painted shall be maintained in metallic contact with the conveyor or other grounded support. Hooks shall be regularly cleaned to ensure adequate grounding of 1 megohm or less. Areas of contact shall be sharp points or knife edges where possible. Points of support of the object shall be concealed from random spray where feasible; and, where the objects being sprayed are supported from a conveyor, the point of attachment to the conveyor shall be located so as to not collect spray material during normal operation.

[x]516-6. Powder Coating. This section shall apply to processes in which combustible dry powders are applied. The hazards associated with combustible dusts are present in such a process to a degree, depending on the chemical composition of the material, particle size, shape, and distribution.

FPN: The hazards associated with combustible dusts are inherent in this process. Generally speaking, the hazard rating of the powders employed depends on the chemical composition of the material, particle size, shape, and distribution.

(a) Electric Equipment and Sources of Ignition. Electric equipment and other sources of ignition shall comply with the requirements of Article 502. Portable electric lamps and other utilization equipment shall not be used within a Class II location during operation of the finishing processes. Where such lamps or utilization equipment are used during cleaning or repairing operations, they shall be of a type approved for Class II, Division 1 locations, and all exposed metal parts shall be effectively grounded.

Exception: Where portable electric lamps are required for operations in spaces not readily illuminated by fixed lighting within the spraying area, they shall be of the type approved for Class II, Division 1 locations where readily ignitible residues may be present.

(b) Fixed Electrostatic Spraying Equipment. The provisions of Section 516-4 and Section 516-6(a) shall apply to fixed electrostatic spraying equipment.

(c) Electrostatic Hand-Spraying Equipment. The provisions of Section 516-5 and Section 516-6(a) shall apply to electrostatic hand-spraying equipment.

(d) Electrostatic Fluidized Beds. Electrostatic fluidized beds and associated equipment shall be of approved types. The high-voltage circuits shall be designed so that any discharge produced when the charging electrodes of the bed are approached or contacted by a grounded object shall not be of sufficient intensity to ignite any powder–air mixture likely to be encountered nor to result in an appreciable shock hazard.

(1) Transformers, power packs, control apparatus, and all other electric portions of the equipment shall be located outside the powder-coating area or shall otherwise comply with the requirements of Section 516-6(a).

Exception: The charging electrodes and their connections to the power supply shall be permitted within the powder-coating area.

(2) All electrically conductive objects within the powder-coating area shall be adequately grounded. The powder-coating equipment shall carry a prominent, permanently installed warning regarding the necessity for grounding these objects.

(3) Objects being coated shall be maintained in electrical contact (less than 1 megohm) with the conveyor or other support in order to ensure proper grounding. Hangers shall be regularly cleaned to ensure effective electrical contact. Areas of electrical contact shall be sharp points or knife edges where possible.

(4) The electric equipment and compressed air supplies shall be interlocked with a ventilation system so that the equipment cannot be operated unless the ventilating fans are in operation.

516-7. Wiring and Equipment Above Class I and II Locations.

(a) Wiring. All fixed wiring above the Class I and II locations shall be in metal raceways, rigid nonmetallic conduit, or electrical nonmetallic tubing, or shall be Type MI, TC, or MC cable. Cellular metal floor raceways shall be permitted only for supplying ceiling outlets or extensions to the area below the floor of a Class I or II location, but such raceways shall have no connections leading into or through the Class I or II location above the floor unless suitable seals are provided.

(b) Equipment. Equipment that may produce arcs, sparks, or particles of hot metal, such as lamps and lampholders for fixed lighting, cutouts, switches, receptacles, motors, or other equipment having make-and-break or sliding contacts, where installed above a Class I or II location or above a location where freshly finished goods are handled, shall be of the totally enclosed type or be constructed so as to prevent the escape of sparks or hot metal particles.

516-16. Grounding. All metal raceways, the metal armor or metallic sheath on cables, and all noncurrent-carrying metal parts of fixed or portable electrical equipment, regardless of voltage, shall be grounded as provided in Article 250. Grounding in Class I locations shall comply with Section 501-16.

NFPA 33-1995, *Standard for Spray Application Using Flammable or Combustible Materials,* covers the spray application of flammable or combustible materials by means of compressed air atomization, "airless" or "hydraulic atomization," or by electrostatic application methods, or any other means in continuous or intermittent processes. It also covers the application of combustible powders where applied by powder spray guns, electrostatic powder spray guns, and fluidized bed application method or electrostatic fluidized bed application method.

NFPA 33 also contains requirements for the maintenance of safe conditions as well as personal safety.

The proper maintenance and operation of processes and process areas where flammable and combustible materials are handled and applied is critical with respect to the protection of life and property from fire or explosion. An analysis of actual experience in industry demonstrates that the largest fire losses and frequency of fires have occurred where the proper codes and standards have not been used or properly applied.

The following definitions are found in Section 1-6 of NFPA 33-1995, *Standard for Spray Application Using Flammable or Combustible Materials.*

> *Spray area.* Any area in which dangerous quantities of flammable or combustible vapors, mists, residues, dusts, or deposits are present due to the operation of spray processes.
>
> A spray area includes:
>
> (a) The interior of spray booths or spray room (with certain exceptions); and
> (b) The interior of any exhaust plenum or any exhaust duct leading from the spraying process; and
> (c) Any area in the direct path of a spray application process.
>
> The authority having jurisdiction may, for the purpose of NFPA 33, define the limits of the spray area in any specific case. The spray area in the vicinity of spraying operations will necessarily vary with the design and arrangement of equipment and with the method of operation. Where spray application operations are strictly confined to predetermined spaces provided with adequate and reliable ventilation, (such as a properly designed and constructed spray booth), the spray area will ordinarily not extend beyond this space. When spray application operations are *not* confined to an adequately ventilated space, then the spray area might extend throughout the room or building area where the spraying is conducted.
>
> *Spray booth.* A power-ventilated structure that encloses a spray application operation or process, and confines and limits the escape of the material being sprayed, including vapors, mists, dusts, and residues that are produced by the spraying operation and conducts or directs these materials to an ex-

haust system. Spray booths are manufactured in a variety of forms, including automotive refinishing, downdraft, open-face, traveling, tunnel, and updraft booths. This definition is not intended to limit the term "spray booth" to any particular design. The entire spray booth is considered part of the spray area. A spray booth is not a spray room.

Spray room. A power-ventilated, fully enclosed room used exclusively for open spraying of flammable or combustible materials. The entire spray room is a spray area. A spray booth is not a spray room.

Waterwash spray booth. A spray booth that is equipped with a water-washing system designed to minimize the concentrations of dusts or residues entering exhaust ducts and to permit the collection of the dusts or residues.

Dry-type spray booth. A spray booth that is not equipped with a water-washing system to remove over spray from the exhaust airstream. A dry spray booth is equipped with one or more of the following: (a) distribution or baffle plates to promote an even flow of air through the booth or to reduce the overspray before it is pulled into the exhaust system; (b) dry media filters, either fired or on rolls, to remove overspray from the exhaust airstream; (c) powder collection systems that capture powder overspray.

Notes on Electrical Installations

The safety of life and property from fire or explosion as a result of spray applications of flammable and combustible materials such as paints, finishes, and adhesives depends on the arrangement and operation of a particular installation. The principal hazards of spray application operations originate from flammable or combustible liquids or powders and their vapors or mists, as well as from highly combustible residues or powders.

Properly constructed spray booths, with adequate mechanical ventilation, may be used to discharge vapors or powder to a safe location and reduce the possibility of an explosion. In like manner, the accumulation of overspray residues, many of which are not only highly combustible but also subject to spontaneous ignition, can be controlled.

The elimination of all sources of ignition in areas where flammable or combustible liquids, vapors, mists, or combustible residues are present, together with constant supervision and maintenance, is essential to the safe operation of spraying.

The human element necessitates careful consideration of the location of the operation and the installation of extinguishing equipment in order to reduce the possibility of fire spreading to other property and minimize the probability of damage to other property.

It is obvious that no open flames or spark-producing equipment should be in any area where, because of inadequate ventilation, explosive vapor–air mixtures or mists are present. It is equally obvious that no open flames or spark-producing equipment should be located where highly combustible spray residues will be deposited on them. Because some residues may be ignited at very low temperatures, additional consideration must be given to operating temperatures of equipment subject to residue deposits. Many deposits may be ignited at temperatures produced by low-pressure steam pipes or incandescent light globes, even fixtures of the explosionproof type.

It should be noted that electrical equipment is generally not permitted inside any spray booth, in the exhaust duct from a spray booth, in the entrained air of an exhaust system from a spraying operation, or in the direct path of spray, unless such equipment is specifically listed for both readily ignitible deposits and flammable vapor.

The determination of the extent of hazardous areas involved in spray application requires an understanding of the multiple hazards of flammable vapors, mists, powders, and highly combustible deposits applied at each individual location.

Where electrical equipment is installed in locations not subject to deposits of combustible residues but, due to inadequate ventilation, subject to explosive concentrations of flammable vapors or mists, only approved explosionproof or other types of equipment approved for Class I, Division 1 locations (for example, purged, pressurized, or intrinsically safe equipment or systems) are permitted.

Where spray areas containing dangerous quantities of flammable or combustible vapors, mists, residues, dusts, or deposits under normal operation have been determined, the adjacent unpartitioned areas, which are safe under normal operating conditions but which may become dangerous due to accident or careless operation, should be given consideration. Equipment known to produce sparks or flames under normal operating conditions should not be installed in these adjacent unpartitioned areas.

Where spraying operations are confined to adequately ventilated spray booths or rooms, there should be no deposits of combustible residues or dangerous concentrations of flammable vapors, mists, or dusts outside the spray booth under normal operating conditions. In the interest of safety, however, it should be noted that unless separated by partitions, an area within a certain distance [see Section 516-2(b)] of the Class I (or Class II), Division 1 spraying area, depending on the arrangement, is classified as Division 2; that is, it should contain no equipment that produces ignition-capable sparks under normal operation. Furthermore, within this distance, electric lamps are required to be enclosed, to prevent hot particles from falling on freshly painted stock or other readily ignitible material and, if subject to physical damage, are required to be properly guarded. See Section 516-7(b).

Even though it is contemplated that areas adjacent to spray booths (particularly where coating-material stocks are located) will be provided with ventilation sufficient to prevent the presence of flammable vapors or deposits, it is nevertheless advisable that electric lamps be totally enclosed to prevent hot particles from falling in any area where there may be freshly painted stock, accidentally spilled flammable or combustible materials, readily ignitible refuse, or flammable or combustible liquid containers that have been left open accidentally. See Section 516-7(b).

Where electric lamps are in areas subject to atmospheres of flammable vapor, lamps should be replaced while electricity is off; otherwise there may be a spark from this source.

Sufficient lighting for coating operations, booth cleaning, and booth repair work should be provided at the time the equipment is installed in order to avoid the use of temporary or emergency lamps connected to ordinary extension cords in this area. See Section 516-3(d). A satisfactory and practical method of lighting is the use of ¼-in.-thick wired or tempered glass panels in the top or sides of the spray booth, with electrical light fixtures located outside the booth, avoiding the direct path of the spray. See Section 516-3(c).

In order to prevent sparks from the accumulation of static electricity, all electrically conductive objects, including metal parts of spray booths, exhaust ducts, piping systems conveying flammable or combustible liquids or paint, solvent tanks, and canisters, should be properly grounded. See Section 4-5 of NFPA 33 and Sections 7-7.3.3 and 7-7.4 of NFPA 77-1993, *Recommended Practice on Static Electricity.*

Automobile undercoating operations in garages, conducted in areas having adequate natural or mechanical ventilation, are exempt from the requirements pertaining to spray-coating operations where (1) undercoating materials not more hazardous than kerosene (as classified by Underwriters Laboratories Inc. with respect to fire hazard rating 30-40) are used, or (2) undercoating materials using only solvents having a flash point in excess of 38°C (100°F) are used, and (3) no open flames are within 20 ft while such operations are conducted.

Article 517 — Health Care Facilities

Contents

A. General

517-1. Scope. The provisions of this article shall apply to electrical construction and installation criteria in health care facilities.

FPN No. 1: This article is not intended to apply to veterinary facilities.

FPN No. 2: For information concerning performance, maintenance, and testing criteria, refer to the appropriate health care facilities documents.

517-2. General. The requirements in Parts B and C apply not only to single-function buildings, but are also intended to be individually applied to their respective forms of occupancy within a multifunction building (e.g., a doctor's examining room located within a limited care facility would be required to meet the provisions of Section 517-10).

The requirements of Article 517 are intended to apply to all types of health care facilities. The requirements for each type of health care facility are nevertheless intended to be applied in a very specific manner. An example of the application of Article 517 would be a suite of doctors' offices within an office building. The doctors' business office would be treated as an ordinary occupancy and would only be required to meet the applicable portion of other parts of this *Code*. However, the examining rooms attached to the doctors' business office would be required to meet the provisions of Part B and Section 517-50 in Article 517.

The scope of Article 517 also includes health care facilities that may be mobile or supply very limited outpatient services. However, the scope does not include animal hospitals.

Other standards referenced in Article 517 are NFPA 99-1996, *Standard for Health Care Facilities;* NFPA *101*®-1997, *Life Safety Code*®; and NFPA 20-1996, *Standard for the Installation of Centrifugal Fire Pumps.*

517-3. Definitions.

Alternate Power Source. One or more generator sets, or battery systems where permitted, intended to provide power during the interruption of the normal electrical services or the public utility electrical service intended to provide power during interruption of service normally provided by the generating facilities on the premises.

As stated in Section 517-30(b)(6), alternate power sources are permitted to serve essential electrical systems of contiguous or same site facilities.

Ambulatory Health Care Center. A building or part thereof used to provide services or treatment to four or more patients at the same time and meeting either (1) or (2).

(1) Those facilities that provide, on an outpatient basis, treatment for patients that would render them incapable of taking action for self-preservation under emergency conditions without assistance from others, such as hemodialysis units or freestanding emergency medical units.

(2) Those facilities that provide, on an outpatient basis, surgical treatment requiring general anesthesia.

The definition of *ambulatory health care center* was added to the 1996 *Code* to define new terms used in new Section 517-45 and to correlate with NFPA 99-1996, *Standard for Health Care Facilities.*

Anesthetizing Location. Any area of a health care facility that has been designated to be used for the administration of any flammable or nonflammable inhalation anesthetic agent in the course of examination or treatment, including the use of such agents for relative analgesia.

The definition of *anesthetizing location* recognizes that in an emergency it may be necessary to administer an anesthetic almost anywhere in a health care facility; however, only those areas in a health care facility set aside specifically for the induction of anesthetics are required to meet the provisions of Part D of Article 517. This definition and the provisions of Part D are not intended to apply to the administering of analgesic anesthetics, such as might be used in a dental office. The term *relative analgesia* is sometimes referred to as "conscious sedation." For guidance on flammable anesthetizing locations, see Annex 2 of NFPA 99-1996, *Standard for Health Care Facilities.*

Critical Branch. A subsystem of the emergency system consisting of feeders and branch circuits supplying energy to task illumination, special power circuits, and selected receptacles serving areas and functions related to patient care, and which are connected to alternate power sources by one or more transfer switches during interruption of the normal power source.

Electrical Life-Support Equipment. Electrically powered equipment whose continuous operation is necessary to maintain a patient's life.

Emergency System. A system of feeders and branch circuits meeting the requirements of Article 700, except as amended by Article 517, and intended to supply alternate power to a limited number of prescribed functions vital to the protection of life and patient safety, with automatic restoration of electrical power within 10 seconds of power interruption.

Revised for the 1999 *Code,* the definition of *emergency system* points out that Article 517, Health Care Facilities, modifies the general requirements of Article 700 for emergency systems. The requirements for sizing a generator and a feeder capacity for the essential electrical system in a health care facility are found in Section 517-30(d). Emergency systems in other occupancies are installed primarily for life safety. The emergency systems in hospitals are for life safety as well as critical equipment.

Equipment System. A system of feeders and branch circuits arranged for delayed, automatic, or manual connection to the alternate power source and that serves primarily 3-phase power equipment.

Essential Electrical System. A system comprised of alternate sources of power and all connected distribution systems and ancillary equipment, designed to ensure continuity of electrical power to designated areas and functions of a health care facility during disruption of normal power sources, and also designed to minimize disruption within the internal wiring system.

Exposed Conductive Surfaces. Those surfaces that are capable of carrying electric current and that are unprotected, unenclosed, or unguarded, permitting personal contact. Paint, anodizing, and similar coatings are not considered suitable insulation, unless they are listed for such use.

Flammable Anesthetics. Gases or vapors, such as fluroxene, cyclopropane, divinyl ether, ethyl chloride, ethyl ether, and ethylene, which may form flammable or explosive mixtures with air, oxygen, or reducing gases such as nitrous oxide.

Flammable Anesthetizing Location. Any area of the facility that has been designated to be used for the administration of any flammable inhalation anesthetic agents in the normal course of examination or treatment.

Hazard Current. For a given set of connections in an isolated power system, the total current that would flow through a low impedance if it were connected between either isolated conductor and ground.

Fault Hazard Current. The hazard current of a given isolated system with all devices connected except the line isolation monitor.

Monitor Hazard Current. The hazard current of the line isolation monitor alone.

Total Hazard Current. The hazard current of a given isolated system with all devices, including the line isolation monitor, connected.

Health Care Facilities. Buildings or portions of buildings that contain, but are not limited to, occupancies such as hospitals; nursing homes; limited care; supervisory care; clinics; medical and dental offices; and ambulatory care, whether permanent or movable.

Hospital. A building or part thereof used for the medical, psychiatric, obstetrical, or surgical care, on a 24-hour basis, of four or more inpatients. *Hospital*, wherever used in this *Code*, shall include general hospitals, mental hospitals, tuberculosis hospitals, children's hospitals, and any such facilities providing inpatient care.

Isolated Power System. A system comprising an isolating transformer or its equivalent, a line isolation monitor, and its ungrounded circuit conductors.

Isolation Transformer. A transformer of the multiple-winding type, with the primary and secondary windings physically separated, which inductively couples its secondary winding to the grounded feeder systems that energize its primary winding.

Life Safety Branch. A subsystem of the emergency system consisting of feeders and branch circuits, meeting the requirements of Article 700 and intended to provide adequate power needs to ensure safety to patients and personnel, and which are automatically connected to alternate power sources during interruption of the normal power source.

Limited Care Facility. A building or part thereof used on a 24-hour basis for the housing of four or more persons who are incapable of self-preservation because of age, physical limitation due to accident or illness, or mental limitations, such as mental retardation/developmental disability, mental illness, or chemical dependency.

Line Isolation Monitor. A test instrument designed to continually check the balanced and unbalanced impedance from each line of an isolated circuit to ground and equipped with a built-in test circuit to exercise the alarm without adding to the leakage current hazard.

Nursing Home. A building or part thereof used for the lodging, boarding, and nursing care, on a 24-hour basis, of four or more persons who, because of mental or physical incapacity, may be unable to provide for their own needs and safety without the assistance of another person. *Nursing home*, wherever used in this *Code*, shall include nursing and convalescent homes, skilled nursing facilities, intermediate care facilities, and infirmaries of homes for the aged.

Nurses' Stations. Areas intended to provide a center of nursing activity for a group of nurses serving bed patients, where the patient calls are received, nurses are dispatched, nurses' notes written, inpatient charts prepared, and medications prepared for distribution to patients. Where such activities are carried on in more than one location within a nursing unit, all such separate areas are considered a part of the nurses' station.

Patient Bed Location. The location of an inpatient sleeping bed; or the bed or procedure table used in a critical patient care area.

Patient Care Area. Any portion of a health care facility wherein patients are intended to be examined or treated. Areas of a health care facility in which patient care is administered are classified as general care areas or critical care areas, either of which may be classified as a wet location. The governing body of the facility designates these areas in accordance with the type of patient care anticipated and with the following definitions of the area classification.

FPN: Business offices, corridors, lounges, day rooms, dining rooms, or similar areas typically are not classified as patient care areas.

(a) General Care Areas. General care areas are patient bedrooms, examining rooms, treatment rooms, clinics, and similar areas in which it is intended that the patient shall come in contact with ordinary appliances such as a nurse call system, electrical beds, examining lamps, telephone, and entertainment devices. In such areas, it may also be intended that patients be connected to electromedical devices (such as heating pads, electrocardiographs, drainage pumps, monitors, otoscopes, ophthalmoscopes, intravenous lines, etc.).

(b) Critical Care Areas. Critical care areas are those special care units, intensive care units, coronary care units, angiography laboratories, cardiac catheterization laboratories, delivery rooms, operating rooms, and similar areas in which patients are intended to be subjected to invasive procedures and connected to line-operated, electromedical devices.

(c) Wet Locations. Wet locations are those patient care areas that are normally subject to wet conditions while patients are present. These include standing fluids on the floor or drenching of the work area, either of which condition is intimate to the patient or staff. Routine housekeeping procedures and incidental spillage of liquids do not define a wet location.

The definition of *patient care area* applies to hospitals as well as patient care areas in outpatient facilities. A patient bed location in a nursing home can be considered a patient care area if a person is examined or treated in that location. However, it excludes such areas as laundry rooms, boiler rooms, and utility areas, which, although routinely wet, are not patient care areas. The governing body of the health care facility may elect to include such areas as hydrotherapy areas, dialysis laboratories, and certain wet laboratories under this definition. Lavatories or bathrooms within a health care facility are not intended to be classified as wet locations. For infection control purposes, many patient and treatment areas have a sink for hand washing, which is not intended to be a wet location either.

Patient Equipment Grounding Point. A jack or terminal bus that serves as the collection point for redundant grounding of electric appliances serving a patient vicinity or for grounding other items in order to eliminate electromagnetic interference problems.

Patient Vicinity. In an area in which patients are normally cared for, the *patient vicinity* is the space with surfaces likely to be contacted by the patient or an attendant who can touch the patient. Typically in a patient room, this encloses a space within the room not less than 6 ft (1.83 m) beyond the perimeter of the bed in its nominal location, and extending vertically not less than 7½ ft (2.29 m) above the floor.

The patient vicinity area is limited to patient beds in their normal position; that is, the position of the bed as called for in the architect's plans, rather than the temporary position of the bed subject to movement by housekeeping staff or the convenience of the medical staff.

Definitions of *hospital, nursing home, ambulatory health care center,* and *limited care facility* are identical to the corresponding definitions found in NFPA *101*-1997, *Life Safety Code.*

Psychiatric Hospital. A building used exclusively for the psychiatric care, on a 24-hour basis, of four or more inpatients.

Reference Grounding Point. The ground bus of the panelboard or isolated power system panel supplying the patient care area.

Selected Receptacles. A minimum number of electric receptacles to accommodate appliances ordinarily required for local tasks or likely to be used in patient care emergencies.

Task Illumination. Provision for the minimum lighting required to carry out necessary tasks in the described areas, including safe access to supplies and equipment, and access to exits.

Therapeutic High-Frequency Diathermy Equipment. Therapeutic high-frequency diathermy equipment is therapeutic induction and dielectric heating equipment.

X-Ray Installations (Long-Time Rating). A rating based on an operating interval of 5 minutes or longer.

X-Ray Installations (Mobile). X-ray equipment mounted on a permanent base with wheels, casters, or a combination of both to facilitate moving the equipment while completely assembled.

X-Ray Installations (Momentary Rating). A rating based on an operating interval that does not exceed 5 seconds.

X-Ray Installations (Portable). X-ray equipment designed to be hand carried.

X-Ray Installations (Transportable). X-ray equipment to be installed in a vehicle or that may be readily disassembled for transport in a vehicle.

B. Wiring and Protection

517-10. Applicability.

(a) Part B shall apply to patient care areas of all health care facilities.

(b) Part B shall not apply to the following:

(1) Business offices, corridors, waiting rooms, and the like in clinics, medical and dental offices, and outpatient facilities

(2) Areas of nursing homes and limited care facilities wired in accordance with Chapters 1 through 4 of this *Code* where these areas are used exclusively as patient sleeping rooms

FPN: See *Life Safety Code®*, NFPA *101®*-1997.

Section 517-10(b) recognizes that patient sleeping rooms in nursing homes are usually not patient care areas. However, if the patient is examined or treated in the patient bed location, it can be considered a patient care area.

517-11. General Installation — Construction Criteria. It is the purpose of this article to specify the installation criteria and wiring methods that will minimize electrical hazards by the maintenance of adequately low-potential differences only between exposed conductive surfaces that are likely to become energized and could be contacted by a patient.

FPN: In a health care facility, it is difficult to prevent the occurrence of a conductive or capacitive path from the patient's body to some grounded object, because that path may be established accidentally or through instrumentation directly connected to the patient. Other electrically conductive surfaces that may make an additional contact with the patient, or instruments that may be connected to the patient, then become possible sources of electric currents that can traverse the patient's body. The hazard is increased as more apparatus is associated with the patient, and, therefore, more intensive precautions are needed. Control of electric shock hazard requires the limitation of electric current that might flow in an electric circuit involving the patient's body by raising the resistance of the conductive circuit that includes the patient, or by insulating exposed surfaces that might become energized, in addition to reducing the potential difference that can appear between exposed conductive surfaces in the patient vicinity, or by combinations of these methods. A special problem is presented by the patient with an externalized direct conductive path to the heart muscle. The patient may be electrocuted at current levels so low that additional protection in the design of appliances, insulation of the catheter, and control of medical practice are required.

This fine print note recognizes the possibility of increased sensitivity to electric shock by patients whose body resistance may be compromised either accidentally or by a necessary medical procedure. Such diverse situations as incontinence or the insertion of a catheter may render a patient much more vulnerable to the effects of an electric current. For these reasons, it is essential that those responsible for the design, installation, and maintenance of the electrical system in patient care areas be well acquainted with at least the rudiments of the hazard as explained in this note.

Since the original recognition of this hazard in the 1971 *Code,* continued clinical evaluation of the problem has provided a better understanding of the extent of the hazard, bringing about the changes in both value and wiring methods now found in the *Code.*

The *Code* clearly assigns responsibility for the designation of the types of patient care areas to the governing body of the health care facility. Both the design and inspection of a patient care area must, therefore, be based on the governing body's designation rather than the superficial appearance of the area.

517-12. Wiring Methods. Except as modified in this article, wiring methods shall comply with the applicable requirements of Chapters 1 through 4 of this *Code.*

517-13. Grounding of Receptacles and Fixed Electric Equipment.

(a) Patient Care Area. In an area used for patient care, the grounding terminals of all receptacles and all noncurrent-carrying conductive surfaces of fixed electric equipment likely to become energized that are subject to personal contact, operating at over 100 volts, shall be grounded by an insulated copper conductor. The grounding conductor shall be sized in accordance with Table 250-122 and installed in metal raceways with the branch-circuit conductors supplying these receptacles or fixed equipment.

Exception No. 1: Metal raceways shall not be required where listed Types MI, MC, or AC cables are used, provided the outer metal armor or sheath of the cable is identified as an acceptable grounding return path.

Exception No. 2: Metal faceplates shall be permitted to be grounded by means of a metal mounting screw(s) securing the faceplate to a grounded outlet box or grounded wiring device.

Exception No. 3: Light fixtures more than 7½ ft (2.2 m) above the floor and switches located outside of the patient vicinity shall not be required to be grounded by an insulated grounding conductor.

Section 517-13(a) applies to the branch circuits in areas used for patient care and is not limited to patient rooms. Additional areas, such as therapy areas, recreational areas, solaria, and certain patient corridors, are also included. It should be clearly understood that Section 517-13(a) requires grounding by means of an insulated copper conductor installed with the branch-circuit conductors. The conductor can be either solid or stranded. A separate insulated equipment grounding conductor is not required to be run to the branch-circuit panelboard with the feeder conductors in a metal raceway.

Exception No. 1 permits certain listed cable types to be used in lieu of metal raceways only where the cable is specifically identified in Section 250-118 as an acceptable grounding return path. The redundant grounding conductor required by Section 517-3(b) is, of course, still required.

Exception No. 2 permits metal faceplates to be grounded by means of the metal mounting screws rather than by having a separate equipment grounding conductor run to the metal plate. See also Section 380-9(b), which requires switches and their metal faceplates to be effectively grounded.

Exception No. 3 exempts lighting fixtures mounted 7½ ft above the floor and switches located outside of the patient vicinity from redundant grounding requirements, since it is unlikely that the patient or attendants will contact these items and the patient at the same time. The patient vicinity space consists of a volume 6 ft horizontally in all directions from the bed and up to a height of 7½ ft above the floor.

(b) Methods. In addition to the requirements of Section 517-13(a), all branch circuits serving patient care areas shall be provided with a ground path for fault current by installation in a metal raceway system or cable assembly. The metal raceway system, or cable armor or sheath assembly, shall itself qualify as an equipment grounding return path in accordance with Section 250-118. Type MC cable and Type MI cable shall have an outer metal armor or sheath that is identified as an acceptable grounding return path.

The purpose of Section 517-13(b) is to clearly point out what can serve as the required additional equipment grounding conductor. This section requires that a redundant ground path be provided through all metal raceways and cables to receptacles and fixed electrical equipment serving or located within patient care areas. This redundant ground path is in addition to the path through the insulated grounding conductor required in Section 517-13(a). The outer metal jacket of interlocking tape of Type MC cable does not qualify as an equipment grounding return path in accordance with Section 250-118 and would not be accepted for this purpose in the patient care area.

Metal-sheathed cable assemblies are not permitted for *emergency circuits* in the patient vicinity because Section 517-30(c)(3) requires such wiring to be protected by installation in a metal raceway.

Patient care areas are not limited to hospitals. They are also found in other health care facilities, such as nursing homes, clinics, medical and dental offices, and so on.

A device that detects and measures leakage current in equipment is illustrated in Figure 517.1.

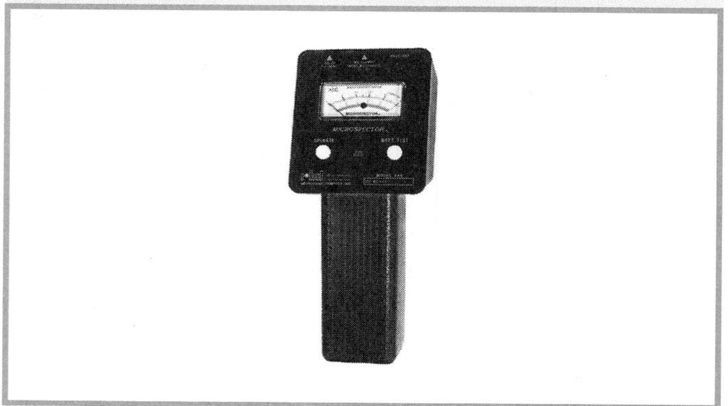

Figure 517.1 *A battery-operated precision instrument for use on all hospital electrical equipment. This instrument measures and detects line voltage and leakage current between the grounding pole of a receptacle and the exposed conductive surfaces of nonelectrical equipment, or between the grounding pole and conductive surfaces of fixed or portable electrical equipment. (Neurodyne-Dempsey, Inc.)*

517-14. Panelboard Bonding. The equipment grounding terminal buses of the normal and essential branch-circuit panelboards serving the same individual patient vicinity shall be bonded together with an insulated continuous copper conductor not smaller than No. 10. Where more than two panels serve the same location, this conductor shall be continuous from panel to panel, but shall be permitted to be broken in order to terminate on the ground bus in each panel.

517-16. Receptacles with Insulated Grounding Terminals. Receptacles with insulated grounding terminals, as permitted in Section 250-146(d), shall be identified; such identification shall be visible after installation.

FPN: Caution is important in specifying such a system with receptacles having insulated grounding terminals, since the grounding impedance is controlled only by the grounding conductors and does not benefit functionally from any parallel grounding paths.

Section 517-16 prevents the indiscriminate use of isolated ground receptacles. Isolated ground receptacles are now required to be identified by an orange triangle located on the face of the receptacle. Older isolated ground receptacles may be identified differently.

517-17. Ground-Fault Protection.

(a) Feeders. Where ground-fault protection is provided for operation of the service disconnecting means or feeder disconnecting means as specified by Sections 230-95 or 215-10, an additional step of ground-fault protection shall be provided in the next level of feeder disconnecting means downstream toward the load. Such protection shall consist of overcurrent devices and current transformers or other equivalent protective equipment that shall cause the feeder disconnecting means to open.

The additional levels of ground-fault protection shall not be installed

(1) On the load side of an essential electrical system transfer switch, or
(2) Between the on-site generating unit(s) described in Section 517-35(b) and the essential electrical system transfer switch(es), or
(3) On electrical systems that are not solidly grounded wye systems with greater than 150 volts to ground, but not exceeding 600 volts phase-to-phase.

(b) Selectivity. Ground-fault protection for operation of the service and feeder disconnecting means shall be fully selective such that the feeder device and not the service device shall open on ground faults on the load side of the feeder device. A six-cycle minimum separation between the service and feeder ground-fault tripping bands shall be provided. Operating time of the disconnecting devices shall be considered in selecting the time spread between these two bands to achieve 100 percent selectivity.

FPN: See Section 230-95, fine print note, for transfer of alternate source where ground-fault protection is applied.

Wherever ground-fault protection (GFP) of equipment is applied to the service providing power to a health care facility, whether by design or by reason of the requirements of Section 230-95, an additional level of ground-fault protection is required downstream. Under this rule, ground-fault protection is required to be applied to every feeder, and additional ground-fault protective devices may be applied farther downstream at the option of the governing body of the health care facility. With proper coordination, this additional ground-fault protection is intended to limit a ground fault to a single feeder and thereby prevent a total outage of the entire health care system. Coordination includes consideration of the trip setting, the time setting, and the time required for operation (opening time) of each level of the ground-fault protection system.

It is not intended that ground-fault protection be installed between the on-site generator and the transfer switch or on the load side of the essential electrical system transfer switch.

However, where a health care installation, such as a doctor's office, is a part of a larger general-use facility (for example, a business office building) that has service ground-fault protection in accordance with Section 230-95, it is not intended that ground-fault protection be required on the health care facility (doctor's office) feeder, since no additional protection will be achieved insofar as the doctor's office is concerned.

Section 230-95(c), FPN No. 3, calls attention to problems that may arise when ground-fault-protected systems are transferred to another supply system.

(c) Testing. When equipment ground-fault protection is first installed, each level shall be performance tested to ensure compliance with Section 517-17(b).

See Section 230-95(c) and its commentary.

517-18. General Care Areas.

(a) Patient Bed Location. Each patient bed location shall be supplied by at least two branch circuits, one from the emergency system and one from the normal system. All branch circuits from the normal system shall originate in the same panelboard.

Exception No. 1: Branch circuits serving only special-purpose outlets or receptacles, such as portable X-ray outlets, shall not be required to be served from the same distribution panel or panels.

Exception No. 2: Requirements of Section 517-18(a) shall not apply to patient bed locations in clinics, medical and dental offices, and outpatient facilities; psychiatric, substance abuse, and rehabilitation hospitals; sleeping rooms of nursing homes and limited care facilities meeting the requirements of Section 517-10(b)(2).

Exception No. 3: A general care patient bed location served from two separate transfer switches on the emergency system shall not be required to have circuits from the normal system.

Patient bed locations in general care areas are prohibited from deriving *all* their branch circuits from the emergency system. At least one branch circuit for each patient bed location is required to originate in a normal system panelboard. This is a reflection of the requirements in Section 517-33.

Revised for the 1999 *Code*, a new Exception No. 3 allows two emergency circuits derived from separate transfer switches, which achieves the same reliability.

(b) Patient Bed Location Receptacles. Each patient bed location shall be provided with a minimum of four receptacles. They shall be permitted to be of the single or duplex types or a combination of both. All receptacles, whether four or more, shall be listed "hospital grade" and so identified. Each receptacle shall be grounded by means of an insulated copper conductor sized in accordance with Table 250-122.

Exception No. 1: Requirements of Section 517-18(b) shall not apply to psychiatric, substance abuse, and rehabilitation hospitals meeting the requirements of Section 517-10(b)(2).

Exception No. 2: Psychiatric security rooms shall not be required to have receptacle outlets installed in the room.

FPN: It is not intended that there be a total, immediate replacement of existing non-hospital grade receptacles. It is intended, however, that non-hospital grade receptacles be replaced with hospital grade receptacles upon modification of use, renovation, or as existing receptacles need replacement.

A significant change was made to Section 517-18(b) in the 1990 *Code*, expanding the requirement for hospital-grade receptacles to include general care patient bed locations. See the commentary following Section 517-19(b).

(c) Pediatric Locations. Fifteen- and 20-ampere, 125-volt receptacles intended to supply patient care areas of pediatric wards, rooms, or areas shall be listed tamper resistant or shall employ a listed tamper resistant cover.

Revised for the 1999 *Code*, Section 517-18(c) now requires that all 15- and 20-ampere, 125-volt receptacles used in pediatric areas must be either listed tamper resistant receptacles or be protected by listed tamper resistant covers. There are no exceptions. The use of locking covers over ordinary receptacles is not a permitted installation method according to the requirements of Section 517-18(c).

517-19. Critical Care Areas.

(a) Patient Bed Location Branch Circuits. Each patient bed location shall be supplied by at least two branch circuits, one or more from the emergency system and one or more circuits from the normal system. At least one branch circuit from the emergency system shall supply an outlet(s) only at that bed location. All branch circuits from the normal system shall be from a single panelboard. Emergency system receptacles shall be identified and shall also indicate the panelboard and circuit number supplying them.

Exception No. 1: Branch circuits serving only special-purpose receptacles or equipment in critical care areas shall be permitted to be served by other panelboards.

Exception No. 2: Critical care locations served from two separate transfer switches on the emergency system shall not be required to have circuits from the normal system.

Exception No. 2 was added in the 1996 *Code* to address a special case where two separate transfer switches supply a single patient care area. Branch circuits supplied from two separate transfer switches provide the same level of redundancy as required by the main requirement.

(b) Patient Bed Location Receptacles.

(1) Each patient bed location shall be provided with a minimum of six receptacles, at least one of which shall be connected to:

(a) The normal system branch circuit required in Section 517-19(a), or
(b) An emergency system branch circuit supplied by a different transfer switch than the other receptacles at the same location.

(2) The above receptacles shall be permitted to be of the single or duplex types, or a combination of both. All receptacles, whether six or more, shall be listed "hospital grade" and so identified. Each receptacle shall be grounded to the reference grounding point by means of an insulated copper equipment grounding conductor.

Section 517-19(a) covers the number and type of branch circuits required for patient bed locations in critical care areas. Section 517-19(b) covers the number of receptacles and the type of circuit supplying them. Each patient bed location must be provided with at least six receptacles. A single receptacle counts as one and a duplex receptacle counts as two receptacles, therefore, three duplex receptacles satisfies the requirement.

Each patient bed location is required to be supplied by at least two branch circuits, one from the *normal* panel and one from the emergency panel, as shown in Figure 517.2. The normal circuits must be supplied from the same panel (L-1). The emergency circuits are permitted to be supplied from different panels (EML-1 and EML-2). However, the emergency branch circuit to patient bed location A cannot supply emergency receptacles for patient bed location B. The patient bed location receptacles can also be supplied

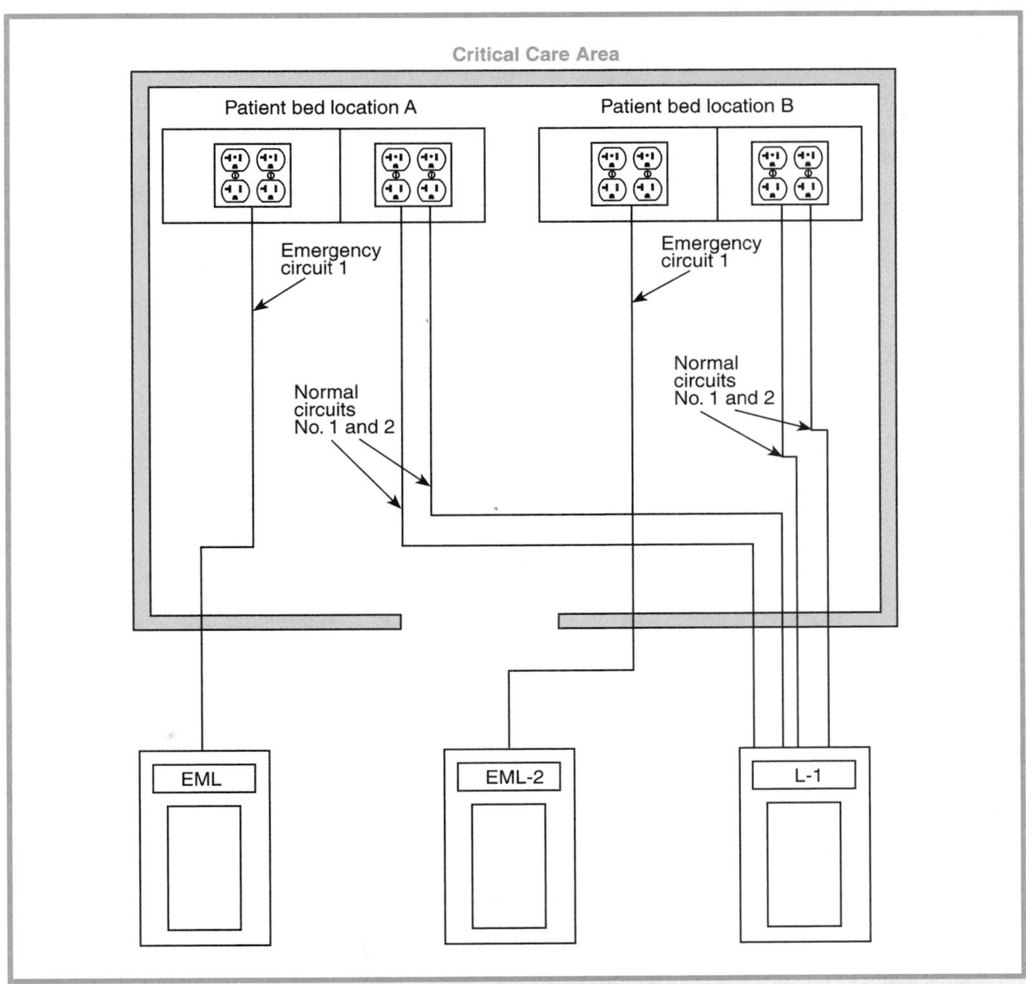

Figure 517.2 *Examples of normal and emergency circuits supplying patient bed locations in a critical care area.*

Figure 517.3 *A hospital-grade receptacle is identified by a green dot on its face. Shown here is a tamper-proof hospital-grade receptacle that fulfills the requirements of Sections 517-18(b) and 517-18(c). (Pass & Seymour/Legrand)*

by two different emergency circuits, instead of one emergency and one normal, provided the emergency circuits are supplied from two different emergency transfer switches.

Receptacles may be of the single or duplex type, provided they are listed hospital-grade type and are identified as such.

A typical method of marking such receptacles is by a green dot on the face of the receptacle. See Figure 517.3. The emergency system receptacles must be marked to indicate the panelboard and circuit number supplying them.

These requirements are intended to ensure that critical care patients will not be without electrical power regardless of whether the equipment, the branch circuits, or the normal system itself is at fault.

(c) Patient Vicinity Grounding and Bonding (Optional). A patient vicinity shall be permitted to have a patient equipment grounding point. The patient equipment grounding point, where supplied, shall be permitted to contain one or more jacks listed for the purpose. An equipment bonding jumper, not smaller than No. 10, shall be used to connect the grounding terminal of all grounding-type receptacles to the patient equipment grounding point. The bonding conductor shall be permitted to be arranged centrically or looped as convenient.

FPN: Where there is no patient equipment grounding point, it is important that the distance between the reference grounding point and the patient vicinity be as short as possible to minimize any potential differences.

(d) Panelboard Grounding. Where a grounded electrical distribution system is used, and metal feeder raceway or Type MC or MI cable is installed, grounding of a panelboard or switchboard shall be ensured by one of the following means at each termination or junction point of the raceway or Type MC or MI cable.

(1) A grounding bushing and a continuous copper bonding jumper, sized in accordance with Section 250-122, with the bonding jumper connected to the junction enclosure or the ground bus of the panel
(2) Connection of feeder raceways or Type MC or MI cable to threaded hubs or bosses on terminating enclosures
(3) Other approved devices such as bonding-type locknuts or bushings

(e) Additional Protective Techniques in Critical Care Areas (Optional). Isolated power systems shall be permitted to be used for critical care areas, and, if used, the isolated power system equipment shall be listed for the purpose and the system designed and installed so that it meets the provisions of and is in accordance with Section 517-160.

Exception: The audible and visual indicators of the line isolation monitor shall be permitted to be located at the nursing station for the area being served.

(f) Isolated Power System Grounding. Where an isolated ungrounded power source is used and limits the first-fault current to a low magnitude, the grounding conductor associated with the secondary circuit shall be permitted to be run outside of the enclosure of the power conductors in the same circuit.

FPN: Although it is permitted to run the grounding conductor outside of the conduit, it is safer to run it with the power conductors to provide better protection in case of a second ground fault.

Running the conductor inside the raceway with the conductors delivering the fault current reduces the impedance of the grounding path.

(g) Special-Purpose Receptacle Grounding. The equipment grounding conductor for special-purpose receptacles, such as the operation of mobile X-ray equipment, shall be extended to the reference grounding points of branch circuits for all locations likely to be served from such receptacles. Where such a circuit is served from an isolated ungrounded system, the grounding conductor shall not be required to be run with the power conductors; however, the equipment grounding terminal of the special-purpose receptacle shall be connected to the reference grounding point.

517-20. Wet Locations.

(a) All receptacles and fixed equipment within the area of the wet location shall have ground-fault circuit-interrupter protection for personnel if interruption of power under fault conditions can be tolerated, or be served by an isolated power system if such interruption cannot be tolerated.

Exception: Branch circuits supplying only listed, fixed, therapeutic and diagnostic equipment shall be permitted to be supplied from a normal grounded service, single- or 3-phase system, provided that

(a) Wiring for grounded and isolated circuits does not occupy the same raceway, and
(b) All conductive surfaces of the equipment are grounded.

(b) Where an isolated power system is utilized, the equipment shall be listed for the purpose and installed so that it meets the provisions of and is in accordance with Section 517-160.

In areas that are designated patient care wet locations by the governing body of the facility, ground-fault circuit-interrupter protection is required for the protection of receptacles and fixed equipment if a circuit interruption can be tolerated. Otherwise, an isolated power system is required. See the commentary following the definition of *patient care area* in Section 517-3.

FPN: For requirements for installation of therapeutic pools and tubs, see Part F of Article 680.

517-21. Ground-Fault Circuit-Interrupter Protection for Personnel. Ground-fault circuit-interrupter protection for personnel shall not be required for receptacles installed in those critical care areas where the toilet and basin are installed within the patient room.

Section 517-21 exempts only receptacles installed in special critical care areas where the toilet and basin are installed within the patient's room. Section 517-21 does not exempt the requirement for GFCI receptacles in bathrooms for patients, staff, or the public, as required by Section 210-8(b)(1).

C. Essential Electrical System

517-25. Scope. The essential electrical system for these facilities shall comprise a system capable of supplying a limited amount of lighting and power service, which is considered essential for life safety and orderly cessation of procedures during the time normal electrical service is interrupted for any reason. This includes clinics, medical and dental offices, outpatient facilities, nursing homes, limited care facilities, hospitals, and other health care facilities serving patients.

FPN: For information as to the need for an essential electrical system, see *Standard for Health Care Facilities,* NFPA 99-1996.

517-30. Essential Electrical Systems for Hospitals.

(a) Applicability. The requirements of Part C, Sections 517-30 through 517-35, shall apply to hospitals where an essential electrical system is required.

FPN No. 1: For performance, maintenance, and testing requirements of essential electrical systems in hospitals, see *Standard for Health Care Facilities,* NFPA 99-1996. For installation of centrifugal fire pumps, see *Standard for the Installation of Centrifugal Fire Pumps,* NFPA 20-1996.

FPN No. 2: For additional information, see *Standard for Health Care Facilities,* NFPA 99-1996.

(b) General.

x**(1)** Essential electrical systems for hospitals shall be comprised of two separate systems capable of supplying a limited amount of lighting and power service, which is considered essential for life safety and effective hospital operation during the time the normal electrical service is interrupted for any reason. These two systems shall be the emergency system and the equipment system.

x**(2)** The emergency system shall be limited to circuits essential to life safety and critical patient care. These are designated the life safety branch and the critical branch.

x**(3)** The equipment system shall supply major electrical equipment necessary for patient care and basic hospital operation.

x**(4)** The number of transfer switches to be used shall be based on reliability, design, and load considerations. Each branch of the essential electrical system shall be served by one or more transfer switches as shown in Figures 517-30(a) and 517-30(b). One transfer switch shall be permitted to serve one or more branches or systems in a facility with a maximum demand on the essential electrical system of 150 kVA as shown in Figure 517-30(c).

FPN: See *Standard for Health Care Facilities,* NFPA 99-1996: 3-4.3.2, Transfer Switch Operation Type I; 3-4.2.1.4, Automatic Transfer Switch Features; and 3-4.2.1.6, Nonautomatic Transfer Device Features.

Figures 517-30(a), 517-30(b), and 517-30(c) in the *Code* indicate possible electrical system connections for hospitals. Figure 517-30(b) illustrates a common situation, where the expansion of the facility necessitates the addition of a normal source to that already in place. As shown in the diagram, this would not necessarily require the addition of another alternate source. Provided the alternate source has a capacity for all the intended load, it may serve multiple services.

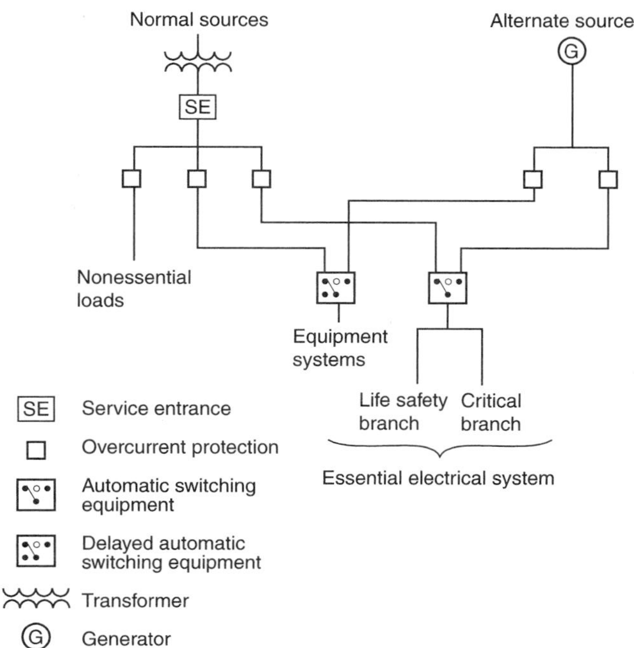

Figure 517-30(a) Small electrical system — hospitals.

For a small electrical system having a maximum demand on the essential electrical system of 150 kVA, see Figure 517-30(c). A small load can be served by a single transfer switch that can handle the loads associated with both the emergency system and the equipment system. This, of course, is based on the assumption that the transfer switch has sufficient capacity to handle the combined loads and that the alternate source of power is sufficiently large to withstand the impact of the simultaneous transfer of both systems in the event of a normal power loss. For further explanation of loads permitted on an emergency system, see NFPA 99-1996, *Standard for Health Care Facilities,* Section 3-4.2.2.

(5) Other Loads. Loads served by the generating equipment not specifically named in Sections 517-32, 517-33, and 517-34 shall be served by their own transfer switches such that these loads

(a) Shall not be transferred if the transfer will overload the generating equipment, and

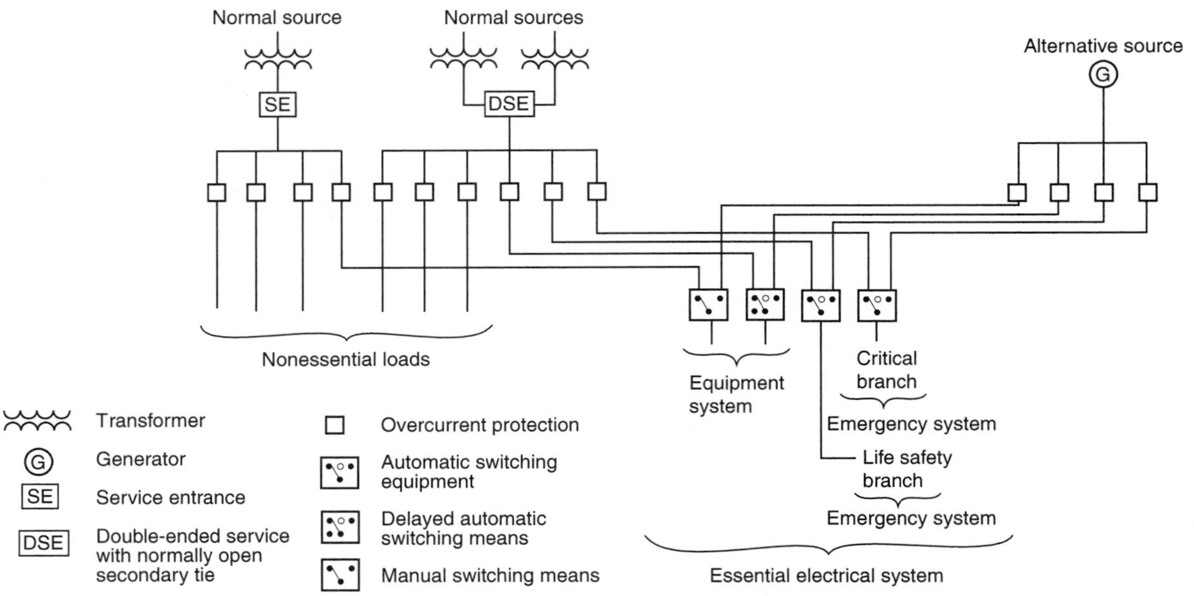

Figure 517-30(b) Typical large electrical system — hospitals.

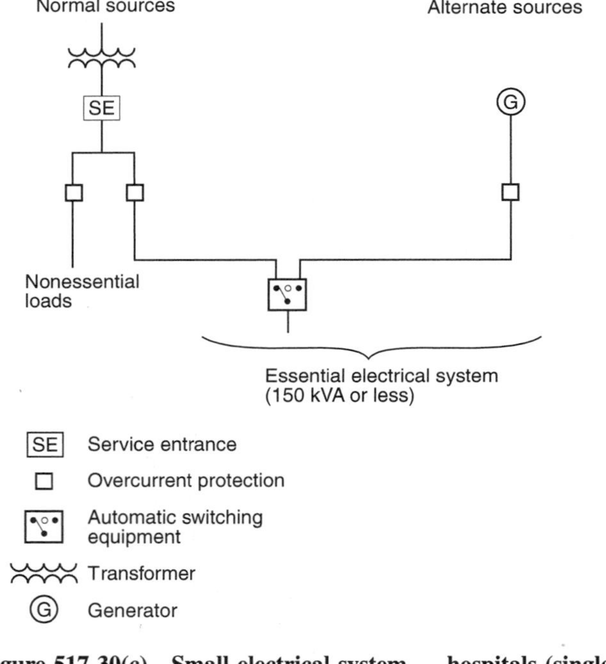

Figure 517-30(c) Small electrical system — hospitals (single transfer switch).

(b) Shall be automatically shed upon generating equipment overloading.

(6) Hospital power sources and alternate power sources shall be permitted to serve the essential electrical systems of contiguous or same site facilities.

Section 517-30(b)(6) was added to the 1999 *Code* and appears in 12-3.3.2 of NFPA 99-1996, *Standard for Health Care Facilities.*

(c) Wiring Requirements.

(1) Separation from Other Circuits. The life safety branch and critical branch of the emergency system shall be kept entirely independent of all other wiring and equipment and shall not enter the same raceways, boxes, or cabinets with each other or other wiring.

Wiring of the life safety branch and the critical branch shall be permitted to occupy the same raceways, boxes, or cabinets of other circuits not part of the branch where such wiring is as follows:

(1) In transfer equipment enclosures, or
(2) In exit or emergency lighting fixtures supplied from two sources, or
(3) In a common junction box attached to exit or emergency lighting fixtures supplied from two sources, or
(4) For two or more emergency circuits supplied from the same branch

Section 517-30(c)(1)(4) permits multiwire branch circuits supplied by the same panelboard.

The wiring of the equipment system shall be permitted to occupy the same raceways, boxes, or cabinets of other circuits that are not part of the emergency system.

(2) Isolated Power Systems. Where isolated power systems are installed in any of the areas in Sections 517-33(a)(1) and (a)(2), each system shall be supplied by an individual circuit serving no other load.

(3) Mechanical Protection of the Emergency System. The wiring of the emergency system of a hospital shall be mechanically protected by installation in nonflexible metal raceways, or shall be wired with Type MI cable. Where installed as branch circuit conductors serving patient care areas, the installation shall comply with the requirements of Section 517-13.

The wiring of emergency systems in hospitals requires additional protection not normally required in other occupancies. Only metal raceways of the nonflexible type and Type MI cable are permitted as a wiring method for hospital emergency systems. (Emergency system circuits can include services, feeders, and branch circuits.) There are five exceptions to this requirement. As described in Section 517-31, the emergency system in a hospital consists of the life safety branch and the critical branch.

Added as a new last sentence to Section 517-30(c)(3) for the 1999 *Code,* the metal raceway (or Type MI cable) requirement may also require redundant grounding where the emergency circuit extends to patient care area receptacles or fixed equipment covered in Section 517-13.

Exception No. 1: Flexible power cords of appliances, or other utilization equipment, connected to the emergency system shall not be required to be enclosed in raceways.

Exception No. 2: Secondary circuits of transformer-powered communications or signaling systems shall not be required to be enclosed in raceways unless otherwise specified by Chapters 7 or 8.

Exception No. 2 exempts nurses' call, telephone, and alarm circuits from being run in metal raceways, provided they comply with their applicable articles elsewhere in the *Code.* Although this provides substantial latitude in the wiring method, it should be noted that the restrictions of Section 300-22 (ducts, plenums, and other air-handling spaces) still apply, unless conductors listed as having adequate fire-resistant and low smoke-producing characteristics are used.

Exception No. 3: Schedule 80 rigid nonmetallic conduit shall be permitted if the branch circuits do not serve patient care areas and it is not prohibited elsewhere in this Code.

Exception No. 4: Where encased in not less than 2 in. (50.8 mm) of concrete, Schedule 40 rigid nonmetallic conduit or electrical nonmetallic tubing shall be permitted if the branch ciruits do not serve patient care areas.

Exception No. 5: Flexible metal raceways and cable assemblies shall be permitted to be used in listed prefabricated medical headwalls, listed office furnishings, or where necessary for flexible connection to equipment.

Exception No. 5 recognizes that flexible connections are sometimes required for equipment and, for practical reasons, permits their limited use.

FPN: See Section 517-13(b) for additional grounding requirements in patient care areas.

(d) Capacity of Systems. The essential electrical system shall have adequate capacity to meet the demand for the operation of all functions and equipment to be served by each system and branch.

Feeders shall be sized in accordance with Articles 215 and 220. The generator set(s) shall have sufficient capacity and proper rating to meet the demand produced by the load of the essential electrical system(s) at any one time.

Demand calculations for sizing of the generator set(s) shall be based on the following:

(1) Prudent demand factors and historical data, or
(2) Connected load, or
(3) Feeder calculation procedures described in Article 220, or
(4) Any combination of the above

The intent of Section 517-30(d) is to mandate the sizing of generators based on actual demand likely to be produced by the connected load of the system at any one time. This method of calculation should help prevent the gross oversizing of generators in health care facilities.

x**517-31. Emergency System.** Those functions of patient care depending on lighting or appliances that are connected to the emergency system shall be divided into two mandatory branches: the life safety branch and the critical branch, described in Sections 517-32 and 517-33.

The branches of the emergency system shall be installed and connected to the alternate power source so that all functions specified herein for the emergency system shall be automatically restored to operation within 10 seconds after interruption of the normal source.

x**517-32. Life Safety Branch.** No function other than those listed in (a) through (f) shall be connected to the life safety branch. The life safety branch of the emergency system shall supply power for the following lighting, receptacles, and equipment.

(a) Illumination of Means of Egress. Illumination of means of egress, such as lighting required for corridors, passageways, stairways, and landings at exit doors, and all necessary ways of approach to exits. Switching arrangements to transfer patient corridor lighting in hospitals from general illumination circuits to night illumination circuits shall be permitted provided only one of two circuits can be selected and both circuits cannot be extinguished at the same time.

FPN: See *Life Safety Code,* NFPA *101*-1997, Sections 5-8 and 5-9.

(b) Exit Signs. Exit signs and exit directional signs.

FPN: See *Life Safety Code,* NFPA *101*-1997, Section 5-10.

(c) Alarm and Alerting Systems. Alarm and alerting systems including the following:

(1) Fire alarms

FPN: See *Life Safety Code,* NFPA *101*-1997, Sections 7-6 and 12-3.4.

(2) Alarms required for systems used for the piping of nonflammable medical gases

FPN: See *Standard for Health Care Facilities,* NFPA 99-1996, 12-3.4.1.

(d) Communications Systems. Hospital communications systems, where used for issuing instructions during emergency conditions.

(e) Generator Set Location. Task illumination battery charger for emergency battery-powered lighting unit(s) and selected receptacles at the generator set location.

(f) Elevators. Elevator cab lighting, control, communications, and signal systems.

ˣ517-33. Critical Branch.

ˣ(a) Task Illumination and Selected Receptacles. The critical branch of the emergency system shall supply power for task illumination, fixed equipment, selected receptacles, and special power circuits serving the following areas and functions related to patient care.

(1) Critical care areas that utilize anesthetizing gases — task illumination, selected receptacles, and fixed equipment
(2) The isolated power systems in special environments
(3) Patient care areas — task illumination and selected receptacles in the following:
   (a) Infant nurseries
   (b) Medication preparation areas
   (c) Pharmacy dispensing areas
   (d) Selected acute nursing areas
   (e) Psychiatric bed areas (omit receptacles)
   (f) Ward treatment rooms
   (g) Nurses' stations (unless adequately lighted by corridor luminaires)
(4) Additional specialized patient care task illumination and receptacles, where needed
(5) Nurse call systems
(6) Blood, bone, and tissue banks
(7) Telephone equipment rooms and closets
(8) Task illumination, selected receptacles, and selected power circuits for the following:
   (a) General care beds (at least one duplex receptacle per patient bedroom)
   (b) Angiographic labs
   (c) Cardiac catheterization labs
   (d) Coronary care units
   (e) Hemodialysis rooms or areas
   (f) Emergency room treatment areas (selected)
   (g) Human physiology labs
   (h) Intensive care units
   (i) Postoperative recovery rooms (selected)
(9) Additional task illumination, receptacles, and selected power circuits needed for effective hospital operation. Single-phase fractional horsepower motors shall be permitted to be connected to the critical branch.

The critical branch is intended to serve a limited number of receptacles and locations, to reduce the load and to minimize the chances of a fault condition. Receptacles in general patient care area corridors are permitted on the critical branch, but they are required to be identified in some manner (color-coded or labeled) as part of the critical branch, in accordance with Section 517-33(c).

ˣ(b) Subdivision of the Critical Branch. It shall be permitted to subdivide the critical branch into two or more branches.

FPN: It is important to analyze the consequences of supplying an area with only critical care branch power when failure occurs between the area and the transfer switch. Some proportion of normal and critical power, or critical power from separate transfer switches, may be appropriate.

ˣ(c) Receptacle Identification. The receptacles or the faceplates for receptacles supplied by the critical branch shall have a distinctive color or marking so as to be readily recognizable.

Section 517-33(c) was added for the 1999 *Code.* As shown in Appendix A, the requirement of this section has been extracted from 3-4.2.2.4(b)(2) of NFPA 99-1996, *Standard for Health Care Facilities.*

ˣ517-34. Equipment System Connection to Alternate Power Source. The equipment system shall be installed and connected to the alternate power source, such that the equipment described in Section 517-34(a) is automatically restored to operation at appropriate time-lag intervals following the energizing of the emergency system. Its arrangement shall also provide for the subsequent connection of equipment described in Section 517-34(b).

ˣ(a) Equipment for Delayed Automatic Connection. The following equipment shall be arranged for delayed automatic connection to the alternate power source.

(1) Central suction systems serving medical and surgical functions, including controls. Such suction systems shall be permitted on the critical branch.
(2) Sump pumps and other equipment required to operate for the safety of major apparatus, including associated control systems and alarms.
(3) Compressed air systems serving medical and surgical functions, including controls. Such air systems shall be permitted on the critical branch.
(4) Smoke control and stair pressurization systems, or both.
(5) Kitchen hood supply or exhaust systems, or both, if required to operate during a fire in or under the hood.

Exception: Sequential delayed automatic connection to the alternate power source to prevent overloading the generator shall be permitted where engineering studies indicate it is necessary.

[x]**(b) Equipment for Delayed Automatic or Manual Connection.** The following equipment shall be arranged for either delayed automatic or manual connection to the alternate power source:

(1) Heating equipment to provide heating for operating, delivery, labor, recovery, intensive care, coronary care, nurseries, infection/isolation rooms, emergency treatment spaces, and general patient rooms

Exception: Heating of general patient rooms and infection/isolation rooms during disruption of the normal source shall not be required under any of the following conditions:

(a) The outside design temperature is higher than +20°F (–6.7°C), or
(b) The outside design temperature is lower than +20°F (–6.7°C) and where a selected room(s) is provided for the needs of all confined patients, then only such room(s) need be heated, or
(c) The facility is served by a dual source of normal power.

FPN No. 1: The design temperature is based on the 97½ percent design value as shown in Chapter 24 of the ASHRAE *Handbook of Fundamentals* (1997).

FPN No. 2: For a description of a dual source of normal power, see Section 517-35(c), FPN.

In some areas, it is common practice to install individual room heating/air conditioners rather than have a central heating/air-conditioning plant. Where these individual units are electrically powered, it may not be practical to apply this high-demand load to the generator. Where the governing body of the nursing home has full-time skilled attendants who can move people to one room(s) that will be heated when this smaller load is picked up by the generator, the intent of the *Code* is satisfied. The provisions for limited heating during emergency conditions are based on consideration of outside design temperature.

(2) An elevator(s) selected to provide service to patient, surgical, obstetrical, and ground floors during interruption of normal power

In instances where interruption of normal power would result in other elevators stopping between floors, throw-over facilities shall be provided to allow the temporary operation of any elevator for the release of patients or other persons who may be confined between floors.

(3) Supply, return, and exhaust ventilating systems for surgical and obstetrical delivery suites, intensive care, coronary care, nurseries, infection/isolation rooms, emergency treatment spaces, and exhaust fans for laboratory fume hoods, nuclear medicine areas where radioactive material is used, ethylene oxide evacuation, and anesthesia evacuation
(4) Hyperbaric facilities
(5) Hypobaric facilities
(6) Automatically operated doors
(7) Minimal electrically heated autoclaving equipment shall be permitted to be arranged for either automatic or manual connection to the alternate source
(8) Controls for equipment listed in Section 517-34
(9) Other selected equipment shall be permitted to be served by the equipment system.

517-35. Sources of Power.

[x]**(a) Two Independent Sources of Power.** Essential electrical systems shall have a minimum of two independent sources of power: a normal source generally supplying the entire electrical system and one or more alternate sources for use when the normal source is interrupted.

(b) Alternate Source of Power. The alternate source of power shall be one of the following:

(1) Generator(s) driven by some form of prime mover(s) and located on the premises
(2) Another generating unit(s) where the normal source consists of a generating unit(s) located on the premises
(3) An external utility service when the normal source consists of a generating unit(s) located on the premises

(c) Location of Essential Electrical System Components. Careful consideration shall be given to the location of the spaces housing the components of the essential electrical system to minimize interruptions caused by natural forces common to the area (e.g., storms, floods, earthquakes, or hazards created by adjoining structures or activities). Consideration shall also be given to the possible interruption of normal electrical services resulting from similar causes as well as possible disruption of normal electrical service due to internal wiring and equipment failures.

FPN: Facilities in which the normal source of power is supplied by two or more separate central station-fed services experience greater than normal electrical service reliability than those with only a single feed. Such a dual source of normal power consists of two or more electrical services fed from separate generator sets or a utility distribution network that has multiple power input sources and is arranged to provide mechanical and electrical separation so that a fault between the facility and the generating sources will not likely cause an interruption of more than one of the facility service feeders.

517-40. Essential Electrical Systems for Nursing Homes and Limited Care Facilities.

x**(a) Applicability.** The requirements of Part C, Sections 517-40(c) through 517-44, shall apply to nursing homes and limited care facilities.

Exception: The requirements of Part C, Section 517-40(c) through 517-44, shall not apply to freestanding buildings used as nursing homes and limited care facilities, provided that:

(a) Admitting and discharge policies are maintained that preclude the provision of care for any patient or resident who may need to be sustained by electrical life-support equipment.
(b) No surgical treatment requiring general anesthesia is offered.
(c) An automatic battery-operated system(s) or equipment is provided that shall be effective for at least 1½ hours and is otherwise in accordance with Section 700-12 and that shall be capable of supplying lighting for exit lights, exit corridors, stairways, nursing stations, medical preparation areas, boiler rooms, and communications areas. This system shall also supply power to operate all alarm systems.

NFPA 99-1996, *Standard for Health Care Facilities,* recognizes two classes of nursing homes or limited care facilities. For the smaller, less complex facility, only a minimum alternate lighting and alarm service need be furnished.

Where treatment of patients is provided, the requirements of Sections 517-41 through 517-44 are required to be applied. The branches of the emergency system for this class of occupancy bear identical titles to their counterparts for hospital-type occupancies.

FPN: See *Life Safety Code,* NFPA 101-1997.

(b) Inpatient Hospital Care Facilities. Nursing homes and limited care facilities that provide inpatient hospital care shall comply with the requirements of Part C, Sections 517-30 through 517-35.

Regardless of the name given to the facility, the type of electrical system depends on the type of patient care provided. Where such care is clearly inpatient hospital care, a hospital-type electrical system is required to be installed.

(c) Facilities Contiguous or Located on the Same Site with Hospitals. Nursing homes and limited care facilities that are contiguous or located on the same site with a hospital shall be permitted to have their essential electrical systems supplied by that of the hospital.

Where a nursing home or limited care facility shares essentially the same building with a hospital, the nursing home is not required to have its own essential electrical system if it derives its power supply from the hospital. It should be noted, however, that this rule applies only to the electrical supply and does not permit the sharing of transfer devices and the like.

FPN: For performance, maintenance, and testing requirements of essential electrical systems in nursing homes and limited care facilities, see *Standard for Health Care Facilities,* NFPA 99-1996.

Section 517-50 and NFPA 99-1996, *Standard for Health Care Facilities,* require an alternate source of power for those types of health care facilities where patients are treated in essentially the same manner as in hospitals, even though the facility may not be called a hospital.

517-41. Essential Electrical Systems.

x**(a) General.** Essential electrical systems for nursing homes and limited care facilities shall be comprised of two separate branches capable of supplying a limited amount of lighting and power service, which is considered essential for the protection of life safety and effective operation of the institution during the time normal electrical service is interrupted for any reason. These two separate branches shall be the life safety branch and the critical branch.

x**(b) Transfer Switches.** The number of transfer switches to be used shall be based on reliability, design, and load considerations. Each branch of the essential electrical system shall be served by one or more transfer switches as shown in Figures 517-41(a) and 517-41(b). One transfer switch shall be permitted to serve one or more branches or systems in a facility with a maximum demand on the essential electrical system of 150 kVa as shown in Figure 517-41(c).

FPN: See *Standard for Health Care Facilities,* NFPA 99-1996, 3-5.3.2, Transfer Switch Operation Type II; 3-4.2.1.4, Automatic Transfer Switch Features; and 3-4.2.1.6, Nonautomatic Transfer Device Features.

(c) Capacity of System. The essential electrical system shall have adequate capacity to meet the demand for the operation of all functions and equipment to be served by each branch at one time.

(d) Separation from Other Circuits. The life safety branch shall be kept entirely independent of all other wiring and equipment and shall not enter the same raceways, boxes, or cabinets with other wiring except as follows:

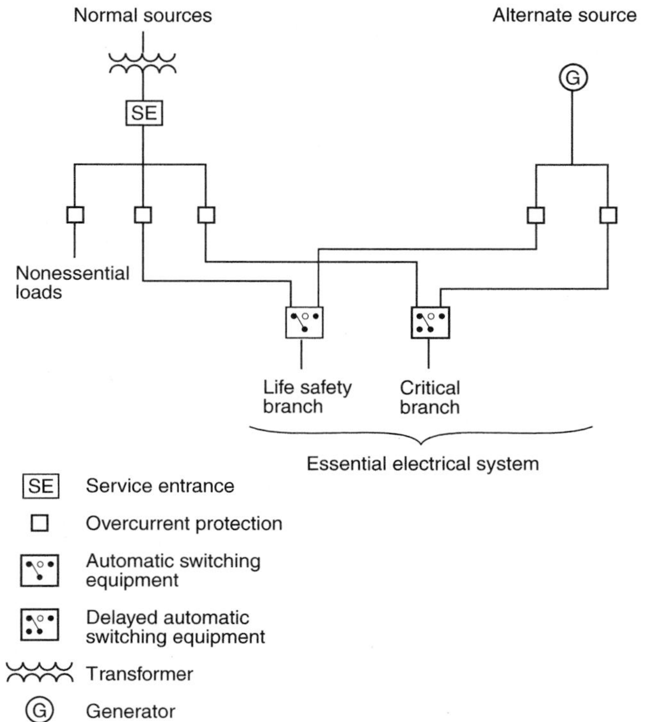

Figure 517-41(a) Small electrical system — nursing homes and limited care facilities.

(1) In transfer switches
(2) In exit or emergency lighting fixtures supplied from two sources, or
(3) In a common junction box attached to exit or emergency lighting fixtures supplied from two sources

The wiring of the critical branch shall be permitted to occupy the same raceways, boxes, or cabinets of other circuits that are not part of the life safety branch.

ˣ517-42. Automatic Connection to Life Safety Branch. The life safety branch shall be installed and connected to the alternate source of power so that all functions specified herein shall be automatically restored to operation within 10 seconds after the interruption of the normal source. No functions other than those listed in (a) through (g) shall be connected to the life safety branch. The life safety branch shall supply power for the following lighting, receptacles, and equipment.

FPN: The life safety branch is called the emergency system in *Standard for Health Care Facilities,* NFPA 99-1996.

(a) Illumination of Means of Egress. Illumination of means of egress as is necessary for corridors, passageways, stairways, landings, and exit doors and all ways of approach to exits. Switching arrangement to transfer patient corridor lighting from general illumination circuits shall be permitted providing only one of two circuits can be selected and both circuits cannot be extinguished at the same time.

FPN: See *Life Safety Code,* NFPA *101*-1997, Sections 5-8 and 5-9.

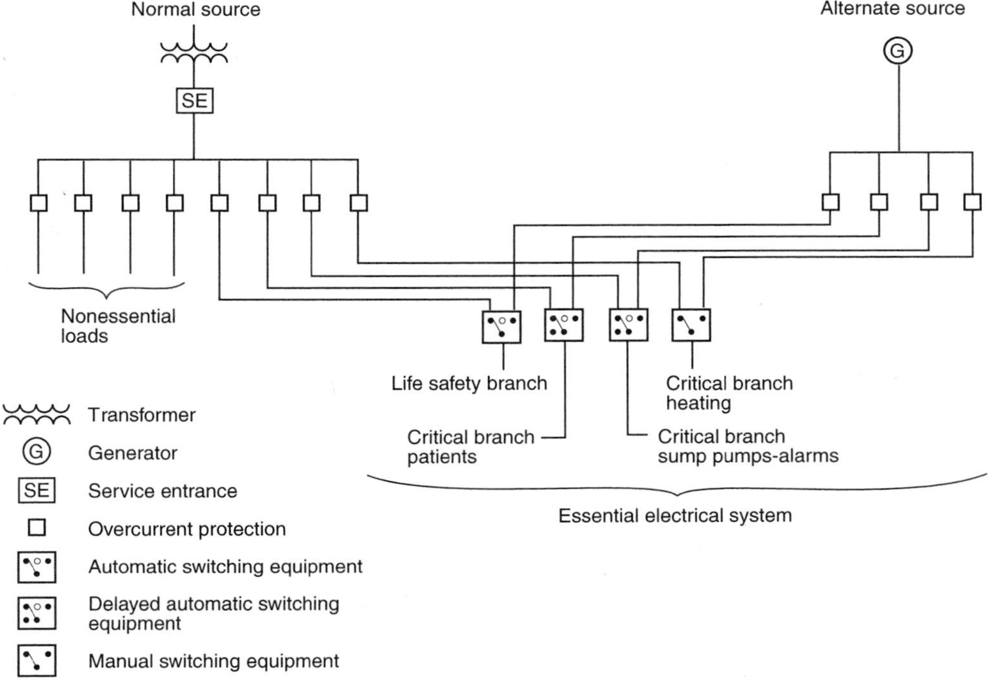

Figure 517-41(b) Typical large electrical system — nursing homes and limited care facilities.

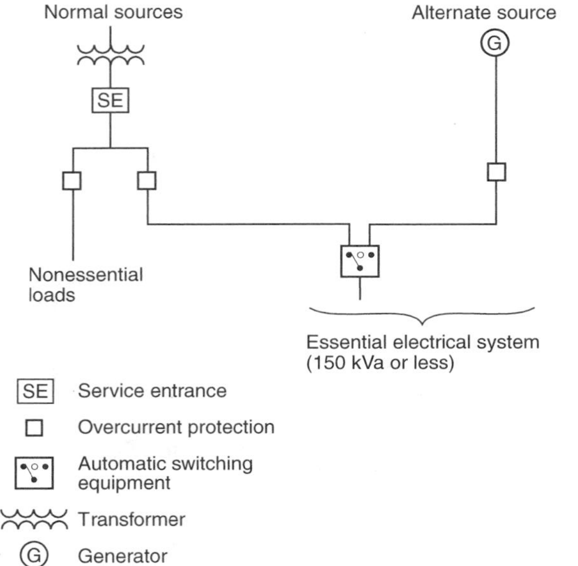

Figure 517-41(c) Small electrical system — nursing homes and limited care facilities (single transfer switch).

(b) Exit Signs. Exit signs and exit directional signs.

FPN: See *Life Safety Code*, NFPA *101*-1997, Section 5-10.

(c) Alarm and Alerting Systems. Alarm and alerting systems, including the following:

(1) Fire alarms

FPN: See *Life Safety Code,* NFPA *101*-1997, Sections 7-6 and 12-3.4.

(2) Alarms required for systems used for the piping of nonflammable medical gases

FPN: See *Standard for Health Care Facilities,* NFPA 99-1996, Section 16-3.4.1.

(d) Communications Systems. Communications systems, where used for issuing instructions during emergency conditions.

(e) Dining and Recreation Areas. Sufficient lighting in dining and recreation areas to provide illumination to exit ways.

(f) Generator Set Location. Task illumination and selected receptacles in the generator set location.

(g) Elevators. Elevator cab lighting, control, communications, and signal systems.

x**517-43. Connection to Critical Branch.** The critical branch shall be installed and connected to the alternate power source so that the equipment listed in Section 517-43(a) shall be automatically restored to operation at appropriate time-lag intervals following the restoration of the life safety branch to operation. Its arrangement shall also provide for the additional connection of equipment listed in Section 517-43(b) by either delayed automatic or manual operation.

(a) Delayed Automatic Connection. The following equipment shall be connected to the critical branch and shall be arranged for delayed automatic connection to the alternate power source.

(1) Patient care areas — task illumination and selected receptacles in the following:
   (a) Medication preparation areas
   (b) Pharmacy dispensing areas
   (c) Nurses' stations (unless adequately lighted by corridor luminaires)
(2) Sump pumps and other equipment required to operate for the safety of major apparatus and associated control systems and alarms
(3) Smoke control and stair pressurization systems
(4) Kitchen hood supply and/or exhaust systems, if required to operate during a fire in or under the hood

(b) Delayed Automatic or Manual Connection. The following equipment shall be connected to the critical branch and shall be arranged for either delayed automatic or manual connection to the alternate power source.

(1) Heating equipment to provide heating for patient rooms.

Exception: Heating of general patient rooms during disruption of the normal source shall not be required under any of the following conditions.

(a) The outside design temperature is higher than +20°F (−6.7°C), or
(b) The outside design temperature is lower than +20°F (−6.7°C) and where a selected room(s) is provided for the needs of all confined patients, then only such room(s) need be heated.
(c) The facility is served by a dual source of normal power as described in Section 517-44(c), FPN.

FPN: The outside design temperature is based on the 97½ percent design values as shown in Chapter 24 of the ASHRAE *Handbook of Fundamentals* (1997).

(2) Elevator Service. In instances where disruption of power would result in elevators stopping between floors, throw-over facilities shall be provided to allow the temporary operation of any elevator for the release of passengers. For elevator cab lighting, control, and signal system requirements, see Section 517-42(g).

(3) Additional illumination, receptacles, and equipment shall be permitted to be connected only to the critical branch.

517-44. Sources of Power.

x**(a) Two Independent Sources of Power.** Essential electrical systems shall have a minimum of two independent sources of power: a normal source generally supplying the entire electrical system and one or more alternate sources for use when the normal source is interrupted.

[x](b) **Alternate Source of Power.** The alternate source of power shall be a generator(s) driven by some form of prime mover(s) and located on the premises.

Exception No. 1: Where the normal source consists of generating units on the premises, the alternate source shall be either another generator set or an external utility service.

Exception No. 2: Nursing homes or limited care facilities meeting the requirements of Section 517-40(a), Exception, shall be permitted to use a battery system or self-contained battery integral with the equipment.

(c) Location of Essential Electrical System Components. Careful consideration shall be given to the location of the spaces housing the components of the essential electrical system to minimize interruptions caused by natural forces common to the area (e.g., storms, floods, earthquakes, or hazards created by adjoining structures or activities). Consideration shall also be given to the possible interruption of normal electrical services resulting from similar causes as well as possible disruption of normal electrical service due to internal wiring and equipment failures.

FPN: Facilities in which the normal source of power is supplied by two or more separate central station-fed services experience greater than normal electrical service reliability than those with only a single feed. Such a dual source of normal power consists of two or more electrical services fed from separate generator sets or a utility distribution network that has multiple power input sources and is arranged to provide mechanical and electrical separation so that a fault between the facility and the generating sources will not likely cause an interruption of more than one of the facility service feeders.

517-45. Essential Electrical Systems for Ambulatory Health Care Centers.

(a) Applicability. The requirements of this section shall apply to those health care facilities described in Section 517-45.

[x]**(b) Connections.** The essential electrical system shall supply power for the following:

(1) Task illumination that is related to the safety of life and that is necessary for the safe cessation of procedures in progress
(2) All anesthesia and resuscitative equipment used in areas where inhalation anesthetics are administered to patients, including alarm and alerting devices

FPN: See *Standard for Health Care Facilities,* NFPA 99-1996, 13-3.4.1.

(3) All electrical life-support equipment in areas where procedures are performed that require such equipment for the support of the patient's life

(c) Alternate Source of Power.

[x]**(1) Power Source.** The alternate source of power for the system shall be specifically designed for this purpose and shall be either a generator, battery system, or self-contained battery integral with the equipment. Where critical care areas are present in the facility, the essential electrical system shall be as required in Sections 517-30 through 517-35.

[x]**(2) System Capacity.** The alternate source of power shall be separate and independent of the normal source and shall have a capacity to sustain its connected loads for a minimum of 1½ hours after loss of the normal source.

[x]**(3) System Operation.** The system shall be arranged so that, in the event of a failure of the normal power source, the alternate source of power shall be automatically connected to the load within 10 seconds.

FPN: See *Standard for Health Care Facilities,* NFPA 99-1996, 3-6.3.2, Transfer Switch Operation for Type III with Generator Sets, and 3-6.3.3, Transfer Switch Operation for Type III with Battery Systems.

517-50. Essential Electrical Systems for Clinics, Medical and Dental Offices, and Other Health Care Facilities Not Covered in Sections 517-30, 517-40, and 517-45.

(a) Applicability. The requirements of this section shall apply to those health care facilities described in Section 517-50.

[x]**(b) Connections.** The essential electrical system shall supply power for the following:

(1) Task illumination that is related to the safety of life and that is necessary for the safe cessation of procedures in progress
(2) All anesthesia and resuscitative equipment used in areas where inhalation anesthetics are administered to patients, including alarm and alerting devices

FPN: See *Standard for Health Care Facilities,* NFPA 99-1996, 14-3.4 and 15-3.4.1.

(c) Alternate Source of Power.

[x]**(1) Power Source.** The alternate source of power for the system shall be specifically designed for this purpose and shall be either a generator, battery system, or self-contained battery integral with the equipment. Where electrical life-support equipment is required, the essential electrical system shall be as required in Sections 517-30 through 517-35.

[x]**(2) System Capacity.** The alternate source of power shall be separate and independent of the normal source and shall have a capacity to sustain its connected loads for a minimum of 1½ hours after loss of the normal source.

[x]**(3) System Operation.** The system shall be arranged so that, in the event of a failure of the normal power source, the alternate source of power shall be automatically connected to the load within 10 seconds.

FPN: See *Standard for Health Care Facilities,* NFPA 99-1996, 3-6.3.2, Transfer Switch Operation for Type III with Generator Sets, and 3-6.3.3, Transfer Switch Operation for Type III with Battery Systems.

D. Inhalation Anesthetizing Locations

FPN: For further information regarding safeguards for anesthetizing locations, see *Standard for Health Care Facilities* NFPA 99-1996.

517-60. Anesthetizing Location Classification.

FPN: If either of the following anesthetizing locations is designated a wet location, refer to Section 517-20.

(a) Hazardous (Classified) Location.

x**(1)** In a location where flammable anesthetics are employed, the entire area shall be considered to be a Class I, Division 1 location that shall extend upward to a level 5 ft (1.52 m) above the floor. The remaining volume up to the structural ceiling is considered to be above a hazardous (classified) location.

(2) Any room or location in which flammable anesthetics or volatile flammable disinfecting agents are stored shall be considered to be a Class I, Division 1 location from floor to ceiling.

For the 1996 edition of NFPA 99, *Health Care Facilities,* substantial changes were made concerning flammable anesthetizing locations. Previously, these requirements were located throughout the standard. Now, all of these requirements are found in Annex 2 of NFPA 99. The reason for this change is explained in the second note of Annex 2 in NFPA 99 and reads as follows.

> The text of this annex is a compilation of requirements included in previous editions of NFPA 99 on safety practices for facilities that used flammable inhalation anesthetics. This material is being retained in this annex by the Technical Committee on Anesthesia Services for the following reasons: 1) the Committee is aware that some countries outside the United States still use this type of anesthetics, and rely on the safety measures herein; and 2) while the Committee is unaware of any medical schools in the U.S. still teaching the proper use of flammable anesthetics or any health care facilities in the U.S. using flammable anesthetics, retaining this material will serve as a reminder of the precautions that would be necessary should the use of this type of anesthetics be re-instituted.

Section 517-60 of the *NEC* divides anesthetizing locations into hazardous (classified) locations, where flammable or nonflammable anesthetics may be interchangeably employed [Section 517-60(a)], and other-than-hazardous (classified) locations, where only nonflammable anesthetics are used [Section 517-60(b)]. In the case of the flammable anesthetizing location, the entire volume of the room, extending upward from a level 5 ft above the floor to the surface of the structural ceiling of the room and including the space between a drop ceiling and the structural ceiling, is considered to be above a hazardous (classified) location.

(b) Other-than-Hazardous (Classified) Location. Any inhalation anesthetizing location designated for the exclusive use of nonflammable anesthetizing agents shall be considered to be an other-than-hazardous (classified) location.

517-61. Wiring and Equipment.

(a) Within Hazardous (Classified) Anesthetizing Locations.

x**(1)** Except as permitted in Section 517-160, each power circuit within, or partially within, a flammable anesthetizing location as referred to in Section 517-60 shall be isolated from any distribution system by the use of an isolated power system.

(2) Isolated power system equipment shall be listed for the purpose and the system designed and installed so that it meets the provisions and is in accordance with Part G.

x**(3)** In hazardous (classified) locations referred to in Section 517-60, all fixed wiring and equipment, and all portable equipment, including lamps and other utilization equipment, operating at more than 10 volts between conductors shall comply with the requirements of Sections 501-1 through 501-15 and Sections 501-16(a) and (b) for Class I, Division 1 locations. All such equipment shall be specifically approved for the hazardous atmospheres involved.

(4) Where a box, fitting, or enclosure is partially, but not entirely, within a hazardous (classified) location(s), the hazardous (classified) location(s) shall be considered to be extended to include the entire box, fitting, or enclosure.

(5) Receptacles and attachment plugs in a hazardous (classified) location(s) shall be listed for use in Class I, Group C hazardous (classified) locations and shall have provision for the connection of a grounding conductor.

(6) Flexible cords used in hazardous (classified) locations for connection to portable utilization equipment, including lamps operating at more than 8 volts between conductors, shall be of a type approved for extra-hard usage in accordance with Table 400-4 and shall include an additional conductor for grounding.

(7) A storage device for the flexible cord shall be provided and shall not subject the cord to bending at a radius of less than 3 in. (76 mm).

(b) Above Hazardous (Classified) Anesthetizing Locations.

(1) Wiring above a hazardous (classified) location referred to in Section 517-60 shall be installed in rigid metal conduit, electrical metallic tubing, intermediate metal conduit, Type MI cable, or Type MC cable that employs a continuous, gas/vaportight metal sheath.

(2) Installed equipment that may produce arcs, sparks, or particles of hot metal, such as lamps and lampholders for fixed lighting, cutouts, switches, generators, motors, or other equipment having make-and-break or sliding contacts, shall be of the totally enclosed type or be constructed so as to prevent escape of sparks or hot metal particles.

Exception: Wall-mounted receptacles installed above the hazardous (classified) location in flammable anesthetizing locations shall not be required to be totally enclosed or have openings guarded or screened to prevent dispersion of particles.

(3) Surgical and other lighting fixtures shall conform to Section 501-9(b).

Exception No. 1: The surface temperature limitations set forth in Section 501-9(b)(2) shall not apply.

Exception No. 2: Integral or pendant switches that are located above and cannot be lowered into the hazardous (classified) location(s) shall not be required to be explosionproof.

(4) Approved seals shall be provided in conformance with Section 501-5, and Section 501-5(a)(4) shall apply to horizontal as well as to vertical boundaries of the defined hazardous (classified) locations.

(5) Receptacles and attachment plugs located above hazardous (classified) anesthetizing locations shall be listed for hospital use for services of prescribed voltage, frequency, rating, and number of conductors with provision for the connection of the grounding conductor. This requirement shall apply to attachment plugs and receptacles of the 2-pole, 3-wire grounding type for single-phase, 120-volt, nominal, ac service.

See the commentary following Section 517-19(b) regarding receptacles listed for hospital use.

(6) Plugs and receptacles rated 250-volts, for connection of 50-ampere, and 60-ampere ac medical equipment for use above hazardous (classified) locations shall be arranged so that the 60-ampere receptacle will accept either the 50-ampere or the 60-ampere plug. Fifty-ampere receptacles shall be designed so as not to accept the 60-ampere attachment plug. The plugs shall be of the 2-pole, 3-wire design with a third contact connecting to the insulated (green or green with yellow stripe) equipment grounding conductor of the electrical system.

(c) Other-than-Hazardous (Classified) Anesthetizing Locations.

(1) Wiring serving other-than-hazardous (classified) locations, as defined in Section 517-60, shall be installed in a metal raceway system or cable assembly. The metal raceway system, or cable armor or sheath assembly, shall qualify as an equipment grounding return path in accordance with Section 250-118. Type MC and Type MI cable shall have an outer metal armor or sheath that is identified as an acceptable grounding return path.

Exception: Pendant receptacle constructions that employ at least Type SJO or equivalent flexible cords suspended not less than 6 ft (1.83 m) from the floor shall not be required to be installed in a metal raceway or cable assembly.

(2) Receptacles and attachment plugs installed and used in other-than-hazardous (classified) locations shall be listed for hospital use for services of prescribed voltage, frequency, rating, and number of conductors with provision for connection of the grounding conductor. This requirement shall apply to 2-pole, 3-wire grounding type for single-phase, 120-, 208-, or 240-volt, nominal, ac service.

See the commentary following Section 517-19(b) regarding receptacles listed for hospital use.

(3) Plugs and receptacles rated 250 volts, for connection of 50-ampere, and 60-ampere ac medical equipment for use in other-than-hazardous (classified) locations shall be arranged so that the 60-ampere receptacle will accept either the 50-ampere or the 60-ampere plug. Fifty-ampere receptacles shall be designed so as not to accept the 60-ampere attachment plug. The plugs shall be of the 2-pole, 3-wire design with a third contact connecting to the insulated (green or green with yellow stripe) equipment grounding conductor of the electrical system.

517-62. Grounding. In any anesthetizing area, all metal raceways and metal-sheathed cables, and all noncurrent-carrying conductive portions of fixed electric equipment, shall be grounded. Grounding in Class I locations shall comply with Section 501-16.

Exception: Equipment operating at not more than 10 volts between conductors shall not be required to be grounded.

It should be noted that the grounding requirements for anesthetizing locations apply only to metal raceways, metal-sheathed cables, and electrical equipment. Carts, tables, and other nonelectrical items are not required to be grounded. In flammable anesthetizing locations, however, portable carts and tables usually have a resistance to ground of not over 1,000,000 ohms, through use of conductive tires and wheels and conductive flooring, to avoid the buildup of static electrical charges. See NFPA 99-1996, *Standard for Health Care Facilities,* Annex 2, 2-6.3.10, for electrostatic safeguards.

517-63. Grounded Power Systems in Anesthetizing Locations.

(a) Battery-Powered Emergency Lighting Units. One or more battery-powered emergency lighting units shall be provided in accordance with Section 700-12(e).

The failure of the emergency circuit feeder that supplies the operating room will ordinarily plunge the room into darkness. Requiring a general-purpose lighting circuit supplied by a normal source feeder will minimize the effect of this kind of failure.

(b) Branch-Circuit Wiring. Branch circuits supplying only listed, fixed, therapeutic and diagnostic equipment, permanently installed above the hazardous (classified) location and in other-than-hazardous (classified) locations, shall be permitted to be supplied from a normal grounded service, single- or three-phase system, provided the following:

(1) Wiring for grounded and isolated circuits does not occupy the same raceway or cable,
(2) All conductive surfaces of the equipment are grounded,
(3) Equipment (except enclosed X-ray tubes and the leads to the tubes) are located at least 8 ft (2.44 m) above the floor or outside the anesthetizing location, and
(4) Switches for the grounded branch circuit are located outside the hazardous (classified) location.

Exception: Sections 517-63(b)(3) and (b)(4) shall not apply in other-than-hazardous (classified) locations.

(c) Fixed Lighting Branch Circuits. Branch circuits supplying only fixed lighting shall be permitted to be supplied by a normal grounded service, provided the following:

(1) Such fixtures are located at least 8 ft (2.44 m) above the floor,
(2) All conductive surfaces of fixtures are grounded,
(3) Wiring for circuits supplying power to fixtures does not occupy the same raceway or cable for circuits supplying isolated power, and
(4) Switches are wall-mounted and located above hazardous (classified) locations.

Exception: Sections 517-63(c)(1) and (c)(4) shall not apply in other-than-hazardous (classified) locations.

(d) Remote-Control Stations. Wall-mounted remote-control stations for remote-control switches operating at 24 volts or less shall be permitted to be installed in any anesthetizing location.

(e) Location of Isolated Power Systems. An isolated power center listed for the purpose and its grounded primary feeder shall be permitted to be located in an anesthetizing location, provided it is installed above a hazardous (classified) location or in an other-than-hazardous (classified) location.

(f) Circuits in Anesthetizing Locations. Except as permitted above, each power circuit within, or partially within, a flammable anesthetizing location as referred to in Section 517-60 shall be isolated from any distribution system supplying other-than-anesthetizing locations.

517-64. Low-Voltage Equipment and Instruments.

(a) Equipment Requirements. Low-voltage equipment that is frequently in contact with the bodies of persons or has exposed current-carrying elements shall be as follows:

(1) Operate on an electrical potential of 10 volts or less, or
(2) Approved as intrinsically safe or double-insulated equipment, or
(3) Moisture resistant

(b) Power Supplies. Power shall be supplied to low-voltage equipment from the following:

(1) An individual portable isolating transformer (autotransformers shall not be used) connected to an isolated power circuit receptacle by means of an appropriate cord and attachment plug, or
(2) A common low-voltage isolating transformer installed in an other-than-hazardous (classified) location, or
(3) Individual dry-cell batteries, or
(4) Common batteries made up of storage cells located in an other-than-hazardous (classified) location

(c) Isolated Circuits. Isolating-type transformers for supplying low-voltage circuits shall have the following:

(1) Approved means for insulating the secondary circuit from the primary circuit, and
(2) Have the core and case grounded

(d) Controls. Resistance or impedance devices shall be permitted to control low-voltage equipment but shall not be used to limit the maximum available voltage to the equipment.

(e) Battery-Powered Appliances. Battery-powered appliances shall not be capable of being charged while in operation unless their charging circuitry incorporates an integral isolating-type transformer.

(f) Receptacles or Attachment Plugs. Any receptacle or attachment plug used on low-voltage circuits shall be of a type that does not permit interchangeable connection with circuits of higher voltage.

> FPN: Any interruption of the circuit, even circuits as low as 10 volts, either by any switch, or loose or defective connections anywhere in the circuit, may produce a spark that is sufficient to ignite flammable anesthetic agents. See 7-5.1.2.3 of *Standard for Health Care Facilities,* NFPA 99-1996.

E. X-Ray Installations

Nothing in this part shall be construed as specifying safeguards against the useful beam or stray X-ray radiation.

> FPN No. 1: Radiation safety and performance requirements of several classes of X-ray equipment are regulated under Public Law 90-602 and are enforced by the Department of Health and Human Services.

FPN No. 2: In addition, information on radiation protection by the National Council on Radiation Protection and Measurements is published as *Reports of the National Council on Radiation Protection and Measurement*. These reports are obtainable from NCRP Publications, P.O. Box 30175, Washington, DC 20014.

517-71. Connection to Supply Circuit.

(a) Fixed and Stationary Equipment. Fixed and stationary X-ray equipment shall be connected to the power supply by means of a wiring method that meets the general requirements of this *Code*.

Exception: Equipment properly supplied by a branch circuit rated at not over 30 amperes shall be permitted to be supplied through a suitable attachment plug and hard-service cable or cord.

(b) Portable, Mobile, and Transportable Equipment. Individual branch circuits shall not be required for portable, mobile, and transportable medical X-ray equipment requiring a capacity of not over 60 amperes.

(c) Over 600-Volt Supply. Circuits and equipment operated on a supply circuit of over 600 volts shall comply with Article 490.

517-72. Disconnecting Means.

(a) Capacity. A disconnecting means of adequate capacity for at least 50 percent of the input required for the momentary rating or 100 percent of the input required for the long-time rating of the X-ray equipment, whichever is greater, shall be provided in the supply circuit.

(b) Location. The disconnecting means shall be operable from a location readily accessible from the X-ray control.

(c) Portable Equipment. For equipment connected to a 120-volt branch circuit of 30 amperes or less, a grounding-type attachment plug and receptacle of proper rating shall be permitted to serve as a disconnecting means.

517-73. Rating of Supply Conductors and Overcurrent Protection.

(a) Diagnostic Equipment.

(1) The ampacity of supply branch-circuit conductors and the current rating of overcurrent protective devices shall not be less than 50 percent of the momentary rating or 100 percent of the long-time rating, whichever is greater.

(2) The ampacity of supply feeders and the current rating of overcurrent protective devices supplying two or more branch circuits supplying X-ray units shall not be less than 50 percent of the momentary demand rating of the largest unit plus 25 percent of the momentary demand rating of the next largest unit plus 10 percent of the momentary demand rating of each additional unit. Where simultaneous biplane examinations are undertaken with the X-ray units, the supply conductors and overcurrent protective devices shall be 100 percent of the momentary demand rating of each X-ray unit.

FPN: The minimum conductor size for branch and feeder circuits is also governed by voltage regulation requirements. For a specific installation, the manufacturer usually specifies minimum distribution transformer and conductor sizes, rating of disconnecting means, and overcurrent protection.

(b) Therapeutic Equipment. The ampacity of conductors and rating of overcurrent protective devices shall not be less than 100 percent of the current rating of medical X-ray therapy equipment.

FPN: The ampacity of the branch-circuit conductors and the ratings of disconnecting means and overcurrent protection for X-ray equipment are usually designated by the manufacturer for the specific installation.

517-74. Control Circuit Conductors.

(a) Number of Conductors in Raceway. The number of control circuit conductors installed in a raceway shall be determined in accordance with Section 300-17.

(b) Minimum Size of Conductors. Size No. 18 or No. 16 fixture wires as specified in Section 725-27 and flexible cords shall be permitted for the control and operating circuits of X-ray and auxiliary equipment where protected by not larger than 20-ampere overcurrent devices.

517-75. Equipment Installations. All equipment for new X-ray installations and all used or reconditioned X-ray equipment moved to and reinstalled at a new location shall be of an approved type.

517-76. Transformers and Capacitors. Transformers and capacitors that are part of X-ray equipment shall not be required to comply with Articles 450 and 460.

Capacitors shall be mounted within enclosures of insulating material or grounded metal.

517-77. Installation of High-Tension X-Ray Cables. Cables with grounded shields connecting X-ray tubes and image intensifiers shall be permitted to be installed in cable trays or cable troughs along with X-ray equipment control and power supply conductors without the need for barriers to separate the wiring.

517-78. Guarding and Grounding.

(a) High-Voltage Parts. All high-voltage parts, including X-ray tubes, shall be mounted within grounded enclosures. Air, oil, gas, or other suitable insulating media shall be used to insulate the high-voltage from the grounded enclosure. The connection from the high-voltage equipment to X-ray tubes and other high-voltage components shall be made with high-voltage shielded cables.

(b) Low-Voltage Cables. Low-voltage cables connecting to oil-filled units that are not completely sealed, such as transformers, condensers, oil coolers, and high-voltage switches, shall have insulation of the oil-resistant type.

(c) Noncurrent-Carrying Metal Parts. Noncurrent-carrying metal parts of X-ray and associated equipment (controls, tables, X-ray tube supports, transformer tanks, shielded cables, X-ray tube heads, etc.) shall be grounded in the manner specified in Article 250, as modified by Sections 517-13(a) and (b).

F. Communications, Signaling Systems, Data Systems, Fire Alarm Systems, and Systems Less than 120 Volts, Nominal

517-80. Patient Care Areas. Equivalent insulation and isolation to that required for the electrical distribution systems in patient care areas shall be provided for communications, signaling systems, data system circuits, fire alarm systems, and systems less than 120 volts, nominal.

FPN: An acceptable alternate means of providing isolation for patient/nurse call systems is by the use of nonelectrified signaling, communications, or control devices held by the patient or within reach of the patient.

517-81. Other-than-Patient-Care Areas. In other-than-patient-care areas, installations shall be in accordance with the appropriate provisions of Articles 640, 725, 760, and 800.

517-82. Signal Transmission Between Appliances.

(a) General. Permanently installed signal cabling from an appliance in a patient location to remote appliances shall employ a signal transmission system that prevents hazardous grounding interconnection of the appliances.

FPN: See Section 517-13(b).

(b) Common Signal Grounding Wire. Common signal grounding wires (i.e., the chassis ground for single-ended transmission) shall be permitted to be used between appliances all located within the patient vicinity, provided the appliances are served from the same reference grounding point.

G. Isolated Power Systems

517-160. Isolated Power Systems.

(a) Installations.

(1) Each isolated power circuit shall be controlled by a switch that has a disconnecting pole in each isolated circuit conductor to simultaneously disconnect all power. Such isolation shall be accomplished by means of one or more transformers having no electrical connection between primary and secondary windings, by means of motor generator sets, or by means of suitably isolated batteries.

x**(2)** Circuits supplying primaries of isolating transformers shall operate at not more than 600 volts between conductors and shall be provided with proper overcurrent protection. The secondary voltage of such transformers shall not exceed 600 volts between conductors of each circuit. All circuits supplied from such secondaries shall be ungrounded and shall have an approved overcurrent device of proper ratings in each conductor. Circuits supplied directly from batteries or from motor generator sets shall be ungrounded, and shall be protected against overcurrent in the same manner as transformer-fed secondary circuits. If an electrostatic shield is present, it shall be connected to the reference grounding point.

(3) The isolating transformers, motor generator sets, batteries and battery chargers, and associated primary or secondary overcurrent devices shall not be installed in hazardous (classified) locations. The isolated secondary circuit wiring extending into a hazardous anesthetizing location shall be installed in accordance with Section 501-4.

x**(4) Isolation Transformers.** An isolation transformer shall not serve more than one operating room except as covered in (a) and (b).

For purposes of this section, anesthetic induction rooms are considered part of the operating room or rooms served by the induction rooms.

(a) *Induction Rooms.* Where an induction room serves more than one operating room, the isolated circuits of the induction room shall be permitted to be supplied from the isolation transformer of any one of the operating rooms served by that induction room.

(b) *Higher Voltages.* Isolation transformers shall be permitted to serve single receptacles in several patient areas where

(1) The receptacles are reserved for supplying power to equipment requiring 150 volts or higher, such as portable X-ray units, and
(2) The receptacles and mating plugs are not interchangeable with the receptacles on the local isolated power system.

(5) The isolated circuit conductors shall be identified as follows:

Isolated Conductor No. 1 — Orange

Isolated Conductor No. 2 — Brown

For 3-phase systems, the third conductor shall be identified as yellow. Where isolated circuit conductors supply 125-volt, single-phase, 15- and 20-ampere receptacles, the orange conductor(s) shall be connected to the terminal(s) on the receptacles that are identified in accordance with Section 200-10(b) for connection to the grounded circuit conductor.

(6) Wire-pulling compounds that increase the dielectric constant shall not be used on the secondary conductors of the isolated power supply.

FPN No. 1: It is desirable to limit the size of the isolation transformer to 10 kVA or less and to use conductor insulation with low leakage to meet impedance requirements.

FPN No. 2: Minimizing the length of branch-circuit conductors and using conductor insulations with a dielectric constant less than 3.5 and insulation resistance constant greater than 6100 megohm-meters (20,000 megohm-ft) at 16°C (60°F) reduces leakage from line to ground reducing the hazard current.

See Figure 517.4 for an example of a hospital isolated power system panel.

Figure 517.4 *An example of a hospital isolated power system panel with built-in isolation transformer, line isolation monitor, load center, and grounded busbar. (Square D Co.)*

(b) Line Isolation Monitor.

(1) In addition to the usual control and overcurrent protective devices, each isolated power system shall be provided with a continually operating line isolation monitor that indicates total hazard current. The monitor shall be designed so that a green signal lamp, conspicuously visible to persons in each area served by the isolated power system, remains lighted when the system is adequately isolated from ground. An adjacent red signal lamp and an audible warning signal (remote if desired) shall be energized when the total hazard current (consisting of possible resistive and capacitive leakage currents) from either isolated conductor to ground reaches a threshold value of 5 mA under nominal line voltage conditions. The line monitor shall not alarm for a fault hazard of less than 3.7 mA or for a total hazard current of less than 5 mA.

Exception: A system shall be permitted to be designed to operate at a lower threshold value of total hazard current. A line isolation monitor for such a system shall be permitted to be approved with the provision that the fault hazard current shall be permitted to be reduced but not to less than 35 percent of the corresponding threshold value of the total hazard current, and the monitor hazard current is to be correspondingly reduced to not more than 50 percent of the alarm threshold value of the total hazard current.

(2) The line isolation monitor shall be designed to have sufficient internal impedance such that, when properly connected to the isolated system, the maximum internal current that can flow through the line isolation monitor, when any point of the isolated system is grounded, shall be 1 mA.

Exception: The line isolation monitor shall be permitted to be of the low-impedance type such that the current through the line isolation monitor, when any point of the isolated system is grounded, will not exceed twice the alarm threshold value for a period not exceeding 5 milliseconds.

FPN: Reduction of the monitor hazard current, provided this reduction results in an increased "not alarm" threshold value for the fault hazard current, will increase circuit capacity.

(3) An ammeter calibrated in the total hazard current of the system (contribution of the fault hazard current plus monitor hazard current) shall be mounted in a plainly visible place on the line isolation monitor with the "alarm on" zone at approximately the center of the scale.

Exception: The line isolation monitor shall be permitted to be a composite unit, with a sensing section cabled to a separate display panel section on which the alarm or test functions are located.

FPN: It is desirable to locate the ammeter so that it is conspicuously visible to persons in the anesthetizing location.

See Figure 517.5 for an example of a line isolation monitor that can monitor several operating rooms from a central location.

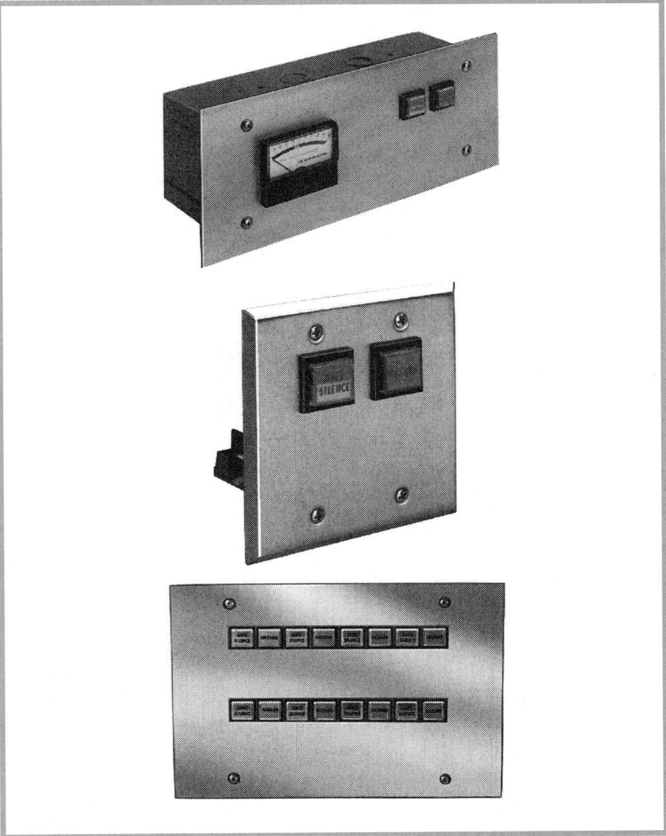

Figure 517.5 *Line isolation monitor (top) for use with isolated power systems, remote indicator alarm (middle), and multiple annunciator panel (bottom), which can monitor several operating rooms from a central location, such as a nurses' station. (Square D Co.)*

Article 518 — Places of Assembly

Contents

518-1. Scope. This article covers all buildings or portions of buildings or structures designed or intended for the assembly of 100 or more persons.

518-2. General Classifications.

(a) Examples. Places of assembly shall include, but not be limited to, the following:

Armories
Assembly halls
Auditoriums
Auditoriums within
  Business establishments
  Mercantile establishments
  Other occupancies
  Schools
Bowling lanes
Church chapels
Club rooms
Conference rooms
Courtrooms
Dance halls
Dining facilities
Exhibition halls
Gymnasiums
Mortuary chapels
Multipurpose rooms
Museums
Places of awaiting transportation
Pool rooms
Restaurants
Skating rinks

(b) Multiple Occupancies. Occupancy of any room or space for assembly purposes by less than 100 persons in a building of other occupancy, and incidental to such other occupancy, shall be classified as part of the other occupancy and subject to the provisions applicable thereto.

(c) Theatrical Areas. Where any such building structure, or portion thereof, contains a projection booth or stage platform or area for the presentation of theatrical or musical productions, either fixed or portable, the wiring for that area, including associated audience seating areas, and all equipment that is used in the referenced area, and portable equipment and wiring for use in the production that will not be connected to permanently installed wiring, shall comply with Article 520.

See the commentary following Section 520-1.

FPN: For methods of determining population capacity, see local building code or, in its absence, the *Life Safety Code*, NFPA *101*-1997.

Article 518 applies to places of assembly designed or intended for the assembly of 100 or more persons. It would apply, for example, to a church chapel or auditorium for occupancy of 100 or more persons, its capacity determined by the methods for occupancy population capacity in accordance with NFPA *101*-1997, *Life Safety Code.* Article 518 does not apply to a supermarket, even though a supermarket may contain 100 or more persons, because a supermarket is not specifically designed or intended for the assembly of persons, nor is it considered to be an auditorium. Article 518 does not apply to an office building or a school building, even though such buildings, as a rule, are designed for occupancy by 100 or more persons. Article 518 does, however, apply to assembly halls, restaurants, and so on, within office or school buildings if these parts of the building are designed or intended for the assembly of 100 or more persons.

The following information for determining new assembly occupancy capacity is contained in NFPA *101*-1997, *Life Safety Code.*

8-1.7 Occupant Load

8-1.7.1 The occupant load permitted in any assembly building, structure, or portion thereof shall be determined on the basis of the following occupant load factors:

(a) An assembly area of concentrated use without fixed seats, such as an auditorium, place of worship, dance floor, discotheque, or lodge hall: one person per 7 net sq ft (0.65 net sq m).

(b) An assembly area of less concentrated use, such as a conference room, dining room, drinking establishment, exhibit room, gymnasium, or lounge: one person per 15 net sq ft (1.4 net sq m).

(c) Bleachers, pews, and similar bench-type seating: one person per 18 linear in. (45.7 linear cm).

(d) Fixed seating. The occupant load of an area having fixed seats shall be determined by the number of fixed seats installed. Required aisle space serving the fixed seats shall not be used to increase the occupant load.

(e) Kitchens. One person per 100 gross sq ft (9.3 gross sq m).

(f) Libraries. In stack areas, one person per 100 gross sq ft (9.3 sq m); in reading rooms, one person per 50 net sq ft (4.6 net sq m).

(g) Swimming Pools. One person per 50 gross sq ft (4.7 sq m) of water surface. Pool decks, one person per 30 gross sq ft (2.8 sq m).

(h) Stages. One person per 15 net sq ft (1.4 net sq m).

(i) Lighting and Access Catwalks, Galleries, and Gridirons. One person per 100 net sq ft (9.3 sq m).

Exception: Larger occupant loads as permitted by 8-1.7.3.

8-1.7.2 The occupant load of a stage area that is part of an assembly area shall be included in determining the occupant load for the assembly area.

8-1.7.3 The occupant load permitted in a building or portion thereof shall be permitted to be increased above that specified in 8-1.7.1 if the necessary aisles and exits are provided. To increase the occupant load, a diagram indicating placement of equipment, aisles, exits, and seating shall be provided to and approved by the authority having jurisdiction prior to any increase in occupant load. In areas not greater than 10,000 sq ft (930 sq m), the occupant load shall not exceed one person in 5 sq ft (0.46 sq m); in areas greater than 10,000 sq ft (930 sq m), the occupant load shall not exceed one person in 7 sq ft (0.65 sq m).

518-3. Other Articles.

(a) Hazardous (Classified) Areas. Electrical installations in hazardous (classified) areas located in places of assembly shall comply with Article 500.

(b) Temporary Wiring. In exhibition halls used for display booths, as in trade shows, the temporary wiring shall be installed in accordance with Article 305. Flexible cables and cords approved for hard or extra-hard usage shall be permitted to be laid on floors where protected from contact by the general public. The ground-fault circuit-interrupter requirements of Section 305-6 shall not apply.

For an illustration of a protection technique for the installation of temporary wiring used in exhibit halls, see Figure 518.1.

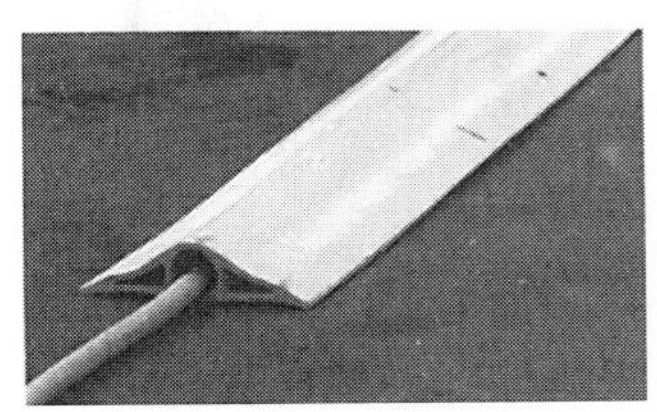

Figure 518.1 *A treadle, such as this one, can be used to protect cords from abuse where they are laid across pedestrian ways; for instance, as part of a temporary wiring scheme such as may be used in exhibition halls. (Daniel Woodhead Co.)*

Exception: Where conditions of supervision and maintenance ensure that only qualified persons will service the installation, flexible cords or cables identified in Table 400-4 for hard usage or extra-hard usage shall be permitted in cable trays used only for temporary wiring. All cords or cables shall be installed in a single layer. A permanent sign shall be attached to the cable tray at intervals not to exceed 25 ft (6.1 m). The sign shall read

CABLE TRAY FOR TEMPORARY WIRING ONLY

(c) Emergency Systems. Control of emergency systems shall comply with Article 700.

518-4. Wiring Methods.

(a) General. The fixed wiring methods shall be metal raceways, flexible metal raceways, nonmetallic raceways encased in not less than 2 in. (50.8 mm) of concrete, Type MI, MC, or AC cable containing an insulated equipment grounding conductor sized in accordance with Table 250-122.

Exception: Fixed wiring methods shall be as provided in

(a) Audio signal processing, amplification, and reproduction equipment — Article 640
(b) Communications circuits — Article 800
(c) Class 2 and Class 3 remote-control and signaling circuits — Article 725
(d) Fire alarm circuits — Article 760

(b) Nonrated Construction. Nonmetallic-sheathed cable, Type AC cable, electrical nonmetallic tubing, and rigid nonmetallic conduit shall be permitted to be installed in those buildings or portions thereof that are not required to be of fire-rated construction by the applicable building code.

FPN: Fire-rated construction is the fire-resistive classification used in building codes.

(c) Spaces with Finish Rating. Electrical nonmetallic tubing and rigid nonmetallic conduit shall be permitted to be installed in restaurants, conference and meeting rooms in hotels or motels, dining facilities and church chapels where

(1) The electrical nonmetallic tubing or rigid nonmetallic conduit is installed concealed within walls, floors, and ceilings where the walls, floors, and ceilings provide a thermal barrier of material that has at least a 15-minute finish rating as identified in listings of fire-rated assemblies.
(2) The electrical nonmetallic tubing or rigid nonmetallic conduit is installed above suspended ceilings where the suspended ceilings provide a thermal barrier of material that has at least a 15-minute finish rating as identified in listings of fire-rated assemblies.

Electrical nonmetallic tubing and rigid nonmetallic conduit are not recognized for use in other space used for environmental air in accordance with Section 300-22(c).

FPN: A finish rating is established for assemblies containing combustible (wood) supports. The finish rating is defined as the time at which the wood stud or wood joist reaches an average temperature rise of 121°C (250°F) or an individual temperature rise of

163°C (325°F) as measured on the plane of the wood nearest the fire. A finish rating is not intended to represent a rating for a membrane ceiling.

Referring to Figure 518.2, the wash rooms and office area of this single-story facility are not places of assembly, as defined in Section 518-2, and therefore require no special wiring methods. Ordinary wiring methods can be used on the inside surface of the storage area walls and on or in the partitions between storage areas, as these also are not places of assembly. Inside any hollow spaces of the fire-rated storage area walls, however, the main requirements of Section 518-4 apply, since the serving corridors are part of the places of assembly as a result of this particular building design. If the hollow spaces of fire-rated walls or ceiling also provide a 15-minute finish rating and are not other spaces for environmental air, as described in Section 300-22(c), then electrical nonmetallic tubing as well as rigid nonmetallic conduit would be permitted for specifically described occupancies.

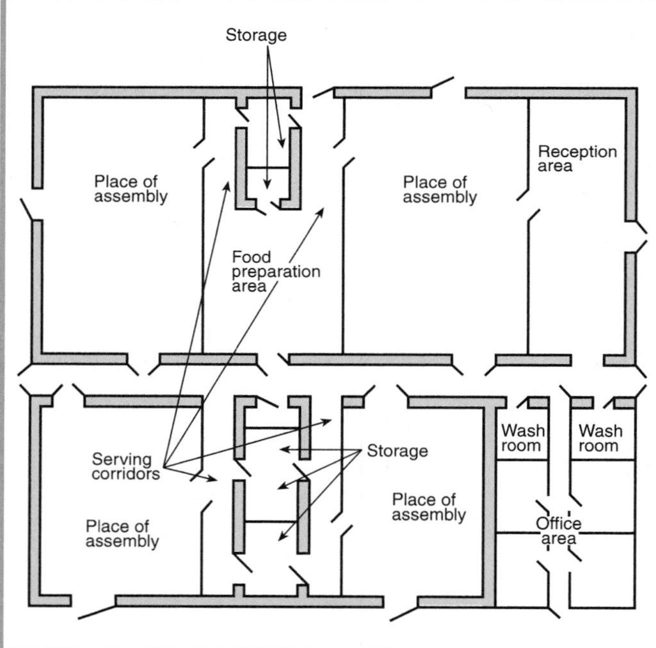

Figure 518.2 The double lines represent walls required by the local building code to be of fire-rated construction. See the commentary text for discussion. The single lines represent walls not required by the local building code to be of fire-rated construction.

In like manner, wiring in ceilings or floors, required to be of fire-rated construction in a place of assembly as defined in Section 518-2, is also required to comply with Section 518-4, except as noted in Section 518-4(a), Exception.

The intent is that, within a place of assembly as defined in Section 518-2, Section 518-4(a) and the exception apply to any wall, floor, or ceiling. Section 518-4(b) applies to those portions of the building and those places of assembly not required to be fire rated. Section 518-4(c) applies only to the specific occupancies described, provided these wiring methods are installed concealed behind a surface that has a 15-minute finish rating.

Places of assembly frequently require emergency wiring and Section 700-9 contains fire protection requirements for emergency circuits.

518-5. Supply. Portable switchboards and portable power distribution equipment shall be supplied only from listed power outlets of sufficient voltage and ampere rating. Such power outlets shall be protected by overcurrent devices. Such overcurrent devices and power outlets shall not be accessible to the general public. Provisions for connection of an equipment grounding conductor shall be provided. The neutral of feeders supplying solid-state, 3-phase, 4-wire dimmer systems shall be considered a current-carrying conductor.

Section 518-5 requires portable switchboards and portable power distribution equipment to be supplied only from listed power outlets rated for the voltage and current they are used for. The overcurrent devices and power outlets used to supply such equip-

Figure 518.3 A listed power outlet for connection of portable switchboards in a place of assembly. (Union Carbide Connector, Inc.)

ment must be located so as not to be accessible to the general public. See Figure 518.3.

Article 520 — Theaters, Audience Areas of Motion Picture and Television Studios, and Similar Locations

Contents

A. General

520-1. Scope. This article covers all buildings or that part of a building or structure designed or used for presentation, dramatic, musical, motion picture projection, or similar purposes and to specific audience seating areas within motion picture or television studios.

The special requirements of Article 520 apply only to that part of a building used as a theater or for a similar purpose and do not necessarily apply to the entire building. For example, the requirements of Article 520 would apply to an auditorium in a school building used for dramatic or other performances. The special requirements of this chapter apply to the stage, auditorium, dressing rooms, and main corridors leading to the auditorium, but not to other parts of the building that are not involved in the use of the auditorium for performances or entertainment. The theater space may be a traditional theater, where the audience sits in the auditorium (house) facing the proscenium arch and views the performance on the stage on the other side of the arch, or other spaces, such as a simple stage platform, either indoors or outdoors, with seats on three or four sides facing the platform.

Article 520 also covers the audience areas only of motion picture and television studios, as defined and covered in Article 530.

520-2. Definitions.

Border Light. A permanently installed overhead strip light.

Breakout Assembly. An adapter used to connect a multipole connector containing two or more branch-circuits to multiple individual branch circuit connectors.

Bundled. Cables or conductors that are physically tied, wrapped, taped, or otherwise periodically bound together.

Connector Strip. A metal wireway containing pendant or flush receptacles.

Drop Box. A box containing pendant- or flush-mounted receptacles attached to a multiconductor cable via strain relief, or a multipole connector.

Footlight. A border light installed on or in the stage.

Grouped. Cables or conductors positioned adjacent to one another but not in continuous contact with each other.

Portable Equipment. Equipment fed with portable cords or cables intended to be moved from one place to another.

Portable Power Distribution Unit. A power distribution box containing receptacles and overcurrent devices.

Proscenium. The wall and arch that separates the stage from the auditorium (house).

Stand Lamp (Work Light). A portable stand that contains a general-purpose lighting fixture or lamp holder with guard for the purpose of providing general illumination on the stage or in the auditorium.

Strip Light. A lighting fixture with multiple lamps arranged in a row.

Two-Fer. An adapter cable containing one male plug and two female cord connectors used to connect two loads to one branch circuit.

An illustration of a two-fer is shown in Figure 520.1.

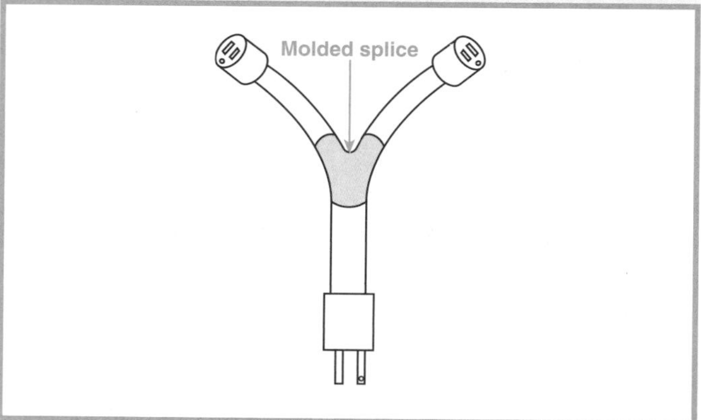

Figure 520.1 *A two-fer consists of two cord connectors on separate cords connected to a single supply cord.*

520-3. Motion Picture Projectors. Motion picture equipment and its installation and use shall comply with Article 540.

520-4. Audio Signal Processing, Amplification, and Reproduction Equipment. Audio signal processing, amplification, and reproduction equipment and its installation shall comply with Article 640.

520-5. Wiring Methods.

(a) General. The fixed wiring method shall be metal raceways, nonmetallic raceways encased in at least 2 in. (50.8 mm) of concrete, Type MI cable, or Type MC cable.

Exception: Fixed wiring methods shall be as provided in Article 640 for audio signal processing, amplification and reproduction equipment, in Article 800 for communication circuits, in Article 725 for Class 2 and Class 3 remote-control and signaling circuits, and in Article 760 for fire alarm circuits.

(b) Portable Equipment. The wiring for portable switchboards, stage set lighting, stage effects, and other wiring not fixed as to location shall be permitted with approved flexible cords and cables as provided elsewhere in Article 520. Fastening such cables and cords by uninsulated staples or nailing shall not be permitted.

(c) Nonrated Construction. Nonmetallic-sheathed cable, Type AC cable, electrical nonmetallic tubing, and rigid nonmetallic conduit shall be permitted to be installed in those buildings or portions thereof that are not required to be of fire-rated construction by the applicable building code.

Theaters and similar buildings are usually required to be of fire-rated construction, as determined by applicable building codes; therefore, the fixed wiring methods are limited. See Section 518-4 for specific wiring methods permitted.

Exceptions to the wiring rules for metal- (or concrete-) enclosed wiring are permitted for the installation of communications circuits, Class 2 and Class 3 remote-control and signaling circuits, sound-reproduction wiring, and fire alarm circuits. Where portability, flexibility, and adjustments are necessary for portable switchboards, stage lighting, and special effects, suitable cords and cables are permitted. Section 520-5(c) permits buildings or portions of buildings not required to be of fire-rated construction to be wired with Type NM cable, Type AC cable, ENT, or RNC.

520-6. Number of Conductors in Raceway. The number of conductors permitted in any metal conduit, rigid nonmetallic conduit as permitted in this article, or electrical metallic tubing for border or stage pocket circuits or for remote-control conductors shall not exceed the percentage fill shown in Table 1 of Chapter 9. Where contained within an auxiliary gutter or a wireway, the sum of the cross-sectional areas of all contained conductors at any cross section shall not exceed 20 percent of the interior cross-sectional area of the auxiliary gutter or wireway. The 30-conductor limitation of Sections 362-5 and 374-5 shall not apply.

520-7. Enclosing and Guarding Live Parts. Live parts shall be enclosed or guarded to prevent accidental contact by persons and objects. All switches shall be of the externally operable type. Dimmers, including rheostats, shall be placed in cases or cabinets that enclose all live parts.

520-8. Emergency Systems. Control of emergency systems shall comply with Article 700.

520-9. Branch Circuits. A branch circuit of any size supplying one or more receptacles shall be permitted to supply stage set lighting. The voltage rating of the receptacles shall not be less than the circuit voltage. Receptacle ampere ratings and branch-circuit conductor ampacity shall not be less than the branch-circuit overcurrent device ampere rating. Table 210-21(b)(2) shall not apply.

The stage set lighting and associated equipment, such as stage effects, both fixed and portable, must be as flexible as possible. Connectors are often used for different purposes and are therefore marked on a show-by-show basis as to the voltage, current, and type of current actually employed. Section 520-9 requires only that connectors be rated sufficiently for the parameters involved, thus permitting connectors with voltage and current ratings higher than the branch-circuit rating to be used.

The intent of Section 520-9 is to exclude the occupancies referenced in Article 520 from all the general requirements relating to connector rating and branch-circuit loading found elsewhere in the *Code*, such as in Table 210-21(b)(2). Section 520-9 modifies several other sections, such as Sections 210-23(c) and (d), which would disallow 40-ampere and larger branch circuits from serving 5000-watt and larger portable lighting fixtures found in the theater.

Stage set lighting is usually planned in advance, and the loads on each receptacle are known. Loads are not casually connected, as they are at a typical general-use wall receptacle. Care is taken to ensure that circuits are not overloaded, thereby avoiding nuisance tripping during a performance.

520-10. Portable Equipment. Portable stage and studio lighting equipment and portable power distribution equipment shall be permitted for temporary use outdoors provided the equipment is supervised by qualified personnel while energized and barriered from the general public.

Section 520-10, added in the 1996 *Code*, permits temporary outdoor use of portable indoor stage or studio equipment not marked suitable for wet or damp locations. If rain occurs, this equipment is typically de-energized, and a protective cover is installed before it is re-energized. At the end of the day, this equipment is either de-energized and protected or dismantled and stored.

B. Fixed Stage Switchboards

520-21. Dead Front. Stage switchboards shall be of the dead-front type and shall comply with Part D of Article 384 unless approved based on suitability as a stage switchboard as determined by a qualified testing laboratory and recognized test standards and principles.

Early stage switchboards were vertical marble or slate slabs mounted on the stage near the proscenium wall, with exposed knife switches and fuseholders mounted on them and with exposed resistance-type dimmer plates across the top. The "dead front," "guarded back," and "metal hood" requirements of the *Code* are intended to provide the operator with some sort of protection from shock, and the heat-producing equipment with some sort of protection from flammable curtains and scenery likely to be above and around the equipment. For these reasons, modern switchboards are totally enclosed.

520-22. Guarding Back of Switchboard. Stage switchboards having exposed live parts on the back of such boards shall be enclosed by the building walls, wire mesh grills, or by other approved methods. The entrance to this enclosure shall be by means of a self-closing door.

520-23. Control and Overcurrent Protection of Receptacle Circuits. Means shall be provided at a stage-lighting switchboard to which load circuits are connected for overcurrent protection of stage-lighting branch circuits, including branch circuits supplying stage and auditorium receptacles used for cord- and plug-connected stage equipment. Where the stage switchboard contains dimmers to control nonstage lighting, the locating of the overcurrent protective devices for these branch circuits at the stage switchboard shall be permitted.

The purpose of Section 520-23 is to ensure that the overcurrent protection devices are readily accessible to stage personnel during the presentation.

520-24. Metal Hood. A stage switchboard that is not completely enclosed dead-front and dead-rear or recessed into a wall shall be provided with a metal hood extending the full length of the board to protect all equipment on the board from falling objects.

Because stages are usually crowded and a great deal of flammable material is often present, a stage switchboard is not permitted to have exposed live parts on its front. Moreover, the space at the rear of a stage switchboard is required to be guarded in order to prevent entrance or contact by unqualified and unauthorized persons. One accepted method of accomplishing this is by enclosing the space between the rear of the switchboard and the wall in a sheet-steel housing with a door at one end.

520-25. Dimmers. Dimmers shall comply with (a) through (d).

Various types of dimmers and a dimmer console are illustrated in Figures 520.2, 520.3, 520.4 and 520.5.

Figure 520.4 shows an autotransformer-type dimmer switchboard. Autotransformer-type dimmers are currently seldom used.

Figure 520.2 *A typical high-density digital SCR dimmer switchboard. (Electronic Theatre Controls, Inc.)*

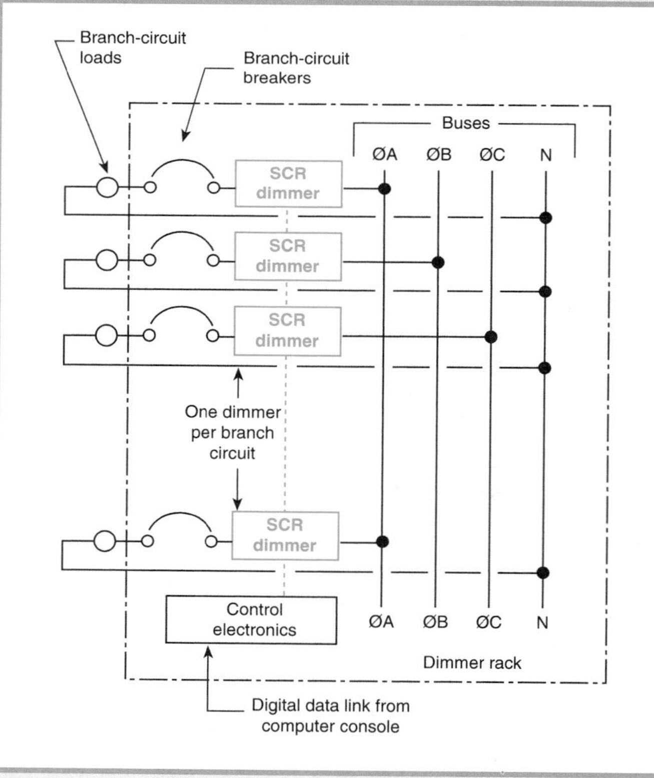

Figure 520.3 *Schematic diagram of a typical high-density SCR dimmer switchboard. (Production Arts Lighting, Inc.)*

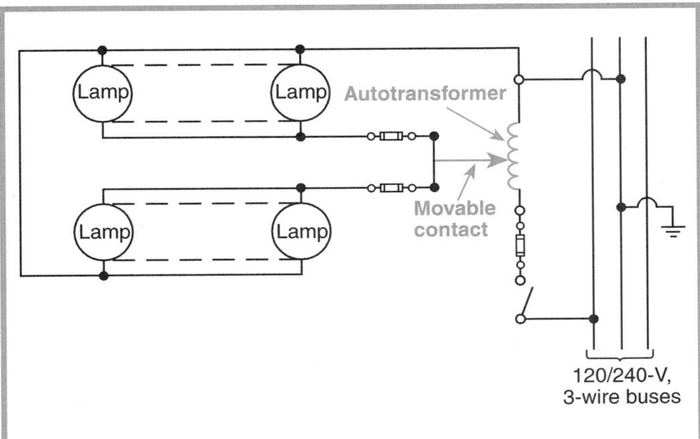

Figure 520.4 Typical connections for an autotransformer-type dimmer. (Kliegl Bros.)

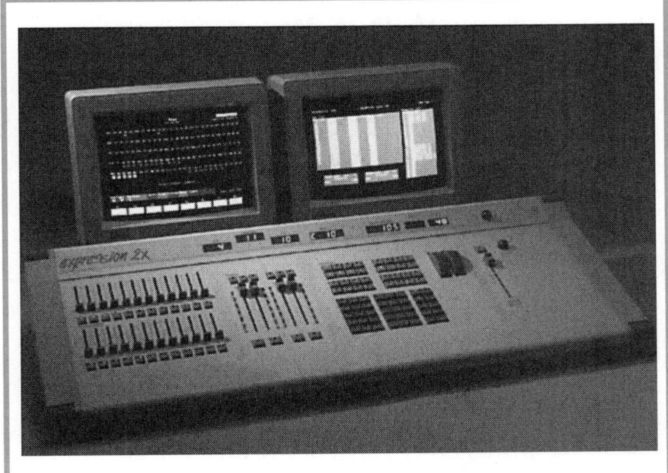

Figure 520.5 An electronic computer lighting control console for remotely controlling solid-state-type dimmers. (Electronic Theatre Controls, Inc.)

(a) Disconnection and Overcurrent Protection. Where dimmers are installed in ungrounded conductors, each dimmer shall have overcurrent protection not greater than 125 percent of the dimmer rating, and shall be disconnected from all ungrounded conductors when the master or individual switch or circuit breaker supplying such dimmer is in the open position.

A modern, high-density digital dimmer rack typically contains one dimmer (usually of 20-, 50-, or 100-ampere capacity) for each branch circuit connected to it. The rack is usually serviced by a 3-phase, 4-wire-plus-ground feeder, which is distributed via buses to all dimmers in the rack. Typical dimmer racks contain between 12 and 96 dimmers and may have total power capacities of up to 288 kW. In large theatrical systems, many racks may be bused together. A central control electronics module drives multiple dimmers in the rack. A digital data link may connect the dimmer rack to the remotely located computer control console.

(b) Resistance- or Reactor-Type Dimmers. Resistance- or series reactor-type dimmers shall be permitted to be placed in either the grounded or the ungrounded conductor of the circuit. Where designed to open either the supply circuit to the dimmer or the circuit controlled by it, the dimmer shall then comply with Section 380-1. Resistance- or reactor-type dimmers placed in the grounded neutral conductor of the circuit shall not open the circuit.

(c) Autotransformer-Type Dimmers. The circuit supplying an autotransformer-type dimmer shall not exceed 150 volts between conductors. The grounded conductor shall be common to the input and output circuits.

Circuits supplying autotransformer-type dimmers are not permitted to exceed 150 volts between conductors. Any desired voltage may be applied to the lamps, from full-line voltage to voltage so low that the lamps provide no illumination, by means of a movable contact tap. This type of dimmer produces very little heat and operates at high efficiency. Its dimming effect, within its maximum rating, is independent of the wattage of the load. See the commentary that follows Section 470-4, which discusses saturable reactors that are sometimes used for stage dimmers.

FPN: See Section 210-9 for circuits derived from autotransformers.

(d) Solid-State-Type Dimmers. The circuit supplying a solid-state dimmer shall not exceed 150 volts between conductors unless the dimmer is listed specifically for higher voltage operation. Where a grounded conductor supplies a dimmer, it shall be common to the input and output circuits. Dimmer chassis shall be connected to the equipment grounding conductor.

Modern stage switchboards are usually of the remote-control type. The switchboard is operated from a remote console, typically a computer system such as the one shown in Figure 520.2 or 520.3. The switchboard or dimmer rack is normally located offstage in a dimmer room, where proper climate control can be furnished and noise from the rack cooling fans will not interfere with the performance onstage. Branch circuits are usually connected to the dimmer rack on a "dimmer per circuit" basis. A digital control cable connects the computer and the dimmer rack, allowing the operator to be positioned on stage or in the auditorium for easy viewing of the performance.

A front view of a typical high-density digital SCR dimmer rack is shown in Figure 520.3 or 520.4. A schematic for this type of dimmer rack is shown in Figure 520.4 or 5. Dimmers for individual circuits are contained in dual plug-in dimmer modules. These modules also contain circuit breakers for overcurrent protection and filter chokes to eliminate acoustic noise from the lamp filaments. The digital control electronics are contained in a plug-in module with front-panel controls for configuration and testing.

520-26. Type of Switchboard. A stage switchboard shall be either one or a combination of the following types.

(a) Manual. Dimmers and switches are operated by handles mechanically linked to the control devices.

Manual-type switchboards usually contain resistance-type or autotransformer-type dimmers. Figure 520.4 is a schematic of a manual autotransformer-type switchboard.

(b) Remotely Controlled. Devices are operated electrically from a pilot-type control console or panel. Pilot control panels shall either be part of the switchboard or shall be permitted to be at another location.

Figure 520.2 illustrates a typical solid-state SCR dimmer switchboard. Figure 520.5 shows a typical control console used to control the dimmer switchboard.

(c) Intermediate. A stage switchboard with circuit interconnections is a secondary switchboard (patch panel) or panelboard remote to the primary stage switchboard. It shall contain overcurrent protection. Where the required branch-circuit overcurrent protection is provided in the dimmer panel, it shall be permitted to be omitted from the intermediate switchboard.

An intermediate stage switchboard, usually called a "patch panel," is located between the dimmer switchboard and the branch circuits. Its purpose is to either break down larger dimmer circuits to smaller branch circuits or to select branch circuits to be controlled by a dimmer, or both.

520-27. Stage Switchboard Feeders.

(a) Type of Feeder. Feeders supplying stage switchboards shall be one of the following.

(1) Single Feeder. A single feeder disconnected by a single disconnect device.

(2) Multiple Feeders to Intermediate Stage Switchboard (Patch Panel). Multiple feeders of unlimited quantity shall be permitted, provided that all multiple feeders are part of a single system. Where combined, neutral conductors in a given raceway shall be of sufficient ampacity to carry the maximum unbalanced current supplied by multiple feeder conductors in the same raceway, but need not be greater than the ampacity of the neutral supplying the primary stage switchboard. Parallel neutral conductors shall comply with Section 310-4.

The feeders to patch panels are often many dimmer-controlled circuits at 100 amperes or less, single phase, so they can be distributed to different combinations of the same size or smaller branch circuits. This type of installation usually requires the use of a common neutral, and because of the quantity of circuits, many installations require several parallel neutrals running in several raceways. Generally, these parallel neutrals are sized as follows:

1. Size the common neutral to the feeder of the primary switchboard.
2. Split this neutral into multiple parallel conductors, one per raceway.
3. Equally divide per phase and size each ungrounded conductor of the many single-phase circuits among the raceways.

In no case is it acceptable to install the ungrounded conductors in one raceway and the common neutral in another.

(3) Separate Feeders to Single Primary Stage Switchboard (Dimmer Bank). Installations with separate feeders to a single primary stage switchboard shall have a disconnecting means for each feeder. The primary stage switchboard shall have a permanent and obvious label stating the number and location of disconnecting means. If the disconnecting means are located in more than one distribution switchboard, the primary stage switchboard shall be provided with barriers to correspond with these multiple locations.

Larger primary stage switchboards usually consist of several sections, often called "dimmer racks," which form a dimmer bank. See Figure 520.2. These dimmer racks may be fed separately or be bused together to accept one or more feeder circuits. If an intermediate stage switchboard is connected to a primary stage switchboard, the primary stage switchboard is usually supplied by a single large feeder, because the intermediate stage switchboard patches only the ungrounded conductors and requires a common neutral. Modern theaters do not use intermediate stage

switchboards, and dimmer banks may have one or several feeders.

(b) Neutral. The neutral of feeders supplying solid-state, 3-phase, 4-wire dimming systems shall be considered a current-carrying conductor.

(c) Supply Capacity. For the purposes of computing supply capacity to switchboards, it shall be permissible to consider the maximum load that the switchboard is intended to control in a given installation, provided that

(1) All feeders supplying the switchboard shall be protected by an overcurrent device with a rating not greater than the ampacity of the feeder.
(2) The opening of the overcurrent device shall not affect the proper operation of the egress or emergency lighting systems.

FPN: For computation of stage switchboard feeder loads, see Section 220-10.

The feeder for single, primary stage switchboards is sized in accordance with the maximum load the switchboard is intended to control for a specific location. The feeder(s) is required to be protected by an overcurrent device that has a rating not greater than the feeder ampacity. Operation of the overcurrent device is not allowed to have any effect on egress or emergency lighting systems. The neutral of feeders supplying solid-state, 3-phase, 4-wire dimming systems will carry third-harmonic currents that are present even under balanced load conditions.

C. Fixed Stage Equipment Other than Switchboards

520-41. Circuit Loads.

(a) Circuits Rated 20 Amperes or Less. Footlights, border lights, and proscenium side lights shall be arranged so that no branch circuit supplying such equipment will carry a load exceeding 20 amperes.

(b) Circuits Rated Greater than 20 Amperes. Where heavy-duty lampholders only are used, such circuits shall be permitted to comply with Article 210 for circuits supplying heavy-duty lampholders.

Sections 210-23(b) and (c) permit 30-, 40-, or 50-ampere branch circuits if heavy-duty lampholders, such as medium- or mogul-base Edison screw-shell types, are used for fixed lighting.

520-42. Conductor Insulation. Foot, border, proscenium, or portable strip light fixtures and connector strips shall be wired with conductors that have insulation suitable for the temperature at which the conductors will be operated, but not less than 125°C (257°F). The ampacity of the 125°C (257°F) conductors shall be that of 60°C (140°F) conductors. All drops from connector strips shall be 90°C (194°F) wire sized to the ampacity of 60°C (140°F) cords and cables with no more than 6 in. (152 mm) of conductor extending into the connector strip. Section 310-15(b)(2)(a) shall not apply.

The 125°C (257°F) minimum temperature rating is based on the heat from the lamps raising the ambient temperature in which the wiring is located. Drops from connector strips are usually flexible cord. While the 90°C-rated cord is also in the higher ambient, it is not in sufficient contact with other circuits that might also heat it. The derating factors of Section 310-15(b)(2) are judged unnecessary because the conductors are not all energized at one time, are not often energized at full intensity (dimmed), and are not energized continuously.

FPN: See Table 310-13 for conductor types.

520-43. Footlights.

(a) Metal Trough Construction. Where metal trough construction is employed for footlights, the trough containing the circuit conductors shall be made of sheet metal not lighter than No. 20 MSG treated to prevent oxidation. Lampholder terminals shall be kept at least ½ in. (12.7 mm) from the metal of the trough. The circuit conductors shall be soldered to the lampholder terminals.

(b) Other-than-Metal Trough Construction. Where the metal trough construction specified in Section 520-43(a) is not used, footlights shall consist of individual outlets with lampholders wired with rigid metal conduit, intermediate metal conduit, or flexible metal conduit, Type MC cable, or mineral-insulated, metal-sheathed cable. The circuit conductors shall be soldered to the lampholder terminals.

(c) Disappearing Footlights. Disappearing footlights shall be arranged so that the current supply will be automatically disconnected when the footlights are replaced in the storage recesses designed for them.

The footlights described in Sections 520-43(a) and (b) are generally obsolete units that were built in the field. Modern footlights are compartmentalized, factory-wired assemblies for field installation, as shown in Figure 520.6. Footlight assemblies may be permanently exposed or be of the disappearing type. Disappearing footlights are arranged so as to automatically disconnect the current supply when the footlights are in the closed position, thereby preventing heat entrapment that could cause a fire. Disconnection is accomplished by mercury switches in the terminal compartment.

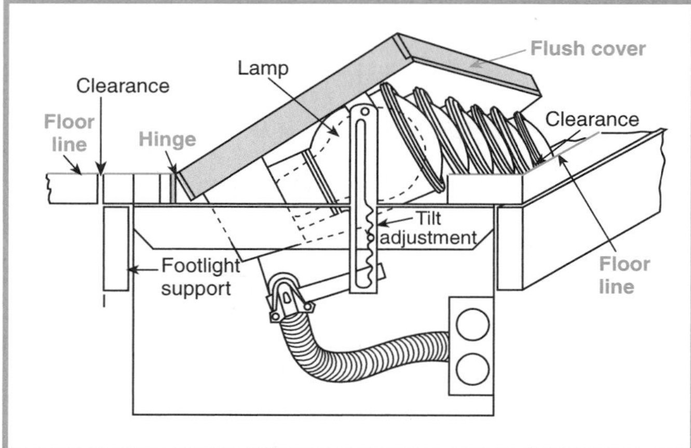

Figure 520.6 To prevent heat entrapment, disappearing footlights are arranged to automatically disconnect the supply circuit when in the closed position.

520-44. Borders and Proscenium Sidelights.

(a) General. Borders and proscenium sidelights shall be as follows:

(1) Constructed as specified in Section 520-43,
(2) Suitably stayed and supported; and
(3) Designed so that the flanges of the reflectors or other adequate guards will protect the lamps from mechanical damage and from accidental contact with scenery or other combustible material.

Figure 520.7 shows a modern border light installed over a stage. Figure 520.8 is a cross-sectional view that illustrates construction details. This particular border light is designed for 200-watt lamps. To obtain the highest illumination efficiency, each lamp is provided with its own reflector. Fitted to each reflector is a glass roundel available in any color. Commonly,

Figure 520.7 A suspended border-light assembly for installation over a stage. (Kliegl Bros.)

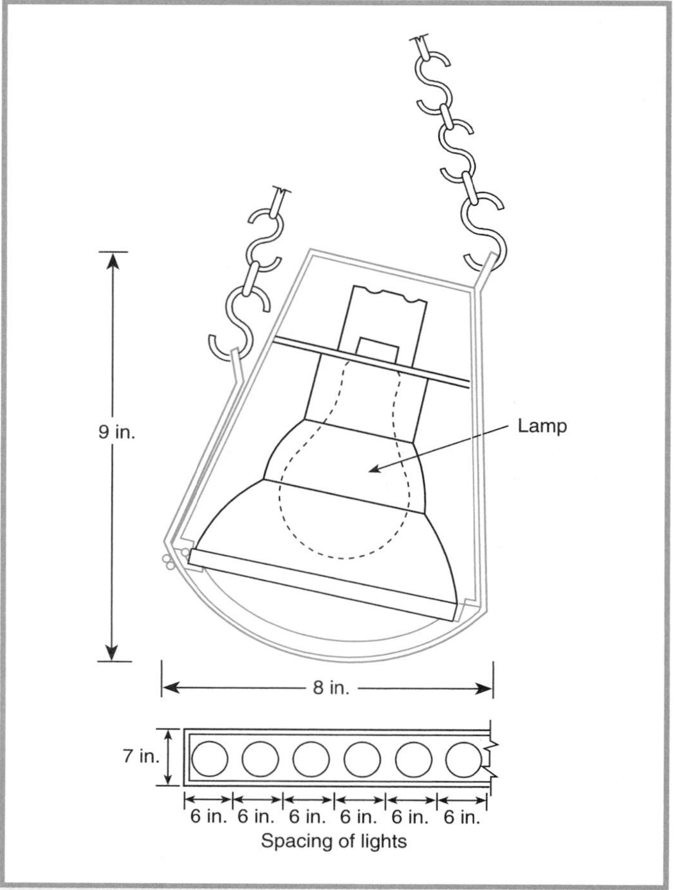

Figure 520.8 A cross-sectional view of a typical light in the border-light assembly shown in Figure 520.7. (Kliegl Bros.)

lampholders are wired alternately on three or four circuits. A splice box is provided on top of the housing for enclosing connections between the cable supplying the border light and the border light's internal wiring, which consists of wiring from the splice box to the lamp sockets in a trough extending the length of the border.

(b) Cords and Cables for Border Lights.

(1) General. Cords and cables for supply to border lights shall be listed for extra-hard usage. The cords and cables shall be suitably supported. Such cords and cables shall be employed only where flexible conductors are necessary. Ampacity of the conductors shall be as provided in Section 400-5.

To facilitate height adjustment for cleaning and lamp replacement, border lights are usually supported by steel cables. Therefore, the circuit conductors supplying the border lights are required to be carried to the border light in a flexible cable. Each of these flexible cables usually contains many circuits; however, its overall size is limited by its ability to travel up and down without getting tangled. See Figure 520.9.

(2) Cords and Cables Not in Contact with Heat-Producing Equipment. Listed multiconductor extra-hard usage-type cords and cables not in direct contact with equipment containing heat-producing elements shall be permitted to have their ampacity determined by Table 520-44. Maximum load current in any conductor with an ampacity determined by Table 520-44 shall not exceed the values in Table 520-44.

Section 520-44(b)(2) permits extra-hard-usage cords not in direct contact with heat-producing equipment to have their ampacity determined by Table 520-44 instead of Section 400-5.

Table 520-44 is based on a minimum 50 percent diversity factor. It includes the fact that not all circuits are on at the same time, not all circuits are at full intensity (dimmed), and not all circuits are on for a long period of time. If the load diversity does not follow this pattern, such as border lights that are all left on at full intensity to light the stage for rehearsal, lecture, or classroom purposes, this table must not be used.

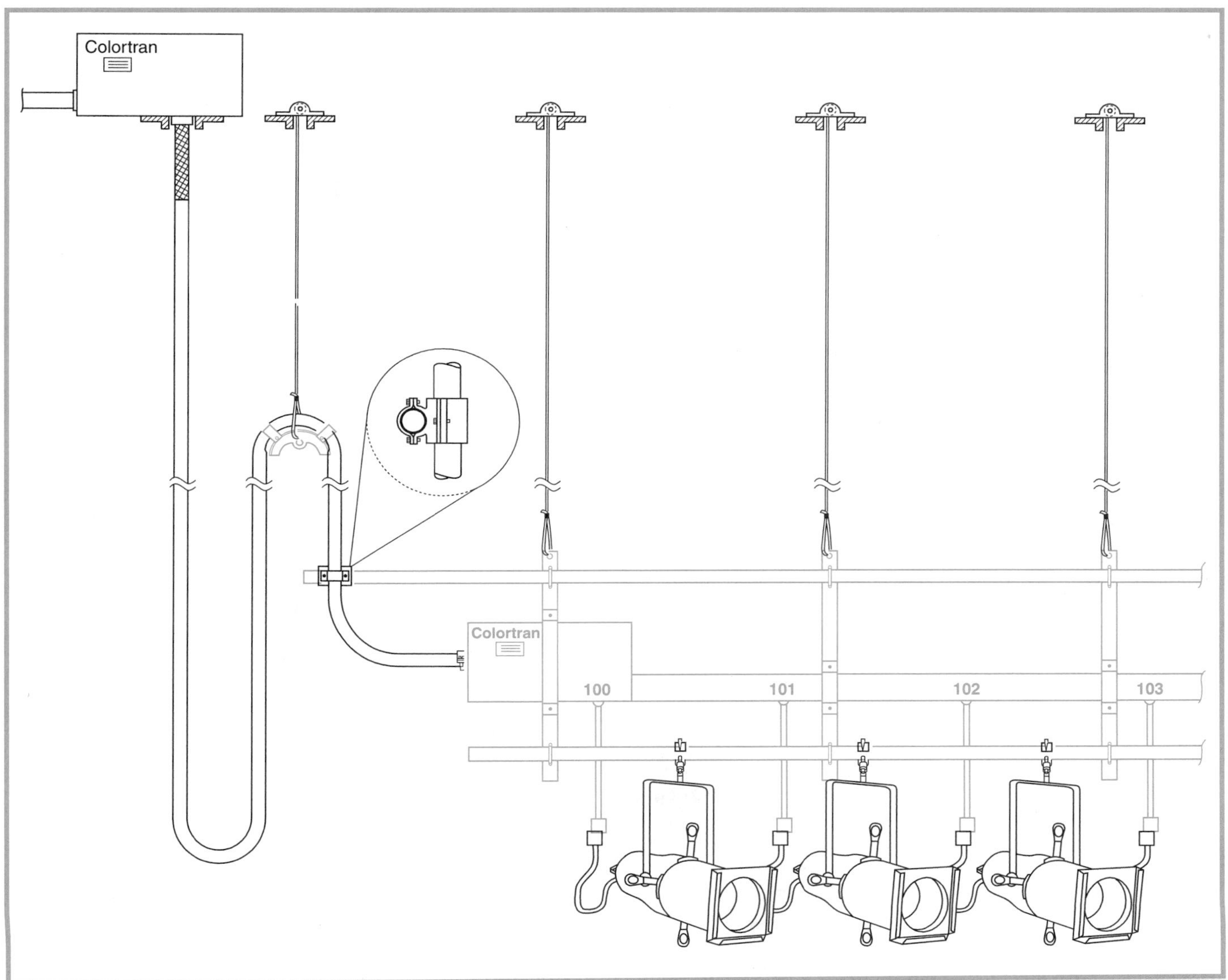

Figure 520.9 *A suspended connector strip with border-light cable attached. (Colortran, Inc.)*

Table 520-44. Ampacity of Listed Extra-Hard Usage Cords and Cables with Temperature Ratings of 75°C (167°F) and 90°C (194°F)*
[Based on Ambient Temperature of 30°C (86°F)]

	Temperature Rating of Cords and Cables		Maximum Rating of Overcurrent Device
Size (AWG)	75°C (167°F)	90°C (194°F)	
14	24	28	15
12	32	35	20
10	41	47	25
8	57	65	35
6	77	87	45
4	101	114	60
2	133	152	80

*Ampacity shown is the ampacity for multiconductor cords and cables where only three copper conductors are current-carrying. If the number of current-carrying conductors in a cord or cable exceeds three and the load diversity factor is a minimum of 50 percent, the ampacity of each conductor shall be reduced as shown in the following table.

Number of Conductors	Percent of Ampacity
4–6	80
7–24	70
25–42	60
43 and above	50

Note: Ultimate insulation temperature. In no case shall conductors be associated together in such a way with respect to the kind of circuit, the wiring method used, or the number of conductors such that the temperature limit of the conductors will be exceeded.

A neutral conductor that carries only the unbalanced current from other conductors of the same circuit need not be considered as a current-carrying conductor.

In a 3-wire circuit consisting of two phase wires and the neutral of a 4-wire, 3-phase, wye-connected system, a common conductor carries approximately the same current as the line-to-neutral currents of the other conductors and shall be considered to be a current-carrying conductor.

On a 4-wire, 3-phase, wye circuit where the major portion of the load consists of nonlinear loads such as electric-discharge lighting, electronic computer/data processing, or similar equipment, there are harmonic currents present in the neutral conductor, and the neutral shall be considered to be a current-carrying conductor.

520-45. Receptacles. Receptacles for electrical equipment or fixtures on stages shall be rated in amperes. Conductors supplying receptacles shall be in accordance with Articles 310 and 400.

520-46. Connector Strips, Drop Boxes, Floor Pockets, and Other Outlet Enclosures. Receptacles for the connection of portable stage-lighting equipment shall be pendant or mounted in suitable pockets or enclosures and shall comply with Section 520-45. Supply cables for connector strips and drop boxes shall be as specified in Section 520-44(b).

Figure 520.9 shows a hanging connector strip with its associated border-light cable. Border lights are hung and supplied in a similar manner. Figures 520.9 through 520.11 illustrate different types of connections for portable stage lighting equipment.

Figure 520.10 A 3-gang, 3-receptacle pin-plug outlet box designed for flush mounting. (Electronic Theatre Controls, Inc.)

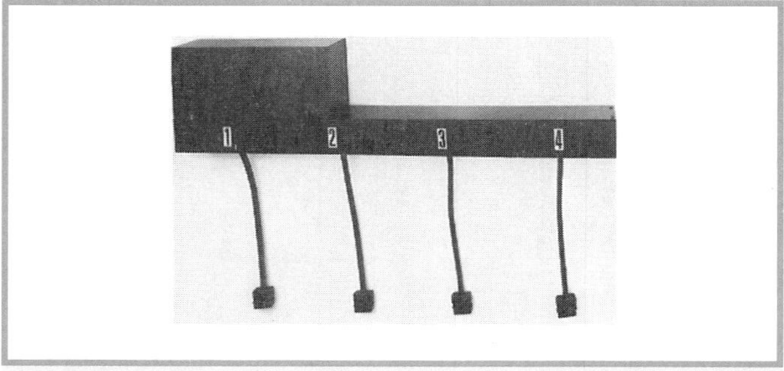

Figure 520.11 A typical four-circuit connector strip designed for wall or pipe mounting. (Electronic Theatre Controls, Inc.)

520-47. Backstage Lamps (Bare Bulbs). Lamps (bare bulbs) installed in backstage and ancillary areas where they can come in contact with scenery shall be located and guarded so as to be free from physical damage and shall provide an air space of not less than 2 in. (50.8 mm) between such lamps and any combustible material.

Exception: Decorative lamps installed in scenery shall not be considered to be backstage lamps for the purpose of this section.

520-48. Curtain Machines. Curtain machines shall be listed.

520-49. Smoke Ventilator Control. Where stage smoke ventilators are released by an electrical device, the circuit operating the device shall be normally closed and shall be controlled by at least two externally operable switches, one switch being placed at a readily accessible location on stage and the other where designated by the authority having jurisdiction. The device shall be designed for the full voltage of the circuit to which it is connected, no resistance being inserted. The device shall be located in the loft above the scenery and shall be enclosed in a suitable metal box having a tight, self-closing door.

In addition to the smoke ventilators being controlled from two externally operable switches at different locations, the design of a normally closed circuit ensures that the smoke ventilators will operate when the circuit opens for any reason, such as a circuit breaker tripping or a fuse blowing.

D. Portable Switchboards on Stage

520-50. Road Show Connection Panel (A Type of Patch Panel). A panel designed to allow for road show connection of portable stage switchboards to fixed lighting outlets by means of permanently installed supplementary circuits. The panel, supplementary circuits, and outlets shall comply with (a) through (d).

Also known as a "road show interconnect" or "intercept panel," a road show connection panel is designed to connect the load side of a portable switchboard to the fixed building branch circuits and associated outlets. It may also provide for the fixed branch circuits to be connected to a fixed switchboard when the portable switchboard is not installed. See Figure 520.12.

Figure 520.12 *A road show interconnect panel (intercept panel) with circuit-breaker-protected pendant male connectors (pigtails) connected either to connector bodies attached to a portable switchboard or to receptacles connected to a fixed switchboard. (Colortran, Inc.)*

(a) Load Circuits. Circuits shall terminate in grounding-type polarized inlets of current and voltage rating that match the fixed-load receptacle.

The grounding-type polarized inlets may be flush or pendant. The fixed-load receptacle is on the other end of the branch circuit that emanates from the panel.

(b) Circuit Transfer. Circuits that are transferred between fixed and portable switchboards shall have all circuit conductors transferred simultaneously.

Section 520-50(b) requires simultaneous transfer of all conductors of the circuit, including any grounded conductors.

(c) Overcurrent Protection. The supply devices of these supplementary circuits shall be protected by branch-circuit overcurrent protective devices. The individual supplementary circuit, within the road show connection panel and theater, shall be protected by branch-circuit overcurrent protective devices of suitable ampacity installed within the road show connection panel.

The branch-circuit overcurrent protection should normally be in the switchboard, but since some older units do not have this protection, backup overcurrent protection is provided by Section 520-50(c).

(d) Enclosure. Panel construction shall be in accordance with Article 384.

520-51. Supply. Portable switchboards shall be supplied only from power outlets of sufficient voltage and ampere rating. Such power outlets shall include only externally operable, enclosed fused switches or circuit breakers mounted on stage or at the permanent switchboard in locations readily accessible from the stage floor. Provisions for connection of an equipment grounding conductor shall be provided. The neutral of feeders supplying solid-state, 3-phase, 4-wire dimmer systems shall be considered a current-carrying conductor.

Power outlets, known in the entertainment industry as "company switches" or "bull switches," are the point in the wiring system where portable feeder cables connect to the fixed building wiring. They may be as simple as an overcurrent-protected multipole receptacle designed to accept the supply cable described in Section 520-53(p), Exception, or they may be multiple sets of parallel single-conductor feeder cables. These single-conductor feeder cables, as described in Section 520-53(h), may be terminated via single-pole separable connectors described in Section 520-53(k), or directly to busbars, fused disconnect switches, or circuit breakers with wire connectors (lugs).

Figure 520.13 shows a typical large company switch. A locking door engaging a safety switch covers the load busbars and trips the main circuit breaker if the door is opened.

The busbars provide for bolted-on wire connectors and for clamp-on wire connectors called "sister lugs."

The single-conductor cables enter through a nonferrous panel in the bottom, provided, in this case, with conventional cable clamps. Often, this strain relief is provided by tying the cable to a tie-off bar with rope. This unit also includes a set of sequential-interlocking-type single-pole separable connectors.

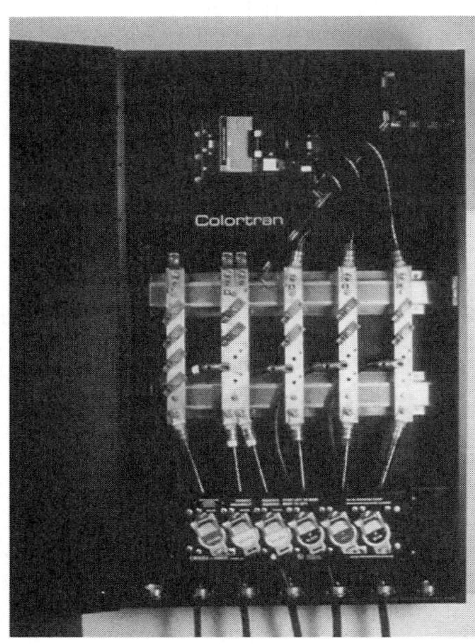

Figure 520.13 *A company switch (bull switch) with a listed sequential-interlocking-type assembly of single-pole separable connectors and with busing to connect single-conductor cables terminated in wire terminal connectors or busbar clamps (sister lugs). (Colortran, Inc.)*

520-52. Overcurrent Protection. Circuits from portable switchboards directly supplying equipment containing incandescent lamps of not over 300 watts shall be protected by overcurrent protective devices having a rating or setting of not over 20 amperes. Circuits for lampholders over 300 watts shall be permitted where overcurrent protection complies with Article 210.

520-53. Construction and Feeders. Portable switchboards and feeders for use on stages shall comply with (a) through (p).

See Figure 520.14 for an example of a portable switchboard.

(a) Enclosure. Portable switchboards shall be placed within an enclosure of substantial construction, which shall be permitted to be arranged so that the enclosure is open during operation. Enclosures of wood shall be completely lined with sheet metal of not less than No. 24 MSG and

Figure 520.14 *A large, portable SCR dimmer switchboard (rolling rack). (Colortran, Inc.)*

shall be well galvanized, enameled, or otherwise properly coated to prevent corrosion or be of a corrosion-resistant material.

(b) Energized Parts. There shall not be exposed energized parts within the enclosure.

(c) Switches and Circuit Breakers. All switches and circuit breakers shall be of the externally operable, enclosed type.

(d) Circuit Protection. Overcurrent devices shall be provided in each ungrounded conductor of every circuit supplied through the switchboard. Enclosures shall be provided for all overcurrent devices in addition to the switchboard enclosure.

(e) Dimmers. The terminals of dimmers shall be provided with enclosures, and dimmer faceplates shall be arranged so that accidental contact cannot be readily made with the faceplate contacts.

(f) Interior Conductors.

(1) Type. All conductors other than busbars within the switchboard enclosure shall be stranded. Conductors shall be approved for an operating temperature at least equal to the approved operating temperature of the dimming devices used in the switchboard and in no case less than the following:

(1) Resistance-type dimmers — 200°C (392°F); or
(2) Reactor-type, autotransformer, and solid-state dimmers — 125°C (257°F).

All control wiring shall comply with Article 725.

(2) Protection. Each conductor shall have an ampacity not less than the rating of the circuit breaker, switch, or fuse that it supplies. Circuit interrupting and bus bracing shall be in accordance with Sections 110-9 and 110-10. The short-circuit current rating shall be marked on the switchboard.

Conductors shall be enclosed in metal wireways or shall be securely fastened in position and shall be bushed where they pass through metal.

(g) Pilot Light. A pilot light shall be provided within the enclosure and shall be connected to the circuit supplying the board so that the opening of the master switch will not cut off the supply to the lamp. This lamp shall be on an individual branch circuit having overcurrent protection rated or set at not over 15 amperes.

The requirement of Section 520-53(g) applies only to switchboards with a main disconnect, if provided, on the switchboard. It serves as a warning at the switchboard as to the presence of power before the main disconnect is activated.

(h) Supply Conductors.

(1) General. The supply to a portable switchboard shall be by means of listed extra-hard usage cords or cables. The supply cords or cable shall terminate within the switchboard enclosure, in an externally operable fused master switch or circuit breaker, or in a connector assembly identified for the purpose. The supply cords or cable (and connector assembly) shall have sufficient ampacity to carry the total load connected to the switchboard and shall be protected by overcurrent devices.

As with the supply end described in Section 520-51, the connection described in Section 520-53(h)(1) may be as simple as permanently terminated multiconductor supply cord or multipole connector assembly (inlet), or as complex as a set of parallel single-conductor feeder cables. These cables may be field-connected to an assembly of single-pole connectors (inlet) or directly connected, with wire connectors, to busbars or a fused switch or breaker.

Section 520-53(h)(1) permits road shows with fixed lighting plans to size the feeder to the actual connected load.

(2) Single-Conductor Cables. Single-conductor portable supply cable sets shall not be smaller than No. 2 conductors. The equipment grounding conductor shall not be smaller than No. 6 conductor. Single conductor grounded neutral cables for a supply shall be sized as per Section 520-53(o)(2). Where single conductors are paralleled for increased ampacity, the paralleled conductors shall be of the same length and size. Single conductor supply cables shall be grouped together but not bundled. The equipment grounding conductor shall be permitted to be of a different type, provided it meets the other requirements of this section, and it shall be permitted to be reduced in size as permitted by Section 250-122. Grounded (neutral) and equipment grounding conductors shall be identified in accordance with Sections 200-6, 250-134(b), and 310-12. Grounded conductors shall be permitted to be identified by marking at least the first 6 in. (152.4 mm) from both ends of each length of conductor with white or natural gray. Equipment grounding conductors shall be permitted to be identified by marking at least the first 6 in. (152.4 mm) from both ends of each length of conductor with green or green with yellow stripes. Where more than one nominal voltage exists within the same premises, each ungrounded conductor shall be identified by system.

(3) Supply Conductors Not Over 10 ft (3.05 m) Long. Where supply conductors do not exceed 10 ft (3.05 m) in length between supply and switchboard or supply and a subsequent overcurrent device, the supply conductors shall be permitted to be reduced in size where all of the following conditions are met.

(a) The ampacity of the supply conductors shall be at least one-quarter of the ampacity of the supply overcurrent protection device.
(b) The supply conductors shall terminate in a single overcurrent protection device that will limit the load to the ampacity of the supply conductors. This single overcurrent device shall be permitted to supply additional overcurrent devices on its load side.
(c) The supply conductors shall not penetrate walls, floors, or ceilings or be run through doors or traffic areas. The supply conductors shall be adequately protected from physical damage.
(d) The supply conductors shall be suitably terminated in an approved manner.
(e) Conductors shall be continuous without splices or connectors.
(f) Conductors shall not be bundled.
(g) Conductors shall be supported above the floor in an approved manner.

(4) Supply Conductors Not Over 20 ft (6.1 m) Long. Where supply conductors do not exceed 20 ft (6.1 m) in length between supply and switchboard or supply and a subsequent overcurrent protection device, the supply conductors shall be permitted to be reduced in size where all of the following conditions are met.

(a) The ampacity of the supply conductors shall be at least one-half of the ampacity of the supply overcurrent protection device.
(b) The supply conductors shall terminate in a single overcurrent protection device that will limit the load to the ampacity of the supply conductors. This single overcurrent device shall be permitted to supply additional overcurrent devices on its load side.
(c) The supply conductors shall not penetrate walls, floors, or ceilings or be run through doors or traffic areas. The

supply conductors shall be adequately protected from physical damage.

(d) The supply conductors shall be suitably terminated in an approved manner.

(e) The supply conductors shall be supported in an approved manner at least 7 ft (2.13 m) above the floor except at terminations.

(f) The supply conductors shall not be bundled.

(g) Tap conductors shall be in unbroken lengths.

Loads of 144 kVA and greater are not uncommon, even on portable switchboard equipment. Installations in the field include lighting for theatrical-type productions with large numbers of lighting fixtures. However, only a fraction of the many fixtures installed are used at any one time. The intent of Sections 520-53(h)(3) and (h)(4) is that the supply conductors are required to be sized according to their overcurrent protection and not by the total connected load. These requirements are similar to the requirements for taps found in Section 240-21.

The tap rules of Sections 520-53(h)(3) and (h)(4) are designed to allow one or more switchboards with smaller feeders to be connected to larger supplies (company switches). If these "rules" are not complied with, proper overcurrent protection devices, either fixed or portable, must be provided for each of the smaller switchboards.

The requirement that the conductors not be bundled is so that Column D of Table 400-5(B) can be employed. If the conductors were bundled, Column F and all applicable derating factors would apply. Most devices used in the theater to terminate single-conductor cables are rated for use at 90°C ampacity. However, if single-conductor cables are terminated directly to a circuit breaker or fused switch, a 75°C ampacity or lower would most likely apply.

(5) Supply Conductors Not Reduced in Ampacity. Supply conductors not reduced in ampacity shall be permitted to pass through holes in walls specifically designed for the purpose. If penetration is through the fire-resistant–rated wall, it shall be in accordance with Section 300-21.

(i) Cable Arrangement. Cables shall be protected by bushings where they pass through enclosures and shall be arranged so that tension on the cable will not be transmitted to the connections. Where power conductors pass through metal, the requirements of Section 300-20 shall apply.

Tension on the connections is removed by using conventional strain relief devices, or often by lashing the cable to the enclosure with rope.

(j) Number of Supply Interconnections. Where connectors are used in a supply conductor, there shall be a maximum number of three interconnections (mated connector pairs) where the total length from supply to switchboard does not exceed 100 ft (30.5 m). In cases where the total length from supply to switchboard exceeds 100 ft (30.5 m), one additional interconnection shall be permitted for each additional 100 ft (30.5 m) of supply conductor.

The addition of excessive numbers of interconnections could jeopardize the mechanical and electrical integrity of the supply conductors.

(k) Single-Pole Separable Connectors. Where single-pole portable cable connectors are used, they shall be listed and of the locking type. Sections 400-10 and 410-56 shall not apply to listed single-pole separable connectors and single-conductor cable assemblies utilizing listed single-pole separable connectors. Where paralleled sets of current-carrying, single-pole separable connectors are provided as input devices, they shall be prominently labeled with a warning indicating the presence of internal parallel connections. The use of single-pole separable connectors shall comply with at least one of the following conditions.

(1) Connection and disconnection of connectors are only possible where the supply connectors are interlocked to the source and it is not possible to connect or disconnect connectors when the supply is energized.

(2) Line connectors are of the listed sequential-interlocking type so that load connectors shall be connected in the following sequence:

(a) Equipment grounding conductor connection
(b) Grounded circuit conductor connection, if provided
(c) Ungrounded conductor connection, and that disconnection shall be in the reverse order

(3) A caution notice shall be provided adjacent to the line connectors indicating that plug connection shall be in the following order:

(a) Equipment grounding conductor connectors
(b) Grounded circuit conductor connectors, if provided
(c) Ungrounded conductor connectors, and that disconnection shall be in the reverse order

Section 520-53(k) provides for a special type of connection device suitable for connecting single-conductor feeder cables. The connection device is to be listed and of the locking type, reducing the likelihood of its separating while under load. The connectors must be used in sets, since they are only single-pole types. It is important that the grounding conductor be connected first and disconnected last, and that the grounded conductor be connected next-to-first and disconnected next-to-last. The connector sets must be arranged so as to reduce the likelihood that the connections will be made in the incorrect order, in accordance with one of the following methods.

1. Provide a scheme whereby the main disconnect cannot be energized until all conductors are connected.
2. Provide a scheme whereby the connectors are precluded from being connected in any order other than the proper one. See Figure 520.12 for an example.
3. Provide a scheme whereby the individual connectors, free of any special electromechanical intervention, are marked with instructions to the user as to proper connection. See Figure 520.13 for an example.

Single-pole separable connectors are quick-connect feeder splicing and terminating devices, not attachment plugs or receptacles. They are designed to be sized, terminated, and inspected by a qualified person before being energized, and are to be guarded from accidental disconnection before being de-energized.

(l) Protection of Supply Conductors and Connectors. All supply conductors and connectors shall be protected against physical damage by an approved means. This protection shall not be required to be raceways.

Rubber mats and commercially available rubber bridges are often used for the protection of supply conductors and connectors.

(m) Flanged Surface Inlets. Flanged surface inlets (recessed plugs) that are used to accept the power shall be rated in amperes.

(n) Terminals. Terminals to which stage cables are connected shall be located so as to permit convenient access to the terminals.

The requirement of Section 520-53(n) facilitates the field connection and disconnection of the large feeder cables as the show travels from place to place.

(o) Neutral.

(1) Neutral Terminal. In portable switchboard equipment designed for use with 3-phase, 4-wire with ground supply, the supply neutral terminal, its associated busbar, or equivalent wiring, or both, shall have an ampacity equal to at least twice the ampacity of the largest ungrounded supply terminal.

Exception: Where portable switchboard equipment is specifically constructed and identified to be internally converted in the field, in an approved manner, from use with a balanced 3-phase, 4-wire with ground supply to a balanced single-phase, 3-wire with ground supply, the supply neutral terminal and its associated busbar, equivalent wiring, or both, shall have an ampacity equal to at least that of the largest ungrounded single-phase supply terminal.

Section 520-53(o)(1) requires careful study, because overlapping concepts are involved. If a 3-phase, 4-wire switchboard of any kind is brought into a space that has only single-phase, 3-wire service, the switchboard will most likely be connected with two phases to one leg and one phase to the other. This connection could double the current flowing through the neutral, so the neutral must be double size to allow for this possibility. The exception to Section 520-53(o)(1) provides for a smaller neutral sized for the single-phase feed where a switchboard contains switching devices that can divide the B-phase load equally between the A-phase and C-phase buses for single-phase operation.

Additionally, 3-phase, 4-wire switchboards that contain solid-state dimming devices must, when connected to a 3-phase, 4-wire supply, be connected to that supply with a multiconductor cable sized by counting the neutral as a current-carrying conductor, or with a set of single-conductor cables where the neutral is sized 130 percent greater than the phases.

For example, a 3-phase, 4-wire switchboard containing six 50-ampere SCR dimmers (100 amperes per phase) without a reassignment switching system would have to have a 200-ampere neutral. (A single-phase, 3-wire-only switchboard would not have to meet this special requirement.) This 200 percent rule would cover all the components making up the neutral conductor system inside or permanently attached to the switchboard, so as to allow for a full-size, single-phase, 3-wire feed when two of the 3-phase, 4-wire phase conductors are terminated to one single-phase, 3-wire leg. Note that the 200 percent neutral already covers the derating requirements (125 percent for a multiconductor feeder system and 130 percent for a single-conductor feeder system) when used in the 3-phase mode. If a reassignment system were added, the neutral would be required to be only 150 amperes. Again, when used in the 3-phase mode, the derating factors would be covered.

Note that the double-neutral requirement covers the terminal and associated busbar or wiring. This requirement begins at the main input terminals or busing, main input inlet connector, or attached main input cord and plug set and includes all wiring on the load side of that point. Power supply feeders easily detached at the terminals or inlet connector need not adhere to the 200 percent neutral rule, because they can easily be sized on a show-by-show basis for the type of supply encountered. These cables must, however, adhere to the requirements of the neutral as a current-carrying conductor, or of the 130 percent single-conductor-cable neutral.

(2) Supply Neutral. The power supply conductors for portable switchboards shall be sized considering the neutral as a current-carrying conductor. Where single-conductor feeder cables, not installed in raceways, are used on multiphase circuits, the grounded neutral conductor shall have an ampacity of at least 130 percent of the ungrounded circuit conductors feeding the portable switchboard.

(p) Qualified Personnel. The routing of portable supply conductors, the making and breaking of supply connectors and other supply connections, and the energization and de-energization of supply services shall be performed by qualified personnel, and portable switchboards shall be so marked, indicating this requirement in a permanent and conspicuous manner.

Exception: A portable switchboard shall be permitted to be connected to a permanently installed supply receptacle by other than qualified personnel, provided that the supply receptacle is protected for its rated ampacity by an overcurrent device of not greater than 150 amperes, and where the receptacle, interconnection, and switchboard further

(a) Employ listed multipole connectors suitable for the purpose for every supply interconnection, and
(b) Prevent access to all supply connections by the general public, and
(c) Employ listed extra-hard usage multiconductor cords or cables with an ampacity suitable for the type of load and not less than the ampere rating of the connectors.

The intent of Section 520-53(p) is to divide the acceptable practices in what are most likely to be professional and professional-grade educational venues from those in amateur or amateur-grade educational venues. The basic requirements allow for such things as single-conductor feeder systems, feeders sized for the current-connected load, tap rules, and so on, and require the services of a qualified person. The exception to Section 520-53(p) provides for a conventional feeder system suitable for use by an untrained person.

E. Portable Stage Equipment Other than Switchboards

520-61. Arc Lamp Fixtures. Arc lamp fixtures, including enclosed arc lamp fixtures and associated ballasts, shall be listed. Interconnecting cord sets and interconnecting cords and cables shall be extra-hard usage type and listed.

520-62. Portable Power Distribution Units. Portable power distribution units shall comply with (a) through (e).

(a) Enclosure. The construction shall be such that no current-carrying part will be exposed.

(b) Receptacles and Overcurrent Protection. Receptacles shall comply with Section 520-45 and shall have branch-circuit overcurrent protection in the box. Fuses and circuit breakers shall be protected against physical damage. Cords or cables supplying pendant receptacles shall be listed for extra-hard usage.

(c) Busbars and Terminals. Busbars shall have an ampacity equal to the sum of the ampere ratings of all the circuits connected to the busbar. Lugs shall be provided for the connection of the master cable.

(d) Flanged Surface Inlets. Flanged surface inlets (recessed plugs) that are used to accept the power shall be rated in amperes.

(e) Cable Arrangement. Cables shall be adequately protected where they pass through enclosures and be arranged so that tension on the cable will not be transmitted to the terminations.

520-63. Bracket Fixture Wiring.

(a) Bracket Wiring. Brackets for use on scenery shall be wired internally, and the fixture stem shall be carried through to the back of the scenery where a bushing shall be placed on the end of the stem. Externally wired brackets or other fixtures shall be permitted where wired with cords designed for hard usage that extend through scenery and without joint or splice in canopy of fixture back and terminate in an approved-type stage connector located, where practical, within 18 in. (457 mm) of the fixture.

(b) Mounting. Fixtures shall be securely fastened in place.

520-64. Portable Strips. Portable strips shall be constructed in accordance with the requirements for border lights and proscenium side lights in Section 520-44(a). The supply cable shall be protected by bushings where it passes through metal and shall be arranged so that tension on the cable will not be transmitted to the connections.

FPN No. 1: See Section 520-42 for wiring of portable strips.

FPN No. 2: See Section 520-68(a)(3) for insulation types required on single conductors.

520-65. Festoons. Joints in festoon wiring shall be staggered. Lamps enclosed in lanterns or similar devices of combustible material shall be equipped with guards.

Festoon wiring is defined in Article 100. Joints in festoon wiring are required to be staggered and properly insulated. This arrangement ensures that connections will not be opposite one another, which could cause sparking due to improper insulation or unraveling of insulation, which, in turn, could ignite lanterns or other combustible material enclosing lamps. Where lampholders have terminals of a type that puncture the conductor insulation and make contact with the conductors, stranded conductors should be used.

520-66. Special Effects. Electrical devices used for simulating lightning, waterfalls, and the like shall be constructed and located so that flames, sparks, or hot particles cannot come in contact with combustible material.

520-67. Multipole Branch-Circuit Cable Connectors. Multipole branch-circuit cable connectors, male and female, for flexible conductors shall be constructed so that tension on the cord or cable will not be transmitted to the connections. The female half shall be attached to the load end of the power supply cord or cable. The connector shall be rated in amperes and designed so that differently rated devices cannot be connected together. Alternating-current multipole connectors shall be polarized and comply with Sections 410-56(g) and 410-58.

FPN: See Section 400-10 for pull at terminals.

520-68. Conductors for Portables.

(a) Conductor Type.

(1) General. Flexible conductors, including cable extensions, used to supply portable stage equipment shall be listed extra-hard usage cords or cables.

(2) Stand Lamps. Reinforced cord shall be permitted to supply stand lamps where the cord is not subject to severe physical damage and is protected by an overcurrent device rated at not over 20 amperes.

See Section 520-2 for the definition of *stand lamp.*

(3) High-Temperature Applications. A special assembly of conductors in sleeving not longer than 3.3 ft (1 m) shall be permitted to be employed in lieu of flexible cord if the individual wires are stranded and rated not less than 125°C (257°F) and the outer sleeve is glass fiber with a wall thickness of at least 0.025 in. (0.635 mm).

Portable stage equipment requiring flexible supply conductors with a higher temperature rating where one end is permanently attached to the equipment shall be permitted to employ alternate, suitable conductors as determined by a qualified testing laboratory and recognized test standards.

Section 520-68(a)(3) provides for connection of stage lighting fixtures, which often operate at elevated temperatures. High-temperature (150°C to 250°C) extra-hard-usage cords are, in general, not available. These less-than-extra-hard-usage cords are limited to 3.3 ft (1 m) in length, to reduce the likelihood that they could be placed on the floor or other area where they might be damaged by traffic or moving scenery.

(4) Breakouts. Listed, hard usage (junior hard service) cords shall be permitted in breakout assemblies where all of the following conditions are met.

(a) The cords are utilized to connect between a single multipole connector containing two or more branch circuits and multiple two-pole, 3-wire connectors.
(b) The longest cord in the breakout assembly does not exceed 20 ft (6.1 m).
(c) The breakout assembly is protected from physical damage by attachment over its entire length to a pipe, truss, tower, scaffold, or other substantial support structure.
(d) All branch circuits feeding the breakout assembly are protected by overcurrent devices rated at not over 20 amperes.

Section 520-68(a)(4) applies to multiconductor cable assemblies with multipole connectors that contain more than one branch circuit. Figure 520.13 shows a multipole connector. The breakout assembly is a multipole connector with several pendant receptacles connected to it, separating the multiple branch circuits into individual branch circuits. It is also possible to use a similar arrangement of pendant plugs to form a "breaking" assembly on the other end of the multiconductor cable.

(b) Conductor Ampacity. The ampacity of conductors shall be as given in Section 400-5, except multiconductor, listed, extra-hard usage portable cords that are not in direct contact with equipment containing heat-producing elements shall be permitted to have their ampacity determined by Table 520-44. Maximum load current in any conductor with an ampacity determined by Table 520-44 shall not exceed the values in Table 520-44.

Section 520-68(b) allows portable, multicircuit, multiconductor cable to be sized in accordance with Table 520-44, in a manner similar to border-light cable. If portable, multicircuit, multiconductor cable is located horizontally directly above heat-producing equipment, in lieu of a connector strip, it should be spaced sufficiently above that equipment to avoid the elevated temperatures, or should be sized in accordance with Section 400-5.

Exception: Where alternate conductors are allowed in Section 520-68(a)(3), their ampacity shall be as given in the appropriate table in this Code for the types of conductors employed.

520-69. Adapters. Adapters, two-fers, and other single and multiple circuit outlet devices shall comply with (a), (b), and (c).

(a) No Reduction in Current Rating. Each receptacle and its corresponding cable shall have the same current and voltage rating as the plug supplying it. It shall not be utilized in a stage circuit with a greater current rating.

(b) Connectors. All connectors shall be wired in accordance with Section 520-67.

Adapters are available where cords and connector bodies of one ampacity are connected to a plug of a larger rating. For example, a No. 12 conductor with an ampacity of 20 amperes could be connected to a 100-ampere circuit. An overload could result in a fire, since the circuit breaker or fuse would not provide adequate protection. Section 520-69(b) requires that both the plug and receptacle be of the same rating.

(c) Conductor Type. Conductors for adapters and two-fers shall be listed, extra-hard usage or listed, hard usage (junior hard service) cord. Hard usage (junior hard service) cord shall be restricted in overall length to 3.3 ft (1 m).

F. Dressing Rooms

520-71. Pendant Lampholders. Pendant lampholders shall not be installed in dressing rooms.

520-72. Lamp Guards. All exposed incandescent lamps in dressing rooms, where less than 8 ft (2.44 m) from the floor, shall be equipped with open-end guards riveted to the outlet box cover or otherwise sealed or locked in place.

Because of the many types of flammable materials present in dressing rooms, such as costumes and wigs, pendant lampholders are not permitted. Lamps are required to be provided with suitable open-end guards that permit relamping and are not easily removed. This makes it difficult to circumvent the guard's intended purpose of preventing contact between the lamps and flammable material.

520-73. Switches Required. All lights and any receptacles adjacent to the mirror(s) and above the dressing table counter(s) installed in dressing rooms shall be controlled by wall switches installed in the dressing room(s). Each switch controlling receptacles adjacent to the mirror(s) and above the dressing table counter(s) shall be provided with a pilot light located outside the dressing room, adjacent to the door to indicate when the receptacles are energized. Other outlets installed in the dressing room shall not be required to be switched.

Revised for the 1999 *Code,* Section 520-73 now only addresses receptacles located adjacent to the mirror and on the countertop. The receptacles located elsewhere in the room are no longer subject to the disconnect and pilot light requirements of Section 520-73. The purpose of the switching requirement is to make sure all the coffee pots, curling irons, hair dryers, and other similar countertop appliances are disconnected at the end of the performance.

G. Grounding

520-81. Grounding. All metal raceways and metal-sheathed cables shall be grounded. The metal frames and enclosures of all equipment, including border lights and portable lighting fixtures, shall be grounded. Grounding, where used, shall be in accordance with Article 250.

Article 525 — Carnivals, Circuses, Fairs, and Similar Events

Contents

A. General Requirements

525-1. Scope. This article covers the installation of portable wiring and equipment for carnivals, circuses, exhibitions, fairs, traveling attractions, and similar functions, including wiring in or on all structures.

Added for the 1996 *Code,* Article 525 addresses the installation of portable wiring and equipment for temporary attractions, such as carnivals, circuses, and fairs. Article 525 is intended to apply to all wiring in or on portable structures, whereas Articles 518 and 520 apply to permanent structures. Installations for portable equipment used at carnivals, circuses, fairs, and the like, were formerly covered under Article 305, Temporary Wiring. Article 525 was developed because the requirements for temporary wiring, as found in Article 305, apply more to construction sites than to events open to the general public. Additionally, Article 525 greatly expands the requirements and scope for installing electrical equipment at these events.

525-3. Other Articles.

(a) Permanent Structures. Articles 518 and 520 shall apply to wiring in permanent structures.

(b) Portable Wiring and Equipment. Wherever the requirements of other articles of this *Code* and Article 525 differ, the requirements of Article 525 shall apply to the portable wiring and equipment.

(c) Audio Signal Processing, Amplification, and Reproduction Equipment. Article 640 shall apply to the wiring and installation of audio signal processing, amplification, and reproduction equipment.

525-6. Protection of Electrical Equipment. Electrical equipment and wiring methods in or on rides, concessions, or other units shall be provided with mechanical protection where such equipment or wiring methods are subject to physical damage.

B. Installation

525-10. Power Sources.

(a) Separately Derived Systems.

(1) Generators. Generators shall comply with the requirements of Article 445.

(2) Transformers. Transformers shall comply with applicable requirements of Sections 240-3(a), (b)(3), and (c); Section 250-30; and Article 450.

(b) Services. Services shall be installed in accordance with applicable requirements of Article 230 and, in addition, shall comply with the following.

(1) Guarding. Service equipment shall not be installed in a location that is accessible to unqualified persons, unless the equipment is lockable.

(2) Mounting and Location. Service equipment shall be mounted on a solid backing and be installed so as to be protected from the weather, unless of weatherproof construction.

Service equipment must be installed in accordance with Article 230 and must not be installed where accessible to unqualified persons unless it is lockable. This includes service equipment connected to separately derived systems, such as generators.

525-12. Overhead Conductor Clearances.

(a) Vertical Clearances. Conductors shall have a vertical clearance to ground in accordance with Section 225-18. These clearances shall apply only to wiring installed outside of tents and concessions.

(b) Clearance to Rides and Attractions. Amusement rides and amusement attractions shall be maintained not less than 15 ft (4.57 m) in any direction from overhead conductors operating at 600 volts or less, except for the conductors supplying the amusement ride or attraction. Amusement rides or attractions shall not be located under or within 15 ft (4.57 m) horizontally of conductors operating in excess of 600 volts.

525-13. Wiring Methods.

(a) Type. Unless otherwise provided for in this article, wiring methods shall comply with the applicable requirements of Chapters 1 through 4 of this *Code*. Where flexible cords or cables are used and are not subject to physical damage, they shall be permitted to be listed for hard usage. When used outdoors, flexible cords and cables shall also be listed for wet locations and shall be sunlight resistant.

(b) Single Conductor. Single conductor cable shall be permitted only in sizes No. 2 or larger.

(c) Open Conductors. Open conductors are prohibited except as part of a listed assembly or festoon lighting installed in accordance with Article 225.

(d) Splices. Flexible cords or cables shall be continuous without splice or tap between boxes or fittings. Cord connectors shall not be laid on the ground unless listed for wet locations. Connectors and cable connections shall not be placed in audience traffic paths or within areas accessible to the public unless guarded.

(e) Support. Wiring for an amusement ride, attraction, tent, or similar structure shall not be supported by any other ride or structure unless specifically designed for the purpose.

(f) Protection. Flexible cords or cables run on the ground, where accessible to the public, shall be covered with approved nonconductive mats. Cables and mats shall be arranged so as not to present a tripping hazard.

(g) Inside Tents and Concessions. Electrical wiring for temporary lighting, where installed inside of tents and concessions, shall be securely installed, and where subject to physical damage, shall be provided with mechanical protection. All temporary lamps for general illumination shall be protected from accidental breakage by a suitable fixture or lampholder with a guard.

525-14. Boxes and Fittings. A box or fitting shall be installed at each connection point, outlet, switchpoint, or junction point.

525-15. Portable Distribution or Termination Boxes. Portable distribution or termination boxes shall comply with (a) through (d).

(a) Construction. Boxes shall be designed so that no live parts are exposed to accidental contact. Where installed outdoors the box shall be of weatherproof construction and mounted so that the bottom of the enclosure is not less than 6 in. (152 mm) above the ground.

Portable distribution or termination equipment must be mounted so that the bottom of the enclosure is at least 6 in. above the ground. This prevents excessive moisture from entering the equipment and allows for proper radius of bend on conductors entering and exiting the equipment from below.

(b) Busbars and Terminals. Busbars shall have an ampere rating not less than the overcurrent device supplying the feeder supplying the box. Where conductors terminate directly on busbars, busbar connectors shall be provided.

(c) Receptacles and Overcurrent Protection. Receptacles shall have overcurrent protection installed within the box. The overcurrent protection shall not exceed the ampere rating of the receptacle, except as permitted in Article 430 for motor loads.

(d) Single-Pole Connectors. Where single-pole connectors are used, they shall comply with Section 530-22.

525-16. Overcurrent Protection. Overcurrent protection of equipment and conductors shall be provided in accordance with Article 240.

525-17. Motors. Motors and associated equipment shall be installed in accordance with Article 430.

525-18. Ground-Fault Circuit-Interrupter Protection for Personnel.

(a) General-Use 15- and 20-Ampere, 125-Volt Receptacles. All 125-volt, single-phase, 15- and 20-ampere receptacle outlets that are in use by personnel shall have listed ground-fault circuit-interrupter protection for personnel. The ground-fault circuit interrupter shall be permitted to be an integral part of the attachment plug or located in the power-supply cord, within 12 in. (305 mm) of the attachment plug. For the purposes of this section, listed cord sets incorporating ground-fault circuit-interrupter protection for personnel shall be permitted. Egress lighting shall not be connected to the load side terminals of a ground-fault circuit-interrupter receptacle.

(b) Appliance Receptacles. Receptacles supplying items, such as cooking and refrigeration equipment, which are incompatible with ground-fault circuit-interrupter devices shall not be required to have ground-fault circuit-interrupter protection.

(c) Other Receptacles. Other receptacle outlets not covered in (a) or (b) shall be permitted to have ground-fault circuit-interrupter protection for personnel, or a written procedure shall be continuously enforced at the site by one or more designated persons to ensure the safety of equipment grounding conductors for all cord sets and receptacles, as described in Section 305-6(b)(2).

C. Grounding and Bonding

525-20. General. All system and equipment grounding shall be in accordance with Article 250.

525-21. Equipment. The following equipment connected to the same source shall be bonded:

(1) Metal raceways and metal sheathed cable
(2) Metal enclosures of electric equipment
(3) Metal frames and metal parts of rides, concessions, trailers, trucks, or other equipment that contain or support electrical equipment

525-22. Equipment Grounding Conductor. All equipment requiring grounding shall be grounded by an equipment grounding conductor of a type and size recognized by Section 250-118 and installed in accordance with Article 250. The equipment grounding conductor shall be bonded to the system grounded conductor at the service disconnecting means, or in the case of a separately derived system such as a generator, at the generator or first disconnecting means supplied by the generator. The grounded circuit conductor shall not be connected to the equipment grounding conductor on the load side of the service disconnecting means or on the load side of a separately derived system disconnecting means.

D. Disconnecting Means

525-30. Type and Location. Each ride and concession shall be provided with a fused disconnect switch or circuit breaker located within sight and within 6 ft (1.83 m) of the operator's station. The disconnecting means shall be readily accessible to the operator, including when the ride is in operation. Where accessible to unqualified persons, the enclosure for the switch or circuit breaker shall be of the lockable type. A shunt trip device that opens the fused disconnect or circuit breaker when a switch located in the ride operator's console is closed shall be a permissible method of opening the circuit.

E. Attractions Utilizing Pools, Fountains, and Similar Installations with Contained Volumes of Water

525-40. Wiring and Equipment. This equipment shall be installed to comply with the applicable requirements of Article 680.

Article 530 — Motion Picture and Television Studios and Similar Locations

Contents

A. General

530-1. Scope. The requirements of this article shall apply to television studios and motion picture studios using either film or electronic cameras, except as provided in Section 520-1, and exchanges, factories, laboratories, stages, or a portion of the building in which film or tape more than ⅞ in. (22 mm) in width is exposed, developed, printed, cut, edited, rewound, repaired, or stored.

FPN: For methods of protecting against cellulose nitrate film hazards, see *Standard for the Storage and Handling of Cellulose Nitrate Motion Picture Film*, NFPA 40-1997.

The requirements for motion picture studios and television studios are virtually the same and are intended to apply only to those locations presenting special hazards. Otherwise, the conditions are similar to theater stages; therefore, the applicable provisions of Article 520 should be observed, such as those for stages and dressing rooms.

The special hazards are temporary structures constructed of wood or other combustible material.

530-2. Definitions.

Alternating-Current Power Distribution Box (Alternating-Current Plugging Box, Scatter Box). An ac distribution center or box that contains one or more grounding-type, polarized receptacles, that may contain overcurrent protection devices.

Bull Switch. An externally-operated wall-mounted safety switch, which may or may not contain overcurrent protection, that is designed for the connection of portable cables and cords.

Location (Shooting Location). A place outside a motion picture studio where a production or part of it is filmed or recorded.

Location Board (Deuce Board). Portable equipment containing a lighting contactor or contactors and overcurrent protection designed for remote control of stage lighting.

Motion Picture Studio (Lot). A building or group of buildings and other structures designed, constructed, or permanently altered for use by the entertainment industry for the purpose of motion picture or television production.

Portable Equipment. Equipment intended to be moved from one place to another.

Plugging Box. A dc device consisting of one or more 2-pole, 2-wire, nonpolarized, nongrounding-type receptacles intended to be used on dc circuits only.

Single-Pole Separable Connector. A device that is installed at the ends of portable, flexible, single-conductor cable that is used to establish connection or disconnection between two cables or one cable and a single-pole, panel-mounted separable connector.

Spider (Cable Splicing Block). A device that contains busbars that are insulated from each other for the purpose of splicing or distributing power to portable cables and cords that are terminated with single-pole busbar connectors.

Stage Effect (Special Effect). An electrical or electromechanical piece of equipment used to simulate a distinctive visual or audible effect such as wind machines, lightning simulators, sunset projectors, and the like.

Stage Property. An article or object used as a visual element in a motion picture or television production, except painted backgrounds (scenery) and costumes.

Stage Set. A specific area set up with temporary scenery and properties designed and arranged for a particular scene in a motion picture or television production.

Stand Lamp (Work Light). A portable stand that contains a general-purpose lighting fixture or lampholder with guard for the purpose of providing general illumination in the studio or stage.

Television Studio or Motion Picture Stage (Sound Stage). A building or portion of a building usually insulated from the outside noise and natural light for use by the entertainment industry for the purpose of motion picture, television, or commercial production.

530-6. Portable Equipment. Portable stage and studio lighting equipment and portable power distribution equipment shall be permitted for temporary use outdoors if the equipment is supervised by qualified personnel while energized and barriered from the general public.

See the commentary following Section 520-10.

B. Stage or Set

530-11. Permanent Wiring. The permanent wiring shall be Type MC cable, Type MI cable, or in approved raceways.

Exception: Communications circuits; audio signal processing, amplification, and reproduction circuits; Class 1, Class 2, and Class 3 remote-control or signaling circuits and power-limited fire alarm circuits shall be permitted to be wired in accordance with Articles 800, 640, 725, and 760.

530-12. Portable Wiring.

(a) Stage Set Wiring. The wiring for stage set lighting and other supply wiring not fixed as to location shall be done with listed hard usage flexible cords and cables. Where subject to physical damage, such wiring shall be listed extra-hard usage flexible cords and cables. Splices or taps in cables shall be permitted if the total connected load does not exceed the maximum ampacity of the cable.

(b) Stage Effects and Electrical Equipment Used as Stage Properties. The wiring for stage effects and electrical equipment used as stage properties shall be permitted to be wired with single- or multi-conductor listed flexible cords or cables if the conductors are protected from physical damage and secured to the scenery by approved cable ties or by insulated staples. Splices or taps shall be permitted where such are made with listed devices and the circuit is protected at not more than 20 amperes.

(c) Other Electrical Equipment. Cords and cables other than extra-hard usage, where supplied as a part of a listed assembly, shall be permitted.

530-13. Stage Lighting and Effects Control. Switches used for studio stage set lighting and effects (on the stages and lots and on location) shall be of the externally operable type. Where contactors are used as the disconnecting means for fuses, an individual externally operable switch, such as a tumbler switch, for the control of each contactor shall be located at a distance of not more than 6 ft (1.83 m) from the contactor, in addition to remote-control switches. A single externally operable switch shall be permitted to simultaneously disconnect all the contactors on any one location board, where located at a distance of not more than 6 ft (1.83 m) from the location board.

530-14. Plugging Boxes. Each receptacle of dc plugging boxes shall be rated at not less than 30 amperes.

530-15. Enclosing and Guarding Live Parts.

(a) Live Parts. Live parts shall be enclosed or guarded to prevent accidental contact by persons and objects.

(b) Switches. All switches shall be of the externally operable type.

(c) Rheostats. Rheostats shall be placed in approved cases or cabinets that enclose all live parts, having only the operating handles exposed.

(d) Current-Carrying Parts. Current-carrying parts of bull switches, location boards, spiders, and plugging boxes shall be enclosed, guarded, or located so that persons cannot accidentally come into contact with them or bring conductive material into contact with them.

530-16. Portable Lamps. Portable lamps and work lights shall be equipped with flexible cords, composition or metal-sheathed porcelain sockets, and substantial guards.

Exception: Portable lamps used as properties in a motion picture set or television stage set, on a studio stage or lot, or on location shall not be considered to be portable lamps for the purpose of this section.

530-17. Portable Arc Lamp Fixtures.

(a) Portable Carbon Arc Lamps. Portable carbon arc lamps shall be substantially constructed. The arc shall be provided with an enclosure designed to retain sparks and carbons and to prevent persons or materials from coming into contact with the arc or bare live parts. The enclosures shall be ventilated. All switches shall be of the externally operable type.

(b) Portable Noncarbon Arc Electric-Discharge Lamp Fixtures. Portable noncarbon arc lamp fixtures, including enclosed arc lamp fixtures, and associated ballasts shall be listed. Interconnecting cord sets, and interconnecting cords and cables, shall be extra-hard usage type and listed.

530-18. Overcurrent Protection.

General. Automatic overcurrent protective devices (circuit breakers or fuses) for motion picture studio stage set lighting and the stage cables for such stage set lighting shall be as given in (a) through (g). The maximum ampacity allowed on a given conductor, cable, or cord size shall be as given in the applicable tables of Articles 310 and 400.

(a) Stage Cables. Stage cables for stage set lighting shall be protected by means of overcurrent devices set at not more than 400 percent of the ampacity given in the applicable tables of Articles 310 and 400.

(b) Feeders. In buildings used primarily for motion picture production, the feeders from the substations to the stages shall be protected by means of overcurrent devices (generally located in the substation) having a suitable ampere rating. The overcurrent devices shall be permitted to be multipole or single-pole gang operated. No pole shall be required in the neutral conductor. The overcurrent device setting for each feeder shall not exceed 400 percent of the ampacity of the feeder, as given in the applicable tables of Article 310.

An overcurrent device setting of up to 400 percent is permissible where the loads are of short duration. Section 530-18(b) permits the use of short-term ratings where the equipment operates for 20 minutes or less. A longer period of operation may pose a fire hazard.

(c) Cable Protection. Cables shall be protected by bushings where they pass through enclosures and shall be arranged so that tension on the cable will not be transmitted to the connections. Where power conductors pass through metal, the requirements of Section 300-20 shall apply.

Portable feeder cables shall be permitted to temporarily penetrate fire-rated walls, floors, or ceilings provided that

(1) The opening be of noncombustible material;
(2) When in use, the penetration is sealed with a temporary seal of a listed firestop material; and
(3) When not in use, the opening shall be capped with a material of equivalent fire rating.

(d) Location Boards. Overcurrent protection (fuses or circuit breakers) shall be provided at the location boards. Fuses in the location boards shall have an ampere rating of not over 400 percent of the ampacity of the cables between the location boards and the plugging boxes.

(e) Plugging Boxes. Cables and cords supplied through plugging boxes shall be of copper. Cables and cords smaller than No. 8 shall be attached to the plugging box by means of a plug containing two cartridge fuses or a 2-pole circuit breaker. The rating of the fuses or the setting of the circuit breaker shall not be over 400 percent of the rated ampacity of the cables or cords as given in the applicable tables of Articles 310 and 400. Plugging boxes shall not be permitted on ac systems.

(f) Alternating-Current Power Distribution Boxes. Alternating-current power distribution boxes used on sound stages and shooting locations shall contain connection receptacles of a polarized, grounding type.

(g) Lighting. Work lights, stand lamps, and fixtures, rated 1000 watts or less and connected to dc plugging boxes, shall be by means of plugs containing two cartridge fuses not larger than 20 amperes, or they shall be permitted to be connected to special outlets on circuits protected by fuses or circuit breakers rated at not over 20 amperes. Plug fuses shall not be used unless they are on the load side of the fuse or circuit breakers on the location boards.

530-19. Sizing of Feeder Conductors for Television Studio Sets.

(a) General. It shall be permissible to apply the demand factors listed in Table 530-19(a) to that portion of the maximum possible connected load for studio or stage set lighting for all permanently installed feeders between substations and stages and to all permanently installed feeders between the main stage switchboard and stage distribution centers or location boards.

Table 530-19(a). Demand Factors for Stage Set Lighting

Portion of Stage Set Lighting Load to Which Demand Factor Applied (volt-amperes)	Feeder Demand Factor
First 50,000 or less at	100%
From 50,001 to 100,000 at	75%
From 100,001 to 200,000 at	60%
Remaining over 200,000 at	50%

(b) Portable Feeders. A demand factor of 50 percent of maximum possible connected load shall be permitted for all portable feeders.

530-20. Grounding. Type MC cable, Type MI cable, metal raceways, and all noncurrent-carrying metal parts of appliances, devices, and equipment shall be grounded as specified in Article 250. This shall not apply to pendant and portable lamps, to stage lighting and stage sound equipment, or to other portable and special stage equipment operating at not over 150 volts dc to ground.

530-21. Plugs and Receptacles.

(a) Rating. Plugs and receptacles shall be rated in amperes. The voltage rating of the plugs and receptacles shall not be less than the circuit voltage. Plug and receptacle ampere ratings for ac circuits shall not be less than the feeder or branch-circuit overcurrent device ampere rating. Table 210-21(b)(2) shall not apply.

(b) Interchangeability. Plugs and receptacles used in portable professional motion picture and television equipment shall be permitted to be interchangeable for ac or dc use on the same premises provided they are listed for ac/dc use and marked in a suitable manner to identify the system to which they are connected.

530-22. Single-Pole Separable Connectors.

(a) General. Where ac single-pole portable cable connectors are used, they shall be listed and of the locking type. Section 400-10 and 410-56 shall not apply to listed single-pole separable connections and single-conductor cable assemblies utilizing listed single-pole separable connectors. Where paralleled sets of current-carrying single-pole separable connectors are provided as input devices, they shall be prominently labeled with a warning indicating the presence of internal parallel connections. The use of single-pole separable connectors shall comply with at least one of the following conditions.

(1) Connection and disconnection of connectors are only possible where the supply connectors are interlocked to the source and it is not possible to connect or disconnect connectors when the supply is energized.

(2) Line connectors are of the listed sequential-interlocking type so that load connectors shall be connected in the following sequence:

(a) Equipment grounding conductor connection
(b) Grounded circuit conductor connection, if provided
(c) Ungrounded conductor connection, and that disconnection shall be in the reverse order

(3) A caution notice shall be provided adjacent to the line connectors indicating that plug connection shall be in the following order:

(a) Equipment grounding conductor connectors
(b) Grounded circuit-conductor connectors, if provided
(c) Ungrounded conductor connectors, and that disconnection shall be in the reverse order

(b) Interchangeability. Single-pole separable connectors used in portable professional motion picture and television equipment shall be permitted to be interchangeable for ac or dc use or for different current ratings on the same premises provided they are listed for ac/dc use and marked in a suitable manner to identify the system to which they are connected.

530-23. Branch Circuits. A branch circuit of any size supplying one or more receptacles shall be permitted to supply stage set lighting loads.

C. Dressing Rooms

530-31. Dressing Rooms. Fixed wiring in dressing rooms shall be installed in accordance with the wiring methods covered in Chapter 3. Wiring for portable dressing rooms shall be approved.

D. Viewing, Cutting, and Patching Tables

530-41. Lamps at Tables. Only composition or metal-sheathed, porcelain, keyless lampholders equipped with suitable means to guard lamps from physical damage and from film and film scrap shall be used at patching, viewing, and cutting tables.

E. Cellulose Nitrate Film Storage Vaults

530-51. Lamps in Cellulose Nitrate Film Storage Vaults. Lamps in cellulose nitrate film storage vaults shall be installed in rigid fixtures of the glass-enclosed and gasketed type. Lamps shall be controlled by a switch having a pole in each ungrounded conductor. This switch shall be located outside of the vault and provided with a pilot light to indicate whether the switch is on or off. This switch shall disconnect from all sources of supply all ungrounded conductors terminating in any outlet in the vault.

530-52. Motors and Other Equipment in Cellulose Nitrate Film Storage Vaults. Except as permitted in Section 530-51, no receptacles, outlets, electric motors, heaters, portable lights, or other portable electric equipment shall be located in cellulose nitrate film storage vaults.

F. Substations

530-61. Substations. Wiring and equipment of over 600 volts, nominal, shall comply with Article 490.

530-62. Portable Substations. Wiring and equipment in portable substations shall conform to the sections applying to installations in permanently fixed substations, but, due to the limited space available, the working spaces shall be permitted to be reduced, provided that the equipment shall be arranged so that the operator can work safely and so that other persons in the vicinity cannot accidentally come into contact with current-carrying parts or bring conducting objects into contact with them while they are energized.

530-63. Overcurrent Protection of Direct-Current Generators. Three-wire generators shall have overcurrent protection in accordance with Section 445-4(e).

530-64. Direct-Current Switchboards.

(a) Switchboards of not over 250 volts dc between conductors, where located in substations or switchboard rooms accessible to qualified persons only, shall not be required to be dead-front.

(b) Frames of dc circuit breakers installed on switchboards shall not be required to be grounded.

G. Separately Derived Systems with 60 Volts to Ground

530-70. General. Use of a separately derived 120-volt, single-phase, 3-wire system with 60 volts on each of two ungrounded conductors to a grounded neutral conductor shall be permitted for the purpose of reducing objectionable noise in audio/video production or other similar sensitive electronic equipment locations provided that its use is restricted to electronic equipment only and that all of the requirements in Sections 530-71 through 530-73 are met.

530-71. Wiring Methods.

(a) Panelboards and Overcurrent Protection. Use of standard single-phase panelboards and distribution equipment with a higher voltage rating shall be permitted. The system shall be clearly marked on the face of the panel or on the inside of the panel doors. Common-trip, two-pole circuit breakers that are identified for operation at the system voltage shall be provided for both ungrounded conductors in all feeders and branch circuits.

(b) Junction Boxes. All junction box covers shall be clearly marked to indicate the distribution panel and the system voltage.

(c) Color Coding. All feeders and branch-circuit conductors installed under this section shall be identified as to system at all splices and terminations by color, marking, tagging, or equally effective means. The means of identification shall be posted at each branch-circuit panelboard and at the disconnecting means for the building.

(d) Voltage Drop. The voltage drop on any branch circuit shall not exceed 1.5 percent. The combined voltage drop of feeder and branch-circuit conductors shall not exceed 2.5 percent.

530-72. Grounding.

(a) General. The system shall be grounded as provided in Section 250-30 as a separately derived single-phase, 3-wire system.

(b) Grounding Conductors Required. Permanently wired utilization equipment and receptacles shall be grounded by means of an equipment grounding conductor run with the circuit conductors to an equipment grounding bus prominently marked "Technical Equipment Ground" in the originating branch-circuit panelboard. The grounding bus shall be connected to the grounded conductor on the line side of the separately derived system's disconnecting means. The grounding conductor shall not be smaller than that specified in Table 250-122 and run with the feeder conductors. The technical equipment grounding bus need not be bonded to the panelboard enclosure. Other grounding methods authorized elsewhere in this *Code* shall be permitted where the impedance of the grounding return path does not exceed the impedance of equipment grounding conductors sized and installed in accordance with Part G of this article.

FPN No. 1: See Section 250-122 for equipment grounding conductor sizing requirements where circuit conductors are adjusted in size to compensate for voltage drop.

FPN No. 2: These requirements limit the impedance of the ground-fault path where only 60 volts applies to a fault condition instead of the usual 120 volts.

530-73. Receptacles.

(a) General. Where receptacles are used as a means of connecting equipment, the following conditions shall be met:

(1) All 15- and 20-ampere receptacle outlets shall be ground-fault circuit-interrupter protected.
(2) All outlet strips, adapters, receptacle covers, and faceplates shall be marked as follows:

WARNING — TECHNICAL POWER
Do not connect to lighting equipment
For electronic equipment use only
60/120 volt 1ø ac
GFCI protected

(3) A 125-volt, single-phase, 15- or 20-ampere rated receptacle having one of its current-carrying poles connected to a grounded circuit conductor shall be located within 6 ft (1.83 m) of all permanently installed 15- or 20-ampere rated 60/120-volt technical power-system receptacles.
(4) All 125-volt receptacles used for 60/120-volt technical power shall be uniquely configured and identified for use with this class of system. All 125-volt, single-phase, 15- or 20-ampere rated receptacle outlets and attachment plugs that are identified for use with grounded circuit conductors shall be permitted in machine rooms, control rooms, equipment rooms, equipment racks, and other

similar locations that are restricted to use by qualified personnel.

(b) Isolated Ground Receptacles. Isolated ground receptacles shall be permitted as described in Section 250-146(d), however, the branch-circuit equipment grounding conductor shall be terminated as required in Section 530-72(b).

Article 540 — Motion Picture Projectors

Contents

A. General

540-1. Scope. The provisions of this article apply to motion picture projection rooms, motion picture projectors, and associated equipment of the professional and nonprofessional types using incandescent, carbon arc, xenon, or other light source equipment that develops hazardous gases, dust, or radiation.

FPN: For further information, see *Standard for the Storage and Handling of Cellulose Nitrate Motion Picture Film,* NFPA 40-1997.

The definitions of hazardous (classified) locations in Article 500 do not include a motion picture projection room, even though some of the older types of film, such as cellulose nitrate film (rarely used now) are highly flammable. In comparison, cellulose acetate film, called "safety film," is in wide use today. Since film is not volatile at ordinary temperatures and no flammable gases are present, the wiring installation is not required to be suitable for hazardous (classified) locations as defined in Article 500, but should be installed with special care to protect against the hazards of fire.

B. Definitions

540-2. Professional Projector. A type of projector using 35- or 70-mm film that has a minimum width of 1⅜ inch (35 mm) and has on each edge 5.4 perforations per inch, or a type using carbon arc, xenon, or other light source equipment that develops hazardous gases, dust, or radiation.

540-3. Nonprofessional Projector. Nonprofessional projectors are those types other than described in Section 540-2.

C. Equipment and Projectors of the Professional Type

540-10. Motion Picture Projection Room Required. Every professional-type projector shall be located within a projection room. Every projection room shall be of permanent construction, approved for the type of building in which the projection room is located. All projection ports, spotlight ports, viewing ports, and similar openings shall be provided with glass or other approved material so as to completely close the opening. Such rooms shall not be considered as hazardous (classified) locations as defined in Article 500.

FPN: For further information on protecting openings in projection rooms handling cellulose nitrate motion picture film, see *Life Safety Code,* NFPA *101*-1997.

540-11. Location of Associated Electrical Equipment.

(a) Motor Generator Sets, Transformers, Rectifiers, Rheostats, and Similar Equipment. Motor generator sets, transformers, rectifiers, rheostats, and similar equipment for the supply or control of current to projection or spotlight equipment shall, where nitrate film is used, be located in a separate room. Where placed in the projection room, they shall be located or guarded so that arcs or sparks cannot come in contact with film, and the commutator end or ends of motor generator sets shall comply with one of the conditions in (1) through (6).

(1) Types. Be of the totally enclosed, enclosed fan-cooled, or enclosed pipe-ventilated type.

(2) Separate Rooms or Housings. Be enclosed in separate rooms or housings built of noncombustible material constructed so as to exclude flyings or lint, and properly ventilated from a source of clean air.

(3) Solid Metal Covers. Have the brush or sliding-contact end of motor-generator enclosed by solid metal covers.

(4) Tight Metal Housings. Have brushes or sliding contacts enclosed in substantial, tight metal housings.

(5) Upper and Lower Half Enclosures. Have the upper half of the brush or sliding-contact end of the motor-generator enclosed by a wire screen or perforated metal and the lower half enclosed by solid metal covers.

(6) Wire Screens or Perforated Metal. Have wire screens or perforated metal placed at the commutator of brush ends. No dimension of any opening in the wire screen or perforated metal shall exceed 0.05 in. (1.27 mm), regardless of the shape of the opening and of the material used.

(b) Switches, Overcurrent Devices, or Other Equipment. Switches, overcurrent devices, or other equipment not normally required or used for projectors, sound reproduction, flood or other special effect lamps, or other equipment shall not be installed in projection rooms.

Exception No. 1: In projection rooms approved for use only with cellulose acetate (safety) film, the installation of appurtenant electrical equipment used in conjunction with the operation of the projection equipment and the control of lights, curtains, and audio equipment, etc., shall be permitted. In such projection rooms, a sign reading "Safety Film Only Permitted in This Room" shall be posted on the outside of each projection room door and within the projection room itself in a conspicuous location.

Exception No. 2: Remote-control switches for the control of auditorium lights or switches for the control of motors operating curtains and masking of the motion picture screen shall be permitted to be installed in projection rooms.

(c) Emergency Systems. Control of emergency systems shall comply with Article 700.

The plan of a projection room of a motion picture theater is illustrated in Figure 540.1. This plan shows one stereopticon or "effect machine" (L), two spot machines (S), and three motion picture projectors (P).

A dc arc lamp is the light source in each of the six machines. The dc supply is furnished by two motor-generator sets, which are usually installed in soundproof areas to avoid interfering with the sound-reproducing equipment and are controlled from the generator panel in the projection room. Two 500-kcmil feeder cables are run from each generator to the dc panelboard.

A branch circuit consisting of two No. 2/0 cables runs from the dc panelboard to each projector (P) and to each spot machine (S). One of the two branch-circuit conductors runs directly to a projector (P); the other passes through an auxiliary gutter to the bank of resistors in the rheostat room and then to the projector. The resistors are equipped with short-circuiting switches, so that the total resistance in series with each arc may be preset to a desired value.

Since the stereopticon or effect machine (L) contains two arc lamps, two circuits with No. 1 conductors are routed to this machine.

Section 540-13 requires that the conductors supplying outlets for arc and xenon lamps of the professional type are not to be smaller than No. 8 and are required to be of sufficient size for the lamps employed. In each case, therefore, the maximum current drawn by the lamps should be determined. In this example, with the arc lamps sized for a large picture, the arc in each projector draws nearly 150 amperes. Four outlets, in addition to the main outlet for supplying the arc, are located at each projector for the following auxiliary circuits.

1. Outlet C supplies a small incandescent lamp inside each lamphouse and/or projector.
2. Outlet G is for the No. 8 equipment grounding conductor, which is connected to each projector frame and to a metal water pipe.
3. Outlet F supplies a foot switch that controls a solenoid-operated shutter behind each lens, for changing from one projector to another.
4. Outlet M supplies the motor used to operate each projector.

Two exhaust fans and two duct systems, one exhausting from the ceiling of the projection room and one connected to the arc lamp housing of each machine, provide ventilation.

540-12. Work Space. Each motion picture projector, floodlight, spotlight, or similar equipment shall have clear working space not less than 30 in. (762 mm) wide on each side and at the rear thereof.

Exception: One such space shall be permitted between adjacent pieces of equipment.

540-13. Conductor Size. Conductors supplying outlets for arc and xenon projectors of the professional type shall not be smaller than No. 8 and shall be of sufficient size for the projector employed. Conductors for incandescent-type projectors shall conform to normal wiring standards as provided in Section 210-24.

540-14. Conductors on Lamps and Hot Equipment. Insulated conductors having a rated operating temperature of not less than 200°C (392°F) shall be used on all lamps or other equipment where the ambient temperature at the conductors as installed will exceed 50°C (122°F).

540-15. Flexible Cords. Cords approved for hard usage, as provided in Table 400-4, shall be used on portable equipment.

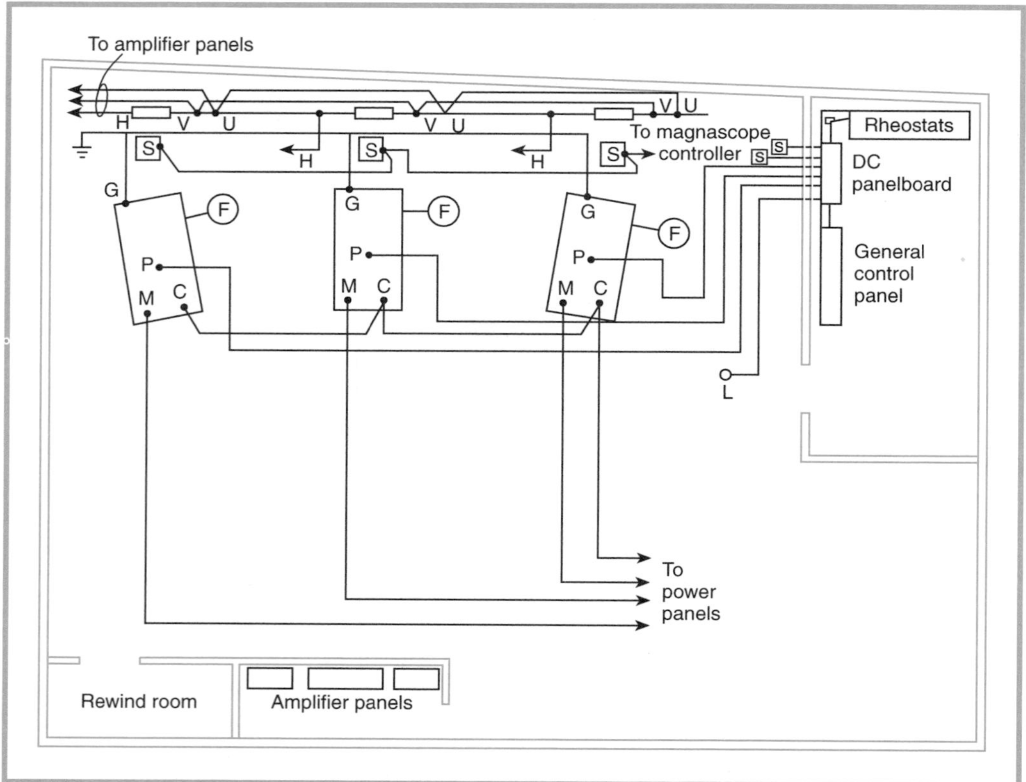

Figure 540.1 A typical layout of a projection room, including associated generator equipment. Modern projectors contain rectifiers as an integral part of their equipment, thereby eliminating generators and other associated equipment.

540-20. Approval. Projectors and enclosures for arc, xenon and incandescent lamps and rectifiers, transformers, rheostats, and similar equipment shall be listed.

540-21. Marking. Projectors and other equipment shall be marked with the manufacturer's name or trademark and with the voltage and current for which they are designed in accordance with Section 110-21.

D. Nonprofessional Projectors

540-31. Motion Picture Projection Room Not Required. Projectors of the nonprofessional or miniature type, where employing cellulose acetate (safety) film, shall be permitted to be operated without a projection room.

540-32. Approval. Projection equipment shall be listed.

E. Audio Signal Processing, Amplification, and Reproduction Equipment

540-50. Audio Signal Processing, Amplification, and Reproduction Equipment. Audio signal processing, amplification, and reproduction equipment shall be installed as provided in Article 640.

Article 545 — Manufactured Buildings

Contents

A. General

545-1. Scope. This article covers requirements for a manufactured building and building components as herein defined.

545-2. Other Articles. Wherever the requirements of other articles of this *Code* and Article 545 differ, the requirements of Article 545 shall apply.

545-3. Definitions.

Building Component. Any subsystem, subassembly, or other system designed for use in or integral with or as part of a structure, which can include structural, electrical, mechanical, plumbing, and fire protection systems, and other systems affecting health and safety.

Building System. Plans, specifications, and documentation for a system of manufactured building or for a type or a system of building components, which can include structural, electrical, mechanical, plumbing, and fire protection systems, and other systems affecting health and safety, and including such variations thereof as are specifically permitted by regulation, and which variations are submitted as part of the building system or amendment thereto.

Closed Construction. Any building, building component, assembly, or system manufactured in such a manner that all concealed parts of processes of manufacture cannot be inspected before installation at the building site without disassembly, damage, or destruction.

Manufactured Building. Any building that is of closed construction and is made or assembled in manufacturing facilities on or off the building site for installation, or assembly and installation on the building site, other than manufactured homes, mobile homes, park trailers, or recreational vehicles.

545-4. Wiring Methods.

(a) Methods Permitted. All raceway and cable wiring methods included in this *Code* and such other wiring systems specifically intended and listed for use in manufactured buildings shall be permitted with listed fittings and with fittings listed and identified for manufactured buildings.

(b) Securing Cables. In closed construction, cables shall be permitted to be secured only at cabinets, boxes, or fittings where No. 10 or smaller conductors are used and protection against physical damage is provided.

545-5. Service-Entrance Conductors. Service-entrance conductors shall meet the requirements of Article 230. Provisions shall be made to route the service-entrance conductors from the service equipment to the point of attachment of the service drop or service lateral.

545-6. Installation of Service-Entrance Conductors. Service-entrance conductors shall be installed after erection at the building site.

Exception: Where point of attachment is known prior to manufacture.

545-7. Service Equipment. Service equipment shall be installed in accordance with Section 230-70.

545-8. Protection of Conductors and Equipment. Protection shall be provided for exposed conductors and equipment during processes of manufacturing, packaging, in transit, and erection at the building site.

545-9. Boxes.

(a) Other Dimensions. Boxes of dimensions other than those required in Table 370-16(a) shall be permitted to be installed where tested, identified, and listed to applicable standards.

(b) Not Over 100 in.^{3} (1640 cm^3). Any box not over 100 in.^{3} (1640 cm^3) in size, intended for mounting in closed construction, shall be affixed with anchors or clamps so as to provide a rigid and secure installation.

545-10. Receptacle or Switch with Integral Enclosure. A receptacle or switch with integral enclosure and mounting means, where tested, identified, and listed to applicable standards, shall be permitted to be installed.

See Figure 545.1 and the commentary following Section 300-15(e) for additional discussion about wiring devices with integral enclosures.

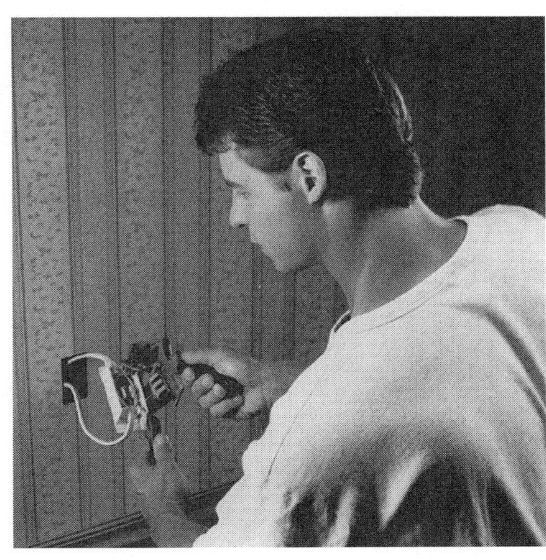

Figure 545.1 *A special type of tool used to connect nonmetallic-sheathed cable to an approved boxless switch or receptacle. (Pass & Seymour/Legrand)*

545-11. Bonding and Grounding. Prewired panels and building components shall provide for the bonding, or bonding and grounding, of all exposed metals likely to become energized, in accordance with Article 250, Parts E, F, and G.

545-12. Grounding Electrode Conductor. Provisions shall be made to route a grounding electrode conductor from the service equipment to the point of attachment to the grounding electrode.

545-13. Component Interconnections. Fittings and connectors that are intended to be concealed at the time of on-site assembly, where tested, identified, and listed to applicable standards, shall be permitted for on-site interconnection of modules or other building components. Such fittings and connectors shall be equal to the wiring method employed in insulation, temperature rise, and fault-current withstand and shall be capable of enduring the vibration and minor relative motions occurring in the components of manufactured building.

Structural components or modules are usually constructed in manufacturing facilities and then transported over the road to a building site for complete assembly of a structure, such as a dwelling unit, motel, or office building. At the on-site location, approved wiring methods are employed to interconnect two or more modules. Figure 545.2 shows a type of nonmetallic-sheathed cable connector permitted for such interconnections.

Figure 545.2 *A type of nonmetallic-sheathed cable connector used for interconnecting modules in a manufacturing building. (Pass & Seymour/Legrand)*

Article 547 — Agricultural Buildings

Contents

547-1. Scope. The provisions of this article shall apply to the following agricultural buildings or that part of a building or adjacent areas of similar or like nature as specified in (a) and (b).

(a) Excessive Dust and Dust with Water. Agricultural buildings where excessive dust and dust with water may accumulate, including all areas of poultry, livestock, and fish confinement systems, where litter dust or feed dust, including mineral feed particles, may accumulate.

(b) Corrosive Atmosphere. Agricultural buildings where a corrosive atmosphere exists. Such buildings include areas:

(1) Where poultry and animal excrement may cause corrosive vapors
(2) Where corrosive particles may combine with water
(3) Where the area is damp and wet by reason of periodic washing for cleaning and sanitizing with water and cleansing agents
(4) Where similar conditions exist

Article 547 applies not only to buildings but also to adjacent areas of similar or like nature.

547-2. Other Articles. For agricultural buildings not having conditions as specified in Section 547-1, the electrical installations shall be made in accordance with the applicable articles in this *Code.*

547-3. Surface Temperatures. Electrical equipment or devices installed in accordance with the provisions of this article shall be installed in a manner such that they will function at full rating without developing surface temperatures in excess of the specified normal safe operating range of the equipment or device.

547-4. Wiring Methods.

(a) Wiring Systems. Types UF, NMC, copper SE cables, rigid nonmetallic conduit, liquidtight flexible nonmetallic conduit, or other cables or raceways suitable for the location, with approved termination fittings, shall be the wiring methods employed. Article 320 and Article 502 wiring methods shall be permitted for areas described in Section 547-1(a).

FPN: See Sections 300-7 and 347-9 for installation of raceway systems exposed to widely different temperatures.

(b) Mounting. All cables shall be secured within 8 in. (203 mm) of each cabinet, box, or fitting. The ¼-in. (6.35-mm) airspace required for nonmetallic boxes, fittings, conduit, and cables in Section 300-6(c) shall not be required in buildings covered by this article.

Cables installed in agricultural buildings are required to be secured within 8 in. of cabinets, boxes, or fittings. This distance is less than that required for cables in other types of occupancies. The requirement for a ¼-in. airspace in Section 300-6(c) is judged unnecessary in agricultural buildings, provided nonmetallic wiring methods are used. Decreasing the support spacing requirements coupled with the elimination of the ¼-in. airspace requirement reduces the potential for mechanical damage to cable-type wiring methods. Locating the wiring methods directly on the interior surface of the building allows a sealant to be placed along the wiring method to facilitate cleaning. See also Section 300-6(c), Exception.

(c) Boxes and Fittings. All boxes and fittings shall comply with Section 547-5.

(d) Flexible Connections. Where necessary to employ flexible connections, dusttight flexible connectors, liquidtight flexible conduit, or flexible cord listed and identified for hard usage shall be used. All connectors and fittings used shall be listed and identified for the purpose.

(e) Physical Protection. All electrical wiring and equipment subject to physical damage shall be protected.

(f) Separate Equipment Grounding Conductor. Noncurrent-carrying metal parts of equipment, raceways, and other enclosures, where required to be grounded, shall be grounded by a copper equipment grounding conductor installed between the equipment and the building disconnecting means. If installed underground, the equipment grounding conductor shall be insulated or covered.

The intent of Section 547-4(f) is to improve the longevity of equipment grounding conductors in the highly corrosive locations that are typical of many farm buildings. This requirement appeared in Section 547-8(c) in the 1996 *Code*.

547-5. Switches, Circuit Breakers, Controllers, and Fuses. Switches, circuit breakers, controllers, and fuses, including pushbuttons, relays, and similar devices, shall be provided with enclosures as specified in (a) and (b).

(a) Excessive Dust and Dust with Water. Buildings described in Section 547-1(a) shall utilize dustproof and weatherproof enclosures.

(b) Corrosive Atmosphere. Buildings described in Section 547-1(b) shall utilize enclosures suitable for the conditions encountered in the application.

FPN No. 1: See Table 430-91 for appropriate enclosure type designations.

FPN No. 2: Cast aluminum and magnetic steels may corrode in agricultural environments.

Watertight enclosures may not provide adequate ventilation for certain types of sensitive equipment, such as electronic equipment. In some cases, it may be necessary to provide ventilation by using other types of enclosures. Section 547-5(b) requires appropriate enclosures for the conditions encountered. This includes consideration of both the type of enclosure and the proper materials used in the construction of the enclosure.

547-6. Motors. Motors and other rotating electrical machinery shall be totally enclosed or designed so as to minimize the entrance of dust, moisture, or corrosive particles.

547-7. Lighting Fixtures. Lighting fixtures shall comply with the following.

(a) Minimize the Entrance of Dust. Lighting fixtures shall be installed to minimize the entrance of dust, foreign matter, moisture, and corrosive material.

(b) Exposed to Physical Damage. Any lighting fixture that may be exposed to physical damage shall be protected by a suitable guard.

(c) Exposed to Water. A fixture that may be exposed to water from condensation, building cleansing water, or solution shall be watertight.

547-8. Service Equipment, Separately Derived Systems, Feeders, Disconnecting Means, and Grounding. Where one or more agricultural buildings are supplied from a distribution point, the disconnecting means and grounding of services and feeders shall comply with (a), (b), or (c).

(a) Disconnecting Means and Overcurrent Protection at a Building(s). Where the disconnecting means and overcurrent protection are located at the load end of the service conductors, service equipment grounding shall meet the requirements of Section 250-24. A disconnecting means shall also be installed at the distribution point when two or more buildings are supplied from that distribution point.

(b) Disconnecting Means and Overcurrent Protection at Distribution Point. Where the disconnecting means and overcurrent protection are located at the distribution point, feeders to buildings shall meet the requirements of Section 250-32 and Article 225, Part B.

(c) Disconnecting Means Without Overcurrent Protection at the Distribution Point. Where the disconnecting means without overcurrent protection is located at the distribution point and a disconnecting means and overcurrent protection is located at the building(s), the grounded circuit conductor connection to the grounding electrode shall not be permitted at the building disconnecting means at each building and all of the following conditions shall be met.

(1) All buildings and premises wiring are under single management.
(2) Disconnecting means suitable for use as service equipment is provided at the distribution point.
(3) An equipment grounding conductor is run with the supply conductors and is of the same size as the largest supply conductor, if of the same material, or is adjusted in size in accordance with the equivalent size columns of Table 250-122 if of different materials.
(4) The equipment grounding conductor is bonded to the grounded circuit conductor at the distribution point or at the source of a separately derived system.
(5) A grounding electrode system is provided and connected to the equipment grounding conductor at the building disconnecting means.

Distribution Point. A centrally located electrical supply structure from which services or feeders to agricultural buildings and other buildings, including the associated farm dwelling, are normally supplied.

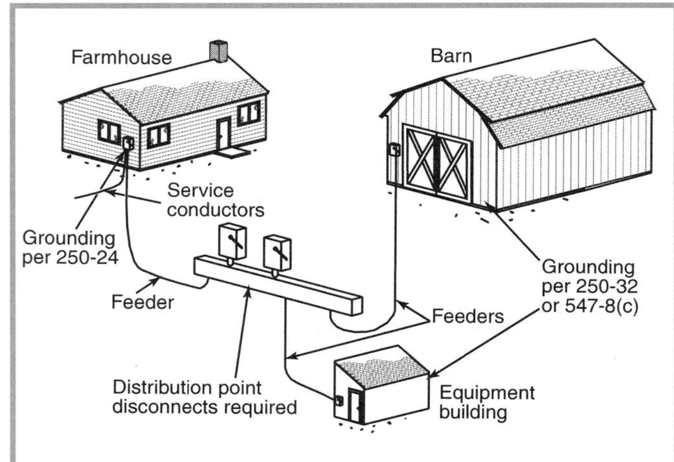

Figure 547.1 *Distribution point located on the load side of the service.*

Section 547-8 was revised for the 1999 *Code*. Three basic arrangements for distribution of power to groups of agricultural buildings from a distribution point are recognized by Section 547-8. They are shown in Figures 547.1, 547.2, and 547.3. These arrangements show a common disconnecting point for this type of multibuilding complex and address the grounding concerns inherent to multiple grounding points.

Figure 547.1 corresponds to Section 547-8(a). In this case, the service disconnect and service overcurrent devices are located at one of the buildings. A feeder is extended from the service to a distribution point. The distribution point, in turn, feeds two additional buildings. An additional disconnecting means is required at the distribution point for each of the buildings that is fed from that location. A disconnecting means may be one or more disconnects. Grounding of the service equipment at the farm house must meet the regular requirements of Section 250-24.

Figure 547.2 corresponds to Section 547-8(b). This example shows the service disconnecting means and overcurrent device(s) located at the distribution point. Feeders extend from the distribution point to the individual buildings. Separate disconnects are then provided at each building in accordance with Section 225-31, and grounding at the separate buildings is in accordance with Section 250-32.

Figure 547.3 corresponds to Section 547-8(c). While the first two arrangements discussed previously fall under established requirements in the *Code*, the third arrangement represents a new requirement. In this case, a main disconnect is installed without overcurrent protection at the distribution point. A disconnecting means with overcurrent protection is installed at each building. Bonding jumpers may not be installed between the grounding electrodes and the grounded conductors at the separate buildings. In addition, the following five other requirements must be met.

1. All buildings must be under single management.
2. The disconnecting means must be suitable for use as service equipment even though it does not include overcurrent devices.
3. An equipment grounding conductor must be run with the supply conductors to each building, and the size of each equipment grounding conductor must be equivalent to the largest supply conductor to each building.
4. The equipment grounding conductor must be bonded to the grounded conductor at the distribution point.

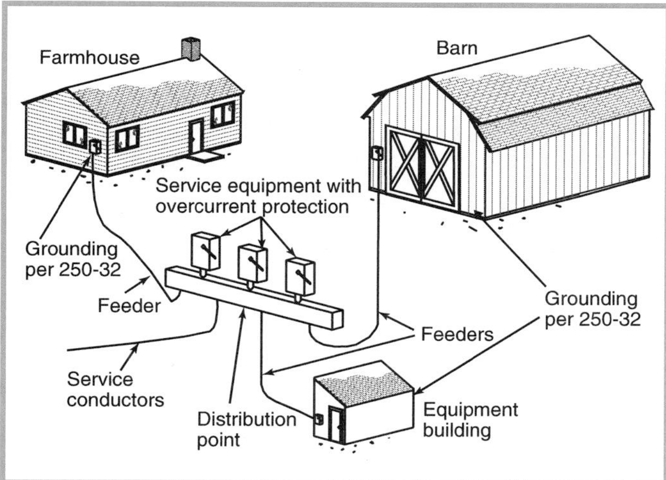

Figure 547.2 Service disconnects and overcurrent protection located at the distribution point.

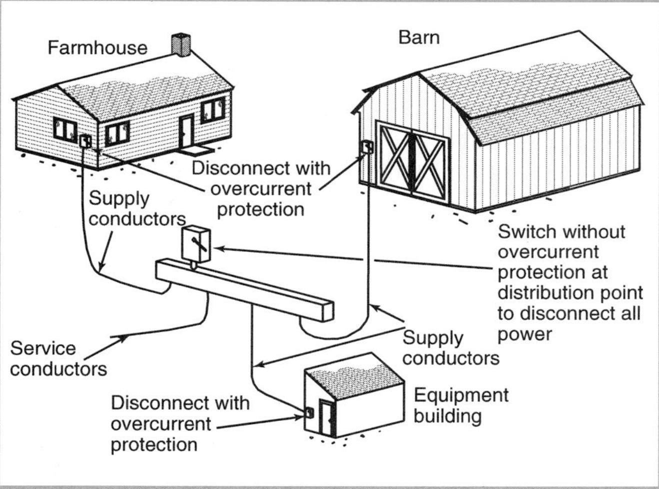

Figure 547.3 Disconnecting means located at the distribution point. It is most important that all grounding comply with Section 547-8(c) for this particular illustration.

5. A grounding electrode system must be provided at each building and must be bonded to the equipment grounding conductor at each building disconnect.

The arrangements shown in Figures 547.1 and 547.2 are not new requirements. Both of these arrangements were previously permitted by existing requirements in the *Code*. However, significant changes were made in Section 547-8(c) and are illustrated in Figure 547.3. The changes in language created some important differences from the previous requirements of the *Code* and created some variations from common practices.

One significant difference can be seen by attempting to label the conductors in Figure 547.3 using defined terms for conductors. Service conductors end at a "service disconnect" by definition. However, according to the definitions of *feeders* and *service equipment*, *feeders* are those conductors between the "service equipment" and the branch-circuit overcurrent devices, and *service equipment* normally includes overcurrent devices in the form of circuit breakers or fuses. The conductors between the distribution point and the building disconnects are not service conductors, so they must be feeders, but they don't quite match the usual definition of feeders either. However, such conductors do fit the description of outside taps of unlimited length as described in Section 240-21(b)(5).

547-9. Bonding and Equipotential Plane.

(a) Definition of Equipotential Plane. An area accessible to livestock where a wire mesh or other conductive elements are embedded in concrete, are bonded to all metal structures and fixed nonelectrical metal equipment that may become energized and are connected to the electrical grounding system to prevent a difference in voltage from developing within the plane. For this section, livestock does not include poultry.

(b) General. Wire mesh or other conductive elements shall be installed in the concrete floor of livestock confinement areas and be bonded to the building grounding electrode system to provide an equipotential plane that may have voltage gradient ramps at entrances and exits that are traversed daily by the same livestock. The bonding conductor shall be copper, insulated, covered or bare, and not smaller than No. 8. The means of bonding to wire mesh or conductive elements shall be by pressure connectors or clamps of brass, copper, copper alloy, or an equally substantial approved means.

Exception No. 1: An equipotential plane shall not be required where there is no electric service to the building nor metal equipment accessible to livestock that is likely to become energized.

Exception No. 2: Slatted floors that are supported by structures that are a part of an equipotential plane shall not be required to be bonded.

FPN No. 1: A natural voltage gradient exists at the edge of an equipotential plane. Typically, voltage gradients exceeding 1 volt per foot at the edge of the equipotential plane will require a voltage gradient ramp.

FPN No. 2: Methods to establish equipotential planes and voltage gradient ramps are described in *Equipo-*

tential Planes in Animal Containment Areas, American Society of Agricultural Engineers (ASAE) EP473-1997.

FPN No. 3: Low grounding electrode system resistances may reduce potential differences in livestock facilities.

(c) Receptacles. All 125-volt, single-phase, 15- and 20-ampere general purpose receptacles in areas having an equipotential plane shall have ground-fault circuit-interrupter protection for personnel.

Article 550 — Mobile Homes, Manufactured Homes, and Mobile Home Parks

Contents

550-23. Service Equipment
(a) Mobile Home Service Equipment
(b) Manufactured Home Service Equipment
(c) Rating
(d) Additional Outside Electrical Equipment
(e) Additional Receptacles
(f) Mounting Height
(g) Marking
550-24. Feeder
(a) Feeder Conductors
(b) Adequate Feeder Capacity

A. General

550-1. Scope. The provisions of this article cover the electrical conductors and equipment installed within or on mobile homes, the conductors that connect mobile homes to a supply of electricity, and the installation of electrical wiring, fixtures, equipment, and appurtenances related to electrical installations within a mobile home park up to the mobile home service-entrance conductors or, if none, the mobile home service equipment.

The *Federal Mobile Home Construction and Safety Standard,* issued by the Federal Housing and Urban Development Administration (HUD), has incorporated many of the provisions of Article 550 of the *NEC.* The federal standard contains the requirements for electrical systems, conductors, and equipment installed within or on mobile homes and the conductors that connect mobile homes to a supply of electricity. Mobile homes are defined as "manufactured homes" in the HUD regulations. For the purposes of this *Code,* and unless otherwise indicated, the term *mobile home* includes manufactured home.

The regulations pertaining to electrical systems are located in the *Code of Federal Regulations,* Title 24, Sections 3280.801 through 3280.816. They require that new mobile homes comply with the federal standard. In some cases, HUD has delegated the enforcement of this standard to state and private inspection agencies and qualified testing laboratories.

See Article 545 for the requirements for manufactured buildings.

550-2. Definitions.

Appliance, Fixed. An appliance that is fastened or otherwise secured at a specific location.

Appliance, Portable. An appliance that is actually moved or can easily be moved from one place to another in normal use.

FPN: For the purpose of this article, the following major appliances, other than built-in, are considered portable if cord connected: refrigerators, range equipment, clothes washers, dishwashers without booster heaters, or other similar appliances.

Appliance, Stationary. An appliance that is not easily moved from one place to another in normal use.

Distribution Panelboard. See definition of *panelboard* in Article 100.

Feeder Assembly. The overhead or under-chassis feeder conductors, including the grounding conductor, together with the necessary fittings and equipment or a power-supply cord listed for mobile home use, designed for the purpose of delivering energy from the source of electrical supply to the distribution panelboard within the mobile home.

Laundry Area. An area containing or designed to contain a laundry tray, clothes washer, or a clothes dryer.

Manufactured Home. A factory-assembled structure or structures that bears a label identifying it as a manufactured home that is transportable in one or more sections, that is built on a permanent chassis and designed to be used as a dwelling with or without a permanent foundation where connected to the required utilities, and that includes the plumbing, heating, air-conditioning, and electrical systems contained therein.

FPN No. 1: See the applicable building code for definition of the term *permanent foundation.*

FPN No. 2: See Part 3280, *Manufactured Home Construction and Safety Standards,* of the federal Department of Housing and Urban Development for additional information on the definition.

For the purpose of this *Code* and unless otherwise indicated, the term *mobile home* includes manufactured homes.

Mobile Home. A factory-assembled structure or structures transportable in one or more sections that is built on a permanent chassis and designed to be used as a dwelling without a permanent foundation where connected to the required utilities, and includes the plumbing, heating, air-conditioning, and electric systems contained therein.

For the purpose of this *Code* and unless otherwise indicated, the term *mobile home* includes manufactured homes.

Manufactured homes (not to be confused with manufactured buildings, covered in Article 545) are covered by Article 550 and are considered mobile homes, even though they are installed on permanent foundations.

It is intended that a mobile home (or manufactured home) constructed under the requirements of Article 550 (or under the HUD regulations) be installed in accordance with the requirements of Article 550. Often, the requirements for mobile homes differ from the requirements of manufactured homes. For example, Section 550-23 was revised for the 1999 *Code* to include a separate section for service equipment for manufactured homes. Also, Section 550-5(a), concerning feeders, includes Exception No. 2 for manufactured homes.

Mobile Home Accessory Building or Structure. Any awning, cabana, ramada, storage cabinet, carport, fence, windbreak, or porch established for the use of the occupant of the mobile home on a mobile home lot.

Mobile Home Lot. A designated portion of a mobile home park designed for the accommodation of one mobile home and its accessory buildings or structures for the exclusive use of its occupants.

Mobile Home Park. A contiguous parcel of land that is used for the accommodation of occupied mobile homes.

Mobile Home Service Equipment. The equipment containing the disconnecting means, overcurrent protective devices, and receptacles or other means for connecting a mobile home feeder assembly.

Park Electrical Wiring Systems. All of the electrical wiring, fixtures, equipment, and appurtenances related to electrical installations within a mobile home park, including the mobile home service equipment.

550-3. Other Articles. Wherever the requirements of other articles of this *Code* and Article 550 differ, the requirements of Article 550 shall apply.

550-4. General Requirements.

(a) Mobile Home Not Intended as a Dwelling Unit. A mobile home not intended as a dwelling unit, for example, those equipped for sleeping purposes only, contractor's on-site offices, construction job dormitories, mobile studio dressing rooms, banks, clinics, mobile stores, or intended for the display or demonstration of merchandise or machinery, shall not be required to meet the provisions of this article pertaining to the number or capacity of circuits required. It shall, however, meet all other applicable requirements of this article if provided with an electrical installation intended to be energized from a 120-volt or 120/240-volt ac power supply system. Where different voltage is required by either design or available power supply system, adjustment shall be made in accordance with other articles and sections for the voltage used.

(b) In Other than Mobile Home Parks. Mobile homes installed in other than mobile home parks shall comply with the provisions of this article.

(c) Connection to Wiring System. The provisions of this article apply to mobile homes intended for connection to a wiring system rated 120/240 volts, nominal, 3-wire ac, with grounded neutral.

(d) Listed or Labeled. All electrical materials, devices, appliances, fittings, and other equipment shall be listed or labeled by a qualified testing agency and shall be connected in an approved manner when installed.

B. Mobile Homes

550-5. Power Supply.

(a) Feeder. The power supply to the mobile home shall be a feeder assembly consisting of not more than one listed 50-ampere mobile home power-supply cord with an integrally molded or securely attached plug cap, or a permanently installed feeder.

Exception No. 1: A mobile home that is factory equipped with gas or oil-fired central heating equipment and cooking appliances shall be permitted to be provided with a listed mobile home power-supply cord rated 40 amperes.

Exception No. 2: Manufactured homes constructed in accordance with Section 550-23(b).

Section 550-23(b) permits the service equipment to be installed in or on a manufactured home.

(b) Power-Supply Cord. If the mobile home has a power-supply cord, it shall be permanently attached to the distribution panelboard or to a junction box permanently connected to the distribution panelboard, with the free end terminating in an attachment plug cap.

Cords with adapters and pigtail ends, extension cords, and similar items shall not be attached to, or shipped with, a mobile home.

A suitable clamp or the equivalent shall be provided at the distribution panelboard knockout to afford strain relief for the cord to prevent strain from being transmitted to the terminals when the power-supply cord is handled in its intended manner.

The cord shall be a listed type with four conductors, one of which shall be identified by a continuous green color or a continuous green color with one or more yellow stripes for use as the grounding conductor.

(c) Attachment Plug Cap. The attachment plug cap shall be a 3-pole, 4-wire, grounding type, rated 50 amperes, 125/250 volts with a configuration as shown in Figure 550-5(c) and intended for use with the 50-ampere, 125/250-volt receptacle configuration shown in Figure 550-5(c). It shall be listed, by itself or as part of a power-supply cord assembly, for the purpose, and shall be molded to or installed on the flexible cord so that it is secured tightly to the cord at the

Receptacle

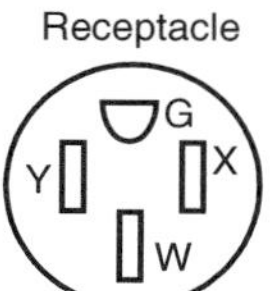

Cap

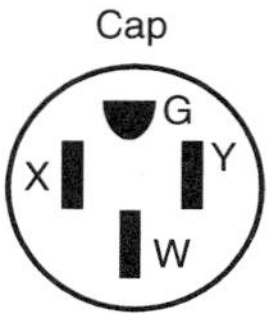

125/250-V, 50-A, 3-pole, 4-wire, grounding type

Figure 550-5(c) 50-ampere, 125/250-volt receptacle and attachment plug cap configurations, 3-pole, 4-wire, grounding-types, used for mobile home supply cords and mobile home parks.

point where the cord enters the attachment plug cap. If a right-angle cap is used, the configuration shall be oriented so that the grounding member is farthest from the cord.

FPN: Complete details of the 50-ampere plug and receptacle configuration can be found in the National Electrical Manufacturers Association *Standard for Dimensions of Attachment Plugs and Receptacles*, ANSI/NEMA WD6-1989, Figure 14-50.

(d) Overall Length of a Power-Supply Cord. The overall length of a power-supply cord, measured from the end of the cord, including bared leads, to the face of the attachment plug cap shall not be less than 21 ft (6.4 m) and shall not exceed 36½ ft (11.13 m). The length of the cord from the face of the attachment plug cap to the point where the cord enters the mobile home shall not be less than 20 ft (6.1 m).

(e) Marking. The power-supply cord shall bear the following marking:

FOR USE WITH MOBILE HOMES — 40 AMPERES.

or

FOR USE WITH MOBILE HOMES — 50 AMPERES.

(f) Point of Entrance. The point of entrance of the feeder assembly to the mobile home shall be in the exterior wall, floor, or roof.

(g) Protected. Where the cord passes through walls or floors, it shall be protected by means of conduits and bushings or equivalent. The cord shall be permitted to be installed within the mobile home walls, provided a continuous raceway having a maximum size of 1¼ in. (31.8 mm) is installed from the branch-circuit panelboard to the underside of the mobile home floor.

(h) Protection Against Corrosion and Mechanical Damage. Permanent provisions shall be made for the protection of the attachment plug cap of the power-supply cord and any connector cord assembly or receptacle against corrosion and mechanical damage if such devices are in an exterior location while the mobile home is in transit.

(i) Mast Weatherhead or Raceway. Where the calculated load exceeds 50 amperes or where a permanent feeder is used, the supply shall be by means of the following:

(1) One mast weatherhead installation, installed in accordance with Article 230, containing four continuous, insulated, color-coded feeder conductors, one of which shall be an equipment grounding conductor; or
(2) A metal raceway or rigid nonmetallic conduit from the disconnecting means in the mobile home to the underside of the mobile home, with provisions for the attachment to a suitable junction box or fitting to the raceway on the underside of the mobile home [with or without conductors as in Section 550-5(i)(1)].

In some locations, mobile homes are permanently connected, as permitted in Sections 550-5(i) and 550-24. Local requirements must be checked for the approved method of installing overhead and underground feeder assemblies.

A raceway is required from the distribution panelboard in the mobile home to the underside of the mobile home. Whether the feeder conductors in this raceway are pulled in by the mobile home manufacturer or by installers in the field is optional. Regardless of who installs them, the conductors should comprise four continuous, insulated, color-coded conductors, as indicated in Sections 550-5(i)(1) and 550-24.

550-6. Disconnecting Means and Branch-Circuit Protective Equipment. The branch-circuit equipment shall be permitted to be combined with the disconnecting means as a single assembly. Such a combination shall be permitted to be designated as a distribution panelboard. If a fused distribution panelboard is used, the maximum fuse size for the mains shall be plainly marked with lettering at least ¼ in. (6.4 mm) high and visible when fuses are changed.

Where plug fuses and fuseholders are used, they shall be tamper-resistant Type S, enclosed in dead-front fuse panelboards. Electrical distribution panelboards containing circuit breakers shall also be dead-front type.

FPN: See Section 110-22 concerning identification of each disconnecting means and each service, feeder, or branch circuit at the point where it originated and the type marking needed.

(a) Disconnecting Means. A single disconnecting means shall be provided in each mobile home consisting of a circuit breaker, or a switch and fuses and its accessories installed in a readily accessible location near the point of entrance of the supply cord or conductors into the mobile home. The main circuit breakers or fuses shall be plainly marked "Main." This equipment shall contain a solderless type of grounding connector or bar for the purposes of grounding, with sufficient terminals for all grounding conductors. The neutral bar termination of the grounded circuit conductors shall be insulated in accordance with Section 550-11(a). The disconnecting equipment shall have a rating suitable for the connected load. The distribution equipment, either circuit breaker or fused type, shall be located a minimum of 24 in. (610 mm) from the bottom of such equipment to the floor level of the mobile home.

FPN: See Section 550-15(b) for information on disconnecting means for branch circuits designed to energize heating or air-conditioning equipment, or both, located outside the mobile home, other than room air conditioners.

A distribution panelboard shall be rated not less than 50 amperes and employ a 2-pole circuit breaker rated 40 amperes for a 40-ampere supply cord, or 50 amperes for a 50-ampere supply cord. A distribution panelboard employing a disconnect switch and fuses shall be rated 60 amperes and shall employ a single 2-pole, 60-ampere fuseholder with 40- or 50-ampere main fuses for 40- or 50-ampere supply cords, respectively. The outside of the distribution panelboard shall be plainly marked with the fuse size.

The distribution panelboard shall be located in an accessible location but shall not be located in a bathroom or a clothes closet. A clear working space at least 30 in. (762 mm) wide and 30 in. (762 mm) in front of the distribution panelboard shall be provided. This space shall extend from the floor to the top of the distribution panelboard.

(b) Branch-Circuit Protective Equipment. Branch-circuit distribution equipment shall be installed in each mobile home and shall include overcurrent protection for each branch circuit consisting of either circuit breakers or fuses.

The branch-circuit overcurrent devices shall be rated as follows:

(1) Not more than the circuit conductors; and
(2) Not more than 150 percent of the rating of a single appliance rated 13.3 amperes or more that is supplied by an individual branch circuit; but
(3) Not more than the overcurrent protection size and of the type marked on the air conditioner or other motor-operated appliance.

A 15-ampere multiple receptacle shall be permitted where connected to a 20-ampere laundry circuit.

(c) Two-Pole Circuit Breakers. Where circuit breakers are provided for branch-circuit protection, 240-volt circuits shall be protected by a 2-pole common or companion trip, or handle-tied paired circuit breakers.

(d) Electrical Nameplates. A metal nameplate on the outside adjacent to the feeder assembly entrance shall read:

THIS CONNECTION FOR 120/240-VOLT,
3-POLE, 4-WIRE, 60 HERTZ, _____ AMPERE
SUPPLY

The correct ampere rating shall be marked in the blank space.

550-7. Branch Circuits. The number of branch circuits required shall be determined in accordance with (a) through (c).

(a) Lighting. Based on 3 volt-amperes/ft^2 (32.26 VA/m^2) times outside dimensions of the mobile home (coupler excluded) divided by 120 volts to determine the number of 15- or 20-ampere lighting area circuits, e.g.,

$$\frac{3 \times \text{length} \times \text{width}}{120 \times 15 \text{ (or 20)}} = \text{No. of 15- (or 20-) ampere circuits}$$

The lighting circuits shall be permitted to serve built-in gas ovens with electric service only for lights, clocks or timers, or listed cord-connected garbage disposal units.

(b) Small Appliances. Small appliance branch circuits shall be installed in accordance with Section 210-52(b).

(c) General Appliances. (Including furnace, water heater, range, and central or room air conditioner, etc.) There shall be one or more circuits of adequate rating in accordance with the following.

FPN No. 1: For the laundry branch circuit, see Section 210-11(c)(2).

FPN No. 2: For central air conditioning, see Article 440.

(1) Ampere rating of fixed appliances not over 50 percent of circuit rating if lighting outlets (receptacles, other than kitchen, dining area, and laundry, considered as lighting outlets) are on the same circuit.

(2) For fixed appliances on a circuit without lighting outlets, the sum of rated amperes shall not exceed the branch-circuit rating. Motor loads or other continuous duty loads shall not exceed 80 percent of the branch-circuit rating.

(3) The rating of a single cord- and plug-connected appliance on a circuit having no other outlets shall not exceed 80 percent of the circuit rating.

(4) The rating of a range branch circuit shall be based on the range demand as specified for ranges in Section 550-13(b)(5).

550-8. Receptacle Outlets.

(a) Grounding-Type Receptacle Outlets. All receptacle outlets (1) shall be of grounding type; (2) shall be installed according to Section 210-7; and (3), except where supplying specific appliances, receptacles shall be 15- or 20-ampere, 125-volt, either single or duplex, and shall accept parallel-blade attachment plugs.

(b) Ground-Fault Circuit Interrupters. All 120-volt, single-phase, 15- and 20-ampere receptacle outlets installed outdoors and in bathrooms, including receptacles in light fixtures, shall have ground-fault circuit-interrupter protection for personnel. Ground-fault circuit-interrupter protection for personnel shall be provided for receptacle outlets located within 6 ft (1.83 m) of any lavatory or sink.

Exception: Receptacles installed for appliances in dedicated spaces, such as for dishwashers, disposals, refrigerators, freezers, and laundry equipment.

No receptacle shall be required in the area occupied by a toilet, shower, tub, or any combination thereof. If a receptacle is installed in such an area, it shall have ground-fault circuit-interrupter protection for personnel.

Feeders supplying branch circuits shall be permitted to be protected by a ground-fault circuit-interrupter in lieu of the provision for such interrupters specified herein.

(c) Cord-Connected Fixed Appliance. A grounding-type receptacle outlet shall be provided for each cord-connected fixed appliance installed.

(d) Required Receptacle Outlets. Receptacle outlets shall be provided in all rooms other than the bath, closet, and hall areas, and shall be installed so that no point along the floor line is more than 6 ft (1.83 m) measured horizontally from an outlet in that space. Countertops shall have receptacles located every 6 ft (1.83 m). The contiguous measurement of countertop and floor line shall be permitted where measured from the required receptacle in rooms requiring small appliance circuits. Receptacle outlets on small appliance circuits shall not be included in determining the spacing for receptacle outlets of other circuits.

Exception No. 1: Where the measured distance is interrupted by an interior doorway, sink, refrigerator, range, oven, or cooktop, an additional receptacle outlet shall be provided where the interrupted space is at least 2 ft (610 mm) wide at the floor line and at least 12 in. (305 mm) wide at the countertop.

Exception No. 2: Receptacles rendered not readily accessible by stationary appliances shall not be considered as the required outlets.

Exception No. 3: The distance along a floor line occupied by a door opened fully against that space shall not be required to be included in establishing the horizontal measurement if the door swing is limited to 90 degrees nominal by that wall space.

Exception No. 4: Receptacle requirements for bar-type counters and for fixed room dividers shall be permitted to be provided by a receptacle outlet in the wall at the nearest point where the counter or room divider attaches to the wall provided the following:

(a) The divider does not exceed 8 ft (2.44 m) in length; and
(b) The divider does not exceed 4 ft (1.22 m) in height; and
(c) The divider is attached to a wall at one end only.

(e) Outdoor Receptacle Outlets. At least one receptacle outlet shall be installed outdoors. A receptacle outlet located in a compartment accessible from the outside of the mobile home shall be considered an outdoor receptacle. Outdoor receptacle outlets shall be protected as required in Section 550-8(b).

(f) Receptacle Outlets Not Permitted.

(1) Shower or Bathtub Space. Receptacle outlets shall not be installed in or within reach [30 in. (762 mm)] of a shower or bathtub space.

(2) Face-Up Position. A receptacle shall not be installed in a face-up position in any countertop.

(g) Pipe Heating Cable Outlet. Where a pipe heating cable outlet is installed, the outlet shall be as follows:

(1) Located within 2 ft (610 mm) of the cold water inlet.
(2) Connected to an interior branch circuit, other than a small appliance branch circuit. It shall be permitted to utilize a bathroom receptacle circuit for this purpose.
(3) On a circuit where all of the outlets are on the load side of the ground-fault circuit-interrupter protection for personnel.
(4) Mounted on the underside of the mobile home and shall not be considered to be the outdoor receptacle outlet required in Section 550-8(e).

Revised for the 1999 *Code,* Section 550-8(g) now requires that if pipe heating cable outlets (sometimes referred to as heat tape outlets) are installed, then they must be located on the underside of the mobile home and be GFCI protected. This protection must be installed according to the four listed provisions of Section 550-8(g).

550-9. Fixtures and Appliances.

(a) Fasten Appliances in Transit. Means shall be provided to securely fasten appliances when the mobile home is in transit. (See Section 550-11 for provisions on grounding.)

(b) Accessibility. Every appliance shall be accessible for inspection, service, repair, or replacement without removal of permanent construction.

(c) Pendants. Pendant-type fixtures or pendant cords shall be listed and identified for the interconnection of building components.

(d) Bathtub and Shower Fixtures. Where a lighting fixture is installed over a bathtub or in a shower stall, it shall be of the enclosed and gasketed type listed for wet locations.

(e) Location of Switches. The switch for shower lighting fixtures and exhaust fans located over a tub or in a shower stall shall be located outside the tub or shower space.

550-10. Wiring Methods and Materials. Except as specifically limited in this section, the wiring methods and materials included in this *Code* shall be used in mobile homes.

(a) Nonmetallic Boxes. Nonmetallic boxes shall be permitted only with nonmetallic cable or nonmetallic raceways.

(b) Nonmetallic Cable Protection. Nonmetallic cable located 15 in. (381 mm) or less above the floor, if exposed, shall be protected from physical damage by covering boards, guard strips, or raceways. Cable likely to be damaged by stowage shall be so protected in all cases.

(c) Metal-Covered and Nonmetallic Cable Protection. Metal-covered and nonmetallic cables shall be permitted to pass through the centers of the wide side of 2 in. by 4 in. studs. However, they shall be protected where they pass through 2 in. by 2 in. studs or at other studs or frames where the cable or armor would be less than 1¼ in. (31.8 mm) from the inside or outside surface of the studs where the wall covering materials are in contact with the studs. Steel plates on each side of the cable, or a tube, with not less than No. 16 MSG wall thickness shall be required to protect the cable. These plates or tubes shall be securely held in place.

(d) Metal Faceplates. Where metal faceplates are used, they shall be effectively grounded.

(e) Installation Requirements. If a range, clothes dryer, or similar appliance is connected by metal-covered cable or flexible metal conduit, a length of not less than 3 ft (914 mm) of free cable or conduit shall be provided to permit moving the appliance. The cable or flexible metal conduit shall be secured to the wall. Type NM or Type SE cable shall not be used to connect a range or dryer. This shall not prohibit the use of Type NM or Type SE cable between the branch-circuit overcurrent-protective device and a junction box or range or dryer receptacle.

(f) Raceways. Where rigid metal conduit or intermediate metal conduit is terminated at an enclosure with a locknut and bushing connection, two locknuts shall be provided, one inside and one outside of the enclosure. Rigid nonmetallic conduit, or electrical nonmetallic tubing, or surface raceway shall be permitted. All cut ends of conduit and tubing shall be reamed or otherwise finished to remove rough edges.

(g) Switches. Switches shall be rated as follows.

(1) For lighting circuits, switches shall be rated not less than 10 amperes, 120 to 125 volts, and in no case less than the connected load.

(2) For motors or other loads, switches shall have ampere or horsepower ratings, or both, adequate for loads controlled. (An ac general-use snap switch shall be permitted to control a motor 2 hp or less with full-load current not over 80 percent of the switch ampere rating.)

(h) Under-Chassis Wiring (Exposed to Weather).

(1) Where outdoor or under-chassis line-voltage (120 volts, nominal, or higher) wiring is exposed to moisture or physical damage, it shall be protected by rigid metal conduit or intermediate metal conduit. The conductors shall be suitable for wet locations.

Exception: Electrical metallic tubing or rigid nonmetallic conduit shall be permitted where closely routed against frames and equipment enclosures.

(2) The cables or conductors shall be Type NMC, Type TW, or equivalent.

(i) Boxes, Fittings, and Cabinets. Boxes, fittings, and cabinets shall be securely fastened in place and shall be supported from a structural member of the home, either directly or by using a substantial brace.

Exception: Snap-in-type boxes. Boxes provided with special wall or ceiling brackets and wiring devices with integral enclosures that securely fasten to walls or ceilings and are identified for the use shall be permitted without support from a structural member or brace. The testing and approval shall include the wall and ceiling construction systems for which the boxes and devices are intended to be used.

(j) Appliance Terminal Connections. Appliances having branch-circuit terminal connections that operate at temperatures higher than 60°C (140°F) shall have circuit conductors as described in (1) or (2).

(1) Branch-circuit conductors having an insulation suitable for the temperature encountered shall be permitted to be run directly to the appliance.

(2) Conductors having an insulation suitable for the temperature encountered shall be run from the appliance terminal connection to a readily accessible outlet box placed at least 1 ft (305 mm) from the appliance. These conductors shall be in a suitable raceway that shall extend for at least 4 ft (1.22 m).

(k) Component Interconnections. Fittings and connectors that are intended to be concealed at the time of assembly shall be listed and identified for the interconnection of building components. Such fittings and connectors shall be equal to the wiring method employed in insulation, temperature rise, and fault-current withstanding and shall be capable of enduring the vibration and shock occurring in mobile home transportation.

550-11. Grounding. Grounding of both electrical and nonelectrical metal parts in a mobile home shall be through connection to a grounding bus in the mobile home distribution panelboard. The grounding bus shall be grounded through the green-colored insulated conductor in the supply cord or the feeder wiring to the service ground in the service-entrance equipment located adjacent to the mobile home location. Neither the frame of the mobile home nor the frame of any appliance shall be connected to the grounded circuit conductor (neutral) in the mobile home.

(a) Insulated Neutral.

(1) The grounded circuit conductor (neutral) shall be insulated from the grounding conductors and from equipment enclosures and other grounded parts. The grounded (neutral) circuit terminals in the distribution panelboard and in ranges, clothes dryers, counter-mounted cooking units, and wall-mounted ovens shall be insulated from the equipment enclosure. Bonding screws, straps, or buses in the distribution panelboard or in appliances shall be removed and discarded.

(2) Connections of ranges and clothes dryers with 120/240-volt, 3-wire ratings shall be made with 4-conductor cord and 3-pole, 4-wire, grounding-type plugs or by Type AC cable, Type MC cable, or conductors enclosed in flexible metal conduit.

The provisions of Section 550-24(a) require that the feeder assembly for a mobile home consist of a listed cord or four color-coded insulated conductors, one of which is the grounded conductor (white) and one of which is used for grounding purposes (green). Thus, the "grounded" and "grounding" conductors are kept independent of each other and are connected only at the service equipment (at the point of connection of the grounding electrode conductor). Grounding of both electrical and nonelectrical metal parts, including the frame of the mobile home or the frame of any appliance, is accomplished by connection to the equipment grounding bus [never to the grounded conductor (neutral bus)].

Bonding screws, straps, or buses, which bond the grounded (neutral) circuit conductors to the noncurrent-carrying metal parts in the mobile home panelboard or in appliances (ranges, clothes dryers), are required to be removed and discarded.

(b) Equipment Grounding Means.

(1) The green-colored insulated grounding wire in the supply cord or permanent feeder wiring shall be connected to the grounding bus in the distribution panelboard or disconnecting means.

(2) In the electrical system, all exposed metal parts, enclosures, frames, lamp fixture canopies, etc., shall be effectively bonded to the grounding terminal or enclosure of the distribution panelboard.

(3) Cord-connected appliances, such as washing machines, clothes dryers, refrigerators, and the electrical system of gas ranges, etc., shall be grounded by means of a cord with grounding conductor and grounding-type attachment plug.

(c) Bonding of Noncurrent-Carrying Metal Parts.

(1) All exposed noncurrent-carrying metal parts that may become energized shall be effectively bonded to the grounding terminal or enclosure of the distribution panelboard. A bonding conductor shall be connected between the distribution panelboard and accessible terminal on the chassis.

(2) Grounding terminals shall be of the solderless type and listed as pressure-terminal connectors recognized for the wire size used. The bonding conductor shall be solid or stranded, insulated or bare, and shall be No. 8 copper minimum, or equivalent. The bonding conductor shall be routed so as not to be exposed to physical damage.

(3) Metallic gas, water, and waste pipes and metallic air-circulating ducts shall be considered bonded if they are connected to the terminal on the chassis [see Section 550-11(c)(1)] by clamps, solderless connectors, or by suitable grounding-type straps.

(4) Any metallic roof and exterior covering shall be considered bonded if (1) the metal panels overlap one another and are securely attached to the wood or metal frame parts by metallic fasteners and (2) if the lower panel of the metallic exterior covering is secured by metallic fasteners at a cross member of the chassis by two metal straps per mobile home unit or section at opposite ends.

The bonding strap material shall be a minimum of 4 in. (102 mm) in width of material equivalent to the skin or a material of equal or better electrical conductivity. The straps shall be fastened with paint-penetrating fittings such as screws and starwashers or equivalent.

550-12. Testing.

(a) Dielectric Strength Test. The wiring of each mobile home shall be subjected to a 1-minute, 900-volt, dielectric strength test (with all switches closed) between live parts (including neutral) and the mobile home ground. Alternatively, the test shall be permitted to be performed at 1080 volts for 1 second. This test shall be performed after branch circuits are complete and after fixtures or appliances are installed.

Exception: Listed fixtures or appliances shall not be required to withstand the dielectric strength test.

(b) Continuity and Operational Tests and Polarity Checks. Each mobile home shall be subjected to the following:

(1) An electrical continuity test to ensure that all exposed electrically conductive parts are properly bonded;
(2) An electrical operational test to demonstrate that all equipment, except water heaters and electric furnaces, is connected and in working order; and
(3) Electrical polarity checks of permanently wired equipment and receptacle outlets to determine that connections have been properly made.

550-13. Calculations. The following method shall be employed in computing the supply-cord and distribution-panelboard load for each feeder assembly for each mobile home in lieu of the procedure shown in Article 220 and shall be based on a 3-wire, 120/240-volt supply with 120-volt loads balanced between the two legs of the 3-wire system.

(a) Lighting and Small Appliance Load.

Lighting Volt-Amperes: Length times width of mobile home floor (outside dimensions) times 3 volt-amperes/ft^2, e.g.,

$$\text{Length} \times \text{width} \times 3 = \text{lighting volt-amperes}$$

Small Appliance Volt-Amperes: Number of circuits times 1500 volt-amperes for each 20-ampere appliance receptacle circuit (see definition of *Appliance, Portable* with note) including 1500 volt-amperes for laundry circuit, e.g.,

$$\text{No. of circuits} \times 1500 = \text{small appliance volt-amperes.}$$

Total: Lighting volt-amperes plus small appliance volt-amperes = total volt-amperes

First 3000 total volt-amperes at 100 percent plus remainder at 35 percent = volt-amperes to be divided by 240 volts to obtain current (amperes) per leg.

(b) Total Load for Determining Power Supply. Total load for determining power supply is the sum of the following.

(1) Lighting and small appliance load as calculated in Section 550-13(a).
(2) Nameplate amperes for motors and heater loads (exhaust fans, air conditioners, electric, gas, or oil heating). Omit smaller of the heating and cooling loads, except include blower motor if used as air-conditioner evaporator motor. Where an air conditioner is not installed and a 40-ampere power-supply cord is provided, allow 15 amperes per leg for air conditioning.
(3) Twenty-five percent of current of largest motor in (2).
(4) Total of nameplate amperes for waste disposer, dishwasher, water heater, clothes dryer, wall mounted oven, cooking units. Where the number of these appliances exceeds three, use 75 percent of total.
(5) Derive amperes for freestanding range (as distinguished from separate ovens and cooking units) by dividing the following values by 240 volts.

Nameplate Rating (watts)	Use (volt-amperes)
0–10,000	80 percent of rating
Over 10,000–12,500	8,000
Over 12,500–13,500	8,400
Over 13,500–14,500	8,800
Over 14,500–15,500	9,200
Over 15,500–16,500	9,600
Over 16,500–17,500	10,000

(6) If outlets or circuits are provided for other than factory-installed appliances, include the anticipated load.

FPN: Refer to Appendix D, Example D11, for an illustration of the application of this calculation.

(c) Optional Method of Calculation for Lighting and Appliance Load. For mobile homes, the optional method for calculating lighting and appliance load shown in Section 220-30 shall be permitted.

550-14. Interconnection of Multiple Section Mobile Home Units. Approved and listed fixed-type wiring methods shall be used to join portions of a circuit that must be electrically joined that are located in adjacent sections of mobile homes after the home is installed on its support foundation. The circuit's junction shall be accessible for disassembly when the home is prepared for relocation.

550-15. Outdoor Outlets, Fixtures, Air-Cooling Equipment, etc.

(a) Listed for Outdoor Use. Outdoor fixtures and equipment shall be listed for outdoor use. Outdoor receptacle or convenience outlets shall be of a gasketed-cover type for use in wet locations.

(b) Outside Heating Equipment, Air-Conditioning Equipment, or Both. A mobile home provided with a branch circuit designed to energize outside heating equipment or air-conditioning equipment, or both, located outside the mobile home, other than room air conditioners, shall have such branch-circuit conductors terminate in a listed outlet box, or disconnecting means, located on the outside of the mobile home. A label shall be permanently affixed adjacent to the outlet box and shall contain the following information:

THIS CONNECTION IS FOR HEATING AND/OR
AIR-CONDITIONING EQUIPMENT. THE
BRANCH CIRCUIT IS RATED AT NOT MORE
THAN _____ AMPERES, AT _____ VOLTS,
60-HERTZ, _____ CONDUCTOR AMPACITY.
A DISCONNECTING MEANS SHALL BE
LOCATED WITHIN SIGHT
OF THE EQUIPMENT.

The correct voltage and ampere rating shall be given. The tag shall be not less than 0.020-in. (508-μm) thick etched brass, stainless steel, anodized or alclad aluminum, or equivalent. The tag shall not be less than 3 in. (76 mm) by 1¾ in. (44.5 mm) minimum size.

C. Services and Feeders

550-21. Distribution System. The mobile home park secondary electrical distribution system to mobile home lots shall be single-phase, 120/240 volts, nominal. For the purpose of Part C, where the park service exceeds 240 volts, nominal, transformers and secondary distribution panelboards shall be treated as services.

Mobile homes are intended for connection to a wiring system nominally rated 120/240 volts, 3-wire ac, with a grounded neutral; therefore, distribution systems at mobile home parks must supply 120/240 volts to the mobile home lot. Because appliances and other equipment are usually installed during the manufacturing process of mobile homes and are rated 120/240 volts, a 120/208-volt supply derived from a 4-wire, 120/208-volt wye system is unsuitable.

Section 550-22 requires park electrical wiring systems to be calculated on the basis of the larger of (1) not less than 16,000 volt-amperes (at 120/240 volts) for each mobile home lot, or (2) the calculated load of the largest typical mobile home the lot will accommodate. However, the ampacity of the feeder-circuit conductors to each mobile home lot is not to be less than 100 amperes (at 120/240 volts), according to Section 550-24(b).

550-22. Minimum Allowable Demand Factors. Park electrical wiring systems shall be calculated (at 120/240 volts) on the larger of (1) 16,000 volt-amperes for each mobile home lot or (2) the load calculated in accordance with Section 550-13 for the largest typical mobile home that each lot will accept. It shall be permissible to compute the feeder or service load in accordance with Table 550-22. No demand factor shall be allowed for any other load, except as provided in this *Code*.

Service and feeder conductors to a mobile home in compliance with Section 310-15(b)(6) shall be permitted.

Table 550-22. Demand Factors for Feeders and Service-Entrance Conductors

Number of Mobile Homes	Demand Factor (percent)
1	100
2	55
3	44
4	39
5	33
6	29
7–9	28
10–12	27
13–15	26
16–21	25
22–40	24
41–60	23
61 and over	22

550-23. Service Equipment.

(a) Mobile Home Service Equipment. The mobile home service equipment shall be located adjacent to the mobile home and not mounted in or on the mobile home. The service equipment shall be located in sight from and not more than 30 ft (9.14 m) from the exterior wall of the mobile home it serves. The service equipment shall be permitted to be located elsewhere on the premises, provided that a disconnecting means suitable for service equipment is located in sight from and not more than 30 ft (9.14 m) from the exterior wall of the mobile home it serves. Grounding at the disconnecting means shall be in accordance with Section 250-32.

(b) Manufactured Home Service Equipment. The manufactured home service equipment shall be permitted to be installed in or on a manufactured home, provided that all of the following conditions are met.

(1) The manufactured home is secured to a permanent foundation that complies with applicable building codes.
(2) The service equipment is installed in a manner acceptable to the authority having jurisdiction.
(3) The installation of the service equipment complies with Article 230.
(4) Means are provided for the connection of a grounding electrode conductor to the service equipment and routing it outside the structure.

Revised for the 1999 *Code,* Section 550-23(b) permits the service equipment to be installed in or on a manufactured home by the manufacturer of the structure or by any qualified individual(s), provided there is compliance with all four provisions of Section 550-23(b). Noticeably absent from these provisions, and different from past editions of the *Code,* is the requirement for approval by the authority having jurisdiction.

(c) Rating. Mobile home service equipment shall be rated at not less than 100 amperes at 120/240 volts, and provisions shall be made for connecting a mobile home feeder assembly by a permanent wiring method. Power outlets used as mobile home service equipment shall also be permitted to contain receptacles rated up to 50 amperes with appropriate overcurrent protection. Fifty-ampere receptacles shall conform to the configuration shown in Figure 550-5(c).

FPN: Complete details of the 50-ampere plug and receptacle configuration can be found in *National Electrical Manufacturers Association Standard for Wiring Devices — Dimensional Requirements,* ANSI/NEMA WD 6-1988, Figure 14-50.

(d) Additional Outside Electrical Equipment. Mobile home service equipment shall also contain a means for connecting a mobile home accessory building or structure, or additional electrical equipment located outside a mobile home by a fixed wiring method.

(e) Additional Receptacles. Additional receptacles shall be permitted for connection of electrical equipment located outside the mobile home, and all such 125-volt, single-phase, 15- and 20-ampere receptacles shall be protected by a listed ground-fault circuit interrupter.

Mobile home service equipment is required to be located in sight of the mobile home, but the equipment can be up to 30 ft from any point on the exterior wall of the mobile home. See Figure 550.1 for an example of mobile home service equipment. This requirement recognizes the use of feeder raceways that are external to the mobile home. Service equipment may be located more than 30 ft from the mobile home if an additional disconnecting means is located within 30 ft of the mobile home and proper grounding and bonding measures are taken. See Section 550-23(a) for the exact requirements.

Figure 550.1 *Mobile home service equipment and power outlet assembly showing metering, disconnecting means, overcurrent protective devices, and receptacles. (Midwest)*

(f) Mounting Height. Outdoor mobile home disconnecting means shall be installed so the bottom of the enclosure containing the disconnecting means is not less than 2 ft (610 mm) above finished grade or working platform. The disconnecting means shall be installed so that the center of the grip of the operating handle, when in the highest position, will not be more than 6 ft 7 in. (2.0 m) above the finished grade or working platform.

(g) Marking. Where a 125/250-volt receptacle is used in mobile home service equipment, the service equipment shall be marked as follows.

TURN DISCONNECTING SWITCH OR CIRCUIT BREAKER OFF BEFORE INSERTING OR REMOVING PLUG. PLUG MUST BE FULLY INSERTED OR REMOVED.

The marking shall be located on the service equipment adjacent to the receptacle outlet.

550-24. Feeder.

(a) Feeder Conductors. Mobile home feeder conductors shall consist of either a listed cord, factory-installed in accordance with Section 550-5(b), or a permanently installed feeder consisting of four, insulated, color-coded conductors that shall be identified by the factory or field marking of the conductors in compliance with Section 310-12. Equipment grounding conductors shall not be identified by stripping the insulation.

Exception: Where a mobile home feeder is installed between service equipment and a mobile home disconnecting means as covered in Section 550-23(a), it shall be permitted to omit the equipment grounding conductor where the grounded circuit conductor is grounded at the disconnecting means as required in Section 250-32(b).

(b) Adequate Feeder Capacity. Mobile home lot feeder circuit conductors shall have adequate capacity for the loads supplied and shall be rated at not less than 100 amperes at 120/240 volts.

Article 551 — Recreational Vehicles and Recreational Vehicle Parks

Contents

A. General

551-1. Scope. The provisions of this article cover the electrical conductors and equipment installed within or on recreational vehicles, the conductors that connect recreational vehicles to a supply of electricity, and the installation of equipment and devices related to electrical installations within a recreational vehicle park.

Laws in many states require a factory inspection by state electrical inspectors. The requirements of such laws follow closely the requirements of NFPA 501C-1996, *Standard on Recreational Vehicles*. Section 1-5 in NFPA 501C specifies that electrical installations in recreational vehicles shall comply with Part A of Article 551 and other applicable sections of the *Code*. This reference to Part A of Article 551 is based on the 1984 edition of the *NEC*. Parts B through F in the 1999 *NEC* were formerly Part A.

551-2. Definitions. (See Article 100 for additional definitions.)

Air-Conditioning or Comfort-Cooling Equipment. All of that equipment intended or installed for the purpose of processing the treatment of air so as to control simultaneously its temperature, humidity, cleanliness, and distribution to meet the requirements of the conditioned space.

Appliance, Fixed. An appliance that is fastened or otherwise secured at a specific location.

Appliance, Portable. An appliance that is actually moved or can easily be moved from one place to another in normal use.

FPN: For the purpose of this article, the following major appliances, other than built-in, are considered portable if cord connected: refrigerators, range equipment, clothes washers, dishwashers without booster heaters, or other similar appliances.

Appliance, Stationary. An appliance that is not easily moved from one place to another in normal use.

Camping Trailer. A vehicular portable unit mounted on wheels and constructed with collapsible partial side walls that fold for towing by another vehicle and unfold at the campsite to provide temporary living quarters for recreational, camping, or travel use. (See Recreational Vehicle.)

Converter. A device that changes electrical energy from one form to another, as from alternating current to direct current.

Dead Front (as applied to switches, circuit breakers, switchboards, and distribution panelboards). Designed, constructed, and installed so that no current-carrying parts are normally exposed on the front.

Disconnecting Means. The necessary equipment usually consisting of a circuit breaker or switch and fuses, and their accessories, located near the point of entrance of supply conductors in a recreational vehicle and intended to constitute the means of cutoff for the supply to that recreational vehicle.

Distribution Panelboard. A single panel or group of panel units designed for assembly in the form of a single panel; including buses, and with or without switches and/or automatic overcurrent-protective devices for the control of light, heat, or power circuits of small individual as well as aggregate capacity; designed to be placed in a cabinet or cutout box placed in or against a wall or partition and accessible only from the front.

Frame. Chassis rail and any welded addition thereto of metal thickness of 16 MSG or greater.

Low Voltage. An electromotive force rated 24 volts, nominal, or less, supplied from a transformer, converter, or battery.

Motor Home. A vehicular unit designed to provide temporary living quarters for recreational, camping, or travel use built on or permanently attached to a self-propelled motor vehicle chassis or on a chassis cab or van that is an integral part of the completed vehicle. (See Recreational Vehicle.)

Power-Supply Assembly. The conductors, including ungrounded, grounded, and equipment grounding conductors, the connectors, attachment plug caps, and all other fittings, grommets, or devices installed for the purpose of delivering energy from the source of electrical supply to the distribution panel within the recreational vehicle.

Recreational Vehicle. A vehicular-type unit primarily designed as temporary living quarters for recreational, camping, or travel use, which either has its own motive power or is mounted on or drawn by another vehicle. The basic entities are travel trailer, camping trailer, truck camper, and motor home.

Recreational Vehicle Park. A plot of land upon which two or more recreational vehicle sites are located, established, or maintained for occupancy by recreational vehicles of the general public as temporary living quarters for recreation or vacation purposes.

Recreational Vehicle Site. A plot of ground within a recreational vehicle park intended for the accommodation of either a recreational vehicle, tent, or other individual camping unit on a temporary basis.

Recreational Vehicle Site Feeder Circuit Conductors. The conductors from the park service equipment to the recreational vehicle site supply equipment.

Recreational Vehicle Site Supply Equipment. The necessary equipment, usually a power outlet, consisting of a circuit breaker or switch and fuse and their accessories, located near the point of entrance of supply conductors to a recreational vehicle site and intended to constitute the disconnecting means for the supply to that site.

Recreational Vehicle Stand. That area of a recreational vehicle site intended for the placement of a recreational vehicle.

Transformer. A device that, when used, will raise or lower the voltage of alternating current of the original source.

Travel Trailer. A vehicular unit, mounted on wheels, designed to provide temporary living quarters for recreational, camping, or travel use, of such size or weight as not to require special highway movement permits when towed by a motorized vehicle, and of gross trailer area less than 320 ft^2 (29.7 m^2). (See Recreational Vehicle.)

Truck Camper. A portable unit constructed to provide temporary living quarters for recreational, travel, or camping use, consisting of a roof, floor, and sides, designed to be loaded onto and unloaded from the bed of a pick-up truck. (See Recreational Vehicle.)

551-3. Other Articles. Wherever the requirements of other articles of this *Code* and Article 551 differ, the requirements of Article 551 shall apply.

551-4. General Requirements.

(a) Not Covered. A recreational vehicle not used for the purposes as defined in Section 551-2 shall not be required to meet the provisions of Part A pertaining to the number or capacity of circuits required. It shall, however, meet all other applicable requirements of this article if the recreational vehicle is provided with an electrical installation intended to be energized from a 120- or 120/240-volt, nominal, ac power-supply system.

(b) Systems. This article covers battery and other low-voltage power systems (24 volts or less), combination electrical systems, generator installations, and 120- or 120/240-volt, nominal, systems.

B. Low-Voltage Systems

551-10. Low-Voltage Systems.

(a) Low-Voltage Circuits. Low-voltage circuits furnished and installed by the recreational vehicle manufacturer, other than automotive vehicle circuits or extensions thereof, are subject to this *Code*. Circuits supplying lights subject to federal or state regulations shall comply with applicable government regulations and this *Code*.

Section 551-10(a) refers to the low-voltage wiring within the recreational vehicle that would be used in place of 120-volt ac supplies and does not refer to the wiring of the automotive vehicle circuits (e.g., those circuits used for ignition).

(b) Low-Voltage Wiring.

(1) Copper conductors shall be used for low-voltage circuits.

It is not the intent of Section 551-10(b)(1) to permit the sidewalls or the roof to serve as the ground return path. See the definition of *frame* in Section 551-2.

Exception: Metal chassis or frame shall be permitted as the return path to the source of supply.

(2) Conductors shall conform to the requirements for Type GXL, HDT, SGT, SGR, or Type SXL or shall have insulation in accordance with Table 310-13 or the equivalent. Conductor sizes No. 6 through No. 18 or SAE shall be listed.

FPN: See SAE Standard J1128-1995 for Types GXL, HDT, and SXL and SAE Standard J1127-1995 for Types SGT and SGR.

(3) Single-wire, low-voltage conductors shall be of the stranded type.

(4) All insulated low-voltage conductors shall be surface marked at intervals not greater than 4 ft (1.22 m) as follows.

(a) Listed conductors shall be marked as required by the listing agency.
(b) SAE conductors shall be marked with the name or logo of the manufacturer, specification designation, and wire gauge.
(c) Other conductors shall be marked with the name or logo of the manufacturer, temperature rating, wire gauge, conductor material, and insulation thickness.

(5) Conductors shall have a minimum insulation rating of 90°C (194°F) for interior installations and 125°C (257°F) for all engine compartment wiring or any under-chassis installations where conductors are located less than 18 in. (457.2 mm) from any component of an internal combustion engine exhaust system.

Section 551-10(b)(5) addresses the location of electrical equipment that is adjacent to hot components of engines and exhaust systems. Its requirements correlate with ANSI/RVIA 12V, *Standard for Low Voltage Systems in Conversion Vehicles*.

(c) Low-Voltage Wiring Methods.

(1) Conductors shall be protected against physical damage and shall be secured. Where insulated conductors are clamped to the structure, the conductor insulation shall be supplemented by an additional wrap or layer of equivalent material, except that jacketed cables shall not be required to be so protected. Wiring shall be routed away from sharp edges, moving parts, or heat sources.

(2) Conductors shall be spliced or joined with splicing devices that provide a secure connection or by brazing, welding, or soldering with a fusible metal or alloy. Soldered splices shall first be spliced or joined so as to be mechanically and electrically secure without solder and then soldered. All splices, joints, and free ends of conductors shall be covered with an insulation equivalent to that on the conductors.

(3) Battery and dc circuits shall be physically separated by at least a ½-in. (12.7-mm) gap or other approved means from circuits of a different power source. Acceptable methods shall be by clamping, routing, or equivalent means that ensure permanent total separation. Where circuits of different power sources cross, the external jacket of the nonmetallic-sheathed cables shall be deemed adequate separation.

(4) Ground connections to the chassis or frame shall be made in an accessible location and shall be mechanically secure. Ground connections shall be by means of copper conductors and copper or copper-alloy terminals of the solderless type identified for the size of wire used. The surface on which ground terminals make contact shall be cleaned and be free from oxide or paint or shall be electrically connected through the use of a cadmium, tin, or zinc-plated internal/external-toothed lockwasher or locking terminals. Ground terminal attaching screws, rivets or bolts, nuts, and lockwashers shall be cadmium, tin, or zinc-plated except rivets shall be permitted to be unanodized aluminum where attaching to aluminum structures.

(5) The chassis-grounding terminal of the battery shall be bonded to the vehicle chassis with a minimum No. 8 copper conductor. In the event the power lead from the battery exceeds No. 8, then the bonding conductor shall be of an equal size.

Sections 551-10(c)(4) and (5) require that the chassis-grounding terminal of the battery be bonded to the vehicle chassis in a mechanically secure manner and

be placed in an accessible location using a minimum No. 8 copper conductor. The ac equipment grounding conductor of the appliance may not have sufficient ampacity to safely conduct the dc fault current. This will necessitate installation of the battery bonding conductor. Some recreational vehicles already have one side of the battery circuit bonded to the frame by a No. 8 or larger copper conductor.

(d) Battery Installations. Storage batteries subject to the provisions of this *Code* shall be securely attached to the vehicle and installed in an area vaportight to the interior and ventilated directly to the exterior of the vehicle. Where batteries are installed in a compartment, the compartment shall be ventilated with openings having a minimum area of 1.7 in.^{2} (1100 mm^2) at both the top and at the bottom. Where compartment doors are equipped for ventilation, the openings shall be within 2 in. (50.8 mm) of the top and bottom. Batteries shall not be installed in a compartment containing spark- or flame-producing equipment, except that they shall be permitted to be installed in the engine generator compartment if the only charging source is from the engine generator.

(e) Overcurrent Protection.

(1) Low-voltage circuit wiring shall be protected by overcurrent protective devices rated not in excess of the ampacity of copper conductors, as follows.

Table 551-10(e)(1). Low-Voltage Overcurrent Protection

Wire Size (AWG)	Ampacity	Wire Type
18	6	Stranded only
16	8	Stranded only
14	15	Stranded or solid
12	20	Stranded or solid
10	30	Stranded or solid

(2) Circuit breakers or fuses shall be of an approved type, including automotive types. Fuseholders shall be clearly marked with maximum fuse size, and both circuit breakers and fuses shall be protected against shorting and physical damage by a cover or equivalent means.

The requirement for protection of fuseholders by a cover or equivalent means is intended to reduce the possibility of the low-voltage system shorting to ground.

FPN: For further information, see *Standard for Electric Fuses (Cartridge Type),* ANSI/SAE J554-1987; *Standard for Blade Type Electric Fuses,* SAE J1284-1988; and *Standard for Automotive Glass Tube Fuses,* UL 275-1993.

(3) Higher current-consuming, dc appliances, such as pumps, compressors, heater blowers, and similar motor-driven appliances, shall be installed in accordance with the manufacturer's instructions.

Motors that are controlled by automatic switching or by latching-type manual switches shall be protected in accordance with Section 430-32(c).

(4) The overcurrent protective device shall be installed in an accessible location on the vehicle within 18 in. (457 mm) of the point where the power supply connects to the vehicle circuits. If located outside the recreational vehicle, the device shall be protected against weather and physical damage.

Exception: External low-voltage supply shall be permitted to have the overcurrent protective device within 18 in. (457 mm) after entering the vehicle or after leaving a metal raceway.

(f) Switches. Switches shall have a dc rating not less than the connected load.

(g) Lighting Fixtures. All low-voltage interior lighting fixtures rated more than 4 watts, employing lamps rated more than 1.2 watts, shall be listed.

(h) Cigarette Lighter Receptacles. Twelve-volt receptacles that will accept and energize cigarette lighters shall be installed in a noncombustible outlet box, or the assembly shall be identified by the manufacturer of the product as thermally protected.

Twelve-volt systems for running and signal lights, similar to those used in a conventional automobile, are covered in Sections 551-10, 551-20, and 551-30. In many recreational vehicles, 12-volt systems are also used for interior lighting and other small loads. The 12-volt system is often supplied from an on-board battery or through a transfer switch from a 120/12-volt transformer in conjunction with a full-wave rectifier.

C. Combination Electrical Systems

551-20. Combination Electrical Systems.

(a) General. Vehicle wiring suitable for connection to a battery or dc supply source shall be permitted to be connected to a 120-volt source, provided the entire wiring system and equipment are rated and installed in full conformity with Parts A, C, D, E, and F requirements of this article covering 120-volt electrical systems. Circuits fed from ac transformers shall not supply dc appliances.

(b) Voltage Converters (120-Volt Alternating Current to Low-Voltage Direct Current). The 120-volt ac side of the voltage converter shall be wired in full conformity with Parts A, C, D, E, and F requirements of this article for 120-volt electrical systems.

Exception: Converters supplied as an integral part of a listed appliance shall not be subject to the above.

All converters and transformers shall be listed for use in recreation vehicles and designed or equipped to provide over-temperature protection. To determine the converter rating, the following formula shall be applied to the total connected load, including average battery charging rate, of all 12-volt equipment:

The first 20 amperes of load at 100 percent, plus

The second 20 amperes of load at 50 percent, plus

All load above 40 amperes at 25 percent

Exception: A low-voltage appliance that is controlled by a momentary switch (normally open) that has no means for holding in the closed position shall not be considered as a connected load when determining the required converter rating. Momentarily energized appliances shall be limited to those used to prepare the vehicle for occupancy or travel.

(c) Bonding Voltage Converter Enclosures. The non-current-carrying metal enclosure of the voltage converter shall be bonded to the frame of the vehicle with a minimum No. 8 copper conductor. The grounding conductor for the battery and the metal enclosure shall be permitted to be the same conductor.

The intent of Section 551-20(c) is to reduce the possibility of damage to the power supply cord by large dc fault currents that may find their way back to the vehicle frame or battery through the ac grounding conductor of the converter. Metal enclosures of listed converters are provided with an external pressure terminal connector for this purpose. See the commentary following Section 551-10(c)(5).

(d) Dual-Voltage Fixtures or Appliances. Fixtures or appliances having both 120-volt and low-voltage connections shall be listed for dual voltage.

In such dual-voltage fixtures, barriers are used to separate the 120-volt and the 12-volt wiring connections.

(e) Autotransformers. Autotransformers shall not be used.

(f) Receptacles and Plug Caps. Where a recreational vehicle is equipped with a 120-volt or 120/240-volt ac system, a low-voltage system, or both, receptacles and plug caps of the low-voltage system shall differ in configuration from those of the 120- or 120/240-volt system. Where a vehicle equipped with a battery or other low-voltage system has an external connection for low-voltage power, the connector shall have a configuration that will not accept 120-volt power.

D. Other Power Sources

551-30. Generator Installations.

(a) Mounting. Generators shall be mounted in such a manner as to be effectively bonded to the recreational vehicle chassis.

(b) Generator Protection. Equipment shall be installed to ensure that the current-carrying conductors from the engine generator and from an outside source are not connected to a vehicle circuit at the same time.

Receptacles used as disconnecting means shall be accessible (as applied to wiring methods) and capable of interrupting their rated current without hazard to the operator.

(c) Installation of Storage Batteries and Generators. Storage batteries and internal-combustion-driven generator units (subject to the provisions of this *Code*) shall be secured in place to avoid displacement from vibration and road shock.

(d) Ventilation of Generator Compartments. Compartments accommodating internal-combustion-driven generator units shall be provided with ventilation in accordance with instructions provided by the manufacturer of the generator unit.

FPN: For generator compartment construction requirements, see *Standard on Recreational Vehicles*, NFPA 501C-1996.

(e) Supply Conductors. The supply conductors from the engine generator to the first termination on the vehicle shall be of the stranded type and be installed in listed flexible conduit or listed liquidtight flexible conduit. The point of first termination shall be in a

(1) Panelboard,
(2) Junction box with a blank cover,
(3) Junction box with a receptacle,
(4) Enclosed transfer switch, or
(5) Receptacle assembly listed in conjunction with the generator.

The panelboard or junction box with a receptacle shall be installed within the vehicle's interior and within 18 in. (457 mm) of the compartment wall but not inside the compartment. If the generator is below the floor level and not in a compartment, the panelboard or junction box with receptacle shall be installed within the vehicle interior within 18 in. (457 mm) of the point of entry into the vehicle. A junction box with a blank cover shall be mounted on the compartment wall and shall be permitted inside or outside the compartment. A receptacle assembly listed in conjunction with the generator shall be mounted in accordance with its listing. If the generator is below floor level and not in a compartment, the junction box with blank cover shall be mounted either to any part of the generator supporting structure (but not to the generator) or to the vehicle floor within 18 in. (457 mm) of any point directly above the generator on either the inside or outside of the floor surface. Overcurrent protection in accordance with Section 240-3 shall be provided for supply conductors as an integral part of a listed generator or shall be located within 18 in. (457 mm) of their point of entry into the vehicle.

551-31. Multiple Supply Source.

(a) Multiple Supply Sources. Where a multiple supply system consisting of an alternate power source and a power-supply cord is installed, the feeder from the alternate power source shall be protected by an overcurrent-protective device. Installation shall be in accordance with Sections 551-30(a) and (b) and 551-40.

(b) Calculation of Loads. Calculation of loads shall be in accordance with Section 551-42.

(c) Multiple Supply Sources Capacity. The multiple supply sources shall not be required to be of the same capacity.

(d) Alternate Power Sources Exceeding 30 Amperes. If an alternate power source exceeds 30 amperes, 120 volts, nominal, it shall be permissible to wire it as a 120-volt, nominal, system or a 120/240-volt, nominal, system, provided an overcurrent-protective device of the proper rating is installed in the feeder.

(e) Power-Supply Assembly Not Less than 30 Amperes. The external power-supply assembly shall be permitted to be less than the calculated load but not less than 30 amperes and shall have overcurrent protection not greater than the capacity of the external power-supply assembly.

551-32. Other Sources. Other sources of ac power, such as inverters or motor generators, shall be listed for use in recreational vehicles and shall be installed in accordance with the terms of the listing. Other sources of ac power shall be wired in full conformity with the requirements in Parts A, C, D, E, and F of this article covering 120-volt electrical systems.

551-33. Alternate Source Restriction. Transfer equipment, if not integral with the listed power source, shall be installed to ensure that the current-carrying conductors from other sources of ac power and from an outside source are not connected to the vehicle circuit at the same time.

E. Nominal 120-Volt or 120/240-Volt Systems

551-40. 120-Volt or 120/240-Volt, Nominal, Systems.

(a) General Requirements. The electrical equipment and material of recreational vehicles indicated for connection to a wiring system rated 120 volts, nominal, 2-wire with ground, or a wiring system rated 120/240 volts, nominal, 3-wire with ground, shall be listed and installed in accordance with the requirements of Parts A, C, D, E, and F of this article.

(b) Materials and Equipment. Electrical materials, devices, appliances, fittings, and other equipment installed, intended for use in, or attached to the recreational vehicle shall be listed. All products shall be used only in the manner in which they have been tested and found suitable for the intended use.

(c) Ground-Fault Circuit-Interrupter Protection. The internal wiring of a recreational vehicle having only one 15- or 20-ampere branch circuit as permitted in Sections 551-42(a) and (b) shall have ground-fault circuit-interrupter protection for personnel. The ground-fault circuit interrupter shall be installed at the point where the power supply assembly terminates within the recreational vehicle. Where a separable cord set is not employed, the ground-fault circuit interrupter shall be permitted to be an integral part of the attachment plug of the power supply assembly. The ground-fault circuit interrupter shall provide protection also under the conditions of an open grounded circuit conductor, interchanged circuit conductors, or both.

551-41. Receptacle Outlets Required.

(a) Spacing. Receptacle outlets shall be installed at wall spaces 2 ft (610 mm) wide or more so that no point along the floor line is more than 6 ft (1.83 m), measured horizontally, from an outlet in that space.

Exception No. 1: Bath and hall areas.

Exception No. 2: Wall spaces occupied by kitchen cabinets, wardrobe cabinets, built-in furniture, behind doors that may open fully against a wall surface, or similar facilities.

(b) Location. Receptacle outlets shall be installed as follows:

(1) Adjacent to countertops in the kitchen [at least one on each side of the sink if countertops are on each side and are 12 in. (305 mm) or over in width]
(2) Adjacent to the refrigerator and gas range space, except where a gas-fired refrigerator or cooking appliance, requiring no external electrical connection, is factory installed
(3) Adjacent to countertop spaces of 12 in. (305 mm) or more in width that cannot be reached from a receptacle required in Section 551-41(b)(1) by a cord of 6 ft (1.83 m) without crossing a traffic area, cooking appliance, or sink

(c) Ground-Fault Circuit-Interrupter Protection. Where provided, each 125-volt, single-phase, 15- or 20-ampere receptacle outlet shall have ground-fault circuit-interrupter protection for personnel in the following locations:

(1) Adjacent to a bathroom lavatory

Section 551-41(c)(1) permits mounting of a bathroom receptacle in the side of a lavatory cabinet where installation of a receptacle is not possible in a thin wall.

(2) Where the receptacles are installed to serve the countertop surfaces, and are within 6 ft (1.83 m) of any lavatory or sink

Exception No. 1: Receptacles installed for appliances in dedicated spaces, such as for dishwashers, disposals, refrigerators, freezers, and laundry equipment.

Exception No. 2: Single receptacles for interior connections of expandable room sections.

Exception No. 3: De-energized receptacles that are within 6 ft (18.3 m) of any sink or lavatory due to the retraction of the expandable room section.

(3) In the area occupied by a toilet, shower, tub, or any combination thereof
(4) On the exterior of the vehicle

Exception: Receptacles that are located inside of an access panel that is installed on the exterior of the vehicle to supply power for an installed appliance shall not be required to have ground-fault circuit-interrupter protection.

The receptacle outlet shall be permitted in a listed lighting fixture. A receptacle outlet shall not be installed in a tub or combination tub–shower compartment.

Clearly, receptacles of any type are not permitted in a tub or tub–shower compartment.

(d) Face-Up Position. A receptacle shall not be installed in a face-up position in any countertop or similar horizontal surfaces within the living area.

551-42. Branch Circuits Required. Each recreational vehicle containing a 120-volt electrical system shall contain one of the following.

(a) One 15-Ampere Circuit. One 15-ampere circuit to supply lights, receptacle outlets, and fixed appliances. Such recreational vehicles shall be equipped with one 15-ampere switch and fuse or one 15-ampere circuit breaker.

(b) One 20-Ampere Circuit. One 20-ampere circuit to supply lights, receptacle outlets, and fixed appliances. Such recreational vehicles shall be equipped with one 20-ampere switch and fuse or one 20-ampere circuit breaker.

(c) Two to Five 15- or 20-Ampere Circuits. A maximum of five 15- or 20-ampere circuits to supply lights, receptacle outlets, and fixed appliances shall be permitted. Such recreational vehicles shall be equipped with a distribution panelboard rated at 120 volts maximum with a 30-ampere rated main power supply assembly. Not more than two 120-volt thermostatically controlled appliances (i.e., air conditioner and water heater) shall be installed in such systems unless appliance isolation switching, energy management systems, or similar methods are used.

Exception: Additional 15- or 20-ampere circuits shall be permitted where a listed energy management system rated at 30-ampere maximum is employed within the system.

FPN: See Section 210-23(a) for permissible loads. See Section 551-45(c) for main disconnect and overcurrent protection requirements.

(d) More than Five Circuits Without a Listed Energy Management System. A 50-ampere, 120/240-volt power-supply assembly shall be used where six or more circuits are employed. The load distribution shall ensure a reasonable current balance between phases.

The calculations for supplies were removed in the 1996 *Code* because experience and field data indicated that calculations were not necessary and, in some cases, mandated oversized power supplies. However, the 50-ampere supply must have six or more circuits, reasonably balanced between phases.

551-43. Branch-Circuit Protection.

(a) Rating. The branch-circuit overcurrent devices shall be rated as follows:

(1) Not more than the circuit conductors, and
(2) Not more than 150 percent of the rating of a single appliance rated 13.3 amperes or more and supplied by an individual branch circuit, but
(3) Not more than the overcurrent protection size marked on an air conditioner or other motor-operated appliances

(b) Protection for Smaller Conductors. A 20-ampere fuse or circuit breaker shall be permitted for protection for fixture leads, cords, or small appliances, and No. 14 tap conductors, not over 6 ft (1.83 m) long for recessed lighting fixtures.

(c) Fifteen-Ampere Receptacle Considered Protected by 20 Amperes. If more than one receptacle or load is on a branch circuit, a 15-ampere receptacle shall be permitted to be protected by a 20-ampere fuse or circuit breaker.

551-44. Power-Supply Assembly.

(a) Fifteen-Ampere Main Power-Supply Assembly. Recreational vehicles wired in accordance with Section 551-42(a) shall use a listed 15-ampere, or larger, main power-supply assembly.

(b) Twenty-Ampere Main Power-Supply Assembly. Recreational vehicles wired in accordance with Section 551-42(b) shall use a listed 20-ampere, or larger, main power-supply assembly.

(c) Thirty-Ampere Main Power-Supply Assembly. Recreational vehicles wired in accordance with Section 551-42(c) shall use a listed 30-ampere, or larger, main power-supply assembly.

(d) Fifty-Ampere Power-Supply Assembly. Recreational vehicles wired in accordance with Section 551-42(d) shall use a listed 50-ampere, 120/240-volt main power-supply assembly.

551-45. Distribution Panelboard.

(a) Listed and Appropriately Rated. A listed and appropriately rated distribution panelboard or other equipment specifically listed for the purpose shall be used. The grounded conductor termination bar shall be insulated from the enclosure as provided in Section 551-54(c). An equipment grounding terminal bar shall be attached inside the metal enclosure of the panelboard.

(b) Location. The distribution panelboard shall be installed in a readily accessible location. Working clearance for the panelboard shall be not less than 24 in. (610 mm) wide and 30 in. (762 mm) deep.

Exception No. 1: Where the panelboard cover is exposed to the inside aisle space, then one of the working clearance dimensions shall be permitted to be reduced to a minimum of 22 in. (559 mm). A panelboard is considered exposed where the panelboard cover is within 2 in. (50.8 mm) of the aisle's finished surface.

Exception No. 2: Compartment doors used for access to a generator shall be permitted to be equipped with a locking system.

(c) Dead-Front Type. The distribution panelboard shall be of the dead-front type and shall consist of one or more circuit breakers or Type S fuseholders. A main disconnecting means shall be provided where fuses are used or where more than two circuit breakers are employed. A main overcurrent protective device not exceeding the power-supply assembly rating shall be provided where more than two branch circuits are employed.

551-46. Means for Connecting to Power Supply.

(a) Assembly. The power-supply assembly or assemblies shall be factory supplied or factory installed and be of one of the types specified herein.

(1) Separable. Where a separable power-supply assembly consisting of a cord with a female connector and molded attachment plug cap is provided, the vehicle shall be equipped with a permanently mounted, flanged surface inlet (male, recessed-type motor-base attachment plug) wired directly to the distribution panelboard by an approved wiring method. The attachment plug cap shall be of a listed type.

(2) Permanently Connected. Each power-supply assembly shall be connected directly to the terminals of the distribution panelboard or conductors within a junction box and provided with means to prevent strain from being transmitted to the terminals. The ampacity of the conductors between each junction box and the terminals of each distribution panelboard shall be at least equal to the ampacity of the power-supply cord. The supply end of the assembly shall be equipped with an attachment plug of the type described in Section 551-46(c). Where the cord passes through the walls or floors, it shall be protected by means of conduit and bushings or equivalent. The cord assembly shall have permanent provisions for protection against corrosion and mechanical damage while the vehicle is in transit.

(b) Cord. The cord exposed usable length shall be measured from the point of entrance to the recreational vehicle or the face of the flanged surface inlet (motor-base attachment plug) to the face of the attachment plug at the supply end.

The cord exposed usable length, measured to the point of entry on the vehicle exterior, shall be a minimum of 23 ft (7.0 m) where the point of entrance is at the side of the vehicle, or shall be a minimum 28 ft (8.5 m) where the point of entrance is at the rear of the vehicle.

Where the cord entrance into the vehicle is more than 3 ft (0.9 m) above the ground, the minimum cord lengths above shall be increased by the vertical distance of the cord entrance heights above 3 ft (0.9 m).

FPN: See Section 551-46(e).

(c) Attachment Plugs.

(1) Recreational vehicles having only one 15-ampere branch circuit as permitted by Section 551-42(a) shall have an attachment plug that shall be 2-pole, 3-wire, grounding type, rated 15 amperes, 125 volts, conforming to the configuration shown in Figure 551-46(c).

FPN: Complete details of this configuration can be found in National Electrical Manufacturers Association's *Standard for Dimensions of Attachment Plugs and Receptacle,* ANSI/NEMA WD 6-1989, Figure 5-15.

(2) Recreational vehicles having only one 20-ampere branch circuit as permitted in Section 551-42(b) shall have an attachment plug that shall be 2-pole, 3-wire, grounding type, rated 20 amperes, 125 volts, conforming to the configuration shown in Figure 551-46(c).

FPN: Complete details of this configuration can be found in National Electrical Manufacturers Association's *Standard for Dimensions of Attachment Plugs and Receptacles*, ANSI/NEMA WD 6-1989, Figure 5-20.

(3) Recreational vehicles wired in accordance with Section 551-42(c) shall have an attachment plug that shall be 2-pole, 3-wire, grounding type, rated 30 amperes, 125 volts, conforming to the configuration shown in Figure 551-46(c) intended for use with units rated at 30 amperes, 125 volts.

FPN: Complete details of this configuration can be found in National Electrical Manufacturers Association's *Standard for Dimensions of Attachment Plugs and Receptacles*, ANSI/NEMA WD 6-1989, Figure TT.

(4) Recreational vehicles having a power-supply assembly rated 50 amperes as permitted by Section 551-42(d) shall have a 3-pole, 4-wire, grounding-type attachment plug rated 50 amperes, 125/250 volts, conforming to the configuration shown in Figure 551-46(c).

The 20- and 50-ampere receptacle and plug configurations used for recreational vehicles appear in Figure 210.6 or 210.7 (NEMA configuration chart); however, the 30-ampere plug and receptacle are special for recreational vehicles and are described in UL 498, *UL Standard for Safety, Attachment Plugs and Receptacles.*

FPN: Complete details of this configuration can be found in National Electrical Manufacturers Association's *Standard for Dimensions of Attachment Plugs and Receptacles*, ANSI/NEMA WD 6-1989, Figure 14-50.

Figure 551-46(c) Configurations for grounding-type receptacles and attachment plug caps used for recreational vehicle supply cords and recreational vehicle lots.

(d) Labeling at Electrical Entrance. Each recreational vehicle shall have permanently affixed to the exterior skin, at or near the point of entrance of the power-supply cord(s), a label 3 in. (76 mm) × 1¾ in. (44.5 mm) minimum size, made of etched, metal-stamped, or embossed brass, stainless steel, or anodized or alclad aluminum not less than 0.020 in. (508 μm) thick, or other suitable material [e.g., 0.005-in. (127-μm) thick plastic laminate] that reads, as appropriate, either

THIS CONNECTION IS FOR 110–125-VOLT AC,
60 HZ, _____ AMPERE SUPPLY.

or

THIS CONNECTION IS FOR 120/240-VOLT AC,
3-POLE, 4-WIRE, 60 HZ, _____ AMPERE SUPPLY.

The correct ampere rating shall be marked in the blank space.

(e) Location. The point of entrance of a power-supply assembly shall be located within 15 ft (4.57 m) of the rear, on the left (road) side or at the rear, left of the longitudinal center of the vehicle, within 18 in. (457 mm) of the outside wall.

Exception No. 1: A recreational vehicle equipped with only a listed flexible drain system or a side-vent drain system shall be permitted to have the electrical point of entrance located on either side, provided the drain(s) for the plumbing system is (are) located on the same side.

Exception No. 2: A recreational vehicle shall be permitted to have the electrical point of entrance located more than 15 ft (4.57 m) from the rear. Where this occurs, the distance beyond the 15-ft (4.57-m) dimension shall be added to the cord's minimum length as specified in Section 551-46(b).

551-47. Wiring Methods.

(a) Wiring Systems. Cables and raceways installed in accordance with Articles 330 through 352 shall be permitted in accordance with their applicable article, except as otherwise specified in this article. An equipment grounding means shall be provided in accordance with Section 250-118.

See the commentary following Section 350-5.

(b) Conduit and Tubing. Where rigid metal conduit or intermediate metal conduit is terminated at an enclosure with a locknut and bushing connection, two locknuts shall be provided, one inside and one outside of the enclosure. All cut ends of conduit and tubing shall be reamed or otherwise finished to remove rough edges.

See the commentary following Sections 348-11 and 300-4(f).

(c) Nonmetallic Boxes. Nonmetallic boxes shall be acceptable only with nonmetallic-sheathed cable or nonmetallic raceways.

(d) Boxes. In walls and ceilings constructed of wood or other combustible material, boxes and fittings shall be flush with the finished surface or project therefrom.

(e) Mounting. Wall and ceiling boxes shall be mounted in accordance with Article 370.

Exception No. 1: Snap-in-type boxes or boxes provided with special wall or ceiling brackets that securely fasten boxes in walls or ceilings shall be permitted.

Exception No. 2: A wooden plate providing a 1½-in. (38-mm) minimum width backing around the box and of a thickness of ½ in. (12.7 mm) or greater (actual) attached directly to the wall panel shall be considered as approved means for mounting outlet boxes.

Exception No. 2 permits mounting of outlet boxes by screws to a wooden plate that is secured directly to the back of the wall panel. This wooden plate is required to be not less than ½ in. thick and is required to extend at least 1½ in. around the box. This recognizes the special construction of recreational vehicle walls, which often make it quite difficult or impossible to attach an outlet box to a structural member, as required by Section 370-23(b).

(f) Sheath Armor. The sheath of nonmetallic-sheathed cable, metal-clad cable, and Type AC cable shall be continuous between outlet boxes and other enclosures.

(g) Protected. Metal-clad, Type AC, or nonmetallic-sheathed cables and electrical nonmetallic tubing shall be permitted to pass through the centers of the wide side of 2 in. by 4 in. wood studs. However, they shall be protected where they pass through 2 in. by 2 in. wood studs or at other wood studs or frames where the cable or tubing would be less than 1¼ in. (31.8 mm) from the inside or outside surface. Steel plates on each side of the cable or tubing, or a steel tube, with not less than No. 16 MSG wall thickness, shall be installed to protect the cable or tubing. These plates or tubes shall be securely held in place. Where nonmetallic-sheathed cables pass through punched, cut, or drilled slots or holes in metal members, the cable shall be protected by bushings or grommets securely fastened in the opening prior to installation of the cable.

(h) Bends. No bend shall have a radius of less than five times the cable diameter.

(i) Cable Supports. Where connected with cable connectors or clamps, cables shall be supported within 12 in. (305 mm) of outlet boxes, distribution panelboards, and splice boxes on appliances. Supports shall be provided every 4½ ft (1.37 m) at other places.

(j) Nonmetallic Box Without Cable Clamps. Nonmetallic-sheathed cables shall be supported within 8 in. (203 mm) of a nonmetallic outlet box without cable clamps. Where wiring devices with integral enclosures are employed with a loop of extra cable to permit future replacement of the device, the cable loop shall be considered as an integral portion of the device.

(k) Physical Damage. Where subject to physical damage, exposed nonmetallic cable shall be protected by covering boards, guard strips, raceways, or other means.

(l) Metal Faceplates. Metal faceplates shall be of ferrous metal not less than 0.030 in. (762 μm) in thickness or of nonferrous metal not less than 0.040 in. (1.016 mm) in thickness. Nonmetallic faceplates shall be listed.

(m) Metal Faceplates Effectively Grounded. Where metal faceplates are used, they shall be effectively grounded.

(n) Moisture or Physical Damage. Where outdoor or under-chassis wiring is 120 volts, nominal, or over and is exposed to moisture or physical damage, the wiring shall be protected by rigid metal conduit, intermediate metal conduit, or by electrical metallic tubing or rigid nonmetallic conduit that is closely routed against frames and equipment enclosures or other raceway or cable identified for the application.

(o) Component Interconnections. Fittings and connectors that are intended to be concealed at the time of assembly shall be listed and identified for the interconnection of building components. Such fittings and connectors shall be equal to the wiring method employed in insulation, temperature rise, and fault-current withstanding, and shall be capable of enduring the vibration and shock occurring in recreational vehicles.

(p) Method of Connecting Expandable Units.

(1) That portion of a branch circuit that is installed in an expandable unit shall be permitted to be connected to the portion of the branch circuit in the main body of the vehicle by means of a flexible cord or attachment plug and cord listed for hard usage. The cord and its connections shall conform to all provisions of Article 400 and shall be considered as a permitted use under Section 400-7. Where the attachment plug and cord are located within the vehicle's interior, use of plastic thermoset or elastomer parallel cord Type SPT-3, SP-3, or SPE shall be permitted.

(2) If the receptacle provided for connection of the cord to the main circuit is located on the outside of the vehicle, it shall be protected with a ground-fault circuit interrupter for personnel and be listed for wet locations. A cord located on the outside of a vehicle shall be identified for outdoor use.

(3) Unless removable or stored within the vehicle interior, the cord assembly shall have permanent provisions for protection against corrosion and mechanical damage while the vehicle is in transit.

(4) If an attachment plug and cord is used, it shall be installed so as not to permit exposed live attachment plug pins.

(q) Prewiring for Air-Conditioning Installation. Prewiring installed for the purpose of facilitating future air-conditioning installation shall conform to the following and other applicable portions of this article. The circuit shall serve no other purpose.

(1) An overcurrent protective device with a rating compatible with the circuit conductors shall be installed in the distribution panelboard and wiring connections completed.

(2) The load end of the circuit shall terminate in a junction box with a blank cover or a device listed for the purpose. Where a junction box with a blank cover is used, the free ends of the conductors shall be adequately capped or taped.

(3) A label conforming to Section 551-46(d) shall be placed on or adjacent to the junction box and shall read

AIR-CONDITIONING CIRCUIT. THIS
CONNECTION IS FOR AIR CONDITIONERS
RATED 110–125-VOLT AC, 60 HZ, ______
AMPERES MAXIMUM. DO NOT EXCEED
CIRCUIT RATING.

An ampere rating, not to exceed 80 percent of the circuit rating, shall be legibly marked in the blank space.

(r) Prewiring for Generator Installation. Prewiring installed for the purpose of facilitating future generator installation shall conform to the following and other applicable portions of this article.

(1) Circuit conductors shall be appropriately sized in relation to the anticipated load and shall be protected by an overcurrent device in accordance with their ampacities.

(2) Where junction boxes are utilized at either of the circuit originating or terminus points, free ends of the conductors shall be adequately capped or taped.

(3) Where devices such as receptacle outlet, transfer switch, etc., are installed, the installation shall be complete, including circuit conductor connections. All devices shall be listed and appropriately rated.

(4) A label conforming to Section 551-46(d) shall be placed on the cover of each junction box containing incomplete circuitry and shall read, as appropriate, either

GENERATOR CIRCUIT. THIS CONNECTION IS FOR GENERATORS RATED 110–125-VOLT AC, 60 HZ, _____ AMPERES MAXIMUM.

or

GENERATOR CIRCUIT. THIS CONNECTION IS FOR GENERATORS RATED 120/240-VOLT AC, 60 HZ, _____ AMPERES MAXIMUM.

The correct ampere rating shall be legibly marked in the blank space.

551-48. Conductors and Boxes. The maximum number of conductors permitted in boxes shall be in accordance with Section 370-16.

551-49. Grounded Conductors. The identification of grounded conductors shall be in accordance with Section 200-6.

551-50. Connection of Terminals and Splices. Conductor splices and connections at terminals shall be in accordance with Section 110-14.

551-51. Switches. Switches shall be rated as follows.

(a) Lighting Circuits. For lighting circuits, switches shall be rated not less than 10 amperes, 120–125 volts and in no case less than the connected load.

(b) Motors or Other Loads. For motors or other loads, switches shall have ampere or horsepower ratings, or both, adequate for loads controlled. (An ac general-use snap switch shall be permitted to control a motor 2 hp or less with full-load current not over 80 percent of the switch ampere rating.)

551-52. Receptacles. All receptacle outlets shall be of the grounding type and installed in accordance with Sections 210-7 and 210-21.

551-53. Lighting Fixtures.

(a) General. Any combustible wall or ceiling finish exposed between the edge of a fixture canopy, or pan and the outlet box, shall be covered with noncombustible material or a material identified for the purpose.

(b) Shower Fixtures. If a lighting fixture is provided over a bathtub or in a shower stall, it shall be of the enclosed and gasketed type and listed for the type of installation, and it shall be ground-fault circuit-interrupter protected.

The switch for shower lighting fixtures and exhaust fans, located over a tub or in a shower stall, shall be located outside the tub or shower space.

Many shower lights have a metal base and, due to low ceilings in recreational vehicles, may be easily reached by most persons under a shower or standing in a bathtub. Section 551-53(b) permits lighting fixtures listed for a wet location to be installed above a tub or shower enclosure only if shock-hazard protection is provided by a ground-fault circuit interrupter.

(c) Outdoor Outlets, Fixtures, Air-Cooling Equipment, etc. Outdoor fixtures and other equipment shall be listed for outdoor use.

551-54. Grounding. (See also Section 551-56 on bonding of noncurrent-carrying metal parts.)

(a) Power-Supply Grounding. The grounding conductor in the supply cord or feeder shall be connected to the grounding bus or other approved grounding means in the distribution panelboard.

(b) Distribution Panelboard. The distribution panelboard shall have a grounding bus with sufficient terminals for all grounding conductors or other approved grounding means.

(c) Insulated Neutral.

(1) The grounded circuit conductor (neutral) shall be insulated from the equipment grounding conductors and from equipment enclosures and other grounded parts. The grounded (neutral) circuit terminals in the distribution panelboard and in ranges, clothes dryers, counter-mounted cooking units, and wall-mounted ovens shall be insulated from the equipment enclosure. Bonding screws, straps, or buses in the distribution panelboard or in appliances shall be removed and discarded.

(2) Connection of electric ranges and electric clothes dryers utilizing a grounded (neutral) conductor, if cord-connected, shall be made with 4-conductor cord and 3-pole, 4-wire, grounding-type plug caps and receptacles.

551-55. Interior Equipment Grounding.

(a) Exposed Metal Parts. In the electrical system, all exposed metal parts, enclosures, frames, lighting fixture canopies, etc., shall be effectively bonded to the grounding terminals or enclosure of the distribution panelboard.

(b) Equipment Grounding Conductors. Bare wires, green-colored wires, or green wires with a yellow stripe(s) shall be used for equipment grounding conductors only.

(c) Grounding of Electrical Equipment. Where grounding of electrical equipment is specified, it shall be permitted as follows.

(1) Connection of metal raceway (conduit or electrical metallic tubing), the sheath of Type MC and Type MI cable where the sheath is identified for grounding, or the armor of Type AC cable to metal enclosures.
(2) A connection between the one or more equipment grounding conductors and a metal box by means of a grounding screw, which shall be used for no other purpose, or a listed grounding device.
(3) The equipment grounding conductor in nonmetallic-sheathed cable shall be permitted to be secured under a screw threaded into the fixture canopy other than a mounting screw or cover screw, or attached to a listed grounding means (plate) in a nonmetallic outlet box for fixture mounting. (Grounding means shall also be permitted for fixture attachment screws.)

(d) Grounding Connection in Nonmetallic Box. A connection between the one or more grounding conductors brought into a nonmetallic outlet box shall be so arranged that a connection can be made to any fitting or device in that box that requires grounding.

(e) Grounding Continuity. Where more than one equipment grounding conductor of a branch circuit enters a box, all such conductors shall be in good electrical contact with each other, and the arrangement shall be such that the disconnection or removal of a receptacle, fixture, or other device fed from the box will not interfere with or interrupt the grounding continuity.

(f) Cord-Connected Appliances. Cord-connected appliances, such as washing machines, clothes dryers, refrigerators, and the electrical system of gas ranges, etc., shall be grounded by means of an approved cord with equipment grounding conductor and grounding-type attachment plug.

551-56. Bonding of Noncurrent-Carrying Metal Parts.

(a) Required Bonding. All exposed noncurrent-carrying metal parts that may become energized shall be effectively bonded to the grounding terminal or enclosure of the distribution panelboard.

(b) Bonding Chassis. A bonding conductor shall be connected between any distribution panelboard and an accessible terminal on the chassis. Aluminum or copper-clad aluminum conductors shall not be used for bonding if such conductors or their terminals are exposed to corrosive elements.

Exception: Any recreational vehicle that employs a unitized metal chassis-frame construction to which the distribution panelboard is securely fastened with a bolt(s) and nut(s) or by welding or riveting shall be considered to be bonded.

(c) Bonding Conductor Requirements. Grounding terminals shall be of the solderless type and listed as pressure terminal connectors recognized for the wire size used. The bonding conductor shall be solid or stranded, insulated or bare, and shall be No. 8 copper minimum, or equal.

(d) Metallic Roof and Exterior Bonding. The metal roof and exterior covering shall be considered bonded where

(1) The metal panels overlap one another and are securely attached to the wood or metal frame parts by metal fasteners, and
(2) The lower panel of the metal exterior covering is secured by metal fasteners at each cross member of the chassis, or the lower panel is bonded to the chassis by a metal strap.

(e) Gas, Water, and Waste Pipe Bonding. The gas, water, and waste pipes shall be considered grounded if they are bonded to the chassis.

(f) Furnace and Metal Air Duct Bonding. Furnace and metal circulating air ducts shall be bonded.

551-57. Appliance Accessibility and Fastening. Every appliance shall be accessible for inspection, service, repair, and replacement without removal of permanent construction. Means shall be provided to securely fasten appliances in place when the recreational vehicle is in transit.

F. Factory Tests

551-60. Factory Tests (Electrical). Each recreational vehicle shall be subjected to the following tests.

(a) Circuits of 120 Volts or 120/240 Volts. Each recreational vehicle designed with a 120-volt or a 120/240-volt electrical system shall withstand the applied potential without electrical breakdown of a 1-minute, 900-volt dielectric strength test, or a 1-second, 1080-volt dielectric strength test, with all switches closed, between ungrounded and grounded conductors and the recreational vehicle ground. During the test, all switches and other controls shall be in the "on" position. Fixtures and permanently installed appliances shall not be required to withstand this test. The test shall be performed after branch circuits are complete prior to energizing the system and again after all outer coverings and cabinetry have been secured.

Each recreational vehicle shall be subjected to the following:

(1) A continuity test to ensure that all metal parts are properly bonded; and
(2) Operational tests to demonstrate that all equipment is properly connected and in working order; and
(3) Polarity checks to determine that connections have been properly made.

(b) Low-Voltage Circuits. An operational test of all low-voltage circuits shall be conducted to demonstrate that all equipment is connected and in electrical working order. This test shall be performed in the final stages of production after all outer coverings and cabinetry have been secured.

G. Recreational Vehicle Parks

551-71. Type Receptacles Provided. Every recreational vehicle site with electrical supply shall be equipped with at least one 20-ampere, 125-volt receptacle. A minimum of 5 percent of all recreational vehicle sites, with electrical supply, shall each be equipped with a 50-ampere, 125/250-volt receptacle conforming to the configuration as identified in Figure 551-46(c). These electrical supplies shall be permitted to include additional receptacles that have configurations in accordance with Section 551-81. A minimum of 70 percent of all recreational vehicle sites with electrical supply shall each be equipped with a 30-ampere, 125-volt receptacle conforming to Figure 551-46(c). This supply shall be permitted to include additional receptacle configurations conforming to Section 551-81. The remainder of all recreational vehicle sites with electrical supply shall be equipped with one or more of the receptacle configurations conforming to Section 551-81. Dedicated tent sites with a 15- or 20-ampere electrical supply shall be permitted to be excluded when determining the percentage of recreational vehicle sites with 30- or 50-ampere receptacles.

Additional receptacles shall be permitted for the connection of electrical equipment outside the recreational vehicle within the recreational vehicle park.

All 125-volt, single-phase, 15- and 20-ampere receptacles shall have listed ground-fault circuit-interrupter protection for personnel.

Section 551-71 requires at least one 20-ampere, 125-volt receptacle at each recreational vehicle campsite. Existing recreational vehicle campgrounds may have some sites that are equipped with 30-ampere receptacles only. Adapter plugs or "cheater" cords are often used to connect a recreational vehicle with a 20-ampere supply cord to a 30-ampere receptacle outlet. This practice does not provide adequate overload protection for the cord or the connected load. Section 551-71 will help ensure availability of the proper receptacle.

Some newer recreational vehicles have a 50-ampere, 120/240-volt supply installed, and Section 551-71 reflects this. Configurations are shown in Figure 551-46(c) and receptacle ratings are described in Section 551-81.

551-72. Distribution System. The recreational vehicle park secondary electrical distribution system to 50-ampere recreational vehicle sites shall be derived from a single-phase, 120/240-volt, 3-wire system. Other recreational vehicle sites with 125-volt, 20- and 30-ampere receptacles may be derived from any grounded distribution system that supplies 120-volt single-phase power. The neutral conductors shall not be reduced in size below the size of the ungrounded conductors for the site distribution. The neutral conductors shall be permitted to be reduced in size below the minimum required size of the ungrounded conductors for 240-volt, line-to-line, permanently connected loads only.

See Section 551-73(b). The reason for requiring single-phase, 120/240-volt, 3-wire systems to the recreational vehicle site is that the recreational vehicles are designed for connection to such systems. If the recreational vehicle park supplies another type of system to the recreational vehicle site (for example, two-phase conductors and a neutral of a 3-phase, 4-wire, 208Y/120-volt system), a hazard could result due to a supply of less than 240 volts, nominal, to motors, for example, or attempts to adapt the recreational vehicle to a different supply system than the one for which it was designed.

551-73. Calculated Load.

(a) Basis of Calculations. Electrical service and feeders shall be calculated on the basis of not less than 9600 volt-amperes per site equipped with 50-ampere, 120/240-volt supply facilities; 3600 volt-amperes per site equipped with both 20-ampere and 30-ampere supply facilities; 2400 volt-amperes per site equipped with only 20-ampere supply facilities; and 600 volt-amperes per site equipped with only 20-ampere supply facilities that are dedicated to tent sites. The demand factors set forth in Table 551-73 shall be the minimum allowable demand factors that shall be permitted in calculating load for service and feeders. Where the electrical supply for a recreational vehicle site has more than one receptacle, the calculated load shall only be computed for the highest rated receptacle.

Dedicated tent sites supplied with electricity do not supply power to an electrical system in a recreational vehicle; therefore, the calculated load can be smaller.

Table 551-73. Demand Factors for Site Feeders and Service-Entrance Conductors for Park Sites

Number of Recreational Vehicle Sites	Demand Factor (percent)
1	100
2	90
3	80
4	75
5	65
6	60
7–9	55
10–12	50
13–15	48
16–18	47
19–21	45
22–24	43
25–35	42
36 plus	41

(b) Transformers and Secondary Distribution Panelboards. For the purpose of this *Code*, where the park service exceeds 240 volts, transformers and secondary distribution panelboards shall be treated as services.

(c) Demand Factors. The demand factor for a given number of sites shall apply to all sites indicated. For example, 20 sites calculated at 45 percent of 3600 volt-amperes results in a permissible demand of 1620 volt-amperes per site or a total of 32,400 volt-amperes for 20 sites.

FPN: These demand factors may be inadequate in areas of extreme hot or cold temperature with loaded circuits for heating or air conditioning.

(d) Feeder-Circuit Capacity. Recreational vehicle site feeder-circuit conductors shall have adequate ampacity for the loads supplied and shall be rated at not less than 30 amperes. The grounded conductors shall have the same ampacity as the ungrounded conductors.

FPN: Due to the long circuit lengths typical in most recreational vehicle parks, feeder conductor sizes found in the ampacity tables of Article 310 may be inadequate to maintain the voltage regulation suggested in the fine print note to Section 210-19. Total circuit voltage drop is a sum of the voltage drops of each serial circuit segment, where the load for each segment is calculated using the load that segment sees and the demand factors of Section 551-73(a).

Loads for other amenities such as, but not limited to, service buildings, recreational buildings, and swimming pools shall be sized separately and then be added to the value calculated for the recreational vehicle sites where they are all supplied by one service.

551-74. Overcurrent Protection. Overcurrent protection shall be provided in accordance with Article 240.

551-75. Grounding. All electrical equipment and installations in recreational vehicle parks shall be grounded as required by Article 250.

551-76. Grounding — Recreational Vehicle Site Supply Equipment.

(a) Exposed Noncurrent-Carrying Metal Parts. Exposed noncurrent-carrying metal parts of fixed equipment, metal boxes, cabinets, and fittings that are not electrically connected to grounded equipment, shall be grounded by a continuous equipment grounding conductor run with the circuit conductors from the service equipment or from the transformer of a secondary distribution system. Equipment grounding conductors shall be sized in accordance with Section 250-122 and shall be permitted to be spliced by listed means.

The arrangement of equipment grounding connnections shall be such that the disconnection or removal of a receptacle or other device will not interfere with, or interrupt, the grounding continuity.

(b) Secondary Distribution System. Each secondary distribution system shall be grounded at the transformer.

(c) Neutral Conductor Not to Be Used as an Equipment Ground. The neutral conductor shall not be used as an equipment ground for recreational vehicles or equipment within the recreational vehicle park.

(d) No Connection on the Load Side. No connection to a grounding electrode shall be made to the neutral conductor on the load side of the service disconnecting means or transformer distribution panelboard.

551-77. Recreational Vehicle Site Supply Equipment.

(a) Location. Where provided, the recreational vehicle site electrical supply equipment shall be located on the left (road) side of the parked vehicle, on a line that is 9 ft (2.74 m), ± 1 ft (0.3 m), from the longitudinal centerline of the stand and shall be located at any point on this line from the rear of the stand to 15 ft (4.57 m) forward of the rear of the stand.

For pull-through sites, the electrical supply equipment shall be permitted to be located at any point along the line from 16 ft (4.88 m) forward of the rear of the stand to 32 ft (9.75 m) forward of the rear of the stand.

Section 551-77(a) is intended to accommodate vehicles towing boats and the like. The location of the site supply equipment permitted for pull-through sites will reduce the use of extension cords.

(b) Disconnecting Means. A disconnecting switch or circuit breaker shall be provided in the site supply equipment for disconnecting the power supply to the recreational vehicle.

(c) Access. All site supply equipment shall be accessible by an unobstructed entrance or passageway not less than 2 ft (610 mm) wide and 6½ ft (1.98 m) high.

(d) Mounting Height. Site supply equipment shall be located not less than 2 ft (610 mm) nor more than 6½ ft (1.98 m) above the ground.

(e) Working Space. Sufficient space shall be provided and maintained about all electrical equipment to permit ready and safe operation, in accordance with Section 110-26.

(f) Marking. Where the site supply equipment contains a 125/250-volt receptacle, the equipment shall be marked as follows: "Turn disconnecting switch or circuit breaker off before inserting or removing plug. Plug must be fully inserted or removed." The marking shall be located on the equipment adjacent to the receptacle outlet.

Partially engaged attachment plugs may result in intermittent neutral (grounded) contact. Loss of the neutral could momentarily apply 250 volts across 125-volt equipment, causing malfunction or damage. See Figure 551.1.

Figure 551.1 *Recreational vehicle site supply equipment with receptacles per Section 551-52 and ground-fault circuit-interrupter protection for the 20-ampere, 125-volt receptacle outlet. (Square D Co.)*

551-78. Protection of Outdoor Equipment.

(a) Wet Locations. All switches, circuit breakers, receptacles, control equipment, and metering devices located in wet locations or outside of a building shall be rainproof equipment.

(b) Meters. If secondary meters are installed, meter sockets without meters installed shall be blanked off with an approved blanking plate.

551-79. Clearance for Overhead Conductors. Open conductors of not over 600 volts, nominal, shall have a vertical clearance of not less than 18 ft (5.49 m) and a horizontal clearance of not less than 3 ft (914 mm) in all areas subject to recreational vehicle movement. In all other areas, clearances shall conform to Sections 225-18 and 225-19.

FPN: For clearances of conductors over 600 volts, nominal, see *National Electrical Safety Code,* ANSI C2-1997.

551-80. Underground Service, Feeder, Branch-Circuit, and Recreational Vehicle Site Feeder-Circuit Conductors.

(a) General. All direct-burial conductors, including the equipment grounding conductor if of aluminum, shall be insulated and identified for the use. All conductors shall be continuous from equipment to equipment. All splices and taps shall be made in approved junction boxes or by use of material listed and identified for the purpose.

(b) Protection Against Physical Damage. Direct-buried conductors and cables entering or leaving a trench shall be protected by rigid metal conduit, intermediate metal conduit, electrical metallic tubing with supplementary corrosion protection, rigid nonmetallic conduit, liquidtight flexible nonmetallic conduit, liquidtight flexible metal conduit, or other approved raceways or enclosures. Where subject to physical damage, the conductors or cables shall be protected by rigid metal conduit, intermediate metal conduit, or Schedule 80 rigid nonmetallic conduit. All such protection shall extend at least 18 in. (457 mm) into the trench from finished grade.

FPN: See Section 300-5 and Article 339 for conductors or Type UF cable used underground or in direct burial in earth.

551-81. Receptacles. A receptacle to supply electric power to a recreational vehicle shall be one of the configurations shown in Figure 551-46(c) in the following ratings.

(a) 50-Ampere. 125/250-volt, 50-ampere, 3-pole, 4-wire, grounding type for 120/240-volt systems

(b) 30-Ampere. 125-volt, 30-ampere, 2-pole, 3-wire, grounding type for 120-volt systems

(c) 20-Ampere. 125-volt, 20-ampere, 2-pole, 3-wire, grounding type for 120-volt systems

FPN: Complete details of these configurations can be found in National Electrical Manufacturers Association's *Standard for Dimensions of Attachment Plugs and Receptacles*, ANSI/NEMA WD 6-1989, Figures 14-50, TT, and 5-20.

Article 552 — Park Trailers

Contents

A. General

552-1. Scope. The provisions of this article cover the electrical conductors and equipment installed within or on park trailers not covered fully under Articles 550 and 551.

Article 552 was new for the 1996 *Code*. The scope covers park trailers that have a single chassis and wheels, not exceeding 400 ft^2 (set up), that are not used as permanent residences. Additionally, Article 552 does not apply to park trailers intended for commercial uses.

It is not uncommon for park trailers to be equipped with electrical loads similar to those used in mobile homes. It is also not uncommon for the park trailer to be located in the same park trailer community for several years without relocation.

This type of equipment is somewhat similar to mobile homes and recreational vehicles, and many requirements contained within Articles 550 and 551 would seem to apply to park trailers. Article 552, therefore, is similar in structure to Articles 550 and 551.

552-2. Definitions. (See Articles 100, 550, and 551 for additional definitions.)

Park Trailer. A unit that meets the following criteria:

(1) Built on a single chassis mounted on wheels, and
(2) Having a gross trailer area not exceeding 400 ft^2 (37.2 m^2) in the set-up mode.

552-3. Other Articles. Wherever the requirements of other articles of this *Code* and Article 552 differ, the requirements of Article 552 shall apply.

552-4. General Requirements. A park trailer as specified in Section 552-2 is intended for seasonal use. It is not intended as a permanent dwelling unit or for commercial uses such as banks, clinics, offices, or similar.

B. Low-Voltage Systems

552-10. Low-Voltage Systems.

(a) Low-Voltage Circuits. Low-voltage circuits furnished and installed by the park trailer manufacturer, other than those related to braking, are subject to this *Code*. Circuits supplying lights subject to federal or state regulations shall comply with applicable government regulations and this *Code*.

Section 552-10(a) refers to the low-voltage wiring within the park trailer that would be used in place of 120-volt ac supplies and does not refer to braking circuits.

(b) Low-Voltage Wiring.

(1) Copper conductors shall be used for low-voltage circuits.

Exception: A metal chassis or frame shall be permitted as the return path to the source of supply.

It is not the intent of Section 552-10(b)(1) to permit the sidewalls or the roof to serve as the ground return path. See the definition of *frame* in Section 551-2.

(2) Conductors shall conform to the requirements for Type GXL, HDT, SGT, SGR, or Type SXL or shall have insulation in accordance with Table 310-13 or the equivalent. Conductor sizes No. 6 through 18 or SAE shall be listed.

FPN: See SAE Standard J1128-1995 for Types GXL, HDT, and SXL and SAE Standard J1127-1995 for Types SGT and SGR.

(3) Single-wire, low-voltage conductors shall be of the stranded type.

(4) All insulated low-voltage conductors shall be surface marked at intervals not greater than 4 ft (1.22 m) as follows.

(a) Listed conductors shall be marked as required by the listing agency.
(b) SAE conductors shall be marked with the name or logo of the manufacturer, specification designation, and wire gauge.
(c) Other conductors shall be marked with the name or logo of the manufacturer, temperature rating, wire gauge, conductor material, and insulation thickness.

(c) Low-Voltage Wiring Methods.

(1) Conductors shall be protected against physical damage and shall be secured. Where insulated conductors are clamped to the structure, the conductor insulation shall be supplemented by an additional wrap or layer of equivalent material, except that jacketed cables shall not be required to be so protected. Wiring shall be routed away from sharp edges, moving parts, or heat sources.

(2) Conductors shall be spliced or joined with splicing devices that provide a secure connection or by brazing, welding, or soldering with a fusible metal or alloy. Soldered splices shall first be spliced or joined to be mechanically and electrically secure without solder and then soldered. All splices, joints, and free ends of conductors shall be covered with an insulation equivalent to that on the conductors.

(3) Battery and other low-voltage circuits shall be physically separated by at least a ½-in. (12.7-mm) gap or other approved means from circuits of a different power source. Acceptable methods shall be by clamping, routing, or equivalent means that ensure permanent total separation. Where circuits of different power sources cross, the external jacket of the nonmetallic-sheathed cables shall be deemed adequate separation.

(4) Ground connections to the chassis or frame shall be made in an accessible location and shall be mechanically secure. Ground connections shall be by means of copper conductors and copper or copper-alloy terminals of the solderless type identified for the size of wire used. The surface on which ground terminals make contact shall be cleaned and be free from oxide or paint or shall be electrically connected through the use of a cadmium, tin, or zinc-plated internal/external-toothed lockwasher or locking terminals. Ground terminal attaching screws, rivets or bolts, nuts, and lockwashers shall be cadmium, tin, or zinc-plated except rivets shall be permitted to be unanodized aluminum where attaching to aluminum structures.

(5) The chassis-grounding terminal of the battery shall be bonded to the unit chassis with a minimum No. 8 copper conductor. In the event the power lead from the battery exceeds No. 8, then the bonding conductor shall be of an equal size.

Sections 552-10(c)(4) and (5) require that the chassis-grounding terminal of the battery be bonded to the vehicle chassis in a mechanically secure manner and be placed in an accessible location using a minimum No. 8 copper conductor. This minimizes the possibility of low-voltage circuit-fault currents passing through the ac panelboard bonding conductor and the equipment grounding conductor of the combination ac/dc appliance and subsequently passing through the negative dc conductor feeding the appliance that also may be bonded to the external metal cover of the appliance. The ac equipment grounding conductor of the appliance may not have sufficient ampacity to safely conduct the dc fault current. This will necessitate installation of the battery bonding conductor. Some recreational vehicles already have one side of the battery circuit bonded to the frame by a No. 8 or larger copper conductor.

(d) Battery Installations. Storage batteries subject to the provisions of this *Code* shall be securely attached to the unit and installed in an area vaportight to the interior and ventilated directly to the exterior of the unit. Where batteries are installed in a compartment, the compartment shall be ventilated with openings having a minimum area of 1.7 in.^{2} (1100 mm^2) at both the top and at the bottom. Where compartment doors are equipped for ventilation, the openings shall be within 2 in. (50.8 mm) of the top and bottom. Batteries shall not be installed in a compartment containing spark- or flame-producing equipment.

(e) Overcurrent Protection.

(1) Low-voltage circuit wiring shall be protected by overcurrent protective devices rated not in excess of the ampacity of copper conductors, as follows.

Table 552-10(e)(1). Low-Voltage Overcurrent Protection

Wire Size (AWG)	Ampacity	Wire Type
18	6	Stranded only
16	8	Stranded only
14	15	Stranded or solid
12	20	Stranded or solid
10	30	Stranded or solid

(2) Circuit breakers or fuses shall be of an approved type, including automotive types. Fuseholders shall be clearly marked with maximum fuse size and shall be protected against shorting and physical damage by a cover or equivalent means.

The requirement for protection of fuseholders by a cover or equivalent means is intended to reduce the possibility of the low-voltage system shorting to ground.

FPN: For further information, see *Standard for Electric Fuses (Cartridge Type),* ANSI/SAE J554-1987; *Standard for Blade Type Electric Fuses,* SAE J1284-1988; and *Standard for Automotive Glass Tube Fuses,* UL 275-1993.

(3) Higher current-consuming, dc appliances such as pumps, compressors, heater blowers, and similar motor-driven appliances shall be installed in accordance with the manufacturer's instructions.

Motors that are controlled by automatic switching or by latching-type manual switches shall be protected in accordance with Section 430-32(c).

(4) The overcurrent protective device shall be installed in an accessible location on the unit within 18 in. (457 mm) of the point where the power supply connects to the unit circuits. If located outside the park trailer, the device shall be protected against weather and physical damage.

Exception: External low-voltage supply shall be permitted to have the overcurrent protective device within 18 in. (457 mm) after entering the unit or after leaving a metal raceway.

(f) Switches. Switches shall have a dc rating not less than the connected load.

(g) Lighting Fixtures. All low-voltage interior lighting fixtures rated more than 4 watts, employing lamps rated more than 1.2 watts, shall be listed.

Twelve-volt systems for running and signal lights, similar to those used in a conventional automobile, are covered in Sections 552-10 and 552-20. In some park trailers, 12-volt systems are also used for interior lighting and other small loads. The 12-volt system is often supplied from an on-board battery or through a transfer switch from a 120/12-volt transformer in conjunction with a full-wave rectifier.

C. Combination Electrical Systems

552-20. Combination Electrical Systems.

(a) General. Unit wiring suitable for connection to a battery or other low-voltage supply source shall be permitted to be connected to a 120-volt source, provided that the entire wiring system and equipment are rated and installed in full conformity with Parts A, C, D, and E requirements of this article covering 120-volt electrical systems. Circuits fed from ac transformers shall not supply dc appliances.

(b) Voltage Converters (120-Volt Alternating Current to Low-Voltage Direct Current). The 120-volt ac side of the voltage converter shall be wired in full conformity with Parts A, C, D, and E requirements of this article for 120-volt electrical systems.

Exception: Converters supplied as an integral part of a listed appliance shall not be subject to the above.

All converters and transformers shall be listed for use in recreation units and designed or equipped to provide over-temperature protection. To determine the converter rating, the following formula shall be applied to the total connected load, including average battery charging rate, of all 12-volt equipment:

The first 20 amperes of load at 100 percent; plus

The second 20 amperes of load at 50 percent; plus

All load above 40 amperes at 25 percent

Exception: A low-voltage appliance that is controlled by a momentary switch (normally open) that has no means for holding in the closed position shall not be considered as a connected load when determining the required converter rating. Momentarily energized appliances shall be limited to those used to prepare the unit for occupancy or travel.

(c) Bonding Voltage Converter Enclosures. The non-current-carrying metal enclosure of the voltage converter shall be bonded to the frame of the unit with a No. 8 copper conductor minimum. The grounding conductor for the battery and the metal enclosure shall be permitted to be the same conductor.

(d) Dual-Voltage Fixtures or Appliances. Fixtures or appliances having both 120-volt and low-voltage connections shall be listed for dual voltage.

In such dual-voltage fixtures, barriers are used to separate the 120-volt and the 12-volt wiring connections.

(e) Autotransformers. Autotransformers shall not be used.

(f) Receptacles and Plug Caps. Where a park trailer is equipped with a 120-volt or 120/240-volt ac system, a low-voltage system, or both, receptacles and plug caps of the low-voltage system shall differ in configuration from those of the 120-volt or 120/240-volt system. Where a unit equipped with a battery or dc system has an external connection for low-voltage power, the connector shall have a configuration that will not accept 120-volt power.

D. Nominal 120-Volt or 120/240-Volt Systems

552-40. 120-Volt or 120/240-Volt, Nominal, Systems.

(a) General Requirements. The electrical equipment and material of park trailers indicated for connection to a wiring system rated 120 volts, nominal, 2-wire with ground, or a wiring system rated 120/240 volts, nominal, 3-wire with ground, shall be listed and installed in accordance with the requirements of Parts A, C, D, and E of this article.

(b) Materials and Equipment. Electrical materials, devices, appliances, fittings, and other equipment installed, intended for use in, or attached to the park trailer shall be listed. All products shall be used only in the manner in which they have been tested and found suitable for the intended use.

552-41. Receptacle Outlets Required.

(a) Spacing. Receptacle outlets shall be installed at wall spaces 2 ft (610 mm) wide or more so that no point along the floor line is more than 6 ft (1.83 m), measured horizontally, from an outlet in that space.

Exception No. 1: Bath and hall areas.

Exception No. 2: Wall spaces occupied by kitchen cabinets, wardrobe cabinets, built-in furniture, behind doors that may open fully against a wall surface, or similar facilities.

(b) Location. Receptacle outlets shall be installed as follows:

(1) Adjacent to countertops in the kitchen [at least one on each side of the sink if countertops are on each side and are 12 in. (305 mm) or over in width]
(2) Adjacent to the refrigerator and gas range space, except where a gas-fired refrigerator or cooking appliance, requiring no external electrical connection, is factory-installed
(3) Adjacent to countertop spaces of 12 in. (305 mm) or more in width that cannot be reached from a receptacle required in Section 552-41(b)(1) by a cord of 6 ft 1.83 m) without crossing a traffic area, cooking appliance, or sink

(c) Ground-Fault Circuit-Interrupter Protection. Where provided, each 125-volt, single-phase, 15- or 20-ampere receptacle outlet shall have ground-fault circuit-interrupter protection for personnel in the following locations:

(1) Adjacent to a bathroom lavatory

Section 552-41(c)(1) permits mounting of a bathroom receptacle in the side of a lavatory cabinet where installation of a receptacle is not possible in a thin wall.

(2) Within 6 ft (1.83 m) of any lavatory or sink

Exception: Receptacles installed for appliances in dedicated spaces, such as for dishwashers, disposals, refrigerators, freezers, and laundry equipment.

(3) In the area occupied by a toilet, shower, tub, or any combination thereof
(4) On the exterior of the unit

Exception: Receptacles that are located inside of an access panel that is installed on the exterior of the unit to supply power for an installed appliance shall not be required to have ground-fault circuit-interrupter protection.

The receptacle outlet shall be permitted in a listed lighting fixture. A receptacle outlet shall not be installed in a tub or combination tub–shower compartment.

(d) Pipe Heating Cable Outlet. Where a pipe heating cable outlet is installed, the outlet shall be as follows:

(1) Located within 2 ft (610 mm) of the cold water inlet
(2) Connected to an interior branch circuit, other than a small appliance branch circuit
(3) On a circuit where all of the outlets are on the load side of the ground-fault circuit-interrupter protection for personnel
(4) Mounted on the underside of the park trailer and shall not be considered to be the outdoor receptacle outlet required in Section 552-41(e)

(e) Outdoor Receptacle Outlets. At least one receptacle outlet shall be installed outdoors. A receptacle outlet located in a compartment accessible from the outside of the park trailer shall be considered an outdoor receptacle. Outdoor receptacle outlets shall be protected as required in Section 552-41(c)(4).

(f) Receptacle Outlets Not Permitted.

(1) Shower or Bathtub Space. Receptacle outlets shall not be installed in or within reach [30 in. (762 mm)] of a shower or bathtub space.

(2) Face-Up Position. A receptacle shall not be installed in a face-up position in any countertop.

552-43. Power Supply.

(a) Feeder. The power supply to the park trailer shall be a feeder assembly consisting of not more than one listed 30-

ampere or 50-ampere park trailer power-supply cord with an integrally molded or securely attached cap, or a permanently installed feeder.

(b) Power-Supply Cord. If the park trailer has a power-supply cord, it shall be permanently attached to the distribution panelboard or to a junction box permanently connected to the distribution panelboard, with the free end terminating in a molded-on attachment plug cap.

Cords with adapters and pigtail ends, extension cords, and similar items shall not be attached to, or shipped with, a park trailer.

A suitable clamp or the equivalent shall be provided at the distribution panelboard knockout to afford strain relief for the cord to prevent strain from being transmitted to the terminals when the power-supply cord is handled in its intended manner.

The cord shall be a listed type with 3-wire, 120-volt or 4-wire, 120/240-volt conductors, one of which shall be identified by a continuous green color or a continuous green color with one or more yellow stripes for use as the grounding conductor.

(c) Mast Weatherhead or Raceway. Where the calculated load exceeds 50 amperes or where a permanent feeder is used, the supply shall be by means of one of the following:

(1) One mast weatherhead installation, installed in accordance with Article 230, containing four continuous, insulated, color-coded feeder conductors, one of which shall be an equipment grounding conductor; or
(2) A metal raceway, rigid nonmetallic conduit, or liquid-tight flexible nonmetallic conduit from the disconnecting means in the park trailer to the underside of the park trailer, with provisions for the attachment to a suitable junction box or fitting to the raceway on the underside of the park trailer [with or without conductors as in Section 550-5(i)(1)]

552-44. Cord.

(a) Permanently Connected. Each power-supply assembly shall be factory supplied or factory installed and connected directly to the terminals of the distribution panelboard or conductors within a junction box and provided with means to prevent strain from being transmitted to the terminals. The ampacity of the conductors between each junction box and the terminals of each distribution panelboard shall be at least equal to the ampacity of the power-supply cord. The supply end of the assembly shall be equipped with an attachment plug of the type described in Section 552-44(c). Where the cord passes through the walls or floors, it shall be protected by means of conduit and bushings or equivalent. The cord assembly shall have permanent provisions for protection against corrosion and mechanical damage while the unit is in transit.

(b) Cord Length. The cord exposed usable length shall be measured from the point of entrance to the park trailer or the face of the flanged surface inlet (motor-base attachment plug) to the face of the attachment plug at the supply end.

The cord exposed usable length, measured to the point of entry on the unit exterior, shall be a minimum of 23 ft (7.0 m) where the point of entrance is at the side of the unit, or shall be a minimum 28 ft (8.5 m) where the point of entrance is at the rear of the unit. The maximum length shall not exceed 36½ ft (11.13 m).

Where the cord entrance into the unit is more than 3 ft (0.9 m) above the ground, the minimum cord lengths above shall be increased by the vertical distance of the cord entrance heights above 3 ft (0.9 m).

(c) Attachment Plugs.

The 50-ampere receptacle and plug configurations used for park trailers appear in Figure 210.6 or 210.7 (NEMA configuration chart); however, the 30-ampere plug and receptacle are special for recreational vehicles and are described in UL 498, *UL Standard for Safety Attachment Plugs and Receptacles.*.

(1) Park trailers wired in accordance with Section 552-46(a) shall have an attachment plug that shall be 2-pole, 3-wire, grounding-type, rated 30 amperes, 125 volts, conforming to the configuration shown in Figure 552-44(c) intended for use with units rated at 30 amperes, 125 volts.

FPN: Complete details of this configuration can be found in National Electrical Manufacturers Association's *Standard for Dimensions of Attachment Plugs and Receptacles*, ANSI/NEMA WD 6-1989, Figure TT.

(2) Park trailers having a power-supply assembly rated 50 amperes as permitted by Section 552-43(b) shall have a 3-pole, 4-wire, grounding-type attachment plug rated 50 amperes, 125/250 volts, conforming to the configuration shown in Figure 552-44(c).

FPN: Complete details of this configuration can be found in National Electrical Manufacturers Association *Standard for Dimensions of Attachment Plugs and Receptacles,* ANSI/NEMA WD 6-1989, Figure 14-50.

Receptacles Caps

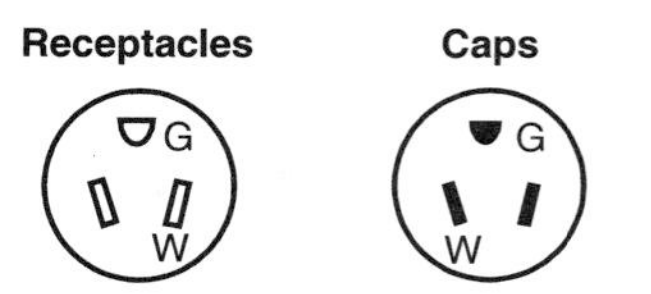

30-A, 125-V, 2-pole, 3-wire, grounding type

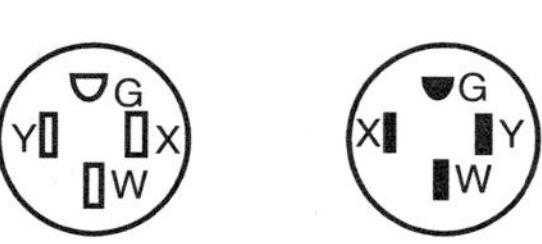

50-A, 125/250-V, 3-pole, 4-wire, grounding type

Figure 552-44(c).

(d) Labeling at Electrical Entrance. Each park trailer shall have permanently affixed to the exterior skin, at or near the point of entrance of the power-supply assembly, a label 3 in. × 1¾ in. (76 mm × 44.5 mm) minimum size, made of etched, metal-stamped, or embossed brass, stainless steel, or anodized or alclad aluminum not less than 0.020-in. (0.508-mm) thick, or other suitable material [e.g., 0.005-in. (0.127-mm) thick plastic laminate], that reads, as appropriate, either

THIS CONNECTION IS FOR 110–125-VOLT AC, 60 HZ, 30 AMPERE SUPPLY.

or

THIS CONNECTION IS FOR 120/240-VOLT AC, 3-POLE, 4-WIRE, 60 HZ, _____ AMPERE SUPPLY.

The correct ampere rating shall be marked in the blank space.

(e) Location. The point of entrance of a power-supply assembly shall be located within 15 ft (4.57 m) of the rear, on the left (road) side or at the rear, left of the longitudinal center of the unit, within 18 in. (457 mm) of the outside wall.

Exception: A park trailer shall be permitted to have the electrical point of entrance located more than 15 ft (4.57 m) from the rear. Where this occurs, the distance beyond the 15-ft (4.57-m) dimension shall be added to the cord's minimum length as specified in Section 551-46(b).

552-45. Distribution Panelboard.

(a) Listed and Appropriately Rated. A listed and appropriately rated distribution panelboard or other equipment specifically listed for the purpose shall be used. The grounded conductor termination bar shall be insulated from the enclosure as provided in Section 552-55(c). An equipment grounding terminal bar shall be attached inside the metal enclosure of the panelboard.

(b) Location. The distribution panelboard shall be installed in a readily accessible location. Working clearance for the panelboard shall be not less than 24 in. (610 mm) wide and 30 in. (762 mm) deep.

Exception: Where the panelboard cover is exposed to the inside aisle space, one of the working clearance dimensions shall be permitted to be reduced to a minimum of 22 in. (559 mm). A panelboard shall be considered exposed where the panelboard cover is within 2 in. (50.8 mm) of the aisle's finished surface.

(c) Dead-Front Type. The distribution panelboard shall be of the dead-front type. A main disconnecting means shall be provided where fuses are used or where more than two circuit breakers are employed. A main overcurrent protective device not exceeding the power-supply assembly rating shall be provided where more than two branch circuits are employed.

552-46. Branch Circuits. Branch circuits shall be determined in accordance with the following.

(a) Two to Five 15- or 20-Ampere Circuits. Two to five 15- or 20-ampere circuits to supply lights, receptacle outlets, and fixed appliances shall be permitted. Such park trailers shall be equipped with a distribution panelboard rated at 120 volts maximum with a 30-ampere rated main power supply assembly. Not more than two 120-volt thermostatically controlled appliances (e.g., air conditioner and water heater) shall be installed in such systems unless appliance isolation switching, energy management systems, or similar methods are used.

Exception: Additional 15- or 20-ampere circuits shall be permitted where a listed energy management system rated at 30 amperes maximum is employed within the system.

(b) More than Five Circuits. Where more than five circuits are needed, they shall be determined in accordance with the following.

(1) Lighting. Based on 3 volt-amperes/ft^2 (32.26 volt-ampere/m^2) multiplied by the outside dimensions of the park trailer (coupler excluded) divided by 120 volts to determine the number of 15- or 20-ampere lighting area circuits, e.g.,

$$\frac{3 \times \text{length} \times \text{width}}{120 \times 15 \text{ (or 20)}} = \text{No. of 15- (or 20-) ampere circuits}$$

The lighting circuits shall be permitted to serve built-in gas ovens with electric service only for lights, clocks or timers, or listed cord-connected garbage disposal units.

(2) Small Appliances. Small appliance branch circuits shall be installed in accordance with Section 210-11(c)(1).

(3) General Appliances. (including furnace, water heater, range, and central or room air conditioner, etc.). There shall be one or more circuits of adequate rating in accordance with the following.

FPN No. 1: For the laundry branch circuit, see Section 210-11(c)(2).

FPN No. 2: For central air conditioning, see Article 440.

(a) Ampere rating of fixed appliances not over 50 percent of circuit rating if lighting outlets (receptacles, other than kitchen, dining area, and laundry, considered as lighting outlets) are on the same circuit.
(b) For fixed appliances on a circuit without lighting outlets, the sum of rated amperes shall not exceed the branch-circuit rating. Motor loads or other continuous duty loads shall not exceed 80 percent of the branch-circuit rating.
(c) The rating of a single cord- and plug-connected appliance on a circuit having no other outlets shall not exceed 80 percent of the circuit rating.
(d) The rating of a range branch circuit shall be based on the range demand as specified for ranges in Section 552-47(b)(5).

552-47. Calculations. The following method shall be employed in computing the supply-cord and distribution-panelboard load for each feeder assembly for each park trailer in lieu of the procedure shown in Article 220 and shall be based on a 3-wire, 120/240-volt supply with 120-volt loads balanced between the two phases of the 3-wire system.

(a) Lighting and Small Appliance Load.

Lighting Volt-Amperes: Length times width of park trailer floor (outside dimensions) times 3 volt-amperes/ft^2, e.g.,

$$\text{Length} \times \text{width} \times 3 = \text{lighting volt-amperes}$$

Small Appliance Volt-Amperes: Number of circuits times 1500 volt-amperes for each 20-ampere appliance receptacle circuit (see definition of *Appliance, Portable* with note) including 1500 volt-amperes for laundry circuit, e.g.,

$$\text{No. of circuits} \times 1500 = \text{small appliance volt-amperes}$$

Total: Lighting volt-amperes plus small appliance volt-amperes = total volt-amperes

First 3000 total volt-amperes at 100 percent plus remainder at 35 percent = volt-amperes to be divided by 240 volts to obtain current (amperes) per leg

(b) Total Load for Determining Power Supply. Total load for determining power supply is the sum of the following:

(1) Lighting and small appliance load as calculated in Section 552-47(a).
(2) Nameplate amperes for motors and heater loads (exhaust fans, air conditioners, electric, gas, or oil heating). Omit smaller of the heating and cooling loads, except include blower motor if used as air-conditioner evaporator motor. Where an air conditioner is not installed and a 50-ampere power-supply cord is provided, allow 15 amperes per phase for air conditioning.
(3) Twenty-five percent of current of largest motor in (2).
(4) Total of nameplate amperes for disposal, dishwasher, water heater, clothes dryer, wall-mounted oven, cooking units. Where the number of these appliances exceeds three, use 75 percent of total.
(5) Derive amperes for freestanding range (as distinguished from separate ovens and cooking units) by dividing the following values by 240 volts.

Nameplate Rating (watts)	Use (volt-amperes)
0–10,000	80 percent of rating
Over 10,000–12,500	8,000
Over 12,500–13,500	8,400
Over 13,500–14,500	8,800
Over 14,500–15,500	9,200
Over 15,500–16,500	9,600
Over 16,500–17,500	10,000

(6) If outlets or circuits are provided for other than factory-installed appliances, include the anticipated load.

FPN: Refer to Appendix D, Example D12, for an illustration of the application of this calculation.

(c) Optional Method of Calculation for Lighting and Appliance Load. For park trailers, the optional method for calculating lighting and appliance load shown in Section 220-30 shall be permitted.

552-48. Wiring Methods.

(a) Wiring Systems. Cables and raceways installed in accordance with Articles 330 through 352 shall be permitted in accordance with their applicable article, except as otherwise specified in this article. An equipment grounding means shall be provided in accordance with Section 250-118.

See the commentary following Section 350-5.

(b) Conduit and Tubing. Where rigid metal conduit or intermediate metal conduit is terminated at an enclosure with a locknut and bushing connection, two locknuts shall be provided, one inside and one outside of the enclosure. All cut ends of conduit and tubing shall be reamed or otherwise finished to remove rough edges.

See the commentary following Sections 348-11 and 373-6(c).

(c) Nonmetallic Boxes. Nonmetallic boxes shall be acceptable only with nonmetallic-sheathed cable or nonmetallic raceways.

(d) Boxes. In walls and ceilings constructed of wood or other combustible material, boxes and fittings shall be flush with the finished surface or project therefrom.

(e) Mounting. Wall and ceiling boxes shall be mounted in accordance with Article 370.

Exception No. 1: Snap-in-type boxes or boxes provided with special wall or ceiling brackets that securely fasten boxes in walls or ceilings shall be permitted.

Exception No. 2: A wooden plate providing a 1½-in. (38-mm) minimum width backing around the box and of a thickness of ½ in. (12.7 mm) or greater (actual) attached directly to the wall panel shall be considered as approved means for mounting outlet boxes.

Exception No. 2 permits the mounting of outlet boxes by screws to a wooden plate that is secured directly to the back of the wall panel. This wooden plate is required to be not less than ½ in. thick and is required to extend at least 1½ in. around the box. This recognizes the special construction of recreational vehicle walls, which often make it quite difficult or impossible to attach an outlet box to a structural member, as required by Section 370-23(b).

(f) Sheath Armor. The sheath of nonmetallic-sheathed cable, metal-clad cable, and Type AC cable shall be continuous between outlet boxes and other enclosures.

(g) Protected. Metal-clad, Type AC, or nonmetallic-sheathed cables and electrical nonmetallic tubing shall be permitted to pass through the centers of the wide side of 2 in. by 4 in. wood studs. However, they shall be protected where they pass through 2 in. by 2 in. wood studs or at other wood studs or frames where the cable or tubing would be less than 1¼ in. (31.8 mm) from the inside or outside surface. Steel plates on each side of the cable or tubing, or a steel tube, with not less than No. 16 MSG wall thickness, shall be installed to protect the cable or tubing. These plates or tubes shall be securely held in place. Where nonmetallic-sheathed cables pass through punched, cut, or drilled slots or holes in metal members, the cable shall be protected by bushings or grommets securely fastened in the opening prior to installation of the cable.

(h) Cable Supports. Where connected with cable connectors or clamps, cables shall be supported within 12 in. (305 mm) of outlet boxes, distribution panelboards, and splice boxes on appliances. Supports shall be provided every 4½ ft (1.37 m) at other places.

(i) Nonmetallic Box Without Cable Clamps. Nonmetallic-sheathed cables shall be supported within 8 in. (203 mm) of a nonmetallic outlet box without cable clamps.

Exception: Where wiring devices with integral enclosures are employed with a loop of extra cable to permit future replacement of the device, the cable loop shall be considered as an integral portion of the device.

(j) Physical Damage. Where subject to physical damage, exposed nonmetallic cable shall be protected by covering boards, guard strips, raceways, or other means.

(k) Metal Faceplates. Metal faceplates shall be of ferrous metal not less than 0.030 in. (0.762 mm) in thickness or of nonferrous metal not less than 0.040 in. (1.016 mm) in thickness. Nonmetallic faceplates shall be listed.

(l) Metal Faceplates Effectively Grounded. Where metal faceplates are used, they shall be effectively grounded.

(m) Moisture or Physical Damage. Where outdoor or under-chassis wiring is 120 volts, nominal, or over and is exposed to moisture or physical damage, the wiring shall be protected by rigid metal conduit, intermediate metal conduit, or by electrical metallic tubing or rigid nonmetallic conduit that is closely routed against frames and equipment enclosures or other raceway or cable identified for the application.

(n) Component Interconnections. Fittings and connectors that are intended to be concealed at the time of assembly shall be listed and identified for the interconnection of building components. Such fittings and connectors shall be equal to the wiring method employed in insulation, temperature rise, and fault-current withstanding, and shall be capable of enduring the vibration and shock occurring in park trailers.

(o) Method of Connecting Expandable Units.

(1) That portion of a branch circuit that is installed in an expandable unit shall be permitted to be connected to the branch circuit in the main body of the vehicle by means of a flexible cord or attachment plug and cord listed for hard usage. The cord and its connections shall conform to all provisions of Article 400 and shall be considered as a permitted use under Section 400-7.

(2) If the receptacle provided for connection of the cord to the main circuit is located on the outside of the unit, it shall be protected with a ground-fault circuit interrupter for personnel and be listed for wet locations. A cord located on the outside of a unit shall be identified for outdoor use.

(3) Unless removable or stored within the unit interior, the cord assembly shall have permanent provisions for protection against corrosion and mechanical damage while the unit is in transit.

(4) If an attachment plug and cord is used it shall be installed so as not to permit exposed live attachment plug pins.

(p) Prewiring for Air-Conditioning Installation. Prewiring installed for the purpose of facilitating future air-conditioning installation shall conform to the following and other applicable portions of this article. The circuit shall serve no other purpose.

(1) An overcurrent protective device with a rating compatible with the circuit conductors shall be installed in the distribution panelboard and wiring connections completed.

(2) The load end of the circuit shall terminate in a junction box with a blank cover or a device listed for the purpose. Where a junction box with a blank cover is used, the free ends of the conductors shall be adequately capped or taped.

(3) A label conforming to Section 552-44(b) shall be placed on or adjacent to the junction box and shall read

AIR-CONDITIONING CIRCUIT. THIS
CONNECTION IS FOR AIR CONDITIONERS
RATED 110–125-VOLT AC, 60 HZ, ____
AMPERES MAXIMUM. DO NOT EXCEED
CIRCUIT RATING.

An ampere rating, not to exceed 80 percent of the circuit rating, shall be legibly marked in the blank space.

552-49. Conductors and Boxes.

(a) Maximum Number of Conductors. The maximum number of conductors permitted in boxes shall be in accordance with Section 370-16.

(b) Free Conductor at Each Box. At least 6 in. (152 mm) of free conductor shall be left at each box except where conductors are intended to loop without joints.

552-50. Grounded Conductors. The identification of grounded conductors shall be in accordance with Section 200-6.

552-51. Connection of Terminals and Splices. Conductor splices and connections at terminals shall be in accordance with Section 110-14.

552-52. Switches. Switches shall be rated as follows.

(a) Lighting Circuits. For lighting circuits, switches shall be rated not less than 10 amperes, 120/125 volts and in no case less than the connected load.

(b) Motors or Other Loads. For motors or other loads, switches shall have ampere or horsepower ratings, or both, adequate for loads controlled. (An ac general-use snap switch shall be permitted to control a motor 2 hp or less with full-load current not over 80 percent of the switch ampere rating.)

552-53. Receptacles. All receptacle outlets shall be of the grounding type and installed in accordance with Sections 210-7 and 210-21.

552-54. Lighting Fixtures.

(a) General. Any combustible wall or ceiling finish exposed between the edge of a fixture canopy, or pan and the outlet box, shall be covered with noncombustible material or a material identified for the purpose.

(b) Shower Fixtures. If a lighting fixture is provided over a bathtub or in a shower stall, it shall be of the enclosed and gasketed type and listed for the type of installation, and it shall be ground-fault circuit-interrupter protected.

The switch for shower lighting fixtures and exhaust fans, located over a tub or in a shower stall, shall be located outside the tub or shower space.

Many shower lights have a metal base and, due to low ceilings in trailers, may be easily reached by most persons under a shower or standing in a bathtub. Section 552-54(b) permits lighting fixtures listed for a wet location to be installed above a tub or shower enclosure if shock-hazard protection is provided by a ground-fault circuit interrupter.

(c) Outdoor Outlets, Fixtures, Air-Cooling Equipment, etc. Outdoor fixtures and other equipment shall be listed for outdoor use.

552-55. Grounding. (See also Section 552-57 on bonding of noncurrent-carrying metal parts.)

(a) Power-Supply Grounding. The grounding conductor in the supply cord or feeder shall be connected to the grounding bus or other approved grounding means in the distribution panelboard.

(b) Distribution Panelboard. The distribution panelboard shall have a grounding bus with sufficient terminals for all grounding conductors or other approved grounding means.

(c) Insulated Neutral.

(1) The grounded circuit conductor (neutral) shall be insulated from the equipment grounding conductors and from equipment enclosures and other grounded parts. The grounded (neutral) circuit terminals in the distribution panelboard and in ranges, clothes dryers, counter-mounted cooking units, and wall-mounted ovens shall be insulated from the equipment enclosure. Bonding screws, straps, or buses in the distribution panelboard or in appliances shall be removed and discarded.

(2) Connection of electric ranges and electric clothes dryers utilizing a grounded (neutral) conductor, if cord-connected, shall be made with 4-conductor cord and 3-pole, 4-wire, grounding-type plug caps and receptacles.

552-56. Interior Equipment Grounding.

(a) Exposed Metal Parts. In the electrical system, all exposed metal parts, enclosures, frames, lighting fixture canopies, etc., shall be effectively bonded to the grounding terminals or enclosure of the distribution panelboard.

(b) Equipment Grounding Conductors. Bare wires, green-colored wires, or green wires with a yellow stripe(s) shall be used for equipment grounding conductors only.

(c) Grounding of Electrical Equipment. Where grounding of electrical equipment is specified, it shall be permitted as follows.

(1) Connection of metal raceway (conduit or electrical metallic tubing), the sheath of Type MC and Type MI cable where the sheath is identified for grounding, or the armor of Type AC cable to metal enclosures.
(2) A connection between the one or more equipment grounding conductors and a metal box by means of a grounding screw, which shall be used for no other purpose, or a listed grounding device.
(3) The equipment grounding conductor in nonmetallic-sheathed cable shall be permitted to be secured under a screw threaded into the fixture canopy other than a mounting screw or cover screw, or attached to a listed grounding means (plate) in a nonmetallic outlet box for fixture mounting (grounding means shall also be permitted for fixture attachment screws).

(d) Grounding Connection in Nonmetallic Box. A connection between the one or more grounding conductors brought into a nonmetallic outlet box shall be arranged so that a connection can be made to any fitting or device in that box that requires grounding.

(e) Grounding Continuity. Where more than one equipment grounding conductor of a branch circuit enters a box, all such conductors shall be in good electrical contact with each other, and the arrangement shall be such that the disconnection or removal of a receptacle, fixture, or other device fed from the box will not interfere with or interrupt the grounding continuity.

(f) Cord-Connected Appliances. Cord-connected appliances, such as washing machines, clothes dryers, refrigerators, and the electrical system of gas ranges, etc., shall be grounded by means of an approved cord with equipment grounding conductor and grounding-type attachment plug.

552-57. Bonding of Noncurrent-Carrying Metal Parts.

(a) Required Bonding. All exposed noncurrent-carrying metal parts that may become energized shall be effectively bonded to the grounding terminal or enclosure of the distribution panelboard.

(b) Bonding Chassis. A bonding conductor shall be connected between any distribution panelboard and an accessible terminal on the chassis. Aluminum or copper-clad aluminum conductors shall not be used for bonding if such conductors or their terminals are exposed to corrosive elements.

Exception: Any park trailer that employs a unitized metal chassis-frame construction to which the distribution panelboard is securely fastened with a bolt(s) and nut(s) or by welding or riveting shall be considered to be bonded.

(c) Bonding Conductor Requirements. Grounding terminals shall be of the solderless type and listed as pressure terminal connectors recognized for the wire size used. The bonding conductor shall be solid or stranded, insulated or bare, and shall be No. 8 copper minimum, or equivalent.

(d) Metallic Roof and Exterior Bonding. The metal roof and exterior covering shall be considered bonded where

(1) The metal panels overlap one another and are securely attached to the wood or metal frame parts by metal fasteners, and
(2) The lower panel of the metal exterior covering is secured by metal fasteners at each cross member of the chassis, or the lower panel is bonded to the chassis by a metal strap.

(e) Gas, Water, and Waste Pipe Bonding. The gas, water, and waste pipes shall be considered grounded if they are bonded to the chassis.

(f) Furnace and Metal Air Duct Bonding. Furnace and metal circulating air ducts shall be bonded.

552-58. Appliance Accessibility and Fastening. Every appliance shall be accessible for inspection, service, repair, and replacement without removal of permanent construction. Means shall be provided to securely fasten appliances in place when the park trailer is in transit.

552-59. Outdoor Outlets, Fixtures, Air-Cooling Equipment, etc.

(a) Listed for Outdoor Use. Outdoor fixtures and equipment shall be listed for outdoor use. Outdoor receptacle or convenience outlets shall be of a gasketed-cover type for use in wet locations.

(b) Outside Heating Equipment, Air-Conditioning Equipment, or Both. A park trailer provided with a branch circuit designed to energize outside heating equipment or air-conditioning equipment, or both, located outside the park trailer, other than room air conditioners, shall have such branch-circuit conductors terminate in a listed outlet box or disconnecting means located on the outside of the park trailer. A label shall be permanently affixed adjacent to the outlet box and shall contain the following information:

THIS CONNECTION IS FOR HEATING AND/OR
AIR-CONDITIONING EQUIPMENT. THE
BRANCH CIRCUIT IS RATED AT NOT MORE
THAN ____ AMPERES, AT ____ VOLTS,
60-Hz, ____ CONDUCTOR AMPACITY.
A DISCONNECTING MEANS SHALL BE
LOCATED WITHIN SIGHT OF THE EQUIPMENT.

The correct voltage and ampere rating shall be given. The tag shall not be less than 0.020-in. (508-μm) thick etched brass, stainless steel, anodized or alclad aluminum, or equivalent. The tag shall not be less than 3 in. × 1¾ in. (76 mm × 44.5 mm) minimum size.

E. Factory Tests

552-60. Factory Tests (Electrical). Each park trailer shall be subjected to the following tests.

(a) Circuits of 120 Volts or 120/240 Volts. Each park trailer designed with a 120-volt or a 120/240-volt electrical system shall withstand the applied potential without electrical breakdown of a 1-minute, 900-volt dielectric strength test, or a 1-second, 1080-volt dielectric strength test, with all switches closed, between ungrounded and grounded conductors and the park trailer ground. During the test, all switches and other controls shall be in the on position. Fixtures and permanently installed appliances shall not be required to withstand this test.

Each park trailer shall be subjected to the following.

(1) A continuity test to ensure that all metal parts are properly bonded;
(2) Operational tests to demonstrate that all equipment is properly connected and in working order;
(3) Polarity checks to determine that connections have been properly made; and
(4) Receptacles requiring GFCI protection shall be tested for correct function by the use of a GFCI testing device.

(b) Low-Voltage Circuits. Low-voltage circuit conductors in each park trailer shall withstand the applied potential without electrical breakdown of a 1-minute, 500-volt or a 1-second, 600-volt dielectric strength test. The potential shall be applied between ungrounded and grounded conductors.

The test shall be permitted on running light circuits before the lights are installed, provided the unit's outer covering and interior cabinetry have been secured. The braking circuit shall be permitted to be tested before being connected to the brakes, provided the wiring has been completely secured.

Article 553 — Floating Buildings

Contents

A. General

553-1. Scope. This article covers wiring, services, feeders, and grounding for floating buildings.

553-2. Definition.

Floating Building. A building unit as defined in Article 100 that floats on water, is moored in a permanent location, and has a premises wiring system served through connection by permanent wiring to an electricity supply system not located on the premises.

553-3. Application of Other Articles. Wiring for floating buildings shall comply with the applicable provisions of other articles of this *Code*, except as modified by this article.

B. Services and Feeders

553-4. Location of Service Equipment. The service equipment for a floating building shall be located adjacent to, but not in or on, the building.

The intent of Section 553-4 is to ensure that supply conductors to a floating building can be disconnected in an emergency, such as during a storm, where the floating dwelling unit has to be moved quickly.

Overcurrent protection for supply conductors is also provided where these conductors may develop excessive leakage, where underwater, or where a short circuit or fault may occur.

553-5. Service Conductors. One set of service conductors shall be permitted to serve more than one set of service equipment.

553-6. Feeder Conductors. Each floating building shall be supplied by a single set of feeder conductors from its service equipment.

Exception: Where the floating building has multiple occupancy, each occupant shall be permitted to be supplied by a single set of feeder conductors extended from the occupant's service equipment to the occupant's panelboard.

553-7. Installation of Services and Feeders.

(a) Flexibility. Flexibility of the wiring system shall be maintained between floating buildings and the supply conductors. All wiring shall be installed so that motion of the water surface and changes in the water level will not result in unsafe conditions.

(b) Wiring Methods. Liquidtight flexible metal conduit or liquidtight flexible nonmetallic conduit with approved fittings shall be permitted for feeders and where flexible connections are required for services. Extra-hard usage portable power cable listed for both wet locations and sunlight resistance shall be permitted for a feeder to a floating building where flexibility is required. Other raceways, suitable for the location, shall be permitted to be installed where flexibility is not required.

Type W cables included in Table 400-4 may or may not be listed for wet locations. Liquidtight flexible nonmetallic conduit with approved fittings is permitted where flexible connections are required.

FPN: See Sections 555-1 and 555-6.

C. Grounding

553-8. General Requirements. Grounding of both electrical and nonelectrical parts in a floating building shall be through connection to a grounding bus in the building panelboard. The grounding bus shall be grounded through a green-colored insulated equipment grounding conductor run with the feeder conductors and connected to a grounding terminal in the service equipment. The grounding terminal in the service equipment shall be grounded by connection through an insulated grounding electrode conductor to a grounding electrode on shore.

553-9. Insulated Neutral. The grounded circuit conductor (neutral) shall be an insulated conductor identified in conformance with Section 200-6. The neutral conductor shall be connected to the equipment grounding terminal in the service equipment, and, except for that connection, it shall be insulated from the equipment grounding conductors, equipment enclosures, and all other grounded parts. The neutral circuit terminals in the panelboard and in ranges, clothes dryers, counter-mounted cooking units, and the like shall be insulated from the enclosures.

553-10. Equipment Grounding.

(a) Electrical Systems. All enclosures and exposed metal parts of electrical systems shall be bonded to the grounding bus.

(b) Cord-Connected Appliances. Where required to be grounded, cord-connected appliances shall be grounded by means of an equipment grounding conductor in the cord and a grounding-type attachment plug.

553-11. Bonding of Noncurrent-Carrying Metal Parts. All metal parts in contact with the water, all metal piping, and all noncurrent-carrying metal parts that may become energized, shall be bonded to the grounding bus in the panelboard.

Article 555 — Marinas and Boatyards

Contents

555-1. Scope. This article covers the installation of wiring and equipment in the areas comprising fixed or floating piers, wharfs, docks, and other areas in marinas, boatyards, boat basins, boathouses, and similar occupancies that are used, or intended for use, for the purpose of repair, berthing, launching, storage, or fueling of small craft and the moorage of floating buildings.

See Article 553 for requirements for floating buildings, including floating dwelling units. Article 555 includes stand-alone boathouses.

555-2. Application of Other Articles. Wiring and equipment for marinas and boatyards shall comply with this article and also with the applicable provisions of other articles of this *Code*.

FPN: For disconnection of auxiliary power from boats, see *Fire Protection Standard for Pleasure and Commercial Motor Craft*, NFPA 302-1994.

555-3. Receptacles. Where shore power is supplied, those accommodations for boats 20 ft (6.1 m) or less in length shall be equipped with shore-power receptacles of a locking and grounding type rated at not less than 20 amperes.

Where shore power is supplied to accommodations for boats longer than 20 ft (6.1 m) in length, shore-power receptacles of a locking and grounding type rated at 30 amperes or more shall be provided.

Fifteen- and 20-ampere, single-phase, 125-volt receptacles other than those supplying shore power to boats located at piers, wharfs, and other locations shall be protected by ground-fault circuit interrupters.

FPN No. 1: For various configurations and ratings of locking- and grounding-type receptacles and caps, see National Electrical Manufacturers Association's *Standard for Dimensions of Attachment Plugs and Receptacles*, ANSI/NEMA 18WD 6-1989.

FPN No. 2: For locking- and grounding-type receptacles for auxiliary power to boats, see *Fire Protection Standard for Marinas and Boatyards*, NFPA 303-1995.

FPN No. 3: In locating receptacles, consideration should be given to the maximum tide level and wave action. See *Fire Protection Standard for Marinas and Boatyards*, NFPA 303-1995, for establishment of datum plane.

Single locking and grounding receptacles are required for providing shore power to boats. See Figures 555.1 and 210.7 for a complete chart of grounding-type locking plug and receptacle configurations. See Figure 555.2 for a power outlet assembly used at marinas and boatyards. See Figure 555.3 for pin and sleeve plug and receptacle configuration.

Each single receptacle that supplies shore power to boats is required to be supplied from an individual branch circuit. Locking- and grounding-type recepta-

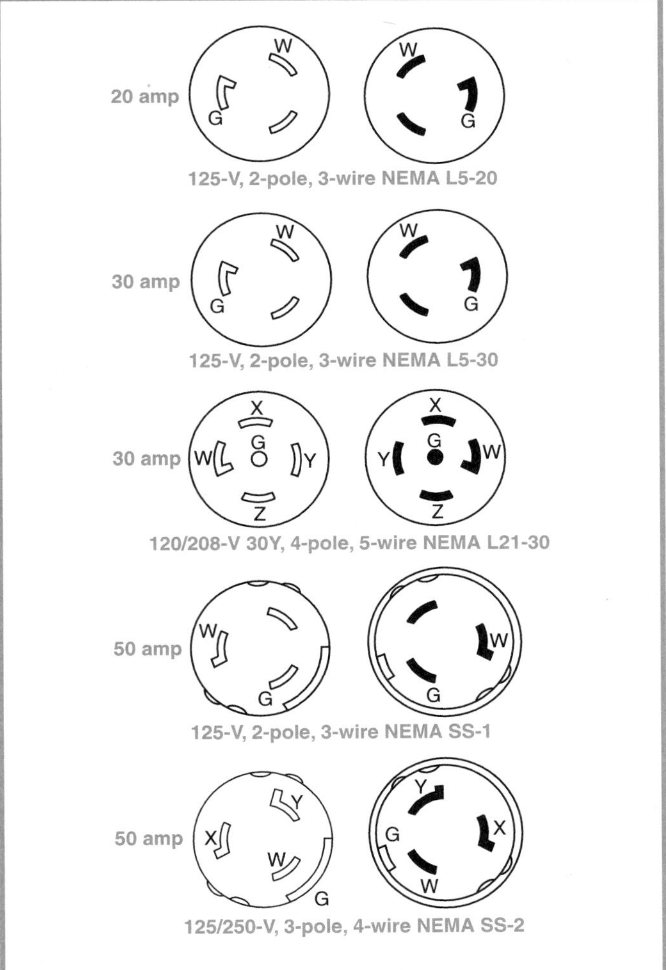

Figure 555.1 *Typical configurations for single locking and grounding receptacles and attachment plug caps used to provide shore power for boats in marinas and boatyards. These configurations are 20 amperes to 50 amperes.*

Figure 555.2 *Power outlet assembly suitable for use at docks, wharves, piers, and other locations that provide shore power to boats. (Hubbell)*

cles and attachment caps are required to ensure proper connections, to prevent unintentional disconnection of on-board equipment such as bilge pumps, refrigerators, and so on. Fifteen- and 20-ampere, single-phase, 125-volt receptacles other than those supplying shore power to boats and used for maintenance or other purposes at piers, wharves, and so on, may be of the general-purpose, nonlocking type and are required to be protected by ground-fault circuit interrupters. See Figure 210.17.

FPN No. 3 calls attention to the possibility of high waves (4 to 5 ft) often created by large cruisers, freighter traffic, and foul weather. This condition may cause outlets and boxes to become filled with water, which results in loss of power and corrosion of the receptacle and attachment plugs.

555-4. Disconnecting Means. A readily accessible disconnecting means shall be provided by which each boat can be isolated from its supply circuit. The disconnecting means shall consist of a circuit breaker or switch, or both, and shall be located within sight of the shore power connection and is intended to constitute the means of cutoff of the supply to the boat.

555-5. Branch Circuits. Each single receptacle that supplies shore power to boats shall be supplied from a power outlet or panelboard by an individual branch circuit of the voltage class and rating corresponding to the rating of the receptacle.

FPN: Supplying receptacles at voltages other than the voltages marked on the receptacle may cause overheating or malfunctioning of connected equipment, for example, supplying single-phase, 120/240-volt, 3-wire loads from a 208Y/120-volt, 3-wire source.

Either an individual or multiwire branch circuit may be used with a common neutral. See the commentary following Section 300-13(b) regarding device removal for multiwire branch circuits.

555-6. Feeders and Services. The load for each feeder and/or service circuit supplying receptacles that supply shore power for boats shall be calculated as follows. These calculations may be modified as indicated in sections (a) and (b).

Number of Receptacles	Sum of the Rating of the Receptacles (percent)
1–4	100
5–8	90
9–14	80
15–30	70
31–40	60
41–50	50
51–70	40
71–plus	30

(a) Where shore power accommodations provide two receptacles specifically for an individual boat slip and these receptacles have different voltages (for example, one 30 ampere, 125 volt and one 50 ampere, 125/250 volt), only the receptacle with the larger kilowatt demand shall be required to be calculated.

(b) If the facility being installed includes individual kilowatt-hour submeters for each slip, and is being calculated using the criteria listed in Section 555-6, the total demand amperes may be multiplied by 0.9 to achieve the final demand amperes.

FPN: These demand factors may be inadequate in areas of extreme hot or cold temperatures with loaded circuits for heating, air-conditioning, or refrigerating equipment.

555-7. Wiring Methods. The wiring method shall be of a type identified for use in wet locations. Extra-hard usage portable power cable listed for both wet locations and sunlight resistance shall be permitted for a feeder where flexibility is required.

Type W cables included in Table 400-4 may or may not be listed for wet locations.

FPN: For further information on wiring methods for various locations and for establishment of datum plane, see *Fire Protection Standard for Marinas and Boatyards,* NFPA 303-1995.

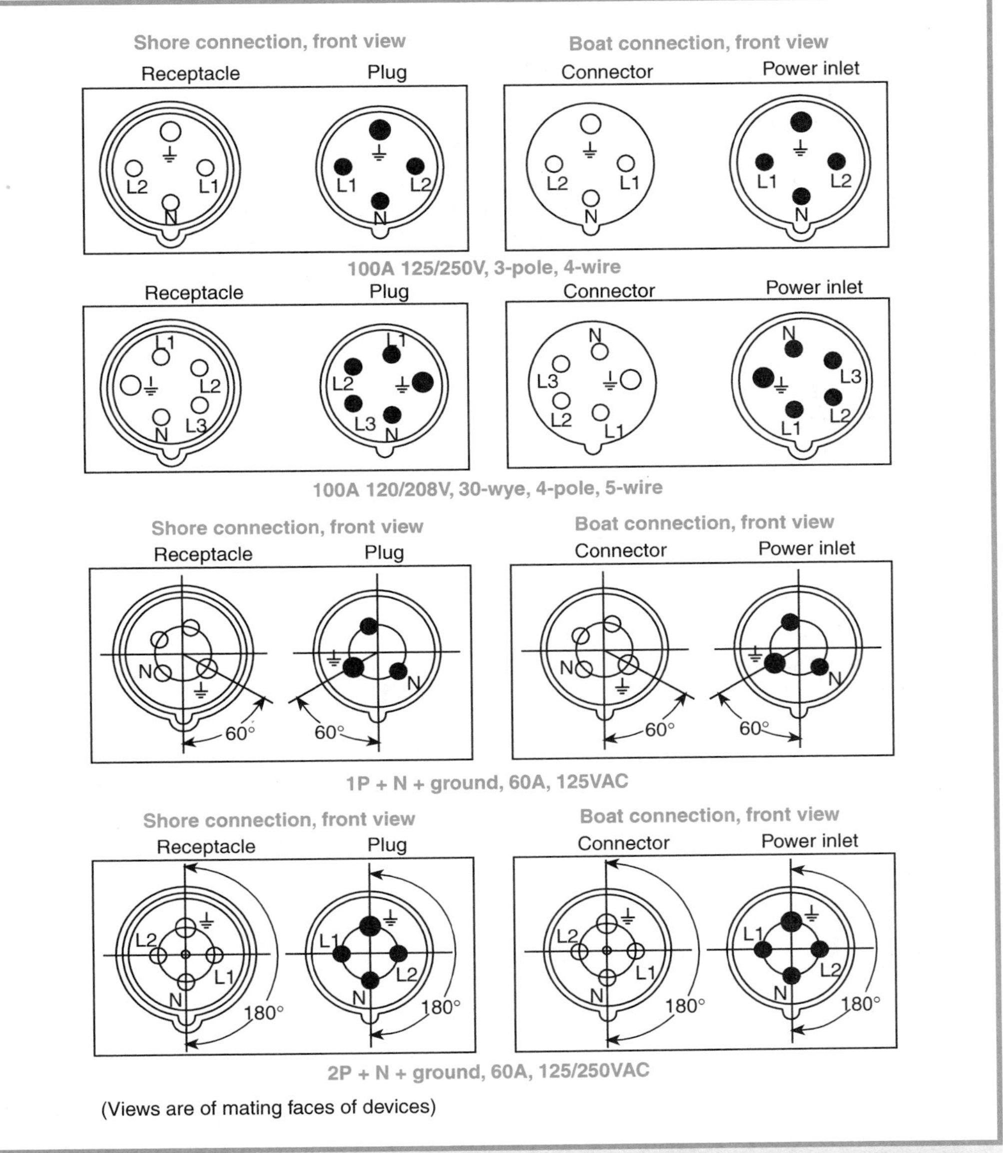

Figure 555.3 *Typical configurations for safety pin-and-sleeve-type receptacles, plugs, connectors, and power inlets used to provide shore power for boats in marinas and boatyards. These configurations are 60 amperes or 100 amperes.*

555-8. Grounding.

(a) Equipment to Be Grounded. The following items shall be connected to an equipment grounding conductor run with the circuit conductors in a raceway or cable:

(1) Boxes, cabinets, and all other metal enclosures
(2) Metal frames of utilization equipment
(3) Grounding terminals of grounding-type receptacles

(b) Type of Equipment Grounding Conductor. The equipment grounding conductor shall be an insulated copper conductor with a continuous outer finish that is either green or green with one or more yellow stripes.

Exception: The equipment grounding conductor of Type MI cable shall be permitted to be identified at terminations.

(c) Size of Equipment Grounding Conductor. The insulated copper equipment grounding conductor shall be sized in accordance with Section 250-122 but not smaller than No. 12.

(d) Branch-Circuit Equipment Grounding Conductor. The insulated equipment grounding conductor for branch circuits shall terminate at a grounding terminal in a remote panelboard or the grounding terminal in the main service equipment.

(e) Feeder Equipment Grounding Conductors. Where a feeder supplies a remote panelboard, an insulated equipment grounding conductor shall extend from a grounding terminal in the service equipment to a grounding terminal in the remote panelboard.

The purpose of Section 555-8 is to require an insulated equipment grounding conductor that will ensure a grounding circuit of high integrity. Because of corrosive conditions present in marinas and boatyards, metal raceways and boxes are not permitted to serve as equipment grounding conductors.

555-9. Wiring Over and Under Navigable Water. Wiring over and under navigable water shall be subject to approval by the authority having jurisdiction.

Some federal and local agencies have specific authority over navigable waterways. Therefore, approval of any proposed installation over or under such a waterway should be obtained from the appropriate authority.

555-10. Gasoline Dispensing Stations — Hazardous (Classified) Locations. Electrical equipment and wiring located in gasoline dispensing stations shall comply with Article 514.

Section 6-3 in NFPA 303-1995, *Fire Protection Standard for Marinas and Boatyards,* and NFPA 30A-1996, *Automotive and Marine Service Station Code,* include requirements pertaining to gasoline dispensing stations. See also Chapter 3 of NFPA 303 for electrical wiring and equipment requirements for marinas and boatyards.

FPN: For further information, see *Automotive and Marine Service Station Code*, NFPA 30A-1996, and *Fire Protection Standard for Marinas and Boatyards*, NFPA 303-1995.

555-11. Location of Service Equipment. The service equipment for floating docks or marinas shall be located adjacent to, but not on or in, the floating structure.

Section 555-11 is similar to Section 553-4 for floating buildings.

CHAPTER 6

Special Equipment

Article 600 — Electric Signs and Outline Lighting

Contents

A. General

600-1. Scope. This article covers the installation of conductors and equipment for electric signs and outline lighting as defined in Article 100.

FPN: As defined in Article 100, electric signs and outline lighting includes all products and installations utilizing neon tubing, such as signs, decorative elements, skeleton tubing, or art forms.

A common misunderstanding arises as to which requirements apply to signs, Article 410 or Article 600. Section 90-3 requires that the requirements in Chapters 1 through 4 be applied to Chapters 5, 6, and 7. Any special requirements for specific occupancies or equipment are found in Chapters 5, 6, or 7. Therefore, the requirements in Article 410 are applicable to signs, except where they are modified or amended in Article 600.

The requirements in Article 600 apply to electric signs of the fixed, stationary, or portable self-contained type. They may have letters or symbols that provide illumination, or they may be illuminated from a source other than a letter or symbol.

600-2. Definitions.

Electric-Discharge Lighting. Systems of illumination utilizing fluorescent lamps, high intensity discharge (HID) lamps, or neon tubing.

Neon Tubing. Electric-discharge tubing manufactured into shapes that form letters, parts of letters, skeleton tubing, outline lighting, other decorative elements, or art forms, and filled with various inert gases.

Sign Body. A portion of a sign that may provide protection from the weather, but is not an electrical enclosure.

Skeleton Tubing. Neon tubing that is itself the sign or outline lighting and not attached to an enclosure or sign body.

600-3. Listing. Electric signs and outline lighting — fixed, mobile, or portable — shall be listed and installed in conformance with that listing, unless otherwise approved by special permission.

(a) Field installed skeleton tubing shall not be required to be listed where installed in conformance with this *Code*.

(b) Outline lighting shall not be required to be listed as a system when it consists of listed lighting fixtures wired in accordance with Chapter 3.

Section 600-3 requires that any electric sign, including outline lighting, be listed, unless it is approved by special permission (i.e., the written consent of the authority having jurisdiction). A listed sign consists of the transformer, channel letter, plastic face, glass tube supports, raceways, glass cups or insulating boots if provided, and disconnecting means when provided by the manufacturer. A listed sign section usually consists of a channel letter, plastic face, glass tube supports, and metal-enclosed electrode receptacles. Sign sections or entire signs are listed for installation indoors or outdoors.

Section 600-3(a) allows skeleton tubing that is not listed to be field installed, provided it is installed in accordance with the *Code*. Section 600-3(b) allows nonlisted outline lighting that is made up of listed lighting fixtures to be installed, provided that it is installed in accordance with Chapter 3.

Large signs are often transported in several parts from the factory to the location where they are assembled. It is at this time that an inspection authority should be present to ensure that the components are assembled in conformity with their listing.

600-4. Markings.

(a) Signs and Outline Lighting Systems. Signs and outline lighting systems shall be marked with the manufacturer's name, trademark, or other means of identification; and, input voltage and current rating.

(b) Signs and outline lighting systems with incandescent lamp holders shall be marked to indicate the maximum allowable wattage of lamps. The markings shall be permanently installed, in letters at least ¼ in. (6.35 mm) high, and shall be located where visible during relamping.

600-5. Branch Circuits.

(a) Required Branch Circuit. Each commercial building and each commercial occupancy accessible to pedestrians shall be provided with at least one outlet in an accessible location at each entrance to each tenant space for sign or outline lighting system use. The outlet(s) shall be supplied by a branch circuit rated at least 20 amperes that supplies no other load. Service hallways or corridors shall not be considered accessible to pedestrians.

(b) Rating.

(1) Branch circuits that supply signs and outline lighting systems containing incandescent and fluorescent forms of illumination shall be rated not to exceed 20 amperes.

(2) Branch circuits that supply neon tubing installations shall not be rated in excess of 30 amperes.

Sections 600-5(b)(1) and (2) were revised for the 1999 *Code* to reflect that the branch-circuit ratings cannot exceed 20 and 30 amperes, respectively. Large signs often require higher ampacities than those permitted by Section 600-5(b). The revised requirement allows a common practice of supplying larger signs with feeders that often require more capacity.

(3) Computed Load. The load for the required branch circuit shall be computed at a minimum of 1200 volt-amperes.

Signs for commercial or industrial occupancies are normally in use for 3 hours or longer and are, therefore, considered continuous loads, which are not permitted to exceed 80 percent of the rating of the branch circuit (16 amperes for a 20-ampere branch circuit).

(c) Wiring Methods.

(1) Supply. The wiring method used to supply signs and outline lighting systems shall terminate within a sign, an outline lighting system enclosure, a suitable box, or a conduit body.

(2) Enclosures as Pull Boxes. Signs and transformer enclosures shall be permitted to be used as pull or junction boxes for conductors supplying other adjacent signs, outline lighting systems, or floodlights that are part of a sign, and shall be permitted to contain both branch and secondary circuit conductors.

(3) Metal poles used to support signs shall be permitted to enclose supply conductors, provided the poles and conductors are installed in accordance with Section 410-15(b).

600-6. Disconnects. Each sign and outline lighting system, or feeder circuit or branch circuit supplying a sign or outline lighting system, shall be controlled by an externally operable switch or circuit breaker that will open all ungrounded conductors. Signs and outline lighting systems located within fountains shall have the disconnect located in accordance with Section 680-12.

Exception No. 1: A disconnecting means shall not be required for an exit directional sign located within a building.

Exception No. 2: A disconnecting means shall not be required for cord-connected signs with an attachment plug.

(a) Location. The disconnecting means shall be within sight of the sign or outline lighting system that it controls. Where the disconnecting means is out of the line of sight from any section that may be energized, the disconnecting means shall be capable of being locked in the open position.

Signs or outline lighting systems operated by electronic or electromechanical controllers located external to the sign or outline lighting system shall be permitted to have a disconnecting means located within sight of the controller or in the same enclosure with the controller. The disconnecting means shall disconnect the sign or outline lighting system and the controller from all ungrounded supply conductors. It shall be designed so that no pole can be operated independently and it shall be capable of being locked in the open position.

Clarified for the 1999 *NEC*, each feeder or branch circuit supplying a sign must have an externally operable switch or circuit breaker to open all the ungrounded conductors. This disconnect must be within sight of the sign or have the capability of being locked in the open position when located so as not to be within sight. See Figures 600.1 and 600.2.

Figure 600.1 *Feeder disconnecting means for multiple supply circuits for a sign.*

(b) Control Switch Rating. Switches, flashers, and similar devices controlling transformers and electronic power supplies shall be rated for controlling inductive loads or have a current rating not less than twice the current rating of the transformer.

FPN: See Section 380-14 for rating of snap switches.

A switching device that controls the primary circuit of a transformer supplying a luminous gas tube is subject to a highly inductive load that causes severe arcing of its contacts. Therefore, the switch or flasher is required to be rated for the inductive load, be a general-use ac snap switch, or have a current rating that is at least twice the current rating of the transformer it controls.

Section 600-6(c) was deleted for the 1999 *Code* because the requirements were more appropriately located in applicable product standards.

600-7. Grounding. Signs and metal equipment of outline lighting systems shall be grounded. Listed flexible metal conduit or listed liquidtight flexible metal conduit that encloses the secondary wiring of a transformer or power supply for use with electric discharge tubing shall be permitted as a bonding means in lengths not exceeding 100 ft (30.5 m). Small metal parts not exceeding 2 in. (50.8 mm) in any dimension, not likely to be energized, and spaced at least ¾ in. (19 mm) from neon tubing shall not require bonding. Where listed nonmetallic conduit is used to enclose the secondary wiring of a transformer or power supply and a bonding conductor is required, the bonding conductor shall be installed separate and remote from the nonmetallic conduit and be spaced at least 1½ in. (38 mm) from the conduit when the circuit is operated at 100 Hz or less or 1¾ in. (44.45 mm) when the circuit is operated at over 100 Hz. Bonding conductors shall be copper and not smaller than No. 14. Metal parts of a building shall not be permitted as a grounded or equipment grounding conductor.

Section 600-7 was revised for the 1999 *Code*. The revised Section 600-7 states that signs and metal equipment of outline lighting systems are to be grounded. All metal parts larger than 2 in. are to be bonded. The revised section also allows flexible metal conduit and liquidtight flexible metal conduit to be used in lengths up to 100 ft as a bonding means. If rigid or flexible nonmetallic conduit is used on secondary wiring of transformers and power supplies, copper bonding conductors must be located external to the nonmetallic raceway. Bonding conductor spacing is required to be at least 1½ in. for circuit frequencies of 100 Hz or less and 1¾ in. for circuit frequencies over 100 Hz. These raceways normally contain only one conductor, which is connected to one side of the neon tube. When rigid nonmetallic conduit or liquidtight flexible nonmetallic conduit are used and any sign parts are required to be bonded, the bonding conductor(s) must be run outside of and be separated from the nonmetallic conduit. Installing bonding conductors inside the nonmetallic conduit with

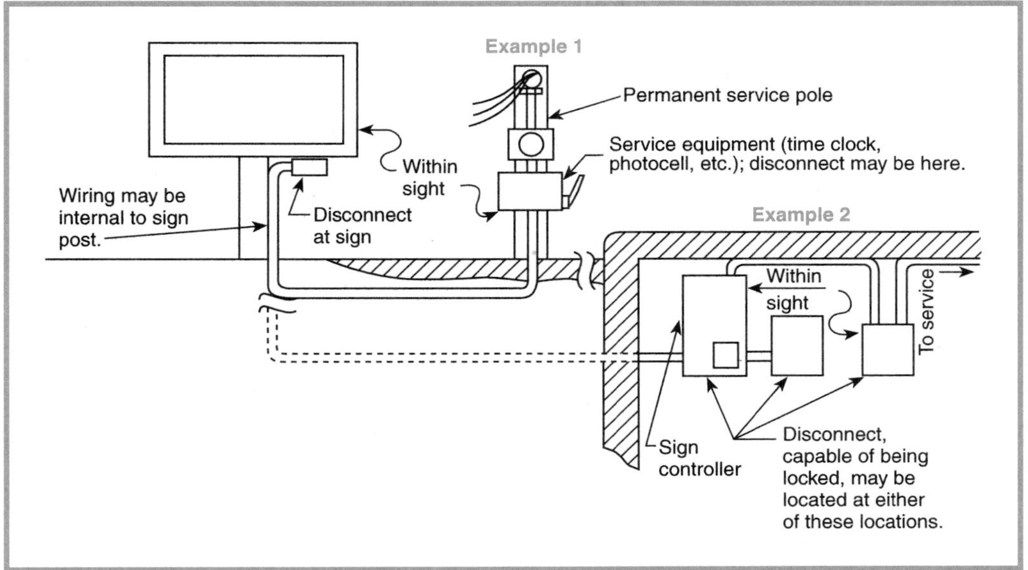

Figure 600.2 *Example 1 shows the disconnecting means placement that satisfies the requirements of Section 600-6(a). Example 2 illustrates the location of a disconnecting means that is not in sight from or more than 50 ft away from the sign.*

secondary power supply conductors is not permitted since it could increase the chance of failure of the conductor or nonmetallic tubing.

FPN: Refer to Section 600-32(j) for additional restrictions on length of high-voltage secondary conductors.

600-8. Enclosures. Live parts other than lamps and neon tubing shall be enclosed.

Exception: A transformer or electronic power supply provided with an integral enclosure, including a primary and secondary circuit splice enclosure, shall not be required to be provided with an additional enclosure.

(a) Strength. Enclosures shall have ample structural strength and rigidity.

(b) Material. Sign and outline lighting system enclosures shall be constructed of metal or shall be listed.

(c) Minimum Thickness of Enclosure Metal. Sheet copper or aluminum shall be at least 0.020 in. (508 μm) thick. Sheet steel shall be at least 0.016 in. (406 μm) (No. 28 MSG) thick.

(d) Protection of Metal. Metal parts of equipment shall be protected from corrosion.

600-9. Location.

(a) Vehicles. Sign or outline lighting system equipment shall be at least 14 ft (4.4 m) above areas accessible to vehicles unless protected from physical damage.

(b) Pedestrians. Neon tubing accessible to pedestrians, other than dry-location portable signs, shall be protected from physical damage.

(c) Adjacent to Combustible Materials. Signs and outline lighting systems shall be installed so that adjacent combustible materials shall not be subjected to temperatures in excess of 90°C (194°F).

The spacing between wood or other combustible materials and an incandescent or HID lamp or lampholder shall not be less than 2 in. (50.8 mm).

(d) Wet Location. Signs and outline lighting system equipment for wet location use, other than listed watertight type, shall be weatherproof and have drain holes, as necessary, in accordance with the following.

(1) Drain holes shall not be larger than ½ in. (12.7 mm) or smaller than ¼ in. (6.35 mm).
(2) Every low point or isolated section of the equipment shall have at least one drain hole.
(3) Drain holes shall be positioned such that there will be no external obstructions.

600-10. Portable or Mobile Signs.

Section 600-10 is not intended to apply to portable or mobile electric signs, such as those mounted on trailers.

(a) Support. Portable or mobile signs shall be adequately supported and readily movable without the use of tools.

(b) Attachment Plug. An attachment plug shall be provided for each portable or mobile sign.

(c) Wet or Damp Location. Portable or mobile signs in wet or damp locations shall meet all of the following.

(1) Cords. All cords shall be junior hard service or hard service types as designated in Table 400-4, and have an equipment grounding conductor.

(2) Ground-Fault Circuit Interrupter. Portable or mobile signs shall be provided with factory-installed ground-fault circuit-interrupter protection for personnel. The ground-fault circuit interrupter shall be an integral part of the attachment plug or shall be located in the power-supply cord within 12 in. (305 mm) of the attachment plug.

(d) Dry Location. Portable or mobile signs in dry locations shall meet the following.

(1) Cords shall be SP-2, SPE-2, SPT-2, or heavier, as designated in Table 400-4.
(2) The cord shall not exceed 15 ft (4.57 m) in length.

It is the intent that a ground-fault circuit interrupter identified for use with portable electric signs be provided with open neutral protection so that the ground-fault circuit interrupter provides protection even if the neutral conductor supplying the sign and ground-fault circuit interrupter opens.

Figure 600.3 A factory-installed GFCI is required as an integral part of the attachment plug or is required to be located in the power supply cord within 12 in. of the attachment plug.

600-21. Ballasts, Transformers, and Electronic Power Supplies.

(a) Accessibility. Ballasts, transformers, and electronic power supplies shall be located where accessible and shall be securely fastened in place.

(b) Location. Ballasts, transformers, and electronic power supplies shall be installed as near to the lamps or neon tubing as practicable to keep the secondary conductors as short as possible.

(c) Wet Location. Ballasts, transformers, and electronic power supplies used in wet locations shall be of the weatherproof type or be of the outdoor type and protected from the weather by placement in a sign body or separate enclosure.

(d) Working Space. A working space at least 3 ft (914 mm) high, 3 ft (914 mm) wide, by 3 ft (914 mm) deep shall be provided at each ballast, transformer, and electronic power supply or its enclosure where not installed in a sign.

(e) Attic Locations. Ballasts, transformers, and electronic power supplies shall be permitted to be located in attics and soffits, provided there is an access door at least 3 ft (914 mm) by 2 ft (610 mm) and a passageway of at least 3 ft (914 mm) high by 2 ft (610 mm) wide with a suitable permanent walkway at least 12 in. (305 mm) wide extending from the point of entry to each component.

(f) Suspended Ceilings. Ballasts, transformers, and electronic power supplies shall be permitted to be located above suspended ceilings, provided their enclosures are securely fastened in place and not dependent on the suspended ceiling grid for support.

600-22. Ballasts.

(a) Type. Ballasts shall be identified for the use and shall be listed.

(b) Thermal Protection. Ballasts shall be thermally protected.

600-23. Transformers and Electronic Power Supplies.

(a) Type. Transformers and electronic power supplies shall be identified for the use and shall be listed.

(b) Secondary-Circuit Ground-Fault Protection. Transformers and electronic power supplies other than the following shall have secondary-circuit ground-fault protection:

(1) Transformers with isolated secondaries and with a maximum open circuit voltage of 7500 volts or less
(2) Transformers with integral porcelain or glass secondary housing for the neon tubing and requiring no field wiring of the secondary circuit

(c) Voltage. Secondary-circuit voltage shall not exceed 15,000 volts, nominal, under any load condition. The voltage to ground of any output terminals of the secondary circuit shall not exceed 7500 volts, under any load conditions.

(d) Rating. Transformers and electronic power supplies shall have a secondary-circuit current rating of not more than 300 mA.

(e) Secondary Connections. Secondary circuit outputs shall not be connected in parallel or in series.

(f) Marking. A transformer or power supply must be marked to indicate that it has secondary fault protection.

B. Field-Installed Skeleton Tubing

600-30. Applicability. Part B of this article shall apply only to field-installed skeleton tubing. These requirements are in addition to the requirements of Part A.

Field-installed skeleton tubing often involves the use of electrically isolated metal components, such as tube supports, fasteners, or decorative channel over the tubing. These metallic components should be located so as to minimize their becoming energized.

600-31. Neon Secondary-Circuit Conductors, 1000 Volts or Less, Nominal.

(a) Wiring Method. Conductors shall be installed using any wiring method included in Chapter 3 suitable for the conditions.

(b) Insulation and Size. Conductors shall be insulated, listed for the purpose, and not smaller than No. 18.

(c) Number of Conductors in Raceway. The number of conductors in a raceway shall be in accordance with Table 1 of Chapter 9.

(d) Installation. Conductors shall be installed so they are not subject to physical damage.

(e) Protection of Leads. Bushings shall be used to protect wires passing through an opening in metal.

600-32. Neon Secondary Circuit Conductors, Over 1000 Volts, Nominal.

(a) Wiring Method. Conductors shall be installed on insulators, in rigid metal conduit, intermediate metal conduit, rigid nonmetallic conduit, liquidtight flexible nonmetallic conduit, flexible metal conduit, liquidtight flexible metal conduit, electrical metallic tubing, metal enclosures, or other equipment listed for the purpose. Wiring methods shall be installed in accordance with the requirements of Chapter 3. Only one conductor shall be installed per length of conduit or tubing. Nonmetallic conduit or flexible nonmetallic conduit, when operated at 100 Hz or less, shall be spaced at least 1½ in. (38 mm) from grounded or bonded parts; and, when operated at over 100 Hz, shall be spaced at least 1¾ in. (44.45 mm) from grounded or bonded parts. Metal parts of a building shall not be used as a grounded or equipment grounding conductor.

See the commentary following Section 600-7.

(b) Insulation and Size. Conductors shall be insulated, listed for the purpose, rated for the voltage, not smaller than No. 18, and have a minimum temperature rating of 105°C (221°F).

(c) Installation. Conductors shall be installed so they are not subject to physical damage.

(d) Bends in Conductors. Sharp bends in insulated conductors shall be avoided.

(e) Spacing. Conductors shall be separated from each other and from all objects other than insulators or neon tubing by a spacing of not less than 1½ in. (38 mm). GTO cable installed in metal conduit or tubing requires no spacing between the cable insulation and the conduit or tubing.

New spacing requirements for GTO cable were added for the 1999 *Code*. GTO cable insulation is sufficient to withstand the stresses caused by locating it near a grounded surface. Also see the commentary following Section 600-7.

(f) Insulators and Bushings. Insulators and bushings for conductors shall be listed for the purpose.

(g) Conductors in Raceways.

(1) Damp or Wet Locations. In damp or wet locations, the insulation on all conductors shall extend not less than 4 in. (102 mm) beyond the metal conduit or tubing.

(2) Dry Locations. In dry locations, the insulation on all conductors shall extend not less than 2½ in. (64 mm) beyond the metal conduit or tubing.

(h) Between Neon Tubing and Grounded Midpoint. Conductors shall be permitted to run from the ends of neon tubing to the grounded midpoint of transformers or electronic power supplies listed for the purpose and provided with terminals at the midpoint. Where such connections are made to the grounded midpoint, the connections between the high-voltage terminals and the line ends of the neon tubing shall be as short as possible.

(i) Dwelling Occupancies. Equipment having an open circuit voltage exceeding 1000 volts shall not be installed in or on dwelling occupancies.

(j) Length of High-Voltage Cable. Not more than 20 ft (7 m) of high-voltage cable shall be permitted in metal conduit or tubing from a high-voltage terminal of a transformer/ power supply to the first neon tube. Not more than 50 ft (15.2 m) of high-voltage cable shall be permitted in nonmetallic conduit from a high-voltage terminal of a transformer/power supply to the first neon tube.

600-41. Neon Tubing.

(a) Design. The length and design of the tubing shall not cause a continuous overcurrent beyond the design loading of the transformer or electronic power supply.

(b) Support. Tubing shall be supported by listed tube supports.

(c) Spacing. A spacing of not less than ¼ in. (6.5 mm) shall be maintained between the tubing and the nearest surface, other than its support.

Electric discharge tubing is required to be of such length and design that it will not cause a continuous overvoltage on the transformer. A tube that is too long or too small in diameter increases the impedance of the load and, thus, stresses the transformer insulation. Gas tube sign transformers are designed to operate at or near short-circuit current. Generally, the primary voltage of the transformers is 120 volts, and proper installation and maintenance of transformers and high-voltage secondary conductors minimizes the possibility of injury or fire. Precautions should be taken to ensure that secondary conductors are

properly terminated to the tube electrodes and that these connections are protected from contact by unauthorized persons or by any flammable or combustible material. Broken tubes should be replaced or de-energized.

600-42. Electrode Connections.

(a) Accessibility. Terminals of the electrode shall not be accessible to unqualified persons.

(b) Electrode Connections. Connections shall be made by use of a connection device, twisting the wires together, or use of an electrode receptacle. Connections shall be electrically and mechanically secure and shall be in an enclosure listed for the purpose.

(c) Support. The neon tubing and conductor shall be supported not more than 6 in. (152 mm) from the electrode connection.

(d) Receptacles. Electrode receptacles shall be listed for the purpose.

(e) Bushings. Where electrodes penetrate an enclosure, bushings listed for the purpose shall be used, unless receptacles are provided.

(f) Wet Locations. A listed cap shall be used to close the opening between neon tubing and a receptacle where the receptacle penetrates a building. Where a bushing or neon tubing penetrates a building, the opening between neon tubing and the bushing shall be sealed.

(g) Electrode Enclosures. Electrode enclosures shall be listed for the purpose.

Article 604 — Manufactured Wiring Systems

Contents

604-1. Scope. The provisions of this article apply to field-installed wiring using off-site manufactured subassemblies for branch circuits, remote-control circuits, signaling circuits, and communications circuits in accessible areas.

604-2. Definition.

Manufactured Wiring System. A system containing component parts that are assembled in the process of manufacture and cannot be inspected at the building site without damage or destruction to the assembly.

604-3. Other Articles. Except as modified by the requirements of this article, all other applicable articles of this *Code* shall apply.

604-4. Uses Permitted. The manufactured wiring systems shall be permitted in accessible and dry locations and in plenums and spaces used for environmental air, where listed for this application and installed in accordance with Section 300-22.

Article 604 covers manufactured wiring systems that are typically comprised of cables and devices with molded plugs and receptacles. These systems greatly reduce installation time and are becoming more widely used and accepted.

For further information, see Article 100 for the definition of *accessible (as applied to wiring methods).*

Exception No. 1: In concealed spaces, one end of tapped cable shall be permitted to extend into hollow walls for direct termination at switch and outlet points.

Exception No. 2: For use in outdoor locations where listed for the purpose.

Exception No. 2 permits the manufacture and installation of listed manufactured wiring systems for use in outdoor locations.

604-5. Uses Not Permitted. Where conductors or cables are limited by the provisions in Articles 333 and 334.

604-6. Construction.

(a) Cable or Conduit Types.

(1) Cable shall be listed armored cable or metal-clad cable containing nominal 600-volt No. 10 or 12 copper-insulated conductors with a bare or insulated copper equipment grounding conductor equivalent in size to the ungrounded conductor.

(2) Conduit shall be listed flexible metal conduit or listed liquidtight flexible conduit containing nominal 600-volt No. 10 or 12 copper-insulated conductors with a bare or insulated copper equipment grounding conductor equivalent in size to the ungrounded conductor.

Exception No. 1 to (1) and (2): A fixture tap, maximum 6 ft (1.83 m) long, intended for connection to a single fixture shall be permitted to contain conductors smaller than No. 12 but not smaller than No. 18.

Exception No. 1 allows metal-clad cable that is not longer than 6 ft to contain conductors smaller than No. 12 but not smaller than No. 18, as illustrated in Figure 604.1.

Figure 604.1 An example of metal-clad cable not longer than 6 ft that contains conductors smaller than No. 12. (AFC Cable Systems, Inc.)

Exception No. 2 to (1) and (2): Conductors smaller than No. 12 shall be permitted for remote-control, signaling, or communications circuits. The assembly shall be listed for the purpose.

Exception No. 3 to (1) and (2): Flexible cord suitable for hard usage, with minimum No. 12 conductors, shall be permitted as part of a listed factory-made assembly not exceeding 6 ft (1.83 m) in length when making a transition between components of a manufactured wiring system and utilization equipment, not permanently secured to the building structure. The cord shall be visible for its entire length and shall not be subject to strain or physical damage.

Exception No. 3 was added to the 1999 *Code* to permit a transition between manufactured wiring systems and utilization equipment found in display cases, merchandise racks, temporary work stations, and the like. This transition is limited, however, to hard-usage cord not over 6 ft in length to minimize damage, as illustrated in Figure 604.2.

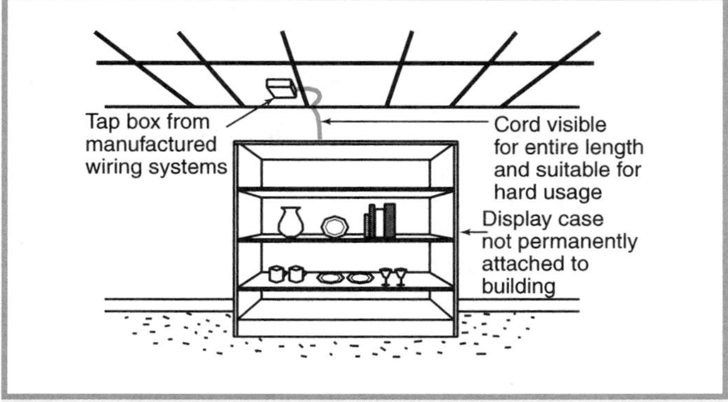

Figure 604.2 Transition wiring between manufactured wiring systems, and utilization equipment.

(3) Each section shall be marked to identify the type of cable or conduit.

(b) Receptacles and Connectors. Receptacles and connectors shall be of the locking type, uniquely polarized and identified for the purpose, and shall be part of a listed assembly for the appropriate system.

Examples of polarized receptacles and connectors are shown in Figure 604.3.

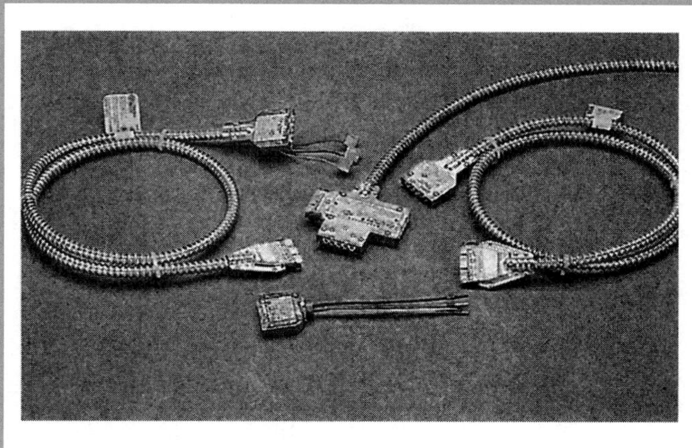

Figure 604.3 Components of a manufactured wiring system. (RELOC, Division of Lithonia Lighting)

(c) Other Component Parts. Other component parts shall be listed for the appropriate system.

604-7. Unused Outlets. All unused outlets shall be capped to effectively close the connector openings.

Article 605 — Office Furnishings (Consisting of Lighting Accessories and Wired Partitions)

Contents

605-8. Freestanding-Type Partitions, Cord and Plug Connected
(a) Flexible Power-Supply Cord
(b) Receptacle Supplying Power
(c) Receptacle Outlets, Maximum
(d) Multiwire Circuits, Not Permitted

605-1. Scope. This article covers electrical equipment, lighting accessories, and wiring systems used to connect, or contained within, or installed on relocatable wired partitions.

605-2. General. Wiring systems shall be identified as suitable for providing power for lighting accessories and appliances in wired partitions. These partitions shall not extend from floor to ceiling.

Exception: Where permitted by the authority having jurisdiction, these relocatable wired partitions shall be permitted to extend to the ceiling but shall not penetrate the ceiling.

(a) Use. These assemblies shall be installed and used only as provided for by this article.

(b) Other Articles. Except as modified by the requirements of this article, all other articles of this *Code* shall apply.

(c) Hazardous (Classified) Locations. Where used in hazardous (classified) locations, these assemblies shall conform with Articles 500 through 517 in addition to this article.

605-3. Wireways. All conductors and connections shall be contained within wiring channels of metal or other material identified as suitable for the conditions of use. Wiring channels shall be free of projections or other conditions that may damage conductor insulation.

A wiring channel may be provided within the system components for the routing of communications and low-voltage wiring, separate from the electrical raceway.

605-4. Partition Interconnections. The electrical connection between partitions shall be a flexible assembly identified for use with wired partitions or shall be permitted to be installed using flexible cord provided all the following conditions are met.

(1) The cord is extra-hard usage type.
(2) The partitions are mechanically contiguous.
(3) The cord is not longer than necessary for maximum positioning of the partitions but is in no case to exceed 2 ft (610 mm).
(4) The cord is terminated at an attachment plug and cord connector with strain relief.

605-5. Lighting Accessories. Lighting equipment listed and identified for use with wired partitions shall comply with all of the following.

(a) Support. A means for secure attachment or support shall be provided.

(b) Connection. Where cord and plug connection is provided, the cord length shall be suitable for the intended application but shall not exceed 9 ft (2.74 m) in length. The cord shall not be smaller than No. 18, shall contain an equipment grounding conductor, and shall be of the hard usage type. Connection by other means shall be identified as suitable for the condition of use.

(c) Receptacle Outlet. Convenience receptacles shall not be permitted in lighting accessories.

605-6. Fixed-Type Partitions. Wired partitions that are fixed (secured to building surfaces) shall be permanently connected to the building electrical system by one of the wiring methods of Chapter 3.

605-7. Freestanding-Type Partitions. Partitions of the freestanding type (not fixed) shall be permitted to be permanently connected to the building electrical system by one of the wiring methods of Chapter 3.

605-8. Freestanding-Type Partitions, Cord and Plug Connected. Individual partitions of the freestanding type, or groups of individual partitions that are electrically connected, mechanically contiguous, and do not exceed 30 ft (9.14 m) when assembled, shall be permitted to be connected to the building electrical system by a single flexible cord and plug, provided all of the following conditions are met.

(a) Flexible Power-Supply Cord. The flexible power-supply cord shall be extra-hard usage type with No. 12 or larger conductors with an insulated equipment grounding conductor and not exceeding 2 ft (610 mm) in length.

(b) Receptacle Supplying Power. The receptacle(s) supplying power shall be on a separate circuit serving only panels and no other loads and shall be located not more than 12 in. (305 mm) from the partition that is connected to it.

(c) Receptacle Outlets, Maximum. Individual partitions or groups of interconnected individual partitions shall not contain more than thirteen 15-ampere, 125-volt receptacle outlets.

(d) Multiwire Circuits, Not Permitted. Individual partitions or groups of interconnected individual partitions shall not contain multiwire circuits.

Partitions (other than the cord- and plug-connected type described in Section 605-8) are permitted to contain multiwire branch circuits.

FPN: See Section 210-4 for circuits supplying partitions in Sections 605-6 and 605-7.

Article 610 — Cranes and Hoists

Contents

A. General

610-1. Scope. This article covers the installation of electrical equipment and wiring used in connection with cranes, monorail hoists, hoists, and all runways.

FPN: For further information, see *Safety Code for Cranes, Derricks, Hoists, Jacks, and Slings*, ANSI B-30.

610-2. Special Requirements for Particular Locations.

(a) Hazardous (Classified) Locations. All equipment that operates in a hazardous (classified) location shall conform to Article 500.

(1) Equipment used in locations that are hazardous because of the presence of flammable gases or vapors shall conform to Article 501.

(2) Equipment used in locations that are hazardous because of combustible dust shall conform to Article 502.

(3) Equipment used in locations that are hazardous because of the presence of easily ignitible fibers or flyings shall conform to Article 503.

See the commentary following Section 503-13(d).

(b) Combustible Materials. Where a crane, hoist, or monorail hoist operates over readily combustible material, the resistors shall be located as permitted in the following:

(1) In a well-ventilated cabinet composed of noncombustible material constructed so that it will not emit flames or molten metal
(2) In a cage or cab constructed of noncombustible material that encloses the sides of the cage or cab from the floor to a point at least 6 in. (152 mm) above the top of the resistors

(c) Electrolytic Cell Lines. See Section 668-32.

Special precautions are necessary on electrolytic cell lines to prevent the introduction of exposed grounded parts, as described in Section 668-32.

B. Wiring

610-11. Wiring Method. Conductors shall be enclosed in raceways or be Type AC cable with insulated grounding conductor, Type MC cable, or Type MI cable unless otherwise permitted in (a) through (e).

Type AC cable with an insulated grounding conductor is permitted as an acceptable wiring method. The insulated equipment grounding conductor in Type AC cable ensures a reliable grounding path.

(a) Contact conductors are not required to be enclosed in raceways.

(b) Short lengths of open conductors at resistors, collectors, and other equipment are not required to be enclosed in raceways.

(c) Where flexible connections are necessary to motors and similar equipment, flexible stranded conductors shall be installed in flexible metal conduit, liquidtight flexible metal conduit, liquidtight flexible nonmetallic conduit, multiconductor cable, or an approved nonmetallic enclosure.

Section 610-11(b) permits the use of short lengths of open wiring on cranes and hoists, where a separately bushed hole from a box or fitting is provided for each conductor and is used for the connection of resistors, collectors, or similar equipment. In addition to other types of raceways and cables, Section 610-11(c) permits flexible metal conduit, liquidtight flexible metal conduit, liquidtight flexible nonmetallic conduit, multiconductor cable, or an approved nonmetallic enclosure (tubing) where flexibility is necessary.

(d) Where multiconductor cable is used with a suspended pushbutton station, the station shall be supported in some satisfactory manner that protects the electric conductors against strain.

(e) Where flexibility is required for power or control to moving parts, a cord suitable for the purpose shall be permitted provided

(1) Suitable strain relief and protection from physical damage is provided, and
(2) In Class 1, Division 2 locations, cord shall be approved for extra-hard usage.

Figure 610.1 is one example of suitable strain relief for the cord.

610-12. Raceway or Cable Terminal Fittings. Conductors leaving raceways or cables shall comply with one of the following.

(a) Separately Bushed Hole. A box or terminal fitting that has a separately bushed hole for each conductor shall be used wherever a change is made from a raceway or cable to open wiring. A fitting used for this purpose shall not contain taps or splices and shall not be used at fixture outlets.

(b) Bushing in Lieu of a Box. A bushing shall be permitted to be used in lieu of a box at the end of a rigid metal conduit, intermediate metal conduit, or electrical metallic

Figure 610.1 *A suitable grip for strain relief with a suspended pushbutton station. (Hubbell, Kellems Division)*

tubing where the raceway terminates at unenclosed controls or similar equipment including contact conductors, collectors, resistors, brakes, power-circuit limit switches, and dc split-frame motors.

610-13. Types of Conductors. Conductors shall comply with Table 310-13 unless otherwise permitted in (a) through (d).

(a) A conductor(s) exposed to external heat or connected to resistors shall have a flame-resistant outer covering or be covered with flame-resistant tape individually or as a group.

(b) Contact conductors along runways, crane bridges, and monorails shall be permitted to be bare, and shall be copper, aluminum, steel, or other alloys or combinations thereof in the form of hard drawn wire, tees, angles, tee rails, or other stiff shapes.

(c) Where flexibility is required, flexible cord or cable shall be permitted to be used and, where necessary, cable reels or take-up devices shall be used.

(d) Conductors for Class 1, Class 2, and Class 3 remote-control, signaling, and power-limited circuits, installed in accordance with Article 725, shall be permitted.

Section 610-13(d) was added for the 1999 *Code* to include a reference to Article 725 for conductors in Class 1, Class 2, and Class 3 control circuits. Since Article 725 does not modify Article 610, this section now permits wiring methods found in Article 725 to be used for control circuits.

610-14. Rating and Size of Conductors.

(a) Ampacity. The allowable ampacities of conductors shall be as shown in Table 610-14(a).

Table 610-14(a). Ampacities of Insulated Copper Conductors Used with Short-Time Rated Crane and Hoist Motors. Based on Ambient Temperature of 30°C (86°F). Up to Four Conductors in Raceway or Cable[1] Up to 3 ac[2] or 4 dc[1] Conductors in Raceway or Cable

Maximum Operating Temperature	75°C (167°F)		90°C (194°F)		125°C (257°F)		Maximum Operating Temperature
	Types MTW, RH, RHW, THW, THWN, XHHW, USE, ZW		Types TA, TBS, SA, SIS, PFA, FEP, FEPB, RHH, THHN, XHHW, Z, ZW		Types FEP, FEPB, PFA, PFAH, SA, TFE, Z, ZW		
Size (AWG or kcmil)	60 Min	30 Min	60 Min	30 Min	60 Min	30 Min	Size (AWG or kcmil)
16	10	12	—	—	—	—	16
14	25	26	31	32	38	40	14
12	30	33	36	40	45	50	12
10	40	43	49	52	60	65	10
8	55	60	63	69	73	80	8
6	76	86	83	94	101	119	6
5	85	95	95	106	115	134	5
4	100	117	111	130	133	157	4
3	120	141	131	153	153	183	3
2	137	160	148	173	178	214	2
1	143	175	158	192	210	253	1
1/0	190	233	211	259	253	304	1/0
2/0	222	267	245	294	303	369	2/0
3/0	280	341	305	372	370	452	3/0
4/0	300	369	319	399	451	555	4/0
250	364	420	400	461	510	635	250
300	455	582	497	636	587	737	300
350	486	646	542	716	663	837	350
400	538	688	593	760	742	941	400
450	600	765	660	836	818	1042	450
500	660	847	726	914	896	1143	500

AMPACITY CORRECTION FACTORS

Ambient Temperature (°C)	For ambient temperatures other than 30°C (86°F), multiply the ampacities shown above by the appropriate factor shown below.						Ambient Temperature (°F)
21–25	1.05	1.05	1.04	1.04	1.02	1.02	70–77
26–30	1.00	1.00	1.00	1.00	1.00	1.00	79–86
31–35	0.94	0.94	0.96	0.96	0.97	0.97	88–95
36–40	0.88	0.88	0.91	0.91	0.95	0.95	97–104
41–45	0.82	0.82	0.87	0.87	0.92	0.92	106–113
46–50	0.75	0.75	0.82	0.82	0.89	0.89	115–122
51–55	0.67	0.67	0.76	0.76	0.86	0.86	124–131
56–60	0.58	0.58	0.71	0.71	0.83	0.83	133–140
61–70	0.33	0.33	0.58	0.58	0.76	0.76	142–158
71–80	—	—	0.41	0.41	0.69	0.69	160–176
81–90	—	—	—	—	0.61	0.61	177–194
91–100	—	—	—	—	0.51	0.51	195–212
101–120	—	—	—	—	0.40	0.40	213–248

Note: Other insulations shown in Table 310-13 and approved for the temperature and location shall be permitted to be substituted for those shown in Table 610-14(a). The allowable ampacities of conductors used with 15-minute motors shall be the 30-minute ratings increased by 12 percent.

[1]For 5 to 8 simultaneously energized power conductors in raceway or cable, the ampacity of each power conductor shall be reduced to a value of 80 percent of that shown in the table.

[2]For 4 to 6 simultaneously energized 125°C (257°F) ac power conductors in raceway or cable, the ampacity of each power conductor shall be reduced to a value of 80 percent of that shown in the table.

FPN: For the ampacities of conductors between controllers and resistors, see Section 430-23.

(b) Secondary Resistor Conductors. Where the secondary resistor is separate from the controller, the minimum size of the conductors between controller and resistor shall be calculated by multiplying the motor secondary current by the appropriate factor from Table 610-14(b) and selecting a wire from Table 610-14(a).

Table 610-14(b). Secondary Conductor Rating Factors

Time in Seconds		Ampacity of Wire in Percent of Full-Load Secondary Current
On	Off	
5	75	35
10	70	45
15	75	55
15	45	65
15	30	75
15	15	85
Continuous Duty		110

(c) Minimum Size. Conductors external to motors and controls shall not be smaller than No. 16 unless otherwise permitted in (1) and (2).

(1) No. 18 wire in multiconductor cord shall be permitted for control circuits at not over 7 amperes.

(2) Wires not smaller than No. 20 shall be permitted for electronic circuits.

(d) Contact Conductors. Contact wires shall have an ampacity not less than that required by Table 610-14(a) for 75°C (167°F) wire, and in no case shall they be smaller than the following:

Distance Between End Strain Insulators or Clamp-Type Intermediate Supports (ft)	Size of Wire (AWG)
Less than 30	6
30–60	4
Over 60	2

Note: For SI units, 1 ft = 0.3048 m.

(e) Calculation of Motor Load.

(1) For one motor, use 100 percent of motor nameplate full-load ampere rating.

(2) For multiple motors on a single crane or hoist, the minimum ampacity of the power supply conductors shall be the nameplate full-load ampere rating of the largest motor or group of motors for any single crane motion, plus 50 percent of the nameplate full-load ampere rating of the next largest motor or group of motors, using that column of Table 610-14(a) that applies to the longest time-rated motor.

(3) For multiple cranes or hoists, or both, supplied by a common conductor system, compute the motor minimum ampacity for each crane as defined in Section 610-14(e), add them together, and multiply the sum by the appropriate demand factor from Table 610-14(e).

Table 610-14(e). Demand Factors

Number of Cranes or Hoists	Demand Factor
2	0.95
3	0.91
4	0.87
5	0.84
6	0.81
7	0.78

(f) Other Loads. Additional loads, such as heating, lighting, and air conditioning, shall be provided for by application of the appropriate sections of this *Code*.

(g) Nameplate. Each crane, monorail, or hoist shall be provided with a visible nameplate marked with the manufacture's name, the rating in volts, frequency, number of phases, and circuit amperes as calculated in Sections 610-14(e) and (f).

610-15. Common Return. Where a crane or hoist is operated by more than one motor, a common-return conductor of proper ampacity shall be permitted.

C. Contact Conductors

610-21. Installation of Contact Conductors. Contact conductors shall comply with (a) through (h).

(a) Locating or Guarding Contact Conductors. Runway contact conductors shall be guarded, and bridge contact conductors shall be located or guarded in a manner that persons cannot inadvertently touch energized current-carrying parts.

(b) Contact Wires. Wires that are used as contact conductors shall be secured at the ends by means of approved strain insulators and shall be mounted on approved insulators so that the extreme limit of displacement of the wire will not bring the latter within less than 1½ in. (38 mm) from the surface wired over.

(c) Supports Along Runways. Main contact conductors carried along runways shall be supported on insulating supports placed at intervals not exceeding 20 ft (6.1 m) unless otherwise permitted in (f).

Such conductors shall be separated not less than 6 in. (152 mm), other than for monorail hoists where a spacing of not less than 3 in. (76 mm) shall be permitted. Where necessary, intervals between insulating supports shall be permitted to be increased up to 40 ft (12.2 m), the separation between conductors being increased proportionately.

(d) Supports on Bridges. Bridge wire contact conductors shall be kept at least 2½ in. (64 mm) apart, and, where the span exceeds 80 ft (24.4 m), insulating saddles shall be placed at intervals not exceeding 50 ft (15.2 m).

(e) Supports for Rigid Conductors. Conductors along runways and crane bridges, that are of the rigid type specified in Section 610-13(b) and not contained within an approved enclosed assembly, shall be carried on insulating supports spaced at intervals of not more than 80 times the vertical dimension of the conductor, but in no case greater than 15 ft (4.57 m), and spaced apart sufficiently to give a clear electrical separation of conductors or adjacent collectors of not less than 1 in. (25.4 mm).

(f) Track as Circuit Conductor. Monorail, tram rail, or crane runway tracks shall be permitted as a conductor of current for one phase of a 3-phase, ac system furnishing power to the carrier, crane, or trolley, provided all of the following conditions are met.

(1) The conductors supplying the other two phases of the power supply are insulated.
(2) The power for all phases is obtained from an insulating transformer.
(3) The voltage does not exceed 300 volts.
(4) The rail serving as a conductor is effectively grounded at the transformer and also shall be permitted to be grounded by the fittings used for the suspension or attachment of the rail to a building or structure.

Crane runway tracks are permitted as a current-carrying conductor where part of a 3-phase system is furnishing power to the crane. Figure 610.2 illustrates a 3-phase, isolated-delta secondary with one phase grounded at the transformer. The track is also permitted to be grounded through the metal supporting means attached to the metal frame of a building.

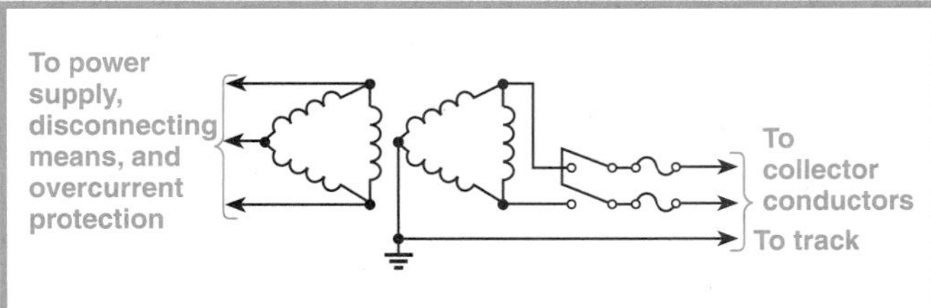

Figure 610.2 *Three-phase, delta-isolating transformer.*

(g) Electrical Continuity of Contact Conductors. All sections of contact conductors shall be mechanically joined to provide a continuous electrical connection.

(h) Not to Supply Other Equipment. Contact conductors shall not be used as feeders for any equipment other than the crane(s) or hoist(s) that they are primarily designed to serve.

610-22. Collectors. Collectors shall be designed so as to reduce to a minimum sparking between them and the contact conductor; and, where operated in rooms used for the storage of easily ignitible combustible fibers and materials, they shall comply with Section 503-13.

D. Disconnecting Means

610-31. Runway Conductor Disconnecting Means. A disconnecting means that has a continuous ampere rating not less than that computed in Sections 610-14(e) and (f) shall be provided between the runway contact conductors and the power supply. Such disconnecting means shall consist of a motor-circuit switch, circuit breaker, or molded case switch. This disconnecting means shall be as follows:

(1) Readily accessible and operable from the ground or floor level
(2) Arranged to be locked in the open position
(3) Open all ungrounded conductors simultaneously
(4) Placed within view of the runway contact conductors

610-32. Disconnecting Means for Cranes and Monorail Hoists. A motor-circuit switch or circuit breaker arranged to be locked in the open position shall be provided in the leads from the runway contact conductors or other power supply on all cranes and monorail hoists. A motor-circuit switch, circuit breaker, or molded case switch arranged to be locked in the open position shall be provided.

Where a monorail hoist or hand-propelled crane bridge installation meets all of the following, the disconnecting means shall be permitted to be omitted.

(1) The unit is controlled from the ground or floor level.
(2) The unit is within view of the power supply disconnecting means.
(3) No fixed work platform has been provided for servicing the unit.

Many crane installations are not arranged so that the unit is within view of the power supply disconnecting means; therefore, a disconnecting means (lock-open type) must be provided in the contact conductors. However, personnel should be aware that while servicing one crane, another unit on the same system could remain energized and could be run into the person performing maintenance on the locked-out unit.

Where the disconnecting means is not readily accessible from the crane or monorail hoist operating station, means shall be provided at the operating station to open the power circuit to all motors of the crane or monorail hoist.

610-33. Rating of Disconnecting Means. The continuous ampere rating of the switch or circuit breaker required by Section 610-32 shall not be less than 50 percent of the combined short-time ampere rating of the motors, nor less than 75 percent of the sum of the short-time ampere rating of the motors required for any single motion.

E. Overcurrent Protection

610-41. Feeders, Runway Conductors.

(a) Single Feeder. The runway supply conductors and main contact conductors of a crane or monorail shall be protected by an overcurrent device(s) that shall not be greater than the largest rating or setting of any branch-circuit protective device plus the sum of the nameplate ratings of all the other loads with application of the demand factors from Table 610-14(e).

(b) More than One Feeder Circuit. Where more than one feeder circuit is installed to supply runway conductors, each feeder circuit shall be sized and protected in compliance with Section 610-41(a).

Multiple feeders are sometimes used to supply long runway conductors in order to minimize voltage drops on the runway conductors. Section 610-41 was revised for the 1999 *Code* to recognize this practice.

610-42. Branch-Circuit Short-Circuit and Ground-Fault Protection. Branch circuits shall be protected as follows.

(a) Fuse or Circuit Breaker Rating. Crane, hoist, and monorail hoist motor branch circuits shall be protected by fuses or inverse-time circuit breakers that have a rating in accordance with Table 430-152. Taps to control circuits shall be permitted to be taken from the load side of a branch-circuit protective device, provided each tap and piece of equipment is properly protected.

Where two or more motors operate a single motion, the sum of their nameplate current ratings shall be considered as a single motor current in the above calculations.

Two or more motors shall be permitted to be connected to the same branch circuit if no tap conductor to an individual motor has an ampacity less than one-third that of the branch circuit and if each motor is protected from overload according to Section 610-43.

(b) Taps to Brake Coils. Taps to brake coils do not require separate overcurrent protection.

610-43. Overload Protection.

(a) Motor and Branch-Circuit Overload Protection. Each motor, motor controller, and branch-circuit conductor shall be protected from overload by one of the following means.

(1) A single motor shall be considered as protected where the branch-circuit overcurrent device meets the rating requirements of Section 610-42.
(2) Overload relay elements in each ungrounded circuit conductor, with all relay elements protected from short circuit by the branch-circuit protection.
(3) Thermal sensing devices, sensitive to motor temperature or to temperature and current, that are thermally in contact with the motor winding(s). A hoist or trolley is considered to be protected if the sensing device is connected in the hoist's upper limit switch circuit so as to prevent further hoisting during an overload condition of either motor.

(b) Manually Controlled Motor. If the motor is manually controlled, with spring return controls, the overload protective device shall not be required to protect the motor against stalled rotor conditions.

(c) Multimotor. Where two or more motors drive a single trolley, truck, or bridge and are controlled as a unit and protected by a single set of overload devices with a rating equal to the sum of their rated full-load currents. A hoist or trolley shall be considered to be protected if the sensing device is connected in the hoist's upper limit switch circuit so as to prevent further hoisting during an overtemperature condition of either motor.

(d) Hoists and Monorail Hoists. Hoists and monorail hoists and their trolleys that are not used as part of an overhead traveling crane shall not require individual motor overload protection, provided the largest motor does not exceed 7½ hp and all motors are under manual control of the operator.

F. Control

610-51. Separate Controllers. Each motor shall be provided with an individual controller unless otherwise permitted in (a) or (b).

(a) Where two or more motors drive a single hoist, carriage, truck, or bridge, they shall be permitted to be controlled by a single controller.

(b) One controller shall be permitted to be switched between motors, provided the following:

(1) The controller has a horsepower rating that is not lower than the horsepower rating of the largest motor.
(2) Only one motor is operated at one time.

610-53. Overcurrent Protection. Conductors of control circuits shall be protected against overcurrent. Control circuits shall be considered as protected by overcurrent devices that are rated or set at not more than 300 percent of the ampacity of the control conductors, unless otherwise permitted in (a) or (b).

(a) Taps to control transformers shall be considered as protected where the secondary circuit is protected by a device rated or set at not more than 200 percent of the rated secondary current of the transformer and not more than 200 percent of the ampacity of the control circuit conductors.

(b) Where the opening of the control circuit would create a hazard, as for example, the control circuit of a hot metal crane, the control circuit conductors shall be considered as being properly protected by the branch-circuit overcurrent devices.

610-55. Limit Switch. A limit switch or other device shall be provided to prevent the load block from passing the safe upper limit of travel of all hoisting mechanisms.

610-57. Clearance. The dimension of the working space in the direction of access to live parts that are likely to require examination, adjustment, servicing, or maintenance while energized shall be a minimum of 2½ ft (762 mm). Where controls are enclosed in cabinets, the door(s) shall either open at least 90 degrees or be removable.

G. Grounding

610-61. Grounding. All exposed noncurrent-carrying metal parts of cranes, monorail hoists, hoists, and accessories, including pendant controls, shall be metallically joined together into a continuous electrical conductor so that the entire crane or hoist will be grounded in accordance with Article 250. Moving parts, other than removable accessories or attachments, that have metal-to-metal bearing surfaces shall be considered to be electrically connected to each other through the bearing surfaces for grounding purposes. The trolley frame and bridge frame shall be considered as electrically grounded through the bridge and trolley wheels and its respective tracks unless local conditions, such as paint or other insulating material, prevent reliable metal-to-metal contact. In this case, a separate bonding conductor shall be provided.

It is not the intent that the trolley frame or bridge frame serve as the equipment grounding conductor for electrical equipment (such as motors, motor controllers, lighting fixtures, transformers, etc.) on a crane. The equipment grounding conductors that are run with the circuit conductors are required to comply with Section 250-118. However, Section 610-61 does not require a separate bar of runway contact conductor to be used as an equipment grounding conductor to serve the mobile portion of the crane, unless reliable metal-to-metal contact through the bridge and trolley wheels and their respective tracks is not available.

Article 620 — Elevators, Dumbwaiters, Escalators, Moving Walks, Wheelchair Lifts, and Stairway Chair Lifts

Contents

A. General

620-1. Scope. This article covers the installation of electrical equipment and wiring used in connection with elevators, dumbwaiters, escalators, moving walks, wheelchair lifts, and stairway chair lifts.

Article 620 was extensively revised for in the 1996 edition of the *Code.* Changes were made to harmonize elevator requirements throughout North America and to keep pace with modern advances in elevator technology. These changes, while continuing to reflect the time-honored safety concerns, more accurately addressed the use of modern equipment.

FPN No. 1: For further information, see *Safety Code for Elevators and Escalators,* ASME/ANSI A17.1-1996.

FPN No. 2: For further information, see *Elevator and Escalator Electrical Equipment Certification Standard,* ASME/ANSI A17.5-1996 (CSA B44.1-1996).

Fine Print Note No. 1 to Section 620-1 points out the existence of the widely recognized elevator code, ASME A17.1, *Safety Code for Elevators and Escalators.* ASME A17.1 provides for many construction and maintenance requirements, from required machine room lighting to seismic considerations.

620-2. Definitions.

Control System. The overall system governing the starting, stopping, direction of motion, acceleration, speed, and retardation of the moving member.

Controller, Motion. The electric device(s) for that part of the control system that governs the acceleration, speed, retardation, and stopping of the moving member.

Controller, Motor. The operative units of the control system comprised of the starter device(s) and power conversion equipment used to drive an electric motor, or the pumping unit used to power hydraulic control equipment.

Controller, Operation. The electric device(s) for that part of the control system that initiates the starting, stopping, and direction of motion in response to a signal from an operating device.

Operating Device. The car switch, push buttons, key or toggle switch(s), or other devices used to activate the operation controller.

Signal Equipment. Includes audible and visual equipment such as chimes, gongs, lights, and displays that convey information to the user.

FPN No. 1: The motor controller, motion controller, and operation controller may be located in a single enclosure or a combination of enclosures.

FPN No. 2: Figure 620-2 is for information only.

Figure 620-2 in the *NEC* provides information to the user of a typical elevator control system. The figure does not intend that every elevator control system should have these exact components.

The definitions in Section 620-2 and Figure 620-2 separate the control system into its functional parts. Article 620 uses these definitions to address and specify the proper safety concerns of each of these controllers: motor, motion, and operation.

620-3. Voltage Limitations. The supply voltage shall not exceed 300 volts between conductors unless otherwise permitted in (a) through (c).

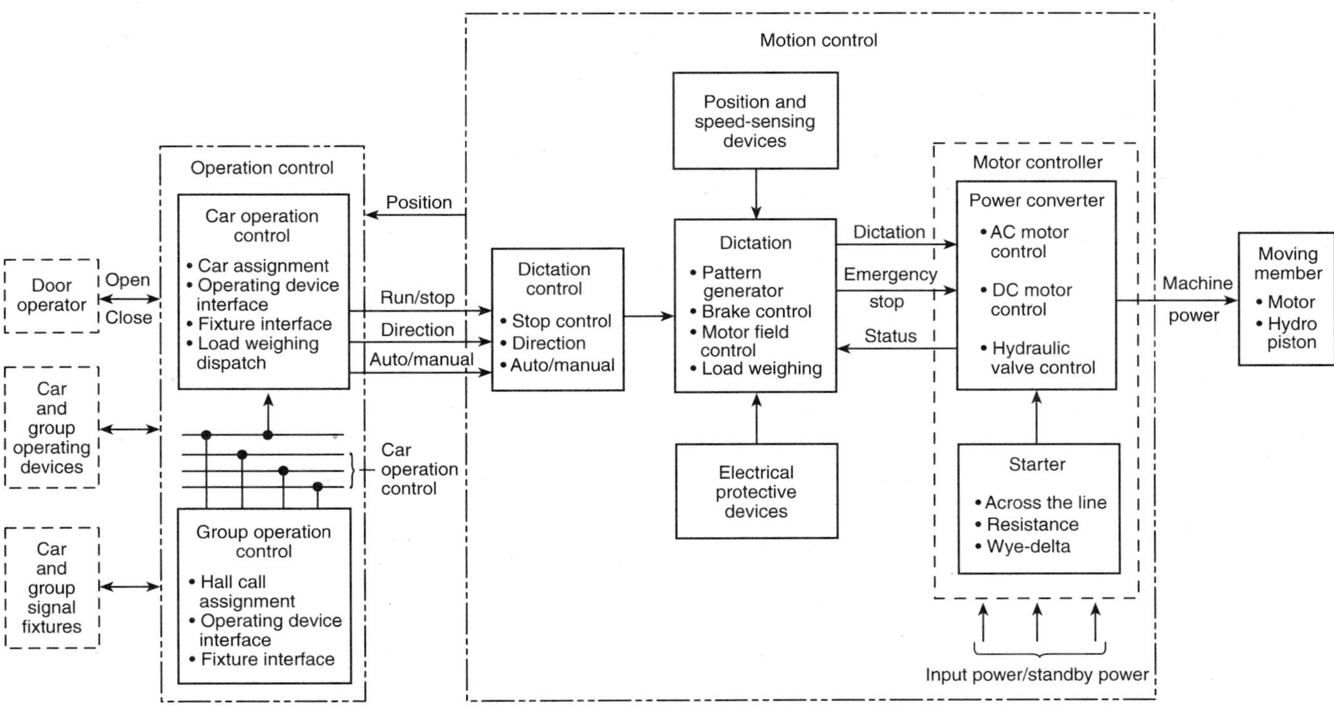

Figure 620-2 Control system.

(a) Power Circuits. Branch circuits to door operator controllers and door motors and branch circuits and feeders to motor controllers, driving machine motors, machine brakes, and motor-generator sets shall not have a circuit voltage in excess of 600 volts. Internal voltages of power conversion and functionally associated equipment, including the interconnecting wiring, shall be permitted to have higher voltages provided that all such equipment and wiring shall be listed for the higher voltages. Where the voltage exceeds 600 volts, warning labels or signs that read "DANGER — HIGH VOLTAGE" shall be attached to the equipment and shall be plainly visible.

(b) Lighting Circuits. Lighting circuits shall comply with the requirements of Article 410.

(c) Heating and Air-Conditioning Circuits. Branch circuits for heating and air-conditioning equipment located on the elevator car shall not have a circuit voltage in excess of 600 volts.

Section 620-3 describes the voltage limitations for control, lighting, and heating and air-conditioning circuits. Voltage limitations for power conversion units and functionally associated equipment have been replaced with a warning sign requirement for voltages above 600 volts.

620-4. Live Parts Enclosed. All live parts of electrical apparatus in the hoistways, at the landings, in or on the cars of elevators and dumbwaiters, in the wellways or the landings of escalators or moving walks, or in the runways and machinery spaces of wheelchair lifts and stairway chair lifts shall be enclosed to protect against accidental contact.

FPN: See Section 110-27 for guarding of live parts (600 volts, nominal, or less).

620-5. Working Clearances. Working space shall be provided about controllers, disconnecting means, and other electrical equipment. The minimum working space shall not be less than that specified in Section 110-26(a).

Where conditions of maintenance and supervision ensure that only qualified persons will examine, adjust, service, and maintain the equipment, the clearance requirements of Section 110-26(a) shall be waived as permitted in (a) through (d).

(a) Flexible Connections to Equipment. Electrical equipment in (1) through (4) is provided with flexible leads to all external connections so that it can be repositioned to meet the clear working space requirements of Section 110-26(a).

Section 620-5(a) was revised for the 1999 *Code* to clarify the intent that flexible leads can be provided on equipment to meet the working clearance requirements of Section 110-26(a) where the equipment is installed in confined locations.

(1) Controllers and disconnecting means for dumbwaiters, escalators, moving walks, wheelchair lifts, and stairway

chair lifts installed in the same space with the driving machine
(2) Controllers and disconnecting means for elevators installed in the hoistway or on the car
(3) Controllers for door operators
(4) Other electrical equipment installed in the hoistway or on the car

(b) Guards. Live parts of the electrical equipment are suitably guarded, isolated, or insulated, and the equipment can be examined, adjusted, serviced, or maintained while energized without removal of this protection.

FPN: See definition of *Exposed* in Article 100.

(c) Examination, Adjusting, and Servicing. Electrical equipment is not required to be examined, adjusted, serviced, or maintained while energized.

(d) Low Voltage. Uninsulated parts are at a voltage not greater than 30 volts rms, 42 volts peak, or 60 volts dc.

B. Conductors

620-11. Insulation of Conductors. The insulation of conductors shall comply with (a) through (d).

FPN: One method of determining that conductors are flame retardant is by testing the conductors to the VW-1 (Vertical-Wire) Flame Test in the *Reference Standard for Electrical Wires, Cables, and Flexible Cords,* ANSI/UL 1581-1991.

(a) Hoistway Door Interlock Wiring. The conductors to the hoistway door interlocks from the hoistway riser shall be flame retardant and suitable for a temperature of not less than 200°C (392°F). Conductors shall be Type SF or equivalent.

(b) Traveling Cables. Traveling cables used as flexible connections between the elevator or dumbwaiter car or counterweight and the raceway shall be of the types of elevator cable listed in Table 400-4 or other approved types.

(c) Other Wiring. All conductors in raceways shall have flame-retardant insulation.

Conductors shall be Type MTW, TF, TFF, TFN, TFFN, THHN, THW, THWN, TW, XHHW, hoistway cable, or any other conductor with insulation designated as flame retardant. Shielded conductors shall be permitted, if such conductors are insulated for the maximum nominal circuit voltage applied to any conductor within the cable or raceway system.

(d) Insulation. All conductors shall have an insulation voltage rating equal to at least the maximum nominal circuit voltage applied to any conductor within the enclosure, cable, or raceway. Insulations and outer coverings that are designated with the suffix "/LS" and are so listed shall be permitted.

Hoistway door interlock wiring is required to be suitable for 200°C (392°F). See Table 310-13, Conductor Application and Insulations. See also Table 310-18.

See Table 400-4 for approved types of elevator cables to be used in hazardous (classified) and nonhazardous locations. See also Note 5 to Table 400-4. A characteristic equally important with respect to safety is the prevention of twisting of cables during their rise and fall with the elevator or dumbwaiter.

620-12. Minimum Size of Conductors. The minimum size of conductors, other than conductors that form an integral part of control equipment, shall be as follows.

(a) Traveling Cables.

(1) For lighting circuits: No. 14 copper. No. 20 copper or larger conductors shall be permitted in parallel, provided the ampacity is equivalent to at least that of No. 14 copper.

(2) For other circuits, No. 20 copper.

(b) Other Wiring. No. 24 copper. Smaller size listed conductors shall be permitted.

The extensive use of electronics with correspondingly lower currents permits the use of smaller wire sizes. Section 620-12(b) permits the use of conductors smaller than No. 24 only if they are listed for the purpose. One application may be the shielded cables interconnecting various microprocessors in an elevator distributive system. All conductors should, of course, have the necessary strength and durability for the conditions to which they will be exposed.

620-13. Feeder and Branch-Circuit Conductors. Conductors shall have an ampacity in accordance with (a) through (d). With generator field control, the conductor ampacity shall be based on the nameplate current rating of the driving motor of the motor-generator set that supplies power to the elevator motor.

Section 620-13 was extensively revised in the 1996 *Code* to reflect a more accurate method of calculating the ampacity of feeder and branch-circuit conductors used to supply intermittent loads. The previous 125 percent requirement was revised because elevator applications are not continuous duty loads, and this requirement now correlates with Sections 430-22(a) and 430-24.

FPN No. 1: The heating of conductors depends on root-mean-square current values, which, with generator field control, are reflected by the nameplate current rating of the motor-generator driving motor rather than by the rating of the elevator motor, which represents actual but short-time and intermittent full-load current values.

FPN No. 2: See Figure 620-13.

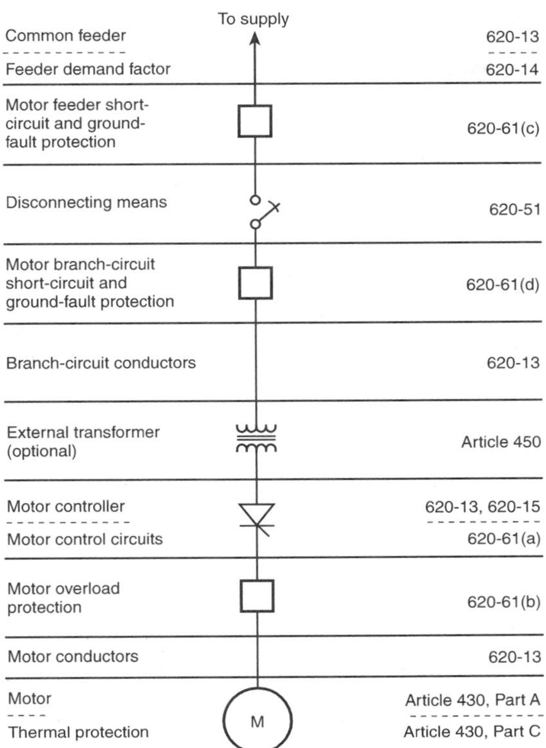

Figure 620-13.

Figure 620-13 in the *Code* depicts the appropriate reference for each part of an elevator circuit.

(a) Conductors Supplying Single Motor. Conductors supplying a single motor shall have an ampacity not less than the percentage of motor nameplate current determined from Sections 430-22(a) and (b).

FPN: Elevator motor currents, or those of similar functions, may exceed the nameplate value, but since they are inherently intermittent duty, and the heating of the motor and conductors is dependent on the root-mean-square (rms) current value, conductors are sized for duty cycle service as shown in Table 430-22(b).

(b) Conductors Supplying a Single Motor Controller. Conductors supplying a single motor controller shall have an ampacity not less than the motor controller nameplate current rating, plus all other connected loads.

FPN: Motor controller nameplate current rating may be derived based on the rms value of the motor current using an intermittent duty cycle and other control system loads, if applicable.

(c) Conductors Supplying a Single Power Transformer. Conductors supplying a single power transformer shall have an ampacity not less than the nameplate current rating of the power transformer plus all other connected loads.

FPN No. 1: The nameplate current rating of a power transformer supplying a motor controller reflects the nameplate current rating of the motor controller at line voltage (transformer primary).

FPN No. 2: See Appendix D, Example No. D10.

(d) Conductors Supplying More than One Motor, Motor Controller, or Power Transformer. Conductors supplying more than one motor, motor controller, or power transformer shall have an ampacity not less than the sum of the nameplate current ratings of the equipment plus all other connected loads. The ampere ratings of motors to be used in the summation shall be determined from Table 430-22(b), and Sections 430-24 and 430-24, Exception No. 1.

FPN: See Appendix D, Examples No. D9 and D10.

620-14. Feeder Demand Factor. Feeder conductors of less ampacity than required by Section 620-13 shall be permitted subject to the requirements of Table 620-14.

FPN: Demand factors are based on 50 percent duty cycle (i.e., half time on and half time off).

Table 620-14. Feeder Demand Factors for Elevators

Number of Elevators on a Single Feeder	Demand Factor
1	1.00
2	0.95
3	0.90
4	0.85
5	0.82
6	0.79
7	0.77
8	0.75
9	0.73
10 or more	0.72

620-15. Motor Controller Rating. The motor controller rating shall comply with Section 430-83. The rating shall be permitted to be less than the nominal rating of the elevator motor, when the controller inherently limits the available power to the motor and is marked as power limited.

Section 620-15 recognizes the inherent power-limiting ability of certain adjustable-speed drive controllers.

FPN: For controller markings, see Section 430-8.

C. Wiring

620-21. Wiring Methods. Conductors and optical fibers located in hoistways, in escalator and moving walk wellways, in wheelchair lifts, stairway chair lift runways, and machinery spaces, in or on cars, and in machine and control rooms, not including the traveling cables connecting the car or counterweight and hoistway wiring, shall be installed in rigid metal conduit, intermediate metal conduit, electrical metallic tubing, rigid nonmetallic conduit, or wireways, or shall be Type MC, MI, or AC cable unless otherwise permitted in (a) through (c).

Section 620-21 permits short lengths of flexible cords and cables if part of listed equipment such as transducers (position, velocity, direction) and if used in low-voltage circuits.

(a) Elevators.

(1) Hoistways.

(a) Flexible metal conduit, liquidtight flexible metal conduit, or liquidtight flexible nonmetallic conduit shall be permitted in hoistways between risers and limit switches, interlocks, operating buttons, and similar devices.

(b) Cables used in Class 2 power-limited circuits shall be permitted to be installed between risers and signal equipment and operating devices provided the cables are supported and protected from physical damage and are of a jacketed and flame-retardant type.

(2) Cars.

(a) Flexible metal conduit, liquidtight flexible metal conduit, or liquidtight flexible nonmetallic conduit of ⅜-in. nominal trade size or larger, not exceeding 6 ft (1.83 m) in length shall be permitted on cars where located so as to be free from oil and if securely fastened in place.

Exception: Liquidtight flexible nonmetallic conduit of ⅜-in. nominal trade size or larger, as defined by Section 351-22(2), shall be permitted in lengths in excess of 6 ft (1.83 m).

(b) Hard-service cords and junior hard-service cords that conform to the requirements of Article 400 (Table 400-4) shall be permitted as flexible connections between the fixed wiring on the car and devices on the car doors or gates. Hard-service cords only shall be permitted as flexible connections for the top-of-car operating device or the car-top work light. Devices or fixtures shall be grounded by means of an equipment grounding conductor run with the circuit conductors. Cables with smaller conductors and other types and thicknesses of insulation and jackets shall be permitted as flexible connections between the fixed wiring on the car and devices on the car doors or gates, if listed for this use.

(c) Flexible cords and cables that are components of listed equipment and used in circuits operating at 30 volts rms or less or 42 volts dc or less shall be permitted in lengths not to exceed 6 ft (1.83 m) provided the cords and cables are supported and protected from physical damage and are of a jacketed and flame-retardant type.

See the commentary following 620-21(a)(1)(b).

(d) Flexible metal conduit, liquidtight flexible metal conduit, liquidtight flexible nonmetallic conduit or flexible cords and cables, or conductors grouped together and taped or corded that are part of listed equipment, a driving machine, or a driving machine brake shall be permitted on the car assembly, in lengths not to exceed 6 ft (1.83 m) without being installed in a raceway and where located to be protected from physical damage and are of a flame-retardant type.

Section 620-21(a)(2)(d) was added to the 1999 *Code.* This section provides the proper wiring methods for the situation where a driving machine or driving machine brake is located on the car. This includes rack and pinion or screw column drives located on cars. See also Section 620-71(b) and see the commentary following 620-21(a)(1)(b).

(3) Machine Room and Machinery Spaces.

(a) Flexible metal conduit, liquidtight flexible metal conduit, or liquidtight flexible nonmetallic conduit of ⅜-in. nominal trade size or larger, not exceeding 6 ft (1.83 m) in length, shall be permitted between control panels and machine motors, machine brakes, motor-generator sets, disconnecting means, and pumping unit motors and valves.

Exception: Liquidtight flexible nonmetallic conduit, as defined in Section 351-22(2), shall be permitted to be installed in lengths in excess of 6 ft (1.83 m).

(b) Where motor-generators, machine motors, or pumping unit motors and valves are located adjacent to or underneath control equipment and are provided with extra-length terminal leads not exceeding 6 ft (1.83 m) in length, such leads shall be permitted to be extended to connect directly to controller terminal studs without regard to the carrying-capacity requirements of Articles 430 and 445. Auxiliary gutters shall be permitted in machine and control rooms between controllers, starters, and similar apparatus.

(c) Flexible cords and cables that are components of listed equipment and used in circuits operating at 30 volts rms or less or 42 volts dc or less shall be permitted in lengths not to exceed 6 ft (1.83 m) provided the cords and cables are supported and protected from physical damage and are of a jacketed and flame-retardant type.

(d) On existing or listed equipment, conductors shall also be permitted to be grouped together and taped or corded without being installed in a raceway. Such cable groups shall be supported at intervals not over 3 ft (914 mm) and located so as to be protected from physical damage.

(4) Counterweight. Flexible metal conduit, liquidtight flexible metal conduit, liquidtight flexible nonmetallic conduit or flexible cords and cables, or conductors grouped together and taped or corded that are part of listed equipment, a driving machine, or a driving machine brake shall be permitted on the counterweight assembly, in lengths not to exceed 6 ft (1.83 m) without being installed in a raceway and where located to be protected from physical damage and are of a flame-retardant type.

See the commentary following 620-21(a)(1)(b).

(b) Escalators.

(1) Flexible metal conduit, liquidtight flexible metal conduit, or liquidtight flexible nonmetallic conduit shall be permitted in escalator and moving walk wellways. Flexible metal conduit or liquidtight flexible conduit of ⅜-in. nominal trade size shall be permitted in lengths not in excess of 6 ft (1.83 m).

Exception: ⅜-in. nominal trade size or larger liquidtight flexible nonmetallic conduit, as defined in Section 351-22(2), shall be permitted to be installed in lengths in excess of 6 ft (1.83 m).

(2) Cables used in Class 2 power-limited circuits shall be permitted to be installed within escalators and moving walkways provided the cables are supported and protected from physical damage and are of a jacketed and flame-retardant type.

See the commentary following 620-21(a)(1)(b).

(3) Hard-service cords that conform to the requirements of Article 400 (Table 400-4) shall be permitted as flexible connections on escalators and moving walk control panels and disconnecting means where the entire control panel and disconnecting means are arranged for removal from machine spaces as permitted in Section 620-5.

(c) Wheelchair Lifts and Stairway Chair Lift Raceways.

(1) Flexible metal conduit or liquidtight flexible metal conduit shall be permitted in wheelchair lifts and stairway chair lift runways and machinery spaces. Flexible metal conduit or liquidtight flexible conduit of ⅜-in. nominal trade size shall be permitted in lengths not in excess of 6 ft (1.83 m).

Exception: ⅜-in. nominal trade size or larger liquidtight flexible nonmetallic conduit, as defined in Section 351-22(2), shall be permitted to be installed in lengths in excess of 6 ft (1.83 m).

(2) Cables used in Class 2 power-limited circuits shall be permitted to be installed within wheelchair lifts and stairway chair lift runways and machinery spaces provided the cables are supported and protected from physical damage and are of a jacketed and flame-retardant type.

See the commentary following 620-21(a)(1)(b).

620-22. Branch Circuits for Car Lighting, Receptacle(s), Ventilation, Heating, and Air Conditioning.

(a) Car Light Source. A separate branch circuit shall supply the car lights, receptacle(s), auxiliary lighting power source, and ventilation on each elevator car. The overcurrent device protecting the branch circuit shall be located in the elevator machine room/machinery space.

(b) Air-Conditioning and Heating Source. A dedicated branch circuit shall supply the air-conditioning and heating units on each elevator car. The overcurrent device protecting the branch circuit shall be located in the elevator machine room/machinery space.

Sections 620-22(a) and 620-22(b) of the 1999 *Code* were revised to specify the location of the overcurrent devices protecting the branch circuits for elevator car lighting and heating/air conditioning in the elevator machine room or machine space. This new requirement will facilitate ease of maintenance and troubleshooting and will also correlate North American code requirements. See Figure 620.1.

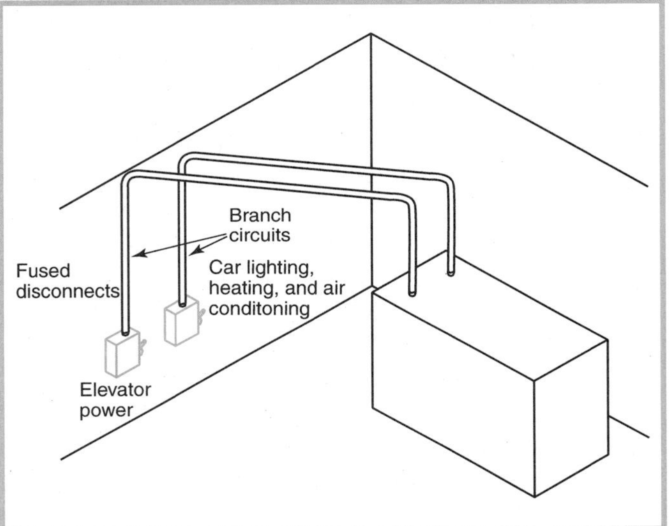

Figure 620.1 *Overcurrent devices for elevator car lighting and heating/air-conditioning branch circuits in the elevator machine room.*

620-23. Branch Circuit for Machine Room/Machinery Space Lighting and Receptacle(s).

(a) A separate branch circuit shall supply the machine room/machinery space lighting and receptacle(s).

Required lighting shall not be connected to the load side of a ground-fault circuit interrupter.

Lighting fixtures are not permitted to be connected to the load side of a GFCI device of any kind because the machine room lighting could be de-energized during certain fault conditions.

(b) The machine room lighting switch shall be located at the point of entry to such machine rooms/machinery spaces.

(c) At least one 125-volt, single-phase, duplex receptacle shall be provided in each machine room and machinery space.

FPN: See *Safety Code for Elevators and Escalators,* ANSI/ASME A17.1-1996, for illumination levels.

620-24. Branch Circuit for Hoistway Pit Lighting and Receptacle(s).

(a) A separate branch circuit shall supply the hoistway pit lighting and receptacle(s).

Required lighting shall not be connected to the load side of a ground-fault circuit interrupter.

(b) The lighting switch shall be located so as to be readily accessible from the pit access door.

(c) At least one 125-volt, single-phase, duplex receptacle shall be provided in the hoistway pit.

FPN: See *Safety Code for Elevators and Escalators,* ANSI/ASME A17.1-1996, for illumination levels.

The receptacles required by Sections 620-23 and 620-24 are also required to be protected by ground-fault circuit-interrupter protection for personnel (see Section 620-85). Lighting fixtures are not permitted to be connected to the load side of GFCI devices of any kind. See also the commentary following Section 620-23(a).

ANSI/ASME A17.1-1996, *Safety Code for Elevators and Escalators,* requires a minimum of 5 foot-candles (54 lux) at the pit floor and requires lighting fixtures to be externally guarded to prevent accidental breakage.

Lighting fixtures in pits should be mounted so that the car or counterweight will not strike them when on fully compressed buffers. It is recommended that lighting fixtures be mounted in the corners of the pit.

D. Installation of Conductors

620-32. Metal Wireways and Nonmetallic Wireways. The sum of the cross-sectional area of the individual conductors in a wireway shall not be more than 50 percent of the interior cross-sectional area of the wireway.

Vertical runs of wireways shall be securely supported at intervals not exceeding 15 ft (4.57 m) and shall have not more than one joint between supports. Adjoining wireway sections shall be securely fastened together to provide a rigid joint.

620-33. Number of Conductors in Raceways. The sum of the cross-sectional area of the individual conductors in raceways shall not exceed 40 percent of the interior cross-sectional area of the raceway, except as permitted in Section 620-32 for wireways.

620-34. Supports. Supports for cables or raceways in a hoistway or in an escalator or moving walk wellway or wheelchair lift and stairway chair lift runway shall be securely fastened to the guide rail; escalator or moving walk truss; or to the hoistway, wellway, or runway construction.

620-35. Auxiliary Gutters. Auxiliary gutters shall not be subject to the restrictions of Section 374-2 as to length or of Section 374-5 as to number of conductors.

620-36. Different Systems in One Raceway or Traveling Cable. Optical fiber cables and conductors for operating devices, operation and motion control, power, signaling, fire alarm, lighting, heating, and air-conditioning circuits of 600 volts or less shall be permitted to be run in the same traveling cable or raceway system if all conductors are insulated for the maximum voltage applied to any conductor within the cables or raceway system and if all live parts of the equipment are insulated from ground for this maximum voltage. Such a traveling cable or raceway shall also be permitted to include shielded conductors and/or one or more coaxial cables, if such conductors are insulated for the maximum voltage applied to any conductor within the cable or raceway system. Conductors shall be permitted to be covered with suitable shielding for telephone, audio, video, or higher frequency communications circuits.

With the use of greater numbers of individual cables and the use of much longer cables in tall buildings, there is a possibility of intertwisting cable loops. In order to eliminate the practice of tying a cable to the traveling cable, one elevator cable or raceway is permitted to enclose optical fiber cables and all the conductors for power, control, lighting, video, fire alarm, and communications circuits, if all conductors are insulated for the maximum voltage applied to any conductor within the cable or raceway and all live parts are also insulated from ground for the maximum voltage present. The revision of Section 620-36 in the 1999 *NEC* clarifies that power-limited and nonpower-limited fire alarm conductors are permitted in the same raceway or traveling cable as power and other types of signaling conductors. All power, signaling, and fire alarm conductors are to be insulated for the maximum voltage applied to any conductor in the raceway or cable.

620-37. Wiring in Hoistways and Machine Rooms.

(a) Uses Permitted. Only such electric wiring, raceways, and cables used directly in connection with the elevator or dumbwaiter, including wiring for signals; for communication with the car; for lighting, heating, air conditioning, and ventilating the elevator car; for fire detecting systems; for pit sump pumps; and for heating, lighting, and ventilating the hoistway, shall be permitted inside the hoistway and the machine room.

(b) Lightning Protection. Bonding of elevator rails (car and/or counterweight) to a lightning protection system grounding down conductor(s), shall be permitted. The lightning protection system grounding down conductor(s) shall not be located within the hoistway. Elevator rails or other hoistway equipment shall not be used as the grounding down conductor for lightning protection systems.

Elevator hoistways are often convenient locations to run wiring from the basement to the roof. However, Section 620-37(b) prohibits equipment and wiring that is not associated with the elevator from being installed in elevator machine rooms and hoistways. Only electrical equipment and wiring used directly in connection with the elevator may be installed inside the hoistway and machine room.

Where a lightning protection system is provided, and if a lightning protection system grounding "down" conductor(s) located outside the hoistway is within a critical horizontal distance of the elevator rails, bonding of the rails to the lightning protection system grounding "down" conductor(s) is required by NFPA 780-1997, *Standard for the Installation of Lightning Protection Systems*. Bonding prevents a dangerous side flash between the lightning protection system grounding "down" conductor(s) and the elevator rails. A lightning strike on the building air terminal will be conducted through the lightning protection system grounding "down" conductor(s), and, if the elevator rails are not at the same potential as the lightning protection system grounding "down" conductor(s), a side flash may occur. Generally, "down" conductors are installed vertically near the perimeter of the structure, so this requirement may have application to "outside" elevators.

FPN: See Section 250-106 for bonding requirements. For further information, see *Standard for the Installation of Lightning Protection Systems*, NFPA 780-1997.

(c) Main Feeders. Main feeders for supplying power to elevators and dumbwaiters shall be installed outside the hoistway unless as follows.

(1) By special permission, feeders for elevators shall be permitted within an existing hoistway if no conductors are spliced within the hoistway.
(2) Feeders shall be permitted inside the hoistway for elevators with driving machine motors located in the hoistway or on the car or counterweight.

620-38. Electrical Equipment in Garages and Similar Occupancies. Electrical equipment and wiring used for elevators, dumbwaiters, escalators, moving walks, and wheelchair lifts and stairway chair lifts in garages shall comply with the requirements of Article 511.

FPN: Garages used for parking or storage and where no repair work is done in accordance with Section 511-2 are not classified.

E. Traveling Cables

620-41. Suspension of Traveling Cables. Traveling cables shall be suspended at the car and hoistways' ends, or counterweight end where applicable, so as to reduce the strain on the individual copper conductors to a minimum.

Traveling cables shall be supported by one of the following means:

(1) By its steel supporting member(s)
(2) By looping the cables around supports for unsupported lengths less than 100 ft (30.5 m)
(3) By suspending from the supports by a means that automatically tightens around the cable when tension is increased for unsupported lengths up to 200 ft (61 m)

FPN: Unsupported length for the hoistway suspension means is that length of cable as measured from the point of suspension in the hoistway to the bottom of the loop, with the elevator car located at the bottom landing. Unsupported length for the car suspension means is that length of cable as measured from the point of suspension on the car to the bottom of the loop, with the elevator car located at the top landing.

Traveling cables between fixed suspension points are not required to be installed in a raceway. If the fixed suspension point is on top of the car, the cable might be exposed on the side of the car. Suitable guards are necessary to protect these cables from damage. If the suspension point is under the car, the cables might be run up the side of the car to the car top junction box. If these runs are over 6 ft, the cables are required to be run in a raceway. Refer to Section 620-44 and Figure 620.2.

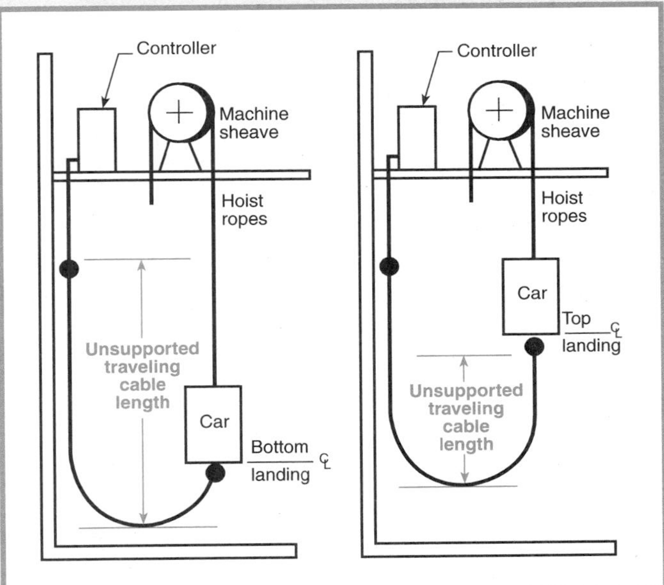

Figure 620.2 *Unsupported length of traveling cable.*

620-42. Hazardous (Classified) Locations. In hazardous (classified) locations, traveling cables shall be of a type approved for hazardous (classified) locations and shall comply with Sections 501-11, 502-12, or 503-10, as applicable.

620-43. Location of and Protection for Cables. Traveling cable supports shall be located so as to reduce to a minimum the possibility of damage due to the cables coming in contact with the hoistway construction or equipment in the hoistway. Where necessary, suitable guards shall be provided to protect the cables against damage.

620-44. Installation of Traveling Cables. Traveling cable shall be permitted to be run without the use of a raceway for a distance not exceeding 6 ft (1.83 m) in length as measured from the first point of support on the elevator car or hoistway wall, or counterweight where applicable, provided the conductors are grouped together and taped or corded, or in the original sheath.

Traveling cables shall be permitted to be continued to elevator controller enclosures and to elevator car and machine room connections, as fixed wiring, provided they are suitably supported and protected from physical damage.

F. Disconnecting Means and Control

620-51. Disconnecting Means. A single means for disconnecting all ungrounded main power supply conductors for each unit shall be provided and be designed so that no pole can be operated independently. Where multiple driving machines are connected to a single elevator, escalator, moving walk, or pumping unit, there shall be one disconnecting means to disconnect the motor(s) and control valve operating magnets.

The disconnecting means for the main power supply conductors shall not disconnect the branch circuit required in Sections 620-22, 620-23, and 620-24.

The branch circuit(s) for the elevator car lighting, receptacles, and HVAC are required to be independent of the control portion of the elevator. Passenger safety, comfort, and elevator maintenance are enhanced by this requirement.

(a) Type. The disconnecting means shall be an enclosed externally operable fused motor circuit switch or circuit breaker capable of being locked in the open position. The disconnecting means shall be a listed device.

FPN: For additional information, see *Safety Code for Elevators and Escalators,* ASME/ANSI A17.1-1996.

(b) Operation. No provision shall be made to open or close this disconnecting means from any other part of the premises. If sprinklers are installed in hoistways, machine rooms, or machinery spaces, the disconnecting means shall be permitted to automatically open the power supply to the affected elevator(s) prior to the application of water. No provision shall be made to automatically close this disconnecting means. Power shall only be restored by manual means.

The *Safety Code for Elevators and Escalators,* ANSI/ASME A17.1-1996, Rule 102.2(c), requires that if sprinklers are installed in hoistways, machine rooms, or machinery spaces, a means must be provided to automatically disconnect the main line power supply to the affected elevator(s) upon or prior to the application of water. Water on elevator electrical equipment can result in hazards such as uncontrolled car movement (wet machine brakes), and movement of elevator with open doors (water on safety circuits bypassing car and/or hoistway door interlocks), and shock hazards.

Automatic disconnection of the main line power supply is not required by ANSI/ASME A17.1-1996, *Safety Code for Elevators and Escalators,* where hoistways and machine rooms are not sprinklered. NFPA 13-1996, *Standard for Sprinkler Systems,* provides requirements for the installation of sprinklers in machine rooms, hoistways, and pits.

Elevator shutdown is generally accomplished through the use of heat detectors located near sprinkler heads. These heat detectors are designed to actuate before the sprinkler heads discharge water and generate an alarm signal. An output control relay powered by the fire alarm system then provides a monitored output to the main line disconnecting means control circuit that activates the shunt trip. This practice ensures that all components have secondary power and are monitored for integrity, as required by NFPA 72, *National Fire Alarm Code®*. Stand-alone heat detectors connected directly to the elevator disconnecting means control circuit are not monitored for integrity, have no secondary power supply, and are not permitted by NFPA 72.

Elevator shutdown can occur even if the car is not at a landing. However, it is highly desirable to recall the car(s) to the designated landing prior to disconnecting the main line power to avoid trapping occupants in the car(s). Most fires produce detectable quantities of smoke before there is sufficient heat to activate a sprinkler head. Therefore, ANSI/ASME A17.1-1996, *Safety Code for Elevators and Escalators,* Rule 102.2(c), requires smoke detectors to be installed in hoistways that are sprinklered for the purposes of recalling the elevator car(s) before the main line power is disconnected. See Section 3-8.15 of NFPA 72, *National Fire Alarm Code,* for additional requirements relating to the fire alarm system and elevator shutdown.

Figure 620.3 illustrates a typical method of super-

vising control power using a fire alarm system. Loss of control power would produce a supervisory signal at the fire alarm control unit that would then be investigated.

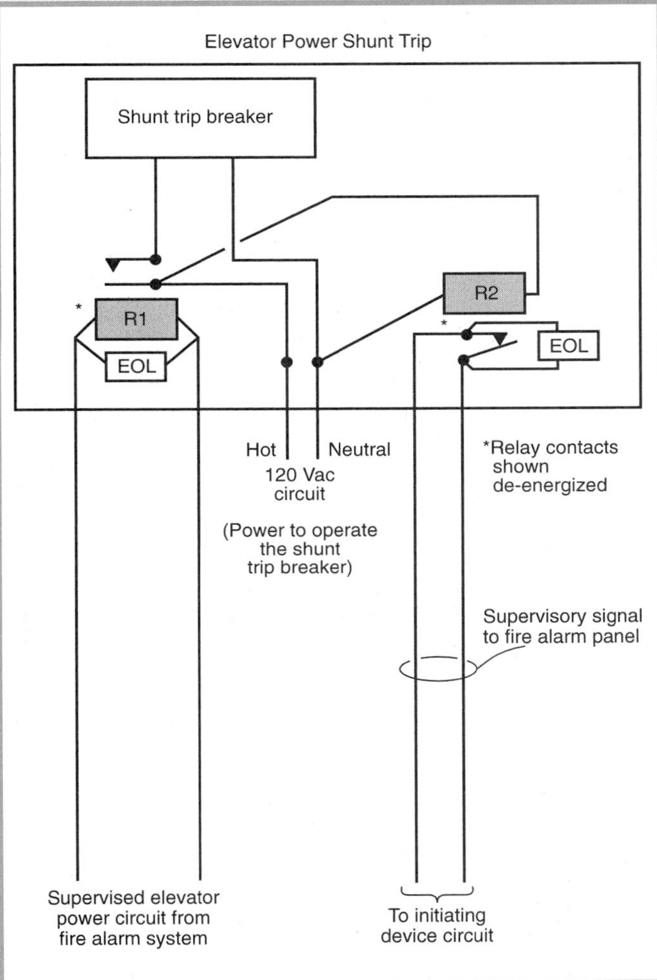

Figure 620.3 Typical method of control power supervision using fire alarm control unit.

FPN: To reduce hazards associated with water on live elevator electrical equipment.

(c) Location. The disconnecting means shall be located where it is readily accessible to qualified persons.

(1) On Elevators Without Generator Field Control. On elevators without generator field control, the disconnecting means shall be located within sight of the motor controller. Driving machines or motion and operation controllers not within sight of the disconnecting means shall be provided with a manually operated switch installed in the control circuit to prevent starting. The manually operated switch(s) shall be installed adjacent to this equipment.

Where the driving machine is located in a remote machinery space, a single means for disconnecting all ungrounded main power supply conductors shall be provided and be capable of being locked in the open position.

See Figure 620.4 for disconnecting means for driving machines or motion and operation controllers not within sight of the main line disconnecting means.

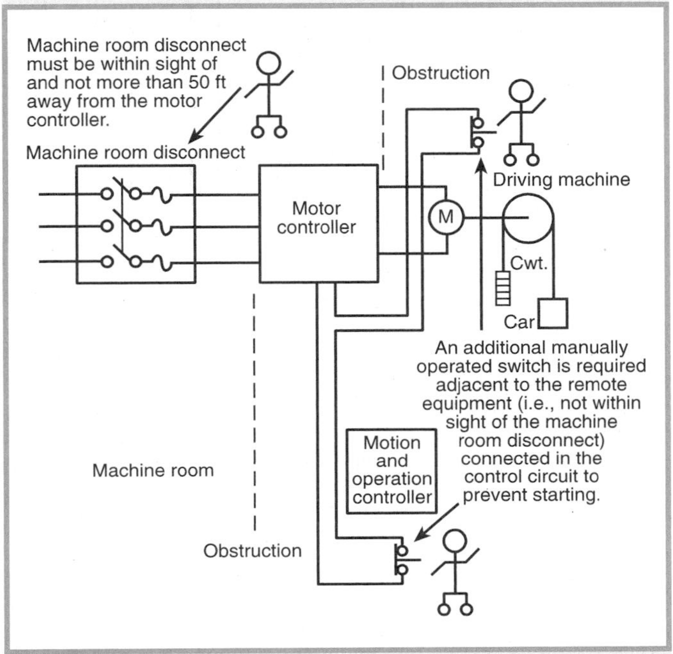

Figure 620.4 Disconnecting means for driving machines or motion and operation controllers not within sight of the main line disconnecting means. (ASME)

(2) On Elevators with Generator Field Control. On elevators with generator field control, the disconnecting means shall be located within sight of the motor controller for the driving motor of the motor-generator set. Driving machines, motor-generator sets, or motion and operation controllers not within sight of the disconnecting means shall be provided with a manually operated switch installed in the control circuit to prevent starting. The manually operated switch(s) shall be installed adjacent to this equipment.

Where the driving machine or the motor-generator set is located in a remote machinery space, a single means for disconnecting all ungrounded main power supply conductors, shall be provided and be capable of being locked in the open position.

See Figures 620.5 and 620.6 for examples of disconnecting means for a motor-generator and for driving machines in a remote location.

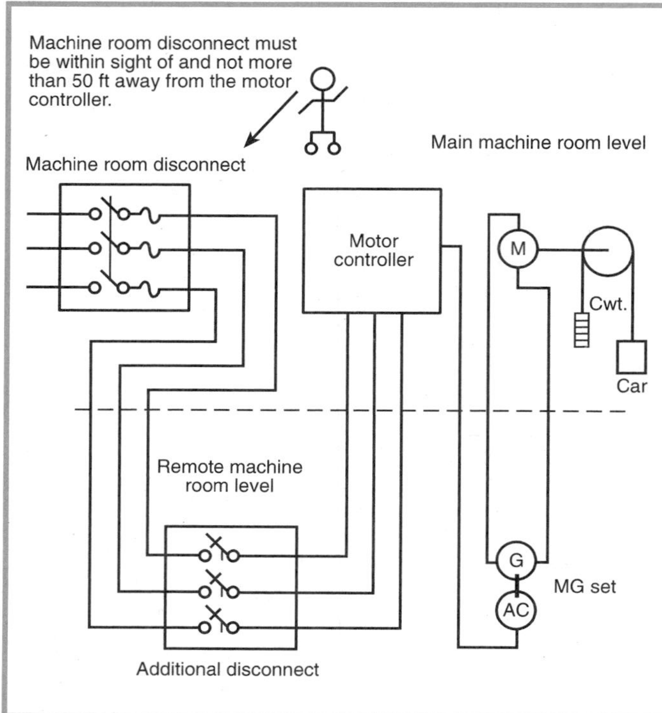

Figure 620.5 Disconnecting means for a motor-generator in a remote location. (ASME)

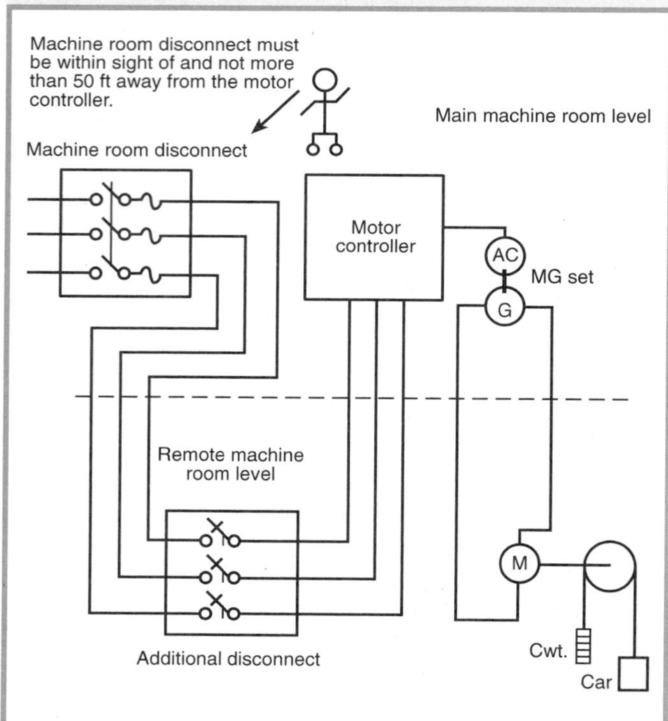

Figure 620.6 Disconnecting means for driving machines in a remote location. (ASME)

(3) On Escalators and Moving Walks. On escalators and moving walks, the disconnecting means shall be installed in the space where the controller is located.

(4) On Wheelchair Lifts and Stairway Chair Lifts. On wheelchair lifts and stairway chair lifts, the disconnecting means shall be located within sight of the motor controller.

(d) Identification and Signs. Where there is more than one driving machine in a machine room, the disconnecting means shall be numbered to correspond to the identifying number of the driving machine that they control.

The disconnecting means shall be provided with a sign to identify the location of the supply side overcurrent protective device.

Sign requirements for the location of supply-side overcurrent devices assist the elevator mechanic in troubleshooting during a power loss.

620-52. Power from More than One Source.

(a) Single-car and Multicar Installations. On single-car and multicar installations, equipment receiving electrical power from more than one source shall be provided with a disconnecting means for each source of electrical power. The disconnecting means shall be within sight of the equipment served.

(b) Warning Sign for Multiple Disconnecting Means. Where multiple disconnecting means are used and parts of the controllers remain energized from a source other than the one disconnected, a warning sign shall be mounted on or next to the disconnecting means. The sign shall be clearly legible and shall read

WARNING — PARTS OF THE CONTROLLER ARE NOT DE-ENERGIZED BY THIS SWITCH.

(c) Interconnection Multicar Controllers. Where interconnections between controllers are necessary for the operation of the system on multicar installations that remain energized from a source other than the one disconnected, a warning sign in accordance with Section 620-52(b) shall be mounted on or next to the disconnecting means.

620-53. Car Light, Receptacle(s), and Ventilation Disconnecting Means. Elevators shall have a single means for disconnecting all ungrounded car light, receptacle(s), and ventilation power-supply conductors for that elevator car.

The disconnecting means shall be capable of being locked in the open position and shall be located in the machine room for that elevator car. Where there is no machine room, the disconnecting means shall be located in the machinery space for that elevator car.

Section 620-53 was revised for the 1999 *Code* to cover disconnecting means for elevators that do not have a machine room. This type of installation includes those designs using drive systems located on the car, counterweight, or in the hoistway. Such designs include screw drive or linear induction motor drives.

See ANSI/ASME A17.1-1996, *Safety Code for Elevators and Escalators,* for more information on this type of arrangement.

Where there is equipment for more than one elevator car in the machine room, the disconnecting means shall be numbered to correspond to the identifying number of the elevator car whose light source they control.

The disconnecting means shall be provided with a sign to identify the location of the supply side overcurrent protective device.

620-54. Heating and Air-Conditioning Disconnecting Means. Elevators shall have a single means for disconnecting all ungrounded car heating and air-conditioning power-supply conductors for that elevator car.

The disconnecting means shall be capable of being locked in the open position and shall be located in the machine room for that elevator car. Where there is no machine room, the disconnecting means shall be located in the machinery space for that elevator car.

Section 620-54 was revised for the 1999 *Code* to cover disconnecting means for elevators that do not have a machine room. This type of installation includes those designs using drive systems located on the car, counterweight, or in the hoistway. Such designs include screw drive or linear induction motor drives. See ANSI/ASME A17.1-1996, *Safety Code for Elevators and Escalators,* for more information on this type of arrangement.

Where there is equipment for more than one elevator car in the machine room, the disconnecting means shall be numbered to correspond to the identifying number of the elevator car whose heating and air conditioning source they control.

The disconnecting means shall be provided with a sign to identify the location of the supply side overcurrent protective device.

G. Overcurrent Protection

620-61. Overcurrent Protection. Overcurrent protection shall be provided as follows.

(a) Operating Devices, Control, and Signaling Circuits. Operating devices, control and signaling circuits shall be protected against overcurrent in accordance with the requirements of Sections 725-23 and 725-24.

Class 2 power-limited circuits shall be protected against overcurrent in accordance with the requirements of Chapter 9, Notes to Tables 11(a) and 11(b).

(b) Overload Protection for Motors.

(1) Duty on elevator and dumbwaiter driving machine motors and driving motors of motor-generators used with generator field control shall be rated as intermittent. Such motors shall be protected against overload in accordance with Section 430-33.

(2) Duty on escalator and moving walk driving machine motors shall be rated as continuous. Such motors shall be protected against overload in accordance with Section 430-32.

(3) Escalator and moving walk driving machine motors and driving motors of motor-generator sets shall be protected against running overload as provided in Table 430-37.

(4) Duty on wheelchair lift and stairway chair lift driving machine motors shall be rated as intermittent. Such motors shall be protected against overload in accordance with Section 430-33.

FPN: For further information, see Section 430-44 for orderly shutdown.

(c) Motor Feeder Short-Circuit and Ground-Fault Protection. Motor feeder short-circuit and ground-fault protection shall be as required in Article 430, Part E.

(d) Motor Branch-Circuit Short-Circuit and Ground-Fault Protection. Motor branch-circuit short-circuit and ground-fault protection shall be as required in Article 430, Part D.

620-62. Selective Coordination. Where more than one driving machine disconnecting means is supplied by a single feeder, the overcurrent protective devices in each disconnecting means shall be selectively coordinated with any other supply side overcurrent protective devices.

Coordination of the overcurrent protective devices is important. For example, if a building contains three elevators and a fault occurs in one, only the overcurrent device ahead of the faulted elevator should open. This leaves the remaining two elevators in operation. This is especially important because elevators are commonly used to carry firefighters and equipment closer to the fire during fire-fighting operations.

If the overcurrent devices in the elevator room do not have proper coordination with the feeder overcurrent device, all three elevators might be disconnected from the power source.

To selectively coordinate overcurrent protective devices, the manufacturer's time-current curves, let-through and withstand capacity data, and the unlatching times data must be used when sizing or setting overcurrent devices.

See Figure 620.7 for an illustration of selective coordination.

H. Machine Room

620-71. Guarding Equipment. Elevator, dumbwaiter, escalator, and moving walk driving machines; motor-generator sets; motor controllers; and disconnecting means shall be installed in a room or enclosure set aside for that purpose

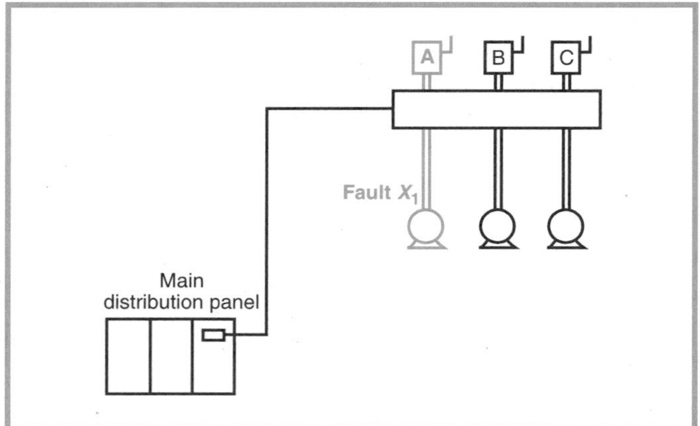

Figure 620.7 *To be selectively coordinated, for a fault at X_1, only the overcurrent devices in switch "A" should open. The feeder overcurrent devices in the main distribution panel should not open. Thus, two elevators will remain in use. See Section 620-51(a) for the requirement of power supply disconnecting means.*

unless otherwise permitted in (a) or (b). The room or enclosure shall be secured against unauthorized access.

(a) Motor Controllers. Motor controllers shall be permitted outside the spaces herein specified, provided they are in enclosures with doors or removable panels that are capable of being locked in the closed position and the disconnecting means is located adjacent to or is an integral part of the motor controller. Motor controller enclosures for escalator or moving walks shall be permitted in the balustrade on the side located away from the moving steps or moving treadway. If the disconnecting means is an integral part of the motor controller, it shall be operable without opening the enclosure.

(b) Driving Machines. Elevators with driving machines located on the car, counterweight, or in the hoistway, and driving machines for dumbwaiters, wheelchair lifts, and stairway lifts shall be permitted outside the spaces herein specified.

J. Grounding

620-81. Metal Raceways Attached to Cars. Metal raceways, Type MC cable, Type MI cable, or Type AC cable attached to elevator cars shall be bonded to grounded metal parts of the car that they contact.

620-82. Electric Elevators. For electric elevators, the frames of all motors, elevator machines, controllers, and the metal enclosures for all electrical equipment in or on the car or in the hoistway shall be grounded in accordance with Article 250.

620-83. Nonelectric Elevators. For elevators other than electric having any electric conductors attached to the car, the metal frame of the car, where normally accessible to persons, shall be grounded in accordance with Article 250.

620-84. Escalators, Moving Walks, Wheelchair Lifts, and Stairway Chair Lifts. Escalators, moving walks, wheelchair lifts, and stairway chair lifts shall comply with Article 250.

620-85. Ground-Fault Circuit-Interrupter Protection for Personnel. Each 125-volt, single-phase, 15- and 20-ampere receptacle installed in pits, on elevator car tops, and in escalator and moving walk wellways shall be of the ground-fault circuit-interrupter type.

All 125-volt, single-phase, 15- and 20-ampere receptacles installed in machine rooms and machinery spaces shall have ground-fault circuit-interrupter protection for personnel.

A single receptacle supplying a permanently installed sump pump shall not require ground-fault circuit-interrupter protection.

The GFCI requirements of Section 620-85 are intended to reduce the shock hazard to maintenance personnel who service elevator equipment using portable hand tools and temporary lighting.

The first paragraph of Section 620-85 requires a GFCI-type receptacle for each 15- and 20-ampere receptacle installed in pits, elevator car tops, and in escalator and moving walk wellways. This requirement is based on the premise that the reset pushbutton for a tripped GFCI receptacle should be within easy reach of an elevator mechanic working in confined spaces.

The second paragraph of Section 620-85 requires that all 15- and 20-ampere receptacles installed in machine rooms and machinery spaces have ground-fault circuit-interrupter protection for personnel. Either a GFCI-type circuit breaker or a GFCI-type receptacle may provide this protection.

The second paragraph of Section 620-85 was revised for the 1999 *Code* to allow GFCI protection, rather than GFCI receptacles, for machinery spaces because machine spaces usually do not involve access hazards.

K. Emergency and Standby Power Systems

620-91. Emergency and Standby Power Systems. An elevator(s) shall be permitted to be powered by an emergency or standby power system.

FPN: See ASME/ANSI A17.1-1996, Rule 211.2, and CAN/CSA-B44-1994, Clause 3.12.13, for additional information.

(a) Regenerative Power. For elevator systems that regenerate power back into the power source, which is unable to absorb the regenerative power under overhauling elevator load conditions, a means shall be provided to absorb this power.

(b) Other Building Loads. Other building loads, such as power and lighting shall be permitted as the energy absorption means required in (a) provided that such loads are automatically connected to the emergency or standby power system operating the elevators and are large enough to absorb the elevator regenerative power.

(c) Disconnecting Means. The disconnecting means required by Section 620-51 shall disconnect the elevator from both the emergency or standby power system and the normal power system.

Where an additional power source is connected to the load side of the disconnecting means, the disconnecting means required in Section 620-51 shall be provided with an auxiliary contact that is positively opened mechanically and the opening shall not be solely dependent on springs. This contact shall cause the additional power source to be disconnected from its load when the disconnecting means is in the open position.

Article 625 — Electric Vehicle Charging System

Contents

A. General

A variety of street- and highway-worthy electric vehicles are expected to be in production. New and proposed legislation in several regions around the United States calls for increasing deployment of electric vehicles as a way to reduce air pollution. In California, the Air Resources Board has mandated that 10 percent of all vehicles produced by major manufacturers and available for sale in that state must be "zero polluting" by 2003. Other states have adopted similar requirements. In addition, the Clean Air Act Amendments of 1990 and the National Energy Policy Act of 1992 have requirements for public and private purchases of clean fuel vehicles and alternatively fueled vehicles, respectively. Electric vehicles fulfill both those requirements. It is apparent that electric vehicle charging will be occurring in all occupancies, including residential, commercial, retail, and public sites.

Article 625 sets forth installation safety requirements for typical hard-wired conductive connections of battery charging equipment, as well as the safety concerns of the new "smart" inductive coupling connections of battery charging equipment. In particular, this article covers the wiring methods, equipment construction, control and protection, and equipment locations for automotive-type vehicle charging equipment. Throughout Article 625, the intent is to prohibit untrained people from being exposed to energized live parts.

625-1. Scope. The provisions of this article cover the electrical conductors and equipment external to an electric vehicle that connect an electric vehicle to a supply of electricity by conductive or inductive means, and the installation of equipment and devices related to electric vehicle charging.

The scope of Article 625 is intended to cover all electrical wiring and equipment installed between the service point and the skin of the "automotive-type" electric vehicle. Automotive-type electric vehicles are emphasized because they are much different from other electric vehicles commonly used today. Most existing electric vehicles are off-road types, such as industrial forklifts, hoists, lifts, transports, golf carts, airport personnel trams, and so on. The charging requirements and other exterior electrical connections are usually serviced and maintained by trained mechanics or technicians. The existing *NEC* has been adequate to allow the authority having jurisdiction to make interpretations that provide the safety levels needed for these installations.

Article 625 specifically excludes off-road vehicles, to avoid conflict with existing articles. Motorcycles are not covered by Article 625 because motorcycles typically have smaller propulsion systems that operate at lower voltages, 12 to 24 volts dc versus 100 to 350 volts dc for electric automotive vehicles. Typically, motorcycles are charged from standard 120-volt, 15-ampere receptacles due to lower battery capacity. GFCI protection is not mandatory for charging electric motorcycles. However, Sections 210-8(a)(2) and (3) typically require GFCI protection for receptacles in locations where an electric motorcycle would normally be charged.

FPN: For industrial trucks, see *Fire Safety Standard for Powered Industrial Trucks Including Type Designations, Areas of Use, Conversions, Maintenance, and Operation,* NFPA 505-1996.

625-2. Definitions.

Several definitions were added to Section 625-2 for the 1999 *Code* to correlate with industry standards such as Society of Automotive Engineers (SAE), J1772 and J1773, and Underwriters Laboratories (UL), 2331.

Electric Vehicle. An automotive-type vehicle for highway use, such as passenger automobiles, buses, trucks, vans, and the like, primarily powered by an electric motor that draws current from a rechargeable storage battery, fuel cell, photovoltaic array, or other source of electric current. For the purpose of this article, electric motorcycles and similar type vehicles and off-road self-propelled electric vehicles, such as industrial trucks, hoists, lifts, transports, golf carts, airline ground support equipment, tractors, boats, and the like, are not included.

The primary difference between electric vehicles as defined in Article 625 and electric vehicles presently covered by other sections in the *NEC* is in their road and highway worthiness. The automotive electric vehicles under consideration are comparable in performance and function to the conventional automobiles and light trucks in use today. See Figure 625.1 for an example of such an electric vehicle. The automotive electric vehicles under development must be capable of complying with the Federal Motor Vehicle Safety Standards and other Department of Transportation, National Highway Traffic Safety Administration, and U.S. Environmental Protection Agency requirements. Special-purpose or limited capability electric vehicles, such as lift trucks, golf carts, and so on, are not covered by Article 625.

Figure 625.1 *Ford ECOSTAR electric vehicle. (Ford Motor Co.)*

Electric Vehicle Connector. A device that by insertion into an electric vehicle inlet, establishes an electrical connection to the electric vehicle for the purpose of charging and information exchange. This is part of the electric vehicle coupler.

Electric Vehicle Coupler. A mating electric vehicle inlet and electric vehicle connector set.

Electric Vehicle Inlet. The device on the electric vehicle into which the electric vehicle connector is inserted for charging and information exchange. This is part of the electric vehicle coupler. For the purposes of this *Code,* the electric vehicle inlet is considered to be part of the electric vehicle and not part of the electric vehicle supply equipment.

Electric Vehicle Nonvented Storage Battery. A hermetically-sealed battery comprised of one or more rechargeable electrochemical cells that has no provision for release of excessive gas pressure, or the addition of water or electrolyte, or for external measurements of electrolyte specific gravity.

Electric Vehicle Supply Equipment. The conductors, including the ungrounded, grounded, and equipment grounding conductors and the electric vehicle connectors, attachment plugs, and all other fittings, devices, power outlets, or apparatuses installed specifically for the purpose of delivering energy from the premises wiring to the electric vehicle.

Electric vehicle supply equipment, as illustrated in Figure 625.2, is comprised of the components between the skin of the electric vehicle and the premises wiring, including any flexible cable, disconnecting means, enclosures, power outlet, and electric vehicle connector. The definition of *electric vehicle* includes all off-vehicle charging equipment and does not include charging equipment installed on the vehicle.

Figure 625.2 *An example of electric vehicle supply equipment. (Square D Co.)*

Personnel Protection System. A system of personnel protection devices and constructional features that when used together provide protection against electric shock of personnel.

625-3. Other Articles. Wherever the requirements of other articles of this *Code* and Article 625 differ, the requirements of Article 625 shall apply.

625-4. Voltages. Unless other voltages are specified, the nominal ac system voltages of 120, 120/240, 208Y/120, 240, 480Y/277, 480, 600Y/347, and 600 volts shall be used to supply equipment covered by this article.

625-5. Listed or Labeled. All electrical materials, devices, fittings, and associated equipment shall be listed or labeled.

B. Wiring Methods

625-9. Electric Vehicle Coupler. The electric vehicle coupler shall comply with (a) through (f).

The electric vehicle connector is the device that inserts into the electric vehicle inlet (charge port) of the vehicle. The electric vehicle inlet is not a premises wiring receptacle or attachment cap. An electric vehicle coupler is the mating set of the electric vehicle connector and electric vehicle inlet. A new requirement was added for the 1999 *Code* to require the coupler to be noninterchangeable to prevent equipment damage or personal injury. Couplers for the electric vehicle charging equipment are not permitted to be standard NEMA-configuration wiring devices.

(a) Polarization. The electric vehicle coupler shall be polarized unless part of a system identified and listed as suitable for the purpose.

(b) Noninterchangeability. The electric vehicle coupler shall have a configuration that is noninterchangeable with wiring devices in other electrical systems. Nongrounding-type electric vehicle couplers shall not be interchangeable with grounding-type electric vehicle couplers.

(c) Construction and Installation. The electric vehicle coupler shall be constructed and installed so as to guard against inadvertent contact by persons with parts made live from the electric vehicle supply equipment or the electric vehicle battery.

Section 625-9(c) prevents a person from coming in contact with energized live parts at the electric vehicle coupler during the connection or disconnection of the vehicle to the charging system.

(d) Unintentional Disconnection. The electric vehicle coupler shall be provided with a positive means to prevent unintentional disconnection.

(e) Grounding Pole. The electric vehicle coupler shall be provided with a grounding pole, unless part of a system identified and listed as suitable for the purpose in accordance with Article 250.

(f) Grounding Pole Requirements. If a grounding pole is provided, the electric vehicle coupler shall be designed so that the grounding pole connection is the first to make and the last to break contact.

C. Equipment Construction

625-13. Electric Vehicle Supply Equipment. Electric vehicle supply equipment rated at 125 volt, single phase, 15 or 20 amperes or part of a system identified and listed as suitable for the purpose and meeting the requirements of Sections 625-18, 625-19, and 625-29 shall be permitted to be cord and plug connected. All other electric vehicle supply equipment shall be permanently connected and fastened in place. This equipment shall have no exposed live parts.

Some manufacturers produce 125-volt, single-phase, 15- or 20-ampere portable charging units for convenience charging. These charging units may be stored in the vehicle. However, Section 625-13 makes it clear that nonportable equipment must be mounted and permanently wired. This equipment may be physically attached to the wall, floor, or ceiling. The provision for no exposed live parts is a safety concern for the general public.

625-14. Rating. Electric vehicle supply equipment shall have sufficient rating to supply the load served. For the purposes of this article, electric vehicle charging loads shall be considered to be continuous loads.

Considering both near-term and long-term requirements for electric vehicle (EV) charging, three methods have been identified for recommended development. Referred to as Level 1, Level 2, and Level 3 EV charging, they cover the range of power levels anticipated for charging EVs.

Level 1. This method, which allows broad access to charge an EV, permits plugging into a common, grounded 120-volt electrical receptacle (NEMA 5-15R or 5-20R). The maximum load on this receptacle is 12 amperes or 1.4 kVA. The minimum circuit and overcurrent rating for this connection is 15 amperes for a 15-ampere receptacle and 20 amperes for a 20-ampere receptacle.

Level 2. This is the primary and preferred method of EV charging at both private and public facilities. It requires special equipment and connection to an electric power supply dedicated to EV charging. The voltage of this connection is either 240 volts or 208 volts. The maximum load is 32 amperes (7.7 kVA at 240 volts or 6.7 kVA at 208 volts). The minimum circuit and overcurrent rating for this connection is 40 amperes (32 × 1.25 = 40 amperes). Electric vehicles are treated as continuous loads. See Section 625-21 for sizing overcurrent protection devices.

Level 3. The EV equivalent of a commercial gasoline dispensing station, this high-speed, high-power method will charge an EV in about the same time it takes to refuel a conventional vehicle. Because of individual supply requirements and available source voltages, exact voltage and load specifications for Level 3 charging have not been defined as in Level 1 and Level 2. These power requirements will be specified by the manufacturer of the equipment.

625-15. Markings. The electric vehicle supply equipment shall comply with (a) through (c).

(a) General. All electric vehicle supply equipment shall be marked by the manufacturer

FOR USE WITH ELECTRIC VEHICLES

(b) Ventilation Not Required. Where marking is required by Section 625-29(c), the electric vehicle supply equipment shall be clearly marked by the manufacturer

VENTILATION NOT REQUIRED

The marking shall be located so as to be clearly visible after installation.

(c) Ventilation Required. Where marking is required by Section 625-29(d), the electric vehicle supply equipment shall be clearly marked by the manufacturer "Ventilation Required." The marking shall be located so as to be clearly visible after installation.

625-16. Means of Coupling. The means of coupling to the electric vehicle shall be either conductive or inductive. Attachment plugs, electric vehicle connectors, and electric vehicle inlets shall be listed or labeled for the purpose.

625-17. Cable. The electric vehicle supply equipment cable shall be Type EV, EVJ, EVE, EVJE, EVT, or EVJT flexible cable as specified in Article 400 and Table 400-4. Ampacities shall be as specified in Table 400-5(A) for No. 10 and smaller and Table 400-5(B) for No. 8 and larger. The overall length of the cable shall not exceed 25 ft (7.63 m). Other cable types and assemblies listed as being suitable for the purpose, including optional hybrid communications, signal, and optical fiber cables, shall be permitted.

The 25-ft maximum cable length is established by adding the 15-ft car length to the 7-ft car width, plus 3 ft to the power outlet securement point. This limits excessive cable lengths that may be exposed to damage. Cable management systems could also be used to retract excessive cable lengths, or cables shorter than 25 ft could be used.

625-18. Interlock. Electric vehicle supply equipment shall be provided with an interlock that de-energizes the electric vehicle connector and its cable whenever the electric connector is uncoupled from the electric vehicle. An interlock shall not be required for portable cord- and plug-connected electric vehicle supply equipment intended for connection to receptacle outlets rated 125 volts, single phase, 15 and 20 amperes.

To reduce shock hazard, a pilot or communications interlock establishes power through the electric vehicle supply equipment. Loss of the pilot or communications circuit locks out power, isolating possible hazardous situations in the electric vehicle supply equipment. See Section 625-29(c) for mechanical ventilation interlock requirements.

For ventilation interlock, see Section 625-29(d)(3) for 125-volt receptacles intended to charge electric vehicles.

625-19. Automatic De-energization of Cable. The electric vehicle supply equipment or the cable-connector combination of the equipment shall be provided with an automatic means to de-energize the cable conductors and electric vehicle connector upon exposure to strain that could result in either cable rupture or separation of the cable from the electric connector and exposure of live parts. Automatic means to de-energize the cable conductors and electric vehicle connector shall not be required for portable cord- and plug-connected electric vehicle supply equipment intended for connection to receptacle outlets rated at 125 volts, single phase, 15 and 20 amperes.

D. Control and Protection

625-21. Overcurrent Protection. Overcurrent protection for feeders and branch circuits supplying electric vehicle supply equipment shall be sized for continuous duty and shall have a rating of not less than 125 percent of the maximum load of the electric vehicle supply equipment. Where noncontinuous loads are supplied from the same feeder or branch circuit, the overcurrent device shall have a rating of not less than the sum of the noncontinuous loads plus 125 percent of the continuous loads.

625-22. Personnel Protection System. The electric vehicle supply equipment shall have a listed system of protection against electric shock of personnel. The personnel protection system shall be composed of listed personnel protection devices and constructional features. Where cord- and plug-connected electric vehicle supply equipment is used, the interrupting device of a listed personnel protection system shall be provided and shall be an integral part of the attachment plug or shall be located in the power supply cable not more than 12 in. (305 mm) from the attachment plug.

The personnel protection system may consist of one or more components that provide protection against electric shock for different portions of the electric vehicle supply equipment circuitry, which may be operating at frequencies other than 50/60 Hz, at direct current potentials, and/or voltages above 150 volts to ground.

Standard GFCI devices do not provide the range of protection needed for the various types of charging systems being developed. Devices or methods that may be used include basic insulation, double insulation, grounding monitors, insulation monitors with interrupters, and leakage current monitors. Many combinations and variations of these devices can be used to provide the personnel protection required. For systems operating at voltages above 150 volts to ground, the protective system may include monitoring systems to ensure that proper grounding is provided and maintained during charging.

625-23. Disconnecting Means. For electric vehicle supply equipment rated more than 60 amperes or more than 150 volts to ground, the disconnecting means shall be provided and installed in a readily accessible location. The disconnecting means shall be capable of being locked in the open position.

625-25. Loss of Primary Source. Means shall be provided such that upon loss of voltage from the utility or other electric system(s), energy cannot be backfed through the electric vehicle supply equipment to the premises wiring system. The electric vehicle shall not be permitted to serve as a standby power supply.

E. Electric Vehicle Supply Equipment Locations

625-28. Hazardous (Classified) Locations. Where electric vehicle supply equipment or wiring is installed in a hazardous (classified) location, the requirements of Articles 500 through 516 shall apply.

Section 625-28 will allow EV charging equipment to be installed in hazardous locations in accordance with the requirements of Chapter 5. This provision is needed in commercial repair garages and combination gasoline/EV charging stations (see Section 511-9).

625-29. Indoor Sites. Indoor sites shall include, but not be limited to, integral, attached, and detached residential garages; enclosed and underground parking structures; repair and nonrepair commercial garages; and agricultural buildings.

(a) Location. The electric vehicle supply equipment shall be located to permit direct connection to the electric vehicle.

(b) Height. Unless specifically listed for the purpose and location, the coupling means of the electric vehicle supply equipment shall be stored or located at a height of not less than 18 in. (457 mm) and not more than 4 ft (1.22 m) above the floor level.

(c) Ventilation Not Required. Where electric vehicle nonvented storage batteries are used or where the electric vehicle supply equipment is listed or labeled as suitable for charging electric vehicles indoors without ventilation and marked in accordance with Section 625-15(b), mechanical ventilation shall not be required.

Major auto manufacturers are taking the necessary steps to make electric vehicle systems safe. Most batteries used in electric vehicles manufactured by major automakers will not emit hydrogen gas in quantities that could cause an explosion. Preventive measures such as mechanical or passive ventilation are not required, because the electric vehicle batteries and charging systems are designed to prevent or limit the emission of hydrogen during charging. The Society of Automotive Engineers (SAE) has developed a recommended practice, J-1718, *Measurement of Hydrogen Gas Emission from Battery-Powered Passenger Cars and Light Trucks During Battery Charging,* that can be used to assess suitability for indoor charging. This standard includes provisions for tests during normal charging operations and potential equipment failure

modes. See Figure 625.3 for an illustration of a garage without ventilation.

(d) Ventilation Required. Where the electric vehicle supply equipment is listed or labeled as suitable for charging electric vehicles that require ventilation for indoor charging and marked in accordance with Section 625-15(c), mechanical ventilation, such as a fan, shall be provided. The ventilation shall include both supply and exhaust equipment and shall be permanently installed and located to intake from, and vent directly to, the outdoors. Positive pressure ventilation systems shall only be permitted in buildings or areas that have been specifically designed and approved for that application. Mechanical ventilation requirements shall be determined by one of the following methods.

The interlock described in Section 625-29(d) prevents a vehicle requiring ventilation from being charged unless ventilation is provided. This interlock feature is included in the pilot connection of the standard electric vehicle supply equipment connection. See Figure 625.4 for an illustration of electric vehicle supply equipment with ventilation.

(1) For supply voltages and currents specified in Table 625-29(d), the minimum ventilation requirements in cubic feet per minute (cfm) shall be as specified in Table 625-29(d) for each of the total number of electric vehicles that can be charged at one time. This table allows for sufficient ventilation for any configuration of electric vehicle supply equipment and electric vehicle charging space.

(2) For supply voltages and currents other than specified in Table 625-29(d), the minimum ventilation requirements in cubic feet per minute (cfm) shall be calculated by means of the following general formulas as applicable.

(a) Single phase:

$$\text{Ventilation}_{\text{single phase}} = \frac{\text{(volts)(amperes)}}{48.7}$$

(b) Three phase:

$$\text{Ventilation}_{\text{three phase}} = \frac{1.732\ \text{(volts)(amperes)}}{48.7}$$

(3) For an electric vehicle supply equipment ventilation system designed by a person qualified to perform such calculations as an integral part of a building's total ventilation system, the minimum ventilation requirements shall be per-

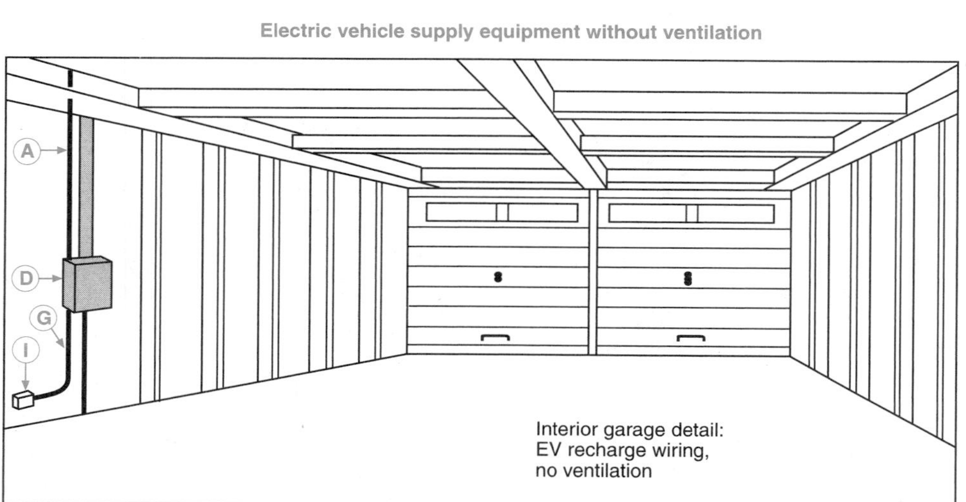

Figure 625.3 *An interior garage detail showing electric vehicle supply equipment where ventilation is not required.*

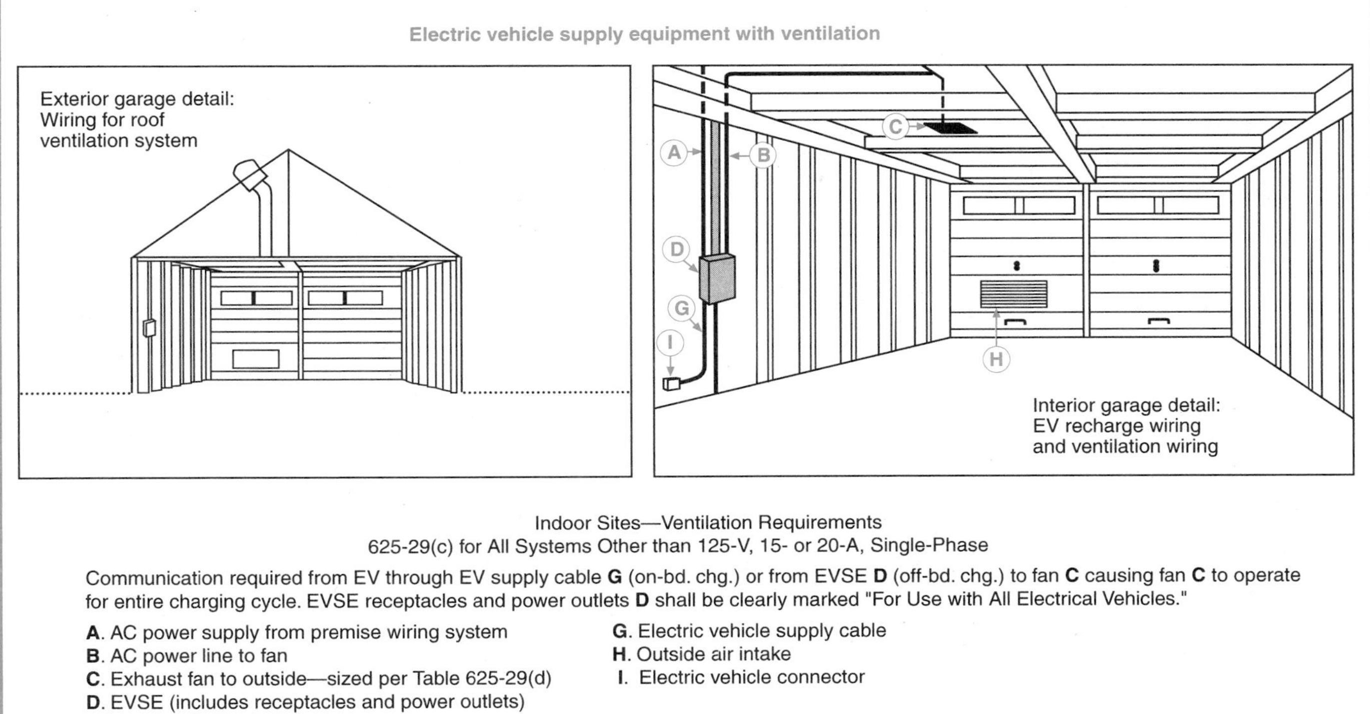

Figure 625.4 *Interior and exterior details for the installation of electric vehicle supply equipment with ventilation.*

mitted to be determined per calculations specified in the engineering study.

The supply circuit to the mechanical ventilation equipment shall be electrically interlocked with the electric vehicle supply equipment and shall remain energized during the entire electric vehicle charging cycle. Electric vehicle supply equipment shall be marked in accordance with Section 625-15. Electric vehicle supply equipment receptacles rated at 125 volts, single phase, 15 and 20 amperes shall be marked in accordance with Section 625-15(c) and shall be switched and the mechanical ventilation system shall be electrically interlocked through the switch supply power to the receptacle.

For receptacles at each electric vehicle charging space identified "For Use with Electric Vehicles," and rated 125 volts, single phase, 15 and 20 amperes, such receptacles shall be switched, and mechanical ventilation shall be electrically interlocked through the switch supply power to the receptacle.

The intent of 625-29(d) is to ensure sufficient diffusion and dilution of hydrogen gas from gas-emitting batteries to prevent a hazardous condition. Certain batteries used in some electric vehicles emit hydrogen gas during the charging process.

Table 625-29(d). Minimum Ventilation Required in Cubic Feet per Minute (cfm) for Each of the Total Number of Electric Vehicles that Can Be Charged at One Time

	Branch-Circuit Voltage						
	Single Phase			3 Phase			
Branch-Circuit Ampere Rating	120 V	208 V	240 V or 120/240 V	208 V or 208 Y/120 V	240 V	480 V or 480 Y/277 V	600 V or 600 Y/347 V
15	37	64	74	—	—	—	—
20	49	85	99	148	171	342	427
30	74	128	148	222	256	512	641
40	99	171	197	296	342	683	854
50	123	214	246	370	427	854	1066
60	148	256	296	444	512	1025	1281
100	246	427	493	740	854	1708	2135
150	—	—	—	1110	1281	2562	3203
200	—	—	—	1480	1708	3416	4270
250	—	—	—	1850	2135	4270	5338
300	—	—	—	2221	2562	5125	6406
350	—	—	—	2591	2989	5979	7473
400	—	—	—	2961	3416	6832	8541

Hydrogen is a colorless, odorless, tasteless, nontoxic flammable gas. At atmospheric pressure, the flammable range for hydrogen is 4 to 75 percent by volume in air.

NFPA 69-1997, *Standard on Explosion Prevention Systems,* establishes requirements to ensure safety with flammable mixtures. Section 3-3, Design and Operating Requirements, requires that combustible gas concentrations be restricted to 25 percent of the lower flammable limit. This design criterion provides a safety margin for personnel working with atmospheres containing hydrogen. Safety is accomplished by keeping the concentration of hydrogen below 25 percent of the lower flammability limit, or 1 percent (25 percent × 4 percent = 1 percent) hydrogen by volume in air; that is, below 10,000 ppm hydrogen.

A ventilation system for a typical residential-type garage includes both supply and mechanical exhaust equipment and is permanently installed. The equipment is located in the space so that it takes in air from outdoors to the space, circulates the air through the space, and exhausts the air directly to the outdoors. Typically, the equipment includes a passive vent for intake on one side of the enclosed space and an exhaust fan vented to the outside on the other side of the space. In enclosed commercial garages and other structures, additional ventilation is not required if the exhaust, as required by the building code for carbon monoxide or other purposes, is greater than the quantity listed in the table. Other engineered electric vehicle ventilation systems are allowed, provided they are designed properly. The electric vehicle charging area is permitted to be ventilated by the building ventilation system. The ventilation system and the charging system must be interlocked to prevent charging if the ventilation is not operating, as shown in Figure 625.5. A qualified person must do the calculation of the ventilation requirements.

625-30. Outdoor Sites. Outdoor sites shall include, but not be limited to, residential carports and driveways, curbside, open parking structures, parking lots, and commercial charging facilities.

Section 625-30 states that mechanical ventilation is not required in open garages, carports, and other structures with two or more open sides.

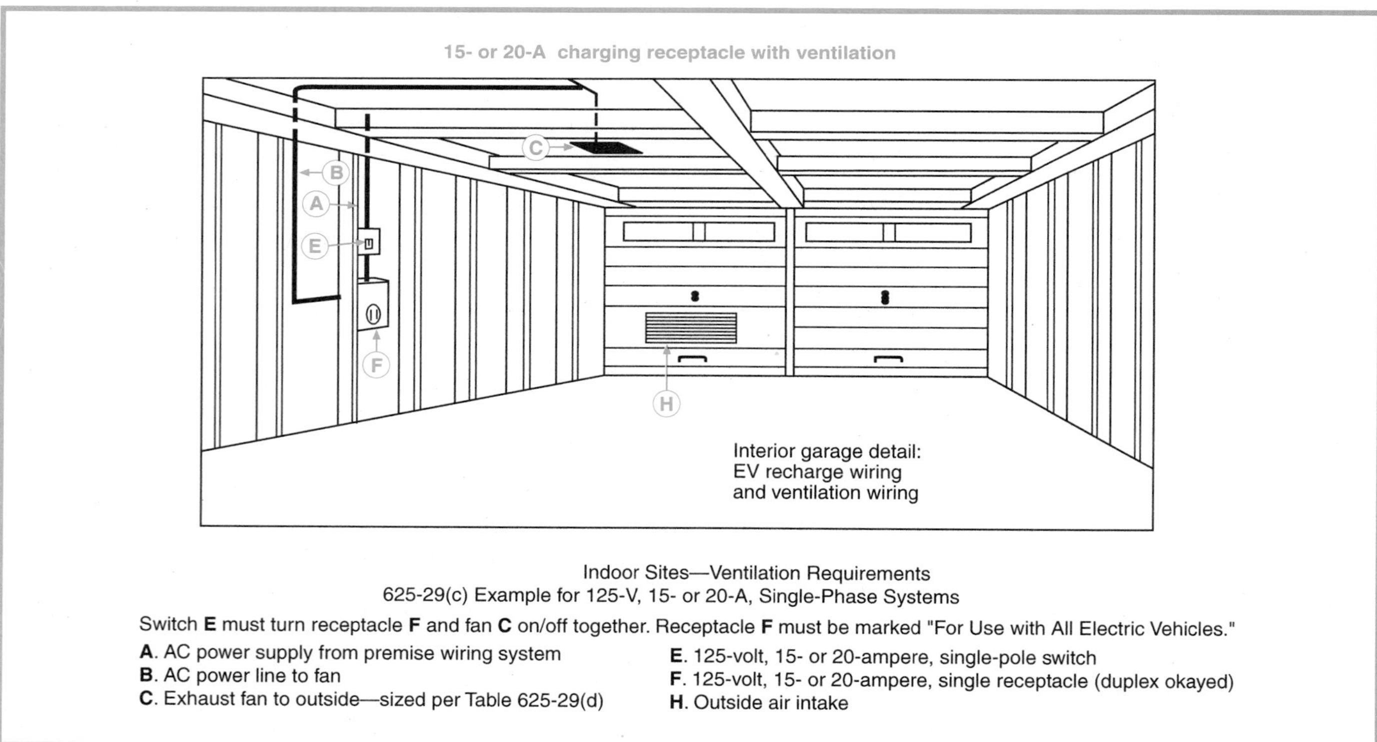

Figure 625.5 *An example showing the ventilation equipment electrically interlocked with the electric vehicle charging equipment receptacle.*

(a) Location. The electric vehicle supply equipment shall be located to permit direct connection to the electric vehicle.

(b) Height. Unless specifically listed for the purpose and location, the coupling means of electric vehicle supply equipment shall be stored or located at a height of not less than 24 in. (610 mm) and not more than 4 ft (1.22 m) above the parking surface.

Article 630 — Electric Welders

Contents

Article 630 has been restructured for the 1999 *Code*. Parts B and C of the 1996 *Code* were combined in the 1999 *Code* to reflect the fact that motor-generator arc welders are in limited use. Additionally, Part C of the 1996 *Code* was deleted for the same reason.

A. General

630-1. Scope. This article covers electric arc welding, resistance welding apparatus, and other similar welding equipment that is connected to an electric supply system.

The two general types of electric welding are resistance welding and arc welding. Resistance welding, or "spot" welding, is the process of joining, or electrically fusing together, two or more metal sheets or parts without any preparation of stock. The metal parts are placed between two electrodes, or welding points, and a heavy current at a low voltage is passed through the electrodes. The metal parts offer resistance to the flow of current such that they heat to a molten state, and a weld is made.

Arc welding is the butting of two metal parts to be welded and then the striking of an arc at this joint with a metal electrode (a flux-coated wire rod). The electrode itself is melted and supplies the extra metal necessary for joining the metal parts.

A transformer supplies current for one ac arc welder, and a generator or rectifier supplies current for one or more dc arc welders.

B. Arc Welders

630-11. Ampacity of Supply Conductors. The ampacity of conductors for arc welders shall be as follows.

(a) Individual Welders. The ampacity of the supply conductors shall not be less than the I_{1eff} value on the rating plate. Alternatively, if the I_{1eff} is not given, the ampacity of the supply conductors shall not be less than the current value determined by multiplying the rated primary current in amperes given on the welder rating plate and the following factor based on the duty cycle of the welder.

	Multiplier for Arc Welders	
Duty Cycle	Nonmotor Generator	Motor Generator
100	1.00	1.00
90	0.95	0.96
80	0.89	0.91
70	0.84	0.86
60	0.78	0.81
50	0.71	0.75
40	0.63	0.69
30	0.55	0.62
20 or less	0.45	0.55

(b) Group of Welders. The ampacity of conductors that supply a group of welders shall be permitted to be less than the sum of the currents, as determined in accordance with (a), of the welders supplied. The conductor rating shall be determined in each case according to the welder loading based on the use to be made of each welder and the allowance permissible in the event that all the welders supplied by the conductors will not be in use at the same time. The load value used for each welder shall take into account both the magnitude and the duration of the load while the welder is in use.

FPN: Conductor ratings based on 100 percent of the current, as determined in accordance with (a), of the two largest welders, 85 percent for the third largest welder, 70 percent for the fourth largest welder, and 60 percent for all remaining welders, can be assumed to provide an ample margin of safety under high-production conditions with respect to the maximum permissible temperature of the conductors. Percentage values lower than those given are permissible in cases where the work is such that a high-operating duty cycle for individual welders is impossible.

Even under high-production conditions, the loads on transformer arc welders are considered intermittent; therefore, it is permissible to reduce the ampacity of feeder conductors supplying several transformers (three or more) to the allowable percentages described in the fine print note. It is obvious that intermittent transformer arc welder loads would be considerably less than a continuous load equal to the sum of the full-load current ratings of all the transformers. See also Section 630-31(b). The ampacity of conductors supplying welders is based on the I_{1eff} rating on the welder rating plate. If the I_{1eff} rating is not available, the size of supply conductors to welders may be calculated. The calculation is done by selecting the appropriate factor from Section 630-12(a) based on the type of welder and duty cycle of the welder. The selected factor is then multiplied by the primary current rating from the welder rating plate to determine the minimum ampacity of the supply conductors.

Obsolete terms and ratings such as *nameplate* and *1-hour duty cycle* were removed from Section 630-12 in the 1999 *NEC*. The terms I_{1eff} and I_{max} were added to the marking requirements. A new FPN includes a formula for I_{1eff}, which is preferred over the value derived from using the factors in Section 630-11. The calculated value is still allowed. I_{max} is basically the same as the rated primary current.

630-12. Overcurrent Protection. Overcurrent protection for arc welders shall be as provided in (a) and (b). Where the values as determined by this section do not correspond with the standard ampere ratings provided in Section 240-6 or the rating or setting specified results in unnecessary opening of the overcurrent device, the next higher standard rating or setting shall be permitted.

(a) For Welders. Each welder shall have overcurrent protection rated or set at not more than 200 percent of I_{1max}. Alternatively, if the I_{1max} is not given, the overcurrent protection shall be rated or set at not more than 200 percent of the rated primary current of the welder.

An overcurrent device shall not be required for a welder that has supply conductors protected by an overcurrent device rated or set at not more than 200 percent of I_{1max} or the rated primary current of the welder.

If the supply conductors for a welder are protected by an overcurrent device rated or set at not more than 200 percent of I_{max} or rated primary current of the welder, a separate overcurrent device shall not be required.

(b) For Conductors. Conductors that supply one or more welders shall be protected by an overcurrent device rated or set at not more than 200 percent of the conductor rating.

Some arc-welding machines have a welding range involving an excess secondary-current output capacity beyond that indicated by the secondary rating marked on the machines. This excess capacity (generally not more than 150 percent of the marked output capacity) is usually supplied by means of one or more secondary taps in addition to the tap, or taps, intended for normal output current; the higher currents thus available are intended to provide for heavier welding work, including the use of larger-size electrodes. This excess capacity is somewhat analogous to the inherent overload capacity of motors and transformers. However, the use of this excess current capacity and the overloading of welding machines, except for relatively short periods of time, may be hazardous and should be undertaken with caution by those involved.

Section 630-12(b) was revised for the 1999 *Code* to require conductor overcurrent protection to be set at not more than 200 percent. This change was made because the calculation of I_{1eff} more accurately reflects the heating effect of the supply conductor while idle as well as while welding.

FPN: I_{1max} is the maximum value of the rated supply current at maximum rated output.

I_{1eff} is the maximum value of the effective supply current, calculated from the rated supply current (I_1), the corresponding duty cycle (duty factor) (X), and the supply current at no-load (I_0) by the following formula.

$$I_{1eff} = \sqrt{I_1^2 X + I_0^2(1 - X)}$$

The ampacity of supply conductors for a welder that is not wired for a specific function — that is, one operated at varying intervals for different applications, such as dissimilar metals or thicknesses — is permitted to be 70 percent of the rated primary current for automatically fed welders and 50 percent of the rated primary current for manually operated welders. The rated primary current can be determined using the following equation and using the values given on the nameplate.

$$\text{Rated primary current} = \frac{\text{kVA} \times 1000}{\text{rated primary voltage}}$$

Where the actual primary current and duty cycle are known, such as for a welder wired for a specific operation, the ampacity of the supply conductors is not permitted to be less than the product of the actual primary current (current drawn during weld operation) and the multiplier, as provided in Section 630-31(a)(2), for the duty cycle at which the welder will be operated. For example, a spot welder is specifically set to perform 300 welds per hour on a 60-Hz system. Each weld draws current for 16 cycles. During the 1-hour period, the welder draws current for 4800 cycles (300 × 16). There are 216,000 cycles per hour (60 × 60 × 60). The duty cycle is calculated as follows:

$$\frac{4800}{216{,}000} \times 100 \text{ percent} = 2.2 \text{ percent (duty cycle)}$$

For a seam welder that draws current for 3 cycles and is off for 4 cycles during every 7-cycle period, the duty cycle is calculated as follows:

$$\frac{3}{7} \times 100 \text{ percent} = 42.9 \text{ percent (duty cycle)}$$

An instrument capable of measuring current impulses for 3 cycles ($^1/_{20}$ second), as shown in the example, is required to measure the actual primary current. The duty cycle is set for a specific operation by adjusting the controller for the welder. When sizing supply conductors, voltage drop should be limited to a value permissible for the satisfactory performance of the welder.

630-13. Disconnecting Means. A disconnecting means shall be provided in the supply circuit for each arc welder that is not equipped with a disconnect mounted as an integral part of the welder.

The disconnecting means shall be a switch or circuit breaker, and its rating shall not be less than that necessary to accommodate overcurrent protection as specified under Section 630-12.

630-14. Marking. A rating plate shall be provided for arc welders giving the following information:

(1) Name of manufacturer
(2) Frequency
(3) Number of phases
(4) Primary voltage
(5) I_{1max} and I_{1eff}, or rated primary current
(6) Maximum open-circuit voltage
(7) Rated secondary current and
(8) Basis of rating, such as the duty cycle

C. Resistance Welders

630-31. Ampacity of Supply Conductors. The ampacity of the supply conductors for resistance welders necessary to limit the voltage drop to a value permissible for the satisfactory performance of the welder is usually greater than that required to prevent overheating as described in (a) and (b).

(a) Individual Welders. The rated ampacity for conductors for individual welders shall comply with the following.

(1) The ampacity of the supply conductors for a welder that may be operated at different times at different values of primary current or duty cycle shall not be less than 70 percent of the rated primary current for seam and automatically fed welders, and 50 percent of the rated primary current for manually operated nonautomatic welders.

(2) The ampacity of the supply conductors for a welder wired for a specific operation for which the actual primary current and duty cycle are known and remain unchanged shall not be less than the product of the actual primary current and the multiplier given below for the duty cycle at which the welder will be operated.

Duty Cycle (percent)	Multiplier
50	0.71
40	0.63
30	0.55
25	0.50
20	0.45
15	0.39
10	0.32
7.5	0.27
5 or less	0.22

(b) Groups of Welders. The ampacity of conductors that supply two or more welders shall not be less than the sum of the value obtained in accordance with (a) for the largest welder supplied and 60 percent of the values obtained for all the other welders supplied.

FPN: **Explanation of Terms**

(1) The *rated primary current* is the rated kilovolt-amperes (kVA) multiplied by 1000 and divided by the rated primary voltage, using values given on the nameplate.
(2) The *actual primary current* is the current drawn from the supply circuit during each welder operation at the particular heat tap and control setting used.
(3) The *duty cycle* is the percentage of the time during which the welder is loaded. For instance, a spot welder supplied by a 60-Hz system (216,000 cycles per hour) making four hundred 15-cycle welds per hour would have a duty cycle of 2.8 percent (400 multiplied by 15, divided by 216,000, multiplied by 100). A seam welder operating 2 cycles "on" and 2 cycles "off" would have a duty cycle of 50 percent.

630-32. Overcurrent Protection. Overcurrent protection for resistance welders shall be as provided in (a) and (b). Where the values as determined by this section do not correspond with the standard ampere ratings provided in Section 240-6 or the rating or setting specified results in unnecessary opening of the overcurrent device, the next higher standard rating or setting shall be permitted.

(a) For Welders. Each welder shall have an overcurrent device rated or set at not more than 300 percent of the rated primary current of the welder. If the supply conductors for a welder are protected by an overcurrent device rated or set at not more than 200 percent of the rated primary current of the welder, a separate overcurrent device shall not be required.

(b) For Conductors. Conductors that supply one or more welders shall be protected by an overcurrent device rated or set at not more than 300 percent of the conductor rating.

630-33. Disconnecting Means. A switch or circuit breaker shall be provided by which each resistance welder and its control equipment can be disconnected from the supply circuit. The ampere rating of this disconnecting means shall not be less than the supply conductor ampacity determined in accordance with Section 630-31. The supply circuit switch shall be permitted as the welder disconnecting means where the circuit supplies only one welder.

630-34. Marking. A nameplate shall be provided for each resistance welder giving the following information:

(1) Name of manufacturer
(2) Frequency
(3) Primary voltage
(4) Rated kilovolt-amperes (kVA) at 50 percent duty cycle
(5) Maximum and minimum open-circuit secondary voltage
(6) Short-circuit secondary current at maximum secondary voltage, and
(7) Specified throat and gap setting

D. Welding Cable

630-41. Conductors. Insulation of conductors intended for use in the secondary circuit of electric welders shall be flame retardant.

630-42. Installation. Cables shall be permitted to be installed in a dedicated cable tray as provided in (a), (b), and (c).

(a) Cable Support. The cable tray shall provide support at not greater than 6-in. (152-mm) intervals.

(b) Spread of Fire and Products of Combustion. The installation shall comply with Section 300-21.

(c) Signs. A permanent sign shall be attached to the cable tray at intervals not greater than 20 ft (6.1 m). The sign shall read

CABLE TRAY FOR WELDING CABLES ONLY

Article 640 — Audio Signal Processing, Amplification, and Reproduction Equipment

Contents

Article 640 was completely revised for the 1999 *Code* to reflect current trends and practices in the industry. New definitions were added to expand understanding of the equipment involved. Separate requirements for permanent and portable equipment were also added. These changes have also closely tied Article 640 with Articles 725 and 760.

A. General

640-1. Scope. This article covers equipment and wiring for audio signal generation, recording, processing, amplification and reproduction; distribution of sound; public address; speech input systems; temporary audio system installations; and electronic organs or other electronic musical instruments. This also includes audio systems subject to Article 517, Part F, and Articles 518, 520, 525, and 530.

Equipment covered by Article 640 includes amplifiers; public address (PA) systems and centralized sound systems used in schools, factories, businesses, stadiums, and similar locations; intercommunications devices and systems; and devices used for recording or reproducing voice or music. The scope remains limited to equipment, which has as its main function the processing, distribution, amplification, and reproduction of audio frequency bandwidth signals. This does not preclude equipment that uses radio frequency or other forms of transmission between equipment components, such as wireless microphone systems.

Electronic organs are synthesizers and synthesizers are also audio signal generators. Electronic organs are still uniquely cited in the scope for clarity, and electronic musical instruments are added to cover all other forms of electronic tone generation. Electronic musical instruments are instruments that create an electronic signal as their sole or primary output, requiring amplification and reproduction equipment to become audible.

FPN No. 1: Examples of permanently installed distributed audio system locations include, but are not limited to, restaurant, hotel, business office, commercial and retail sales environments, churches, and schools. Both portable and permanently installed equipment locations include, but are not limited to, residences, auditoriums, theaters, stadiums, and movie and television studios. Temporary installations include, but are not limited to, auditoriums, theaters, stadiums (which use both temporary and permanently installed systems), and outdoor events such as fairs, festivals, circuses, public events, and concerts.

FPN No. 2: Fire and burglary alarm signaling devices are specifically not encompassed by this article.

640-2. Definitions. For purposes of this article, the following definitions apply.

Audio Amplifier or Pre-Amplifier. Electronic equipment that increases the current or voltage, or both, potential of an audio signal intended for use by another piece of audio equipment. *Amplifier* is the term used to denote an audio amplifier within this article.

Audio Autotransformer. A transformer with a single winding and multiple taps intended for use with an amplifier loudspeaker signal output.

Audio Signal Processing Equipment. Electrically operated equipment that produces or processes, or both, electronic signals that, when appropriately amplified and reproduced by a loudspeaker, produce an acoustic signal within the range of normal human hearing (typically 20-20 kHz). Within this article, the terms *equipment* and *audio equipment* are assumed to be equivalent to audio signal processing equipment.

FPN: This equipment includes, but is not limited to, loudspeakers; headphones; pre-amplifiers; microphones and their power supplies; mixers; MIDI (musical instrument digital interface) equipment or other digital control systems; equalizers, compressors, and other audio signal processing equipment; audio media recording and playback equipment including turntables, tape decks and disk players (audio and multimedia), synthesizers, tone generators, and electronic organs. Electronic organs and synthesizers may have integral or separate amplification and loudspeakers. With the exception of amplifier outputs, virtually all such equipment is used to process signals (utilizing analog or digital techniques) that have nonhazardous levels of voltage or current potential.

The definition of *audio signal processing equipment* clarifies the limits of signal processing (frequency bandwidth), which falls under Article 640.

The FPN enumerates the breadth of equipment that is considered to fall within the defined scope of Article 640. It is intended to provide a sufficiently broad list of current technology equipment to assist in determining the applicability of Article 640 to the equipment under review.

"MIDI equipment or other digital control systems" is mentioned specifically because while MIDI or similar digital control signals may issue from an electronic musical instrument, such signals may also be obtained from a computer that is appropriately configured to perform similar controlling functions.

Audio System. Within this article, the term *audio system* means the totality of all equipment and interconnecting wiring used to fabricate a fully functional audio signal processing, amplification, and reproduction system.

Audio Transformer. A transformer with two or more electrically isolated windings and multiple taps intended for use with an amplifier loudspeaker signal output.

The definition of *audio transformer* is included to clearly state that such transformers are only intended for use with audio signals, not light and power.

Equipment Rack. A framework for the support enclosure, or both, of equipment. May be portable or stationary. See ANSI/EIA/310-D-1992, *Cabinets, Racks, Panels and Associated Equipment.*

ANSI/EIA/310-D-92, *Cabinets, Racks, Panels and Associated Equipment,* defines *commercial equipment racks.* However, custom-fabricated racks are also used utilizing equipment mounting hole patterns that generally comply with this standard, but may not be constructed of steel. Within Article 640, the terms *equipment rack* and *rack* are both used to refer to equipment enclosures that are conceptually similar in intended use to those defined by the ANSI/EIA standard.

Loudspeaker. Equipment that converts an ac electric signal into an acoustic signal. The term *speaker* is commonly used to mean loudspeaker.

See Figure 640.1 for one example of a surface-mounted loudspeaker.

Figure 640.1 *Surface-mounted loudspeaker for indoor or outdoor use. (Bose Corp.)*

Maximum Output Power. The maximum output power delivered by an amplifier into its rated load as determined under specified test conditions. This may exceed the manufacturer's rated output power for the same amplifier.

Mixer. Equipment used to combine and level match a multiplicity of electronic signals, such as from microphones, electronic instruments, and recorded audio.

Typical peak signal operating voltages for such equipment varies from a few millivolts for microphones to 2 to 4 volts for disk players. A mixer's purpose is to balance these inputs to provide (typically) a 1-volt peak output signal to an amplifier.

Mixer–Amplifier. Equipment that combines the functions of a mixer and amplifier within a single enclosure.

Portable Equipment. Equipment fed with portable cords or cables intended to be moved from one place to another.

Powered Loudspeaker. Equipment that consists of a loudspeaker and amplifier within the same enclosure. Other signal processing may also be included.

Rated Load Impedance. The amplifier manufacturer's stated or marked speaker impedance into which an amplifier will deliver its rated output power. 2Ω, 4Ω, and 8Ω are typical ratings.

Rated Output Power. The amplifier manufacturer's stated or marked output power capability into its rated load.

Rated Output Voltage. For audio amplifiers of the constant-voltage type, this is the nominal output voltage when the amplifier is delivering full rated power. Rated output voltage is used for determining approximate acoustic output in distributed speaker systems that typically employ impedance matching transformers. Typical ratings are 25 volts, 70.7 volts, and 100 volts.

Technical Power System. An electrical distribution system with grounding in accordance with Section 250-146(d), where the equipment grounding conductor is isolated from the premises grounded conductor except at a single grounded termination point within a branch circuit panelboard, the originating (main breaker) branch-circuit panelboard, or at the premises grounding electrode.

The terms *technical power* and *technical ground* are commonly used by audio/video technicians and electricians to designate a wiring system that is in compliance with Section 250-146(d). Currently, the term *technical power* is only used in Article 530, Part G, Separately Derived Systems with 60 Volts to Ground. Including the definition of *technical power system* in Article 640 is intended to broaden the scope of this term to include the commonly employed distribution systems fabricated in compliance with Section 250-146(d).

Temporary Equipment. Portable wiring and equipment intended for use with events of a transient or temporary nature where all equipment is presumed to be removed at the conclusion of the event.

Temporary equipment may be used in facilities of a permanent or temporary nature, or with no facilities other than a source of electrical power. Locations would include both indoor and outdoor areas such as (indoor) athletic facilities, halls, auditoriums and (outdoor) concert "shells," athletic fields, beaches, or other places designated for public assembly.

640-3. Locations and Other Articles. Circuits and equipment shall comply with (a) through (k), as applicable.

(a) Spread of Fire or Products of Combustion. See Section 300-21.

(b) Ducts, Plenums, and Other Air-Handling Spaces. See Section 300-22, where installed in ducts or plenums or other space used for environmental air.

FPN: *Standard for the Installation of Air Conditioning and Ventilation Systems,* NFPA 90A-1996, 2-3.10.1(a), Exception No. 3, permits loudspeakers, loudspeaker assemblies, and their accessories listed in accordance with *Fire Test for Heat and Visible Smoke Release for Discrete Products and Their Accessories Installed in Air-Handling Spaces,* UL 2043-1996, to be installed in other spaces used for environmental air (ceiling cavity plenums).

(c) Cable Trays. Cable trays shall be used in accordance with Article 318.

FPN: See Section 725-61(c) for the use of Class 2, Class 3, and Type PLTC cable in cable trays.

(d) Hazardous (Classified) Locations. Equipment used in hazardous (classified) locations shall comply with the applicable requirements of Chapter 5.

(e) Places of Assembly. Equipment used in places of assembly shall comply with Article 518.

Places of assembly, as described in Section 518-2, are some of the most common locations for the installation of both distributed audio systems (e.g., background music) and centralized systems (permanently installed sound reinforcement systems for meeting rooms, auditoriums, gymnasiums, etc.).

(f) Theaters, Audience Areas of Motion Picture and Television Studios, and Similar Locations. Equipment used in theaters, audience areas of motion picture and television studios, and similar locations shall comply with Article 520.

(g) Carnivals, Circuses, Fairs, and Similar Events. Equipment used in carnivals, circuses, fairs, and similar events shall comply with Article 525.

(h) Motion Picture and Television Studios. Equipment used in motion picture and television studios shall comply with Article 530.

(i) Swimming Pools, Fountains, and Similar Locations. Audio equipment used in or near swimming pools, fountains, and similar locations shall comply with Article 680.

The underwater installation of audio equipment is found in Section 680-23. Section 640-10 covers the acceptable placement, wiring, and use of equipment used near (rather than immersed within) bodies of water, both natural and artificial. Section 640-10 covers natural bodies of water. Article 680 does not cover natural bodies of water, such as lakes, rivers, and streams.

(j) Combination Systems. Where the authority having jurisdiction permits audio systems for paging or music, or both, to be combined with fire alarm systems, the wiring shall comply with Article 760.

Fire alarm systems frequently use loudspeakers for verbal announcements as well as alarm tones. All such systems need to comply with Article 760 and

NFPA 72-1996, *National Fire Alarm Code*®. Audio systems that use the paging or background music system are also permitted to be used as part of a fire alarm warning system, but require compliance with Article 760. The installation of fire alarm systems is governed by NFPA 72-1996, *National Fire Alarm Code*. Also, refer to NFPA *101*®-1997, *Life Safety Code*®, for multipurpose systems.

FPN: For installation requirements for such combination systems, refer to *National Fire Alarm Code*®, NFPA 72-1996, and *Life Safety Code*®, NFPA *101*®, 1997.

(k) Antennas. Equipment used in audio systems that contain an audio or video tuner and an antenna input shall comply with Article 810. Wiring other than antenna wiring that connects such equipment to other audio equipment shall comply with this article.

The term *receiver* is commonly used in the consumer market to mean an amplifier combined with a radio tuner (typically AM/FM) and other signal processing and/or switching functions. Except for the tuner function and the antenna input, the signal processing functions are the same as those provided by equipment in Article 640. Article 810, Part B, covers the antenna installation for such equipment, and Section 810-2 references Article 640 as appropriate for wiring requirements (other than the antenna).

640-4. Protection of Electrical Equipment. Amplifiers, loudspeakers, and other equipment shall be so located or protected so as to guard against environmental exposure or physical damage, such as might result in fire, shock, or personal hazard.

640-5. Access to Electrical Equipment Behind Panels Designed to Allow Access. Access to equipment shall not be denied by an accumulation of wires and cables that prevents removal of panels, including suspended ceiling panels.

640-6. Mechanical Execution of Work. Equipment and cabling shall be installed in a neat and workmanlike manner. Cables for installed systems shall be supported by the building structure in such a manner that the cable will not be damaged by normal building use.

FPN: One way to determine accepted industry practice is to refer to nationally recognized standards such as the following: *Commercial Building Telecommunications Wiring Standard*, ANSI/EIA/TIA 568-A-1995; *Commercial Building Standard for Telecommunications Pathways and Spaces*, ANSI/EIA/TIA 569-A-1997; and *Residential and Light Commercial Telecommunications Wiring Standard*, ANSI/ EIA/TIA 570-1991, or other ANSI-approved installation standards.

640-7. Grounding.

(a) General. Wireways and auxiliary gutters shall be grounded and bonded in accordance with the requirements of Article 250. Where the wireway or auxiliary gutter does not contain power-supply wires, the equipment grounding conductor shall not be required to be larger than No. 14 copper or its equivalent. Where the wireway or auxiliary gutter contains power-supply wires, the equipment grounding conductor shall not be smaller than specified in Section 250-122.

(b) Separately Derived Systems with 60 Volts to Ground. Grounding of separately derived systems with 60 volts to ground shall be in accordance with Section 530-72.

(c) Isolated Ground Receptacles. Isolated grounding-type receptacles shall be permitted as described in Section 250-146(d), and for the implementation of other technical power systems in compliance with Article 250. For separately derived systems with 60 volts to ground, the branch-circuit equipment grounding conductor shall be terminated as required in Section 530-72(b).

Section 640-7(b) references Section 530-72 for the proper technique for grounding separately derived systems with 60 volts to ground (Article 530, Part G). This separately derived system is valid as a technique for the reduction of electromagnetic noise in audio systems and in video systems.

Section 640-7(c) is intended to clarify the proper use of isolated ground receptacles when used with technical power systems of the separately-derived-systems-with-60-volts-to-ground type.

FPN: See Section 410-56(c) for grounding-type receptacles and required identification.

640-8. Grouping of Conductors. Insulated conductors of different systems grouped or bundled so as to be in close physical contact with each other in the same raceway or other enclosure, or in portable cords or cables, shall comply with Section 300-3(c)(1).

640-9. Wiring Methods.

(a) Wiring to and Between Audio Equipment.

(1) Wiring and equipment from source of power to and between devices connected to the premises wiring systems shall comply with the requirements of Chapters 1 through 4, except as modified by this article.

(2) Separately derived systems shall comply with the applicable articles of this *Code*, except as modified by this article. Separately derived systems with 60 volts to ground shall be permitted for use in audio system installations as specified in Article 530, Part G.

(3) All wiring not connected to the premises wiring system or to a wiring system separately derived from the premises wiring system shall comply with Article 725.

(b) Auxiliary Power Supply Wiring. Equipment that has a separate input for an auxiliary power supply shall be wired in compliance with Article 725. Battery installation shall be in accordance with Article 480.

FPN No. 1: This section does not apply to the use of uninterruptable power supply (ups) equipment, or other sources of supply, that are intended to act as a direct replacement for the primary circuit power source and are connected to the primary circuit input.

Audio equipment with a separate input for an auxiliary power supply is typically used for emergency paging or fire alarm systems. These auxiliary power supply inputs typically range from 12 to 48 volts dc. Article 480 adequately covers installation and overcurrent protection of battery circuits of this type. The term *auxiliary* is used to indicate that the equipment is also capable of being powered by the premises wiring system through an independent input connector, cord, or cable.

FPN No. 1 clarifies that while the equipment can be powered by a replacement source for the premises wiring system, such a source (a UPS or standby generator) is not considered to be supplying the equipment auxiliary power supply unless it is directly connected to the auxiliary power supply input and supplying a dc voltage.

FPN No. 2: Refer to *National Fire Alarm Code,* NFPA 72-1996, where equipment is used for a fire alarm system.

(c) Output Wiring and Listing of Amplifiers. Amplifiers with output circuits carrying audio program signals shall be permitted to employ Class 1, Class 2, or Class 3 wiring where the amplifier is listed and marked for use with the specific class of wiring method. Such listing shall ensure the energy output is equivalent to the shock and fire risk of the same class as stated in Article 725. Overcurrent protection shall be provided and shall be permitted to be inherent to the amplifier.

Audio circuits wired using Class 1 wiring methods shall not occupy the same raceway or enclosure with other than audio circuits wired using Class 1 wiring methods.

Audio circuits wired using Class 2 wiring methods shall not occupy the same raceway or enclosure with other than audio circuits wired using Class 2 wiring methods.

Audio circuits wired using Class 3 wiring methods shall not occupy the same raceway or enclosure with other than audio circuits wired using Class 3 wiring methods.

FPN No. 1: *Amplifiers for Fire Protective Signaling Systems,* ANSI/UL 1711-1994, contains requirements for the listing of amplifiers used for fire alarm systems in compliance with *National Fire Alarm Code,* NFPA 72-1996.

FPN No. 2: Examples of requirements for listing amplifiers used in residential, commercial, and professional use are found in *Commercial Audio Equipment,* ANSI/UL 813-1996, *Professional Video and Audio Equipment,* ANSI/UL 1419-1997, *Audio-Video Products and Accessories,* ANSI/ UL 1492-1996, or *Audio/Video and Musical Instrument Apparatus for Household, Commercial, and Similar Use,* ANSI/UL 6500-1996.

(d) Use of Audio Transformers and Autotransformers. Audio transformers and autotransformers shall only be used for audio signals in a manner so as not to exceed the manufacturer's stated input or output voltage, impedance, or power limitations. The input or output wires of an audio transformer or autotransformer shall be allowed to connect directly to the amplifier or loudspeaker terminals. No electrical terminal or lead shall be required to be grounded or bonded.

Audio transformers and autotransformers are commonly used between the amplifier output and the loudspeaker input for the following reasons:

1. **At the output of the amplifier to change the amplifier's operating voltage to match the design impedance of the loudspeaker**
2. **At the loudspeaker, where the inherently low voice coil impedance is raised to match the output voltage of the amplifier (or autotransformer)**
3. **Between the amplifier output and loudspeaker input as an attenuating device (volume control)**

Audio autotransformers are only similar in concept to autotransformers used for light and power but they are not designed for such use. Audio transformers are commonly used for providing electrical isolation of the speakers from the signal source. Either type of audio transformer (two winding or autotransformer) is frequently referred to as an "impedance matching transformer." The last sentence of Section 640-9(d) specifically addresses the fact that electrical terminals are not to be treated in the same manner as transformers that are used for light and power might be (e.g., grounding the common terminal of an autotransformer). Some amplifier outputs are deliberately isolated from equipment ground, in which case such a connection could damage the amplifier and violate the manufacturer's recommended use. The frame of the transformer may or may not require bonding, depending on the manufacturer's installation instructions.

640-10. Audio Systems Near Bodies of Water. Audio systems near bodies of water, either natural or artificial, shall be subject to the following restrictions.

Exception: This section does not include audio systems intended for use on boats, yachts, or other forms of land or water transportation used near bodies of water, whether or not supplied by branch-circuit power.

FPN: See Section 680-23 for installation of underwater audio equipment.

(a) Equipment Supplied by Branch-Circuit Power. Audio system equipment supplied by branch-circuit power shall not be placed laterally within 5 ft (1.52 m) of the inside wall of a pool, spa, hot tub, or fountain, nor within 5 ft (1.52 m) of the prevailing or tidal high water mark. The equipment shall be provided with branch-circuit power protected by a ground-fault circuit interrupter where required by other articles.

Article 680 is limited by its scope to the use of underwater loudspeakers. This particular application for audio equipment is unique in construction and wiring to pools and is appropriate for Article 680.

Other locations where audio equipment might be used "near bodies of water" are not addressed in Article 680. Section 640-10(a), combined with the contents of the rest of Article 640, is intended to address locations where audio equipment is used near bodies of water.

The exception to Section 640-10(a) excludes sound systems on vehicles used in or near the water, such as amphibious vehicles and boats of all sizes.

The term "prevailing or tidal high water mark" recognizes that natural bodies of water can have the "edge" of the water fluctuate. This phrase clarifies that such changes can be anticipated.

The requirement for use of a ground-fault circuit interrupter is placed here to ensure that personnel are properly protected. Where the equipment (specifically an amplifier or receiver) is installed in an electrical cabinet or a room not near bodies of water, the requirement for GFCI protection does not apply, unless required by other sections of the *Code.*

(b) Equipment Not Supplied by Branch-Circuit Power. Audio system equipment powered by a listed Class 2 power supply or by the output of an amplifier listed as permitting the use of Class 2 wiring shall only be restricted in placement by the manufacturer's recommendations.

FPN: Placement of the power supply or amplifier, if supplied by branch-circuit power, is still subject to Section 640-10(a).

B. Permanent Audio System Installations

Permanent audio systems are characterized by the fixed locations for the wiring, signal processing equipment, and reproduction equipment. Wiring is attached to the building structure and is frequently concealed. Speakers in commercial, hospital, school, and restaurant areas are commonly recessed into ceiling or wall surfaces, or mounted to structure surfaces using brackets, usually beyond the reach of a standing person.

640-21. Use of Flexible Cords and Cables.

(a) Between Equipment and Branch-Circuit Power. Power supply cords for audio equipment shall be suitable for the use and shall be permitted to be used where the interchange, maintenance, or repair of such equipment is facilitated through the use of a power supply cord.

(b) Between Loudspeakers and Amplifiers, or Between Loudspeakers. Cables used to connect loudspeakers to each other or to an amplifier shall comply with Article 725. Other listed cable types and assemblies, including optional hybrid communications, signal, and optical fiber cables, shall be permitted.

Some loudspeakers are specifically identified for outdoor use. Section 110-11 requires equipment as well as the supplying conductors to be identified for use in the operating environment. Figure 640.2 shows a loudspeaker identified for outdoor use and located partially in ground. The conductors supplying this outdoor speaker must also be identified for the environment.

Figure 640.2 *Loudspeaker for outdoor use above ground or partially in ground, as shown. (Bose Corp.)*

(c) Between Equipment. Cables used for the distribution of audio signals between equipment shall comply with Article 725. Other listed cable types and assemblies, including optional hybrid communications, signal, and optical fiber cables, shall be permitted. Other cable types and assemblies specified by the equipment manufacturer as acceptable for the use shall be permitted in accordance with Section 110-3(b).

(d) Between Equipment and Power Supplies Other than Branch-Circuit Power.

(1) Storage batteries shall be installed and wired in accordance with the requirements of this *Code* for the voltage and power delivered.

(2) Transformers, transformer rectifiers, and other ac or dc power supplies shall be installed and wired in accordance with the requirements of this *Code* for the voltage and power delivered.

FPN: For some equipment, sources such as in (1) and (2) will serve as the only source of power. These could, in turn, be supplied with intermittent or continuous branch-circuit power.

(e) Between Equipment Racks and Premises Wiring System. Flexible cords and cables shall be permitted for the electrical connection of permanently installed equipment racks to the premises wiring system to facilitate access to equipment or for the purpose of isolating the technical ground of the rack from the premises ground. Connection shall be made using either approved plugs and receptacles or by direct connection within an approved enclosure. Flexible cords and cables shall not be subjected to physical manipulation or abuse while the rack is in use.

640-22. Wiring of Equipment Racks. Equipment racks shall be fabricated of metal and grounded. Bonding shall not be required if the rack is connected to a technical power ground.

Equipment racks shall be wired in a neat and workmanlike manner. Wires, cables, structural components, or other equipment shall not be placed in such a manner as to prevent reasonable access to equipment power switches and resettable or replaceable circuit overcurrent protection devices.

Supply cords or cables, if used, shall terminate within the equipment rack enclosure in an identified connector assembly. The supply cords or cable (and connector assembly, if used) shall have sufficient ampacity to carry the total load connected to the equipment rack and shall be protected by overcurrent devices.

640-23. Conduit or Tubing.

(a) Number of Conductors. The number of conductors permitted in a single conduit or tubing shall not exceed the percentage fill specified in Table 1, Chapter 9.

(b) Nonmetallic Conduit or Tubing and Insulating Bushings. The use of nonmetallic conduit or tubing and insulating bushings shall be permitted where a technical ground system is employed and shall comply with applicable articles.

640-24. Wireways, Gutters, and Auxiliary Gutters. The use of metallic and nonmetallic wireways, gutters, and auxiliary gutters shall be permitted for use with audio signal conductors and shall comply with applicable articles with respect to permitted locations, construction, and fill.

640-25. Loudspeaker Installation in Fire Resistance-Rated Partitions, Walls, and Ceilings. Loudspeakers installed in a fire resistance-rated partition, wall, or ceiling shall be listed for the purpose or installed in an enclosure or recess that maintains the fire resistance rating.

Section 640-25, revised for the 1999 *NEC*, clarifies the requirement of an enclosure that maintains the requisite fire resistance rating of the surface within which a flush-mount loudspeaker is installed. Since a very limited number of listed enclosures are manufactured, site-built enclosures installed with the approval of the authority having jurisdiction are a common solution to this installation problem.

The reference to NFPA 251-1995, *Standard Methods of Tests of Fire Endurance of Building Construction and Materials,* is intended to clarify a typical classification method for a component of the ceiling, or components installed therein.

FPN: Fire-rated construction is the fire-resistive classification used in building codes. One method of determining fire rating is testing in accordance with *Standard Methods of Tests of Fire Endurance of Building Construction and Materials*, NFPA 251-1995.

C. Portable and Temporary Audio System Installations

Portable and temporary audio system installations are characterized by the portable nature of the signal processing equipment and the reproduction equipment. While the equipment may not fundamentally differ from that used in permanent installations, the enclosures that serve as portable equipment racks provide both transit protection and mechanical protection from abuse while the equipment is in use. Such enclosures may accommodate one or multiple pieces of equipment, and may be of metal, wood, plastic, or reinforced plastic construction. The nonmetal construction enclosures frequently do not comply with ANSI/EIA/310-D-92, *Cabinets, Racks, Panels and Associated Equipment,* except for the mounting attachment points for the equipment.

Three conditions distinguish the difference between a portable and a temporary installation. The first is the nature of use and possession. For example, a theater might employ portable equipment that is moved during the course of a performance, but is never used outside of a theater event. The second condition is the transitory nature of the event. A performer could use his or her personal portable equipment for a performance in the same theater, but because the presence of the equipment within the theater terminates with the end of the performance, the equipment's use qualifies as temporary. Finally, an event may be considered temporary and obviously uses temporary audio systems. Examples would include fairs, circuses, outdoor concerts, and folk festivals. Part C of Article 640 covers portable and temporary installations, the differences being primarily ones of intent and, for outdoor temporary use, adequate ability to protect the equipment from environmental hazards.

640-41. Multipole Branch-Circuit Cable Connectors. Multipole branch-circuit cable connectors, male and female, for power supply cords and cables shall be constructed so that tension on the cord or cable will not be transmitted to the connections. The female half shall be attached to the load end of the power supply cord or cable. The connector shall be rated in amperes and designed so that differently rated devices cannot be connected together. Alternating-current multipole connectors shall be polarized and comply with Sections 410-56(g) and 410-58. Alternating-current or direct-current multipole connectors utilized for connection between loudspeakers and amplifiers, or between loudspeakers, shall not be compatible with nonlocking 15- or 20-ampere rated connectors intended for branch-circuit power, nor with connectors rated 250 volts or greater of either the locking or nonlocking type. Signal cabling not intended for such loudspeaker and amplifier interconnection shall not be permitted to be compatible with multipole branch-circuit cable connectors of any accepted configuration.

FPN: See Section 400-10 for pull at terminals.

640-42. Use of Flexible Cords and Cables.

(a) Between Equipment and Branch-Circuit Power. Power supply cords for audio equipment shall be listed and shall be permitted to be used where the interchange, maintenance, or repair of such equipment is facilitated through the use of a power supply cord.

(b) Between Loudspeakers and Amplifiers, or Between Loudspeakers. Flexible cords and cables used to connect loudspeakers to each other or to an amplifier shall comply with Article 400 and Article 725, respectively. Cords and cables listed for portable use, either hard or extra-hard usage as defined by Article 400, shall also be permitted. Other listed cable types and assemblies, including optional hybrid communications, signal, and optical fiber cables, shall be permitted.

(c) Between Equipment and/or Between Equipment Racks. Flexible cords and cables used for the distribution of audio signals between equipment shall comply with Article 400 and Article 725, respectively. Cords and cables listed for portable use, either hard or extra-hard service as defined by Article 400, shall also be permitted. Other listed cable types and assemblies, including optional hybrid communications, signal, and optical fiber cables, shall be permitted.

(d) Between Equipment, Equipment Racks, and Power Supplies Other than Branch-Circuit Power.

(1) Storage batteries shall be installed and wired in accordance with the requirements of this *Code* for the voltage and power delivered.

(2) Transformers, transformer rectifiers, and other ac or dc power supplies shall be installed and wired in accordance with the requirements of this *Code* for the voltage and power delivered.

(3) Generators shall be installed in accordance with Article 445.

Portable generators have been added to reference Article 445. Article 445 does not strictly mention "portable," but it references Article 250. Section 250-34 addresses the special issues of portable and vehicle-mounted generators.

Portable generator use is common. A festival at the beach and a parade or carnival float are some examples. Storage batteries and power inverters are a quieter way to achieve "mobile" audio on a parade float. This is NOT the same thing as an automobile sound system. This would be (typically) the type of equipment used for a stationary (temporary or permanent) installation, but set up on a flatbed truck, for instance.

(e) Between Equipment Racks and Branch-Circuit Power. The supply to a portable equipment rack shall be by means of listed extra-hard usage cords or cables, as defined in Article 400. For outdoor portable or temporary use, the cords or cables shall be further listed as being suitable for wet locations and sunlight resistant.

(1) Where equipment racks include audio and lighting and/or power equipment, Articles 520 and 525 shall apply as appropriate.

(2) The usage and construction of cable extensions, adapters, and breakout assemblies shall be in accordance with Article 520 or 525, as appropriate.

640-43. Wiring of Equipment Racks. Equipment racks fabricated of metal shall be grounded. Nonmetallic racks with covers (if provided) removed shall not allow access to Class 1, Class 3, or primary circuit power without the removal of covers over terminals or the use of tools.

Equipment racks shall be wired in a neat and workmanlike manner. Wires, cables, structural components, or other equipment shall not be placed in such a manner as to prevent reasonable access to equipment power switches and resettable or replaceable circuit overcurrent protection devices.

Wiring that exits the equipment rack for connection to other equipment or to a power supply shall be relieved of strain or otherwise suitably terminated such that a pull on the flexible cord or cable shall not increase the risk of damage to the cable or connected equipment such as to cause an unreasonable risk of fire or electric shock.

See Figure 640.3 for an example of portable equipment racks used for a temporary installation.

Figure 640.3 Portable equipment racks of composite wood/metal construction used for temporary audio installations. (Bose Corp.)

640-44. Environmental Protection of Equipment. Temporary outdoor, unsheltered placement or use of portable equipment not listed for the purpose shall be permitted only where appropriate protection of such equipment from adverse weather conditions is provided to prevent risk of fire or electrical shock. Where the system is intended to remain operable during adverse weather, arrangements shall be made for maintaining operation and ventilation of heat dissipating equipment.

Section 640-44 recognizes that most audio equipment used in temporary audio systems is not listed for use in an outdoor environment. Section 640-44 allows the use of such equipment provided an appropriate means of protection can be provided to protect the equipment from anticipated adverse weather conditions.

Figure 640.4 depicts the initial stage of a temporary audio system installation. Temporary audio system installation requirements are covered in detail in Part C of Article 640.

Figure 640.4 Temporary installation of loudspeakers and a portable equipment rack. (Bose Corp.)

640-45. Protection of Wiring. Where accessible to the public, flexible cords and cables laid or run on the ground or on the floor shall be covered with approved nonconductive mats. Cables and mats shall be arranged so as not to present a tripping hazard.

640-46. Equipment Access. Equipment likely to present a risk of fire, electrical shock, or physical injury to the public shall be protected by barriers or supervised by qualified personnel so as to prevent public access.

Article 645 — Information Technology Equipment

Contents

645-1. Scope. This article covers equipment, power-supply wiring, equipment interconnecting wiring, and grounding of information technology equipment and systems, including terminal units, in an information technology equipment room.

The term *information technology equipment* replaces other terms that describe computer-based business, personal, and industrial equipment. This terminology is also used by UL 1950, as well as international standards, as a more inclusive term for the equipment being addressed by Article 645.

FPN: For further information, see *Standard for the Protection of Electronic Computer/Data Processing Equipment,* NFPA 75-1995.

645-2. Special Requirements for Information Technology Equipment Room. This article applies, provided all the following conditions are met.

(a) Disconnecting means complying with Section 645-10 are provided.

(b) A separate heating/ventilating/air-conditioning (HVAC) system is provided that is dedicated for information technology equipment use and is separated from other areas of occupancy. Any HVAC system that serves other occupancies shall be permitted to also serve the information technology equipment room if fire/smoke dampers are provided at the point of penetration of the room boundary. Such dampers shall operate on activation of smoke detectors and also by operation of the disconnecting means required by Section 645-10.

FPN: For further information, see *Standard for the Protection of Electronic Computer/Data Processing Equipment,* NFPA 75-1995.

(c) Listed information technology equipment is installed.

FPN: For further information, see *Standard for the Protection of Electronic Computer/Data Processing Equipment,* NFPA 75-1995.

(d) Occupied only by those personnel needed for the maintenance and functional operation of the installed information technology equipment.

FPN: For further information, see *Standard for the Protection of Electronic Computer/Data Processing Equipment,* NFPA 75-1995.

(e) The room is separated from other occupancies by fire-resistant-rated walls, floors, and ceilings with protected openings.

(f) The building construction, rooms, or areas and occupancy comply with the applicable building code.

The requirements in Article 645 are based on the assumption that the room complies with NFPA 75-1999, *Standard for the Protection of Electronic Computer/Data Processing Equipment.*

Article 645 applies only to equipment and wiring located within the information technology equipment room. An information technology equipment room is an enclosed area, with one or more means of entry, that contains computer-based business and industrial equipment. It is designed to comply with the special construction and fire protection provisions of NFPA 75-1999, *Standard for the Protection of Electronic Computer/Data Processing Equipment,* as well as Section 645-2 of the *Code.*

Small terminals, such as remote telephone terminal units, remote data terminals, personal computers, and cash registers in stores and supermarkets, are not covered by Article 645.

645-5. Supply Circuits and Interconnecting Cables.

(a) Branch-Circuit Conductors. The branch-circuit conductors supplying one or more units of a data processing system shall have an ampacity not less than 125 percent of the total connected load.

(b) Connecting Cables. The data processing system shall be permitted to be connected to a branch circuit by any of the following means listed for the purpose.

(1) Computer/data processing cable and attachment plug cap.
(2) Flexible cord and an attachment plug cap.
(3) Cord set assembly. Where run on the surface of the floor, they shall be protected against physical damage.

(c) Interconnecting Cables. Separate data processing units shall be permitted to be interconnected by means of cables and cable assemblies listed for the purpose. Where run on the surface of the floor, they shall be protected against physical damage.

(d) Under Raised Floors. Power cables, communications cables, connecting cables, interconnecting cables, and receptacles associated with the information technology equipment shall be permitted under a raised floor, provided the following.

(1) The raised floor is of suitable construction and the area under the floor is accessible.

FPN: See *Standard for the Protection of Electronic Computer/Data Processing Equipment,* NFPA 75-1995.

(2) The branch-circuit supply conductors to receptacles or field-wired equipment are in rigid metal conduit, rigid nonmetallic conduit, intermediate metal conduit, electrical metallic tubing, metal wireway, surface metal raceway with metal cover, flexible metal conduit, liquidtight flexible metal or nonmetallic conduit, Type MI cable, Type MC cable, or Type AC cable. These supply conductors shall be installed in accordance with the requirements of Section 300-11.

Branch-circuit conductors installed under the raised floor of an information technology equipment room by any of the wiring methods listed in Section 645-5(d)(2) are required to conform to the specific article for the wiring method used. In addition, Article 300 applies, except where modified by Article 645. For example, Section 300-11 requires raceways and boxes to be securely fastened in place, even though they are installed below a raised floor.

(3) Ventilation in the underfloor area is used for the information technology equipment room only.
(4) Openings in raised floors for cables protect cables against abrasions and minimize the entrance of debris beneath the floor.
(5) Cables, other than those covered in (2) and those complying with (a), (b), and (c) below, shall be listed as Type DP cable having adequate fire-resistant characteristics suitable for use under raised floors of an information technology equipment room.
    (a) Interconnecting cables enclosed in a raceway.
    (b) Interconnecting cables listed with equipment manufactured prior to July 1, 1994, being installed with that equipment.
    (c) Cable type designations Type TC (Article 340); Types CL2, CL3, and PLTC (Article 725); Types NPLF and FPL (Article 760); Types OFC and OFN (Article 770); Types CM and MP (Article 800); Type CATV (Article 820). These designations shall be permitted to have an additional letter P or R or G. Green insulated single conductor cables, No. 4 and larger, marked "for use in cable trays" or "for CT use" shall be permitted for equipment grounding.

Section 645-5(d)(5) requires interconnecting cables used under raised floors (other than branch-circuit conductors) to be listed as Type DP cables. Cables listed as part of equipment manufactured before the effective date of July 1, 1994, were not required to be listed. Cables in raceways are also exempt. Section 645-5(d)(5)(c) allows cables that pass the *Vertical Tray Flame Test,* ANSI/UL 1581-1991, or the *Vertical Flame Test — Cables in Cable Trays,* CSA C22.2 No. 0.3-M-1992, where not more than 4 ft of cable is damaged during the CSA test, to be installed under raised floors of computer rooms. Section 645-5(d)(5)(c) now recognizes Type DP cables that satisfy the tests specified above.

FPN: One method of defining fire resistance is by establishing that the cables do not spread fire to the top of the tray in the "Vertical Tray Flame Test" referenced in the *Standard for Electrical Wires, Cables, and Flexible Cords,* ANSI/UL 1581-1991. Another method of defining fire resistance is for the damage (char length) not to exceed 4 ft 11 in. (1.5 m) when performing the CSA "Vertical Flame Test — Cables in Cable Trays," as described in *Test Methods for Electrical Wires and Cables,* CSA C22.2 No. 0.3-M-1985.

(e) Securing in Place. Power cables; communications cables; connecting cables; interconnecting cables; and associated boxes, connectors, plugs, and receptacles that are listed as part of, or for, information technology equipment shall not be required to be secured in place.

645-6. Cables Not in Information Technology Equipment Room. Cables extending beyond the information technology equipment room shall be subject to the applicable requirements of this *Code.*

FPN: For signaling circuits, refer to Article 725; for fiber optic circuits, refer to Article 770; and for communications circuits, refer to Article 800. For fire alarm systems, refer to Article 760.

645-7. Penetrations. Penetrations of the fire-resistant room boundary shall be in accordance with Section 300-21.

645-10. Disconnecting Means. A means shall be provided to disconnect power to all electronic equipment in the information technology equipment room. There shall also be a similar means to disconnect the power to all dedicated HVAC systems serving the room and cause all required fire/ smoke dampers to close. The control for these disconnecting means shall be grouped and identified and shall be readily accessible at the principal exit doors. A single means to control both the electronic equipment and HVAC systems shall be permitted.

Section 645-10 requires two separate disconnecting means, but permits a single control, such as one pushbutton, to electrically operate both disconnecting means. The disconnecting means is required to disconnect the conductors of each circuit from their supply source and close all required fire/smoke dampers. See the definition of *disconnecting means* in Article 100. The disconnecting means is permitted to be remote-controlled switching devices, such as relays, with pushbutton stations at the principal exit doors.

The requirements of Sections 645-10 and 645-7 for sealing penetrations are intended to minimize the passage of smoke or fire to other parts of the building.

Exception: Installations qualifying under the provisions of Article 685.

645-11. Uninterruptible Power Supplies (UPS). Unless otherwise permitted in (a) or (b), UPS systems installed within the information technology room, and their supply and output circuits, shall comply with Section 645-10. The disconnecting means shall also disconnect the battery from its load.

(a) Installations qualifying under the provisions of Article 685.

(b) A disconnecting means complying with Section 645-10 shall not be required for power sources capable of supplying 750 volt-amperes or less derived either from UPS equipment or from battery circuits integral to electronic equipment, provided all other requirements of Section 645-11 are met.

645-15. Grounding. All exposed noncurrent-carrying metal parts of an information technology system shall be grounded in accordance with Article 250 or shall be double insulated. Power systems derived within listed information technology equipment that supply information technology systems through receptacles or cable assemblies supplied as part of this equipment shall not be considered separately derived for the purpose of applying Section 250-20(d). Where signal reference structures are installed, they shall be bonded to the equipment grounding system provided for the information technology equipment.

The last sentence of Section 645-15 was added for the 1999 *Code* to recognize that properly bonded high-frequency signal reference structures provide additional safety measures.

FPN No. 1: The bonding and grounding requirements in the product standards governing this listed equipment ensure that it complies with Article 250.

FPN No. 2: Where isolated grounding-type receptacles are used, see Sections 250-146(d) and 410-56(c).

645-16. Marking. Each unit of an information technology system supplied by a branch circuit shall be provided with a manufacturer's nameplate, which shall also include the input power requirements for voltage, frequency, and maximum rated load in amperes.

Article 650 — Pipe Organs

Contents

650-1. Scope. This article covers those electrical circuits and parts of electrically operated pipe organs that are employed for the control of the sounding apparatus and keyboards.

650-2. Other Articles. Electronic organs shall comply with the appropriate provisions of Article 640.

650-3. Source of Energy. The source of power shall be a transformer-type rectifier, the dc potential of which shall not exceed 30 volts dc.

650-4. Grounding. The rectifier shall be grounded according to the provisions in Section 250-112(b).

650-5. Conductors. Conductors shall comply with (a) through (d).

(a) Size. Not less than No. 28 for electronic signal circuits and not less than No. 26 for electromagnetic valve supply and the like. A main common-return conductor in the electromagnetic supply shall not be less than No. 14.

Note that No. 14 wire is required only for a main common return conductor in the electromagnetic supply.

(b) Insulation. Conductors shall have thermoplastic or thermosetting insulation.

(c) Conductors to Be Cabled. Except for the common-return conductor and conductors inside the organ proper, the organ sections and the organ console conductors shall be cabled. The common-return conductors shall be permitted under an additional covering enclosing both cable and return conductor, or shall be permitted as a separate conductor and shall be permitted to be in contact with the cable.

(d) Cable Covering. Each cable shall be provided with an outer covering, either overall or on each of any subassemblies of grouped conductors. Tape shall be permitted in place of a covering. Where not installed in metal raceway, the covering shall be flame retardant or the cable or each cable subassembly shall be covered with a closely wound fireproof tape.

650-6. Installation of Conductors. Cables shall be securely fastened in place and shall be permitted to be attached directly to the organ structure without insulating supports. Cables shall not be placed in contact with other conductors.

Insulating supports are not required; however, measures should be taken to prevent contact between cables and conductors of other systems.

650-7. Overcurrent Protection. Circuits shall be arranged so that all conductors shall be protected from overcurrent by an overcurrent device rated at not more than 6 amperes.

Exception: The main supply conductors and the common-return conductors.

Article 660 — X-Ray Equipment

Contents

A. General

660-1. Scope. This article covers all X-ray equipment operating at any frequency or voltage for industrial or other nonmedical or nondental use.

FPN: See Article 517, Part E, for X-ray installations in health care facilities.

Nothing in this article shall be construed as specifying safeguards against the useful beam or stray X-ray radiation.

FPN No. 1: Radiation safety and performance requirements of several classes of X-ray equipment are regulated under Public Law 90-602 and are enforced by the Department of Health and Human Services.

FPN No. 2: In addition, information on radiation protection by the National Council on Radiation Protection and Measurements is published as *Reports of the National Council on Radiation Protection and Measurement.* These reports can be obtained from NCRP Publications, 7910 Woodmont Ave., Suite 1016, Bethesda, MD 20814.

660-2. Definitions.

Long-Time Rating. A rating based on an operating interval of 5 minutes or longer.

Mobile. X-ray equipment mounted on a permanent base with wheels and/or casters for moving while completely assembled.

Momentary Rating. A rating based on an operating interval that does not exceed 5 seconds.

Portable. X-ray equipment designed to be hand-carried.

Transportable. X-ray equipment to be installed in a vehicle or that may be readily disassembled for transport in a vehicle.

660-3. Hazardous (Classified) Locations. Unless approved for the location, X-ray and related equipment shall not be installed or operated in hazardous (classified) locations.

FPN: See Article 517, Part D.

X-ray equipment in industrial establishments or similar locations is usually commonly used for the purpose of inspecting a process or product. This method permits nondestructive testing without dismantling or applying stress to detect cracks, flaws, or structural defects. Welded joints are frequently inspected with X-ray equipment to detect hidden defects that may cause failure under stress.

Among the industrial applications of X-rays, the most common is radiography, where shadow pictures of the subject matter are produced on photographic film. The type and thickness of the material involved govern the voltage to be employed, which may range from a few thousand volts (kilovolts) to millions of volts (megavolts). It is possible to X-ray metal objects that are 20 in. (508 mm) thick.

Fluoroscopy is another X-ray technique used for industrial or commercial applications. This method is similar to radiography but operates at a much lower voltage range (less than 250 kilovolts) and, instead of producing a film, a shadow picture is projected on a screen, similar to those used for security checks of luggage at airport terminals. Fluoroscopy is capable of detecting minute flaws or defects.

660-4. Connection to Supply Circuit.

(a) Fixed and Stationary Equipment. Fixed and stationary X-ray equipment shall be connected to the power supply by means of a wiring method meeting the general requirements of this *Code*. Equipment properly supplied by a branch circuit rated at not over 30 amperes shall be permitted to be supplied through a suitable attachment plug cap and hard-service cable or cord.

(b) Portable, Mobile, and Transportable Equipment. Individual branch circuits shall not be required for portable, mobile, and transportable X-ray equipment requiring a capacity of not over 60 amperes. Portable and mobile types of X-ray equipment of any capacity shall be supplied through a suitable hard-service cable or cord. Transportable X-ray equipment of any capacity shall be permitted to be connected to its power supply by suitable connections and hard-service cable or cord.

(c) Over 600 Volts, Nominal. Circuits and equipment operated at more than 600 volts, nominal, shall comply with Article 490.

660-5. Disconnecting Means. A disconnecting means of adequate capacity for at least 50 percent of the input required for the momentary rating or 100 percent of the input required for the long-time rating of the X-ray equipment, whichever is greater, shall be provided in the supply circuit. The disconnecting means shall be operable from a location readily accessible from the X-ray control. For equipment connected to a 120-volt, nominal, branch circuit of 30 amperes or less, a grounding-type attachment plug cap and receptacle of proper rating shall be permitted to serve as a disconnecting means.

660-6. Rating of Supply Conductors and Overcurrent Protection.

(a) Branch-Circuit Conductors. The ampacity of supply branch-circuit conductors and the overcurrent protective devices shall not be less than 50 percent of the momentary rating or 100 percent of the long-time rating, whichever is greater.

(b) Feeder Conductors. The rated ampacity of conductors and overcurrent devices of a feeder for two or more branch circuits supplying X-ray units shall not be less than 100 percent of the momentary demand rating [as determined by (a)] of the two largest X-ray apparatus plus 20 percent of the momentary ratings of other X-ray apparatus.

FPN: The minimum conductor size for branch and feeder circuits is also governed by voltage regulation requirements. For a specific installation, the manufacturer usually specifies minimum distribution transformer and conductor sizes, rating of disconnect means, and overcurrent protection.

660-7. Wiring Terminals. X-ray equipment not provided with a permanently attached cord or cord set shall be provided with suitable wiring terminals or leads for the connection of power-supply conductors of the size required by the rating of the branch circuit for the equipment.

660-8. Number of Conductors in Raceway. The number of control circuit conductors installed in a raceway shall be determined in accordance with Section 300-17.

660-9. Minimum Size of Conductors. Size No. 18 or 16 fixture wires, as specified in Section 725-27, and flexible cords shall be permitted for the control and operating circuits of X-ray and auxiliary equipment where protected by not larger than 20-ampere overcurrent devices.

660-10. Equipment Installations. All equipment for new X-ray installations and all used or reconditioned X-ray equipment moved to and reinstalled at a new location shall be of an approved type.

B. Control

660-20. Fixed and Stationary Equipment.

(a) Separate Control Device. A separate control device, in addition to the disconnecting means, shall be incorporated in the X-ray control supply or in the primary circuit to the high-voltage transformer. This device shall be a part of the X-ray equipment but shall be permitted in a separate enclosure immediately adjacent to the X-ray control unit.

(b) Protective Device. A protective device, which shall be permitted to be incorporated into the separate control device, shall be provided to control the load resulting from failures in the high-voltage circuit.

660-21. Portable and Mobile Equipment. Portable and mobile equipment shall comply with Section 660-20, but the manually controlled device shall be located in or on the equipment.

660-23. Industrial and Commercial Laboratory Equipment.

(a) Radiographic and Fluoroscopic Types. All radiographic- and fluoroscopic-type equipment shall be effectively enclosed or shall have interlocks that de-energize the equipment automatically to prevent ready access to live current-carrying parts.

(b) Diffraction and Irradiation Types. Diffraction- and irradiation-type equipment or installations not effectively enclosed or provided with interlocks to prevent access to live current-carrying parts during operation shall be provided with a positive means to indicate when they are energized. The indicator shall be a pilot light, readable meter deflection, or equivalent means.

660-24. Independent Control. Where more than one piece of equipment is operated from the same high-voltage circuit, each piece or each group of equipment as a unit shall be provided with a high-voltage switch or equivalent disconnecting means. This disconnecting means shall be constructed, enclosed, or located so as to avoid contact by persons with its live parts.

A control device provides means for initiating and terminating X-ray exposures and automatically times their duration.

C. Transformers and Capacitors

660-35. General. Transformers and capacitors that are part of an X-ray equipment shall not be required to comply with Articles 450 and 460.

High-ratio step-up transformers that are an integral part of an X-ray are not required to comply with Article 450 and are generally used to provide the high voltage necessary for X-ray tubes. There is a lesser degree of fire hazard due to the low primary voltage; therefore, X-ray transformers are not required to be installed in fire-resistant vaults.

660-36. Capacitors. Capacitors shall be mounted within enclosures of insulating material or grounded metal.

D. Guarding and Grounding

660-47. General.

(a) High-Voltage Parts. All high-voltage parts, including X-ray tubes, shall be mounted within grounded enclosures. Air, oil, gas, or other suitable insulating media shall be used to insulate the high voltage from the grounded enclosure. The connection from the high-voltage equipment to X-ray tubes and other high-voltage components shall be made with high-voltage shielded cables.

(b) Low-Voltage Cables. Low-voltage cables connecting to oil-filled units that are not completely sealed, such as transformers, condensers, oil coolers, and high-voltage switches, shall have insulation of the oil-resistant type.

Grounded enclosures are required to be provided for all high-voltage X-ray equipment, including X-ray tubes. High-voltage shielded cables are required to be used to connect high-voltage equipment to X-ray tubes, and the shield is required to be grounded, as specified in Section 660-48.

660-48. Grounding. Noncurrent-carrying metal parts of X-ray and associated equipment (controls, tables, X-ray tube supports, transformer tanks, shielded cables, X-ray tube heads, etc.) shall be grounded in the manner specified in Article 250. Portable and mobile equipment shall be provided with an approved grounding-type attachment plug cap.

Exception: Battery-operated equipment.

Article 665 — Induction and Dielectric Heating

Contents

A. General

665-1. Scope. This article covers the construction and installation of induction and dielectric heating equipment and accessories for industrial and scientific applications, but not for medical or dental applications, appliances, or line frequency pipelines and vessels heating.

FPN No. 1: See Article 422 for appliances.

FPN No. 2: See Article 427, Part E, for line frequency pipelines and vessels heating.

To prevent spurious radiation caused by induction and dielectric heating equipment and to ensure that the frequency spectrum is utilized equitably, the Federal Communications Commission (FCC) has established rules (*Code of Federal Regulations,* Title 47, Part 18) that govern the use of this type of industrial heating equipment operating above 10 kHz.

665-2. Definitions.

Dielectric Heating. Dielectric heating is the heating of a nominally insulating material due to its own dielectric losses when the material is placed in a varying electric field.

Heating Equipment. The term *heating equipment* as used in this article includes any equipment used for heating purposes whose heat is generated by induction or dielectric methods.

Induction Heating. The heating of a nominally conductive material due to its own I^2R losses when the material is placed in a varying electromagnetic field.

Induction and dielectric heating are used for ovens, furnaces, and industrial equipment where pieces of material are heated by a rapidly alternating magnetic or electric field. For further information on electric heating systems using an induction heater or a dielectric heater in ovens and furnaces, see NFPA 86-1995, *Standard for Ovens and Furnaces,* and NFPA 86D-1995, *Standard for Industrial Furnaces Using Vacuum as an Atmosphere.*

Theory of Operation — Solid-State Converter Power Circuit

The solid-state converter power circuit consists of three sections: the rectifier section, the inverter section, and the tank circuit, which includes the load coil located on the outside. The rectifier section input power is 480 volts ac nominal, 3 phase, 60 Hz (see Figure 665.1). The output of the rectifier section is generally a fixed 600 volts dc nominal. This feeds the inverter section, which is a tuned variable frequency switching network that converts the dc power to single-phase, nominal, 180-Hz, 1-kHz, 3-kHz, or 10-kHz square wave or sinusoidal power. The operating frequency is determined by the control section. The single-phase output drives the tank circuit, which consists of a load coil and tuning capacitors.

Variable output power is achieved by frequency variation of the inverter section. Since the tank circuit is a tuned load, as the output approaches the frequency of the tank circuit, the power into the load approaches the maximum output. At minimum frequency, the output power is very low. As the frequency increases, the power increases, until a maximum is reached. This is the resonant frequency of the tank circuit.

Induction heating loads that are magnetic (carbon steel) exhibit a dynamic change as the work piece passes through the Curie temperature point (where

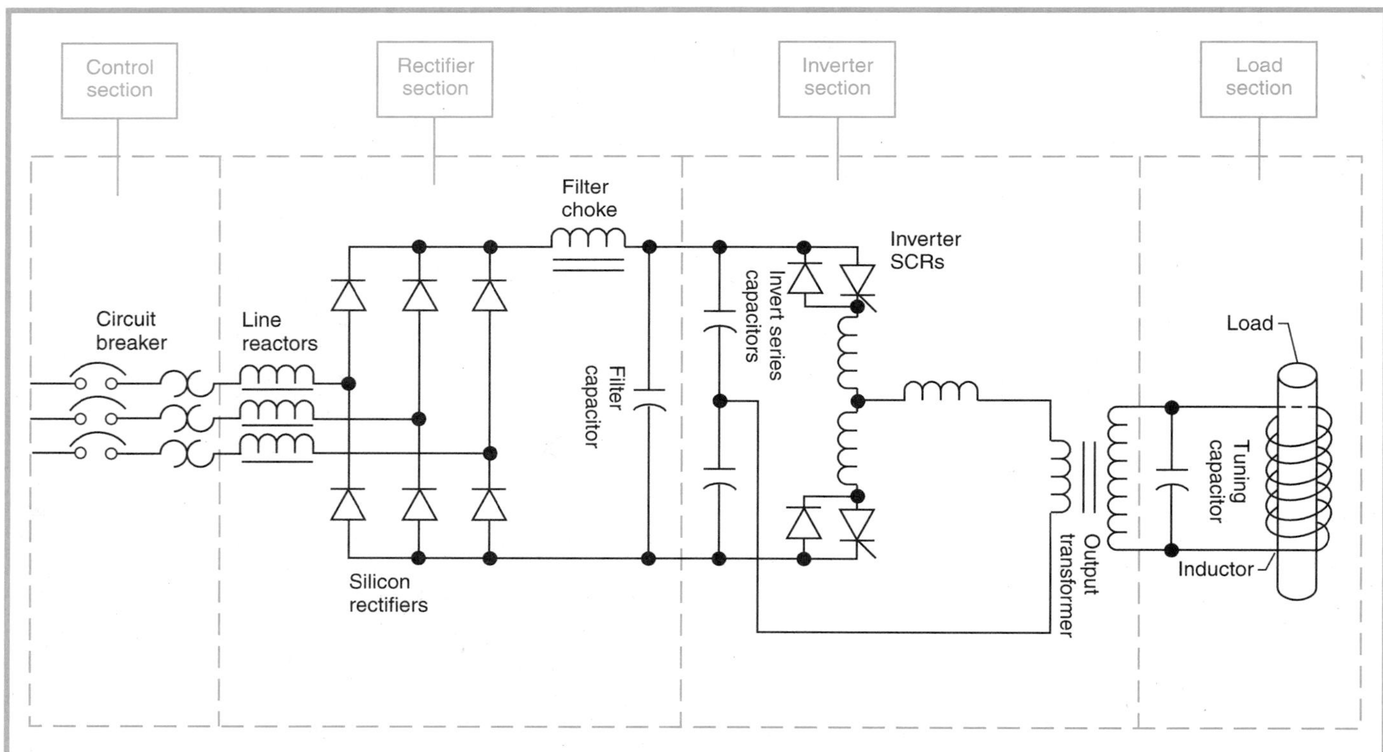

Figure 665.1 Simplified diagram of the components of a solid-state converter used for induction heating. (Tocco Division, Park Ohio Industries)

the work load changes from the magnetic to the nonmagnetic state). The converter automatically adjusts to maintain a constant output power during this dynamic load change.

Induction Heating

Induction heating is accomplished with the aid of a current-carrying coil that induces an eddy current in the work piece. Induction heating, in general, involves frequencies ranging from 50 Hz to approximately 500 kHz, and power outputs from a few hundred watts to several thousand kilowatts.

Where induction heating is employed, a nominally conductive material is placed in an inductor coil. The effective intensity of inductance is caused by a current flow in the coil at a high frequency, which produces a rapidly alternating magnetic field. This induces a voltage in the material to be heated and causes a current to flow through the resistance of the material (I_2R loss), producing induction heating. See Figure 665.2.

Dielectric Heating

Dielectric heating equipment is similar to induction heating equipment; however, the frequencies are generally higher (on the order of 3 MHz or more) than those in induction heating. Dielectric heating is useful for heating materials commonly thought of as being nonconductive; for instance, heating plastic preforms before molding, curing glue and plywood, drying rayon cakes, and many similar applications. Frequencies for this type of equipment range from 1 to 200 MHz, and they fall especially within the 1 to 50 MHz range. Vacuum-tube generators are used exclusively to supply dielectric heating power, and outputs range from a few hundred watts to several hundred kilowatts.

A typical wiring diagram of a vacuum-tube genera-

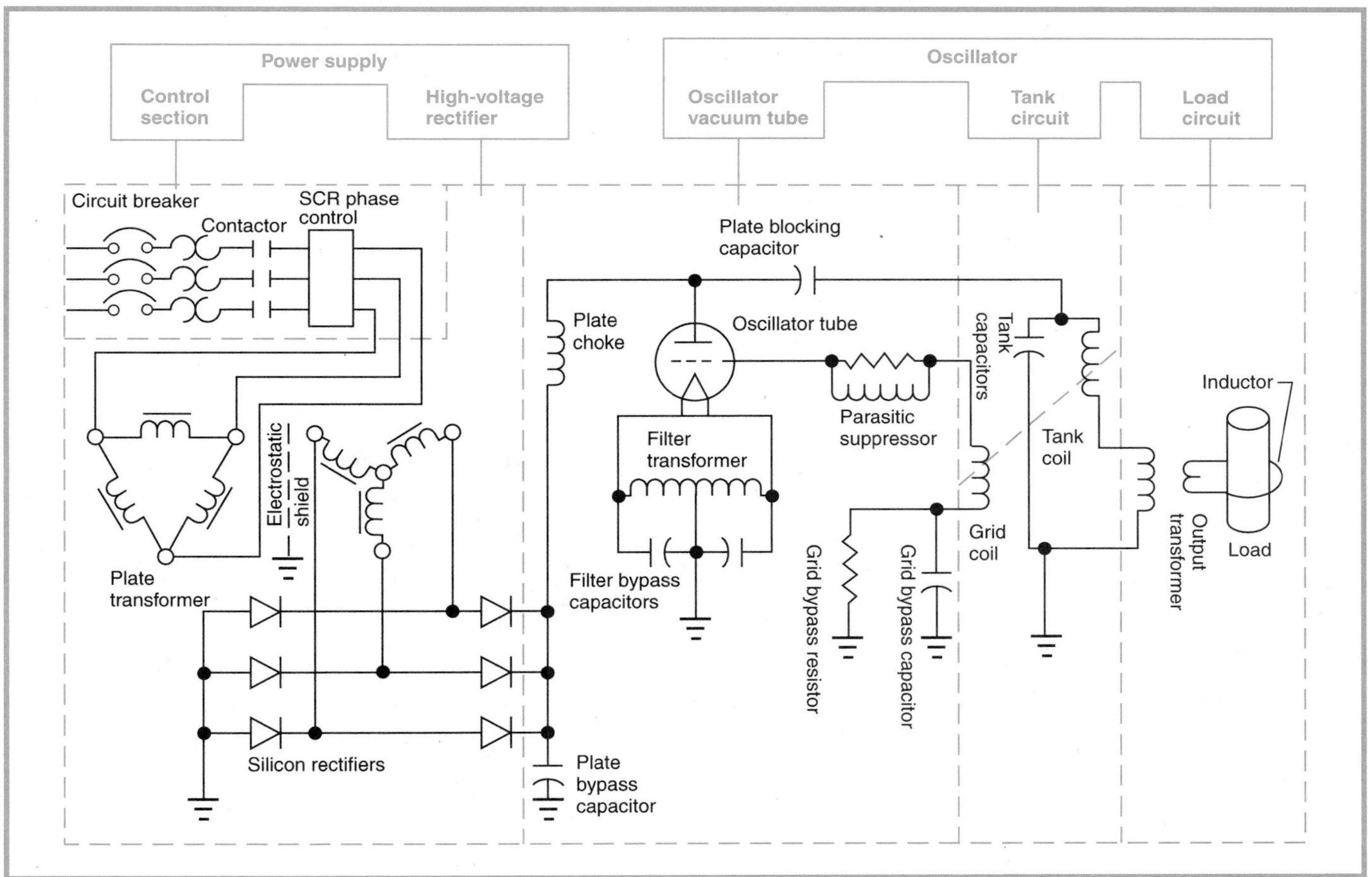

Figure 665.2 *Simplified diagram of the components of a vacuum-tube generator used for induction heating. (Tocco Division, Park Ohio Industries)*

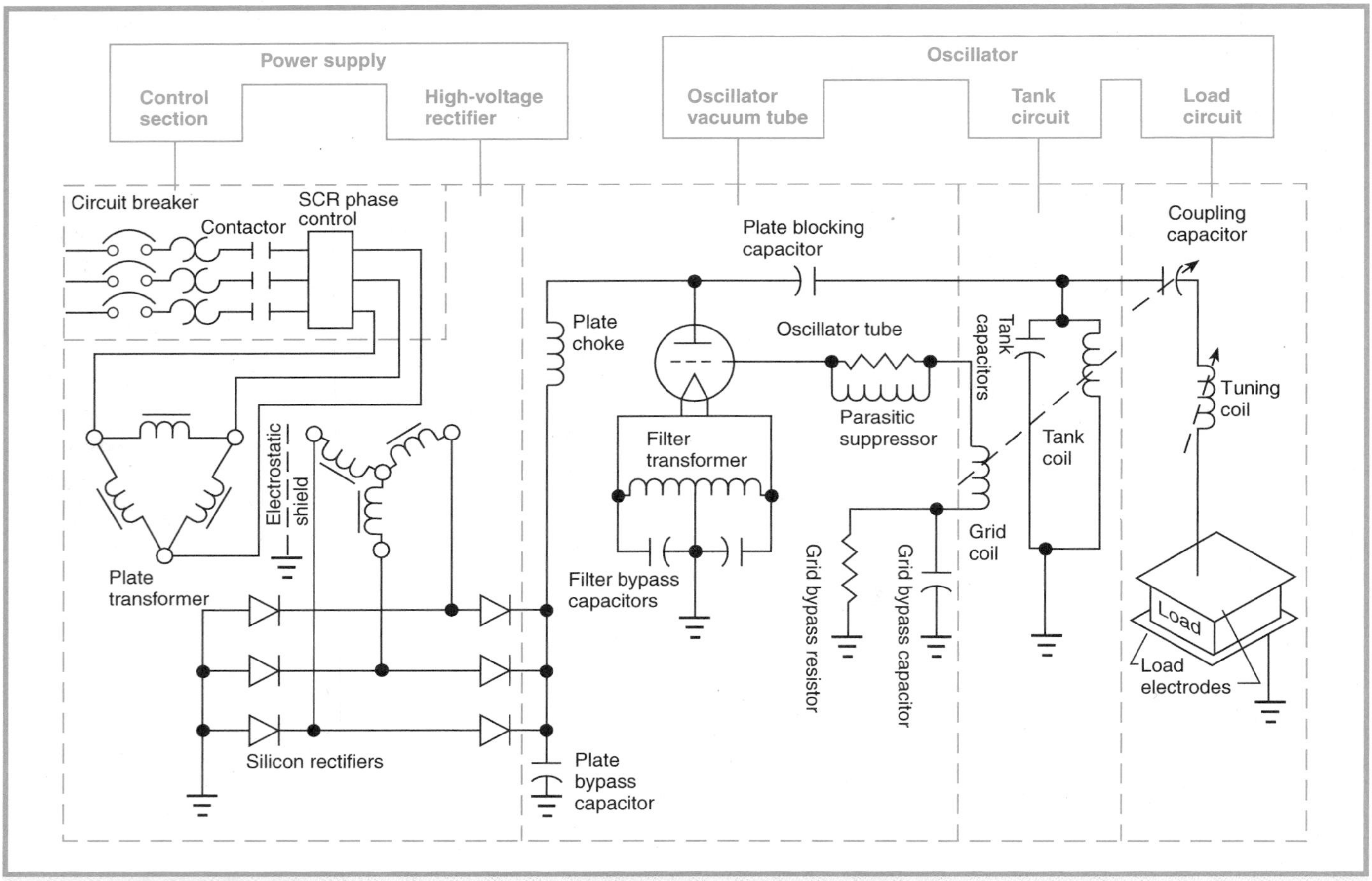

Figure 665.3 *Simplified diagram of the components of a vacuum-tube generator used for dielectric heating. (Tocco Division, Park Ohio Industries)*

tor is shown in Figure 665.3. Whereas induction heating uses a varying magnetic field, dielectric heating employs a varying electric field. This is done by placing the material to be heated between a pair of metal plates, called electrodes, in the output circuit of the generator. When high-frequency voltage is applied to the electrodes, a rapidly alternating electric field is set up between them, passing through the material to be heated. Because of the electrical charges within the molecules of this material, the field causes the molecules to vibrate in proportion to its frequency. This internal molecular action generates the heat used for dielectric heating.

See also Figure 665.4 for an example of an 80-kW oscillator used for heating.

665-3. Other Articles. Wiring from the source of power to the heating equipment shall comply with Chapters 1 through 4. Circuits and equipment operated at more than 600 volts, nominal, shall comply with Article 490.

Figure 665.4 *An 80-kW oscillator used to harden the exhaust valve seats of an automotive engine. (Tocco Division, Park Ohio Industries)*

665-4. Hazardous (Classified) Locations. Induction and dielectric heating equipment shall not be installed in hazardous (classified) locations as defined in Article 500 unless the equipment and wiring are designed and approved for the hazardous (classified) locations.

B. Guarding, Grounding, and Labeling

665-20. Enclosures. The converting apparatus (including the dc line) and high-frequency electric circuits (excluding the output circuits and remote-control circuits) shall be completely contained within an enclosure or enclosures of noncombustible material.

665-21. Panel Controls. All panel controls shall be of dead-front construction.

665-22. Access to Internal Equipment. Doors or detachable panels shall be employed for internal access. Where doors are used giving access to voltages from 500 to 1000 volts ac or dc, either door locks shall be provided or interlocking shall be installed. Where doors are used giving access to voltages of over 1000 volts ac or dc, either mechanical lockouts with a disconnecting means to prevent access until voltage is removed from the cubicle, or both door interlocking and mechanical door locks, shall be provided. Detachable panels not normally used for access to such parts shall be fastened in a manner that will make them inconvenient to remove.

665-23. Warning Labels or Signs. Warning labels or signs that read "DANGER — HIGH VOLTAGE — KEEP OUT" shall be attached to the equipment and shall be plainly visible where unauthorized persons might come in contact with energized parts, even when doors are open or when panels are removed from compartments containing over 250 volts ac or dc.

665-24. Capacitors. Where capacitors in excess of 0.1 microfarad are used in dc circuits, either as rectifier filter components or suppressers, etc., having circuit voltages of over 240 volts to ground, bleeder resistors or grounding switches shall be used as grounding devices. The time of discharge shall be in accordance with Section 460-6(a).

Where capacitors are individually switched out of a circuit, a bleeder resistor or automatic switch shall be used as a discharge means.

Where auxiliary rectifiers are used with filter capacitors in the output for bias supplies, tube keyers, etc., bleeder resistors shall be used, even though the dc voltage may not exceed 240 volts.

665-25. Work Applicator Shielding. Protective cages or adequate shielding shall be used to guard work applicators other than induction heating coils. Induction heating coils shall be permitted to be protected by insulation or refractory materials, or both. Interlock switches shall be used on all hinged access doors, sliding panels, or other easy means of access to the applicator. All interlock switches shall be connected in such a manner as to remove all power from the applicator when any one of the access doors or panels is open. Interlocks on access doors or panels shall not be required if the applicator is an induction heating coil at dc ground potential or operating at less than 150 volts ac.

665-26. Grounding and Bonding. Grounding or inter-unit bonding, or both, shall be used wherever required for circuit operation, for limiting to a safe value radio frequency potentials between all exposed noncurrent-carrying parts of the equipment and earth ground, between all equipment parts and surrounding objects, and between such objects and earth ground. Such grounding and bonding shall be installed in accordance with Article 250.

Because of stray currents flowing between units of the equipment or to the ground, bonding presents special problems at radio frequencies. These special bonding requirements are needed particularly at dielectric heating frequencies (100 to 200 MHz) due to the differences in radio-frequency potential that can exist between the equipment and surrounding metal units or other units of the installation. Satisfactory bonding can be accomplished by placing all units of the equipment on a flooring or base consisting of a copper or aluminum sheet and by thoroughly bonding, if necessary, by soldering, welding, or bolting.

By use of such special bonding, the radio-frequency resistance and reactance between units is held to a minimum, and any stray circulating currents flowing through this bonding will not cause a dangerous voltage drop.

It is necessary to provide operator protection from high radio-frequency potentials by shielding at dielectric heating frequencies. Interference with radio communications systems at such high frequencies may be eliminated by totally enclosing all components in a shielding of copper or aluminum.

665-27. Marking. Each heating equipment shall be provided with a nameplate giving the manufacturer's name and model identification and the following input data: line volts, frequency, number of phases, maximum current, full-load kilovolt-amperes (kVA), and full-load power factor.

665-28. Control Enclosures. Direct-current or low-frequency ac shall be permitted in the control portion of the heating equipment. This shall be limited to not over 150 volts. Solid or stranded wire No. 18 or larger shall be used. A step-down transformer with proper overcurrent protection shall be permitted in the control enclosure to obtain an ac voltage of less than 150 volts. The higher-voltage terminals shall be guarded to prevent accidental contact. 60-Hz components shall be permitted to control high frequency where properly rated by the induction heating equipment manufacturer. Electronic circuits utilizing solid-state devices and tubes shall be permitted printed circuits or wires smaller than No. 18.

C. Motor-Generator Equipment

665-40. General. Motor-generator equipment shall include all rotating equipment designed to operate from an ac or dc motor or by mechanical drive from a prime mover, producing an alternating current of any frequency for induction or dielectric heating, or both.

665-41. Ampacity of Supply Conductors. The ampacity of supply conductors shall be determined in accordance with Article 430.

665-42. Overcurrent Protection. Overcurrent protection shall be provided as specified in Article 430 for the electric supply circuit.

665-43. Disconnecting Means. The disconnecting means shall be provided as specified in Article 430.

A readily accessible disconnecting means shall be provided by which each heating equipment can be isolated from its supply circuit. The ampere rating of this disconnecting means shall not be less than the nameplate current rating of the equipment. The supply circuit disconnecting means shall be permitted as a heating equipment disconnecting means where the circuit supplies only one equipment.

665-44. Output Circuit. The output circuit shall include all output components external to the generator, including contactors, transformers, busbars, and other conductors, and shall comply with (a) and (b).

(a) Generator Output. The output circuit shall be isolated from ground unless otherwise permitted as follows:

(1) Where the capacitive coupling inherent in the generator causes the generator terminals to have voltages from terminal to ground that are equal or

(2) Where a vacuum or controlled atmosphere is used with a coil in a tank or chamber, the center point of the coil shall be grounded to maintain an equal potential between each terminal and ground.

Where rated at over 500 volts, the output circuit shall incorporate a dc ground protector unit. The dc impressed on the output circuit shall not exceed 30 volts and shall not exceed a current capability of 5 mA.

An isolating transformer for matching the load and the source shall be permitted in the output circuit if the output secondary is not at dc ground potential.

(b) Component Interconnections. The various components required for a complete induction heating equipment installation shall be connected by properly protected multiconductor cable, busbar, or coaxial cable. Cables shall be installed in nonferrous raceways. Busbars shall be protected, where required, by nonferrous enclosures.

665-47. Remote Control.

(a) Selector Switch. Where remote controls are used for applying power, a selector switch shall be provided and interlocked to provide power from only one control point at a time.

(b) Foot Switches. Switches operated by foot pressure shall be provided with a shield over the contact button to avoid accidental closing of a switch.

D. Equipment Other than Motor-Generators

665-60. General. Equipment other than motor-generators shall consist of all static multipliers and oscillator-type units utilizing vacuum tubes or solid-state devices, or both. The equipment shall be capable of converting ac or dc to an ac frequency suitable for induction or dielectric heating, or both.

665-61. Ampacity of Supply Conductors. The ampacity of supply conductors shall be determined in accordance with (a) and (b).

(a) Nameplate Rating. The ampacity of the circuit conductors shall not be less than the nameplate current rating of the equipment.

(b) Two or More. The ampacity of conductors supplying two or more equipments shall not be less than the sum of the nameplate current ratings on all equipments.

If simultaneous operation of two or more equipments supplied from the same feeder is not possible, the ampacity of the feeder shall not be less than the sum of the nameplate ratings for the largest group of machines capable of simultaneous operation, plus 100 percent of the stand-by currents of the remaining machines supplied.

665-62. Overcurrent Protection. Overcurrent protection shall be provided as specified in Article 240 for the equipment as a whole. This overcurrent protection shall be provided separately or as a part of the equipment.

665-63. Disconnecting Means. A readily accessible disconnecting means shall be provided by which each heating equipment can be isolated from its supply circuit. The disconnecting means shall be located within sight from the controller or be capable of being locked in the open position. The rating of this disconnecting means shall not be less than the nameplate rating of the equipment. The supply circuit disconnecting means shall be permitted for disconnecting the heating equipment where the circuit supplies only one equipment.

665-64. Output Circuit. The output circuit shall include all output components external to the converting device, including contactors, transformers, busbars, and other conductors, and shall comply with (a) and (b).

(a) Converter Output. The output circuit shall be isolated from ground unless a dc voltage can exist at the terminals from an internal component failure; then the output circuit (direct or coupled) shall be at dc ground potential.

(b) Converter and Applicator Connection. Where the connections between the converter and the work applicator exceed 2 ft (610 mm) in length, the connections shall be enclosed or guarded with nonferrous, noncombustible material.

665-66. Line Frequency in Converter Equipment Output. Commercial frequencies of 25- to 60-Hz ac output shall be permitted to be coupled for control purposes, but shall be limited to not over 150 volts during periods of circuit operation.

665-67. Keying. Where high-speed keying circuits dependent on the effect of "oscillator blocking" are employed, the peak radio frequency output voltage during the blocked portion of the cycle shall not exceed 100 volts in units employing radio frequency converters.

665-68. Remote Control.

(a) Selector Switch. Where remote controls are used for applying power, a selector switch shall be provided and interlocked to provide power from only one control point at a time.

(b) Foot Switches. Switches operated by foot pressure shall be provided with a shield over the contact button to avoid accidental closing of the switch.

Article 668 — Electrolytic Cells

Contents

668-1. Scope. The provisions of this article apply to the installation of the electrical components and accessory equipment of electrolytic cells, electrolytic cell lines, and process power supply for the production of aluminum, cadmium, chlorine, copper, fluorine, hydrogen peroxide, magnesium, sodium, sodium chlorate, and zinc.

Not covered by this article are cells used as a source of electric energy and for electroplating processes and cells used for the production of hydrogen.

FPN No. 1: In general, any cell line or group of cell lines operated as a unit for the production of a particular metal, gas, or chemical compound may differ from any other cell lines producing the same product because of variations in the particular raw materials used, output capacity, use of proprietary methods or process practices, or other modifying factors to the extent that detailed *Code* requirements become overly restrictive and do not accomplish the stated purpose of this *Code*.

FPN No. 2: For further information, see *Standard for Electrical Safety Practices in Electrolytic Cell Line Working Zones,* IEEE 463-1993.

An electrolytic cell line and its dc process power supply circuit, both within a cell line working zone, are treated as an individual machine supplied from a single source, even though they may cover acres of space, have a load current in excess of 400,000 amperes dc, or have a circuit voltage in excess of 1000 volts dc. The cell line process current passes through each cell in a series connection, and the load current cannot be subdivided, as it can in the heating circuit of a resistance-type electric furnace.

Because a cell line is supplied by its individual dc rectifier system, the rectifier or the entire cell line circuit is de-energized by removing its source of primary power.

In some electrolytic cell systems, the terminal voltage of the process supply can be appreciable. The voltage-to-ground of exposed live parts from one end of a cell line to the other is variable between the limits of the terminal voltage. Hence, operating and maintenance personnel and their tools are required to be insulated from ground. See Figure 668.1 for an example of a pot room in an aluminum reduction plant.

Figure 668.1 *A typical pot room in an aluminum reduction plant. (Alcoa)*

668-2. Definitions.

Cell Line. An assembly of electrically interconnected electrolytic cells supplied by a source of direct-current power.

Cell Line Attachments and Auxiliary Equipment. As applied to this article, cell line attachments and auxiliary equipment include, but are not limited to, auxiliary tanks; process piping; duct work; structural supports; exposed cell line conductors; conduits and other raceways; pumps, positioning equipment, and cell cutout or bypass electrical devices. Auxiliary equipment includes tools, welding machines, crucibles, and other portable equipment used for operation and maintenance within the electrolytic cell line working zone.

In the cell line working zone, auxiliary equipment includes the exposed conductive surfaces of ungrounded cranes and crane-mounted cell-servicing equipment.

Electrically Connected. A connection capable of carrying current as distinguished from connection through electromagnetic induction.

Electrolytic Cell. A tank or vat in which electrochemical reactions are caused by applying electrical energy for the purpose of refining or producing usable materials.

Electrolytic Cell Line Working Zone. The cell line working zone is the space envelope wherein operation or maintenance is normally performed on or in the vicinity of exposed energized surfaces of electrolytic cell lines or their attachments.

668-3. Other Articles.

(a) Lighting, Ventilating, Material Handling. Chapters 1 through 4 shall apply to services, feeders, branch circuits, and apparatus for supplying lighting, ventilating, material handling, and the like, that are outside the electrolytic cell line working zone.

(b) Systems Not Electrically Connected. Those elements of a cell line power-supply system that are not electrically connected to the cell supply system, such as the primary winding of a two-winding transformer, the motor of a motor-generator set, feeders, branch circuits, disconnecting means, motor controllers, and overload protective equipment shall be required to comply with all applicable provisions of this *Code*.

(c) Electrolytic Cell Lines. Electrolytic cell lines shall comply with the provisions of Chapters 1, 2, 3, and 4 except as amended in (1), (2), (3), or (4).

(1) The electrolytic cell line conductors shall not be required to comply with the provisions of Articles 110, 210, 215, 220, and 225. See Section 668-11.

(2) Overcurrent protection of electrolytic cell dc process power circuits shall not be required to comply with the requirements of Article 240.

(3) Equipment located or used within the electrolytic cell line working zone or associated with the cell line dc power circuits shall not be required to comply with the provisions of Article 250.

(4) The electrolytic cells, cell line attachments, and the wiring of auxiliary equipments and devices within the cell line working zone shall not be required to comply with the provisions of Articles 110, 210, 215, 220, and 225. See Section 668-30.

FPN: See Section 668-15 for equipment, apparatus, and structural component grounding.

668-10. Cell Line Working Zone.

(a) Area Covered. The space envelope of the cell line working zone shall encompass any space

(1) Within 96 in. (2.44 m) above energized surfaces of electrolytic cell lines or their energized attachments.
(2) Below energized surfaces of electrolytic cell lines or their energized attachments, provided the headroom in the space beneath is less than 96 in. (2.44 m).
(3) Within 42 in. (1.07 m) horizontally from energized surfaces of electrolytic cell lines or their energized attachments or from the space envelope described in Sections 668-10(a)(1) or (a)(2).

(b) Area Not Covered. The cell line working zone shall not be required to extend through or beyond walls, floors, roofs, partitions, barriers, or the like.

668-11. Direct-Current Cell Line Process Power Supply.

(a) Not Grounded. The dc cell line process power-supply conductors shall not be required to be grounded.

(b) Metal Enclosures Grounded. All metal enclosures of dc cell line process power-supply apparatus operating at a power-supply potential between terminals of over 50 volts shall be grounded as follows:

(1) Through protective relaying equipment, or
(2) By a minimum 2/0 copper grounding conductor or a conductor of equal or greater conductance

(c) Grounding Requirements. The grounding connections required by Section 668-11(b) shall be installed in accordance with Sections 250-8, 250-10, 250-12, 250-68, and 250-70.

668-12. Cell Line Conductors.

(a) Insulation and Material. Cell line conductors shall be either bare, covered, or insulated and of copper, aluminum, copper-clad aluminum, steel, or other suitable material.

(b) Size. Cell line conductors shall be of such cross-sectional area that the temperature rise under maximum load conditions and at maximum ambient shall not exceed the safe operating temperature of the conductor insulation or the material of the conductor supports.

(c) Connections. Cell line conductors shall be joined by bolted, welded, clamped, or compression connectors.

668-13. Disconnecting Means.

(a) More than One Process Power Supply. Where more than one dc cell line process power supply serves the same cell line, a disconnecting means shall be provided on the cell line circuit side of each power supply to disconnect it from the cell line circuit.

(b) Removable Links or Conductors. Removable links or removable conductors shall be permitted to be used as the disconnecting means.

668-14. Shunting Means.

(a) Partial or Total Shunting. Partial or total shunting of cell line circuit current around one or more cells shall be permitted.

(b) Shunting One or More Cells. The conductors, switches, or combination of conductors and switches used for shunting one or more cells shall comply with the applicable requirements of Section 668-12.

668-15. Grounding. For equipment, apparatus, and structural components that are required to be grounded by provisions of Article 668, the provisions of Article 250 shall apply, except a water pipe electrode shall not be required to be used. Any electrode or combination of electrodes described in Sections 250-50 and 250-52 shall be permitted.

668-20. Portable Electrical Equipment.

(a) Portable Electrical Equipment Not to Be Grounded. The frames and enclosures of portable electrical equipment used within the cell line working zone shall not be grounded.

Exception No. 1: Where the cell line voltage does not exceed 200 volts dc, these frames and enclosures shall be permitted to be grounded.

Exception No. 2: These frames and enclosures shall be permitted to be grounded where guarded.

(b) Isolating Transformers. Electrically powered, hand-held, cord-connected portable equipment with ungrounded frames or enclosures used within the cell line working zone shall be connected to receptacle circuits that have only ungrounded conductors such as a branch circuit supplied by an isolating transformer with an ungrounded secondary.

(c) Marking. Ungrounded portable electrical equipment shall be distinctively marked and shall employ plugs and receptacles of a configuration that prevents connection of this equipment to grounding receptacles and that prevents inadvertent interchange of ungrounded and grounded portable electrical equipments.

668-21. Power Supply Circuits and Receptacles for Portable Electrical Equipment.

(a) Isolated Circuits. Circuits supplying power to ungrounded receptacles for hand-held, cord-connected equipments shall be electrically isolated from any distribution system supplying areas other than the cell line working zone and shall be ungrounded. Power for these circuits shall be supplied through isolating transformers. Primaries of such transformers shall operate at not more than 600 volts between conductors and shall be provided with proper overcurrent protection. The secondary voltage of such transformers shall not exceed 300 volts between conductors, and all circuits supplied from such secondaries shall be ungrounded and shall have an approved overcurrent device of proper rating in each conductor.

(b) Noninterchangeability. Receptacles and their mating plugs for ungrounded equipment shall not have provision for a grounding conductor and shall be of a configuration that prevents their use for equipment required to be grounded.

(c) Marking. Receptacles on circuits supplied by an isolating transformer with an ungrounded secondary shall be a distinctive configuration, distinctively marked, and shall not be used in any other location in the plant.

668-30. Fixed and Portable Electrical Equipment.

(a) Electrical Equipment Not Required to Be Grounded. Alternating-current systems supplying fixed and portable electrical equipments within the cell line working zone shall not be required to be grounded.

(b) Exposed Conductive Surfaces Not Required to Be Grounded. Exposed conductive surfaces, such as electrical equipment housings, cabinets, boxes, motors, raceways, and the like, that are within the cell line working zone shall not be required to be grounded.

(c) Wiring Methods. Auxiliary electrical equipment such as motors, transducers, sensors, control devices, and alarms, mounted on an electrolytic cell or other energized surface, shall be connected to premises wiring systems by any of the following means.

(1) Multiconductor hard usage cord.
(2) Wire or cable in suitable raceways or metal or nonmetallic cable trays. If metal conduit, cable tray, armored cable, or similar metallic systems are used, they shall be installed with insulating breaks such that they will not cause a potentially hazardous electrical condition.

(d) Circuit Protection. Circuit protection shall not be required for control and instrumentation that are totally within the cell line working zone.

(e) Bonding. Bonding of fixed electrical equipment to the energized conductive surfaces of the cell line, its attachments, or auxiliaries shall be permitted. Where fixed electrical equipment is mounted on an energized conductive surface, it shall be bonded to that surface.

668-31. Auxiliary Nonelectric Connections. Auxiliary nonelectric connections, such as air hoses, water hoses, and the like, to an electrolytic cell, its attachments, or auxiliary equipments shall not have continuous conductive reinforcing wire, armor, braids, and the like. Hoses shall be of a nonconductive material.

668-32. Cranes and Hoists.

(a) Conductive Surfaces to Be Insulated from Ground. The conductive surfaces of cranes and hoists that enter the cell line working zone shall not be required to be grounded. The portion of an overhead crane or hoist that contacts an energized electrolytic cell or energized attachments shall be insulated from ground.

(b) Hazardous Electrical Conditions. Remote crane or hoist controls that may introduce hazardous electrical conditions into the cell line working zone shall employ one or more of the following systems:

(1) Isolated and ungrounded control circuit in accordance with Section 668-21(a)
(2) Nonconductive rope operator
(3) Pendant pushbutton with nonconductive supporting means and having nonconductive surfaces or ungrounded exposed conductive surfaces
(4) Radio

668-40. Enclosures. General-purpose electrical equipment enclosures shall be permitted where a natural draft ventilation system prevents the accumulation of gases.

Article 669 — Electroplating

Contents

669-1. Scope. The provisions of this article apply to the installation of the electrical components and accessory equipment that supply the power and controls for electroplating, anodizing, electropolishing, and electrostripping. For purposes of this article, the term *electroplating* shall be used to identify any or all of these processes.

Because of the extremely high currents and low voltages normally involved, conventional wiring methods cannot be used in electroplating, anodizing, electropolishing, and electrostripping processes. Note the permission to use bare conductors, even in systems exceeding 50 volts dc. See Figures 669.1 and 669.2. Some systems in the aluminum anodizing process have potentials up to 240 volts. Warning signs are required to be posted to indicate the presence of bare conductors.

Figure 669.1 A typical electroplating line.

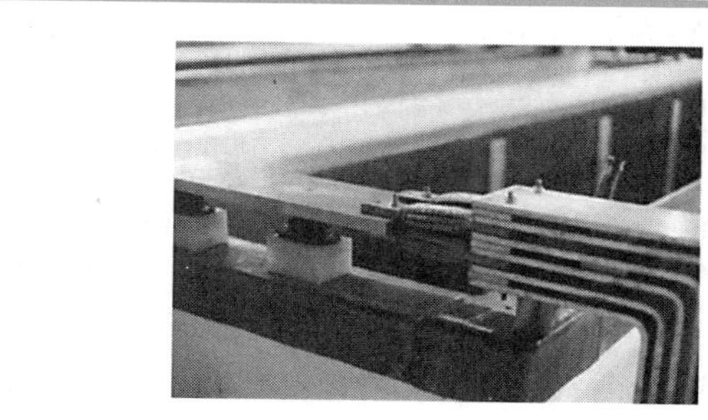

Figure 669.2 Bare busbar to supply an electroplating tank.

669-2. Other Articles. Except as modified by this article, wiring and equipment used for electroplating processes shall comply with the applicable requirements of Chapters 1 through 4.

669-3. General. Equipment for use in electroplating processes shall be identified for such service.

669-5. Branch-Circuit Conductors. Branch-circuit conductors supplying one or more units of equipment shall have an ampacity of not less than 125 percent of the total connected load. The ampacities for busbars shall be in accordance with Section 374-6.

669-6. Wiring Methods. Conductors connecting the electrolyte tank equipment to the conversion equipment shall be as follows.

(a) Systems Not Exceeding 50 Volts Direct Current. Insulated conductors shall be permitted to be run without insulated support provided they are protected from physical damage. Bare copper or aluminum conductors shall be permitted where supported on insulators.

(b) Systems Exceeding 50 Volts Direct Current. Insulated conductors shall be permitted to be run on insulated supports, provided they are protected from physical damage. Bare copper or aluminum conductors shall be permitted where supported on insulators and guarded against accidental contact up to the point of termination in accordance with Section 110-27.

669-7. Warning Signs. Warning signs shall be posted to indicate the presence of bare conductors.

669-8. Disconnecting Means.

(a) More than One Power Supply. Where more than one power supply serves the same dc system, a disconnecting means shall be provided on the dc side of each power supply.

(b) Removable Links or Conductors. Removable links or removable conductors shall be permitted to be used as the disconnecting means.

669-9. Overcurrent Protection. Direct-current conductors shall be protected from overcurrent by one or more of the following:

(1) Fuses or circuit breakers,
(2) A current-sensing device that operates a disconnecting means, or
(3) Other approved means

Article 670 — Industrial Machinery

Contents

670-1. Scope. This article covers the definition of, the nameplate data for, and the size and overcurrent protection of supply conductors to industrial machinery.

FPN: For further information, see *Electrical Standard for Industrial Machinery*, NFPA 79-1997.

670-2. Definitions.

Industrial Machinery (Machine). A power-driven machine (or a group of machines working together in a coordinated manner), not portable by hand while working, that is used to process material by cutting; forming; pressure; electrical, thermal, or optical techniques; lamination; or a combination of these processes. It can include associated equipment used to transfer material or tooling (including fixtures), assemble/disassemble, inspect or test, or package. [The associated electrical equipment including the logic controller(s) and associated software or logic together with the machine actuators and sensors are considered as part of the industrial machine.]

See Figure 670.1 for one example of an industrial machine. Sections 670-1 and 670-2 permit the inclusion of other types of industrial machines without the need to continuously modify the scope of Article 670 and NFPA 79-1997, *Electrical Standard for Industrial Machinery*. Also, the scope and definitions are more in harmony with NFPA 79 and IEC publication 60204-1.

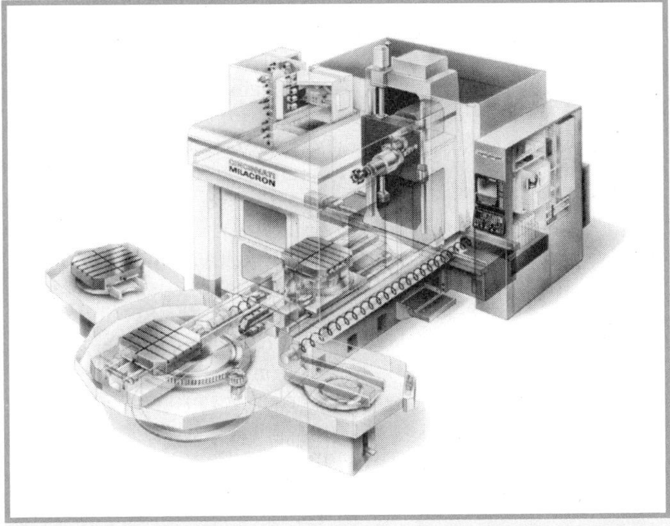

Figure 670.1 A horizontal machining center with touch screen computer control. (Cincinnati Milicron)

Industrial Manufacturing System. A systematic array of one or more industrial machines not portable by hand and that includes any associated material handling, manipulating, gauging, measuring, or inspection equipment.

670-3. Machine Nameplate Data.

(a) Permanent Nameplate. A permanent nameplate that lists supply voltage, phase, frequency, full-load current, the maximum ampere rating of the short-circuit and ground-fault protective device, ampere rating of largest motor or load, short-circuit interrupting capacity of the machine overcurrent-protective device, if furnished, and diagram number shall be attached to the control equipment enclosure or machine where plainly visible after installation.

The full-load current shown on the nameplate shall not be less than the sum of the full-load currents required for all motors and other equipment that may be in operation at the same time under normal conditions of use. Where unusual type loads, duty cycles, etc., require oversized conductors, the required capacity shall be included in the marked "full-load current." Where more than one incoming supply circuit is to be provided, the nameplate shall state the above information for each circuit.

(b) Overcurrent Protection. Where overcurrent protection is provided in accordance with Section 670-4(b), the machine shall be marked "overcurrent protection provided at machine supply terminals."

670-4. Supply Conductors and Overcurrent Protection.

(a) Size. The size of the supply conductor shall be such as to have an ampacity not less than 125 percent of the full-load current rating of all resistance heating loads plus 125 percent of the full-load current rating of highest rated motor plus the sum of the full-load current ratings of all other connected motors and apparatus that may be in operation at the same time.

FPN: See the 0–2000-volt ampacity tables of Article 310 for ampacity of conductors rated 600 volts and below.

(b) Overcurrent Protection. A machine shall be considered as an individual unit and therefore shall be provided with a disconnecting means. The disconnecting means shall be permitted to be supplied by branch circuits protected by either fuses or circuit breakers. The disconnecting means shall not be required to incorporate overcurrent protection. Where furnished as part of the machine, overcurrent protection shall consist of a single circuit breaker or set of fuses, the machine shall bear the marking required in Section 670-3, and the supply conductors shall be considered either as feeders or taps as covered by Section 240-21.

The rating or setting of the overcurrent protective device for the circuit supplying the machine shall not be greater than the sum of the largest rating or setting of the branch-circuit short-circuit and ground-fault protective device provided with the machine, plus 125 percent of the full-load current rating of all resistance heating loads, plus the sum of the full-load currents of all other motors and apparatus that may be in operation at the same time.

Clause 7.10 in NFPA 79-1997, *Electrical Standard for Industrial Machinery,* states, in part:

> The operating handle of the disconnecting means shall be readily accessible. The center of the grip of the operating handle of the disconnecting means, when in its highest position, shall not be more than 6 ft 7 in. (2 m) above the floor. A permanent operating platform, readily accessible by means of a permanent stair or ladder, shall be considered as the floor for the purpose of this requirement. The operating handle shall be capable of being locked only in the open (off) position. When the control enclosure door is closed, the operating handle shall positively indicate whether the disconnecting means is in the open (off) or closed position.

Exception: Where one or more instantaneous trip circuit breakers or motor short-circuit protectors are used for motor branch-circuit short-circuit and ground-fault protection as permitted by Section 430-52(c), the procedure specified above for determining the maximum rating of the protective device for the circuit supplying the machine shall apply with the following provision. For the purpose of the calculation, each instantaneous trip circuit breaker or motor short-circuit protector shall be assumed to have a rating not exceeding the maximum percentage of motor full-load current permitted by Table 430-152 for the type of machine supply circuit protective device employed.

Where no branch-circuit short-circuit and ground-fault protective device is provided with the machine, the rating or setting of the overcurrent protective device shall be based on Sections 430-52 and 430-53, as applicable.

A commonly asked question is, "When does the *NEC* apply to wiring for machinery defined in Article 670?" The equipment and wiring of industrial machinery, where different component parts may be purchased and assembled at the location where they will be used, must be installed in accordance with the applicable articles in the *NEC.*

Machinery assembled by the manufacturer, in accordance with NFPA 79, then disassembled for shipping and reassembled at its place of use, comes only under Article 670 and any *NEC* sections referenced therein. In this case, the machinery is treated as a package unit. The nameplate provides the necessary information to size the branch-circuit conductors, disconnecting means, and overcurrent protection. The computation for motor and nonmotor loads is reflected on the nameplate as full-load amperes, and no further calculation is necessary.

670-5. Clearance. Where the conditions of maintenance and supervision ensure that only qualified persons will service the installation, the dimensions of the working space in the direction of access to live parts operating at not over 150 volts line-to-line or line-to-ground that are likely to require examination, adjustment, servicing, or maintenance while energized shall be a minimum of 2½ ft (762 mm). Where controls are enclosed in cabinets, the door(s) shall open at least 90 degrees or be removable.

Exception: Where the enclosure requires a tool to open, and where only diagnostic and troubleshooting testing is involved on live parts, the clearances shall be permitted to be less than 2½ ft (762 mm).

It is essential that *only qualified persons* perform the service or maintenance on energized live parts. See the definitions of *energized* and *live parts* in Article 100.

Article 675 — Electrically Driven or Controlled Irrigation Machines

Contents

A. General

675-1. Scope. The provisions of this article apply to electrically driven or controlled irrigation machines, and to the branch circuits and controllers for such equipment.

Electric pump motors used to supply water to irrigation machines are governed by the general requirements of the *Code,* not by Article 675.

675-2. Definitions.

Center Pivot Irrigation Machines. A multimotored irrigation machine that revolves around a central pivot and employs alignment switches or similar devices to control individual motors.

Collector Rings. An assembly of slip rings for transferring electrical energy from a stationary to a rotating member.

Irrigation Machines. An electrically driven or controlled machine, with one or more motors, not hand portable, and used primarily to transport and distribute water for agricultural purposes.

675-3. Other Articles. These provisions are in addition to, or amendatory of, the provisions of Article 430 and other articles in this *Code* that apply except as modified in this article.

The requirements of Article 675 apply to special equipment for a particular condition; they supplement or modify the general rules. See Section 90-3, Code Arrangement.

675-4. Irrigation Cable.

(a) Construction. The cable used to interconnect enclosures on the structure of an irrigation machine shall be an assembly of stranded, insulated conductors with nonhygroscopic and nonwicking filler in a core of moisture- and flame-resistant nonmetallic material overlaid with a metallic covering and jacketed with a moisture-, corrosion-, and sunlight-resistant nonmetallic material.

The conductor insulation shall be of a type listed in Table 310-13 for an operating temperature of 75°C (167°F) and for use in wet locations. The core insulating material thickness shall not be less than 30 mils (762 μm), and the metallic overlay thickness shall not be less than 8 mils (203 μm). The jacketing material thickness shall not be less than 50 mils (1.27 mm).

A composite of power, control, and grounding conductors in the cable shall be permitted.

(b) Alternate Wiring Methods. Other cables listed for the purpose.

(c) Supports. Irrigation cable shall be secured by straps, hangers, or similar fittings identified for the purpose and installed as not to damage the cable. Cable shall be supported at intervals not exceeding 4 ft (1.22 m).

(d) Fittings. Fittings shall be used at all points where irrigation cable terminates. The fittings shall be designed for use with the cable and shall be suitable for the conditions of service.

675-5. More than Three Conductors in a Raceway or Cable. The signal and control conductors of a raceway or cable shall not be counted for the purpose of derating the conductors as required in Section 310-15(b)(2)(a).

675-6. Marking on Main Control Panel. The main control panel shall be provided with a nameplate that shall give the following information:

(1) The manufacturer's name, the rated voltage, the phase, and the frequency
(2) The current rating of the machine, and
(3) The rating of the main disconnecting means and size of overcurrent protection required

675-7. Equivalent Current Ratings. Where intermittent duty is not involved, the provisions of Article 430 shall be used for determining ratings for controllers, disconnecting means, conductors, and the like. Where irrigation machines have inherent intermittent duty, the following determinations of equivalent current ratings shall be used.

(a) Continuous-Current Rating. The equivalent continuous-current rating for the selection of branch-circuit conductors and overcurrent protection shall be equal to 125 percent of the motor nameplate full-load current rating of the largest motor plus a quantity equal to the sum of each of the motor nameplate full-load current ratings of all remaining motors on the circuit multiplied by the maximum percent duty cycle at which they can continuously operate.

(b) Locked-Rotor Current. The equivalent locked-rotor current rating shall be equal to the numerical sum of the locked-rotor current of the two largest motors plus 100 percent of the sum of the motor nameplate full-load current ratings of all the remaining motors on the circuit.

675-8. Disconnecting Means.

(a) Main Controller. A controller that is used to start and stop the complete machine shall meet all of the following requirements:

(1) An equivalent continuous current rating not less than specified in Sections 675-7(a) or 675-22(a)
(2) A horsepower rating not less than the value from Tables 430-151(A) and (B) based on the equivalent locked-rotor current specified in Sections 675-7(b) or 675-22(b)

(b) Main Disconnecting Means. The main disconnecting means for the machine shall provide overcurrent protection and shall be at the point of connection of electrical power to the machine or shall be visible and not more than 50 ft (15.2 m) from the machine and shall be readily accessible and capable of being locked in the open position. This disconnecting means shall have a horsepower and current rating not less than required for the main controller.

Section 675-8(b) permits the main disconnecting means to be up to 50 ft from the machine if readily accessible and capable of being locked in the open position. This eliminates one set of overcurrent protective devices and one disconnecting means where the circuit originates at the motor control panel for the irrigation pump and this panel is within 50 ft of the center pivot machine. It also alleviates some potential problems with machines designed to be towed to a second site.

Exception: Circuit breakers without marked horsepower ratings shall be permitted in accordance with Section 430-109.

(c) Disconnecting Means for Individual Motors and Controllers. A disconnecting means shall be provided to simultaneously disconnect all ungrounded conductors for each motor and controller and shall be located as required by Article 430, Part J. The disconnecting means shall not be required to be readily accessible.

Article 430, Part I, provides for safety during maintenance and inspection shutdown periods. See the commentary following Section 430-103.

675-9. Branch-Circuit Conductors. The branch-circuit conductors shall have an ampacity not less than specified in Sections 675-7(a) or 675-22(a).

675-10. Several Motors on One Branch Circuit.

(a) Protection Required. Several motors, each not exceeding 2-hp rating, shall be permitted to be used on an irrigation machine circuit protected at not more than 30 amperes at 600 volts, nominal, or less, provided all of the following conditions are met.

Section 430-53 provides for motor branch-circuit short-circuit and ground-fault protection for several motors on one branch circuit. A combination of these requirements, which are modified for this special equipment application, is found in Section 675-10.

(1) The full-load rating of any motor in the circuit shall not exceed 6 amperes.
(2) Each motor in the circuit shall have individual overload protection in accordance with Section 430-32.
(3) Taps to individual motors shall not be smaller than No. 14 copper and not more than 25 ft (7.62 m) in length.

(b) Individual Protection Not Required. Individual branch-circuit short-circuit protection for motors and motor controllers shall not be required where the requirements of Section 675-10(a) are met.

675-11. Collector Rings.

(a) Transmitting Current for Power Purposes. Collector rings shall have a current rating not less than 125 percent of the full-load current of the largest device served plus the full-load current of all other devices served, or as determined from Sections 675-7(a) or 675-22(a).

(b) Control and Signal Purposes. Collector rings for control and signal purposes shall have a current rating not less than 125 percent of the full-load current of the largest device served plus the full-load current of all other devices served.

(c) Grounding. The collector ring used for grounding shall have a current rating of not less than that sized in accordance with Section 675-11(a).

(d) Protection. Collector rings shall be protected from the expected environment and from accidental contact by means of a suitable enclosure.

675-12. Grounding. The following equipment shall be grounded:

(1) All electrical equipment on the irrigation machine
(2) All electrical equipment associated with the irrigation machine
(3) Metal junction boxes and enclosures, and
(4) Control panels or control equipment that supply or control electrical equipment to the irrigation machine

Exception: Grounding shall not be required on machines where all of the following provisions are met.

(a) The machine is electrically controlled but not electrically driven.
(b) The control voltage is 30 volts or less.
(c) The control or signal circuits are current limited as specified in Chapter 9, Tables 11(a) and 11(b).

675-13. Methods of Grounding. Machines that require grounding shall have a noncurrent-carrying equipment grounding conductor provided as an integral part of each cord, cable, or raceway. This grounding conductor shall be sized not less than the largest supply conductor in each cord, cable, or raceway. Feeder circuits supplying power to irrigation machines shall have an equipment grounding conductor sized according to Table 250-122.

675-14. Bonding. Where electrical grounding is required on an irrigation machine, the metallic structure of the machine, metallic conduit, or metallic sheath of cable shall be bonded to the grounding conductor. Metal-to-metal contact with a part that is bonded to the grounding conductor and the noncurrent-carrying parts of the machine shall be considered as an acceptable bonding path.

675-15. Lightning Protection. If an irrigation machine has a stationary point, a grounding electrode system in accordance with Article 250, Part C, shall be connected to the machine at the stationary point for lightning protection.

If the electrical power supply to irrigation machine equipment is a service, provisions in Sections 250-50 and 250-52 require a made grounding electrode, such as a driven ground rod. Consideration should be given to Section 250-60 and NFPA 780-1995, *Lightning Protection Code,* in areas where lightning protection is critical.

675-16. Energy from More than One Source. Equipment within an enclosure receiving electrical energy from more than one source shall not be required to have a disconnecting means for the additional source, provided that its voltage is 30 volts or less and it meets the requirements of Part C of Article 725.

675-17. Connectors. External plugs and connectors on the equipment shall be of the weatherproof type.

Unless provided solely for the connection of circuits meeting the requirements of Part C of Article 725, external plugs and connectors shall be constructed as specified in Section 250-124(a).

B. Center Pivot Irrigation Machines

675-21. General. The provisions of Part B are intended to cover additional special requirements that are peculiar to center pivot irrigation machines. See Section 675-2 for definition *of Center Pivot Irrigation Machines.*

675-22. Equivalent Current Ratings. In order to establish ratings of controllers, disconnecting means, conductors, and the like, for the inherent intermittent duty of center pivot irrigation machines, the following determination shall be used.

The ratings of electrical components of any circuit should be selected so as to avoid extensive damage to the equipment during a short circuit or ground fault. Section 675-22 contains requirements for establishing ratings of components of special equipment for inherent intermittent duty. Also see the commentary following Sections 110-10 and 430-52. See Figure 675.1 for an example of a center pivot irrigation machine.

Figure 675.1 *A center pivot irrigation machine. (Lockwood Corp.)*

(a) Continuous-Current Rating. The equivalent continuous-current rating for the selection of branch-circuit conductors and branch-circuit devices shall be equal to 125 percent of the motor nameplate full-load current rating of the largest motor plus 60 percent of the sum of the motor nameplate full-load current ratings of all remaining motors on the circuit.

(b) Locked-Rotor Current. The equivalent locked-rotor current rating shall be equal to the numerical sum of two times the locked-rotor current of the largest motor plus 80 percent of the sum of the motor nameplate full-load current ratings of all the remaining motors on the circuit.

Article 680 — Swimming Pools, Fountains, and Similar Installations

Contents

A. General

680-1. Scope. The provisions of this article apply to the construction and installation of electrical wiring for and equipment in or adjacent to all swimming, wading, therapeutic, and decorative pools, fountains, hot tubs, spas, and hydromassage bathtubs, whether permanently installed or storable, and to metallic auxiliary equipment, such as pumps, filters, and similar equipment.

Article 680 applies to decorative pools and fountains; swimming, wading, and wave pools; therapeutic tubs and tanks; hot tubs; spas; hydromassage bathtubs; and similar installations. These installations can be indoors or outdoors, permanent or storable, whether or not served by electrical circuits of any nature. Studies conducted by Underwriters Laboratories Inc., various manufacturers, and others indicate that a person in a swimming pool can receive a severe electric shock by reaching over and touching the energized casing of a faulty appliance, such as a radio, hair dryer, and so on, as the person's body establishes a conductive path through the water and pool to earth. Also, a person not in contact with a faulty appliance or any grounded object may receive an electric shock and be rendered immobile by a potential gradient in the water itself. Accordingly, the requirements of Article 680 covering effective bonding and grounding, installation of receptacles and lighting fixtures, use of ground-fault circuit interrupters, modified wiring methods, and so on, apply not only to the installation of the pool but also to installations and equipment adjacent to or associated with the pool.

The definitions of *pool* and *fountain* were contained in the fine print note following Section 680-1 in the 1996 *Code*. For the 1999 *Code*, these definitions were moved to Section 680-4.

680-2. Approval of Equipment. All electrical equipment installed in the water, walls, or decks of pools, fountains, and similar installations shall comply with the provisions of this article.

680-3. Other articles.

(a) Except as modified by this section, wiring and equipment in or adjacent pools and fountains shall comply with the applicable requirements of Chapters 1 through 4.

FPN: See Section 370-23 for junction boxes and Section 347-3 for rigid nonmetallic conduit.

Note that Section 370-23(d) specifies the requirements for the support of threaded boxes that do not contain devices, and Section 347-3(b) does not permit equipment to be supported by rigid nonmetallic conduit. Figure 680.1 shows a properly supported junction box for a wet-niche fixture. Also see the commentary following Section 370-23(d).

(b) The installation and wiring of audio equipment adjacent to pools and fountains shall comply with the applicable requirements of Article 640. Underwater loudspeakers shall be installed in accordance with Section 680-23.

680-4. Definitions.

Cord- and Plug-Connected Lighting Assembly. A lighting assembly consisting of a lighting fixture intended for installation in the wall of a spa, hot tub, or storable pool, and a cord- and plug-connected transformer.

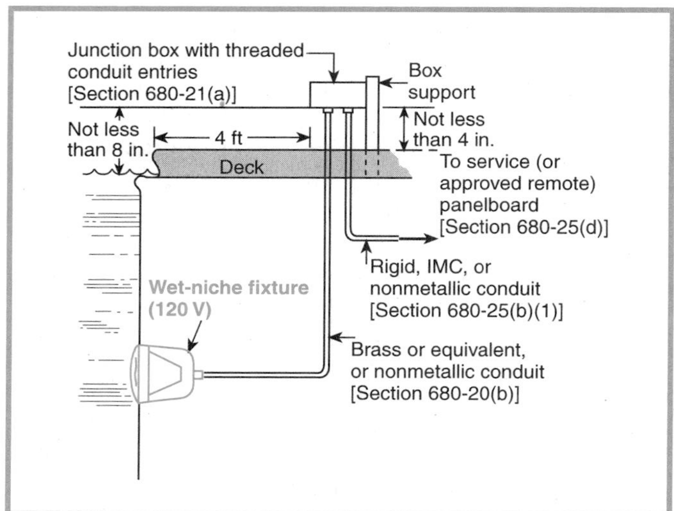

Figure 680.1 *Wet-niche fixture installation with junction box supported above pool deck.*

Dry-Niche Lighting Fixture. A lighting fixture intended for installation in the wall of a pool or fountain in a niche that is sealed against the entry of pool water.

Forming Shell. A structure designed to support a wet-niche lighting fixture assembly and intended for mounting in a pool or fountain structure.

Fountain. As used in this article, the term includes fountains, ornamental pools, display pools, and reflection pools. It does not include drinking fountains.

Hydromassage Bathtub. A permanently installed bathtub equipped with a recirculating piping system, pump, and associated equipment. It is designed so it can accept, circulate, and discharge water upon each use.

See the commentary following Part G, Hydromassage Bathtubs.

No-Niche Lighting Fixture. A lighting fixture intended for installation above or below the water without a niche.

Packaged Spa or Hot Tub Equipment Assembly. A factory-fabricated unit consisting of water-circulating, heating, and control equipment mounted on a common base, intended to operate a spa or hot tub. Equipment may include pumps, air blowers, heaters, lights, controls, sanitizer generators, etc.

The definition of *packaged spa or hot tub equipment assembly* clarifies which assemblies are subject to the requirements of Section 680-42.

Packaged Therapeutic Tub or Hydrotherapeutic Tank Equipment Assembly. A factory-fabricated unit consisting of water-circulating, heating, and control equipment mounted on a common base, intended to operate a therapeutic tub or hydrotherapeutic tank. Equipment may include pumps, air blowers, heaters, lights, controls, sanitizer generators, etc.

Permanently Installed Decorative Fountains and Reflection Pools. Those that are constructed in the ground, on the ground, or in a building in such a manner that the fountain cannot be readily disassembled for storage, whether or not served by electrical circuits of any nature. These units are primarily constructed for their aesthetic value and are not intended for swimming or wading.

Permanently Installed Swimming, Wading, and Therapeutic Pools. Those that are constructed in the ground or partially in the ground, and all others capable of holding water in a depth greater than 42 in. (1.07 m), and all pools installed inside of a building, regardless of water depth, whether or not served by electrical circuits of any nature.

See the commentary following Part F, Pools and Tubs for Therapeutic Use.

Pool. As used in this article, the term includes swimming, wading, and permanently installed therapeutic pools.

Pool Cover, Electrically Operated. Motor-driven equipment designed to cover and uncover the water surface of a pool by means of a flexible sheet or rigid frame.

The requirements for electrically operated pool covers are found in Section 680-26.

Self-Contained Spa or Hot Tub. Factory-fabricated unit consisting of a spa or hot tub vessel with all water-circulating, heating, and control equipment integral to the unit. Equipment may include pumps, air blowers, heaters, lights, controls, sanitizer generators, etc.

The definition of *self-contained spa or hot tub,* clarifies which assemblies are subject to the requirements of Section 680-42.

Self-Contained Therapeutic Tubs or Hydrotherapeutic Tanks. A factory-fabricated unit consisting of a therapeutic tub or hydrotherapeutic tank with all water-circulating, heating, and control equipment integral to the unit. Equipment may include pumps, air blowers, heaters, light controls, sanitizer generators, etc.

Spa or Hot Tub. A hydromassage pool, or tub for recreational or therapeutic use, not located in health care facilities, designed for immersion of users, and usually having a filter, heater, and motor-driven blower. It may be installed indoors or outdoors, on the ground or supporting structure, or in the ground or supporting structure. Generally, a spa or hot tub is not designed or intended to have its contents drained or discharged after each use.

See the commentary following Part D, Spas and Hot Tubs.

Storable Swimming or Wading Pool. Those that are constructed on or above the ground and are capable of holding water to a maximum depth of 42 in. (1.07 m), or a pool with nonmetallic, molded polymeric walls or inflatable fabric walls regardless of dimension.

See the commentary following Section 680-31 and Figure 680.16.

Originally, storable pools were not specifically addressed in the *NEC.* Article 680 was written to provide guidance relative to permanent, in-ground pools and their unique construction requirements because of the unusual earth-water-electricity-human body environment created in the finished product. The conductivity of moist concrete or metal walls buried in the ground, the incorporation of large masses of reinforcing steel, and the inclusion of stainless-steel handrails and diving-board stands as well as 120-volt lights into the pool structure all called for the strict wiring, bonding, and grounding requirements of Article 680.

Storable pools, on the other hand, are intended to be temporary structures, without the need for special wiring or modification to the pool site. They are most often sold as a complete package, consisting of the pool walls, vinyl liner, plumbing kit, and pump/filter device. A storable pool is likely to be disassembled and stored during the winter months. Regional preferences, weather patterns, economic considerations, and design characteristics of the pool are all factors influencing this action. The original Article 680 definition of a storable pool was, "One that is so constructed that it may be readily disassembled for storage and reassembled to its original integrity."

Part C of Article 680 was created to address the special equipment specifications of storable pools, and Underwriters Laboratories Inc. developed testing and labeling criteria for listing the pump/filter units designed especially for these pools. This equipment has the following four notable aspects:

1. It must have an approved system of double insulation or its equivalent.
2. It is permitted to have a flexible cord equipped with a parallel-blade, grounding-type attachment plug for electrical connection.
3. It must have a grounding conductor included in the flexible cord.
4. The flexible cord is not limited to 3 ft, as permitted in Section 680-7, and is specified by UL to be 25 ft long. This length was chosen to discourage the use of extension cords.

The UL labeling requirement for these listed units includes the wording "For use with storable pools only." In recent years, many consumers and members of the swimming pool industry have found it desirable to use these pump/filter units on any aboveground or on-ground pool, regardless of the pool's dimensions or the "storability" of the pool.

Storable pools are supplied in two distinct versions. One version is intended to be disassembled at the end of each swimming season. The second version, by the nature of its construction, can be disassembled, but manufacturers do not recommend this. The pools in the latter category frequently require special modification to and preparation of the pool site, making them impractical to disassemble. Draining these pools, especially large ones, increases the likelihood of costly damage due to shrinkage of the vinyl liner material.

The main factor differentiating the two types of pools is wall height. Generally, pools intended to be disassembled at season's end have wall heights of 42 in. or less, while those not intended for disassembly have wall heights of 48 in. or more. The surface area of the pools is not a factor.

Wet-Niche Lighting Fixture. A lighting fixture intended for installation in a forming shell mounted in a pool or fountain structure where the fixture will be completely surrounded by water.

680-5. Transformers and Ground-Fault Circuit Interrupters.

(a) Transformers. Transformers used for the supply of underwater fixtures, together with the transformer enclosure, shall be identified for the purpose. The transformer shall be an isolated winding type that has a grounded metal barrier between the primary and secondary windings.

Unless marked otherwise, UL-listed swimming pool and spa transformers are not suitable for connection to a conduit that extends directly to an underwater pool light forming shell. Swimming pool and spa transformers are not permitted to be used outdoors unless marked "For Outdoor Use" or in an equivalent manner that signifies that they have been found acceptable for both outdoor and indoor use. See Section 110-3(b).

(b) Ground-Fault Circuit Interrupters. Ground-fault circuit interrupters shall be self-contained units, circuit-breaker types, receptacle types, or other approved types.

See the definition of *ground-fault circuit interrupter* in Article 100.

A ground-fault circuit interrupter is intended to be used only in a circuit that has a solidly grounded conductor; however, an equipment grounding conductor is not necessary in order for the GFCI to function. A Class A GFCI trips if the current to ground has a value in the range of 4 through 6 mA; it is suitable for use in swimming pool circuits. It should be noted, however, that swimming pool circuits installed before local adoption of the 1965 edition of the *Code* may have sufficient leakage current to cause a Class A GFCI to trip. A Class B GFCI trips if the current to ground exceeds 20 mA; it is suitable for use with underwater swimming pool lighting fixtures only.

(c) Wiring. Conductors on the load side of a ground-fault circuit interrupter or of a transformer, used to comply with the provisions of Section 680-20(a)(1), shall not occupy raceway, boxes, or enclosures containing other conductors unless the other conductors are protected by ground-fault circuit interrupters or are grounding conductors. Supply conductors to a feed-through type ground-fault circuit interrupter shall be permitted in the same enclosure.

Ground-fault circuit interrupters shall be permitted in a panelboard that contains circuits protected by other than ground-fault circuit interrupters.

680-6. Receptacles, Lighting Fixtures, Lighting Outlets, Switching Devices, and Ceiling-Suspended (Paddle) Fans.

The term *ceiling fan* has been changed to clarify that the *Code* refers to a fan listed to UL 507, which refers to this type of fan as a "paddle fan."

(a) Receptacles.

(1) A receptacle(s) that provides power for a water-pump motor(s) for, or other loads directly related to the circulation and sanitation system, a permanently installed pool or fountain, as permitted in Section 680-7, shall be permitted between 5 ft and 10 ft (1.52 m and 3.05 m) from the inside walls of the pool or fountain, and, where so located, shall be single and of the locking and grounding types and shall be protected by a ground-fault circuit interrupter(s).

Other receptacles on the property shall be located at least 10 ft (3.05 m) from the inside walls of a pool or fountain.

(2) Where a permanently installed pool is installed at a dwelling unit(s), at least one 125-volt 15- or 20-ampere receptacle on a general-purpose branch circuit shall be located a minimum of 10 ft (3.05 m) from and not more than 20 ft (6.08 m) from the inside wall of the pool. This receptacle shall be located not more than 6 ft 6 in. (1.98 m) above the floor, platform, or grade level serving the pool.

(3) All 125-volt receptacles located within 20 ft (6.08 m) of the inside walls of a pool or fountain shall be protected by a ground-fault circuit interrupter.

FPN: In determining the above dimensions, the distance to be measured is the shortest path the supply cord of an appliance connected to the receptacle would follow without piercing a floor, wall, ceiling, doorway with hinged or sliding door, window opening, or other effective permanent barrier.

Section 680-6(a) applies to receptacles located near a permanently installed pool or fountain. It does not apply to direct-connected equipment. Section 680-6(a)(1) makes it clear that a single locking- and grounding-type receptacle used for a recirculation pump motor is permitted where it will be not less than 5 ft from the inside walls of the pool or fountain, is of the single-locking type, and is protected by a GFCI.

Section 680-6(a)(2) requires each permanently installed pool in a residential setting to have at least one receptacle located near the pool and be within easy reach. Distance limits are clearly indicated. The intent of this requirement is to permit ordinary appliances to be safely plugged in and used near the pool, but avoid the need for extension cords in the vicinity of the pool. The 10-ft minimum dimension was chosen since an appliance with a 6-ft cord could not be accidentally knocked into the pool.

Section 680-6(a)(3) requires all receptacles located within 20 ft of a pool or fountain to be protected by GFCIs. This requirement applies to pools located outdoors or indoors, permanently installed or storable, as well as for residential or commercial use. Since people in this area are normally subjected to dampness and moisture, the GFCI requirements within the 20-ft space is most reasonable.

Examples of receptacles meeting the requirements of Section 680-6(a) are shown in Figures 680.2 and 680.5. Figure 680.4 illustrates how to correctly make the distance measurements required by this section.

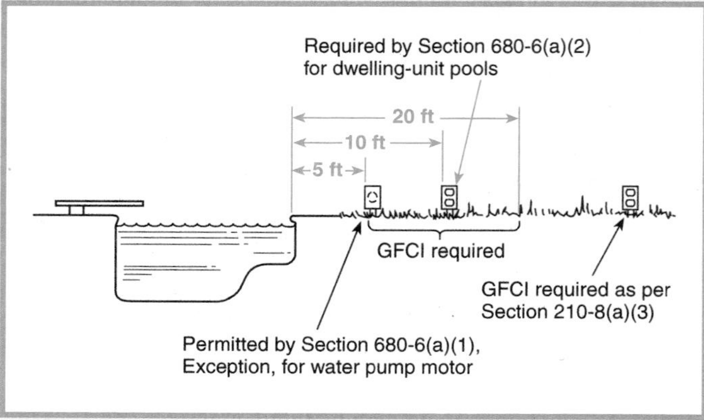

Figure 680.2 *For permanently installed pools at a dwelling unit(s), it is mandatory to install a 125-volt receptacle between 10 ft and 20 ft from the inside wall of the pool.*

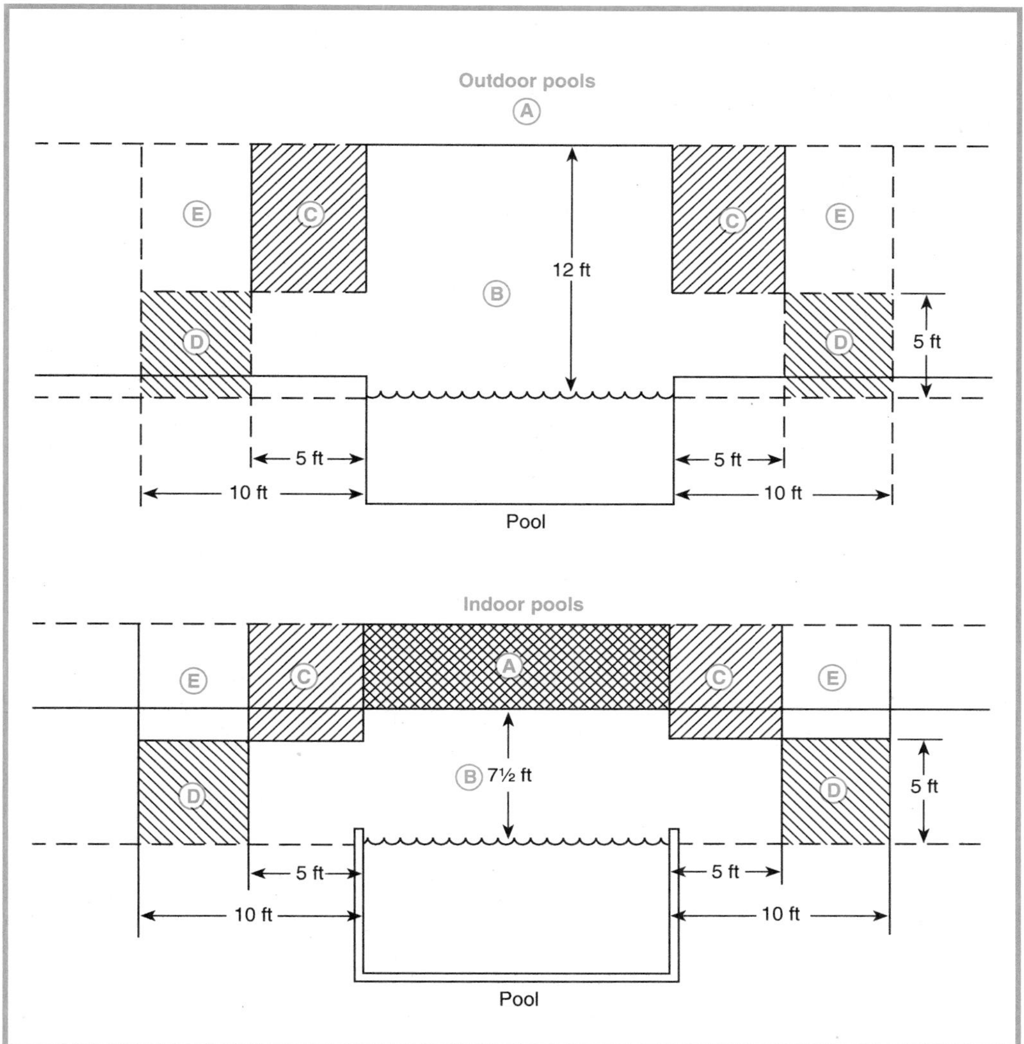

Figure 680.3 *Diagrams clarifying the limitations applicable to certain zones surrounding outdoor and indoor pools.*

See Figure 680.3 for diagrams that clarify the limitations applicable to certain zones surrounding outdoor and indoor pools.

Outdoor Pools

Area A. Lighting fixtures, lighting outlets, and ceiling-suspended (paddle) fans permitted above 12 ft. Ceiling-suspended (paddle) fans are common where the pool is located in a cabana-type structure. See Section 680-6(b)(1).

Area B. Lighting fixtures, lighting outlets, and ceiling-suspended (paddle) fans not permitted below 12 ft. See Section 680-6(b)(1).

Area C. Existing lighting fixtures and lighting outlets permitted in this space if rigidly attached to existing structure (GFCI required). See Section 680-6(b)(2).

Area D. Lighting fixtures and lighting outlets permitted if protected by a GFCI. See Section 680-6(b)(4).

Area E. Lighting fixtures and lighting outlets permitted if rigidly attached. See Section 680-6(b)(4).

Indoor Pools

Area A. Totally enclosed lighting fixtures protected by a GFCI and ceiling-suspended (paddle) fans protected by a GFCI permitted above $7^1/_2$ ft. See Section 680-6(b)(3).

Area B. Lighting fixtures, lighting outlets, and

ceiling-suspended (paddle) fans not permitted below 5 ft. See Section 680-6(b)(3).

Areas C, D, and E. Same as outdoor pools.

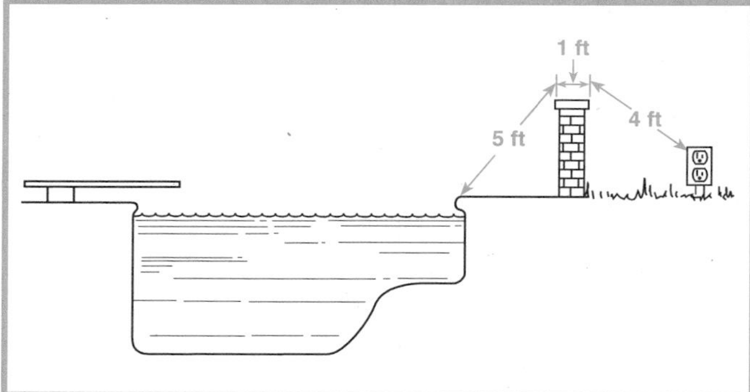

Figure 680.4 The shortest path a supply cord of an appliance connected to the receptacle could follow without piercing the wall, as described in Section 680-6(a)(3), FPN.

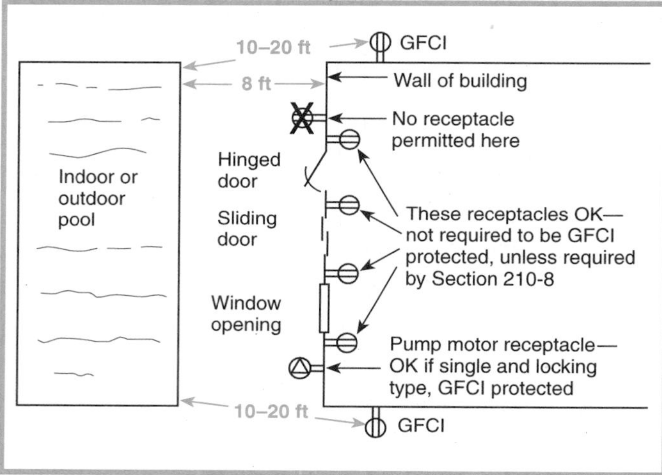

Figure 680.5 Acceptable receptacle locations within 20 ft of a swimming pool.

(b) Lighting Fixtures, Lighting Outlets, and Ceiling-Suspended (Paddle) Fans.

(1) In outdoor pool areas, lighting fixtures, lighting outlets, and ceiling-suspended (paddle) fans shall not be installed over the pool or over the area extending 5 ft (1.52 m) horizontally from the inside walls of a pool unless no part of the lighting fixture or ceiling-suspended (paddle) fan is less than 12 ft (3.66 m) above the maximum water level.

(2) Existing lighting fixtures and lighting outlets located less than 5 ft (1.52 m) measured horizontally from the inside walls of a pool shall be at least 5 ft (1.52 m) above the surface of the maximum water level, shall be rigidly attached to the existing structure, and shall be protected by a ground-fault circuit interrupter.

(3) In indoor pool areas, the limitations of Section 680-6(b)(1) shall not apply if all of the following conditions are complied with

(a) Fixtures are of a totally enclosed type,
(b) A ground-fault circuit interrupter is installed in the branch circuit supplying the fixture(s) or ceiling-suspended (paddle) fans, and
(c) The distance from the bottom of the fixture or ceiling-suspended (paddle) fan to the maximum water level is not less than 7 ft 6 in. (2.29 m).

(4) Lighting fixtures and lighting outlets installed in the area extending between 5 ft (1.52 m) and 10 ft (3.05 m) horizontally from the inside walls of a pool shall be protected by a ground-fault circuit interrupter unless installed 5 ft (1.52 m) above the maximum water level and rigidly attached to the structure adjacent to or enclosing the pool.

(5) Cord-connected lighting fixtures shall meet the same specifications as other cord- and plug-connected equipment as set forth in Section 680-7 where installed within 16 ft (4.88 m) of any point on the water surface, measured radially.

(c) Switching Devices. Switching devices on the property shall be located at least 5 ft (1.52 m) horizontally from the inside walls of a pool unless separated from the pool by a solid fence, wall, or other permanent barrier.

Panelboards, time clocks, pool light switches, and so on, where located not less than 5 ft horizontally from the inside walls of a pool without a solid fence, wall, or other permanent barrier, will be out of reach of persons who are in a pool, preventing contact and possible shock hazards.

(d) Motors in Other than Dwelling Units. Wiring supplying pool pump motors rated 15 and 20 amperes, 125 volt or 240 volt, single phase, whether by receptacle or direct connection, shall be provided with ground-fault circuit-interrupter protection for personnel.

New for the 1999 *NEC*, all single-phase, pool pump motor circuits, rated 15 and 20 amperes for nondwelling installations are required to have GFCI protection. This requirement applies to municipal swimming pools, hotel and motel swimming pools, water park pools, school swimming pools, and similar nondwelling pools. This requirement does not apply to swimming pools that are part of a dwelling occupancy such as a single-family home, multifamily residential occupancy, apartment complex or condominium complex. Refer to the definition of *dwelling* in Article 100.

680-7. Cord- and Plug-Connected Equipment. Fixed or stationary equipment rated 20 amperes or less, other than an underwater lighting fixture for a permanently installed pool, shall be permitted to be connected with a flexible cord to facilitate the removal or disconnection for maintenance or repair. For other than storable pools, the flexible cord shall not exceed 3 ft (914 mm) in length and shall have a copper equipment grounding conductor not smaller than No. 12 with a grounding-type attachment plug.

FPN: See Section 680-25(e) for connection with flexible cords.

In some climates, it is preferable to disconnect a permanent pool's filter pump during cold-weather months. A 3-ft cord is permitted to facilitate the removal of fixed or stationary equipment for maintenance and storage. The 3-ft (914-mm) cord limitation does not apply to cord- and plug-connected filter pumps used with storable-type pools (covered in Part C of Article 680), since these pumps are neither fixed nor stationary. Listed filter pumps for use with storable pools are considered portable and are permitted to be equipped with cords that are longer than 3 ft.

680-8. Overhead Conductor Clearances. The following parts of pools shall not be placed under existing service-drop conductors or any other open overhead wiring; nor shall such wiring be installed above the following:

(1) Pools and the area extending 10 ft (3.05 m) horizontally from the inside of the walls of the pool,
(2) Diving structure, or
(3) Observation stands, towers, or platforms unless the installations provide the clearances in Table 680-8

Table 680-8. Clearances

	Insulated Supply or Service Drop Cables, 0–750 Volts to Ground, Supported on and Cabled Together with an Effectively Grounded Bare Messenger or Effectively Grounded Neutral Conductor	All Other Supply or Service-Drop Conductors Voltage to Ground	
		0–15 kV	Greater than 15–50 kV
A Clearance in any direction to the water level, edge of water surface, base of diving platform, or permanently anchored raft	22 ft (6.7 m)	25 ft (7.62 m)	27 ft (8.23 m)
B Clearance in any direction to the diving platform or tower	14 ft (4.27 m)	17 ft (5.2 m)	18 ft (5.49 m)
C Horizontal limit of clearance measured from inside wall of the pool	This limit shall extend to the outer edge of the structures listed in (1) and (2) but not less than 10 ft (3.05 m).		

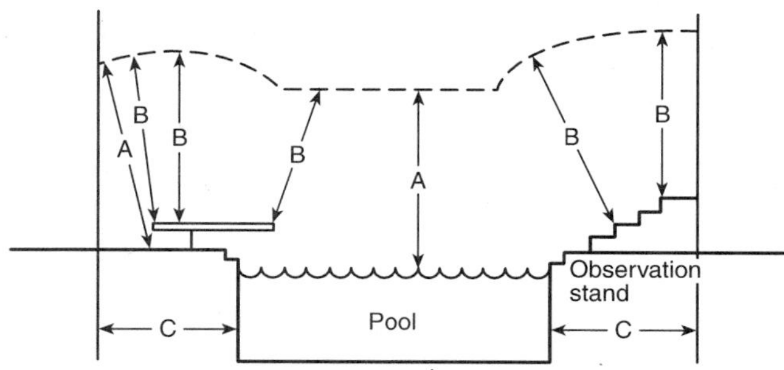

Figure 680-8.

Community antenna system coaxial cables complying with Article 820 and the supporting messengers shall be permitted at a height of not less than 10 ft (3.05 m) above swimming and wading pools, diving structures, and observation stands, towers, or platforms.

The clearances for conductors from pools and pool structures were increased for the 1999 *Code*. These changes harmonize the *NEC* with the *National Electrical Safety Code* (NESC).

FPN: See Sections 225-18 and 225-19 for clearances for conductors not covered by this section.

680-9. Electric Pool Water Heaters. All electric pool water heaters shall have the heating elements subdivided into loads not exceeding 48 amperes and protected at not more than 60 amperes.

The ampacity of the branch-circuit conductors and the rating or setting of overcurrent protective devices shall not be less than 125 percent of the total load of the nameplate rating.

680-10. Underground Wiring Location. Underground wiring shall not be permitted under the pool or within the area extending 5 ft (1.52 m) horizontally from the inside wall of the pool unless this wiring is necessary to supply pool equipment permitted by this article. Where space limitations prevent wiring from being routed 5 ft (1.52 m) or more from the pool, such wiring shall be permitted where installed in rigid metal conduit, intermediate metal conduit, or a nonmetallic raceway system. All metal conduit shall be corrosion resistant and suitable for the location. The minimum burial depth shall be as follows.

Wiring Method	Minimum Burial (in.)
Rigid metal conduit	6
Intermediate metal conduit	6
Nonmetallic raceways listed for direct burial without concrete encasement	18
Other approved raceways*	18

Note: For SI units, 1 in. = 25.4 mm.

*Raceways approved for burial only where concrete encased shall require a concrete envelope not less than 2 in. (50.8 mm) thick.

680-11. Equipment Rooms and Pits. Electric equipment shall not be installed in rooms or pits that do not have adequate drainage to prevent water accumulation during normal operation or filter maintenance.

680-12. Disconnecting Means. A disconnecting means shall be provided and be accessible, located within sight from all pools, spas, and hot tub equipment, and shall be located at least 5 ft (1.52 m) from the inside walls of the pool, spa, or hot tub.

A disconnecting means is required to be installed within sight of the pool, spa, and hot tub equipment to allow service personnel to disconnect the power while servicing these units. This requirement ensures that a disconnect is available to workers servicing pool, spa, and hot tub equipment such as motors, heaters, and lighting fixtures. See Figure 680.6.

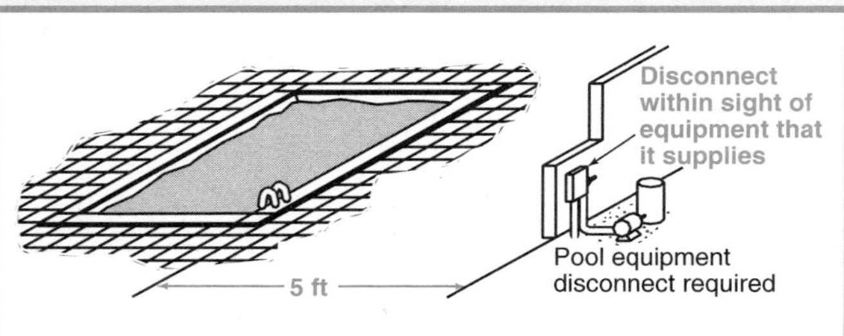

Figure 680.6 Disconnect for pool, spa, and hot tub equipment located within sight and within 5 ft of the pool.

B. Permanently Installed Pools

680-20. Underwater Lighting Fixtures. Paragraphs (a) through (d) apply to all lighting fixtures installed below the normal water level of the pool.

(a) General.

(1) The design of an underwater lighting fixture supplied from a branch circuit either directly or by way of a transformer meeting the requirements of Section 680-5(a) shall be such that, where the fixture is properly installed without a ground-fault circuit interrupter, there is no shock hazard with any likely combination of fault conditions during normal use (not relamping).

In addition, a ground-fault circuit interrupter shall be installed in the branch circuit supplying fixtures operating at more than 15 volts, so that there is no shock hazard during relamping. The installation of the ground-fault circuit interrupter shall be such that there is no shock hazard with any likely fault-condition combination that involves a person in a conductive path from any ungrounded part of the branch circuit or the fixture to ground.

Compliance with this requirement shall be obtained by the use of a listed underwater lighting fixture and by installation of a listed ground-fault circuit interrupter in the branch circuit.

(2) No lighting fixtures shall be installed for operation on supply circuits over 150 volts between conductors.

(3) Lighting fixtures mounted in walls shall be installed with the top of the fixture lens at least 18 in. (457 mm) below the normal water level of the pool, unless the lighting fixture is listed and identified for use at a depth of not less than 4 in. (102 mm) below the normal water level of the pool.

The reason for the 18-in. minimum submergence requirement is to reduce the likelihood that a person in the water, hanging onto the side of the pool directly over the fixture, will have his or her chest cavity in line with the fixture. This section covers fixtures that have been investigated and found acceptable for use where the chest cavity may be directly in front of the fixture. The highest level of leakage current in the pool coming from a wet-niche fixture with a broken lens and bulb is found directly in front of the fixture.

(4) A lighting fixture facing upward shall have the lens adequately guarded to prevent contact by any person.

(5) Fixtures that depend on submersion for safe operation shall be inherently protected against the hazards of overheating when not submerged.

Fixtures that depend on submersion for safe operation are required to be inherently protected against the hazards of overheating when not submerged, such as during a relamping process. Protection against overheating is required to be built into the fixture or to be a part of it. A remotely located low-water cutoff switch does not provide the intended protection.

(b) Wet-Niche Fixtures.

(1) Forming shells shall be installed for the mounting of all wet-niche underwater fixtures and shall be equipped with provisions for conduit entries.

Conduit shall extend from the forming shell to a suitable junction box or other enclosure located as provided in Section 680-21. Conduit shall be rigid metal, intermediate metal, liquidtight flexible nonmetallic, or rigid nonmetallic.

Metal conduit shall be of brass or other approved corrosion-resistant metal.

Where a nonmetallic conduit is used, a No. 8 insulated copper conductor shall be installed in this conduit with provisions for terminating in the forming shell, junction box or transformer enclosure, or ground-fault circuit-interrupter enclosure unless a listed low-voltage lighting system is used, not requiring grounding. The termination of the No. 8 conductor in the forming shell shall be covered with, or encapsulated in, a listed potting compound to protect such connection from the possible deteriorating effect of pool water. Metal parts of the fixture and forming shell in contact with the pool water shall be of brass or other approved corrosion-resistant metal.

Where rigid nonmetallic conduit or liquidtight flexible nonmetallic conduit is used between a forming shell for a wet-niche fixture and a junction box or other enclosure, a No. 8 insulated copper bonding conductor is required to be installed within this conduit to provide electrical continuity between the forming shell and the junction box or other enclosure. The conduit should be sized large enough to enclose both the No. 8 insulated copper bonding conductor and the approved flexible cord that supplies the wet-niche fixture, to facilitate easy withdrawal and insertion of the bonding conductor and the cord. However, the 1999 *Code* was revised to exempt low-voltage lighting systems from this requirement.

(2) The end of the flexible-cord jacket and the flexible-cord conductor terminations within a fixture shall be covered with, or encapsulated in, a suitable potting compound to prevent the entry of water into the fixture through the cord or its conductors. In addition, the grounding connection within a fixture shall be similarly treated to protect such connection from the deteriorating effect of pool water in the event of water entry into the fixture.

(3) The fixture shall be bonded to and secured to the forming shell by a positive locking device that ensures a low-resistance contact and requires a tool to remove the fixture from the forming shell. Bonding is not required for fixtures listed for the application, having no noncurrent-carrying metal parts.

(c) Dry-Niche Fixtures. A dry-niche lighting fixture shall be provided with a provision for drainage of water and a means for accommodating one equipment grounding conductor for each conduit entry.

Approved rigid metal conduit, intermediate metal conduit, liquidtight flexible nonmetallic conduit, or rigid nonmetallic conduit shall be installed from the fixture to the service equipment or panelboard. Where installed on buildings, electrical metallic tubing shall be permitted to be used to protect conductors. Where installed within buildings, electrical nonmetallic tubing or electrical metallic tubing shall be permitted to be used to protect conductors. A junction box shall not be required but, if used, shall not be required to be elevated or located as specified in Section 680-21(a)(5), if the fixture is specifically identified for the purpose.

Underwater lighting fixtures of either dry-niche, no-niche, or wet-niche types operating at more than 15 volts require ground-fault circuit-interrupter protection. See the commentary following Section 680-5(b).

Branch-circuit conductors for dry-niche fixtures are required to be installed in approved rigid metal conduit, intermediate metal conduit, or rigid nonmetallic conduit from the fixture to a panelboard or the service equipment. Branch-circuit conductors for wet-niche fixtures leaving the pool junction box are required to be enclosed in rigid metal conduit, intermediate metal conduit, liquidtight flexible nonmetallic conduit, or rigid nonmetallic conduit, except where located in or on buildings, where the conductors are permitted to be installed in electrical metallic tubing or electrical nonmetallic tubing. However, unlike wet-niche fixtures, a junction box is not required for dry-niche fixtures. If one is used, it is not required to be elevated or located as specified in Section 680-21(a)(5). See Figures 680.1 and 680.5.

(d) No-Niche Fixtures. A no-niche fixture shall be as follows:

(1) Listed for the purpose
(2) Installed in accordance with the requirements of Section 680-20(b)

Where connection to a forming shell is specified, the connection shall be to the mounting bracket.

680-21. Junction Boxes and Enclosures for Transformers or Ground-Fault Circuit Interrupters.

(a) Junction Boxes. A junction box connected to a conduit that extends directly to a forming shell or mounting bracket of a no-niche fixture shall be as follows.

(1) Listed and labeled for the purpose.

Section 680-21(a)(1) was added to the 1999 *Code* to ensure that the integral grounding terminals necessary for grounding and bonding of the underwater fixtures will be available. Boxes used for the supply wiring to wet-niche and no-niche underwater lighting fixtures must be listed for the purpose by a recognized testing laboratory. A box that is listed, but not specifically for use with swimming pools, does not provide the correct number of integral grounding and bonding terminals. See Figure 680.7 for an illustration of the installation requirements for a wet-niche forming shell and flush deck box. See Figure 680.1 for surface deck boxes.

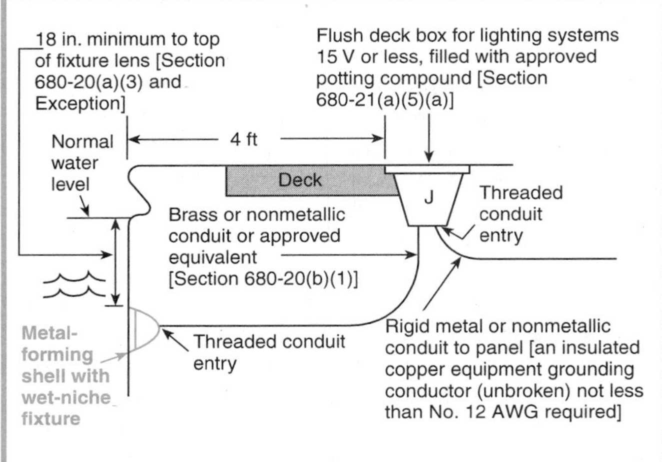

Figure 680.7 *Installation requirements for a wet-niche forming shell and flush deck box installed in accordance with Section 680-21(a).*

(2) Equipped with threaded entries or hubs or a nonmetallic hub listed for the purpose.

(3) Of copper, brass, suitable plastic, or other approved corrosion-resistant material.

(4) Provided with electrical continuity between every connected metal conduit and the grounding terminals by means of copper, brass, or other approved corrosion-resistant metal that is integral with the box.

(5) Located not less than 4 in. (102 mm), measured from the inside of the bottom of the box, above the ground level, or pool deck, or not less than 8 in. (203 mm) above the maximum pool water level, whichever provides the greater elevation, and located not less than 4 ft (1.22 m) from the inside wall of the pool, unless separated from the pool by a solid fence, wall, or other permanent barrier. If used on a lighting system operating at 15 volts or less, a flush deck box shall be permitted provided

(a) An approved potting compound is used to fill the box to prevent the entrance of moisture, and
(b) The flush deck box is located not less than 4 ft (1.22 m) from the inside wall of the pool.

(b) Other Enclosures. An enclosure for a transformer, ground-fault circuit interrupter, or a similar device connected to a conduit that extends directly to a forming shell or mounting bracket of a no-niche fixture shall be as follows:

(1) Listed and labeled for the purpose
(2) Equipped with threaded entries or hubs or a nonmetallic hub listed for the purpose
(3) Provided with an approved seal, such as duct seal at the conduit connection, that prevents circulation of air between the conduit and the enclosures
(4) Provided with electrical continuity between every connected metal conduit and the grounding terminals by means of copper, brass, or other approved corrosion-resistant metal that is integral with the enclosures
(5) Located not less than 4 in. (102 mm), measured from the inside bottom of the enclosure, above the ground level or pool deck, or not less than 8 in. (203 mm) above the maximum pool water level, whichever provides the greater elevation, and located not less than 4 ft (1.22 m) from the inside wall of the pool, unless separated from the pool by a solid fence, wall, or other permanent barrier

(c) Protection. Junction boxes and enclosures mounted above the grade of the finished walkway around the pool shall not be located in the walkway unless afforded additional protection, such as by location under diving boards, adjacent to fixed structures, and the like.

(d) Grounding Terminals. Junction boxes, transformer enclosures, and ground-fault circuit-interrupter enclosures connected to a conduit that extends directly to a forming shell or mounting bracket of a no-niche fixture shall be provided with a number of grounding terminals that shall be at least one more than the number of conduit entries.

(e) Strain Relief. The termination of a flexible cord of an underwater lighting fixture within a junction box, transformer enclosure, ground-fault circuit interrupter, or other enclosure shall be provided with a strain relief.

680-22. Bonding. It shall not be the intent of this section to require that the No. 8 or larger solid copper bonding conductor be extended or attached to any remote panelboard, service equipment, or any electrode, but only that it shall be employed to eliminate voltage gradients in the pool area as prescribed.

The primary purpose of bonding is to ensure that voltage gradients in the pool area are eliminated. The revised wording of Section 680-22 in the 1999 *Code* makes it clear that the No. 8 conductor is only for elimination of the voltage gradient in the pool area and is not required to provide a path for fault current that may occur as a result of electrical equipment failure. See Figure 680.8.

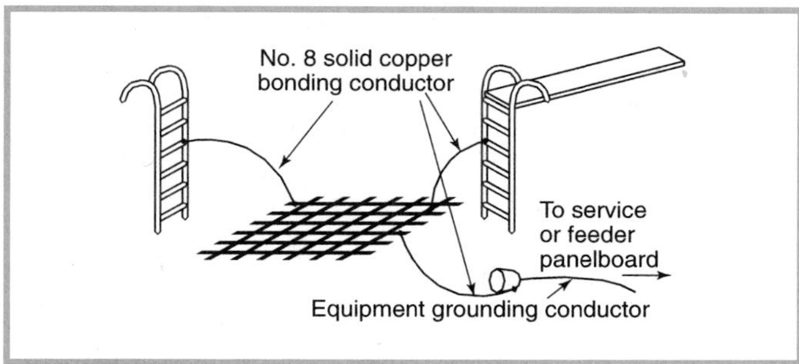

Figure 680.8 *Bonding at swimming pools.*

(a) Bonded Parts. The following parts shall be bonded together.

(1) All metallic parts of the pool structure, including the reinforcing metal of the pool shell, coping stones, and deck. The usual steel tie wires shall be considered suitable for bonding the reinforcing steel together, and welding or special clamping shall not be required. These tie wires shall be made tight. Where reinforcing steel is effectively insulated by a listed encapsulating nonconductive compound, at the time of manufacture and installation, it shall not be required to be bonded.

The 1999 *Code* recognizes that encapsulated reinforcing steel may not provide the conductivity necessary to establish the required common bonding grid. A common bonding grid will not be formed if the steel is effectively encapsulated by a listed compound during installation and manufacturing. Therefore, a bonding connection is not required for this type of application. See Figure 680.9.

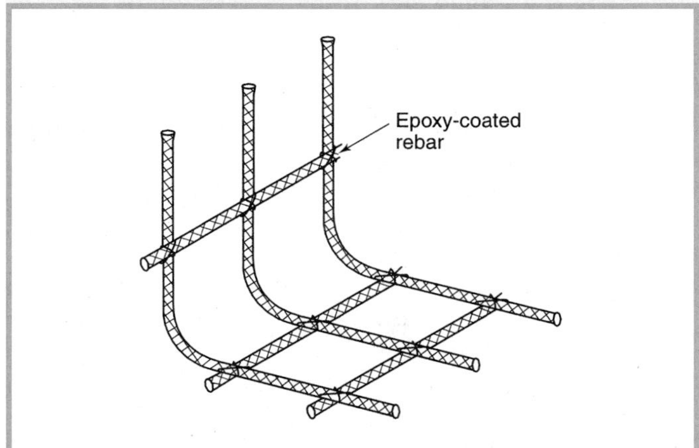

Figure 680.9 *Bonding not required for epoxy-coated rebar.*

The structural reinforcing steel serves as the common bonding grid, to which all metal appurtenances associated with the pool are connected. Safety-rope hooks are not required to be bonded, as specified in Section 680-22(a). The flush deck box meets the provisions of Section 680-21(a).

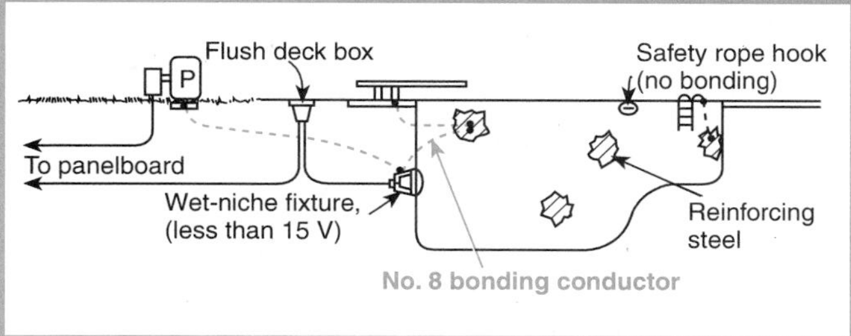

Figure 680.10 *A poured-concrete pool.*

(2) All forming shells and mounting brackets of a no-niche fixture unless a listed low-voltage lighting system is used, not requiring bonding.

(3) All metal fittings within or attached to the pool structure. Isolated parts that are not over 4 in. (102 mm) in any dimension and do not penetrate into the pool structure more than 1 in. (25.4 mm) shall not require bonding.

(4) Metal parts of electrical equipment associated with the pool water circulating system, including pump motors and metal parts of equipment associated with pool covers, including electric motors. Metal parts of listed equipment incorporating an approved system of double insulation and providing a means for grounding internal nonaccessible, noncurrent-carrying metal parts shall not be bonded.

(5) Metal-sheathed cables and raceways, metal piping, and all fixed metal parts that are within 5 ft (1.52 m) horizontally of the inside walls of the pool, and within 12 ft (3.66 m) above the maximum water level of the pool, or any observation stands, towers, or platforms, or from any diving structures, and that are not separated from the pool by a permanent barrier.

Section 680-22(a)(5) would also include, for example, a metal door frame or metal window frame.

(b) Common Bonding Grid. The parts specified in (a) shall be connected to a common bonding grid with a solid copper conductor, insulated, covered, or bare, not smaller than No. 8. Connection shall be made by exothermic welding or by pressure connectors or clamps that are labeled as being suitable for the purpose and are of the following material: stainless steel, brass, copper, or copper alloy. The common bonding grid shall be permitted to be any of the following:

Section 680-22(b) in the 1999 *Code* was changed to recognize exothermic welding as a means of connecting to a common bonding grid. This change also recognizes pressure connectors and clamps that are

specifically listed for the purpose. Connections in pool areas must be suitable for wet conditions and high levels of chlorine. See Figure 680.11 for an illustration of bonding connections.

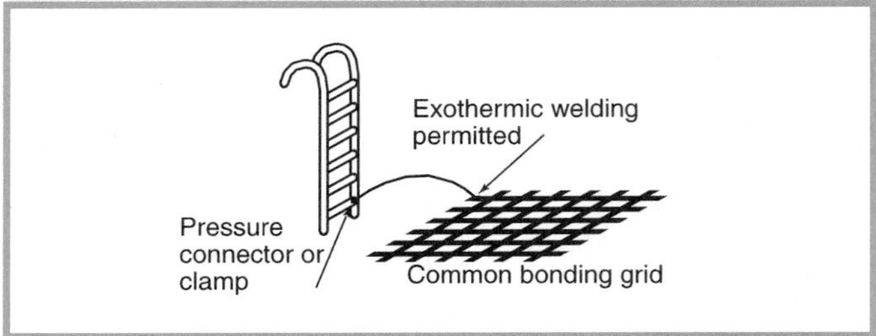

Figure 680.11 *Bonding connections at swimming pools.*

(1) The structural reinforcing steel of a concrete pool where the reinforcing rods are bonded together by the usual steel tie wires or the equivalent
(2) The wall of a bolted or welded metal pool
(3) A solid copper conductor, insulated, covered, or bare, not smaller than No. 8
(4) Rigid metal conduit or intermediate metal conduit of brass or other identified corrosion-resistant metal conduit

Section 680-22(b)(4) was added for the 1999 *Code* to recognize that metal conduits provide an effective bonding path. Corrosion-resistant conduits can be used as part of the common bonding grid, particularly where installed above ground and in accessible locations. See Figure 680.12.

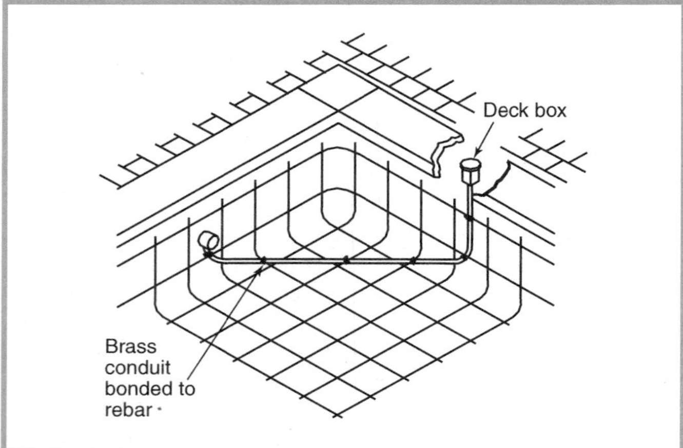

Figure 680.12 *Corrosion-resistant conduits used as part of the common bonding grid.*

Structural reinforcing steel or the walls of bolted or welded metal pool structures shall be permitted as a common bonding grid for nonelectrical parts where connections can be made in accordance with Section 250-8.

Section 250-8 requires that the connection be made by exothermic welding, listed clamps, connectors, or other listed means. If a pool is built with structural reinforcing steel (rebar) encapsulated by a nonconductive coating, such as epoxy or plastic, the rebar is not required to be bonded to the other metal parts of the pool system. See the commentary following 680-22(a)(1).

(c) Pool Water Heaters. For pool water heaters rated at more than 50 amperes that have specific instructions regarding bonding and grounding, only those parts designated to be bonded shall be bonded, and only those parts designated to be grounded shall be grounded.

It is important to know the difference between the terms *bonding* and *grounding* as they apply to Article 680. As defined in Article 100, *bonding* is the permanent joining of metallic parts to form an electrically conductive path that will ensure electrical continuity and the capacity to conduct safely any current likely to be imposed. See the commentary following Section 680-25(f) for further information.

Section 680-22(c) identifies the metal parts required to be bonded, including all metal parts of electrical equipment associated with the pool water circulating system, all metal parts of the pool structure, and all fixed metal parts, which include conduit and piping, metal door frames and metal window frames, and so on, within 5 ft of the inside walls of the pool and not separated by a permanent barrier. It should be understood that the bonding of these parts does not mean that they are required to be connected to each other; it means that they are required to be connected to a common bonding grid with an insulated, covered, or bare solid copper conductor not smaller than No. 8. See Figure 680.13. Connections are required to be made by exothermic welds or by listed pressure connectors, clamps, or other listed means, in accordance with Section 250-8.

The reason for connecting metal parts (ladders, handrails, water-circulating equipment, forming shells, diving boards, etc.) to a common bonding grid (pool reinforcing steel, pool metal wall, or a No. 8 solid conductor) is to ensure that all such metal parts will be at the same electrical potential. This grid reduces any possible shock hazard created by stray currents in the ground or piping connected to the swimming pool. Stray currents can also exist in nonmetallic piping because of the low resistivity of chlorinated water.

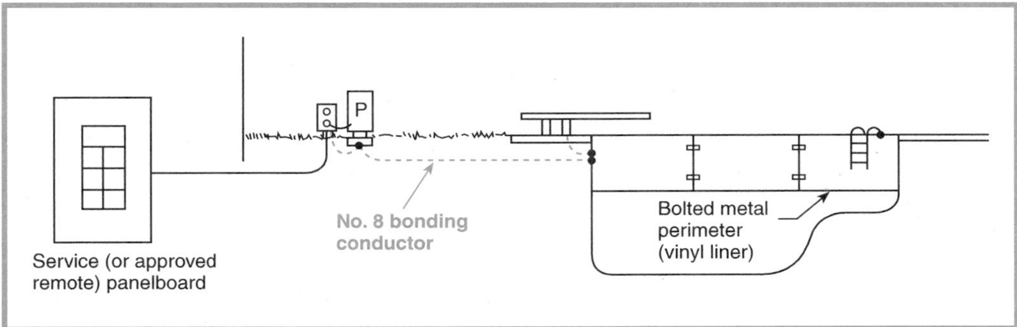

Figure 680.13 *A metal-perimeter (e.g., steel or aluminum) pool with bolted or welded sections. The metal perimeter serves as the common bonding grid to which the metal ladder, metal diving board, and pump motor are connected.*

Since corrosion is normally associated with the wet conditions of swimming pool areas, wiring and connections should be checked periodically, especially bonding connections between the No. 8 copper conductor and, for instance, an aluminum (or other dissimilar metal) ladder.

680-23. Underwater Audio Equipment. All underwater audio equipment shall be identified for the purpose.

(a) Speakers. Each speaker shall be mounted in an approved metal forming shell, the front of which is enclosed by a captive metal screen, or equivalent, that is bonded to and secured to the forming shell by a positive locking device that ensures a low-resistance contact and requires a tool to open for installation or servicing of the speaker. The forming shell shall be installed in a recess in the wall or floor of the pool.

(b) Wiring Methods. Rigid metal conduit or intermediate metal conduit of brass or other identified corrosion-resistant metal or rigid nonmetallic conduit shall extend from the forming shell to a suitable junction box or other enclosure as provided in Section 680-21. Where rigid nonmetallic conduit is used, a No. 8 insulated copper conductor shall be installed in this conduit with provisions for terminating in the forming shell and the junction box. The termination of the No. 8 conductor in the forming shell shall be covered with, or encapsulated in, a suitable potting compound to protect such connection from the possible deteriorating effect of pool water.

(c) Forming Shell and Metal Screen. The forming shell and metal screen shall be of brass or other approved corrosion-resistant metal.

680-24. Grounding. The following equipment shall be grounded:

(1) Wet-niche and no-niche underwater lighting fixtures, other than those low-voltage systems listed for the application without a grounding conductor
(2) Dry-niche underwater lighting fixtures
(3) All electrical equipment located within 5 ft (1.52 m) of the inside wall of the pool
(4) All electrical equipment associated with the recirculating system of the pool
(5) Junction boxes
(6) Transformer enclosures
(7) Ground-fault circuit interrupters
(8) Panelboards that are not part of the service equipment and that supply any electrical equipment associated with the pool

680-25. Methods of Grounding.

(a) General. The following provisions shall apply to the grounding of underwater lighting fixtures, junction boxes, metal transformer enclosures, panelboards, motors, and other electrical enclosures and equipment.

(b) Pool Lighting Fixtures and Related Equipment.

(1) Wet-niche, dry-niche, or no-niche lighting fixtures shall be connected to an equipment grounding conductor sized in accordance with Table 250-122 but not smaller than No. 12.

Exception: An equipment grounding conductor between the wiring chamber of the secondary winding of a transformer and a junction box shall be sized in accordance with the overcurrent device in this circuit.

(2) The equipment grounding conductor shall be an insulated copper conductor and shall be installed with the circuit conductors in rigid metal conduit, intermediate metal conduit, liquidtight flexible nonmetallic conduit, or rigid nonmetallic conduit.

(3) Where installed on buildings, electrical metallic tubing shall be permitted to be used to protect conductors. Where installed within buildings, electrical nonmetallic tubing or electrical metallic tubing shall be permitted to be used to protect conductors.

For circuits supplying a pool, electrical nonmetallic tubing is now a permissible wiring method for wiring located within a building.

FPN: For requirements of electrical nonmetallic tubing, see Article 331.

Exception: Where connecting to transformers for pool lights, liquidtight flexible metal conduit or liquidtight flexible nonmetallic conduit shall be permitted to be used when installed in accordance with Article 351 and does not exceed 6 ft (1.83 m) for any one length or 10 ft (3.05 m) of total length used.

(4) The junction box, transformer enclosure, or other enclosure in the supply circuit to a wet-niche or no-niche lighting fixture and the field-wiring chamber of a dry-niche lighting fixture shall be grounded to the equipment grounding terminal of the panelboard. This terminal shall be directly connected to the panelboard enclosure. The equipment grounding conductor shall be installed without joint or splice except as permitted in (a) and (b).

(a) Where more than one underwater lighting fixture is supplied by the same branch circuit, the equipment grounding conductor, installed between the junction boxes, transformer enclosures, or other enclosures in the supply circuit to wet-niche fixtures, or between the field-wiring compartments of dry-niche fixtures, shall be permitted to be terminated on grounding terminals.

(b) Where the underwater lighting fixture is supplied from a transformer, ground-fault circuit interrupter, clock-operated switch, or a manual snap switch that is located between the panelboard and a junction box connected to the conduit that extends directly to the underwater lighting fixture, the equipment grounding conductor shall be permitted to terminate on grounding terminals on the transformer, ground-fault circuit interrupter, clock-operated switch enclosure, or an outlet box used to enclose a snap switch.

See the commentary following Section 680-5(a).

(5) Wet-niche or no-niche lighting fixtures that are supplied by a flexible cord or cable shall have all exposed noncurrent-carrying metal parts grounded by an insulated copper equipment grounding conductor that is an integral part of the cord or cable. This grounding conductor shall be connected to a grounding terminal in the supply junction box, transformer enclosure, or other enclosure. The grounding conductor shall not be smaller than the supply conductors and not smaller than No. 16.

(c) Motors. Pool-associated motors shall be connected to an equipment grounding conductor sized in accordance with Table 250-122 but not smaller than No. 12. It shall be an insulated copper conductor and shall be installed with the circuit conductors in rigid metal conduit, intermediate metal conduit, rigid nonmetallic conduit, or Type MC cable listed for the application. Where installed on or within buildings, electrical metallic tubing shall be permitted to be used to protect the conductors.

Where necessary to employ flexible connections at or adjacent to the motor, liquidtight flexible metal or nonmetallic conduit with approved fittings shall be permitted.

In the interior of a one-family dwelling or in the interior of another building or structure associated with a one-family dwelling, any of the wiring methods recognized in Chapter 3 of this *Code* that contain a copper equipment grounding conductor that is insulated or covered by the outer sheath of the wiring method and is not smaller than No. 12 shall be permitted to be used for the connection of pool-associated motors. Flexible cord shall be permitted in accordance with Section 680-7.

Type MC cable listed for the application has been added to the list of wiring methods permitted for swimming pool motor circuits.

Section 680-25(c) permits wiring methods such as nonmetallic-sheathed cable, MC cable, and armored cable in the interior of one-family dwelling units. It does not permit less than a No. 12 copper conductor for the equipment grounding conductor. The equipment grounding conductor must be an insulated conductor or a bare conductor within an overall sheath, such as the bare conductor in nonmetallic-sheathed cable. The use of conduit or Type MC cable would mandate the use of an insulated equipment grounding conductor. See Figure 680.7.

(d) Panelboards. A panelboard and, where installed, a disconnecting means, that are not part of the service equipment or source of a separately derived system, shall have an equipment grounding conductor installed between its grounding terminal and the grounding terminal of the applicable service equipment or source of a separately derived system. This conductor shall be sized in accordance with Table 250-122 but not smaller than No. 12. On separately derived systems, this conductor shall be sized in accordance with Table 250-66 but not smaller than No. 8. It shall be an insulated conductor and shall be installed with the feeder conductors in rigid metal conduit, intermediate metal conduit, liquidtight flexible nonmetallic conduit, or rigid nonmetallic conduit. Electrical metallic tubing shall be permitted to be used to protect conductors where installed on or within the building in accordance with Article 348. Electrical nonmetallic tubing shall be permitted to be used to enclose the conductors where installed within the building in accordance with Article 331. The equipment grounding conductor shall be connected to an equipment grounding terminal of the panelboard and, where installed, to the enclosure for a disconnecting means.

(1) Where the equipment grounding conductor between an existing remote panelboard and the service equipment is connected by means of a flexible metal conduit or an approved cable assembly with an insulated or covered equipment grounding conductor, the conduits listed above shall not be required.

(2) A panelboard at a separate building shall be permitted to supply swimming pool equipment if the feeder meets the requirements for grounding in Section 250-32. Where installed, an equipment grounding conductor shall be an insulated conductor.

The insulated equipment grounding conductor may be aluminum or copper and may be installed in a raceway. It should be understood that for an existing remote panelboard, Section 680-25(d) permits an approved cable assembly with an insulated or covered aluminum or copper equipment grounding conductor. See Figure 680.14.

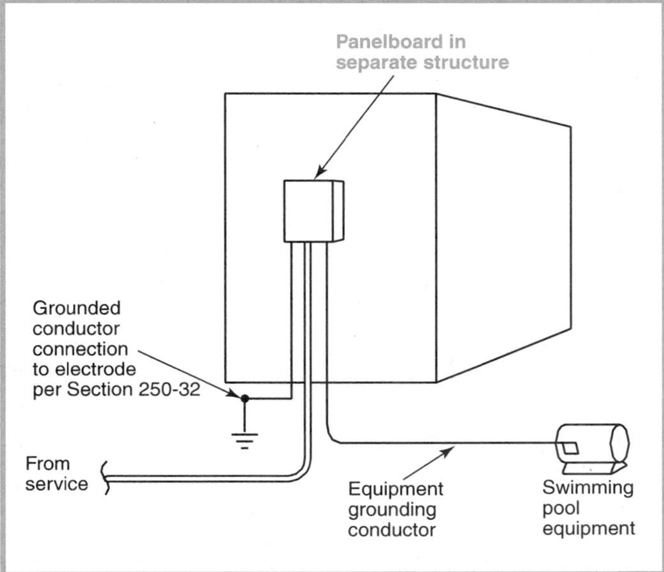

Figure 680.14 *A remote panelboard supplying pool equipment can have its supply from any wiring method permitted by Chapter 3, provided the equipment grounding conductor is No. 12 copper or larger.*

Section 680-25(d) recognizes pool equipment supplied by a separately derived system. If a remote panelboard supplying a pool is supplied by a separately derived system, the rules covering the grounding conductor only apply to the feeder between the separately derived system and the panelboard, and not all the way back to the service, which might be high voltage.

The general rule in Section 680-25(d) requires an equipment grounding conductor to be installed between a panelboard serving swimming pool equipment and the service or the source of a separately derived system. Section 680-25(d)(2) was added for the 1999 *Code* to allow pool equipment to be supplied from a panelboard in a separate building when no equipment grounding conductor is extended from the service (or derived system) to the panelboard. Grounding at the separate building must meet the requirements of Section 250-32. See Figure 680.15.

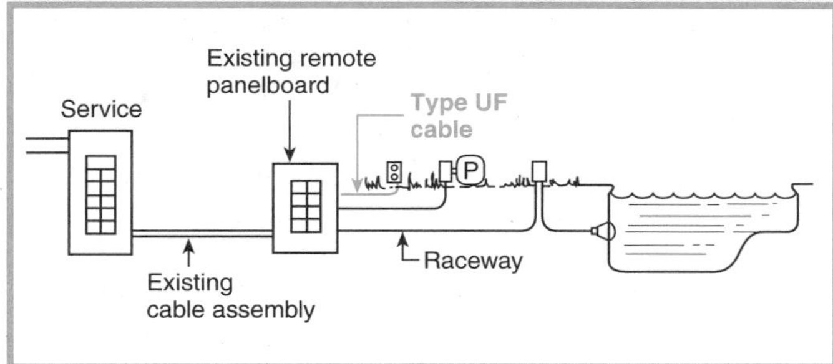

Figure 680.15 *Pool equipment supplied from a panelboard in a separate building as per Section 680-25(d)(2).*

Section 680-25(d) was also revised for the 1999 *Code* to require a minimum No. 8 equipment grounding conductor for separately derived systems. Section 250-30(a)(1) references Section 250-102(c) for the sizing of bonding jumpers for separately derived systems. Section 250-102(c) references Table 250-66, which specifies a minimum size No. 8 conductor.

(e) Cord-Connected Equipment. Where fixed or stationary equipment is connected with a flexible cord to facilitate removal or disconnection for maintenance, repair, or storage as provided in Section 680-7, the equipment grounding conductors shall be connected to a fixed metal part of the assembly. The removable part shall be mounted on or bonded to the fixed metal part.

(f) Other Equipment. Other electrical equipment shall be grounded in accordance with Article 250 and connected by wiring methods of Chapter 3.

Equipment other than underwater lighting fixtures and pool-associated motors is required to be connected by the wiring methods of Chapter 3 and grounded in accordance with Article 250. For example, an outdoor receptacle is permitted to be wired with Type UF cable containing an insulated or bare conductor for equipment grounding purposes. Circuits for pools may be derived from an existing remote panelboard supplied by an approved cable assembly, as specified in Section 680-25(d). Section 680-25(f) permits Type UF cable to be used for the receptacle required by Section 680-6(a)(2) and other equipment, but circuit conductors for underwater lighting fixtures are required to be run in raceways. Circuit conductors for pool-associated motors other than flexible cord, as permitted by Section 680-7, are required to be run in raceway. An exception is the interior of one-family dwelling units, where any wiring method permitted by Chapter 3 is acceptable if the equipment grounding conductor is at least No. 12 copper. See Figure 680.14.

Sections 680-24 and 680-25 provide requirements for equipment grounding. These sections require equipment grounding conductors to be connected to noncurrent-carrying metal parts of the specified equipment. These equipment grounding conductors are required to be run with the circuit conductors in rigid metal conduit, intermediate conduit, listed MC cable for motors, or rigid nonmetallic conduit (electrical nonmetallic tubing is permitted inside buildings), and they must be terminated at the grounding terminal bus of the service panelboard, source of the separately derived system, or subpanel. This equipment grounding conductor provides a path of low impedance that limits the voltage to ground and facilitates operation of the circuit overcurrent protective device(s). The equipment grounding conductor is required to be an insulated copper conductor not smaller than No. 12. If installed on or in buildings, the conductor may be run with the circuit conductors in EMT. See also Section 680-25(c).

The requirements in Sections 680-24 and 680-25 are in addition to the bonding requirements in Section 680-22. The intent of the bonding requirements is to establish an equipotential plane to limit the voltage between all noncurrent-carrying parts of electrical and nonelectrical equipment in the pool area.

Bonding conductors may be insulated, covered, or bare and are required to be No. 8 solid copper or larger. They may be direct buried, and, if connected to metal parts of the pool structure or metal parts of electrical equipment, they may be externally clamped or attached and are not required to be accessible. All these parts form a common bonding grid that establishes an equipotential grounding system, and they do not have to be run to the equipment grounding terminals of panelboards or service equipment.

680-26. Electrically Operated Pool Covers.

(a) Motors and Controllers. The electric motors, controllers, and wiring shall be located at least 5 ft (1.52 m) from the inside wall of the pool unless separated from the pool by a wall, cover, or other permanent barrier. Electric motors installed below grade level shall be of the totally enclosed type.

FPN No. 1: For cabinets installed in damp and wet locations, see Section 373-2(a).

FPN No. 2: For switches or circuit breakers installed in wet locations, see Section 380-4.

FPN No. 3: For protection against liquids, see Section 430-11.

(b) Wiring Methods. The electric motor and controller shall be connected to a circuit protected by a ground-fault circuit interrupter.

680-27. Deck Area Heating. The provisions of this section shall apply to all pool deck areas, including a covered pool, where electrically operated comfort heating units are installed within 20 ft (6.1 m) of the inside wall of the pool.

(a) Unit Heaters. Unit heaters shall be rigidly mounted to the structure and shall be of the totally enclosed or guarded types. Unit heaters shall not be mounted over the pool or within the area extending 5 ft (1.52 m) horizontally from the inside walls of a pool.

(b) Permanently Wired Radiant Heaters. Radiant electric heaters shall be suitably guarded and securely fastened to their mounting device(s). Heaters shall not be installed over a pool or within the area extending 5 ft (1.52 m) horizontally from the inside walls of the pool and shall be mounted at least 12 ft (3.66 m) vertically above the pool deck unless otherwise approved.

(c) Radiant Heating Cables Not Permitted. Radiant heating cables embedded in or below the deck shall not be permitted.

Only unit heaters and permanently connected radiant heaters are permitted in the area that extends 5 ft to 20 ft horizontally from the inside walls of a pool. Radiant heat cables embedded in the deck are not permitted.

680-28. Double Insulated Pool Pumps. A permanently installed pool shall be permitted to be provided with listed cord- and plug-connected pool pumps incorporating an approved system of double insulation that provides a means for grounding only the internal and nonaccessible, noncurrent-carrying metal parts of the pump.

Cord- and plug-connected double-insulated swimming pool filter pumps have been used with aboveground and some storable pools, regardless of the pool's size, for many years without any known field-related problems. The internal metal parts of a swimming pool filter pump incorporating a system of double insulation are not bonded, since the act of bonding compromises the double-insulation system.

C. Storable Pools

680-30. Pumps. A cord-connected pool filter pump shall incorporate an approved system of double insulation or its equivalent and shall be provided with means for grounding only the internal and nonaccessible noncurrent-carrying metal parts of the appliance.

The means for grounding shall be an equipment grounding conductor run with the power-supply conductors in the flexible cord that is properly terminated in a grounding-type attachment plug having a fixed grounding contact member.

680-31. Ground-Fault Circuit Interrupters Required. All electrical equipment, including power-supply cords, used with storable pools shall be protected by ground-fault circuit interrupters.

A storable pool may be readily disassembled and has a maximum wall height of 42 in. Pools of any dimension with inflatable walls are considered storable. See the definition of *storable swimming or wading pool* in Section 680-4. This type of pool and its associated equipment does not require bonding conductors. However, the filter pump is required to be double insulated, and grounding means consisting of an equipment grounding conductor that is an integral part of the flexible cord is required to be provided. There are portable filter pumps for use with storable pools listed by Underwriters Laboratories Inc.

The receptacle is required to be located at least 10 ft from the pool [see Section 680-6(a)], and all electrical equipment is required to have ground-fault circuit-interrupter protection for personnel. See Figure 680.16 for the requirements for a storable-type pool.

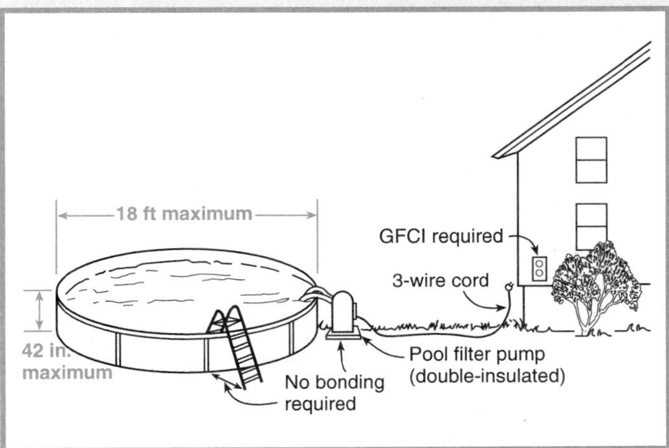

Figure 680.16 *The requirements for a storable-type pool. Metal appurtenances are not required to be bonded. The 3-wire cord may be longer than 3 ft. (Some listed filter pumps are equipped with cords 25 ft long.) The receptacle shown can be a GFCI-type receptacle, a receptacle supplied through a GFCI-type receptacle, or a receptacle protected by a GFCI-type circuit breaker.*

FPN: Where flexible cords are used, see Section 400-4.

680-32. Lighting Fixtures.

(a) 15 Volts or Less. A lighting fixture installed in or on the wall of a storable pool shall be part of a cord- and plug-connected lighting assembly. This assembly shall be as follows:

(1) Have no exposed metal parts
(2) Have a fixture lamp that operates at 15 volts or less
(3) Have an impact-resistant polymeric lens, fixture body, and transformer enclosure
(4) Have a transformer meeting the requirements of Section 680-5(a) with a primary rating not over 150 volts, and
(5) Be listed as an assembly for the purpose

(b) Not Over 150 Volts. A lighting assembly without a transformer, and with the fixture lamp(s) operating at not over 150 volts, shall be permitted to be cord- and plug-connected where the assembly complies with all of the following.

(1) It has no exposed metal parts.
(2) It has an impact-resistant polymeric lens and fixture body.
(3) A ground-fault circuit interrupter with open neutral protection is provided as an integral part of the assembly.
(4) The fixture lamp is permanently connected to the ground-fault circuit interrupter with open-neutral protection.
(5) It complies with the requirements of Section 680-20(a).
(6) It is listed as an assembly for the purpose.

Section 680-32(b)(6) recognizes lighting fixtures installed in or on storable pools. It should be understood that these cord- and plug-connected fixtures are required to be listed as an assembly.

D. Spas and Hot Tubs

680-38. Emergency Switch for Spas and Hot Tubs. A clearly labeled emergency shutoff or control switch for the purpose of stopping the motor(s) that provide power to the recirculation system and jet system shall be installed readily accessible to the users and at least 5 ft (1.52 m) away, adjacent to, and within sight of the spa or hot tub. This requirement shall not apply to single-family dwellings.

Added for the 1999 *Code,* Section 680-38 requires a local disconnecting device for spas and hot tubs that is capable of being used in an emergency. This requirement was added to address entrapment hazards associated with spas and hot tubs. A definitive publication on this issue entitled *Guideline for Entrapment Hazards: Making Pools and Spas Safer* (Publication Number 363) is available from the U.S. Consumer Product Safety Commission, Washington, D.C. 20207.

Accordingly, the emergency shutoff switch must be installed within sight of and at least 5 ft from the spa or hot tub. This shutoff switch must be clearly labeled as an "Emergency Shutoff." See Figure 680.17 for an illustration of the switch location. The shutoff switch can be either a line-operated device or a remote-controlled circuit that causes the pump circuit

to open. This requirement does not apply to one-family dwellings.

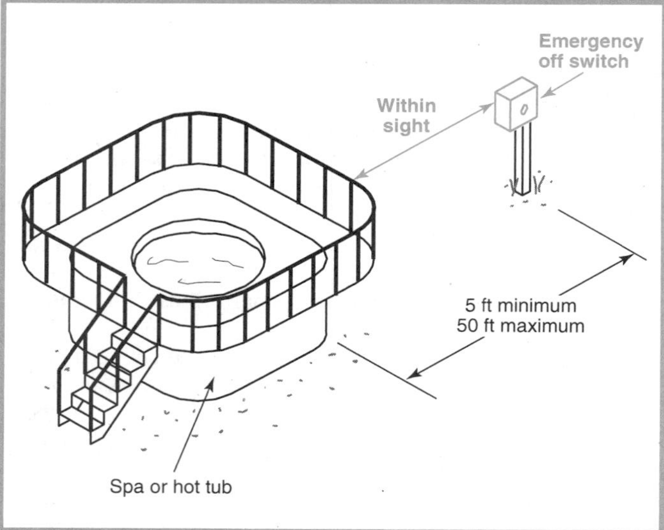

Figure 680.17 Location of the emergency shutoff device as required by Section 680-38.

680-40. Outdoor Installations. A spa or hot tub installed outdoors shall comply with the provisions of Parts A and B of this article except as permitted in (a) and (b).

(a) Flexible Connections. Listed packaged units utilizing a factory-installed remote panelboard shall be permitted to be connected with not more than 6 ft (1.828 m) of liquidtight flexible conduit or be cord and plug connected with a cord not longer than 15 ft (4.57 m) if protected by a ground-fault circuit interrupter.

Section 680-40(a) permits the use of liquidtight flexible metal or nonmetallic conduit. This modifies the requirements for wiring methods in Section 680-25(d).

(b) Bonding. Bonding by metal-to-metal mounting on a common frame or base shall be permitted. The metal bands or hoops used to secure wooden staves shall not be required to be bonded as required in Section 680-22.

680-41. Indoor Installations. A spa or hot tub installed indoors shall comply with the provisions of Parts A and B except as modified by this section, and shall be connected by the wiring methods of Chapter 3.

Listed spa and hot tub packaged units rated 20 amperes or less shall be permitted to be cord and plug connected to facilitate the removal or disconnection of the unit for maintenance and repair.

(a) Receptacles. At least one 125-volt, 15- or 20-ampere receptacle on a general-purpose branch circuit shall be located a minimum of 5 ft (1.52 m) from and not more than 10 ft (3.05 m) from the inside wall of the spa or hot tub.

(1) Receptacles on the property shall be located at least 5 ft (1.52 m) measured horizontally from the inside walls of the spa or hot tub.

(2) Receptacles of 125 volts located within 10 ft (3.05 m) of the inside walls of a spa or hot tub shall be protected by a ground-fault circuit interrupter.

FPN: In determining the above dimensions, the distance to be measured is the shortest path the supply cord of an appliance connected to the receptacle would follow without piercing a floor, wall, or ceiling of a building or other effective permanent barrier.

See Section 680-6(a)(3), FPN, and Figure 680.4.

(3) Receptacles that provide power for a spa or hot tub shall be ground-fault circuit-interrupter protected.

(b) Mounting Height of Lighting Fixtures, Lighting Outlets, and Ceiling-Suspended (Paddle) Fans.

(1) Lighting fixtures, lighting outlets, and ceiling-suspended (paddle) fans located over the spa or hot tub or within 5 ft (1.52 m) from the inside walls of the spa or hot tub shall be a minimum of 7 ft 6 in. (2.29 m) above the maximum water level and shall be protected by a ground-fault circuit interrupter.

Lighting fixtures, lighting outlets, and ceiling-suspended (paddle) fans that are located 12 ft (3.66 m) or more above the maximum water level shall not require a ground-fault circuit interrupter for protection.

(2) Lighting fixtures meeting the requirements of (a) or (b) and protected by a ground-fault circuit interrupter shall be permitted to be installed less than 7 ft 6 in. (2.29 m) over a spa or hot tub.

(a) Recessed fixtures with a glass or plastic lens and nonmetallic or electrically isolated metal trim, suitable for use in damp locations.

(b) Surface-mounted fixtures with a glass or plastic globe and a nonmetallic body or a metallic body isolated from contact. Such fixtures shall be suitable for use in damp locations.

(c) Wall Switches. Switches shall be located at least 5 ft (1.52 m), measured horizontally, from the inside walls of the spa or hot tub.

Receptacles, wall switches, and electrical devices and controls not associated with a spa or hot tub are required to be located at least 5 ft away. Receptacles within 10 ft are required to be protected by a GFCI. Receptacles supplying power to a spa or hot tub are also required to be protected by a GFCI unless the

unit is a listed package unit with integral GFCI protection.

Lighting fixtures, lighting outlets, and ceiling-suspended (paddle) fans located over a spa or hot tub or within 5 ft horizontally are required to be protected by a GFCI.

(d) Bonding. The following parts shall be bonded together.

(1) All metal fittings within or attached to the spa or hot tub structure.

(2) Metal parts of electrical equipment associated with the spa or hot tub water circulating system, including pump motors.

(3) Metal conduit and metal piping within 5 ft (1.52 m) of the inside walls of the spa or hot tub and that are not separated from the spa or hot tub by a permanent barrier.

(4) All metal surfaces that are within 5 ft (1.52 m) of the inside walls of the spa or hot tub and not separated from the spa or hot tub area by a permanent barrier.

Exception: Small conductive surfaces not likely to become energized, such as air and water jets and drain fittings, where not connected to metallic piping, towel bars, mirror frames, and similar nonelectrical equipment, shall not be required to be bonded.

(5) Electrical devices and controls not associated with the spas or hot tubs shall be located a minimum of 5 ft (1.52 m) away from such units or be bonded to the spa or hot tub system.

Bonding and grounding requirements are similar to those in Parts A and B of Article 680, except that metal-to-metal mounting on a common frame or base is an acceptable bonding method.

Small conductive surfaces such as air and water jets, drain fittings, towel bars, and so on, are not required to be bonded. See Section 680-41(d)(4), Exception.

Listed packaged units may be cord connected.

(e) Methods of Bonding. All metal parts associated with the spa or hot tub shall be bonded by any of the following methods:

(1) The interconnection of threaded metal piping and fittings
(2) Metal-to-metal mounting on a common frame or base
(3) The provisions of a copper bonding jumper, insulated, covered, or bare, not smaller than No. 8 solid

(f) Grounding. The following equipment shall be grounded:

(1) All electric equipment located within 5 ft (1.52 m) of the inside wall of the spa or hot tub
(2) All electric equipment associated with the circulating system of the spa or hot tub

(g) Methods of Grounding.

(1) All electrical equipment shall be grounded in accordance with Article 250 and be connected by the wiring methods of Chapter 3.

(2) Where equipment is connected with a flexible cord, the equipment grounding conductor shall be connected to a fixed metal part of the assembly.

(h) Electric Water Heaters. All electric spa or hot tub water heaters shall be listed and shall have the heating elements subdivided into loads not exceeding 48 amperes and protected at not more than 60 amperes.

The ampacity of the branch-circuit conductors, and the rating or setting of overcurrent protective devices, shall not be less than 125 percent of the total load of the nameplate rating.

(i) Underwater Audio Equipment. Underwater audio equipment shall comply with the provisions of Parts B or C of this article.

680-42. Protection. The outlet(s) that supplies:

(a) A self-contained spa or hot tub, or
(b) A packaged spa or hot tub equipment assembly, or
(c) A field-assembled spa or hot tub with a heater load of 50 amperes or less shall be protected by a ground-fault circuit interrupter.

A listed self-contained unit or listed packaged equipment assembly marked to indicate that integral ground-fault circuit-interrupter protection is provided for all electrical parts within the unit or assembly (pumps, air blowers, heaters, lights, controls, sanitizer generators, wiring, etc.) shall not require that the outlet supply be protected by a ground-fault circuit interrupter.

A field-assembled spa or hot tub rated greater than 250 volts or rated 3 phase shall not require the supply to be protected by a ground-fault circuit interrupter.

A combination pool/hot tub or spa assembly commonly bonded need not be protected by a ground-fault circuit interrupter.

FPN: See Section 680-4 for definitions of *self contained spa or hot tub,* and for *packaged spa or hot equipment assembly.*

Section 680-42 has been revised for the 1999 *Code* to require field assembled spas and hot tubs with heater loads of 50 amperes or less to be GFCI protected. Spas and hot tubs utilizing voltages over 250 volts or 3-phase power are not required to have GFCI protection because GFCI devices are not available in all voltage, amperage, and phasing arrangements. Combination spa/pool or hot tub/pool arrangements are not required to have GFCI protection if they share a common bonding grid.

E. Fountains

Part E applies to permanently installed decorative fountains and reflecting pools in the ground, partially in the ground, or in a building. These units are primarily for aesthetic value and are not intended for swimming or wading.

Part E does not cover installations in natural lakes, rivers, or ponds. However, it may be used in conjunction with the rest of the *Code* where electrical equipment is installed in a natural body of water.

680-50. General. The provisions of Part E shall apply to all permanently installed fountains as defined in Section 680-4. Self-contained, portable fountains not larger than 5 ft (1.52 m) in any dimension are not covered by Part E. Fountains that have water common to a pool shall comply with the pool requirements of this article.

680-51. Lighting Fixtures, Submersible Pumps, and Other Submersible Equipment.

(a) Ground-Fault Circuit Interrupter. A ground-fault circuit interrupter shall be installed in the branch circuit supplying fountain equipment unless the equipment is listed for operation at 15 volts or less and is supplied by a transformer that complies with Section 680-5(a).

(b) Operating Voltage. No lighting fixtures shall be installed for operation on supply circuits over 150 volts between conductors. Submersible pumps and other submersible equipment shall operate at 300 volts or less between conductors.

(c) Lighting Fixture Lenses. Lighting fixtures shall be installed with the top of the fixture lens below the normal water level of the fountain unless approved for above-water locations. A lighting fixture facing upward shall have the lens adequately guarded to prevent contact by any person.

(d) Overheating Protection. Electrical equipment that depends on submersion for safe operation shall be protected against overheating by a low-water cutoff or other approved means when not submerged.

(e) Wiring. Equipment shall be equipped with provisions for threaded conduit entries or be provided with a suitable flexible cord. The maximum length of exposed cord in the fountain shall be limited to 10 ft (3.05 m). Cords extending beyond the fountain perimeter shall be enclosed in approved wiring enclosures. Metal parts of equipment in contact with water shall be of brass or other approved corrosion-resistant metal.

(f) Servicing. All equipment shall be removable from the water for relamping or normal maintenance. Fixtures shall not be permanently imbedded into the fountain structure so that the water level must be reduced or the fountain drained for relamping, maintenance, or inspection.

(g) Stability. Equipment shall be inherently stable or be securely fastened in place.

680-52. Junction Boxes and Other Enclosures.

(a) General. Junction boxes and other enclosures used for other than underwater installation shall comply with Sections 680-21(a), (b), (c), (d), and (e).

(b) Underwater Junction Boxes and Other Underwater Enclosures. Junction boxes and other underwater enclosures shall be submersible and

(1) Be equipped with provisions for threaded conduit entries or compression glands or seals for cord entry;
(2) Be of copper, brass, or other approved corrosion-resistant material;
(3) Be filled with an approved potting compound to prevent the entry of moisture; and
(4) Be firmly attached to the supports or directly to the fountain surface and bonded as required.

Where the junction box is supported only by the conduit, the conduit shall be of copper, brass, or other approved corrosion-resistant metal. Where the box is fed by nonmetallic conduit, it shall have additional supports and fasteners of copper, brass, or other approved corrosion-resistant material.

FPN: See Section 370-23 for support of enclosures.

680-53. Bonding. All metal piping systems associated with the fountain shall be bonded to the equipment grounding conductor of the branch circuit supplying the fountain.

FPN: See Section 250-122 for sizing of these conductors.

680-54. Grounding. The following equipment shall be grounded:

(1) All electrical equipment located within the fountain or within 5 ft (1.52 m) of the inside wall of the fountain
(2) All electrical equipment associated with the recirculating system of the fountain
(3) Panelboards that are not part of the service equipment and that supply any electrical equipment associated with the fountain

680-55. Methods of Grounding.

(a) Applied Provisions. The provisions of Section 680-25 shall apply, excluding paragraph (e).

(b) Supplied by a Flexible Cord. Electrical equipment that is supplied by a flexible cord shall have all exposed non-current-carrying metal parts grounded by an insulated copper equipment grounding conductor that is an integral part of this cord. This grounding conductor shall be connected to a grounding terminal in the supply junction box, transformer enclosure, or other enclosure.

680-56. Cord- and Plug-Connected Equipment.

(a) Ground-Fault Circuit Interrupter. All electrical equipment, including power-supply cords, shall be protected by ground-fault circuit interrupters.

(b) Cord Type. Flexible cord immersed in or exposed to water shall be of the hard-service type as designated in Table 400-4 and shall be marked water resistant.

(c) Sealing. The end of the flexible cord jacket and the flexible cord conductor termination within equipment shall be covered with, or encapsulated in, a suitable potting compound to prevent the entry of water into the equipment through the cord or its conductors. In addition, the ground connection within equipment shall be similarly treated to protect such connections from the deteriorating effect of water that may enter into the equipment.

(d) Terminations. Connections with flexible cord shall be permanent, except that grounding-type attachment plugs and receptacles shall be permitted to facilitate removal or disconnection for maintenance, repair, or storage of fixed or stationary equipment not located in any water-containing part of a fountain.

680-57. Signs.

(a) General. Includes only fixed, stationary electrically illuminated utilization equipment with words or symbols designed to convey information or attract attention.

(b) Ground-Fault Circuit-Interrupter Protection for Personnel. All circuits supplying the sign shall have ground-fault circuit-interrupter protection for personnel.

(c) Location. Any sign installed inside a fountain shall be at least 5 ft (1.52 m) inside the fountain measured from the outside edges of the fountain.

(d) Disconnect. Shall comply with Section 600-6.

(e) Bonding. Shall comply with Section 600-7.

(f) Grounding. Any equipment associated with the sign shall be grounded as per Article 250.

The use of signs in fountains has become increasingly popular. Section 680-57 was added for the 1999 *Code* to address signs located in fountains. Signs were previously addressed by Article 600 and fountains were addressed by Article 680, but this new Section 680-57 now addresses them where used together. Signs in fountains are required to have GFCI protection for personnel. This protection may be provided in the feeder or branch circuit. To prevent contact by persons around the fountain, the sign must be at least 5 ft from the edge of the fountain (see Figure 680.18). Disconnecting and bonding requirements in Article 600 apply, and grounding is to be provided in accordance with Article 250.

F. Pools and Tubs for Therapeutic Use

Part F recognizes therapeutic equipment in other locations under the same or similar conditions of use, such as athletic training rooms and the like. Portable therapeutic appliances are covered by the provisions of Article 422. They are required to provide protection from electrocution while in the on or off position. The device used is an immersion-detection circuit interrupter (IDCI).

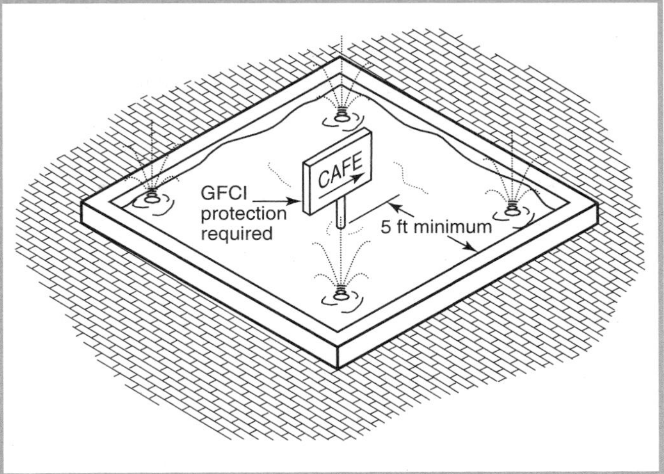

Figure 680.18 *Signs located in fountains as described in Section 680-57.*

Permanently installed therapeutic pools that cannot be readily disassembled are required to comply with Parts A and B of Article 680. The limitations regarding lighting fixtures over and around a swimming pool do not apply to therapeutic pools and tubs. The lighting fixtures in the tub area are required to be totally enclosed.

Therapeutic tubs that cannot easily be moved are subject to the same basic requirements.

Bonding and grounding requirements are similar to those in Parts A and B of Article 680, except that metal-to-metal mounting on a common frame or base is permitted. If equipment is connected by a flexible cord, the equipment grounding conductor is required to be connected to a fixed metal part of the assembly.

680-60. General. The provisions of Part F shall apply to pools and tubs for therapeutic use in health care facilities, gymnasiums, athletic training rooms, and similar areas. See Section 517-3 for definition of *health care facilities*. Portable therapeutic appliances shall comply with Article 422.

680-61. Permanently Installed Therapeutic Pools. Therapeutic pools that are constructed in the ground, on the ground, or in a building in such a manner that the pool cannot be readily disassembled shall comply with Parts A and B of this article.

Exception: The limitations of Sections 680-6(b)(1) and (2) shall not apply where all lighting fixtures are of the totally enclosed type.

680-62. Therapeutic Tubs (Hydrotherapeutic Tanks). Therapeutic tubs, used for the submersion and treatment of patients, that are not easily moved from one place to another in normal use or that are fastened or otherwise secured at a specific location, including associated piping systems, shall conform to this part.

(a) Protection. The outlet(s) that supplies the following shall be protected by a ground-fault circuit interrupter:

(1) A self-contained therapeutic tub or hydrotherapeutic tank, or
(2) A packaged therapeutic tub or hydrotherapeutic tank, or
(3) A field-assembled therapeutic tub or hydrotherapeutic tank with a heater load of 50 amperes or less

A listed self-contained unit or listed packaged equipment assembly marked to indicate that integral ground-fault circuit-interrupter protection is provided for all electrical parts within the unit or assembly (pumps, air blowers, heaters, lights, controls, sanitizer generators, wiring, etc.) shall not require that the outlet supply be protected by a ground-fault circuit interrupter.

A field-assembled therapeutic tub or hydrotherapeutic tank rated greater than 250 volts or rated 3 phase shall not require the supply to be protected by a ground-fault circuit interrupter.

Section 680-62(a) was revised for the 1999 *Code* to clarify the rules for GFCI protection of large hydrotherapeutic tanks and large therapeutic tubs. The requirements in Section 680-62(a) are similar to the requirements for large hot tubs and spas. Large field-assembled therapeutic tubs are not required to have GFCI protection in their electrical supply when the heater load is over 50 amperes.

(b) Ground-Fault Circuit Interrupter. A ground-fault circuit interrupter shall protect all therapeutic equipment.

Exception: Portable therapeutic appliances shall comply with Section 250-114.

(c) Bonding. The following parts shall be bonded together.

(1) All metal fittings within or attached to the tub structure.

(2) Metal parts of electrical equipment associated with the tub water circulating system, including pump motors.

(3) Metal-sheathed cables and raceways and metal piping that are within 5 ft (1.52 m) of the inside walls of the tub and not separated from the tub by a permanent barrier.

(4) All metal surfaces that are within 5 ft (1.52 m) of the inside walls of the tub and not separated from the tub area by a permanent barrier.

(5) Electrical devices and controls not associated with the therapeutic tubs shall be located a minimum of 5 ft (1.52 m) away from such units or be bonded to the therapeutic tub system.

(d) Methods of Bonding. All metal parts associated with the tub shall be bonded by any of the following methods:

(1) The interconnection of threaded metal piping and fittings,
(2) Metal-to-metal mounting on a common frame or base,
(3) Connections by suitable metal clamps, or
(4) By the provisions of a solid copper bonding jumper, insulated, covered, or bare, not smaller than No. 8

(e) Grounding. The following equipment shall be grounded:

(1) All electrical equipment located within 5 ft (1.52 m) of the inside wall of the tub
(2) All electrical equipment associated with the circulating system of the tub

(f) Methods of Grounding.

(1) All electrical equipment shall be grounded in accordance with Article 250 and connected by wiring methods of Chapter 3.

(2) Where equipment is connected with a flexible cord, the equipment grounding conductor shall be connected to a fixed metal part of the assembly.

(g) Receptacles. All receptacles within 5 ft (1.52 m) of a therapeutic tub shall be protected by a ground-fault circuit interrupter.

(h) Lighting Fixtures. All lighting fixtures used in therapeutic tub areas shall be of the totally enclosed type.

G. Hydromassage Bathtubs

Hydromassage bathtubs (see definition in Section 680-4) are required to be protected by a GFCI. In addition, all 15- and 20-ampere receptacles within 5 ft of the inside wall of the hydromassage bathtub are required to be GFCI protected. Except for this protection, hydromassage bathtubs are treated the same as ordinary bathtubs. See Section 410-4(d) for special requirements relating to cord-connected fixtures, hanging fixtures, and pendants near bathtubs. Also see Sections 210-8(a)(1) and 210-8(b)(1) for requirements for GFCI protection of bathroom receptacles.

680-70. Protection. Hydromassage bathtubs and their associated electrical components shall be protected by a ground-fault circuit interrupter. All 125-volt, single-phase receptacles within 5 ft (1.52 m) measured horizontally of the inside walls of a hydromassage tub shall be protected by a ground-fault circuit interrupter(s).

680-71. Other Electrical Equipment. Lighting fixtures, switches, receptacles, and other electrical equipment located in the same room, and not directly associated with a hydromassage bathtub, shall be installed in accordance with the requirements of Chapters 1 through 4 in this *Code* covering the installation of that equipment in bathrooms.

680-72. Accessibility. Hydromassage bathtub electrical equipment shall be accessible without damaging the building structure or building finish.

This new Section 680-72 was added for the 1999 *Code* to require access to electrical equipment associated with the hydromassage tub. Building codes and plumbing codes may not require access to this equipment. This requirement is intended to ensure that the electrical equipment associated with hydromassage bathtubs can be accessed for maintenance and repair without damaging the finish or structure of the building. The access may be either an integral part of the tub or one that is provided in the finish that encloses the tub. See Figure 680.19.

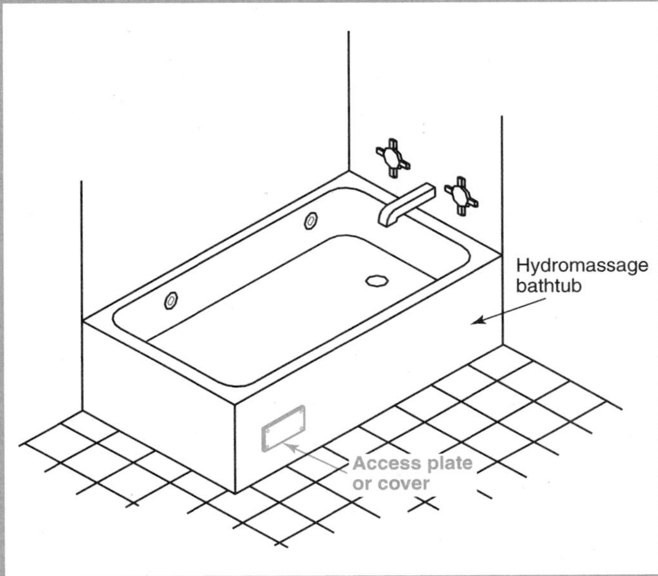

Figure 680.19 *Access plate located as described in Section 680-72.*

680-73. Bonding. All metal piping systems, metal parts of electrical equipment, and pump motors associated with the hydromassage tub shall be bonded together using a copper bonding jumper, insulated, covered, or bare, not smaller than No. 8 solid.

Metal parts of listed equipment incorporating an approved system of double insulation and providing a means for grounding internal nonaccessible, noncurrent-carrying metal parts shall not be bonded.

Article 685 — Integrated Electrical Systems

Contents

A. General

685-1. Scope. This article covers integrated electrical systems, other than unit equipment, in which orderly shutdown is necessary to ensure safe operation. An *integrated electrical system* as used in this article is a unitized segment of an industrial wiring system where all of the following conditions are met:

(1) An orderly shutdown is required to minimize personnel hazard and equipment damage,
(2) The conditions of maintenance and supervision ensure that qualified persons will service the system, and
(3) Effective safeguards, acceptable to the authority having jurisdiction, are established and maintained

Integrated electrical systems are commonly used in large industrial establishments where the electrical system and equipment are designed, installed, and operated by engineering work forces. The control equipment, including overcurrent devices, is located to be accessible to qualified personnel but may not be readily accessible (as defined in Article 100).

Orderly shutdown is sometimes required to prevent equipment damage or personal injury due to sudden loss of electrical power to the equipment. Orderly shutdown might be commonly employed in nuclear power generating facilities, paper mills, or other areas with hazardous processes.

685-2. Application of Other Articles. In the following other articles applying to particular cases of installation of conductors and equipment, there are orderly shutdown requirements that are in addition to those of this article or are modifications of them.

	Section
More than one building or other structure	225, Part B
Ground-fault protection of equipment	230-95, Exception No. 1
Protection of conductors	240-3
Electrical system coordination	240-12
Ground-fault protection of equipment	240-13(1)
Grounding ac systems of 50 to 1000 volts	250-21(c)
Equipment protection	427-22
Orderly shutdown	430-44

	Section
Disconnection	430-74, Exception Nos. 1 and 2
Disconnecting means in sight from controller	430-102(a), Exception No. 2
Energy from more than one source	430-113, Exception Nos. 1 and 2
Disconnecting means	645-10, Exception
Point of connection	705-12(a)

B. Orderly Shutdown

685-10. Location of Overcurrent Devices in or on Premises. Location of overcurrent devices that are critical to integrated electrical systems shall be permitted to be accessible, with mounting heights permitted to ensure security from operation by nonqualified personnel.

685-12. Direct-Current System Grounding. Two-wire dc circuits shall be permitted to be ungrounded.

685-14. Ungrounded Control Circuits. Where operational continuity is required, control circuits of 150 volts, or less, from separately derived systems shall be permitted to be ungrounded.

Article 690 — Solar Photovoltaic Systems

Contents

A. General

690-1. Scope. The provisions of this article apply to solar photovoltaic electrical energy systems including the array circuit(s), inverter(s), and controller(s) for such systems. [See Figures 690-1(a) and (b).] Solar photovoltaic systems covered by this article may be interactive with other electrical power production sources or stand alone, with or without electrical energy storage such as batteries. These systems may have ac or dc output for utilization.

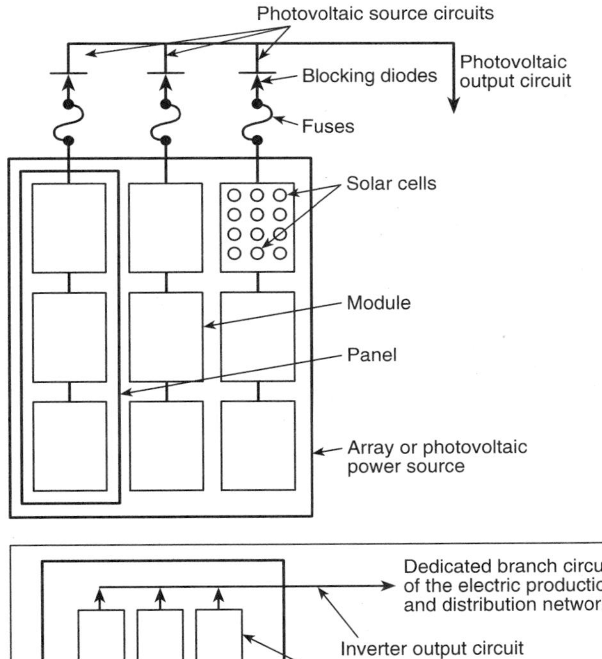

Note 1: These diagrams are intended to be a means of identification for photovoltaic system components, circuits, and connections.
Note 2: Disconnecting means required by Article 690, Part C are not shown.
Note 3: System grounding and equipment grounding are not shown. See Article 690, Part E.

Figure 690-1(a) Identification of solar photovoltaic system components.

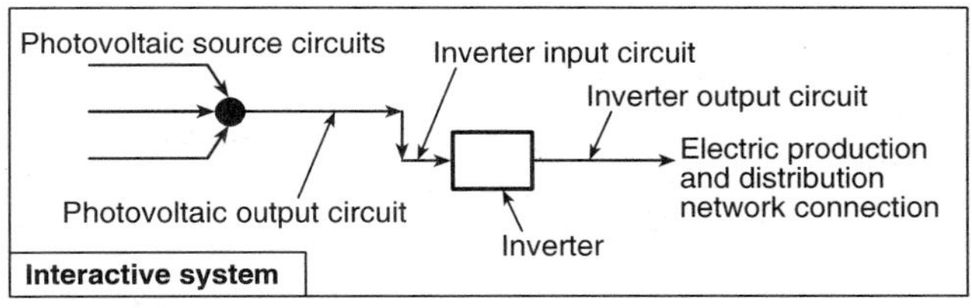

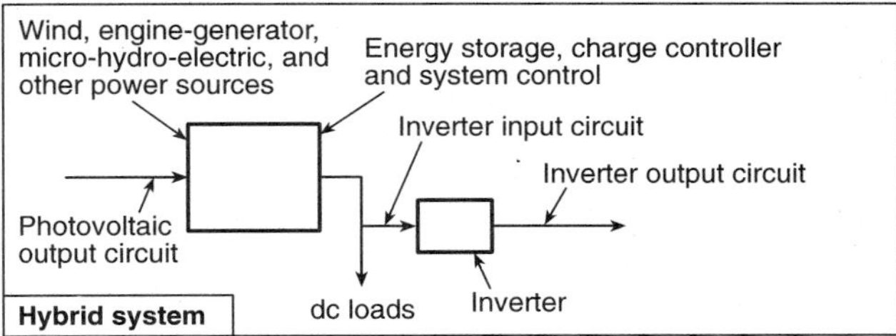

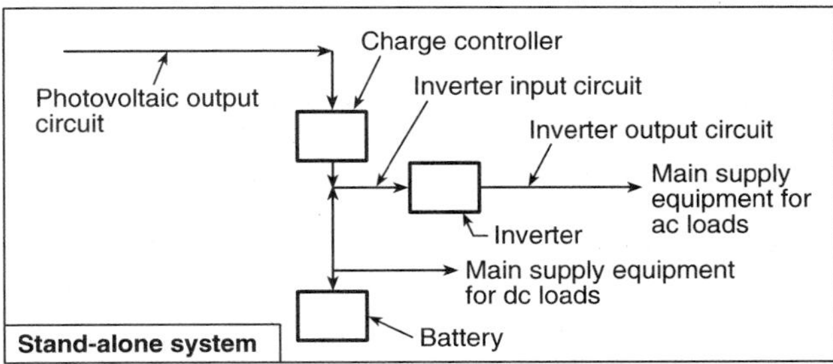

Note 1: These diagrams are intended to be a means of identification for photovoltaic system components, circuits, and connections.
Note 2: Disconnecting means and overcurrent protection required by Article 690 are not shown.
Note 3: System grounding and equipment grounding are not shown. See Article 690, Part E.
Note 4: Custom designs occur in each configuration and some components are optional.

Figure 690-1(b) Identification of solar photovoltaic system components in common system configurations.

In accordance with Section 705-3, Article 705 is also applicable to a solar photovoltaic system that is interconnected with another power production source, which may be an electric utility or an on-site generating system, such as a wind turbine, hydroelectric generator, or the like.

For an example of a custom designed home using a solar photovoltaic electrical system, see Figure 690.1.

Typical solar photovoltaic systems are illustrated in Figures 690.2 through 690.5. Other circuit arrangements are permissible.

690-2. Definitions.

Alternating-Current Module (Alternating-Current Photovoltaic Module). A complete, environmentally protected unit consisting of solar cells, optics, inverter, and

Figure 690.1 The south roof of this solar house is an integrated array of glass panels, most of which generate electricity directly from the sun using photovoltaics. (Solar Design Associates, Inc.)

other components, exclusive of tracker, designed to generate ac power when exposed to sunlight.

An alternating-current photovoltaic module (ACPV module) consists of a single, mechanical unit. Since there is no accessible, field-installed direct-current wiring in this single unit, the dc photovoltaic source circuit requirements in this *Code* are not applicable to the dc wiring in an ACPV module.

Array. A mechanically integrated assembly of modules or panels with a support structure and foundation, tracker, and other components, as required, to form a direct-current power-producing unit.

The building blocks of an array are illustrated in Figure 690.6.

Blocking Diode. A diode used to block reverse flow of current into a photovoltaic source circuit.

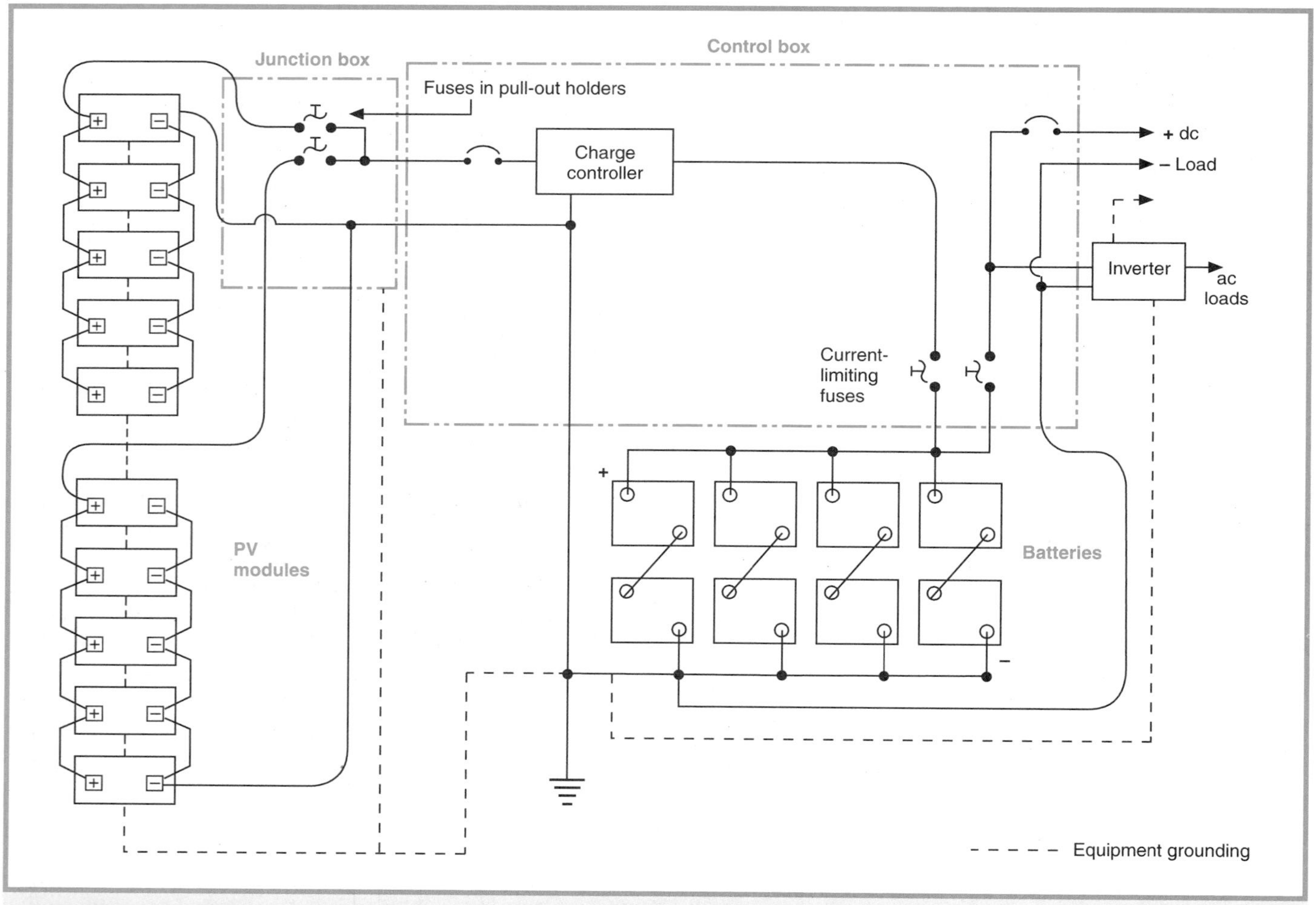

Figure 690.2 Small residential stand-alone system.

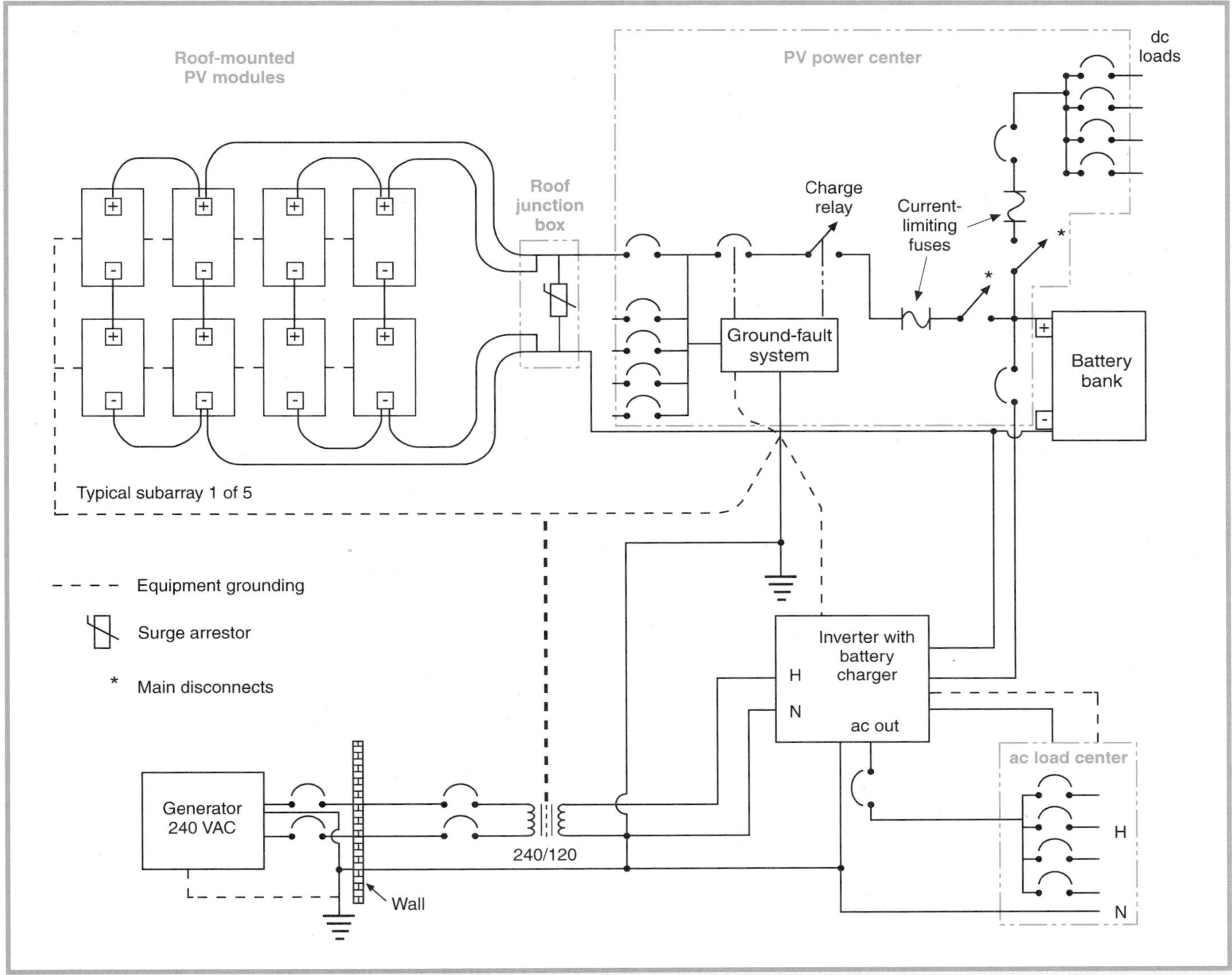

Figure 690.3 *Medium-sized residential hybrid system.*

Blocking diodes are not required by this *Code* although they may be required in the instructions or labels supplied with the photovoltaic module.

Charge Controller. Equipment that controls dc voltage or dc current, or both, used to charge a battery.

Electrical Production and Distribution Network. A power production, distribution, and utilization system, such as a utility system and connected loads, that is external to and not controlled by the photovoltaic power system.

Hybrid System. A system comprised of multiple power sources. These power sources may include photovoltaic, wind, micro-hydro generators, engine-driven generators, and others, but do not include electrical production and distribution network systems. Energy storage systems, such as batteries, do not constitute a power source for the purpose of this definition.

Interactive System. A solar photovoltaic system that operates in parallel with and may deliver power to an electrical production and distribution network. For the purpose of this definition, an energy storage subsystem of a solar photovoltaic system, such as a battery, is not another electrical production source.

Inverter. Equipment that is used to change voltage level or waveform, or both, of electrical energy. Commonly, an inverter [also known as a power conditioning unit (PCU) or power conversion system (PCS)] is a device that changes dc input to an ac output. Inverters may also function as

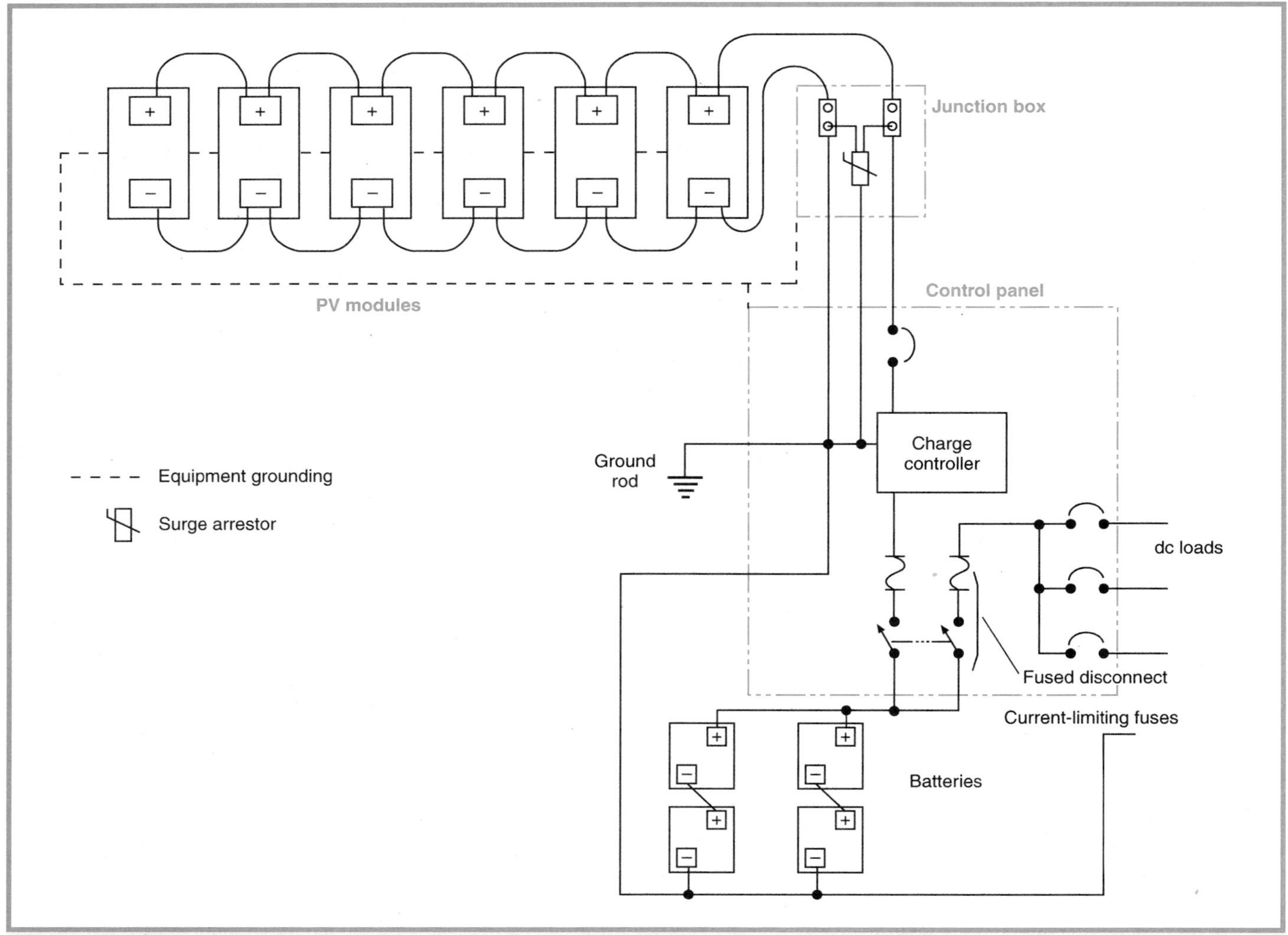

Figure 690.4 Remote cabin dc-only system.

battery chargers that use alternating current from another source and convert it into direct current for charging batteries.

Inverter Input Circuit. Conductors between the inverter and the battery in stand-alone systems or the conductors between the inverter and the photovoltaic output circuits for electrical production and distribution network.

Inverter Output Circuit. Conductors between the inverter and an ac load center for stand-alone systems or the conductors between the inverter and the service equipment or another electric power production source, such as a utility, for electrical production and distribution network.

Module. A complete, environmentally protected unit consisting of solar cells, optics, and other components, exclusive of tracker, designed to generate dc power when exposed to sunlight.

Panel. A collection of modules mechanically fastened together, wired, and designed to provide a field-installable unit.

Photovoltaic Output Circuit. Circuit conductors between the photovoltaic source circuit(s) and the inverter or dc utilization equipment.

Photovoltaic Power Source. An array or aggregate of arrays that generates dc power at system voltage and current.

Photovoltaic Source Circuit. Circuits between modules and from modules to the common connection point(s) of the dc system.

Solar Cell. The basic photovoltaic device that generates electricity when exposed to light.

Solar Photovoltaic System. The total components and subsystems that, in combination, convert solar energy into electrical energy suitable for connection to a utilization load.

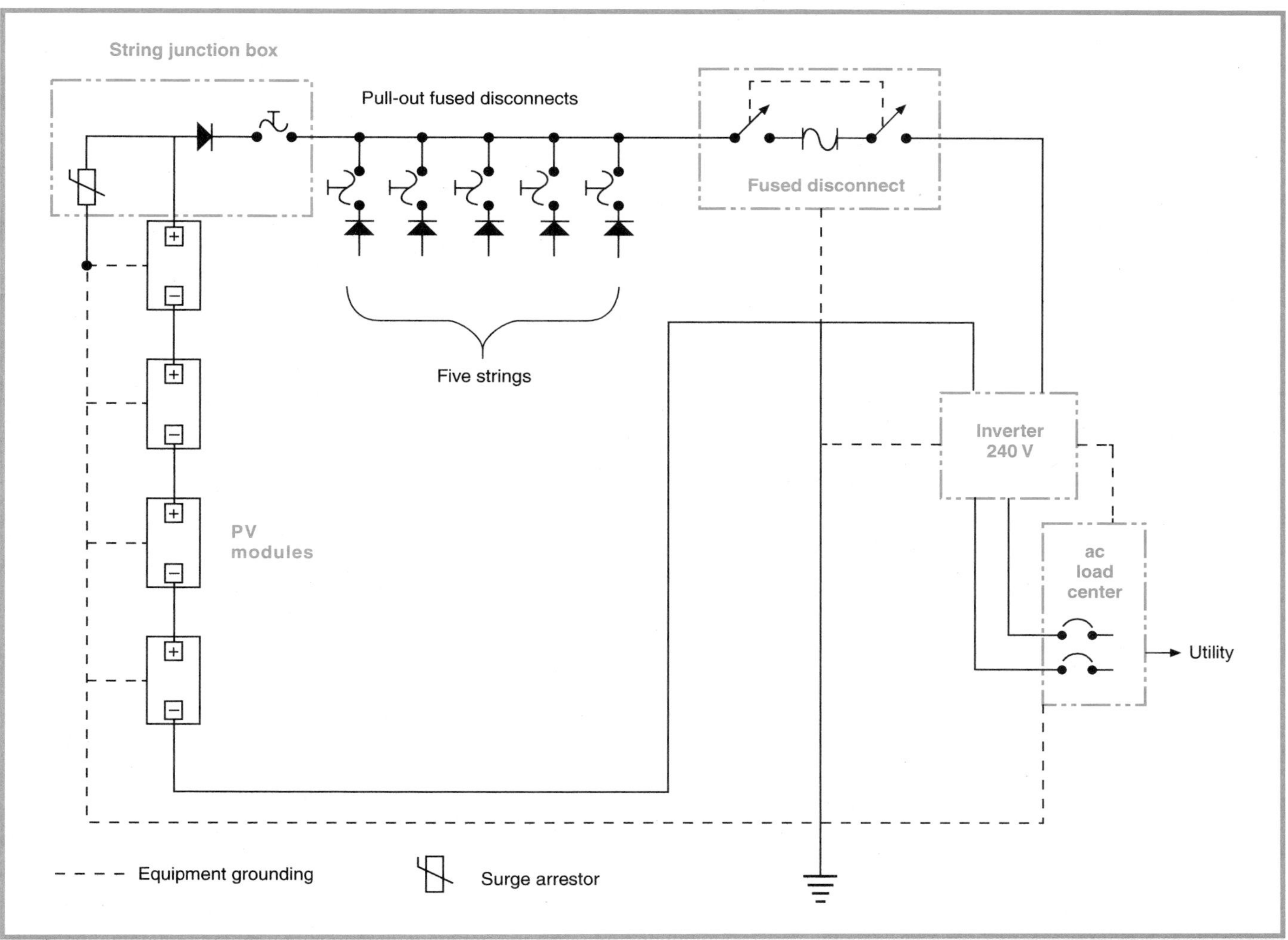

Figure 690.5 *Rooftop grid-connected system.*

Stand-Alone System. A solar photovoltaic system that supplies power independently of an electrical production and distribution network.

The simplified circuit diagrams in Figures 690.2 through 690.5 demonstrate the use of various components in a photovoltaic system. Specific requirements for overcurrent protection, disconnecting means, and grounding are covered in other sections of Article 690 and should not be inferred from these diagrams. Instructions and/or labels on the photovoltaic module may require additional overcurrent devices that are not shown.

System Voltage. The dc voltage of any photovoltaic source circuit or output circuit. For 3-wire or multiwire installations, including 2-wire circuits connected to 3-wire systems, the system voltage shall be the highest voltage between any two conductors.

690-3. Other Articles. Wherever the requirements of other articles of this *Code* and Article 690 differ, the requirements of Article 690 shall apply.

690-4. Installation.

(a) Solar Photovoltaic System. A solar photovoltaic system shall be permitted to supply a building or other structure in addition to any service(s) of another electricity supply system(s).

(b) Conductors of Different Systems. Photovoltaic source circuits and photovoltaic output circuits shall not be contained in the same raceway, cable tray, cable, outlet box, junction box, or similar fitting as feeders or branch circuits

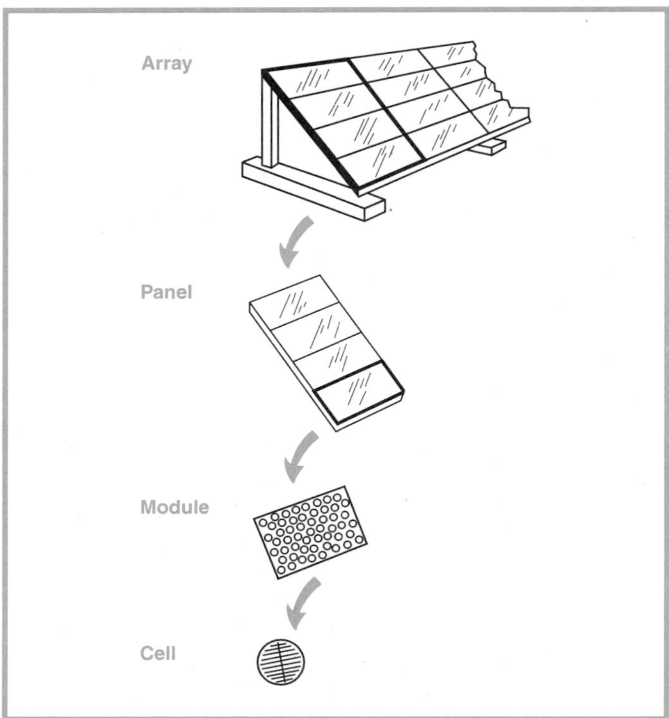

Figure 690.6 Array components.

of other systems, unless the conductors of the different systems are separated by a partition or are connected together.

As an example, Section 690-4(b) requires the conductors of a yard light located near a roof-mounted photovoltaic (PV) array to be in a separate raceway or cable from the conductors of PV source circuits or PV output circuits.

(c) Module Connection Arrangement. The connections to a module or panel shall be arranged so that removal of a module or panel from a photovoltaic source circuit does not interrupt a grounded conductor to another photovoltaic source circuit. Sets of modules interconnected as systems rated at 50 volts or less, with or without blocking diodes, and having a single overcurrent device shall be considered as a single-source circuit. Supplementary overcurrent devices used for the exclusive protection of the photovoltaic modules are not considered as overcurrent devices for the purpose of this section.

In general, Section 690-4(c) requires the installation of a jumper between a module terminal or lead and the connection point to the grounded photovoltaic source circuit conductor so that a module can be removed without interrupting the grounded conductor to other photovoltaic source circuits. If interrupted, such conductors, although identified as grounded, would be operating at the system potential with respect to ground, and a shock hazard could result.

(d) Equipment. Inverters or motor generators shall be identified for use in solar photovoltaic systems.

690-5. Ground-Fault Protection. Roof-mounted dc photovoltaic arrays located on dwellings shall be provided with dc ground-fault protection to reduce fire hazards.

Ground-fault detection and interruption for solar photovoltaic systems should not be confused with the requirements for ground-fault circuit-interrupter (GFCI) protection, as defined in Article 100. A GFCI is intended for the protection of personnel in single-phase ac systems. It functions to open the ungrounded conductor when a 5-mA fault current is detected. In contrast, Section 690-5 is intended to prevent fires in dc photovoltaic circuits due to ground faults.

(a) Ground-Fault Detection and Interruption. The ground-fault protection device or system shall be capable of detecting a ground fault, interrupting the flow of fault current, and providing an indication of the fault.

(b) Disconnection of Conductors. The ungrounded conductors of the faulted source circuit shall be automatically disconnected. If the grounded conductors of the faulted source circuit are disconnected to comply with the requirements of Section 690-5(a), all conductors of the faulted source circuit shall be opened automatically and simultaneously. Interrupting the ground-fault current path to open the grounded conductor of the array or faulted sections of the array shall be permitted.

The last sentence of Section 690-5(b) allows grounded conductors to be automatically opened by a device that meets the requirements of Section 690-5(b).

(c) Labels and Markings. Labels and markings shall be applied near the ground-fault indicator at a visible location stating that if a ground fault is indicated, the normally grounded conductors may be energized and ungrounded.

690-6. Alternating-Current Modules.

(a) Photovoltaic Source Circuits. The requirements of Article 690 pertaining to photovoltaic source circuits shall not apply to ac modules. The photovolaic source circuit, conductors, and inverters shall be considered as internal wiring of an ac module.

(b) Inverter Output Circuit. The output of an ac module shall be considered an inverter output circuit.

(c) Disconnecting Means. A single disconnecting means, in accordance with Sections 690-15 and 690-17, shall be permitted for the combined ac output of one or more ac modules. Additionally, each ac module in a multiple ac-module system shall be provided with a connector, bolted, or terminal-type disconnecting means.

Alternating-current photovoltaic modules, as utility-interactive devices, are designed to produce ac power only when connected to an external source of ac power at the correct voltage and frequency. A single disconnecting means will remove the external source and turn off all ac photovoltaic modules connected to that disconnecting device.

(d) Ground-Fault Detection. Alternating-current-module systems shall be permitted to use a single detection device to detect only ac ground faults and to disable the array by removing ac power to the ac module(s).

The requirement of Section 690-6(d) for ac photovoltaic modules replaces the requirements of Section 690-5 that applies only to conventional dc photovoltaic modules. As in Section 690-5, this is a fire prevention device and is not intended to be a shock prevention device. The detection device can be a 25 mA to 30 mA, or higher, equipment ground-fault protection circuit breaker on the required dedicated circuit.

(e) Overcurrent Protection. The output circuits of ac modules shall be permitted to have overcurrent protection and conductor sizing in accordance with Section 240-4(b)(2).

B. Circuit Requirements

690-7. Maximum Voltage.

(a) Maximum System Voltage. In a dc photovoltaic source circuit or output circuit, the maximum system voltage for that circuit shall be computed as the sum of the rated open-circuit voltage of the series-connected photovoltaic modules corrected for the lowest expected ambient temperature. For crystalline and multi-crystalline silicon modules, the rated open-circuit voltage shall be multiplied by the correction factor provided in Table 690-7. This voltage shall be used to determine the voltage rating of cables, disconnects, overcurrent devices, and other equipment. Where the lowest expected ambient temperature is below −40°C (−40°F), or where other than crystalline or multi-crystalline silicon photovoltaic modules are used, the system voltage adjustment shall be made in accordance with the manufacturer's instructions.

Table 690-7. Voltage Correction Factors for Crystalline and Multi-Crystalline Silicon Modules

Ambient Temperature (°C)	Correction factors for ambient temperatures below 25°C (77°F), multiply the rated open-circuit voltage by the appropriate correction factor shown below.	Ambient Temperature (°F)
25 to 10	1.06	77 to 50
9 to 0	1.10	49 to 32
−1 to −10	1.13	31 to 14
−11 to −20	1.17	13 to −4
−21 to −40	1.25	−5 to −40

A photovoltaic source is not a constant-voltage source, and the difference between the rated operating voltage determined under controlled laboratory conditions and the open-circuit voltage under field-installed conditions may be significant. Consequently, the higher-rated open-circuit voltage must be used to select circuit components with proper voltage ratings.

The voltage potential (both open circuit and operating) of a photovoltaic power source increases with decreasing temperature. The installer should note the temperature conditions under which the photovoltaic device was rated. If the anticipated lowest daytime temperatures at the installation site are lower than the rating conditions (25°C), Table 690-7 should be used to adjust the maximum open-circuit voltage of the system before selecting conductors, overcurrent devices, and switchgear.

Previous to the 1999 edition of this *Code,* the temperature adjustment requirement was included in the instructions provided with listed photovoltaic modules and was established at a fixed 125 percent, regardless of temperature.

Bipolar photovoltaic systems (with positive and negative voltages) must use the sum of the absolute values of the open-circuit voltages to determine the rated open-circuit system voltage. For example, a system with open-circuit voltages of plus and minus 480 volts would have a system open-circuit voltage of 960 volts. This voltage should be multiplied by a temperature-dependent factor from Table 690-7, yielding a system design voltage of up to 1200 volts. This system design voltage should be used in selecting cables and other equipment. Also see the definition of *system voltage* in Section 690-2.

(b) Direct-Current Utilization Circuits. The voltage of dc utilization circuits shall conform with Section 210-6.

Section 690-7(b) covers installations where the photovoltaic output is connected to direct-current utilization circuits.

(c) Photovoltaic Source and Output Circuits. In one- and two-family dwellings, photovoltaic source circuits and photovoltaic output circuits that do not include lampholders, fixtures, or receptacles shall be permitted to have a maximum system voltage up to 600 volts. Other installations with a maximum system voltage over 600 volts shall comply with Article 690, Part I.

Photovoltaic direct-current circuits in buildings are permanently connected using wiring systems recognized by this *Code.* Requirements for protecting unqualified persons from contact with these circuits are included in Sections 690-7(b) and (d). Unqualified persons are not likely to service equipment in these circuits due to its complexity. There is a significant difference between the rated open-circuit voltage and the operating voltage in photovoltaic direct-current circuits. In order for the photovoltaic system to perform its intended function, rated direct-current open-circuit voltages of up to 600 volts may be present.

(d) Circuits Over 150 Volts to Ground. In one- and two-family dwellings, live parts in photovoltaic source circuits and photovoltaic output circuits over 150 volts to ground shall not be accessible to other than qualified persons while energized.

FPN: See Section 110-27 for guarding of live parts, and Section 210-6 for voltage to ground and between conductors.

Where direct-current circuitry over 150 volts to ground is present in one- and two-family dwellings, additional protection for unqualified persons is needed. Protection may be in the form of a locked cabinet or an enclosure that requires tools to open and permits entry only for servicing purposes.

690-8. Circuit Sizing and Current.

(a) Computation of Maximum Circuit Current. The maximum current for the specific circuit shall be computed as follows.

(1) Photovoltaic Source Circuit Currents. The maximum current shall be the sum of parallel module rated short-circuit currents multiplied by 125 percent.

The use of the array short-circuit current will allow for proper sizing of conductors to handle the current generated during extended periods of operation under a short-circuit current operating point.

The 125 percent factor mentioned in Section 690-8(a)(1) is required because photovoltaic modules, photovoltaic source circuits, and photovoltaic output circuits deliver output currents higher than the rated short-circuit currents for more than 3 hours near solar noon. Prior to the 1999 Edition of this *Code,* this requirement was specified in the instructions provided with each listed module. It is not related to the additional 125 percent factor required by Section 690-8(b).

Photovoltaic modules in hot climates operate at temperatures of 60°C to 80°C due to solar heating. Conductors with insulation types rated at least 90°C should be used, and these conductors should have the ampacity derated in accordance with Table 310-16 or 310-17.

(2) Photovoltaic Output Circuit Currents. The maximum current shall be the sum of parallel source circuit maximum currents as calculated in (1).

(3) Inverter Output Circuit Current. The maximum current shall be the inverter continuous output current rating.

Both stand-alone and utility-interactive inverters are power-limited devices. Output circuits connected to these devices are sized on the continuous rated outputs of these devices and are not based on load calculations.

(4) Stand-Alone Inverter Input Circuit Current. The maximum current shall be the stand-alone continuous inverter input current rating when the inverter is producing rated power at the lowest input voltage.

Stand-alone inverters are constant-output-voltage devices. As the input battery voltage decreases, the input battery current increases to maintain a constant ac output. The input current for such inverters is calculated by taking the rated full-power output of the inverter in watts and dividing it by the lowest operating battery voltage and then by the rated efficiency of the inverter. For example, the input current for a 4000-watt, 24-volt inverter that is 85 percent efficient at 22 volts would be calculated as follows:

$$\begin{aligned}\text{Ampere input} &= \frac{\text{watt output}}{\text{voltage} \times \text{efficiency}}\\ &= \frac{4000}{22 \times 0.85}\\ &= 214 \text{ amperes}\end{aligned}$$

Ripple currents may be present in the dc-input circuits of single-phase, stand-alone inverters. These ripple currents may cause nuisance operation of overcurrent devices at continuous high inverter outputs. In these cases, the measured maximum rms (root mean square) value of the input current, which will be greater than the average current calculated above, should be used in determining conductor sizes and overcurrent device ratings.

(b) Ampacity and Overcurrent Device Ratings. Photovoltaic system currents shall be considered continuous. The circuit conductors and overcurrent devices shall be sized to carry not less than 125 percent of the maximum currents as computed in (a). The rating or setting of overcurrent devices shall be permitted in accordance with Sections 240-3(b) and (c).

Exception: Circuits containing an assembly together with its overcurrent device(s) that is listed for continuous operation at 100 percent of its rating shall be permitted to be utilized at 100 percent of its rating.

The exception to Section 690-8(b) permits use at the full rating of assemblies, such as panelboards, incorporating overcurrent devices listed for continuous operation at 100 percent of the rating.

(c) Systems with Multiple Direct-Current Voltages. For a photovoltaic power source that has multiple output circuit voltages and employs a common-return conductor, the ampacity of the common-return conductor shall not be less than the sum of the ampere ratings of the overcurrent devices of the individual output circuits.

690-9. Overcurrent Protection.

(a) Circuits and Equipment. Photovoltaic source circuit, photovoltaic output circuit, inverter output circuit, and storage battery circuit conductors and equipment shall be protected in accordance with the requirements of Article 240. Circuits connected to more than one electrical source shall have overcurrent devices located so as to provide overcurrent protection from all sources.

Exception: An overcurrent device shall not be required for circuit conductors sized in accordance with Section 690-8(b) and located where

(a) There are no external sources such as parallel-connected source circuits, batteries, or backfeed from inverters, or
(b) The short-circuit currents from all sources do not exceed the ampacity of the conductors.

FPN: Possible backfeed of current from any source of supply, including a supply through an inverter into the photovoltaic output circuit and photovoltaic source circuits, is a consideration in determining whether adequate overcurrent protection from all sources is provided for conductors and modules.

In the circuit illustrated in Figure 690.7, the photovoltaic source circuit overcurrent devices are required to be rated so that the source circuit conductors are protected in accordance with Article 240 and the overcurrent device ratings do not exceed the maximum overcurrent device rating marked on the modules. Possible backfeed from the other photovoltaic source circuits, other supply sources through the power conditioning unit, and storage battery circuits, if any, have to be considered.

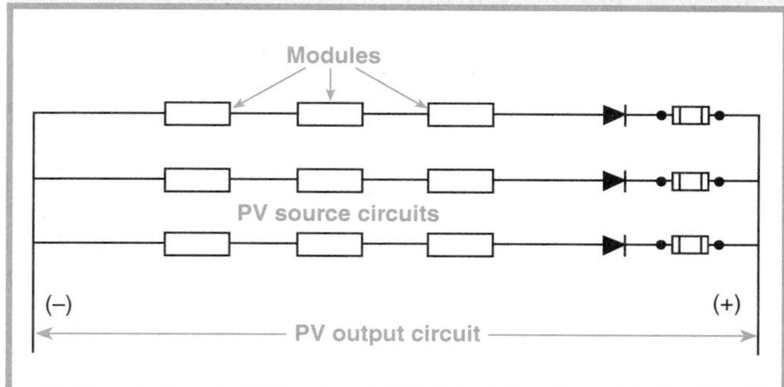

Figure 690.7 *Application of blocking diodes and source circuit overcurrent devices.*

Diodes (possibly required by the module manufacturer for performance reasons) may lose their blocking ability (due to overtemperature conditions or internal breakdown). Therefore, overcurrent protection has to be considered with a condition of shorted blocking diodes if they are used in the circuit.

At the inverter or battery/charge controller end of the photovoltaic output circuit, the need for overcurrent protection has to be considered with respect to the maximum backfeed fault current available from other sources.

(b) Power Transformers. Overcurrent protection for a transformer with a source(s) on each side shall be provided in accordance with Section 450-3 by considering first one side of the transformer, then the other side of the transformer, as the primary.

Exception: A power transformer with a current rating on the side connected toward the photovoltaic power source not less than the short-circuit output current rating of the inverter shall be permitted without overcurrent protection from that source.

(c) Photovoltaic Source Circuits. Branch-circuit or supplementary-type overcurrent devices shall be permitted to provide overcurrent protection in photovoltaic source circuits. The overcurrent devices shall be accessible, but shall not be required to be readily accessible.

By considering the overcurrent protection of photovoltaic source circuits as supplementary overcurrent protection, the use of overcurrent device types and ratings other than those suitable for branch-circuit protection can be permitted. Use of such devices permits module protection closer to the specified ratings required on the labels attached to listed modules. It is anticipated that replacement or resetting of overcurrent devices in photovoltaic source circuits will be done by qualified service personnel. Consequently, ready access to the user need not be provided. These supplemental overcurrent devices must be listed for dc operation and have appropriate voltage and current ratings.

(d) Direct-Current Rating. Overcurrent devices, either fuses or circuit breakers, used in any dc portion of a photovoltaic power system shall be listed for use in dc circuits and shall have the appropriate voltage, current, and interrupt ratings.

Direct-current (dc) faults are considerably harder to interrupt than alternating-current (ac) faults. Overcurrent devices marked or listed only for ac use should not be used in dc circuits. Automotive-type fuses, although listed for use in automotive dc systems, are not suitable for use in electrical power systems meeting the requirements of this *Code.*

690-10. Stand-Alone Systems. The premises wiring system shall be adequate to meet the requirements of this *Code* for a similar installation connected to a service. The wiring on the supply side of the building or structure disconnecting means shall comply with this *Code* except as modified by (a), (b), and (c).

(a) Inverter Output. The ac inverter output from a stand-alone system shall be permitted to supply ac power to the building or structure disconnecting means at current levels below the rating of that disconnecting means.

A stand-alone residential or commercial photovoltaic installation may have an alternating-current output and be connected to a building wired in full compliance with all articles of this *Code.* Even though such an installation may have service-entrance equipment rated at 100 or 200 amperes at 120/240 volts, there is no requirement that the photovoltaic source provide either the rated full current or the dual voltages of the service equipment. While safety requirements dictate full compliance with the ac wiring sections of this *Code,* a photovoltaic installation is usually designed so that the actual ac demands on the system are sized to the output rating of the photovoltaic system.

(b) Sizing and Protection. The circuit conductors between the inverter output and the building or structure disconnecting means shall be sized based on the output rating of the inverter. These conductors shall be protected from overcurrents in accordance with Article 240. The overcurrent protection shall be located at the output of the inverter.

(c) Single 120-Volt Supply. The inverter output of a stand-alone solar photovoltaic system shall be permitted to supply 120 volts to single-phase, 3-wire 120/240-volt service equipment or distribution panels where there are no 240-volt outlets and where there are no multiwire branch circuits. In all installations, the rating of the overcurrent device connected to the output of the inverter shall be less than the rating of the neutral bus in the service equipment. This equipment shall be marked

WARNING — SINGLE 120-VOLT SUPPLY — DO NOT CONNECT MULTIWIRE BRANCH CIRCUITS!

Multiwire branch circuits are common in single- and two-family dwelling units. When connected to a normal 120/240-volt ac service, the currents in the neutral conductors of these multiwire branch circuits (typically 14–3 AWG) cancel or are, at most, no larger than the rating of the branch-circuit overcurrent device. When these electrical systems are connected to a single 120-volt photovoltaic power system inverter by paralleling the two ungrounded conductors in the load center, the currents in the neutral conductor for each multiwire branch circuit add and do not cancel. The currents in the neutral conductor may be as high as twice the rating of the branch-circuit overcurrent device. Neutral conductor overloading is possible.

C. Disconnecting Means

690-13. All Conductors. Means shall be provided to disconnect all current-carrying conductors of a photovoltaic power source from all other conductors in a building or other structure. Where a circuit grounding connection is not designed to be automatically interrupted as part of the ground-fault protection system required by Section 690-5, a switch or circuit breaker used as a disconnecting means shall not have a pole in the grounded conductor.

FPN: The grounded conductor may have a bolted or terminal disconnecting means to allow maintenance or troubleshooting by qualified personnel.

690-14. Additional Provisions. The provisions of Article 230, Part F, as modified by (a) and (b), shall apply to the photovoltaic power source disconnecting means.

(a) Disconnecting Means. The disconnecting means shall not be required to be suitable as service equipment and shall be rated in accordance with Section 690-17.

A disconnecting means rated in accordance with Section 690-17 may be used, but is not required to be marked suitable for use as service equipment.

(b) Equipment. Equipment such as photovoltaic source circuit isolating switches, overcurrent devices, and blocking diodes shall be permitted on the photovoltaic side of the photovoltaic disconnecting means.

In general, there is a need to disconnect equipment from sources of supply for servicing. However, in a solar photovoltaic system, some equipment, as indicated, is permitted to be located on the photovoltaic power source side of the disconnecting means. See Figure 690.8. In order to service the exempted equipment, the array or portions of the array may have to be disabled, as explained in the commentary following Section 690-18.

There is no intent or requirement to have a disconnecting means located in each photovoltaic source circuit or located physically at each photovoltaic module location. Unlike load circuits (e.g., roof-top air conditioners), photovoltaic source-circuit conductors may be energized at any time from the photovoltaic modules. A centrally located disconnect near the inverter or batteries will serve to disconnect the photovoltaic source circuits from the other portions of the electrical power system.

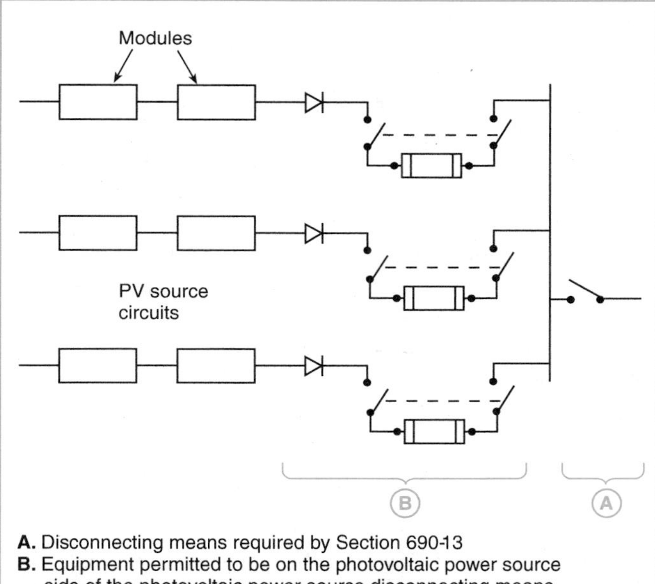

Figure 690.8 *Equipment permitted on the photovoltaic power source side of the photovoltaic power source disconnecting means.*

690-15. Disconnection of Photovoltaic Equipment. Means shall be provided to disconnect equipment, such as inverters, batteries, charge controllers, and the like, from all ungrounded conductors of all sources. If the equipment is energized from more than one source, the disconnecting means shall be grouped and identified.

A single disconnecting means in accordance with Section 690-17 shall be permitted for the combined ac output of one or more inverters or ac modules in an interactive system.

690-16. Fuses. Disconnecting means shall be provided to disconnect a fuse from all sources of supply if the fuse is energized from both directions and is accessible to other than qualified persons. Such a fuse in a photovoltaic source circuit shall be capable of being disconnected independently of fuses in other photovoltaic source circuits.

Switches, pullouts, or similar devices that have suitable ratings may serve as means to disconnect fuses from all sources of supply.

690-17. Switch or Circuit Breaker. The disconnecting means for ungrounded conductors shall consist of a manually operable switch(es) or circuit breaker(s)

(1) Located where readily accessible,
(2) Externally operable without exposing the operator to contact with live parts,
(3) Plainly indicating whether in the open or closed position, and
(4) Shall have an interrupting rating sufficient for the nominal circuit voltage and the current that is available at the line terminals of the equipment.

Where all terminals of the disconnecting means may be energized in the open position, a warning sign shall be mounted on or adjacent to the disconnecting means. The sign shall be clearly legible and shall read substantially:

WARNING — ELECTRIC SHOCK HAZARD —
DO NOT TOUCH TERMINALS —
TERMINALS ON BOTH THE LINE AND
LOAD SIDES MAY BE ENERGIZED
IN THE OPEN POSITION.

Exception No. 1: A disconnecting means located on the dc side shall be permitted to have an interrupting rating less than the current-carrying rating where the system is designed so that the dc switch cannot be opened under load.

Exception No. 2: A connector shall be permitted to be used as an ac or dc disconnecting means provided that it complies with the requirements of Section 690-33 and is listed and identified for the use.

690-18. Installation and Service of an Array. Open circuiting, short circuiting, or opaque covering shall be used to disable an array or portions of an array for installation and service.

FPN: Photovoltaic modules are energized while exposed to light. Installation, replacement, or servicing of array components while a module(s) is irradiated may expose persons to electric shock.

To prevent contact with energized parts for installation, servicing, or other functions, a number of methods may accomplish disablement of an array or portions of an array.

One method, used infrequently due to the expense in time and materials, is to cover all or portions of an array with an opaque material. Care must be taken that all of the area to be covered is shielded from light.

Another method is to provide for the array to be divided into nonhazardous segments. This may be accomplished by switches or connectors. Also see Section 690-33.

Short circuiting all or portions of an array by means of switches or plug-in connectors in conjunction with bypass diodes may also provide the necessary disablement. (Bypass diodes are incorporated in photovoltaic power sources for performance purposes.)

D. Wiring Methods

690-31. Methods Permitted.

(a) Wiring Systems. All raceway and cable wiring methods included in this *Code* and other wiring systems and fittings specifically intended and identified for use on photovoltaic arrays shall be permitted. Where wiring devices with integral enclosures are used, sufficient length of cable shall be provided to facilitate replacement.

(b) Single Conductor Cable. Types SE, UF, and USE single-conductor cable shall be permitted in photovoltaic source circuits where installed in the same manner as a Type UF multiconductor cable in accordance with Article 339. Where exposed to direct rays of the sun, Type UF cable identified as sunlight-resistant or Type USE cable shall be used.

Some photovoltaic modules are designed for a direct series connection by having terminations located at both ends. To accommodate such a direct series connection without the waste of one or more conductors in a multiconductor cable, use of a single-conductor Type USE, SE, or UF cable is permitted in photovoltaic source circuits. The Article 339 reference to installation as a multiconductor cable permits the single-conductor cable to be routed separately, not necessarily with the other conductors of a circuit. Since photovoltaic source circuits are direct-current circuits, increased impedance due to separated circuit conductors is not a problem. Since photovoltaic modules may operate at high temperatures and are installed in outdoor, exposed locations, the use of high-temperature, wet-rated conductors such as USE-2 is suggested. UF conductors may not have the necessary temperature ratings required.

(c) Flexible Cords and Cables. Flexible cords and cables, where used to connect the moving parts of tracking PV modules, shall comply with Article 400 and shall be of a type identified as a hard service cord or portable power cable; shall be suitable for extra-hard usage, listed for outdoor use, water resistant, and sunlight resistant. Allowable ampacities shall be in accordance with Section 400-5. For ambient temperatures exceeding 30°C (86°F), the ampacities shall be derated by the appropriate factors given in Table 690-31(c).

(d) Small Conductor Cables. Single-conductor cables listed for outdoor use that are sunlight resistant and moisture resistant with sizes No. 16 and No. 18 shall be permitted for module interconnections where such cables meet the ampacity requirements of Section 690-8. Section 310-15 shall be used to determine the cable ampacity and temperature derating factors.

Since these smaller cables are not normally marked with standard *Code*-recognized markings (e.g., USE, UF, SE), information should be provided by the photovoltaic module manufacturer or installer establishing that these cables have the necessary sunlight and moisture resistance and are suitable for exposed, outdoor use.

Section 200-6(a), allows grounded conductors smaller than No. 6 to be marked with a white marking when used in photovoltaic source circuits.

690-32. Component Interconnections. Fittings and connectors that are intended to be concealed at the time of on-site assembly, where listed for such use, shall be permitted for on-site interconnection of modules or other array components. Such fittings and connectors shall be equal to the wiring method employed in insulation, temperature rise, and fault-current withstand, and shall be capable of resisting the effects of the environment in which they are used.

690-33. Connectors. The connectors permitted by Article 690 shall comply with (a) through (e).

(a) Configuration. The connectors shall be polarized and shall have a configuration that is noninterchangeable with receptacles in other electrical systems on the premises.

(b) Guarding. The connectors shall be constructed and installed so as to guard against inadvertent contact with live parts by persons.

(c) Type. The connectors shall be of the latching or locking type.

(d) Grounding Member. The grounding member shall be the first to make and the last to break contact with the mating connector.

(e) Interruption of Circuit. The connectors shall be capable of interrupting the circuit current without hazard to the operator.

Connectors that are not load-break rated will be marked with a warning label.

Table 690-31(c). Correction Factors

Ambient Temperature (°C)	Temperature Rating of Conductor 60°C (140°F)	75°C (167°F)	90°C (194°F)	105°C (221°F)	Ambient Temperature (°F)
30	1.00	1.00	1.00	1.00	86
31–35	0.91	0.94	0.96	0.97	87–95
36–40	0.82	0.88	0.91	0.93	96–104
41–45	0.71	0.82	0.87	0.89	105–113
46–50	0.58	0.75	0.82	0.86	114–122
51–55	0.41	0.67	0.76	0.82	123–131
56–60	—	0.58	0.71	0.77	132–140
61–70	—	0.33	0.58	0.68	141–149
71–80	—	—	0.41	0.58	150–158

The connectors discussed in Section 690-33(e) are those concealed connectors used in the installation of the photovoltaic modules. Section 690-33(e) does not refer to other connectors that might be used elsewhere in the system.

690-34. Access to Boxes. Junction, pull, and outlet boxes located behind modules or panels shall be installed so that the wiring contained in them can be rendered accessible directly or by displacement of a module(s) or panel(s) secured by removable fasteners and connected by a flexible wiring system.

E. Grounding

690-41. System Grounding. For a photovoltaic power source, one conductor of a 2-wire system rated over 50 volts and a neutral conductor of a 3-wire system shall be solidly grounded.

Exception: Other methods that accomplish equivalent system protection and that utilize equipment listed and identified for the use shall be permitted.

FPN: See Section 250-2(a).

Low-voltage systems that are not grounded must have overcurrent protection in each of the ungrounded conductors, as required by Section 240-21.

With respect to the exception to Section 690-44, other methods that employ available equipment may be used to achieve objectives contained in Section 250-2(a), thereby providing protection for the photovoltaic power source circuits equivalent to solid grounding.

690-42. Point of System Grounding Connection. The dc circuit grounding connection shall be made at any single point on the photovoltaic output circuit.

FPN: Locating the grounding connection point as close as practicable to the photovoltaic source will better protect the system from voltage surges due to lightning.

If other-than-solid grounding is utilized, as permitted by Section 690-41, Exception, the connections should be made in accordance with the markings on the equipment or in the installation instructions.

Stand-alone photovoltaic systems may require the grounding connection point to be located close to the high-current conductors associated with the battery and the inverter.

Photovoltaic systems on the roofs of dwellings that require ground-fault protection of equipment (see Section 690-5), may require that the single-point grounding connection be made inside the ground-fault protection equipment. Connections should be made in accordance with markings on the equipment or in the installation instructions.

690-43. Equipment Grounding. Exposed noncurrent-carrying metal parts of module frames, equipment, and conductor enclosures shall be grounded regardless of voltage.

Equipment grounding is required even in low-voltage (12- and 24-volt) systems not otherwise required to have a system ground. A grounding electrode must be added to an ungrounded system to accommodate the equipment grounds.

690-45. Size of Equipment Grounding Conductor. The equipment grounding conductor shall be not smaller than the required size of the circuit conductors in systems where the available photovoltaic power source short-circuit current is less than twice the current rating of the overcurrent device.

In other systems, the equipment grounding conductor shall be sized in accordance with Section 250-122.

Where fault-current sources may be significant in photovoltaic systems with batteries and/or multiple parallel strings of photovoltaic modules, the equipment grounding requirements of Section 250-122 will generally apply. In systems with only one or two strings of parallel-connected modules and no other source of fault currents, the equipment-grounding conductor must be as large as the circuit conductors.

690-47. Grounding Electrode System. A grounding electrode system shall be provided in accordance with Sections 250-50 through 250-60.

The ac and dc grounding systems for a photovoltaic system, where both are present, should tie to a common grounding electrode or grounding system. If separate electrodes are used, they should be bonded together.

F. Marking

690-51. Modules. Modules shall be marked with identification of terminals or leads as to polarity, maximum overcurrent device rating for module protection, and with rated

(1) Open-circuit voltage
(2) Operating voltage
(3) Maximum permissible system voltage
(4) Operating current
(5) Short-circuit current, and
(6) Maximum power

690-52. Alternating-Current Photovoltaic Modules. Alternating-current modules shall be marked with identification of terminals or leads, and with identification of the rated

(1) Nominal operating ac voltage
(2) Nominal operating ac frequency
(3) Maximum ac power
(4) Maximum ac current, and
(5) Maximum overcurrent device rating for ac module protection

690-53. Photovoltaic Power Source. A marking, specifying the photovoltaic power source rated as follows shall be provided by the installer at the site at an accessible location at the disconnecting means for the photovoltaic power source:

(1) Operating current
(2) Operating voltage
(3) Maximum system voltage, and
(4) Short-circuit current

FPN: Reflecting systems used for irradiance enhancement may result in increased levels of output current and power.

After installation of photovoltaic arrays, it may be difficult to determine the system's rated voltage and current. These ratings, along with the open-circuit voltage and short-circuit current, are necessary to size the remainder of the system components, as specified elsewhere in Article 690.

Generally, the marking described in Section 690-53 is required to be provided by the installer. The rated values for the photovoltaic power source can be calculated by adding voltage ratings of series-connected modules and adding current ratings of parallel-connected modules or photovoltaic source circuits.

With respect to the fine print note, a deliberate increase in the level of irradiance by reflectors or the like can cause the power source to operate at levels above those recommended by the manufacturer. See Section 110-3.

690-54. Interactive System Point of Interconnection. All interactive system(s) points of interconnection with other sources shall be marked at an accessible location at the disconnecting means as a power source with the maximum ac output operating current and the operating ac voltage.

G. Connection to Other Sources

690-60. Identified Interactive Equipment. Only inverters and ac modules listed and identified as interactive shall be permitted in interactive systems.

690-61. Loss of Interactive System Power. An inverter or an ac module in an interactive solar photovoltaic system shall automatically de-energize its output to the connected electrical production and distribution network upon loss of voltage in that system and shall remain in that state until the electrical production and distribution network voltage has been restored.

A normally interactive solar photovoltaic system shall be permitted to operate as a stand-alone system to supply loads that have been disconnected from electrical production and distribution network sources.

The requirement of Section 690-61 prevents energizing of otherwise de-energized system conductors or output conductors of other off-site sources (e.g., an electrical utility) and is intended to prevent electric shock. This feature is normally provided as part of the power conditioning unit.

690-62. Ampacity of Neutral Conductor. If a single-phase, 2-wire inverter output is connected to the neutral and one ungrounded conductor (only) of a 3-wire system or of a 3-phase, 4-wire, wye-connected system, the maximum load connected between the neutral and any one ungrounded conductor plus the inverter output rating shall not exceed the ampacity of the neutral conductor.

690-63. Unbalanced Interconnections.

(a) Single Phase. Single-phase inverters for photovoltaic systems and ac modules in interactive solar photovoltaic systems shall not be connected to 3-phase power systems unless the interconnected system is designed so that significant unbalanced voltages cannot result.

(b) Three Phase. Three-phase inverters and 3-phase ac modules in interactive systems shall have all phases automatically de-energized upon loss of, or unbalanced, voltage in one or more phases unless the interconnected system is designed so that significant unbalanced voltages will not result.

690-64. Point of Connection. The output of a photovoltaic power source shall be connected as specified in (a) or (b).

(a) Supply Side. A photovoltaic power source shall be permitted to be connected to the supply side of the service disconnecting means as permitted in Section 230-82(5).

(b) Load Side. A photovoltaic power source shall be permitted to be connected to the load side of the service disconnecting means of the other source(s) at any distribution equipment on the premises provided that all of the following conditions are met.

(1) Each source interconnection shall be made at a dedicated circuit breaker or fusible disconnecting means.

(2) The sum of the ampere ratings of overcurrent devices in circuits supplying power to a busbar or conductor shall not exceed the rating of the busbar or conductor.

Exception: For a dwelling unit, the sum of the ampere ratings of the overcurrent devices shall not exceed 120 percent of the rating of the busbar or conductor.

(3) The interconnection point shall be on the line side of all ground-fault protection equipment.

Exception: Connection shall be permitted to be made to the load side of ground-fault protection, provided that there is ground-fault protection for equipment from all ground-fault current sources.

(4) Equipment containing overcurrent devices in circuits supplying power to a busbar or conductor shall be marked to indicate the presence of all sources.

Exception: Equipment with power supplied from a single point of connection.

(5) Equipment such as circuit breakers, if backfed, shall be identified for such operation.

H. Storage Batteries

690-71. Installation.

(a) General. Storage batteries in a solar photovoltaic system shall be installed in accordance with the provisions of Article 480. The interconnected battery cells shall be considered grounded where the photovoltaic power source is installed in accordance with Section 690-41, Exception.

Batteries in photovoltaic power systems are usually grounded when the photovoltaic power system is grounded in accordance with Article 690, Part E.

Grounded metal trays and cases or containers (as required by Section 250-110) in flooded, lead-acid battery systems operating over 50 volts have been shown to be a contributing factor to ground faults. Nonconductive trays and cases are recommended to alleviate this problem.

(b) Dwellings.

(1) Storage batteries for dwellings shall have the cells connected so as to operate at less than 50 volts.

Exception: Where live parts are not accessible during routine battery maintenance, a battery system voltage in accordance with Section 690-7 shall be permitted.

(2) Live parts of battery systems for dwellings shall be guarded to prevent accidental contact by persons or objects, regardless of voltage or battery type.

FPN: Batteries in solar photovoltaic systems are subject to extensive charge-discharge cycles and typically require frequent maintenance, such as checking electrolyte and cleaning connections.

At any voltage, a primary safety concern in battery systems is that a fault (e.g., a metal tool dropped onto terminals) may create a fire or explosive hazard. *Guarded,* as defined in Article 100, describes the best method to reduce this hazard.

(c) Current Limiting. A listed, current-limiting, overcurrent device shall be installed in each circuit adjacent to the batteries where the available short-circuit current from a battery or battery bank exceeds the interrupting or withstand ratings of other equipment in that circuit. The installation of current-limiting fuses shall comply with Section 690-16.

Large banks of storage batteries can deliver significant amounts of short-circuit current. Current-limiting overcurrent devices should be used if necessary.

690-72. Charge Control. Equipment shall be provided to control the charging process of the battery. Charge control shall not be required where the design of the photovoltaic source circuit is matched to the voltage rating and charge current requirements of the interconnected battery cells, and the maximum charging current multiplied by 1 hour is less than 3 percent of the rated battery capacity expressed in ampere-hours or as recommended by the battery manufacturer. All adjusting means for control of the charging process shall be accessible only to qualified persons.

FPN: Certain battery types such as valve-regulated lead acid or nickel cadmium can experience thermal failure when overcharged.

690-74. Battery Interconnections. Flexible cables, as identified in Article 400, in sizes No. 2/0 and larger shall be permitted within the battery enclosure from battery terminals to nearby junction box where they shall be connected to an approved wiring method. Flexible battery cables shall also be permitted between batteries and cells within the battery enclosure. Such cables shall be listed for hard service use and identified as moisture resistant.

Battery plates and terminals are frequently constructed of relatively soft lead and lead alloys encased in plastics that are often sealed with asphalt. Large-size, low-stranding, stiff copper conductors, when attached to these components, may cause them to be distorted. The use of flexible cables (see Article 400) will reduce such distortions. Listed cables with the appropriate physical and chemical resistant properties should be used. Welding and "battery" cables are not allowed or described in the *NEC* for this use. Flexible "building wire"-type cables (Chapter 3) are also available and suitable.

I. Systems Over 600 Volts

690-80. General. Solar photovoltaic systems with a maximum system voltage over 600 volts dc shall comply with Article 490 and other requirements applicable to installations rated over 600 volts.

690-85. Definitions. For the purposes of Part I of this article, the voltages used to determine cable and equipment ratings shall be defined as follows.

Battery Circuits. In battery circuits, the voltage shall be the highest voltage experienced under charging conditions.

Photovoltaic Circuits. In dc photovoltaic source circuits and photovoltaic output circuits, the voltage shall be the maximum system voltage.

Article 695 — Fire Pumps

Contents

695-1. Scope.

(a) Covered. This article covers the installation of the following:

(1) Electric power sources and interconnecting circuits
(2) Switching and control equipment dedicated to fire pump drivers

(b) Not Covered. This article does not cover the following:

(1) The performance, maintenance, and acceptance testing of the fire pump system, and the internal wiring of the components of the system
(2) Pressure maintenance (jockey or makeup) pumps

FPN: See *Standard for the Installation of Centrifugal Fire Pumps,* NFPA 20-1996, for further information.

Article 695 was extensively rewritten for the 1999 *Code.* The revision involves some important changes such as provisions to allow for power sources that do not directly originate from a utility. These revisions also harmonize the *NEC* and NFPA 20-1996, *Standard for the Installation of Centrifugal Fire Pumps.* Another major issue that was addressed was the issue of providing power for electric fire pumps installed in multibuilding campus-style arrangements.

Testing and performance criteria for a fire pump system remain within the domain of NFPA 20.

Article 695 does not apply to pumps used to supply sprinkler systems in one- and two-family dwellings. NFPA 13D-1996, *Standard for the Installation of Sprinkler Systems in One- and Two-Family Dwellings and Manufactured Homes,* does not require the use of a listed pump; thus, neither NFPA 20 nor Article 695 is applicable.

Although jockey pumps are not covered by Article 695, they are permitted to be connected to the fire pump feeder.

695-2. Application of Other Articles. Except as required or permitted by this article, the installation of electrical conductors and equipment for fire pumps shall comply with all applicable requirements in Chapters 1 through 4 of this *Code.*

695-3. Power Source(s) for Electric Motor-Driven Fire Pumps. Electric motor-driven fire pumps shall have a reliable source of power.

The power source for an electric motor-driven fire pump must be reliable and have adequate capacity to carry the locked-rotor currents of the fire pump motor and accessory equipment. These two main requirements ensure that the fire pump will operate in the event of a fire without being accidentally disconnected and that the fire pump will continue to operate until the fire is extinguished, the fire pump is purposely shut down, or the pump itself is destroyed.

One or more of the following three basic types of power sources are permitted. If only one source is used, that source must be capable of supplying locked-rotor current to the fire pump and accessory equipment.

1. Power may be supplied by a separate utility service or connection ahead of the main disconnecting means.
2. Power may be supplied by an on-site power production system.
3. Power may be supplied by multiple sources (two or more feeders) on a multibuilding campus where utility or power produced on site are not reasonably available and approved by the authority having jurisdiction.

If none of these options is individually capable of providing reliable power with adequate capacity, a combination (two or more) of the above sources must be used.

ˣ**(a) Individual Sources.** Where reliable, and where capable of carrying indefinitely the sum of the locked-rotor current of the fire pump motor(s) and the pressure maintenance pump motor(s) and the full-load current of the associated fire pump accessory equipment when connected to this power supply, the power source for an electric motor-driven fire pump shall be one or more of the following.

(1) Electric Utility Service Connection. A fire pump shall be permitted to be supplied by a separate service, or by a tap located ahead of and not within the same cabinet, enclosure, or vertical switchboard section as the service disconnecting means. The connection shall be located and arranged so as to minimize the possibility of damage by fire from within the premises and from exposing hazards. A tap ahead of the service disconnecting means shall comply with Section 230-82(4). The service equipment shall comply with the labeling requirements in Section 230-2 and the location requirements in Section 230-72(b).

(2) On-Site Power Production Facility. A fire pump shall be permitted to be supplied by an on-site power production facility. The source facility shall be located and protected to minimize the possibility of damage by fire.

Examples of power sources are illustrated in Figures 695.1 and 695.2.

The determination of whether the serving electric utility may be judged a reliable source of power is an issue for the authority having jurisdiction. Appendix A of NFPA 20 provides guidance for evaluating the reliability of the power supply.

Performance requirements for an alternate source of electric power may be found in *Emergency and Standby Power Systems,* NFPA 110-1996.

Most important in Section 695-3(a)(2) is the requirement that a fire in one source not affect the reliability of another source. For example, if an emergency feeder were physically routed above the normal switchboard and a fire occurred in the normal switchboard, the reliability of the emergency feeder during the fire would be questionable. There are many other examples for such interruption scenarios. Section 695-6(a) and (b) contains requirements for fire protection of the fire pump supply conductors.

ˣ**(b) Multiple Sources.** Where reliable power cannot be obtained from a source described in Section 695-3(a), power shall be supplied from an approved combination of two or more of either of such sources or from an approved combination of one or more of such sources in combination with an on-site standby generator complying with (1) and (3) or two or more feeders complying with (2) or (3) or from two or more sets of feeders complying with (2) and (3).

(1) Generator Capacity. An on-site generator(s) used to comply with this section shall be of sufficient capacity to allow normal starting and running of the motor(s) driving the fire pump(s) while supplying all other simultaneously

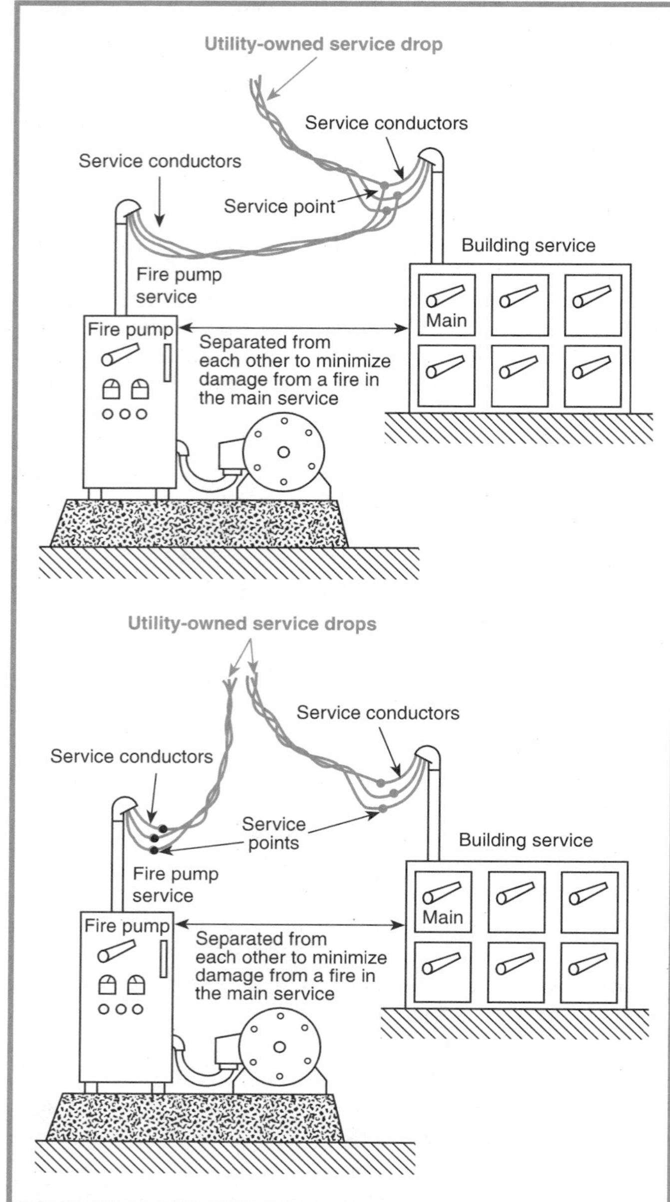

Figure 695.1 *Two different configurations permitted for connecting to electric utility-owned service drops.*

operated load. Automatic shedding of one or more optional standby loads in order to comply with this capacity requirement shall be permitted. A tap ahead of the on-site generator disconnecting means shall not be required.

If the alternate source of power is an on-site generator, the alternate source disconnecting means and the alternate source overcurrent protective device(s) for the electric-drive fire pump are not required to be sized for locked-rotor current of the fire pump motor(s). Rather, the alternate source circuit components are permitted to be sized according to Article 430, provided they are "selected or set to allow instantaneous pickup and running of the fire pump load." See Section 445-4(a).

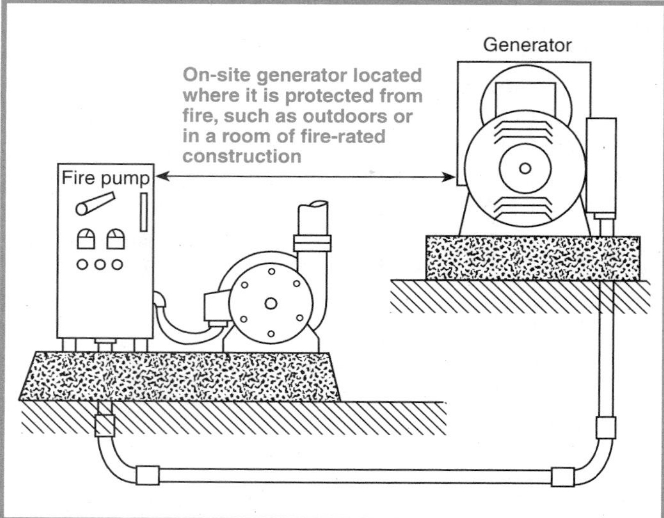

Figure 695.2 *On-site generator as a power source for a fire pump installation.*

(2) Feeder Sources. This section applies to multi-building campus-style complexes with fire pumps at one or more buildings. Where sources in Section 695-3(a) are not practicable, and with the approval of the authority having jurisdiction, two or more feeder sources shall be permitted as one power source or as more than one power source where such feeders are connected to or derived from separate utility services. The connection(s), overcurrent protective device(s), and disconnecting means for such feeders shall meet the requirements of Section 695-4(b).

Section 695-3(b)(2) was added for the 1999 *Code* to permit the use of feeder sources for campus-style applications. Chapters 6 and 7 of NFPA 20-1996, *Standard for the Installation of Centrifugal Fire Pumps,* permit the use of a reliable feeder to supply a fire pump if acceptable to the authority having jurisdiction. See Figure 695.3. In NFPA 20, the use of a *reliable* feeder as a power supply for a fire pump is based on performance criteria, such as the following:

1. Redundant power supply features
2. Special requirements for generators
3. Special requirements for transfer switches
4. Generator overcurrent protection

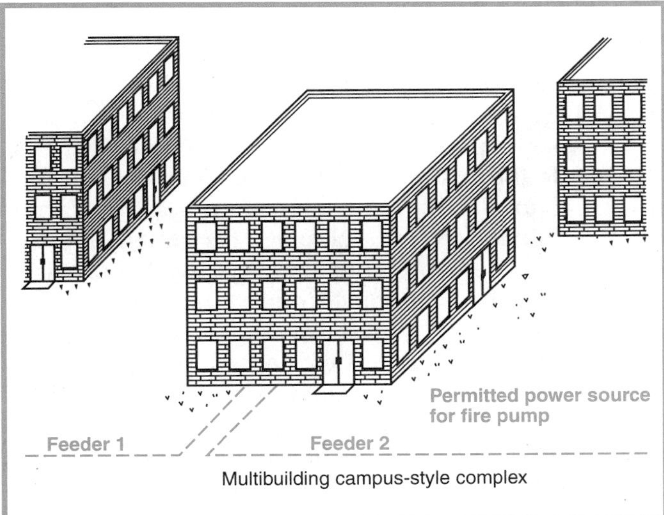

Figure 695.3 *Multiple feeder sources as described in Section 695-3(b)(2).*

(3) Arrangement. The power sources shall be arranged so that a fire at one source will not cause an interruption at the other source.

ˣ**695-4. Continuity of Power.** Circuits that supply electric motor-driven fire pumps shall be supervised from inadvertent disconnection as covered in (a) or (b).

(a) Direct Connection. The supply conductors shall directly connect the power source to either a listed fire pump controller or listed combination fire pump controller and power transfer switch.

(b) Supervised Connection. A single disconnecting means and associated overcurrent protective device(s) shall be permitted to be installed between a remote power source and one of the following:

(a) A listed fire pump controller
(b) A listed fire pump power transfer switch, or
(c) A listed combination fire pump controller and power transfer switch

For systems installed under the provisions of Section 695-3(b)(2) only, such additional disconnecting means and associated overcurrent protective device(s) shall be permitted as required to comply with other provisions of this *Code*. For on-site generators installed under the provisions of Section 695-3(b)(1), the overcurrent devices shall not be required. All disconnecting means and overcurrent protective devices that are unique to the fire pump loads shall comply with the following.

(1) Overcurrent Device Selection. The overcurrent protective device(s) shall be selected or set to carry indefinitely the sum of the locked-rotor current of the fire pump motor(s) and the pressure maintenance pump motor(s) and the full-load current of the associated fire pump accessory equipment when connected to this power supply.

(2) Disconnecting Means. The disconnecting means shall be

(a) Identified as suitable for use as service equipment,
(b) Lockable in the closed position, and
(c) Located sufficiently remote from other building or other fire pump source disconnecting means that inadvertent contemporaneous operation would be unlikely.

(3) Disconnect Marking. The disconnecting means shall be marked "Fire Pump Disconnecting Means." The letters shall be at least 1 in. (25.4 mm) in height, and they shall be visible without opening enclosure doors or covers.

(4) Controller Marking. A placard shall be placed adjacent to the fire pump controller stating the location of this disconnecting means and the location of the key (if the disconnecting means is locked).

Typically, the supply conductors of the power source are directly connected to the fire pump controller or combination fire pump controller and power transfer switch without a disconnecting means or overcurrent protection. However, a single disconnecting means is allowed provided all of the conditions in Section 695-4(b), Supervised Connection, are met.

In addition to the fire pump controller, where a separate service disconnect with overcurrent protection is provided for the fire pump circuit, it will be as follows:

1. The overcurrent device will be sized to carry the locked-rotor currents of all the fire pump motors.
2. The fire pump disconnection means will be located away from the other service disconnecting means. This disconnecting means is rated as service equipment, and is lockable in the on position.
3. The disconnect and controller will be marked and placarded as described.
4. The circuit will be supervised in the closed position, in accordance with Section 695-4(b)(4).

Specifically, the size of the overcurrent protective device is the sum of the locked-rotor currents of all the permitted motors plus the sum of any other fire pump auxiliary loads.

Example

If a fusible service disconnect switch supplies power to a 100-hp, 460-volt, 3-phase fire pump and a 1½-hp, 460-volt, 3-phase jockey pump, what would be the smallest standard-size service disconnect switch and overcurrent protective device permitted for this service?

According to the nameplate of the motors, the locked-rotor current is 725 amperes for the 100-hp motor and 20 amperes for the 1½-hp motor. If the locked-rotor amperes are not on the nameplate, use Table 430-151(B) would be used.

The size is calculated by summing the locked-rotor current of both motors and then going to the next larger standard-size overcurrent device. The motor locked-rotor currents (LRC) are taken from the motor nameplates and are summed as follows:

100-hp, 3-phase LRC	= 725 amperes
1½-hp, 3-phase LRC	= 20 amperes
Total LRC	= 745 amperes

Accordingly, the next larger standard-size disconnect switch and overcurrent device is 800 amperes. An adjustable-trip circuit breaker of 750 amperes is also permitted, since it, too, will carry the locked-rotor current indefinitely.

Even though the disconnect switch, fuse, and circuit breakers are sized according to locked-rotor currents, the feeder conductors to the fire pump controller "shall have an ampacity at least equal to the sum of the full-load current rating as determined by Section 430-6(a)(1) of all the motors, plus 25 percent of the highest rated motor in the group, plus the ampere rating of other loads," according Section 430-24. (Section 695-2 permits using Article 430 to size conductors, since the sizing of conductors is not specifically addressed in Article 695.) Using the same motors as above, the feeder to the fire pump controller would be sized as follows, using Section 430-6(a) and Table 430-150 for full-load currents (FLC) of the motors.

100-hp, 3-phase FLC	= 124 amperes
1½-hp, 3-phase FLC	= 3 amperes
25% of 100-hp motor FLC	= 31 amperes
Total FLC	= 158 amperes

Using Table 310-16, a 2/0 copper conductor is the minimum size under the 75°C column A.

(5) Supervision. The disconnecting means shall be supervised in the closed position by one of the following methods:

(a) Central station, proprietary, or remote station signal device
(b) Local signaling service that will cause the sounding of an audible signal at a constantly attended point
(c) Locking the disconnecting means in the closed position
(d) Sealing of disconnecting means and approved weekly recorded inspections when the disconnecting means are located within fenced enclosures or in buildings under the control of the owner

Supervision of the disconnecting means by a local (protected premises) fire alarm system, central station, proprietary supervising station, or remote supervising station requires a connection to the premises fire alarm system. The connection is generally made through the use of dry contacts in the fire pump controller that are connected to a fire alarm system initiating device circuit. This circuit is programmed to generate a supervisory signal at the fire alarm control unit upon loss of voltage to the fire pump controller. A supervisory signal indicates that the suppression system is off normal. Connection of this device must cause a supervisory signal and may not cause a trouble or alarm signal. For more information, see NFPA 72-1996, *National Fire Alarm Code®*.

695-5. Transformers. Where the service or system voltage is different from the utilization voltage of the fire pump motor, transformer(s) protected by disconnecting means and overcurrent protective devices shall be permitted to be installed between the system supply and the fire pump controller in accordance with (a) and (b), or (c). Only transformers covered in (c) shall be permitted to supply loads not directly associated with the fire pump system.

(a) Size. Where a transformer supplies an electric motor-driven fire pump, it shall be rated at a minimum of 125 percent of the sum of the fire pump motor(s) and pressure maintenance pump(s) motor loads, and 100 percent of the associated fire pump accessory equipment supplied by the transformer.

(b) Overcurrent Protection. The primary overcurrent protective device(s) shall be selected or set to carry indefinitely the sum of the locked-rotor current of the fire pump motor(s) and the pressure maintenance pump motor(s) and the full-load current of the associated fire pump accessory equipment when connected to this power supply. Secondary overcurrent protection shall not be permitted.

See Figure 695.4.

Dedicated transformer and overcurrent protection sizing may be broken down into three basic requirements. Generally stated, they are as follows.

1. The transformer must be sized to at least 125 percent of the sum of the loads.
2. The transformer primary overcurrent device's size must be at least a specified minimum size.
3. The transformer secondary must not contain any overcurrent devices whatsoever.

Example

If a 4160/480-volt, 3-phase, dedicated transformer supplies power to a 100-hp, 460-volt, 3-phase, code

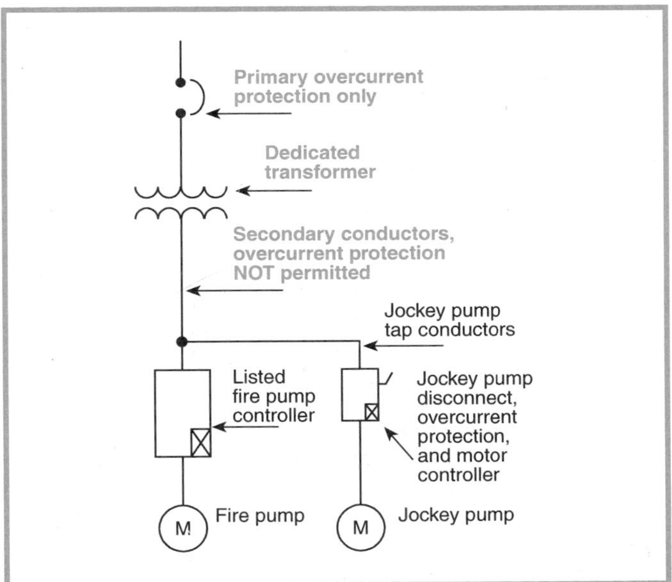

Figure 695.4 *The overcurrent device in the primary of a dedicated transformer supplying a fire pump installation is required to be sized to carry the locked-rotor current motor(s) and associated fire pump accessory equipment indefinitely.*

letter "G" fire pump and a 1½-hp, 460-volt, 3-phase, code letter "H" jockey pump, what would be the minimum standard-size transformer required? What would be the minimum-size primary overcurrent protective device permitted for this transformer?

Step 1. According to Section 695-5(a), the size of this transformer is calculated by summing the load on the transformer, increasing the load of the fire and jockey pumps to 125 percent, and then, using this calculation as a minimum, selecting a transformer size. The motor full-load amperes (FLA) are from Table 430-150.

100-hp, 3-phase FLA	= 124 amperes
1½-hp, 3-phase FLA	= 3 amperes
Total FLA	= 127 amperes

According to Section 695-5(a), the sum of the fire pump motor and the jockey pump motor is then increased to 125 percent.

$$127 \text{ amperes} \times 1.25 = 158.75 \text{ amperes}$$

Then, the transformer is sized as follows:

$$\text{Transformer kVA} = \frac{\text{volts} \times \text{amperes} \times \sqrt{3}}{1000}$$

$$\text{Transformer kVA} = \frac{480 \times 158.75 \times \sqrt{3}}{1000} = 131.98 \text{ kVA}$$

The minimum-size transformer permitted by Section 695-5(a) would be 131.98 kVA. The next larger standard-size transformer available is 150 kVA, but any larger size is permitted.

Step 2. Calculate the minimum-size primary overcurrent protection device permitted for this transformer.

According to Section 695-5(b), the minimum primary overcurrent protection device must allow the transformer to carry the locked-rotor current of the fire pump and, in this case, the jockey pump.

The total locked-rotor currents of both motors must be individually calculated. This data may be available directly from the motor nameplate, if the actual locked-rotor current is indicated. In this example, however, we are assuming that only the kVA code letters are available.

Then, according to Section 430-7(b) and using the maximum values for the individual code letters according to Table 430-7(b), the maximum locked-rotor current is calculated as follows:

For the 100-hp motor, code letter "G" locked-rotor amperes, the locked-rotor amperes = motor horsepower × maximum code letter value kVA × 1000 × $\sqrt{3}$

$$= 100 \text{ hp} \times 6.29 \frac{\text{kVA}}{\text{hp}} \times \frac{1000}{460 \times \sqrt{3}}$$

The 100-hp motor locked-rotor amperes is 789.46 amperes.

The 1½-hp motor, code letter "H" locked-rotor amperes can be calculated using the same formula.

$$\text{Locked-rotor amperes} = 1\tfrac{1}{2} \text{ hp} \times 7.09 \frac{\text{kVA}}{\text{hp}} \times \frac{1000}{460 \times \sqrt{3}}$$

The 1½-hp motor locked-rotor amperes is 13.35 amperes.

Both motors are totalled as follows:

100-hp LRA	= 789.49 amperes
1½-hp LRA	= 13.35 amperes
Total LRA	= 802.81 or 803 amperes for both fire pump and jockey pump

The equivalent locked-rotor current on the primary side of the transformer, based on the calculated locked-rotor current of the secondary of the transformer, can be calculated as follows:

Locked-rotor $\text{amperes}_{\text{primary}}$

$$= \frac{\text{secondary voltage}}{\text{primary voltage}} \times \text{locked-rotor amperes}_{\text{secondary}}$$

$$= \frac{480 \text{ volts}}{4160} \times 803 \text{ amperes}$$

$$= 92.65 \text{ or } 93 \text{ amperes}$$

This value of 93 amperes represents the secondary locked-rotor amperes reflected to the primary side of this transformer. Since this value is the absolute smallest overcurrent protective device permitted, the next larger standard size, according to Section 240-6, is 100 amperes. The minimum-size overcurrent protective device is 100 amperes.

Summary

The calculation for a 4160/480-volt, 3-phase transformer supplying a 100-hp fire pump and a $1^1/_2$-hp jockey pump, both at 460 volts, 3 phase may be summarized as follows:

1. The smallest standard-size transformer permitted is 150 kVA.
2. The smallest standard-size overcurrent protective device permitted on the primary of the transformer is 100 amperes.
3. A secondary overcurrent protective device is not permitted.

(c) Feeder Source. Where a feeder source is provided in accordance with Section 695-3(b)(2), transformers supplying the fire pump system shall be permitted to supply other loads. All other loads shall be calculated in accordance with Article 220, including demand factors as applicable.

(1) Size. Transformers shall be rated at a minimum of 125 percent of the sum of the fire pump motor(s) and pressure maintenance pump(s) motor loads, and 100 percent of the remaining load supplied by the transformer.

(2) Overcurrent Protection. The transformer size, the feeder size, and the overcurrent protective device(s) shall be coordinated such that overcurrent protection is provided for the transformer in accordance with Section 450-3 and for the feeder in accordance with Section 215-3, and such that the overcurrent protective device(s) is selected or set to carry indefinitely the sum of the locked-rotor current of the fire pump motor(s), the pressure maintenance pump motor(s), the full-load current of the associated fire pump accessory equipment, and 100 percent of the remaining loads supplied by the transformer.

695-6. Power Wiring. Power circuits and wiring methods shall comply with the requirements in (a) through (g), and as permitted in Section 230-90(a), Exception No. 4; Section 230-94, Exception No. 4; Section 230-95, Exception No. 2; Section 240-13(3); Section 230-208; Section 240-3(a); and Section 430-31.

(a) Service Conductors. Supply conductors shall be physically routed outside a building(s) and shall be installed as service entrance conductors in accordance with Article 230. Where supply conductors cannot be physically routed outside buildings, they shall be permitted to be routed through buildings where installed in accordance with Condition No. 1 or Condition No. 2 of Section 230-6. Where a fire pump is wired under the provisions of Section 695-3(b)(2), this requirement shall apply to all supply conductors on the load side of the service disconnecting means that constitute the normal source of supply to that fire pump.

Exception: Where there are multiple sources of supply with means for automatic connection from one source to the other, the requirement shall only apply to those conductors on the load side of that point of automatic connection between sources.

(b) Circuit Conductors. Fire pump supply conductors on the load side of the final disconnecting means and overcurrent device(s) permitted by Section 695-4(b) shall be kept entirely independent of all other wiring. They shall only supply loads that are directly associated with the fire pump system, and they shall be protected to resist potential damage by fire, structural failure, or operational accident. They shall be permitted to be routed through a building(s) encased in 2 in. (50.8 mm) of concrete or within enclosed construction dedicated to the fire pump circuit(s) and having a minimum of a 1-hour fire resistance rating, or they shall be permitted to be within listed electrical circuit protective systems with a minimum of 1-hour fire resistance. The installation shall comply with any restrictions provided in the listing of the electrical circuit protective system used.

It is important to understand the difference between a 1-hr fire rating of an electrical circuit, such as a conduit with wires, and a 1-hr fire-resistance rating of a structural member, such as a wall. Simply stated, at the end of a 1-hr fire test on an electrical conduit with wires, the circuit must still function electrically (no short circuits, grounds, or opens are permitted). The circuit and its insulation must still be intact and electrically functioning. However, a wall subjected to a 1-hr fire-resistant test must only prevent a fire from passing through or past the wall, without regard to damage to the wall. All fire ratings and fire-resistance ratings are based on the assumption that the structural supports for the assembly are not impaired by the effects of the fire.

Various Electrical Circuit Protective Systems (FHIT), Electrical Circuit Protective Materials (FHIY), and Fire Resistive Cables (FHJR) used in the fire protection of electrical circuits for fire pumps are described in the UL 1998 *Building Materials Directory.* The four-letter code (shown in parentheses) following each of the category headings is the UL product category guide designation. For information on electrical circuit protective systems, see *Fire Tests for Electrical Circuit Protective Systems,* UL Subject 1724.

Exception: The supply conductors located in the electrical equipment room where they originate and in the fire pump room shall not be required to have the minimum 1-hour fire separation or fire resistance rating, unless otherwise required by Section 700-9(d) of this Code.

(c) Conductor Size.

(1) Conductors supplying a fire pump motor(s), pressure maintenance pumps, and associated fire pump accessory equipment shall have a minimum rating of 125 percent of the sum of the fire pump motor(s) and pressure maintenance motor(s) full-load current(s), and 100 percent of the associated fire pump accessory equipment.

(2) Conductors supplying only a fire pump motor(s) shall have a minimum rating of 125 percent of the fire pump motor(s) full-load current(s).

(d) Overload Protection. Power circuits shall not have automatic protection against overloads. Except as provided in Section 695-5(c)(2), branch-circuit and feeder conductors shall be protected against short circuit only. Where a tap is made to supply a fire pump, and the tap wiring is run in accordance with Section 230-6, the applicable distance and size restrictions in Section 240-21 shall not apply.

The requirements for ground-fault protection are not permitted to apply to the fire pump power wiring. [Ground-fault protection (GFP) is equipment protection and should not be confused with GFCI protection for personnel.] See Section 230-95, Exception No. 2, and Section 240-13(3).

Exception: Conductors between storage batteries and the engine shall not require overcurrent protection or disconnecting means.

(e) Pump Wiring. All wiring from the controllers to the pump motors shall be in rigid metal conduit, intermediate metal conduit, liquidtight flexible metal conduit, or liquidtight flexible nonmetallic conduit Type LFNC-B, or Type MI cable.

The requirement of Section 695-6(e) does not apply to light switches, convenience receptacles, telephone outlets, fire detectors, and similar equipment located within the fire pump room.

(f) Junction Points. Where wire connectors are used in the fire pump circuit, the connectors shall be listed. A fire pump controller or fire pump power transfer switch, where provided, shall not be used as a junction box to supply other equipment including a pressure maintenance (jockey) pump(s). A fire pump controller and fire pump power transfer switch, where provided, shall not serve any load other than the fire pump for which it is intended.

(g) Mechanical Protection. All wiring from engine controllers and batteries shall be protected against mechanical injury, and shall be installed in accordance with the controller and engine manufacturer's instructions.

x695-7. Voltage Drop. The voltage at the controller line terminals shall not drop more than 15 percent below normal (controller-rated voltage) under motor starting conditions. The voltage at the motor terminals shall not drop more than 5 percent below the voltage rating of the motor when the motor is operating at 115 percent of the full-load current rating of the motor.

Exception: This limitation shall not apply for emergency run mechanical starting.

x695-10. Listed Equipment. Diesel engine fire pump controllers, electric fire pump controllers, electric motors, fire pump power transfer switches, foam pump controllers, and limited service controllers shall be listed for fire pump service.

Section 695-10 was added for the 1999 *Code* to harmonize with NFPA 20-1996, *Standard for the Installation of Centrifugal Fire Pumps,* which requires components used in fire pump systems to be listed.

695-12. Equipment Location.

(a) Controllers and Transfer Switches. Electric motor-driven fire pump controllers and power transfer switches shall be located as close as practicable to the motors that they control and shall be within sight of the motors.

(b) Engine-Drive Controllers. Engine-drive fire pump controllers shall be located as close as is practical to the engines that they control and shall be within sight of the engines.

(c) Storage Batteries. Storage batteries for diesel engine drives shall be rack supported above the floor, secured against displacement, and located where they will not be subjected to excessive temperature, vibration, mechanical injury, or flooding with water.

(d) Energized Equipment. All energized equipment parts shall be located at least 12 in. (305 mm) above the floor level.

(e) Protection Against Pump Water. Fire pump controllers and power transfer switches shall be located or protected so that they will not be damaged by water escaping from pumps or pump connections.

(f) Mounting. All fire pump control equipment shall be mounted in a substantial manner on noncombustible supporting structures.

Neither NFPA 20-1996, *Standard for the Installation of Centrifugal Fire Pumps,* nor this *Code* mandates a dedicated room for the fire pump. However, a suitable space for this equipment is specified within NFPA 20.

The phrase "as close as practicable" may require that additional space be available to achieve the minimum maintenance working space set forth in Section 110-26.

Generally, fire pump controllers are housed in substantial enclosures suitable to protect the contents against limited amounts of falling water and dirt. In addition, all energized parts within the enclosure must be mounted at least 12 in. above the floor. Typically, the floor space for this area is equipped with a floor drain.

Section 695-12(f) does not allow fire pump control equipment to be mounted on combustible backboards (such as plywood).

695-14. Control Wiring.

[x]**(a) Control Circuit Failures.** External control circuits shall be arranged so that failure of any external circuit (open or short circuit) shall not prevent the operation of a pump(s) from all other internal or external means. Breakage, disconnecting, shorting of the wires, or loss of power to these circuits may cause continuous running of the fire pump, but shall not prevent the controller(s) from starting the fire pump(s) due to causes other than these external control circuits.

[x]**(b) Sensor Functioning.** No undervoltage, phase-loss, frequency-sensitive, or other sensor(s) shall be installed that automatically or manually prohibit actuation of the motor contactor.

[x]**(c) Remote Device(s).** No remote device(s) shall be installed that will prevent automatic operation of the transfer switch.

[x]**(d) Engine-Drive Control Wiring.** All wiring between the controller and the diesel engine shall be stranded and sized to continuously carry the charging or control currents as required by the controller manufacturer. Such wiring shall be protected against mechanical injury. Controller manufacturer's specifications for distance and wire size shall be followed.

(e) Electric Fire Pump Control Wiring Methods. All electric motor-driven fire pump control wiring shall be in rigid metal conduit, intermediate metal conduit, liquidtight flexible metal conduit, or Type MI cable.

The wiring methods described in Section 695-14(e) apply only to the control wiring for electric motor-driven fire pumps. They do not apply to the control wiring for engine-driven fire pumps, since there are no similar requirements presently in NFPA 20-1996, *Standard for the Installation of Centrifugal Fire Pumps.*

CHAPTER 7

Special Conditions

Article 700 — Emergency Systems

Contents

A. General

700-1. Scope. The provisions of this article apply to the electrical safety of the installation, operation, and maintenance of emergency systems consisting of circuits and equipment intended to supply, distribute, and control electricity for illumination or power, or both, to required facilities when the normal electrical supply or system is interrupted.

Emergency systems are those systems legally required and classed as emergency by municipal, state, federal, or other codes, or by any governmental agency having jurisdiction. These systems are intended to automatically supply illumination or power, or both, to designated areas and equipment in the event of failure of the normal supply or in the event of accident to elements of a system intended to supply, distribute, and control power and illumination essential for safety to human life.

This scope statement includes the definition of *emergency system.* This article applies to the installation of emergency systems essential for safety to human life where such systems are legally required by municipal, state, federal, or other codes or a governmental agency having jurisdiction.

Article 700 does not determine whether emergency systems are required nor does this article specify the location of emergency or exit lights. These determinations may be made by using NFPA *101*®-1997, *Life Safety Code*®.

FPN No. 1: For further information regarding wiring and installation of emergency systems in health care facilities, see Article 517.

FPN No. 2: For further information regarding performance and maintenance of emergency systems in health care facilities, see *Standard for Health Care Facilities,* NFPA 99-1996.

FPN No. 3: Emergency systems are generally installed in places of assembly where artificial illumination is required for safe exiting and for panic control in buildings subject to occupancy by large numbers of persons, such as hotels, theaters, sports arenas, health care facilities, and similar institutions. Emergency systems may also provide power for such functions as ventilation where essential to maintain life, fire detection and alarm systems, elevators, fire pumps, public safety communications systems, industrial processes where current interruption would produce serious life safety or health hazards, and similar functions.

FPN No. 4: For specification of locations where emergency lighting is considered essential to life safety, see *Life Safety Code,*® NFPA *101*®-1997.

FPN No. 5: For further information regarding performance of emergency and standby power systems, see *Standard for Emergency and Standby Power Systems,* NFPA 110-1996.

Emergency systems are designed and installed to maintain a specific degree of illumination or provide power for essential equipment if the normal power supply fails. Essential equipment might be fire pumps or operating room and life-support equipment in hospitals, for example.

If authorities determine that emergency lighting, including the proper placement of exit signs, is required for safe egress from various classes of buildings or parts of buildings, then sufficient illumination is required for corridors, passageways, stairways, lobbies, and so on.

700-2. Application of Other Articles. Except as modified by this article, all applicable articles of this *Code* shall apply.

700-3. Equipment Approval. All equipment shall be approved for use on emergency systems.

700-4. Tests and Maintenance.

(a) Conduct or Witness Test. The authority having jurisdiction shall conduct or witness a test of the complete system upon installation and periodically afterward.

(b) Tested Periodically. Systems shall be tested periodically on a schedule acceptable to the authority having jurisdiction to ensure the systems are maintained in proper operating condition.

(c) Battery Systems Maintenance. Where battery systems or unit equipments are involved, including batteries used for starting, control, or ignition in auxiliary engines, the authority having jurisdiction shall require periodic maintenance.

(d) Written Record. A written record shall be kept of such tests and maintenance.

(e) Testing Under Load. Means for testing all emergency lighting and power systems during maximum anticipated load conditions shall be provided.

Testing of the emergency system may be broken down into two general categories. The first category is acceptance testing. Acceptance testing is performed after the system has been installed but before the system is used. Acceptance testing ensures that the system meets or exceeds the original installation specification.

The second category is operational testing, which is performed during the life of the system. Operational testing ensures that the system remains functional and that maintenance has been performed adequately.

Section 700-4 clearly requires both types of testing. Written records of both types of testing and maintenance are required.

Further information on tests and maintenance may be found in NFPA 72-1996, *National Fire Alarm Code®;* NFPA 99-1996, *Standard for Health Care Facilities;* NFPA *101*-1997, *Life Safety Code;* NFPA 110-1996, *Standard for Emergency and Standby Power Systems;* and NFPA 111-1996, *Standard on Stored Electrical Energy Emergency and Standby Power Systems.*

700-5. Capacity.

(a) Capacity and Rating. An emergency system shall have adequate capacity and rating for all loads to be operated simultaneously. The emergency system equipment shall be suitable for the maximum available fault current at its terminals.

It is essential that the emergency system be designed with adequate capacity and rating to safely carry the entire load connected to the emergency system at one time. The emergency system is required to be capable of restarting emergency loads, such as motors, that may have stopped. Emergency system equipment is required to be suitable for the available fault current. Using devices that limit the available fault current is one method of achieving suitability.

(b) Selective Load Pickup, Load Shedding, and Peak Load Shaving. The alternate power source shall be permitted to supply emergency, legally required standby and optional standby system loads where automatic selective load pickup and load shedding is provided as needed to ensure adequate power to (1) the emergency circuits, (2) the legally required standby circuits, and (3) the optional standby circuits, in that order of priority. The alternate power source shall be permitted to be used for peak load shaving, provided the above conditions are met.

Section 700-5(b) permits a generator to serve more than one level of emergency, standby, or other loads. It also permits the use of a generator for peak load shaving, supplying back-up power, and other uses. However, assurance is required that priority loads will be properly and reliably served. To provide the necessary assurance, such systems must be maintained and periodically tested.

Peak load-shaving operation shall be permitted for satisfying the test requirement of Section 700-4(b), provided all other conditions of Section 700-4 are met.

A portable or temporary alternate source shall be available whenever the emergency generator is out of service for major maintenance or repair.

If a generator is used for peak load shaving or in a cogeneration system, major down time for maintenance of the generator must be anticipated. On the other hand, using the emergency generator on a regular basis for nonemergency loads provides assurance that the emergency generator will supply emergency power when it's needed. The requirement for a portable or temporary alternate source is to provide emergency power when the generator set is out of service for a long time. A major maintenance or repair procedure is one that keeps the generator set out of service for more than a few hours.

700-6. Transfer Equipment.

(a) Transfer equipment, including automatic transfer switches, shall be automatic, identified for emergency use, and approved by the authority having jurisdiction. Transfer equipment shall be designed and installed to prevent the inadvertent interconnection of normal and emergency sources of supply in any operation of the transfer equipment.

(b) Means shall be permitted to bypass and isolate the transfer equipment. Where bypass isolation switches are used, inadvertent parallel operation shall be avoided.

(c) Automatic transfer switches shall be electrically operated and mechanically held.

Section 700-6(c) was added in the 1999 *Code* to ensure that relay contacts will be mechanically held in the event of coil failure. This change correlates the *NEC* with the requirements found in NFPA 110-1996, *Standard for Emergency and Standby Power Systems.*

(d) Transfer equipment shall supply only emergency loads.

Moved to Section 700-6(d) in the 1999 *Code,* this requirement first appeared as Exception No. 1 to Section 700-9(b) in the 1996 *Code.*

Although the alternate power source is permitted to supply emergency loads as well as other loads, the transfer switch used for the emergency system is strictly limited to emergency loads only, that is, loads classed as emergency in accordance with Section 700-1. Other loads, such as legally required standby loads or optional standby loads (covered by Articles 701 and 702) are not permitted to be supplied from the emergency system transfer switch. If a single generator is used to supply both emergency and nonemergency loads, then multiple transfer switches are required.

700-7. Signals. Audible and visual signal devices shall be provided, where practicable, for the following purposes.

(a) Derangement. To indicate derangement of the emergency source.

(b) Carrying Load. To indicate that the battery is carrying load.

(c) Not Functioning. To indicate that the battery charger is not functioning.

(d) Ground Fault. To indicate a ground fault in solidly grounded wye emergency systems of more than 150 volts to ground and circuit-protective devices rated 1000 amperes or more. The sensor for the ground-fault signal devices shall be located at, or ahead of, the main system disconnecting means for the emergency source, and the maximum setting of the signal devices shall be for a ground-fault current of 1200 amperes. Instructions on the course of action to be taken in event of indicated ground fault shall be located at or near the sensor location.

FPN: For signals for generator sets, see *Standard for Emergency and Standby Power Systems,* NFPA 110-1996.

The major causes of emergency equipment failure are improper testing or lack of testing, inadequate maintenance, and the failure of attendants to see the signals that indicate malfunctioning battery-charging equipment.

Signal devices should be located where attendants or other personnel who are familiar with the operation of the emergency equipment can see or hear them.

In locations such as theaters or assembly halls, audible signal bells or horns that annunciate the functions specified in Section 700-7 should be located where their sounding will not cause panic.

Battery-operated unit equipment generally has a test switch to simulate a failure of the normal system and an indicating light that glows brightly while charging and dims when ready. Transparent cases for lead-acid batteries allow easy viewing of electrolyte levels.

A storage battery system is normally capable of delivering 12 volts, 24 volts, 32 volts, or 120 volts and consists of monitoring and distribution cabinets and a console with battery and charger. It generally includes audio, visual, and remote signal devices and a test switch, and may include a trouble bell and silence switch.

Although Section 700-26 indicates that ground-fault protection of equipment is not required on the alternate source for emergency systems, ground faults can occur on such systems. They can result in equipment burndown. Because of the emergency nature of such systems, automatic disconnect in the event of a ground fault is not appropriate. Detection of such a fault, however, is desirable.

700-8. Signs.

(a) Emergency Sources. A sign shall be placed at the service entrance equipment indicating type and location of on-site emergency power sources.

Exception: A sign shall not be required for individual unit equipment as specified in Section 700-12(e).

(b) Grounding. Where the grounded circuit conductor connected to the emergency source is connected to a grounding electrode conductor at a location remote from the emergency source, there shall be a sign at the grounding location that shall identify all emergency and normal sources connected at that location.

Section 700-8(b) requires a sign at the grounding location, if the emergency source is a separately derived system and is connected to a grounding electrode conductor at a location that is remote from the emergency source.

B. Circuit Wiring

700-9. Wiring, Emergency System.

(a) Identification. All boxes and enclosures (including transfer switches, generators, and power panels) for emergency circuits shall be permanently marked so they will be readily identified as a component of an emergency circuit or system.

The marking may be a color code, the identification "emergency system," or any other type of identification that will identify the box or enclosure as a component of the emergency system.

(b) Wiring. Unless otherwise permitted in (1) through (4), wiring from an emergency source or emergency source distribution overcurrent protection to emergency loads shall be kept entirely independent of all other wiring and equipment. Wiring of two or more emergency circuits supplied from the same source shall be permitted in the same raceway, cable, box, or cabinet.

(1) The normal power source wiring shall be permitted to be located in transfer equipment enclosures.

(2) In exit or emergency lighting fixtures, wiring supplied from two sources shall be permitted.

(3) In a common junction box, attached to exit or emergency lighting fixtures, wiring supplied from two sources shall be permitted.

(4) The wiring within a common junction box attached to unit equipment, containing only the branch circuit supplying the unit equipment and the emergency circuit supplied by the unit equipment shall be permitted.

Emergency circuit wiring is not permitted to enter the same raceway, cable, box, or cabinet with the regular or normal wiring of the building concerned. Wiring for the emergency circuits is required to be completely independent of all other wiring and equipment. This ensures that any fault on the normal wiring circuits will not affect the performance of the emergency wiring or equipment.

To effect an immediate transfer from one system to the other, both the normal source and the emergency source must be present within a transfer switch enclosure per Section 700-9(b)(1).

Sections 700-9(b)(2) and (3) permit the use of two-lamp exit or two-lamp emergency fixtures where one lamp is connected to the normal supply and one lamp is connected to the alternate supply. Both lamps may be illuminated as part of the regular lighting operation.

Wiring on the load side of a transfer switch serves as both the emergency circuit wiring and the normal circuit wiring. It is not intended that two sets of wiring supply emergency loads from the transfer switch to the emergency load distribution panel, as shown in Figures 700.3 and 700.4, or from these emergency distribution panels to the emergency loads.

(c) Wiring Design and Location. Emergency wiring circuits shall be designed and located to minimize the hazards that might cause failure due to flooding, fire, icing, vandalism, and other adverse conditions.

The purpose of this requirement is to minimize the likelihood of impairment of the emergency system due to flood, fire, icing, vandalism, and other adverse conditions. The same requirement applies to sources of power covered in Section 700-12.

(d) Fire Protection. Emergency systems shall meet the following additional requirements in assembly occupancies greater than 1000 persons or in buildings above 75 ft (23 m) in height with any of the following occupancy classes: assembly, educational, residential, detention and correctional, business, and mercantile.

(1) Feeder-circuit wiring shall meet one of the following conditions:

(a) Be installed within buildings that are fully protected by an approved automatic fire suppression system
(b) Be a listed electrical circuit protective system with a minimum 1-hour fire rating

(c) Be protected by a listed thermal barrier system for electrical system components
(d) Be protected by a fire-rated assembly having a minimum fire rating of 1 hour
(e) Be embedded in a minimum of 2 in. (50.8 mm) of concrete
(f) Be a cable listed to maintain circuit integrity for a minimum of 1 hour when installed in accordance with the listing requirements

(2) Equipment for feeder circuits (including transfer switches, transformers, panelboards, etc.) shall be located either in spaces fully protected by approved automatic fire suppression systems (sprinklers, carbon dioxide systems, etc.) or in spaces with a 1-hour fire resistance rating.

FPN: For the definition of *occupancy class*, see Section 4-1 of *Life Safety Code*, NFPA *101*-1997.

Proper operation of emergency electrical systems is critical for densely populated occupancies and for high-rise occupancies. Therefore, fire protection requirements for both emergency system feeder circuits and equipment ensure the integrity as well as the performance of the emergency electrical system. These fire protection requirements were added in the 1996 *Code* and the methods of protection have been expanded for the 1999 *Code*.

If feeders and equipment are located within buildings that are fully protected by an approved fire suppression system, then no further fire protection techniques are generally required.

Sprinkler systems are the most common fire suppression systems and they are covered in NFPA 13-1996, *Standard for the Installation of Sprinkler Systems*. Buildings that are fully protected by automatic sprinkler systems meet the requirements of this section. Additional fire suppression systems are included in the following standards:

1. NFPA 11-1998, *Standard for Low-Expansion Foam*
2. NFPA 12-1998, *Standard on Carbon Dioxide Extinguishing Systems*
3. NFPA 12A-1997, *Standard on Halon 1301 Fire Extinguishing Systems*
4. NFPA 15-1996, *Standard for Water Spray Fixed Systems for Fire Protection*
5. NFPA 17-1998, *Standard for Dry Chemical Extinguishing Systems*
6. NFPA 2001-1996, *Standard on Clean Agent Fire Extinguishing Systems*

If feeders and equipment are not located within buildings that are fully protected by an approved fire suppression system, many other methods and protection techniques are available to comply with these fire protection requirements.

Further information about these fire protection methods and techniques are as follows.

*1. **Listed electrical circuit protective systems.*** These systems are described in the 1998 UL *Building Materials Directory*. The four-letter code (shown in parentheses) following each of the category headings is the UL product category guide designation. Examples of these systems include Electrical Circuit Protective Systems (FHIT), Electrical Circuit Protective Materials (FHIY), and Fire Resistive Cables (FHJR).

*2. **Listed thermal barrier systems.*** These systems are described in the 1998 UL *Building Materials Directory* as Thermal Barrier Systems (XCLF). Examples of this protection technique include batts and blankets (XCLR), packing material (XCMD), and performed mineral and fiber units (XCMK) wrapped or otherwise formed over the conduit to achieve a predetermined fire rating.

*3. **Fire-rated assembly.*** These systems are described in the 1998 *UL Fire Resistance Directory*, Volumes 1 and 2. The assemblies found in Volume 1 include hourly ratings for beams, floors, roofs, columns, and walls and partitions. Volume 2 of this directory includes hourly ratings for joint systems and through-penetration firestop systems.

All fire ratings and fire resistance ratings are based on the assumption that the structural supports for the assembly are not impaired by the fire.

*4. **Embedded in concrete.*** Embedding a conduit in concrete is most effective when implemented during original construction. This method has been successful for many years in protecting premises from service conductors. According to Section 230-6, conductors embedded in not less than 2 in. of concrete are considered outside of the building.

*5. **Cables listed to maintain circuit integrity.*** Circuit integrity cables are classified by Underwriters Laboratories 1998 *Building Materials Directory* under the existing product category Fire Resistive Cables (FHJR).

It is important to understand the difference between a 1-hour fire rating of an electrical cable and a 1-hour fire resistance rating of a structural member, such as a wall. Simply stated, at the end of a 1-hour fire rating test on an electrical cable, the circuit and its insulation must be intact and electrically functioning. (No short circuits, grounds, or opens are permitted.) However, a wall subjected to a 1-hour fire resistance test must only prevent a fire from passing through or past the wall, without regard to damage to the wall.

C. Sources of Power

700-12. General Requirements. Current supply shall be such that, in the event of failure of the normal supply to, or within, the building or group of buildings concerned, emergency lighting, emergency power, or both shall be available within the time required for the application but not to exceed 10 seconds. The supply system for emergency purposes, in addition to the normal services to the building and meeting the general requirements of this section, shall be one or more of the types of systems described in (a) through (d). Unit equipment in accordance with (e) shall satisfy the applicable requirements of this article.

In selecting an emergency source of power, consideration shall be given to the occupancy and the type of service to be rendered, whether of minimum duration, as for evacuation of a theater, or longer duration, as for supplying emergency power and lighting due to an indefinite period of current failure from trouble either inside or outside the building.

Equipment shall be designed and located to minimize the hazards that might cause complete failure due to flooding, fires, icing, and vandalism.

Equipment for sources of power as described in (a) through (d) where located within assembly occupancies greater than 1000 persons or in buildings above 75 ft (23 m) in height with any of the following occupancy classes — assembly, educational, residential, detention and correctional, business, and mercantile — shall be installed either in spaces fully protected by approved automatic fire suppression systems (sprinklers, carbon dioxide systems, etc.), or in spaces with a 1-hour fire rating.

FPN No. 1: For the definition of *occupancy class,* see Section 4-1 of *Life Safety Code,* NFPA *101*-1997.

FPN No. 2: Assignment of degree of reliability of the recognized emergency supply system depends on the careful evaluation of the variables at each particular installation.

(a) Storage Battery. Storage batteries used as a source of power for emergency systems shall be of suitable rating and capacity to supply and maintain the total load for a period of 1½ hours minimum, without the voltage applied to the load falling below 87½ percent of normal.

Batteries, whether of the acid or alkali type, shall be designed and constructed to meet the requirements of emergency service and shall be compatible with the charger for that particular installation.

For a sealed battery, the container shall not be required to be transparent. However, for the lead acid battery that requires water additions, transparent or translucent jars shall be furnished. Automotive-type batteries shall not be used.

An automatic battery charging means shall be provided.

(b) Generator Set.

(1) A generator set driven by a prime mover acceptable to the authority having jurisdiction and sized in accordance with Section 700-5. Means shall be provided for automatically starting the prime mover on failure of the normal service and for automatic transfer and operation of all required electrical circuits. A time-delay feature permitting a 15-minute setting shall be provided to avoid retransfer in case of short-time reestablishment of the normal source.

(2) Where internal combustion engines are used as the prime mover, an on-site fuel supply shall be provided with an on-premise fuel supply sufficient for not less than 2 hours full-demand operation of the system. Where power is needed for the operation of the fuel transfer pumps to deliver fuel to a generator set day tank, this pump shall be connected to the emergency power system.

Engine-driven generators that require a fuel pump may not start or continue operating if the fuel pump is not operating. The last sentence of Section 700-12(b)(2) was added to the 1999 *Code* to ensure that fuel transfer pumps will have power when needed.

(3) Prime movers shall not be solely dependent on a public utility gas system for their fuel supply or municipal water supply for their cooling systems. Means shall be provided for automatically transferring from one fuel supply to another where dual fuel supplies are used.

Exception: Where acceptable to the authority having jurisdiction, the use of other than on-site fuels shall be permitted where there is a low probability of a simultaneous failure of both the off-site fuel delivery system and power from the outside electrical utility company.

(4) Where a storage battery is used for control or signal power, or as the means of starting the prime mover, it shall be suitable for the purpose and shall be equipped with an automatic charging means independent of the generator set. Where the battery charger is required for the operation of the generator set, it shall be connected to the emergency system. Where power is required for the operation of dampers used to ventilate the generator set, the dampers shall be connected to the emergency system.

Engine-driven generators that are not equipped with an alternator to provide battery charging require a separate battery charging system. Failure of the power supply to a battery charging system or failure of the battery charging system itself could render the engine-driven generator inoperable. The second sentence of Section 700-12(b)(4) was added to the 1999 *Code* to ensure that the charging system will always have power.

Ventilation dampers may require electrical power in order to operate. If this is the case, this new requirement also requires them to be powered by the emergency power source to ensure operation when the engine-driven generator operates. These additional loads require the system designer to ensure that the engine-driven generator is of adequate capacity in accordance with Section 700-5(a).

(5) Generator sets that require more than 10 seconds to develop power shall be acceptable, provided an auxiliary power supply will energize the emergency system until the generator can pick up the load.

Figure 700.1 illustrates a typical generator installation supplying standby power in ratings from 55 kW to 930 kW, 60 Hz.

(c) Uninterruptible Power Supplies. Uninterruptible power supplies used to provide power for emergency systems shall comply with the applicable provisions of Section 700-12(a) and (b).

(d) Separate Service. Where acceptable to the authority having jurisdiction as suitable for use as an emergency source, a second service shall be permitted. This service shall be in accordance with Article 230, with separate service drop or lateral, widely separated electrically and physically from the normal service to minimize the possibility of simultaneous interruption of supply.

(e) Unit Equipment. Individual unit equipment for emergency illumination shall consist of the following:

(1) A rechargeable battery
(2) A battery charging means
(3) Provisions for one or more lamps mounted on the equipment, or shall be permitted to have terminals for remote lamps, or both, and
(4) A relaying device arranged to energize the lamps automatically upon failure of the supply to the unit equipment

The batteries shall be of suitable rating and capacity to supply and maintain at not less than 87½ percent of the nominal battery voltage for the total lamp load associated with the unit for a period of at least 1½ hours, or the unit equipment shall supply and maintain not less than 60 percent of the initial emergency illumination for a period of at least 1½ hours. Storage batteries, whether of the acid or alkali type, shall be designed and constructed to meet the requirements of emergency service.

Unit equipment shall be permanently fixed in place (i.e., not portable) and shall have all wiring to each unit installed in accordance with the requirements of any of the wiring methods in Chapter 3. Flexible cord and plug connection shall be permitted, provided that the cord does not exceed 3 ft (914 mm) in length. The branch circuit feeding the unit equipment shall be the same branch circuit as that serving the normal lighting in the area and connected ahead of any local switches. The branch circuit that feeds unit equipment shall be clearly identified at the distribution panel. Emergency illumination fixtures that obtain power from a unit equipment and are not part of the unit equipment shall be wired to the unit equipment as required by Section 700-9 and by one of the wiring methods of Chapter 3.

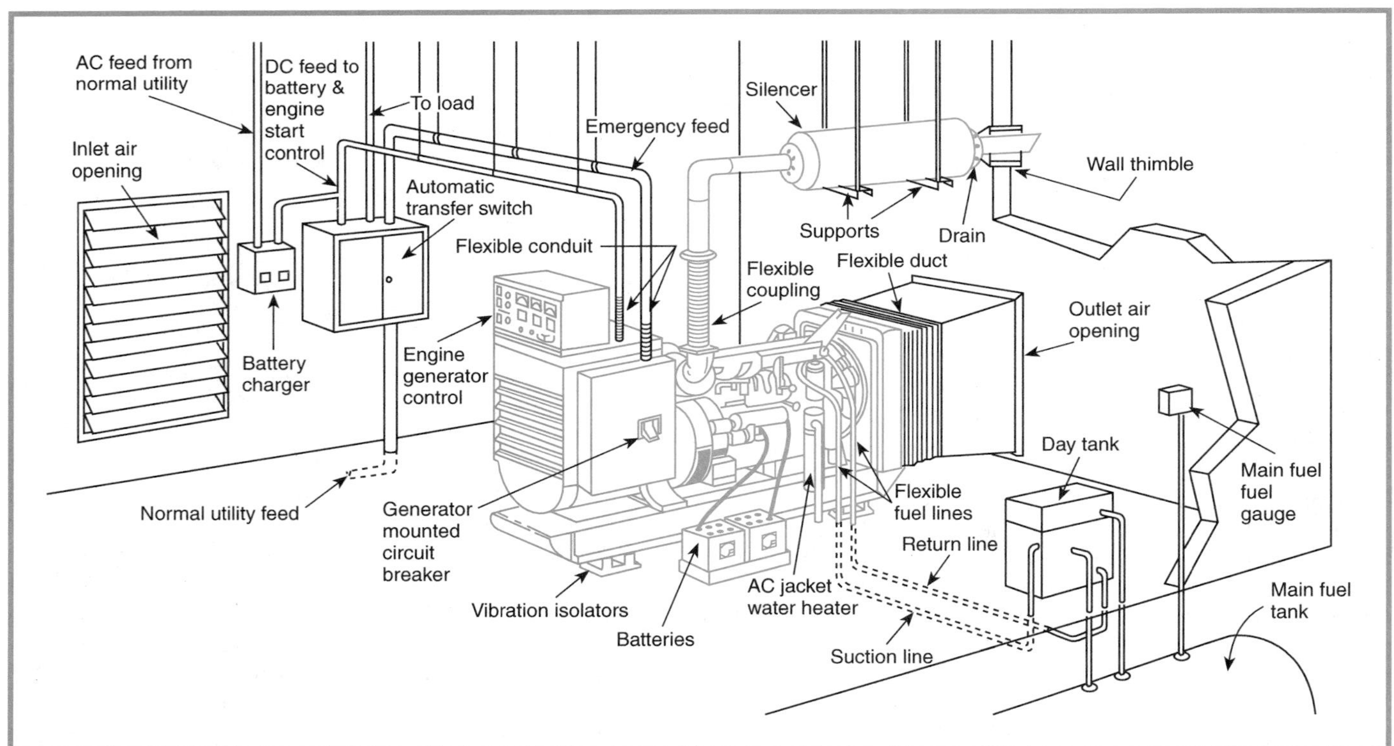

Figure 700.1 *A typical generator installation supplying standby power in ratings from 55 kW to 930 kW, 60 Hz. (Caterpillar)*

Exception: In a separate and uninterrupted area supplied by a minimum of three normal lighting circuits, a separate branch circuit for unit equipment shall be permitted if it originates from the same panelboard as that of the normal lighting circuits and is provided with a lock-on feature.

When designing emergency systems, whether for lighting or power, or both, the *type* of service that is needed must be considered.

Supply systems for emergency systems can be designed as one or more of the following.

1. One storage battery or a group of storage batteries provided with an automatic battery-charging means. (See also Article 480, Storage Batteries.)

2. A generator set driven by a prime mover, acceptable to the authority having jurisdiction, and with adequate capacity to carry the maximum load connected. Prime movers may be internal-combustion engines, steam or gas turbines, or other approved types. A storage battery used to start the prime mover is required to be provided with an automatic battery-charging means. An on-site fuel supply sufficient to operate internal-combustion engines at full load for 2 hours is required to be available.

Off-site fuel supplies may be used where experience has demonstrated their reliability. Off-site fuel supplies may also be used where they will provide greater reliability than gasoline or diesel engines or in isolated areas where maintenance or refueling could be a problem.

Some types of drivers, particularly large ones, may take longer than 10 seconds to accelerate and develop generator voltage. Gas and steam turbines and large internal-combustion engines may have prolonged starting times. Depending on the specific loads, short-time supply could be provided by an uninterruptible power supply; a generator shared with other loads; or a generator with limited emergency supply, such as an expander, steam turbine, or waste heat system.

3. Uninterruptible power supplies (UPS), which generally include a rectifier, storage battery, and inverter to ac. These may be very complex systems with redundant components and high-speed solid-state switching. It is common practice to include an automatic bypass for UPS malfunction to permit maintenance.

4. The use of a separate service requires a judgment by the authority having jurisdiction. Such judgment should be based on the nature of the emergency loads and the expected reliability of the other available sources.

5. Unit equipment may be wired with a flexible cord (not longer than 3 ft) and attachment plug cap. This equipment must be permanently fixed in place, usually by mounting screws that are accessible only from within the unit. One or more lamps may be mounted on or remote from the unit. The unit should be located where it can be readily checked or tested for proper performance. See Figure 700.2.

Figure 700.2 *Self-contained, fully automatic unit equipment for operating emergency lighting located on the unit or for remotely located exit signs or lighting heads. (Dual-Lite, Inc.)*

Unit equipment is intended to provide illumination for the area where it is installed. For instance, if a unit is located in a corridor, it is required to be connected to the branch circuit supplying the normal corridor lights (on the line side of any switching arrangements). If normal power fails, the unit would automatically energize the unit lamps, restoring illumination to the corridor. A separate circuit is not permitted for unit equipment (except as noted in the exception) because, if applied to the above example, failure of the normal corridor circuit would not affect the unit equipment, and the corridor would remain dark. The branch circuit feeding the unit must be identified at the panelboard.

Notes on General Requirements for Emergency Lighting Systems

At least two sources of power must be provided — one normal supply and one or more emergency systems described in Section 700-12. The sources may be (1) two services, one normal supply and one emergency supply (preferably from separate utility stations), (2) one normal service and a storage battery (or unit equipment) system, or (3) one normal service and a generator set. (See Figures 700.3 and 700.4.)

A transfer means (or throw-over switch) must be provided to energize the emergency equipment from the alternate supply when the normal source of supply is interrupted.

If a separate service is used, both may operate normally, but equipment for emergency lighting and power must be arranged to be energized from either service.

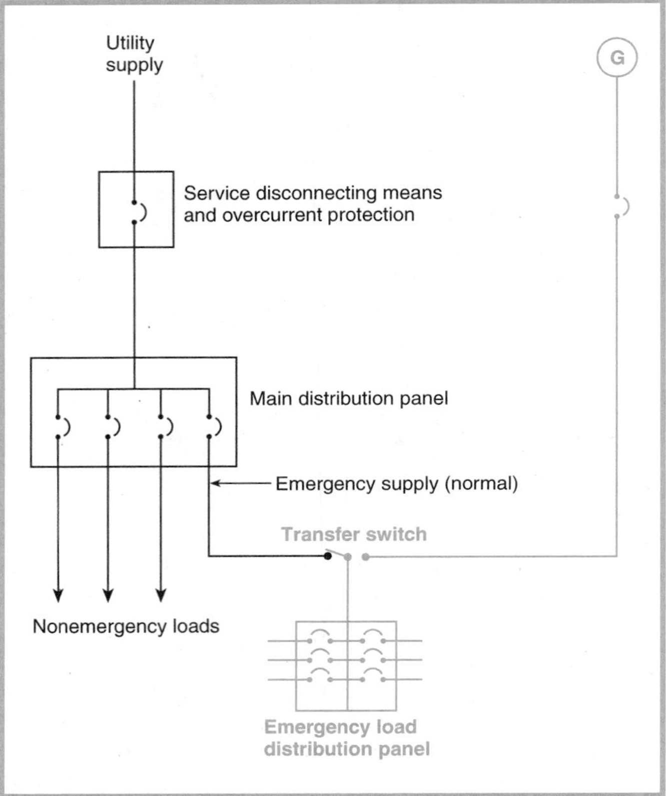

Figure 700.3 *Emergency load arranged to be supplied from a generator, as permitted by Section 700-12(b).*

If the alternate or emergency source of supply is a storage battery or generator set, then the single emergency system is usually operated on the normal service, and the battery (or batteries) or generator operates only if the normal service fails. A generator may be used, however, for peak load shaving and the like in accordance with Section 700-5(b).

Two or more separate and complete systems may provide power for emergency lighting, but means must be provided for energizing one system if the other one fails.

It should be noted that provisions for disconnecting means and overcurrent protection (see Figures 700.3 and 700.4) are to be provided for emergency systems as required by Article 230. (See also Section 230-83.)

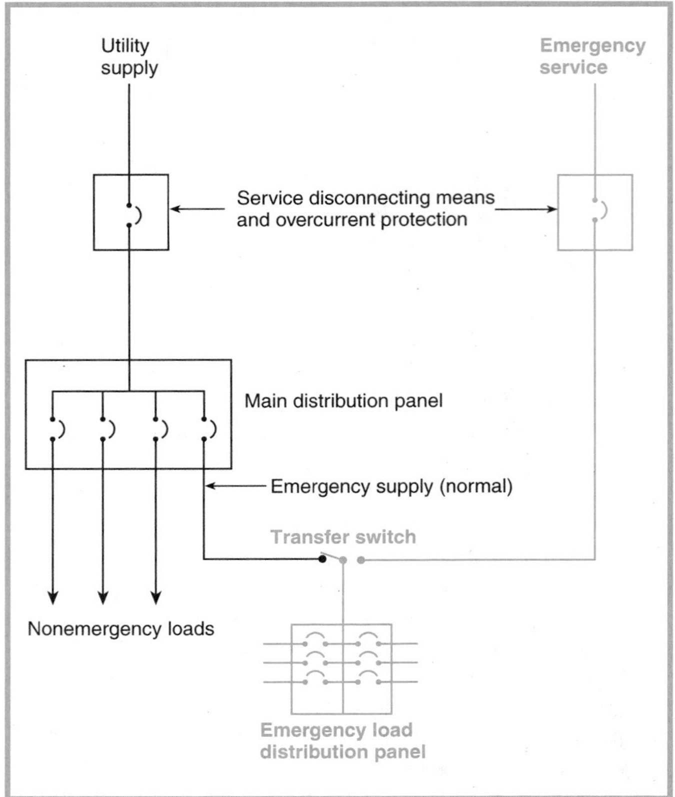

Figure 700.4 *Emergency load arranged to be supplied from two widely separated services, as permitted by Section 700-12(d). When one service fails, the emergency load will be transferred to the other service.*

D. Emergency System Circuits for Lighting and Power

700-15. Loads on Emergency Branch Circuits. No appliances and no lamps, other than those specified as required for emergency use, shall be supplied by emergency lighting circuits.

700-16. Emergency Illumination. Emergency illumination shall include all required means of egress lighting, illuminated exit signs, and all other lights specified as necessary to provide required illumination.

Emergency lighting systems shall be designed and installed so that the failure of any individual lighting element, such as the burning out of a light bulb, cannot leave in total darkness any space that requires emergency illumination.

Where high-intensity discharge lighting such as high- and low-pressure sodium, mercury vapor, and metal halide is used as the sole source of normal illumination, the emergency lighting system shall be required to operate until normal illumination has been restored.

Exception: Alternative means that ensure emergency lighting illumination level is maintained shall be permitted.

High-intensity discharge (HID) fixtures take some time to start once they are energized. Therefore, if they are the sole source of normal illumination in an area, the *Code* requires that the emergency lighting system operate not only until the normal system is returned to service, but also until the HID fixtures provide illumination. This may require a timing circuit, photoelectric monitoring system, or the equivalent.

700-17. Circuits for Emergency Lighting. Branch circuits that supply emergency lighting shall be installed to provide service from a source complying with Section 700-12 when the normal supply for lighting is interrupted. Such installations shall provide either one of the following: (1) an emergency lighting supply, independent of the general lighting supply, with provisions for automatically transferring the emergency lights upon the event of failure of the general lighting system supply, or (2) two or more separate and complete systems with independent power supply, each system providing sufficient current for emergency lighting purposes. Unless both systems are used for regular lighting purposes and are both kept lighted, means shall be provided for automatically energizing either system upon failure of the other. Either or both systems shall be permitted to be a part of the general lighting system of the protected occupancy if circuits supplying lights for emergency illumination are installed in accordance with other sections of this article.

General Considerations for Transfer Switches

Automatic transfer switches of double-throw construction are used primarily for emergency and standby power generation systems rated 600 volts or less. These transfer switches do not normally incorporate overcurrent protection and are designed and applied in accordance with the *Code,* particularly Articles 700 and 701. They are available in ratings from 30 to 3000 amperes. For reliability, most automatic transfer switches rated above 100 amperes are mechanically held and electrically operated from the power source to which the load is to be transferred.

An automatic transfer switch is usually located in the main or secondary distribution bus that feeds the branch circuits. Because of its location in the system, the capabilities that must be designed into the transfer switch are unique and extensive as compared with the design requirements for other branch-circuit and feeder devices. For example, special consideration should be given to the following characteristics of an automatic transfer:

1. Its ability to close against high inrush currents
2. Its ability to carry full-rated current continuously from normal and emergency sources
3. Its ability to withstand fault currents
4. Its ability to interrupt six times the full-load current

In addition to considering each of the above characteristics individually, it is also necessary to consider the effect each has on the other.

In arrangements that provide protection against failure of the utility service, consideration should also be given to the following:

1. An open circuit within the building area on the load side of the incoming service
2. Overload or fault condition
3. Electrical or mechanical failure of the electric power distribution system within the building

It is, therefore, desirable to locate transfer switches close to the load and to keep the operation of the transfer switches independent of overcurrent protection. It is often desirable to use multiple transfer switches of lower current rating located near the load rather than one large transfer switch at the point of incoming service.

Location of Overcurrent Devices

The location of overcurrent devices for both normal and emergency power is covered by Section 240-21 and is not affected by the installation of an automatic transfer switch. Transfer switches should be rated for continuous duty and have low contact temperature rise.

Solid Neutral on Alternating-Current and Direct-Current Systems

If automatic ac-to-ac transfer switches are used, then solid neutrals can be used with the grounding connections, as required in Section 250-24. If multiple grounding creates objectionable ground current, then corrective action, as specified in Section 250-6(b), must be made.

Section 230-95 requires ground-fault protection of equipment. Because the normal source and the emergency source are typically grounded at their locations, the multiple neutral-to-ground connections usually require some additional means or devices to ensure proper ground-fault sensing by the ground-fault protection device. Additional means or devices are generally required because the usual alterations used to stop objectionable current, per Section 250-26(b), do not apply if the objectionable current is a ground-fault current. [See Section 250-26(c).] Rather, solutions often include adding an overlapping neutral transfer pole or conventional fourth pole to the transfer switch. Other solutions include using isolation transformers and special ground-fault circuits, or using the service ground to also ground the generator neutral with 3-pole transfer switches.

On ac-to-dc automatic transfer switches, a solid neutral tie between the ac and dc neutrals is not permitted where both sources of supply are exterior distribution systems. Section 250-164, which addresses the location of grounds for dc exterior systems, clearly specifies that the dc system can be grounded only at the supply station.

If the dc system is an interior isolated system, such as a storage battery, then solid neutral connection between the ac system neutral and the dc source is acceptable.

On an ac-to-dc automatic transfer switch where the neutral must be switched, the size of the neutral switching pole must be considered. A 4-pole, double-throw switch must be used where a 3-phase, 4-wire normal source and a 2-wire dc emergency source are transferred. Because the neutral is switched, a 4-pole, double-throw transfer switch is required. In this instance, one pole of the dc emergency source carries three times the current of the other poles.

Close Differential Voltage Supervision of Normal Source

Most often, the normal source is an electric utility company whose power is transmitted many miles to the point of utilization. The automatic transfer switch control panel continuously monitors the voltage of all phases. (Because utility frequency is, for all practical purposes, constant, only the voltage needs to be monitored.) For single-phase power systems, the line-to-line voltage is monitored. For 3-phase power systems, all three line-to-line voltages should be monitored to provide full phase protection.

Monitoring protects against operation at reduced voltage, such as during brownouts, which can damage loads such as motors. Since the voltage sensitivity of loads varies, the pickup (acceptable) voltage setting and dropout (unacceptable) voltage setting of the monitors should be adjustable. Typical range of adjustment for the pickup is 85 percent to 100 percent of nominal, while the dropout setting, which is a function of the pickup setting, is 75 to 98 percent of the pickup selected. Typical settings for most loads are 95 percent of nominal for pickup and 85 percent of nominal for dropout (90 percent of pickup).

Consideration must be given to voltage supervision at closer differential for many installations where the load circuits are critical to voltage.

Electronic equipment load is frequently voltage critical. These installations include patient care equipment in health care facilities, X-ray equipment, television stations, cable TV centers, microwave communications, telephone communications, computers, computer-operated equipment, computer centers, and similar applications.

Polyphase motors operating at low load have a tendency to single phase, despite the loss of voltage in one phase, leading to burnout of the motor. A close differential of voltage supervision should be applied to automatic transfer switches for motor installations of the polyphase type. Differential voltage relays with a close adjustment of 2 percent for transfer and retransfer values aid in the detection of phase outages and provide protection from single phasing.

Automatic Transfer Switches with Emergency Source on Automatically Started Power Plant

In these installations, the normal source is usually a utility power line, and the emergency source is an automatically started engine generator set that starts when the normal source fails. To ensure maximum reliability, a minimum installation should be arranged to do the following.

1. Initiate engine starting of the power plant from a contact on the automatic transfer switch control panel (see Figure 700.5).

2. Sustain connection of load circuits to the normal

Figure 700.5 *Automatic emergency transfer switch and control panel. (Automatic Switch Co.)*

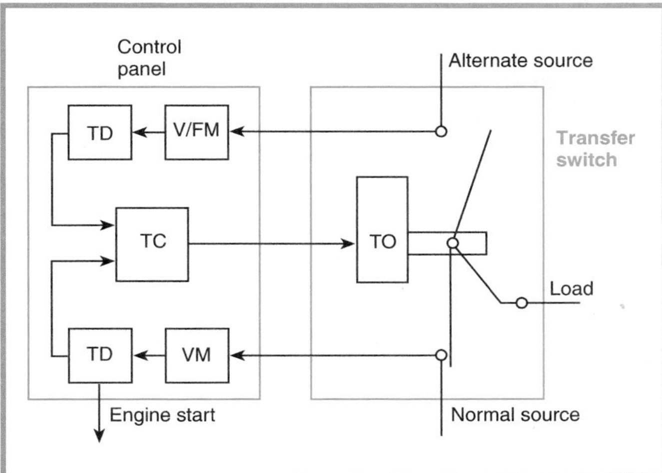

Figure 700.6 *Diagram of connections of the automatic transfer switch shown in Figure 700.5. TD, time delay; TO, transfer; TC, transfer controls; V/FM, voltage-frequency-sensitive monitor; VM, voltage monitor. (Automatic Switch Co.)*

source during the starting period, to provide utilization of any existing service from the normal source.

3. Measure output voltage and frequency of emergency source through the use of a voltage-frequency-sensitive monitor (see Figure 700.6) and effect transfer of the load circuits to the power plant only when both voltage and frequency of the power plant are approximately normal. Sensing of the emergency source need only be single phase, since most applications involve an on-site engine generator with a relatively short line run to the automatic transfer switch. In addition to monitoring voltage, the emergency source's frequency should be monitored. Unlike the utility power, the engine generator frequency can vary during start-up. Frequency monitoring will avoid overloading the engine generator while it is starting and can thus prevent stalling the engine. Combined frequency and voltage monitoring will prevent loads from being transferred to an engine generator set with an unacceptable output.

4. Provide visual signal and auxiliary contact for remote indication when the power plant is feeding the load.

Time-Delay Devices on Automatic Transfer Switches

Time delays are provided to program operation of the automatic transfer switch. To avoid unnecessary starting and transfer to the alternate supply, a nominal 1-second time delay, adjustable up to 6 seconds, can override momentary interruptions and reductions in normal source voltage but still allow starting and transfer if the reduction or outage is sustained. (See Figure 700.6.) The advantages of this feature are realized in all types of automatic transfer installations. In standby plant installations, the reduced number of false starts is especially important to minimize wear on the starting gear, battery, and associated equipment. This delay is generally set at 1 second but may be set higher if reclosers on the high lines take longer to operate or if momentary power dips exceed 1 second. If longer delay settings are used, care must be taken to ensure that sufficient time remains to meet 10-second power restoration requirements.

Once the load is transferred to the alternate source, another timer delays retransfer to the normal source until that source has time to stabilize. See commentary following the exception to Section 700-16. This timer is required by Section 700-12(b)(1) and is controlled by the preferred source voltage monitors. The timer is adjustable from 0 to 30 minutes and is normally set at 30 minutes. Another important function of this retransfer timer is to allow an engine generator to operate under load for a preselected minimum time to ensure continued good performance of the set and its starting system. This delay should be automatically nullified if the alternate source fails and the normal source is available, as determined by the voltage monitors.

Engine generator manufacturers often recommend a cool-down period for their sets that allows them to run unloaded after the load is retransferred to the normal source. A third time delay, usually 5 minutes, is provided for this purpose. Running an unloaded engine for more than 5 minutes is neither necessary nor recommended, because it can cause deterioration in engine performance.

If more than one automatic transfer switch is connected to the same engine generator, it is sometimes recommended that transfer of the loads be purposely sequenced to the alternate source. Using such a sequencing scheme can reduce starting kVA capacity requirements of the generator. A fourth timer, adjustable from 0 to 5 minutes, will delay transfer to the emergency supply source for this and other similar requirements.

700-18. Circuits for Emergency Power. For branch circuits that supply equipment classed as emergency, there shall be an emergency supply source to which the load will be transferred automatically upon the failure of the normal supply.

E. Control — Emergency Lighting Circuits

700-20. Switch Requirements. The switch or switches installed in emergency lighting circuits shall be arranged so that only authorized persons will have control of emergency lighting.

Exception No. 1: Where two or more single-throw switches are connected in parallel to control a single circuit, at least one of these switches shall be accessible only to authorized persons.

Exception No. 2: Additional switches that act only to put emergency lights into operation but not disconnect them shall be permissible.

Switches connected in series or 3- and 4-way switches shall not be used.

700-21. Switch Location. All manual switches for controlling emergency circuits shall be in locations convenient to authorized persons responsible for their actuation. In places of assembly, such as theaters, a switch for controlling emergency lighting systems shall be located in the lobby or at a place conveniently accessible thereto.

In no case shall a control switch for emergency lighting in a theater, or motion-picture theater or place of assembly, be placed in a motion-picture projection booth or on a stage or platform.

Exception: Where multiple switches are provided, one such switch shall be permitted in such locations where arranged so that it can energize the circuit only, but it cannot de-energize the circuit.

700-22. Exterior Lights. Those lights on the exterior of a building that are not required for illumination when there is sufficient daylight shall be permitted to be controlled by an automatic light-actuated device.

F. Overcurrent Protection

700-25. Accessibility. The branch-circuit overcurrent devices in emergency circuits shall be accessible to authorized persons only.

FPN: Fuses and circuit breakers for emergency circuit overcurrent protection, where coordinated to ensure selective clearing of fault currents, increase overall reliability of the system.

700-26. Ground-Fault Protection of Equipment. The alternate source for emergency systems shall not be required to have ground-fault protection of equipment with automatic disconnecting means. Ground-fault indication of the emergency source shall be provided per Section 700-7(d).

Article 701 — Legally Required Standby Systems

Contents

A. General

701-1. Scope. The provisions of this article apply to the electrical safety of the installation, operation, and maintenance of legally required standby systems consisting of circuits and equipment intended to supply, distribute, and control electricity to required facilities for illumination or power, or both, when the normal electrical supply or system is interrupted.

The systems covered by this article consist only of those that are permanently installed in their entirety, including the power source.

FPN No. 1: For additional information, see *Standard for Health Care Facilities,* NFPA 99-1996.

FPN No. 2: For further information regarding performance of emergency and standby power systems, see *Standard for Emergency and Standby Power Systems,* NFPA 110-1996.

FPN No. 3: For further information, see *Recommended Practice for Emergency and Standby Power Systems for Industrial and Commercial Applications,* ANSI/IEEE 446-1995.

Legally required standby systems are intended to provide electric power to aid in fire fighting, rescue operations, control of health hazards, and similar operations. In comparison, emergency systems (see Article 700) are those that are essential for safety to life. Optional standby systems (see Article 702) are those in which failure can cause physical discomfort, serious interruption of an industrial process, damage to process equipment, or disruption of business, for example.

The requirements for legally required standby systems are much the same as for emergency systems. There are, however, some differences. When normal power is lost, legally required systems must be able to supply standby power in 60 seconds or less, instead of the 10 seconds or less required of emergency systems. Wiring for legally required standby systems may occupy the same raceways, cables, boxes, and cabinets as other general wiring. Wiring for emergency systems is required to be kept entirely independent of other wiring. Legally required standby systems take second priority to emergency systems if they are involved in sharing an alternate supply and/or load shedding or peak shaving schemes.

701-2. Legally Required Standby Systems. Legally required standby systems are those systems required and so classed as legally required standby by municipal, state, federal, or other codes or by any governmental agency having jurisdiction. These systems are intended to automatically supply power to selected loads (other than those classed as emergency systems) in the event of failure of the normal source.

FPN: Legally required standby systems are typically installed to serve loads, such as heating and refrigeration systems, communications systems, ventilation and smoke removal systems, sewerage disposal, lighting systems, and industrial processes, that, when stopped during any interruption of the normal electrical supply, could create hazards or hamper rescue or fire-fighting operations.

701-3. Application of Other Articles. Except as modified by this article, all applicable articles of this *Code* shall apply.

701-4. Equipment Approval. All equipment shall be approved for the intended use.

701-5. Tests and Maintenance for Legally Required Standby Systems.

(a) Conduct or Witness Test. The authority having jurisdiction shall conduct or witness a test of the complete system upon installation.

(b) Tested Periodically. Systems shall be tested periodically on a schedule and in a manner acceptable to the authority having jurisdiction to ensure the systems are maintained in proper operating condition.

(c) Battery Systems Maintenance. Where batteries are used for control, starting, or ignition of prime movers, the authority having jurisdiction shall require periodic maintenance.

(d) Written Record. A written record shall be kept on such tests and maintenance.

(e) Testing Under Load. Means for testing legally required standby systems under load shall be provided.

701-6. Capacity and Rating. A legally required standby system shall have adequate capacity and rating for the supply of all equipment intended to be operated at one time. Legally required standby system equipment shall be suitable for the maximum available fault current at its terminals.

The alternate power source shall be permitted to supply legally required standby and optional standby system loads where automatic selective load pickup and load shedding is provided as needed to ensure adequate power to the legally required standby circuits.

701-7. Transfer Equipment.

(a) Transfer equipment, including automatic transfer switches, shall be automatic and identified for standby use and approved by the authority having jurisdiction. Transfer equipment shall be designed and installed to prevent the inadvertent interconnection of normal and alternate sources of supply in any operation of the transfer equipment.

(b) Means to bypass and isolate the transfer switch equipment shall be permitted. Where bypass isolation switches are used, inadvertent parallel operation shall be avoided.

(c) Automatic transfer switches shall be electrically operated and mechanically held.

Section 701-7(c) was added to the 1999 *Code* to ensure that relay contacts will be mechanically held in the event of coil failure. This change also correlates the *NEC* with NFPA 110-1996, *Standard for Emergency and Standby Power Systems.*

701-8. Signals. Audible and visual signal devices shall be provided, where practicable, for the following purposes.

(a) Derangement. To indicate derangement of the standby source.

(b) Carrying Load. To indicate that the standby source is carrying load.

(c) Not Functioning. To indicate that the battery charger is not functioning.

FPN: For signals for generator sets, see *Standard for Emergency and Standby Power Systems,* NFPA 110-1996.

701-9. Signs.

(a) Mandated Standby. A sign shall be placed at the service entrance indicating type and location of on-site legally required standby power sources.

Exception: A sign shall not be required for individual unit equipment as specified in Section 701-11(f).

(b) Grounding. Where the grounded circuit conductor connected to the emergency source is connected to a grounding electrode conductor at a location remote from the emergency source, there shall be a sign at the grounding location that shall identify all emergency and normal sources connected at that location.

B. Circuit Wiring

701-10. Wiring Legally Required Standby Systems. The legally required standby system wiring shall be permitted to occupy the same raceways, cables, boxes, and cabinets with other general wiring.

C. Sources of Power

701-11. Legally Required Standby Systems. Current supply shall be such that, in the event of failure of the normal supply to, or within, the building or group of buildings concerned, legally required standby power will be available within the time required for the application but not to exceed 60 seconds. The supply system for legally required standby purposes, in addition to the normal services to the building, shall be permitted to comprise one or more of the types of systems described in (a) through (e). Unit equipment in accordance with (f) shall satisfy the applicable requirements of this article.

In selecting a legally required standby source of power, consideration shall be given to the type of service to be rendered, whether of short-time duration or long duration.

Consideration shall be given to the location or design, or both, of all equipment to minimize the hazards that might cause complete failure due to floods, fires, icing, and vandalism.

FPN: Assignment of degree of reliability of the recognized legally required standby supply system depends on the careful evaluation of the variables at each particular installation.

(a) Storage Battery. A storage battery shall be of suitable rating and capacity to supply and maintain at not less than 87½ percent of system voltage the total load of the circuits supplying legally required standby power for a period of at least 1½ hours.

Batteries, whether of the acid or alkali type, shall be designed and constructed to meet the service requirements of emergency service and shall be compatible with the charger for that particular installation.

For a sealed battery, the container shall not be required to be transparent. However, for the lead acid battery that requires water additions, transparent or translucent jars shall be furnished. Automotive-type batteries shall not be used.

An automatic battery charging means shall be provided.

(b) Generator Set.

(1) A generator set driven by a prime mover acceptable to the authority having jurisdiction and sized in accordance with Section 701-6. Means shall be provided for automatically starting the prime mover upon failure of the normal service and for automatic transfer and operation of all required electrical circuits. A time-delay feature permitting a 15-minute setting shall be provided to avoid retransfer in case of short-time re-establishment of the normal source.

(2) Where internal combustion engines are used as the prime mover, an on-site fuel supply shall be provided with an on-premise fuel supply sufficient for not less than 2 hours full-demand operation of the system.

(3) Prime movers shall not be solely dependent on a public utility gas system for their fuel supply or municipal water supply for their cooling systems. Means shall be provided for automatically transferring one fuel supply to another where dual fuel supplies are used.

Exception: Where acceptable to the authority having jurisdiction, the use of other than on-site fuels shall be permitted where there is a low probability of a simultaneous failure of both the off-site fuel delivery system and power from the outside electrical utility company.

(4) Where a storage battery is used for control or signal power, or as the means of starting the prime mover, it shall be suitable for the purpose and shall be equipped with an automatic charging means independent of the generator set.

(c) Uninterruptible Power Supplies. Uninterruptible power supplies used to provide power for legally required standby systems shall comply with the applicable provisions of Section 701-11(a) and (b).

(d) Separate Service. Where acceptable to the authority having jurisdiction, a second service shall be permitted. This service shall be in accordance with Article 230, with separate service drop or lateral widely separated electrically and physically from the normal service to minimize the possibility of simultaneous interruption of supply.

(e) Connection Ahead of Service Disconnecting Means. Where acceptable to the authority having jurisdiction, connections ahead of, but not within, the main service disconnecting means shall be permitted. The legally required standby service shall be sufficiently separated from the normal main service disconnecting means to prevent simultaneous interruption of supply through an occurrence within the building or groups of buildings served.

FPN: See Section 230-82 for equipment permitted on the supply side of a service disconnecting means.

Section 230-82 provides requirements for service equipment. These requirements provide safe interruption of available fault current from the utility.

(f) Unit Equipment. Individual unit equipment for legally required standby illumination shall consist of the following:

(1) A rechargeable battery
(2) A battery charging means
(3) Provisions for one or more lamps mounted on the equipment and shall be permitted to have terminals for remote lamps, and
(4) A relaying device arranged to energize the lamps automatically upon failure of the supply to the unit equipment

The batteries shall be of suitable rating and capacity to supply and maintain at not less than 87½ percent of the nominal battery voltage for the total lamp load associated with the unit for a period of at least 1½ hours, or the unit equipment shall supply and maintain not less than 60 percent of the initial legally required standby illumination for a period of at least 1½ hours. Storage batteries, whether of the acid or alkali type, shall be designed and constructed to meet the requirements of emergency service.

Unit equipment shall be permanently fixed in place (i.e., not portable) and shall have all wiring to each unit installed in accordance with the requirements of any of the wiring methods in Chapter 3. Flexible cord and plug connection shall be permitted, provided that the cord does not exceed 3 ft (914 mm) in length. The branch circuit feeding the unit equipment shall be the same branch circuit as that serving the normal lighting in the area and connected ahead of any local switches. Legally required standby illumination fixtures that obtain power from a unit equipment and are not part of the unit equipment shall be wired to the unit equipment by one of the wiring methods of Chapter 3.

Exception: In a separate and uninterrupted area supplied by a minimum of three normal lighting circuits, a separate branch circuit for unit equipment shall be permitted if it originates from the same panelboard as that of the normal lighting circuits and is provided with a lock-on feature.

D. Overcurrent Protection

701-15. Accessibility. The branch-circuit overcurrent devices in legally required standby circuits shall be accessible to authorized persons only.

701-17. Ground-Fault Protection of Equipment. The alternate source for legally required standby systems shall not be required to have ground-fault protection of equipment.

Article 702 — Optional Standby Systems

Contents

A. General

702-1. Scope. The provisions of this article apply to the installation and operation of optional standby systems.

The systems covered by this article consist only of those that are permanently installed in their entirety, including prime movers.

702-2. Optional Standby Systems. Optional standby systems are intended to protect public or private facilities or property where life safety does not depend on the performance of the system. Optional standby systems are intended to supply on-site generated power to selected loads either automatically or manually.

FPN: Optional standby systems are typically installed to provide an alternate source of electric power for such facilities as industrial and commercial buildings, farms, and residences, and to serve loads such as heating and refrigeration systems, data processing and communications systems, and industrial processes that, when stopped during any power outage, could cause discomfort, serious interruption of the process, damage to the product or process, or the like.

Optional standby systems are those in which failure can cause physical discomfort, serious interruption of an industrial process, damage to process equipment, or disruption of business, for example.

702-3. Application of Other Articles. Except as modified by this article, all applicable articles of this *Code* shall apply.

702-4. Equipment Approval. All equipment shall be approved for the intended use.

702-5. Capacity and Rating. An optional standby system shall have adequate capacity and rating for the supply of all equipment intended to be operated at one time. Optional standby system equipment shall be suitable for the maximum available fault current at its terminals. The user of the optional standby system shall be permitted to select the load connected to the system.

702-6. Transfer Equipment. Transfer equipment shall be suitable for the intended use and designed and installed so as to prevent the inadvertent interconnection of normal and alternate sources of supply in any operation of the transfer equipment.

Transfer equipment, located on the load side of branch-circuit protection, shall be permitted to contain supplementary overcurrent protection having an interrupting rating sufficient for the available fault current that the generator can deliver.

702-7. Signals. Audible and visual signal devices shall be provided, where practicable, for the following purposes.

(1) Derangement. To indicate derangement of the optional standby source.

(2) Carrying Load. To indicate that the optional standby source is carrying load.

702-8. Signs.

(a) Standby. A sign shall be placed at the service-entrance equipment that indicates the type and location of on-site optional standby power sources. A sign shall not be required for individual unit equipment for standby illumination.

(b) Grounding. Where the grounded circuit conductor connected to the emergency source is connected to a grounding electrode conductor at a location remote from the emergency source, there shall be a sign at the grounding location that shall identify all emergency and normal sources connected at that location.

B. Circuit Wiring

702-9. Wiring Optional Standby Systems. The optional standby system wiring shall be permitted to occupy the same raceways, cables, boxes, and cabinets with other general wiring.

Article 705 — Interconnected Electric Power Production Sources

Contents

705-1. Scope. This article covers installation of one or more electric power production sources operating in parallel with a primary source(s) of electricity.

FPN: Examples of the types of primary sources are a utility supply, on-site electric power source(s), or other sources.

Article 705 sets forth basic safety requirements for the installation of generators and other types of power production sources that are interconnected and operate in parallel. Power sources include any system that produces electric power. They include not only electric utility sources, but also on-premises sources ranging from rotating generators (see Article 445) to solar photovoltaic systems (see Article 690) to fuel cells.

Article 705 addresses the basic safety requirements, specifically related to parallel operation, for the generators and other power sources, the power system that interconnects the power sources, and equipment that is connected to these systems. The proper application of these systems requires a thorough review of the entire power system.

705-2. Definition. For purposes of this article, the following definition applies.

Interactive System. An electric power production system that is operating in parallel with and capable of delivering energy to an electric primary source supply system.

705-3. Other Articles. Interconnected electric power production sources shall comply with this article and also the applicable requirements of the following articles:

	Article
Generators	445
Emergency systems	700
Legally required standby systems	701
Optional standby systems	702

Exception: Installation of solar photovoltaic systems operated as interconnected power sources shall be in accordance with Article 690.

705-10. Directory. A permanent plaque or directory, denoting all electrical power sources on or in the premises, shall be installed at each service equipment location and at locations of all electric power production sources capable of being interconnected.

Exception: Installations with large numbers of power production sources shall be permitted to be designated by groups.

705-12. Point of Connection. The outputs of electric power production systems shall be interconnected at the premises service disconnecting means.

(a) The outputs shall be permitted to be interconnected at a point or points elsewhere on the premises where the system qualifies as an integrated electric system and incorporates protective equipment in accordance with all applicable sections of Article 685.

(b) The outputs shall be permitted to be interconnected at a point or points elsewhere on the premises where all of the following conditions are met.

(1) The aggregate of nonutility sources of electricity has a capacity in excess of 100 kW, or the service is above 1000 volts;
(2) The conditions of maintenance and supervision ensure that qualified persons will service and operate the system; and
(3) Safeguards and protective equipment are established and maintained.

The point of interconnection is required to be at the premises service disconnecting means. This requirement may be difficult to meet for systems of 100 kW or smaller. It is intended to prevent the indiscriminate interconnection of small generators or other sources of power without proper protection against fire and electric shock. (See Figure 705.1.) It is important to use disconnect devices (switches, etc.) that are suitable for the purpose.

The requirement specifying "at the premises service disconnecting means" permits connection ahead of the disconnect or on the load side. This is to accommodate the safe work practices of many utilities that provide a readily accessible disconnect for dispersed generation. It is a contract matter between the utility and the customer and does not adversely affect the safety of the premises wiring; therefore, it is not within the scope of the *Code.*

Sections 705-12(a) and (b) recognize that generators and other power sources can be safely connected elsewhere on the premises system. These locations include where the premises has an integrated electrical system as set forth in Article 685, where the total generator capacity on premises is greater than 100 kW, or where the service is greater than 1000 volts.

Experience has shown that a higher level of design input and more responsible installation are used for larger systems than are commonly used for small systems (100-kW or less, 600-volt or less services).

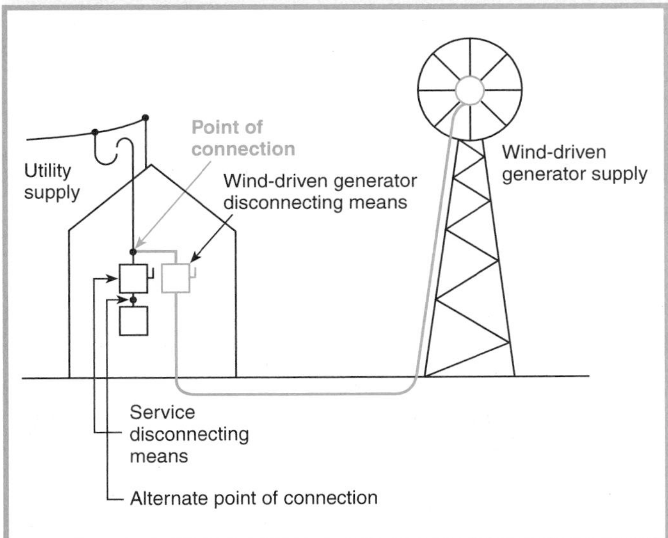

Figure 705.1 *The point of interconnection is required to be at the premises service disconnecting means.*

705-14. Output Characteristics. The output of a generator or other electric power production source operating in parallel with an electric supply system shall be compatible with the voltage, wave shape, and frequency of the system to which it is connected.

FPN: The term *compatible* does not necessarily mean matching the primary source wave shape.

The level of output voltage and frequency must be controlled to permit real power and reactive power to flow in the intended amount and proper direction. Control of the driver speed causes real power (kW) to flow from a rotating generator. Control of voltage causes reactive power (kVAr) to flow to or from a synchronous generator. The parallel operation of generators is a complex balance of several variables. These are design parameters and are beyond the scope of the *Code.* A considerable amount of data is available for equipment application and design.

The output characteristics of a rotating generator are significantly different from those of a solid-state power source. Their compatibility with other sources and with different types of loads will be limited in different ways.

Where either the power source or the loads have solid-state equipment, such as inverters, uninterruptible power supplies (UPS), or solid-state variable speed drives, harmonic currents will flow in the system. (See Figures 705.2, 705.3, and 705.4.) These multiples of the basic supply frequency (usually 60 Hz) cause additional heating, which may require derating of generators, transformers, cables, and motors. Special generator voltage control systems are required to avoid erratic operation or destruction of control devices. Circuit breakers may require derating if the higher harmonics become significant.

Significant magnitudes of harmonics may be inad-

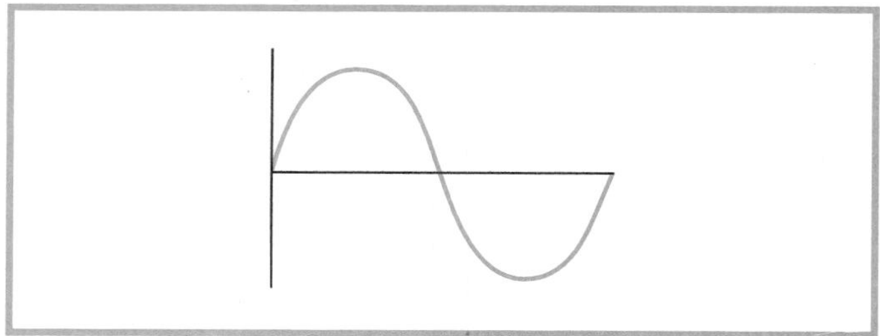

Figure 705.2 *Typical output wave shape of rotating generator and system wave shape normally encountered with motor, lighting, and heating loads.*

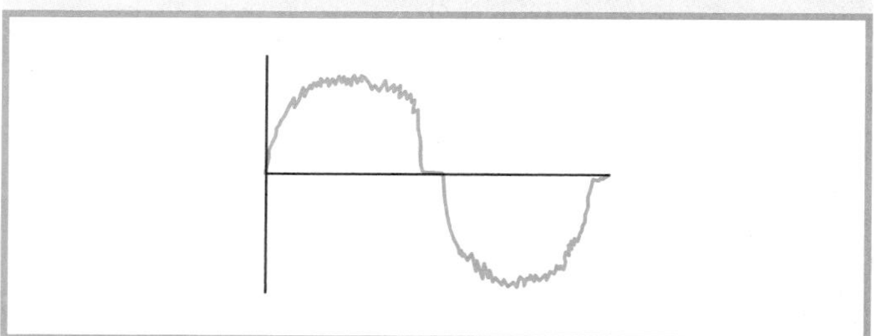

Figure 705.3 *Typical output wave shape with inverter source. Motors and transformers will be driven by harmonic-rich voltage and may require derating.*

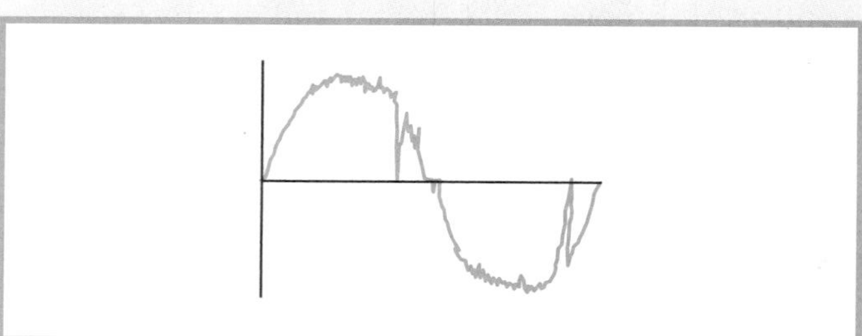

Figure 705.4 *Wave shape typical of system with variable speed drive, rectifier, elevator, and uninterruptible power to supply loads. Source generator may require derating, and special voltage control may be needed.*

vertently matched to system resonance and result in the opening of capacitor fuses, overheating of circuits, and erratic operation of controls. The usual solution is to detune the system by rearrangement or installation of reactors, or both.

705-16. Interrupting and Short-Circuit Current Rating. Consideration shall be given to the contribution of fault currents from all interconnected power sources for the interrupting and short-circuit current ratings of equipment on interactive systems.

705-20. Disconnecting Means, Sources. Means shall be provided to disconnect all ungrounded conductors of an electric power production source(s) from all other conductors. See Article 230.

705-21. Disconnecting Means, Equipment. Means shall be provided to disconnect equipment, such as inverters or transformers, associated with a power production source, from all ungrounded conductors of all sources of supply. Equipment intended to be operated and maintained as an integral part of a power production source exceeding 1000 volts shall not be required to have a disconnecting means.

An example of equipment that is covered under the second sentence of Section 705-21 is a generator designed for 4160 volts and connected to a 13,800-volt system through a transformer. The transformer and generator are to be operated as a unit. A disconnecting means is not required between the generator and transformer. A disconnecting means, however, is required between the transformer and the point of connection to the power system. Added for the 1999 *Code,* Section 445-10 addresses the general requirements for disconnecting means associated with generators.

705-22. Disconnect Device. The disconnecting means for ungrounded conductors shall consist of a manually or power operable switch(es) or circuit breaker(s)

(1) Located where accessible;
(2) Externally operable without exposing the operator to contact with live parts and, if power operable, of a type that can be opened by hand in the event of a power supply failure;
(3) Plainly indicating whether in the open or closed position; and
(4) Having ratings not less than the load to be carried and the fault current to be interrupted.

For disconnect equipment energized from both sides, a marking shall be provided to indicate that all contacts of the disconnect equipment may be energized.

FPN No. 1: In parallel generation systems, some equipment, including knife blade switches and fuses, are likely to be energized from both directions. See Section 240-40.

FPN No. 2: Interconnection to an off-premises primary source could require a visibly verifiable disconnecting device.

The requirements for disconnects in Section 705-22 are probably the most important requirements in Article 705. A disconnecting means is to serve each generating source. This device, or another, will be the service-entrance disconnect. Still another may be applied to separate the generating systems.

The basic requirement in Section 705-22 recognizes the success of applying switches as well as molded-case circuit breakers in this service. Most safe work practices on the premises use these disconnect devices.

The disconnect at the service entrance is required for disconnecting the premises wiring system from the utility. The utility safe work practices may also use this disconnect device. Utility work practices may require a visibly verifiable disconnect device. For this reason, some utility contracts require that a visible break be provided. The second fine print note brings attention to this common utility requirement.

705-30. Overcurrent Protection. Conductors shall be protected in accordance with Article 240. Equipment overcurrent protection shall be in accordance with the articles referenced in Article 240. Equipment and conductors connected to more than one electrical source shall have a sufficient number of overcurrent devices located so as to provide protection from all sources.

(a) Generators shall be protected in accordance with Section 445-4.

(b) Solar photovoltaic systems shall be protected in accordance with Article 690.

(c) Overcurrent protection for a transformer with a source(s) on each side shall be provided in accordance with Section 450-3 by considering first one side of the transformer, then the other side of the transformer, as the primary.

705-32. Ground-Fault Protection. Where ground-fault protection is used, the output of an interactive system shall be connected to the supply side of the ground-fault protection.

Exception: Connection shall be permitted to be made to the load side of ground-fault protection provided that there is ground-fault protection for equipment from all ground-fault current sources.

Larger systems have become very sophisticated and involve many special relays. The more common relays are under- and over-voltage, under- and over-frequency, voltage-restrained overcurrent, anti-motoring, loss-of-field, over-temperature, and shutdown for derangement of the mechanical driver. Small generator installations cannot justify the cost of these switchgear-type relays. Small generator protection is comprised of more common devices with fewer features. Application guides for relay protection are available in technical literature and manufacturers' data.

It is important to note that the requirements in Article 705 are only for protection against conditions that may occur because generating sources are being operated in parallel. The installation must also meet the requirements of the referenced articles to provide the basic protection and safeguards for all equipment, whether or not the equipment is involved with more than one source.

705-40. Loss of Primary Source. Upon loss of primary source, an electric power production source shall be automatically disconnected from all ungrounded conductors of the primary source and shall not be reconnected until the primary source is restored.

FPN No. 1: Risks to personnel and equipment associated with the primary source could occur if an interactive electric power production source can operate as an island. Special detection methods can be required to determine that a primary source supply system outage has occurred, and whether there should be automatic disconnection. When the primary source supply system is restored, special detection methods can be required to limit exposure of power production sources to out-of-phase reconnection.

FPN No. 2: Induction-generating equipment on systems with significant capacitance can become self-excited upon loss of primary source and experience severe over-voltage as a result.

When two interconnected power systems separate, they will drift out of synchronism. When the utility separates, there is a risk of damage to the system if restoration of the utility occurs out of phase. If the timing of the reconnection is random, there will be violent electromechanical stresses. These can destroy mechanical components such as gears, couplings, and shafts, and can displace coils. It is essential to disconnect the premises wiring system or generator from the primary source. Many technical guides are available from which to select appropriate protective systems and equipment for large systems. A limited choice of low-cost devices is still available for application to small systems.

Induction generators are commonly used today. They have characteristics quite different from synchronous machines. They are more rugged because of the construction of the rotor. They are less expensive because of their basic design, availability, and type of starting and control equipment. Theoretically, an induction machine can continue to run on an isolated system if a large capacitor bank provides excitation. In reality, an induction machine will lose stability and be shut down quickly by one of the protective devices.

705-42. Unbalanced Interconnections. A 3-phase electric power production source shall be automatically disconnected from all ungrounded conductors of the interconnected systems when one of the phases of that source opens. This requirement shall not be applicable to an electric power production source providing power for an emergency or legally required standby system.

705-43. Synchronous Generators. Synchronous generators in a parallel system shall be provided with the necessary equipment to establish and maintain a synchronous condition.

705-50. Grounding. Interconnected electric power production sources shall be grounded in accordance with Article 250.

Exception: For direct-current systems connected through an inverter directly to a grounded service, other methods that accomplish equivalent system protection and that utilize equipment listed and identified for the use shall be permitted.

Article 720 — Circuits and Equipment Operating at Less than 50 Volts

Contents

720-1. Scope. This article covers installations operating at less than 50 volts, direct current or alternating current.

Exception: As covered in Articles 411, 551, 650, 669, 690, 725, and 760.

Article 411 was added in the 1996 *Code* to cover lighting systems operating at 30 volts or less.

720-2. Hazardous (Classified) Locations. Installations coming within the scope of this article and installed in hazardous (classified) locations shall also comply with the appropriate provisions of Articles 500 through 517.

It should be understood that low voltage alone does not render a circuit incapable of igniting flammable atmospheres. Under some conditions, ordinary flashlights using two $1^1/_2$-volt D cells can be a source of ignition in some hazardous (classified) locations.

720-4. Conductors. Conductors shall not be smaller than No. 12 copper or equivalent. Conductors for appliance branch circuits supplying more than one appliance or appliance receptacle shall not be smaller than No. 10 copper or equivalent.

720-5. Lampholders. Standard lampholders that have a rating of not less than 660 watts shall be used.

720-6. Receptacle Rating. Receptacles shall have a rating of not less than 15 amperes.

720-7. Receptacles Required. Receptacles of not less than 20-ampere rating shall be provided in kitchens, laundries, and other locations where portable appliances are likely to be used.

720-8. Overcurrent Protection. Overcurrent protection shall comply with Article 240.

720-9. Batteries. Installations of storage batteries shall comply with Article 480.

720-10. Grounding. Grounding shall be as provided in Article 250.

720-11. Mechanical Execution of Work. Circuits operating at less than 50 volts shall be installed in a neat and workmanlike manner. Cables shall be supported by the building structure in such a manner that the cable will not be damaged by normal building use.

Section 720-11 requires these cables to be installed in a manner that is consistent with standard industry practice.

Article 725 — Class 1, Class 2, and Class 3 Remote-Control, Signaling, and Power-Limited Circuits

Contents

A. General

725-1. Scope. This article covers remote-control, signaling, and power-limited circuits that are not an integral part of a device or appliance.

FPN: The circuits described herein are characterized by usage and electrical power limitations that differentiate them from electric light and power circuits; therefore, alternative requirements to those of Chapters 1 through 4 are given with regard to minimum wire sizes, derating factors, overcurrent protection, insulation requirements, and wiring methods and materials.

Article 725 was substantially revised and reformatted for the 1996 *Code*. Definitions of Class 1, Class 2, and Class 3 circuits were added to help users understand the nature of these circuits. Many sections were combined, simplified, and relocated in order to group the installation requirements and the listing requirements. Requirements for mechanical execution of work were added. Definitive listing requirements of Class 2 and Class 3 power supplies were added. The previous tables associated with sizing the class of power supply were relocated to Tables 11(a) and 11(b) of Chapter 9.

Included within the scope of Article 725 may be such systems as burglar alarm circuits, access control circuits, sound circuits, nurse call circuits, intercom circuits, some computer network systems, some control circuits for lighting dimmer systems, and some low-voltage control circuits that originate from listed appliances or originate from listed computer equipment. Fire alarm circuits and systems are covered by Article 760. Communications circuits are covered by Article 800. Network-powered broadband communications circuits are covered by new Article 830. (See Figure 725.1.)

Figure 725.1 Typical burglar alarm system keypad. (Napco)

The installation requirements for the low-voltage wiring of information technology equipment (electronic data processing and computer equipment) located within the confines of a room that is constructed according to the requirements of NFPA 75-1999, *Standard for the Protection of Electronic Computer/Data Processing Equipment,* is not covered by this article. Low-voltage wiring within these specially constructed rooms is covered in Article 645.

Also, if listed computer equipment is interconnected and in close proximity, the wiring is considered an integral part of the equipment and is not subject to the requirements of Article 725. If the wiring leaves the group of equipment to connect to other devices in the same room or elsewhere in the building, the wiring is considered "wiring within buildings" and is subject to the requirements of this article.

The format for this article is similar to the format for Articles 760, 770, 800, and 820, and 830.

General Discussion of Remote-Control, Signaling, and Power-Limited Circuits

The wiring methods required by Chapters 1 through 4 of the *Code* apply to remote-control, signaling, and power-limited circuits, except as amended by Article 725 for specified conditions.

A remote-control, signaling, or power-limited circuit is the portion of the wiring system between the load side of the overcurrent device or the power-limited supply and all connected equipment. It is separated into Class 1, Class 2, or Class 3 circuits.

Class 1 circuits are not permitted to exceed 600 volts. In many cases, Class 1 circuits are extensions of power systems and are subject to the requirements of the power systems, except under the following conditions.

1. Conductors sized No. 16 and 18 may be used if properly protected against overcurrent (see Section 725-23).
2. Where damage to the circuit would introduce a hazard, the circuit must be mechanically protected by a suitable means [see Section 725-8(b)].
3. The adjustment factors of Section 310-15(b)(2) apply only if such conductors carry a continuous load (see Section 725-28 for the exact requirements for adjustment factors affecting ampacity).

Class 1 remote-control circuits are commonly used to operate motor controllers in conjunction with moving equipment or mechanical processes, elevators, conveyors, and other such equipment. They may also be employed as shunt trip circuits for circuit breakers. Class 1 signaling circuits often operate at 120 volts but are not limited to this value.

Conductors and equipment on the supply side of overcurrent protection, transformers, or current-limiting devices of Class 2 and Class 3 circuits must be installed according to the applicable requirements of Chapter 3. Load-side conductors and equipment must comply with Article 725. They are required to be separated from, and not occupy, the same raceways, cable trays, cables, or enclosures as electric light, power, and Class 1 conductors. Exceptions are noted in Section 725-54(a).

Dry-cell batteries are considered Class 2 power supplies, provided the voltage is 30 volts or less and the capacity is equal to or less than that available from series-connected No. 6 carbon-zinc cells. (A No. 6 dry cell is a cylindrical-shaped battery with nominal dimensions of 2.5 in. in diameter by 6 in. tall and weighing just over 2 lb. It is about 10 times the volume of a standard D-cell battery commonly used in flashlights.) Circuits originating from thermocouples are classified Class 2 circuits also. Neither dry-cell batteries nor thermocouples are required to be listed.

725-2. Definitions. For purposes of this article, the following definitions apply.

Class 1 Circuit. The portion of the wiring system between the load side of the overcurrent device or power-limited supply and the connected equipment. The voltage and power limitations of the source are in accordance with Section 725-21.

Class 2 Circuit. The portion of the wiring system between the load side of a Class 2 power source and the connected equipment. Due to its power limitations, a Class 2 circuit considers safety from a fire initiation standpoint and provides acceptable protection from electric shock.

Class 3 Circuit. The portion of the wiring system between the load side of a Class 3 power source and the connected equipment. Due to its power limitations, a Class 3 circuit considers safety from a fire initiation standpoint. Since higher levels of voltage and current than Class 2 are permitted, additional safeguards are specified to provide protection from an electric shock hazard that could be encountered.

725-3. Locations and Other Articles. Circuits and equipment shall comply with (a) through (e). Only those sections of Article 300 referenced in this article shall apply to Class 1, Class 2, and Class 3 circuits.

(a) Spread of Fire or Products of Combustion. Section 300-21.

(b) Ducts, Plenums, and Other Air-Handling Spaces. Section 300-22, where installed in ducts or plenums or other space used for environmental air.

Exception to (b): As permitted in Section 725-61(a).

See the commentary following Sections 300-22(b), 300-22(c), and 725-71(a), FPN.

(c) Hazardous (Classified) Locations. Articles 500 through 516 and Article 517, Part D, where installed in hazardous (classified) locations.

(d) Cable Trays. Article 318, where installed in cable tray.

(e) Motor Control Circuits. Article 430, Part F, where tapped from the load side of the motor branch-circuit protective device(s) as specified in Section 430-72(a).

725-5. Access to Electrical Equipment Behind Panels Designed to Allow Access. Access to equipment shall not be denied by an accumulation of conductors and cables that prevents removal of panels, including suspended ceiling panels.

The excess accumulation of conductors and cables can limit access to electrical equipment by preventing the removal of access panels. To service, rearrange, or install electrical equipment, the worker must have an accessible work space for working safely on energized equipment. See Section 300-11(a), which contains requirements relating to securing and supporting. It permits the use of support wires and approved fittings that are independent of the suspended ceiling support wires. (See Figures 725.2 and 725.3; also see Section 725-7.)

725-6. Class 1, Class 2, and Class 3 Circuit Grounding. Class 1, Class 2, and Class 3 circuits and equipment shall be grounded in accordance with Article 250.

725-7. Mechanical Execution of Work. Class 1, Class 2, and Class 3 circuits shall be installed in a neat and workmanlike manner. Cables shall be supported by the building structure in such a manner that the cable will not be damaged by normal building use.

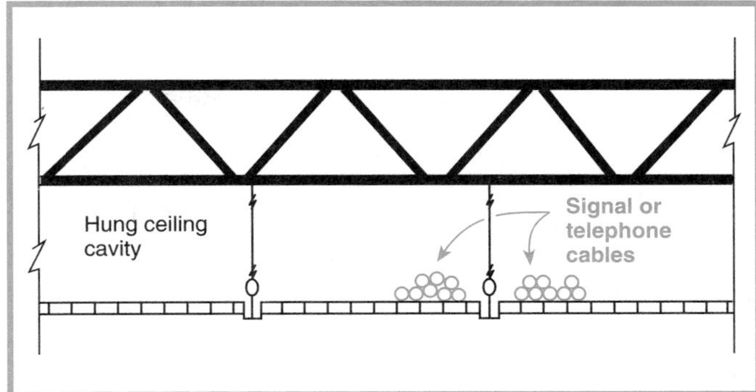

Figure 725.2 *The excess accumulation of conductors and cables prevents access to equipment or cables. Incorrect methods shown.*

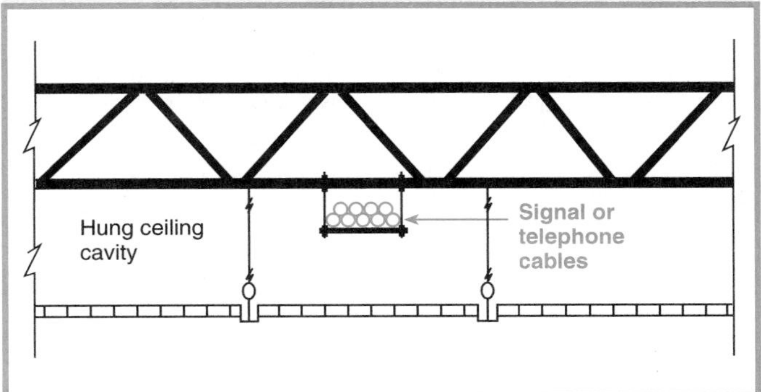

Figure 725.3 *The excess accumulation of conductors and cables prevents access to equipment or cables. Correct methods shown.*

FPN: One way to determine accepted industry practice is to refer to nationally recognized standards such as *Commercial Building Telecommunications Wiring Standard,* ANSI/EIA/TIA 568-1991; *Commercial Building Standard for Telecommunications Pathways and Spaces,* ANSI/EIA/TIA 569-1990; and *Residential and Light Commercial Telecommunications Wiring Standard,* ANSI/EIA/TIA 570-1991.

This requirement makes it clear that the *Code* requires these cables to be installed in a neat and workmanlike manner. The fine print note refers to industry standards that provide guidance on good workmanship. [See Figures 725.2 and 725.3 and Section 725-54(d).]

725-8. Safety-Control Equipment.

(a) Remote-Control Circuits. Remote-control circuits for safety-control equipment shall be classified as Class 1 if the failure of the equipment to operate introduces a direct fire or life hazard. Room thermostats, water temperature regulating devices, and similar controls used in conjunction with electrically controlled household heating and air conditioning shall not be considered safety-control equipment.

(b) Physical Protection. Where damage to remote-control circuits of safety control equipment would introduce a hazard, as covered in Section 725-8(a), all conductors of such remote-control circuits shall be installed in rigid metal conduit, intermediate metal conduit, rigid nonmetallic conduit, electrical metallic tubing, Type MI cable, Type MC cable, or be otherwise suitably protected from physical damage.

The remote-control circuits to safety-control devices may be required to be classified as Class 1 if failure of the safety-control circuit could cause a direct fire or life hazard. One example of a direct fire hazard could be a boiler explosion caused by the failure of the low-water cutoff circuit. See Figure 725.4. This is an example of a direct link between failure and the initiation of a hazard, that is, where the failure of the equipment itself causes the hazard.

Generally, fire alarm systems and nurses' call systems do not fit this category. These systems do not have a direct link to the initiation of fire or the initiation of a life hazard but, rather, serve as the reporting or warning link of a hazard initiated by some other (indirect) cause.

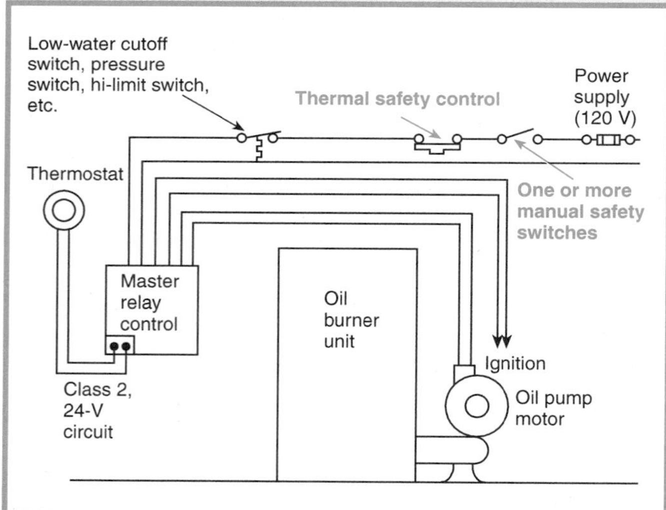

Figure 725.4 *Typical installation of an automatic oil burner unit for a boiler employing a safety shutdown circuit, according to the requirements of Section 725-8.*

725-15. Class 1, Class 2, and Class 3 Circuit Requirements. A remote-control, signaling, or power-limited circuit shall comply with the following parts of this article:

(1) Class 1 Circuits, Parts A and B
(2) Class 2 and Class 3 Circuits, Parts A and C

B. Class 1 Circuits

725-21. Class 1 Circuit Classifications and Power Source Requirements. Class 1 circuits shall be classified as either Class 1 power-limited circuits where they comply with the power limitations of (a), or as Class 1 remote-control and signaling circuits where they are used for remote control or signaling purposes and comply with the power limitations of (b).

(a) Class 1 Power-Limited Circuits. These circuits shall be supplied from a source that has a rated output of not more than 30 volts and 1000 volt-amperes.

(1) Class 1 Transformers. Transformers used to supply power-limited Class 1 circuits shall comply with Article 450.

(2) Other Class 1 Power Sources. Power sources other than transformers shall be protected by overcurrent devices rated at not more than 167 percent of the volt-ampere rating of the source divided by the rated voltage. The overcurrent devices shall not be interchangeable with overcurrent devices of higher ratings. The overcurrent device shall be permitted to be an integral part of the power supply.

To comply with the 1000 volt-ampere limitation of (a), the maximum output (VA_{max}) of power sources other than transformers shall be limited to 2500 volt-amperes, and the product of the maximum current (I_{max}) and maximum voltage (V_{max}) shall not exceed 10,000 volt-amperes. These ratings shall be determined with any overcurrent-protective device bypassed.

VA_{max} is the maximum volt-ampere output after one minute of operation regardless of load and with overcurrent protection bypassed, if used. Current-limiting impedance shall not be bypassed when determining VA_{max}.

I_{max} is the maximum output current under any noncapacitive load, including short circuit, and with overcurrent protection bypassed, if used. Current-limiting impedance should not be bypassed when determining I_{max}. Where a current-limiting impedance, listed for the purpose, or as part of a listed product, is used in combination with a stored energy source, e.g., storage battery, to limit the output current, I_{max} limits apply after 5 seconds.

V_{max} is the maximum output voltage regardless of load with rated input applied.

(b) Class 1 Remote-Control and Signaling Circuits. These circuits shall not exceed 600 volts. The power output of the source shall not be required to be limited.

Often, remote-control and signaling circuits that do not meet the requirements of a Class 2 or a Class 3 circuit are classified Class 1 circuits by default. For example, a listed nurses' call system may contain a power supply with an output of 500 watts at 24 volts. Since this power supply obviously exceeds the maximum permitted output of a Class 2 power supply and the output terminals are not marked to indicate this equipment as suitable for a Class 2 power supply,

the output circuit wiring is classified as Class 1 and subject to all the Class 1 circuit requirements.

725-23. Class 1 Circuit Overcurrent Protection. Overcurrent protection for conductors No. 14 and larger shall be provided in accordance with the conductor ampacity, without applying the derating factors of Section 310-15 to the ampacity calculation. Overcurrent protection shall not exceed 7 amperes for No. 18 conductors and 10 amperes for No. 16.

Exception: Where other articles of this Code permit or require other overcurrent protection.

FPN: For example, see Section 430-72 for motors, Section 610-53 for cranes and hoists, and Sections 517-74(b) and 660-9 for X-ray equipment.

Section 240-4(b)(2) gives alternate sizes for overcurrent protective devices.

725-24. Class 1 Circuit Overcurrent Device Location. Overcurrent devices shall be located at the point where the conductor to be protected receives its supply.

Exception No. 1: Where the overcurrent device protecting the larger conductor also protects the smaller conductor.

Exception No. 2: Transformer secondary conductors. Class 1 circuit conductors supplied by the secondary of a single-phase transformer having only a 2-wire (single-voltage) secondary shall be permitted to be protected by overcurrent protection provided on the primary (supply) side of the transformer, provided this protection is in accordance with Section 450-3 and does not exceed the value determined by multiplying the secondary conductor ampacity by the secondary-to-primary transformer voltage ratio. Transformer secondary conductors other than 2 wire shall not be considered to be protected by the primary overcurrent protection.

Exception No. 2 correlates with Section 240-3(f) and Section 430-72(b), Exception No. 2.

Exception No. 3: Electronic power source output conductors. Class 1 circuit conductors supplied by the output of a single-phase, listed electronic power source, other than a transformer, having only a 2-wire (single-voltage) output for connection to Class 1 circuits shall be permitted to be protected by overcurrent protection provided on the input side of the electronic power source, provided this protection does not exceed the value determined by multiplying the Class 1 circuit conductor ampacity by the output-to-input voltage ratio. Electronic power source outputs, other than 2 wire (single voltage), connected to Class 1 circuits shall not be considered to be protected by overcurrent protection on the input of the electronic power source.

Exception No. 3 was added to the 1999 *Code*. Electronic power supplies that do not supply energy directly through the use of transformers are now covered by this exception. This exception permits overcurrent protection for the Class 1 circuit conductors to be installed on the input side of the power source rather than the output side.

FPN: A single-phase, listed electronic power supply whose output supplies a 2-wire (single-voltage) circuit is an example of a Class 1 power source meeting the requirements of Section 725-23.

Exception No. 4: Class 1 circuit conductors No. 14 and larger that are tapped from the load side of the overcurrent-protective device(s) of a controlled light and power circuit shall require only short-circuit and ground-fault protection and shall be permitted to be protected by the branch-circuit overcurrent-protective device(s) where the rating of the protective device(s) is not more than 300 percent of the ampacity of the Class 1 circuit conductor.

Exception No. 4 correlates with Section 240-3(g).

725-25. Class 1 Circuit Wiring Methods. Installations of Class 1 circuits shall be in accordance with the appropriate articles in Chapter 3.

Exception No. 1: As provided in Sections 725-26 through 725-28.

Exception No. 2: Where other articles of this Code permit or require other methods.

725-26. Conductors of Different Circuits in Same Cable, Enclosure, or Raceway.

(a) Two or More Class 1 Circuits. Class 1 circuits shall be permitted to occupy the same cable, enclosure, or raceway without regard to whether the individual circuits are alternating current or direct current, provided all conductors are insulated for the maximum voltage of any conductor in the cable, enclosure, or raceway.

(b) Class 1 Circuits with Power Supply Circuits. Class 1 circuits and power supply circuits shall be permitted to occupy the same cable, enclosure, or raceway only where the equipment powered is functionally associated.

Exception No. 1: Where installed in factory- or field-assembled control centers.

Exception No. 2: Underground conductors in a manhole where one of the following conditions is met.

(a) The power-supply or Class 1 circuit conductors are in a metal-enclosed cable or Type UF cable.
(b) The conductors are permanently separated from the power-supply conductors by a continuous firmly fixed nonconductor, such as flexible tubing, in addition to the insulation on the wire.
(c) The conductors are permanently and effectively separated from the power supply conductors and securely fastened to racks, insulators, or other approved supports.

Section 725-26(b) permits the installation of Class 1 power-limited circuit conductors in manholes with wiring of nonpower-limited systems where suitable separation provides the same degree of protection as required for the following:

1. Class 2 and Class 3 power-limited circuits in Section 725-54(a)
2. Communications circuits in Section 800-52(a)
3. Radio/television antennas and lead-in conductors in Section 810-18(a)
4. CATV conductors in Section 820-52(a)

725-27. Class 1 Circuit Conductors.

(a) Sizes and Use. Conductors of sizes No. 18 and 16 shall be permitted to be used, provided they supply loads that do not exceed the ampacities given in Section 402-5 and are installed in a raceway, an approved enclosure, or a listed cable. Conductors larger than No. 16 shall not supply loads greater than the ampacities given in Section 310-15. Flexible cords shall comply with Article 400.

(b) Insulation. Insulation on conductors shall be suitable for 600 volts. Conductors larger than No. 16 shall comply with Article 310. Conductors in sizes No. 18 and 16 shall be Type FFH-2, KF-2, KFF-2, PAF, PAFF, PF, PFF, PGF, PGFF, PTF, PTFF, RFH-2, RFHH-2, RFHH-3, SF-2, SFF-2, TF, TFF, TFFN, TFN, ZF, or ZFF. Conductors with other types and thicknesses of insulation shall be permitted if listed for Class 1 circuit use.

Section 725-27 requires all Class 1 circuit conductors to be rated at 600 volts. This effectively requires Class 1 circuits to be wired using the wiring methods found in Chapter 3.

725-28. Number of Conductors in Cable Trays and Raceway, and Derating.

(a) Class 1 Circuit Conductors. Where only Class 1 circuit conductors are in a raceway, the number of conductors shall be determined in accordance with Section 300-17. The derating factors given in Section 310-15(b)(2)(a) shall apply only if such conductors carry continuous loads in excess of 10 percent of the ampacity of each conductor.

(b) Power-Supply Conductors and Class 1 Circuit Conductors. Where power-supply conductors and Class 1 circuit conductors are permitted in a raceway in accordance with Section 725-26, the number of conductors shall be determined in accordance with Section 300-17. The derating factors given in Section 310-15(b)(2)(a) shall apply as follows:

(1) To all conductors where the Class 1 circuit conductors carry continuous loads in excess of 10 percent of the ampacity of each conductor and where the total number of conductors is more than three

(2) To the power-supply conductors only, where the Class 1 circuit conductors do not carry continuous loads in excess of 10 percent of the ampacity of each conductor and where the number of power-supply conductors is more than three

(c) Class 1 Circuit Conductors in Cable Trays. Where Class 1 circuit conductors are installed in cable trays, they shall comply with the provisions of Sections 318-9 through 318-11.

725-29. Circuits Extending Beyond One Building. Class 1 circuits that extend aerially beyond one building shall also meet the requirements of Article 225.

C. Class 2 and Class 3 Circuits

725-41. Power Sources for Class 2 and Class 3 Circuits.

(a) Power Source. The power source for a Class 2 or a Class 3 circuit shall be as specified in (1), (2), (3), or (4):

FPN No. 1: Figure 725-41 illustrates the relationships between Class 2 or Class 3 power sources, their supply, and the Class 2 or Class 3 circuits.

FPN No. 2: Tables 11(a) and 11(b) in Chapter 9 provide the requirements for listed Class 2 and Class 3 power sources.

(1) A listed Class 2 or Class 3 transformer
(2) A listed Class 2 or Class 3 power supply
(3) Other listed equipment marked to identify the Class 2 or Class 3 power source

FPN: Examples of other listed equipment are as follows:

(1) A circuit card listed for use as a Class 2 or Class 3 power source where used as part of a listed assembly
(2) A current-limiting impedance, listed for the purpose, or part of a listed product, used in conjunction with a nonpower-limited transformer or a stored energy source, e.g., storage battery, to limit the output current
(3) A thermocouple

Exception: Thermocouples shall not require listing as a Class 2 power source.

(4) Listed information technology (computer) equipment limited power circuits

FPN: One way to determine applicable requirements for listing of information technology (computer) equipment is to refer to *Standard for Safety of Information Technology Equipment, Including Electrical Business Equipment,* UL 1950-1995. Typically such circuits are used to interconnect information technology equipment for the purpose of exchanging information (data).

(5) A dry cell battery shall be considered an inherently limited Class 2 power source, provided the voltage is

30 volts or less and the capacity is equal to or less than that available from series connected No. 6 carbon zinc cells.

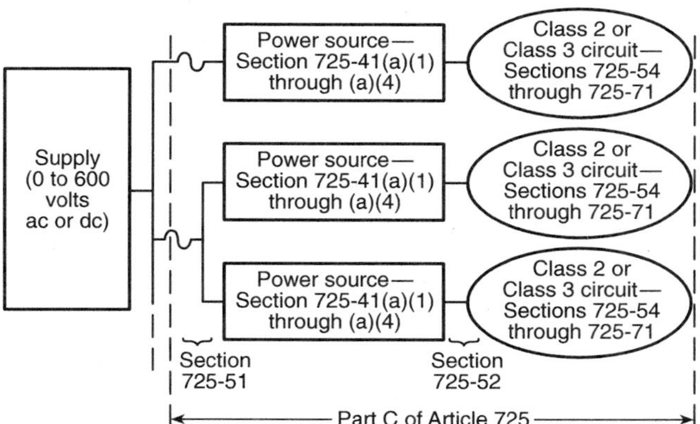

Figure 725-41 Class 2 and Class 3 circuits.

(b) Interconnection of Power Sources. Class 2 or Class 3 power sources shall not have the output connections paralleled or otherwise interconnected unless listed for such interconnection.

Except for dry-cell batteries and thermocouples, Section 725-41(b) now requires listed power sources for Class 2 and Class 3 circuits.

725-51. Wiring Methods on Supply Side of the Class 2 or Class 3 Power Source. Conductors and equipment on the supply side of the power source shall be installed in accordance with the appropriate requirements of Chapters 1 through 4. Transformers or other devices supplied from electric light or power circuits shall be protected by an overcurrent device rated not over 20 amperes.

Exception: The input leads of a transformer or other power source supplying Class 2 and Class 3 circuits shall be permitted to be smaller than No. 14, but not smaller than No. 18 if they are not over 12 in. (305 mm) long and if they have insulation that complies with Section 725-27(b).

Listed Class 2 and Class 3 transformers must be protected by an overcurrent device not exceeding 20 amperes, unless the transformers are fed from circuits other than power or lighting.

725-52. Wiring Methods and Materials on Load Side of the Class 2 or Class 3 Power Source. Conductors on the load side of the power source shall be insulated at not less than the requirements of Section 725-71 and shall be installed in accordance with Sections 725-54 and 725-61.

Section 725-52 requires listed cable for all applications, separation from other circuits, and specially designated cables (i.e., plenum rated or tray rated) for certain areas within buildings. However, this section does not prohibit the use of Chapter 3 wiring methods.

725-54. Installation of Conductors and Equipment.

(a) Separation from Electric Light, Power, Class 1, Nonpower-Limited Fire Alarm Circuit Conductors, and Medium Power Network-Powered Broadband Communications Cables.

(1) In Cables, Compartments, Cable Trays, Enclosures, Manholes, Outlet Boxes, Device Boxes, and Raceways. Cables and conductors of Class 2 and Class 3 circuits shall not be placed in any cable, cable tray, compartment, enclosure, manhole, outlet box, device box, raceway, or similar fitting with conductors of electric light, power, Class 1, nonpower-limited fire alarm circuits, and medium power network-powered broadband communications cables.

Section 725-54(a)(1) was revised for the 1999 *Code* to specifically include cables of Class 2 and Class 3 circuits. Jackets of listed Class 2 and Class 3 cables do not have sufficient construction specifications to permit them to be installed with electric light power, Class 1, nonpower-limited fire alarm circuits, and medium power network-powered broadband communications cables. Failure of the cable insulation due to a fault could lead to hazardous voltages being imposed on the Class 2 or Class 3 circuit conductors.

Exception No. 1: Where the conductors of the electric light, power, Class 1, nonpower-limited fire alarm, and medium power network-powered broadband communications circuits are separated by a barrier from the Class 2 and Class 3 circuits. In enclosures, Class 2 or Class 3 circuits shall be permitted to be installed in a raceway within the enclosure to separate them from Class 1, electric light, power, nonpower-limited fire alarm, and medium power network-powered broadband communications circuits.

Exception No. 2: Conductors in compartments, enclosures, device boxes, outlet boxes, or similar fittings, where electric light, power, Class 1, nonpower-limited fire alarm, and medium power network-powered broadband communications circuit conductors are introduced solely to connect to the equipment connected to Class 2 or Class 3 circuits to which the other conductors are connected, and

(a) The electric light, power, Class 1, nonpower-limited fire alarm, and medium power network-powered broadband communications circuit conductors are routed to maintain a minimum of 0.25 in. (6.35 mm) separation from the conductors and cables of Class 2 and Class 3 circuits, or

(b) The circuit conductors operate at 150 volts or less to ground and also comply with one of the following:

(1) The Class 2 and Class 3 circuits are installed using Type CL3, CL3R, or CL3P or permitted substitute cables, provided these Class 3 cable conductors extending beyond the jacket are separated by a minimum of 0.25 in. (6.35 mm) or by a nonconductive sleeve or nonconductive barrier from all other conductors, or

(2) The Class 2 and Class 3 circuit conductors are installed as a Class 1 circuit in accordance with Section 725-21.

Exception No. 3: Conductors entering compartments, enclosures, device boxes, outlet boxes, or similar fittings, where electric light, power, Class 1, nonpower-limited fire alarm, and medium power network-powered communications circuit conductors are introduced solely to connect the equipment connected to Class 2 or Class 3 circuits to which the other conductors in the enclosure are connected. If the conductors must enter an enclosure that is provided with a single opening, they shall be permitted to enter through a single fitting (such as a tee) provided the conductors are separated from the conductors of the other circuits by a continuous and firmly fixed nonconductor, such as flexible tubing.

An example of Exception No. 2 and Exception No. 3 is power circuit and Class 2 circuit conductors in the same motor-starter enclosure, where the Class 2 circuit source is the secondary of a control transformer in the same motor-starter enclosure. In such an installation, the Class 2 conductor insulation is not required to have the same voltage rating as the insulation on the power conductors in the same enclosure.

Exception No. 4: Underground conductors in a manhole where one of the following conditions is met.

(a) The electric light, power, Class 1, nonpower-limited fire alarm, and medium power network-powered broadband communications circuit conductors are in a metal-enclosed cable or Type UF cable.

(b) The conductors are permanently and effectively separated from the conductors of the other circuits by a continuous and firmly fixed nonconductor, such as flexible tubing, in addition to the insulation or covering on the wire.

(c) The conductors are permanently and effectively separated from conductors of the other circuits and securely fastened to racks, insulators, or other approved supports.

Exception No. 4 permits the installation of Class 2 and Class 3 power-limited circuit conductors in manholes having wiring for electric light, power, Class 1, nonpower-limited fire alarm circuits, and from medium power network-powered broadband communications cables where suitable separation provides the same degree of protection as required in Section 725-54(a). See the commentary following Section 725-26(b), Exception No. 2(c).

Exception No. 5: As permitted by Section 780-6(a) and installed in accordance with Article 780.

Exception No. 6: In cable trays, where the conductors of the electric light, power, Class 1, and nonpower-limited fire alarm circuits are separated by a solid fixed barrier of a material compatible with the cable tray, or where the Class 2 and Class 3 circuits are installed in Type MC cable.

Exception No. 6 was added to the 1999 *Code* to permit mixing of circuits in some cable tray applications because, under these circumstances, jacket damage that results in faults between circuits is very unlikely.

Section 725-54 was revised in the 1999 *Code* cycle to include separation of Class 2 and Class 3 cables and conductors from medium power network-powered broadband communications cables covered by new Article 830.

(2) In Hoistways. Class 2 or Class 3 circuit conductors shall be installed in rigid metal conduit, rigid nonmetallic conduit, intermediate metal conduit, or electrical metallic tubing in hoistways.

Exception: As provided for in Section 620-21 for elevators and similar equipment.

(3) Other Applications. Conductors of Class 2 and Class 3 circuits shall be separated by at least 2 in. (50.8 mm) from conductors of any electric light, power, Class 1, nonpower-limited fire alarm, or medium power network-powered broadband communications circuits.

Exception No. 1: Where either (1) all of the electric light, power, Class 1, nonpower-limited fire alarm, and medium power network-powered broadband communications circuit conductor or (2) all of the Class 2 and Class 3 circuit conductors are in raceway or in metal-sheathed, metal-clad, nonmetallic-sheathed, or Type UF cables.

Exception No. 2: Where all of the electric light, power, Class 1, nonpower-limited fire alarm, and medium power network-powered broadband communications circuit conductors are permanently separated from all of the Class 2 and Class 3 circuit conductors by a continuous and firmly fixed nonconductor, such as porcelain tubes or flexible tubing, in addition to the insulation on the conductors.

(b) Conductors of Different Circuits in Same Cable, Enclosure, or Raceway.

(1) Two or More Class 2 Circuits. Conductors of two or more Class 2 circuits shall be permitted within the same cable, enclosure, or raceway.

(2) Two or More Class 3 Circuits. Conductors of two or more Class 3 circuits shall be permitted within the same cable, enclosure, or raceway.

(3) Class 2 Circuits with Class 3 Circuits. Conductors of one or more Class 2 circuits shall be permitted within the same cable, enclosure, or raceway with conductors of Class 3 circuits, provided that the insulation of the Class 2 circuit conductors in the cable, enclosure, or raceway is at least that required for Class 3 circuits.

(4) Class 2 and Class 3 Circuits with Communications Circuits. Class 2 and Class 3 circuit conductors shall be permitted in the same cable with communications circuits, in which case the Class 2 and Class 3 circuits shall be classified as communications circuits and shall meet the requirements of Article 800. The cables shall be listed as communications cables or multipurpose cables.

Data circuits between computers are Class 2 circuits. In a typical office environment consisting of a group of computers in a local area network, the data wiring is as prevalent as telephone wiring. A common way to minimize the amount of cabling is to run the telephone and data circuits in the same cable. Figure 725.5 illustrates such an arrangement. Section 725-54(b)(4) requires that either a communications or multipurpose cable be used for this purpose.

Exception: Cables constructed of individually listed Class 2, Class 3, and communications cables under a common jacket shall not be required to be classified as communications cables. The fire-resistance rating of the composite cable shall be determined by the performance of the composite cable.

(5) Class 2 or Class 3 Cables with Other Circuit Cables. Jacketed cables of Class 2 or Class 3 circuits shall be permitted in the same enclosure or raceway with jacketed cables of any of the following:

(a) Power-limited fire alarm systems in compliance with Article 760
(b) Nonconductive and conductive optical fiber cables in compliance with Article 770
(c) Communications circuits in compliance with Article 800
(d) Community antenna television and radio distribution systems in compliance with Article 820
(e) Low power network-powered broadband communications in compliance with Article 830

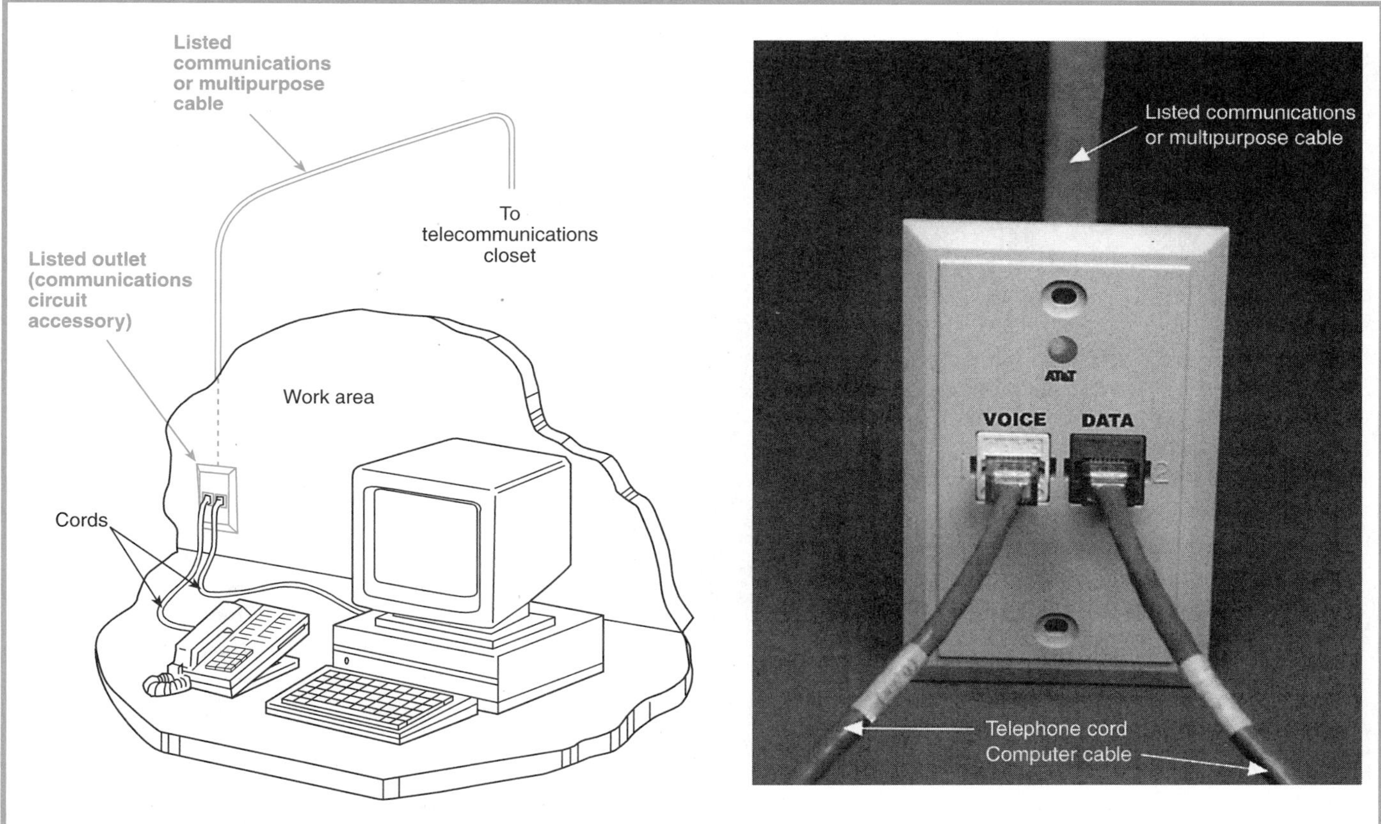

***Figure 725.5** Telephone and data circuits in the same cable.*

(c) Circuit Conductors Extending Beyond One Building. Where Class 2 or Class 3 circuit conductors extend beyond one building and are run so as to be subject to accidental contact with electric light or power conductors operating at over 300 volts to ground, or are exposed to lightning on interbuilding circuits on the same premises, the requirements of the following shall also apply:

(1) Sections 800-10, 800-12, 800-13, 800-30, 800-31, 800-32, 800-33, and 800-40 for other than coaxial conductors
(2) Sections 820-10, 820-33, and 820-40 for coaxial conductors

(d) Support of Conductors. Class 2 or Class 3 circuit conductors shall not be strapped, taped, or attached by any means to the exterior of any conduit or other raceway as a means of support.

Exception: Except as permitted by Section 300-11(b)(2).

See the commentary following Section 725-7.

725-61. Applications of Listed Class 2, Class 3, and PLTC Cables. Class 2, Class 3, and PLTC cables shall comply with (a) through (g).

(a) Plenum. Cables installed in ducts, plenums, and other spaces used for environmental air shall be Type CL2P or CL3P.

Exception: Listed wires and cables installed in compliance with Section 300-22.

(b) Riser. Cables installed in vertical runs and penetrating more than one floor, or cables installed in vertical runs in a shaft, shall be Type CL2R or CL3R. Floor penetrations requiring Type CL2R or CL3R shall contain only cables suitable for riser or plenum use.

Exception No. 1: Other cables as covered in Table 725-61 and other listed wiring methods as covered in Chapter 3, where installed in metal raceways or located in a fireproof shaft having firestops at each floor.

Exception No. 2: Types CL2, CL3, CL2X, and CL3X cable in one- and two-family dwellings.

FPN: See Section 300-21 for firestop requirements for floor penetrations.

(c) Cable Trays. Cables installed in cable trays outdoors shall be Type PLTC. Cables installed in cable trays indoors shall be Types PLTC, CL3P, CL3R, CL3, CL2P, CL2R, and CL2.

FPN: See Section 800-52(d) for cables permitted in cable trays.

(d) Hazardous (Classified) Locations. Cables installed in hazardous (classified) locations shall be Type PLTC. Where the use of Type PLTC cable is permitted in Sections 501-4(b), 502-4(b), and 504-20, the cable shall be installed in cable trays; in raceways; supported by messenger wire, or otherwise adequately supported and mechanically protected by angles, struts, channels, other mechanical means; or directly buried where the cable is listed for this use.

Exception No. 1: For Class 2 circuits as permitted by Section 501-4(b), Exception.

Exception No. 2: Conductors in Type PLTC cables used for Class 2 thermocouple circuits shall be permitted to be any of the materials used for thermocouple extension wire.

Exception No. 3: In industrial establishments where the conditions of maintenance and supervision ensure that only qualified persons will service the installation, and where the cable is not subject to physical damage, Type PLTC cable that complies with the crush and impact requirements of Type MC cable and is identified for such use shall be permitted as open wiring between cable tray and utilization equipment in lengths not to exceed 50 ft (15.24 m), where the cable is supported and protected against physical damage using mechanical protection such as dedicated struts, angles, or channels. The cable shall be supported and secured at intervals not exceeding 6 ft (1.83 m).

Exception No. 3 was added to Section 725-61(d) in the 1999 *Code* to allow limited use of a special Type PLTC cable in open wiring applications between a cable tray and utilization equipment in industrial establishments. This offers the installer an alternative to Type MC cable. If used in this manner, Type PLTC cable must meet the crush and impact requirements of Type MC cable. Type PLTC cable must also be mechanically protected and installed in runs of 50 ft or less.

(e) Other Wiring Within Buildings. Cables installed in building locations other than the locations covered in (a) through (d) shall be Type CL2 or CL3.

Exception No. 1: Type CL2X or CL3X where installed in a raceway, or other wiring methods as covered in Chapter 3.

Exception No. 2: In nonconcealed spaces where the exposed length of cable does not exceed 10 ft (3.05 m).

Exception No. 3: Listed Type CL2X, Class 2 cables less than 0.25 in. (6.4 mm) in diameter and listed Type CL3X, Class 3 cables less than 0.25 in. (6.4 mm) in diameter installed in one- or two-family dwellings.

Exception No. 4: Listed Type CL2X, Class 2 cables less than 0.25 in. (6.4 mm) in diameter and listed Type CL3X, Class 3 cables less than 0.25 in. (6.4 mm) installed in nonconcealed spaces in multifamily dwellings.

Exception No. 5: Type CMUC undercarpet communications wires and cables installed under carpet.

(f) Cross-Connect Arrays. Type CL2 or CL3 conductors or cable shall be used.

Cross-connect arrays and patch panels located within building spaces must be cross wired, jumpered, or interconnected using listed cables.

(g) Class 2 and Class 3 Cable Uses and Permitted Substitutions. The uses and permitted substitutions for Class 2 and Class 3 cables listed in Table 725-61 shall be considered suitable for the purpose and shall be permitted.

FPN: For information on Types CMP, CMR, CMG, CM, and CMX cables, see Section 800-50.

725-71. Listing and Marking of Class 2, Class 3, and Type PLTC Cables. Class 2, Class 3, and Type PLTC cables installed as wiring within buildings shall be listed as being resistant to the spread of fire and other criteria in accordance with (a) through (g) and shall be marked in accordance with (h).

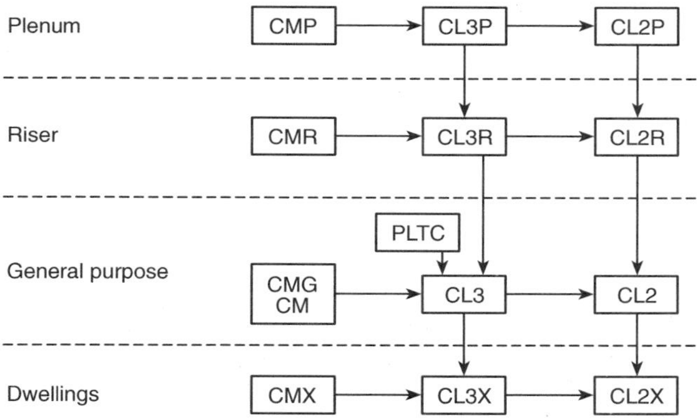

Figure 725-61 Cable substitution hierarchy.

(a) Types CL2P and CL3P. Types CL2P and CL3P plenum cables shall be listed as being suitable for use in ducts, plenums, and other space used for environmental air and shall also be listed as having adequate fire-resistant and low smoke-producing characteristics.

FPN: One method of defining low smoke-producing cable is by establishing an acceptable value of the smoke produced when tested in accordance with *Standard Method of Test for Fire and Smoke Characteristics of Wires and Cables,* NFPA 262-1994, to a maximum peak optical density of 0.5 and a maximum average optical density of 0.15. Similarly, one method of defining fire-resistant cables is by establishing a maximum allowable flame travel distance of 5 ft (1.52 m) when tested in accordance with the same test.

NFPA 262-1994, *Standard Method of Test for Fire and Smoke Characteristics of Wires and Cables,* is a test method for electrical wires and cables that are to be installed without raceways in plenums and other spaces used for environmental air. It was originally developed as UL 910, which is an adaptation of the Steiner Tunnel Test (NFPA 255/ASTM E84/UL 723).

The test is conducted in a 25-ft-long horizontal duct lined with insulating masonry faced with a row of refractory fire brick. See Figure 725.6. One side of the duct is provided with a row of double-paned observation windows that permit monitoring of the

Table 725-61. Cable Uses and Permitted Substitutions

Cable Type	Use	References	Permitted Substitutions
CL3P	Class 3 plenum cable	725-61(a)	CMP
CL2P	Class 2 plenum cable	725-61(a)	CMP, CL3P
CL3R	Class 3 riser cable	725-61(b)	CMP, CL3P, CMR
CL2R	Class 2 riser cable	725-61(b)	CMP, CL3P, CL2P, CMR, CL3R
PLTC	Power-limited tray cable	725-61(c) and (d)	
CL3	Class 3 cable	725-61(b), (e), and (f)	CMP, CL3P, CMR, CL3R, CMG, CM, PLTC
CL2	Class 2 cable	725-61(b), (e), and (f)	CMP, CL3P, CL2P, CMR, CL3R, CL2R, CMG, CM, PLTC, CL3
CL3X	Class 3 cable, limited use	725-61(b) and (e)	CMP, CL3P, CMR, CL3R, CMG, CM, PLTC, CL3, CMX
CL2X	Class 2 cable, limited use	725-61(b) and (e)	CMP, CL3P, CL2P, CMR, CL3R, CL2R, CMG, CM, PLTC, CL3, CL2, CMX, CL3X

Table 725-61 was extensively revised for the 1999 *Code* to reflect the more realistic use of field applications for various cable types.

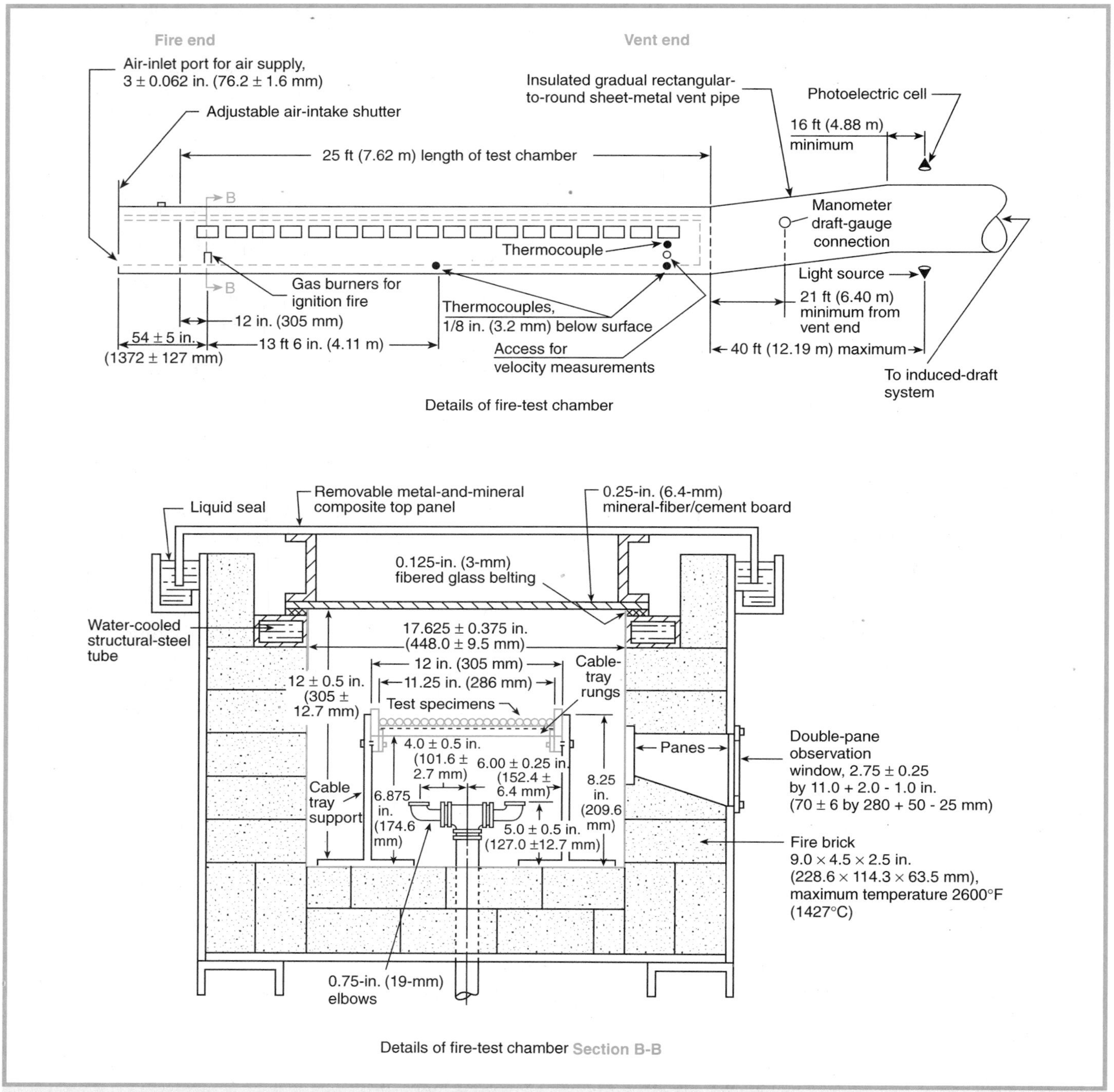

***Figure 725.6** Steiner Tunnel Test chamber used in NFPA 262/UL 910-1994, Standard Method of Test for Fire and Smoke Characteristics of Wires and Cables.*

progress of the test. One end of the chamber, designated the "fire end," is provided with two gas burners that deliver flames upward to engulf the test specimen. The burners are positioned on each side of the centerline of the furnace so that flame is evenly distributed across the sample. Test specimens are installed in a cable tray 23.9 ft long.

Smoke output is measured in a metal vent pipe at the vent end of the furnace, which is opposite the fire end. Where smoke passes through the vent pipe, it passes between a light source and a photocell. The output signal of the photocell is directly proportional to the amount of light received.

The cable specimens are exposed to a fire source with an output of 300,000 Btu/hour for a period of 20 minutes. A graph is plotted, illustrating the flame distance beyond $4^1/_2$ ft (measured from the end of the flame delivered by the burners) versus time taken for the duration of the test. This graph is plotted against a flame spread curve representative of a red oak specimen. A graph is also plotted for optical density versus time for the sample. Optical density is calculated as follows:

$$\text{Optical density} = \log_{10} \frac{T_0}{T}$$

where:

T_0 = initial light transmission

T = light transmission during test

NFPA 262 does not list pass/fail criteria. The criteria for acceptance of a given application are given in the appropriate sections of the *Code*. See the fine print note following Section 725-71(a) and Sections 760-31(d), FPN; 770-51(a), FPN; 800-51(a), FPN; and 820-51(a), FPN.

A Class 2 or Class 3 cable that has passed the requirements of this test may be used in ducts, plenums, or other air-handling spaces. In addition, such cable may be used anywhere in a building where Class 2 or Class 3 cable is permitted. (See Table 725-61.)

(b) Types CL2R and CL3R. Types CL2R and CL3R riser cables shall be listed as being suitable for use in a vertical run in a shaft or from floor to floor and shall also be listed as having fire-resistant characteristics capable of preventing the carrying of fire from floor to floor.

FPN: One method of defining fire-resistant characteristics capable of preventing the carrying of fire from floor to floor is that the cables pass the requirements of *Test for Flame Propagation Height of Electrical and Optical-Fiber Cable Installed Vertically in Shafts*, ANSI/UL 1666-1997.

In the fire test in UL 1666, cables are arranged in a simulated vertical shaft and subjected to an ignition source. The shaft is a 19-ft-high concrete shaft divided into two compartments at the 12-ft level. There is a 1 ft by 2 ft opening between the two compartments. The ignition source is a burner with a heat output rate of 495,000 Btu/hr. See Figure 725.7.

The individual cable lengths are suspended from a support system on the second floor and held in place just below the support system and at the first-floor slot.

The support frame consists of a steel bar located on the second floor, above the slot opening. The individual cable lengths are suspended from the support frame by one of the following methods.

1. The cables are draped over the support frame.
2. The cables are arranged in a clamping device that is suspended from the support frame by two hooks.
3. In the case of large-diameter cables, each individual cable is placed in a wire-mesh grip. Each grip is attached to a hook that is suspended from the support frame.

In order to pass, cables must not propagate flame to the top of the 12-ft-high compartment during the 30-minute test.

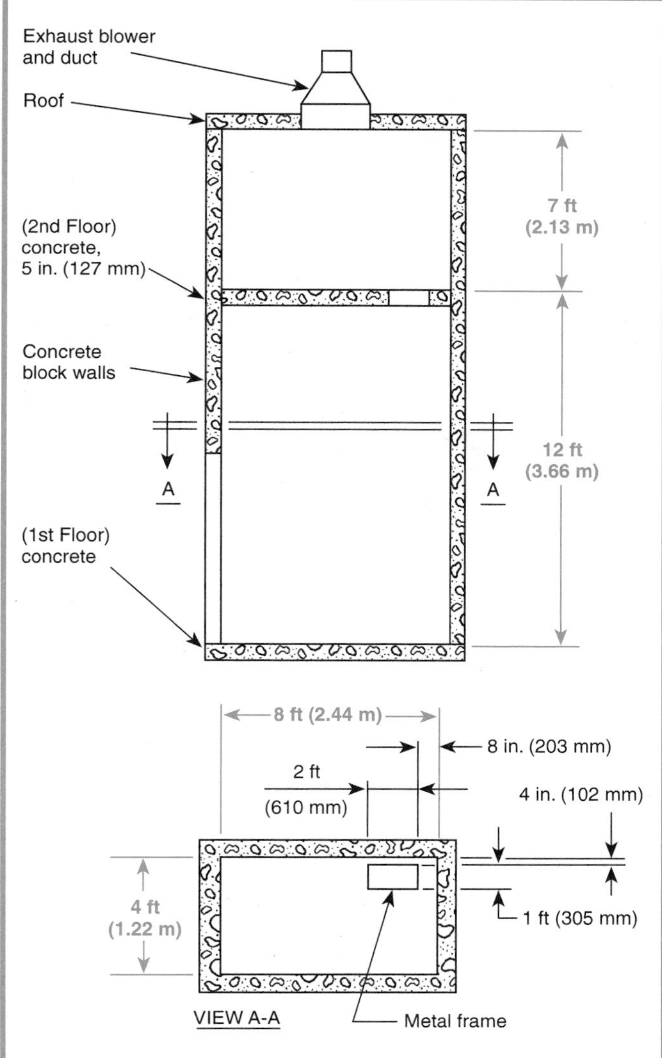

Figure 725.7 *Fire test chamber for Test for Flame Propagation Height of Electrical and Optical-Fiber Cable Installed Vertically in Shafts, ANSI/UL 1666-1997.*

(c) Types CL2 and CL3. Types CL2 and CL3 cables shall be listed as being suitable for general-purpose use, with the exception of risers, ducts, plenums, and other space used for environmental air, and shall also be listed as being resistant to the spread of fire.

FPN: One method of defining resistant to the spread of fire is that the cables do not spread fire to the top of the tray in the vertical tray flame test in *Reference Standard for Electrical Wires, Cables and Flexible Cords,* ANSI/UL 1581-1991.

Another method of defining resistant to the spread of fire is for the damage (char length) not to exceed 4 ft 11 in. (1.5 m) when performing the CSA vertical flame test for cables in cable trays, as described in *Test Methods for Electrical Wires and Cables,* CSA C22.2 No. 0.3-M-1985.

The UL Vertical Tray Flame Test determines whether cables installed in a ladder-type cable tray will propagate fire from a given exposure. The test flame is supplied by a strip or ribbon-type propane gas burner.

The cables are installed in a steel ladder-type cable tray that is 12 in. wide, 3 in. deep, and 96 in. long. Cables are installed in the tray spaced half a diameter apart, filling 75 percent of the tray width. A burner supplying 70,000 Btu/hr (20.32 kW/hr) is positioned 18 in. above the bottom of the tray, midway between two rungs. The burner flame is applied to the samples for a period of 20 minutes. See Figure 725.8. The samples are considered to have passed the test if flame has not propagated to the top of the cable tray by the test's conclusion. See also the commentary following Section 725-71(d), FPN.

A cable that has passed this testing may be listed as a Type CL2 or CL3 cable.

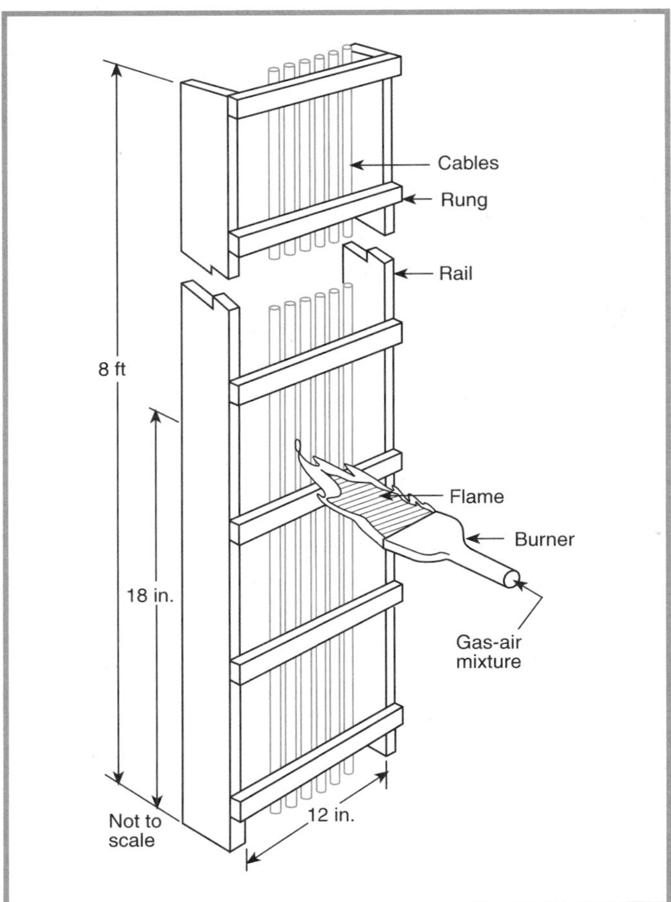

Figure 725.8 *UL Vertical Tray Flame Test.*

(d) Types CL2X and CL3X. Types CL2X and CL3X limited-use cables shall be listed as being suitable for use in dwellings and for use in raceway and shall also be listed as being flame retardant.

FPN: One method of determining that cable is resistant to flame spread is by testing the cable to the VW-1 (vertical-wire) flame test in the *Reference Standard for Electrical Wires, Cables and Flexible Cords,* ANSI/UL 1581-1991.

UL 1581, *Reference Standard for Electrical Wires, Cables, and Flexible Cords,* contains basic requirements for conductors, insulation, jackets, and other coverings, and the methods of sample preparation, specimen selection and conditioning, and measurements and calculations required in the standards for UL 44, *Rubber-Insulated Wires and Cables;* UL 83, *Thermoplastic-Insulated Wires and Cables;* and UL 62, *Flexible Cord and Fixture Wire.* The flame test methods of these standards include the Vertical Wire Flame Test (VW-1) and the Vertical Tray Flame Test.

The VW-1 test is a small-scale test of the flammability of insulating material. It uses a single specimen, secured in a vertical position in a three-sided metal enclosure that has a layer of untreated surgical cotton on the bottom. See Figure 725.9. For thermoplastic- or rubber-insulated wire and cable, the specimen used is an 18-in. insulated No. 14 copper or No. 12 aluminum conductor. A small kraft-paper indicator flag is attached to the specimen near its top. A flame is impinged 10 in. below the bottom of the flag for a period of 15 seconds. It is then removed for 15 seconds and reapplied for 15 seconds. This procedure is repeated for five 15-second flame applications. The specimen is considered to be capable of propagating fire if

1. More than 25 percent of the indicator flag is burned away or charred,

2. Flaming or glowing particles are emitted that ignite the surgical cotton, or
3. The specimen continues to flame longer than 60 seconds after removal of the burner.

A sample that has passed this test may be listed as Type CL2X or CL3X, suitable for use as a limited-use, power-limited cable. This cable may be used in one- and two-family and multifamily dwellings.

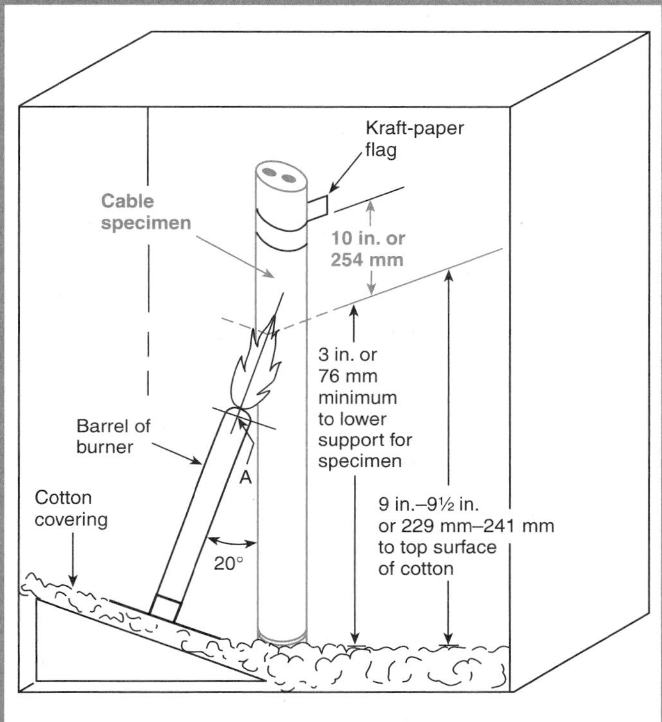

Figure 725.9 *UL Vertical Wire Flame Test.*

(e) Type PLTC. Type PLTC nonmetallic-sheathed, power-limited tray cable shall be listed as being suitable for cable trays and shall consist of a factory assembly of two or more insulated conductors under a nonmetallic jacket, marked in accordance with Section 310-11 and Table 725-71. The insulated conductors shall be Nos. 22 through 12. The conductor material shall be copper (solid or stranded). Insulation on conductors shall be suitable for 300 volts. The cable core shall be either (1) two or more parallel conductors, (2) one or more group assemblies of twisted or parallel conductors, or (3) a combination thereof. A metallic shield or a metallized foil shield with drain wire(s) shall be permitted to be applied either over the cable core, over groups of conductors, or both. The cable shall be listed as being resistant to the spread of fire. The outer jacket shall be a sunlight- and moisture-resistant nonmetallic material.

Exception No. 1: Where a smooth metallic sheath, welded and corrugated metallic sheath, or interlocking tape armor is applied over the nonmetallic jacket, an overall nonmetallic jacket shall not be required. On metallic-sheathed cable without an overall nonmetallic jacket, the information required in Section 310-11 shall be located on the nonmetallic jacket under the sheath.

Exception No. 2: Conductors in PLTC cables used for Class 2 thermocouple circuits shall be permitted to be any of the materials used for thermocouple extension wire.

FPN: One method of defining resistant to the spread of fire is that the cables do not spread fire to the top of the tray in the vertical tray flame test in the *Reference Standard for Electrical Wires, Cables and Flexible Cords,* ANSI/UL 1581-1991.

Another method of defining resistant to the spread of fire is for the damage (char length) not to exceed 4 ft 11 in. (1.5 m) when performing the CSA vertical flame test for cables in cable trays, as described in *Test Methods for Electrical Wires and Cables,* CSA C22.2 No. 0.3-M-1985.

(f) Class 3 Cable Voltage Rating. Class 3 cables shall have a voltage rating of not less than 300 volts.

(g) Class 3 Single Conductors. Class 3 single conductors used as other wiring within buildings shall not be smaller than No. 18 and shall be Type CL3. Conductor types described in Section 725-27(b) that are also listed as Type CL3 shall be permitted.

Section 725-71(g) was revised for the 1999 *Code* to clarify that single conductors must also be resistant to the spread of fire.

FPN: One method of defining resistant to the spread of fire is that the cables do not spread fire to the top of the tray in the Vertical Tray Flame Test in *Reference Standard for Electrical Wires, Cables and Flexible Cords,* ANSI/UL 1581-1991.

Another method of defining resistant to the spread of fire is for the damage (char length) not to exceed 4 ft 11 in. (1.5 m) when performing the CSA Vertical Flame Test — Cables in Cable Trays as described in *Test Methods for Electrical Wires and Cables*, CSA C22.2 No. 0.3-M-1985.

(h) Marking. Cables shall be marked in accordance with Table 725-71. Voltage ratings shall not be marked on the cables.

FPN: Voltage markings on cables may be misinterpreted to suggest that the cables may be suitable for Class 1 electric light and power applications.

Exception: Voltage markings shall be permitted where the cable has multiple listings and a voltage marking is required for one or more of the listings.

Table 725-71. Cable Markings

Cable Marking	Type	Listing References
CL3P	Class 3 plenum cable	725-71(a), (f), and (h)
CL2P	Class 2 plenum cable	725-71(a) and (h)
CL3R	Class 3 riser cable	725-71(b), (f), and (h)
CL2R	Class 2 riser cable	725-71(b) and (h)
PLTC	Power-limited tray cable	725-71(e) and (h)
CL3	Class 3 cable	725-71(c), (f), and (h)
CL2	Class 2 cable	725-71(c), (f), and (h)
CL3X	Class 3 cable, limited use	725-71(d), (f), and (h)
CL2X	Class 2 cable, limited use	725-71(d), (f), and (h)

FPN: Class 2 and Class 3 cable types are listed in descending order of fire resistance rating, and Class 3 cables are listed above Class 2 cables, because Class 3 cables can substitute for Class 2 cables.

Article 727 — Instrumentation Tray Cable: Type ITC

Contents

727-1. Scope. This article covers the use, installation, and construction specifications of instrumentation tray cable for application to instrumentation and control circuits operating at 150 volts or less and 5 amperes or less.

Article 727 permits an alternate wiring method for circuits that do not exceed 5 amperes and 150 volts. It is particularly suited for instrumentation circuits in industrial establishments where qualified persons perform service and maintenance.

727-2. Definition. *Type ITC instrumentation tray cable* is a factory assembly of two or more insulated conductors, with or without a grounding conductor(s), enclosed in a nonmetallic sheath.

727-3. Other Articles. In addition to the provisions of this article, installation of Type ITC cable shall comply with other applicable articles of this *Code*, such as Articles 240, 250, 300, and 318.

727-4. Uses Permitted. Type ITC cable shall be permitted to be used in industrial establishments where the conditions of maintenance and supervision ensure that only qualified persons will service the installation

(1) In cable trays
(2) In raceways
(3) In hazardous locations as permitted in Articles 501, 502, 503, 504, and 505
(4) As open wiring where equipped with a smooth metallic sheath, continuous corrugated metallic sheath, or interlocking tape armor applied over the nonmetallic sheath in accordance with Section 727-6

The cable shall be supported and secured at intervals not exceeding 6 ft (1.83 m).

Exception No. 1: Type ITC cable without a metallic sheath or armor shall be permitted to be installed as open wiring between cable tray and equipment in lengths not to exceed 50 ft (15.24 m), where the cable is supported and protected against physical damage using mechanical protection, such as dedicated struts, angles, or channels. The cable shall be supported and secured at intervals not exceeding 6 ft (1.83 m).

Exception No. 2: Type ITC cable that complies with the crush and impact requirements of Type MC cable and is identified for such use shall be permitted as open wiring between the cable tray and the equipment in lengths not to exceed 50 ft (15.24 m). The cable shall be supported and secured at intervals not exceeding 6 ft (1.83 m).

(5) As aerial cable on a messenger
(6) Direct buried where identified for the use
(7) Under raised floors in control rooms and rack rooms where arranged to prevent damage to the cable

727-5. Uses Not Permitted. Type ITC cable shall not be installed on circuits operating at more than 150 volts or more than 5 amperes.

Installation of Type ITC cable with other cables shall be subject to the stated provisions of the specific articles for the other cables. Where the governing articles do not contain stated provisions for installation with Type ITC cable, the installation of Type ITC cable with the other cables shall not be permitted.

Type ITC cable shall not be installed with power, lighting, Class 1, or nonpower-limited circuits.

Exception No. 1: Where terminated within equipment or junction boxes and separations are maintained by insulating barriers or other means.

Exception No. 2: Where a metallic sheath or armor is applied over the nonmetallic sheath of the Type ITC cable.

727-6. Construction. The insulated conductors of Type ITC cable shall be in sizes No. 22 through No. 12. The conductor material shall be copper or thermocouple alloy. Insulation on the conductors shall be rated for 300 volts. Shielding shall be permitted.

The cable shall be listed as being resistant to the spread of fire. The outer jacket shall be sunlight and moisture resistant.

Where a smooth metallic sheath, continuous corrugated metallic sheath, or interlocking tape armor is applied over the nonmetallic sheath, an overall nonmetallic jacket shall be permitted to be applied, but shall not be required.

727-7. Marking. The cable shall be marked in accordance with Section 310-11.

727-8. Allowable Ampacity. The allowable ampacity of the conductors shall be 5 amperes, except for No. 22 conductors that shall have an allowable ampacity of 3 amperes.

727-9. Overcurrent Protection. Overcurrent protection shall not exceed 5 amperes for Nos. 20 and larger conductors, and 3 amperes for No. 22 conductors.

727-10. Bends. Bends in Type ITC cables shall be made so as not to damage the cable.

Article 760 — Fire Alarm Systems

Contents

760-61. Applications of Listed PLFA Cables
(a) Plenum
(b) Riser
(c) Other Wiring Within Buildings
(d) Cable Uses and Permitted Substitutions
760-71. Listing and Marking of PLFA Cables and Insulated Continuous Line-Type Fire Detectors
(a) Conductor Materials
(b) Conductor Size
(c) Ratings
(d) Type FPLP
(e) Type FPLR
(f) Type FPL
(g) Fire Alarm Circuit Integrity (CI) Cable
(h) Coaxial Cables
(i) Cable Marking
(j) Insulated Continuous Line-Type Fire Detectors

A. General

760-1. Scope. This article covers the installation of wiring and equipment of fire alarm systems including all circuits controlled and powered by the fire alarm system.

FPN No. 1: Fire alarm systems include fire detection and alarm notification, guard's tour, sprinkler waterflow, and sprinkler supervisory systems. Circuits controlled and powered by the fire alarm system include circuits for the control of building systems safety functions, elevator capture, elevator shutdown, door release, smoke doors and damper control, fire doors and damper control and fan shutdown, but only where these circuits are powered by and controlled by the fire alarm system. For further information on the installation and monitoring for integrity requirements for fire alarm systems, refer to the *National Fire Alarm Code,®* NFPA 72-1996.

FPN No. 2: Class 1, 2, and 3 circuits are defined in Article 725.

Article 760 has been revised for the 1999 *Code* to change the permitted cable uses and substitutions and to include the addition of a new circuit integrity (CI) cable type. See the commentary following Section 760-31(f).

FPN No. 2 explains that fire alarm circuits include those powered and controlled from the fire alarm system power source. Article 760 covers only those circuits powered and controlled by the fire alarm system. Fire safety features such as smoke door control, damper control, fan shutdown, and elevator recall powered and controlled by the fire alarm system are covered by Article 760. Circuits powered and controlled by other building systems such as heating, ventilating, and air conditioning (HVAC); security; lighting controls; and time recording are covered by Article 725.

The format for Article 760 is similar to that for Articles 725, 770, 800, 820, and 830.

Article 760 covers the wiring between the devices and equipment required by NFPA 72-1996, *National Fire Alarm Code*. NFPA 72 provides the requirements for the selection, installation, performance, use, and testing and maintenance of the fire alarm system components. The requirement for whether an occupancy is required to have a fire alarm system is found in NFPA *101, Life Safety Code,* or other local codes.

Examples of fire alarm devices and equipment are shown in Figures 760.1 and 760.2. The installation of these system component items is covered by NFPA 72, but the circuit wiring associated with these components must be according to Article 760.

Paragraph 1-5.5.4 of NFPA 72 requires that "all wiring, cable, and equipment shall be in accordance with NFPA 70, *National Electrical Code,* and specifically with Article 760." Additionally, 1-5.1.2 of NFPA 72 requires all equipment to be listed for the purpose for which it is used.

NFPA 1221-1994, *Standard for the Installation, Maintenance, and Use of Public Fire Service Communication*

Figure 760.1 Typical fire alarm control unit. (Fire Control Instruments)

Figure 760.2 *Typical spot-type smoke detector. (System Sensor)*

Systems, covers the installation, maintenance, and use of all public fire service communication facilities. These facilities include public reporting, dispatching, telephone, and both two-way and microwave radio systems, all of which fulfill two principal functions: the receipt of fire alarms or other emergency calls from the public and the retransmission of these alarms and emergency calls to fire companies and other appropriate agencies.

760-2. Definitions. For purposes of this article, the following definitions apply.

Fire Alarm Circuit. The portion of the wiring system between the load side of the overcurrent device or the power-limited supply and the connected equipment of all circuits powered and controlled by the fire alarm system. Fire alarm circuits are classified as either nonpower-limited or power-limited.

Fire Alarm Circuit Integrity (CI) Cable. Cable used in fire alarm systems to ensure continued operation of critical circuits during a specified time under fire conditions.

Nonpower-Limited Fire Alarm Circuit (NPLFA). A fire alarm circuit powered by a source that complies with Sections 760-21 and 760-23.

Power-Limited Fire Alarm Circuit (PLFA). A fire alarm circuit powered by a source that complies with Section 760-41.

760-3. Locations and Other Articles. Circuits and equipment shall comply with (a) through (f). Only those sections of Article 300 referenced in this article shall apply to fire alarm systems.

(a) Spread of Fire or Products of Combustion. See Section 300-21.

(b) Ducts, Plenums, and Other Air-Handling Spaces. Section 300-22, where installed in ducts or plenums or other spaces used for environmental air.

Exception to (b): As permitted in Sections 760-30(b)(1) and (2) and Section 760-61(a).

See the commentary following Sections 300-22(b) and 300-22(c).

(c) Hazardous (Classified) Locations. Articles 500 through 516 and Article 517, Part D, where installed in hazardous (classified) locations.

(d) Corrosive, Damp, or Wet Locations. Sections 110-11, 300-6, and 310-9 where installed in corrosive, damp, or wet locations.

Section 760-3(d) requires cables and equipment that is used in wet or damp locations, high ambient temperature areas, or corrosive locations to be identified as suitable for the particular use. Underground installations are considered wet locations.

(e) Building Control Circuits. Article 725 where building control circuits (e.g., elevator capture, fan shutdown) are associated with the fire alarm system.

(f) Optical Fiber Cables. Where optical fiber cables are utilized for fire alarm circuits, the cables shall be installed in accordance with Article 770.

760-5. Access to Electrical Equipment Behind Panels Designed to Allow Access. Access to equipment shall not be denied by an accumulation of conductors and cables that prevents removal of panels, including suspended ceiling panels.

The excess accumulation of wires and cables can limit access to equipment by preventing the removal of access panels. This section also prohibits installation where cables may be damaged by ordinary building use. See Section 300-11(a), which contains requirements relating to securing and supporting. Section 300-11(a) permits the use of support wires and approved fittings that are independent of the suspended ceiling support wires. See commentary following Section 760-8, FPN, and Figures 760.3 and 760.4.

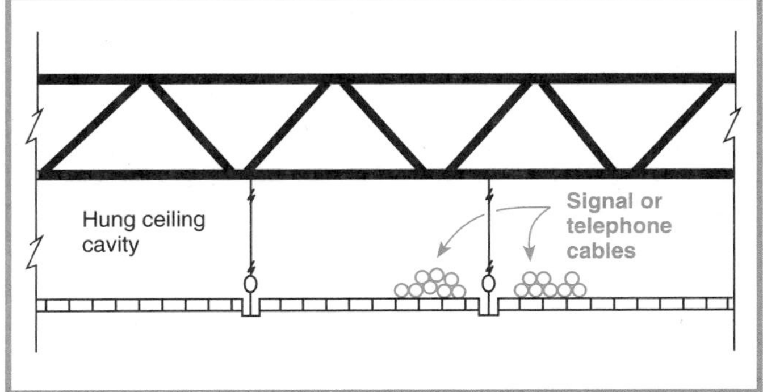

Figure 760.3 *The excess accumulation of conductors and cables prevents access to equipment or cables. Incorrect methods shown.*

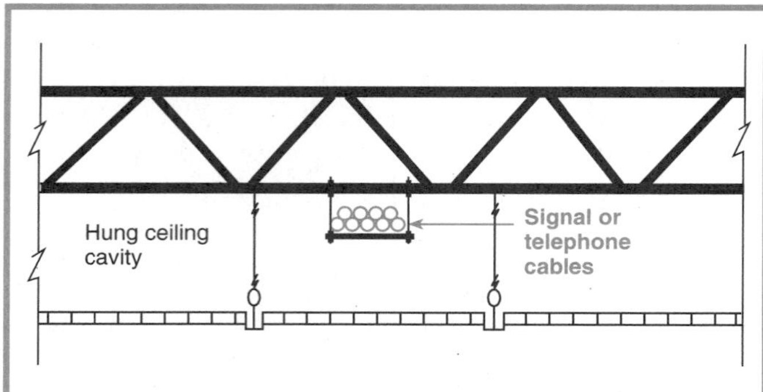

Figure 760.4 The excess accumulation of conductors and cables prevents access to equipment or cables. Correct methods shown.

760-6. Fire Alarm Circuit and Equipment Grounding. Fire alarm circuits and equipment shall be grounded in accordance with Article 250.

760-7. Fire Alarm Circuits Extending Beyond One Building. Power-limited fire alarm circuits that extend beyond one building and run outdoors shall either meet the installation requirements of Parts B, C, and D of Article 800, or shall meet the installation requirements of Article 225. Nonpower-limited fire alarm circuits that extend beyond one building and run outdoors shall meet the requirements of Article 225.

760-8. Mechanical Execution of Work. Fire alarm circuits shall be installed in a neat and workmanlike manner. Cables shall be supported by the building structure in such a manner that the cable will not be damaged by normal building use.

FPN: One way to determine accepted industry practice is to refer to nationally recognized standards such as *Commercial Building Telecommunications Wiring Standard*, ANSI/EIA/TIA 568-A-1995; *Commercial Building Standard for Telecommunications Pathways and Spaces*, ANSI/EIA/TIA 569-1990; and *Residential and Light Commercial Telecommunications Wiring Standard*, ANSI/EIA/TIA 570-1991.

Section 760-8 makes it clear that the *Code* requires these cables to be installed in a neat and workmanlike manner. The fine print note refers to industry standards that provide guidance on good workmanship.

760-10. Fire Alarm Circuit Identification. Fire alarm circuits shall be identified at terminal and junction locations, in a manner that will prevent unintentional interference with the signaling circuit during testing and servicing.

760-15. Fire Alarm Circuit Requirements. Fire alarm circuits shall comply with the following parts of this article.

(a) Nonpower-Limited Fire Alarm (NPLFA) Circuits. See Parts A and B.

(b) Power-Limited Fire Alarm (PLFA) Circuits. See Parts A and C.

Exact power source limitations for power-limited fire alarm circuits used by testing laboratories are found in Chapter 9, Tables 12(a) and 12(b). Table 12(a) covers alternating-current source limitations and Table 12(b) covers direct-current source limitations.

B. Nonpower-Limited Fire Alarm (NPLFA) Circuits

760-21. NPLFA Circuit Power Source Requirements. The power source of nonpower-limited fire alarm circuits shall comply with Chapters 1 through 4, and the output voltage shall not be more than 600 volts, nominal.

760-23. NPLFA Circuit Overcurrent Protection. Overcurrent protection for conductors No. 14 and larger shall be provided in accordance with the conductor ampacity without applying the derating factors of Section 310-15 to the ampacity calculation. Overcurrent protection shall not exceed 7 amperes for No. 18 conductors and 10 amperes for No. 16 conductors.

Exception: Where other articles of this Code permit or require other overcurrent protection.

760-24. NPLFA Circuit Overcurrent Device Location. Overcurrent devices shall be located at the point where the conductor to be protected receives its supply.

Exception No. 1: Where the overcurrent device protecting the larger conductor also protects the smaller conductor.

Exception No. 2: Transformer secondary conductors. Nonpower-limited fire alarm circuit conductors supplied by the secondary of a single-phase transformer that has only a 2-wire (single-voltage) secondary shall be permitted to be protected by overcurrent protection provided by the primary (supply) side of the transformer, provided the protection is in accordance with Section 450-3 and does not exceed the value determined by multiplying the secondary conductor ampacity by the secondary-to-primary transformer voltage ratio. Transformer secondary conductors other than 2-wire shall not be considered to be protected by the primary overcurrent protection.

Exception No. 3: Electronic power source output conductors. Nonpower-limited circuit conductors supplied by the output of a single-phase, listed electronic power source, other than a transformer, having only a 2 wire (single-voltage) output for connection to nonpower-limited circuits shall be permitted to be protected by overcurrent protection provided on the input side of the electronic power source, provided this protection does not exceed the value determined by multiplying the nonpower-limited circuit conductor ampacity by the output-to-input voltage ratio. Electronic power source outputs, other than 2 wire (single voltage), connected to nonpower-limited circuits shall not be considered to be protected by overcurrent protection on the input of the electronic power source.

Exception No. 3 was added to the 1999 *Code*. Nonpower-limited electronic power supplies that do not supply energy directly through the use of transformers are now covered by this exception. This exception permits overcurrent protection for the nonpower-limited circuit conductors to be installed on the input side of the electronic power source rather than on the output side for 2-wire circuits only.

FPN: A single-phase, listed electronic power supply whose output supplies a 2-wire (single-voltage) circuit is an example of a nonpower-limited power source that meets the requirements of Section 760-21.

760-25. NPLFA Circuit Wiring Methods. Installation of nonpower-limited fire alarm circuits shall be in accordance with Sections 110-3(b), 300-11(a), 300-15, 300-17, and other appropriate articles of Chapter 3.

Exception No. 1: As provided in Sections 760-26 through 760-30.

Exception No. 2: Where other articles of this Code require other methods.

Section 760-25 requires the appropriate wiring methods in Chapter 3 to be used for nonpower-limited circuits. However, Section 760-25, Exception No. 1, permits special nonpower-limited cable types to be used in place of Chapter 3 wiring methods.

Section 760-25 was revised for the 1999 *Code* to require devices be mounted in accordance with Chapter 3. Section 300-11(a) requires devices and equipment to be securely mounted. Section 300-15(b) is also referenced to require nonpower-limited circuit terminations to be made in a box or conduit body. However, Exception No. 4 to Section 300-15(b) permits devices with integral terminal enclosures and mounting brackets to be used without a box. Devices are required to be mounted on a box or conduit body where the instructions or listing indicate the use of a box. Fire alarm system components such as manual fire alarm boxes are frequently tested. Secure mounting is necessary to ensure that they will remain in place. (See Figure 760-5.)

760-26. Conductors of Different Circuits in Same Cable, Enclosure, or Raceway.

(a) Class 1 with NPLFA Circuits. Class 1 and nonpower-limited fire alarm circuits shall be permitted to occupy the same cable, enclosure, or raceway without regard to whether the individual circuits are alternating current or direct current, provided all conductors are insulated for the maximum voltage of any conductor in the enclosure or raceway.

Figure 760.5 *Typical manual fire alarm box. (Edwards/EST)*

(b) Fire Alarm with Power-Supply Circuits. Power-supply and fire alarm circuit conductors shall be permitted in the same cable, enclosure, or raceway only where connected to the same equipment.

760-27. NPLFA Circuit Conductors.

(a) Sizes and Use. Only copper conductors shall be permitted to be used for fire alarm systems. No. 18 conductors and No. 16 conductors shall be permitted to be used, provided they supply loads that do not exceed the ampacities given in Table 402-5 and are installed in a raceway, an approved enclosure, or a listed cable. Conductors larger than No. 16 shall not supply loads greater than the ampacities given in Section 310-15, as applicable.

The minimum size conductors permitted to be used on nonpower-limited fire protective signaling circuits is No. 18. The load must not exceed the conductor ampacities specified in Table 402-5.

NFPA 72-1996, *National Fire Alarm Code,* requires fire alarm device and appliance voltages to be between 85 and 110 percent of nominal rated voltage. Calculations should be made to ensure that all devices or appliances will be operating within these limits at full circuit load. Where future circuit extensions are anticipated, larger conductors should be considered. Some manufacturers specify maximum circuit loop resistances. The equipment specifications should be consulted to ensure that maximum allowable loop resistances are not exceeded.

(b) Insulation. Insulation on conductors shall be suitable for 600 volts. Conductors larger than No. 16 shall comply with Article 310. Conductors in sizes No. 18 and 16 shall be Type KF-2, KFF-2, PAFF, PTFF, PF, PFF, PGF, PGFF, RFH-2, RFHH-2, RFHH-3, SF-2, SFF-2, TF, TFF, TFN, TFFN, ZF, or ZFF. Conductors with other types and thickness of insulation shall be permitted if listed for nonpower-limited fire alarm circuit use.

FPN: For application provisions, see Table 402-3.

(c) Conductor Materials. Conductors shall be solid or stranded copper.

Exception to (b) and (c): Wire Types PAF and PTF shall be permitted only for high-temperature applications between 90°C (194°F) and 250°C (482°F).

The special stranding requirements for fire alarm circuit conductors were deleted in the 1996 *Code*. Either solid or stranded conductors are permitted. In addition, some commonly available flexible fixture wires in sizes No. 16 or 18 (such as Types TFF and TFFN) were added to the list of permitted conductor insulations for fire alarm circuits.

760-28. Number of Conductors in Cable Trays and Raceways, and Derating.

(a) NPLFA Circuits and Class 1 Circuits. Where only nonpower-limited fire alarm circuit and Class 1 circuit conductors are in a raceway, the number of conductors shall be determined in accordance with Section 300-17. The derating factors given in Section 310-15(b)(2)(a) shall apply if such conductors carry continuous load in excess of 10 percent of the ampacity of each conductor.

(b) Power-Supply Conductors and Fire Alarm Circuit Conductors. Where power-supply conductors and fire alarm circuit conductors are permitted in a raceway in accordance with Section 760-26, the number of conductors shall be determined in accordance with Section 300-17. The derating factors given in Section 310-15(b)(2)(a), shall apply as follows:

(1) To all conductors where the fire alarm circuit conductors carry continuous loads in excess of 10 percent of the ampacity of each conductor and where the total number of conductors is more than three

(2) To the power-supply conductors only, where the fire alarm circuit conductors do not carry continuous loads in excess of 10 percent of the ampacity of each conductor and where the number of power-supply conductors is more than three

(c) Cable Trays. Where fire alarm circuit conductors are installed in cable trays, they shall comply with Sections 318-9 through 318-11.

760-30. Multiconductor NPLFA Cables. Multiconductor nonpower-limited fire alarm cables that meet the requirements of Section 760-31 shall be permitted to be used on fire alarm circuits operating at 150 volts or less and shall be installed in accordance with (a) and (b).

(a) NPLFA Wiring Method. Multiconductor nonpower-limited fire alarm circuit cables shall be installed as follows.

(1) In raceway or exposed on surface of ceiling and sidewalls or fished in concealed spaces. Cable splices or terminations shall be made in listed fittings, boxes, enclosures, fire alarm devices, or utilization equipment. Where installed exposed, cables shall be adequately supported and installed in such a way that maximum protection against physical damage is afforded by building construction such as baseboards, door frames, ledges, etc. Where located within 7 ft (2.13 m) of the floor, cables shall be securely fastened in an approved manner at intervals of not more than 18 in. (457 mm).

(2) In metal raceway or rigid nonmetallic conduit where passing through a floor or wall to a height of 7 ft (2.13 m) above the floor unless adequate protection can be afforded by building construction such as detailed in (1), or unless an equivalent solid guard is provided.

(3) In rigid metal conduit, rigid nonmetallic conduit, intermediate metal conduit, or electrical metallic tubing where installed in hoistways.

Exception: As provided for in Section 620-21 for elevators and similar equipment.

(b) Applications of Listed NPLFA Cables. The use of nonpower-limited fire alarm circuit cables shall comply with (1) through (4).

(1) Ducts and Plenums. Multiconductor nonpower-limited fire alarm circuit cables, Types NPLFP, NPLFR, and NPLF, shall not be installed exposed in ducts or plenums. See Section 300-22(b).

Wiring methods for nonpower-limited circuits in ducts and plenums must be in accordance with the Chapter 3 wiring methods covered by Section 300-22(b).

(2) Other Spaces Used for Environmental Air. Cables installed in other spaces used for environmental air shall be Type NPLFP.

Other spaces used for environmental air are covered by Section 300-22(c) and the related fine print note. Spaces over suspended ceilings used as an environmental air-handling return are considered as other spaces used for environmental air by Section 300-22(c). Nonpower-limited cables used in other spaces used for environmental air must, however, be plenum rated. [See Section 760-31(d).]

Exception No. 1: Types NPLFR and NPLF cables installed in compliance with Section 300-22(c).

Exception No. 2: Other wiring methods in accordance with Section 300-22(c) and conductors in compliance with Section 760-27(c).

(3) Riser. Cables installed in vertical runs and penetrating more than one floor or cables installed in vertical runs in a shaft shall be Type NPLFR. Floor penetrations requiring Type NPLFR shall contain only cables suitable for riser or plenum use.

Exception No. 1: Type NPLF or other cables specified in Chapter 3 that are in compliance with Section 760-27(c) and encased in metal raceway.

Exception No. 2: Type NPLF cables located in a fireproof shaft having firestops at each floor.

FPN: See Section 300-21 for firestop requirements for floor penetrations.

(4) Other Wiring Within Buildings. Cables installed in building locations other than the locations covered in Section 760-30(b)(1), (2), and (3) shall be Type NPLF.

Exception No. 1: Chapter 3 wiring methods with conductors in compliance with Section 760-27(c).

Exception No. 2: Type NPLFP or Type NPLFR cables shall be permitted.

760-31. Listing and Marking of NPLFA Cables. Nonpower-limited fire alarm cables installed as wiring within buildings shall be listed in accordance with (a) and (b) and as being resistant to the spread of fire in accordance with (c) through (f), and shall be marked in accordance with (g).

(a) NPLFA Conductor Materials. Conductors shall be No. 18 or larger solid or stranded copper.

See the commentary following Section 760-27(c), Exception.

(b) Insulated Conductors. Insulated conductors shall be suitable for 600 volts. Insulated conductors No. 14 and larger shall be one of the types listed in Table 310-13 or one that is identified for this use. Insulated conductors No. 18 and No. 16 shall be in accordance with Section 760-27.

(c) Type NPLFP. Type NPLFP nonpower-limited fire alarm cable for use in other space used for environmental air shall be listed as being suitable for use in other space used for environmental air as described in Section 300-22(c) and shall also be listed as having adequate fire-resistant and low smoke-producing characteristics.

FPN: One method of defining low smoke-producing cable is by establishing an acceptable value of the smoke produced when tested in accordance with the *Standard Method of Test for Fire and Smoke Characteristics of Wires and Cables*, NFPA 262-1994, to a maximum peak optical density of 0.5 and a maximum average optical density of 0.15. Similarly, one method of defining fire-resistant cables is by establishing a maximum allowable flame travel distance of 5 ft (1.52 m) when tested in accordance with the same test.

For further information on the fire test method for plenum cables, see the commentary following Section 725-71(a), FPN. Also see the commentary following Section 760-30(b)(2).

(d) Type NPLFR. Type NPLFR nonpower-limited fire alarm riser cable shall be listed as being suitable for use in a vertical run in a shaft or from floor to floor and shall also be listed as having fire-resistant characteristics capable of preventing the carrying of fire from floor to floor.

FPN: One method of defining fire-resistant characteristics capable of preventing the carrying of fire from floor to floor is that the cables pass the *Test for Flame Propagation Height of Electrical and Optical-Fiber Cables Installed Vertically in Shafts*, ANSI/UL 1666-1997.

For further information on the fire test method for riser cables, see the commentary following Section 725-71(b), FPN.

(e) Type NPLF. Type NPLF nonpower-limited fire alarm cable shall be listed as being suitable for general-purpose fire alarm use, with the exception of risers, ducts, plenums, and other space used for environmental air, and shall also be listed as being resistant to the spread of fire.

FPN: One method of defining resistant to the spread of fire is that the cables do not spread fire to the top of the tray in the vertical-tray flame test in the *Reference Standard for Electrical Wires, Cables and Flexible Cords,* ANSI/UL 1581-1991.

Another method of defining resistant to the spread of fire is for the damage (char length) not to exceed 4 ft 11 in. (1.5 m) when performing the CSA vertical flame test — cables in cable trays, as described in *Test Methods for Electrical Wires and Cables,* CSA C22.2 No. 0.3-M-1985.

For further information on the fire test method for cables used as other wiring within buildings, see the commentary following Section 725-71(c), FPN.

(f) Fire Alarm Circuit Integrity (CI) Cable. Cables suitable for use in fire alarm systems to ensure survivability of critical circuits during a specified time under fire conditions shall be listed as circuit integrity (CI) cable. Cables identified in Sections 760-31(c), (d), and (e) meeting the requirements for circuit integrity shall have the additional classification using the suffix CI (for example, NPLFP-CI, NPLFR-CI and NPLF-CI).

FPN No. 1: This cable may be used for fire alarm circuits to comply with the survivability requirements of *National Fire Alarm Code,*® NFPA 72-1996, subsections 3-2.4, 3-4.4, 3-12.4, and 3-12.4.3, that the cable maintain its electrical function during fire conditions for a defined period of time.

Circuit integrity (CI) cable was added in the 1999 *Code* to meet the performance requirements for survivability required by 3-2.4, 3-4.4, and 3-12.4 of NFPA 72-1996, *National Fire Alarm Code*. This type of cable is designed to retain vital electrical performance during and immediately after fire exposure. CI cable carries the CI suffix, and is considered as a 2-hour-rated cable assembly. This type of cable provides an alternative to fire-rated mineral insulated (MI) cable.

FPN No. 2: One method of defining circuit integrity (CI) cable is by establishing a minimum 2-hour fire resistance rating for the cable when tested in accordance with the *Standard for Tests of Fire Resistive Cables,* UL 2196-1995.

(g) NPLFA Cable Markings. Multiconductor nonpower-limited fire alarm cables shall be marked in accordance with Table 760-31(g). Nonpower-limited fire alarm circuit cables shall be permitted to be marked with a maximum usage voltage rating of 150 volts. Cables that are listed for circuit integrity shall be identified with the suffix CI as defined in (f).

Table 760-31(g). NPLFA Cable Markings

Cable Marking	Type	Reference
NPLFP	Nonpower-limited fire alarm circuit cable for use in other space used for environmental air	760-31(c) and (g)
NPLFR	Nonpower-limited fire alarm circuit riser cable	760-31(d) and (g)
NPLF	Nonpower-limited fire alarm circuit cable	760-31(e) and (g)

Note: Cables identified in (c), (d), and (e) meeting the requirements for circuit integrity shall have the additional classification using the suffix "CI" (for example, NPLFP-CI, NPLFR-CI, and NPLF-CI).

FPN: Cable types are listed in descending order of fire-resistance rating.

C. Power-Limited Fire Alarm (PLFA) Circuits

760-41. Power Sources for PLFA Circuits. The power source for a power-limited fire alarm circuit shall be as specified in (a), (b), or (c).

FPN: Tables 12(a) and 12(b) in Chapter 9 provide the listing requirements for power-limited fire alarm circuit sources.

(a) Transformers. A listed PLFA or Class 3 transformer.

(b) Power Supplies. A listed PLFA or Class 3 power supply.

(c) Listed Equipment. Listed equipment marked to identify the PLFA power source.

FPN: Examples of listed equipment are a fire alarm control panel with integral power source; a circuit card listed for use as a PLFA source, where used as part of a listed assembly; a current-limiting impedance, listed for the purpose or part of a listed product, used in conjunction with a nonpower-limited transformer or a stored energy source, e.g., storage battery, to limit the output current.

760-42. Circuit Marking. The equipment shall be durably marked where plainly visible to indicate each circuit that is a power-limited fire alarm circuit.

FPN: See Section 760-52(a), Exception No. 3, where a power-limited circuit is to be reclassified as a nonpower-limited circuit.

760-51. Wiring Methods on Supply Side of the PLFA Power Source. Conductors and equipment on the supply side of the power source shall be installed in accordance with the appropriate requirements of Part B and Chapters 1 through 4. Transformers or other devices supplied from power-supply conductors shall be protected by an overcurrent device rated not over 20 amperes.

Exception: The input leads of a transformer or other power source supplying power-limited fire alarm circuits shall be permitted to be smaller than No. 14, but not smaller than No. 18, if they are not over 12 in. (305 mm) long and if they have insulation that complies with Section 760-27(b).

760-52. Wiring Methods and Materials on Load Side of the PLFA Power Source. Fire alarm circuits on the load side of the power source shall be permitted to be installed using wiring methods and materials in accordance with either (a) or (b).

Section 760-52 permits individual power-limited circuits to be installed using Chapter 3 wiring methods, nonpower-limited fire alarm circuit wiring methods, or power-limited circuit wiring methods. If it is desirable to run power-limited circuits in the same cable or raceway with nonpower-limited circuits, they may be reclassified as permitted by Section 760-52(a), Exception No. 3.

(a) NPLFA Wiring Methods and Materials. Installation shall be in accordance with Section 760-25, and conductors shall be solid or stranded copper.

Section 760-52(a) was revised for the 1999 *Code* to require devices and equipment be mounted in accordance with Chapter 3. Section 300-11(a) requires devices and equipment to be securely mounted. Section

300-15(b) is also referenced to require power-limited circuits to be terminated in a box or conduit body. However, Section 300-15(b), Exception No. 4, permits devices with integral terminal enclosures and mounting brackets to be used without a box. Devices are required to be mounted on a box or conduit body where the instructions or listing indicate the use of a box. Fire alarm system components such as manual fire alarm boxes are frequently tested. Secure mounting is necessary to ensure that they will remain in place.

Exception No. 1: The derating factors given in Section 310-15(b)(2)(a) shall not apply.

Exception No. 2: Conductors and multiconductor cables described in and installed in accordance with Sections 760-27 and 760-30 shall be permitted.

Exception No. 3: Power-limited circuits shall be permitted to be reclassified and installed as nonpower-limited circuits if the power-limited fire alarm circuit markings required by Section 760-42 are eliminated and the entire circuit is installed using the wiring methods and materials in accordance with Part B, Nonpower-Limited Fire Alarm Circuits.

Section 760-52(a) permits any of the wiring methods in Chapter 3 to be used for power-limited circuits. In addition, Section 760-52(a), Exception No. 3, allows power-limited circuits to be reclassified and installed in accordance with the requirements for nonpower-limited circuits. Where installed as nonpower-limited circuits, separation from power-limited circuits must be maintained, in accordance with Sections 760-26 and 760-54.

FPN: Power-limited circuits reclassified and installed as nonpower-limited circuits are no longer power-limited circuits, regardless of the continued connection to a power-limited source.

(b) PLFA Wiring Methods and Materials. Power-limited fire alarm conductors and cables described in Section 760-71 shall be installed as detailed in (1), (2), or (3) of this section. Devices shall be installed in accordance with Sections 110-3(b), 300-11(a), and 300-15.

(1) In raceway or exposed on the surface of ceiling and sidewalls or fished in concealed spaces. Cable splices or terminations shall be made in listed fittings, boxes, enclosures, fire alarm devices, or utilization equipment. Where installed exposed, cables shall be adequately supported and installed in such a way that maximum protection against physical damage is afforded by building construction such as baseboards, door frames, ledges, etc. Where located within 7 ft (2.13 m) of the floor, cables shall be securely fastened in an approved manner at intervals of not more than 18 in. (457 mm).

Section 760-52(b)(1) requires mechanical protection at splices and termination points. Since failure of a circuit often occurs at splices or termination points, this requirement offers more protection and strain relief for these cable connections.

(2) In metal raceways or rigid nonmetallic conduit where passing through a floor or wall to a height of 7 ft (2.13 m) above the floor, unless adequate protection can be afforded by building construction such as detailed in (1), or unless an equivalent solid guard is provided.

(3) In rigid metal conduit, rigid nonmetallic conduit, intermediate metal conduit, or electrical metallic tubing where installed in hoistways.

Exception No. 1: As provided for in Section 620-21 for elevators and similar equipment.

Section 760-52(b), Exception No. 1, permits the mixing of different wiring methods on a single circuit only for circuit extensions, replacement of conductors and cables, or in certain locations or areas covered by Section 760-3. Mixing of different wiring methods on a single circuit would be permitted when the circuit passes through a classified area, a wet location, a plenum, or extends outdoors to other buildings. See the commentary following Section 760-52.

Exception No. 2: Other wiring methods and materials installed in accordance with the requirements of Section 760-3 shall be permitted to extend or replace the conductors and cables described in Section 760-71 and permitted by Section 760-52(b).

760-54. Installation of Conductors and Equipment.

(a) Separation from Electric Light, Power, Class 1, NPLFA, and Medium Power Network-Powered Broadband Communications Circuit Conductors.

(1) In Cables, Compartments, Enclosures, Outlet Boxes, or Raceways. Power-limited circuit cables and conductors shall not be placed in any cable, cable tray, compartment, enclosure, outlet box, raceway, or similar fitting with conductors of electric light, power, Class 1, nonpower-limited fire alarm circuit conductors, or medium power network-powered broadband communications circuits.

Section 760-54(a)(1) was revised for the 1999 *Code* to specifically include cables of power-limited fire alarm circuits. Jackets of listed power-limited fire alarm cables do not have sufficient construction specifications to permit them to be installed with electric light, power, Class 1, nonpower-limited fire alarm circuits, and medium power network-powered broadband communications cables. Failure of the cable insulation due to a fault could lead to hazardous voltages being imposed on the power-limited fire alarm circuit conductors. This change also adds cable trays to the list of places requiring separation.

Section 725-54(a) was revised for the 1999 *Code* to include separation of power-limited cables and conductors from medium power network-powered broadband communications cables covered by new Article 830.

Exception No. 1: Where the conductors of the electric light, power, Class 1, nonpower-limited fire alarm, and medium power network-powered broadband communications circuit conductors are separated by a barrier from the power-limited fire alarm circuits. In enclosures, power-limited fire alarm circuits shall be permitted to be installed in a raceway within the enclosure to separate them from Class 1, electric light, power, nonpower-limited fire alarm, and medium power network-powered broadband communications circuits.

Exception No. 2: Conductors in compartments, enclosures, device boxes, outlet boxes, or similar fittings, where electric light, power, Class 1, nonpower-limited fire alarm, or medium power network-powered broadband communications circuit conductors are introduced solely to connect to the equipment connected to power-limited circuits to which the other conductors are connected, and

(a) The electric light, power, Class 1, nonpower-limited fire alarm, and medium power network-powered broadband communications circuit conductors are routed to maintain a minimum of 0.25 in. (6.35 mm) separation from the conductors and cables of power-limited fire alarm circuits, or

(b) The circuit conductors operate at 150 volts or less to ground and also comply with one of the following.

(1) The fire alarm power-limited circuits are installed using Types FPL, FPLR, FPLP or permitted substitute cables, provided these power-limited cable conductors extending beyond the jacket are separated by a minimum of 0.25 in. (6.35 mm) or by a nonconductive sleeve or nonconductive barrier from all other conductors, or

(2) The fire alarm power-limited circuit conductors are installed as nonpower-limited fire alarm circuits in accordance with Section 760-25.

Exception No. 3: Conductors entering compartments, enclosures, device boxes, outlet boxes, or similar fittings, where electric light, power, Class 1, nonpower-limited fire alarm, or medium power network-powered broadband communications circuit conductors are introduced solely to connect to the equipment connected to power-limited fire alarm circuits or to other circuits controlled by the fire alarm system to which the other conductors in the enclosure are connected. If the conductors must enter an enclosure that is provided with a single opening, they shall be permitted to enter through a single fitting (such as a tee) provided the conductors are separated from the conductors of the other circuits by a continuous and firmly fixed nonconductor, such as flexible tubing.

(2) In Hoistways. Power-limited fire alarm circuit conductors shall be installed in rigid metal conduit, intermediate metal conduit, or electrical metallic tubing in hoistways.

Exception: As provided for in Section 620-21 for elevators and similar equipment.

(3) Other Applications. Power-limited fire alarm circuit conductors shall be separated at least 2 in. (50.8 mm) from conductors of any electric light, power, Class 1, nonpower limited fire alarm, or medium power network-powered broadband communications circuits.

Exception No. 1: Where either (1) all of the electric light, power, Class 1, nonpower-limited fire alarm, and medium power network-powered broadband communications circuit conductors or (2) all of the power-limited fire alarm circuit conductors are in raceway or in metal-sheathed, metal-clad, nonmetallic-sheathed, or Type UF cables.

Exception No. 2: Where all of the electric light, power, Class 1, nonpower-limited fire alarm, and medium power network-powered broadband communications circuit conductors are permanently separated from all of the power-limited fire alarm circuit conductors by a continuous and firmly fixed nonconductor, such as porcelain tubes or flexible tubing in addition to the insulation on the wire.

(b) Conductors of Different PLFA Circuits, Class 2, Class 3, and Communications Circuits in Same Cable, Enclosure, or Raceway.

(1) Two or More PLFA Circuits. Cable and conductors of two or more power-limited fire alarm circuits, communications circuits, or Class 3 circuits shall be permitted in the same cable, enclosure, or raceway.

(2) Class 2 Circuits with PLFA Circuits. Conductors of one or more Class 2 circuits shall be permitted within the same cable, enclosure, or raceway with conductors of power-limited fire alarm circuits, provided that the insulation of the Class 2 circuit conductors in the cable, enclosure, or raceway is at least that required by the power-limited fire alarm circuits.

(3) Low Power Network-Powered Broadband Communications Cables and PLFA Cables. Low power network-powered broadband communications circuits shall be permitted in the same enclosure or raceway with PLFA cables.

(c) Support of Conductors. Power-limited fire alarm circuit conductors shall not be strapped, taped, or attached by any means to the exterior of any conduit or other raceway as a means of support.

See the commentary following Section 760-8.

(d) Conductor Size. Conductors of No. 26 shall only be permitted where spliced with a connector listed as suitable for No. 26 to No. 24 or larger conductors that are terminated on equipment or where the No. 26 conductors are terminated on equipment listed as suitable for No. 26 conductors. Single conductors shall not be smaller than No. 18.

Due to a signaling method called *multiplexing*, power-limited fire alarm cable may contain circuit conductors as small as No. 26. In the past, these small conductors were typically reserved for communications circuits, but due to recent technological advances, they now have application within the fire alarm industry. Of course, these small circuit conductors are only permitted to be used as specified in Section 760-54(d) and as permitted by the listing or installation instructions of specific fire alarm equipment.

760-55. Current-Carrying Continuous Line-Type Fire Detectors.

(a) Application. Listed continuous line-type fire detectors, including insulated copper tubing of pneumatically operated detectors, employed for both detection and carrying signaling currents shall be permitted to be used in power-limited circuits.

(b) Installation. Continuous line-type fire detectors shall be installed in accordance with Sections 760-42 through 760-52 and Section 760-54.

760-61. Applications of Listed PLFA Cables. Power-limited fire alarm cables shall comply with (a) through (c) or, where cable substitution is being made, with (d).

(a) Plenum. Cables installed in ducts, plenums, and other spaces used for environmental air shall be Type FPLP.

Exception: Types FPLP, FPLR, and FPL cable installed in compliance with Section 300-22.

(b) Riser. Cables installed in vertical runs and penetrating more than one floor or cables installed in vertical runs in a shaft shall be Type FPLR. Floor penetrations requiring Type FPLR shall contain only cables suitable for riser or plenum use.

Exception No. 1: Where the cables are installed in metal raceway or are located in a fireproof shaft having firestops at each floor.

Exception No. 2: Type FPL in one- and two-family dwellings.

FPN: See Section 300-21 for firestop requirements for floor penetrations.

(c) Other Wiring Within Buildings. Cables installed in building locations other than the locations covered in (a) and (b) shall be Type FPL.

Exception No. 1: Where the cables are enclosed in raceway.

Exception No. 2: Cables specified in Chapter 3 that meet the requirements of Sections 760-71(a) and (b) and are installed in nonconcealed spaces where the exposed length of cable does not exceed 10 ft (3.05 m).

Exception No. 3: A portable fire alarm system provided to protect a stage or set when not in use shall be permitted to use wiring methods in accordance with Section 530-12.

(d) Cable Uses and Permitted Substitutions. The uses and permitted substitutions for power-limited fire alarm circuit cables listed in Table 760-61 shall be considered suitable for the purpose and shall be permitted.

FPN: For information on multipurpose cables (Types MPP, MPR, MPG, MP) and communications cables (Types CMP, CMR, CMG, CM), see Section 800-50.

760-71. Listing and Marking of PLFA Cables and Insulated Continuous Line-Type Fire Detectors. Type FPL cables installed as wiring within buildings shall be listed as being resistant to the spread of fire and other criteria in accordance with (a) through (h) and shall be marked in accordance with (i). Insulated continuous line-type fire detectors shall be listed in accordance with (j).

(a) Conductor Materials. Conductors shall be solid or stranded copper.

Table 760-61. Cable Uses and Permitted Substitutions

Cable Type	Use	References	Permitted Substitutions: Multiconductor	Permitted Substitutions: Coaxial
FPLP	Power-limited fire alarm plenum cable	760-61(a)	CMP	MPP
FPLR	Power-limited fire alarm riser cable	760-61(b)	CMP, FPLP, CMR	MPP, MPR
FPL	Power-limited fire alarm cable	760-61(c)	CMP, FPLP, CMR, FPLR, CMG, CM	MPP, MPR, MPG, MP

Table 760-61 has been extensively revised for the 1999 *Code* to reflect current field applications for various cable types.

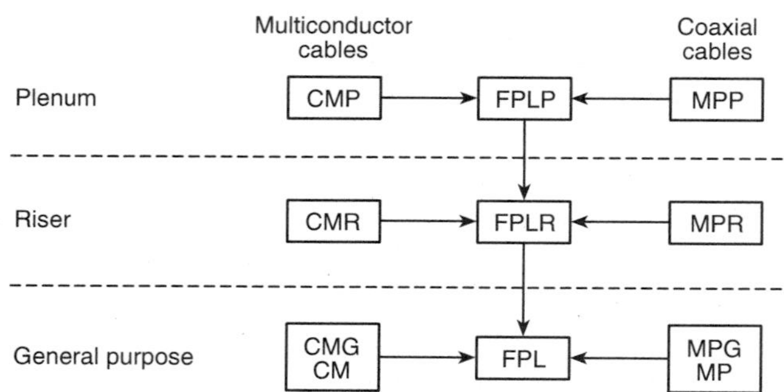

Figure 760-61 Cable substitution hierarchy.

Some line-type fire detectors may not be made exclusively of copper, but are listed for the application. Also see the commentary following Section 760-27(c), Exception.

(b) Conductor Size. The size of conductors in a multiconductor cable shall not be smaller than No. 26. Single conductors shall not be smaller than No. 18.

(c) Ratings. The cable shall have a voltage rating of not less than 300 volts.

(d) Type FPLP. Type FPLP power-limited fire alarm plenum cable shall be listed as being suitable for use in ducts, plenums, and other space used for environmental air and shall also be listed as having adequate fire-resistant and low smoke-producing characteristics.

FPN: One method of defining low smoke-producing cable is by establishing an acceptable value of the smoke produced when tested in accordance with the *Standard Method of Test for Fire and Smoke Characteristics of Wires and Cables,* NFPA 262-1994, to a maximum peak optical density of 0.5 and a maximum average optical density of 0.15. Similarly, one method of defining fire-resistant cables is by establishing maximum allowable flame travel distance of 5 ft (1.52 m) when tested in accordance with the same test.

For further information on the fire test method for plenum cables, see the commentary following Section 725-71(a), FPN.

(e) Type FPLR. Type FPLR power-limited fire alarm riser cable shall be listed as being suitable for use in a vertical run in a shaft or from floor to floor and shall also be listed as having fire-resistant characteristics capable of preventing the carrying of fire from floor to floor.

FPN: One method of defining fire-resistant characteristics capable of preventing the carrying of fire from floor to floor is that the cables pass the requirements of the *Standard Test for Flame Propagation Height of Electrical and Optical-Fiber Cable Installed Vertically in Shafts,* ANSI/UL 1666-1997.

For further information on the fire test method for riser cables, see the commentary following Section 725-71(b), FPN.

(f) Type FPL. Type FPL power-limited fire alarm cable shall be listed as being suitable for general-purpose fire alarm use, with the exception of risers, ducts, plenums, and other spaces used for environmental air and shall also be listed as being resistant to the spread of fire.

FPN: One method of defining resistant to the spread of fire is that the cables do not spread fire to the top of the tray in the vertical-tray flame test in the *Reference Standard for Electrical Wires, Cables and Flexible Cords,* ANSI/UL 1581-1991. Another method of defining resistant to the spread of fire is for the damage (char length) not to exceed 4 ft 11 in. (1.5 m) when performing the CSA vertical flame test — cables in cable trays, as described in *Test Methods for Electrical Wires and Cables,* CSA C22.2 No. 0.3-M-1985.

For further information on the fire test method for cables used as other wiring within buildings, see the commentary following Section 725-71(c), FPN.

(g) Fire Alarm Circuit Integrity (CI) Cable. Cables suitable for use in fire alarm systems to ensure survivability of critical circuits during a specified time under fire conditions shall be listed as circuit integrity (CI) cable. Cables identified in Sections 760-71(d), (e), and (f) meeting the requirements for circuit integrity shall have the additional classification using the suffix CI (for example, FPLP-CI, FPLR-CI and FPL-CI).

FPN No. 1: This cable is used for fire alarm circuits as one method of complying with the survivability requirements of the *National Fire Alarm Code,* NFPA 72-1996, 3-2.4, 3-4.4, 3-12.4, and 3-12.4.3, that the cable maintain its electrical function during fire conditions for a defined period of time.

FPN No. 2: One method of defining circuit integrity (CI) cable is by establishing a minimum 2-hour fire resistance rating for the cable when tested in accordance with the *Standard for Tests of Fire Resistive Cables*, UL 2196-1995.

There are provisions in the *National Fire Alarm Code®*, NFPA 72-1996, that require continued operation of the fire alarm system, including circuit wiring, under severe conditions such as attack by fire. To provide this integrity, NFPA 72 recognizes the use of 2-hour fire-rated cable assemblies. FPN No. 2 refers to cables tested in accordance with UL 2196 as an example of the type of wiring method that would qualify as circuit integrity (CI) cable. For one such example of CI cable, see Figure 760.6.

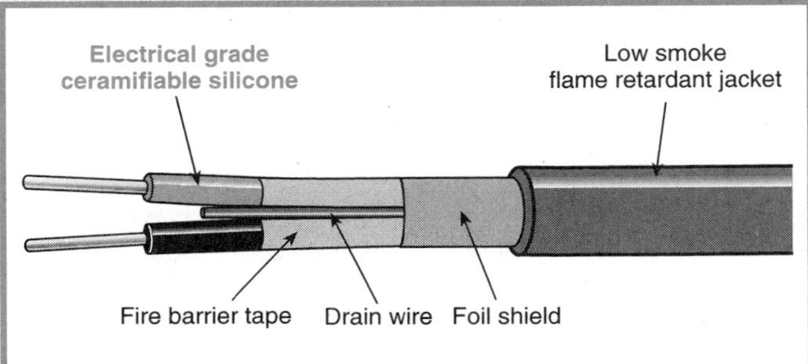

Figure 760.6 *Circuit integrity cable. (Rockbestos-Surprenant Cable Corp.)*

(h) Coaxial Cables. Coaxial cables shall be permitted to use 30 percent conductivity copper-covered steel center conductor wire and shall be listed as Type FPLP, FPLR, or FPL cable.

(i) Cable Marking. The cable shall be marked in accordance with Table 760-71(i). The voltage rating shall not be marked on the cable. Cables that are listed for circuit integrity shall be identified with the suffix CI as defined in Section 760-71(g).

FPN: Voltage ratings on cables may be misinterpreted to suggest that the cables may be suitable for Class 1, electric light, and power applications.

Exception: Voltage markings shall be permitted where the cable has multiple listings and voltage marking is required for one or more of the listings.

FPN: Cable types are listed in descending order of fire-resistance rating.

Table 760-71(i). Cable Markings

Cable Marking	Type	Listing References
FPLP	Power-limited fire alarm plenum cable	760-71(d) and (i)
FPLR	Power-limited fire alarm riser cable	760-71(e) and (i)
FPL	Power-limited fire alarm cable	760-71(f) and (i)

Note: Cables identified in (d), (e), and (f) meeting the requirements for circuit integrity shall have the additional classification using the suffix "CI" (for example, FPLP-CI, FPLR-CI, and FPL-CI).

(j) Insulated Continuous Line-Type Fire Detectors. Insulated continuous line-type fire detectors shall be rated in accordance with (c), listed as being resistant to the spread of fire in accordance with (d) through (f), marked in accordance with (i), and the jacket compound shall have a high degree of abrasion resistance.

Article 770 — Optical Fiber Cables and Raceways

Contents

770-53. Applications of Listed Optical Fiber Cables and Raceways
(a) Plenum
(b) Riser
(c) Other Wiring Within Buildings
(d) Hazardous (Classified) Locations
(e) Cable Trays
(f) Cable Substitutions

A. General

770-1. Scope. The provisions of this article apply to the installation of optical fiber cables and raceways. This article does not cover the construction of optical fiber cables and raceways.

Article 770 is included to permit the orderly development and usage of optical fiber technology where used in conjunction with electrical conductors for communications, signaling, and control circuits in lieu of metallic conductors. The most common optical fiber cable used in buildings is nonconductive. (See Figure 770.1.)

Optical fiber cables may be desirable in some circumstances where electrical noise is a problem, as they are not affected by electrical noise. Optical fiber cables may be nonconductive or may contain electrical conductors. See Figures 770.1 and 770.2.

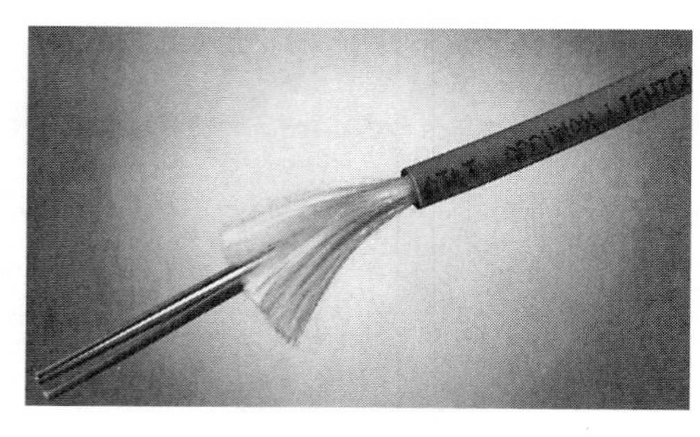

Figure 770.1 *An example of a nonconductive optical fiber cable. (AT&T Network Cable Systems)*

770-2. Definitions.

Exposed. The circuit is in such a position that, in case of failure of supports and insulation, contact with another circuit may result.

FPN: See Article 100 for two other definitions of *Exposed.*

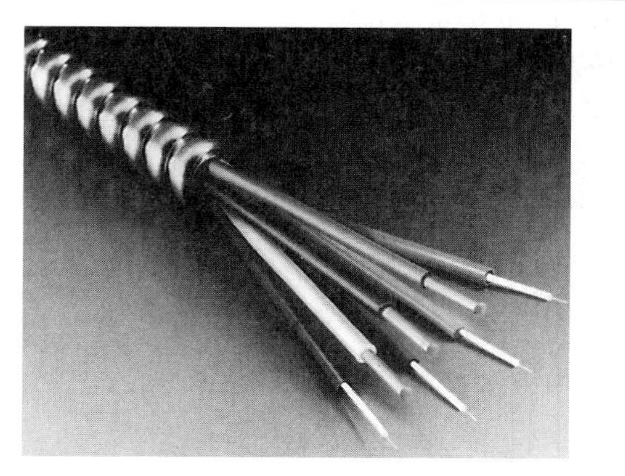

Figure 770.2 *An example of a composite optical fiber cable. This cable also meets the requirements of Article 334 and is referred to as Type MC cable. (AFC Cable Systems)*

Optical Fiber Raceway. A raceway designed for enclosing and routing nonconductive optical fiber cables.

Point of Entrance. The point at which the wire or cable emerges from an external wall, from a concrete floor slab, or from a rigid metal conduit or an intermediate metal conduit grounded to an electrode in accordance with Section 800-40(b).

770-3. Locations and Other Articles. Circuits and equipment shall comply with (a) and (b). Only those sections of Article 300 referenced in this article shall apply to optical fiber cables and raceways.

See the commentary following Sections 300-22(b) and 300-22(c).

(a) Spread of Fire or Products of Combustion. See Section 300-21.

(b) Ducts, Plenums, and Other Air-Handling Spaces. Section 300-22, where installed in ducts or plenums or other space used for environmental air.

Exception to (b): As permitted in Section 770-53(a).

770-4. Optical Fiber Cables. Optical fiber cables transmit light for control, signaling, and communications through an optical fiber.

770-5. Types. Optical fiber cables can be grouped into three types.

(a) Nonconductive. These cables contain no metallic members and no other electrically conductive materials.

(b) Conductive. These cables contain noncurrent-carrying conductive members such as metallic strength members, metallic vapor barriers, and metallic armor or sheath.

(c) Composite. These cables contain optical fibers and current-carrying electrical conductors, and shall be permitted to contain noncurrent-carrying conductive members such as metallic strength members and metallic vapor barriers. Composite optical fiber cables shall be classified as electrical cables in accordance with the type of electrical conductors.

770-6. Raceways for Optical Fiber Cables. The raceway shall be of a type permitted in Chapter 3 and installed in accordance with Chapter 3.

Exception: Listed nonmetallic optical fiber raceway identified as general purpose, riser, or plenum optical fiber raceway in accordance with Section 770-51 and installed in accordance with Sections 331-7 through 331-14, where the requirements applicable to electrical nonmetallic tubing shall apply. Unlisted underground or outside plant construction plastic innerduct shall be terminated at the point of entrance.

Section 770-6 was revised for the 1999 *Code* to emphasize that conduit fill requirements apply where the optical fiber is installed in a raceway with electrical conductors.

FPN: For information on listing requirements for optical fiber raceways, see *Standard for Optical Fiber Raceways,* UL 2024.

Where optical fiber cables are installed within the raceway without current-carrying conductors, the raceway fill tables of Chapter 3 and Chapter 9 shall not apply.

Where nonconductive optical fiber cables are installed with electric conductors in a raceway, the raceway fill tables of Chapter 3 and Chapter 9 shall apply.

770-7. Access to Electrical Equipment Behind Panels Designed to Allow Access. Access to equipment shall not be denied by an accumulation of cables that prevents removal of panels, including suspended ceiling panels.

See the commentary following Section 725-5.

770-8. Mechanical Execution of Work. Optical fiber cables shall be installed in a neat and workmanlike manner. Cables shall be supported by the building structure in such a manner that the cable will not be damaged by normal building use.

FPN: One way to determine accepted industry practice is to refer to nationally recognized standards such as *Commercial Building Telecommunications Wiring Standard,* ANSI/EIA/TIA 568A-1995; *Commercial Building Standard for Telecommunications Pathways and Spaces,* ANSI/EIA/TIA 569-A-1997; and *Residential and Light Commercial Telecommunications Wiring Standard,* ANSI/EIA/TIA 570-1991, or other ANSI-approved installation standards.

Section 770-8 makes it clear that the *Code* requires these cables to be installed in a neat and workmanlike manner. The fine print note refers to industry standards that provide guidance on good workmanship.

B. Protection

770-33. Grounding of Entrance Cables. Where exposed to contact with electric light or power conductors, the non-current-carrying metallic members of optical fiber cables entering buildings shall be grounded as close to the point of entrance as practicable or shall be interrupted as close to the point of entrance as practicable by an insulating joint or equivalent device.

C. Cables Within Buildings

770-49. Fire Resistance of Optical Fiber Cables. Optical fiber cables installed as wiring within buildings shall be listed as being resistant to the spread of fire in accordance with Sections 770-50 and 770-51.

770-50. Listing, Marking, and Installation of Optical Fiber Cables. Optical fiber cables in a building shall be listed as being suitable for the purpose, and cables shall be marked in accordance with Table 770-50.

Table 770-50. Cable Markings

Cable Marking	Type	Reference
OFNP	Nonconductive optical fiber plenum cable	770-51(a) and 770-53(a)
OFCP	Conductive optical fiber plenum cable	770-51(a) and 770-53(a)
OFNR	Nonconductive optical fiber riser cable	770-51(b) and 770-53(b)
OFCR	Conductive optical fiber riser cable	770-51(b) and 770-53(b)
OFNG	Nonconductive optical fiber general-purpose cable	770-51(c) and 770-53(c)
OFCG	Conductive optical fiber general-purpose cable	770-51(c) and 770-53(c)
OFN	Nonconductive optical fiber general-purpose cable	770-51(d) and 770-53(c)
OFC	Conductive optical fiber general-purpose cable	770-51(d) and 770-53(c)

Exception No. 1: Optical fiber cables shall not be required to be listed and marked where the length of the cable within the building, measured from its point of entrance, does not exceed 50 ft (15.2 m) and the cable enters the building from the outside and is terminated in an enclosure.

FPN: Splice cases or terminal boxes, both metallic and plastic types, are typically used as enclosures for splicing or terminating optical fiber cables.

Exception No. 2: Conductive optical fiber cable shall not be required to be listed and marked where the cable enters the building from the outside and is run in rigid metal conduit or intermediate metal conduit and such conduits are grounded to an electrode in accordance with Section 800-40(b).

Exception No. 3: Nonconductive optical fiber cables shall not be required to be listed and marked where the cable enters the building from the outside and is run in raceway installed in compliance with Chapter 3.

FPN No. 1: Cable types are listed in descending order of fire resistance rating. Within each fire resistance rating, nonconductive cable is listed first, since it may substitute for the conductive cable.

FPN No. 2: See the referenced sections for requirements and permitted uses.

770-51. Listing Requirements for Optical Fiber Cables and Raceways. Optical fiber cables shall be listed in accordance with (a) through (d), and optical fiber raceways shall be listed in accordance with (e) through (g).

(a) Types OFNP and OFCP. Types OFNP and OFCP nonconductive and conductive optical fiber plenum cables shall be listed as being suitable for use in ducts, plenums, and other space used for environmental air and shall also be listed as having adequate fire-resistant and low smoke-producing characteristics.

FPN: One method of defining low smoke-producing cables is by establishing an acceptable value of the smoke produced when tested in accordance with the *Standard Method of Test for Fire and Smoke Characteristics of Wires and Cables,* NFPA 262-1994, to a maximum peak optical density of 0.5 and a maximum average optical density of 0.15. Similarly, one method of defining fire-resistant cables is by defining maximum allowable flame travel distance of 5 ft (1.52 m) when tested in accordance with the same test.

For further information on the fire test method for plenum cables, see the commentary following Section 725-71(a), FPN.

(b) Types OFNR and OFCR. Types OFNR and OFCR nonconductive and conductive optical fiber riser cables shall be listed as being suitable for use in a vertical run in a shaft or from floor to floor and shall also be listed as having fire-resistant characteristics capable of preventing the carrying of fire from floor to floor.

FPN: One method of defining fire-resistant characteristics capable of preventing the carrying of fire from floor to floor is that the cables pass the requirements of the *Standard Test for Flame Propagation Height of Electrical and Optical-Fiber Cable Installed Vertically in Shafts,* ANSI/UL 1666-1997.

For further information on the fire test method for riser cables, see the commentary following Section 725-71(b), FPN.

(c) Types OFNG and OFCG. Types OFNG and OFCG nonconductive and conductive general-purpose optical fiber cables shall be listed as being suitable for general-purpose use, with the exception of risers and plenums, and shall also be listed as being resistant to the spread of fire.

FPN: One method of defining resistance to the spread of fire is for the damage (char length) not to exceed 4 ft 11 in. (1.5 m) when performing the vertical flame test — cables in cable trays, as described in *Test Methods for Electrical Wires and Cables,* CSA C22.2 No. 0.3-M 1985.

For further information on the fire test method for cables used as other wiring within buildings, see the commentary following Section 725-71(c), FPN.

(d) Types OFN and OFC. Types OFN and OFC nonconductive and conductive optical fiber cables shall be listed as being suitable for general-purpose use, with the exception of risers, plenums, and other space used for environmental air, and shall also be listed as being resistant to the spread of fire.

FPN: One method of defining resistant to the spread of fire is that the cables do not spread fire to the top of the tray in the vertical-tray flame test in the *Reference Standard for Electrical Wires, Cables and Flexible Cords,* ANSI/UL 1581-1991.

Another method of defining resistant to the spread of fire is for the damage (char length) not to exceed 4 ft 11 in. (1.5 m) when performing the vertical flame test — cables in cable trays, as described in *Test Methods for Electrical Wires and Cables*, CSA C22.2 No. 0.3-M-1985.

(e) Plenum Optical Fiber Raceway. Plenum optical fiber raceways shall be listed as having adequate fire-resistant and low smoke-producing characteristics.

(f) Riser Optical Fiber Raceway. Riser optical fiber raceways shall be listed as having fire-resistant characteristics capable of preventing the carrying of fire from floor to floor.

(g) General-Purpose Optical Fiber Cable Raceway. General-purpose optical fiber cable raceway shall be listed as being resistant to the spread of fire.

The optical fiber raceways indicated in Sections 770-51(e) and (g) are listed raceways used in plenum, riser, or general-purpose applications. This listing includes raceways and fittings for installations of nonconductive optical fiber cables in accordance with this article. These raceways are not suitable for installation of current-carrying conductors, cords, or cables. Nor are these raceways suitable for installations of hybrid cables that contain optical fiber members and current-carrying conductors.

A raceway marked "plenum" is suitable for use in ducts, plenums, or other spaces used for environmental air in accordance with Section 770-53(a) when used to enclose optical fiber cables marked OFNP. This raceway exhibits a maximum peak optical density of 0.5, a maximum average optical density of 0.15, and a maximum flame-spread distance of 5 ft when tested in accordance with the *Test For Flame Propagation and Smoke-Density Values for Electrical and Optical-Fiber Cables Used in Spaces Transporting Environmental Air,* UL 910. This raceway is identified by a marking on the surface of the raceway or on a marker tape indicating "plenum." A raceway marked "plenum" is also suitable for installation in risers when used to enclose optical fiber cables marked OFNP or OFNR, and general purpose use when used to enclose optical fiber cables marked OFNP, OFNR, OFNG, or OFN.

A raceway marked "riser" is suitable for installation in risers in accordance with Section 770-53(b) when used to enclose optical fiber cable marked OFNP or OFNR. This raceway has fire-resistant characteristics capable of preventing the carrying of fire from floor to floor. This raceway meets the test requirements of the *Standard Test for Flame Propagation Height of Electrical and Optical-Fiber Cable Installed Vertically in Shafts,* UL 1666. This raceway is identified by a marking on the surface of the raceway or on a marker tape indicating "riser." A raceway marked "riser" is also suitable for general-purpose use when used to enclose optical fiber cable marked OFNP, OFNR, OFNG, or OFN.

A raceway marked "general purpose" is suitable for installation in general-purpose areas in accordance with Section 770-53(c) when used to enclose optical fiber cable marked OFNP, OFNR, OFNG, or OFN. This raceway has fire-resistant characteristics that are capable of preventing the spread of fire.

Pliable raceway is raceway that can be bent by hand without the use of tools. The smallest radius of the curve of the inner edge of any bend to which the raceway may be bent without cracking either on the outer surface or internally is not less than 2½ times the outside diameter of the raceway.

770-52. Installation of Optical Fibers and Electrical Conductors.

(a) With Conductors for Electric Light, Power, Class 1, Nonpower-Limited Fire Alarm, or Medium Power Network-Powered Broadband Communications Circuits. Optical fibers shall be permitted within the same composite cable for electric light, power, Class 1, nonpower-limited fire alarm, or medium power network-powered broadband communications circuits operating at 600 volts or less only where the functions of the optical fibers and the electrical conductors are associated.

Nonconductive optical fiber cables shall be permitted to occupy the same cable tray or raceway with conductors for electric light, power, Class 1, nonpower-limited fire alarm, or medium power network-powered broadband communications circuits operating at 600 volts or less. Conductive optical fiber cables shall not be permitted to occupy the same cable tray or raceway with conductors for electric light, power, Class 1, nonpower-limited fire alarm, or medium power network-powered broadband communications circuits.

Composite optical fiber cables containing only current-carrying conductors for electric light, power, Class 1 circuits rated 600 volts or less shall be permitted to occupy the same cabinet, cable tray, outlet box, panel, raceway, or other termination enclosure with conductors for electric light, power, or Class 1 circuits operating at 600 volts or less.

Nonconductive optical fiber cables shall not be permitted to occupy the same cabinet, outlet box, panel, or similar enclosure housing the electrical terminations of an electric light, power, Class 1, nonpower-limited fire alarm, or medium power network-powered broadband communications circuit.

Exception No. 1: Occupancy of the same cabinet, outlet box, panel, or similar enclosure shall be permitted where nonconductive optical fiber cable is functionally associated with the electric light, power, Class 1, nonpower-limited fire alarm, or medium power network-powered broadband communications circuit.

Exception No. 2: Occupancy of the same cabinet, outlet box, panel, or similar enclosure shall be permitted where nonconductive optical fiber cables are installed in factory- or field-assembled control centers.

Exception No. 3: In industrial establishments only, where conditions of maintenance and supervision ensure that only qualified persons will service the installation, nonconductive optical fiber cables shall be permitted with circuits exceeding 600 volts.

Exception No. 4: In industrial establishments only, where conditions of maintenance and supervision ensure that only qualified persons will service the installation, composite optical fiber cables shall be permitted to contain current-carrying conductors operating over 600 volts.

Installations in raceway shall comply with Section 300-17.

Section 770-52 was revised for the 1999 *Code* to include separation of optical fibers from medium power network-powered broadband communications cables covered by new Article 830.

(b) With Other Conductors. Optical fibers shall be permitted in the same cable, and conductive and nonconductive optical fiber cables shall be permitted in the same cable tray, enclosure, or raceway with conductors of any of the following:

(1) Class 2 and Class 3 remote-control, signaling, and power-limited circuits in compliance with Article 725
(2) Power-limited fire alarm systems in compliance with Article 760
(3) Communications circuits in compliance with Article 800
(4) Community antenna television and radio distribution systems in compliance with Article 820
(5) Low power network-powered broadband communications circuits in compliance with Article 830

(c) Grounding. Noncurrent-carrying conductive members of optical fiber cables shall be grounded in accordance with Article 250.

770-53. Applications of Listed Optical Fiber Cables and Raceways. Nonconductive and conductive optical fiber cables shall comply with (a) through (f) as applicable.

(a) Plenum. Cables installed in ducts, plenums, and other spaces used for environmental air shall be Type OFNP or OFCP.

Also, listed plenum optical fiber raceways shall be permitted to be installed in ducts and plenums as described in Section 300-22(b) and in other space used for environmental air as described in Section 300-22(c). Only Type OFNP cable shall be permitted to be installed in these raceways.

Exception: Types OFNR, OFCR, OFNG, OFN, OFCG, and OFC cables installed in compliance with Section 300-22.

(b) Riser. Cables installed in vertical runs and penetrating more than one floor or cables installed in vertical runs in a shaft shall be Types OFNR or OFCR. Floor penetrations requiring Types OFNR or OFCR shall contain only cables suitable for riser or plenum use.

Also, listed riser optical fiber raceways shall be permitted to be installed in vertical runs in a shaft or from floor to floor. Only Types OFNR and OFNP cables shall be permitted to be installed in these raceways.

Exception No. 1: Where Types OFNG, OFN, OFCG, and OFC cables are encased in metal raceway or are located in a fireproof shaft having firestops at each floor.

Exception No. 2: Type OFNG, OFN, OFCG, or OFC cable in one- and two-family dwellings.

FPN: See Section 300-21 for firestop requirements for floor penetrations.

(c) Other Wiring Within Buildings. Cables installed in building locations other than the locations covered in (a) and (b) shall be Type OFNG, OFN, OFCG, or OFC. Such cables shall be permitted to be installed in listed general-purpose optical fiber raceways.

(d) Hazardous (Classified) Locations. Cables installed in hazardous (classified) locations shall be any type indicated in Table 770-53.

(e) Cable Trays. Optical fiber cables of the types listed in Table 770-50 shall be permitted to be installed in cable trays.

FPN: It is not the intent to require that these optical fiber cables be listed specifically for use in cable trays.

(f) Cable Substitutions. The substitutions for optical fiber cables listed in Table 770-53 shall be permitted.

Table 770-53. Cable Substitutions

Cable Type	Permitted Substitutions
OFNP	None
OFCP	OFNP
OFNR	OFNP
OFCR	OFNP, OFCP, OFNR
OFNG, OFN	OFNP, OFNR
OFCG, OFC	OFNP, OFCP, OFNR, OFCR, OFNG, OFN

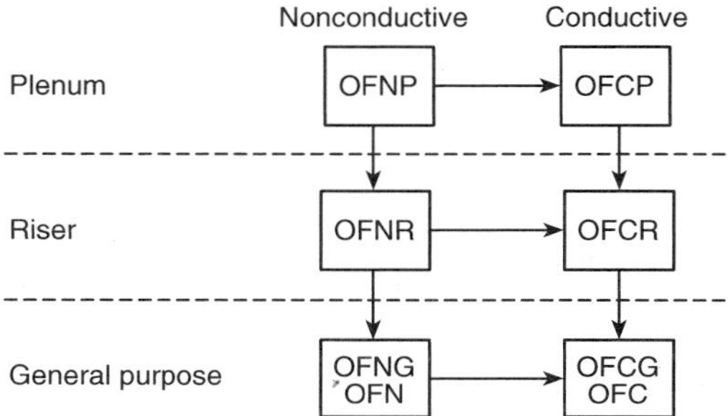

Figure 770-53 Cable substitution hierarchy.

Article 780 — Closed-Loop and Programmed Power Distribution

Contents

780-1. Scope. The provisions of this article apply to premise power distribution systems jointly controlled by a signaling between the energy controlling equipment and utilization equipment.

Article 780 provides requirements for the "smart house" concept, which involves universal cable terminating in universal outlets.

Buildings wired by conventional methods require separate sets of conductors for different systems. In the smart house, however, multiple conductors for 120-volt ac power, 24-volt dc UPS, telephone, remote-control and signaling, as well as coaxial cable, are combined in a single construction known as *hybrid cabling.*

This hybrid cabling serves multipurpose receptacle outlets known as *convenience centers,* which are capable of supplying different types of energy and signals to specific appliances or equipment.

The smart house utilizes an energy safety technique called *closed-loop control* to reduce shock hazard. In conventional wiring, receptacles are energized at all times under normal operating conditions. In the closed-loop configuration, receptacles are not energized until the insertion of an attachment plug generates a characteristic electrical identification.

Figure 780.1 illustrates a typical smart house installation. The 1987 *Handbook* illustrated a system distributing power at both 120/240-volts ac and 48-volts dc, with 12-volts dc UPS. Present smart house technology

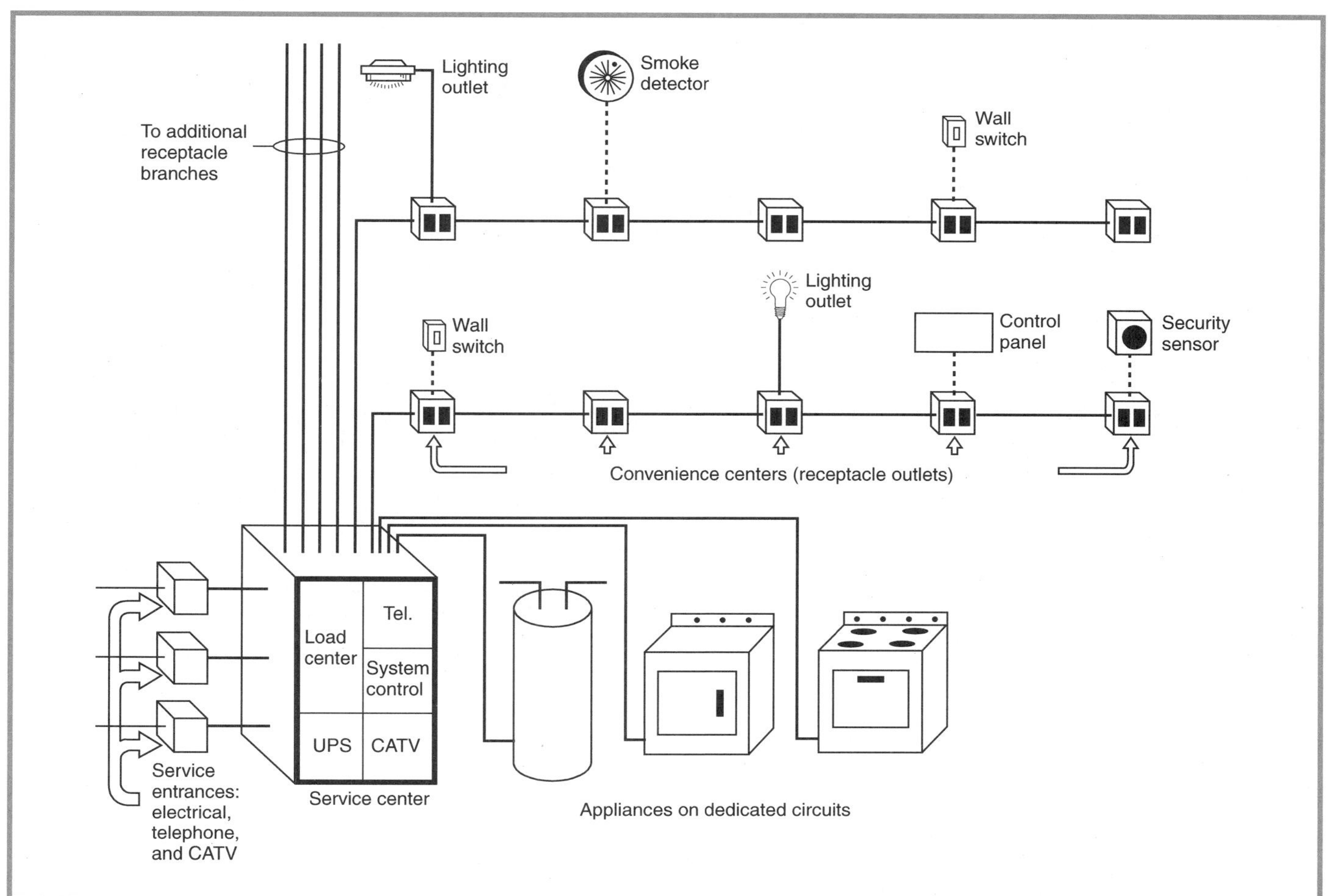

Figure 780.1 *A typical smart house installation. (Smart House)*

utilizes only 120/240-volts ac, with 24-volts dc UPS to maintain system electronics in the event of a transient or utility power outage.

780-2. General.

(a) Other Articles. Except as modified by the requirements of this article, all other applicable articles of this *Code* shall apply.

(b) Component Parts. All equipment and conductors shall be listed and identified.

780-3. Control. The control equipment and all power switching devices operated by the control equipment shall be listed and identified. The system shall operate as follows.

(a) Characteristic Electrical Identification Required. Outlets of a closed-loop power distribution system shall not be energized unless the utilization equipment first exhibits a characteristic electrical identification.

Receptacles are energized with 120-volts ac power only when electronic circuitry in the convenience center receives this characteristic identification.

(b) Conditions for De-energization. Outlets shall be de-energized when any of the following conditions occur.

(1) A nominal-operation acknowledgement signal is not being received from the utilization equipment connected to the outlet in a closed-loop power distribution system.

Convenience center receptacles are de-energized when the characteristic electrical identification ceases (when the attachment plug is withdrawn). In addition, appliances with built-in smart house communications chips transmit a continuous nominal-operation signal to the convenience center electronics. If this signal is interrupted, indicating a possible malfunction or safety problem, the receptacle is automatically de-energized.

(2) A ground-fault condition exists.
(3) An overcurrent condition exists.

(c) Additional Conditions for De-energization When an Alternate Source of Power is Used. In addition to the requirements in (b), outlets shall be de-energized when any of the following conditions occur.

(1) The grounded conductor is not properly grounded.
(2) Any ungrounded conductor is not at nominal voltage.

(d) Controller Malfunction. In the event of a controller malfunction, all associated outlets shall be de-energized.

780-5. Power Limitation in Signaling Circuits. For signaling circuits not exceeding 24 volts, the current required shall not exceed 1 ampere where protected by an overcurrent device or an inherently limited power source.

780-6. Cables and Conductors.

(a) Hybrid Cable. Listed hybrid cable consisting of power, communications, and signaling conductors shall be permitted under a common jacket. The jacket shall be applied so as to separate the power conductors from the communications and signaling conductors. An optional outer jacket shall be permitted to be applied. The individual conductors of a hybrid cable shall conform to the *Code* provisions applicable to their current, voltage, and insulation rating. The signaling conductors shall not be smaller than No. 24 copper.

(b) Cables and Conductors in the Same Cabinet, Panel, or Box. The power, communications, and signaling conductors of listed hybrid cable are permitted to occupy the same cabinet, panel, or outlet box (or similar enclosure housing the electrical terminations of electric light or power circuits) only if connectors specifically listed for hybrid cable are employed.

780-7. Noninterchangeability. Receptacles, cord connectors, and attachment plugs used on closed-loop power distribution systems shall be constructed so that they are not interchangeable with other receptacles, cord connectors, and attachment plugs.

Convenience center receptacles are constructed so that they will not accept an attachment plug with a different voltage or current rating than that for which the device is intended. Attachment plugs for use with closed-loop power distribution systems will not fit into conventional receptacles, to ensure that "smart" appliances are not used on other power distribution systems that lack closed-loop control features.

CHAPTER 8

Communications Systems

Article 800 — Communications Circuits

Contents

A. General

800-1. Scope. This article covers telephone, telegraph (except radio), outside wiring for fire alarm and burglar alarm, and similar central station systems; and telephone systems not connected to a central station system but using similar types of equipment, methods of installation, and maintenance.

FPN No. 1: For further information for fire alarm, guard tour, sprinkler waterflow, and sprinkler supervisory systems, see Article 760.

FPN No. 2: For installation requirements of optical fiber cables, see Article 770.

FPN No. 3: For installation requirements for network-powered broadband communications circuits, see Article 830.

Section 90-3, Code Arrangement, states that Chapter 8, which includes Articles 800, 810, 820, and 830, covers communications systems and is independent of the other chapters, except where they are specifically referenced therein. For instance, Section 800-10(a)(3) references Section 225-14(d), Section 800-30(c) references Article 500, and Section 800-52(c) references Section 300-22(c).

Although information technology equipment systems are often used for or with communications systems, Article 800 does not cover wiring of this equipment. Article 645 provides wiring requirements for wiring contained solely within an information technology equipment (computer) room. See Section 645-2 for the definition of *information technology equipment room*. Article 725 provides wiring requirements for wiring that extends beyond a computer room. See Article 760 for wiring requirements for a fire alarm system.

In some cases, the telephone system wiring is also used for data transmission; this use is covered by Article 800. Telephone company central offices are exempt from the requirements of Article 800. [See Section 90-2(b)(4).]

New Article 830 has been added for the 1999 *Code* to cover network-powered broadband communications systems.

The format for Article 800 is similar to that for Articles 725, 760, 770, and 820.

800-2. Definitions. See Article 100. For purposes of this article, the following additional definitions apply.

Block. A square or portion of a city, town, or village enclosed by streets and including the alleys so enclosed, but not any street.

Cable. A factory assembly of two or more conductors having an overall covering.

Cable Sheath. A covering over the conductor assembly that may include one or more metallic members, strength members, or jackets.

Exposed. A circuit that is in such a position that, in case of failure of supports and insulation, contact with another circuit may result.

FPN: See Article 100 for two other definitions of *Exposed.*

Point of Entrance. The point of entrance within a building is the point at which the wire or cable emerges from an external wall, from a concrete floor slab, or from a rigid metal conduit or an intermediate metal conduit grounded to an electrode in accordance with Section 800-40(b).

See the commentary following Sections 800-10(c) and 800-11(c).

Premises. The land and buildings of a user located on the user side of the utility-user network point of demarcation.

Wire. A factory assembly of one or more insulated conductors without an overall covering.

See Article 100 for definitions of *conductor, equipment,* and *raceway.*

800-3. Hybrid Power and Communications Cables. The provisions of Section 780-6 shall apply for listed hybrid power and communications cables in closed-loop and programmed power distribution.

See Section 800-51(i) for listing requirements and use of hybrid power and communications cables in one- and two-family residences for applications other than closed-loop and programmed power distribution.

FPN: See Section 800-51(i) for hybrid power and communications cable in other applications.

800-4. Equipment. Equipment intended to be electrically connected to a telecommunications network shall be listed for the purpose. Installation of equipment shall also comply with Section 110-3(b).

FPN: One way to determine applicable requirements is to refer to the *Standard for Safety of Information Technology Equipment, Including Electrical Business Equipment,* UL 1950-1993, third edition; the *Standard for Safety, Telephone Equipment,* UL 1459-1995, third edition; or the *Standard for Safety, Communications Circuit Accessories,* UL 1863- 1995, second edition. For information on listing requirements for communications raceways, see UL 2024-1995, *Standard for Optical Fiber Raceways.*

UL 1459 and UL 1863 are examples of safety standards that contain requirements for determining that equipment connected to the telecommunications network is suitable for the purpose. Listed equipment that is connected to the telecommunications network and is evaluated to other U.S. safety standards is also subject to telecommunications requirements appropriate for the equipment. Examples of this equipment are information technology equipment, audio–video equipment, and signaling equipment connected to a central station. The appropriate requirements are either contained within the applicable safety standard or are extracted from UL 1459 or UL 1863, or both.

Exception: This listing requirement shall not apply to test equipment that is intended for temporary connection to a telecommunications network by qualified persons during the course of installation, maintenance, or repair of telecommunications equipment or systems.

Except for test equipment, all permanently installed electrical components of the communications network are subject to the listing requirements of Section 800-4.

800-5. Access to Electrical Equipment Behind Panels Designed to Allow Access. Access to equipment shall not be denied by an accumulation of wires and cables that prevents removal of panels, including suspended ceiling panels.

The excess accumulation of wires and cables can limit access to equipment by preventing the removal of access panels. (See Figures 800.1 and 800.2.)

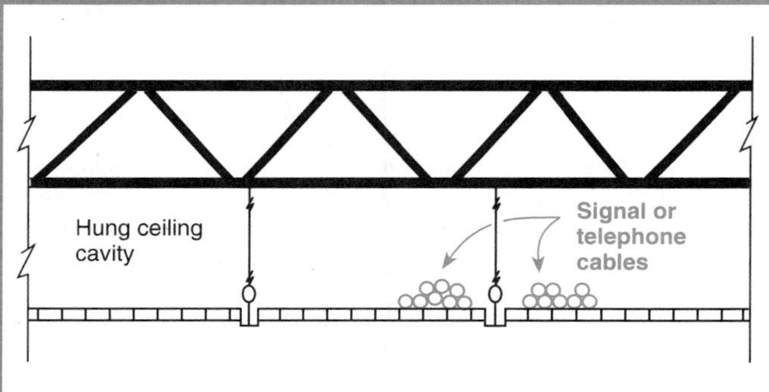

Figure 800.1 *Excess accumulation of conductors and cables prevents access to equipment or cables. Incorrect methods shown.*

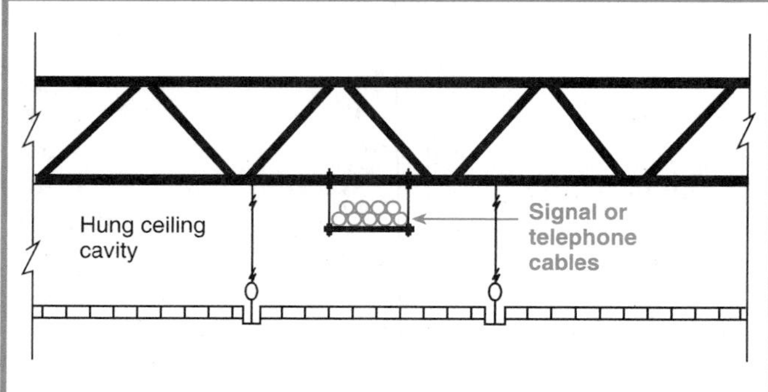

Figure 800.2 *Excess accumulation of conductors and cables prevents access to equipment or cables. Correct methods shown.*

800-6. Mechanical Execution of Work. Communications circuits and equipment shall be installed in a neat and workmanlike manner. Cables shall be supported by the building structure in such a manner that the cable will not be damaged by normal building use.

FPN: One way to determine accepted industry practice is to refer to nationally recognized standards such as *Commercial Building Telecommunications Cabling Standard*, ANSI/ EIA/TIA 568-A-1995; *Commercial Building Standard for Telecommunications Pathways and Spaces*, ANSI/TIA/EIA 569-A-1997; and *Residential and Light Commercial Telecommunications Wiring Standard,* ANSI/EIA/TIA 570-1991, or other ANSI-approved installation standards.

The fine print note following Section 800-6 makes it clear that there are industry standards.

800-7. Hazardous (Classified) Locations. Communications circuits and equipment installed in a location that is classified in accordance with Article 500 shall comply with the applicable requirements of Chapter 5.

Section 800-7 was new in the 1996 *Code*. It alerts users that communications circuits installed in locations classified in accordance with Article 500 must conform to the applicable requirements of Chapter 5.

B. Conductors Outside and Entering Buildings

800-10. Overhead Communications Wires and Cables. Overhead communications wires and cables entering buildings shall comply with (a) and (b).

(a) On Poles and In-Span. Where communications wires and cables and electric light or power conductors are supported by the same pole or run parallel to each other in-span, the following conditions shall be met.

(1) Relative Location. Where practicable, the communications wires and cables shall be located below the electric light or power conductors.

(2) Attachment to Crossarms. Communications wires and cables shall not be attached to a crossarm that carries electric light or power conductors.

(3) Climbing Space. The climbing space through communications wires and cables shall comply with the requirements of Section 225-14(d).

(4) Clearance. Supply service drops of 0–750 volts running above and parallel to communications service drops shall have a minimum separation of 12 in. (30.48 cm) at any point in the span, including the point of and at their attachment to the building, provided the nongrounded conductors are insulated and that a clearance of not less than 40 in. (1.02 m) is maintained between the two services at the pole.

(b) Above Roofs. Communications wires and cables shall have a vertical clearance of not less than 8 ft (2.44 m) from all points of roofs above which they pass.

Exception No. 1: Auxiliary buildings, such as garages and the like.

Exception No. 2: A reduction in clearance above only the overhanging portion of the roof to not less than 18 in. (457 mm) shall be permitted if (1) not more than 4 ft (1.22 mm) of communications service-drop conductors pass above the roof overhang and (2) they are terminated at a through- or above-the-roof raceway or approved support.

Exception No. 3: Where the roof has a slope of not less than 4 in. (102 mm) in 12 in. (305 mm), a reduction in clearance to not less than 3 ft (914 mm) shall be permitted.

800-11. Underground Circuits Entering Buildings. Underground communications wires and cables entering buildings shall comply with (a) through (c).

(a) With Electric Light or Power Conductors. Underground communications wires and cables in a raceway, handhole, or manhole containing electric light, power, Class 1, or nonpower-limited fire alarm circuit conductors shall be in a section separated from such conductors by means of brick, concrete, or tile partitions or by means of a suitable barrier.

(b) Underground Block Distribution. Where the entire street circuit is run underground and the circuit within the block is placed so as to be free from likelihood of accidental contact with electric light or power circuits of over 300 volts to ground, the insulation requirements of Sections 800-12(a) and (c) shall not apply, insulating supports shall not be required for the conductors, and bushings shall not be required where the conductors enter the building.

(c) Point of Entry. The point of entry for communications wiring and cables shall be within 20 ft (7.0 m) of the electrical service entry point.

Exception: Where it is impracticable to install the communications service in this manner, a separate grounding electrode, installed in compliance with Section 800-40(b)(3) and bonded in accordance with Sections 800-40(c) and (d).

Section 800-10(c) is new for the 1999 *Code*. Proper bonding is usually not accomplished where the point of entry of communications cables and wiring is remote from the electrical service. Improper bonding can result in a dangerous voltage potential being superimposed on the communications circuit grounding conductor or metallic cable sheath.

FPN: Under certain conditions, the length of the bonding conductor has a direct relationship to the difference in potential between the communications and the power circuits.

800-12. Circuits Requiring Primary Protectors. Circuits that require primary protectors as provided in Section 800-30 shall comply with the following.

(a) Insulation, Wires, and Cables. Communications wires and cables without a metallic shield, running from the last outdoor support to the primary protector, shall be listed as being suitable for the purpose and shall have current-carrying capacity as specified in Section 800-30(a)(1)(b) or 800-30(a)(1)(c).

(b) On Buildings. Communications wires and cables in accordance with Section 800-12(a) shall be separated at least 4 in. (102 mm) from electric light or power conductors not in a raceway or cable, or be permanently separated from conductors of the other system by a continuous and firmly fixed nonconductor in addition to the insulation on the wires, such as porcelain tubes or flexible tubing. Communications wires and cables in accordance with Section 800-12(a) exposed to accidental contact with electric light and power conductors operating at over 300 volts to ground and attached to buildings shall be separated from woodwork by being supported on glass, porcelain, or other insulating material.

Exception: Separation from woodwork shall not be required where fuses are omitted as provided for in Section 800-30(a)(1), or where conductors are used to extend circuits to a building from a cable having a grounded metal sheath.

(c) Entering Buildings. Where a primary protector is installed inside the building, the communications wires and cables shall enter the building either through a noncombustible, nonabsorbent insulating bushing or through a metal raceway. The insulating bushing shall not be required where the entering communications wires and cables (1) are in metal-sheathed cable, (2) pass through masonry, (3) meet the requirements of Section 800-12(a) and fuses are omitted as provided in Section 800-30(a)(1), or (4) meet the requirements of Section 800-12(a) and are used to extend circuits to a building from a cable having a grounded metallic sheath. Raceways or bushings shall slope upward from the outside or, where this cannot be done, drip loops shall be formed in the communications wires and cables immediately before they enter the building.

Raceways shall be equipped with an approved service head. More than one communications wire and cable shall be permitted to enter through a single raceway or bushing. Conduits or other metal raceways located ahead of the primary protector shall be grounded.

800-13. Lightning Conductors. Where practicable, a separation of at least 6 ft (1.83 m) shall be maintained between communications wires and cables on buildings and lightning conductors.

C. Protection

800-30. Protective Devices.

(a) Application. A listed primary protector shall be provided on each circuit run partly or entirely in aerial wire or aerial cable not confined within a block. Also, a listed primary protector shall be provided on each circuit, aerial or underground, located within the block containing the building served so as to be exposed to accidental contact with electric light or power conductors operating at over 300 volts to ground. In addition, where there exists a lightning exposure, each interbuilding circuit on a premises shall be protected by a listed primary protector at each end of the interbuilding circuit. Installation of primary protectors shall also comply with Section 110-3(b).

FPN No. 1: On a circuit not exposed to accidental contact with power conductors, providing a listed primary protector in accordance with this article will help protect against other hazards, such as lightning and above-normal voltages induced by fault currents on power circuits in proximity to the communications circuit.

FPN No. 2: Interbuilding circuits are considered to have a lightning exposure unless one or more of the following conditions exist.

(1) Circuits in large metropolitan areas where buildings are close together and sufficiently high to intercept lightning.
(2) Interbuilding cable runs of 140 ft (42.7 m) or less, directly buried or in underground conduit, where a continuous metallic cable shield or a continuous metallic conduit containing the cable is bonded to each building grounding electrode system.
(3) Areas having an average of five or fewer thunderstorm days per year and earth resistivity of less than 100 ohm-meters. Such areas are found along the Pacific coast.

Telephone utility companies ordinarily provide primary protectors where telephone lines are exposed to lightning. Installers of private networks that include interbuilding cables should also install primary protectors where cables are exposed to lightning. Generally, cables are considered to be exposed to lightning unless one or more of the conditions in FPN No. 2 exists. A primary protector is required at each end of an interbuilding communications circuit where lightning exposure exists.

(1) Fuseless Primary Protectors. Fuseless-type primary protectors shall be permitted under any of the following conditions:

(a) Where conductors enter a building through a cable with grounded metallic sheath member(s) and if the conductors in the cable safely fuse on all currents greater than the current-carrying capacity of the primary protector and of the primary protector grounding conductor
(b) Where insulated conductors in accordance with Section 800-12(a) are used to extend circuits to a building from a cable with an effectively grounded metallic sheath member(s) and if the conductors in the cable or cable stub, or the connections between the insulated conductors and the exposed plant, safely fuse on all currents greater than the current-carrying capacity of the primary protector, or the associated insulated conductors and of the primary protector grounding conductor
(c) Where insulated conductors in accordance with Section 800-12(a) or (b) are used to extend circuits to a building from other than a cable with a metallic sheath member(s) if (1) the primary protector is listed for this purpose, and (2) the connections of the insulated conductors to the exposed plant or the conductors of the exposed plant safely fuse on all currents greater than the current-carrying capacity of the primary protector, or the associated insulated conductors and of the primary protector grounding conductor
(d) Where insulated conductors in accordance with Section 800-12(a) are used to extend circuits aerially to a building from an unexposed buried or underground circuit
(e) Where insulated conductors in accordance with Section 800-12(a) are used to extend circuits to a building from cable with an effectively grounded metallic sheath member(s) and if (1) the combination of the primary protector and insulated conductors is listed for this purpose, and (2) the insulated conductors safely fuse on all currents greater than the current-carrying capacity of the primary protector and of the primary protector grounding conductor

The term *effectively grounded* is defined in Article 100.

(2) Fused Primary Protectors. Where the requirements listed under Sections 800-30(a)(1)(a) through (e) are not met, fused-type primary protectors shall be used. Fused-type primary protectors shall consist of an arrester connected between each line conductor and ground, a fuse in series with each line conductor, and an appropriate mounting arrangement. Primary protector terminals shall be marked to indicate line, instrument, and ground, as applicable.

(b) Location. The primary protector shall be located in, on, or immediately adjacent to the structure or building served and as close as practicable to the point at which the exposed conductors enter or attach.

See Figure 800.3 for an example of a primary protector unit typically installed in commercial buildings.

See Figure 800.4 for an example of applications of listed communications and multipurpose cables.

For purposes of this section, the point at which the exposed conductors enter shall be considered to be the point of emergence through an exterior wall, a concrete floor slab, or from a rigid metal conduit or an intermediate metal conduit grounded to an electrode in accordance with Section 800-40(b).

For purposes of this section, primary protectors located at mobile home service equipment located in sight from and not more than 30 ft (9.14 m) from the exterior wall of the mobile home it serves, or at a mobile home disconnecting means grounded in accordance with Section 250-32 and located in sight from and not more than 30 ft (9.14 m) from

Figure 800.3 *A primary protector unit typically installed in commercial buildings. This is the interface to the outside plant cable. (AT&T)*

the exterior wall of the mobile home it serves, shall be considered to meet the requirements of this section.

FPN: Selecting a primary protector location to achieve the shortest practicable primary protector grounding conductor will help limit potential differences between communications circuits and other metallic systems.

(c) Hazardous (Classified) Locations. The primary protector shall not be located in any hazardous (classified) location as defined in Article 500, nor in the vicinity of easily ignitible material.

Exception: As permitted in Sections 501-14, 502-14, and 503-12.

800-31. Primary Protector Requirements. The primary protector shall consist of an arrester connected between each line conductor and ground in an appropriate mounting. Primary protector terminals shall be marked to indicate line and ground as applicable.

FPN: One way to determine applicable requirements for a listed primary protector is to refer to the *Standard for Protectors for Paired Conductor Communications Circuits,* ANSI/UL 497-1995.

800-32. Secondary Protector Requirements. Where a secondary protector is installed in series with the indoor communications wire and cable between the primary protector and the equipment, it shall be listed for the purpose. The secondary protector shall provide means to safely limit currents to less than the current-carrying capacity of listed indoor communications wire and cable, listed telephone set line cords, and listed communications terminal equipment having ports for external wire line communications circuits. Any overvoltage protection, arresters, or grounding connection shall be connected on the equipment terminals side of the secondary protector current-limiting means.

FPN No. 1: One way to determine applicable requirements for a listed secondary protector is to refer to the *Standard for Secondary Protectors for Communications Circuits,* UL 497A-1996.

FPN No. 2: Secondary protectors on exposed circuits are not intended for use without primary protectors.

800-33. Cable Grounding. The metallic sheath of communications cables entering buildings shall be grounded as close as practicable to the point of entrance or shall be interrupted as close to the point of entrance as practicable by an insulating joint or equivalent device.

For purposes of this section, the point of entrance shall be considered to be at the point of emergence through an exterior wall, a concrete floor slab, or from a rigid metal conduit or an intermediate metal conduit grounded to an electrode in accordance with Section 800-40(b).

D. Grounding Methods

800-40. Cable and Primary Protector Grounding. The metallic member(s) of the cable sheath, where required to be grounded by Section 800-33, and primary protectors shall be grounded as specified in (a) through (d).

(a) Grounding Conductor.

(1) Insulation. The grounding conductor shall be insulated and shall be listed as suitable for the purpose.

(2) Material. The grounding conductor shall be copper or other corrosion-resistant conductive material, stranded or solid.

(3) Size. The grounding conductor shall not be smaller than No. 14.

(4) Run in Straight Line. The grounding conductor shall be run to the grounding electrode in as straight a line as practicable.

(5) Physical Damage. Where necessary, the grounding conductor shall be guarded from physical damage. Where the grounding conductor is run in a metal raceway, both ends of the raceway shall be bonded to the grounding conductor or the same terminal or electrode to which the grounding conductor is connected.

(b) Electrode. The grounding conductor shall be connected as follows.

(1) To the nearest accessible location on the following:

(a) The building or structure grounding electrode system as covered in Section 250-50

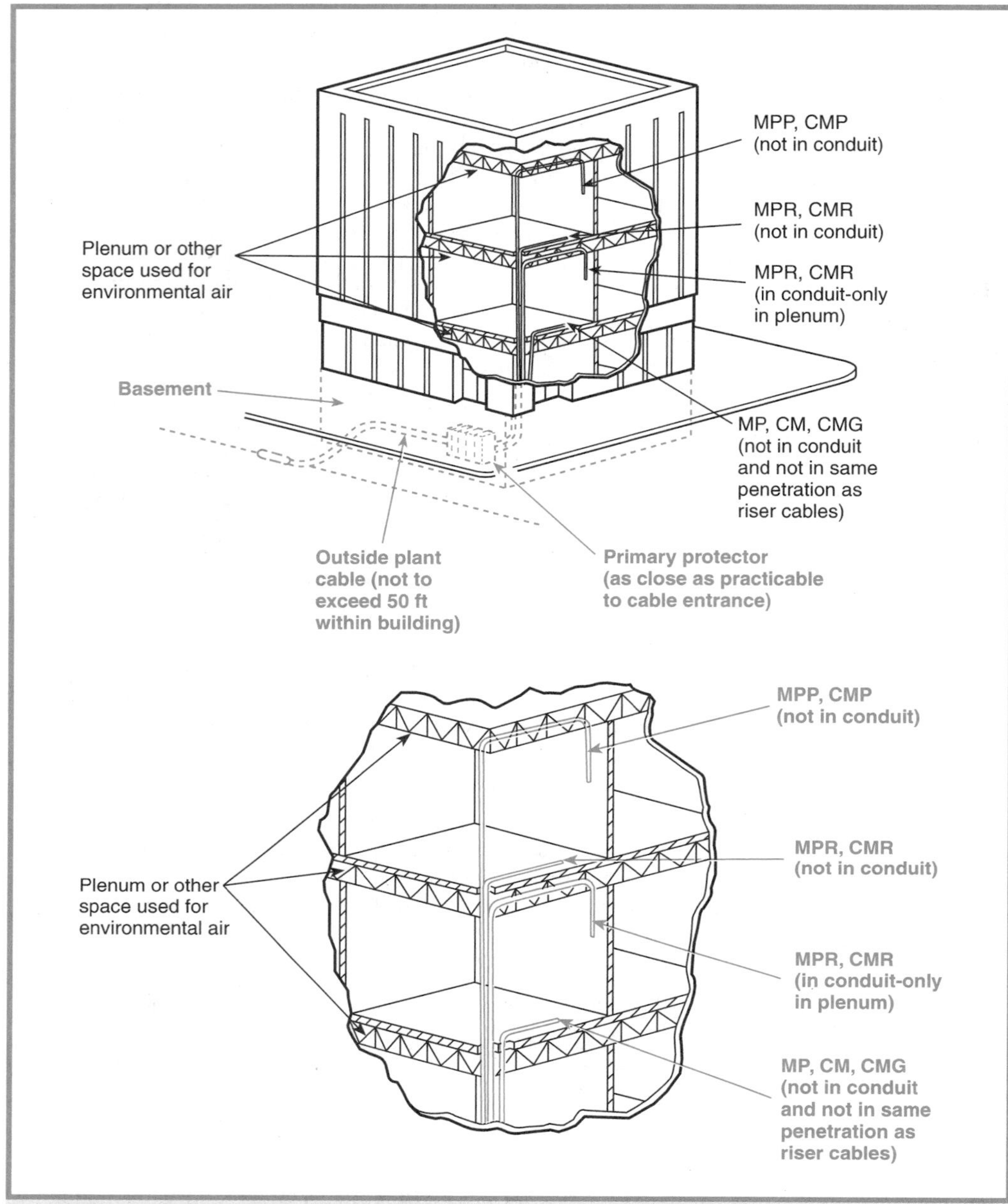

Figure 800.4 *An example of applications of listed communications and multipurpose cables.*

(b) The grounded interior metal water piping system as covered in Section 250-104(a)
(c) The power service accessible means external to enclosures as covered in Section 250-92(b)
(d) The metallic power service raceway
(e) The service equipment enclosure
(f) The grounding electrode conductor or the grounding electrode conductor metal enclosure; or
(g) To the grounding conductor or the grounding electrode of a building or structure disconnecting means that is grounded to an electrode as covered in Section 250-32.

For purposes of this section, the mobile home service equipment or the mobile home disconnecting means, as described in Section 800-30(b), shall be considered accessible.

(2) If the building or structure served has no grounding means, as described in (b)(1), to any one of the individual electrodes described in Section 250-50; or

(3) If the building or structure served has no grounding means, as described in (b)(1) or (b)(2), to an effectively grounded metal structure or to a ground rod or pipe not less than 5 ft (1.52 m) in length and ½ in. (12.7 mm) in diameter, driven, where practicable, into permanently damp earth and separated from lightning conductors as covered in Section 800-13 and at least 6 ft (1.83 m) from electrodes of other systems. Steam or hot water pipes or lightning-rod conductors shall not be employed as electrodes for protectors.

(c) Electrode Connection. Connections to grounding electrodes shall comply with Section 250-70. Connectors, clamps, fittings, or lugs used to attach grounding conductors and bonding jumpers to grounding electrodes or to each other that are to be concrete-encased or buried in the earth shall be suitable for its application.

(d) Bonding of Electrodes. A bonding jumper not smaller than No. 6 copper or equivalent shall be connected between the communications grounding electrode and power grounding electrode system at the building or structure served where separate electrodes are used. Bonding together of all separate electrodes shall be permitted.

Exception: At mobile homes as covered in Section 800-41.

FPN No. 1: See Section 250-60 for use of air terminals (lightning rods).

FPN No. 2: Bonding together of all separate electrodes will limit potential differences between them and between their associated wiring systems.

Section 800-40(d) requires bonding of communications and power grounding electrodes at the same building or structure. Section 800-11(c), Exception, may require a separate grounding electrode for communications systems.

800-41. Primary Protector Grounding and Bonding at Mobile Homes.

(a) Grounding. Where there is no mobile home service equipment located in sight from and not more than 30 ft (9.14 m) from the exterior wall of the mobile home it serves, or there is no mobile home disconnecting means grounded in accordance with Section 250-32 and located within sight from and not more than 30 ft (9.14 m) from the exterior wall of the mobile home it serves, the primary protector ground shall be in accordance with Sections 800-40(b)(2) and (3).

(b) Bonding. The primary protector grounding terminal or grounding electrode shall be bonded to the metal frame or available grounding terminal of the mobile home with a copper grounding conductor not smaller than No. 12 under any of the following conditions:

(1) Where there is no mobile home service equipment or disconnecting means as in (a), or
(2) The mobile home is supplied by cord and plug.

E. Communications Wires and Cables Within Buildings

The 1996 *Code* clarified that data circuits between computers are Class 2 circuits. In a typical office environment that consists of a group of computers in a local area network, data wiring is as prevalent as telephone wiring. A common way to minimize the amount of cabling is to run the telephone and data circuits in the same cable. Figure 800.5 illustrates such an arrangement. Section 725-54(b)(4) requires that either a communications or multipurpose cable be used for this purpose.

800-48. Raceways for Communications Wires and Cables. Where communications wire and cables are installed in a raceway, the raceway shall be of a type permitted in Chapter 3 and installed in accordance with Chapter 3.

Exception: Listed nonmetallic communications raceway identified as general purpose, riser, or plenum in accordance with Section 800-51 and installed in accordance with Sections 331-7 through 331-14, where the requirements applicable to electrical nonmeallic tubing shall apply.

800-49. Fire Resistance of Communications Wires and Cables. Communications wires and cables installed as wiring within a building shall be listed as being resistant to the spread of fire in accordance with Sections 800-50 and 800-51.

800-50. Listing, Marking, and Installation of Communications Wires and Cables. Communications wires and cables installed as wiring within buildings shall be listed as being suitable for the purpose and installed in accordance with Section 800-52. Communications cables and undercarpet communications wires shall be marked in accordance with Table 800-50. The cable voltage rating shall not be marked on the cable or on the under-carpet communications wire.

FPN: Voltage markings on cables may be misinterpreted to suggest that the cables may be suitable for Class 1, electric light, and power applications.

Exception No. 1: Voltage markings shall be permitted where the cable has multiple listings and voltage marking is required for one or more of the listings.

Exception No. 2: Listing and marking shall not be required where the cable enters the building from the outside and is continuously enclosed in a rigid metal conduit system or an intermediate metal conduit system and such conduit systems are grounded to an electrode in accordance with Section 800-40(b).

Exception No. 3: Listing and marking shall not be required where the length of the cable within the the building, measured from its point of entrance, does not exceed 50 ft (15.2 m) and the cable enters the building from the outside

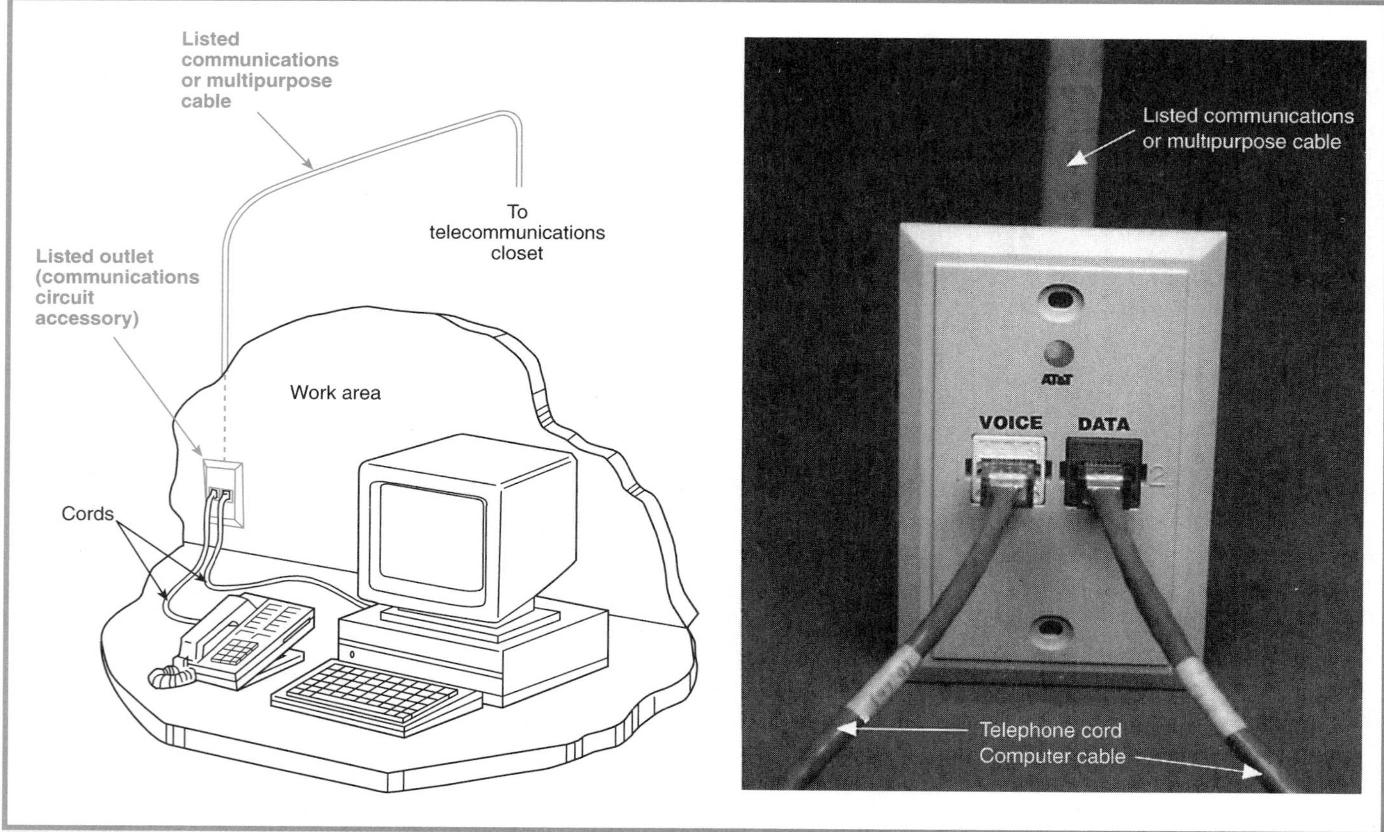

Figure 800.5 An example of telephone and data circuits in the same cable.

and is terminated in an enclosure or on a listed primary protector.

FPN No. 1: Splice cases or terminal boxes, both metallic and plastic types, are typically used as enclosures for splicing or terminating telephone cables.

FPN No. 2: This exception limits the length of unlisted outside plant cable to 50 ft (15.2 m), while Section 800-30(b) requires that the primary protector shall be located as close as practicable to the point at which the cable enters the building. Therefore, in installations requiring a primary protector, the outside plant cable may not be permitted to extend 50 ft (15.2 m) into the building if it is practicable to place the primary protector closer than 50 ft (15.2 m) to the entrance point.

Exception No. 4: Multipurpose cables shall be considered as being suitable for the purpose and shall be permitted to substitute for communications cables as provided for in Section 800-53(f).

800-51. Listing Requirements for Communications Wires and Cables and Communications Raceways. Communications wires and cables shall have a voltage rating of not less than 300 volts and shall be listed in accordance with (a) through (i), and communications raceways shall be listed in accordance with (j) through (l). Conductors in communications cables, other than in a coaxial cable, shall be copper.

A voltage rating of 300 volts is necessary for the following reasons:

1. To coordinate with protector installation requirements (i.e., protectors are not required within a block unless the cable is exposed to over 300 volts)
2. To recognize the fact that primary protectors are designed to allow voltages below 300 to pass
3. To accommodate the voltages ordinarily found on a telephone line (48 volts dc plus ringing voltage up to 130 volts rms)
4. To permit communications cables to substitute for 300-volt power-limited fire-protective signaling cables

FPN: See Section 800-4 for listing requirement for equipment.

Table 800-50. Cable Markings

Cable Marking	Type	Reference
MPP	Multipurpose plenum cable	800-51(g) and 800-53(a)
CMP	Communications plenum cable	800-51(a) and 800-53(a)
MPR	Multipurpose riser cable	800-51(g) and 800-53(b)
CMR	Communications riser cable	800-51(b) and 800-53(b)
MPG	Multipurpose general-purpose cable	800-51(g) and 800-53(d)
CMG	Communications general-purpose cable	800-51(c) and 800-53(d)
MP	Multipurpose general-purpose cable	800-51(g) and 800-53(d)
CM	Communications general-purpose cable	800-51(d) and 800-53(d)
CMX	Communications cable, limited use	800-51(e) and 800-53(d), Exception Nos. 1, 2, 3, and 4
CMUC	Under-carpet communications wire and cable	800-51(f) and 800-53(d), Exception No. 5

FPN No. 1: Cable types are listed in descending order of fire resistance rating, and multipurpose cables are listed above communications cables because multipurpose cables may substitute for communications cables.

FPN No. 2: See the referenced sections for permitted uses.

(a) Type CMP. Type CMP communications plenum cable shall be listed as being suitable for use in ducts, plenums, and other spaces used for environmental air and shall also be listed as having adequate fire-resistant and low smoke-producing characteristics.

FPN: One method of defining low smoke-producing cables is by establishing an acceptable value of the smoke produced when tested in accordance with the *Standard Method of Test for Fire and Smoke Characteristics of Wires and Cables,* NFPA 262-1994, to a maximum peak optical density of 0.5 and a maximum average optical density of 0.15. Similarly, one method of defining fire-resistant cables is by establishing a maximum allowable flame travel distance of 5 ft (1.52 m) when tested in accordance with the same test.

See the commentary following Section 725-71(a), FPN.

(b) Type CMR. Type CMR communications riser cable shall be listed as being suitable for use in a vertical run in a shaft or from floor to floor and shall also be listed as having fire-resistant characteristics capable of preventing the carrying of fire from floor to floor.

FPN: One method of defining fire-resistant characteristics capable of preventing the carrying of fire from floor to floor is that the cables pass the requirements of the *Standard Test for Flame Propagation Height of Electrical and Optical-Fiber Cable Installed Vertically in Shafts,* ANSI/UL 1666-1997.

See the commentary following Section 725-71(b), FPN.

(c) Type CMG. Type CMG general-purpose communications cable shall be listed as being suitable for general-purpose communications use, with the exception of risers and plenums, and shall also be listed as being resistant to the spread of fire.

FPN: One method of defining resistant to the spread of fire is for the damage (char length) not to exceed 4 ft 11 in. (1.5 m) when performing the vertical flame test — cables in cable trays, as described in *Test Methods for Electrical Wires and Cables,* CSA C22.2 No. 0.3-M 1985.

See the commentary following Section 725-71(c), FPN.

(d) Type CM. Type CM communications cable shall be listed as being suitable for general-purpose communications use, with the exception of risers and plenums, and shall also be listed as being resistant to the spread of fire.

FPN: One method of defining resistant to the spread of fire is that the cables do not spread fire to the top of the tray in the vertical-tray flame test in the *Reference Standard for Electrical Wires, Cables and Flexible Cords,* ANSI/UL 1581-1991.

Another method of defining resistant to the spread of fire is for the damage (char length) not to exceed 4 ft 11 in. (1.5 m) when performing the vertical flame test — cables in cable trays, as described in *Test Methods for Electrical Wires and Cables,* CSA C22.2 No. 0.3-M-1985.

See the commentary following Section 725-71(d), FPN.

(e) Type CMX. Type CMX limited use communications cable shall be listed as being suitable for use in dwellings and for use in raceway and shall also be listed as being resistant to flame spread.

FPN: One method of determining that cable is resistant to flame spread is by testing the cable to the VW-1 (vertical-wire) flame test in the *Reference Standard for Electrical Wires, Cables and Flexible Cords,* ANSI/UL 1581-1991.

(f) Type CMUC Under-Carpet Wire and Cable. Type CMUC under-carpet communications wire and cable shall be listed as being suitable for under-carpet use and shall also be listed as being resistant to flame spread.

FPN: One method of determining that cable is resistant to flame spread is by testing the cable to the VW-1 (vertical-wire) flame test in the *Reference Standard for Electrical Wires, Cables and Flexible Cords,* ANSI/UL 1581-1991.

(g) Multipurpose (MP) Cables. Cables that meet the requirements for Types CMP, CMR, CMG, and CM and also satisfy the requirements of Section 760-71(b) for multiconductor cables and Section 760-71(h) for coaxial cables shall be permitted to be listed and marked as multipurpose cable Types MPP, MPR, MPG, and MP, respectively.

More copper communications cables, such as Types MPP, MPR, MPG, and MP, now qualify for listing as multipurpose cables, since the stranding requirements for fire alarm cables have been deleted.

(h) Communications Wires. Communications wires, such as distributing frame wire and jumper wire, shall be listed as being resistant to the spread of fire.

FPN: One method of defining resistant to the spread of fire is that the cables do not spread fire to the top of the tray in the vertical-tray flame test in the *Reference Standard for Electrical Wires, Cables and Flexible Cords,* ANSI/UL 1581-1991.

Another method of defining resistant to the spread of fire is for the damage (char length) not to exceed 4 ft 11 in. (1.5 m) when performing the vertical flame test — cables in cable trays, as described in *Test Methods for Electrical Wires and Cables,* CSA C22.2 No. 0.3-M-1985.

(i) Hybrid Power and Communications Cable. Listed hybrid power and communications cable shall be permitted where the power cable is a listed Type NM or NM-B conforming to the provisions of Article 336, and the communications cable is a listed Type CM, and the jackets on the listed NM or NM-B and listed CM cables are rated for 600 volts minimum, and the hybrid cable is listed as being resistant to the spread of fire.

FPN: One method of defining resistant to the spread of fire is that the cables do not spread fire to the top of the tray in the vertical-tray flame test in the *Reference Standard for Electrical Wires, Cables and Flexible Cords,* ANSI/UL 1581-1991.

Another method of defining resistant to the spread of fire is for the damage (char length) not to exceed 4 ft 11 in. (1.5 m) when performing the vertical flame test — cables in cable trays, as described in *Test Methods for Electrical Wires and Cables,* CSA C22.2 No. 0.3-M-1985.

(j) Plenum Communications Raceways. Plenum communications raceways listed as plenum optical fiber raceways shall be permitted for use in ducts, plenums, and other spaces used for environmental air and shall also be listed as having adequate fire-resistant and low-smoke producing characteristics.

(k) Riser Communications Raceway. Riser communications raceways shall be listed as having adequate fire-resistant characteristics capable of preventing the carrying of fire from floor to floor.

(l) General-Purpose Communications Raceway. General-purpose communications raceways shall be listed as having adequate fire-resistant characteristics.

The communications raceways indicated in Section 800-51(j) through (l) are listed raceways used in plenum, riser, or general-purpose applications. This listing includes raceways and fittings for the installation of communications cables in accordance with this article. The raceways are not suitable for installation of wires, cords, or cables with or without communication members.

A raceway marked "plenum" is suitable for use in ducts, plenums, or other spaces used for environmental air in accordance with Section 800-53(a) when used to enclose communications cables marked CMP. This raceway exhibits a maximum peak optical density of 0.5, a maximum average optical density of 0.15, and a maximum flame-spread distance of 5 ft when tested in accordance with the *Test for Flame Propagation and Smoke-Density Values for Electrical and Optical-Fiber Cables Used in Spaces Transporting Environmental Air,* UL 910. This raceway is identified by a marking on the surface of the raceway or on a marker tape indicating "plenum." A raceway marked "plenum" is also suitable for installation in risers when used to enclose communications cables marked CMP or CMR; general-purpose use when used to enclose communications cables marked CMP, CMR, CMG, or CM; and dwellings use when used to enclose communications cables marked CMP, CMR, CMG, CM, or CMX.

A raceway marked "riser" is suitable for installation in risers in accordance with Section 800-53(b) when used to enclose communications cables marked CMP or CMR. This raceway has fire-resistant characteristics capable of preventing the carrying of fire from floor to floor. This raceway meets the test requirements of the *Standard Test for Flame Propagation Height of Electrical and Optical-Fiber Cable Installed Vertically in Shafts,* UL 1666. This raceway is identified by a marking on the surface of the raceway or on a marker tape indicating "riser." A raceway marked "riser" is also suitable for general-purpose use when used to enclose communications cables marked CMP, CMR, CMG, or CM and dwellings use when used to enclose communications cables marked CMP, CMR, CMG, CM, or CMX.

A raceway marked "general purpose" is suitable for installation in general-purpose areas in accordance with Section 800-53(d) when used to enclose communications cables marked CMP, CMR, CMG, or CM and dwellings use when used to enclose communications cables marked CMP, CMR, CMG, CM, or CMX.

Pliable raceway is raceway that can be bent by hand without the use of tools. The smallest radius of the curve of the inner edge of any bend to which the raceway may be bent without cracking either on the outer surface or internally is not less than 2½ times the outside diameter of the raceway.

800-52. Installation of Communications Wires, Cables, and Equipment. Communications wires and cables from the protector to the equipment or, where no protector is required, communications wires and cables attached to the outside or inside of the building shall comply with (a) through (e).

Section 800-52 was revised for the 1999 *Code* to include nonpower-limited fire alarm circuits covered by Article 760 and network-powered broadband communications circuits covered by Article 830.

(a) Separation from Other Conductors.

(1) In Raceways, Boxes, and Cables.

(a) *Other Power-Limited Circuits.* Communications cables shall be permitted in the same raceway or enclosure with cables of any of the following:

(1) Class 2 and Class 3 remote-control, signaling, and power-limited circuits in compliance with Article 725
(2) Power-limited fire alarm systems in compliance with Article 760
(3) Nonconductive and conductive optical fiber cables in compliance with Article 770
(4) Community antenna television and radio distribution systems in compliance with Article 820
(5) Low power network-powered broadband communications circuits in compliance with Article 830

(b) *Class 2 and Class 3 Circuits.* Class 1 circuits shall not be run in the same cable with communications circuits. Class 2 and Class 3 circuit conductors shall be permitted in the same cable with communications circuits, in which case the Class 2 and Class 3 circuits shall be classified as communications circuits and shall meet the requirements of this article. The cables shall be listed as communications cables or multipurpose cables.

Exception: Cables constructed of individually listed Class 2, Class 3, and communications cables under a common jacket shall not be required to be classified as communications cable. The fire-resistance rating of the composite cable shall be determined by the performance of the composite cable.

(c) *Electric Light, Power, Class 1, Nonpower-Limited Fire Alarm, and Medium Power Network-Powered Broadband Communications Circuits.*

1. In Raceways, Compartments, and Boxes. Communications conductors shall not be placed in any raceway, compartment, outlet box, junction box, or similar fitting with conductors of electric light, power, Class 1, nonpower-limited fire alarm or medium power network-powered broadband communications circuits.

Exception No. 1: Where all of the conductors of electric light, power, Class 1, nonpower-limited fire alarm, and medium power network-powered broadband communications circuits are separated from all of the conductors of communications circuits by a barrier.

Exception No. 2: Power conductors in outlet boxes, junction boxes, or similar fittings or compartments where such conductors are introduced solely for power supply to communications equipment. The power circuit conductors shall be routed within the enclosure to maintain a minimum of 0.25-in. (6.35-mm) separation from the communications circuit conductors.

(2) Other Applications. Communications wires and cables shall be separated at least 2 in. (50.8 mm) from conductors of any electric light, power, Class 1, nonpower-limited fire alarm, or medium power network-powered broadband communications circuits.

Exception No. 1: Where either (1) all of the conductors of the electric light, power, Class 1, nonpower-limited fire alarm, and medium power network-powered broadband communications circuits are in a raceway or in metal-sheathed, metal-clad, nonmetallic-sheathed, Type AC, or Type UF cables, or (2) all of the conductors of communications circuits are encased in raceway.

Exception No. 2: Where the communications wires and cables are permanently separated from the conductors of electric light, power, Class 1, nonpower-limited fire alarm, and medium power network-powered broadband communications circuits by a continuous and firmly fixed nonconductor, such as porcelain tubes or flexible tubing, in addition to the insulation on the wire.

(b) Spread of Fire or Products of Combustion. Installations in hollow spaces, vertical shafts, and ventilation or air-handling ducts shall be made so that the possible spread of fire or products of combustion will not be substantially increased. Openings around penetrations through fire resistance-rated walls, partitions, floors, or ceilings shall be firestopped using approved methods.

(c) Equipment in Other Space Used for Environmental Air. Section 300-22(c) shall apply.

(d) Cable Trays. Types MPP, MPR, MPG, and MP multipurpose cables and Types CMP, CMR, CMG, and CM communications cables shall be permitted to be installed in cable trays.

(e) Support of Conductors. Raceways shall be used for their intended purpose. Communications cables or wires shall not be strapped, taped, or attached by any means to the exterior of any conduit or raceway as a means of support.

See Sections 800-5 and 800-6. These sections require communications cables to be supported by the building structure in such a manner that they will not be damaged by ordinary building use.

Exception: Overhead (aerial) spans of communications cables or wires shall be permitted to be attached to the exterior of a raceway-type mast intended for the attachment and support of such conductors.

There are some instances where the only way to achieve the proper clearance above roadways, driveways, or structures is to use a mast. The exception added to Section 800-52(e) for the 1999 *Code* permits overhead spans of communications cables to be attached to the exterior of a raceway-type mast only if the mast is installed to support communications cables. Communications cables are prohibited from being attached to a service mast according to Section 230-28.

800-53. Applications of Listed Communications Wires and Cables, and Communications Raceways. Communications wires and cables shall comply with (a) through (f).

Note that the length of unlisted outside-plant cable permitted in the building depends on the location of the primary protector, in accordance with Sections 800-30(b) and 800-50, Exception No. 3.

(a) Plenum. Cables installed in ducts, plenums, and other spaces used for environmental air shall be Type CMP. Also, listed plenum communications raceways shall be permitted to be installed in ducts and plenums as described in Section 300-22(b) and in other spaces used for environmental air as described in Section 300-22(c). Only Type CMP cable shall be permitted to be installed in these raceways.

Section 800-53(a) was revised for the 1999 *Code* to recognize listed plenum communications raceways. These raceways provide limited mechanical protection and ease of installation, but they are limited to Type CMP plenum-rated cable if installed in ducts and plenums.

Exception: Types CMP, CMR, CMG, CM, and CMX and communications wire installed in compliance with Section 300-22.

(b) Riser. Cables installed in vertical runs and penetrating more than one floor, or cables installed in vertical runs in a shaft, shall be Type CMR. Floor penetrations requiring Type CMR shall contain only cables suitable for riser or plenum use. Also, listed riser communications raceways shall be permitted to be installed in vertical riser runs in a shaft from floor to floor. Only Type CMR and CMP cables shall be permitted to be installed in these raceways.

Section 800-53(b) was revised for the 1999 *Code* to recognize listed plenum communications raceways. These raceways provide limited mechanical protection and ease of installation, but they are limited to Type CMP plenum-rated cable or Type CMR riser-rated cable where installed in risers.

FPN: See Section 800-52(b) for firestop requirements for floor penetrations.

Exception No. 1: Where the listed cables are encased in metal raceway or are located in a fireproof shaft having firestops at each floor.

Exception No. 2: Types CM and CMX cable in one- and two-family dwellings.

(c) Distributing Frames and Cross-Connect Arrays. Communications wires shall be used in distributing frames and cross-connect arrays.

Exception: Types CMP, CMR, CMG, and CM cables shall be permitted to be used.

(d) Other Wiring Within Buildings. Cables installed in building locations other than the locations covered in (a), (b), and (c) shall be Type CMG or Type CM. Only Type CMG, CM, CMR, or CMP cables shall be permitted to be installed in general purpose communications raceways.

Exception No. 1: Where listed communications wires and cables are enclosed in raceway of a type included in Chapter 3.

Exception No. 2: Type CMX communications cable in nonconcealed spaces where the exposed length of cable does not exceed 10 ft (3.05 m).

Exception No. 3: Type CMX communications cables that are less than 0.25 in. (6.35 mm) in diameter and installed in one- or two-family dwellings.

Exception No. 4: Type CMX communications cables that are less than 0.25 in. (6.35 mm) in diameter and installed in nonconcealed spaces in multifamily dwellings.

Exception No. 5: Type CMUC under-carpet communications wires and cables installed under carpet.

(e) Hybrid Power and Communications Cable. Hybrid power and communications cable listed in accordance with Section 800-51(i) shall be permitted to be installed in one- and two-family dwellings.

(f) Cable Substitutions. Substitutions for communications cables listed in Table 800-53 shall be considered suitable for the purpose and shall be permitted.

Table 800-53. Cable Uses and Permitted Substitutions

Cable Type	Use	References	Permitted Substitutions
CMP	Communications plenum cable	800-53(a)	MPP
CMR	Communications riser cable	800-53(b)	MPP, CMP, MPR
CMG, CM	Communications general-purpose cable	800-53(d)	MPP, CMP, MPR, CMR, MPG, MP
CMX	Communications cable, limited use	800-53(d)	MPP, CMP, MPR, CMR, MPG, MP, CMG, CM

Note: See Figure 800-53, Cable substitution hierarchy.

Table 800-53 was extensively revised for the 1999 *Code* to reflect the more realistic use of field applications for various cable types.

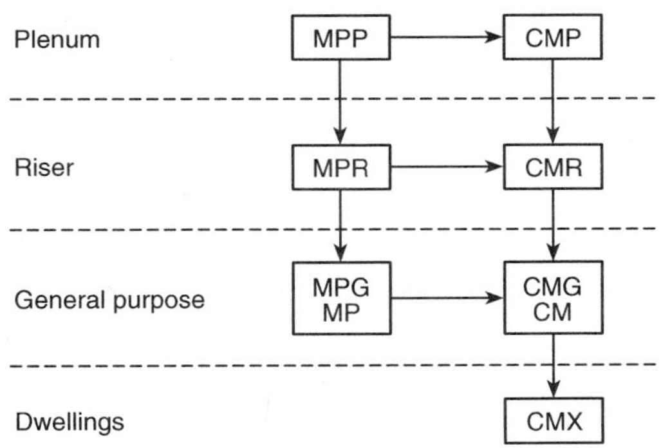

Figure 800-53 Cable substitution hierarchy.

Article 810 — Radio and Television Equipment

Contents

A. General

810-1. Scope. This article covers antenna systems for radio and television receiving equipment, amateur radio transmitting and receiving equipment, and certain features of transmitter safety. This article covers antennas such as multi-element, vertical rod, and dish, and also covers the wiring and cabling that connects them to equipment. This article does not cover equipment and antennas used for coupling carrier current to power line conductors.

Article 810 covers wiring requirements for equipment used to receive television and radio equipment, specifically including digital satellite receiving equipment for television signals. The wiring for the power supply is covered by Chapters 1 through 4. Article 640 contains requirements for sound distribution systems. The interior wiring of coaxial cable is covered by Article 820.

810-2. Other Articles. Wiring from the source of power to and between devices connected to the interior wiring system shall comply with Chapters 1 through 4 other than as modified by Parts A and B of Article 640. Wiring for audio signal processing, amplification, and reproduction equipment shall comply with Article 640. Coaxial cables that connect antennas to equipment shall comply with Article 820.

810-3. Community Television Antenna. The antenna shall comply with this article. The distribution system shall comply with Article 820.

810-4. Radio Noise Suppressors. Radio interference eliminators, interference capacitors, or noise suppressors connected to power-supply leads shall be of a listed type. They shall not be exposed to physical damage.

810-5. Definitions. See Article 100.

B. Receiving Equipment — Antenna Systems

810-11. Material. Antennas and lead-in conductors shall be of hard-drawn copper, bronze, aluminum alloy, copper-clad steel, or other high-strength, corrosion-resistant material.

Exception: Soft-drawn or medium-drawn copper shall be permitted for lead-in conductors where the maximum span between points of support is less than 35 ft (10.67 m).

810-12. Supports. Outdoor antennas and lead-in conductors shall be securely supported. The antennas or lead-in conductors shall not be attached to the electric service mast. They shall not be attached to poles or similar structures carrying open electric light or power wires or trolley wires of over 250 volts between conductors. Insulators supporting the antenna conductors shall have sufficient mechanical strength to safely support the conductors. Lead-in conductors shall be securely attached to the antennas.

810-13. Avoidance of Contacts with Conductors of Other Systems. Outdoor antennas and lead-in conductors from an antenna to a building shall not cross over open conductors of electric light or power circuits and shall be kept well away from all such circuits so as to avoid the possibility of accidental contact. Where proximity to open electric light or power service conductors of less than 250 volts between conductors cannot be avoided, the installation shall be such as to provide a clearance of at least 2 ft (610 mm).

Where practicable, antenna conductors shall be installed so as not to cross under open electric light or power conductors.

One of the leading causes of electrical shock and electrocution, according to statistical reports, is the accidental contact of radio, television, and amateur radio transmitting and receiving antennas and equipment with light or power conductors. Extreme caution should therefore be exercised during this type of installation, and periodic visual inspections conducted thereafter.

810-14. Splices. Splices and joints in antenna spans shall be made mechanically secure with approved splicing devices or by such other means as will not appreciably weaken the conductors.

Conductor spans from antennas should be of sufficient size and strength to maintain clearances and avoid possible contact with light or power conductors. Splices and joints should be made with approved connectors or other means that provide sufficient mechanical strength so that conductors are not weakened appreciably, which could cause them to break and come into contact with higher-voltage conductors.

810-15. Grounding. Masts and metal structures supporting antennas shall be grounded in accordance with Section 810-21.

810-16. Size of Wire-Strung Antenna — Receiving Station.

(a) Size of Antenna Conductors. Outdoor antenna conductors for receiving stations shall be of a size not less than given in Table 810-16(a).

Table 810-16(a). Size of Receiving Station Outdoor Antenna Conductors

	Minimum Size of Conductors Where Maximum Open Span Length Is		
Material	Less than 35 ft	35 ft to 150 ft	Over 150 ft
Aluminum alloy, hard-drawn copper	19	14	12
Copper-clad steel, bronze, or other high-strength material	20	17	14

Note: For SI units, 1 ft = 0.3048 m.

(b) Self-Supporting Antennas. Outdoor antennas, such as vertical rods, dishes, or dipole structures, shall be of corrosion-resistant materials and of strength suitable to withstand ice and wind loading conditions, and shall be located well away from overhead conductors of electric light and power circuits of over 150 volts to ground, so as to avoid the possibility of the antenna or structure falling into or making accidental contact with such circuits.

Section 810-16(b) includes dish-type (parabolic) antennas.

810-17. Size of Lead-in — Receiving Station. Lead-in conductors from outside antennas for receiving stations shall, for various maximum open span lengths, be of such size as to have a tensile strength at least as great as that of the conductors for antennas as specified in Section 810-16. Where the lead-in consists of two or more conductors that are twisted together, are enclosed in the same covering, or are concentric, the conductor size shall, for various maximum open span lengths, be such that the tensile strength of the combination will be at least as great as that of the conductors for antennas as specified in Section 810-16.

810-18. Clearances — Receiving Stations.

(a) Outside of Buildings. Lead-in conductors attached to buildings shall be installed so that they cannot swing closer than 2 ft (610 mm) to the conductors of circuits of 250 volts or less between conductors, or 10 ft (3.05 m) to the conductors of circuits of over 250 volts between conductors, except that in the case of circuits not over 150 volts between conductors, where all conductors involved are supported so as to ensure permanent separation, the clearance shall be permitted to be reduced but shall not be less than 4 in. (102 mm). The clearance between lead-in conductors and any conductor forming a part of a lightning rod system shall not be less than 6 ft (1.83 m) unless the bonding referred to in Section 250-60 is accomplished. Underground conductors shall be separated at least 12 in. (305 mm) from conductors of any light or power circuits or Class 1 circuits.

Exception: Where the electric light or power conductors, Class 1 conductors, or lead-in conductors are installed in raceways or metal cable armor.

(b) Antennas and Lead-ins — Indoors. Indoor antennas and indoor lead-ins shall not be run nearer than 2 in. (50.8 mm) to conductors of other wiring systems in the premises.

Exception No. 1: Where such other conductors are in metal raceways or cable armor.

Exception No. 2: Where permanently separated from such other conductors by a continuous and firmly fixed nonconductor, such as porcelain tubes or flexible tubing.

(c) In Boxes or Other Enclosures. Indoor antennas and indoor lead-ins shall be permitted to occupy the same box or enclosure with conductors of other wiring systems where separated from such other conductors by an effective permanently installed barrier.

810-19. Electric Supply Circuits Used in Lieu of Antenna — Receiving Stations. Where an electric supply circuit is used in lieu of an antenna, the device by which the radio receiving set is connected to the supply circuit shall be listed.

The approved device is usually a small, fixed capacitor connecting the antenna terminal of the receiver and one wire of the supply circuit. As is the case with most receivers, the capacitor should be designed for operation at not less than 300 volts. This ensures a high degree of safety and minimizes the possibility of a breakdown in the capacitor, thereby avoiding a short circuit to ground through the antenna coil of the set.

810-20. Antenna Discharge Units — Receiving Stations.

(a) Where Required. Each conductor of a lead-in from an outdoor antenna shall be provided with a listed antenna discharge unit.

Exception: Where the lead-in conductors are enclosed in a continuous metallic shield that is either permanently and effectively grounded or is protected by an antenna discharge unit.

(b) Location. Antenna discharge units shall be located outside the building or inside the building between the point of entrance of the lead-in and the radio set or transformers, and as near as practicable to the entrance of the conductors to the building. The antenna discharge unit shall not be located near combustible material or in a hazardous (classified) location as defined in Article 500.

A lightning arrester is not required if the lead-in conductors are enclosed in a continuous metal shield, such as rigid or intermediate metal conduit, electrical metallic tubing, or any metal raceway or metal-shielded cable that is effectively grounded. A lightning discharge will take the path of lower impedance

and jump from the lead-in conductors to the metal raceway or shield rather than take the path through the antenna coil of the receiver.

(c) Grounding. The antenna discharge unit shall be grounded in accordance with Section 810-21.

810-21. Grounding Conductors — Receiving Stations. Grounding conductors shall comply with (a) through (j).

(a) Material. The grounding conductor shall be of copper, aluminum, copper-clad steel, bronze, or similar corrosion-resistant material. Aluminum or copper-clad aluminum grounding conductors shall not be used where in direct contact with masonry or the earth or where subject to corrosive conditions. Where used outside, aluminum or copper-clad aluminum shall not be installed within 18 in. (457 mm) of the earth.

(b) Insulation. Insulation on grounding conductors shall not be required.

If metal enclosures are used to enclose the grounding conductor, then bonding must be provided to ensure an adequate low-impedance current path.

(c) Supports. The grounding conductors shall be securely fastened in place and shall be permitted to be directly attached to the surface wired over without the use of insulating supports.

Exception: Where proper support cannot be provided, the size of the grounding conductors shall be increased proportionately.

(d) Mechanical Protection. The grounding conductor shall be protected where exposed to physical damage, or the size of the grounding conductors shall be increased proportionately to compensate for the lack of protection. Where the grounding conductor is run in a metal raceway, both ends of the raceway shall be bonded to the grounding conductor or to the same terminal or electrode to which the grounding conductor is connected.

(e) Run in Straight Line. The grounding conductor for an antenna mast or antenna discharge unit shall be run in as straight a line as practicable from the mast or discharge unit to the grounding electrode.

(f) Electrode. The grounding conductor shall be connected as follows.

(1) To the nearest accessible location on the following:

(a) The building or structure grounding electrode system as covered in Section 250-50
(b) The grounded interior metal water piping system as covered in Section 250-104(a)
(c) The power service accessible means external to enclosures as covered in Section 250-92(b)
(d) The metallic power service raceway
(e) The service equipment enclosure, or
(f) The grounding electrode conductor or the grounding electrode conductor metal enclosures; or

(2) If the building or structure served has no grounding means, as described in (f)(1), to any one of the individual electrodes described in Section 250-50; or

(3) If the building or structure served has no grounding means, as described in (f)(1) or (f)(2), to an effectively grounded metal structure or to any of the individual electrodes described in Section 250-52.

(g) Inside or Outside Building. The grounding conductor shall be permitted to be run either inside or outside the building.

(h) Size. The grounding conductor shall not be smaller than No. 10 copper, No. 8 aluminum, or No. 17 copper-clad steel or bronze.

(i) Common Ground. A single grounding conductor shall be permitted for both protective and operating purposes.

(j) Bonding of Electrodes. A bonding jumper not smaller than No. 6 copper or equivalent shall be connected between the radio and television equipment grounding electrode and the power grounding electrode system at the building or structure served where separate electrodes are used.

The requirements for grounding are in accordance with Article 250. Antenna masts must be grounded to the same grounding electrode used for the electrical system of the building. This is necessary to ensure that all exposed, noncurrent-carrying metal parts are at the same potential. In many cases, masts are incorrectly connected to conveniently located vent pipes, metal gutters, and downspouts. This could create potential differences between lead-in conductors and various metal parts located in or on buildings, resulting in possible shock and fire hazards. An underground gas piping system is not permitted to be used as a grounding electrode.

Section 810-21(j) clarifies that the bonding requirement applies only to electrodes at the same building or structure. The use of separate radio/television grounding electrodes is not required.

C. Amateur Transmitting and Receiving Stations — Antenna Systems

810-51. Other Sections. In addition to complying with Part C, antenna systems for amateur transmitting and receiving stations shall also comply with Sections 810-11 through 810-15.

The communications raceways indicated in Section 800-51(j) through (l) are listed raceways used in plenum, riser, or general-purpose applications. This list-

ing includes raceways and fittings for the installation of communications cables in accordance with this article. The raceways are not suitable for installation of wires, cords, or cables with or without communication members.

A raceway marked "plenum" is suitable for use in ducts, plenums, or other spaces used for environmental air in accordance with Section 800-53(a) when used to enclose communications cables marked CMP. This raceway exhibits a maximum peak optical density of 0.5, a maximum average optical density of 0.15, and a maximum flame-spread distance of 5 ft when tested in accordance with the *Test for Flame Propagation and Smoke-Density Values for Electrical and Optical-Fiber Cables Used in Spaces Transporting Environmental Air,* UL 910. This raceway is identified by a marking on the surface of the raceway or on a marker tape indicating "plenum." A raceway marked "plenum" is also suitable for installation in risers when used to enclose communications cables marked CMP or CMR; general-purpose use when used to enclose communications cables marked CMP, CMR, CMG, or CM; and dwellings use when used to enclose communications cables marked CMP, CMR, CMG, CM, or CMX.

A raceway marked "riser" is suitable for installation in risers in accordance with Section 800-53(b) when used to enclose communications cables marked CMP or CMR. This raceway has fire-resistant characteristics capable of preventing the carrying of fire from floor to floor. This raceway meets the test requirements of the *Standard Test for Flame Propagation Height of Electrical and Optical-Fiber Cable Installed Vertically in Shafts,* UL 1666. This raceway is identified by a marking on the surface of the raceway or on a marker tape indicating "riser." A raceway marked "riser" is also suitable for general-purpose use when used to enclose communications cables marked CMP, CMR, CMG, or CM and dwellings use when used to enclose communications cables marked CMP, CMR, CMG, CM, or CMX.

A raceway marked "general purpose" is suitable for installation in general-purpose areas in accordance with Section 800-53(d) when used to enclose communications cables marked CMP, CMR, CMG, or CM and dwellings use when used to enclose communications cables marked CMP, CMR, CMG, CM, or CMX.

Pliable raceway is raceway that can be bent by hand without the use of tools. The smallest radius of the curve of the inner edge of any bend to which the raceway may be bent without cracking either on the outer surface or internally is not less than 2½ times the outside diameter of the raceway.

810-52. Size of Antenna. Antenna conductors for transmitting and receiving stations shall be of a size not less than given in Table 810-52.

Table 810-52. Size of Amateur Station Outdoor Antenna Conductors

	Minimum Size of Conductors Where Maximum Open Span Length Is	
Material	**Less than 150 ft**	**Over 150 ft**
Hard-drawn copper	14	10
Copper-clad steel, bronze, or other high-strength material	14	12

Note: For SI units, 1 ft = 0.3048 m.

810-53. Size of Lead-in Conductors. Lead-in conductors for transmitting stations shall, for various maximum span lengths, be of a size at least as great as that of conductors for antennas as specified in Section 810-52.

810-54. Clearance on Building. Antenna conductors for transmitting stations, attached to buildings, shall be firmly mounted at least 3 in. (76 mm) clear of the surface of the building on nonabsorbent insulating supports, such as treated pins or brackets equipped with insulators having not less than 3-in. (76-mm) creepage and airgap distances. Lead-in conductors attached to buildings shall also comply with these requirements.

Exception: Where the lead-in conductors are enclosed in a continuous metallic shield that is permanently and effectively grounded, they shall not be required to comply with these requirements. Where grounded, the metallic shield shall also be permitted to be used as a conductor.

The creepage distance is measured from the conductor across the face of the supporting insulator to the building surface. The air gap distance is measured from the conductor (at its closest point) across the air space (not necessarily in a straight line) to the surface of the building. The exception covers coaxial cable with the shield permanently and effectively grounded.

810-55. Entrance to Building. Except where protected with a continuous metallic shield that is permanently and effectively grounded, lead-in conductors for transmitting stations shall enter buildings by one of the following methods:

(1) Through a rigid, noncombustible, nonabsorbent insulating tube or bushing
(2) Through an opening provided for the purpose in which

the entrance conductors are firmly secured so as to provide a clearance of at least 2 in. (50.8 mm), or
(3) Through a drilled window pane

810-56. Protection Against Accidental Contact. Lead-in conductors to radio transmitters shall be located or installed so as to make accidental contact with them difficult.

810-57. Antenna Discharge Units — Transmitting Stations. Each conductor of a lead-in for outdoor antennas shall be provided with an antenna discharge unit or other suitable means that will drain static charges from the antenna system.

If an antenna discharge unit is not installed at a transmitting station, then protection against lightning may be provided by a switch that connects the lead-in to ground during the time the station is not in operation.

Exception No. 1: Where protected by a continuous metallic shield that is permanently and effectively grounded.

Exception No. 2: Where the antenna is permanently and effectively grounded.

810-58. Grounding Conductors — Amateur Transmitting and Receiving Stations. Grounding conductors shall comply with (a) through (c).

(a) Other Sections. All grounding conductors for amateur transmitting and receiving stations shall comply with Sections 810-21(a) through (j).

(b) Size of Protective Grounding Conductor. The protective grounding conductor for transmitting stations shall be as large as the lead-in, but not smaller than No. 10 copper, bronze, or copper-clad steel.

(c) Size of Operating Grounding Conductor. The operating grounding conductor for transmitting stations shall not be less than No. 14 copper or its equivalent.

D. Interior Installation — Transmitting Stations

810-70. Clearance from Other Conductors. All conductors inside the building shall be separated at least 4 in. (102 mm) from the conductors of any electric light, power, or signaling circuit.

Exception No. 1: As provided in Article 640.

Exception No. 2: Where separated from other conductors by raceway or some firmly fixed nonconductor, such as porcelain tubes or flexible tubing.

810-71. General. Transmitters shall comply with (a) through (c).

(a) Enclosing. The transmitter shall be enclosed in a metal frame or grille, or separated from the operating space by a barrier or other equivalent means, all metallic parts of which are effectively connected to ground.

(b) Grounding of Controls. All external metal handles and controls accessible to the operating personnel shall be effectively grounded.

(c) Interlocks on Doors. All access doors shall be provided with interlocks that will disconnect all voltages of over 350 volts between conductors when any access door is opened.

Article 820 — Community Antenna Television and Radio Distribution Systems

Contents

A. General

820-1. Scope. This article covers coaxial cable distribution of radio frequency signals typically employed in community antenna television (CATV) systems.

Article 820 covers the installation of coaxial cables for the distribution of radio frequency (RF) signals associated with closed-circuit television, cable television, and security television cameras. Interior coaxial cable for radio and television receiving equipment is also covered by this article. New Article 830 was added to the 1999 *Code* to cover network-powered broadband system installations.

820-2. Definitions. See Article 100. For the purposes of this article, the following additional definitions apply.

Exposed. An exposed cable is one that is in such a position that, in case of failure of supports and insulation, contact with another circuit may result.

FPN: See Article 100 for two other definitions of *Exposed*.

Point of Entrance. The point within a building at which the cable emerges from an external wall, from a concrete floor slab, or from a rigid metal conduit or an intermediate metal conduit grounded to an electrode in accordance with Section 820-40(b).

Premises. The land and buildings of a user located on the user side of utility-user network point of demarcation.

820-3. Locations and Other Articles. Circuits and equipment shall comply with (a) through (g).

The addition of Section 820-3 in the 1999 *Code* permits Article 830 wiring methods to substitute for Article 820 wiring methods. The substitution of these wiring methods facilitates an upgrade of Article 820 installations to network-powered broadband applications.

(a) Spread of Fire or Products of Combustion. Section 300-21.

(b) Ducts, Plenums, and Other Air-Handling Spaces. Section 300-22, where installed in ducts or plenums or other spaces used for environmental air.

Exception: As permitted in Section 820-53(a).

(c) Installation and Use. Section 110-3(b) shall apply.

(d) Installations of Conductive and Nonconductive Optical Fiber Cables. Article 770.

(e) Communications Circuits. Article 800.

(f) Network-Powered Broadband Communications Systems. Article 830.

(g) Alternate Wiring Methods. The wiring methods of Article 830 shall be permitted to substitute for the wiring methods of Article 820.

FPN: Use of Article 830 wiring methods will facilitate the upgrading of Article 820 installations to network-powered broadband applications.

820-4. Energy Limitations. The coaxial cable shall be permitted to deliver low-energy power to equipment that is directly associated with the radio frequency distribution system if the voltage is not over 60 volts and if the current supply is from a transformer or other device that has energy-limiting characteristics.

820-5. Access to Electrical Equipment Behind Panels Designed to Allow Access. Access to equipment shall not be denied by an accumulation of cables that prevents removal of panels, including suspended ceiling panels.

The excess accumulation of wires and cables can limit access to equipment by preventing the removal of access panels. (See Figures 820.1 and 820.2.)

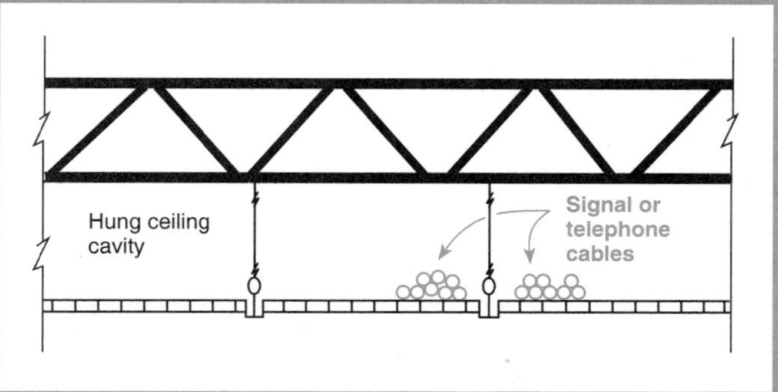

Figure 820.1 *Excess accumulation of conductors and cables prevents access to equipment or cables. Incorrect methods shown.*

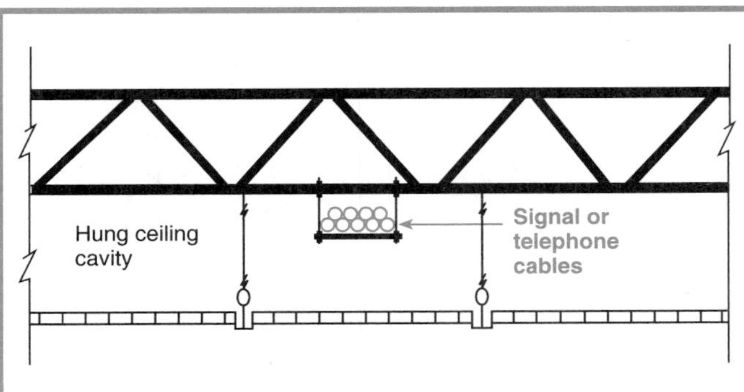

Figure 820.2 *Excess accumulation of conductors and cables prevents access to equipment or cables. Correct methods shown.*

820-6. Mechanical Execution of Work. Community antenna television and radio distribution systems shall be installed in a neat and workmanlike manner. Cables shall be supported by the building structure in such a manner that the cable will not be damaged by normal building use.

FPN: One way to determine accepted industry practice is to refer to nationally recognized standards such as *Commercial Building Telecommunications Cabling Standard,* ANSI/ EIA/TIA 568-A-1995; *Commercial Building Standard for Telecommunications Pathways and Spaces,* ANSI/EIA/TIA 569-1990; and *Residential and Light Commercial Telecommunications Wiring Standard,* ANSI/EIA/TIA 570-1991.

This fine print note makes it clear that there are industry standards.

B. Cables Outside and Entering Buildings

820-10. Outside Cables. Coaxial cables, prior to the point of grounding, as defined in Section 820-33, shall comply with (a) through (f).

(a) On Poles. Where practicable, conductors on poles shall be located below the electric light, power, Class 1, or nonpower-limited fire alarm circuit conductors and shall not be attached to a crossarm that carries electric light or power conductors.

(b) Lead-in Clearance. Lead-in or aerial-drop cables from a pole or other support, including the point of initial attachment to a building or structure, shall be kept away from electric light, power, Class 1, or nonpower-limited fire alarm circuit conductors so as to avoid the possibility of accidental contact.

Exception: Where proximity to electric light, power, Class 1, or nonpower-limited fire alarm circuit service conductors cannot be avoided, the installation shall be such as to provide clearances of not less than 12 in. (305 mm) from light, power, Class 1, or nonpower-limited fire alarm circuit service drops.

(c) On Masts. Aerial cable shall be permitted to be attached to an above-the-roof raceway mast that does not enclose or support conductors of electric light or power circuits.

(d) Above Roofs. Cables shall have a vertical clearance of not less than 8 ft (2.44 m) from all points of roofs above which they pass.

Exception No. 1: Auxiliary buildings such as garages and the like.

Exception No. 2: A reduction in clearance above only the overhanging portion of the roof to not less than 18 in. (457 mm) shall be permitted if (1) not more than 4 ft (1.22 m) of communications service drop conductors pass above the roof overhang, and (2) they are terminated at a raceway mast or other approved support.

Exception No. 3: Where the roof has a slope of not less than 4 in. (102 mm) in 12 in. (305 mm), a reduction in clearance to not less than 3 ft (914 mm) shall be permitted.

(e) Between Buildings. Cables extending between buildings and also the supports or attachment fixtures shall be acceptable for the purpose and shall have sufficient strength to withstand the loads to which they may be subjected.

Wind and ice loads can sometimes be excessive and should be considered.

Exception: Where a cable does not have sufficient strength to be self-supporting, it shall be attached to a supporting messenger cable that, together with the attachment fixtures or supports, shall be acceptable for the purpose and shall have sufficient strength to withstand the loads to which they may be subjected.

(f) On Buildings. Where attached to buildings, cables shall be securely fastened in such a manner that they will be separated from other conductors as follows.

(1) Electric Light or Power. The coaxial cable shall have a separation of at least 4 in. (102 mm) from electric light, power, Class 1, or nonpower-limited fire alarm circuit conductors not in raceway or cable, or be permanently separated from conductors of the other system by a continuous and firmly fixed nonconductor in addition to the insulation on the wires.

(2) Other Communications Systems. Coaxial cable shall be installed so that there will be no unnecessary interference in the maintenance of the separate systems. In no case shall the conductors, cables, messenger strand, or equipment of one system cause abrasion to the conductors, cable, messenger strand, or equipment of any other system.

(3) Lightning Conductors. Where practicable, a separation of at least 6 ft (1.83 m) shall be maintained between any coaxial cable and lightning conductors.

820-11. Entering Buildings.

(a) Underground Systems. Underground coaxial cables in a duct, pedestal, handhole, or manhole that contains electric light or power conductors or Class 1 circuits shall be in a section permanently separated from such conductors by means of a suitable barrier.

(b) Direct-Buried Cables and Raceways. Direct-buried coaxial cable shall be separated at least 12 in. (305 mm) from conductors of any light or power or Class 1 circuit.

Exception No. 1: Where electric service conductors or coaxial cables are installed in raceways or have metal cable armor.

Exception No. 2: Where electric light or power branch-circuit or feeder conductors or Class 1 circuit conductors are installed in a raceway or in metal-sheathed, metal-clad, or Type UF or Type USE cables; or the coaxial cables have metal cable armor or are installed in a raceway.

See the commentary following Section 800-10(c), Exception.

C. Protection

820-33. Grounding of Outer Conductive Shield of a Coaxial Cable. Where coaxial cable is exposed to lightning or to accidental contact with lightning arrester conductors or power conductors operating at a voltage of over 300 volts to ground, the outer conductive shield of the coaxial cable shall be grounded at the building premises as close to the point of cable entry as practicable. For purposes of this section, the point at which the exposed cable enters shall be considered to be the point of emergence through an exterior wall, a concrete floor slab, or from a rigid or intermediate metal conduit grounded to an electrode in accordance with Section 820-40(b).

For purposes of this section, grounding located at mobile home service equipment located in sight from and not more than 30 ft (9.14 m) from the exterior wall of the mobile home it serves, or at a mobile home disconnecting means grounded in accordance with Section 250-32 and located in sight from and not more than 30 ft (9.14 m) from the exterior wall of the mobile home it serves, shall be considered to meet the requirements of this section.

FPN: Selecting a grounding location to achieve the shortest practicable grounding conductor will help limit potential differences between CATV and other metallic systems.

(a) Shield Grounding. Where the outer conductive shield of a coaxial cable is grounded, no other protective devices shall be required.

(b) Shield Protection Devices. Grounding of a coaxial drop cable shield by means of a protective device that does not interrupt the grounding system within the premises shall be permitted.

Section 820-33(b) permits the use of a shield protection device that does not interrupt the grounding system within the premises. This permits protection against overheating for the CATV service-drop cable. Overheating can occur due to neutral fault currents in the power and lighting systems. Such a protective device would have to maintain the integrity of the coaxial system to prevent RF leakage. An ordinary fuse, for example, would not be suitable.

D. Grounding Methods

820-40. Cable Grounding. Where required by Section 820-33, the shield of the coaxial cable shall be grounded as specified in (a) through (d).

(a) Grounding Conductor.

(1) Insulation. The grounding conductor shall be insulated and shall be listed as suitable for the purpose.

(2) Material. The grounding conductor shall be copper or other corrosion-resistant conductive material, stranded or solid.

(3) Size. The grounding conductor shall not be smaller than No. 14. It shall have a current-carrying capacity approximately equal to that of the outer conductor of the coaxial cable.

(4) Run in Straight Line. The grounding conductor shall be run to the grounding electrode in as straight a line as practicable.

(5) Physical Protection. Where subject to physical damage, the grounding conductor shall be adequately protected. Where the grounding conductor is run in a metal raceway, both ends of the raceway shall be bonded to the grounding conductor or the same terminal or electrode to which the grounding conductor is connected.

(b) Electrode. The grounding conductor shall be connected as follows.

(1) To the nearest accessible location on the following:

(a) The building or structure grounding electrode system as covered in Section 250-50
(b) The grounded interior metal water piping system as covered in Section 250-104(a)
(c) The power service accessible means external to enclosures as covered in Section 250-92(b)
(d) The metallic power service raceway
(e) The service equipment enclosure
(f) The grounding electrode conductor or the grounding electrode conductor metal enclosure, or
(g) To the grounding conductor or to the grounding electrode of a building or structure disconnecting means that is grounded to an electrode as covered in Section 250-32; or

(2) If the building or structure served has no grounding means as described in (b)(1), to any one of the individual electrodes described in Section 250-50; or

(3) If the building or structure served has no grounding means as described in (b)(1) or (b)(2), to an effectively grounded metal structure or to any one of the individual electrodes described in Section 250-52.

(c) Electrode Connection. Connections to grounding electrodes shall comply with Section 250-70.

(d) Bonding of Electrodes. A bonding jumper not smaller than No. 6 copper or equivalent shall be connected between the antenna systems grounding electrode and the power grounding electrode system at the building or structure served where separate electrodes are used.

Section 820-40(d) requires bonding of CATV and power grounding electrodes at the same building or structure. Section 820-11(c), Exception, may require a separate grounding electrode for communications systems.

Exception: At mobile homes as covered in Section 820-42.

FPN No. 1: See Section 250-60 for use of air terminals (lightning rods).

FPN No. 2: Bonding together of all separate electrodes will limit potential differences between them and between their associated wiring systems.

The most common error made in grounding CATV systems is to connect the coaxial cable sheath to a ground rod driven by the CATV installer at a convenient location near the point of cable entry to the building and not bonding it to the electrical service grounding electrode system, service raceway, or other components that make up the grounding electrode system. A separate grounding electrode is permitted by the *Code* only if the building or structure has none of the grounding means described in Section 820-40(b)(1) or 820-40(b)(2), which is rare. The *Code* requires that some means be provided that are accessible and external to the service equipment for making the bonding and grounding connection for other systems. One of the following means must be provided:

1. Exposed service raceways
2. Exposed grounding electrode conductor
3. An approved means for the external connection of a conductor (A No. 6 copper conductor with one end bonded to the service raceway or equipment with about 6 in. exposed is acceptable.)

Proper bonding of the CATV system coaxial cable sheath to the electrical power ground is needed to prevent potential fire and shock hazards. The earth cannot be used as an equipment grounding conductor or bonding conductor, as it does not have the low-impedance path required. (See Section 250-54.)

Both CATV systems and power systems are subject to current surges as a result, for example, of induced voltages from lightning in the vicinity of the usually extensive outside distribution systems. Surges also result from switching operations on power systems. If the grounded conductors and parts of the two systems are not bonded by a low-impedance path, such line surges can raise the potential difference between the two systems to many thousands of volts. This can result in arcing between the two systems, for example, wherever the coaxial cable jacket contacts a grounded part, such as a metal water pipe or metal structural member, inside the building.

If an individual is the interface between the two systems and the bonding has not been done in accordance with the *Code*, the high-voltage surge could result in electric shock. More common, however, is burnout of the television tuner, as this part is almost always an interface between the two systems. The tuner is connected to the power system ground through the grounded neutral of the power supply, even if the television itself is not provided with an equipment grounding conductor.

Also see the commentary following Sections 250-92(b), FPN No. 2, and 820-33(b).

820-41. Equipment Grounding. Unpowered equipment and enclosures or equipment powered by the coaxial cable shall be considered grounded where connected to the metallic cable shield.

820-42. Bonding and Grounding at Mobile Homes.

(a) Grounding. Where there is no mobile home service equipment located in sight from and not more than 30 ft (9.14 m) from the exterior wall of the mobile home it serves, or there is no mobile home disconnecting means grounded in accordance with Section 250-32 and located within sight from and not more than 30 ft (9.14 m) from the exterior wall of the mobile home it serves, the coaxial cable shield ground, or surge arrester ground, shall be in accordance with Sections 820-40(b)(2) and (3).

(b) Bonding. The coaxial cable shield grounding terminal, surge arrester grounding terminal, or grounding electrode shall be bonded to the metal frame or available grounding terminal of the mobile home with a copper grounding conductor not smaller than No. 12 under any of the following conditions:

(1) Where there is no mobile home service equipment or disconnecting means as in (a), or
(2) Where the mobile home is supplied by cord and plug

E. Cables Within Buildings

820-49. Fire Resistance of CATV Cables. Coaxial cables installed as wiring within buildings shall be listed as being resistant to the spread of fire in accordance with Sections 820-50 and 820-51.

820-50. Listing, Marking, and Installation of Coaxial Cables. Coaxial cables in a building shall be listed as being suitable for the purpose, and cables shall be marked in accordance with Table 820-50. The cable voltage rating shall not be marked on the cable.

FPN: Voltage markings on cables may be misinterpreted to suggest that the cables may be suitable for Class 1, electric light, and power applications.

Exception No. 1: Voltage markings shall be permitted where the cable has multiple listings and voltage marking is required for one or more of the listings.

Exception No. 2: Listing and marking shall not be required where the cable enters the building from the outside and is run in rigid metal conduit or intermediate metal conduit, and such conduits are grounded to an electrode in accordance with Section 820-40(b).

Exception No. 3: Listing and marking shall not be required where the length of the cable within the building, measured from its point of entrance, does not exceed 50 ft (15.2 m) and the cable enters the building from the outside and is terminated at a grounding block.

Table 820-50. Cable Markings

Cable Marking	Type	Reference
CATVP	CATV plenum cable	820-51(a) and 820-53(a)
CATVR	CATV riser cable	820-51(b) and 820-53(b)
CATV	CATV cable	820-51(c) and 820-53(c)
CATVX	CATV cable, limited use	820-51(d) and 820-53(c), Exception Nos. 1, 2, 3, and 4

FPN No. 1: Cable types are listed in descending order of fire-resistance rating.

FPN No. 2: See the referenced sections for listing requirements and permitted uses.

820-51. Additional Listing Requirements. Cables shall be listed in accordance with (a) through (d).

(a) Type CATVP. Type CATVP community antenna television plenum cable shall be listed as being suitable for use in ducts, plenums, and other spaces used for environmental air and shall also be listed as having adequate fire-resistant and low smoke-producing characteristics.

FPN: One method of defining low smoke-producing cables is by establishing an acceptable value of the smoke produced when tested in accordance with the *Standard Method for Test for Fire and Smoke Characteristics of Wires and Cables,* NFPA 262 -1994, to a maximum peak optical density of 0.5 and a maximum average optical density of 0.15. Similarly, one method of defining fire-resistant cables is by establishing maximum allowable flame travel distance of 5 ft (1.52 m) when tested in accordance with the same test.

See the commentary following Section 725-71(a), FPN.

(b) Type CATVR. Type CATVR community antenna television riser cable shall be listed as being suitable for use in a vertical run in a shaft or from floor to floor and shall also be listed as having fire-resistant characteristics capable of preventing the carrying of fire from floor to floor.

FPN: One method of defining fire-resistant characteristics capable of preventing the carrying of fire from floor to floor is that the cables pass the requirements of the *Standard Test for Flame Propagation Height of Electrical and Optical-Fiber Cable Installed Vertically in Shafts,* ANSI/UL 1666-1997.

See the commentary following Section 725-71(b), FPN.

(c) Type CATV. Type CATV community antenna television cable shall be listed as being suitable for general purpose CATV use, with the exception of risers and plenums, and shall also be listed as being resistant to the spread of fire.

FPN: One method of defining resistant to the spread of fire is that the cables do not spread fire to the top of the tray in the vertical-tray flame test in the *Reference Standard for Electrical Wires, Cables and Flexible Cords,* ANSI/UL 1581-1991.

Another method of defining resistant to the spread of fire is for the damage (char length) not to exceed 4 ft 11 in. (1.5 m) when performing the vertical flame test — cables in cable trays, as described in *Test Methods for Electrical Wires and Cables,* CSA C22.2 No. 0.3-M-1985.

See the commentary following Section 725-71(c), FPN.

(d) Type CATVX. Type CATVX limited-use community antenna television cable shall be listed as being suitable for use in dwellings and for use in raceway and shall also be listed as being resistant to flame spread.

FPN: One method of determining that cable is resistant to flame spread is by testing the cable to the VW-1 (vertical-wire) flame test in the *Reference Standard for Electrical Wires, Cables and Flexible Cords,* ANSI/UL 1581-1991.

See the commentary following Section 725-71(d), FPN.

820-52. Installation of Cables and Equipment. Beyond the point of grounding, as defined in Section 820-33, the cable installation shall comply with (a) through (e).

Section 820-52 was revised for the 1999 *Code* to specifically include separation from network-powered

broadband communications circuits. Jackets of coaxial cables do not have sufficient construction specifications to permit them to be installed with electric light, power, Class 1, nonpower-limited fire alarm circuits, and medium and high power network-powered broadband communications cables. Failure of the cable insulation due to a fault could lead to hazardous voltages being imposed on the Class 2 or Class 3 circuit conductors.

(a) Separation from Other Conductors.

(1) In Raceways and Boxes.

(a) *Other Circuits.* Coaxial cables shall be permitted in the same raceway or enclosure with jacketed cables of any of the following:

(1) Class 2 and Class 3 remote-control, signaling, and power-limited circuits in compliance with Article 725
(2) Power-limited fire alarm systems in compliance with Article 760
(3) Communications circuits in compliance with Article 800
(4) Nonconductive and conductive optical fiber cables in compliance with Article 770
(5) Low power network-powered broadband communications circuits in compliance with Article 830

(b) *Electric Light, Power, Class 1, Nonpower-Limited Fire Alarm, and Medium Power Network-Powered Broadband Communications Circuits.* Coaxial cable shall not be placed in any raceway, compartment, outlet box, junction box, or other enclosures with conductors of electric light, power, Class 1, nonpower-limited fire alarm, or medium power network-powered broadband communications circuits.

Exception No. 1: Where all of the conductors of electric light, power, Class 1, nonpower-limited fire alarm, and medium power network-powered broadband communications circuits are separated from all of the coaxial cables by a barrier.

Exception No. 2: Power circuit conductors in outlet boxes, junction boxes, or similar fittings or compartments where such conductors are introduced solely for power supply to the coaxial cable system distribution equipment. The power circuit conductors shall be routed within the enclosure to maintain a minimum 0.25-in. (6.35-mm) separation from coaxial cables.

(2) Other Applications. Coaxial cable shall be separated at least 2 in. (50.8 mm) from conductors of any electric light, power, Class 1, nonpower-limited fire alarm, or medium power network-powered broadband communications circuits.

Exception No. 1: Where either (1) all of the conductors of electric light, power, Class 1, nonpower-limited fire alarm, and medium power network-powered broadband communications and circuits are in a raceway, or in metal-sheathed, metal-clad, nonmetallic-sheathed, Type AC, or Type UF cables, or (2) all of the coaxial cables are encased in raceway.

Exception No. 2: Where the coaxial cables are permanently separated from the conductors of electric light, power, Class 1, nonpower-limited fire alarm, and medium power network-powered broadband communications circuits by a continuous and firmly fixed nonconductor, such as porcelain tubes or flexible tubing, in addition to the insulation on the wire.

(b) Spread of Fire or Products of Combustion. Installations in hollow spaces, vertical shafts, and ventilation or air-handling ducts shall be made so that the possible spread of fire or products of combustion will not be substantially increased. Openings around penetrations through fire resistance-rated walls, partitions, floors, or ceilings shall be firestopped using approved methods.

(c) Equipment in Other Space Used for Environmental Air. Section 300-22(c) shall apply.

(d) Hybrid Power and Coaxial Cabling. The provisions of Section 780-6 shall apply for listed hybrid power and coaxial cabling in closed-loop and programmed power distribution.

(e) Support of Conductors. Raceways shall not be used as a means of support for coaxial cables.

820-53. Applications of Listed CATV Cables. CATV cables shall comply with (a) through (d).

Table 820-53 was extensively revised for the 1999 *Code* to reflect the current field applications for various cable types.

(a) Plenum. Cables installed in ducts, plenums, and other spaces used for environmental air shall be Type CATVP.

Exception: Types CATVP, CATVR, CATV, and CATVX cables installed in compliance with Section 300-22.

(b) Riser. Cables installed in vertical runs and penetrating more than one floor, or cables installed in vertical runs in a shaft, shall be Type CATVR. Floor penetrations requiring Type CATVR shall contain only cables suitable for riser or plenum use.

Exception No. 1: Types CATV and CATVX cables encased in metal raceway or located in a fireproof shaft having firestops at each floor.

Exception No. 2: Types CATV and CATVX cables in one- and two-family dwellings.

FPN: See Section 820-52(b) for the firestop requirements for floor penetrations.

(c) Other Wiring Within Buildings. Cables installed in building locations other than the locations covered in (a) and (b) shall be Type CATV.

Exception No. 1: Type CATVX cable enclosed in raceway.

Exception No. 2: Type CATVX cable in nonconcealed spaces where the exposed length of cable does not exceed 10 ft (3.05 m).

Exception No. 3: Type CATVX cables that are less than 0.375 in. (9.52 mm) in diameter and installed in one- or two-family dwellings.

Exception No. 4: Type CATVX cables that are less than 0.375 in. (9.52 mm) in diameter and installed in nonconcealed spaces in multifamily dwellings.

(d) Cable Substitutions. The substitutions for community antenna television cables listed in Table 820-53 shall be considered suitable for the purpose and shall be permitted.

Table 820-53. Coaxial Cable Uses and Permitted Substitutions

Cable Type	Use	References	Permitted Substitutions
CATVP	Coaxial plenum cable	820-53(a)	CMP
CATVR	Coaxial riser cable	820-53(b)	CATVP, CMP, CMR
CATV	Coaxial general-purpose cable	820-53(c)	CATVP, CMP, CATVR, CMR, CMG, CM
CATVX	Coaxial cable, limited use	820-53(c)	CATVP, CMP, CATVR, CMR, CATV, CMG, CM

Note: See Figure 820-53, Cable substitution hierarchy.

FPN: The substitute cables in Table 820-53 are only coaxial-type cables.

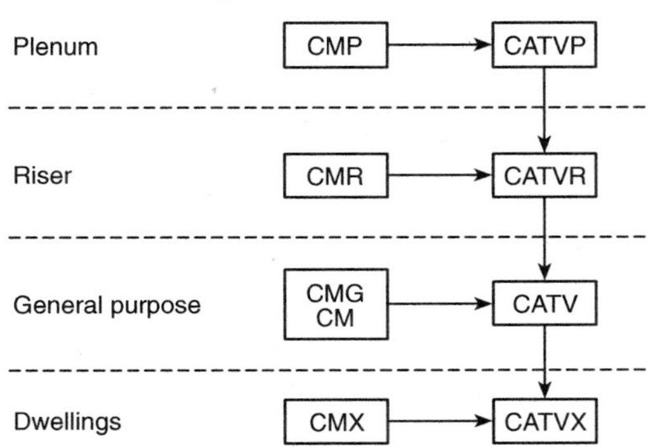

Figure 820-53 Cable substitution hierarchy.

Article 830 — Network-Powered Broadband Communications Systems

Contents

A. General

830-1. Scope. This article covers network-powered broadband communications systems that provide any combination of voice, audio, video, data, and interactive services through a network interface unit.

Article 830 was added to the 1999 *Code* to permit network-powered broadband communications circuits that provide a wide array of services to a subscriber. Among these services are voice, data (such as internal access), interactive services, and television signals.

FPN No. 1: A typical basic system configuration includes a cable supplying power and broadband signal to a network interface unit that converts the broadband signal to the component signals. Typical cables are coaxial cable with both broadband signal and power on the center conductor, composite metallic cable with a coaxial member for the broadband signal and a twisted pair for power, and composite optical fiber cable with a pair of conductors for power. Larger systems may also include network components such as amplifiers that require network power.

FPN No. 2: See Section 90-2(b)(4) for installations of broadband communications systems that are not covered.

830-2. Definitions. See Article 100. For purposes of this article, the following additional definitions apply.

Block. A square or portion of a city, town, or village enclosed by streets and including the alleys so enclosed, but not any street.

Exposed to Accidental Contact with Electrical Light or Power Conductors. The circuit is in such a position that, in case of failure of supports or insulation, contact with another circuit may result.

Fault Protection Device. An electronic device intended for the protection of personnel that functions under fault conditions, such as network-powered broadband communications cable short or open circuit, to limit the current or voltage, or both, for a low power network-powered broadband communications circuit and provide acceptable protection from electric shock.

Network Interface Unit (NIU). A device that converts a broadband signal into component voice, audio, video, data, and interactive services signals. The NIU provides isolation between the network power and the premises signal circuits. The NIU may also contain primary and secondary protectors.

Network-Powered Broadband Communications Circuit. The circuit extending from the communications utility's serving terminal or tap up to and including the NIU.

FPN: A typical single-family network-powered communications circuit consists of a communications drop or communications service cable and an NIU, and includes the communications utility's serving terminal or tap where it is not under the exclusive control of the communications utility.

Point of Entrance. The point within a building at which the cable emerges from an external wall, from a concrete floor slab, or from a rigid metal conduit or an intermediate metal conduit grounded to an electrode in accordance with Section 830-40(b).

Premises Wiring. The circuits located on the user side of the network interface unit.

830-3. Locations and Other Articles. Circuits and equipment shall comply with (a) through (d).

Article 830 contains provisions for wiring both inside and outside of buildings. However, other articles cover the wiring derived from the network interface unit (NIU) into the premises. For example, Article 725 covers wiring of Class 2 and Class 3 circuits, Article 760 covers wiring of fire alarm systems, Article 770 covers the installation of optical fiber cables, Article 800 covers telephone wiring, and Article 820 covers coaxial cable installations for television signals. The major difference between Article 820 and Article 830 is the voltage present on the circuit conductors. Article 820 systems are limited to 60 volts, but Article 830 systems are permitted to have voltages as high as 150 volts. Higher voltages allows the system to power more sophisticated electronics and provide a wider variety of services. (See Figure 830.1.)

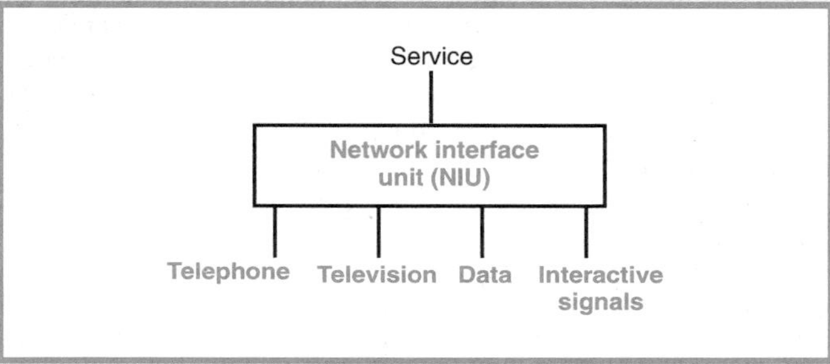

Figure 830.1 *A network interface unit (NIU) diagram showing derived circuits.*

(a) Spread of Fire or Products of Combustion. Section 300-21.

(b) Ducts, Plenums, and Other Air-Handling Spaces. Section 300-22, where installed in ducts or plenums or other spaces used for environmental air.

Exception: As permitted in Section 830-55(a).

(c) Installation and Use. Section 110-3(b).

(d) Output Circuits. As appropriate for the services provided, the output circuits derived from the network interface unit shall comply with the requirements of the following:

(1) Installations of communications circuits — Article 800
(2) Installations of community antenna television and radio distribution circuits — Article 820

Exception: Section 830-30(b)(3) shall apply where protection is provided in the output of the NIU.

(3) Installations of optical fiber cables — Article 770
(4) Installations of Class 2 and Class 3 circuits — Article 725
(5) Installations of power-limited fire alarm circuits — Article 760

830-4. Power Limitations. Network-powered broadband communications systems shall be classified as having low or medium power sources as defined in Table 830-4.

Table 830-4. Limitations for Network-Powered Broadband Communications Systems

Network Power Source	Low	Medium
Circuit voltage, V_{max} (volts) (Note 1)	0–100	0–150
Power limitation, VA_{max} (volt-amperes) (Note 1)	250	250
Current limitation, I_{max} (amperes) (Note 1)	$1000/V_{max}$	$1000/V_{max}$
Maximum power rating (volt-amperes)	100	100
Maximum voltage rating (volts)	100	150
Maximum overcurrent protection (amperes) (Note 2)	$100/V_{max}$	NA

Notes:

1. V_{max}, I_{max}, and VA_{max} are determined with the current-limiting impedance in the circuit (not bypassed) as follows:

V_{max}—Maximum system voltage regardless of load with rated input applied.

I_{max}—Maximum system current under any noncapacitive load, including short circuit, and with overcurrent protection bypassed if used.

I_{max} limits apply after 1 minute of operation.

VA_{max}—Maximum volt-ampere output after 1 minute of operation regardless of load and overcurrent protection bypassed if used.

2. Overcurrent protection is not required where the current-limiting device provides equivalent current limitation and the current-limiting device does not reset until power or the load is removed.

High power network-powered broadband communications circuits are not covered by this *Code.*

830-5. Network-Powered Broadband Communications Equipment and Cables. Network-powered broadband communications equipment and cables shall be listed as suitable for the purpose.

Exception No. 1: This listing requirement shall not apply to community antenna television and radio distribution system coaxial cables that were installed prior to January 1, 2000, in accordance with Article 820 and are used for low power network-powered broadband communications circuits. See Section 830-9.

Exception No. 2: Substitute cables for network-powered broadband communications cables shall be permitted as shown in Table 830-58.

Article 830 permits three types of listed cables including coaxial, coaxial with a twisted pair of conductors, and fiber optical with a twisted pair of conductors. In the coaxial configuration, power and signal are carried on the coaxial center conductor. With coaxial and a twisted pair of conductors, the signal is carried on the coaxial center conductor, and the power is carried on the twisted pair. With fiber optical and twisted pair conductors, the signal is carried on the fiber, and the power is carried on the twisted pair.

(a) Listing and Marking. Listing and marking of network-powered broadband communications cables shall comply with (1) or (2).

A voltage rating of 300 volts is necessary for the following reasons:

1. To coordinate with protector installation requirements (i.e., protectors are not required within a block unless the cable is exposed to over 300 volts)
2. To recognize the fact that primary protectors are designed to allow voltages below 300 to pass
3. To accommodate the voltages ordinarily found on a network-powered broadband communications circuit (voltage up to 150 volts rms)

(1) Type BMU, Type BM, and Type BMR Cables. Network-powered broadband communications medium power underground cable, Type BMU; network-powered broadband communications medium power cable, Type BM; and network-powered broadband communications medium power riser cable, Type BMR, shall be factory-assembled cables consisting of a jacketed coaxial cable, a jacketed combination of coaxial cable and multiple individual conductors, or a jacketed combination of an optical fiber cable and multiple individual conductors. The insulation for the individual conductors shall be rated for 300 volts minimum. Cables intended for outdoor use shall be listed as suitable for the application. Cables shall be marked in accordance with Section 310-11. Type BMU cables shall be jacketed and listed as being suitable for outdoor underground use. Type BM cables shall be listed as being suitable for general purpose use, with the exception of risers and plenums, and shall also be listed as being resistant to the spread of fire. Type BMR cables shall be listed as being suitable for use in a vertical run in a shaft or from floor to floor and shall also be listed as having fire-resistant characteristics capable of preventing the carrying of fire from floor to floor.

FPN No. 1: One method of defining resistant to spread of fire is that the cables do not spread fire to the top of the tray in the vertical tray flame test in the *Reference Standard for Electrical Wires, Cables and Flexible Cords,* ANSI/UL 1581-1991. Another method of defining resistant to the spread of fire is for the damage (char length) not to exceed 4 ft. 11 in. (1.5 m) when performing the CSA vertical flame test for cables in cable trays, as described in *Test Methods for Electrical Wires and Cables,* CSA C22.2 No. 0.3-M-1985.

See the commentary following Section 725-71(c), FPN.

FPN No. 2: One method of defining fire-resistant characteristics capable of preventing the carrying of fire from floor to floor is that the cables pass the requirements of the *Standard Test for Flame Propagation Height of Electrical and Optical-Fiber Cable Installed Vertically in Shafts,* ANSI/UL 1666-1997.

See the commentary following Section 725-71(b), FPN.

(2) Type BLU, Type BLX, and Type BLP Cables. Network-powered broadband communications low power underground cable, Type BLU; limited use network-powered broadband communications low power cable, Type BLX; and network-powered broadband communications low power plenum cable, Type BLP, shall be factory assembled cables consisting of a jacketed coaxial cable, a jacketed combination of coaxial cable and multiple individual conductors, or a jacketed combination of an optical fiber cable and multiple individual conductors. The insulation for the individual conductors shall be rated for 300 volts minimum. Cables intended for outdoor use shall be listed as suitable for the application. Cables shall be marked in accordance with Section 310-11. Type BLU cables shall be jacketed and listed as being suitable for outdoor underground use. Type BLX limited-use cables shall be listed as being suitable for use outside, for use in dwellings, and for use in raceways, and shall also be listed as being flame retardant. Type BLP cables shall be listed as being suitable for use in ducts, plenums, and other spaces for environmental air and shall also be listed as having adequate fire-resistant and low smoke-producing characteristics.

FPN No. 1: One method of determining that cable is flame retardant is by testing the cable to VW-1 (vertical-wire) flame test in the *Reference Standard for Electrical Wires, Cables and Flexible Cords,* ANSI/UL 1581-1991.

See the commentary following Section 725-71(d), FPN.

FPN No. 2: One method of defining low smoke-producing cable is by establishing an acceptable value of the smoke produced when tested in accordance with *Standard Method of Test for Fire and Smoke Characteristics of Wires and Cables,* NFPA 262-1994, to a maximum peak optical density of 0.5 and a maximum average optical density of 0.15. Similarly, one method of defining fire-resistant cables is by establishing maximum allowable flame travel distance of 5 ft (1.52 m) when tested in accordance with the same test.

See the commentary following Section 725-71(a), FPN.

830-6. Access to Electrical Equipment Behind Panels Designed to Allow Access. Access to equipment shall not be denied by an accumulation of cables that prevents removal of panels, including suspended ceiling panels.

The excess accumulation of wires and cables can limit access to equipment by preventing the removal of access panels. (See Figures 830.2 and 830.3.)

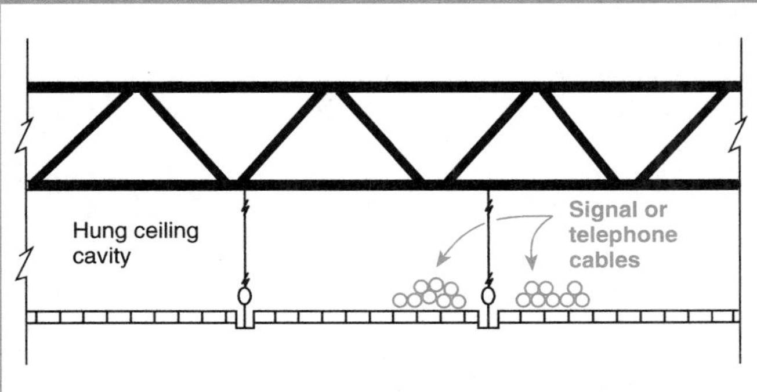

Figure 830.2 Excess accumulation of conductors and cables prevents access to equipment or cables. Incorrect methods shown.

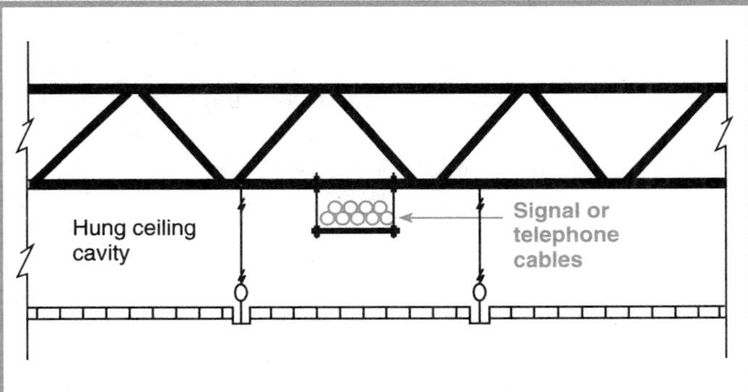

Figure 830.3 Excess accumulation of conductors and cables prevents access to equipment or cables. Correct methods shown.

830-7. Mechanical Execution of Work. Network-powered broadband communications circuits and equipment shall be installed in a neat and workmanlike manner. Cables shall be supported by the building structure in such a manner that the cable will not be damaged by normal building use.

FPN: One way to determine accepted industry practice is to refer to nationally recognized standards such as *Commercial Building Telecommunications Cabling Standard,* ANSI/ EIA/TIA 568-A-1995; *Commercial Building Standard for Telecommunications Pathways and Spaces,* ANSI/TIA/EIA 569-A-1997; and *Residential and Light Commercial Telecommunications Wiring Standard,* ANSI/EIA/TIA 570-1991, or other ANSI-approved installation standards.

The fine print note to Section 830-7 makes it clear that there are industry standards.

830-8. Hazardous (Classified) Locations. Network-powered broadband communications circuits and equipment installed in a location that is classified in accordance with Article 500 shall comply with the applicable requirements of Chapter 5.

B. Cables Outside and Entering Buildings

830-9. Entrance Cables. Cables installed outdoors shall be listed as suitable for the application. In addition, network-powered broadband communications cables located outside and entering buildings shall comply with (a) and (b).

(a) Medium Power Circuits. Medium power network-powered broadband communications circuits located outside and entering buildings shall be installed using Type BMU, Type BM, or Type BMR network-powered broadband communications medium power cables.

(b) Low Power Circuits. Low power network-powered broadband communications circuits located outside and entering buildings shall be installed using Type BLU or Type BLX low power network-powered broadband communications cables. Cables shown in Table 830-58 shall be permitted to substitute.

Exception: Outdoor community antenna television and radio distribution system coaxial cables installed prior to January 1, 2000, and installed in accordance with Article 820, shall be permitted for low-power type, network-powered broadband communications circuits.

830-10. Aerial Cables. Aerial network-powered broadband communications cables shall comply with (a) through (g).

(a) On Poles. Where practicable, network-powered broadband communications cables on poles shall be located below the electric light, power, Class 1, or nonpower-limited fire alarm circuit conductors and shall not be attached to a crossarm that carries electric light or power conductors.

(b) Climbing Space. The climbing space through network-powered broadband communications cables shall comply with the requirements of Section 225-14(d).

(c) Lead-in Clearance. Lead-in or aerial-drop network-powered broadband communications cables from a pole or other support, including the point of initial attachment to a building or structure, shall be kept away from electric light, power, Class 1, or nonpower-limited fire alarm circuit conductors so as to avoid the possibility of accidental contact.

Exception: Where proximity to electric light, power, Class 1, or nonpower-limited fire alarm circuit service conductors cannot be avoided, the installation shall be such as to provide clearances of not less than 12 in. (305 mm) from light, power, Class 1, or nonpower-limited fire alarm circuit service drops.

(d) Clearance from Ground. Overhead spans of network-powered broadband communication cables shall conform to not less than the following:

9.5 ft (2.89 m) — above finished grade, sidewalks, or from any platform or projection from which they might be reached and accessible to pedestrians only

11.5 ft (3.50 m) — over residential property and driveways, and those commercial areas not subject to truck traffic

15.5 ft (4.72 m) — over public streets, alleys, roads, parking areas subject to truck traffic, driveways on other than residential property, and other land traversed by vehicles such as cultivated, grazing, forest, and orchard

FPN: These clearances have been specifically chosen to correlate with the *National Electrical Safety Code,* ANSI C2-1997, Table 232-1, which provides for clearances of wires, conductors, and cables above ground and roadways, rather than using the clearances referenced in Section 225-18. Because Article 800 and Article 820 have had no required clearances, the communications industry has used the clearances from the NESC for their installed cable plant.

Network-powered broadband communications systems contain sufficient energy to shock and kill. For that reason, they are subjected to requirements similar to those for other high-powered circuits.

Conductor spans should be of sufficient size and strength to maintain clearances and avoid possible contact with light or power conductors. Splices and joints should be made with approved connectors or other means that provide sufficient mechanical strength so that conductors are not weakened appreciably, which could cause them to break and come into contact with higher-voltage conductors.

(e) Over Pools. Clearance of network-powered broadband communications cable in any direction from the water level, edge of pool, base of diving platform, or anchored raft shall be not less than 22 ft (6.71 m). Clearance in any direction to the diving platform or tower shall be not less than 14 ft (4.27 m).

Exception No. 1: Where the pool is fully enclosed by a solid or screened permanent structure.

FPN: These clearances have been specifically chosen to correlate with Section 680-8 and to incorporate the more stringent clearance requirements of the *National Electrical Safety Code,* ANSI C2-1997, Table 234-3, which provides for clearances of wires, conductors, and cables over pools.

Exception No. 2: Where the cables are located more than 10 ft (3.05 m) horizontally from the following:

(1) The edge of the pool
(2) Diving structure or diving tower, or
(3) Observation stands, towers, or platforms

(f) Above Roofs. Network-powered broadband communications cables shall have a vertical clearance of not less than 8 ft (2.44 m) from all points of roofs above which they pass.

Exception No. 1: Auxiliary buildings such as garages and the like.

Exception No. 2: A reduction in clearance above only the overhanging portion of the roof to not less than 18 in. (457 mm) shall be permitted if (1) not more than 4 ft (1.22 m) of the broadband communications drop cables pass above the roof overhang, and (2) they are terminated at a through-the-roof raceway or support.

Exception No. 3: Where the roof has a slope of not less than 4 in. (102 mm) in 12 in. (305 mm), a reduction in clearance to not less than 3 ft (914 mm) shall be permitted.

(g) Final Spans. Final spans of network-powered broadband communications cables without an outer jacket shall be permitted to be attached to the building, but they shall be kept not less than 3 ft (914 mm) from windows that are designed to be opened, doors, porches, balconies, ladders, stairs, fire escapes, or similar locations.

Exception: Conductors run above the top level of a window shall be permitted to be less than the 3-ft (914-mm) requirement above.

Overhead network-powered broadband communications cables shall not be installed beneath openings through which materials may be moved, such as openings in farm and commercial buildings, and shall not be installed where they will obstruct entrance to these building openings.

(h) Between Buildings. Network-powered broadband communications cables extending between buildings and also the supports or attachment fixtures shall be acceptable for the purpose and shall have sufficient strength to withstand the loads to which they may be subjected.

Wind and ice loads can sometimes be excessive and should be considered.

Exception: Where a network-powered broadband communications cable does not have sufficient strength to be self-supporting, it shall be attached to a supporting messenger cable that, together with the attachment fixtures or supports, shall be acceptable for the purpose and shall have sufficient strength to withstand the loads to which they may be subjected.

(i) On Buildings. Where attached to buildings, network-powered broadband communications cables shall be securely fastened in such a manner that they will be separated from other conductors as follows.

(1) Electric Light or Power. The network-powered broadband communications cable shall have a separation of at least 4 in. (102 mm) from electric light, power, Class 1, or nonpower-limited fire alarm circuit conductors not in raceway or cable, or be permanently separated from conductors of the other system by a continuous and firmly fixed nonconductor in addition to the insulation on the wires.

(2) Other Communications Systems. Network-powered broadband communications cables shall be installed so that there will be no unnecessary interference in the maintenance of the separate systems. In no case shall the conductors, cables, messenger strand, or equipment of one system cause abrasion to the conductors, cables, messenger strand, or equipment of any other system.

(3) Lightning Conductors. Where practicable, a separation of at least 6 ft (1.83 m) shall be maintained between any network-powered broadband communications cable and lightning conductors.

(4) Protection from Damage. Network-powered broadband communications cables attached to buildings and located within 8 ft (2.44 m) of finished grade shall be protected by enclosures, raceways, or other approved means.

Exception: A low power network-powered broadband communications circuit that is equipped with a listed fault protection device, appropriate to the network-powered broadband communications cable used, and located on the network side of the network-powered broadband communications cable being protected.

830-11. Underground Circuits Entering Buildings.

(a) Underground Systems. Underground network-powered broadband communications cables in a duct, pedestal, handhole, or manhole that contains electric light, power conductors, nonpower-limited fire alarm circuit conductors or Class 1 circuits shall be in a section permanently separated from such conductors by means of a suitable barrier.

(b) Direct-Buried Cables and Raceways. Direct-buried network-powered broadband communications cables shall be separated at least 12 in. (305 mm) from conductors of any light, power, nonpower-limited fire alarm circuit conductors or Class 1 circuit.

Exception No. 1: Where electric service conductors or network-powered broadband communications cables are installed in raceways or have metal cable armor.

Exception No. 2: Where electric light or power branch-circuit or feeder conductors, nonpower-limited fire alarm circuit conductors, or Class 1 circuit conductors are installed in a raceway or in metal-sheathed, metal-clad, or Type UF or Type USE cables; or the network-powered broadband communications cables have metal cable armor or are installed in a raceway.

(c) Mechanical Protection. Direct-buried cable, conduit, or other raceways shall be installed to meet the minimum cover requirements of Table 830-11. In addition, direct-buried cables emerging from the ground shall be protected by enclosures, raceways, or other approved means extending from the minimum cover distance required by Table 830-11 below grade to a point at least 8 ft (2.44 m) above finished grade. In no case shall the protection be required to exceed 18 in. (457 mm) below finished grade. Type BMU and BLU direct-buried cables emerging from the ground shall be installed in rigid metal conduit, intermediate metal conduit, rigid nonmetallic conduit, or other approved means extending from the minimum cover distance required by Table 830-11 below grade to the point of entrance.

Exception: A low power network-powered broadband communications circuit that is equipped with a listed fault protection device, appropriate to the network-powered broadband communications cable used, and located on the network side of the network-powered broadband communications cable being protected.

(d) Pools. Underground cables shall not be permitted under the pool or within the area extending 5 ft (1.52 m) horizontally from the inside wall of the pool.

Exception: Where space limitations prevent cables from being routed 5 ft (1.52 m) or more from the pool, such wiring shall be permitted where installed in rigid metal conduit, intermediate metal conduit, or a nonmetallic raceway system. All metal conduit shall be corrosion resistant and suitable for the location. The minimum burial depth shall be as follows:

Wiring Method	*Minimum Burial (in.)*
Rigid metal conduit	*6*
Intermediate metal conduit	*6*
Nonmetallic raceways listed for direct burial	*18*
*Other approved raceways**	*18*

Note: For SI units, 1 in. = 25.4 mm.

**Raceways approved for burial only where concrete encased shall require a concrete envelope not less than 2 in. (50.8 mm) thick.*

C. Protection

830-30. Primary Electrical Protection.

(a) Application. Primary electrical protection shall be provided on all network-powered broadband communications conductors that are neither grounded nor interrupted and are run partly or entirely in aerial cable not confined within a block. Also, primary electrical protection shall be provided on all aerial or underground network-powered broadband communications conductors that are neither grounded nor interrupted and are located within the block containing the building served so as to be exposed to lightning or accidental contact with electric light or power conductors operating at over 300 volts to ground.

Exception: Where electrical protection is provided on the derived circuit(s) (output side of the NIU) in accordance with Section 830-30(b)(3).

FPN No. 1: On network-powered broadband communications conductors not exposed to lightning or accidental contact with power conductors, providing primary electrical protection in accordance with this article will help protect against other hazards, such as ground potential rise caused by power fault cur-

Table 830-11. Network-Powered Broadband Communications Systems Minimum Cover Requirements, Burial in Inches (*Cover* is the shortest distance measured between a point on the top surface of any direct-buried cable, conduit, or other raceway and the top surface of finished grade, concrete, or similar cover.)

Location of Wiring Method or Circuit	Direct Burial Cables (in.)	Rigid Metal Conduit or Intermediate Metal Conduit (in.)	Nonmetallic Raceways Listed for Direct Burial; Without Concrete Encasement or Other Approved Raceways (in.)
All locations not specified below	18	6	12
In trench below 2-in. thick concrete or equivalent	12	6	6
Under a building (in raceway only)	0	0	0
Under minimum of 4-in. thick concrete exterior slab with no vehicular traffic and the slab extending not less than 6 in. beyond the underground installation	12	4	4
One- and two-family dwelling driveways and outdoor parking areas and used only for dwelling related purposes	12	12	12

Notes:

1. For SI units, 1 in. = 25.4 mm.
2. Raceways approved for burial only where concrete encased shall require a concrete envelope not less than 2 in. (50.8 mm) thick.
3. Lesser depths shall be permitted where cables rise for terminations or splices or where access is otherwise required.
4. Where solid rock is encountered, all wiring shall be installed in metal or nonmetallic raceway permitted for direct burial. The raceways shall be covered by a minimum of 2 in. (50.8 mm) of concrete extending down to rock.
5. Low power network-powered broadband communications circuits using directly buried community antenna television and radio distribution system coaxial cables that were installed outside and entering buildings prior to January 1, 2000, in accordance with Article 820 shall be permitted where buried to a minimum depth of 12 in. (30.48 cm).

rents, and above-normal voltages induced by fault currents on power circuits in proximity to the network-powered broadband communications conductors.

FPN No. 2: Network-powered broadband communications circuits are considered to have a lightning exposure unless one or more of the following conditions exist.

(1) Circuits in large metropolitan areas where buildings are close together and sufficiently high to intercept lightning.
(2) Interbuilding cable runs of 140 ft (42.7 m) or less, directly buried or in underground conduit, where a continuous metallic cable shield or a continuous metallic conduit containing the cable is bonded to each building grounding electrode system.
(3) Areas having an average of five or fewer thunderstorm days per year and earth resistivity of less than 100 ohmmeters. Such areas are found along the Pacific coast.

The utility companies may provide primary protectors if conductors are exposed to lightning. Generally, cables are not considered to be exposed to lightning unless one or more of the conditions in FPN No. 2 exists. A primary protector is required at each end of a communications circuit if lightning exposure exists, unless protection is provided on the output side of the network interface unit (NIU).

(1) Fuseless Primary Protectors. Fuseless-type primary protectors shall be permitted where power fault currents on all protected conductors in the cable are safely limited to a value no greater than the current-carrying capacity of the primary protector and of the primary protector grounding conductor.

(2) Fused Primary Protectors. Where the requirements listed under (1) are not met, fused-type primary protectors shall be used. Fused-type primary protectors shall consist of an arrester connected between each conductor to be protected and ground, a fuse in series with each conductor to be protected, and an appropriate mounting arrangement. Fused primary protector terminals shall be marked to indicate line, instrument, and ground, as applicable.

(b) Location. The location of the primary protector, where required, shall comply with (1), (2), or (3).

(1) A listed primary protector shall be applied on each network-powered broadband communications cable external to and on the network side of the network interface unit.

(2) The primary protection function shall be an integral part of and contained in the network interface unit. The network interface unit shall be listed for the purpose and shall have an external marking indicating that it contains primary electrical protection.

(3) The primary protector(s) shall be provided on the derived circuit(s) (output side of the NIU), and the combination of the NIU and the protector(s) shall be listed for the purpose.

A primary protector, whether provided integrally or external to the network interface unit, shall be located as close as practicable to the point of entrance.

For purposes of this section, a network interface unit and any externally provided primary protectors located at mobile home service equipment located in sight from and not more than 30 ft (9.14 m) from the exterior wall of the mobile home it serves, or at a mobile home disconnecting means grounded in accordance with Section 250-32 and located in sight from and not more than 30 ft (9.14 m) from the exterior wall of the mobile home it serves, shall be considered to meet the requirements of this section.

FPN: Selecting a network interface unit and primary protector location to achieve the shortest practicable primary protector grounding conductor will help limit potential differences between communications circuits and other metallic systems.

(c) Hazardous (Classified) Locations. The primary protector or equipment providing the primary protection function shall not be located in any hazardous (classified) location as defined in Article 500 or in the vicinity of easily ignitible material.

Exception: As permitted in Sections 501-14, 502-14, and 503-12.

830-33. Grounding or Interruption of Metallic Members of Network-Powered Broadband Communications Cables. The shields of network-powered broadband communications cables used for communications or powering shall be grounded at the building as close to the point of entrance as practicable. Metallic cable members not used for communications or powering shall be grounded or interrupted by an insulating joint or equivalent device as close to the point of entrance as practicable.

For purposes of this section, grounding or interruption of network-powered broadband communications cable metallic members installed at mobile home service equipment located in sight from and no more than 30 ft (9.14 m) from the exterior wall of the mobile home it serves, or at a mobile home disconnecting means grounded in accordance with Section 250-24 and located in sight from and not more than 30 ft (9.14 m) from the exterior wall of the mobile home it serves, shall be considered to meet the requirements of this section.

FPN: Selecting a grounding location to achieve the shortest practicable grounding conductor will help limit potential differences between the network-powered broadband communications circuits and other metallic systems.

D. Grounding Methods

830-40. Cable, Network Interface Unit, and Primary Protector Grounding. Network interface units containing protectors, NIUs with metallic enclosures, primary protectors, and the grounded metallic members of the network-powered broadband communications cable shall be grounded as specified in (a) through (d).

(a) Grounding Conductor.

(1) Insulation. The grounding conductor shall be insulated and shall be listed as suitable for the purpose.

(2) Material. The grounding conductor shall be copper or other corrosion-resistant conductive material, stranded or solid.

(3) Size. The grounding conductor shall not be smaller than No. 14, and shall have a current-carrying capacity approximately equal to that of the grounded metallic member(s) and protected conductor(s) of the network-powered broadband communications cable. The grounding conductor shall not be required to exceed No. 6.

(4) Run in Straight Line. The grounding conductor shall be run to the grounding electrode in as straight a line as practicable.

(5) Physical Protection. Where subject to physical damage, the grounding conductor shall be adequately protected. Where the grounding conductor is run in a metal raceway, both ends of the raceway shall be bonded to the grounding conductor or the same terminal or electrode to which the grounding conductor is connected.

(b) Electrode. The grounding conductor shall be connected as follows.

(1) To the nearest accessible location on the following:

(a) The building or structure grounding electrode system as covered in Section 250-50
(b) The grounded interior metal water piping system as covered in Section 250-104(a)
(c) The power service accessible means external to enclosures as covered in Section 250-92(b)
(d) The metallic power service raceway
(e) The service equipment enclosure
(f) The grounding electrode conductor or the grounding electrode conductor metal enclosure, or
(g) To the grounding conductor or to the grounding electrode of a building or structure disconnecting means that

is grounded to an electrode as covered in Section 250-32

For purposes of this section, the mobile home service equipment or the mobile home disconnecting means, as described in Section 830-33, shall be considered accessible.

(2) If the building or structure served has no grounding means as described in (b)(1), to any one of the individual electrodes described in Section 250-50; or

(3) If the building or structure served has no grounding means, as described in (b)(1) or (b)(2), to an effectively grounded metal structure or to a ground rod or pipe not less than 5 ft (1.52 m) in length and ½ in. (12.7 mm) in diameter, driven, where practicable, into permanently damp earth and separated from lightning conductors as covered in Section 800-13 and at least 6 ft (1.83 m) from electrodes of other systems. Steam or hot water pipes or lightning rod conductors shall not be employed as electrodes for protectors, NIUs with integral protection, grounded metallic members, NIUs with metallic enclosures, and other equipment.

(c) Electrode Connection. Connections to grounding electrodes shall comply with Section 250-70. Connectors, clamps, fittings, or lugs used to attach grounding conductors and bonding jumpers to grounding electrodes or to each other that are to be concrete encased or buried in the earth shall be suitable for its application.

(d) Bonding of Electrodes. A bonding jumper not smaller than No. 6 copper or equivalent shall be connected between the network-powered broadband communications system grounding electrode and the power grounding electrode system at the building or structure served where separate electrodes are used.

Exception: At mobile homes as covered in Section 830-42.

FPN No. 1: See Section 250-60 for use of lightning rods.

FPN No. 2: Bonding together of all separate electrodes will limit potential differences between them and between their associated wiring systems.

The most common error made in grounding network-powered broadband communications systems is to connect the cable sheath to a ground rod driven by the utility installer at a convenient location near the point of cable entry to the building and not bonding it to the electrical service grounding electrode system, service raceway, or other components that make up the grounding electrode system. A separate grounding electrode is permitted by the *Code* only if the building or structure has none of the grounding means described in Section 820-40(b). The *Code* requires that some means be provided that are accessible and external to the service equipment for making the bonding and grounding connection for other systems. One of the following means must be provided:

1. Exposed service raceways
2. Exposed grounding electrode conductor
3. An approved means for the external connection of a conductor (A No. 6 copper conductor with one end bonded to the service raceway or equipment with about 6 in. exposed is acceptable.)

Proper bonding of the network-powered broadband communications system cable sheath to the electrical power ground is needed to prevent potential fire and shock hazards. The earth cannot be used as an equipment grounding conductor or bonding conductor, as it does not have the low-impedance path required. (See Section 250-54.)

Both network-powered broadband communications systems and power systems are subject to current surges as a result, for example, of induced voltages from lightning in the vicinity of the usually extensive outside distribution systems. Surges also result from switching operations on power systems. If the grounded conductors and parts of the two systems are not bonded by a low-impedance path, such line surges can raise the potential difference between the two systems to many thousands of volts. This can result in arcing between the two systems, for example, wherever the coaxial cable jacket contacts a grounded part, such as a metal water pipe or metal structural member, inside the building.

If an individual is the interface between the two systems and the bonding has not been accomplished in accordance with the *Code,* the high-voltage surge could result in electric shock. More common, however, is burnout of the television tuner, as this part is almost always an interface between the two systems. The tuner is connected to the power system ground through the grounded neutral of the power supply, even if the television itself is not provided with an equipment grounding conductor.

Also see the commentary following Sections 250-92(b), FPN No. 2, and 820-33(b).

830-42. Bonding and Grounding at Mobile Homes.

(a) Grounding. Where there is no mobile home service equipment located in sight from and not more than 30 ft (9.14 m) from the exterior wall of the mobile home it serves, or there is no mobile home disconnecting means grounded in accordance with Section 250-32 and located within sight from and not more than 30 ft (9.14 m) from the exterior wall of the mobile home it serves, the network-powered broadband communications cable, network interface unit, and primary protector ground, shall be installed in accordance with Sections 830-40(b)(2) and (3).

(b) Bonding. The network-powered broadband communications cable grounding terminal, network interface unit grounding terminal, if present, and primary protector grounding terminal shall be bonded together with a copper bonding conductor not smaller than No. 12. The network-powered broadband communications cable grounding terminal, network interface unit grounding terminal, primary protector grounding terminal, or the grounding electrode shall be bonded to the metal frame or available grounding terminal of the mobile home with a copper bonding conductor not smaller than No. 12 under any of the following conditions:

(1) Where there is no mobile home service equipment or disconnecting means as in (a), or
(2) Where the mobile home is supplied by cord and plug

E. Network-Powered Broadband Communications Systems Wiring Methods Within Buildings

830-54. Medium Power Network-Powered Broadband Communications System Wiring Methods. Medium power network-powered broadband communications systems shall be installed within buildings using listed Type BM or Type BMR, network-powered broadband communications medium power cables.

(a) Ducts, Plenums, and Other Air-Handling Spaces. Section 300-22 shall apply.

Section 300-22 requires the appropriate methods of Chapter 3 to be used for the installation of medium power network-powered broadband communications circuits in plenums and other spaces used for environmental air within buildings. There is no plenum-rated cable for medium power network-powered broadband communications circuits. However, if the wiring downstream of the NIU is covered by other articles, then listed cables may be available.

(b) Riser. Cables installed in vertical runs and penetrating more than one floor, or cables installed in vertical runs in a shaft, shall be Type BMR. Floor penetrations requiring Type BMR shall contain only cables suitable for riser or plenum use.

Exception No. 1: Type BM cables encased in metal raceway or located in a fireproof shaft that has firestops at each floor.

Exception No. 2: Type BM cables in one- and two-family dwellings.

(c) Other Wiring. Cables installed in locations other than the locations covered in (a) and (b) shall be Type BM.

Exception: Type BMU cable where the cable enters the building from the outside and is run in rigid metal conduit or intermediate metal conduit, and such conduits are grounded to an electrode in accordance with Section 830-40(b).

830-55. Low Power Network-Powered Broadband Communications System Wiring Methods. Low power network-powered broadband communications systems shall be installed within buildings using listed Type BLX or Type BLP network-powered broadband communications low power cables.

(a) Ducts, Plenums, and Other Air-Handling Spaces. Cables installed in ducts, plenums, and other spaces used for environmental air shall be Type BLP.

Section 830-55(a) permits the use of listed Type BLP cable for low power network-powered broadband communications circuits within plenums and other spaces used for environmental air within buildings.

Exception: Type BLX cables installed in compliance with Section 300-22.

(b) Riser. Cables installed in vertical runs and penetrating more than one floor, or cables installed in vertical runs in a shaft, shall be Type BLP or Type BMR. Floor penetrations requiring Type BMR shall contain only cables suitable for riser or plenum use.

Exception No. 1: Type BLX cables encased in metal raceway or located in a fireproof shaft that has firestops at each floor.

Exception No. 2: Type BLX cable that is less than 0.375 in. (9.52 mm) in diameter in one- and two-family dwellings.

(c) Other Wiring. Cables installed in locations other than the locations covered in (a) and (b) shall be Type BLP or Type BM.

Exception No. 1: Type BLX cable enclosed in raceway.

Exception No. 2: Type BLU cables where the cable enters the building from the outside and is run in rigid metal conduit or intermediate metal conduit, and such conduits are grounded to an electrode in accordance with Section 830-40(b).

Exception No. 3: Type BLX cable that is less than 0.375 in. (9.52 mm) in diameter in one- and two-family dwellings.

Exception No. 4: Type BLX cable where the length of cable within the building does not exceed 50 ft (15.2 m) and the cable enters the building from outside and is terminated at a grounding block or a primary protection location.

FPN: This exception limits the length of Type BLX cable to 50 ft (15.2 m), while Section 830-30(b) requires that the primary protector, or NIU with integral protection, be located as close as practicable to the point at which the cable enters the building. Therefore, in installations requiring a primary protector, or NIU with integral protection, Type BLX cable may not be permitted to extend 50 ft (15.2 m) into the building if it is practicable to place the primary protector closer than 50 ft (15.2 m) to the entrance point.

830-56. Protection Against Physical Damage. Section 300-4 shall apply.

830-57. Bends. Bends in network broadband cable shall be made so as not to damage the cable.

830-58. Installation of Network-Powered Broadband Communications Cables and Equipment. Cable and equipment installations within buildings shall comply with (a) through (e), as applicable.

(a) Separation of Conductors.

(1) In Raceways and Enclosures.

(a) *Low and Medium Power Network-Powered Broadband Communications Circuit Cables.* Low and medium power network-powered broadband communications cables shall be permitted in the same raceway or enclosure.

(b) *Low Power Network-Powered Broadband Communications Circuit Cables.* Low power network-powered broadband communications cables shall be permitted in the same raceway or enclosure with jacketed cables of any of the circuits shown below:

(1) Class 2 and Class 3 remote-control, signaling, and power-limited circuits in compliance with Article 725
(2) Power-limited fire alarm systems in compliance with Article 760
(3) Communications circuits in compliance with Article 800
(4) Nonconductive and conductive optical fiber cables in compliance with Article 770
(5) Community antenna television and radio distribution systems in compliance with Article 820

(c) *Medium Power Network-Powered Broadband Communications Circuit Cables.* Medium power network-powered broadband communications cables shall not be permitted in the same raceway or enclosure with conductors of any of the circuits shown below:

(1) Class 2 and Class 3 remote-control, signaling, and power-limited circuits in compliance with Article 725
(2) Power-limited fire alarm systems in compliance with Article 760
(3) Communications circuits in compliance with Article 800
(4) Conductive optical fiber cables in compliance with Article 770
(5) Community antenna television and radio distribution systems in compliance with Article 820

(d) *Electric Light, Power, Class 1, Nonpower-Limited Fire Alarm Circuits.* Network-powered broadband communications cable shall not be placed in any raceway, compartment, outlet box, junction box, or similar fittings with conductors of electric light, power, Class 1, or nonpower-limited fire alarm circuit cables.

Exception No. 1: Where all of the conductors of electric light, power, Class 1, nonpower-limited fire alarm circuits are separated from all of the network-powered broadband communications cables by a barrier.

Exception No. 2: Power circuit conductors in outlet boxes, junction boxes, or similar fittings or compartments where such conductors are introduced solely for power supply to the network-powered broadband communications system distribution equipment. The power circuit conductors shall be routed within the enclosure to maintain a minimum 0.25-in. (6.35-mm) separation from network-powered broadband communications cables.

(2) Other Applications. Network-powered broadband communications cable shall be separated at least 2 in. (50.8 mm) from conductors of any electric light, power, Class 1, and nonpower-limited fire alarm circuits.

Exception No. 1: Where either (1) all of the conductors of electric light, power, Class 1, and nonpower-limited fire alarm circuits are in a raceway, or in metal-sheathed, metal-clad, nonmetallic-sheathed, Type AC, or Type UF cables, or (2) all of the network-powered broadband communications cables are encased in raceway.

Exception No. 2: Where the network-powered broadband communications cables are permanently separated from the conductors of electric light, power, Class 1, and nonpower-limited fire alarm circuits by a continuous and firmly fixed nonconductor, such as porcelain tubes or flexible tubing, in addition to the insulation on the wire.

(b) Spread of Fire or Products of Combustion. Installations in hollow spaces, vertical shafts, and ventilation or air-handling ducts shall be made so that the possible spread of fire or products of combustion will not be substantially increased. Openings around penetrations through fire–resistance-rated walls, partitions, floors, or ceilings shall be firestopped using approved methods.

(c) Equipment in Other Space Used for Environmental Air. Section 300-22(c) shall apply.

(d) Support of Conductors. Raceways shall be used for their intended purpose. Network-powered broadband communications cables shall not be strapped, taped, or attached by any means to the exterior of any conduit or raceway as a means of support.

See the commentary following Section 300-11.

(e) Cable Substitutions. The substitutions for network-powered broadband cables listed in Table 830-58 shall be permitted. All cables in Table 830-58, other than network-powered broadband cables, shall be coaxial cables.

Table 830-58. Cable Substitutions

Cable Type	Permitted Cable Substitutions
BM	BMR
BLP	MPP, CMP, CL3P
BLX	MPP, CMP, CL3P, MPR, CMR, CL3R, MPG, MP, CMG, CM, CL3, CMX, CL3X, BMR, BM, BLP

FPN: The substitute multipurpose, communications, and Class 3 cables in Table 830-58 are only coaxial-type cables.

CHAPTER 9

Tables

Contents

Substantial changes were made to these tables in the 1996 *Code,* but very few changes were made for the 1999 edition.

Because most conduit or tubing presently manufactured has a different internal diameter, and in order to make the tables as accurate as possible, revised tables for the 1996 *Code* were developed that provide the diameter and the actual area of different raceways for 31, 40, 53, and 100 percent fill. In addition, the tabular format of the tables for dimensions of conductors has been changed to provide greater accuracy.

Appendix C contains 12 raceway conductor fill tables. These tables were expanded to provide specific and accurate raceway conductor fill information, with separate tables for both metallic- and nonmetallic-type conduit and tubing raceways, as well as flexible raceways. Many examples of how to use them have been included, both here and in Appendix C.

Table 1. Percent of Cross Section of Conduit and Tubing for Conductors

Number of Conductors	All Conductor Types
1	53
2	31
Over 2	40

FPN No. 1: Table 1 is based on common conditions of proper cabling and alignment of conductors where the length of the pull and the number of bends are within reasonable limits. It should be recognized that, for certain conditions, a larger size conduit or a lesser conduit fill should be considered.

Table 1 sets forth the absolute maximum fill permitted for conduit or tubing. FPN No. 1 is a warning that, in some cases, underfilling or oversizing of the conduit may be necessary.

FPN No. 2: When pulling three conductors or cables into a raceway, if the ratio of the raceway (inside diameter) to the conductor or cable (outside diameter) is between 2.8 and 3.2, jamming can occur. While jamming can occur when pulling four or more conductors or cables into a raceway, the probability is very low.

This note changed for the 1999 *Code*. Previously, it appeared as Note 10. New in the 1999 *Code,* this note now appears as a fine print note (FPN), as it contains explanatory text rather than mandatory requirements.

FPN No. 2 alerts the user that conductor jamming may occur during the installation (pulling) of conductors into a conduit even though 40 percent fill allowances are observed. During the installation of three conductors or cables into the raceway, it is possible that one conductor may slip between the other two conductors. This is more likely to take place at bends, where the raceway may be slightly oval.

Appendix C, Table C1, permits three No. 8 AWG conductors in a ½-in. EMT. Using this example, ½-in. EMT has an internal diameter (ID) of 0.622 in. (from Table 4), and a No. 8 THHN conductor has an outside diameter (OD) of 0.216 in. (from Table 5).

While the EMT has an internal diameter of 0.622 in. in straight runs, it may not be round at a bend, where one conductor may slip between the other two and cause a jam as the conductors exit the bend. In a straight run, assuming no variation in the EMT's internal diameter or the conductor's outside diameter, one conductor usually cannot slip between the other two, since 3 × 0.216 in. = 0.648 in., which is greater than the EMT's internal diameter of 0.622 in. However, at a bend, the major internal diameter of the raceway may increase due to bending, particularly in tubing, to a diameter slightly larger than 0.648 in. permitting the middle conductor to be pulled between the outer two conductors. As the conductors exit the bend and the raceway's internal diameter returns to a normal circle of 0.622 in., the conductors may jam. This can also occur in straight runs where the ratio of the raceway's internal diameter to the conductor's outside diameter approaches 3. The jam ratio is calculated as follows:

$$\text{Jam ratio} = \frac{\text{ID of raceway}}{\text{OD of conductor}} = \frac{0.622}{0.216} = 2.88$$

Industry representatives recommend that a jam ratio of 2.8 to 3.2 should be avoided.

Notes to Tables

(1) See Appendix C for the maximum number of conductors and fixture wires, all of the same size (total cross-sectional area including insulation) permitted in trade sizes of the applicable conduit or tubing.

The percentage fill allowed for conduit and tubing has not changed in the 1996 or the 1999 *Code*. What was changed significantly in the 1996 edition was the addition of many new conduit fill tables for 12 types of conduit. In lieu of Table 3A, 3B, or 3C from editions of the *Code* prior to 1996, the user is offered Appendix C, asked to select the appropriate table, then cross reference the correct size and quantity of conductors in order to select the minimum trade size conduit or tubing. See the commentary in Appendix C for examples of how to use these tables.

(2) Table 1 applies only to complete conduit or tubing systems and is not intended to apply to sections of conduit or tubing used to protect exposed wiring from physical damage.

The maximum fill requirements do not apply to short sections of conduit or tubing used for the physical protection of conductors and cables. Cables are commonly protected from physical damage by conduit or tubing sleeves sized to enable the cable to be passed through with relative ease without injuring or abrading the protective jacket of the cable. However, a fitting is required on the end(s) of the conduit or tubing to protect the conductors or cables from abrasion. [See Section 300-15(c).]

(3) Equipment grounding or bonding conductors, where installed, shall be included when calculating conduit or tubing fill. The actual dimensions of the equipment grounding or bonding conductor (insulated or bare) shall be used in the calculation.

All conductors, whether insulated, covered, or bare, occupy space within a raceway. Therefore all conductors, including equipment grounding conductors, bonding conductors, and grounded conductors, must be counted when calculating the percent fill of a conduit or tubing. The only exception to this rule is the addition of an uninsulated grounding conductor permitted within $^3/_8$-in. flexible metal conduit (see footnote to Table 350-12). The dimensions of bare conductors are given in Table 8.

(4) Where conduit or tubing nipples having a maximum length not to exceed 24 in. (610 mm) are installed between boxes, cabinets, and similar enclosures, the nipples shall be permitted to be filled to 60 percent of their total cross-sectional area, and Section 310-15(b)(2)(a) adjustment factors need not apply to this condition.

(5) For conductors not included in Chapter 9, such as multiconductor cables, the actual dimensions shall be used.

For conductors not included in Chapter 9, such as high-voltage types, the cross-sectional area may be calculated in the following manner, using the actual dimensions of each conductor:

$$d^2\text{cm} = \text{cross-sectional area}$$

where:

d = outside diameter of a conductor (including insulation) [1 in. = 1000 mils (1 mil = 0.001 in.)]

cm = circular mils (a unit measure of area) = $\pi(3.1416)/4 = 0.7854$ square mils

Example

Three 15-kV single conductors are to be installed in conduit. The outside diameter of each conductor measures $1^5/_8$ in. ($1^5/_8$ in. = 1.625 in.).

$$1.625 \times 1.625 \times 0.7854 \times 3 = 6.2218 \text{ in.}^2$$

Therefore, the area within the cross section of the conduit that will be displaced by these three conductors is 6.222 in.^{2}

Table 1 allows 40 percent conduit fill for three or more conductors. Table 4 indicates that 40 percent of a 5-in. rigid metal conduit is 8.085 in.^{2}, thus accommodating three 15-kV single conductors.

(6) For combinations of conductors of different sizes, use Tables 5 and 5A for dimensions of conductors and Table 4 for the applicable conduit or tubing dimensions.

Two examples follow for computing the minimum trade size conduit or tubing required for mixed conductor sizes.

Example 1

A 200-ampere feeder is routed in various wiring methods [EMT, PVC (Schedule 40), and rigid metal conduit] from the main switchboard in one building to a second building. The circuit consists of four No. 4/0 XHHW conductors and one No. 6 XHHW. Select the proper trade size for the various types of conduit and tubing to be used for this feeder.

Answer

All the raceways for this example require conduit fill to be calculated according to Table 1 of Chapter 9. (See Section 346-7 for RMC, Section 347-11 for PVC, and Section 348-8 for EMT.) Table 1 of Chapter 9 permits a conduit fill to a maximum of 40 percent for over two conductors in a raceway. According to Note 6, the user is directed to Table 5 for the area required for each insulated conductor. Multiplying the quantity of conductors by the area of each conductor results in the total conductor-occupied space within the conduit. Then, according to Note 6, the user is directed to Table 4 to select the appropriate trade size conduits or tubing. Table 4 contains the applicable cross-sectional area of conduit and tubing based on conductor-occupied space (40 percent maximum in this example).

Step 1. No. 4/0 XHHW = 0.3197 in.^{2} in area (Table 5)

Step 2. No. 6 XHHW = 0.059 in.^{2} in area (Table 5)

Step 3. Calculate the total area occupied by the wires, as follows:

For four No. 4/0 XHHW,	4 × 0.3197 =	1.2788 in.^{2}
For one No. 6 XHHW,	1 × 0.059 =	0.059 in.^{2}
Total Area		1.3378 in.^{2}

The total area is 1.338 in.^{2} (rounded off).

Step 4. According to Table 4, the EMT portion of this feeder will require a minimum 2-in. trade size. Since 2-in. EMT has 1.342 in.^{2} of available space for over two conductors, and the minimum required space is 1.338 in.^{2}, this is within the 2-in. EMT limit.

Rigid metal conduit also requires a minimum trade size of 2 in., since 2-in. RMC has 1.363 in.^{2} of available space for over two conductors.

But, PVC (Schedule 40) will require a minimum $2^1/_2$-in. trade size conduit, since the 2-in. PVC has only 1.316 in.^{2} available space for over two conductors, and the circuit under consideration requires at least 1.338 in.^{2}

Example 2

Determine the minimum size rigid metal conduit allowed for the following mixed conductor sizes and types.

Quantity	Wire Size and Type	Table 5 Cross-Sectional Area (Each)	Subtotal Cross-Sectional Area
4	12 THWN	0.0133	0.0532
3	8 TW	0.0437	0.1311
3	6 THW	0.0726	0.2178
		Total cross-sectional area:	0.4021

The "Over 2 Wires" column of Table 4 indicates that 40 percent of a $1^1/_4$-in. rigid metal conduit is 0.610 in.^{2} Therefore, $1^1/_4$ in. is the minimum permitted conduit size for these 10 conductors.

(7) When calculating the maximum number of conductors permitted in a conduit or tubing, all of the same size (total cross-sectional area including insulation), the next higher whole number shall be used to determine the maximum number of conductors permitted when the calculation results in a decimal of 0.8 or larger.

Example

Determine how many No. 10 THHN conductors are permitted in a $1^1/_4$-in. rigid metal conduit.

Answer

Table 1 permits a 40 percent fill for over 2 conductors. Using Table 4, 40 percent fill for a $1^1/_4$-in. rigid metal conduit is 0.610 in., and, from Table 5, the cross-sectional area of a No. 10 THHN conductor is 0.0211 in.^{2}

$$\frac{0.610 \text{ in.}^2}{0.0211 \text{ in.}^2 \text{ per conductor}} = 28.910 \text{ conductors}$$

Without Note 7, the maximum quantity of conductors could not exceed 28. However, 29 conductors are permitted in a $1^1/_4$-in. rigid metal conduit. This is because the decimal 0.910 is more than 0.8, and the quantity of conductors may be increased to the next higher whole number. Appendix C, Table C8, verifies that 29 No. 10 THWN conductors are permitted in a $1^1/_4$-in. rigid metal conduit.

If the decimal is less than 0.8, the decimal is dropped.

(8) Where bare conductors are permitted by other sections of this *Code*, the dimensions for bare conductors in Table 8 shall be permitted.

(9) A multiconductor cable of two or more conductors shall be treated as a single conductor for calculating percentage conduit fill area. For cables that have elliptical cross sections, the cross-sectional area calculation shall be based on using the major diameter of the ellipse as a circle diameter.

Table 4. Dimensions and Percent Area of Conduit and Tubing (Areas of Conduit or Tubing for the Combinations of Wires Permitted in Table 1, Chapter 9)

Electrical Metallic Tubing

Trade Size (in.)	Internal Diameter (in.)	Total Area 100% ($in.^2$)	2 Wires 31% ($in.^2$)	Over 2 Wires 40% ($in.^2$)	1 Wire 53% ($in.^2$)
½	0.622	0.304	0.094	0.122	0.161
¾	0.824	0.533	0.165	0.213	0.283
1	1.049	0.864	0.268	0.346	0.458
1¼	1.380	1.496	0.464	0.598	0.793
1½	1.610	2.036	0.631	0.814	1.079
2	2.067	3.356	1.040	1.342	1.778
2½	2.731	5.858	1.816	2.343	3.105
3	3.356	8.846	2.742	3.538	4.688
3½	3.834	11.545	3.579	4.618	6.119
4	4.334	14.753	4.573	5.901	7.819

Electrical Nonmetallic Tubing

Trade Size (in.)	Internal Diameter (in.)	Total Area 100% ($in.^2$)	2 Wires 31% ($in.^2$)	Over 2 Wires 40% ($in.^2$)	1 Wire 53% ($in.^2$)
½	0.560	0.246	0.076	0.099	0.131
¾	0.760	0.454	0.141	0.181	0.240
1	1.000	0.785	0.243	0.314	0.416
1¼	1.340	1.410	0.437	0.564	0.747
1½	1.570	1.936	0.600	0.774	1.026
2	2.020	3.205	0.994	1.282	1.699
2½	—	—	—	—	—
3	—	—	—	—	—
3½	—	—	—	—	—
4	—	—	—	—	—

Flexible Metal Conduit

Trade Size (in.)	Internal Diameter (in.)	Total Area 100% ($in.^2$)	2 Wires 31% ($in.^2$)	Over 2 Wires 40% ($in.^2$)	1 Wire 53% ($in.^2$)
⅜	0.384	0.116	0.036	0.046	0.061
½	0.635	0.317	0.098	0.127	0.168
¾	0.824	0.533	0.165	0.213	0.282
1	1.020	0.817	0.253	0.327	0.433
1¼	1.275	1.277	0.396	0.511	0.677
1½	1.538	1.857	0.576	0.743	0.984
2	2.040	3.269	1.013	1.307	1.732
2½	2.500	4.909	1.522	1.964	2.602
3	3.000	7.069	2.191	2.827	3.746
3½	3.500	9.621	2.983	3.848	5.099
4	4.000	12.566	3.896	5.027	6.660

Intermediate Metal Conduit

Trade Size (in.)	Internal Diameter (in.)	Total Area 100% ($in.^2$)	2 Wires 31% ($in.^2$)	Over 2 Wires 40% ($in.^2$)	1 Wire 53% ($in.^2$)
⅜	—	—	—	—	—
½	0.660	0.342	0.106	0.137	0.181
¾	0.864	0.586	0.182	0.235	0.311
1	1.105	0.959	0.297	0.384	0.508
1¼	1.448	1.646	0.510	0.658	0.872
1½	1.683	2.223	0.689	0.889	1.178
2	2.150	3.629	1.125	1.452	1.923
2½	2.557	5.135	1.592	2.054	2.722
3	3.176	7.922	2.456	3.169	4.199
3½	3.671	10.584	3.281	4.234	5.610
4	4.166	13.631	4.226	5.452	7.224

Liquidtight Flexible Nonmetallic Conduit (Type LFNC-B*)

Trade Size (in.)	Internal Diameter (in.)	Total Area 100% ($in.^2$)	2 Wires 31% ($in.^2$)	Over 2 Wires 40% ($in.^2$)	1 Wire 53% ($in.^2$)
⅜	0.494	0.192	0.059	0.077	0.102
½	0.632	0.314	0.097	0.125	0.166
¾	0.830	0.541	0.168	0.216	0.287
1	1.054	0.872	0.270	0.349	0.462
1¼	1.395	1.528	0.474	0.611	0.810
1½	1.588	1.979	0.614	0.792	1.049
2	2.033	3.245	1.006	1.298	1.720

*Corresponds to Section 351-22(2).

Liquidtight Flexible Nonmetallic Conduit (Type LFNC-A*)

Trade Size (in.)	Internal Diameter (in.)	Total Area 100% ($in.^2$)	2 Wires 31% ($in.^2$)	Over 2 Wires 40% ($in.^2$)	1 Wire 53% ($in.^2$)
⅜	0.495	0.192	0.060	0.077	0.102
½	0.630	0.312	0.097	0.125	0.165
¾	0.825	0.535	0.166	0.214	0.283
1	1.043	0.854	0.265	0.341	0.452
1¼	1.383	1.501	0.465	0.600	0.796
1½	1.603	2.017	0.625	0.807	1.069
2	2.063	3.341	1.036	1.336	1.771

*Corresponds to Section 351-22(1).

Table 4. (Continued)

Liquidtight Flexible Metal Conduit

Trade Size (in.)	Internal Diameter (in.)	Total Area 100% (in.^{2})	2 Wires 31% (in.^{2})	Over 2 Wires 40% (in.^{2})	1 Wire 53% (in.^{2})
⅜	0.494	0.192	0.059	0.077	0.102
½	0.632	0.314	0.097	0.125	0.166
¾	0.830	0.541	0.168	0.216	0.287
1	1.054	0.872	0.270	0.349	0.462
1¼	1.395	1.528	0.474	0.611	0.810
1½	1.588	1.979	0.614	0.792	1.049
2	2.033	3.245	1.006	1.298	1.720
2½	2.493	4.879	1.513	1.952	2.586
3	3.085	7.475	2.317	2.990	3.962
3½	3.520	9.731	3.017	3.893	5.158
4	4.020	12.692	3.935	5.077	6.727
5	—	—	—	—	—
6	—	—	—	—	—

Rigid Metal Conduit

Trade Size (in.)	Internal Diameter (in.)	Total Area 100% (in.^{2})	2 Wires 31% (in.^{2})	Over 2 Wires 40% (in.^{2})	1 Wire 53% (in.^{2})
⅜	—	—	—	—	—
½	0.632	0.314	0.097	0.125	0.166
¾	0.836	0.549	0.170	0.220	0.291
1	1.063	0.888	0.275	0.355	0.470
1¼	1.394	1.526	0.473	0.610	0.809
1½	1.624	2.071	0.642	0.829	1.098
2	2.083	3.408	1.056	1.363	1.806
2½	2.489	4.866	1.508	1.946	2.579
3	3.090	7.499	2.325	3.000	3.975
3½	3.570	10.010	3.103	4.004	5.305
4	4.050	12.883	3.994	5.153	6.828
5	5.073	20.213	6.266	8.085	10.713
6	6.093	29.158	9.039	11.663	15.454

Rigid PVC Conduit, Schedule 80

Trade Size (in.)	Internal Diameter (in.)	Total Area 100% (in.^{2})	2 Wires 31% (in.^{2})	Over 2 Wires 40% (in.^{2})	1 Wire 53% (in.^{2})
½	0.526	0.217	0.067	0.087	0.115
¾	0.722	0.409	0.127	0.164	0.217
1	0.936	0.688	0.213	0.275	0.365
1¼	1.255	1.237	0.383	0.495	0.656
1½	1.476	1.711	0.530	0.684	0.907
2	1.913	2.874	0.891	1.150	1.1523
2½	2.290	4.119	1.277	1.647	2.183
3	2.864	6.442	1.997	2.577	3.414
3½	3.326	8.688	2.693	3.475	4.605
4	3.786	11.258	3.490	4.503	5.967
5	4.768	17.855	5.535	7.142	9.463
6	5.709	25.598	7.935	10.239	13.567

Rigid PVC Conduit, Schedule 40, and HDPE Conduit

Trade Size (in.)	Internal Diameter (in.)	Total Area 100% (in.^{2})	2 Wires 31% (in.^{2})	Over 2 Wires 40% (in.^{2})	1 Wire 53% (in.^{2})
½	0.602	0.285	0.088	0.114	0.151
¾	0.804	0.508	0.157	0.203	0.269
1	1.029	0.832	0.258	0.333	0.441
1¼	1.360	1.453	0.450	0.581	0.770
1½	1.590	1.986	0.616	0.794	1.052
2	2.047	3.291	1.020	1.316	1.744
2½	2.445	4.695	1.455	1.878	2.488
3	3.042	7.268	2.253	2.907	3.852
3½	3.521	9.737	3.018	3.895	5.161
4	3.998	12.554	3.892	5.022	6.654
5	5.016	19.761	6.126	7.904	10.473
6	6.031	28.567	8.856	11.427	15.141

Type A, Rigid PVC Conduit

Trade Size (in.)	Internal Diameter (in.)	Total Area 100% (in.^{2})	2 Wires 31% (in.^{2})	Over 2 Wires 40% (in.^{2})	1 Wire 53% (in.^{2})
½	0.700	0.385	0.119	0.154	0.204
¾	0.910	0.650	0.202	0.260	0.345
1	1.175	1.084	0.336	0.434	0.575
1¼	1.500	1.767	0.548	0.707	0.937
1½	1.720	2.324	0.720	0.929	1.231
2	2.155	3.647	1.131	1.459	1.933
2½	2.635	5.453	1.690	2.181	2.890
3	3.230	8.194	2.540	3.278	4.343
3½	3.690	10.694	3.315	4.278	5.668
4	4.180	13.723	4.254	5.489	7.273
5	—	—	—	—	—
6	—	—	—	—	—

Type EB, PVC Conduit

Trade Size (in.)	Internal Diameter (in.)	Total Area 100% (in.^{2})	2 Wires 31% (in.^{2})	Over 2 Wires 40% (in.^{2})	1 Wire 53% (in.^{2})
½	—	—	—	—	—
¾	—	—	—	—	—
1	—	—	—	—	—
1¼	—	—	—	—	—
1½	—	—	—	—	—
2	2.221	3.874	1.201	1.550	2.053
2½	—	—	—	—	—
3	3.330	8.709	2.700	3.484	4.616
3½	3.804	11.365	3.523	4.546	6.024
4	4.289	14.448	4.479	5.779	7.657
5	5.316	22.195	6.881	8.878	11.764
6	6.336	31.530	9.774	12.612	16.711

Table 5. Dimensions of Insulated Conductors and Fixture Wires

Type: AF, FFH-2, RFH-1, RFH-2, RH, RHH*, RHW*, RHW-2*, RHH, RHW, RHW-2, SF-1, SF-2, SFF-1, SFF-2, TF, TFF, THHW, THW, THW-2, TW, XF, XFF

Type	Size (AWG or kcmil)	Approximate Diameter (in.)	Approximate Area (in.^{2})
RFH-2,	18	0.136	0.0145
FFH-2	16	0.148	0.0172
RH	14	0.163	0.0209
	12	0.182	0.0260
RHW-2, RHH, RHW	14	0.193	0.0293
	12	0.212	0.0353
RH, RHH, RHW, RHW-2	10	0.236	0.0437
	8	0.326	0.0835
	6	0.364	0.1041
	4	0.412	0.1333
	3	0.440	0.1521
	2	0.472	0.1750
	1	0.582	0.2660
	1/0	0.622	0.3039
	2/0	0.668	0.3505
	3/0	0.720	0.4072
	4/0	0.778	0.4754
	250	0.895	0.6291
	300	0.950	0.7088
	350	1.001	0.7870
	400	1.048	0.8626
	500	1.133	1.0082
	600	1.243	1.2135
	700	1.314	1.3561
	750	1.348	1.4272
	800	1.380	1.4957
	900	1.444	1.6377
	1000	1.502	1.7719
	1250	1.729	2.3479
	1500	1.852	2.6938
	1750	1.966	3.0357
	2000	2.072	3.3719
SF-2, SFF-2	18	0.121	0.0115
	16	0.133	0.0139
	14	0.148	0.0172
SF-1, SFF-1	18	0.091	0.0065
RFH-1, AF, XF, XFF	18	0.106	0.0080
AF, TF, TFF, XF, XFF	16	0.118	0.0109
AF, TW, XF, XFF	14	0.133	0.0139
TW	12	0.152	0.0181
	10	0.176	0.0243
	8	0.236	0.0437
RHH*, RHW*, RHW-2*, THHW, THW, THW-2	14	0.163	0.0209

Table 5. (Continued)

Type: AF, RHH*, RHW*, RHW-2*, THHN, THHW, THW, THW-2, TFN, TFFN, THWN, THWN-2, XF, XFF

Type	Size (AWG or kcmil)	Approximate Diameter (in.)	Approximate Area (in.^{2})
RHH*, RHW*, RHW-2*,	12	0.182	0.0260
THHW, THW, AF, XF, XFF,	10	0.206	0.0333
RHH*, RHW*, RHW-2*, THHW, THW, THW-2	8	0.266	0.0556
TW, THW, THHW, THW-2, RHH*, RHW*, RHW-2*	6	0.304	0.0726
	4	0.352	0.0973
	3	0.380	0.1134
	2	0.412	0.1333
	1	0.492	0.1901
	1/0	0.532	0.2223
	2/0	0.578	0.2624
	3/0	0.630	0.3117
	4/0	0.688	0.3718
	250	0.765	0.4596
	300	0.820	0.5281
	350	0.871	0.5958
	400	0.918	0.6619
	500	1.003	0.7901
	600	1.113	0.9729
	700	1.184	1.1010
	750	1.218	1.1652
	800	1.250	1.2272
	900	1.314	1.3561
	1000	1.372	1.4784
	1250	1.539	1.8602
	1500	1.662	2.1695
	1750	1.776	2.4773
	2000	1.882	2.7818
TFN, TFFN	18	0.084	0.0055
	16	0.096	0.0072
THHN, THWN, THWN-2	14	0.111	0.0097
	12	0.130	0.0133
	10	0.164	0.0211
	8	0.216	0.0366
	6	0.254	0.0507
	4	0.324	0.0824
	3	0.352	0.0973
	2	0.384	0.1158
	1	0.446	0.1562
	1/0	0.486	0.1855
	2/0	0.532	0.2223
	3/0	0.584	0.2679
	4/0	0.642	0.3237
	250	0.711	0.3970
	300	0.766	0.4608

Table 5. (Continued)

Type: FEP, FEPB, PAF, PAFF, PF, PFA, PFAH, PFF, PGF, PGFF, PTF, PTFF, TFE, THHN, THWN, THWN-2, Z, ZF, ZFF			
Type	**Size (AWG or kcmil)**	**Approximate Diameter (in.)**	**Approximate Area ($in.^2$)**
THHN, THWN, THWN-2	350	0.817	0.5242
	400	0.864	0.5863
	500	0.949	0.7073
	600	1.051	0.8676
	700	1.122	0.9887
	750	1.156	1.0496
	800	1.188	1.1085
	900	1.252	1.2311
	1000	1.310	1.3478
PF, PGFF, PGF, PFF, PTF, PAF, PTFF, PAFF	18	0.086	0.0058
	16	0.098	0.0075
PF, PGFF, PGF, PFF, PTF, PAF, PTFF, PAFF, TFE, FEP, PFA, FEPB, PFAH	14	0.113	0.010
TFE, FEP, PFA, FEPB, PFAH	12	0.132	0.0137
	10	0.156	0.0191
	8	0.206	0.0333
	6	0.244	0.0468
	4	0.292	0.0670
	3	0.320	0.0804
	2	0.352	0.0973
TFE, PFAH	1	0.422	0.1399
TFE,PFA, PFAH, Z	1/0	0.462	0.1676
	2/0	0.508	0.2027
	3/0	0.560	0.2463
	4/0	0.618	0.3000
ZF, ZFF	18	0.076	0.0045
	16	0.088	0.0061
Z, ZF, ZFF	14	0.103	0.0083
Z	12	0.122	0.0117
	10	0.156	0.0191
	8	0.196	0.0302
	6	0.234	0.0430
	4	0.282	0.0625
	3	0.330	0.0855
	2	0.362	0.1029
	1	0.402	0.1269

Table 5. (Continued)

Type: KF-1, KF-2, KFF-1, KFF-2, XHH, XHHW, XHHW-2, ZW			
Type	**Size (AWG or kcmil)**	**Approximate Diameter (in.)**	**Approximate Area ($in.^2$)**
XHHW, ZW, XHHW-2, XHH	14	0.133	0.0139
	12	0.152	0.0181
	10	0.176	0.0243
	8	0.236	0.0437
	6	0.274	0.0590
	4	0.322	0.0814
	3	0.350	0.0962
	2	0.382	0.1146
XHHW, XHHW-2, XHH	1	0.442	0.1534
	1/0	0.482	0.1825
	2/0	0.528	0.2190
	3/0	0.58	0.2642
	4/0	0.638	0.3197
	250	0.705	0.3904
	300	0.76	0.4536
	350	0.811	0.5166
	400	0.858	0.5782
	500	0.943	0.6984
	600	1.053	0.8709
	700	1.124	0.9923
	750	1.158	1.0532
	800	1.190	1.1122
	900	1.254	1.2351
	1000	1.312	1.3519
	1250	1.479	1.7180
	1500	1.602	2.0157
	1750	1.716	2.3127
	2000	1.822	2.6073
KF-2, KFF-2	18	0.063	0.0031
	16	0.075	0.0044
	14	0.090	0.0064
	12	0.109	0.0093
	10	0.133	0.0139
KF-1, KFF-1	18	0.057	0.0026
	16	0.069	0.0037
	14	0.084	0.0055
	12	0.103	0.0083
	10	0.127	0.0127

*Types RHH, RHW, and RHW-2 without outer covering.

Table 5A. Compact Aluminum Building Wire Nominal Dimensions* and Areas

Size (AWG or kcmil)	Bare Conductor		Types THW and THHW		Type THHN		Type XHHW		Size (AWG or kcmil)
	Number of Strands	Diameter (in.)	Approximate Diameter (in.)	Approximate Area (in.^{2})	Approximate Diameter (in.)	Approximate Area (in.^{2})	Approximate Diameter (in.)	Approximate Area (in.^{2})	
8	7	0.134	0.255	0.0510	—	—	0.224	0.0394	8
6	7	0.169	0.290	0.0660	0.240	0.0452	0.260	0.0530	6
4	7	0.213	0.335	0.0881	0.305	0.0730	0.305	0.0730	4
2	7	0.268	0.390	0.1194	0.360	0.1017	0.360	0.1017	2
1	19	0.299	0.465	0.1698	0.415	0.1352	0.415	0.1352	1
1/0	19	0.336	0.500	0.1963	0.450	0.1590	0.450	0.1590	1/0
2/0	19	0.376	0.545	0.2332	0.495	0.1924	0.490	0.1885	2/0
3/0	19	0.423	0.590	0.2733	0.540	0.2290	0.540	0.2290	3/0
4/0	19	0.475	0.645	0.3267	0.595	0.2780	0.590	0.2733	4/0
250	37	0.520	0.725	0.4128	0.670	0.3525	0.660	0.3421	250
300	37	0.570	0.775	0.4717	0.720	0.4071	0.715	0.4015	300
350	37	0.616	0.820	0.5281	0.770	0.4656	0.760	0.4536	350
400	37	0.659	0.865	0.5876	0.815	0.5216	0.800	0.5026	400
500	37	0.736	0.940	0.6939	0.885	0.6151	0.880	0.6082	500
600	61	0.813	1.050	0.8659	0.985	0.7620	0.980	0.7542	600
700	61	0.877	1.110	0.9676	1.050	0.8659	1.050	0.8659	700
750	61	0.908	1.150	1.0386	1.075	0.9076	1.090	0.9331	750
1000	61	1.060	1.285	1.2968	1.255	1.2370	1.230	1.1882	1000

*Dimensions are from industry sources.

Most aluminum building wire in Types THW, THWN/THHN, and XHHW conductors is compact stranded. Table 5A provides appropriate dimensions for these types of wire.

Table 8. Conductor Properties

		Conductors				Direct-Current Resistance at 75°C (167°F)		
		Stranding		Overall		Copper		Aluminum
Size (AWG or kcmil)	Area (Circular Mils)	Quantity	Diameter (in.)	Diameter (in.)	Area (in.^{2})	Uncoated (ohm/1000 ft)	Coated (ohm/1000 ft)	(ohm/1000 ft)
18	1620	1	—	0.040	0.001	7.77	8.08	12.8
18	1620	7	0.015	0.046	0.002	7.95	8.45	13.1
16	2580	1	—	0.051	0.002	4.89	5.08	8.05
16	2580	7	0.019	0.058	0.003	4.99	5.29	8.21
14	4110	1	—	0.064	0.003	3.07	3.19	5.06
14	4110	7	0.024	0.073	0.004	3.14	3.26	5.17
12	6530	1	—	0.081	0.005	1.93	2.01	3.18
12	6530	7	0.030	0.092	0.006	1.98	2.05	3.25
10	10380	1	—	0.102	0.008	1.21	1.26	2.00
10	10380	7	0.038	0.116	0.011	1.24	1.29	2.04
8	16510	1	—	0.128	0.013	0.764	0.786	1.26
8	16510	7	0.049	0.146	0.017	0.778	0.809	1.28
6	26240	7	0.061	0.184	0.027	0.491	0.510	0.808
4	41740	7	0.077	0.232	0.042	0.308	0.321	0.508
3	52620	7	0.087	0.260	0.053	0.245	0.254	0.403
2	66360	7	0.097	0.292	0.067	0.194	0.201	0.319
1	83690	19	0.066	0.332	0.087	0.154	0.160	0.253
1/0	105600	19	0.074	0.372	0.109	0.122	0.127	0.201
2/0	133100	19	0.084	0.418	0.137	0.0967	0.101	0.159
3/0	167800	19	0.094	0.470	0.173	0.0766	0.0797	0.126
4/0	211600	19	0.106	0.528	0.219	0.0608	0.0626	0.100
250	—	37	0.082	0.575	0.260	0.0515	0.0535	0.0847
300	—	37	0.090	0.630	0.312	0.0429	0.0446	0.0707
350	—	37	0.097	0.681	0.364	0.0367	0.0382	0.0605
400	—	37	0.104	0.728	0.416	0.0321	0.0331	0.0529
500	—	37	0.116	0.813	0.519	0.0258	0.0265	0.0424
600	—	61	0.099	0.893	0.626	0.0214	0.0223	0.0353
700	—	61	0.107	0.964	0.730	0.0184	0.0189	0.0303
750	—	61	0.111	0.998	0.782	0.0171	0.0176	0.0282
800	—	61	0.114	1.030	0.834	0.0161	0.0166	0.0265
900	—	61	0.122	1.094	0.940	0.0143	0.0147	0.0235
1000	—	61	0.128	1.152	1.042	0.0129	0.0132	0.0212
1250	—	91	0.117	1.289	1.305	0.0103	0.0106	0.0169
1500	—	91	0.128	1.412	1.566	0.00858	0.00883	0.0141
1750	—	127	0.117	1.526	1.829	0.00735	0.00756	0.0121
2000	—	127	0.126	1.632	2.092	0.00643	0.00662	0.0106

Notes:

1. These resistance values are valid **only** for the parameters as given. Using conductors having coated strands, different stranding type, and, especially, other temperatures changes the resistance.
2. Formula for temperature change: $R_2 = R_1[1 + \alpha(T_2 - 75)]$ where: $\alpha_{cu} = 0.00323$, $\alpha_{AL} = 0.00330$
3. Conductors with compact and compressed stranding have about 9 percent and 3 percent, respectively, smaller bare conductor diameters than those shown. See Table 5A for actual compact cable dimensions.
4. The IACS conductivities used: bare copper = 100%, aluminum = 61%.
5. Class B stranding is listed as well as solid for some sizes. Its overall diameter and area is that of its circumscribing circle.

Traditionally, wire sizes have been expressed as either American Wire Gage (AWG), circular mil (cmil) area, or thousands of circular mil area (kcmil). Today, wire is available with its cross-sectional area expressed in square millimeters (mm^2) as well.

The 1999 *NEC* specifically requires that insulated conductors be marked with their sizes expressed in either American Wire Gage or in circular mil area. [See Sections 110-6 and 310-11(a)(4).] There are no exceptions to either of these two requirements. Since Article 310 does not specifically prohibit optional marking on insulated conductors, the *Code* would permit square millimeter (mm^2) markings on conductors, but only if it were in addition to the required traditional marking of AWG or circular mil area.

According to American National Standard (ANSI) *Standard for Use of the International System of Units (SI): The Modern Metric System*, IEEE/ASTM SI 10-1997, to convert from circular mils to square meters, multiply circular mils by 5.067075×10^{-10}. Since square millimeters, rather than square meters, is the standard marking for wire size, and because the reciprocal is more appropriate for this conversion, a simpler conversion factor to convert from square millimeters to circular mils (approximately) follows:

$$k = 1973.53\ \frac{\text{circular mils}}{\text{mm}^2}$$

In order to compare the square millimeter wire gauge to traditional wire sizes, the following example is provided.

Example

What traditional wire size does the size 125 mm^2 approximately represent?

Answer

$$\begin{aligned}\text{Circular mil area} &= \text{wire size (mm}^2) \\ &\quad \times \text{ conversion factor} \\ &= 125\ \text{mm}^2 \times 1973.53\ \frac{\text{circular mils}}{\text{mm}^2} \\ &= 246{,}691 \text{ circular mils, or} \\ &\quad 246.691 \text{ kcmil}\end{aligned}$$

Therefore, the 125 mm^2 wire is larger than No. 4/0 but smaller than a 250-kcmil conductor.

If a 125 mm^2 wire is determined to be the minimum or recommended sized conductor, it is important to understand that a size 250 kcmil would be the only Table 8 conductor with equivalent cross-sectional area. Size No. 4/0 is simply not enough metal, so 250 kcmil would be the choice for minimum equivalency.

FPN: The construction information is per NEMA WC8-1992. The resistance is calculated per National Bureau of Standards Handbook 100, dated 1966, and Handbook 109, dated 1972.

Table 9. Alternating-Current Resistance and Reactance for 600-Volt Cables, 3-Phase, 60 Hz, 75°C (167°F) — Three Single Conductors in Conduit

	Ohms to Neutral per 1000 Feet														
	X_L (Reactance) for All Wires		Alternating-Current Resistance for Uncoated Copper Wires			Alternating-Current Resistance for Aluminum Wires			Effective Z at 0.85 *PF* for Uncoated Copper Wires			Effective Z at 0.85 *PF* for Aluminum Wires			
Size (AWG or kcmil)	PVC, Aluminum Conduits	Steel Conduit	PVC Conduit	Aluminum Conduit	Steel Conduit	PVC Conduit	Aluminum Conduit	Steel Conduit	PVC Conduit	Aluminum Conduit	Steel Conduit	PVC Conduit	Aluminum Conduit	Steel Conduit	Size (AWG or kcmil)
14	0.058	0.073	3.1	3.1	3.1	—	—	—	2.7	2.7	2.7	—	—	—	14
12	0.054	0.068	2.0	2.0	2.0	3.2	3.2	3.2	1.7	1.7	1.7	2.8	2.8	2.8	12
10	0.050	0.063	1.2	1.2	1.2	2.0	2.0	2.0	1.1	1.1	1.1	1.8	1.8	1.8	10
8	0.052	0.065	0.78	0.78	0.78	1.3	1.3	1.3	0.69	0.69	0.70	1.1	1.1	1.1	8
6	0.051	0.064	0.49	0.49	0.49	0.81	0.81	0.81	0.44	0.45	0.45	0.71	0.72	0.72	6
4	0.048	0.060	0.31	0.31	0.31	0.51	0.51	0.51	0.29	0.29	0.30	0.46	0.46	0.46	4
3	0.047	0.059	0.25	0.25	0.25	0.40	0.41	0.40	0.23	0.24	0.24	0.37	0.37	0.37	3
2	0.045	0.057	0.19	0.20	0.20	0.32	0.32	0.32	0.19	0.19	0.20	0.30	0.30	0.30	2
1	0.046	0.057	0.15	0.16	0.16	0.25	0.26	0.25	0.16	0.16	0.16	0.24	0.24	0.25	1
1/0	0.044	0.055	0.12	0.13	0.12	0.20	0.21	0.20	0.13	0.13	0.13	0.19	0.20	0.20	1/0
2/0	0.043	0.054	0.10	0.10	0.10	0.16	0.16	0.16	0.11	0.11	0.11	0.16	0.16	0.16	2/0
3/0	0.042	0.052	0.077	0.082	0.079	0.13	0.13	0.13	0.088	0.092	0.094	0.13	0.13	0.14	3/0
4/0	0.041	0.051	0.062	0.067	0.063	0.10	0.11	0.10	0.074	0.078	0.080	0.11	0.11	0.11	4/0
250	0.041	0.052	0.052	0.057	0.054	0.085	0.090	0.086	0.066	0.070	0.073	0.094	0.098	0.10	250
300	0.041	0.051	0.044	0.049	0.045	0.071	0.076	0.072	0.059	0.063	0.065	0.082	0.086	0.088	300
350	0.040	0.050	0.038	0.043	0.039	0.061	0.066	0.063	0.053	0.058	0.060	0.073	0.077	0.080	350
400	0.040	0.049	0.033	0.038	0.035	0.054	0.059	0.055	0.049	0.053	0.056	0.066	0.071	0.073	400
500	0.039	0.048	0.027	0.032	0.029	0.043	0.048	0.045	0.043	0.048	0.050	0.057	0.061	0.064	500
600	0.039	0.048	0.023	0.028	0.025	0.036	0.041	0.038	0.040	0.044	0.047	0.051	0.055	0.058	600
750	0.038	0.048	0.019	0.024	0.021	0.029	0.034	0.031	0.036	0.040	0.043	0.045	0.049	0.052	750
1000	0.037	0.046	0.015	0.019	0.018	0.023	0.027	0.025	0.032	0.036	0.040	0.039	0.042	0.046	1000

Notes:

1. These values are based on the following constants: UL-type RHH wires with Class B stranding, in cradled configuration. Wire conductivities are 100 percent IACS copper and 61 percent IACS aluminum, and aluminum conduit is 45 percent IACS. Capacitive reactance is ignored, since it is negligible at these voltages.
These resistance values are valid only at 75°C (167°F) and for the parameters as given, but are representative for 600-volt wire types operating at 60 Hz.

2. *Effective Z* is defined as $R \cos(\theta) + X \sin(\theta)$, where θ is the power factor angle of the circuit. Multiplying current by effective impedance gives a good approximation for line-to-neutral voltage drop. Effective impedance values shown in this table are valid only at 0.85 power factor.
For another circuit power factor (*PF*), effective impedance (*Ze*) can be calculated from *R* and X_L values given in this table as follows:
$Ze = R \times PF + X_L \sin[\arccos(PF)]$.

Voltage-drop calculations using the dc-resistance formula are not always accurate for ac circuits, especially for those with a less-than-unity power factor or for those that use conductors larger than No. 2 AWG. Table 9 allows the *Code* user to perform simple ac voltage-drop calculations. Table 9 was compiled using the Neher-McGrath ac-resistance calculation method; the values presented are both reliable and conservative. This table contains completed calculations of effective impedance *(Z)* for the average ac circuit with an 85 percent power factor. If calculations with a different power factor are necessary, the table also contains the appropriate values of inductive reactance and ac resistance. An example of each calculation follows.

The basic assumptions and the limitations of Table 9 are as follows.

1. Capacitive reactance is ignored.
2. There are three conductors in a raceway.
3. The voltage-drop values calculated are approximate.
4. For circuits with other parameters, the Neher-McGrath ac-resistance calculation method is used.

Example 1

Determine the voltage drop for the following.

A feeder has a 100-ampere continuous load. The system source is 240 volts, 3 phase, and the supplying circuit breaker is 125 amperes. The feeder is a $1^1/_4$-in. aluminum conduit with three No. 1 THHN copper conductors operating at their maximum temperature rating of 75°C. The circuit length is 150 ft, and the power factor is 85 percent. What is the approximate voltage drop of this circuit, using Table 9?

Answer

Step 1. Using the Table 9 column "Effective Z (Impedance) at 0.85 *PF* for Uncoated Copper Wires," select aluminum conduit and size No. 1 copper wire. A value of 0.16 ohms per 1000 ft is given. Use 0.16 ohms in the formula, as follows.

Step 2. Find the approximate line-to-neutral voltage drop.

$$\text{Voltage drop}_{\text{(line-to-neutral)}}$$
$$= \text{table value} \times \frac{\text{circuit length}}{1000\text{ ft}} \times \text{circuit load}$$
$$= 0.16\text{ ohms} \times \frac{150\text{ ft}}{1000\text{ ft}} \times 100\text{ amperes}$$
$$= 2.40\text{ volts}$$

Step 3. Find the line-to-line voltage drop.

$$\text{Voltage drop}_{\text{(line-to-line)}}$$
$$= \text{voltage drop}_{\text{(line-to-neutral)}} \times \sqrt{3}$$
$$= 2.40\text{ volts} \times 1.732$$
$$= 4.157\text{ volts}$$

Step 4. Find the voltage drop expressed as a percentage of the circuit voltage.

$$\text{Percentage voltage drop}_{\text{(line-to-line)}}$$
$$= \frac{4.157\text{ volts}}{240\text{ volts}} \times 100$$
$$= 1.73\%$$

Step 5. Find the voltage present at the load end of the circuit.

$$240\text{ volts} - 4.157\text{ volts} = 235.84\text{ volts}$$

Example 2

Determine the voltage drop for the following.

A 270-ampere continuous load is present on a feeder. The circuit consists of a single 4-in. PVC conduit with three 600 kcmil XHHW/USE aluminum conductors fed from a 480-volt, 3-phase, 3-wire source. The conductors are operating at their maximum rated temperature of 75°C. If the power factor is 0.7 and the circuit length is 250 ft, is the voltage drop excessive?

Answer

Step 1. Using the table column "X_L (Reactance) for All Wires," select PVC conduit and the row for size 600 kcmil. A value of 0.039 ohms per 1000 ft is given as this X_L. Next, using the column "Alternating-Current Resistance for Aluminum Wires," select PVC conduit and the row for size 600 kcmil. A value of 0.036 ohms per 1000 ft is given as this R.

Step 2. Find the angle representing a power factor of 0.7.

Using a calculator with trigonometric functions or a trigonometric function table, find the arccosine ($\cos^{-1}$) of 0.7, which is 45.57°. For this example, we will call this angle ∅. Continuing to use the table or a calculator, find the sin of 45.57°, which is 0.7141.

Step 3. Find the impedance (Z_c) corrected to 0.7 power factor (Z_c).

$$Z_c = (R \times \cos ø) + (X_L \times \sin ø)$$
$$= (0.036 \times 0.7) + (0.039 \times 0.7141)$$
$$= 0.0252 + 0.0279$$
$$= 0.0531\text{ ohms to neutral}$$

Step 4. As in Example No. 1, find the approximate line-to-neutral voltage drop.

$$\text{Voltage drop}_{\text{(line-to-neutral)}}$$
$$= Z_c \times \frac{\text{circuit length}}{1000\text{ ft}} \times \text{circuit load}$$
$$= 0.0530 \times \frac{250\text{ ft}}{1000\text{ ft}} \times 270\text{ amperes}$$
$$= 3.577\text{ volts}$$

Step 5. Find the approximate line-to-line voltage drop.

$$\text{Voltage drop}_{\text{(line-to-line)}}$$
$$= \text{voltage drop}_{\text{(line-to-neutral)}} \times \sqrt{3}$$
$$= 3.577\text{ volts} \times 1.732$$
$$= 6.196\text{ volts}$$

Step 6. **Find the approximate voltage drop expressed as a percentage of the circuit voltage.**

$$\text{Percentage voltage drop}_{\text{(line-to-line)}} = \frac{6.196 \text{ volts}}{480 \text{ volts}} \times 100 = 1.29\%$$

Step 7. **Find the voltage present at the load end of the circuit.**

$$480 \text{ volts} - 6.196 \text{ volts} = 473.8 \text{ volts}$$

Conclusion

According to Section 215-2(d), FPN No. 2, this voltage drop does not appear to be excessive.

Tables 11(a) and 11(b)

For listing purposes, Tables 11(a) and 11(b) provide the required power source limitations for Class 2 and Class 3 power sources. Table 11(a) applies for alternating-current sources, and Table 11(b) applies for direct-current sources.

The power for Class 2 and Class 3 circuits shall be either (1) inherently limited, requiring no overcurrent protection, or (2) not inherently limited, requiring a combination of power source and overcurrent protection. Power sources designed for interconnection shall be listed for the purpose.

As part of the listing, the Class 2 or Class 3 power source shall be durably marked where plainly visible to indicate the class of supply and its electrical rating. A Class 2 power source not suitable for wet location use shall be so marked.

Exception: Limited power circuits used by listed information technology equipment.

Overcurrent devices, where required, shall be located at the point where the conductor to be protected receives its supply and shall not be interchangeable with devices of higher ratings. The overcurrent device shall be permitted as an integral part of the power source.

Moved in the 1996 *Code,* Tables 11(a) and 11(b) were previously Tables 725-31(a) and (b). Due to the listing requirements for Class 2 and 3 power supplies, this information is no longer useful to the average user. It has been retained in Chapter 9 as Tables 11(a) and 11(b) to provide direction for organizations that are properly equipped and qualified to evaluate these products.

Table 11(a). Class 2 and Class 3 Alternating-Current Power Source Limitations

		Inherently Limited Power Source (Overcurrent Protection Not Required)				Not Inherently Limited Power Source (Overcurrent Protection Required)			
Power Source		**Class 2**			**Class 3**	**Class 2**		**Class 3**	
Source voltage V_{max} (volts) (see Note 1)		0 through 20*	Over 20 and through 30*	Over 30 and through 150	Over 30 and through 100	0 through 20*	Over 20 and through 30*	Over 30 and through 100	Over 100 and through 150
Power limitations VA_{max} (volt-amperes) (see Note 1)		—	—	—	—	250 (see Note 3)	250	250	N.A.
Current limitations I_{max} (amperes) (see Note 1)		8.0	8.0	0.005	$150/V_{max}$	$1000/V_{max}$	$1000/V_{max}$	$1000/V_{max}$	1.0
Maximum overcurrent protection (amperes)		—	—	—	—	5.0	$100/V_{max}$	$100/V_{max}$	1.0
Power source maximum nameplate rating	VA (volt-amperes)	$5.0 \times V_{max}$	100	$0.005 \times V_{max}$	100	$5.0 \times V_{max}$	100	100	100
	Current (amperes)	5.0	$100/V_{max}$	0.005	$100/V_{max}$	5.0	$100/V_{max}$	$100/V_{max}$	$100/V_{max}$

*Voltage ranges shown are for sinusoidal ac in indoor locations or where wet contact is not likely to occur. For nonsinusoidal or wet contact conditions, see Note 2.

Table 11(b). Class 2 and Class 3 Direct-Current Power Source Limitations

Power Source		Inherently Limited Power Source (Overcurrent Protection Not Required)					Not Inherently Limited Power Source (Overcurrent Protection Required)			
		Class 2				Class 3	Class 2		Class 3	
Source voltage V_{max} (volts) (see Note 1)		0 through 20*	Over 20 and through 30*	Over 30 and through 60*	Over 60 and through 150	Over 60 and through 100	0 through 20*	Over 20 and through 60*	Over 60 and through 100	Over 100 and through 150
Power limitations VA_{max} (volt-amperes) (see Note 1)		—	—	—	—	—	250 (see Note 3)	250	250	N.A.
Current limitations I_{max} (amperes) (see Note 1)		8.0	8.0	$150/V_{max}$	0.005	$150/V_{max}$	$1000/V_{max}$	$1000/V_{max}$	$1000/V_{max}$	1.0
Maximum overcurrent protection (amperes)		—	—	—	—	—	5.0	$100/V_{max}$	$100/V_{max}$	1.0
Power source maximum nameplate rating	VA (volt-amperes)	$5.0 \times V_{max}$	100	100	$0.005 \times V_{max}$	100	$5.0 \times V_{max}$	100	100	100
	Current (amperes)	5.0	$100/V_{max}$	$100/V_{max}$	0.005	$100/V_{max}$	5.0	$100/V_{max}$	$100/V_{max}$	$100/V_{max}$

*Voltage ranges shown are for continuous dc in indoor locations or where wet contact is not likely to occur.For interrupted dc or wet contact conditions, see Note 4.

Notes for Table 11(a) and 11(b)

1. V_{max}, I_{max}, and VA_{max} are determined with the current limiting impedance in the circuit (not bypassed) as follows.

   V_{max}: Maximum output current under any noncapacitive load, including short circuit, and with overcurrent protection bypassed if used. Where a transformer limits the output current, I_{max} limits apply voltage regardless of load with rated input applied.

   I_{max}: Maximum output after 1 minute of operation. Where a current-limiting impedance, listed for the purpose, or as part of a listed product, is used in combination with a nonpower-limited transformer or a stored energy source, e.g., storage battery, to limit the output current, I_{max} limits apply after 5 seconds.

   VA_{max}: Maximum volt-ampere output after 1 minute of operation regardless of load and overcurrent protection bypassed if used.

2. For nonsinusoidal ac, V_{max} shall not be greater than 42.4 volts peak. Where wet contact (immersion not included) is likely to occur, Class 3 wiring methods shall be used or V_{max} shall not be greater than 15 volts for sinusoidal ac and 21.2 volts peak for nonsinusoidal ac.
3. If the power source is a transformer, VA_{max} is 350 or less when V_{max} is 15 or less.
4. For dc interrupted at a rate of 10 to 200 Hz, V_{max} shall not be greater than 24.8 volts peak. Where wet contact (immersion not included) is likely to occur, Class 3 wiring methods shall be used or V_{max} shall not be greater than 30 volts for continuous dc; 12.4 volts peak for dc that is interrupted at a rate of 10 to 200 Hz.

Tables 12(a) and 12(b)

For listing purposes, Tables 12(a) and 12(b) provide the required power source limitations for power-limited fire alarm sources. Table 12(a) applies for alternating-current sources, and Table 12(b) applies for direct-current sources.

The power for power-limited fire alarm circuits shall be either (1) inherently limited, requiring no overcurrent protection, or (2) not inherently limited, requiring the power to be limited by a combination of power source and overcurrent protection.

As part of the listing, the PLFA power source shall be durably marked where plainly visible to indicate that it is a power-limited fire alarm power source.

The overcurrent device, where required, shall be located at the point where the conductor to be protected receives its supply and shall not be interchangeable with devices of higher ratings. The overcurrent device shall be permitted as an integral part of the power source.

Moved in the 1996 *Code,* Tables 12(a) and 12(b) were previously Tables 760-21(a) and (b). Due to the listing requirements for these power supplies, this information is no longer useful to the average user. These tables have been moved to Chapter 9 as Tables 12(a) and 12(b) to provide direction for organizations that are properly equipped and qualified to evaluate these products.

Table 12(a). PLFA Alternating-Current Power Source Limitations

Power Source		Inherently Limited Power Source (Overcurrent Protection Not Required)			Not Inherently Limited Power Source (Overcurrent Protection Required)		
Circuit voltage V_{max} (volts) (see Note 1)		0 through 20	Over 20 and through 30	Over 30 and through 100	0 through 20	Over 20 and through 100	Over 100 and through 150
Power limitations VA_{max} (volt-amperes) (see Note 1)		—	—	—	250 (see Note 2)	250	N.A.
Current limitations I_{max} (amperes) (see Note 1)		8.0	8.0	$150/V_{max}$	$1000/V_{max}$	$1000/V_{max}$	1.0
Maximum overcurrent protection (amperes)		—	—	—	5.0	$100/V_{max}$	1.0
Power source maximum nameplate ratings	VA (volt-amperes)	$5.0 \times V_{max}$	100	100	$5.0 \times V_{max}$	100	100
	Current (amperes)	5.0	$100/V_{max}$	$100/V_{max}$	5.0	$100/V_{max}$	$100/V_{max}$

Table 12(b). PLFA Direct-Current Power Source Limitations

Power Source		Inherently Limited Power Source (Overcurrent Protection Not Required)				Not Inherently Limited Power Source (Overcurrent Protection Required)		
Circuit voltage V_{max} (volts) (see Note 1)		0 through 20	Over 20 and through 30	Over 30 and through 100	Over 100 and through 250	0 through 20	Over 20 and through 100	Over 100 and through 150
Power limitations VA_{max} (volt-amperes) (see Note 1)		—	—	—	—	250 (see Note 2)	250	N.A.
Current limitations I_{max} (amperes) (see Note 1)		8.0	8.0	$150/V_{max}$	0.030	$1000/V_{max}$	$1000/V_{max}$	1.0
Maximum overcurrent protection (amperes)		—	—	—	—	5.0	$100/V_{max}$	1.0
Power source maximum nameplate ratings	VA (volt-amperes)	$5.0 \times V_{max}$	100	100	$0.030 \times V_{max}$	$5.0 \times V_{max}$	100	100
	Current (amperes)	5.0	$100/V_{max}$	$100/V_{max}$	0.030	5.0	$100/V_{max}$	$100/V_{max}$

Notes for Tables 12(a) and 12(b)

1. V_{max}, I_{max}, and VA_{max} are determined as follows.

   V_{max}: Maximum output voltage regardless of load with rated input applied.

   I_{max}: Maximum output current under any noncapacitive load, including short circuit, and with overcurrent protection bypassed if used. Where a transformer limits the output current, I_{max} limits apply after 1 minute of operation. Where a current-limiting impedance, listed for the purpose, is used in combination with a nonpower-limited transformer or a stored energy source, e.g., storage battery, to limit the output current, I_{max} limits apply after 5 seconds.

   VA_{max}: Maximum volt-ampere output after 1 minute of operation regardless of load and overcurrent protection bypassed if used. Current limiting impedance shall not be bypassed when determining I_{max} and VA_{max}.

2. If the power source is a transformer, VA_{max} is 350 or less when V_{max} is 15 or less.

APPENDIX A

Extract References

(See Section 90-3, Paragraph 4)

A-250-104(b)	*National Fuel Gas Code,* NFPA 54-1996	Paragraph 3.14(a)
A-500-5(a)(1)	*Recommended Practice for the Classification of Flammable Liquids, Gases, or Vapors and of Hazardous (Classified) Locations for Electrical Installations in Chemical Process Areas,* NFPA 497-1997	Section 1-3, Definitions, Combustible Material
A-500-5(a)(2)	*Recommended Practice for the Classification of Flammable Liquids, Gases, or Vapors and of Hazardous (Classified) Locations for Electrical Installations in Chemical Process Areas,* NFPA 497-1997	Section 1-3, Definitions, Combustible Material
A-500-5(a)(3)	*Recommended Practice for the Classification of Flammable Liquids, Gases, or Vapors and of Hazardous (Classified) Locations for Electrical Installations in Chemical Process Areas,* NFPA 497-1997	Section 1-3, Definitions, Combustible Material
A-500-5(a)(4)	*Recommended Practice for the Classification of Flammable Liquids, Gases, or Vapors and of Hazardous (Classified) Locations for Electrical Installations in Chemical Process Areas,* NFPA 497-1997	Section 1-3, Definitions, Combustible Material
A-500-5(b)(1)	*Recommended Practice for the Classification of Combustible Dusts and of Hazardous (Classified) Locations for Electrical Installations in Chemical Process Areas,* NFPA 499-1997	Section 1-3, Definitions, Combustible Dusts
A-500-5(b)(2)	*Recommended Practice for the Classification of Combustible and of Hazardous (Classified) Locations for Electrical Installations in Chemical Process Areas,* NFPA 499-1997	Section 1-3, Definitions, Combustible Dusts
A-505-7(a)	*Recommended Practice for the Classification of Flammable Liquids, Gases, or Vapors and of Hazardous (Classified) Locations in Chemical Process Areas,* NFPA 497-1997	Section 1-3, Definitions, Combustible Material
A-505-7(b)	*Recommended Practice for the Classification of Flammable Liquids, Gases, or Vapors and of Hazardous (Classified) Locations in Chemical Process Areas,* NFPA 497-1997	Section 1-3, Definitions, Combustible Material
A-505-7(c)	*Recommended Practice for the Classification of Flammable Liquids, Gases, or Vapors and of Hazardous (Classified) Locations in Chemical Process Areas,* NFPA 497-1997	Section 1-3, Definitions, Combustible Material
A-514-2	*Automotive and Marine Service Station Code,* NFPA 30A-1996	Sections 7-1 and 7-3
A-Table 514-2	*Automotive and Marine Service Station Code,* NFPA 30A-1996	Table 7
A-Fig. 514-2	*Automotive and Marine Service Station Code,* NFPA 30A-1996	Figure 7-1
A-514-5(b)	*Automotive and Marine Service Station Code,* NFPA 30A-1996	Paragraph 9-4.5
A-514-5(c)	*Automotive and Marine Service Station Code,* NFPA 30A-1996	Paragraph 9-5.3
A-515-2	*Flammable and Combustible Liquids Code,* NFPA 30-1996	Paragraphs 5-9.5.1, 5-9.5.3
A-Table 515-2	*Flammable and Combustible Liquids Code,* NFPA 30-1996	Table 5-9.5.3
A-Fig. 515-2	*Flammable and Combustible Liquids Code,* NFPA 30-1996	Figure 5-7.16
A-516-2(a)(1)	*Standard for Spray Application Using Flammable or Combustible Materials,* NFPA 33-1995	Section 1-6, Definitions, Spray Area
A-516-2(a)(2)	*Standard for Spray Application Using Flammable or Combustible Materials,* NFPA 33-1995	Section 1-6, Definitions, Spray Area
A-516-2(a)(3)	*Standard for Spray Application Using Flammable or Combustible Materials,* NFPA 33-1995	Section 1-6, Definitions, Spray Area
A-516-2(a)(4)	*Standard for Dipping and Coating Processes Using Flammable or Combustible Liquids,* NFPA 34-1995	Section 1-6, Definitions, Vapor Source, and Paragraph 4-2.2
A-516-2(a)(5)	*Standard for Dipping and Coating Processes Using Flammable or Combustible Liquids,* NFPA 34-1995	Paragraph 4-2.1
A-516-2(a)(6)	*Standard for Dipping and Coating Processes Using Flammable or Combustible Liquids,* NFPA 34-1995	Paragraph 4-3.1
A-516-2(b)(1)	*Standard for Spray Application Using Flammable or Combustible Materials,* NFPA 33-1995	Paragraph 4-3.1

A-516, Fig. 516-2(b)(1)	*Standard for Spray Application Using Flammable or Combustible Materials,* NFPA 33-1995	Figure 4-3.1
A-516-2(b)(2)	*Standard for Spray Application Using Flammable or Combustible Materials,* NFPA 33-1995	Paragraph 4-3.2
A-516, Fig. 516-2(b)(2)	*Standard for Spray Application Using Flammable or Combustible Materials,* NFPA 33-1995	Figures 4-3.2(a), 4-3.2(b)
A-516-2(b)(3)	*Standard for Spray Application Using Flammable or Combustible Materials,* NFPA 33-1995	Paragraph 4-3.3
A-516-2(b)(4)	*Standard for Spray Application Using Flammable or Combustible Materials,* NFPA 33-1995	Paragraph 4-3.4
A-516, Fig. 516-2(b)(4)	*Standard for Spray Application Using Flammable or Combustible Materials,* NFPA 33-1995	Figure 4-3.4
A-516-2(b)(5)	*Standard for Dipping and Coating Processes Using Flammable or Combustible Liquids,* NFPA 34-1995	Paragraph 4-2.3
A-516-2(b)(6)	*Standard for Dipping and Coating Processes Using Flammable or Combustible Liquids,* NFPA 34-1995	Paragraph 4-2.4
A-516, Fig. 516-2(b)(5)	*Standard for Dipping and Coating Processes Using Flammable or Combustible Liquids,* NFPA 34-1995	Figure 4-2(a)
A-516-2(c)	*Standard for Dipping and Coating Processes Using Flammable or Combustible Liquids,* NFPA 34-1995	Section 4-3.2
A-516-3(b)	*Standard for Spray Application Using Flammable or Combustible Materials,* NFPA 33-1995	Section 4-2
A-516-3(d)	*Standard for Spray Application Using Flammable or Combustible Materials,* NFPA 33-1995	Section 4-7
A-516-4	*Standard for Spray Application Using Flammable or Combustible Materials,* NFPA 33-1995	Chapter 9
A-516-5	*Standard for Spray Application Using Flammable or Combustible Materials,* NFPA 33-1995	Chapter 10
A-516-6	*Standard for Spray Application Using Flammable or Combustible Materials,* NFPA 33-1995	Chapter 13
A-517-30(b)(1)	*Standard for Health Care Facilities,* NFPA 99-1996	Paragraph 3-4.2.2.1
A-517-30(b)(2)	*Standard for Health Care Facilities,* NFPA 99-1996	Paragraph 3-4.2.2.1
A-517-30(b)(3)	*Standard for Health Care Facilities,* NFPA 99-1996	Paragraph 3-4.2.2.1
A-517-30(b)(4)	*Standard for Health Care Facilities,* NFPA 99-1996	Paragraph 12-3.3.2
A-517-31	*Standard for Health Care Facilities,* NFPA 99-1996	Paragraphs 3-4.2.2.2(a), 3-5.2.2.2
A-517-32	*Standard for Health Care Facilities,* NFPA 99-1996	Paragraph 3-4.2.2.2(b)
A-517-33	*Standard for Health Care Facilities,* NFPA 99-1996	Paragraph 3-4.2.2.2(c)
A-517-33(c)	*Standard for Health Care Facilities,* NFPA 99-1996	Paragraph 3-4.2.2.4(b)2.
A-517-34	*Standard for Health Care Facilities,* NFPA 99-1996	Paragraph 3-4.2.2.3(b)
A-517-34(a)	*Standard for Health Care Facilities,* NFPA 99-1996	Paragraph 3-4.2.2.3(c)
A-517-34(b)	*Standard for Health Care Facilities,* NFPA 99-1996	Paragraph 3-4.2.2.3(d)
A-517-35(a)	*Standard for Health Care Facilities,* NFPA 99-1996	Paragraph 3-4.1.1.2
A-517-40(a)	*Standard for Health Care Facilities,* NFPA 99-1996	Paragraphs 16-3.3.2, Exception; 17-3.3.2, Exception
A-517-41(a)	*Standard for Health Care Facilities,* NFPA 99-1996	Paragraph 3-5.2.2.1
A-517-41(b)	*Standard for Health Care Facilities,* NFPA 99-1996	Paragraph 3-5.2.2.1
A-517-42	*Standard for Health Care Facilities,* NFPA 99-1996	Paragraphs 3-5.2.2.2, 3-5.3.1
A-517-43	*Standard for Health Care Facilities,* NFPA 99-1996	Paragraph 3-5.2.2.3
A-517-44(a)	*Standard for Health Care Facilities,* NFPA 99-1996	Paragraphs 3-5.1, 3-4.1.1.2
A-517-44(b)	*Standard for Health Care Facilities,* NFPA 99-1996	Paragraphs 3-5.1, 3-4.1.1.3, 16-3.3.2.1, 17-3.3.2.1
A-517-45(b)	*Standard for Health Care Facilities,* NFPA 99-1996	Paragraphs 3-6.2.2.2, 13-3.4.1
A-517-45(c)(1)	*Standard for Health Care Facilities,* NFPA 99-1996	Paragraph 3-6.1
A-517-45(c)(2)	*Standard for Health Care Facilities,* NFPA 99-1996	Paragraph 3-6.3.1.1
A-517-45(c)(3)	*Standard for Health Care Facilities,* NFPA 99-1996	Paragraph 3-6.3.1.2
A-517-50(b)	*Standard for Health Care Facilities,* NFPA 99-1996	Paragraphs 3-6.2.2.2, 14-3.4.1, 15-3.4.1
A-517-50(c)(1)	*Standard for Health Care Facilities,* NFPA 99-1996	Paragraph 3-6.1
A-517-50(c)(2)	*Standard for Health Care Facilities,* NFPA 99-1996	Paragraph 3-6.3.1.1

A-517-50(c)(3)	*Standard for Health Care Facilities,* NFPA 99-1996	Paragraph 3-6.3.1.2
A-517-60(a)(1)	*Standard for Health Care Facilities,* NFPA 99-1996	ANNEX 2, Sections 2-1 and 2-2
A-517-61(a)(1)	*Standard for Health Care Facilities,* NFPA 99-1996	ANNEX 2, Paragraph 2-6.3.2
A-517-61(a)(3)	*Standard for Health Care Facilities,* NFPA 99-1996	ANNEX 2, Para graphs 2-2.1, 2-4.5, 2-4.6, 2-4.7
A-517-160(a)(2)	*Standard for Health Care Facilities,* NFPA 99-1996	Paragraph 3-3.2.2.1
A-517-160(a)(4)	*Standard for Health Care Facilities,* NFPA 99-1996	Paragraphs 12-4.1.2.6(d), 12-4.1.2.6(e)
A-695-3(a)	*Standard for the Installation of Centrifugal Fire Pumps,* NFPA 20-1996	Section 6-2 and Paragraphs 6-2.1, 6-2.2
A-695-3(b)	*Standard for the Installation of Centrifugal Fire Pumps,* NFPA 20-1996	Paragraphs 6-2.3.1, 6-2.4.1
A-695-4	*Standard for the Installation of Centrifugal Fire Pumps,* NFPA 20-1996	Paragraphs 6-2.4.3, 6-3.2.2
A-695-7	*Standard for the Installation of Centrifugal Fire Pumps,* NFPA 20-1996	Paragraph 6-3.1.2
A-695-10	*Standard for the Installation of Centrifugal Fire Pumps,* NFPA 20-1996	Paragraphs 6-4.1.1, 7-1.1.1, 7-7, 7-8.1, 7-9.1, 9-1.1.1.
A-695-14(a)	*Standard for the Installation of Centrifugal Fire Pumps,* NFPA 20-1996	Paragraph 7-5.2.5
A-695-14(b)	*Standard for the Installation of Centrifugal Fire Pumps,* NFPA 20-1996	Paragraph 7-4.5.6
A-695-14(c)	*Standard for the Installation of Centrifugal Fire Pumps,* NFPA 20-1996	Paragraph 7-8.1.3
A-695-14(d)	*Standard for the Installation of Centrifugal Fire Pumps,* NFPA 20-1996	Paragraph 9-3.5.1

Appendix A identifies those *NEC* figures and tables that are identified in the *NEC* by a superscript letter "x" and the NFPA document and section, or equivalent, from which the material in the *NEC* was extracted. For example, Section [x]515-2 in Article 515 corresponds with A-515-2 in the first column of Appendix A. The second and third columns in Appendix A list the "parent" NFPA document and the section or area of the NFPA document from which the material was extracted.

Requests for interpretations or proposed revisions of the extracted text are referred to the NFPA committee that is responsible for the document from which the text was extracted.

APPENDIX B

This appendix is not a part of the requirements of this Code but is included for informational purposes only.

B-310-15(b)(1). Formula Application Information. This appendix provides application information for ampacities calculated under engineering supervision.

The data in Appendix B is based on calculations using the Neher-McGrath formula, which is found in Section 310-15(c). See the commentary following Section 310-15(c).

B-310-15(b)(2). Typical Applications Covered by Tables. Typical ampacities for conductors rated 0 through 2000 volts are shown in Tables B-310-1 through B-310-10. Underground electrical duct bank configurations, as detailed in Figures B-310-3, B-310-4, and B-310-5, are utilized for conductors rated 0 through 5000 volts. In Figures B-310-2 through B-310-5, where adjacent duct banks are used, a separation of 5 ft (1.52 m) between the centerlines of the closest ducts in each bank or 4 ft (1.22 m) between the extremities of the concrete envelopes is sufficient to prevent derating of the conductors due to mutual heating. These ampacities were calculated as detailed in the basic ampacity paper, *The Calculation of the Temperature Rise and Load Capability of Cable Systems,* by J. H. Neher and M. H. McGrath, AIEE Paper 57-660. For additional information concerning the application of these ampacities, see *Power Cable Ampacities,* IEEE/ICEA Standard S-135/P-46-426 and IEEE Standard 835-1994, *Standard Power Cable Ampacity Tables.*

Typical values of thermal resistivity (Rho) are as follows:

Average soil (90 percent of USA) = 90

Concrete = 55

Damp soil (coastal areas, high water table) = 60

Paper insulation = 550

Polyethylene (PE) = 450

Polyvinyl chloride PVC = 650

Rubber and rubber-like = 500

Very dry soil (rocky or sandy) = 120

Thermal resistivity, as used in this appendix, refers to the heat transfer capability through a substance by conduction. It is the reciprocal of thermal conductivity and is normally expressed in the units °C-cm/watt. For additional information on determining soil thermal resistivity (Rho), see *Guide for Soil Thermal Resistivity Measurements,* ANSI/IEEE Standard 442-1996.

A soil resistivity higher than 90 will reduce the ampacities below those in the underground ampacity tables B-310-5, B-310-6, B-310-7, B-310-8, B-310-9, and B-310-10 if other factors remain the same. Conversely, a load factor of less than 100 percent will increase ampacities if other factors remain the same. See Section B-310-15(b)(7) for allowable adjustments if the load factor is less than 100 percent. Reduced load factors are used in Figures B-310-3, B-310-4, and B-310-5.

B-310-15(b)(3). Criteria Modifications. Where values of load factor and Rho are known for a particular electrical duct bank installation and they are different from those shown in a specific table or figure, the ampacities shown in the table or figure can be modified by the application of factors derived from the use of Figure B-310-1.

Where two different ampacities apply to adjacent portions of a circuit, the higher ampacity can be used beyond the point of transition, a distance equal to 10 ft (3.05 m) or 10 percent of the circuit length figured at the higher ampacity, whichever is less.

The information given in Section B-310-15(b)(3) is the same as that found in the exception to Section 310-15(a)(2). See the commentary following this exception.

Where the burial depth of direct burial or electrical duct bank circuits are modified from the values shown in a figure or table, ampacities can be modified as shown in (1) and (2) as follows.

(1) Where burial depths are increased in part(s) of an electrical duct run to avoid underground obstructions, no decrease in ampacity of the conductors is needed, provided the total length of parts of the duct run increased in depth to avoid obstructions is less than 25 percent of the total run length.

See Figure B.1.

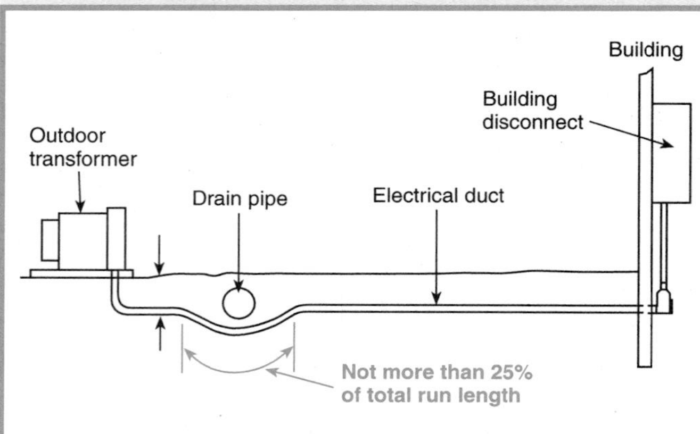

Figure B.1 If that portion deeper than 30 in. does not exceed 25 percent of the total run length, no decrease in ampacity is required, even though part of the run is more than 30 in. deep, which is the maximum depth assumed in Figure B-310-2, Note 1, in order to maintain the accuracy of the tables.

(2) Where burial depths are deeper than shown in a specific underground ampacity table or figure, an ampacity derating factor of 6 percent per increased foot (305 mm) of depth for all values of Rho can be utilized. No rating change is needed where the burial depth is decreased.

For example, in accordance with Table B-310-7 and Figure B-310-2, the ampacity of six paralleled runs of 500 kcmil, Type XHHW copper conductors (Rho = 90, Detail 3) is 6 × 273 = 1638 amperes at a depth measuring no greater than 30 in. to the top duct in the bank. If the burial depth is 6 ft, the ampacity is 1638 − (3.5 × 0.06 × 1638) = 1294 amperes.

B-310-15(b)(4). Electrical Ducts. The term *electrical duct(s)* is defined in Section 310-60.

B-310-15(b)(5). Tables B-310-6 and B-310-7.

(1) To obtain the ampacity of cables installed in two electrical ducts in one horizontal row with 7.5-in. (191-mm) center-to-center spacing between electrical ducts, similar to Figure B-310-2, Detail 1, multiply the ampacity shown for one duct in Tables B-310-6 and B-310-7 by 0.88.

(2) To obtain the ampacity of cables installed in four electrical ducts in one horizontal row with 7.5-in. (191-mm) center-to-center spacing between electrical ducts, similar to Figure B-310-2, Detail 2, multiply the ampacity shown for three electrical ducts in Tables B-310-6 and B-310-7 by 0.94.

The underground ampacity tables are based on the 7.5-in. center-to-center spacing illustrated in Figure B-310-2. Although moving directly buried cables or electrical ducts further apart will increase ampacities, the effect is surprisingly small. One calculation indicates that two side-by-side electrical ducts buried 30 in. below grade would have to be spaced about 5 ft apart before they could be considered single electrical ducts as shown in Detail 1, Figure B-310-2. Decreasing the burial depth, decreasing the thermal resistivity of the earth or other surrounding medium, and decreasing the load factor have a much greater effect in increasing ampacity than does increasing the horizontal spacing.

B-310-15(b)(6). Electrical Ducts Utilized in Figure B-310-2. If spacing between electrical ducts, as shown in Figure B-310-2, is less than specified in Figure B-310-2, where electrical ducts enter equipment enclosures from underground, the ampacity of conductors contained within such electrical ducts need not be reduced.

B-310-15(b)(7). Examples Showing Use of Figure B-310-1 for Electrical Duct Bank Ampacity Modifications. Figure B-310-1 is used for interpolation or extrapolation for values of Rho and load factor for cables installed in electrical ducts. The upper family of curves shows the variation in ampacity and Rho at unity load factor in terms of I_1, the ampacity for Rho = 60, and 50 percent load factor. Each curve is designated for a particular ratio I_2/I_1, where I_2 is the ampacity at Rho = 120 and 100 percent load factor.

The lower family of curves shows the relationship between Rho and load factor that will give substantially the same ampacity as the indicated value of Rho at 100 percent load factor.

As an example, to find the ampacity of a 500 kcmil copper cable circuit for six electrical ducts as shown in Table B-310-5: At the Rho = 60, LF = 50, I_1 = 583; for Rho = 120 and LF = 100, I_2 = 400. The ratio I_2/I_1 = 0.686. Locate Rho = 90 at the bottom of the chart and follow the 90 Rho line to the intersection with 100 percent load factor where the equivalent Rho = 90. Then follow the 90 Rho line to I_2/I_1 ratio of 0.686 where F = 0.74. The desired ampacity = 0.74 × 583 = 431, which agrees with the table for Rho = 90, LF = 100.

To determine the ampacity for the same circuit where Rho = 80 and LF = 75, using Figure B-310-1, the equivalent Rho = 43, F = 0.855, and the desired ampacity = 0.855 × 583 = 498 amperes. Values for using Figure B-310-1 are found in the electrical duct bank ampacity tables of this appendix.

Where the load factor is less than 100 percent and can be verified by measurement or calculation, the ampacity of electrical duct bank installations can be modified as shown. Different values of Rho can be accommodated in the same manner.

Table B-310-1. Ampacities of Two or Three Insulated Conductors, Rated 0 through 2000 Volts, Within an Overall Covering (Multiconductor Cable), in Raceway in Free Air Based on Ambient Air Temperature of 30°C (86°F)

Size	Temperature Rating of Conductor. See Table 310-13.						Size
	60°C (140°F)	75°C (167°F)	90°C (194°F)	60°C (140°F)	75°C (167°F)	90°C (194°F)	
AWG or kcmil	Types TW, UF	Types RH, RHW, THHW, THW, THWN, XHHW, ZW	Types THHN, THHW, THW-2, THWN-2, RHH, RWH-2, USE-2, XHHW, XHHW-2, ZW-2	Type TW	Types RH, RHW, THHW, THW, THWN, XHHW	Types THHN, THHW, THW-2, THWN-2, RHH, RWH-2, USE-2, XHHW, XHHW-2, ZW-2	AWG or kcmil
	COPPER			ALUMINUM OR COPPER-CLAD ALUMINUM			
14	16*	18*	21*	—	—	—	14
12	20*	24*	27*	16*	18*	21*	12
10	27*	33*	36*	21*	25*	28*	10
8	36	43	48	28	33	37	8
6	48	58	65	38	45	51	6
4	66	79	89	51	61	69	4
3	76	90	102	59	70	79	3
2	88	105	119	69	83	93	2
1	102	121	137	80	95	106	1
1/0	121	145	163	94	113	127	1/0
2/0	138	166	186	108	129	146	2/0
3/0	158	189	214	124	147	167	3/0
4/0	187	223	253	147	176	197	4/0
250	205	245	276	160	192	217	250
300	234	281	317	185	221	250	300
350	255	305	345	202	242	273	350
400	274	328	371	218	261	295	400
500	315	378	427	254	303	342	500
600	343	413	468	279	335	378	600
700	376	452	514	310	371	420	700
750	387	466	529	321	384	435	750
800	397	479	543	331	397	450	800
900	415	500	570	350	421	477	900
1000	448	542	617	382	460	521	1000
Ambient Temp. (°C)	For ambient temperatures other than 30°C (86°F), multiply the ampacities shown above by the appropriate factor shown below.						Ambient Temp. (°F)
21–25	1.08	1.05	1.04	1.08	1.05	1.04	70–77
26–30	1.00	1.00	1.00	1.00	1.00	1.00	79–86
31–35	0.91	0.94	0.96	0.91	0.94	0.96	88–95
36–40	0.82	0.88	0.91	0.82	0.88	0.91	97–104
41–45	0.71	0.82	0.87	0.71	0.82	0.87	106–113
46–50	0.58	0.75	0.82	0.58	0.75	0.82	115–122
51–55	0.41	0.67	0.76	0.41	0.67	0.76	124–131
56–60	—	0.58	0.71	—	0.58	0.71	133–140
61–70	—	0.33	0.58	—	0.33	0.58	142–158
71–80	—	—	0.41	—	—	0.41	160–176

*Unless otherwise specifically permitted elsewhere in this *Code,* the overcurrent protection for these conductor types shall not exceed 15 amperes for No. 14, 20 amperes for No. 12, and 30 amperes for No. 10 copper; or 15 amperes for No. 12 and 25 amperes for No. 10 aluminum and copper-clad aluminum.

Table B-310-2 (1996 *Code*) was relocated from Appendix B to Article 310 and renumbered Table 310-20 for the 1999 *Code*. See the commentary following Table 310-20.

Table B-310-3. Ampacities of Multiconductor Cables with Not More than Three Insulated Conductors, Rated 0 Through 2000 Volts, in Free Air Based on Ambient Air Temperature of 40°C (104°F) (For Types TC, MC, MI, UF, and USE Cables)

Size	Temperature Rating of Conductor. See Table 310-13.								Size
	60°C (140°F)	75°C (167°F)	85°C (185°F)	90°C (194°F)	60°C (140°F)	75°C (167°F)	85°C (185°F)	90°C (194°F)	
AWG or kcmil	COPPER				ALUMINUM OR COPPER-CLAD ALUMINUM				AWG or kcmil
18	—	—	—	11*	—	—	—	—	18
16	—	—	—	16*	—	—	—	—	16
14	18*	21*	24*	25*	—	—	—	—	14
12	21*	28*	30*	32*	18*	21*	24*	25*	12
10	28*	36*	41*	43*	21*	28*	30*	32*	10
8	39	50	56	59	30	39	44	46	8
6	52	68	75	79	41	53	59	61	6
4	69	89	100	104	54	70	78	81	4
3	81	104	116	121	63	81	91	95	3
2	92	118	132	138	72	92	103	108	2
1	107	138	154	161	84	108	120	126	1
1/0	124	160	178	186	97	125	139	145	1/0
2/0	143	184	206	215	111	144	160	168	2/0
3/0	165	213	238	249	129	166	185	194	3/0
4/0	190	245	274	287	149	192	214	224	4/0
250	212	274	305	320	166	214	239	250	250
300	237	306	341	357	186	240	268	280	300
350	261	337	377	394	205	265	296	309	350
400	281	363	406	425	222	287	317	334	400
500	321	416	465	487	255	330	368	385	500
600	354	459	513	538	284	368	410	429	600
700	387	502	562	589	306	405	462	473	700
750	404	523	586	615	328	424	473	495	750
800	415	539	604	633	339	439	490	513	800
900	438	570	639	670	362	469	514	548	900
1000	461	601	674	707	385	499	558	584	1000
Ambient Temp. (°C)	For ambient temperatures other than 40°C (104°F), multiply the ampacities shown above by the appropriate factor shown below.								Ambient Temp. (°F)
21–25	1.32	1.20	1.15	1.14	1.32	1.20	1.15	1.14	70–77
26–30	1.22	1.13	1.11	1.10	1.22	1.13	1.11	1.10	79–86
31–35	1.12	1.07	1.05	1.05	1.12	1.07	1.05	1.05	88–95
36–40	1.00	1.00	1.00	1.00	1.00	1.00	1.00	1.00	97–104
41–45	0.87	0.93	0.94	0.95	0.87	0.93	0.94	0.95	106–113
46–50	0.71	0.85	0.88	0.89	0.71	0.85	0.88	0.89	115–122
51–55	0.50	0.76	0.82	0.84	0.50	0.76	0.82	0.84	124–131
56–60	—	0.65	0.75	0.77	—	0.65	0.75	0.77	133–140
61–70	—	0.38	0.58	0.63	—	0.38	0.58	0.63	142–158
71–80	—	—	0.33	0.44	—	—	0.33	0.44	160–176

*Unless otherwise specifically permitted elsewhere in this *Code,* the overcurrent protection for these conductor types shall not exceed 15 amperes for No. 14, 20 amperes for No. 12, and 30 amperes for No. 10 copper; or 15 amperes for No. 12 and 25 amperes for No. 10 aluminum and copper-clad aluminum.

Table B-310-4 (1996 *Code*) was relocated from Appendix B to Article 310 and renumbered Table 310-21 for the 1999 *Code*. See the commentary following Table 310-21.

Table B-310-5. Ampacities of Single Insulated Conductors, Rated 0 through 2000 Volts, in Nonmagnetic Underground Electrical Ducts (One Conductor per Electrical Duct), Based on Ambient Earth Temperature of 20°C (68°F), Electrical Duct Arrangement per Figure B-310-2, Conductor Temperature 75°C (167°F)

Size (kcmil)	3 Electrical Ducts (Fig. B-310-2, Detail 2)			6 Electrical Ducts (Fig. B-310-2, Detail 3)			9 Electrical Ducts (Fig. B-310-2, Detail 4)			3 Electrical Ducts (Fig. B-310-2, Detail 2)			6 Electrical Ducts (Fig. B-310-2, Detail 3)			9 Electrical Ducts (Fig. B-310-2, Detail 4)			Size (kcmil)
	Types RHW, THHW, THW, THWN, XHHW, USE			Types RHW, THHW, THW, THWN, XHHW, USE			Types RHW, THHW, THW, THWN, XHHW, USE			Types RHW, THHW, THW, THWN, XHHW, USE			Types RHW, THHW, THW, THWN, XHHW, USE			Types RHW, THHW, THW, THWN, XHHW, USE			
	COPPER									ALUMINUM OR COPPER-CLAD ALUMINUM									
	RHO 60 LF 50	RHO 90 LF 100	RHO 120 LF 100	RHO 60 LF 50	RHO 90 LF 100	RHO 120 LF 100	RHO 60 LF 50	RHO 90 LF 100	RHO 120 LF 100	RHO 60 LF 50	RHO 90 LF 100	RHO 120 LF 100	RHO 60 LF 50	RHO 90 LF 100	RHO 120 LF 100	RHO 60 LF 50	RHO 90 LF 100	RHO 120 LF 100	
250	410	344	327	386	295	275	369	270	252	320	269	256	302	230	214	288	211	197	250
350	503	418	396	472	355	330	446	322	299	393	327	310	369	277	258	350	252	235	350
500	624	511	484	583	431	400	545	387	360	489	401	379	457	337	313	430	305	284	500
750	794	640	603	736	534	494	674	469	434	626	505	475	581	421	389	538	375	347	750
1000	936	745	700	864	617	570	776	533	493	744	593	557	687	491	453	629	432	399	1000
1250	1055	832	781	970	686	632	854	581	536	848	668	627	779	551	508	703	478	441	1250
1500	1160	907	849	1063	744	685	918	619	571	941	736	689	863	604	556	767	517	477	1500
1750	1250	970	907	1142	793	729	975	651	599	1026	796	745	937	651	598	823	550	507	1750
2000	1332	1027	959	1213	836	768	1030	683	628	1103	850	794	1005	693	636	877	581	535	2000
Ambient Temp. (°C)	For ambient temperatures other than 20°C (68°F), multiply the ampacities shown above by the appropriate factor shown below.																		Ambient Temp. (°F)
6–10	1.09			1.09			1.09			1.09			1.09			1.09			43–50
11–15	1.04			1.04			1.04			1.04			1.04			1.04			52–59
16–20	1.00			1.00			1.00			1.00			1.00			1.00			61–68
21–25	0.95			0.95			0.95			0.95			0.95			0.95			70–77
26–30	0.90			0.90			0.90			0.90			0.90			0.90			79–86

Table B-310-6. Ampacities of Three Insulated Conductors, Rated 0 through 2000 Volts, Within an Overall Covering (Three-Conductor Cable) in Underground Electrical Ducts (One Cable per Electrical Duct) Based on Ambient Earth Temperature of 20°C (68°F), Electrical Duct Arrangement per Figure B-310-2, Conductor Temperature 75°C (167°F)

Size (AWG or kcmil)	1 Electrical Duct Fig. B-310-2, Detail 1)			3 Electrical Ducts (Fig. B-310-2, Detail 2)			6 Electrical Ducts (Fig. B-310-2, Detail 3)			1 Electrical Duct Fig. B-310-2, Detail 1)			3 Electrical Ducts (Fig. B-310-2, Detail 2)			6 Electrical Ducts (Fig. B-310-2, Detail 3)			Size (AWG or kcmil)
	Types RHW, THHW, THW, THWN, XHHW, USE			Types RHW, THHW, THW, THWN, XHHW, USE			Types RHW, THHW, THW, THWN, XHHW, USE			Types RHW, THHW, THW, THWN, XHHW, USE			Types RHW, THHW, THW, THWN, XHHW, USE			Types RHW, THHW, THW, THWN, XHHW, USE			
	COPPER									ALUMINUM OR COPPER-CLAD ALUMINUM									
	RHO 60 LF 50	RHO 90 LF 100	RHO 120 LF 100	RHO 60 LF 50	RHO 90 LF 100	RHO 120 LF 100	RHO 60 LF 50	RHO 90 LF 100	RHO 120 LF 100	RHO 60 LF 50	RHO 90 LF 100	RHO 120 LF 100	RHO 60 LF 50	RHO 90 LF 100	RHO 120 LF 100	RHO 60 LF 50	RHO 90 LF 100	RHO 120 LF 100	
8	58	54	53	56	48	46	53	42	39	45	42	41	43	37	36	41	32	30	8
6	77	71	69	74	63	60	70	54	51	60	55	54	57	49	47	54	42	39	6
4	101	93	91	96	81	77	91	69	65	78	72	71	75	63	60	71	54	51	4
2	132	121	118	126	105	100	119	89	83	103	94	92	98	82	78	92	70	65	2
1	154	140	136	146	121	114	137	102	95	120	109	106	114	94	89	107	79	74	1
1/0	177	160	156	168	137	130	157	116	107	138	125	122	131	107	101	122	90	84	1/0
2/0	203	183	178	192	156	147	179	131	121	158	143	139	150	122	115	140	102	95	2/0
3/0	233	210	204	221	178	158	205	148	137	182	164	159	172	139	131	160	116	107	3/0
4/0	268	240	232	253	202	190	234	168	155	209	187	182	198	158	149	183	131	121	4/0
250	297	265	256	280	222	209	258	184	169	233	207	201	219	174	163	505	144	132	250
350	363	321	310	340	267	250	312	219	202	285	252	244	267	209	196	245	172	158	350
500	444	389	375	414	320	299	377	261	240	352	308	297	328	254	237	299	207	190	500
750	552	478	459	511	388	362	462	314	288	446	386	372	413	314	293	374	254	233	750
1000	628	539	518	579	435	405	522	351	321	521	447	430	480	361	336	433	291	266	1000
Ambient Temp. (°C)	For ambient temperatures other than 20°C (68°F), multiply the ampacities shown above by the appropriate factor shown below.																		Ambient Temp. (°F)
6–10	1.09			1.09			1.09			1.09			1.09			1.09			43–50
11–15	1.04			1.04			1.04			1.04			1.04			1.04			52–59
16–20	1.00			1.00			1.00			1.00			1.00			1.00			61–68
21–25	0.95			0.95			0.95			0.95			0.95			0.95			70–77
26–30	0.90			0.90			0.90			0.90			0.90			0.90			79–86

Table B-310-7. Ampacities of Three Single Insulated Conductors, Rated 0 Through 2000 Volts, in Underground Electrical Ducts (Three Conductors per Electrical Duct) Based on Ambient Earth Temperature of 20°C (68°F), Electrical Duct Arrangement per Figure B-310-2, Conductor Temperature 75°C (167°F)

Size (AWG or kcmil)	1 Electrical Duct Fig. B-310-2, Detail 1)			3 Electrical Ducts (Fig. B-310-2, Detail 2)			6 Electrical Ducts (Fig. B-310-2, Detail 3)			1 Electrical Duct Fig. B-310-2, Detail 1)			3 Electrical Ducts (Fig. B-310-2, Detail 2)			6 Electrical Ducts (Fig. B-310-2, Detail 3)			Size (AWG or kcmil)
	Types RHW, THHW, THW, THWN, XHHW, USE			Types RHW, THHW, THW, THWN, XHHW, USE			Types RHW, THHW, THW, THWN, XHHW, USE			Types RHW, THHW, THW, THWN, XHHW, USE			Types RHW, THHW, THW, THWN, XHHW, USE			Types RHW, THHW, THW, THWN, XHHW, USE			
	COPPER									ALUMINUM OR COPPER-CLAD ALUMINUM									
	RHO 60 LF 50	RHO 90 LF 100	RHO 120 LF 100	RHO 60 LF 50	RHO 90 LF 100	RHO 120 LF 100	RHO 60 LF 50	RHO 90 LF 100	RHO 120 LF 100	RHO 60 LF 50	RHO 90 LF 100	RHO 120 LF 100	RHO 60 LF 50	RHO 90 LF 100	RHO 120 LF 100	RHO 60 LF 50	RHO 90 LF 100	RHO 120 LF 100	
8	63	58	57	61	51	49	57	44	41	49	45	44	47	40	38	45	34	32	8
6	84	77	75	80	67	63	75	56	53	66	60	58	63	52	49	59	44	41	6
4	111	100	98	105	86	81	98	73	67	86	78	76	79	67	63	77	57	52	4
3	129	116	113	122	99	94	113	83	77	101	91	89	83	77	73	84	65	60	3
2	147	132	128	139	112	106	129	93	86	115	103	100	108	87	82	101	73	67	2
1	171	153	148	161	128	121	149	106	98	133	119	115	126	100	94	116	83	77	1
1/0	197	175	169	185	146	137	170	121	111	153	136	132	144	114	107	133	94	87	1/0
2/0	226	200	193	212	166	156	194	136	126	176	156	151	165	130	121	151	106	98	2/0
3/0	260	228	220	243	189	177	222	154	142	203	178	172	189	147	138	173	121	111	3/0
4/0	301	263	253	280	215	201	255	175	161	235	205	198	219	168	157	199	137	126	4/0
250	334	290	279	310	236	220	281	192	176	261	227	218	242	185	172	220	150	137	250
300	373	321	308	344	260	242	310	210	192	293	252	242	272	204	190	245	165	151	300
350	409	351	337	377	283	264	340	228	209	321	276	265	296	222	207	266	179	164	350
400	442	376	361	394	302	280	368	243	223	349	297	284	321	238	220	288	191	174	400
500	503	427	409	460	341	316	412	273	249	397	338	323	364	270	250	326	216	197	500
600	552	468	447	511	371	343	457	296	270	446	373	356	408	296	274	365	236	215	600
700	602	509	486	553	402	371	492	319	291	488	408	389	443	321	297	394	255	232	700
750	632	529	505	574	417	385	509	330	301	508	425	405	461	334	309	409	265	241	750
800	654	544	520	597	428	395	527	338	308	530	439	418	481	344	318	427	273	247	800
900	692	575	549	628	450	415	554	355	323	563	466	444	510	365	337	450	288	261	900
1000	730	605	576	659	472	435	581	372	338	597	494	471	538	385	355	475	304	276	1000
Ambient Temp. (°C)	For ambient temperatures other than 20°C (68°F), multiply the ampacities shown above by the appropriate factor shown below.																		Ambient Temp. (°F)
6–10	1.09			1.09			1.09			1.09			1.09			1.09			43–50
11–15	1.04			1.04			1.04			1.04			1.04			1.04			52–59
16–20	1.00			1.00			1.00			1.00			1.00			1.00			61–68
21–25	0.95			0.95			0.95			0.95			0.95			0.95			70–77
26–30	0.90			0.90			0.90			0.90			0.90			0.90			79–86

Table B-310-8. Ampacities of Two or Three Insulated Conductors, Rated 0 Through 2000 Volts, Cabled Within an Overall (Two- or Three-Conductor) Covering, Directly Buried in Earth, Based on Ambient Earth Temperature of 20°C (68°F), Arrangement per Figure B-310-2, 100 Percent Load Factor, Thermal Resistance (Rho) of 90

Size (AWG or kcmil)	1 Cable (Fig. B-310-2, Detail 5)		2 Cables (Fig. B-310-2, Detail 6)		1 Cable (Fig. B-310-2, Detail 5)		2 Cables (Fig. B-310-2, Detail 6)		Size (AWG or kcmil)
	60°C (140°F)	75°C (167°F)	60°C (140°F)	75°C (167°F)	60°C (140°F)	75°C (167°F)	60°C (140°F)	75°C (167°F)	
	TYPES				TYPES				
	UF	RHW, THHW, THW, THWN, XHHW, USE	UF	RHW, THHW, THW, THWN, XHHW, USE	UF	RHW, THHW, THW, THWN, XHHW, USE	UF	RHW, THHW, THW, THWN, XHHW, USE	
	COPPER				ALUMINUM OR COPPER-CLAD ALUMINUM				
8	64	75	60	70	51	59	47	55	8
6	85	100	81	95	68	75	60	70	6
4	107	125	100	117	83	97	78	91	4
2	137	161	128	150	107	126	110	117	2
1	155	182	145	170	121	142	113	132	1
1/0	177	208	165	193	138	162	129	151	1/0
2/0	201	236	188	220	157	184	146	171	2/0
3/0	229	269	213	250	179	210	166	195	3/0
4/0	259	304	241	282	203	238	188	220	4/0
250	—	333	—	308	—	261	—	241	250
350	—	401	—	370	—	315	—	290	350
500	—	481	—	442	—	381	—	350	500
750	—	585	—	535	—	473	—	433	750
1000	—	657	—	600	—	545	—	497	1000
Ambient Temp. (°C)	For ambient temperatures other than 20°C (68°F), multiply the ampacities shown above by the appropriate factor shown below.								Ambient Temp. (°F)
6–10	1.12	1.09	1.12	1.09	1.12	1.09	1.12	1.09	43–50
11–15	1.06	1.04	1.06	1.04	1.06	1.04	1.06	1.04	52–59
16–20	1.00	1.00	1.00	1.00	1.00	1.00	1.00	1.00	61–68
21–25	0.94	0.95	0.94	0.95	0.94	0.95	0.94	0.95	70–77
26–30	0.87	0.90	0.87	0.90	0.87	0.90	0.87	0.90	79–86

Note: For ampacities of Type UF cable in underground electrical ducts, multiply the ampacities shown in the table by 0.74.

Table B-310-9. Ampacities of Three Triplexed Single Insulated Conductors, Rated 0 Through 2000 Volts, Directly Buried in Earth Based on Ambient Earth Temperature of 20°C (68°F), Arrangement per Figure B-310-2, 100 Percent Load Factor, Thermal Resistance (Rho) of 90

Size (AWG or kcmil)	See Fig. B-310-2, Detail 7		See Fig. B-310-2, Detail 8		See Fig. B-310-2, Detail 7		See Fig. B-310-2, Detail 8		Size (AWG or kcmil)
	60°C (140°F)	75°C (167°F)	60°C (140°F)	75°C (167°F)	60°C (140°F)	75°C (167°F)	60°C (140°F)	75°C (167°F)	
	TYPES				TYPES				
	UF	USE	UF	USE	UF	USE	UF	USE	
	COPPER				ALUMINUM OR COPPER-CLAD ALUMINUM				
8	72	84	66	77	55	65	51	60	8
6	91	107	84	99	72	84	66	77	6
4	119	139	109	128	92	108	85	100	4
2	153	179	140	164	119	139	109	128	2
1	173	203	159	186	135	158	124	145	1
1/0	197	231	181	212	154	180	141	165	1/0
2/0	223	262	205	240	175	205	159	187	2/0
3/0	254	298	232	272	199	233	181	212	3/0
4/0	289	339	263	308	226	265	206	241	4/0
250	—	370	—	336	—	289	—	263	250
350	—	445	—	403	—	349	—	316	350
500	—	536	—	483	—	424	—	382	500
750	—	654	—	587	—	525	—	471	750
1000	—	744	—	665	—	608	—	544	1000
Ambient Temp. (°C)	For ambient temperatures other than 20°C (68°F), multiply the ampacities shown above by the appropriate factor shown below.								Ambient Temp. (°F)
6–10	1.12	1.09	1.12	1.09	1.12	1.09	1.12	1.09	43–50
11–15	1.06	1.04	1.06	1.04	1.06	1.04	1.06	1.04	52–59
16–20	1.00	1.00	1.00	1.00	1.00	1.00	1.00	1.00	61–68
21–25	0.94	0.95	0.94	0.95	0.94	0.95	0.94	0.95	70–77
26–30	0.87	0.90	0.87	0.90	0.87	0.90	0.87	0.90	79–86

Table B-310-10. Ampacities of Three Single Insulated Conductors, Rated 0 Through 2000 Volts, Directly Buried in Earth Based on Ambient Earth Temperature of 20°C (68°F), Arrangement per Figure B-310-2, 100 Percent Load Factor, Thermal Resistance (Rho) of 90

Size (AWG or kcmil)	See Fig. B-310-2, Detail 9		See Fig. B-310-2, Detail 10		See Fig. B-310-2, Detail 9		See Fig. B-310-2, Detail 10		Size (AWG or kcmil)
	60°C (140°F)	75°C (167°F)	60°C (140°F)	75°C (167°F)	60°C (140°F)	75°C (167°F)	60°C (140°F)	75°C (167°F)	
	TYPES				TYPES				
	UF	USE	UF	USE	UF	USE	UF	USE	
	COPPER				ALUMINUM OR COPPER-CLAD ALUMINUM				
8	84	98	78	92	66	77	61	72	8
6	107	126	101	118	84	98	78	92	6
4	139	163	130	152	108	127	101	118	4
2	178	209	165	194	139	163	129	151	2
1	201	236	187	219	157	184	146	171	1
1/0	230	270	212	249	179	210	165	194	1/0
2/0	261	306	241	283	204	239	188	220	2/0
3/0	297	348	274	321	232	272	213	250	3/0
4/0	336	394	309	362	262	307	241	283	4/0
250	—	429	—	394	—	335	—	308	250
350	—	516	—	474	—	403	—	370	350
500	—	626	—	572	—	490	—	448	500
750	—	767	—	700	—	605	—	552	750
1000	—	887	—	808	—	706	—	642	1000
1250	—	979	—	891	—	787	—	716	1250
1500	—	1063	—	965	—	862	—	783	1500
1750	—	1133	—	1027	—	930	—	843	1750
2000	—	1195	—	1082	—	990	—	897	2000
Ambient Temp. (°C)	For ambient temperatures other than 20°C (68°F), multiply the ampacities shown above by the appropriate factor shown below.								Ambient Temp. (°F)
6–10	1.12	1.09	1.12	1.09	1.12	1.09	1.12	1.09	43–50
11–15	1.06	1.04	1.06	1.04	1.06	1.04	1.06	1.04	52–59
16–20	1.00	1.00	1.00	1.00	1.00	1.00	1.00	1.00	61–68
21–25	0.94	0.95	0.94	0.95	0.94	0.95	0.94	0.95	70–77
26–30	0.87	0.90	0.87	0.90	0.87	0.90	0.87	0.90	79–86

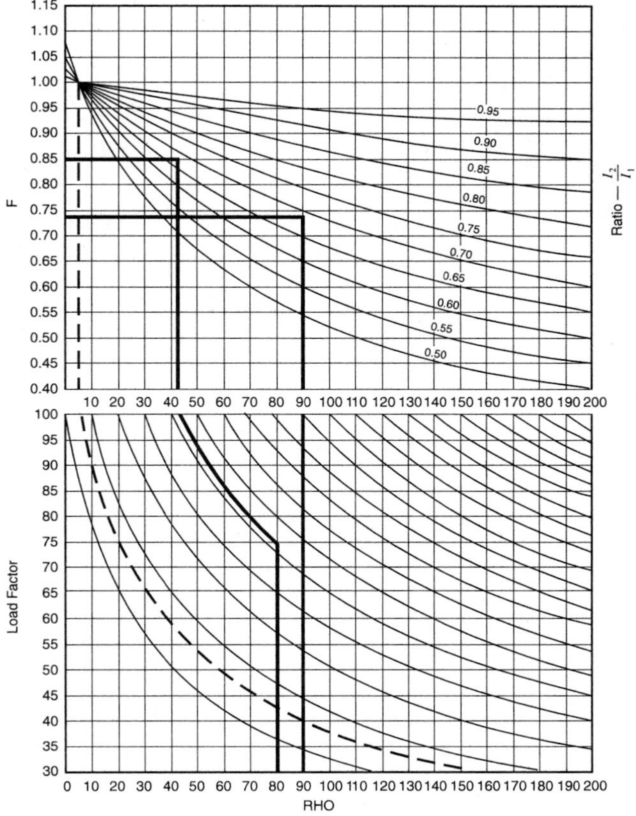

Figure B-310-1 Interpolation chart for cables in a duct bank I_1 = **ampacity for Rho = 60, 50 LF;** I_2 = **ampacity for Rho = 120, 100 LF (load factor); desired ampacity =** $F \times I_1$.

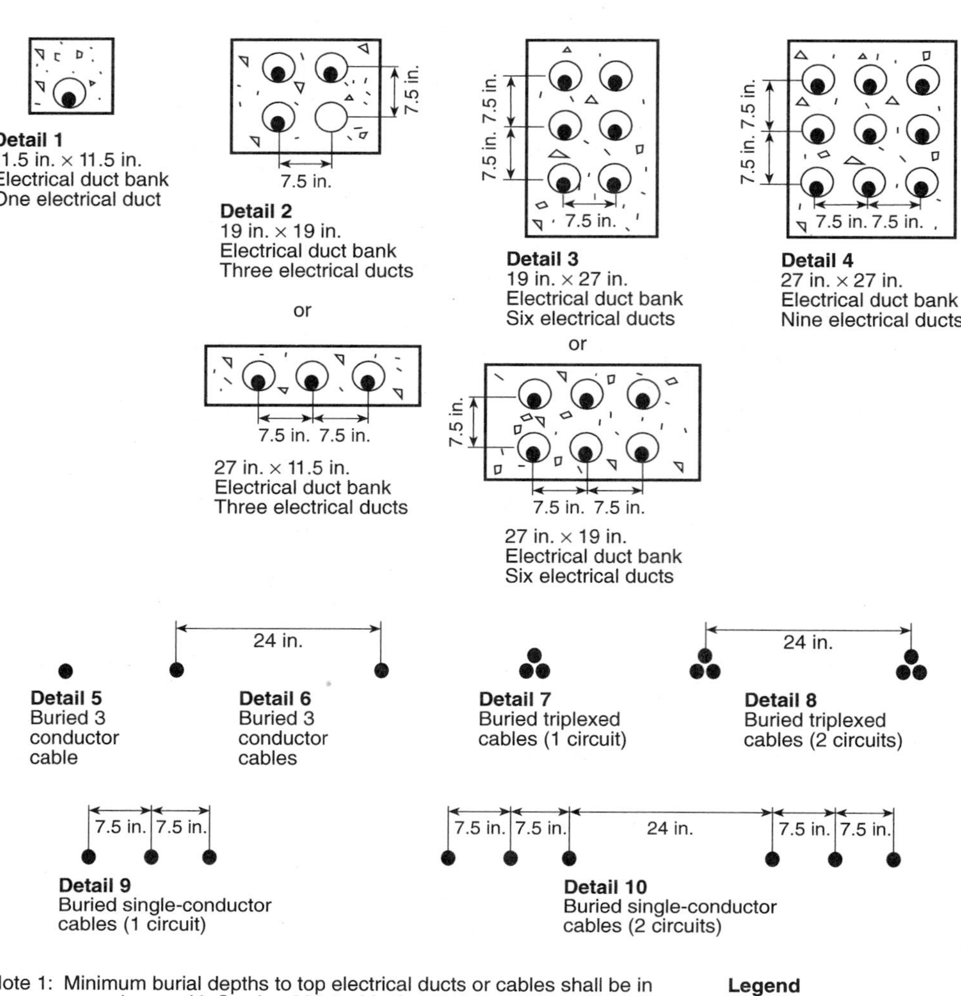

Note 1: Minimum burial depths to top electrical ducts or cables shall be in accordance with Section 300-5. Maximum depth to the top of electrical duct banks shall be 30 in. and maximum depth to the top of direct buried cables shall be 36 in.

Note 2: For two and four electrical duct installations with electrical ducts installed in a single row, see Section B-310-15(b)(5).

Note 3: For SI units: 1 in. = 25.4 mm; 1 ft = 0.3048 m.

Legend

- Backfill (earth or concrete)
- Electrical duct
- Cable or cables

Figure B-310-2 Cable installation dimensions for use with Tables B-310-5 through B-310-10.

Figure B-310-3 Ampacities of single insulated conductors rated 0 through 5000 volts in underground electrical ducts (three conductors per electrical duct) nine single-conductor cables per phase based on ambient earth temperature of 20°C (68°F), conductor temperature 75°C (167°F).

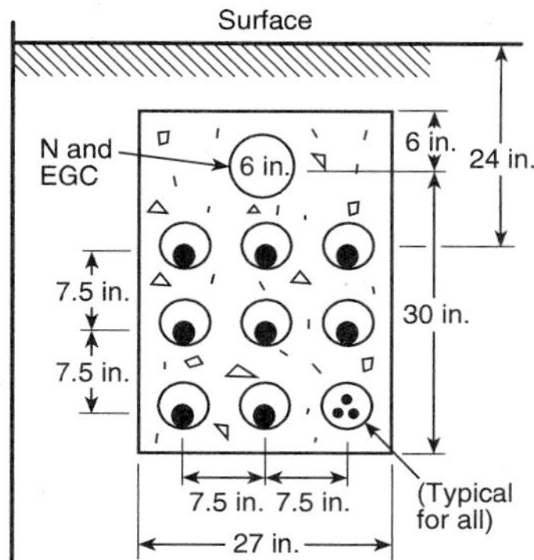

Design Criteria
Neutral and Electrical Grounding Conductor (EGC) Duct = 6 in.
Phase Ducts = 3 to 5 in.
Conductor Material = Copper
Number of Cables per Duct = 3

Number of Cables per Phase = 9
Rho Concrete = Rho Earth 5
Rho PVC Duct = 650
Rho Cable Insulation = 500
Rho Cable Jacket = 650

For SI units: 1 in. = 25.4 mm.

Notes:
1. Neutral configuration per Section 300-5(i), Exception No. 2, for isolated phase installations in nonmagnetic ducts.
2. Phasing is A, B, C in rows or columns. Where magnetic electrical ducts are used, conductors are installed A, B, C per electrical duct with the neutral and all equipment grounding conductors in the same electrical duct. In this case, the 6-in. trade size neutral duct is eliminated.
3. Maximum harmonic loading on the neutral conductor cannot exceed 50 percent of the phase current for the ampacities shown in the table.
4. Metallic shields of Type MV-90 cable shall be grounded at one point only where using A, B, C phasing in rows or columns.

Size	TYPES RHW, THHW, THW, THWN, XHHW, USE, OR MV-90*			Size
	Total per Phase Ampere Rating			
kcmil	RHO EARTH 60 LF 50	RHO EARTH 90 LF 100	RHO EARTH 120 LF 100	kcmil
250	2340 (260A/Cable)	1530 (170A/Cable)	1395 (155A/Cable)	250
350	2790 (310A/Cable)	1800 (200A/Cable)	1665 (185A/Cable)	350
500	3375 (375A/Cable)	2160 (240A/Cable)	1980 (220A/Cable)	500

Ambient Temp. (°C)	For ambient temperatures other than 20°C (68°F), multiply the ampacities shown above by the appropriate factor shown below.					Ambient Temp. (°F)
6–10	1.09	1.09	1.09	1.09	1.09	43–50
11–15	1.04	1.04	1.04	1.04	1.04	52–59
16–20	1.00	1.00	1.00	1.00	1.00	61–68
21–25	0.95	0.95	0.95	0.95	0.95	70–77
26–30	0.90	0.90	0.90	0.90	0.90	79–86

*Limited to 75°C conductor temperature.

Figure B-310-4 Ampacities of single insulated conductors rated 0 through 5000 volts in nonmagnetic underground electrical ducts (one conductor per electrical duct) four single-conductor cables per phase based on ambient earth temperature of 20°C (68°F), conductor temperature 75°C (167°F).

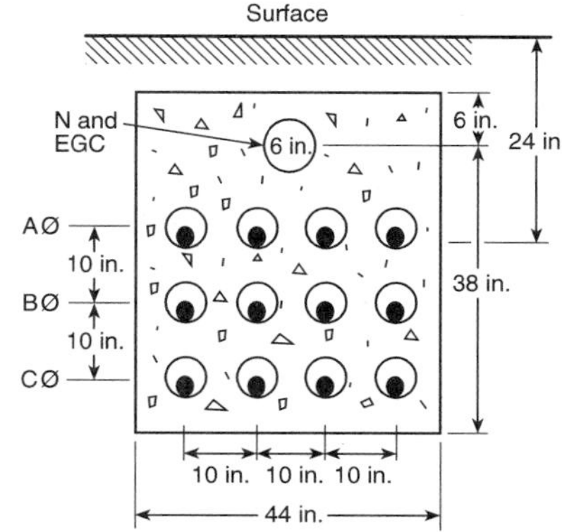

Design Criteria
Neutral and Electrical Grounding Conductor (EGC) Duct = 6 in.
Phase Ducts = 3 in. (min.)
Conductor Material = Copper
Number of Cables per Duct = 1

Number of Cables per Phase = 4
Rho Concrete = Rho Earth – 5
Rho PVC Duct = 650
Rho Cable Insulation = 500
Rho Cable Jacket = 650

Notes:
1. For SI units, 1 in. = 25.4 mm.
2. Neutral configuration per Section 300-5(i), Exception No. 2.
3. Maximum harmonic loading on the neutral conductor cannot exceed 50 percent of the phase current for the ampacities shown in the table.
4. Metallic shields of Type MV-90 cable shall be grounded at one point only.

Size	TYPES			Size
	RHW, THHW, THW, THWN, XHHW, USE, OR MV-90*			
	Total Per Phase Ampere Rating			
kcmil	RHO EARTH 60 LF 50	RHO EARTH 90 LF 100	RHO EARTH 120 LF 100	kcmil
750	2520 (705A/Cable)	1860 (465A/Cable)	1680 (420A/Cable)	750
1000	3300 (825A/Cable)	2140 (535A/Cable)	1920 (480A/Cable)	1000
1250	3700 (925A/Cable)	2380 (595A/Cable)	2120 (530A/Cable)	1250
1500	4060 (1015A/Cable)	2580 (645A/Cable)	2300 (575A/Cable)	1500
1750	4360 1090A/Cable)	2740 (685A/Cable)	2460 (615A/Cable)	1750

Ambient Temp. (°C)	For ambient temperatures other than 20°C (68°F), multiply the ampacities shown above by the appropriate factor shown below.					Ambient Temp. (°F)
6–10	1.09	1.09	1.09	1.09	1.09	43–50
11–15	1.04	1.04	1.04	1.04	1.04	52–59
16–20	1.00	1.00	1.00	1.00	1.00	61–68
21–25	0.95	0.95	0.95	0.95	0.95	70–77
26–30	0.90	0.90	0.90	0.90	0.90	79–86

*Limited to 75°C conductor temperature.

Figure B-310-5 Ampacities of single insulated conductors rated 0 through 5000 volts in nonmagnetic underground electrical ducts (one conductor per electrical duct) five single-conductor cables per phase based on ambient earth temperature of 20°C (68°F), conductor temperature 75°C (167°F).

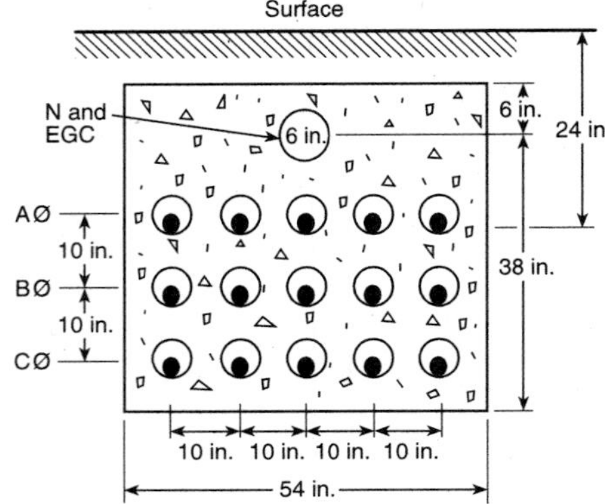

Design Criteria	
Neutral and Electrical Grounding Conductor (EGC) Duct = 6 in.	Number of Cables per Phase = 5
Phase Ducts = 3 in. (min.)	Rho Concrete = Rho Earth − 5
Conductor Material = Copper	Rho PVC Duct = 650
Number of Cables per Duct = 1	Rho Cable Insulation = 500
	Rho Cable Jacket = 650

Notes:
1. For SI units, 1 in. = 25.4 mm.
2. Neutral configuration per Section 300-5(i), Exception No. 2.
3. Maximum harmonic loading on the neutral conductor cannot exceed 50 percent of the phase current for the ampacities shown in the table.
4. Metallic shields of Type MV-90 cable shall be grounded at one point only.

Size	TYPES RHW, THHW, THW, THWN, XHHW, USE, OR MV-90*			Size
	Total Per Phase Ampere Rating			
kcmil	RHO EARTH 60 LF 50	RHO EARTH 90 LF 100	RHO EARTH 120 LF 100	kcmil
2000	5575 (1115A/Cable)	3375 (675A/Cable)	3000 (600A/Cable)	2000

Ambient Temp. (°C)	For ambient temperatures other than 20°C (68°F), multiply the ampacities shown above by the appropriate factor shown below.					Ambient Temp. (°F)
6–10	1.09	1.09	1.09	1.09	1.09	43–50
11–15	1.04	1.04	1.04	1.04	1.04	52–59
16–20	1.00	1.00	1.00	1.00	1.00	61–68
21–25	0.95	0.95	0.95	0.95	0.95	70–77
26–30	0.90	0.90	0.90	0.90	0.90	79–86

*Limited to 75°C conductor temperature.

Table B-310-11. Adjustment Factors for More than Three Current-Carrying Conductors in a Raceway or Cable with Load Diversity

Number of Current-Carrying Conductors	Percent of Values in Tables as Adjusted for Ambient Temperature if Necessary
4–6	80
7–9	70
10–24	70*
25–42	60*
43–85	50*

*These factors include the effects of a load diversity of 50 percent.

FPN: The ampacity limit for the number of current-carrying conductors in 10 through 85 is based on the formula shown below. For greater than 85 conductors, special calculations are required that are beyond the scope of this table.

$$A_2 = \sqrt{\frac{0.5N}{E}} \times (A_1) \text{ or } A_1, \text{ whichever is less}$$

Where:

A_1 = ampacity from Tables 310-16; 310-18; B-310-1; B-310-6; and B-310-7 multiplied by the appropriate factor from Table B-310-11.

N = total number of conductors used to obtain multiplying factor from Table B-310-11.

E = desired number of current-carrying conductors in the raceway or cable.

A_2 = ampacity limit for the current-carrying conductors in the raceway or cable.

Example No. 1

Calculate the ampacity limit for 12 No. 14 THWN current-carrying conductors (75°C) in a raceway that contains 24 conductors.

$$A_2 = \sqrt{\frac{(0.5)(24)}{12}} \times 20(0.7)$$
$$= 14 \text{ amperes (i.e., 50 percent diversity)}$$

Example No. 2

Calculate the ampacity limit for 18 No. 14 THWN current-carrying conductors (75°C) in a raceway that contains 24 conductors.

$$A_2 = \sqrt{\frac{(0.5)(24)}{18}} \times 20(0.7) = 11.5 \text{ amperes}$$

APPENDIX C

Conduit and Tubing Fill Tables for Conductors and Fixture Wires of the Same Size

Contents

C12(A). Maximum Number of Compact Conductors in Type EB, PVC Conduit (Based on Table 1, Chapter 9)

Appendix C is not a part of the requirements of this *Code* and is included for information only. However, by using the tables in this appendix, one is afforded very accurate calculations without having to perform the calculations according to Chapter 9, Table 1.

These new Appendix C tables became necessary because of the different internal dimensions of the many types of conduit and tubing. The previous Chapter 9 Tables 3A, 3B, and 3C are no longer used in calculating conduit fill. The Chapter 9 tables were becoming restricted and less accurate as new wiring methods were introduced.

As in the past editions of the *Code,* Chapter 9, Table 1, sets forth the percentage fill required, and Chapter 9, Tables 4, 5, and 5A, list the accurate conduit, tubing, and wire dimensions. The user may calculate the percent fill, as permitted in past editions of the *Code,* or may use Appendix C, since all of Appendix C data was generated using Tables 1, 4, 5, and 5A.

At first glance, the number of tables contained in this appendix may appear daunting. However, after closer inspection, the reader will realize that there are only 12 sets (or groups) of tables, to correspond with the 12 wiring methods.

Within each set of tables are three types of tables. First, there are the conductors for general wiring. They appear in an order similar to previous Chapter 9, Tables 3A, 3B, and 3C. Next are fixture wires, in an order similar to the previous Chapter 9, Table 2. Finally, compact stranded conductors are listed, similar to former Chapter 9, Table 5A. Tables using compact stranding are listed as "A" tables.

In order to select the correct trade size conduit or tubing, the user is encouraged to proceed as follows.

Step 1. Select the appropriate wiring method from Tables C1 through C12 using the following lists of metallic and nonmetallic wiring methods.

Metallic Wiring Methods

Type of Wiring	Appropriate Table
Electrical metallic tubing	Table C1
Flexible metal conduit	Table C3
Intermediate metal conduit	Table C4
Liquidtight flexible metal conduit	Table C7
Rigid metal conduit	Table C8

Nonmetallic Wiring Methods

Type of Wiring	Appropriate Table
Electrical nonmetallic tubing	Table C2
Liquidtight flexible nonmetallic conduit (FNMC-B)	Table C5
Liquidtight flexible nonmetallic conduit (FNMC-A)	Table C6
Rigid PVC conduit, Schedule 80	Table C9
Rigid PVC conduit, Schedule 40 and HDPE conduit	Table C10
Type A, rigid PVC conduit	Table C11
Type EB, rigid PVC conduit	Table C12

Step 2. Choose the appropriate conductors (general wiring conductors, fixture wires, or compact conductors).

Step 3. Choose the appropriate insulation.

Step 4. Select the correct trade size conduit or tubing for the given quantity and size of conductors required.

See the following four examples of how to determine the correct trade size conduit or tubing.

Example 1

A circuit will require ten No. 10 THWN-2 conductors in an underground conduit across a parking lot for exterior lighting. What size PVC conduit will be required?

Answer

Appendix C lists various types of PVC conduit, each of which is included below.

Wiring Method	Table	Trade Size Conduit or Tubing
PVC Schedule 40 and HDPE	Table C10	1 in.
PVC Type A	Table C11	¾ in.
PVC Schedule 80	Table C9	1 in.
PVC Type EB	Table C12	This conduit is only available in trade sizes 2 in. and larger.

Example 2

An underground service lateral requires four 600-kcmil XHHW compact-stranded aluminum conductors. What trade size conduit will be required for RMC, IMC, PVC Schedule 40, PVC Schedule 80, and PVC Type EB?

Answer

Appendix C lists the following:

Wiring Method	Table	Trade Size Conduit or Tubing
RMC	Table C8A	3 in.
PVC Schedule 40	Table C10A	3 in.
PVC Schedule 80	Table C9A	3½ in.
PVC Type EB	Table C12A	3 in.
IMC	Table C4A	3 in.

Most aluminum building wire in Types THW, THWN/THHN, and XHHW is compact stranded.

Example 3

A 40-hp motor will be supplied by three No. 4 TW conductors. What size metal conduit or tubing will be required? What size metal flex will be required at the motor termination?

Answer

Appendix C lists various types of metal conduit and metal flex, each of which is included below.

Wiring Method	Table	Trade Size Conduit or Tubing
EMT	Table C1	1 in.
IMC	Table C4	1 in.
RMC	Table C8	1 in.
FMC	Table C3	1 in.
Liquidtight, FMC	Table C7	1 in.

Example 4

A fire alarm system will require the riser to contain 21 No. 16 TFF conductors. What size electrical metallic tubing will be required?

Answer

According to Table C1, a 1-in. EMT is required.

Table C1. Maximum Number of Conductors and Fixture Wires in Electrical Metallic Tubing *(Based on Table 1, Chapter 9)*

CONDUCTORS											
	Conductor Size (AWG/ kcmil)	Trade Size (in.)									
Type		½	¾	1	1¼	1½	2	2½	3	3½	4
RH	14	6	10	16	28	39	64	112	169	221	282
	12	4	8	13	23	31	51	90	136	177	227
RHH, RHW, RHW,	14	4	7	11	20	27	46	80	120	157	201
	12	3	6	9	17	23	38	66	100	131	167
RH, RHH, RHW, RHW-2	10	2	5	8	13	18	30	53	81	105	135
	8	1	2	4	7	9	16	28	42	55	70
	6	1	1	3	5	8	13	22	34	44	56
	4	1	1	2	4	6	10	17	26	34	44
	3	1	1	1	4	5	9	15	23	30	38
	2	1	1	1	3	4	7	13	20	26	33
	1	0	1	1	1	3	5	9	13	17	22
	1/0	0	1	1	1	2	4	7	11	15	19
	2/0	0	1	1	1	2	4	6	10	13	17
	3/0	0	0	1	1	1	3	5	8	11	14
	4/0	0	0	1	1	1	3	5	7	9	12
	250	0	0	0	1	1	1	3	5	7	9
	300	0	0	0	1	1	1	3	5	6	8
	350	0	0	0	1	1	1	3	4	6	7
	400	0	0	0	1	1	1	2	4	5	7
	500	0	0	0	0	1	1	2	3	4	6
	600	0	0	0	0	1	1	1	3	4	5
	700	0	0	0	0	0	1	1	2	3	4
	750	0	0	0	0	0	1	1	2	3	4
	800	0	0	0	0	0	1	1	2	3	4
	900	0	0	0	0	0	1	1	1	3	3
	1000	0	0	0	0	0	1	1	1	2	3
	1250	0	0	0	0	0	0	1	1	1	2
	1500	0	0	0	0	0	0	1	1	1	1
	1750	0	0	0	0	0	0	1	1	1	1
	2000	0	0	0	0	0	0	1	1	1	1
TW	14	8	15	25	43	58	96	168	254	332	424
	12	6	11	19	33	45	74	129	195	255	326
	10	5	8	14	24	33	55	96	145	190	243
	8	2	5	8	13	18	30	53	81	105	135
RHH*, RHW*, RHW-2*, THHW, THW THW-2,	14	6	10	16	28	39	64	112	169	221	282
RHH*, RHW*, RHW-2*, THHW, THW	12	4	8	13	23	31	51	90	136	177	227
	10	3	6	10	18	24	40	70	106	138	177
RHH*, RHW*, RHW-2*, THHW, THW, THW-2	8	1	4	6	10	14	24	42	63	83	106

*Types RHH, RHW, and RHW-2 without outer covering.

Table C1. Continued

CONDUCTORS											
	Conductor Size (AWG/ kcmil)	Trade Size (in.)									
Type		½	¾	1	1¼	1½	2	2½	3	3½	4
RHH*, RHW*, RHW-2*, TW, THW, THHW, THW-2	6	1	3	4	8	11	18	32	48	63	81
	4	1	1	3	6	8	13	24	36	47	60
	3	1	1	3	5	7	12	20	31	40	52
	2	1	1	2	4	6	10	17	26	34	44
	1	1	1	1	3	4	7	12	18	24	31
	1/0	0	1	1	2	3	6	10	16	20	26
	2/0	0	1	1	1	3	5	9	13	17	22
	3/0	0	1	1	1	2	4	7	11	15	19
	4/0	0	0	1	1	1	3	6	9	12	16
	250	0	0	1	1	1	3	5	7	10	13
	300	0	0	1	1	1	2	4	6	8	11
	350	0	0	0	1	1	1	4	6	7	10
	400	0	0	0	1	1	1	3	5	7	9
	500	0	0	0	1	1	1	3	4	6	7
	600	0	0	0	1	1	1	2	3	4	6
	700	0	0	0	0	1	1	1	3	4	5
	750	0	0	0	0	1	1	1	3	4	5
	800	0	0	0	0	1	1	1	3	3	5
	900	0	0	0	0	0	1	1	2	3	4
	1000	0	0	0	0	0	1	1	2	3	4
	1250	0	0	0	0	0	1	1	1	2	3
	1500	0	0	0	0	0	1	1	1	1	2
	1750	0	0	0	0	0	0	1	1	1	2
	2000	0	0	0	0	0	0	1	1	1	1
THHN, THWN, THWN-2	14	12	22	35	61	84	138	241	364	476	608
	12	9	16	26	45	61	101	176	266	347	443
	10	5	10	16	28	38	63	111	167	219	279
	8	3	6	9	16	22	36	64	96	126	161
	6	2	4	7	12	16	26	46	69	91	116
	4	1	2	4	7	10	16	28	43	56	71
	3	1	1	3	6	8	13	24	36	47	60
	2	1	1	3	5	7	11	20	30	40	51
	1	1	1	1	4	5	8	15	22	29	37
	1/0	1	1	1	3	4	7	12	19	25	32
	2/0	0	1	1	2	3	6	10	16	20	26
	3/0	0	1	1	1	3	5	8	13	17	22
	4/0	0	1	1	1	2	4	7	11	14	18
	250	0	0	1	1	1	3	6	9	11	15
	300	0	0	1	1	1	3	5	7	10	13
	350	0	0	1	1	1	2	4	6	9	11
	400	0	0	0	1	1	1	4	6	8	10
	500	0	0	0	1	1	1	3	5	6	8
	600	0	0	0	1	1	1	2	4	5	7
	700	0	0	0	1	1	1	2	3	4	6
	750	0	0	0	0	1	1	1	3	4	5
	800	0	0	0	0	1	1	1	3	4	5
	900	0	0	0	0	1	1	1	3	3	4
	1000	0	0	0	0	1	1	1	2	3	4
FEP, FEPB, PFA, PFAH, TFE	14	12	21	34	60	81	134	234	354	462	590
	12	9	15	25	43	59	98	171	258	337	430
	10	6	11	18	31	42	70	122	185	241	309
	8	3	6	10	18	24	40	70	106	138	177
	6	2	4	7	12	17	28	50	75	98	126
	4	1	3	5	9	12	20	35	53	69	88
	3	1	2	4	7	10	16	29	44	57	73
	2	1	1	3	6	8	13	24	36	47	60
PFA, PFAH, TFE	1	1	1	2	4	6	9	16	25	33	42
PFA, PFAH, TFE, Z	1/0	1	1	1	3	5	8	14	21	27	35
	2/0	0	1	1	3	4	6	11	17	22	29
	3/0	0	1	1	2	3	5	9	14	18	24
	4/0	0	1	1	1	2	4	8	11	15	19
Z	14	14	25	41	72	98	161	282	426	556	711
	12	10	18	29	51	69	114	200	302	394	504
	10	6	11	18	31	42	70	122	185	241	309
	8	4	7	11	20	27	44	77	117	153	195
	6	3	5	8	14	19	31	54	82	107	137
	4	1	3	5	9	13	21	37	56	74	94
	3	1	2	4	7	9	15	27	41	54	69
	2	1	1	3	6	8	13	22	34	45	57
	1	1	1	2	4	6	10	18	28	36	46

*Types RHH, RHW, and RHW-2 without outer covering.

Table C1. Continued

CONDUCTORS											
	Conductor Size (AWG/ kcmil)	Trade Size (in.)									
Type		½	¾	1	1¼	1½	2	2½	3	3½	4
XHH, XHHW, XHHW-2, ZW	14	8	15	25	43	58	96	168	254	332	424
	12	6	11	19	33	45	74	129	195	255	326
	10	5	8	14	24	33	55	96	145	190	243
	8	2	5	8	13	18	30	53	81	105	135
	6	1	3	6	10	14	22	39	60	78	100
	4	1	2	4	7	10	16	28	43	56	72
	3	1	1	3	6	8	14	24	36	48	61
	2	1	1	3	5	7	11	20	31	40	51
XHH, XHHW, XHHW-2	1	1	1	1	4	5	8	15	23	30	38
	1/0	1	1	1	3	4	7	13	19	25	32
	2/0	0	1	1	2	3	6	10	16	21	27
	3/0	0	1	1	1	3	5	9	13	17	22
	4/0	0	1	1	1	2	4	7	11	14	18
	250	0	0	1	1	1	3	6	9	12	15
	300	0	0	1	1	1	3	5	8	10	13
	350	0	0	1	1	1	2	4	7	9	11
	400	0	0	0	1	1	1	4	6	8	10
	500	0	0	0	1	1	1	3	5	6	8
	600	0	0	0	1	1	1	2	4	5	6
	700	0	0	0	0	1	1	2	3	4	6
	750	0	0	0	0	1	1	1	3	4	5
	800	0	0	0	0	1	1	1	3	4	5
	900	0	0	0	0	1	1	1	3	3	4
	1000	0	0	0	0	0	1	1	2	3	4
	1250	0	0	0	0	0	1	1	1	2	3
	1500	0	0	0	0	0	1	1	1	1	3
	1750	0	0	0	0	0	0	1	1	1	2
	2000	0	0	0	0	0	0	1	1	1	1

FIXTURE WIRES							
	Conductor Size (AWG/ kcmil)	Trade Size (in.)					
Type		½	¾	1	1¼	1½	2
FFH-2, RFH-2, RFHH-3	18	8	14	24	41	56	92
	16	7	12	20	34	47	78
SF-2, SFF-2	18	10	18	30	52	71	116
	16	8	15	25	43	58	96
	14	7	12	20	34	47	78
SF-1, SFF-1	18	18	33	53	92	125	206
AF, RFH-1, RFHH-2, TF, TFF, XF, XFF	18	14	24	39	68	92	152
AF, RFHH-2, TF, TFF, XF, XFF	16	11	19	31	55	74	123
AF, XF, XFF	14	8	15	25	43	58	96
TFN, TFFN	18	22	38	63	108	148	244
	16	17	29	48	83	113	186
PF, PFF, PGF, PGFF, PAF, PTF, PTFF, PAFF	18	21	36	59	103	140	231
	16	16	28	46	79	108	179
	14	12	21	34	60	81	134
ZF, ZFF, ZHF, HF, HFF	18	27	47	77	133	181	298
	16	20	35	56	98	133	220
	14	14	25	41	72	98	161
KF-2, KFF-2	18	39	69	111	193	262	433
	16	27	48	78	136	185	305
	14	19	33	54	93	127	209
	12	13	23	37	64	87	144
	10	8	15	25	43	58	96
KF-1, KFF-1	18	46	82	133	230	313	516
	16	33	57	93	161	220	362
	14	22	38	63	108	148	244
	12	14	25	41	72	98	161
	10	9	16	27	47	64	105
AF, XF, XFF	12	4	8	13	23	31	51
	10	3	6	10	18	24	40

Note: This table is for concentric stranded conductors only. For compact stranded conductors, Table C1(A) should be used.

Table C1(A). Maximum Number of Compact Conductors in Electrical Metallic Tubing *(Based on Table 1, Chapter 9)*

COMPACT CONDUCTORS											
	Conductor Size (AWG/ kcmil)	Trade Size (in.)									
Type		½	¾	1	1¼	1½	2	2½	3	3½	4
THW, THW-2, THHW	8	2	4	6	11	16	26	46	69	90	115
	6	1	3	5	9	12	20	35	53	70	89
	4	1	2	4	6	9	15	26	40	52	67
	2	1	1	3	5	7	11	19	29	38	49
	1	1	1	1	3	4	8	13	21	27	34
	1/0	1	1	1	3	4	7	12	18	23	30
	2/0	0	1	1	2	3	5	10	15	20	25
	3/0	0	1	1	1	3	5	8	13	17	21
	4/0	0	1	1	1	2	4	7	11	14	18
	250	0	0	1	1	1	3	5	8	11	14
	300	0	0	1	1	1	3	5	7	9	12
	350	0	0	1	1	1	2	4	6	8	11
	400	0	0	0	1	1	1	4	6	8	10
	500	0	0	0	1	1	1	3	5	6	8
	600	0	0	0	1	1	1	2	4	5	7
	700	0	0	0	1	1	1	2	3	4	6
	750	0	0	0	0	1	1	1	3	4	5
	1000	0	0	0	0	1	1	1	2	3	4
THHN, THWN, THWN-2	8	—	—	—	—	—	—	—	—	—	—
	6	2	4	7	13	18	29	52	78	102	130
	4	1	3	4	8	11	18	32	48	63	81
	2	1	1	3	6	8	13	23	34	45	58
	1	1	1	2	4	6	10	17	26	34	43
	1/0	1	1	1	3	5	8	14	22	29	37
	2/0	1	1	1	3	4	7	12	18	24	30
	3/0	0	1	1	2	3	6	10	15	20	25
	4/0	0	1	1	1	3	5	8	12	16	21
	250	0	1	1	1	1	4	6	10	13	16
	300	0	0	1	1	1	3	5	8	11	14
	350	0	0	1	1	1	3	5	7	10	12
	400	0	0	1	1	1	2	4	6	9	11
	500	0	0	0	1	1	1	4	5	7	9
	600	0	0	0	1	1	1	3	4	6	7
	700	0	0	0	1	1	1	2	4	5	7
	750	0	0	0	1	1	1	2	4	5	6
	1000	0	0	0	0	1	1	1	3	3	4
XHHW, XHHW-2	8	3	5	8	15	20	34	59	90	117	149
	6	1	4	6	11	15	25	44	66	87	111
	4	1	3	4	8	11	18	32	48	63	81
	2	1	1	3	6	8	13	23	34	45	58
	1	1	1	2	4	6	10	17	26	34	43
	1/0	1	1	1	3	5	8	14	22	29	37
	2/0	1	1	1	3	4	7	12	18	24	31
	3/0	0	1	1	2	3	6	10	15	20	25
	4/0	0	1	1	1	3	5	8	13	17	21
	250	0	1	1	1	2	4	7	10	13	17
	300	0	0	1	1	1	3	6	9	11	14
	350	0	0	1	1	1	3	5	8	10	13
	400	0	0	1	1	1	2	4	7	9	11
	500	0	0	0	1	1	1	4	6	7	9
	600	0	0	0	1	1	1	3	4	6	8
	700	0	0	0	1	1	1	2	4	5	7
	750	0	0	0	1	1	1	2	3	5	6
	1000	0	0	0	0	1	1	1	3	4	5

Definition: Compact stranding is the result of a manufacturing process where the standard conductor is compressed to the extent that the interstices (voids between strand wires) are virtually eliminated.

Table C2. Maximum Number of Conductors and Fixture Wires in Electrical Nonmetallic Tubing *(Based on Table 1, Chapter 9)*

CONDUCTORS							
	Conductor Size (AWG/ kcmil)	Trade Size (in.)					
Type		½	¾	1	1¼	1½	2
RH	14	4	8	15	27	37	61
	12	3	7	12	21	29	49
RHH, RHW, RHW-2	14	3	6	10	19	26	43
	12	2	5	9	16	22	36
RH, RHH, RHW, RHW-2	10	1	4	7	13	17	29
	8	1	1	3	6	9	15
	6	1	1	3	5	7	12
	4	1	1	2	4	6	9
	3	1	1	1	3	5	8
	2	0	1	1	3	4	7
	1	0	1	1	1	3	5
	1/0	0	0	1	1	2	4
	2/0	0	0	1	1	1	3
	3/0	0	0	1	1	1	3
	4/0	0	0	1	1	1	2
	250	0	0	0	1	1	1
	300	0	0	0	1	1	1
	350	0	0	0	1	1	1
	400	0	0	0	1	1	1
	500	0	0	0	0	1	1
	600	0	0	0	0	1	1
	700	0	0	0	0	0	1
	750	0	0	0	0	0	1
	800	0	0	0	0	0	1
	900	0	0	0	0	0	1
	1000	0	0	0	0	0	1
	1250	0	0	0	0	0	0
	1500	0	0	0	0	0	0
	1750	0	0	0	0	0	0
	2000	0	0	0	0	0	0
TW	14	7	13	22	40	55	92
	12	5	10	17	31	42	71
	10	4	7	13	23	32	52
	8	1	4	7	13	17	29
RHH*, RHW*, RHW-2*, THHW, THW, THW-2	14	4	8	15	27	37	61
RHH*, RHW*, RHW-2*, THHW, THW	12	3	7	12	21	29	49
	10	3	5	9	17	23	38
RHH*, RHW*, RHW-2*, THHW, THW, THW-2	8	1	3	5	10	14	23
RHH*, RHW*, RHW-2*, TW, THW, THHW, THW-2	6	1	2	4	7	10	17
	4	1	1	3	5	8	13
	3	1	1	2	5	7	11
	2	1	1	2	4	6	9
	1	0	1	1	3	4	6
	1/0	0	1	1	2	3	5
	2/0	0	1	1	1	3	5
	3/0	0	0	1	1	2	4
	4/0	0	0	1	1	1	3
	250	0	0	1	1	1	2
	300	0	0	0	1	1	2
	350	0	0	0	1	1	1
	400	0	0	0	1	1	1
	500	0	0	0	1	1	1
	600	0	0	0	0	1	1
	700	0	0	0	0	1	1
	750	0	0	0	0	1	1
	800	0	0	0	0	1	1
	900	0	0	0	0	0	1
	1000	0	0	0	0	0	1
	1250	0	0	0	0	0	1
	1500	0	0	0	0	0	0
	1750	0	0	0	0	0	0
	2000	0	0	0	0	0	0

*Types RHH, RHW, and RHW-2 without outer covering.

Table C2. Continued

CONDUCTORS							
Type	Conductor Size (AWG/ kcmil)	Trade Size (in.) ½	¾	1	1¼	1½	2
THHN, THWN, THWN-2	14	10	18	32	58	80	132
	12	7	13	23	42	58	96
	10	4	8	15	26	36	60
	8	2	5	8	15	21	35
	6	1	3	6	11	15	25
	4	1	1	4	7	9	15
	3	1	1	3	5	8	13
	2	1	1	2	5	6	11
	1	1	1	1	3	5	8
	1/0	0	1	1	3	4	7
	2/0	0	1	1	2	3	5
	3/0	0	1	1	1	3	4
	4/0	0	0	1	1	2	4
	250	0	0	1	1	1	3
	300	0	0	1	1	1	2
	350	0	0	0	1	1	2
	400	0	0	0	1	1	1
	500	0	0	0	1	1	1
	600	0	0	0	1	1	1
	700	0	0	0	0	1	1
	750	0	0	0	0	1	1
	800	0	0	0	0	1	1
	900	0	0	0	0	1	1
	1000	0	0	0	0	0	1
FEP, FEPB, PFA, PFAH, TFE	14	10	18	31	56	77	128
	12	7	13	23	41	56	93
	10	5	9	16	29	40	67
	8	3	5	9	17	23	38
	6	1	4	6	12	16	27
	4	1	2	4	8	11	19
	3	1	1	4	7	9	16
	2	1	1	3	5	8	13
PFA, PFAH, TFE	1	1	1	1	4	5	9
PFA, PFAH, TFE, Z	1/0	0	1	1	3	4	7
	2/0	0	1	1	2	4	6
	3/0	0	1	1	1	3	5
	4/0	0	1	1	1	2	4
Z	14	12	22	38	68	93	154
	12	8	15	27	48	66	109
	10	5	9	16	29	40	67
	8	3	6	10	18	25	42
	6	1	4	7	13	18	30
	4	1	3	5	9	12	20
	3	1	1	3	6	9	15
	2	1	1	3	5	7	12
	1	1	1	2	4	6	10
XHH, XHHW, XHHW-2, ZW	14	7	13	22	40	55	92
	12	5	10	17	31	42	71
	10	4	7	13	23	32	52
	8	1	4	7	13	17	29
	6	1	3	5	9	13	21
	4	1	1	4	7	9	15
	3	1	1	3	6	8	13
	2	1	1	2	5	6	11
XHH, XHHW, XHHW-2	1	1	1	1	3	5	8
	1/0	0	1	1	3	4	7
	2/0	0	1	1	2	3	6
	3/0	0	1	1	1	3	5
	4/0	0	0	1	1	2	4
	250	0	0	1	1	1	3
	300	0	0	1	1	1	3
	350	0	0	1	1	1	2
	400	0	0	0	1	1	1
	500	0	0	0	1	1	1
	600	0	0	0	1	1	1
	700	0	0	0	0	1	1
	750	0	0	0	0	1	1
	800	0	0	0	0	1	1
	900	0	0	0	0	1	1
	1000	0	0	0	0	0	1
	1250	0	0	0	0	0	1
	1500	0	0	0	0	0	1
	1750	0	0	0	0	0	0
	2000	0	0	0	0	0	0

Table C2. Continued

FIXTURE WIRES							
Type	Conductor Size (AWG/ kcmil)	Trade Size (in.) ½	¾	1	1¼	1½	2
FFH-2, RFH-2, RFHH-3	18	6	12	21	39	53	88
	16	5	10	18	32	45	74
SF-2, SFF-2	18	8	15	27	49	67	111
	16	7	13	22	40	55	92
	14	5	10	18	32	45	74
SF-1, SFF-1	18	15	28	48	86	119	197
AF, RFH-1, RFHH-2, TF, TFF, XF, XFF	18	11	20	35	64	88	145
AF, RFHH-2, TF, TFF, XF, XFF	16	9	16	29	51	71	117
AF, XF, XFF	14	7	13	22	40	55	92
TFN, TFFN	18	18	33	57	102	141	233
	16	13	25	43	78	107	178
PF, PFF, PGF, PGFF, PAF, PTF, PTFF, PAFF	18	17	31	54	97	133	221
	16	13	24	42	75	103	171
	14	10	18	31	56	77	128
ZF, ZFF, ZHF, HF, HFF	18	22	40	70	125	172	285
	16	16	29	51	92	127	210
	14	12	22	38	68	93	154
KF-2, KFF-2	18	31	58	101	182	250	413
	16	22	41	71	128	176	291
	14	15	28	49	88	121	200
	12	10	19	33	60	83	138
	10	7	13	22	40	55	92
KF-1, KFF-1	18	38	69	121	217	298	493
	16	26	49	85	152	209	346
	14	18	33	57	102	141	233
	12	12	22	38	68	93	154
	10	7	14	24	44	61	101
AF, XF, XFF	12	3	7	12	21	29	49
	10	3	5	9	17	23	38

Note: This table is for concentric stranded conductors only. For compact stranded conductors, Table C2(A) should be used.

Table C2(A). Maximum Number of Compact Conductors in Electrical Nonmetallic Tubing *(Based on Table 1, Chapter 9)*

	COMPACT CONDUCTORS						
	Conductor Size (AWG/ kcmil)	Trade Size (in.)					
Type		½	¾	1	1¼	1½	2
THW, THW-2, THHW	8	1	3	6	11	15	25
	6	1	2	4	8	11	19
	4	1	1	3	6	8	14
	2	1	1	2	4	6	10
	1	0	1	1	3	4	7
	1/0	0	1	1	3	4	6
	2/0	0	1	1	2	3	5
	3/0	0	1	1	1	3	4
	4/0	0	0	1	1	2	4
	250	0	0	1	1	1	3
	300	0	0	1	1	1	2
	350	0	0	0	1	1	2
	400	0	0	0	1	1	1
	500	0	0	0	1	1	1
	600	0	0	0	1	1	1
	700	0	0	0	0	1	1
	750	0	0	0	0	1	1
	1000	0	0	0	0	0	1
THHN, THWN, THWN-2	8	—	—	—	—	—	—
	6	1	4	7	12	17	28
	4	1	2	4	7	10	17
	2	1	1	3	5	7	12
	1	1	1	2	4	5	9
	1/0	1	1	1	3	5	8
	2/0	0	1	1	3	4	6
	3/0	0	1	1	2	3	5
	4/0	0	1	1	1	2	4
	250	0	0	1	1	1	3
	300	0	0	1	1	1	3
	350	0	0	1	1	1	2
	400	0	0	0	1	1	2
	500	0	0	0	1	1	1
	600	0	0	0	1	1	1
	700	0	0	0	1	1	1
	750	0	0	0	1	1	1
	1000	0	0	0	0	1	1
XHHW, XHHW-2	8	2	4	8	14	19	32
	6	1	3	6	10	14	24
	4	1	2	4	7	10	17
	2	1	1	3	5	7	12
	1	1	1	2	4	5	9
	1/0	1	1	1	3	5	8
	2/0	0	1	1	3	4	7
	3/0	0	1	1	2	3	5
	4/0	0	1	1	1	3	4
	250	0	0	1	1	1	3
	300	0	0	1	1	1	3
	350	0	0	1	1	1	3
	400	0	0	1	1	1	2
	500	0	0	0	1	1	1
	600	0	0	0	1	1	1
	700	0	0	0	1	1	1
	750	0	0	0	1	1	1
	1000	0	0	0	0	1	1

Definition: Compact stranding is the result of a manufacturing process where the standard conductor is compressed to the extent that the interstices (voids between strand wires) are virtually eliminated.

Table C3. Maximum Number of Conductors and Fixture Wires in Flexible Metal Conduit *(Based on Table 1, Chapter 9)*

	CONDUCTORS										
	Conductor Size (AWG/ kcmil)	Trade Size (in.)									
Type		½	¾	1	1¼	1½	2	2½	3	3½	4
RH	14	6	10	15	24	35	62	94	135	184	240
	12	5	8	12	19	28	50	75	108	148	193
RHH, RHW, RHW-2	14	4	7	11	17	25	44	67	96	131	171
	12	3	6	9	14	21	37	55	80	109	142
RH, RHH, RHW, RHW-2	10	3	5	7	11	17	30	45	64	88	115
	8	1	2	4	6	9	15	23	34	46	60
	6	1	1	3	5	7	12	19	27	37	48
	4	1	1	2	4	5	10	14	21	29	37
	3	1	1	1	3	5	8	13	18	25	33
	2	1	1	1	3	4	7	11	16	22	28
	1	0	1	1	1	2	5	7	10	14	19
	1/0	0	1	1	1	2	4	6	9	12	16
	2/0	0	1	1	1	1	3	5	8	11	14
	3/0	0	0	1	1	1	3	5	7	9	12
	4/0	0	0	1	1	1	2	4	6	8	10
	250	0	0	0	1	1	1	3	4	6	8
	300	0	0	0	1	1	1	2	4	5	7
	350	0	0	0	1	1	1	2	3	5	6
	400	0	0	0	0	1	1	1	3	4	6
	500	0	0	0	0	1	1	1	3	4	5
	600	0	0	0	0	1	1	1	2	3	4
	700	0	0	0	0	0	1	1	1	3	3
	750	0	0	0	0	0	1	1	1	2	3
	800	0	0	0	0	0	1	1	1	2	3
	900	0	0	0	0	0	1	1	1	2	3
	1000	0	0	0	0	0	1	1	1	1	3
	1250	0	0	0	0	0	0	1	1	1	1
	1500	0	0	0	0	0	0	1	1	1	1
	1750	0	0	0	0	0	0	1	1	1	1
	2000	0	0	0	0	0	0	0	1	1	1
TW	14	9	15	23	36	53	94	141	203	277	361
	12	7	11	18	28	41	72	108	156	212	277
	10	5	8	13	21	30	54	81	116	158	207
	8	3	5	7	11	17	30	45	64	88	115
RHH*, RHW*, RHW-2*, THHW, THW, THW-2	14	6	10	15	24	35	62	94	135	184	240
RHH*, RHW*, RHW-2*, THHW, THW	12	5	8	12	19	28	50	75	108	148	193
	10	4	6	10	15	22	39	59	85	115	151
RHH*, RHW*, RHW-2*, THHW, THW, THW-2	8	1	4	6	9	13	23	35	51	69	90
RHH*, RHW*, RHW-2*, TW, THW, THHW, THW-2	6	1	3	4	7	10	18	27	39	53	69
	4	1	1	3	5	7	13	20	29	39	51
	3	1	1	3	4	6	11	17	25	34	44
	2	1	1	2	4	5	10	14	21	29	37
	1	1	1	1	2	4	7	10	15	20	26
	1/0	0	1	1	1	3	6	9	12	17	22
	2/0	0	1	1	1	3	5	7	10	14	19
	3/0	0	1	1	1	2	4	6	9	12	16
	4/0	0	0	1	1	1	3	5	7	10	13
	250	0	0	1	1	1	3	4	6	8	11
	300	0	0	1	1	1	2	3	5	7	9
	350	0	0	0	1	1	1	3	4	6	8
	400	0	0	0	1	1	1	3	4	6	7
	500	0	0	0	1	1	1	2	3	5	6
	600	0	0	0	0	1	1	1	3	4	5
	700	0	0	0	0	1	1	1	2	3	4
	750	0	0	0	0	1	1	1	2	3	4
	800	0	0	0	0	1	1	1	1	3	4
	900	0	0	0	0	0	1	1	1	3	3
	1000	0	0	0	0	0	1	1	1	2	3
	1250	0	0	0	0	0	1	1	1	1	2
	1500	0	0	0	0	0	0	1	1	1	1
	1750	0	0	0	0	0	0	1	1	1	1
	2000	0	0	0	0	0	0	1	1	1	1

*Types RHH, RHW, and RHW-2 without outer covering.

Table C3. Continued

CONDUCTORS											
Type	Conductor Size (AWG/ kcmil)	Trade Size (in.) ½	¾	1	1¼	1½	2	2½	3	3½	4
THHN, THWN, THWN-2	14	13	22	33	52	76	134	202	291	396	518
	12	9	16	24	38	56	98	147	212	289	378
	10	6	10	15	24	35	62	93	134	182	238
	8	3	6	9	14	20	35	53	77	105	137
	6	2	4	6	10	14	25	38	55	76	99
	4	1	2	4	6	9	16	24	34	46	61
	3	1	1	3	5	7	13	20	29	39	51
	2	1	1	3	4	6	11	17	24	33	43
	1	1	1	1	3	4	8	12	18	24	32
	1/0	1	1	1	2	4	7	10	15	20	27
	2/0	0	1	1	1	3	6	9	12	17	22
	3/0	0	1	1	1	2	5	7	10	14	18
	4/0	0	1	1	1	1	4	6	8	12	15
	250	0	0	1	1	1	3	5	7	9	12
	300	0	0	1	1	1	3	4	6	8	11
	350	0	0	1	1	1	2	3	5	7	9
	400	0	0	0	1	1	1	3	5	6	8
	500	0	0	0	1	1	1	2	4	5	7
	600	0	0	0	0	1	1	1	3	4	5
	700	0	0	0	0	1	1	1	3	4	5
	750	0	0	0	0	1	1	1	2	3	4
	800	0	0	0	0	1	1	1	2	3	4
	900	0	0	0	0	0	1	1	1	3	4
	1000	0	0	0	0	0	1	1	1	3	3
FEP, FEPB, PFA, PFAH, TFE	14	12	21	32	51	74	130	196	282	385	502
	12	9	15	24	37	54	95	143	206	281	367
	10	6	11	17	26	39	68	103	148	201	263
	8	4	6	10	15	22	39	59	85	115	151
	6	2	4	7	11	16	28	42	60	82	107
	4	1	3	5	7	11	19	29	42	57	75
	3	1	2	4	6	9	16	24	35	48	62
	2	1	1	3	5	7	13	20	29	39	51
PFA, PFAH, TFE	1	1	1	2	3	5	9	14	20	27	36
PFA, PFAH, TFE, Z	1/0	1	1	1	3	4	8	11	17	23	30
	2/0	1	1	1	2	3	6	9	14	19	24
	3/0	0	1	1	1	3	5	8	11	15	20
	4/0	0	1	1	1	2	4	6	9	13	16
Z	14	15	25	39	61	89	157	236	340	463	605
	12	11	18	28	43	63	111	168	241	329	429
	10	6	11	17	26	39	68	103	148	201	263
	8	4	7	11	17	24	43	65	93	127	166
	6	3	5	7	12	17	30	45	65	89	117
	4	1	3	5	8	12	21	31	45	61	80
	3	1	2	4	6	8	15	23	33	45	58
	2	1	1	3	5	7	12	19	27	37	49
	1	1	1	2	4	6	10	15	22	30	39
XHH, XHHW, XHHW-2, ZW	14	9	15	23	36	53	94	141	203	277	361
	12	7	11	18	28	41	72	108	156	212	277
	10	5	8	13	21	30	54	81	116	158	207
	8	3	5	7	11	17	30	45	64	88	115
	6	1	3	5	8	12	22	33	48	65	85
	4	1	2	4	6	9	16	24	34	47	61
	3	1	1	3	5	7	13	20	29	40	52
	2	1	1	3	4	6	11	17	24	33	44
XHH, XHHW, XHHW-2	1	1	1	1	3	5	8	13	18	25	32
	1/0	1	1	1	2	4	7	10	15	21	27
	2/0	0	1	1	2	3	6	9	13	17	23
	3/0	0	1	1	1	3	5	7	10	14	19
	4/0	0	1	1	1	2	4	6	9	12	15
	250	0	0	1	1	1	3	5	7	10	13
	300	0	0	1	1	1	3	4	6	8	11
	350	0	0	1	1	1	2	4	5	7	9
	400	0	0	0	1	1	1	3	5	6	8
	500	0	0	0	1	1	1	3	4	5	7
	600	0	0	0	0	1	1	1	3	4	5
	700	0	0	0	0	1	1	1	3	4	5
	750	0	0	0	0	1	1	1	2	3	4
	800	0	0	0	0	1	1	1	2	3	4
	900	0	0	0	0	0	1	1	1	3	4
	1000	0	0	0	0	0	1	1	1	3	3
	1250	0	0	0	0	0	1	1	1	1	3
	1500	0	0	0	0	0	1	1	1	1	2
	1750	0	0	0	0	0	0	1	1	1	1
	2000	0	0	0	0	0	0	1	1	1	1

Table C3. (Continued)

FIXTURE WIRES							
Type	Conductor Size (AWG/ kcmil)	Trade Size (in.) ½	¾	1	1¼	1½	2
FFH-2, RFH-2, RFHH-3	18	8	14	22	35	51	90
	16	7	12	19	29	43	76
SF-2, SFF-2	18	11	18	28	44	64	113
	16	9	15	23	36	53	94
	14	7	12	19	29	43	76
SF-1, SFF-1	18	19	32	50	78	114	201
AF, RFH-1, RFHH-2, TF, TFF, XF, XFF	18	14	24	37	58	84	148
AF, RFHH-2, TF, TFF, XF, XFF	16	11	19	30	47	68	120
AF, XF, XFF	14	9	15	23	36	53	94
TFN, TFFN	18	23	38	59	93	135	237
	16	17	29	45	71	103	181
PF, PFF, PGF, PGFF, PAF, PTF, PTFF, PAFF	18	22	36	56	88	128	225
	16	17	28	43	68	99	174
	14	12	21	32	51	74	130
ZF, ZFF, ZHF, HF, HFF	18	28	47	72	113	165	290
	16	20	35	53	83	121	214
	14	15	25	39	61	89	157
KF-2, KFF-2	18	41	68	105	164	239	421
	16	28	48	74	116	168	297
	14	19	33	51	80	116	204
	12	13	23	35	55	80	140
	10	9	15	23	36	53	94
KF-1, KFF-1	18	48	82	125	196	285	503
	16	34	57	88	138	200	353
	14	23	38	59	93	135	237
	12	15	25	39	61	89	157
	10	10	16	25	40	58	103
AF, XF, XFF	12	5	8	12	19	28	50
	10	4	6	10	15	22	39

Note: This table is for concentric stranded conductors only. For compact stranded conductors, Table C3(A) should be used.

Table C3(A). Maximum Number of Compact Conductors in Flexible Metal Conduit ***(Based on Table 1, Chapter 9)***

COMPACT CONDUCTORS											
Type	Conductor Size (AWG/ kcmil)	Trade Size (in.) ½	¾	1	1¼	1½	2	2½	3	3½	4
THW, THHW, THW-2	8	2	4	6	10	14	25	38	55	75	98
	6	1	3	5	7	11	20	29	43	58	76
	4	1	2	3	5	8	15	22	32	43	57
	2	1	1	2	4	6	11	16	23	32	42
	1	1	1	1	3	4	7	11	16	22	29
	1/0	1	1	1	2	3	6	10	14	19	25
	2/0	0	1	1	1	3	5	8	12	16	21
	3/0	0	1	1	1	2	4	7	10	14	18
	4/0	0	1	1	1	1	4	6	8	11	15
	250	0	0	1	1	1	3	4	7	9	12
	300	0	0	1	1	1	2	4	6	8	10
	350	0	0	1	1	1	2	3	5	7	9
	400	0	0	0	1	1	1	3	5	6	8
	500	0	0	0	1	1	1	3	4	5	7
	600	0	0	0	0	1	1	1	3	4	6
	700	0	0	0	0	1	1	1	3	4	5
	750	0	0	0	0	1	1	1	2	3	5
	1000	0	0	0	0	0	1	1	1	3	4
THHN, THWN, THWN-2	8	—	—	—	—	—	—	—	—	—	—
	6	3	4	7	11	16	29	43	62	85	111
	4	1	3	4	7	10	18	27	38	52	69
	2	1	1	3	5	7	13	19	28	38	49
	1	1	1	2	3	5	9	14	21	28	37
	1/0	1	1	1	3	4	8	12	17	24	31
	2/0	1	1	1	2	4	6	10	14	20	26
	3/0	0	1	1	1	3	5	8	12	17	22
	4/0	0	1	1	1	2	4	7	10	14	18
	250	0	1	1	1	1	3	5	8	11	14
	300	0	0	1	1	1	3	5	7	9	12
	350	0	0	1	1	1	3	4	6	8	10
	400	0	0	1	1	1	2	3	5	7	9
	500	0	0	0	1	1	1	3	4	6	8
	600	0	0	0	1	1	1	2	3	5	6
	700	0	0	0	0	1	1	1	3	4	6
	750	0	0	0	0	1	1	1	3	4	5
	1000	0	0	0	0	0	1	1	1	3	4
XHHW, XHHW-2	8	3	5	8	13	19	33	50	71	97	127
	6	2	4	6	9	14	24	37	53	72	95
	4	1	3	4	7	10	18	27	38	52	69
	2	1	1	3	5	7	13	19	28	38	49
	1	1	1	2	3	5	9	14	21	28	37
	1/0	1	1	1	3	4	8	12	17	24	31
	2/0	1	1	1	2	4	7	10	15	20	26
	3/0	0	1	1	1	3	5	8	12	17	22
	4/0	0	1	1	1	2	4	7	10	14	18
	250	0	1	1	1	1	4	5	8	11	14
	300	0	0	1	1	1	3	5	7	9	12
	350	0	0	1	1	1	3	4	6	8	11
	400	0	0	1	1	1	2	4	5	7	10
	500	0	0	0	1	1	1	3	4	6	8
	600	0	0	0	1	1	1	2	3	5	6
	700	0	0	0	0	1	1	1	3	4	6
	750	0	0	0	0	1	1	1	3	4	5
	1000	0	0	0	0	1	1	1	2	3	4

Definition: Compact stranding is the result of a manufacturing process where the standard conductor is compressed to the extent that the interstices (voids between strand wires) are virtually eliminated.

Table C4. Maximum Number of Conductors and Fixture Wires in Intermediate Metal Conduit ***(Based on Table 1, Chapter 9)***

CONDUCTORS											
Type	Conductor Size (AWG/ kcmil)	Trade Size (in.) ½	¾	1	1¼	1½	2	2½	3	3½	4
RH	14	6	11	18	31	42	69	98	151	202	261
	12	5	9	14	25	34	56	79	122	163	209
RHH, RHW, RHW-2	14	4	8	13	22	30	49	70	108	144	186
	12	4	6	11	18	25	41	58	89	120	154
RH, RHH, RHW, RHW-2	10	3	5	8	15	20	33	47	72	97	124
	8	1	3	4	8	10	17	24	38	50	65
	6	1	1	3	6	8	14	19	30	40	52
	4	1	1	3	5	6	11	15	23	31	41
	3	1	1	2	4	6	9	13	21	28	36
	2	1	1	1	3	5	8	11	18	24	31
	1	0	1	1	2	3	5	7	12	16	20
	1/0	0	1	1	1	3	4	6	10	14	18
	2/0	0	1	1	1	2	4	6	9	12	15
	3/0	0	0	1	1	1	3	5	7	10	13
	4/0	0	0	1	1	1	3	4	6	9	11
	250	0	0	1	1	1	1	3	5	6	8
	300	0	0	0	1	1	1	3	4	6	7
	350	0	0	0	1	1	1	2	4	5	7
	400	0	0	0	1	1	1	2	3	5	6
	500	0	0	0	1	1	1	1	3	4	5
	600	0	0	0	0	1	1	1	2	3	4
	700	0	0	0	0	1	1	1	2	3	4
	750	0	0	0	0	1	1	1	1	3	4
	800	0	0	0	0	0	1	1	1	3	3
	900	0	0	0	0	0	1	1	1	2	3
	1000	0	0	0	0	0	1	1	1	2	3
	1250	0	0	0	0	0	1	1	1	1	2
	1500	0	0	0	0	0	0	1	1	1	1
	1750	0	0	0	0	0	0	1	1	1	1
	2000	0	0	0	0	0	0	1	1	1	1
TW	14	10	17	27	47	64	104	147	228	304	392
	12	7	13	21	36	49	80	113	175	234	301
	10	5	9	15	27	36	59	84	130	174	224
	8	3	5	8	15	20	33	47	72	97	124
RHH*, RHW*, RHW-2*, THHW, THW, THW-2	14	6	11	18	31	42	69	98	151	202	261
RHH*, RHW*, RHW-2*, THHW, THW	12	5	9	14	25	34	56	79	122	163	209
	10	4	7	11	19	26	43	61	95	127	163
RHH*, RHW*, RHW-2*, THHW, THW, THW-2	8	2	4	7	12	16	26	37	57	76	98
RHH*, RHW*, RHW-2*, TW, THW, THHW, THW-2	6	1	3	5	9	12	20	28	43	58	75
	4	1	2	4	6	9	15	21	32	43	56
	3	1	1	3	6	8	13	18	28	37	48
	2	1	1	3	5	6	11	15	23	31	41
	1	1	1	1	3	4	7	11	16	22	28
	1/0	1	1	1	3	4	6	9	14	19	24
	2/0	0	1	1	2	3	5	8	12	16	20
	3/0	0	1	1	1	3	4	6	10	13	17
	4/0	0	1	1	1	2	4	5	8	11	14
	250	0	0	1	1	1	3	4	7	9	12
	300	0	0	1	1	1	2	4	6	8	10
	350	0	0	1	1	1	2	3	5	7	9
	400	0	0	0	1	1	1	3	4	6	8
	500	0	0	0	1	1	1	2	4	5	7
	600	0	0	0	1	1	1	1	3	4	5
	700	0	0	0	0	1	1	1	3	4	5
	750	0	0	0	0	1	1	1	2	3	4
	800	0	0	0	0	1	1	1	2	3	4
	900	0	0	0	0	1	1	1	2	3	4
	1000	0	0	0	0	0	1	1	1	3	3
	1250	0	0	0	0	0	1	1	1	1	3
	1500	0	0	0	0	0	1	1	1	1	2
	1750	0	0	0	0	0	0	1	1	1	1
	2000	0	0	0	0	0	0	1	1	1	1

*Types RHH, RHW, and RHW-2 without outer covering.

Table C4. Continued

CONDUCTORS											
Type	Conductor Size (AWG/ kcmil)	Trade Size (in.) ½	¾	1	1¼	1½	2	2½	3	3½	4
THHN, THWN, THWN-2	14	14	24	39	68	91	149	211	326	436	562
	12	10	17	29	49	67	109	154	238	318	410
	10	6	11	18	31	42	68	97	150	200	258
	8	3	6	10	18	24	39	56	86	115	149
	6	2	4	7	13	17	28	40	62	83	107
	4	1	3	4	8	10	17	25	38	51	66
	3	1	2	4	6	9	15	21	32	43	56
	2	1	1	3	5	7	12	17	27	36	47
	1	1	1	2	4	5	9	13	20	27	35
	1/0	1	1	1	3	4	8	11	17	23	29
	2/0	1	1	1	3	4	6	9	14	19	24
	3/0	0	1	1	2	3	5	7	12	16	20
	4/0	0	1	1	1	2	4	6	9	13	17
	250	0	0	1	1	1	3	5	8	10	13
	300	0	0	1	1	1	3	4	7	9	12
	350	0	0	1	1	1	2	4	6	8	10
	400	0	0	1	1	1	2	3	5	7	9
	500	0	0	0	1	1	1	3	4	6	7
	600	0	0	0	1	1	1	2	3	5	6
	700	0	0	0	1	1	1	1	3	4	5
	750	0	0	0	1	1	1	1	3	4	5
	800	0	0	0	0	1	1	1	3	4	5
	900	0	0	0	0	1	1	1	2	3	4
	1000	0	0	0	0	1	1	1	2	3	4
FEP, FEPB, PFA, PFAH, TFE	14	13	23	38	66	89	145	205	317	423	545
	12	10	17	28	48	65	106	150	231	309	398
	10	7	12	20	34	46	76	107	166	221	285
	8	4	7	11	19	26	43	61	95	127	163
	6	3	5	8	14	19	31	44	67	90	116
	4	1	3	5	10	13	21	30	47	63	81
	3	1	3	4	8	11	18	25	39	52	68
	2	1	2	4	6	9	15	21	32	43	56
PFA, PFAH, TFE	1	1	1	2	4	6	10	14	22	30	39
PFA, PFAH, TFE, Z	1/0	1	1	1	4	5	8	12	19	25	32
	2/0	1	1	1	3	4	7	10	15	21	27
	3/0	0	1	1	2	3	6	8	13	17	22
	4/0	0	1	1	1	3	5	7	10	14	18
Z	14	16	28	46	79	107	175	247	381	510	657
	12	11	20	32	56	76	124	175	271	362	466
	10	7	12	20	34	46	76	107	166	221	285
	8	4	7	12	21	29	48	68	105	140	180
	6	3	5	9	15	20	33	47	73	98	127
	4	1	3	6	10	14	23	33	50	67	87
	3	1	2	4	7	10	17	24	37	49	63
	2	1	1	3	6	8	14	20	30	41	53
	1	1	1	3	5	7	11	16	25	33	43
XHH, XHHW, XHHW-2, ZW	14	10	17	27	47	64	104	147	228	304	392
	12	7	13	21	36	49	80	113	175	234	301
	10	5	9	15	27	36	59	84	130	174	224
	8	3	5	8	15	20	33	47	72	97	124
	6	1	4	6	11	15	24	35	53	71	92
	4	1	3	4	8	11	18	25	39	52	67
	3	1	2	4	7	9	15	21	33	44	56
	2	1	1	3	5	7	12	18	27	37	47
XHH, XHHW, XHHW-2	1	1	1	2	4	5	9	13	20	27	35
	1/0	1	1	1	3	5	8	11	17	23	30
	2/0	1	1	1	3	4	6	9	14	19	25
	3/0	0	1	1	2	3	5	7	12	16	20
	4/0	0	1	1	1	2	4	6	10	13	17
	250	0	0	1	1	1	3	5	8	11	14
	300	0	0	1	1	1	3	4	7	9	12
	350	0	0	1	1	1	3	4	6	8	10
	400	0	0	1	1	1	2	3	5	7	9
	500	0	0	0	1	1	1	3	4	6	8
	600	0	0	0	1	1	1	2	3	5	6
	700	0	0	0	1	1	1	1	3	4	5
	750	0	0	0	1	1	1	1	3	4	5
	800	0	0	0	0	1	1	1	3	4	5
	900	0	0	0	0	1	1	1	2	3	4
	1000	0	0	0	0	1	1	1	2	3	4
	1250	0	0	0	0	0	1	1	1	2	3
	1500	0	0	0	0	0	1	1	1	1	2
	1750	0	0	0	0	0	1	1	1	1	2
	2000	0	0	0	0	0	0	1	1	1	1

Table C4. Continued

FIXTURE WIRES							
Type	Conductor Size (AWG/ kcmil)	Trade Size (in.) ½	¾	1	1¼	1½	2
FHH-2, RFH-2, RFHH-3	18	9	16	26	45	61	100
	16	8	13	22	38	51	84
SF-2, SFF-2	18	12	20	33	57	77	126
	16	10	17	27	47	64	104
	14	8	13	22	38	51	84
SF-1, SFF-1	18	21	36	59	101	137	223
AF, RFH-1, RFHH-2, TF, TFF, XF, XFF	18	15	26	43	75	101	165
AF, RFH-2, TF, TFF, XF, XFF	16	12	21	35	60	81	133
AF, XF, XFF	14	10	17	27	47	64	104
TFN, TFFN	18	25	42	69	119	161	264
	16	19	32	53	91	123	201
PF, PFF, PGF, PGFF, PAF, PTF, PTFF, PAFF	18	23	40	66	113	153	250
	16	18	31	51	87	118	193
	14	13	23	38	66	89	145
ZF, ZFF, ZHF, HF, HFF	18	30	52	85	146	197	322
	16	22	38	63	108	145	238
	14	16	28	46	79	107	175
KF-2, KFF-2	18	44	75	123	212	287	468
	16	31	53	87	149	202	330
	14	21	36	60	103	139	227
	12	14	25	41	70	95	156
	10	10	17	27	47	64	104
KF-1, KFF-1	18	52	90	147	253	342	558
	16	37	63	103	178	240	392
	14	25	42	69	119	161	264
	12	16	28	46	79	107	175
	10	10	18	30	52	70	114
AF, XF, XFF	12	5	9	14	25	34	56
	10	4	7	11	19	26	43

Note: This table is for concentric stranded conductors only. For compact stranded conductors, Table C4(A) should be used.

Table C4(A). Maximum Number of Compact Conductors in Intermediate Metal Conduit ***(Based on Table 1, Chapter 9)***

COMPACT CONDUCTORS											
	Conductor Size (AWG/ kcmil)	Trade Size (in.)									
Type		½	¾	1	1¼	1½	2	2½	3	3½	4
THW, THW-2, THHW	8	2	4	7	13	17	28	40	62	83	107
	6	1	3	6	10	13	22	31	48	64	82
	4	1	2	4	7	10	16	23	36	48	62
	2	1	1	3	5	7	12	17	26	35	45
	1	1	1	1	4	5	8	12	18	25	32
	1/0	1	1	1	3	4	7	10	16	21	27
	2/0	0	1	1	3	4	6	9	13	18	23
	3/0	0	1	1	2	3	5	7	11	15	20
	4/0	0	1	1	1	2	4	6	9	13	16
	250	0	0	1	1	1	3	5	7	10	13
	300	0	0	1	1	1	3	4	6	9	11
	350	0	0	1	1	1	2	4	6	8	10
	400	0	0	1	1	1	2	3	5	7	9
	500	0	0	0	1	1	1	3	4	6	8
	600	0	0	0	1	1	1	2	3	5	6
	700	0	0	0	1	1	1	1	3	4	5
	750	0	0	0	1	1	1	1	3	4	5
	1000	0	0	0	0	1	1	1	2	3	4
THHN, THWN, THWN-2	8	—	—	—	—	—	—	—	—	—	—
	6	3	5	8	14	19	32	45	70	93	120
	4	1	3	5	9	12	20	28	43	58	74
	2	1	1	3	6	8	14	20	31	41	53
	1	1	1	3	5	6	10	15	23	31	40
	1/0	1	1	2	4	5	9	13	20	26	34
	2/0	1	1	1	3	4	7	10	16	22	28
	3/0	0	1	1	3	4	6	9	14	18	24
	4/0	0	1	1	2	3	5	7	11	15	19
	250	0	1	1	1	2	4	6	9	12	15
	300	0	0	1	1	1	3	5	7	10	13
	350	0	0	1	1	1	3	4	7	9	11
	400	0	0	1	1	1	2	4	6	8	10
	500	0	0	1	1	1	2	3	5	7	9
	600	0	0	0	1	1	1	2	4	5	7
	700	0	0	0	1	1	1	2	3	5	6
	750	0	0	0	1	1	1	1	3	4	6
	1000	0	0	0	0	1	1	1	2	3	4
XHHW, XHHW-2	8	3	6	9	16	22	37	52	80	107	138
	6	2	4	7	12	16	27	38	59	80	103
	4	1	3	5	9	12	20	28	43	58	74
	2	1	1	3	6	8	14	20	31	41	53
	1	1	1	3	5	6	10	15	23	31	40
	1/0	1	1	2	4	5	9	13	20	26	34
	2/0	1	1	1	3	4	7	11	17	22	29
	3/0	0	1	1	3	4	6	9	14	18	24
	4/0	0	1	1	2	3	5	7	11	15	20
	250	0	1	1	1	2	4	6	9	12	16
	300	0	0	1	1	1	3	5	8	10	13
	350	0	0	1	1	1	3	4	7	9	12
	400	0	0	1	1	1	3	4	6	8	11
	500	0	0	1	1	1	2	3	5	7	9
	600	0	0	0	1	1	1	2	4	5	7
	700	0	0	0	1	1	1	2	3	5	6
	750	0	0	0	1	1	1	1	3	4	6
	1000	0	0	0	0	1	1	1	2	3	4

Definition: Compact stranding is the result of a manufacturing process where the standard conductor is compressed to the extent that interstices (voids between strand wires) are virtually eliminated.

Table C5. Maximum Number of Conductors and Fixture Wires in Liquidtight Flexible Nonmetallic Conduit (Type FNMC-B*) ***(Based on Table 1, Chapter 9)***

CONDUCTORS								
	Conductor Size	Trade Size (in.)						
Type	**(AWG/kcmil)**	⅜	½	¾	1	1¼	1½	2
RH	14	3	6	10	16	29	38	62
	12	3	5	8	13	23	30	50
RHH, RHW, RHW-2,	14	2	4	7	12	21	27	44
	12	1	3	6	10	17	22	36
RH, RHH, RHW, RHW-2	10	1	3	5	8	14	18	29
	8	1	1	2	4	7	9	15
	6	1	1	1	3	6	7	12
	4	0	1	1	2	4	6	9
	3	0	1	1	1	4	5	8
	2	0	1	1	1	3	4	7
	1	0	0	1	1	1	3	5
	1/0	0	0	1	1	1	2	4
	2/0	0	0	1	1	1	1	3
	3/0	0	0	0	1	1	1	3
	4/0	0	0	0	1	1	1	2
	250	0	0	0	0	1	1	1
	300	0	0	0	0	1	1	1
	350	0	0	0	0	1	1	1
	400	0	0	0	0	1	1	1
	500	0	0	0	0	1	1	1
	600	0	0	0	0	0	1	1
	700	0	0	0	0	0	0	1
	750	0	0	0	0	0	0	1
	800	0	0	0	0	0	0	1
	900	0	0	0	0	0	0	1
	1000	0	0	0	0	0	0	1
	1250	0	0	0	0	0	0	0
	1500	0	0	0	0	0	0	0
	1750	0	0	0	0	0	0	0
	2000	0	0	0	0	0	0	0
TW	14	5	9	15	25	44	57	93
	12	4	7	12	19	33	43	71
	10	3	5	9	14	25	32	53
	8	1	3	5	8	14	18	29
RHH†, RHW†, RHW-2†, THHW, THW, THW-2	14	3	6	10	16	29	38	62
RHH†, RHW†, RHW-2†, THHW, THW	12	3	5	8	13	23	30	50
	10	1	3	6	10	18	23	39
RHH†, RHW†, RHW-2†, THHW, THW, THW-2	8	1	1	4	6	11	14	23
RHH†, RHW†, RHW-2†, TW, THW, THHW, THW-2	6	1	1	3	5	8	11	18
	4	1	1	1	3	6	8	13
	3	1	1	1	3	5	7	11
	2	0	1	1	2	4	6	9
	1	0	1	1	1	3	4	7
	1/0	0	0	1	1	2	3	6
	2/0	0	0	1	1	2	3	5
	3/0	0	0	1	1	1	2	4
	4/0	0	0	0	1	1	1	3
	250	0	0	0	1	1	1	3
	300	0	0	0	1	1	1	2
	350	0	0	0	0	1	1	1
	400	0	0	0	0	1	1	1
	500	0	0	0	0	1	1	1
	600	0	0	0	0	1	1	1
	700	0	0	0	0	0	1	1
	750	0	0	0	0	0	1	1
	800	0	0	0	0	0	1	1
	900	0	0	0	0	0	0	1
	1000	0	0	0	0	0	0	1
	1250	0	0	0	0	0	0	1
	1500	0	0	0	0	0	0	0
	1750	0	0	0	0	0	0	0
	2000	0	0	0	0	0	0	0

*Corresponds to Section 351-22(2).

†Types RHH, RHW, and RHW-2 without outer covering.

Table C5. Continued

CONDUCTORS								
Type	Conductor Size (AWG/kcmil)	Trade Size (in.) ⅜	½	¾	1	1¼	1½	2
THHN, THWN, THWN-2	14	8	13	22	36	63	81	133
	12	5	9	16	26	46	59	97
	10	3	6	10	16	29	37	61
	8	1	3	6	9	16	21	35
	6	1	2	4	7	12	15	25
	4	1	1	2	4	7	9	15
	3	1	1	1	3	6	8	13
	2	1	1	1	3	5	7	11
	1	0	1	1	1	4	5	8
	1/0	0	1	1	1	3	4	7
	2/0	0	0	1	1	2	3	6
	3/0	0	0	1	1	1	3	5
	4/0	0	0	1	1	1	2	4
	250	0	0	0	1	1	1	3
	300	0	0	0	1	1	1	3
	350	0	0	0	1	1	1	2
	400	0	0	0	0	1	1	1
	500	0	0	0	0	1	1	1
	600	0	0	0	0	1	1	1
	700	0	0	0	0	1	1	1
	750	0	0	0	0	0	1	1
	800	0	0	0	0	0	1	1
	900	0	0	0	0	0	1	1
	1000	0	0	0	0	0	0	1
FEP, FEPB, PFA, PFAH, TFE	14	7	12	21	35	61	79	129
	12	5	9	15	25	44	57	94
	10	4	6	11	18	32	41	68
	8	1	3	6	10	18	23	39
	6	1	2	4	7	13	17	27
	4	1	1	3	5	9	12	19
	3	1	1	2	4	7	10	16
	2	1	1	1	3	6	8	13
PFA, PFAH, TFE	1	0	1	1	2	4	5	9
PFA, PFAH, TFE, Z	1/0	0	1	1	1	3	4	7
	2/0	0	1	1	1	3	4	6
	3/0	0	0	1	1	2	3	5
	4/0	0	0	1	1	1	2	4
Z	14	9	15	26	42	73	95	156
	12	6	10	18	30	52	67	111
	10	4	6	11	18	32	41	68
	8	2	4	7	11	20	26	43
	6	1	3	5	8	14	18	30
	4	1	1	3	5	9	12	20
	3	1	1	2	4	7	9	15
	2	0	1	1	3	6	7	12
	1	0	1	1	2	5	6	10
XHH, XHHW, XHHW-2, ZW	14	5	9	15	25	44	57	93
	12	4	7	12	19	33	43	71
	10	3	5	9	14	25	32	53
	8	1	3	5	8	14	18	29
	6	1	1	3	6	10	13	22
	4	1	1	2	4	7	9	16
	3	1	1	1	3	6	8	13
	2	1	1	1	3	5	7	11
XHH, XHHW, XHHW-2	1	0	1	1	1	4	5	8
	1/0	0	1	1	1	3	4	7
	2/0	0	0	1	1	2	3	6
	3/0	0	0	1	1	1	3	5
	4/0	0	0	1	1	1	2	4
	250	0	0	0	1	1	1	3
	300	0	0	0	1	1	1	3
	350	0	0	0	1	1	1	2
	400	0	0	0	0	1	1	1
	500	0	0	0	0	1	1	1
	600	0	0	0	0	1	1	1
	700	0	0	0	0	1	1	1
	750	0	0	0	0	0	1	1
	800	0	0	0	0	0	1	1
	900	0	0	0	0	0	1	1
	1000	0	0	0	0	0	0	1
	1250	0	0	0	0	0	0	1
	1500	0	0	0	0	0	0	1
	1750	0	0	0	0	0	0	0
	2000	0	0	0	0	0	0	0

Table C5. Continued

FIXTURE WIRES								
Type	Conductor Size (AWG/kcmil)	Trade Size (in.) ⅜	½	¾	1	1¼	1½	2
FFH-2, RFH-2	18	5	8	15	24	42	54	89
	16	4	7	12	20	35	46	75
SF-2, SFF-2	18	6	11	19	30	53	69	113
	16	5	9	15	25	44	57	93
	14	4	7	12	20	35	46	75
SF-1, SFF-1	18	11	19	33	53	94	122	199
AF, RFH-1, RFHH-2, TF, TFF, XF, XFF	18	8	14	24	39	69	90	147
AF, RFHH-2, TF, TFF, XF, XFF	16	7	11	20	32	56	72	119
AF, XF, XFF	14	5	9	15	25	44	57	93
TFN, TFFN	18	14	23	39	63	111	144	236
	16	10	17	30	48	85	110	180
PF, PFF, PGF, PGFF, PAF, PTF, PTFF, PAFF	18	13	21	37	60	105	136	223
	16	10	16	29	46	81	105	173
	14	7	12	21	35	61	79	129
HF, HFF, ZF, ZFF, ZHF	18	17	28	48	77	136	176	288
	16	12	20	35	57	100	129	212
	14	9	15	26	42	73	95	156
KF-2, KFF-2	18	24	40	70	112	197	255	418
	16	17	28	49	79	139	180	295
	14	12	19	34	54	95	123	202
	12	8	13	23	37	65	85	139
	10	5	9	15	25	44	57	93
KF-1, KFF-1	18	29	48	83	134	235	304	499
	16	20	34	58	94	165	214	350
	14	14	23	39	63	111	144	236
	12	9	15	26	42	73	95	156
	10	6	10	17	27	48	62	102
AF, XF, XFF	12	3	5	8	13	23	30	50
	10	1	3	6	10	18	23	39

Note: This table is for concentric stranded conductors only. For compact stranded conductors, Table C5(A) should be used.

Table C5(A). Maximum Number of Compact Conductors in Liquidtight Flexible Nonmetallic Conduit (Type FNMC-B*) *(Based on Table 1, Chapter 9)*

COMPACT CONDUCTORS								
	Conductor Size	Trade Size (in.)						
Type	(AWG/kcmil)	3/8	1/2	3/4	1	1¼	1½	2
THW, THW-2, THHW	8	1	2	4	7	12	15	25
	6	1	1	3	5	9	12	19
	4	1	1	2	4	7	9	14
	2	1	1	1	3	5	6	11
	1	0	1	1	1	3	4	7
	1/0	0	1	1	1	3	4	6
	2/0	0	0	1	1	2	3	5
	3/0	0	0	1	1	1	3	4
	4/0	0	0	1	1	1	2	4
	250	0	0	0	1	1	1	3
	300	0	0	0	1	1	1	2
	350	0	0	0	1	1	1	2
	400	0	0	0	0	1	1	1
	500	0	0	0	0	1	1	1
	600	0	0	0	0	1	1	1
	700	0	0	0	0	1	1	1
	750	0	0	0	0	0	1	1
	1000	0	0	0	0	0	1	1
THHN, THWN, THWN-2	8	—	—	—	—	—	—	—
	6	1	2	4	7	13	17	28
	4	1	1	3	4	8	11	17
	2	1	1	1	3	6	7	12
	1	0	1	1	2	4	6	9
	1/0	0	1	1	1	4	5	8
	2/0	0	1	1	1	3	4	6
	3/0	0	0	1	1	2	3	5
	4/0	0	0	1	1	1	3	4
	250	0	0	1	1	1	1	3
	300	0	0	0	1	1	1	3
	350	0	0	0	1	1	1	2
	400	0	0	0	1	1	1	2
	500	0	0	0	0	1	1	1
	600	0	0	0	0	1	1	1
	700	0	0	0	0	1	1	1
	750	0	0	0	0	1	1	1
	1000	0	0	0	0	0	1	1
XHHW, XHHW-2	8	1	3	5	9	15	20	33
	6	1	2	4	6	11	15	24
	4	1	1	3	4	8	11	17
	2	1	1	1	3	6	7	12
	1	0	1	1	2	4	6	9
	1/0	0	1	1	1	4	5	8
	2/0	0	1	1	1	3	4	7
	3/0	0	0	1	1	2	3	5
	4/0	0	0	1	1	1	3	4
	250	0	0	1	1	1	1	3
	300	0	0	0	1	1	1	3
	350	0	0	0	1	1	1	3
	400	0	0	0	1	1	1	2
	500	0	0	0	0	1	1	1
	600	0	0	0	0	1	1	1
	700	0	0	0	0	1	1	1
	750	0	0	0	0	1	1	1
	1000	0	0	0	0	0	1	1

*Corresponds to Section 351-22(2).

Definition: Compact stranding is the result of a manufacturing process where the standard conductor is compressed to the extent that the interstices (voids between strand wires) are virtually eliminated.

Table C6. Maximum Number of Conductors and Fixture Wires in Liquidtight Flexible Nonmetallic Conduit (Type FNMC-A*) *(Based On Table 1, Chapter 9)*

CONDUCTORS								
	Conductor Size	Trade Size (in.)						
Type	(AWG/kcmil)	3/8	1/2	3/4	1	1¼	1½	2
RH	14	3	6	10	16	28	38	64
	12	3	4	8	13	23	31	51
RHH, RHW, RHW-2	14	2	4	7	11	20	27	45
	12	1	3	6	9	17	23	38
RH, RHH, RHW, RHW-2	10	1	3	5	8	13	18	30
	8	1	1	2	4	7	9	16
	6	1	1	1	3	5	7	13
	4	0	1	1	2	4	6	10
	3	0	1	1	1	4	5	8
	2	0	1	1	1	3	4	7
	1	0	0	1	1	1	3	5
	1/0	0	0	1	1	1	2	4
	2/0	0	0	1	1	1	1	4
	3/0	0	0	0	1	1	1	3
	4/0	0	0	0	1	1	1	3
	250	0	0	0	0	1	1	1
	300	0	0	0	0	1	1	1
	350	0	0	0	0	1	1	1
	400	0	0	0	0	1	1	1
	500	0	0	0	0	0	1	1
	600	0	0	0	0	0	1	1
	700	0	0	0	0	0	0	1
	750	0	0	0	0	0	0	1
	800	0	0	0	0	0	0	1
	900	0	0	0	0	0	0	1
	1000	0	0	0	0	0	0	1
	1250	0	0	0	0	0	0	0
	1500	0	0	0	0	0	0	0
	1750	0	0	0	0	0	0	0
	2000	0	0	0	0	0	0	0
TW	14	5	9	15	24	43	58	96
	12	4	7	12	19	33	44	74
	10	3	5	9	14	24	33	55
	8	1	3	5	8	13	18	30
RHH†, RHW†, RHW-2†, THHW, THW, THW-2	14	3	6	10	16	28	38	64
RHH†, RHW†, RHW-2†, THHW, THW	12	3	4	8	13	23	31	51
	10	1	3	6	10	18	24	40
RHH†, RHW†, RHW-2†, THHW, THW, THW-2	8	1	1	4	6	10	14	24
RHH†, RHW†, RHW-2†, TW, THW, THHW, THW-2	6	1	1	3	4	8	11	18
	4	1	1	1	3	6	8	13
	3	1	1	1	3	5	7	11
	2	0	1	1	2	4	6	10
	1	0	1	1	1	3	4	7
	1/0	0	0	1	1	2	3	6
	2/0	0	0	1	1	1	3	5
	3/0	0	0	1	1	1	2	4
	4/0	0	0	0	1	1	1	3
	250	0	0	0	1	1	1	3
	300	0	0	0	1	1	1	2
	350	0	0	0	0	1	1	1
	400	0	0	0	0	1	1	1
	500	0	0	0	0	1	1	1
	600	0	0	0	0	1	1	1
	700	0	0	0	0	0	1	1
	750	0	0	0	0	0	1	1
	800	0	0	0	0	0	1	1
	900	0	0	0	0	0	0	1
	1000	0	0	0	0	0	0	1
	1250	0	0	0	0	0	0	1
	1500	0	0	0	0	0	0	1
	1750	0	0	0	0	0	0	0
	2000	0	0	0	0	0	0	0

*Corresponds to Section 351-22(1)

†Types RHH, RHW, and RHW-2 without outer covering.

Table C6. Continued

CONDUCTORS								
Type	Conductor Size (AWG/kcmil)	Trade Size (in.) ⅜	½	¾	1	1¼	1½	2
THHN, THWN, THWN-2	14	8	13	22	35	62	83	137
	12	5	9	16	25	45	60	100
	10	3	6	10	16	28	38	63
	8	1	3	6	9	16	22	36
	6	1	2	4	6	12	16	26
	4	1	1	2	4	7	9	16
	3	1	1	1	3	6	8	13
	2	1	1	1	3	5	7	11
	1	0	1	1	1	4	5	8
	1/0	0	1	1	1	3	4	7
	2/0	0	0	1	1	2	3	6
	3/0	0	0	1	1	1	3	5
	4/0	0	0	1	1	1	2	4
	250	0	0	0	1	1	1	3
	300	0	0	0	1	1	1	3
	350	0	0	0	1	1	1	2
	400	0	0	0	0	1	1	1
	500	0	0	0	0	1	1	1
	600	0	0	0	0	1	1	1
	700	0	0	0	0	1	1	1
	750	0	0	0	0	0	1	1
	800	0	0	0	0	0	1	1
	900	0	0	0	0	0	1	1
	1000	0	0	0	0	0	0	1
FEP, FEPB, PFA, PFAH, TFE	14	7	12	21	34	60	80	133
	12	5	9	15	25	44	59	97
	10	4	6	11	18	31	42	70
	8	1	3	6	10	18	24	40
	6	1	2	4	7	13	17	28
	4	1	1	3	5	9	12	20
	3	1	1	2	4	7	10	16
	2	1	1	1	3	6	8	13
PFA, PFAH, TFE	1	0	1	1	2	4	5	9
PFA, PFAH, TFE, Z	1/0	0	1	1	1	3	5	8
	2/0	0	1	1	1	3	4	6
	3/0	0	0	1	1	2	3	5
	4/0	0	0	1	1	1	2	4
Z	14	9	15	25	41	72	97	161
	12	6	10	18	29	51	69	114
	10	4	6	11	18	31	42	70
	8	2	4	7	11	20	26	44
	6	1	3	5	8	14	18	31
	4	1	1	3	5	9	13	21
	3	1	1	2	4	7	9	15
	2	1	1	1	3	6	8	13
	1	1	1	1	2	4	6	10
XHH, XHHW, XHHW-2, ZW	14	5	9	15	24	43	58	96
	12	4	7	12	19	33	44	74
	10	3	5	9	14	24	33	55
	8	1	3	5	8	13	18	30
	6	1	1	3	5	10	13	22
	4	1	1	2	4	7	10	16
	3	1	1	1	3	6	8	14
	2	1	1	1	3	5	7	11
XHH, XHHW, XHHW-2	1	0	1	1	1	4	5	8
	1/0	0	1	1	1	3	4	7
	2/0	0	0	1	1	2	3	6
	3/0	0	0	1	1	1	3	5
	4/0	0	0	1	1	1	2	4
	250	0	0	0	1	1	1	3
	300	0	0	0	1	1	1	3
	350	0	0	0	1	1	1	2
	400	0	0	0	0	1	1	1
	500	0	0	0	0	1	1	1
	600	0	0	0	0	1	1	1
	700	0	0	0	0	1	1	1
	750	0	0	0	0	0	1	1
	800	0	0	0	0	0	1	1
	900	0	0	0	0	0	1	1
	1000	0	0	0	0	0	0	1
	1250	0	0	0	0	0	0	1
	1500	0	0	0	0	0	0	1
	1750	0	0	0	0	0	0	0
	2000	0	0	0	0	0	0	0

Table C6. Continued

FIXTURE WIRES								
Type	Conductor Size (AWG/kcmil)	Trade Size (in.) ⅜	½	¾	1	1¼	1½	2
FFH-2, RFH-2, RFHH-3	18	5	8	14	23	41	55	92
	16	4	7	12	20	35	47	77
SF-2, SFF-2	18	6	11	18	29	52	70	116
	16	5	9	15	24	43	58	96
	14	4	7	12	20	35	47	77
SF-1, SFF-1	18	12	19	33	52	92	124	205
AF, RFH-1, RFHH-2, TF, TFF, XF, XFF	18	8	14	24	39	68	91	152
AF, RFHH-2, TF, TFF, XF, XFF	16	7	11	19	31	55	74	122
AF, XF, XFF	14	5	9	15	24	43	58	96
TFN, TFFN	18	14	22	39	62	109	146	243
	16	10	17	29	47	83	112	185
PF, PFF, PGF, PGFF, PAF, PTF, PTFF, PAFF	18	13	21	37	59	103	139	230
	16	10	16	28	45	80	107	178
	14	7	12	21	34	60	80	133
HF, HFF, ZF, ZFF, ZHF	18	17	27	47	76	133	179	297
	16	12	20	35	56	98	132	219
	14	9	15	25	41	72	97	161
KF-2, KFF-2	18	25	40	69	110	193	260	431
	16	17	28	48	77	136	183	303
	14	12	19	33	53	94	126	209
	12	8	13	23	36	64	86	143
	10	5	9	15	24	43	58	96
KF-1, KFF-1	18	29	48	82	131	231	310	514
	16	21	33	57	92	162	218	361
	14	14	22	39	62	109	146	243
	12	9	15	25	41	72	97	161
	10	6	10	17	27	47	63	105
AF, XF, XFF	12	3	4	8	13	23	31	51
	10	1	3	6	10	18	24	40

Note: This table is for concentric stranded conductors only. For compact stranded conductors, Table C6(A) should be used.

Table C6(A). Maximum Number of Compact Conductors in Liquidtight Flexible Nonmetallic Conduit (Type FNMC-A*) ***(Based on Table 1, Chapter 9)***

		COMPACT CONDUCTORS						
	Conductor Size	Trade Size (in.)						
Type	(AWG/kcmil)	3/8	1/2	3/4	1	1 1/4	1 1/2	2
THW, THW-2, THHW	8	1	2	4	6	11	16	26
	6	1	1	3	5	9	12	20
	4	1	1	2	4	7	9	15
	2	1	1	1	3	5	6	11
	1	0	1	1	1	3	4	8
	1/0	0	1	1	1	3	4	7
	2/0	0	0	1	1	2	3	5
	3/0	0	0	1	1	1	3	5
	4/0	0	0	1	1	1	2	4
	250	0	0	0	1	1	1	3
	300	0	0	0	1	1	1	3
	350	0	0	0	1	1	1	2
	400	0	0	0	0	1	1	1
	500	0	0	0	0	1	1	1
	600	0	0	0	0	1	1	1
	700	0	0	0	0	1	1	1
	750	0	0	0	0	0	1	1
	1000	0	0	0	0	0	1	1
THHN, THWN, THWN-2	8	—	—	—	—	—	—	—
	6	1	2	4	7	13	18	29
	4	1	1	3	4	8	11	18
	2	1	1	1	3	6	8	13
	1	0	1	1	2	4	6	10
	1/0	0	1	1	1	3	5	8
	2/0	0	1	1	1	3	4	7
	3/0	0	0	1	1	2	3	6
	4/0	0	0	1	1	1	3	5
	250	0	0	1	1	1	1	3
	300	0	0	0	1	1	1	3
	350	0	0	0	1	1	1	3
	400	0	0	0	1	1	1	2
	500	0	0	0	0	1	1	1
	600	0	0	0	0	1	1	1
	700	0	0	0	0	1	1	1
	750	0	0	0	0	1	1	1
	1000	0	0	0	0	0	1	1
XHHW, XHHW-2	8	1	3	5	8	15	20	34
	6	1	2	4	6	11	15	25
	4	1	1	3	4	8	11	18
	2	1	1	1	3	6	8	13
	1	0	1	1	2	4	6	10
	1/0	0	1	1	1	3	5	8
	2/0	0	1	1	1	3	4	7
	3/0	0	0	1	1	2	3	6
	4/0	0	0	1	1	1	3	5
	250	0	0	1	1	1	2	4
	300	0	0	0	1	1	1	3
	350	0	0	0	1	1	1	3
	400	0	0	0	1	1	1	2
	500	0	0	0	0	1	1	1
	600	0	0	0	0	1	1	1
	700	0	0	0	0	1	1	1
	750	0	0	0	0	1	1	1
	1000	0	0	0	0	0	1	1

*Corresponds to Section 351-22(1).

Definition: Compact stranding is the result of a manufacturing process where the standard conductor is compressed to the extent that the interstices (voids between strand wires) are virtually eliminated.

Table C7. Maximum Number of Conductors and Fixture Wires in Liquidtight Flexible Metal Conduit ***(Based on Table 1, Chapter 9)***

		CONDUCTORS									
	Conductor Size	Trade Size (in.)									
Type	(AWG/kcmil)	1/2	3/4	1	1 1/4	1 1/2	2	2 1/2	3	3 1/2	4
RH	14	6	10	16	29	38	62	93	143	186	243
	12	5	8	13	23	30	50	75	115	149	195
RHH, RHW, RHW-2	14	4	7	12	21	27	44	66	102	133	173
	12	3	6	10	17	22	36	55	84	110	144
RH, RHH, RHW, RHW-2,	10	3	5	8	14	18	29	44	68	89	116
	8	1	2	4	7	9	15	23	36	46	61
	6	1	1	3	6	7	12	18	28	37	48
	4	1	1	2	4	6	9	14	22	29	38
	3	1	1	1	4	5	8	13	19	25	33
	2	1	1	1	3	4	7	11	17	22	29
	1	0	1	1	1	3	5	7	11	14	19
	1/0	0	1	1	1	2	4	6	10	13	16
	2/0	0	1	1	1	1	3	5	8	11	14
	3/0	0	0	1	1	1	3	4	7	9	12
	4/0	0	0	1	1	1	2	4	6	8	10
	250	0	0	0	1	1	1	3	4	6	8
	300	0	0	0	1	1	1	2	4	5	7
	350	0	0	0	1	1	1	2	3	5	6
	400	0	0	0	1	1	1	1	3	4	6
	500	0	0	0	1	1	1	1	3	4	5
	600	0	0	0	0	1	1	1	2	3	4
	700	0	0	0	0	0	1	1	1	3	3
	750	0	0	0	0	0	1	1	1	2	3
	800	0	0	0	0	0	1	1	1	2	3
	900	0	0	0	0	0	1	1	1	2	3
	1000	0	0	0	0	0	1	1	1	1	3
	1250	0	0	0	0	0	0	1	1	1	1
	1500	0	0	0	0	0	0	1	1	1	1
	1750	0	0	0	0	0	0	1	1	1	1
	2000	0	0	0	0	0	0	0	1	1	1
TW	14	9	15	25	44	57	93	140	215	280	365
	12	7	12	19	33	43	71	108	165	215	280
	10	5	9	14	25	32	53	80	123	160	209
	8	3	5	8	14	18	29	44	68	89	116
RHH*, RHW*, RHW-2*, THHW, THW, THW-2	14	6	10	16	29	38	62	93	143	186	243
RHH*, RHW*, RHW-2*, THHW, THW	12	5	8	13	23	30	50	75	115	149	195
	10	3	6	10	18	23	39	58	89	117	152
RHH*, RHW*, RHW-2*, THHW, THW, THW-2	8	1	4	6	11	14	23	35	53	70	91
RHH*, RHW*, RHW-2*, TW, THW, THHW, THW-2	6	1	3	5	8	11	18	27	41	53	70
	4	1	1	3	6	8	13	20	30	40	52
	3	1	1	3	5	7	11	17	26	34	44
	2	1	1	2	4	6	9	14	22	29	38
	1	1	1	1	3	4	7	10	15	20	26
	1/0	0	1	1	2	3	6	8	13	17	23
	2/0	0	1	1	2	3	5	7	11	15	19
	3/0	0	1	1	1	2	4	6	9	12	16
	4/0	0	0	1	1	1	3	5	8	10	13
	250	0	0	1	1	1	3	4	6	8	11
	300	0	0	1	1	1	2	3	5	7	9
	350	0	0	0	1	1	1	3	5	6	8
	400	0	0	0	1	1	1	3	4	6	7
	500	0	0	0	1	1	1	2	3	5	6
	600	0	0	0	1	1	1	1	3	4	5
	700	0	0	0	0	1	1	1	2	3	4
	750	0	0	0	0	1	1	1	2	3	4
	800	0	0	0	0	1	1	1	2	3	4
	900	0	0	0	0	0	1	1	1	3	3
	1000	0	0	0	0	0	1	1	1	2	3
	1250	0	0	0	0	0	1	1	1	1	2
	1500	0	0	0	0	0	0	1	1	1	2
	1750	0	0	0	0	0	0	1	1	1	1
	2000	0	0	0	0	0	0	1	1	1	1

*Types RHH, RHW, and RHW-2 without outer covering.

Table C7. Continued

CONDUCTORS

Type	Conductor Size (AWG/kcmil)	½	¾	1	1¼	1½	2	2½	3	3½	4
		Trade Size (in.)									
THHN, THWN, THWN-2	14	13	22	36	63	81	133	201	308	401	523
	12	9	16	26	46	59	97	146	225	292	381
	10	6	10	16	29	37	61	92	141	184	240
	8	3	6	9	16	21	35	53	81	106	138
	6	2	4	7	12	15	25	38	59	76	100
	4	1	2	4	7	9	15	23	36	47	61
	3	1	1	3	6	8	13	20	30	40	52
	2	1	1	3	5	7	11	17	26	33	44
	1	1	1	1	4	5	8	12	19	25	32
	1/0	1	1	1	3	4	7	10	16	21	27
	2/0	0	1	1	2	3	6	8	13	17	23
	3/0	0	1	1	1	3	5	7	11	14	19
	4/0	0	1	1	1	2	4	6	9	12	15
	250	0	0	1	1	1	3	5	7	10	12
	300	0	0	1	1	1	3	4	6	8	11
	350	0	0	1	1	1	2	3	5	7	9
	400	0	0	0	1	1	1	3	5	6	8
	500	0	0	0	1	1	1	2	4	5	7
	600	0	0	0	1	1	1	1	3	4	6
	700	0	0	0	1	1	1	1	3	4	5
	750	0	0	0	0	1	1	1	3	3	5
	800	0	0	0	0	1	1	1	2	3	4
	900	0	0	0	0	1	1	1	2	3	4
	1000	0	0	0	0	0	1	1	1	3	3
FEP, FEPB, PFA, PFAH, TFE	14	12	21	35	61	79	129	195	299	389	507
	12	9	15	25	44	57	94	142	218	284	370
	10	6	11	18	32	41	68	102	156	203	266
	8	3	6	10	18	23	39	58	89	117	152
	6	2	4	7	13	17	27	41	64	83	108
	4	1	3	5	9	12	19	29	44	58	75
	3	1	2	4	7	10	16	24	37	48	63
	2	1	1	3	6	8	13	20	30	40	52
PFA, PFAH, TFE	1	1	1	2	4	5	9	14	21	28	36
PFA, PFAH, TFE, Z	1/0	1	1	1	3	4	7	11	18	23	30
	2/0	1	1	1	3	4	6	9	14	19	25
	3/0	0	1	1	2	3	5	8	12	16	20
	4/0	0	1	1	1	2	4	6	10	13	17
Z	14	20	26	42	73	95	156	235	360	469	611
	12	14	18	30	52	67	111	167	255	332	434
	10	8	11	18	32	41	68	102	156	203	266
	8	5	7	11	20	26	43	64	99	129	168
	6	4	5	8	14	18	30	45	69	90	118
	4	2	3	5	9	12	20	31	48	62	81
	3	2	2	4	7	9	15	23	35	45	59
	2	1	1	3	6	7	12	19	29	38	49
	1	1	1	2	5	6	10	15	23	30	40
XHH, XHHW, XHHW-2, ZW	14	9	15	25	44	57	93	140	215	280	365
	12	7	12	19	33	43	71	108	165	215	280
	10	5	9	14	25	32	53	80	123	160	209
	8	3	5	8	14	18	29	44	68	89	116
	6	1	3	6	10	13	22	33	50	66	86
	4	1	2	4	7	9	16	24	36	48	62
	3	1	1	3	6	8	13	20	31	40	52
	2	1	1	3	5	7	11	17	26	34	44
XHH, XHHW, XHHW-2	1	1	1	1	4	5	8	12	19	25	33
	1/0	1	1	1	3	4	7	10	16	21	28
	2/0	0	1	1	2	3	6	9	13	17	23
	3/0	0	1	1	1	3	5	7	11	14	19
	4/0	0	1	1	1	2	4	6	9	12	16
	250	0	0	1	1	1	3	5	7	10	13
	300	0	0	1	1	1	3	4	6	8	11
	350	0	0	1	1	1	2	3	5	7	10
	400	0	0	0	1	1	1	3	5	6	8
	500	0	0	0	1	1	1	2	4	5	7
	600	0	0	0	1	1	1	1	3	4	6
	700	0	0	0	1	1	1	1	3	4	5
	750	0	0	0	0	1	1	1	3	3	5
	800	0	0	0	0	1	1	1	2	3	4
	900	0	0	0	0	1	1	1	2	3	4
	1000	0	0	0	0	0	1	1	1	3	3
	1250	0	0	0	0	0	1	1	1	1	3
	1500	0	0	0	0	0	1	1	1	1	2
	1750	0	0	0	0	0	0	1	1	1	2
	2000	0	0	0	0	0	0	1	1	1	2

Table C7. Continued

FIXTURE WIRES

Type	Conductor Size (AWG/kcmil)	½	¾	1	1¼	1½	2
		Trade Size (in.)					
FFH-2, RFH-2, RFHH-3	18	8	15	24	42	54	89
	16	7	12	20	35	46	75
SF-2, SFF-2	18	11	19	30	53	69	113
	16	9	15	25	44	57	93
	14	7	12	20	35	46	75
SF-1, SFF-1	18	19	33	53	94	122	199
AF, RFH-1, RFHH-2, TF, TFF, XF, XFF	18	14	24	39	69	90	147
AF, RFHH-2, TF, TFF, XF, XFF	16	11	20	32	56	72	119
AF, XF, XFF	14	9	15	25	44	57	93
TFN, TFFN	18	23	39	63	111	144	236
	16	17	30	48	85	110	180
PF, PFF, PGF, PGFF, PAF, PTF, PTFF, PAFF	18	21	37	60	105	136	223
	16	16	29	46	81	105	173
	14	12	21	35	61	79	129
HF, HFF, ZF, ZFF, ZHF	18	28	48	77	136	176	288
	16	20	35	57	100	129	212
	14	15	26	42	73	95	156
KF-2, KFF-2	18	40	70	112	197	255	418
	16	28	49	79	139	180	295
	14	19	34	54	95	123	202
	12	13	23	37	65	85	139
	10	9	15	25	44	57	93
KF-1, KFF-1	18	48	83	134	235	304	499
	16	34	58	94	165	214	350
	14	23	39	63	111	144	236
	12	15	26	42	73	95	156
	10	10	17	27	48	62	102
AF, XF, XFF	12	5	8	13	23	30	50
	10	3	6	10	18	23	39

Note: This table is for concentric stranded conductors only. For compact stranded conductors, Table C7(A) should be used.

Table C7(A). Maximum Number of Compact Conductors in Liquidtight Flexible Metal Conduit *(Based on Table 1, Chapter 9)*

COMPACT CONDUCTORS												
	Conductor Size	Trade Size (in.)										
Type	(AWG/kcmil)	3/8	1/2	3/4	1	1 1/4	1 1/2	2	2 1/2	3	3 1/2	4
THW, THW-2, THHW	8	1	2	4	7	12	15	25	38	58	76	99
	6	1	1	3	5	9	12	19	29	45	59	77
	4	1	1	2	4	7	9	14	22	34	44	57
	2	1	1	1	3	5	6	11	16	25	32	42
	1	0	1	1	1	3	4	7	11	17	23	30
	1/0	0	1	1	1	3	4	6	10	15	20	26
	2/0	0	0	1	1	2	3	5	8	13	16	21
	3/0	0	0	1	1	1	3	4	7	11	14	18
	4/0	0	0	1	1	1	2	4	6	9	12	15
	250	0	0	0	1	1	1	3	4	7	9	12
	300	0	0	0	1	1	1	2	4	6	8	10
	350	0	0	0	1	1	1	2	3	5	7	9
	400	0	0	0	0	1	1	1	3	5	6	8
	500	0	0	0	0	1	1	1	3	4	5	7
	600	0	0	0	0	1	1	1	1	3	4	6
	700	0	0	0	0	1	1	1	1	3	4	5
	750	0	0	0	0	0	1	1	1	3	3	5
	1000	0	0	0	0	0	1	1	1	1	3	4
THHN, THWN, THWN-2	8	—	—	—	—	—	—	—	—	—	—	—
	6	1	2	4	7	13	17	28	43	66	86	112
	4	1	1	3	4	8	11	17	26	41	53	69
	2	1	1	1	3	6	7	12	19	29	38	50
	1	0	1	1	2	4	6	9	14	22	28	37
	1/0	0	1	1	1	4	5	8	12	19	24	32
	2/0	0	1	1	1	3	4	6	10	15	20	26
	3/0	0	0	1	1	2	3	5	8	13	17	22
	4/0	0	0	1	1	1	3	4	7	10	14	18
	250	0	0	1	1	1	1	3	5	8	11	14
	300	0	0	0	1	1	1	3	4	7	9	12
	350	0	0	0	1	1	1	2	4	6	8	11
	400	0	0	0	1	1	1	2	3	5	7	9
	500	0	0	0	0	1	1	1	3	5	6	8
	600	0	0	0	0	1	1	1	2	4	5	6
	700	0	0	0	0	1	1	1	1	3	4	6
	750	0	0	0	0	1	1	1	1	3	4	5
	1000	0	0	0	0	0	1	1	1	2	3	4
XHHW, XHHW-2	8	1	3	5	9	15	20	33	49	76	98	129
	6	1	2	4	6	11	15	24	37	56	73	95
	4	1	1	3	4	8	11	17	26	41	53	69
	2	1	1	1	3	6	7	12	19	29	38	50
	1	0	1	1	2	4	6	9	14	22	28	37
	1/0	0	1	1	1	4	5	8	12	19	24	32
	2/0	0	1	1	1	3	4	7	10	16	20	27
	3/0	0	0	1	1	2	3	5	8	13	17	22
	4/0	0	0	1	1	1	3	4	7	11	14	18
	250	0	0	1	1	1	1	3	5	8	11	15
	300	0	0	0	1	1	1	3	5	7	9	12
	350	0	0	0	1	1	1	3	4	6	8	11
	400	0	0	0	1	1	1	2	4	6	7	10
	500	0	0	0	0	1	1	1	3	5	6	8
	600	0	0	0	0	1	1	1	2	4	5	6
	700	0	0	0	0	1	1	1	1	3	4	6
	750	0	0	0	0	1	1	1	1	3	4	5
	1000	0	0	0	0	0	1	1	1	2	3	4

Definition: Compact stranding is the result of a manufacturing process where the standard conductor is compressed to the extent that the interstices (voids between strand wires) are virtually eliminated.

Table C8. Maximum Number of Conductors and Fixture Wires in Rigid Metal Conduit *(Based on Table 1, Chapter 9)*

CONDUCTORS													
	Conductor Size	Trade Size (in.)											
Type	(AWG/kcmil)	1/2	3/4	1	1 1/4	1 1/2	2	2 1/2	3	3 1/2	4	5	6
RH	14	6	10	17	29	39	65	93	143	191	246	387	558
	12	5	8	13	23	32	52	75	115	154	198	311	448
RHH, RHW, RHW-2	14	4	7	12	21	28	46	66	102	136	176	276	398
	12	3	6	10	17	23	38	55	85	113	146	229	330
RH, RHH, RHW, RHW-2	10	3	5	8	14	19	31	44	68	91	118	185	267
	8	1	2	4	7	10	16	23	36	48	61	97	139
	6	1	1	3	6	8	13	18	29	38	49	77	112
	4	1	1	2	4	6	10	14	22	30	38	60	87
	3	1	1	2	4	5	9	12	19	26	34	53	76
	2	1	1	1	3	4	7	11	17	23	29	46	66
	1	0	1	1	1	3	5	7	11	15	19	30	44
	1/0	0	1	1	1	2	4	6	10	13	17	26	38
	2/0	0	1	1	1	2	4	5	8	11	14	23	33
	3/0	0	0	1	1	1	3	4	7	10	12	20	28
	4/0	0	0	1	1	1	3	4	6	8	11	17	24
	250	0	0	0	1	1	1	3	4	6	8	13	18
	300	0	0	0	1	1	1	2	4	5	7	11	16
	350	0	0	0	1	1	1	2	4	5	6	10	15
	400	0	0	0	1	1	1	1	3	4	6	9	13
	500	0	0	0	1	1	1	1	3	4	5	8	11
	600	0	0	0	0	1	1	1	2	3	4	6	9
	700	0	0	0	0	1	1	1	1	3	4	6	8
	750	0	0	0	0	0	1	1	1	3	3	5	8
	800	0	0	0	0	0	1	1	1	2	3	5	7
	900	0	0	0	0	0	1	1	1	2	3	5	7
	1000	0	0	0	0	0	1	1	1	1	3	4	6
	1250	0	0	0	0	0	0	1	1	1	1	3	5
	1500	0	0	0	0	0	0	1	1	1	1	3	4
	1750	0	0	0	0	0	0	1	1	1	1	2	4
	2000	0	0	0	0	0	0	0	1	1	1	2	3
TW	14	9	15	25	44	59	98	140	216	288	370	581	839
	12	7	12	19	33	45	75	107	165	221	284	446	644
	10	5	9	14	25	34	56	80	123	164	212	332	480
	8	3	5	8	14	19	31	44	68	91	118	185	267
RHH*, RHW*, RHW-2*, THHW, THW, THW-2	14	6	10	17	29	39	65	93	143	191	246	387	558
RHH*, RHW*, RHW-2*, THHW, THW	12	5	8	13	23	32	52	75	115	154	198	311	448
	10	3	6	10	18	25	41	58	90	120	154	242	350
RHH*, RHW*, RHW-2*, THHW, THW, THW-2	8	1	4	6	11	15	24	35	54	72	92	145	209
RHH*, RHW*, RHW-2, TW, THW, THHW, THW-2	6	1	3	5	8	11	18	27	41	55	71	111	160
	4	1	1	3	6	8	14	20	31	41	53	83	120
	3	1	1	3	5	7	12	17	26	35	45	71	103
	2	1	1	2	4	6	10	14	22	30	38	60	87
	1	1	1	1	3	4	7	10	15	21	27	42	61
	1/0	0	1	1	2	3	6	8	13	18	23	36	52
	2/0	0	1	1	2	3	5	7	11	15	19	31	44
	3/0	0	1	1	1	2	4	6	9	13	16	26	37
	4/0	0	0	1	1	1	3	5	8	10	14	21	31
	250	0	0	1	1	1	3	4	6	8	11	17	25
	300	0	0	1	1	1	2	3	5	7	9	15	22
	350	0	0	0	1	1	1	3	5	6	8	13	19
	400	0	0	0	1	1	1	3	4	6	7	12	17
	500	0	0	0	1	1	1	2	3	5	6	10	14
	600	0	0	0	1	1	1	1	3	4	5	8	12
	700	0	0	0	0	1	1	1	2	3	4	7	10
	750	0	0	0	0	1	1	1	2	3	4	7	10
	800	0	0	0	0	1	1	1	2	3	4	6	9
	900	0	0	0	0	1	1	1	1	3	4	6	8
	1000	0	0	0	0	0	1	1	1	2	3	5	8
	1250	0	0	0	0	0	1	1	1	1	2	4	6
	1500	0	0	0	0	0	1	1	1	1	2	3	5
	1750	0	0	0	0	0	0	1	1	1	1	3	4
	2000	0	0	0	0	0	0	1	1	1	1	3	4

*Types RHH, RHW, and RHW-2 without outer covering.

Table C8. Continued

CONDUCTORS

Type	Conductor Size (AWG/ kcmil)	½	¾	1	1¼	1½	2	2½	3	3½	4	5	6
		Trade Size (in.)											
THHN, THWN, THWN-2	14	13	22	36	63	85	140	200	309	412	531	833	1202
	12	9	16	26	46	62	102	146	225	301	387	608	877
	10	6	10	17	29	39	64	92	142	189	244	383	552
	8	3	6	9	16	22	37	53	82	109	140	221	318
	6	2	4	7	12	16	27	38	59	79	101	159	230
	4	1	2	4	7	10	16	23	36	48	62	98	141
	3	1	1	3	6	8	14	20	31	41	53	83	120
	2	1	1	3	5	7	11	17	26	34	44	70	100
	1	1	1	1	4	5	8	12	19	25	33	51	74
	1/0	1	1	1	3	4	7	10	16	21	27	43	63
	2/0	0	1	1	2	3	6	8	13	18	23	36	52
	3/0	0	1	1	1	3	5	7	11	15	19	30	43
	4/0	0	1	1	1	2	4	6	9	12	16	25	36
	250	0	0	1	1	1	3	5	7	10	13	20	29
	300	0	0	1	1	1	3	4	6	8	11	17	25
	350	0	0	1	1	1	2	3	5	7	10	15	22
	400	0	0	1	1	1	2	3	5	7	8	13	20
	500	0	0	0	1	1	1	2	4	5	7	11	16
	600	0	0	0	1	1	1	1	3	4	6	9	13
	700	0	0	0	1	1	1	1	3	4	5	8	11
	750	0	0	0	0	1	1	1	3	4	5	7	11
	800	0	0	0	0	1	1	1	2	3	4	7	10
	900	0	0	0	0	1	1	1	2	3	4	6	9
	1000	0	0	0	0	1	1	1	1	3	4	6	8
FEP, FEPB, PFA, PFAH, TFE	14	12	22	35	61	83	136	194	300	400	515	808	1166
	12	9	16	26	44	60	99	142	219	292	376	590	851
	10	6	11	18	32	43	71	102	157	209	269	423	610
	8	3	6	10	18	25	41	58	90	120	154	242	350
	6	2	4	7	13	17	29	41	64	85	110	172	249
	4	1	3	5	9	12	20	29	44	59	77	120	174
	3	1	2	4	7	10	17	24	37	50	64	100	145
	2	1	1	3	6	8	14	20	31	41	53	83	120
PFA, PFAH, TFE	1	1	1	2	4	6	9	14	21	28	37	57	83
PFA, PFAH, TFE, Z	1/0	1	1	1	3	5	8	11	18	24	30	48	69
	2/0	1	1	1	3	4	6	9	14	19	25	40	57
	3/0	0	1	1	2	3	5	8	12	16	21	33	47
	4/0	0	1	1	1	2	4	6	10	13	17	27	39
Z	14	15	26	42	73	100	164	234	361	482	621	974	1405
	12	10	18	30	52	71	116	166	256	342	440	691	997
	10	6	11	18	32	43	71	102	157	209	269	423	610
	8	4	7	11	20	27	45	64	99	132	170	267	386
	6	3	5	8	14	19	31	45	69	93	120	188	271
	4	1	3	5	9	13	22	31	48	64	82	129	186
	3	1	2	4	7	9	16	22	35	47	60	94	136
	2	1	1	3	6	8	13	19	29	39	50	78	113
	1	1	1	2	5	6	10	15	23	31	40	63	92
XHH, XHHW, XHHW-2, ZW	14	9	15	25	44	59	98	140	216	288	370	581	839
	12	7	12	19	33	45	75	107	165	221	284	446	644
	10	5	9	14	25	34	56	80	123	164	212	332	480
	8	3	5	8	14	19	31	44	68	91	118	185	267
	6	1	3	6	10	14	23	33	51	68	87	137	197
	4	1	2	4	7	10	16	24	37	49	63	99	143
	3	1	1	3	6	8	14	20	31	41	53	84	121
	2	1	1	3	5	7	12	17	26	35	45	70	101
XHH, XHHW, XHHW-2	1	1	1	1	4	5	9	12	19	26	33	52	76
	1/0	1	1	1	3	4	7	10	16	22	28	44	64
	2/0	0	1	1	2	3	6	9	13	18	23	37	53
	3/0	0	1	1	1	3	5	7	11	15	19	30	44
	4/0	0	1	1	1	2	4	6	9	12	16	25	36
	250	0	0	1	1	1	3	5	7	10	13	20	30
	300	0	0	1	1	1	3	4	6	9	11	18	25
	350	0	0	1	1	1	2	3	6	7	10	15	22
	400	0	0	1	1	1	2	3	5	7	9	14	20
	500	0	0	0	1	1	1	2	4	5	7	11	16
	600	0	0	0	1	1	1	1	3	4	6	9	13
	700	0	0	0	1	1	1	1	3	4	5	8	11
	750	0	0	0	0	1	1	1	3	4	5	7	11
	800	0	0	0	0	1	1	1	2	3	4	7	10
	900	0	0	0	0	1	1	1	2	3	4	6	9
	1000	0	0	0	0	1	1	1	1	3	4	6	8
	1250	0	0	0	0	0	1	1	1	2	3	4	6
	1500	0	0	0	0	0	1	1	1	1	2	4	5
	1750	0	0	0	0	0	0	1	1	1	1	3	5
	2000	0	0	0	0	0	0	1	1	1	1	3	4

Table C8. Continued

FIXTURE WIRES

Type	Conductor Size (AWG/ kcmil)	½	¾	1	1¼	1½	2
		Trade Size (in.)					
FFH-2, RFH-2, RFHH-3	18	8	15	24	42	57	94
	16	7	12	20	35	48	79
SF-2, SFF-2	18	11	19	31	53	72	118
	16	9	15	25	44	59	98
	14	7	12	20	35	48	79
SF-1, SFF-1	18	19	33	54	94	127	209
AF, RFH-1, RFHH-2, TF, TFF, XF, XFF	18	14	25	40	69	94	155
AF, RFHH-2, TF, TFF, XF, XFF	16	11	20	32	56	76	125
AF, XF, XFF	14	9	15	25	44	59	98
TFN, TFFN	18	23	40	64	111	150	248
	16	17	30	49	84	115	189
PF, PFF, PGF, PGFF, PAF, PTF, PTFF, PAFF	18	21	38	61	105	143	235
	16	16	29	47	81	110	181
	14	12	22	35	61	83	136
HF, HFF, ZF, ZFF, ZHF	18	28	48	79	135	184	303
	16	20	36	58	100	136	223
	14	15	26	42	73	100	164
KF-2, KFF-2	18	40	71	114	197	267	439
	16	28	50	80	138	188	310
	14	19	34	55	95	129	213
	12	13	23	38	65	89	146
	10	9	15	25	44	59	98
KF-1, KFF-1	18	48	84	136	235	318	524
	16	34	59	96	165	224	368
	14	23	40	64	111	150	248
	12	15	26	42	73	100	164
	10	10	17	28	48	65	107
AF, XF, XFF	12	5	8	13	23	32	52
	10	3	6	10	18	25	41

Note: This table is for concentric stranded conductors only. For compact stranded conductors, Table C8(A) should be used.

Table C8(A). Maximum Number of Compact Conductors in Rigid Metal Conduit *(Based on Table 1, Chapter 9)*

COMPACT CONDUCTORS													
	Conductor Size (AWG/ kcmil)	Trade Size (in.)											
Type		**½**	**¾**	**1**	**1¼**	**1½**	**2**	**2½**	**3**	**3½**	**4**	**5**	**6**
THW, THW-2, THHW	8	2	4	7	12	16	26	38	59	78	101	158	228
	6	1	3	5	9	12	20	29	45	60	78	122	176
	4	1	2	4	7	9	15	22	34	45	58	91	132
	2	1	1	3	5	7	11	16	25	33	43	67	97
	1	1	1	1	3	5	8	11	17	23	30	47	68
	1/0	1	1	1	3	4	7	10	15	20	26	41	59
	2/0	0	1	1	2	3	6	8	13	17	22	34	50
	3/0	0	1	1	1	3	5	7	11	14	19	29	42
	4/0	0	1	1	1	2	4	6	9	12	15	24	35
	250	0	0	1	1	1	3	4	7	9	12	19	28
	300	0	0	1	1	1	3	4	6	8	11	17	24
	350	0	0	1	1	1	2	3	5	7	9	15	22
	400	0	0	1	1	1	1	3	5	7	8	13	20
	500	0	0	0	1	1	1	3	4	5	7	11	17
	600	0	0	0	1	1	1	1	3	4	6	9	13
	700	0	0	0	1	1	1	1	3	4	5	8	12
	750	0	0	0	0	1	1	1	3	4	5	7	11
	1000	0	0	0	0	1	1	1	1	3	4	6	9
THHN, THWN, THWN-2	8	—	—	—	—	—	—	—	—	—	—	—	—
	6	2	5	8	13	18	30	43	66	88	114	179	258
	4	1	3	5	8	11	18	26	41	55	70	110	159
	2	1	1	3	6	8	13	19	29	39	50	79	114
	1	1	1	2	4	6	10	14	22	29	38	60	86
	1/0	1	1	1	4	5	8	12	19	25	32	51	73
	2/0	1	1	1	3	4	7	10	15	21	26	42	60
	3/0	0	1	1	2	3	6	8	13	17	22	35	51
	4/0	0	1	1	1	3	5	7	10	14	18	29	42
	250	0	1	1	1	2	4	5	8	11	14	23	33
	300	0	0	1	1	1	3	4	7	10	12	20	28
	350	0	0	1	1	1	3	4	6	8	11	17	25
	400	0	0	1	1	1	2	3	5	7	10	15	22
	500	0	0	0	1	1	1	3	5	6	8	13	19
	600	0	0	0	1	1	1	2	4	5	6	10	15
	700	0	0	0	1	1	1	1	3	4	6	9	13
	750	0	0	0	1	1	1	1	3	4	5	9	13
	1000	0	0	0	0	1	1	1	2	3	4	6	9
XHHW, XHHW-2	8	3	5	9	15	21	34	49	76	101	130	205	296
	6	2	4	6	11	15	25	36	56	75	97	152	220
	4	1	3	5	8	11	18	26	41	55	70	110	159
	2	1	1	3	6	8	13	19	29	39	50	79	114
	1	1	1	2	4	6	10	14	22	29	38	60	86
	1/0	1	1	1	4	5	8	12	19	25	32	51	73
	2/0	1	1	1	3	4	7	10	16	21	27	43	62
	3/0	0	1	1	2	3	6	8	13	17	22	35	51
	4/0	0	1	1	1	3	5	7	11	14	19	29	42
	250	0	1	1	1	2	4	5	8	11	15	23	34
	300	0	0	1	1	1	3	5	7	10	13	20	29
	350	0	0	1	1	1	3	4	6	9	11	18	25
	400	0	0	1	1	1	2	4	6	8	10	16	23
	500	0	0	0	1	1	1	3	5	6	8	13	19
	600	0	0	0	1	1	1	2	4	5	7	10	15
	700	0	0	0	1	1	1	1	3	4	6	9	13
	750	0	0	0	1	1	1	1	3	4	5	8	12
	1000	0	0	0	0	1	1	1	2	3	4	7	10

Definition: Compact stranding is the result of a manufacturing process where the standard conductor is compressed to the extent that the interstices (voids between strand wires) are virtually eliminated.

Table C9. Maximum Number of Conductors and Fixture Wires in Rigid PVC Conduit, Schedule 80 *(Based on Table 1, Chapter 9)*

CONDUCTORS													
	Conductor Size (AWG/ kcmil)	Trade Size (in.)											
Type		**½**	**¾**	**1**	**1¼**	**1½**	**2**	**2½**	**3**	**3½**	**4**	**5**	**6**
RH	14	4	8	13	23	32	55	79	123	166	215	341	490
	12	3	6	10	19	26	44	63	99	133	173	274	394
RHH, RHW, RHW-2	14	3	5	9	17	23	39	56	88	118	153	243	349
	12	2	4	7	14	19	32	46	73	98	127	202	290
RH, RHH, RHW, RHW-2	10	1	3	6	11	15	26	37	59	79	103	163	234
	8	1	1	3	6	8	13	19	31	41	54	85	122
	6	1	1	2	4	6	11	16	24	33	43	68	98
	4	1	1	1	3	5	8	12	19	26	33	53	77
	3	0	1	1	3	4	7	11	17	23	29	47	67
	2	0	1	1	3	4	6	9	14	20	25	41	58
	1	0	1	1	1	2	4	6	9	13	17	27	38
	1/0	0	0	1	1	1	3	5	8	11	15	23	33
	2/0	0	0	1	1	1	3	4	7	10	13	20	29
	3/0	0	0	1	1	1	3	4	6	8	11	17	25
	4/0	0	0	0	1	1	2	3	5	7	9	15	21
	250	0	0	0	1	1	1	2	4	5	7	11	16
	300	0	0	0	1	1	1	2	3	5	6	10	14
	350	0	0	0	1	1	1	1	3	4	5	9	13
	400	0	0	0	0	1	1	1	3	4	5	8	12
	500	0	0	0	0	1	1	1	2	3	4	7	10
	600	0	0	0	0	0	1	1	1	3	3	6	8
	700	0	0	0	0	0	1	1	1	2	3	5	7
	750	0	0	0	0	0	1	1	1	2	3	5	7
	800	0	0	0	0	0	1	1	1	2	3	4	7
	1000	0	0	0	0	0	1	1	1	1	2	4	5
	1250	0	0	0	0	0	0	1	1	1	1	3	4
	1500	0	0	0	0	0	0	1	1	1	1	2	4
	1750	0	0	0	0	0	0	0	1	1	1	2	3
	2000	0	0	0	0	0	0	0	1	1	1	1	3
TW	14	6	11	20	35	49	82	118	185	250	324	514	736
	12	5	9	15	27	38	63	91	142	192	248	394	565
	10	3	6	11	20	28	47	67	106	143	185	294	421
	8	1	3	6	11	15	26	37	59	79	103	163	234
RHH*, RHW*, RHW-2*, THHW, THW, THW-2	14	4	8	13	23	32	55	79	123	166	215	341	490
RHH*, RHW*, RHW-2*, THHW, THW	12	3	6	10	19	26	44	63	99	133	173	274	394
	10	2	5	8	15	20	34	49	77	104	135	214	307
RHH*, RHW*, RHW-2*, THHW, THW, THW-2	8	1	3	5	9	12	20	29	46	62	81	128	184
RHH*, RHW*, RHW-2*, TW, THW, THHW, THW-2	6	1	1	3	7	9	16	22	35	48	62	98	141
	4	1	1	3	5	7	12	17	26	35	46	73	105
	3	1	1	2	4	6	10	14	22	30	39	63	90
	2	1	1	1	3	5	8	12	19	26	33	53	77
	1	0	1	1	2	3	6	8	13	18	23	37	54
	1/0	0	1	1	1	3	5	7	11	15	20	32	46
	2/0	0	1	1	1	2	4	6	10	13	17	27	39
	3/0	0	0	1	1	1	3	5	8	11	14	23	33
	4/0	0	0	1	1	1	3	4	7	9	12	19	27
	250	0	0	0	1	1	2	3	5	7	9	15	22
	300	0	0	0	1	1	1	3	5	6	8	13	19
	350	0	0	0	1	1	1	2	4	6	7	12	17
	400	0	0	0	1	1	1	2	4	5	7	10	15
	500	0	0	0	1	1	1	1	3	4	5	9	13
	600	0	0	0	0	1	1	1	2	3	4	7	10
	700	0	0	0	0	1	1	1	2	3	4	6	9
	750	0	0	0	0	0	1	1	1	3	4	6	8
	800	0	0	0	0	0	1	1	1	3	3	6	8
	900	0	0	0	0	0	1	1	1	2	3	5	7
	1000	0	0	0	0	0	1	1	1	2	3	5	7
	1250	0	0	0	0	0	1	1	1	1	2	4	5
	1500	0	0	0	0	0	0	1	1	1	1	3	4
	1750	0	0	0	0	0	0	1	1	1	1	3	4
	2000	0	0	0	0	0	0	0	1	1	1	2	3

*Types RHH, RHW, and RHW-2 without outer covering.

Table C9. Continued

CONDUCTORS													
Type	Conductor Size (AWG/kcmil)	Trade Size (in.) ½	¾	1	1¼	1½	2	2½	3	3½	4	5	6
THHN, THWN, THWN-2	14	9	17	28	51	70	118	170	265	358	464	736	1055
	12	6	12	20	37	51	86	124	193	261	338	537	770
	10	4	7	13	23	32	54	78	122	164	213	338	485
	8	2	4	7	13	18	31	45	70	95	123	195	279
	6	1	3	5	9	13	22	32	51	68	89	141	202
	4	1	1	3	6	8	14	20	31	42	54	86	124
	3	1	1	3	5	7	12	17	26	35	46	73	105
	2	1	1	2	4	6	10	14	22	30	39	61	88
	1	0	1	1	3	4	7	10	16	22	29	45	65
	1/0	0	1	1	2	3	6	9	14	18	24	38	55
	2/0	0	1	1	1	3	5	7	11	15	20	32	46
	3/0	0	1	1	1	2	4	6	9	13	17	26	38
	4/0	0	0	1	1	1	3	5	8	10	14	22	31
	250	0	0	1	1	1	3	4	6	8	11	18	25
	300	0	0	0	1	1	2	3	5	7	9	15	22
	350	0	0	0	1	1	1	3	5	6	8	13	19
	400	0	0	0	1	1	1	3	4	6	7	12	17
	500	0	0	0	1	1	1	2	3	5	6	10	14
	600	0	0	0	0	1	1	1	3	4	5	8	12
	700	0	0	0	0	1	1	1	2	3	4	7	10
	750	0	0	0	0	1	1	1	2	3	4	7	9
	800	0	0	0	0	1	1	1	2	3	4	6	9
	900	0	0	0	0	0	1	1	1	3	3	6	8
	1000	0	0	0	0	0	1	1	1	2	3	5	7
FEP, FEPB, PFA, PFA, PFAH, TFE	14	8	16	27	49	68	115	164	257	347	450	714	1024
	12	6	12	20	36	50	84	120	188	253	328	521	747
	10	4	8	14	26	36	60	86	135	182	235	374	536
	8	2	5	8	15	20	34	49	77	104	135	214	307
	6	1	3	6	10	14	24	35	55	74	96	152	218
	4	1	2	4	7	10	17	24	38	52	67	106	153
	3	1	1	3	6	8	14	20	32	43	56	89	127
	2	1	1	3	5	7	12	17	26	35	46	73	105
PFA, PFAH, TFE	1	1	1	1	3	5	8	11	18	25	32	51	73
PFA, PFAH, TFE, Z	1/0	0	1	1	3	4	7	10	15	20	27	42	61
	2/0	0	1	1	2	3	5	8	12	17	22	35	50
	3/0	0	1	1	1	2	4	6	10	14	18	29	41
	4/0	0	0	1	1	1	4	5	8	11	15	24	34
Z	14	10	19	33	59	82	138	198	310	418	542	860	1233
	12	7	14	23	42	58	98	141	220	297	385	610	875
	10	4	8	14	26	36	60	86	135	182	235	374	536
	8	3	5	9	16	22	38	54	85	115	149	236	339
	6	2	4	6	11	16	26	38	60	81	104	166	238
	4	1	2	4	8	11	18	26	41	55	72	114	164
	3	1	2	3	5	8	13	19	30	40	52	83	119
	2	1	1	2	5	6	11	16	25	33	43	69	99
	1	0	1	2	4	5	9	13	20	27	35	56	80
XHH, XHHW, XHHW-2, ZW	14	6	11	20	35	49	82	118	185	250	324	514	736
	12	5	9	15	27	38	63	91	142	192	248	394	565
	10	3	6	11	20	28	47	67	106	143	185	294	421
	8	1	3	6	11	15	26	37	59	79	103	163	234
	6	1	2	4	8	11	19	28	43	59	76	121	173
	4	1	1	3	6	8	14	20	31	42	55	87	125
	3	1	1	3	5	7	12	17	26	36	47	74	106
	2	1	1	2	4	6	10	14	22	30	39	62	89
XHH, XHHW, XHHW-2	1	0	1	1	3	4	7	10	16	22	29	46	66
	1/0	0	1	1	2	3	6	9	14	19	24	39	56
	2/0	0	1	1	1	3	5	7	11	16	20	32	46
	3/0	0	1	1	1	2	4	6	9	13	17	27	38
	4/0	0	0	1	1	1	3	5	8	11	14	22	32
	250	0	0	1	1	1	3	4	6	9	11	18	26
	300	0	0	1	1	1	2	3	5	7	10	15	22
	350	0	0	0	1	1	1	3	5	6	8	14	20
	400	0	0	0	1	1	1	3	4	6	7	12	17
	500	0	0	0	1	1	1	2	3	5	6	10	14
	600	0	0	0	0	1	1	1	3	4	5	8	11
	700	0	0	0	0	1	1	1	2	3	4	7	10
	750	0	0	0	0	1	1	1	2	3	4	6	9
	800	0	0	0	0	1	1	1	1	3	4	6	9
	900	0	0	0	0	0	1	1	—	3	3	5	8
	1000	0	0	0	0	0	1	1	1	2	3	5	7
	1250	0	0	0	0	0	1	1	1	1	2	4	6
	1500	0	0	0	0	0	0	1	1	1	1	3	5
	1750	0	0	0	0	0	0	1	1	1	1	3	4
	2000	0	0	0	0	0	0	1	1	1	1	2	4

Table C9. Continued

FIXTURE WIRES							
Type	Conductor Size (AWG/kcmil)	Trade Size (in.) ½	¾	1	1¼	1½	2
FFH-2, RFH-2, RFHH-3	18	6	11	19	34	47	79
	16	5	9	16	28	39	67
SF-2, SFF-2	18	7	14	24	43	59	100
	16	6	11	20	35	49	82
	14	5	9	16	28	39	67
SF-1, SFF-1	18	13	25	42	76	105	177
AF, RFH-1, RFHH-2, TF, TFF, XF, XFF	18	10	18	31	56	77	130
AF, RFHH-2, TF, TFF, XF, XFF	16	8	15	25	45	62	105
AF, XF, XFF	14	6	11	20	35	49	82
TFN, TFFN	18	16	29	50	90	124	209
	16	12	22	38	68	95	159
PF, PFF, PGF, PGFF, PAF, PTF, PTFF, PAFF	18	15	28	47	85	118	198
	16	11	22	36	66	91	153
	14	8	16	27	49	68	115
HF, HFF, ZF, ZFF, ZHF	18	19	36	61	110	152	255
	16	14	27	45	81	112	188
	14	10	19	33	59	82	138
KF-2, KFF-2	18	28	53	88	159	220	371
	16	19	37	62	112	155	261
	14	13	25	43	77	107	179
	12	9	17	29	53	73	123
	10	6	11	20	35	49	82
KF-1, KFF-1	18	33	63	106	190	263	442
	16	23	44	74	133	185	310
	14	16	29	50	90	124	209
	12	10	19	33	59	82	138
	10	7	13	21	39	54	90
AF, XF, XFF	12	3	6	10	19	26	44
	10	2	5	8	15	20	34

Note: This table is for concentric stranded conductors only. For compact stranded conductors, Table C9(A) should be used.

Table C9(A). Maximum Number of Compact Conductors in Rigid PVC Conduit, Schedule 80 *(Based on Table 1, Chapter 9)*

COMPACT CONDUCTORS													
Type	Conductor Size (AWG/kcmil)	Trade Size (in.) ½	¾	1	1¼	1½	2	2½	3	3½	4	5	6
THW, THW-2, THHW	8	1	3	5	9	13	22	32	50	68	88	140	200
	6	1	2	4	7	10	17	25	39	52	68	108	155
	4	1	1	3	5	7	13	18	29	39	51	81	116
	2	1	1	1	4	5	9	13	21	29	37	60	85
	1	0	1	1	3	4	6	9	15	20	26	42	60
	1/0	0	1	1	2	3	6	8	13	17	23	36	52
	2/0	0	1	1	1	3	5	7	11	15	19	30	44
	3/0	0	0	1	1	2	4	6	9	12	16	26	37
	4/0	0	0	1	1	1	3	5	8	10	13	22	31
	250	0	0	1	1	1	2	4	6	8	11	17	25
	300	0	0	0	1	1	2	3	5	7	9	15	21
	350	0	0	0	1	1	1	3	5	6	8	13	19
	400	0	0	0	1	1	1	3	4	6	7	12	17
	500	0	0	0	1	1	1	2	3	5	6	10	14
	600	0	0	0	0	1	1	1	3	4	5	8	12
	700	0	0	0	0	1	1	1	2	3	4	7	10
	750	0	0	0	0	1	1	1	2	3	4	7	10
	1000	0	0	0	0	0	1	1	1	2	3	5	8
THHN, THWN, THWN-2	8	—	—	—	—	—	—	—	—	—	—	—	—
	6	1	3	6	11	15	25	36	57	77	99	158	226
	4	1	1	3	6	9	15	22	35	47	61	98	140
	2	1	1	2	5	6	11	16	25	34	44	70	100
	1	1	1	1	3	5	8	12	19	25	33	53	75
	1/0	0	1	1	3	4	7	10	16	22	28	45	64
	2/0	0	1	1	2	3	6	8	13	18	23	37	53
	3/0	0	1	1	1	3	5	7	11	15	19	31	44
	4/0	0	0	1	1	2	4	6	9	12	16	25	37
	250	0	0	1	1	1	3	4	7	10	12	20	29
	300	0	0	1	1	1	3	4	6	8	11	17	25
	350	0	0	0	1	1	2	3	5	7	9	15	22
	400	0	0	0	1	1	1	3	5	6	8	13	19
	500	0	0	0	1	1	1	2	4	5	7	11	16
	600	0	0	0	1	1	1	1	3	4	6	9	13
	700	0	0	0	0	1	1	1	3	4	5	8	12
	750	0	0	0	0	1	1	1	3	4	5	8	11
	1000	0	0	0	0	0	1	1	1	3	3	5	8
XHHW, XHHW-2	8	1	4	7	12	17	29	42	65	88	114	181	260
	6	1	3	5	9	13	21	31	48	65	85	134	193
	4	1	1	3	6	9	15	22	35	47	61	98	140
	2	1	1	2	5	6	11	16	25	34	44	70	100
	1	1	1	1	3	5	8	12	19	25	33	53	75
	1/0	0	1	1	3	4	7	10	16	22	28	45	64
	2/0	0	1	1	2	3	6	8	13	18	24	38	54
	3/0	0	1	1	1	3	5	7	11	15	19	31	44
	4/0	0	0	1	1	2	4	6	9	12	16	26	37
	250	0	0	1	1	1	3	5	7	10	13	21	30
	300	0	0	1	1	1	3	4	6	8	11	17	25
	350	0	0	1	1	1	2	3	5	7	10	15	22
	400	0	0	0	1	1	1	3	5	7	9	14	20
	500	0	0	0	1	1	1	2	4	5	7	11	17
	600	0	0	0	1	1	1	1	3	4	6	9	13
	700	0	0	0	0	1	1	1	3	4	5	8	12
	750	0	0	0	0	1	1	1	2	3	5	7	11
	1000	0	0	0	0	0	1	1	1	3	3	6	8

Definition: Compact stranding is the result of a manufacturing process where the standard conductor is compressed to the extent that the interstices (voids between strand wires) are virtually eliminated.

Table C10. Maximum Number of Conductors and Fixture Wires in Rigid PVC Conduit, Schedule 40 and HDPE Conduit *(Based on Table 1, Chapter 9)*

CONDUCTORS													
Type	Conductor Size (AWG/kcmil)	Trade Size (in.) ½	¾	1	1¼	1½	2	2½	3	3½	4	5	6
RH	14	5	9	16	28	38	63	90	139	186	240	378	546
	12	4	8	12	22	30	50	72	112	150	193	304	439
RHH, RHW, RHW-2	14	4	7	11	20	27	45	64	99	133	171	269	390
	12	3	5	9	16	22	37	53	82	110	142	224	323
RH, RHH, RHW, RHW-2	10	2	4	7	13	18	30	43	66	89	115	181	261
	8	1	2	4	7	9	15	22	35	46	60	94	137
	6	1	1	3	5	7	12	18	28	37	48	76	109
	4	1	1	2	4	6	10	14	22	29	37	59	85
	3	1	1	1	4	5	8	12	19	25	33	52	75
	2	1	1	1	3	4	7	10	16	22	28	45	65
	1	0	1	1	1	3	5	7	11	14	19	29	43
	1/0	0	1	1	1	2	4	6	9	13	16	26	37
	2/0	0	0	1	1	1	3	5	8	11	14	22	32
	3/0	0	0	1	1	1	3	4	7	9	12	19	28
	4/0	0	0	1	1	1	2	4	6	8	10	16	24
	250	0	0	0	1	1	1	3	4	6	8	12	18
	300	0	0	0	1	1	1	2	4	5	7	11	16
	350	0	0	0	1	1	1	2	3	5	6	10	14
	400	0	0	0	1	1	1	1	3	4	6	9	13
	500	0	0	0	0	1	1	1	3	4	5	8	11
	600	0	0	0	0	1	1	1	2	3	4	6	9
	700	0	0	0	0	0	1	1	1	3	3	6	8
	750	0	0	0	0	0	1	1	1	2	3	5	8
	800	0	0	0	0	0	1	1	1	2	3	5	7
	900	0	0	0	0	0	1	1	1	2	3	5	7
	1000	0	0	0	0	0	1	1	1	1	3	4	6
	1250	0	0	0	0	0	0	1	1	1	1	3	5
	1500	0	0	0	0	0	0	1	1	1	1	3	4
	1750	0	0	0	0	0	0	1	1	1	1	2	3
	2000	0	0	0	0	0	0	0	1	1	1	2	3
TW	14	8	14	24	42	57	94	135	209	280	361	568	822
	12	6	11	18	32	44	72	103	160	215	277	436	631
	10	4	8	13	24	32	54	77	119	160	206	325	470
	8	2	4	7	13	18	30	43	66	89	115	181	261
RHH*, RHW*, RHW-2*, THHW, THW, THW-2	14	5	9	16	28	38	63	90	139	186	240	378	546
RHH*, RHW*, RHW-2*, THHW, THW*	12	4	8	12	22	30	50	72	112	150	193	304	439
	10	3	6	10	17	24	39	56	87	117	150	237	343
RHH*, RHW*, RHW-2*, THHW, THW, THW-2	8	1	3	6	10	14	23	33	52	70	90	142	205
RHH*, RHW*, RHW-2*, TW, THW, THHW, THW-2	6	1	2	4	8	11	18	26	40	53	69	109	157
	4	1	1	3	6	8	13	19	30	40	51	81	117
	3	1	1	3	5	7	11	16	25	34	44	69	100
	2	1	1	2	4	6	10	14	22	29	37	59	85
	1	0	1	1	3	4	7	10	15	20	26	41	60
	1/0	0	1	1	2	3	6	8	13	17	22	35	51
	2/0	0	1	1	1	3	5	7	11	15	19	30	43
	3/0	0	1	1	1	2	4	6	9	12	16	25	36
	4/0	0	0	1	1	1	3	5	8	10	13	21	30
	250	0	0	1	1	1	3	4	6	8	11	17	25
	300	0	0	1	1	1	2	3	5	7	9	15	21
	350	0	0	0	1	1	1	3	5	6	8	13	19
	400	0	0	0	1	1	1	3	4	6	7	12	17
	500	0	0	0	1	1	1	2	3	5	6	10	14
	600	0	0	0	0	1	1	1	3	4	5	8	11
	700	0	0	0	0	1	1	1	2	3	4	7	10
	750	0	0	0	0	1	1	1	2	3	4	6	10
	800	0	0	0	0	1	1	1	2	3	4	6	9
	900	0	0	0	0	0	1	1	1	3	3	6	8
	1000	0	0	0	0	0	1	1	1	2	3	5	7
	1250	0	0	0	0	0	1	1	1	1	2	4	6
	1500	0	0	0	0	0	1	1	1	1	1	3	5
	1750	0	0	0	0	0	0	1	1	1	1	3	4
	2000	0	0	0	0	0	0	1	1	1	1	3	4

*Types RHH, RHW, and RHW-2 without outer covering.

Table C10. Continued

CONDUCTORS													
	Conductor Size (AWG/ kcmil)	Trade Size (in.)											
Type		½	¾	1	1¼	1½	2	2½	3	3½	4	5	6
THHN, THWN, THWN-2	14	11	21	34	60	82	135	193	299	401	517	815	1178
	12	8	15	25	43	59	99	141	218	293	377	594	859
	10	5	9	15	27	37	62	89	137	184	238	374	541
	8	3	5	9	16	21	36	51	79	106	137	216	312
	6	1	4	6	11	15	26	37	57	77	99	156	225
	4	1	2	4	7	9	16	22	35	47	61	96	138
	3	1	1	3	6	8	13	19	30	40	51	81	117
	2	1	1	3	5	7	11	16	25	33	43	68	98
	1	1	1	1	3	5	8	12	18	25	32	50	73
	1/0	1	1	1	3	4	7	10	15	21	27	42	61
	2/0	0	1	1	2	3	6	8	13	17	22	35	51
	3/0	0	1	1	1	3	5	7	11	14	18	29	42
	4/0	0	1	1	1	2	4	6	9	12	15	24	35
	250	0	0	1	1	1	3	4	7	10	12	20	28
	300	0	0	1	1	1	3	4	6	8	11	17	24
	350	0	0	1	1	1	2	3	5	7	9	15	21
	400	0	0	0	1	1	1	3	5	6	8	13	19
	500	0	0	0	1	1	1	2	4	5	7	11	16
	600	0	0	0	1	1	1	1	3	4	5	9	13
	700	0	0	0	0	1	1	1	3	4	5	8	11
	750	0	0	0	0	1	1	1	2	3	4	7	11
	800	0	0	0	0	1	1	1	2	3	4	7	10
	900	0	0	0	0	1	1	1	2	3	4	6	9
	1000	0	0	0	0	0	1	1	1	3	3	6	8
FEP, FEPB, PFA, PFAH, TFE	14	11	20	33	58	79	131	188	290	389	502	790	1142
	12	8	15	24	42	58	96	137	212	284	366	577	834
	10	6	10	17	30	41	69	98	152	204	263	414	598
	8	3	6	10	17	24	39	56	87	117	150	237	343
	6	2	4	7	12	17	28	40	62	83	107	169	244
	4	1	3	5	8	12	19	28	43	58	75	118	170
	3	1	2	4	7	10	16	23	36	48	62	98	142
	2	1	1	3	6	8	13	19	30	40	51	81	117
PFA, PFAH, TFE	1	1	1	2	4	5	9	13	20	28	36	56	81
PFA, PFAH, TFE, Z	1/0	1	1	1	3	4	8	11	17	23	30	47	68
	2/0	0	1	1	3	4	6	9	14	19	24	39	56
	3/0	0	1	1	2	3	5	7	12	16	20	32	46
	4/0	0	1	1	1	2	4	6	9	13	16	26	38
Z	14	13	24	40	70	95	158	226	350	469	605	952	1376
	12	9	17	28	49	68	112	160	248	333	429	675	976
	10	6	10	17	30	41	69	98	152	204	263	414	598
	8	3	6	11	19	26	43	62	96	129	166	261	378
	6	2	4	7	13	18	30	43	67	90	116	184	265
	4	1	3	5	9	12	21	30	46	62	80	126	183
	3	1	2	4	6	9	15	22	34	45	58	92	133
	2	1	1	3	5	7	12	18	28	38	49	77	111
	1	1	1	2	4	6	10	14	23	30	39	62	90
XHH, XHHW, XHHW-2, ZW	14	8	14	24	42	57	94	135	209	280	361	568	822
	12	6	11	18	32	44	72	103	160	215	277	436	631
	10	4	8	13	24	32	54	77	119	160	206	325	470
	8	2	4	7	13	18	30	43	66	89	115	181	261
	6	1	3	5	10	13	22	32	49	66	85	134	193
	4	1	2	4	7	9	16	23	35	48	61	97	140
	3	1	1	3	6	8	13	19	30	40	52	82	118
	2	1	1	3	5	7	11	16	25	34	44	69	99
XHH, XHHW, XHHW-2	1	1	1	1	3	5	8	12	19	25	32	51	74
	1/0	1	1	1	3	4	7	10	16	21	27	43	62
	2/0	0	1	1	2	3	6	8	13	17	23	36	52
	3/0	0	1	1	1	3	5	7	11	14	19	30	43
	4/0	0	1	1	1	2	4	6	9	12	15	24	35
	250	0	0	1	1	1	3	5	7	10	13	20	29
	300	0	0	1	1	1	3	4	6	8	11	17	25
	350	0	0	1	1	1	2	3	5	7	9	15	22
	400	0	0	0	1	1	1	3	5	6	8	13	19
	500	0	0	0	1	1	1	2	4	5	7	11	16
	600	0	0	0	1	1	1	1	3	4	5	9	13
	700	0	0	0	0	1	1	1	3	4	5	8	11
	750	0	0	0	0	1	1	1	2	3	4	7	11
	800	0	0	0	0	1	1	1	2	3	4	7	10
	900	0	0	0	0	1	1	1	2	3	4	6	9
	1000	0	0	0	0	0	1	1	1	3	3	6	8
	1250	0	0	0	0	0	1	1	1	1	3	4	6
	1500	0	0	0	0	0	1	1	1	1	2	4	5
	1750	0	0	0	0	0	0	1	1	1	1	3	5
	2000	0	0	0	0	0	0	1	1	1	1	3	4

Table C10. Continued

FIXTURE WIRES							
	Conductor Size	Trade Size (in.)					
Type	(AWG/kcmil)	½	¾	1	1¼	1½	2
FFH-2, RFH-2, RFHH-3	18	8	14	23	40	54	90
	16	6	12	19	33	46	76
SF-2, SFF-2	18	10	17	29	50	69	114
	16	8	14	24	42	57	94
	14	6	12	19	33	46	76
SF-1, SFF-1	18	17	31	51	89	122	202
AF, RFH-1, RFHH-2, TF, TFF, XF, XFF	18	13	23	38	66	90	149
AF, RFHH-2, TF, TFF, XF, XFF	16	10	18	30	53	73	120
AF, XF, XFF	14	8	14	24	42	57	94
TFN, TFFN	18	20	37	60	105	144	239
	16	16	28	46	80	110	183
PF, PFF, PGF, PGFF, PAF, PTF, PTFF, PAFF	18	19	35	57	100	137	227
	16	15	27	44	77	106	175
	14	11	20	33	58	79	131
HF, HFF, ZF, ZFF, ZHF	18	25	45	74	129	176	292
	16	18	33	54	95	130	216
	14	13	24	40	70	95	158
KF-2, KFF-2	18	36	65	107	187	256	424
	16	26	46	75	132	180	299
	14	17	31	52	90	124	205
	12	12	22	35	62	85	141
	10	8	14	24	42	57	94
KF-1, KFF-1,	18	43	78	128	223	305	506
	16	30	55	90	157	214	355
	14	20	37	60	105	144	239
	12	13	24	40	70	95	158
	10	9	16	26	45	62	103
AF, XF, XFF	12	4	8	12	22	30	50
	10	3	6	10	17	24	39

Note: This table is for concentric stranded conductors only. For compact stranded conductors, Table C10(A) should be used.

Table C10(A). Maximum Number of Compact Conductors in Rigid PVC Conduit, Schedule 40 and HDPE Conduit *(Based on Table 1, Chapter 9)*

COMPACT CONDUCTORS													
	Conductor Size (AWG/ kcmil)	Trade Size (in.)											
Type		½	¾	1	1¼	1½	2	2½	3	3½	4	5	6
THW, THW-2, THHW	8	1	4	6	11	15	26	37	57	76	98	155	224
	6	1	3	5	9	12	20	28	44	59	76	119	173
	4	1	1	3	6	9	15	21	33	44	57	89	129
	2	1	1	2	5	6	11	15	24	32	42	66	95
	1	1	1	1	3	4	7	11	17	23	29	46	67
	1/0	0	1	1	3	4	6	9	15	20	25	40	58
	2/0	0	1	1	2	3	5	8	12	16	21	34	49
	3/0	0	1	1	1	3	5	7	10	14	18	29	42
	4/0	0	1	1	1	2	4	5	9	12	15	24	35
	250	0	0	1	1	1	3	4	7	9	12	19	27
	300	0	0	1	1	1	2	4	6	8	10	16	24
	350	0	0	1	1	1	2	3	5	7	9	15	21
	400	0	0	0	1	1	1	3	5	6	8	13	19
	500	0	0	0	1	1	1	2	4	5	7	11	16
	600	0	0	0	1	1	1	1	3	4	5	9	13
	700	0	0	0	0	1	1	1	3	4	5	8	12
	750	0	0	0	0	1	1	1	2	3	5	7	11
	1000	0	0	0	0	1	1	1	1	3	4	6	9
THHN, THWN, THWN-2	8	—	—	—	—	—	—	—	—	—	—	—	—
	6	2	4	7	13	17	29	41	64	86	111	175	253
	4	1	2	4	8	11	18	25	40	53	68	108	156
	2	1	1	3	5	8	13	18	28	38	49	77	112
	1	1	1	2	4	6	9	14	21	29	37	58	84
	1/0	1	1	1	3	5	8	12	18	24	31	49	72
	2/0	0	1	1	3	4	7	9	15	20	26	41	59
	3/0	0	1	1	2	3	5	8	12	17	22	34	50
	4/0	0	1	1	1	3	4	6	10	14	18	28	41
	250	0	0	1	1	1	3	5	8	11	14	22	32
	300	0	0	1	1	1	3	4	7	9	12	19	28
	350	0	0	1	1	1	3	4	6	8	10	17	24
	400	0	0	1	1	1	2	3	5	7	9	15	22
	500	0	0	0	1	1	1	3	4	6	8	13	18
	600	0	0	0	1	1	1	2	4	5	6	10	15
	700	0	0	0	1	1	1	1	3	4	5	9	13
	750	0	0	0	1	1	1	1	3	4	5	8	12
	1000	0	0	0	0	1	1	1	2	3	4	6	9
XHHW, XHHW-2	8	3	5	8	14	20	33	47	73	99	127	200	290
	6	1	4	6	11	15	25	35	55	73	94	149	215
	4	1	2	4	8	11	18	25	40	53	68	108	156
	2	1	1	3	5	8	13	18	28	38	49	77	112
	1	1	1	2	4	6	9	14	21	29	37	58	84
	1/0	1	1	1	3	5	8	12	18	24	31	49	72
	2/0	1	1	1	3	4	7	10	15	20	26	42	60
	3/0	0	1	1	2	3	5	8	12	17	22	34	50
	4/0	0	1	1	1	3	5	7	10	14	18	29	42
	250	0	0	1	1	1	4	5	8	11	14	23	33
	300	0	0	1	1	1	3	4	7	9	12	19	28
	350	0	0	1	1	1	3	4	6	8	11	17	25
	400	0	0	1	1	1	2	3	5	7	10	15	22
	500	0	0	0	1	1	1	3	4	6	8	13	18
	600	0	0	0	1	1	1	2	4	5	6	10	15
	700	0	0	0	1	1	1	1	3	4	5	9	13
	750	0	0	0	1	1	1	1	3	4	5	8	12
	1000	0	0	0	0	1	1	1	2	3	4	6	9

Definition: Compact stranding is the result of a manufacturing process where the standard conductor is compressed to the extent that the interstices (voids between strand wires) are virtually eliminated.

Table C11. Maximum Number of Conductors and Fixture Wires in Type A, Rigid PVC Conduit *(Based on Table 1, Chapter 9)*

CONDUCTORS											
	Conductor Size	Trade Size (in.)									
Type	(AWG/kcmil)	½	¾	1	1¼	1½	2	2½	3	3½	4
RH	14	7	12	20	34	44	70	104	157	204	262
	12	6	10	16	27	35	56	84	126	164	211
RHH, RHW, RHW-2	14	5	9	15	24	31	49	74	112	146	187
	12	4	7	12	20	26	41	61	93	121	155
RH, RHH, RHW, RHW-2	10	3	6	10	16	21	33	50	75	98	125
	8	1	3	5	8	11	17	26	39	51	65
	6	1	2	4	6	9	14	21	31	41	52
	4	1	1	3	5	7	11	16	24	32	41
	3	1	1	3	4	6	9	14	21	28	36
	2	1	1	2	4	5	8	12	18	24	31
	1	0	1	1	2	3	5	8	12	16	20
	1/0	0	1	1	2	3	5	7	10	14	18
	2/0	0	1	1	1	2	4	6	9	12	15
	3/0	0	1	1	1	1	3	5	8	10	13
	4/0	0	0	1	1	1	3	4	7	9	11
	250	0	0	1	1	1	1	3	5	7	8
	300	0	0	1	1	1	1	3	4	6	7
	350	0	0	0	1	1	1	2	4	5	7
	400	0	0	0	1	1	1	2	4	5	6
	500	0	0	0	1	1	1	1	3	4	5
	600	0	0	0	0	1	1	1	2	3	4
	700	0	0	0	0	1	1	1	2	3	4
	750	0	0	0	0	1	1	1	1	3	4
	800	0	0	0	0	1	1	1	1	3	3
	900	0	0	0	0	0	1	1	1	2	3
	1000	0	0	0	0	0	1	1	1	2	3
	1250	0	0	0	0	0	1	1	1	1	2
	1500	0	0	0	0	0	0	1	1	1	1
	1750	0	0	0	0	0	0	1	1	1	1
	2000	0	0	0	0	0	0	1	1	1	1
TW	14	11	18	31	51	67	105	157	235	307	395
	12	8	14	24	39	51	80	120	181	236	303
	10	6	10	18	29	38	60	89	135	176	226
	8	3	6	10	16	21	33	50	75	98	125
RHH*, RHW*, RHW-2*, THHW, THW, THW-2	14	7	12	20	34	44	70	104	157	204	262
RHH*, RHW*, RHW-2*, THHW, THW	12	6	10	16	27	35	56	84	126	164	211
	10	4	8	13	21	28	44	65	98	128	165
RHH*, RHW*, RHW-2*, THHW, THW, THW-2	8	2	4	8	12	16	26	39	59	77	98
RHH*, RHW*, RHW-2*, TW, THW, THHW, THW-2	6	1	3	6	9	13	20	30	45	59	75
	4	1	2	4	7	9	15	22	33	44	56
	3	1	1	4	6	8	13	19	29	37	48
	2	1	1	3	5	7	11	16	24	32	41
	1	1	1	1	3	5	7	11	17	22	29
	1/0	1	1	1	3	4	6	10	14	19	24
	2/0	0	1	1	2	3	5	8	12	16	21
	3/0	0	1	1	1	3	4	7	10	13	17
	4/0	0	1	1	1	2	4	6	9	11	14
	250	0	0	1	1	1	3	4	7	9	12
	300	0	0	1	1	1	2	4	6	8	10
	350	0	0	1	1	1	2	3	5	7	9
	400	0	0	1	1	1	1	3	5	6	8
	500	0	0	0	1	1	1	2	4	5	7
	600	0	0	0	1	1	1	1	3	4	5
	700	0	0	0	1	1	1	1	3	4	5
	750	0	0	0	1	1	1	1	3	3	4
	800	0	0	0	0	1	1	1	2	3	4
	900	0	0	0	0	1	1	1	2	3	4
	1000	0	0	0	0	1	1	1	1	3	3
	1250	0	0	0	0	0	1	1	1	1	3
	1500	0	0	0	0	0	1	1	1	1	2
	1750	0	0	0	0	0	0	1	1	1	1
	2000	0	0	0	0	0	0	1	1	1	1

*Types RHH, RHW, and RHW-2 without outer covering.

Table C11. Continued

		CONDUCTORS									
	Conductor Size	Trade Size (in.)									
Type	(AWG/kcmil)	½	¾	1	1¼	1½	2	2½	3	3½	4
THHN, THWN, THWN-2	14	16	27	44	73	96	150	225	338	441	566
	12	11	19	32	53	70	109	164	246	321	412
	10	7	12	20	33	44	69	103	155	202	260
	8	4	7	12	19	25	40	59	89	117	150
	6	3	5	8	14	18	28	43	64	84	108
	4	1	3	5	8	11	17	26	39	52	66
	3	1	2	4	7	9	15	22	33	44	56
	2	1	1	3	6	8	12	19	28	37	47
	1	1	1	2	4	6	9	14	21	27	35
	1/0	1	1	2	4	5	8	11	17	23	29
	2/0	1	1	1	3	4	6	10	14	19	24
	3/0	0	1	1	2	3	5	8	12	16	20
	4/0	0	1	1	1	3	4	6	10	13	17
	250	0	1	1	1	2	3	5	8	10	14
	300	0	0	1	1	1	3	4	7	9	12
	350	0	0	1	1	1	2	4	6	8	10
	400	0	0	1	1	1	2	3	5	7	9
	500	0	0	1	1	1	1	3	4	6	7
	600	0	0	0	1	1	1	2	3	5	6
	700	0	0	0	1	1	1	1	3	4	5
	750	0	0	0	1	1	1	1	3	4	5
	800	0	0	0	1	1	1	1	3	4	5
	900	0	0	0	0	1	1	1	2	3	4
	1000	0	0	0	0	1	1	1	2	3	4
FEP, FEPB, PFA, PFAH, TFE	14	15	26	43	70	93	146	218	327	427	549
	12	11	19	31	51	68	106	159	239	312	400
	10	8	13	22	37	48	76	114	171	224	287
	8	4	8	13	21	28	44	65	98	128	165
	6	3	5	9	15	20	31	46	70	91	117
	4	1	4	6	10	14	21	32	49	64	82
	3	1	3	5	8	11	18	27	40	53	68
	2	1	2	4	7	9	15	22	33	44	56
PFA, PFAH, TFE	1	1	1	3	5	6	10	15	23	30	39
PFA, PFAH, TFE, Z	1/0	1	1	2	4	5	8	13	19	25	32
	2/0	1	1	1	3	4	7	10	16	21	27
	3/0	1	1	1	3	3	6	9	13	17	22
	4/0	0	1	1	2	3	5	7	11	14	18
Z	14	18	31	52	85	112	175	263	395	515	661
	12	13	22	37	60	79	124	186	280	365	469
	10	8	13	22	37	48	76	114	171	224	287
	8	5	8	14	23	30	48	72	108	141	181
	6	3	6	10	16	21	34	50	76	99	127
	4	2	4	7	11	15	23	35	52	68	88
	3	1	3	5	8	11	17	25	38	50	64
	2	1	2	4	7	9	14	21	32	41	53
	1	1	1	3	5	7	11	17	26	33	43
XHH, XHHW, XHHW-2, ZW	14	11	18	31	51	67	105	157	235	307	395
	12	8	14	24	39	51	80	120	181	236	303
	10	6	10	18	29	38	60	89	135	176	226
	8	3	6	10	16	21	33	50	75	98	125
	6	2	4	7	12	15	24	37	55	72	93
	4	1	3	5	8	11	18	26	40	52	67
	3	1	2	4	7	9	15	22	34	44	57
	2	1	1	3	6	8	12	19	28	37	48
XHH, XHHW, XHHW-2,	1	1	1	3	4	6	9	14	21	28	35
	1/0	1	1	2	4	5	8	12	18	23	30
	2/0	1	1	1	3	4	6	10	15	19	25
	3/0	0	1	1	2	3	5	8	12	16	20
	4/0	0	1	1	1	3	4	7	10	13	17
	250	0	1	1	1	2	3	5	8	11	14
	300	0	0	1	1	1	3	5	7	9	12
	350	0	0	1	1	1	3	4	6	8	10
	400	0	0	1	1	1	2	3	5	7	9
	500	0	0	1	1	1	1	3	4	6	8
	600	0	0	0	1	1	1	2	3	5	6
	700	0	0	0	1	1	1	1	3	4	5
	750	0	0	0	1	1	1	1	3	4	5
	800	0	0	0	1	1	1	1	3	4	5
	900	0	0	0	0	1	1	1	2	3	4
	1000	0	0	0	0	1	1	1	2	3	4
	1250	0	0	0	0	0	1	1	1	2	3
	1500	0	0	0	0	0	1	1	1	1	2
	1750	0	0	0	0	0	1	1	1	1	2
	2000	0	0	0	0	0	0	1	1	1	1

Table C11. Continued

		FIXTURE WIRES					
	Conductor Size	Trade Size (in.)					
Type	(AWG/kcmil)	½	¾	1	1¼	1½	2
FFH-2, RFH-2, RFHH-3	18	10	18	30	48	64	100
	16	9	15	25	41	54	85
SF-2, SFF-2	18	13	22	37	61	81	127
	16	11	18	31	51	67	105
	14	9	15	25	41	54	85
SF-1, SFF-1	18	23	40	66	108	143	224
AF, RFH-1, RFHH-2, TF, TFF, XF, XFF	18	17	29	49	80	105	165
AF, RFHH-2, TF, TFF, XF, XFF	16	14	24	39	65	85	134
AF, XF, XFF	14	11	18	31	51	67	105
TFN, TFFN	18	28	47	79	128	169	265
	16	21	36	60	98	129	202
PF, PFF, PGF, PGFF, PAF, PTF, PTFF, PAFF	18	26	45	74	122	160	251
	16	20	34	58	94	124	194
	14	15	26	43	70	93	146
HF, HFF, ZF, ZFF, ZHF	18	34	58	96	157	206	324
	16	25	42	71	116	152	239
	14	18	31	52	85	112	175
KF-2, KFF-2	18	49	84	140	228	300	470
	16	35	59	98	160	211	331
	14	24	40	67	110	145	228
	12	16	28	46	76	100	157
	10	11	18	31	51	67	105
KF-1, KFF-1	18	59	100	167	272	357	561
	16	41	70	117	191	251	394
	14	28	47	79	128	169	265
	12	18	31	52	85	112	175
	10	12	20	34	55	73	115
AF, XF, XFF	12	6	10	16	27	35	56
	10	4	8	13	21	28	44

Note: This table is for concentric stranded conductors only. For compact stranded conductors, Table C11(A) should be used.

Table C11(A). Maximum Number of Compact Conductors in Type A, Rigid PVC Conduit ***(Based on Table 1, Chapter 9)***

COMPACT CONDUCTORS											
	Conductor Size	Trade Size (in.)									
Type	(AWG/kcmil)	½	¾	1	1¼	1½	2	2½	3	3½	4
THW, THW-2, THHW	8	3	5	8	14	18	28	42	64	84	107
	6	2	4	6	10	14	22	33	49	65	83
	4	1	3	5	8	10	16	24	37	48	62
	2	1	1	3	6	7	12	18	27	36	46
	1	1	1	2	4	5	8	13	19	25	32
	1/0	1	1	1	3	4	7	11	16	21	28
	2/0	1	1	1	3	4	6	9	14	18	23
	3/0	0	1	1	2	3	5	8	12	15	20
	4/0	0	1	1	1	3	4	6	10	13	17
	250	0	1	1	1	1	3	5	8	10	13
	300	0	0	1	1	1	3	4	7	9	11
	350	0	0	1	1	1	2	4	6	8	10
	400	0	0	1	1	1	2	3	5	7	9
	500	0	0	1	1	1	1	3	4	6	8
	600	0	0	0	1	1	1	2	3	5	6
	700	0	0	0	1	1	1	1	3	4	5
	750	0	0	0	1	1	1	1	3	4	5
	1000	0	0	0	0	1	1	1	2	3	4
THHN, THWN, THWN-2	8	—	—	—	—	—	—	—	—	—	—
	6	3	5	9	15	20	32	48	72	94	121
	4	1	3	6	9	12	20	30	45	58	75
	2	1	2	4	7	9	14	21	32	42	54
	1	1	1	3	5	7	10	16	24	31	40
	1/0	1	1	2	4	6	9	13	20	27	34
	2/0	1	1	1	3	5	7	11	17	22	28
	3/0	1	1	1	3	4	6	9	14	18	24
	4/0	0	1	1	2	3	5	8	11	15	19
	250	0	1	1	1	2	4	6	9	12	15
	300	0	1	1	1	1	3	5	8	10	13
	350	0	0	1	1	1	3	4	7	9	11
	400	0	0	1	1	1	2	4	6	8	10
	500	0	0	1	1	1	2	3	5	7	9
	600	0	0	0	1	1	1	3	4	5	7
	700	0	0	0	1	1	1	2	3	5	6
	750	0	0	0	1	1	1	2	3	4	6
	1000	0	0	0	0	1	1	1	2	3	4
XHHW, XHHW-2	8	4	6	11	18	23	37	55	83	108	139
	6	3	5	8	13	17	27	41	62	80	103
	4	1	3	6	9	12	20	30	45	58	75
	2	1	2	4	7	9	14	21	32	42	54
	1	1	1	3	5	7	10	16	24	31	40
	1/0	1	1	2	4	6	9	13	20	27	34
	2/0	1	1	1	3	5	7	11	17	22	29
	3/0	1	1	1	3	4	6	9	14	18	24
	4/0	0	1	1	2	3	5	8	12	15	20
	250	0	1	1	1	2	4	6	9	12	16
	300	0	1	1	1	1	3	5	8	10	13
	350	0	0	1	1	1	3	5	7	9	12
	400	0	0	1	1	1	3	4	6	8	11
	500	0	0	1	1	1	2	3	5	7	9
	600	0	0	0	1	1	1	3	4	5	7
	700	0	0	0	1	1	1	2	3	5	6
	750	0	0	0	1	1	1	2	3	4	6
	1000	0	0	0	0	1	1	1	2	3	4

Definition: Compact stranding is the result of a manufacturing process where the standard conductor is compressed to the extent that the interstices (voids between strand wires) are virtually eliminated.

Table C12. Maximum Number of Conductors in Type EB, PVC Conduit ***(Based on Table 1, Chapter 9)***

CONDUCTORS							
	Conductor Size	Trade Size (in.)					
Type	(AWG/kcmil)	2	3	3½	4	5	6
RH	14	74	166	217	276	424	603
	12	59	134	175	222	341	485
RHH, RHW, RHW-2	14	53	119	155	197	303	430
	12	44	98	128	163	251	357
RH, RHH, RHW, RHW-2	10	35	79	104	132	203	288
	8	18	41	54	69	106	151
	6	15	33	43	55	85	121
	4	11	26	34	43	66	94
	3	10	23	30	38	58	83
	2	9	20	26	33	50	72
	1	6	13	17	21	33	47
	1/0	5	11	15	19	29	41
	2/0	4	10	13	16	25	36
	3/0	4	8	11	14	22	31
	4/0	3	7	9	12	18	26
	250	2	5	7	9	14	20
	300	1	5	6	8	12	17
	350	1	4	5	7	11	16
	400	1	4	5	6	10	14
	500	1	3	4	5	9	12
	600	1	3	3	4	7	10
	700	1	2	3	4	6	9
	750	1	2	3	4	6	9
	800	1	2	3	4	6	8
	900	1	1	2	3	5	7
	1000	1	1	2	3	5	7
	1250	1	1	1	2	3	5
	1500	0	1	1	1	3	4
	1750	0	1	1	1	3	4
	2000	0	1	1	1	2	3
TW	14	111	250	327	415	638	907
	12	85	192	251	319	490	696
	10	63	143	187	238	365	519
	8	35	79	104	132	203	288
RHH*, RHW*, RWH-2*, THHW, THW, THW-2	14	74	166	217	276	424	603
RHH*, RHW*, RHW-2*, THHW, THW	12	59	134	175	222	341	485
	10	46	104	136	173	266	378
RHH*, RHW*, RHW-2*, THHW, THW THW-2	8	28	62	81	104	159	227
RHH*, RHW*, RHW-2*, TW, THW, THHW, THW-2	6	21	48	62	79	122	173
	4	16	36	46	59	91	129
	3	13	30	40	51	78	111
	2	11	26	34	43	66	94
	1	8	18	24	30	46	66
	1/0	7	15	20	26	40	56
	2/0	6	13	17	22	34	48
	3/0	5	11	14	18	28	40
	4/0	4	9	12	15	24	34
	250	3	7	10	12	19	27
	300	3	6	8	11	17	24
	350	2	6	7	9	15	21
	400	2	5	7	8	13	19
	500	1	4	5	7	11	16
	600	1	3	4	6	9	13
	700	1	3	4	5	8	11
	750	1	3	4	5	7	11
	800	1	3	3	4	7	10
	900	1	2	3	4	6	9
	1000	1	2	3	4	6	8
	1250	1	1	2	3	4	6
	1500	1	1	1	2	4	6
	1750	1	1	1	2	3	5
	2000	0	1	1	1	3	4

*Types RHH, RHW, and RHW-2 without outer covering.

Table C12. Continued

CONDUCTORS							
Type	Conductor Size (AWG/kcmil)	Trade Size (in.) 2	3	3½	4	5	6
THHN, THWN, THWN-2	14	159	359	468	595	915	1300
	12	116	262	342	434	667	948
	10	73	165	215	274	420	597
	8	42	95	124	158	242	344
	6	30	68	89	114	175	248
	4	19	42	55	70	107	153
	3	16	36	46	59	91	129
	2	13	30	39	50	76	109
	1	10	22	29	37	57	80
	1/0	8	18	24	31	48	68
	2/0	7	15	20	26	40	56
	3/0	5	13	17	21	33	47
	4/0	4	10	14	18	27	39
	250	4	8	11	14	22	31
	300	3	7	10	12	19	27
	350	3	6	8	11	17	24
	400	2	6	7	10	15	21
	500	1	5	6	8	12	18
	600	1	4	5	6	10	14
	700	1	3	4	6	9	12
	750	1	3	4	5	8	12
	800	1	3	4	5	8	11
	900	1	3	3	4	7	10
	1000	1	2	3	4	6	9
FEP, FEPB, PFA, PFAH, TFE	14	155	348	454	578	888	1261
	12	113	254	332	422	648	920
	10	81	182	238	302	465	660
	8	46	104	136	173	266	378
	6	33	74	97	123	189	269
	4	23	52	68	86	132	188
	3	19	43	56	72	110	157
	2	16	36	46	59	91	129
PFA, PFAH, TFE	1	11	25	32	41	63	90
PFA, PFAH, TFE, Z	1/0	9	20	27	34	53	75
	2/0	7	17	22	28	43	62
	3/0	6	14	18	23	36	51
	4/0	5	11	15	19	29	42
Z	14	186	419	547	696	1069	1519
	12	132	297	388	494	759	1078
	10	81	182	238	302	465	660
	8	51	115	150	191	294	417
	6	36	81	105	134	206	293
	4	24	55	72	92	142	201
	3	18	40	53	67	104	147
	2	15	34	44	56	86	122
	1	12	27	36	45	70	99
XHH, XHHW, XHHW-2, ZW	14	111	250	327	415	638	907
	12	85	192	251	319	490	696
	10	63	143	187	238	365	519
	8	35	79	104	132	203	288
	6	26	59	77	98	150	213
	4	19	42	56	71	109	155
	3	16	36	47	60	92	131
	2	13	30	39	50	77	110
XHH, XHHW, XHHW-2	1	10	22	29	37	58	82
	1/0	8	19	25	31	48	69
	2/0	7	16	20	26	40	57
	3/0	6	13	17	22	33	47
	4/0	5	11	14	18	27	39
	250	4	9	11	15	22	32
	300	3	7	10	12	19	28
	350	3	6	9	11	17	24
	400	2	6	8	10	15	22
	500	1	5	6	8	12	18
	600	1	4	5	6	10	14
	700	1	3	4	6	9	12
	750	1	3	4	5	8	12
	800	1	3	4	5	8	11
	900	1	3	3	4	7	10
	1000	1	2	3	4	6	9
	1250	1	1	2	3	5	7
	1500	1	1	1	3	4	6
	1750	1	1	1	2	4	5
	2000	0	1	1	1	3	5

NOTE: This table is for concentric stranded conductors only. For compact stranded conductors, Table C12(A) should be used.

Table C12(A). Maximum Number of Compact Conductors in Type EB, PVC Conduit *(Based on Table 1, Chapter 9)*

CONDUCTORS							
Type	Conductor Size (AWG/kcmil)	Trade Size (in.) 2	3	3½	4	5	6
THW, THW-2, THHW	8	30	68	89	113	174	247
	6	23	52	69	87	134	191
	4	17	39	51	65	100	143
	2	13	29	38	48	74	105
	1	9	20	26	34	52	74
	1/0	8	17	23	29	45	64
	2/0	6	15	19	24	38	54
	3/0	5	12	16	21	32	46
	4/0	4	10	14	17	27	38
	250	3	8	11	14	21	30
	300	3	7	9	12	19	26
	350	3	6	8	11	17	24
	400	2	6	7	10	15	21
	500	1	5	6	8	12	18
	600	1	4	5	6	10	14
	700	1	3	4	6	9	13
	750	1	3	4	5	8	12
	1000	1	2	3	4	7	9
THHN, THWN, THWN-2	8	—	—	—	—	—	—
	6	34	77	100	128	196	279
	4	21	47	62	79	121	172
	2	15	34	44	57	87	124
	1	11	25	33	42	65	93
	1/0	9	22	28	36	56	79
	2/0	8	18	23	30	46	65
	3/0	6	15	20	25	38	55
	4/0	5	12	16	20	32	45
	250	4	10	13	16	25	35
	300	4	8	11	14	22	31
	350	3	7	9	12	19	27
	400	3	6	8	11	17	24
	500	2	5	7	9	14	20
	600	1	4	6	7	11	16
	700	1	4	5	6	10	14
	750	1	4	5	6	9	14
	1000	1	3	3	4	7	10
XHHW, XHHW-2	8	39	88	115	146	225	320
	6	29	65	85	109	167	238
	4	21	47	62	79	121	172
	2	15	34	44	57	87	124
	1	11	25	33	42	65	93
	1/0	9	22	28	36	56	79
	2/0	8	18	24	30	47	67
	3/0	6	15	20	25	38	55
	4/0	5	12	16	21	32	46
	250	4	10	13	17	26	37
	300	4	8	11	14	22	31
	350	3	7	10	12	19	28
	400	3	7	9	11	17	25
	500	2	5	7	9	14	20
	600	1	4	6	7	11	16
	700	1	4	5	6	10	14
	750	1	3	5	6	9	13
	1000	1	3	4	5	7	10

Definition: Compact stranding is the result of a manufacturing process where the standard conductor is compressed to the extent that the interstices (voids between strand wires) are virtually eliminated.

APPENDIX D

Examples

Contents

These examples were moved from Chapter 9 following the tables to this new Appendix D for the 1999 *Code*. They are now more appropriately located in an explanatory appendix.

Numerous but slight corrections and enhancements were made to Examples D1(a), D1(b), D2(a), D2(b), D2(c), D4(a), and D10. Example D8 was reformatted and contains numerous changes. And, Examples D10, Mobile Home, and D12, Park Trailer, were moved from their respective Articles 550 and 552 to Appendix D.

Selection of Conductors. In the following examples, the results are generally expressed in amperes (A). To select conductor sizes, refer to the 0 through 2000 volt (V) ampacity tables of Article 310 and Section 310-5 that pertains to such tables.

Voltage. For uniform application of Articles 210, 215, and 220, a nominal voltage of 120, 120/240, 240, and 208Y/120 V is used in computing the ampere load on the conductor.

Fractions of an Ampere. Except where the computations result in a major fraction of an ampere (0.5 or larger), such fractions are permitted to be dropped.

Power Factor. Calculations in the following examples are based, for convenience, on the assumption that all loads have the same power factor (PF).

Ranges. For the computation of the range loads in these examples, Column A of Table 220-19 has been used. For optional methods, see Columns B and C of Table 220-19. Except where the computations result in a major fraction of a kilowatt (0.5 or larger), such fractions are permitted to be dropped.

SI Units. For metric conversions, 1 ft^2 = 0.093 m^2 and 1 ft = 0.3048 m.

In the following examples, loads are assumed to be properly balanced on the system. If loads are not properly balanced, additional feeder capacity may be required. These calculations are based on the standard method.

Example No. D1(a). One-Family Dwelling

The dwelling has a floor area of 1500 ft^2, exclusive of an unfinished cellar not adaptable for future use, unfinished attic, and open porches. Appliances are a 12-kW range and a 5.5-kW, 240-V dryer. Assume range and dryer kW ratings equivalent to kVA ratings in accordance with Sections 220-18 and 220-19.

Computed Load *[see Section 220-10]*

General Lighting Load: 1500 ft^2 at 3 VA per ft^2 = 4500 VA

Minimum Number of Branch Circuits Required *[see Section 210-11(e)(1)]*

General Lighting Load: 4500 VA ÷ 120 V = 37.5 A
This requires three 15-A, 2-wire or two 20-A, 2-wire circuits
Small Appliance Load: Two 2-wire, 20-A circuits *[see Section 210-11(c)(1)]*
Laundry Load: One 2-wire, 20-A circuit *[see Section 210-11(c)(2)]*
Bathroom Branch Circuit: One 2-wire, 20-A circuit (no additional load calculation is required for this circuit) *[see Section 210-11(c)(3)]*

Minimum Size Feeder Required *[see Section 220-10]*

General Lighting	4500 VA
Small Appliance	3000 VA
Laundry	1500 VA
Total	9000 VA
3000 VA at 100%	3000 VA
9000 VA − 3000 VA = 6000 VA at 35%	2100 VA
Net Load	5100 VA
Range *(see Table 220-19)*	8000 VA
Dryer *(see Table 220-18)*	5500 VA
Net Computed Load	18,600 VA

Net Computed Load for 120/240-V, 3-wire, single-phase service or feeder

18,600 VA ÷ 240 V = 77.5 A

Net computed load exceeds 10 kVA. Section 230-42(b) would require service conductors to be 100 A.

Calculation for Neutral for Feeder and Service

Lighting and Small Appliance Net Load	5100 VA
Range: 8000 VA at 70% *(see Section 220-22)*	5600 VA
Dryer: 5500 VA at 70% *(see Section 220-22)*	3850 VA
Total	14,550 VA

Computed Load for Neutral

14,550 VA ÷ 240 V = 60.6 A

The general lighting and general-use receptacle load is computed from the outside dimensions of the building, apartment, or other area involved. For a dwelling unit, the computed floor area is not to include open porches, garages, or as stated in the heading, "an unfinished cellar not adaptable for future use." See Sections 220-3(a) and 220-3(b)(10) for the requirements on how to calculate the lighting and general-use receptacle load for this occupancy.

Example

A two-story dwelling measures 30 ft × 30 ft for the first floor and 30 ft × 20 ft for the second floor.

30 ft × 30 ft = 900 ft² (first floor)
30 ft × 20 ft = 600 ft² (second floor)
Total area = 1500 ft²

Example No. D1(b). One-Family Dwelling

Assume same conditions as Example No. D1(a), plus addition of one 6-A, 230-V, room air-conditioning unit and one 12-A, 115-V, room air-conditioning unit,* one 8-A, 115-V rated waste disposer, and one 10-A, 120-V, rated dishwasher.* See Article 430 for general motors and Article 440, Part G, for air-conditioning equipment. Motors have nameplate ratings of 115 V and 230 V for use on 120-V and 240-V nominal voltage systems.

From Example No. D1(a), feeder current is 78 A (3-wire, 240 V).

*(For feeder neutral, use larger of the two appliances for unbalance.)

	Line A	Neutral	Line B
Amperes from Example No. D1(a)	78	61	78
One 230-V air conditioner	6	—	6
One 115-V air conditioner and 120-V dishwasher	12	10	10
One 115-V disposer	—	8	8
25% of largest motor *(see Section 430-24)*	3	2	2
Total amperes per line	99	81	104

Therefore, the service would be rated 110 A

Example No. D2(a). Optional Calculation for One-Family Dwelling, Heating Larger than Air Conditioning *[see Section 220-30]*

The dwelling has a floor area of 1500 ft², exclusive of an unfinished cellar not adaptable for future use, unfinished attic, and open porches. It has a 12-kW range, a 2.5-kW water heater, a 1.2-kW dishwasher, 9 kW of electric space heating installed in five rooms, a 5-kW clothes dryer, and a 6-A, 230-V, room air-conditioning unit. Assume range, water heater, dishwasher, space heating, and clothes dryer kW ratings equivalent to kVA.

Air Conditioner kVA Calculation

6 A × 230 V ÷ 1000 = 1.38 kVA

The air-conditioning load is calculated at 100 percent and is calculated separately to comply with the requirements of Section 220-30(c)(1).

This 1.38 kVA (*Section 220-30(c)(1)*) is less than 40% of 9 kVA of separately controlled electric heat (*Section 220-3(c)(6)*), so the 1.38 kVA need not be included in the service calculation.

General Load

1500 ft² at 3 VA	4500 VA
Two 20-A appliance outlet circuits at 1500 VA each	3000 VA
Laundry circuit	1500 VA
Range (at nameplate rating)	12,000 VA
Water heater	2500 VA
Dishwasher	1200 VA
Clothes dryer	5000 VA
Total	29,700 VA

Application of Demand Factor *(see Section 220-30(b))*

First 10 kVA of general load at 100%	10,000 VA
Remainder of general load at 40% (19.7 kVA × 0.4)	7880 VA
Total of general load	17,880 VA
9 kVA of heat at 40% (9000 VA × 0.4) =	3600 VA
Total	21,480 VA

Calculated Load for Service Size

21,480 VA ÷ 240 V = 89.5 A

The minimum service size would be 100 A.

Feeder Neutral Load, per Section 220-22

1500 ft² at 3 VA	4500 VA
Three 20-A circuits at 1500 VA	4500 VA
Total	9000 VA
3000 VA at 100%	3000 VA
9000 VA − 3000 VA = 6000 VA at 35%	2100 VA
Subtotal	5100 VA
Range: 8 kVA at 70%	5600 VA
Clothes dryer: 5 kVA at 70%	3500 VA
Dishwasher	1200 VA
Total	15,400 VA

Calculated Load for Neutral

15,400 VA ÷ 240 V = 64.2 A

Example No. D2(b). Optional Calculation for One-Family Dwelling, Air Conditioning Larger than Heating *[see Section 220-30]*

The dwelling has a floor area of 1500 ft², exclusive of an unfinished cellar not adaptable for future use, unfinished attic, and open porches. It has two 20-A small appliance circuits, one 20-A laundry circuit, two 4-kW wall-mounted ovens, one 5.1-kW counter-mounted cooking unit, a 4.5-kW water heater, a 1.2-kW dishwasher, a 5-kW combination clothes washer and dryer, six 7-A, 230-V room air-conditioning units, and a 1.5-kW permanently installed bathroom space heater. Assume wall-mounted ovens, counter-mounted cooking unit, water heater, dishwasher, and combination clothes washer and dryer kW ratings equivalent to kVA.

Air Conditioning kVA Calculation

Total amperes = 6 units × 7 A = 42 A

42 A × 240 V ÷ 1000 = 10.08 kVA (assume PF = 1.0)

Load Included at 100%

Air Conditioning: Included below *(Section 220-30(c)(1))*
Space Heater: Omit *(Section 220-30(c)(5))*

General Load

1500 ft² at 3 VA	4500 VA
Two 20-A small appliance circuits at 1500 VA each	3000 VA
Laundry circuit	1500 VA
Two ovens	8000 VA
One cooking unit	5100 VA
Water heater	4500 VA
Dishwasher	1200 VA
Washer/dryer	5000 VA
Total general load	32,800 VA
First 10 kVA at 100%	10,000 VA
Remainder at 40% (22.8 kVA × 0.4 × 1000)	9120 VA
Subtotal general load	19,120 VA
Air conditioning	10,080 VA
Total	29,200 VA

Calculated Load for Service

29,200 VA ÷ 240 V = 122 A (service rating)

Feeder Neutral Load, per Section 220-22

Assume that the two 4-kVA wall-mounted ovens are supplied by one branch circuit, the 5.1-kVA counter-mounted cooking unit by a separate circuit.

1500 ft² at 3 VA	4500 VA
Three 20-A circuits at 1500 VA	4500 VA
Subtotal	9000 VA
3000 VA at 100%	3000 VA
9000 VA − 3000 VA = 6000 VA at 35%	2100 VA
Subtotal	5100 VA

Two 4-kVA ovens plus one 5.1-kVA cooking unit = 13.1 kVA. Table 220-19 permits 55% demand factor or 13.1 kVA × 0.55 = 7200 VA feeder capacity.

Subtotal from above	5100 VA
Ovens and cooking unit: 7200 VA × 70% for neutral load	5040 VA
Clothes washer/dryer: 5 kVA × 70% for neutral load	3500 VA
Dishwasher	1200 VA
Total	14,840 VA

Calculated Load for Neutral

14,840 VA ÷ 240 V = 61.83 A (use 62 A)

Example No. D2(c). Optional Calculation for One-Family Dwelling with Heat Pump (Single-Phase, 240/120-Volt Service)
(see Section 220-30)

The dwelling has a floor area of 2000 ft², exclusive of an unfinished cellar not adaptable for future use, unfinished attic, and open porches. It has a 12-kW range, a 4.5-kW water heater, a 1.2-kW dishwasher, a 5-kW clothes dryer, and a 2½-ton (24-A) heat pump with 15 kW of backup heat.

Heat Pump kVA Calculation

24 A × 240 V ÷ 1000 = 5.76 kVA

This 5.76 kVA is less than 15 kVA of the backup heat; therefore, the heat pump load need not be included in the service calculation *(see Section 220-30(c))*.

General Load

2000 ft² at 3 VA	6000 VA
Two 20-A appliance outlet circuits at 1500 VA each	3000 VA
Laundry circuit	1500 VA
Range (at nameplate rating)	12,000 VA
Water heater	4500 VA
Dishwasher	1200 VA
Clothes dryer	5000 VA
Subtotal general load	33,200 VA
First 10 kVA of general load at 100%	10,000 VA
Remainder of general load at 40% (23,200 VA × 0.4)	9280 VA
Total net general load	19,280 VA

Heat Pump and Supplementary Heat*

240 V × 24 A = 5760 VA

15-kW Electric Heat:

5760 VA + 15,000 VA = 20,760 VA = 20.76 kVA
20.76 kVA × 100% = 20.76 kVA

*If supplementary heat is not on at same time as heat pump, heat pump kVA need not be added to total.

Totals

Net general load	19,280 VA
Heat pump and supplementary heat	20,760 VA
Total	40,040 VA

Calculated Load for Service

40.04 kVA × 1000 ÷ 240 V = 166.8 A

This dwelling unit would be permitted to be served by a 175-A service.

Example No. D3. Store Building

A store 50 ft by 60 ft, or 3000 ft², has 30 ft of show window. There are a total of 80 duplex receptacles. The service is a 120/240 V, single phase 3-wire service. Actual connected lighting load is 8500 VA.

Computed Load *(see Section 220-10)*

Noncontinuous Loads:	
Receptacle Load *(see Section 220-13)*	
80 receptacles at 180 VA	14,400 VA
10,000 VA at 100%	10,000 VA
14,400 VA – 10,000 VA = 4400 VA at 50%	2,200 VA
Subtotal	12,200 VA
Continuous Loads:	
General Lighting* 3000 ft² at 3 VA per ft²	9000 VA
Show Window Lighting 30 ft at 200 VA per ft	6000 VA
Outside Sign Circuit *[see Section 600-5(b)(3)]*	1200 VA
Subtotal	16,200 VA
Subtotal from noncontinuous	12,200 VA
Total noncontinuous + continuous loads =	28,400 VA

*In the example, 125% of the actual connected lighting load (8500 VA × 1.25 = 10,625 VA) is less than 125% of the load from Table 220-3(a), so the minimum lighting load from Table 220-3(a) is used in the calculation. Had the actual lighting load been greater than the value computed from Table 220-3(a), 125% of the actual connected lighting load would have been used.

Minimum Number of Branch Circuits Required

General Lighting: Branch circuits need only be installed to supply the actual connected load *[see Section 210-11(d)]*.

8500 VA × 1.25 = 10,625 VA

10,625 VA ÷ 240 V = 44 A for 3-wire, 120/240 V

The lighting load would be permitted to be served by 2-wire or 3-wire, 15- or 20-A circuits with combined capacity equal to 44 A or greater for

3-wire circuits or 88 A or greater for 2-wire circuits. The feeder capacity as well as the number of branch-circuit positions available for lighting circuits in the panelboard must reflect the full calculated load of 9000 VA × 1.25 = 11,250 VA.

Show Window

6000 VA × 1.25 = 7500 VA

7500 VA ÷ 240 V = 31 A for 3-wire, 120/240 V

The show window lighting is permitted to be served by 2-wire or 3-wire circuits with a capacity equal to 31 A or greater for 3-wire circuits or 62 A or greater for 2-wire circuits.

Receptacles required by Section 210-62 are assumed to be included in the receptacle load above if these receptacles do not supply the show window lighting load.

Receptacles

Receptacle Load: 14,400 VA ÷ 240 V = 60 A for 3-wire, 120/240 V

The receptacle load would be permitted to be served by 2-wire or 3-wire circuits with a capacity equal to 60 A or greater for 3-wire circuits or 120 A or greater for 2-wire circuits.

Minimum Size Feeder (or Service) Overcurrent Protection *[see Sections 215-3 or 230-90(a)]*

Subtotal noncontinuous loads	12,200 VA
Subtotal continuous load at 125% (16,200 VA × 1.25)	20,250 VA
Total	32,450 VA

32,450 VA ÷ 240 V = 135 A

The next higher standard size is 150 A *(see Section 240-6).*

Minimum Size Feeders (or Service Conductors) Required *[see Sections 215-2, 215-3, and 230-42(a)]*

For 120/240-V, 3-wire system,

32,450 VA ÷ 240 V = 135 A

Service or feeder conductor is 1/0 Cu per Section 215-3 and Table 310-16 (with 75°C terminations).

Example No. D4(a). Multifamily Dwelling

A multifamily dwelling has 40 dwelling units.

Meters are in two banks of 20 each with individual feeders to each dwelling unit.

One-half of the dwelling units are equipped with electric ranges not exceeding 12 kW each. Assume range kW rating equivalent to kVA rating in accordance with Section 220-19. Other half of ranges are gas ranges.

Area of each dwelling unit is 840 ft^2.

Laundry facilities on premises are available to all tenants. Add no circuit to individual dwelling unit. Add 1500 VA for each laundry circuit to house load and add to the example as a "house load."

Computed Load for Each Dwelling Unit *(see Article 220)*

General Lighting: 840 ft^2 at 3 VA per ft^2 = 2520 VA
Special Appliance Load: Electric range *(see Section 220-19)* = 8000 VA

Minimum Number of Branch Circuits Required for Each Dwelling Unit *(see Section 220-4)*

General Lighting Load: 2520 VA ÷ 120 V = 21 A or two 15-A, 2-wire or two 20-A, 2-wire circuits
Small Appliance Load: Two 2-wire circuits of No. 12 wire *[see Section 210-11(c)(1)]*
Range Circuit: 8000 VA ÷ 240 V = 33 A. A circuit of two No. 8 conductors and one No. 10 conductor as permitted by Section 220-22 *[see Section 210-19(c)]*

Minimum Size Feeder Required for Each Dwelling Unit *(see Section 215-2)*

Computed Load *(see Article 220)*:	
General Lighting	2520 VA
Small Appliance (two 20-ampere circuits)	3000 VA
Subtotal Computed Load (without ranges)	5520 VA

Application of Demand Factor *(see Table 220-11)*

First 3000 VA at 100%	3000 VA
5520 VA – 3000 VA = 2520 VA at 35%	882 VA
Net Computed Load (without ranges)	3882 VA
Range Load	8000 VA
Net Computed Load (with ranges)	11,882 VA

Size of Each Feeder *(see Section 215-2)*

For 120/240-V, 3-wire system (without ranges)

Net computed load of 3882 VA ÷ 240 V = 16.2 A

For 120/240-V, 3-wire system (with ranges)

Net computed load of 11,882 VA ÷ 240 V = 49.5 A

Feeder Neutral

Lighting and Small Appliance	3882 VA
Range: 8000 VA at 70% *(see Section 220-22)* (only for apartments with electric range)	5600 VA
Net Computed Load (neutral)	9482 VA

Calculated Load for Neutral

9482 VA ÷ 240 V = 39.5 A

Minimum Size Feeder Required from Service Equipment to Meter Bank (For 20 Dwelling Units — 10 with Ranges)

Total Computed Load:	
Lighting and Small Appliance 20 units × 5520 VA	110,400 VA
Application of Demand Factor	
First 3000 VA at 100%	3000 VA
110,400 VA – 3000 VA = 107,400 VA at 35%	37,590 VA
Net Computed Load	40,590 VA
Range: 10 ranges (less than 12 kVA) *(see Col. A, Table 220-19)*	25,000 VA
Net Computed Load (with ranges)	65,590 VA

Net computed load for 120/240-V, 3-wire system,

65,590 VA ÷ 240 V = 273 A

Feeder Neutral

Lighting and Small Appliance	40,590 VA
Range: 25,000 VA at 70% *(see Section 220-22)*	17,500 VA
Net Computed Load (neutral)	58,090 VA

Calculated Load for Neutral

58,090 VA ÷ 240 V = 242 A

Further Demand Factor *(Section 220-22)*

200 A at 100%	200 A
242 A – 200 A = 42 A at 70%	29 A
Net Computed Load (neutral)	229 A

Minimum Size Main Feeders (or Service Conductors) Required (Less House Load) (For 40 Dwelling Units — 20 with Ranges)

Total Computed Load:	
Lighting and Small Appliance Load 40 units × 5520 VA	220,800 VA

Application of Demand Factor (from Table 220-11)

First 3000 VA at 100%	3000 VA
Next 120,000 VA − 3000 VA = 117,000 VA at 35%	40,950 VA
Remainder 220,800 VA − 120,000 VA = 100,800 VA at 25%	25,200 VA
Net Computed Load	69,150 VA
Range: 20 ranges (less than 12 kVA) *(see Col. A, Table 220-19)*	35,000 VA
Net Computed Load	104,150 VA

For 120/240-V, 3-wire system

Net computed load of 104,150 VA ÷ 240 V = 434 A

Feeder Neutral

Lighting and Small Appliance Load	69,150 VA
Range: 35,000 VA at 70% *(see Section 220-22)*	24,500 VA
Computed Load (neutral)	93,650 VA

93,650 VA ÷ 240 V = 390 A

Further Demand Factor *(see Section 220-22)*

200 A at 100%	200 A
390 A – 200 A = 190 A at 70%	133 A
Net Computed Load (neutral)	333 A

[See Tables 310-16 through 310-21, and Section 310-15(b)(2) and (b)(4).]

Example No. D4(b). Optional Calculation for Multifamily Dwelling

A multifamily dwelling equipped with electric cooking and space heating or air conditioning has 40 dwelling units.

Meters are in two banks of 20 each with house metering and individual feeders to each dwelling unit.

Each dwelling unit is equipped with an electric range of 8-kW nameplate rating, four 1.5-kW separately controlled 240-V electric space heaters, and a 2.5-kW, 240-V electric water heater. Assume range, space heater, and water heater kW ratings equivalent to kVA.

A common laundry facility is available to all tenants *[see Section 210-52(f), Exception No. 1]*.

Area of each dwelling unit is 840 ft^2.

Computed Load for Each Dwelling Unit *(see Article 220)*

General Lighting Load:	
840 ft^2 at 3 VA per ft^2	2520 VA
Other loads:	
Electric range	8000 VA
Electric heat: 6 kVA (or air conditioning if larger)	6000 VA
Electric water heater	2500 VA

Minimum Number of Branch Circuits Required for Each Dwelling Unit

General Lighting: 2520 VA ÷ 120 V = 21 A
Two 15-A, 2-wire, or two 20-A, 2-wire circuits
Small Appliance Load: Two 2-wire circuits of No. 12 *[see Section 210-11(c)(1)]*
Range Circuit: 8000 VA × 80% ÷ 240 V = 27 A
Three No. 10 conductors as permitted in Column C of Table 220-19
Space Heating: 6000 VA ÷ 240 V = 25 A
Number of circuits *(see Section 210-11)*

Minimum Size Feeder Required for Each Dwelling Unit *(see Section 215-2)*

Computed Load *(see Article 220):*	
General Lighting	2520 VA
Small Appliance (two 20-A circuits)	3000 VA
Subtotal Computed Load (without range and space heating)	5520 VA

Application of Demand Factor

First 3000 VA at 100%	3000 VA
5520 VA – 3000 VA = 2520 VA at 35%	882 VA
Net Computed Load (without range and space heating)	3882 VA
Range	6400 VA
Space Heating *(see Section 220-15)*	6000 VA
Water Heater	2500 VA
Net Computed Load (for individual dwelling unit)	18,782 VA

Size of Each Feeder

For 120/240-V, 3-wire system,

Net computed load of 18,782 VA ÷ 240 V = 78 A

Feeder Neutral *(see Section 220-22)*

Lighting and Small Appliance	3882 VA
Range: 6400 VA at 70% *(see Section 220-22)*	4480 VA
Water and Space Heating (no neutral): 240 V	0 VA
Net Computed Load (neutral)	8362 VA

Calculated Load for Neutral

8362 VA ÷ 240 V = 35 A

Minimum Size Feeder Required from Service Equipment to Meter Bank (for 20 Dwelling Units)

Total Computed Load:	
Lighting and Small Appliance Load	
20 units × 5520 VA	110,400 VA
Water and Space Heating Load	
20 units × 8500 VA	170,000 VA
Range: 20 × 8000 VA	160,000 VA
Net Computed Load (20 dwelling units)	440,400 VA
Net Computed Load Using Optional Calculation *(see Table 220-32)*	
440,400 VA × 0.38	167,352 VA

167,352 VA ÷ 240 V = 697 A

Minimum Size Main Feeder Required (Less House Load) (For 40 Dwelling Units)

Computed Load:	
Lighting and Small Appliance Load	
40 units × 5520 VA	220,800 VA
Water and Space Heating Load	
40 units × 8500 VA	340,000 VA
Range: 40 ranges × 8000 VA	320,000 VA
Net Computed (40 dwelling units)	880,800 VA

Net Computed Load Using Optional Calculation *(see Table 220-32)*

880,800 VA × 0.28 = 246,624 VA

246,624 VA ÷ 240 V = 1028 A

Feeder Neutral Load for Feeder from Service Equipment to Meter Bank (For 20 Dwelling Units)

Lighting and Small Appliance Load	
20 units × 5520 VA	110,400 VA
First 3000 VA at 100%	3000 VA
110,400 VA – 3000 VA = 107,400 VA at 35%	37,590 VA
Net Computed Load	40,590 VA
20 ranges: 35,000 VA at 70% *(see Table 220-19 and Section 220-22)*	24,500 VA
Total	65,090 VA

65,090 VA ÷ 240 V = 271 A

Further Demand Factor *(see Section 220-22)*

First 200 A at 100%	200 A
Balance: 271 A – 200 A = 71 A at 70%	50 A
Total	250 amperes

Feeder Neutral Load for Main Feeder (Less House Load) (For 40 Dwelling Units)

Lighting and Small Appliance Load	
40 units × 5520 VA	220,800 VA
First 3000 VA at 100%	3000 VA
Next 120,000 VA – 3000 VA = 117,000 VA at 35%	40,950 VA
Remainder 220,800 VA – 120,000 VA = 100,800 VA at 25%	25,200 VA
Net Computed Load	69,150 VA
40 ranges: 55,000 VA at 70% *(see Table 220-19 and Section 220-22)*	38,500 VA
Total	107,650 VA

107,650 VA ÷ 240 V = 449 A

Further Demand Factor *(see Section 220-22)*

First 200 A at 100%	200 A
Balance: 449 A − 200 A = 249 A at 70%	174 A
Total	374 A

Example No. D5(a). Multifamily Dwelling Served at 208Y/120 Volts, Three Phase

All conditions and calculations are the same as for the multifamily dwelling of Example No. D4(a) served at 120/240 V, single phase except service to each dwelling unit would be two phase legs and neutral.

Minimum Number of Branch Circuits Required for Each Dwelling Unit *(see Section 210-11)*

Range Circuit:

8000 VA ÷ 208 V = 38 A

A circuit of two No. 8 conductors and one No. 10 conductor as permitted by Section 210-19(c)

Minimum Size Feeder Required for Each Dwelling Unit *(see Section 215-2)*

For 120/208-V, 3-wire system (without ranges),
Net Computed Load of 3882 VA ÷ 2 legs ÷ 120 V/leg = 16.2 A
For 120/208-V, 3-wire system (with ranges),

Net Computed Load (range) of 8000 VA ÷ 208 V = 38.5 A
Total Load (range + lighting) = 38.5 A + 16.2 A = 54.7 A

Feeder Neutral: (range) of 8000 VA × 70% =
5600 VA ÷ 208 V = 26.9 A

Total Load: (range + lighting) = 26.9 A + 16.2 A = 43.1 A

Minimum Size Feeder Required from Service Equipment to Meter Bank (for 20 Dwelling Units — 10 with Ranges)

For 208Y/120-V, 3-phase, 4-wire system,
Ranges:
Maximum number between any two phase legs = 4
2 × 4 = 8.
Table 220-19 demand = 23,000 VA
Per phase demand = 23,000 VA ÷ 2 = 11,500 VA
Equivalent 3-phase load = 34,500 VA
Net Computed Load (total):

40,590 VA + 34,500 VA = 75,090 VA
75,090 VA ÷ (208 V)(1.732) = 208.4 A

Feeder Neutral Size:

Net Computed Lighting and Appliance Load & Equivalent Range Load

40,590 VA + (34,500 VA × 70%) = 64,740 VA

Net Computed Neutral Load:

64,740 VA ÷ (208 V)(1.732) = 179.6 A

Minimum Size Main Feeder Required (Less House Load) (For 40 Dwelling Units — 20 with Ranges)

For 208Y/120-V, 3-phase, 4-wire system,
Ranges:
Maximum number between any two phase legs = 7
2 × 7 = 14.
Table 220-19 demand = 29,000 VA
Per phase demand = 29,000 VA ÷ 2 = 14,500 VA
Equivalent 3-phase load = 43,500 VA
Net Computed Lighting and Appliance Load & Equivalent Range Load:

69,150 VA + 43,500 VA = 112,650 VA
112,650 VA ÷ (208 V)(1.732) = 312.7 A

Main Feeder Neutral Size:

69,150 VA + (43,500 VA × 70%) = 99,600 VA
99,600 VA ÷ (208 V)(1.732) = 276.5 A

Further Demand Factor *(see Section 220-22)*

200 A at 100%	200.0 A
276.5 A – 200 A = 76.5 A at 70%	53.6 A
Net Computed Load (neutral)	253.6 A

Example No. D5(b). Optional Calculation for Multifamily Dwelling Served at 208Y/120 Volts, Three Phase

All conditions and calculations are the same as for Optional Calculation for the multifamily dwelling of Example No. D4(b) served at 120/240 V, single phase except service to each dwelling unit would be two phase legs and neutral.

Minimum Number of Branch Circuits Required for Each Dwelling Unit *(see Section 210-11)*

Range Circuit:

8000 VA × 80% ÷ 208 V = 30.7 A

A circuit of two No. 8 conductors and one No. 10 conductor as permitted by Section 210-19(c)
Space Heating: 6000 VA ÷ 208 V = 28.8 A
Two 20-ampere, 2-pole circuits using No. 12 conductors

Minimum Size Feeder Required for Each Dwelling Unit

120/208-V, 3-wire circuit,
Net computed load of 18,782 VA ÷ 208 V = 90.3 A
Net computed lighting load (line to neutral):

3882 VA ÷ 2 legs ÷ 120 V per leg = 16.2 amperes

Line to line = 14,900 VA ÷ 208 V = 71.6 A
Total Load = 16.2 A + 71.6 A = 87.8 A

Minimum Size Feeder Required from Service Equipment to Meter Bank (For 20 Dwelling Units)

Net Computed Load

167,352 VA ÷ (208 V)(1.732) = 464.9 A

Feeder Neutral Load:

65,080 VA ÷ (208 V)(1.732) = 180.65 A

Minimum Size Main Feeder Required (Less House Load) (For 40 Dwelling Units)

Net computed load:

246,624 VA ÷ (208 V)(1.732) = 684.6 A

Main Feeder Neutral Load:

107,650 VA ÷ (208 V)(1.732) = 298.8 A

Further Demand Factor *(see Section 220-22)*

200 A at 100%	200.0 A
298.8 A – 200 A = 98.8 A at 70%	69.2 A
Net Computed Load (neutral)	269.2 A

Example No. D6. Maximum Demand for Range Loads

Table 220-19, Column A applies to ranges not over 12 kW. The application of Note 1 to ranges over 12 kW (and not over 27 kW) and Note 2 to ranges over 8¾ kW (and not over 27 kW) is illustrated in the following two examples:

Ranges All the Same Rating *(see Table 220-19, Note 1)*

Assume 24 ranges, each rated 16 kW.
From Table 220-19 Column A, the maximum demand for 24 ranges of 12-kW rating is 39 kW.
16 kW exceeds 12 kW by 4.
5% × 4 = 20% (5% increase for each kW in excess of 12)
39 kW × 20% = 7.8 kW increase
39 kW + 7.8 kW = 46.8 kW (value to be used in selection of feeders)

Ranges of Unequal Rating *(see Table 220-19, Note 2)*

Assume 5 ranges, each rated 11 kW; 2 ranges, each rated 12 kW; 20 ranges, each rated 13.5 kW; 3 ranges, each rated 18 kW.

5 ranges × 12 kW = 60 kW (use 12 kW for range rated less than 12)
2 ranges × 12 kW = 24 kW
20 ranges × 13.5 kW = 270 kW
3 ranges × 18 kW = 54 kW
30 ranges total kW = 408 kW
408 kW ÷ 30 ranges = 13.6 kW (average to be used for computation)

From Table 220-19 Column A, the demand for 30 ranges of 12-kW rating is 15 kW + (1 kW × 30 ranges) = 45 kW. 13.6 kW exceeds 12 kW by 1.6 kW (use 2 kW).

5% × 2 = 10% (5% increase for each kW in excess of 12 kW)
45 kW × 10% = 4.5 kW increase
45 kW + 4.5 kW = 49.5 kW (value to be used in selection of feeders)

Example No. D8. Motor Circuit Conductors, Overload Protection, and Short-Circuit and Ground-Fault Protection

(see Sections 240-6, 430-6, 430-22, 430-23, 430-24, 430-32, 430-34, 430-52, and 430-62, Tables 430-150 and 430-152)

Determine the minimum required conductor ampacity, the motor overload protection, the branch-circuit short-circuit and ground-fault protection, and the feeder protection, for three induction-type motors on a 480-V, 3-phase feeder, as follows:

(a) One 25-hp, 460-V, 3-phase, squirrel-cage motor, nameplate full-load current 32 A, Design B, Service Factor 1.15

(b) Two 30-hp, 460-V, 3-phase, wound-rotor motors, nameplate primary full-load current 38 A, nameplate secondary full-load current 65 A, 40°C rise.

Conductor Ampacity

The full-load current value used to determine the minimum required conductor ampacity is obtained from Table 430-150 *[see Section 430-6(a)]* for the squirrel-cage motor and the primary of the wound-rotor motors. To obtain the minimum required conductor ampacity, the full-load current is multiplied by 1.25 *[see Sections 430-22 and 430-23(a)]*.

For the 25-hp motor,

34 A × 1.25 = 42.5 A

For the 30-hp motors,

40 A × 1.25 = 50 A
65 A × 1.25 = 81.25 A

Motor Overload Protection

Where protected by a separate overload device, the motors are required to have overload protection rated or set to trip at not more than 125% of the nameplate full-load current *[see Sections 430-6(a) and 430-32(a)(1)]*.

For the 25-hp motor,

32 A × 1.25 = 40.0 A

For the 30-hp motors,

38 A × 1.25 = 47.5 A

Where the separate overload device is an overload relay (not a fuse or circuit breaker), and the overload device selected at 125% is not sufficient to start the motor or carry the load, the trip setting is permitted to be increased in accordance with Section 430-34.

Branch-Circuit Short-Circuit and Ground-Fault Protection

The selection of the rating of the protective device depends on the type of protective device selected, in accordance with Section 430-52 and Table 430-152. The following is for the 25-hp motor.

Nontime-Delay Fuse: The fuse rating is 300% × 34 A = 102 A.
The next larger standard fuse is 110 A *[see Sections 240-6 and 430-52(c)(1), Exception No. 1]*. If the motor will not start with a 110-A nontime-delay fuse, the fuse rating is permitted to be increased to 125 A because this rating does not exceed 400% *[see Section 430-52(c)(1), Exception No. 2a]*.

Time-Delay Fuse: The fuse rating is 175% × 34 A = 59.5 A.
The next larger standard fuse is 60 A *[see Sections 240-6 and 430-52(c)(1), Exception No. 1]*. If the motor will not start with a 60-A time-delay fuse, the fuse rating is permitted to be increased to 70 A because this rating does not exceed 225% *[see Section 430-52(c)(1), Exception No. 2b]*.

Feeder Short-Circuit and Ground-Fault Protection

The rating of the feeder protective device is based on the sum of the largest branch-circuit protective device (example is 100 A) plus the sum of the full-load currents of the other motors, or 100 A + 40 A + 40 A = 180 A. The nearest standard fuse that does not exceed this value is 175 A *[see Sections 240-6 and 430-62(a)]*.

Example No. D9. Feeder Ampacity Determination for Generator Field Control

[see Sections 220-10, 430-24, 430-24 Exception No. 1, 620-13, 620-14, 620-61, and Tables 430-22(b) and 620-14]

Determine the conductor ampacity for a 460-V 3-phase, 60-Hz ac feeder supplying a group of six elevators. The 460-V ac drive motor nameplate rating of the largest MG set for one elevator is 40 hp and 52 A, and the remaining elevators each have a 30-hp, 40-A, ac drive motor rating for their MG sets. In addition to a motor controller, each elevator has a separate motion/operation controller rated 10 A continuous to operate microprocessors, relays, power supplies, and the elevator car door operator. The MG sets are rated continuous.

Conductor Ampacity

Conductor ampacity is determined as follows:

(a) Per Sections 620-13(d) and 620-61(b)(1), use Table 430-22(b) for intermittent duty (elevators). For intermittent duty using a continuous rated motor, the percentage of nameplate current rating to be used is 140%.

(b) For the 30-hp ac drive motor,

140% × 40 A = 56 A

For the 40-hp ac drive motor,

140% × 52 A = 73 A

(c) The total conductor ampacity is the sum of all the motor currents.

(1 motor × 73 A) + (5 motors × 56 A) = 353 A

(d) Per Section 620-14 and Table 620-14, the conductor (feeder) ampacity would be permitted to be reduced by the use of a demand factor. Constant loads are not included *(see Section 620-14, FPN)*. For six elevators, the demand factor is 0.79.
Therefore, feeder diverse ampacity = 0.79 × 353 A = 279 A.

(e) Per Sections 430-24 and 215-3

Controller continuous current = 125% × 10 A = 12.5 A

(f) The total feeder ampacity is the sum of the diverse current and all the controller continuous current.

I_{total} = 279 A + (6 elevators × 12.5 A) = 354 A

(g) This ampacity would be permitted to be used to select the wire size. See Figure D9.

Example No. D10. Feeder Ampacity Determination for Adjustable Speed Drive Control

[see Sections 215-3, 430-24, 430-24(b), 620-13, 620-14, 620-61, and Tables 430-22(b) and 620-14]

Determine the conductor ampacity for a 460-V, 3-phase, 60-Hz ac feeder supplying a group of six identical elevators. The system is adjustable-speed SCR dc drive. The power transformers are external to the drive (motor controller) cabinet. Each elevator has a separate motion/operation controller connected to the load side of the main line disconnect switch rated 10 A continuous to operate microprocessors, relays, power supplies, and the

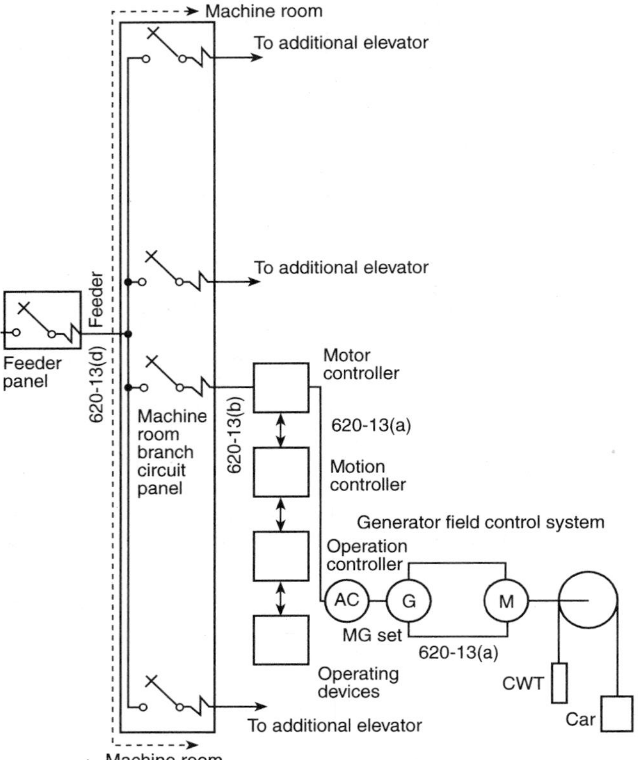

Figure D9.

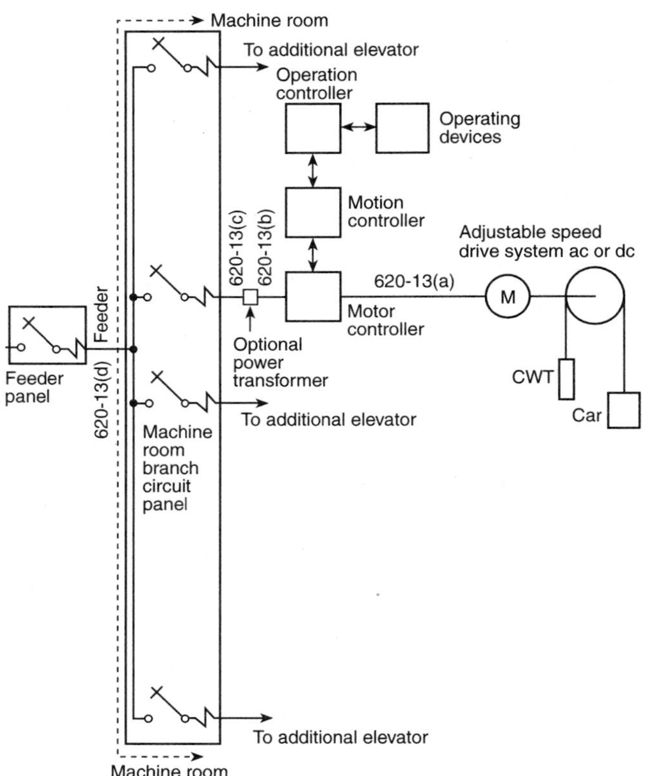

Figure D10.

elevator car door operator. Each transformer is rated 95 kVA with an efficiency of 90%.

Conductor Ampacity. Conductor ampacity is determined as follows:

(a) Calculate the nameplate rating of the transformer:

$$I = \frac{95\text{ kVA} \times 1000}{\sqrt{3} \times 460\text{ V} \times 0.90\text{ eff.}} = 133\text{ A}$$

(b) Per Section 620-13(d) for six elevators, the total conductor ampacity is the sum of all the currents.

$$6\text{ elevators} \times 133\text{ A} = 798\text{ A}$$

(c) Per Section 620-14 and Table 620-14, the conductor (feeder) ampacity would be permitted to be reduced by the use of a demand factor. Constant loads are not included *(see Section 620-13, FPN No. 2)*. For six elevators, the demand factor is 0.79.
Therefore, feeder diverse ampacity = 0.79 × 798 A = 630 A

(d) Per Sections 430-24 and 215-3,

$$\text{Controller continuous current} = 125\% \times 10\text{ A} = 12.5\text{ A}$$

(e) The total feeder ampacity is the sum of the diverse current and all the controller constant current.

$$I_{\text{total}} = 630\text{ A} + (6\text{ elevators} \times 12.5\text{ A}) = 705\text{ A}$$

(f) This ampacity would be permitted to be used to select the wire size. See Figure D10.

Example No. D11. Mobile Home

Example D11 was relocated from Section 550-13(b) to Appendix D for the 1999 *Code* since it is considered explanatory material and is therefore more appropriately located in an appendix.

A mobile home floor is 70 ft by 10 ft and has two small appliance circuits, a 1000-VA, 240-V heater, a 200-VA, 120-V exhaust fan, a 400-VA, 120-V dishwasher, and a 7000-VA electric range.

Lighting and Small Appliance Load

Lighting (70 ft × 10 ft × 3 VA per ft^2)	2100 VA
Small appliance (1500 VA × 2 circuits)	3000 VA
Laundry (1500 VA × 1 curcuit)	1500 VA
Subtotal	6600 VA
First 3000 VA at 100%	3000 VA
Remainder (6600 VA – 3000 VA = 3600 VA) × 35%	1260 VA
Total	4260 VA

$$4260\text{ VA} \div 240\text{ V} = 17.75\text{ A per leg}$$

Amperes per Leg

	Leg A	Leg B
Lighting and appliances	17.75	17.75
Heater (1000 VA ÷ 240 V)	4.20	4.20
Fan (200 VA × 125% ÷ 120 V)	2.08	—
Dishwasher (400 VA ÷ 120 V)	—	3.30
Range (7000 VA × 0.8 ÷ 240 V)	23.30	23.30
Total amperes per leg	47.33	48.55

Based on the higher current calculated for either leg, a minimum 50-A supply cord would be required.

For SI units, 1 ft^2 = 0.093 m^2 and 1 ft = 0.3048 m

Example No. D12. Park Trailer

Example D12 was relocated from Section 552-47(b) to Appendix D for the 1999 *Code* since it is considered explanatory material and is therefore more appropriately located in an appendix.

A park trailer floor is 40 ft by 10 ft and has two small appliance circuits, a 1000-VA, 240-V heater, a 200-VA, 120-V exhaust fan, a 400-VA, 120-V dishwasher, and a 7000-VA electric range.

Lighting and Small Appliance Load

Lighting (40 ft × 10 ft × 3 VA per ft^2)	1200 VA
Small appliance (1500 VA × 2 curcuits)	3000 VA
Laundry (1500 VA × 1 circuit)	1500 VA
Subtotal	5700 VA
First 3000 VA at 100%	3000 VA
Remainder (5700 VA – 3000 VA = 2700 VA) × 35%	945 VA
Total	3945 VA

3945 VA ÷ 240 V = 16.44 A per leg

Amperes per Leg

	Leg A	Leg B
Lighting and appliances	16.44	16.44
Heater (1000 VA ÷ 240 V)	4.20	4.20
Fan (200 VA × 125% ÷ 120 V)	2.08	—
Dishwasher (400 VA ÷ 120 V)	—	3.3
Range (7000 VA × 0.8 ÷ 240 V)	23.30	23.30
Total amperes per leg	46.02	47.24

Based on the higher current calculated for either leg, a minimum 50-A supply cord would be required.

For SI units, 1 ft^2 = 0.093 m^2 and 1 ft = 0.3048 m

APPENDIX E

Article 250 Cross Reference

This appendix is not a part of the recommendations of this NFPA document but is included for informational purposes only.

E1. Cross Reference from the 1999 NEC to the 1996 NEC

Topic	1999 NEC	1996 NEC
Part A. General		
Scope	−1	−1
General Requirements for Grounding and Bonding	−2	−51, −1 FPN 1 & 2
Application of Other Articles	−4	−2
Objectionable Current Over Grounding Conductors	−6	−21
Connection of Grounding and Bonding Equipment	−8	−113
Protection of Ground Clamps and Fittings	−10	−117
Clean Surfaces	−12	−118
Part B. Circuit and System Grounding		
Alternating-Current Circuits and Systems to Be Grounded	−20	−5
Alternating-Current Systems of 50 Volts to 1000 Volts Not Required to be Grounded	−21	−5
Circuits Not to Be Grounded	−22	−7
Grounding Service-Supplied Alternating-Current Systems	−24	−23, −53(a)
Conductor to Be Grounded — Alternating-Current Systems	−26	−25
Main Bonding Jumpers	−28	−79, −53(b)
Grounding Separately Derived Alternating-Current Systems	−30	−26
Two or More Buildings or Structures Supplied from a Common Service	−32	−24
Portable and Vehicle-Mounted Generators	−34	−6
High-Impedance Grounded Neutral Systems	−36	−27
Part C. Grounding Electrode System and Grounding Electrode Conductor		
Grounding Electrode System	−50	−81
Made and Other Electrodes	−52	−83
Supplementary Grounding Electrodes	−54	−91(c)
Resistance of Made Electrodes	−56	−84
Common Grounding Electrode	−58	−54
Use of Air Terminals	−60	−86
Grounding Electrode Conductor Material	−62	−91(a)
Grounding Electrode Conductor Installation	−64	−91(a), −92
Size of Alternating-Current Grounding Electrode Conductor	−66	−94
Grounding Electrode Conductor Connection to Electrodes	−68	−112
Methods of Grounding Conductor Connection to Electrodes	−70	−115
Part D. Enclosure, Raceway, and Service Cable Grounding		
Service Raceways and Enclosures	−80	−32
Underground Service Cable or Conduit	−84	−55
Other Conductor Enclosures and Raceways	−86	−33
Part E. Bonding		
General	−90	−70
Services	−92	−71
Method of Bonding at the Service	−94	−72
Bonding Other Enclosures	−96	−75
Bonding for Over 250 Volts	−97	−76
Bonding Loosely Jointed Metal Raceways	−98	−77
Bonding in Hazardous Classified) Locations	−100	−78
Equipment Bonding Jumpers	−102	−79
Bonding of Piping Systems and Exposed Structural Steel	−104	−80
Lightning Protection Systems	−106	−46
Part F. Equipment Grounding and Equipment Grounding Conductors		
Equipment Fastened in Place or Connected by Permanent Wiring Methods (Fixed)	−110	−42
Fastened in Place or Connected by Permanent Wiring Methods (Fixed) — Specific	−112	−43
Equipment Connected by Cord and Plug	−114	−45
Nonelectric Equipment	−116	−44
Types of Equipment Grounding Conductors	−118	−91(b)

Topic	1999 NEC	1996 NEC
Identification of Equipment Grounding Conductors	−119	−57(b)
Equipment Grounding Conductor Installation	−120	−92(c)
Size of Equipment Grounding Conductors	−122	−95
Equipment Grounding Conductor Continuity	−124	−99
Identification of Wiring Device Terminals	−126	−119
Part G. Methods of Equipment Grounding		
Equipment Grounding Conductor Connections	−130	−50
Short Sections of Raceway	−132	−56
Equipment Fastened in Place or Connected by Permanent Wiring Methods (Fixed) — Grounding	−134	−57
Equipment Considered Effectively Grounded	−136	−58
Cord- and Plug-Connected Equipment	−138	−59
Frames of Ranges and Clothes Dryers	−140	−60
Use of Grounded Circuit Conductor for Grounding Equipment	−142	−61
Multiple Circuit Connections	−144	−62
Connecting Receptacle Grounding Terminal to Box	−146	−74
Continuity and Attachment of Equipment Grounding Conductors to Boxes	−148	−114
Part H. Direct-Current Systems		
General	−160	(new)
Direct-Current Circuits and Systems to Be Grounded	−162	−3
Point of Connection for Direct-Current Systems	−164	−22
Size of Direct-Current Grounding Electrode Conductor	−166	−93
Direct-Current Bonding Jumper	−168	−79(d)
Ungrounded Direct-Current Separately Derived Systems	−169	(new)
Part J. Instruments, Meters, and Relays		
Instrument Transformer Circuits	−170	−121
Instrument Transformer Cases	−172	−122
Cases of Instruments, Meters, and Relays Operating at Less than 1000 Volts	−174	−123
Cases of Instruments, Meters, and Relays Operating Voltage 1 kV and Over	−176	−124
Instrument Grounding Conductor	−178	−125
Part K. Grounding of Systems and Circuits of 1 kV and Over (High Voltage)		
General	−180	−150
Derived Neutral Systems	−182	−151
Solidly Grounded Neutral Systems	−184	−152
Impedance Grounded Neutral Systems	−186	−153
Grounding of Systems Supplying Portable or Mobile Equipment	−188	−154
Grounding of Equipment	−190	−155

E2. Cross Reference from the 1996 NEC to the 1999 NEC

Topic	1996 NEC	1999 NEC
Part A. General		
Scope	−1	−1
Application of Other Articles	−2	−4
Part B. Circuit and System Grounding		
Direct-Current Systems	−3	−162
Alternating-Current Circuits and Systems to Be Grounded	−5	−20, −21
Portable and Vehicle-Mounted Generators	−6	−34
Circuits Not to Be Grounded	−7	−22
Part C. Location of System Grounding Connections		
Objectionable Current Over Grounding Conductors	−21	−6
Point of Connection for Direct-Current Systems	−22	−164
Grounding Service-Supplied Alternating-Current Systems	−23	−24
Two or More Buildings or Structures Supplied from a Common Service	−24	−32
Conductor to Be Grounded — Alternating-Current Systems	−25	−26
Grounding Separately Derived Alternating-Current Systems	−26	−30
High-Impedance Grounded Neutral System Connections	−27	−36
Part D. Enclosure and Raceway Grounding		
Service Raceways and Enclosures	−32	−80
Other Conductor Enclosures and Raceways	−33	−86
Part E. Equipment Grounding		
Equipment Fastened in Place or Connected by Permanent Wiring Methods (Fixed)	−42	−110
Fastened in Place or Connected by Permanent Wiring Methods (Fixed) Specific	−43	−112
Nonelectric Equipment	−44	−116
Equipment Connected by Cord and Plug	−45	−114
Spacing from Lightning Rods	−46	−106
Part F. Methods of Grounding		
Equipment Grounding Conductor Connections	−50	−130
Effective Grounding Path	−51	−2
Grounding Path to Grounding Electrode at Services	−53	−24, −28
Common Grounding Electrode	−54	−58

Topic	1996 NEC	1999 NEC
Underground Service Cable	−55	−84
Short Sections of Raceway	−56	−132
Equipment Fastened in Place or Connected by Permanent Wiring Methods (Fixed) — Grounding	−57	−119, −134
Equipment Considered Effectively Grounded	−58	−136
Cord- and Plug-Connected Equipment	−59	−138
Frames of Ranges and Clothes Dryers	−60	−140
Use of Grounded Circuit Conductor for Grounding Equipment	−61	−142
Multiple Circuit Connections	−62	−144
Part G. Bonding		
General	−70	−90
Service Equipment	−71	−92
Method of Bonding Service Equipment	−72	−94
Metal Armor or Tape of Service Cable	−73	(removed)
Connecting Receptacle Grounding Terminal to Box	−74	−146
Bonding Other Enclosures	−75	−96
Bonding for Over 250 Volts	−76	−97
Bonding Loosely Jointed Metal Raceways	−77	−98
Bonding in Hazardous (Classified) Locations	−78	−100
Main and Equipment Bonding Jumpers	−79	−28, −102, −168
Bonding of Piping Systems and Exposed Structural Steel	−80	−104
Part H. Grounding Electrode System		
Grounding Electrode System	−81	−50
Made and Other Electrodes	−83	−52
Resistance of Made Electrodes	−84	−56
Use of Lightning Rods	−86	−60
Part J. Grounding Conductors		
Material	−91	−54, −62, −64, −118
Installation	−92	−64, −120
Size of Direct-Current Grounding Electrode Conductor	−93	−166
Size of Alternating-Current Grounding Electrode Conductor	−94	−66
Size of Equipment Grounding Conductors	−95	−122
Outline Lighting	−97	(600-7)
Equipment Grounding Conductor Continuity	−99	−124
Part K. Grounding Conductor Connections		
To Grounding Electrode	−112	−68
To Conductors and Equipment	−113	−8
Continuity and Attachment of Equipment Grounding Conductors to Boxes	−114	−148
Connection to Electrodes	−115	−70
Protection of Attachment	−117	−10
Clean Surfaces	−118	−12
Identification of Wiring Device Terminals	−119	−126
Part L. Instrument Transformers, Relays, Etc.		
Instrument Transformer Circuits	−121	−170
Instrument Transformer Cases	−122	−172
Cases of Instruments, Meters, and Relays — Operating at Less than 1000 Volts	−123	−174
Cases of Instruments, Meters, and Relays — Operating Voltage 1 kV and Over	−124	−176
Instrument Grounding Conductor	−125	−178
Part M. Grounding of Systems and Circuits of 1 kV and Over (High Voltage)		
General	−150	−180
Derived Neutral Systems	−151	−182
Solidly Grounded Neutral Systems	−152	−184
Impedance Grounded Neutral Systems	−153	−186
Grounding of Systems Supplying Portable or Mobile Equipment	−154	−188
Grounding of Equipment	−155	−190

Tentative Interim Amendment

NFPA 70

National Electrical Code®

1996 Edition

Reference: 90-1(d) (New)
TIA 96-2 (NFPA 70)

Pursuant to Section 5 of the NFPA Regulations Governing Committee Projects, the National Fire Protection Association has issued the following Tentative Interim Amendment to NFPA 70, *National Electrical Code®,* 1996 edition. The TIA was processed by the National Electrical Code Committee, and was issued by the Standards Council on January 14, 1999, with an effective date of February 3, 1999.

A Tentative Interim Amendment is tentative because it has not been processed through the entire standards-making procedures. It is interim because it is effective only between editions of the standard. A TIA automatically becomes a proposal of the proponent for the next edition of the standard; as such, it then is subject to all of the procedures of the standards-making process.

1. Add a new 90-1(d) and Fine Print Note (FPN) to read as follows:

(d) **Relation to International Standards.** The requirements in this *Code* address the fundamental principles of protection for safety contained in International Standard "Electrical Installations of Buildings, IEC 60364-1, Section 131.

FPN: IEC 60364-1, Section 131 contains fundamental principles of protection for safety that encompass: protection against electric shock, protection against thermal effects, protection against overcurrent, protection against fault currents, and protection against overvoltage. All of the above potential hazards are addressed by the requirements in this *Code*.

Index

-B-

-E-

-G-

-H-

-N-

-O-

-T-

-U-

-V-

-W-

-X-

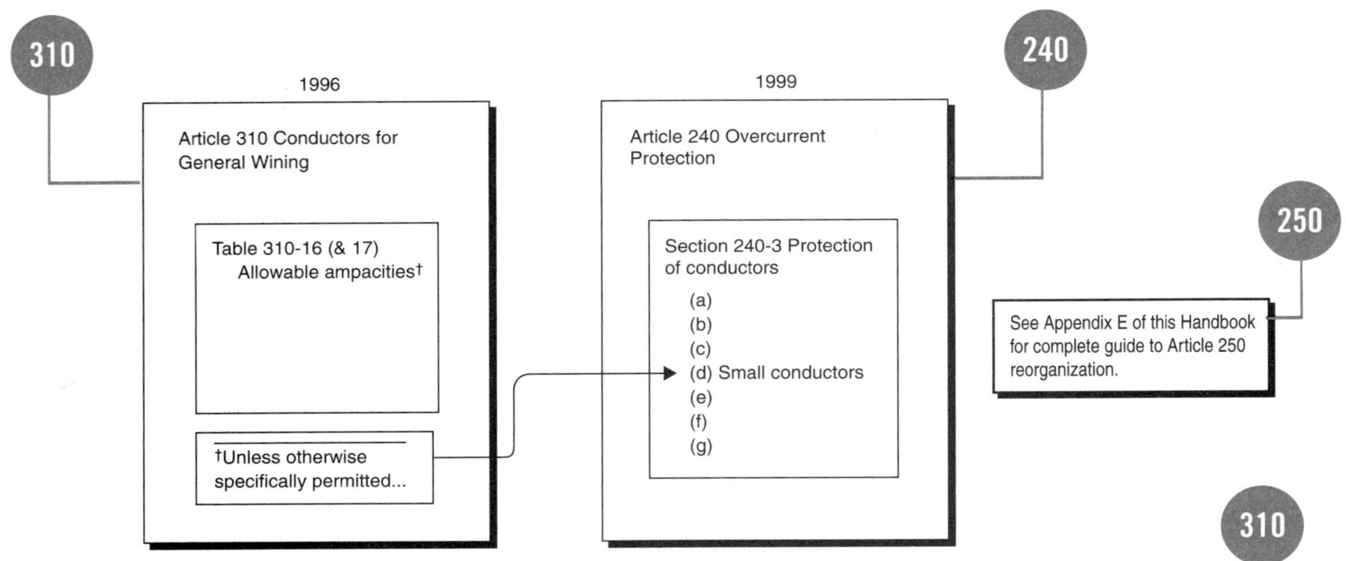

Article 310 Reorganization

1996 NEC Section	*Section Heading or Description*	*1999 NEC Section*
310-15	Ampacities for Conductors Rated 0-2000 Volts	310-15
310-15 (part)	General—Tables or Engineering Supervision	310-15(a)(1)
310-15(FPN)	Tables do not account for voltage drop	310-15(a)(1)(FPN No. 1)
Tables Note 7 FPN	Type MTW Machine Tool Wire Ampacities	310-15(1)(FPN No. 2)
310-15(c)	General—Selection of Ampacity	310-15(a)(2)
310-15(a) (part)	Tables	310-15(b)
310-15(a)(FPN)	Load based on Article 220—ampacity considerations	310-15(b)(FPN)
Tables Note 1	Tables—General—insulation-type letters	310-15(b)(1)
Tables Note 9(a)	Adjustment Factors and Table	310-15(b)(2)(a)
Tables Note 8(b)	More Than One Conduit, Tube, or Raceway	310-15(b)(2)(b)
Tables Note 5	Bare or Covered Conductors	310-15(b)(3)
Tables Note 10	Neutral Conductor—when counted as current-carrying	310-15(b)(4)
Tables Note 11	Grounding or Bonding Conductor	310-15(b)(5)
Tables Note 3	Dwelling Units—feeder and service conductor ampacities	310-15(b)(6)
Tables Note 6	Mineral-Insulated, Metal-Sheathed Cable	310-15(b)(7)
310-15(b)	Engineering Supervision	310-15(c)
Tables Note 9	Overcurrent Protection	Deleted—Sec 240-3
Obelisk Notes at bottom of Tables 310-16 & 310-17	Limitations on overcurrent protection for small conductors	240-3(d)

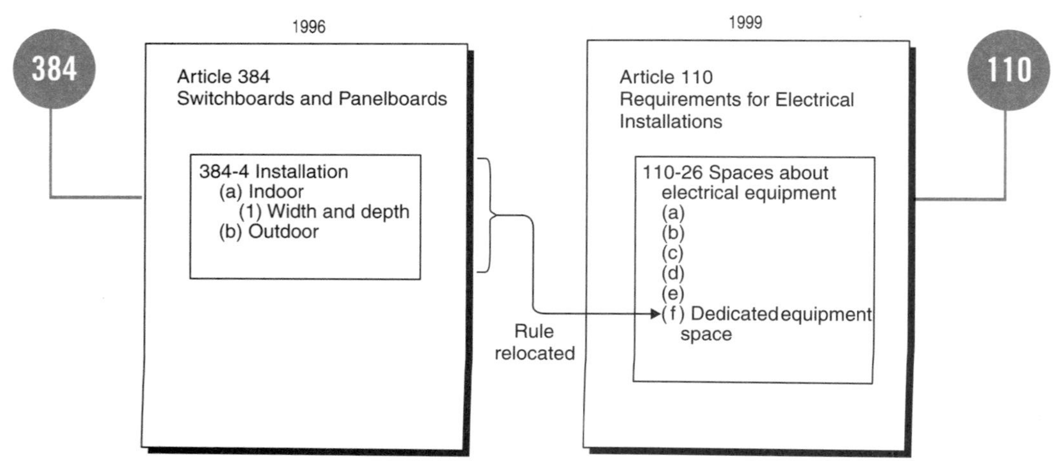